Nutrition in the Prevention and Treatment of Disease

Second Edition

NUTRITION IN THE PREVENTION AND TREATMENT OF DISEASE

SECOND EDITION

Edited by

ANN M. COULSTON
Nutrition Consultant
Mountain View, CA

CAROL J. BOUSHEY
Department of Foods and Nutrition
Purdue University
West Lafayette, IN

ELSEVIER

AMSTERDAM • BOSTON • HEIDELBERG • LONDON • NEW YORK • OXFORD
PARIS • SAN DIEGO • SAN FRANCISCO • SINGAPORE • SYDNEY • TOKYO
Academic Press is an imprint of Elsevier

Elsevier Academic Press
30 Corporate Drive, Suite 400, Burlington, MA 01803, USA
525 B Street, Suite 1900, San Diego, California 92101-4495, USA
84 Theobald's Road, London WC1X 8RR, UK

This book is printed on acid-free paper. ♾

Library of Congress Cataloging-in-Publication Data
Nutrition in the prevention and treatment of disease / edited by Ann M. Coulston, Carol J. Boushey. – 2nd ed.
p. ; cm.
Includes bibliographical references and index.
ISBN-13: 978-0-12-374118-9 (hardcover : alk. paper) 1. Dietetics. 2. Nutrition.
3. Diet in disease. 4. Diet therapy. I. Coulston, Ann M. II. Boushey, Carol. J.
[DNLM: 1. Diet Therapy–methods. 2. Nutrition Disorders–diet therapy.
3. Nutrition Disorders–epidemiology. 4. Nutrition Disorders–prevention & control.
5. Nutritional Support. WB 400 N97534 2008]
RM216.N875 2008
615.8′54–dc22
2007043040

British Library Cataloguing in Publication Data
A catalogue record for this book is available from the British Library

ISBN 13: 978-012-374118-9

For all information on all Elsevier Academic Press publications visit our Web site at www.books.elsevier.com

Printed in China
08 09 10 9 8 7 6 5 4 3 2 1

We dedicate this book to our teachers and colleagues,
women and men who have dedicated their
professional lives to advance the field of nutrition science.

Contents

CHAPTER 6

Overview of Nutritional Epidemiology

Marian L. Neuhouser and Ruth E. Patterson

CHAPTER 7

Analysis, Presentation, and Interpretation of Dietary Data

Debra Coward-McKenzie and Rachel K. Johnson

CHAPTER 8

Current Theoretical Bases for Nutrition Intervention and Their Uses

Karen Glanz

CHAPTER 9

Nutrition Intervention: Lessons from Clinical Trials

Linda G. Snetselaar

CHAPTER 10

Tools and Techniques to Facilitate Nutrition Intervention

Linda M. Delahanty and Joan M. Heins

CHAPTER 11

Evaluation of Nutrition Interventions

Alan R. Kristal

CHAPTER 12

Biomarkers and Their Use in Nutrition Intervention

Johanna W. Lampe and Cheryl L. Rock

B. Overweight and Obesity

CHAPTER 21

Genetics of Human Obesity

Janis S. Fisler and Nancy A. Schonfeld-Warden

CHAPTER 22

Obesity: Overview of Treatments and Interventions

Helen M. Seagle, Holly R. Wyatt, and James O. Hill

CHAPTER 23

Obesity: The Role of Physical Activity in Adults

Marcia L. Stefanick

CHAPTER 24

Macronutrient Intake and the Control of Body Weight

David A. Levitsky

CHAPTER 25

Behavioral Risk Factors for Overweight and Obesity: Diet and Physical Activity

Nancy E. Sherwood, Mary Story, Dianne Neumark-Sztainer, and Kate Bauer

CHAPTER 26

Dietary Approaches to Exploit Energy Balance Utilities for Body Weight Control

Richard D. Mattes

List of Contributors

MERLIN W. ARIEFDJOHAN, PhD, Department of Foods and Nutrition, Purdue University, West Lafayette, IN

MARION TAYLOR BAER, PhD, RD, Department of Community Health Sciences, School of Public Health, University of California, Los Angeles, and USC UCEDD, Department of Pediatrics University of Southern California, Los Angeles, CA

BRYAN C. BATCH, MD, Department of Medicine, Duke University Medical Center, Durham, NC

KATE BAUER, MS, Division of Epidemiology and Community Health, School of Public Health, University of Minnesota, Minneapolis, MN

PETER L. BEYER, MS, RD, Department of Dietetics and Nutrition, University of Kansas Medical Center, Kansas City, KS

CAROL J. BOUSHEY, PhD, MPH, RD, Department of Foods and Nutrition, Purdue University, West Lafayette, IN

BETTE J. CAAN, Dr PH, Kaiser Permanente, Division of Research, Oakland, CA

MONA S. CALVO, PhD, Center for Food Safety and Applied Nutrition, Food and Drug Administration, Laurel, MD

REBECCA B. COSTELLO, PhD, Office of Dietary Supplements, National Institutes of Health, Bethesda, MD

ANN M. COULSTON, MS, RD, Nutrition Consultant, Mountain View, CA

DEBRA COWARD-MCKENZIE, MS, Department of Nutrition, The University of North Carolina at Chapel Hill, Chapel Hill, NC

LINDA M. DELAHANTY, MS, RD, LDN, MGH Diabetes Center, Massachusetts General Hospital, Boston, MA

WENDY DEMARK-WAHNEFRIED, PhD, RD, Department of Behavioral Science, MD Anderson Cancer Center, The University of Texas, Houston, TX

MARIA DUARTE-GARDEA, PhD, RD, College of Health Sciences, University of Texas at El Paso, El Paso, TX

JOHANNA DWYER, DSc, RD, Office of Dietary Supplements, National Institutes of Health, Bethesda, MD, and Jean Mayer Human Nutrition Research Center on Aging at Tufts University, Boston, MA

PHILIP M. FARRELL, MD, PhD, UW School of Medicine and Public Health, Madison, WI

JANIS S. FISLER, PhD, Department of Nutrition, University of California at Davis, Davis, CA

JO L. FREUDENHEIM, PhD, RD, Department of Social and Preventive Medicine, State University of New York at Buffalo, Buffalo, NY

KAREN GLANZ, PhD, MPH, Department of Behavioral Sciences and Health Education, Rollins School of Public Health, Emory University, Atlanta, GA

REJEANNE GOUGEON, PhD, Crabtree Laboratory, Royal Victoria Hospital, Montreal, QC, Canada

JESSICA GRIEGER, PhD, Department of Nutrition, Pennsylvania State University, University Park, PA

ANNE BRADFORD HARRIS, PhD, MPH, RD, Waisman Center, Department of Pediatrics, University of Wisconsin-Madison, WI

ROBERT P. HEANEY, MD, John A. Creighton University Professor, Creighton University, Omaha, NE

JOAN M. HEINS, MS, RD, LD, CDE, Division of Health Behavior Research, Washington University School of Medicine, St. Louis, MO

STEVE HERTZLER, PhD, RD, Ross Products Division, Abbott Laboratories, Columbus, OH

JAMES O. HILL, PhD, Health Sciences Center, University of Colorado at Denver and Health Sciences Center, Denver, CO

KARRY A. JACKSON, BS, Bartlett, IL

ASKER E. JEUKENDRUP, PhD, Human Performance Laboratory, School of Sport and Exercise Sciences, University of Birmingham, Edgbaston, United Kingdom

RACHEL K. JOHNSON, PhD, MPH, RD, Cals Deans Office, The University of Vermont, Burlington, VT

JAMES A. JOSEPH, PhD, USDA Human Nutrition Research Center on Aging, Tufts University, Boston, MA

LAURENCE N. KOLONEL, MD, PhD, Epidemiology Program, Cancer Research Center of Hawaii, University of Hawaii, Honolulu, HI

PENNY M. KRIS-ETHERTON, PhD, RD, Department of Nutrition, Pennsylvania State University, University Park, PA

ALAN R. KRISTAL, PhD, Cancer Prevention Program, Fred Hutchinson Cancer Research Center, Seattle, WA

HUICHUAN J. LAI, PhD, RD, School of Medicine and Public Health, Madison, WI

JOHANNA W. LAMPE, PhD, RD, Cancer Prevention Program, Fred Hutchinson Cancer Research Center, Seattle, WA

JENNY H. LEDIKWE, PhD, Department of Nutrition, Pennsylvania State University, University Park, PA

DAVID A. LEVITSKY, PhD, Department of Psychology, Division of Nutritional Sciences (DNS), Cornell University, Ithaca, NY

PAO-HWA LIN, PhD, Department of Medicine, Duke University Medical Center, Durham, NC

ROBERT MARCUS, MD, Professor Emeritus, Department of Medicine, Stanford University, Stanford, CA

JULIE A. MARES, PhD, Department of Ophthalmology and Visual Sciences, University of Wisconsin School of Medicine and Public Health, Madison, WI

RICHARD D. MATTES, PhD, MPH, RD, Department of Foods and Nutrition, Purdue University, West Lafayette, IN

MEGAN A. MCCRORY, PhD, Department of Foods and Nutrition and Department of Psychological Sciences, Purdue University, West Lafayette, IN

AMY E. MILLEN, Social and Preventive Medicine, State University of New York at Buffalo, Buffalo, NY

JOHN A. MILNER, PhD, National Cancer Institute, Division of Cancer Prevention, National Institutes of Health, US Department of Health and Human Services, Bethesda, MD

SUZANNE P. MURPHY, PhD, RD, Cancer Research Center of Hawaii, University of Hawaii, Honolulu, HI

MAUREEN A. MURTAUGH, PhD, RD, Department of Internal Medicine, Division of Epidemiology, University of Utah, Salt Lake City, UT

MARIAN L. NEUHOUSER, PhD, RD, Cancer Prevention Program, Fred Hutchinson Cancer Research Center, Seattle, WA

DIANNE NEUMARK-SZTAINER, PhD, MPH, RD, Division of Epidemiology and Community Health, School of Public Heath, University of Minnesota, Minneapolis, MN

MIHAI D. NICULESCU, MD, PhD, Department of Nutrition, School of Public Health, University of North Carolina at Chapel Hill, Chapel Hill, NC

JOSE M. ORDOVAS, PhD, Nutrition and Genomics Laboratory, Jean Mayer USDA HNRCA, Tufts University, Boston, MA

SONG-YI PARK, PhD, Epidemiology Program, Cancer Research Center of Hawaii, University of Hawaii, Honolulu, HI

RUTH E. PATTERSON, PhD, RD, Cancer Prevention Program, Fred Hutchinson Cancer Research Center, Seattle, WA

GEORGE PERRY, PhD, College of Sciences, University of Texas at San Antonio, San Antonio, TX

MICHELLE PIETZAK, MD, Department of Pediatrics, University of Southern California Keck School of Medicine, Los Angeles, CA

CHERYL L. ROCK, PhD, RD, Department of Family and Preventive Medicine, School of Medicine, University of California at San Diego, La Jolla, CA

SARAH TAYLOR ROLLER, JD, RD, MPH, Covington & Burling LLP, Washington, DC

BARBARA J. ROLLS, PhD, Department of Nutrition, Pennsylvania State University, University Park, PA

MICHAELL ROUSSELL, BS, Department of Nutrition, Pennsylvania State University, University Park, PA

EDWARD SALTZMAN, MD, Energy Metabolism Laboratory, Jean Mayer USDA HNRCA, Tufts University, Boston, MA

DENNIS A. SAVAIANO, PhD, Department of Foods and Nutrition, College of Consumer and Family Sciences, Purdue University, West Lafayette, IN

NANCY A. SCHONFELD-WARDEN, MD, Department of Pediatrics, University of California at Davis, Davis, CA

HELEN M. SEAGLE, MS, RD, Weight Management, Kaiser Permanente, Denver, CO

HAROLD E. SEIFRIED, PhD, DABT, National Cancer Institute, Division of Cancer Prevention, National Institutes of Health, US Department of Health and Human Services, Rockville, MD

NANCY E. SHERWOOD, PhD, Division of Epidemiology and Community Health, School of Public Health, University of Minnesota HealthPartners Research Foundation, Minneapolis, MN

BARBARA SHUKITT-HALE, PhD, Human Nutrition Research Center on Aging, Jean Mayer USDA, Tufts University, Boston, MA

MARTHA L. SLATTERY, Division of Epidemiology, Department of Medicine, University of Utah, Salt Lake City, UT

MARK A. SMITH, PhD, School of Medicine, Pathology, Case Western Reserve University, Cleveland, OH

LINDA G. SNETSELAAR, PhD, RD, Department of Epidemiology, Department of Community and Behavioral Health, Preventive Nutrition Education, College of Public Health, University of Iowa, Iowa City, IA

MARCIA L. STEFANICK, PhD, Stanford Prevention Research Center, Department of Medicine, Stanford University School of Medicine, Stanford, CA

CHARLES B. STEPHENSEN, PhD, USDA Western Human Nutrition Research Center, University of California at Davis, Davis, CA

MARY STORY, PhD, RD, Division of Epidemiology and Community Health, School of Public Health, University of Minnesota, Minneapolis, MN

FABRIZIS L SUAREZ, MD, PhD, FACN, Ross Products Division, Medical Safety & Surveillance Abbott Laboratories, Columbus, OH

AMY F. SUBAR, PhD, MPH, RD, Division of Cancer Control and Population Sciences, National Cancer Institute, Bethesda, MD

LAURA P. SVETKEY, MD, MHS, Department of Medicine, Division of Nephrology, Duke University Medical Center, Durham, NC

FRANCES E. THOMPSON, PhD, MPH, Division of Cancer Control and Population Sciences, National Cancer Institute, Bethesda, MD

CRISTINE M. TRAHMS, MS, RD, FADA, Center on Human Development and Disability, Department of Pediatrics, University of Washington, Seattle, WA

CONNIE M. WEAVER, Department of Foods and Nutrition, Purdue University, West Lafayette, IN

JENNIFER WEBB, MS, RD, Childrens Hospital, Los Angeles, CA

SUSAN J. WHITING, PhD, College of Pharmacy and Nutrition, University of Saskatchewan, Saskatoon, SK, Canada

HOLLY R. WYATT, MD, Health Sciences Center, University of Colorado at Denver and Health Sciences Center, Denver, CO

STEVEN H. ZEISEL, MD, PhD, Director of the UNC Nutrition Institute, School of Public Health, University of North Carolina at Chapel Hill, Chapel Hill, NC

Preface to the First Edition

The purpose of this text is to provide an update of current knowledge in clinical nutrition and an overview of the rationale and science base of its application to practice in the treatment and prevention of disease. The text addresses basic principles and concepts that are central to the major clinical nutrition-related activities, such as nutritional assessment and monitoring, current theoretical base and knowledge of efficacious interventions, interactions between genetic and nutritional factors, and the use and interpretation of population-based or clinical epidemiological evidence. The various roles of clinical nutrition and current knowledge of nutrition in the prevention and treatment of major disease-specific conditions are also reviewed, with an emphasis on past and current scientific evidence that supports these roles. New areas of interest and study are also discussed, with the perspective that the application of the scientific method is by definition an evolutionary process.

Treatment of the disease diabetes mellitus provides an excellent and current example of treatment evolution. In the early part of the 20th century, before the discovery of insulin by F. G. Banting and C. H. Best in 1921, the treatment of choice for individuals with diabetes mellitus was morphine for pain abatement along with a very restricted, starvation diet. When insulin injections became available, dietary protocols were developed. Initially, dietary treatment was based on food exchange lists that encouraged prescribed intakes of carbohydrate, protein, and fat. Recent research from the Diabetes Control and Complications Trial and a similar research trial in the United Kingdom has been the base for the current dietary management emphasizing blood glucose monitoring throughout the day and individualized adjustment of carbohydrate ingestion and insulin injection in individuals who require insulin therapy. Nutrition intervention plays a major role in the management of the patient with diabetes mellitus and in the treatment of the disease and the prevention or delay of complications.

Another essential role for nutrition intervention is in the prevention of cancer. Cancer represents a disease continuum, and at all stages, from primary prevention to treatment, nutrition is a key factor. As discussed in the section highlighting nutrition and major cancer types, an explosion of new knowledge has identified nutrition as a major factor in the etiology and progression of disease.

Nutrition is the process by which the human body utilizes food for the production of energy, for the maintenance of health, for growth, and for the normal functioning of every organ and tissue. Clinical nutrition is the study of nutrition and diet as related to the development and treatment of human disease. Nutrition is an interdisciplinary field of study, built on a foundation of biomedical and behavioral sciences. Clinical nutrition is the aspect of nutrition science that is related to the development, progression, or management of disease, as differentiated from the issues of normal requirements, cellular functions and activities, and various topics that must be addressed in meeting basic requirements to enable normal growth and development.

Areas of study that contribute to knowledge in clinical nutrition include the disease-relevant biochemistry, metabolism, and activities of nutrients and dietary factors within the tissues and cells; the bioavailability and utilization of nutrients from various food sources, as disease risk or diagnosis may influence these factors; the regulation and compartmentalization of nutrients in the body; the attitudes about food and the eating patterns and behaviors of the targeted individual or group; the technology of food science and specialized or modified food products; and the technology involved in providing adequate and appropriate nutrients or foods to individuals and various community-based or institutionalized groups. Other aspects of clinical nutrition include the development and evaluation of nutrition education efforts; the development of nutrition policies, guidelines, and practice standards that affect the goals and objectives of government and private health agencies, professional practice groups, and health-related organizations; and the design and implementation of individual, clinical, and community-based nutrition and diet interventions. Clinical nutrition interventions range in scope from efforts to maintain health during short-term illness, to optimization of health status in individuals at risk for or diagnosed with chronic diseases, to major nutritional and diet modifications as specific or adjuvant treatments for disease. Clinical nutrition encompasses primary, secondary, and tertiary disease prevention, in addition to management of disease.

Dietary intake or nutritional status may be altered as a result of disease or by the treatment modalities that are utilized, such as surgical treatments, or medical management strategies, such as drugs. The altered needs must be met by dietary or nutritional interventions in order to prevent malnutrition and the associated consequences, which would contribute to overall morbidity and mortality. Also, nutrition intervention can be a critical component of disease prevention, an important aspect of disease management, or the primary treatment for disease. A complicating factor is that people generally eat food, rather than nutrients, so that the practical and psychosocial aspects of diet modification and food or food product availability must be considered in any nutrition intervention, whether individual or community based and irrespective of whether the goal is primary prevention or disease treatment.

As in any area of the biomedical sciences, the importance of science-based activities and practices cannot be understated. Clinical nutrition concepts and practices that can become popular with either the lay public or professionals are sometimes based on the type of scientific evidence that cannot truly support the rationales and practices, regardless of how standard and common they might be. Popular theories may be generated by observational epidemiological studies, case series, or anecdotal reports, all of which lack the capability of truly demonstrating a causative or efficacious role for the nutritional factor. Such studies are useful for generating hypotheses, but the apparent associations between diets and disease may be confounded by uncontrolled or unmeasured factors and other determinants of health and disease. Unproven diet therapies exist for the treatment of numerous conditions, and many aspects of common nutrition interventions are sorely in need of testing in an appropriate research design. As in any other aspect of disease prevention and treatment, the use of nutrition interventions or diet therapies should be based on a scientific rationale and sound data, not on anecdotal experience. The scientific basis for clinical nutrition needs to expand considerably in order to fully support claims for the efficacy of many of the common activities and interventions, and progress in this area is being made.

Our definitions of diseases need to further evolve to bring greater clarity and improve precision of treatment. As gene–diet interactions are scientifically delineated, laser-sharp therapies may be applied to specific individuals. For the public, however, generating and analyzing data that summarize dietary intake and its association with disease will be valuable tasks in both treating disease and developing disease prevention strategies. Well-designed focused screening will be an aid in disease detection, and well-founded medical nutrition therapies can minimize disease development and related complications. Providing scientifically sound, creative, and effective nutrition interventions can be challenging. In so doing, however, we will serve the public good.

It is our goal to update our knowledge and its application through updated editions of this text. In addition, we plan to provide online access to relevant new findings and their import to nutrition in the prevention and treatment of disease. It is our goal to raise the bar for both understanding and treatment.

Ann M. Coulston
Cheryl L. Rock
Elaine R. Monsen

Preface to the Second Edition

Like with the first edition of this text, our purpose is to provide a compilation of current knowledge in clinical nutrition and an overview of the rationale and science base of its application to practice in the prevention and treatment of disease. The first section addresses basic principles and concepts that are central to clinical nutrition research methodology. Because nutrition information is gathered from a variety of study designs, research methodology, epidemiology, and intervention studies are reviewed, coupled with data analysis, intervention techniques, and application of behavioral principles to nutrition intervention. The use of biomarkers to monitor nutrition intervention is an example of a rapidly expanding field in research methodology. Throughout these chapters, new areas of study are discussed with the perspective that the application of the scientific method is by definition an evolutionary process. Specific examples, drawn from recently published reports, bring the principles to life.

The second section covers areas of study that contribute to knowledge in clinical nutrition, including disease-relevant biochemistry, metabolism, dietary factors within tissues and cells, and attitudes about food and the eating patterns and behaviors of targeted individuals or groups. This section presents a rich array of topics that cover areas of general interest with several new topics of nutrition relationships.

Clinical nutrition is the aspect of nutrition science that is related to the development, progression, or management of disease, as differentiated from the issues of normal requirements, cellular functions, and activities. Interventions range from efforts to maintain health during short-term illness, to optimization of health status in individuals at risk for or diagnosed with chronic diseases, to major nutritional and dietary modifications as specific or adjuvant treatments for disease. The first condition addressed is the ever-growing concern with overweight and obesity. As with many of the following disease groups, this grouping begins with a chapter on the genetics of human obesity and moves on to issues related to treatment, role of physical activity, nutrient-related considerations, childhood and adolescent issues, and environmental queues controlling energy intake.

Cardiovascular disease, also a condition closely related to nutrition, is summarized in three chapters that examine genetic considerations, lipid disorders, and hypertension. Closely related to obesity and cardiovascular disease is diabetes mellitus. It is interesting how many of the clinical nutrition areas interrelate—obesity is a risk factor for cardiovascular disease and diabetes, whereas diabetes is an independent risk factor for cardiovascular disease. Dietary intake or nutritional status may be altered as a result of disease or by the treatment modalities that are used, such as surgical treatments or medical management strategies, including prescription medications. The altered needs must be met by dietary or nutrition interventions in order to prevent malnutrition and the associated consequences that contribute to morbidity and mortality.

Nutrition intervention can be a critical component of disease prevention, an important aspect of disease management, or the primary treatment for disease. This is exemplified by the chapters dealing with cancer, beginning again with a discussion of the genetic components, followed by a discussion of malignancies that have connections to nutrition and specific nutrients. Gastrointestinal diseases, especially the newer knowledge about diet and microflora of the gastrointestinal tract, demonstrate the importance of food choices in disease prevention, treatment, and management. The bone health chapters cover three important topics linked by the nutrients calcium and vitamin D and tell an important story of the value of early nutrition on health in later years.

Generating and analyzing data that summarize dietary intake and its association with disease are valuable tasks in treating disease and developing disease prevention strategies. Well-founded medical nutrition therapies can minimize disease development and related complications. Providing scientifically sound, creative, and effective nutrition interventions is challenging and rewarding. We plan to update our knowledge and its application through future editions of this text.

Ann M. Coulston
Carol J. Boushey

SECTION I

Research Methodology

A. Assessment Methods for Research and Practice

CHAPTER **1**

Dietary Assessment Methodology

FRANCES E. THOMPSON AND AMY F. SUBAR
National Cancer Institute, Bethesda, Maryland

Contents

I. INTRODUCTION

This chapter is a revision of the similarly named chapter in the 2001 edition of this book [1], which itself was based on the "Dietary Assessment Resource Manual" [2] by Frances E. Thompson and Tim Byers, adapted with permission from the *Journal of Nutrition.* Dietary assessment encompasses food consumption at the national level (e.g., food supply and production), household level, and individual level. This review focuses only on individual-level food intake assessment. It is intended as a resource for those who wish to assess diet in a research study. The first section reviews major dietary assessment methods, their advantages and disadvantages and validity. The next sections describe which dietary assessment methods are most appropriate for different types of studies and for various types of populations. Finally, specific issues that relate to all methods are discussed. The intent of this chapter is to contribute to an understanding of various dietary assessment methods so that the most appropriate method for a particular need is chosen.

II. DIETARY ASSESSMENT METHODS

A. Dietary Records

For the dietary record approach, the respondent records the foods and beverages and the amounts of each consumed over one or more days. The amounts consumed may be measured, using a scale or household measures (such as cups, tablespoons), or estimated, using models, pictures, or no particular aid. Typically, if multiple days are recorded, they are consecutive, and no more than 3 or 4 days are included. Recording periods of more than 4 consecutive days are usually unsatisfactory, as reported intakes decrease [3] because of respondent fatigue. Theoretically, the information is recorded at the time of the eating occasion, but it need not be done on paper. Dictaphones, computer recording, and self-recording scales have been used [4–7] and hold special promise for low-literacy groups and other difficult-to-assess populations because of their ease of administration and potential accuracy, though tape recording has not been shown to be useful among school-aged children [8]. A recently developed prototype of a computer-administered instrument illustrates the potential benefits of technology, particularly for low-literacy groups: the respondent selects the food consumed and subsequently the appropriate portion size of the selected foods via food photographs on the computer screen [6]. Computerized programs for recording food intake may be delivered on the Internet [9], a CD-ROM [6], or on a hand-held personal digital assistant (PDA) [5, 10]. A PDA itself can be coupled with a camera that photographs foods selected [11]. When these programs are linked with appropriate databases (see Section V.D), the burden of coding data can be dramatically relieved. However, response problems, particularly accurate estimation of portion size, remain [10].

To complete a dietary record, the respondent must be trained in the level of detail required to adequately describe the foods and amounts consumed, including the name of the food (brand name, if possible), preparation methods, recipes for food mixtures, and portion sizes. In some studies, this is enhanced by contact and review of the report after 1 day of recording. At the end of the recording period, a trained interviewer should review the records with the respondent to clarify entries and to probe for forgotten foods. Someone other than the subject can also record dietary records. This is often the method used with children or people in institutions.

While intake data using dietary records are typically collected in an open-ended form, close-ended forms have also been developed [5, 12–15]. These forms consist of listings of food groups; the respondent indicates whether that food group has been consumed. Portion size can also be asked, either in an open-ended manner or in categories. In format, these "checklist" forms resemble the food frequency questionnaire (FFQ) (see Section II.C). Unlike FFQs, which generally query about intake over a specified time period such as the past year or month, they are filled out either concurrently with actual intake (for precoded records) or at the end of a day for that day's intake (daily recall).

The dietary record method has the potential for providing quantitatively accurate information on food consumed during the recording period [16]. Recording foods as they are consumed lessens the problem of omission and the foods are more fully described. Further, the measurement of amounts of food consumed at each occasion should provide more accurate portion sizes than if the respondents were recalling portion sizes of foods previously eaten.

A major disadvantage of the dietary record method is that it is subject to bias both in the selection of the sample and in the measurement of the diet. Dietary record keeping requires that respondents or respondent proxies be both motivated and literate (if done on paper), which can potentially limit use of the method in some population groups (e.g., those of low socioeconomic status, the poorly educated, recent immigrants, children, and some elderly groups). The requirements for cooperation in keeping records can limit the generalizability of the findings from the dietary records to the broader population from which the study sample was drawn. Research indicates that there is a significant increase in incomplete records as more days of records are kept, and the validity of the collected information decreases in the later days of a 7-day recording period, in contrast to collected information in the earlier days [3]. Part of this decrease may occur because many respondents develop the practice of filling out the record at one time for a previous period.

When respondents record only once per day, the record method approaches the 24-hour recall in terms of respondents relying on memory rather than concurrent recording. More important, recording foods as they are being eaten can affect both the types of food chosen and the quantities consumed [17]. The knowledge that food requires recording and the demanding task of doing it, therefore, may alter the dietary behaviors the tool is intended to measure [18]. This effect is a weakness when the aim is to measure unaltered dietary behavior. However, when the aim is to enhance awareness of dietary behavior and change that behavior, as in some intervention studies, this effect can be seen as an advantage [19]. Recording, by itself, is an effective weight-loss technique [20]. Recent interest in "real time" assessment [21] has led to the development and testing of a dietary intake self-monitoring system delivered through a personal digital assistant that enables concurrent recording and immediate, automated feedback. A pilot study testing this approach found improved self-monitoring and adherence to dietary goals [19].

As is true with all quantitative dietary information, the information collected on dietary records can be burdensome to code and can lead to high personnel costs. Dietary assessment software that allows for easier data entry using common spellings of foods can save considerable time in data coding. Even with high-quality data entry, maintaining overall quality control for dietary records can be difficult because information is often not recorded consistently from respondent to respondent.

These weaknesses may be less pronounced for the hybrid method of the "checklist" form, because checking off a food item may be easier than recording a complete description of the food, and the costs of data processing can be minimal, especially if the form is machine scannable. The checklist can be developed to assess particular "core foods," which contribute substantially to intakes of some nutrients. However, as the comprehensiveness of the nutrients to be assessed increases, the length of the form also increases and becomes more burdensome to complete at each eating occasion. The checklist method may be most appropriate in settings with limited diets or for assessment of a limited set of foods or nutrients.

Several studies indicate that reported energy and protein intakes on diet records for selected small samples of adults are underestimated in the range of 4% to 37% when compared to energy expenditure as measured by doubly-labeled water or protein intake as measured by urinary nitrogen [20, 22–34]. Because of these findings, the record is considered an imperfect gold standard. Underreporting on food records is probably a result of the combined effects of incomplete recording and the impact of the recording process on dietary choices leading to undereating [20, 31]. The highest levels of underreporting on food records have been found among individuals with higher body mass indexes (BMIs) [24, 26, 27, 35, 36], particularly women [24, 26, 27, 37–39]. This relationship has been found among elderly individuals also [40]. This effect, however, may arise, in part, because heavier individuals are more likely to be dieting on any given individual day [41]. Other research shows that demographic or psychological indices such as education, employment grade, social desirability, body image, or dietary restraint may also be important factors related to underreporting on diet records [24, 31, 38, 39, 42–44]. The research evidence for the psychosocial factors related to energy misreporting is reviewed in Mauer *et al.* [45]. A few studies suggest that low-energy reporters compared to non-low-energy reporters have intakes that are lower in absolute intake of most nutrients [36], higher in percentage of energy from protein [36, 39], and lower in percentage of energy as carbohydrate [36, 39, 46, 47] and in percentage of energy from fat [47]. Underreporters may also report lower intakes of desserts, sweet baked goods, butter, and alcoholic beverages [36, 47] but more grains, meats, salads, and vegetables [36].

Some approaches have been suggested to overcome the underreporting in the record approach. Some suggest enhanced training of respondents. A different approach is to incorporate psychosocial questions known to be related to underreporting in order to estimate the level of underreporting [45]. Another suggested approach is to calibrate dietary records to doubly-labeled water (DLW), a biological indicator of energy expenditure, including covariates of gender, weight, and height, to more accurately predict individuals' energy intakes [48]. Further research is needed to develop and test these and other ideas.

B. 24-Hour Dietary Recall

For the 24-hour dietary recall, the respondent is asked to remember and report all the foods and beverages consumed in the preceding 24 hours or in the preceding day. The recall typically is conducted by interview, in person or by telephone [49, 50], either computer assisted [51] or using a paper-and-pencil form. Well-trained interviewers are crucial in administering a 24-hour recall because much of the dietary information is collected by asking probing questions. Ideally, interviewers would be dietitians with education in foods and nutrition; however, non-nutritionists who have been trained in the use of a standardized instrument can be effective. All interviewers should be knowledgeable about foods available in the marketplace and about preparation practices, including prevalent regional or ethnic foods.

The interview is often structured, usually with specific probes, to help the respondent remember all foods consumed throughout the day. An early study found that respondents with interviewer probing reported 25% higher dietary intakes than did respondents without interviewer probing [52]. Probing is especially useful in collecting necessary details, such as how foods were prepared. It is also useful in recovering many items not originally reported, such as common additions to foods (e.g., butter on toast) and eating occasions not originally reported (e.g., snacks and beverage breaks). However, interviewers should be provided with standardized neutral probing questions so as to avoid leading the respondent to specific answers when the respondent really does not know or remember.

The current state-of-the-art 24-hour dietary recall instrument is the United States Department of Agriculture's (USDA) Automated Multiple Pass Method (AMPM) [51], and it is used in the U.S. National Health and Nutrition Examination Survey (NHANES), the only nationally representative dietary survey in the United States. In the AMPM, intake is reviewed more than once in an effort to retrieve forgotten eating occasions and foods. The process consists of (1) an initial "quick list," where the respondent reports all the foods and beverages consumed without interruption from the interviewer; (2) a forgotten foods list of 9 food categories commonly omitted in 24-hour recall reporting; (3) time and occasion, where the respondent reports the time each eating occasion began and names the occasion; (4) a detail pass, where probing questions ask for more detailed information about each food and the portion size, in addition to review of the eating occasions and times between the eating occasions; and (5) final review, where questions about any other item not already reported are asked [51]. Research at the USDA allowed development of the Food Model Booklet [53], a portion size booklet used in the NHANES to facilitate more accurate portion size estimation. A 24-hour recall interview using the multiple pass approach typically requires between 30 and 45 minutes.

A quality control system to minimize error and increase reliability of interviewing and coding 24-hour recalls is essential. Such a system should include a detailed protocol for administration, training, and retraining sessions for interviewers, duplicate collection and coding of some of the recalls throughout the study period, and the use of a computerized database system for nutrient analysis. A research study among girls evaluated the marginal gains in accuracy of the estimates of mean and variance with increasing levels of quality control [54]. The authors recommended that the extent of quality control procedures adopted for a particular study should be carefully considered in light of that study's desired accuracy and precision and its resource constraints.

There are many advantages to the 24-hour recall. An interviewer administers the tool and records the responses, so literacy of the respondent is not required. Because of the immediacy of the recall period, respondents are generally able to recall most of their dietary intakes. Because there is relatively little burden on the respondents, those who agree to give 24-hour dietary recalls are more likely to be representative of the population than are those who agree to keep food records. Thus, the 24-hour recall method is useful across a wide range of populations. In addition, interviewers can be trained to capture the detail necessary so that the foods eaten by any population can be researched later by the coding staff and coded appropriately. Finally, in contrast to record methods, dietary recalls occur after the food has been consumed, so there is less potential for the assessment method to interfere with dietary behavior.

Computerized software systems allow direct coding of the foods reported during the interview. The potential benefits of automated software include substantial cost reductions for processing dietary data, less missing data, and greater standardization of interviews [55, 56]. However, a potential problem in direct coding of interview responses is the loss of the respondent's reported name and description of the food, in contrast to paper records of the interview, which are then available for later review and editing. If direct coding is used for the interview, methods for the interviewer to easily enter those foods not found in the system should be available, and these methods should be reinforced by interviewer training and quality control procedures.

Another technological advance in 24-hour dietary recall methodology is the increasing development of automated data collection systems [57–62]. These systems vary in the number of foods in their databases, the approach to asking about portion size, and their inclusion of probes regarding details of foods consumed and possible additions. One system, designed to assess heavy metal and pesticide use in children, has been developed for a hand-held device with wireless Internet access [61]. The National Cancer Institute (NCI) is currently developing a Web-based automated self-administered 24-hour dietary recall [62]. The goal is to create software that respondents can use to complete a dietary recall with the aid of multimedia visual cues, prompts, and animated characters, versus standard methods that require a trained interviewer. The system design relies on the most current USDA survey database [63] and includes many elements of the AMPM 24-hour interview developed by the USDA [64] and currently used in the NHANES. Respondents will be asked about portion sizes with the help of digital photographs depicting up to eight sizes. The instrument will be delivered via the Internet and will be available to researchers at a nominal cost.

The main weakness of the 24-hour recall approach is that individuals may not report their food consumption accurately for various reasons related to knowledge, memory, and the interview situation. These cognitive influences are discussed in more detail in Section V.F. Because most individuals' diets vary greatly from day to day, it is not appropriate to use data from a single 24-hour recall to characterize an individual's usual diet. Neither should a single day's intake, whether it is determined by a recall or food record, be used to estimate the proportion of the population that has adequate or inadequate diets (e.g., the proportion of individuals with less than 30% of energy from fat, or who are deficient in vitamin C intake) [65]. This is because the true distribution of usual diets is much narrower than is the distribution of daily diets (there is variation not only among people in usual diet but also in the day-to-day for each person). The principal use of a single 24-hour recall is to describe the average dietary intake of a group because the means are robust and unaffected by within-person variation. Multiple days of recalls or records can better assess the individual's usual intake and population distributions but require special statistical procedures designed for that purpose (see Section V.C).

The validity of the 24-hour dietary recall has been studied by comparing respondents' reports of intake either with intakes unobtrusively recorded/weighed by trained observers or with biological markers. Numerous observational studies of the effectiveness of the 24-hour recall have been conducted with children (see Section IV.C). In some studies with adults, group mean nutrient estimates from 24-hour recalls have been found to be similar to observed intakes [3, 66], although respondents with lower observed intakes have tended to overreport, and those with higher observed intakes have tended to underreport their intakes [66]. One observational study found energy underreporting during a self-selected eating period in both men and women, similar underreporting during a controlled diet period in men, and accurate reporting during this controlled diet in women; underestimates of portion sizes accounted for much of the underreporting [67]. Similar to findings for food records, studies with biological markers such as doubly-labeled water and urinary nitrogen generally have found underreporting using 24-hour dietary recalls, for energy in the range of 3% to 26% [6, 22, 25, 31, 68–72] and for protein in the range of 11% to 28% [69, 72, 73]. However, underreporting is not always found. Some have found overreporting of energy from 24-hour dietary recalls compared to DLW in the proxy reports for young children [74]. One study found overreporting of protein from 13% to 25% depending on level of BMI [75]. In addition, it is likely that the commonly reported phenomenon of underreporting in Western countries may not occur in all cultures; for example, Harrison *et al.* reported that 24-hour recalls collected from Egyptian women were well within expected amounts [76]. Finally, energy adjustment has been found in many studies to reduce error. For example, for protein density (i.e., percentage energy from protein), 24-hour dietary recall reports in the large NCI-funded Observing Protein and Energy Nutrition (OPEN) study were in close agreement to the biomarkers-based measure [72]. Evaluation of the USDA AMPM in two small observational studies indicated good agreement in mean intakes of macronutrients among men [77] and among obese women [78] and some overreporting of mean energy and carbohydrate intakes among normal and overweight women [78]. In a small, highly selected group of normal-weight women in energy balance, mean intake of energy using AMPM agreed with energy expenditure measured by DLW [79]. A large DLW study using recalls collected with the AMPM is currently being analyzed.

In past national dietary surveys using multiple-pass methods, data suggested that underreporting may have affected up to 15% of all 24 hour recalls [80, 81]. Underreporters compared to nonunderreporters tended to report fewer numbers of foods, fewer mentions of foods consumed, and smaller portion sizes across a wide range of food groups and tended to report more frequent intakes of low-fat/diet foods and less frequent intakes of fat added to foods [80]. As was found for records, factors such as obesity, gender, social desirability, restrained eating, education, literacy, perceived health status, and race/ethnicity have been shown in various studies to be related to underreporting in recalls [31, 41, 42, 70, 80–84].

C. Food Frequency

The food frequency approach [85, 86] asks respondents to report their usual frequency of consumption of each food from a list of foods for a specific period of time. Information is collected on frequency and sometimes portion size, but little detail is collected on other characteristics of the foods as eaten, such as the methods of cooking or the combinations of foods in meals. To estimate relative or absolute nutrient intakes, many FFQs also incorporate portion size questions, or specify portion sizes as part of each question. Overall nutrient intake estimates are derived by summing, over all foods, the products of the reported frequency of each food by the amount of nutrient in a specified (or assumed) serving of that food to produce an estimated daily intake of nutrients, dietary constituents, and food groups.

Many FFQs are available, and many continue to be adapted and developed for different populations and different purposes. Among those evaluated and commonly used for U.S. adults are the Health Habits and History Questionnaire (HHHQ) or Block questionnaires [87–97], the Fred Hutchinson Cancer Research Center Food Frequency Questionnaire (a revised HHHQ) [98, 99], the Harvard University Food Frequency Questionnaires or Willett questionnaires [85, 95–97, 100, 101–107], and the NCI's Diet History Questionnaire [72, 97, 108, 109]. The latter was designed with an emphasis on cognitive ease for respondents [110–112]. Other instruments have been developed for specific populations. Two FFQs have been developed by researchers at the University of Arizona, the University of Arizona Food Frequency Questionnaire and the Southwest Food Frequency Questionnaire, to capture the diverse diets of Latinos and Native Americans [113–115]. Other investigators have developed FFQs for Hispanic adults [116, 117]. Investigators at the University of Hawaii have developed a questionnaire for assessing the diverse diets of Hawaiian, Japanese, Chinese, Filipino, and non-Hispanic white ethnic groups [118, 119]. This instrument was adapted for use in a multiethnic cohort study conducted in Hawaii and Los Angeles [120]. In Europe, a number of FFQs have been developed within Western European countries for the European Prospective Investigation into Cancer and Nutrition (EPIC) [30, 121–126]. In addition, abbreviated FFQs attempting to assess total diet have been developed composed of shorter lists of 40 to 60 line items from the original 100 or so items [127–131]. "Brief" FFQs that assess a limited number of dietary exposures are discussed in the next section. Because of the number of FFQs available, investigators need to carefully consider which best suits their research needs.

The appropriateness of the food list is crucial in the food frequency method [88]. The full variability of an individual's diet, which includes many foods, brands, and preparation practices, cannot be fully captured with a finite food list. Obtaining accurate reports for foods eaten both as single items and in mixtures is particularly problematic. FFQs can ask the respondent to report either a combined frequency for a particular food eaten both alone and in mixtures or separate frequencies for each food use (e.g., one could ask about beans eaten alone and in mixtures, or one could ask separate questions about refried beans, bean soups, beans in burritos). The first approach is cognitively complex for the respondent, but the second approach may lead to double counting (e.g., burritos with beans may be reported both as beans and as a Mexican mixture). Often FFQs include similar foods in a single question (e.g., beef, pork, or lamb). However, such grouping can create a cognitively complex question (e.g., for someone who often eats beef and occasionally eats pork and lamb). In addition, when a single question is applied to a group of foods, assumptions about the relative frequencies of intake of the foods constituting the group are made in the assignment of values in the nutrient database. These assumptions are generally based on information from an external study population (such as from a national survey sample) even though true eating patterns may differ considerably across population subgroups and over time.

Each quantitative FFQ must be associated with a database to allow for the estimation of nutrient intakes for an assumed or reported portion size of each food queried. For example, the FFQ item of macaroni and cheese encompasses a wide variety of recipes with different nutrient compositions, yet the FFQ database must have a single nutrient composition profile. Several approaches are used to construct such a database [85]. A database approach uses quantitative dietary intake information from the target population to define the typical nutrient density of a particular food group category. For example, for the food group macaroni and cheese, all reports of the individual food codes reported in a population survey can be collected and a mean or median nutrient composition (by portion size if necessary) estimated. Values can also be calculated by gender and age. Dietary analyses software, specific to each FFQ, is then used to compute nutrient intakes for individual respondents. These analyses are available commercially for the Block, Willett, and Hutchinson FFQs and are publicly available for the NCI FFQ.

In pursuit of improving the validity of the FFQ, investigators have addressed a variety of frequency questionnaire design issues such as length, closed versus open-ended response categories, portion size, seasonality, and time frame. Frequency instruments designed to assess total diet generally list more than 100 individual line items, many with additional portion size questions, requiring 30 to 60 minutes to complete. This raises concern about length and its effect on response rates. Though respondent burden is a factor in obtaining reasonable response rates for studies in general, a few studies have shown this not to be a decisive factor for FFQs [111, 132–136]. This tension between length and specificity highlights the difficult issue of how to define a closed-ended list of foods for a food frequency instrument. The increasing use of optically scanned instruments has necessitated the use of closed-ended response categories forcing a loss in specificity [137].

Although the amounts consumed by individuals are considered an important component in estimating dietary intakes, for the FFQ instrument it is controversial as to whether or not portion size questions should be included. Frequency has been found to be a greater contributor than typical serving size to the variance in intake of most foods [138]; therefore, some prefer to use FFQs without the additional respondent burden of reporting serving sizes [85]. Others cite small improvements in the performance of FFQs that ask the respondents to report a usual serving size for each food [90, 91]. Some incorporate portion size and frequency into one question, asking how often a particular portion of the food is consumed [85]. Although some research has been conducted to determine the best ways to ask about portion size on FFQs [110, 139, 140], the marginal benefit of such information in a particular study may depend on the study objective and population characteristics.

Another design issue is the time frame about which intake is queried. Many instruments inquire about usual intakes over the past year [88, 101], but it is possible to ask about the past week or month [141] depending on specific research situations. Even when respondents are asked about intake over the past year, some studies indicate that the season in which the questionnaire is administered influences reporting over the entire year [142, 143].

Finally, analytical decisions are required in how food frequency data are processed. In research applications where there are no automated quality checks to assure that all questions are asked, decisions about how to handle missing data are needed. In particular, in self-administered situations, there are usually many initial frequency questions that are not answered. One approach is to assign null values, as some research indicates that respondents selectively omit answering questions about foods they seldom or never eat [144, 145]. Another approach is to assign the median value from those who did provide valid answers. One study compared these two approaches within a case-control setting and found that the two were equivalent in terms of introducing bias into the relative risk estimates [146].

Strengths of the FFQ approach are that it is inexpensive to administer and process and aims to estimate the respondent's usual intake of foods over an extended period of time. Unlike other methods, the FFQ can be used to circumvent recent changes in diet (e.g., changes resulting from disease) by obtaining information about individuals' diets as recalled about a prior time period. Retrospective reports about diet nearly always use a food frequency approach. Food frequency responses are used to rank individuals according to their usual consumption of nutrients, foods, or groups of foods. Nearly all food frequency instruments are designed to be self-administered, require 30 to 60 minutes to complete depending on the instrument and the respondent, and are either optically scanned paper versions or automated to be administered electronically [87, 98, 113, 147, 148]. Because the costs of data collection and processing and the respondent burden are typically much lower for FFQs than for multiple diet records or recalls, FFQs have become a common way to estimate usual dietary intake in large epidemiological studies.

The major limitation of the food frequency method is that it contains a substantial amount of measurement error [72, 109]. Many details of dietary intake are not measured, and the quantification of intake is not as accurate as with recalls or records. Inaccuracies result from an incomplete listing of all possible foods and from errors in frequency and usual serving size estimations. The estimation tasks required for an FFQ are complex and difficult [149]. As a result, the scale for nutrient intake estimates from an FFQ may be shifted considerably, yielding inaccurate estimates of the average intake for the group. Research suggests that longer food frequency lists may overestimate, whereas shorter lists may underestimate intake of fruits and vegetables [150], but it is unclear as to whether or how this applies to nutrients and other food groups.

The serving size of foods consumed is difficult for respondents to evaluate and is thus problematic for all assessment instruments (see Section V.A). However, the inaccuracies involved in respondents attempting to estimate usual serving size in FFQs may be even greater because a respondent is asked to estimate an average for foods that may have highly variable portion sizes across eating occasions [151].

Because of the error inherent in the food frequency approach, it is generally considered inappropriate to use FFQ data to estimate quantitative parameters, such as the mean and variance, of a population's usual dietary intake [101, 152–156]. Although some FFQs seem to produce estimates of population average intakes that are reasonable [152], different FFQs will perform in often unpredictable ways in different populations, so the levels of nutrient intakes estimated by FFQs should best be regarded as only approximations [153]. FFQs are generally used for ranking subjects according to food or nutrient intake rather than for estimating absolute levels of intake, and they are used widely in case-control or cohort studies to assess the association between dietary intake and disease risk [157–159]. For estimating relative risks, the degree of misclassification of subjects is more important than is the quantitative scale on which the ranking is made [160].

The definitive validity study for a food frequency–based estimate of long-term usual diet would require nonintrusive observation of the respondent's total diet over a long time. No such studies have ever been done. One early feeding study, with three defined 6-week feeding cycles (in which all intakes were known), showed some significant differences in known absolute nutrient intakes as compared to the Willett FFQ for several fat components, mostly in the direction of underestimation by the FFQ [161]. The most practical approach to examining the concordance of food frequency responses and usual diet is to use multiple food recalls or records over a period as an indicator of usual diet. This approach has been used in many studies examining various FFQs (see [162] for register of such studies). In these studies, the correlations between the methods for most foods and nutrients are in the range of 0.4 to 0.7. However, recalls and records cannot be considered as accurate reference instruments as they themselves suffer from error that may be correlated with error in the FFQ and, in addition, may not represent the time period of interest. Biomarkers that do represent usual intake without bias are available for energy (doubly-labeled water) [163] and protein (urinary nitrogen) [164]. Validation studies of various FFQs using these biomarkers have found large underestimates of self-reported energy intake [25, 31, 34, 69, 71, 72] and some underestimation of protein intake [29, 30, 69, 72, 126, 165–168]. Correlations of FFQs and the biomarkers have ranged from 0.1 to 0.5 for energy [25, 69, 72] and from 0.2 to 0.7 for protein [29, 30, 69, 72, 126, 165–168]. One study showed that protein density (kcal of protein as a percentage of total kcal) was less problematic—a slight overestimation among women and similar estimates among men [72], and correlations of 0.3 to 0.4 [109]—indicating that energy adjustment may alleviate some of the error inherent in food frequency instruments. Various statistical methods employing measurement error models and energy adjustment are used not only to assess the validity of FFQs but also to adjust estimates of relative risks for disease outcomes [169–179]. However, analyses indicate that correlations between an FFQ and a reference instrument, such as the 24-hour recall, may be overestimated because of correlated errors [109]. Furthermore, other analyses comparing relative risk estimation from FFQs to dietary records [180, 181] in prospective cohort studies indicate that observed relationships using an FFQ are severely attenuated, thereby obscuring associations that might exist. Accordingly, some epidemiologists have suggested that the error in FFQs is a serious enough problem that alternative means (such as food records) of collecting dietary data in large-scale prospective studies be considered [182, 183].

D. Brief Dietary Assessment Instruments

Many brief dietary assessment instruments have been developed. These instruments can be useful in situations that do not require either assessment of the total diet or quantitative accuracy in dietary estimates. For example, a brief diet assessment of some specific components might be used to triage large numbers of individuals into groups to allow more focused attention on those at greatest need for intervention or education. Measurement of dietary intake, no matter how crude, can also activate interest in the respondent to facilitate nutrition education. These brief instruments may, therefore, have utility in clinical settings or in situations where health promotion and health education are the goals. In the intervention setting, brief instruments focusing on specific aspects of a dietary intervention have also been used to track changes in diet, although there is concern that responses to questions of intake that directly evolve from intervention messages may be biased [184] and that these instruments lack sensitivity to detect change [185]. Brief instruments of specific dietary components such as fruits and vegetables are used often for population surveillance at the state or local level, for example in the Centers for Disease Control and Prevention's (CDC) Behavioral Risk Factor Surveillance System [186] and the California Health Interview Survey [187] (see Section III.A). Brief instruments can be used to examine relationships between some specific aspects of diet and other exposures, as in the National Health Interview Survey [188]. Finally, some groups use short screeners to evaluate the effectiveness of policy initiatives [187].

Brief instruments can be simplified/targeted FFQs or questionnaires that focus on specific eating behaviors other than the frequency of consuming specific foods. Complete FFQs typically contain 100 or more food items to capture the range of foods contributing to the many nutrients in the diet. If an investigator is interested only in estimating the intake of a single nutrient or food group, however, then fewer foods need to be assessed. Often, only 15 to 30 foods might be required to account for most of the intake of a particular nutrient [189, 190].

Numerous short questionnaires using a food frequency approach have been developed and compared with multiple days of food records, dietary recalls, complete FFQs, or biological indicators of diet. Single-exposure abbreviated FFQs have been developed and tested for protein [191], calcium [90, 192–194], iron [195], isoflavones [196], phytoestrogens [197], soy foods [196, 198], folate [199–201], sugar snacks [202], heterocyclic aromatic amines [203], and alcohol [204, 205]. Much of the focus in brief instrument development has been on fruits and vegetables and fats.

Food frequency type instruments to measure fruit and vegetable consumption range from a single overall question to 45 or more individual questions [206–210]. An early seven-item tool developed by the NCI and private grantees for the NCI's 5 A Day for Better Health Program effort has been used widely in the United States [211–213]. The tool is similar to one used in CDC's Behavioral Risk Factor Surveillance System (BRFSS) [186, 214, 215]. Validation studies of the CDC and 5 A Day brief instruments to assess fruit and vegetable intake have suggested that, without portion size adjustments, they often underestimate actual intake [206, 215–217] (see also [211]). Using cognitive interviewing findings (see Section V.F), NCI has revised the tool, including adding portion size questions. Using the revised tool, some studies indicate improved performance [218] and utility in surveillance studies. However, its performance in community interventions was mixed; in six of eight site/gender comparisons, fruit and vegetable consumption was significantly overestimated relative to multiple 24-hour recalls [219]. More important, the screener indicated change in consumption in both men and women when none was seen with the 24-hour recalls [220].

A fat screener, originally developed by Block [221] and currently composed of 17 items [87], was designed to account for most of the intake of fat using information about sources of fat intake in the U.S. population. The fat screener was useful as an initial screen for high fat intake in the Women's Health Trial [221] and in the CDC's Behavioral Risk Factor and Surveillance System for nutritional surveillance [222]. However, the screener did not perform well among young, low-income Hispanic women [222], and it substantially underestimated percentage energy from fat and was only modestly correlated ($r = 0.36$) with multiple 24-hour recalls in a sample of medical students [223]. In samples of men participating in intervention trials, the screener was not as precise [185] or as sensitive [224] as complete FFQs, possibly because the foods were not selected to preserve between-person variability [185]. The MEDFICTS (meats, eggs, dairy, fried foods, fat in baked goods, convenience foods, fats added at the table, and snacks) questionnaire, initially developed to assess adherence to low-fat (≤30% energy from fat) diets [225], asks about frequency of intake and portion size of 20 individual foods, major food sources of fat and saturated fat in the U.S. diet. Its initial evaluation showed high correlations with food records [225]. In additional cross-sectional studies, the MEDFICTS underestimated percentage energy from fat; it was effective in identifying individuals with very high fat intakes, but it was not effective in identifying individuals with moderately high fat diets [226] or correctly identifying those individuals consuming low-fat diets [227]. In a longitudinal setting, positive changes in the MEDFICTS score have been correlated with improvements in serum lipids and waist circumference among cardiac rehabilitation patients [228]. Other fat screeners have been developed to preserve the between-person variability of intake [229–231]. A 20-item screener developed and tested in the German site of the EPIC study correlated with a complete FFQ [229, 230]. A 16-item percentage energy from fat screener correlated 0.6 with 24-hour recalls in an older U.S. population [231]; however, its performance in intervention studies was variable [232].

Often, interventions are designed to target specific food preparation or consumption behaviors rather than frequency of consuming specific foods. Examples of such behaviors might be trimming the fat from red meats, removing the skin from chicken, or choosing low-fat dairy products. Many questionnaires have been developed in various populations to measure these types of dietary behaviors [222, 233–239], and many have been found to correlate with fat intake estimated from other more detailed dietary instruments [240, 241] or with blood lipids [237, 242, 243]. In addition, some studies have found that changes in dietary behavior scores have correlated with changes in blood lipids [242, 244, 245]. The Kristal Food Habits Questionnaire, sometimes also called the Eating Behaviors Questionnaire, was originally developed in 1990 [246]. It measures five dimensions of fat-related behavior: avoid fat as a spread or flavoring, substitute low-fat foods, modify meats, replace high-fat foods with fruits and vegetables, and replace high-fat foods with lower-fat alternatives. The instrument has been updated and modified for use in different settings and populations [243, 247, 248]. A modification tested in African-American adolescent girls was correlated with multiple 24-hour dietary recalls [249]. In another modification developed for African-American women [250], a subset of 30 items from the *SisterTalk* Food Habits Questionnaire correlated with change in BMI as strongly as did the original 91 items [251].

Recognizing the utility of assessing a few dimensions of diet simultaneously, several multifactor short instruments have been developed and evaluated, many combining fruits and vegetables with fiber or fat components [252–257]. Others assess additional components of the diet. For example, Prime-Screen is composed of 18 FFQ items asking about the respondents' consumption of fruits and vegetables, whole and low-fat dairy products, whole grains, fish and red meat, and sources of saturated and trans fatty acids (and seven supplement questions); the average correlation with nutrient estimates from a full FFQ was 0.6 [258]. The 5-Factor Screener used in the 2005 National Health Interview Survey Cancer Control Supplement assessed fruits and vegetables, fiber, added sugar, calcium, and dairy servings [259], and the dietary screener used in the 2005 California Health Interview Survey assessed fruits and vegetables and added sugar [260].

Some multicomponent behavioral questionnaires have also been developed. The Kristal Food Habits Questionnaire was expanded to not only measure the five fat factors (described above) but also to measure three factors related to fiber: consumption of cereals and grains, consumption of fruits and vegetables, and substitution of high-fiber for low-fiber foods [261]. This fat- and fiber-related eating behavior questionnaire correlated with food frequency measures of fat and fiber among participants from a health maintenance organization in Seattle, Washington [261]. Schlundt *et al.* have developed a 51-item Eating Behavior Patterns Questionnaire targeted at fat and fiber assessment in African-American women [262]. Newly incorporated in this questionnaire are questions to reflect emotional eating and impulsive snacking.

Some instruments combine aspects of food frequency and behavioral questions to assess multiple dietary patterns. For example, the Rapid Eating and Activity Assessment for Patients (REAP), composed of 27 items assessing consumption of whole grains, calcium-rich foods, fruits and vegetables, fats, sugary beverages and foods, sodium, and alcohol, correlated moderately with records and an FFQ [263].

Because the cognitive processes for answering food frequency–type questions can be complex, some attempts have been made to reduce respondent burden by asking questions that require only yes or no answers. Kristal *et al.* developed a questionnaire to assess total fat, saturated fat, fiber, and percent energy from fat; the questionnaire contains 44 food items for which respondents are asked whether they eat the items at a specified frequency. A simple index based on the number of yes responses was found to correlate well with diet as measured by 4-day records and with an FFQ assessing total diet [264]. This same yes-no approach to questioning for a food list has also been used as a modification of the 24-hour recall [265]. These "targeted" 24-hour recall instruments aim to assess particular foods, not the whole diet [266–269]. They present a precoded close-ended food list and ask whether the respondent ate each food on the previous day; portion size questions may also be asked.

The brevity of these instruments and their correspondence with dietary intake as estimated by more extensive methods create a seductive option for investigators who would like to measure dietary intake at a low cost. Although brief instruments have many applications, they have several limitations. First, they do not capture information about the entire diet. Most measures are not quantitatively meaningful and, therefore, estimates of dietary intake for the population usually cannot be made. Even when the goal is to estimate total intake, the estimates are not precise and have large measurement error. Finally, the specific dietary behaviors found to correlate with dietary intake in a particular population may not correlate similarly in another population, or even in the same population in another time period. For example, behavioral questionnaires developed and tested in middle-class, middle-aged U.S. women [246] were found to perform very differently when applied to Canadian male manual laborers [270]; to a low-income, low-education adult Canadian population [271]; and to participants in a worksite intervention program in Nevada [272]. Investigators should carefully consider the needs of their study and their target population's dietary patterns before choosing an off-the-shelf instrument designed to briefly measure either food frequency or specific dietary behaviors.

E. Diet History

The term "diet history" is used in many ways. In the most general sense, a dietary history is any dietary assessment that asks the respondent to report about past diet. Originally, as coined by Burke, the term "dietary history" referred to the collection of information not only about the frequency of intake of various foods but also about the typical makeup of meals [273, 274]. Many now imprecisely use the term "dietary history" to refer to the food frequency method of dietary assessment. However, several investigators have developed diet history instruments that provide information about usual food intake patterns beyond simply food frequency data [275–278]. Some of these instruments characterize foods in more detail than is allowed in food frequency lists (e.g., preparation methods and foods eaten in combination), and some of these instruments ask about foods consumed at every meal [277, 279]. The term "diet history" is therefore probably best reserved for dietary assessment methods that are designed to ascertain a person's usual food intake in which many details about characteristics of foods as usually consumed are assessed in addition to the frequency and amount of food intake.

The Burke diet history included three elements: a detailed interview about usual pattern of eating, a food list asking for amount and frequency usually eaten, and a 3-day diet record [273, 274]. The detailed interview (which sometimes includes a 24-hour recall) is the central feature of the Burke dietary history, with the food frequency checklist and the 3-day diet record used as cross-checks of the history. The original Burke diet history has not often been exactly reproduced, because of the effort and expertise involved in capturing and coding the information if it is collected by an interviewer. However, many variations of the Burke method have been developed and used in a variety of settings [275–278, 280–283]. These variations attempt to ascertain the usual eating patterns for an extended period of time, including type, frequency, and amount of foods consumed; many include a cross-check feature [284, 285]. Some diet history instruments have been automated and adapted for self-administration, thus eliminating the need for an interviewer to ask the questions [277, 286]. Other diet histories have been automated but still continue to be administered by an interviewer [287]. Short-term recalls or records are often used for validation or calibration rather than as a part of the tool.

The major strength of the diet history method is its assessment of meal patterns and details of food intake rather than intakes for a short period of time (as in records or recalls) or only frequency of food consumption. Details of the means of preparation of foods can be helpful in better characterizing nutrient intake (e.g., frying vs. baking), as well as exposure to other factors in foods (e.g., charcoal broiling). When the information is collected separately for each meal, analyses of the joint effects of foods eaten together are possible (e.g., effects on iron absorption of concurrent intake of tea or foods containing vitamin C). Although a meal-based approach often requires more time from the respondent than does a food-based approach, it may provide more cognitive support for the recall process. For example, the respondent may be better able to report total bread consumption by reporting bread as consumed at each meal.

A weakness of the approach is that respondents are asked to make many judgments both about the usual foods and the amounts of those foods eaten. These subjective tasks may be difficult for many respondents. Burke cautioned that nutrient intakes estimated from these data should be interpreted as relative rather than absolute. All of these limitations are also shared with the food frequency method. The meal-based approach is not useful for individuals who have no particular eating pattern and may be of limited use for individuals who "graze" (i.e., eat small bits throughout the day, rather than eat at defined meals). The approach, when conducted by interviewers, requires trained nutrition professionals.

The validity of diet history approaches is difficult to assess because we lack independent knowledge of the individual's usual long-term intake. Nutrient estimates from diet histories have often been found to be higher than nutrient estimates from tools that measure intakes over short periods, such as recalls or records [288–293]. However, results for these types of comparisons depend on both the approach used and study characteristics. Validation studies that estimate correlations between reference data from recalls, records, or observations and diet histories are limited and show correlations in ranges similar to those for FFQs [278, 292, 294]. There are few validations of diet history questionnaires using biological markers as a basis of comparison. One study showed that, on average, 12 adults completing a diet history underreported by 12% in comparison to energy expenditure (measured by doubly-labeled water) [295]; another showed that, in comparison to protein intake as measured by urinary nitrogen (UN), 64 respondents completing a diet history questionnaire underreported by 3% [296].

F. Blended Instruments

Better understanding of various instruments' strengths and weaknesses has led to creative blending of approaches, in hopes of maximizing the strengths of each instrument. For example, a record-assisted 24-hour dietary recall has been used in several studies with children [297, 298]. The child keeps notes of what she or he has eaten and then uses these notes as memory prompts in a later 24-hour dietary recall. Several researchers have combined elements of a 24-hour dietary recall and FFQ, often to assess specific dietary components. For example, in the assessment of fruits and vegetables, a limited set of questions is asked about yesterday's intake and is combined with usual frequency of consumption of common fruits and vegetables [299, 300]. Similarly, the Nutritionist Five Collection Form combines a 2-day dietary recall with food frequency questions [301]. Thompson *et al.* have combined information from a series of daily checklists (i.e., precoded records) with frequency reports from a food frequency instrument to form checklist-adjusted estimates of intake. In a validation study of this approach, validity improved for energy and protein, but it was unchanged for protein density [302].

One advance is the development of statistical methods that seek to better estimate usual intake of episodically consumed foods. A two-part statistical model uses information from two or more 24-hour recalls, allowing for the inclusion of daily frequency estimates derived from an FFQ, as well as any other potentially contributing characteristic (such as age, race/ethnicity), as covariates [303]. Frequency information contributes to the model by providing additional information about an individual's propensity to consume a food—information not available from only a few recalls. The recalls, however, provide information about the nature and amount of the food consumed. Such methods are used to better measure usual intakes (see Section V.C).

The development of these hybrid instruments in addition to developing new analytical techniques combining information from different assessment methods may hold great promise for furthering our ability to accurately assess diets. Table 1 summarizes the information related to the dietary assessment methods outlined in this section.

III. DIETARY ASSESSMENT IN SPECIFIC SITUATIONS

The choice of the most appropriate dietary assessment method for a specific research question requires careful consideration. The primary research question must be clearly formed and questions of secondary interest should be recognized as such. Projects can fail to achieve their primary goal because of too much attention to secondary goals. The choice of the most appropriate dietary assessment tool depends on many factors. Questions that must be answered in evaluating which dietary assessment tool is most appropriate for a particular research need include [158] the following: (1) Is information needed about foods, nutrients, other food constituents or specific dietary behaviors? (2) Is the average intake of a group or the intake of each individual needed? (3) Is absolute or relative intake needed? (4) What level of accuracy is needed? (5) What time period is of interest? (6) What are the research constraints in terms of money, time, staff, and respondent characteristics?

A. Cross-Sectional Surveys

One of the most common types of population-level studies is the simple cross-sectional survey, a set of measurements of a population at a particular point in time. When measurements are collected on a cross-section at two or more times, the data can be used for purposes of monitoring or surveillance. Data collected in cross-sectional surveys are used at the national, state, and local levels as the bases for assessment of risk of deficiency, toxicity, overconsumption, adherence to dietary guidelines and public health programs, and food and nutrition policy. These cross-sectional surveys assess food and nutrient intakes of individuals as well as

TABLE 1 Advantages and Disadvantages of Dietary Assessment Instruments

Instrument	Advantages	Disadvantages
Food record	1. Intake quantified 2. Could enhance self-monitoring for weight control or other behavior change 3. Does not require recall of foods eaten	1. High investigator cost 2. High respondent burden 3. Extensive respondent training and motivation required 4. Many days needed to capture individual's usual intake 5. Affects eating behavior 6. Intake often underreported 7. Reports of intake decrease with time 8. Attrition increases with number of daily records requested 9. May lead to nonrepresentative sample and subsequent nonresponse bias
24-hour dietary recall	1. Intake quantified 2. Appropriate for most populations, thus less potential for nonresponse bias 3. Relatively low respondent burden 4. Does not affect eating behavior	1. High investigator cost 2. Many days needed to capture individual's usual intake 3. Intake often underreported
Food frequency questionnaire	1. Usual individual intake asked 2. Information on total diet obtained 3. Low investigator cost 4. Does not affect eating behavior	1. Not quantifiably precise 2. Difficult cognitive task for respondent 3. Intake often misreported
Brief instruments	1. Usual individual intake often asked 2. Low investigator cost 3. Low respondent burden 4. Does not affect eating behavior	1. Not quantifiably precise 2. Difficult cognitive task for respondent 3. Assessment limited to small number of nutrients/ foods 4. Intake often misreported
Diet history	1. Usual individual intake asked 2. Information on total diet obtained 3. Information often available on foods consumed by meal 4. Can have low investigator cost 5. Does not affect eating behavior	1. Not quantifiably precise 2. Difficult cognitive task for respondent 3. Intake often misreported 4. Can have high investigator burden

intake distributions of populations (see Section V.C) at one point in time. To assess trends in intakes over time, it would be ideal for the data collection methods, sampling procedures, and food composition databases used in dietary surveillance to be similar from survey to survey. As a practical matter, however, this is difficult. The dietary assessment method used consistently over the years in national dietary surveillance is the interviewer-administered 24-hour dietary recall. However, recall methodology has changed over time based on cognitive research, the addition of multiple interviewing passes, standardization of probes, and automation of the interview [304]. Another issue that affects the assessment of trends over time is changes in the nutrient/food grouping databases and specification of default foods. Changes in the food supply are reflected in additions or subtractions to food databases, whereas changes in consumption trends lead to subsequent reassignment of default choices for some foods (e.g., type of milk or fat addition). Food composition databases, too, are modified over time because of true changes in food composition or improved analytic methods for particular nutrients. More recently, databases have included values for food groups. The first of these for a public use dataset was the Pyramid Servings Database, produced by the USDA [305]. This database is available in national dietary surveys conducted in 1994–1996, 1998, and 1999–2002 and translates quantities of all foods (disaggregating ingredients of mixed dishes at the commodity level) into servings of food groups consistent with the 2000 Dietary Guidelines for Americans [306]. This has recently been updated to the MyPyramid Equivalents Database [305] modified to reflect the recommendations in the 2005 Dietary Guidelines [307].

In the past, there were two major cross-sectional surveillance surveys, the NHANES and the Continuing Survey of Food Intakes by Individuals (CSFII), conducted by the National Center for Health Statistics (NCHS) and the USDA, respectively [308–327]. Starting in 1999, these two surveys were merged into a single national dietary surveillance survey called What We Eat in America, NHANES [51]. The 24-hour dietary recalls are collected using USDA's AMPM and the data are analyzed and processed by the USDA. The 24-hour recalls in NHANES also query the intake of dietary supplements. In the NHANES 2003–2004 and 2005–2006, two 24-hour dietary recalls were conducted along with an extensive FFQ without portion size (called the Food Propensity Questionnaire or FPQ) [328]. Frequency data from the FPQ are intended to be used as covariates in a new model to assess usual dietary intakes [303] (see Section V.C). Information about the NHANES surveys is available on both the USDA and NCHS websites [329, 330].

The type of information required for a surveillance or monitoring system can vary. For some purposes, quantitative estimates of intake are needed, whereas for other purposes, only qualitative estimates of intake, like food frequency or behavioral indicators, are needed. There is a particular need to monitor dietary trends at the local level. To help provide local data, brief FFQs to assess fruit and vegetable intake developed by the CDC for telephone administration have been used periodically within BRFSS [186]. The California Department of Health, in its California Dietary Practices Survey, has assessed dietary practices since 1989 [331]. The California Health Interview Survey has assessed fruit and vegetable intake in 2001 and 2005 [187].

B. Case-Control Studies

A case-control study design classifies individuals with regard to disease status currently (as cases or controls) and relates this to past (retrospective) exposures. For dietary exposure, the period of interest could be either the recent past (e.g., the year before diagnosis) or the distant past (e.g., 10 years ago or in childhood). Because of the need for information about diet before the onset of disease, dietary assessment methods that focus on current behavior, such as the 24-hour recall, are not useful in retrospective studies. The food frequency and diet history methods are well suited for assessing past diet and are therefore the only good choices for case-control studies.

In any food frequency or diet history interview, the respondent is not asked to call up specific memories of each eating occasion but to respond on the basis of general perceptions of how frequently he or she ate a food. In case-control studies, the relevant period is often the year before diagnosis of disease or onset of symptoms, or at particular life stages, such as adolescence and childhood. Thus, in assessing past diet, an additional requirement is to orient the respondent to the appropriate period.

The validity of recalled diet from the distant past is difficult to assess, because definitive recovery biomarker information (e.g., DLW or UN), is not available for large samples from long ago. Instead, relative validity and long-term reproducibility of various FFQs have been assessed in various populations by asking participants from past dietary studies to recall their diets from that earlier time [332, 333]. These studies have found that correlations between past and current reports about the past vary by nutrient and by food group [85], with higher correspondence for very frequently consumed and rarely consumed foods compared to that for moderately consumed foods [334]. Correspondence of retrospective diet reports with the diet as measured in the original study has usually been greater than correspondence with diet reported by subjects for the current (later) period. This observation implies that if diet from years in the past is of interest, it is usually preferable to ask respondents to recall it than to simply consider current diet as a proxy for past diet. The current diets of respondents may affect their retrospective reports about past diets. In particular, retrospective diet reports from seriously ill individuals may be biased by recent dietary changes [332, 335]. Studies of groups in whom diet was previously measured indicate no consistent differences in the accuracy of retrospective reporting between those who recently became ill and others [336, 337].

C. Prospective (Cohort) Studies

In the prospective study design, exposures of interest are assessed at baseline in a group (cohort) of people and disease outcomes occurring over time (prospectively) are then related to the baseline exposure levels. In prospective dietary studies, dietary status at baseline is measured and related to later incidence of disease. In studies of many chronic diseases, large numbers of individuals need to be followed for years before enough new cases with that disease accrue for statistical analyses. A broad assessment of diet is usually desirable in prospective studies, as many dietary exposures and many disease end points will ultimately be investigated and areas of interest may not even be recognized at the beginning of a prospective study.

To relate diet at baseline to the eventual occurrence of disease, a measure of the usual intake of foods by study subjects is needed. Although a single 24-hour recall or a food record for a single day would not adequately characterize the usual diet of study subjects in a cohort study, such information could be later analyzed at the group level for contrasting the average dietary intakes of subsequent cases with those who did not acquire the disease. Multiple dietary recalls, records, diet histories, and food frequency methods have all been used effectively in prospective studies. Cost and logistic issues tend to favor food frequency methods, as many prospective studies require thousands of respondents. However, because of concern about significant measurement error and attenuation attributed to the FFQ [180, 181, 183, 338, 339], other approaches are being sought. One potential approach is the use of automated self-administered 24-hour recall instruments (see Section II.B). Another approach is collection of multiple days of food records at baseline, with later coding and analysis of records for those respondents selected for analysis, using a nested case-control design.

Even in large studies using FFQs, it is desirable to include multiple recalls or records in subsamples of the population (preferably before beginning the study) to construct or modify the food frequency instrument and to calibrate it (see Section V.G). Information on the foods consumed could be used to ensure that the questionnaire includes the major food sources of key nutrients, with reasonable portion sizes. Because the diets of individuals change over time, it is desirable to measure diet throughout the follow-up period rather than just at baseline. One study revealed that data from annual administrations of FFQs showed only small dietary changes over time and that repeat administrations more than 5 years apart would be acceptable to assess dietary change over time [340]. If diet is measured repeatedly over the years, repeated calibration is also desirable. Information from calibration studies can be used for three purposes: to give design information (e.g., the sample size needed [341]), to show how values from the food frequency tool (or a brief food list thus derived) relate to values from the recalls/records [156, 160], and to determine the degree of attenuation/measurement error in the estimates of association observed in the study (e.g., between diet and disease) [172, 173, 175, 177, 179, 342–344] (see Section V.G).

D. Intervention Studies

Intervention studies range from relatively small, highly controlled clinical studies of targeted participants to large trials of population groups. Intervention studies may use dietary assessment for two purposes: initial screening for inclusion (or exclusion) into the study and measurement of dietary changes resulting from an intervention. Not all intervention trials require initial screening. For those that do, screening can be performed using either detailed instruments or less burdensome instruments. For example, food frequency instruments have been used in the Women's Health Trial [221] and in the Women's Health Initiative Dietary Modification Trial [345] to identify groups with high fat intake.

Measurement of the effects of a dietary intervention requires a valid measure of diet before, during, and after the intervention period. In small, intense clinical studies, the expected intervention effect is usually relatively large, whereas in larger community dietary intervention trials, the expected intervention effect may be relatively small and thus difficult to detect. Some work has been done to examine the validity of methods to measure dietary change in individuals or in populations [248, 346–348]. In relatively small studies, dietary records, multiple 24-hour recalls, and diet history questionnaires have been used. Large intervention studies, because of resource constraints, usually rely on less precise measures of diet, including FFQs and brief instruments. However, these resource constraints may become less important as automated self-administered instruments become available. Measurement of specific dietary behaviors in addition to, or even in place of, dietary intake could be considered in intervention evaluations when the nature of the intervention involves education about specific behaviors. If, for instance, a community-wide campaign to choose low-fat dairy products were to be evaluated, food selection and shopping behaviors specific to choosing those items could be measured.

Intentional behavior change is a complex and sequential phenomenon, as has been shown for tobacco cessation [349]. A complex sequence of events may also lead to dietary change [350]. The effects of educational interventions might also be assessed by measuring knowledge, attitudes, beliefs, barriers, and perceptions of readiness for dietary change, although the reliability of these types of questions has not been well assessed.

Whether an intervention is targeting individuals or the entire population, repeated measures of diet among study subjects can reflect reporting bias in the direction of the change being promoted. Even though not intending to be deceptive, some respondents may tend to report what they think investigators want to hear. Social desirability bias and social approval bias can be measured and the resulting scales incorporated into intervention analyses. Behavioral questions and the food frequency method, because of their greater subjectivity, may be more susceptible to social desirability biases than is the 24-hour recall method [49, 184]. On the other hand, greater awareness of diet because of the intervention may enhance accuracy of report. Nonetheless, because self-reports of diet are subject to bias in the context of an intervention study, an independent assessment of dietary change should be considered. For example, food availability or sales in worksite cafeterias, school cafeterias, or vending machines could be monitored. One such method useful in community-wide interventions is monitoring food sales [351]. Often, cooperation can be obtained from food retailers [352]. Because of the large number of food items, only a small number should be monitored, and the large effects on sales of day-to-day pricing fluctuations should be carefully considered. Another method to consider is measuring changes in biomarkers of diet, for example, serum carotenoids [353] or serum cholesterol [354], in the population. Consistency of changes in self-reported diet and appropriate biomarkers provides further evidence for real changes in the diet. See Chapter 11 for a more in-depth

TABLE 2 Dietary Assessment in Different Study Situations

Study Situation	Methods Commonly Used
Cross-sectional/ surveillance	24-hour recall; FFQ; brief instruments
Case-control (retrospective)	FFQ; diet history
Cohort (prospective)	FFQ; diet history; 24-hour recall; record
Intervention	FFQ; brief instruments; 24-hour recall

discussion of evaluation of nutrition interventions and Chapter 12 for the use of biomarkers in intervention studies. A quick guide summarizing preferred dietary assessment methods based on study design is shown in Table 2.

IV. DIETARY ASSESSMENT IN SPECIAL POPULATIONS

A. Respondents Unable to Self-Report

In many situations, respondents are unavailable or unable to report about their diets. For example, in case-control studies, surrogate reports may be obtained for cases who have died or who are too ill to interview. Although the accuracy of surrogate reports has not been examined, comparability of reports by surrogates and subjects has been studied in hopes that surrogate information might be used interchangeably with information provided by subjects [355]. Common sense indicates that individuals who know most about a subject's lifestyle would make the best surrogate reporters. Adult siblings provide the best information about a subject's early life, and spouses or children provide the best information about a subject's adult life. When food frequency instruments are used, the level of agreement between subject and surrogate reports of diet varies with the food and possibly with other variables such as number of shared meals, interview situation, case status, and sex of the surrogate reporter. Mean frequencies of use computed for individual foods and food groups between surrogate reporters and subject reporters tend to be similar [356–358], but agreement is much lower when detailed categories of frequency are compared. Several studies have shown that agreement is better for alcoholic beverages, coffee, and tea than for other foods.

Although subjects reporting themselves in the extremes of the distribution are seldom reported by their surrogates in the opposite extreme, many subjects who report they are in an extreme are reported by their surrogates in the middle of the distribution [359]. This may limit the usefulness of surrogate information for analyses at the individual level that rely on proper ranking. Furthermore, there may be a substantial difference in the quality of surrogate reports between spouses of deceased subjects and spouses of surviving subjects [360]. Thus far, however, there is little evidence that dietary intakes are systematically overreported or underreported depending on case status of the subject [361–363]. Nonetheless, use of surrogate respondents should be minimized for obtaining dietary information in analytical studies. When used, analyses excluding the surrogate reports should be done to examine the sensitivity of the reported associations to possible errors or biases in the surrogate reports. If planning a study using surrogate reports, sample size should be inflated to account for the higher frequency of missing data, inability to recruit surrogates for some number of cases, and reduced precision of dietary estimates.

B. Ethnic Populations

Special modifications are needed in the content of dietary assessment methods when the study population is composed of individuals whose cuisine or cooking practices are not mainstream [364]. If the method requires an interview, interviewers of the same ethnic or cultural background are preferable so that dietary information can be more effectively communicated. If dietary information is to be quantified into nutrient estimates, examination of the nutrient composition database is necessary to ascertain whether ethnic foods are included and whether those foods and their various preparation methods represent those consumed by the target population [365]. It is also necessary to examine the recipes and assumptions underlying the nutrient composition of certain ethnic foods. Very different foods may be called the same name, or similar foods may be called by different names [366]. For these reasons, it may be necessary to obtain detailed recipe information for all ethnic mixtures reported.

To examine the suitability of the initial database, baseline recalls or records with accompanying interviews should be collected from individuals in the ethnic groups. These interviews should focus on all the kinds of food eaten and the ways in which foods are prepared in that culture. Recipes and alternative names of the same food should be collected, and interviewers should be familiarized with the results of these interviews. Recipes and food names that are relatively uniform should be included in the nutrient composition database. Even with these modifications, it may be preferable to collect detailed descriptions of ethnic foods reported rather than to directly code these foods using preselected lists most common in computer-assisted methods. This would prevent the detail of food choice and preparation from being lost by *a priori* coding.

Use of FFQs developed for the majority population may be suboptimal for many individuals with ethnic eating patterns. Many members of ethnic groups consume both foods common in the mainstream culture and foods that are specific to their own ethnic group. A food list can be developed either by modifying an existing food list based on expert judgment of the diet of the target population or, preferably, by examining the frequency of reported foods in the population from a set of dietary records or recalls. FFQs for specific groups including Navajos [367], Chinese Americans [368], individuals in Northern India [369], Hispanics [117, 370], and Israelis [371] have been developed using these approaches.

Besides the food list, however, there are other important issues to consider when adapting existing FFQs for use in other populations. The relative intake of different foods within a food group line item may differ, thus requiring a change in the nutrient database associated with each line item. For example, Latino populations may consume more tropical fruit nectars and less apple and grape juice than the general U.S. population and therefore would require a different nutrient composition standard for juices. In addition, the portion sizes generally used may differ [372]. For example, rice may be consumed in larger quantities in Latino and Asian populations; the amount attributed to a large portion for the general population may be substantially lower than the amount typically consumed by Latino and Asian populations. Adaptation of an existing FFQ considering all of these factors is illustrated for an elderly Puerto Rican population [373], for white and African-American adults in the Lower Mississippi Delta [374], and for the Hawaii–Los Angeles Multiethnic Cohort Study [120].

The performance of FFQs varies across ethnic groups [375]. Questionnaires aimed at allowing comparison of intakes across multiple cultures have been developed; however, studies done thus far indicate that there are validity differences among the various cultural groups [116, 120, 373, 376–378]. Understanding these validity differences is crucial to the appropriate interpretation of study results.

C. Children and Adolescents

Assessing the diets of children is considered to be even more challenging than assessing the diets of adults. Children tend to have diets that are highly variable from day-to-day, and their food habits can change rapidly. Younger children are less able to recall, estimate, and cooperate in usual dietary assessment procedures, so much information by necessity has to be obtained by surrogate reporters. Adolescents, although more able to report, may be less interested in giving accurate reports. Baranowski and Domel have posited a cognitive model of how children report dietary information [379].

The literature about dietary assessment in children and adolescents has been reviewed [380–386]. The 24-hour dietary recall, food records (including precoded checklists [15]), dietary histories, food frequency instruments, brief instruments [193, 387, 388], and blended instruments, such as a record-assisted 24-hour dietary recall [297], have all been used to assess children's intakes. The use of direct observation of children's diets has also been used extensively, often as a reference method to compare with self-reported instruments [389]. As predicted from the model posited earlier, children's estimates of portion size had large errors [390], and children were found to be less able than adults to estimate portion sizes [391] (also see Section V.A).

For preschool-aged children, information is obtained from surrogates, usually the primary caretaker(s), who may typically be a parent or an external caregiver. If information can be obtained only from one surrogate reporter, the reports are likely to be less complete. Even for periods when the caregiver and child are together, foods tend to be underestimated [392]. A "consensus" recall method, in which the child and parents report as a group on a 24-hour dietary recall, has been shown to give more accurate information than a recall from either parent alone [393]. Sobo and Rock [394] have described such interviews and suggested tips for interviewers to maximize data accuracy. For older children, a blended instrument, the record assisted 24-hour recall (in which the children record only the names of foods and beverages consumed throughout a 24-hour period, serving as a cue for the later 24-hour recall interview) has been developed and tested [297, 298]. However, for school-aged and adolescents, there is no consensus of which dietary assessment method is most accurate.

Adaptation of food frequency instruments originally developed for adults requires consideration of the instrument itself (food list, question wording and format, portion size categories) and the database for converting responses to nutrient intakes. Food frequency instruments have been especially developed and tested for use in child and adolescent populations [13, 395, 396]. Generally correlations between the criterion instrument and food frequency instruments have been lower in child and adolescent populations than in adult populations.

D. Elderly

Measuring diets among the elderly can, but does not necessarily, present special problems [397, 398]. Both recall and food frequency techniques are inappropriate if memory is impaired or if the use of medications impairs cognitive functioning, Similarly, self-administered tools may be inappropriate if physical disabilities including poor vision are present. Direct observation in institutional care facilities or shelf inventories for elders who live at home can be useful. Even when cognitive integrity is not impaired, other factors can affect the assessment of diet among the elderly. Because of the frequency of chronic illness in this age group, special diets (e.g., low sodium, low fat, high fiber) have often been recommended. Such recommendations could not only affect actual dietary intake but could also bias reporting, as individuals may report what they should

TABLE 3 Optimal Strategies for Special Populations

Special Population	Optimal Strategies
Respondents unable to self-report	Use best-informed surrogate. Analyze effect of potential bias on study results.
Ethnic populations	Use interviewers of same ethnic background. Use nutrient composition database reflective of foods consumed. For FFQs, use appropriate food list and nutrient composition database.
Children	For young children, use caretakers in conjunction with child. For older children and adolescents, blended instrument and other creative ways of engagement and motivation may work best.
Elderly	Assess any special considerations, including memory, special diets, dentition, use of supplements, and so on, and adapt methods accordingly.

eat rather than what they do eat. Alternatively, respondents on special diets may be more aware of their diets and may more accurately report them. When dentition is poor, the interviewer should probe for foods that are prepared or consumed in different ways. Elderly individuals frequently take multiple types of nutritional supplements, which present special problems in dietary assessment (see Chapter 2). Because of the concern of malnutrition among the elderly, specific instruments to detect risk of malnutrition, such as the Nutrition Screening Initiative [399] and the Mini Nutritional Assessment [400], have been developed.

Adaptations of standard dietary assessment methods have been suggested and evaluated, including the use of memory strategies, prior notification of a dietary interview [401], combining methods [300], adapting existing instruments [402], and developing new approaches. Research suggests that under many circumstances the validity of dietary information collected from the elderly is comparable to that collected from younger adults [403].

The principles discussed in this section are summarized in Table 3.

V. SELECTED ISSUES IN DIETARY ASSESSMENT METHODS

A. Estimation of Portion Size

Research has shown that untrained individuals have difficulty estimating portion sizes of foods, both when examining displayed foods and when reporting about foods previously consumed [391, 404–415], and that children are worse than adults [391]. Further, respondents appear to be relatively insensitive to changes made in portion size amounts shown in reference categories asked on FFQs [416]. Portion sizes of foods that are commonly bought or consumed in defined units (e.g., bread by the slice, pieces of fruit, beverages in cans or bottles) may be more easily reported than amorphous foods (e.g., steak, lettuce, pasta) or poured liquids. Other studies indicate that small portion sizes tend to be overestimated and large portion sizes underestimated [406, 417].

Aids are commonly used to help respondents estimate portion size. NHANES, What We Eat in America, uses an extensive set of three-dimensional models for an initial in-person 24-hour dietary recall; respondents are then given a Food Model Booklet developed by the USDA [53] along with a limited number of three-dimensional models (such as measuring cups and spoons) for recalls collected by telephone. Studies that have compared three-dimensional food models to two-dimensional photographs have shown that there is little difference in the reporting accuracy between methods [390, 415, 418, 419]. The accuracy of reporting using either models or household measures can be improved with training [420–423], but the effects deteriorate with time [424].

B. Mode of Administration

For interviewer-administered instruments, one way that the costs of collecting dietary information may be reduced is to administer the instrument by telephone. Costs can be further reduced by use of self-administered questionnaires by mail. Both telephone and mail surveys are less invasive than are face-to-face interviews. The use of telephone surveys to collect dietary information has been reviewed [425]. Telephone surveys have higher response rates than do mail surveys [426] and have been used in a variety of public health research settings [427]. However, there is increasing concern about response rates in telephone surveys given the prevalence of telemarketing and technology, which allows for the screening of calls. Nevertheless, interviews by telephone can be substantially less expensive than face-to-face interviews. The difficulty of reporting serving sizes by telephone can be eased by mailing picture booklets or other portion size estimation aids to the participants before the interview.

Many studies have evaluated the comparability of data from telephone versus in-person 24-hour dietary recall interviews. Several have found substantial but imperfect agreement between dietary data collected by telephone and that estimated by other methods, including face-to-face interviews [50, 428–433], expected intakes [434], or observed intakes [435]. Accuracy of portion size estimates for known quantities of foods consumed assessed by telephone and by in-person interviews was examined and found to be similarly accurate [415, 436]. One study found comparability of telephone and in-person interviews among urban African-American women [437]. However, some segments of the population do not have telephones, and some persons will not answer their telephones under certain circumstances. In addition, an increasing proportion of the population has only wireless phones; this is particularly prevalent among young adults [438]. Therefore, it is important to consider sampling schemes that account for these concerns so that potential respondents without landline telephones are included [427].

For FFQs or brief instruments, self-administration is less costly than interviewer-administration. However, self-administration may be unfeasible for large portions of the population who have low literacy levels or limited motivation. Thus, selection bias is a potential problem.

Self-administered tools can be completed on paper or electronically. Various FFQs [147] and dietary history questionnaires [439] are available for completion on the computer. Web-administered dietary assessment with wireless phone and PDA technology is also an area of great potential. As this mode of administration becomes more prevalent, it will be important to examine the comparability with in-person and telephone-administered modes, as well as the potential for selection effects. One study in Sweden found a lower initial response rate to a Web questionnaire compared to a mailed printed questionnaire, but there was a greater compliance in answering follow-up questions with Web respondents [440].

C. Estimating Usual Intakes of Nutrients and Foods

Knowledge of the usual intakes of foods and nutrients are of interest to assess both individual intakes and population distributions. To assess risk of chronic disease for epidemiological research or in clinical settings, it is individual level data that are of interest. To assess the proportion of the population at risk for deficiency or toxicity, however, usual intake distributions are necessary. In theory, usual intake is defined as the long-run average intake of a food or nutrient. A major problem is that true usual intake is not observable.

Data from FFQs, 24-hour recalls, and records have all been used as surrogates to estimate usual intake. Dietary recalls are most often used in surveillance, and FFQs, which ask about usual intake over a specified time period, are primarily used in epidemiological research investigating diet and disease relationships. FFQs, however, are limited in their ability to estimate usual intake well and are known to contain a substantial amount of measurement error [72, 109]. Recalls or records, which also contain error, focus on short time periods but provide richer detail about types of foods and amounts consumed. Importantly, intakes reported for only a few days do not capture day-to-day variation in intakes, an important factor when attempting to estimate usual intakes. Although usual intake estimations can be improved with additional days of data collection, averaging intakes across a few days does not adequately represent usual intakes [441]. Thus, more sophisticated methods based on statistical modeling have evolved [442]. Without such methods, as many as 7 to 14 days of data collection might be necessary for most nutrients and food groups [443], a number that is impractical in most large-scale nutrition research.

When only a few days are used to represent usual intake distributions, the lack of information regarding within-person variation is considerable, leading to biased estimates. Distributions generated from averaging only a few days of data are generally substantially wider than true usual intakes, thereby overestimating the proportion of the population above or below a certain cut point. Statistical modeling mitigates some of the limitation of having only a few days of intake by analytically estimating and removing the effects of within-person variation in dietary intake [441]. The earliest efforts at statistical modeling were developed by the Institute of Medicine [444] and then extended by researchers at Iowa State University [445, 446]. Both of these methods work best for estimating usual intakes of nutrients, most of which are consumed nearly every day by most everyone. For foods or food groups that are more episodically consumed (e.g., dark green vegetables), there is a revised version of the modeling by Iowa State University [447]. The NCI recently developed a new usual intake model that is an improvement over previous statistical methods [303]. This method uses two 24-hour recalls to estimate intake of both nutrients and episodically consumed foods. Unlike previous methods, this model allows for covariates such as sex, age, race, or information from an FFQ to supplement the model. The use of frequency information from an FFQ as a covariate in a statistical model is novel. Such data may add important information about an individual's propensity to consume a food—information that is often missing with just a few dietary recalls [328]. A frequency instrument can substantially improve the power to detect relationships between dietary intakes and other variables. The amount of improvement depends on the specific food or nutrient in question and the population under study. Extensive frequency data were collected in 2003–2004 and 2005–2006 for use in estimating usual intakes (see Section III.A).

D. Choice of Nutrient/Food Database

When dietary data are to be converted to nutrient intake data, it is necessary to use a nutrient composition database. Typically, such a database includes the description of the food, a food code, and the nutrient composition per 100 grams of the food. The number of foods and nutrients included varies with the database.

Some values in nutrient databases are obtained from laboratory analysis; however, because of the high cost of laboratory analyses, many values are estimated based on conversion factors or other knowledge about the food [448].

In addition, accepted analytical methods are not yet available for some nutrients of interest [449], analytical quality of the information varies with nutrient [449, 450], and the variances or ranges of nutrient composition of individual foods is in most cases unknown [451]. Rapid growth in the food processing sector and the global nature of the food supply add further challenges to estimating the mean and variability in the nutrient composition of foods.

One of the USDA's primary missions is to provide nutrient composition data for foods in the U.S. food supply, accounting for various types of preparation [452]. USDA produces and maintains the "Nutrient Database for Standard Reference," which includes information on up to 140 food components for 7293 foods. Values for individual carotenoids, selected *trans*-fatty acids, individual sugars, and vitamin K have been incorporated into Release 19 of the Nutrient Database for Standard Reference (NDSR) [453]. In addition, information on about 50 traditional or subsistence foods from the American Indian/Alaska Native Foods Database have been added to this release. Interest in nutrients and food components potentially associated with diseases has led to development of databases for limited numbers of foods. These include databases for flavonoids, proanthocyanidins, choline, and fluoride [454]. Information regarding USDA's nutrient composition databases is available at the USDA's Nutrient Data Laboratory home page [455]. Another database, the USDA Food and Nutrient Database for Dietary Studies, is used for the analysis of survey data in NHANES [63]. It includes information describing each of the approximately 7000 foods, food portions and weights, and nutrient information (derived from the NDSR). The International Network of Food Data Systems (INFOODS) maintains an international directory of nutrient composition data [456]. The recent compilation of some databases throughout the world is found on its website [457].

Research on nutrients (or other dietary constituents) and foods is ongoing. Constant interest in updating current values and providing new values for a variety of dietary constituents remains a priority for researchers. In addition, in the United States, methods that relate dietary intake to dietary guidance have been developed [458, 459]. The 2005 MyPyramid Food Guidance System is based on the 2005 Dietary Guidelines for Americans [307] and produces estimates of servings for 30 components of MyPyramid guidance (e.g., dairy, fruits, vegetables) [460]. Each food not only has a nutrient profile but also has a MyPyramid profile. One limitation in all nutrient databases is the variability in the nutrient content of foods within a food category [461, 462] and the volatility of nutrient composition in the manufactured foods; these limitations are particularly problematic for estimating fatty acids. Depending on the level of detail queried on the dietary assessment instrument, the respondent's knowledge of specific brand names, and the specificity of a particular nutrient database, estimating accurate fatty acid intake can be problematic. For FFQs, collapsing foods into categories that might have highly variable nutrient contents compounds this problem.

Many other databases are available in the United States for use in analyzing records and recalls, but most are based fundamentally on the USDA database, often with added foods and specific brand names. One prominent such database in the United States is the Nutrition Data System for Research (NDS-R) developed by the Nutrition Coordinating Center (NCC) housed at the University of Minnesota [463]. This database includes information on 144 food components for more than 18,000 foods including 8000 brand name products. The NCC is constantly updating its database to reflect values in the latest release of the NDSR.

Estimates of nutrient intake from dietary recalls and records are often affected by the nutrient composition database that is used to process the data [464–466]. Any differences may be due to the number of food items in the database, the recency of nutrient data, and the number of missing or imputed nutrient composition values. Therefore, before choosing a nutrient composition database, a prime factor to consider is the completeness and accuracy of the data for the nutrients of interest. For some purposes, it may be useful to choose a database in which each nutrient value for each food also contains a code for the quality of the data (e.g., analytical value, calculated value, imputed value, or missing). Investigators need to be aware that a value of zero is assigned to missing values in some databases. The nutrient database should also include weight/volume equivalency information for each food item. Many foods are reported in volumetric measures (e.g., 1 cup) and must be converted to weight in grams. The number of common mixtures (e.g., spaghetti with sauce) available in the database is another important factor. If the study requires precision of nutrient estimates, then procedures for calculating the nutrients in various mixtures must be developed and incorporated into nutrient composition calculations. Another key consideration is how the database is maintained and supported.

Developing a nutrient database for an FFQ presents additional challenges as nutrient composition values need to be assigned for a food grouping instead of an individual food item. Various approaches which rely on 24-hour recall data, either from a national population sample or a sample similar to the target population, have been used [88, 112, 467]. Generally, individual food codes from 24-hour recall data are grouped into FFQ food groupings, and a composite nutrient profile for each food grouping is estimated based on the individual foods' relative consumption. Again, for this approach to be effective, the 24-hour data need to be connected to a trustworthy nutrient database.

E. Choice of Dietary Analysis Software

Computerized data processing requires creating a file that includes a food code and an amount consumed for each food reported. Computer software then links the nutrient composition of each food on the separate nutrient composition database file, converts the amount reported to multiples of 100 g, multiplies by that factor, stores that information, and sums across all foods for each nutrient for each individual. Many computer packages have been developed that include both a nutrient composition database and software to convert individual responses to specific foods and, ultimately, to nutrients. A listing of many commercial dietary analysis software products was made available in 2006 [457].

Software should be chosen on the basis of the research needs, the level of detail necessary, the quality of the nutrient composition database, and the hardware and software requirements [468]. If precise nutrient information is required, it is important that the system is able to expand to incorporate information about newer foods in the marketplace and to integrate detailed information about food preparation (e.g., homemade stew) by processing recipe information. Sometimes the study purpose requires analysis of dietary data to derive intake estimates not only for nutrients but also for food groups (e.g., fruits and vegetables) food components other than standard nutrients (e.g., nitrites) or food characteristics (e.g., fried foods). These additional requirements limit the choice of appropriate software.

The automated food coding system used for the NHANES is the USDA's AMPM [56]. The AMPM is a network dietary coding system that provides online coding, recipe modification and development, data editing and management, and nutrient analysis of dietary data with multiple user access to manage the survey activities. It is available to government agencies and the general public only through special arrangement with USDA. A similar program is available in a commercial software program called the Food Intake Analysis System [469] available from the University of Texas.

Many diet history and food frequency instruments have also been automated. Users of these software packages should be aware of the source of information in the nutrient database and the assumptions about the nutrient content of each food item listed in the questionnaire.

F. Cognitive Testing Research Related to Dietary Assessment

Nearly all studies using dietary information about subjects rely on the subjects' own reports of their diets. Because such reports are based on complex cognitive processes, it is important to understand and take advantage of what is known about how respondents remember dietary information and how that information is retrieved and reported to the investigator. Several investigators have discussed the need for and importance of such considerations in the assessment of diet [332, 379, 416, 470–472], and research using cognitive testing methods in dietary assessment has been reported [110, 218, 263, 279, 416, 473–478]. A thorough description of cognitive interviewing methods is found in Willis [479].

There is an important distinction between specific and generic memories of diet. Specific memory relies on particular memories about episodes of eating and drinking, whereas generic memory relies on general knowledge about the respondent's typical diet. A 24-hour recall relies primarily on specific memory of all actual events in the very recent past, whereas an FFQ that directs a respondent to report the usual frequency of eating a food over the previous year relies primarily on generic memory. As the time between the behavior and the report increases, respondents may rely more on generic memory and less on specific memory [471].

What can the investigator do to enhance retrieval and improve reporting of diet? Research indicates that the amount of dietary information retrieved from memory can be enhanced by the context in which the instrument is administered and by use of specific memory cues and probes. For example, for a 24-hour dietary recall, foods that the respondent did not initially report can be recovered by interviewer probes. The effectiveness of these probes is well established and is therefore part of the interviewing protocols for all standardized high-quality 24-hour dietary recalls including those administered in the NHANES. Probes can be useful in improving generic memory, too, when subjects are asked to report their usual diets from periods in the past [332, 472]. Such probes can feature questions about past living situations and related eating habits.

The way in which questions are asked can affect responses. Certain characteristics of the interviewing situation may impact the social desirability of particular responses for foods seen as "good" or "bad." For example, the presence of other family members during the dietary interview may enhance social desirability bias, especially for certain foods like alcoholic beverages. An interview in a health setting such as a clinic may also enhance biases related to the social desirability of foods tied to compliance with dietary recommendations previously made for health reasons. In all instances, interviewers should be trained to refrain from either positive or negative feedback and should repeatedly encourage subjects to accurately report all foods.

G. Validation/Calibration Studies

It is important and desirable that any new dietary assessment method be validated or calibrated against other more established methods [176, 177, 179, 480]. Furthermore, even if an instrument has been evaluated, its proposed use in a different population may warrant additional validation research in that population. The purpose of such studies is to better understand how the method works in the particular research setting and to use that information to better interpret results from the overall study. Before a new FFQ or brief assessment questionnaire is used in the main study, for example, it should be evaluated in a validation/calibration study that compares the questionnaire to another dietary assessment method, for example, 24-hour dietary recalls or a more detailed FFQ, obtained from the same individuals, and, preferably, to biological markers such as DLW or UN. The NCI maintains a register of validation/calibration studies and publications on the Web [481].

Validation studies yield information about how well the new method is measuring what it is intended to measure, and calibration studies use the same information to relate (calibrate) the new method to a reference method using a regression model. Validation/calibration studies are challenging because of the difficulty and expense in collecting independent dietary information. Some researchers have used observational techniques to establish true dietary intake [83, 392, 482, 483]. Others have used laboratory measures, such as the 24-hour urine collection to measure protein, sodium, and potassium intakes and the DLW technique to measure energy expenditure [24–30, 70, 75, 126, 165, 166, 168, 484]. However, the high cost of this latter technique can make it impractical for most studies. The overall validity of energy intake estimates from the dietary assessment can be roughly checked by comparing weight data to reported energy intakes, in conjunction with use of equations to estimate basal energy expenditure [27, 36, 38, 39, 43, 46, 81, 82, 484–486] (also discussed in Chapter 4).

Because they are relatively expensive to conduct, validation/calibration studies are done on subsamples of the total study sample. However, the sample should be sufficiently large to estimate the relationship between the study instrument and a reference method with reasonable precision. Increasing the numbers of individuals sampled and decreasing the number of repeat measures per individual (e.g., two nonconsecutive 24-hour recalls on 100 people rather than four recalls on 50 people) can often help to increase precision without extra cost [487]. To the extent possible, the sample should be chosen randomly.

The resulting statistics, which quantify the relationship between the new method and the reference method, can be used for a variety of purposes. Because, in most cases, the reference method (usually records or recalls) is itself imperfect and subject to within-person error (day-to-day variability), measures such as correlation coefficients may underestimate the level of agreement with the actual usual intake. This phenomenon, referred to as "attenuation bias," can be corrected for using measurement error models that allow for within-person error in the reference instrument, resulting in estimates that more nearly reflect the correlation between the diet measure and true diet [341, 344]. The corrected correlation coefficients also give guidance as to the sample size required in a study, as the less precise the diet measure, the more individuals will be needed to attain the desired statistical power [341]. The estimated regression relationship between the new method and the reference method can also be used to adjust the relationships between diet and outcome as assessed in the larger study [160]. For example, the mean amounts of foods or nutrients, and their distributions, as estimated by a brief method, can be adjusted according to the calibration study results [488]. In addition, methods to adjust estimates of relationships measured in studies (e.g., relative risk of disease for subjects with high nutrient intake compared to those with low intake) have been described [172, 173, 344, 489, 490]. Many of these adjustments require the assumption that the reference method is unbiased [172, 342]. There is much evidence, however, that, at least for some nutrients, the reported intakes from recalls and records are also biased in a manner correlated with the tool of interest (such as an FFQ) [109], violating this assumption. Violation of this assumption would lead to overestimates of validity. For these reasons, researchers have sometimes used as reference measures biomarkers such as UN that have been shown in feeding studies to be unbiased measures of intake. Currently, however, only a few such biomarkers are known. Another area in need of further study is the effect of measurement error in a multivariate context, as most research thus far has been limited to the effect on univariate relationships [175, 179, 491, 492].

Acknowledgments

We gratefully acknowledge the contributions of Susan M. Krebs-Smith, Rachel Ballard-Barbash, Douglas Midthune, and Gordon B. Willis in reviewing and editing portions of this chapter. We also thank Meredith A. Morrissette and Penny Randall-Levy for invaluable research assistance.

References

1. Thompson, F. E., and Subar, A. F. (2001). Chapter 1. Dietary assessment methodology. *In* "Nutrition in the Prevention and Treatment of Disease" (A. M. Coulston, C. L. Rock, and E. R. Monsen, Eds.). Academic Press, San Diego, CA, pp. 3–30.
2. Thompson, F. E., and Byers, T. (1994). Dietary assessment resource manual. *J. Nutr.* **124**, 2245S–2318S.
3. Gersovitz, M., Madden, J. P., and Smiciklas-Wright, H. (1978). Validity of the 24-hr. dietary recall and seven-day record for group comparisons. *J. Am. Diet. Assoc.* **73**, 48–55.
4. Todd, K. S., Hudes, M., and Calloway, D. H. (1983). Food intake measurement: problems and approaches. *Am. J. Clin. Nutr.* **37**, 139–146.
5. Kretsch, M. J., and Fong, A. K. (1993). Validity and reproducibility of a new computerized dietary assessment method: effects of gender and educational level. *Nutr. Res.* **13**, 133–146.
6. Di, N. J., Contento, I. R., and Schinke, S. P. (2007). Criterion validity of the Healthy Eating Self-monitoring Tool (HEST) for black adolescents. *J. Am. Diet. Assoc.* **107**, 321–324.
7. Bingham, S. A., Gill, C., Welch, A., Day, K., Cassidy, A., Khaw, K. T., Sneyd, M. J., Key, T. J., Roe, L., and Day, N. E. (1994). Comparison of dietary assessment methods in nutritional epidemiology: weighed records v. 24 h recalls,

food-frequency questionnaires and estimated-diet records. *Br. J. Nutr.* **72**, 619–643.

8. Lindquist, C. H., Cummings, T., and Goran, M. I. (2000). Use of tape-recorded food records in assessing children's dietary intake. *Obes. Res.* **8**, 2–11.
9. My Pyramid, Tracker. (2007). Center for Nutrition Policy and Promotion, U.S. Department of Agriculture. Available at www.mypyramidtracker.gov.
10. Beasley, J., Riley, W. T., and Jean-Mary, J. (2005). Accuracy of a PDA-based dietary assessment program. *Nutrition* **21**, 672–677.
11. Wang, D. H., Kogashiwa, M., and Kira, S. (2006). Development of a new instrument for evaluating individuals' dietary intakes. *J. Am. Diet. Assoc.* **106**, 1588–1593.
12. Johnson, N. E., Sempos, C. T., Elmer, P. J., Allington, J. K., and Matthews, M. E. (1982). Development of a dietary intake monitoring system for nursing homes. *J. Am. Diet. Assoc.* **80**, 549–557.
13. Hammond, J., Nelson, M., Chinn, S., and Rona, R. J. (1993). Validation of a food frequency questionnaire for assessing dietary intake in a study of coronary heart disease risk factors in children. *Eur. J. Clin. Nutr.* **47**, 242–250.
14. Johnson, N. E., Nitzke, S., and VandeBerg, D. L. (1974). A reporting system for nutrition adequacy. *Home. Econ. Res. J.* **2**, 210–221.
15. Lillegaard, I. T., Loken, E. B., and Andersen, L. F. (2007). Relative validation of a pre-coded food diary among children, under-reporting varies with reporting day and time of the day. *Eur. J. Clin. Nutr.* **61**, 61–68.
16. Gibson, R. S. (2005). "Principles of Nutritional Assessment." Oxford University Press, New York.
17. Rebro, S. M., Patterson, R. E., Kristal, A. R., and Cheney, C. L. (1998). The effect of keeping food records on eating patterns. *J. Am. Diet. Assoc.* **98**, 1163–1165.
18. Vuckovic, N., Ritenbaugh, C., Taren, D. L., and Tobar, M. (2000). A qualitative study of participants' experiences with dietary assessment. *J. Am. Diet. Assoc.* **100**, 1023–1028.
19. Glanz, K., Murphy, S., Moylan, J., Evensen, D., and Curb, J. D. (2006). Improving dietary self-monitoring and adherence with hand-held computers: a pilot study. *Am. J. Health Promot.* **20**, 165–170.
20. Goris, A. H., Westerterp-Plantenga, M. S., and Westerterp, K. R. (2000). Undereating and underrecording of habitual food intake in obese men: selective underreporting of fat intake. *Am. J. Clin. Nutr.* **71**, 130–134.
21. Glanz, K., and Murphy, S. (2007). Dietary assessment and monitoring in real time. *In* "The Science of Real-Time Data Capture: Self-Reports in Health Research" (A. A. Stone, S. Schiffman, A. A. Atienza, and L. Nebeling, Eds.), pp. 151–168. Oxford University Press, New York.
22. Trabulsi, J., and Schoeller, D. A. (2001). Evaluation of dietary assessment instruments against doubly labeled water, a biomarker of habitual energy intake. *Am. J. Physiol. Endocrinol. Metab.* **281**, E891–E899.
23. Hill, R. J., and Davies, P. S. (2001). The validity of self-reported energy intake as determined using the doubly labelled water technique. *Br. J. Nutr.* **85**, 415–430.
24. Taren, D. L., Tobar, M., Hill, A., Howell, W., Shisslak, C., Bell, I., and Ritenbaugh, C. (1999). The association of energy intake bias with psychological scores of women. *Eur. J. Clin. Nutr.* **53**, 570–578.
25. Sawaya, A. L., Tucker, K., Tsay, R., Willett, W., Saltzman, E., Dallal, G. E., and Roberts, S. B. (1996). Evaluation of four methods for determining energy intake in young and older women: comparison with doubly labeled water measurements of total energy expenditure. *Am. J. Clin. Nutr.* **63**, 491–499.
26. Black, A. E., Prentice, A. M., Goldberg, G. R., Jebb, S. A., Bingham, S. A., Livingstone, M. B., and Coward, W. A. (1993). Measurements of total energy expenditure provide insights into the validity of dietary measurements of energy intake. *J. Am. Diet. Assoc.* **93**, 572–579.
27. Black, A. E., Bingham, S. A., Johansson, G., and Coward, W. A. (1997). Validation of dietary intakes of protein and energy against 24 hour urinary N and DLW energy expenditure in middle-aged women, retired men and post-obese subjects: comparisons with validation against presumed energy requirements. *Eur. J. Clin. Nutr.* **51**, 405–413.
28. Martin, L. J., Su, W., Jones, P. J., Lockwood, G. A., Tritchler, D. L., and Boyd, N. F. (1996). Comparison of energy intakes determined by food records and doubly labeled water in women participating in a dietary-intervention trial. *Am. J. Clin. Nutr.* **63**, 483–490.
29. Rothenberg, E. (1994). Validation of the food frequency questionnaire with the 4-day record method and analysis of 24-h urinary nitrogen. *Eur. J. Clin. Nutr.* **48**, 725–735.
30. Bingham, S. A., Gill, C., Welch, A., Cassidy, A., Runswick, S. A., Oakes, S., Lubin, R., Thurnham, D. I., Key, T. J., Roe, L., Khaw, K. T., and Day, N. E. (1997). Validation of dietary assessment methods in the UK arm of EPIC using weighed records, and 24-hour urinary nitrogen and potassium and serum vitamin C and carotenoids as biomarkers. *Int. J. Epidemiol.* **26**(suppl 1), S137–S151.
31. Bathalon, G. P., Tucker, K. L., Hays, N. P., Vinken, A. G., Greenberg, A. S., McCrory, M. A., and Roberts, S. B. (2000). Psychological measures of eating behavior and the accuracy of 3 common dietary assessment methods in healthy postmenopausal women. *Am. J. Clin. Nutr.* **71**, 739–745.
32. Seale, J. L., and Rumpler, W. V. (1997). Comparison of energy expenditure measurements by diet records, energy intake balance, doubly labeled water and room calorimetry. *Eur. J. Clin. Nutr.* **51**, 856–863.
33. Seale, J. L., Klein, G., Friedmann, J., Jensen, G. L., Mitchell, D. C., and Smiciklas-Wright, H. (2002). Energy expenditure measured by doubly labeled water, activity recall, and diet records in the rural elderly. *Nutrition* **18**, 568–573.
34. Mahabir, S., Baer, D. J., Giffen, C., Subar, A., Campbell, W., Hartman, T. J., Clevidence, B., Albanes, D., and Taylor, P. R. (2006). Calorie intake misreporting by diet record and food frequency questionnaire compared to doubly labeled water among postmenopausal women. *Eur. J Clin. Nutr.* **60**, 561–565.
35. Lichtman, S. W., Pisarska, K., Berman, E. R., Pestone, M., Dowling, H., Offenbacher, E., Weisel, H., Heshka, S., Matthews, D. E., and Heymsfield, S. B. (1992). Discrepancy between self-reported and actual caloric intake and exercise in obese subjects. *N. Engl. J. Med.* **327**, 1893–1898.
36. Pryer, J. A., Vrijheid, M., Nichols, R., Kiggins, M., and Elliott, P. (1997). Who are the "low energy reporters" in the

dietary and nutritional survey of British adults? *Int. J. Epidemiol.* **26**, 146–154.

37. Johnson, R. K., Goran, M. I., and Poehlman, E. T. (1994). Correlates of over- and underreporting of energy intake in healthy older men and women. *Am. J. Clin. Nutr.* **59**, 1286–1290.
38. Hirvonen, T., Mannisto, S., Roos, E., and Pietinen, P. (1997). Increasing prevalence of underreporting does not necessarily distort dietary surveys. *Eur. J. Clin. Nutr.* **51**, 297–301.
39. Lafay, L., Basdevant, A., Charles, M. A., Vray, M., Balkau, B., Borys, J. M., Eschwege, E., and Romon, M. (1997). Determinants and nature of dietary underreporting in a free-living population: the Fleurbaix Laventie Ville Sante (FLVS) Study. *Int. J. Obes. Relat. Metab. Disord.* **21**, 567–573.
40. Bazelmans, C., Matthys, C., De Henauw, S., Dramaix, M., Kornitzer, M., De Backer, G., and Leveque, A. (2007). Predictors of misreporting in an elderly population: the "Quality of life after 65" study. *Public Health Nutr.* **10**, 185–191.
41. Ballard-Barbash, R., Graubard, I., Krebs-Smith, S. M., Schatzkin, A., and Thompson, F. E. (1996). Contribution of dieting to the inverse association between energy intake and body mass index. *Eur. J. Clin. Nutr.* **50**, 98–106.
42. Hebert, J. R., Clemow, L., Pbert, L., Ockene, I. S., and Ockene, J. K. (1995). Social desirability bias in dietary self-report may compromise the validity of dietary intake measures. *Int. J. Epidemiol.* **24**, 389–398.
43. Stallone, D. D., Brunner, E. J., Bingham, S. A., and Marmot, M. G. (1997). Dietary assessment in Whitehall II: the influence of reporting bias on apparent socioeconomic variation in nutrient intakes. *Eur. J. Clin. Nutr.* **51**, 815–825.
44. Champagne, C. M., Bray, G. A., Kurtz, A. A., Monteiro, J. B., Tucker, E., Volaufova, J., and DeLany, J. P. (2002). Energy intake and energy expenditure: a controlled study comparing dietitians and non-dietitians. *J. Am. Diet. Assoc.* **102**, 1428–1432.
45. Maurer, J., Taren, D. L., Teixeira, P. J., Thomson, C. A., Lohman, T. G., Going, S. B., and Houtkooper, L. B. (2006). The psychosocial and behavioral characteristics related to energy misreporting. *Nutr. Rev.* **64**, 53–66.
46. Kortzinger, I., Bierwag, A., Mast, M., and Muller, M. J. (1997). Dietary underreporting: validity of dietary measurements of energy intake using a 7-day dietary record and a diet history in non-obese subjects. *Ann. Nutr. Metab.* **41**, 37–44.
47. Lafay, L., Mennen, L., Basdevant, A., Charles, M. A., Borys, J. M., Eschwege, E., and Romon, M. (2000). Does energy intake underreporting involve all kinds of food or only specific food items? Results from the Fleurbaix Laventie Ville Sante (FLVS) study. *Int. J. Obes. Relat. Metab. Disord.* **24**, 1500–1506.
48. Seale, J. L. (2002). Predicting total energy expenditure from self-reported dietary records and physical characteristics in adult and elderly men and women. *Am. J. Clin. Nutr.* **76**, 529–534.
49. Buzzard, I. M., Faucett, C. L., Jeffery, R. W., McBane, L., McGovern, P., Baxter, J. S., Shapiro, A. C., Blackburn, G. L., Chlebowski, R. T., Elashoff, R. M., and Wynder, E. L. (1996). Monitoring dietary change in a low-fat diet intervention study: advantages of using 24-hour dietary recalls vs food records. *J. Am. Diet. Assoc.* **96**, 574–579.
50. Casey, P. H., Goolsby, S. L., Lensing, S. Y., Perloff, B. P., and Bogle, M. L. (1999). The use of telephone interview methodology to obtain 24-hour dietary recalls. *J. Am. Diet. Assoc.* **99**, 1406–1411.
51. What we eat in America, NHANES. (2007). Agricultural Research Service, U.S. Department of Agriculture, www.ars.usda.gov/Services/docs.htm?docid=9098.
52. Campbell, V. A., and Dodds, M. L. (1967). Collecting dietary information from groups of older people. *J. Am. Diet. Assoc.* **51**, 29–33.
53. McBride, J. (2001). Was it a slab, a slice, or a sliver? High-tech innovations take food survey to new level. *Agric. Res.*, 4–7.
54. Cullen, K. W., Watson, K., Himes, J. H., Baranowski, T., Rochon, J., Waclawiw, M., Sun, W., Stevens, M., Slawson, D. L., Matheson, D., and Robinson, T. N. (2004). Evaluation of quality control procedures for 24-h dietary recalls: results from the Girls Health Enrichment Multisite Studies. *Prev. Med.* **38**(suppl), S14–S23.
55. Probst, Y. C., and Tapsell, L. C. (2005). Overview of computerized dietary assessment programs for research and practice in nutrition education. *J. Nutr. Educ. Behav.* **37**, 20–26.
56. USDA automated multiple-pass method. (2005) . Agricultural Research Service, U.S. Department of Agriculture, www.ars.usda.gov/Services/docs.htm?docid=7710.
57. Mennen, L. I., Bertrais, S., Galan, P., Arnault, N., Potier de Couray, G., and Hercberg, S. (2002). The use of computerised 24 h dietary recalls in the French SU.VI.MAX Study: number of recalls required. *Eur. J. Clin. Nutr.* **56**, 659–665.
58. Baranowski, T., Islam, N., Baranowski, J., Cullen, K. W., Myres, D., Marsh, T., and de Moor, C. (2002). The food intake recording software system is valid among fourth-grade children. *J. Am. Diet. Assoc.* **102**, 380–385.
59. Vereecken, C. A., Covents, M., Matthys, C., and Maes, L. (2005). Young adolescents' nutrition assessment on computer (YANA-C). *Eur. J. Clin. Nutr.* **59**, 658–667.
60. Zoellner, J., Anderson, J., and Gould, S. M. (2005). Comparative validation of a bilingual interactive multimedia dietary assessment tool. *J. Am. Diet. Assoc.* **105**, 1206–1214.
61. Lu, C., Pearson, M., Renker, S., Myerburg, S., and Farino, C. (2006). A novel system for collecting longitudinal self-reported dietary consumption information: the Internet data logger (iDL). *J. Expo. Sci. Environ. Epidemiol.* **16**, 427–433.
62. Subar, A. F., Thompson, F. E., Potischman, N., Forsyth, B. H., Buday, R., Richards, D., McNutt, S., Hull, S. G., Guenther, P. M., Schatzkin, A., and Baranowski, T. (2007). Formative research of a quick list for an automated self-administered 24-hour dietary recall. *J. Am. Diet. Assoc.* **107**, 1002–1007.
63. USDA food and nutrient database for dietary studies. Agricultural Research Service, U.S. Department of Agriculture, www.ars.usda.gov/Services/docs.htm?docid=12089.
64. Moshfegh, A. J., Borrud, L. G., Perloff, B., and LaComb, R. P. (1999). Improved method for the 24-hour dietary recall for use in national surveys. *FASEB. J.* **13**, A603.
65. National Research Council. (1986). "Nutrient Adequacy. Assessment Using Food Consumption Surveys." National Academy Press, Washington, DC.

66. Madden, J. P., Goodman, S. J., and Guthrie, H. A. (1976). Validity of the 24-hr. recall: analysis of data obtained from elderly subjects. *J. Am. Diet. Assoc.* **68**, 143–147.
67. Jonnalagadda, S. S., Mitchell, D. C., Smiciklas-Wright, H., Meaker, K. B., Van Heel, N., Karmally, W., Ershow, A. G., and Kris-Etherton, P. M. (2000). Accuracy of energy intake data estimated by a multiple-pass, 24-hour dietary recall technique. *J. Am. Diet. Assoc.* **100**, 303–308.
68. Tran, K. M., Johnson, R. K., Soultanakis, R. P., and Matthews, D. E. (2000). In-person vs telephone-administered multiple-pass 24-hour recalls in women: validation with doubly labeled water. *J. Am. Diet. Assoc.* **100**, 777–783.
69. Kroke, A., Klipstein-Grobusch, K., Voss, S., Moseneder, J., Thielecke, F., Noack, R., and Boeing, H. (1999). Validation of a self-administered food-frequency questionnaire administered in the European Prospective Investigation into Cancer and Nutrition (EPIC) Study: comparison of energy, protein, and macronutrient intakes estimated with the doubly labeled water, urinary nitrogen, and repeated 24-h dietary recall methods. *Am. J. Clin. Nutr.* **70**, 439–447.
70. Johnson, R. K., Soultanakis, R. P., and Matthews, D. E. (1998). Literacy and body fatness are associated with underreporting of energy intake in US low-income women using the multiple-pass 24-hour recall: a doubly labeled water study. *J. Am. Diet. Assoc.* **98**, 1136–1140.
71. Hebert, J. R., Ebbeling, C. B., Matthews, C. E., Hurley, T. G., Ma, Y., Druker, S., and Clemow, L. (2002). Systematic errors in middle-aged women's estimates of energy intake: comparing three self-report measures to total energy expenditure from doubly labeled water. *Ann. Epidemiol.* **12**, 577–586.
72. Subar, A. F., Kipnis, V., Troiano, R. P., Midthune, D., Schoeller, D. A., Bingham, S., Sharbaugh, C. O., Trabulsi, J., Runswick, S., Ballard-Barbash, R., Sunshine, J., and Schatzkin, A. (2003). Using intake biomarkers to evaluate the extent of dietary misreporting in a large sample of adults: the OPEN study. *Am. J. Epidemiol.* **158**, 1–13.
73. Slimani, N., Bingham, S., Runswick, S., Ferrari, P., Day, N. E., Welch, A. A., Key, T. J., Miller, A. B., Boeing, H., Sieri, S., Veglia, F., Palli, D., Panico, S., Tumino, R., Bueno-De-Mesquita, B., Ocke, M. C., Clavel-Chapelon, F., Trichopoulou, A., van Staveren, W. A., and Riboli, E. (2003). Group level validation of protein intakes estimated by 24-hour diet recall and dietary questionnaires against 24-hour urinary nitrogen in the European Prospective Investigation into Cancer and Nutrition (EPIC) calibration study. *Cancer Epidemiol. Biomarkers Prev.* **12**, 784–795.
74. Montgomery, C., Reilly, J. J., Jackson, D. M., Kelly, L. A., Slater, C., Paton, J. Y., and Grant, S. (2005). Validation of energy intake by 24-hour multiple pass recall: comparison with total energy expenditure in children aged 5–7 years. *Br. J. Nutr.* **93**, 671–676.
75. Heerstrass, D. W., Ocke, M. C., Bueno-de-Mesquita, H. B., Peeters, P. H., and Seidell, J. C. (1998). Underreporting of energy, protein and potassium intake in relation to body mass index. *Int. J. Epidemiol.* **27**, 186–193.
76. Harrison, G. G., Galal, O. M., Ibrahim, N., Khorshid, A., Stormer, A., Leslie, J., and Saleh, N. T. (2000). Underreporting of food intake by dietary recall is not universal: a comparison of data from Egyptian and American women. *J. Nutr.* **130**, 2049–2054.
77. Conway, J. M., Ingwersen, L. A., and Moshfegh, A. J. (2004). Accuracy of dietary recall using the USDA five-step multiple-pass method in men: an observational validation study. *J. Am. Diet. Assoc.* **104**, 595–603.
78. Conway, J. M., Ingwersen, L. A., Vinyard, B. T., and Moshfegh, A. J. (2003). Effectiveness of the US Department of Agriculture 5-step multiple-pass method in assessing food intake in obese and nonobese women. *Am. J. Clin. Nutr.* **77**, 1171–1178.
79. Blanton, C. A., Moshfegh, A. J., Baer, D. J., and Kretsch, M. J. (2006). The USDA Automated Multiple-Pass Method accurately estimates group total energy and nutrient intake. *J. Nutr.* **136**, 2594–2599.
80. Krebs-Smith, S. M., Graubard, B. I., Kahle, L. L., Subar, A. F., Cleveland, L. E., and Ballard-Barbash, R. (2000). Low energy reporters vs others: a comparison of reported food intakes. *Eur. J. Clin. Nutr.* **54**, 281–287.
81. Briefel, R. R., McDowell, M. A., Alaimo, K., Caughman, C. R., Bischof, A. L., Carroll, M. D., and Johnson, C. L. (1995). Total energy intake of the US population: the third National Health and Nutrition Examination Survey, 1988–1991. *Am. J. Clin. Nutr.* **62**, 1072S–1080S.
82. Klesges, R. C., Eck, L. H., and Ray, J. W. (1995). Who underreports dietary intake in a dietary recall? Evidence from the Second National Health and Nutrition Examination Survey. *J. Consult. Clin. Psychol.* **63**, 438–444.
83. Poppitt, S. D., Swann, D., Black, A. E., and Prentice, A. M. (1998). Assessment of selective under-reporting of food intake by both obese and non-obese women in a metabolic facility. *Int. J. Obes. Relat. Metab. Disord.* **22**, 303–311.
84. Tooze, J. A., Subar, A. F., Thompson, F. E., Troiano, R., Schatzkin, A., and Kipnis, V. (2004). Psychosocial predictors of energy underreporting in a large doubly labeled water study. *Am. J. Clin. Nutr.* **79**, 795–804.
85. Willett, W. C. (1998). "Nutritional Epidemiology." Oxford University Press, New York.
86. Zulkifli, S. N., and Yu, S. M. (1992). The food frequency method for dietary assessment. *J. Am. Diet. Assoc.* **92**, 681–685.
87. NutritionQuest: questionnaires & screeners. (2007). NutritionQuest, www.nutritionquest.com/products/questionnaires_screeners.htm.
88. Block, G., Hartman, A. M., Dresser, C. M., Carroll, M. D., Gannon, J., and Gardner, L. (1986). A data-based approach to diet questionnaire design and testing. *Am. J. Epidemiol.* **124**, 453–459.
89. Block, G., Woods, M., Potosky, A. L., and Clifford, C. (1990). Validation of a self-administered diet history questionnaire using multiple diet records. *J. Clin. Epidemiol.* **43**, 1327–1335.
90. Cummings, S. R., Block, G., McHenry, K., and Baron, R. B. (1987). Evaluation of two food frequency methods of measuring dietary calcium intake. *Am. J. Epidemiol.* **126**, 796–802.
91. Sobell, J., Block, G., Koslowe, P., Tobin, J., and Andres, R. (1989). Validation of a retrospective questionnaire assessing diet 10–15 years ago. *Am. J. Epidemiol.* **130**, 173–187.
92. Block, G., Thompson, F. E., Hartman, A. M., Larkin, F. A., and Guire, K. E. (1992). Comparison of two dietary questionnaires validated against multiple dietary records collected during a 1-year period. *J. Am. Diet. Assoc.* **92**, 686–693.

93. Mares-Perlman, J. A., Klein, B. E., Klein, R., Ritter, L. L., Fisher, M. R., and Freudenheim, J. L. (1993). A diet history questionnaire ranks nutrient intakes in middle- aged and older men and women similarly to multiple food records. *J. Nutr.* **123**, 489–501.
94. Coates, R. J., Eley, J. W., Block, G., Gunter, E. W., Sowell, A. L., Grossman, C., and Greenberg, R. S. (1991). An evaluation of a food frequency questionnaire for assessing dietary intake of specific carotenoids and vitamin E among low-income black women. *Am. J. Epidemiol.* **134**, 658–671.
95. Caan, B. J., Slattery, M. L., Potter, J., Quesenberry, C. P. J., Coates, A. O., and Schaffer, D. M. (1998). Comparison of the Block and the Willett self-administered semiquantitative food frequency questionnaires with an interviewer-administered dietary history. *Am. J. Epidemiol.* **148**, 1137–1147.
96. McCann, S. E., Marshall, J. R., Trevisan, M., Russell, M., Muti, P., Markovic, N., Chan, A. W., and Freudenheim, J. L. (1999). Recent alcohol intake as estimated by the Health Habits and History Questionnaire, the Harvard Semiquantitative Food Frequency Questionnaire, and a more detailed alcohol intake questionnaire. *Am. J. Epidemiol.* **150**, 334–340.
97. Subar, A. F., Thompson, F. E., Kipnis, V., Midthune, D., Hurwitz, P., McNutt, S., McIntosh, A., and Rosenfeld, S. (2001). Comparative validation of the Block, Willett, and National Cancer Institute food frequency questionnaires: the Eating at America's Table Study. *Am. J. Epidemiol.* **154**, 1089–1099.
98. Nutrition assessment. (2007). Fred Hutchinson Cancer Research Center, www.fhcrc.org/science/shared_resources/nutrition.
99. Patterson, R. E., Kristal, A. R., Tinker, L. F., Carter, R. A., Bolton, M. P., and Agurs-Collins, T. (1999). Measurement characteristics of the Women's Health Initiative food frequency questionnaire. *Ann. Epidemiol.* **9**, 178–187.
100. HSPH nutrition department's file download site: directory listing of /health/FFQ/files. (2007). Nutrition Department, Harvard School of Public Health, https://regepi.bwh. harvard .edu/health/FFQ/files.
101. Rimm, E. B., Giovannucci, E. L., Stampfer, M. J., Colditz, G. A., Litin, L. B., and Willett, W. C. (1992). Reproducibility and validity of an expanded self-administered semi-quantitative food frequency questionnaire among male health professionals. *Am. J. Epidemiol.* **135**, 1114–1126.
102. Willett, W. C., Sampson, L., Stampfer, M. J., Rosner, B., Bain, C., Witschi, J., Hennekens, C. H., and Speizer, F. E. (1985). Reproducibility and validity of a semiquantitative food frequency questionnaire. *Am. J. Epidemiol.* **122**, 51–65.
103. Willett, W. C., Reynolds, R. D., Cottrell-Hoehner, S., Sampson, L., and Browne, M. L. (1987). Validation of a semi-quantitative food frequency questionnaire: comparison with a 1-year diet record. *J. Am. Diet. Assoc.* **87**, 43–47.
104. Salvini, S., Hunter, D. J., Sampson, L., Stampfer, M. J., Colditz, G. A., Rosner, B., and Willett, W. C. (1989). Food based validation of a dietary questionnaire: the effects of week-to-week variation in food consumption. *Int. J. Epidemiol.* **18**, 858–867.
105. Feskanich, D., Rimm, E. B., Giovannucci, E. L., Colditz, G. A., Stampfer, M. J., Litin, L. B., and Willett, W. C. (1993). Reproducibility and validity of food intake measurements from a semiquantitative food frequency questionnaire. *J. Am. Diet. Assoc.* **93**, 790–796.
106. Suitor, C. J., Gardner, J., and Willett, W. C. (1989). A comparison of food frequency and diet recall methods in studies of nutrient intake of low-income pregnant women. *J. Am. Diet. Assoc.* **89**, 1786–1794.
107. Wirfalt, A. K. E., Jeffery, R. W., and Elmer, P. J. (1998). Comparison of food frequency questionnaires: the reduced Block and Willett questionnaires differ in ranking on nutrient intakes. *Am. J. Epidemiol.* **148**, 1148–1156.
108. Diet history questionnaire. (2007). National Cancer Institute, http://riskfactor.cancer.gov/DHQ.
109. Kipnis, V., Subar, A. F., Midthune, D., Freedman, L. S., Ballard-Barbash, R., Troiano, R. P., Bingham, S., Schoeller, D. A., Schatzkin, A., and Carroll, R. J. (2003). Structure of dietary measurement error: results of the OPEN biomarker study. *Am. J. Epidemiol.* **158**, 14–21.
110. Subar, A. F., Thompson, F. E., Smith, A. F., Jobe, J. B., Ziegler, R. G., Potischman, N., Schatzkin, A., Hartman, A., Swanson, C., Kruse, L., Hayes, R. B., Riedel-Lewis, D., and Harlan, L. C. (1995). Improving food frequency questionnaires: a qualitative approach using cognitive interviewing. *J. Am. Diet. Assoc.* **95**, 781–788.
111. Subar, A. F., Ziegler, R. G., Thompson, F. E., Johnson, C. C., Weissfeld, J. L., Reding, D., Kavounis, K. H., and Hayes, R. B. (2001). Is shorter always better? Relative importance of dietary questionnaire length and cognitive ease on response rates and data quality for two dietary questionnaires. *Am. J. Epidemiol.* **153**, 404–409.
112. Subar, A. F., Midthune, D., Kulldorff, M., Brown, C. C., Thompson, F. E., Kipnis, V., and Schatzkin, A. (2000). An evaluation of alternative approaches to assign nutrient values to food groups in food frequency questionnaires. *Am. J. Epidemiol.* **152**, 279–286.
113. Questionnaires. (2004). The Arizona Diet, Behavioral, and Quality of Life Assessment Center **12**, http://azdiet-behavior .azcc.arizona.edu/questions.htm.
114. Ritenbaugh, C., Aickin, M., Taren, D., Teufel, N., Graver, E., Woolf, K., and Alberts, D. S. (1997). Use of a food frequency questionnaire to screen for dietary eligibility in a randomized cancer prevention phase III trial. *Cancer Epidemiol. Biomarkers Prev.* **6**, 347–354.
115. Garcia, R. A., Taren, D., and Teufel, N. I. (2000). Factors associated with the reproducibility of specific food items from the Southwest Food Frequency Questionnaire. *Ecol. Food Nutr.* **38**, 549–561.
116. Kristal, A. R., Feng, Z., Coates, R. J., Oberman, A., and George, V. (1997). Associations of race/ethnicity, education, and dietary intervention with the validity and reliability of a food frequency questionnaire: the Women's Health Trial Feasibility Study in Minority Populations [published erratum appears in *Am. J. Epidemiol.*, October 15, 1998 **148**(8): 820]. *Am. J. Epidemiol.* **146**, 856–869.
117. Block, G., Wakimoto, P., Jensen, C., Mandel, S., and Green, R. R. (2006). Validation of a food frequency questionnaire for Hispanics. *Prev. Chronic. Dis.* **3**, A77.

118. Hankin, J. H., Yoshizawa, C. N., and Kolonel, L. N. (1990). Reproducibility of a diet history in older men in Hawaii. *Nutr. Cancer* **13**, 129–140.
119. Hankin, J. H., Wilkens, L. R., Kolonel, L. N., and Yoshizawa, C. N. (1991). Validation of a quantitative diet history method in Hawaii. *Am. J. Epidemiol.* **133**, 616–628.
120. Stram, D. O., Hankin, J. H., Wilkens, L. R., Pike, M. C., Monroe, K. R., Park, S., Henderson, B. E., Nomura, A. M., Earle, M. E., Nagamine, F. S., and Kolonel, L. N. (2000). Calibration of the dietary questionnaire for a multiethnic cohort in Hawaii and Los Angeles. *Am. J. Epidemiol.* **151** 358–370.
121. Ocke, M. C., Bueno-de-Mesquita, H. B., Goddijn, H. E., Jansen, A., Pols, M. A., van Staveren, W. A., and Kromhout, D. (1997). The Dutch EPIC food frequency questionnaire. I. Description of the questionnaire, and relative validity and reproducibility for food groups. *Int. J. Epidemiol.* **26**(suppl 1), S37–S48.
122. Katsouyanni, K., Rimm, E. B., Gnardellis, C., Trichopoulos, D., Polychronopoulos, E., and Trichopoulou, A. (1997) Reproducibility and relative validity of an extensive semi-quantitative food frequency questionnaire using dietary records and biochemical markers among Greek schoolteachers. *Int. J. Epidemiol.* **26**(suppl 1), S118–S127.
123. Bohlscheid-Thomas, S., Hoting, I., Boeing, H., and Wahrendorf, J. (1997). Reproducibility and relative validity of food group intake in a food frequency questionnaire developed for the German part of the EPIC project. European Prospective Investigation into Cancer and Nutrition. *Int. J. Epidemiol.* **26**(suppl 1), S59–S70.
124. Bohlscheid-Thomas, S., Hoting, I., Boeing, H., and Wahrendorf, J. (1997). Reproducibility and relative validity of energy and macronutrient intake of a food frequency questionnaire developed for the German part of the EPIC project. European Prospective Investigation into Cancer and Nutrition. *Int. J. Epidemiol.* **26**(suppl 1), S71–S81.
125. Riboli, E., Elmstahl, S., Saracci, R., Gullberg, B., and Lindgarde, F. (1997). The Malmo Food Study: validity of two dietary assessment methods for measuring nutrient intake. *Int. J. Epidemiol.* **26**(suppl 1), S161–S173.
126. Pisani, P., Faggiano, F., Krogh, V., Palli, D., Vineis, P., and Berrino, F. (1997). Relative validity and reproducibility of a food frequency dietary questionnaire for use in the Italian EPIC centres. *Int. J. Epidemiol.* **26**(suppl 1), S152–S160.
127. Block, G., Hartman, A. M., and Naughton, D. (1990). A reduced dietary questionnaire: development and validation. *Epidemiology* **1**, 58–64.
128. Harlan, L. C., and Block, G. (1990). Use of adjustment factors with a brief food frequency questionnaire to obtain nutrient values. *Epidemiology* **1**, 224–231.
129. Feskanich, D., Marshall, J., Rimm, E. B., Litin, L. B., and Willett, W. C. (1994). Simulated validation of a brief food frequency questionnaire. *Ann. Epidemiol.* **4**, 181–187.
130. Potischman, N., Carroll, R. J., Iturria, S. J., Mittl, B., Curtin, J., Thompson, F. E., and Brinton, L. A. (1999). Comparison of the 60- and 100-item NCI-block questionnaires with validation data. *Nutr. Cancer* **34**, 70–75.
131. Rockett, H. R., Berkey, C. S., and Colditz, G. A. (2007). Comparison of a short food frequency questionnaire with the Youth/Adolescent Questionnaire in the Growing Up Today Study. *Int. J. Pediatr. Obes.* **2**, 31–39.
132. Kristal, A. R., Glanz, K., Feng, Z., Hebert, J. R., Probart, C., Eriksen, M., and Heimendinger, J. (1994). Does using a short dietary questionnaire instead of a food frequency improve response rates to a health assessment survey? *J. Nutr. Educ.* **26**, 224–226.
133. Eaker, S., Bergstrom, R., Bergstrom, A., Adami, H. O., and Nyren, O. (1998). Response rate to mailed epidemiologic questionnaires: a population-based randomized trial of variations in design and mailing routines. *Am. J. Epidemiol.* **147**, 74–82.
134. Morris, M. C., Colditz, G. A., and Evans, D. A. (1998). Response to a mail nutritional survey in an older bi-racial community population. *Ann. Epidemiol.* **8**, 342–346.
135. Johansson, L., Solvoll, K., Opdahl, S., Bjorneboe, G. E., and Drevon, C. A. (1997). Response rates with different distribution methods and reward, and reproducibility of a quantitative food frequency questionnaire. *Eur. J. Clin. Nutr.* **51**, 346–353.
136. Kuskowska-Wolk, A., Holte, S., Ohlander, E. M., Bruce, A., Holmberg, L., Adami, H. O., and Bergstrom, R. (1992). Effects of different designs and extension of a food frequency questionnaire on response rate, completeness of data and food frequency responses. *Int. J. Epidemiol.* **21**, 1144–1150.
137. Tylavsky, F. A., and Sharp, G. B. (1995). Misclassification of nutrient and energy intake from use of closed-ended questions in epidemiologic research. *Am. J. Epidemiol.* **142**(3), 342–352.
138. Heady, J. A. (1961). Diets of bank clerks. Development of a method of classifying the diets of individuals for use in epidemiological studies. *J. R. Statist. Soc. A.* **124**, 336–371.
139. Kumanyika, S., Tell, G. S., Fried, L., Martel, J. K., and Chinchilli, V. M. (1996). Picture-sort method for administering a food frequency questionnaire to older adults. *J. Am. Diet. Assoc.* **96**, 137–144.
140. Kumanyika, S. K., Tell, G. S., Shemanski, L., Martel, J., and Chinchilli, V. M. (1997). Dietary assessment using a picture-sort approach. *Am. J. Clin. Nutr.* **65**, 1123S–1129S.
141. Eck, L. H., Klesges, L. M., and Klesges, R. C. (1996). Precision and estimated accuracy of two short-term food frequency questionnaires compared with recalls and records. *J. Clin. Epidemiol.* **49**, 1195–1200.
142. Subar, A. F., Frey, C. M., Harlan, L. C., and Kahle, L. (1994). Differences in reported food frequency by season of questionnaire administration: the 1987 National Health Interview Survey. *Epidemiology* **5**, 226–253.
143. Tsubono, Y., Nishino, Y., Fukao, A., Hisamichi, S., and Tsugane, S. (1995). Temporal change in the reproducibility of a self-administered food frequency questionnaire. *Am. J. Epidemiol.* **142**, 1231–1235.
144. Caan, B., Hiatt, R. A., and Owen, A. M. (1991). Mailed dietary surveys: response rates, error rates, and the effect of omitted food items on nutrient values. *Epidemiology* **2**, 430–436.
145. Holmberg, L., Ohlander, E. M., Byers, T., Zack, M., Wolk, A., Bruce, A., Bergstrom, R., Bergkvist, L., and Adami, H.O. (1996). A search for recall bias in a case-control study of diet and breast cancer. *Int. J. Epidemiol.* **25**, 235–244.

146. Hansson, L. M., and Galanti, M. R. (2000). Diet-associated risks of disease and self-reported food consumption: how shall we treat partial nonresponse in a food frequency questionnaire? *Nutr. Cancer* **36**, 1–6.
147. Diet history questionnaire: web-based DHQ. (2007). National Cancer Institute, http://riskfactor.cancer.gov/DHQ/webquest/index.html.
148. Matthys, C., Pynaert, I., De Keyzer, W., and De Henauw, S. (2007). Validity and reproducibility of an adolescent web-based food frequency questionnaire. *J. Am. Diet. Assoc.* **107**, 605–610.
149. Smith, A. F. (1993). Cognitive psychological issues of relevance to the validity of dietary reports. *Eur. J. Clin. Nutr.* **47** (suppl 2), S6–18.
150. Krebs-Smith, S. M., Heimendinger, J., Subar, A. F., Patterson, B. H., and Pivonka, E. (1994). Estimating fruit and vegetable intake using food frequency questionnaires: a comparison of instruments. *Am. J. Clin. Nutr.* **59**, 283s.
151. Hunter, D. J., Sampson, L., Stampfer, M. J., Colditz, G. A., Rosner, B., and Willett, W. C. (1988). Variability in portion sizes of commonly consumed foods among a population of women in the United States. *Am. J. Epidemiol.* **127**, 1240–1249.
152. Block, G., and Subar, A. F. (1992). Estimates of nutrient intake from a food frequency questionnaire: the 1987 National Health Interview Survey. *J. Am. Diet. Assoc.* **92**, 969–977.
153. Briefel, R. R., Flegal, K. M., Winn, D. M., Loria, C. M., Johnson, C. L., and Sempos, C. T. (1992). Assessing the nation's diet: limitations of the food frequency questionnaire. *J. Am. Diet. Assoc.* **92**, 959–962.
154. Sempos, C. T. (1992). Invited commentary: some limitations of semiquantitative food frequency questionnaires. *Am. J. Epidemiol.* **135**, 1127–1132.
155. Rimm, E. B., Giovannucci, E. L., Stampfer, M. J., Colditz, G. A., Litin, L. B., and Willett, W. C. (1992). Authors' response to "Invited commentary: some limitations of semiquantitative food frequency questionnaires." *Am. J. Epidemiol.* **135**, 1133–1136.
156. Carroll, R. J., Freedman, L. S., and Hartman, A. M. (1996). Use of semiquantitative food frequency questionnaires to estimate the distribution of usual intake. *Am. J. Epidemiol.* **143**, 392–404.
157. Kushi, L. H. (1994). Gaps in epidemiologic research methods: design considerations for studies that use food-frequency questionnaires. *Am. J. Clin. Nutr.* **59**, 180s–184s.
158. Beaton, G. H. (1994). Approaches to analysis of dietary data: relationship between planned analyses and choice of methodology. *Am. J. Clin. Nutr.* **59**, 253s–261s.
159. Sempos, C. T., Liu, K., and Ernst, N. D. (1999). Food and nutrient exposures: what to consider when evaluating epidemiologic evidence. *Am. J. Clin. Nutr.* **69**, 1330S–1338S.
160. Freedman, L. S., Schatzkin, A., and Wax, Y. (1990). The impact of dietary measurement error on planning sample size required in a cohort study. *Am. J. Epidemiol.* **132**, 1185–1195.
161. Schaefer, E. J., Augustin, J. L., Schaefer, M. M., Rasmussen, H., Ordovas, J. M., Dallal, G. E., and Dwyer, J. T. (2000). Lack of efficacy of a food-frequency questionnaire in assessing dietary macronutrient intakes in subjects consuming diets of known composition. *Am. J. Clin. Nutr.* **71**, 746–751.
162. Dietary assessment calibration/validation register: studies and their associated publications. (2007). National Cancer Institute, www-dacv.ims.nci.nih.gov.
163. Livingstone, M. B., and Black, A. E. (2003). Markers of the validity of reported energy intake. *J. Nutr.* **133**(suppl 3), 895S–920S.
164. Bingham, S. A. (2003). Urine nitrogen as a biomarker for the validation of dietary protein intake. *J. Nutr.* **133**(suppl 3), 921S–924S.
165. Bingham, S. A., Cassidy, A., Cole, T. J., Welch, A., Runswick, S. A., Black, A. E., Thurnham, D., Bates, C., Khaw, K. T., and Key, T. J. (1995). Validation of weighed records and other methods of dietary assessment using the 24 h urine nitrogen technique and other biological markers. *Br. J. Nutr.* **73**, 531–550.
166. Pijls, L. T., De Vries, H., Donker, A. J., and van Eijk, J. T. (1999). Reproducibility and biomarker-based validity and responsiveness of a food frequency questionnaire to estimate protein intake. *Am. J. Epidemiol.* **150**, 987–995.
167. Ocke, M. C., Bueno-de-Mesquita, H. B., Pols, M. A., Smit, H. A., van Staveren, W. A., and Kromhout, D. (1997). The Dutch EPIC food frequency questionnaire. II. Relative validity and reproducibility for nutrients. *Int. J. Epidemiol.* **26**(suppl 1), S49–S58.
168. Bingham, S. A. (1997). Dietary assessments in the European prospective study of diet and cancer (EPIC). *Eur. J. Cancer Prev.* **6**, 118–124.
169. Bingham, S. A., and Day, N. E. (1997). Using biochemical markers to assess the validity of prospective dietary assessment methods and the effect of energy adjustment. *Am. J. Clin. Nutr.* **65**, 1130S–1137S.
170. Flegal, K. M. (1999). Evaluating epidemiologic evidence of the effects of food and nutrient exposures. *Am. J. Clin. Nutr.* **69**, 1339S–1344S.
171. Burack, R. C., and Liang, J. (1987). The early detection of cancer in the primary-care setting: factors associated with the acceptance and completion of recommended procedures. *Prev. Med.* **16**, 739–751.
172. Prentice, R. L. (1996). Measurement error and results from analytic epidemiology: dietary fat and breast cancer. *J. Natl. Cancer Inst.* **88**, 1738–1747.
173. Kipnis, V., Freedman, L. S., Brown, C. C., Hartman, A. M., Schatzkin, A., and Wacholder, S. (1997). Effect of measurement error on energy-adjustment models in nutritional epidemiology. *Am. J. Epidemiol.* **146**, 842–855.
174. Hu, F. B., Stampfer, M. J., Rimm, E., Ascherio, A., Rosner, B. A., Spiegelman, D., and Willett, W. C. (1999). Dietary fat and coronary heart disease: a comparison of approaches for adjusting for total energy intake and modeling repeated dietary measurements. *Am. J. Epidemiol.* **149**, 531–540.
175. Carroll, R. J., Freedman, L. S., Kipnis, V., and Li, L. (1998). A new class of measurement-error models, with applications to dietary data. *Can. J. Stat.* **26**, 467–477.
176. Kohlmeier, L., and Bellach, B. (1995). Exposure assessment error and its handling in nutritional epidemiology. *Annu. Rev. Public Health* **16**, 43–59.

177. Kaaks, R., Riboli, E., and van Staveren, W. (1995). Calibration of dietary intake measurements in prospective cohort studies. *Am. J. Epidemiol.* **142**, 548–556.
178. Bellach, B., and Kohlmeier, L. (1998). Energy adjustment does not control for differential recall bias in nutritional epidemiology. *J. Clin. Epidemiol.* **51**, 393–398.
179. Kipnis, V., Carroll, R. J., Freedman, L. S., and Li, L. (1999). Implications of a new dietary measurement error model for estimation of relative risk: application to four calibration studies. *Am. J. Epidemiol.* **150**, 642–651.
180. Bingham, S. A., Luben, R., Welch, A., Wareham, N., Khaw, K. T., and Day, N. (2003). Are imprecise methods obscuring a relation between fat and breast cancer? *Lancet* **362**, 212–214.
181. Freedman, L. S., Potischman, N., Kipnis, V., Midthune, D., Schatzkin, A., Thompson, F. E., Troiano, R. P., Prentice, R., Patterson, R., Carroll, R., and Subar, A. F. (2006). A comparison of two dietary instruments for evaluating the fat-breast cancer relationship. *Int J. Epidemiol.* **35**, 1011–1021.
182. Kristal, A. R., Peters, U., and Potter, J. D. (2005). Is it time to abandon the food frequency questionnaire? *Cancer Epidemiol. Biomarkers Prev.* **14**, 2826–2828.
183. Kristal, A. R., and Potter, J. D. (2006). Not the time to abandon the food frequency questionnaire: counterpoint. *Cancer Epidemiol. Biomarkers Prev.* **15**, 1759–1760.
184. Kristal, A. R., Andrilla, C. H., Koepsell, T. D., Diehr, P. H., and Cheadle, A. (1998). Dietary assessment instruments are susceptible to intervention-associated response set bias. *J. Am. Diet. Assoc.* **98**, 40–43.
185. Neuhouser, M. L., Kristal, A. R., McLerran, D., Patterson, R. E., and Atkinson, J. (1999). Validity of short food frequency questionnaires used in cancer chemoprevention trials: results from the Prostate Cancer Prevention Trial. *Cancer Epidemiol. Biomarkers Prev.* **8**, 721–725.
186. Behavioral Risk Factor Surveillance System (BRFSS) (2007). Centers for Disease Control and Prevention, www.cdc.gov/brfss.
187. California Health Interview Survey. (2005). California Department of Health Services and the Public Health Institute, www.chis.ucla.edu.
188. Thompson, F. E., Midthune, D., Subar, A. F., McNeel, T., Berrigan, D., and Kipnis, V. (2005). Dietary intake estimates in the National Health Interview Survey, 2000: methodology, results, and interpretation. *J. Am. Diet. Assoc.* **105**, 352–363.
189. Pickle, L. W., and Hartman, A. M. (1985). Indicator foods for vitamin A assessment. *Nutr. Cancer* **7**, 3–23.
190. Byers, T., Marshall, J., Fiedler, R., Zielezny, M., and Graham, S. (1985). Assessing nutrient intake with an abbreviated dietary interview. *Am. J. Epidemiol.* **122**, 41–50.
191. Morin, P., Herrmann, F., Ammann, P., Uebelhart, B., and Rizzoli, R. (2005). A rapid self-administered food frequency questionnaire for the evaluation of dietary protein intake. *Clin. Nutr.* **24**, 768–774.
192. Hertzler, A. A., and Frary, R. B. (1994). A dietary calcium rapid assessment method (RAM). *Topics Clin. Nutr.* **9**, 76–85.
193. Harnack, L. J., Lytle, L. A., Story, M., Galuska, D. A., Schmitz, K., Jacobs, D. R., Jr., and Gao, S. (2006). Reliability and validity of a brief questionnaire to assess calcium intake of middle-school-aged children. *J. Am. Diet. Assoc.* **106**, 1790–1795.
194. Sebring, N. G., Denkinger, B. I., Menzie, C. M., Yanoff, L. B., Parikh, S. J., and Yanovski, J. A. (2007). Validation of three food frequency questionnaires to assess dietary calcium intake in adults. *J. Am. Diet. Assoc.* **107**, 752–759.
195. Hertzler, A. A., and McAnge, T. R., Jr. (1986). Development of an iron checklist to guide food intake. *J. Am. Diet. Assoc.* **86**, 782–786.
196. Kirk, P., Patterson, R. E., and Lampe, J. (1999). Development of a soy food frequency questionnaire to estimate isoflavone consumption in US adults. *J. Am. Diet. Assoc.* **99**, 558–563.
197. Horn-Ross, P. L., Barnes, S., Lee, V. S., Collins, C. N., Reynolds, P., Lee, M. M., Stewart, S. L., Canchola, A. J., Wilson, L., and Jones, K. (2006). Reliability and validity of an assessment of usual phytoestrogen consumption (United States). *Cancer Causes Control* **17**, 85–93.
198. Williams, A. E., Maskarinec, G., Hebshi, S., Oshiro, C., Murphy, S., and Franke, A. A. (2003). Validation of a soy questionnaire with repeated dietary recalls and urinary isoflavone assessments over one year. *Nutr. Cancer* **47**, 118–125.
199. Pufulete, M., Emery, P. W., Nelson, M., and Sanders, T. A. (2002). Validation of a short food frequency questionnaire to assess folate intake. *Br. J. Nutr.* **87**, 383–390.
200. Hickling, S., Knuiman, M., Jamrozik, K., and Hung, J. (2005). A rapid dietary assessment tool to determine intake of folate was developed and validated. *J. Clin. Epidemiol.* **58**, 802–808.
201. Owens, J. E., Holstege, D. M., and Clifford, A. J. (2007). Comparison of two dietary folate intake instruments and their validation by RBC folate. *J. Agric. Food Chem.* **55**, 3737–3740.
202. Kiwanuka, S. N., Astrom, A. N., and Trovik, T. A. (2006). Sugar snack consumption in Ugandan schoolchildren: validity and reliability of a food frequency questionnaire. *Community Dent. Oral Epidemiol.* **34**, 372–380.
203. Rohrmann, S., and Becker, N. (2002). Development of a short questionnaire to assess the dietary intake of heterocyclic aromatic amines. *Public Health Nutr.* **5**, 699–705.
204. Serra-Majem, L., Santana-Armas, J. F., Ribas, L., Salmona, E., Ramon, J. M., Colom, J., and Salleras, L. (2002). A comparison of five questionnaires to assess alcohol consumption in a Mediterranean population. *Public Health Nutr.* **5**, 589–594.
205. Herran, O. F., and Ardila, M. F. (2006). Validity and reproducibility of two semi-quantitative alcohol frequency questionnaires for the Colombian population. *Public Health Nutr.* **9**, 763–770.
206. Field, A. E., Colditz, G. A., Fox, M. K., Byers, T., Serdula, M., Bosch, R. J., and Peterson, K. E. (1998). Comparison of 4 questionnaires for assessment of fruit and vegetable intake. *Am. J. Public Health* **88**, 1216–1218.
207. Cullen, K. W., Baranowski, T., Baranowski, J., Hebert, D., and de Moor, C. (1999). Pilot study of the validity and reliability of brief fruit, juice and vegetable screeners among inner city African-American boys and 17 to 20 year old adults. *J. Am. Coll. Nutr.* **18**, 442–450.

208. Resnicow, K., Odom, E., Wang, T., Dudley, W. N., Mitchell, D., Vaughan, R., Jackson, A., and Baranowski, T. (2000). Validation of three food frequency questionnaires and 24-hour recalls with serum carotenoid levels in a sample of African-American adults. *Am. J. Epidemiol.* **152**, 1072–1080.
209. Prochaska, J. J., and Sallis, J. F. (2004). Reliability and validity of a fruit and vegetable screening measure for adolescents. *J. Adolesc. Health* **34**, 163–165.
210. Andersen, L. F., Veierod, M. B., Johansson, L., Sakhi, A., Solvoll, K., and Drevon, C. A. (2005). Evaluation of three dietary assessment methods and serum biomarkers as measures of fruit and vegetable intake, using the method of triads. *Br. J. Nutr.* **93**, 519–527.
211. Campbell, M. K., Polhamus, B., McClelland, J. W., Bennett, K., Kalsbeek, W., Coole, D., Jackson, B., and Demark-Wahnefried, W. (1996). Assessing fruit and vegetable consumption in a 5 a day study targeting rural blacks: the issue of portion size. *J. Am. Diet. Assoc.* **96**, 1040–1042.
212. Baranowski, T., Smith, M., Baranowski, J., Wang, D. T., Doyle, C., Lin, L. S., Hearn, M. D., and Resnicow, K. (1997). Low validity of a seven-item fruit and vegetable food frequency questionnaire among third-grade students. *J. Am. Diet. Assoc.* **97**, 66–68.
213. Hunt, M. K., Stoddard, A. M., Peterson, K., Sorensen, G., Hebert, J. R., and Cohen, N. (1998). Comparison of dietary assessment measures in the Treatwell 5 A Day worksite study. *J. Am. Diet. Assoc.* **98**, 1021–1023.
214. Serdula, M., Coates, R., Byers, T., Mokdad, A., Jewell, S., Chavez, N., Mares-Perlman, J., Newcomb, P., Ritenbaugh, C., Treiber, F., and Block, G. (1993). Evaluation of a brief telephone questionnaire to estimate fruit and vegetable consumption in diverse study populations. *Epidemiology* **4**, 455–463.
215. Smith-Warner, S. A., Elmer, P. J., Fosdick, L., Tharp, T. M., and Randall, B. (1997). Reliability and comparability of three dietary assessment methods for estimating fruit and vegetable intakes. *Epidemiology* **8**, 196–201.
216. Armstrong, B. (1979). Diet and hormones in the epidemiology of breast and endometrial cancers. *Nutr. Cancer* **1**, 90–95.
217. Kristal, A. R., Vizenor, N. C., Patterson, R. E., Neuhouser, M. L., Shattuck, A. L., and McLerran, D. (2000). Precision and bias of food frequency-based measures of fruit and vegetable intakes. *Cancer Epidemiol. Biomarkers Prev.* **9**, 939–944.
218. Thompson, F. E., Subar, A. F., Smith, A. F., Midthune, D., Radimer, K. L., Kahle, L. L., and Kipnis, V. (2002). Fruit and vegetable assessment: performance of 2 new short instruments and a food frequency questionnaire. *J. Am. Diet. Assoc.* **102**, 1764–1772.
219. Green, G. W., Resnicow, K., Thompson, F. E., Peterson, K. E., Hurley, T. G., Hebert, J. R., Toobert, D. J., Williams, G., Elliot, D. L., Sher, T. G., Domas, A., Midthune, D., Stacewicz-Sapuntzakis, M., Yaroch, A. L., and Nebling, L. (2008) Correspondence of the NCI fruit and vegetable screener to repeat 24-H recalls and serum carotenoids in behavioral intervention trials. *J. Nutr.* **138**, 2005–2045.
220. Peterson, K E, Hebert, J. R., Hurley, T. G., Resnicow, K., Thompson, F. E., Greene, G. W., Shaikh, A. R., Yaroch, A. L., Williams, G. C., Salkeld, J., Toobert, D. J., Domas, A., Elliot, D. L., Hardin, J., and Nebeling, L. (2008). Accuracy and precision of two short screeners to assess change in fruit and vegetable consumption among diverse populations participating in health promotion intervention trials. *J. Nutr.* **138**, 2185–2255.
221. Block, G., Clifford, C., Naughton, M. D., Henderson, M., and McAdams, M. (1989). A brief dietary screen for high fat intake. *J. Nutr. Educ.* **21**, 199–207.
222. Coates, R. J., Serdula, M. K., Byers, T., Mokdad, A., Jewell, S., Leonard, S. B., Ritenbaugh, C., Newcomb, P., Mares-Perlman, J., and Chavez, N. (1995). A brief, telephone-administered food frequency questionnaire can be useful for surveillance of dietary fat intakes. *J. Nutr.* **125**, 1473–1483.
223. Spencer, E. H., Elon, L. K., Hertzberg, V. S., Stein, A. D., and Frank, E. (2005). Validation of a brief diet survey instrument among medical students. *J. Am. Diet. Assoc.* **105**, 802–806.
224. Caan, B., Coates, A., and Schaffer, D. (1995). Variations in sensitivity, specificity, and predictive value of a dietary fat screener modified from Block et al. *J. Am. Diet. Assoc.* **95**, 564–568.
225. Kris-Etherton, P., Eissenstat, B., Jaax, S., Srinath, U., Scott, L., Rader, J., and Pearson, T. (2001). Validation for MEDFICTS, a dietary assessment instrument for evaluating adherence to total and saturated fat recommendations of the National Cholesterol Education Program Step 1 and Step 2 diets. *J. Am. Diet. Assoc.* **101**, 81–86.
226. Taylor, A. J., Wong, H., Wish, K., Carrow, J., Bell, D., Bindeman, J., Watkins, T., Lehmann, T., Bhattarai, S., and O'Malley, P. G. (2003). Validation of the MEDFICTS dietary questionnaire: a clinical tool to assess adherence to American Heart Association dietary fat intake guidelines. *Nutr. J.* **2**, 4.
227. Teal, C. R., Baham, D. L., Gor, B. J., and Jones, L. A. (2007). Is the MEDFICTS Rapid Dietary Fat Screener valid for premenopausal African-American women? *J. Am. Diet. Assoc.* **107**, 773–781.
228. Holmes, A. L., Sanderson, B., Maisiak, R., Brown, A., and Bittner, V. (2005). Dietitian services are associated with improved patient outcomes and the MEDFICTS dietary assessment questionnaire is a suitable outcome measure in cardiac rehabilitation. *J. Am. Diet. Assoc.* **105**, 1533–1540.
229. Rohrmann, S., and Klein, G. (2003). Validation of a short questionnaire to qualitatively assess the intake of total fat, saturated, monounsaturated, polyunsaturated fatty acids, and cholesterol. *J. Hum. Nutr. Diet.* **16**, 111–117.
230. Rohrmann, S., and Klein, G. (2003). Development and validation of a short food list to assess the intake of total fat, saturated, mono-unsaturated, polyunsaturated fatty acids and cholesterol. *Eur. J. Public Health* **13**, 262–268.
231. Thompson, F. E., Midthune, D., Subar, A. F., Kipnis, V., Kahle, L. L., and Schatzkin, A. (2007). Development and evaluation of a short instrument to estimate usual dietary intake of percentage energy from fat. *J. Am. Diet. Assoc.* **107**, 760–767.
232. Thompson, F E, Midthune, D., Williams, G., Yaroch, A. L., Hurley, T. G., Resnicow, K., Hebert, J. R., Toobert, D. J., Greene, G. W., Peterson, K., and Nebeling, L. C. (2008).

Evaluation of a short dietary assessment instrument for percentage energy from fat in an intervention study. *J. Nutr.* **138**, 1935–1995.

233. Yaroch, A. L., Resnicow, K., and Khan, L. K. (2000). Validity and reliability of qualitative dietary fat index questionnaires: a review. *J. Am. Diet. Assoc.* **100**, 240–244.
234. van Assema, P., Brug, J., Kok, G., and Brants, H. (1992). The reliability and validity of a Dutch questionnaire on fat consumption as a means to rank subjects according to individual fat intake. *Eur. J. Cancer Prev.* **1**, 375–380.
235. Ammerman, A. S., Haines, P. S., DeVellis, R. F., Strogatz, D. S., Keyserling, T. C., and Simpson, R. J., Jr., and Siscovick, D. S., (1991). A brief dietary assessment to guide cholesterol reduction in low-income individuals design and validation. *J. Am. Diet. Assoc.* **91**, 1385–1390.
236. Hopkins, P. N., Williams, R. R., Kuida, H., Stults, B. M., Hunt, S. C., Barlow, G. K., and Ash, K. O. (1989). Predictive value of a short dietary questionnaire for changes in serum lipids in high-risk Utah families. *Am. J. Clin. Nutr.* **50**, 292–300.
237. Kemppainen, T., Rosendahl, A., Nuutinen, O., Ebeling, T., Pietinen, P., and Uusitupa, M. (1993). Validation of a short dietary questionnaire and a qualitative fat index for the assessment of fat intake. *Eur. J. Clin. Nutr.* **47**, 765–775.
238. Retzlaff, B. M., Dowdy, A. A., Walden, C. E., Bovbjerg, V. E., and Knopp, R. H. (1997). The Northwest Lipid Research Clinic Fat Intake Scale: validation and utility. *Am. J. Public Health* **87**, 181–185.
239. Little, P., Barnett, J., Margetts, B., Kinmonth, A. L., Gabbay, J., Thompson, R., Warm, D., Warwick, H., and Wooton, S (1999). The validity of dietary assessment in general practice. *J. Epidemiol. Community Health* **53**, 165–172.
240. Kinlay, S., Heller, R. F., and Halliday, J. A. (1991). A simple score and questionnaire to measure group changes in dietary fat intake. *Prev. Med.* **20**, 378–388.
241. Beresford, S. A., Farmer, E. M., Feingold, L., Graves, K. L., Sumner, S. K., and Baker, R. M. (1992). Evaluation of a self-help dietary intervention in a primary care setting. *Am. J. Public Health* **82**, 79–84.
242. Connor, S. L., Gustafson, J. R., Sexton, G., Becker, N., Artaud-Wild, S., and Connor, W. E. (1992). The Diet Habit Survey: a new method of dietary assessment that relates to plasma cholesterol changes. *J. Am. Diet. Assoc.* **92**, 41–47.
243. Glasgow, R. E., Perry, J. D., Toobert, D. J., and Hollis, J. F. (1996). Brief assessments of dietary behavior in field settings. *Addict. Behav.* **21**, 239–247.
244. Heller, R. F., Pedoe, H. D., and Rose, G. (1981). A simple method of assessing the effect of dietary advice to reduce plasma cholesterol. *Prev. Med.* **10**, 364–370.
245. Retzlaff, R. M., Dowdy, A. A., Walden, C. E., Bovbjerg, V. E., and Knopp, R. H. (1997). The northwest lipid research clinic fat intake scale: validation and utility. *Am. J. Public Health* **87**, 181–185.
246. Kristal, A. R., Shattuck, A. L., and Henry, H. J. (1990). Patterns of dietary behavior associated with selecting diets low in fat: reliability and validity of a behavioral approach to dietary assessment. *J. Am. Diet. Assoc.* **90**, 214–220.
247. Kristal, A. R., White, E., Shattuck, A. L., Curry, S., Anderson, G. L., Fowler, A., and Urban, N. (1992). Long-term maintenance of a low-fat diet: durability of fat-related dietary habits in the Women's Health Trial. *J. Am. Diet. Assoc.* **92**, 553–559.
248. Kristal, A. R., Beresford, S. A., and Lazovich, D. (1994). Assessing change in diet-intervention research. *Am. J. Clin. Nutr.* **59**, 185S–189S.
249. Yaroch, A. L., Resnicow, K., Petty, A. D., and Khan, L. K. (2000). Validity and reliability of a modified qualitative dietary fat index in low-income, overweight, African American adolescent girls. *J. Am. Diet. Assoc.* **100**, 1525–1529.
250. Risica, P M, Burkholder, G. B., Gans, K. M., Lasater, T. M., Acharyya S, Davis, C., and Kirtania, U. (2007). Assessing fat-related dietary behaviors among black women: reliability and validity of a new Food Habits Questionnaire. *J. Nutr. Educ. Behav.* **39**, 197–204.
251. Anderson, C. A., Kumanyika, S. K., Shults, J., Kallan, M. J., Gans, K. M., and Risica, P. M. (2007). Assessing change in dietary-fat behaviors in a weight-loss program for African Americans: a potential short method. *J. Am. Diet. Assoc.* **107**, 838–842.
252. Block, G., Gillespie, C., Rosenbaum, E. H., and Jenson, C. (2000). A rapid food screener to assess fat and fruit and vegetable intake. *Am. J. Prev. Med.* **18**, 284–288.
253. Andersen, L. F., Johansson, L., and Solvoll, K. (2002). Usefulness of a short food frequency questionnaire for screening of low intake of fruit and vegetable and for intake of fat. *Eur. J. Public Health* **12**, 208–213.
254. Buzzard, I. M., Stanton, C. A., Figueiredo, M., Fries, E. A., Nicholson, R., Hogan, C. J., and Danish, S. J. (2001). Development and reproducibility of a brief food frequency questionnaire for assessing the fat, fiber, and fruit and vegetable intakes of rural adolescents. *J. Am. Diet. Assoc.* **101**, 1438–1446.
255. Thompson, F. E., Midthune, D., Subar, A. F., Kahle, L. L., Schatzkin, A., and Kipnis, V. (2004). Performance of a short tool to assess dietary intakes of fruits and vegetables, percentage energy from fat and fibre. *Public Health Nutr.* **7**, 1097–1105.
256. Svilaas, A., Strom, E. C., Svilaas, T., Borgejordet, A., Thoresen, M., and Ose, L. (2002). Reproducibility and validity of a short food questionnaire for the assessment of dietary habits. *Nutr. Metab. Cardiovasc. Dis.* **12**, 60–70.
257. Laviolle, B., Froger-Bompas, C., Guillo, P., Sevestre, A., Letellier, C., Pouchard, M., Daubert, J. C., and Paillard, F. (2005). Relative validity and reproducibility of a 14-item semi-quantitative food frequency questionnaire for cardiovascular prevention. *Eur. J. Cardiovasc. Prev. Rehabil.* **12**, 587–595.
258. Rifas-Shiman, S. L., Willett, W. C., Lobb, R., Kotch, J., Dart, C., and Gillman, M. W. (2001). PrimeScreen, a brief dietary screening tool: reproducibility and comparability with both a longer food frequency questionnaire and biomarkers. *Public Health Nutr.* **4**, 249–254.
259. Five-factor screener in the 2005 NHIS cancer control supplement. (2007). National Cancer Institute, http://appliedresearch.cancer.gov/surveys/nhis/5factor.
260. The diet screener in the 2005 California Health Interview Survey. (2007). National Cancer Institute, http://appliedresearch.cancer.gov/surveys/chis/dietscreener.

261. Shannon, J., Kristal, A. R., Curry, S. J., and Beresford, S. A. (1997). Application of a behavioral approach to measuring dietary change: the fat-and fiber-related diet behavior questionnaire. *Cancer Epidemiol. Biomarkers Prev.* **6**, 355–361.
262. Schlundt, D. G., Hargreaves, M. K., and Buchowski, M. S. (2003). The Eating Behavior Patterns Questionnaire predicts dietary fat intake in African American women. *J. Am. Diet. Assoc.* **103**, 338–345.
263. Gans, K. M., Risica, P. M., Wylie-Rosett, J., Ross, E. M., Strolla, L. O., McMurray, J., and Eaton, C. B. (2006). Development and evaluation of the nutrition component of the Rapid Eating and Activity Assessment for Patients (REAP): a new tool for primary care providers. *J. Nutr. Educ. Behav.* **38**, 286–292.
264. Kristal, A. R., Shattuck, A. L., Henry, H. J., and Fowler, A. S. (1989). Rapid assessment of dietary intake of fat, fiber, and saturated fat: validity of an instrument suitable for community intervention research and nutritional surveillance. *Am. J. Health Promot.* **4**, 288–295.
265. Kristal, A. R., Abrams, B. F., Thornquist, M. D., Disogra, L., Croyle, R. T., Shattuck, A. L., and Henry, H. J. (1990). Development and validation of a food use checklist for evaluation of community nutrition interventions. *Am. J. Public Health* **80**, 1318–1322.
266. Neuhouser, M. L., Patterson, R. E., Kristal, A. R., Eldridge, A. L., and Vizenor, N. C. (2001). A brief dietary assessment instrument for assessing target foods, nutrients and eating patterns. *Public Health Nutr.* **4**, 73–78.
267. Yen, J., Zoumas-Morse, C., Pakiz, B., and Rock, C. L. (2003). Folate intake assessment: validation of a new approach. *J. Am. Diet. Assoc.* **103**, 991–1000.
268. Zhou, S. J., Schilling, M. J., and Makrides, M. (2005). Evaluation of an iron specific checklist for the assessment of dietary iron intake in pregnant and postpartum women. *Nutrition* **21**, 908–913.
269. Haraldsdottir, J., Thorsdottir, I., de Almeida, M. D., Maes, L., Perez, R. C., Elmadfa, I., and Frost, A. L. (2005). Validity and reproducibility of a precoded questionnaire to assess fruit and vegetable intake in European 11- to 12-year-old schoolchildren. *Ann. Nutr. Metab.* **49**, 221–227.
270. Birkett, N. J., and Boulet, J. (1995). Validation of a food habits questionnaire: poor performance in male manual laborers. *J. Am. Diet. Assoc.* **95**, 558–563.
271. Gray-Donald, K., O'Loughlin, J., Richard, L., and Paradis, G. (1997). Validation of a short telephone administered questionnaire to evaluate dietary interventions in low income communities in Montreal, Canada. *J. Epidemiol. Community Health* **51**, 326–331.
272. Spoon, M. P., Devereux, P. G., Benedict, J. A., Leontos, C., Constantino, N., Christy, D., and Snow, G. (2002). Usefulness of the food habits questionnaire in a worksite setting. *J. Nutr. Educ. Behav.* **34**, 268–272.
273. Burke, B. S. (1947). The dietary history as a tool in research. *J. Am. Diet. Assoc.* **23**, 1041–1046.
274. Burke, B. S., and Stuart, H. C. (1938). A method of diet analysis: applications in research and pediatric practice. *J. Pediatr.* **12**, 493–503.
275. McDonald, A., Van Horn, L., Slattery, M., Hilner, J., Bragg, C., Caan, B., Jacobs, D., Jr., Liu, K., Hubert, H., and Gernhofer, N. (1991). The CARDIA dietary history: development, implementation, and evaluation. *J. Am. Diet. Assoc.* **91**, 1104–1112.
276. Visser, M., De Groot, L. C., Deurenberg, P., and van Staveren, W. A. (1995). Validation of dietary history method in a group of elderly women using measurements of total energy expenditure. *Br. J. Nutr.* **74**, 775–785.
277. Kohlmeier, L., Mendez, M., McDuffie, J., and Miller, M. (1997). Computer-assisted self-interviewing: a multimedia approach to dietary assessment. *Am. J. Clin. Nutr.* **65**, 1275S–1281S.
278. Landig, J., Erhardt, J. G., Bode, J. C., and Bode, C. (1998). Validation and comparison of two computerized methods of obtaining a diet history. *Clin. Nutr.* **17**, 113–117.
279. Kohlmeier, L. (1994). Gaps in dietary assessment methodology: meal vs list-based methods. *Am. J. Clin. Nutr.* **59**, 175s–179s.
280. van Staveren, W. A., de Boer, J. O., and Burema, J. (1985). Validity and reproducibility of a dietary history method estimating the usual food intake during one month. *Am. J. Clin. Nutr.* **42**, 554–559.
281. Jain, M. (1989). Diet history: questionnaire and interview techniques used in some retrospective studies of cancer. *J. Am. Diet. Assoc.* **89**, 1647–1652.
282. Kune, S., Kune, G. A., and Watson, L. F. (1987). Observations on the reliability and validity of the design and diet history method in the Melbourne Colorectal Cancer Study. *Nutr. Cancer* **9**, 5–20.
283. Tapsell, L. C., Brenninger, V., and Barnard, J. (2000). Applying conversation analysis to foster accurate reporting in the diet history interview. *J. Am. Diet. Assoc.* **100**, 818–824.
284. van Beresteyn, E. C., van't Hof, M. A., van der Heiden-Winkeldermaat, H. J., ten Have-Witjes, A., and Neeter, R. (1987). Evaluation of the usefulness of the cross-check dietary history method in longitudinal studies. *J. Chronic. Dis.* **40**, 1051–1058.
285. Bloemberg, B. P., Kromhout, D., Obermann-De Boer, G. L., and Van Kampen-Donker, M. (1989). The reproducibility of dietary intake data assessed with the cross-check dietary history method. *Am. J. Epidemiol.* **130**, 1047–1056.
286. Mensink, G. B., Haftenberger, M., and Thamm, M. (2001). Validity of DISHES 98, a computerised dietary history interview: energy and macronutrient intake. *Eur. J. Clin. Nutr.* **55**, 409–417.
287. Slattery, M. L., Caan, B. J., Duncan, D., Berry, T. D., Coates, A., and Kerber, R. (1994). A computerized diet history questionnaire for epidemiologic studies. *J. Am. Diet. Assoc.* **94**, 761–766.
288. Jain, M., Howe, G. R., Johnson, K. C., and Miller, A. B. (1980). Evaluation of a diet history questionnaire for epidemiologic studies. *Am. J. Epidemiol.* **111**, 212–219.
289. Nes, M., van Staveren, W. A., Zajkas, G., Inelmen, E. M., and Moreiras-Varela, O. (1991). Validity of the dietary history method in elderly subjects. Euronut SENECA investigators. *Eur. J. Clin. Nutr.* **45**(suppl 3), 97–104.
290. Jain, M., Howe, G. R., and Rohan, T. (1996). Dietary assessment in epidemiology: comparison of a food frequency and a diet history questionnaire with a 7-day food record. *Am. J. Epidemiol.* **143**, 953–960.

291. EPIC Group of Spain. (1997). Relative validity and reproducibility of a diet history questionnaire in Spain. I. Foods. *Int. J. Epidemiol.* **26**(suppl 1), S91–S99.
292. EPIC Group of Spain. (1997). Relative validity and reproducibility of a diet history questionnaire in Spain. II. Nutrients. *Int. J. Epidemiol.* **26**(suppl 1), S100–S109.
293. van Liere, M. J., Lucas, F., Clavel, F., Slimani, N., and Villeminot, S. (1997). Relative validity and reproducibility of a French dietary history questionnaire. *Int. J. Epidemiol.* **26**(suppl 1), S128–S136.
294. Liu, K., Slattery, M., Jacobs, D. J., Cutter, G., McDonald, A., Van Horn, L., Hilner, J. E., Caan, B., Bragg, C., and Dyer, A. (1994). A study of the reliability and comparative validity of the Cardia Dietary History. *Ethn. Dis.* **4**, 15–27.
295. Rothenberg, E., Bosaeus, I., Lernfelt, B., Landahl, S., and Steen, B. (1998). Energy intake and expenditure: validation of a diet history by heart rate monitoring, activity diary and doubly labeled water. *Eur. J. Clin. Nutr.* **52**, 832–838.
296. EPIC Group of Spain. (1997). Relative validity and reproducibility of a diet history questionnaire in Spain. III. Biochemical markers. *Int. J. Epidemiol.* **26**(suppl 1), S110–S117.
297. Lytle, L. A., Nichaman, M. Z., Obarzanek, E., Glovsky, E., Montgomery, D., Nicklas, T., Zive, M., and Feldman, H. (1993). Validation of 24-hour recalls assisted by food records in third-grade children. *J. Am. Diet. Assoc.* **93**, 1431–1436.
298. Weber, J. L., Lytle, L., Gittelsohn, J., Cunningham-Sabo, L., Heller, K., Anliker, J. A., Stevens, J., Hurley, J., and Ring, K. (2004). Validity of self-reported dietary intake at school meals by American Indian children: the Pathways Study. *J. Am. Diet. Assoc.* **104**, 746–752.
299. Andersen, L. F., Bere, E., Kolbjornsen, N., and Klepp, K. I. (2004). Validity and reproducibility of self-reported intake of fruit and vegetable among 6th graders. *Eur. J. Clin. Nutr.* **58**, 771–777.
300. Kristjansdottir, A. G., Andersen, L. F., Haraldsdottir, J., de Almeida, M. D., and Thorsdottir, I. (2006). Validity of a questionnaire to assess fruit and vegetable intake in adults. *Eur. J. Clin. Nutr.* **60**, 408–415.
301. Amend, A., Melkus, G. D., Chyun, D. A., Galasso, P., and Wylie-Rosett, J. (2007). Validation of dietary intake data in black women with type 2 diabetes. *J. Am. Diet. Assoc.* **107**, 112–117.
302. Thompson, F., Subar, A., Potischman, N., Midthune, D., Kipnis, V., Troiano, R. P., and Schatzkin, A. (2006). A checklist-adjusted food frequency method for assessing dietary intake. *In* "Sixth International Conference on Dietary Assessment Methods: Complementary Advances in Diet and Physical Activity Assessment Methodologies" Diet Research Foundation, the Danish Network of Nutritional Epidemiologists, Copenhagen, Denmark .
303. Tooze, J. A., Midthune, D., Dodd, K. W., Freedman, L. S., Krebs-Smith, S. M., Subar, A. F., Guenther, P. M., Carroll, R. J., and Kipnis, V. (2006). A new statistical method for estimating the usual intake of episodically consumed foods with application to their distribution. *J. Am. Diet. Assoc.* **106**, 1575–1587.
304. Raper, N., Perloff, B., Ingwersen, L., Steinfeldt, L., and Anand, J. (2004). An overview of USDA's dietary intake data system. *J. Food Compost. Anal.* **17**, 545–555.
305. Foodlink archives. (2007). Agricultural Research Services, U.S. Department of Agriculture, www.ars.usda.gov/Services/docs.htm?docid=8634.
306. U.S. Department of Health and Human Services (2000). "Nutrition and Your Health: Dietary Guidelines for Americans," 5th ed. U.S. Government Printing Office, Washington, DC.
307. U.S. Department of Health and Human Services, and U.S. Department of Agriculture. (2005). "Dietary Guidelines for Americans, 2005." U.S. Government Printing Office, Washington, DC.
308. National Center for Health Statistics. (1992). Dietary methodology workshop for the Third National Health and Nutrition Examination Survey. Vital and Health Statistics 4(27), Publication No: 92–1464. Department of Health and Human Services, Hyattsville, MD.
309. National Center for Health Statistics. (1994). Plan and operation of the Third National Center for Health and Nutrition Examination Survey, 1988–1994. Vital Health Stat 1(32), Publ No. 94–1308. Department of Health and Human Services, Hyattsville, MD.
310. National Center for Health Statistics (1985). "Plan of Operation of the Hispanic Health and Nutrition Examination Survey 1982–84." U.S. Dept. of Health Human Services, Hyattsville, MD.
311. U.S. Department of Agriculture, H.N.I.S. (1984). Nutrient Intakes: Individuals in 48 States, Year 1977–78. Nationwide Food Consumption Survey 1977–78. Report No. I-2. U.S. Department of Agriculture, Hyattsville, MD.
312. U.S. Department of Agriculture, H.N.I.S.C.N.D. (1983). Food Intakes: Individuals in 48 States, Year 1977–78. Nationwide Food Consumption Survey 1977–78. Report No. I-1. U.S. Department of Agriculture, Hyattsville, MD.
313. Peterkin, B. B., Rizek, R. L., and Tippett, K. S. (1988). Nationwide food consumption survey, 1987. *Nutr. Today,* 18–23.
314. U.S. Department of Agriculture, H.N.I.S. (1993). Food and nutrient intakes by individuals in the United States, 1 day, 1987-88: Nationwide food consumption survey 1987–88. NFCS Rep. No. 87-I-1. U.S. Department of Agriculture, Hyattsville, MD.
315. U.S. Department of Agriculture, H.N.I.S. (1987). CSFII. Nationwide Food Consumption Survey, Continuing Survey of Food Intakes by Individuals. Women 19–50 Years and Their Children 1–5 Years, 1 Day. CSFII Report No. 86-1. U.S. Department of Agriculture, Hyattsville, MD.
316. U.S. Department of Agriculture, H.N.I.S. (1986). CSFII. Nationwide Food Consumption Survey. Continuing Survey of Food Intakes by Individuals. Women 19–50 Years and their Children 1–5 Years, 4 Days. NFCS, CSFII Report No. 86-3. U.S. Department of Agriculture, Hyattsville, MD.
317. U.S. Department of Agriculture, H.N.I.S.N.M.D. (1985). Nationwide food consumption survey continuing survey of food intakes by individuals: women 19–50 years and their children 1–5 years, 1 day. NFCS, CSFII Report No. 85-1. U.S. Department of Agriculture, Hyattsville, MD.
318. U.S. Department of Agriculture, H.N.I.S.N.M.D. (1985). Nationwide food consumption survey continuing survey of food intakes by individuals: low income women 19–50 years

and their children 1–5 years, 1 day. NFCS, CSFII Report No. 85-2. U.S. Department of Agriculture, Hyattsville, MD.
319. U.S. Department of Agriculture, H.N.I.S.N.M.D. (1985). Nationwide food consumption survey continuing survey of food intakes by individuals: low-income women 19–50 years and their children 1–5 years, 4 days. NFCS, CSFII Report No. 85-5. U.S. Department of Agriculture, Hyattsville, MD.
320. U.S. Department of Agriculture, H.N.I.S.N.M.D. (1986). Nationwide food consumption survey continuing of food intakes by individuals: low- income women 19–50 years and their children 1–5 years, 1 day. NFCS, CSFII Report No. 86-2. U.S. Department of Agriculture, Hyattsville, MD.
321. U.S. Department of Agriculture, H.N.I.S.N.M.D. (1986). Nationwide food consumption continuing survey of food intakes by individuals: low-income women 19–50 years and their children 1–5 years, 4 days. NFCS, CSFII, Report No. 86–4. U.S. Department of Agriculture, Hyattsville, MD.
322. U.S. Department of Agriculture, H.N.I.S.N.M.D. (1985). Nationwide food consumption survey continuing survey of food intakes by individuals: men 19–50 years, 1 day. NFCS, CSFII Rep. No. 85-3. U.S. Department of Agriculture, Hyattsville, MD.
323. U.S. Department of Agriculture, H.N.I.S. (1993). NFCS continuing survey of food intakes by individuals, 1989. 5285 Port Royal Rd, Springfield, VA 22161: Accession No. PB93-500411. Computer Tape. U.S. Dept. Commerce, National Technical Information Service, Springfield, VA.
324. U.S. Department of Agriculture, H.N.I.S. (1993). NFCS continuing survey of food intakes by individuals, 1990. 5285 Port Royal Rd, Springfield, VA 22161. Accession No. PB93-504843. Computer Tape. U.S. Department of Commerce, National Technical Information Service, Springfield, VA.
325. U.S. Department of Agriculture, H.N.I.S. (1994). NFCS continuing survey of food intakes by individuals, 1991. 5285 Port Royal Rd, Springfield, VA 22161, Accession No. PB94-500063, Computer Tape. U.S. Dept. of Commerce, National Technical Information Service, Springfield, VA.
326. Interagency Board for Nutrition Monitoring and Related Research. (1992). Nutrition Monitoring in the United States: the Directory of Federal and State Nutrition Monitoring Activities. (PHS) 92-1255-1. Department of Health and Human Services, Hyattsville, MD.
327. U.S. Department of Agriculture, A.R.S. (1998). Food and nutrient intakes by individuals in the United States, by sex and age, 1994–96, Nationwide Food Surveys. NFS Report No. 96-2. U.S. Department of Agriculture, Hyattsville, MD.
328. Subar, A. F., Dodd, K. W., Guenther, P. M., Kipnis, V., Midthune, D., McDowell, M., Tooze, J. A., Freedman, L. S., and Krebs Smith, S. M. (2006). The food propensity questionnaire: concept, development, and validation for use as a covariate in a model to estimate usual food intake. *J. Am. Diet. Assoc.* **106**, 1556–1563.
329. Food surveys research group. (2007). Agricultural Research Services, U.S. Department of Agriculture, http://www.ars.usda.gov/main/site _main.htm?modecode=12-35-50-00.
330. National Health and Nutrition Examination Survey. (2007). National Center for Health Statistics, www.cdc.gov/nchs/about/major/nhanes/survey_results_and_products.htm.
331. California statewide surveys. (2004). California Department of Health Services, www.dhs.ca.gov/ps/cdic/cpns/research/rea_surveys.htm.
332. Friedenreich, C. M., Slimani, N., and Riboli, E. (1992). Measurement of past diet: review of previous and proposed methods. *Epidemiol. Rev.* **14**, 177–196.
333. Maruti, S. S., Feskanich, D., Rockett, H. R., Colditz, G. A., Sampson, L. A., and Willett, W. C. (2006). Validation of adolescent diet recalled by adults. *Epidemiology* **17**, 226–229.
334. Thompson, F. E., Lamphiear, D. E., Metzner, H. L., Hawthorne, V. M., and Oh, M. S. (1987). Reproducibility of reports of frequency of food use in the Tecumseh diet methodology study. *Am. J. Epidemiol.* **125**, 658–671.
335. Malila, N., Virtanen, M., Pietinen, P., Virtamo, J., Albanes, D., Hartman, A. M., and Heinonen, O. P. (1998). A comparison of prospective and retrospective assessments of diet in a study of colorectal cancer. *Nutr. Cancer* **32**, 146–153.
336. Friedenreich, C. M., Howe, G. R., and Miller, A. B. (1991). An investigation of recall bias in the reporting of past food intake among breast cancer cases and controls. *Ann. Epidemiol.* **1**, 439–453.
337. Friedenreich, C. M., Howe, G. R., and Miller, A. B. (1991). The effect of recall bias on the association of calorie-providing nutrients and breast cancer. *Epidemiology* **2**, 424–429.
338. Willett, W. C., and Hu, F. B. (2007). The food frequency questionnaire. *Cancer Epidemiol. Biomarkers Prev.* **16**, 182–183.
339. Freedman, L. S., Schatzkin, A., Thiebaut, A. C., Potischman, N., Subar, A. F., Thompson, F. E., and Kipnis, V. (2007). Abandon neither the food frequency questionnaire nor the dietary fat-breast cancer hypothesis [letter to the editor]. *Cancer Epidemiol. Biomarkers Prev.* **16**, 1321–1322.
340. Goldbohm, R. A., van't Veer, P., van den Brandt, P. A., van't, H., Brants, H. A., Sturmans, F., and Hermus, R. J. (1995). Reproducibility of a food frequency questionnaire and stability of dietary habits determined from five annually repeated measurements. *Eur. J. Clin. Nutr.* **49**, 420–429.
341. Freedman, L. S., Carroll, R. J., and Wax, Y. (1991). Estimating the relation between dietary intake obtained from a food frequency questionnaire and true average intake. *Am. J. Epidemiol.* **134**, 310–320.
342. Rosner, B., Willett, W. C., and Spiegelman, D. (1989). Correction of logistic regression relative risk estimates and confidence intervals for systematic within-person measurement error. *Stat. Med.* **8**, 1051–1069.
343. Kaaks, R., Plummer, M., Riboli, E., Esteve, J., and van Staveren, W. (1994). Adjustment for bias due to errors in exposure assessments in multicenter cohort studies on diet and cancer: a calibration approach. *Am. J. Clin. Nutr.* **59**, 245s–250s.
344. Carroll, R. J., Freedman, L. S., and Kipnis, V. (1998). Measurement error and dietary intake. *Adv. Exp. Med. Biol.* **445**, 139–145.
345. Ritenbaugh, C., Patterson, R. E., Chlebowski, R. T., Caan, B., Fels Tinker, L., Howard, B., and Ockene, J. (2003). The Women's Health Initiative Dietary Modification trial: overview and baseline characteristics of participants. *Ann. Epidemiol.* **13**, S87–S97.

346. Briefel, R. R. (1994). Assessment of the US diet in national nutrition surveys: national collaborative efforts and NHANES. *Am. J. Clin. Nutr.* **59**, 164s–167s.
347. Georgiou, C. C. (1993). Saturated fat intake of elderly women reflects perceived changes in their intake of foods high in saturated fat and complex carbohydrate. *J. Am. Diet. Assoc.* **93**, 1444–1445.
348. Srinath, U., Shacklock, F., Scott, L. W., Jaax, S., and Kris-Etherton, P. M. (1993). Development of MEDFICTS—a dietary assessment instrument for evaluating fat, saturated fat, and cholesterol intake. *J. Am. Diet. Assoc.* October 28, A105(abstract).
349. Prochaska, J. O., DiClemente, C. C., and Norcross, J. C. (1992). In search of how people change: applications to addictive behaviors. *Am. Psychol.* **47**, 1102–1114.
350. Glanz, K., Patterson, R. E., Kristal, A. R., DiClemente, C. C., Heimendinger, J., Linnan, L., and McLerran, D. F. (1994). Stages of change in adopting healthy diets: fat, fiber, and correlates of nutrient intake [published erratum appears in *Health Educ. Q.,* May 1995, **22**(2), 261]. *Health Educ. Q.* **21**, 499–519.
351. Cheadle, A., Psaty, B. M., Diehr, P., Koepsell, T., Wagner, E., Curry, S., and Kristal, A. (1995). Evaluating community-based nutrition programs: comparing grocery store and individual-level survey measures of program impact. *Prev. Med.* **24**, 71–79.
352. Cheadle, A., Psaty, B. M., Curry, S., Wagner, E., Diehr, P., Koepsell, T., and Kristal, A. (1993). Can measures of the grocery store environment be used to track community-level dietary changes? *Prev. Med.* **22**, 361–372.
353. Smith-Warner, S. A., Elmer, P. J., Tharp, T. M., Fosdick, L., Randall, B., Gross, M., Wood, J., and Potter, J. D. (2000). Increasing vegetable and fruit intake: randomized intervention and monitoring in an at-risk population. *Cancer Epidemiol. Biomarkers Prev.* **9**, 307–317.
354. Sasaki, S., Ishikawa, T., Yanagibori, R., and Amano, K. (1999). Responsiveness to a self-administered diet history questionnaire in a work-site dietary intervention trial for mildly hypercholesterolemic Japanese subjects: correlation between change in dietary habits and serum cholesterol levels. *J. Cardiol.* **33**, 327–338.
355. Samet, J. M., and Alberg, A. J. (1998). Surrogate sources of dietary information. *In* "Nutritional Epidemiology" (W Willett, Ed.), Oxford University Press, New York.
356. Kolonel, J. N., Hirohata, T., and Nomura, A. M. Y. (1977) Adequacy of survey data collected from substitute respondents. *Am. J. Epidemiol.* **106**(6), 476–484.
357. Marshall, J., Priore, R., Haughey, B., Rzepka, T., and Graham, S. (1980). Spouse-subject interviews and the reliability of diet studies. *Am. J. Epidemiol.* **112**(6), 675–683.
358. Humble, C. G., Samet, J. M., and Skipper, B. E. (1984) Comparison of self-and surrogate-reported dietary information. *Am. J. Epidemiol.* **119**(1) 86–98.
359. Metzner, H. L., Lamphiear, D. E., Thompson, F. E., Oh M. S., and Hawthorne, V. M. (1989). Comparison of surrogate and subject reports of dietary practices, smoking habits and weight among married couples in the Tecumseh Diet Methodology Study. *J. Clin. Epidemiol.* **42** 367–375.
360. Hislop, T. G., Coldman, A. J., Zheng, Y. Y., Ng, V. T., and Labo, T. (1992). Reliability of dietary information from surrogate respondents. *Nutr. Cancer* **18**, 123–129.
361. Herrmann, N. (1985). Retrospective information from questionnaires. I. Comparability of primary respondents and their next-of-kin. *Am. J. Epidemiol.* **121**, 937–947.
362. Petot, G. J., Debanne, S. M., Riedel, T. M., Smyth, K. A., Koss, E., Lerner, A. J., and Friedland, R. P. (2002). Use of surrogate respondents in a case control study of dietary risk factors for Alzheimer's disease. *J. Am. Diet. Assoc.* **102**, 848–850.
363. Fryzek, J. P., Lipworth, L., Signorello, L. B., and McLaughlin, J. K. (2002). The reliability of dietary data for self and next-of-kin respondents. *Ann. Epidemiol.* **12**, 278–283.
364. Hankin, J. H., and Wilkens, L. R. (1994). Development and validation of dietary assessment methods for culturally diverse populations. *Am. J. Clin. Nutr.* **59**, 198s–200s.
365. Lyons, G. K., Woodruff, S. I., Candelaria, J. I., Rupp, J. W., and Elder, J. P. (1996). Development of a protocol to assess dietary intake among Hispanics who have low literacy skills in English. *J. Am. Diet. Assoc.* **96**, 1276–1279.
366. Loria, C. M., McDowell, M. A., Johnson, C. L., and Woteki, C. E. (1991). Nutrient data for Mexican-American foods: are current data adequate? *J. Am. Diet. Assoc.* **91**, 919–922.
367. Indian Health Service. (1992). Navajo Health and Nutrition Survey Manual. Navajo Area Indian Health Service, Nutrition and Dietetics Branch, Health Promotion/Disease Prevention Program.
368. Lee, M. M., Lee, F., Wang-Ladenla, S., and Miike, R. (1994). A semiquantitative dietary history questionnaire for Chinese Americans. *Ann. Epidemiol.* **4**, 188–197.
369. Hebert, J. R., Gupta, P. C., Bhonsle, R. B., Sinor, P. N., Mehta, H., and Mehta, F. S. (1999). Development and testing of a quantitative food frequency questionnaire for use in Gujarat, India. *Public Health Nutr.* **2**, 39–50.
370. Taren, D., de Tobar, M., Ritenbaugh, C., Graver, E., Whitacre, R., and Aiken, M. (2000). Evaluation of the Southwest Food Frequency Questionnaire. *Ecol. Food Nutr.* **38**, 515–547.
371. Shahar, D., Shai, I., Vardi, H., Brener-Azrad, A., and Fraser, D. (2003). Development of a semi-quantitative Food Frequency Questionnaire (FFQ) to assess dietary intake of multiethnic populations. *Eur. J. Epidemiol.* **18**, 855–861.
372. Sharma, S., Cade, J., Landman, J., and Cruickshank, J. K. (2002). Assessing the diet of the British African-Caribbean population: frequency of consumption of foods and food portion sizes. *Int. J. Food Sci. Nutr.* **53**, 439–444.
373. Tucker, K. L., Bianchi, L. A., Maras, J., and Bermudez, O. I. (1998). Adaptation of a food frequency questionnaire to assess diets of Puerto Rican and non-Hispanic adults. *Am. J. Epidemiol.* **148**, 507–518.
374. Tucker, K. L., Maras, J., Champagne, C., Connell, C., Goolsby, S., Weber, J., Zaghloul, S., Carithers, T., and Bogle, M. L. (2005). A regional food-frequency questionnaire for the US Mississippi Delta. *Public Health Nutr.* **8**, 87–96.
375. Coates, R. J., and Monteilh, C. P. (1997). Assessments of food frequency questionnaires in minority populations. *Am. J. Clin. Nutr.* **65**, 1108S–1115S.
376. Mayer-Davis, E. J., Vitolins, M. Z., Carmichael, S. L., Hemphill, S., Tsaroucha, G., Rushing, J., and Levin, S.

(1999). Validity and reproducibility of a food frequency interview in a multi-cultural epidemiologic study. *Ann. Epidemiol.* **9**, 314–324.

377. Baumgartner, K. B., Gilliland, F. D., Nicholson, C. S., McPherson, R. S., Hunt, W. C., Pathak, D. R., and Samet, J. M. (1998). Validity and reproducibility of a food frequency questionnaire among Hispanic and non-Hispanic white women in New Mexico. *Ethn. Dis.* **8**, 81–92.
378. Cullen, K. W., and Zakeri, I. (2004). The youth/adolescent questionnaire has low validity and modest reliability among low-income African-American and Hispanic seventh-and eighth grade-youth. *J. Am. Diet. Assoc.* **104**, 415–419.
379. Baranowski, T., and Domel, S. B. (1994). A cognitive model of children's reporting of food intake. *Am. J. Clin. Nutr.* **59**, 212s–217s.
380. Hertzler, A. A. (1990). A review of methods to research nutrition knowledge and dietary intake of preschoolers. *Topics Clin. Nutr.* **6**, 1–9.
381. Rockett, H. R., and Colditz, G. A. (1997). Assessing diets of children and adolescents. *Am. J. Clin. Nutr.* **65**, 1116S–1122S.
382. McPherson, R. S., Hoelscher, D. M., Alexander, M., Scanlon, K. S., and Serdula, M. K. (2000). Dietary assessment methods among school-aged children: validity and reliability. *Prev. Med.* **31**, S11–S33.
383. Livingstone, M. B., and Robson, P. J. (2000). Measurement of dietary intake in children. *Proc. Nutr. Soc.* **59**, 279–293.
384. Serdula, M. K., Alexander, M. P., Scanlon, K. S., and Bowman, B. A. (2001). What are preschool children eating? A review of dietary assessment. *Annu. Rev. Nutr.* **21**, 475–498.
385. Rockett, H. R., Berkey, C. S., and Colditz, G. A. (2003). Evaluation of dietary assessment instruments in adolescents. *Curr. Opin. Clin. Nutr. Metab. Care* **6**, 557–562.
386. NCS dietary assessment literature review. (2007). National Cancer Institute, http://riskfactor.cancer.gov/tools/children/review.
387. Dennison, B. A., Jenkins, P. L., and Rockwell, H. L. (2000). Development and validation of an instrument to assess child dietary fat intake. *Prev. Med.* **31**, 214–224.
388. Koehler, K. M., Cunningham-Sabo, L., Lambert, L. C., McCalman, R., Skipper, B. J., and Davis, S. M. (2000). Assessing food selection in a health promotion program: validation of a brief instrument for American Indian children in the southwest United States. *J. Am. Diet. Assoc.* **100**, 205–211.
389. Simons-Morton, B. G., and Baranowski, T. (1991). Observation in assessment of children's dietary practices. *J. Sch. Health* **61**, 204–207.
390. Matheson, D. M., Hanson, K. A., McDonald, T. E., and Robinson, T. N. (2002). Validity of children's food portion estimates: a comparison of 2 measurement aids. *Arch. Pediatr. Adolesc. Med.* **156**, 867–871.
391. Frobisher, C., and Maxwell, S. M. (2003). The estimation of food portion sizes: a comparison between using descriptions of portion sizes and a photographic food atlas by children and adults. *J. Hum. Nutr. Diet.* **16**, 181–188.
392. Baranowski, T., Sprague, D., Baranowksi, J. H., and Harrison, J. A. (1991). Accuracy of maternal dietary recall for preschool children. *J. Am. Diet. Assoc.* **91**, 669–674.
393. Eck, L. H., Klesges, R. C., and Hanson, C. L. (1989). Recall of a child's intake from one meal: are parents accurate? *J. Am. Diet. Assoc.* **89**, 784–789.
394. Sobo, E. J., and Rock, C. L. (2001). "You ate all that!?" Caretaker-child interaction during children's assisted dietary recall interviews. *Med. Anthropol. Q.* **15**, 222–244.
395. Rockett, H. R., Breitenbach, M., Frazier, A. L., Witschi, J., Wolf, A. M., Field, A. E., and Colditz, G. A. (1997). Validation of a youth/adolescent food frequency questionnaire. *Prev. Med.* **26**, 808–816.
396. Klohe, D. M., Clarke, K. K., George, G. C., Milani, T. J., Hanss Nuss, H., and Freeland Graves, J. (2005). Relative validity and reliability of a food frequency questionnaire for a triethnic population of 1-year-old to 3-year-old children from low-income families. *J. Am. Diet. Assoc.* **105**, 727–734.
397. McDowell, M. A., Harris, T. B., and Briefel, R. R. (1991). Dietary surveys of older persons. *Clin. Appl. Nutr.* **1**, 51–60.
398. van Staveren, W. A., de Groot, L. C., Blauw, Y. H., and van der Wielen, R. P. (1994). Assessing diets of elderly people: problems and approaches. *Am. J. Clin. Nutr.* **59**, 221s–223s.
399. Mitchell, D. C., Smiciklas-Wright, H., Friedmann, J. M., and Jensen, G. (2002). Dietary intake assessed by the Nutrition Screening Initiative Level II Screen is a sensitive but not a specific indicator of nutrition risk in older adults. *J. Am. Diet. Assoc.* **102**, 842–844.
400. Guigoz, Y. (2006). The Mini Nutritional Assessment (MNA) review of the literature—What does it tell us? *J. Nutr. Health Aging* **10**, 466–485.
401. Chianetta, M. M., and Head, M. K. (1992). Effect of prior notification on accuracy of dietary recall by the elderly. *J. Am. Diet. Assoc.* **92**, 741–743.
402. Klipstein-Grobusch, K., den Breeijen, J. H., Goldbohm, R. A., Geleijnse, J. M., Hofman, A., Grobbee, D. E., and Witteman, J. C. (1998). Dietary assessment in the elderly: validation of a semiquantitative food frequency questionnaire. *Eur. J. Clin. Nutr.* **52**, 588–596.
403. Goldbohm, R. A., van den Brandt, P. A., Brants, H. A., van't Veer, P. A., Al, M., Sturmans, F., and Hermus, R. J. (1994). Validation of a dietary questionnaire used in a large-scale prospective cohort study on diet and cancer. *Eur. J. Clin. Nutr.* **48**, 253–265.
404. Thompson, C. H., Head, M. K., and Rodman, S. M. (1987). Factors influencing accuracy in estimating plate waste. *J. Am. Diet. Assoc.* **87**, 1219–1220.
405. Guthrie, H. A. (1984). Selection and quantification of typical food portions by young adults. *J. Am. Diet. Assoc.* **84**, 1440–1444.
406. Nelson, M., Atkinson, M., and Darbyshire, S. (1996). Food photography II: use of food photographs for estimating portion size and the nutrient content of meals. *Br. J. Nutr.* **76**, 31–49.
407. Hebert, J. R., Gupta, P. C., Bhonsle, R., Verghese, F., Ebbeling, C., Barrow, R., Ellis, S., and Ma, Y. (1999). Determinants of accuracy in estimating the weight and volume of commonly used foods: a cross-cultural comparison. *Ecol. Food Nutr.* **37**, 475–502.
408. Young, L. R., and Nestle, M. S. (1995). Portion sizes in dietary assessment: issues and policy implications. *Nutr. Rev.* **53**, 149–158.

409. Young, L. R., and Nestle, M. (1998). Variation in perceptions of a medium food portion: implications for dietary guidance. *J. Am. Diet. Assoc.* **98**, 458–459.
410. Cypel, Y. S., Guenther, P. M., and Petot, G. J. (1997). Validity of portion-size measurement aids: a review. *J. Am. Diet. Assoc.* **97**, 289–292.
411. Hernandez, T., Wilder, L., Kuehn, D., Rubotzky, K., Veillon, P. M., Godwin, S., Thompson, C., and Wang, C. (2006). Portion size estimation and expectation of accuracy. *J. Food Compost. Anal.* **19**, S14–S21.
412. Chambers, E., McGuire, B., Godwin, S., McDowell, M., and Vecchio, F. (2000). Quantifying portion sizes for selected snack foods and beverages in 24-hour dietary recalls. *Nutr. Res.* **20**, 315–326.
413. Robinson, F., Morritz, W., McGuiness, P., and Hackett, A. F (1997). A study of the use of a photographic food atlas to estimate served and self-served portion sizes. *J. Hum. Nutr. Diet.* **10**, 117–124.
414. Robson, P. J., and Livingstone, M. B. (2000). An evaluation of food photographs as a tool for quantifying food and nutrient intakes. *Public Health Nutr.* **3**, 183–192.
415. Godwin, S. L., Chambers, E., and Cleveland, L. (2004) Accuracy of reporting dietary intake using various portion-size aids in-person and via telephone. *J. Am. Diet. Assoc* **104**, 585–594.
416. Smith, A. F., Jobe, J. B., and Mingay, D. J. (1991). Question induced cognitive biases in reports of dietary intake by college men and women. *Health. Psychol.* **10**, 244–251.
417. Harnack, L., Steffen, L., Arnett, D. K., Gao, S., and Luepker R. V. (2004). Accuracy of estimation of large food portions *J. Am. Diet. Assoc.* **104**, 804–806.
418. Pao, E.M. (1987). Validation of food intake reporting by men. Administrative Report No: 382. U.S. Department of Agriculture, Human Nutrition Information Service, Hyattsville, MD.
419. Posner, B. M., Smigelski, C., Duggal, A., Morgan, J. L., Cobb, J., and Cupples, L. A. (1992). Validation of two-dimensional models for estimation of portion size in nutrition research. *J. Am. Diet. Assoc.* **92**, 738–741.
420. Bolland, J. E., Yuhas, J. A., and Bolland, T. W. (1988). Estimation of food portion sizes: effectiveness of training. *J. Am. Diet. Assoc.* **88**, 817–821.
421. Howat, P. M., Mohan, R., Champagne, C., Monlezun, C., Wozniak, P., and Bray, G. A. (1994). Validity and reliability of reported dietary intake data. *J. Am. Diet. Assoc.* **94**, 169–173.
422. Weber, J. L., Tinsley, A. M., Houtkooper, L. B., and Lohman, T. G. (1997). Multimethod training increases portion-size estimation accuracy. *J. Am. Diet. Assoc.* **97**, 176–179.
423. Williamson, D. A., Allen, H. R., Martin, P. D., Alfonso, A. J., Gerald, B., and Hunt, A. (2003). Comparison of digital photography to weighed and visual estimation of portion sizes. *J. Am. Diet. Assoc.* **103**, 1139–1145.
424. Bolland, J. E., Ward, J. Y., and Bolland, T. W. (1990). Improved accuracy of estimating food quantities up to 4 weeks after training. *J. Am. Diet. Assoc.* **90**, 1402–4, 1407.
425. Fox, T. A., Heimendinger, J., and Block, G. (1992). Telephone surveys as a method for obtaining dietary information: a review. *J. Am. Diet. Assoc.* **92**, 729–732.
426. Dillman, D. A. (1978). "Mail and Telephone Surveys: The Total Design Method." John Wiley & Sons, New York.
427. Marcus, A. C., and Crane, L. A. (1986). Telephone surveys in public health research. *Med. Care* **24**, 97–112.
428. Morgan, K. J., Johnson, S. R., Rizek, R. L., Reese, R., and Stampley, G. L. (1987). Collection of food intake data: an evaluation of methods. *J. Am. Diet. Assoc.* **87**, 888–896.
429. Leighton, J., Neugut, A. I., and Block, G. (1988). A comparison of face-to-face frequency interviews and self-administered questionnaires. *Am. J. Epidemiol.* **128**, 891(Abstract).
430. Lyu, L. C., Hankin, J. H., Liu, L. Q., Wilkens, L. R., Lee, J. H., Goodman, M. T., and Kolonel, L. N. (1998). Telephone vs face-to-face interviews for quantitative food frequency assessment. *J. Am. Diet. Assoc.* **98**, 44–48.
431. Moshfegh, A. M., Borrud, L. G., Perloff, P., and LaComb, R. (1999). Improved method for the 24-hour dietary recall for use in National Surveys. *FASEB. J.* **13**, A603.
432. Bogle, M., Stuff, J., Davis, L., Forrester, I., Strickland, E., Casey, P. H., Ryan, D., Champagne, C., McGee, B., Mellad, K., Neal, E., Zaghloul, S., Yadrick, K., and Horton, J. (2001). Validity of a telephone-administered 24-hour dietary recall in telephone and non-telephone households in the rural Lower Mississippi Delta region. *J. Am. Diet. Assoc.* **101**, 216–222.
433. Brustad, M., Skeie, G., Braaten, T., Slimani, N., and Lund, E. (2003). Comparison of telephone vs face-to-face interviews in the assessment of dietary intake by the 24 h recall EPIC SOFT program the Norwegian calibration study. *Eur. J. Clin. Nutr.* **57**, 107–113.
434. Posner, B. M., Borman, C. L., Morgan, J. L., Borden, W. S., and Ohls, J. C. (1982). The validity of a telephone-administered 24-hour dietary recall methodology. *Am. J. Clin. Nutr.* **36**, 546–553.
435. Krantzler, N. J., Mullen, B. J., Schutz, H. G., Grivetti, L. E., Holden, C. A., and Meiselman, H. L. (1982). Validity of telephoned diet recalls and records for assessment of individual food intake. *Am. J. Clin. Nutr.* **36**, 1234–1242.
436. Gruber, L., Shadle, M., and Polich, C. (1988). From movement to industry: the growth of HMOs. *Health Aff.*, 197–208.
437. Yanek, L. R., Moy, T. F., Raqueno, J. V., and Becker, D. M. (2000). Comparison of the effectiveness of a telephone 24-hour dietary recall method vs an in-person method among urban African-American women. *J. Am. Diet. Assoc.* **100**, 1172–1177.
438. Blumberg, S. J., Luke, J. V., and Cynamon, M. L. (2006). Telephone coverage and health survey estimates: evaluating the need for concern about wireless substitution. *Am. J. Public Health* **96**, 926–931.
439. Matthys, C., Pynaert, I., Roe, M., Fairweather-Tait, S. J., Heath, A. L., and De Henauw, S. (2004). Validity and reproducibility of a computerised tool for assessing the iron, calcium and vitamin C intake of Belgian women. *Eur. J. Clin. Nutr.* **58**, 1297–1305.
440. Balter, K. A., Balter, O., Fondell, E., and Lagerros, Y. T. (2005). Web-based and mailed questionnaires: a comparison of response rates and compliance. *Epidemiology* **16**, 577–579.
441. Dodd, K. W., Guenther, P. M., Freedman, L. S., Subar, A. F., Kipnis, V., Midthune, D., Tooze, J. A., and Krebs-Smith,

S. M. (2006). Statistical methods for estimating usual intake of nutrients and foods: a review of the theory. *J. Am. Diet. Assoc.* **106**, 1640–1650.

442. Carriquiry, A. L. (2003). Estimation of usual intake distributions of nutrients and foods. *J. Nutr.* **133**, 601S–608S.
443. Hartman, A. M., Brown, C. C., Palmgren, J., Pietinen, P., Verkasalo, M., Myer, D., and Virtamo, J. (1990). Variability in nutrient and food intakes among older middle-aged men: implications for design of epidemiologic and validation studies using food recording. *Am. J. Epidemiol.* **132**, 999–1012.
444. Dietary reference intakes: applications in dietary planning. (2003). Institute of Medicine, Food and Nutrition Board, www.nap.edu/books/0309088534/html.
445. Nusser, S. M., Carriquiry, A. L., Dodd, K. W., and Fuller, W. A. (1996). A semiparametric transformation approach to estimating usual daily intake distributions. *J. Am. Stat. Assoc.* **91**, 1440–1449.
446. Guenther, P. M., Kott, P. S., and Carriquiry, A. L. (1997). Development of an approach for estimating usual nutrient intake distributions at the population level. *J. Nutr.* **127**, 1106–1112.
447. Nusser, S. M., Fuller, W. A., and Guenther, P. M. (1997). Estimating usual dietary intake distributions: adjusting for measurement error and non-normality in 24-hour food intake data. *In* "Survey Measurement and Process Quality" (L. Lyberg, P. Biemer, M. Collins, E. deLeeuw, C. Dippo, N. Schwartz, and D. Trewin, Eds.), pp. 689–709. Wiley, New York.
448. Schakel, S. F., Buzzard, I. M., and Gebhard, S. E. (1997). Procedures for estimating nutrient values for food composition databases. *J. Food Compost. Anal.* **10**, 102–114.
449. Beecher, G. R., and Matthews, R. H. (1990). Nutrient composition of foods. *In* "Present Knowledge in Nutrition" (M. L. Brown, Ed.) 6th ed., pp. 430–439. International Life Sciences Institute, Nutrition Foundation, Washington, DC.
450. Interagency Board for Nutrition Monitoring and Related Research. Third report on nutrition monitoring in the United States (Volume 1). U.S. Government Printing Office Washington, DC.
451. (1997). What are the variances of food composition data? *J. Food Compost. Anal.* **10**, 89.
452. Perloff, B. P. (1989). Analysis of dietary data. *Am. J. Clin. Nutr.* **50**, 1128–1132.
453. U.S. Department of Agriculture (2006). "Composition of Foods Raw, Processed, Prepared. USDA National Nutrient Database for Standard Reference, Release 19." Beltsville, MD.
454. Haytowitz, D.B., Patterson, K.Y., Pehrsson, P.R., Exler, J., and Holden, J.M. (2005). The National Food and Nutrient Analysis Program: accomplishments and lessons learned. U.S. Department of Agriculture, Beltsville, MD.
455. USDA nutrient data laboratory. (2007). National Agricultural Library, U.S. Department of Agricultural, http://fnic.-nal.usda.gov/nal_display/index.php?info_center=4&tax_level=2&tax_subject=279&topic_id=1387.
456. The International Network of Food Data Systems (INFOODS): International food composition tables directory. (2007). Agricultural and Consumer Protection Department, Food and Agricultural Organization of the United Nations, www.fao.org/infoods/directory_en.stm.
457. 2006 International nutrient databank directory. (2006). Steering Commitee of the National Nutrient databank Conference, www.medicine.uiowa.edu/gcrc/nndc/2006%20International%20Nutrient%20Databank%20Directory.pdf.
458. Smith, S. A., Campbell, D. R., Elmer, P. J., Martini, M. C., Slavin, J. L., and Potter, J. D. (1995). The University of Minnesota Cancer Prevention Research Unit vegetable and fruit classification scheme (United States). *Cancer Causes Control* **6**, 292–302.
459. Cleveland, L. E., Cook, D. A., Krebs-Smith, S. M., and Friday, J. (1997). Method for assessing food intakes in terms of servings based on food guidance. *Am. J. Clin. Nutr.* **65**, 1254S–1263S.
460. Pyramid servings database for USDA survey food codes version 2.0. (2007). Community Nutrition Research Group. U.S. Department of Agricultured, www.ba.ars.usda.gov/cnrg/services/deload.html.
461. Byers, T., and Gieseker, K. (1997). Issues in the design and interpretation of studies of fatty acids and cancer in humans. *Am. J. Clin. Nutr.* **66**, 1541S–1547S.
462. Innis, S. M., Green, T. J., and Halsey, T. K. (1999). Variability in the trans fatty acid content of foods within a food category: implications for estimation of dietary trans fatty acid intakes. *J. Am. Coll. Nutr.* **18**, 255–260.
463. Food and nutrient database. (2007). University of Minnesota, www.ncc.umn.edu/products/database.html.
464. Jacobs, D. R., Jr., Elmer, P. J., Gorder, D., Hall, Y., and Moss, D. (1985). Comparison of nutrient calculation systems. *Am. J. Epidemiol.* **121**, 580–592.
465. Lee, R. D., Nieman, D. C., and Rainwater, M. (1995). Comparison of eight microcomputer dietary analysis programs with the USDA Nutrient Data Base for Standard Reference. *J. Am. Diet. Assoc.* **95**, 858–867.
466. McCullough, M. L., Karanja, N. M., Lin, P. H., Obarzanek, E., Phillips, K. M., Laws, R. L., Vollmer, W. M., O'Connor, E. A., Champagne, C. M., and Windhauser, M. M. (1999). Comparison of 4 nutrient databases with chemical composition data from the Dietary Approaches to Stop Hypertension trial. DASH Collaborative Research Group. *J. Am. Diet. Assoc.* **99**, S45–S53.
467. Kristal, A. R., Shattuck, A. L., and Williams, A. E. (1992). Current issues and concerns on the uses of food composition data: food frequency questionnaires for diet intervention research. *In* "17th National Nutrient Databank Conference Proceedings, June 7–10, 1992, Baltimore, MD" pp 110–125. International Life Sciences Institute, Washington, DC.
468. Buzzard, I. M., Price, K. S., and Warren, R. A. (1991). Considerations for selecting nutrient calculation software: evaluation of the nutrient database. *Am. J. Clin. Nutr.* **54**, 7–9.
469. The Dell Center for Healthy Living: FIAS millennium. (2004). The University of Texas School of Public Health, www.sph.uth.tmc.edu/DellHealthyLiving/default.asp?id=4008.
470. Dwyer, J. T., Krall, E. A., and Coleman, K. A. (1987). The problem of memory in nutritional epidemiology research. *J. Am. Diet. Assoc.* **87**, 1509–1512.

471. Smith, A. F., Jobe, J. B., and Mingay, D. J. (1991). Retrieval from memory of dietary information. *Appl. Cogn. Psychol.* **5**, 269–296.
472. Friedenreich, C. M. (1994). Improving long-term recall in epidemiologic studies. *Epidemiology* **5**, 1–4.
473. Satia, J. A., Patterson, R. E., Taylor, V. M., Cheney, C. L., Shiu-Thornton, S., Chitnarong, K., and Kristal, A. R. (2000). Use of qualitative methods to study diet, acculturation, and health in Chinese-American women. *J. Am. Diet. Assoc.* **100**, 934–940.
474. Wolfe, W. S., Frongillo, E. A., and Cassano, P. A. (2001). Evaluating brief measures of fruit and vegetable consumption frequency and variety: cognition, interpretation, and other measurement issues. *J. Am. Diet. Assoc.* **101**, 311–318.
475. Thompson, F. E., Subar, A. F., Brown, C. C., Smith, A. F., Sharbaugh, C. O., Jobe, J. B., Mittl, B., Gibson, J. T., and Ziegler, R. G. (2002). Cognitive research enhances accuracy of food frequency questionnaire reports: results of an experimental validation study. *J. Am. Diet. Assoc.* **102**, 212–225.
476. Chambers, E., Godwin, S. L., and Vecchio, F. A. (2000). Cognitive strategies for reporting portion sizes using dietary recall procedures. *J. Am. Diet. Assoc.* **100**, 891–897.
477. Johnson-Kozlow, M., Matt, G. E., and Rock, C. L. (2006). Recall strategies used by respondents to complete a food frequency questionnaire: an exploratory study. *J. Am. Diet. Assoc.* **106**, 430–433.
478. Matt, G. E., Rock, C. L., and Johnson-Kozlow, M. (2006). Using recall cues to improve measurement of dietary intakes with a food frequency questionnaire in an ethnically diverse population: an exploratory study. *J. Am. Diet. Assoc.* **106**, 1209–1217.
479. Willis, G. (2005). "Cognitive Interviewing: a Tool of Improving Questionnaire Design." Sage Publications, Thousand Oaks, CA.
480. Buzzard, I. M., and Sievert, Y. A. (1994). Research priorities and recommendations for dietary assessment methodology. *Am. J. Clin. Nutr.* **59**, 275s–280s.
481. Dietary Assessment Calibration/Validation Register, (2007). National Cancer Institute, http://appliedresearch.cancer.gov/cgi-bin/dacv/index.pl.
482. Baranowski, T., Dworkin, R., Henske, J. C., Clearman, D. R., Dunn, J. K., Nader, P. R., and Hooks, P. C. (1986). The accuracy of children's self-reports of diet: Family Health Project. *J. Am. Diet. Assoc.* **86**, 1381–1385.
483. Domel, S. B., Baranowski, T., Leonard, S. B., Davis, H., Riley, P., and Baranowski, J. (1994). Accuracy of fourth and fifth-grade students' food records compared with school-lunch observations. *Am. J. Clin. Nutr.* **59**, 218s–220s.
484. Bingham, S. A. (1994). The use of 24-h urine samples and energy expenditure to validate dietary assessments. *Am. J. Clin. Nutr.* **59**, 227s–231s.
485. Samaras, K., Kelly, P. J., and Campbell, L. V. (1999). Dietary underreporting is prevalent in middle-aged British women and is not related to adiposity (percentage body fat). *Int. J. Obes. Relat. Metab. Disord.* **23**, 881–888.
486. Ajani, U. A., Willett, W. C., and Seddon, J. M. (1994). Reproducibility of a food frequency questionnaire for use in ocular research. Eye Disease Case-Control Study Group. *Invest. Ophthalmol. Vis. Sci.* **35**, 2725–2733.
487. Hartman, A. M., and Block, G. (1992). Dietary assessment methods for macronutrients. *In* "Macronutrients: Investigating Their Role in Cancer" (M. S. Micozzi, and T. E. Moon, Eds.), pp. 87–124. Marcel Dekker, New York.
488. Thompson, F. E., Kipnis, V., Subar, A. F., Krebs-Smith, S. M., Kahle, L. L., Midthune, D., Potischman, N., and Schatzkin, A. (2000). Evaluation of 2 short instruments and a food-frequency questionnaire to estimate daily number of servings of fruit and vegetables. *Am. J. Clin. Nutr.* **71**, 1503–1510.
489. Rosner, B., Spiegelman, D., and Willett, W. C. (1990). Correction of logistic regression relative risk estimates and confidence intervals for measurement error: the case of multiple covariates measured with error. *Am. J. Epidemiol.* **132**, 734–745.
490. Paeratakul, S., Popkin, B. M., Kohlmeier, L., Hertz-Picciotto, I., Guo, X., and Edwards, L. J. (1998). Measurement error in dietary data: implications for the epidemiologic study of the diet-disease relationship. *Eur. J. Clin. Nutr.* **52**, 722–727.
491. Plummer, M., and Clayton, D. (1993). Measurement error in dietary assessment: an investigation using covariance structure models. Part II. *Stat. Med.* **12**, 937–948.
492. Plummer, M., and Clayton, D. (1993). Measurement error in dietary assessment: an investigation using covariance structure models. Part I. *Stat. Med.* **12**, 925–935.

CHAPTER 2

Assessment of Dietary Supplement Use

JOHANNA DWYER AND REBECCA B. COSTELLO
Office of Dietary Supplements, National Institutes of Health, Bethesda, Maryland

Contents

I. INTRODUCTION

A. Rationale

Nutritional status assessment is incomplete without assessing intake of dietary supplements as well as food, because supplements provide many essential nutrients and other bioactive substances that affect health outcomes and disease. For those suffering from health problems, specific dietary supplements may be recommended or self-prescribed. Therefore, it is essential to assess intakes of dietary supplements because Americans so commonly use them today [1, 2].

Dietary supplements come in a variety of product formulations, but for the purposes of discussions in this chapter, dietary supplements are defined using the legal definitions in the Dietary Supplement Health and Education Act (DSHEA), which became law in 1994. A dietary supplement is a product (other than tobacco) that is intended to supplement the diet; contains one or more dietary ingredients (including vitamins, minerals, herbs or other botanicals, amino acids, and other substances) or their constituents; is intended to be taken by mouth as a pill, capsule, tablet, or liquid; and is labeled on the front panel as being a dietary supplement [3].

B. Purposes of Dietary Supplement Intake Assessment

1. Obtain Total Intakes of Nutrients

Dietary supplements contribute substantial amounts to total intakes of some nutrients, such as calcium in postmenopausal women, or folic acid, vitamin B_6 and vitamin B_{12} in patients receiving hemodialysis. Failure to include these nutrient sources would lead to serious underestimation of intakes and overall diet quality. Half of American adults report use of dietary supplements. Among the supplement users, nearly half use multivitamin-mineral supplements, and the remainder use single or multiple vitamin or mineral preparations that also contribute substantial amounts of nutrients to intakes [2]. Therefore, it is essential to assess dietary supplement intakes among users of these products.

2. Assess Risk of Toxicities

Supplements are often highly concentrated sources of nutrients, and when combined with frequent use of highly fortified foods, nutrient intakes may exceed the tolerable upper levels (UL), placing the individual at risk of toxicity. Some of the bioactive compounds in supplements can also have toxic effects, as was the case with the botanical ephedra, which was used in many weight loss and performance enhancing dietary supplements until the Food and Drug Administration (FDA) banned its use in 2004. Also, some dietary supplements of low quality may be contaminated by heavy metals or other ingredients that are toxic, and noting their use on medical records may be helpful.

3. Assess Nutrient-Nutrient and Nutrient-Drug Interactions

These interactions are important to document not only in order to assess and plan dietary intakes but to avoid drug-supplement interactions. Supplement-nutrient and supplement-drug interactions are of particular concern. Some dietary supplements, particularly botanicals, contain many bioactive substances that may interact with other drugs or nutrients [4].

4. Clarify Associations between Dietary Supplement Intake and Health Status

Data on dietary supplement use are also important to clarify associations between the supplements used and risk of various diseases. There may also be a relationship between possible changes in a supplement with changing patterns of disease.

5. Assess Conformity with Health Promotion and Disease Prevention Recommendations and Guidelines

The Food and Nutrition Board (FNB) of the Institute of Medicine, the National Academy of Sciences, the U.S. Preventive Health Services Task Force, Healthy People 2010, and numerous other bodies such as the Committee on Nutrition of the American Academy of Pediatrics and consensus statements issued on behalf of other professional societies make recommendations on the use of dietary supplements. Health screening tools may include queries to ascertain if individuals are following these guidelines. For example, the FNB recommends that women in the childbearing years who are at risk of becoming pregnant should take a dietary supplement of folic acid as well as assure that their food intakes of folate are adequate.

C. Health Profiles of Dietary Supplement Users

The health profiles of dietary supplement users differ from those of nonusers in some respects that affect health, making causative associations between supplement use and health status difficult to establish without first correcting for them. These factors include physical activity, smoking status, age, income, education, and prior health status. For example, multivitamin mineral supplement users tend to have better diets and to be healthier than nonusers, probably because of the differential presence of these other factors. However, generalization is hazardous. Cancer survivors often have high levels of dietary supplement use, in the hopes that it will stave off a return of the malignancy [5–7]. Users of "condition-specific" supplements often suffer from specific health problems. For example, those who are already ill or suffering from joint pain or osteoarthritis often use glucosamine and chondroitin or chondroitin sulphate as medicines [8]. Those with prostate problems or prostate cancer survivors commonly use saw palmetto [9]. There are many illnesses that consumers believe will be alleviated or cured by use of such condition-specific supplements.

D. Prevalence of Dietary Supplement Use

The prevalence of dietary supplement use has increased since 1994 when supplements became more widely available after the passage of the DSHEA. Today prevalence is high, with roughly half of all adults and substantial but somewhat lesser numbers of children using dietary supplements. The best data are from population-based samples, such as the National Health and Nutrition Examination Survey (NHANES). On the basis of such studies, it is evident that use is particularly high in certain subgroups. For example, it is very high among elders, somewhat less but also high among young children, and lower in adolescents and adults [2, 10]. Supplement use is also positively associated with educational status, income, and in most cases with health status. However, those who are in poor current health status are often also heavy users of particular supplements, perhaps hoping that these products will mitigate or cure their conditions.

Limited surveys of dietary supplement use have also been done of health professionals, such as dietitians, but the response rates in existing surveys are too low to provide data that can be extrapolated to the profession as a whole [11].

The *Nutrition Business Journal* publishes a list of the most popular dietary supplements each year, and the latest data are summarized in Table 1. Although the sales data on which these lists are based do not conform precisely to prevalence of dietary supplement use, they do give some idea of what dietary supplements Americans are purchasing today. The dietary supplement marketplace is constantly changing, and as it does, consumption patterns also change, so it is important to keep up with trends in the industry [12].

TABLE 1 Dietary Supplements Ranked by 2005 Sales

Product	Consumer Sales, $Millions
Multivitamins	$4157
Meal replacements	2300
Sports nutrition supplements	2217
Calcium	1010
B vitamins	937
Vitamin C	836
Glucosamine/chondroitin	810
Homeopathics	619
Vitamin E	440
Fish/animal oils	359
CoQ 10	339
Vitamin A/carotenoids	258
Noni juice	250
Probiotics	243
Iron	219
Magnesium	202
Plant oils	197
Digestive enzymes	166
Garlic	166
Echinacea	154
Green tea	148
Saw palmetto	137
Chromium	125
Mangosteen juice	122
Ginkgo biloba	109

Source: As reported in *Nutrition Business Journal*, pp. 2–91, Nutrition Business Journal Supplement Business Report, New Hope Natural Medicines Penton Media Inc., San Diego, CA, 2006.

According to proprietary data collected by the Natural Marketing Institute using survey data of primary grocery shoppers and heads of household in the United States who were surveyed by National Family Opinion, vitamin supplements have stayed steady, minerals have also done so with only a slight downward trend, whereas herbal supplements have decreased, and the so-called condition-specific supplements for certain conditions (such as memory, weight loss, bone and joint health) have increased or been repackaged to appeal to a wider audience [13].

E. Patterns of Dietary Supplement Use among the Ill

Table 2 provides some questions that are useful to ask when querying people about their actual use or plans to use dietary supplements. Nutrition professionals need to be aware that people use dietary supplements in various ways when they fall ill. They are not always forthcoming about what they are doing when discussing their health problems with their physicians or other health professionals. In part, this lack of candor may be due to the law. DSHEA classified dietary supplements as foods, with less stringent standards for quality and efficacy than would be true if they were categorized as drugs. By definition, under the law, dietary supplements are not to be used for the prevention, mitigation, or cure of disease, although many people use dietary supplements for these purposes [14]. The bottom line is that dietary supplements are inappropriate as substitutes for evidence-based medical therapies prescribed by physicians.

The vast majority, perhaps 65%, of the American public turns first to prescription drugs if they are ill. A small proportion of the public turns first to dietary supplements or other alternative medicines for treating their illnesses, and as a result they may delay obtaining care from conventional practitioners. In spite of the limitation of the legitimate uses of dietary supplements to health promotion under DSHEA, dietary supplements are often used in ways that are different than the law intends. However, caution is needed, especially for those who are undergoing medical treatment or who are ill. Many consumers still view dietary supplements as helpful in the prevention and treatment of many conditions, including arthritis, colds and flu, osteoporosis, lack of energy, memory problems, and cancer, and therefore they often self-medicate with dietary

TABLE 2 Guidelines for Health Professionals When Considering Recommendations for Dietary Supplements

Is there a shortfall in nutrient intakes from recommendations when all sources of nutrients (food, beverages, medications like antacids that contain nutrients, and dietary supplements) are taken into account? If so, try to satisfy needs first by the use of usual foods or fortified foods, or are dietary supplements in order?
Does the supplement contain more than 100% Daily Value and especially more than the upper level (UL) of nutrient intake recommended? If so, choose a supplement that contains lesser amounts; there is no known benefit to consuming more than the RDA (or AI) of a particular nutrient.
Is the dietary supplement standardized and certified by an accredited source, such as the U.S. Phamacopoeia (USP) or National Formulary (NF) showing that the USP standards for identity, purity, packaging, and labeling were followed? If so, they are labeled with the USP/NF symbol. The USP website is at www.usp.org.
Is the product fresh? If so, it has not exceeded its expiration date on the label.
Does the supplement contain labeled amounts of nutrients or other constituents? There is no single source at present that provides verifications for label claims for nutrients and other constituents in supplements. Some useful information may be available if the product has been tested by Consumerlab.com and found to contain amounts of constituents claimed on the label (see www.consumerlab.com).
Is the dietary supplement reasonable in cost and within the patient's economic means?
Has the patient read the patient package insert and cautionary directions and dosage limits on the label? If so, patients should be alert to adverse reactions specified and report them to their physician if they arise.
Does the dietary supplement claim to prevent or cure a condition or disease? By law, dietary supplements cannot be claimed to prevent or cure disease, so the claim is not supportable and conventional methods for doing so should be sought.
Is the dietary supplement safe? Certain dietary supplements such as ephedra (ma huang) have been declared unsafe by the Food and Drug Administration (FDA) and the U.S. Department of Health and Human Services. Check the Center for Food Science and Nutrition (CFSAN) website at FDA for updates on other products.
Is the dietary supplement efficacious? If so, authoritative bodies such as the Food and Nutrition Board, National Academy of Sciences, the Consensus Conferences of the Agency for Healthcare Research on Quality, or others should have indicated that such uses are safe and effective. The Office of Dietary Supplements within the National Institutes of Health summarizes information on many studies of safety and efficacy at http://dietary-supplements.info.nih.gov.

Source: Adopted and modified from Costello, Leser, and Coates [33].

supplements for these purposes. They also think they may have a role in the prevention or treatment of depression, stress, heartburn, high cholesterol, and vision, heart, and blood pressure problems. Vitamin mineral supplements are viewed as somewhat more effective than herbals and botanicals, not only in affecting overall health and wellness but in prevention and treatment. Many consumers use a combination of prescribed, over-the-counter drugs and dietary supplements all at the same time. It is this group that runs the highest risks of potential supplement-drug interactions, particularly if they are taking many medications and if they are taking medications or supplements that are especially likely to interact adversely with each other. For example, those on Coumadin, a commonly used anticoagulant, may experience adverse reactions if they self-medicate with various herbal and botanical drugs that affect blood clotting [15,16].

Websites that are helpful in assessing possible interactions of dietary supplements with drugs include the National Center for Complementary and Alternative Medicine (NCCAM) website at http://www.nccam.nih.gov, the Natural Medicines Comprehensive Database at www.naturaldatabase.com, Consumer Lab at www.Consumerlab.com, the Natural Standard database at www.naturalstandard.com, and the Center for Education and Research on Therapeutics at www.QTdrugs.org.

People who have poor reported health status, both physical and emotional, tend to be users of herbal and condition-specific as well as other types of supplements. Therefore, it is important to check dietary supplement use in the ill. If patients are undergoing medical treatment and prescription drugs have been prescribed, they should be encouraged to use them as directed *first* and by themselves to gain the full effects of the therapy. If the patient insists on continuing dietary supplement use, possible adverse interactions with the drug regimen the patient is taking should be investigated and the patient counseled on how to avoid potential adverse interactions. Any interactions identified should be entered into the patient's chart.

F. Motivations for Dietary Supplement Use

People have many different motivations for dietary supplement use. Motivations vary not only from person to person by such factors as age, sex, income, education, and ethnicity, but also by attitudes such as concerns about deficiencies, readiness to engage in preventive behaviors, and health status. Motivations also vary within and between individuals depending on the type of product and whether it is a nutrient or non-nutrient supplement, over time, and depending on the definition of dietary supplement that is employed.

Motivations and use of dietary supplements are related, but probably in complex ways that may differ from one individual to another. One theory is that knowledge and attitudes (motivations) cause supplement use. It is also possible that people use the supplements and then attitudes and knowledge (motivations) follow, perhaps to rationalize or justify use. Probably some people get into the habit and then find reasons for their behavior, often perhaps due to social influence, whereas others operate in a more deliberate manner, gathering knowledge and attitudes. The implications for nutrition and other health professionals are that they must consider both ways to influence behavior when collecting information about dietary supplement use. For example, this may have utility in developing methods for persuading women in the reproductive age group who might become pregnant to increase their use of folic acid supplements.

II. METHODS FOR ASSESSING DIETARY SUPPLEMENT INTAKE

A. Dietary Supplement Intake (Exposure)

Most of the methods for assessing intake of dietary supplements are similar to those discussed in Chapter 1 on dietary assessment methodology. The strengths and limitations of the various dietary assessment methods covered in Chapter 1 apply to capturing intakes of dietary supplements. A method applicable to dietary supplements rather than foods is the pill inventory, which is widely used in obtaining information about other medications. For some supplements, inferences about use can be made from blood or urine biomarkers, if these are available, although they provide only qualitative rather than quantitative information. The unique features of the methods for collecting dietary supplement information are detailed in Table 3.

B. Assessing Supplement Intake in Clinical Settings

1. In-Patient Settings

Most hospitals prohibit self-medication with dietary supplements or other over-the-counter medications without the written permission of the physician, so that supplement use in the hospital is usually limited. However, prior use may be of interest, as some botanicals may take days or weeks to be excreted from the body. The patient or a family member/caregiver should be asked to provide a list of the types and amounts of dietary supplements that the patient uses or has used in the recent past, and this information should be entered into the chart and electronic medical records as part of the dietary assessment. It is important for all health professionals who see the patient to query him or her about dietary supplement use; some patients are reluctant to share their usage patterns with the doctor but are willing to share this information with dietitians. It is

TABLE 3 Dietary Supplement Assessment Methods

Method	Advantages	Disadvantages	Comments
Pill inventories	Actual labels are available and can be examined, doses recorded.	Patients/clients may forget or refuse to bring supplements when requested, or they may only produce socially acceptable/legal products. May change dietary supplement use reporting. UPC codes for dietary supplements are not unique, and the formulations in a given specific UPC coded product may change.	Technique is commonly used in studies of medication use. A variant is to ask the patient/respondent to provide grocery receipts and if these include UPC codes it may be possible to identify the supplement.
Diet records	May be useful in clinical settings with willing patients to improve adherence and obtain specific details about usage patterns. They provide actual record of intake going forward. Respondent has the bottle from which he or she records information, and recall of items used is not necessary. May provide useful contextual information for improving adherence.	May change eating or dietary supplement use behavior, especially if supplements are taken as a prescription. Extremely time consuming for client/patient, usual intake may only be revealed with many days of recording. Forgetting to record is common.	Usually the record includes food and drink as well as dietary supplement use; can be expanded to also cover other medications when drug-supplement or food-supplement interactions are suspected.
Food frequency questionnaires	Retrospective, so they do not affect food consumption. Lists may help to prod memory and make recall easier. May provide an estimate of usual intake. Quick to fill out. The standardized format makes them useful for large-scale studies. If write-ins are accommodated, then information can be more complete.	Lists not usually complete and may be nonspecific. For some condition-specific and other supplements, use is infrequent and may not show up if the window of recall is 30 days or so. Even semiquantitative dietary supplement intake forms are not quantitatively precise.	Frequency questionnaires for dietary supplement use range from simple checklists for categories of dietary supplements (such as multivitamin mineral supplements, single vitamin or mineral supplements, and others) or specific supplements to semiquantitative questionnaires that tap both frequency and amount used.
24-hour dietary recalls	Retrospective, so they do not affect food consumption and usually are easy for patient to recall. Quantifies intake. May point to problems with timing or other aspects of supplement use. Some computerized dietary assessment programs now include a module for ascertainment of both food and supplement intakes.	Rely on memory, and some items may be forgotten or individual may not be able to provide sufficient detail about the exact supplement name, dose, and so on. Many days are needed to estimate usual intake. May be useful clinically but more difficult to use in large studies in which standardization is necessary.	Individual is usually asked to provide his or her intake of dietary supplements as well as food and drink that has been consumed in the previous 24 hours.
Diet histories	Food intake is not changed because method is retrospective. Permit obtaining information on total diet.	Recall is involved and memory may be faulty. Respondent may not recall usual pattern of supplement intake or may not have a usual pattern. The amounts are usually not precise. Time consuming for both investigator and respondent.	Individual provides the professional with information on usual food intake patterns.
Brief dietary supplement assessment forms	Do not change eating behavior because they are retrospective. Focus solely on dietary supplement use. Easy to fill out and inexpensive.	Only a small number of supplements or foods or both can be asked about. Often information on dose, type of supplement, timing, and so on are not provided.	Individual is asked to respond about how often and how much dietary supplements are taken.

(continues)

TABLE 3 *(continued)*

Method	Advantages	Disadvantages	Comments
Blood and urine	Do not change supplement use because they are retrospective. If the only source of the biomarker is the dietary supplement, it is possible to state with certainty that the product was consumed.	Not all nutrients or botanicals have easily identifiable biomarkers. Method is not quantifiably precise.	A blood (e.g., folic acid) or urine (e.g., creatine) biomarker is used to ascertain intake, either in conjunction with or instead of usage data on dietary supplements.

everyone's responsibility to collect this information, and dietitians, nurses, and pharmacists as well as physicians should be alert to the possibility of dietary supplement use. When electronic medical records are developed, a question on the patient's use of dietary supplements should be included, and the supplements that are used should be named in the medical record.

2. Outpatient Settings

During nutrition assessment, all patients should be asked about their use of dietary supplements: what supplements they use, how much they use, and how often. Replies should be written in the medical record. Dietary supplement use should be included in diet history taking and included in calculations of nutrient intakes. Some food frequency questionnaires and food checklists include items on intake of the most commonly consumed dietary supplements, but these may not include less commonly used products or supplements that are used only occasionally that may also be important to health. Therefore, it is wise to probe for additional supplement use and to encourage patients to write these dietary supplements into their questionnaires. If questions remain or there is need for further documentation, the patient can be asked to keep a supplement intake record, which he or she can bring to the next visit. One useful way to elicit further information about dietary supplement use from ambulatory patients who report very high use but cannot remember details of what they take at home is to use the "brown bag" technique. The patient is asked to bring all of the dietary supplements and medications that he or she uses to the next visit. The doses and types of dietary supplements and other medications can be recorded in the chart, and their potential impacts taken into account in dietary assessment and assessment of possible supplement-nutrient interactions. Another helpful tool is a simple diary for documenting dietary supplement use that can be given to patients. One example is the NIH's brochure "What Supplements Are You Taking?" which can be downloaded and printed from the Office of Dietary Supplements website (http://ods.od.nih.gov/pubs/partnersbrochure. asp) and includes not only a diary, but some questions about dietary supplement use and medication. Patients can fill this out at their leisure, and then the health professional can review it. Sources such as the Health Professional's Guide to Dietary Supplements [17] and A Healthcare Professional's Guide to Evaluating Dietary Supplements [18] or the American Pharmacists Association at www.aphanet.org may be helpful in obtaining estimates of ingredients and content.

C. Estimating Dietary Supplement Intake

Once the patient's reported intake has been elicited, the information must be evaluated. If nutrient intakes are needed, nutrients from food, beverages, and nutrient-containing supplements must be added together to get an estimate of total dietary intake in the individual user. For some, this may include capturing intakes from food bars or sports drinks that are highly fortified.

D. Assessment of Dietary Supplement Intake in Some Large-Scale National Surveys

1. National Health and Nutrition Examination Survey (NHANES)

The National Health and Nutrition Examination Survey (NHANES) is the nation's population-based survey to assess the dietary intakes, health, and nutritional status of noninstitutionalized adults and children in the United States. About 5000 people are surveyed each year in five communities nationwide. In addition to food intake, information on dietary supplement use (frequency, amount, and duration) is collected from NHANES respondents during the interviews in their households. The dietary supplements include vitamins, minerals, other prescription and nonprescription dietary supplements, and antacids (a major source of calcium) taken in the last month. During the household interview, details on supplement use, such as how long the product has been used, how often it was taken over the past month, and how much was taken, are established. Those respondents who say that they have taken dietary supplements are then asked to provide the supplement containers. About two-thirds of them do so. Supplement containers are viewed, and the interviewer records the product label's name, strength of the ingredient (for certain vitamins and minerals), and other information. During the household interview, details on supplement use, such as how long the product has been used, how often it was taken over the last month, and how much was taken, are established [19].

2. National Health Interview Survey (NHIS)

The National Health Interview Survey has periodically obtained information on dietary supplement intake, and this has been useful in assessing such issues as use of vitamin-mineral supplements among cancer survivors [20]. Also, questions were asked in depth about use of certain supplements as part of the NCCAM, which has sponsored additional modules on two occasions. The latest survey, which queries respondents in depth about use of vitamin-mineral supplements and herbal and other nonvitamin supplements, is currently in the field. The advantage is that the survey obtains detailed information on motivations, associations with conventional medical treatments, and cost, as well as dietary supplement use. The disadvantage is that food intake and health indices are not included.

3. National Cancer Institute Diet History Questionnaire: A Public Use Semiquantitative Food Frequency Questionnaire

The National Cancer Institute (NCI) has developed a semiquantitative food frequency questionnaire called the Diet History Questionnaire (DHQ), which is publicly available and is described in Chapter 1. It can be reprogrammed to add specific dietary supplements. As issued, it includes specific questions on fiber and fiber supplements; multivitamins; herbals; antioxidant supplements; beta carotene; vitamins A, C, and E; and calcium over the past year. For calcium users, there are additional questions on how long calcium containing supplements or antacids have been used. There are also specific queries on approximately 10 vitamins and minerals as well as fatty acids, and about 24 different herbals and botanicals. The questionnaire includes both foods and supplements, and it can be used for many purposes, not simply cancer therapy. The 2006 version of the DHQ was validated using a checklist approach and was found to be an improvement over earlier versions of the NCI questionnaire known as the 1992 NCI/Block questionnaire [21]. The DHQ was also concurrently validated against four 24-hour dietary recalls, and the DHQ performed better than the 1995 Block and Willett (purple version) food frequency questionnaires [22]. The DHQ can be accessed at http://www.riskfactor.cancer.gov/DHQ.

4. Other Instruments for Assessing Dietary Supplement Intake in Epidemiological Studies

Many proprietary questionnaires that are variations of food frequency questionnaires with additional questions on dietary supplements exist for assessing dietary supplement intakes in epidemiological studies of various large cohorts. However, the questions vary from one questionnaire to another, depending on the focus of the study. When data on dietary supplement composition are available as part of these questionnaires, the source of the information is generally from manufacturers or self-entered values. Total intakes reported reflect not only intakes from food but from specific dietary supplements that were queried.

There are many examples of tools for adults that query dietary supplements as well as foods; those for children and infants are fewer in number. The questionnaires require specific food composition and dietary supplement databases to be analyzed, and these are not generally in the public domain. Adult questionnaires that query some dietary supplements include the Harvard (Willett) semiquantitative food frequency questionnaire, different versions of which have been used in a number of studies; the University of Hawaii Cancer Center's multiethnic cohort questionnaire; the Women's Health Initiative semiquantitative food frequency questionnaire; the Women's Health Study questionnaire; the National Institute of Environmental Health Science's Sisters Study Questionnaire; and the American Cancer Society's food frequency questionnaire. The availability of these questionnaires varies, and depends on obtaining permission from the owners of the questionnaires and their willingness to collaborate with other investigators. Some university research groups, such as the dietary assessment groups at the University of Washington-Fred Hutchinson Cancer Research Center, Harvard University, and the University of Hawaii Cancer Center, will permit their questionnaires to be used and will analyze results for a fee.

The Fred Hutchinson Cancer Research Center at the University of Washington was the Data Coordinating Center for the Women's Health Initiative (WHI). Based on the work with the WHI, a module was refined for use with two other NIH sponsored clinical trials of adults: the Selenium and Vitamin E Cancer Prevention Trial (SELECT) and the Vitamin and Lifestyle Cohort Study (Vital). These tools are available to outside groups and can be processed for a fee. The questionnaires have both nutrient supplements and herbal/botanical supplements, and they come in both male and female versions. The center has a website that describes the services it provides to outside users at www.fhcrc.org/science/shared_resources/nutrition.

The University of Hawaii Cancer Center questionnaire has 180 foods and 10 supplements on the current questionnaire that is available outside the center—investigators purchase the questionnaire from the Cancer Center (for a fee) and then send completed questionnaires for processing into nutrient intakes (also for a fee). The analysis provides separate variables for intake of 54 nutrients from food and intake of 22 nutrients from supplements. The tool includes many supplements used by Asian Americans. The questionnaire has detailed questions and many defaults to permit more precise and accurate information on the dietary supplements that are used. A more extensive list of supplements is currently being developed.

Other food frequency questionnaires are available from commercial services. Perhaps the best known one is the

Nutritionquest group, which offers the Nutritionquest or Block semiquantitative food frequency questionnaire for sale, and it can be accessed at www.nutritionquest.com. The basic 2005 food frequency questionnaire has about 110 foods and also items on multiple vitamin supplements, single vitamins and minerals, and one item on herbals. Upon request and for a fee, the Block questionnaires can be tailored for individual research purposes. The Block Nutritionquest group has also developed a brief calcium/vitamin D intake assessment screening tool that includes 19 foods and three supplements as well as questions to adjust for food fortification practices. Another Block intake assessment screening tool is the Block Folic Acid/Dietary Folate Equivalents Screener, based on NHANES 1999–2001 dietary recall data. It includes 21 questions and provides separate estimates of total, supplement, and food-only intakes [23]. The Block Soy Foods Screener focuses on 10 food and supplement items; it is designed to measure intakes of daidzein, genistein, coumestrol, and total isoflavones. The problem is that the lists of dietary supplements are usually short.

E. Other Instruments Used for Assessing Dietary Supplement Intake in Clinical Research Studies

Many other techniques have been used to obtain information on dietary supplement intakes. Serial random 24-hour dietary recalls that included foods as well as supplement intakes were used successfully in the Women's Intervention Nutrition Study (WINS), a large randomized trial of diet as adjuvant therapy in women who had been treated for breast cancer [24]. In the Hemodialysis (HEMO) study of patients undergoing renal replacement therapy, food and dietary supplement intake was assessed using a 2-day diet diary assisted recall technique, in addition to medication inventories and other techniques to check on adherence to use of high dose B vitamin supplements [25].

III. DIETARY SUPPLEMENT COMPOSITION DATABASES FOR ANALYSIS OF DIETARY SUPPLEMENT INTAKE

The analysis of dietary supplement intakes ideally requires complete analytically verified tables of dietary supplement composition by chemical analysis, which still does not exist. Therefore, results tend to be imprecise and inaccurate, particularly for intakes of some of the botanical ingredients in dietary supplements. The situation is slowly changing, but dietary supplement databases are still incomplete both with respect to how representative they are of the universe of dietary supplements marketed and sold in the United States and how well documented the levels of ingredients are. At present, the majority of the currently available databases rely on label claims rather than analytically verified data. For dietary supplements and many highly fortified processed foods that lack analytical data on micronutrients, intake estimates obtained from product label declarations are likely to be biased; overages are likely, especially for vitamins because regulations require that the actual nutrient content of products be equal to or greater than the declared level on the label, after taking into account processing effects and shelf life losses [26].

A. Dietary Supplement Label Databases

1. NHANES Label Database

The composition of dietary supplements consumed in NHANES is available at the National Center for Health Statistics website, although the primary purpose of the database is to store information on nutrients taken from the dietary supplement labels collected from NHANES respondents. NHANES research nutritionists obtain additional label data for the dietary supplement database by contacting manufacturers and distributors, company websites, and other Internet sources. Changes in supplement composition are tracked and entered into the database when reformulations are identified. The NHANES label database is publicly available and permits nutrition scientists to better assess total intakes of nutrients from all sources than ever before. However, it has its limitations. Only supplements that were used by respondents in the survey are provided in the database. Only about 10,000 respondents are included in each NHANES data release. Although this may seem like a large number of respondents, for rarely used supplements, there may be few or no users who respond. Only levels of nutrients are noted, although the names of other ingredients are recorded as well. The quantitative data on nutrients that it provides rely on nutrient content declarations on the labels and are not analytically verified. Because it is a violation of the law to declare levels of a nutrient on the label as being more than what is provided, manufacturers tend to add more than the declared label value to many products. The amount added depends on the nutrient in question, its stability, cost, bulk, and other characteristics; there is no single "correction factor" that can be used. The supplements that are reported in NHANES during the most recent interview cycle are released every 2 years. Unfortunately, supplements change rapidly, and many of the products may not be on the market at the time the database is accessed. Default values are also included in the database because many respondents are unable to supply the exact supplement or strength that was consumed, although some information is available. Since NHANES uses a nationally representative sampling procedure, defaults developed with the NHANES data may be useful in other surveys as well, particularly if it is not possible to collect data with this level of detail. The defaults are based on the frequency of supplements reported in the latest 2-year NHANES release that is available, as well as on manufacturer information on sales. For example, default matches for adults include matching multivitamins to multivitamin minerals, vitamin A to 8000 IU, vitamin C to 500 mg, vitamin B_6 to 100 mg, vitamin D to 400 IU, vitamin E to 400 IU, folic acid to 400 mcg, calcium to 500 mg, iron to 65 mg, and zinc to 50 mg.

2. Natural Standard Database

Natural Standard is an international research collaboration that aggregates and synthesizes data on dietary supplements and other complementary and alternative therapies to provide objective and reliable information for clinicians using an evidence-based, consensus-based, and peer-reviewed procedure with reproducible grading scales. For more information, the website is www.naturalstandard.com.

The group has compiled a compendium of evidence-based reviews on herbs and dietary supplements that is available on line as well as in print. The compendium is also available via the National Library of Medicine at www.nlm.nih.gov/medlineplus/druginfo/herb_All.html.

3. Natural Products Association Database

The Natural Products Association (NPA) is a trade association (formerly the National Nutritional Foods Association) that has a foundation that operates a two-part quality assurance program that includes a third-party certification program for Good Manufacturing Practice standards as determined by NPA based on a dialogue between suppliers and others. Those who meet the standard and pass audits can use the NPA logo. Members can also participate in the TruLabel program, which includes data on ingredients in 23,000 product labels.

4. Other Label Databases

Several other private compilations of dietary supplement label information may be purchased, such as the Natural Medicines Database at www.NaturalDatabase.com and HealthNotes Clinical Essentials (Portland, Oregon) at www.healthnotes.com.

B. Chemical Analytically Verified Dietary Supplement Databases

Several chemically analyzed dietary supplement databases now exist, but they contain only a few products, are not always based on representative numbers of products, and some are proprietary. The major ones are described here:

1. *U.S. Department of Agriculture Dietary Supplement Ingredient Database.* An initiative is under way with the U.S. Department of Agriculture to develop an analytically substantiated dietary supplement ingredient database (DSID) for nutrients and eventually for other constituents as well. Initial efforts are focusing on multivitamin-mineral supplements, as Americans commonly consume these types [26]. The database will be publicly available in late 2008 or 2009.

2. *Consumerlabs.com.* ConsumerLab.com, LLC is a provider of independent test results and information to help consumers and health care professionals evaluate dietary supplements and other health, wellness, and nutrition products. Data are published only on products that have been tested. The products are bought off the shelf in consumer outlets and chemically analyzed. ConsumerLab.com does not publish a comprehensive database that is publicly available. However, a subscription to their reports is available at www.consumerlab.com for a reasonable cost.

3. *NSF.* NSF is an independent, not-for-profit testing organization offering product testing of dietary supplements in its NSF/American National Standards Institute. It does not simply evaluate test data submitted by manufacturers or analyze a single sample of a product and approve it, but NSF conducts its own product testing in its accredited laboratories. The three main components of the NSF Dietary Supplements Certification Program are verification that the contents of the supplement actually match what is printed on the label, assurance that there are no ingredients present in the supplement that are not openly disclosed on the label, and assurance that there are no unacceptable levels of contaminants present in the supplement. The major disadvantage of the values published by NSF is that they do not constitute a comprehensive database. Only products that have been certified are included in the database. At present there are about 400 products from 24 companies that are certified on the NSF website, www.nsf.org/consumer.

C. Computerized Dietary Assessment Programs That Include Dietary Supplements

1. University of Minnesota Dietary Supplement Module

Some computerized dietary assessment programs include dietary supplements in the interview and also have databases on dietary supplement composition. For example, the University of Minnesota's Nutrient Data System (NDS) is developing a dietary supplement module (DSM), which can be used in conjunction with existing software to obtain information about food intake; the new module should be completed and available to researchers by 2008.

2. Other Computerized Dietary Assessment Programs

Other computerized dietary assessment programs permit the addition of supplement information to the database even if it is not included in the food composition database, but none yet provide complete lists of the most commonly used supplements in the software package.

IV. THE DIETARY SUPPLEMENT LABEL

FDA regulations require that certain information appear on dietary supplement labels. Information that must be on a dietary supplement label includes a descriptive name of the product stating that it is a "supplement"; the name and place of business of the manufacturer, packer, or distributor; a complete list of ingredients; and the net contents of the product. The regulations are described in depth in the FDA's Dietary Supplement Labeling Guide (accessible at www.cfsan.fda.gov). The FDA has recently issued regulations for good manufacturing practices (GMP) that touch upon such topics as verification of identity, purity, strength, and supplement composition and will go into effect in the next few years. More information is available at the Center for Food Safety and Applied Nutrition (CFSAN) website and at http://www.fda.gov/bbs/topics/NEWS/2007/NEW01657.html.

A. Dietary Supplement Label: Ingredients

1. Differences between Food and Dietary Supplement Labels

The Supplement Facts panel on dietary supplements must list dietary ingredients as well as those ingredients that do not have recommended daily intakes (RDIs) or daily reference values (DRVs). It is optional to list the source of a dietary ingredient on the Supplement Facts panel whereas sources of a dietary ingredient and ingredients without RDIs or DVs are not permitted on the food Nutrition Facts label. Also the part of the plant from which a dietary ingredient is derived must be listed on the dietary supplement either on the Supplement Facts panel or the "other ingredients" area below the panel. In contrast, this information cannot be listed on the food label. The Supplement Facts panel does not permit listing of "zero" amounts of nutrients although the Nutrition Facts panel for food requires it. The percentage Daily Value (%DV) or the Reference Daily Intake or Daily Reference Value of a dietary ingredient that is in a serving of a dietary supplement product must be declared for all ingredients for which there are DVs except protein. Dietary supplements for infants, children younger than 4 years old, and pregnant and lactating women do not require this, however.

2. Supplement Facts Label

The Supplement Facts panel must list the names and amounts of the dietary ingredients present in the product, the serving size, and servings per container. A serving for a dietary supplement is the maximum amount recommended as appropriate on the label for consumption at one time or if recommendations are not given 1 unit (e.g., tablet, capsule, packet, teaspoon). Thus, if the label says to take one to three tablets with breakfast, the serving size is three tablets.

3. Ingredient List

Other dietary ingredients that do not have DVs are also listed in the Supplement Facts panel after the ingredients that do have them, in addition to their correct botanical (Latin) names. They are also listed by their common or usual names and must be accompanied by their weight per serving.

B. Dietary Supplement Label: Claims

Table 4 describes the three categories of claims that can be used on dietary supplements: health claims, nutrient content claims, and structure/function claims. These claims also apply to food labels.

TABLE 4 Claims That Can Be Used on Dietary Supplements

1. Health Claims

Health claims describe a relationship between a food, food component, or dietary supplement ingredient, and reducing risk of a disease or health-related condition. A "health claim" definition has two essential components: (1) a substance (whether a food, food component, or dietary ingredient) and (2) a disease or health-related condition. A statement lacking either one of these components does not meet the regulatory definition of a health claim.

The FDA has oversight in determining which health claims may be used on a dietary supplement label. Its authority comes from several laws:

NLEA Authorized Health Claims

The Nutrition Labeling and Education Act (NLEA) of 1990, the Dietary Supplement Act of 1992, and the Dietary Supplement Health and Education Act of 1994 (DSHEA) provide for health claims used on labels that characterize a relationship between a food, a food component, dietary ingredient, or dietary supplement and risk of a disease provided the claims meet certain criteria and are authorized by a FDA regulation. The FDA authorizes these types of health claims based on an extensive review of the scientific literature, generally as a result of the submission of a health claim petition, using the significant scientific agreement standard to determine that the nutrient/disease relationship is well established. For an explanation of the significant scientific agreement standard, visit www.cfsan.fda.gov/~dms/ssaguide.html.

(continues)

TABLE 4 *(continued)*

Qualified Health Claims

The FDA's 2003 *Consumer Health Information for Better Nutrition Initiative* provides for the use of qualified health claims when there is emerging evidence for a relationship between a food, food component, or dietary supplement and reduced risk of a disease or health-related condition. In this case, the evidence is not well enough established to meet the significant scientific agreement standard required for FDA to issue an authorizing regulation. Qualifying language is included as part of the claim to indicate that the evidence supporting the claim is limited. Both conventional foods and dietary supplements may use qualified health claims. The FDA uses its enforcement discretion for qualified health claims after evaluating and ranking the quality and strength of the totality of the scientific evidence. Although The FDA's "enforcement discretion" letters are issued to the petitioner requesting the qualified health claim, the qualified claims are available for use on any food or dietary supplement product meeting the enforcement discretion conditions specified in the letter. The FDA has prepared a guide on interim procedures for qualified health claims and on the ranking of the strength of evidence supporting a qualified claim (visit www.cfsan.fda.gov/~dms/hclmgui3.html). Qualified health claim petitions that are submitted to the FDA will be available for public review and comment. A listing of petitions open for public comment is at the FDA Dockets Management website. A summary of the qualified health claims authorized by the FDA may be found at www.cfsan.fda.gov/~dms/qhc-sum.html. For more information on Qualified Health Claims, visit www.cfsan.fda.gov/~dms/lab-qhc.html.

2. Nutrient Content Claims

Most nutrient content claim regulations apply only to those nutrients or dietary substances that have an established daily value and are expressed as percentage Daily Value (see www.cfsan.fda.gov/~dms/flg-7a.html). Percentage claims for dietary supplements are another category of nutrient content claims used to describe a percentage level of a dietary ingredient for which there is no established Daily Value.

3. Structure/Function Claims

Statements that address a role of a specific substance in maintaining normal healthy structures or functions of the body are considered to be structure/function claims. Structure/function claims may not explicitly or implicitly link the relationship to a disease or health-related condition.

Structure/function claims on dietary supplements describe the role of a nutrient or dietary ingredient intended to affect normal structure or function in humans, for example, "calcium builds strong bones." In addition, they may characterize the means by which a nutrient or dietary ingredient acts to maintain such structure or function, for example, "fiber maintains bowel regularity" or "antioxidants maintain cell integrity," or they may describe general well-being from consumption of a nutrient or dietary ingredient. Structure/function claims may also describe a benefit related to a nutrient deficiency disease (like vitamin C and scurvy), as long as the statement also tells how widespread such a disease is in the United States. If a dietary supplement label includes such a claim, it must state in a "disclaimer" that the FDA has not evaluated the claim. The disclaimer must also state that the dietary supplement product is not intended to "diagnose, treat, cure or prevent any disease," because only a drug can legally make such a claim.

V. AUTHORITATIVE INFORMATION AND RESOURCES ABOUT DIETARY SUPPLEMENTS

A. Office of Dietary Supplements, National Institutes of Health

The Office of Dietary Supplements at the National Institutes of Health has as its mission to strengthen knowledge and understanding of dietary supplements by evaluating scientific information, stimulating and supporting research, disseminating research results, and educating the public about the efficacy and safety of dietary supplements in order to foster an enhanced quality of life and health for the U.S. population. Its website contains much useful information for health professionals and can be accessed at http://ods.od.nih.gov.

1. International Bibliography of Information on Dietary Supplements and Clinical IBIDS

The search engine on the ODS website assists health professionals in finding research studies on dietary supplements. The clinical version provides useful searches for commonly used dietary supplements encountered in clinical practice. The International Bibliographic Information on Dietary Supplements (IBIDS) is available at http://ods.od.nih.gov/databases/ibids.html. Special topics of interest to clinicians are featured on the web page.

2. Computer Access to Research on Dietary Supplements

For those who wish to view research on dietary supplements that is currently being supported by the federal government, Computer Access to Research on Dietary Supplements (CARDS) is an invaluable resource [27]. This resource is located at http://ods.od.nih.gov/databases/cards.html.

3. Other Resources

ODS also provides a great deal of other authoritative health information on its website. Of particular use to health professionals are dietary supplement fact sheets and the annual bibliographies of significant advances in dietary supplement research (access at http://dietary/supplements.info.nih.gov/health_information/health_information.aspx).

B. Food and Drug Administration

1. Center for Food Safety and Applied Nutrition for Health Claims

The FDA's CFSAN's website has a variety of materials on dietary supplements, including recent recalls, frequently asked questions, and some materials for consumers. It can be accessed at http://www.cfsan.fda.gov/~dms/supplmnt.html.

2. Center for Food Safety and Applied Nutrition Adverse Events Reporting System and Medwatch

CFSAN has developed a new Adverse Events Reporting System (CAERS), which replaces the patchwork of existing adverse event systems that were maintained previously by individual Offices within CFSAN. The FDA will use the CAERS system as a monitoring tool to identify potential public health issues that may be associated with the use of a particular product already in the marketplace. Information gathered in CAERS will also assist the FDA in the formulation and dissemination of CFSAN's post-marketing policies and procedures. At present, adverse event reports from the dietary supplement industry, consumers, and health professionals should be submitted to MedWatch.

3. Food and Drug Administration Electronic Newsletter

The FDA produces a newsletter on dietary supplements and other updates, which is available free at www.cfsan.fda .gov/~dms/infonet.html#fda-dsfl.

C. National Center for Complementary and Alternative Medicine

The NIH's NCCAM sponsors research on dietary supplements and also provides fact sheets on a number of products, especially those that are being used for the prevention or treatment of disease. The "herbs at a glance" series contains authoritative fact sheets on a number of different herbs and botanicals including common names, uses, potential side effects, and resources. Information can be obtained at http://nccam.nih.gov/health/herbsataglance.htm.

The premier source for medical research is the National Library of Medicine's MEDLINE database of life science journals. Free access to this resource is available online using PubMed. When searching PubMed, an additional filter labeled "CAM" has been made available through the NCCAM to limit searches to references using alternative and complementary treatments. Access this resource at www.nlm.nih.gov/nccam/camonpubmed.html.

D. National Cancer Institute

The NIH's NCI operates a number of research programs that involve dietary supplements. It also occasionally produces fact sheets and papers on cancer treatment and prevention measures that include dietary supplements. The NCI's Division of Cancer Prevention and the Division of Cancer Control and Population Sciences continue to develop and maintain a website called the Dietary Assessment Calibration and Validation Register discussed in Chapter 1. The website is accessible at www.dacv.ims.nci. nih.gov.

Health care providers who are treating cancer patients may wish to consult the NCI website for information on dietary supplements and other alternative and complementary therapies for cancer patients. It can be accessed at www.cancer.gov/cancertopics/treatment/cam.

Health professionals and patients who are seeking to enroll in clinical trials of dietary supplements or other therapies for cancers or other diseases should consult the federal government's list of registered clinical trials available at www.clinicaltrials.gov.

E. Agency for Healthcare Research and Quality, U.S. Department of Health and Human Services

The Agency for Healthcare Research and Quality (AHRQ) works closely with the NIH and other federal agencies to develop systematic evidence based reviews of the health literature on topics of public health significance. This agency also operates state-of-the-science and consensus conferences on critical topics and publishes the deliberations from them. Several recent evidence-based reviews and conferences have involved dietary supplements. These include multivitamin mineral supplements, omega-3 fatty acids, ephedra and ephedrine for weight loss and athletic performance, antioxidants, vitamin C, vitamin E and CoQ10 and cardiovascular disease and cancer, B vitamins and berries, and neurodegenerative diseases and other topics. The web address is www.ahrq.gov.

F. National Library of Medicine

1. PubMed and Medline

As mentioned previously, this is a world famous computerized bibliography, which includes biomedical information on dietary supplements; it is freely available to the public over the Web. Access it at www.pubmed.gov.

2. Medline Plus (Subscription)

The National Library of Medicine sponsors a database by subscription service for specialized searches that can be accessed at MedLine Plus, www.nlm.nih.gov/medlineplus/druginformation.html.

3. Bibliographies

Occasionally the National Library of Medicine publishes bibliographies of various topics dealing with dietary supplements under its current bibliographies in medicine series.

G. U.S. Department of Agriculture National Agricultural Library/Food and Nutrition Information Center

The Food and Nutrition Information Center (FNIC) compiles and disseminates authoritative bibliographies for laypersons and generalist practitioners on various topics, including dietary supplements, with partial support for these efforts from the ODS at NIH. These are available free of cost at www.nal.usda.gov/fnic. Among the recent materials available for consumers are the following: Dietary Supplements: General Resources for Consumers, www.nal.usda.gov/fnic/pubs/bibs/gen/dietsupp.html. Specialized bibliographies are also available, such as these on specific supplements:

Dietary Supplements: Consumer Resources on Vitamins Minerals and Antioxidants, www.nal.usda.gov/fnic/pubs/bibs/gen/diet_supp_antioxident.html

Dietary Supplements: Consumer Resources on Herbs Botanicals, Phytochemicals, and Other Specific Dietary Supplements, www.nal.usda.gov/fnic/pubs/bibs/gen/diet_supp_other.html

Dietary Supplements and Heart Health Resources for Consumers, www.nal.usda.gov/fnic/pubs/bibs/gen/diet_supp_heart.html

H. Department of Defense: U.S. Army Center for Health Promotion and Preventive Medicine

The U.S. Army Center for Health Promotion and Preventive Medicine is a resource center for members of the armed forces as well as the general public on various health issues. Some excellent materials for laypersons are available particularly on reasonable dietary supplement use and on performance at http://chppm-www.apgea.army.mil/dhpw/wellness/PPNC.

I. Canadian Government Resources

The Canadian government's Natural Health Product Ingredients Database includes a display of toxicity restrictions, registry numbers for the chemicals by Chemical Abstracts Service (CAS) and other registry numbers, herbals and hyperlinks to the Canadian Natural Health Products Directorate (NHPD) and the Therapeutic Products Directorate (TPD) monographs. The website can be accessed at http://cpe0013211b4c6d-cm0014e88ee7a4.cpe.net.cable.rogers.com/IngredientDAtabaseV1_1_1/homeRequst.do.

J. U.S. Pharmacopoeia

The U.S. Pharmacopoeial Convention is an independent science-based public health organization and official public standards-setting authority for all prescription and over-the-counter medicines, dietary supplements, and other healthcare products manufactured and sold in the United States. The standards are legally enforceable for drugs and a dietary supplement program also exists. Quality standards are determined by a voluntary expert committee and products which are submitted for evaluation and pass audits are listed on their website; those products that fail the evaluation are not listed. At present, there are about 100 certified products. The data can be accessed at www.usp.org/USPNF.

K. American Dietetic Association (ADA)

1. Position Papers and Other Materials

The American Dietetic Association (ADA) is the professional association for dietetic and nutrition professionals. Members of the association have developed a number of useful position papers and other materials on dietary supplements, and the association continues to publish useful articles on dietary supplements [28].

2. Evidence Analysis Library

The ADA has recently created an evidence analysis library (EAL) that provides authoritative evaluation of the evidence on various clinical topics, including some that involve dietary supplements. Members receive access to the library as part of their dues; access to it by others is by subscription. To learn more, access the ADA website at www.eatright.org.

3. Practice Groups

The Complementary and Alternative Medicine Practice Group (CAM) of the American Dietetic Association focuses specifically on dietary supplements. It produces an excellent newsletter, and members also receive free or reduced prices on many professional resources that are useful in assessing dietary supplement intakes.

L. Books

Among the useful reference books is the *Physician's Desk Reference for Nonprescription Drugs, Dietary Supplements, and Herbs* [29], which covers the full spectrum of nutritional supplements including: vitamins, minerals, amino acids, probiotics, metabolites, hormones, enzymes, and cartilage products. For each supplement there are listed precautions, contraindications, side effects, and possible interactions with medications. The *Commission E Monographs* [30] summarize the German Commission E monographs on various herbal medicines. *Herbs of Commerce* [31] is a comprehensive listing of more than 2000 botanicals that have current and historical uses as therapeutic agents. Botanical synonyms are included, so that older botanical names that are no longer accepted can be cross-referenced. Also included are the Ayurvedic names and the Chinese names for more than 500 herbs. The book contains the Latin binomials, the standardized common names, the

Ayurvedic names, the Pinyin name (simplified Chinese name), and other common names. The *Encyclopedia of Dietary Supplements* [32] reviews many over-the-counter supplements carried in today's nutritional products marketplace and presents peer-reviewed, objective entries that review the most significant scientific research, including basic chemical, preclinical, and clinical studies.

VI. HOW TO REPORT PROBLEMS WITH DIETARY SUPPLEMENT INTAKE

A. Food and Drug Administration

The MedWatch program allows health care providers to report problems possibly caused by FDA-regulated products such as drugs, medical devices, medical foods, and dietary supplements. The identity of the patient is kept confidential. Reported adverse effects and drug interactions are also posted on the FDA Dietary Supplement Information page of its website. If a consumer or health care provider thinks a patient has suffered a serious harmful effect or illness from a dietary supplement, she or he can report it by calling the FDA's MedWatch hotline at 1-800-FDA-1088 or by accessing the website at www.fda.gov/medwatch/report/hcp.htm. Consumers may also report an adverse event or illness they believe to be related to the use of a dietary supplement by calling the FDA at 1-800-FDA-1088 or using the website, www.fda.gov/medwatch/report/consumer/consumer.htm.

B. Federal Trade Commission

The Federal Trade Commission (FTC) has authority over the advertising of dietary supplements. This agency has issued advertising guidelines for the supplement industry that explain how truth in advertising applies to this industry and the kinds of claims manufacturers can and cannot make. The guidelines, titled "Dietary Supplements: An Advertising Guide for Industry," can be accessed at www.ftc.gov/bcp/conline/pubs/buspubs/dietsupp.htm. The FTC can take action against supplement manufacturers who make claims that lack "sound scientific evidence" in their advertising or that they deem false or misleading. FTC consumer protection can be accessed at www.ftc.gov/bcp/index.shtml.

C. Poison Control Centers

The American Association of Poison Control Centers operates a hotline for suspected poisonings from drugs or dietary supplements at 1-800-222-1222. The site can be accessed at www.aapc.org.

VII. CONCLUSION

Best practices today for dietitians and health care providers include a careful assessment of dietary supplement use by consumers and patients in order to better assess the health effects of dietary supplements. In some cases, health professionals will find it useful to encourage the use of specific supplements and in other cases they will not, but in all cases use should be documented.

References

1. Ervin, R. B., Wright, J. D., and Kennedy, S. J. (1999). Use of dietary supplements in the United States, 1988–94. National Center for Health Statistics, Hyattsville, MD.
2. Radimer, K., Bindewald, B., Hughes, J., Ervin, B., Swanson, C., and Picciano, M. F. (2004). Dietary supplement use by US adults: data from the National Health and Nutrition Examination Survey, 1999–2000. *Am. J. Epidemiol.* **160**, 339–349.
3. Center for Food Science and Nutrition, Office of Nutritional Products, Labeling and Dietary Supplements. (2005). "Guidance for Industry: A Dietary Supplement Labeling Guide." U.S. Food and Drug Administration, Washington, DC.
4. Timbo, B. B., Ross, M. P., McCarthy, P. V., and Lin, C. T. J. (2006). Dietary supplements in a national survey: prevalence of use and reports of adverse events. *J. Am. Diet Assoc.* **106**, 1966–1974.
5. Rock, C. L., Newman, V. A., Neuhouser, M. L., Major, J., and Carnett, M. J. (1994). Antioxidant supplement use in cancer survivors and the general population. *J. Nutr.* **134** (suppl), 3194S–3195S.
6. Rock, C. L., Newman, V., Flatt, S. W., Faerber, S., Wright, F. A., and Pierce, J. P. (1997). Nutrient intakes from foods and dietary supplements in women at risk for breast cancer recurrence. *Nutr. Cancer* **29**, 133–139.
7. Newman, V., Rock, C. L., Faerber, S., Flatt, S. W., Wright, F. A., and Pierce, J. P. (1998). Dietary supplement use by women at risk for breast cancer recurrence. *J. Am. Diet Assoc.* **98**, 285–292.
8. Felson, D. T., Lawrence, R. C., Hochberg, M. C., McAlindon, T., Dieppe, P., Minor, M. A., Blair, S. N., Berman, B. M., Fries, J. F., Weinberger, M., Lorig, K. R., Jacobs, J. J., and Goldberg, V. (2000). Osteoarthritis: new insights. *Ann. Intern. Med.* **133**, 726–737.
9. Wiygul, J. B., Evans, B. R., Peterson, B. L., Polaseik, T., Walther, P., Robertson, C., Alboai, D., and Denmark-Wahnefried, W. (2005). Supplement use among men with prostate cancer. *Urology* **66**, 161–166.
10. Picciano, M. F., Dwyer, J. T., Radimer, K. L., Wilson, D. H., Fisher, K. D., Thomas, P. R., Yetley, E. A., Swanson, C. A., Moshfegh, A. J., Levy, P. S., Nielson, S. J., and Marriott, B. M. (in press). Dietary supplement use among infants, children, and adolescents in the United States (US): 1999–2002. *Arch. Pediatrics.*
11. White, J. V., Pitman, S., and Blumberg, J. B. (2007). Dietitians and multivitamin use: personal and professional practices. *Nutrition Today* **42**, 62–68.
12. Saldanha, L. G. (2007). The dietary supplement marketplace. *Nutrition Today* **42**, 52–54.

13. Sloan, E. (2007). Why people use vitamin and mineral supplements. *Nutrition Today* **42**, 55–61.
14. Hoffman, F. A. (2001). Regulation of dietary supplements in the United States: understanding the Dietary Supplement and Health Education Act. *Clin Obstet Gynecol* **44**, 780–788.
15. Brazier, N. C., and Levine, M. A. (2003). Drug-herb interaction among commonly used conventional medicines: a compendium for health care professionals. *Am. J. Therapeutics* **10**, 163–169.
16. Couris, R. R. (2005). Vitamins and minerals that affect hemostasis and antithrombotic therapies. *Thrombosis Res.* **117**, 25–31.
17. American Dietetic Association/Allison Sarubin. (2003). "Health Professional's Guide to Popular Dietary Supplements from the Joint Working Group on Dietary Supplements." American Dietetic Association, Chicago.
18. ADA/APHA. (2000). Special report from the Joint Working Group on Dietary Supplements: a healthcare professional's guide to evaluating dietary supplements available at www.aphanet.org
19. Dwyer, J. T., Picciano, M. F., Raiten, D. J., and Members of the Steering Committee: National Health and Nutrition Examination Survey. (2003). Collection of food and dietary supplement intake data: what we eat in America. *NHANES J. Nutr.* **133**,575S–635S.
20. McDavid, K., Breslow, R. A., and Radimer, K. (2001). Vitamin/mineral supplementation among cancer survivors 1987 and 1992 National Health Interview Surveys. *Nutr. Cancer* **41**, 29–32.
21. Thompson, F. E., Subar, A. F., Brown, C. C., Smith, A. F., Sharbaugh, C. O., Jobe, J. B., Mittl, B., Gibson, J. T., and Ziegler, R. G. (2002). Cognitive research enhances accuracy of food frequency questionnaire reports: results of an experimental validation study. *J. Am. Diet. Assoc.* **102**, 212–25.
22. Subar, A. F., Thompson, F. E., Kipnis, V., Midthune, D., Hurwitz, P., McNutt, S., McIntosh, A., and Rosenfeld. S. (2001). Comparative validation of the Block, Willett, and National Cancer Institute food frequency questionnaires. *Am. J. Epidemiol.* **154**, 1089–1099.
23. Clifford, A. J., Noceti, E. M., Block-Jay, A., Block, T., and Block, G. (2005). Erythrocyte folate and its response to folic acid supplementation is assay dependent in women. *J. Nutr.* **135**, 137–143.
24. Chlebowski, R. T., Blackburn, G. L., Thomson, C. A., Nixon, D. W., Shapiro, A., Hoy, M. K., Goodman, M. T., Giuliano, A. E., Karanja, N., McAndrew, P., Hudis, C., Butler, J., Merkel, D., Kristal, A., Caan, B., Michaelson, R., Vinciguerra, V., Del Prete, S., Winker, M., Hall, R., Simon, M., Winters, B. L., and Elashoff, R. M. (2007). Dietary fat reduction and breast cancer outcome: interim efficacy results from the women's intervention nutrition study. *J. Natl. Cancer Inst.* **98**, 1767–1776.
25. Dwyer, J.T, Cunniff, P. J., Maroni, B. J., Kopple, J. D., Burrowes, J. D., Powers, S. N., Cockram, D. B., Chumlea, W. C., Kusek, J. W., Makoff, R., Goldstein, J., and Paranandi, L. (1998). The hemodialysis (hemo) pilot study: nutrition program and participant characteristics at baseline. *J. Renal Nutr.* **8**, 11–20.
26. Dwyer, J. T., Holden, J., Andrews, K., Roseland, J., Zhao, C., Schweitzer, A., Perry, C. R., Harnly, J., Wolf, W. R., Picciano, M. F., Fisher, K. D., Saldanha, L. G., Yetley, E. A., Betz, J. M., Coates, P. M., Milner, J. A., Whitted, J., Burt, V., Radimer, K., Wilger, J., Sharpless, K. E., and Hardy, C. J. (2007). Measuring vitamins and minerals in dietary supplements for nutrition studies in the USA Press. *Anal. Bioanal. Chem.* (suppl) **389**: 37–46.
27. Haggans, C. J., Regan, K. S., Brown, L. M., Wang, C., Krebs-Smith, J., Coates, P. M., and Swanson, C. A. (2005). Computer access to research on dietary supplements: a database of federally funded dietary supplement research. *J. Nutr.* **135**, 1796–1798.
28. Mathieu, J. (2007). Sifting through the research on supplements. *JADA* **107**, 912–914.
29. Thomson, P. D. R. (2007). "PDR® for Nonprescription Drugs, Dietary Supplements, and Herbs: The Definitive Guide to Over the Counter Medications." Thomson PDR, New York.
30. Blumenthal, M., Busse, W. R., Goldberg, A., Gruenwald, J., Hall, T., Riggins, C. W., and Rister, R. S., eds. (1998). "The Complete German Commission E Monographs: Therapeutic Guide to Herbal Medicines." American Botanical Council Integrative Medicine Communications. Austin, TX.
31. McGuffin, M., Leung, A., and Tucker, A. P. (2000). "Herbs of Commerce," 2nd ed. American Herbal Products Association, Denver, CO.
32. Coates, P. M., Blackman, M. R., Cragg, G. M., Levine, M., Moss, J., and White, J. D. (editors) (2005). "Encyclopedia of Dietary Supplements." Marcel Dekker, New York.
33. Costello, R. B., Leser, M., and Coates, P. M. Dietary Supplements for Health Maintenance and Disease Prevention, In: *Handbook of Clinical Nutrition and Aging*. Humana Press Inc., 2008.

CHAPTER 3

Physical Assessment of Nutritional Status

EDWARD SALTZMAN[1,*] AND MEGAN A. McCRORY[2]

[1]*Jean Mayer USDA Human Nutrition Research Center on Aging, Tufts University, Boston, Massachusetts*

[2]*Department of Foods and Nutrition and Department of Psychological Sciences, Purdue University, West Lafayette, Indiana*

Contents

I. INTRODUCTION

Physical assessment contributes to the determination of nutritional status, the identification of nutrition-related health problems, the risk of further nutrition-related morbidity, and the identification of factors resulting in malnutrition. The mutual influence of nutrition and disease may make distinguishing between the ill effects of disease and malnutrition difficult or impossible, but physical assessment allows for the evaluation of manifestations of, or risk for, nutritional and medical problems.

On a population level, or in apparently healthy persons with low risk, screening may be used in lieu of a more detailed and resource-intensive assessment. The results of screening may indicate the need for further assessment, and several screening tools have such staged approaches. Individuals who already have existing nutritional compromise or are at high risk because of illness or other factors should undergo a more detailed assessment. Indications and tools for screening in contrast to detailed assessment have been reviewed in detail elsewhere [1–3].

The term *malnutrition* is often used to describe protein energy malnutrition (PEM), but micronutrient malnutrition occurs in multiple disease states and must also be assessed. Nutritional disorders of excess should also be considered malnutrition and are of increasing importance. Such disorders include obesity, diets characterized by imbalanced macronutrient intake, and micronutrient toxicity induced by food faddism or supplement use.

II. COMPONENTS OF CLINICAL ASSESSMENT

Clinical assessment of nutritional status includes the clinical and dietary history, assessment of anthropometric parameters, physical examination, assessment of functional status, and laboratory evaluation (Table 1). Although

TABLE 1 Components of Assessment

Component	Examples
History	Current and past health Medications and dietary supplements Alcohol and tobacco use Family health history Social history
Diet	Ability to shop and prepare food 24-hour recall Food diary Food preferences Food sensitivities
Anthropometrics	Weight and weight change Height Skinfold thickness Circumferences
Physical Examination	See Table 4
Functional Assessment	Hand grip strength Activities of daily living Walking
Laboratory	Blood and urine tests Dual x-ray absorptiometry

* Corresponding author

This material is based upon work supported by the U.S. Department of Agriculture, Agricultural Research Service, under agreement No. 58-1950-7-707. Any opinions, findings, conclusion, or recommendations expressed in this publication are those of the author(s) and do not necessarily reflect the view of the U.S. Dept. of Agriculture.

TABLE 2 Common Medications and Potential Effects on Nutrient Status

Class (Specific Example)	Effect
Amphetamines	↓ Appetite and weight
Antibiotics (*N*-methylthiotetrazole side chains)	↓ Vitamin K function
Anticonvulsants (phenytoin, phenobarbital)	↓ Calcium absorption
	↓ Vitamin D and folate
	Bone loss
Antipsychotics (clozapine)	↑ Appetite and weight
Bile acid sequestrants	↓ Absorption vitamins A, D, E, K
Corticosteroids	↑ Appetite, fat mass and weight, ↓ lean mass
	Hyperglycemia
	↓ Vitamin D and calcium
	Bone loss
	↓ Vitamin B_6 levels (significance unclear)
Diuretics	↓ Potassium, magnesium, sodium
	↑ Calcium (thiazides)
	↓ Thiamin
Ethanol	↓ Thiamin, folate, riboflavin, vitamins B_6, A, D
	PEM
Insulin	↑ Appetite and weight
Isoniazid	↓ Vitamin B_6, niacin
	↓ Vitamin D (significance unclear)
Lithium	↑ Appetite and weight
Methotrexate	↓ Folate
Orlistat	↓ Vitamins A, D, E, K
Progesterone	↑ Appetite and weight
Selective serotonin reuptake inhibitors	↓ or ↑ Appetite and weight
Sulfasalazine	↓ Folate
Sulfonylureas	↑ Appetite and weight
Theophylline	↓ Appetite and weight
Thiazolidinediones	↑ Weight and subcutaneous fat
Topiramate	↓ Appetite and weight
Tricyclic antidepressants	↑ Appetite and weight

some elements of the clinical evaluation are relatively insensitive or nonspecific indicators of nutritional disorders, it is these clinical observations that stimulate further confirmatory measures, such as detailed assessment of diet and appropriate diagnostic tests.

Dietary assessment and important elements of the dietary history are discussed in other chapters of this volume. Medical histories address details of the present complaint or illness, body weight change, and the past medical history. A review of organ systems to elicit relevant clinical factors not directly related to the present illness should also be conducted. The history must also address the spectrum of behaviors and physiological functions necessary to maintain adequate nutritional status, including appetite and thirst, and the abilities to procure, prepare, and ingest food. Socioeconomic and psychosocial factors may figure prominently in nutritional status and should not be neglected.

Medication use should be reviewed, and in hospitalized patients, prehospitalization medications and dietary supplements should also be included. Relevant medications include prescription and over-the-counter medications, vitamin and mineral supplements, and herbal preparations. Although approximately 50% of the U.S. population reports recent use of at least one vitamin, mineral, or dietary supplement [4], many do not consider nutritional supplements to be medications, and direct questioning may be necessary to elicit this history. Medications interfere with nutritional status by multiple mechanisms, including alterations in appetite, taste, thirst, nutrient absorption, nutrient metabolism, or excretion. Many medications induce changes in bowel function, which in turn may influence intake. Table 2 describes nutritional effects of commonly used medications.

III. ANTHROPOMETRIC ASSESSMENT

Anthropometric measurements quantify basic physical characteristics and include height, weight, circumference of body parts, and skinfold thickness. Assessment of these parameters allows comparison to population norms or to values collected over time in the same individual.

A. Height

Measurement of height is necessary to estimate ideal body weight or desirable body weight, body mass index (BMI), and body surface area, and it is used to calculate energy requirements as well as body composition. Height should be directly measured when possible and is best obtained by a stadiometer. Vertical height decreases with age as a result of vertebral bone loss, vertebral compression fractures, and thinning of intervertebral disks and weight-bearing cartilage. Height begins to decline at approximately age 30 years for both men and women, and this decline accelerates with age; in one longitudinal series, between the ages of 30 and 80 years, women lost 8 cm and men lost 5 cm [5]. Loss of vertebral mass and disk compression may induce kyphosis (curvature with backward convexity of the spine), which will reduce measured height.

When height cannot be accurately measured, such as in acutely ill or immobilized patients or in those with severe osteoporotic changes, alternatives include self-reported height, estimated height, or surrogate methods to estimate height. Self-reported height is less accurate than measured height as men tend to overreport and women tend to underreport [6]. Self-reported height is more accurate, however, than estimation of height by visualization of supine patients, which has been found to overestimate height [7]. In addition, accuracy of visual estimation of height was better for taller patients compared to shorter patients, possibly because taller patients were closer to the length of the beds in which they were lying, which provided a frame of reference for estimation [7].

Surrogate measures for height include arm span, knee height, and seated height. Use of knee height or arm span to estimate vertical height may be useful in clinical as well as research situations for individuals who cannot stand, who are debilitated, or who have experienced loss of height [5, 8]. These measures correlate with vertical height but are influenced less by age-related changes in stature and impediments to the measurement of vertical height such as disability or frailty [8–11]. Surrogates of height have been used to predict both current height as well as previous adult maximal height. Arm span, which is the entire distance from the tip of the middle finger of one hand to the other, can be measured with arms stretched at right angles to the body, with the measuring tape crossing in front of the clavicles. Frail or debilitated persons may require assistance to maintain the correct position for measurement [9, 10]. Half arm span (the distance from the sternal notch to the tip of the middle finger of one hand) can also be measured and then doubled to determine arm span. Knee height is best measured with specialized calipers, which can be performed either in sitting or recumbent positions, making this useful in most ambulatory and hospital settings.

Prediction equations for the estimation of height from arm span or knee height are available for specific age, gender, racial, and ethnic groups [10, 12–18]. In several trials that directly compared measured height to surrogates, disagreement between measured height and surrogates increased when the measurement was conducted in ill patients instead of healthy subjects; compared to measured height, mean differences were 0 to 2 cm for self-reported height, −0.6 to 4 cm for knee height, and 0 to 7 cm for arm span [14–16, 19]. Despite a reported lack of agreement with measured height, errors in using these surrogates are smaller than those for visual estimation and are not likely to result in errors that are clinically meaningful.

B. Weight

Body weight can be utilized as a single measurement at a point in time or as multiple measurements reflecting a trend over time. Loss of body weight in the setting of starvation or illness suggests PEM and is associated with increased risk of morbidity and mortality [20–24]. In hospitalized patients with a variety of gastrointestinal, infectious, and neoplastic diseases, PEM at admission was associated with an approximately twofold risk of subsequent complications [25]. Involuntary change in weight, as opposed to assessment of a single static measure of weight, may better predict risk for PEM-related complications [22, 26, 27]. In patients with cancer who were undergoing chemotherapy, a loss of 5% or more of usual body weight was associated with impaired functional status and significantly decreased median survival compared to patients without weight loss [20]. More than 60 years ago, Studley [23] recognized that unintentional weight loss of 20% or more of usual body weight before surgery for peptic ulcer significantly increased the risk of postoperative mortality. Others have confirmed that PEM preceding surgery increases risk of postoperative complications [22, 24, 28]. Patients who have lost 10% to 20% of initial body weight over 6 months and have associated physiological defects, or those who have lost 20% or more over 6 months, should be considered at high risk [26, 29].

In obese persons, "adjusted" weight or ideal body weight is used in some prediction equations to guide provision of energy and medications [30–34]. This adjustment is based on the assumption that in obese persons approximately 25% of weight gained is fat free mass. Stores of lean and fat mass may act as buffers against chronic PEM, but even in obese patients they do not prevent PEM in acute illness. Despite high levels of energy reserves and expanded lean body mass, obese patients still experience PEM when acutely ill [30].

Ideally, body weight should be measured by use of calibrated beam-type or electronic scales. Alternatives are home scales, calibrated bed scales, chair scales, or wheelchair scales. To monitor changes in weight over time, the use of the same scale is recommended given variability between scales. Technological advances now allow automatic remote monitoring of home scales via the Internet, a method gaining popularity in clinical practice as well as in research.

Self-reported weights are often inaccurate. Overweight women and men tend to underestimate weight, whereas lower weight men tend to overestimate [6]. Use of a single self-reported weight is also an insensitive measure of weight change in ill patients, because weight loss in approximately one-third of patients may be missed [35]. In cases where a person cannot be weighed or provide a self-reported weight, weight may be estimated, an inaccurate practice that does, however, improve with experience [7].

Gains and losses in weight should be documented, and related factors including illness, medication changes, and psychosocial contributors should be sought to explain weight change. Precipitous changes in weight are usually due to alterations in body water because of conditions such as congestive heart failure, cirrhosis, renal failure, or treatments for these conditions. Thus, large weight fluctuations should stimulate investigation for volume overload, shifts, or dehydration.

C. Weight for Height

Weight is often expressed as a function of height to facilitate comparison of individuals of varied heights. Historically, ideal body weight or desirable body weight was defined by actuarial data of weight for height, often with adjustment for frame size, and published as tables. As with all predictive methods, these tables best apply to members of the population from which the data were derived, and thus they may have reduced applicability to diverse ethnic groups, older adults, or those with chronic illnesses [36]. Frame size can be determined by measurement of elbow or wrist breadth or of wrist circumference, which requires the use of specialized calipers or measuring tape. Percentage of ideal body weight was previously used to classify underweight and overweight, and today it is still, on occasion, utilized for these purposes or to estimate energy needs or drug dosing.

Body weight expressed as a function of height takes the general form weight/height^x and is called body mass index (BMI). Whereas x may be any number, Quetelet's index, or kg/m^2, has become synonymous with BMI. The use of the BMI to assess weight for height in individuals, with the classifications found in Table 3, reflects recommendations of the National Institutes of Health [37] and World Health Organization [38].

BMI correlates with body fat, although a linear relationship is not observed throughout the range of BMI (Fig. 1). Although BMI correlates highly with body fat for populations, interpretation in individuals must include consideration of clinical and other factors. For example, BMI may be elevated despite relatively low levels of body fat in those with edema or in bodybuilders. Similarly, the same low BMI may be observed in a patient who has experienced significant loss of weight and in a long-distance runner who is healthy and weight stable. The relationship between BMI and body fat differs between sexes, varies among racial and ethnic groups, and also changes over the life span [39].

TABLE 3 Classification of Weight by Body Mass Index

			Disease Risk Relative to Normal Weight and Waist Circumference (Inches)	
	Grade	**BMI (kg/m^2)**	**Men ≤ 40 Women ≤ 35**	**>40 >35**
Underweight	III	<16		
	II	16–16.99		
	I	17–18.49		
Normal		18.5–24.9		
Overweight		25–29.9	Increased	High
Obesity	I	30–34.9	High	Very high
	II	35–39.9	Very high	Very high
	III	≥40	Extremely high	Extremely high

Source: Adapted from Ferro-Luzzi, A., Sette, S., Franklin, M., and James, W. P. (1992). A simplified approach of assessing adult chronic energy deficiency. *Eur. J. Clin. Nutr.* **46,** 173–186; and NHLBI Obesity Education Initiative Expert Panel on the Identification, Evaluation, and Treatment of Overweight and Obesity in Adults. (1998). Clinical guidelines on the identification, evaluation, and treatment of overweight and obesity in adults: the evidence report. *Obes. Res.* **6** (suppl 2), 51S–209S.

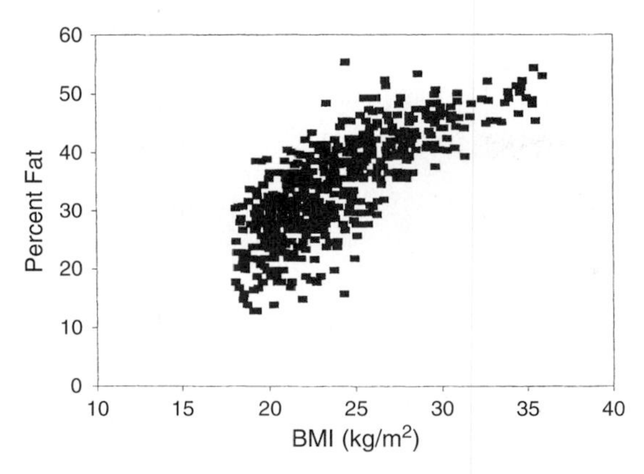

FIGURE 1 Relationship between BMI and fat mass in women (Reproduced with permission from Heysmfield, S. B., Tighe, A. and Wang, Z.-M. (1994). Nutritional assessment by anthropometric and biochemical methods. *In* "Modern Nutrition in Health and Disease" 8th ed., 1994 (M. E. Shils, J. A. Olson, and M. Shike, Eds.), p. 824. Lea and Febiger, Malvern, PA.)

A single BMI classification for the entire adult age range does not reflect the loss of lean mass and gain in fat mass that accompany aging. Despite these potential problems, BMI remains an easily calculated and useful method of classifying weight relative to height, especially for populations. In individuals, BMI can be used as one of several indicators of nutritional status and should be interpreted in light of physical examination and other findings.

Both low and high BMI correlate with morbidity and mortality [40–44]. In addition to susceptibility to infection and multiple other diseases, low levels of BMI are also associated with lethargy and diminished work productivity [38]. The lowest average survivable BMI, as observed in starvation, famine, anorexia nervosa, or by theoretical models, has been estimated to be 12 to 13 kg/m^2 [45, 46]. However, when weight loss is rapid or associated with illness, morbidity and mortality can occur at any level of BMI.

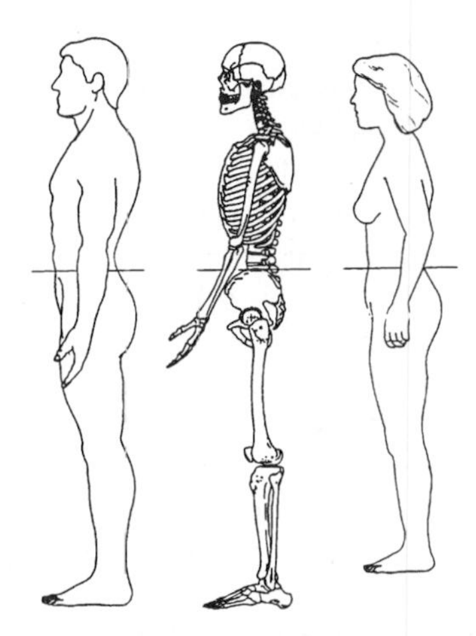

FIGURE 2 Landmarks for measurement of waist circumference. (From National Heart, Lung, and Blood Institute. [1998]. Clinical guidelines on the identification, evaluation, and treatment of overweight and obesity in adults: the evidence report. *Obes. Res.* **6**(suppl 2), 51S–209S.)

Obesity has deleterious effects on every organ system as well as quality of life and productivity, and the health effects of obesity are discussed in detail in this volume and elsewhere [47, 48]. Of note, a significant proportion of weight-related comorbidities, such as type 2 diabetes or obstructive sleep apnea, remain undiagnosed in obese persons [49, 50]. The reasons for this are complex, including contributions of both patients and providers, but this observation underscores then need for careful assessment of obese persons in whom weight-related comorbidity might be present.

D. Body Fat Distribution

Central distribution of body fat increases risk for type 2 diabetes, coronary heart disease and its risk factors including hypertension, and dyslipidemia [51–53]. Central obesity is a predictor of risk for diabetes and other cardiovascular risk facts beyond BMI, and when contrasted to BMI, in some cases central obesity has been a better predictor of disease risk than BMI [51]. Central adiposity reflects accumulation of fat in the intraabdominal and subcutaneous compartments. Measures of central adiposity include waist-to-hip ratio and single measurements of the abdomen at its greatest or narrowest circumference. Current guidelines suggest that abdominal adiposity be assessed by waist circumference measured at the level of the top of the iliac crest (Fig. 2) [37]. Anthropometric measures of central adiposity correlate with subcutaneous and intra-abdominal fat measured by computed tomography (CT) or magnetic resonance imaging (MRI), but circumferences do not distinguish between these compartments. Like BMI, the relationship between anthropometric measures of central obesity and disease risk varies with sex, age, race, and ethnic group [51]. In populations, measures of central obesity generally increase with increasing BMI; thus the use of a single cutoff for waist circumference to indicate central obesity for the entire range of BMI is problematic. The need for BMI-specific cutoffs, especially for BMI in ranges less than 30 kg/m^2, has been proposed [51].

E. Circumferences and Skinfold Thickness Measurements

Circumferences of the trunk or limbs reflect amounts of underlying lean and fat mass. Skinfold thickness describes the amount of subcutaneous fat present when the skin at various sites is pinched by specialized calipers, and it provides information about fat stores. The sites at which these measurements are conducted are illustrated in Figure 3. Because measurements of circumference or skinfold thickness

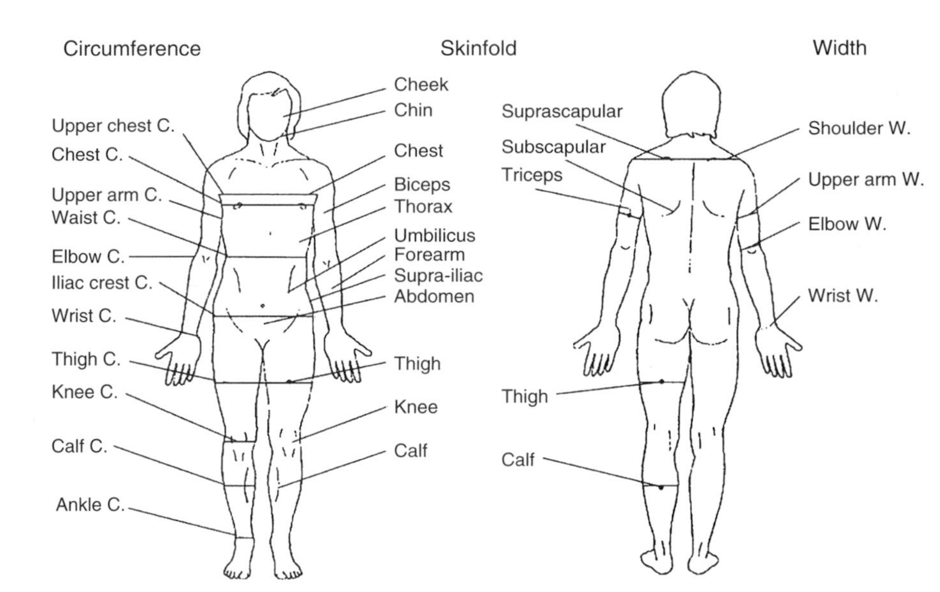

FIGURE 3 Body sites for measurement of circumferences, skinfold thickness, and widths. (From Wang, J., Thornton, J. C., Kolesnik, S., and Pierson, R. N. Jr. (2000). Anthropometry in body composition: an overview. *Ann. N.Y. Acad. Sci.* **904,** 317–326. Used with permission of Wiley-Blackwell.)

from a single body site are likely to be influenced by the interindividual variability in body composition, measurements at multiple sites are conducted to reduce the potential contribution of this variability. Single site measurements may, however, provide some data regarding changes over time in the same individual. Combinations of circumference and skinfold thickness measurements at multiple sites have been related to reference measures of body composition such as hydrodensitometry, dual energy x-ray absorptiometry (DXA), or CT to develop prediction equations for lean and fat mass for the whole body or specific segments [54, 55].

Circumferences and skinfold thickness may be influenced by several factors, including age, sex, race or ethnicity, and state of hydration [56]. Specific reference data for the individual or population should be used when possible. Most reference data have been developed in healthy populations. Limited reference data exist for hospitalized patients; however, acute illness is associated with widely fluctuating perturbations in body water such that generalizable reference data may never be obtainable. Upper body measurements may be preferable in those individuals with evidence of edema or ascites, because the upper body is less likely to accumulate excess body water. Measurements obtained in reduced-obese persons are likely to be of diminished predictive value because of error introduced by redundant skin and other factors, especially when large amounts of weight have been lost.

IV. BODY COMPOSITION ASSESSMENT

Body composition describes and quantifies various compartments within the body and is most commonly conducted to assess of the amount of fat tissue and lean tissue (or fat-free tissue) in the body. Measurement of skinfold thicknesses and circumferences is one such method that is used to predict lean and fat compartments based on equations derived from relationships between anthropometric measurements and body composition from reference methods. When the amount of fat tissue is assessed, it is usually expressed as a percentage of total body mass or body weight. Thus, "percentage body fat" is calculated as (fat mass/body mass) $\times$ 100%. If desired, more detailed body composition assessments can quantify the amount of different components of lean tissue, such as bone mineral, water, and protein [57].

Body composition can be assessed at the level of the body as a whole (e.g., weight or BMI), by division into lean and fat tissues; by division into molecules such as water, protein, and fat; or at an atomic level into elements such as carbon and potassium (Fig. 4). Methods to assess body composition vary by the compartments being measured. Some commonly employed methods include dual energy x-ray absorptiometry (DXA), which can divide the body into fat, fat-free, and bone compartments, and density methods such as air displacement plethysmography (ADP, using a device known as the BOD POD) and hydrostatic weighing (HW) or underwater weighing (UWW). These methods utilize body density, body volume, and weight to estimate fat and fat-free compartments. Hydrostatic weighing is traditionally known as the gold standard for percentage body fat determination. It is difficult for some individuals to perform, as it involves whole body submersion and full lung exhalation; however, it is considered highly accurate in persons who are comfortable performing this maneuver. Bioelectrical impedance (BIA) measures body water, from which fat-free mass can be estimated. ADP and DXA may be the most favorable due to their combined ease of use, accuracy, and precision. The primary use for DXA on a clinical basis is to provide a measure of bone density in order

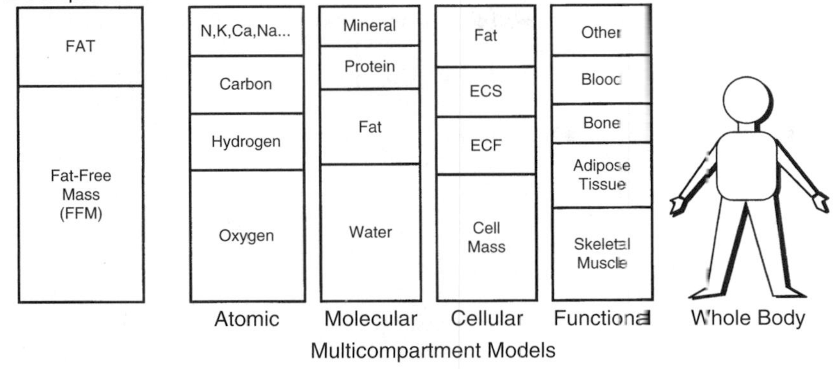

FIGURE 4 Models of body composition compartments. ECS = extracellular solid, and ECF = extracellular fluid. (From Ellis, K. J. (2000). Human body composition: in vivo methods. *Physiol. Rev.* **80,** 649–680. Used with permission)

to assess osteoporosis risk. Therefore, it is not widely available for routine measurement of percentage body fat. Also, DXA is not appropriate for use in pregnant women and small children because of the small amount of radiation dose associated with the measurement. Of the methods discussed, BIA is the least accurate for individuals but may be useful for monitoring changes over time. Although not yet routinely used clinically, these methods can be sought out at some hospitals, universities, and research institutions, and health clubs.

A single body composition assessment can be used to help determine chronic disease risk due to excess body fat particularly in cases where assessment from BMI alone may be in question. For example, a suspected high BMI due to having an exceptionally high muscle mass, such as in a bodybuilder or highly competitive athlete, can be confirmed with body composition testing. In another case, an extremely sedentary person may have a normal-weight BMI that would classify him or her as having low disease risk. However, if upon body composition testing this person has a high percentage of body fat due to having low muscle and/ or bone mass, he or she is therefore at higher risk for chronic disease than previously thought.

A high percentage of body fat may be associated with established chronic disease risk factors such as elevated serum cholesterol and triglycerides, glucose, and blood pressure. Specific cutoffs for chronic disease risk associated with different percentage body fat levels have not been established; however, it has been suggested that women not exceed 30% body fat and men not exceed 20% body fat [58]. As discussed earlier, body fat may vary substantially with age, gender, and other factors; thus these are crude guidelines that should be interpreted with caution. Perhaps more useful than assessing specific disease risk by a single measurement at one point in time is monitoring percentage body fat changes over time in various situations such as in response to lifestyle modifications (e.g., adoption of consistent exercise habits or healthful eating patterns), with aging, or with wasting disease processes themselves (e.g., cachexia associated with AIDS or cancer).

For optimal accuracy and precision, body composition should be assessed under the appropriate conditions. These include in the fasting state or a minimum of 4 hours after a small meal, under normal hydration levels, and while at rest (not after moderate or heavy exercise).

V. PHYSICAL MANIFESTATIONS OF MALNUTRITION

Although physical examination may reveal manifestations of malnutrition, these findings often occur late in the course of malnutrition. The specificity of physical exam is limited because deficiencies of several micronutrients may result in similar manifestations, and micronutrient deficiencies may not occur in isolation such that physical findings may reflect multiple deficiencies. Nonetheless, physical findings suggestive of malnutrition elicit relevant medical and historical elements and stimulate biochemical tests to assess deficiencies.

The following sections describe physical findings in organ systems or disease states. Because physical findings may be nonspecific, associated historical, anthropometric, and functional and biochemical findings are discussed. Table 4 summarizes selected physical findings found in nutrient deficiency or excess, some of which are depicted in Figure 5.

A. Head and Mouth

The temporalis muscles should be visualized for evidence of wasting, a classic sign of PEM. Dull hair that is easily plucked may indicate PEM, and brittle hair is associated with several micronutrient deficiencies.

TABLE 4 Physical Signs of Malnutrition

System	Sign	Nutrient or Condition
Mouth	Glossitis	Deficiencies of riboflavin, niacin, biotin, vitamin B_6, vitamin B_{12}, folate, iron, zinc
	Angular stomatitis	Deficiencies of riboflavin, niacin, biotin, vitamin B_6, iron,
	Cheilosis	zinc
	Gingivitis	Vitamin C deficiency
	Gingival bleeding	
	Parotid hyperplasia	Bulimia nervosa
	Dental erosions	
Eyes	Xerophthalmia	Deficiency of vitamin A
	Night blindness	
	Photophobia	
	Xerosis	
	Bitot's spots	
	Corneal ulceration	
	Corneal scarring	
	Diplopia	Toxicity of vitamin A
	Nystagmus	Thiamin deficiency
	Lateral gaze deficit	
	Optic nerve atrophy	Vitamin B_{12} deficiency
	Blindness	
	Retinitis pigmentosa	Vitamin E deficiency
	Visual deficits	
	Kayser-Fleischer ring	Copper toxicity
	Sunflower cataract	
Skin	Seborrheic-like dermatitis	Deficiencies of vitamin B_6, zinc
	Impaired wound healing	Deficiencies of vitamin C, zinc
	Erythematous or scaly rash at sun-exposed areas (e.g., extremities and neck) (Casal's necklace)	Niacin deficiency
	Perifollicular petechiae	Vitamin C deficiency
	Ecchymosis (bruising)	
	Easy bruising	Vitamin K deficiency
	Dry flaky skin	Essential fatty acid deficiency
	Depigmentation	Protein-energy malnutrition
	Yellow or orange discoloration	Carotenoid excess
	Pallor	Anemia due to deficiencies of iron, vitamin B_{12}, folate
Nails	Koilonychia (spoon-shaped nails)	Iron deficiency
	Discolored or thickened nails	Selenium toxicity
Hair	Swan-neck deformity	Vitamin C deficiency
	Discoloration	Protein energy malnutrition
	Dullness	
	Easy pluckability	
	Alopecia	Vitamin A toxicity, biotin deficiency
Cardiovascular	High output congestive heart failure	Thiamin deficiency
	Cardiomyopathy and heart failure	Selenium deficiency
Gastrointestinal	Stomatitis	Niacin deficiency
	Proctitis	
	Esophagitis	
	Hepatomegaly	Hepatic steatosis due to diabetes, obesity, deficiency of choline, carnitine
Musculoskeletal	Generalized or proximal weakness	Vitamin D deficiency
	Bone tenderness	

(continues)

TABLE 4 *(continued)*

System	Sign	Nutrient or Condition
	Fracture	
	Weakness	PEM, hypophosphatemia, hypokalemia, hypomagnesemia, vitamin D deficiency, iron deficiency
	Muscle wasting	PEM
	Carpopedal spasm	Hypocalcemia
Neurologic	Peripheral neuropathy	Deficiencies of vitamins B_6, B_{12}, E, thiamin; vitamin B_6 toxicity
	Mental status changes	Deficiencies of thiamin, vitamins B_6, B_{12}, niacin, biotin,
	Delirium	hypophosphatemia, hypermagnesemia
	Dementia	Deficiencies of vitamin B_{12}, thiamin, niacin

The mouth provides considerable information about nutrition status (see Fig. 5). Cracking or ulceration of the lips (cheilosis), or cracking or ulceration at the corners of the mouth (angular stomatitis), is seen in several B vitamin, zinc, and other deficiencies. Cheilosis may be a more specific indicator of malnutrition, but many non-nutritional factors cause both of these findings. Angular stomatitis may also be due to poor fitting dentures, which is associated with reduced food intake. The tongue may become sore and may appear beefy red or magenta because of deficiencies of several micronutrients and zinc, whereas loss of papilla leading to atrophy may accompany PEM. Pale gums may indicate anemia. Bleeding from the gums may indicate coagulopathy resulting from vitamin K deficiency or scurvy. Cancer chemotherapy may result in pain or ulcers in the mouth and throat (mucositis), a common cause of poor intake in patients undergoing chemotherapy and in some patients undergoing radiation therapy.

Dental health directly impacts dietary intake and eating enjoyment [59, 60]. Those who are edentulous and without dentures, or with only one denture, are at higher risk for poor intake than are those with two dentures [60]. Acutely ill edentulous patients are often hospitalized without dentures, which may predispose to in-hospital malnutrition by limiting ability to chew.

Although now far more rare than previously, goiter as a result of iodine deficiency may be apparent by visualization or palpation of the thyroid.

B. Skin

The skin may reveal signs of acute or chronic nutritional problems (Fig. 5). Tenting of the skin, seen with dehydration, is persistence of a tentlike fold after pinching the skin. Pallor of the skin suggests anemia. Dermatitis accompanies many micronutrient deficiencies as well as essential fatty acid deficiency (Table 4 and Fig. 5), but it may only occur in advanced states of deficiency. Dermatitis may present in classic patterns, for example, the dermatitis of pellagra in sun-exposed areas such as the neck and extremities or the erythematous perioral and perianal dermatitis of zinc deficiency. The skin should be examined for evidence of decubitus ulcer and impaired wound healing, which are more likely to occur in setting of PEM or micronutrient deficiencies.

Skin lesions are not only the result of nutritional deficiencies but may represent metabolic disorders or sensitivities to components of food. For example, hyperinsulimia may cause acanthosis nigricans, a gray discoloration around the base of the neck and on extensor surfaces, whereas lipid disorders may lead to lesions on the extremities and around the eyes. Gluten sensitivity causes dermatitis herpetaformis, an itchy eruptive rash. Dermatitis herpetaformis may occur in isolation or may be accompanied by subclinical or symptomatic celiac sprue; even if diarrhea and weight loss are absent, consideration should be made of subclinical intestinal disease and screening for vitamin deficiency. Yellow or orange discoloration of the skin may indicate carotenoid excess, which can be distinguished from jaundice since the former spares the sclerae.

C. Cardiovascular System

Examination of the cardiovascular system is unlikely to detect specific nutritional problems, although nutrition has a central role in the development of and is compromised by several cardiovascular diseases.

Cardiac cachexia may occur in patients with chronic congestive heart failure (CHF) and is manifest by progressive loss of weight and lean mass. It is not only due to loss of appetite, which may often accompany CHF, but reflects a complex disorder made up of imbalances between anabolic and catabolic processes, increased resting energy expenditure, disturbances in neurohumoral factors and gut hormones, and edema of the gut [61]. Because increasing food intake alone may not influence loss of lean mass, therapies are directed to the underlying pathophysiological processes as well as increasing food intake to prevent weight loss [61]. In patients with CHF, body daily weight may vary by several kilograms because of disease decompensation, sodium intake, or diuretic treatment. Large and rapid changes in

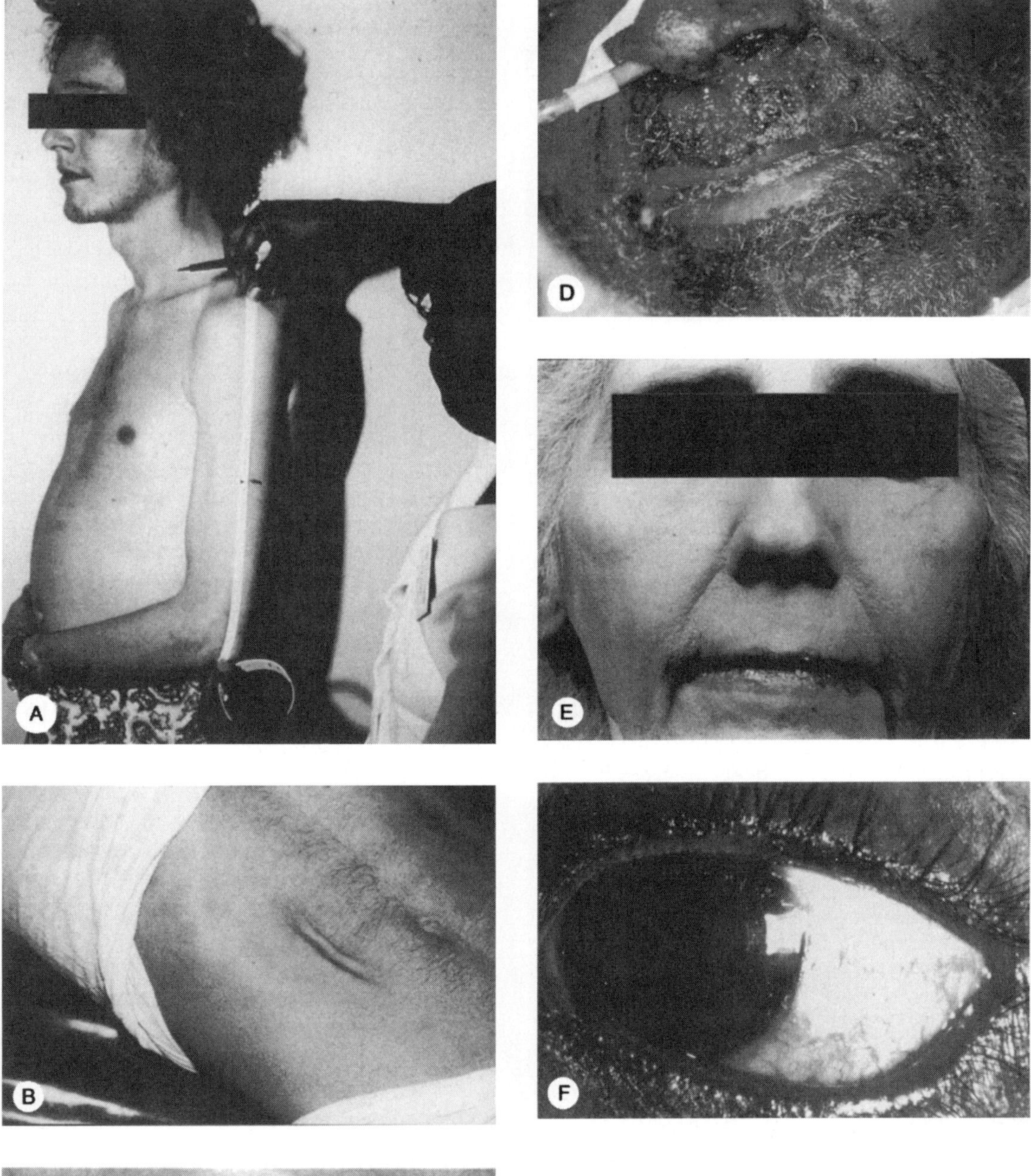

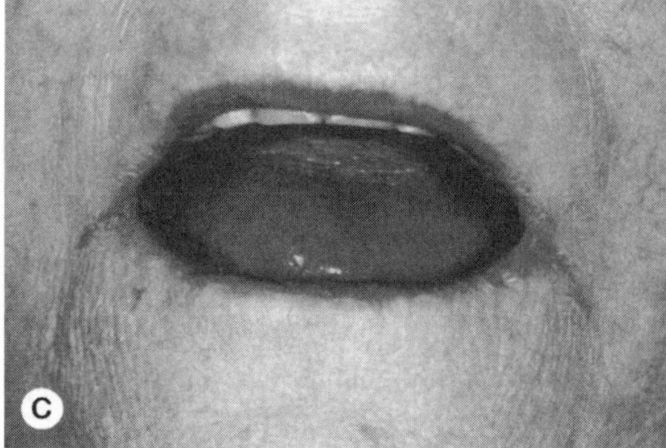

FIGURE 5 Physical signs associated with nutrient deficiencies. (A) Muscle wasting in severe PEM. (B) Tenting of skin in dehydration; the skin retains the tented shape after being pinched. (C) Glossitis and angular stomatitis associated with multiple B vitamin deficiencies. (D) Dermatitis associated with zinc deficiency. (E) Cheilosis, or vertical fissuring of the lips, associated with multiple B vitamin deficiencies. (F) Bitot's spot accompanying vitamin A deficiency. (Photos courtesy of Dr. Robert Russell and Dr. Joel Mason.) See color plate.

body weight should be interpreted cautiously; as discussed previously, home remote monitoring of body weight in CHF patients now allows improved tracking of body weight patterns.

Thiamin deficiency may result in congestive heart failure ("wet" beriberi). This high-output heart failure is characterized by rapid heart rate and pulmonary and peripheral edema. Symptomatic thiamin deficiency has typically been observed in those with alcohol abuse, with very poor intake, or with carbohydrate refeeding. However, it is now recognized that patients who have undergone bariatric surgery who experience frequent vomiting or poor intake, despite the absence of traditional risk factors, may experience symptomatic thiamin deficiency.

D. Pulmonary System

Like the cardiac exam, physical findings of the pulmonary system seldom implicate specific nutritional factors, but pulmonary pathology is both caused by and contributes to malnutrition. Respiratory muscle strength may be diminished in PEM and predisposes to respiratory complications in those with chronic pulmonary disease or in acutely ill patients. Respiratory muscle strength and spirometry have been utilized as components of functional assessment for PEM. Hypophosphatemia may result in acute respiratory failure or need for prolonged mechanical ventilation. Hypophosphatemia may occur in settings of PEM, alcohol abuse, during correction of diabetic ketoacidosis, and with refeeding syndrome.

Chronic obstructive lung disease (COPD), like CHF, is associated with PEM because of a constellation of factors including diminished intake, increased resting energy expenditure, systemic inflammation, and muscle dysfunction [62, 63]. In patients with severe COPD, dyspnea induced simply by eating can limit intake. Also, acute exacerbations or severe COPD may be accompanied by aerophagia (swallowing of air), which is evidenced by tympanitic sounds on percussion of the abdomen, and may cause gastric distention, abdominal bloating, and decreased intake. Chronic treatment of COPD with corticosteroids predisposes to further loss of lean mass and gains in fat mass characterized by a Cushingoid appearance, thin skin with easy bruising, glucose intolerance, and bone loss.

E. Gastrointestinal System

Examination of the abdomen may suggest gastrointestinal or liver conditions, many of which promote PEM and micronutrient malnutrition. In addition to nutritional disorders caused by disease, dietary factors may result in gastrointestinal symptoms; some of these are uncomfortable but harmless, whereas others may progress to serious or life-threatening diseases. Included in this spectrum and discussed in greater detail elsewhere in this text are lactose intolerance, consumption of osmotic cathartics such as sorbitol (in food, candy, or medications), and gluten-sensitive enteropathy (celiac disease). Following gastric resection or gastric bypass for obesity, dumping syndromes may occur in response to hyperosmolar meals and are characterized by varying degrees of lightheadedness or near-syncope, nausea, diaphoresis, chest pain, and abdominal pain or cramps.

Digestive and absorptive processes are site specific in the gastrointestinal system. Accordingly, the extent and location of both disease activity and resection should be noted. Inflammatory bowel disease, celiac disease, resection of the intestinal tract, and short bowel syndrome are associated with PEM and deficiencies of water- and fat-soluble vitamins, as well as iron, calcium, other minerals, and essential fatty acids [64–67]. Geerling *et al.* (66) found that patients with Crohn's disease in remission had persistent deficiencies in several water- and fat-soluble vitamins as well as zinc. Thus, signs or symptoms of micronutrient deficiency in the absence of gastrointestinal symptoms should not be ignored.

End-stage liver disease or cirrhosis is frequently associated with PEM and deficiencies of fat-soluble vitamins [68]. Manifestations of deficiency states may be masked by malaise associated with chronic illness or be asymptomatic and may only become apparent when specifically elicited, as was observed in patients with primary sclerosing cholangitis who experienced problems such as night blindness, bone pain, or easy bleeding [68].

Chronic pancreatitis or pancreatic insufficiency (e.g., with cystic fibrosis) may lead to maldigestion of macronutrients and fat-soluble vitamins, and postprandial pain in pancreatitis may inhibit food intake.

Atrophic gastritis, which predisposes to vitamin B_{12} deficiency, increases in prevalence with advancing age. Not surprisingly, more than 12% of a free-living elderly population was found to be deficient in vitamin B_{12} [69].

F. Musculoskeletal System

Protein energy malnutrition may result in reductions in muscular size and strength, as well as in functional changes such as diminution of work capacity or endurance. Muscle wasting may be most apparent at the temporalis muscle (temporal atrophy or wasting), the shoulder girdle, and between the bones of the dorsum of the hand (interosseus wasting). Generalized weakness is a highly nonspecific symptom and is observed in dehydration, iron deficiency (even before the development of significant anemia), hypophosphatemia, and in multiple vitamin and mineral deficiencies. It is now recognized that vitamin D deficiency may also result in generalized or proximal weakness, as well as frank myopathies [70].

Children with rickets may demonstrate deformation of long bones or prominence of costochrondral joints (the rachitic rosary). Exam findings of osteomalacia in adults are subtler and may include tenderness of the bones and sternum. A history of fractures and bone pain may indicate metabolic bone disease or osteoporosis and should stimulate evaluation of the skeleton as well as of calcium and vitamin D intake and status. Persons with diseases known to influence calcium and vitamin D metabolism (such as those with malabsorptive disorders, chronic renal failure, and the institutionalized elderly) are at high risk for metabolic bone disease, and appropriate monitoring and treatment should be undertaken. An emerging risk factor for osteopenia as well as osteomalacia is malabsorptive bariatric surgery. Hypovitaminosis D and secondary hyperparathyroidism may occur in patients who have undergone malabsorptive procedures such as gastric bypass or biliopancreatic diversion, and symptomatic hypocalcaemia may occur in those with the more malabsorptive biliopancreatic diversion [71].

G. Hematological System

Physical examination of the hematological system is not possible, but signs and symptoms of hematological disorders may be easily visualized in areas such as skin and mucous membranes. Physical findings of anemia include pallor of the skin and mucous membranes, and iron deficiency may cause nails to be spoon shaped. Classical nutritional anemias result from deficiencies of vitamin B_{12}, folate, and iron. Populations at increased risk include alcoholics (vitamin B_{12} and folate), the elderly (vitamin B_{12}), women with menometorrhagia (iron), and vegans (vitamin B_{12}). However, anemia as a result of PEM and deficiencies of vitamin C, vitamin B_6, riboflavin, and copper has also been reported [72].

Abnormal bleeding as a result of vitamin K deficiency may be observed. Most vitamin K–related bleeding usually occurs in the setting of oral warfarin use, which antagonizes vitamin K action. Dietary vitamin K deficiency must be severe before coagulopathy occurs, but it is observed in alcoholics, those with malabsorptive disorders or poor intake, and those taking some medications (Table 2). Vitamin C deficiency may cause gum bleeding, splinter hemorrhages of the nails, petechial hemorrhages of the skin and larger hemorrhages of the muscles and skin that are manifest as ecchymoses (bruises).

H. Renal System

Chronic renal failure is associated with PEM, risk for several micronutrient deficiencies, and risk for hypervitaminosis A. Physical findings of PEM may be masked by volume overload and body weight variation resulting from dialysis. Chronic renal failure and dialysis patients are at increased risk for PEM, and an inverse relationship exists between body mass index and mortality [73]. Micronutrient issues in chronic renal failure include deficiency of vitamin D, hyperphosphatemia and related bone disease, deficiencies of several water-soluble vitamins, and toxicity of vitamin A.

I. Neurological and Psychiatric Systems

Dementia and neurological disorders, such as stroke, Parkinson's disease, and head injury, may impair feeding skills and swallowing. A history of difficulty initiating a swallow, choking or gagging, wet cough, and retained food in the mouth are among the signs that should stimulate further evaluation for dysphagia [74].

Signs of bulimia may include dental erosions and parotid hyperplasia because of frequent vomiting. Nutritional problems associated with anorexia nervosa and bulimia include PEM, electrolyte abnormalities, vitamin and mineral deficiencies, and, in the longer term, osteopenia.

Multiple psychiatric or neurological syndromes related to nutrient deficiency (e.g., thiamin, niacin, vitamin B_6, vitamin B_{12}, vitamin E, essential fatty acids) or excess because of supplementation or faddism have been described (Table 4). In the United States, common predisposing factors to deficiency syndromes are alcoholism and malabsorptive disorders, as well as aging in the case of vitamin B_{12}. Of particular importance is that deficiency of vitamin B_{12} may be manifest by neurological or psychiatric symptoms in the absence of anemia or macrocytosis [75]. Vitamin B_{12} deficiency may result in subtle neuropsychiatric symptoms as well as the more classical signs of combined systems degeneration. Thiamin deficiency results in progressive deficits manifest by cognitive changes, cerebellar dysfunction, gaze palsy, sensory and motor manifestations, and eventually dementia.

VI. FUNCTIONAL ASSESSMENT

Functional assessment is based on the premise that PEM or other forms of malnutrition may be reflected by impairment in strength, mobility, and other functions [29, 76]. Functional impairment may also signal risk for future poor nutrition given the complex set of behaviors needed to adequately obtain, prepare, and consume adequate intake. Tools developed for functional assessment vary in complexity from simple measures of handgrip strength or respiratory muscle strength to batteries of multistage tasks requiring complex physical and cognitive processes.

A common simple functional test is handgrip strength. Handgrip strength as measured by handgrip dynamometry correlates with lean body mass, and reductions in handgrip strength are associated with PEM [77–79]. In patients undergoing surgery, preoperative handgrip strength has also been found to predict risk of postoperative complications [80–82].

Handgrip dynamometry requires the cooperation of a conscious patient and can be hampered by factors such as neuromuscular disease or arthritis. Handgrip strength is useful in the serial assessment of an individual, but it can also be used for reference to age- and sex-specific norms [82, 83]. There are two caveats regarding comparison of individual values to population norms. First, two different types of devices (strain-gauged versus mechanical) have been utilized to measure handgrip strength and values may differ based on the device used. Second, in the normative data of Bassey and Harries [83], longitudinal changes in grip strength exceeded cross-sectional differences with age, the significance of which is unclear.

When serial assessments are conducted in malnourished patients, handgrip strength may improve early after nutrition support, likely indicating effects of repletion of intracellular energy, micronutrients, and hydration long before significant accretion of protein [64, 78]. After this initial improvement, more gradual gains in handgrip strength occur and reflect repletion of lean mass.

VII. MULTICOMPONENT ASSESSMENT TOOLS

Attempts have been made to improve sensitivity and specificity in detecting malnutrition or nutrition-related risk by combining various assessment parameters. Multicomponent tools include variable combinations of historical, anthropometric, physical examination, and biochemical elements. Some tools are intended for initial screening, with a more in-depth assessment to follow if certain criteria are met, whereas other tools are utilized for more extensive assessment. Tools have been developed specifically for children, community-dwelling adults, older persons, and hospitalized patients. Several of the tools include measures that may be influenced by illness as well as by malnutrition. Because the ultimate value of assessment tools is to detect persons at risk for adverse outcomes, it may not be necessary to differentiate between factors that are strictly nutritional. In studies where these tools where contrasted, none has been consistently superior [84, 85]. Table 5 contrasts the components of some multicomponent assessment tools, and some examples are discussed in greater depth next.

A. Subjective Global Assessment

Subjective global assessment (SGA) combines historical elements with examination for edema and loss of subcutaneous fat, to stratify individuals into three categories: well nourished, moderately (or suspected of being) malnourished, or severely malnourished [86] (Table 6). Patients with a variety of illnesses who have been classified as moderately or severely malnourished have been shown to have greater postoperative complications, longer hospital stays, and accrued greater hospital charges compared to well-nourished patients [87]. Covinsky *et al.* [88] used SGA to assess 369 hospitalized older adults (≥70 years old) and then followed patient outcomes at 3 months and 1 year. Patients classified as severely malnourished were more likely to be dependent in activities of daily living at 3 months after discharge and were more likely to have spent time in a nursing home during the year after hospitalization. Both moderately and severely malnourished patients were more likely to have died at the 3-month and 1-year follow-up after discharge compared to the well-nourished group. When the SGA was compared to the Nutritional Risk Screening (NRS) 2002, an assessment tool utilized more commonly in Europe that has more objectively defined criteria for some elements, there was close agreement with SGA in patients at nutritional risk [89]. Although other multicomponent indices include biochemical tests such as visceral protein concentration, addition of these tests to SGA did not improve predictive performance and are not included.

TABLE 5 Multicomponent Assessment Tools (Yes Indicates Inclusion)

	Population	Health Status	Dietary Intake	Weight or BMI	Weight Change	Skinfold Thickness or Circumferences	Physical Exam	Visceral Protein	Immunity	Inflammation
PNI	Hospitalized					Yes		Yes	Yes	
NRI	Hospitalized				Yes			Yes		
PINI	Hospitalized							Yes		Yes
SGA	Hospitalized	Yes	Yes		Yes		Yes			
MNA	Elderly	Yes	Yes	Yes	Yes					
NSI	Elderly		Yes							
NRS	Hospitalized	Yes	Yes	Yes	Yes					
MUST	Community	Yes		Yes	Yes					

PNI, Prognostic Nutritional Index; NRI, Nutritional Risk Index; PINI, Prognostic Inflammatory and Nutritional Index; SGA, Subjective Global Assessment; NSI, Nutrition Screening Initiative; NRS, Nutritional Risk Screening 2002; MUST, Malnutrition Universal Screening.

TABLE 6 Components of Subjective Global Assessment

Patient history:
- Weight change
  - Overall loss in past 6 months
  - Change in past 2 weeks (increase, stable, or decrease)
- Dietary intake change relative to normal
  - No change
  - Change
  - Duration (number of weeks)
  - Types of change
    - Suboptimal solid diet
    - Full liquid diet
    - Hypocaloric liquids
    - Starvation
- Gastrointestinal symptoms (that persisted >2 weeks)
  - No symptoms
  - Nausea
  - Vomiting
  - Diarrhea
  - Anorexia
- Disease and its relation to nutritional requirements
  - Primary diagnosis
  - Metabolic demand (stress)
    - No stress
    - Mild stress
    - High stress

Physical exam:
- For each trait specify: 0 = normal, 1+ = mild, 2+ = moderate, 3+ = severe
  - Loss of subcutaneous fat (triceps, chest)
  - Muscle wasting (quadriceps, deltoids)
  - Ankle edema
  - Sacral edema
  - Ascites

Source: Reprinted from Destsky AS, McLaughlin JR, Baker JP et al., What is subjective global assessment of nutritional status? *JPEN J Parenter Enteral Nutr.* 1987;**11**:8–13 with permission from the American Society for Parenteral and Enteral Nutrition (A.S.P.E.N.). A.S.P.E.N. does not endorse the use of this material in any form other than its entirety.

B. Mini Nutritional Assessment

The mini nutritional assessment was initially developed for the frail elderly, but it has been validated for use in other elderly populations [90]. Designed to be administered by a health care professional, the mini nutritional assessment consists of 18 questions relating to anthropometrics (including weight loss, BMI, midarm circumference, and calf circumference), dietary intake (including change in appetite, number of meals/day, and autonomy of feeding), global assessment (including mobility, lifestyle, and medication use), and subjective assessment (self-perception of health and nutritional status). Each area of assessment receives a score, which is tallied at the end of the assessment, to a maximum score of 30. A score of 24 to 30 points indicates no risk of malnutrition, 17 to 23.5 indicates risk of malnutrition, and a score of <17 points indicates existing malnutrition [90, 91].

VIII. SUMMARY

Physical assessment is an important component of nutritional assessment and provides data regarding body weight, body composition, and micronutrient status. Anthropometry is the foundation for assessment of PEM, but it must be interpreted in light of physical exam findings. Although physical findings of micronutrient malnutrition lack sensitivity and specificity, they provide data regarding the extent and chronicity of deficiencies. Functional assessment and multicomponent assessment tools improve our ability to detect malnutrition and to predict nutrition-related adverse events.

References

1. ASPEN Board of Directors, The Clinical Guidelines Task Force. (2002). Guidelines for the use of parenteral and enteral nutrition in adult and pediatric patients. *JPEN* **26**.
2. Kondrup, J., Allison, S. P., Elia, M., Vellas, B., and Plauth, M. (2003). ESPEN guidelines for nutrition screening 2002. *Clin. Nutr.* **22**, 415–421.
3. Hensrud, D. D. (1999). Nutrition screening and assessment. *Med. Clin. N. Am.* **83**, 1525–1546.
4. Radimer, K., Bindewald, B., Hughes, J., Ervin, B., Swanson, C., and Picciano, M. F. (2004). Dietary supplement use by US adults: data from the National Health and Nutrition Examination Survey, 1999–2000. *Am. J. Epidemiol.* **160**, 339–349.
5. Sorkin, J. D., Muller, D. C., and Andres, R. (1999). Longitudinal change in height of men and women: implications for the interpretation of the body mass index. *Am. J. Epidemiol.* **150**, 969–977.
6. Pirie, P., Jacobs, D., Jeffery, R., and Hannan, P. (1981). Distortion in self-reported height and weight. *J. Am. Diet. Assoc.* **78**, 601–606.
7. Coe, T. R., Halkes, M., Houghton, K., and Jefferson, D. (1999). The accuracy of visual estimation of weight and height in pre-operative surgical patients. *Anaesthesia* **54**, 582–586.
8. Roubenoff, R., and Wilson, P. W. F. (1993). Advantage of knee height over height as an index of stature in expression of body composition in adults. *Am. J. Clin. Nutr.* **57**, 609–613.
9. Kwok, T., and Whitelaw, M. N. (1991). The use of armspan in nutritional assessment of the elderly. *J. Am. Geriatr. Soc.* **39**, 494–496.
10. Mitchell, C. O., and Lipschitz, D. A. (1982). Arm length measurement as alternative to height in nutritional assessment of the elderly. *JPEN* **6**, 226–229.
11. Chumlea, W. C., and Guo, S. (1992). Equations for predicting stature in white and black elderly individuals. *J Gerontol* **47**, M197–M203.
12. Chumlea, W. C., Guo, S. S., Wholihan, K., Cockram, D., Kuczmarski, R. J., and Johnson, C. L. (1998). Stature prediction equations for elderly non-Hispanic white, non-Hispanic black, and Mexican-American persons developed from NHANES III data. *J. Am. Diet. Assoc.* **98**, 137–142.

13. Weinbrenner, T., Vioque, J., Barber, X., and Asensio, L. (2006). Estimation of height and body mass index from demi-span in elderly individuals. *Gerontology* **52**, 275–281.
14. Hickson, M., and Frost, G. (2003). A comparison of three methods for estimating height in the acutely ill elderly population. *J. Hum. Nutr. Diet.* **16**, 13–20.
15. Brown, J. K., Feng, J. Y., and Knapp, T. R. (2002). Is self-reported height or arm span a more accurate alternative measure of height? *Clin. Nurs. Res.* **11**, 417–432.
16. Manonai, J., Khanacharoen, A., Theppisai, U., and Chittacharoen, A. (2001). Relationship between height and arm span in women of different age groups. *J. Obstet. Gynaecol. Res.* **27**, 325–327.
17. Reeves, S. L., Varakamin, C., and Henry, C. J. (1996). The relationship between arm-span measurement and height with special reference to gender and ethnicity. *Eur. J. Clin. Nutr.* **50**, 398–400.
18. Bassey, E. J. (1986). Demi-span as a measure of skeletal size. *Ann. Hum. Biol.* **13**, 499–502.
19. Beghetto, M. G., Fink, J., Luft, V. C., and de Mello, E. D. (2006). Estimates of body height in adult inpatients. *Clin. Nutr.* **25**, 438–443.
20. Dewys, W. D., Begg, C., Lavin, P. T., Band, P. R., Bennett, J. M., Bertino, J. R., Cohen, M. H., Douglass, H. O., Engstrom, P. F., Ezdinli, E. Z., Horton, J., Johnson, G. J., Moertel, C. G., Oken, M. M., Perlia, C., Rosenbaum, C., Silverstein, M. N., Skeel, R. T., Sponzo, R. W., and Tormey, D. C. (1980). Prognostic effect of weight loss prior to chemotherapy in cancer patients. *Am. J. Med.* **69**, 491–497.
21. Reynolds, M. W., Fredman, L., Langenberg, P., and Magaziner, J. (1999). Weight, weight change, and mortality in a random sample of older community-dwelling women. *J. Am. Geriatr. Soc.* **47**, 1409–1414.
22. Seltzer, M. H., Slocum, B. A., Cataldi-Bethcher, E. L., Fileti, C., Gerson, N. (1982). Instant nutritional assessment: absolute weight loss and surgical mortality. *JPEN* **6**, 218–221.
23. Studley, H. O. (1936). Percentage of weight loss: a basic indicator of surgical risk in patients with chronic peptic ulcer disease. *JAMA* **106**, 458–460.
24. Windsor, J. A., and Hill, G. L. (1988). Weight loss with physiologic impairment: a basic indicator of surgical risk. *Ann. Surg.* **207**, 290–296.
25. Naber, T. H., de Bree, A., Schermer, T. R., Bakkeren, J., Bar, B., dc Wild, G., and Katan, M. B. (1997). Specificity of indexes of malnutrition when applied to apparently healthy people: the effect of age. *Am. J. Clin. Nutr.* **65**, 1721–1725.
26. Windsor, J. A. (1993). Underweight patients and the risks of major surgery. *World J. Surg.* **17**, 165–172.
27. Fischer, J., and Johnson, M. A. (1990). Low body weight and weight loss in the aged. *J. Am. Diet. Assoc.* **90**, 1697–1706
28. Engelman, D. T., Adams, D. H., Byrne, J. G., Aranki, S. F., Collins, J. J., Couper, G. S., Allred, E. N., Cohn, L. H., and Rizzo, R. J. (1999). Impact of body mass index and albumin on morbidity and mortality after cardiac surgery. *J. Thorac. Cardiovasc. Surg.* **118**, 866–873.
29. Hill, G. L. (1992). Body composition research: implications for the practice of clinical nutrition. *JPEN* **16**, 197–218.
30. Choban, P. S., Burge, J. C., Scales, D., and Flancbaum, L. (1997). Hypoenergetic nutrition support in hospitalized obese patients; a simplified method for clinical application. *Am. J. Clin. Nutr.* **66**, 546–550.
31. Cutts, M. E., Dowdy, R. P., Ellersieck, M. R., and Edes, T. E. (1997). Predicting energy needs in ventilator-dependent critically ill patients: effect of adjusting for edema or adiposity. *Am. J. Clin. Nutr.* **66**, 1250–1256.
32. Wurtz, R., Itokazu, F., and Rodvold, K. (1997). Antimicrobial dosing in obese patients. *Clin. Infect. Dis.* **25**, 112–118.
33. Frankenfield, D., Roth-Yousey, L., and Compher, C. (2005). Comparison of predictive equations for resting metabolic rate in healthy nonobese and obese adults: a systematic review. *J. Am. Diet. Assoc.* **105**, 775–789.
34. Berger, M. M., and Chiolero, R. L. (2007). Hypocaloric feeding: pros and cons. *Curr. Opin. Crit. Care.* **13**, 80–186.
35. Morgan, D. B., Hill, G. L., and Burkinshaw, L. (1980). The assessment of weight loss from a single measurement of body weight: the problems and limitations. *Am. J. Clin. Nutr.* **33**, 2101–2105.
36. Robinett-Weiss, N., Hixson, M. L., Keir, B., and Sieberg, J. (1984). The metropolitan height-weight tables: perspectives for use. *J. Am. Diet. Assoc.* **84**, 1480–1481.
37. NHLBI Obesity Education Initiative Expert Panel on the Identification Evaluation and Treatment of Overweight and Obesity in Adults. (1998). Clinical guidelines on the identification, evaluation, and treatment of overweight and obesity in adults: the evidence report. *Obes. Res.* **6** (suppl 2), 51S–209S.
38. World Health Organization. (1995). "Physical status: the use and interpretation of anthropometry." World Health Organization, Geneva.
39. Prentice, A. M., and Jebb, S. A. (2001). Beyond body mass index. *Obes. Rev.* **2**, 141–147.
40. Calle, E. E., Rodriguez, C., Walker-Thurmond, K., and Thun, M. J. (2003). Overweight, obesity, and mortality from cancer in a prospectively studied cohort of U.S. adults. *N. Engl. J. Med.* **348**, 1625–1638.
41. World Health Organization. (1990). Diet, nutrition, and the prevention of chronic diseases. Report of a WHO Study Group. *World Health Organ. Tech. Rep. Ser.* **797**, 1–204.
42. Landi, F., Zuccala, G., Gambassi, G., Incalzi, R.A., Manigrasso, L., Pagano, F., Carbonin, P., and Bernabei, R. (1999). Body mass index and mortality among older people living in the community. *J. Am. Geriatr. Soc.* **47**, 1072–1076.
43. Visscher, T. L. S., Seidell, J. C., Menotti, A., Blackburn, H., Nissinen, A., Feskens, E. J. M., and Kromhout, D. (2000). Underweight and overweight in relation to mortality among men aged 40–59 and 50–69. *Am. J. Epidemiol.* **151**, 660–666.
44. Must, A., Spadano, J., Coakley, E. H., Field, A. E., Colditz, G., and Dietz, W. H. (1999). The disease burden associated with overweight and obesity. *JAMA* **282**, 1523–1529.
45. Heymsfield, S. B. (1999). Nutritional assessment of malnutrition by anthropometric methods. *In* "Modern Nutrition in Health and Disease." (M. E. Shils, J. A. Olson, M. Shike, and A. C. Ross, Eds.). Williams & Wilkins, Baltimore, pp. 903–921.
46. Henry, C. J. (1990). Body mass index and the limits of human survival. *Eur. J. Clin. Nutr.* **44**, 329–335.
47. Saltzman, E. (2006). Obesity as a health issue. *In* "Present Knowledge in Nutrition" (B. A. Bowman, R. M. Russell, Eds.) International Life Sciences Institute, Washington, DC, pp. 637–648.
48. Kopelman, P. (2007). Health risks associated with overweight and obesity. *Obes. Rev.* **8** (suppl 1), 13–17.

49. Centers for Disease Control and Prevention. (2003). Prevalence of diabetes and impaired fasting glucose in adults—United States, 1999–2000. *MMWR* **52**, 833–837.
50. Young, T., Evans, L., Finn, L., and Palta, M. (1997). Estimation of the clinically diagnosed proportion of sleep apnea syndrome middle-aged men and women. *Sleep* **20**, 705–706.
51. Klein, S., Allison, D. B., Heymsfield, S. B., Kellcy, D. E., Leibel R. L., Nonas, C., and Kahn, R. (2007). Waist circumference and cardiometabolic risk: a consensus statement from shaping America's health: Association for Weight Management and Obesity Prevention; NAASO, the Obesity Society; the American Society for Nutrition; and the American Diabetes Association. *Diabetes Care* **30**, 1647–1652.
52. Despres, J. (1998). The insulin resistance-dyslipidemic syndrome of visceral obesity: effect on patients' risk. *Obes. Res.* **6** (suppl), 8s–17s.
53. Molarius, A., and Siedell, J. C. (1998). Selection of anthropometric indicators for classification of abdominal fatness: a critical review. *Int. J. Obes.* **22**, 719–727.
54. Durnin, J. V. G. A., Lonergan, M. E., Good, J., and Ewan, A. (1974). A cross-sectional nutritional and anthropometric study, with an interval of 7 years, on 611 young adolescent schoolchildren. *Brit. Med. J.* **32**, 169–178.
55. Jackson, A. S., and Pollock, M. L. (1978). Generalized equations for predicting body density of men. *Br. J. Nutr.* **40**, 497–504.
56. Wang, J., Thornton, J. C., Kolesnik, S., and Pierson, R. N, Jr. (2000). Anthropometry in body composition: an overview. *Ann. N. Y. Acad. Sci.* **904**, 317–326.
57. Wang, Z. M., Pierson, R. N., Jr., and Heymsfield, S. B. (1992). The five-level model: a new approach to organizing body-composition research. *Am. J. Clin. Nutr.* **56**, 19–28.
58. McArdle, W. D., Katch, F. I., and Katch, V. L. (1996). Body Composition Assessment: "Exercise Physiology: Energy, Nutrition, and Human Performance." Williams & Wilkins, Baltimore.
59. Lamy, M., Mojon, P., Kalykakis, G., Legrand, R., and Butz-Jorgensen, E. (1999). Oral status and nutrition in the institutionalized elderly. *J. Dentistry* **27**, 443–448.
60. Papas, A. S., Palmer, C. A., Rounds, M. C., and Russell, R. M. (1998). The effects of denture status on nutrition. *Spec. Care Dentist.* **18**, 17–25.
61. von Haehling, S., Doehner, W., and Anker, S. D. (2007). Nutrition, metabolism, and the complex pathophysiology of cachexia in chronic heart failure. *Cardiovasc. Res.* **73**, 298–309.
62. Agusti, A. G. (2005). Systemic effects of chronic obstructive pulmonary disease. *Proc. Am. Thorac. Soc.* **2**, 367–370.
63. Schols, A. M., Soeters, P. B., Mostert, R., Saris, W. H., and Wouters, E. F. (1991). Energy balance in chronic obstructive pulmonary disease. *Am. Rev. Resp. Dis.* **143**, 1248–1252.
64. Christie, P. M., and Hill, G. L. (1990). Effect of intravenous nutrition on nutrition and function in acute attacks of inflammatory bowel disease. *Gastroenterology* **99**, 730–736.
65. Bousvaros, A., Zurakowski, D., Duggan, C., Law, T., Rifai, N., Goldberg, N. E., and Leichtner, A. M. (1998). Vitamins A and E serum levels in children and young adults with inflammatory bowel disease: effect of disease activity. *J. Ped. Gastroenterol. Nutr.* **26**, 129–135.
66. Geerling, B. J., Badart-Smook, A., Stockbrugger, R. W., and Brummer, R. J. (1998). Comprehensive nutritional status in patients with long-standing Crohn disease currently in remission. *Am. J. Clin. Nutr.* **67**, 919–926.
67. Siguel, E. N., and Lerman, R. H. (1996). Prevalence of essential fatty acid deficiency in patients with chronic gastrointestinal disorders. *Metabolism* **45**, 12–23.
68. Lee, Y. M., and Kaplan, M. M. (1995). Primary sclerosing cholangitis. *N. Engl. J. Med.* **332**, 924–933.
69. Lindenbaum, J., Rosenberg, I. H., Wilson, P. W., Stabler, S. P., and Allen, R. H. (1994). Prevalence of cobalamin deficiency in the Framingham elderly population. *Am. J. Clin. Nutr.* **60**, 2–11.
70. Prabhala, A., Garg, R., and Dandona, P. (2000). Severe myopathy associated with vitamin D deficiency in western New York. *Arch. Int. Med.* **160**, 1199–1203.
71. De Prisco, C., and Levine, S. N. (2005). Metabolic bone disease after gastric bypass surgery for obesity. *Am. J. Med. Sci.* **329**, 57–61.
72. Chanarin, I. (1999). Nutritional aspects of hematologic disorders. *In* "Modern Nutrition in Health and Disease" (M. E. Shils, J. A. Olson, M. Shike, and A. C. Ross, Eds.) . Williams & Wilkins, Baltimore, pp. 1419–1439.
73. Kopple, J. D., Zhu, X., Lew, N. L., and Lowrie, E. G. (1999). Body weight-for-height relationships predict mortality in maintenance hemodialysis patients. *Kidney Intl.* **56**, 1136–1148.
74. Perlman, A. L. (1999). Dysphagia: populations at risk and methods of diagnosis. *Nutr. Clin. Prac.* **14**, S2–S9.
75. Lindenbaum, J., Healton, E. B., Savage, D. G., Brust, J. C., Garrett, T. J., Podell, E. R., Marcell, P. D., Stabler, S. P., and Allen, R. H. (1988). Neuropsychiatric disorders caused by cobalamin deficiency in the absence of anemia or macrocytosis. *N. Engl. J. Med.* **318**, 1720–1728.
76. Jeejeebhoy, K. N. (1998). Nutritional assessment. *Gastroenterol. Clin. N. Am.* **27**, 347–369.
77. Payette, H., Hanussaik, N., Boutier, V., Morais, J. A., and Gray-Donald, K. (1998). Muscle strength and functional mobility in relation to lean body mass in free-living frail elderly women. *Eur. J. Clin. Nutr.* **52**, 45–53.
78. Russell, D. M., Prendergast, P. J., Darby, P. L., Garfinkel, P. E., Whitwell, J., and Jeejeebhoy, K. N. (1983). A comparison between muscle function and body composition in anorexia nervosa: the effect of refeeding. *Am. J. Clin. Nutr.* **38**, 229–237.
79. Vaz, M., Thangam, S., Prabhu, A., and Shetty, P. S. (1996). Maximal voluntary contraction as a functional indicator of adult chronic undernutrition. *Brit. J. Nutr.* **76**, 9–15.
80. Hunt, D. R., Rowlands, B. J., and Johnston, D. (1985). Hand grip strength: a simple prognostic indicator in surgical patients. *JPEN* **9**, 701–704.
81. Kalfarentzos, F., Spiliotis, J., Velimezis, G., Dougenis, D., and Androulakis, J. (1989). Comparison of forearm muscle dynamometry with nutritional prognostic index, as a preoperative indicator in cancer patients. *JPEN* **13**, 34–36.
82. Webb, A. R., Newman, L. A., Taylor, M., and Keogh, J. B. (1989). Hand grip dynamometry as a predictor of postoperative complications reappraisal using age standardized grip strengths. *JPEN* **13**, 30–33.
83. Bassey, E. J., and Harries, U. J. (1993). Normal values for handgrip strength in 920 men and women aged over 65 years,

and longitudinal changes over 4 years in 620 survivors. *Clin. Sci.* **84**, 331–337.

84. Alberda, C., Graf, A., and McCargar, L. (2006). Malnutrition: etiology, consequences, and assessment of a patient at risk. *Best Pract. Res. Clin. Gastroenterol.* **20**, 419–439.
85. Alvares-da-Silva, M. R., and Reverbel da Silveira, T. (2005). Comparison between handgrip strength, subjective global assessment, and prognostic nutritional index in assessing malnutrition and predicting clinical outcome in cirrhotic outpatients. *Nutrition* **21**, 113–117.
86. Detsky, A., McLaughlin, J., Baker, J., Johnston, N., Whittaker, S., Mendelson, R., and Jeejeebhoy, K. (1987). What is subjective global assessment of nutritional status? *JPEN* **11**, 8–13.
87. Schneider, S. M., and Hebuterne, X. (2000). Use of nutritional scores to predict outcomes in chronic diseases. *Nutr. Rev.* **58**, 31–38.
88. Covinsky, K., Martin, G., Beyth, R., Justice, A., Sehgal, A., and Landefeld, C. (1999). The relationship between clinical assessments of nutritional status and adverse outcomes in older hospitalized medical patients. *J. Am. Geriatr. Soc.* **47**, 532–538.
89. Valero, M. A., Diez, L., El Kadaoui, N., Jimenez, A. E., Rodriguez, H., and Leon, M. (2005). Are the tools recommended by ASPEN and ESPEN comparable for assessing the nutritional status? *Nutricion Hospitalaria* **20**, 259–267.
90. Garry, P. J., and Vellas, B. J. (1999). Practical and validated use of the mini nutritional assessment in geriatric evaluation. *Nutr. Clin. Care* **2**, 146–154.
91. Omran, M. L., and Morley, J. E. (2000). Assessment of protein energy malnutrition in older persons, part I: history, examination, body composition, and screening tools. *Nutrition* **16**, 50–63.

CHAPTER 4

Energy Requirement Methodology

DEBRA COWARD-McKENZIE[1] AND RACHEL K. JOHNSON[2]
[1]*University of North Carolina at Chapel Hill, Chapel Hill, North Carolina*
[2]*University of Vermont, Burlington, Vermont*

Contents

I. INTRODUCTION

Knowledge of energy requirements throughout the life cycle and during various physiological conditions and disease states is essential to the promotion of optimal human health. In the past, the measurement of energy intake served as an important tool from which to base energy requirements among all age groups [1]. The heavy reliance on such subjective forms of measurement, however, prompted research for more valid techniques to estimate energy needs [2]. The aim of this chapter is to familiarize the reader with contemporary techniques to measure energy expenditure. These include the sophisticated technique of doubly-labeled water (DLW), which allows nutrition scientists to accurately estimate energy requirements based on the measurement of total energy expenditure in free-living subjects.

II. COMPONENTS OF ENERGY EXPENDITURE

The human body expends energy in the form of resting energy expenditure (REE), the thermic effect of food (TEF), and energy expended in physical activity (EEPA) [3]. These three components make up a person's daily total energy expenditure (TEE). Except in extremely active subjects, the REE constitutes the largest portion (50% to 60%) of the TEE. The TEF represents approximately 10% of the daily TEE. The contribution of physical activity is the most variable component of TEE, which may be as low as 100 kcal/day in sedentary people or as high as 3000 kcal/day in very active people (Fig. 1).

A. Resting Energy Expenditure

Resting energy expenditure is the energy cost of the physiological functions necessary to maintain homeostasis. These involuntary functions include respiration, cardiac output, body temperature regulation, and other functions of the sympathetic nervous system.

The term "basal energy expenditure" (BEE) is also used to describe this component of daily energy expenditure. BEE can be defined as the minimal amount of energy expended that is compatible with life. BEE is the amount of energy used in 24 hours by a person who is lying at physical and mental rest, at least 12 hours after the last meal, in a thermo-neutral environment that prevents the activation of heat generating processes, such as shivering. Basal metabolic rate (BMR) measurements are made early in the morning, before the person has engaged in any physical activity, and with no ingestion of tea or coffee or inhalation of nicotine-containing tobacco smoke for at least

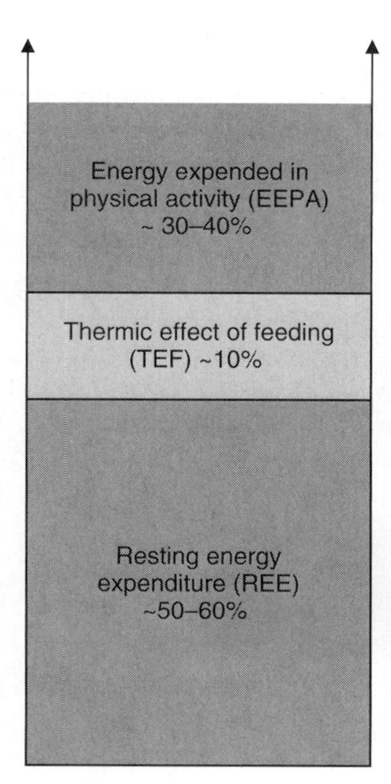

FIGURE 1 The components of total energy expenditure.

12 hours before the measurement. If any of the conditions for BMR are not met, the energy expenditure should be termed the resting metabolic rate (RMR). For practical reasons, the BMR is rarely measured. In its place, RMR measurements are used, which, in most cases, are higher than the BMR.

1. Determinants of Resting Energy Expenditure

The determinants of resting energy expenditure are well established in both adults and children. The principal factors contributing to individual variation in REE include body size and composition, gender, age, physical fitness, hormonal status, genetics, and environmental influences [4–6].

a. Body Size

Both height and weight are important in determining the amount of energy a person expends and therefore requires to maintain body weight. Larger people have higher metabolic rates than do people of smaller size because additional body tissue requires additional basal metabolic activity. For example, a difference in weight of 10 kg would lead to a difference in RMR of approximately 120 kcal/day in adult men or women, or a difference in total daily energy expenditure of approximately 200 kcal/day in people with low levels of physical activity.

b. Body Composition

Fat-free mass (FFM) or lean body mass is the primary determinant of TEE in all age groups [7]. FFM is the metabolically active tissue in the body. Hence, most, about 73 percent [8], of the variation in REE between people can be accounted for by the variation in their FFM. FFM is in turn affected by other factors such as age, gender, and physical fitness. FFM can be accurately measured using a number of techniques. These include underwater weighing, measuring total body water using stable isotopes of deuterium or oxygen-18, and dual-energy x-ray absorptiometry (DXA). DXA is a scanning technique that accurately estimates fat-free mass, lean body mass, and bone mineral mass. Subjects lie supine on a padded table for 20 to 45 minutes during which time two very low energy (6.4 and 11.2 fJ) x-ray beams are passed through the body (Fig. 2; see the color plate at the back of the book). The total x-ray dose is less than 1 mrem, on the order of a single day's background radiation, for a whole-body composition analysis. Svendsen *et al.* [9] discuss in detail the calculation of fat mass and lean body mass using these data. Thus, DXA provides an accurate measure of lean body mass and fat mass in a short time with a minimal radiation dose. Pregnancy tests should be performed before administering the DXA to further ensure safety.

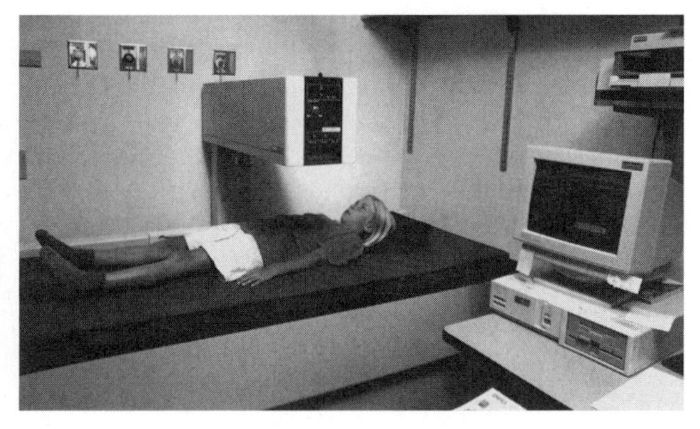

FIGURE 2 Dual-energy x-ray absorptiometry is a scanning technique that accurately estimates bone mineral, fat, and fat-free soft tissue. See color plate.

Because of the expense or impractical nature of research techniques for body composition, other, less accurate methods are often used in practice to estimate body composition. These include skinfold anthropometry, bioelectrical impedance, and near-infrared interactance. See Chapter 3 for a detailed description of body composition measurement techniques.

It has long been questioned whether people who are obese or at risk of developing obesity have reduced metabolic rates. Several studies have produced no evidence that children at high risk of obesity have lower metabolic rates than lean children when the measurements are adjusted for lean body mass [10, 11]. The Dietary Reference Intakes (DRI) database of TEE measurements [6] and more recent studies [12] show that heavier women have greater TEE values than women of lower weight but of the same height. Although Saltzman and Roberts [12] found that RMR is consistently depressed during active weight loss beyond loss of FFM, other cross-sectional studies comparing post-obese with never-obese individuals report no difference in RMR, suggesting that RMR stabilizes with weight [13–15]. Hence, there is little evidence that a greater metabolic efficiency leads to the development of obesity.

c. Gender

Gender is another factor that affects REE. The values for REE and 24-hour energy expenditure in a metabolic chamber are lower in female than in male subjects by approximately 50 kcal/day [16, 17]. In general, women have metabolic rates that are 5% to 10% lower than men of the same weight and height. Much of the differences between the sexes can be explained by differences in body composition. Women typically have more fat in proportion to muscle than men. Also, some of the difference found between males and females could be due to variation in the lean tissues. It is well documented [18] that different body tissues have different metabolic rates, with adipose and muscle having a lower rate than brain and organ tissues. It is possible that lower RMR seen in women is due to a different balance of organ, brain, and skeletal muscle compared to men.

While cyclical changes are believed to impact RMR (see the Hormonal Status section), the reduced energy expenditure in female subjects is not fully explained by the confounding effects of the menstrual cycle because the effect is consistent in postmenopausal women as well as prepubertal girls [4, 5].

d. Age

There is a well-documented age-related reduction in resting metabolic rate [19], with the suggested decline occurring around age 40 years in men and 50 years in women [20]. Researchers have shown that older healthy people have RMRs that are significantly lower than younger people, even when adjusted for differences in fat mass and fat-free mass. This decline in REE can be partly explained by a reduction in the quantity, as well as the metabolic activity, of lean body mass [19], including changes in the relative size of organs and tissues [21]. If individuals gain weight as they age, RMR may actually increase because of gains of FFM and fat mass.

e. Physical Fitness

Athletes with greater muscular development show approximately a 5% increase in RMR over that in nonathletic individuals owing to their greater FFM. Habitual exercise has been shown to cause an 8% to 14% higher metabolic rate in men who were moderately and highly active, respectively, owing to increased FFM [22].

f. Hormonal Status

Hormonal status can impact metabolic rate, particularly in endocrine disorders, such as hyperthyroidism and hypothyroidism, when energy expenditure is increased or decreased, respectively. Stimulation of the sympathetic nervous system, such as occurs during emotional excitement or stress, increases cellular activity by the release of epinephrine, which acts directly to promote glycogenolysis. Other hormones, such as cortisol, growth hormone, and insulin, also influence metabolic rate. Researchers have shown that serum leptin concentrations are a positive determinant of RMR [23].

The metabolic rate of adult females fluctuates with the menstrual cycle. An average of 359 kcal/day difference in the BMR has been measured between its low point, about 1 week before ovulation at day 14, and its high point, just before the onset of menstruation. The mean increase in energy expenditure is about 150 kcal/day during the second half of the menstrual cycle [24]. Stable RMR cannot be assumed in women [25]. During pregnancy, RMR decreases in the early stages, whereas later in pregnancy, the metabolic rate is increased by the processes of uterine, placental, and fetal growth and by the mother's increased cardiac work [26].

g. Ethnicity and Genetics

REE can also be impacted by both ethnic origin and genetic inheritance. Most studies comparing African-American and non-Hispanic white adults and children have reported that RMRs adjusted for body composition are significantly lower in African Americans by about 10% [6]. In addition, other ethnic groups have been investigated for potential differences in energy requirements, with no studies confirming differences in RMR. In Pima Indians, a group believed to have a form of genetic obesity, RMR was not found to differ from that of non-Hispanic whites after adjustment for body composition [27, 28]. Mohawk Indian children were reported to have higher values of TEE than non-Hispanic white children, but the difference was due to higher levels of EEPA [29].

The most obvious impact of genetic inheritance on RMR is due to differences in body composition, with genetics accounting for 25% to 50% of interindividual variability [30]. There also appears to be genetic influence beyond body composition. Bogardus and coworkers [31] reported a significant intrafamily influence on RMR independent of FFM, age, and gender. Although the origin of this association is not fully understood, it may potentially be due to differences in the relative size of FFM components, such as muscle, brain, organs, as organ size determined by magnetic resonance imaging strongly predicts RMR [32]. The question of which genes are responsible for these differences is being studied, but little is known at this time. Further work in this area is needed. Currently, insufficient data exist to create prediction equations for differences in energy requirements among specific genetic or ethnic groups that would be accurate for both males and females throughout the life stages.

h. Environmental Influences

REE is affected by extremes in environmental temperature. People living in tropical climates usually have RMRs that are 5% to 20% higher than those living in a temperate area. Exercise in temperatures greater than 86°F also imposes a small additional metabolic load of about 5% owing to increased sweat gland activity. The extent to which energy metabolism increases in extremely cold environments depends on the insulation available from body fat and protective clothing.

A summary of changes in BMR among individuals migrating between tropic and temperate climates found that changes in ambient temperature do not produce a long-term change in metabolic rate [33]. Instead, it appears to be confined to the period of time during which the change in temperature occurs.

Studies have found that changes in altitude can also impact BMR [34–37]. High altitude increases BMR, but it is not clear at which heights the effect becomes prominent. A study of men at 14,100 ft found an increase in BMR of about 200 to 500 kcal/d when energy intakes were maintained [38] while studies on women found the effect to be less definitive [39].

2. Measuring Resting Energy Expenditure: Indirect Calorimetry

The technique of indirect calorimetry has become the method of choice in most circumstances when a

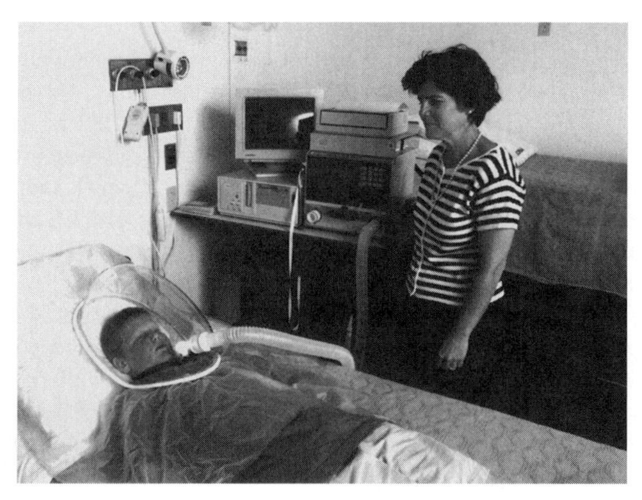

FIGURE 3 Measurement of resting metabolic rate using indirect calorimetry. See color plate.

measurement of REE is needed. This method does not directly measure heat transfer. To determine energy (heat) production, O_2 consumption and CO_2 production are measured. Thus, the term *"indirect"* has been adopted. The equipment varies but a ventilated hood system is most commonly used. A clear plastic hood is placed over the subject's head and made airtight around the neck (Fig. 3; see the color plate at the back of the book). Indirect calorimetry has the advantage of mobility and low equipment cost and is frequently used in clinical settings to assess patients' energy requirements. Indirect calorimetry also provides quantitative information about the types of substrates that are oxidized [40]. The pretesting environment impacts the measurement of RMR. Food, ethanol, caffeine, nicotine, and physical activity impact RMR for a variable number of hours and should be controlled before measurements are taken [41]. Outpatient-test experimental conditions have been shown to overestimate RMR by approximately 8% compared with inpatient measurements of RMR [42]. This factor should be taken into account when results are compared between laboratories and when daily energy requirements based on measures of REE are being evaluated.

3. Estimating Resting Energy Expenditure: Prediction Equations

While numerous equations have been developed to estimate REE, only two will be discussed here: The Harris–Benedict equations and those developed by the Institute of Medicine/Food and Nutrition Board to determine DRIs.

a. Harris–Benedict Equations

The Harris–Benedict equations remain the most commonly used tool by clinicians when estimating people's REE. The equations are often used as a basis for prescribing energy intake for hospitalized patients and to formulate energy intake goals for weight loss. A review of the data used in the formulation of the Harris–Benedict equations in the early 1930s deduced that the methods and conclusions of Harris and Benedict appear valid but not error free [43]. Only between 50% and 75% of the variability in REE is explained by the equations and subsequent equations have not generally improved on this level of error [44–46]. The equations in use today are shown in Table 1. Although equations have been developed for obese populations, standard equations developed for nonobese populations have been shown to be more accurate in this population (mean BMI 48.9 kg/(m^2) [12]. Currently, the identification of error in equations (e.g., Harris–Benedict, Mifflin-St Jeor, Cunningham) available for use in practice has become an

TABLE 1 Examples of Equations for Predicting Resting and Basal Energy Expenditure

	Harris–Benedict Equations
Women	REE (kcal/day) = 655 + 9.56 (weight) + 1.85 (height) – 4.68 (age) $r^2 = 0.53$, $F = 37.8$, $P < 0.001$
Men	REE (kcal/day) = 66.5 + 13.75 (weight) + 5.0 (height) – 6.76 (age) $r^2 = 0.75$, $F = 135.2$, $P < 0.001$
	DRI Equations
Women	BEE (kcal/d) = 255 – (2.35 × age) + 361.6 × height + 9.39 × weight Residual = ± 125, $R^2 = 0.39$
Men	BEE (kcal/d) = 204 – (4 x age) + 450.5 × height + 11.69 × weight Residual = ± 149, $R^2 = 0.46$

Sources: The Harris–Benedict Equations: Frankenfield, D. C., Muth, E. R., and Rowe, W. A. (1998). The Harris–Benedict studies of human basal metabolism: history and limitations. *J. Am. Diet. Assoc.* **98**, 439–445.

DRI Equations for Basal Energy Expenditure: Institute of Medicine of The National Academies. (2005). "Dietary Reference Intakes: Energy, Carbohydrate, Fiber, Fat, Fatty Acids, Cholesterol, Protein, and Amino Acids." The National Academy Press, Washington, DC.

Note: Weight is in kilograms, height is in centimeters, and age is in years.

active area of research [44]. The reader is encouraged to follow this literature as it applies to populations differing in body size, health status, sex, and age.

b. DRI Equations

The BEE prediction equations developed for the DRIs are shown in Table 1. These equations were derived from observed BEE values found in the Doubly-Labeled Water database [6]. This database was developed by searching published research that involved use of the DLW method. The investigators associated with the identified publications were solicited to contribute to the database. Twenty investigators responded and submitted individual TEE data and ancillary data including age, gender, height, weight, BEE (both observed and estimated), and descriptors for each individual in the data set. This collected information comprises the database [6]. These data were not obtained from randomly selected individuals and thus are not a representative sample of the U.S. population. However, because the measurements were obtained from men, women, and children, over a wide range of weights, heights, and ages, it is believed that the data still offer the best currently available information [6].

c. Predicting REE in Disease and Physiological Conditions

REE has been characterized for a variety of various disease states and physiological conditions. These include burns [47], anorexia nervosa [46, 48], severe central nervous system impairment [49], cerebral palsy [50], pregnancy [51], and lactation [52]. In addition, REE has been studied in both children [4] and the elderly [53, 54]. Clinicians should not assume that prediction equations, which were developed in normal, healthy people, are valid in special populations [44].

B. Thermic Effect of Food

The thermic effect of food (TEF) is the increase in energy expenditure associated with the consumption of food. It accounts for approximately 10% of TEE. TEF is also termed "diet-induced thermogenesis." TEF can be divided into obligatory and facultative thermogenesis. Obligatory thermogenesis is the energy needed to digest, absorb, and metabolize nutrients. Facultative thermogenesis is the excess energy expended above the obligatory thermogenesis attributable to metabolic inefficiency. TEF is not measured in clinical settings but rather estimated as 10% of TEE. Hence, the following equation:

$$\text{TEE} = (\text{RMR} + \text{EEPA}) \times 1.10$$

where TEE = total energy expenditure, RMR = resting energy expenditure, EEPA = energy expended in physical activity, and the factor of 1.10 accounts for the TEF in the equation.

1. Determinants of Thermic Effect of Food

The composition of the diet affects the TEF, with TEF being greater after carbohydrate and protein consumption in comparison with fat. This is accounted for by the differences in metabolic efficiency when metabolizing carbohydrate and protein versus fat. Fat is efficiently metabolized and stored with only 4% wastage, compared with 25% inefficiency when carbohydrate is converted to fat for storage. These differences have been hypothesized to contribute to the obesity-promoting characteristics of high-fat diets [55].

Other dietary factors are known to impact TEF. For example, spicy foods such as chili and mustard have been shown to increase the metabolic rate significantly in comparison with unspiced meals. The effect was prolonged, lasting up to 3 hours [56]. Caffeine has been shown to increase TEF by 8% to 11% [57]. Other factors such as nicotine and cold also stimulate the TEF [58].

C. Energy Expended in Physical Activity

Energy expended in physical activity (EEPA) is the most variable component of total energy expenditure. EEPA includes energy expended in voluntary exercise, which include activities of daily living (bathing, feeding, and grooming, for example), sports and leisure, and occupational activities. EEPA also includes the energy expended in nonexercise activity thermogenesis (NEAT), which is associated with fidgeting, maintenance of posture, and other physical activities of daily life. The share of total energy expenditure accounted for by physical activity is obviously greater for active individuals [59]. It can vary from 10% in a bedridden person to as high as 50% or more of TEE in athletes. Because of the alarmingly high rates of obesity in the United States, increasing EEPA through voluntary physical activity is being stressed as an effective way of achieving healthy weight [60].

NEAT can add significantly to a person's energy expenditure. Researchers found that when nonobese volunteers were fed 1000 kcal/day in excess of weight-maintenance requirements for 8 weeks, two-thirds of the increase in total daily energy expenditure was due to increased NEAT. Changes in NEAT accounted for 10-fold differences in fat storage that occurred and directly predicted resistance to fat gain with overfeeding. These results suggest that as humans overeat, activation of NEAT dissipates excess energy to preserve leanness and that failure to activate NEAT may result in fat gain [61]. NEAT has not been found to decrease with age [62].

There is some controversy over whether NEAT is impacted by intentional exercise. Some studies have found that planned physical activities had a minimal effect on energy expenditure, presumably due to decreased activity, including NEAT at other times of the day [63, 64]. In contrast, Blaak *et al.* [65] found that obese boys enrolled in an exercise training program reported a significant increase in spontaneous activity.

A novel study suggested that gum chewing is sufficiently exothermic that it can lead to a meaningful increase in energy expenditure over time. Chewing gum led to a mean increase in energy expenditure of 11 ± 3 kcal/hour in seven nonobese volunteers with stable weight. The researchers speculated that if a person chewed gum during waking hours and changed no other components of energy balance, he or she could realize a yearly loss of more than 5 kg of body fat [66].

1. Determinants of Energy Expended in Physical Activity

Differences in EEPA are due both to patterns of activity (both voluntary and involuntary) as well as to the body composition that results from that pattern of activity [67]. Hence, the level of a person's fitness will affect EEPA because of the increased lean body mass and metabolically active tissue. In addition, physical activity can also affect RMR in the postexercise period by 5% or more up to 24 hours after exercise [68].

In general, EEPA declines with age. Because the decline in EEPA is often disproportionately greater than the decline in energy intake, positive energy balance results. This often leads to increased total and central body fatness, a loss of muscle mass, and a greater predisposition to comorbidities associated with obesity and physical inactivity [69]. Fortunately, it has been demonstrated that regular aerobic exercise can successfully increase EEPA in middle-aged men and hence raise the daily energy requirements for weight maintenance [70]. Thus, regular exercise may counter the age-related tendency toward obesity. Table 2 shows average energy costs of typical activities, with the values expressed as multiples of RMR.

TABLE 2 Examples of Activity Energy Costs

Activity	Energy Cost (Multiple of Basal Metabolism)
Lying quietly	1.0
Riding in a vehicle	1.0
Light activity while sitting	1.5
Walking (2 mph)	2.5
Watering plants	2.5
Walking the dog	3.0
Cycling leisurely and household tasks (moderate effort)	3.5
Raking the lawn	4.0
Golfing (no cart) and gardening (no lifting)	4.4
Walking (4 mph) and slow swimming	4.5
Dancing, ballroom (fast) or square	5.5
Dancing, aerobic or ballet	6.0
Walking (5 mph)	8.0
Jogging (10 min miles)	10.2
Skipping rope	12.0

Source: Reprinted with permission from the National Academies Press, Copyright 2005, National Academy of Sciences, Institute of Medicine of the National Academies. (2005). "Dietary Reference Intakes for Energy, Carbohydrate, Fiber, Fat, Fatty Acids, Cholesterol, Protein, and Amino Acids." The National Academy Press, Washington, DC. Adapted from Table 12-2.

2. Measuring Energy Expended in Physical Activity

Obtaining a valid and appropriate measurement of EEPA is a challenging task. Measures fall into three general categories: direct observation; subjective reports, such as physical activity questionnaires; and objective assessment tools, including doubly-labeled water technique, movement counters, and heart rate monitoring. The objective activity assessment tools are often used to validate the subjective activity measures.

a. Objective Measures of EEPA

i. Doubly-Labeled Water EEPA can be estimated by determining the difference between total daily energy expenditure as measured by doubly-labeled water and resting metabolic rate and thermic effect of food measured by indirect calorimetry. The advantages of the DLW approach are that it requires little subject cooperation, is unobtrusive, and measures free-living activity throughout a person's daily routine. Unfortunately, the high price of the isotopes and mass spectrometer instrumentation as well as the need for technical expertise in sample preparation and measurement have limited its widespread application. Nevertheless, DLW has been used as a gold standard for validating other methods to measure EEPA in free-living individuals [71]. DLW measurement of total energy expenditure is discussed in detail in the TEE section of this chapter.

ii. Movement Counters Accelerometers are movement counters that can measure the occurrence of body movement and its intensity. Accelerometers cannot be used to measure the static component in exercises, like weight lifting or carrying loads. However, in normal daily life, it is assumed that the effect of static exercise on the total level of physical activity is negligible [72]. Since 1999, there has been a push to improve accelerometers as it was concluded that those available were not practical for large-scale studies because of high cost, uncertain reliability, and difficulties in interpreting data [73]. The National Health and Nutrition Examination Survey (NHANES) 2003–2004 began using accelerometers on survey participants ages 6 years and older who agreed to wear them, representing the largest implementation of objective physical activity monitoring [73].

One-axial accelerometers, such as the Caltrac, measure vertical movement. Energy expenditure is estimated by entering the subject's age, height, weight, and gender. Activity counts are displayed when predetermined constants are entered in place of the subject's personal data.

The Caltrac was not found to be a meaningful predictor of EEPA in a group of free-living, school-aged children [74] or in older men and women [69]. The Caltrac may be useful, however, in supervised settings in which researchers want to assess children's levels of physical activity [75].

The triaxial accelerometer measures movement on three planes and possesses the ability to store extensive data within its internal computer for retrieval at a later time. The evidence to date suggests that a triaxial accelerometer provides a better estimate of usual physical activity or energy expenditure than a single-plane accelerometer [76]. Studies have demonstrated that these instruments can be used to distinguish differences in activity levels between individuals and to assess the effect of interventions on physical activity within individuals [72].

iii. Heart Rate Monitoring The use of heart rate monitoring as a proxy measure for EEPA is based on the principle that heart rate and oxygen consumption (Vo_2) tend to be linearly related throughout a large portion of the aerobic work range. When this relationship is known, the exercise heart rate can be used to estimate Vo_2 [77]. Heart rate monitoring is relatively inexpensive, can assess people in a free-living state, and has the potential for providing information on the pattern as well as the total level of energy expenditure [78]. The technique has been successfully applied in small groups of children, lactating women, normal adults, athletes, and in remote indigenous populations [77]. Wareham and colleagues have confirmed its feasibility for assessing the pattern and total level of energy expenditure in the epidemiological context [78]. However, users of the technique must be cognizant of the uncertainties and likely sources of error in its use.

b. Physical Activity Questionnaires

Physical activity questionnaires have been used in many studies because they are easy to administer to large numbers of people and do not intrude on people's everyday activities. Although questionnaires do not provide precise estimates of EEPA, they may be helpful in ranking groups of subjects from the least to the most active. The ranking can then be used to correlate activity levels with disease outcomes [61]. An accurate questionnaire is both reliable and valid. A reliable questionnaire consistently provides similar results under the same circumstances, whereas a valid questionnaire truly measures what it was designed to measure. The validity of a physical activity questionnaire should be determined by comparing it with an objective measure of EEPA such as the doubly-labeled water method. Starling and colleagues [71] found that the Yale Physical Activity Survey estimates of EEPA compared favorably with DLW on a group basis. However, its use as a proxy measure for individual EEPA is limited. In the same study, the Minnesota Leisure Time Physical Activity Questionnaire significantly underestimated EEPA in free-living older men and women [71]. This points out the importance of ascertaining the validity of a questionnaire before applying it in large epidemiological studies. Multiple validity studies on various questionnaires have been done [71, 79–81]. A collection of the most used physical activity questionnaires, along with descriptions for their use, is available [82].

III. TOTAL ENERGY EXPENDITURE

Total energy expenditure (TEE) is composed of the energy required for both unconscious and conscious activities. As already mentioned, it consists of three distinct parts: resting energy expenditure, thermic effect of food, and the energy expended in physical activity. TEE estimates energy requirements, assuming that people are in energy balance [6]. The state of energy balance requires that, to maintain a specific weight, body energy stores must remain constant, that is, energy intake must match total energy expenditure. Thus, in significant weight gain or loss, when body composition is changing, metabolizable energy intake is not an accurate predictor of energy expenditure. Therefore, changes in body composition must be monitored for the accurate determination of energy expenditure.

A. Measuring Total Energy Expenditure

1. Direct Calorimetry

Direct calorimetry obtains a direct measurement of the amount of heat generated by the body within a structure large enough to permit moderate amounts of activity. These structures are called whole-room calorimeters. Direct calorimetry provides a measure of energy expended in the form of heat. The technique of direct calorimetry has several disadvantages. The structure is costly, requires complex engineering, and appropriate facilities are scarce around the world. Subjects must remain in a physically confined environment for long periods. In addition, direct calorimetry does not provide any information about the nature of substrates that are being oxidized to generate energy within the body [40].

2. Doubly-Labeled Water

The introduction of the DLW technique for human use in 1982 by Schoeller and van Santen [83] provided a scientific breakthrough in the measurement of TEE in free-living humans. The method was originally described by Lifson *et al.* [84] in the 1940s. It is based on the principle that carbon dioxide production can be estimated by the difference in elimination rates of body hydrogen and oxygen. Through his observations, Lifson concluded that the oxygen in expired carbon dioxide was derived from total body water [85]. This results from the equilibrium between the oxygen in body water and the oxygen in respiratory carbon dioxide [86]. With this finding, Lifson and colleagues predicted that carbon dioxide production could be indirectly measured by separately labeling both the hydrogen and oxygen pool of the body water with naturally occurring, stable isotopes. Based on the theory that the hydrogen of body water will exit as water and the oxygen of body water as carbon dioxide and water, the differential elimination rates of hydrogen and oxygen will indicate the amount of carbon dioxide expired during a given period. Carbon dioxide production can then be equated with energy expenditure under the assumption that carbon dioxide is an end product of substrate metabolism. Therefore, carbon dioxide production can be converted to an estimate of oxygen production using a food quotient estimated from total dietary intake.

The term "doubly-labeled water" indicates that the water consists of isotopic forms of hydrogen and oxygen. Deuterium oxide (^2H_2O) is the isotope that labels the hydrogen component of body water, and oxygen–18 ($H_2{}^{18}O$) is the isotope that labels the oxygen component of body water. Together, these isotopes label the hydrogen and oxygen of body water and trace their path within the body during the course of the study. The levels of the isotopes can be determined by periodic sampling of body water through urine, saliva, or plasma. Before a preweighed loading dose of DLW is orally administered, a baseline measurement of body water is obtained to detect the amount of deuterium oxide and oxygen–18 already present within the body water. After the baseline sample collection and administration of the DLW, two samples of body water are collected several days apart (usually 10 to 14 days) to determine the elimination rates of the isotopes. This two-point method has been shown to minimize the errors encountered from daily variations in the rate of carbon dioxide production [86].

a. Assumptions of the Doubly-Labeled Water Method

As with all scientific methodology, assumptions are made that need to be understood when using the DLW technique. A few of the fundamental assumptions are examined here. First, it is assumed that there is a constant total body water pool in which the isotopes turn over. However, it is known that with certain conditions such as infancy or intense exercise, the body water pool may change because of increases or decreases in body mass. This change in body mass must be greater than 15%, however, to produce an error in the total energy expenditure measure of less than 1% to 2% [87].

Second, the rate of flux for carbon dioxide and water should remain constant under steady-state kinetics. However, because water intake and physical activity are episodic in nature, this rate of flux is not constant. By implementing the previously mentioned two-point sample collection method, an average carbon dioxide production rate can be obtained over the measurement period without encountering the daily fluctuations that will contribute to error [87].

Another assumption states that no carbon dioxide or water enters the body via the skin or lungs. However, when a subject is exposed to higher-than-normal levels of carbon dioxide, such as cigarette smoke, there may be an apparent increase in carbon dioxide production. Nonetheless, to produce a 3% to 6% error in the estimation of true carbon dioxide production, the person would have to smoke roughly three packs of cigarettes a day [87]. In addition, water vapor can easily penetrate the lungs and skin to be directly absorbed into the body. However, Pinson and Langham [88] demonstrated that this type of exchange does not affect the calculation of carbon dioxide production, because the elimination rates of hydrogen and oxygen are proportional. Other assumptions of the method are associated with the isotopic dilution spaces and isotopic fractionations. These are beyond the scope of this chapter but are thoroughly reviewed by Schoeller [87] and Wolf [89].

b. Advantages and Disadvantages of Doubly-Labeled Water

The DLW technique has numerous advantages that make it the ideal method for measuring total daily energy expenditure in a variety of populations. The advantages to the researcher and the subject are the ease of administration and the ability of the subject to engage in free-living activities during the measurement period. This is extremely advantageous in infants, young children, the elderly, and populations with disabilities that cannot be subjected to the rigorous testing involved in the measurement of oxygen consumption during various activities. The DLW technique provides the objective criterion measure necessary to validate more subjective estimates of energy expenditure such as activity logs and diet recalls. Most important, the method is accurate and has a precision of between 2% and 8% [87].

Although there are many advantages to the DLW technique, there are also drawbacks, namely, the expense of the stable isotope (approximately $500 to dose an average weight adult), periodic worldwide shortages of oxygen–18, and the expertise required to operate the highly sophisticated and costly mass spectrometer for analysis of the enrichments

FIGURE 4 Mass spectrometer used for the analysis of isotope enrichments in the DLW method. See color plate.

TABLE 3 Advantages and Disadvantages of the Doubly-Labeled Water Technique to Measure Total Energy Expenditure

Advantages	Disadvantages
Noninvasive, unobstrusive, and easily administered	Availability and expense of oxygen-18 (approximately $900 for 70-kg adult)
Measurement performed under free-living conditions over extended time period (7 to 14 days)	Reliance on isotope ratio mass spectrometry for analysis of samples
Accurate and precise (2% to 8%)	A direct measure of CO_2 production, not energy expenditure, does not measure substrate oxidation
Can be used to estimate activity energy expenditure when combined with measurement of resting metabolic rate	Not suitable for large-scale epidemiological studies

(Fig. 4; see the color plate at the back of the book). The advantages and disadvantages of the DLW technique are highlighted in Table 3.

B. Estimating Total Energy Expenditure

The introduction of the DLW technique in humans has produced a large and robust database of TEE measurements in a variety of populations. A meta-analysis of 574 DLW measurements helped to establish the average and range of habitual energy expenditures in different age and sex groups [90]. Since this study, many DLW studies have been completed. The Food and Nutrition Board used data from these DLW studies to compile a worldwide TEE database [6]. Table 4 is a summary of data from this database. The database provides a frame of reference for energy needs in the general population and can be used to evaluate other estimates of energy expenditure. Special circumstances such as illness or enforced exertion were excluded from this database.

Because of the high cost and small numbers of laboratories with the capacity to do DLW studies, clinicians continue to rely heavily on prediction equations to estimate energy requirements. The TEE database compiled for the DRIs was used to derive equations for predicting energy expenditure requirements. These equations are provided in Table 5.

TABLE 4 Doubly-Labeled Water Data for Individuals with a Body Mass Index (BMI) in the Range from 18.5 to 25 kg/m^2

Age Group (yr)	n	Mean Body Mass Index (kg/m^2)	TEE Mean (kcal/day)	BEE Mean (kcal/day)	Mean Physical Activity Level Factor (TEE/BEE)
Females					
3–8	227	15.6	1487	1035	1.57
9–13	89	17.4	1907	1320	1.68
14–18	42	20.4	2302	1729	1.73
19–30	82	21.4	2436	1769	1.70
31–50	61	21.6	2404	1675	1.68
51–70	71	22.2	2066	1524	1.69
71+	24	21.8	1564	1480	1.62
Males					
3–8	129	15.4	1441	1004	1.64
9–13	28	17.2	2079	1186	1.74
14–18	10	20.4	3116	1361	1.75
19–30	48	22.0	3081	1361	1.85
31–50	59	22.6	3021	1322	1.77
51–70	24	23.0	2469	1226	1.64
70+	38	22.8	2238	1183	1.61

Source: Reprinted with permission from the National Academies Press, Copyright 2005, National Academy of Sciences, Institute of Medicine of the National Academies. (2005). "Dietary Reference Intakes for Energy, Carbohydrate, Fiber, Fat, Fatty Acids, Cholesterol, Protein, and Amino Acids." The National Academy Press, Washington, DC. From Table 5–10.

TABLE 5 Equations to Predict Energy Requirements in Females and Males 19 Years and Older

For Females

EER = 354.1 − 6.91 × age + PA × (9.36 × weight + 726 × height)
Where PA is the physical activity coefficient:
PA = 1.00 if PAL is estimated to be ≥1.0<1.4 (sedentary)
PA = 1.12 if PAL is estimated to be ≥1.4<1.6 (low active)
PA = 1.27 if PAL is estimated to be ≥1.6<1.9 (active)
PA = 1.45 if PAL is estimated to be ≥1.9<2.5 (very active)

For Males

EER = 661.8 − 9.53 × age + PA × (15.91 × weight + 539.6 × height)
Where PA is the physical activity coefficient:
PA = 1.00 if PAL is estimated to be ≥1.0<1.4 (sedentary)
PA = 1.12 if PAL is estimated to be ≥1.4<1.6 (low active)
PA = 1.27 if PAL is estimated to be ≥1.6<1.9 (active)
PA = 1.45 if PAL is estimated to be ≥1.9<2.5 (very active)

Source: Institute of Medicine of the National Academies (2005). "Dietary Reference Intake for Energy, Carbohydrate, Fiber, Fat, Fatty Acids, Cholesterol, Protein, and Amino Acids." The National Academy Press, Washington, DC.

Note: Weight is in kilograms, height is in centimeters, and age is in years.

C. Total Energy Expenditure in Special Populations

The number of studies using DLW to examine TEE in various disease states, physiological conditions, and across the life cycle have proliferated during the past decade. Hence, data are now available on energy expenditure during infancy [91], childhood [92], adolescence [93], and in the elderly [94]. TEE during pregnancy and lactation has been well characterized in some elegant longitudinal studies [26,95]. In addition, TEE has been examined in adults and children with obesity [96,97], burns [98], cerebral palsy [99], Down syndrome [100], HIV/AIDs [101], and Alzheimer's disease [102]. The effects of these various conditions on energy requirements are highlighted in Table 6.

IV. RECOMMENDED ENERGY INTAKES

Classically the Recommended Dietary Allowances (RDAs) have been used as a guide to determine energy intakes for groups of normal, healthy people. The last RDA for energy, which was labeled the recommended energy intake (REI), was set in 1989. RDAs for all nutrients except energy are set at levels well above those estimated to minimize the occurrence of deficiency syndromes. For energy, this obviously is not the case as there are adverse effects to individuals who consume energy above their requirements over time resulting in weight gain. Recommendations for energy have always been set as an average of energy requirements for a population group. In the past, recommended energy intakes have relied heavily on data from dietary surveys that estimate energy intake. This is based on the assumption that people are in energy balance at the time of measurement and that the estimates of energy intake are valid. A large body of evidence now demonstrates that self-reported estimates of food intake do not provide accurate or unbiased estimates of people's energy intake and that underreporting of food intake is pervasive in dietary studies [103]. Underreporting is discussed in detail in Chapter 7.

The DRIs [6] adopted an alternative approach and instead summarized data from multiple DLW studies to estimate energy requirements. There is no RDA for energy, as it would be inappropriate to recommend levels that would exceed requirements of 97% to 98% of individuals. Instead, the requirement for energy for individuals of

TABLE 6 Effects of Disease States and Physiological Conditions on Energy Requirements: Results from Doubly-Labeled Water Studies

Disease or Condition	Effect on Energy Requirements	Explanation
Obesity	Increased	Increased fat-free mass coupled with decreased physical activity
Burned children	No change	Increased resting metabolic rate counteracts decreased physical activity
Anorexia	Increased	To counteract increased physical activity with underweight and hypometabolism
HIV/AIDS	No change	Energy expenditure not elevated, reduced energy intake causes weight loss
Cerebral palsy	Relative to individual	High interindividual variation in energy expended in physical activity; ambulation status an important predictor
Alzheimer's	No change	Energy expenditure not elevated, low-energy intake predisposes to weight loss
Spinal cord injury	Decreased	Lower energy expended in physical activity, resting metabolic rate, and thermic effect of food
Pregnancy	Relative to individual	No prediction of metabolic response
Lactation	Increased	Energy cost of milk production partially offset by reduced physical activity

normal weight is expressed as an estimated energy requirement (EER), which reflects the energy expenditure based on the individual's sex, age, height, weight, and physical activity (see Table 5 for equations). For overweight individuals, these equations estimate total energy expenditure (TEE), rather than the EER, which is reserved for normal-weight individuals. The equations are estimates of energy needs based on maintaining current weight and activity level and therefore were not designed to lead to weight loss in overweight individuals [6].

As the prevalence of obesity has reached epidemic proportions in the United States, people's energy expended in physical activity has become so low that it may become increasingly difficult to meet the micronutrient needs on the energy intakes required to keep people in energy balance. Hence, it will become increasingly imperative that emphasis be placed on increased energy expended in physical activity for the maintenance of optimal health.

References

1. World Health Organization. (1985). Energy and protein requirements. *WHO Tech. Rep. Ser.* p. 724.
2. Schoeller, D. A., and Racette, S. B. (1990). A review of field techniques for the assessment of energy expenditure. *J. Nutr.* **120**, 1492–1495.
3. Hildreth, H. G., and Johnson, R. K. (1995). The doubly labeled water technique for the determination of human energy requirements. *Nutr. Today* **30**, 254–260.
4. Goran, M. I., Kaskoun, M., and Johnson, R. K. (1994). Determinants of resting energy expenditure in young children. *J. Pediatr.* **125**, 362–367.
5. Ravussin, E., Lilloja, S., Anderson, T. E., Christin, L., and Bogardus, C. (1986). Determinants of 24-hour energy expenditure in man. *J. Clin. Invest.* **78**, 1568–1578.
6. Institute of Medicine and the National Academies (2005). "Dietary Reference Intakes for Energy, Carbohydrate, Fiber, Fat, Fatty Acids, Cholesterol, Protein, and Amino Acids." National Academy Press, Washington, DC.
7. Weinsier, R. L., Schutz, Y., and Bracco, D. (1992). Reexamination of the relationship of resting metabolic rate to fat-free mass and to the metabolically active components of fat-free mass in humans. *Am. J. Clin. Nutr.* **55**, 790–794.
8. Nelson, K. M., Weinsier, R. L., Long, C. L., and Schutz, Y. (1992). Prediction of resting energy expenditure from fat-free mass and fat mass. *Am. J. Clin. Nutr.* **56**, 848–856.
9. Svendsen, O. L., Haarbo, J., Hassager, C., and Christiansen, C. (1993). Accuracy of measurements of body composition by dual-energy-ray absorptiometry *in vivo*. *Am. J. Clin. Nutr.* **57**, 605–608.
10. Wurmser, H., Laessle, R., Jacob, K., Langhard, S., Uhl, H., Angst, A., Muller, A., and Pirke, K. M. (1998). Resting metabolic rate in preadolescent girls at high risk of obesity. *Int. J. Obes. Relat. Metab. Disord.* **22**, 793–799.
11. Goran, M. I., Nagy, T. R., Gower, B. A., Mazariegos, M., Solomons, N., Hood, V., and Johnson, R. K. (1998). Influence of sex, seasonality, ethnicity, and geographic location on the components of total energy expenditure in young children: implications for energy requirements. *Am. J. Clin. Nutr.* **68**, 675–682.
12. Das, S. K., Saltzman,E., McCrory, M. A., Hsu, L. K., Shikora, S. A., Dolnikowski, G., Kehayias, J. J., and Roberts, S. B. (2004). Energy expenditure is very high in extremely obese women. *J. Nutr.* **134**, 1412–1416.
13. Saltzman, E., and Roberts, S. B. (1995). The role of energy expenditure in energy regulation: findings from a decade of research. *Nutr. Rev.* **53**, 209–220.
14. Larson, D. E., Ferraro, R. T., Robertson, D. S., and Ravussin, E. (1995). Energy metabolism in weight-stable postobese individuals. *Am. J. Clin. Nutr.* **62**, 735–739.
15. Weinsier, R. L., Nagy, T. R., Hunter, G. R., Darnell, B. E., Hensrud, D. D., and Weiss, H. L. (2000). Do adaptive changes in metabolic rate favor weight regain in weight-reduced individuals? An examination of the set-point theory. *Am. J. Clin. Nutr.* **72**, 1088–1094.
16. Ferraro, R. T., Lillioja, S., Fontvielle, A. M., Rising, R., Bogardus, C., and Ravussin, E. (1992). Lower sedentary metabolic rate in women compared to men. *J. Clin. Invest.* **90**, 780–784.
17. Arciero, P. J., Goran, M. I., and Poehlman, E. T. (1993). Resting metabolic rate is lower in women compared to men. *J. Appl. Physiol.* **75**, 2514–2520.
18. FAO, WHO, UNU. (1985). Energy and protein requirements. Report of a joint FAO/WHO/UNU expert consultation. *Technical Report Series* 724. World Health Organization, Geneva.
19. Piers, L. S., Soare, M. J., McCormack, L. M., and O'Dea, K. (1998). Is there evidence for an age-related reduction in metabolic rate? *J. Appl. Physiol.* **85**, 2196–2204.
20. Keys, A., Taylor, H. L., and Grande, F. (1973). Basal metabolism and age of adult man. *Metabolism* **22**, 579–587.
21. Henry, C. J. (2000). Mechanisms of changes in basal metabolism during aging. *Eur. J. Clin. Nutr.* **54**, 577–591.
22. Horton, T., and Geissler, C. (1994). Effect of habitual exercise on daily energy expenditure and metabolic rate during standardized activity. *Am. J. Clin. Nutr.* **59**, 13–19.
23. Jorgensen, J. O., Vahl, N., Dall, R., and Christiansen, J. S. (1998). Resting metabolic rate in healthy adults: relation to growth hormone status and leptin levels. *Metabolism* **47**, 1134–1139.
24. Webb, P. (1986). 24-Hour energy expenditure and the menstrual cycle. *Am. J. Clin. Nutr.* **44**, 614–661.
25. Henry, J., Lightowler, H. J., and Marchini, J. (2003). Intra-individual variation in resting metabolic rate during the menstrual cycle. *Br. J. Nutr.* **89**, 811–817.
26. Goldberg, G. R., Prentice, A. M., Coward, W. A., Davies, H. L., Murgatroyd, P. R., Sawyer, M. B., Ashford, J., and Black, A. E. (1993). Longitudinal assessment of energy expenditure in pregnancy by the doubly labeled water method. *Am. J. Clin. Nutr.* **57**, 494–505.
27. Fontvieille, A. M., Dwyer, J., and Ravussin, E. (1992). Resting metabolic rate and body composition of Pima Indians and Caucasian children. *Int. J. Obes. Relat. Metab. Disord.* **16**, 535–542.
28. Weyer, C., Snitker, S., Rising, R., Bogardus, C., and Ravussin, E. (1999). Determinants of energy expenditure and fuel utilization in man: effects of body composition, age, sex, ethnicity and glucose tolerance in 916 subjects. *Int. J. Obes. Relat. Metab. Disord.* **23**, 715–722.

29. Goran, M. I., Kaskoun, M., Johnson, R., Martinez, C., Kelly, B., and Hood, V. (1995). Energy expenditure and body fat distribution in Mohawk children. *Pediatrics* **95**, 89–95.
30. Bouchard, C., and Perusse, L. (1993). Genetics of obesity. *Annu. Rev. Nutr.* **13**, 337–354.
31. Bogardus, C., Lillioja, S., Ravussin, E., Abbott, W., Zawadzki, J. K., Young, A., Knowler, W. C., Jacobowitz, R., and Moll, P.P. (1986). Familial dependence of the resting metabolic rate. *N. Engl. J. Med.* **315**, 96–100.
32. Illner, K., Brinkman, G., Heller, M., Bosy-Westphal, A., and Mulier, M. J. (2000). Metabolically active components of fat free mass and resting energy expenditure in nonobese adults. *Am. J. Physiol.* **278**, E308–E315.
33. Hayter, J. E., and Henry, C. J. (1993). Basal metabolic rate in human subjects migrating between tropical and temperate regions: a longitudinal study and review of previous work. *Eur. J. Clin. Nutr.* **47**, 724–734.
34. Cartee, G. D., Douen, A. G., Ramlal, T., Klip, A., and Holloszy, J. O. (1991). Stimulation of glucose transport in skeletal muscle by hypoxia. *J. Appl. Physiol.* **70**, 1593–1600.
35. Zinker, B. A., Wilson, R. D., and Wasserman, D. H. (1995). Interaction of decreased arterial PO_2 and exercise on carbohydrate metabolism in the dog. *Am. J. Physiol.* **269**, E409–E417.
36. Brooks, G. A., Butterfield, G. E., Wolfe, R. R., Groves, B. M., Mazzeo, R. S., Sutton, J. R., Wolfel, E. E., and Reeves, J. T. (1991). Increased dependence on blood glucose after acclimatization to 4,300 m. *J. Appl. Physiol.* **70**, 919–927.
37. Brooks, G. A., Wolfel, E. E., Groves, B. M., Bender, P. R., Butterfield, G. E., Cymerman, A., Mazzeo, R. S., Sutton, J. R., Wolfe, R. R., and Reeves, J. T. (1992). Muscle accounts for glucose disposal but not lactate appearance during exercise after acclimatization to 4,300 m. *J. Appl. Physiol.* **72**, 2435–2445.
38. Butterfield, G. E., Gates, J., Fleming, S., Brooks, G. A., Sutton, J. R., and Reeves, J. T. (1992). Increased energy intake minimizes weight loss in men at high altitude. *J. Appl. Physiol.* **72**, 1741–1748.
39. Mawson, J. T., Braun, B., Rock, P. B., Moore, L. G., Mazzeo, R., and Butterfield, G. E. (2000). Women at high altitude: energy requirement at 4300 m. *J. Appl. Physiol.* **88**, 272–281.
40. Simonson, D. C., and DeFronzo, R. A. (1990). Indirect calorimetry: methodological and interpretative problems. *Am. J. Physiol.* **258**, E399–E412.
41. Compher, C., Frankenfield, D., Keim, N., and Roth-Yousey, L. (2006). Best practice methods to apply to measurement of testing metabolic rate in adults: a systematic review. *J. Am. Diet. Assoc.* **106**, 881–903.
42. Berke, E. M., Gardner, A. W., Goran, M. I., and Poehlman, E. T. (1992). Resting metabolic rate and the influence of the pretesting environment. *Am. J. Clin. Nutr.* **55**, 626–629.
43. Frankenfield, D. C., Muth, E. R., and Rowe, W. A. (1998). The Harris–Benedict studies of human basal metabolism: history and limitations. *J. Am. Diet. Assoc.* **98**, 439–445.
44. Boullata, J., Williams, J., Cottrell, F., Hudson, L., and Compher, C. (2007). Accurate determination of energy needs in hospitalized patients. *J. Am. Diet. Assoc.* **107**, 393–401.
45. Frankenfield, D., Roth-Yousey, L., and Compher, C. (2005). Comparison of predictive equations for resting metabolic rate in healthy nonobese and obese adults: systematic review. *J. Am. Diet. Assoc.* **105**, 75–89.
46. Cuerda, C., Ruiz, A., Velasco, C., Breton, I., Camblor, M., and Garcia-Peris, P. (2006). How accurate are predictive formulas calculating energy expenditure in adolescent patients with anorexia nervosa? *Clin. Nutr.* **26**, 100–106.
47. Goran, M. I., Broemeling, L., Herndon, D. N., Peters, D. J., and Wolfe, R. R. (1991). Estimating energy requirements in burned children: a new approach derived from measurements of resting energy expenditure. *Am. J. Clin. Nutr.* **54**, 35–40.
48. Krahn, D. D., Rock, C., Dechert, R. E., Nairn, K. K., and Hasse, S. A. (1993). Changes in resting energy expenditure and body composition in anorexia nervosa patients during refeeding. *J. Am. Diet. Assoc.* **93**, 434–438.
49. Bandini, L. G., Puelzi-Quinn, H. M., Morelli, J. A., and Fukagawa, N. K. (1995). Estimation of energy requirements in persons with severe central nervous system impairment. *J. Pediatr.* **126**, 828–832.
50. Johnson, R. K., Goran, M. I., Ferrar, M. S., and Poehlman, E. T. (1996). Athetosis increases resting metabolic rate in adults with cerebral palsy. *J. Am. Diet. Assoc.* **96**, 145–148.
51. Goldberg, G. R., Prentice, A. M., Coward, W. A., Davies, H. I., Murgatroyd, P. R., Wensing, C., Black, A. E., Harding, M., and Sawyer, M. (1993). Longitudinal assessment of energy expenditure in pregnancy by the doubly labeled water method. *Am. J. Clin. Nutr.* **57**, 494–505.
52. Spaaij, C., van Raaij, J., de Groot, L., ven der Heijden, L., Boeholt, H. A., and Hautvast, J. C. (1994). Effect of lactation on resting metabolic rate and on diet- and work-induced thermogenesis. *Am. J. Clin. Nutr.* **59**, 42–47.
53. Arciero, P. J., Goran, M. I., Gardner, A. M., Ades, P. A., Tyzbir, R. S., and Poehlman, E. T. (1993). A practical equation to predict resting metabolic rate in older females. *J. Am. Geriatr. Soc.* **41**, 389–395.
54. Arciero, P. J., Goran, M. I., Gardner, A. M., Ades, P. A., and Poehlman, E. T. (1993). A practical equation to predict resting metabolic rate in older men. *Metabolism.* **42**, 950–957.
55. Prentice, A. M. (1995). All calories are not equal. *Int. Dialogue Carbohydr.* **5**, 1–3.
56. McCrory, P. (1994). Energy balance, food intake and obesity. *In* "Exercise and Obesity" (A. P. Hills, and M. L. Wahlqvist, Eds.). Smith-Gordon & Co., London.
57. Dulloo, A. G., Geissler, C. A., Horton, T., Collins, A., and Miller, D. S. (1989). Normal caffeine consumption: influence on thermogenesis and daily energy expenditure in lean and post-obese human volunteers. *Am. J. Clin. Nutr.* **49**, 44–50.
58. Hofstetter, A. (1986). Increased 24-hour energy expenditure in cigarette smokers. *N. Engl. J. Med.* **314**, 79–82.
59. Kriska, A. M., and Caspersen, C. J. (1997). Introduction to a collection of physical activity questionnaires. *Med. Sci. Sports Exer.* **29**, S5–S9.
60. Pate, R. R., Prat, M., Blair, S. N., Haskell, W. L., Macera, C. A., Bouchard, C., Buchner, D., Ettinger, W., Heath, G. W., King, A. C., Kirska, A., Leon, A. S., Marcus, B. H., Morris, J., Paffenbarger, R. S., Patrick, K., Pollack, M. L., Rippe, J. M., Sallis, J., and Wilmore, J. H. (1995). Physical activity and public health: a recommendation from the Centers for Disease Control and Prevention and the American College of Sports Medicine. *J. Am. Med. Assoc.* **273**, 402–407.

61. Levine, J. A., Eberhardt, N. L., and Jensen, M. D. (1999). Role of nonexercise activity thermogenesis in resistance to fat gain in humans. *Science* **283**, 212–214.
62. Harriss, A. M., Lanningham-Foster, L. M., McCrady, S. K., and Levine, J. A. (2007). Nonexercise movement in elderly compared with young people. *Am. J. Physiol. Endrocrinol Metab.* **292**, E1207–E1212.
63. Shah, M., Geissler, C. A., and Miller, D. S. (1988). Metabolic rate during and after aerobic exercise in post-obese and lean women. *Eur. J. Clin. Nutr.* **42**, 455–464.
64. Geissler, C. A., and Aldouri, M. S. (1985). Racial differences in the energy cost of standardized activities. *Ann. Nutr. Metab* **29**, 40–47.
65. Blaak, E. E., Westerterp, K. R., Bar-Or, O., Wouters, L. J., and Saris, W. H. (1992). Total energy expenditure and spontaneous activity in relation to training in obese boys. *Am. J. Clin. Nutr* **55**, 777–782.
66. Levine, J., Baukol, P., and Pavlidis, I. (1999). The energy expended in chewing gum. *N. Engl. J. Med.* **341**, 2100.
67. Food and Nutrition Board, National Research Council, National Academy of Sciences. (1989). "Recommended Dietary Allowances," 10th ed. National Academy Press, Washington, DC.
68. Bielinski, R., Schutz, Y., and Jequier, E. (1985). Energy metabolism during the postexercise recovery in man. *Am. J. Clin. Nutr.* **42**, 69–82.
69. Vaughan, L., Zurlo, F., and Ravessin, E. (1991). Aging and energy expenditure. *Am. J. Clin. Nutr.* **53**, 821–825.
70. Bunyard, L. B., Katzel, L. I., Busby-Whitehead, J., Wu, Z., and Goldberg, A. P. (1998). Energy requirements of middle-aged men are modifiable by physical activity. *Am. J. Clin. Nutr.* **68**, 1136–1142.
71. Starling, R. D., Matthews, D. E., Ades, P. A., and Poehlman, E. T. (1999). Assessment of physical activity in older individuals: a doubly labeled water study. *J. Appl. Physiol.* **86**, 2090–2096.
72. Westerterp, K. R. (1999). Physical activity assessment with accelerometers. *Int. J. Obes. Relat. Metab. Disord.* **23**, S45–S49.
73. Troiano, R. P. (2005). A timely meeting: objective measurement of physical activity. *Med. Scien. Sports Exercise* **37**, S582–S588.
74. Johnson, R. K., Russ, J., and Goran, M. I. (1998). Physical activity related energy expenditure in children by doubly labeled water as compared with the Caltrac accelerometer. *Int. J. Obes. Relat. Metab. Disord.* **22**, 1046–1052.
75. Sallis, J. F., Buono, J. F., Roby, J. L., Carlson, D., and Nelson, J. (1990). The Caltrac accelerometer as a physical activity monitor for school-age children. *Med. Sci. Sports Exer.* **22**, 698–703.
76. Plasqui, G., Joosen, A., Kester, A. D., Goris, A. H. C., and Westererp, K. R. (2005). Measuring free-living energy expenditure and physical activity with triaxial accelerometry. *Obes. Res.* **13**, 1363–1369.
77. Livingstone, M. B. E. (1997). Heart-rate monitoring: the answer for assessing energy expenditure and physical activity in population studies? *Br. J. Nutr.* **78**, 869–871.
78. Wareham, N. J., Hennings, S. J., Prentice, A. M., and Day, N. E. (1997). Feasibility of heart-rate monitoring to estimate total level and pattern of energy expenditure in a population-based epidemiological study: the Ely young cohort feasibility study 1994–5. *Br. J. Nutr.* **87**, 889–900.
79. Mahabir, S., Baer, D. J., Giffen, C., Clevidence, B. A., Campbell, W. S., Taylor, P. R., and Hartman, T. J. (2006). Comparison of energy expenditure estimates from 4 physical activity questionnaires with double labeled water estimates in postmenopausal women. *Am. J. Clin. Nutr.* **84**, 230–236.
80. Conway, J. M., Irwin, M. L., and Ainsworth, B. E. (2002). Estimating energy expenditure from the Minnesota Leisure Time Physical Activity and Tecumseh Occupational Activity questionnaires: a doubly labeled water validation. *J. Clin. Epid.* **55**, 392–399.
81. Bonnefoy, M., Normand, S., Pachiaudi, C., Lacour, J. R., Laville, M., and Kostka, T. (2001). Simultaneous validation of ten physical activity questionnaires in older men: a doubly labeled water study. *J. Am. Geriatr. Soc* **49**, 28–35.
82. Kriska, A. M., and Caspersen, C. J., Eds. (1997). A collection of physical activity questionnaires for health-related research. *Med. Sci. Sports Exer.* **29**, S5–S205.
83. Schoeller, D. A., and van Santen, E. (1982). Measurement of energy expenditure in humans by the doubly labeled water method. *J. Appl. Physiol.* **53**, 955–959.
84. Lifson, N., Gordon, G. B., Visscher, M. B., and Nier, A. O. (1949). The fate of utilized molecular oxygen and the source of heavy oxygen of respiratory carbon dioxide, studied with the aid of heavy oxygen. *J. Biol. Chem.* **180**, 803–811.
85. Lifson, N., Gordon, G. B., and McClintock, R. (1997). Measurement of total carbon dioxide production by means of D_2O^{18}. *Obes Res.* **5**, 78–84.
86. Welles, S. (1990). Two-point vs. multipoint sample collection for the analysis of energy expenditure by use of the double labeled water method. *Am. J. Clin. Nutr.* **52**, 1134–1138.
87. Schoeller, D. A. (1988). Measurement of energy expenditure in free-living humans by using doubly labeled water. *J. Nutr.* **118**, 1278–1289.
88. Pinson, E. A., and Langham, W. H. (1957). Physiology and toxicology of tritium in man. *J. Appl. Physiol.* **10**, 108–126.
89. Wolf, R. R. (1992). "Radioactive and Stable Isotope Tracers in Biomedicine Principles and Practice of Kinetic Analysis," pp. 207–234. Wiley-Liss, New York.
90. Black, A. E., Coward, W. A., Cole, R. J., and Prentice, A. M. (1996). Human energy expenditure in affluent societies: an analysis of 574 doubly-labeled water measurements. *Eur. J. Clin. Nutr.* **50**, 72–92.
91. Davies, P. S. W. (1998). Energy requirements for growth and development in infancy. *Am. J. Clin. Nutr.* **68**, 939S–943S.
92. Goran, M. I., Gower, B. A., Nagy, T. R., and Johnson, R. K. (1998). Developmental changes in energy expenditure and physical activity in children: evidence for a decline in physical activity in girls before puberty. *Pediatrics* **101**, 887–891.
93. Bratteby, L. E., Snadhagen, B., Fan, H., Enghardt, H., and Samuelson, G. (1998). Total energy expenditure and physical activity as assessed by the doubly labeled water method in Swedish adolescents in whom energy intake was underestimated by 7-d records. *Am. J. Clin. Nutr.* **67**, 905–911.
94. Roberts, S. B., Young, V. R., Fuss, P., Heyman, M. B., Fiatarone, M., Dallal, G. E., Cortiella, J., and Evans, W. J.

(1992). What are the dietary energy needs of elderly adults? *Int. J. Obes. Relat. Metab. Disord.* **16**, 969–976.

95. Goldberg, G. R., Prentice, A. M., Coward, W. A., Davies, H. L., Murgatroyd, P. R., Wensing, C., Black, A. E., Harding, M., and Sawyer, M. (1993). Longitudinal assessment of energy expenditure in pregnancy by the doubly labeled water method. *Am. J. Clin. Nutr.* **57**, 494–505.
96. Lichtman, S. W., Krystyna, P., Berman, E. R., Pestone, M., Dowling, H., Offenbacher, E., Weisel, H., Heshka, S., Matthews, D. E., and Heymsfield, S. B. (1992). Discrepancy between self-reported and actual caloric intake and exercise in obese subjects. *N. Engl. J. Med.* **327**, 1893–1998.
97. Goran, M. I., Shewchuk, R., Bower, B. A., Nagy, T. R., Carpenter, W. H., and Johnson, R. K. (1998). Longitudinal changes in fatness in white children: no effect of childhood energy expenditure. *Am. J. Clin. Nutr.* **67**, 309–316.
98. Goran, M. I., Peters, E. J., Herndon, D. N., and Wolfe, R. R. (1990). Total energy expenditure in burned children using the doubly labeled water technique. *Am. J. Physiol.* **259**, E576–E585.
99. Johnson, R. K., Hildreth, H. G., Contompasis, S. H., and Goran, M. I. (1997). Total energy expenditure in adults with cerebral palsy as assessed by double labeled water. *J. Am. Diet. Assoc.* **97**, 966–970.
100. Luke, A., Roizen, N. J., Sutton, M., and Schoeller, D. A. (1994). Energy expenditure in children with Down syndrome: correcting metabolic rate for movement. *J. Pediatr.* **125**, 829–838.
101. Macallan, D. C., Noble, C., Baldwin, C., Jebb, S. A., Prentice, A. M., and Coward, A. (1995). Energy expenditure and wasting in human immunodeficiency virus infection. *N. Engl. J. Med.* **333**, 83–88.
102. Toth, M. J., Goran, M. I., Carpenter, W. H., Newhouse, P., Rosen, C. J., and Poehlman, E. T. (1997). Daily energy expenditure in free-living non-institutionalized Alzheimer's patients: a doubly labeled water study. *Neurolology* **48**, 997–1002.
103. Johnson, R. K. (2000). What are people *really* eating? *Nutr. Today* **35**, 40–46.

B. Research and Applied Methods for Observational and Intervention Studies

CHAPTER 5

Application of Research Paradigms to Nutrition Practice

CAROL J. BOUSHEY
Department of Foods and Nutrition, Purdue University, West Lafayette, Indiana

Contents

I. INTRODUCTION

The importance of health-related research was eloquently stated by Professor Atkinson from the Department of Pediatrics at McMaster University while presenting the Ryley-Jeffs Memorial Lecture. In her lecture, she stated [1]:

> The 21st-century model of health research is founded on a broad base of multidisciplinary research that is expeditiously and effectively translated into evidence-based practice, education, policy, and advocacy. The key objective is to improve the health of populations.

Given that nutrition is involved in almost all of the metabolic processes of human life, there would be no argument that nutrition research clearly comprises an important part of health-related research. Nutrition plays critical roles in the etiology, progression, or treatment of the majority of chronic degenerative diseases contributing to the largest proportion of morbidity and mortality among the population of the United States. By virtue of the multidisciplinary role of nutrition, a field requiring knowledge in cellular and molecular biology, biochemistry, physiology, genetics, food science, and the social sciences, individuals trained in the field of nutrition are in an ideal position to discover and use health research.

To assist with the application of research to the practice of clinical and community nutrition, this chapter reviews several paradigms of research. Other chapters in this book cover specific research methods and study designs (see other chapters in Section 1). There is an emphasis on extracting information from all areas of research including the basic sciences. This chapter provides conceptual models and frameworks that can guide the practice of nutrition.

II. BROAD RESEARCH AREAS

In the field of nutrition, research has been traditionally referred to as "basic" or "applied" research [2]. This dichotomous designation of nutrition research may be too simplistic. Goldstein and Brown [3] elaborated on a biomedical model that included basic science, disease-oriented research (DOR), and patient-oriented research (POR), particularly in reference to medical doctors. In the biomedical model, basic science involves the exploration of living systems at the molecular or tissue level. The DOR deliberately focuses on the pathogenesis or treatment of specific diseases, but it does not depend on patients. The POR would involve physicians treating the whole patient and observing, analyzing, and managing a patient. Recognition was made that all three areas are needed to achieve full understanding of the application of new discoveries to human health [3]. Although this model is not ideal with regard to referring to nutrition research, it does point to the difficulty of using the word "applied." In reality, this entire model is "applied" if it refers to the ultimate end point of improving the human condition. On the other hand, if "applied" refers to interaction with patients that would be only the POR. In other words, the term "applied" may not fully communicate the research activities to scientists or the public.

One metaphor for research to consider is the three-legged stool as shown in Figure 1. In this model, the "legs" represent broad research areas. The "seat" represents research-based conclusions. The combined evidence from each broad research area is needed to keep the seat stable or keep the seat from falling down. An unbalanced stool would represent an incomplete picture from which to draw conclusions and recommendations. In this model, the

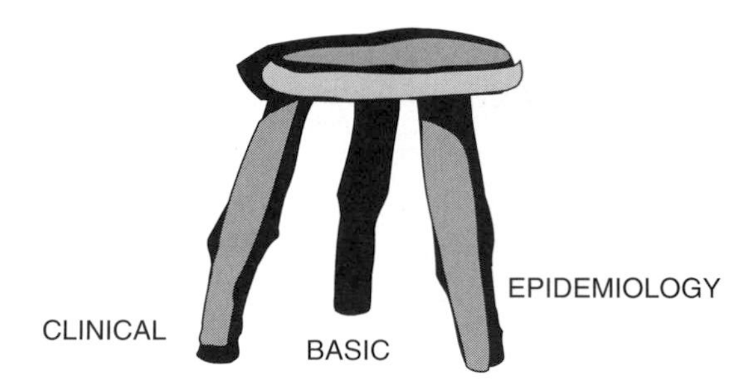

FIGURE 1 Research model: A three-legged stool.

basic research would involve examination of cellular models, biological specimens, or food components. This research is set apart in that no direct interaction occurs with human subjects. The clinical research refers to human studies that are under investigator control with regard to an intervention or treatment. The epidemiology studies would refer to observational studies using descriptive, prospective, or case-control study designs [4].

A. Broad Research Areas Example

An example that demonstrates the importance of these three broad areas is the conundrum surrounding the consumption of nuts and the subsequent effect on body weight. Tree nuts and peanuts are considered to be high-fat, energy-dense foods, thus their incorporation into a diet promoting energy balance would seem counterintuitive. However, evidence from the research areas of epidemiology, the clinical sciences, and the basic sciences suggests that the relationship between tree nuts and peanuts and weight maintenance bears reconsideration [5]. Using the broad areas represented by the three-legged stool, some examples of research regarding nuts will be outlined below.

An epidemiological study design that insures that the exposure measure (in this case, nut consumption) occurred before the outcome measure (in this case, body mass index) is the prospective cohort study [4]. Several reports from prospective analysis have been published with regard to nut consumption and body weight [6–8]. The Adventist Health Study enrolled 31,208 non-Hispanic white California Seventh-Day Adventists and collected extensive dietary data at baseline. In this cohort of adult men and women, a statistically significant negative association between consumption of nuts and obesity as assessed by body mass index was found [6]. This was probably the earliest report indicating that those individuals with higher nut consumption were less likely to be obese compared to those individuals with lower nut consumption. Another well-known prospective analysis, the Nurses' Health Study [7], also reported a negative association between nut consumption and body mass index. Using data from the Physicians' Health Study, Albert *et al.* [8] reported no association between nut consumption and body mass index. The results from the studies noted here and additional cross-sectional studies [9–11] have not reported a higher body mass index among frequent nut and peanut consumers compared to low consumers or nonconsumers. Collectively, these findings are compelling; however, to make a final consensus, one would want to see if this relationship would hold under more controlled study conditions or the clinical "leg."

A quasi-experimental study was conducted with a sample of 15 healthy normal-weight adults [12]. The participants were provided additional energy as 500 kcal of peanuts to consume daily for 8 weeks on top of their usual diets. The sample size allowed the investigators to monitor the subjects closely with regard to following the protocol and collecting background dietary intake. With no compensation, the predicted weight gain was an average of 3.6 kg for the group. Measured change was only 1 kg or about 28% of predicted level. These results would suggest that additional energy from nuts may have less of influence on weight than expected; however, the lack of a control group limits the conclusions that can be drawn. A study with an experimental design would be considered more useful. A prospective design using a randomized, crossover approach was completed among 20 healthy overweight adult women [13]. In this study, the women received 344 kcal of almonds daily for one 10-week period or no almonds for another 10 weeks with a 3-week washout between these two treatments. The theoretical weight gain was estimated to be 3.4 kg. However, the average weight of the participants was 70.4 ± 9 kg at the start of the almond arm and 70.3 ± 9 kg at the end, indicating no change. Other clinical studies have corroborated these results [5].

How might the basic science arm help with explaining these observations? Early work on this topic suggested greater energy loss from whole peanuts compared to peanut butter [14] and elevated fecal fat loss with nut consumption. As a result, a research team started to examine the nonstarch polysaccharides and phenolic compounds in the cell walls of almonds to possibly explain the epidemiological and clinical observations [15]. Using gas-liquid chromatography, HPLC, and microscopy, it was observed that bioaccessbility is limited in nuts. This inefficiency stems from resistance of the parenchyma cells constituting the cell walls of nuts to microbial and enzymatic degradation. Thus, cells that are not ruptured during mastication may pass through the gastrointestinal tract without releasing the lipid they contain. Although insufficient information is available to estimate the amount of energy that may be unavailable, the findings provide an explanation for the epidemiological and clinical results.

The studies highlighted here represent only a sample of the varied research activities addressing the original observation made in 1992 from the Adventist Health Study [6]. Whereas some individuals may have been quick to change practice after reading the results of the Adventist Health Study, and other individuals may have been quick to discount the results, the lesson is that no one research area can provide a clear picture for which application to practice. Another caveat with this example is the lack of describing the contribution of results from animal models, which does not imply less importance to this research area. Admittedly, public health decisions can be made without full knowledge of biological plausibility [16]; however, practice is enhanced and more effective with a complete picture.

III. EVIDENCE-BASED PRACTICE

In the 1990s, a movement within the medical profession made a concerted effort to steer away from reliance on expert opinion and start the process of informing practice decisions through a systematic examination of the scientific literature. An extensive series of manuscripts were published in *JAMA* starting with the Evidence-Based Medicine Working Group [17] and championed by two medical doctors, Gordon Guyatt and David Sackett [18]. Evidence-based medicine is a model for clinical decision making that integrates three elements: "best research evidence with clinical expertise and patient values" [19]. The research evidence includes clinically relevant research from the three areas of research covered under Section I.A. of this chapter (epidemiology, clinical, and basic science). Clinical expertise recognizes that practitioners have clinical skills and past knowledge that can clarify consistent and unique patterns in patients. Each patient holds his or her own set of expectations, concerns, and preferences, thus the inclusion of patient values. These elements guide a patient's plan of care to optimize clinical outcomes and quality of life. Evidence-based medicine has broadened its scope to all health care practices to which the term "evidence-based practice" is applied.

A. Steps to Achieve Evidence-Based Practice

The steps for evidence-based practice first include the recognition of a patient problem and constructing a structured clinical question [19, 20]. For example, from the Evidence Analysis Library (EAL) of the American Dietetic Association, there is the question, "What gastric volume level should be reached before stopping/holding enteral nutrition?"

The second step is to efficiently and effectively search information resources to retrieve the best available evidence to answer the clinical question. This involves accessing available databases such as MEDLINE using the publicly available PubMed portal through the National Library of Medicine (www.nlm.nih.gov). Efficiently finding the best evidence may require learning about MeSH terms and appropriate filters. It would be important to use correct nomenclature. For example, prealbumin is not a precursor to albumin as implied by the name. Prealbumin is more accurately termed "transthyretin." The name selected by the Joint Commission on Biochemical Nomenclature best describes transthyretin's role as a serum transport protein for thyroxin and retinal-binding protein. A PubMed search of prealbumin does not come up empty; however, a search of transthyretin generates three times the number of publications. One creates in advance some definitions of the type of study that would be most useful based on study design, patient outcomes, and exposures. Also, one decides in advance if reviews will be used.

Once applicable resources are gathered, the third step involves the critical appraisal of the evidence assembled. This step would involve assessing whether the selected report is relevant to the question and meets predetermined criteria of quality, such as study design, length of time, sample size, characteristics of sample, blinding, and use of current measures. Table 1 outlines the study designs of published manuscripts of original data as having an order of strongest to weakest empirical evidence when properly executed. Methods that synthesize primary research have an order of strength of evidence also. The quantitative synthesis methods, such as meta-analysis [21], are especially useful when sample sizes of individual studies tend to be small [22] or results tend to be inconsistent between studies [23]. The Agency for Healthcare and Research Quality with the U.S. Department of Health and Human Services has a number of modules for completing an evidence-based practice review, as well as completed reports in a variety of areas (www.ahrq.gov/clinic/epcix.htm).

The fourth step is directed to gaining a full understanding of the study results through summarizing the evidence. A table outlining the results as often used with meta-analysis [22, 23] can be useful at this stage. Various grading schemes

TABLE 1 Study Types Listed in Order from Strongest to Weakest Evidence by Primary Research (Left Column) and Synthesis Research (Right Column)

Primary Research of Original Data	Synthesis of Primary Research
Experimental Studies	***Meta-Analysis***
Randomized controlled clinical trial (RCT)	Decision analysis
Randomized cross-over clinical trial	Cost-benefit analysis
Randomized controlled laboratory study	Cost-effectiveness study
Observational Studies	***Systematic Review***
Prospective (cohort, longitudinal) study	
Case-control study	Narrative review
Ecological study	Consensus statement
Cross-sectional (prevalence) study	Consensus report
Population based	
Convenience sample	Medical opinion
Case series	
Case report	

are available for summarizing the strength of the collective evidence. Values range from 1 being good/strong, 2 being fair, 3 being limited/weak, 4 being expert opinion only, and 5 being grade not assignable. The number of studies meeting quality criteria would influence the final conclusion or recommendation.

In reality, the fifth and final step can only be made through integrating the evidence with the clinical information about the individual patient and the patient's desires and goals for care. Consideration of the best clinical care option for a patient is still the decision-making process of the health care practitioner with the patient. This can probably be best appreciated when one considers the number of options available for renal replacement therapy. The final decision is a complex process that ultimately will influence the patient's diet and the use of dietary supplements.

B. Examples of Evidence-Based Practice in Nutrition

The American Dietetic Association has created an Evidence Analysis Library (EAL) that is available free to members of the association and available for a fee to others (www.adaevidencelibrary.com). Working groups have completed numerous evidence analyses and accompanying practice guidelines. This represents a rich resource for nutrition; however, the EAL does not release practitioners from actively engaging in the process. The final step still involves the application of professional judgment within the range of the patient's expressed priorities.

Other systematic reviews have been completed and these reviews can be consulted for application to practice. The National Cholesterol Education Program (NCEP) has used a systematic review process to develop guidelines for cardiovascular disease [24]. Extensive systematic reviews of the influence of omega-3 fatty acids across a wide range of clinical conditions were completed [25]. A number of limitations were identified regarding the published studies of omega-3 fatty acids. Many of the limitations were common to all health sciences research, such as poor study design, lack of information about an intervention, and nondiverse samples. However, problems unique to omega-3 fatty acids and nutrition, in general, were identified. When dealing with a pharmaceutical drug, there is little difference in the active substance other than dose. In contrast, omega-3 fatty acids and other nutrients may be available in different forms that vary in bioavailability or metabolic activity. Further, food sources of nutrients add the complexity of variation in concentration and the possibility of other constituents in the food interfering with absorption. The circulating levels of a nutrient, which may be a biomarker of compliance, may be influenced by other metabolic factors. One of the most serious problems is the lack of attention to the baseline dietary intake or the background diet. Many studies don't even communicate that the study subjects, whether in the intervention or control groups, may have been independently consuming the nutrient of interest. Thus, the systematic compilation of evidence for nutrition will continue to present challenges not faced by other health fields.

The individual in clinical nutrition practice may question the need to be concerned with results from basic research or even research in general. A well-meaning guide for the busy clinician even suggested that just scanning an abstract and the results section of a research article would provide maximum information for the reader [26]. However, this couldn't be further from the truth [27]. To base practice on scientific principles, original research by necessity deserves more than a quick scan. Otherwise, a full understanding of any results cannot be fully appreciated. The serious practitioner will make time to become familiar with each "leg" of science related to his or her area of expertise or special practice. A study of registered dietitians found that those individuals who read professional publications weekly compared to monthly had higher perceptions, attitudes, and knowledge toward evidence-based practice [28].

IV. TRANSLATIONAL RESEARCH

The most recent research paradigm is "translational research." This paradigm grew from the desire to link the discoveries of molecular networks in health and disease to the patient and public benefit [29]. The challenge calls for novel interdisciplinary approaches to advance science and improve the health of the nation. The Translational Research Working Group of the National Cancer Institute reached consensus on an operational definition of translational research as "research that transforms scientific discoveries arising in the lab, clinic, or population into new clinical tools and applications that reduce cancer incidence, morbidity, and mortality." Education and practice in nutrition require knowledge and skills in the biological and social sciences [2]. As a result, nutrition is well positioned to be a player in interdisciplinary research.

Figure 2 depicts translational research and signifies that discovery may occur anywhere in the cycle (i.e., at the bench, clinical, or population level). As noted with the three-legged stool model, all of these areas of research interact and depend on one another. The aspect of the translational model that separates this model from the others is the additional step of developing new tools and applications for use as new chemotherapeutic drugs, new devices, nontraditional approaches [30], behavioral interventions, new screening assays, and educational training. Dissemination to health care providers, patients, and the public is considered a part of the research process. Further, these new tools feed back into the discovery cycle. Should the new intervention not perform as expected, then this

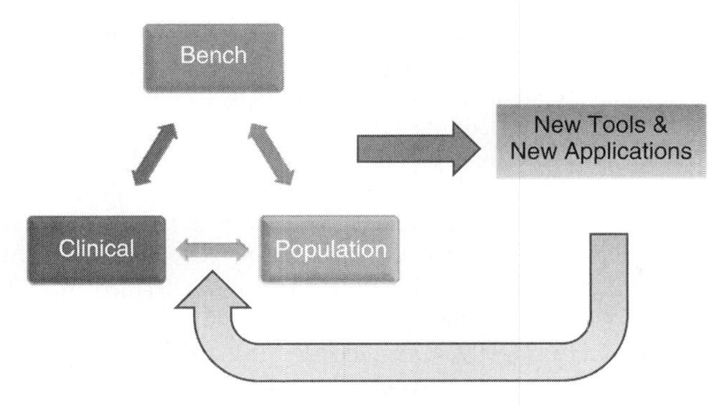

FIGURE 2 Translational research model.

observation will generate new research to ensure that dissemination brings improvement to the nation's health.

The combined results from the bench, clinical, and population sciences regarding calcium and bone in youth is thoroughly covered in Chapter 44. This information is briefly summarized here to show the application of translational research in nutrition. Dietary intake data from the USDA's "What We Eat in America, 2001–2002 NHANES" [31] show that the average calcium intakes of adolescent girls and boys aged 9 to 13 years are well below the 1300 mg adequate intake (AI) level for calcium set by the Institute of Medicine for girls and boys 9 through 18 years [32, 33]. One of the major lines of evidence to establish the AI in this age group was calcium retention to meet peak bone mineral accretion. Furthermore, older teens tend to drink more carbonated beverages, fruit drinks, and citrus juice, but less milk and noncitrus juices [34]. Because of the increase in noncalcium-rich beverages, there is a concern that the lack of calcium intake during adolescence can contribute to poor bone formation, which may potentially increase risk for osteoporosis in later life [35]. Because bone mass accrues until the peak is reached by early adulthood, maximizing peak bone mass within an individual's genetic potential requires optimal calcium intakes throughout growth. This is especially important during adolescence when the rate of calcium accretion is highest because approximately half an adult's bone mass accumulates during these few short years [36].

Clearly the preceding information fed to the application of a targeted intervention to young girls. French and colleagues [37] developed a theoretically based intervention that incorporated behavior-changing activities into Girl Scout programs called the CAL-Girls study. The messages were directed to girls 9 to 11 years old. At the end of the intensive program that was uniformly delivered significant changes were observed for increases in dietary calcium intake; however, no significant intervention effects were observed for bone mineral content at any specific bone site. Further, no gains were made in increasing weight-bearing physical activity. Therefore, the results fed back into the research cycle to initiate new research questions. Specifically, more data are needed on optimal dosages of weight-bearing physical activity and level of behavior change for consumption of calcium-rich foods needed to effect bone mass gains in early adolescent girls. Further, the question is raised, how can behavioral models be modified to better reflect positive changes in adolescents?. Completely unrelated to the dosage is the concern that the methods of measuring diet and physical activity cannot adequately distinguish changes in intake or output. This type of feedback about inadequate assessment techniques has generated a number of research programs directed to improving the measurements of diet and physical activity.

V. SUMMARY

This chapter introduced several research paradigms used to assist researchers and practitioners in reaching a common goal of enhancing the health of the nation. All individuals in the field of nutrition engage with research either through the discovery process or the learning process. The latter constitutes the synthesis of research studies to identify current best practices. With the advent of technologies, nutrition practitioners have available numerous resources to form work groups, access original published literature, and summarize evidence within the context of their working environments.

References

1. Atkinson, S. A. (2006). A nutrition odyssey: knowledge discovery, translation, and outreach. 2006 Ryley-Jeffs Memorial Lecture. *Can. J. Diet. Pract. Res.* **67**, 150–156.
2. Opportunities in the nutritions and food sciences: research challenges and the next generation of investigators. Special Committee of the Food and Nutrition Board of the Institute of Medicine, National Academy of Sciences. *J. Nutr.* **124**, 763–769.
3. Goldstein, J. L., and Brown, M. S. (1997). The clinical investigator: bewitched, bothered, and bewildered—but still beloved. *J. Clin. Invest.* **99**, 2803–2812.
4. Boushey, C., Harris, J., Bruemmer, V., Archer, S. L., and Van Horn, L. (2006). Publishing nutrition research: a review of study design, statistical analyses, and other key elements of manuscript preparation, Part 1. *J. Am. Diet. Assoc.* **106**, 89–96.
5. St-Onge, M. (2005). Dietary fats, teas, dairy, and nuts: potential functional foods for weight control? *Am. J. Clin. Nutr.* **81**, 7–15.
6. Fraser, G. E., Sabate, J., Beeson, W. L., and Strahan, T. M. (1992). A possible protective effect of nut consumption on risk of coronary heart disease. *Arch. Intern. Med.* **152**, 1416–1424.
7. Hu, F. B., Stampfer, M. J., Manson, J. E., Rimm, E. B., Colditz, G. A., Rosner, B. A., Speizer, F. E., Hennekens, C. H., and Willett, W. C. (1998). Frequent nut consumption and risk of coronary heart disease in women: prospective cohort study. *BMJ* **317**, 1341–1345.

8. Albert, C. M., Gaziano, J. M., Willett, W. C., and Manson, J. E. (2002). Nut consumption and decreased risk of sudden cardiac death in the Physicians' Health Study. *Arch. Intern. Med.* **162**, 1382–1387.
9. Ellsworth, J. L., Kushi, L. H., and Folsom, A. R. (2001). Frequent nut intake and risk of death from coronary heart disease and all causes in postmenopausal women: the Iowa Women's Health Study. *Nutr. Metab. Cardiovasc. Dis.* **11**, 372–377.
10. Sabate, J. (2003). Nut consumption and body weight. *Am. J. Clin. Nutr.* **78**(3 suppl), 647S–650S.
11. Griel, A. E., Eissenstat, B., Juturu, V., Hsieh, G., and Kris-Etherton, P. M. (2004). Improved diet quality with peanut consumption. *J. Am. Coll. Nutr.* **23**, 660–668.
12. Alper, C. M., and Mattes, R. D. (2002). The effects of chronic peanut consumption on energy balance and hedonics. *Int. J. Obes.* **26**, 1129–1137.
13. Hollis, J. H., and Mattes, R. D. (2007). Effect of chronic consumption of almonds on body weight in healthy humans. *Br. J. Nutr.* **98**, 651–656.
14. Levine, A. S., and Silvis, S. E. (1980). Absorption of whole peanuts, peanut oil, and peanut butter. *N. Engl. J. Med.* **303**, 917–918.
15. Ellis, P. R., Kendall, C. W. C., Ren, Y., Parker, C., Pacy, J. F., Waldron, K. W., and Jenkins, D. J. A. (2004). Role of cell walls in the bioaccessibility of lipids in almond seeds. *Am. J. Clin. Nutr.* **80**, 604–613.
16. Rothman, K. J. (2002). "Epidemiology: An Introduction," pp 1–223. Oxford University Press, New York.
17. Evidence-based medicine: a new approach to teaching the practice of medicine. (1992). Evidence-Based Medicine Working Group. *JAMA* **268**, 2420–2425.
18. Oxman, A. D., Sackett, D. L., and Guyatt, G. H. (1993). Users' guides to the medical literature. I. How to get started. The Evidence-Based Medicine Working Group. *JAMA* **270**, 2093–2095.
19. Centre for Evidence-Based Medicine. Available at www.cebm.utoronto.ca. Accessed on September 8, 2007.
20. Hatala, R., Keitz, S. A., Wilson, M. C., and Guyatt, G. (2006). Moving toward an integrated evidence-based medicine curriculum. *J. Gen. Intern. Med.* **21**, 538–541.
21. Petitti, D. B. (2000). "Meta-Analysis, Decision Analysis, and Cost-Effectiveness Analysis," 2nd ed., pp 1–303. Oxford University Press, New York. Monographs in Epidemiology and Biostatistics, Volume 31. Kesley, J. L., Marmot, M.G., Stolley, P. D., and Vessey, M. P.
22. Savaiano, D., Boushey, C. J., and McCabe, G. P. (2006). Lactose intolerance symptoms assessed by meta-analysis: a grain of truth that leads to exaggeration. *J. Nutr.* **136**, 1107–1113.
23. Boushey, C. J., Beresford, S. A. A., Omenn, G. S., and Motulsky, A. G. (1995). A quantitative assessment of plasma homocysteine as a risk factor for vascular disease: probable benefits of increasing folic acid intakes. *JAMA* **274**, 1049–1057.
24. Van Horn, L., and Ernst, N. (2001). A summary of the science supporting the new National Cholesterol Education Program dietary recommendations: what dietitians should know. *J. Am. Diet. Assoc.* **101**, 1148–1154.
25. Balk, E. M., Horsley, T. A., Newberry, S. J., Lichtenstein, A. H., Yetley, E. A., Schachter, H. M., Moher, D., MacLean, C. H., and Lau, J. (2007). A collaborative effort to apply the evidence-based review process to the field of nutrition: challenges, benefits, and lessons learned. *Am. J. Clin. Nutr.* **85**, 1448–1456.
26. Crist, E. A. (1999). Mapping a practice route through the journal. *J. Am. Diet. Assoc.* **99**, 1041–1042.
27. Mattes, R. D., and Boushey, C. J. (2000). To read or not to read original research articles: it should not be a question. *J. Am. Diet. Assoc.* **100**, 171–174.
28. Byham-Gray, L. D., Gilbride, J. A., Dixon, L. B., and Stage, F. K. (2005). Evidence-based practice: what are dietitians' perceptions, attitudes, and knowledge? *J. Am. Diet. Assoc.* **105**, 1574–1581.
29. Zerhouni, E. A. (2005). US biomedical research: basic, translational, and clinical sciences. *JAMA* **294**, 1352–1358.
30. Szczurko, O., Cooley, K., Busse, J. W., Seely, D., Bernhardt, B., Guyatt, G. H., Zhou, Q., and Mills, E. J. (2007). Naturopathic care for chronic low back pain: a randomized trial. *PLoS One* **2**, e919.
31. Moshfegh, A., Goldman, J., and Cleveland, L. (2005). What we eat in America, NHANES 2001–2002: Usual Nutrient Intakes from Food Compared to Dietary Reference Intakes. U.S. Department of Agriculture, Agricultural Research Service, Washington, DC.
32. Institute of Medicine, National Academy of Sciences. (1997). "Dietary Reference Intakes for Calcium, Phosphorus, Magnesium, Vitamin D, and Fluoride," National Academy Press, Washington, DC.
33. U.S. Department of Agriculture, Agricultural Research Service. (1999). "Food and Nutrient Intakes by Children 1994–96, 1998," ARS Food Surveys Research Group.
34. Weinbery, L. G., Berner, L. A., and Groves, J. E. (2004). Nurtrient contributions of dairy foods in the United States, continuing survey of food intakes by individuals, 1994–1996, 1998. *J. Am. Diet. Assoc.* **104**, 895–902.
35. Bowman, S. A. (2002). Beverage choices of young females: changes and impact on nutrient intakes. *J. Am. Diet. Assoc.* **102**, 1234–1239.
36. Weaver, C. M. (2000). The growing years and prevention of osteoporosis later in life. *Proceedings of the Nutrition Society* **59**, 303–306.
37. French, S. A., Storey, M., Fulkerson, J. A., and Himes J. H, Hannan, P., Neumark-Sztainer, D., and Ensurd, K. (2005). Increasing weight-bearing physical activity and calcium-rich foods to promote bone mass among 9–11 year old girls: outcome of the Cal-Girls Study. *Int. J. Behav. Nutr. Phys. Act.* **2**, 1–11.

CHAPTER 6

Overview of Nutritional Epidemiology

MARIAN L. NEUHOUSER AND RUTH E. PATTERSON
Cancer Prevention Program, Fred Hutchinson Cancer Research Center, Seattle, Washington

Contents

I. INTRODUCTION

Epidemiology is the science of public health whose laboratory setting is human populations. More specifically, epidemiology is the study of the distribution and determinants of disease frequency in human populations [1]. Epidemiology addresses such questions as who gets a disease, when, and why. This information is then used to create strategies for prevention or treatment. Nutritional epidemiology is the study of how dietary factors relate to the occurrence of disease in human populations. For example, observations that persons who consumed fish at least twice a week had a lower risk of incident cardiovascular disease and stroke as well as a lower risk of sudden cardiac death than persons who ate no fish [2, 3] led to widespread recommendations to include fish in the diet for the prevention of cardiovascular disease (CVD). More recently, population-based studies demonstrating that consumption of *trans*-fatty acids may be associated with unfavorable blood lipid profiles have led to a decline in the use of these fats by food manufacturers and outright bans by some local health departments [4–7]. Another major finding by nutritional epidemiologists that has had major public health policy implications was the observation that the use of multivitamins containing folic acid reduced the risk of neural tube defect–affected pregnancies. Consistent evidence from numerous epidemiological studies led to the fortification of the U.S. food supply with folic acid as a primary prevention strategy in 1998 [8, 9]. Thus, some of the major nutrition-related public health contributions since the 1950s have been made possible by nutritional epidemiology.

One factor that distinguishes the discipline of nutritional epidemiology is the extraordinary challenge of the exposure assessment. In epidemiology, exposure is defined as participant characteristics, lifestyle behaviors, or agents (e.g., food, medications such as hormone replacement therapy, tobacco, sun) with which a participant comes into contact that may relate to disease risk (see Table 1) [10]. Measuring dietary intake is particularly complicated for many reasons.

TABLE 1 Examples of Exposures Relevant to Nutritional Epidemiological Studies

Exposure	Diet-Related Example	Other Example
Agent that may cause or protect from disease	Vegetable consumption may be protective for colon cancer.	Physical activity may be protective for colon cancer.
Constitutional host factors	Genetic predisposition to nutrition-related disease.	Older adults are more predisposed to chronic disease.
Other host factors	Food preferences that determine food choices.	More educated adults may have better disease screening.
Agents that may confound the association between another agent and disease	Correlation between dietary constituents (e.g., a diet high in fruits and vegetables is usually low in fat).	Smokers are less likely to engage in physical activity.
Agents that may modify effects of other agents	Fruits and vegetables may protect against lung cancer among smokers.	Alcohol results in increased risk of lung cancer from smoking.
Agents that may determine outcome of disease	Malnutrition.	Medical treatment.

For purposes of illustration, we can compare the assessment challenges for two common exposures: cigarette smoking and diet. Smoking is a single (yes/no) activity—people are either smokers or nonsmokers, so individuals usually report with accuracy whether or not they smoke. Because smoking is addictive, it tends to be a consistent long-term behavior. Further, because smoking is a habit, most people smoke roughly the same number of cigarettes per day (e.g., one or two packs per day). In comparison, over the course of even 1 week, an individual can consume hundreds, even thousands of distinct food items in various combinations, making it cognitively challenging for respondents to accurately report on their intakes. Meals can be prepared by others (e.g., in a restaurant, by a spouse, as prepackaged food) so that the respondent may not be cognizant of preparation details such as fat or salt used in cooking or portion size. Food choices typically vary with seasons and other life activities (e.g., weekends, holidays, vacations). The day-to-day variability in food intake can be so large that it is difficult to identify any underlying consistent pattern. In addition, foods themselves are often a surrogate for the exposure of interest (e.g., dietary fat or fiber), which means that investigators must rely on food composition databases to calculate the exposure variable. Given even this superficial summary of the problems inherent in assessing dietary intake, it is not surprising that, despite some of the important contributions noted, it has often been difficult to obtain consistent and strong evidence regarding how diet affects disease risk.

The majority of nutritional epidemiological research since the early 1980s has focused on the identification of foods or specific nutrients in foods that prevent or promote the occurrence of chronic diseases, such as cancer, diabetes, and cardiovascular disease, in the general population. Therefore, the tools and methods of nutritional epidemiology were developed to address scientific issues unique to the biology of chronic diseases. In particular, epidemiological methods were designed to consider these important methodological issues: (1) the extensive time for disease development, (2) the multifactorial nature of chronic diseases, and (3) research conducted in free-living humans, which establishes associations only and precludes direct tests of cause and effect. Next, each of these issues in relation to nutritional epidemiology is discussed.

A. Extensive Time for Disease Development

Chronic diseases develop over many years and even decades. This lengthy progression has important implications for the field of nutritional epidemiology. For example, the currently accepted model of colon cancer assumes that it is a multistep pathogenic pathway beginning with mutations (germline or somatic) leading to growth of polyps (preneoplastic growths), which become adenomas and progress to carcinoma [11]. Even after an adenoma finally develops, many years may elapse before it is clinically detected. Upon clinical detection of colon cancer, it is clear that a meal consumed that day, or the month before, could have no significant effect on the disease for two reasons. First, the critical period of cancer initiation and promotion likely occurred many years before disease diagnosis. Second, the biologically relevant exposure is the long-term or usual diet, rather than any single eating occasion. Therefore, the exposure of interest in the development of cancer occurred throughout the previous 10 to 20 years and perhaps the previous 5 years for cardiovascular disease and diabetes.

This time lag between dietary exposure and disease occurrence presents significant difficulties in the study of diet and chronic disease. These difficulties have been addressed in two major ways. First, dietary assessment instruments for epidemiology were developed to capture information on usual, long-term dietary intake [12]. The most common are the food frequency and diet habits questionnaires. Second, study designs that were already in use in other areas of epidemiology/population sciences were applied to studies of diet-disease risk, namely, case-control and cohort studies. These designs were already in use to gather data on remote exposures (e.g., use of oral contraceptives, age at menarche) or life events (e.g., age at first pregnancy, number of full-term pregnancies), which proved to be very useful for establishing associations of various risk factors with disease risk [1]. For example, asking detailed questions about reproductive history of women with and without breast cancer has provided extremely useful and reliable data about breast cancer risk factors. Using the same approach, nutritional epidemiologists have asked people about retrospective diet in case-control studies or assessed current diet in cohort studies and followed participants over time to monitor occurrence of disease. The challenge with the latter approach is that it may take many years for disease to develop and in the meantime, people may change their dietary habits.

B. Multifactorial Nature of Chronic Diseases

Chronic diseases such as cancer and cardiovascular disease have complex etiologies that are multifactorial in nature. In addition to diet, other determinants of chronic diseases include host susceptibility factors (e.g., genetic susceptibility) and other lifestyle habits (e.g., cigarette smoking, body habitus, physical activity, alcohol intake). These other factors may confound our ability to find an association of dietary intake with disease risk. For example, individuals with an interest in health may eat a diet high in fruits and vegetables (which also tends to be low in fat) and have high physical activity levels. If vegetable intake is found to be associated with a reduced risk of colon cancer, it can be difficult to disentangle whether it is the high vegetable intake or the low-fat intake that affects disease risk. It is also possible that diet itself is not related to disease risk but is merely serving as a marker for some other healthful behavior (such as physical activity or adherence to cancer screening guidelines such as getting routine mammograms and colonoscopies) [13].

Another illustration of the multifactorial nature of chronic disease is the complex relationship between dietary factors and a person's genetic constitution. For example, the phase II enzymes, glutathione-*S*-transferases, are up-regulated by vegetables from the *Brassica* family (broccoli, Brussels' sprouts, cauliflower); in other words, compounds in these vegetables increase the activity of this family of enzymes [14–16]. There is also growing evidence that the health benefits of *Brassica* vegetables may vary depending on whether someone has any of the common genetic variations in these enzymes. Humans who are GSTM1-null (approximately 50% of U.S. non-Hispanic whites and 35% of African Americans) or GSTT1-null (approximately 15% of U.S. non-Hispanic whites and 25% of U.S. African Americans) have greatly reduced (heterozygote) or no (homozygote) enzyme activity [14]. Among those with the null variants, compounds in the *Brassica* vegetable, such as isothiocyanates, may be metabolized and eliminated more slowly, so they may be available for a longer period of time in target tissues and organs [17]. Several studies have found stronger associations of *Brassica* vegetables with reduced risks of lung [18], breast [19], and colon cancer [20] among persons GSTM1- or GSTT1-null. This example of diet-gene interaction illustrates the complexity and multifactorial nature of diet-related chronic disease development.

C. Research in Human Beings

There are several important considerations with regard to conducting research in humans. Notably, population-based nutrition research or nutritional epidemiology is often conducted among healthy persons, at least at the time of initial assessment. For studies that involve dietary interventions, federal guidelines for the protection of human subjects dictate that study participants (both healthy and diseased) cannot (knowingly) be exposed to potentially dangerous dietary regimens that restrict essential nutrients or introduce known carcinogens at high levels over long periods of time. However, short-term controlled feeding studies can reasonably test the effect of potentially harmful food constituents for the purpose of learning about nutrient metabolism and potential disease mechanisms in relation to food intake [21]. For example, participants can be fed charred meat containing nitrosamines to learn how humans metabolize or eliminate such potential carcinogens [22, 23], or they can be fed restricted diets devoid of certain nutrients to learn about nutrient absorption, metabolism, and excretion [24, 25]. Indeed, such, information is critical for public health recommendations regarding nutrient requirements. However, tightly controlled human feeding or intervention studies are expensive and laborious and not feasible to conduct in a large-scale manner, especially when studying general population risk. For these reasons, nutritional epidemiology is primarily an observational discipline that consists largely of (1) measuring an exposure (e.g., usual dietary intake), (2) measuring an outcome (e.g., disease occurrence), and (3) using statistical techniques to quantify the magnitude of the association between these two observations.

Epidemiology encompasses three major topic areas: (1) exposure assessment, (2) study design, and (3) interpretation of the associations. Next we discuss each of these in relation to nutritional epidemiology.

II. PRINCIPLES OF EXPOSURE MEASUREMENT IN NUTRITIONAL EPIDEMIOLOGY

A variety of dietary assessment methods are used in nutritional epidemiology. The choice of instrument used depends on the hypothesis being tested, the population, and cost considerations. The three primary tools are food frequency questionnaires, 24-hour dietary recalls, and food records or food diaries.

A. Food Frequency Questionnaires

Chronic disease risk develops over many years. As such, the biologically relevant exposure is usual or long-term diet, consumed many years before the appearance of clinical symptoms and disease diagnosis. Therefore, assessment instruments that only capture data on recent dietary intake (e.g., food records or recall) may not be as useful for studies examining chronic diseases that evolve over a longer period of time (years). For these reasons, food frequency questionnaires (FFQs) have been regarded as the dietary assessment instrument best suited for most epidemiological applications (12). Although the design of FFQs can vary somewhat, they typically contain the following sections: (1) adjustment questions, (2) the food checklist, and (3) summary questions.

FFQs often contain *adjustment questions* that assess the nutrient content of specific food items. For example, participants are asked what type of milk they usually drink and are given several options (e.g., whole, skim, soy), which saves space and reduces participant burden compared to asking for the frequency of consumption and usual portion sizes of many different types of milk. Adjustment questions also permit more refined analyses of fat intake by asking about food preparation practices (e.g., removing fat from red meat) and types of added fats (e.g., use of butter versus margarine on bread).

The main section consists of a *food* or *food group checklist*, with questions on usual frequency of intake and portion size. The foods are selected to capture data on (1) major sources of energy and nutrients for most people,

(2) between-person variability in food intake, and (3) major hypotheses regarding diet and disease. Frequency of consumption is typically categorized from "never or less than once per month" to "2+ per day" for foods and "6+ per day" for beverages. Portion sizes are often assessed by asking respondents to mark "small," "medium," or "large" in comparison to a given medium portion size. However, some questionnaires only ask about the frequency of intake of a "usual" portion size (e.g., 1 cup milk). In the latter instances, respondents are asked to calculate the frequency of the amount given, rather than actual serving size consumed.

Summary *questions* ask about usual intake of fruits and vegetables because the long lists of these foods (needed to capture micronutrient intake) lead to overreporting of intake [26].

Note that development of an FFQ is a daunting, complex task requiring considerable understanding of exposure measurement in nutritional epidemiology, food composition knowledge, formatting and questionnaire design expertise, and computer programming resources [27, 28]. Other limitations of FFQs are that they may be fraught with considerable measurement error, which limits their usefulness [29, 30].

B. Other Dietary Assessment Tools Used in Nutritional Epidemiology: Recalls and Records

Food records and 24-hour dietary recalls are the two other common tools used in nutritional epidemiology [12]. Briefly, food records or diaries require individuals to record all foods and beverages consumed over a specified period of time, usually 3 to 7 days. Participants are asked to carry the record with them and to record foods as eaten in real time. Some protocols require participants to weigh or measure foods before eating and review the record with a registered dietitian, whereas less stringent protocols use food models or photographs and other aids to instruct respondents on estimating serving sizes and do not engage in extensive review or documentation [31]. Regardless of the data collection protocol, ultimately the food consumption information from records/diaries is entered into a specialized software program for calculation of nutrient intakes. This data-entry step is a time-consuming task and requires trained data technicians or registered dietitians. Food records are somewhat burdensome for study participants to complete. In the foreseeable future, the use of personal digital assistants (PDAs), digital cameras, mobile telephones, and other electronic devices both to record and transmit food record data will likely alleviate some of the participant burden [32]. Food records are also expensive to administer, so for large studies with tens of thousands of study participants they are usually cost prohibitive.

Twenty-four-hour dietary recalls are frequently used in population-based studies, including the National Health and Nutrition Examination (NHANES) surveys and "What We Eat in America" (33). Data from these recalls provide snapshots of U.S. eating patterns and are frequently used for formulations of dietary recommendations. A 24-hour dietary recall is a 20- to 30-minute interview in which the respondent is asked to recall all foods and beverages consumed over the previous 24 hours. Ideally the interview is conducted with real-time direct data entry into a software program for analysis. It is very important that the interviewer be well trained; tone of voice, body posture (when in person), and reactions to participant descriptions of foods consumed can influence the quality of the data.

A single recall is suitable for group data, such as "What We Eat in America" [33]. However, for individual assessment, day-to-day variability of food intake is so high that for both records and recalls several days of data are required to characterize usual intake for an individual. Using data on variability in intake from food records completed by 194 participants in the Nurses Health Study [12], the number of days needed to estimate the mean intakes for individuals within 10% of "true" means would be 57 days for fat, 117 days for vitamin C, and 67 days for calcium. For estimating food consumption for individuals, variability can be even greater. For example, the number of days needed to estimate the following foods within 10% of "true" means would be 55 days for white fish and 217 days for carrots. Unfortunately, research has shown that reported energy intake, nutrient intake, and recorded numbers of foods decrease with as few as 4 days of recording dietary intake [34]. These changes may reflect reduced accuracy and completeness of recording intake or actual changes in dietary intake to reduce the burden of recording intake. For group data, however, recalls have consistently provided good population-level food intake data that characterize American dietary habits. For example, NHANES 24-hour recall data have shown that portion sizes of all foods, as well as frequency of snacking and beverage consumption, have increased over time, possibly contributing to the obesity epidemic since the 1970s [35–37]. These findings are now being used to formulate public health recommendations about meal size and beverage consumption frequency [36].

C. Vitamin/Mineral Supplement Assessment

Historically, less attention has been paid to measuring vitamin/mineral supplement use compared to food intake. However, assessing vitamin/mineral supplement use is important because supplement use per se is an exposure of interest for the risk of several chronic diseases [38–42]. In addition, supplements are used by about half of all Americans, so they contribute a large proportion of total (diet plus supplement) micronutrient intake [27, 43–45].

Epidemiological studies typically use personal interviews or self-administered questionnaires to obtain information on three to five general classes of multiple vitamins (multivitamins with or without minerals, stress supplements, antioxidant mixtures, and other mixtures, including multivitamins with herbals), single supplements, the dose of single supplements, and sometimes frequency or duration of use [46].

In a validity study comparing a self-administered assessment method to label transcription among 104 supplement users, correlation coefficients ranging from 0.1 for iron to 0.8 for vitamin C were found [46]. The principal sources of error were investigator error in assigning the micronutrient composition of multiple vitamins and respondent confusion regarding the distinction between multiple vitamins and single supplements. These results suggest that commonly used epidemiological methods of assessing supplement use may incorporate significant amounts of error in estimates of some nutrients. In a subsequent study, we found that a similar inventory reporting system captured supplement use when compared to blood, toenail, and urine biomarkers [44, 47]. In a marketplace that is rapidly becoming more complex, with vitamins, minerals, and botanical compounds combined in unusual mixtures at highly variable doses, the association of dietary supplements with disease risk is becoming increasingly difficult, but important, to assess [48].

D. Use of Biomarkers in Nutritional Epidemiology

Dietary biomarkers are critical to the advancement of nutritional epidemiology [29, 49]. Because biomarkers from blood, urine, or stool specimens may serve as objective measures of diet, in theory, some of the problems associated with self-reported diet such as measurement errors, underreporting, and missed or forgotten foods may be avoided. The drawbacks of biomarkers are that biological specimens are expensive to collect, store, and analyze in large studies, and not all foods or nutrients have suitable biomarkers. For example, biomarkers fall into one of two general categories: recovery biomarkers and concentration biomarkers. Recovery biomarkers are those that have a known quantitative time-associated relationship between dietary intake and excretion or recovery in human waste (e.g., urine or feces) [50, 51]. On the other hand, concentration biomarkers (blood concentrations of vitamins, minerals, and carotenoids) are responsive to diet and generally have a linear association with diet [52, 53], but they cannot be used to estimate absolute intake in the same manner as recovery biomarkers [50, 51].

Studies using doubly-labeled water as a recovery biomarker to estimate energy expenditure have found significant underreporting and person-specific biases in nutrient estimates, such as the tendency for obese persons to underestimate dietary intake [54–56]. Identification and understanding of the effect of these person-specific biases are the major challenges now facing nutrition studies. Movement in this field, however, has been hampered by the lack of practical and available biomarkers. For example, there is no established biomarker for total fat or carbohydrate intake. In addition, many biomarkers (e.g., serum β-carotene for total fruits and vegetables) are concentration (qualitative) biomarkers, as opposed to quantitative biomarkers, and, therefore, are of limited usefulness in assessing overall or person-specific biases in self-report.

However, there is a growing awareness of the importance of biomarker substudies for the interpretation of observational nutritional epidemiological studies. For example, the Women's Health Initiative has a sizable substudy that includes doubly-labeled water measures to assess total energy expenditure and 24-hour urinary nitrogen to assess protein intake [57, 58]. These types of studies are an important step toward strengthening the reliability and interpretability of epidemiological studies of diet and disease.

III. STUDY DESIGNS USED IN NUTRITIONAL EPIDEMIOLOGY

Epidemiological studies can be divided into two general types: observational and experimental. The three primary observational study designs are ecological, case-control, and cohort studies. In human studies, the main experimental studies are called intervention trials, or randomized controlled trials. An overview of these study designs is given here in relation to nutritional epidemiology.

A. Observational Studies

1. Ecological and Migrant Studies

Important hypothesis-generating studies have examined the relationship, between countries, of national estimates of per capita supply of foods (e.g., dietary fat) with time-lagged rates of cancer or heart disease incidence or mortality [59–62]. These analyses strongly suggest that dietary fat intake increases risk, whereas plant foods decrease risk of these major diseases. However, it must be noted that (1) estimates of per capita intake from food disappearance data are extremely imprecise and include nonhuman consumption uses such as livestock feed and manufacturing use of food or food end products (e.g., corn biofuel, soybean-based inks used in newsprint), (2) it is generally not feasible to control for other differences between the countries (e.g., differences in physical activity levels or smoking prevalence), and (3) it is unknown whether the individuals within the countries that are exposed to specific dietary factors are the same individuals experiencing the disease. Migrant studies have often shown that with a single move from less-developed to Westernized countries, large and significant increases occur in risk of several chronic diseases, such as breast cancer [63, 64]. These changes occur rapidly, often after just one generation as immigrants become acculturated to the diet and other habits of their new country (65). Migrant studies offer strong evidence to support an important role for lifestyle and environment exposures as disease risk factors; however, few such studies have included pertinent dietary assessment to be able to address these questions properly [65].

2. Case-Control Studies

In a case-control study, individuals are identified and studied according to a single disease outcome. Specifically, individuals who have recently been diagnosed with a disease (e.g., colon cancer) are asked about their past exposure to diet and other risk factors and often provide a blood sample. A comparable set of control individuals, usually drawn from the same population, is also enrolled in the study and the individuals are asked about their past exposures. The two groups (those with and without the disease) are compared for differences in dietary intake and other exposures. The major advantage of this design is that an entire study can be completed in just a few years with a smaller sample size than is needed for other study designs (could be as small as 200 cases and controls). However, this study design can only answer questions about a single disease outcome. In addition, these studies can introduce potentially serious biases. Two major concerns with case-control studies are recall bias and selection bias. In studies of chronic disease, investigators typically ask participants to recall behavior and other exposures (e.g., dietary intake) from 5 to 10 years, or earlier in the past. Bias can occur when cases recall exposure to potential risk factors differently than controls. Selection bias occurs when controls agree to join the study because of an interest in health and are therefore more likely to exhibit healthy behavior (e.g., eat healthful diets, are physically active). The higher prevalence of healthy behavior in the controls appears to be associated with reduced risk of disease when actually it is associated with willingness to participate in a research study on health. Thus, control selection is an extremely important part of study design [1, 66]. Further, because cases are usually recruited relatively soon after diagnosis, unless remote diet is assessed, the dietary habits reported over the previous year or more recent time frame actually represent dietary intake in the preclinical phase of disease. Inferences from such data are not clear with respect to understanding diet-disease relationships. Another problem with case-control studies is that biomarkers of diet (e.g., serum micronutrient concentrations) are potentially affected by the disease process and therefore may not be reliable indicators of long-term status (e.g., risk) in cases. This problem is partially overcome in nested case-control studies described next.

3. Prospective Cohort Studies

The prospective cohort study typically enrolls people who are free of disease, assesses baseline risk factors, and follows the participants over time to monitor disease occurrence. The major advantage of cohort studies is that exposure to potential risk factors is assessed before the development of disease [1, 66]. Therefore, exposures such as self-reported dietary intake or nutritional biomarkers from blood samples cannot be influenced by the disease process. In addition, cohort studies can examine many exposures in relation to different disease outcomes. A cohort study is generally a large enterprise because most diseases affect only a small proportion of a population, even if the population is followed for many years. These studies typically have sample sizes exceeding 50,000, can have a total cost in excess of $100 million, and require that the cohort be followed for 10 or more years [1, 66]. Despite the cost and logistics, cohort studies have been useful in establishing diet-disease risk factors with ensuing recommendations for public health [67–72].

Because of the large size of these studies, the analysis of biological markers (e.g., serum micronutrient concentrations) for all participants is prohibitively expensive. Therefore, cohort studies often archive (e.g., bank) serum or plasma, white blood cells, toenails, DNA, or other biological specimens for the purpose of conducting nested case-control studies in the future [73, 74]. In a nested case-control study, a sample of cohort participants who develop a disease such as breast cancer (e.g., cases) are matched with other individuals in the cohort who do not develop the disease (e.g., controls). Biological samples from cases and controls are retrieved and analyses are performed to determine whether there are differences in prevalence of exposures (e.g., diet) between the cases and the controls [75, 76]. This can be an efficient and powerful study design that avoids many of the pitfalls of the classic case-control studies.

B. Intervention Trials

Intervention trials prospectively examine the effect of an exposure randomly assigned by the investigators, such as a low-fat diet or a particular dietary supplement, on an outcome such as disease occurrence, risk factor for a disease, or a biomarker. An important consideration when designing these studies is the degree of dietary control needed: controlled diet provided by the investigators versus vitamin supplementation versus dietary counseling. The stringency of dietary control is determined in part by the expected size of the response (e.g., change in disease risk) and the length of the treatment period required. For example, if the required dietary treatment period exceeds several months, a controlled feeding study is usually not logistically or financially viable. It is also important to note that, with the exception of dietary supplement intervention trials, most dietary interventions are not double-blinded. If the study compares a low-fat eating pattern to usual diet, for example, each participant will know to which arm he or she has been randomized, because they are being asked to make specific dietary changes. Further, as with any intervention trial, one must account for "drop-in" and "drop-out" rates. Some study participants may find the required intervention activities too burdensome, so they may drop out or be less than 100% adherent to study activities [77, 78]. Control participants, on the other hand, even if they are asked not to change their diet or take any new dietary supplements, may begin new dietary patterns that could be similar to the intervention. Both drop-in and drop-out phenomena can diminish the amount of contrast between the intervention and comparison groups, thereby attenuating any actual effects of the intervention.

In an intervention trial, the random assignment of participants to the control versus the intervention group means that participants with predisposing conditions or unmeasured factors that might influence the outcome are equally likely to be randomized into the intervention or the control group. As such, there is usually less uncontrolled confounding in randomized intervention studies [66]. In addition, random allocation of the exposure eliminates the possibilities of selection bias and recall bias. However, these projects are usually expensive and labor intensive and, therefore, are only conducted for important public health questions where the observational data are suggestive, but a true experiment is needed before issuing public health recommendations. Unlike observational studies, randomized trials are the only epidemiological study design where cause and effect may be inferred [1, 66].

IV. INTERPRETATION OF CAUSE AND EFFECT IN NUTRITIONAL EPIDEMIOLOGY

Given that nutritional epidemiology is the study of dietary intake and its association with disease risk, we must use careful scientific judgment in determining when the strength of the evidence supports a causal link between the exposure and the outcome. When assessing causality, important considerations include (1) the main measure of association used in epidemiological studies; (2) the major alternative explanation for an observed association in observational studies, which is confounding; and (3) methods for assessing causality in studies of associations. Other important considerations include (1) biological plausibility; (2) temporal association; (3) the strengths of the association; (4) dose-response relationship; and (5) consistency with other studies [1, 66].

A. Measures of Association

The most commonly used measure of association between dietary intake and disease risk is relative risk. The relative risk (RR) estimates the magnitude of an association between the dietary exposure and disease and indicates the probability of developing the disease in the exposed group relative to those who are not exposed [1, 66]. For example, an RR of 1.0 indicates that the incidence of disease in both the exposed and unexposed groups is the same. An RR greater than 1.0 indicates a positive association. For example, an RR of 2.0 between dietary fat and colon cancer indicates that individuals eating a high-fat diet are twice as likely to develop colon cancer as those eating a low-fat diet. RRs less than 1.0 are typically considered protective. An RR of 0.5 for the association of vegetable intake with colon cancer risk indicates that among individuals with diets high in vegetables, the risk of colon cancer is approximately half compared to those with diets low in vegetables. Often RRs are given for the highest category of intake (e.g., highest quartile of fat or vegetable intake) in comparison to the lowest category of intake.

Given the degree of measurement error in dietary intake estimates, RRs in nutritional epidemiology rarely exceed 3.0. RRs are typically presented with their associated confidence interval (e.g., RR 2.0, 95% confidence interval of 1.3 to 2.9), which provides information on the precision of the point estimate (e.g., the RR). Specifically, it is the range within which the true point estimates lie with a certain degree of assurance. Typically 95% confidence intervals (CIs) are given, which corresponds to the traditional test of statistical significant, $p < 0.05$, meaning that the there is less than a 5% probability that the findings occurred by chance. A 95% CI that does not include the null value (1.0) is, by definition, statistically significant at the $p = 0.05$ level. The width of the CI also provides information about the variability in the point estimate, which is a function of sample size. Therefore, the wider the CI, the more variability in the measure, the smaller the sample size, and the less confidence we can have that the observed point estimate is the true point estimate.

It is important to separate the strength of an RR from its public health relevance. For example, a large RR (e.g., RR 5.0) might be observed between a certain food, dietary supplement, or food component, such as an additive, and a risk of disease. However, if consumption of that food is rare, then its overall influence on the population's total morbidity or mortality will be minimal. Conversely, an RR of 1.5 might be very important from a public health perspective if the dietary exposure is common. Once RR estimates are used to determine the strength of association, then projections of the consequence of an exposure on public health (termed "population attributable risk") become important in the development of policy and allocation of resources. For example, the consistent observations from observational studies that *trans* fats were associated with unfavorable serum lipid profiles and cardiovascular disease led to new food labeling laws requiring that *trans* fats be listed among the "Nutrition Facts" [5].

B. Confounding

Confounding occurs when an observed association between the dietary intake and disease is actually due to other differences in the exposed versus the nonexposed groups but not due to the dietary factor itself [1, 66]. Confounding is a critical concept in nutritional epidemiology because it is plausible that people who choose one behavior (e.g., healthful diets) might differ from those who did not choose that behavior with regard to other exposures (such as physical activity) [12].

For example, a population-based study of participant characteristics associated with dietary supplement use among 1449 adults confirmed that supplement users were more likely to be female, older, better educated, nonsmokers, regular exercisers, and to consume diets higher in fruits and vegetables and lower in fat [13]. In this sample, other behaviors observed were previously unreported associations of supplement use with cancer screening, use of other chemopreventive agents (hormone replacement therapy and aspirin), and a psychosocial factor: belief in a diet–cancer connection. These relationships could confound studies of supplement use and cancer risk in complex ways. For example, male supplement users were more likely to have had a prostate-specific antigen test, which is associated with increased diagnosis of prostate cancer [79]. Therefore, supplement use could spuriously appear to be associated with increased incidence of prostate cancer.

The observed relationship between supplement use and belief in a connection between diet and cancer is especially interesting. Health beliefs influence cancer risk through behavior such as diet and exercise. For example, in a previous prospective study, it was observed that belief in a connection between diet and cancer was a statistically significant predictor of changes to more healthful diets over time [80]. In cohort studies, the increasing healthfulness of supplement users' diets and other health practices over time could result in a spurious positive association between supplement use and chronic disease risk.

It is important to note that in studies in which nutrient intake is summed from foods and supplements, the intake of micronutrients in the highest exposure category often appears too high to be obtained from food alone and probably reflects supplement use (see Chapter 2). Therefore, studies of nutrient intake may also be confounded by the relationship between supplement use and healthful lifestyle. In these studies, consistency of findings for the nutrient from foods and vitamin supplements separately would increase our confidence that an observed association was not confounded by supplement users' healthful lifestyles.

In theory, statistical adjustment in analyses for participant characteristics and major health-related behavior can control for some of the effects of confounding factors. However, the absence of residual confounding cannot be assured, especially if other important confounding factors are unknown, not assessed at all, assessed with error, or not included in the analyses

C. Evidence of Causality

Epidemiology is the study of associations, and statistical methods provide the means to conduct hypothesis testing to quantify the association. However, it is important to note that the existence of a statistically significant association does not indicate that the observed relationship is one of cause and effect. For any observed association, the following questions should be considered:

- How likely is it that the observed association is due to chance?
- Could this association be the result of poor study design, poor implementation, or inappropriate analysis?
- How well do these results meet other criteria of causality, as given in Table 2 [1]? Specifically, is the association weak or strong? Is there a plausible biological mechanism? Did the exposure precede the outcome? Is there a dose-response relationship?
- How well do these results fit in the context of all available evidence on this association? Causality is supported when a number of studies, conducted at different times, using different methods, among different populations, show similar results. Note that true causality can only be inferred within the context of an experimental study, a randomized controlled trial.

In a field characterized by as much uncertainty as nutritional epidemiology, it is rare for a cause-and-effect relationship to be considered unequivocal. However, lack of complete certainty does not mean that we should ignore the information that we have or postpone action that appears needed at a given time [1, 81, 82]. It merely means that we

TABLE 2 Criteria for Judging Whether Observed Associations between Diet and Disease Risk Are Causal [1, 66]

Criteria	
Strength of the association	The stronger the association, the less likely that it is due to the effect of an unsuspected or uncontrolled confounding variable.
Biological credibility	A known or postulated biological mechanism supports causality. However, an association that does not appear biologically credible at one time may eventually prove to be so. Implausible associations may be the beginning of the advancement of knowledge regarding mechanisms.
Time sequence	The exposure of interest must precede the disease outcome by a time span consistent with known biological mechanisms.
Dose–response	Evidence for a dose–response relationship (i.e., increased risk associated with increased exposure) is considered supportive of causality.

exercise prudence and thoughtful consideration before acting on epidemiological evidence.

V. OBSTACLES TO FINDING ASSOCIATIONS OF DIETARY INTAKE AND DISEASE RISK

Here we review the major obstacles to epidemiological research, including error in exposure assessment and limitations of study designs.

A. Sources of Error in Food Frequency Questionnaires

Table 3 gives a summary of the potential sources of error and bias in estimating dietary intake using an FFQ. Many of these errors are respondent based, including problems with memory, errors in frequency judgments and portion size estimation, and social desirability bias. The form itself is a major source of error because of limitations inherent in closed-ended scannable response options, the use of a limited food list (generally about 100 items) to minimize respondent burden, inadequate food composition information, and the requirement that respondents mentally average intake over long periods of time. Finally research indicates that dietary interventions themselves introduce reporting bias toward the more desirable responses [83].

1. Assessing the Reliability and Validity of Food Frequency Questionnaires

Reliability is generally used to refer to reproducibility, or whether an instrument will measure an exposure (e.g., nutrient intake) in the same way twice on the same respondents. Validity, which refers to the accuracy of an instrument, is a considerably higher standard. A convergent validity study compares a practical, epidemiological measurement method (e.g., an FFQ) with a more accurate but more burdensome method (e.g., food records). Reliability

TABLE 3 Sources of Error or Bias in Dietary Intake Estimates from a Food Frequency Questionnaire

Source of Error	Type of Error	Reason for Error
Participant	Memory	Unable to recall food consumption. This error increases with interval of memory required.
	Frequency judgments	Respondent has cognitive difficulty accurately providing this information. May be a particular problem in low-literacy respondents.
	Question comprehension	Respondent may not understand what foods are being asked about, understand the frequency categories, or be able to estimate portion sizes.
	Response errors	Respondent mistakenly codes incorrect frequency or skips questions.
	Portion size errors	Respondent cannot conceptualize reference portion size or his or her own portion sizes.
	Social desirability bias	Respondent unintentionally (or intentionally) misrepresents dietary intake in order to please investigators. For example, obese participants may underestimate intake.
Questionnaire (investigator)	Food list	Food list is too short or not appropriate for population being studied, and therefore dietary intake data are incomplete.
	Food groups	Food groups may not appropriately group foods by nutrient composition.
	Portion sizes	The reference portion size may be too large or small for the population such that they consistently over- or underestimate amounts of food consumed.
	Categorization of frequencies	Information is lost by using closed-ended categories (e.g., 2 to 4 times/week) instead of using an open-ended format.
	Poor design	Font is too small, skip patterns are not clear, or instructions are not clear.
	Database	Database may have incorrect nutrient values, incomplete nutrient values, or be missing important exposures altogether (e.g., isoflavones).
	Data collection and programming errors	Scanning errors may occur. Nutrient analysis program may contain errors. Data from incomplete FFQs are used in analysis.
	Seasonal variation	It may not be possible to adequately report intake of foods where intake varies markedly over seasons.
	Unusual dietary patterns	Respondents with unusual eating patterns (e.g., liquid diets) may not be able to accurately report dietary patterns.
Other	Intervention-associated bias	Respondents in an intervention are more likely to report socially desirable responses.

and convergent validity are typically investigated by means of statistical measures of bias and precision [10].

In a reliability study, reproducibility is assessed by comparing mean intake estimates from two administrations of the FFQ in the same group of respondents. If an instrument is reliable, the mean intake estimates should not vary substantially between the two administrations. In a convergent validity study, bias is generally assessed by comparing the mean estimates from an FFQ to those from food records or recalls in the same respondents. This comparison allows us to determine whether nutrient intake estimates from an FFQ appear to be generally under- or overreported in comparison to the criterion measure [10, 12]. Bias is especially important when the objective is to measure absolute intakes for comparison to dietary recommendations or some other objective criteria. For example, bias is critical when estimating how close Americans are to meeting the dietary recommendation to eat five servings of fruits and vegetables per day.

Precision is concerned with whether an FFQ accurately ranks individuals from low to high nutrient intakes, which is typically the analytical approach used to assess associations of dietary intake with risk of disease [12]. In this situation, bias in the estimate of absolute intake is not important as long as precision is good. In a reliability study, reproducibility is assessed as the correlation coefficient between nutrient intakes estimated from two administrations of the FFQ in the same group of respondents. In a convergent validity study, precision is the correlation coefficient between nutrient intake estimates from the FFQ in comparison to a criterion measure (usually dietary recalls or records). Often FFQ studies also assess convergent validity by ranking nutrient intake estimates, dividing them into categories (e.g., quartiles) and comparing these to similar categories calculated from another instrument. However, classifying a continuous exposure into a small number of categories does not reduce the effects of measurement error [10] and, therefore, this analysis does not provide additional information about correlation coefficients.

It is important to know that an instrument can be reliable without being accurate. That is, it can yield the same nutrient estimates two times and be wrong (e.g., biased upward or downward) both times. Alternatively, an instrument can be very reliable and consistently yield an accurate group mean (e.g., unbiased), but have poor precision such that it does not accurately rank individuals in the group from low to high in nutrient intake. Reliability is easy to measure, and nutrient correlation coefficients between two administrations of the same FFQ are generally in the range of 0.6–0.7. Estimates of reliability give an upper bound to the accuracy of an instrument. Whereas a high reliability coefficient does not imply a high convergent validity coefficient, a low reliability coefficient clearly means poor validity.

Studies comparing FFQs with records or recalls are often called ***convergent*** validation studies. The theory behind this type of study is that the major sources of error associated with FFQs are independent of those associated with short-term dietary recall and recording methods, which avoids spuriously high estimates of ***convergent*** validity resulting from correlated errors. As summarized by Willett [12], the errors associated with FFQs are the restrictions imposed by a fixed list of foods, perception of portion sizes, and the cognitive challenge of assessing frequency of food consumption over a broad time frame. These sources of error are only minimally shared by diet records, which are open-ended, do not depend on memory, and permit measurement of portion sizes. Biases in food records result from coding errors and changes in eating habits while keeping the records. Like food records, dietary recalls are open-ended. However, recalls are usually collected without advance notification. Therefore, participants cannot change what they eat retroactively and the instrument itself should not affect food intake. Bias in recalls results from estimation of portion sizes, participant memory, and coding errors. Nonetheless, it is apparent that there are correlated errors between FFQs and records or recalls [30, 56, 84]. Social desirability could influence how participants record or recall food intake across all types of dietary assessment instruments [83, 84]. Participant error in estimating portion sizes could bias recall and FFQ estimates of intake in similar ways. There are also correlated errors in nutrient databases. For example, estimates of selenium intake from FFQs and food records are correlated, which is merely the result of correlated errors in the nutrient database. Finally, research using doubly-labeled water to determine energy requirements has demonstrated significant underreporting of energy intakes from FFQs and recalls that may vary by participant characteristics [55, 56]. It is important to be aware of limitations of records and recalls as criterion measures of dietary intake and to interpret cautiously results based on these measures [30].

2. Effects of Error in Food Frequency Questionnaires

Error in dietary assessment can be of two types, with markedly different consequences. Random error refers to mistakes such as inadvertently marking the wrong frequency column, skipping questions, and lapses in judgment. These errors introduce noise into nutrient estimates such that our ability to find the "signal" (e.g., an association of dietary fat and breast cancer) is masked or attenuated (biased toward no association).

Systematic error refers to under- or overreporting of intake across the population (e.g., bias), but also person-specific sources of bias. For example, studies indicate that obese women are more likely to underestimate dietary intake than normal-weight women [54, 55, 85]. Systematic error may result in either null associations or spurious associations. In one report, Prentice used data from FFQs collected in a low-fat dietary intervention trial to simulate the effects of random and systematic error on an association of dietary fat and breast cancer, where the true RR was assumed to be 4.0 [86]. Assuming only random error exists in the estimate of fat intake, the projected (i.e., observed) RR for fat and breast cancer would be 1.4. Assuming both random error and systematic error exists, the projected RR would be 1.1, similar to that reported in a pooled analysis of cohort studies [87]. These results clearly suggest that FFQs may not be adequate to detect many associations of diet with disease, even if a strong relationship exists [88, 89]. Therefore, it is not surprising that results from diet–chronic disease studies are often null or conflicting, given the error in our dietary assessment methods [58, 86].

TABLE 4 Estimates of the Observed Association[a] between Dietary Fat Intake (per 10 Grams of Fat) and BMI after Adjustment for Random Measurement Error in the Measure of Dietary Fat

Correlation Coefficient[b] between the FFQ Estimate and "True" Fat Intake	Observed Increase in BMI for Every 10 g of Fat Consumed[c]
1.00 (FFQ is perfect measure of fat intake)	4.0[c]
0.70 (FFQ is a good measure of fat intake)	2.8
0.50 (FFQ is a weak measure of fat intake)	2.0
0.30 (FFQ is a poor measure of fat intake)	1.2

[a] $\beta_{observed} = \beta_{true} \times$ validity coefficient.

[b] Correlation coefficient from validity study comparing FFQ to multiple 24-hour recalls.

[c] Assume true regression coefficient from a multivariate model predicting BMI equals 4.0.

B. Limitations in Research Designs

1. Observational Studies

In studies of nutritional epidemiology, unique obstacles exist to finding clear and interpretable relationships between dietary intake and disease risk [58, 81]. In roughly increasing order of importance, these obstacles include the following:

- Current or recent dietary intake may differ from intake over the time frame relevant to the development of disease, which will reduce our ability to find associations between diet and disease.
- Certain nutrient intakes within a population may not be highly variable. For example, energy from dietary fat in a population of postmenopausal women may only vary from 25% to 40%, resulting in inadequate range of disease risk to find an association with breast cancer. This situation is akin to assessing whether smoking causes cancer by studying men who smoke one pack per day in comparison to men who smoke 1.5 packs per day. Minimal heterogeneity in exposures provides insufficient contrasts.
- Diet is a complex mixture of foods and nutrients, including many highly correlated compounds, making it difficult to separate the effects of any one compound from other dietary factors.
- Dietary intake may relate in a complicated manner to other risk factors such as hormonal status, obesity, or hypertension. These relationships (some of which may be in the causal pathway) make it difficult to appropriately control for confounding factors.
- Measurement properties of existing dietary self-report instruments are largely unknown, although it is clear that there are many sources of random error and systematic error, both of which obscure our ability to find associations between dietary intake and disease risk.

An important point to consider is that most of the obstacles listed above will limit or attenuate our ability to find associations between dietary intake and disease. For example, as shown in Table 4, an observed association of dietary fat intake with body mass index (BMI) might appear too small to be clinically important. However, if we assume that significant measurement error exists in our estimate of fat intake (e.g., a correlation of 0.30 between our measure and "true" intake), then the real association would be 4.0 BMI points per 10 g of fat intake, which is considerably more important. Therefore, studies showing weak or no associations between dietary intake and disease (e.g., null results) need to be interpreted cautiously.

Even this cursory review of the obstacles to interpretation of observational studies of diet and disease makes it clear that these studies alone may not provide reliable information on the associations of dietary intake and disease, regardless of their size or duration.

2. Limitations of Clinical Trials of Dietary Intake and Disease Risk

In spite of the many desirable features of dietary intervention trials, unique obstacles are present in these types of studies, as summarized below.

The costs of a long-term dietary intervention trial can be formidable. For example, the National Institutes of Health–sponsored Women's Health Initiative (WHI) tested whether a "low-fat eating pattern" would reduce the incidence of breast cancer, colorectal cancer, and coronary heart disease among 48,837 postmenopausal women in the United States [90–92]. The dietary intervention required participants to attend monthly sessions (run by specially trained nutritionists) for the first 18 months and then quarterly classes for the remainder of the trial, about 8.5 years in total [93].

In addition, new intervention components were added to the trial to encourage adherence. The costs of implementing this type of intervention far exceed those required for comparatively simple pill-placebo trials or observational studies.

Maintenance of dietary adherence for a sufficient period of time to be able to ascertain clinical outcomes (e.g., disease risk) can be a formidable task. On one hand, the greater the contrast in dietary intake between the intervention and control groups, the more likely the study will be able to detect an effect on the outcome. However, it is clearly more difficult to get participants to adhere to very strict or limited regimes, which can result in such poor adherence that the trial becomes futile [94]. Monitoring of dietary adherence typically requires use of self-reported dietary instruments, with their attendant weaknesses (discussed earlier).

VI. FUTURE RESEARCH DIRECTIONS

As is apparent from this overview of nutritional epidemiology, the biggest challenge is that of addressing random, systematic, and person-specific sources of error in dietary assessment. Only when well-designed true validity studies clarify these sources of error will we be able to markedly improve our ability to draw valid inferences from epidemiological studies of diet and disease.

An exciting area of research concerns diet-gene and diet-environment interactions in the etiology and pathogenesis of many chronic diseases. Despite the vigorous investigation of environmental causes of disease, it has long been recognized that not all persons exposed to the same risk factors will develop the associated disease [95, 96]. For example, although it is well accepted that smoking causes lung cancer, only 10% to 15% of smokers will be diagnosed with the disease in their lifetime. More and more, we are beginning to understand the impact of differential genetic susceptibility in the etiology and pathogenesis of common diseases such as coronary heart disease and cancer. If only a subgroup of individuals is sensitive to certain dietary exposures, the effect will be diluted and the association will be undetectable when the entire population is the focus of study. Better understanding of these individual susceptibilities has the potential to bring considerable clarity to nutritional epidemiological research. Another exciting area is the use of new technology to assess the influence of diet on the various "omics," such as proteomics, metabolomics [97]. These small molecules may prove to be more informative biomarkers of diet and diet-disease relationships than simple assessment of blood nutrients.

To summarize, despite the challenges in nutritional epidemiology and the measurement error issues that have impeded progress in the field, nutritional epidemiology studies have made important scientific contributions that have shaped public health policy and practice. To move the field forward, we must investigate strategies to improve methods of dietary assessment and reduce measurement error. While the complete elimination of error in dietary assessment methods is probably not a realistic objective, a better understanding of these errors (based on objective biomarkers), combined with statistical methods to address these errors, may be a reachable goal. It is the combined contribution of many different study types (e.g., observational, intervention, biomarker, mechanistic feeding studies, genetic susceptibility studies) that offers the greatest potential for identification of lifestyle strategies for disease prevention.

References

1. Hennekens, C. H., and Buring, J. E. (1987). "Epidemiology in Medicine." Little, Brown & Company, Boston/Toronto.
2. Mozaffarian, D., Gottdiener, J. S., and Siscovick, D. S. (2006). Intake of tuna or other broiled or baked fish versus fried fish and cardiac structure, function, and hemodynamics. *Am. J. Cardiol.* **97**, 216–222.
3. Mozaffarian, D., Katan, M. B., Ascherio, A., Stampfer, M., and Willett, W. C. (2006). Trans fatty acids and cardiovascular disease [review]. *N. Engl. J. Med.* **354**, 1601–1613.
4. Willett, W. C. (2006). Trans fatty acids and cardiovascular disease-epidemiological data [review]. *Atheroscler. Suppl.* **7**, 5–8.
5. Moss, J. (2006). Labeling of trans fatty acid content in food, regulations and limits: the FDA view [review]. *Atheroscler. Suppl.* **7**, 57–59.
6. Korver, O., and Katan, M. B. (2006). The elimination of trans fats from spreads: how science helped to turn an industry around. *Nutr, Rev.* **64**, 275–279.
7. Okie, S. (2007). New York to trans fats: you're out! *N. Engl. J. Med.* **356**, 2017–2021.
8. Khoury, M. J., Shaw, G. M., Moore, C. A., Lammer, E. J., and Mulinare, J. (1996). Does periconceptional multivitamin use reduce the risk of neural tube defects associated with other birth defects? Data from two population-based case-control studies. *Am. J. Med. Genet.* **61**, 30–36.
9. Anonymous (1999). Folic acid for the prevention of neural tube defects. American Academy of Pediatrics. *Pediatrics* **104**, 325–327.
10. Armstrong, B. K., White, E., and Saracci, R. (1992). "Principles of exposure measurement in epidemiology." Oxford University Press, Oxford.
11. Potter, J. D., Slattery, M. L., Bostick, R. M., and Gapstur, S. M. (1993). Colon cancer: a review of the epidemiology. *Epidemiol. Rev.* **15**, 499–545.
12. Willett, W. (1998). "Nutritional Epidemiology." Oxford University Press, New York.
13. Patterson, R. E., Neuhouser, M. L., White, E., Hunt, J. R., and Kristal, A. R. (1998). Cancer-related behavior of vitamin supplement users. *Cancer Epidemiol. Biomarkers Prev.* **7**, 79–81.
14. Lampe, J. W., Chen, C., Li, S., Prunty, J., Grate, M. T., Meehan, D. E., Barale, K. V., Dightman, D. A., Feng, Z., and Potter, J. D. (2000). Modulation of human glutathione *S*-transferases by

botanically defined vegetable diets. *Cancer Epidemiol. Biomarkers Prev.* **9**, 787–793.

15. Lampe, J. W., and Peterson, S. (2002). Brassica, biotransformation and cancer risk: genetic polymorphisms after the preventive effects of cruciferous vegetables. *J. Nutr.* **132**, 2991–2994.
16. Fowke, J. H., Shu, X. O., Dai, Q., Shintani, A., Conaway, C. C., Chung, F. L., Cai, Q., Gao, Y. T., and Zheng, W. (2003). Urinary isothiocyanate excretion, brassica consumption, and gene polymorphisms among women living in Shanghai, China. *Cancer Epidemiol. Biomarkers Prev.* **12**, 1536–1539.
17. Gasper, A. V., Al-janobi, A., Smith, J. A., Bacon, J. R., Fortun, P., Atherton, C., Taylor, M. A., Hawkey, C. J., Barrett, D. A., and Mithen, R. F. (2005). Glutathione S-transferase M1 polymorphism and metabolism of sulforaphane from standard and high-glucosinolate broccoli *Am. J. Clin. Nutr.* **82**, 1283–1291.
18. Brennan, P., Hsu, C. C., Moullan, N., Szeszenia-Dabrowska, N., Lissowska, J., Zaridze, D., Rudnai, P., Fabianova, E., Mates, D., Bencko, V., Foretova, L., Janout, V., Gemignani, F., Chabrier, A., Hall, J., Hung, R. J., Boffetta, P., and Canzian, F. (2005). Effect of cruciferous vegetables on lung cancer patients stratified by genetic status: a Mendelian randomisation approach. *Lancet* **366**, 1558–1560.
19. Fowke, J. H., Chung, F. L., Jin, F., Qi, D., Cai, Q., Conaway, C. C., Cheng, J-R., Shu, X-O., Gao, Y-T., and Zheng, W. (2003). Urinary isothiocyanate levels, brassica, and human breast cancer. *Cancer Res* **63**, 3980–3986.
20. Turner, F., Smith, G., Sachse, C., Lightfoot, T., Garner, R. C., Wolf, C. R., Forman, D., Bishop, D. T., and Barrett, J. H. (2004). Vegetable, fruit and meat consumption and potential risk modifying genes in relation to colorectal cancer. *Int. J. Cancer* **112**, 259–264.
21. Lampe, J. (2004). Nutrition and cancer prevention: small-scale human studies for the 21st century. *Cancer Epidemiol. Biomarkers Prev.* **13**, 1987–1997.
22. Sinha, R., Rothman, N., Brown, E. D., Mark, S. D., Hoover, R. N., Caporaso, N. E., Orville, L. A., Knize, M. G., Land, N. P., and Kadlubar, F. F. (1994). Pan-fried meat containing high levels of heterocyclic aromatic amines but low levels of aromatic hydrocarbons induces cytochrome P451A2 activity in humans. *Cancer Res.* **54**, 6154–6159.
23. Cross, A. J., Pollock, J. R., and Bingham, S. A. (2003). Haem, not protein or inorganic iron, is responsible for endogenous intestinal N-nitrosation arising from red meat. *Cancer Res.* **63**, 1258–1260.
24. Davis, S. R., Quinlivan, E. P., Shelnutt, K. P., Maneval, D. R., Ghandour, H., Capdevila, A., Coats, B. S., Wagner, C., Selhub, J., and Bailey, L. B. (2005). The methylenetetrahydrofolate reductase 677C→T polymorphism and dietary folate restriction affect plasma one-carbon metabolites and red blood cell folate concentrations and distribution in women. *J. Nutr.* **135**, 1040–1044.
25. Davis, S. R., Quinlivan, E. P., Stacpoole, P. W., and Gregory, J. F. (2006). Plasma glutathione and cystathionine concentrations are elevated but cysteine flux is unchanged by dietary vitamin B_6 restriction in young men and women. *J. Nutr.* **136**, 373–378.
26. Kristal, A. R., Vizenor, N. C., Patterson, R. E., Neuhouser, M. L., Shattuck, A. L., and McLerran, D. (2000). Precision and bias of food frequency-based measures of fruit and vegetable intakes. *Cancer Epidemiol. Biomarkers Prev.* **9**, 939–944.
27. Patterson, R. E., Kristal, A. R., Carter, R. A., Fels-Tinker, L., Bolton, M. P., and Agurs-Collins, T. (1999). Measurement characteristics of the Women's Health Initiative food frequency questionnaire. *Ann. Epidemiol.* **9**, 178–187.
28. Block, G., Hartman, A. M., Dresser, C. M., Carroll, M. D., Gannon, J., and Gardner, L. (1986). A data-based approach to diet questionnaire design and testing. *Am. J. Epidemiol.* **124**, 453–469.
29. Subar, A. F., Kipnis, V., Troiano, R. P., Midthune, D., Schoeller, D. A., Bingham, S., Sharbaugh, C. O., Trabulsi, J., Runswick, S., Ballard-Barbash, R., Sunshine, J., and Schatzkin, A. (2003). Using intake biomarkers to evaluate the extent of dietary misreporting in a large sample of adults: the OPEN Study. *Am. J. Epidemiol.* **158**, 1–13.
30. Kipnis, V., Midthune, D., Freedman, L. S., Bingham, S., Schatzkin, A., Subar, A. F., and Carroll, R. J. (2001). Empirical evidence of correlated biases in dietary assessment instruments and its implications. *Am. J. Epidemiol.* **153**, 394–403.
31. Kolar, A. S., Patterson, R. E., White, E., Neuhouser, M. L., Frank, L. L., Standley, J., Potter, J. D., and Kristal, A. R. (2005). A practical method for collecting 3-day food records in a large cohort. *Epidemiology* **16**, 579–583.
32. Wang, D.-H., Kogashiwa, M., Ohta, S., and Kira, S. (2002). Validity and reliability of a dietary assessment method: the application of a digital camera with a mobile phone card attachment. *J. Nutr. Sci. Vitaminol.* **48**, 498–504.
33. Dwyer, J., Picciano, M. F., and Raiten, D. J. (2003). Collection of food and dietary supplement intake data: what we eat in America-NHANES. *J. Nutr.* **133**, 590–600.
34. Rebro, S., Patterson, R. E., Kristal, A. R., and Cheney, C. (1998). The effect of keeping food records on eating patterns. *J. Am. Diet. Assoc.* **98**, 1163–1165.
35. Nielsen, S. J., Siega-Riz, A. M., and Popkin, B. M. (2002). Trends in energy intake in US between 1977 and 1996: similar shifts seen across age groups. *Obesity Res.* **10**, 370–378.
36. Nielsen, S. J., and Popkin, B. M. (2003). Patterns and trends in food portion sizes, 1977–1998. *JAMA* **289**, 450–453.
37. Zizza, C., Siega-Riz, A. M., and Popkin, B. M. (2001). Significant increase in young adults' snacking between 1977–1978 and 1994–1996 represents a cause for concern. *Prev. Med.* **32**, 303–310.
38. Satia-Abouta, J., Kristal, A. R., Patterson, R. E., Littman, A. J., Stratton, K. L., and White, E. (2003). Dietary supplement use and medical conditions: the VITAL study. *Am. J. Prev. Med.* **24**, 43–51.
39. Brennan, L. A., Morris, G. M., Wasson, G. R., Hannigan, B. M., and Barnett, Y. A. (2000). The effect of vitamin C or vitamin E supplementation on basal and H_2O_2-induced DNA damage in human lymphocytes. *Br. J. Nutr.* **84**, 195–202.
40. Neuhouser, M. L., Patterson, R. E., and Kristal, A. (2005). Dietary supplements and cancer risk: epidemiological research and recommendations. *In* "Preventive Nutrition" (A. Bendich, and R. J. Deckelbaum, Eds.), pp. 89–121. Humana Press, Totowa, NJ.
41. Kristal, A. R., Stanford, J. L., Cohen J. H., Wicklund, K., and Patterson, R. E. (1999). Vitamin and mineral supplement use is associated with reduced risk of prostate cancer. *Cancer Epidemiol Biomarkers Prev.* **8**, 887–892.

42. Leitzmann, M., Stampfer, M., Colditz, G. A., Willett, C. G., and Giovannucci, E. (2003). Zinc supplement use and risk of prostate cancer. *J. Natl. Cancer Inst.* **95**, 1004–1007.
43. Neuhouser, M. L., Kristal, A. R., Patterson, R. E., Goodman, P. T., and Thompson, I. M. (2001). Dietary supplement use in the Prostate Cancer Prevention Trial: implications for prevention trials. *Nutr. Cancer* **39**, 12–18.
44. Satia-Abouta, J., Patterson, R. E., King, I. B., Stratton, K. L., Shattuck, A. L., Kristal, A. R., Potter, J. D., Thornquist, M. D., and White, E. (2003). Reliability and validity of self-report of vitamin and mineral supplement use in the VITamins and Lifestyle Study. *Am. J. Epidemiol.* **157**, 944–954.
45. Ervin, R. B., Wright, J. D., and Kennedy-Stephenson, J. J. (1999). Use of dietary supplements in the United States, 1988–1994. *Vital Health Stat* **244**, 1–14.
46. Patterson, R. E., Kristal, A. R., Levy, L., McLerran, D., and White, E. (1998). Validity of methods used to assess vitamin and mineral supplement use. *Am. J. Epidemiol.* **148**, 643–649.
47. Satia, J. A., King, I. B., Morris, J. S., Stratton, K., and White, E. (2006). Toenail and plasma levels as biomarkers of selenium exposure. *Ann. Epidemiol.* **16**, 53–58.
48. Gunther, S., Patterson, R. E., Kristal, A. R., Stratton, K. L., and White, E. (2004). Demographic and health-related correlates of herbal and specialty supplement use. *J. Am. Diet. Assoc.* **104**, 27–34.
49. Bingham, S. A. (2003). Urine nitrogen as a biomarker for the validation of dietary protein intake. *J. Nutr.* **133**, 921S–924S.
50. Kaaks, R. (1997). Biochemical markers as additional measurements in studies of the accuracy of dietary questionnaire measurements: conceptual issues. *Am. J. Clin. Nutr.* **65**, 1232S–1239S.
51. Kipnis, V., Subar, A. F., Midthune, D., Freedman, L. S., Ballard-Barbash, R., Troiano, R. P., Bingham, S., Schoeller, D. A., Schatzkin, A., and Carroll, R. J. (2003). Structure of dietary measurement error: results of the OPEN biomarker study. *Am. J. Epidemiol.* **158**, 14–21.
52. Campbell, D. R., Gross, M. D., Martini, M. C., Grandits, G. A., Slavin, J. L., and Potter, J. D. (1994). Plasma carotenoids as biomarkers of vegetable and fruit intake. *Cancer Epidemiol. Biomarkers Prev.* **3**, 493–500.
53. Neuhouser, M. L., Patterson, R. E., King, I. B., Horner, N. K., and Lampe, J. W. (2003). Selected nutritional biomarkers predict diet quality. *Public Health Nutr.* **6**, 703–709.
54. Black, A., Bingham, S., Johansson, G., and Coward, W. (1997). Validation of dietary intakes of protein and energy against 24 hour urinary N and DLW energy expenditures in middle-aged women, retired men and post-obese subjects: comparisons with validation against presumed energy requirements. *Eur. J. Clin. Nutr.* **51**, 405–413.
55. Horner, N. K., Patterson, R. E., Neuhouser, M. L., Lampe, J. W., Beresford, S. A., and Prentice, R. L. (2002). Participant characteristics associated with errors in self-reported energy intake from the Women's Health Initiative food-frequency questionnaire. *Am. J. Clin. Nutr.* **76**, 766–773.
56. Johnson, R. K., Soultanakis, R. P., and Matthers, D. E. (1998). Literacy and body fatness are associated with underreporting of energy intake in US low-income women using the multiple-pass 24-hour recall: a doubly labeled water experiment. *J. Am. Diet. Assoc.* **98**, 1136–1140.
57. Neuhouser, M. L., Tinker, L. F., Schoeller, D., Beresford, S., Van Horn, L., Caan, B., Bingham, S., Assaf, A., Heiss, G., Kuller, L., Okene, J., Sarto, G., Satterfield, S., Stefanick, M., Thomson, C., Wactawski-Wende, J., and Prentice, R. L. (2006). Use of nutritional biomarkers to describe participant-related measurement error from dietary self-report in the Women's Health Initiative. *In* "Sixth International Conference on Dietary Assessment Methods: Complementary Advances in Diet and Physical Activity Assessment Methodologies" (B. L. Heitmann, L. Lissner, and A. Winkvist, Eds.), pp. AY01–01. Diet Research Foundation; The Danish Network of Nutritional Epidemiologists, Swedish Network for Nutritional Epidemiology, Copenhagen.
58. Prentice, R. L., Sugar, E., Wang, C. Y., Neuhouser, M. L., and Patterson, R. E. (2002). Research strategies and the use of nutrient biomarkers in studies of diet and chronic disease. *Public Health Nutr.* **5**, 977–984.
59. World Cancer Research Fund. (1997). Food, nutrition and the prevention of cancer: a global perspective. American Institute for Cancer Research.
60. Aldercreutz, H. (1990). Western diet and Western diseases: some hormonal and biochemical mechanisms and associations. *Scand. J. Clin. Lab. Invest.* **50**, 3–23.
61. Yu, H., Harris, R. E., Gao, Y.-T., Gao, R., and Wynder, E. L. (1991). Comparative epidemiology of cancers of the colon, rectum, prostate and breast in Shanghai, China versus the United States. *Int. J. Epidemiol.* **20**, 76–81.
62. Lee, M., Gomez, S., Chang, J., Wey, M., Wang, R., and Hsing, A. W. (2003). Soy and isoflavone consumption in relation to prostate cancer risk in China. *Cancer Epidemiol. Biomarkers Prev.* **12**, 665–668.
63. Shimizu, H., Ross, R. K., Bernstein, L., Yatani, R., Henderson, B. E., and Mack, T. M. (1991). Cancers of the prostate and breast among Japanese and white immigrants in Los Angeles County. *Br. J. Cancer* **63**, 963–966.
64. Pineda, M. D., White, E., Kristal, A. R., and Taylor, V. (2001). Asian breast cancer survival in the US: a comparison between Asian immigrants, US-born Asian Americans and Caucasians. *Int. J. Epidemiol.* **30**, 976–982.
65. Neuhouser, M. L., Thompson, B., Coronado, G. D, and Solomon, C. C. (2004). Higher fat intake and lower fruit and vegetable intakes are associated with greater acculturation among Mexicans living in Washington State. *J. Am. Diet. Assoc.* **104**, 51–57.
66. Rothman, K. J. (1986). "Modern Epidemiology." Little, Brown & Company, Boston/Toronto.
67. Willett, W. C., Stampfer, M. J., Colditz, G. A., Rosner, B. A., and Speizer, F. E. (1990). Relation of meat, fat, and fiber intake to the risk of colon cancer in a prospective study among women. *N. Engl. J. Med.* **323**, 1664–1672.
68. Leitzmann, M. F., Stampfer, M. J., Michaud, D. S., Augustsson, K., Colditz, G. C., Willett, W. C., and Giovannucci, E. L. (2004). Dietary intake of n-3 and n-6 fatty acids and the risk of prostate cancer. *Am. J. Clin. Nutr.* **80**, 204–216.
69. Gonzalez, A. J., White, E., Kristal, A., and Littman, A. J. (2006). Calcium intake and 10-year weight change in middle-aged adults. *J. Am. Diet. Assoc.* **106**, 1066–1073; quiz 1082.
70. Augustsson, K., Michaud, D. S., Rimm, E. B., Leitzmann, M. F., Stampfer, M. J., Willett, W. C., and Giovannucci, E. (2003).

A prospective study of intake of fish and marine fatty acids and prostate cancer. *Cancer Epidemiol. Biomarkers Prev.* **12**, 64–77

71. Calle, R., Rodriguez, C., Walker-Thurmond, K., and Thun, M (2003). Overweight, obesity, and mortality from cancer in a prospectively studied cohort of U.S. adults. *N. Engl. J. Med.* **348**, 1625–1638.
72. McCullough, M. L., Robertson, A. S., Rodriguez, C., Jacobs E. J., Chao, A., Carolyn, J., Calle, E. E., Willett, W. C., and Thun, M. J. (2003). Calcium, vitamin D, dairy products, and risk of colorectal cancer in the Cancer Prevention Study II Nutrition Cohort (United States). *Cancer Causes Control* **14** 1–12.
73. Kristal, A. R., King, I. B., Albanes, D., Pollak, M. N., Stanzyk, F. Z., Santella, R. M., and Hoque, A. (2005). Centralized blood processing for the selenium and vitamin E cancer prevention trial: effects of delayed processing on carotenoids, tocopherols, insulin-like growth factor-I, insulin-like growth factor binding protein 3, steroid hormones, and lymphocyte viability. *Cancer Epidemiol. Biomarkers Prev.* **14**, 727–730.
74. King, I. B., Satia-Abouta, J., Thornquist, M. D., Bigler, J., Patterson, R. E., Kristal, A. R., Shattuck, A. L., Potter, J. D., and White, E. (2003). Buccal cell DNA yield, quality, and collection costs: comparison of methods for large-scale studies. *Cancer Epidemiol. Biomarkers Prev.* **11**, 1130–1133.
75. King, I. B., Kristal, A. R., Schaffer, S., Thornquist, M., and Goodman, G. E. (2005). Serum trans-fatty acids are associated with risk of prostate cancer in beta-carotene and retinol efficacy trial. *Cancer Epidemiol. Biomarkers Prev.* **14**, 988–992.
76. Helzlsouer, K. J., Huang, H.-Y., Alberg, A. J., *et al.* (2000). Association between α-tocopherol, γ-tocopherol, selenium and subsequent prostate cancer. *J. Natl. Cancer Inst.* **92**, 2018–2023.
77. Women's Health Initiative Study Group. (2004). Dietary adherence in the Women's Health Initiative Dietary Modification Trial. *J. Am. Diet. Assoc.* **104**, 654–658.
78. Tinker, L., Patterson, R., Kristal, A., Bowen, D., and Taylor, V. (2001). Accuracy of two self-monitoring tools used in a low-fat intervention trial. *J. Am. Diet. Assoc.* **101**, 1031–1040.
79. Etzioni, R., Penson, D. F., Legler, J. M., di Tommaso, D., Boer, R., Gann, P. H., and Feuer, E. J. (2002). Overdiagnosis due to prostate-specific antigen screening: lessons from U.S. prostate cancer incidence trends. *J. Natl. Cancer Inst.* **94**, 981–990.
80. Patterson, R. E., Kristal, A. R., and White, E. (1996). Do beliefs, knowledge, and perceived norms about diet and cancer predict dietary change? *Am. J. Public Health* **86**, 1394–1400.
81. Prentice, R. L., Willett, W. C., Greenwald, P., Alberts, D., Bernstein, L., Boyd, N. F., Byers, T., Clinton, S. K., Fraser, G., Freedman, L., Hunter, D., Kipnis, V., Kolonel, L. N., Kristal, B. S., Kristal, A., Lampe, J. W., McTiernan, A., Milner, J., Patterson, R. E., Potter, J. D., Riboli, E., Schatzkin, A., Yates, A., and Yetley, E. (2004). Nutrition and physical activity and chronic disease prevention: research strategies and recommendations. *J. Natl. Cancer Inst.* **96**, 1276–1287.
82. Neuhouser, M. L. (2006). The long and winding road of diet and breast cancer prevention. *Cancer Epidemiol. Biomarkers Prev.* **15**, 1755–1756.
83. Kristal, A. R., Andrilla, C. H., Koepsell, T. D., Diehr, P. H., and Cheadle, A. (1998). Dietary assessment instruments are susceptible to intervention-associated response set bias. *J. Am. Diet. Assoc.* **98**, 40–43.
84. Hebert, J. R., Clemow, L., Pbert, L., Ockene, I. S., and Ockene, J. K. (1995). Social desirability bias in dietary self-report may compromise the validity of dietary intake measures. *Int. J. Epidemiol.* **24**, 389–398.
85. Heitmann, B. L., and Lissner, L. (1995). Dietary underreporting by obese individuals: is it specific or non-specific? *Br. Med. J.* **311**, 986–989.
86. Prentice, R. L. (1996). Measurement error and results from analytic epidemiology: dietary fat and breast cancer. *J. Natl. Cancer. Inst.* **88**, 1738–1747.
87. Smith-Warner, S. A., Spiegelman, D., Adami, H. O., Beeson, W. L., van den Brandt, P. A., Folsom, A. R., Fraser, G. E., Freudenheim, J. L., Goldbohm, R. A., and Graham, S. (2001). Types of dietary fat and breast cancer: a pooled analysis of cohort studies. *Int. J. Cancer.* **92**, 767–774.
88. Bingham, S., Luben, R., Welch, A., Wareham, N., Khaw, K. T., and Day, N. (2003). Are imprecise methods obscuring a relation between fat and breast cancer? *Lancet* **362**, 212–214.
89. Freedman, L. S., Potischman, N. A., Kipnis, V., Midthune, D., Schatzkin, A., Thompson, F. E., Troiano, R. P., Prentice, R. L., Patterson, R., Carroll R., and Subar, A. F. (2006). A comparison of two dietary instruments for evaluating the fat-breast cancer relationship. Int. *J. Epidemiol.* **35**, 1011–1021.
90. Prentice, R. L., Caan, B., Chlebowski, R. T., Patterson, R., Kuller, L. H., Ockene, J. K., Margolis, K. L., Limacher, M. C., Manson, J. E., Parker, L. M., Paskett, E., Phillips, L., Robbins, J., Rossouw, J. E., Sarto, G. E., Shikany, J. M., Stefanick, M. L., Thomson, C. A., Van Horn, L., Vitolins, M. Z., Wactawski-Wende, J., Wallace, R. B., Wassertheil-Smoller, S., Whitlock, E., Yano, K., Adams-Campbell, L., Anderson, G. L., Assaf, A. R., Beresford, S. A., Black, H. R., Brunner, R. L., Brzyski, R. G., Ford, L., Gass, M., Hays, J., Heber, D., Heiss, G., Hendrix, S. L., Hsia, J., Hubbell, F. A., Jackson, R. D., Johnson, K. C., Kotchen, J. M., LaCroix, A. Z., Lane, D. S., Langer, R. D., Lasser, N. L., and Henderson, M. M. (2006). Low-fat dietary pattern and risk of invasive breast cancer: the Women's Health Initiative Randomized Controlled Dietary Modification Trial [see comment]. *JAMA* **295**, 629–642.
91. Beresford, S. A., Johnson, K. C., Ritenbaugh, C., Lasser, N. L., Snetselaar, L. G., Black, H. R., Anderson, G. L., Assaf, A. R., Bassford, T., Bowen, D., Brunner, R. L., Brzyski, R. G., Caan, B., Chlebowski, R. T., Gass, M., Harrigan, R. C., Hays, J., Heber, D., Heiss, G., Hendrix, S. L., Howard, B. V., Hsia, J., Hubbell, F. A., Jackson, R. D., Kotchen, J. M., Kuller, L. H., LaCroix, A. Z., Lane, D. S., Langer, R. D., Lewis, C. E., Manson, J. E., Margolis, K. L., Mossavar-Rahmani, Y., Ockene, J. K., Parker, L. M., Perri, M. G., Phillips, L., Prentice, R. L., Robbins, J., Rossouw, J. E., Sarto, G. E., Stefanick, M. L., Van Horn, L., Vitolins, M. Z., Wactawski-Wende, J., Wallace, R. B., and Whitlock, E. (2006). Low-fat dietary pattern and risk of colorectal cancer: the Women's Health Initiative Randomized Controlled Dietary Modification Trial. *JAMA* **295**, 643–654.
92. Howard, B. V., Van Horn, L., Hsia, J., Manson, J. E., Stefanick, M. L., Wassertheil-Smoller, S., Kuller, L. H., LaCroix, A. Z., Langer, R. D., Lasser, N. L., Lewis, C. E., Limacher, M. C., Margolis, K. L., Mysiw, W. J., Ockene, J. K., Parker, L. M., Perri,

M. G., Phillips, L., Prentice, R. L., Robbins, J., Rossouw, J. E., Sarto, G. E., Schatz, I. J., Snetselaar, L. G., Stevens, V. J., Tinker, L. F., Trevisan, M., Vitolins, M. Z., Anderson, G. L., Assaf, A. R., Bassford, T., Beresford, S. A., Black, H. R., Brunner, R. L., Brzyski, R. G., Caan, B., Chlebowski, R. T., Gass, M., Granek, I., Greenland, P., Hays, J., Heber, D., Heiss, G., Hendrix, S. L., Hubbell, F. A., Johnson, K. C., and Kotchen, J. M. (2006). Low-fat dietary pattern and risk of cardiovascular disease: the Women's Health Initiative Randomized Controlled Dietary Modification Trial [see comment]. *JAMA* **295**, 655–666.

93. Ritenbaugh, C., Patterson, R. E., Chlebowski, R. T., Caan, C., Fels-Tinker, L., Howard, B., and Ockene, J. (2003). The Women's Health Initiative Dietary Modification Trial: overview and baseline characteristics of participants. *Ann. Epidemiol.* **13**, A87–A97.
94. Tinker, L. F., Perri, M. G., Patterson, R. E., Bowen, D. J., McIntosh, M., Parker, L. M., Sevick, M. A., and Wodarski, L. A. (2002). The effects of physical and emotional status on adherence to a low-fat dietary pattern in the Women's Health Initiative. *J. Am. Diet. Assoc.* **102**, 799–800.
95. Slattery, M. L., Kampman, E., Samowitz, W., Caan, B. J., and Potter, J. D. (2000). Interplay between dietary inducers of GST and the *GSTM-1* genotype in colon cancer. *Int. J. Cancer* **87**, 728–733.
96. Goode, E. L., Ulrich, C. M., and Potter, J. D. (2002). Polymorphisms in DNA repair genes and associations with cancer risk. Cancer Epidemiol. *Biomarkers Prev.* **11**, 1513–1530.
97. Milner, J. A., McDonald, S. S., Anderson, D. E., and Greenwald, P. (2001). Molecular targets for nutrients involved with cancer prevention. *Nutr. Cancer* **41**, 1–16.

CHAPTER 7

Analysis, Presentation, and Interpretation of Dietary Data

DEBRA COWARD-McKENZIE[1] AND RACHEL K. JOHNSON[2]
[1]*University of North Carolina at Chapel Hill, Chapel Hill, North Carolina*
[2]*University of Vermont, Burlington, Vermont*

Contents

I. INTRODUCTION

Nutritional epidemiological studies, while not an exact science, play a critical role in relating dietary intake to risk of disease. These investigations often require the gathering of dietary intake data from various samples, which must then be translated into a usable form. This chapter discusses what is done with the dietary data once collected. This includes *analysis*, the examination of the dietary data to determine the nutritional composition of the subjects' diets; *presentation*, the communication of the data and results in a logical format, such as comparing the results to a standard; and *interpretation*, the translation of the data and results—what do the data really tell us?

II. ANALYSIS OF DIETARY DATA

The methods most often used to obtain dietary intake information for research investigations include 24-hour dietary recalls, dietary records or diaries, and food frequency questionnaires (FFQs). The 24-hour dietary recalls and dietary records provide detailed descriptions of the types and amounts of foods and beverages consumed throughout a specified period of time, normally 1 to 7 days. The FFQ provides a less detailed list of selected foods and the frequency of their consumption in the past. (See Chapter 1 for further description of these methods.) The data received must then be analyzed to determine the total intake of nutrients or food components consumed by each subject.

A. Computer-Based Analysis

In the past, analysis was performed manually. This process was painstakingly tedious and expensive, requiring a highly trained person to code and enter data. Coding included looking up every food in a table to find a code number to be entered. Amounts were entered by unit and a multiplier. These coding techniques required many calculations to be performed by hand, leaving numerous possibilities for error. Now, a variety of computer-based food composition databases and nutrient computation systems are available in which the foods can be entered directly by name and computation of nutrient values is automated. The accuracy of the data obtained from these systems will differ, depending on several factors:

1. *Updating of the database.* New foods are constantly being introduced in the market, so the best databases are updated often to keep up with these changes. Virtually all databases use the U.S. Department of Agriculture (USDA) Nutrient Database for Standard Reference (SR) as their primary source of nutrient data. The SR contains information once published in the Agriculture Handbook 8, but it is no longer available in the printed form [1]. The 2006 release of the SR or number 19 [2] contains more than 130 nutrients for more than 7000 foods. Although the information is not complete, specific criteria have been established for evaluating foods to ensure the data are as accurate as possible [3, 4]. Many databases also add information from specific food manufacturers to provide information on name brand foods not available in the SR.
2. *The numbers and types of food items available.* This is particularly important for recalls and dietary records or for FFQs containing write-in sections where all foods must be assigned nutrient values. In regions with ethnocultural

diversity, special care must be taken when selecting databases [5]. Databases that contain a variety of ethnic foods will provide greater accuracy and will require less manual entry of nutrient values for foods.

3. *The ability to add foods or nutrients.* This is most important for those investigations in areas with multiethnicity or when there is a high tendency for the subjects to include restaurant foods that may not be included in the database. The ability to adapt or add recipe information should also be available. For example, if a subject had homemade beef stew for lunch, the database should allow the coder to either add or delete ingredients from an existing recipe or add a new recipe to the file.
4. *The ease of data entry and analysis.* It is no longer necessary to code diets manually. Systems should be easy to use to avoid unnecessary coding errors that can occur. Entry of products by name, particularly brand names, should be available. Some research databases, such as the Food Intake Analysis System [6], offer default options. These choices provide average estimations for foods for which exact information is not known. For example, if a subject had chicken breast but was not sure of the cooking method or the serving size, the coder has the default option to choose from instead of making guesses. These options can help decrease differences in nutrient intake values caused by multiple coders or data entry technicians.
5. *The nutrients available.* Not every database contains all nutrients and some contain more accurate data for particular nutrients. Systems should be evaluated for the accuracy of the nutrient values that are being studied. Analysis should include the option of choosing nutrient calculations for each food as well as summaries for an individual meal or day.
6. *The handling of missing nutrient values.* If a specific nutrient value is unknown for a particular food, the way the database handles the missing information may affect the accuracy of the nutrient information. Some systems impute values, whereas others simply use a value of zero. An imputed value is almost always a better estimation [7]. However, imputing nutrient values is a labor-intensive task and requires nutritionists with knowledge of data evaluation and imputing procedures. Therefore, caution must be taken when using databases with imputed values, because few database developers have access to qualified nutrition-trained personnel required for accurately estimating values [8].

When computing nutrient intake from food consumption data, it is assumed that the nutrient quality and content of certain foods are virtually constant and that what is consumed is available for use. However, we know that this assumption is not totally correct. There are various reasons why the actual value of a consumed nutrient may differ from the calculated value. The level of certain nutrients in foods may be affected by differences in growing and harvesting conditions (e.g., selenium [9, 10]), storage, processing, and cooking (e.g., vitamin C [11, 12]). Databases have tried to account for some of the differences by increasing the databanks to include preparation methods, cuts of meat, and specific manufacturers for processed food. For example, if chicken is entered into the database, the coder may have approximately 455 items from which to choose. This large number includes name brand foods, particular pieces of chicken available (e.g., breast or thigh), and cooking methods (e.g., baked or fried, cooked with skin on or off, skin eaten or not). Because so many choices are offered, recalls and records should be as detailed as possible to provide enough information to make an accurate selection.

The use of controlled feeding trials in a study, such as the Dietary Approaches to Stop Hypertension (DASH) trial [13], can help alleviate some of the differences between the calculated and actual nutrient values of food. The DASH trial was a multicenter study designed to compare the effects of dietary patterns on blood pressure. The subjects were asked to consume only foods prepared by the centers. Food procurement, production, and distribution guidelines were set and strictly adhered to at all sites to ensure that menus consistently met nutrient goals. For example, food items were given specific purchasing sizes, detailed descriptions, or defined brand names to ensure that all site recipes were of uniform composition [14]. Menu items were analyzed in a laboratory to obtain nutrient content values [15]. When possible, foods can be obtained from central suppliers to further eliminate any differences in nutritional content of foods due to regional variations in a study of this type.

The diet as a whole can also affect the availability of some nutrients. For example, high-fiber diets may decrease the availability of certain nutrients, such as zinc and iron [16, 17]. Computer-based analysis programs do not generally examine the overall diet and cannot determine how nutrient-nutrient interactions may affect availability. Iron, for example, is a mineral for which intake is not a good marker for availability. The absorption of iron is influenced by many components: (1) the source of iron (more heme iron is absorbed than nonheme iron); (2) the iron status of the individual (decreased iron stores increase absorption); and (3) the overall composition of the meal. These components play a role in determining how much of the iron consumed is available to the body [17]. In turn, iron consumption can also affect the absorption of other nutrients, such as zinc. Nutrient-nutrient interactions can greatly determine how well a calculated nutrient value represents the actual available amount of a nutrient.

Other factors that should be taken into account are drug-nutrient interactions and those people who may be malnourished or suffer from malabsorption. For example, the elderly are more likely to have a decreased ability to absorb vitamin B_{12} than do younger adults. The elderly population is also at higher risk for drug-nutrient interactions because they are often prescribed many medications. Researchers must be aware of any illnesses or medications taken by subjects that could interfere with nutrient absorption.

The development of fat-blocking drugs, such as Orlistat [18], which has been approved for the treatment of obesity, and fat replacers, such as Olestra [19], which has been approved by the U.S. Food and Drug Administration (FDA) in savory snacks, may decrease absorption of fat. Reduced absorption of fatty acids and other fat-soluble nutrients, such as vitamins A, D, E, and K may also occur. Although fat-soluble vitamins are added to Olestra-containing foods, actual absorption may still be affected for constituents not fortified in Olestra-containing foods such as carotenoids.

Although food composition databases are increasingly becoming more accurate and may be closer to actual values of energy intake than laboratory analysis [15], they cannot provide exact measurements for all nutrient intakes. Furthermore, even if these values are determined to be accurate, intake does not necessarily mean the nutrient is available for use. To obtain more accurate information on nutrient status, other methods, such as biomarkers (see Chapter 12), should be employed. Also, familiarity with the subjects' diets is essential for more accurate calculations. This includes, but is not limited to, factors such as supplement use, medications used, presence of diseases or illness, as well as special diets (e.g., vegetarian).

B. Total Diet Analysis

Because accurate measures of nutrient values based on food intake are difficult to calculate, some investigators choose to evaluate the overall diet. Although using these techniques will not provide exact numbers for specific nutrients, they will give investigators a better picture of overall diet quality and health risk. With the increasing evidence that other non-nutrient constituents, such as phytochemicals, may play a role in disease prevention [20], these indexes could prove to be very useful.

In reviewing the indexes of overall diet quality, Kant [21] found that there were three major approaches to the development of indexes: (1) derived from nutrients only, (2) based on foods or food groups, and (3) based on a combination of nutrients and foods. The definition of "diet quality" differs based on the attributes chosen by the investigators of each index [21], so the index chosen will depend on the needs of the study. These indexes based on nutrients only tend to look at consumption as a percentage of one of the nutrient-based reference values of the dietary reference intakes (DRI), such as the Recommended Dietary Allowance (RDA) as a marker for diet quality. Those based on foods and food groups examine the intake patterns of foods to identify patterns associated with adequacy [21].

Although numerous tools are available for examining the overall diet quality, those most commonly applied are based on the combination of nutrients and foods. These indexes, including the Healthy Eating Index (HEI) [22] and the Diet Quality Index (DQI) [23, 24], use the Dietary Guidelines for Americans [25] and the Food Guide Pyramid [26] to score the overall diet. These are based on the premise that if the USDA Food Guide Pyramid and guidelines are followed, including a variety of foods within each food group, the resulting diet will be adequate in nutrients and promote optimal health [27].

Patterson *et al.* [23] were among the first to relate diet quality to the Dietary Guidelines for Americans. The DQI included measures of eight food groups and the recommendations from the Committee on Diet and Health of the National Research Council Food and Nutrition Board, published in 1989 [28]. Haines *et al.* revised the index in 1999, now called the Dietary Quality Index–Revised (DQI-R) [24], to reflect the updated guidelines. However, no revision has been made to date to reflect the 2005 guidelines. The DQI has been altered for specific populations, including preschool-age children [29] and pregnant women [30], to allow for more accurate evaluation of diet quality based on intake recommendations for these populations.

The DQI-R incorporates both nutrients and food components to determine diet quality. It is based on 10 components, with a 100-point scale, each component worth 10 points (Table 1). Components are based on total fat and saturated fat as a percentage of energy; milligrams of cholesterol consumed; recommended servings for fruit, vegetables, and grains; adequacy of calcium and iron intake; dietary diversity; and dietary moderation. The dietary diversity score was developed to show differences in intake across 23 broad food group categories including seven grain-based products, seven vegetable components, two fruit and juice categories, and seven animal-based products [24]. Dietary moderation scores added sugars, discretionary fat, sodium intake, and alcohol intake. The DQI-R was designed to monitor dietary changes in populations but can provide an estimate of diet quality for an individual relative to the national guidelines and can note improvement or decline of diet quality with multiple calculations [24].

The HEI [22, 27] was first developed in 1995 by the USDA Center for Nutrition Policy and Promotion (CNPP) to assess and monitor the dietary status of Americans using the 1989 data from the Continuing Survey of Food Intake of Individuals (CSFII) [27]. It has undergone revision to reflect updated guidelines, including the most recent revision in 2006 [22]. HEI also uses both nutrients and food components to determine overall diet quality. The new HEI uses a density approach with the standards expressed as a percentage of calories or per 1000 calories (Table 2). Whereas the old HEI resembled the DQI-R, the new HEI differs in several areas. The HEI now contains 12 components, looking at not only the total intake of fruit, vegetables, and grains, but also taking into account whole fruits and grains and specific vegetables consumed. Cholesterol

TABLE 1 Diet Quality Index, Revised

Component	Max. Score Criteria[a]	Min. Score Criteria[a]
Total fat (% of energy intake)	≤30%	>40%
Saturated fat (% of energy intake)	≤10%	>13%
Dietary cholesterol	≤300 mg	>400 mg
% Recommended servings of fruit per day (2 to 4 based on energy intake)	≥100%	<50%
% Recommended servings of vegetables/day (3 to 5 based on energy intake)	≥100%	<50%
% Recommended servings of bread per day (6 to 11 based on energy intake)	≥100%	<50%
Calcium (% adequate intake for age)	≥100%	<50%
Iron intake (% 1989 RDA for age)	≥100%	<50%
Dietary diversity score	≥6	<3
Dietary moderation score	≥7	<4

Source: Adapted from Haines, P. S., Siega-Riz, A. M., and Popkin, B. M. (1999). The Diet Quality Index revised: a measurement instrument for populations. Copyright © The American Dietetic Association. Reprinted by permission from *Journal of the American Dietetic Association*, Vol. 99: 697–704.

[a]Scoring range for each component is 0 (min.) to 10 (max.).

TABLE 2 Healthy Eating Index, 2005—Components and Standards for Scoring[a]

Component	Maximum Points	Standard for Maximum Score	Standard for Minimum Score (0)
Total fruit (includes 100% juice)	5	≥0.8 cup equiv. per 1000 kcal	No fruit
Whole fruit (not juice)	5	≥0.4 cup equiv. per 1000 kcal	No whole fruit
Total vegetables	5	≥1.1 cup per 1000 kcal	No vegetables
Dark green and orange vegetables and legumes[b]	5	≥0.4 cup equiv. per 1000 kcal	No dark green or orange vegetables or legumes
Total grains	5	≥3.0 oz equiv. per 1000 kcal	No grains
Whole grains	5	≥1.5 oz equiv. per 1000 kcal	No whole grains
Milk[c]	10	≥1.3 cup equiv. per 1000 kcal	No milk
Meat and beans	10	≥2.5 oz equiv. per 1000 kcal	No meat or beans
Oils[d]	10	≥12 grams per 1000 kcal	No oil
Saturated fat	10	≤7% of energy	≥15% of energy
Sodium	10	≤0.7 gram per 1000 kcal[e]	≥2.0 grams per 1000 kcal
Calories from solid fat, alcohol, and added sugar (SoFAAS)	20	≤20% of energy[e]	≥50% of energy

Source: Reprinted from Guenther, P. M., Krebs-Smith, S. M., and Reedy, J., et al. (2006). The Healthy Eating Index-2005, CNPP- Fact Sheet No. 1. U.S. Department of Agriculture, Center for Nutritional Policy and Promotion, Alexandria, VA.

[a]Intakes between the minimum and maximum levels are scored proportionately, except for saturated fat and sodium (see note *e*).

[b]Legumes counted as vegetables only after meat and beans standard is met.

[c]Includes all milk products, such as fluid milk, yogurt, and cheese.

[d]Includes nonhydrogenated vegetable oils and oils in fish, nuts, and seeds.

[e]Saturated fat and sodium get a score of 8 for intake levels that reflect the 2005 Dietary Guidelines, <10% of calories from saturated fat and 1.1 grams of sodium/1000 kcal, respectively.

intake, total fat, and food variety components have been removed and replaced with "Oils" and "Calories from Solid Fat, Alcohol, and Added Sugars." Reliability analyses suggests that the new individual components provide additional insight to the summary score [22].

Categorizing foods into appropriate groups, particularly combination foods, can be a problem when using these analysis techniques. Cleveland *et al.* [31] developed a method for assessing food intakes in terms of food servings. These guidelines help overcome two major obstacles when assessing food intake with respect to the dietary guidelines, including dealing with food mixtures and differing units of measurement used. Because many foods are eaten as mixtures and are difficult to categorize into food groups, Cleveland *et al.* [31] developed a recipe file that helps break down food mixtures into ingredients so they can be assigned their respective groups more easily. Standard serving sizes were assigned gram weights to help overcome the units problem, allowing for the use of only one unit of measure. The 1994 USDA-CSFII database data file [32] incorporates these guidelines. Although combined foods must still be separated and coded before entering, this file does provide a more refined method for developing reproducible data [24].

III. PRESENTATION OF DATA

Once the data gathered from the dietary assessment methods have been analyzed, they must then be presented in a useful fashion. This is often done by comparing the analyzed data to a standard. These standards may include the Dietary Reference Intake (DRI) or comparison to a national average, such as the National Health and Nutrition Examination Surveys (NHANES). However, when comparing data for analyses, researchers must keep in mind differences that may exist between survey methods, questionnaire wording, data processing, and databases that could impact comparisons, and these differences should be taken into account.

The DRIs are a set of nutrient-based reference values. This set includes an estimated average requirement (EAR), a Recommended Dietary Allowance (RDA), and an adequate intake (AI), which are defined by nutrient adequacy and may relate to the reduction of the risk of chronic disease [33]. Once the EARs have been established they are used to set the new RDAs, which should be used as a daily intake goal by healthy individuals and should be sufficient to meet the needs of 97% to 98% of all healthy individuals. If there is not sufficient evidence to determine an EAR, then an AI is set, once again based on groups of healthy individuals. A tolerable upper intake level (UL) is also set where information is available as an indicator of excess for nutrients [33].

Each value has a specific goal and use [33] (Table 3). For example, the EAR is the estimate that is believed to meet the nutrient needs of half of the healthy individuals in a gender or life-stage group. When assessing nutrient intake of healthy groups, it is highly recommended that the EAR be used instead of the RDA [33].

DRIs are set for specific subgroups based on age and gender. They are to be applied to healthy populations and may not be adequate for those who are or have been malnourished or have certain diseases or conditions that increase nutrient requirements. For individuals, the RDA and AI can serve as a goal for nutrient intake. A more complete description of the DRIs can be found in Chapter 13 and the complete DRI tables are in the appendix.

Some researchers use national survey data as a standard when presenting dietary data. What We Eat in America (WWEIA) is the dietary intake component of NHANES. This survey is a joint effort between the U.S. Department of Agriculture (USDA) and the U.S. Department of Health and Human Services (DHHS), with data released in 2-year intervals. NHANES provides medical history, physical measurements, biochemical evaluation, physical signs and symptoms, and diet information from two 24-hour recalls. Table 4 provides an overview of the dietary information collected. Researchers may wish to compare results to the information obtained from these surveys to determine how their study sample compares to the national average. Although the data from these surveys may be applied to certain subgroups, such as specific age groups, gender, socioeconomic levels, education levels, and some ethnic groups, they cannot be used as guides for others, such as malnourished or specific disease states.

TABLE 3 Uses of Dietary Reference Intakes for Healthy Individuals and Groups

Type of Use	For the Individual	For a Group
Planning	**RDA**: Aim for this intake.	**EAR**: Use in conjunction with a measure of variability of the group's intake to set goals for the mean intake of a specific population.
	AI: Aim for this intake.	
	UL: Use as a guide to limit intake; chronic intake of higher amounts may increase risk of adverse effects.	
Assessment[a]	**EAR**: Use to examine the possibility of inadequacy; evaluation of true status requires clinical, biochemical, or anthropometric data.	**EAR**: Use in the assessment of the prevalence of inadequate intakes within a group.
	UL: Use to examine the possibility of overconsumption; evaluation of true status requires clinical, biochemical, or anthropometric data.	

Source: Reprinted with permission from the National Academies Press, Copyright 2005, National Academy of Sciences, Institute of Medicine and the National Academies (2005). "Dietary Reference Intakes for Energy, Carbohydrate, Fiber, Fat, Fatty Acids, Cholesterol, Protein, and Amino Acids." The National Academy Press, Washington, DC.

Key: EAR, estimated average requirement; RDA, recommended dietary allowance; AI, adequate intake; UL, tolerable upper intake level.

[a]Requires statistically valid approximation of usual intake.

TABLE 4 Overview of What We Eat in America (WWEIA)

Two days of 24-hour dietary recalls are collected using the USDA's computerized dietary data collection instrument, the Automated Multiple-Pass Method (AMPM).	Day 1: In-person interview in the mobile Examination Center (MEC); three-dimensional food models are available.
	Day 2: Telephone interview on a different day of the week than the MEC interview; USDA's Food Model Booklet and a limited number of three-dimensional models are provided.
Information Collected during the Interviews	
For each food and beverage consumed during previous 24-hour period	Detailed description
	Additions to the food
	Amount consumed
	What foods were eaten in combination
	Time eating occasion began
	Name of eating occasion
	Food source (where obtained)
	Whether food was eaten at home
	Amounts of food energy and 60+ nutrients/food components provided by the amount of food (calculated)
For each respondent on each day	Day of the week
	Amount and type of water consumed, including total plain water, tap water, and plain carbonated water
	Source of tap water
	Daily intake usual, much more or much less than usual
	Use and type of salt at table and in preparation
	Whether on a special diet and type of diet
	Frequency of fish and shellfish consumption (children 1 to 5 and women 16 to 49 years)
	Daily total intakes of food energy and 60+ nutrients/food components (calculated)

Researchers who choose to use the HEI or the DQI-R report their data by using the 100-point scale. For example, the HEI scores are based on the assumption that a "good" diet has a score greater than or equal to 80, a diet that "needs improvement" has a score of 51 to 80, and diets that score 50 or below are considered "poor" [34]. When this method was applied to the NHANES 1999–2000 data, the HEI mean score was 63.8 [34]. Researchers found that 74% of the population had diets that needed improvement, 10% had good diets, and 16% had poor diets. Researchers may also compare the scores from their study sample to the national scores for that time period, when available, to determine if their study population falls in line with the national average.

IV. INTERPRETATION OF DATA

Once the dietary intake data have been analyzed and compared to a standard, researchers must then look at the results to determine what the data really mean. How the data are interpreted can depend on the assessment method used and the nutrient being studied, the study type, and the accuracy of the subject's responses.

A. Assessment Methods

The assessment method chosen for use in a study can determine how the data collected can be interpreted. Recalls and records gather present intake data, while FFQs provide data based on past intake. It is known that an individual's nutrient and energy intake varies not only from day to day, but season to season as well. So if past intake is needed, FFQs may be the better choice.

The number of days of food intake records or recalls available can also affect the interpretation of the data. If high levels of accuracy are needed for an individual's intake of a nutrient, a greater number of days will be required than if a group average could be used. Care must be taken when determining the number of days to use in a study. For example, researchers using data from a single 24-hour recall from 832 men found that saturated fat intake was inversely associated with stroke [35]. Because of the day-to-day variability, dietary changes or recommendations for an individual should not be based on a single day's intake. Basiotis *et al.* [36] determined the number of days of food intake data needed to estimate individual intake as well as group intake for food energy and 18 nutrients. They found that for a female, an average of 35 days is needed to determine a true average of energy intake for each individual, whereas 3 days of food records from each subject are required to determine a group average.

It is also important that data obtained from dietary assessment methods be utilized properly. The FFQ was developed to rank nutrient intakes from low to high. It was not intended to be used to determine and develop levels of nutrient intake to prevent disease. However, many researchers have chosen to use the FFQ for this purpose. For example, studies based on FFQ data have recommended levels of vitamin E intake to reduce risk for heart disease [37, 38]. Studies have concluded that these levels may not have the effect on heart disease development as previously suggested [39–41]. When choosing a method for dietary assessment, care must be taken to ensure that the data are properly interpreted.

B. The Hierarchy of Scientific Evidence

To make a judgment about dietary changes, the hierarchy of scientific evidence should be kept in mind (Fig. 1). The most rigorous evaluation of a dietary hypothesis is the randomized feeding trial, which should ideally be conducted as a double-blind experiment. Large, randomized feeding trials may eventually provide definitive answers to some of the questions we have regarding the relationships between dietary factors and the major illnesses of our society. Unfortunately, these are extremely costly and may be of long duration if the disease studied has a long latency period. For now, our knowledge of many of these relationships has been derived mainly from observational studies.

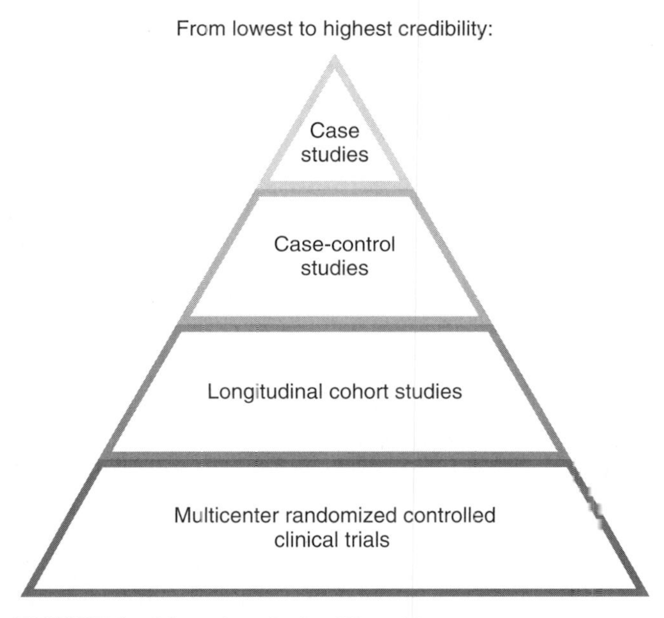

FIGURE 1 Hierarchy of scientific evidence.

C. Data Validity

One major concern when interpreting dietary data is the accuracy of the information reported. Because most of the information gathered is self-reported, the reliability and validity of the data depend on the reporter's motivation, memory, ability to communicate, and awareness of the foods consumed. Most methods have been proven to be generally reliable; that is, they will provide the same estimate on different occasions. However, do the techniques gather true and accurate measurements, or valid measurements, of what people are really eating? In the past, assumptions were made that the information was indeed valid. The validity of the techniques was often verified by comparing the different methods to each other. For example, an FFQ may be "validated" by comparing results to food records. If the food records are 100% accurate, this method of convergent validation would be fine. However, because both methods require self-reported intake, the validity of both methods should be questioned. Therefore, external independent markers of intake are needed for true validation [42].

During the 1980s, the search began for biochemical indicators that closely reflect dietary intake. These indicators, a type of biomarker, measure specific variables in body fluids or tissues to reflect intake of a food component [43]. Doubly-labeled water (DLW) is the most widely used biomarker for energy intake at this time (see Chapter 4 for further details). Although this method can only determine the accuracy of a dietary intake method with respect to total energy, it is reasonable to assume that if that method is accurate (or not) for energy, it will also be so for specific macro- and micronutrients [44, 45]. Therefore, it is accepted that if group estimates of energy intake are found to be truly valid using a method such as DLW, the estimates of macro- and micronutrient intakes may also be considered valid [46]; however, this conclusion needs further research.

Other biomarkers that have been used to validate nutrient intake include fatty acid patterns in blood to reflect fatty acid intake [47], urinary nitrogen to validate protein intake [48–50], and serum carotenoids and vitamin C concentrations as markers of fruit and vegetable consumption [51, 52]. With the increased use of these biomarkers, particularly DLW, to determine the accuracy of dietary intake data in a variety of subjects, it has become clear that the traditional dietary assessment methods are not completely accurate. It is now well accepted that misreporting of food intake—over- or underreporting—is widespread and that no data-gathering method is immune [42].

1. Overreporting

Overreporting occurs when reported intakes are higher than the measured energy expenditure levels. Overreporting has not been found to be as big of a problem as underreporting, but it still has the potential to interfere with results and conclusions. Johansson *et al.* [53] found that individuals who overreport tend to be younger and, with lower body mass indexes, often considered lean. The highest proportion of overreporters was found among those subjects who wanted to increase body weight. A study of the elderly population found that overreporting was higher in men [54]. Although overreporters are not common, care should be taken when obtaining data from subjects that are at highest risk.

2. Underreporting

Underreporting occurs when reported intakes are much lower than measured energy expenditure levels. These reports are often so low that basal metabolic needs could not be met and are not biologically plausible. Depending on the age, gender, and body composition of a given sample, underreporters may compose anywhere from 2% to 85% of the total sample [55]. It is now understood that underreporting tends to be associated with certain groups. It is well accepted that the obese are at greatest risk of underreporting dietary intake. Many studies have shown that the obese underreport more often and to a greater degree—30% to 47%—than the lean [56, 57]. Women have also been found to underreport more often than men [58–61]. Other characteristics of low energy intake reporters include those associated with low socioeconomic status—that is, low income and low education levels [58–61]. Table 5 provides a summary of groups who have been found to be underreporters. These factors are further complicated by the possibility that they are risk factors for many chronic diseases.

3. Reasons for Underreporting

Once underreporting became well accepted and documented, researchers began, and continue, to look for reasons why people underreport. We know that being obese is not the cause of underreporting, but it is most likely the psychological and behavioral characteristics associated with obesity that lead to underreporting [55, 62, 63]. A need for social acceptance, a desire to be liked or accepted by the interviewer, may cause the subject to underreport "sinful" foods. A high level of body dissatisfaction—that is, if a person sees a leaner physique as being healthier or more desirable than his or her own—may cause him or her to misreport foods. Also, researchers have found that women who scored higher on restrained eating scales, those who feel they are making a conscious effort to avoid certain foods, tend to underreport as well [55, 62, 63].

Another explanation for underreporting may be related more to the meal size than body size. Wansink and Chandon [64] asked overweight and normal weight adults to estimate the calories in both small and large fast food meals. They found that both groups underestimated the number of calories in the larger meals. Therefore, greater underestimation of calories by obese individuals may be, in part, a result of their tendency to consume larger meals.

TABLE 5 Summary of Underreporting

	Populations Most Likely to Underreport
Women	Smokers
Obese	Lower education
Low socioeconomic status	Older adults
Psychosocial Characteristics Associated with Underreporting	
High scorers on restrained eating scales	A history of dieting behavior
Discrepancy between perceived and ideal body size	
Foods Most Likely to Be Underreported	
U.S. Survey[a]	
Cake/pie	Meat, fish, poultry, egg sandwiches or mixtures
Savory snacks: chips, popcorn, pretzels	Regular soft drinks
Cheese	Fat-type spreads
White potatoes	Condiments
British Survey[b]	
Cake	Breakfast cereal
Sugars	Milk
Fats	

[a]U.S. Survey from Krebs-Smith, S. M., Graubard, B. I., Kahle, L. L., Subar, A. F., Cleveland, L. E., and Ballard-Barbash, R. (2000). Low energy reporters vs others: a comparison of reported food intakes. *Eur. J. Clin. Nutr.* **54**, 281–287.

[b]British Survey from Bingham, S. A., Cassidy, A., Cole, T. J., Welch, A., Runswick, S. A., Black, A. E., Thurnham, D., Bates, C., Khaw, K. T., Key, T. J. A., Day, N. E. (1995). Validation of weighed records and other methods of dietary assessment using the 24 h urine nitrogen technique and other biological markers. *Br. J. Nutr.* **73**, 531–550.

4. Foods Most Often Underreported

If underreporting occurred in all foods and nutrients to the same degree, the solution would be to add a correction factor to the data of the underreporters. This would bring their reported intake of all nutrients into line with that of the valid reporters. Unfortunately, it has become clear that underreporters often fail to report those foods that are seen as "bad" or "sinful" [45]. In a U.S. survey of 8334 adults, 1224 were found to be low energy reporters [65]. Some of the foods that were found to be most often underreported included cakes/pies, savory snacks, cheese, white potatoes, meats, regular soft drinks, fat-type spreads, and condiments [65]. British researchers found that little difference was seen between underreporters and normal energy reporters with regard to bread, potatoes, meat, vegetables, or fruit, but a significant difference was seen with cakes, sugars, fat, and breakfast cereal [49]. Table 5 gives a summary of underreported foods. Some researchers have found that underreporters tend to report lower intakes of fat, and higher intakes of protein and carbohydrates as a percentage of total energy [66, 67], whereas others show that reports of added sugar intake are significantly lower [68]. No agreement has been reached as to how much, if at all, specific macronutrients are misreported. Therefore, adding a correction factor would not be effective.

5. The Problems with Underreporting

A major problem with underreporting occurs when researchers begin to classify dietary intake information to determine diet and disease associations. This is often done by ranking nutrient intakes from low to high and then looking for any associations between nutrient intake and occurrence of disease. There is a real danger of misclassification of subjects if this ranking is based on false or underreported intakes. As pointed out earlier, those who tend to be at higher risk for underreporting are also those who are at greater risk for many chronic diseases. For example, obesity is a known risk factor for coronary heart disease as well as underreporting. Because bias in measuring dietary intake has the potential of removing as well as creating associations, it can generate misleading conclusions about the impact of diet and disease [69, 70].

Underreporting is now understood to be a real and potentially misleading problem in nutrition research. Although the use of biomarkers can help validate and interpret dietary intake data and, ideally, should become routine in nutritional epidemiology, they cannot totally replace the collection of estimated dietary intakes. Food is made up of many components, not simply the nutrients that we are aware of and that can be found in a database. Records of foods consumed may help researchers find non-nutrient constituents that help reduce risks for diseases. More evidence is accumulating to support the belief that the entire food, not just specific components, leads to health [71, 72]. Also, there is not a biomarker for every dietary constituent that is of interest to researchers. Finally, self-report collection methods can be performed on large populations at a much lower cost than the use of biomarkers and requires less technology and fewer skills.

6. Identifying Underreporters

To help identify underreporters, researchers can apply methods such as the Goldberg cutoff, extensively described by Goldberg and colleagues [73] and Black [74] and colleagues [75]. The Goldberg cutoff identifies the most obviously implausible intake values by evaluating the energy intake against estimated energy requirements. Basal metabolic rate (BMR) can be measured using methods described in Chapter 4, or height and weight measurements can be used to predict BMR from a standard formula (the Schofield equation is recommended by Goldberg *et al.* [73]). A ratio of the estimated energy intake (EI) to measured or predicted BMR is calculated as EI/BMR. This ratio can then be compared with a study-specific cutoff value (see Black [74] for a practical guide to using the Goldberg cutoff). This cutoff represents the lowest value of EI/BMR that could reasonably reflect the energy expenditure based on physical activity level if the information is available or assuming a sedentary lifestyle if no activity information is gathered. A summary of the principles of the cutoffs can be found in Table 6. Studies using the Goldberg cutoff classified 28% to 39% of the women and 18% to 27% of the men as low energy reporters [59, 65].

The Goldberg cutoff has several limitations [74]. It has poor sensitivity for defining invalid reports at the individual level as it identifies only the extreme underreporters. It also does not distinguish between varying degrees of underreporting. The major limitation is that the cutoff depends on knowledge of energy requirements or energy expenditure. If no physical activity information is available, the cutoff assumes a sedentary lifestyle and will therefore underestimate the underreporters. If researchers can gather information on lifestyle, occupation, leisure time, and particularly information regarding physical activity of the subjects, calculations can be more specific and improve the chances of identifying the underreporters [76].

Researchers have also proposed a method [77, 78], based on the principles of the Goldberg cutoff, to screen for implausible reports by comparing reported energy intake with predicted total energy requirements. Researchers offer that because total energy requirements are predicted, not just basal metabolic rate, this new method eliminates the potential error caused by assigning inaccurate physical activity levels when there is limited or no information of activity levels available [77]. Using this method to analyze CSFII 1994 to 1996 data, Huang *et al.* [78] found that implausible reported energy intakes reduced the overall validity of the sample and including misreporters could lead to inappropriate conclusions about the impact of diet on health outcomes.

TABLE 6 Principles of the Goldberg Cutoff

Principle 1: Validation of reported energy intake rests on the following:

$$\text{Energy intake (EI)} = \text{Energy expenditure (EE)} \pm \text{Changes in body stores}$$

Principle 2: Assumes subjects are weight stable and therefore are in energy balance:

$$EI = EE$$

Principle 3: Express energy requirements as multiples of basal metabolic rate (BMR):

$$EE = \text{Physical activity level (PAL)} \times BMR$$

Principle 4: Since $EI = EE$ and $EE = PAL \times BMR$, then the following can be assumed:

$$EI = PAL \times BMR \text{ or } EI{:}BMR = PAL$$

Principle 5: Reported energy intake expressed as EI_{rep}: BMR can be compared with expected PAL.

Principle 6: Since error exists in all the measured elements of the equation, absolute agreement cannot be expected.

Principle 7: Confidence limits (cutoffs) of the agreement between EI_{rep}: BMR and PAL must be determined to establish if reported values within the subjects are acceptable. (See Black, 2000, for cutoff equations.)

Source: Data found in Black, A. E. (2000). Critical evaluation of energy intake using the Goldberg cut-off for energy intake:basal metabolic rate: a practical guide to its calculation, use and limitations. *Inter. J. Obes.* 24, 1119–1130.

7. Handling Underreporting in Dietary Data

Researchers are still not sure how to handle data sets containing large numbers of underreporters. Several approaches have been suggested, but none is ideal. One technique would be to exclude anyone who is found to report implausible energy intakes. The problem with this method is that the underreporters tend to fall into specific subgroups (i.e., obese, smokers) and, as stated previously, will seriously alter the sample. Some investigators have analyzed their data with all the subjects and then again after the underreporters have been removed [79]. If there are no significant discrepancies between the finding, this can improve the confidence in the results and conclusions.

Some have suggested that once the difference between the low reporters and valid reporters is noted, upward adjustments of all the nutrients could be made. Because underreporting does not appear to occur equally for all nutrients, this method is not advisable. Research has shown that underreporters tend to report micronutrient-rich diets when compared to valid reporters [58]. Adjusting would give a false impression of the nutrient status and could mask possible risks of disease.

Other researchers have suggested adjusting nutrient intakes for energy intake using the regression of nutrient versus energy [80]. This would be feasible only if portion sizes were underestimated, yet the actual foods were all reported accurately. Otherwise, this method could make the reports worse [81]. As noted earlier, it is most likely that foods are systematically omitted from recalls. So if, for example, fat-containing foods (e.g., desserts) are often underreported, whereas vitamin A–containing foods (e.g., cantaloupe) are not, energy adjustments would provide lower than actual measures of fat intake but a higher measure of vitamin A intake. Many researchers have recognized that adjustments cannot eliminate the bias caused by selective underreporting [82].

V. CONCLUSION

Although dietary assessment methods do not provide true nutritional intake values, their importance in nutritional research studies still remains. Nutritional health cannot be evaluated without also examining the total diet composition. Improving dietary intake methodology is critical to the credibility of nutrition research. Identifying underreporters and improving dietary database validity through analytical approaches should remain in the forefront of dietary assessment until methodology improvements can be found. Researchers as well as practitioners need to keep in mind that bias does exist, and they should interpret dietary intake data with skepticism to prevent any misleading diet and health associations.

References

1. U.S. Department of Agriculture, Agricultural Research Service, Nutrient Data Laboratory. (2007). "About the Nutrient Data Lab," www.ars.usda.gov/Aboutus/docs.htm?docid=4441.
2. U.S. Department of Agriculture, Agricultural Research Service. (2006). USDA National Nutrient Database for Standard Reference, Release 19. Nutrient Data Laboratory home page, www.ars.usda.gov/ba/bhnrc/ndl.
3. Haytowitz, D. B., Pehrsson, P. R., and Holden, J. M. (2002). The identification of key foods for food composition research. *J. Food Comp. Anal.* **15**, 183–194.

4. Phillips, K. M., Patterson, K. Y., Rasor, A. S., Exler, J., Haytowitz, D. B., Holden, J. M., and Pehrsson, P. R. (2006). Quality-control materials in the USDA National Food and Nutrient Analysis Program (NFNAP). *Anal. Bioanal. Chem.* **384**(6), 1341–1355.
5. Akinyele, I. O., and Aminu, F. T. (1997). Computerized database of ethnocultural foods commonly eaten in Nigeria. *Am. J. Clin. Nutr.* **65**(suppl), 1331S.
6. Food Intake Analysis System developed by the University of Texas, Health Science Center at Houston, School of Public Health and the U.S. Department of Agriculture. Human Nutrition Information Service, Washington, DC.
7. Cowin, I., and Emmett, P. (1999). The effect of missing data in the supplements to McCance and Widdowson's food tables on calculated nutrient intakes. *Eur. J. Clin. Nutr.* **53**(11), 891–894.
8. Buzzard, I. M., Price, K. S., and Warren, R. A. (1991). Considerations for selecting nutrient-calculation software: evaluation of the nutrient database. *Am. J. Clin. Nutr.* **54**, 7–9.
9. Diplock, A. T. (1987). Trace elements in human health with special reference to selenium. *Am. J. Clin. Nutr.* **45**, 1313–1322.
10. Levander, O. A. (1991). Scientific rationale for the 1989 Recommended Dietary Allowance for selenium. *J. Am. Diet. Assoc.* **91**, 1572–1576.
11. Sinha, R., Block, G., and Taylor, P. (1993). Problems with estimating vitamin C intakes. *Am. J. Clin. Nutr.* **57**, 547–550.
12. Snyder, P. O., and Matthews, E. (1983). Percent retention of vitamin C in whipped potatoes after pre-service holding. *J. Am. Diet. Assoc.* **83**, 454–458.
13. Vogt, T. M., Appel, L. J., Obarzanek, E., Moore, T. J., Vollmer, W. M., Svetkey, L. P., Sacks, F. M., Bray, G. A., Cutler, J. A., Windhauser, W. M., Lin, P.-H., and Karanja, N. M., for the DASH Collaborative Research Group. (1999). Dietary approaches to stop hypertension: rationale, design, and methods. *J. Am. Diet. Assoc.* **99**(suppl), S12–S18.
14. Swain, J. F., Windhauser, M. M., Hoben, K. P., Evans, M. A., McGee, B. B., and Steele, P. D., for the DASH Collaborative Research Group. (1999). Menu design and selection for multicenter controlled feeding studies: process used in the dietary approaches to stop hypertension trial. *J. Am. Diet. Assoc.* **99** (suppl), S54–S59.
15. McCullough, M. L., Karanja, N. M., Lin, P.-H., Obarzanek, E., Phillips, K. M., Laws, R. L., Vollmer, W. M., O'Connor, E. A., Champagne, C. M., and Windhauser, M. M., for the DASH Collaborative Research Group. (1999). Comparison of 4 nutrient databases with chemical composition data from the dietary approaches to stop hypertension trial. *J. Am. Diet. Assoc.* **99** (suppl), S45–S53.
16. O'Dell, B. L. (1984). Bioavailability of trace elements. *Nutr. Rev.* **42**, 301–308.
17. Lynch, S. R. (1997). Interaction of iron with other nutrients. *Nutr. Rev.* **55**, 102–110.
18. Cahill, A., and Lean, M. E. (1999). Review article: malnutrition and maltreatment: a comment on orlistat for the treatment of obesity. *Aliment. Pharmacol. Ther.* **13**(8), 997–1002.
19. Prince, D. M., and Welschenbach, M. A. (1998). Olestra: a new food additive. *J. Am. Diet. Assoc.* **98**(5), 565–569.
20. Bloch, A., and Thomson, C. A. (2004). Position of the American Dietetic Association: functional foods. *J. Am. Diet. Assoc.* **104**, 814–826.
21. Kant, A. K. (1996). Indexes of overall diet quality: a review. *J. Am. Diet. Assoc.* **96**, 785–791.
22. Guenther, P. M., Krebs-Smith, S. M., Reedy, J., Britten, P., Juan, W. Y., Lino, M., Carlson, A., Hiza, H. A., and Basiotis, P. P. (2006). The Healthy Eating Index-2005, CNPP-Fact Sheet No. 1. U.S. Department of Agriculture, Center for Nutritional Policy and Promotion, Alexandria, VA.
23. Patterson, R. E., Haines, P. S., and Popkin, B. M. (1994). Diet Quality Index: capturing a multidimensional behavior. *J. Am. Diet. Assoc.* **94**, 57–64.
24. Haines, P. S., Siega-Riz, A. M., and Popkin, B. M. (1999). The Diet Quality Index Revised: a measurement instrument for populations. *J. Am. Diet. Assoc.* **99**, 697–704.
25. U.S. Department of Agriculture, Agricultural Resource Service. (2005). "Dietary Guidelines for Americans 2005," www.health.gov/dietaryguidelines/dga2005/document.
26. U.S. Department of Agriculture. "MyPyramid.gov," www.mypyramid.gov.
27. Kennedy, E. T., Ohls, J., Carlson, S., and Fleming, K. (1995). The Healthy Eating Index: design and applications. *J. Am. Diet. Assoc.* **95**, 1103–1108.
28. Food and Nutrition Board. (1989). "Diet and Health: Implications for Reducing Chronic Disease Risk." National Academy Press, Washington, DC.
29. Kranz, S., Hartman, T., Siega-Riz, A. M., and Herring, A. H. (2006). A diet quality index for American preschoolers based on current dietary intake recommendations and an indicator of energy balance. *J. Am. Diet. Assoc.* **106**, 1594–1604.
30. Bodnar, L. M., and Siega-Riz, A. M. (2002). A diet quality index for preganancy detects variation in diet and differences by sociodemographic factors. *Public Health Nutr.* **5**, 801–809.
31. Cleveland, L. E., Cook, D. A., Krebs-Smith, S. M., and Friday, J. (1997). Method for assessing food intakes in terms of servings based on food guidance. *Am. J. Clin. Nutr.* **65** (suppl), 1254S–1263S.
32. Food Surveys Research Group, ARS. (1994). "Continuing Survey Food Intakes by Individuals (CSFII), 1994, Recipe and Pyramid Servings" [database on CD-ROM]. Department of Agriculture, Agricultural Research Service, Washington, DC.
33. Institute of Medicine and the National Academies. (2005). "Dietary Reference Intakes for Energy, Carbohydrate, Fiber, Fat, Fatty Acids, Cholesterol, Protein, and Amino Acids." National Academy Press, Washington, DC.
34. Basiotis, P. P., Carlson, A., Gerrior, S. A., Juan, N. Y., and Lino, M. (2002). "The Healthy Eating Index: 1999–2000," CNPP-12. U.S. Department of Agriculture, Center for Nutritional Policy and Promotion, Washington, DC.
35. Gillman, W. W., Cupples, L. A., Millen, B. E., Ellison, R. C., and Wolf, P. A. (1997). Inverse association of dietary fat with development of ischemic stroke in men. *JAMA* **278**, 2145–2150.
36. Basiotis, P. P., Welsh, S. O., Cronin, F. J., Kelsay, J. L., and Mertz, W. (1987). Number of days of food intake records required to estimate individual and group nutrient intakes with defined confidence. *J. Nutr.* **117**, 1638–1641.
37. Rimm, E. B., Stampfer, M. J., Ascherio, A., Giovannucci, E., Colditz, G. A., and Willett, W. C. (1993). Vitamin E consumption and the risk of coronary disease in women. *N. Engl. J. Med.* **328**, 1450–1456.

38. Stampfer, M. J., Hennekens, C. H., Manson, J. E., Colditz, G. A., Rosner, B., and Willett, W. C. (1993). Vitamin E consumption and the risk of coronary disease in women. *N. Engl. J. Med.* **328**, 1444–1449.
39. Yusuf, S., Dagenais, G., Pogue, J., Bosch, J., and Sleight, P. (2000). Vitamin E supplementation and cardiovascular events in high-risk patients: the Heart Outcomes Prevention Evaluation Study Investigators. *N. Engl. J. Med.* **342**(3), 154–160.
40. Lonn, E., Bosch, J., Yusuf, S., Sheridan, P., Pogue, J., Arnold, J. M., Ross, C., Arnold, A., Sleight, P., Probstfield, J., and Dagenias, G. R.; HOPE and HOPE-TOO Trail Investigators. (2005). Effects of long term vitamin E supplementation on cardiovascular events and cancer: a randomized controlled trial. *JAMA* **293**, 1338–1347.
41. Chiabrando, C., Avanzini, F., Rivalta, C., Colombo, F., Fanelli, R., Palumbo, G., and Roncaglioni, M. C.; PPP Collaborative Group on the antioxidant effect of vitamin E (2002). Long term vitamin E supplementation fails to reduce lipid peroxidation in people at cardiovascular risk: analysis of underlying factors. *Curr. Control Trails Cardiovasc. Med.* **3**, 5–11.
42. Black, A. E., Prentice, A. M., Goldberg, G. R., Jebb, S. A., Bingham, S. A., Livingstone, B. E., and Coward, A. W. (1993). Measurements of total energy expenditure provide insights into the validity of dietary measurements of energy intake. *J. Am. Diet. Assoc.* **93**, 572–579.
43. Katan, M. B. (1998). Biochemical indicators of dietary intake. *Eur. J. Clin. Nutr.* **52**, S5 (abstract).
44. Schoeller, D. A., and Van Santen, E. (1982). Measurement of energy expenditure in humans by the doubly labeled water method. *J. Appl. Physiol.* **53**, 955–959.
45. Mertz, W. (1992). Food intake measurements: is there a "gold standard"? *J. Am. Diet. Assoc.* **92**, 1463–1465.
46. Johnson, R. K., Driscoll, P., and Goran, M. I. (1996). Comparison of multiple-pass 24-hour recall estimates of energy intake with total energy expenditure determined by the doubly labeled water method in young children. *J. Am. Diet. Assoc.* **96**, 1140–1144.
47. Andersen, L. F., Solvoll, K., and Drevon, D. A. (1996). Very-long-chain n-3 fatty acids as biomarkers for intake of fish and n-3 fatty acid concentrates. *Am. J. Clin. Nutr.* **64**, 305–311.
48. Bingham, S. A., and Cummings, J. (1985). Urine nitrogen as an independent validatory measure of dietary intake: a study of nitrogen balance in individuals consuming their normal diet. *Am. J. Clin. Nutr.* **42**, 1276–1289.
49. Bingham, S. A., Cassidy, A., Cole, T. J., Welch, A., Runswick, S. A., Black, A. E., Thurnham, D., Bates, C., Khaw, K. T., Key, T. J. A., and Day, N. E. (1995). Validation of weighed records and other methods of dietary assessment using the 24 h urine nitrogen technique and other biological markers. *Br. J. Nutr.* **73**, 531–550.
50. Bingham, S. A. (1997). Dietary assessments in the European Prospective Study of diet and Cancer (EPIC). *Eur. J. Cancer Prev.* **6**, 118–124.
51. Le Marchand, L., Hankin, J. H., Carter, F. S., Essling, C., Luffey, D., Franke, A. A., Wilkens, L. R., Cooney, R. V., and Kolonel, L. N. (1994). A pilot study on the use of plasma carotenoids and ascorbic acid as markers of compliance to a high fruit and vegetable dietary intervention. *Cancer Epidemiol. Biomarkers Prev.* **3**, 245–251.
52. Pierce, J. P., Faerber, S., Wright, F. A., Newman, V., Flatt, S. W., Kealey, S., Rock, C. L., Hryniuk, W., and Greenberg, E. R. (1997). Feasibility of a randomized trial of a high-vegetable diet to prevent breast cancer recurrence. *Nutr. Can.* **289**, 282–288.
53. Johansson, L., Solvoll, K., Bjorneboe, G.-E. A., and Drevon, C. A. (1998). Under- and overreporting of energy intake related to weight status and lifestyle in a nationwide sample. *Am. J. Clin. Nutr.* **68**, 266–274.
54. Bazelmans, C., Matthys, C., De Henauw, S., Dramaix, M., Kornitzer, M., DeBacker, G., and Leveque, A. (2007). Predictors of misreporting in an elderly population: the "Quality of Life after 65" study. *Public Health Nutr.* **10**, 185–191.
55. Maurer, J., Taren, D. L., Teixeira, P. J., Thomson, C. A., Lohman, T. G., Going, S. B., and Houtkooper, L.B. (2006). The psychosocial and behavioral characteristics related to energy misreporting. *Nutr. Rev.* **64**, 53–66.
56. Prentice, A. M., Black, A. E., Coward, W. A., Davies, H. L., Goldberg, G. R., Murgatroyd, P. R., Ashford, J., Sawyer, M., and Whitehead, R. G. (1986). High levels of energy expenditure in obese women. *Br. Med. J.* **292**, 983–987.
57. Lichtman, S. W., Pisarska, K., Berman, E. R., Pestones, M., Dowling, H., Offenbacher, E., Weisel, H., Heshka, S., Matthews, D. E., and Heymsfield, S. B. (1992). Discrepancy between self-reported and actual caloric intake and exercise in obese subjects. *N. Engl. Med. J.* **327**, 1893–1898.
58. Price, G. M., Paul, A. A., Cole, T. J., and Wadsworth, M. E. J. (1997). Characteristics of the low-energy reporters in a longitudinal national dietary survey. *Br. J. Nutr.* **77**, 833–851.
59. Pryer, J. A., Vrijheid, M., Nichols, R., Kiggins, M., and Elliot, P. (1997). Who are the "low energy reporters" in the dietary and nutritional survey of British adults? *Int. J. Epidemiol.* **26**, 146–154.
60. Johnson, R. K., Soultanakis, R. P., and Matthews, D. E. (1998). Literacy and body fatness are associated with underreporting of energy intake in U.S. low-income women using the multiple-pass 24-hour recall: a doubly labeled water study. *J. Am. Diet. Assoc.* **98**, 1136–1140.
61. Dwyer, J., Picciano, M. F., and Raiten, D. J. (2003). Estimation of usual intakes: What We Eat in America–NHANES. *J. Nutr.* **133**, 609S–623S.
62. Taren, D., Tobar, M., Hill, A., Howell, W., Shisslack, C., Bell, I., and Ritenbaugh, C. (1999). The association of energy intake bias with psychological score of women. *Eur. J. Clin Nutr.* **53**, 570–578.
63. Johnson, R. K., Soultanakis, R. P., and Matthews, D. E. (1999). Psychological factors and energy intake underreporting in women. *FASEB J.* **13**, A695.
64. Wansink, B., and Chandon, P. (2006). Meal size, not body size, explains errors in estimating the calorie content of meals. *Ann. Intern. Med.* **145**, 326–332.
65. Krebs-Smith, S. M., Graubard, B. I., Kahle, L. L., Subar, A. F., Cleveland, L. E., and Ballard-Barbash, R. (2000). Low energy reporters vs others: a comparison of reported food intakes. *Eur. J. Clin. Nutr.* **54**, 281–287.

66. Briefel, R. R., Sempos, C. T., McDowell, M. A., Chien, S. C. Y., and Alaimo, K. (1997). Dietary methods research in the third National Health and Nutrition Examination Survey: underreporting of energy intake. *Am. J. Clin. Nutr.* **65**, 1203S–1209S.
67. Voss, S., Kroke, A., Lipstein-Grobusch, K., and Boeing, H. (1998). Is macronutrient composition of dietary intake data affected by underreporting? Results from the EPIC-Potsdam study. *Eur. J. Clin. Nutr.* **52**, 119–126.
68. Poppitt, S. D., Swann, D., Black, A. E., and Prentice, A. M. (1998). Assessment of selective underreporting of food intake by obese and non-obese women in a metabolic facility. *Int. J. Obes. Relat. Metab. Disord.* **22**, 303–311.
69. Johnson, R. K., Black, A. E., and Cole, T. J. (1998). Dietary fat intake and the risk of coronary heart disease in women [letter to the editor]. *N. Engl. J. Med.* **338**, 918–919.
70. Livingstone, M. B. E., Prentice, A. M., Strain, J. J., Coward, E. A., Black, A. E., Barker, M. E., McKenna, P. G., and Whitehead, R. G. (1990). Accuracy of weighed dietary records in studies of diet and health. *Br. Med. J.* **300**, 708–712.
71. Thomson, C., Bloch, A. S., and Hasler, C. M. (1999). Position of the American Dietetic Association: functional foods. *J. Am. Diet. Assoc.* **99**, 1278–1285.
72. Craig, W. J. (1999). Health-promoting properties of common herbs. *Am. J. Clin. Nutr.* **70**, 491S–499S.
73. Goldberg, G. R., Black, A. E., Jebb, S. A., Cole, T. J., Murgatroyd, P. R., Coward, W. A., and Prentice, A. M. (1991). Critical evaluation of energy intake data using fundamental principles of energy physiology: 1. Derivation of cut-off limits to identify under-recording. *Eur. J. Clin. Nutr.* **45**, 569–581.
74. Black, A. E. (2000). Critical evaluation of energy intake using the Goldberg cut-off for energy intake:basal metabolic rate. A practical guide to its calculation, use and limitations. *Inter. J. Obes.* **24**, 1119–1130.
75. Black, A. E., Goldberg, G. R., Jebb, S. A., Livingston, M. B. E., Cole, T. J., and Prentice, A. M. (1991). Critical evaluation of energy intake data using fundamental principles of energy physiology: 2. Evaluating the results of published surveys. *Eur. J. Clin. Nutr.* **45**, 583–599.
76. Black, A. E. (1998). Poor validity of dietary assessment: what have we learnt? *Eur. J. Clin. Nutr.* **52**, S17.
77. McCrory, M. A., Hajduk, C. L., and Roberts, S. B. (2002). Procedures for screening out inaccurate reports of dietary energy intake. *Public Health Nutr* **5**, 873–882.
78. Huang, T. T.-K., Roberts, S. B., Howarth, N. C., and McCrory, M. A. (2005). Effect of screening out implausible energy intake reports on relationships between diet and BMI. *Obes. Res.* **13**, 1205–1217.
79. Munoz, K., Krebs-Smoth, S., Ballard-Barbash, R., and Cleveland, L. (1997). Food intakes of U.S. children and adolescents compared with recommendations. *Pediatrics* **100**(3), 323–329.
80. Willett, W., and Stampfer, M. J. (1986). Total energy intake: implications for epidemiologic analyses. *Am. J. Epidemiol.* **124**, 17–27.
81. Carter, L. M., and Whiting, S. J. (1998). Underreporting of energy intake, socioeconomic status, and expression of nutrient intake. *Nutr. Rev.* **56**, 179–182.
82. Stallone, D. D., Brunner, E. J., Bingham, S. A., and Marmot, M. G. (1997). Dietary assessment in Whitehall II: the influence of reporting bias on apparent socioeconomic variation in nutrient intakes. *Eur. J. Clin. Nutr.* **51**, 815–825.

CHAPTER **8**

Current Theoretical Bases for Nutrition Intervention and Their Uses

KAREN GLANZ

Rollins School of Public Health, Emory University, Atlanta, Georgia

Contents

I. INTRODUCTION

This chapter discusses contemporary theoretical bases for nutrition intervention for disease prevention and management and their applications in practice. Other chapters in this text provide specific recommendations regarding dietary advice for disease prevention and for nutritional management of patients. This chapter (1) introduces key concepts related to the application of theory in understanding and improving dietary behavior, (2) reviews behavioral issues related to healthful diets, (3) describes several current theoretical models that can be helpful in planning and conducting nutrition intervention, and (4) highlights important issues and constructs that cut across theories.

Nutritional intervention is a central component of disease prevention and management. Health professionals' roles in nutritional intervention are pivotal because of their centrality in health care and their credibility as patient educators [1]. The most recently published evidence-based recommendations in the United States advise including intensive behavioral dietary counseling for adult patients with known risk factors for chronic disease, noting that this counseling can be provided by physicians or other clinicians such as dietitians [2, 3]. Although the evidence for brief, low-intensity counseling of unselected patients is not sufficient to recommend routine dietary counseling [3], people who report receiving advice about dietary change during health care visits report more health-enhancing diet changes than those who received no such advice [4]. Further, nonclinical community sites such as worksites, churches, and community centers are becoming increasingly important as settings for nutrition interventions, especially with the growing and widespread nature of obesity and diabetes [5].

II. THE IMPORTANCE OF UNDERSTANDING INFLUENCES ON DIETARY BEHAVIOR

Successful nutrition interventions take many forms. Interventions to yield desirable changes in eating patterns can be best designed with an understanding of relevant theories of dietary behavior change and the ability to use them skillfully [6]. Although many early reports of nutrition interventions did not cite a particular theory or model as the basis for the strategies they employ [7, 8], the application of sound behavioral science theory in nutrition interventions is becoming increasingly common [9]. Also, emerging evidence suggests that interventions developed with an explicit theoretical foundation are more effective than those lacking a theoretical base [10] and that some studies combine multiple theories [11].

We have identified six theoretical models that are in current use and can be particularly useful for understanding the processes of changing eating patterns in health care and community settings: social cognitive theory, the stages of change construct from the transtheoretical model, consumer information processing, the theory of planned behavior, multiattribute utility theory, and social ecological models [6, 12, 13]. The central elements of each theory and how they can be used to help formulate nutrition interventions are described in this chapter.

A. Multiple Determinants of Food Choice

Many social, cultural, and economic factors contribute to the development, maintenance, and change of dietary patterns. No single factor or set of factors has been found to adequately account for why people eat as they do. Physiological and psychological factors, acquired food preferences, and knowledge about foods are important individual determinants of food intake. Families, social relationships, socioeconomic status, culture, and geography are also important influences on food choices. A broad understanding of some of the key factors and models for understanding food choice can provide a foundation for well-informed clinical nutrition intervention, help identify the most influential factors for a particular patient, and enable clinicians to focus on issues that are most salient for their patients.

B. Multiple Levels of Influence

Common wisdom holds that nutrition interventions are most likely to be effective if they embrace an ecological perspective for health promotion [13, 14]. That is, they should not only be targeted at individuals but should also affect interpersonal, organizational, and environmental factors that influence dietary behavior [15, 16]. This is most clearly illustrated when one thinks of the context of selecting and purchasing food. Consumers learn about foods through advertising and promotion in the media, by labels on food packages, and via product information in grocery stores, cafeterias, and restaurants [17]. Their actual purchases are influenced by personal preferences, family habits, medical advice, availability, cost, packaging, placement, and intentional meal planning. The foods they consume may be further changed in the preparation process, either at home or while eating out. The process is complex and clearly determined not only by multiple factors but by factors at multiple levels. Still, much food choice can be represented by routines and simple, internalized rules.

Traditionally, health/patient educators focus on intraindividual factors such as a person's beliefs, knowledge, and skills. Contemporary thinking suggests that thinking beyond the individual to the social milieu and environment can enhance the chance of successful health promotion and patient education [14, 16]. Health providers can and should work toward understanding the various levels of influence that affect the patient's behavior and health status. This will be discussed and illustrated with examples later in this chapter.

III. WHAT IS THEORY?

A theory is a set of interrelated concepts, definitions, and propositions that present a systematic view of events or situations by specifying relations among variables in order to explain and predict the events or situations. The notion of generality, or broad application, is important [6]. Even though various theoretical models of health behavior may reflect the same general ideas, each theory employs a unique vocabulary to articulate the specific factors considered to be important. Theories vary in the extent to which they have been conceptually developed and empirically tested.

Theory can be helpful during the various stages of planning, implementing, and evaluating interventions [6]. Theories can be used to guide the search for reasons why people are or are not consuming a healthful diet or adhering to a therapeutic dietary regimen. They can help pinpoint what you need to know before working effectively with a client, group, or patient. They also help to identify what should be monitored, measured, or compared in evaluating the efficacy of nutrition intervention.

IV. EXPLANATORY AND CHANGE THEORIES

Theories can guide the search to understand *why* people do or do not follow medical advice, help identify *what* information is needed to design an effective intervention strategy, and provide insight into *how* to design an educational program so it is successful [6]. Thus, theories and models help *explain* behavior as well as suggest how to develop more effective ways to influence and *change* behavior. These types of theories often have different emphases but are quite complementary [6]. For example, understanding why people choose the foods they eat is one step toward successful nutrition management, but even the best explanations will not be enough by themselves to fully guide change to improve health. Some type of change model is also needed. All of the theories and models described here have some potential as both explanatory and change models, although they might be better for one or the other purpose. For example, the theory of planned behavior was originally developed as an explanatory model, whereas in contrast the stages of change construct were conceived to help guide planned change efforts.

V. UNIQUE FEATURES OF DIETARY BEHAVIOR TO CONSIDER WHEN USING THEORY

Dietary behavior changes are most likely to be effective for preventing or managing disease when they are sustained over the long term and in people's natural environments, outside the clinical setting. To be effective in nutritional intervention, health care providers need to understand both the principles of clinical nutrition management and a variety of behavioral and educational issues [18].

There are several core issues about nutritional change that should be recognized. First, most diet-related risk factors are asymptomatic and do not present immediate or dramatic symptoms. And, by the time symptoms are recognized, the changes needed are often very challenging. Second, health-enhancing dietary changes require qualitative change, not just modification of the amount of food consumed, and cessation is not a viable option (as with smoking or other addictive behaviors). Third, both the act of making changes and self-monitoring require accurate knowledge about the nutrient composition of foods or a convenient, practical reference source. Thus, information acquisition and processing may be more complex for dietary change than for changes in some other health behaviors, such as smoking and exercise. Because of this, consumer information processing models (described later) are more important for nutrition intervention than for other types of health-related behaviors. Other important issues include long-term maintenance, the format and medium for providing dietary advice, nutritional adequacy, options for initiating the change process, the changing food supply, fad diets, and special populations.

A. Long-Term Dietary Change

Because nutritional intervention leads to meaningful improvements in health only when long-term change is achieved, both providers and patients both need to "look down the road" when formulating expectations and setting goals [19]. For example, for most patients without other major risk factors who follow recommended dietary changes for cholesterol reduction, significant reductions are seen within 4 to 6 weeks, and cholesterol reduction goals can be reached within 4 to 6 months [20]. Even after goals are achieved, new dietary habits must be maintained. Thus, if it takes several weeks or even months to adjust to the new dietary regimen, patience and persistence by both physician and patient may be worthwhile in the long run. Maintenance of new eating patterns occurs mainly outside any clinical or therapeutic setting. Different skills are required to make initial changes and to maintain them over the long term, so follow-up consultations and advice should address new issues, not merely repeat or rehash old information.

B. Restrictive and Additive Recommendations: Typical Reactions

Traditionally, nutrition intervention has focused on advice to *restrict* intake of certain foods or nutrients (e.g., reducing fat and saturated fat intake, limiting calorie intake, and limiting sodium/salt). Yet the most often-mentioned obstacle to achieving a healthful diet is not wanting to give up the foods we like [21]. Basic psychological principles hold that when people are faced with a restriction, or loss of a choice, that choice or commodity becomes more attractive. In other words, focusing mainly on what *not* to eat or on eating less of some types of foods may evoke conscious or unconscious negativism in some people. In contrast, emphasizing *additive* recommendations, such as increasing intake of fruits and vegetables or eating more fiber-rich foods, often appeals to people because it sanctions their doing more of something. The challenge is to make these recommendations attractive to patients, and to ensure that they are presented in the context of an overall healthful diet.

C. Implications of Counseling for Gradual Change or Very Strict Diets

A generally held view is that the chances of long-term dietary change are greater when efforts to change occur in a gradual, stepwise manner. This might involve attempting changes within specific food groups one at a time, until the total diet comes close to recommendations. A basic principle involved is that small successes (i.e., recognition of each successful behavioral change) increase confidence and motivation for each successive change. Although this is effective for many people, others become impatient or even lose their enthusiasm for changes that are minimally recognizable. An alternative is to begin with a highly restrictive diet such as a very low-fat diet or a very low-calorie diet for weight loss. These types of programs, with strict dietary regimens, may be useful for patients who are highly motivated, postsurgically, after a coronary incident, or for those who have not been successful in making gradual changes. In some cases, a strict diet for an initial short time period will yield visible or clinical changes that help motivate patients to continue adhering to a less extreme regimen. Such diets require careful supervision, however.

D. Special Populations

Ideally, each patient should be treated as an individual with unique circumstances and health history. Still, epidemiological research indicates that certain demographic subgroups differ in terms of both risk factors and diet. Understanding these population trends can help prepare a provider to work with various types of patients. Minority and lower socioeconomic status individuals are disproportionately affected by obesity [22] and related risk factors. Targeted interventions are needed for disadvantaged groups [23]. Age may also make a difference: younger persons may feel invulnerable to coronary events, and older adults may be managing multiple chronic conditions and using both prescribed and over-the-counter medications that could interact with foods. These are just a few examples of how population subgroups may differ, and these serve as a reminder to be sensitive to group patterns but to avoid stereotyping in the absence of firsthand evidence about an individual.

VI. IMPORTANT THEORIES AND THEIR KEY CONSTRUCTS

There are several available and widely used models and theories of behavior change that are applicable to nutrition intervention. This section describes six theoretical models that are in current use and make unique contributions to the interventionist's tool kit. They are social cognitive theory, the stages of change construct from the transtheoretical model, consumer information processing, the theory of planned behavior, multiattribute utility theory, and social ecological models [6, 9, 12, 24]. The central elements of each theory and how they can be used to help formulate nutrition interventions are described in this chapter. Table 1 provides illustrative statements showing the application of each theory.

A. Social Cognitive Theory

Social cognitive theory (SCT), the cognitive formulation of social learning theory that has been best articulated by Bandura [25, 26], explains human behavior in terms of a three-way, dynamic, reciprocal model in which personal factors, environmental influences, and behavior continually interact. SCT synthesizes concepts and processes from cognitive, behavioristic, and emotional models of behavior change, so it can be readily applied to nutritional intervention for disease prevention and management. A basic premise of SCT is that people learn not only through their own experiences but also by observing the actions of others and the results of those actions [6]. Key constructs of social cognitive theory that are relevant to nutritional intervention include observational learning, reinforcement, self-control, and self-efficacy [9].

Principles of behavior modification, which have often been used to promote dietary change, are derived from SCT. Some elements of behavioral dietary interventions based on SCT constructs of self-control, reinforcement, and self-efficacy include goal-setting, self-monitoring, and behavioral contracting [9, 18]. As will be discussed later, goal setting and self-monitoring seem to be particularly useful components of effective interventions.

Self-efficacy, or a person's confidence in his or her ability to take action and to persist in that action despite obstacles or challenges, seems to be especially important for influencing health behavior and dietary change efforts [26]. Health providers can make deliberate efforts to increase patients' self-efficacy using three types of strategies: (1) setting small incremental and achievable goals; (2) using formalized behavioral contracting to establish goals and specify rewards; and (3) monitoring and reinforcement, including patient self-monitoring by keeping records [6]. In group nutrition programs, it is possible to easily incorporate activities such as cooking demonstrations, problem-solving discussions, and self-monitoring that are rooted in social cognitive theory.

The key social cognitive theory construct of reciprocal determinism means that a person can be both an agent for change and a responder to change. Thus, changes in the environment, the examples of role models, and reinforcements can be used to promote healthier behavior.

This core construct is also central to social ecological models and more important today than ever before.

TABLE 1 Statements Representing Theoretical Approaches to Understanding and Changing Dietary Behavior

Theory	Statement(s)
Social cognitive theory	Overeating at holidays is triggered by food advertisements, store displays, and party buffets that people encounter.
Stages of change	If someone feels that the time is right and she or he is "ready to change," that person will probably be more successful with a nutrition intervention.
Consumer information processing	The information on nutrition labels sometimes overloads consumers. Consumers who are concerned about nutrition tend to look at nutrient labels before deciding which food to buy.
Theory of planned behavior	An individual who plans to change how she or he eats and specifies what, when, how, and where the changes will happen is more likely to follow through than will someone with a more general plan. Further, if the individual thinks that his or her spouse will be supportive, motivation will be higher.
Multiattribute utility theory	If taste and convenience are foremost in someone's mind when deciding what to eat, a nutrition intervention that zeroes in on these factors has the greatest promise of success.
Social ecological model	If healthful food is easily available and low in cost, and the company provides good cooking and food preparation facilities, then workers will be more likely to follow healthy eating patterns.

Source: Adapted in part from Glanz, K., Rimer, B. K., and Lewis, F. M. (Eds.). (2002). "Health Behavior and Health Education: Theory, Research and Practice," 3rd ed. Jossey-Bass, San Francisco; and Glanz, K., and Rudd, J. (1993). Views of theory, research, and practice: a survey of nutrition education and consumer behavior professionals. *J. Nutr. Educ.* **25**, 269–273.

B. Stages of Change

Long-term dietary change for disease prevention or risk reduction involves multiple actions and adaptations over time. Some people may not be ready to attempt changes whereas others may have already begun implementing diet modifications. The construct of "stage of change" is a key element of the transtheoretical model of behavior change and proposes that people are at different stages of readiness to adopt healthful behaviors [27, 28]. The notion of readiness to change, or stage of change, has been examined in dietary behavior research and found useful in explaining and predicting eating habits [29–33].

Stages of change is a heuristic model that describes a sequence of steps in successful behavior change: precontemplation (no recognition of need for or interest in change), contemplation (thinking about changing), preparation (planning for change), action (adopting new habits), and maintenance (ongoing practice of new, healthier behavior) [27]. People do not always move through the stages of change in a linear manner—they often recycle and repeat certain stages; for example, individuals may relapse and go back to an earlier stage depending on their level of motivation and self-efficacy.

The stages of change model can be used both to help understand why patients might not be ready to undertake dietary change, and to improve the success of nutritional intervention [18]. Patients can be classified according to their stage of change by asking a few simple questions: Are they interested in trying to change their eating patterns, thinking about changing their diet, ready to begin a new eating plan, already making dietary changes, or trying to sustain changes they have been following for some time? By knowing their current stage, you can help determine how much time to spend with the patient, whether to wait until he or she is more ready to attempt active changes, whether referral for in-depth nutritional counseling is warranted, and so on. Knowledge of the patient's current stage of change can also lead to appropriate follow-up questions about past efforts to change, obstacles, and challenges, and available strategies for overcoming barriers or obstacles to change [18].

C. Consumer Information Processing

People require information about how to choose nutritious foods in order to follow guidelines for healthy eating. A central premise of consumer information-processing theory is that individuals can process only a limited amount of information at one time [34]. People tend to seek only enough information to make a satisfactory choice. They develop heuristics, or rules of thumb, to help them make choices quickly within their limited information-processing capacity [34]. The nutrition information environment is often complex and confusing, especially when programs rely heavily on print nutrition-education materials that may be written at too sophisticated a level in terms of wording and concepts. Recently, introduction of the new food "Nutrition Facts" food labels have simplified generally available nutrition information somewhat and contributed to an increase in usual label use and satisfaction with their content [35].

There are several implications of consumer information-processing theory for nutrition intervention. Information that is provided should be made easily accessible, not confusing, and understandable with limited effort. Messages that are food focused rather than nutrient focused may be particularly helpful [36]. Nutrition information should be tailored to the comprehension level of the audience, matched to their lifestyles and experience, and either portable or available at or near the point of food selection [34].

Knowledge about which foods to choose and how much to consume on a therapeutic diet are the *sine qua non* of dietary adherence. However, knowledge of how to use nutrition information and the skills to choose or prepare healthful foods are not enough without motivation and support. Further, patients with low literacy skills may require more explanations and fewer printed materials, thus posing important challenges [37]. One community-based dietary fat intervention that used few print materials, emphasized interactive experiences, and was targeted to the cultural backgrounds of participants was successful in promoting desirable dietary changes [38].

D. Theory of Planned Behavior

Often people's food choices are influenced by how they view the actions they are considering and whether they believe important others such as family members or peers would approve or disapprove of their behavior. The theory of planned behavior (TPB), which evolved from its predecessor, the theory of reasoned action (TRA), focuses on the relationships between behavior and beliefs, attitudes, subjective norms, and intentions [39]. The concept of perceived behavioral control involves the belief about whether one can control his or her performance of a behavior [39]—that is, people may feel motivated if they feel they "can do it." A central assumption of TPB is that "behavioral intentions" are the most important determinants of behavior [40].

TPB has been applied widely to help understand and explain many types of behaviors, including eating behavior [41]. As predicted, the core constructs of TPB have been found to predict healthy eating such as fruit and vegetable intake [42] and to help explain low-fat milk consumption [43]. Attitudes, beliefs, and behavioral beliefs about eating have also been examined within community interventions in schools [42] and communities [43] and found to mediate outcomes of the interventions. That is, the theoretical constructs help explain why some people changed and others did not after the nutrition education program or communication campaign.

Although the idea that behavioral intentions are important is central to TPB, there has been some concern that they are still too far removed to be good predictors of actual behavior. The concept of implementation intentions involves encouraging patients or people receiving an intervention to be specific about how they would change. One study found that providing implementation intention prompts led to greater weight reduction in a commercial weight loss program [44].

E. Multiattribute Utility Theory

Both health professionals and marketers recognize that people seek the things they like and that give them pleasure and that they take action to obtain these things. Identifying those concerns that are most important to a person's decision about performing a specific behavior can lead to the development of effective interventions and decision aides to promote desirable behaviors [6]. Multiattribute utility theory (MAU) is a form of value expectancy theory that aims to specify how people define and evaluate the elements of decision making about performing a specific behavior. Key elements of value expectancy theory are the valence, or importance, of a particular feature of a behavior or product; and the expectancy, or subjective probability, that a given consequence will occur if the behavior is performed [45].

MAU is a form of value expectancy theory with particular relevance to understanding influences on food choice and changes in eating habits. MAU posits that people evaluate decisions based on *multiple attributes* and somehow consciously or unconsciously weight the alternatives before deciding what actions to take. The literature on food choice has identified several key factors that appear to be important in food selection: taste, nutrition, cost, convenience, and weight control. For the general public, taste has been reported to be the most important influence on food choice, followed by cost [12]. Understanding the relative importance of various concerns to individuals can guide the design of nutrition counseling and nutrition education programs. For example, by designing and promoting a nutritious diet as tasty, an intervention might be more successful than if it is presented primarily as nutritious or inexpensive. A project that used tailored messages based on alternative food choices and the attributes among them was found to be effective for increasing dietary fiber consumption; it was based on the behavioral alternatives model, which has many similar features to MAU theory [46].

F. Social Ecological Model

The last conceptual model is the social ecological model, which helps to understand factors affecting behavior and also provides guidance for developing successful programs through social environments. Social ecological models emphasize multiple levels of influence (such as individual, interpersonal, organizational, community, and public policy) and the idea that behaviors both shape and are shaped by the social environment [13, 14].

The principles of social ecological models are consistent with social cognitive theory concepts that suggest initially creating an environment conducive to change is important to making it easier to adopt healthy behaviors [25]. Given the widespread problems of overnutrition in developed countries, more attention is being focused toward increasing the health-promoting features of communities and neighborhoods and reducing the ubiquity of high-calorie, high-fat food choices [15, 16].

G. Selecting an Appropriate Theoretical Model or Models

Effective nutrition intervention depends on marshaling the most appropriate theory and practice strategies for a given situation [6]. Different theories are best suited to different individuals and situations. For example, when attempting to overcome patients' personal barriers to changing their diet to reduce their cholesterol level, the theory of planned behavior may be useful. The stages of change model may be especially useful in developing diabetes education interventions. When trying to teach low-literacy patients how to choose and prepare healthy foods, consumer information processing may be more suitable. The choice of the most fitting theory or theories should begin with identifying the problem, goal, and units of practice [6], *not* with simply selecting a theoretical framework because it is intriguing or familiar.

When it comes to practical application, theories are often judged in the context of activities of fellow practitioners. To apply the criterion of *usefulness* to a theory, most providers are concerned with whether it is consistent with everyday observations [6]. In contrast, researchers usually make scientific judgments of how well a theory conforms to observable reality *when empirically tested.* Patient educators should review the research literature periodically to supplement their firsthand experience and that of their colleagues. A central premise in applying an understanding of the influences on health behavior to patient education is that you can gain an understanding of a patient through an interview or written assessment, and better focus in on that individual's readiness, self-efficacy, knowledge level, and so on. Clearly, it is necessary to select a short list of factors to evaluate, and this may differ depending on clinical risk factors or a patient's history. Once there is a good understanding of that person's cognitive or behavioral situation, the intervention can be personalized, or tailored. Tailored messages and feedback have been found to be promising strategies for encouraging healthful behavior changes in primary care, community, and home-based settings [47–49].

The challenge of successfully applying theoretical frameworks in nutrition programs involves evaluating the frameworks and their key concepts in terms of both conceptual relevance and practical value [6]. Also, the integration of multiple theories into a comprehensive model tailored for a given individual or community group requires careful analysis of the audience and frequent reexamination during program design and implementation.

VII. FINDINGS REGARDING APPLICATIONS OF THEORY TO NUTRITIONAL BEHAVIOR

Since the 1980s, there has been an increase in published research applying theoretical models to the study of nutritional behavior [8–11]. Numerous studies have examined the determinants of eating behavior using coherent theoretical frameworks and constructs. Most of this research has been somewhat limited by the use of cross-sectional designs. There continues to be a need for more studies using longitudinal designs, studying families and changing food roles in families, and studying the relationships among various eating behaviors and not just a few nutrients or types of foods.

Since the early 1990s, there has been a substantial increase in research applying the stages of change model to dietary behavior [29, 30, 33, 51]. Several intervention trials have explicitly used this model to help shape their nutrition promotion programs [52–54]. Prospective intervention research examining employees' readiness to change their eating patterns has revealed forward movement across the stage continuum in worksite nutrition studies and shown that changes in stage of change for healthy eating is significantly associated with dietary improvements [55].

Also, people's initial stage of change may influence their participation in nutritional intervention. People who are initially in the later stages of change (preparation, action, and maintenance) tend to spend more time on dietary change [32] and to report making more healthful changes in their food choices [48].

Several large worksite nutrition programs have applied constructs from social cognitive theory (SCT) [52, 53, 56]. Another multisite study, the Working Well Trial, used an intervention rooted in SCT, consumer information processing, and the stages of change [8, 57]. Several of these nutrition interventions have been found effective compared to a control condition, though they have done little to test the elements of social cognitive theory that might be most associated with observed changes.

An extensive evidence review of nutrition interventions for chronic disease risk reduction identified goal setting—and the associated monitoring—as promising elements in most effective interventions [58]. Similarly, developments in theory-based nutrition interventions have emphasized the role of self-monitoring in many forms [59]. Self-monitoring can be seen as emerging from social cognitive theory and also as a self-regulation approach. The Women's Health Initiative (WHI) dietary modification trial [60] incorporated new self-monitoring tools as this large clinical trial progressed over multiple years [61]. New technologies including personal digital assistants (PDAs) were also tried on a small scale in the WHI, with impressive short-term success [62]. Internet-based interventions for weight loss and weight maintenance have shown encouraging results, especially when combined with email counseling [63–66]. Those programs typically include ongoing self-monitoring including daily weighing, which seems to have been proven to be a positive and helpful element of weight loss maintenance [67].

A widely cited report used multiattribute utility (MAU) theory as the framework for analyzing surveys of a national sample of 2967 adults. The study examined the relative importance of taste, nutrition, cost, convenience, and weight control on personal eating choices. Taste was reported to be the most important overall influence on food choices, followed by cost. The importance of nutrition and weight control was best predicted by respondents' being within a particular health lifestyle cluster [12]. No published reports have explicitly applied MAU to design a nutrition intervention, but a tailored message intervention program used the behavioral alternatives model, which has important similarities to MAU [46].

Although social ecological models are often discussed, especially in the context of the obesity epidemic [16], only a handful of studies explicitly use the ecological model as a foundation for interventions. One study used an intervention aimed at multiple levels of influence (institutional, individual, and group) for low-income postpartum women [68]. Another report used the social ecological model as a theoretical perspective to describe a project to promote systems change in school lunches, including food service meals and vending machines [69]. A multilevel community-wide childhood obesity prevention program in Massachusetts that included school, home, and community environments showed promising results after the first year [70]. Interestingly, this program explicitly focused on community participation in developing the program and evaluation, but it implicitly used a social ecological approach.

VIII. CONSTRUCTS AND ISSUES ACROSS THEORIES

It is important to bear in mind that the various theories that can be used for nutrition intervention are not mutually exclusive. Not surprisingly, they share several constructs and common issues across theories. It is often challenging to sort out the key issues in various models. This section focuses on important issues and constructs across models. The first of these is that successful dietary behavior change depends on a sound understanding of the patient's, or consumer's, view of the world.

A. The Patient's View of the World: Perceptions, Cognitions, Emotions, and Habits

For health professionals who work with patients and provide them with advice on health and lifestyle, adherence to treatment is often disappointingly poor, even in response to relatively simple medical advice. Such poor adherence often arises because patients do not have the necessary behavioral skills to make changes to their diet. Following a heart attack, for example, patients might well understand the importance of adopting such changes to their lifestyle but be unable to make those changes. There will be other circumstances where patients might not understand the importance of such changes and may even believe that such changes pose an additional risk to their health. In other circumstances still, a patient might be experiencing depression or anxiety, such that emotional dysfunction will be a major barrier to compliance.

Traditionally, it has been assumed that the relationship between knowledge, attitudes, and behavior is a simple and direct one. Indeed, over the years, many prevention and patient education programs have been based on the premise that if people understand the health consequences of a particular behavior, they will modify it accordingly. Moreover, the argument goes, if people have a negative attitude toward an existing lifestyle practice and a positive attitude toward change, they will make healthful changes. However, we now know from research conducted since the 1970s that the relationships between knowledge, awareness of the need to change, intention to change, and an actual change in behavior are complex indeed.

Ideally, each patient should be treated as an individual with unique circumstances and health history. Still, epidemiological research indicates that certain demographic subgroups differ in terms of risk factors and health behaviors [22]. Understanding these population trends can help prepare a provider to work with various types of patients. For example, younger persons may feel invulnerable to coronary events, and older adults may be managing multiple chronic conditions. An active middle-aged professional may place returning to his previous level of activity above important health protective actions. These examples serve as a reminder to be sensitive to group patterns, yet to avoid stereotyping in the absence of firsthand evidence about an individual. Within this general context, various theories and models can guide the search for effective ways to reach and positively motivate patients.

B. Behavior Change as a Process

Sustained health behavior change involves multiple actions and adaptations over time. Some people may not be ready to attempt changes, some may be thinking about attempting change, and others may have already begun implementing behavioral modifications. One central issue that has gained wide acceptance is the simple notion that *behavior change is a process, not an event.* Rather, it is important to think of the change process as one that occurs in stages. It is not a question of people deciding one day to change their diet and the next day becoming low-fat eaters for life! Likewise, most people will not be able to dramatically change their eating patterns all at once. The idea that behavior change occurs in a number of steps is not particularly new, but it has gained wider recognition. Indeed, various multistage theories of behavior change date back more than 50 years to the work of Lewin, McGuire, Weinstein, Marlatt and Gordon, and others [6, 71–74].

The notion of readiness to change, or stage of change, has been examined in health behavior research and found useful in explaining and predicting a variety of types of behaviors. Prochaska, Velicer, DiClemente, and their colleagues have been leaders in beginning to formally identify the dynamics and structure of change that underlie both self-mediated and clinically facilitated health behavior change. The construct of stage of change (described earlier in this chapter) is a key element of their transtheoretical model of behavior change and proposes that people are at different stages of readiness to adopt healthful behaviors [27, 28].

While the stages of change construct cuts across various circumstances of individuals who need to change or want to change, other theories also address these processes. Here we look across various models to illustrate four key concerns in understanding the process of behavior change: (1) motivation versus intention, (2) intention versus action, (3) changing behavior versus maintaining behavior change, and (4) the role of biobehavioral factors.

1. Motivation versus Intention

Behavior change is challenging for most people even if they are highly motivated to change. As has already been noted in this chapter, the set of relationships between knowledge, awareness of the need to change, intention to change, and an actual change in behavior is complex indeed. For individuals who are coping with disease and illness and who often have to make significant changes to their lifestyle and other aspects of their lives, this challenge is even greater. According to the transtheoretical model, people in precontemplation are neither motivated nor planning to change, those in contemplation intend to change, and those in preparation are acting on their intentions by taking specific steps toward the action of change [28].

2. Intention versus Action

The transtheoretical model makes a clear distinction between the stages of contemplation and preparation, and overt action [27, 28]. A further application of this distinction comes from the theory of planned behavior (TPB) [39, 40]. The TPB proposes that intentions are the best predictor of behavior [39]. However, researchers are increasingly focusing attention on "implementation intentions" as being more proximal and even better predictors of behavior and behavior change [44].

3. Changing Behaviors versus Maintaining Behavior Change

Even where this is good initial compliance to a lifestyle change program, such as changing diet, relapse is common. It is widely recognized that many overweight persons are able to lose weight, only to regain it within a year. Thus, it has become clear to researchers and clinicians that undertaking initial behavior changes and maintaining behavior change require different types of strategies. The Transtheoretical Method (TTM) distinction between "action" and "maintenance" stages implicitly addresses this phenomenon [27, 28]. Another model that is not described in detail here, Marlatt and Gordon's relapse prevention model, specifically focuses on strategies for dealing with maintenance of a recently changed behavior [74]. It involves developing self-management and coping strategies and establishing new behavior patterns that emphasize perceived control, environmental management, and improved self-efficacy. These strategies are an eclectic mix drawn from social cognitive theory [25], the theory of planned behavior [39, 40], applied behavioral analysis, and the forerunners of the stages of change model.

C. Biobehavioral Factors

The behavioral and social theories described thus far have some important limitations, many of which are only now beginning to be understood. Notably, for some health behaviors—especially addictive or addiction-like behaviors—there are other important determinants of behavior, which may be physiological or metabolic. Among the best known are the addictive effects of nicotine, alcohol, and some drugs. Physiological factors increase psychological cravings and create withdrawal syndromes that may impede even highly motivated persons from changing their behaviors (quitting smoking, not consuming alcoholic beverages, etc.). Some behavior changes—for example, weight loss—also affect energy metabolism and make long-term risk factor reduction an even greater challenge than it would be if it depended on cognitive-behavioral factors alone. Research into the psychobiology of appetite offers intriguing possibilities for understanding biobehavioral models of food intake.

D. Barriers to Actions, Pros and Cons, and Decisional Balance

According to social cognitive theory [25], a central determinant of behavior involves the interaction between individuals and their environments. Behavior and environment are said to continuously interact and influence one another, which is known as the principle of *reciprocal determinism*. The concept of barriers to action, or perceived barriers, can be found in several theories of health behavior, either explicitly or as an application. It is part of social cognitive theory [25] and the theory of planned behavior [39]. In the transtheoretical model, there are parallel constructs labeled as the "pros" (the benefits of change) and "cons" (the costs of change) [27, 28]. Taken together, these constructs are known as "decisional balance."

The idea that individuals engage in relative weighing of the pros and cons has its origins in Janis and Mann's model of decision making, published in their seminal study released in the 1970s [75], although the idea had emerged much earlier in social psychological discourse. Lewin's idea of force field analysis [72] and other work on persuasion and decision counseling by Janis and Mann predated that important work. Indeed, this notion is basic to models of rational decision making, in which people intellectually think about the advantages and disadvantages, obstacles and facilitators, barriers and benefits, or pros and cons of engaging in a particular action.

E. Control over Behavior and Health: Control Beliefs and Self-Efficacy

Sometimes, "control beliefs" and self-efficacy hold people back from achieving better health. These deterrents to positive health behavior change are common and can be found in several models of health behavior, including social cognitive theory, the theory of planned behavior, and relapse prevention. One of the most important challenges for these models—and ultimately for health professionals who apply them—is to enhance perceived behavioral control and increase self-efficacy, thereby improving patients' motivation and persistence in the face of obstacles.

IX. IMPLICATIONS AND OPPORTUNITIES

Theory and research suggest that the most effective nutrition interventions are those that use multiple strategies and aim to achieve multiple goals of awareness, information transmission, skill development, and supportive environments and policies [58]. The range of nutrition intervention tools and techniques is extensive and varied. Programs differ based on their goals and objectives, the needs of clients, and the available resources, staff, and expertise.

Nutrition programs can stand alone or be part of broader, multicomponent and multiple-focus health promotion and patient education programs.

What can be expected? Program design relates closely to what one can expect in terms of results. Generally speaking, minimally intensive intervention efforts such as one-time group education sessions can reach large audiences but seldom lead to behavior changes. More intensive programs typically appeal to at-risk or motivated groups, cost more to offer, and can achieve relatively greater changes in knowledge, attitudes, and eating patterns [3].

Nutrition interventions must be sensitive to audience and contextual factors. Food selection decisions are made for many reasons other than just nutrition: taste, cost, convenience, and cultural factors all play significant roles [12]. Dietary change strategies must take these issues into consideration. The health promotion motto "know your audience" has a true and valuable meaning. Planning processes can consider multiple theories in a systematic way through approaches such as intervention mapping [76].

Further, change is incremental. Many people have practiced a lifetime of less than optimal nutrition behaviors. It is unreasonable to expect that significant and lasting changes will occur during the course of a program that lasts only a few months. Programs need to pull participants along the continuum of change, being sure to be just in front of those most ready to change with attractive, innovative offerings.

In population-focused programs, it appears to be of limited value to adopt a program solely oriented toward modifying individual choice (e.g., teaching and persuading individuals to choose low-fat dairy products). A more productive strategy would also include environmental change efforts (e.g., expanding the availability of more nutritious food choices) [13–15]. When this is done along with individual skill training, long-lasting and meaningful changes can be achieved.

Finally, when planning interventions we should strive to be creative. Nutrition interventions should be as entertaining and engaging as the other activities they are competing with. People will want to participate if they can have fun with the nutrition programs. Communication technologies are opening many different channels for engaging people's interest in better nutrition. The communication of nutrition information, now matter how important it is to good health, is secondary to attracting and retaining the interest and enthusiasm of the audience.

References

1. Glanz, K., and Gilboy, M. B. (1992). Physicians, preventive, care, and applied nutrition: selected literature. *Acad. Med.* **67**, 776–781.
2. U.S. Preventive Services Task Force. (2003). Behavioral counseling in primary care to promote a healthy diet: recommendations and rationale. *Am. J. Prev. Med.* **24**, 93–100.
3. Pignone, M. P., Ammerman, A., Fernandez, L., Orleans, C. T., Pender, N., Woolf, S., Lohr, K. N., and Sutton, S. (2003). Counseling to promote a healthy diet in adults: a summary of the evidence for the U.S. Preventive Services Task Force. *Am. J. Prev. Med.* **24**, 75–92.
4. Hunt, J. R., Kristal, A. R., White, E., Lynch, J. C., and Fries, E. (1995). Physician recommendations for dietary change: their prevalence and impact in a population-based sample. *Am. J. Public Health* **185**, 722–726.
5. Katz, D. L., O'Connell, M., Yeh, M. C., Nawaz, H., Njike, V., Anderson, L. M., Cory, S., and Dietz, W.; Task Force on Community Preventive Services. (2005). Public health strategies for preventing and controlling overweight and obesity in school and worksite settings: a report on recommendations of the Task Force on Community Preventive Services. *MMWR Recommend. Rep.* **54**, 1–12.
6. Glanz, K., Rimer, B. K., and Lewis, F. M. (2002). "Health Behavior and Health Education: Theory, Research and Practice," 3rd ed. Jossey-Bass, San Francisco.
7. Glanz, K., and Seewald-Klein, T. (1986). Nutrition at the worksite: an overview. *J. Nutr. Educ.* **18**(1, suppl), S1–S12.
8. Glanz, K., and Eriksen, M. P. (1993). Individual and community models for dietary behavior change. *J. Nutr. Educ.* **25**, 80–86.
9. Glanz, K. (1997). Behavioral research contributions and needs in cancer prevention and control: dietary change. *Prev. Med.* **26**, S43–S55.
10. Contento, I. (1995). Nutrition education and implications. *J. Nutr. Educ.* **27**, special issue.
11. Achterberg, C., and Miller, C. (2004). Is one theory better than another in nutrition education? A viewpoint: more is better. *J. Nutr. Educ. Behav.* **36**, 40–42.
12. Glanz, K., Basil, M., Maibach, E., Goldberg, J., and Snyder, D. (1998). Why Americans eat what they do: taste, nutrition, cost, convenience, and weight control as influences on food consumption. *J. Am. Diet. Assoc.* **98**, 1118–1126.
13. McLeroy, K., Bibeau, D., Steckler, A., and Glanz, K. (1988). An ecological perspective on health promotion programs. *Health Educ. Q.* **15**, 351–377.
14. Sallis, J., and Owen, N. (2002). Ecological models. *In* "Health Behavior and Health Education: Theory, Research, and Practice," 3rd ed. (K. Glanz, B. K. Rimer, and F. M. Lewis, Eds.), pp. 462–484. Jossey-Bass.
15. Glanz, K., Lankenau, B., Foerster, S., Temple, S., Mullis, R., and Schmid, T. (1995). Environmental and policy approaches to cardiovascular disease prevention through nutrition: opportunities for state and local action. *Health Educ. Q.* **22**, 512–527.
16. Glanz, K., Sallis, J. F., Saelens, B. E., and Frank, L. D. (2005). Healthy nutrition environments: concepts and measures. *Am. J. Health Promot.* **19**, 330–333.
17. Glanz, K., Hewitt, A. M., and Rudd, J. (1992). Consumer behavior and nutrition education: an integrative review. *J. Nutr. Educ.* **24**, 267–277.
18. Glanz, K. (1992). Nutritional intervention: a behavioral and educational perspective. *In* "Prevention of Coronary Heart Disease" (I. S. Ockene, and J. K. Ockene, Eds.), pp. 231–265. Little, Brown, Boston.
19. Glanz, K. (1988). Patient and public education for cholesterol reduction: a review of strategies and issues. *Patient Educ. Couns.* **12**, 235–257.

20. Third Report of the National Cholesterol Education Program (NCEP) Expert Panel on Detection, Evaluation and Treatment of High Blood Cholesterol in Adults (Adult Treatment Panel III): Executive Summary. 2001. National Institutes of Health National Heart, Lung, and Blood Institute (NIH Publication 01-3670).
21. Morreale, S. J., and Schwartz, N. E. (1995). Helping Americans eat right: developing practical and actionable public nutrition education messages based on the ADA Survey of American Dietary Habits. *J. Am. Diet. Assoc.* **95** 305–308.
22. Wang, Y., and Beydoun, M. (2007). The obesity epidemic in the United States—gender, age, socioeconomic, racial/ethnic, and geographic characteristics: a systematic review and meta-regression analysis. *Epidemiol. Rev.* **29**, 6–28.
23. Kumanyika, S., and Grier, S. (2006). Targeting interventions for ethnic minority and low-income populations. *Future Child* **16**, 187–207.
24. Glanz, K., and Rudd, J. (1993). Views of theory, research, and practice: a survey of nutrition education and consumer behavior professionals. *J. Nutr. Educ.* **25**, 269–273.
25. Bandura, A. (1986). "Social Foundations of Thought and Action: A Social Cognitive Theory." Prentice-Hall, Englewood Cliffs, NJ.
26. Bandura, A. (1997). "Self-Efficacy: The Exercise of Control." W. H. Freeman, New York.
27. Prochaska, J. O., DiClemente, C. C., and Norcross, J. (1992). In search of how people change: applications to addictive behaviors. *Am. Psychol.* **47**, 1102–1114.
28. Prochaska, J. O., Redding, C., and Evers, K. (2002). The transtheoretical model of behavior change. *In* "Health Behavior and Health Education: Theory, Research, and Practice," 3rd ed. (K. Glanz, B. K. Rimer, and F. M. Lewis, Eds.), pp. 99–120. Jossey-Bass, San Francisco.
29. Glanz, K., Patterson, R., Kristal, A., DiClemente, C., Heimendinger, J., Linnan, L., and McLerran, D. (1994). Stages of change in adopting healthy diets: fat, fiber and correlates of nutrient intake. *Health Educ. Q.* **21**, 499–519.
30. Greene, G. W., Rossi, S. R., Reed, G. R., Willey, C., and Prochaska, J. O. (1994). Stage of change for reducing dietary fat to 30% of energy or less. *J. Am. Diet. Assoc.* **94**, 1105–1110.
31. Glanz, K., Kristal, A. R., Tilley, B. C., and Hirst, K. (1998). Psychosocial correlates of healthful diets among male auto workers. *Cancer Epid. Biomarkers Prev.* **7**, 119–126.
32. Glanz, K., Patterson, R. E., Kristal, A. R., Feng, Z., Linnan, L., Heimendinger, J., and Hebert, J. R. (1998). Impact of work site health promotion on stages of dietary change: the Working Well Trial. *Health Educ. Behav.* **25**, 448–463.
33. Kristal, A. R., Glanz, K., Curry, S. J., and Patterson, R. E. (1999). How can stages of change be best used in dietary interventions? *J. Am. Diet. Assoc.* **99**, 679–684.
34. Rudd, J., and Glanz, K. (1990). How consumers use information for health action: consumer information processing. *In* "Health Behavior and Health Education: Theory, Research, and Practice," (K. Glanz, F. M. Lewis, and B. K. Rimer, Eds.), pp. 115–139. Jossey-Bass, San Francisco.
35. Kristal, A., Levy, L., Patterson, R., Li, S., and White, E. (1998). Trends in food label use associated with new nutrition labeling regulations. *Am. J. Public Health* **88**, 1212–1215.
36. Gehling, R. K., Magarey, A. M., and Daniels, L. A. (2004). Food-based recommendations to reduce fat intake: an evidence-based approach to the development of a family-focused child weight management programme. *J. Paediatr. Child Health* **41**, 112–118.
37. Macario, E., Emmons, K.M., Sorensen, G., Hunt, M. K., and Rudd, R. E. (1998). Factors influencing nutrition education for patients with low literacy skills. *J. Am. Diet. Assoc.* **98**, 559–564.
38. Howard-Pitney, B., Winkleby, M., Albright, C. L., Bruce, B., and Fortmann, S. P. (1997). The Stanford Nutrition Action Program: a dietary fat intervention for low-literacy adults. *Am. J. Public Health* **87**, 1971–1976.
39. Montano, D. E., and Kasprzyk, D. (2002). The theory of reasoned action and the theory of planned behavior. *In* "Health Behavior and Health Education: Theory, Research, and Practice," 3rd ed. (K. Glanz, B. K. Rimer, and F. M. Lewis, Eds.), pp. 67–98. Jossey-Bass, San Francisco.
40. Ajzen, I. (1991). The theory of planned behavior. *Org. Behav. Hum. Decis. Proc.* **50**, 179–211.
41. Anderson, A. S., Cox, D.N., McKellar, S., Reynolds, J., Lean, M. E., and Mela, D. J. (1998). Take Five, a nutrition education intervention to increase fruit and vegetable intakes: impact on attitudes toward dietary change. *Br. J. Nutr.* **80**, 133–140.
42. Lien, N., Lytle, L. A., and Komro, K. A. (2002). Applying theory of planned behavior to fruit and vegetable consumption of young adults. *Am. J. Health Promot.* **16**, 189–197.
43. Butterfield-Booth, S., and Reger, B. (2004). The message changes belief and the rest is theory: the "1% or less" milk campaign and reasoned action. *Prev. Med.* **39**, 581–588.
44. Luszczynska, A., Sobczyk, A., and Abraham, C. (2007). Planning to lose weight: randomized controlled trial of an implementation intention prompt to enhance weight reduction among overweight and obese women. *Health Psychol.* **26**, 507–512.
45. Carter, W. B. (1990). Health behavior as a rational process: theory of reasoned action and multiattribute utility theory. *In* "Health Behavior and Health Education: Theory, Research, and Practice," 3rd ed. (K. Glanz, B. K. Rimer, and F. M. Lewis, Eds.), pp. 63–91. Jossey-Bass, San Francisco.
46. Brinberg, D., Axelson, M .L., and Price, S. (2000). Changing food knowledge, food choice, and dietary fiber consumption by using tailored messages. *Appetite* **35**, 35–43.
47. Campbell, M. K., DeVellis, B. M., Strecher, V. J., Ammerman, A. S., DeVellis, R. F., and Sandler, R. S. (1994). Improving dietary behavior: the effectiveness of tailored messages in primary care settings. *Am. J. Public Health* **84**, 783–787.
48. Beresford, S. A., Curry, S. J., Kristal, A. R., Lazovich, D., Feng, Z., and Wagner, E. H. (1997). A dietary intervention in primary care practice: The Eating Patterns Study. *Am. J. Public Health* **87**, 610–616.
49. Brug, J., Glanz, K., Van Assema, P., Kok, G., and Van Breukelen, G. (1998). The impact of computer-tailored feedback and iterative feedback on fat, fruit, and vegetable intake. *Health Educ. Behav.* **25**, 517–531.
50. Kristal, A., Patterson, R., Glanz, K., Heimendinger, J., Hebert, J., Feng, Z., and Probart, C. (1995). Psychosocial correlates of healthful diets: baseline results from the Working Well Study. *Prev. Med.* **24**, 221–228.

51. Campbell, M. K., Reynolds, K. D., Havas, S., Curry, S., Bishop, D., Nicklas, T., Palombo, R., Buller, D., Feldman, R., Topor, M., Johnson, C., Beresford, S., Motsinger, B., Morrill, C., and Heimendinger, J. (1999). Stages of change for increasing fruit and vegetable consumption among adults and young adults participating in the National 5 A Day for Better Health Community Studies. *Health Educ. Behav.* **26**, 513–534.
52. Sorensen, G., Thompson, B., Glanz, K., Feng, Z., Kinne, S., DiClemente, C., Emmons, K., Heimendinger, J., Probart, C., and Lichtenstein, E. (1996). Working well: results from a worksite-based cancer prevention trial. *Am. J. Public Health* **86**, 939–947.
53. Tilley, B., Glanz, K., Kristal, A., Hirst, K., Li, S., Vernon, S., and Myers, R. (1999). Nutrition intervention for high-risk auto workers: results of the Next Step Trial. *Preventive Medicine* **28**, 284–292.
54. Sorensen, G., Stoddard, A., and Macario, E. (1998). Social support and readiness to make dietary changes. *Health Educ. Behav.* **25**, 586–598.
55. Kristal, A., Glanz, K., Tilley, B. C., and Li, S. (2000). Mediating factors in dietary change: understanding the impact of a worksite nutrition intervention. *Health Educ. Behav.* **27**, 112–125.
56. Sorensen, G., Hunt, M. K., Cohen, N., Stoddard, A., Stein, E., Phillips, J., Baker, F., Combe, C., Hebert, J., and Palombo, R. (1998). Worksite and family education for dietary change: the Treatwell 5 A Day Program. *Health Educ. Res.* **13**, 577–591.
57. Abrams, D., Boutwell, W. B., Grizzle, J., Heimendinger, J., Sorensen, G., and Varnes, J. (1994). Cancer control at the workplace: the Working Well Trial. *Prev. Med.* **23**, 15–27.
58. Ammerman, A. S., Lindquist, C. H., Lohr, K. N., and Hersey, J. (2002). The efficacy of behavioral interventions to modify dietary fat and fruit and vegetable intake: a review of the evidence. *Prev. Med.* **35**, 25–41.
59. Burke, L. E., Warziski, M., Starrett, T., Choo, J., Music, E., Sereika, S., Stark, S., and Sevick, M. A. (2005). Self-monitoring dietary intake: current and future practices. *J. Renal Nutr.* **15**, 281–290.
60. Ritenbaugh, C., Patterson, R., Chlebowski, R., Caan, B., Fels-Tinker, L., Howard, B., and Ockene, J. (2003). The Women's Health Initiative dietary modification trial: overview and baseline characteristics of participants. *Ann. Epidemiol.* **13**, 87–97.
61. Mossavar-Rahmani, Y., Henry, H., Rodabaugh, R., Bragg, C., Brewer, A., Freed, T., Kinzel, L., Pedersen, M., Soule, C. O., and Vosburg, S. (2004). Additional self-monitoring tools in the dietary modification component of the Women's Health Initiative. *J. Am. Diet. Assoc.* **104**, 76–85.
62. Glanz, K., Murphy, S., Moylan, J., Evensen, D., and Curb, J. D. (2006). Improving dietary self-monitoring and adherence with hand-held computers: a pilot study. *Am. J. Health Promot.* **20**, 165–170.
63. Tate, D. F., Jackvony, E. H., and Wing, R. R. (2003). Effects of Internet behavioral counseling on weight loss in adults at risk for type 2 diabetes: a randomized trial. *JAMA* **289**, 1833–1836.
64. Tate, D. F., Jackvony, E. H., and Wing, R. R. (2006). A randomized trial comparing human e-mail counseling, computer-automated tailored counseling, and no counseling in an Internet weight loss program. *Arch. Intern. Med.* **166**, 1620–1625.
65. Wing, R. R., Tate, D. F., Gorin, A. A., Raynor, H. A., and Fava, J. L. (2006). A self-regulation program for maintenance weight loss. *N. Eng. J. Med.* **355**, 1563–1571.
66. Saperstein, S. L., Atkinson, N. L., and Gold, R. S. (2007). The impact of Internet use for weight loss. *Obes. Rev.* **8**, 459–465.
67. Wing, R. R., Tate, D. F., Gorin, A. A., Raynor, H. A., Fava, J. L., and Machan, J. (2007). STOP regain: are there negative effects of daily weighing? *J. Consult. Clin. Psychol.* **75**, 652–656.
68. Peterson, K. E., Sorensen, G., Pearson, M., Hebert, J. R., Gottlieb, B. R., and McCormick, M. C. (2002). Design of an intervention addressing multiple levels of influence on dietary and activity patterns of low-income, postpartum women. *Health Educ. Res.* **17**, 531–540.
69. Suarez-Balcazar, Y., Redmond, L., Kouba, J., Hellwig, M., Davis, R., Martinez, L. I., and Jones, L. (2007). Introducing systems change in the schools: the case of luncheons and vending machines. *Am. J. Commun. Psychol.* **39**, 335–354.
70. Economos, C. D., Hyatt, R., Goldberg, J. P., Must, A., Naumova, E. N., Collins, J. J., and Nelson, M. E. (2007). A community intervention reduces BMI z-score in children: Shape-Up Somerville first year results. *Obesity* **15**, 1325–1336.
71. Weinstein, N. D. (1993). Testing four competing theories of health-protective behavior. *Health Psychol.* **12**, 324–333.
72. Lewin, K. (1935). "A Dynamic Theory of Personality." McGraw Hill, New York.
73. McGuire, W. J. (1984). Public communication as a strategy for inducing health promoting behavioral change. *Prev. Med.* **13**, 299–313.
74. Marlatt, A. G., and Gordon, J. R. (1985). "Relapse Prevention: Maintenance Strategies in the Treatment of Addictive Behaviors." Guilford Press, New York.
75. Janis, I., and Mann, L. (1977). "Decision Making: A Psychological Analysis of Conflict." Free Press, New York.
76. Brug, J., Oenema, A., and Ferriera, I. (2005). Theory, evidence and intervention mapping to improve behavior nutrition and physical activity interventions. *Int. J. Behav. Nutr. Phys. Act.* **2**, 2.

CHAPTER 9

Nutrition Intervention: Lessons from Clinical Trials

LINDA G. SNETSELAAR
University of Iowa, Iowa City, Iowa

Contents

I. INTRODUCTION

The modification of dietary patterns to prevent and optimize the management of chronic disease traditionally has been perceived as a difficult and challenging task. However, much has been learned since the 1980s about how to successfully modify eating patterns. For example, in clinical trials several diet intervention studies focused on the prevention of cancer or cardiovascular disease have demonstrated the feasibility of reducing dietary fat intake in targeted groups. Additionally, complex dietary modifications testing the effect of diet on the progression of renal disease have also been successfully achieved.

II. CONCEPTUAL MODELS OF MOTIVATION

A. Self-Regulation Theory

This theory, originally described by Kanfer, states that behavior is regulated by cycles that involve self-monitoring, comparing goal achievement with expectations, and correcting the course of action when the goal is not met [1, 2]. To change dietary behavior, a person seeks to increase knowledge of the discrepancy between current status and the identified goal. Two ways to accomplish this are (1) to increase the awareness of current status (e.g., through feedback such as dietary self-monitoring) or (2) to change the goal to make it more attainable. In conflict situations, when a goal is desired and yet not seen as important enough to strive to attain, ambivalence (feeling at least two different ways about something) is a normal, key obstacle to dietary change.

B. Rokeach's Value Theory

Studies in persons who have undergone sudden transformation shifts in behavior show that personality is organized around concentric layers [3]. An individual's attitudes, numbering in the thousands, represent an organizational series of steps inward. More central are our beliefs and even more central are our core personal values. The most central is the sense of personal identity. The more central the shift occurs, the more likely that the resulting behavior change will be maintained over time.

C. Health Belief Model

The health belief model is attempts to explain and predict health behaviors by focusing on the attitudes and beliefs of individuals. The key variables of the health belief model are as follows [4]:

1. *Degree of perceived risk of a disease.* This variable includes perceived susceptibility of contracting a health condition associated with lack of a healthy diet and its perceived severity once the disease is contracted.
2. *Perceived benefits of diet adherence.* A second benefit is the believed effectiveness of dietary strategies designed to help reduce the threat of disease.
3. *Perceived barriers to diet adherence.* This variable includes potential negative consequences that may result from changing dietary patterns, including physical (weight gain or loss), psychological (lack of spontaneity in food selection), and financial demands (cost of new foods).
4. *Cues to action.* Events that motivate people to take action in changing their dietary habits are crucial determinants of change.

5. *Self-efficacy*. A very important variable is the belief in being able to successfully execute the dietary behavior required to produce the desired outcomes [5–7].
6. *Other variables.* Demographic, sociopsychological, and structural variables affect an individual's perceptions of dietary change and thus indirectly influence his or her ability to sustain new eating behaviors.

Motivation for change depends on the presence of a sufficient degree of perceived risk in combination with sufficient self-efficacy relative to achieving dietary change. Perceived risk without self-efficacy tends to result in defensive cognitive coping, such as denial and rationalization rather than behavior change.

D. Decisional Balance

The classic Janis and Mann decisional balance model [8] is a rational view, describing a decision as a process of weighing cognitively the pros and cons of change. Change depends on the pros of change outweighing the cons. Determining the strengths and weaknesses of a decision can be facilitated by the counselor who is adept at reflective listening skills designed to assist the client in reviewing those ideas that result in an optimum final choice.

E. Interaction

According to Miller and Rollnick [9], motivation can be thought of not as a client trait but as an interpersonal process between nutrition counselor and participant. Rather than seeing motivational change as something the client achieves, this process is one that the nutrition counselor and the client experience in tandem.

III. THEORIES USED IN ACHIEVING DIETARY BEHAVIOR CHANGE IN CLINICAL TRIALS

The nutrition components of clinical trials require skills in long-term dietary maintenance. These skills go beyond educating participants and instead involve strategies designed to reinitiate participants who no longer comply with the recommended eating plan. The studies described here provide research data collected when the theories presented earlier are initiated in a clinical trial setting.

A. Women's Health Initiative

The Women's Health Initiative (WHI) [10, 11, 12] is a randomized controlled clinical trial designed to look at prevention of breast and colorectal cancer. The dietary arm of this study focused on a diet with 20% energy from fat plus five servings of fruits and vegetables and six servings of grain products per day.

To accomplish this change in dietary habits, nutritionists in the study used a variety of behavior change techniques based on the models discussed here. The stages of change model drove efforts to increase compliance in the WHI. The Prochaska–DiClemente model includes six designated stages of change: precontemplation, contemplation, determination, action, maintenance, and relapse [13]. In an effort to simplify and accommodate different levels of adherence, WHI investigators chose to use only three levels of readiness to change: ready to change, unsure, and not ready to change. The decision to simplify levels is based on work with study participants showing that strategies to modify behavior fall within these three categories.

To test the effectiveness of using motivational strategies targeted at these three levels of change, a small research study was devised. Results of that study showed a positive change in dietary behavior following its implementation [14]. In this pilot study, researchers evaluated an intensive intervention program with diet. The basis of the program was use of motivational interviewing with participants in the WHI. The goal was to meet the study nutrition goal of 20% energy from fat.

WHI dietary intervention participants ($n = 175$) from three clinical centers were randomized to intervention or control status. Those randomized to the intensive intervention program participated in three individual motivational interviewing contacts from a nutritionist, plus the usual WHI dietary intervention. Those randomized to the control group continued with the usual WHI dietary management intervention. Percentage energy from fat was estimated at intensive intervention program baseline and intensive intervention program follow-up (1 year later) using the WHI food frequency questionnaire (FFQ).

The change in percentage energy from fat between the intensive intervention program and control group at baseline and 1-year follow-up was -1.2 percentage points from total fat for intensive intervention program participants and $+1.4$ percentage points for control participants. The result was an overall difference of 2. 6 percentage points ($p \leq 0.001$).

Table 1 presents summary statistics on the intensive intervention program effects comparing baseline levels of fat consumption. The changes in fat consumed varied by intensive intervention program baseline fat intake as a percentage of energy intake. Participants having the highest intensive intervention program baseline fat intake ($\geq$30% energy) showed the largest overall change in percentage energy from fat between intensive intervention program baseline and intensive intervention program follow-up. As might be expected, the smallest change was found in participants who consumed between 25% and 30% of energy from fat at intensive intervention program baseline. These participants were closer to their goal of 20% energy from fat at baseline, allowing for less overall change.

TABLE 1 Effect of IIP Intervention on FFQ Percentage Engergy from Fat Stratified by Baseline Percentage Energy from Fat[a]

	n	Baseline X (SD)	Follow-up X (SD)	Difference
% Energy from fat: <20.0				
Intervention control	23	17.75(1.8)	17.86(3.9)	0.1
	25	17.35(2.3)	19.70(4.5)	2.3
Difference[b]		−0.4	1.8	2.2
% Energy from fat: =20.0 and <25.0				
Intervention control	25	22.72(1.4)	21.68(4.6)	−1.0
	26	23.17(1.7)	25.29(4.8)	2.1
Difference[b]		0.5	3.6	3.1
% Energy from fat: =25.0 and <30.0				
Intervention control	21	27.42(1.6)	26.3(4.6)	−1.1
	15	26.94(1.2)	26.89(4.6)	0.0
Difference[b]		−0.5	0.6	1.1
% Energy from fat: =30.0				
Intervention control	13	34.24(2.5)	30.11(6.5)	−4.1
	16	33.81(3.1)	33.82(5.0)	0.0
Difference[b]		−0.4	3.7	4.1

Source: Modified from Bowen, D., Ehret, C., Pedersen, M., Snetselaar, L., and Johnson, M. (2002). Results of an adjunct dietary intervention program in the Women's Health Initiative. *J. Am. Diet. Assoc.* **102**, 1631–1637.

[a]Participants with missing FFQ data were excluded.

[b]$p<0.05$ using paired *t*-test.

The results of this study show that a protocol based on motivational interviewing and delivered through contacts with trained nutritionists is effective. Those subjects who participated in the intervention arm of the study further lowered their dietary fat intake to achieve study goals.

B. Diet Intervention Study in Children

A similar protocol was used in the Diet Intervention Study in Children (DISC) [15]. DISC was a randomized, multicenter clinical trial assessing the efficacy and safety of lowering dietary fat to decrease low-density lipoprotein cholesterol in children at high risk for cardiovascular disease [16, 17]. Children began this study between ages 7 and 10 and participated in group dietary intervention programs. As they moved into adolescence (ages 13 to 17) and encountered added obstacles to dietary adherence and retention, researchers in the study designed and implemented an individual-level motivational intervention. The diet prescription in the DISC study required providing 28% energy from total fat, less than 8% energy from saturated fat, up to 9% energy from polyunsaturated fat, and less than 75 mg/1000 kcal/day of cholesterol. The diet met the age- and sex-specific Recommended Dietary Allowance for energy, protein, and micronutrients.

Researchers used a pre- to postintervention design among a subset of the total intervention cohort ($n = 334$). The first 127 participants who appeared for regularly scheduled intervention visits after implementation of the new intervention method were considered part of the study. These participants ranged from 13 to 17 years of age with equal numbers of boys and girls. Nutrition interventionists asked all of the 127 participants to return in 4 to 8 weeks for a follow-up session. Initial sessions were conducted in person, and follow-up sessions were conducted either in person or by telephone.

Three 24-hour dietary recalls were collected within 2 weeks after the follow-up session. These dietary data were compared to three baseline 24-hour dietary recalls collected in the year preceding initial exposure to the motivational intervention method.

Self-reported data were also collected. At initial and follow-up intervention sessions, participants were shown "assessment rulers" (see Fig. 1) numbered 1 to 12 and asked to rate their adherence to dietary guidelines and their readiness to make new or additional dietary changes.

Results from the study show that the mean energy from total fat decreased from 27.7% to 25.6% ($p < 0.001$) (Table 2) and the mean energy from saturated fat decreased from 9.5% to 8.6% of total energy intake ($p < 0.001$). Additionally, dietary cholesterol decreased from 182.9 to 157.3 mg/1000 kcal ($p < 0.003$). A comparison of males and females showed no differences in gender relative to study results. Note that for this preliminary test, no control group was randomly assigned or examined. Therefore, the researchers cannot predict if significant reductions in consumption of dietary fat and cholesterol are attributable to the intervention.

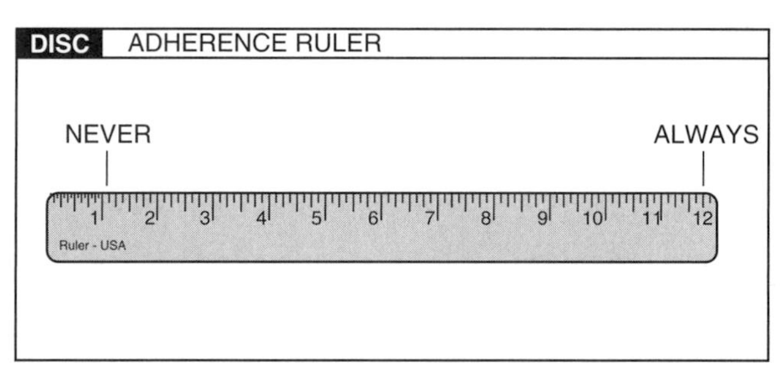

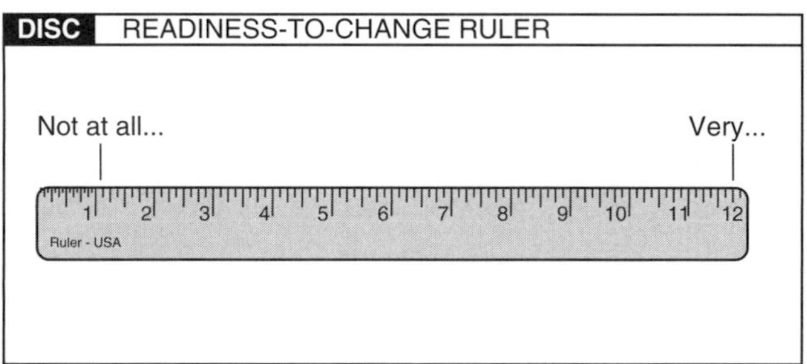

FIGURE 1 Assessment rulers. (Data from Berg-Smith, S. M., Stevens, V. J., Brown, K. M., VanHorn, L., Gernhofer, N., Peters, E., Greenberg, R., Snetselaar, L., Ahrens, L., and Smith, K., for the Dietary Intervention Study in Children [DISC] Research Group. [1999]. A brief motivational intervention to improve dietary adherence in adolescents. *Health Educ. Res.* **14**, 399–410.)

The self-reported adherence rating score and readiness to change score increased by approximately 1 point on a scale from 1 to 12 (both $p < 0.001$). To help accomplish goals, action plans were also made. The study results show that 94% of the participants made action plans and 89% successfully implemented them.

TABLE 2 Changes in Total Fat Intake, Saturated Fat Intake, and Dietary Cholesterol after Two Intervention Sessions

	Mean	SD	*p* Level
Total Fat Intake			
Baseline	27.7	6.1	—
Follow-up	25.6	6.1	—
Change	−2.1	7.0	<0.001
Saturated Fat Intake			
Baseline	9.5	2.7	—
Follow-up	8.6	2.4	—
Change	−0.9	3.1	<0.001
Dietary Cholesterol			
Baseline	182.9	97.6	—
Follow-up	157.3	87.6	—
Change	−25.6	92.3	<0.003

Source: Data from Berg-Smith, S. M., Stevens, V. J., Brown, K. M., VanHorn, L., Gernhofer, N., Peters, E., Greenberg, R., Snetselaar, L., Ahrens, L., and Smith, K., for the Dietary Intervention Study in Children (DISC) Research Group. (1999). A brief motivational intervention to improve dietary adherence in adolescents. *Health Educ. Res.* **14**, 399–410.

TABLE 3 Nutrition Counselor Satisfaction with the Motivational Intervention Method

Level of Satisfaction	Percentage of the Intervention Sessions
Very satisfying	39%
Satisfying	35%
Somewhat satisfying	19%
Slightly or not satisfying	7%

Source: Data from Berg-Smith, S. M., Stevens, V. J., Brown, K. M., VanHorn, L., Gernhofer, N., Peters, E., Greenberg, R., Snetselaar, L., Ahrens, L., and Smith, K., for the Dietary Intervention Study in Children (DISC) Research Group. (1999). A brief motivational intervention to improve dietary adherence in adolescents. *Health Educ. Res.* **14**, 399–410.

This study also examined counselor satisfaction. The results showed that nearly three-quarters of the nutrition counselors were satisfied or very satisfied with using the motivational intervention methods (Table 3).

C. Motivational Intervention Method

Figure 2 provides a method for establishing rapport before tailoring the intervention to the readiness to change level: ready to change, unsure, and not ready to change. Figure 3 provides specific strategies for each level of change.

1. First Level: Not Ready to Change

The main goal for this level of intervention is to raise awareness of the need to continue meeting goals (e.g., fat grams, carbohydrate grams, energy intake). Additionally, to achieve this goal it is necessary to reduce resistance and barriers to meeting goals (e.g., decreasing cues to eat high-fat foods). Also, importantly, focus is placed on increasing interest in considering behavioral steps toward meeting the goals noted earlier.

Throughout the initial interview, when working with a patient in this level, it is important to ask open-ended questions, listen reflectively, affirm, summarize, and elicit self-motivational statements. Figure 3 provides examples of questions designed to facilitate the participant's ability to make motivational statements.

a. Ask Open-Ended Questions

Initially for a participant at this level, it is important to ask questions that require explaining or discussing. Questions focus on requiring more than one-word answers.

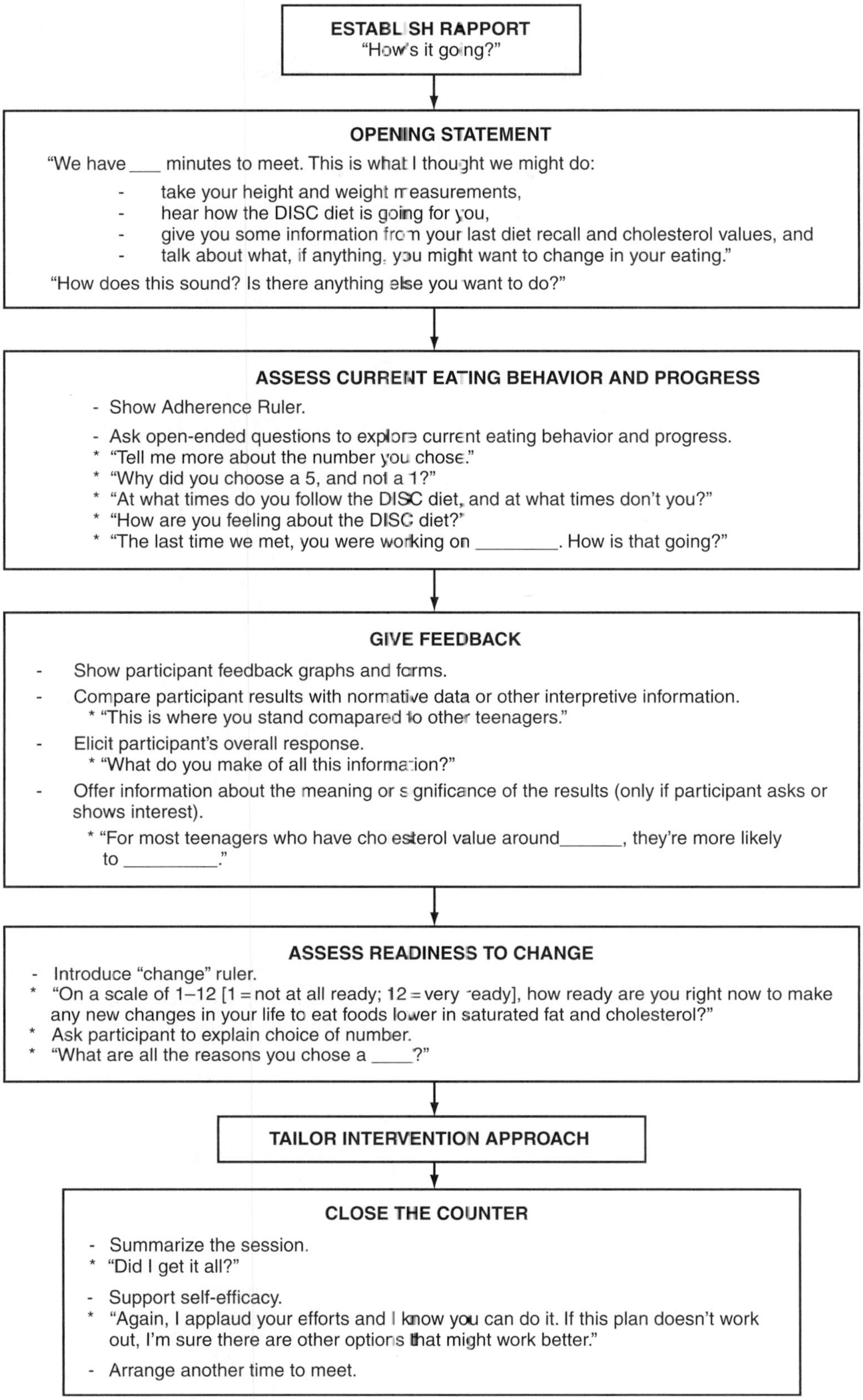

ESTABLISH RAPPORT
"How's it going?"

↓

OPENING STATEMENT

"We have ___ minutes to meet. This is what I thought we might do:

- take your height and weight measurements,
- hear how the DISC diet is going for you,
- give you some information from your last diet recall and cholesterol values, and
- talk about what, if anything, you might want to change in your eating."

"How does this sound? Is there anything else you want to do?"

↓

ASSESS CURRENT EATING BEHAVIOR AND PROGRESS

- Show Adherence Ruler.
- Ask open-ended questions to explore current eating behavior and progress.
  * "Tell me more about the number you chose."
  * "Why did you choose a 5, and not a 1?"
  * "At what times do you follow the DISC diet, and at what times don't you?"
  * "How are you feeling about the DISC diet?"
  * "The last time we met, you were working on ________. How is that going?"

↓

GIVE FEEDBACK

- Show participant feedback graphs and forms.
- Compare participant results with normative data or other interpretive information.
  * "This is where you stand comapared to other teenagers."
- Elicit participant's overall response.
  * "What do you make of all this information?"
- Offer information about the meaning or significance of the results (only if participant asks or shows interest).
  * "For most teenagers who have cholesterol value around______, they're more likely to _________."

↓

ASSESS READINESS TO CHANGE

- Introduce "change" ruler.
  * "On a scale of 1–12 [1 = not at all ready; 12 = very ready], how ready are you right now to make any new changes in your life to eat foods lower in saturated fat and cholesterol?"
  * Ask participant to explain choice of number.
  * "What are all the reasons you chose a ____?"

↓

TAILOR INTERVENTION APPROACH

↓

CLOSE THE COUNTER

- Summarize the session.
  * "Did I get it all?"
- Support self-efficacy.
  * "Again, I applaud your efforts and I know you can do it. If this plan doesn't work out, I'm sure there are other options that might work better."
- Arrange another time to meet.

FIGURE 2 Motivational intervention model. (From Berg-Smith, S. M., Stevens, V. J., Brown, K. M., VanHorn, L., Gernhofer, N., Peters, E., Greenberg, R., Snetselaar, L., Ahrens, L., and Smith, K., for the Dietary Intervention Study in Children [DISC] Research Group. [1999]. A brief motivational intervention to improve dietary adherence in adolescents. *Health Educ. Res.* **14**, 399–410.)

TAILOR INTERVENTION APPROACH

Level 1

Not ready

Goal: to raise awareness

Major task: inform and encourage

- Ask key open-ended questions.
  * "That's interesting—why did you give yourself a 3 and not a 1?"
  * "What would need to be different for you to consider making new additional changes in your eating?"
  * "You say you are a __ on the ruler. What would have to happen to you for you to move from a ___ to a ___?" "How could I help get you there?"
- Respectfully acknowledge their decisions.
  * "I respect your decision to not make any new or additional changes in your eating. You're the best judge of what's right for you."
- Offer professional advice.
  * "As you might guess, my recommendation is that you ___. But of course, it's your decision. If there comes a time soon when you decide to make any new or additional changes with your eating, I'm always available to help. In the meantime, I'd like to stay in touch."

Level 2

Unsure

Goal: to buid motivation and confidence

Major task: explore ambivalence

- Explore ambivalence.
  * "What are some of the things you like (and dislike) about your current eating habits?"
  * "What are some of the good (and less good) things about making new or additional changes?"
- Look into the future.
  * "I can see why you're unsure about making new or additional changes in your eating. Let's just stand back for a moment and imagine that you decided to change. What would it be like? Why would you want to do this?"
- Refer to other teens.
  * "What do your friends like to eat?"
  * "What would your friends think if you ate this way?"
- Ask about the next step.
  * "Where does this leave you now?" (Let the patient raise the topic of change.)

Level 3

Ready

Goal: to negotiate a plan

Major task: facilitate decision making

- Identify change options.
  * "What do you think needs to change?"
  * "What are your ideas for making a change?"
  * "Which option makes the most sense to you?"
- Help the participant set a realistic and achievable short-term goal.
- Develop an action plan.
- Choose a reward.
- Summarize the plan.
- Complete an "action plan."

FIGURE 3 Motivational intervention components for three specific levels. (Modified from Berg-Smith, S. M., Stevens, V. J., Brown, K. M., VanHorn, L., Gernhofer, N., Peters, E., Greenberg, R., Snetselaar, L., Ahrens, L., and Smith, K., for the Dietary Intervention Study in Children [DISC] Research Group. [1999]. A brief motivational intervention to improve dietary adherence in adolescents. *Health Educ. Res.* **14**, 399–410.)

The goal is to guide the participants to talk about their dietary change progress and difficulties. Figure 2 provides some opening questions. Other questions and statements are presented here:

"Let's discuss your experience with diet up to now. Tell me how changing your diet has been for you."

"What things would you like to discuss about your experiences with dietary change and your progress with changes? What do you like about these changes? What don't you like about these changes?"

---

b. Listen Reflectively

Listening goes beyond hearing what a person has said and acknowledging those words. Crucial in responding to a patient or participant is the understanding of what is meant beyond the words. Reflective listening involves a guess at what the person feels and is phrased as a statement rather than a question. Stating the feeling behind the statement serves two purposes. It allows the participant to tell you if your judgment of the feeling is on target. It also shows that you really are trying to understand more than just words and do care about feelings also. Here are some participant–nutritionist interactions that illustrate reflective listening:

Scenario 1

Participant: "There are times when I do a wonderful job of meeting my fat gram goal, but sometimes I don't do so well. I keep trying though."

Nutritionist: "You seem to feel badly that you don't always meet your fat gram goal."

Scenario 2

Participant: "I am so tired of trying to follow this diet. It seems that I have put hours into following it, and I have little to show for it. I certainly have not lost weight."

Nutritionist: "You feel frustrated and angry about trying so hard and still getting nowhere."

Scenario 3

Participant: "When I don't fill in a food diary, I am not sure that I am doing well or not."

Nutritionist: "You are worried on days when you do not fill in a food diary."

Scenario 4

Participant: "I really don't want to continue following this diet. I have other things that are more of a priority now."

Nutritionist: "You seem hassled by these other competing desires and feel that following a new eating pattern is getting in the way."

c. Affirm

Communicating support to participants is an excellent way of letting them know that you appreciate what they are doing. Affirmations are statements that indicate alignment and normalization of the participant's issues. Alignment means telling participants that you understand them and are with them in their difficulties. Normalization means telling the participants that they are perfectly within reason and "normal" to have such reactions and feelings. Examples of affirmations include the following:

"It is very hard to struggle with competing priorities. You've done amazingly well."

"That is an insightful idea."

"Thank you for telling me that. It must have been hard for you to tell me."

"I can see why you would have this difficulty. Many people have the same problem."

d. Summarize

Periodically, and at the point when you begin to elicit self-motivational statements, summarize the content of what the participant has said. Cover key points even if they involve negative feelings. You can discuss conflicting ideas that the participant has brought up by using the strategy "on the one hand you . . . and on the other hand you. . . ." This reminds both of you about the issues and ensures clarity.

e. Elicit Self-Motivational Statements

The most important part of self-motivational statements is that they help participants realize that a problem exists, that they are concerned about the problem, that they intend to correct the problem, and that they think they can do better in the future. Figure 4 provides questions to elicit self-motivational statements. These statements fall into four categories: problem recognition, concern, intention to change, and optimism.

It is important to respectfully acknowledge decisions that participants make. These decisions may mean that a participant has decided not to make changes immediately (see Fig. 3).

It is appropriate to offer professional advice but still leave the actual decision to make a change up to the participant. Figure 3 provides some ideas on how to approach the participant.

Close the discussion with another summary. Concentrate on any self-motivational statements that the participant has made. End the session with the idea that both of you should think about what has been discussed and that you can revisit the issues next time.

2. Second Level: Unsure about Change

The main goal for this intervention is to tip the balance toward working to meet the goals. Four steps are important in meeting the goals: (1) regroup, (2) ask key questions, (3) negotiate a plan, and (4) conclude the work.

a. Regroup

The first step in dealing with a participant who is unsure about changing dietary habits is regrouping to focus on the transition from not being ready to deal with the problems of change to moving toward a reinitiating of behavior adjustment. This process of regrouping can serve as a reminder of what has happened in previous sessions. Below are four ways to regroup:

1. *Summarize* the participant's perceptions of what is going on. The summary might include self-motivational statements that the participant has made.
2. *Identify* ambivalence or other conflicting issues.
3. *Review* any self-monitoring related to dietary intake.
4. *Restate* intentions or plans to change or do better in the future.

b. Ask Key Questions

Ask questions that focus on the participant's statements regarding future plans to make dietary changes. The goal is to ask open-ended questions that cause the participant to think about what you have just summarized and come to the conclusion that action is necessary. The goal is for the

1. **Problem Recognition**
   What things make you think that this is a problem?
   What difficulties have you had in relation to your diet?
   In what ways do you think you or other people have been inconvenienced by your diet?
   In what ways has this been a problem for you?
   How has your diet stopped you from doing what you want to do?

2. **Concern**
   What is there about your diet that you or other people might see as reasons for concern?
   What worries you about your diet?
   How do you feel about your dietary problems?
   How much does that concern you?
   In what ways does this concern you?
   What do you think will happen if you don't make a change?

3. **Intention to Change**
   The fact that you're here indicates that at least a part of you thinks it's time to do something. What are the reasons you see for making a change?
   What makes you think that you need to make a change?
   If you were 100% successful and things worked out exactly as you would like, what would be different?
   What things make you think that you should stop following your diet? . . . And what about the other side? What makes you think it's time for a change?
   What are you thinking about your diet at this point?
   What would be the advantages of making a change?
   I can see that you're feeling stuck at the moment. What's going to have to change?

4. **Optimism**
   What makes you think that if you decide to make a change, you could do it?
   What encourages you that you can change if you want to?
   What do you think would work for you, if you decided to change?

FIGURE 4 Examples of questions designed to elicit self-motivational statements. (Modified from Bowen, D., Ehret, C., Pedersen, M., *et al.* [2002]. Results of an adjunct dietary intervention program in the Women's Health Initiative. *J. Am. Diet. Assoc.* **102**, 1631–1637.)

participant to provide a statement showing the desire to change. Here are some examples of questions that facilitate positive participant statements:

"How might we work together to proceed from here?"
"Hearing my summary of how things have gone in the past, what do you want to do?"
"How might you become more involved in dietary change?
"What are the good parts and the bad parts about continuing to change?"
"You are currently unsure of what to do. How might we work together to resolve the issue?"

c. Negotiate a Plan

There are three parts to the negotiation process. The first involves setting goals; the second, considering the options; and finally, arriving at a plan.

1. *Set goals.* Past wisdom dictated that setting goals meant being specific and behavioral. "I will eat candy bars only one time per week, on Sunday." Motivational interviewing dictates that goal setting may start out broadly at first and then move to behavioral goals that are specific. To elicit broadly stated goals, the following questions might be used:

"What about your diet would you like to change?"

"How would you like things to be different from how they are now?"

2. *Consider options.* Make a list of things that might be changed to bring the participant closer to the dietary goal. Ask the participant to choose among the options. If the first one does not work, the participant has many choices as backups.
3. *Arrive at a plan.* Ask the participant to arrive at a plan. Include in the plan the specific behavioral goals with potential problems that may serve as barriers to making these changes.

d. Conclude the Work

Always end the session with an encouraging statement and a reflection on the participant's resourcefulness in identifying the plan. Follow this statement with the idea that he is the best expert relative to behavior change. Indicate that you will stay in touch to check on how the behavior change is going.

3. Third Level: Ready to Change

The main goal for this third intervention is to reengage the participant in meeting the dietary goals.

a. Review

Cover the previous discussions with the participant. Focus on the statements the participant made that show interest and a willingness to change. Use statements that show that it was the participant's idea to meet dietary goals previously set, for example, "You said that you were interested in trying again."

b. Encourage Choices and Activities

Ask the participant what she would like to do to reengage. Encourage the participant to make her own choice. Collaborate and negotiate short-term, easily attained goals initially, for example, "I will drink 1% milk for a week in place of 2% milk and gradually reduce to fat-free milk."

c. Summarize

Review the discussion, the issues, and difficulties on both sides. Remind the participant to keep trying and to believe in herself.

D. Modification of Diet and Renal Disease Study (MDRD)

The Modification of Diet and Renal Disease Study (MDRD) [18] used the self-management approach to modify dietary behavior. In this population of persons diagnosed with renal disease, research nutritionists counseled participants on diets low in protein and phosphorus. Although the diets were difficult to follow, the study participants showed great motivation based on their desires to avoid renal dialysis [19]. The following strategies were used successfully in the MDRD [20]:

1. *Single-nutrient approach to dietary behavior change.* The focus in this study was on reducing protein content of the diet. With this reduction, dietary phosphorus was also reduced. When other nutrients were modified, specific food groups were identified. Even with changes in other nutrients, protein was still a primary focus.
2. *Self-monitoring.* The participant's ability to self-monitor was crucial to keeping protein intake down to goal levels. Weighing and measuring was used as a means of matching dietary change on a daily basis with the biological marker of urinary nitrogen. Further study in self-monitoring matched with the biological marker showed that problems occur in knowing how to best represent dietary intake [21]. For example, often it is difficult to closely mirror the exact amount of protein in a cut of meat if that cut is not precisely specified.
3. *Staging changes.* In the MDRD, nutritionists also staged changes in dietary protein by gradually reducing the dietary protein intake. This gradual change made day-to-day modifications easier.
4. *Modeling.* Nutritionists modeled dietary changes by offering both recipes low in protein and taste-testing sessions. Group sessions were held at which a special meal was offered with food preparation techniques modeled.

E. Diabetes Control and Complications Trial (DCCT)

The Diabetes Control and Complications Trial (DCCT) [22] used techniques similar to those of the MDRD study [23]:

1. *Single-nutrient approach.* Investigators focused on carbohydrate as a single nutrient, where it was matched with insulin to achieve normalized blood glucose.
2. *Self-monitoring.* Monitoring consisted of following blood glucose concentrations and dietary intakes to verify where problems might be occurring. If dietary intake was high along with blood glucose concentrations, dietary intake or insulin was modified to achieve normal blood glucose levels.
3. *Staging changes.* Changes were staged by working on specific times of day that were most problematic. If lunchtimes were most often high, we focused on dietary intake modifications to alter blood glucose levels. Also insulin and exercise often played a role.
4. *Modeling.* Dietary modifications were facilitated by providing recipes, modeling, and going to restaurants to identify and anticipate blood glucose levels after eating a favorite lunch or other meals out of the home.

IV. SUMMARY

Considerable experience in clinical trials suggests that dietary modification requires a process of making changes on an individual basis with constant negotiation with the patient or participant. Working as a team, the nutritionist and participant can achieve dietary change that alters biological markers and may reduce disease risk and optimize management.

References

1. Agostinelli, G., Brown, J. M., and Miller, W. (1995). Effects of normative feedback on consumption among heavy drinking college attendants. *J. Drug Educ.* **25**, 31–40.
2. Brown, J. M., and Miller, W. R. (1993). Impact of motivational interviewing on participation in residential alcoholism treatment. *Psychol. Addict. Behav.* **7**, 211–218.

3. Rokeach, M. (1973). "The Nature of Human Values." Free Press, New York.
4. Rosenstock, I., Strecher, V., and Becker, M. (1994). The health belief model and HIV risk behavior change. *In* "Preventing AIDS: Theories and Methods of Behavioral Interventions" (R. J. DiClemente and J. L. Peterson, Eds.), pp. 5–24. Plenum Press, New York.
5. Bandura, A. (1989). Perceived self-efficacy in the exercise of control over AIDS infection. *In* "Primary Prevention of AIDS: Psychological Approaches," pp. 128–141. Sage, London.
6. Bandura, A. (1977). Self-efficacy: toward a unifying theory of behavioral change. *Psychol. Rev.* **84**, 191–215.
7. Bandura, A. (1991). Self-efficacy mechanism in physiological activation and health-promoting behavior. *In* "Neurobiology of Learning, Emotion and Affect" (J. Madden, Ed.), pp. 229–270. Raven Press, New York.
8. Janis, J. L., and Mann, L. (1977). "Decision-Making: A Psychological Analysis of Conflict, Choice and Commitment," Free Press, New York.
9. Miller, W. R., and Rollnick, S. (1991). "Motivational Interviewing: Preparing People to Change Addictive Behavior." Guilford Press, New York.
10. Women's Health Initiative. (1994). WHI Protocol for Clinical Trial and Observational Components, NIH Publication no. N01-WH-2-2110. Fred Hutchinson Cancer Research Center, Seattle, WA.
11. Prentice, R. L., Caan, B., Chlebowski, R. T., Patterson, R., and Kuller, L. H., *et al.* (2006). Low-fat dietary pattern and risk of invasive breast cancer: the Women's Health Initiative Randomized Controlled Dietary Modification Trial. *JAMA* **295**, 629–642.
12. Beresford, S. A., Johnson, K. C., Ritenbaugh, C., Lasser, N. L., and Snetselaar, L. G., *et al.* (2006). Low-fat dietary pattern and risk of colorectal cancer: the Women's Health Initiative Randomized Controlled Dietary Modification Trial. *JAMA* **295**, 643–654.
13. Prochaska, J., and DiClemente, C. (1982). Transtheoretical therapy: toward a more integrative model of change. *Psychother. Theory Res. Prac.* **19**, 276–288.
14. Bowen, D., Ehret, C., Pedersen, M., and Johnson, M. (2002). Results of an adjunct dietary intervention program in the Women's Health Initiative. *J. Am. Diet. Assoc.* **102**, 1631–1637.
15. Berg-Smith, S. M., Stevens, V. J., Brown, K. M., VanHorn, L., Gernhofer, N., Peters, E., Greenberg, R., Snetselaar, L., Ahrens, L., and Smith, K., for the Dietary Intervention Study in Children (DISC) Research Group. (1999). A brief motivational intervention to improve dietary adherence in adolescents. *Health Educ. Res.* **14**, 399–410.
16. DISC Collaborative Research Group. (1995). Efficacy and safety of lowering dietary intake of fat and cholesterol in children with elevated low-density lipoprotein cholesterol: the Dietary Intervention Study in Children (DISC). *JAMA* **273**, 1429–1435.
17. DISC Collaborative Research Group. (1993). Dietary Intervention Study in Children (DISC) with elevated LDL cholesterol: design and baseline characteristics. *Ann. Epidemiology* **3**, 393–402.
18. Klahr, S., Levey, A. S., Beck, G. J., Caggiula, A. W., Hunsicker, L., Kusek, J. W., and Striker, G., for the Modification of Diet in Renal Disease Study Group. (1994). The effects of dietary protein restriction and blood-pressure control on the progression of chronic renal disease. *N. Engl. J. Med.* **330**, 877–884.
19. Milas, C., Norwalk, M. P., Akpele, L., Castaldo, L., Coyne, T., Doroshenko, L., Kigawa, L., Korzec-Ramirez, D., Scherch, L., and Snetselaar, L. (1995). Factors associated with adherence to the dietary protein intervention in the modification of diet in renal disease. *J. Am. Diet. Assoc.* **95**(11), 1295–1300.
20. Snetselaar, L. (1992). Dietary compliance issues in patients with early stage renal disease. *Clin. Appl. Nutr.* **2**(3), 47–52.
21. Snetselaar, L., Chenard, C. A., Hunsicker, L. G., and Stumbo, P. J. (1995). Protein calculation from food diaries underestimates biological marker. *J. Nutr.* **125**, 2333–2340.
22. The Diabetes Control and Complications Trial Research Group. (1993). The effect of intensive treatment of diabetes on the development and progression of long-term complications in insulin-dependent diabetes mellitus. *N. Engl. J. Med.* **329**, 977–986.
23. Greene, T., Bourgorgnie, J., Hawbe, V., Kusek, J., Snetselaar, L., Soucie, J., and Yamamoto, M. (1993). Baseline characteristics in the modification of diet in renal disease study. *J. Am. Soc. Neprhol.* **3**(11), 1819–1834.

CHAPTER **10**

Tools and Techniques to Facilitate Nutrition Intervention

LINDA M. DELAHANTY[1] AND JOAN M. HEINS[2]

[1] *Massachusetts General Hospital, Boston, Massachusetts*

[2] *Washington University School of Medicine, St. Louis, Missouri*

Contents

I. INTRODUCTION

To effectively facilitate nutrition interventions using medical nutrition therapy (MNT), dietitians must understand the teaching–learning process and conduct individualized assessments of learning style and motivational readiness so they can determine which tools and techniques are most appropriate to help patients make healthy changes in eating behavior.

The first section of this chapter reviews the various factors that affect the teaching–learning process, including age, literacy, culture, and learner style, and reviews the different domains in which people learn. The next section discusses nutrition education techniques and focuses on the process that the nutrition counselor uses to assess learning needs, determine the level of education, and select the educational method and tools that match the patient's needs. The final section reviews the stages of change model and how it can be used to guide the nutrition counselor in selecting behavior change techniques that are tailored to an individual patient's level of motivational readiness. The nutrition counselor can use the behavioral counseling techniques of consciousness raising and motivational interviewing to address patients in the preaction stages of change (precontemplation and contemplation) and use behavioral change techniques that focus on skills training such as goal setting, self-monitoring, stimulus control, problem-solving barriers, coping skills and stress management, and increasing social support for patients in the action stages (preparation, action, and maintenance). This chapter includes case scenarios demonstrating how the counselor might apply these approaches and provides examples of how to phrase the counseling dialogue.

II. THE NUTRITION EDUCATION AND COUNSELING PROCESS

The potential effect of nutrition interventions and medical nutrition therapy to improve health outcomes from cardiovascular, cancer, diabetes, obesity, gastrointestinal, and other health conditions is clearly described in other chapters in this text. However, the process of implementing medical nutrition therapy is not as straightforward and is based on thorough assessment of each client's lifestyle, capabilities, and motivation to change.

At a glance, one might perceive the nutrition education and counseling process to be routine, where nutrition information and recommended food choices are discussed as they pertain to particular health concerns of a client. For a patient with a high cholesterol level, teach the National Cholesterol Education Program's (NCEP) multifaceted therapeutic lifestyle changes (TLC) approach; for high blood pressure, teach weight loss and dietary approaches to stop hypertension (DASH) diet; for type 2 diabetes, teach weight loss, exercise, and possibly carbohydrate counting. If only changing eating behavior was that straightforward! The truth is that each person who is referred for nutrition counseling presents with varying levels of knowledge and motivation for changing eating habits. Today's nutrition counselor must draw on knowledge from the biomedical and behavioral sciences to define the nutrition prescription and design an intervention that will truly impact eating.

The traditional model for delivery of nutrition interventions has been individual consultations in health care settings. This paradigm is changing. Increased focus on chronic disease prevention has resulted in an enhanced need for nutrition education at a time when economic constraints in health care have led to a decline in traditional hospital-based nutrition counseling services. Thus, dietitians are providing counseling services in shopping centers, offering cholesterol education classes in health clubs, and communicating with clients via telephone and the Internet. Although individualized counseling will continue as an important component of clinical nutrition, the use of alternative methods, such as group sessions or guided self-study, has been shown to be both efficient and effective. Regardless of the setting, nutrition counseling has a common goal: to help people make healthy changes in eating behaviors. This chapter discusses tools and techniques for applying education and behavior change theories in the practice of medical nutrition therapy.

III. THE TEACHING/LEARNING PROCESS

Knowledge is not sufficient but is essential for behavior change [1]. To be effective, nutrition counselors need to understand the elements that influence learning and the different domains in which people learn.

A. Factors Influencing Learning

Learning is influenced by many factors including age, literacy, culture, and individual learner style.

1. Age

Much of our understanding of the differences in the way adults and children learn comes from the work of Knowles [2], who identified concepts of *need to know, performance centered*, and *experiential learning* as important to the adult learner. These concepts have been expanded by the research of others who describe self-directed learning [3] and critical thinking techniques [4] as elements that enhance adult learning. A common theme that emerges across learning theories and studies of adult learning is the importance of active involvement of the individual in the learning process [5].

2. Literacy

The term "literacy" includes not only the ability to read and write but also to process information. With the wide use of printed materials to support nutrition education, client literacy and the reading level of teaching materials are important issues. Assessing level of formal education is easy, but unfortunately it is not always an accurate indicator of literacy. Tests such as ABLE [6], WRAT [7], and the Cloze procedure [8] can be used to evaluate an individual's literacy level. These tests, however, take time to administer and can be embarrassing to the client. Their best use may be to assess reading levels in targeted groups before developing educational materials rather than for individual assessment in clinical practice. The steps for evaluating nutrition education materials are relatively simple, and dietitians can use them when information on the reading level of material is not provided. Methods such as the SMOG readability formula [9], Gunning's FOG Index [10], and the Fry Readability Graph [11] are described in the reading and health education literature [12–14]. These formulas use the number of words, the number of syllables, and similar criteria to assign a grade or reading level. They can be scored by hand or with computer programs. One of the difficulties of using readability formulas in clinical nutrition is that core words such as "calories," "carbohydrate," "vitamins," "minerals," and "cholesterol" contain three or more syllables and consequently raise the reading level of materials. These words may be sufficiently common, however, so that people with reading abilities lower than the assigned grade level can process and comprehend the materials based on their familiarity with the words as well as the organization of the written text [15]. Recognition that poor health outcomes are not related exclusively to reading levels has led to an emerging field of health literacy. Health literacy is defined as "the degree to which individuals have the capacity to obtain, process and understand basic health information and services needed to make appropriate health decisions" [16]. A simple technique such as giving a client a nutrition tool to read and then asking what the information means to him or her can identify important limitations in the individual's health literacy. The U.S. Department of Health and Human Services website for Health Literacy, www.hrsa.gov/healthliteracy, provides basic information and practical tips for health professionals to use in communicating with patients as well as links to additional resources.

3. Culture

Ethnicity and culture influence the way people learn. Addressing cultural difference is not simply translating educational materials into the primary language of the client. Fundamentals of values, health beliefs, and communication styles also vary by culture. The changing demographics of the American population have resulted in an increased need for nutrition education materials appropriate for a wide range of cultures. One response to this need is the *Ethnic and Regional Food Practice* series developed by the Diabetes Care and Education practice group of the American Dietetic Association [17]. The series, written for health professionals, discusses traditions and beliefs influencing health behaviors such as eating habits, provides nutrient analysis of foods specific to the culture, outlines sample recipes, and includes a reproducible master of a client teaching tool. Other sources for ethnically appropriate teaching materials include government printing agencies, volunteer health organizations, ethnic special interest groups, and vendors of health education materials. Even when specific guidelines are not available, an appreciation that food and health hold different meanings to people of different cultures can aid dietitians in counseling individuals from diverse ethnic backgrounds. Dietitians can use a model for multicultural nutrition counseling competencies, developed by Harris-Davis and Haughton, as a basis for self-evaluation and selection of opportunities to enhance their skills [18].

4. Learner Style

Learner style is a phrase that has been used to describe the way people cognitively process information [19] and the interaction of the individual with the learning environment [20]. Drawing from the work of Carl Jung, Osterman has identified four types of learners: *feelers, thinkers, sensors*, and *intuitors* [21]. Table 1 summarizes key characteristics that differentiate these styles. Teaching strategies can be selected to match the learning style or characteristic of the individual. When working with groups, an education session can be constructed to include components that will appeal to all learning styles [21]. Walker, in her review of adult learner characteristics [20], contrasts different environmental methods of learning, such as group versus individual, computer versus print versus video, didactic versus emotional appeal in messages, and directed learning versus self-directed approaches. A number of studies have examined the effectiveness of different methods for delivering education and counseling for health behavior change. Findings show that group settings are both clinically and cost effective and an efficient use of clinicians' time [22–25]. A comparison of group versus self-directed education for blood cholesterol reduction found that the self-directed approach was a viable alternative to group diet instruction [26]. Studies of telephone, e-mail, and Internet counseling show these approaches can change health behaviors in diverse patient populations [27–29]. These findings support use of alternatives to individual counseling sessions to match client learning styles and provide efficient methods for dietitians to use to extend nutrition services to more people.

B. Domains of Learning

Learning includes knowledge, attitude, and skill. The education literature describes these areas of learning as domains: cognitive (knowledge), affective (attitude), and psychomotor (skill) [30, 31]. Within each domain there is a range or level of learning that can be achieved.

The cognitive domain concentrates on knowledge outcome and includes six hierarchical levels: knowledge, comprehension, application, analysis, synthesis, and evaluation. The affective domain encompasses an individual's feelings or attitudes associated with a particular topic. The five levels of the affective domain progress from receiving to responding, valuing, organizing, and characterizing within a value or a value complex. The psychomotor domain looks at skill development including perception set (prepared for learning), guided response (performance), mechanism (response more habitual), complex overt response (response is effective and routine), and adaptation (response continues in new situations.)

Detailed classification systems or taxonomies have been developed that include action verbs to define levels of learning [32, 33]. The taxonomies are used to set learning objectives. Objectives can be written for any domain and for any level within that domain, depending on the desired outcome of the counseling session. A learning plan can be developed for an individual that systematically advances him or her from a low level to a high level of competence.

Application of the taxonomies of learning to nutrition counseling offers a framework for integrating learning and behavior change theories. Learning objectives are written in language that clearly identifies *who* will do *what, when, where*, and *how. Who* is the client, *what* is the information, *how* is the measurable behavior, and *when* and *where* define the situation. For example, a behavioral objective for a patient with hyperlipidemia, targeted at the application level of the cognitive domain, could be as follows: At the end of the counseling session (*when*), the client (*who*) will be able to identify (*how*) low-fat entrée options (*what*) from sample menus (*where*).

The value of setting behavioral objectives for nutrition interventions is well supported in the literature. A review by Contento and colleagues found that programs that were more behaviorally focused were more effective [5]. This extensive review identified education and behavior change strategies that were successful in changing eating habits. Educational strategies of self-evaluation or self-assessment and active participation worked well for individual or group

TABLE 1 Learning Styles

	Feelers	Thinkers	Sensors	Intuitors
Looks for:	Meaning, clarity	Facts and information	Practical application	Alternatives
Learns by:	Listening and sharing	Thinking through ideas	Problem solving	Trial and error
Best format:	Discussion	Lecture	Demonstration	Self-discovery
Favorite question:	Why?	What?	How?	If?

Source: Adapted from Osterman, D. N. (1984). The feedback lecture: matching teaching and learning styles. *J. Am. Diet. Assoc.* **84**, 1221–1222.

interventions. Effective behavioral strategies included the use of a systematic behavior change process, tailoring the intervention to the specific needs of the individual, involvement of others to provide social support, and an empowerment approach that enhances personal control.

In summary, for nutrition counseling to be successful, learning principles suggest that as much attention must be given to selecting the most effective way to communicate the diet to the individual as is given to assessing what would be the most effective diet for the individual. Most health professionals have strong grounding in the biological sciences but less exposure to the behavioral sciences. For this reason, the comfort level for determining the appropriate nutrient intake is greater than for evaluating learner needs and setting behavioral objectives. The current demand in all areas of health care to measure effectiveness in terms of patient outcomes requires clinicians to look beyond the diagnosis and intervention phases of care and evaluate the ultimate results. Nutrition interventions that do not show measurable improvement in patient status are not effective. The argument that treatment failure is due to patient noncompliance does not negate the lack of effect. Dietitians and other health care providers must be adept at combining nutrition, learning, and behavior change principles in the process of nutrition education.

IV. NUTRITION EDUCATION TECHNIQUES

Length of time since diagnosis, acuity of disease condition, complexity of the nutrition intervention, preferred client learning style, and readiness to change behaviors are factors to consider in developing a client's education plan. The counselor needs to assess learning needs, decide on the level of education, select an optimal method, and choose appropriate nutrition education tools.

A. Assessing Learning Needs

Assessment of learning needs should be an integral part of nutrition counseling. Table 2 provides a basic list of variables that should be assessed. Additional items are added to gather information pertinent to the individual and the clinical condition. The amount of time spent in assessment can limit time spent in counseling. Some studies report up to 55% of a counseling session devoted to the assessment phase [34]. However, time spent on a comprehensive assessment is regained by the effectiveness of a nutrition counseling session that has been tailored to the individual's needs.

A variety of approaches can be used to reduce time spent on assessment. Before the counseling session, data can be collected from medical records, and patients can be asked to submit information. Questionnaires can be sent and received by mail, via fax, through the Internet, or completed by the client in the waiting room before the visit. If advanced data collection has not been successful, the counselor may ask the patient. Although clients may not recount a comprehensive description of their referring physician's intent, their perception of what the interaction should achieve provides a basis for assessing learning needs and willingness to make behavior changes.

TABLE 2 Nutrition Educational Assessment

Demographic	**Clinical**
Age	Medical history
Gender	Medication
Occupation	Height/weight
Education	Food allergies/intolerances
Social	**Health Habits**
Family status	Eating patterns
Living environment	Physical activity
Social network	Smoking status
Cultural factors	Alcohol intake
Religious practices	Health practices
Health beliefs	Use of health services
Learner Characteristics	
Previous health education	
Expectations for current education	
Preferred learning methods	
Learning style	
Health literacy	
Readiness for change	

B. Levels of Education

Nutrition education should be planned as a continuum of learning that starts with fundamental guidelines and then incrementally adds more complex information as basic applications are mastered. The terms "initial/survival," "practical," and "continuing" have been used to differentiate three levels of education [34].

1. Survival Level

The first level focuses on essential information that the client needs to make important fundamental adjustments in health behaviors. Ideally, initial education will occur shortly after diagnosis. The extent of information included at the survival level differs by disease condition and learner characteristics. A person with diabetes treated with insulin needs enough information to understand the association between food, activity, and insulin so that he or she can select appropriate meals to avoid hyper- and hypoglycemia. For the patient with congestive heart failure with frequent hospital admissions, the survival information would focus on the sodium content of foods and avoidance of fluid retention. Survival education needs to be simple and directive; the dietitian serves as a teacher by providing concrete guidelines on what the patient should or should not do.

2. Practical Level

The practical level of education can occur as follow-up to initial counseling or as a new encounter, with the patient having had initial instruction some time before. Information should expand on the fundamentals learned for survival by applying them to a variety of situations. New topics can be introduced as well. Clients often will identify "need to know" information they have found important to learn, such as how to eat in restaurants, modify recipes, or interpret food labels. At this level, the dietitian serves as a counselor by providing guidelines for patients to use in making decisions.

3. Continuing Education

Once a client has mastered the basic skills and can apply them successfully in his or her life, continuing education can be used to reinforce learning, update information, and achieve higher levels of knowledge. In-depth knowledge of the relationship between nutrition and the disease process, nutrition principles, food preparation, and eating behaviors can enable patients to "take charge" of their disease management. The dietitian at this level serves as a consultant, helping the client synthesize and personalize information.

C. Educational Methods

In addition to the content of the educational intervention, the process or method offers different techniques to make nutrition education more effective. In-person individual or group sessions are the most common formats used for nutrition counseling. Although decisions on group or individual counseling sessions may be made by feasibility criteria (i.e., time, money), the client's learning preference should be considered as well. Using Osterman's classification of learning styles (Table 1), *feelers* would like group sessions that have discussion opportunities, whereas *thinkers* could be frustrated by this method because they "just want the facts." *Sensors* would respond to either method as long as an application exercise is included. A combination of individual and group sessions offers practical advantages. Information presented during group sessions can be tailored to the individual in a one-on-one discussion either in-person or by telephone or e-mail.

Another method for nutrition education is the use of self-study materials. This approach has been used more often in population or community nutrition interventions than in group or individual counseling sessions. Self-study modules offer advantages in terms of convenience, pace of learning, active involvement of the individual, and economics. Modules range in size and scope. Two classic examples are the 28-page pamphlet of the Shape Up America program [35], now available online [36], and the 312-page Learn Program manual [37]. The American Dietetic Association introduced a 56-page workbook, "Real Solutions: Weight Loss Workbook in 2004" [38]. Self-study modules generally include self-assessment exercises, general information on the health topic, steps for identifying personal behaviors to modify, guidelines for making behavioral changes, methods for monitoring behavior change, and tips on sustaining the new behaviors in special situations (e.g., eating out, stressful days). Self-study programs are now available on CD-ROM and over the Internet. In addition to Shape Up America, the Food Guide Pyramid is available in a self-help module via the Internet at www.mypyramid.gov. This method of learning offers something for all learning styles, especially *intuitors* who will appreciate the opportunity for self-discovery.

Caban and colleagues have developed a model for using patient assessment to guide selection of education methods for weight management (Fig. 1) [39]. They use a comprehensive set of physical and psychological indicators to

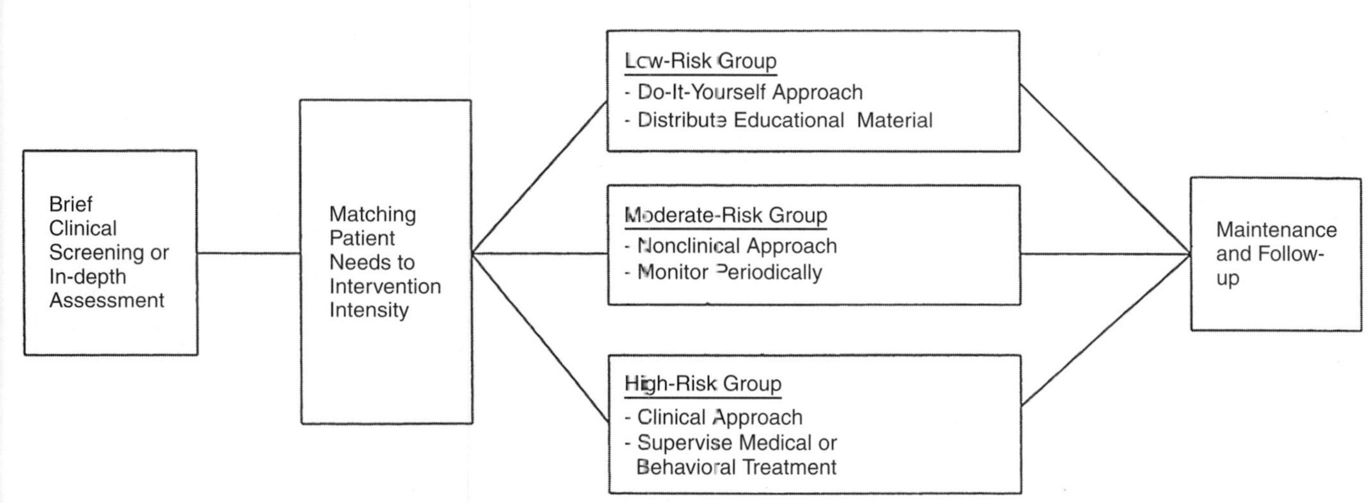

FIGURE 1 Model for individualizing treatment approaches. (Adapted with permission from Caban, A., Johnson, P., Marseille, D., and Wylie-Rosett, J. [1999]. Tailoring a lifestyle change approach and resources to the patient. *Diab. Spectrum* **12**, 33–38.)

classify individuals by risk status, and then they select interventions with an intensity that matches individual needs. People who are in stable clinical condition and highly motivated with few behaviors to change are classified as low risk and are candidates for self-directed learning methods. Those with clinical conditions targeted for improvement and who are less motivated with several behaviors to change require more structure and support. Group interaction, periodic monitoring, and structured activities are appropriate for this type of client. The high-risk client, who has multiple health problems that need to be addressed, has little motivation, and requires a great deal of support will need individualized care. In their program, computers are used to access assessment data, tailor questionnaires to individual profiles, and provide feedback to the client. The authors recognize that all practice settings do not have these resources, but they encourage counselors to use even a brief assessment to collect information that will determine the best type of program for the individual.

D. Nutrition Educational Tools

Nutrition counseling does not suffer from a lack of tools. Information on nutrition education materials can be found in the catalogs of publishing houses, volunteer health and professional organizations, government agencies, reviews published in professional journals, and by professional networking. Education materials also are available from companies manufacturing health products (e.g., pharmaceutical companies) and associations that promote a food or food group (e.g., National Dairy Council). Although there may not be a teaching tool that matches all aspects of a patient's learning profile, availability of educational materials is not the prominent problem. Resources for acquisition and storage are more common deterrents to dietitians' being able to use a teaching tool tailored to a client's specific needs.

1. Meal Planning Tools

Several formats for nutrition education tools have been applied in multiple areas of clinical nutrition. These include guidelines, menu approaches, counting methods, and exchange systems.

Guidelines such as the Food Guide Pyramid are tools that provide basic information to help people make healthy food choices. They may include some information on servings and food preparation but not the specificity of nutrient information that can be found with other methods. Guidelines work well as a tool for initial education if precision in nutrient intake is not required. They may contain sufficient information for some people to be able to make eating behavior changes that reduce health risks and improve clinical indicators.

Menus are a tool to give clients specific direction on what to eat, including food type, preparation method, and serving size. For survival education, several days of menus can be written using familiar foods that ensure appropriate nutrition. The number of patients who carefully collect their tray menus during a hospital stay attests to the popularity of this approach. Menu planning tools generally rely on input from the patient so that food likes and dislikes can be taken into consideration. Menu planning can be combined with other instructional formats and used at the practical and ongoing education levels as well. Computer programs are becoming available that plan menus, taking into account the nutrient prescription and individual food preferences.

Counting methods have become popular in recent years. Calorie counting is a standard approach for weight management. Fat gram counting is utilized in teaching materials for cholesterol reduction, cancer prevention, and weight management. Carbohydrate counting, initially considered a technique for intensive insulin therapy, is now being used with all types of diabetes for initial as well as continuing education.

Exchange approaches focus on food groups versus individual foods to teach nutrition principles. A popular example is the *Exchange Lists for Meal Planning* [40] that has been used for decades to instruct patients with diabetes. The system provides a tool for teaching patients how to select a diet that meets a macronutrient prescription. The exchange lists concept has been adapted for weight management education and is sometimes applied to nutrient information provided with recipes and manufactured food products.

The plate model, a common teaching tool in Europe, was used in the Diabetes Atherosclerosis Intervention Study (DIAS) and found to be as effective as the more quantitative exchange lists [41]. The method uses a visual tool of a plate divided into three sections: one covers one-half of the plate and the other two are one-quarter sections. The guidelines place vegetables including salads in the one-half plate section, starches in one of the one-quarter sections and meat and alternatives in the other one-quarter section. Dairy and fruits are included as side dish servings. The Plate Model method has been adapted for diabetes education and offers a simple method for encouraging portion control as well as healthy food choices.

Different versions of these tools can be found to match many idiosyncratic learner characteristics. Simplified versions appropriate for initial education or for low-literacy clients and versions that have been translated into Spanish are the most common. Purchasing reproducible tools offers nutrition counselors an efficient way to have a wide range of teaching materials available to meet individual client needs. Even more promising are computer software programs providing nutrition education tools. Computer programs allow the provider to tailor materials to patient characteristics and to change content if it becomes obsolete.

2. Skills Training

Meal planning tools, even those carefully selected and matched to individual learner characteristics, are not sufficient for nutrition education. People may learn what they need to do but lack the "skill" or ability to translate knowledge into practice. Teaching an individual how to read a nutrition label provides the knowledge necessary to interpret the information. Applying that knowledge to decisions made at the point of purchase or in meal preparation requires skill. Comparing nutrition labels on two products requires analytical skills if the products have different serving sizes. After using the nutrition label to select a product, additional skills are required to integrate the selected food into the total diet. Tools are available in print, on video, on CD-ROM, and as three-dimensional models for teaching clients a wide range of nutrition skills.

Some clients, however, cannot or choose not to master certain skills. Meal planning is a prime example. No matter how well educated and trained, there are some people who find meal planning a real barrier to following their diet. "Just tell me what to eat" is their common complaint. Preprinted menus offer a solution in some cases. The American Diabetes Association introduced the first of their *Month of Meals* publications in 1989 and quickly added several more editions in response to the demand [42]. Books offering low fat and/or low calorie menus are abundant as well.

Menus, however, do not meet the needs of people who lack the skills and interest to prepare meals from menus. The Cardiovascular Risk Reduction Dietary Intervention Trial addressed the difficulty many people have in translating complex dietary recommendations into meals. The trial compared a nutritionally complete prepared meals program with (1) a diet prescribed to meet the same nutrient levels established for the prepared meals and (2) usual-care dietary therapy [43, 44]. The prepared meals program resulted in significantly greater weight loss and reduction in blood pressure than the other two interventions, and improvement in lipid and glucose levels compared with the usual-care therapy.

This "need" for more specific information that will help people choose healthy meals is not limited to clinical nutrition. The food industry, recognizing that each generation of Americans spends less time cooking, has responded with carry-out and order-in meals from restaurants and deli, salad bar, and complete "meals to go" in supermarkets. Chef-prepared meals for home delivery can be ordered online from a long list of vendors, and meal prep franchises are flourishing. Today's nutrition counselor must be just as adept at helping clients develop the skills to select healthy prepared meals as teaching the fundamentals of reducing dietary fat intake.

To change eating habits, nutrition counseling needs to include more than well-selected education techniques. A review has found that many studies based on dissemination of information and teaching of skills were not effective in changing eating behavior [5]. Even studies supposedly following the knowledge, attitude, behavior communications model were not successful if the behavioral components were simply skills training (e.g., label reading, food preparation). Only in studies where participants were self-selected and already motivated were they able to show changes in eating patterns with an intervention that did not have a true behavior change technique. Unfortunately, a diagnosis of a disease or health risk does not automatically result in a motivated patient; therefore, nutrition counselors need to include behavior change techniques in their practice.

V. BEHAVIOR CHANGE TECHNIQUES

Theories for behavior change are extensive and are described in Chapter 8. Certain theories offer techniques that are practical for use in nutrition counseling. Some techniques such as self-monitoring, behavioral contracting, and goal setting are patient focused. They allow the individual to "personalize" the nutrition intervention, to be an active participant in the process, and to receive feedback on progress made toward identified goals. Counselors typically implement other techniques such as motivational interviewing, consciousness raising, and assessing readiness for change. The stages of change transtheoretical model [45] offers dietitians a systematic framework for applying behavior change techniques to nutrition education and counseling. Although the model can be used with all education methods, the following section will describe its application in nutrition counseling sessions.

A. Stages of Change

The stages of change model was described by Prochaska and colleagues. It postulates that both cessation of high-risk behaviors and the acquisition of healthy behaviors involve progression through five stages of change: precontemplation, contemplation, preparation, action, and maintenance (Table 3) [45]. The precontemplation stage is characterized by having no intention of changing the behavior in question in the foreseeable future. People in this stage tend to be unaware that they have a problem and are resistant to efforts to modify the behavior in question. Contemplation is characterized by awareness that the particular behavior is a problem and serious consideration about resolving the problem but having no commitment to take action in the near future. The preparation stage is the stage of decision making. The person has made a commitment to take action within the next 30 days and is already making small behavioral changes. In the action stage, clients make notable overt efforts to change. Maintenance involves working to stabilize the particular behavior change and avoid relapse for the next 6 months. People do not simply progress through the stages in a straight line; they may recycle by relapsing and repeating stage progressions and they can enter or exit at any point [46, 47]. This transtheoretical model has been shown to generalize across a broad range of problem behaviors (including diet, exercise, and weight control) and populations [46].

TABLE 3 Stages of Change

Stage	Definition	Client Characteristics	Expressions Client May Say
Precontemplation	Person is not interested in making changes within the next 6 months.	May be unaware a problem exists.	It's not my problem.
		Sees no reason or need to change.	I can't change.
		Not interested in talking about the situation or behavior.	Leave me alone.
			Quit nagging me.
Contemplation	Person is thinking about making changes within the next 6 months.	Has limited knowledge about the problem.	I'll change when the time is better—someday.
		Is weighing risks/benefits of changing a behavior.	Why should I? I don't understand.
		Is waiting for the "magic moment" to start.	I know I should, but __________.
		Is wishing problem behavior would solve itself.	It takes too much effort.
Preparation	Person is planning to make changes within the next 30 days or has tried making some small changes but is not consistent with new behavior yet.	Is motivated and ready to learn.	I am ready to __________.
		Knows what to do, but not sure how to start.	I want to __________.
		May have tried small changes, but now ready to try more.	How can I start?
			Where can I go to learn __________?
Action	Person has changed a behavior within the past 6 months and is becoming more consistent with it.	Efforts to change are more visible to others and are more consistent.	I can __________.
		Believes change is possible.	It's getting easier for me to __________.
		Started making changes in his or her environment to support the changed behavior.	I'm doing it, but sometimes __________.
			I was amazed to learn that __________.
Maintenance	Person has maintained a new behavior longer than 6 months.	Is maintaining the new behavior—the change has become a part of his or her daily routine.	I just do it now.
		Is trying to avoid slipping back into old habits.	It's not a problem for me to __________.
		Is confident about ability to maintain change.	I feel good about __________.
			I don't have to think about it much anymore.

Source: Reprinted with permission from HealthPartners, Inc., Bloomington, MN. © 2000. All other rights reserved. For more information, contact HealthPartners, Center for Health Promotion.

Integral to the stages of change model is a "standard" or goal for the behavior that has been proven to produce results. For example, the goal of the NCEP TLC diet for a dietary saturated fat intake of less than 7% of energy and less than 200 mg cholesterol per day was established based on research showing associated reduction in serum LDL cholesterol concentration. Stages of change techniques help individuals evaluate current behaviors against a defined standard and then assess their readiness to change. The objective is to help patients achieve the standard for the behavior; however, incremental goals are often required. Fortunately, many nutrition interventions in clinical nutrition show benefits associated with stepwise progression to the goal.

If dietitians use the stages of change model as a basic technique for approaching nutrition counseling sessions, it often requires a modification in their teaching and counseling style [48]. The focus of attention shifts from establishing an agenda to merely educate the individual and shifts toward an approach that assesses the status of the individual as a basis for tailoring the counseling session. For example, instructing a client with hypercholesterolemia who is clearly in the precontemplation stage on an NCEP TLC diet will be less effective than discussing the risks associated with high cholesterol levels and the benefits of reducing the saturated fat content of the diet. Although the counselor may feel an obligation to impart specific information, mismatching stages and interventions could break rapport and may lead the client to avoid further follow-up sessions.

B. Effective Use of Stages of Change Technique in Nutrition Counseling

Nutrition assessments are typically done at the initial counseling session. It is important for dietitians to use the stages of change model in each session to assess attitudes toward nutrition and health, readiness to learn, and willingness to change. Assessment of these attitudes requires proficient use of open-ended questions and listening skills.

Open-ended questions begin with "what," "how," "why," and "could." The first clue about a client's stage of change is in the response to opening questions such as "How can I help you?" "What are your goals for our meeting today?" Attentive listening is an important tool to assess the client's responses to these questions. Attentive listening involves not only allowing sufficient silence to hear the client's verbal response but also paying attention to facial expression, voice tone, and body language as the person is speaking. The nutrition counselor can use the listening technique of paraphrasing to see if he or she has accurately understood the content of the client's statements or the listening technique of reflection to find out if he or she understands the emotional feeling that the client is trying to convey [49]. It is the combination of open-ended questioning and proficient listening skills that builds rapport.

As the counselor collects assessment information in the first part of the counseling session, there are repeated opportunities to evaluate a client's stage of change through the use of open-ended questions and effective listening skills.

C. Consciousness Raising: A Technique for Precontemplators

1. A Case of Precontemplation

Dietitian: "How can I help you?" "What are your goals for today's session?"

Patient: "I don't know. The doctor sent me." "I feel fine." "I already have so many problems, so why bother?" (arms crossed, sitting back in the chair, stiff body posture, voice tone may display resignation, anger, upset, or denial concerning necessity for changing habits or coming to see the dietitian).

Dietitian: "Did your doctor say why he wanted you to come?" "What did your doctor tell you about your ________ (e.g., cholesterol levels/diabetes control/blood pressure)?"

Patients in the precontemplation stage tend to focus more on the difficulties or disadvantages of changing eating behavior—the diet restriction, the inconvenience, or the expense of making healthy food choices or keeping food records. They tend to put much less emphasis on the benefits of changing eating behavior. Patients can be in the precontemplation stage for various reasons including lack of knowledge, lack of skills, lack of resources, distorted health beliefs, or competing priorities [50]. The counselor needs to explore and assess the reasons particular to each client before proceeding further with the session.

QUESTIONS TO ASSESS KNOWLEDGE, SKILLS, RESOURCES, HEALTH BELIEFS, AND COMPETING PRIORITIES

"What did your doctor tell you about your ________ (e.g., cholesterol level/diabetes control/blood pressure)?"

"What did your doctor tell you about your laboratory results?"

"Do you know what the goals are for ________ (cholesterol levels/hemoglobinA1c levels/blood glucose levels)?"

"What are your personal goals?"

"What do you know about how your diet affects your ________ (e.g., cholesterol level/hemoglobin A1c/ blood glucose patterns)?"

"What do you think is most important when reading a nutrition label if you have __________ (e.g., diabetes/ high cholesterol)?"
"Do you have a blood glucose meter?"
"Do you test your blood glucose levels at home?" "How often and what times of day?"
"What do you think that you would need to do to improve your __________ (e.g., cholesterol levels/hemoglobinA1c levels/blood glucose levels/activity level)?"
"Are there any factors that you feel make it difficult for you to focus on improving your __________ (e.g., cholesterol levels/hemoglobinA1c levels/blood glucose levels/ activity level)?"

2. Consciousness Raising

Consciousness raising is an important strategy for dietitians to use if a client presents to a session in the precontemplation stage [47]. The following steps can be used to facilitate consciousness raising (Table 4):

1. Discuss the medical problem or condition of concern.
2. Review lab results related to the condition versus target/normal values for the test results.
3. Review the relationship of diet, exercise, and other self-care habits to the medical condition and personal lab data.
4. Use visual aids (e.g., test tubes of fat, 1-lb or 5-lb fat models), audiovisuals, and personalized profile sheets to enhance the message.
5. Elicit feedback from the client regarding this information.

As the interaction between the counselor and the patient proceeds, the counselor can watch for changes in body language (leaning forward, changing voice tone, arms uncrossing) and in level of interaction (asking more questions). Once this information-sharing process is complete, the counselor has provided the client with the information necessary to make an informed choice about changing eating behavior. The clients who are in the precontemplation stage because of lack of knowledge or skills may move toward contemplation or preparation stages once these issues are addressed. On the other hand, some clients may be fully informed about their medical condition, lab data, and the impact of eating, activity, and self-care habits on their health and still do not view changing eating habits as a priority. These clients are less likely to progress to contemplation or preparation within one session. In these cases, it is important to avoid judgment and show understanding by acknowledging their feelings and choices. For example, a client may feel too depressed to consider change at the time of the session and a competing priority may be to seek counseling and treatment for depression first. The counselor's responsibility is to provide information and then reassess the client's stage of change. It's the client's responsibility to make an informed choice.

D. Motivational Interviewing: A Technique for Contemplators

1. A Case of Contemplation

Dietitian: "How can I help you?" "What would you like to accomplish at our meeting today?"
Patient: "I'm not sure. I know what to do. I just need motivation." "I know that I should _______ (e.g., lose weight/eat less fat), but __________."

Patients in the contemplation stage view the pros and cons of changing eating, exercise, or self-care habits as about equal, with the cons slightly greater than the pros. They may focus on the short-term costs of changing eating behaviors (e.g., limiting food choices, adjusting food purchases when shopping or menu selections when eating out, investing time in keeping food records or nutrition appointments) and pay less attention to the long-term health benefits [50]. Clients can be in the contemplation stage because of limited knowledge about the problem, wishing that someone else would fix the problem or that the problem would solve itself, competing priorities, or low self-efficacy regarding ability to change eating habits.

If limited knowledge is the problem, then the counselor can use the technique of consciousness raising to provide information, clarify any misconceptions, and respond to specific questions. Before proceeding further with the session, the counselor needs to develop a clear understanding of the client's ambivalence about changing eating behavior. To do this, the counselor needs to explore the client's concerns and perceived barriers to making changes as well as the perceived benefits. This technique of exploring the pros and cons of behavior change is called *decisional balance*.

The counselor also needs to evaluate each client's self-efficacy for changing eating behavior because perceived self-efficacy influences the acquisition of new behaviors and the inhibition of existing behaviors. It also affects people's choices of behavioral settings, the amount of effort they will expend on a task, and the length of time they will persist in the face of obstacles. Finally, self-efficacy affects people's emotional reactions and thought patterns [51]. The process of using decisional balance and other techniques to increase motivation and self-efficacy is often referred to as *motivational interviewing*.

For clients to move from the contemplation stage toward the preparation stage of change, they must believe that the advantages of changing their eating behavior outweigh the disadvantages. The counselor can help reduce or minimize barriers to change by assisting the client with practical problem solving (some barriers may be related to access to treatment and resources and other barriers will be more attitudinal, such as fear of change or the results of change).

TABLE 4 Stage-Matched Counseling Techniques

Stage	Counseling Strategies	Do's/Don'ts That Motivate/Hinder Change
Precontemplation	Raise self-awareness of their behavior.	Don't be judgmental.
	Raise awareness of health concern and implications.	Don't rush them into action.
	Show how their behavior may affect others around them.	Don't ignore their emotional reaction to idea of change.
	Encourage them to express their feelings.	Do listen and acknowledge their feelings.
		Do help them understand effects of their behavior.
		Do help them become aware a problem exists.
Contemplation	Identify and discuss their concerns, beliefs, and perceived barriers toward their behavior.	Don't ignore the potential impact the change may have on their family.
	Show how benefits of change outweigh the risks of not changing.	Don't nag or preach to them demanding they change.
	Clarify any ambivalence felt toward changing.	Don't provide "how to" information.
	Provide or suggest resources for learning more about the solution.	Do listen to their concerns.
		Do help them identify benefits they'll receive if they decide to change a behavior.
		Do provide "facts" answering the "why" questions they have.
Preparation	Help them develop a plan of action.	Don't recommend general behavior change—be specific.
	Teach specific "how to" skills.	Don't refer to small changes as "not good enough."
	Help build their self-confidence so they can form a new behavior.	Don't create new barriers for them to overcome.
	Help them access an educational program and obtain resources needed for change.	Do provide specific ideas on how to change.
		Do remind them that "change" is hard work.
		Do help them realize that "any change" is better than "no change."
Action	Reinforce their decision to change.	Don't nag or preach.
	Provide emotional support to continue behavior change they've started.	Don't make assumptions that they're not having lapses.
	Explain the difference between lapse and relapse.	Don't overpraise so that they only report successful outcomes.
	Help them evaluate their progress, and teach additional skills, as needed.	Do explore their feelings about changes they've made.
		Do celebrate "milestones," big and small.
		Do compliment positive behavior changes you observe.
Maintenance	Offer ideas or ways to maintain change.	Don't assume that initial action means permanent change.
	Help them build a supportive environment around them to maintain the change.	Don't be discouraged or disappointed when lapses happen.
	Continue teaching relapse prevention techniques.	Don't underestimate environmental barriers.
	Help validate rationale for change.	Do remind them of their overall progress (from the start).
		Do review benefits they've received from changing their behavior.
		Do ask open-ended questions about how they are coping with change and how they feel about their changed behavior.

Source: Reprinted with permission from HealthPartners, Inc., Bloomington, MN. © 2000. All other rights reserved. For more information, contact Health Partners, Center for Health Promotion.

It is also helpful to identify the client's positive incentives for continuing with current behaviors (i.e., the payoff for staying the same) and use strategies to decrease the perceived desirability of the behavior (e.g., increase awareness of the negative consequences of the behavior using facts and audiovisuals) [50].

Motivational interviewing is a client-centered technique designed to build commitment and reach a decision to change [52]. It involves an interviewing process that assesses decisional balance and self-efficacy beliefs and focuses on increasing the client's intrinsic motivation to change and self-confidence in ability to do so. The following types of questions can be used to assess decisional balance [50].

2. Decisional Balance Questions

Explore the advantages of not changing (staying the same):

Dietitian: "What are some of the positive aspects of not _________ (e.g., reducing your fat intake/losing weight)?"
Patient: "I can eat what I want and enjoy my food and not feel deprived."
Dietitian: "What else is good about it?"
Patient: "I don't have to risk failing at this again."
Dietitian: "What else?"
Patient: "Those are the main things."

Explore the disadvantages of not changing:

Dietitian: "What are some of the not so good things about _________ (e.g., eating a high-fat diet/maintaining your current weight)?"
Patient: "My cholesterol is too high. My clothes don't fit and I'm sluggish."
Dietitian: "Anything else?"
Patient: "I have a higher risk for a heart attack."
Dietitian: "Any other disadvantages?"
Patient: "My family worries about me."

Summarize by paraphrasing:

Dietitian: "It sounds like you like to eat and enjoy your food without feeling deprived. I also get the sense that it's important to you not to feel like a failure when you try to change your eating habits. On the other hand, it sounds as though there are some disadvantages to your current way of eating: your high cholesterol level, the risk for heart disease, and feeling sluggish. Also, it sounds as though you think your family worries about you and your health."
Patient: "That's right."
Dietitian: "How do you feel about this now? What would you like to do about this?"
Patient: "I'm not sure."
Dietitian: "It's important to enjoy your food and not feel deprived. If we could come up with some ideas to help lower your cholesterol that are doable and that won't make you feel deprived, would you be interested? If you'd like, we can explore some of the possibilities together."

Each client's motivation to change is influenced by two components: importance and confidence [53]. Assessing importance and confidence is a technique that dietitians can use to help decide what steps to take with patients who are in the contemplation stage of change.

3. Importance and Confidence Technique*

INTRODUCE THE DISCUSSION

"I'm not really sure how you feel about __________ (e.g., reducing your fat intake/losing weight/exercising more). Can you help me by answering two simple questions and then we can see where to go from there?"

ASSESS IMPORTANCE AND CONFIDENCE

"How do you feel *right now* about __________ (eating less fat/losing weight/exercising more)? On a scale from 0 to 10, with 0 meaning 'not important at all' and 10 meaning 'very important,' how *important* is it to you personally to __________ (eat less fat/lose weight/exercise more)?"

"If you decided *right now* to __________ (eat less fat/lose weight/exercise more), how *confident* do you feel that you would succeed? If 0 is 'not confident at all' and 10 is 'very confident,' what number would you give yourself?"

Summarize the answers:

SELECTING THE FOCUS

- If importance is low (<3), focus on importance first.
- If both are the same, focus on importance first.
- If one number is distinctly lower than the other, focus on the lower number first.
- If both are very low (<3), explore feelings about participating further in the session.

EXPLORING IMPORTANCE

"This is (very/pretty/somewhat/a little bit) important to you. What made you choose _______ and not 1 or 2?"

- Reflect reasons given (self-motivational statements).
- Ask for elaboration. ("What makes that important to you?" "Tell me more about that.")

*This section is reprinted with permission of Allan Zuckoff, M. A., University of Pittsburgh Medical Center; adapted from S. Rollnick, P. Mason, and C. Butler (1999). "Health Behavior Change: A Guide for Practitioners." Churchill Livingstone, London.

- Ask for more reasons. ("What else makes it [very/pretty/somewhat/a little] important?")
- Summarize. ("So what makes this [very/pretty/somewhat/a little] important right now is . . . ." "What would have to happen for you to move up to a _______ [score plus 3–4]?" "What stops you from being at a _________ [score plus 3–4]?")
- Reflect, ask for elaboration, ask "What else?"
- Summarize. ("So what makes this [very/pretty/somewhat/a little] important is . . . ." "It would become more important to you if . . . ." "What has kept it from being more important is . . . .")

EXPLORING CONFIDENCE

"You feel (very/pretty/somewhat/a little) confident that you could do this if you tried. What made you choose _________ and not 1 or 2?"

- Reflect reasons given (self-motivational statements).
- Ask for elaboration. ("Tell me more about that.")
- Ask for more reasons. ("What else makes you feel [very/pretty/somewhat/a little] confident?")
- Summarize ("So what makes you [very/pretty/somewhat/a little] confident right now is . . . .") "What would help you move up to a _______ [score plus 3–4]?" "What stops you from being at a ________ [score plus 3–4]?")
- Reflect, ask for elaboration, ask "What else?"
- Summarize. ("So what makes you [very /pretty/somewhat/a little] confident is . . . ." "You would feel more confident if . . . ." "What has kept you from feeling more confident is . . . .")

The following types of questions can be used in the motivational interviewing process to reinforce confidence/self-efficacy. These types of questions are referred to as *competency-focused interviewing* [49].

QUESTIONS THAT ARE COMPETENCY FOCUSED AND SUCCESS ORIENTED

- "What do you know about __________ (e.g., diabetes, high cholesterol)?"
- "What kind of changes have you made in your diet so far?"
- "How have you fit exercise into your schedule in the past?"
- "What is the most important thing that you were able to learn?"
- "What do you think you would do differently this time to enable you to __________ (lose weight/lower your cholesterol/improve your blood glucose levels?)"
- "How do you think you would feel when you reach your goal?"
- "What do you see as the next step?"

4. FRAMES Strategy

Miller and Rollnick have suggested that the following specific motivational techniques can be combined to achieve an effective motivational interviewing strategy (FRAMES) [50].

1. **F**eedback. Clearly discuss the client's current health situation and risks and explain results of objective tests; share observations based on food records and weight trends. Clarify goals by comparing feedback on the patient's current situation to some standard and set goals toward that standard that are realistic and attainable.
2. **R**esponsibility. Emphasize that it is the client's responsibility to change.
3. **A**dvice. Clearly identify the problem or risk area, explain why change is important, and advocate specific change.
4. **M**enu. Offer the client a menu of alternative strategies for changing eating habits. Offering each client a range of options allows the individual to select strategies that match his or her particular situation and enhance the sense of perceived personal choice.
5. **E**mpathy. Show warmth, respect, support, caring, concern, understanding, commitment, and active interest through attentive listening skills.
6. **S**elf-efficacy. Reinforce the client's self-efficacy via competency-based questions and statements. Research has shown that the counselor's belief in the client's ability to change can be a significant determinant of outcome [50].

The dietitian's responsibility in counseling patients in the contemplation stage is to help the client reduce the barriers to making changes, focus more on the benefits, simplify the "to do" steps, and enhance self-efficacy. Once the dietitian has completed this process, it is the client's responsibility and choice to make a decision about changing eating habits.

E. Goal Setting: A Technique for the Preparation Stage

1. A Case of Preparation

Dietitian: "How can I help you?" "What are your goals for today's session?"
Patient: "I want to __________ (improve my blood glucose control/lower my cholesterol level/lose weight)."

Patients in the preparation stage feel that the advantages of changing their eating habits outweigh the disadvantages. They may have already tried making small changes, but they are looking for more specific guidance and support.

WEEKLY GOALS

NUTRITION /BEHAVIOR
1.______________________
2.______________________

EXERCISE
1.______________________
2.______________________

The following roadblocks could interfere with my ability to achieve these goals.
Therefore, I have devised these coping strategies.

Roadblock #1:

Plan:

Roadblock #2:

Plan:

Roadblock #3:

Plan:

Evaluation

How did it go? ______________________

What did you learn? ______________________

Would you do anything differently next time? ______________________

FIGURE 2 Weekly goals form.

Their ambivalence may not have disappeared. However, they show more interest in change by making self-motivational statements and by asking more questions about change and experimenting with small changes [50]. The dietitian's responsibility is to strengthen the client's commitment to change and to assist the client in making realistic plans to modify his or her lifestyle and eating habits. It is the client's responsibility to participate in goal setting by considering the options and selecting a strategy that provides sufficient direction to prevent floundering, but not so many goals that it undermines self-efficacy and success.

2. Goal Setting

For clients to move toward the action stage, they must resolve their ambivalence and establish a firm commitment to a plan of action. The counselor can use the FRAMES technique to guide the client toward a specific action plan. In addition, the counselor can ask questions and make statements that imply competency and build self-efficacy. It is important for the dietitian to accentuate and reinforce behaviors that the client is doing right and then set goals from there. Goals that are specific, realistic, positive, short term, and measurable are best. It is also important to help the client anticipate obstacles to success and problem-solve strategies to deal with those barriers (Fig. 2). Finally, the counselor can ask questions to be sure that the client has set reasonable achievable goals. If clients are not at least 80% confident in their ability to achieve their goals, it is important to help them reset their goals to a level at which they feel they can be successful.

QUESTIONS TO FACILITATE GOAL SETTING

- "What do you think you would like to do?"
- "What would you say is the first step that you need to take?"

- "What other changes would you like to make?"
- "What options would you consider trying?"
- "What obstacles can you anticipate that might interfere with your ability to accomplish your goals?"
- "How do you think you will handle these obstacles?"
- "How confident are you that you will be able to accomplish your goals?"

Patient: "I'd like to start to increase my activity to improve my blood sugars."

Dietitian: "That's great! It's good that you have so many options for exercise—a treadmill, outdoor walking, an exercise bike. You've also already got a great start on increasing your activity by walking with your coworker at lunch two times per week for 20 minutes each time. If you could increase your activity minutes toward 150 minutes per week, that is a level that would be a good target. What would you say would be a reasonable next step to increase your activity?"

Patient: "I'd like to try to walk four times per week for 20 minutes at lunch."

Dietitian: "Are there any problems or roadblocks that you can anticipate that would get in the way of accomplishing this goal?"

Patient: "If it rains, we might not go out."

Dietitian: "How will you handle that situation?"

Patient: "I'll either suggest that we walk indoors, or I'll make a plan to walk on my own on the weekend."

Dietitian: "Are there any other barriers or obstacles that you can anticipate?"

Patient: "No, not right now."

Dietitian: "Out of 100% confidence, how confident are you that you can do 80 minutes of activity in the next week?"

Patient: "I'm 80% confident."

Dietitian: "Good, because if you were less than 80% confident, that would mean that we should consider changing the goal so that you feel that the likelihood of success is fairly high."

F. Self-Management Skills Training: Techniques for Action

1. A Case of Action

Dietitian: "How can I help you?" "What would you like to accomplish in our meeting today?"

Patient: "I'm doing well with my food choices and exercise during the week, but on weekends it can be harder if I go out to eat."

Patients in the action stage have started making changes in their environment to support changes in eating habits. They need positive reinforcement for making behavioral changes and assistance strengthening self-management skills. The counselor's responsibility is to provide continued praise and support for positive behavioral changes and offer ongoing information and advice to enhance self-management skills. The client's responsibility is to actively participate in the session by sharing feelings about changes that he or she has made and discussing questions and concerns about maintaining the behavior changes.

2. Self-Management Skills Training

The core techniques that dietitians can combine to help clients in the action stage to strengthen self-management skills are concrete nutrition information and advice, self-monitoring, stimulus control, and exercise [1].

1. *Concrete nutrition information and advice.* Dietitians' expertise in translating nutrition recommendations into food choices that are meaningful and satisfying is key in supporting clients in the action stage. In particular, dietitians can assist patients in trying new recipe ideas/modifying favorite recipes, finding healthier food choices, and suggesting new and interesting food combinations at meals and snacks. In this way, dietitians can help patients learn that dietary change can occur without disrupting family food patterns or personal enjoyment of food.
2. *Self-monitoring.* Dietitians can encourage clients to use self-monitoring as a tool to enhance behavior change. When clients keep track of their eating habits by recording the amounts and types of food eaten, they become more aware of how their food choices affect their health outcomes (e.g., weight, cholesterol, blood sugar). Patients who are overweight or have hypercholesterolemia may focus on food records that track their weight, fat gram, or energy intake, whereas patients with diabetes may focus on self-monitoring food, carbohydrate intake, activity, and blood sugar patterns and then use the data to learn a problem-solving approach for understanding food–activity–blood sugar relationships.
3. *Stimulus control.* Dietitians can discuss with patients how to set up their environment for success. If patients remove problem foods from their environment and follow a shopping list that includes only healthy food choices, then they can create an environment conducive to successful dietary change. Clients can also learn strategies to reduce the temptations for undesired foods by minimizing exposure to these food items at parties or buffets.
4. *Exercise.* Increasing exercise is a positive lifestyle change that can improve high-density lipoprotein (HDL) cholesterol, blood glucose control, blood pressure, weight loss and weight maintenance, self-esteem, and motivation. Increasing physical activity therefore can enhance the impact of dietary change on health outcomes and may also strengthen the self-esteem, self-efficacy, and motivation necessary to maintain diet behavior change.

G. Problem-Solving Skills and Coping Strategies: Techniques for Maintenance

1. A Case of Maintenance

Dietitian: "How can I help you?" "What are your goals for today's session?"

Patient: "I feel good about the way that I'm eating now. It has become more of a habit to _______ (eat less fat/exercise more/count carbohydrates)."

Patients in the maintenance stage have been actively working on changing eating habits for at least 6 months. Although the changes in eating habits may have become part of a patient's routine, there is still a risk of lapse or relapse. The dietitian's responsibility is to help clients plan ahead for high-risk situations and develop the problem-solving and coping skills necessary to avoid relapsing. The client's responsibility is to share any feelings or concerns related to the changes that he or she has made, discuss the particular situations that challenge his or her ability to continue with eating behavior change, and actively participate in the problem-solving process.

2. Problem-Solving Skills and Coping Strategies

Some examples of the high-risk situations that can lead to lapse or relapse of eating behavior change are eating out, stress and other emotions (feeling anxious or depressed), hunger, and vacations. If clients do not develop coping strategies to deal with high-risk situations, then they are likely to interpret an experience of overeating as a failure. This can diminish self-efficacy and undermine long-term success. Alternatively, if clients can respond to high-risk situations with effective problem-solving and coping strategies, then the experience of managing the situation improves self-efficacy and increases the likelihood of sustaining behavior change [1].

When patients can identify barriers to achieving their goals and anticipate high-risk situations, then the nutrition counselor can teach them the following steps to the problem-solving approach:

- Describe the problem or barrier in detail.
- Brainstorm options to address it.
- Pick an option to try.
- Make a positive action plan or goal (see Fig. 2).
- Anticipate and plan to handle roadblocks.
- Identify ways to increase the likelihood of success.
- Try the plan and see how it goes.

When the nutrition counselor models the problem-solving approach with the client and then the client practices this skill successfully, then self-efficacy improves [54].

The best way to help clients prevent relapse is to focus on both cognitive and behavioral techniques to appropriately respond to lapses. The behavioral steps to help clients include the following: (1) anticipate and identify high-risk situations, (2) facilitate a problem-solving approach to determine possible solutions, (3) select a coping strategy, and (4) evaluate the effectiveness of the plan. The cognitive techniques that are important in preventing relapse are directed at how clients think and feel in response to a relapse. Cognitive restructuring techniques include the following: (1) listen to self-talk associated with a lapse and evaluate if thoughts are logical, reasonable, or helpful; (2) counter any negative self-talk with positive statements; and (3) stay focused on the progress so far and the advantages of making changes in eating behavior [37].

In the process of discussing lapses in high-risk situations, it is important for dietitians to remind clients that lapses are normal and to ask open-ended questions about how clients feel regarding changes they have made in eating habits and respond to them with empathy and not judgment [50].

QUESTIONS TO FACILITATE DISCUSSION OF HIGH-RISK SITUATIONS

- "How are you feeling about the changes that you have made in your eating habits so far?"
- "Are there any situations that make it more challenging for you to sustain your _______ (activity level/reduced fat intake/weight loss)?"
- "What strategies have you tried so far to deal with the situation?"
- "How did they work?"
- "Are there any other ideas that you might try?"
- "Would you like to hear about some strategies that have worked for other people in the same situation?"
- "If we take a moment to review the various options that we have discussed, are there any that you would like to try?"
- "Which ones?" (See Fig. 2.)

The problem-solving skills and coping strategies that are important for working with clients in the maintenance stage often incorporate the techniques used to help move the client forward through the stages of change. Note that patients can recycle by relapsing and repeating stage progressions, and they can exit and reenter at any point. In fact, many new clients that dietitians see are relapsers who are coming to reenter and recycle through the eating behavior change process. In these cases, there are some important questions that the counselor can ask to assess each client's experience.

QUESTIONS TO ASSESS PRIOR EXPERIENCE OF RELAPSERS

- "What were your three most important reasons for _______ (losing weight/eating less fat/exercising more)?"
- "How long did you sustain the behavior change?"
- "Who supported you at that time?"
- "How did you handle temptations?"

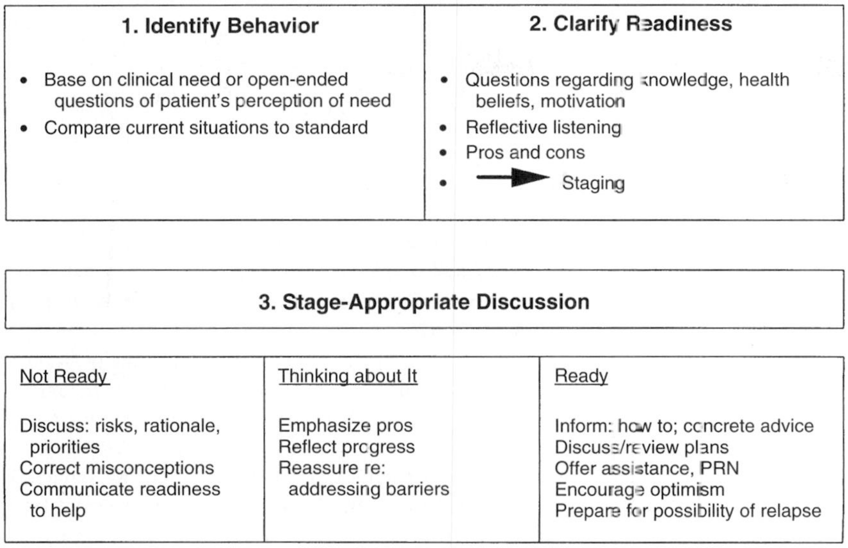

1. Identify Behavior	2. Clarify Readiness
• Base on clinical need or open-ended questions of patient's perception of need • Compare current situations to standard	• Questions regarding knowledge, health beliefs, motivation • Reflective listening • Pros and cons • → Staging

3. Stage-Appropriate Discussion

Not Ready	Thinking about It	Ready
Discuss: risks, rationale, priorities Correct misconceptions Communicate readiness to help	Emphasize pros Reflect progress Reassure re: addressing barriers	Inform: how to; concrete advice Discuss/review plans Offer assistance, PRN Encourage optimism Prepare for possibility of relapse

FIGURE 3 Three-phase model within 15-minute encounter. (Adapted with verbal permission from E. B. Fisher, Jr.)

- "What coping strategies did you use?"
- "What was going on in your life and how were you feeling when you started slipping?"
- "What did you learn from that experience?"

In sum, the stages of change model is a useful technique for approaching nutrition counseling sessions. At first glance, however, adding a new component to counseling sessions may appear difficult when allocation of time already is an issue [34]. Fisher uses a three-phase model to show that stages of change techniques can be applied in sessions as brief as 15 minutes (Fig. 3). The model expedites identification of the behavior, clarification of patient's readiness to change, and selection of stage-appropriate tools and techniques. The time the dietitian allocates to tailor counseling to information that the patient is receptive to learning will result in a more productive session.

H. Social Support

Social support influences all stages of change. A number of studies have shown the benefit of support from family, friends, and coworkers [1, 5, 55]. When spouses or significant others understand the nutritional advice given and the importance of self-monitoring, stimulus control, and exercise in managing eating behaviors, then they are better able to provide support. Support might include cooking and shopping for appropriate food items, keeping tempting food out of sight, modeling a slower eating speed, exercising together, showing a positive attitude, and offering praise.

Support from clinical staff (e.g., nurses, dietitians) was studied in the very successful Diabetes Control and Complications Trial (DCCT) [56]. The success of the DCCT was due in part to participants in the intensively treated group being able to make multiple behavior changes that resulted in their achieving improved glycemic control [57]. A study examining the effect of diet behaviors found that adherence to meal plans and appropriate treatment of hypoglycemia were associated with better diabetes control [58]. After the close of DCCT, telephone interviews were used to ask a sample of participants about the types of staff support that helped them adhere to all aspects of their treatment plan. Nondirective types of support (suggests, willing to help but does not take over) were mentioned more often than directive types of support (tells, assumes responsibility) [55]. Studies of staff support conducted with patients with acute as well as chronic illnesses indicate that stage or phase of the patient's clinical condition may influence the type of staff support that is most helpful [59]. Patients newly diagnosed or in an acute phase of their illness appear to appreciate directive support (gives me great solutions), whereas those in stable conditions value nondirective support (gives me suggestions but lets me make up my own mind). Applying type of social support to stages of change techniques, nondirective support is most appropriate for the precontemplation, contemplation, and maintenance phases, whereas directive support would be appropriate for problem solving in the preparation and action stages.

VI. CONCLUSION

The challenge in facilitating nutrition interventions is to prevent or treat disease by changing people's eating habits. Research on nutrition education is extensive but not conclusive. Although a definitive model has not been identified, a variety of strategies have shown success and can be incorporated into nutrition education programs. There is consensus on two elements: (1) education needs to be tailored to the individual's learning needs, and (2) both education and behavior change techniques are necessary. Understanding factors that influence the learning and behavior change processes enables better selection of educational tools and techniques to make counseling more effective. The stages of change model is a useful technique for tailoring education to the individual and applying behavior change strategies.

References

1. Brownell, K. D., and Cohen, L. R. (1995). Adherence to dietary regimens 2: components of effective interventions. *Behav. Med.* **20**, 155–164.
2. Knowles, M. (1990). "The Adult Learner: A Neglected Species." Gulf Publishing, Houston, TX.
3. Tough, A. (1985). How adults learn and change. *Diabetes Educator* **11**, 12–25.
4. Brookfield, S. D. (1987). "Developing Critical Thinkers: Challenging Adults to Explore Alternative Ways of Thinking and Acting." Jossey-Bass, San Francisco.
5. Contento, I., Bronner, Y. I., Paige, D. M., Gross, S. M., Bisignani, L., Lytle, L. A., Maloney, S. K., White, S. L., Olson, C. M., and Swadener, S. S. (1995). The effectiveness of nutrition education and implications for nutrition education policy, programs, and research: a review of research. *J. Nutr. Educ.* **27**, 355–364.
6. Karlsen, B., and Gardner, E. F. (1986). "Adult Basic Learning Examination Norms Booklet." Harcourt Brace, San Antonio, TX.
7. Wilkinson, G. S. (1993). "Wide Range Achievement Test Administration Manual." Wide Range, Wilmington, DE.
8. Taylor, S. C. (1953). Cloze procedure: a new test for measuring readability. *Journalism Q.* **10**, 425–433.
9. McLaughlin, G. H. (1969). SMOG grading: a new readability formula. *J. Reading* **12**, 639–646.
10. Gunning, R. (1952). "The Technique of Clear Writing." McGraw-Hill, New York.
11. Fry, E. B. (1977). Fry's readability graph: clarifications, validity, and extension to level 17. *J. Reading* **21**, 242–252.
12. Powers, R. D., Summer, W. A., and Kearl, B. E. (1958). A recalculation of four readability formulas. *J. Educ. Psychol.* **48**, 99–105.
13. Vaughn, J. Jr. (1976). Interpreting readability assessments. *J. Reading* **19**, 635–639.
14. Pitchert, J., and Elam, P. (1985). Readability formulas may mislead you. *Patient Educ. Couns.* **7**, 181–191.
15. Nitzke, S. V. J. (1992). Overview of reading and literacy research and applications in nutrition education. *J. Nutr. Educ.* **24**, 261–265.
16. U.S. Department of Health and Human Services. "Healthy People 2010: Understanding and Improving Health," November 2000, www.healthypeople.gov/Document/pdf/uih/2010uih.pdf (accessed April 24, 2007).
17. "Ethnic and Regional Food Practices: A Series (1989–2002)." The American Dietetic Association, Chicago.
18. Harris-Davis, E., and Haughton, B. (2000). Model for multicultural nutrition counseling competencies. *J. Am Diet. Assoc.* **100**, 1178–1185.
19. Achterberg, C. (1988). Factors that influence learner readiness. *J. Am. Diet. Assoc.* **88**, 1426–1429.
20. Walker, E. (1999). Characteristics of the adult learner. *Diabetes Educator* **25**, 16–24.
21. Osterman, D. N. (1984). The feedback lecture: matching teaching and learning styles. *J. Am. Diet. Assoc.* **84**, 1221–1222.
22. Norris, S. L., Engelgau, M. M., and Narayan, K. M. (2001). Effective of self-management training in type 2 diabetes: a systematic review of randomized controlled trials. *Diabetes Care* **24**, 561–587.
23. Deakin, T., McShane, C. E., Cade, J. E., and Williams, R. D. R. (2006). Group based training for self-management strategies in people with type 2 diabetes mellitus. Cochrane LIBR; (4) (CD003417).
24. Goldfield, G. S., Epstein, L. H., Kilanowski, C. K., Paluch, R. A., and Kogut-Bossler, B. (2001). Cost-effectiveness of group and mixed family-based treatment for childhood obesity. *Int. J. Obes.* **25**, 1843–1849.
25. Murphy, A., Guilar, A., and Donat, D. (2004). Nutrition education for women with newly diagnosed gestational diabetes mellitus: small-group vs individual counseling. *Can. J. Diabetes* **28**, 1–5.
26. Johnston, J. M., Jansen, G. R., Anderson, J., and Kendell, P. (1994). Comparison of group diet instruction to a self-directed education program for cholesterol reduction. *J. Nutr. Educ.* **26**, 140–145.
27. Tate, D. F., Jackvony, E. H., and Wing, R. R. (2003). Effects of Internet counseling on weight loss in adults at risk for type 2 diabetes: a randomized trial. *JAMA* **289**, 1833–1836.
28. Williamson, D. A., Martin, P., Davis, M., White, A., Newton, R., Walden, H., York-Crowe, E., Alfonso, A., Gordon, S., and Ryan, D. (2005). Efficacy of an Internet-based behavioral weight loss program for overweight adolescent African-American girls. *Eating & Weight Disorders* **10**, 193–203.
29. Tate, D. F., Jackvony, E. H., and Wing, R. R. (2006). A randomized trial comparing human email counseling, computer-automated tailored counseling, and no counseling in an Internet weight loss program. *Arch. Intern. Med.* **166**, 1620–1625.
30. Mager, R. F. (1975). "Preparing Instructional Objectives." Fearon, Belmont, CA.
31. Houston, C., and Haire-Joshu, D. (1995). Application of health behavior models. *In* "Management of Diabetes Mellitus: Perspectives Across the Lifespan" (D. Haire-Joshu, Ed.). Mosby-Year Book, St. Louis.
32. Bloom, B. S. (1956). "A Taxonomy of Educational Objectives. Handbook 1: Cognitive Domain." David McKay, New York.
33. Harrow, A. (1971). "A Taxonomy of the Psychomotor Domain." David McKay, New York.
34. Pichert, J. W. (1987). Teaching strategies for effective nutrition counseling. *In* "Handbook of Diabetes Nutritional Counseling" (M. A. Powers, Ed.). Aspen, Rockville, MD.
35. Glass, W. (1996). "On Your Way to Fitness." Shape Up America, Bethesda, MD.

36. Shape Up America. "On Your Way to Fitness, 1996," www.shapeup.org/publications/on.your.way.to.fitness/index.html (accessed April 24, 2007).
37. Brownell, K. D. (2004). "The Learn Program for Weight Control." American Health Publishing, Dallas, TX.
38. Piechota, T. (2004). "Real Solutions: Weight Loss Workbook." American Dietetic Association, Chicago.
39. Caban, A., Johnson, P., Marseille, D., and Wylie-Rosett, J. (1999). Tailoring a lifestyle change approach and resources to the patient. *Diab. Spectrum* **12**, 33–38.
40. "Exchange Lists for Meal Planning." (2002). The American Dietetic Association, Chicago, and American Diabetes Association, Alexandria, VA.
41. Camelon, K. M., Hadell, K., Jamsen, P. T., Ketonen, K. J. Kohtamaki, H. M., Makimatilla, S., Tormala, M. L., and Valve, R. H. (1998). The plate model: a visual method of teaching meal planning. *J. Am. Diet. Assoc.* **98**, 1155–1162.
42. Month of Meals series. (1989–1994). American Diabetes Association, Alexandria, VA.
43. Metz, J. A., Kris-Etherton, P. M., Morris, C. D., Mustad, V. A., Stern, J. S., Oparil, S., Chait, A., Haynes, R. B., Resnick, L. M., Clark, S., Hatton, D. C., McMahon, M., Holcomb, S., Snyder, G. W., Pi-Sunyer, F. X., and McCarron, D. A. (1997). Dietary compliance and cardiovascular risk reduction with a prepared meal plan compared to a self-selected diet. *Am. J. Clin. Nutr.* **66**, 373–385.
44. Haynes, R. B., Kris-Etherton, P., McCarron, D. A., Oparil, S., Chait, A., Resnick, L. M., Morris, C. D., Clark, S., Hatton, D. C., Metz, J. A., McMahon, M., Holcomb, S., Snyder, G. W., Pi-Sunyer, F. X., and Stern, J. S. (1999). Nutritionally complete prepared meal plan to reduce cardiovascular risk factors: a randomized clinical trial. *J. Am. Diet. Assoc.* **99**, 1077–1083.
45. Prochaska, J. O., and DiClementi, C. C. (1983). Stages and processes of self-change in smoking: toward an integrative model of change. *J. Consult. Clin. Psychol.* **51**, 390–395.
46. Prochaska, J. O., Velicier, W. F., Rossi, J. S., Goldstein, M. G., Marcus, B. H., Rakowski, W., Fiore, C., Harlow, L. L., Redding, C. A., Rosenbloom, D., and Rossi, S. R. (1994). Stages of change and decisional balance for 12 problem behaviors. *Health Psychol.* **13**, 39–46.
47. Prochaska, J. O., DiClementi, C. C., and Norcross, J. C. (1993). In search of how people change: applications to addictive behaviors. *Diab. Spectrum* **6**, 25–33.
48. Gehling, E. (1999). The next step: changing us or changing them? *Diab. Care Educ. Newsflash* **20**, 31–33.
49. Powers, M. J. (1996). Counseling skills for improved behavior change. *In* "Handbook of Diabetes Nutritional Counseling" (M. A. Powers, Ed.). Aspen, Gaithersburg, MD.
50. Miller, W. R., and Rollnick, S. (2002). "Motivational Interviewing: Preparing People for Change." 2nd ed. Guilford Press, New York.
51. Strecher, V. J., McEvoy Devellis, B., Becker, M. H., and Rosenstock, I. M. (1986). The role of self-efficacy in achieving health behavior change. *Health Educ. Q.* **13**, 73–91.
52. DiLillo, V., Siegfried, N. J., and Smith, West D. (2003). Incorporating motivational interviewing into behavioral obesity treatment. *Cogn. Behav. Pract.* **10**, 120–130.
53. Rollnick, S., Mason, P., and Butler, C. (1999). "Health Behavior Change: A Guide for Practitioners." Churchill Livingstone, London.
54. D'Zurilla, T. J., and Nezu, A. C. (1982). Social problem solving skills in adults. *In* "Advances in Cognitive-Behavioral Research and Therapy" (P. C. Kendall, Ed.), pp. 201–274. Academic Press, New York.
55. Fisher, E. B., Jr., La Greca, A. M., Arfken, C., and Schneiderman, N. (1997). Directive and nondirective support in diabetes management. *Int. J. Behav. Med.* **4**, 131–144.
56. Davis, K., Heins, J., and Fisher, E. B. Jr. (1997). Types of social support deemed important by participants in the DCCT. Diabetes ***46***(suppl 1), 89A.
57. The Diabetes Control and Complications Trial Research Group. (1993). The effect of intensive treatment of diabetes on the development and progression of long-term complications in insulin-dependent diabetes mellitus. *N. Engl. J. Med.* **329**, 977–994.
58. Delahanty, L. M., and Halford, B. N. (1993). The role of diet behaviors in achieving improved glycemic control in intensively treated patients in the Diabetes Control and Complications Trial. *Diab. Care* **16**, 1453–1458.
59. Walker, M. S., Zona, D. M., and Fisher, E. B. (2006). Depressive symptoms after lung cancer surgery: their relation to coping style and social support. *Psychooncology* **15**, 684–693.

CHAPTER 11

Evaluation of Nutrition Interventions

ALAN R. KRISTAL
Fred Hutchinson Cancer Research Center, Seattle, Washington

Contents

I. INTRODUCTION

Nutrition interventions include a broad array of programs and activities with many different goals. Interventions may be designed for treatment of acute or chronic disease, prevention of specific diseases, or simply improvement of nutritional status. Interventions can focus on changing an individual's dietary behavior, both directly and indirectly, or they can target the composition, manufacture, and availability of food. Research in dietary intervention can be either behavioral, to test whether a dietary intervention program can promote dietary change, or clinical/epidemiological, to test whether dietary change can affect a disease endpoint or disease risk. Nutrition intervention programs can be delivered as services to individual clients, groups, or entire communities. The evaluation needs of each nutrition intervention program will differ, based on the program content, design, and goals.

The optimal way to evaluate a nutrition intervention is to complete a hypothesis-driven, randomized trial. This means that an *a priori* hypothesis should be used to evaluate whether or not the intervention was effective, an experimental design created, and careful attention paid to factors that contribute to the overall validity of a scientific experiment, such as protocol development, measurement, and statistical analysis. Although expertise from many scientific disciplines is required to complete such a trial, it is important for nutritional scientists to understand the methodological issues that underlie the design of a valid intervention evaluation.

This chapter provides a general overview of quantitative evaluation, with an emphasis on those aspects likely to be the responsibility of a nutritional scientist. The focus will be on quantitative outcome and, in particular, on whether or not an intervention is effective in achieving change in dietary behavior.

II. OVERVIEW: TYPES OF NUTRITION INTERVENTION PROGRAM EVALUATIONS

A well-designed and clearly articulated evaluation plan is a key aspect of a successful nutrition intervention program. An evaluation plan requires an intervention program to have clearly defined and realistic objectives. An evaluation plan can also give timely feedback at each stage of program implementation, allowing modifications to improve program effectiveness. The three types of evaluations that are most suitable for nutrition intervention programs are (1) formative evaluation, focusing on program design; (2) process evaluation, emphasizing program implementation, quality assurance, and participant reaction; and (3) outcome evaluation, measuring the achievement of program objectives.

A. Formative Evaluation

One challenge for nutrition intervention programs is matching the content of the interventions to the interests and needs of the intended audience. Nutrition information is inherently complex, and it must balance between being scientifically correct and still comprehensible and useful to the intended audience. Intervention activities should also be reasonable for the context in which they are delivered. At the stage of formative evaluation, a nutritionist assesses whether or not materials and programs are appropriately intensive, scientifically coherent, convenient, and otherwise consistent with their intended uses.

B. Process Evaluation

Once in place, it is important to know if the intervention program is reaching its audience and how it is being received. There is a tendency for persons who are already interested in nutrition and motivated to change to participate in intervention trials. Thus, if an intervention is to be generalizable, it is important to ensure that program components successfully reach men, younger persons, racial and ethnic minorities, and other groups less likely to be drawn to programs in nutrition. This is also the stage at which to evaluate whether the audiences' reactions to the program are favorable or whether changes are needed to have a broader impact.

C. Outcome Evaluation

Ultimately, the effectiveness of a nutrition intervention will be judged based on its ability to achieve program objectives. The authors judge the specification and collection of outcome measures, and their correct analysis, to be essential to any nutrition intervention evaluation.

III. OUTCOMES OR ENDPOINTS USED TO ASSESS INTERVENTION EFFECTIVENESS

The most obvious intervention outcomes or endpoints are based on changes in nutrient intake, but for comprehensive evaluations of dietary interventions, this is too limited. Often indirect measures of intervention effectiveness, for example, changes in supermarket sales or implementation of a worksite catering policy, can serve as meaningful outcomes. In addition, to understand how an intervention did or did not work, it is necessary to measure intermediate or mediating factors for dietary change, such as beliefs, attitudes, or nutrition knowledge [1].

Before discussing outcomes in detail, two overarching points need to be emphasized. First, the single most important consideration for selecting outcomes is that they must have clear interpretations that relate directly to the intervention you are evaluating. Examples of poor evaluation outcomes include "compliance with the USDA MyPyramid," "dietary adequacy," or "compliance with the Recommended Dietary Allowances." These types of outcomes are not useful as evaluation endpoints because they are too multidimensional, cannot be precisely defined, or their interpretations are too subjective. Second, if you will not measure dietary change per se, then the outcomes you select should have a known and reasonably strong relationship to dietary behavior. For example, increased nutrition knowledge or awareness of relationships between diet and disease are not sufficient in themselves as intervention endpoints, because they have low or no predictive value for dietary behavior change. Experience suggests that it is best to carefully formulate and define intervention outcomes as a part of the overall intervention design. This results in a more focused intervention program and yields evaluation data that are optimally informative.

A. Types of Outcomes or Endpoints Used to Assess Intervention Effectiveness

Outcomes can be classified most broadly into four types: (1) Physiological or biological measures, (2) behavioral measures based on self-report, (3) diet-related psychosocial measures, and (4) environmental or surrogate measures of dietary behavior.

Biological or physiological measures are objective indicators of dietary change. For well-funded and relatively small clinical intervention trials, measures based on serological concentration of nutrients or metabolic changes associated with dietary change are optimal approaches to evaluation. The strength of biological measures is that they are objective and unbiased. Their weaknesses are that they do not exist for many outcomes of interest (e.g., reduced percentage of energy from fat in the diet) and they are rarely sufficiently sensitive to detect the relatively modest dietary changes one can expect from low-intensity health promotion interventions. It can also be procedurally difficult and prohibitively expensive to add biological measures to large, health-promotion intervention trials.

Self-reported dietary behavior is the most often used basis for intervention evaluation. There are two conceptually distinct types of measures based on self-reported diet. The most common measures are the intakes of specific nutrients in an individual's diet, such as percentage of energy from fat or milligrams of beta-carotene, or measures of food use, such as servings per day of fruits and vegetables. One can also characterize dietary habits—for example, removing the skin before eating chicken or using low-fat instead of regular salad dressings. The primary weakness of all self-reported behavioral outcomes is that persons exposed to an intervention may bias their reports of behavior to exaggerate true behavior change [2, 3]. The strengths of self-reported behavioral outcomes are that they are easy to interpret and often reflect an intervention's specified goals.

Psychosocial outcomes consist of theoretical constructs that relate to diet or dietary change. These include nutrition knowledge, attitudes, and beliefs about diet and intentions and self-efficacy to change diet. These constructs are best interpreted in the context of structured theoretical models of behavior change. For example, an intervention based on the precede/proceed model [4] would assess changes in predisposing, enabling, and reinforcing factors for dietary change. The primary weakness of psychosocial outcomes is that they do not measure dietary change; rather they measure factors that relate, often quite weakly, to dietary behavior. Their strength is that they are often the actual target or focus of the intervention. For example, an intervention designed to increase awareness of the benefits of eating more fruits and vegetables could be evaluated by measuring changes in perceived benefits. Psychosocial measures are best considered as mediating factors for dietary change—that is, factors that explain how an intervention ultimately results in changed behavior [5–7]. Collecting information on psychosocial factors can yield valuable insights into how to improve the design and content of dietary interventions [8] and thus should be given high priority in research designs.

Environmental measures assess characteristics of communities, organizations, or the physical environment that in some way reflect dietary behavior or dietary change. Measures some researchers have used are the percentage of supermarket shelf space used for healthful versions of staple foods (e.g., low-fat milk or whole-grain breads) and the percentage of foods in vending machines that are low in fat. The weakness of these measures is that there is relatively little research to support their validity as measures of dietary change, and the available evidence suggests that associations between change in environmental measures and individual dietary behavior are modest. However, the strengths of these measures are that they are objective and unbiased, and they are frequently inexpensive to collect.

Selection from among these various types of intervention outcomes is based on many criteria. However, the primary criterion is that they should be meaningful measures of the desired intervention outcome. This requires judgment and thoughtful consideration of both the goals of the intervention and of the evaluation. The following examples illustrate the diversity of options available for evaluating different types of interventions. For an intensive, clinical intervention with a goal of lowering fat intake from 35% to 20% of energy, outcomes could include self-reported diet and measured body weight. For a worksite-based intervention to increase the availability of healthful foods to workers during the workday, outcomes could include foods offered at cafeterias and in vending machines. For a worksite-based intervention designed to test different approaches to promoting healthy dietary patterns, outcomes could include self-reported diet as well as mediating psychosocial factors such as knowledge of fat in foods and stage of change to adopt a low-fat diet. For a public health campaign to increase the use of lower-fat milk products, outcomes could include supermarket sales data and random digit dial surveys to assess consumption of low-fat milk.

IV. DESIGN OF NUTRITION INTERVENTION EVALUATIONS

Evaluation design encompasses the protocols for participant recruitment, measuring intervention delivery and outcomes, and analyzing and presenting results. Choices made during the development of an evaluation design are primarily guided by two considerations: (1) the content, type, or design of the nutrition intervention, and (2) the purpose of the evaluation.

By type of intervention, the broadest distinctions are between *clinical* interventions and *public health* interventions. Clinical interventions target high-risk individuals and generally consist of intensive, multiple individual or group sessions that address both dietary behavior (nutrition education) and psychological support for maintaining dietary change. In contrast, public health interventions target large groups of individuals, usually not limited to persons at high risk, and generally consist of a broad range of low-intensity and low-cost components such as media messages and self-help materials. Some interventions fall between these two extremes (e.g., worksite-based health promotion programs may offer a series of intensive nutrition education classes) and some interventions may fall outside of this classification altogether (e.g., the decision to fortify cereal-grain products with folic acid). For purposes of evaluation, the two most important distinctions between clinical and public health interventions are the timing and amount of expected change: intensive clinical interventions produce rapid and dramatic change, whereas public health interventions yield slow change over long periods of time.

For the purposes of the evaluation, a broad distinction is made between *research* that contributes to scientific knowledge and *documentation* that serves the needs of practitioners and administrators. Comprehensive scientific evaluations are generally beyond the financial means of any but the most well-funded research trials, because the costs of evaluation generally far exceed those for intervention design and delivery. Practitioners and administrators must assume that the interventions being delivered are at least somewhat effective. As a result, evaluation is focused on evidence that the program is reaching the intended target population and is benefiting those who participate.

A. Design Components of Nutrition Intervention Evaluations

Five components characterize the design of an intervention evaluation:

1. A representative sample of persons who would be likely program participants or targets of an intervention
2. One or more measures of the evaluation outcome at preintervention
3. One or more comparison groups, most often a control group not receiving an intervention
4. Randomized assignment to treatment (intervention or control) groups
5. One or more measures of the evaluation outcome at postintervention

The most robust research evaluations include all of these design characteristics. However, not all of these components need be present to have a scientifically valid design. The only necessary characteristics are that there be postintervention outcome measures and a comparison group. Intervention evaluations not based on a randomized trial can be either quasi-experimental, in which treatment is assigned in a fashion other than randomization, or observational, in which epidemiological methods are used to statistically model differences between those receiving and not receiving an intervention. Because of their complexity, nutritionists should consult with appropriate experts before choosing such design alternatives.

Administrative evaluation of program effectiveness should strive to incorporate as many of these evaluation components as feasible. In practice, however, an administrative evaluation may simply consist of documenting the number of persons receiving the intervention and measuring changes among those exposed. One improvement in this design would be to calculate participation rates, based, for example, on the number of persons offered the intervention or the number eligible. This would give administrators insight into the acceptability and penetration of the intervention into eligible populations. Another improvement would be to document level of exposure to or participation in the intervention and to correlate level of exposure to changes in outcome. This type of dose–response analysis can suggest whether more intensive intervention would be cost effective.

1. Representative Sample

Evaluation of a representative sample means that evaluation participants include persons from demographic and socioeconomic groups who would be targets for the intervention. Representativeness in dietary interventions is often difficult, because participation in nutrition interventions tends to be higher among women and older people [9]. However, representativeness is not always important or even desirable. If the purpose of the evaluation is to test whether an intervention can work at all (e.g., in the best of all possible circumstances), it will be preferable to recruit only highly motivated volunteers using highly selective enrollment criteria. It may even be appropriate to require participants to complete a prerandomization run-in activity, for example, completing a 4-day food record, to eliminate participants not likely to complete the trial [10]. Alternatively, if an evaluation is designed to examine how an intervention will work as it is to be delivered in practice, then participants are best recruited from representative samples from defined populations in an attempt to achieve as high a recruitment rate as possible [11].

2. Preintervention Measures

Measures before intervention are desirable for two reasons. First, even in randomized experiments, there might be differences in baseline measures across treatment groups. Second, pre- and postintervention measures allow evaluators to calculate change from baseline. Basing evaluation on differences in change between treatment groups rather than simply on differences in outcome measures postintervention almost always yields superior statistical power. If possible, preintervention assessments should be completed before treatment group assignment, so that neither the evaluation staff nor the participant can be biased by knowing which intervention they will receive.

3. Comparison Group

The choice of comparison group(s) depends on the goal of the intervention. The most common design is to compare outcomes in a group receiving an intervention to one not receiving an intervention. Options include comparisons between a new and a standard intervention, or between groups receiving different levels of the same intervention. It is rarely satisfactory to have no comparison group, because it is not possible to determine whether any observed changes can be attributed to the intervention or to other factors outside of an investigator's control or not directly associated with the intervention itself.

4. Randomized Assignment to Treatment Groups

Randomization is the best way to ensure comparability between treatment groups, but it is not always possible. In this case, it may be feasible to devise an unbiased approach to assign individuals to contrasting treatments, for example, based on health care practitioner, day of the week, or hospital clinic. One entirely unacceptable design is to offer an intervention and compare results in self-selected participants to those refusing to participate. Participation in a nutrition intervention is strongly associated with characteristics that *a priori* predict dietary change, such as sex, age, and interest in nutrition and health. With this approach, comparisons between participants and nonparticipants are so strongly biased that they should be excluded from consideration for any type of evaluation design. Examination of change among participants alone would be a better design, because no inferences can be made beyond documentation that change occurred in those who received the intervention.

5. Postintervention Assessments

It is optimal to complete two postintervention measures, one after the intervention is complete and one delayed by at least several months. The first measure assesses the immediate impact of the intervention, whereas the latter gives insight into whether effects are durable and whether there is continued change over time. Basing an evaluation on an early measure alone can be misleading: Some interventions may have no long-term effect, because behavior change is not sustained. Some interventions may yield continual, gradual change over time, in which case only a long-term assessment will demonstrate intervention effectiveness.

B. Analysis of Intervention Effects

Statistical analysis of nutrition interventions can pose many challenges. Some of these statistical issues are described here, and nontechnical recommendations for the most commonly used designs and statistical models are given.

1. Level of Measurement and Units of Analysis

There are many choices for how outcomes are assessed and analyzed. Unit of measurement describes whether the outcomes are assessed on individual participants (e.g., self-reported diet) or on a group-level or environmental characteristic (e.g., supermarket sales or availability of fresh fruit in worksite cafeterias). Unit of analysis describes whether analyses of outcomes are based on individual observations (e.g., individual changes in fat intake), on aggregated measures of individual observations (e.g., percentage of population drinking low-fat milk in the previous day), or on group-level or environmental outcomes. For clinical interventions and for public health interventions in which outcome assessments are based on measurements of individuals, the units of measurement and analysis are almost always at the individual level. For some public health interventions, especially those that target large groups or communities, there are many options for combinations of unit of measurement and unit of analysis. The most common is to aggregate measures on a sample of individuals and interpret these as measures of the community [12]. One can also measure outcomes at the community level, such as availability of healthful foods in supermarkets and restaurants, existence of nutrition programs in the community, or media coverage of nutrition-related information.

It is of utmost importance to match the evaluation design, in particular, the randomization scheme, to the unit of analysis. In grouped randomized designs, for example, a trial randomizing work sites or schools to different intervention treatments, the analysis must be based on the unit of randomization, not the individuals participating in the intervention. Unlike individually randomized designs, the sample size for analysis of grouped randomized designs is a function both of the number of groups and the number of participants in each group. Consider an intervention trial that is evaluating whether a school nutrition unit can affect students' lunch choices. If 12 schools with 2000 students were each randomized such that 6 implemented the curriculum and 6 did not, the number of experimental units is 12, not 24,000. Readers will find an excellent and nontechnical overview of these issues in reviews by Koepsell [13, 14].

2. Calculating the Simple Intervention Effect

The best and most comprehensible measure of intervention effectiveness is the difference between the change in intervention group participants minus the change in control (or alternative treatment) group participants. This measure can be considered the *intervention effect*. As an example, in an intervention with a goal to decrease the percentage of energy from fat the intervention effect is defined as

$$[\text{Fat}(\%\text{en})_b - \text{Fat}(\%\text{en})_f]_I - [\text{Fat}(\%\text{en})_b - \text{Fat}(\%\text{en})_F]_C$$

where Fat (%en) is the percentage of energy from dietary fat, subscripts b and f refer to baseline and follow-up, and subscripts I and C refer to intervention and control groups. The statistical test of whether or not the intervention effect is different from zero is based on the standard error of the intervention effect, which is defined as

$$\sqrt{\frac{\text{var}[\text{Fat}(\%\text{en})_b - \text{Fat}(\%\text{en})_f]_I}{n_I} + \frac{\text{var}[\text{Fat}(\%\text{en})_b - \text{Fat}(\%\text{en})_f]_C}{n_C}},$$

where n_I is the sample size in the intervention group and n_c is the sample size in the control group.

Most measures of dietary intake and serum nutrient concentrations have log-normal or other non-normal distributions; however, changes in these measures are characteristically normally distributed. It is therefore rarely necessary to transform measures before analysis, and for simplicity of interpretation it should be avoided.

3. Calculating Intervention Effects Adjusted for Sociodemographic and Confounding Factors

The intervention effect is best estimated after adjustment for characteristics that are associated with dietary behavior. This is because (1) randomization may not have resulted in these characteristics being evenly divided between treatment groups, and (2) there may be increased statistical power to detect a statistically significant intervention effect if variance associated with these factors is controlled for in the analyses.

Here are two approaches to calculating adjusted intervention effects, which differ depending on the scale of measurement of the outcome variable. Most outcome measures are either continuous (e.g., percentage of energy from fat) or ordered categories (e.g., motivation to change measured on a scale of 1 to 10), and thus multiple linear regression can be used to calculate adjusted intervention effects. For a simple, two-treatment randomized design, the best approach is to build a regression model as follows: The dependent variable is calculated as the change from baseline to follow-up in the outcome measure; the covariates are the baseline value of the outcome measure plus the characteristics and diet-related measures you wish to control for in the analysis; the independent variable representing treatment group is an indicator variable coded 0 = control and 1 = treatment. In this model, the regression coefficient for the treatment indicator variable is the adjusted treatment effect, and the standard error of this regression coefficient is

TABLE 1 Example of Statistical Analyses Used to Report Outcome of a Dietary Intervention to Reduce Fat and Increase Fruit and Vegetable Intakes

	n	Baseline	Change at 3 Months	Change at 12 Months
Fat-related diet habits[a]				
Intervention ($x \pm$ SD)	601	2.29 ± 0.49	−0.09 ± 0.37	−0.09 ± 0.38
Control ($x \pm$ SD)	604	2.30 ± 0.49	−0.01 ± 0.36	−0.00 ± 0.40
Intervention effect ($x \pm$ SE)		Unadjusted	−0.08 ± 0.02	−0.09 ± 0.02
		Adjusted[b]	−0.09 ± 0.02	−0.10 ± 0.02
		p-Value	<0.0001	<0.0001
Fruit and vegetables (svgs/day)				
Intervention ($x \pm$ SD)	601	3.62 ± 1.49	0.41 ± 1.88	0.47 ± 1.83
Control ($x \pm$ SD)	604	3.47 ± 1.41	0.08 ± 1.63	0.14 ± 1.80
Intervention effect ($x \pm$ SE)		Unadjusted	0.33 ± 0.09	0.33 ± 0.10
		Adjusted[b]	0.39 ± 0.10	0.46 ± 0.10
		p-Value	<0.0001	<0.0001

Source: From Kristal, A., Curry, S., Shattuck, A., Feng, Z., and Li, S. (2000). A randomized trial of a tailored, self-help dietary intervention: the Puget Sound Eating Patterns Study. *Prev. Med.* **31**, 380–389.

[a]Score from 21-item scale, scored from 1.0 (low fat) to 4.0 (high fat).

[b]Adjusted for baseline value, age, sex, race, body mass index, and income.

used to test the statistical significance of the adjusted intervention effect. Table 1 gives an example of how this approach is used for the primary analysis of a two-group randomized trial of a dietary intervention.

If an outcome measure is categorical, for example, whether or not a participant lost 10 pounds or more, it is then appropriate to use logistic regression models. For a simple, two-treatment design, the following model would be appropriate: The dependent variable is an indicator variable, coded 0 or 1, representing whether or not the outcome is absent or present; the covariates are variables you wish to control in the analysis; and the independent variable representing treatment group is an indicator variable coded 0 = control and 1 = treatment. The exponentiated regression coefficient of the treatment group indicator variable (e^{β}) is the relative odds of the outcome comparing the control to treatment group. The standard error of the regression coefficient is used to determine whether the relative odds are statistically different from 1.0. Multiple categorical outcomes, for example, movement through stages of dietary change, pose considerable statistical challenges [15, 16]. Consultation with a biostatistician is important in modeling these types of outcomes, and some approach to simplifying the analysis may prove to yield more interpretable and useful results.

C. Statistical Power

A final aspect of an evaluation design is to choose an appropriate sample size. This requires making a judgment on what size intervention effect is worth detecting. Even a trivially small intervention effect can be found statistically significant given a large enough sample size, and a clinically meaningful intervention effect may not reach statistical significance if the sample size is too small. For clinical interventions, it is reasonable to choose an effect size that is meaningful for an individual. For a public health intervention, the effect size will be much smaller and needs only be meaningful in terms of changes in population distributions of the outcome measure. When choosing minimum effect sizes, it is worthwhile to review what other interventions have achieved. The general recommendation is to never set a minimum detectable intervention effect larger than 1.5 times that observed in other interventions unless there is strong reason to believe that the new intervention being evaluated is far superior.

V. MEASUREMENT ISSUES WHEN ASSESSING DIETARY CHANGE AND OTHER INTERVENTION OUTCOMES

Once outcomes are well defined and an evaluation design is established, one must select, modify, or develop measures. The two standard characteristics of dietary assessment methods, validity and reliability, are described in detail elsewhere in this text (see Chapter 1). Here the discussion extends to cover measures of psychosocial factors and aspects of measurement that are relevant to measuring dietary change, as well as practical considerations that have important implications for trial design and feasibility. Section VI describes how these measurement issues influence selection from among alternative measures of intervention outcomes.

A. Validity

Simply stated, validity is the extent to which your assessment tool measures what you want it to measure. Validity is not necessarily an intrinsic aspect of a particular tool or assessment instrument, because validity can vary as a function of method of administration, participants, or time. There are many types of validity, each with implications for intervention evaluation. Content validity is the extent to which you have sampled the domain of what you are trying to measure. For example, if your intervention goal were to lower total fat intake, high content validity would mean that you have measured all foods with meaningful amounts of fat. A limited but nevertheless important part of content validity is commonly referred to as *face validity*. Face validity is a judgment made by experts about whether or not a completed questionnaire measures what it is supposed to measure [17]. Construct validity is primarily a consideration for psychosocial measures. In this context, high construct validity means that the items used to form a scale are a good measure of some meaningful, underlying, or latent construct. An example is a scale that measures enabling factors for dietary change, consisting of six items on barriers, norms, and social support [18]. Criterion validity considers how well one measure correlates with another measure of the same construct. A type of criterion validity most important for nutrition intervention evaluations is predictive validity, which is based on whether a measure made at one time point predicts change at a later point in time. For example, based on the theory of reasoned action [19], one would validate a measure of intention to eat more fruits and vegetables by examining how it predicts increased intake.

B. Reliability

Two types of reliability are important for evaluating intervention trials. Test–retest reliability measures agreement between multiple assessments. In practice, this means that a measure taken on one day would be strongly correlated with a measure taken on another day. Although no measures have perfect reliability, measures of daily nutrient intake or specific dietary behavior have particularly low reliability because of the variability in the amounts and types of foods people eat from day to day. This type of variability, termed intraindividual or within-persons variability, makes even a perfect assessment of a single day's diet not informative for evaluating whether or not a person has changed his/her usual diet as a response to an intervention.

A second and entirely different type of reliability, which is relevant primarily to measures of psychosocial factors, is internal consistency reliability. Most psychosocial factors cannot be assessed directly (e.g., social support for eating low-fat foods), and they are generally measured using a set of items that taken together characterize the construct (e.g., "How much support do you get from your coworkers to select healthy foods from the cafeteria at lunch?"). The statistic called Cronbach's alpha, which ranges from 0 to 1, is a measure of how well the mean of scale items measures an underlying construct. High internal consistency reliability is a function of two factors, the average correlation among items in the scale *and* the number of items in the scale. Most scientists suggest a minimum of 0.7 for internal consistency; however, this ignores the practical problem that in applied evaluations it is not feasible to use lengthy scales. When scales are restricted to three or four items, a Cronbach's alpha of 0.50 is satisfactory.

C. Intervention-Associated Recall Bias

Bias is a measure of the extent to which an instrument under- or overestimates what it is attempting to measure. Ample evidence indicates that person-specific biases exist in self-reports of diet; for example, overweight persons tend to systematically underestimate energy intake [20]. If person-specific biases are constant, they will have little impact on evaluation because analyses will be based on differences between measures assessed at baseline and follow-up. However, two unique sources of bias are introduced whenever evaluation of a dietary intervention is based on self-reported behavior. First, repeated monitoring results in changed responses to assessment instruments. For example, the number of different foods reported on a 4-day diet record decreases from day 1 to day 4, suggesting that study participants simplify their diets [21]. There is also some evidence that the quality of dietary intake data improves with repeated assessments, and this may differ by intervention treatment group [21]. Second, although well-designed dietary intervention trials randomize participants to intervention and control groups, the delivery of the behavioral intervention cannot be blinded as in conventional placebo-controlled trials [22]. Thus, if intervention group participants report eating diets that match the goals of the intervention program rather than what they actually ate, there will be a bias toward overestimating intervention effectiveness. This bias can be substantial [2, 3, 23], and it appears to be larger in women than men [24–26].

D. Responsiveness

A measure only recently introduced into the nutrition literature, related specifically to measuring change, is termed "responsiveness." Conceptually, responsiveness is a measure of whether an instrument captures information on intervention-related dietary patterns [27]. In intensive clinical interventions, in which a successful intervention results in large changes in foods consumed and food-preparation techniques, a sensitive instrument will capture information on both the most common foods and dietary practices of the sample at baseline and on new practices adopted because of the intervention. In public health interventions, in which there are only modest changes in dietary behavior, a sensitive measure will detect very small changes in behavior targeted by the intervention. Statistically, responsiveness is the ratio of the intervention effect divided by the standard deviation of the intervention effect. Responsiveness is thus a function of several aspects of a measure: (1) how well an instrument measures the intervention outcome both pre- and postintervention, (2) the magnitude of the intervention effect, and (3) the variance in the dietary or other measure used as the intervention outcome.

E. Participant Burden

If outcome assessments are too long or complicated, require biological samples, or must be repeated many times, high participant burden results. High participant burden can significantly compromise an evaluation. During recruitment, participation rates will be poor and participants will be less representative of the intervention target population. Once entered into the study, long or repeated follow-up assessments will contribute to high dropout rates. Once recruited and assigned to a treatment group, all participants should be included in the evaluation whether or not they complete the intervention or subsequent follow-up assessments. Thus, high dropout rates (larger than 25%) make an evaluation suspect.

F. Instrument Complexity

Accurate assessment of nutrient intake requires complex instrumentation, regardless of whether instruments are interviewer- or self-administered. Some diet-related psychosocial factors may also require many questions with complicated skip patterns. It is important to remember that nutrition knowledge and literacy may be poor in some populations. Intervention participants may have difficulty answering detailed questions about food preparation or serving sizes, or they may not be able to understand complex questions about attitudes and beliefs [28].

Complexity of analysis is an issue that is often overlooked. Before using an evaluation instrument, understand how it will be analyzed, evaluate the underlying nutrient database and associated software, and make sure that these analyses produce the variables you wish to measure. Dietary records, food frequency questionnaires, or scales to measure diet-related psychosocial factors are not useful unless there are means to transform data from these instruments into interpretable measures of the evaluation outcomes. Similarly, although it is simple to collect blood for serological outcomes, an analysis of many diet-related serological measures is possible only in specialized research laboratories. These laboratories are primarily in academic research centers, and they are rarely capable of or interested in processing the large numbers of samples required for evaluating a dietary intervention.

G. Costs

Randomized trials are expensive, and the costs of evaluating a nutrition intervention are often far greater than the costs of intervention delivery. A large proportion of evaluation cost can be attributed to outcome assessments, especially if they are based on dietary records or recalls or on serological measures of micronutrient concentrations. Not surprisingly, the high cost of dietary assessments strongly motivates the use of alternative measures.

VI. DIETARY ASSESSMENT INSTRUMENTS AND THEIR APPLICABILITY FOR INTERVENTION EVALUATION

Most research on dietary assessment is motivated by the needs of nutritional epidemiologists and is focused on how to best understand relationships between diet and health outcomes or on surveillance of population-level nutritional status. Dietary intervention studies have tended to use dietary assessment methods developed for other purposes, with the hope that they will serve the current intended purpose. However, as described earlier in this chapter, evaluating dietary interventions and measuring dietary change have many special nuances. Here, the available tools for measuring intervention effectiveness are reviewed with a focus on the characteristics most relevant for outcome evaluation.

A. Anthropometric and Biochemical Measures

Many anthropometric and biochemical measures correlate with nutrient intake, but relatively few of these are useful or appropriate for the evaluation of interventions to change dietary behavior (Table 2). The exceptions will be interventions that are designed to increase intake of a specific micronutrient, for example, the fortification of cereal-grain products in the United States with folic acid or the fortification of sugar with vitamin A in Guatemala [29].

The most useful anthropometric measure is change in body weight, which can either be the goal of the nutritional intervention or a marker of decreased energy intake relative to expenditure, or fat intakes. Given the lack of objective markers of dietary fat reduction, this relationship with weight deserves more comment. Randomized trials of fat reduction show intervention effects for weight of about 3 kg associated with a 10 percentage point decrease in the percentage energy from fat [10, 30] and of about 0.25 kg with decreases of 1 percentage point [11]. Weight loss associated with fat reduction is likely due to incomplete substitution of nonfat sources of energy in a fat-restricted diet [31].

Biochemical measures are useful for assessing changes in foods containing unique constituents that can be measured easily in blood or urine. The most often used measure

TABLE 2 Anthropometric and Biochemical Measures Suitable for Intervention Evaluation

Measure	Intervention(s)	Comments/References
Weight	Energy or possibly fat	[10, 30]
Serum carotenoids	Fruits and vegetables	Fasting blood samples are optimal. Control for serum cholesterol, body mass index, and smoking necessary. Confounded by beta-carotene supplements. [32–34, 36]
Urinary isoflavonoids	Soy products	Spot sample may be adequate. Hidden sources of soy (e.g., processed meats) may confound measure. [99, 100]
Red cell or phospholipid	Fish	Confounded by fish oil supplements.
Fatty acid	Fat modification	[101–103]
Plasma/urinary isothiocyanates	Cruciferous vegetables	[104, 105]
Serum cholesterol	Fat modification	[106, 107]

is change in total serum carotenoids, which reflects usual intakes of carotenoid-containing fruits and vegetables [32–35]; however, total carotenoids minus lycopene is a superior measure [36] because of the low correlation of serum lycopene with other serum carotenoids. Biochemical measures are also useful to assess metabolic changes that result from dietary change. Lipids, such as low-density lipoprotein and serum cholesterol, can be used as a measure of polyunsaturated and decreased saturated fatty acid intake in some target groups [37, 38]. Because biochemical measures are expensive and require invasive collection procedures, they are impractical to use as primary outcome measures in public health interventions. An alternative is to collect biochemical measures on a small subsample of volunteers and use these data as secondary outcomes to confirm results based on dietary self-report.

B. Self-Reported Dietary Behavior

Table 3 gives an overview of measures used to collect data on self-reported diet. Selection from among these many choices requires balancing the costs and participant burden of "gold standard" measures such as multiple 24-hour dietary recalls with the practical benefits of using short, self-administered questionnaires that assess specific dietary patterns.

The primary distinctions among types of instruments are based on whether diet is measured as foods actually consumed, foods "usually" consumed, or patterns of food consumption. Note that there is an inherent hierarchy across these types of instruments; one can measure dietary patterns based on any dietary assessment instrument or foods usually consumed based on foods actually consumed. Measuring nutrients from foods actually consumed is a gold standard for intervention evaluation because any changes in foods, portion sizes, and preparation techniques can be captured.

TABLE 3 Measures of Self-Reported Diet Suitable for Intervention Evaluation

Nutrients from foods consumed
24-hour dietary recalls
Nutrients from "usual diet"
Food frequency questionnaires
Dietary patterns
Diet behavior questionnaires
Short food frequency questionnaires
"Focused" 24-hour dietary recalls

However, because many days of foods must be assessed to characterize an individual's usual diet, any measure based on actual foods consumed will be expensive and have high participant burden. The benefit of assessing nutrients from foods usually consumed is that only a single measure is needed at each time point, but it is not possible to capture details on all foods and their preparation methods in precise detail. The benefits of assessing dietary patterns alone are that it is a less burdensome task and is often a more direct measure of whether new dietary behaviors were adopted. The limitation is that changes in dietary patterns cannot be directly interpreted as changes in nutrient intake and are thus less well accepted in the nutritional science community.

1. Nutrients from Foods Actually Consumed

The best approach to measuring nutrients from foods actually consumed is unannounced (unscheduled), interviewer-administered 24-hour dietary recalls. Unannounced recalls are administered by telephone, which in practice is facilitated by collecting information at the beginning of an evaluation on convenient days, places, and times to call. Participants can be given serving size booklets so that they can refer to specific pictures when reporting amounts of foods consumed. The protocol or script used for collecting 24-hour recalls can also be modified to focus on assessing the intervention outcomes (e.g., for an intervention to decrease saturated fat, type of margarine will be important) and de-emphasize details that are uninformative (e.g., for a fruit and vegetable intervention, probes for added salt are not relevant). Finally, it is important to consider the limitations to the accuracy of 24-hour recalls, because respondents often do not know answers to questions about food composition, preparation, or portion size.

High costs (between $35 and $55 per day) and participant burden are serious drawbacks to any study wishing to evaluate intervention effectiveness using dietary recalls. This is especially true when the evaluation is at the individual level and therefore many days of intake must be captured for each participant. For randomized designs, it may be sufficient to calculate intervention effects based on one pre- and one postintervention recall per participant. This measure of the intervention effect will be unbiased, but there are problems because of the low reliability of a single day's measure of nutrient intake. Most important, because the group-level intervention effect will have high variance, there will be low statistical power to detect differences across treatment groups. Further, it is statistically inappropriate to adjust this measure of the intervention effect for individuals' characteristics or other diet-related factors [39]. Intervention effects calculated from a single 24-hour recall can be used in much the same way as biochemical measures as a secondary outcome are used to corroborate a more practical but less valid measure of nutrient intake.

Other methods of capturing foods actually consumed, including food records and scheduled 24-hour recalls, are inappropriate for intervention evaluation. This is because both record keeping and anticipating a dietary recall can substantially change dietary behavior [2]. Although unannounced 24-hour recalls may be subject to intervention-associated bias through differential recall of foods consumed, food records or announced recalls are much more likely to be biased because of their effects on food choice. For example, three multipass 24-hour recalls were used at baseline, 6, and 12 months to generate Diet Quality Index scores [40, 41] using the Diet Quality Index-Revised (DQI-R) [42] among women cancer survivors (>65 years of age) participating in an intervention to improve diet quality. Investigators reported significant improvements among the intervention group at 6 months, but not at 12 months. These results could be due to the reasons mentioned earlier in this paragraph or a diminishing effect of the intervention over time. The true reason explaining these results may have been elucidated had an appropriate biomarker been available for use (see Chapter 12).

2. Nutrients from Foods Usually Consumed

Since the mid-1980s, food frequency questionnaires (FFQs) have become a standard measure in nutritional epidemiology. They have also become quite common for intervention evaluations because they can be self-administered, processed using mark-sense technology, and cost less than $15 per administration.

Two characteristics of any FFQ detract from their use for measuring dietary change. First, whereas FFQs are convenient for investigators, they are burdensome to participants because they require time, high literacy, and knowledge about food. Second, the cognitive processes required to complete an FFQ might make them highly subject to intervention-associated bias. Respondents must construct answers to FFQ items using knowledge about their characteristic diet, because memory about actual eating and drinking episodes erodes after only a few days [43]. Thus, not knowing the true answer to most questions, it is likely that perceptions of behavior deemed desirable because of the intervention would bias FFQ responses [26]. Nevertheless, practical considerations often require using FFQs. For these evaluations, it is necessary to incorporate methods to minimize response burden and increase response rates, such as limiting total questionnaire length and giving incentives.

Standard FFQs developed for epidemiological studies are not necessarily good for intervention evaluation. An intervention-focused FFQ must usually collect detailed information on food choices that reflect the intervention goals. This may require any of the following: (1) regrouping foods into categories relevant to the intervention (e.g., grouping soups into "creamed," "vegetable," "bean," and "broth" to capture fat, carotenoids, and fiber); (2) assessing relatively fine distinctions between similar foods (e.g., nonfat versus regular mayonnaise, refried versus baked beans, or plain lettuce salads versus mixed salads with carrots and tomatoes); (3) assessing preparation methods (e.g., chicken with or without skin, shellfish boiled or fried); and (4) collecting information on portion sizes. It is also important to examine the nutrient database and computer algorithms underlying analysis of the FFQ. Several approaches are available for assigning nutrient values to FFQ items [44, 45], and some may not produce a nutrient database that reflects substantial differences in relevant nutrients between foods targeted by the intervention. There are also different approaches in how FFQ analysis software incorporates information about types of foods (e.g., type of milk) or preparations (e.g., chicken with or without skin) when computing nutrients. It may be necessary to modify these algorithms to better reflect changes targeted by an intervention. Unfortunately, developing a new FFQ is time consuming and complex, and most evaluations must choose wisely from among those available based on how well they capture the dietary behavior of interest. Some commercially available FFQ software packages do allow minor modifications that can be helpful in developing FFQs for intervention trials [46], and some FFQs are available that were developed specifically for evaluating interventions [22, 44, 47].

3. Dietary Patterns

The many practical and scientific benefits to using short, simple instruments to measure dietary patterns targeted by an intervention were described earlier in this chapter. A modest amount of research has been done in this area, and it has yielded four approaches to short dietary assessment: (1) short FFQs; (2) prediction equations, based on regression models; (3) diet-habits questionnaires; and (4) focused 24-hour recalls. These approaches differ substantially in their underlying statistical assumptions and their use of behavioral theory related to dietary behavior change. Understanding these differences can help nutritionists to select and modify available instruments or to develop instruments suited to the intervention under evaluation.

a. Short Food Frequency Questionnaires

These instruments typically contain between 10 and 15 FFQ items and are designed to assess intake of a specific nutrient or frequency of eating a specific group of foods. Examples in the literature include instruments for fat [48], calcium [49], and fruits and vegetables [50–53]. For intervention evaluation, a short FFQ should be based on knowledge of dietary patterns in the sample receiving the intervention [54, 55] and the behavioral targets of the intervention. There is modest evidence that an approach based solely on statistical criteria can be valid in epidemiological studies [45], but this approach lacks face validity when evaluating an intervention. Short FFQs are appropriate for intervention evaluation only if a small number of foods are being targeted by the intervention or if the outcome is a nutrient that is concentrated in very few foods. Thus, short FFQs are best for nutrients such as calcium [56] or for food groups such as fruits and vegetables, and they are suspect for nutrients such as fat or sodium that are spread throughout the food supply.

b. Prediction Equations

Many nutritional scientists have proposed using regression models to predict nutrient intake from a short set of questions. These models are built typically on data from FFQs, in which the frequencies of eating specific foods are entered into a stepwise linear regression model predicting the nutrient of interest. A relatively small number of foods will predict a large amount of variance in most nutrients [57], and on the surface this appears to be a useful way to simplify dietary assessment. However, these models have poor face validity because these prediction equations assign coefficients (weights) to food items that often have little to do with the nutrient of interest [58]. The models also have poor criterion validity because models developed from one sample typically do not predict nutrient intake in a different, independent sample [59, 60].

A related approach is based on using factor analysis to identify patterns of association among foods on an FFQ. These patterns are given descriptive names (e.g., "junk food" or "plain home cooking") based on the interpretation of the foods in the factor, and these factors are treated as meaningful measures of an underlying dietary pattern. These patterns lack both face validity and criterion validity and are not reproducible between samples. One of the reasons both regression and factor analysis do not yield useful measures is that they are based on correlations among FFQ items in a specific sample. There is little consistency in the correlations among FFQ items across time or across samples and even less consistency if a dietary intervention changes dietary patterns. Measures based on regression models or factor analysis are entirely inappropriate for intervention evaluation.

c. Diet-Habits Questionnaires

Diet-habits questionnaires are unique because they have been developed specifically to measure intervention outcomes. These questionnaires typically contain between 15 and 30 questions, with response options that are qualitative (e.g., rarely/never to usually/always) or ordered frequency categories. The most robust of these measures is based on theoretical or at least explicit models of dietary behavior change [61, 62], and their development is facilitated by an understanding of nutrition, psychometric theory, and cognitive psychology, as well as skill in item construction and questionnaire design. There is ample evidence that these measures can be valid measures of dietary change [61–69], and they tend to have higher responsiveness than alternatives such as short FFQs, full FFQs, or repeated 24-hour recalls [9, 11, 27, 70, 71].

Diet behavior questionnaires are well suited to evaluations of public health interventions because of their low participant burden and high sensitivity to change, and they are less well suited for clinical interventions requiring a measure of changes in nutrient intake. Before beginning an evaluation, it is important to pilot these measures in the target population to make sure that the questionnaire format language and diet patterns being measured are appropriate. Nutritionists should take the liberty of modifying existing questionnaires to better suit their target population, as long as the basic structure and content of the instrument remains intact.

d. Focused 24-Hour Recalls

A focused recall uses techniques similar to those used to collect standard 24-hour recalls, but it collects only information related to the evaluation outcome. For example, if the intervention goal is to increase servings of fruits and vegetables, then the only information captured during the interview is about fruits and vegetables. The amount of detail collected can also vary. A focused recall could collect details about the type and serving size of all targeted foods, or an interviewer could simply read a list of foods and ask respondents whether of not they ate them in the previous day [72]. Focused recalls are far simpler to administer and analyze than standard 24-hour recalls, yet they share the characteristic of low bias because of their reliance on episodic rather than general memory [43]. Similar to a 24-hour recall, only repeated assessments can be used to characterize behavior of an individual, but a single measure is fine for characterizing a group. For example, one could contrast the percentages of participants in the intervention and control groups who drank skim milk or ate french fried potatoes on the previous day. More complex focused recalls have been developed to measure carotenoid intake [73], and more research on developing these measures is warranted. Nutritionists should consider developing instruments based on this approach when the 24-hour recalls are the desired evaluation tool but cannot be used because of costs or participant burden.

4. Diet-Related Psychosocial Factors

Table 4 gives an overview of psychosocial factors that can be used to assess intervention outcomes. Comprehensive behavioral models are built from many constructs [74], but in the context of an intervention evaluation it is generally both feasible and necessary to measure only those constructs that are central to the model. Thus, for each of the behavioral models used to design nutrition interventions, those psychosocial factors that are key to their evaluation are selected.

Little research exists on the validity and responsiveness of diet-related psychosocial measures. Measures developed and validated for one behavioral domain—for example, for smoking cessation—are often adapted for dietary intervention research without consideration of their face or construct validity. For some diet-related psychosocial constructs, measurement is simple and little validation is necessary. For example, one could measure intention to change diet with a single item such as "In the next 6 months, how likely is it that you will change your diet to eat less fat?" Alternatively, measuring a construct such as stage of dietary change can be complex and requires considerable developmental research. When adopting psychosocial measures from other behavioral domains, it is important to remember that dietary behavior is unique. First, dietary behavior is not a single behavior but a composite of many behaviors consisting of food choice, preparation, and frequency of consumption. Psychosocial factors related to one aspect of dietary behavior may differ from those relating to another. Second, dietary behavior is not discrete, such as smoking; rather it occurs on a continuum in which any discrete definition of desirable dietary behavior is necessarily arbitrary (e.g., <30% energy from fat). Thus, diet-related psychosocial factors related to a discrete criterion, such as eating five or more servings a day of fruits and vegetables, may not capture the intent of an intervention to simply increase servings regardless of baseline intake.

One of the most popular approaches to organizing dietary intervention programs is based on using the "stage of change" construct from the transtheoretical model [75]. Interventions are designed to move participants from preaction stages of dietary change (precontemplation, contemplation, and decision) into action and then maintenance. There is increasing evidence that interventions can move people through stages of dietary change [76], and that movement through stages of change is associated with dietary behavior change [15, 16]. However, much controversy surrounds the conceptualization of stages of change when applied to dietary behavior and thus little agreement exists in the literature on how it should be assessed [76–78].

A second popular approach for organizing nutrition interventions is precede/proceed, which is not a behavioral model but rather a planning model for intervention development and delivery. Practical measures of the main constructs from this model—predisposing, enabling, and reinforcing factors—have been developed and validated [18, 79]. One study has demonstrated that change in these factors, in addition to stage of dietary change, explained up to 55% of the intervention effect in a large randomized trial [16].

TABLE 4 Diet-Related Psychosocial Factors Suitable for Intervention Evaluation

Theoretical Model	Key Constructs	References
Utilization of health services	Predisposing (knowledge, attitudes, beliefs) Enabling (barriers, norms) Reinforcing (social support)	[79, 108, 109]
The theory of reasoned action	Intention	[109–111]
Health belief model	Perceived benefits, susceptibility, and severity	[109, 112, 113]
Social learning theory	Self-efficacy	[114]
Transtheoretical model	Stages of change	[77, 115–117]

C. Environmental and Surrogate Measures of Diet

Table 5 gives an overview of environmental indicators and other surrogate measures that can be used to assess outcomes of dietary interventions. With the exception of household food inventories, these measures are only useful to evaluate intervention outcomes at the group or community level, and thus they are best suited to evaluate environmental-level interventions.

Household food inventories consist of asking a set of questions about whether or not specific foods are currently available in the household. Characteristically the list consists of 10 to 15 foods that relate to the intervention. Examples of

TABLE 5 Surrogate and Environmental Measures Suitable for Intervention Evaluation

	References
Individual or household level	[81, 84]
Household food (pantry) inventory	
Group or community level	
Cafeteria plate observation	[118]
Cafeteria sales	[91, 92]
Vending machine sales	[90]
Supermarket sales	[86]
Supermarket environment	

foods that could be included are types of staple foods (e.g., regular versus low-fat mayonnaise) or the types of foods available for snacks (potato chips, fresh fruit, or baby carrots). There are no studies that formally evaluate household inventories as intervention outcome measures, but there is evidence that study participants can recall foods in their households accurately [80] and that the types of foods correlate with individually assessed nutrient intake [81, 82]. Because of the simplicity of this approach, further efforts to evaluate these measures are well motivated.

Monitoring food sales can be used as a direct measure of an intervention effect, but it is extraordinarily challenging [83–86]. Relating changes in supermarket sales to nutrition interventions is difficult, and most interventions attempting this approach have failed. Reasons are many, including the complexity of the food supply, the large number of food items that are sold in typical supermarkets, business confidentiality of sales data, and poor match between the data needs for business and the needs of nutrition researchers. Careful planning, pilot studies, and ongoing contact between researchers and persons responsible for data collection are necessary to make supermarket sales data useful for outcome evaluation. Monitoring sales in food services such as worksite and school cafeterias or vending machines is much simpler and has been used successfully to evaluate interventions [87–92]. One good approach is to devise a simple scheme for unobtrusively observing and recording food choices as customers move through a cafeteria line [93].

Changes in the food environment, such as supermarket signage or distribution of supermarket shelf space, are also potential surrogate measures of intervention outcomes. A series of studies has shown that it is possible to reliably measure supermarket environments, that measures of shelf space correlate with community-level measures of diet, and that changes in supermarket shelf space correlate weakly with changes in community-level diet [94–98]. It is possible that changes in supermarket signage or foods offered in restaurants could reflect an effective community-level intervention, as businesses adjust to demands from consumers for information about and access to healthier foods. These are not likely to be sensitive measures, because the supermarket environment is saturated with signage, and restaurant menus are difficult to evaluate objectively.

VII. CONCLUSION

Evaluations of nutrition interventions can be both scientifically and operationally challenging. Nutritionists can and should take the lead in conceptualizing and interpreting the evaluation of a nutrition intervention, but they should also seek collaborations or consultations with scientists in other disciplines to ensure that methods are optimal. When planning an evaluation, make sure that the timeline allows you to test, pilot, and, if necessary, refine measures and procedures, even if you are using previously developed instruments and methods. Know that your measures will have sufficient responsiveness to detect an intervention effect. Make sure that your measures assess the behaviors targeted by the intervention and that the effect sizes you expect are reasonable. These steps are often expensive and slow, but the ultimate result of using appropriate evaluation methods is that you will obtain clearly interpretable and valid results.

References

1. Baranowski, T. (2006). Advances in basic behavioral research will make the most important contributions to effective dietary change programs at this time. *J. Am. Diet. Assoc.* **106**, 808–811.
2. Caan, B., Ballard-Barbash, R., Slattery, M. L., Pinsky, J. L., Iber, F. L., Mateski, D. J., Marshall, J. R., Paskett, E. D., Shike, M., Weissfeld, J. L., Schatzkin, A., and Lanza, E. (2004). Low energy reporting may increase in intervention participants enrolled in dietary intervention trials. *J. Am. Diet. Assoc.* **104**, 357–366.
3. Baranowski, T., Allen, D. D., Masse, L. C., and Wilson, M. (2006). Does participation in an intervention affect responses on self-reported questionnaires? *Health Educ. Res.* **21**, i98–i109.
4. Bandura, A. (1986). "Social Foundations of Thought and Action: A Social Cognitive Theory," pp. 1–617. Prentice-Hall, Englewood Cliffs, NJ.
5. MacKinnon, D. P., and Dwyer, J. H. (1993). Estimating mediated effect in prevention studies. *Eval. Rev.* **17**, 144–148.
6. Hansen, W. B., and McNeal, R. B. (1996). The law of maximum expected potential effect: constraints placed on program effectiveness by mediator relationships. *Health Educ. Res.* **11**, 501–507.
7. Baranowski, T., Lin, L. S., Wetter, D. W., Resnicow, K., and Hearn, M. D. (1997). Theory as mediating variables: why aren't community interventions working as desired. *Ann. Epidemiol.* **S7**, S89–S95.
8. Glanz, K., and Steffen, A. (in press). Development and reliability testing for measures of psychosocial constructs associated with adolescent girls' calcium intake. *J. Am. Diet. Assoc.*

9. Beresford, S. A., Curry, S. J., Kristal, A. R., Lazovich, D., Feng, Z., and Wagner, E. H. (1997). A dietary intervention in primary care practice: the eating patterns study. *Am. J. Public Health* **87**, 610–616.
10. Kristal, A. R., Shattuck, A. L., Bowen, D. J., Sponzo, R. W., and Nixon, D. W. (1997). Feasibility of using volunteer research staff to deliver and evaluate a low-fat dietary intervention: the American Cancer Society Breast Cancer Dietary Intervention project. *Cancer Epidemiol. Biomarkers Prev.* **6**, 459–467.
11. Kristal, A. R., Curry, S. J., Shattuck, A. L., Feng, Z., and Li, S. (2000). A randomized trial of a tailored, self-help dietary intervention: the Puget Sound Eating Patterns Study. *Prev. Med.* **31**, 380–389.
12. French, S. A., Story, M., Fulkerson, J. A., Himes, J. H., Hannan, P., Neumark-Sztainer, D., and Ensrud, K. (2005). Increasing weight-bearing physical activity and calcium-rich foods to promote bone mass gains among 9-11 year old girls: outcomes of the Cal-Girls study. *Int. J. Behav. Nutr. Phys. Act.* **2**, 1–11.
13. Koepsell, T. D., Diehr, P. H., Cheadle, A., and Kristal, A. (1995). Invited commentary: symposium at community intervention trials. *Am. J. Epidemiol.* **142**, 594–599.
14. Koepsell, T. D., Wagner, E. H., Cheadle, A. C., Patrick, D. L., Martin, D. C., Diehr, P. H., Perrin, E. B., Kristal, A. R., Allan-Andrilla, C. H., and Dey, L. J. (1992). Selected methodological issues in evaluating community-based health promotion and disease prevention programs. *Annu. Rev. Nutr.* **13**, 31–57.
15. Glanz, K., Patterson, R. E., Kristal, A. R., Feng, Z., Linnan, L., Heimendinger, J., and Hebert, J. R. (1998). Impact of work site health promotion on stages of dietary change: the working well trial. *Health Educ. Behav.* **25**, 448–463.
16. Kristal, A. R., Glanz, K., Tilley, B. C., and Li, S. (2000). Mediating factors in dietary change: understanding the impact of a worksite nutrition intervention. *Health Educ. Behav.* **27**, 112–125.
17. Nunally, J. C., and Bernstein, I. H. (1994). "Psychometric Theory." McGraw-Hill, New York.
18. Glanz, K., Kristal, A. R., Sorensen, G., Palombo, R., Heimendinger, J., and Probart, C. (1993). Development and validation of measures of psychosocial factors influencing fat- and fiber-related dietary behavior. *Prev. Med.* **22**, 373–387.
19. Ajzen, I., and Fishbein, M. (1980). "Understanding Attitudes and Predicting Social Behavior." Prentice Hall, Englewood Cliffs, NJ.
20. Johnson, R. K., Soultanakis, R. P., and Matthews, D. E. (1998). Literacy and body fatness are associated with underreporting of energy intake in US low-income women using the multiple-pass 24-hour recall: a doubly labeled water study. *J. Am. Diet. Assoc.* **98**, 1136–1140.
21. Rebro, S. M., Patterson, R. E., Kristal, A. R., and Cheney, C. L. (1998). The effect of keeping food records on eating patterns. *J. Am. Diet. Assoc.* **98**, 1163–1165.
22. Kristal, A. R., Feng, Z., Coates, R. J., Oberman, A., and George, V. (1997). Associations of race, ethnicity, education and dietary intervention on validity and reliability of a food frequency questionnaire in the Women's Health Initiative Feasibility Study in Minority Populations. *Am. J. Epidemiol.* **146**, 856–869.
23. Forster, J., Jeffrey, R., VanNatta, M., and Pirie, P. (1990). Hypertension prevention trial: do 24-h food records capture usual eating behavior in a dietary change study? *Am. J. Clin. Nutr.* **51**, 253–257.
24. Hebert, J. R., Clemow, L., Pbert, L., and Ockene, J. (1995). Social desirability bias in dietary self-report may compromise the validity of dietary intake measures. *Int. J. Epidemiol.* **24**, 389–398.
25. Herbert, J. R., Ma, Y., Clemow, L., Ockene, I. S., Saperia, G., Stanek, E. J., Merriam, P. A., and Ockene, J. K. (1997). Gender differences in social desirability and social approval bias in dietary self-report. *Am. J. Epidemiol.* **146**, 1046–1055.
26. Kristal, A. R., Andrilla, C. H. A., Koepsell, T. D., Diehr, P. H., and Cheadle, A. (1998). Dietary assessment instruments are susceptible to intervention-associated response set bias. *J. Am. Diet. Assoc.* **98**, 40–43.
27. Kristal, A. R., Beresford, S. A. A., and Lazovich, D. (1994). Assessing change in diet-intervention research. *Am. J. Clin. Nutr.* (Suppl) **59**, 185S–189S.
28. Weinrich, S., Boyde, M., and Pow, B. (1997). Tool adaptation for socioeconomically disadvantaged populations. *In* "Instruments for Clinical Health Care Research" (M. Stromburd and S. Olson, Eds.), pp. 20–29. Jones & Bartlett, Norwalk, CT.
29. Arroyave, G., Mejia, L. A., and Aguilar, J. (1981). The effect of vitamin A fortification of sugar on the serum vitamin A levels of preschool Guatemalan children. *Am. J. Clin. Nutr.* **34**, 41–49.
30. Sheppard, L., Kristal, A. R., and Kushi, L. H. (1991). Weight loss in women participating in randomized low-fat diets. *Am. J. Clin. Nutr.* **54**, 821–828.
31. Hill, J. O., and Peters, J. C. (1998). Environmental contributions to the obesity epidemic. *Science* **280**, 1371–1375.
32. Drewnowski, A., Rock, C. L., Henderson, S. A., Shore, A. B., Fischler, C., Galan, P., Preziosi, P., and Herceberg, S. (1997). Serum beta-carotene and vitamin C as biomarkers of vegetable and fruit intakes in a community-based sample of French adults. *Arch. Intern. Med.* **65**, 1796–1802.
33. Rock, C. L., Flatt, S. W., Wright, F. A., Faerber, S., Newman, V., and Pierce, J. P. (1997). Responsiveness of carotenoids to a high vegetable diet intervention designed to prevent breast cancer recurrence. *Cancer Epidem. Biomar.* **6**, 617–623.
34. Rock, C. L., Thornquist, M. D., Kristal, A. R., Patterson, R. E., Cooper, D. A., Newhouser, M. L., Neumark-Sztainer, D., and Cheskin, L. J. (1999). Demographic, dietary and lifestyle factors differentially explain variability in serum carotenoids and fat soluble vitamins: baseline results form the Olestra Post-Marketing Surveillance Study. *J. Nutr.* **129**, 855–864.
35. Murphy, S. P., Kaiser, L. L., Townsend, M. S., and Allen, L. H. (2001). Evaluation of validity of items for a food behavior checklist. *J. Am. Diet. Assoc.* **101**, 761.
36. Campbell, D. R., Gross, M. D., Martini, M. C., Grandits, G. A., Slavin, J. L., and Potter, J. D. (1994). Plasma carotenoids as biomarkers of vegetable and fruit intake. *Cancer Epidem. Biomar.* **3**, 493–500.
37. Mazier, M. J., and Jones, P. J. (1999). Dietary fat saturation, but not the feeding state, modulates rates of cholesterol esterification in normolipidemic men. *Metabolism* **48**, 1210–1215.
38. Holmes, A. L., Sanderson, B., Maisiak, R., Brown, A., and Bittner, V. (2005). Dietitian services are associated with improved patient outcomes and the MEDFICTS dietary assessment questionnaire is a suitable outcome measure in cardiac rehabilitation. *J. Am. Diet. Assoc.* **105**, 1533–1540.
39. Lui, K. (1988). Measurement error and its impact on partial correlation and multiple regression analysis. *Am. J. Epidemiol.* **127**, 864–874.

40. Haines, P. S., Siega-Riz, A. M., and Popkin, B. M. (1999). The Diet Quality Index revised: a measurement instrument for populations. *J. Am. Diet. Assoc.* **99**, 697–704.
41. Patterson, R. E., Haines, P. S., and Popkin, B. M. (1994). Diet Quality Index: capturing a multidimensional behavior. *J. Am. Diet. Assoc.* **94**, 57–64.
42. Snyder, D. C., Sloane, R., Haines, P. S., Miller, P., Clipp, E. C., Morey, M. C., Pieper, C., Cohen, H., and Demark-Wahnefried, W. (2007). The Diet Quality Index-Revised: a tool to promote and evaluate dietary change among older cancer survivors enrolled in a home-based intervention trial. *J. Am. Diet. Assoc.* **107**, 1519–1529.
43. Smith, A. F., Jobe, J. B., and Mingay, D. J. (1991). Retrieval from memory of dietary information. *Appl. Cogn. Psychol.* **5**, 269–296.
44. Kristal, A. R., Shattuck, A. L., and Williams, A. E. (1992). Food frequency questionnaires for diet intervention research. *In* "17th National Nutrient Databank Conference," pp. 110–125. International Life Sciences Institute, Baltimore.
45. Block, G., Hartman, A. M., Dresser, C. M., Carroll, M. D., Gannon, J., and Gardner, L. (1986). A data-based approach to diet questionnaire design and testing. *Am. J. Epidemiol.* **124**, 453–469.
46. Block, G., Coyle, L. M., Hartman, A. M., and Scoppa, S. M. (1994). Revision of dietary analysis software for the health habits and history questionnaire. *Am. J. Epidemiol.* **139**, 1190–1196.
47. Patterson, R. E., Kristal, A. R., Carter, R. A., Fels-Tinker, L., Bolton, M. P., and Agurs-Collins, T. (1999). Measurement characteristics of the Women's Health Initiative food frequency questionnaire. *Ann. Epidemiol.* **9**, 178–187.
48. Block, G., Clifford, C., Naughton, M. P., Henderson, M., and McAdams, M. (1989). A brief dietary screen for high fat intake. *J. Nutr. Educ.* **21**, 199–207.
49. Cummings, S. R., Block, G., McHenry, K., and Baron, R. B. (1987). Evaluation of two food frequency methods of measuring dietary calcium intake. *Am. J. Epidemiol.* **126**, 796–802.
50. Serdula, M., Coates, R., Byers, T., Mokdad, A., Jewell, S., Chavez, N., Mares-Perlman, J., Newcomb, P., Ritenbaugh, C., Treiber, F., and Block, G. (1993). Evaluation of a brief telephone questionnaire to estimate fruit and vegetable consumption in diverse study populations. *Epidemiology* **4**, 455–463.
51. Field, A. E., Colditz, G. A., Fox, M. K., Byers, T., Serdula, M., Bosch, R. J., and Peterson, K. E. (1998). Comparison of 4 questionnaires for assessment of fruit and vegetable intake. *Am. J. Public Health* **88**, 1216–1218.
52. Hunt, M. K., Stoddard, A. M., Peterson, K., Sorensen, G., Herbert, J. R., and Cohen, N. (1998). Comparison of dietary assessment measures in the Treatwell 5 A Day worksite study. *J. Am. Diet. Assoc.* **98**, 1021–1023.
53. Kristal, A. R., Vizenor, N. C., Patterson, R. E., Neuhouser, M. L., and Shattuck, A. L. (2000). Validity of food frequency based measures of fruit and vegetable intakes. *Cancer Epidem. Biomar.* **9**, 939–944.
54. Block, G., Dresser, C. M., Hartman, A. M., and Carroll, M. D. (1985). Nutrient sources in the American diet: quantitative data from the NHANES II survey. II. Macronutrients and fats. *Am. J. Epidemiol.* **122**, 27–38.
55. Block, G., Dresser, C. M., Hartman, A. M., and Carroll, M. D. (1985). Nutrient sources in the American diet: quantitative data from the NHANES II survey. I. Vitamins and minerals. *Am. J. Epidemiol.* **122**, 13–26.
56. Jensen, J. K., Gustafson, D., Boushey, C. J., Auld, G., Bock, M. A., Bruhn, C. M., Gabel, K., Misner, S., Novotny, R., Peck, L., and Read, M. (2004). Development of a food frequency questionnaire to measure calcium intake of Asian, Hispanic, and White youth. *J. Am. Diet. Assoc.* **104**, 762–769.
57. Byers, T., Marshall, J., Fielder, R., Zielezny, M., and Graham, S. (1985). Assessing nutrient intake with an abbreviated dietary interview. *Am. J. Epidemiol.* **122**, 41–50.
58. Gray, G. E., Paganini-Hill, A., Ross, R. K., and Henderson, B. E. (1984). Assessment of three brief methods of estimation of vitamin A and C intakes for a prospective study of cancer: comparison with dietary history. *Am. J. Epidemiol.* **119**, 581–590.
59. Hankin, J. H., Messinger, H. B., and Stallones, R. A. (1968). A short dietary method for epidemiological studies. IV. Evaluation of questionnaire. *Am. J. Epidemiol.* **91**, 562–567.
60. Hankin, J. H., Rawlings, V., and Nomura, A. (1978). Assessment of a short dietary method for a prospective study on cancer. *Am. J. Clin. Nutr.* **31**, 355–359.
61. Kristal, A. R., Shattuck, A. L., and Henry, H. J. (1990). Patterns of dietary behavior associated with selecting diets low in fat: reliability and validity of a behavioral approach to dietary assessment. *J. Am. Diet. Assoc.* **90**, 214–220.
62. Shannon, J., Kristal, A. R., Curry, S. J., and Beresford, S. A. (1997). Application of a behavioral approach to measuring dietary change: the fat- and fiber-related diet behavior questionnaire. *Cancer Epidem. Biomar.* **6**, 355–361.
63. Conner, S. L., Gustafson, J. R., Sexton, G., Becker, N., Artaud-Wild, S., and Conner, W. E. (1992). The Diet Habit Survey: a new method of dietary assessment that relates to plasma cholesterol changes. *J. Am. Diet. Assoc.* **92**, 41–47.
64. Gans, K., Sundaram, S., McPhillips, J., Hixson, M., Linnan, L., and Carleson, R. (1993). Rate your plate: an eating pattern assessment and educational tool that relates to plasma cholesterol changes. *J. Nutr. Educ.* **25**, 29–36.
65. Peters, J. R., Quiter, E. S., Brekke, M. L., Admire, J., Brekke, M. J., Mullins, R. M., and Hunninghake, D. B. (1994). The eating pattern assessment tool: a simple instrument for assessing dietary fat and cholesterol intake. *J. Am. Diet. Assoc.* **94**, 1008–1013.
66. Kristal, A. R., White, E., Shattuck, A. L., Curry, S., Anderson, G. L., Fowler, A., and Urban, N. (1992). Long-term maintenance of a low-fat diet: durability of fat-related dietary habits in the Women's Health Trial. *J. Am. Diet. Assoc.* **92**, 553–559.
67. Kristal, A. R., Shattuck, A. L., and Patterson, R. E. (1999). Differences in fat related dietary patterns between black, Hispanic, and white women: results from the women's Health Trial Feasibility Study in minority populations. *Public Health Nutr.* **2**, 273–276.
68. Kinlay, S., Heller, R. F., and Halliday, J. A. (1991). A simple score and questionnaire to measure group changes in dietary fat intake. *Prev. Med.* **20**, 378–388.
69. Hartman, T. J., McCarthy, P. R., and Himes, J. H. (1993). Use of eating-pattern messages to evaluate changes in eating behaviors in a worksite cholesterol education program. *J. Am. Diet. Assoc.* **93**, 1119–1123.

70. Hartman, T. J., McCarthy, P. R., Park, R. J., Schuster, E., and Kushi, L. (1997). Results of a community-based low-literacy nutrition education program. *J. Community Health* **22**, 325–341.
71. Glasgow, R. E., Perry, J. D., Toobert, D. J., and Hollins, J. F. (1996). Brief assessments of dietary behavior in field settings. *Addict. Behav.* **21**, 239–247.
72. Kristal, A. R., Abrams, B. F., Thornquist, M. D., DiSorga, L., Croyle, R. T., Shattuck, A. L., and Henry, H. J. (1990). Development and validation of a food use checklist for evaluation of community nutrition interventions. *Am. J. Public Health* **80**, 1318–1322.
73. Neuhouser, M. L., Patterson, R. E., Kristal, A. R., Eldridge, A. L., and Vizenor, N. C. (2000). A brief dietary assessment instrument for assessing target foods, nutrients and eating patterns. *Public Health Nutr.* **4**, 73–78.
74. Glanz, K., Lewis, F. M., and Rimer, B. K. (1997). "Health Behavior and Health Education Theory, Research, and Practice." Jossey-Bass, San Francisco.
75. Prochaska, J. O., and Velicer, W. F. (1997). The transtheoretical model of health behavior change. *Am. J. Health Promot.* **12**, 38–48.
76. Kristal, A. R., Glanz, K., Curry, S. J., and Patterson, R. E. (1999). How can stages of change be best used in dietary interventions? *J. Am. Diet. Assoc.* **99**, 679–684.
77. Povey, R., Conner, M., Sparks, P., James, R., and Shepherd, R. (1999). A critical examination of the application of the transtheoretical model's stages of change to dietary behaviours. *Health Educ. Res.* **14**, 641–651.
78. Greene, G. W., Rossi, S. R., Rossi, J. S., Velicer, W. F., Fava, J. L., and Prochaska, J. O. (1999). Dietary applications of the stages of change model. *J. Am. Diet. Assoc.* **99**, 673–678.
79. Glanz, K., Kristal, A. R., Tilley, B. C., and Hirst, K. (1998). Psychosocial correlates of healthful diets among male auto workers. *Cancer Epidemiol. Biomarkers Prev.* **7**, 119–126.
80. Crocket, S. J., Potter, J. D., Wright, M. S., and Bacheller, A. (1992). Validation of a self-reported shelf inventory to measure food purchases behavior. *J. Am. Diet. Assoc.* **92**, 694–697.
81. Satia, J. A., Patterson, R. E., Kristal, A., Pineda, M., and Hislop, T. G. (2001). A household food inventory for North American Chinese. *Public Health Nutr.* **2**, 241–247.
82. Patterson, R. E., Kristal, A. R., Shannon, J., Hunt, J. R., and White, E. (1997). Using a brief household inventory as an environmental indicator of individual dietary practices. *Am. J. Public Health* **87**, 272–275.
83. Shucker, R., Levy, A., Tenny, J., and Mathews, O. (1992). Nutrition self-labeling and consumer purchase behavior. *J. Nutr.* **24**, 553–559.
84. Patterson, B. H., Kessler, L. G., Wax, Y., Bernstein, A., Light, L., Midthune, D. N., Portnoy, B., Tenny, J., and Tuckermanty, E. (2002). Evaluation of a supermarket intervention. *Eval. Res.* **16**, 464–490.
85. Kristal, A. R., Goldenhar, L., Muldoon, J., and Morton, R. F. (1997). A randomized trial of a supermarket intervention to increase consumption of fruits and vegetables. *Am. J. Health Promot.* **11**, 422–425.
86. Odenkirchen, J., Portnoy, B., Blair, J., Rodgers, A., Light, L., and Tenny, J. (1992). In-store monitoring of a supermarket nutrition intervention. *Fam. Community Health* **14**, 1–9.
87. Wilbur, C. S., Zifferblatt, S. M., Pinsky, J. L., and Zifferblatt, S. (1981). Healthy vending: a cooperative pilot research program to stimulate good health in the marketplace. *Prev. Med.* **10**, 85–89.
88. Schmitz, M. F., and Fielding, J. E. (1986). Point-of-choice nutrition labeling: evaluation in a worksite cafeteria. *J. Nutr. Educ.* **19**, 85–92.
89. Cincirpini, P. M. (1984). Changing food selections in a public cafeteria. *Behav. Modif.* **8**, 520–539.
90. Hoerr, S. M., and Louden, V. A. (1993). Can nutrition information increase sales of healthful vended snacks? *J. Sch. Health* **63**, 386–390.
91. Jeffery, R. W., French, S. A., Raether, C., and Baxter, J. E. (1994). An environmental intervention to increase fruit and salad purchases in a cafeteria. *Prev. Med.* **23**, 788–792.
92. Perlmutter, C. C., Canter, D. D., and Gregoire, M. B. (1997). Profitability and acceptability of fat- and sodium-modified hot entrees in a worksite cafeteria. *J. Am. Diet. Assoc.* **97**, 391–395.
93. Mayer, J., Brown, T., Heins, J., and Bishop, D. (1987). A multi-component intervention for modifying food selection in a worksite cafeteria. *J. Nutr. Educ.* **6**, 277–280.
94. Cheadle, A., Psaty, B., Wagner, E., Diehr, P., Koepsell, T., Curry, S., and Von Korff, M. (1990). Evaluating community-based nutrition programs. Assessing reliability of a survey of grocery store product displays. *Am. J. Public Health* **80**, 709–711.
95. Cheadle, A., Psaty, B. M., Curry, S., Wagner, E., Diehr, P., Koepsell, T., and Kristal, A. (1991). Community-level comparisons between the grocery store environment and individual dietary practices. *Prev. Med.* **20**, 250–261.
96. Cheadle, A., Wagner, E., Koepsell, T., Kristal, A., and Patrick, D. (1992). Environmental indicators: a tool for evaluating community-based health-promotion programs. *Am. J. Prev. Med.* **9**, 78–84.
97. Cheadle, A., Psaty, B. M., Curry, S., Wagner, E., Diehr, P., Koepsell, T., and Kristal, A. (1993). Can measures of the grocery story environment be used to track community-level dietary changes? *Prev. Med.* **22**, 361–372.
98. Cheadle, A., Psaty, B., Diehr, P., Koepsell, T., Wagner, E., Curry, S., and Kristal, A. (1995). Evaluating community-based nutrition programs: comparing grocery store and individual-level survey measures of program impact. *Prev. Med.* **24**, 71–79.
99. Karr, S. C., Lampe, J. W., Hutchins, A. M., and Slavin, J. L. (1997). Urinary isoflavonoid excretion in humans is dose-dependent at low to moderate levels of soy protein consumption. *Am. J. Clin. Nutr.* **66**, 46–51.
100. Seow, A., Shi, C. C., Franke, A. A., Hankin, J. H., Lee, H. P., and Yu, M. C. (1998). Isoflavonoid levels in spot urine are associated with frequency of dietary soy intake in a population-based sample of middle-aged and older Chinese in Singapore. *Cancer Epidem. Biomar.* **7**, 135–140.
101. Stanford, J. L., King, I., and Kristal, A. R. (1991). Long-term storage of red blood cells and correlations between red cell and dietary fatty acids: results from a pilot study. *Nutr. Cancer* **16**, 183–188.
102. Agren, J., Hanninen, O., Julkunen, A., Fogelholm, L., Vidgred, H., Schwab, U., Pynnonen, O., and Uusitupa, M.

(1996). Fish diet, fish oil and docosahexaenoic acid rich oil lower fasting and postprandial plasma lipid levels. *Eur. J. Clin. Nutr.* **50**, 765–771.

103. Burr, M. L., Fehily, A. M., Gilbert, J. F., Rogers, S., Holliday, R. M., Sweetnam, P. M., Elwood, P. C., and Deadman, N. M. (1989). Effects of changes in fat, fish, and fibre intakes on death and myocardial reinfarction: diet and reinfarction trial (DART). *Lancet* **2**, 757–761.
104. Chung, F., Jiao, D., Getahun, S., and Yu, M. (1998). A urinary biomarker for uptake of dietary isothiocyanates in humans. *Cancer Epidem. Biomar.* **7**, 103–108.
105. Seow, A., Shi, C., Chung, F., Jioa, D., Hankin, J., Lee, H., Coetzec, G., and Yu, M. (1998). Urinary total isothiocyanate (ITC) in a population-based sample of middle-aged and older Chinese in Singapore: relationship with dietary total ITC and glutathione S-transferase MI/TI/PI genotypes. *Cancer Epidem. Biomar.* **7**, 775–781.
106. McDougall, J., Litzau, K., Haver, E., Saunders, V., and Spiller, G. (1995). Rapid reduction of serum cholesterol and blood pressure by a twelve-day, very low fat, strictly vegetarian diet. *J. Am. Coll. Nutr.* **14**, 491–496.
107. Howell, W. H., McNamara, D. J., Tosca, M. A., Smith, B. T., and Gaines, J. A. (1997). Plasma lipid and lipoprotein responses to dietary fat and cholesterol: a meta-analysis. *Am. J. Clin. Nutr.* **65**, 1747–1764.
108. Green, L. W., and Kreuter, M. W. (1991). "Health Promotion Planning: An Educational and Environmental Approach." Mayfield, Mountain View, CA.
109. Paradis, G., O'Loughlin, J., Elliot, M., Masson, P., Renaud, L., Sacks-Silver, G., and Lampron, G. (1995). Coeur en sante St-Henri—a heart health promotion programme in a low income, low education neighborhood in Montreal, Canada: theoretical model and early field experience. *J. Epidemiol. Community Health* **49**, 503–512.
110. Richardson, N. J., Shepherd, R., and Elliman, N. A. (1993). Current attitudes and future influences on meat consumption in the U.K. *Appetite* **1993**, 41–51.
111. Brewer, J. L., Blake, A. J., Rankin, S. A., and Douglass, L. W. (1999). Theory of reasoned action predicts milk consumption in women. *J. Am. Diet. Assoc.* **99**, 33–44.
112. Kloeblen, A. S., and Batish, S. S. (1999). Understanding the intention to permanently follow a high folate diet among a sample of low-income pregnant women according to the health belief model. *Health Educ. Res.* **14**, 327–388.
113. Schafer, R. B., Keith, P. M., and Schafer, E. (1995). Predicting fat in diets of marital partners using the health belief model. *J. Behav. Med.* **18**, 419–433.
114. Ling, A., and Howarth, C. (1999). Self-efficacy and consumption of fruit and vegetables: validation of a summated scale. *Am. J. Health Promot.* **13**, 290–298.
115. Glanz, K., Patterson, R. E., Kristal, A. R., DiClemente, C. C., Heimendinger, J., Linnan, L., and McLerran, D. F. (1994). Stages of change in adopting healthy diets: fat, fiber, and correlates of nutrient intake. *Health Educ. Q.* **21**, 499–519.
116. Campbell, M. K., Symons, M., Demark-Wahnefried, W., Polhamus, B., Bernhardt, J. M., McClelland, J. W., and Washington, C. (1998). Stages of change and psychosocial correlates of fruit and vegetable consumption among rural African-American church members. *Am. J. Health Promot.* **12**, 185–191.
117. Nitzke, S., Auld, G., McNulty, J., Bock, M., Bruhn, C., Gabel, K., Lauritzen, G., Lee, Y., Medeiros, D., Newman, R., Ortiz, M., Read, M., Schutz, H., and Sheehan, E. (1999). Stages of change for reducing fat and increasing fiber among dietitians and adults with diet-related chronic disease. *J. Am. Diet. Assoc.* **99**, 728–731.
118. Graves, K., and Shannon, B. (1983). Using visual plate waste measurement to assess school lunch behavior. *J. Am. Diet. Assoc.* **83**, 163–165.

CHAPTER **12**

Biomarkers and Their Use in Nutrition Intervention

JOHANNA W. LAMPE[1] AND CHERYL L. ROCK[2]
[1]Fred Hutchinson Cancer Research Center, Seattle, Washington
[2]University of California at San Diego, La Jolla, California

Contents

This chapter presents the basic concepts and key issues related to the various uses of biomarkers in nutrition intervention research; it is not intended to be a comprehensive review. Identification and use of biomarkers is continuously evolving with the growing understanding of biological processes and the improved sensitivity of laboratory assays. Consequently, our examples of existing biomarkers are mere snapshots in the greater scheme of biomarker development and application.

I. INTRODUCTION

A biomarker or biological indicator is a characteristic that is measured and evaluated as a marker of normal biological processes, pathogenic processes, or responses to an intervention. In theory, almost any measurement that reflects a change in a biochemical process, structure, or function can be used as a biomarker. In addition, an exogenous compound that, as a result of ingestion, inhalation, or absorption, can be measured in tissues or body fluids can also be considered a biomarker.

Biomarkers can be classified broadly into markers of exposure, effect, and susceptibility and have numerous applications in nutrition. They can be used to assess dietary intakes (exposure), biochemical or physiological responses to a dietary behavior or nutrition intervention (effect), and predisposition to a particular disease or response to treatment (genetic susceptibility).

Although clinicians have used certain biological markers, such as serum cholesterol and glucose, for generations, the use of biomarkers has taken on new importance with the dramatic advances in various fields of biology and desire for objective measures in large-scale, population-based, descriptive and intervention nutrition research. Exquisitely sensitive laboratory techniques can detect subtle alterations in molecular processes that reflect events known or believed to occur along the continuum between health and disease.

II. BIOMARKERS OF DIETARY INTAKE OR EXPOSURE

Biomarkers are used to monitor dietary exposure and for nutritional assessment for several reasons. One reason is to provide biochemical data on nutritional status by generating objective evidence that enables the evaluation of dietary adequacy or ranking of individuals on exposure to particular nutrients or dietary constituents. Biochemical or biological measurements may also be collected to characterize objectively a dietary pattern, such as fruit and vegetable consumption, or to validate dietary assessment instruments or self-reported dietary data, or to monitor compliance to a dietary intervention. Another purpose for obtaining these biological measures is to establish the biological link between the nutritional factor and a physiological or biochemical process—often a hallmark of nutrition intervention studies—when the concentration of the micronutrient or dietary constituent is measured in a peripheral tissue.

A. Biomarkers of Nutrient Intake

Biochemical measures of nutrients can be a valuable component of nutritional assessment and monitoring. Overall, the usefulness of biochemical indicators of nutritional status or exposure is based on knowledge of the physiological and other determinants of the measure. For several micronutrients, the concentration of the nutrient in the circulating body pool (i.e., serum) appears to be a reasonably accurate reflection of overall status for the nutrient (Table 1). In contrast, the amount of some micronutrients in the circulating pool may be homeostatically regulated when the storage pool is adequate, or may be unrelated to intake, and thus has little relationship with total body reserves or overall status. Figure 1 illustrates the relation between various compartments or body pools that may be sampled in the measurement of biological indicators.

TABLE 1 Biomarkers of Nutrient Intake[a]

Nutrient	Biomarkers of Dietary Exposure	Possible Functional Markers
Dietary fiber, nonstarch polysaccharides	Fecal hemicellulose	Fecal weight
	Fecal short-chain fatty acids	
Thiamin		Erythrocyte transketolase activation
Biotin	Urinary 3-hydroxy-isovalerate with a loading dose of leucine	Erythrocyte pyruvate carboxylase activity
Riboflavin	Plasma FAD Erythrocyte FAD	EGRAC
Niacin	Erythrocyte NAD Urinary metabolites of niacin	Erythrocyte nicotinate-nucleotide: pyrophosphate phosphoribosyltransferase activity (not very responsive)
Vitamin B_6	Plasma pyridoxal 5-phosphate Urinary 4-pyridoxic acid	Erythrocyte aspartate or alanine aminotransferase
Folate	Plasma folate Erythrocyte folate	Plasma homocysteine
Vitamin B_{12}	Plasma B_{12}	
Vitamin C	Plasma vitamin C Erythrocyte, lymphocyte, or platelet vitamin C	Urinary deoxypyridinoline:total collagen cross-links Urinary carnitine
Vitamin A	Plasma retinol:retinol-binding protein	
Vitamin E	α- and/or γ-tocopherol in serum or plasma, erythrocytes, lymphocytes, lipoproteins, adipose tissue, or buccal mucosal cells	LDL oxidation Breath pentane and ethane Platelet adhesion and aggregation
Vitamin D	Serum 25-hydroxyvitamin D	
Vitamin K	Plasma vitamin K	Plasma prothrombin concentrations
Phosphorus	Serum inorganic phosphate	
Magnesium	Erythrocyte or lymphocyte magnesium	
Calcium	Calcium retention	Bone mass Serum osteocalcin Serum levels of skeletal alkaline phosphatase Urinary and serum measures of collagen turnover
Iron		Serum ferritin[b] Transferrin saturation Erythrocyte protoporphyrin Mean corpuscular volume Serum transferrin receptor Hemoglobin or packed cell volume
Copper	Platelet copper	Erythrocyte SOD Platelet cytochrome *c* oxidase activity Serum peptidylglycine α-aminating monooxygenase activity? Plasma diamine oxidase

(continues)

TABLE 1 *(continued)*

Nutrient	Biomarkers of Dietary Exposure	Possible Functional Markers
Zinc		Erythrocyte metallothionein Erythrocyte SOD Monocyte metallothionein mRNA Serum thymulin activity Plasma 5-nucleotidase activity
Manganese	Serum manganese	Lymphocyte Mn-SOD activity Blood arginase activity
Molybdenum		Urinary levels of sulfate, uric acid, sulfite, hypoxanthine, xanthine, and other sulfur-containing compounds
Iodine	Plasma iodine Urinary iodine	Plasma TSH, T_4, and T_3 (total and free)
Selenium	Plasma or whole-blood selenium Hair or toenail selenium	Plasma GSH peroxidase activities Erythrocyte GSH peroxidase activities Blood cell selenoperoxidase activities Plasma T_4 and T_3

Key: FAD, flavin adenine dinucleotide; EGRAC, FAD-dependent erythrocyte glutathione reductase activation coefficient; NAD, nicotinamide adenine dinucleotide; LDL, low-density lipoprotein; GSH, glutathione; SOD, superoxide dismutase; TSH, thyroid-stimulating hormone; T_3, triiodothyronine; T_4, thyroxine.

[a]Direct measures of dietary exposure and nutrient-specific functional markers. This table includes both established markers and additional markers that show promise.

[b]In approximate order of increasing severity of iron shortage. See Cook, J. D. (1999). Defining optimal body iron. *Proc. Nutr. Soc.* **58**, 489–495.

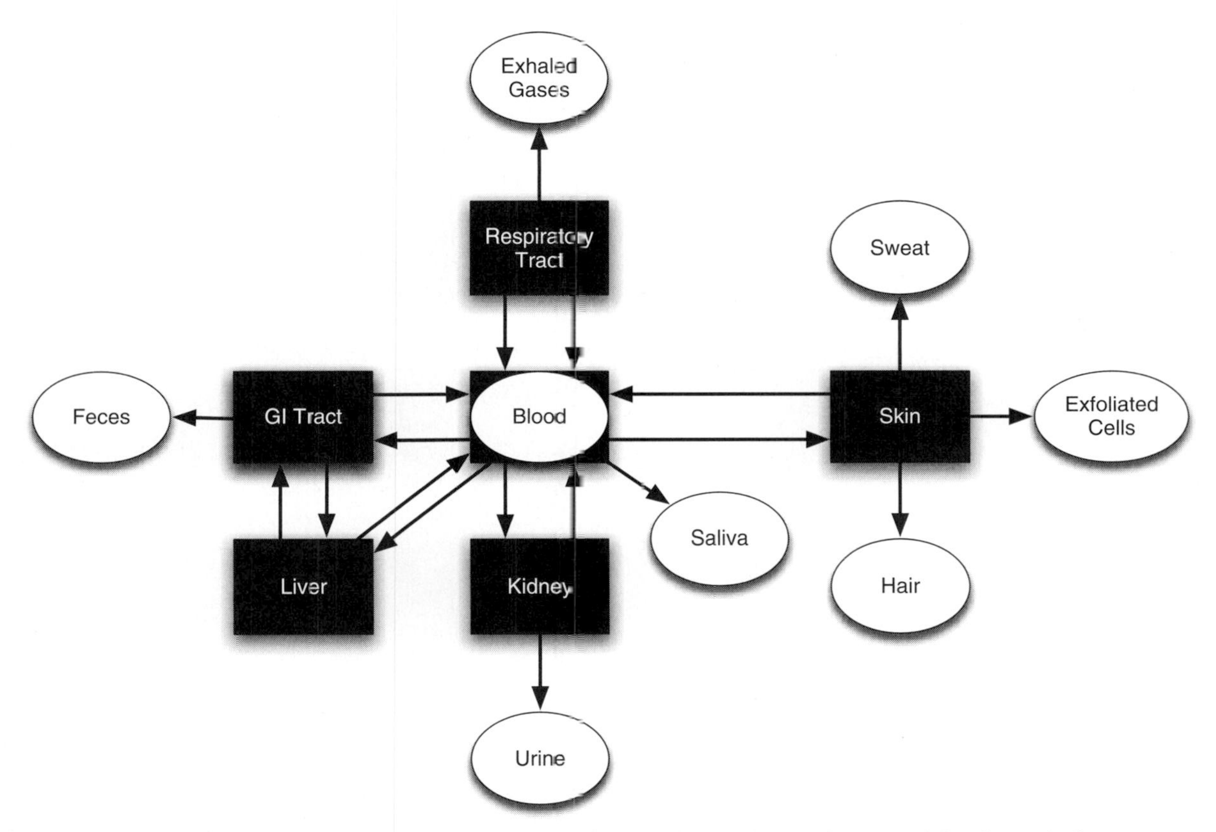

FIGURE 1 The relation between body compartments and biological specimens that can be assayed for dietary indicators.

Knowledge of the influencing nondietary factors is particularly important for accurate interpretation of the nutrient concentration in tissues. For example, tocopherols and carotenoids are transported in the circulation nonspecifically by the cholesterol-rich lipoproteins [1, 2], so higher concentrations of these lipoproteins are predictive of higher concentrations of the associated micronutrients in the circulation, independent of dietary intake or total body pool. Smoking and alcohol consumption need to be considered in the interpretation of serum and other tissue concentrations of several micronutrients, particularly compounds that may be subject to oxidation (e.g., vitamin C, tocopherols, carotenoids, folate). Knowledge of the relationship between the indicator and the risk of nutrient depletion, in addition to the responsiveness of the indicator to interventions or change, is also necessary [3]. For some nutrients, such as calcium and zinc, a specific sensitive exposure marker of diet simply has not yet been identified.

Table 1 lists examples of biochemical measures of nutrients that may serve as useful biomarkers in nutritional assessment or monitoring of dietary intake. For more details, the reader is referred to in-depth reviews addressing the use of biomarkers for assessing nutrient exposure [4–7]. Unfortunately, a static measurement (i.e., a tissue concentration) is typically not as sensitive as a functional marker in the assessment of status; however, good functional measures are still lacking in many instances.

B. Biomarkers of Other Dietary Exposures

Numerous dietary constituents, particularly of plant origin, although not recognized as essential for life, have demonstrated biological activity and are thought to play an important role in the prevention of chronic disease [8, 9]. These phytochemicals are absorbed to various degrees, often metabolized in the intestinal epithelium and liver, and excreted; thus, the metabolites can be monitored in serum or plasma or urine.

Some classes of compounds such as flavonoids are found in many plant foods, whereas others such as isoflavones are limited to select sources (Table 2). The isoflavones, daidzein and genistein, are highly concentrated in soybeans and soy products [10, 11]. Urinary isoflavone excretion is associated strongly and directly with soy protein intake under controlled dietary conditions [12]. In intervention studies of soy foods or isoflavone supplementation, urinary isoflavonoid excretion, and serum or plasma isoflavone concentrations are useful markers of study compliance [13–14]. However, because the plasma half-lives of the isoflavones genistein and daidzein are short (6 to 8 hours) [15], the timing of soy consumption in relation to urine or blood sampling may under- or overestimate isoflavone exposure. Metabolism of isoflavones is also linked to the health of colonic bacterial populations, and therefore the effects of diet and drugs on the colonic environment may influence plasma and urinary levels.

Dietary exposure to flavonoids and other polyphenols can be monitored by measuring parent compounds and metabolites in urine or plasma [16, 17]. Several compounds in cruciferous vegetables, such as sulforaphane and other isothiocyanates, have been of interest because of their potential chemopreventive effects. Concentrations of these compounds and the metabolic derivatives of them can be measured in plasma and urine by liquid chromatography/mass spectrometry [18]. In addition, dithiocarbamates (conversion products of isothiocyanates and their metabolites) can be quantified readily in urine, following extraction and measurement by high-performance liquid

TABLE 2 Phytochemical Content of Plant Food Families and Select Plant Foods[a]

Plant Foods	Flavonoids	Isoflavones	Lignans	Carotenoids	Organosulfides	Isothiocyanates	Terpenes	Phytates
Cruciferae[b]	✓			✓	✓	✓	✓	
Rutaceae[c]	✓			✓			✓	
Alliaceae[d]	✓			✓	✓		✓	✓
Solanaceae[e]	✓			✓			✓	
Umbelliferae[f]	✓			✓			✓	
Curcurbitaceae[g]	✓			✓			✓	
Cereals	✓		✓				✓	✓
Soybeans	✓	✓		✓			✓	✓
Flaxseed	✓		✓					
Measurable in biological samples[h]	U, P, S	U, P, S	P	P	P, B	U, P	P, S	U, P

Source: Adapted from Caragay, A. B. (1992). Cancer-preventive foods and ingredients. *Food Technol.* **46**, 65–68; and Fahey, J. W., Clevidence, B. A., and Russell, R. M. (1999). Methods for assessing the biological effects of specific plant components. *Nutr. Rev.* **57**, S34–S40.

[a]Some phytochemicals are present in most plant foods; others are restricted to particular botanical families or even particular plant species.

[b]Cabbage family; [c]citrus; [d]onion family; [e]tomato family; [f]carrot family; [g]squash family.

[h]U, urine; P, plasma or serum; S, stool; B, breath.

chromatography. Both of these approaches provide a way to monitor cruciferous-vegetable exposure during an intervention [19].

Biomarkers also exist for monitoring exposure to less desirable food constituents, such as mycotoxins (e.g., aflatoxin) in mold-contaminated grain products and pyrolysis products that result from cooking meat at high temperatures (e.g., heterocyclic amines). Because of the nature of these compounds, exposure to potentially carcinogenic compounds can be determined by measuring the presence of adducts—the result of covalent binding of the chemicals to proteins or to nucleic acids in DNA. The rationale for using measurements of carcinogen-DNA adducts is based on the assumption that DNA adducts formed *in vivo* are responsible for genetic alterations in genes critical for carcinogenesis and that protein adducts formed through the same processes reflect the formation of DNA adducts [20]. Because adducts represent an integration of exposure and interindividual variability in carcinogen metabolism and DNA repair, they may provide a more relevant measure of exposure (i.e., a biologically effective dose [20]). Some adducts are specific for dietary exposure; aflatoxin-albumin adducts result from ingestion of aflatoxin. Other adducts, such as benzo[*a*]pyrene-DNA adducts, are nonspecific because benzo[*a*]pyrene comes from a variety of sources besides diet, including air pollution, tobacco, and occupational exposures. Adducts can be used to monitor exposure within individuals. They can also serve as early markers of the efficacy of interventions designed to prevent exposure to genotoxic agents or to modify the metabolism of procarcinogens once exposure has occurred. An example of this latter use is an intervention to reduce aflatoxin-DNA adducts using a broccoli sprout supplement [21].

C. Biomarkers of Energy Intake

To date, few biological measures are available that objectively monitor energy intake, and those that are available are cumbersome in free-living populations or expensive. Under steady-state conditions, indirect calorimetry provides an estimate of energy expenditure and some insight about intake. Indirect calorimetry estimates the rate of oxidation or energy expenditure from the rate of oxygen consumption (VO_2) and the rate of carbon dioxide production (VCO_2). This technique is relatively inexpensive and portable, although some participant cooperation is required. These traits make the technique attractive primarily for clinical applications [22].

Energy expenditure can also be measured using a doubly-labeled water technique [23], as discussed in detail in Chapter 4. This method uses nonradioactive isotopes of hydrogen (^2H) and oxygen (^{18}O) to measure free-living total energy expenditure by monitoring urinary isotope excretion. Energy expenditures determined by room calorimetry, indirect calorimetry, and doubly-labeled water measures are not significantly different within the calorimeter environment; however, in free-living individuals, doubly-labeled, water-derived energy expenditures are found to be 13% to 15% higher than those for other methods [24]. The doubly-labeled water method has the distinct advantage of allowing the study participants to go about their usual activities with energy expenditure calculated after a study period of 7 to 14 days. Unfortunately, the ^{18}O isotope required to conduct doubly-labeled water studies is expensive and is often in short supply. Although doubly-labeled water methodology is suited to nutrition research aimed at quantifying total energy expenditure for specific groups, the cost for large samples limits broad use.

D. Biomarkers as General Dietary Indicators

Monitoring changes in patterns in response to dietary interventions presents additional challenges. The goal in this case is to monitor the intake of certain types of foods or food groups rather than specific nutrients; therefore, these dietary indicators ideally should be distributed generally within certain types of foods.

Plasma carotenoids provide a good example of the use of biomarkers as a dietary indicator when the goal is to assess and monitor dietary patterns. Vegetables and fruits contribute the vast majority of carotenoids in the diet, and plasma carotenoid concentrations have been shown to be useful biomarkers of vegetable and fruit intakes in cross-sectional descriptive studies, controlled feeding studies, and clinical trials [25–28]. The consistency of this relationship across diverse groups and involving various concurrent diet manipulations (with differences in amounts of dietary factors that could alter carotenoid bioavailability) is notable, although considerable interindividual variation in the degree of response is typically observed. Also, nondietary factors that are among the determinants of plasma carotenoid concentrations (e.g., body mass, plasma cholesterol concentration) will influence the absolute concentration that is observed in response to dietary intake, so these characteristics must be used as adjustment factors.

Although vitamin C also is provided predominantly by fruits and vegetables in the diet, this measure is much less useful as a biomarker of this dietary pattern because the relationship between vitamin C intake and plasma concentration is linear only up to a certain threshold [29]. The use of vitamin C supplements (which is common in the U.S. population) often increases the intake level beyond the range in which linearity between intake and plasma concentration occurs and thus obscures the relationship between food choices and tissue concentrations.

Lignans are a group of compounds present in high-fiber foods, particularly cereals and fruits and vegetables [30]. These compounds are not found in animal products and, similar to carotenoids, may be useful markers of a plant-based diet [31]. Lignans provide an example of how using dietary constituents as biomarkers requires an understanding of the metabolism of the compounds. Lignans in plant foods are altered by intestinal microflora, so that the specific compounds, enterodiol and enterolactone, monitored in plasma or urine are actually bacterial metabolites. Because of this bacterial conversion, lignan concentrations in urine or plasma in response to a similar dietary dose vary significantly among individuals. In addition, nondietary factors (e.g., use of oral antibiotics) reduce enterolactone and enterodiol production [32].

As another example, the fatty acid composition of membrane phospholipids is in part determined by the omega-6 and omega-3 fatty acid composition of the diet. Thus, the fatty acid pattern of serum phospholipids or plasma aliquots has been used as a biomarker of compliance with omega-3 fatty acid supplementation in clinical trials [33, 34]. Although enzyme selectivity and other physiological factors are also important determinants of the fatty acid composition of phospholipids, a diet high in omega-3 polyunsaturated fats will result in increased amounts of eicosapentaenoic and docosahexaenoic acids in circulating tissue pools.

Specific fatty acids also can be associated with certain types of foods. Pentadecanoic acid (15:0) and heptadecanoic acid (17:0) are fatty acids produced by bacteria in the rumen of ruminant animals. These fatty acids, with uneven numbers of carbon atoms, are not synthesized by humans; therefore, their presence in human biological samples can indicate dietary exposure to milk fat. Proportions of 15:0 and 17:0 in adipose tissue and concentrations of 15:0 in serum have been found to correlate with milk fat intake in men and women [35, 36].

III. FUNCTIONAL BIOMARKERS

If a nutrient or dietary constituent has an identified impact on physiological, biochemical, or genetic factors, measuring markers of those functional effects can be extremely useful. Such functional indices can be classified as those that are measures of *discrete* functions of a nutritional factor and those that are measures of more *general* functions [37]. A discrete functional index often relates to the first limiting biochemical system, for example, a particular enzymatic pathway [38]. These markers can be used to identify the dosage or concentration of a nutritional factor necessary to achieve a clinically meaningful response or to define optimum nutrient status (Table 1). Unfortunately, for many nutritional factors, the first limiting biochemical system is unknown or not readily measured or accessible in humans. A general functional index is less specific but may be more directly linked to the pathogenesis of disease or ill health. Often a panel of markers, rather than one specific measure, provides a better picture. Examples of general functional indices or markers are oxidative stress, immune function, bone health, and cell turnover—processes that have been shown to play roles in the risk of various diseases.

For a functional index to be an effective nutritional biomarker for intervention studies related to disease risk, a cause-and-effect relationship must be established between (1) nutritional status and the functional index, (2) between the functional index and ill health, and (3) between nutrient status and ill health. Such an undertaking is a daunting and time-consuming task, but it is especially important if a functional biomarker is going to be used as a proxy, or surrogate, for a clinical end point or disease outcome. A clinical end point is a characteristic or variable that measures how a patient feels, functions, or survives. A surrogate end point biomarker is an index whose modulation has been shown to indicate the progression or reversal of the disease process; it is a biomarker that is intended to substitute for a clinical end point. In an intervention trial, the use of surrogate end point biomarkers (rather than the frank diagnosis of disease) requires substantially less time and fewer resources in the evaluation of efforts aimed toward reducing risk for chronic diseases such as cancer, cardiovascular disease, and osteoporosis [39].

To date, few markers have been established as true surrogate end point biomarkers (i.e., they cannot accurately substitute for a clinical end point [40]). The evidence supporting the linkage of a biomarker to a clinical end point may be derived from epidemiological studies, clinical trials, *in vitro* analyses, animal models, and simulated biological systems. Many biomarkers have been proposed as potential surrogate end points, but relatively few are likely to achieve this status because of the complexity of disease mechanisms and the limited capability of a single biomarker to reflect the collective impact of multiple therapeutic effects on ultimate outcome.

A. Biomarkers of Enzyme Function

Understanding how diet influences enzyme systems is important in developing strategies for disease prevention and treatment. For example, dietary modulation of enzymes involved in carcinogen metabolism may be important in reducing cancer risk, and dietary intervention that reduces expression of rate-limiting enzymes in cholesterol synthesis may alter cardiovascular disease risk. Enzymes that require micronutrients as cofactors are also used as biomarkers of nutritional status (Table 1).

Components of diet have the capacity to modulate protein synthesis and function. An ideal discrete functional marker would be one that reflects the direct effect of a dietary constituent—for example, mRNA amount when the dietary factor regulates gene expression or level of enzyme activity when the factor acts as a competitive inhibitor of the enzyme (Fig. 2). Unfortunately, at present, monitoring at these levels in the pathway in an intact human

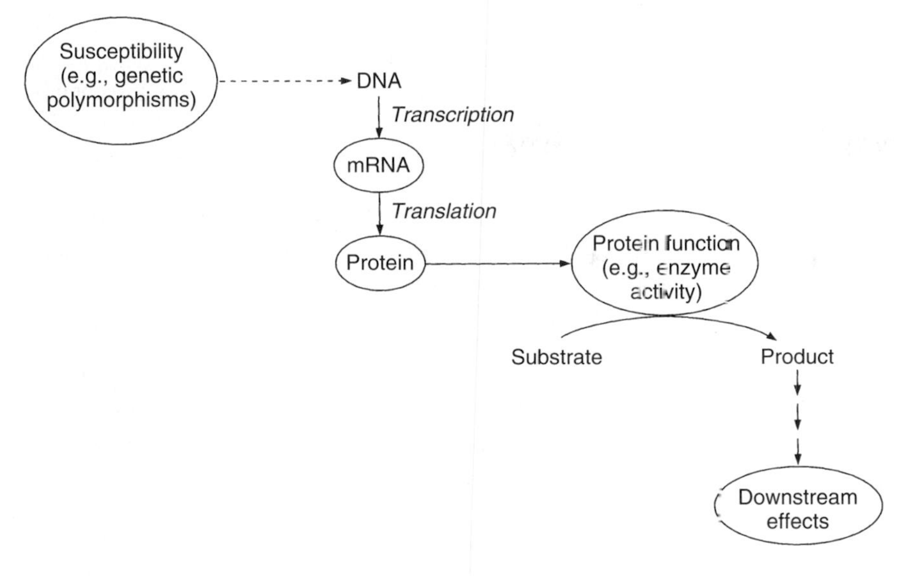

FIGURE 2 Direct functional markers of dietary exposure.

is not always feasible. Often, we rely primarily on a downstream marker, whose measurement may be influenced by subsequent or parallel pathways and may give a diluted signal.

Often the enzymes of interest are located primarily in tissues that are not readily accessible (e.g., liver, intestine, lung). One approach to meeting this challenge is to measure the enzymes in more accessible tissue; for example, enzymes that are present in high levels in the liver can often be measured in plasma or serum as a result of normal hepatocyte turnover. Enzyme activity of glutathione *S*-transferase (GST), a biotransformation enzyme important in carcinogen detoxification, can be measured spectrophotometrically in serum [41] or concentrations of the enzyme itself can be determined in serum by immunoassay [42]. Serum concentration of the GST isoenzyme GST-α has been shown to increase when cruciferous vegetables are added to the diet [42]. A limitation of using serum measures of a hepatic enzyme is that the assumption is made that liver function is normal. Thus, including other measures of liver function in the data collection is important to verify that no underlying hepatic disease is resulting in spurious GST values. Additionally, some enzymes are present in isoforms in various tissues. GST-μ, another GST isoenzyme, is present in lymphocytes as well as in liver; therefore, for this isoenzyme, GST activity or protein concentration can be measured in cells extracted from blood samples [42].

Another approach to monitor enzyme activity *in vivo* is to use a drug probe. Many of the same xenobiotic metabolizing enzymes that metabolize carcinogens also metabolize and are modulated by commonly used drugs. The metabolites of these drugs can be monitored in serum, plasma, or urine and used to determine enzyme activities. For example, measuring caffeine metabolites in urine samples collected 4 hours after consumption of a defined caffeine dose allows determination of cytochrome P450 1A2, *N*-acetyltransferase, and xanthine oxidase activities [43], and urinary concentrations of the glucuronide and sulfate conjugates of acetaminophen (paracetamol) are used to measure UDP-glucuronosyltransferase and sulfotransferase activities [44]. Drugs can be administered as probes during a nutrition intervention to determine the degree of change in enzyme activity in response to diet and to examine gene–diet interactions [45–47].

Measurement of arachidonic acid metabolism, which involves measuring the concentration of prostaglandins or leukotrienes (metabolic products) or enzymes in the eicosanoid metabolic pathway (i.e., cyclooxygenase), provides another example. Altered arachidonic acid metabolism is among the biochemical activities of nonsteroidal anti-inflammatory agents and may also be influenced by antioxidant micronutrients, such as vitamin E [48], and quantitative changes in these products or enzymes in tissues serve as biomarkers of this activity [49]. A reasonable amount of biological evidence suggests some role for this enzymatic pathway in colon carcinogenesis [50], but the overall relationship with the disease process is still under investigation.

Similarly, endogenous compounds can serve as probes to monitor enzyme activity. Serum concentrations of the amino acid homocysteine, compared to serum and red blood cell folate concentration, are a more sensitive systemic measure of cellular folate depletion [51, 52]. Serum concentrations of homocysteine increase with folate inadequacy because the remethylation of homocysteine requires *N*-5-methyltetrahydrofolate as a co-substrate [53], which therefore provides a functional marker for folate status. Nonetheless, because homocysteine is at the intersection of two metabolic pathways, remethylation and transsulfuration, deficiencies in other nutrients in these pathways, namely, vitamin B_{12}, vitamin B_6, and possibly riboflavin, can contribute to elevated serum homocysteine concentrations [51].

B. Biomarkers of Oxidative Stress

Oxidative stress has been suggested to play a role in the pathophysiological disease process in cancer, atherosclerotic cardiovascular disease, and many other acute and chronic conditions [54], although the specific relationship with the disease process remains to be established in most instances. Cellular damage caused by reactive oxygen species, which are generated from cellular respiration, co-oxidation during metabolism, and the activity of phagocytic cells of the immune system, is controlled by antioxidant defense mechanisms that involve several micronutrients. Oxidative stress describes the condition of oxidative damage resulting when the balance between free-radical generation and antioxidant defenses is unfavorable. Direct measurement of active oxygen and related species in biological samples is challenging, mainly because these compounds have short half-lives. Thus, the oxidative stress biomarkers used in human studies are typically adducts or end products that reflect reactions that have occurred between free radicals and compounds such as lipids, proteins, carbohydrates, DNA, and other molecules that are potential targets [55].

One frequently described assay used as an oxidative stress biomarker is the thiobarbituric acid reactive substances (TBARS) assay. The TBARS assay basically quantifies a product of malondialdehyde, which presumably reflects lipid hydroperoxides in the sample. Direct measurement of malondialdehyde in biological samples using high-performance liquid chromatography (HPLC) has also been examined as an alternative approach, and plasma malondialdehyde (MDA) concentrations have been reported to respond to alterations in antioxidant nutrient status, albeit not consistently [55].

Measurement of breath pentane is another biomarker of oxidative stress that has been utilized in human studies [56]. The approach basically involves collecting exhaled air for the measurement of the products of peroxidation of unsaturated fatty acids, a portion of which are volatile and released in the breath, using gas chromatography methods. However, the specific measurement methodologies vary a great deal and are not always reliable, and standardization of the procedure and knowledge of various influencing factors are needed to improve the usefulness of this approach [55].

Another biomarker of oxidative stress is urinary 8-hy droxydeoxyguanosine (8OHdG), which can be measured by several different methods with high sensitivity [57]. The 8-hydroxylation of the guanine base is a frequent type of oxidative DNA damage, and 8OHdG is subsequently excreted without further metabolism in the urine after repair *in vivo* by exonucleases. In previous studies, certain demographic factors and physiological characteristics such as gender and body mass [58] have been observed to influence urinary 8OHdG concentration, so these factors may need to be considered in interpretation. Urinary 8OHdG is increased in association with conditions known to be characterized by increased oxidative stress, such as smoking, whole-body irradiation, and cytotoxic chemotherapy [57–59]. Urinary 8OHdG (unadjusted or adjusted for urinary creatinine) has also been observed to decline in response to a high-vegetable and fruit diet intervention in human subjects [60], which is particularly interesting because this type of diet has been suggested to promote reduced oxidative stress [8, 60, 61].

Prostaglandin-like compounds produced by nonenzymatic free-radical-catalyzed peroxidation of arachidonic acid, termed F_2 isoprostanes, are currently of great interest as useful biomarkers of oxidative damage [62]. Specific gas chromatography/mass spectrometry assays for the measurement of some of these compounds, such as iPF2a-III (also called 8-iso-$PGPF_2$) and iPF_2-VI, have been developed and used to quantify the compounds in urine and blood samples. These markers have been shown to be less variable than 8OHdG [63]. Elevated levels have been observed in plasma and urine samples from subjects under a wide variety of conditions of enhanced oxidative stress [64], and the measure can be altered in dietary intervention studies designed to reduce oxidative damage [65].

Another approach to measuring DNA oxidative damage that has been evaluated for use in human nutrition intervention research is the measurement of 5-hydroxymethyluracil levels in DNA in blood. 5-Hydroxymethyluracil is produced when DNA is exposed to oxidants, is relatively stable when compared to other oxidation products, and can be quantified with gas chromatography/mass spectrometry [66, 67]. At the same time, its high intraindividual variation reduces its utility as a biomarker [68].

Oxidative damage to low-density lipoprotein (LDL) has been specifically linked to atherogenesis, and in an application of this biological activity, measurement of LDL oxidation *ex vivo* has been used in clinical studies as a biomarker of oxidative stress [69]. Basically, this process involves isolating the LDL fraction from a blood sample, exposing this fraction to oxidants such as Cu^{2+} and measuring the lag time before oxidation. Also, various specific methodologies are used across laboratories, and the lack of standardization in the approaches in use constrains the ability to make comparisons across studies.

Several other approaches to measuring biomarkers of oxidative stress have been proposed and are under study, and the reader is referred to a review of this topic [55]. Because of their inherent variability and uncertain responsiveness to dietary manipulations, clinical researchers often employ a panel of potential measures of oxidative stress, rather than relying on a single indicator [60].

C. Biomarkers of Immune Function

The human immune system is a complex and highly interactive network of cells and their products that has a central role in protecting against various external disease-promoting factors and perhaps against malignant cells. Many of the components of the immune system can serve as biomarkers and are monitored *in vivo* or *ex vivo* [70, 71]. Because of the complexity of the system, the selection of assays should be closely aligned with the research question being asked. Furthermore, multiple parameters need to be measured: one single biomarker is inadequate to monitor immune function.

Cell-mediated immune variables include the absolute amounts or ratios of various white blood cell (WBC) types (e.g., total counts, WBC differentials, T-cell subsets) and measures of T-cell function (e.g., lymphocyte proliferation, cytokine release from mitogen-stimulated cultures, cytotoxic capacities, delayed-type hypersensitivity). Both number and activity of natural killer cells (one of the cell types that plays an important role in immune surveillance) are used as biomarkers in nutrition intervention trials [72, 73]. Cytokines (e.g., the interleukins) are soluble factors released by immune cells, which control and direct the function of other immune effectors. Some of these have been used as markers of immune response in randomized trials of vitamin supplementation [73, 74].

An *in vivo* functional test, the delayed-type hypersensitivity (DTH) skin test, is widely used to monitor the immune system in humans, including in studies of dietary modulation of immune function [75]. It measures the capacity of an individual's immune system to mount a response to antigenic stimulation. The DTH test typically involves the simultaneous intradermal application of one or several DTH antigens. These antigens elicit an immunological reaction involving the release of lymphokines by antigen-sensitized T cells. These compounds, in turn, activate macrophages, which release inflammatory mediators, resulting in measurable skin induration.

D. Biomarkers of Bone Health

Bone mass measurements and biomarkers of bone turnover are used as functional indices of bone health and, to a certain extent, can also be used as markers of the adequacy of calcium intake. Measures of bone mass include bone mineral content (BMC; i.e., the amount of mineral at a particular skeletal site, such as the femoral neck, lumbar spine, or total body) and bone mineral density (BMD; i.e., bone mineral content divided by the area of the scanned region); both are strong predictors of fracture risk [76–78]. Controlled calcium intervention trials that measure change in BMD provide evidence for the intake requirement for calcium [79] (see Chapters 44 and 45 for more details).

Biochemical markers of bone turnover predict bone mass changes and fracture risk and respond to dietary calcium intake [79]; thus, they provide some promise for a biochemical indicator of calcium status. Unlike BMD, they reflect more subtle changes in bone metabolism. Bone turnover is the cyclical process by which the skeleton undergoes renewal by a coupled, but time-separated, sequence of bone resorption and bone formation [80]. Markers of bone turnover rely on the measurement in serum or urine of enzymes or matrix proteins synthesized by osteoblasts or osteoclasts that spill over into body fluids, or of osteoclast-generated degradation products of the bone matrix [79]. Currently, serum levels of skeletal alkaline phosphatase and osteocalcin are used as markers of bone formation, and products of collagen degradation measured in urine are used to measure bone resorption. These markers exhibit substantial short-term and long-term fluctuations related to time of day, phase of menstrual cycle, and season of the year, as well as other factors that alter bone remodeling (e.g., exercise) [81].

E. Biomarkers of Cell Turnover

Cellular markers of proliferation, differentiation, and apoptosis (i.e., programmed cell death) can be useful as biomarkers in research focused on nutritional factors and cancer, although the measured effect is a general indicator of an altered cell growth regulation effect. Use of such markers is severely restricted by the difficulty accessing the tissue of interest. Consequently, research in this area has been limited primarily to tissue available via endoscopic or fine-needle biopsy procedures (e.g., gastrointestinal tract, breast, prostate).

As a general rule, increased proliferation of undifferentiated cells defines one aspect or characteristic of carcinogenesis, and in colon cancer, this relationship has been well established. For example, cell proliferation occurs at the base of the colonic crypts, and as cells migrate from the crypts to the luminal surface, they become increasingly differentiated and mature and lose their proliferative capabilities [82]. The shift in which the proliferative zone extends to the surface, so that cells on the luminal surface retain proliferative capabilities and are immature and underdifferentiated, may be considered a field defect that sets the stage for current and future neoplastic changes [82–84]. Early work in this area relied on the incorporation of tritiated-thymidine or bromodeoxyuridine into the DNA of dividing cells during incubation of a biopsy specimen. These methods required that the tissue be freshly obtained, so the cells were viable and replicating. Often, label incorporation was incomplete. Now, with increased sensitivity in immunohistochemical techniques, proteins present in proliferating cells (e.g., proliferating cell nuclear antigen [PCNA] and Ki67) are used more widely to quantify proliferative activity in tissue specimens. Labeling indices involving tritiated thymidine and PCNA have been used to quantify the proliferative activity in colonic mucosal samples from human subjects [85] and have been used successfully as end points in several nutrition intervention studies to prevent colon cancer [86]. These indices are being further refined by staining for proteins present during apoptosis (e.g., Bax, Bcl–2) and in differentiated cells in order to provide a more complete picture of cell dynamics.

Adoption of aberrant crypt foci (early morphological changes in colonic epithelium that are considered potential precursors of adenomatous polyps) as biomarkers in humans is an example of how improvements in technology have led to the adoption of a biomarker that until recently could only be used in animal studies. Development of magnifying endoscopes with improved resolution now allows investigators to monitor aberrant crypt foci in colon tissue samples from healthy humans [87].

IV. BIOMARKERS OF GENETIC SUSCEPTIBILITY

The health of individuals and the population in general is the result of interaction between genetic and environmental factors. For the great majority of human diseases, purely genetic or purely environmental etiologies are insufficient to explain individual variability in occurrence, prognosis, or outcome [88]. This is especially the case with chronic diseases, such as heart disease and cancer. Genetically determined susceptibility factors alter disease frequency or treatment response through variations in the DNA coding sequences of genes. As a result, genetically susceptible individuals produce proteins that are structurally different, or produce them in greater or lesser amounts than individuals who are not at increased risk of disease.

By genetic standards, traits with a frequency of between 1/100 and 1/10,000 in a population are considered uncommon, and rare traits are those with a frequency of less than 1/10,000 [89]. Typically, these are low-prevalence and high-penetrance genes (e.g., genes associated with familial cancers). Common genetic traits are those in which the least common allele is present in at least 1% of the population [89]. Traits with this characteristic are known as genetic polymorphisms. They include the high-prevalence, low-penetrance genes—"susceptibility genes"—thought to contribute to disease risk.

In cases where specific genetic mutations or variations may indicate disease risk or progression or may be modified by nutritional factors, genetic markers can also be useful biomarkers. Various molecular techniques have been developed to help characterize genetic abnormalities or differences. Genetic factors are important to consider in nutrition research for several reasons. One reason is that it is increasingly evident that genetic polymorphisms may contribute substantially to differences in the response to environmental and dietary exposures [90]. For example, genetic variations in the expression of the xenobiotic metabolizing enzymes may mediate the potentially mutagenic effect of heterocyclic amines (obtained from meat cooked at high temperature) [91] (see also Chapter 34). Also, results from laboratory animal studies suggest that dietary modifications can promote alterations in genetic factors [92], so that measuring genetic abnormalities may be considered an approach to demonstrating a biological link between dietary factors and disease risk.

Some polymorphic traits may only be important in the presence of a particular dietary exposure. For example, carrying the 5,10-methylene-tetrahydrofolate reductase (MTHFR) thermal-labile variant has been shown to be a risk factor for colorectal adenomatous polyps, but only in the context of low-folate, vitamin B_{12}, and vitamin B_6 intake [93].

Given that a goal of the field of nutrition is the prevention and treatment of disease, genetic markers may aid in this effort by identifying population subgroups at high risk of disease in the presence of particular dietary exposures. Genetic susceptibility markers may also strengthen our understanding of disease by focusing attention on possible pathways of disease causation and progression. There is considerable heterogeneity in disease risk within populations; thus, markers of susceptibility may also help to clarify associations between dietary exposures and diseases within population subgroups [94].

V. CRITERIA FOR SELECTING AND USING BIOMARKERS

When a candidate biomarker is identified, certain basic considerations need to be addressed before it can be adopted for use in research or in a clinical setting [95]. These considerations relate to the reliability of the laboratory assay itself, the biological relevance of the marker, and the characteristics of the marker within a population. Whether or not a particular established marker is used also depends on the purpose the marker will serve.

Development of a biomarker usually builds on scientific knowledge from various types of laboratory studies, including tissue culture and animal studies. In the laboratory, an initial priority is to determine a marker's reliability or reproducibility. Assay performance can be evaluated using coefficients of variation (CV%; $[SD/mean] \times 100$) to estimate within- and between-batch precision; these are measures of analytical, laboratory performance and do not reflect intra- or interindividual variation. The within-batch precision is determined by dividing single samples into multiple aliquots and analyzing them together. Between-batch precision is determined by analyzing multiple aliquots on separate days. It is difficult to generalize about acceptable numerical values for the laboratory CV% because the degree of error acceptable depends on the use of the biomarker data. For epidemiological studies, if the goal is to establish a stable estimate of the group mean, an acceptable CV% may depend on the number of samples available, the mean concentration of the biomarker, and between-individual variation [96]. Techniques of quality control [97, 98] and statistical methodologies for managing quality control data [99] have been established and are used in clinical laboratories.

Biomarkers should be relatively easy to measure and require relatively noninvasive techniques of tissue sampling. This requires that the biological relevance or validity—that is, the relationship between biomarkers in tissues readily available for human monitoring (e.g., peripheral blood) to those in target tissue (e.g., lung, liver) and the relationship between the biomarker and the disease or exposure being studied—of the measure be established. One example is the use of serum ferritin as a marker of body iron stores. Serum ferritin was validated as a measure of iron stores against bone-marrow examination for stainable iron, the criterion—but very invasive—method for measuring iron stores [100]. As a result, serum ferritin has been adopted as a simple, quantitative biomarker of iron stores in otherwise healthy individuals.

If a particular biomarker is to be used as a measure of dietary exposure, it must be evaluated with respect to its sensitivity to that intake. Several approaches, both observational and interventional, can be used to define the relationship between long-term dietary intake and biological levels. Investigators can rely on geographical differences in exposure: tissue samples from areas of known nutrient deficiency of a specific nutrient can be compared with samples from average and high-exposure areas. This approach has the advantage that it can reflect the long-term intake of a settled group of individuals; however, identifying and controlling confounding factors is a major challenge [96]. Another observational approach is to establish within individuals the relation between a dietary exposure and the biochemical marker. Participants for such a study can be selected randomly or can be selected specifically to maximize the range of intakes. For example, in one study designed to test the use of plasma carotenoids as markers of vegetable and fruit intake, study participants were selected on the basis of their reported vegetable and fruit intake—only those who had intakes ≤2 or ≥5 of servings per day were recruited into the study [26]. Rigorous testing of the relationship between intake of the dietary factor and a biomarker under controlled dietary conditions is also valuable to establish dose–response relations; however, these trials are usually limited to weeks or months and, if they involve extensive changes to usual diet, blinding of participants may not be feasible.

Depending on the biomarker, significant variability can be seen in a biomarker. Sources of variation can be internal (e.g., age, sex, genetics, body build, biological rhythm) or environmental (e.g., diet, season, time of day, immobilization, exercise, drugs). These can contribute to both within- and interindividual variation. Additional external sources of variation, beyond laboratory accuracy and precision, can include an individual's posture during sample collection and sample handling and storage; protocols should be established to minimize these latter sources of variability.

Selection of a biomarker is dependent in part on its use. A biological indicator that is going to be used as a measure of a dietary exposure in an epidemiological study needs to be a valid representation of long-term intake [96]. Repeated sampling and measurement of a biomarker over time can provide some estimate of the within-individual variability and, therefore, the likelihood that the biomarker is a stable estimate of long-term intake. If repeated measures of a biochemical indicator vary substantially over time in the same individuals, then a single measure will not reflect true, long-term intake [96]. This lack of consistency may occur because diet has changed over the sampling interval or because the measure is overly sensitive to short-term influences, such as recent intake. When using dietary constituents or their metabolites as biomarkers, an understanding of the metabolism and pharmacokinetics of the compound and the frequency of exposure will help to establish the utility of the measure as a biomarker of long-term intake.

A nutritional intervention study may require a biomarker that is a short-term measure of response to treatment. A biomarker that is to serve as a short-term measure of response needs to change within the time frame of the intervention. For example, serum folate provides a measure of recent folic acid exposure; however, erythrocyte concentrations are dependent on the life span of the cells and therefore will not reflect short-term changes in dietary folate. Serum folate concentrations decline within 3 weeks after the initiation of a low-folate diet, whereas erythrocyte folate concentration remains in the normal range for at least 17 weeks [101].

Additional practical considerations in the use of established biomarkers include the ability to conveniently access the body compartment for measurement, the procedures necessary to collect and process the sample, the burden to study participants or patients, and the resources for laboratory analysis. For example, multiday collections of feces or urine can be a major burden for many individuals. In addition, they can result in incomplete collections, which also compromise the final results. An accurate quantification of vitamin C or folate in a circulating body pool requires processing steps that must be conducted immediately after blood collection to preserve the sample appropriately and prevent degradation that would otherwise make the resulting measurement inaccurate. These extra steps can add time and effort to the labor of blood processing. Furthermore, the complexity of an assay method can vary from the ability to analyze hundreds of samples a day at a cost of a few dollars per sample to a labor-intensive, weeklong process that costs hundreds of dollars per sample.

The ability to measure particular biomarkers is also often linked to technological challenges and existing capabilities. For example, the development of HPLC in the 1970s, and improved separation and detection technologies that are currently emerging, facilitates the quantification of many micronutrients and other dietary constituents that are present in very low concentrations in biological samples. Similarly, immunoassays allow for quantitation of phytochemicals, proteins, and so on in small volumes of serum or plasma, where previous methods required substantial quantities of sample. The development of microarray technology has provided the ability to analyze the expression profiles for thousands of genes in parallel [102, 103]. This technique will rapidly advance knowledge regarding the mechanisms by which nutrition and diet affect disease risk; however, its application in intact humans will still be limited by access to the tissue of interest and the capacity to detect small but relevant changes in gene expression in response to diet.

VI. SUMMARY

The use of biomarkers in humans is an integral component of nutrition intervention research. Biochemical measurements of dietary constituents in blood or other tissues can provide a useful assessment of the intake of certain dietary factors. However, for some nutrients, functional markers, or direct functional indices, provide a better estimate of the significance of the true status for a nutrient. More general functional indices relating to processes associated with disease risk are important for establishing the relationships between diet and disease prevention and response to treatment.

The development of biomarkers continues at a rapid pace. New types of markers are being proposed constantly and analytical techniques for existing markers are improved. This advancement requires establishing the accuracy, reliability, and interpretability of the biomarkers; obtaining data on marker distributions within different age and sex groupings in normal populations; determining the extent of intraindividual variation in markers with respect to tissue localization and persistence; and assessing the contribution of genetic and acquired susceptibility factors to interindividual variability.

References

1. Clevidence, B. A., and Bieri, J. G. (1993). Association of carotenoids with human plasma lipoproteins. *Methods Enzymol.* **214**, 33–46.
2. Romanchik, J. E., Morel, D. W., and Harrison, E. H. (1995). Distribution of carotenoids and α-tocopherol among lipoproteins do not change when human plasma is incubated in vitro. *J. Nutr.* **125**, 2610–2617.
3. Habicht, J. P., and Pelletier, D. L. (1990). The importance of context in choosing nutritional indicators. *J. Nutr.* **120**(suppl), 1519–1524.
4. Shils, M. E., Olson, J. A., Shike, M., and Ross, A. C. Eds. (1999). "Modern Nutrition in Health and Disease," 9th ed. Williams & Wilkins, Philadelphia.
5. Ziegler, E. E., and Filer, L. J., Eds. (1996). "Present Knowledge in Nutrition," 7th ed. ILSI Press, Washington, DC.
6. Board, F. A. N. (1998). "Dietary Reference Intakes for Thiamin, Riboflavin, Niacin, Vitamin B6, Folate, Vitamin B12, Pantothenic Acid, Biotin, and Choline." National Academy Press, Washington, DC.
7. Kaaks, R., Riboli, E., and Sinha, R. (1997). Biochemical markers of dietary intake. *In* "Application of Biomarkers in Cancer Epidemiology" (P. Toniolo, P. Boffetta, D. E. G. Shuker, N. Rothman, B. Hulka, and N. Pearce, Eds.), pp. 103–126. International Agency for Research on Cancer, Lyon.
8. IARC. (2003). "Fruit and Vegetables, IARC Handbook of Cancer Prevention," vol. 8. IARC Press, Lyon, France.
9. Ignarro, L. J., Balestrieri, M. L., and Napoli, C. (2007). Nutrition, physical activity, and cardiovascular disease: an update. *Cardiovasc. Res.* **73**, 326–340.
10. Coward, L., Barnes, N., Setchell, K. D. R., and Barnes, S. (1993). Genistein, daidzein and their β-glycoside conjugates: antitumor isoflavones in soybean foods from American and Asian diets. *J. Agric. Food Chem.* **41**, 1961–1967.
11. Franke, A. A., Custer, L. J., Cerna, C. M., and Narala, K. (1994). Quantitation of phytoestrogens in legumes by HPLC. *J. Agric. Food Chem.* **42**, 1905–1913.
12. Karr, S. C., Lampe, J. W., Hutchins, A. M., and Slavin, J. L. (1997). Urinary isoflavonoid excretion in humans is dose-dependent at low to moderate levels of soy protein consumption. *Am. J. Clin. Nutr.* **66**, 46–51.
13. Maskarinec, G., Franke, A. A., Williams, A. E., Hebshi, S., Oshiro, C., Murphy, S., and Stanczyk, F. Z. (2004). Effects of a 2-year randomized soy intervention on sex hormone levels in premenopausal women. *Cancer Epidemiol. Biomarkers Prev.* **13**, 1736–1744.
14. Marini, H., Minutoli, L., Polito, F., Bitto, A., Altavilla, D., Atteritano, M., Gaudio, A., Mazzaferro, S., Frisina, A., Frisina, N., Lubrano, C., Bonaiuto, M., D'Anna, R., Cannata, M. L., Corrado, F., Adamo, E. B., Wilson, S., and Squadrito, F. (2007). Effects of the phytoestrogen genistein on bone metabolism in osteopenic postmenopausal women: a randomized trial. *Ann. Intern. Med.* **146**, 839–847.
15. Watanabe, S., Yamaguchi, M., Sobue, T., Takahashi, T., Miura, T., Arai, Y., Mazur, W., Wähälä, K., and Adlercreutz, H. (1998). Pharmacokinetics of soybean isoflavones in plasma, urine and feces of men after ingestion of 60 g baked soybean powder (kinako). *J. Nutr.* **128**, 1710–1715.
16. Gross, M. D., Pfeiffer, M., Martini, M., Campbell, D., Slavin, J., and Potter, J. (1996). The quantitation of metabolites of quercetin flavonols in human urine. *Cancer Epidemiol. Biomarkers Prev.,* 711–720.
17. Noroozi, M., Burns, J., Crozier, A., Kelly, I. E., and Lean, M. E. J. (2000). Prediction of dietary flavonol consumption from fasting plasma concentration or urinary excretion. *Eur. J. Clin. Nutr.* **54**, 143–149.
18. Gasper, A. V., Al-Janobi, A., Smith, J. A., Bacon, J. R., Fortun, P., Atherton, C., Taylor, M. A., Hawkey, C. J., Barrett, D. A., and Mithen, R. F. (2005). Glutathione S-transferase M1 polymorphism and metabolism of sulforaphine from standard and high-glucosinolate broccoli. *Am. J. Clin. Nutr.* **82**, 1283–1291.
19. Shapiro, T. A., Fahey, J. W., Dinkova-Kostova, A. T., Holtzclaw, W. D., Stephenson, K. K., Wade, K. L., Ye, L.,

and Talalay, P. (2006). Safety, tolerance, and metabolism of broccoli sprout glucosinolates and isothiocyanates: a clinical phase I study. *Nutr. Cancer.* **55**, 53–62.

20. Wild, C. P., and Pisani, P. (1997). Carcinogen—DNA and carcinogen—protein adducts in molecular epidemiology. *In* "Application of Biomarkers in Cancer Epidemiology" (P. Toniolo, P. Boffetta, D. E. G. Shuker, N. Rothman B. Hulka, and N. Pearce, Eds.), pp. 143–158. International Agency for Research on Cancer, Lyon.
21. Kensler, T. W., Chen, J. G., Egner, P. A., Fahey, J. W., Jacobson, L. P., Stephenson, K. K., Ye, L., Coady, J. L., Wang, J. B., Wu, Y., Sun, Y., Zhang, Q. N., Zhang, B. C., Zhu, Y. R., Qian, G. S., Carmella, S. G., Hecht, S. S., Benning, L., Gange, S. J., Groopman, J. D., and Talalay, P. (2005). Effects of glucosinolate-rich broccoli sprouts on urinary levels of aflatoxin-DNA adducts and phenanthrene tetraols in a randomized clinical trial in He Zuo township, Qidong, People's Republic of China. *Cancer Epidemiol. Biomarkers Prev.* **14**, 2605–2613.
22. McClave, S. A., and Snider, H. L. (1992). Use of indirect calorimetry in clinical nutrition. *Nutr. Clin. Prac.* **7**, 207–221.
23. Speakman, J. R. (1998). The history and theory of the doubly labeled water technique. *Am. J. Clin. Nutr.* **68**(suppl), 932S–938S.
24. Seale, J. (1995). Energy expenditure measurements in relation to energy requirements. *Am. J. Clin. Nutr.* **62**(suppl), 1042S–1046S.
25. Martini, M. C., Campbell, D. R., Gross, M. D., Grandits, G. A., Potter, J. D., and Slavin, J. L. (1995). Plasma carotenoids as biomarkers of vegetable intake: the Minnesota CPRU feeding studies. *Cancer Epidemiol. Biomarkers Prev.* **4**, 491–495.
26. Campbell, D. R., Gross, M. D., Martini, M. C., Grandits, G. A., Slavin, J. L., and Potter, J. D. (1994). Plasma carotenoids as biomarkers of vegetable and fruit intake. *Cancer Epidemiol. Biomarkers Prev.* **3**, 493–500.
27. Rock, C. L., Flatt, S. W., Wright, F. A., Faerber, S., Newman, V., Kealey, S., and Pierce, J. P. (1997). Responsiveness of carotenoids to a high vegetable diet intervention designed to prevent breast cancer recurrence. *Cancer Epidemiol. Biomarkers Prev.* **6**, 617–623.
28. Le Marchand, L., Hankin, J. H., Carter, F. S., Essling, C., Luffey, D., Franke, A. A., Wilkens, L. R., Cooney, R. V., and Kolonel, L. N. (1994). A pilot study on the use of plasma carotenoids and ascorbic acid as markers of compliance to a high fruit and vegetable diet intervention. *Cancer Epidemiol. Biomarkers Prev.* **3**, 245–251.
29. Blanchard, J., Toxer, T. N., and Rowland, M. (1997). Pharmacokinetic perspectives on megadoses of ascorbic acid. *Am. J. Clin. Nutr.* **66**, 1165–1171.
30. Milder, I. E. J., Arts, I. C. W., van de Putte, B., Venema, D. P., and Hollman, P. C. H. (2005). Lignan contents of Dutch plant foods: a database including lariciresinol, pinoresinol, secoisolariciresinol and matairesinol. *Br. J. Nutr.* **93**, 393–402.
31. Kirkman, L. M., Lampe, J. W., Campbell, D. R., Martini, M. C., and Slavin, J. L. (1995). Urinary lignan and isoflavonoid excretion in men and women consuming vegetable and soy diets. *Nutr. Cancer* **24**, 1–12.
32. Kilkkinen, A., Pietinen, P., Klaukka, T., Virtamo, J., Korhonen, P., and Adlercreutz, H. (2002). Use of oral antimicrobials decreases serum enterolactone concentrations. *Am. J. Epidemiol.* **155**, 472–477.
33. Meydani, S. N., Endres, S., Woods, M. M., Goldin, B. R., Soo, C., Morrill-Labrode, A., Dinarello, C. A., and Gorbach, S. L. (1991). Oral (n-3) fatty acid supplementation suppresses cytokine production and lymphocyte proliferation: comparison between young and older women. *J. Nutr.* **121**, 547–555.
34. Soyland, E., Funk, J., Rajka, G., Sandberg, M., Thune, P., Rustad, L., Hellend, S., Middefat, K., Odu, S., Falk, E. S., Solvoll, K., Bjorneboe, G. A., and Drevaon, C. A. (1993). Effect of dietary supplementation with very-long-chain n-3 fatty acids in patients with psoriasis. *N. Engl. J. Med.* **328**, 1812–1816.
35. Wolk, A., Vessby, B., Ljung, H., and Barrefors, P. (1998). Evaluation of a biologic marker for dairy fat intake. *Am. J. Clin. Nutr.* **68**, 291–295.
36. Smedman, A. E. M., Gustafsson, I. B., Berglund, L. G. T., and Vessby, B. O. H. (1999). Pentadecanoic acid in serum as a marker for intake of milk fat: relations between intake of milk fat and metabolic risk factors. *Am. J. Clin. Nutr.* **69**, 22–29.
37. Turnlund, J. R. (1994). Future directions for establishing mineral/trace element requirements. *J. Nutr.* **124**, 1765S–1770S.
38. Strain, J. J. (1999). Optimal nutrition: an overview. *Proc. Nutr. Soc.* **58**, 395–396.
39. Kelloff, G. J., Sigman, C. C., Johnson, K. M., Boone, C. W., Greenwald, P., Crowell, J. A., Hawk, E. T., and Doody, L. A. (2000). Perspectives on surrogate end points in the development of drugs that reduce the risk of cancer. *Cancer Epidemiol. Biomarkers Prev.* **9**, 127–137.
40. Fleming, T. R., and DeMets, D. L. (1996). Surrogate end points in clinical trials: are we being misled? *Ann. Int. Med.* **125**, 605–613.
41. Habig, W. H., Pabst, M. J., and Jakoby, W. B. (1974). Glutathione *S*-transferases: the first enzymatic step in mercapturic acid formation. *J. Biol. Chem.* **249**, 7130–7139.
42. Bogaards, J. J. P., Verhagen, H., Willems, M. I., van Poppel, G., and van Bladeren, P. J. (1994). Consumption of Brussels sprouts results in elevated a-class glutathione S-transferase levels in human blood plasma. *Carcinogenesis* **15**, 1073–1075.
43. Kashuba, A. D. M., Bertino, J. S., Kearns, G. L., Leeder, J. S., James, A. W., Gotschall, R., and Nafziger, A. N. (1998). Quantitation of three-month intraindividual variability and influence of sex and menstrual cycle phase on CYP1A2, N acetyltransferase-2, and xanthine oxidase activity determined with caffeine phenotyping. *Clin. Pharmacol. Ther.* **63**, 540–551.
44. Pantuck, E. J., Pantuck, C. B., Anderson, K. E., Wattenberg, L. W., Conney, A. H., and Kappas, A. (1984). Effect of Brussels sprouts and cabbage on drug conjugation. *Clin. Pharmacol. Ther.* **35**, 161–169.
45. Sinha, R., Rothman, N., Brown, E. D., Mark, S. D., Hoover, R. N., Capraso, N. E., Levander, O. A., Knize, M. G., Lang, N. P., and Kadlubar, F. F. (1994). Pan-fried meat containing high levels of heterocyclic aromatic amines but low levels of

polycyclic aromatic hydrocarbons induces cytochrome P4501A2 activity in humans. *Cancer Res.* **54**, 6154–6159.
46. Kall, M. A., Vang, O., and Clausen, J. (1997). Effects of dietary broccoli on human drug metabolising activity. *Cancer Lett.* **114**, 169–170.
47. Lampe, J. W., Chen, C., Li, S., Prunty, J., Grate, M. T., Meehan, D. E., Barale, K. V., Dightman, D. A., Feng, Z., and Potter, J. D. (2000). Modulation of human glutathione *S*-transferases by botanically defined vegetable diets. *Cancer Epidemiol. Biomarkers Prev.* **9**, 787–793.
48. Lauritsen, K., Laursen, L. S., Bukhave, K., and Rask-Madsen, J. (1987). Does vitamin E supplementation modulate *in vivo* arachidonate metabolism in human inflammation? *Pharmacol. Toxicol.* **61**, 246–249.
49. Ruffin, M. T., Krishnan, K., Rock, C. L., Normolle, D., Vaerten, M. A., Peters-Golden, M., Crowell, J., Kelloff, G., Boland, C. R., and Brenner, D. E. (1997). Suppression of human colorectal mucosal prostaglandins: determining the lowest effective aspirin dose. *J. Natl. Cancer Inst.* **89**, 1152–1160.
50. Krishnan, K., Ruffin, M. T., and Brenner, D. E. (1998). Clinical models of chemoprevention for colon cancer. *Hem. Onc. Clin. N. Am.* **12**, 1079–1113.
51. Selhub, J., and Miller, J. W. (1992). The pathogenesis of homocysteinemia: interruption of the coordinate regulation by *S*-adenosylmethionine of the remethylation and transsulfuration of homocysteine. *Am. J. Clin. Nutr.* **55**, 131–138.
52. Kim, Y. I., Fawaz, K., Knox, T., Lee, Y. M., Norton, R., Arora, S., Paiva, L., and Mason, J. B. (1998). Colonic mucosal concentrations of folate correlate well with blood measurements of folate status in persons with colorectal polyps. *Am. J. Clin. Nutr.* **68**, 866–872.
53. Stabler, S. P., Marcell, P. D., Podell, E. R., Allen, R. H., Savage, D. G., and Lindenbaum, J. (1988). Elevation of total homocysteine in the serum of patients with cobalamin or folate deficiency detected by capillary gas chromatography-mass spectrometry. *J. Clin. Invest.* **81**, 466–474.
54. Rock, C. L., Jacob, R. A., and Bowen, P. A. (1996). Update on the biological characteristics of the antioxidant micronutrients: vitamin C, vitamin E, and the carotenoids. *J. Am. Diet. Assoc.* **96**, 693–702.
55. Mayne, ST. (2003). Antioxidant nutrients and chronic disease: use of biomarkers of exposure and oxidative stress status in epidemiologic research. *J. Nutr.* **133**(suppl 3), 933S–940S.
56. Lemoyne, M., Gossum, A. V., Kurian, R., Ostro, M., Azler, J., and Jeejeebhoy, K. N. (1987). Breath pentane analysis as an index of lipid peroxidation: a functional test of vitamin E status. *Am. J. Clin. Nutr.* **46**, 267–272.
57. Mei, S., Yao, Q., Wu, C., and Xu, G. (2005). Determination of urinary 8-hydroxy-2′-deoxyguanosine by two approaches-capillary electrophoresis and GC/MS: an assay for in vivo oxidative DNA damage in cancer patients. *J. Chromatogr. B Analyt. Technol. Biomed. Life Sci.* **827**, 83–87.
58. Loft, S., Vistisen, K., Ewertz, M., Tjonneland, A., Overvad, K., and Poulsen, H. E. (1992). Oxidative DNA damage estimated by 8-hydroxydeoxyguanosine excretion in humans: influence of smoking, gender and body mass index. *Carcinogenesis* **13**, 2241–2247.
59. Tagesson, C., Kallberg, M., Klintenberg, C., and Starkhammar, H. (1995). Determination of urinary 8-hydroxydeoxy guanosine by automated coupled-column high performance liquid chromatography: a powerful technique for assaying *in vivo* oxidative DNA damage in cancer patients. *Eur. J. Cancer* **31A**, 934–940.
60. Thompson, H. J., Heimendinger, J., Haegele, A., Sedlacek, S. M., Gillette, C., O'Neill, C., Wolfe, P., and Conry, C. (1999). Effect of increased vegetable and fruit consumption on markers of oxidative cellular damage. *Carcinogenesis* **20**, 2261–2266.
61. Johansson, G., Holmen, A., Persson, L., Hogstedt, R., Wassen, C., Ottova, L., and Gustafsson, J. A. (1992). The effect of a shift from a mixed diet to a lacto-vegetarian diet on human urinary and fecal mutagenic activity. *Carcinogenesis* **13**, 153–157.
62. Witztum, J. L. (1998). To E or not to E—how do we tell? *Circulation* **98**, 2785–2787.
63. Morrow, J. D., Harris, T. M., and Roberts, L. J. (1990). Noncyclooxygenase oxidative formation of a series of novel prostaglandins: analytical ramifications for measurement of eicosanoids. *Anal. Biochem.* **184**, 1–10.
64. Morrow, J. D., and Roberts, L. J. (1997). The isoprostanes: unique bioactive products of lipid peroxidation. *Prog. Lipid Res.* **36**, 1–21.
65. Atteritano, M., Marini, H., Minutoli, L., Polito, F., Bitto, A., Altavilla, D., Mazzaferro, S., Anna, R. D., Cannata, M. L., Gaudio, A., Frisina, A., Frisina, N., Corrado, F., Cancellieri, F., Lubrano, C., Bonaiuto, M., Adamo, E. B., and Squadrito, F. (2007). Effects of the phytoestrogen genistein on some predictors of cardiovascular risk in osteopenic, post-menopausal women: a 2-years randomized, double-blind, placebo-controlled study. *J. Clin. Endocrinol. Metab.* **92**, 3068–3075.
66. Djuric, Z., Lu, M. H., Lewis, S. M., Luongo, D. A., Chen, X. W., Heilbrun, L. K., Reading, B. A., Duffy, P. H., and Hart, R. W. (1992). Oxidative DNA damage levels in rats fed low-fat, high-fat, or calorie-restricted diets. *Toxicol. Appl. Pharmacol.* **115**, 156–160.
67. Djuric, Z., Heilbrun, L. K., Reading, B. A., Boomer, A., Valeriote, F. A., and Martino, S. (1991). Effects of a low-fat diet on levels of oxidative damage to DNA in human peripheral nucleated blood cells. *J. Natl. Cancer Inst.* **83**, 766–769.
68. Kato, I., Ren, J., Heilbrun, L. K., and Djuric, Z. (2006). Intra- and inter-individual variability in measurements of biomarkers for oxidative damage in vivo: Nutrition and Breast Health Study. *Biomarkers* **11**, 143–152.
69. Mosca, L., Rubenfire, M., Mandel, C., Rock, C., Tarshis, T., Tsai, A., and Pearson, T. (1997). Antioxidant nutrient supplementation reduces the susceptibility of low density lipoprotein to oxidation in patients with coronary artery disease. *J. Am. Coll. Cardiol.* **30**, 392–399.
70. Field, C. J. (2000). Use of T cell function to determine the effect of physiologically active food components. *Am. J. Clin. Nutr.* **71**(6 suppl), 1720S–1750S.
71. Lourd, B., and Mazari, L. (1999). Nutrition and immunity in the elderly. *Proc. Nutr. Soc.* **58**, 685–695.
72. Murata, T., Tamai, H., Morinobu, T. M. M., Takenaka, H., Hayashi, K., and Mino, M. (1994). Effect of long-term administration of β-carotene on lymphocyte subsets in humans. *Am. J. Clin. Nutr.* **60**, 597–602.
73. Santos, M. S., Meydani, S. N., Leka, L., Wu, D., Fotouhi, N., Meydani, M., Hennekens, C. H., and Gaziano, J. M. (1996).

Natural killer cell activity in elderly men is enhanced by β-carotene supplementation. *Am. J. Clin. Nutr.* **64**, 772–777.
74. Jeng, K.-C. G., Yang, C.-S., Siu, W.-Y., Tsai, Y.-S., Liao, W. J., and Kuo, J.-S. (1996). Supplementation of vitamins C and E enhances cytokine production by peripheral blood mononuclear cells in healthy adults. *Am. J. Clin. Nutr.* **64**, 960–965.
75. Bogden, J. D., Bendich, A., Kemp, F. W., Bruening, K. S., Skurnick, J. H., Denny, T., Baker, H., and Louria, D. B. (1994). Daily micronutrient supplements enhance delayed-hypersensitivity skin test responses in older people. *Am. J. Clin. Nutr.* **60**, 437–447.
76. Black, D. M., Cummings, S. R., Genant, H. K., Nevitt, M. C., Palermo, L., and Browner, W. (1992). Axial and appendicular bone density predict fractures in older women. *J. Bone Mineral Res.* **7**, 633–638.
77. Cummings, S. R., Black, D. M., Nevitt, M. C., Browner, W. Cauley, J., Ensrud, K., Genant, H. K., Palermo, L., Scott, J. and Vogt, T. M. (1993). Bone density at various sites for prediction of hip fracture. The Study of Osteoporotic Fractures Research Group. *Lancet* **341**, 72–75.
78. Melton, L. J. I., Atkinson, E. J., O'Fallon, W. M., Wahner, H. W., and Riggs, B. L. (1993). Long-term fracture prediction by bone mineral assessed at different skeletal sites. *J. Bone Mineral Res.* **8**, 1227–1233.
79. Cashman, K. D., and Flynn, A. (1999). Optimal nutrition: Calcium, magnesium and phosphorus. *Proc. Nutr. Soc.* **58**, 477–487.
80. Kanis, J. A. (1991). Calcium requirements for optimal skeletal health in women. *Calcified Tissue Int.* **49**, S33–S41.
81. Watts, N. B. (1999). Clinical utility of biochemical markers of bone remodeling. *Clin. Chem.* **45**, 1359–1368.
82. Boland, C. R. (1993). The biology of colorectal cancer. *Cancer* **71**(suppl), 4181–4186.
83. Einspahr, J. G., Alberts, D. S., Gapstur, S. M., Bostick, R. M., Emerson, S. S., and Gerner, E. W. (1997). Surrogate end-point biomarkers as measures of colon cancer risk and their use in cancer chemoprevention trials. *Cancer Epidemiol. Biomarkers Prev.* **6**, 37–48.
84. Lipkin, M., and Newmark, H. (1985). Effect of added dietary calcium on colonic epithelial-cell proliferation in subjects at high risk for familial colonic cancer. *N. Engl. J. Med.* **313**, 1381–1384.
85. Bostick, R. M., Fosdick, L., Lillemoe, T. J., Overn, P., Wood, J. R., Grambsch, P., Elmer, P., and Potter, J. D. (1997). Methodological findings and considerations in measuring colorectal epithelial cell proliferation in humans. *Cancer Epidemiol. Biomarkers Prev.* **6**, 931–942.
86. Vargas, P. A., and Alberts, D. S. (1992). Primary prevention of colorectal cancer through dietary modification. *Cancer* **70**, 1229–1235.
87. Takayama, T., Katsuki, S., Takahashi, Y., Ohi, M., Nojiri, S., Sakamaki, S., Kato, J., Kogawa, K., Miyake, H., and Niitsu, Y. (1998). Aberrant crypt foci of the colon as precursors of adenoma and cancer. *N. Engl. J. Med.* **339**, 1277–1284.
88. Garte, S., Zocchetti, C., and Taioli, E. (1997). Gene-environment interactions in the application of biomarkers of cancer susceptiblity in epidemiology. *In* "Application of Biomarkers in Cancer Epidemiology" (P. Toniolo, P. Boffetta, D. E. G. Shuker, N. Rothman, B. Hulka, and N. Pearce, Eds.), pp. 251–264. International Agency for Research on Cancer, Lyon.
89. Murray, R. F. (1986). Tests of so-called susceptibility. *J. Occup. Med.* **28**, 1103–1107.
90. Lai, C., and Shields, P. G. (1999). The role of interindividual variation in human carcinogenesis. *J. Nutr.* **129**(suppl), 552S–555S.
91. Peters, U., Sinha, R., Bell, D. A., Rothman, N., Grant, D. J., Watson, M. A., Kulldorff, M., Brooks, L. R., Warren, S. H., and DeMarini, D. M. (2004). Urinary mutagenesis and fried red meat intake: influence of cooking temperature, phenotype, and genotype of metabolizing enzymes in a controlled feeding study. *Environ. Mol. Mutagen.* **43**, 53–74.
92. Kim, Y. I., Pogribney, I. P., Basnakian, A. G., Miller, J. W., Selhub, J., James, S. J., and Mason, J. B. (1997). Folate deficiency in rats induces DNA strand breaks and hypomethylation within the p53 tumor suppressor gene. *Am. J. Clin. Nutr.* **65**, 46–52.
93. Ulrich, C. M., Kampman, E., Bigler, J., Schwartz, S. M., Chen, C., Bostick, R., Fosdick, L., Beresford, S. A. A., Yasui, Y., and Potter, J. D. (1999). Colorectal adenomas and the C677T *MTHFR* polymorphism: evidence for gene-environment interaction? *Cancer Epidemiol. Biomarkers Prev.* **8**, 659–668.
94. Vine, M. F., and McFarland, L. T. (1990). Markers of susceptibility. *In* "Biological Markers in Epidemiology" (B. S. Hulka, T. C. Wilcosky, and J. D. Griffith, Eds.), pp. 196–213. Oxford University Press, New York.
95. Blanck, H. M., Bowman, B. A., Cooper, G. R., Myers, G. L., and Miller, D. T. (2003). Laboratory issues: use of nutritional biomarkers. *J Nutr.* **133**(suppl 3):888S–894S.
96. Hunter, D. (1998). Biochemical indicators of dietary intake. *In* "Nutritional Epidemiology" (W. Willett, Ed.), 2nd ed, pp. 174–243. Oxford University Press, New York.
97. Whitehead, T. P. (1977). "Quality Control in Clinical Chemistry." John Wiley, New York.
98. Aitio, A., and Apostoli, P. (1995). Quality assurance in biomarker measurement. *Toxicol. Lett.* **77**, 195–204.
99. Westgard, J. O., Barry, P. L., Hunt, M. R., and Groth, T. (1981). A multi-rule Shewhart chart for quality control in clinical chemistry. *Clin. Chem.* **27**, 493–501.
100. Lipschitz, D. A., Cook, J. D., and Finch, C. A. (1974). A clinical evaluation of serum ferritin. *N. Engl. J. Med.* **290**, 1213–1216.
101. Herbert, V. (1987). Recommended dietary intakes (RDI) of folate in humans. *Am. J. Clin. Nutr.* **45**, 661–670.
102. Trujillo, E., Davis, C., and Milner, J. (2006). Nutrigenomics, proteomics, metabolomics, and the practice of dietetics. *J. Am. Diet. Assoc.* **106**, 403–413.
103. Niculescu, M. D., da Costa, K. A., Fischer, L. M., and Zeisel, S. H. (2007). Lymphocyte gene expression in subjects fed a low-choline diet differs between those who develop organ dysfunction and those who do not. *Am. J. Clin. Nutr.* **86**, 230–239.

SECTION II

Nutrition for Health Maintenance, Prevention, and Disease-Specific Treatment

---

A. Food and Nutrient Intake for Health

CHAPTER **13**

Nutrition Guidelines to Maintain Health

SUZANNE P. MURPHY
Cancer Research Center of Hawaii, University of Hawaii, Honolulu, Hawaii

Contents

I. INTRODUCTION

Nutrition guidelines for Americans fall into two broad categories: those that focus primarily on nutrient intakes and those that are primarily oriented to food choices. In addition, clinicians and researchers have come to realize that healthy food and nutrient intakes are inevitably linked to healthy activity levels, so that physical activity guidance is often included with dietary guidance.

The Food and Nutrition Board of the National Academy of Sciences has periodically released recommendations for nutrient intakes by Americans. The 1989 Recommended Dietary Allowances (RDAs) [1] have been replaced by Dietary Reference Intakes (DRIs) [2–10]. DRI is an umbrella term for four types of nutrient recommendations:

- Estimated Average Requirement (EAR)
- Recommended Dietary Allowance (RDA)
- Adequate Intake (AI)
- Tolerable Upper Intake Level (UL)

DRIs have been set for energy, 14 vitamins, 15 minerals (with ULs for an additional 3), and 7 macronutrients. A summary of all the DRI values is given in Appendix 13A.

Federal government agencies have developed three types of guidance to help consumers make healthy food and nutrient choices. The U.S. Department of Agriculture (USDA) and the U.S. Department of Health and Human Services (USDHHS) have jointly issued the Dietary Guidelines for Americans every 5 years since 1980. The guidelines for 2005 were the most recent release [11, 12]. To help apply the new guidelines, MyPyramid was developed as a tool to be used in nutrition education efforts, and it can be particularly useful in helping the public make healthy food choices [13]. MyPyramid updates the recommendations from the 1992 Food Guide Pyramid (FGP) [14]. It translates the guidance offered by both the DRIs and the dietary guidelines into a graphic format, which illustrates the recommended number of servings from each of six food groups. Finally, the Nutrition Facts label [15] and the Supplement Facts label [16] are now required by law, and they provide consumers with useful information at the point of purchase.

Physical activity guidance has been offered by several organizations and is also included in the 2005 dietary guidelines. Advice to increase physical activity now frequently accompanies dietary guidance because health professionals recognize that maintenance of a healthy body size and sustained cardiopulmonary fitness can only be achieved through an active lifestyle coupled with healthy dietary choices.

II. GUIDELINES FOR NUTRIENT ADEQUACY AND SAFETY

A. Dietary Reference Intakes for Nutrient Adequacy

The Dietary Reference Intakes (DRIs) offer guidance on the level of nutrient intake that will promote health and reduce the risk of chronic disease. The Food and Nutrition Board developed the process for setting the new DRIs after substantial feedback from the nutrition professionals [17] and several new concepts were incorporated:

- An increased focus on reduction of chronic disease risk as well as the prevention of nutritional deficiencies.
- Determination of an estimated average requirement (EAR) that would be the intake that would be adequate for approximately 50% of a healthy population.
- Calculation of a Recommended Dietary Allowance (RDA) from the EAR, by adding two standard deviations

of the requirement distribution. The RDA is thus the level of intake that would be adequate for 97.5% of the population.
- Use of an adequate intake (AI) when the scientific database is not sufficient to set an EAR (and its associated RDA).
- Determination of a tolerable upper intake level (UL) that represents the upper level of intake that poses a low risk of adverse effects. Usual intakes above this level are not recommended.

DRI reports have been issued for sets of nutrients: bone-related nutrients (calcium, phosphorus, magnesium, vitamin D, and fluoride) [2]; B vitamins (thiamin, riboflavin, niacin, vitamin B_6, vitamin B_{12}, folate, and choline) [3]; antioxidant nutrients (vitamin C, vitamin E, and selenium) [4]; micronutrients (vitamins A and K, iron, zinc, copper, iodine, molybdenum, and several other minerals) [5]; macronutrients (energy, protein, fat, carbohydrates, and dietary fiber) [6]; and electrolytes (potassium, sodium, and chloride) [7].

Because there are many new uses of the DRIs, a subcommittee on uses and interpretation of the DRIs was convened and has published two reports: one on assessing intakes [8] and the other on planning intakes [9]. A summary report on the DRIs is now available [10] and provides a concise reference to the values for the DRIs, as well as their appropriate uses.

When offering guidance to consumers on healthy nutrient intakes, health professionals should suggest the RDA (or the AI if an RDA is not available) as the appropriate target. Because an individual's actual requirement is almost never known, the goal is to reduce to a very low level the *risk* that an intake is inadequate. By definition, usual intake at the level of the RDA or AI has a low risk of inadequacy (2% to 3% for the RDA). For example, the appropriate target for magnesium intake for a woman 31 to 50 years of age is the RDA of 320 mg/d. Her target for calcium intake should be the AI of 1000 mg/d.

To reflect newer information on bioavailability, the RDAs for two nutrients are in different forms than were used in the past: folate is in mcg of dietary folate equivalents (DFE) rather than total mcg, whereas vitamin E is in mg of α-tocopherol rather than in mg of α-tocopherol *equivalents*. For folate, the new DFE unit reflects an increased availability of fortification and supplemental forms of folate and thus will tend to increase estimated intakes of this nutrient.

The situation is reversed for vitamin E. Forms of tocopherol other than alpha-tocopherol (such as β-tocopherol and γ-tocopherol) are not considered active forms of the vitamin because of poor transport from the liver. Furthermore, the all rac-α-tocopherol form that is commonly used for fortification and in dietary supplements has a lower activity. Therefore, intakes measured in the older units (α-tocopherol equivalents) will overestimate intakes of the 2D-stereoisomers of α-tocopherol.

B. The Tolerable Upper Intake Level: A New DRI to Reduce the Risk of Adverse Effects

The DRIs from the Food and Nutrition Board also include a level of intake that should not be exceeded. Health professionals and consumers may use the tolerable upper intake level (UL) to ensure that nutrient intakes are not too high. Appendix 13A shows the ULs that have been set for eight vitamins and 16 minerals. For example, the UL for calcium is 2500 mg/d for children and adults. It is unlikely that this level of intake could be achieved from unfortified foods alone. However, through the use of heavily fortified foods or dietary supplements, intake exceeding 2500 mg/d is possible. Intakes above the UL for calcium carry an increased risk of milk-alkali syndrome, a potentially serious metabolic disorder.

The UL for magnesium for adults is 350 mg/d from pharmacological forms. Because magnesium salts can cause osmotic diarrhea, intakes above 350 mg/d are not recommended. Food sources of magnesium do not cause diarrhea and thus are not included in this UL. Indeed, the UL for magnesium is below the RDA for magnesium for adult men (420 mg/d), reflecting the different forms of the nutrient in each of these DRIs. Several other nutrient DRIs also use different forms for the RDA and the UL: niacin and folate (for which the UL is from fortification or supplemental forms only), and vitamin E (where the UL is for all forms of α-tocopherol, whereas the RDA is only for the 2D-stereoisomers of α-tocopherol).

ULs are not available for all nutrients, not because intake at any level in considered safe, but because there was not sufficient scientific data to set an upper level. ULs should never be considered a target intake; the RDA and AI are the appropriate targets. In some instances, controlled trials or feeding studies, or therapeutic prescriptions, may utilize nutrient levels above the UL. In these cases, when medical supervision is provided, intakes above the UL may be appropriate.

C. How Do Current Nutrient Intakes Compare to the DRIs?

Evaluating nutrient intakes of either individuals or of groups presents many challenges, as noted by a number of authors [18–22]. The National Health and Nutrition Examination Survey (NHANES) is conducted continuously by the U.S. Department of Agriculture and the U.S. Department of Health and Human Services. Table 1 summarizes a report on the adequacy of dietary intakes of Americans using data from the NHANES 2001–2002 survey for adults 19 years of age and older [23]. When evaluating intakes from the national surveys, it is desirable to know the percentage of the population below the EAR, because that percentage is an accurate estimate of the prevalence of inadequate intakes for most nutrients (this approach cannot

TABLE 1 How Adequate Are the Dietary Nutrient Intakes of American Adults? Selected Nutrients for Men and Women 19 Years of Age and Older

Nutrient	Gender	Recommended Intake[a]	Reported Mean Intake[b]	Prevalence of Dietary Inadequacy[b,c]
Vitamin A (RAE/d)	Men	900	656	57%
	Women	700	564	48%
Vitamin C (mg/d)[d]	Men	90	105	36%
	Women	75	84	32%
Vitamin E (mg α-tocopherol/d)	Men	15	8.2	89%
	Women	15	6.3	97%
Dietary fiber (g/d)	Men	30–38	18.0	Unknown
	Women	21–25	14.3	Unknown
Vitamin B_6 (mg/d)	Men	1.3–1.7	2.23	7%
	Women	1.3–1.5	1.53	28%
Folate (ug DFE/d)	Men	400	636	6%
	Women	400	483	16%
Vitamin B_{12} (ug/d)[e]	Men	2.4	6.45	<3%
	Women	2.4	4.33	7–9%
Iron (mg/d)	Men	8	18.0	<3%
	Women	8–18	13.1	10%
Zinc (mg/d)	Men	11	14.2	11%
	Women	8	9.7	17%
Calcium (mg/d)	Men	1000–1200	984	Unknown
	Women	1000–1200	735	Unknown
Magnesium (mg/d)	Men	400–420	322	64%
	Women	310–320	240	67%

[a]Adequate intake (AI) or Recommended Dietary Allowance (RDA) as shown in Appendix 13A.

[b]From What We Eat in America, NHANES, 2001–2002 [23]; intake from food only. Mean intakes would increase if supplements were included.

[c]The percentage of intakes below the EAR [8]. All intake data were adjusted for day-to-day variation in intakes before examining the proportion below the EAR. The prevalence of usual intakes below the EAR approximates the percentage of the population with inadequate nutrient intakes. For iron, a full probability approach was used to estimate the prevalence of inadequacy because the requirements are not normally distributed. For nutrients with an AI (calcium and fiber), the prevalence of inadequacy is assumed to be low if the mean intake is above the AI but the prevalence of inadequacy is unknown if the mean intake is below the AI.

[d]Vitamin C recommendations are for non-smokers, and are 35 mg/d higher for smokers. The prevalences of vitamin C inadequacy are for non-smokers. Corresponding prevalences are 69% for men smokers and 76% for women smokers.

[e]Prevalence for participants 19 to 50 years of age. Prevalence was not calculated for participants age 50 and older because 10% to 30% of older adults may not absorb vitamin B_{12} in foods, so this age group should take a supplement or consume foods that are fortified with vitamin B_{12}.

RAE = Retinol activity equivalents; DFE = Dietary folate equivalents.

be used if requirements are skewed, however, such as with iron requirements for menstruating women) [8]. The analyses in Table 1 used this theoretically correct approach and included adjustment of the distributions to remove the effect of day-to-day variation in intakes. Thus, the prevalence of dietary inadequacy should approximate the prevalence of persons who demonstrate low values for the functional indicator that was chosen for the nutrient. For nutrients with an adequate intake (AI) rather than an EAR, it is not possible to quantify the prevalence of inadequacy, but mean intakes above the AI indicate that a low prevalence of inadequacy is likely.

The prevalence of dietary inadequacy is 10% or less for vitamin B_{12} and iron for both men and women and for vitamin B_6 and folate for men. Inadequacy is particularly high for vitamin E (89% for men and 97% for women), although clinical deficiencies are rare [4]. Given the relatively high prevalence of inadequacy for several other nutrients (vitamin A, vitamin C, and magnesium), it appears that nutrient intakes need to be improved.

Because ULs have been set for many nutrients, it is now possible to examine the prevalence of intakes at risk of being excessive (Table 2). Only sodium intakes are above the UL for a substantial proportion of the adult population (94% of men and 74% of women). For children up to 8 years, however, high zinc intakes are a concern, as are retinol, copper, and selenium intakes for children 1 to 3 years of age. Sodium intakes are too high for all ages of children and adolescents. Because many of the ULs are extrapolated from data for adults, it is possible that they are unnecessarily low for some nutrients.

An important issue to consider when evaluating intakes is whether all sources of a nutrient are measured and included in total intake. In particular, many Americans

TABLE 2 What Is the Prevalence of Intakes from Food That Are above the UL?

Nutrient	Prevalence for Children, 1–18 Years	Prevalence for Adults, 19 and Older
Retinol	1–3 y: 12% Other ages: < 3%	< 3%
Folic acid	1–3 y: 5% 4–8 y: 4% Other ages: < 3%	< 3%
Vitamin B_6	< 3%	< 3%
Vitamin C	< 3%	< 3%
Calcium	< 3%	< 3%
Phosphorus	< 3%	< 3%
Iron	< 3%	< 3%
Zinc	1–3 y: 69% 4–8 y: 22% Other ages: < 3%	< 3%
Copper	1–3 y: 15% Other ages: < 3%	< 3%
Selenium	1–3 y: 8% Other ages: < 3%	< 3%
Sodium	1–3 y: 83% 4–8 y: 94% 9–18 y: 74%–97%	Men: 94% Women: 74%

Source: What We Eat in America, NHANES, 2001–2002 [23]; intake from food only. Prevalence of intakes above the UL would increase if supplements were included.

take dietary supplements, and the contribution of supplements to intakes should be considered before evaluating the adequacy of an individual diet or the prevalence of inadequacy for a group of individuals. Although NHANES 2001–2002 collected and quantified nutrient intakes from supplements, the data in Table 1 do not reflect intakes from supplements. Therefore, the mean intakes for many of these nutrients are lower than they would have been with supplements included, and the percentage below the EAR is probably an overestimate of the true prevalence of dietary inadequacy. Furthermore, the prevalence of intakes above the UL may be underestimated for nutrients that are common in vitamin and mineral supplements.

III. GUIDELINES FOR HEALTHY FOOD CHOICES

A. Dietary Guidelines for Americans

By law, the secretaries of Agriculture and Health and Human Services are required to jointly issue dietary guidelines for Americans every 5 years. The recently released Dietary Guidelines for Americans [11, 12] fulfill that requirement for the year 2005. Rather than just considering how the previous guidelines (from 2000) should be changed, the 2005 Dietary Guidelines Advisory Committee (DGAC) was asked to conduct an evidence-based review of diet and health [12]. This process led to nine main messages outlined in detail in the report from the DGAC [12]. Subsequently, the Dietary Guidelines were released in a 70-page report from the U.S. Departments of Agriculture and Health and Human Services, which concisely stated the key recommendations within each of the nine messages, as well as key recommendations for specific population groups for most of the messages [11]. The messages and key recommendations are summarized in Table 3. A shorter consumer brochure is also available and gives an abbreviated version of the guidelines [24]. These consumer messages are also shown in Table 3. Finally, a consumer book from the USDHHS offers suggestions on "Everyday healthy eating and physical activity for life, based on the Dietary Guidelines." [25]. It includes healthy eating plans and worksheets, nearly 100 heart-healthy recipes, and helpful websites and tips.

1. Adequate Nutrients within Calorie Needs

This guideline encompasses key recommendations about food choices to ensure nutrient adequacy, as well as how to balance energy intakes within energy needs. A specific suggestion is to follow either the USDA Food Guide or the Dietary Approaches to Stop Hypertension (DASH) eating plan. Shortly after the release of the 2005 Dietary Guidelines, the USDA released a new food guide, called MyPyramid [26]. More information on MyPyramid is covered in the next section of this chapter.

TABLE 3 Dietary Guidelines for Americans, 2005

Focus Areas	Consumer Messages[a]	Professional Recommendations[b]
Adequate nutrients within calorie needs	Make smart choices from every food group. Get the most nutrition out of your calories.	Consume a variety of nutrient-dense foods and beverages within and among the basic food groups while choosing foods that limit the intake of saturated and *trans* fats, cholesterol, added sugars, salt, and alcohol. Meet recommended intakes within energy needs by adopting a balanced eating pattern, such as the USDA Food Guide or the DASH Eating Plan.
Weight management	Find your balance between food and physical activity.	To maintain body weight in a healthy range, balance calories from foods and beverages with calories expended. To prevent gradual weight gain over time, make small decreases in food and beverage calories and increase physical activity.
Physical activity		Engage in regular physical activity and reduce sedentary activities to promote health, psychological well-being, and a healthy body weight. To reduce the risk of chronic disease in adulthood, engage in at least 30 minutes of moderate-intensity physical activity, above usual activity, at work or at home on most days of the week. For most people, greater health benefits can be obtained by engaging in physical activity of more vigorous intensity or longer duration. To help manage body weight and prevent gradual, unhealthy body weight gain in adulthood, engage in approximately 60 minutes of moderate- to vigorous-intensity activity on most days of the week while not exceeding caloric intake requirements. To sustain weight loss in adulthood, participate in at least 60 to 90 minutes of daily moderate-intensity physical activity while not exceeding caloric intake requirements. Some people may need to consult with a health care provider before participating in this level of activity. Achieve physical fitness by including cardiovascular conditioning, stretching exercises for flexibility, and resistance exercises or calisthenics for muscle strength and endurance.
Food groups to encourage		Consume a sufficient amount of fruits and vegetables while staying within energy needs. Two cups of fruit and $2\frac{1}{2}$ cups of vegetables per day are recommended for a reference 2000-calorie intake, with higher or lower amounts depending on the calorie level. Choose a variety of fruits and vegetables each day. In particular, select from all five vegetable subgroups (dark green, orange, legumes, starchy vegetables, and other vegetables) several times a week. Consume 3 or more ounce equivalents of whole-grain products per day, with the rest of the recommended grains coming from enriched or whole-grain products. In general, at least half the grains should come from whole grains. Consume 3 cups per day of fat-free or low-fat milk or equivalent milk products.
Fats	Know your fats.	Consume less than 10% of calories from saturated fatty acids and less than 300 mg/d of cholesterol, and keep *trans*-fatty acid consumption as low as possible. Keep total fat intake between 20% and 35% of calories, with most fats coming from sources of polyunsaturated and monounsaturated fatty acids, such as fish, nuts, and vegetable oils. When selecting and preparing meat, poultry, dry beans, and milk or milk products, make choices that are lean, low fat, or fat free. Limit intake of fat and oils high in saturated or *trans*-fatty acids, and choose products low in such fats and oils.
Carbohydrates	Don't sugarcoat it.	Choose fiber-rich fruits, vegetables, and whole grains often. Choose and prepare foods and beverages with little added sugars or caloric sweeteners, such as amounts suggested by the USDA Food Guide and the DASH eating plan.

(continues)

TABLE 3 *(continued)*

Focus Areas	Consumer Messages[a]	Professional Recommendations[b]
		Reduce the incidence of dental caries by practicing good oral hygiene and consuming sugar- and starch-containing foods and beverages less frequently.
Sodium and potassium	Reduce sodium (salt), increase potassium.	Consume less than 2300 mg (approximately 1 tsp of salt) of sodium per day. Choose and prepare foods with little salt. At the same time, consume potassium-rich foods, such as fruits and vegetables.
Alcoholic beverages	About alcohol: If you choose to drink alcohol, do so in moderation.	Those who choose to drink alcoholic beverages should do so sensibly and in moderation—defined as the consumption of up to one drink per day for women and up to two drinks per day for men. Alcoholic beverages should not be consumed by some individuals, including those who cannot restrict their alcohol intake, women of childbearing age who may become pregnant, pregnant and lactating women, children and adolescents, individuals taking medications that can interact with alcohol, and those with specific medical conditions. Alcoholic beverages should be avoided by individuals engaging in activities that require attention, skill, or coordination, such as driving or operating machinery.
Food safety	Play it safe with food.	To avoid microbial foodborne illness: Clean hands, food contact surfaces, and fruit and vegetables. Meat and poultry should *not* be washed or rinsed. Separate raw, cooked, and ready-to-eat foods while shopping, preparing, or storing foods. Cook foods to a safe temperature to kill microorganisms. Chill (refrigerate) perishable food promptly and defrost foods properly. Avoid raw (unpasteurized) milk or any products made from unpasteurized milk, raw or partially cooked eggs or foods containing raw eggs, raw or undercooked meat and poultry, unpasteurized juices, and raw sprouts.

[a] From USDHHS, USDA. (2005). Finding your way to a healthier you: Based on the Dietary Guidelines for Americans [24].
[b] From USDHHS, USDA. (2005). Dietary Guidelines for Americans [11].

The concept of variety is presented in this guideline by suggesting consumption of a variety of nutrient-dense foods and beverages. Thus, the broader recommendation to consume a variety of foods (in versions of the Dietary Guidelines before 2000) has been clarified to emphasize selection of nutrient-dense foods as well as the importance of limiting food components such as saturated and *trans*-fats, cholesterol, added sugars, salt, and alcohol. Research on the role of dietary variety in ensuring nutrient adequacy indicates that variety can contribute to improved nutrient intake, even within a fixed energy intake [27, 28].

Considerations for specific populations include recommendations on the use of dietary supplements: persons over 50 years of age should obtain the vitamin B_{12} RDA by eating fortified foods or taking a supplement; women of childbearing age should consume 400 mcg/d of synthetic folic acid (from fortified foods or supplements); and the elderly and individuals with dark skin, as well as anyone with insufficient exposure to ultraviolet radiation, should consume 25 mcg/d of vitamin D from foods and supplements.

2. Weight Management

The remarkable increase in the rates of obesity among Americans has added urgency to the need to find better ways to help individuals avoid accumulating body fat [29, 30]. Using the range for healthy weight as a body mass index between 19.0 and 25.0, only 41% of all individuals 20 years of age and older were classified as having a healthy weight (39% of men and 44% of women) [29]. Guidelines to maintain weight include balancing energy intake and expenditure and making small decreases in energy intake (and increases in physical activity) over time to prevent an increase in weight with aging.

3. Physical Activity

Furthermore, the low rates of physical activity among both youth and adults are a cause for concern not only as a contributing factor to the percentage of the population that is overweight and obese, but also to a declining level of muscular and cardiopulmonary fitness that is likely to be associated with cardiovascular disease, diabetes, and osteoporosis, especially among older adults [31]. The physical activity guideline recommends at least 30 minutes of moderate-intensity exercise on most days of the week to reduce the risk of chronic disease; 60 minutes on most days to prevent gradual, unhealthy weight gain; and 60 to 90 minutes daily to sustain weight loss. Children and adolescents are encouraged to engage in at least 60 minutes of physical activity on most days.

4. Food Groups to Encourage

The key recommendations within this guideline suggest specific amounts of fruits, vegetables, whole grains, and fat-free/low-fat milk to consume. These recommendations are reflected in the guidance that is summarized by MyPyramid [26]. Recommendations for children and adolescents are also included. The importance of choosing a variety of fruits and vegetables is emphasized.

Fruits, vegetables, and grains are the base of MyPyramid and form the foundation of a diet that is nutritionally adequate. In addition, high consumption of these foods has been shown to reduce the risk of a variety of chronic diseases, including cardiovascular disease and cancer [32–34]. Indeed, the association between fruit, vegetable, and whole-grain intakes and reduced risk of disease is often stronger than the association between intakes of their nutrient components (such as fiber, vitamins and minerals, or carotenoids) and risk. The reasons for this are not well understood, but they might include unmeasured food components, interactions among the components within a food, and inaccurate composition data for the components. Furthermore, fruits, vegetables, and whole grains tend to be low in both fat and energy and thus can be consumed in greater volume than more energy-dense foods without exceeding energy needs. There is evidence that a diet that is based on foods with a low energy density can both promote weight loss and prevent weight gain [35, 36].

The concept of discretionary calories was introduced in the 2005 Dietary Guidelines to help consumers quantify how much of their energy "budget" would remain after consuming the recommended amounts from all of the food groups. As shown later in Figure 2, discretionary calories are low (less than 300 kcal/d) for many of the age and gender categories. This is an important concept to communicate to consumers, to emphasize that a healthy diet does not have much room for foods that are high in fat and sugar.

5. Fats

The fat guideline represents research on the role of fat intake in overall health. First, the evidence is strong that intake of both saturated fat and cholesterol is positively associated with cardiovascular disease [37, 38]. Thus, the guideline continues to promote a maximum of 10% of energy from saturated fat and a maximum of 300 mg/d of dietary cholesterol, even though the DRIs did not specify these limits. Table 4 shows that over 70% of Americans are meeting the cholesterol goal, but only 36% meet the saturated fat goal. The guideline also recommends that 20% to 35% of energy intake comes from fat in diets of adults, as was specified by the DRIs for total fat. Very low-fat diets (those below 20% of energy from fat) may not be ideal for all individuals; indeed, diets that are high in carbohydrates may actually contribute to cardiovascular disease for some people [39, 40]. Advice regarding the distribution of macronutrients in a diet should be tailored, whenever possible, to an individual's specific risk factors (such as low-density lipoprotein and high-density lipoprotein cholesterol levels).

6. Carbohydrates

The carbohydrate guideline focuses on consuming fiber-rich foods, avoiding added sugars and caloric sweeteners, and reducing dental caries. The guidelines discuss the concept of discretionary calories and include an example of a

TABLE 4 How Well Are Americans Following the Food Guide Pyramid and the Dietary Guidelines for Fat Intake?

	Recommended Intake[a]	Average Intake[b]	Percentage Meeting Recommendation[c]
Grain group	6–11 servings/day	6.8 svg/d	34
Vegetable group	3–5 servings/day	3.0 svg/d	34
Fruit group	2–4 servings/day	1.6 svg/d	24
Dairy group	2–3 servings/day	1.7 svg/d	26[d]
Meat group	5–7 ounce equiv	5.3 ounce equiv	38
Discretionary fat	Use sparingly	25.3% of energy	(not quantified)
Added sugars	Use sparingly	16.6% of energy	(not quantified)
Total fat	20–35% of energy	34% of energy	53
Saturated fat	<10% of energy	11% of energy	41
Cholesterol	<300 mg/d	273 mg/d	69

[a] From [11] and [14]. Food group intakes are compared to the Food Guide Pyramid recommendations because intake data for MyPyramid food groups are not currently available.

[b] Intake data are for all individuals aged 2 years and older (excluding breast-fed children), 1-day averages based on data reported in the 1999–2002 NHANES [56] except for the following: data for percentage meeting the recommendations for saturated fat and cholesterol are from NHANES 1999–2000 [59]; data for total fat and saturated fat intakes and percentage meeting recommendations for total fat are for adults aged 20 years and older, two-day averages from NHANES 2003–2004 [58]; data for cholesterol intakes are for ages 2 years and older, NHANES 2001–2002 [60].

[c] Based on servings recommended for a specific caloric intake (for example, 6 svg/d of grains if 2200 kcal/d or less; 9 svg/d if 2200 to 2800 kcal/d; and 11 svg/d if greater than 2800 kcal/d) [14].

[d] Uses three servings per day as the recommendation for women who are pregnant or lactating, and for all individuals over age 50.

typical healthy 2000-calorie diet that would afford about 32 grams of added sugars (less than the 33 grams in most soft drinks). Many fruits, vegetables, and dairy products are high in sugars such as fructose and lactose, but these foods are also important sources of many vitamins and minerals. In contrast, foods such as nondiet carbonated beverages or fruit-flavored punches usually supply few nutrients other than energy. Thus, the guideline urges Americans to avoid foods that are high in added sugars (and a table lists examples of these foods: soft drinks, sugars and candy, cakes, cookies, pies, fruit-flavored drinks and fruit punch, and dairy desserts). High intakes of sugars are also associated with an increased incidence of dental caries, providing yet another reason to moderate sugar intake [41]. Dietary fiber intakes would increase if consumption of whole grains increased, so the guideline emphasizes the importance of consuming at least three servings of whole grains daily.

7. Sodium and Potassium

Because most Americans consume more salt than they need, this guideline reinforces the DRI recommendation to consume less than 2300 mg of sodium per day (about 1 teaspoon of salt). Another key recommendation within this guideline targets specific population groups: those who already have hypertension, blacks, and middle-aged and older adults. For these groups of people, sodium intake should be limited to 1500 mg/d, while ensuring that potassium intake meets the adequate intake of 4700 mg/d [7].

8. Alcoholic Beverages

Moderate intake of alcohol is defined as no more than one drink per day for women and two drinks per day for men, where 12 ounces of beer, 5 ounces of wine, or 1.5 ounces of distilled spirits count as a serving. The alcohol guideline also lists individuals who should not drink (see Table 3).

9. Food Safety

Concerns about foodborne illness led to the inclusion of a guideline on food safety in both 2000 and 2005. The U.S. Government Accounting Office has estimated a range of 6.5 to 33 million cases of foodborne illness a year [42]. More recently, the CDC estimated that foodborne diseases cause approximately 76 million illnesses, 325,000 hospitalizations, and 5000 deaths in the United States each year [43]. The estimates are difficult to derive because the majority of cases are not reported to health care providers. Even the lower end of the range raises substantial concern about the burden of illness (and the cost of lost productivity) attributable to these illnesses. Although consumers cannot control all sources of food contamination, there is reason to believe that high-risk behaviors by consumers contribute to a substantial portion of foodborne illness [44]. Therefore, an important objective of this guideline is to improve food-handling practices, and the recommendations follow the same four topics that are promoted by the Fight-BAC! campaign by the Partnership for Food Safety Education [45]: clean, separate, cook, and chill (see Table 3 for more details). In addition, a specific recommendation has been added to avoid raw milk and eggs, undercooked meat and poultry, unpasteurized juices, and raw sprouts.

B. MyPyramid

The original Food Guide Pyramid (FGP) was released by the USDA in 1992 [14] and has been widely distributed to both health professionals and consumers. In 2005, it was replaced by MyPyramid, which reflects the 2005 Dietary Guidelines for Americans (Fig. 1) [13]. The MyPyramid image is simple, illustrating foods that should be included in a healthy diet and the relative amounts by the width of the colored bands. The importance of physical activity is emphasized by the figure of a person climbing the steps of the pyramid. A single-sheet consumer handout shows the graphic image on one side and provides more specific advice on the other side (Fig. 1). Amounts to eat from each of the food groups are illustrated for a 2000-calorie diet, as well as messages about balancing food with physical activity and limiting fats, sugars, and salt. However, the handout is not intended to provide advice that is tailored to a particular person. For this purpose, consumers are urged to go to www.MyPyramid.gov to obtain information on the exact amounts that are appropriate, based on energy needs [46]. Food intake patterns have been developed for 12 energy intake levels, ranging from 1000 kcal/d to 3200 kcal/d (Fig. 2). The appropriate energy intake category is determined at the MyPyramid website based on age, gender, weight, and level of physical activity. Figure 3 shows the approximate energy intake ranges for different age and gender groups. A children's version of MyPyramid is also available [47] (see Fig. 4). It is visually similar to MyPyramid except for the figure climbing the stairs.

Both the form and the content of MyPyramid underwent extensive testing, and a series of papers describe the process [48]. Numerous focus groups were conducted to ensure that consumers understood the messages being conveyed and that the pyramid was a meaningful graphic for offering dietary guidance [49]. In addition, extensive analyses were conducted to determine what guidance would ensure adequacy (provision of recommended levels of nutrients) and moderation (low-fat, low-energy choices from each of the groups) [50]. Finally, typical dietary patterns in the United States were considered, which led to the selection of the food groups and the number of servings of each that would provide the recommended levels of nutrients at the various daily energy levels shown in Figure 2. Because the food groups are broadly defined, they can be adapted to meet specific cultural and personal preferences. Thus, MyPyramid combines the dietary guidelines and the DRIs into a single tool that is both scientifically based and consumer friendly.

GRAINS Make half your grains whole	VEGETABLES Vary your veggies	FRUITS Focus on fruits	MILK Get your calcium-rich foods	MEAT & BEANS Go lean with protein
Eat at least 3 oz. of whole-grain cereals, breads, crackers, rice, or pasta every day 1 oz. is about 1 slice of bread, about 1 cup of breakfast cereal, or ½ cup of cooked rice, cereal or pasta	Eat more dark-green veggies like broccoli, spinach, and other dark leafy greens Eat more orange vegetables like carrots and sweetpotatoes Eat more dry beans and peas like pinto beans, kidney beans, and lentils	Eat a variety of fruit Choose fresh, frozen, canned, or dried fruit Go easy on fruit juices	Go low-fat or fat-free when you choose milk, yogurt, and other milk products If you don't or can't consume milk, choose lactose-free products or other calcium sources such as fortified foods and beverages	Choose low-fat or lean meats and poultry Bake it, broil it, or grill it Vary your protein routine – choose more fish, beans, peas, nuts, and seeds
For a 2,000-calorie diet, you need the amounts below from each food group. To find the amounts that are right for you, go to MyPyramid.gov.				
Eat 6 oz. every day	Eat 2½ cups every day	Eat 2 cups every day	Get 3 cups every day; for kids aged 2 to 8, it's 2	Eat 5½ oz. every day

Find your balance between food and physical activity

- Be sure to stay within your daily calorie needs.
- Be physically active for at least 30 minutes most days of the week.
- About 60 minutes a day of physical activity may be needed to prevent weight gain.
- For sustaining weight loss, at least 60 to 90 minutes a day physical activity may be required.
- Children and teenagers should be physically active for 60 minutes every day, or most days.

Know the limits on fats, sugars, and salt (sodium)

- Make most of your fat sources from fish, nuts, and vegetable oils.
- Limit solid fats like butter, margarine, shortening, and lard, as well as foods that contain these.
- Check the Nutrition Facts label to keep saturated fats, *trans* facts, and sodium low.
- Choose food and beverages low in added sugars. Added sugars contribute calories with few, if any, nutrients.

MyPyramid.gov
STEPS TO A HEALTHIER YOU

U.S. Department of Agriculture
Center for Nutrition Policy and Promotion
April 2005
CNPP-15
USDA

USDA is an equal opportunity provider and employer.

FIGURE 1 MyPyramid: Steps to a healthier you. See color plate.

Like the FGP, MyPyramid consists of five basic food groups: fruits, vegetables, grains, meat/beans, and milk. However, MyPyramid also has an "oils" group. The daily amount to consume from each of the groups is shown in Figure 2. Rather than specify amounts as "servings," MyPyramid gives amounts in cups, ounces, and teaspoons. The goal of the additional specificity for portions is to reduce consumer confusion about the size of a serving.

MyPyramid

Food Intake Patterns

The suggested amounts of food to consume from the basic food groups, subgroups, and oils to meet recommended nutrient intakes at 12 different calorie levels. Nutrient and energy contributions from each group are calculated according to the nutrient-dense forms of foods in each group (e.g., lean meats and fat-free milk). The table also shows the discretionary calorie allowance that can be accommodated within each calorie level, in addition to the suggested amounts of nutrient-dense forms of foods in each group.

Daily Amount of Food From Each Group

Calorie Level[1]	1,000	1,200	1,400	1,600	1,800	2,000	2,200	2,400	2,600	2,800	3,000	3,200
Fruits[2]	1 cup	1 cup	1.5 cups	1.5 cups	1.5 cups	2 cups	2 cups	2 cups	2 cups	2.5 cups	2.5 cups	2.5 cups
Vegetables[3]	1 cup	1.5 cups	1.5 cups	2 cups	2.5 cups	2.5 cups	3 cups	3 cups	3.5 cups	3.5 cups	4 cups	4 cups
Grains[4]	3 oz-eq	4 oz-eq	5 oz-eq	5 oz-eq	6 oz-eq	6 oz-eq	7 oz-eq	8 oz-eq	9 oz-eq	10 oz-eq	10 oz-eq	10 oz-eq
Meat and Beans[5]	2 oz-eq	3 oz-eq	4 oz-eq	5 oz-eq	5 oz-eq	5.5 oz-eq	6 oz-eq	6.5 oz-eq	6.5 oz-eq	7 oz-eq	7 oz-eq	7 oz-eq
Milk[6]	2 cups	2 cups	2 cups	3 cups	3 cups	3 cups	3 cups	3 cups	3 cups	3 cups	3 cups	3 cups
Oils[7]	3 tsp	4 tsp	4 tsp	5 tsp	5 tsp	6 tsp	6 tsp	7 tsp	8 tsp	8 tsp	10 tsp	11 tsp
Discretionary calorie allowance[8]	165	171	171	132	195	267	290	362	410	426	512	648

1 Calorie Levels are set across a wide range to accommodate the needs of different individuals. The attached table "Estimated Daily Calorie Needs" can be used to help assign individuals to the food intake pattern at a particular calorie level.

2 Fruit Group includes all fresh, frozen, canned, and dried fruits and fruit juices. In general, 1 cup of fruit or 100% fruit juice, or 1/2 cup of dried fruit can be considered as 1 cup from the fruit group.

3 Vegetable Group includes all fresh, frozen, canned, and dried vegetables and vegetable juices. In general, 1 cup of raw or cooked vegetables or vegetable juice, or 2 cups of raw leafy greens can be considered as 1 cup from the vegetable group.

Vegetable Subgroup Amounts Are Per Week

Calorie Level	1,000	1,200	1,400	1,600	1,800	2,000	2,200	2,400	2,600	2,800	3,000	3,200
Dark green veg.	1 c/wk	1.5 c/wk	1.5 c/wk	2 c/wk	3 c/wk	3 c/wk	3 c/wk	3 c/wk	3 c/wk	3 c/wk	3 c/wk	3 c/wk
Orange veg.	.5 c/wk	1 c/wk	1 c/wk	1.5 c/wk	2 c/wk	2 c/wk	2 c/wk	2 c/wk	2.5 c/wk	2.5 c/wk	2.5 c/wk	2.5 c/wk
Legumes	.5 c/wk	1 c/wk	1 c/wk	2.5 c/wk	3 c/wk	3 c/wk	3 c/wk	3 c/wk	3.5 c/wk	3.5 c/wk	3.5 c/wk	3.5 c/wk
Starchyveg.	1.5 c/wk	2.5 c/wk	2.5 c/wk	2.5 c/wk	3 c/wk	3 c/wk	6 c/wk	6 c/wk	7 c/wk	7 c/wk	9 c/wk	9 c/wk
Other veg.	3.5 c/wk	4.5 c/wk	4.5 c/wk	5.5 c/wk	6.5 c/wk	6.5 c/wk	7 c/wk	7 c/wk	8.5 c/wk	8.5 c/wk	10 c/wk	10 c/wk

4 Grains Group includes all foods made from wheat, rice, oats, cornmeal, barley, such as bread, pasta, oatmeal, breakfast cereals, tortillas, and grits. In general, 1 slice of bread, 1 cup of ready-to-eat cereal, or 1/2 cup of cooked rice, pasta, or cooked cereal can be considered as 1 ounce equivalent from the grains group. At least half of all grains consumed should be whole grains.

5 Meat & Beans Group includes, in general, 1 ounce of leanmeat, poultry, orfish,1 egg, 1 Tbsp. peanut butter, 1/4 cup cooked dry dry beans, or 1/2 ounce of nuts or seeds can be considered as 1 ounce equivalent from the meat and beans group

6 Milk Group includes all fluid milk products and foods made from milk that retain their calcium content, such as yogurt and cheese. Foods made from milk that have little to no calcium, such as cream cheese, cream, and butter, are not part of the group. Most milk group choices should be fat-free or low-fat. In general, 1 cup of milk or yogurt, 1 1/2 ounces of natural cheese, or 2 ounces of processed cheese can be considered as 1 cup from the milk group.

7 Oils include fats from many different plants and from fish that are liquid at room temperature, such as canola, corn, olive, soybean, and sunflower oil. Some foods are naturally high in oils, like nuts, olives, some fish, and avocados. Foods that are mainly oil include mayonnaise, certain salad dressings, and soft margarine.

8 Discretionary Calorie Allowance is the remaining amount of calories in a food intake pattern after accounting for the calories needed for all food groups—using forms of foods that are fat-free or low-fat and with no added sugars.

FIGURE 2 MyPyramid food intake patterns.

Estimated Daily Calorie Needs

To determine which food intake pattern to use for an individual, the following chart gives an estimate of individual calorie needs. The calorie range for each age/sex group is based on physical activity level, from sedentary to active.

	Calorie Range		
Children	Sedentary	→	Active
2–3 years	1,000	→	1,400
Females			
4–8 years	1,200	→	1,800
9–13	1,600	→	2,200
14–18	1,800	→	2,400
19–30	2,000	→	2,400
31–50	1,800	→	2,200
51+	1,600	→	2,200
Males			
4–8 years	1,400	→	2,000
9–13	1,800	→	2,600
14–18	2,200	→	3,200
19–30	2,400	→	3,000
31–50	2,200	→	3,000
51+	2,000	→	2,800

Sedentary means a lifestyle that includes only the light physical activity associated with typical day-to-day life.

Active means a lifestyle that includes physical activity equivalent to walking more than 3 miles per day at 3 to 4 miles per hour, in addition to the light physical activity associated with typical day-to-day life.

U.S. Department of Agriculture
Center for Nutrition Policy and Promotion
April 2005

FIGURE 3 Estimated daily calorie needs.

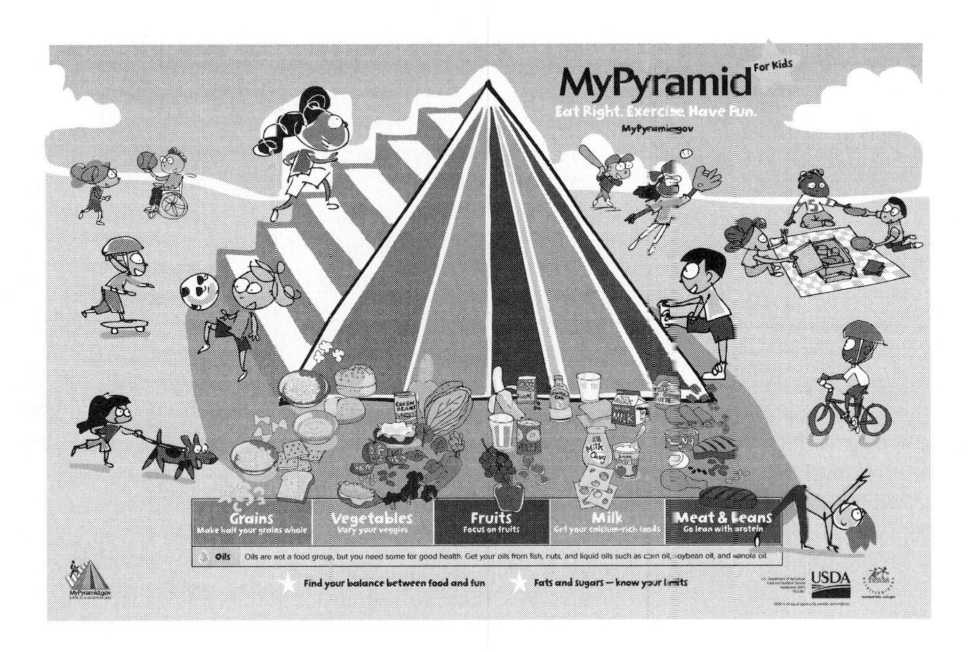

FIGURE 4 MyPyramid for kids. See color plate.

Ounce equivalents are specified for grains because a half-cup of cooked grains is equivalent to a 1-ounce slice of bread. Likewise, ounce equivalents are used for meat and beans because a tablespoon of peanut butter, an egg, and a quarter cup of beans are equivalent to 1 ounce of meat, fish, or poultry. It is important to note that the portion sizes are smaller than many typical portions. For example, an ounce equivalent of grains is a slice of bread or one-half cup of pasta, although typical portions are substantially larger than these (adults typically consume about two slices of bread per eating occasion, and about one cup of pasta [51]).

Additional specificity is suggested for the grain and vegetable groups. At least half of all grains consumed should be whole grains. Vegetables are divided into five subcategories, and the number of cups per weeks from each is specified as part of each food intake pattern (Fig. 2).

The recommended amounts of fruit and vegetables are higher than in the previous FGP. For example, 5 cups of fruits and vegetables are recommended for the 2200-calorie pattern, compared to 3.5 cups (seven half-cup servings) used for the FGP. The amount of milk that should be consumed has increased from 2 cups to 3 cups for all the patterns at 1600 kcal/d and higher; previously 3 cups was recommended only for teenagers and older adults. However, the amount of grains has been decreased slightly, by 1- or 2-ounce equivalents per day. The amount of oils to be consumed ranges from 3 to 11 teaspoons across the food patterns.

To help consumers better understand how many calories are available for solid fats, sweets, and alcohol (the "tip" of the FGP), the concept of discretionary calories has been introduced with MyPyramid (Fig. 2). These are the calories remaining after consuming the recommended amount from each of the food groups (including the oils). For the patterns for 1800 kcal/d or less, the discretionary calorie allowance is less than 200 kcal/d. At higher energy intake patterns, the number of discretionary calories increases, up to 648 kcal/d in the 3200-calorie pattern. Thus, individuals with greater energy needs, perhaps as a result of greater physical activity, have more calories that may be used for foods that contain solid fats, sweets, and alcohol. Of course, this additional energy need could also be met by consuming more of the foods in the various MyPyramid food groups.

Consumers may use a website called MyPyramid Tracker to evaluate their diets relative to the recommendations in MyPyramid [52]. This evolving website has many of the features of the former Interactive Healthy Eating Index website [53]. The user enters the foods and amounts consumed for a day, and the tracker calculates the intakes of nutrient and of the MyPyramid food groups. These are compared to recommendations, so the individual can see where improvements are needed. If the user also specifies physical activities, the tracker will estimate energy expenditure and graphically illustrate the balance between intake and expenditure. Intakes may be tracked for up to 1 year. A computer game is also available for children (*MyPyramid Blast Off*) at the MyPyramid website [47]. Many other resources and materials are also available at www.MyPyramid.gov [46].

Many food guides have been proposed in the United States, but most are specific interpretations of the more general guidance offered by the FGP and MyPyramid. Although pyramids for various cultural or ethnic groups have been proposed, there is seldom the same level of analytical research for these variations as was conducted for the original, more general, FGP and MyPyramid. Health professionals should be cautious about recommending food guides that have not undergone the rigorous testing of MyPyramid.

C. Food Exchanges to Design Meal Plans

Food exchanges were originally developed by the American Dietetic Association and the American Diabetes Association as a consumer-friendly tool that dietitians could use to plan meals for diabetic patients. Starting in 1989, the exchange lists were adapted for use in weight management as well, and they have been updated regularly [54]. Dietitians have widely used these exchanges to provide simple guidelines for their clients who wished to control macronutrient intake. Although the food exchanges were developed many years before the Food Guide Pyramid was published in 1992, the exchanges are remarkably similar to the food groups used in both the FGP and MyPyramid. Following are the exchange list food groups:

- Starch (includes bread, cereals, grains, starchy vegetables, crackers, snacks, and legumes)
- Meat and meat substitutes: lean, medium fat, and high fat plus plant-based proteins
- Non-starchy vegetables
- Fruits
- Milk: fat free, low fat/1%, reduced fat/2%, whole
- Fat
- Alcohol

Each of the exchanges is assigned an approximate value for energy and for grams of carbohydrate, protein, and fat. A dietitian can then plan a diet for a patient to include a specified number of exchanges from each group. As a guide, a healthy diet for teenagers and adults should include 2–3 servings of non-starchy vegetables, 2 servings of fruits, 6 servings of starch, 2 servings of low-fat or fat-free milk, about 6 ounces of meat or meat substitutes, and *small* amounts of fat and sugar. Once the number of exchanges is determined, the grams of the macronutrients and the energy content of the diet can be easily estimated.

The patient then uses the booklet to decide what food selections fit within each exchange. For example, if a patient's plan includes six starch exchanges, he or she may choose among a variety of cereals, grains, pasta, beans, starchy vegetables such as corn or potatoes, bread, and crackers. High-fat sweets and desserts like frosted cake count as a starch exchange *and* a fat exchange (for a 2-inch square piece). Portion sizes for one exchange of each food are specified, such as one (1 ounce) slice of bread or 1 cup of milk.

When using exchanges, health professionals should be aware that while the exchange food groups and the MyPyramid food group are similar, there are some important differences that may lead to confusion if a patient is also familiar with MyPyramid. For example, as mentioned earlier, potatoes count as a starch exchange, not as a vegetable. Corn and green beans are also considered starches, while legumes count as one starch plus one lean meat exchange. Some high-fat foods like cream and avocados count as fat exchanges only, while regular cheese counts as a high-fat meat exchange, not as a milk exchange. Many vegetables are considered "free" foods rather than vegetables because they contain few calories per serving (e.g., salad greens, cabbage, cucumber, green beans, and carrots).

A primary appeal of the exchange lists continues to be the flexibility offered to the health professional in planning macronutrient-controlled diets for patients. If combined with the minimum recommended number of servings from MyPyramid, the exchange lists provide an attractive alternative for a health professional who wishes to work with a client to design a meal plan that is nutritionally adequate as well as individually tailored for the client's specific macronutrient goals.

HOW TO READ A NUTRITION FACTS LABEL

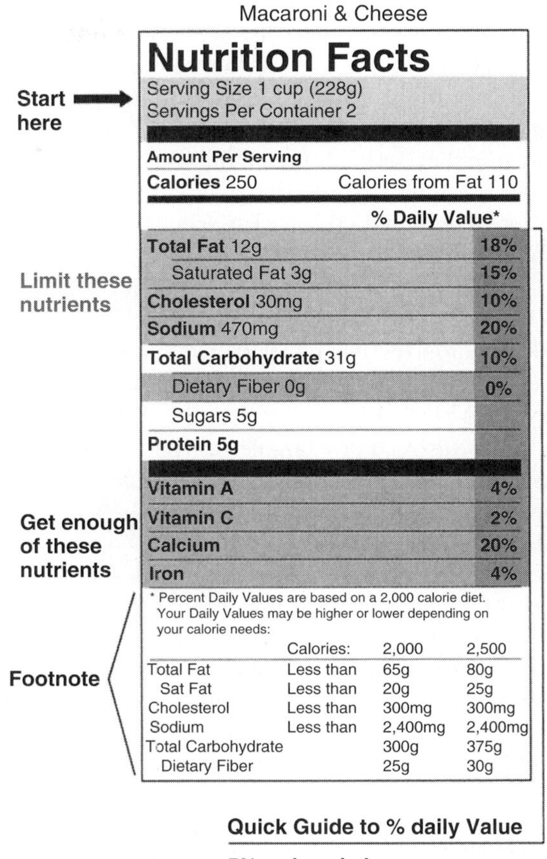

FIGURE 5 How to read a Nutrition Facts label. Adapted from USDHHS, USDA. (2005). Finding your way to a healthier you: based on the Dietary Guidelines for Americans [24].

D. The Role of Food and Supplement Labels in Helping Consumers Follow the Dietary Guidelines

The Nutrition Facts on the food label provide information that can help consumers follow the dietary guidelines (Fig. 5). The Nutrition Labeling and Education Act (NLEA) of 1990 requires a Nutrition Facts panel on most packaged food products [15]. Consumers can use this information to monitor their consumption of energy, fat, saturated fat, sugar, and sodium, as recommended in the body weight, fat, sugar, and sodium guidelines. Persons who wish to increase their intakes—for example, of vitamins, minerals, and dietary fiber—also will find the Nutrition Facts useful. For example, tofu may be a good source of calcium if it is precipitated with a calcium salt, but some tofu products have little calcium. By examining either the ingredient list (for a calcium salt) or the Nutrition Facts, it is possible to quickly tell if the product provides substantial amounts of calcium.

For most nutrients, amounts in weights such as grams, milligrams, or micrograms would not be meaningful to most consumers. As a result, the nutrients are shown as a percentage of a daily value (DV). For vitamins and minerals, these DVs are based on reference daily intakes, which in turn are based on the RDAs from 1968 [55]. For macronutrients, cholesterol, fiber, sodium, and potassium, the DVs are based on Daily Reference Values. The DRVs user older recommendations regarding fat intake (less than 30%, rather than 35%, of energy intake) and the current recommendation for saturated fat intake (less than 10% of energy intake). The DV for dietary fiber intake is 11.5 grams per 1000 kcal, for cholesterol it is 300 mg/d, for sodium it is 2400 mg/d, and for potassium it is 3500 mg/d. For example, for a 2000-calorie diet, the DV for fat is 65 g/d, the DV for saturated fat is 20 g/d, and the DV for dietary fiber is 25 g. The DVs for two calorie intake levels (2000 kcal/d and 2500 kcal/d) are specified on the label for total fat, saturated fat, cholesterol, sodium, total carbohydrate, and dietary fiber. The percentage of the 2000-kcal/d DV that is contained in the specific product is also given in larger type in the center of the panel. For calories, calories from fat, total fat, saturated fat, cholesterol, and sodium, the actual amount per serving is also given. For example, the macaroni and cheese Nutrition Facts shown in Figure 5 indicate that one serving of this product (1 cup) contains 18% of the DV for total fat (12 g) and 15% of the DV for saturated fat (3 g). Nutrition information on unpackaged fruits and vegetables must be posted in the produce department of grocery stores.

Consumers can readily scan the %DVs to see if this food item is a high or low source of specific nutrients, although 100% of the DV does not always mean 100% of that person's RDA. Not only are the DVs based on much older nutrient standards (from 1968), but they are also based on the RDA for the age and sex group with the highest RDA (often, teenage boys).

Because they serve different purposes, the serving sizes used for the food label do not always correspond to those used for MyPyramid. The NLEA specifies the servings on the food label, and they are intended to reflect the usual portion sizes that a consumer might typically select. In contrast, the MyPyramid amounts were not necessarily usual portion sizes but rather were portions that allowed flexibility in providing dietary guidance. In most cases, the MyPyramid portion is smaller than the corresponding food label serving size. For example, a MyPyramid serving of pasta is one-half cup, although a typical portion is closer to 1 cup. Thus, the serving size of pasta on the food label is 1 cup to correspond to this typical portion.

Starting in March, 1999, a Supplement Facts label was required for most dietary supplements [16]. The format of this label is similar to that for the Nutrition Facts label on foods, but it also allows for more flexibility in reporting non-nutrient components. For example, a ginseng supplement would indicate the number of micrograms in the supplement, but no daily value is given.

E. How Do Current Food Group Intakes Compare to Recommendations?

Because the supporting databases for calculating intakes of MyPyramid food groups have only recently been available, an evaluation of intakes relative to the MyPyramid recommendations has not been published for the national survey data. However, Table 4 shows the recommended servings from the five major food groups of the FGP and the actual intakes reported by a national sample of Americans aged 2 years and older in 1999–2002 [56]. Although mean intakes of grains and vegetables exceed the minimum recommended number of servings (six and three, respectively), approximately two-thirds of the population is not meeting the calorie-adjusted recommendations. The minimum number of servings applies to persons who are consuming a diet of 1600 kcal (or less), so those with higher energy intakes should also have a higher number of servings from each of the FGP food groups. Intakes of fruit and dairy are even lower, with about three-quarters of Americans reporting diets that do not meet these recommendations. The majority of people do not even meet the recommendation for meat intake.

As Table 4 also shows, Americans consume a substantial portion of their energy intake as discretionary fat and added sugar: together these items contribute 42% of the calories in a typical American diet [56]. Although the discretionary fat shown in the table includes both oils and solid fats, it is clear that Americans are consuming far more discretionary calories than MyPyramid would recommend (Fig. 2). For food intake patterns in the range of 2000 to 2400 kcal/d, discretionary calories are only 13% to 15% of energy intake. Thus, individuals who are consuming over 40% of their energy from discretionary fat and added sugars are either substituting these items for foods that are more nutrient dense or are adding these items to their diets at the risk of overconsumption of energy. Although the contribution of fat and sugar to obesity has been difficult to quantify (probably in large part because survey subjects underreport these foods [57]), if energy consumption exceeds energy expenditure (and, for children and adolescents, the energy requirements for growth), fat storage will occur. Until the role of these foods as a factor in obesity can be better understood, it remains appropriate to encourage a reduction in the consumption of foods with a low nutrient-to-calorie ratio.

The current recommended range for fat intake is 20% to 35% of energy intake for adults [6]. Over half of the adult population reported diets that were within this range [58], whereas 41% of diets for all Americans aged 2 years and older were under the recommended limit for saturated fat (less than 10% of energy intake) [58, 59]. The cholesterol goal (less than 300 mg/d) was being met by almost 70% of the sample [59, 60].

Overall, Americans are not choosing diets that follow either the dietary guidelines or the Food Guide Pyramid. Only two recommendations are being met by at least half the population, those for total fat and cholesterol. The Healthy Eating Index (HEI) provides a tool for evaluating overall conformance with the dietary guidelines [53, 59]. An individual is scored on 10 dietary parameters, most of which are shown in Table 4 (consumption of the correct number of food group servings and meeting the guidelines for fat and cholesterol; in addition, a diet is scored based on one's level of sodium intake and also for a measure of dietary variety based on the number of different food commodities that were consumed). Out of a possible score of 100, the average was 64 in 1999–2000 [59]. The score has improved slightly from 1989, when the average was 61.5, but is exactly the same as the score in 1996. By this measure as well, there is room for substantial improvement in the pattern of the American diet.

IV. BEYOND FOOD AND NUTRIENT GUIDELINES: PHYSICAL ACTIVITY GUIDELINES

A. The 2005 Dietary Guideline on Physical Activity

Increasing concerns about both the fitness level and the obesity rates among Americans led to a dietary guideline in 2000 that focused specifically on physical activity. As noted previously, the 2005 Dietary Guidelines also include one for physical activity, with similar advice: at least 30 minutes of moderate physical activity most days of the week for adults and 60 minutes for children (Table 3). However, in 2005, additional advice was offered for preventing weight gain among adults (60 minutes of moderate- to vigorous-intensity activity) and for sustaining weight loss among adults (60 to 90 minutes of daily moderate-intensity activity) [11].

In addition, an active lifestyle can promote a healthy diet by increasing energy requirements. With a larger energy budget, individuals can spare calories for optional foods containing added sugars and discretionary fats without displacing foods from the healthy foundation of their diets. The ability to choose a larger variety of foods, without gaining weight, also increases the enjoyment of eating, and decreases the guilt associated with occasional choices of foods of low nutrient density. The calories per hour expended for several types of moderate and vigorous physical activities are given in the Dietary Guidelines booklet [11].

The text in the Dietary Guidelines booklet [11] also discusses the many benefits of regular physical activity:

- Increases physical fitness and enables one to meet the physical demands of work and leisure comfortably
- Decreases risk for overweight and obesity
- Decreases risk for chronic diseases such as high blood pressure, stroke, coronary artery disease, type 2 diabetes, and colon cancer
- Can aid in managing mild to moderate depression and anxiety

Different types of physical activity may have different benefits. For example, resistance exercise can increase muscular strength and endurance, if performed on 2 or more days per week. Vigorous physical activity provides better fitness and thus a lower risk of developing chronic diseases. Weight-bearing exercise reduces the risk of osteoporosis by increasing peak bone mass (during growth), maintaining bone mass (during adulthood), and reducing the rate of bone loss (during aging).

A later chapter in this book examines the role of inactivity in the etiology of obesity. Inactivity not only predisposes an individual to weight gain but also can reduce the efficacy of weight loss attempts [61]. Obesity in turn is a risk factor for a variety of chronic diseases. Furthermore, an inactive lifestyle is a direct risk factor for several chronic diseases, beyond the indirect role of inactivity in promoting obesity, which in turn is a risk factor [61]. The role of both obesity and inactivity as risk factors for chronic diseases (e.g., osteoporosis, colon cancer, cardiovascular disease, and diabetes) is discussed elsewhere in this book.

B. Other Recommendations

The physical activity dietary guideline is based on guidance offered in 1995 by the Centers for Disease Control and Prevention and the American College of Sports Medicine: "Every US adult should accumulate 30 minutes or more of moderate-intensity physical activity on most, preferably all, days of the week" [31, p. 402]. This message has been widely disseminated and provides an achievable goal for most Americans. One of the appealing concepts for the public has been the ability to count short periods of exercise toward the daily goal of 30 minutes. Thus, walking up stairs or walking an extra few blocks to work can accumulate toward the total, as long as the activity is carried out at a moderate pace. Moderate activity is defined as an activity that requires three to six metabolic equivalents (METs), where a MET is the ratio of the metabolic rate while performing the activity to the resting metabolic rate (measured in kcal/min). Examples of moderate activity include walking briskly (3 to 4 miles per hour), cycling for pleasure or transportation, and swimming [31]. An increased emphasis on resistance training to promote strength and balance is based on a position stand of the American College of Sports Medicine [62].

Physical activity guidance for both adults (18–65 years of age) and older adults (over age 65) were updated by the American College of Sports Medicine and the American Heart Association in 2007 [63, 64]. For both of these age groups, the newer guidelines continue to recommend moderate-intensity aerobic (endurance) physical activity for a minimum of 30 minutes on five days each week or vigorous-intensity aerobic activity for a minimum of 20 minutes on three days each week. In addition, all adults would benefit by performing activities using the major muscles of the body that maintain or increase muscular strength and endurance at least twice a week. Specifically, it is recommended that 8–10 exercises be performed on each of these two days, and that the weight that is used should result in substantial fatigue after 8–12 repetitions of each exercise. Other recommendations that are specifically for adults over the age of 65 include performing exercises that maintain or improve balance in order to reduce the risk of injury from falls and having a plan for obtaining sufficient physical activity. The guidance for children's physical activity is based primarily on recommendations from the Centers for Disease Control and Prevention [65], which suggest that 60 minutes of moderate activity per day is appropriate. In addition, evidence linking television watching and body weight and level of fatness in children [6] has led to a recommendation from the American Academy of Pediatrics to limit television and video viewing to a maximum of 2 hours per day to reduce the risk of overweight among children [67].

C. How Do Current Activity Levels Compare to These Guidelines?

Goals for physical activity and fitness for Americans are specified as part of the Healthy People 2010 objectives, along with the observed levels at baseline (between 1994

TABLE 5 How Well Are Americans Following the Physical Activity Guidelines? Percentage at Baseline and Year 2010 Targets

Healthy People 2010 Objective	Age Group	Percentage at Baseline (Year)	Target for 2010
Reduce the proportion of adults who engage in no leisure-time physical activity.	18 y and older	40% (1997)	20%
Increase the proportion of adults who engage regularly in moderate physical activity for at least 30 min. per day.	18 y and older	15% (1997)	30%
Increase the proportion of adults who engage in vigorous activity 3 or more days per week for 20 min. or more.	18 y and older	23% (1997)	30%
Increase the proportion of adults who enhance and maintain muscular strength and endurance.	18 y and older	18% (1998)	30%
Increase the proportion of adults who enhance and maintain flexibility.	18 y and older	30% (1998)	43%
Increase the proportion of adolescents who engage in moderate physical activity for at least 30 min. on 5 or more days per week.	Grades 9–12	27% (1999)	35%
Increase the proportion of adolescents who engage in vigorous physical activity at least 20 min. on 3 or more days per week.	Grades 9–12	65% (1999)	85%
Increase the proportion of schools that require daily physical education for all students.	Middle schools	17% (1994)	25%
	High schools	2% (1994)	5%
Increase the proportion of adolescents who participate in daily school physical education.	Grades 9–12	29% (1999)	50%
Increase the proportion of adolescents who spend at least 50% of school physical education class time being physically active.	Grades 9–12	38% (1999)	50%
Increase the proportion of adolescents who view television 2 or fewer hours on a school day.	Grades 9–12	57% (1999)	75%

Source: Adapted from U.S. Department of Health and Human Services. (2007). Healthy people 2010: objectives for improving health, www.healthypeople.gov/document/pdf/Volume2/22Physical.pdf. Accessed on May 1, 2007.

and 1999) [68]. These objectives and targets are summarized in Table 5. Few American adults (only 15%) engage in moderate physical activity for at least 30 minutes per day, and 40% engage in no leisure-time physical activity. Clearly, it would take a substantial change in lifestyle to expand daily moderate activities to most American adults. Engagement in other types of physical activity is also low: 23% engage in vigorous activity at least 3 days per week, 18% try to enhance muscular strength, and 30% try to enhance flexibility.

For adolescents in grades 9 to 12, physical activity in the schools is lower than the objectives (Table 5). The percentage engaging in daily school physical education is only 29%, and the proportion that spends at least 50% of physical education class time being active is only 38%. Furthermore, only 17% of middle schools and 2% of high schools require daily physical education for all students. No recent data have been compiled for younger children, but in 1997, only one state required daily school physical education from kindergarten through 12th grade [65].

However, 65% of adolescents engaged in vigorous activity at least three times per week, and 27% engaged in moderate physical activity on 5 or more days per week. Because physical activity in schools appears to be limited, these levels of vigorous and moderate activity may be attributable to after-school sports or similar nonschool activities.

Thus, a public health priority is to identify ways to provide safe and enjoyable opportunities to engage in physical activity for both adults and children. Physical education in schools provides an obvious opportunity to increase the activity level of younger children. For older children, team sports may offer more appeal. Community-based activities may be the best way to reach the large majority of sedentary adults.

A midcourse review of the objectives shows progress toward the goals for seven of the physical activity and fitness objectives, but none had met the target [69]. It will take a concerted effort by health professionals if activity levels in America are to increase substantially.

V. SUMMARY

There are currently a variety of consumer-friendly tools that promote the consumption of healthful diets and the need for daily physical activity. In general, these interact smoothly to provide incentives for the selection of nutritious foods and at least moderately strenuous activity levels.

However, despite these educational tools, consumers do not appear to be motivated to make the dietary changes recommended by these nutrient and food guidelines or to select more frequent or more strenuous physical activities. As the survey summaries in Tables 1, 4, and 5 show, many improvements could be made to American food and activity choices. The burden falls on nutrition educators and behavioral scientists to provide approaches that will inspire the public to change their food practices and activity levels in a positive direction. The rewards are many, both at the individual level (in reduced rates of chronic disease) and the societal level (in reduced medical care costs and lost productivity). One estimate suggests that 16.6% of all deaths in the United States in 2000 (400,000 deaths) were due to poor diet and activity patterns [70]. The challenge for health professionals is to provide the motivation and the environment that will facilitate more healthful diet and activity patterns by the American public. Successful and practical intervention programs are greatly needed.

References

1. National Research Council (NRC), Food and Nutrition Board. (1989). Recommended Dietary Allowances. National Academy Press, Washington, DC.
2. Institute of Medicine (IOM). (1997). Dietary Reference Intakes for calcium, phosphorus, magnesium, vitamin D, and fluoride. National Academy Press, Washington, DC.
3. Institute of Medicine (IOM). (1998). Dietary Reference Intakes for thiamin, riboflavin, niacin, vitamin B_6, folate, vitamin B_{12}, pantothenic acid, biotin, and choline. National Academy Press, Washington, DC.
4. Institute of Medicine (IOM). (2000). Dietary Reference Intakes for vitamin C, vitamin E, selenium, and carotenoids. National Academy Press, Washington, DC.
5. Institute of Medicine (IOM). (2001). Dietary Reference Intakes for vitamin A, vitamin K, arsenic, boron, chromium, copper, iodine, iron, manganese, molybdenum, nickel, silicon, vanadium, and zinc. National Academy Press, Washington, DC.
6. Institute of Medicine (IOM). (2002). Dietary Reference Intakes for energy, carbohydrate, fiber, fat, fatty acids, cholesterol, protein, and amino acids. National Academy Press, Washington, DC.
7. Institute of Medicine (IOM). (2005). Dietary Reference Intakes for water, potassium, sodium, chloride, and sulfate. National Academies Press, Washington, DC.
8. Institute of Medicine (IOM). (2000). Dietary Reference Intakes: applications in dietary assessment. National Academy Press, Washington, DC.
9. Institute of Medicine (IOM). (2003). Dietary Reference Intakes: applications in dietary planning. National Academies Press, Washington, DC.
10. Institute of Medicine (IOM). (2006). Dietary Reference Intakes: the essential guide to nutrient requirements. National Academies Press, Washington, DC.
11. USDHHS, USDA. (2005). Dietary Guidelines for Americans 2005, 6th ed. HHS Pub. No.: HHS-ODPDP-2005-01-DGA-A. USDA Pub. No.: Home and Garden Bulletin No. 232. U.S. Government Printing Office, Washington, DC.
12. Dietary Guidelines Advisory Committee. (2005). Dietary Guidelines for Americans, www.health.gov/dietaryguidelines/dga2005/report. Accessed September 7, 2004.
13. MyPyramid. Available at www.MyPyramid.gov. Accessed June 2005.
14. U.S. Department of Agriculture (USDA). (1992). The Food Guide Pyramid. Home and Garden Bulletin No. 252. U.S. Government Printing Office, Washington, DC.
15. Kurtzweil, P. (1993). "Nutrition Facts" to help consumers eat smart. *FDA Consumer* **27**(4), 22–27.
16. Kurtzweil, P. (1998). An FDA guide to dietary supplements. *FDA Consumer* **32**(5), 28–35.
17. Institute of Medicine (IOM). (1994). How should the Recommended Dietary Allowances be revised? National Academy Press, Washington, DC.
18. National Research Council (NRC), Food and Nutrition Board. (1986). Nutrient adequacy: assessment using food consumption surveys. National Academy Press, Washington, DC.
19. Thompson, F. E., and Beyers, T. (1994). Dietary assessment resource manual. *J. Nutr.* **124**, 2245S–2317S.
20. Buzzard, I. M., and Willett, W. C., Eds. (1994). Dietary assessment methods. *Am. J. Clin. Nutr.* **59**, 143S–306S.
21. Willett, WC., and Sampson, L., Eds. (1997). Dietary assessment methods. *Am. J. Clin. Nutr.* **65**, 1097S–1368S.
22. Taren, D., Ritenbaugh, C., and Freedman, L., Eds. (2002). Fourth international conference on dietary assessment methods. *Public Health Nutrition* **5**(6A): 815–1109.
23. Food Surveys Research Group, U.S. Department of Agriculture. (2005). What we eat in America, NHANES 2001–2002: usual nutrient intakes from food compared to Dietary Reference Intakes, www.usda.gov/fsrg. Accessed on April 15, 2007.
24. USDHHS, USDA. (2005). Finding your way to a healthier you: based on the Dietary Guidelines for Americans. HHS Pub. No.: HHS-ODPDP-2005-01-DGA-B. USDA Pub. No.: Home and Garden Bulletin No. 232-CP. U.S. Government Printing Office, Washington, DC.
25. U.S. Department of Health and Human Services. (2005). A healthier you. Based on the Dietary Guidelines for Americans. U.S. Government Printing Office, Washington, DC.
26. Center for Nutrition Policy and Promotion, U.S. Department of Agriculture. (2005) MyPyramid. Steps to a healthier you, www.mypyramid.gov. Accessed February 14. 2007.
27. Foote, J. A., Murphy, S. P., Wilkens, L. R., Basiotis, P. P., and Carlson, A. (2004). Dietary variety increases the probability of nutrient adequacy among adults. *J. Nutr.* **134**(7), 1779–1785.
28. Murphy, S. P., Foote, J. A., Wilkens, L. R., Basiotis, P. P., Carlson, A., White, K. L., and Yonemori, K. (2006). Simple measures of dietary variety are associated with improved dietary quality. *J. Am. Diet. Assoc.* **106**, 425–439.
29. Kuczmarski, R. J., Flegal, K. M., Campbell, S. M., and Johnson, C. L. (1994). Increasing prevalence of overweight among US adults. *J. Am. Med. Assoc.* **272**, 205–211.
30. National Center for Health Statistics, CDC. (1997). Update: prevalence of overweight among children, adolescents, and adults–United States 1988–1994. *Morb. Mortal. Wkly. Rep.* **46**, 199–202.
31. Pate, R. R., Pratt, M., Blair, S. N., Haskell, W. L., Macera, C. A., Bouchard, C., Buchner, D., Ettinger, W., Heath, G. W., King, A. C., Kriska, A., Leon, A. S., Marcus, B. H., Morris, J., Paffenbarger, R. S., Patrick, K., Pollack, M. L., Rippe, J. M., Sallis, J., and Wilmore, J. H. (1995). Physical activity and public health—a recommendation from the Centers for Disease Control and Prevention and the American College of Sports Medicine. *J. Am. Med. Assoc.* **273**, 402–407.

32. Steinmetz, K. A., and Potter, J. D. (1996). Vegetables, fruit, and cancer prevention: a review. *J. Am. Diet. Assoc.* **96**, 1027–1039.
33. Ness, A. R., and Powles, J. W. (1997). Fruit and vegetables, and cardio-vascular disease: a review. *Intl. J. Epi.* **26**, 1–13.
34. Jacobs, D. R. Jr., Meyer, K. A., Kushi, L. H., and Folsom, A. R. (1998). Whole grain intake may reduce the risk of ischemic heart disease death in postmenopausal women: the Iowa Women's Health Study. *Am. J. Clin. Nutr.* **68**, 248–257.
35. Rolls, B. J., Bell, E. A., Castellanos, V. H., Chow, M., Pelkman, C. L., and Thorwart, M. L. (1999). Energy density but not fat content of foods affected energy intake in lean and obese women. *Am. J. Clin. Nutr.* **69**(5), 863–871.
36. Rolls, B. J., Castellanos, V. H., Halford, J. C., Kilara, A., Panyam, D., Pelkman, C. L., Smith, G. P., and Thorwart, M. L. (1998). Volume of food consumed affects satiety in men. *Am. J. Clin. Nutr.* **67**, 1170–1177.
37. Gordon, D. J. (1995). Cholesterol and mortality: what can meta-analyses tell us? *In* "Cardiovascular Disease" (L. L. Gallo, Ed.), pp. 333–340. Plenum Press, New York.
38. Gordon, D. J. (1995). Cholesterol lowering and total mortality. *In* "Lowering Cholesterol in High Risk Individuals and Populations" (B. M. Rifkind, Ed.), pp. 33–48. Marcel Dekker, New York.
39. Grundy, S. M. (1998). Overview: Second International Conference on Fats and Oil Consumption in Health and Disease: how we can optimize dietary composition to combat metabolic complications and decrease obesity. *Am. J. Clin. Nutr.* **67**(3), 497S–499S.
40. Krauss, R. M. (1998). Triglycerides and atherogenic lipoproteins: rationale for lipid management. *Am. J. Med.* **105**(1A), 58S–62S.
41. Depaola, D. P., Faine, M. P., and Palmer, C. A. (1999). Nutrition in relation to dental medicine. *In* "Modern Nutrition in Health and Disease" (M. E. Shils, J. A. Olson, M. Shike, and A. C. Ross, Eds.), 9th ed., pp. 1099–1124. Williams & Wilkins, Baltimore.
42. U.S. General Accounting Office. (1996). Food safety: information on foodborne illnesses: report to Congressional Committees. Washington, DC, GAO/RCED-96-96.
43. Mead, P. S., Slutsker, L., Dietz, V., McCaig, L. F., Bresee, J. S., Shapiro, C., Griffin, P. M., and Tauxe, R. V. (1999). Food-related illness and death in the United States. *Emerging Infect. Dis.* **5**, 607–625.
44. Yang, S., Leff, M. G., McTague, D., Horvath, K. A., Jackson-Thompson, J., Murayi, T., Boeselager, G. K., Melnik, T. A., Gildemaster, M. C., Ridings, D. L., Altekruse, S. F., and Angulo, F. J. (1998). Multistate surveillance for food-handling, preparation, and consumption behaviors associated with food-borne diseases: 1995 and 1996 BRFSS food-safety questions. *Morb. Mortal. Wkly. Rep.* **47**, 33–57.
45. Kurtzweil, P. (1998). A year of food safety accomplishments. *FDA Consumer* **32**(5), 8–9.
46. Haven, J., Burn, A., Herring, D., and Britten, P. (2006). MyPyramid.gov provides consumers with practical nutrition information at their fingertips. *J. Nutr. Educ. Behav.* **38**, S153–S154.
47. French, L., Howell, G., Haven, J., and Britten, P. (2006) Designing MyPyramid for Kids materials to help children eat right, exercise, have fun. *J. Nutr. Educ. Behav.* **38,** S158–S159.
48. Murphy, S. Ed. (2006). Development of the MyPyramid food guidance system. Supplement to the *J. Nutr. Educ. Behav.* **38,** S77–S161.
49. Britten, P., Haven, J., and Davis, C. (2006). Consumer research for development of educational messages for the MyPyramid food guidance system. *J. Nutr. Educ. Behav.* **38**, S108–S123.
50. Britten, P., Marcoe, K., Yamini, S., and Davis, C. (2006). Development of food intake patterns for the MyPyramid food guidance system. *J. Nutr. Educ. Behav.* **38**, S78–S92.
51. Krebs-Smith, S. M., Guenther, P. M., Cook, A., Thompson, F. E., Cucinelli, J., and Ulder, J. (1997). Foods commonly consumed per eating occasion and in a day, 1989-91. NFS Report No. 91–3. USGPO, Washington, DC.
52. Juan, W. Y., Gerrior, S., and Hiza, H. (2006). MyPyramid Tracker assesses food consumption, physical activity, and energy balance status interactively. *J. Nutr. Educ. Behav.* **38,** S155–S157.
53. U.S. Department of Agriculture, Center for Nutrition Policy and Promotion. (1995). The Healthy Eating Index. CNPP-1. USDA/CNPP, Washington, DC.
54. American Dietetic Association, American Diabetes Association. (2007). Choose Your Foods: Weight Management. American Dietetic Association,. Chicago, IL.
55. Kurtzweil, P. (1993). "Daily Values" encourage healthy diet. *FDA Consumer* **27**(4), 28–32.
56. Cook, A. J., and Friday, J. E. (2007). Pyramid servings intakes in the United States 1999-2002, 1 Day. Beltsville, MD: USDA, Agricultural Research Service, Community Nutrition Research Group, CNRG Table Set 3.0, www.ba.ars.usda.gov/cnrg. Accessed May 2, 2007.
57. Briefel, R. R., Sempos, C. T., McDowell, M. A., Chien, S., and Alaimo, K. (1997). Dietary methods research in the Third National Health and Nutrition Examination Survey: Under-reporting of energy intake. *Am. J. Clin. Nutr.* **65**, 1203S–1209S.
58. Moshfegh, A. J., Goldman, J. D., and Lacomb, R. P. (2007). Levels and sources of fat in the diets of Americans. *FASEB J.* **21**, A1062 (abstract).
59. Basiotis, P. P., Carlson, A., Gerrior, S. A., Juan, W. Y., and Lino, M. (2002). The Healthy Eating Index: 1999–2000. Center for Nutrition Policy and Promotion, Alexandria, VA.
60. Food Surveys Research Group, U.S. Department of Agriculture. Nutrient intakes: mean amount consumed per individual, one day, 2001–2002, www.ars.usda.gov/SP2UserFiles/Place/12355000/pdf/Table_1_BIA.pdf. Accessed on May 1, 2007.
61. U.S. Department of Health and Human Services. (1996). Physical Activity and Health: A Report of the Surgeon General. U.S. Department of Health and Human Services, Centers for Disease Control and Prevention, National Center for Chronic Disease Prevention and Health Promotion, Atlanta, GA.
62. American College of Sports Medicine. Position Stand. (2002). Progression models in resistance training for healthy adults. *Med. Sci. Sports Exerc.* **34**(2), 364–380.
63. Haskell, W. L., Lee, I-M., Pate, R. R., Powell, K. E., Blair, S. N., Franklin, B. A., Macera, C. A., Heath, G. W., Thompson, P. D., and Bauman, A. (2007). Physical activity and public health: updated recommendation for adults from the American College

of Sports Medicine and the American Heart Association. *Med. Sci. Sports Exerc.* **39**, 1423–1434.

64. Nelson, M. E., Rejeski, W. J., Blair, S. N., Duncan, P. W., Judge, J. O., King, A. C., Macera, C. A., and Castaneda-Sceppa, C. (2007). Physical activity and public health in older adults: recommendation from the American College of Sports Medicine and the American Heart Association. *Med. Sci. Sports Exerc.* **39**, 1435–1445.
65. Centers for Disease Control and Prevention (CDC). (1997). Guidelines for school and community health programs to promote lifelong physical activity among young people. *Morb. Mortal. Wkly. Rep.* **46,** RR-6, 1–34.
66. Andersen, R. E., Crespo, C. J., Bartlett, S. J., Cheskin, L. J., and Pratt, M. (1998). Relationship of physical activity and television watching with body weight and level of fatness among children: results from the Third National Health and Nutrition Examination Survey. *J. Am. Med. Assoc.* **279**, 938–942, 1998.
67. American Academy of Pediatrics Policy Statement. (2003). Prevention of pediatric overweight and obesity. *Pediatrics* **112**(2), 424–430.
68. U.S. Department of Health and Human Services. (2000). Healthy people 2010: objectives for improving health, www.healthypeople.gov/document/pdf/Volume2/22Physical.pdf. Accessed on May 1, 2007.
69. U.S. Department of Health and Human Services. (2007). Healthy people 2010, www.healthypeople.gov/data/midcourse/pdf/FA22.pdf. Accessed on May 15, 2007.
70. Mokdad, A. H., Marks, J. S., Stroup, D. F., and Gerberding, J. L. (2004). Actual causes of death in the United States, 2000. *JAMA* **291**, 1238–1245.

CHAPTER **14**

Nutrition, Health Policy, and the Problem of Proof

ROBERT P. HEANEY[1] AND SARAH TAYLOR ROLLER[2]
[1]*Creighton University, Omaha, Nebraska*
[2]*Covington & Burling LLP, Washington, D.C.*

Contents

I. BACKGROUND CONSIDERATIONS

In the early days of molecular biology there was a dictum that went "one gene, one protein," expressing, first, the insight that the blueprints for proteins were encoded in the genome and, second, the speculation that, as each protein was unique, so it had a unique genetic blueprint. The latter belief had to be abandoned quickly when it became apparent that there were a great many more proteins than there were genes. Rather than a setback, the resulting revisions in the blueprint model greatly enriched the science of cell biology. Nutrition today faces a similar need to reformulate the basic approach to its own science.

Although less explicitly articulated than the one-gene-one-protein principle, nutrition implicitly holds to a "one nutrient, one disease" conceptual model. As commonly taught, *the* disease of thiamin deficiency is beriberi; *the* disease of niacin deficiency is pellagra; *the* disease of vitamin D deficiency is rickets (or osteomalacia in adults); and so on. Although there is no expressed objection by nutritional scientists to recognizing multiple systemic consequences of nutrient inadequacy, the hold of the one-nutrient-one-disease model continues to dominate nutritional policy and regulation of health claims. It is expressed, for example, in the reluctance of the field, in the case of vitamin D, to label as "deficiency" the osteoporosis, fracture risk, propensity to falls, immune defects, hypertension, and cancer risk of low vitamin D status. Clinical scientists call all this morbidity "vitamin D insufficiency." It cannot be "deficiency," so the thinking goes, because such patients do not have rickets or osteomalacia.

A. Scientific Limits and Policy Issues

When such scientific tensions become integrated into the fabric of nutrition policy, peculiar outcomes for personal and public health can be the unwitting result. For example, in the context of food and drug regulation, the reluctance to attribute a broader spectrum of morbidity to nutritional "deficiency" has helped perpetuate strained distinctions between the scope of disease-related benefits attributable to the nutritional value of "foods" and those that are the presumptive therapeutic benefits of "drugs." More specifically, under the governing legal framework of the Federal Food, Drug, and Cosmetic Act ("FDCA"),[1] any "food" that is marketed in a manner suggesting that it may help prevent disease is subject to regulation as a "drug,"[2] except under two limited conditions. The first is when the benefit of the food concerns a classical essential nutrient deficiency disease (e.g., pellagra, scurvy); and the second is when the benefit is represented in conformance with a "health claim" authorized by the Food and Drug Administration (FDA).

The room for health claims is limited, however, because a claim will not be authorized for food unless the benefit the claim would convey is supported by scientific evidence concerning the relationship between a specifically identified food "substance" and a particular disease condition. That evaluation is based on an "evidence-based" ranking system[3]

[1] 21 U.S.C. 301 *et seq.*

[2] 21 U.S.C. 321(g)(1)(B)(defining "drug" to mean "articles intended for use in the diagnosis, cure, mitigation, treatment, or prevention of disease in man or other animals)".

[3] Health claims that are authorized through the submission of a premarket notification under procedures established by the Food and Drug Administration Modernization Act of 1997 for health claims based on the U.S. Dietary Guidelines for Americans and certain other authoritative statements of federal government agencies and the National Academy of Science are not evaluated under the evidence-based ranking system, but instead upon FDA procedural guidelines designed to ensure health claims are supported by "significant scientific agreement" through equivalent systems of scientific evaluation. Food and Drug Modernization Act, P.L. No. 105–115, §303, 111 Stat. 2296 (1997); Food and Drug Administration, Guidance for Industry: Notification of a Health Claim or Nutrient Content Claim Based on an Authoritative Statement of a Scientific Body (1998).

that was adapted from systems designed for medical therapies (see The Matter of Proof section). Under this system, the evidence, in addition to statistical significance, must show that the magnitude of the defined substance-disease benefit would be "physiologically meaningful and achievable in the general population" through the particular foods that the health claim would represent as beneficial.[4–6]

An illustration of the consequences of this system is found in the fact that, in order for a food product that is rich in calcium and vitamin D to make a health claim and still qualify as a "food" (rather than a "drug"), representations concerning disease-related benefits would need to be confined either to the classical deficiency disease, rickets,[7] or to osteoporosis, in compliance with an authorized health claim.[8] The same food would be subject to regulation as a "drug," however, if the disease-related benefits were represented to extend to other disease conditions, such as fracture risk, propensity to falls, immune defects, hypertension, or cancer risks that have been shown to be related to low vitamin D status. The distinctive regulatory standards governing "food" products that revolve around whether or not a disease condition results from an essential nutrient "deficiency" effectively integrates the scientific limitations of the one-nutrient-one-disease model into the fabric of FDCA regulatory policies distinguishing "foods" from "drugs" and thus strongly influences the extent to which the disease-related benefits of foods achieve recognition and acceptance.

Although research into the workings of nutrients in ways distinct from those classically associated with their nominal deficiency diseases certainly exists and is even flourishing, the conceptualization of nutrition remains hobbled by the limitations imposed by the one-nutrient-one-disease model and is further frustrated, particularly at a regulatory and nutritional policy level, by the reductionistic and positivistic biases of evidence-based ranking systems, which unduly constrain the consideration of relevant scientific evidence and make no presumption of disease-related benefit from the simple satisfaction of nutritional needs in the absence of what is judged affirmative proof.

B. Characterizing Nutritional Relationships between Chronic Undernutrition and Systemic Disease Manifestations

Fundamental in the endeavor to reformulate the scientific framework for nutrition is the need to develop alternatives to the one-nutrient-one-disease model for purposes of characterizing the manifestations of disease that result from chronic, suboptimal intake of essential nutrients.

Several years ago an alternative conceptual framework was proposed, one that explicitly recognized two main classes of mechanisms of nutrient action and made explicit provision for a spectrum of latency periods for disease expression. Examples of its application to three nutrients, calcium, vitamin D, and folate, were given [1]. This model is set forth diagrammatically in Figures 1 and 2. (Fig. 1 is applicable to all nutrients, whereas Fig. 2 exemplifies the model for one nutrient, vitamin D.) The model is arbitrary in that it makes an artificial distinction between so-called index and nonindex mechanisms and diseases. Doing so simply pays homage to the diseases classically associated with each nutrient (their "index" diseases) and to the mechanisms elucidated for their pathogenesis. But it is artificial in that it gives special place to the disorder first associated with the nutrient, which, as it turns out, may not always be the most important expression of deficiency. Moreover, it lumps all of the other disorders into a catch-all category of "nonindex" diseases. The principal value of the model is that it forces attention to the multiplicity of effects and gives explicit

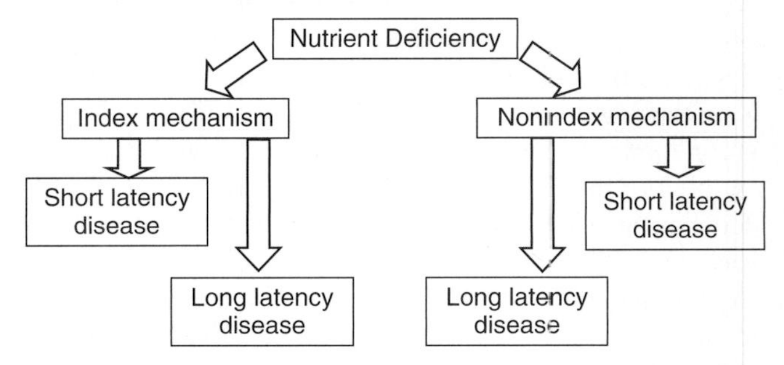

FIGURE 1 Scheme setting forth broad classes of latencies and mechanisms by which deficiency of a given nutrient produces disease or dysfunction. (Copyright by Robert P. Heaney, 2007. Used with permission.)

---

[4] DHHS, FDA, Interim Evidence Based Ranking System for Scientific Data (July 10, 2003) (modeling the FDA evidence based ranking system on that of the Institute for Clinical Systems Improvement as adapted by the American Dietetic Association) (www.cfsan.fda.gov/~dms/hclmgui4.html); see Greer, N., Mosser G., Logan G., Wagstrom Halaas, G. (2000). A practical approach to evidence grading. *Jt. Comm. J. Qual. Improv.* **26**, 700–712; Myers, E. F., Pritchett, E., Johnson, E. Q. (2001). Evidence-based practice guides vs. protocols: what's the difference? *J. Am. Diet. Assoc.* **101**, 1085–1090.

[5] DHHS, FDA, Interim Evidence Based Ranking System for Scientific Data (July 10, 2003) weighting studies by design type in descending order from Type One to Type Four:
Type One: Randomized, controlled intervention trials;
Type Two: Prospective observational cohort studies;
Type Three: Nonrandomized intervention trials with concurrent or historical controls; case-control studies;
Type Four: Cross-sectional studies; analyses of secondary disease end points in intervention trials; case series.

[6] www.cfsan.fda.gov/~dms/hclmgui4.html.

[7] 21 U.S.C. 321(g)(1)(C) (excluding from "drug" status foods represented to have benefits with respect to supporting "the structure or any function of the body"); see also 21 U.S.C. 343(r)(6) (permitting representations of benefit "related to a classical nutrient deficiency disease and disclos[ing] the prevalence of such disease in the United States"); 21 C.F.R. 101.14(a)(5) (excluding from requirements for "health claims" representations of benefit concerning "diseases resulting from essential nutrient deficiencies (e.g., scurvy, pellagra))."

[8] See 21 C.F.R. 101.72.

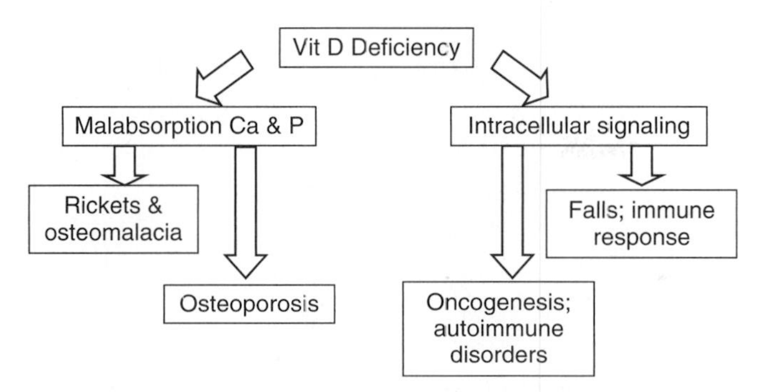

FIGURE 2 The scheme shown in Figure 1 made specific for vitamin D. (Copyright by Robert P. Heaney, 2007. Used with permission.)

recognition to long latency disorders, in contrast to the short latency that is characteristic of all the index diseases.

With vitamin D, for example, it is now clear that non-index processes certainly account for the great majority of the nutrient's actions in the body, literally dwarfing its effects on bone and the calcium economy (which, at the same time, remain real and important in their own right).

Behind this revised conceptualization lies the fact that all cells and tissues need all nutrients. They all need energy, they all need raw materials and building blocks, they all need cofactors, catalysts, and minerals. When an organism's diet is deficient in any given nutrient, all cells and tissues are to some extent functionally impaired. In the early days of nutritional science, we recognized only the most rapidly developing disorders involving only the most vulnerable tissues or systems. We can do better today.

There are two important considerations that flow from this broader understanding: (1) nutrient deficiency states, even when one defect is more prominent than others, tend to be pluriform; and (2) although their public health impact may be large, measurable effects of nutrients in isolated, single tissues or systems are often small (see the Effect Size section). The problems created by these features can be exemplified by two observations relating specifically to calcium.

Many papers describing trials of calcium conclude somewhat as follows: "While the reduction in blood pressure [or fracture or obesity or colon cancer or PMS] was statistically significant, the effect was too small to warrant recommending a change in calcium intake for the general population" [2, 3]. Such a conclusion is usually incorrect, because it ignores the public health impact of even small changes (see Effect Size) and also because it ignores all the other real, but sometimes equally small, measurable effects involving other body systems and diseases.

The problem is further exemplified by a study analyzing diet quality in women using a diet score based on nine essential nutrients with low covariance in foods [4]. Diets poor in calcium were typically poor in five of the nine selected nutrients, whereas diets adequate in calcium were adequate in at least eight of the nine. In other words, calcium deficiency is rarely the only problem with a given diet.

Both of these features have implications, first, for nutritional policy, since manifestly multiple system effects have to be factored into the decision process, and, second, for nutritional interventions, since mononutrient supplementation or fortification will be an inadequate response to what is often a polynutrient problem.

II. THE MATTER OF PROOF

Both current clinical science and regulatory policy are eagerly pursuing an approach to certainty described as "evidence based," despite frequently overlooked shortcomings in this approach as practiced [5, 6]. Evidence-based medicine (EBM), even before it got that name, had typically been applied to evaluation of procedural or pharmacological interventions in individual patients, such as "Does radical mastectomy for breast cancer produce better outcomes than lumpectomy?" or "Do beta blockers improve survival after myocardial infarction?" Such questions tend to be discrete and narrowly defined, and indeed every attempt is made to make them so. EBM attempts to rank the studies addressing such questions according to study design (or type), assigning greatest weight to the randomized controlled trial (RCT), then cohort studies, then case-control studies, then case reports, and finally, little or no weight to expert opinion.[9]

Despite its many problems [5, 6], this approach is generally held to be the best way to evaluate new medical treatments. Unfortunately, because nutrition has no corresponding approach of its own, EBM and its criteria, developed for medical treatments, are today being applied to nutritional questions and to issues of nutritional policy [7]. We say "unfortunately" because the drug model is poorly suited to the nutrient context. Table 1 summarizes several of the principal inherent differences between the two, which we now explore in more detail.

A. Contrast Groups

It is relatively easy to measure drug effects, first, because drugs are added to a drug-free state. Drugs are not normally present in the body except when administered by a physician or investigator, and when introduced, their concentration is not homeostatically regulated. Nutrients, by contrast, can essentially never have a nutrient-free state, either in the diet or in the body. Even when deficient, they are always still present in both, and often their serum concentration is tightly regulated (e.g., calcium and magnesium), which is why that concentration is often a poor index of nutrient status.

[9] See note 5.

TABLE 1 Structural Differences between Drugs and Nutrients Related to Tests of Efficacy

Characteristic	Drug	Nutrient
Contrast groups	Drug-free Drug-added	Regular diet (or low) Augmented diet
Scope of action	Largely single system	Usually multisystem
Effect size	Measurably large	Measurably small, system-by-system
Response characteristic	Linear	Threshold
Adjuvants	Minimize co-therapy	Optimize total nutrition

B. Scope

It is easy to measure drug effects for yet another reason: those effects are narrow in scope. The primary outcome variable for an antihypertensive is blood pressure. But the primary outcome variable for any given nutrient is not easy to define adequately, as it will usually be multiple. Despite a vast literature of nutrient efficacy studies with unitary end points, the fact is that nutrients should almost never be evaluated using single-system outcomes. They require some sort of global index, which nutrition today lacks.

C. Effect Size

It is easy to observe drug effects for another reason as well: They are, by design, measurably large in size. An antihypertensive agent that lowers blood pressure by 4 mm Hg in patients with hypertension would not be considered useful and would not be tested. Instead, agents would be designed to lower blood pressure by, perhaps, 40 mm Hg. Such drugs could be tested satisfactorily and economically in sample sizes of less than 50 patients. By contrast, a nutrient that lowers blood pressure at a population level by 4 mm Hg would have a profound effect on public health but would require sample sizes in excess of 1000 healthy individuals simply to demonstrate that it had any effect at all.

It is important to stress that the apparent smallness of nutrient effects does not mean that they are thereby unimportant. Figure 3 makes this point graphically for a typical distribution of systolic blood pressure values in adult women. A downward shift of the distribution by just 4 mm Hg reduces the fraction of the population above 140 mm Hg (i.e., the cutoff for the diagnosis, hypertension) by nearly 50%—a profoundly important public health benefit from a change that, in individuals, may seem trivially small. Further examples of large cumulative impacts of small changes are seen in the fact that negative calcium balance of just 30 mg/d translates to 1% bone loss/year or

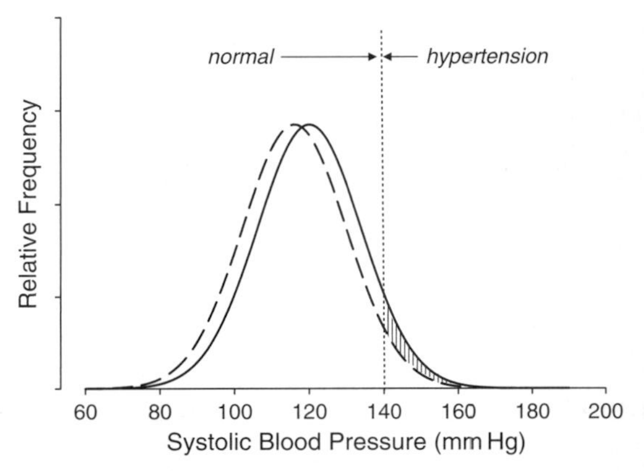

FIGURE 3 Typical frequency distribution for systolic blood pressure in perimenopausal women (with mean 120 mm Hg and SD 14 mm Hg)—the solid line. Approximately 8% of this population has a value above 140 mm Hg and hence would be considered hypertensive. Shifting the distribution downwards (i.e., to the left) by just 4 mm Hg (the dashed curve) lowers the proportion above 140 mm Hg to about 4.3%, or a reduction of nearly 50%. The shaded zone is the graphic representation of the difference in population proportions that are hypertensive, which is produced by the lowering of the distribution of blood pressure values. (Copyright by Robert P. Heaney, 2007. Used with permission.)

30% in 30 years (i.e., a skeletal deterioration from normal status to osteoporosis). Weight gain presents a similar contrast. Energy imbalance of as little as 70 kCal/d (less than one-half can of a typical carbonated beverage) leads to a weight gain of 70 lbs in just 10 years (i.e., it accounts for a transition from normal weight to obesity). Such small changes underlie many of the long latency disorders, yet they are effectively imperceptible when viewed up close.

D. Response Characteristic

Drug effects are usually linear across a broad range of doses, whereas nutrient effects typically exhibit threshold behavior (Fig. 4). The drug response is monotonic—that is, it is directionally the same (and hence is detectable) everywhere along the intake range (Fig. 5). By contrast, a perceptible nutrient effect occurs only if at least one of the contrasted intakes is below the response threshold (Fig. 6). If both groups have intakes above the threshold, no difference in response to the two intakes will occur or should be expected.

E. Adjuvants

An additional and very important difference between the two investigative challenges lies in the role of helper agents in enabling or obscuring the proper effects of an intervention. When testing drugs, co-therapies are usually excluded, so as to allow clearer definition of precisely what the drug is doing. However, in evaluating individual nutrients, intakes

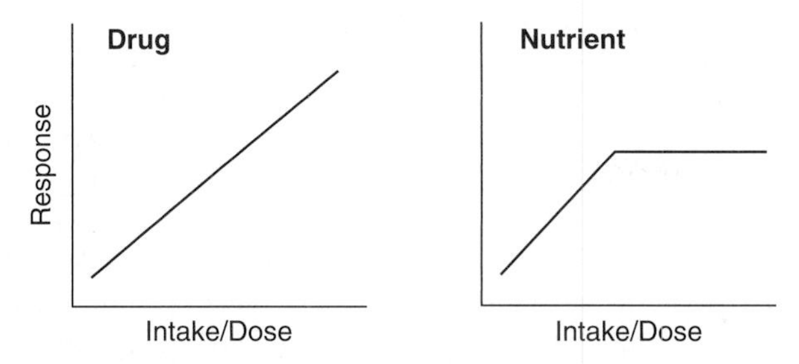

FIGURE 4 Response curves typical of drugs (*left*) and nutrients (*right*). The drug response tends to be monotonic, so that higher doses produce larger effects across the range of plausible doses, whereas for nutrients, there is usually a response threshold, above which the benefit of the nutrient is constant, irrespective of intake. (Copyright by Robert P. Heaney, 2007. Used with permission.)

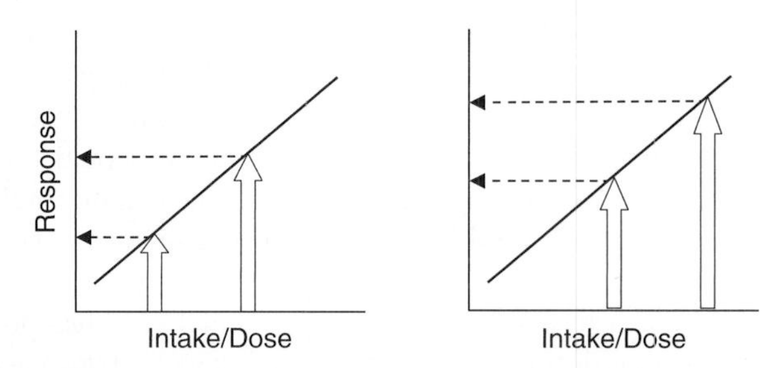

FIGURE 5 For drugs, the response can be detected along much or all of the response range. (*Left*) Low dose. (*Right*) High dose. (Copyright by Robert P. Heaney, 2007. Used with permission.)

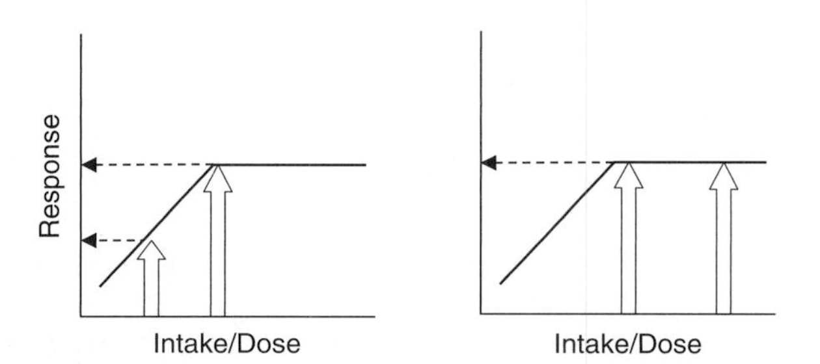

FIGURE 6 For nutrients, response occurs only if one or both of the contrasting intakes is below the response threshold (*left*). If both are at or above the threshold (*right*) no difference in response occurs. (Copyright by Robert P. Heaney, 2007. Used with permission.)

of other nutrients not only must continue, but ensuring adequacy of their intakes can become critically important if the test nutrient is to exert a measurable effect. Calcium, vitamin D, and protein provide a useful example of this interdependency.

It can easily be shown that practicable calcium intakes will not produce useful quantities of absorbed calcium in the absence of vitamin D and, conversely, that inadequate calcium intake will not permit adequate absorption even in the presence of full vitamin D repletion [8]. If, as is generally recognized, the effect of either calcium or vitamin D is considered to be increased calcium delivery into the body, then it follows that one needs both nutrients to obtain (and observe) the effect of either. Using a drug model would typically require studying each nutrient alone, as some of the published meta-analyses have mistakenly done [9–11].

But to take the example a step further, studies have shown that simply getting more calcium into the body is not itself enough to ensure protection of bone. The effect of calcium on bone mass turns out to be dependent on protein intake. Dawson-Hughes and Harris [12] showed that the bone protective effect of supplementation with calcium and vitamin D was confined to individuals in the highest tertile of protein intakes; and in the Omaha Nun database on midlife women, the positive association of calcium intake with skeletal calcium retention was confined to women with protein intakes above the median for the group [13]. The results from these two studies are entirely concordant, and they ought not be surprising, as protein makes up nearly 50% of the volume of bone, and bone remodeling requires a continuing supply of fresh dietary protein if bone replacement is to keep pace with bone removal. Finally, data suggest that magnesium may be crucial also, inasmuch as vitamin D deficiency seems to be associated with subclinical magnesium deficiency [14]. In brief, as already noted, diets are seldom deficient in just one nutrient, and tests of the efficacy of single nutrients may fail, as many have done [15], if all other intake shortfalls are not corrected as well.

Efforts can be made to overcome or compensate for many of these structural problems, once recognized, but contextual issues are more difficult to deal with. Table 2 lists three important contextual differences between drugs and nutrients in the testing of efficacy.

TABLE 2 Contextual Differences between Drugs and Nutrients Related to Tests of Efficacy

Characteristic	Drug	Nutrient
Profit margin (access to funds for testing)	Large	Small
Patent protection	Usually available	Usually not available
Ethics	Placebo controls often acceptable	Placebo controls usually unacceptable when outcome involves serious disease

F. Profits and Patents

With drugs, the cost of establishing efficacy (estimated at about $1 billion for an average drug today) will be recovered from the "profits" derived from drug sales for the number of years for which patent protection prevents unfair competition. Neither recourse is usually available for foods or nutrients. Profit margins are low, and patent protection is limited or nonexistent. Thus, the food industry is precluded from funding the kinds of EBM-recommended tests needed to show population-level benefits.

G. Ethics

Because of the difference in response characteristic between drugs and nutrients, a valid RCT for a nutrient requires that the control group be placed on an inadequate intake. With new drugs, one can always summon equipoise. One does not know, in advance, whether the agent being tested will confer any net benefit; and it is not, in fact, a deprivation to be "deprived" of a drug that is ineffective. But that posture is never possible for nutrients. All nutrients are, ultimately, essential, and to test the putative benefit of any single nutrient requires, in effect, that our trial measure what happens to someone who clearly does not get enough. Once that necessity is recognized, it is immediately apparent why there is an ethical problem, because to show a benefit, the investigators must harm the control group [16]. Such a trial becomes even more clearly indefensible when the outcome concerned is a serious health problem such as hip fracture or preeclampsia.[10]

III. APPROACHES

The validity of the foregoing analysis must be judged on its own merits, not on the feasibility or attractiveness of any solutions the authors may propose. Nevertheless, suggesting some solutions seems appropriate, if only to illustrate how the problem of establishing efficacy for nutrients might be approached or to evoke alternative suggestions. For the purposes of this chapter, we shall confine ourselves to two distinct approaches that together attempt to address most of the problems of the drug model listed in Tables 1 and 2. One is a global index, and the other is a design alternative to the RCT.

[10] The CPEP trial (calcium in preeclampsia prevention) is a case in point [17]. Calcium supplementation in this trial did not significantly reduce preeclampsia incidence, in contrast to results from previous RCTs. But there was no low-calcium contrast group in CPEP. The control group received a calcium intake averaging *above* the current recommendations for pregnancy. It would not, in fact, have been ethically permissible to design the trial with a low calcium intake group because, for the trial to be successful, there would have had to have been an excess of well over 20 cases of preeclampsia in the control group, relative to the intervention group.

A. A Global Index

As nutrients affect many systems and organs, it seems desirable to test nutrient effects in a way that captures that multiplicity. Such an approach calls for a global index that aggregates the multiple, sometimes small, measurable changes produced in as many systems. Unfortunately, no such index exists for nutrients. There have been promising attempts to define food quality in a positive way—that is, by what a food contains rather than by what it does not have (or has less of) [18, 19]. An analogous approach might be taken to health promotion by individual nutrients. Although it is unlikely that a single index could be developed for all nutrients, the components of the index for a specific nutrient would logically be made up of those functions or outcomes clinically linked to the nutrient concerned. An illustration of the value of capturing multiple end points is provided in Figure 7, showing important health benefits for vitamin D in four distinct systems. Although each benefit had been established individually, the required sample sizes and costs were large. Manifestly, an aggregate outcome measure that incorporated all four end points would both more fairly and more accurately represent the health benefit of vitamin D and could have rendered detection of benefit less difficult and less costly.

Also, with respect to vitamin D, Cannell [21] has proposed recognition of a "vitamin D deficiency syndrome" consisting of low serum 25(OH)D levels together with some combination of osteoporosis, osteomalacia, poor calcium absorption, heart disease, hypertension, autoimmune disease, certain cancers, depression, and chronic pain, or any of the other disorders that have been plausibly linked to

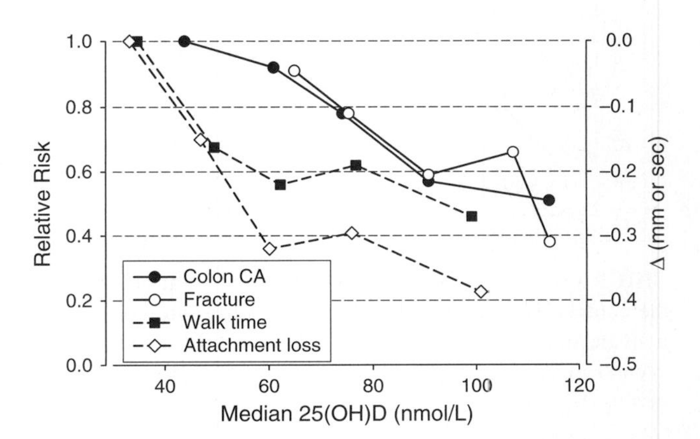

FIGURE 7 Health status improvement in several clinical studies of four health outcomes (periodontal disease, neuromuscular function, fractures, and colon cancer), plotted as a function of median achieved vitamin D status in the various studies assembled in this analysis, as assessed by serum 25(OH)D concentration. The solid lines are for relative risk of colon cancer and fracture (left-hand axis) and the dashed lines for reductions in walk time in the elderly and tooth attachment loss in periodontal disease (right-hand axis). (Redrawn from Bischoff-Ferrari *et al.* [20]. Used with permission.)

low vitamin D status. This approach explicitly recognizes the number of ways that vitamin D functions and therefore the ways deficiency may manifest itself. The features of this syndrome definition could simply be inverted, thus characterizing the *benefit* of improved vitamin D status. A global index could be fabricated, constructing it so that high values would be indicative of greater benefit, and low values of less benefit (or increased risk of frank disease). A certain number of points might be given for every incremental lowering in systolic blood pressure, a certain number of points for every incremental improvement in calcium absorption, a certain number of points for every incremental lowering in frequency of falls, a certain number of points for every incremental reduction in prevalence of key cancers (e.g., breast, colon, prostate, lung, and marrow-lymphoma, among others), and a certain number of points for insulin responsiveness (expressed, for example, in response to a standard glucose challenge).

This example is intended solely to illustrate the type of approach that might be taken and not to suggest specific components or the values attached to them. Issues that would need to be resolved would be the relative importance of the various end points to the individual concerned, to the public health, or to the health care budget of the nation. Such relative importance would be expressed in the index by the weights (number of points) given each component of the index. However the score may be confected, and however its various components may be weighted, the score itself would be the primary outcome variable of an appropriately designed clinical investigation, explicitly comparing the aggregate of system effects in individuals having better vitamin D status with those with poorer status.

Comparable scores could surely be developed for other nutrients. For example, the components of a calcium score would include blood pressure, colon cancer risk, bone remodeling rate, bone mass, fracture risk, kidney stone risk, and body composition, among others. Much work would have to be done in order to elicit a sufficient consensus in the scientific community to make the use of such indices persuasive, but lacking that effort, the field has no good way to assess the multiplicity of system effects that is characteristic of nutrients. Such indices for vitamin D and calcium seem a good place to start since their components represent already established end points. Thus, experimentation with the index will be a test of the *concept*, not a test of the *nutrient* (i.e., how well does the index work in identifying a health benefit, not how well does vitamin D or calcium work).

B. An Alternative to the Randomized Controlled Trial

Before describing a design that might solve some of the problems implicit in Tables 1 and 2, it will be useful to review the major design types, with particular emphasis on their propensity to bias (i.e., what makes them, structurally, more or less persuasive).

1. Minimizing Bias

One inescapable fact of clinical research is that individuals respond differently to any given nutritional state, disease context, or treatment intervention. These differences are based in countless genetic variations in receptor binding, enzyme efficiency, gene promotion, and all of the other signaling and response mechanisms of cells and tissues of the body. Additionally, individuals bring different life experiences, different morbidities, differing nutritional status, differing motivations, and differing abilities to adhere to or sustain a given therapeutic intervention. The result is that, when one envisions an investigation of a nutrient or drug intervention in a given sample of the population, one knows in advance that some will do better than others. Some will have gotten better or worse by themselves, irrespective of the intervention. In brief, some will show improvement or worsening of a particular physiological function for reasons unrelated to the intervention. The problem lies in the fact that we cannot tell who these persons are in advance, nor do we know who will actually get better or worse until we have done the investigation. Then the challenge is to figure out whether it was the intervention that was responsible for a measured difference between treated groups, or some of the other myriad factors that vary among the subjects, factors that may, by chance, have been disproportionately represented in the contrast groups of the investigation.

2. The Randomized Controlled Trial (RCT)

The preferred approach to this issue is the true experimental design (i.e., a randomized controlled trial). Its strength lies in the fact that, in randomly allocating the members of the sample to one intervention or the other, we actually randomly allocate the unrecognizable "improvers" and "worseners" to the treatment or control groups. From centuries of experience with the laws of probability we are able to estimate the likelihood that a substantial disproportion of the improvers were by chance allocated to the active agent, and correspondingly a substantial fraction of the worseners were allocated to the control group. When we make a probability statement about the results (e.g., $p < 0.05$ or $p < 0.01$), we are simply saying that the chances are less than 1 in 20 or less than 1 in 100 that the difference we observed between the two treatment groups was due to this luck-of-the-draw disproportion in responders rather than to the effect of the intervention we are studying.

Although the RCT permits strong causal inference, the problems enumerated in Tables 1 and 2 and in the foregoing discussion make the RCT unsuitable (or even unacceptable) for much research involving nutrients. Additionally, the RCT has serious generalizability problems, too often ignored [22].

Nevertheless, an RCT is favored by EBM, because it allows us to quantify our chances of being wrong when we impute causality to the intervention. It is sobering to bear in mind that we can never truly prove causality; all we can do is state our chances of being wrong when we make that imputation. There is always the possibility that we will have been unlucky, that we had gotten that 1 time in 100 when pure random chance "put" an excess of the improvers in the treatment group and an excess of the worseners in the control group.[11]

3. Observational Studies

By contrast, all other trial designs (generally termed "observational" studies) have to deal with the fact that the assignment of one group of individuals to a given intervention and another to no treatment at all (or to low or high intakes of the nutrient being investigated) will not be due to random chance but to other factors not under the control of the investigator, factors that may be related to the measured outcome. With such trial designs, although one can assess the chances that a given difference is greater than might have been expected from random chance, one has no way of knowing whether that difference was due to the intervention or to other, unrelated factors that influenced who received the intervention and who did not, simply because those factors were not distributed randomly to the contrast groups.

The now classic example of how one can be misled by such factors is found in the relationship between postmenopausal estrogen use and coronary artery disease in women. Before the Women's Health Initiative (WHI), most observational studies had shown a seemingly protective effect of estrogen and, in fact, the Institute of Medicine had issued a report to the effect that WHI was probably unnecessary because we already knew the answer [23]. As it turned out, we did not, and WHI showed that estrogen use in postmenopausal women, rather than being protective, actually increased the risk of coronary artery disease [24] at least in older women [25]. This discordance is usually explained by invoking the likelihood that women voluntarily taking estrogens were also involved in myriad other health-promoting behaviors and that estrogen use, therefore, was simply a marker for a generally healthier lifestyle. Whether that explanation is ultimately correct or complete will perhaps never be known, but the important point is that factors other than estrogen use—factors that could not be randomized—were responsible for the seemingly better outcomes in the estrogen users.

4. Investigative Interference

An additional and crucial feature of a properly designed RCT is the double blind. This is important because the placebo effect is a powerful one, and individuals in a trial, knowing that they are receiving a potentially efficacious treatment for a serious health problem, respond differently than do individuals outside the investigative context. This difference in response is known technically as "interference" of the investigation in the outcome. In any concurrent investigation, whether a randomized controlled trial or a concurrent cohort study, the placebo effect will always be present and may be responsible for a substantial portion of the response. The purpose of the blind, in an RCT, is to equalize that interference—that is, both the placebo and the treatment group experience a placebo response, and any residual difference will, therefore, likely be due to the intervention. But with a concurrent cohort study (often termed a "prospective" design) we must contend not only with the influence of other, unrecognized factors, but with unbalanced interference. Because both the study participants and the investigative staff are aware of who is a member of the active and placebo arms of the study, there is unequal interference.

There are two observational designs that avoid this unequal interference problem: One is the case control study, and another is the nonconcurrent cohort study. In the former, the contrast groups are assembled from individuals who have and have not actually developed the outcome under study, and then their exposure to the factor being tested is determined after the fact. In the latter, the exposure is also determined after the fact, but the study is basically prospective, inasmuch as one assembles the contrast groups by exposure, not outcomes, using various previously recorded databases or personal recall. In both cases, the investigation occurs after the outcome has developed, and hence no interference of the investigation in the outcome is possible.

The weakness of case-control studies lies in their tendency to "admission rate bias" [26], a problem that can be avoided only by either obtaining one's samples from the population itself, using random methods, or by total sampling of the outcomes concerned. (Then, with either approach, admission into the study is not influenced by factors related to the outcome.) But such stratagems are rarely possible and even more rarely implemented. Hence, case-control studies, as commonly reported, rank low on EBM's scale of persuasiveness.

5. The Nonconcurrent Cohort Design

There is yet another design that needs to be described—the nonconcurrent cohort study. Like its concurrent cousin, the nonconcurrent cohort study assembles its contrast groups on the basis of exposure to the intervention (or in the case of nutrients, to high or low intakes of the nutrient concerned). However, it does so after the fact. An example would be a study of the effects of smoking. Contrast groups can be accurately assembled many years after the onset of smoking.

[11] Moreover, given the multiplicity of randomized controlled trials that are reported every week in the literature, one can be certain that some of them will have fallen victim to that chance.

Nonconcurrent cohort studies circumvent some of the problems of the RCT, and when feasible, may well be the preferred nonexperimental design and perhaps the only one reasonably suited to the study of nutrients.[12] Nonconcurrent cohort designs sidestep the ethical problem, because the low-intake group is self-determined, and the exposure (or lack thereof) is not under investigative control. We may deplore the inadequate intake, but we neither cause it nor tolerate it.

The principal problem with a nonconcurrent cohort study lies not in its design per se, but in its execution. This is because one may not be able to establish with any certainty, after the fact, to which contrast group an individual might have belonged. How can we be certain that one group has had a high intake of vitamin B_6 for the past 10 years and a contrasting group had a low intake? It will not always be easy. However, there are now many large databases, both in Europe and in North America, that do permit such kinds of assessments.

One example is the demonstration by Munger *et al.* [27] of an association between low vitamin D status and risk of development of multiple sclerosis (MS). That study was possible because the U.S. military has a huge repository of frozen serum samples that could be analyzed, in this case for 25(OH)D, in order to ascertain the level of vitamin D nutrition before the development of MS. This study, as executed, used a case-control design, rather than a nonconcurrent cohort design, but that was purely for reasons of cost. (Because MS is a fortunately rare disorder, starting with the cases greatly reduced the required investigative and analytical work.) However, there would have been no theoretical barrier to ascertaining 25(OH)D concentrations in all of the millions of frozen specimens or some reasonable sample thereof and thus constructing the study as a nonconcurrent design. Just 2 years earlier, Munger *et al.* [28] had done precisely that for the Nurses' Health Study (however, using recorded vitamin D intakes rather than serum 25[OH]D to define the exposure cohorts) and had found a similar apparent protection of vitamin D with respect to risk of MS. In both cases it was possible to estimate the actual vitamin D status before the development of MS. However, as that status was unknown to both the participants and the investigators until after development of the outcome, no interference could have occurred. That leaves only the possibility of some extraneous, but unrelated, factor being responsible for allocation of the improvers and worseners to the two contrast groups. This problem can, in many cases, be mitigated.

There will never be a method of estimating the chance that the observed effect was caused by unrelated factors, which is as certain as random allocation, but that does not mean that one cannot take steps to reduce disproportions in the distribution of these other factors or to match the contrast groups for them. For example, in the issue of postmenopausal estrogen use and coronary artery disease risk, in addition to taking estrogen use as a determinant of the contrast groups, hindsight instructs us that we could have matched the groups for other markers of health-promoting behaviors such as smoking, alcohol use, regular dental visits, weekly exercise, body mass index, family history, and undoubtedly many others. In other words, in assembling the contrast groups that constitute the two cohorts of a nonconcurrent cohort study, one needs to use not only exposure to the agent being tested, but all other factors that we know, at the time of the investigation, may be associated with the outcome of interest. Manifestly, this is not easy and will not always be possible. But, where possible, such an approach, particularly when combined with a global index, offers the best possible chance of skirting both the ethical problem of RCTs and the inferential weakness often faced by observational studies of the efficacy of nutrients.

Many reports from observational studies attempt to evaluate the impact of these factors by "adjusting" for various confounders. In reports from such studies, the method of adjustment is almost never stated and the reader (and often the investigator as well) does not know the basis for the adjustment. Rather than adjusting, it would seem to be far preferable to match the contrast groups for the confounding factors in advance of the analysis.

IV. CONCLUSION

This chapter has attempted to address an issue that is both a challenge and an opportunity for nutritional science at the beginning of its second century as a scientific discipline. It is a challenge in that the drug model, widely used as a basis for claims of efficacy, is ill suited for the task of establishing nutrient effects, yet it is increasingly being applied to nutrients, particularly for the framing of nutritional policy. It is an opportunity in that it gives nutrition a chance to break out of its implicit one-nutrient-one-disease mold and to focus more explicitly on total body health, not as a sop to a romanticized "holistic" approach to science, but simply because total body health is precisely the proper object of nutrition per se.

For nutrition to live up to its promise in reducing the substantial burden of diet-related morbidity and mortality, both the conceptualization of nutrition and the systems for evaluating relevant scientific evidence will need to be reframed to characterize more fully and accurately the human relationship of dependency on the food environment that nutrition, ultimately, endeavors to understand. Given the ecological nature and evolutionary origins of this relationship, this reframing will need to consider features of the

[12] Such designs are sometimes mistakenly called "retrospective" cohort studies, but all cohort studies are basically prospective inasmuch as the contrast groups are assembled on the basis of exposure occurring *before* the development of the outcome of interest.

past food environments that have shaped the evolution of human nutritional needs, as well as those of contemporary food environments that are shaping current food consumption patterns and nutritional inadequacies. These, in turn will have to be integrated into a system of evaluation that places clinical, observational, and other research findings into an appropriate environmental health context, both to expose those disease relationships with greater biological plausibility and significance and to minimize the risk of overlooking subtler relationships of substantial importance to public health. Ultimately, a reformulated approach for evaluating a more diverse range of relevant scientific evidence will be necessary if the nutritional relationships to disease that have greatest importance for improving both personal and public health are to be recognized and woven into the fabric of nutrition policy.

This chapter is little more than a first step. Although the investigational design that avoids most of the problems presented by randomized trials (i.e., the nonconcurrent cohort study) would seem to offer promise, much more thought and testing need to be done to define better the probably nutrient-specific confounding factors that ought to be factored into assembling the intake cohorts needed to discern total body effects of nutrients. Additionally, better tools, such as global effect indices, need to be developed and tested. Unfortunately, the EBM juggernaut is rapidly rolling downhill and threatens to crush public health nutrition with its reductionist methods. Hence, the challenge is both real and urgent.

References

1. Heaney, R. P. (2003). Long-latency deficiency disease: insights from calcium and vitamin D. *Am. J. Clin. Nutr.* **78**, 912–919.
2. Griffith, L. E., Guyatt, G. H., Cook, R. J., Bucher, H. C., and Cook, D. J. (1999). The influence of dietary and nondietary calcium supplementation on blood pressure. An updated meta-analysis of randomized controlled trials. *Am. J. Hypertens.* **12**, 84–92.
3. Allender, P. S., Cutler, J. A., Follmann, D., Cappuccio, F. P., Pryer, J., and Elliott, P. (1996). Dietary calcium and blood pressure: a meta-analysis of randomized clinical trials. *Ann. Intern. Med.* **124**, 825–831.
4. Barger-Lux, M. J., Heaney, R. P., Packard, P. T., Lappe, J. M., and Recker, R. R. (1992). Nutritional correlates of low calcium intake. *Clinics Appl. Nutr.* **2**, 39–44.
5. Service, F. J. (2002). Idle thoughts from an addled mind. *Endocr. Prac.* **8**, 135–136.
6. Heaney, R. P. (2007). Evidence-based medicine and common sense: practical and ethical issues in clinical trials for osteoporosis. *Future Rheumatol.* **2**, 104–110.
7. Roller, S. T., Voorhees, T., and Lunkenheimer, Jr., A. K. (2006). Obesity, food marketing and consumer litigation: threat or opportunity? *Food Drug Law J* **61**, 419–444.
8. Heaney, R. P. (2005). Vitamin D: Role in the calcium economy. *In* "Vitamin D," 2nd ed. (D. Feldman, F. H. Glorieux, and J. W. Pike, Eds.), pp. 773–787. Academic Press, San Diego.
9. Shea, B., Wells, G., Cranney, A., Zytaruk, N., Robinson, V., Griffith, L., Ortiz, Z., Peterson, J., Adachi, J., Tugwell, P., and Guyatt, G. (2002). VII: Meta-analysis of calcium supplementation for the prevention of postmenopausal osteoporosis. *Endocr. Rev.* **23**, 552–559.
10. Papadimitropoulos, E., Wells, G., Shea, B., Gillespie, W., Weaver, B., Zytaruk, N., Cranney, A., Adachi, J., Tugwell, P., Josse, R., Greenwood, C., and Guyatt, G. (2002). VIII: Meta-analysis of the efficacy of vitamin D treatment in preventing osteoporosis in postmenopausal women. *Endocr. Rev.* **23**, 560–569.
11. Cranney, A., Guyatt, G., Griffith, L., Wells, G., Tugwell, P., and Rosen, C. (2002). IX: Summary of meta-analyses of therapies for postmenopausal osteoporosis. *Endocr. Rev.* **23**, 570–578.
12. Dawson-Hughes, B., and Harris, S. S. (2002). Calcium intake influences the association of protein intake with rates of bone loss in elderly men and women. *Am. J. Clin. Nutr.* **75**, 773–779.
13. Heaney, R. P. (2007). Effects of protein on the calcium economy. *In* "Nutritional Aspects of Osteoporosis" (P. Burckhardt, B. Dawson-Hughes, and R. P. Heaney, Eds.), pp. 191–197. Elsevier, Amsterdam.
14. Sahota, O., Mundey, M. K., San, P., Godber, I. M., and Hosking, D. J. (2006). Vitamin D insufficiency and the blunted PTH response in established osteoporosis: the role of magnesium deficiency. *Osteoporos. Int.* **17**, 1013–1021.
15. Omenn, G. S., Goodman, G. E., Thornquist. M. D., Balmes, J., Cullen, M. R., Glass, A., Keogh, J. P., Meyskens, F. L., Valanis, B., Williams, J. H., Jr., Barnhart, S., and Hammar, S. (1996). Effects of a combination of beta carotene and vitamin A on lung cancer and cardiovascular disease. *N. Engl. J. Med.* **334**, 1150–1155.
16. Levine, R. J. (2003). Placebo controls in clinical trials of new therapies for osteoporosis. *J. Bone Miner. Res.* **18**, 1154–1159.
17. Levine, R. J., Hauth, J. C., Curet, L. B., Sibai, B. M., Catalano, P. M., Morris, C. D., DerSimonian, R., Esterlitz, J. R., Raymond, E. G., Bild, D. E., Clemens, J. D., and Cutler, J. A. (1997). Trial of calcium to prevent preeclampsia. *N. Engl. J. Med.* **337**, 69–76.
18. Drewnowski, A. (2005). Concept of a nutritious food: toward a nutrient density score. *Am. J. Clin. Nutr.* **82**, 721–732.
19. Heaney, R. P., and Rafferty, K. (2006). Assessing nutritional quality. *Am. J. Clin. Nutr.* **83**, 722–723.
20. Bischoff-Ferrari, H. A., Giovannucci, E., Willett, W. C., Dietrich, T., and Dawson-Hughes, B. (2006). Estimation of optimal serum concentrations of 25-hydroxyvitamin D for multiple health outcomes. *Am. J. Clin. Nutr.* **84**, 18–28.
21. www.vitamindcouncil.com/vdds.shtml. Accessed March 23, 2007.
22. Thaul, S., and Hotra D., Eds. (1993). An assessment of the NIH Women's Health Initiative. Washington, DC: National Academy of Science Press.
23. Feinstein, A. R. (1989). Epidemiologic analyses of causation: the unlearned scientific lessons of randomized trials. *J. Clin. Epidemiol.* **42**, 481–489.

24. Manson, J., Hsia, J., Johnson, K. C., Rossouw, J. E., Assaf, A. R. Lasser, N. L., Trevisan, M., Black, H. R., Heckbert, S. R., Detrano R., Strickland, O. L., Wong, N. D., Crouse, J. R., Stein, E., and Cushman, M. (2003). Estrogen plus progestin and the risk of coronary heart disease. *N. Engl. J. Med.* **349**, 523–534.
25. Rossouw, J. E., Prentice, R. L., Manson, J. E., Wu, L., Barad D., Barnaberi, V. M., Ko, M., LaCroix, A. Z., Margolis, K. L. and Stefanick, M. L. (2007). Postmenopausal hormone therapy and risk of cardiovascular disease by age and years since menopause. *J. Am. Med. Assoc.* **297**, 1465–1477.
26. Heaney, R. P., and Dougherty, C. J. (1981, 1986, 1998). "Research for Health Professionals: Design, Analysis, and Ethics." Creighton University, Varsity Press, 1981, 1986; Iowa State University Press, 1988.
27. Munger, K. L., Levin, L. I., Hollis, B. W., Howard, N. S., and Ascherio, A. (2006). Serum 25-hydroxyvitamin D levels and risk of multiple sclerosis. *JAMA* **296**, 2832–2838.
28. Munger, K. L., Zhang, S. M., O'Reilly, E., Hernan, M. A., Olek, M. J., Willett, W. C., and Ascherio, A. (2004). Vitamin D intake and incidence of multiple sclerosis. *Neurology* **62**, 60–65.

CHAPTER 15

Choline and Neural Development

MIHAI D. NICULESCU[1] AND STEVEN H. ZEISEL[2]
[1]*Department of Nutrition, University of North Carolina, Chapel Hill*
[2]*UNC Nutrition Research Institute at Kannapolis, North Carolina*

Contents

I. INTRODUCTION

An increasing amount of evidence supports the hypothesis that chronic illness in adult life may partially have its origins before birth [1]. In such cases prevention rather than treatment becomes the active principle in establishing long-term public health policies that will enable an overall improvement in the health status of the general population. Various nutrient deficiencies (docosahexaenoic acid, iron, protein/amino acids, energy restriction, folate, etc.), occurring during pregnancy or perinatally, are associated with defects in brain development that range from impaired physiological functions, such as decreased visual acuity, to severe birth defects [1–5].

The importance of adequate nutrition during brain development has been debated extensively for many years. A growing body of evidence indicates that the relationship between various nutrients and the development of the nervous system is complex and not necessarily confined to one specific period of gestation, but there is no doubt that specific nutrients play essential roles in neural development [6–15]. This chapter discusses the role choline has in brain development and the subsequent implications in the physiology of memory and brain aging.

II. CHOLINE METABOLISM AND BIOCHEMISTRY

A. Intestinal Absorption

Dietary free choline or choline-derived compounds are first metabolized in the intestine. Part of free choline is metabolized by the gut bacteria to betaine and methylamines [16], whereas the choline-derived compounds are hydrolyzed by enzymes from the pancreatic secretions and from intestinal mucosal cells, such as phospholipases A_1, A_2, and B [17]. The free choline is absorbed by the enterocytes via carrier-mediated transport [18, 19], whereas betaine is absorbed most probably by active Na^+ or Cl^- coupled, and passive Na^+-independent transport systems (reviewed in [20]), at a faster rate than choline [21, 22].

The bioavailability of choline-derived compounds is different in infants than in adults, as a consequence of differences in both the physiology of their digestive system [17] and in the choline content of the milk [23].

B. Transport and Tissue Uptake

After intestinal absorption, choline is transported to the liver via the portal circulation mainly as phosphatidylcholine [24], or it is incorporated into chylomicrons and released into the systemic circulation via the lymphatic system [25]. Choline accumulates in all tissues [19] by diffusion and mediated transport (reviewed in [26, 27]), via three distinct systems: low-affinity facilitated diffusion, high-affinity Na^+-dependent transport, and a Na^+-independent transport with intermediate affinity [26, 27]. Based on the affinity for choline, three types of transporters have been identified: cation transporters (OCTs) of low affinity, choline-transporter-like (CTLs) with intermediate affinity, and choline transporters (CHTs) with high affinity [26, 27]. In addition, phosphatidylcholine (PC) is trafficked via the ATP-binding cassette transporters (especially ABCA1 and ABCG1) [28–30]. Choline is transported across the blood-brain barrier by the CHT high-affinity system [31–33].

Choline is excreted in the primary urine by glomerular filtration, but only 2% of the filtered choline is found in the final urine because of intense reabsorption mainly in the proximal tubules [34] by an organic cation transport system [35].

C. Metabolism

Figure 1 presents a general overview of choline metabolism. Of special importance is the accumulation of choline by liver, kidney, brain, mammary gland, and placenta [19, 26].

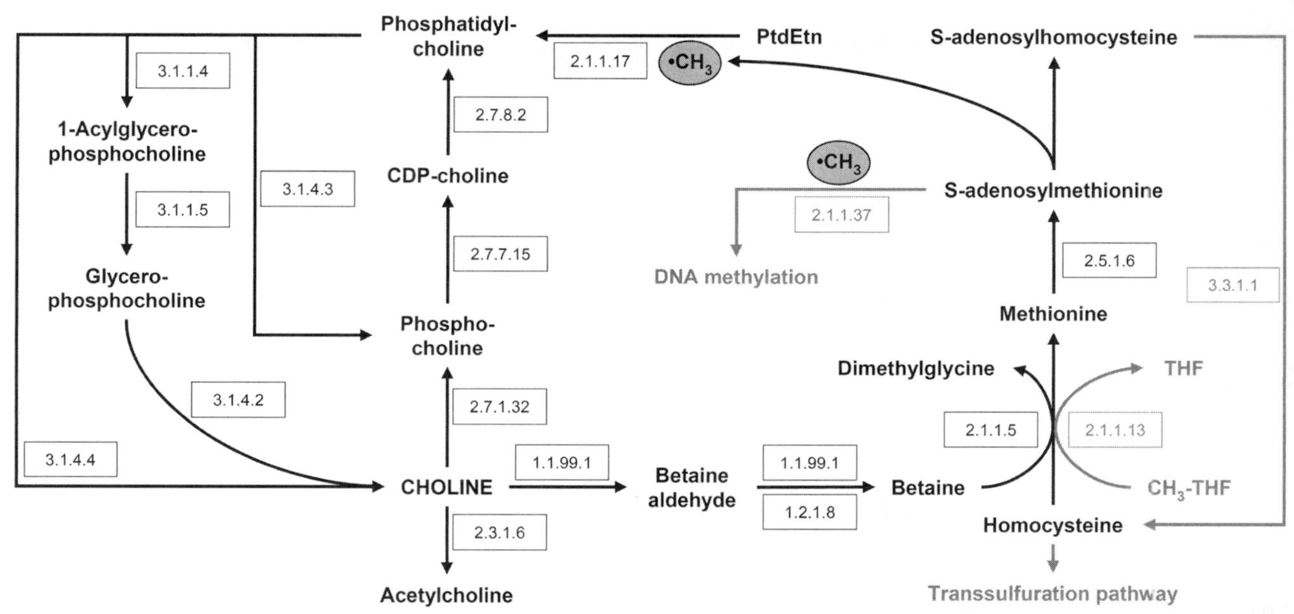

FIGURE 1 Choline metabolism. Enzymes are represented by EC numbers (1.1.99.1, choline dehydrogenase; 1.2.1.8, betaine-aldehyde dehydrogenase; 2.1.1.5, betaine-homocysteine *S*-methyltransferase; 2.1.1.13, methionine synthase; 2.1.1.17, phosphatidylethanolamine *N*-methyltransferase; 2.1.1.37, DNA (cytosine-5-)-methyltransferase; 2.3.1.6, choline *O*-acetyltransferase; 2.5.1.6, methionine adenosyltransferase; 2.7.1.32, choline kinase; 2.7.7.15, choline-phosphate cytidylyltransferase; 2.7.8.2, diacylglycerol cholinephosphotransferase; 3.1.1.4, phospholipase A_2; 3.1.1.5, lysophospholipase; 3.1.4.2, glycerophosphocholine phosphodiesterase; 3.1.4.3, phospholipase C; 3.1.4.4, phospholipase D; 3.3.1.1, adenosylhomocysteinase). THF, tetrahahydrofolate; PtdEtn, phosphatidylethanolamine.

Choline is involved in three major pathways: acetylcholine synthesis, methyl donation via betaine, and phosphatidylcholine synthesis; the latter two have a special importance in brain development [36, 37]. Phosphatidylcholine synthesis occurs by two independent pathways (Fig. 1). Choline is phosphorylated by choline kinase and converted subsequently to cytidine diphosphocholine (CDP-choline). In combination with diacylglycerol, CDP-choline forms phosphatidylcholine (reaction catalyzed by diacylglycerol cholinephosphotransferase). In an alternate pathway, choline is synthesized *de novo* by the methylation of phosphatidylethanolamine (PtdEtn) to phosphatidylcholine in a reaction catalyzed by phosphatidylethanolamine-*N*-methyltransferase (Pemt, EC 2.1.1.17) using *S*-adenosylmethionine as the methyl donor [38, 39]. This pathway, although most active in liver, has been identified in other tissues like fetal brain and mammary gland [40–42].

Choline is also involved in the one-carbon metabolism via its irreversible oxidation to betaine [43], which methylates homocysteine to form methionine, thus contributing to the *S*-adenosylmethionine synthesis and linking choline with folate metabolism (Fig. 1).

During pregnancy, choline is transported across the placenta by active mechanisms, against a concentration gradient [44]. The fetus is exposed to a high concentration of choline, and plasma choline concentration progressively declines after the first weeks of life [45]. In human newborns, plasma free choline can reach concentrations of around 70 μmol/L [46], whereas in adults, plasma free choline is much lower (7–20 μmol/L) [47]. The majority of choline in blood circulates as phosphatidylcholine (1–1.5 mmol/L) [47].

III. CHOLINE IN FOODS AND DIETARY REQUIREMENTS

A. Dietary Sources

Most of the foods we eat contain various amounts of free choline, choline esters, and betaine [48]. In 2004, the U.S. Department of Agriculture (USDA) released its first database on choline content in common foods (www.ars.usda.gov/Services/docs.htm?docid=6232). In most foods, choline and phosphatidylcholine are the most abundant compounds. The foods most abundant in choline are of animal origin, especially eggs and liver, but some vegetables have also significant amounts of free choline and phosphatidylcholine (Brussels sprouts, cauliflower, nuts, etc.). Cereals and many baked products contain high amounts of betaine as well.

In human breast milk, free choline and all main choline esters are abundant, with total choline levels around 1.5 mmol/L [49]. Manufacturers of infant formulas have modified the content of choline compounds to levels similar to human breast milk [23, 49].

B. Dietary Requirements

Although choline has not been initially considered an essential nutrient, because of the presence of its endogenous *de novo* synthesis from phosphatidylethanolamine [50], several human studies demonstrated that dietary choline is required (reviewed in [37]). In 1998, the U.S. Institute of Medicine (Food and Nutrition Board) established for the first time an adequate intake (AI) and tolerable upper intake limit (UL) values for choline, based on limited human studies [51] (Table 1). UL values range from 1000 mg/d in children to 3500 mg/d in adults [51]. However, for some age categories for which adequate data were missing, AI values have been set by extrapolating from adult values (for ages 1 through 18 years) and for infants (for ages 7 through 12 months) [51]. Although a previous controlled study has indicated that choline intakes may be adequate [52], a later epidemiological study using participants from the Framingham Offspring Study indicated that the mean intake for total choline (energy adjusted) is below the AI values, with a mean intake of 313 mg/d; moreover, there was an inverse association between choline intake and plasma total homocysteine concentration in subjects with low folate intakes [53]. This later study strengthens previous similar findings in pregnant women, where a significant percentage of individuals had lower choline intakes than the recommended AI values [54, 55].

C. Consequences of Dietary Choline Deficiency in Humans

Studies performed in humans have categorized the effects induced by modulating dietary choline in three groups: (1) changes in acetylcholine synthesis and release in brain, (2) organ dysfunction/metabolic changes in adults eating a low choline diet for a relatively short time, and (3) birth defects (abnormal neural tube closure, cleft palate) in newborns from mothers eating diets lower in choline content as well as effects of choline on brain progenitor cell proliferation and survival in rodents.

Historically, the first reports on the effects induced by changing dietary choline focused on acetylcholine release in human brain and the consequences this had on memory, small motor movements, and the release of other neurotransmitters, like γ-aminobutyric acid (reviewed in [56]). In this context, the effectiveness of choline supplementation in improving the acetylcholine release was, for most of the studies, limited to individuals with neurological disorders like tardive dyskinesia and Alzheimer's disease [57]. Using rodent models, dietary choline has been shown to modulate the density of the benzodiazepine receptors and the function of the GABA-ergic receptors in the cortex [56], but whether these findings can be applied to healthy humans is unclear.

Human dietary requirements for choline have been studied in subjects fed low choline diets under controlled conditions; they develop reversible fatty liver as well as liver and muscle damage [58, 59]. These clinical outcomes are associated with increased apoptosis in lymphocytes [60]. Premenopausal women were less likely to develop organ dysfunction when fed a low-choline diet than were men or postmenopausal women, because estrogen induces the gene that makes endogenous synthesis of choline possible (phosphatidylethanolamine *N*-methyltransferase gene [PEMT]). Interestingly, the risk of developing such clinical signs is associated with the presence of several polymorphisms within genes involved in folate and choline metabolism, such as the *PEMT* gene, choline dehydrogenase gene (*CHDH*), and 5,10-methylenetetrahydrofolate dehydrogenase (*MTHFD1*) gene [61, 62], suggesting that dietary choline requirements may differ with genotype, gender, and estrogen status [61, 62].

The third category of effects of manipulating dietary choline intake all relate to critical periods when choline is needed for fetal and infant development.

TABLE 1 Adequate Intakes for Choline

Group	Age	AI
Infants and children	0–6 mo	125 mg/d
	7–12 mo	150 mg/d
	1–3 y	200 mg/d
	4–8 y	250 mg/d
	9–13 y	375 mg/d
Boys	14–18 y	550 mg/d
Girls	14–18 y	400 mg/d
Pregnant women	All ages	450 mg/d
Lactating women	All ages	550 mg/d
Other men		550 mg/d
Other women		425 mg/d

Source: Adapted from Institute of Medicine and National Academy of Sciences USA. (1998). Choline. In "Dietary Reference Intakes for Folate, Thiamin, Riboflavin, Niacin, Vitamin B12, Pantothenic Acid, Biotin, and Choline," pp. 390–422. National Academy Press, Washington, DC.

IV. CHOLINE AND NEURAL DEVELOPMENT

Dietary choline is essential in at least two distinct stages of brain development: neurulation (when the neural tube forms and subsequently closes) and a later stage (late pregnancy) during the maturation of hippocampus and other brain areas. Although the mechanisms responsible for these outcomes have been partially elucidated in animal models, confirmation in human studies is available only for early pregnancy at present.

A. Choline Deficiency in Early Pregnancy

When day 9 gestation mouse embryos were exposed *in vitro* to either an inhibitor of choline uptake (2-dimethylaminoethanol, DMAE) or to an inhibitor of phosphatidylcholine synthesis (1-O-octadecyl-2-O-methyl-rac-glycero-3-phosphocholine, ET-18-OCH[3]), they developed craniofacial hypoplasia and open neural tube defects in the forebrain, midbrain, and hindbrain regions [63], alterations that are similar to those induced by folate deficiency [64]. Increased cell death was associated especially with the inhibition of phosphatidylcholine synthesis [63]. A subsequent study using the same mouse embryo model revealed that the mechanisms responsible for these outcomes were related to alterations in choline metabolism pathways [65], stressing the importance of dietary choline. These two animal studies were enhanced by data from two human epidemiological studies describing similar outcomes in newborns from mothers who ate lower choline diets during pregnancy. In these studies, Shaw *et al.* reported that maternal dietary choline and betaine intakes during pregnancy inversely correlated with the risk of having a baby with neural tube defects and orofacial clefts [54, 55]. Women in the lowest quartile of total choline intake during the periconceptional period (≤290.41 mg/d) and eating diets low in folate content had a fourfold higher risk of having a baby with neural tube defects (spina bifida and anencephaly) than the women in the highest quartile for choline intake and a low folate diet [55]. A similar relationship was also found between maternal choline intake and the increased risk of cleft lip with cleft palate [54].

Independent of the maternal choline intake, the presence in infants of single nucleotide polymorphisms of two genes involved in choline metabolism choline kinase A (*CHKA*) and CTP:phosphocholine cytidylyltransferase1 (*PCYT1A*) increased the risk of spina bifida in babies, regardless the maternal choline intake [66]. Altogether, these data suggest that the risk for neural tube defects is related to the maternal dietary choline and folate intakes and to the presence in the offspring of genetic polymorphisms in genes related to the metabolism of these two nutrients.

B. Choline Deficiency in Late Pregnancy

During later gestation, rat and mouse models have been used to explore the role of choline in fetal brain development.

1. Choline Deficiency Inhibits Cell Proliferation

Rats and mice fed a low choline diet in late pregnancy (gestational days 12 to 17 in mice, days 12 to 18 or 20 in rats) had reduced neural progenitor cell proliferation and increased apoptosis in fetal hippocampus and cortex [67–69]. Similar outcomes were reported when pregnant mice are fed a low-folate diet [70], suggesting, again, potential synergistic mechanisms of action between folate and choline. These alterations were confirmed in cell culture studies using primary neurons, pheochromocytoma cells, and human neuroblastoma cells. Dividing neuron-like cells (PC12) exposed to a choline-deficient medium had decreased cell division and increased apoptosis [69] and had diminished concentrations of phosphatidylcholine in their membranes [71]. The apoptosis that is induced is caspase 3-dependent in both cell culture and fetal brain models [68, 71].

These changes are associated with important alterations of proteins involved in cell signaling, neuronal differentiation, and in the regulation of cell-cycle progression. Choline deficiency altered the expression of structural and signaling proteins like TGFβ1, vimentin, and MAP1 in rat hippocampus [72]. Decreased cell proliferation was also associated with increased expression of neuronal and glial differentiation marker proteins like vimentin, TOAD-64 (dihydropyriminidase-like 2, Dpysl2), and calretinin [67, 72, 73]. Moreover, choline deficiency altered the protein expression of netrin and DCC, two proteins required for axonal growth and guidance in the developing nervous system [74]. These data suggest that choline deficiency induces a net trend toward the differentiation of neural progenitors and a subsequent reduction of the pool of the available precursor cells. Some differences persist for the lifetime of the offspring of treated mothers; prenatal choline supplementation decreased calretinin protein levels in the adult (24-month) rat hippocampus [73].

The mechanisms associating choline deficiency with decreased cell proliferation are, in part, related to the overexpression of cyclin-dependent kinase inhibitors (Cdkn) like p27Kip1 [75], p15Ink4b [75, 76], and Cdkn3 [76, 77], suggesting that choline deficiency inhibits cell proliferation by inducing G_1 arrest because of the inhibition of the interaction between cyclin-dependent kinases and cyclins. Moreover, in human neuroblastoma cells (IMR-32), choline deficiency decreased the phosphorylation of the retinoblastoma protein (p110, Rb) [77]. This interaction model between p27Kip1 and p15Ink4b cyclin-dependent kinase inhibitors, TGFβ, and the Rb proteins, fits the previously described model of cell-cycle regulation (reviewed in [78]), where the net outcome is cell-cycle arrest in the G_1 phase of the cell cycle. These findings were reinforced by a study using mouse hippocampal and cortical progenitor cells exposed to choline deficiency for 48 hours. Using oligonucleotide arrays, the authors reported extensive changes in more than a thousand genes, of which 331 were related to cell division, apoptosis, neuronal and glial differentiation, methyl metabolism, and calcium-binding protein ontology classes [79], where, again, the net result was toward reduced cell proliferation, increased apoptosis, and increased differentiation.

2. Choline Deficiency Alters Gene Expression via Epigenetic Mechanisms

a. DNA Methylation Regulates Gene Expression

Methylation of DNA can modify gene expression and is a mechanism for genomic imprinting [80]. It is one of several epigenetic mechanisms that regulate gene expression. In mammalian cells, virtually all methylation occurs at the position 5 of the carbon ring of cytosine (C), within cytosine-guanine (CpG) sites, such that 60% to 90% of these sites are methylated, and this corresponds to 3% to 8% of all cytosine residues [80]. DNA methylation is catalyzed by DNA-methyl transferases (DNMTs) using methyl groups donated by *S*-adenosylmethionine (Fig. 1) [80].

DNA methylation is not a random process. Many promoter regions have a higher incidence of CpG repeats than expected (called CpG islands), and these islands are the main targets for changes in methylation [81]. Such CpG islands are defined as DNA sequences with a C + G content of more than 55%, and CpG observed/expected ratio of more than 0.65 [82]. Three biologically active DNA-methyltransferases (DNMT1, DNMT3a, and DNMT3b) are capable of adding methyl groups to the cytosine in CpG dinucleotides, and all are essential for embryonic development [83]. DNMT3a and 3b are *de novo* methylases, whereas DNMT1 is a maintenance methylase with preferential activity for hemi-methylated DNA [84, 85].

If methylation occurs at a CpG island in the promoter region of a gene, then expression is usually repressed [86] because methylated CpG sites prevent the binding of transcription factors [87]. Methylated CpG sites bind to methyl-CpG binding domain proteins (MBDs and MeCP2) and recruit histone deacetylases, and the resulting deacetylated histones induce chromatin compaction and gene silencing [88].

b. DNA Methylation, Fetal Development, and Cell Differentiation

DNA methylation is very important during embryogenesis and late fetal development. Although the original DNA methylation pattern is germline-specific (sperm DNA is hypermethylated compared to oocyte DNA), almost immediately after fertilization (within one to two cell divisions) there is a dramatic erasure of methylation, which continues until blastocyst implantation [89]. Following implantation, mouse embryonic stem cells (ES) are subjected to *de novo* methylation catalyzed by Dnmt3a and Dnmt3b genes (*de novo* methylases), with the exception of certain tissue-specific genes that remain unmethylated [90]. Once established, the new methylation pattern is conserved during cell replication (catalyzed by Dnmt1 in the S-phase of the cell division [90]). These methylation patterns can be changed as cells differentiate (cell differentiation is associated with a genome-wide DNA hypo-methylation followed by remethylation [91–93]) and can be altered by dietary intake of methyl donors (discussed next).

c. DNA Methylation and Neural Development

Neural development is influenced by DNA methylation. Overall levels of methylation decrease as neuronal differentiation proceeds [94] and the treatment of neural progenitor cells with demethylating agents induces them to differentiate into cholinergic and adrenergic neurons [95]. These methylation patterns are cell-type specific: Whereas mature neurons express DNA methyltransferases, these genes have much lower levels of expression in oligodendrocytes and astrocytes in the white matter [96]. The expression of DNA methyltransferases has a different importance, based on the differentiation stage of the cell: Deletion of Dnmt1 in postmitotic neurons does not affect overall levels of DNA methylation, whereas the same deletion in neural progenitors markedly decreases methylation levels and causes severe defects in neurogenesis [97]. Astrocyte differentiation is also dependent on the methylation status of glial fibrillary acidic protein (Gfap) promoter at the binding site of STAT3 transcription factor. This promoter site becomes hypo-methylated before differentiation, and its methylation prevents gene activation by STAT3 and astrocyte differentiation [98].

d. Choline Deficiency Alters DNA Methylation in Fetal Brain

Although the relationship between nutrition and epigenetics has been firmly established [99], less is known about the role nutrition has in the epigenetic regulation of fetal brain development. However, available data allow us to identify an important role for choline in DNA methylation during brain development [100]. Global DNA methylation is decreased in the neuroepithelial layer of the hippocampus in choline-deficient mouse fetal brains [76], whereas opposite effects were reported in the fetal brains from Pemt −/− mice, which also had increased *S*-adenosylmethionine levels [101]. Interestingly, Pemt −/− mice also had altered methylation of the lysine 4 and 9 within histone 3, suggesting that alterations in the *S*-adenosylmethionine availability in the fetal brain may be crucial for DNA and histone methylation [101]. Along with decreased global methylation, changes in gene-specific methylation were reported, where a cyclin-dependent kinase (*Cdkn3*) was hypomethylated in its promoter by choline deficiency [76, 77] in the progenitor layer of the hippocampus and in human neuroblastoma cells. These alterations were associated with increased protein expression of this cyclin-dependent kinase inhibitor [76], and this model is consistent with previous findings regarding the epigenetic regulation of cyclin-dependent kinase inhibitors and their roles in cell proliferation [102].

Because dietary choline is an important player in the maintenance of the *S*-adenosylmethionine pool (the methyl donor for DNA methylation), along with folate and methionine (Fig. 1), it is attractive to hypothesize that choline influences the epigenetic status of the developing brain

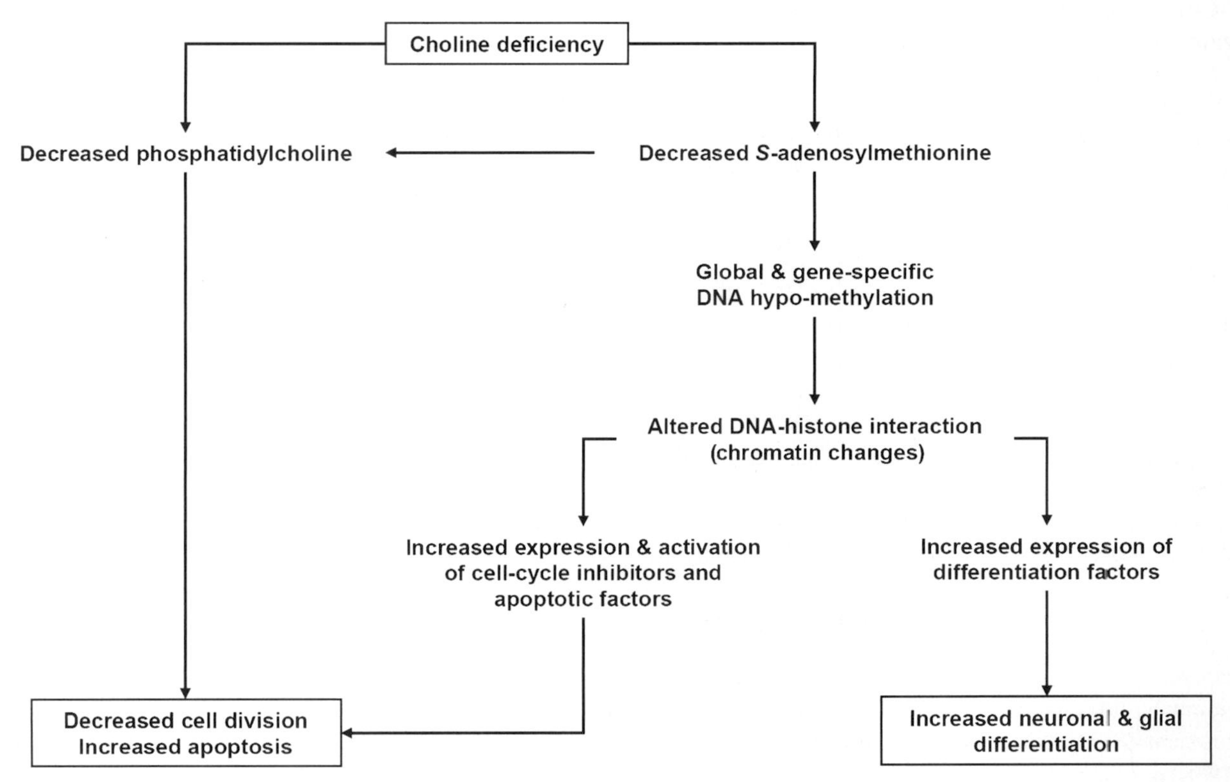

FIGURE 2 Choline deficiency alters brain development.

and thereby permanently alters brain function [36]. Figure 2 summarizes the hypothesized role that choline may have in the epigenetic regulation of brain development.

V. LONG-LASTING CONSEQUENCES OF PRENATAL CHOLINE AVAILABILITY

A. Molecular and Functional Changes

The changes induced by dietary choline in fetal brain have long-lasting effects that alter the neuronal function throughout the adult life. When pregnant rats were choline supplemented during late gestation, basal and receptor-stimulated phospholipase D activity was up-regulated in the hippocampus of the offspring during postnatal development [103]. Acetylcholine metabolism and choline uptake mechanisms were also permanently altered in the adult brain. Choline acetyltransferase (ChAT) and acetylcholinesterase (AChE) activities were increased in the adult hippocampus of rats exposed to choline deficiency while *in utero*, and choline incorporation into acetylcholine was more dependent on high-affinity choline uptake (HACU) mechanisms, compared to controls or to rats that were choline-supplemented while *in utero* [104]. The increase in AChE activity was later found to be due to increased AChE protein synthesis [105], strongly suggesting that gene expression or posttranscriptional regulation was permanently altered by choline deficiency. This hypothesis was confirmed recently in a study showing that the gene expression of the choline transporter Cht was increased in the adult hippocampus from rats exposed to prenatal choline deficiency, and that this correlated with an increased number of neurons that were CHT-immunoreactive [106]. In addition, many other changes in gene expression were described to occur in the adult hippocampus and cortex, initiated by prenatal choline manipulation [107].

Cell signaling is also influenced by prenatal choline availability. In juvenile rats, the phosphorylation and activation of hippocampal mitogen-activated protein kinase (MAPK) and cAMP-response element binding protein (CREB) in response to glutamate, *N*-methyl-D-aspartate stimulation or to depolarizing concentrations of K^+ were increased by choline supplementation and reduced by choline deficiency while *in utero* [108]. Choline supplementation while *in utero* also increased the levels of nerve growth factor (NGF) in the adult rat hippocampus and cortex [109], suggesting that prenatal choline availability had an important role in promoting neurogenesis in the adult hippocampus, mediated by the nootropic action of NGF [110]. Opposite changes were reported for choline deficiency while *in utero* in other areas of the forebrain, such as medial septal nucleus, nucleus of the diagonal band, and the nucleus basalis of Meynert [111].

Although confirmation in humans is needed, these findings provide evidence that prenatal choline availability initiates a pattern of permanent metabolic alterations (metabolic imprinting) that, once established, plays an important role in later life [112].

B. Choline Deficiency Induces Cognitive and Memory Deficits

The functional and molecular changes described earlier are, at least in part, responsible for behavioral and memory changes initiated by prenatal variation in the availability of choline to developing brain (reviewed in [100]). Prenatal choline supplementation protects against the neurotoxicity induced by the administration of the NMDA receptor antagonist dizocilpine (MK-801) to female adolescent rats [113, 114]. When status epilepticus is induced in adult rats using kainic acid, rats receiving supplemental choline between gestational day 11 and postnatal day 7, performed better in the water maze tests than did the deficient and control groups, while the activity of hippocampal ChAT activity was 18% lower in the choline-deficient animals compared with the other two groups [115].

Maternal dietary choline availability during late pregnancy was associated with long-lasting changes in the hippocampal function of the adult offspring. Choline supplementation during this period enhanced visuospatial and auditory memory in the adult rats throughout their life span [116–120]. It also enhanced a property of the hippocampus, long-term potentiation (LTP) [121–123]. The offspring from mothers fed a choline-deficient diet manifested opposite outcomes [117, 121].

VI. IMPLICATIONS FOR HUMAN BRAIN DEVELOPMENT

It is always difficult to extrapolate to humans the findings reported using animal models. However, limited data are available to support the hypothesis that similar mechanisms are involved in humans. Because of ethical constraints, no studies are available in children or pregnant mothers to validate the rodent model. Because pregnant women are at risk of becoming choline deficient [45], and thus having increased risk of giving birth to babies with neural tube defects [55], the recommendation that pregnant women should attempt to consume diets adequate in choline content seems reasonable. In addition, because half of the population has gene polymorphisms that affect choline and folate metabolisms [61, 62], it is likely that different individuals may have different dietary requirements for choline and may need to pay special attention to choline intake during pregnancy.

Acknowledgments

This work was funded by grants from the National Institutes of Health (DK55865, AG09525, ES012997) and the U.S. Department of Agriculture (2004-01833). Support for this work was also provided by grants from the National Institutes of Health to the University of North Carolina Clinical Nutrition Research Unit (DK56350), the University of North Carolina General Clinical Research Center (RR00046), and the Center for Environmental Health and Susceptibility (ES10126).

References

1. Gordon, N. (1997). Nutrition and cognitive function. *Brain Dev.* **19**, 165–170.
2. Guesry, P. (1998). The role of nutrition in brain development. *Prev. Med.* **27**, 189–194.
3. Mattson, M. P. (2003). Gene-diet interactions in brain aging and neurodegenerative disorders. *Ann. Intern. Med.* **139**, 441–444.
4. Mattson, M. P., and Shea, T. B. (2003). Folate and homocysteine metabolism in neural plasticity and neurodegenerative disorders. *Trends Neurosci.* **26**, 137–146.
5. McNamara, R. K., and Carlson, S. E. (2006). Role of omega-3 fatty acids in brain development and function: potential implications for the pathogenesis and prevention of psychopathology. *Prostaglandins Leukot Essent. Fatty Acids* **75**, 329–349.
6. Georgieff, M. K. (2007). Nutrition and the developing brain: nutrient priorities and measurement. *Am. J. Clin. Nutr.* **85**, 614S–620S.
7. Bourre, J. M. (2006). Effects of nutrients (in food) on the structure and function of the nervous system: update on dietary requirements for brain. Part 2 : macronutrients. *J. Nutr. Health Aging* **10**, 386–399.
8. Bourre, J. M. (2006). Effects of nutrients (in food) on the structure and function of the nervous system: update on dietary requirements for brain. Part 1: micronutrients. *J. Nutr. Health Aging* **10**, 377–385.
9. Beard, J. (2003). Iron deficiency alters brain development and functioning. *J. Nutr.* **133**, 1468S–1472S.
10. Beard, J. (2007). Recent evidence from human and animal studies regarding iron status and infant development. *J. Nutr.* **137**, 524S–530S.
11. Golub, M. S., Keen, C. L., Gershwin, M. E., Styne, D. M., Takeuchi, P. T., Ontell, F., Walter, R. M., and Hendrickx, A. G. (1996). Adolescent growth and maturation in zinc-deprived rhesus monkeys [see comment]. *Am. J. Clin. Nutr.* **64**, 274–282.
12. Golub, M. S., Takeuchi, P. T., Keen, C. L., Hendrickx, A. G., and Gershwin, M. E. (1996). Activity and attention in zinc-deprived adolescent monkeys. *Am.. J. Clin. Nutr.* **64**, 908–915.
13. Tian, D., Su, M., Song, T., Li, G., and Xu, X. (2002). Studies on survival and outgrowth of processes of cultured rat hippocampus neurons in containing selenium and free serum medium. *Zhonghua yu fang yi xue za zhi [Chinese journal of preventive medicine].* **36**, 167–171.

14. Tian, D., Su, M., Xu, X., Wu, X., Li, Q., and Zheng, R. (2002). Effects of selenium and iodine on c-fos and c-jun mRNA and their protein expressions in cultured rat hippocampus cells. *Biol. Trace Elem. Res.* **90**, 175–186.
15. Shaw, G. M., Schaffer, D., Velie, E. M., Morland, K., and Harris, J. A. (1995). Periconceptional vitamin use, dietary folate, and the occurrence of neural tube defects. *Epidemiology* **6**, 219–226.
16. Zeisel, S. H., Wishnok, J. S., and Blusztajn, J. K. (1983). Formation of methylamines from ingested choline and lecithin. *J. Pharmacol. Exp. Ther.* **225**, 320–324.
17. Jones, P. J. H., and Kubow, S. (2006). Lipids, sterols, and their metabolites. *In* "Modern Nutrition in Health and Disease" (M. E. Shils, M. Shike, A. C. Ross, B. Caballero, and R. J. Cousins, Eds.), pp. 92–122. Lippincott Williams & Wilkins, Philadelphia.
18. Kamath, A. V., Darling, I. M., and Morris, M. E. (2003). Choline uptake in human intestinal Caco-2 cells is carrier-mediated. *J. Nutr.* **133**, 2607–2611.
19. Zeisel, S. H., and Blusztajn, J. K. (1994). Choline and human nutrition. *Annu. Rev. Nutr.* **14**, 269–296.
20. Craig, S. A. S. (2004). Betaine in human nutrition. *Am. J. Clin. Nutr.* **80**, 539–549.
21. Kettunen, H., Peuranen, S., Tiihonen, K., and Saarinen, M. (2001). Intestinal uptake of betaine in vitro and the distribution of methyl groups from betaine, choline, and methionine in the body of broiler chicks. *Comp. Biochem. Physiol.—Part A: Molecular & Integrative Physiology* **128**, 269–278.
22. Schwab, U., Torronen, A., Meririnne, E., Saarinen, M., Alfthan, G., Aro, A., and Uusitupa, M. (2006). Orally administered betaine has an acute and dose-dependent effect on serum betaine and plasma homocysteine concentrations in healthy humans. *J. Nutr.* **136**, 34–38.
23. Holmes-McNary, M. Q., Cheng, W. L., Mar, M. H., Fussell, S., and Zeisel, S. H. (1996). Choline and choline esters in human and rat milk and in infant formulas. *Am. J. Clin. Nutr.* **64**, 572–576.
24. Savendahl, L., Mar, M. H., Underwood, L. E., and Zeisel, S. H. (1997). Prolonged fasting in humans results in diminished plasma choline concentrations but does not cause liver dysfunction. *Am. J. Clin. Nutr.* **66**, 622–625.
25. Schlierf, C., Falor, W. H., Wood, P. D., Lee, Y.-L., and Kinsell, L. W. (1969). Composition of human chyle chylomicrons following single fat feedings. *Am. J. Clin. Nutr.* **22**, 79–86.
26. Lockman, P. R., and Allen, D. D. (2002). The transport of choline. *Drug Dev. Ind. Pharm.* **28**, 749–771.
27. Michel, V., Yuan, Z., Ramsubir, S., and Bakovic, M. (2006). Choline transport for phospholipid synthesis. *Exp. Biol. Med. (Maywood)* **231**, 490–504.
28. Kobayashi, A., Takanezawa, Y., Hirata, T., Shimizu, Y., Misasa, K., Kioka, N., Arai, H., Ueda, K., and Matsuo, M. (2006). Efflux of sphingomyelin, cholesterol, and phosphatidylcholine by ABCG1. *J. Lipid Res.* **47**, 1791–1802.
29. Schmitz, G., and Langmann, T. (2001). Structure, function and regulation of the ABC1 gene product. *Curr. Opin. Lipidol.* **12**, 129–140.
30. Schmitz, G., Langmann, T., and Heimerl, S. (2001). Role of ABCG1 and other ABCG family members in lipid metabolism. *J. Lipid Res.* **42**, 1513–1520.
31. Allen, D. D., and Lockman, P. R. (2003). The blood-brain barrier choline transporter as a brain drug delivery vector. *Life Sci.* **73**, 1609–1615.
32. Allen, D. D., Lockman, P. R., Roder, K. E., Dwoskin, L. P., and Crooks, P. A. (2003). Active transport of high-affinity choline and nicotine analogs into the central nervous system by the blood-brain barrier choline transporter. *J. Pharmacol. Exp. Ther.* **304**, 1268–1274.
33. Lockman, P. R., McAfee, J. H., Geldenhuys, W. J., and Allen, D. D. (2004). Cation transport specificity at the blood-brain barrier. *Neurochem Res.* **29**, 2245–2250.
34. Acara, M., Roch-Ramel, F., and Rennick, B. (1979). Bidirectional renal tubular transport of free choline: a micropuncture study. *Am. J. Physiol. Renal Physiol.* **236**, F112–F118.
35. Pietruck, F., Horbelt, M., Feldkamp, T., Engeln, K., Herget-Rosenthal, S., Philipp, T., and Kribben, A. (2006). Digital fluorescence imaging of organic cation transport in freshly isolated rat proximal tubules. *Drug Metab. Dispos.* **34**, 339–342.
36. Zeisel, S. H., and Niculescu, M. D. (2006). Perinatal choline influences brain structure and function *Nutr. Rev.* **64**, 197–203.
37. Zeisel, S. H., and Niculescu, M. D. (2006). Choline and phosphatidylcholine. *In* "Modern Nutrition in Health and Disease" (M. E. Shils, M. Shike, A. C. Ross, B. Caballero, and R. J. Cousins, Eds.), pp. 525–536. Lippincott Williams & Wilkins, Philadelphia.
38. Blusztajn, J. K., Zeisel, S. H., and Wurtman, R. J. (1979). Synthesis of lecithin (phosphatidylcholine) from phosphatidylethanolamine in bovine brain. *Brain Res.* **179**, 319–327.
39. Kaneshiro, T., and Law, J. H. (1964). Phosphatidylcholine synthesis in agrobacterium tumefaciens. I. Purification and properties of a phosphatidylethanolamine N-methyltransferase. *J. Biol. Chem.* **239**, 1705–1713.
40. Blusztajn, J. K., Zeisel, S. H., and Wurtman, R. J. (1985). Developmental changes in the activity of phosphatidylethanolamine N-methyltransferases in rat brain. *Biochem. J.* **232**, 505–511.
41. Vance, J. E., Stone, S. J., and Faust, J. R. (1997). Abnormalities in mitochondria-associated membranes and phospholipid biosynthetic enzymes in the mnd/mnd mouse model of neuronal ceroid lipofuscinosis. *Biochim. Biophys. Acta.* **1344**, 286–299.
42. Yang, E. K., Blusztajn, J. K., Pomfret, E. A., and Zeisel, S. H. (1988). Rat and human mammary tissue can synthesize choline moiety via the methylation of phosphatidylethanolamine. *Biochem. J.* **256**, 821–828.
43. Niculescu, M. D., and Zeisel, S. H. (2002). Diet, methyl donors and DNA methylation: interactions between dietary folate, methionine and choline. *J. Nutr.* **132**, 2333S–2335S.
44. Sweiry, J. H., Page, K. R., Dacke, C. G., Abramovich, D. R., and Yudilevich, D. L. (1986). Evidence of saturable uptake mechanisms at maternal and fetal sides of the perfused human placenta by rapid paired-tracer dilution: studies with calcium and choline. *J. Dev. Physiol.* **8**, 435–445.
45. McMahon, K. E., and Farrell, P. M. (1985). Measurement of free choline concentrations in maternal and neonatal blood by micropyrolysis gas chromatography. *Clinica Chimica Acta; Intl. J. Clin. Chem.* **149**, 1–12.

46. Buchman, A. L., Sohel, M., Moukarzel, A., Bryant, D., Schanler, R., Awal, M., Burns, P., Dorman, K., Belfort, M., and Jenden, D. J. (2001). Plasma choline in normal newborns, infants, toddlers, and in very-low-birth-weight neonates requiring total parenteral nutrition. *Nutrition* **17**, 18–21.
47. Zeisel, S. H. (2000). Choline: an essential nutrient for humans. *Nutrition* **16**, 669–671.
48. Zeisel, S. H., Mar, M.-H., Howe, J. C., and Holden, J. M. (2003). Concentrations of choline-containing compounds and betaine in common foods. *J. Nutr.* **133**, 1302–1307.
49. Ilcol, Y. O., Ozbek, R., Hamurtekin, E., and Ulus, I. H. (2005). Choline status in newborns, infants, children, breast-feeding women, breast-fed infants and human breast milk. *J. Nutr. Biochem.* **16**, 489–499.
50. Bremer, J., and Greenberg, D. M. (1960). Biosynthesis of choline in vitro. *Biochim. Biophys. Acta* **37**, 173–175.
51. Institute of Medicine and National Academy of Sciences USA (1998). Choline. *In* "Dietary Reference Intakes for Folate Thiamin, Riboflavin, Niacin, Vitamin B_{12} Panthothenic Acid Biotin, and Choline," pp. 390–422. National Academy Press Washington, DC.
52. Fischer, L. M., Scearce, J. A., Mar, M. H., Patel, J. R. Blanchard, R. T., Macintosh, B. A., Busby, M. G., and Zeisel S. H. (2005). Ad libitum choline intake in healthy individuals meets or exceeds the proposed adequate intake level. *J. Nutr* **135**, 826–829.
53. Cho, E., Zeisel, S. H., Jacques, P., Selhub, J., Dougherty, L. Colditz, G. A., and Willett, W. C. (2006). Dietary choline and betaine assessed by food-frequency questionnaire in relation to plasma total homocysteine concentration in the Framingham Offspring Study. *Am. J. Clin. Nutr.* **83**, 905–911.
54. Shaw, G. M., Carmichael, S. L., Laurent, C., and Rasmussen, S. A. (2006). Maternal nutrient intakes and risk of orofacial clefts. *Epidemiology* **17**, 285–291.
55. Shaw, G. M., Carmichael, S. L., Yang, W., Selvin, S., and Schaffer, D. M. (2004). Periconceptional dietary intake of choline and betaine and neural tube defects in offspring. *Am. J. Epidemiol.* **160**, 102–109.
56. Miller, L. G. (1990). Dietary choline alteration. Implications for gamma-aminobutyric acid and other neurotransmitter receptors. *Biochem. Pharmacol.* **40**, 1179–1182.
57. Wurtman, R. J., and Growdon, J. H. (1978). Dietary enhancement of CNS neurotransmitters. *Hosp. Pract.* **13**, 71–77.
58. da Costa, K. A., Badea, M., Fischer, L. M., and Zeisel, S. H. (2004). Elevated serum creatine phosphokinase in choline-deficient humans: mechanistic studies in C2C12 mouse myoblasts. *Am. J. Clin. Nutr.* **80**, 163–170.
59. da Costa, K. A., Gaffney, C. E., Fischer, L. M., and Zeisel, S. H. (2005). Choline deficiency in mice and humans is associated with increased plasma homocysteine concentration after a methionine load. *Am. J. Clin Nutr.* **81**, 440–444.
60. da Costa, K. A., Niculescu, M. D., Craciunescu, C. N., Fischer, L. M., and Zeisel, S. H. (2006). Choline deficiency increases lymphocyte apoptosis and DNA damage in humans. *Am. J. Clin. Nutr.* **84**, 88–94.
61. da Costa, K. A., Kozyreva, O. G., Song, J., Galanko, J. A., Fischer, L. M., and Zeisel, S. H. (2006). Common genetic polymorphisms affect the human requirement for the nutrient choline. *FASEB J.* **20**, 1336–1344.
62. Kohlmeier, M., da Costa, K. A., Fischer, L. M., and Zeisel, S. H. (2005). Genetic variation of folate-mediated one-carbon transfer pathway predicts susceptibility to choline deficiency in humans. *Proc. Natl. Acad. Sci. USA* **102**, 16025–16030.
63. Fisher, M. C., Zeisel, S. H., Mar, M. H., and Sadler, T. W. (2001). Inhibitors of choline uptake and metabolism cause developmental abnormalities in neurulating mouse embryos. *Teratology* **64**, 114–122.
64. Antony, A. C. (2007). In utero physiology: role of folic acid in nutrient delivery and fetal development. *Am. J. Clin. Nutr.* **85**, 598S–603S.
65. Fisher, M. C., Zeisel, S. H., Mar, M. H., and Sadler, T. W. (2002). Perturbations in choline metabolism cause neural tube defects in mouse embryos in vitro. *FASEB J.* **16**, 619–621.
66. Enaw, J. O., Zhu, H., Yang, W., Lu, W., Shaw, G. M., Lammer, E. J., and Finnell, R. H. (2006). CHKA and PCYT1A gene polymorphisms, choline intake and spina bifida risk in a California population. *BMC Med.* **4**, p.36.
67. Albright, C. D., Friedrich, C. B., Brown, E. C., Mar, M. H., and Zeisel, S. H. (1999). Maternal dietary choline availability alters mitosis, apoptosis and the localization of TOAD-64 protein in the developing fetal rat septum. *Brain Res. Dev. Brain Res.* **115**, 123–129.
68. Craciunescu, C. N., Albright, C. D., Mar, M. H., Song, J., and Zeisel, S. H. (2003). Choline availability during embryonic development alters progenitor cell mitosis in developing mouse hippocampus. *J. Nutr.* **133**, 3614–3618.
69. Holmes-McNary, M. Q., Loy, R., Mar, M. H., Albright, C. D., and Zeisel, S. H. (1997). Apoptosis is induced by choline deficiency in fetal brain and in PC12 cells. *Brain Res. Dev. Brain Res.* **101**, 9–16.
70. Craciunescu, C. N., Brown, E. C., Mar, M. H., Albright, C. D., Nadeau, M. R., and Zeisel, S. H. (2004). Folic acid deficiency during late gestation decreases progenitor cell proliferation and increases apoptosis in fetal mouse brain. *J. Nutr.* **134**, 162–166.
71. Yen, C. L., Mar, M. H., and Zeisel, S. H. (1999). Choline deficiency-induced apoptosis in PC12 cells is associated with diminished membrane phosphatidylcholine and sphingomyelin, accumulation of ceramide and diacylglycerol, and activation of a caspase. *FASEB J.* **13**, 135–142.
72. Albright, C. D., Tsai, A. Y., Mar, M. H., and Zeisel, S. H. (1998). Choline availability modulates the expression of TGFbeta1 and cytoskeletal proteins in the hippocampus of developing rat brain. *Neurochem. Res* . **23**, 751–758.
73. Albright, C. D., Siwek, D. F., Craciunescu, C. N., Mar, M. H., Kowall, N. W., Williams, C. L., and Zeisel, S. H. (2003). Choline availability during embryonic development alters the localization of calretinin in developing and aging mouse hippocampus. *Nutr. Neurosci.* **6**, 129–134.
74. Albright, C. D., Mar, M. H., Craciunescu, C. N., Song, J., and Zeisel, S. H. (2005). Maternal dietary choline availability alters the balance of netrin-1 and DCC neuronal migration proteins in fetal mouse brain hippocampus. *Brain Res. Dev. Brain Res.* **159**, 149–154.
75. Albright, C. D., Mar, M. H., Friedrich, C. B., Brown, E. C., and Zeisel, S. H. (2001). Maternal choline availability alters the localization of p15Ink4B and p27Kip1 cyclin-dependent kinase inhibitors in the developing fetal rat brain hippocampus. *Dev. Neurosci.* **23**, 100–106.

76. Niculescu, M. D., Craciunescu, C. N., and Zeisel, S. H. (2006). Dietary choline deficiency alters global and gene-specific DNA methylation in the developing hippocampus of mouse fetal brains. *FASEB J.* **20**, 43–49.
77. Niculescu, M. D., Yamamuro, Y., and Zeisel, S. H. (2004). Choline availability modulates human neuroblastoma cell proliferation and alters the methylation of the promoter region of the cyclin-dependent kinase inhibitor 3 gene. *J. Neurochem.* **89**, 1252–1259.
78. Ravitz, M. J., and Wenner, C. E. (1997). Cyclin-dependent kinase regulation during G1 phase and cell cycle regulation by TGF-beta. *Adv Cancer Res.* **71**, 165–207.
79. Niculescu, M. D., Craciunescu, C. N., and Zeisel, S. H. (2005). Gene expression profiling of choline-deprived neural precursor cells isolated from mouse brain. *Brain Res. Mol. Brain Res.* **134**, 309–322.
80. Jeltsch, A. (2002). Beyond Watson and Crick: DNA methylation and molecular enzymology of DNA methyltransferases. *Chembiochem.* **3**, 382.
81. Robertson, K. D., and Wolffe, A. P. (2000). DNA methylation in health and disease. *Nat. Rev. Genet.* **1**, 11–19.
82. Wang, Y., and Leung, F. C. (2004). An evaluation of new criteria for CpG islands in the human genome as gene markers. *Bioinformatics.* **20**, 1170–1177.
83. Bestor, T. H. (2000). The DNA methyltransferases of mammals. *Hum. Mol. Genet.* **9**, 2395–2402.
84. Lei, H., Oh, S. P., Okano, M., Juttermann, R., Goss, K. A., Jaenisch, R., and Li, E. (1996). De novo DNA cytosine methyltransferase activities in mouse embryonic stem cells. *Development* **122**, 3195–3205.
85. Okano, M., Xie, S., and Li, E. (1998). Cloning and characterization of a family of novel mammalian DNA (cytosine-5) methyltransferases. *Nat. Genet.* **19**, 219–220.
86. Oligny, L. L. (2001). Human molecular embryogenesis: an overview. *Pediatr. Dev. Pathol.* **4**, 324–343.
87. Jones, P. A., and Takai, D. (2001). The role of DNA methylation in mammalian epigenetics. *Science* **293**, 1068–1070.
88. Malone, C. S., Miner, M. D., Doerr, J. R., Jackson, J. P., Jacobsen, S. E., Wall, R., and Teitell, M. (2001). CmC(A/T)GG DNA methylation in mature B cell lymphoma gene silencing. *Proc. Natl. Acad. Sci. USA* **98**, 10404–10409.
89. Monk, M., Boubelik, M., and Lehnert, S. (1987). Temporal and regional changes in DNA methylation in the embryonic, extraembryonic and germ cell lineages during mouse embryo development. *Development* **99**, 371–382.
90. Cerny, J., and Quesenberry, P. J. (2004). Chromatin remodeling and stem cell theory of relativity. *J. Cell Physiol.* **201**, 1–16.
91. Boyd, A. W., and Schrader, J. W. (1982). Derivation of macrophage-like lines from the pre-B lymphoma ABLS 8.1 using 5-azacytidine. *Nature* **297**, 691–693.
92. Jost, J. P., Oakeley, E. J., Zhu, B., Benjamin, D., Thiry, S., Siegmann, M., and Jost, Y. C. (2001). 5-Methylcytosine DNA glycosylase participates in the genome-wide loss of DNA methylation occurring during mouse myoblast differentiation. *Nucleic Acids Res.* **29**, 4452–4461.
93. Zhu, B., Benjamin, D., Zheng, Y., Angliker, H., Thiry, S., Siegmann, M., and Jost, J. P. (2001). Overexpression of 5-methylcytosine DNA glycosylase in human embryonic kidney cells EcR293 demethylates the promoter of a hormone-regulated reporter gene. *Proc. Natl. Acad. Sci. USA* **98**, 5031–5036.
94. Costello, J. F. (2003). DNA methylation in brain development and gliomagenesis. *Front Biosci.* **8**, s175–s184.
95. Mattson, M. P. (2003). Methylation and acetylation in nervous system development and neurodegenerative disorders. *Ageing Res. Rev.* **2**, 329–342.
96. Goto, K., Numata, M., Komura, J. I., Ono, T., Bestor, T. H., and Kondo, H. (1994). Expression of DNA methyltransferase gene in mature and immature neurons as well as proliferating cells in mice. *Differentiation* **56**, 39–44.
97. Fan, G., Beard, C., Chen, R. Z., Csankovszki, G., Sun, Y., Siniaia, M., Biniszkiewicz, D., Bates, B., Lee, P. P., Kuhn, R., Trumpp, A., Poon, C., Wilson, C. B., and Jaenisch, R. (2001). DNA hypomethylation perturbs the function and survival of CNS neurons in postnatal animals. *J.Neurosci.* **21**, 788–797.
98. Takizawa, T., Nakashima, K., Namihira, M., Ochiai, W., Uemura, A., Yanagisawa, M., Fujita, N., Nakao, M., and Taga, T. (2001). DNA methylation is a critical cell-intrinsic determinant of astrocyte differentiation in the fetal brain. *Dev. Cell.* **1**, 749–758.
99. Feil, R. (2006). Environmental and nutritional effects on the epigenetic regulation of genes. *Mut. Res./Fundamental and Molecular Mechanisms of Mutagenesis* **600**, 46–57.
100. Zeisel, S. H. (2006). The fetal origins of memory: The role of dietary choline in optimal brain development. *J. Pediatr.* **149**, S131–S136.
101. Zhu, X., Mar, M. H., Song, J., and Zeisel, S. H. (2004). Deletion of the Pemt gene increases progenitor cell mitosis, DNA and protein methylation and decreases calretinin expression in embryonic day 17 mouse hippocampus. *Brain Res. Dev. Brain Res.* **149**, 121–129.
102. Fukai, K., Yokosuka, O., Imazeki, F., Tada, M., Mikata, R., Miyazaki, M., Ochiai, T., and Saisho, H. (2005). Methylation status of p14ARF, p15INK4b, and p16INK4a genes in human hepatocellular carcinoma. *Liver International.* **25**, 1209–1216.
103. Holler, T., Cermak, J. M., and Blusztajn, J. K. (1996). Dietary choline supplementation in pregnant rats increases hippocampal phospholipase D activity of the offspring. *FASEB J.* **10**, 1653–1659.
104. Cermak, J. M., Holler, T., Jackson, D. A., and Blusztajn, J. K. (1998). Prenatal availability of choline modifies development of the hippocampal cholinergic system. *FASEB J.* **12**, 349–357.
105. Cermak, J. M., Blusztajn, J. K., Meck, W. H., Williams, C. L., Fitzgerald, C. M., Rosene, D. L., and Loy, R. (1999). Prenatal availability of choline alters the development of acetylcholinesterase in the rat hippocampus. *Dev. Neurosci.* **21**, 94–104.
106. Mellott, T. J., Kowall, N. W., Lopez-Coviella, I., and Blusztajn, J. K. (2007). Prenatal choline deficiency increases choline transporter expression in the septum and hippocampus during postnatal development and in adulthood in rats. *Brain Res.* **1151**, 1–11.
107. Mellott, T. J., Follettie, M. T., Diesl, V., Hill, A. A., Lopez-Coviella, I., and Krzysztof Blusztajn, J. (2007). Prenatal choline availability modulates hippocampal and cerebral cortical gene expression. *FASEB J.* **21**(7), 1311–1323.

108. Mellott, T. J., Williams, C. L., Meck, W. H., and Blusztajn, J. K. (2004). Prenatal choline supplementation advances hippocampal development and enhances MAPK and CREB activation. *FASEB J.* **18**, 545–547.
109. Sandstrom, N. J., Loy, R., and Williams, C. L. (2002). Prenatal choline supplementation increases NGF levels in the hippocampus and frontal cortex of young and adult rats. *Brain Res.* **947**, 9–16.
110. Frielingsdorf, H., Simpson, D. R., Thal, L. J., and Pizzo, D. P. (2007). Nerve growth factor promotes survival of new neurons in the adult hippocampus. *Neurobiol. Dis.* **26**, 47–55.
111. McKeon-O'Malley, C., Siwek, D., Lamoureux, J. A., Williams, C. L., and Kowall, N. W. (2003). Prenatal choline deficiency decreases the cross-sectional area of cholinergic neurons in the medial septal nucleus. *Brain Res.* **977**, 278–283.
112. Meck, W. H., and Williams, C. L. (2003). Metabolic imprinting of choline by its availability during gestation: Implications for memory and attentional processing across the lifespan. *Neurosci. Biobehav. Rev.* **27**, 385–399.
113. Guo-Ross, S. X., Clark, S., Montoya, D. A., Jones, K. H., Obernier, J., Shetty, A. K., White, A. M., Blusztajn, J. K., Wilson, W. A., and Swartzwelder, H. S. (2002). Prenatal choline supplementation protects against postnatal neurotoxicity. *J. Neurosci.* **22**, RC195.
114. Guo-Ross, S. X., Jones, K. H., Shetty, A. K., Wilson, W. A., and Swartzwelder, H. S. (2003). Prenatal dietary choline availability alters postnatal neurotoxic vulnerability in the adult rat. *Neurosci. Lett.* **341**, 161–163.
115. Holmes, G. L., Yang, Y., Liu, Z., Cermak, J. M., Sarkisian, M. R., Stafstrom, C. E., Neill, J. C., and Blusztajn, J. K. (2002). Seizure-induced memory impairment is reduced by choline supplementation before or after status epilepticus. *Epilepsy Res.* **48**, 3–13.
116. Meck, W. H., and Williams, C. L. (1997). Perinatal choline supplementation increases the threshold for chunking in spatial memory. *Neuroreport.* **8**, 3053–3059.
117. Meck, W. H., and Williams, C. L. (1997). Simultaneous temporal processing is sensitive to prenatal choline availability in mature and aged rats. *Neuroreport.* **8**, 3045–3051.
118. Meck, W. H., and Williams, C. L. (1997). Characterization of the facilitative effects of perinatal choline supplementation on timing and temporal memory. *Neuroreport.* **8**, 2831–2835.
119. Meck, W. H., and Williams, C. L. (1999). Choline supplementation during prenatal development reduces proactive interference in spatial memory. *Brain Res. Dev. Brain Res.* **118**, 51–59.
120. Williams, C. L., Meck, W. H., Heyer, D. D., and Loy, R. (1998). Hypertrophy of basal forebrain neurons and enhanced visuospatial memory in perinatally choline-supplemented rats. *Brain Res.* **794**, 225–238.
121. Jones, J. P., Meck, W., Williams, C. L., Wilson, W. A., and Swartzwelder, H. S. (1999). Choline availability to the developing rat fetus alters adult hippocampal long-term potentiation. *Brain Res. Dev. Brain Res.* **118**, 159–167.
122. Montoya, D. A., White, A. M., Williams, C. L., Blusztajn, J. K., Meck, W. H., and Swartzwelder, H. S. (2000). Prenatal choline exposure alters hippocampal responsiveness to cholinergic stimulation in adulthood. *Brain Res. Dev. Brain Res.* **123**, 25–32.
123. Pyapali, G., Turner, D., Williams, C., Meck, W., and Swartzwelder, H. S. (1998). Prenatal choline supplementation decreases the threshold for induction of long-term potentiation in young adult rats. *J. Neurophysiol.* **79**, 1790–1796.

CHAPTER **16**

Antioxidants in Health and Disease

HAROLD E. SEIFRIED AND JOHN A. MILNER
National Cancer Institute, Rockville, Maryland

Contents

I. INTRODUCTION

Antioxidants are a loosely characterized group of compounds that are defined by their general ability to decrease or delay oxidation. Dietary antioxidants are recognized to have the ability to inhibit the formation of both reactive oxygen (ROS) and nitrogen species (RNS), which can adversely affect normal cellular processes and physiological functions [1–3]. Under normal conditions, the balance between production and elimination of radicals is maintained by enzymes (such as glutathione peroxidases, catalase and superoxide dismutases, thioredoxin reductase, and heme oxygenase) and a host of nonenzymatic (some metals, glutathione, thiols, vitamins, and phytochemicals such as isoflavones, flavonoids, and polyphenols) components, which can be influenced by eating behaviors [1–3]. Table 1 includes a number of these antioxidants with their respective structures and properties.

This range of chemical classes within the antioxidant domain is unprecedented and part of the reason that the grouping of compounds by their ability to interact with reactive species (ROS and RNS) can lead to an oversimplification and generalization. Hence, descriptors such as "conundrum" and "double-edged sword" are often used to characterize the relationship between antioxidants and health. Part of this controversy stems from the innate properties of ROS, as this class of compounds can influence both disease prevention and its promotion. Although the generation of ROS had been viewed as primarily, or solely, detrimental to health, more recent advances in research have shown they can have crucial roles in normal physiological processes including being growth factors, influencing immunocompetence, and initiating apoptosis in damaged cells.

In spite of these beneficial functions of ROS, abnormal production or nonhomeostatic regulation of ROS is positively linked to the development of common diseases and associated conditions, including cancer and cardiovascular disease (CVD), as well as a number of neurological and metabolic diseases. As a result, antioxidants have been deemed promising for the prevention and possibly treatment of several of these diseases. Some of this optimism is partly based on the frequently observed case-control association between diets high in fruits and vegetables (and presumably antioxidant exposures) and decreased disease risk, including cancer. Although there is evidence that antioxidants may offer health benefits in populations that are at increased risk because of environmental or medical conditions, there are many inconsistencies in the literature. Because evidence does exist that several antioxidants at "normal" exposure levels can regulate signal transduction and thus regulate proliferation and the immune response, both normal physiological processes and mechanisms other than their antioxidant properties may be functioning. Overall, the physiological or pharmacological importance of antioxidants as regulators of radicals as a possible means for promoting health continues to receive widespread scientific attention and debate.

A. Antioxidant Usage and Measures of Oxidative Stress

The attention of public and scientific literature to the purported health benefits of antioxidants [4–7] has paralleled the increased use of antioxidant supplements by the U.S. adult population. Based on 1999–2000 National Health and Nutrition Examination Survey (NHANES) data, more than half the population are using dietary supplements; approximately 33% take multivitamins; and over 12% use vitamin E or C supplements [6]. This increased interest in effects of vitamins in general and specifically the class of "antioxidants" as protecting agents was sparked by a series of articles beginning in the 1970s. In the decades since, an expanded knowledge base in the nutritional sciences about

TABLE 1 Some Antioxidants and Their Biological Activity

Antioxidant	Structure	Detected or Effective Level	Cancer Activity	CVD Activity	Hormonal Response	Deficiency Symptoms
Carotenoids	β carotene 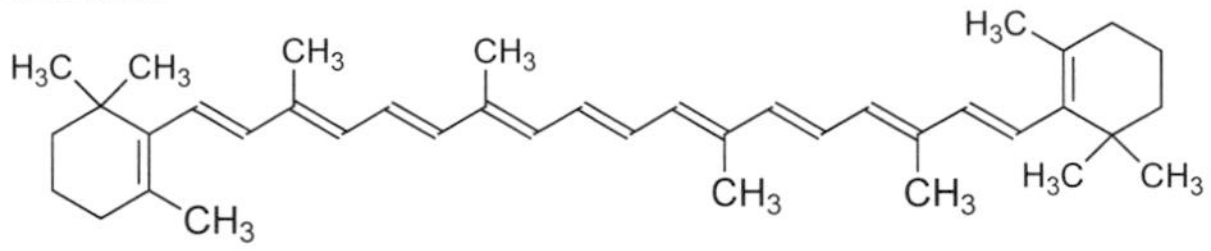	5–50 μg/dL	Induces connexin–differentiation, immune function	Food sources lower CVD	No	Only vitamin A—xeroderma, night vision
Curcumin	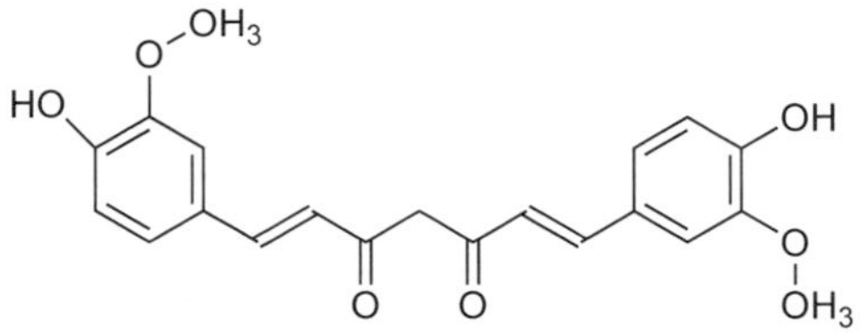	> 2 μmol/l at 8 g/day	Anti-inflammatory; increases apoptosis, interferes with chemotherapeutics and cell signaling; interferes with Phase I enzymes, stimulates Phase II	??	No	No
Flavonoids	genistein 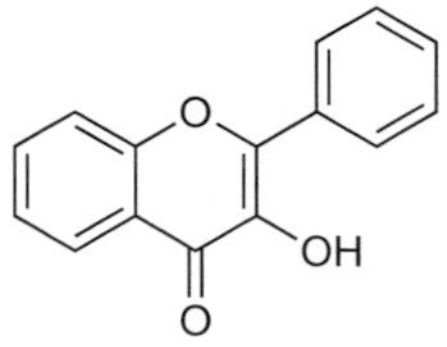	1–10 μmol/l	Modulates cell signaling, inhibits angiogenesis, increases apoptosis, much data from cell studies	Food sources lower CVD, minimal cholesterol lowering	Some are estrogenic, some anti, some both (dose related)	No
Glutathione	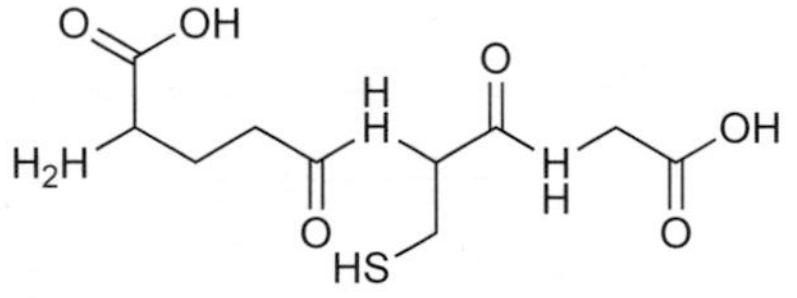		Substrate of GSH glutathione peroxidase reduction of H_2O_2 and lipid hydroperoxides and substrate for Phase II carcinogen metabolism	??	No	??
Oleanic acid	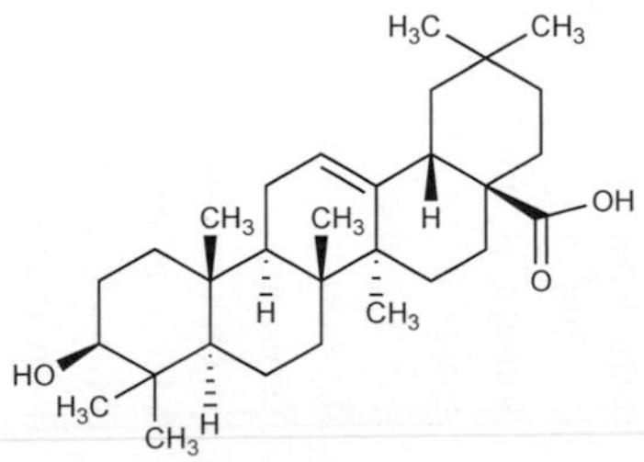		Inhibits proliferation angiogenesis, metastasis; induces apoptosis and cell differentiation; anti-inflammatory	Antiatherosclerotic, hypolioidemic	No	No

TABLE 1 *(continued)*

Antioxidant	Structure	Detected or Effective Level	Cancer Activity	CVD Activity	Hormonal Response	Deficiency Symptoms
Resveratrol	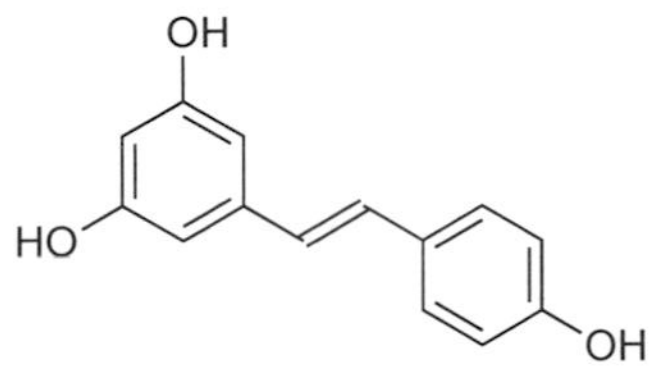	Mostly in cell culture	Inhibits proliferation angiogenesis, induces apoptosis and cell cycle arrest in cancer cells in culture	Inhibits platelet aggregation cell adhesion in culture	Both estrogen (E2) agonist without E2 and antagonist with E2	No
Selenium	Se	45–80 μg/l 40–100 μg/d	Increases glutathione peroxidase, stimulates immunity	Weakly cardio-protective	No	Muscle wasting, cardiomyopathy
Vitamin C	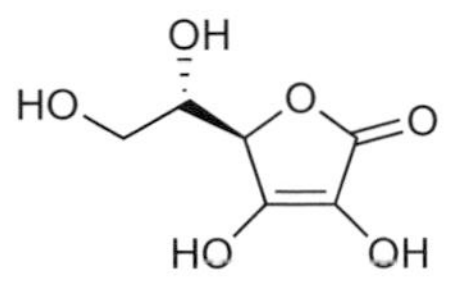	Saturated blood levels at 400 mg per day ~70 μmol/l	Reduces lung, breast, oral, upper GI, gastric, colorectal CA with oral C; IV doses increase survival	Mixed results on CVD reduction, more + than −, lowers blood pressure, coronary artery dilation	No	Scurvy, impaired collagen synthesis
Vitamin D	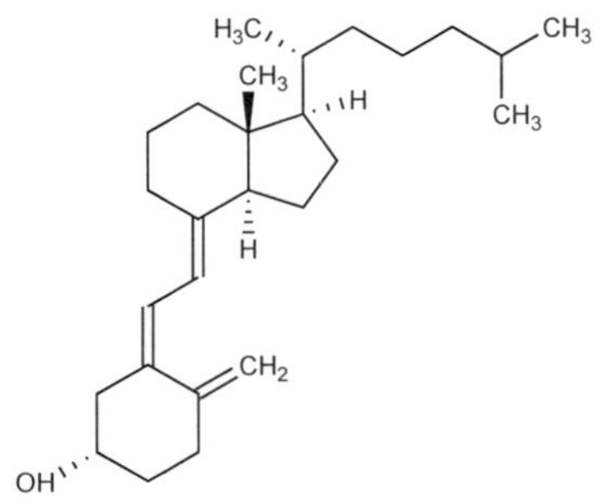		Induces differentiation, lowers proliferation	Lowers blood pressure	No	Rickets, seizures, osteomalacia
Vitamin E	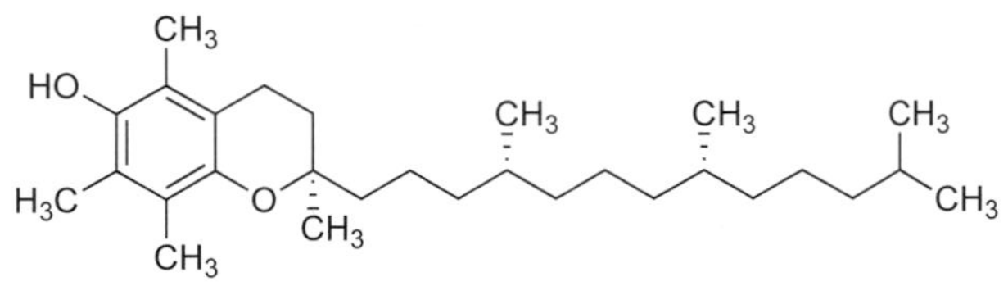	20 μmol/l	Affects cell signaling, boosts immunity, inflammation; minimal CA effect	Platelet aggregation, lower MI, carotid atherosclerosis	No	27 to 40+% of population below 20 μmol/l

the potential molecular targets and interactions that may account for the health benefits of antioxidants has surfaced from a wide range of preclinical, clinical, and population studies [7, 8].

The major cellular targets of ROS include membrane lipids, proteins, nucleic acids, and carbohydrates. Several biomarkers for oxidative damage and antioxidant defense have been introduced as potential indicators of alterations in normal homeostatic mechanism. Basically, oxidative stress biomarkers can be separated into those that (1) reflect modified molecules caused by ROS and (2) reflect shifts in biological measures of small molecular weight compounds or the induction of enzymes (Table 2). The ability to monitor these biomarkers at multiple time points in blood, urine, ductal lavage, and so on allows for repeated greater sensitivity in detecting and understanding stress status, which is not always possible for multiple reasons, including ethics, with more invasive procedures. Unfortunately, at present, it is not easy to link these biomarkers to a specific clinical outcome and the field is expanding and maturing. The uncertainty about what constitutes normal and abnormal values what are the appropriate target tissues, and the unsubstantiated relationship of a change in a biomarker to a specific phenotypic response also raises significant concerns about many of these measures. It is possible that eventually an "oxidative stress profile" that incorporates changes in multiple biomarkers may be useful in establishing who will benefit, or who will potentially be placed at risk, by the exaggerated use of antioxidants through foods or dietary supplements.

Two types of methods have been primarily used to evaluate the antioxidant properties of foods: these assays are based on hydrogen atom transfer (HAT) reactions or on electron transfer (ET). Most HAT-based assays use a competitive reaction scheme, in which antioxidant and substrate compete for thermally generated peroxyl radicals through the decomposition of azo-compounds. These assays include inhibition of induced low-density lipoprotein auto-oxidation, oxygen radical absorbance capacity (ORAC), total radical trapping antioxidant parameter (TRAP), and crocin bleaching assays. The ET-based assays measure the capacity of an antioxidant in the reduction of an oxidant and corresponding change in color when reduced. ET-based assays include the total phenols assay by Folin-Ciocalteu reagent (FCR), Trolox equivalence antioxidant capacity (TEAC), ferric ion reducing antioxidant power (FRAP), "total antioxidant potential" assay using a Cu(II) complex as an oxidant, and DPPH, the oxygen radical absorbance capacity [9, 10]. Although each of the methods has value for a relative comparison across a variety of food items, it remains unclear if the measures truly reflect their physiological value after consumption. The lack of detailed information about multiple exposures and with multiple durations makes the interpretation of existing studies with sometimes subtle changes in fluids and cells in humans extremely challenging. The merits of these analyses may also be questionable because possible postprandial or diurnal variations that are not directly related to the intake of dietary antioxidants per se are not considered. Finally, it should be noted that plasma antioxidant capacity may be significantly affected by nonantioxidant dietary constituents, which influence uptake, tissue mobilization, or metabolism of endogenous or exogenous antioxidants [2, 8, 11].

The beneficial health effects of fruits and vegetables have been attributed, at least in part, to their antioxidant content. Significant and generally transient increases in plasma total antioxidant capacity are frequently observed following the human consumption of flavonoid-rich foods [11]. This concept that these flavonoids or possibly other bioactive food components function by modifying oxidative stress has been challenged by observations that typical

TABLE 2 Some Biomarkers of Oxidative Damage

Total Antioxidant Potential	Total Radical Trapping Antioxidant
Lipid peroxidation	Malondialdehyde-lysine, 4-hydroxy-2-nonenallysine, acrolein-lysine, F2-isoprostane, thiobarbituric acid reactive substances
DNA oxidation	8-hydroxy-2′-deoxyguanosine
Glyco-oxidation	Carboxymethyl-lysine, pentosidine, argpyrimidine, methylglyoxal
Nitro-oxidation	Nitrotyrosine, nitrite/nitrate
Protein oxidation	o,o′-dityrosine, ortho-tyrosine, bilirubin oxidative metabolites, oxidized glutathione
Enzyme activities	Superoxide dismutase, catalase, glutathione peroxidase, glutathione reductase, glutathione-*S*-transferase, thioredoxin reductase, heme oxygenase
Protein concentrations	Albumin, ferritin, transferrin, lactoferrin, ceruloplasmin, thioredoxin
Concentrations of low-molecular-weight molecules	Bilirubin, tocopherols, carotenoids, ubiquinol/ubiquinone, ascorbate, glutathione, cysteine, urate, selenium

intakes only result in minimal shifts in circulating concentrations in plasma, and that extensively metabolism of the agents likely diminishes their *in vivo* antioxidant capacity. Lotito and Frei [11] concluded that the large increase in plasma total antioxidant capacity observed after the consumption of flavonoid-rich foods is not caused by the flavonoids per se but is likely the consequence of increased uric acid levels.

The possible usefulness of antioxidants in disease prevention, particularly for cardiovascular diseases and cancer, stems from a number of epidemiological findings and a number of follow-up intervention studies. However, a substantive compendium of negative cardiovascular and cancer effects of antioxidant use, especially relating to dietary supplements, has also emerged as will be discussed subsequently. A resulting concern is the potentially deleterious effects of antioxidant supplements on normal ROS levels. This is especially important as precise modulation of ROS levels is needed to allow normal cell function or to promote apoptotic cell death of precancerous or transformed cells [12]. Conflicting findings on risks and benefits have led to the "Evidence Report from the Agency for Healthcare Research and Quality (AHRQ) on Vitamin C, E and Coenzyme Q10" [13]. This AHRQ report included a broad search of the literature and concluded that the evidence did not support a positive benefit of vitamin E supplementation for cardiovascular events. It did not support significant potential for harm either. Conclusions about vitamin C and coenzyme Q were mixed. It should also be noted that their influence on cancer was not evaluated, but Taylor and Greenwald have reviewed a number of completed nutritional cancer prevention trials [14], which helps fill out the background picture

This collection of potential positive and negative antioxidant effects on ROS deserves further examination because of the molecular evidence for the multiple role(s) of ROS in development and progression of cancer, cardiovascular disease, and other diseases [2, 8]. This increased attention is timely as there have been studies [2] that link some nutritional antioxidants with increased mortality from cancer and CVD, as well as some clinical studies that do not support the cancer prevention efficacy of some antioxidants.

B. Reactive Oxygen Species and Normal Physiology

ROS typically arise as by-products of cellular metabolism and ionizing radiation, usually reflected in the formation of the following four species: superoxide anion (O_2^-), hydrogen peroxide (H_2O_2), hydroxyl radical ($OH^\cdot$), and singlet oxygen ($^1O_2$). Although H_2O_2 and $^1O_2$ are not free radicals per se, these species behave similarly to free radicals. The reactivity of O_2^- or H_2O_2 with other molecules is not appreciable, but the presence of trace amounts of transition metals fosters their conversion to OH *via* the Fenton or Haber-Weiss reactions. ROS formation is a natural consequence of aerobic metabolism and is integral for maintaining tissue oxygen homeostasis [15].

Oxygen homeostasis—the balance between constitutive oxidants and antioxidants—is maintained through a natural series of reduction-oxidation (redox) reactions involving the transfer of electrons between two chemical species: compounds that lose electrons (oxidized) and those that gain electrons (reduced). When oxygen homeostasis is not maintained, the cellular environment becomes oxidatively stressed. Approximately 1% to 3% of oxygen consumed by the body is converted into ROS [16]. Three of the major ROS—superoxide radical O_2^-, hydrogen peroxide, and hydroxyl radical $OH^\cdot$—are normal metabolic by-products that are generated continuously by the mitochondria in growing cells [15, 17, 18]. Other significant intracellular sources of ROS include microsomal cytochrome P450 enzymes, flavoprotein oxidases, and peroxisomal enzymes involved in fatty acid metabolism [15]. The potentially damaging oxidative stress can be caused by excess ROS, which are kept in check by endogenous cellular antioxidant mechanisms. Oxidative stress-related enzymes include superoxide dismutases, for eliminating the superoxide radical as well as catalase and glutathione peroxidases for removing hydrogen peroxide and organic peroxides [15, 17]. Polymorphisms have been observed in these enzymes, which can affect an individual's capacity to respond to changes in ROS; this is discussed in more detail at the end of this chapter.

Transient fluctuations of ROS levels can influence activity of signal transduction pathways leading to cell proliferation, or to apoptosis or necrosis, depending on the dosage and duration of ROS changes and also on cell type. Typically, low levels of ROS can be mitogenic, whereas medium (normal homeostatic) levels lead to temporary or permanent growth arrest (replicative senescence), and elevated levels usually result in cell death either by apoptosis or necrosis [18–20]. Although necrosis and apoptosis may be viewed as negative events in terms of cell loss, these processes also have positive roles in the down-regulation of immune responses (more detail in [3]) and the elimination of transformed cells ("tumor suppression" via apoptosis).

C. Reactive Oxygen Species in Disease Conditions

Imbalanced ROS homeostasis has been linked to increased risk of several diseases including cancer, CVD, atherosclerosis diabetes, and neurodegenerative diseases including Alzheimer's. Understanding the molecular effects of ROS on these different roles should assist in unraveling the varied and sometimes contradictory evidence about these diseases and assist in evaluating the importance and safety of antioxidants.

1. Cancer

Cancer is a number of somewhat distinct diseases that can be characterized on the basis of uncontrolled cellular growth resulting from a series of altered sets of genetic and epigenetic manifestations. Hanahan and Weinberg indicated the "hallmark capabilities" necessary for tumorigenesis: (1) self-sufficiency in growth signals, (2) insensitivity to antigrowth signals, (3) evasion of apoptosis, (4) limitless potential for replication, (5) sustained angiogenesis, and (6) tissue invasion and metastasis [21]. Excessive ROS and RNS are involved in all these processes and thus contribute to cancer progression either positively by promoting cell division or negatively by stimulating apoptosis and slowing the growth. Belief in the protective effects of antioxidant supplements has led to their widespread use by cancer patients [22]. The expanded use of nutritional aides is not limited to the United States, as many other countries are reporting increased used of various alternative and complementary approaches and strategies [23]. A greater understanding of both the negative and positive consequences of ROS and antioxidants in the etiology and progression of carcinogenesis is crucial to making clear advances in cancer prevention and treatment. At present, the two faces (benefit/risk) of ROS/RNS in malignant diseases make it difficult to present a true, clear, and concise message to consumers.

2. Cardiovascular Disease

Cardiovascular disease (CVD), encompassing atherosclerosis and its associated vascular disorders, is the leading cause of mortality in developed countries [24, 25]. In addition, other vascular insults such as those associated with cigarette smoking, diabetes mellitus, hypertension, and hyperlipidemia can trigger an inflammatory response in blood vessels. Chronic low-grade inflammation is generally accepted to accompany atherogenesis [26, 27]. This inflammatory state, which has been linked in part to ROS mediation, can result in damage to smooth muscle and vascular endothelial cells. This in turn leads to a dysfunctional endothelium, which can be characterized by pathological alterations in the endothelial cell's anticoagulant, anti-inflammatory, and vascular-relaxation properties, which can promote the recruitment of monocytes, macrophages, growth factors, and cellular hypertrophy. All of these factors can contribute to atherosclerotic plaque formation. In summary, increased ROS activity helps drive many of these processes involved in the development of CVD if left unchecked.

II. ANTIOXIDANTS IN DISEASE ETIOLOGY, TREATMENT, AND PREVENTION

Several studies have reported that diets high in fruits and vegetables can be associated with a markedly decreased risk of CVD and cancer. This has been frequently attributed to high levels of antioxidants present in these foods. Other studies, however, do not provide clear and unequivocal support for this assumption. For example, data from historical food frequency questionnaires collected during the cohort Nurses' Health and Health Professionals Follow-up Studies indicate that cancer incidence was not influenced by increased fruit or vegetable consumption; however, modest reductions in CVD were detected [28] (Table 3). Similarly, studies of dietary supplements and individual antioxidants are not consistent; observed effects have ranged from benefit to possible harm [29, 30]. A number of factors may contribute to these contradictory findings, including participant baseline health and ROS levels, exposure to environmental carcinogens, and genetic differences in ROS metabolism [31]. Some argue that higher doses of some antioxidants have a prooxidant effect [2], and there is a high probability that absorption, distribution, metabolism, and excretion of a simple ingested supplement is quite different from the complex in fruits and vegetables, which contains many different types of antioxidants as well as other phytochemicals.

Because many examples of "J" and "U"-shaped dose-response curves have been reported among nutrients and food additives, a more critical question should be, what is the potential interaction of ROS and antioxidants with chemotherapy and radiation therapy? The interaction occurs sometimes in a positive manner, enhancing the efficacy of the treatment, but also sometimes negatively interfering with the agent or treatment. Similarly, individual antioxidants can inhibit or stimulate tumor cell growth or survival depending on the tumor, the antioxidant, and the oxidative status; this means that there is no simple answer for the overall effectiveness of antioxidants, and each must be considered on a case-by-case basis. The following sections provide the results of some major clinical and epidemiological trials on antioxidant effects on cancer and observed side effects, particularly on CVD, focusing on the antioxidants selenium, vitamin E, vitamin C, and β-carotene, all of which are readily available in the marketplace as dietary supplements. Key study findings and some experimental details are presented Table 3.

A. Antioxidants and Cancer

The potential benefits of the dietary antioxidants selenium, vitamin E, vitamin C, and β-carotene in numerous observational and clinical trials have been examined for several years [7, 32]. Since the 1990s, however, evidence has emerged which suggests that some antioxidants, in certain patients, perhaps at untoward high doses or in the presence of specific conditions or cancer treatments, may modulate normal protective benefits whereas others may bring about deleterious effects, such as interfering with the efficacy of cancer drug treatment or increasing an individual's risk for cancer or heart disease [33, 34].

TABLE 3 Human Intervention Studies on Antioxidants

Study and Publication Date	Study Details, Size, And Duration	Intervention Details	Study Results
Linxian study China (1993) [35]	29,584 men (5 years)	βC (15 mg) vita E (30 mg) and Selenium (50 μg) daily	Protective effects Cancer: −13% Mortality: −9%
ATBC study Finland (1994) [42, 44, 45, 52, 53, 54, 55, 56, 74]	29,133 male smokers (5–8 years) with up to 19-year follow-up	βC (20 mg) ± vita E (50 mg) daily	Lung cancer: +18% but mortality mixed Prostate cancer: −32% Colorectal cancer: −22% No decrease CVD or angina Slight excess hemorrhagic stroke and βC related mortality +7%
CARET study USA (1996) [30, 43, 46]	18,314 male smokers or asbestos exposed (4 years + 6-year follow-up)	βC (30 mg) ± vita A (25,000 IU) daily	Lung cancer: +28% Mortality: +17% Baseline levels after 10 y
CHAOS United Kingdom (1996) [77]	2002 atherosclerosis patients ~80% males (510 days)	535 or 263 mg vita E daily	Decreased non-fatal myocardial infarction (77%)
PHS study United States (1996) [36, 38, 86]	22,071 physicians (12 years) (11% smokers, 39% former smokers)	50 mg βC every other day	No significant effect on cancer or CVD Decreased prostate cancer −32%
NPC Trial USA (1996) [61, 62, 63]	1312 men and women with dermal basal or squamous cell carcinoma (4.5 years + 6.4-year follow-up)	200 μg Selenium daily	No effect on skin cancer Lung (−46%), colorectal (−58%), prostate (−63%), mortality (−50%)
WHS study USA (1999) [37, 87]	39,876 female 45+ years old health professionals (2.1 years + 2-year follow-up) (13% smokers)	50 mg βC or vita E (400 mg) every other day + aspirin	No βC effect on cancer rates nor on CVD but high fruit and veggie intake associated with lower MI
GISSI Italy (1999) [90]	11,324 myocardial infarction survivors ~85% male (3–5 years)	vita E (300 mg) ± PUFA (1 g) daily	No effect of vita E on CVD, PUFA effects protective Cardiovascular deaths: −30% Mortality: −20%
ASAP (2002) [83]	520 high-cholesterol male and female (postmenopausal) smokers and nonsmokers (6 years)	vita C (250 mg) and 100 mg α-tocopherol acetate twice daily	Carotid atherosclerosis −25% combined, −30% in men, decreased plaque size in >50%
IVUS (2002) [84]	40 cardiac transplant patients (1 year)	vita C (500 mg) and vita E (263 mg) twice daily	Decreased atherosclerotic plaque size
British Heart Protection Study (2002) [85]	20,536 increased risk of MI 15,454 males, 5082 females (5 years)	vita C (250 mg) vita E (600 mg) and 20 mg βC	No significant effect on CVD, mortality, or cardiac events (MI) No effect on cancer, stroke, or dementia
SU.VI.MAX France (2003) [57–60]	12,741 men (aged 45–60) and women (aged 35–60) (7.1 years)	βC (6 mg), vita E (30 mg), vita C (120 μg) + Zn (20 mg) and Selenium (100 μg) daily dosing	Protective in men Prostate cancer: −~58% All Cancer: −31% and mortality: −37% No supplement effect in women but lower CVD with high fruits and vegetables
HOPE & HOPE-TOO Canada, United States, Europe, South America (2005) [91, 92]	1138 men and women (over 55) with left ventricular dysfunction or diabetes (4 years, follow-up to 7 years)	vita E (400 mg) daily	No significant effect on cancer incidence or mortality No significant effect on CVD events but heart failure +13%
E3N Prospective Cohort Study France (2006) [49]	59,910 women—selected 700 with smoking-related cancers (7.4 years)	βC intake divided into quartiles from diet reports	Nonsmokers 20% to 60+% reduction lung CA Smokers 1.5 to 2X increase

1. Antioxidant Supplementation and Cancer Prevention

On initial examination of Table 3, large-scale trials do not consistently demonstrate a definite benefit of antioxidant supplements in healthy individuals. However, closer examination suggests that in individuals with initially low background antioxidant status, supplementation to achieve a normal range may bring about some health benefit [35]. However, in some cases, high-dose supplementation may lead to potentially harmfully elevated blood levels, especially in oxidatively stressed individuals, such as smokers who consume large amounts of β-carotene.

a. β-Carotene

A major issue in assessing antioxidant efficacy is whether any antioxidant interventions, especially at elevated doses above the normal nutritional level, result in a dose-dependent benefit to individuals at risk for disease. For example, two large-scale, randomized intervention trials evaluated the effect of β-carotene supplementation over 12 years on the primary prevention of cancer and CVD in male physicians (PHS) [36] and 2+ years in female health professionals (WHS) [37]. The dose administered was 50 mg β-carotene every other day. Upon initial evaluation, no evidence of either stimulation or inhibition of either diseases was noted when placebo and treated groups were compared. However, subgroup analyses of the PHS data later found that men with the lowest quartile for plasma β-carotene at initial baseline had a lower risk for total cancer and in particular prostate cancer when β-carotene supplement users were compared with placebo [38]. It has been reported that the average blood levels achieved during the PHS study were 120 μg/dL [39], whereas normal serum β-carotene is in the 5- to 50-μg/dL range [1]. An effect of smoking was not observable in these two studies, possibly because only 11% of the physicians and 13% of the women were smokers so the study was not adequately powered to examine this aspect. It should be noted that there is no reason to suspect that these individuals were nutritionally compromised either, so the results may not be universally applicable to the general population. Additional β-carotene findings were derived from cohort subsets of randomized clinical trials of β-carotene. These studies found no evidence that β-carotene supplementation prevented recurrence of basal or squamous carcinomas of the skin (50 mg/d) in the United States [40] or of colorectal adenomas (20 mg/d) in Australia [41].

Similarly several randomized trials in high-risk populations have reported reduced disease risk using antioxidants, at least in baseline deficient populations when subsets of the populations are examined. The NCI trials, conducted in Linxian, China, with a population at high risk for esophageal cancer, noted a significant benefit for those receiving a β-carotene/vitamin E/selenium combination—a 13% decrease in the cancer mortality rate, a 21% decrease in stomach cancer mortality, a 4% decrease in esophageal cancer mortality, a 10% decrease in deaths from strokes, and a 9% decrease in deaths from all causes [35]. The generalizability of these findings from Linxian may not be universally appropriate because individuals in this study appear to have limited intakes of several micronutrients. Thus, these trial findings may not be applicable to well-nourished populations.

2. Effects on Smokers

When smokers were selected as the target population, very different views surfaced about the health consequences of β-carotene supplements. Two highly publicized trials provided evidence for potentially adverse effects of antioxidants, particularly in smokers and in individuals exposed to certain environmental hazards, such as asbestos. These two independent, randomized, clinical trials, the 5- to 8-year ATBC (α-tocopherol β-carotene) [42] and the 4-year CARET (β-CArotene and Retinol Efficacy Trial) [43], reported adverse effects of 20- and 30-mg β-carotene daily supplementation on lung cancer risk in high-risk populations. The ATBC studied effects primarily in smokers, whereas the Physicians Health Study (PHS) included over half who were nonsmokers [36]. Similar to the PHS study results for β-carotene, the ATBC study data demonstrated a reduction (32%) in prostate cancer incidence in the α-tocopherol (vitamin E)-supplemented group [44] and 22% fewer cases of colorectal cancer [45] upon reanalysis. Although lung cancer incidence was elevated in both ATBC and CARET smokers with β-carotene, after a 12-year follow-up of CARET participants, a significant decrease in lung cancer incidence in the placebo arm was observed linked to fruit and vegetable intake when the lowest versus highest quintiles were compared [46]. An earlier case-control study of lung cancer in nonsmokers conducted in New York concluded that dietary β-carotene, raw fruits and vegetables, and vitamin E supplements reduce the risk of lung cancer in nonsmoking men and women [47]. A subsequent evaluation of the Nurse's Health Study from Harvard implicated carrots, but not β-carotene, further adding to the confusion [48]. Interestingly, in the French Etude Epidemiologique de Femmes de la Mutuelle Generale de l'Education Nationale (E3N) prospective investigation, the nonsmoking women showed a dose-dependent lowering of lung cancer risk when intakes were considered. The largest reduction was observed in those women taking β-carotene supplements, but, of course, smokers showed the commonly observed increased risk [49]. Research has implicated β-carotene cleavage products as likely factors in the increased cancer activity of high β-carotene doses, exacerbated by smoking-induced alterations in β-carotene metabolism [39, 50, 51].

3. Follow-ups of Existing Studies

Recent follow-up examinations of a subset of the ATBC cohort has reported a marked association between elevated serum levels of α-tocopherol and reduced prostate cancer risk in a 6-year prospective study with 100 prostate cancer patients and 200 controls [52]; however, a separate post-intervention study has reported that the excess risk for lung and the beneficial effect for prostate cancers were no longer significantly different from controls [53]. In addition, with the exception of a slight carotene protective effect on early-stage laryngeal cancer, other upper aerodigestive cancers were not affected [54], although a dose-responsive decrease in mortality with increasing vitamin E serum levels was noted [55]. Interestingly, evaluation of dietary records in the ATBC study indicated that consumption of fruits and vegetables was associated with a lower lung cancer risk, as was comparison of the levels of dietary lycopene, and other carotenoids, as well as actual serum levels of β-carotene and retinol, when the highest and lowest baseline quintiles were compared [56]. A 6-year follow-up of the CARET cohort has reported that the increased risk was no longer significant for both lung cancer and cardiovascular disease, but subgroup analysis suggests excess risk in heretofore unreported susceptible groups (females and former smokers), a difference from the ATBC study, which may be due to the higher β-carotene dose in CARET [30] or perhaps the presence of vitamin E (α-tocopherol). The return to normal risk in the smokers suggests that subtle changes were introduced such that this group actually achieved protection from prior treatment. Thus, it is conceivable that the increased risk observed in the ATBC study may have reflected an increase in those with a preexisting precancerous lesion or tumor but protection in those without. Thus, the question of benefits or merits of β-carotene in smokers remains controversial, as it may depend on the presence or absence of precancerous lesions.

Further follow-up and reexamination of subgroup analysis may also help explain these unexpected results. For example, the recently reported French study, SUppléments en VItamines et Minéraux AntioXydants (Antioxidant Vitamin and Mineral Supplements [SU.VI.MAX.]), has examined the effects of supplementation with vitamins C and E (120 and 30 mg, respectively), 6-mg β-carotene, 100-μg selenium, and 20-mg zinc on the health of ~13,000 men (45 to 60 years) and women (35 to 60 years) [57–60]. In men only, antioxidant supplementation reduced the risk of developing all types of cancers by 31%. This effect was most pronounced in men with low baseline levels of β-carotene who had the greatest increased risk of developing cancer. The most recent evaluation of the study data has shown there was a nonsignificant reduction in prostate cancer rate associated with the supplementation in all men, but there was a significant difference between men with normal baseline prostate specific antigen (PSA) and those with elevated PSA. Among men with normal PSA, there was a statistically significant reduction in the prostate cancer rate among men receiving the supplements. Surprisingly, the supplementation was associated with an increased incidence of prostate cancer, with borderline statistical significance in men with elevated PSA at baseline [60]. It is interesting to note that the average β-carotene blood levels both at the beginning of the study (~39 μg/dL in women and ~25 μg/dL in men) and at the end (~90 μg/dL for women and ~50 μg/dL for men) were below those seen in the PHS, whereas no such change was observed for serum levels of vitamins E or C, selenium, or zinc.

a. Selenium

Dietary selenium is another potentially important antioxidant and potential chemopreventive agent because of its importance in the functionality of glutathione peroxidase. Strong evidence linking selenium supplementation and decreased cancer risk comes from the secondary analysis of the Nutritional Prevention of Cancer (NPC) Trial, a study designed for patients with a history of basal or squamous cell carcinomas [61]. Although no benefit was observed for skin cancer, the primary end point of the study, secondary end point analyses showed significant reductions in relative risk for total cancer mortality (50%), total cancer incidence (37%), as well as incidences of lung (46%), colorectal (58%), and prostate (63%) cancer for patients who received selenium supplements. Further reanalysis of the NPC data confirmed a 49% reduction in prostate cancer incidence in men receiving selenium supplements as compared to controls, although more recent follow-up suggests the benefits may be decreasing with time [62]. Another reanalysis that included 3 additional years of follow-up data found a nonsignificant decrease in lung cancer risk for all patients who received selenium supplements; however, the analysis suggested a significant risk reduction following supplementation among those with the lowest baseline plasma selenium [63]. A separate study by Li *et al.* [64] also suggested a reduction in prostate cancer incidence occurred following selenium supplementation [64]. The authors of these various studies differ in their conclusions concerning whether selenium prevents initiation of carcinogenesis or inhibits tumor progression; nevertheless, the benefit of selenium for prostate cancer prevention is fairly consistent. A wealth of preclinical studies supports the anticancer properties of selenium. Much of the evidence concerning selenium's anticancer/antitumor properties suggests mechanisms of action other than those associated with its antioxidant properties [65].

4. Antioxidant Combinations

More broadly, a meta-analysis of antioxidant supplementation in patients at elevated risk of gastrointestinal cancers revealed no protective effect for vitamins A, C, and E on esophageal, gastric, colorectal, pancreatic, and liver cancers. However, selenium supplementation was noted as possibly a cancer deterrent. The authors have noted that the findings may not be easily extrapolated to the effects of fruits and vegetables, which are rich sources of only some of these antioxidants [29]. Likewise, the applicability of these findings to healthy individuals is unclear. The follow-up examination of fruit and vegetable use and breast cancer risk within the European Prospective Investigation into Cancer and Nutrition (EPIC) indicated that fruits and vegetables had no effect on cancer when the highest versus the lowest quintiles were compared [66]. These findings add to the confusion in light of other studies that suggest fruits and vegetables have benefits, particularly when associated with their antioxidant content [46, 47, 49]. These inconsistencies may reflect the variation in the intake of individual fruits and vegetables, their interactions with environmental insults, or the genetic background of the consumer.

Completion of the human genome sequence and the advent of DNA microarrays using cDNAs enhanced the detection and identification of hundreds of differentially expressed genes in response to antioxidants including flavonoids, selenium, zinc, and several vitamins [67]. Phenolic antioxidant resveratrol found in berries and grapes has been reported to inhibit the growth formation of prostate tumor cells by acting on the regulatory genes such as p53 while activating a cascade of genes involved in cell cycle and apoptosis including p300, Apaf-1, cdk inhibitor p21, p57 (KIP2), p53-induced Pig 7, Pig 8, Pig 10, cyclin D, and DNA fragmentation factor 45, although some findings were not confirmed by Rt-PCR. [68]. Likewise, the expression of a host of genes is influenced by selenium supplementation [69].

B. Antioxidants and Cardiovascular Disease

In spite of the general acceptance of antioxidants' safety and potential efficacy [70], clinically based studies involving meta-analysis of vitamin E supplementation showing increased mortality above a 400-mg dose [71], versus a protective effect of fruit and vegetable consumption in EPIC [66] and a meta-analysis of fruit and vegetable cohort studies [72], provide mixed evidence. A study by Kushi in 1996 examined the role of dietary antioxidants in randomly chosen members of the prospective IOWA Women's Health Study [73]. A dose-dependent inverse relationship between vitamin E consumption and coronary heart disease mortality was detected. The strongest inverse association was observed among the approximately 22,000 women who did not consume vitamin supplements compared to those who did [73]. The more recent meta-analysis of high-dose vitamin E supplementation by Miller, however, suggested a positive association of all-cause mortality when supplementation was above 400 IU/day [71]. In spite of a number of issues with the choice of studies to be included and excluded in the meta-analysis, as well as the failure to consider a dose-response relationship in the physiological range, Wright's earlier analysis of continuous serum α-tocopherol values in the ATBC study indicated a dose-dependent reduction in mortality because of chronic disease with increasing concentrations up to approximately 13 to 14 mg/L (30–33 μmol/L), after which no further benefit was noted [55]. The use of super-physiological exposures with questionable health benefits, especially for those at the highest status, raises serious concerns about the wisdom of megadoses of vitamin E supplements. A recent meta-analysis of more than 60+ clinical trials that focused on antioxidant use (including vitamin E) reached similar conclusions about slightly increased mortality but again may have introduced some biases in the evaluation [74]. Another study, the Canadian Heart Outcomes Prevention Evaluation (HOPE) randomized controlled prospective investigation of vitamin E supplementation (400 IU/day) in approximately 4700 patients with diabetes and other cardiovascular risk factors, linked supplementation with an increased risk of heart failure but no protection from cancer in the 2.6-year follow-up portion of the study; no differences in cardiovascular effects were noted during the first 4.6 years [75].

Research evidence from large-scale trials also indicated no clear benefit of antioxidant supplements in healthy individuals when considered *in toto*. Upon closer examination, supplementing individuals with low background levels to a normal range may be beneficial, but high-dose supplementation leading to markedly elevated blood levels may be harmful, especially in oxidatively stressed individuals, such as smokers in the case of β-carotene or in unhealthy or older patients in the case of vitamin E.

1. Targets of Antioxidant-Related Disease Protection

Vitamin E, vitamin C, and β-carotene intake, whether by supplementation or as components of foods, has been extensively studied for their potential to reduce CVD or conditions related to the sequelae of CVD and other vascular diseases. A number of older studies suggest that vitamin E, alone or in combination with other antioxidants, may be protective against atherosclerosis in at-risk populations. However, more recent studies question the overall benefits [42, 76–78]. In addition to a potential role in minimizing oxidative DNA damage and lipid peroxidation, vitamin E may modify several other functions. Several of these functions specifically protect against processes known to contribute to atherosclerosis, including inhibiting protein kinase C activity and smooth muscle cell proliferation, inhibiting cell adhesion and platelet aggregation, counteracting inflammation, and enhancing bioavailability of nitric oxide to improve endothelium-dependent vasodilator function [79–81]. Vitamin E also may improve insulin-mediated glucose uptake, thus possibly decreasing risk for type 2 diabetes, a condition that also contributes to atherosclerosis [82].

2. Clinical Trial Interventions for CVD Amelioration

Antioxidants have been tested for the ability to ameliorate development of CVD in some at-risk populations. The

Antioxidant Supplementation in Atherosclerosis Prevention (ASAP) study, a 6-year randomized trial, supplemented 520 hypercholesterolemic men and postmenopausal women (45 to 69 years old) with vitamin C and vitamin E [83]. A significant decrease (26%) in the progression of carotid atherosclerosis was observed in men, although no significant effect was seen in women. In the Harvard Intravascular UltraSonography (IVUS) study, a combination of vitamin C and vitamin E was given to 40 (35 males) cardiac transplant patients. Cardiac transplantation leads to oxidative stress, which may contribute to atherosclerosis; supplementation of patients with these vitamins slowed progress of coronary atherosclerosis [84].

Two large, recently completed studies have reported preliminary results. The British Heart Protection Study, a multicenter, randomized, double-blind, placebo-controlled trial with a 2 × 2 factorial design, enrolled 20,536 patients aged 55 to 75 years who were at an increased 5-year risk of myocardial infarction [85]. Participants received antioxidant vitamins (combination of 600 IU vitamin E, 250 mg vitamin C, and 20 mg β-carotene) or simvastatin or both. No adverse or beneficial effect on vascular or nonvascular morbidity or mortality could be attributed to supplementation with vitamins. The American PHS II enrolled 15,000 physicians aged 55 years and older in a randomized, double-blind, placebo-controlled trial to test β-carotene, vitamin E (400 IU synthetic on alternate days), vitamin C (500 mg/d), and multivitamin (RDA of most vitamins and minerals) in a 2 × 2 × 2 × 2 factorial design [86]. This study will be completed in December 2007. The effect of supplementation on the prevention of total and prostate cancer, CVD, and age-related eye diseases (cataracts and macular degeneration) are being examined. Preliminary analysis reported no significant effect of any of the vitamins on cardiovascular outcomes. The WHS tested the ability of aspirin, vitamin E, and β-carotene to prevent cancer and CVD [37]. Analysis of β-carotene supplementation showed no effect of β-carotene on risk of CVD. Despite the lack of evidence for an effect of discrete nutrients on CVD, high fruit and vegetable intake was associated with lower risk for myocardial infarction in the WHS study [87].

As seen in the WHS, dietary habits sometimes may be linked to decreased CVD risk, despite a lack of effectiveness for individual supplements. Similarly, Mennen *et al.* [88] found that a diet rich in flavonoids reduced cardiovascular disease risk in a subset of women participating in the SU.VI.MAX. study [57–60]. Dietary intakes were estimated using six 24-hour dietary records collected during the course of 1 year. In women, flavonoid-rich food consumption was associated with decreased systolic blood pressure and a decreased risk for CVD; this relationship was not observed in men. The lack of an effect in men could be attributed to the men's higher risk for CVD; inconsistencies in dietary reporting and measuring of flavonoid consumption also could contribute to this discrepancy. However, this study found that, after 7.5 years of follow-up, no protective effect against ischemic heart disease attributable to antioxidant supplementation could be discerned in either men or women [88].

Four large, randomized clinical trials specifically tested the ability of vitamin E supplementation to slow the progression of CVD in individuals at increased risk for CVD death (reviewed by Salonen *et al.*, 2002) [89]. The ATBC trial, originally designed to test cancer prevention abilities of vitamins C and E and β-carotene, tested supplementation with 50 mg α-tocopherol acetate daily for 5 to 8 years [42]. In this study of male smokers, α-tocopherol supplementation was associated with a modest trend toward decreased incidence of angina pectoris but no decrease in CVD mortality. β-carotene supplementation had no preventive effect and was associated with a slight increase in the occurrence of angina [76]. An apparent excess of hemorrhagic stroke in the treatment group complicated this study. The Cambridge Heart Antioxidant Study (CHAOS) tested RRR-α-tocopherol from "natural sources" in 2002 participants with clinical evidence of CVD [77]. Trial analysis identified a 77% decrease in nonfatal myocardial infarction but also a nonsignificant increase in early deaths from CVD and total mortality, although no increase in risk of hemorrhagic stroke was observed. Interpretation of this study was complicated by unbalanced randomization, incomplete follow-up, and a midstudy change in vitamin E dose (800 IU per day to 400 IU per day) [78]. The Italian Gruppo Italiano per lo Studio della Supravvivenza nell Infarto Miocardio (GISSI) study supplemented 11,324 participants with previous myocardial infarction with 300 mg all-rac α-tocopherol daily, with or without 1 gm polyunsaturated fatty acids (PUFAs) [90]. No effect was observed in the group receiving both supplements, but a significant 20% decrease in cardiovascular death was observed in the group receiving only PUFAs [90]. In the HOPE study, 2545 women and 6996 men at high risk for CVD received either 400 IU per day "natural source" vitamin E or ramipril; this study found no effect of supplementation on any parameters related to CVD [91, 92]. As noted earlier, however, the follow-up study suggests an increase in CVD after an additional 2.6 years [75].

ROS can influence the inflammatory process. Since inflammation is thought to be a significant cause of damage to blood vessels, contributing to CVD, it may be a target for treatment with antioxidants. Vascular smooth muscle cell accumulation and hypertrophy, and nitric oxide regulation of endothelial vasorelaxation and vasodilation, may also be processes that could be therapeutically modulated by antioxidants. Although individual supplements have not proven to be strongly effective, diets high in fruit and vegetables, and their attendant antioxidant and phytochemical combinations, generally may have been shown to be somewhat protective [72]. Of course, diets high in fruits and vegetables may indirectly reduce the risk of CVD by promoting healthy body weight, decreasing the risk of developing conditions contributing to CVD such as hyperlipidemia and type 2 diabetes.

Studies with defined populations have generally failed to demonstrate marked prevention potential for cancer or heart disease, yet animal studies suggest benefits should occur. There are a number of possible explanations for this conundrum. Foremost among these is that significant nutrient–nutrient interactions are likely occurring along with the importance of environmental insults as determinants of the overall response. Some of the case-control studies may actually be selecting participants with a general interest in the pursuit of a healthy lifestyle. It is known that volunteers for observational antioxidant studies tend to have better diets, exercise more, use less alcohol, and come from higher socioeconomic backgrounds, all of which all may contribute to a decreased baseline risk for CVD and other disease conditions [93]. Additionally, supplement users tend to have healthy lifestyles and diets favorable for disease prevention [93].

C. Polymorphism: An Additional Risk Factor in Cancer

Evidence is increasingly surfacing that genetic polymorphisms can influence the response to an arsenal of agents used in the battle against cancer, including both drugs and dietary components. However, these investigations are often conflicting because the cancer process is not a simple process but involves multiple cellular events, many of which are likely influenced by genetic polymorphisms at the site of the target or how the agent is modified through absorption, metabolism, or excretion.

Although reactive oxygen species are integral to many cellular and biomolecular processes associated with acute coronary syndromes, the relationship to specific genetic polymorphisms has not been overly compelling and remains controversial. It may be that current eating behaviors prevent the easy identification of diet–gene–health interactions in this disease condition. However, mechanistically, the linkage with free radical–related gene polymorphisms is logical and deserves additional attention as a potential subtle long-term regulator of both cancer and heart disease risk. Little attention has been given to genes associated with oxidative stress and heart disease; the primary focus has been on cholesterol homeostasis and control [94].

A number of scientists have been exploring the effects of genetic polymorphisms on oxidative stress or the ability of antioxidants to influence the cancer process. Several examples about the possible utility of using specific polymorphisms as predictors of cancer risk have surfaced. For example, the manganese superoxide dismutase (MnSOD) protein is involved in decreasing the levels of superoxide anion generated during normal biochemical processes. A polymorphism that changes an alanine residue to valine in position 9 in the protein has been shown to be a risk factor in prostate carcinogenesis and is found in a statistically higher frequency in individuals with prostate cancer [95]. Likewise, breast cancer risk was slightly increased (odds ratio 1.3) in women with Ala/Ala genotype compared with those with Val/Val genotype [96]. Interestingly, patients carrying the Val allele may have a higher prevalence of cardiomyopathy [97]. However, these findings need to be confirmed.

The catalase (CAT) gene polymorphism at 262C –> T may also have a role in breast cancer development. This antioxidant enzyme neutralizes hydrogen peroxide and is known to be induced by oxidative challenges. A -262C –> T polymorphism in the promoter region of CAT is associated with risk of several conditions related to oxidative stress. Interestingly, the CC genotype relationship with breast cancer reduction was only observed among Caucasians and not in African Americans [98]. CAT polymorphism at codon 262 does not appear to be real to diabetes and the risk of heart disease [99]. The reason for the disconnect among genetic polymorphisms (CAT and MnSOD), oxidative damage, and risk of cancer and heart disease warrants additional attention.

Still another gene involved in antioxidant activity and cancer risk is glutathione peroxidase 1 (GPX-1) (100). Although codon 198 can lead to leucine or proline, it was determined that the leucine-containing allele was more frequently associated with breast cancer than the proline-containing allele [100]; however, there are inconsistencies in the literature [101]. Combinations of gene polymorphism may offer additional insights into risk. Cox *et al.* observed an increased breast cancer relative risk (OR = 1.87) in individuals with the MnSOD Ala16Ala genotype and the Leu198Leu genotype of GPX-1, while neither surfaced as risk modifiers when considered independently [102].

Environmental factors may also dictate the importance of genetic polymorphisms in determining cancer risk. For example, the T-allele in the GPX-1 gene at position 198 is considered to be protective in smokers. In a study of smokers with 432 lung cancer cases and 366 controls, those possessing the variant T-allele were significantly less likely to develop lung cancer than individuals without it [103]. Likewise, patients with alcohol-induced cirrhosis who had the genotype consistent with low MnSOD activity and also with high-GPX-1 activity had much lower levels of potentially toxic levels of iron. In individuals with both polymorphisms, none developed hepatocellular carcinoma, as opposed to other polymorphism combinations where anywhere from 16% to 32% of the individuals developed liver cancer [104]. In addition, consumption of a number of food items may influence the relationship between individual polymorphisms and health. Both aldehyde dehydrogenase-2 (ALDH2) and x-ray repair cross-complementing 1 (XRCC1) genes have surfaced as having a role in the response to dietary selenium. The glutamic acid 487 lysine polymorphism in the ALDH2 gene and the arginine 399 glutamine polymorphism in the XRCC1 gene are both associated with an increased risk of esophageal cancer in individuals consuming a low-selenium diet. Additionally, this risk becomes even more pronounced in individuals who smoke or consume alcohol [105]. This may reflect the individual pharmokinetics in handling supplemental selenium. Likewise, the protective effect of the catalase CC genotype was even more pronounced among women who used dietary supplements, as well as those with high fruit and vegetable intakes [106]. In another study, an inverse association between fruit and vegetable consumption and breast cancer risk was observed among women with the wild-type genotype for codon 84 of their O(6)-methylguanine DNA methyl-transferase (MGMT) gene [107]. In this group, the observed OR was 0.8, while for individuals with other polymorphisms the effects of a varied intake were not statistically significant. The association between fruit and vegetable consumption and reduced breast cancer risk was also seen in individuals with a variant allele for codon 143 in their mgMT gene. In one study, the intake of α- and β-carotene as modifiers of skin cancer was found to relate to MnSOD V16A polymorphism, such that an inverse association of intake was limited to the Val carriers, whereas no association was observed among women with the AA genotype [108].

D. Antioxidants: Prevention versus Treatment in Clinical Applications

As has been indicated, antioxidants can have both beneficial and deleterious effects on disease. In the case of cardiovascular disease in general and atherosclerosis specifically, the effects of antioxidants would be expected to be primarily beneficial as lipid oxidation and inflammation are important factors in the sequelae involving atherosclerotic plaque development. The overall consensus is that dietary antioxidant interventions seem to have a minor effect on improving atherosclerosis prevention [109]. However, reviews on the relationships of the statin drugs with free radical generation and inflammation provide evidence that statin drugs have a strong protective effect over and above their more well-known cholesterol-lowering activity [110]. Some of these protective effects may, however, be due to increased expression of antioxidant enzymes such as catalases and a suppression of pro-oxidant enzyme systems which reduce the production of the free radicals superoxide and peroxynitrite [111, 112].

Antioxidant and ROS effects on the cancer process are in part dependent on the status and behaviors of the consumer as well as the stage of development and type of cancer involved. The ability to predict and distinguish between the positive and negative effects of antioxidants is especially crucial in those receiving cancer treatments. Many cancer patients take vitamin supplements for protection and health promotion, especially during these times when eating behaviors are unusual. A survey of patients at a comprehensive cancer center found that 60% of patients used vitamins and the majority combined them with conventional therapy [113]. Similarly, among a Massachusetts cohort of women with early-stage breast cancer, 60% used megavitamin therapy along with surgery, chemotherapy, and/or radiation therapy [114]. Given the large numbers of cancer patients taking antioxidant vitamins, with or without the knowledge of their oncologists, greater understanding of the actions of antioxidants within the cancer disease and treatment milieu is crucial. Consumer awareness of the issues between antioxidants and cancer appear to be increasing but are far behind other health-related dietary practices [115].

Radiotherapy, as well as many chemotherapeutic agents, eliminated cancer cells by inducing apoptosis via generation of ROS. Thus, it was feared that antioxidants, especially in megadoses, may decrease ROS production and thus decrease the benefits of these treatments. Nevertheless, a number of *in vitro* and some clinical studies have shown that antioxidant treatment during standard cancer therapy does not always interfere with these therapies. The use of high-dose levels of antioxidants appears to be critical to achieve a beneficial outcome when antioxidants are used with more conventional treatments. In several reported studies, low doses of antioxidants stimulated the growth of human cancer cells in culture, and a single low dose of antioxidant before radiation therapy protected cells against radiation damage, suggesting that low doses of antioxidants might have detrimental effects on cancer treatment [116, 117]. High-dose levels have been defined in humans as up to 10 g vitamin C/day, up to 1,000 IU vitamin E/day, and up to 60 mg β-carotene/day; in tissue culture, high levels have been considered up to 200 μg vitamin C/ml, up to 20 μg vitamin E/ml, and up to 15 μg β-carotene/ml [118]. Low-dose levels have been defined in humans as approximately the recommended dietary allowance values and in tissue culture as up to 50 μg vitamin C/ml, up to 5 μg vitamin E/ml, and up to 1 μg β-carotene/ml [116].

Although antioxidants may cause some inhibition of ROS production caused by chemo- or radiation therapy, at high levels the antioxidant can directly inhibit tumor growth and can appear to stimulate the drug effects. Prasad *et al.* described the ability of the α-tocopherol succinate form of vitamin E to enhance the cytotoxic effects of adriamycin in human prostate cancer, glioma, and HeLa cells; enhance cisplatin, tamoxifen, and decarbazine on human melanoma and parotid acinar carcinoma cells; and promote doxorubicin to inhibit murine leukemia cell lines [118].

Data from large-scale trials with antioxidants are not available, although some smaller studies provide evidence that antioxidant supplementation can increase the efficacy of standard chemotherapies in humans. In 18 nonrandomized patients with small cell lung cancer, supplementation with multiple antioxidants along with chemotherapy or radiation therapy resulted in increased survival times. Retinoic acid and interferon enhanced the effect of radiation on locally advanced cervical adenoma, again improving survival times [119]. Two patients with advanced epithelial ovarian carcinoma (stage IIIC) received antioxidant therapy (1200 IU vitamin E, 300 mg coenzyme Q_{10}, 9000 mg vitamin C, 25 mg mixed carotenoids, and 10,000 IU vitamin A) before conventional chemotherapy and 60 g ascorbic acid twice weekly at the end of therapy. Both women had normal CA-125 levels and remained disease-free more than 3 years after initial diagnosis [120].

In an open trial in patients with small cell lung cancer, patients who received individualized daily supplements of trace elements, fatty acids, and vitamins (including 15–40,000 IU vitamin A, 10–20,000 IU β-carotene, 300–800 IU vitamin E, 2–5 g vitamin C, and 1600–3400 μg selenium) had a greater 2-year survival rate than historical controls (33% versus <15%) [121, 122]. Furthermore, an analysis of 385 breast cancer patients asked to recall antioxidant supplement use showed that users of supplements were less likely to have a breast cancer recurrence or breast cancer–related death than nonusers [123]. In contrast, an analysis of outcomes for 90 women with unilateral nonmetastatic breast cancer who received megadoses of vitamins and minerals (β-carotene, vitamin C, niacin, selenium, coenzyme Q_{10}, and zinc) showed no improvement in survival compared to matched controls (5-year survival rate was 72% for cases, 81% for controls) [124]. Although definitive conclusions cannot be drawn from this limited number of studies improvement in survival of patients receiving antioxidants in addition to chemotherapy has been observed. The use of antioxidant vitamins in conjunction with standard therapy warrants additional attention [125].

III. OVERALL CONCLUSION AND DISCUSSION

ROS are a potential double-edged sword in disease promotion and prevention. Whereas generation of ROS once was viewed as detrimental to the overall health of the organism, advances in research have shown that ROS play crucial roles in normal physiological processes including response to growth factors, the immune response, and apoptotic elimination of damaged cells. Notwithstanding these beneficial functions, aberrant production or regulation of ROS activity has been demonstrated to contribute to the development of some prevalent diseases and conditions, including cancer and CVD. Antioxidant supplementation has historically been viewed as a promising therapy for prevention and treatment of these diseases, especially given the tantalizing but sometimes conflicting links observed between diets high in fruits and vegetables (and presumably antioxidants) and decreased risks for cancer and CVD. Trials of individual antioxidants, however, have rarely shown strongly beneficial effects. In most healthy individuals, endogenous antioxidant defenses may be sufficient, and extra supplementation may have little effect on disease susceptibility. In some populations, particularly those with underlying illness or compromised nutritional status, dietary or other supplementation may be helpful. A better understanding of the impact of supplementation on disease risk also may be furthered by increased knowledge of individual genetic polymorphisms related to the metabolism and detoxification of ROS and antioxidants, as well as interactions of metabolic intermediates. A more critical issue is the interaction of antioxidants and orthodox cancer therapy.

Advances in tools used to determine oxidative status may allow for the estimating of an individual antioxidant profile. Whether supplements have positive or negative effects may depend on an individual's baseline antioxidant and ROS levels. Antioxidant supplementation in those with low baseline ROS could be detrimental, because it may impair normal immune function and prevent ROS-mediated apoptotic elimination of precancerous or cancerous cells. Similarly, better measurements of ROS status may help predict the potential for antioxidants in individuals exposed to specific environmental toxins, including but not limited to cigarette smoke, which may influence the magnitude and direction of the response.

Current measures of antioxidant status in an individual vary greatly among different research settings making comparison of results problematic. Establishment of "gold standard" biomarkers of oxidative stress and establishment of guidelines to accurately and precisely determine levels of a given marker in normal, healthy individuals should help resolve the uncertainties in the literature. Better measurements also may help determine why a diet high in fruits and vegetables often is more beneficial than specific antioxidant supplements. Additional research focusing on critical events that contribute to disease progression will help in defining the most efficient times for intervention with antioxidant supplementation.

Future research must focus on defining molecular mechanisms that engender oxidative stress that transforms healthy conditions to diseased states and the identification of timelines for effective interventions with antioxidants as either preventive and therapeutic agents. Although much of the current literature about antioxidants, including administration of pharmacological or dietary amounts, focuses on enhancing ROS elimination and the inhibiting ROS generation, many other cellular processes are likely involved. An improved understanding of these cellular processes and how they relate to measures of oxidative stress will provide important clues about those who will benefit most or be placed at risk from antioxidant usage, whether provided as foods or supplements. The identification of the reliable and sensitive biomarkers of antioxidant exposures, of their biological effects or consequences, and of susceptibility factors including genetic and environmental modifiers will have far-reaching implications for the monitoring and treatment of oxidative-stress related to health and a number of diseased conditions.

References

1. Panel on Dietary Antioxidants and Related Compounds. (2000). A report by the Panel on Dietary Antioxidants and Related Compounds: Dietary Reference Intakes for Vitamin C, Vitamin E, Selenium, and Carotenoids. Food and Nutrition Board, Institute of Medicine, Washington, D.C.
2. Halliwell, B. (2007). Oxidative stress and cancer: have we moved forward? *Biochem. J.* **401**, 1–11.
3. Seifried, H. E., Anderson, D. E., Milner, J. A., and Greenwald, P. (2006). Reactive oxygen species and dietary antioxidants: double-edged swords? *In* "New Developments in Antioxidant Research" (H. Panglossi, Ed.), pp. 1–25. Nova Science, Hauppauge (NY).
4. U.S. Surgeon General. (1979). Healthy People: "The Surgeon General's Report on Health Promotion and Disease Prevention," Chapter 11, United States Public Health Service; Report No.: DHEW (PHS) Publication No. 79-55071, Washington, DC.
5. American Institute for Cancer Research/World Cancer Research Fund. (1997). *In* "Food, Nutrition and the Prevention of Cancer: A Global Perspective" (World Cancer Research Fund, Ed.). American Institute for Cancer Research, Washington, DC.
6. Radimer, K. L., Bindewald, B., Hughes, J., Ervin, B., Swanson, C., and Picciano, M. F. Dietary supplement use by US adults: data from the National Health and Nutrition Examination Survey, 1999–2000. *Am. J. Epidemiol.* **160**(4) 339–349.
7. Mishra, V. (2007). Oxidative stress and role of antioxidant supplementation in critical illness. *Clin. Lab.* **53**, 199–209.
8. Seifried, H. E., Anderson, D. E., Fisher, E. I., and Milner, J. A. (2007). A review of the interaction among dietary antioxidants and reactive oxygen species. *J. Nutr. Biochem.*, Mar. 13 [Epub ahead of print].
9. Prior, R. L., Wu, X., and Schaich, K. (2005). Standardized methods for the determination of antioxidant capacity and phenolics in foods and dietary supplements. *J. Agric. Food Chem.* **53**(10), 4290–4302.
10. Huang, D., Ou, B., and Prior, R. L. (2005). The chemistry behind antioxidant capacity assays. *J. Agric. Food Chem.* **53**(6), 1841–1856.
11. Lotito, S. B., and Frei, B. (2006). Consumption of flavonoid-rich foods and increased plasma antioxidant capacity in humans: cause, consequence, or epiphenomenon? *Free Radic. Biol. Med.* **41**(12), 1727–1746.
12. Martindale, J. L., and Holbrook, N. J. (2002). Cellular response to oxidative stress: signaling for suicide and survival. *J. Cell Physiol.* **192**, 1–15.
13. Shekelle, P., Hardy, M. L., Coulter, I., Udani, J., Spar, M., Oda, K., Jungvig, L. K., Tu, W., Suttorp, M. J., Valentine, D., Ramirez, L., Shanman, R., and Newberry, S. J. (2003). Effects of the supplemental use of antioxidant vitamin C, vitamin E and coenzyme Q10 for the prevention and treatment of cancer. *Evid. Rep. Technol. Assess. (Summ.)* **75**, 1–3.
14. Taylor, P. R., and Greenwald, P. (2005). Nutritional interventions in cancer prevention. *J. Clin. Oncol.* **23**, 333–345.
15. Castro, L., and Freeman, B. A. (2001). Reactive oxygen species in human health and disease. *Nutrition* **17**, 161, 163–165.
16. Sohal, R. S., and Weindruch, R. (1996). Oxidative stress, caloric restriction, and aging. *Science* **273**, 59–63.
17. McCord, J. M. (2000). The evolution of free radicals and oxidative stress. *Am. J. Med.* **108**, 652–659.
18. Lopaczynski, W., and Zeisel, S. H. (2004). Antioxidants, programmed cell death, and cancer. *Nutr. Res.* **21**, 295–307.
19. Finkel, T., and Holbrook, N. J. (2000). Oxidants, oxidative stress and the biology of ageing. *Nature* **408**, 239–247.
20. Holbrook, N. J., and Ikeyama, S. (2002). Age-related decline in cellular response to oxidative stress: links to growth factor signaling pathways with common defects. *Biochem. Pharmacol.* **64**, 999–1005.
21. Hanahan, D., and Weinberg, R. A. (2000). The hallmarks of cancer. *Cell* **100**, 57–70.
22. Dy, G. K., Bekele, L., Hanson, L. J., Furth, A., Mandrekar, S., Sloan, J. A. and Adjei, A. A. (2004). Complementary and alternative medicine use by patients enrolled onto phase I clinical trials. *J. Clin. Oncol.* **22**, 4810–4815.
23. Gerson-Cwilich, R., Serrano-Olvera, A., and Villalobos-Prieto, A. (2006). Complementary and alternative medicine (CAM) in Mexican patients with cancer. *Clin. Transl. Oncol.* **8**(3), 200–207.
24. Itabe, H. (2003). Oxidized low-density lipoproteins: what is understood and what remains to be clarified. *Biol. Pharm. Bull.* **26**, 1–9.
25. Duval, C., Cantero, A. V., Auge, N., Mabile, L., Thiers, J. C., Negre-Salvayre, A., and Salvayre, R. (2003). Proliferation and wound healing of vascular cells trigger the generation of extracellular reactive oxygen species and LDL oxidation. *Free Radic. Biol. Med.* **35**, 1589–1598.
26. Droge, W. (2002). Free radicals in the physiological control of cell function. *Physiol. Rev.* **82**, 47–95.
27. Sullivan, G. W., Sarembock, I. J., and Linden, J. (2000). The role of inflammation in vascular diseases. *J. Leukoc. Biol.* **67**, 591–602.
28. Hung, H. C., Joshipura, K. J., Jiang, R., Hu, F. B., Hunter, D., Smith-Warner, S. A., Colditz, G. A., Rosner, B., Spiegelman, D., and Willett, W. C. (2004). Fruit and vegetable intake and risk of major chronic disease. *J. Natl. Cancer Inst.* **96**, 1577–1584.
29. Bjelakovic, G., Nikolova, D., Simonetti, R. G., and Gluud, C. (2004). Antioxidant supplements for prevention of gastrointestinal cancers: a systematic review and meta-analysis. *Lancet* **364**, 1219–1228.
30. Goodman, G. E., Thornquist, M. D., Balmes, J., Cullen, M. R., Meyskens, F. L. Jr., Omenn, G. S., Valanis, B., and Williams, J. H. Jr. (2004). The Beta-Carotene and Retinol Efficacy Trial: incidence of lung cancer and cardiovascular disease mortality during 6-year follow-up after stopping beta-carotene and retinol supplements. *J. Natl. Cancer Inst.* **96**, 1743–1750.
31. Salganik, R. I. (2001). The benefits and hazards of antioxidants: controlling apoptosis and other protective mechanisms

in cancer patients and the human population. *J. Am. Coll. Nutr.* **20**, 464S–472S.

32. Food and Drug Administration Office of Special Nutritionals. (1993). Conference on Antioxidant Vitamins and Cancer and Cardiovascular Disease, November 1-3. Food and Drug Administration, Washington, DC.
33. Seifried, H. E., McDonald, S. S., Anderson, D. E., Greenwald, P., and Milner, J. A. (2003). The antioxidant conundrum in cancer. *Cancer Res.* **63**, 4295–4298.
34. Seifried, H. E., Anderson, D. E., Sorkin, B. C., and Costello, R. B. (2004). Free radicals: the pros and cons of antioxidants. Executive summary report. *J. Nutr.* **134**, 3143S–3163S.
35. Blot, W. J., Li, J.-Y., Taylor, P. R., Guo, W., Dawsey, S., Wang, G.-Q., Yang, C. S., Zheng, S.-F., Gail, M., Li, G.-Y., Yu, Y., Liu, B., Tangrea, J., Sun, Y., Liu, F., Fraumeni, J. F., Zhang, Y.-H., and Jr., Li, B. (1993). Nutrition intervention trials in Linxian, China: supplementation with specific vitamin/mineral combinations, cancer incidence, and disease-specific mortality in the general population. *J. Natl. Cancer Inst.* **85**, 1483–1492.
36. Hennekens, C. H., Buring, J. E., Manson, J. E., Stampfer, M., Rosner, B., Cook, N. R., Belanger, C., LaMotte, F., Gaziano, J. M., Ridker, P. M., Willett, W., and Peto, R. (1996). Lack of effect of long-term supplementation with beta carotene on the incidence of malignant neoplasms and cardiovascular disease. *N. Engl. J. Med.* **334**, 1145–1149.
37. Lee, I. M., Cook, N. R., Manson, J. E., Buring, J. E., and Hennekens, C. H. (1999). Beta-carotene supplementation and incidence of cancer and cardiovascular disease: the Women's Health Study. *J. Natl. Cancer Inst.* **91**, 2102–2106.
38. Cook, N. R., Stampfer, M. J., Ma, J., Manson, J. E., Sacks, F. M., Buring, J. E., and Hennekens, C. H. (1999). Beta-carotene supplementation for patients with low baseline levels and decreased risks of total and prostate carcinoma. *Cancer* **86**, 1783–1792.
39. Russell, R. M. (2004). The enigma of beta-carotene in carcinogenesis: what can be learned from animal studies. *J. Nutr.* **134**, 262S–268S.
40. Greenberg, E. R., Baron, J. A., Stukel, T. A., Stevens, M. M., Mandel, J. S., Spencer, S. K., Elias, P. M., Lowe, N., Nierenberg, D. W., and Bayrd, G. (1990). A clinical trial of beta carotene to prevent basal-cell and squamous-cell cancers of the skin. The Skin Cancer Prevention Study Group. *N. Engl. J. Med.* **323**, 789–795.
41. MacLennan, R., Macrae, F., Bain, C., Battistutta, D., Chapuis, P., Gratten, H., Lambert, J., Newland, R. C., Ngu, M., Russell, A., Ward, M., and Wahlqvist, M. L. (1995). Randomized trial of intake of fat, fiber, and beta carotene to prevent colorectal adenomas. The Australian Polyp Prevention Project. *J. Natl. Cancer Inst.* **87**, 1760–1766.
42. Albanes, D., Heinonen, O. P., Huttunen, J. K., Taylor, P. R., Virtamo, J., Edwards, B. K., Haapakoski, J., Rautalahti, M., Hartman, A. M., Palmgren, J., and Greenwald, P. (1995). Effects of alpha-tocopherol and beta-carotene supplements on cancer incidence in the Alpha-Tocopherol Beta-Carotene Cancer Prevention Study. *Am. J. Clin. Nutr.* **62**(suppl 6), 1427S–1430S.
43. Omenn, G. S., Goodman, G. E., Thornquist, M. D., Balmes, J., Cullen, M. R., Glass, A., Keogh, J. P., Meyskens, F. L. Jr., Valanis, B., Williams, J. H. Jr., Barnhart, S., Cherniack, M. G., Brodkin, C. A., and Hammar, S. (1996). Risk factors for lung cancer and for intervention effects in CARET, the Beta-Carotene and Retinol Efficacy Trial. *J. Natl. Cancer Inst.* **88**(21), 1550–1559.
44. Heinonen, O. P., Albanes, D., Virtamo, J., Taylor, P. R., Huttunen, J. K., Hartman, A. M., Haapakoski, J., Malila, N., Rautalahti, M., Ripatti, S., Maenpaa, H., Teerenhovi, L., Koss, L., Virolainen, M., and Edwards, B. K. (1998). Prostate cancer and supplementation with alpha-tocopherol and beta-carotene: incidence and mortality in a controlled trial. *J. Natl. Cancer Inst.* **90**, 440–446.
45. Albanes, D., Malila, N., Taylor, P. R., Huttunen, J. K., Virtamo, J., Edwards, B. K., Rautalahti, M., Hartman, A. M., Barrett, M. J., Pietinen, P., Hartman, T. J., Sipponen, P., Lewin, K., Teerenhovi, L., Paivi Hietanen, P., Tangrea, J. A., Virtanen, M., and Heinonen, O. P. (2000). Effects of supplemental alpha-tocopherol and beta-carotene on colorectal cancer: results from a controlled trial (Finland). *Cancer Causes Control* **11**, 197–205.
46. Neuhouser, M. L., Patterson, R. E., Thornquist, M. D., Omenn, G. S., King, I. B., and Goodman, G. E. (2003). Fruits and vegetables are associated with lower lung cancer risk only in the placebo arm of the beta-carotene and retinol efficacy trial (CARET). *Cancer Epidemiol. Biomarkers Prev.* **12**(4), 350–358.
47. Mayne, S. T., Janerich, D. T., Greenwald, P., Chorost, S., Tucci, C., Zaman, M. B., Melamed, M. R., Kiely, M., and McKneally, M. F. (1994). Dietary beta carotene and lung cancer risk in U.S. nonsmokers. *J. Natl. Cancer Inst.* **86**(1), 33–38.
48. Speizer, F. E., Colditz, G. A., Hunter, D. J., Rosner, B., and Hennekens, C. (1999). Prospective study of smoking, antioxidant intake, and lung cancer in middle-aged women (USA). *Cancer Causes Control* **10**(5), 475–482.
49. Touvier, M., Kesse, E., Clavel-Chapelon, F., and Boutron-Ruault, M. C. (2005). Dual association of beta-carotene with risk of tobacco-related cancers in a cohort of French women. *J. Natl. Cancer Inst.* **97**(18), 1338–1344.
50. Liu, C., Russell, R. M., and Wang, X. D. (2004). Alpha-tocopherol and ascorbic acid decrease the production of beta-apo-carotenals and increase the formation of retinoids from beta-carotene in the lung tissues of cigarette smoke-exposed ferrets in vitro. *J. Nutr.* **134**(2), 426–430.
51. Alija, A. J., Bresgen, N., Sommerburg, O., Langhans, C. D., Siems, W., and Eckl, P. M. (2006). Beta-carotene breakdown products enhance genotoxic effects of oxidative stress in primary rat hepatocytes. *Carcinogenesis* **27**(6), 1128–1133.
52. Weinstein, S. J., Wright, M. E., Pietinen, P., King, I., Tan, C., Taylor, P. R., Virtamo, J., and Albanes, D. (2005). Serum alpha-tocopherol and gamma-tocopherol in relation to prostate cancer risk in a prospective study. *J. Natl. Cancer Inst.* **97**, 396–399.
53. Virtamo, J., Pietinen, P., Huttunen, J. K., Korhonen, P., Malila, N., Virtanen, M. J., Albanes, D., Taylor, P. R., and Albert, P., and ATBC Study Group. (2003). Incidence of cancer and mortality following alpha-tocopherol and beta-carotene supplementation: a postintervention follow-up. *JAMA* **290**, 476–485.
54. Wright, M. E., Virtamo, J., Hartman, A. M., Pietinen, P., Edwards, B. K., Taylor, P. R., Huttunen, J. K., and Albanes, D.

(2007). Effects of alpha-tocopherol and beta-carotene supplementation on upper aerodigestive tract cancers in a large, randomized controlled trial. *Cancer* **109**(5), 891–898.
55. Wright, M. E., Lawson, K. A., Weinstein, S. J., Pietinen, P., Taylor, P. R., Virtamo, J., and Albanes, D. (2006). Higher baseline serum concentrations of vitamin E are associated with lower total and cause-specific mortality in the Alpha-Tocopherol, Beta-Carotene Cancer Prevention Study. *Am. J. Clin. Nutr.* **84**(5), 1200–1207.
56. Holick, C. N., Michaud, D. S., Stolzenberg-Solomon, R., Mayne, S. T., Pietinen, P., Taylor, P. R., Virtamo, J., and Albanes, D. (2002). Dietary carotenoids, serum beta-carotene, and retinol and risk of lung cancer in the alpha-tocopherol, beta-carotene cohort study. *Am. J. Epidemiol.* **156**, 536–547.
57. Hercberg, S., Preziosi, P., Galan, P., Faure, H., Arnaud, J., Duport, N., Malvy, D., Roussel, A. M., Briançon, S., and Favier, A. (1999). "The SU.VI.MAX Study": a primary prevention trial using nutritional doses of antioxidant vitamins and minerals in cardiovascular diseases and cancers. SUpplementation on VItamines et Mineraux AntioXydants. *Food Chem. Toxicol.* **37**, 925–930.
58. Hercberg, S., Galan, P., Preziosi, P., Malvy, M., Briancon, S., Ait, H. M., Rahim, B., and Favier, A. (2002). The SU.VI.-MAX trial on antioxidants. *IARC Sci. Publ.* **156**, 451–455.
59. Hercberg, S., Galan, P., Preziosi, P., Bertrais, S., Mennen, L., Malvy, D., Roussel, A. M., Favier, A., and Briançon, S. (2004). The SU.VI.MAX Study: a randomized, placebo-controlled trial of the health effects of antioxidant vitamins and minerals. *Arch. Intern. Med.* **164**, 2335–2342.
60. Meyer, F., Galan, P., Douville, P., Bairati, I., Kegle, P., Bertrais, S., Estaquio, C., and Hercberg, S. (2005). Antioxidant vitamin and mineral supplementation and prostate cancer prevention in the SU.VI.MAX trial. *Int. J. Cancer* **116**(2), 182–186.
61. Clark, L. C., Combs G. F. Jr., Turnbull, B. W., Slate, E. H., Chalker, D. K., Chow, J., Davis, L. S., Glover, R. A., Graham, G. F., Gross, E. G., Krongrad, A., Lesher, J. L., Park, H. K., Sanders, B. B., Smith, C. L., and Taylor, J. R. (1996). Effects of selenium supplementation for cancer prevention in patients with carcinoma of the skin: a randomized controlled trial. Nutritional Prevention of Cancer Study Group. *JAMA* **276**, 1957–1963.
62. Duffield-Lillico, A. J., Dalkin, B. L., Reid, M. E., Turnbull, B. W., Slate, E. H., Jacobs, E. T., Marshall, J. R., Clark. L. C., and the Nutritional Prevention of Cancer Study Group. (2003). Selenium supplementation, baseline plasma selenium status and incidence of prostate cancer: an analysis of the complete treatment period of the Nutritional Prevention of Cancer Trial. *BJU Int.* **91**, 608–612.
63. Reid, M. E., Duffield-Lillico, A. J., Garland, L., Turnbull, B. W., Clark, L. C., and Marshall, J. R. (2002). Selenium supplementation and lung cancer incidence: an update of the nutritional prevention of cancer trial. *Cancer Epidemiol. Biomarkers Prev.* **11**, 1285–1291.
64. Li, H., Stampfer, M. J., Giovannucci, E. L., Morris, J. S., Willett, W. C., Gaziano, J. M., and Ma, J. (2004). A prospective study of plasma selenium levels and prostate cancer risk. *J. Natl. Cancer Inst.* **96**, 696–703.
65. Fan, T. W., Higashi, R. M., and Lane, A. N. (2006). Integrating metabolomics and transcriptomics for probing SE anticancer mechanisms. *Drug Metab. Rev.* **38**(4), 707–732.
66. van Gils, C. H., Peeters, P. H., Bueno-de-Mesquita, H. B., Boshuizen, H. C., Lahmann, P. H., Clavel-Chapelon, F., Thiebaut, A., Kesse, E., Sieri, S., Palli, D., Tumino, R., Panico, S., Vineis, P., Gonzalez, C. A., Ardanaz, E., Sanchez, M. J., Amiano, P., Navarro, C., Quiros, J. R., Key, T. J., Allen, N., Khaw, K. T., Bingham, S. A., Psaltopoulou, T., Koliva, M., Trichopoulou, A., Nagel, G., Linseisen, J., Boeing, H., Berglund, G., Wirfalt, E., Hallmans, G., Lenner, P., Overvad, K., Tjonneland, A., Olsen, A., Lund, E., Engeset, D., Alsaker, E., Norat, T., Kaaks, R., Slimani, N., and Riboli, E. (2005). Consumption of vegetables and fruits and risk of breast cancer. *JAMA* **293**, 183–193.
67. Narayanan, B. A. (2006). Chemopreventive agents alter global gene expression pattern: predicting their mode of action and targets. *Curr. Cancer Drug Targets* **6**(8), 711–727.
68. Narayanan, B. A., Narayanan, N. K., Re, G. G., and Nixon, D. W. (2003). Differential expression of genes induced by resveratrol in LNCaP cells: P53-mediated molecular targets. *Int. J. Cancer* **104**(2), 204–212.
69. Goulet, A. C., Watts, G., Lord, J. L., and Nelson, M. A. (2007). Profiling of selenomethionine responsive genes in colon cancer by microarray analysis. *Cancer Biol. Ther.* **6**(4) [Epub print].
70. Food and Drug Administration Office of Special Nutritionals, (1993). Conference on Antioxidant Vitamins and Cancer and Cardiovascular Disease, November 1-3. Food and Drug Administration, Washington, DC.
71. Miller, E. R. III, Pastor-Barriuso, R., Dalal, D., Riemersma, R. A., Appel, L. J., and Guallar, E. (2005). Meta-analysis: high-dosage vitamin E supplementation may increase all-cause mortality. *Ann. Intern. Med.* **142**, 37–46.
72. Dauchet, L., Amouyel, P., Hercberg, S., and Dallongeville, J. (2006). Fruit and vegetable consumption and risk of coronary heart disease: a meta-analysis of cohort studies. *J. Nutr.* **136**(10), 2588–2593.
73. Kushi, L. H., Folsom, A. R., Prineas, R. J., Mink, P. J., Wu, Y., and Bostick, R. M. (1996). Dietary antioxidant vitamins and death from coronary heart disease in postmenopausal women. *N. Engl. J. Med.* **334**, 1156–1162.
74. Bjelakovic, G., Nikolova, D., Gluud, L. L., Simonetti, R. G., and Gluud, C. (2007). Mortality in randomized trials of antioxidant supplements for primary and secondary prevention: systematic review and meta-analysis. *JAMA* **297**(8), 842–857.
75. Lonn, E., Bosch, J., Yusuf, S., Sheridan, P., Pogue, J., Arnold, J. M., Ross, C., Arnold, A., Sleight, P., Probstfield, J., and Dagenais, G. R. (2005). Effects of long-term vitamin E supplementation on cardiovascular events and cancer: a randomized controlled trial. *JAMA* **293**, 1338–1347.
76. Rapola, J. M., Virtamo, J., Haukka, J. K., Heinonen, O. P., Albanes, D., Taylor, P. R., and Huttunen, J. K. (1996). Effect of vitamin E and beta carotene on the incidence of angina pectoris: a randomized, double-blind, controlled trial. *JAMA* **275**, 693–698.
77. Stephens, N. G., Parsons, A., Schofield, P. M., Kelly, F., Cheeseman, K., and Mitchinson, M. J. (1996). Randomised controlled trial of vitamin E in patients with coronary disease: Cambridge Heart Antioxidant Study (CHAOS). *Lancet* **347**, 781–786.
78. Morris, C. D., and Carson, S. (2003). Routine vitamin supplementation to prevent cardiovascular disease: a summary of the

evidence for the U.S. Preventive Services Task Force. *Ann. Intern. Med.* **139**, 56–70.

79. Azzi, A., Breyer, I., Feher, M., Pastori, M., Ricciarelli, R., Spycher, S., Staffieri, M., Stocker, A., Zimmer, S., and Zingg, J. M. (2000). Specific cellular responses to alpha-tocopherol. *J. Nutr.* **130**, 1649–1652.
80. Azzi, A., Gysin, R., Kempna, P., Ricciarelli, R., Villacorta, L., Visarius, T., and Zingg, J. M. (2003). The role of alpha-tocopherol in preventing disease: from epidemiology to molecular events. *Mol. Aspects Med.* **24**, 325–336.
81. Sung, L., Greenberg, M. L., Koren, G., Tomlinson, G. A., Tong, A., Malkin, D., and Feldman, B. M. (2003). Vitamin E: the evidence for multiple roles in cancer. *Nutr. Cancer* **46**, 1–14.
82. Gokkusu, C., Palanduz, S., Ademoglu, E., and Tamer, S. (2001). Oxidant and antioxidant systems in niddm patients: influence of vitamin E supplementation. *Endocr. Res.* **27**, 377–386.
83. Salonen, R. M., Nyyssonen, K., Kaikkonen, J., Porkkala-Sarataho, E., Voutilainen, S., Rissanen, T. H., Tuomainen, T. P., Valkonen, V. P., Ristonmaa, U., Lakka, H. M., Vanharanta, M., Salonen, J. T., and Poulsen, H. E. (2003). Six-year effect of combined vitamin C and E supplementation on atherosclerotic progression: the Antioxidant Supplementation in Atherosclerosis Prevention (ASAP) Study. *Circulation* **107**, 947–953.
84. Fang, J. C., Kinlay, S., Beltrame, J., Hikiti, H., Wainstein, M., Behrendt, D., Suh, J., Frei, B., Mudge, G. H., Selwyn, A. P., and Ganz, P. (2002). Effect of vitamins C and E on progression of transplant-associated arteriosclerosis: a randomized trial. *Lancet* **359**, 1008–1013.
85. Heart Protection Study Collaborative Group. (2002). MRC/BHF Heart Protection Study of antioxidant vitamin supplementation in 20,536 high-risk individuals: a randomized placebo-controlled trial. *Lancet* **360**, 23–33.
86. Christen, W. G., Gaziano, J. M., and Hennekens, C. H. (2000). Design of Physicians' Health Study II–a randomized trial of beta-carotene, vitamins E and C, and multivitamins, in prevention of cancer, cardiovascular disease, and eye disease, and review of results of completed trials. *Ann. Epidemiol.* **10**, 125–134.
87. Liu, S., Manson, J. E., Lee, I. M., Cole, S. R., Hennekens, C. H., Willett, W. C., and Buring, J. E. (2000). Fruit and vegetable intake and risk of cardiovascular disease: the Women's Health Study. *Am. J. Clin. Nutr.* **72**, 922–928.
88. Mennen, L. I., Sapinho, D., de Bree, A., Arnault, N., Bertrais, S., Galan, P., and Hercberg, S. (2004). Consumption of foods rich in flavonoids is related to a decreased cardiovascular risk in apparently healthy French women. *J. Nutr.* **134**, 923–926.
89. Salonen, J. T. (2002). Clinical trials testing cardiovascular benefits of antioxidant supplementation. *Free Radic. Res.* **36**, 1299–1306.
90. Gruppo Italiano per lo Studio della Sopravvivenza nell Infarto miocardico. (1999). Dietary supplementation with n-3 polyunsaturated fatty acids and vitamin E after myocardial infarction: results of the GISSI-Prevenzione trial. Gruppo Italiano per lo Studio della Sopravvivenza nell'Infarto miocardico. *Lancet* **354**, 447–455.
91. Yusuf, S., Sleight, P., Pogue, J., Bosch, J., Davies, R., and Dagenais, G. (2000). Effects of an angiotensin-converting-enzyme inhibitor, ramipril, on cardiovascular events in high-risk patients. The Heart Outcomes Prevention Evaluation Study Investigators. *N. Engl. J. Med.* **342**, 145–153.
92. Dagenais, G. R., Yusuf, S., Bourassa, M. G., Yi, Q., Bosch, J., Lonn, E. M., Kouz, S., and Grover, J. (2001). Effects of ramipril on coronary events in high-risk persons: results of the Heart Outcomes Prevention Evaluation Study. *Circulation* **104**, 522–526.
93. Harrison, R. A., Holt, D., Pattison, D. J., and Elton, P. J. (2004). Are those in need taking dietary supplements? A survey of 21,923 adults. *Br. J. Nutr.* **91**, 617–623.
94. Ordovas, J. M. (2007). Identification of a functional polymorphism at the adipose fatty acid binding protein gene (FABP4) and demonstration of its association with cardiovascular disease: a path to follow. *Nutr. Rev.* **65**(3), 130–134.
95. Ergen, H. A., Narter, F., Timirci, O., and Isbir, T. (2007). Effects of manganese superoxide dismutase Ala-9Val polymorphism on prostate cancer: a case-control study. *Anticancer Res.* **27**(2), 1227–1230.
96. Cai, Q., Shu, X. O., Wen, W., Cheng, J. R., Dai, Q., Gao, Y. T., and Zheng, W. (2004). Genetic polymorphism in the manganese superoxide dismutase gene, antioxidant intake, and breast cancer risk: results from the Shanghai Breast Cancer Study. *Breast Cancer Res.* **6**(6), R647–R655.
97. Valenti, L., Conte, D., Piperno, A., Dongiovanni, P., Fracanzani, A. L., Fraquelli, M., Vergani, A., Gianni, C., Carmagnola, L., and Fargion, S. (2004). The mitochondrial superoxide dismutase A16V polymorphism in the cardiomyopathy associated with hereditary haemochromatosis. *J. Med. Genet.* **41**(12), 946–950.
98. Ahn, J., Nowell, S., McCann, S. E., Yu, J., Carter, L., Lang, N. P., Kadlubar, F. F., Ratnasinghe, L. D., and Ambrosone, C. B. (2006). Associations between catalase phenotype and genotype: modification by epidemiologic factors. 98.1. *Cancer Epidemiol, Biomarkers Prev.* **15**(6), 1217–1222.
99. Ukkola, O., Erkkila, P. H., Savolainen, M. J., and Kesaniemi, Y. A. (2001). Lack of association between polymorphisms of catalase, copper-zinc superoxide dismutase (SOD), extracellular SOD and endothelial nitric oxide synthase genes and macroangiopathy in patients with type 2 diabetes mellitus. *J. Intern. Med.* **249**(5), 451–459.
100. Hu, Y. J., and Diamond, A. M. (2003). Role of glutathione peroxidase 1 in breast cancer: loss of heterozygosity and allelic differences in the response to selenium. *Cancer Res.* **63**(12), 3347–3351.
101. Ahn, J., Gammon, M. D., Santella, R. M., Gaudet, M. M., Britton, J. A., Teitelbaum, S. L., Terry, M. B., Neugut, A. I., and Ambrosone, C. B. (2005). No association between glutathione peroxidase Pro198 Leu polymorphism and breast cancer risk. *Cancer Epidemiol. Biomarkers Prev.* **14**(10), 2459–2461.
102. Cox, D. G., Tamimi, R. M., and Hunter, D. J. (2006). Gene × Gene interaction between MnSOD and GPX-1 and breast cancer risk: a nested case-control study. *BMC Cancer* **6**, 217.

103. Raaschou-Nielsen, O., Sorensen, M., Hansen, R. D., Frederiksen, K., Tjonneland, A., Overvad, K., and Vogel, U. (2007). GPX1 Pro198Leu polymorphism, interactions with smoking and alcohol consumption, and risk for lung cancer. *Cancer Lett.* **247**(2), 293–300.
104. Sutton, A., Nahon, P., Pessayre, D., Rufa, t P., Poire, A., Ziol, M., Vidaud, D., Barget, N., Ganne-Carrie, N., Charnaux, N., Trinchet, J. C., Gattegno, L., and Beaugrand, M. (2006). Genetic polymorphisms in antioxidant enzymes modulate hepatic iron accumulation and hepatocellular carcinoma development in patients with alcohol-induced cirrhosis. *Cancer Res.* **66**(5), 2844–2852.
105. Cai, L., You, N. C., Lu, H., Mu, L. N., Lu, Q. Y., Yu, S. Z., Le, A. D., Marshall, J., Heber, D., and Zhang, Z. F. (2006). Dietary selenium intake, aldehyde dehydrogenase-2 and X-ray repair cross-complementing 1 genetic polymorphisms, and the risk of esophageal squamous cell carcinoma. *Cancer* **106**(11), 2345–2354.
106. Ahn, J., Gammon, M. D., Santella, R. M., Gaudet, M. M., Britton, J. A., Teitelbaum, S. L., Terry, M. B., Nowell, S., Davis, W., Garza, C., Neugut, A. I., and Ambrosone, C. B. (2005). Associations between breast cancer risk and the catalase genotype, fruit and vegetable consumption, and supplement use. *Am. J. Epidemiol.* **162**(10), 943–952.
107. Shen, J., Terry, M. B., Gammon, M. D., Gaudet, M. M., Teitelbaum, S. L., Eng, S. M., Sagiv, S. K., Neugut, A. I., and Santella, R. M. (2005). MGMT genotype modulates the associations between cigarette smoking, dietary antioxidants and breast cancer risk. *Carcinogenesis* **26**(12), 2131–2137.
108. Han, J., Colditz, G. A., and Hunter, D. J. (2007). Manganese superoxide dismutase polymorphism and risk of skin cancer (United States). *Cancer Causes Control* **18**(1), 79–89.
109. Puddu, G. M., Cravero, E., Arnone, G., Muscari, A., and Puddu, P. (2005). Molecular aspects of atherogenesis: new insights and unsolved questions. *J. Biomed Sci.* **12**, 839–853.
110. Stoll, L. L.,, McCornick, M. L., Denning, G. M., and Weintraub, N. L. (2004). Antioxidant effects of statins. *Drugs Today* (Barc) **40**, 975–990.
111. Drinitsina, S. V., and Zateishchikova, D. A. (2005). Antioxidant properties of statins [Russian]. *Kardiologia* **45**, 65–72.
112. Guzik, T. J., and Harrion, D. G. (2006). Vascular NADPH oxidases as drug targets for novel antioxidant strategies. *Drug Discovery Today* **11**, 524–533.
113. Richardson, M. A., Sanders, T., Palmer, J. L., Greisinger, A., and Singletary, S. E. (2000). Complementary/alternative medicine use in a comprehensive cancer center and the implications for oncology. *J. Clin. Oncol.* **18**, 2505–2514.
114. Burstein, H. J., Gelber, S., Guadagnoli, E., and Weeks, J. C. (1999). Use of alternative medicine by women with early-stage breast cancer. *N. Engl. J. Med.* **340**, 1733–1739.
115. Toner, C. (2004). Consumer perspectives about antioxidants. *J. Nutr.* **134**, 3192S–3193S.
116. Prasad, K. N., Cole, W. C., Kumar, B., and Che, P. K. (2002). Pros and cons of antioxidant use during radiation therapy. *Cancer Treat. Rev.* **28**, 79–91.
117. Prasad, K. N. (2004). Rationale for using high-dose multiple dietary antioxidants as an adjunct to radiation therapy and chemotherapy. *J. Nutr.* **134**, 3182S–3183S.
118. Prasad, K. N., Kumar, B., Yan, X. D., Hanson, A. J., and Cole, W. C. (2003). Alpha-tocopheryl succinate, the most effective form of vitamin E for adjuvant cancer treatment: a review. *J. Am. Coll. Nutr.* **22**, 108–117.
119. Prasad, K. N., Cole, W. C., Kumar, B., and Prasad, K. C. (2001). Scientific rationale for using high-dose multiple micronutrients as an adjunct to standard and experimental cancer therapies. *J. Am. Coll. Nutr.* **20**, 450S–463S.
120. Drisko, J. A., Chapman, J., and Hunter, V. J. (2003). The use of antioxidants with first-line chemotherapy in two cases of ovarian cancer. *J. Am. Coll. Nutr.* **22**, 118–123.
121. Lamson, D. W., and Brignall, M. S. (1999). Antioxidants in cancer therapy; their actions and interactions with oncologic therapies. *Altern. Med. Rev.* **4**, 304–329.
122. Jaakkola, K., Lahteenmaki, P., Laakso, J., Harju, E., Tykka, H., and Mahlberg, K. (1992). Treatment with antioxidant and other nutrients in combination with chemotherapy and irradiation in patients with small-cell lung cancer. *Anticancer Res.* **12**, 599–606.
123. Fleischauer, A. T., Simonsen, N., and Arab, L. (2003). Antioxidant supplements and risk of breast cancer recurrence and breast cancer-related mortality among postmenopausal women. *Nutr. Cancer* **46**, 15–22.
124. Lesperance, M. L., Olivotto, I. A., Forde, N., Zhao, Y., Speers, C., Foster, H., Tsao, M., MacPherson, N., and Hoffer, A. (2002). Mega-dose vitamins and minerals in the treatment of non-metastatic breast cancer: an historical cohort study. *Breast Cancer Res. Treat.* **76**, 137–143.
125. Conklin, K. A. (2000). Dietary antioxidants during cancer chemotherapy: impact on chemotherapeutic effectiveness and development of side effects. *Nutr. Cancer* **37**, 1–18.

CHAPTER 17

Nutrients and Food Constituents in Cognitive Decline and Neurodegenerative Disease

JAMES A. JOSEPH,[1] MARK A. SMITH,[2] GEORGE PERRY,[3] AND BARBARA SHUKITT-HALE[4]

[1]*USDA, HNRCA at Tufts University, Boston, Massachusetts*

[2]*Case Western Reserve University, Cleveland, Ohio*

[3]*University of Texas at San Antonio, San Antonio, Texas*

[4]*USDA, HNRCA at Tufts University, Boston, Massachusetts*

Contents

I. INTRODUCTION

According to Wikipedia: "Pollyanna tells the story of Pollyanna Whittier, a young girl who goes to live with her wealthy Aunt Polly after her father's death. Pollyanna's philosophy of life centers around what she calls 'The Glad Game': she always tries to find something to be glad about in every situation." Given the increasing proportion of aged individuals in the United States and other countries, with all of their attendant ills, if there was ever a time when a "Pollyanna" is needed, it is now. The purpose of this review is to provide some additional "Pollyannas" in the form of polyphenols contained in berries, Concord grape juice, curcumin, and other natural products such as tea catechins, and the polyunsaturated fatty acids contained in such commodities as fish oils and nuts that may act as harbingers of good news for healthy aging. Note, however, that since there have already been multiple reviews of vitamins E and C (e.g., [1, 2]) and such supplements as ginkgo biloba (e.g., [3]), these topics will not be covered here.

As is well known, with aging there is an increase in the number of multiple co-occurring chronic conditions, including cognitive decline and dementia. Because the proportion of the population in the United States and other nations that are aged continues to increase, cognitive and motor deficits are growing rapidly. Cognitive impairment and dementia are major causes of disability in our nation, and their financial impact and long-term care costs are enormous. The major cause of dementia is Alzheimer's disease (AD). Aging clearly results in declines in brain size, weight, and function [4–6]. However, to date the cellular and morphological substrates underlying these changes remain poorly characterized. Although dogma suggests that aging is associated with a significant loss of neurons in the brain, careful studies have uncovered only modest, if any, change in cell number and size in a variety of brain regions with aging, including the neocortex and hippocampus [7–10]. A lack of evident morphological changes associated with neurodegeneration in the entorhinal cortex with aging further supports the concept that there are fundamental differences between the types of changes that occur in normal "healthy" aging and the pathological changes that occur in age-related neurodegenerative processes, such as AD.

Among the clear functional changes that occur in the brain with aging are declines in various aspects of cognition and memory. In particular, short-term memory [11], memory acquisition and early retrieval [5], working memory [12], recognition memory [13, 14], reasoning [15], and processing speed [16, 17] are affected with aging. In fact, a great deal of research has shown, in both humans and animal models, the occurrence of numerous neuronal and behavioral deficits during aging in the absence of neurodegenerative disease. These changes may include decrements in receptor sensitivity, most notably adrenergic [18], dopaminergic [19, 20], muscarinic [21, 22], and opioid [23]. These decrements, and those involving neuronal signaling [24] and decreases in neurogenesis [25], can be expressed, ultimately, as alterations in both motor [26, 27] and cognitive behaviors [28]. The alterations in motor function may include decreases in balance, muscle strength, and coordination [26], whereas cognitive deficits are seen primarily with respect to spatial learning and memory [29, 30]. Indeed, these characterizations have been supported by a great deal of research both in animals [28–30] and humans [31]. Age-related deficits in motor performance are thought to be the result of alterations either in the striatal dopamine (DA) system (as the striatum shows marked structural and functional changes with age in Parkinson's disease [PD]) or in the cerebellum, which also shows age-related alterations [32, 33].

Memory alterations appear to occur primarily in secondary memory systems and are reflected in the storage of newly acquired information [22, 34]. It is thought that the hippocampus mediates allocentric spatial navigation (i.e., place learning), and that the prefrontal cortex is critical to acquiring the rules that govern performance in particular tasks (i.e., procedural knowledge), whereas the dorsomedial striatum mediates egocentric spatial orientation (i.e., response and cue learning) [35–38]. More importantly data from a variety of experiments suggest that the contributing factors to the behavioral decrements seen in aging involve oxidative stress (OS) [39] and inflammation (INF) [40, 41]. This review discusses some of the nutritional interventions in aging and their putative utility in neurodegenerative disease.

II. GENDER DIFFERENCES IN DEMENTIA

Epidemiological observations and evidence of gender-related differences in cognition and behavior suggest that there may be important genetic or biological factors related to gender that are operating in the pathogenesis of neurological disease, particularly in AD. Clinicians who diagnose and treat patients with AD recognize that there is heterogeneity in its cognitive and behavioral manifestations. Research suggests that gender may be an important modifying factor in AD development and expression. One of the most intriguing aspects concerning the epidemiology of AD is that the prevalence rate in women is roughly twice that in men, and this skewed sex ratio is specific for AD but not for other dementias. Age is the most important risk factor associated with dementia. Males tend to have shorter life spans than females and even though the life-span gap narrows as men live longer, still at the age of 75 and older, there are significantly more women with AD than men.

Long-term effects of the metabolic and hormonal differences between men and women may play a relevant role on the observed age-associated cognitive impairment and behavioral changes. Some studies have considered metabolic differences in cerebral glucose between men and women as an important factor in cognitive decline. These studies have only shown a decreased parietal activity in early-onset dementia of AD, independent of a gender effect [42]. Another aspect regarding differences in prevalence of AD among men and women focuses on the roles of estrogen and testosterone in disease pathogenesis, and there are a number of lines of evidence suggesting that estrogen deficiency, following menopause, may contribute to the etiology of AD in women [43, 44]. The decreased incidence [45] and a delay in the onset [46] of AD among women on hormone replacement therapy following menopause [47] has also contributed to a belief that these agents may play a relevant role in brain function and cognitive decline associated with aging [48]. However, a decline in estrogen or testosterone does not explain why males with Down's syndrome are at significantly higher risk of developing AD-type changes and at an earlier age than their female counterparts [49]. Indeed, the concentration of estrogen and testosterone in both sexes is similar in patients with Down's syndrome compared to those in the general population. Studies have also cast doubt on estrogen replacement therapy as being protective against AD [50–53].

There are a number of other hormones involved in the hypothalamic-pituitary-gonadal axis that regulate reproductive function and, importantly, receptors for these other hormones are expressed in many nonreproductive tissues including the brain. Supporting evidence indicates that other hormones of the hypothalamic-pituitary-gonadal axis may be playing a central role in the pathogenesis of AD [54].

Several studies of gender differences in cognition have pointed to greater language deficits in women with AD as compared with men [55, 56]. However, other studies have reported absence of gender-related language differences or other measures of cognition, including memory and perception in AD [57, 58]. Although the most prominent change noted in patients with AD is decreased cognition, behavioral disturbances also frequently occur. Interestingly, although several reports have suggested that increased behavioral disturbance in AD is related to dementia severity across gender, qualitative differences between men and women in the manifestation of the disturbances also have been reported. Female patients with AD exhibit tendencies to be more reclusive and emotionally labile. In comparison, men with AD show more psychomotor and vegetative changes and aggressive behaviors [59]. Male patients exhibit greater problems than female patients in wandering, abusiveness, and social impropriety, particularly in the more advanced stages of the disorder. In addition, male patients with AD have increased physical, verbal, and sexual aggression than women [60–63]. Depression, on the other hand, does appear to be more prevalent in female than in male patients with AD [63]. Thus, several observations suggest that there may be important genetic factors related to gender that are operative in the pathogenesis of AD. However, it remains controversial whether men and women differ in the incidence of AD and whether there are clearly recognizable sex disparities operating in the cognitive and behavioral changes among those afflicted.

III. OXIDATIVE STRESS IN AGING

Oxidative stress results from the shift toward reactive oxygen species (ROS) production in the equilibrium between ROS generation and the antioxidant defense system [64]. In the brain, this is particularly important, because studies have found indications of increased OS in brain aging, including reductions in redox active iron [65, 66], as well as increases in Bcl-2 [67] and membrane lipid peroxidation [68]. Studies have also shown that there are significant increases in cellular hydrogen peroxide [69]. Additionally, there is significant lipofuscin accumulation [65] along with alterations in membrane lipids [70]. Studies have also suggested the involvement of lipid rafts with oxidative stress sensitivity [71]. Importantly, the consequences of these increases in oxidative stress at several levels may result in reduced calcium homeostasis, alterations in cellular signaling cascades, and changes in gene expression [72–77], which combine to contribute to the increased vulnerability to OS seen in the aging population [78, 79] and which is elevated in neurodegenerative diseases, such as AD [80–82] and PD [83, 84].

Oxidative stress vulnerability in aging also may be the result of microvasculature changes and increases in oxidized proteins and lipids [85], as well as alterations in (1) membrane microenvironment and structure [86, 87] (2) calcium buffering ability, and (3) the vulnerability of neurotransmitter receptors to OS (discussed later). Additional "vulnerability factors" include critical declines in endogenous antioxidant protection, involving alterations in the ratio of oxidized to total glutathione [88] and reduced glutamine synthetase [89]. Taken together, these findings indicate that there are increases in OS in aging, that the CNS may be particularly vulnerable to these increases (see [87, 90] for review), and the efficacy of antioxidants may be reduced in aging.

Calcium buffering has been shown to be significantly reduced in senescence [91–93]. The consequences of such long-lasting increases in cytosolic calcium may involve cell death induced by several mechanisms (e.g., xanthine oxidase activation [94]), with subsequent pro-oxidant generation and loss of functional capacity of the cell.

However, it is important to note here that OS may only be a partial contributor to neuronal and behavioral changes in senescence. For example, OS may contribute to these age-related diseases by inducing the expression of proinflammatory cytokines through activation of the oxidative stress-sensitive nuclear factor kappa B (NF-κB) [95, 96]. NF-κB in turn up-regulates the inflammatory response leading to a further increase in ROS [97], which results in a continuous increase in oxidative stress and inflammation and thus vulnerability to further stressors.

A. Oxidative Stress in Alzheimer's Disease

Various reports have shown increased reactive carbonyls in association with AD [98]. These changes have been identified in senile plaques [99, 100], neurofibrillary tangles (NFT) [100, 101], and the primary component of the latter, tau protein [101, 102]. The significance of these findings was initially questioned by suggestions that the lesions of AD, such as those that occur in vessel walls [103, 104], accumulate damage through low protein turnover [105]. What was missing from this criticism was not the accumulative nature of carbonyl modification but that the products first identified, advanced glycation end products (AGE), are "active-modifications," by which we mean they are the result of metal-catalyzed redox chemistry and are continuing sites of redox chemistry [106]. Also, we have demonstrated that the lesions not only are sites of AGE accumulation but also continuing sites of glycation, because the initial Amadori product is closely associated with NFT [107].

Early reports of oxidative modifications were followed in close succession by the identification in NFT of reactive carbonyls [108, 109] and protein adducts of the lipid peroxidation product, hydroxynonenal [110, 111]. What was remarkable in using these different markers either resulting from carbonyl adduction or, in the case of reactive carbonyls, from direct protein oxidation is that although highly stable modifications involving cross-linked proteins are predominantly associated with the lesions, metastable modifications are more commonly associated with the neuronal cytoplasm. Specifically, populations of neurons involved in AD, and not others, show this change, suggesting that the most active site of oxidative damage is the neuronal cytoplasm.

Studies analyzing certain physical properties of the oxidized proteins forming cross-linking compounds, specifically those properties that make these biological molecules refractory to light, have shown the presence of modified proteins in the brain of AD patients [112]. In addition, oxidation of the modified proteins not only renders the modified protein more resistant to degradation but also appears to competitively inhibit the proteosome [112]. These changes may underlie the accumulation of ubiquitin conjugates observed in the neurons in AD [113]. Protein nitration is a non-cross-link-related oxidative modification of protein resulting from either peroxynitrite attack or peroxidative nitration. In investigating the distribution of nitrotyrosine in AD, we found that the major site of nitrotyrosine was in the cytoplasm of non-NFT-containing neurons [114] and that neurons containing NFT actually showed lower levels of nitrotyrosine than similar neurons lacking NFT. These relationships were confirmed when we examined RNA, a cellular component with a relatively rapid turnover rate. A major oxidation product of RNA, 8-hydroxyguanosine (8OHG), has a distribution similar to nitrotyrosine, except that it is absent from NFT and reduced in the surrounding cytoplasm [115], even though NFT contain associated RNA [116]. The concurrence of RNA and protein damage suggests that the major site of oxidative damage in AD is localized predominantly in the neuronal cytoplasm.

B. Source of Reactive Oxygen Species

Both location and type of damage are important to understand the source of oxidative damage. First, the location of damage, which involves every category of biomacromolecules, appears to be restricted to neurons. Classically, nitrotyrosine is considered the product of peroxynitrite attack of tyrosine, and 8OHG is considered the product of –OH attack of guanosine. However, the separation is not simple; nitrotyrosine can be formed from peroxidative nitration by nitrite and H_2O_2 and peroxynitrite is produced by the reaction of nitric oxide (NO–) with superoxide (O_2–). In the case of peroxidative nitration, treating tissue sections with nitrite and H_2O_2 yields increased nitrotyrosine of the same distribution found during the disease in AD, but not control, cases [117]. An issue with peroxynitrite is diffusibility, being the result of the fusion of nitric oxide (NO) and O_2–; it can diffuse several cell diameters from its source to attack vulnerable target proteins [118]. In AD, one of the most striking findings is the restriction of damage to the cell bodies of vulnerable neurons. Although amyloid-β deposits and NFT contain redox-active iron, like oxidative damage, the most conspicuous changes in iron are within the cytoplasm of vulnerable neurons [119, 120]. Significantly, cytoplasmic redox-active iron is barely detectable in controls. Redox-active iron is the critical element for Fenton chemistry generation of –OH from H_2O_2. Ultrastructural localization of iron shows it is diffusely associated with the cytoplasm, primarily in the endoplasmic reticulum but also in granules identified as lipofuscin as well as their associated vacuoles.

Lipofuscin is thought to represent the terminal phase of autophagic lysosomes that involve iron-rich mitochondria [121]. Therefore, the increased redox-active iron in such lysosomes in AD lends credence to the notion of mitochondrial abnormalities in AD. Mitochondrial DNA, as well as the protein cytochrome oxidase-I, is increased several-fold in vulnerable neurons in AD. Ultrastructural examination showed, although both markers were in mitochondria, that in AD the increased levels were in the cytoplasm and, in the case of mDNA, in vacuoles associated with lipofuscin, the same sites that showed increased redox-active iron. The majority of iron is in the endoplasmic reticulum, suggesting that the role for mitochondria is probably not to directly supply –OH but instead to supply its precursors, H_2O_2 and redox-active metals. Although the proposed mechanism is distinct from nonmetal-catalyzed peroxynitrite formation, it does not discount an important role for NO. Neurons in AD show activation of NO synthetase as well as its modulator, dimethylargininase [122]. Nitric oxide has strong antioxidant activity (see the next section) as well as inhibitor activity for cytochromes. The latter could play a role in the hypometabolism consistently found in AD [123] as well as the altered mitochondrial dynamics noted here.

C. Relationship to Lesions

In AD, the putative source of the reactive oxygen species was supposed to be the lesions. Amyloid-β by itself was proposed to generate reactive oxygen species [124]. This mechanism has fallen into question for both chemical and biological issues [125]. Nevertheless, amyloid-β, under some circumstances, can bind iron and promote catalytic redox cycling, yielding reactive oxygen [126]. *In vivo* oxidative damage is inversely correlated to amyloid-β load, indicating that rather than being a source of the reactive oxygen, amyloid-β may be a modulator that can either increase or decrease reactive oxygen production [86, 127, 128]. Further, the relative paucity of short-lived oxidative changes surrounding amyloid-β deposits [115], rather than those that accumulate in long-lived proteins [100], also puts into question the idea that reactive oxygen resulting from inflammation is an important mechanism for oxidative damage. In fact, although the notion of inflammation in AD is well established [129], this appears to be a secondary response to the underlying pathological changes.

IV. INFLAMMATION

As mentioned previously, evidence also suggests that in addition to oxidative stress, CNS inflammatory events may have an important role in affecting neuronal and behavioral deficits in aging [130]. It has been shown that activated glial cells increase in the normal aging brain, which exhibits greater immunoreactivity in markers for both microglia and astrocytes [131–133]. Additionally, increased glial fibrillary acid protein expression is observed by middle age [131], and in the elderly this increase even occurs in the absence of a defined stimulus [134]. Glial cells mediate the endogenous immune system within the microenvironment in the CNS [135], and their activation is the hallmark of inflammation in the brain [136]. Activated microglia produce inflammatory molecules such as cytokines, growth factors, and complement proteins [134, 137, 138]. These proinflammatory mediators in turn activate other cells to produce additional signaling molecules that further activate microglia in a positive feedback loop to perpetuate and amplify the inflammatory signaling cascade [139]. Activated microglia produce proinflammatory cytokines such as interleukin-1 (IL-1), interleukin-6 (IL-6), and tumor necrosis factor-α (TNF-α) [140, 141].

Increases in TNF-α have also been reported as a function of age [142], as well as associated inhibition of glia [143]. Similarly, research in both aged mice and humans has found increases in TNF-α, IL-6 [142–145], and C reactive protein [146]. All of these changes appear to be accompanied by up-regulations in downstream indicators of inflammation (e.g., complement C1q) in microarray studies [147].

Additionally, studies indicate that the expression of cyclooxygenase (COX) 2 appears to be associated with amyloid-β deposition in the hippocampus [148, 149], and inflammatory prostaglandins (PG) such as PGE show increases in the hippocampus, as well as other areas in aging [150]. Because the PG synthesis pathway appears to be a major source of ROS in brain [151], and in other organ systems, these findings indicate that inflammation may be accompanied by and even generate its "evil twin," OS, in producing the deleterious effects of aging. Thus, such factors as cytokines, cyclooxygenases, prostaglandins, and others may act as extracellular signals in generating additional ROS that are associated with decrements in neuronal function or glial neuronal interactions [152–156] and ultimately the deficits in behavior that have been observed in aging.

If this is the case, it should be possible to induce behavioral (cognitive and motor) deficits similar to those seen in aging using procedures that induce oxidative or inflammatory stressors. Indeed, these changes have been induced in several experiments. Rodent studies have suggested that young animals exposed to OS show similar neuronal and behavioral changes to those seen in aged animals. The results have shown that young animals irradiated with particles of high energy and charge show behavioral deficits paralleling those observed in aging [157–159]. High energy and charged particles (specifically 600 MeV or 1 GeV ^{56}Fe) also disrupt the functioning of the dopamine-mediated behaviors, such as motor behavior [160], spatial learning and memory behavior [161], and amphetamine-induced conditioned taste aversion [162].

Inflammatory mediators have been shown to produce similar deficits in behavior [41]. For example, the administration of lipopolysaccharide (LPS) intrahippocampally was found to up-regulate inflammatory mediators, inducing degeneration of hippocampal pyramidal neurons, and produced decrements in working memory [40, 163, 164]. Similarly, the chronic ventricular infusion of LPS into young rats produces many of the same alterations in behavior that have been reported in AD. These changes are accompanied by inflammatory, neurochemical, and neuropathological alterations [40, 41, 163, 165].

Thus, these studies and those reviewed above suggest that one method to forestall or perhaps even reverse the behavioral/declines that have been observed in aging might be to increase endogenous antioxidant/anti-inflammatory protection.

V. AGE–ALZHEIMER DISEASE PARALLELS

As discussed previously, there are increases in oxidative and inflammatory stressors as a function of age that appear to be involved in the decrements seen in both cognitive and motor behaviors. If this is the case, then it might be postulated that neurodegenerative diseases, which are age-dependent, would be superimposed upon an environment already vulnerable to these insults. Indeed, OS plays a major role in the cascade of effects associated with AD (e.g., damage to DNA, protein oxidation, lipid peroxidation [90], and abnormal sequestration of metals [166–168]) that may be independent of amyloid-β deposition. Thus, the free radical perturbations would have an even greater effect in an aged organism, because, as pointed out in the previous sections, there is increased vulnerability to OS and inflammatory insults in senescence. These inflammatory mediators are prominent in the AD brain and they also have been observed in lower concentrations in nondemented brains from aged individuals. As pointed out in the previous section, multiple endogenous sources including microglia, astrocytes, and brain endothelial cells can produce these inflammatory mediators in AD [169–172]. Glial cells play important roles in supporting survival of neurons [173–176] and are extraordinarily sensitive to changes in the brain microenvironment. Brain astrocytes (RA) in particular show reactive gliosis to several forms of central nervous system (CNS) lesions [177, 178]. Additionally, gliosis, which can lead to brain damage by several mechanisms [179], is a feature common to virtually every neurodegenerative disease (e.g., multiple sclerosis, AD, tumor, HIV encephalitis, or prion disease) [180–184].

VI. POLYPHENOL SUPPLEMENTATION AND REDUCTIONS OF OXIDATIVE STRESS AND INFLAMMATION

There have been numerous studies in which antioxidants have been examined with respect to reducing the deleterious effects of brain aging, with mixed results. However, our research suggests that the combinations of antioxidant/anti-inflammatory polyphenolics found in fruits and vegetables may show efficacy in aging. Plants, including food plants (fruits and vegetables), synthesize a vast array of chemical compounds that are not involved in their primary metabolism. These "secondary compounds" instead serve a variety of ecological functions, ultimately, to enhance the plant's survivability. Interestingly, these compounds also may be responsible for the multitude of beneficial effects of fruits and vegetables on an array of health-related bioactivities; two of the most important may be their antioxidant and anti-inflammatory properties. Because OS appears to be involved in the signaling and behavioral losses seen in senescence, an important question becomes whether increasing antioxidant or anti-inflammatory intake would forestall or prevent these changes, and the literature is replete with studies (e.g., vitamins E and C [185]) in which a large variety of dietary agents have been employed to alter behavioral and neuronal deficits with aging. Instead, since recent studies have indicated that they have been shown to have considerable efficacy in reducing the deleterious effects of neuronal aging, this chapter will focus more on the antioxidant/anti-inflammatory potential of green tea catechins, curcumin, and berry fruits.

A. Green Tea Catechins

Catechins are derived from a number of sources, including green tea, red wine, and dark chocolate (see [186]). The most extensively studied have been those from green tea. (−) Epigallocatechin-3-gallate (EGCG) is the primary compound in green tea that is thought to provide the numerous beneficial effects that have been shown in many studies to provide a number of health benefits ranging from cancer treatment to cardiovascular function. Youdim, Mandel, and colleagues have provided several extensive reviews of the properties and molecular mechanisms involved in the health benefits of EGCG [187]. Thus, an extensive review will not be provided here except as these effects relate to neuroprotection, where it appears that the strongest evidence suggests that the primary beneficial properties of EGCG may be its antioxidant, anti-inflammatory, and metal chelating abilities. In addition, these catechins appear to enhance prosurvival genes and, as described later for blueberries, act to enhance neuroprotection and reduce stress signaling. This multiplicity of effects appears to provide a significant protection against oxidative and inflammatory stressors. By far the bulk of the data concerning neuroprotection have been provided by studies showing reduced ischemic-induced neuronal degeneration in various models of cerebral artery occlusion [188–194]. Green tea catechins also have been found to provide significant neuroprotection against *N*-methyl-4-phenyl-1,2,3,6-tetrahydropyridine (MPTP)-induced neurotoxicity in mice in several experiments (e.g., see [188]). Significant protection has also been seen in numerous *in vitro* experiments in several cell models (as reviewed in [194, 195]). Clearly, there appear to be numerous beneficial properties of green tea catechins on several oxidative stress- or inflammatory-mediated conditions. However, the possible benefits in a human population and the amounts of green tea necessary to produce beneficial effects remain to be determined.

B. Curcumin

One of the most exciting polyphenolic-containing dietary products that has emerged into the neuroscience literature in recent years is curcumin. Largely as a result of the groundbreaking work of Dr. Greg Cole and his colleagues [196], this spice, which has been used to treat illnesses for hundreds of years, has recently been shown to possess some putative important beneficial effects for neuroprotection and possible treatment in AD. In a manner similar to that seen with respect to the green tea catechins discussed earlier and the berries in the next section, curcumin, which is derived from *Curcuma longa* Linn (aka turmeric), appears to have potent anti-inflammatory/antioxidant properties. Moreover, it has also been shown to have potent anticancer effects (see [196] for review). However, of interest in this present review is curcumin's putative effect on AD. Curcumin supplementation prevented extensive damage following transient forebrain ischemia in CA1 neurons in the rat [197], suggesting that it may have important neuroprotective properties. Additionally, *in vitro* studies have demonstrated that curcumin reduces inflammatory activity of microglial cells [198, 199]. Several studies in AD transgenic mice also showed that curcumin down-regulated amyloid expression and reduced inflammatory markers [200–202]. Data in humans are still forthcoming, but from the animal and cell experiments it appears that curcumin may be important in altering the course of plaque deposition and expression of AD.

C. Berry Fruits

In our first study, we utilized fruits and vegetables identified as being high in antioxidant activity via the Oxygen Radical Absorbance Capacity Assay (ORAC) [203–205] and showed that long-term (from age 6 to 15 months in F344 rats) feeding with a supplemented American Institute of Nutrition (AIN)-93 diet (strawberry extract or spinach extract [1% to 2% of the diet] or vitamin E [500 IUD]) retarded age-related decrements in cognitive or neuronal function. Results indicated that the supplemented diets could prevent the onset of age-related deficits in several indices (e.g., cognitive behavior, Morris water-maze performance) [206].

In a subsequent experiment [91], we found that dietary supplementation (for 8 weeks) with spinach, strawberry, or blueberry (BB) extracts in an AIN-93 diet was effective in reversing age-related deficits in neuronal and behavioral (cognitive) function in aged (19 mo) F344 rats. However, only the BB-supplemented group exhibited improved performance on tests of motor function. Specifically, the BB-supplemented group displayed improved performance on two motor tests that rely on balance and coordination, rod walking, and the accelerating rotarod, whereas none of the other supplemented groups differed from control on these tasks [91]. The rodents in all diet groups, but not the control group, showed improved working memory (short-term memory) performance in the Morris water maze, demonstrated as one-trial learning following the 10-minute retention interval [91]. We also observed significant increases in several indices of neuronal signaling (e.g., muscarinic receptor [MAChR] sensitivity) and found that the BB diet reversed age-related "dysregulation" in Ca^{45} buffering capacity. Examinations of ROS in the brain tissue obtained from animals in the various diet groups indicated that the striata obtained from all of the supplemented groups exhibited significantly lower ROS levels (by assaying DCF;2′,7′-dichlorofluorescein diacetate) than the controls. A subsequent study using a BB-supplemented NIH-31 diet replicated the previous findings [207]. However, it was clear from these supplementation studies [91, 207] that the significant effects of BBs on both motor and cognitive behavior were due to a multiplicity of actions, in addition to those involving antioxidant and anti-inflammatory activity. We also have shown that BB-supplemented senescent animals show increased neurogenesis [208].

With respect to AD, we have shown that BB-supplemented (from age 1 to 12 mo) mice transgenic for amyloid precursor protein and presenilin-1 mutations (which show the formation of numerous plaques in the brain, similar to those seen in AD) do not show behavioral deficits in Y-maze performance as seen by those given a control diet [283]. The supplemented mice also showed enhancements in several signaling molecules associated with cognitive function (e.g., extracellular signal-regulated kinase activity). These findings suggest that it is possible to delay or prevent cognitive dysfunction despite the pathological changes in this mouse model and further suggest that the inclusion in the diet of fruits high in antioxidant activity may help prevent the deleterious effects of this disease later in life (see Fig. 1).

D. Polyunsaturated Fatty Acids

The major source of omega-3 (eicosapentaenoic acid, EPA) and omega-6 (docosahexaenoic acid DHA) fatty acids is fish oil. Numerous studies to date have suggested that dietary supplementation with EPA and DHA has a host of beneficial effects in many of the diseases that increase as a function of aging such as heart disease [209–211], hypertriglyceridemia [212], cancer [213, 214], and neurodegenerative disease [215, 216]. Studies suggest, for example, that aging mice, which have reduced levels of brain polyunsaturated fatty acids, appear to show alterations in neuronal membranes, such that the mice show memory loss, learning disabilities, cognitive alterations, and even decrements in visual acuity, which can be reduced by supplementing the mice with DHA containing fish oil or DHA [217]. Similar findings have also been seen in the rat [218]. It also appears that AD patients exhibit lower amounts of DHA in plasma [219] and brain [220]. Epidemiologically, it appears that increased DHA or dietary fatty fish intake reduce the risk of AD [221]. Importantly, however, the mechanisms involved in the putative beneficial effects of DHA in these models are not well understood. Florent and colleagues [222] showed that protection against amyloid beta involved activation of ERK 1,2 survival pathways. They showed that cortical neurons pretreated with DHA showed less cell death, reduced apoptosis, caspase

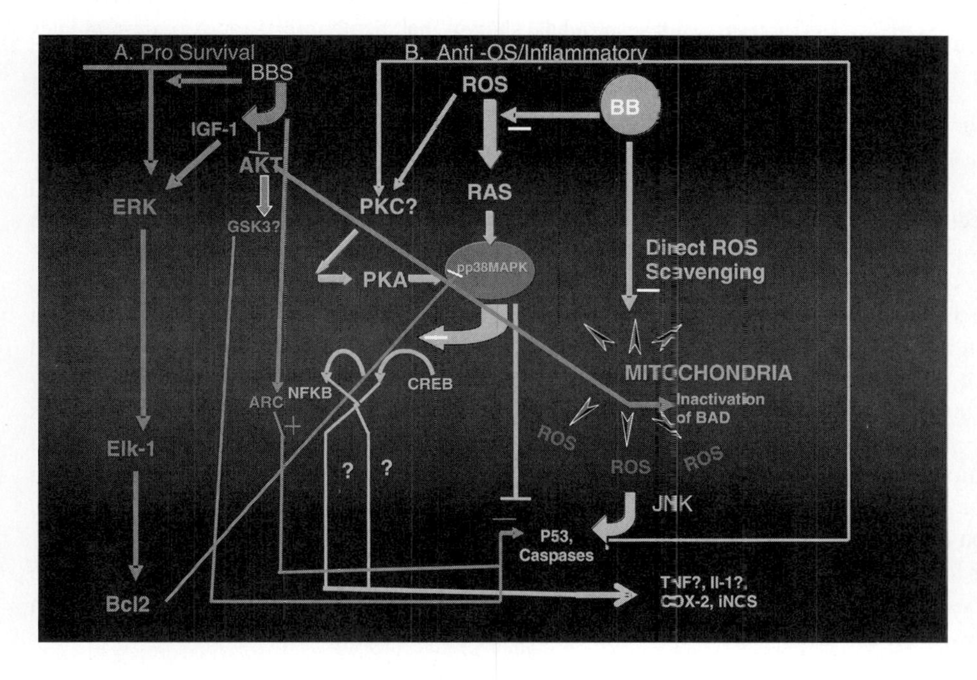

FIGURE 1 Possible direct and indirect effects of blueberry supplementation (BBS) that reduce stress signaling and increase survival.

activity, and arachidonic acid activity. The study by Florent and colleagues [222] supports previous studies in a variety of cell types showing DHA protective effects [223]. In addition to its effects on ERK protective signaling, DHA has also been shown to be the derivative of the docosanoid neuroprotectin D1 (NPD1) that was shown to provide protection against oxidative stress in retinal cells [224]. It also appears that DHA alters membrane lipid rafts to induce phosphatidylserine accumulation in cell membranes, to impinge on additional points in the cell survival pathways (e.g., Akt and Raf-1 [225, 226]). Thus, it appears that the protective effects of polyunsaturated fatty acids, much as the fruit polyphenols, may involve enhancing protective signaling pathways.

Results from several studies have indicated that the regular consumption of foods containing omega-3 fatty acid including soybean oil, fish oil, and nuts may lower mortality from cardiovascular disease [227–234]. Importantly, evidence also indicates that tree nuts such as walnuts may be beneficial in cardiovascular disease for their effects on serum lipids (reviewed in [235]). Although studies of nuts are not nearly as extensive as studies on the effects of fish oil, and few studies have focused on the brain effects of tree nuts such as walnuts, one could surmise based on the cardiovascular findings that there may be secondary or primary benefits on neuronal function. Moreover, because tree nuts such as walnuts also contain flavonoids similar to those found in fruits and vegetables, additional benefits might occur from synergistic interactions with the nut-derived fatty acids. Preliminary research from our lab, for example, has shown that senescent rats maintained on a diet containing walnuts showed enhanced performance in both cognitive and motor behaviors relative to the animals maintained on the control diets [unpublished].

E. Putative Signaling Mechanisms Involved in Polyphenol Regulation of Oxidative Stress and Inflammation

There are multiple sources of OS in the cell that result from food metabolism, ionizing radiation, smoking, and so on. There are also mitochondrial sources of ROS that emerge from the energy metabolism of the cell. Importantly, ROS are also generated from inflammatory processes. Subsequently, a great deal of research has shown that there is a cascade of stress signals that are generated from ROS. For example, ROS are believed to play an important role in the pathophysiology of neurodegenerative diseases such as AD or Parkinson's disease (PD), involving the production of inflammatory mediators [236, 237].

One of the first steps in the production of these mediators is the generation of the protein kinase C (PKC) family. Of particular importance is the generation of the protein kinase Cγ (PKCγ) isoform, one of the major forms that is found in memory control brain areas such as the hippocampus. Of the different isoforms of PKC present in the brain, the γ-subtype is the most abundant representative in the rat hippocampus [238, 239]. The PKC pathway is part of a major signal transduction system in inflammation [240] and is activated by several inflammatory agents, including the tumor-promoting phorbol ester PMA (phorbol 12-myristate 13-acetate). The exact mechanism of the involvement of PKC in the stress pathway remains to be determined, but it has been suggested that ROS-induced PKCγ may target lipid rafts [241]. Importantly, PKC isoforms have been associated with the LPS-generated increased production of nitric oxide (NO) and inducible NO synthase (iNOS) [242]. Moreover, Hall and colleagues [243] found that the induction of PKC from microglial cells via amyloid-β 25–35 induced COX-2 (cyclooxygenase-2) expression.

In this respect, microglial cells are a major source of inflammation increasing the production of cytokines and nitric oxide, among other compounds, that can induce cell death and decrements in neuronal activity [134, 244]. It has been shown that p38 MAPK is intimately involved with microglia activation, the stress response [245], and c-Jun N-terminal kinase (JNK) [246]. Importantly, activation of neuronal p38 MAPK and JNK has been shown to directly disrupt long-term potentiation (LTP), and inhibition of microglial activation was found to prevent LTP disruption [247–250].

An integral part of the stress pathway is the activation of nuclear factor kappa B (NFκB). It is present in the cytoplasm in an inhibitory form and attached to an inhibitory protein, IκB, where it is tightly controlled [251]. During stimulation with the uncoupling of IκB, NFκB translocates to the nucleus and mediates the transcription of many "inflammatory" genes (e.g., COX-2, TNF-α, interleukin 1-beta [IL-1β] and inducible nitric oxide synthase [iNOS]) to further promulgate inflammatory signals and neuronal degeneration [252]. NFκB usually acts in concert with cyclic AMP response element binding protein (CREB), and research has shown that acute mild hypoxia up-regulates CREB at serine 133 (see [253, 254] for reviews). It has also been shown that CREB is activated by hydrogen peroxide in Jurkat T lymphocytes [255] and by cadmium in mouse neuronal cells [256], as well as during stroke [257].

Thus, this brief discussion shows that inflammatory and oxidative stressors can elicit a cascade of signals that result ultimately in the generation of additional stressors, loss of cell function, and, in the case of neurodegenerative disease, reductions in the protective capacity of the organism in senescence. However, it appears from our findings and others that polyphenols similar to those contained in berry fruits such as blueberries can activate protective pathways to reduce the deleterious effects of inflammation and oxidative stress. Additionally, previous research has shown that under OS or inflammatory conditions, polyphenols similar to those contained in blueberries, tea, red wine, or ginkgo biloba altered signaling in ERK activity (e.g., see [258, 259], as well as PKC [260, 261] and CREB [262] in several models described next).

1. BV2 Mouse Microglial Cells

As mentioned previously, accumulating evidence indicates that inflammation in the CNS increases during normal aging and age-related neurodegenerative diseases augment neuroinflammation. Neuroinflammation is largely mediated through the activation of microglial cells. Microglial activation has been attributed to enhanced signal transduction leading to the induction of inflammatory enzymes such as iNOS and COX-2, as well as cytokines such as IL-1β and TNF-α. In an earlier study, we showed that blueberries were effective in attenuating the production of these inflammatory mediators in LPS-activated murine BV2 microglia [263]. To extend these findings, we also examined a purified extract of blueberries (post-C18) and showed a suppression in the LPS-induced increases in iNOS, p38 MAPK and NF-kB in the BV2 mouse microglial cells.

2. Muscarinic-Transfected Receptors

We and others have shown that there are increases in vulnerability to OS in aging that include striatal muscarinic receptor (MAChR) sensitivity to hydrogen peroxide application [264]. Given their importance in a variety of functions including memory [265], amyloid precursor protein (APP) processing [266], and vascular functioning [267], OS and age-sensitive deficits in MAChR may result in the cognitive, behavioral, and neuronal aberrations observed in aging that are exacerbated in AD and vascular dementia. In this regard, findings have indicated that COS-7 cells transfected with one of the five MAChRs and exposed to dopamine (DA) [258] showed differences in OS sensitivity expressed as a function of $Ca2^+$ buffering (i.e., the ability to extrude or sequester $Ca2^+$ following oxotremorine-induced depolarization). The loss of calcium buffering in these experiments is similar to that reported in many studies with respect to aging (see [268, 269]), and such losses can have a profound effect on the functioning and viability of the cell [270–272], further increasing OS [273] and leading ultimately to decrements in motor and memory function in senescent rats [30, 274]. It is also important to note here that there are significant differences in the rates of aging among various brain regions, with areas such as the hippocampus [274, 275], cerebellum [276, 277], and striatum [264, 278] showing profound alterations with aging in such factors as morphology, electrophysiology, and receptor sensitivity. However, we showed [279] that COS-7 cells transfected with M1 muscarinic receptors (which show increased vulnerability to DA-induced OS) are protected from these changes if pretreated with blueberry extract. Additional analyses suggested that mechanistically the protective effects of the blueberry may be derived from their ability to reduce stress signaling. These analyses revealed that blueberry treatment decreased both CREB- and PKCγ-induced signaling increases induced by DA, while increasing protective MAPK signals.

3. Primary Hippocampal Cells

In a subsequent experiment, we showed that deficits in Ca^{2+} buffering induced by DA or amyloid beta $(A\beta)_{42}$ in primary hippocampal neuronal cells were antagonized by blueberry extract. The results indicated that Aβ-induced increases in *p*-MAPK were suppressed by blueberries while blueberries further enhanced DA-induced increases in *p*-MAPK. However, blueberries antagonized both DA- and Aβ(42)-induced increases in PKCγ, *p*-CREB, *p*-p38 MAPK, *p*-JNK, and IGF-1. Previous studies have shown that OS/INF stressors such as Aβ can increase transcription factors (e.g., *p*-CREB) associated with OS/INF and possibly decrease Ca^{2+} homeostasis, but it appears that the beneficial effects of blueberry polyphenols may involve reductions in stress signaling.

VII. CONCLUSION

From the previous sections, it should be clear that there are a number of sources of oxidative stress and inflammation and that these insults are superimposed on an increasingly vulnerable environment in aging. Moreover, in genetic aberrations in conditions such as AD or Parkinson's disease this vulnerability increases even further. Because this is the case, it is critical that methods be explored to reduce this vulnerability. What we have tried to show in this review is that one method of accomplishing this may be through diets containing polyphenols and polyunsaturated fatty acids. An abundance of epidemiological data indicates that diets rich in these compounds, which have antioxidant and anti-inflammatory activities, may play a pivotal role in maintaining human health [280–282].

Therefore, it is important for the diets to contain fruits and vegetables, and this appears to be especially true in fostering healthy aging and possibly in preventing the onset of AD. We have reviewed studies that have shown reversals in age-related cognitive and motor behaviors with fruit or vegetable supplementation [91] and have increased signaling and prevented cognitive decline in APP/PSI mice [283]. In the case of AD, there is an inverse correlation between the intake of wine flavonoids [284] or fruit and vegetable intake [285] and the development of dementia. Thus, these studies, as well as those reviewed previously, suggest a positive role for dietary polyphenols and polyunsaturated fatty acids in both the prevention and delay of the deleterious effects of aging and AD. Finally, it should be pointed out that studies in cell models indicate that green tea extracts may also be of some benefit in reducing the neurotoxicity associated with Parkinson's disease [286–288]. Given these considerations, it is evident that polyphenols and polyunsaturated fatty acids that have antioxidant and anti-inflammatory properties may be critical elements in a diet to maintain motor and cognitive health throughout the life span and should increase the likelihood of achieving successful aging.

Acknowledgments

The authors would like to thank Dr. Donna Bielinski and Vivian Cheng for their help in the preparation and editing of this chapter.

References

1. Martin, A., Janigian, D., Shukitt-Hale, B., Prior, R.L, and Joseph, J. A. (1999). Effect of vitamin E intake on levels of vitamins E and C in the central nervous system and peripheral tissues: implications for health recommendations. *Brain Res.* **845**, 50–59.
2. Martin, A., Cherubini, A., Andres-Lacueva, C., Paniagua, M., and Joseph, J. A. (2002). Effects of fruits and vegetables on levels of vitamins E and C in the brain and their association with cognitive performance. *J. Nutr. Health Aging* **6**, 392–404.
3. Christen, Y. (2004). Ginkgo biloba and neurodegenerative disorders. *Front. Biosci.* **9**, 3091–3104.
4. Cabeza, R., Grady, C. L., Nyberg, L., *et al.* (1997). Age-related differences in neural activity during memory encoding and retrieval: a positron emission tomography study. *J. Neurosci.* **17**, 391–400.
5. Small, S. A., Stern, Y., Tang, M., and Mayeux, R. (1999). Selective decline in memory function among healthy elderly. *Neurology* **52**, 1392–1396.
6. Murphy, D. G., DeCarli, C., Schapiro, M. B., *et al.* (1992). Age-related differences in volumes of subcortical nuclei, brain matter, and cerebrospinal fluid in healthy men as measured with magnetic resonance imaging. *Arch Neurol.* **49**, 839–845.
7. Morrison, J. H., and Hof, P. R. (1997). Life and death of neurons in the aging brain. *Science* **278**, 412–419.
8. West, M. J. (1993). Regionally specific loss of neurons in the aging human hippocampus. *Neurobiol. Aging* **14**, 287–293.
9. Ball, M. J., and West, M. J. (1998). Aging in the human brain: A clarion call to stay the course. *Neurobiol. Aging* **19**, 1.
10. Peters, A., Morrison, J. H., Rosene, D. L., and Hyman, B. T. (1998). Feature article: Are neurons lost from the primate cerebral cortex during normal aging? *Cereb. Cortex* **8**, 295–300.
11. Bartus, R. T., Fleming, D., and Johnson, H. R. (1978). Aging in the rhesus monkey: debilitating effects on short-term memory. J. *Gerontol.* **33**, 858–871.
12. Grady, C. L., McIntosh, A. R., Bookstein, F., *et al.* (1998). Age-related changes in regional cerebral blood flow during working memory for faces. *Neuroimage* **8**, 409–425.
13. Moss, M. B., Killiany, R. J., Lai, Z. C., *et al.* (1997). Recognition memory span in rhesus monkeys of advanced age. *Neurobiol Aging* **18**, 13–19.
14. Rapp, P. R., and Amaral, D. G. (1991). Recognition memory deficits in a subpopulation of aged monkeys resemble the effects of medial temporal lobe damage. *Neurobiol Aging* **12**, 481–486.
15. Gilinsky, A. S., and Judd, B. B. (1994). Working memory and bias in reasoning across the life span. *Psychol. Aging* **9**, 356–371.
16. Kail, R., and Salthouse, T. A. (1994). Processing speed as a mental capacity. *Acta Psychol.* (Amst.). and 199–225.
17. Robbins, T. W., James, M., Owen, A. M., *et al.* (1994). Cambridge Neuropsychological Test Automated Battery (CANTAB): a factor analytic study of a large sample of normal elderly volunteers. *Dementia* **5**, 266–281.
18. Gould, N., Chadman, K., and Bickford, PC. (1998). Antioxidant protection of cerebellar beta-adrenergic receptor function in aged F344 rats. *Neurosci Left.* **250**, 165–168.
19. Joseph, J. A., Berger, R. E., Engel, B. T., and Roth, G. S. (1978). Age-related changes in the nigrostriatum: a behavioral and biochemical analysis. *J. Gerontology* **33**, 643–649.
20. Cepeda, C., Colwell, C. S., Itri, I. N., *et al.* (1998). Dopaminergic modulation of NMDA-induced whole cell currents in neostriatal neurons in slices: contribution of calcium conductances. *J. Neurophysiol.* **79**, 82–94.
21. Egashira, T. (2000). Effects of breeding conditions on neurochemical cholinergic and monoaminergic markers in aged rat brain. *Nippon Ronen Igakkai Zasshi* **37**, 233–238.
22. Joseph, J. A. (1992). The putative role of free radicals in the loss of neuronal functioning in senescence. *Integr. Physiol. Behav. Sci.* **27**, 216–227.
23. Kornhuber, J., Schoppmeyer, K., Bendig, C., and Riederer, P. (1996). Characterization of [3H]pentazocine binding sites in postmortem human frontal cortex. Neural Transm. **103**, 45–53.
24. Galli, R. L., Shukitt-Hale, B., Youdim, K. A., and Joseph, J. A. (2002). Fmit polyphenolics and brain aging: Nutritional interventions targeting age-related neuronal and behavioral deficits. *Ann. NY Acad. Sci.* **959**, 128–132.
25. Kuhn, H. G., Dickinson-Anson, H., and Gage, F. H. (1996). Neurogenesis in the dentate gyrus of the adult rat: Age-related decrease of neuronal progenitor proliferation. *J. Neurosci.* **16**, 2027–2033.
26. Joseph, J. A., Bartus, R. T., Clody, D., *et al.* (1983). Psychomotor performance in the senescent rodent: Reduction of deficits via striatal dopamine receptor up-regulation. *Neurobiol. Aging* **4**, 313–319.
27. Kluger, A., Gianutsos, J. G., Golomb, J., *et al.* (1997). Motor/psychomotor dysfunction in normal aging, mild cognitive decline, and early Alzheimer's disease: Diagnostic and differential diagnostic features. *Int. Psychogeriatr.* **9**(suppl), 307–316, discussion 317–321.
28. Bartus, R. T. (1990). Drugs to treat age-related neurodegenerative problems. The final frontier of medical science? *J. Am. Geriatr. Soc.* **38**, 680–695.
29. Ingram, D. K., Spangler, E. L., Iijima, S., *et al.* (1994). New pharmacological strategies for cognitive enhancement using a rat model of age-related memory impairment. *Ann NY Acad Sci.* **717**, 16–32.
30. Shukitt-Hale, B., Mouzakis, G., and Joseph, J. A. (1998). Psychomotor and spatial memory performance in aging male Fischer 344 rats. *Exp. Gerontal.* **33**, 615–624.
31. Muir, J. L. (1997). Acetylcholine, aging, and Alzheimer's disease. *Pharmacal. Biochem. Behav.* **56**, 687–696.
32. Bickford, P, Heron, C, and Young D. A., (1992). Impaired acquisition of novel locomotor tasks in aged and norepinephrine-depleted F344 rats. *Neurobiol Aging.* **13**, 475–481.
33. Bickford, P. (1993). Motor leaning deficits in aged rats are con-elated with loss of cerebellar noradrenergic function. *Brain Res.* **620**, 133–138.

34. Bartus, R. T., Dean, R. L., and Beer, B. (1982). Neuropeptide effects on memory in aged monkeys. *Neurobiol. Aging* **3**, 61–68.
35. Devan, B. D., Goad, E. H., and Petri, H. L. (1996). Dissociation of hippocampal and striatal contributions to spatial navigation in the water maze. *Neurobiol. Learn Mem.* **66**, 305–323.
36. McDonald, R. J., and White, N. M. (1994). Parallel information processing in the water maze: evidence for independent memory systems involving dorsal striatum and hippocampus. *Behav. Neural. Bioi.* **61**, 260–270.
37. Oliveira, M. G., Bueno, O. F., Pomarico, A. C., and Gugliano, E. B. (1997). Strategies used by hippocampal- and caudate putamen-lesioned rats in a learning task. *Neurobioi. Learn. Mem.* **68**, 32–41.
38. Zyzak, D. R., Otto, T., Eichenbaum, H., and Gallagher, M. (1995). Cognitive decline associated with normal aging in rats: a neuropsychological approach. *Learn Mem.* **2**, 1–16.
39. Shukitt-Hale, B., Smith, D. E., Meydani, M., and Ioseph, J. A. (1999). The effects of dietary antioxidants on psychomotor performance in aged mice. *Exp. Gerontal.* **34**, 797–808.
40. Hauss-Wegrzyniak, B., Willard, L. B., Del Soldato, P., *et al.* (1999). Peripheral administration of novel anti-inflammatories can attenuate the effects of chronic inflammation within the CNS. *Brain Res.* **815**, 36–43.
41. Hauss-Wegrzyniak, B., Vannucchi, M. G., and Wenk, G. L. (2000). Behavioral and ultrastructural changes induced by chronic neuroinflammation in young rats. *Brain Res.* **859**, 157–166.
42. Small, G. W., Kuhl, D. E., Riege, W. H., *et al.* (1989). Cerebral glucose metabolic patterns in Alzheimer's disease: effect of gender and age at dementia onset. *Arch. Gen. Psychiatry* **46**, 527–532.
43. Jorm, A. F., Korten, A. E., and Henderson, A. S. (1987). The prevalence of dementia: a quantitative integration of the literature. *Acta Psychiatr. Scand.* **76**, 465–479.
44. McGonigal, G., Thomas, B., McQuade, C., *et al.* (1993). Epidemiology of Alzheimer's presenile dementia in Scotland, 1974-88. *BMl* **306**, 680–683.
45. Henderson, V. W., Paganini-Hill, A., Emanuel, C. K., *et al.* (1994). Estrogen replacement therapy in older women. Comparisons between Alzheimer's disease cases and nondemented control subjects. *Arch Neurol.* **51**, 896–900.
46. Tang, M. X., Jacobs, D., Stem, Y., *et al.* (1996). Effect of oestrogen during menopause on risk and age at onset of Alzheimer's disease. *Lancet* **348**, 429–432.
47. Kawas, C., Resnick, S., Morrison, A., *et al.* (1997). A prospective study of estrogen replacement therapy and the risk of developing Alzheimer's disease: the Baltimore Longitudinal Study of Aging. *Neurology* **48**, 1517–1521.
48. Stam, F. C., Wigboldus, J. M., and Smeulders, A. W. (1986). Age incidence of senile brain amyloidosis. *Pathol. Res. Pract.* **181**, 558–562.
49. Schupf, N., Kapell, D., Nightingale, B., *et al.* (1998). Earlier onset of Alzheimer's disease in men with Down syndrome. *Neurology* **50**, 991–995.
50. Mulnard, R. A. (2000). Estrogen as a treatment for Alzheimer's disease. *JAMA* **284**, 307–308.
51. Mulnard, R. A., Cotmau, C. W., Kawas, C., *et al.* (2000). Estrogeu replacement therapy for treatment of mild to moderate Alzheimer disease: a randomized controlled trial. Alzheimer's Disease Cooperative Study. *JAMA* **283**, 1007–1015.
52. Wang, P. N., Liao, S. Q., Liu, R. S., *et al.* (2000). Effects of estrogen on cognition, mood, and cerebral blood flow in AD: a controlled study. *Neurology* **54**, 2061–2066.
53. Seshadri, S., Zornberg, G. L., Derby, L. E., *et al.* (2001). Postmenopausal estrogen replacement therapy and the risk of Alzheimer disease. *Arch. Neurol.* **58**, 435–440.
54. Genazzani, A. R., Gastaldi, M., Bidzinska, B., *et al.* (1992). The brain as a target organ of gonadal steroids. *Psychoneuroendocrinology* **17**, 385–390.
55. Ripich, D. N., Petrill, S. A., Whitehouse, P. J., and Ziol, E. W. (1995). Gender differences in language of AD patients: a longitudinal study. *Neurology* **45**, 299–302.
56. Buckwalter, J. G., Sobel, E., Dunn, M. E., *et al.* (1993). Gender differences on a brief measure of cognitive functioning in Alzheimer's disease. *Arch. Neurol.* **50**, 757–760.
57. Hebert, L. E., Wilson, R. S., Gilley, D. W., *et al.* (2000). Decline of language among women and men with Alzheimer's disease. *J. Gerontol. B. Psychol. Sci. Soc. Sci.* **55**, P354–P360.
58. Bayles, K. A., Azuma, T., Cruz, R. F., *et al.* (1999). Gender differences in language of Alzheimer disease patients revisited. *Alzheimer Dis. Assoc. Disord.* **13**, 138–146.
59. Ott, B. R., Tate, C. A., Gordon, N. M., and Heindel, W. E. (1996). Gender differences in the behavioral manifestations of Alzheimer's disease. *J. Am. Geriatr. Soc.* **44**, 583–587.
60. Drachman, D. A., Swearer, J. M., O'Donnell, B. F., *et al.* (1992). The Caretaker Obstreperous-Behavior Rating Assessment (COBRA) Scale. *J. Am. Geriatr. Soc.* **40**, 463–470.
61. Lyketsos, C. G., Steele, C., Galik, E., *et al.* (1999). Physical aggression in dementia patients and its relationship to depression. *Am. J. Psychiatry* **156**, 66–71.
62. Lyketsos, C. G., Chen, L. S., and Authony, J. C. (1999). Cognitive decline in adulthood: an 11.5-year follow-up of the Baltimore Epidemiologic Catchment Area Study. *Am. J. Psychiatry* **156**, 58–65.
63. Cohen, D, Eisdorfer, C., and Gorelick, P. (1993). Sex differences in the psychiatric manifestations of Alzheimer's disease. *J. Am. Geriatr. Soc.* **41**, 229–232.
64. Halliwell, B., and Gutteridge, J. M. (1985). Oxygen radicals and the nervous system. *Trends Neurosci.* **8**, 22–26.
65. Gilissen, E. P., Jacobs, R. E., and Allman, J. M. (1999). Magnetic resonance microscopy of iron in the basal forebrain cholinergic structures of the aged mouse lemur. *J. Neurol. Sci.* **168**, 21–27.
66. Savory, J., Rao, J. K., Huang, Y., Letada, P. R., and Herman, M. M. (1999). Age-related hippocampal changes in Bcl-2:Bax ratio, oxidative stress, redox-active iron and apoptosis associated with aluminum-induced neurodegeneration: increased susceptibility with aging. *Neurotoxicology* **20**, 805–817.
67. Sadoul, R. (1998). Bcl-2 family members in the development and degenerative pathologies of the nervous system. *Cell Death Differ.* **5**, 805–815.
68. Yu, B. P. (1994). Cellular defenses against damage from reactive oxygen species [published erratum appears in

Physiol. Rev. 1995 Jan, 75(1): preceding 1]. *Physiol. Rev.* **74**, 139–162.

69. Cavazzoni, M., Barogi, S., Baracca, A., Parenti Castelli, G., and Lenaz, G. (1999). The effect of aging and an oxidative stress on peroxide levels and the mitochondrial membrane potential in isolated rat hepatocytes. *FEBS Lett.* **449**, 53–56.
70. Denisova, N. A., Erat, S. A., Kelly, J. F., and Roth, G. S. (1998). Differential effect of aging on cholesterol modulation of carbachol-stimulated low-K(m) GTPase in striatal synaptosomes. *Exp. Gerontol.* **33**, 249–265.
71. Shen, H. M., Lin, Y., Choksi, S., Tran, J., Jin, T., Chang, L., Karin, M., Zhang, J., and Liu, Z. G. (2004). Essential roles of receptor-interacting protein and TRAF2 in oxidative stress-induced cell death. *Mol. Cell Biol.* **24**, 5914–5922.
72. Annunziato, L., Pannaccione, A., Cataldi, M., Secondo, A., Castaldo, P., Di Renzo, G., and Taglialatela, M. (2002). Modulation of ion channels by reactive oxygen and nitrogen species: a pathophysiological role in brain aging? *Neurobiol. Aging* **23**, 819–834.
73. Dalton, T. P., Shertzer, H. G., and Puga, A. (1999). Regulation of gene expression by reactive oxygen. *Annu. Rev. Pharmacol. Toxicol.* **39**, 67–101.
74. Davies, K. J. (2000). Oxidative stress, antioxidant defenses, and damage removal, repair, and replacement systems. *IUBMB Life* **50**, 279–289.
75. Hughes, K. A., and Reynolds, R. M. (2005). Evolutionary and mechanistic theories of aging. *Annu. Rev. Entomol.* **50**, 421–445.
76. Perez-Campo, R., Lopez-Torres, M., Cadenas, S., Rojas, C., and Barja, G. (1998). The rate of free radical production as a determinant of the rate of aging: evidence from the comparative approach. *J. Comp. Physiol.* [B] **168**, 149–158.
77. Waring, P. (2005). Redox active calcium ion channels and cell death. *Arch. Biochem. Biophys.* **434**, 33–42.
78. Halliwell, B. (2001). Role of free radicals in the neurodegenerative diseases: therapeutic implications for antioxidant treatment. *Drugs Aging* **18**, 685–716.
79. Rego, A. C., and Oliveira, C. R. (2003). Mitochondrial dysfunction and reactive oxygen species in excitotoxicity and apoptosis: implications for the pathogenesis of neurodegenerative diseases. *Neurochem. Res.* **28**, 1563–1574.
80. Lovell, M. A., Ehmann, W. D., Butler, S. M., and Markesbery, W. R. (1995). Elevated thiobarbituric acid-reactive substances and antioxidant enzyme activity in the brain in Alzheimer's disease. *Neurology* **45**, 1594–1601.
81. Marcus, D. L., Thomas, C., Rodriguez, C., Simberkoff, K., Tsai, J. S., Strafaci, J. A., and Freedman, M. L. (1998). Increased peroxidation and reduced antioxidant enzyme activity in Alzheimer's disease. *Exp. Neurol.* **150**, 40–44.
82. Smith, C. D., Carney, J. M., Starke-Reed, P. E., Oliver, C. N., Stadtman, E. R., Floyd, R. A., and Markesbery, W. R. (1991). Excess brain protein oxidation and enzyme dysfunction in normal aging and in Alzheimer disease. *Proc. Natl. Acad. Sci.* **88**, 10540–10543.
83. Dexter, D. T., Holley, A. E., Flitter, W. D., Slater, T. F., Wells, F. R., Daniel, S. E., Lees, A. J., Jenner, P., and Marsden, C. D. (1994). Increased levels of lipid hydroperoxides in the parkinsonian substantia nigra: an HPLC and ESR study. *Mov. Disord.* **9**, 92–97.
84. Spencer, J. P., Jenner, P., Daniel, S. E., Lees, A. J., Marsden, D. C., and Halliwell, B. (1998). Conjugates of catecholamines with cysteine and GSH in Parkinson's disease: possible mechanisms of formation involving reactive oxygen species. *J. Neurochem.* **71**, 2112–2222.
85. Floyd, R. A., and Hensley, K. (2002). Oxidative stress in brain aging. Implications for therapeutics of neurodegenerative diseases. *Neurobiol. Aging* **23**, 795–807.
86. Joseph, J., Shukitt-Hale, B., Denisova, N. A., Martin, A., Perry, G., and Smith, M. A. (2001). Copernicus revisited: amyloid beta in Alzheimer's disease. *Neurobiol. Aging* **22**, 131–146.
87. Joseph, J. A., Denisova, N., Fisher, D., Shukitt-Hale, B., Bickford, P., Prior, R., and Cao, G. (1998). Membrane and receptor modifications of oxidative stress vulnerability in aging. Nutritional considerations. *Ann. NY Acad. Sci.* **854**, 268–276.
88. Olanow, C. W. (1992). An introduction to the free radical hypothesis in Parkinson's disease. *Ann. Neurol.* **32**(suppl), S2–S9.
89. Carney, J. M., Smith, C. D., Carney, A. M., and Butterfield, D. A. (1994). Aging- and oxygen-induced modifications in brain biochemistry and behavior. *Ann. NY Acad. Sci.* **738**, 44–53.
90. Joseph, J. A., Denisova, N., Fisher, D., Bickford, P., Prior, R., and Cao, G. (1998). Age-related neurodegeneration and oxidative stress: putative nutritional intervention. *Neurol. Clin.* **16**, 747–755.
91. Joseph, J. A., Shukitt-Hale, B., Denisova, N. A., Bielinski, D., Martin, A., McEwen, J. J., and Bickford, P. C. (1999). Reversals of age-related declines in neuronal signal transduction, cognitive, and motor behavioral deficits with blueberry, spinach, or strawberry dietary supplementation. *J. Neurosci.* **19**, 8114–8121.
92. Landfield, P. W., and Eldridge, J. C. (1994). The glucocorticoid hypothesis of age-related hippocampal neurodegeneration: role of dysregulated intraneuronal Ca2+. *Ann. NY Acad. Sci.* **746**, 308–321.
93. Toescu, E. C., and Verkhratsky, A. (2004). Ca2+ and mitochondria as substrates for deficits in synaptic plasticity in normal brain ageing. *J. Cell Mol. Med.* **8**, 181–190.
94. Cheng, Y., Wixom, P., James-Kracke, M. R., and Sun, A. Y. (1994). Effects of extracellular ATP on Fe2+-induced cytotoxicity in PC-12 cells. *J. Neurochem.* **66**, 895–902.
95. Durany, N., Munch, G., Michel, T., and Riederer, P. (1999). Investigations on oxidative stress and therapeutical implications in dementia. *Eur. Arch. Psychiatry Clin. Neurosci.* **249**(suppl 3), 68–73.
96. Munch, G., Schinzel, R., Loske, C., Wong, A., Durany, N., Li, J. J., Vlassara, H., Smith, M. A., Perry, G., and Riederer, P. (1998). Alzheimer's disease—synergistic effects of glucose deficit, oxidative stress and advanced glycation endproducts. *J. Neural. Transm.* **105**, 439–461.
97. Lane, N. (2003). A unifying view of ageing and disease: the double-agent theory. *J. Theor. Biol.* **225**, 531–540.
98. Smith, C. D., Carney, J. M., Tatsumo, T., *et al.* (1992). Protein oxidation in aging brain. *Ann.* NY Acad. Sci. ***663**, 110–119.*
99. Vitek, M. P., Bhattacharya, K., Glendening, J. M., *et al.* (1994). Advanced glycation end products contribute to amyloidosis in Alzheimer disease. *Proc. Natl. Acad. Sci. USA* **91**, 4766–4770.

100. Smith, M. A., Taneda, S., Richey, P. I., *et al.* (1994). Advanced Maillard reaction end products are associated with Alzheimer disease pathology. *Proc. Natl. Acad. Sci. USA* **91**, 5710–5714.
101. Yan, S. D., Chen, X., Schmidt, A. M., *et al.* (1994). Glycated tau protein in Alzheimer disease: a mechanism for induction of oxidant stress. *Proc. Natl. Acad. Sci. USA* **91**, 7787–7791.
102. Ledesma, M. D., Bonay, P., Colaco, C., and Avila, J. (1994). Analysis of microtubule associated protein tau glycation in paired helical filaments. *J. Bioi. Chem.* **269**, 21614–21619.
103. Salomon, R. G., Subbanagounder, G., O'Neil, J., *et al.* (1997). LevuglandinE2-protein adducts in human plasma and vasculature. *Chem. Res. Toxicol.* **10**, 536–545.
104. Sayre, L. M., Perry, G., and Smith, M. A. (1999). In situ methods for detection and localization of markers of oxidative stress: application in neurodegenerative disorders. *Methods Enzymol.* **309**, 133–152.
105. Mattson, M. P., Carney, J. W., and Butterfield, D. A. (1995). A tombstone in Alzheimer's? *Nature* **373**, 481.
106. Smith, M. A., Sayre, L. M., Vitek, M. P., *et al.* (1995). Early AGEing and Alzheimer's. *Nature* **374**, 316.
107. Castellani, R. J., Harris, P. I., Sayre, L. M., *et al.* (2001). Active glycation in neurofibrillary pathology of Alzheimer disease: N(epsilon)-(carboxymethyl) lysine and hexitol-lysine. *Free Radic. Bioi. Med.* **31**, 175–180.
108. Smith, M. A., Perry, G., Richey, P. I., *et al.* (1996). Oxidative damage in Alzheimer's. *Nature* **382**, 120–121.
109. Smith, M. A., Sayre, L. M., Anderson, V. E., *et al.* (1998). Cytochemical demonstration of oxidative damage in Alzheimer disease by immunochemical enhancement of the carbonyl reaction with 2,4-dinitrophenylhydrazine. *J. Histochem. Cytochem.* **46**, 731–735.
110. Montine, T. J., Amarnath, V., Martin, M. E., *et al.* (1996). E-4-hydroxy-2-nonenal is cytotoxic and cross-links cytoskeletal proteins in P19 neuroglial cultures. *Am. J. Pathol.* **148**, 8993.
111. Sayre, L. M., Zelasko, D. A., Harris, P. I., *et al.* (1997). 4-Hydroxynoueual-derived advanced lipid peroxidation end products are increased in Alzheimer's disease. *J. Neurochem.* **68**, 2092–2097.
112. Friguet, B., Stadtman, E. R., and Szweda, L. I. (1994). Modification of glucose-6-phosphate dehydrogenase by 4-hydroxy-2-nonenal. Formation of cross-linked protein that inhibits the multicatalytic protease. *J. Bioi. Chern.* **269**, 21639–21643.
113. Mori, H., Kondo, J., and lhara, Y. (1987). Ubiquitin is a component of paired helical filaments in Alzheimer's disease. *Science* **235**, 1641–1644.
114. Smith, M. A., Richey Harris, P. I., Sayre, L.M, *et al.* (1997). Widespread peroxynitrite-mediated damage in Alzheimer's disease. *J. Neurosci.* **17**, 2653–2657.
115. Nunomura, A., Perry, G., Pappolla, M.A, *et al.* (1999). RNA oxidation is a prominent feature of vulnerable neurons in Alzheimer's disease. *J. Neurosci.* **19**, 1959–1964.
116. Ginsberg, S. D., Crino, P. B., Lee, V. M., *et al.* (1997). Sequestration of RNA in Alzheimer's disease neurofibrillary tangles and senile plaques. *Ann. Neurol.* **41**, 200–209.
117. Nunomura, A., Perry, G., Aliev, G., *et al.* (2001). Oxidative damage is the earliest event in Alzheimer disease. *J. Neuropathol. Exp. Neurol.* **60**, 759–767.
118. Sampson, J. B., Ye, Y., Rosen, H., and Beckman, J. S. (1998). Myeloperoxidase and horseradish peroxidase catalyze tyrosine nitration in proteins from nitrite and hydrogen peroxide. *Arch. Bioehern. Biophys.* **356**, 207–213.
119. Smith, M. A., Harris, P. I., Sayre, L. M., and Pelty, G. (1997). Iron accumulation in Alzheimer disease is a source of redox-generated free radicals. *Proc. Natl. Acad. Sci. USA* **94**, 9866–9868.
120. Sayre, L. M., Perry, G., Hatris, P. I., *et al.* (2000). In situ oxidative catalysis by neurofibrillary tangles and senile plaques in Alzheimer's disease: a central role for bound transition metals. *J. Neurochem.* **74**, 270–279.
121. Brunk, U. T., Jones, C. B., and Sohal, R. S. (1992). A novel hypothesis of lipofuscinogenesis and cellular aging based on interactions between oxidative stress and autophagocytosis. *Mutat. Res.* **275**, 395–403.
122. Smith, M. A., Vasak, M., Knipp, M., *et al.* (1998). Dimethylargininase, a nitric oxide regulatory protein, in Alzheimer disease. *Free Radic. Bioi. Med.* **25**, 898–902.
123. Small, G. W., Mazziotta, J. C., Collins, M. T., *et al.* (1995). Apolipoprotein E type 4 allele and cerebral glucose metabolism in relatives at risk for familial Alzheimer disease. *JAMA* **273**, 942–947.
124. Hensley, K., Carney, J. M., Mattson, M. P., *et al.* (1994). A model for beta-amyloid aggregation and neurotoxicity based on free radical generation by the peptide: relevance to Alzheimer disease. *Proc. Natl. Acad. Sci. USA* **91**, 3270–3274.
125. Sayre, L. M., Zagorski, M. G., Surewicz, W. K., *et al.* (1997). Mechanisms of neurotoxicity associated with amyloid beta deposition and the role of free radicals in the pathogenesis of Alzheimer's disease: a critical appraisal. *Chem. Res. Toxicol.* **10**, 518–526.
126. Rottkamp, C. A., Raina, A. K., Zhu, X., *et al.* (2001). Redox-active iron mediates amyloid-beta toxicity. *Free Radic. Bio. Med.* **30**, 447–450.
127. Perry, G., Nunomura, A., Raina, A. K., and Smith, M. A. (2000). Amyloid-beta junkies. *Lancet* **355**, 757.
128. Smith, M. A., Joseph, J. A., and Peny, G. (2000). Arson: tracking the culprit in Alzheimer's disease. *Ann. NY Acad. Sci.* **924**, 35–38.
129. Lukiw, W. J., and Bazan, N. G. (2000). Neuroinflammatory signaling upregulation in Alzheimer's disease. *Neurochem. Res.* **25**, 1173–1184.
130. Bodles, A. M., and Barger, S. W. (2004). Cytokines and the aging brain: what we don't know might help us. *Trends Neurosci.* **27**, 621–626.
131. Rozovsky, I., Finch, C. E., and Morgan, T. E. (1998). Age-related activation of microglia and astrocytes: *in vitro* studies show. *Neurobiol. Aging* **19**, 97–103.
132. Sheng, J. G., Mrak, R. E., and Griffin, W. S. (1998). Enlarged and phagocytic, but not primed, interleukin-1 alpha-immunoreactive microglia increase with age in normal human brain. *Acta Neuropathol.* (Berl.) **95**, 229–234.
133. Sloane, J. A., Hollander, W., Moss, M. B., Rosene, D. L., and Abraham, C. R. (1999). Increased microglial activation and protein nitration in white matter of the aging monkey. *Neurobiol. Aging* **20**, 395–405.
134. McGeer, P. L., and McGeer, E. G. (1995). The inflammatory response system of brain: implications for therapy of Alzheimer

and other neurodegenerative diseases. *Brain Res. Rev.* **21**(2), 195–218.

135. Kreutzberg, G. W. (1996). Microglia: a sensor for pathological events in the CNS. *Trends Neurosci.* **19**, 312–318.
136. Orr, C. F., Rowe, D. B., and Halliday, G. M. (2002). An inflammatory review of Parkinson's disease. *Prog. Neurobiol.* **68**, 325–340.
137. Chen, S., Frederickson, R. C., and Brunden, K. R. (1996). Neuroglial-mediated immunoinflammatory responses in Alzheimer's disease: complement activation and therapeutic approaches. *Neurobiol. Aging* **17**, 781–787.
138. Darley-Usmar, V., Wiseman, H., and Halliwell, B. (1995). Nitric oxide and oxygen radicals: a question of balance. *FEBS Lett.* **369**, 131–135.
139. Floyd, R. A. (1999). Neuroinflammatory processes are important in neurodegenerative diseases: an hypothesis to explain the increased formation of reactive oxygen and nitrogen species as major factors involved in neurodegenerative disease development. *Free Radic. Biol. Med.* **26**, 1346–1355.
140. Luterman, J. D., Haroutunian, V., Yemul, S., Ho, L., Purohit, D., Aisen, P. S., Mohs, R., and Pasinetti, G. M. (2000). Cytokine gene expression as a function of the clinical progression of Alzheimer disease dementia. *Arch. Neurol.* **57**, 1153–1160.
141. Tarkowski, E., Liljeroth, A. M., Minthon, L., Tarkowski, A., Wallin, A., and Blennow, K. (2003). Cerebral pattern of pro- and anti-inflammatory cytokines in dementias. *Brain Res. Bull.* **61**, 255–260.
142. Chang, H. N., Wang, S. R., Chiang, S. C., Teng, W. J., Chen, M. L., Tsai, J. J., Huang, D. F., Lin, H. Y., and Tsai, Y. Y. (1996). The relationship of aging to endotoxin shock and to production of TNF-α. *J. Gerontology* **51, M**, 220–M222.
143. Chang, R. C., Chen, W., Hudson, P., Wilson, B., Han, D. S., and Hong, J. S. (2001). Neurons reduce glial responses to lipopolysaccharide (LPS) and prevent injury of microglial cells from over-activation by LPS. *J. Neurochem.* **76**, 1042–1049.
144. Spaulding, C. C., Walford, R. L., and Effros, R. B. (1997). Calorie restriction inhibits the age-related dysregulation of the cytokines TNF-alpha and IL-6 in C3B10RF1 mice. *Mech. Ageing Dev.* **93**, 87–94.
145. Volpato, S., Guralnik, J. M., Ferrucci, L., Balfour, J., Chaves, P., Fried, L. P., and Harris, T. B. (2001). Cardiovascular disease, interleukin-6, and risk of mortality in older women: the women's health and aging study. *Circulation* **103**, 947–953.
146. Kushner, I. (2001). C-reactive protein elevation can be caused by conditions other than inflammation and may reflect biologic aging. *Cleve. Clin. J. Med.* **68**, 535–537.
147. Weindruch, R., Kayo, T., Lee, C. K., and Prolla, T. A. (2001). Microarray profiling of gene expression in aging and its alteration by caloric restriction in mice. *J. Nutr.* **131**, 918S–923S.
148. Ho, L., Pieroni, C., Winger, D., Purohit, D. P., Aisen, P. S., and Pasinetti, G. M. (1999). Regional distribution of cyclooxygenase-2 in the hippocampal formation in Alzheimer's disease. *J. Neurosci. Res.* **57**, 295–303.
149. Hoozemans, J. J., Bruckner, M. K., Rozemuller, A. J., Veerhuis, R., Eikelenboom, P., and Arendt, T. (2002). Cyclin D1 and cyclin E are co-localized with cyclo-oxygenase 2 (COX-2) in pyramidal neurons in Alzheimer disease temporal cortex. *J. Neuropathol. Exp. Neurol.* **61**, 678–688.
150. Casolini, P., Catalani, A., Zuena, A. R., and Angelucci, L. (2002). Inhibition of COX-2 reduces the age-dependent increase of hippocampal inflammatory markers, corticosterone secretion, and behavioral impairments in the rat. *J. Neurosci. Res.* **68**, 337–343.
151. Baek, B. S., Kim, J. W., Lee, J. H., Kwon, H. J., Kim, N. D., Kang, H. S., Yoo, M. A., Yu, B. P., and Chung, H. Y. (2001). Age-related increase of brain cyclooxygenase activity and dietary modulation of oxidative status. *J. Gerontol. A. Biol. Sci. Med. Sci.* **56,** B426–B431.
152. Rosenman, S., Shrikant, P., Dubb, L., Benveniste, E., and Ransohoff, R. (1995). Cytokine-induced expression of vascular cell adhesion molecule-1 (VCAM-1) by astrocytes and astrocytoma cell lines. *J. Immunol.* **154**, 1888–1899.
153. Schipper, H. (1996). Astrocytes, brain aging, and neurodegeneration. *Neurob. Aging* **17**, 467–480.
154. Steffen, B., Breier, G., Butcher, E., Schulz, M., and Engelhardt, B. (1996). ICAM-1, VCAM-1, and MAdCAM-1 are expressed on choroid plexus epithelium but not endothelium and mediate binding of lymphocytes *in vitro*. *Am. J. Pathol.* **148**, 1819–1838.
155. Stella, N., Estelles, A., Siciliano, J., Tence, M., Desagher, S., Piomelli, D., Glowinski, J., and Premont. J. (1997). Interleukin-1 enhances the ATP-evoked release of arachidonic acid from mouse astrocytes. *J. Neurosci.* **17**(9), 2939–2946.
156. Woodroofe, M. N. (1995). Cytokine production in the central nervous system. *Neurology* **45**(6 suppl 6), S6–S10.
157. Joseph, J. A., Erat, S., and Rabin, B. M. (1998). CNS effects of heavy particle irradiation in space: behavioral implications. *Adv. Space Res.* **22**, 209–216.
158. Joseph, J. A., Shukitt-Hale, B., McEwen, J., and Rabin, B. M. (2000). CNS-induced deficits of heavy particle irradiation in space: the aging connection. *Adv. Space Res.* **25**, 2057–2064.
159. Shukitt-Hale, B., Casadesus, G., McEwen, J. J., Rabin, B. M., and Joseph, J. A. (2000). Spatial learning and memory deficits induced by exposure to iron-56-particle radiation. *Radiat. Res.* **154**, 28–33.
160. Joseph, J. A., Hunt, W. A., Rabin, B. M., and Dalton, T. K. (1992). Possible "accelerated striatal aging" induced by 56Fe heavy-particle irradiation: implications for manned space flights. *Radiat. Res.* **130**, 88–93.
161. Rabin, B. M., Joseph, J. A., and Erat, S. (1998). Effects of exposure to different types of radiation on behaviors mediated by peripheral or central systems. *Adv. Space Res.* **22**, 217–225.
162. Bickford, P. C., Shukitt-Hale, B., and Joseph, J. A. (1999). Effects of aging on cerebellar noradrenergic function and motor learning: nutritional interventions. *Mech. Ageing Dev.* **111**, 141–154.
163. Hauss-Wegrzyniak, B., Dobrzanski, P., Stoehr, J. D., and Wenk, G. L. (1998). Chronic neuroinflammation in rats reproduces components of the neurobiology of Alzheimer's disease. *Brain Res.* **780**, 294–303.
164. Yamada, K., Komori, Y., Tanaka, T., Senzaki, K., Nikai, T., Sugihara, H., Kameyama, T., and Nabeshima, T. (1999).

Brain dysfunction associated with an induction of nitric oxide synthase following an intracerebral injection of lipopolysaccharide in rats. *Neuroscience* **88**, 281–294.
165. Hauss-Wegrzyniak, B., Vraniak, P., and Wenk, G. L. (1999). The effects of a novel NSAID on chronic neuroinflammation are age dependent. *Neurobiol. Aging* **20**, 305–313.
166. Christen, Y. (2000). Oxidative stress and Alzheimer disease. *Am. J. Clin. Nutr.* **71**, 621S–629S.
167. Markesbery, W. R., and Carney, J. M. (1999). Oxidative alterations in Alzheimer's disease. *Brain Pathol.* **9**, 133–146.
168. Ham, D., and Schipper, H. M. (2000). Heme oxygenase-I induction and mitochondrial iron sequestration in astroglia exposed to amyloid peptides. *Cell Mol. Bioi.* (Noisy-Ie-grand) **46**, 587–596.
169. Agrimi, U., and Di Guardo, G. (1993). Amyloid, amyloid-inducers, cytokines and heavy metals in scrapie and other human and animal subacute spongiform encephalopathies: some hypotheses. *Med. Hypotheses* **40**, 113–116.
170. Paris, D., Townsend, K. P., Obregon, D. F., *et al.* (2002). Pro-inflammatory effect of freshly solubilized beta-amyloid peptides in the brain. *Prostaglandins Other Lipid Mediat.* **70**, 1–12.
171. Meda, L, Baron, P., and Scarlato, G. (2001). Glial activation in Alzheimer's disease: the role of Abeta and its associated proteins. *Neurobiol. Aging* **22**, 885–893.
172. McGeer, E., and McGeer, P. (1998). The importance of inflammatory mechanisms in Alzheimer disease. *Exp. Gerontal.* **33**, 371–378.
173. Yoshida, M., Saito, H., and Katsuki, H. (1995). Neurotrophic effects of conditioned media of astrocytes isolated from different brain regions on hippocampal and cortical neurons. *Experientia* **51**, 133–136.
174. Wiese, S., Metzger, F., Holtmann, B., and Sendtner, M. (1999). Mechanical and excitotoxic lesion of motoneurons: effects of neurotrophins and ciliary neurotrophic factor on survival and regeneration. *Acta Neurochir. Suppl.* (Wien) **73**, 31–39.
175. Paratcha, G., Ledda, F., Baars, L., *et al.* (2001). Released GFR alpha 1 potentiates downstream signaling, neuronal survival, and differentiation via a novel mechanism of recruitment of c-Ret to lipid rafts. *Neuron* **29**, 171–184.
176. Rakowicz, W. P., Staples, C. S., Milbrandt, J., *et al.* (2002). Glial cell line-derived neurotrophic factor promotes the survival of early postnatal spinal motor neurons in the lateral and medial motor columns in slice culture. *J. Neurosci.* **22**, 3953–3962.
177. Garcia-Ovejero, D., Veiga, S., Garcia-Segura, L. M., and Doncarlos, L. L. (2002). Glial expression of estrogen and androgen receptors after rat brain injury. *J. Camp Neural.* **450**, 256–271.
178. Malhotra, S. K., Shnitka, T. K., and Elbrink, J. (1990). Reactive astrocytes: a review. *Cytobios* **61**, 133–160.
179. McGraw, J., Hiebert, G. W., and Steeves, J. D. (2001). Modulating astrogliosis after neurotrauma. *J. Neurosci. Res.* **63**, 109–115.
180. Brown, D. R. (2001). Microglia and prion disease. *Microsc. Res. Tech.* **54**, 71–80.
181. Persidsky, Y., Limoges, J., Rasmussen, J., *et al.* (2001). Reduction in glial immunity and neuropathology by a PAF antagonist and an MMP and TNFalpha inhibitor in SCID mice with HIV-I encephalitis. *J. Neuroimmunol.* **114**, 57–68.
182. Gendron, F. P., Neary, J. T., Theiss, P. M., *et al.* (2003). Mechanisms of P2X7 receptor-mediated ERKI/2 phosphorylation in human astrocytoma cells. *Am. J. Physiol. Cell. Physiol.* **284, C**, 571–C581.
183. Irizarry, M. C., and Hyman, B. T. (2001). Alzheimer disease therapeutics. *J. Neuropathol. Exp. Neural.* **60**, 923–928.
184. Kobayashi, K., Hayashi, M., Nakano, H., *et al.* (2002). Apoptosis of astrocytes with enhanced lysosomal activity and oligodendrocytes in white matter lesions in Alzheimer's disease. *Neuropathol. Appl. Neurobiol.* **28**, 238–251.
185. Martin, A., Prior, R., Shukitt-Hale, B., Cao, G., and Joseph, J. A. (2000). Effect of fruits, vegetables, or vitamin E—rich diet on vitamins E and C distribution in peripheral and brain tissues: implications for brain function. *J. Gerontol. A. Biol. Sci. Med. Sci.* **55**, B144–B151.
186. Sutherland, B. A., Rahman, R. M., and Appleton, I. (2006). Mechanisms of action of green tea catechins, with a focus on ischemia-induced neurodegeneration. *J. Nutr. Biochem.* **17**, 291–306.
187. Mandel, S. A., Avramovich-Tirosh, Y., Reznichenko, L., Zheng, H., Weinreb, O., Amit, T., and Youdim, M. B. (2005). Multifunctional activities of green tea catechins in neuroprotection. Modulation of cell survival genes, iron-dependent oxidative stress and PKC signaling pathway. *Neurosignals* **14**, 46–60.
188. Dajas, F., Rivera, F., Blasina, F., Arredondo, F., Echeverry, C., Lafon, L., *et al.* (2003). Cell culture protection and *in vivo* neuroprotective capacity of flavonoids. *Neurotox. Res.* **5**, 425–432.
189. Rivera, F., Urbanavicius, J., Gervaz, E., Morquio, A., and Dajas, F. (2004). Some aspects of the *in vivo* neuroprotective capacity of flavonoids: bioavailability and structure–activity relationship. *Neurotox. Res.* **6**, 543–553.
190. Hong, J. T., Ryu, S. R., Kim, H. J., Lee, J. K., Lee, S. H., Kim, D. B., *et al.* (2000). Neuroprotective effect of green tea extract in experimental ischemia-reperfusion brain injury. *Brain Res. Bull.* **53**, 743–749.
191. Choi, Y. B., Kim, Y. I., Lee, K. S., Kim, B. S., and Kim, D. J. (2004). Protective effect of epigallocatechin gallate on brain damage after transient middle cerebral artery occlusion in rats. *Brain Res.* **1019**, 47–54.
192. Matsuoka, Y., Hasegawa, H., Okuda, S., Muraki, T., Uruno, T., and Kubota, K. (1995). Ameliorative effects of tea catechins on active oxygen-related nerve cell injuries. *J. Pharmacol. Exp. Ther.* **274**, 602–608.
193. Suzuki, M., Tabuchi, M., Ikeda, M., Umegaki, K., and Tomita, T. (2004). Protective effects of green tea catechins on cerebral ischemic damage. *Med. Sci. Monit.* **10,** BR166–BR174.
194. Sutherland, B. A., Shaw, O. M., Clarkson, A. N., Jackson, D. N., Sammut, I. A., and Appleton, I. (2005). Neuroprotective effects of (−)-epigallocatechin gallate following hypoxia–ischemia-induced brain damage: novel mechanisms of action. *FASEB J.* **19**, 258–260.
195. Levites, Y., Weinreb, O., Maor, G., Youdim, M. B. H., and Mandel, S. (2001). Green tea polyphenol (-)-epigallocatechin-3-gallate prevents N-methyl-4-phenyl-1,2,3,6-tetrahydropyridine-induced dopaminergic neurodegeneration. *J. Neurochem.* **78**, 1073–1082.

196. Ringman, J. M., Frautschy, S. A., Cole, G. M., Masterman, D. L., and Cummings, J. L. (2005). A potential role of the curry spice curcumin in Alzheimer's disease. *Curr. Alzheimer Res.* **2**, 131–136.
197. Al-Omar, F. A., Nagi, M. N., Abdulgadir, M. M., Al Joni, K. S., and Al-Majed, A. A. (2006). Immediate and delayed treatments with curcumin prevents forebrain ischemia-induced neuronal damage and oxidative insult in the rat hippocampus. *Neurochem. Res.* **31**, 611–618.
198. Kim, H. Y., Park, E. J., Joe, E. H., and Jou, I (2003). Curcumin suppresses Janus kinase-STAT inflammatory signaling through activation of Src homology 2 domain-containing tyrosine phosphatase 2 in brain microglia. *J. Immunol.* **171**, 6072–6079.
199. Jung, K. K., Lee, H. S., Cho, J. Y., Shin, W. C., Rhee, M. H., Kim, T. G., Kang, J. H., Kim, S. H., Hong, S., and Kang, S. Y. (2006). Inhibitory effect of curcumin on nitric oxide production from lipopolysaccharide-activated primary microglia. *Life Sci.* **79**, 2022–2031.
200. Garcia-Alloza, M., Borrelli, L. A., Rozkalne, A., Hyman, B. T., and Bacskai, B. J. (2007). Curcumin labels amyloid pathology *in vivo*, disrupts existing plaques, and partially restores distorted neurites in an Alzheimer mouse model. *J. Neurochem.* **102**, 1095–1104.
201. Yang, F., Lim, G. P., Begum, A. N., Ubeda, O. J., Simmons, M. R., Ambegaokar, S. S., Chen, P. P., Kayed, R., Glabe, C. G., Frautschy, S. A., and Cole, G. M. (2005). Curcumin inhibits formation of amyloid beta oligomers and fibrils, binds plaques, and reduces amyloid *in vivo*. *J. Biol. Chem.* **280**, 5892–5901.
202. Holt, P. R., Katz, S., and Kirshoff, R. (2005). Curcumin therapy in inflammatory bowel disease: a pilot study. *Dig. Dis. Sci.* **50**, 2191–2193.
203. Cao, G., Giovanoni, M., and Prior, R. L. (1996). Antioxidant capacity in different tissues of young and old rats. *Proc. Soc. Exp. Bioi. Med.* **211**, 359–365.
204. Prior, RL, and Cao, G. (1999). Antioxidant capacity and polyphenolic components of teas: Implications for altering in vivo antioxidant status. *Proc. Soc. Exp. Bioi. Med.* **220**, 255–261.
205. Wang, S. Y., and Lin, H. S. (2000). Antioxidant activity in fruits and leaves of blackberry, raspberry, and strawberry varies with cultivar and developmental stage. *J. Agric. Food Chem.* **48**, 140–146.
206. Joseph, J. A., Shukitt-Hale, B., Denisova, N. A., Prior, R. L., Cao, G., Martin, A., Taglialatela, G., and Bickford, P. C. (1998). Long-term dietary strawberry, spinach, or vitamin E supplementation retards the onset of age-related neuronal signal-transduction and cognitive behavioral deficits. *J. Neurosci.* **18**, 8047–8055.
207. Youdim, K. A., Shukitt-Hale, B., MacKinnon, S., *et al.* (2000). Polyphenolics enhance red blood cell resistance to oxidative stress: in vitro and in vivo. *Biochim. Biophys. Acta* **1523**, 117–122.
208. Casadesus, G., Stellwagen, H., Szprengiel, A., *et al.* (2002). Modulation of hippocampal neurogenesis and cognitive performance in the aged rat: the blueberry effect. *Soc. Neurosci. Abs.* **28**, 294.
209. Billman, G. E., Kang, J. X., and Leaf, A. (1999). Prevention of sudden cardiac death by dietary pure omega-3 polyunsaturated fatty acids in dogs. *Circulation* **99**, 2452–2457.
210. Grimminger, F., Grimm, H., Fuhrer, D., Papavassilis, C., Lindemann, G., Blecher, C., Mayer, K., Tabesch, F., Kramer, H. J., Stevens, J., and Seeger, W. (1996). Omega-3 lipid infusion in a heart allotransplant model: shift in fatty acid and lipid mediator profiles and prolongation of transplant survival. *Circulation* **93**, 365–371.
211. Biscione, F., Pignalberi, C., Totteri, A., Messina, F., and Altamura, G. (2007). Cardiovascular effects of omega-3 free fatty acids. *Curr. Vasc. Pharmacol.* **5**, 163–172.
212. Oh, R. C., and Lanier, J. B. (2007). Management of hypertriglyceridemia. *Am. Fam. Physician.* **75**, 1365–1371.
213. Chapkin, R. S., Davidson, L. A., Ly, L., Weeks, B. R., Lupton, J. R., and McMurray, D. N. (2007). Immunomodulatory effects of (n-3) fatty acids: putative link to inflammation and colon cancer. *J. Nutr.* **137**, 200S–204S.
214. Simopoulos, A. P. (2006). Evolutionary aspects of diet, the omega-6/omega-3 ratio and genetic variation: nutritional implications for chronic diseases. *Biomed. Pharmacother.* **60**, 502–507.
215. Cole, G. M., and Frautschy, S. A. (2006). Docosahexaenoic acid protects from amyloid and dendritic pathology in an Alzheimer's disease mouse model. *Nutr. Health.* **18**, 249–259.
216. Cole, G. M., Lim, G. P., Yang, F., Teter, B., Begum, A., Ma, Q., Harris-White, M. E., and Frautschy, S. A. (2005). Prevention of Alzheimer's disease: omega-3 fatty acid and phenolic anti-oxidant interventions. *Neurobiol. Aging* **26**(suppl 1), 133–136.
217. Yehuda, S., .Rabinovitz, S, Carasso, R. L., and Mostofsky, D. I. (2002). The role of polyunsaturated fatty acids in restoring the aging neuronal membrane. *Neurobiol. Aging* **23**, 843–853.
218. Ikemoto, A., Ohishi, M., Sato, Y., Hata, N, Misawa, Y., Fujii, Y., and Okuyama, H. (2001). Reversibility of n-3 fatty acid deficiency-induced alterations of learning behavior in the rat: level of n-6 fatty acids as another critical factor. *J. Lipid Res.* **42**, 1655–1663.
219. Conquer, J. A., Tierney, M. C., Zecevic, J., Bettger, W. J., and Fisher, R. H. (2000). Fatty acid analysis of blood plasma of patients with Alzheimer's disease, other types of dementia, and cognitive impairment. *Lipids* **35**, 1305–1312.
220. Soderberg, M, Edlund, C, Kristensson, K, and Dallner, G. (1991). Fatty acid composition of brain phospholipids in aging and in Alzheimer's disease. *Lipids* **26**, 421–425.
221. Kalmijn, S., van Boxtel, M. P., Ocke, M., Verschuren, W. M., Kromhout, D., and Launer, L. J. (2004). Dietary intake of fatty acids and fish in relation to cognitive performance at middle age. *Neurology* **62**, 275–280.
222. Florent, S., Malaplate-Armand, C., Youssef, I., Kriem, B., Koziel, V., Escanye, M. C., Fifre, A., Sponne, I., Leininger-Muller, B., Olivier, J. L., Pillot, T., and Oster, T. (2006). Docosahexaenoic acid prevents neuronal apoptosis induced by soluble amyloid-beta oligomers. *J. Neurochem.* **96**, 385–395.
223. Stillwell, W., and Wassall, S. R. (2003). Docosahexaenoic acid: membrane properties of a unique fatty acid. *Chem. Phys. Lipids* **126**, 1–27.

224. Bazan, N. G. (2006). The onset of brain injury and neurodegeneration triggers the synthesis of docosanoid neuroprotective signaling. *Cell Mol. Neurobiol.* **26**, 901–913.
225. Akbar, M., and Kim, H. Y. (2002). Protective effects of docosahexaenoic acid in staurosporine-induced apoptosis: involvement of phosphatidylinositol-3 kinase pathway. *J. Neurochem.* **82**, 655–665.
226. Akbar, M., Calderon, F., Wen, Z., and Kim, H. Y. (2005). Docosahexaenoic acid: a positive modulator of Akt signaling in neuronal survival. *Proc. Natl. Acad. Sci. USA* **102**, 10858–10863.
227. Erkkila, A. T., Lehto, S., Pyorala, K., and Uusitupa, M. I. (2003). n-3 Fatty acids and 5-y risks of death and cardiovascular disease events in patients with coronary artery disease. *Am. J. Clin. Nutr.* **78**, 65–71.
228. He, K., Song, Y., Daviglus, M. L., Liu, K., Van Horn, L., Dyer, A. R., and Greenland, P. (2004). Accumulated evidence on fish consumption and coronary heart disease mortality: a meta-analysis of cohort studies. *Circulation* **109**, 2705–2711.
229. Hu, F. B., Bronner, L., Willett, W. C., Stampfer, M. J., Rexrode, K. M., Albert, C. M., Hunter, D., and Manson, J. E. (2002). Fish and omega-3 fatty acid intake and risk of coronary heart disease in women. *JAMA* **287**, 1815–1821.
230. He, K., Song, Y., Daviglus, M. L., Liu, K., Van Horn, L., Dyer, A. R., Goldbourt, U., and Greenland, P. (2004). Fish consumption and incidence of stroke: a meta-analysis of cohort studies. *Stroke* **35**, 1538–1542.
231. de Lorgeril, M., Salen, P., Martin, J. L., Monjaud, I., Delaye, J., and Mamelle, N. (1999). Mediterranean diet, traditional risk factors, and the rate of cardiovascular complications after myocardial infarction: final report of the Lyon Diet Heart Study. *Circulation* **99**, 779–785.
232. Bucher, H.C., Hengstler, P., Schindler, C., and Meier, G. (2002). N-3 polyunsaturated fatty acids in coronary heart disease: a meta-analysis of randomized controlled trials. *Am. J. Med.* **112**, 298–304.
233. Burr, M. L. (1992). Fish food, fish oil and cardiovascular disease. *Clin. Exp. Hypertens. A.* **14**, 181–192.
234. Dolecek, T. A., and Granditis, G. (1991). Dietary polyunsaturated fatty acids and mortality in the Multiple Risk Factor Intervention Trial (MRFIT). *World Rev. Nutr. Diet.* **66**, 205–216.
235. Feldman, E. B. (2002). The scientific evidence for a beneficial health relationship between walnuts and coronary heart disease. *J. Nutr.* **132**, 1062S–1101S.
236. Akiyama-Oda, Y., Hotta, Y., Tsukita, S., and Oda, H. (2000). Mechanism of glia-neuron cell-fate switch in the Drosophila thoracic neuroblast 6-4 lineage. *Development* **127**, 3513–3522.
237. Barnham, K. J., Masters, C. L., and Bush, A. I. (2004). Neurodegenerative diseases and oxidative stress. *Nat. Rev. Drug Discov.* **3**, 205–214.
238. Huang, F. L., Yoshida, Y., Nakabayashi, H., Young, W. S. 3rd, and Huang, K. P. (1988). Immunocytochemical localization of protein kinase C isozymes in rat brain. *J. Neurosci* **8**, 4734–4744.
239. Saito, N., Kose, A., Ito, A., Hosoda, K., Mori, M., Hirata, M., Ogita, K., Kikkawa, U., Ono, Y., Igarashi, K., *et al.* (1989) Immunocytochemical localization of beta II subspecies of protein kinase C in rat brain. *Proc. Natl. Acad. Sci. USA* **86**, 3409–3413.
240. Spitaler, M., and Cantrell, D. A. (2004). Protein kinase C and beyond. *Nat. Immunol.* **5**, 785–790.
241. Lin, D., and Takemoto, D. J. (2005). Oxidative activation of protein kinase C gamma through the C1 domain. Effects on gap junctions. *J. Biol. Chem.* **280**, 13682–13693.
242. Salonen, T., Sareila, O., Jalonen, U., Kankaanranta, H., Tuominen, R., and Moilanen, E. (2006). Inhibition of classical PKC isoenzymes downregulates STAT1 activation and iNOS expression in LPS-treated murine J774 macrophages. *Br. J. Pharmacol.* **147**, 790–799.
243. Hall, A. J., Tripp, M., Howell, T., Darland, G., Bland, J. S., and Babish, J. G. (2006). Gastric mucosal cell model for estimating relative gastrointestinal toxicity of non-steroidal anti-inflammatory drugs. *Prostaglandins Leukot. Essent. Fatty Acids* **75**, 9–17.
244. McGeer, E. G., and McGeer, P. L. (2003). Inflammatory processes in Alzheimer's disease. *Prog Neuropsychopharmacol Biol. Psychiatry* **27**, 741–749.
245. Zhu, X., Raina, A. K., Lee, H. G., Casadesus, G., Smith, M. A., and Perry, G. (2004). Oxidative stress signalling in Alzheimer's disease. *Brain Res.* **1000**, 32–39.
246. Hensley, K., Floyd, R. A., Zheng, N. Y., Nael, R., Robinson, K. A., Nguyen, X., Pye, Q. N., Stewart, C. A., Geddes, J., Markesbery, W. R., Patel, E., Johnson, G. V., and Bing, G. (1999). p38 kinase is activated in the Alzheimer's disease brain. *J. Neurochem.* **72,** 2053–2058.
247. Yrjanheikki, J., Keinanen, R., Pellikka, M., Hokfelt, T., and Koistinaho, J. (1998). Tetracyclines inhibit microglial activation and are neuroprotective in global brain ischemia. *Proc. Natl. Acad. Sci. USA* **95**, 15769–15774.
248. Tikka, T., Fiebich, B. L., Goldsteins, G., Keinanen, R., and Koistinaho, J. (2001). Minocycline, a tetracycline derivative, is neuroprotective against excitotoxicity by inhibiting activation and proliferation of microglia. *J. Neurosci.* **21**, 2580–2588.
249. Zhu, S., Stavrovskaya, I. G., Drozda, M., Kim, B. Y., Ona, V., Li, M., Sarang, S., Liu, A. S., Hartley, D. M., Wu, D. C., Gullans, S., Ferrante, R. J., Przedborski, S., Kristal, B. S., and Friedlander, R. M. (2002). Minocycline inhibits cytochrome c release and delays progression of amyotrophic lateral sclerosis in mice. *Nature* **417**, 74–78.
250. Dong, X. X., Hui, Z. J., Xiang, W. X., Rong, Z. F., Jian, S., and Zhu, C. J. (2007). Ginkgo biloba extract reduces endothelial progenitor-cell senescence through augmentation of telomerase activity. *J. Cardiovasc. Pharmacol.* **49**, 111–115.
251. Yamamoto, Y., and Gaynor, R. B. (2004). IkappaB kinases: key regulators of the NF-kappaB pathway. *Trends Biochem. Sci.* **29**, 72–79.
252. Farooqui, A. A., Horrocks, L. A., and Farooqui, T. (2007). Modulation of inflammation in brain: a matter of fat. *J. Neurochem.* **101**, 577–599.
253. Beitner-Johnson, D., and Millhorn, D. E. (1998). Hypoxia induces phosphorylation of the cyclic AMP response element-binding protein by a novel signaling mechanism. *J. Biol. Chem.* **273**, 19834–19839.
254. Cummins, E. P., and Taylor, C. T. (2005). Hypoxia-responsive transcription factors. *Pflugers Arch.* **450**, 363–371.

255. Rodriguez-Mora, O. G., Howe, C. J., Lahair, M. M., McCubrey, J. A., and Franklin, R. A. (2005). Inhibition of CREB transcriptional activity in human T lymphocytes by oxidative stress. *Free Radic. Biol. Med.* **38**, 1653–1661.
256. Rockwell, P., Martinez, J., Papa, L., and Gomes, E. (2004). Redox regulates COX-2 upregulation and cell death in the neuronal response to cadmium. *Cell Signal* **16**, 343–353.
257. Gerzanich, V., Ivanova, S., and Simard, J. M. (2003). Early pathophysiological changes in cerebral vessels predisposing to stroke. *Clin. Hemorheol. Microcirc.* **29**, 291–294.
258. Joseph, J. A., Fisher, D. R., and Strain, J. (2002). Muscarinic receptor subtype determines vulnerability to oxidative stress in COS-7 cells. *Free Radic. Biol. Med.* **32**, 153–161.
259. Joseph, J. A., Fisher, D. R., and Bielinski, D. (2006). Blueberry extract alters oxidative stress-mediated signaling in COS-7 cells transfected with selectively vulnerable muscarinic receptor subtypes. *J. Alzheimers. Dis.* **9**, 35–42.
260. He, L. M., Chen, L. Y., Lou, X. L., Qu, A. L., Zhou, Z., and Xu, T. (2002). Evaluation of beta-amyloid peptide 25–35 on calcium homeostasis in cultured rat dorsal root ganglion neurons. *Brain Res.* **939**, 65–75.
261. Stutzmann, G. E., Smith, I., Caccamo, A., Oddo, S., Laferla, F. M., and Parker, I. (2006). Enhanced ryanodine receptor recruitment contributes to Ca2+ disruptions in young, adult, and aged Alzheimer's disease mice. *J. Neurosci.* **26**, 5180–5189.
262. O'Neill, C., Cowburn, R. F., Bonkale, W. L., Ohm, T. G., Fastbom, J., Carmody, M., and Kelliher, M. (2001). Dysfunctional intracellular calcium homoeostasis: a central cause of neurodegeneration in Alzheimer's disease. *Biochem. Soc. Symp.* 177–194.
263. Lau, F. C., Bielinski, D. F., and Joseph, J. A. (2007). Inhibitory effects of blueberry extract on the production of inflammatory mediators in lipopolysaccharide-activated BV2 microglia. *J. Neurosci. Res.* **85**, 1010–1017.
264. Joseph, J. A., Villalobos-Molina, R., Denisova, N., Erat, S., Cutler, R., and Strain, J. G. (1996). Age differences in sensitivity to H_2O_2-or NO-induced reductions in K+-evoked dopamine release from superfused striatal slices: reversals by PBN or Trolox. *Free Rad. Biol. Med.* **20**, 821–830.
265. Bartus, R. T., Dean, R. L., Beer, B., and Lippa, A. S. (1982). The cholinergic hypothesis of geriatric memory dysfunction. *Science* **217**, 408–417.
266. Rossner, S., Ueberham, U., Schliebs, R., Perez-Polo, J. R., and Bigl, V. (1998). The regulation of amyloid precursor protein metabolism by cholinergic mechanisms and neurotrophin receptor signaling. *Prog. Neurobiol.* **56**, 541–569.
267. Elhusseiny, A., Cohen, Z., Olivier, A., Stanimirovic, D. B., and Hamel, E. (1999). Functional acetylcholine muscarinic receptor subtypes in human brain microcirculation: identification and cellular localization. *J. Cereb. Blood Flow Metab.* **19**, 794–802.
268. Toescu, E. C., and Verkhratsky, A. (2000). Parameters of calcium homeostasis in normal neuronal ageing. *J. Anatomy* **197**, 563–569.
269. Herman, J. P., Chen, K. C., Booze, R., and Landfield, P. W. (1998). Upregulation of alpha1D Ca2+ channel subunit mRNA expression in the hippocampus of aged 344 rats. *Neurobiol. Aging* **19**, 581–587.
270. Lynch, D. R., and Dawson, T. M. (1994). Secondary mechanisms in neuronal trauma. *Curr. Opin. Neurobiol.* **7**, 510–516.
271. Vannucci, R. C., Brucklacher, R. M., and Vannucci, S. J. (2001). Intracellular calcium accumulation during the evolution of hypoxicischemic brain damage in the immature rat. *Brain Res. Dev.* **126**, 117–120.
272. Mattson, M. P. (2000). Emerging apoptosis in neurodegenerative disorders. Nature Reviews. *Mol. Cell Biol.* **1**, 120–129.
273. DeSarno, P., Shestopal, S. A., King, T. D., Zmijewska, A., Songand, L., and Jope, R. S. (2003). Muscarinic receptor activation protects cells from apoptotic effects of DNA damage, oxidative stress and mitochondrial inhibition. *J. Biol. Chem.* **278**, 11086–11093.
274. Huidobro, A., Blanco, P., Villalba, M, Gomez-Puertas, P., Villa, A., Pereira, R., Bogonez, E., Martinez-Serrano, A., Aparicio, J. J., and Satrustegui, J. (1993.). Age-related changes in calcium homeostatic mechanisms in synaptosomes in relation with working memory deficiency. *Neurobiol. Aging* **14**, 479–486.
275. Nyakas, C., Oosterink, B. J., Keijser, J., Felszeghy, K., de Jong, G. I., Korf, J., and Luiten, D. G. (1997). Selective decline of 5-HT1A receptor binding sites in rat cortex, hippocampus and cholinergic basal forebrain nuclei during aging. *J. Chem. Neuroanatomy* **13**, 53–61.
276. Kaufmann, J. L., Bickford, P. C., and Taglialatela, G. (2001). Oxidative stress dependent up-regulation of Bcl-2 expression in the central nervous system of aged Fisher 344 rats. *J. Neurochem.* **76**, 1099–1108.
277. Hartmann, H., Velbinger, K., Eckert, A., and Muller, W. E. (1996). Region-specific downregulation of free intracellular calcium in the aged rat brain. *Neurobiol. Aging* **17**, 557–563.
278. Kaasinen, V., Vilkman, H., Hietala, J., Nagren, K., Helenius, H., Olsson, H., Farde, L., and Rinne, J. (2000). Age-related dopamine D2/D3 receptor loss in extrastriatal regions of the human brain. *Neurobio. Aging* **21**, 683–688.
279. Joseph, J. A., Fisher, D. R., Carey, A. N., Neuman, A., and Bielinski, D. F. (2006). Dopamine-induced stress signaling in COS-7 cells transfected with selectively vulnerable muscarinic receptor subtypes is partially mediated via the i3 loop and antagonized by blueberry extract. *J. Alzheimers Dis.* **10**, 423–437.
280. Miquel, J. (2001). Nutrition and ageing. *Public Health Nutr.* **4**, 1385–1388.
281. Kris-Etherton, P. M., and Keen, C. L. (2002). Evidence that the antioxidant flavonoids in tea and cocoa are beneficial for cardiovascular health. *Curr. Opin. Lipidol.* **13**, 41–49.
282. Kris-Etherton, P. M., Hecker, K. D., Bonanome, A., *et al.* (2002). Bioactive compounds in foods: Their role in the prevention of cardiovascular disease and cancer. *Am. J. Med.* I **13**(suppl 9B), 71S–88S.
283. Joseph, J. A., Arendash, G., Gordon, M., Diamond, D., Shukitt-Hale, B., and Morgan, D. (2003). Blueberry supplementation enhances signaling and prevents behavioral deficits in an Alzheimer disease model. *Nutr. Neurosci.* **6**, 153–162.

284. Commenges, D., Scotet, V., Renaud, S., Jacqmin-Gadda, H., Barberger-Gateau, P., and Dartigues, J. F. (2000). Intake of flavonoids and risk of dementia. *Eur. J. Epidemiol.* **16**, 357–363.
285. Grant, W. B., Campbell, A., Itzhaki, R. F., and Savory, J. (2002). The significance of environmental factors in the etiology of Alzheimer's disease. *J. Alzheimers Dis.* **4**, 179–189.
286. Levites, Y., Amit, T., Youdim, M. B., and Mandel, S. (2002). Involvement of protein kinase C activation and cell survival/cell cycle genes in green tea polyphenolepigallocatechin 3-gallate neuroprotective action. *J. Bioi. Chem.* **277**, 30574–30580.
287. Levites, Y., Weinreb, O., Maor, G., *et al.* (2001). Green tea polyphenol-epigallocatechin-3-gallate prevents N-methyl-4-phenyl-I,2,3,6-tetrahydropyridine induced dopaminergic neurodegeneration. *J. Neurochem.* **78**, 1073–1082.
288. Levites, Y., Youdim, M. B., Maor, G., and Mandel, S. (2002). Attenuation of 6-hydroxydopamine (6-0HDA)-induced nuclear factor-kappaB (NF-kappaB) activation and cell death by tea extracts in neuronal cultures. *Biochem. Pharmacol.* **63**, 21–29.

CHAPTER **18**

Diet and Supplements in the Prevention and Treatment of Eye Diseases

JULIE A. MARES[1] AND AMY E. MILLEN[2]

[1]*University of Wisconsin at Madison, Madison, Wisconsin*

[2]*State University at Buffalo, New York*

Contents

I. INTRODUCTION

A. Nutrition and Common Age-Related Eye Diseases

The deterioration of human vision advances with age. Over 80% of blindness worldwide occurs in people over age 50. This chapter addresses the influence of diet on the most common causes of vision loss in middle-aged and older people: age-related cataract, age-related macular degeneration, and diabetic retinopathy.

Significant advances in the understanding of how nutrition may influence eye diseases of the aging population have been made since the 1980s. The aging public's awareness of the decline in vision with age, and of the possibility that nutrition may influence this decline, has driven the marketing of nutritional supplements, which are sometimes costly and of uncertain benefit. In this chapter, we consider the existing evidence for the benefits of certain diets and supplements in slowing age-related visual problems associated with cataracts, macular degeneration, and diabetic retinopathy (summarized in Tables 1–3).

B. Nutrition and Other Aspects of Vision

The reader is referred to other texts and reviews that describe the role of nutrition in other aspects of vision that are briefly summarized in the following paragraphs. The role of vitamin A in preventing night blindness and xerophthalmia, which remain common problems in developing countries, has been widely discussed [1]. Food shortages, as occurred in Cuba in 1991 to 1994 and among the allied prisoners of World War II, or chronic alcohol use can result in a condition broadly referred to as nutritional amblyopia, which results in blurred vision and reduced visual acuity (recently discussed [2]). This may be the result of poor intake and absorption of B vitamins or antioxidants, alcohol and tobacco toxicity, or a combination of these factors.

Nutritional supplements may be of benefit to some patients with visual disorders. One such condition is retinitis pigmentosa, an inherited condition of progressive visual loss. It begins with the loss of night vision in childhood or adolescence. This is followed by the loss of peripheral vision because of the degeneration of rods, and finally the loss of central vision because of the degeneration of cones (recently reviewed [3]). Vitamin A supplements have been demonstrated to improve some aspects of retina function in patients with retinitis pigmentosa. In addition, diet or blood levels of omega-3 fatty acids have been directly associated with slower rates of retina degeneration and of visual decline in retinitis pigmentosa patients.

Nutrition may also be important to the development of the visual system in newborns. Some studies (but not all) have observed better visual development in infants who were breastfed, as opposed to those who were bottle fed. This has led to a search for the nutritional differences between breast milk and infant formulas that may explain better vision in breastfed infants. In two recent studies, breastfeeding was associated with better visual acuity at age 3.5 [4] and 4 to 6 years [5]. Breast milk contains high levels of the long-chain fatty acid, docosahexaenoic acid (DHA), which rapidly accumulates in retinal photoreceptor membranes neonatally. DHA supplementation has been reported in some, but not all, studies to improve visual functions in some preterm and term infants (reviewed in [6]). Some suggest that improvements may only be transient. One study reported that DHA supplementation for 6 months postnatally did not improve vision in later childhood [5]. This suggests the possibility that other components

TABLE 1 Summary of Evidence Relating Nutritional Exposures to Cataract

Nutritional Exposure	Strength of Evidence	Comment
Carbohydrate	**Possible increased risk associated with high levels of specific or overall refined carbohydrates:** Animal studies suggest several mechanisms. Population studies are limited in number and conflicting.	Animal studies suggest the effect of carbohydrates may be influenced by dietary antioxidants, but this has not been considered in population studies.
Antioxidants	**Benefit of food antioxidants is likely:** Animal studies prove that oxidative stress leads to lens opacities and that antioxidants lower indicators of oxidative stress or damage.	In population studies, diets rich in specific antioxidant nutrients are likely to be markers for diets rich in plant foods (fruits, vegetables, whole grains), which contribute a wide range of nutritive and non-nutritive antioxidants.
	Population studies in many samples indicate lower risk of cataract with higher intake or blood levels of various antioxidants (vitamins C, E, or lutein and zeaxanthin); data are most consistent for diets rich in lutein and zeaxanthin.	Clinical trials do not generally support benefit of one or two specific antioxidant nutrients or a combination of high-dose antioxidants.
Lead	**Exposure possibly increases risk:** Evidence is limited to one study. This risk factor is poorly investigated.	
Diet patterns	**Benefit of following U.S. Dietary Guidelines (or other micronutrient-rich diet patterns) is likely:** Evidence for a direct association with Healthy Eating index score (reflecting adherence to 2000 U.S. Dietary Guidelines) is limited to one study in women, but results are supported by a large overall body of evidence for lower rates of cataract among people associated with nutrient-rich diets.	
Multivitamin supplements	**Benefit is likely:** Results of numerous population studies are supportive, particularly for intake over 10 or more years.	Protective associations in population studies may reflect better diets or other healthy lifestyles in multivitamin supplement users. Data from clinical trials over 6 years in length are lacking but soon forthcoming.

TABLE 2 Summary of Evidence Relating Diet to Age-Related Macular Degeneration (AMD)

Nutritional Exposure	Strength of Evidence	Comment
Antioxidants	**Benefits of a combination antioxidant supplement in slowing progression has been proven:** A high-dose combination antioxidant supplement (400 IU vitamin E, 500 mg vitamin C, 15 mg beta-carotene, 80 mg zinc, and 2 mg copper) lowered progression of intermediate to advanced AMD by 25% over 6 years and reduced moderate vision loss by 19% in a large, multicenter, placebo-controlled trial (AREDS). This result and observation that AREDS supplements reduce the oxidation of cysteine in the blood in people supports the idea that antioxidants play a role in protecting against AMD that is also suggested by retinal abnormalities in experimental animals deficient in antioxidants.	Caveats regarding AREDS-tested supplements: — There is no evidence that high-dose antioxidants lower the onset of AMD among people with early signs of the disease or who merely have a family history of AMD. — The longer-term risks and benefits are unknown. — Whether lower doses or different combinations of nutrients may have more benefit or lower long-term risk is unknown (and is being studied).

(continues)

TABLE 2 *(continued)*

Nutritional Exposure	Strength of Evidence	Comment
	Benefits of antioxidant-rich diets in slowing the onset of AMD is strongly suggested by the above and because of the following: Diets that are higher in one or more antioxidant nutrients of foods (vegetables and fruits, for example) are often related to lower risk for AMD (even though the specific single nutrient[s] related to AMD differ across samples and type of AMD). The intake of foods that were high in nutrients contained in AREDS supplements, but at much lower doses, was associated with lower incidence of AMD in one study, suggesting that the intake of foods that are high in several antioxidants may protect against earlier stages of AMD.	Trials of one or two high-dose antioxidants have not shown benefit.
Lutein and zeaxanthin	**Benefit in slowing AMD or improving vision is possible:** Supplementation in animal studies reduces light damage to retina. The biological plausibility that lutein and zeaxanthin could lower risk by protection against oxidative stress or reduction of damage caused by light exposure is strong, but impact in humans has not been proven. Diets high in foods that contain lutein and zeaxanthin are sometimes, but not always, associated with lower risk of AMD in population studies. Several small and short-term clinical trials provide preliminary evidence to suggest that lutein and zeaxanthin supplementation may improve vision in people with AMD.	Diets rich in lutein and zeaxanthin may reflect the overall benefit of high intakes of many micronutrient-rich foods. The benefit of lutein and zeaxanthin supplements in slowing or preventing AMD is currently being studied in clinical trials. The ability to accumulate lutein and zeaxanthin (and therefore, the potential benefit) varies across people.
Zinc	**Benefit in slowing progression is proven:** High-dose zinc supplements slowed the progression of intermediate to advanced AMD in a large, multicenter, placebo-controlled clinical trial over 6 years. In combination with antioxidant supplements, this supplement also reduced moderate visual acuity loss. Benefit of adequate zinc intake from foods in slowing development of AMD is likely. Zinc deficiency impairs retinal function in animals and humans. Diets high in zinc have been related to lower AMD in some but not all epidemiological studies, but this may reflect other aspects of consuming foods rich in zinc (milk, beans, meats, and shellfish).	The long-term benefits and risks of high-dose zinc supplementation are unknown (and are being further studied). See additional caveats about high-dose antioxidant supplements (above).
Dietary fat	**High intake of total fat is likely to encourage development of AMD:** Overall fat intake is associated with lower prevalence or progression of AMD in most population studies (although not always statistically significant).	Higher risk for AMD among people with diets high in fat might reflect lower overall nutrient density of high-fat diets or other aspects of lifestyle associated with the intake of high-fat diets.
	Benefit of foods rich in long-chain omega-3 fatty acids is likely: High intake of long-chain omega-3 fatty acids or fish is associated with lower risk for AMD in 7 of 8 study samples.	Diets high in long-chain omega-3 fats or fish may reflect other aspects of diet (intake of vitamin D or selenium) or lifestyle that were unmeasured or controlled for. The benefit of fish oil supplements is unknown and being tested.
Vitamin D	**Benefit of good vitamin D status (from adequate sunlight, foods or supplements) is possible but only recently studied:** Animal and cell studies suggest anti-inflammatory properties of vitamin D. Lower risk for early AMD was observed among Americans with lower blood levels of vitamin D in one large, cross-sectional study in a representative sample of the U.S. population.	Higher blood levels of vitamin D could be related to other aspects of diet or lifestyle that could protect against AMD.

(continues)

TABLE 2 *(continued)*

Nutritional Exposure	Strength of Evidence	Comment
Glycemic index	**Diets with low glycemic index score might reduce risk of developing AMD:** An association between glycemic index score and AMD was observed in only one population study to date.	A low glycemic index score is likely to be a marker for a plant-food and nutrient-rich diet, which might protect against AMD, especially over many years. The impact of overall diet patterns on AMD is poorly studied.
Multivitamin supplements	**Benefit is unknown:** The use of multivitamin supplements has not been associated with lower risk for AMD in population studies (except in Americans who did not report drinking milk daily).	The impact of multivitamins on the onset or worsening of AMD has not been tested in clinical trials.

TABLE 3 Summary of Evidence Relating Diet to Diabetic Retinopathy (DR)

Nutritional Exposure	Strength of Evidence	Comment
Dietary fat and fiber	**A benefit of high-fiber diets that are low or moderate in fat may be that they help to maintain blood glucose control, which is strongly related to the development of DR.** Some observational studies support lower risk of DR among people whose diets are high in fiber and low in fat.	High-fiber diets that are low or moderate in fat are also likely to be rich in other micronutrients, which may explain benefit in studies to date. Existing studies are limited by their cross-sectional designs, short-term assessment of dietary intake, small sample sizes, and lack of adjustment for confounding factors. No conclusions can be made thus far regarding intake of fiber and fat and risk of DR development and progression.
Antioxidants	**Benefit of foods rich in antioxidants is likely but inadequately studied:** Data from cell culture and animal studies strongly support a protective effect of antioxidant intake and protection against development of DR. Only three observational studies have investigated the associations between antioxidants, in the serum or diet, and risk of prevalent DR. These studies do not support a protective effect of antioxidants on DR. One study suggested that long-term intake of antioxidant supplements or multivitamins may protect against prevalent DR. Short-term trials of single antioxidants have not demonstrated benefit on blood glucose control.	Population studies are limited in number and design, and results may be biased by changes in diet and supplementation practices upon development of diabetic complications. Long-term, prospective studies are needed.
Other potential dietary factors	**The benefit of consuming diets rich in many nutrients is suggested:** Potential roles of magnesium, B-vitamins, and vitamin D in DR development are suggested by emerging evidence. New studies also suggest that genetic differences in enzymes involved in the metabolism of these nutrients may modify associations between diet and DR relationships. Certain dietary patterns may lower risk for the onset or worsening of DR. Results of one small case-control study in India suggest that plant-based diets may be protective.	Long-term, prospective studies are needed to investigate the putative role of newly implicated nutrients in protection against the development of DR, particularly in the context of specific diet patterns that are linked in existing studies to control of blood glucose and that are antioxidant rich.

of breast milk missing from infant formulas, such as carotenoids, may be responsible for better vision in breastfed infants, but this idea has not been adequately studied.

Overall nutritional intake in infancy, childhood, and adolescence might influence chronic age-related eye diseases that develop in later life. Evidence suggests that childhood diet influences the risk of cardiovascular disease later in life [7], but the influence on age-related eye diseases, which are the focus of this chapter, has not been investigated.

II. CATARACT

Cataract is the leading cause of blindness worldwide, accounting for almost half of all blindness [8]. The visual burden of cataract is largest in developing countries, where the relatively simple surgical excision of cataract is less available. However, the economic burden of cataract in developing countries is high. In the United States, an estimated 17% of Americans older than 40 years have cataract in either eye, and 5% have had cataracts extracted [9]. Cataract surgery accounted for more than 12% of the Medicare budget the last time this was evaluated in 1992 [10]. The occurrence of cataract will dramatically increase in the next 20 years as the aging U.S. population grows [9]. The large increase in cataract surgical procedures predicted for the U.S. population is expected to have a substantial effect on health care spending and, potentially, on the fiscal stability of the Medicare system [11]. Nutrition appears to be a primary means to substantially reduce the burden of age-related cataract; it has been estimated that if preventive measures could delay cataract by only 10 years, the visual and surgical burden would be cut in half [12].

Cataracts develop as opaque regions in the lens of the eye (Fig. 1) as a result of either acute metabolic insults or, slowly, as a result of the gradual accumulation of damage with age. The lens must remain clear in order to collect light and focus it on the retina. However, lens opacities scatter light, blurring vision. When light transmission is blocked substantially enough to reduce visual acuity, a cataract exists; if opacities become severe, blindness can occur.

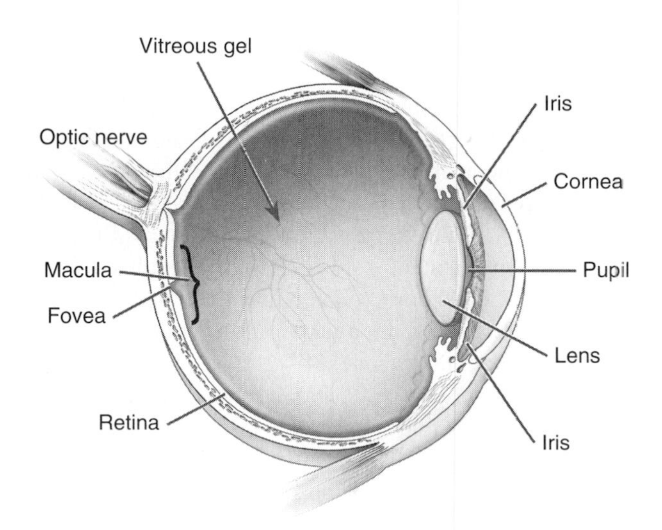

FIGURE 1 Anatomical features of the human eye. (Courtesy of National Eye Institute, National Institutes of Health.)

Cataracts generally occur in three regions of the lens: (1) the nucleus, or center, of the lens (nuclear cataract), (2) the outside edge of the lens, just under the lens capsule that separates the lens from the vitreous humor (posterior subcapsular cataracts, or PSC), and (3) the fiber layer between the outside edges of the lens and the nucleus (cortical cataract). Cataracts sometimes occur exclusively in one or the other of these regions; however, sometimes opacities occur in multiple regions, particularly as cataracts become more severe, and along with the advancement of age.

Nuclear cataract may reflect cumulative insults that have occurred since early childhood, because nuclear lens fibers do not have mitochondria, nuclei, or other cytoplasmic organelles and thus lack the capacity to repair damage over one's lifetime. Nuclear cataracts are the most common type of cataract in whites [13–15], while cortical cataracts are most common in blacks [15, 16]. Least frequently, cataracts develop in the PSC region of the lens; but when they do, they frequently necessitate lens extraction [17] because opacities in this region often severely limit vision.

A. Causes

The pathogenesis of cataract is known to involve the accumulated stresses resulting from inability of the eye to sufficiently defend against, or repair the damage, caused by physiological and environmental stressors that arise from such things as the photochemical formation of free radicals and osmotic imbalances. Smoking is the most common modifiable risk factor for cataract in population studies [18–26]. Other probable risk factors include exposure to ultraviolet (UV)-B light (particularly in the cortical region of the lens) [27–29], myopia [30], and diabetes (especially for cortical cataract and sometimes nuclear cataract and PSC) [26, 31–34]. Risk is sometimes, but not always, higher among people who have a higher body mass index [35–37], use alcoholic beverages heavily [38–40], and have elevated markers of systemic and local inflammation [41] or arthritis [26]. Risk has been reported to be lower among users of cholesterol-lowering statin drugs [42] and among women who have used postmenopausal estrogens [43] or have reproductive health histories that heighten exposure to endogenous estrogens [44]. Several aspects of diet could modify these risk factors or independently influence the risk of cataracts.

Animal studies have indicated that cataract can develop in certain species as a result of extreme metabolic perturbations, which have been induced by high galactose diets, experimental diabetes, nutrients deficiencies (riboflavin, calcium, zinc, and selenium), or nutrient excesses (selenium) (reviewed previously [45, 46]). Nutritional influences on the development of cataract are also suggested by the high prevalence of cataract in developing countries [8], although this may also reflect the higher concentration of sunlight to which people living near the equator are exposed, or the increased prevalence of diarrheal diseases. In samples of people from a few developing countries, a higher occurrence of cataract has been reported among those with low intakes of protein-containing foods [47], fruits and vegetables [48], or other selected nutrients [49]. In a region of China in which chronic deficiencies of nutrients is common, multivitamin supplementation for 5 years reduced the prevalence of one type of cataract [50].

More often, lens opacities develop slowly over many years, beginning in the second decade in life and progressing more rapidly in middle and older age. A gradual influence of nutrition on the development of cataract over adult life is suspected and is supported by studies in the Emory mouse, a type that develops cataracts in adulthood. In these mice, calorie restriction, which is generally associated with reduced oxidative stress and improved glucoregulation (recently reviewed [51]), slows the development of lens opacities [52]. Since the 1980s, a body of evidence has accumulated to indicate lower rates of cataracts among populations of people who eat micronutrient-rich diets or take multivitamins (summarized in Table 1). These findings, discussed later, support the idea that nutritional status over adult life influences the development of cataract with age. Yet most epidemiological studies of relationships of diet to cataract have been conducted in relatively well-nourished and older populations in developed countries, and over short time periods. Therefore, these studies may underestimate the role of diet in the development of cataracts over time and accompanied by conditions of extreme limitations of food intake.

B. Carbohydrates

In experimental animals, sugar cataracts are easily developed by feeding monosaccharide-rich diets or agents that promote diabetes (reviewed in [46]), which itself is a risk factor for cataract. Sugars do not require insulin to enter the lens. When the metabolic pathways to utilize them are overwhelmed, sugar alcohols are formed by aldose reductase, accumulate in the lens, and can cause osmotic cataracts. Moreover, prolonged exposure to monosaccharides can lead to nonenzymatic glycosylation and the accumulation of advanced glycation end products in the lens [53], or cataracts [54]. In people with diabetes, poor glycemic control has been associated with higher lens density [55].

However, the overall impact of blood glucose on cataract risk is uncertain, on the basis of epidemiological studies. In two studies, risk for cataract was associated with either high levels of dietary carbohydrate [56] or diets with high levels of refined carbohydrates, as indicated by a high glycemic index [56] score. Yet in two other populations [57], no associations with the intake of refined carbohydrates were observed. It may be that the impact of dietary or blood carbohydrates on cataractogenesis is modified by other aspects of diet that have not been adequately accounted for in population studies, such as the antioxidant content of the diet. In experimental animals and in experiments in cultured lenses, vitamins C and E protect against the development of sugar cataracts (previously reviewed [58]).

C. Antioxidants

It is well known that oxidative stress increases lens damage. Relationships of antioxidant nutrients to cataract development in animals, and to the occurrence of cataract in populations, have been extensively reviewed [45, 46, 58–60]. Deficiencies of riboflavin, selenium, and zinc, cofactors for enzymes that play important roles in oxidative defense, cause cataracts in some species (previously reviewed [46]). Deficiencies of vitamins C or E and major water and lipid soluble antioxidants, respectively, have not been reported to cause cataracts independently, but they do protect against oxidative damage in a variety of animal and cell systems. Vitamin C is abundant in human lenses, and its levels in the lens reflect those in the diet [61]. Lipid-soluble antioxidants in lenses include vitamin E and the carotenoids lutein and zeaxanthin [62], which have both been demonstrated to protect against UV-light induced peroxidation in lens cells [63].

In population studies, blood levels of antioxidants or diets rich in one or more antioxidants (vitamins C, E, or lutein and zeaxanthin) are often related to lower prevalence or incidence of cataracts or cataract extractions [64–71]. Associations are less frequently found in cross-sectional studies, in which diet or blood antioxidants and cataracts are assessed concurrently [70, 72], but these studies are likely to be biased by recent dietary improvements among older people with chronic disease diagnoses. Additionally, the oldest people sampled, who may have high rates of cataracts by virtue of having lived longer, are likely to have better diets than members of their birth cohorts who had died.

There is inconsistency across studies with respect to the relationship between cataract and any single antioxidant nutrient. This is not surprising, because the influence of any single antioxidant is likely to be modulated by the abundance of other antioxidants and their opportunity for synergy—a factor not often taken into account in analyses. In order to attempt to account for this, some researchers have noted lower rates of cataract among persons with a combination of high levels of antioxidants in diet or serum [71, 73].

An exception to the inconsistency in associations of single nutrients to cataract in population studies is the association observed with lutein and zeaxanthin. Diets and serum rich in lutein and zeaxanthin are consistently related to lower incidence or prevalence of nuclear cataract or cataract extraction in longitudinal studies [64–69]. However, foods rich in these carotenoids, such as green vegetables and dark green leafy vegetables, are rich in many antioxidants, so that the consistency of this association may simply reflect antioxidant-rich diets in general.

Clinical trials of only one or two antioxidant nutrients for up to 6 years have not been generally effective in slowing cataract [50, 74–76]. One exception was the lower prevalence of nuclear cataract among persons aged 65 to 74 years in a poorly nourished region of China, who had received supplements of riboflavin and niacin for 5 years [50]. In one clinical trial, the rate of progression of lens opacities was lower among 81 persons who took three antioxidant nutrients (18 mg of beta-carotene, 750 mg of vitamin C, and 600 IU of vitamin E) for 3 years, compared to 77 people who did not take these supplements. However, in a large (4629 subjects) double-masked, placebo-controlled trial, the Age-Related Eye Disease Study (AREDS), a similar formulation of high-dose antioxidants (containing 15 mg beta-carotene, 500 mg vitamin C, and 400 IU vitamin E) taken for an average of 6.3 years did not influence progression of age-related lens opacities.

Thus, the overall body of evidence from animal and population studies suggests that antioxidant components of the diet may contribute to protection against cataract development. However, there is little evidence to suggest that short-term supplementation with one or a few antioxidants is likely to have an important impact on cataracts, which develop over many years and are influenced by a wide variety of dietary, health, and lifestyle factors.

D. Lead

Lead exposure may be one contributor to higher risk of cataract among people of lower socioeconomic status. In one study, there was a threefold higher estimated risk of cataract among men in the highest quintile for level of lead in the tibia, after adjusting for other known and measured risk factors (smoking, diabetes, dietary antioxidants) [77]. However, additional studies that confirm this association are needed. Reductions in environmental lead exposure in the United States since the 1980s, through controls on gasoline emissions and pipes that supply drinking water, could reduce the potential impact of lead to cataract development. The extent to which lead might contribute to cataract, worldwide, is unknown.

Lead exposures early in life can accumulate in bones. As bones thin with age, they not only release calcium but also other stored minerals like lead. Therefore, if lifetime lead accumulation is high, anything that increases osteoporosis could increase lead exposure. This suggests the possibility that risk factors for osteoporosis might also increase risk for cataract and that some shared risk factors (weight, smoking) could be explained by this mechanism.

E. Diet Patterns

Only one investigation has directly evaluated the overall impact of a healthy diet on the occurrence of cataract. This study was conducted in a subsample of participants of the Nurse's Health Study. They observed that adherence to 1990 Dietary Guidelines for Americans, as reflected by Healthy Eating Index (HEI) scores for diets over a 10-year time period, lowered the estimated risk for early nuclear lens opacities [78]. Women in the highest two quartiles for the HEI score had more than twofold lower prevalence of nuclear opacities than women in the lowest quartile. The prevalence of nuclear opacities was lower among women in highest versus lowest quartiles for intake of several foods (fruits, vegetables, whole grains, milk), but results were not always statistically significant. However, associations were strongest when all aspects of diet were considered together. Given the large magnitude of impact of healthy diets on early lens opacities in this relatively homogeneous and well-nourished sample of women, the overall impact of micronutrient dense diets on cataract occurrence over longer time periods, and in populations with wider ranges of intake, could be much stronger.

F. Supplements

Numerous studies have evaluated relationships of supplement use to the occurrence of cataract and are the subject of several recent reviews [79–81]. As discussed earlier, most clinical trials of supplements that contain high doses of antioxidant nutrients have not observed that their use lowers rates of cataract. Differently, there is evidence from eight observational studies [16, 68, 82–87] and one nonrandom clinical trial analysis [88] that indicate lower rates of some types of cataracts or cataract extractions among people who use multiple vitamin supplements, compared with only two studies whose designs were limited: one by use of cataract extraction as an outcome [89] (which may reflect factors other than degree of opacity) and the other by a sample in which supplement use was not prevalent [90]. In many of these studies, the protection was strongest among people who used multivitamin supplements for more than 10 years [68, 83, 85–87]. However, people who use supplements generally have healthier diets and lifestyles [91], and this may explain the protective relationships, in full or in part. Analyses in two studies that stratify by diet quality or propensity to use supplements based on measured demographic and medical characteristics [83] [88] suggest that these are not the pertinent factors explaining the associations, at least in these samples. Results of one clinical trial conducted in the Linxian region of China, where malnutrition is common, supports the benefit of supplements in preventing nuclear cataract: multivitamin supplements administered for 5 to 6 years in a double-masked placebo-controlled fashion resulted in a lower prevalence of nuclear cataract among persons 65 to 74 years of age (but not among persons 45 to 64 years of age, in whom the rate of cataract development was likely to be slower). Results of a second clinical study, expected to be completed in 2007, will provide additional insights. This 13-year trial involving about 1000 people in Italy will be, at its completion, the longest clinical trial conducted to evaluate the impact of nutritional supplements on cataract [92].

III. AGE-RELATED MACULAR DEGENERATION

A. Overview

Age-related macular degeneration (AMD) is a result of deterioration of the part of the retina (the macula, see Fig. 1) that is responsible for central vision and reading fine detail. It is the leading cause of blindness in developed countries and the third leading cause worldwide [8]. In the United States, by age 65, roughly 8% of people have an intermediate form of AMD that has a high risk of progressing to advanced AMD [93], which is associated with central vision loss; 12% of Americans over 80 years have advanced AMD [93]. There is no cure for this condition, and medical and surgical treatments are limited to people with one type of advanced AMD (wet AMD) and are of limited long-term effectiveness. Because of the steep increase in AMD with age, prevention is likely to have a large impact on the social and economic burden of this condition.

The early stages of AMD are signaled by yellowish white deposits, or drusen (Fig. 2), which can vary in size and area; more extensive drusen development is predictive of greater eventual progression to advanced stages. In intermediate stages, there are sometimes areas of hyper- or hypopigmentation (retinal pigment abnormalities) in the retina pigment epithelial (RPE) cells, a single layer of cells that support the rods and cones (photoreceptors), caused by disturbances in the distribution of melanin pigments. Extensive drusen and retinal pigment abnormalities are thought to signal the existence of pathological processes associated with a distressed central retina. These changes are theorized to compromise the ability of nutrients and oxygen to flow from the choroidal blood supply, through Bruch's membrane, to the RPE cells and the photoreceptors they support. Drusen may also reflect the existence of inflammatory processes that contribute to degradation of this area.

When deterioration proceeds to the extent that the RPE cells and the photoreceptors they support die, advanced AMD occurs, which interferes with vision in the center of the visual field (such as that needed to view a person's face straight on) and the ability to read fine detail needed, for example, to read newspapers. This type of advanced AMD is sometimes referred to as "dry" AMD. If growth of new blood vessels occurs, then the advanced AMD is referred to as "wet" AMD. Bleeding or leaking of these vessels can also cause acute limits in vision.

B. Causes

AMD appears to develop as a result of a complex interplay of multiple dietary, environmental, and genetic factors that influence oxidative stress [94], inflammation [95–97], and light damage [98]. Smoking is the most commonly reported risk factor [99] for AMD in epidemiological studies. Although it is commonly thought that damage by light, especially in the blue range, promotes AMD, epidemiological studies have not consistently observed higher rates of

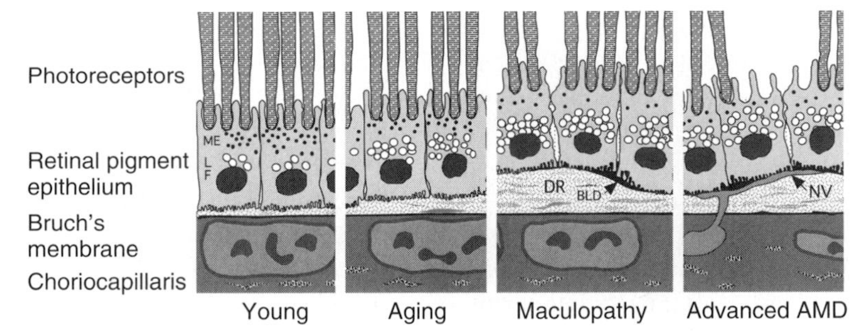

FIGURE 2 Changes to layers of retina in the macular region during aging, and progression of age-related macular degeneration (AMD) in early and advanced stages. With age, there is an accumulation of lipofuscin (LF), a decline in melanin (ME) in the retinal pigment epithelium, and a thickening and lipid enrichment of Bruch's membrane. Basal laminar deposits (BLD) are seen in early maculopathy but may also reflect aging. Drusen (DR) accumulation also characterizes early maculopathy. In intermediate AMD, drusen becomes more extensive, and there are areas of the retinal pigment epithelium that are hypopigmented or hyperpigmented (because of melanin clumping). In more advanced AMD, there is atrophy of the photoreceptors and there may be neovascularization (NV)—the growth of new blood vessels and narrowing of the choriocapillaris. (Courtesy of François Delori, Schepens Eye Research Institute/Harvard Medical School, Boston, Massachusetts.)

AMD among people with high levels of sunlight exposure. (This might be due to the difficulty in assessing the amount, type, and timing of sun exposure.) Some studies indicate high risk for AMD among people with cardiovascular disease [100, 101] or its risk factors [102]. Aspects of diet that may explain these risk factors or be independently related are discussed later.

Having a first-degree relative with AMD increases one's risk for the disease as well. Variants in genes that have roles in complement activation in the inflammatory response (complement factor H, factor B) or others that are close to regions that involve immunological response, are more common in people who have AMD [103]. The genetic predisposition to AMD may involve a propensity for inflammation or exacerbation of an inflammatory response [97]. Some studies find high prevalence of AMD among people with elevated C-reactive protein [104, 105]—an indicator of systemic inflammation—or among people with inflammatory diseases, such as gout [106], or who have used anti-inflammatory medicines, which may signal the presence of inflammatory disease [107]. The evidence that diet and supplements may prevent or slow AMD is discussed below and summarized in Table 2.

C. Antioxidants

Oxidative damage to proteins, lipids, and DNA, by free radicals within the photoreceptor outer segments or RPE, can be the result of photo-oxidation of lipids from light exposure; they can also be the by-product of metabolic events, such as oxidative metabolism or enzyme reactions that use oxygen (such as xanthine oxidase). The retina is particularly susceptible to oxidative stress, because of its high rates of oxidative metabolism and light exposure, and its high concentration of lipids with double bonds that are vulnerable to this type of attack. Free radicals that propagate oxidative damage may also be produced directly as a mechanism for biological defense, as is the case when white blood cells respond immunologically to pathogens with an oxidant boost.

The retina is generally rich in antioxidants and antioxidant enzymes that quench free radicals, including vitamins C and E [108], carotenoids [109, 110], and zinc [111], a cofactor for superoxide dismutase. Eighty percent or more of carotenoids in the retina are lutein and zeaxanthin (discussed later). Lycopene, an efficient quencher of singlet oxygen, is also found in the RPE/choroidal area [110]. Although other carotenoids are not commonly found in human eye tissues, there is the possibility that they may have systemic effects on oxidative damage, which could modify processes that might promote AMD (such as cardiovascular disease) or spare carotenoids that accumulate in the retina. In studies with experimental animals, deficiencies of vitamins E and C, carotenoids, or nutrients such as zinc, which are components in enzymes that protect against oxidative stress, result in pathological changes to the retina (as previously reviewed [94, 112–114]).

Results of clinical trials of antioxidant supplements have been varied. Single or dual high-dose antioxidant supplements have not been associated with lower risk or progression of AMD [115, 116]. This is consistent with evidence that oxidant defense is carried out by a highly complex system of mechanisms, in which antioxidant effects of several different individual antioxidant molecules have been found to be synergistic [117–119]. Differently, in the Age-Related Eye Disease Study (AREDS), a 6.2-year randomized, placebo-controlled clinical trial [120], high doses of the antioxidants (500 mg of vitamin C, 400 IU of vitamin E, 15 mg of beta-carotene, and 80 mg of zinc along with 2 mg of copper) slowed the progression of AMD from intermediate to more advanced stages by 28%. There is also evidence that this treatment reduces the age-related oxidation of cysteine in the blood, which supports the possibility that the benefit is due to an improvement in oxidative stress [121]. A small short-term study of men with atrophic AMD, who received antioxidant supplements along with lutein, reported improved visual function [122]. Men randomized to the control group did not report improved vision.

Although these studies suggest that nutrition, and in particular dietary antioxidants, may be important in slowing the progression of AMD, questions remain regarding the optimal dose and combination of nutrients in supplements, their long-term benefit and safety, and whether other aspects of diet are equally or more important. Moreover, there is no evidence that people who have early AMD or who are at risk for AMD because of a family history will benefit from high-dose antioxidant supplements.

Epidemiological studies also generally support the idea that the intake of foods that are rich in antioxidants are associated with lower risk for AMD. Although the level of individual antioxidants in foods is usually much lower than levels provided by current high-dose antioxidant supplements, foods provide a more varied array of antioxidants than supplements, and the impact of decades of consuming dietary antioxidants on AMD may be substantial and has not been adequately studied. In epidemiological studies, when associations with antioxidant nutrients are considered one at a time, low levels of one or more antioxidants in the blood or diet have often, but not always, been related to higher prevalence or incidence of certain age-related changes in the macula. Dietary antioxidant nutrients related to lower occurrence of AMD include vitamin E [123–125] or one or more carotenoids (reviewed [126] and subsequently reported [127–130]) or zinc [124, 125, 127–129, 131, 132]).

The potential impact of a combination of dietary antioxidants was estimated in the Rotterdam Eye Study [125]. They observed a 35% lower risk of incident AMD among the 10% of participants who were consuming antioxidant-rich diets (i.e., diets with above the median intake of all four antioxidants in AREDS supplements). Risk in persons with antioxidant-poor diets (with all four antioxidant nutrients at levels below the median) was higher, although this was not statistically significant. Overall, there is a clear trajectory of lower risk for AMD that is associated with eating diets higher in several antioxidants. Lowering the prevalence of early or intermediate stages of AMD by eating antioxidant-rich foods has the potential to prevent even more cases of advanced AMD than does beginning supplements once intermediate AMD becomes manifest [133].

Although the level of antioxidant vitamins in foods is usually lower than in high-dose antioxidant supplements, the number of different known and unknown antioxidants in foods is likely to exceed those in supplements. Foods high in antioxidants, such as vegetables, might lower oxidative stress to a greater degree than supplements. In a recent randomized, crossover trial, eating two or more cups of brassica vegetables (such as broccoli) lowered a urinary marker of oxidative stress, whereas moderate levels of supplementation with antioxidants in multivitamins did not [134]. However, it is not known whether the non-nutritive antioxidants in foods reach the retina (besides lutein and zeaxanthin, which are discussed next) or whether lowering the systemic level of oxidative stress by consuming such foods influences age-related macular degeneration. The impact of decades of consuming dietary antioxidants on AMD may be substantial and has not been adequately studied.

D. Lutein and Zeaxanthin

Lutein and its structural isomer zeaxanthin are selectively concentrated into the retina and other ocular tissues [109, 110] over all other known carotenoids. Evidence for the existence of specific xanthophyll-binding proteins in the vertebrate retina has been reported which could explain this selective uptake [135, 136]. These carotenoids must be provided by the diet, because the body does not synthesize them. They are contained in a wide variety of fruits and vegetables, seeds, and eggs and are especially concentrated in green vegetables [137, 138]. They have also been available in dietary supplements since 1995. From 1988 through 1994, Americans, on average, consumed 1 to 2 mg of these carotenoids daily from foods [139]. Their structure is similar to pro-vitamin A carotenoids, like beta-carotene, but with additional hydroxyl groups, which makes them more polar and unable to be used in vitamin A synthesis.

The highest density of these carotenoids is found in the inner retina, in the Henle fiber and inner plexiform layers [140], where they comprise a yellow pigment referred to as macular pigment. At these locations, they are likely to function as an optical filter that absorbs short-wavelength visible (blue) light [141]. This might protect against AMD or simply enhance vision. Removal of blue light by carotenoids is thought to remove the effects of chromatic aberration, resulting in better resolution of visual images, and is theorized to reduce glare [142].

One mechanism by which macular pigment could protect against the development and progression of AMD is by blocking light-induced formation of a toxic di-retinal conjugate, A2E. Drusen deposits, characteristic of early AMD, which may be promoted by inflammatory processes [96], contain lipofuscin, the principal component of which is a toxic compound, A2E, which forms as a consequence of light-related vitamin A cycling in the retina. A2E accumulates in the RPE during phagocytosis of the rods and cones, and it is taken up by the lysosomes. A2E, in excess levels, has a variety of potential toxic effects, one of which is sensitivity to blue light damage [143]. When a critical intracellular level of this compound has been reached, cell damage to DNA occurs, induced by blue light irradiation [144]. Because macular pigment could absorb 40% to 90% of incident blue light [141], it could reduce A2E toxicity.

The macular pigment density reduces about twofold between the central macula, where the zeaxanthin isomer predominates, and the periphery of the fovea where the lutein isomer predominates [109]. The total concentration reduces 100-fold from the center of the cone-dominated fovea to the rod-dominated peripheral retina [145]. An antioxidant role is supported by the presence of oxidation products of lutein and zeaxanthin in the retina [146]. These carotenoids might also influence membrane stability by their unique alignment in biological membranes [147].

Animal models to study the impact of macular pigment on AMD or related photoreceptor health are limited to species that accumulate these carotenoids in their ocular tissues: primates, and some birds. Primates fed diets deficient in these xanthophyll carotenoids [114, 148] suffer a loss of retinal pigment epithelial (RPE) cells and increased photoreceptor cell death [149]. Retinal zeaxanthin has been demonstrated to prevent light-induced photoreceptor death in quail [150].

Lower levels of lutein and zeaxanthin have been found in autopsy specimens of donor eyes with AMD [145]. However, to date, cross-sectional relationships of AMD to macular pigment density, measured noninvasively in living persons, are inconsistent [151–154] and likely to reflect relationships of AMD to short-term diet rather than diet over decades of adult life when AMD was likely to develop. Prospective studies of the relationships of macular pigment to the onset of AMD are needed in order to understand whether low macular pigment is a consequence of, or a contributor to, AMD.

Associations between the intake and blood levels of lutein and zeaxanthin and the occurrence of age-related macular degeneration are inconsistent as well. Lower risk for one or more types of AMD was associated with levels of these carotenoids in the diet or blood in some studies [120, 127, 129, 155, 156] but not others [124, 157–161]. The inconsistency might reflect that the intake of lutein and zeaxanthin are simply markers for diets rich in antioxidants.

They may be two of the many components in fruits and vegetables that may slow earlier stages of AMD. Several studies suggest that high intakes of fruits [157], vegetables [130, 162], or fruits and vegetables combined [163] are associated with lower risk for AMD.

However, it is too early to draw conclusions from population studies about the potential protective influence of lutein and zeaxanthin on AMD [126]. One reason is that the time period over which lutein and zeaxanthin intake has been measured in such studies often reflects only a few years. Diet may have changed, relative to diets over the longer periods of adult life, which influenced the development of AMD. In the Carotenoids in Age-Related Eye Disease Study, high intake of lutein and zeaxanthin was associated with lower odds for intermediate AMD, only after excluding persons who were likely to have had unstable diets [132]. Selective mortality bias might also explain the inconsistent results in population studies. Older participants in observational studies may be more likely to have eaten diets rich in fruits and vegetables over their adult lifetime than people in their birth cohort who are no longer living, making it more difficult to discern the effects of those diets on disease outcomes. Longer-term prospective studies, particularly of the youngest persons at risk for AMD, will provide more insights in years to come.

The benefit and risks of taking lutein and zeaxanthin supplements has only been studied in a limited number of small clinical trials. Visual acuity improved slightly in the fewer than 30 subjects who received exclusively lutein supplements for 1 year [122] and in another receiving lutein and zeaxanthin in combination with other antioxidants [164]. In another trial, data suggest that lutein or zeaxanthin supplementation may improve visual performance at low illumination [165]. However, in another small trial, supplementation did not improve contrast sensitivity over 9 months [166]. Larger and longer-term trials are needed to address potential benefits (and risks) of supplementation with lutein and zeaxanthin more conclusively. A large multicenter trial of these carotenoids (and omega-3 fatty acids) on progression of advanced AMD (AREDS II) is currently under way [167].

The benefits of lutein and zeaxanthin supplements on preventing or slowing AMD may also depend on a person's ability to accumulate carotenoids from the diet or supplements. This appears to vary among individuals. Between 20% and 50% of subjects in previous investigations have low serum or retinal response to oral supplementation with these carotenoids [168–170]. In general, blood responses to oral carotenoids vary between individuals as well [171, 172]. The influences on the ability to take up carotenoids by the intestinal tract and the eye are largely unknown. Having high levels of body fat and diabetes are conditions that are related to lower macular density [173] and may influence the ability to absorb and accumulate lutein and zeaxanthin in the eye. These conditions might reflect levels of oxidative stress or lipoprotein distributions, which could, in turn, influence the ability to accumulate carotenoids.

Despite the possibility that vegetables, particularly green vegetables and dark leafy greens, may lower risk for AMD (and cataract as discussed previously), some older people limit their intake because they contain high levels of vitamin K, which could interfere with warfarin that has been prescribed to prevent blood clotting. A sudden increase in vitamin K intake from these foods can reduce the effectiveness of the drug. However, patients can be advised to consult with the physician who manages their warfarin dose to adjust it to the highest daily green vegetable intake that the patient can consistently eat.

E. Zinc

Zinc may be particularly important to the retina because concentrations of zinc in the retina exceed those elsewhere in the body, with the exception of the prostate [111]. Deficiency of zinc in both animals and people impairs retinal functioning, as previously reviewed [112]. There is evidence for numerous mechanisms (catalytic, regulatory, and structural [174]) by which zinc could influence retinal integrity. Zinc catalyzes enzymatic reactions and is a cofactor of more than 100 enzymes, some of which are involved in oxidant defense. Zinc depletion in RPE cells has reduced levels of catalase, glutathione peroxidase, and metallothionein and has reduced ability to phagocytize photoreceptor outer segments [175]. Zinc performs structural roles; it facilitates protein folding to produce biologically active molecules (zinc fingers). Zinc is also involved in immune responses [176, 177]. Both zinc deficiency and excesses impair immunity [177]. Zinc depletion may also trigger apoptosis of RPE cells or increase the vulnerability of RPE cells to photic injury [178]. However, zinc supplementation can also enhance stress-induced effects in RPE cells [179].

The longer-term benefit of zinc on risk for AMD is not clear from observational studies. A high intake of zinc was related to lower prevalence of early AMD [131], to incidence of one type of AMD in one population [124], and to lower risk for overall incident AMD in another [125]. In other studies, zinc intake has been unrelated to AMD [158, 180].

By contrast, the use of high-dose zinc supplements (80 mg as zinc oxide, along with 2 mg of cupric oxide) for 6 years with or without antioxidants was associated with modestly lower progression from intermediate to advanced AMD in the AREDS [181]. One smaller zinc supplementation trial had previously reported a benefit of zinc supplementation on vision loss in patients with AMD [182], whereas another did not [183]. No serious safety issues with zinc supplementation were identified in the 6-year AREDS study (aside from more frequent hospitalization for genitourinary problems in men and more frequent anemia, unsupported by differences in hematocrit), and zinc supplementation was related to lower mortality in this sample [184]. However, the longer-term benefits and risks of zinc supplementation at the high levels tested in AREDS are unknown.

F. Dietary Fat

There are at least three broad mechanisms by which dietary fats might either enhance or slow age-related macular degeneration: first, because of the high caloric density of fats, eating high-fat foods can displace other nutrient-dense foods that may have otherwise protected against AMD. Second, eating high-fat and low nutrient density foods may contribute to high body mass, which is sometimes reported to be a risk factor for AMD [185–187]. Third, fatty acids themselves have numerous biological effects as components of biological membranes and regulators of biochemical pathways. Some dietary fats increase risk for atherosclerosis, which is related to AMD risk in some studies [100, 101]. Certain fatty acids can also have direct physiological effects on the retina by modulating oxidative stress or by the inflammatory response, which can promote AMD pathogenesis (discussed later).

In laboratory studies of mouse models of atherosclerosis, feeding high-fat diets resulted in the accumulation of lipid-like droplets in the retina and degenerative changes in RPE cells and Bruch's membrane [188–191]. In epidemiological studies, high dietary fat levels have been generally associated with increased risk for early and late AMD [185–196], even though these associations have not always been statistically significant. Some exceptions to this trend include prevalence or short-term incidence studies with low power to evaluate associations with either early AMD [196] or advanced AMD [192–194].

However, there is inconsistency across studies in the type of fat that was most related to AMD. AMD was more strongly related to high intake of saturated fats in some studies [193–195, 197, 198] and to high intake of polyunsaturated fatty acids or monounsaturated fatty acids in other studies [185, 197, 198]. The intake of *trans*-fatty acids, provided in diets by margarines and other processed foods, was related to higher risk for AMD in two studies [185, 197]. Because fat intake often changes, particularly in relation to the common diagnosis for cardiovascular diseases, which can be related to AMD, these relationships are difficult to interpret. Moreover, there is limited ability in such studies to adjust for the numerous other protective aspects of diet that accompany a more moderate, as compared to high, intake of fat.

In contrast, in seven of eight published observational studies, lower risk for early or late AMD is reported among people with higher intake of fish or long-chain omega-3 polyunsaturated fatty acids (LC Omega-3 PUFAs) [195–201]. Only in one population-based sample from Wisconsin with low fish intake was there no association between fish intake and AMD observed [193]. However, curiously, the rates of intermediate and advanced AMD (characterized by pigmentary abnormalities and geography atrophy) in a sample of older adults from Iceland are markedly higher than rates in three other populations of European ancestry [202, 203], despite having the highest per capita fish intake among Europeans and the fact that 56% report to use cod liver oil, a rich source of LC omega-3 fatty acids.

The protective influence of omega-3 fatty acids could be somewhat dependent on the omega-6 content of the diet. A higher ratio of omega-3 to -6 fatty acids could increase formation of anti-inflammatory eicosanoids from omega-3 fatty acids, because the omega-6 fatty acids compete for the desaturase enzyme that creates them [204] or replace the omega-6 content of membranes. In three past studies, the risk reduction associated with a high intake of LC omega-3 PUFAs was stronger among subjects who had low intake of omega-6 PUFAs [185, 197, 198].

LC omega-3 PUFAs, such as docosahexaenoic acid (DHA) or eicosapentaenoic acid (EPA), may be particularly important to the health of the retina. The pathogenesis of AMD may be influenced by atherosclerosis or involve parallel processes [102]. There is a large body of evidence to suggest that EPA and DHA lower CVD mortality by several mechanisms: by an improvement of blood lipids by lowering blood triglycerides; decreasing inflammation, blood pressure, and platelet aggregation; and improving vascular reactivity (reviewed in [205]). These may influence AMD pathogenesis directly or indirectly.

DHA is the most abundant LC omega-3 PUFA in rod outer segment membranes [206, 207] at a concentration that exceeds levels found elsewhere in the body (reviewed in [208]). Its presence in membranes affects their biophysical properties and may influence membrane-bound enzymes, receptors, and transport. This is important in visual transduction [209], but it may also influence the pathogenesis of AMD. LC omega-3 PUFAs may protect against AMD by direct influence on retinal cell survival [210]. DHA has also been demonstrated to protect RPE cells from oxidative stress [210, 211]. High intake of this type of fat might also lower risk for AMD because of its anti-inflammatory properties [204]. Numerous cell culture studies provide clues for possible mechanisms by which LC omega-3 PUFAs could enhance the integrity of vascular and basement membranes and prevent neovascularization (recently reviewed [212]). The benefit of supplementing with 1 gm per day of LC omega-3 PUFAS is unknown and is currently being tested in a large clinical trial (AREDS II).

G. Vitamin D

Data are emerging to suggest that vitamin D might play a role in protection against AMD. Evidence for the high prevalence of complement H polymorphisms among people with AMD suggests that inflammation may play an important role in the development of AMD [213–215]. This is further supported by the observation of immunocompetent proteins associated with inflammation entrapped in drusen, suggesting that localized inflammatory response plays a role in drusen biogenesis [216]. Additionally, high levels of C-reactive protein (a systemic marker of inflammation) [104, 105] and the use of anti-inflammatory medications [107] were significantly related to AMD, independent of other established risk factors, in some but not all [217, 218] previous studies. Nutrients that attenuate the inflammatory response or enhance immunity to pathogens might protect against AMD.

Vitamin D is an example of one such food component with anti-inflammatory properties and immune-enhancing properties [219–221]. Moreover, the vitamin D receptor is found on activated T-lymphocytes [222] and macrophages [223]. Vitamin D may suppress inflammation by regulating the transcription of certain cytokine genes [224, 225].

Additionally, there is some evidence to suggest that vitamin D possesses antiangiogenic properties. This is also relevant to AMD because angiogenesis and the breaking of weak blood vessels within the retina lead to wet AMD. Antiangiogenic properties of vitamin D have been suggested in mice and rat models [226, 227].

A recently published study identified the first epidemiological associations between high blood levels of 25-hydroxyvitamin D and the prevalence of early AMD in the Third National Health and Nutrition Examination Survey [199]. Consistent findings in other samples are required to better evaluate this as one of many food components that could protect against AMD. There were too few cases of advanced AMD to reliably evaluate relationships with blood vitamin D in this study, but vitamin D might protect against neovascular AMD by virtue of its antiangiogenic properties (discussed earlier). This remains to be investigated.

This finding suggests that vitamin D might be another of the food components that protects the aging retina. If confirmed, vitamin D status could have a large impact on risk for AMD, given that a large proportion of the U.S. population is estimated to be at risk for poor vitamin D status [228, 229]. Because sunlight is an important source of vitamin D, the elderly may be at particularly high risk for poor vitamin D status because they have a reduced ability to produce this nutrient in the skin, in response to UV light [230, 231]. The avoidance of excessive incident sunlight to minimize risk for skin cancers, cataract, and macular degeneration might further jeopardize vitamin D status in older people. Interestingly, previously reported dietary associations between AMD and fish and zinc (for which milk is an important source) might reflect, in part, a protective effect from vitamin D, as both fish and milk (fortified with vitamin D in the United States) are also important sources of vitamin D in the diet.

H. Glycemic Index

The glycemic index of foods was recently introduced to be another possible aspect of diet that could influence the development of AMD [232]. Although advanced glycation end products have been found in drusen, it is not yet known whether they are a cause or consequence of degenerative changes. Degeneration of the retinal vasculature is a well-known complication of diabetes mellitus; yet the presence of diabetes has sometimes, but not always, been related to AMD in epidemiological studies. The biological plausibility that elevation in blood sugar promotes AMD, particularly in the absence of diabetes, remains untested. Nevertheless, diets with a low glycemic index often include few refined grains and sugars and plenty of fruits, vegetables, whole grains, legumes, and milk, which have numerous ingredients that could protect against AMD. Thus, high glycemic index diets, like high-fat diets, may be related to higher rates of AMD, in part or in whole, because they are poorer in a wide variety of protective nutrients and other diet components. Relationships of overall diet patterns to AMD have not yet been studied.

I. Herbal Supplements

The use of herbal supplements has increased in the United States. Several herbal supplements, such as those containing ginkgo biloba and bilberry, have been promoted to benefit the health of the retina. However, there are no scientific studies that support their benefit except one very small (20 persons) study of ginkgo biloba in patients with AMD, in which improvement in visual acuity was indicated in a preliminary report (recently reviewed [233]).

IV. DIABETIC RETINOPATHY

A. Overview

Diabetic retinopathy (DR) is a disease among people with diabetes that is considered to be the result of damage to the microvasculature of the retina. It is the leading cause of new cases of blindness in working age U.S. adults (20–74 years) [234]. Other diabetic complications include cataract and other diseases of the micro- and macrovasculature, such as atherosclerosis, nephropathy, and neuropathy.

Approximately 9.6% (20.6 million) of Americans aged 20 years and older have diabetes [235]. Diabetes, and thus its complications, is projected to increase in the coming years [236], particularly with the increasing number of overweight and obese individuals in the United States [237]. Diabetes is especially burdensome in minority populations, such as African Americans, Mexican Americans [238], and Native Americans [239, 240], in whom the prevalence and incidence are higher than the national average.

Preclinical stages of DR include changes in retinal blood flow. Nonproliferative diabetic retinopathy (NPDR) consists of the formation of clinical lesions (microaneurysms and intraretinal hemorrhages), the appearance of retinal exudates (lipid deposits from leaky blood vessels) and cotton wool spots (resulting from localized ischemia), and the appearance of venous bleeding and loops [241]. Blindness can result at theses stages if macular edema occurs [242]. In the later proliferative stage (PDR), new vessels and fibrous tissue can originate from the optic disc or elsewhere in the retina. Problems arise if the vessels grow through the inner limiting membrane into the vitreous humor of the eye, which is constantly contracting and condensing. Often this movement can lead to vessel tear, hemorrhage, and blindness [241, 243].

The annual economic and social burden of medical care and disability due to vision loss and blindness caused by DR is quite large [244] and the only effective medical treatment is laser photocoagulation [241], in which a laser is used to cauterize leaking, newly developed blood vessels that have grown as a result of DR. Photocoagulation does not necessarily prevent vision loss in all individuals, and it involves investing time in treatment that may cause discomfort.

B. Causes

Randomized controlled trials have demonstrated that maintenance of tight blood glucose control via intensive insulin therapy is associated with lower incidence and progression of DR in individuals with types 1 and 2 diabetes [245–247]. Even so, intensive insulin therapy did not prevent the occurrence of DR in all individuals and was often associated with increased bouts of hypoglycemia [245, 246]. Other risk factors for DR are duration of diabetes [248], hypertension [249], and elevated serum cholesterol or triglycerides [250–252]. Because DR is not 100% preventable or curable [253], there is still a need to continue to investigate possible preventive measures for its incidence and progression.

Nutrition may also have a significant affect on DR, and its role has been relatively unexplored (summarized in Table 3). Dietary intake may contribute to risk for DR either (1) by contributing to maintenance of tight blood glucose control, (2) by lowering blood pressure or serum lipid levels, or (3) via mechanisms independent of such factors. However, few observational studies have investigated the overall impact of differing diets and supplementation on the risk for DR in human populations [254–260]. Medical nutrition therapy for patients with diabetes could be targeted at preventing DR or its progression and may be less costly than current treatment.

C. Dietary Fat and Fiber

Some of the original investigations of the associations between dietary intake and DR were conducted in the 1980s and involved intervention studies of linoleic acid, an omega-6 PUFA [261, 262]. These studies, conducted among individuals with type 2 diabetes, suggested that intake of diets enriched with linoleic acid decreased development and progression of DR, especially among those with poor glycemic control. The authors hypothesized that linoleic acid enrichment of cell membranes would increase the sensitivity of the insulin receptor and thus improve blood glucose control [261]. However, these earlier intervention studies' effects could have been due to linoleic acid consumption displacing consumption of other more harmful fatty acids, such as saturated fat. Two studies conducted a decade later, one cross-sectional [256] and one with retrospectively collected dietary data [254], observed no statistically significant differences in dietary fat consumption between persons with and without prevalent DR. Differently, results of the Diabetes Control and Complications Trial (DCCT) of 1041 patients with type 1 diabetes indicated that baseline intake of total fat (% kcals) was directly correlated with prestudy DR and overall progression of DR [260]. Data from one ecological study also support the hypothesis that intake of cholesterol, total fat, and saturated fat increases risk for DR [263].

Diets rich in fiber, especially soluble fiber, have been shown to improve glycemic control and insulin sensitivity and to lower plasma lipids among persons with type 2 diabetes [264]. The DCCT also observed that baseline intake of dietary fiber (% kcals) was inversely correlated with prestudy DR, as well as overall progression of DR [260]. One [256] of two [254, 256] earlier conducted studies also observed lower prevalence of DR to be associated with high fiber intake.

Results of another small case-control study of people with type 2 diabetes, conducted in India, indicated that diets high in energy, animal proteins and fats were more common among persons with PDR than controls [255]. Controls tended to eat more pulses and vegetables, sources of dietary fiber. No published studies have investigated relationships of dietary patterns to risk of DR. This study suggests that overall dietary patterns, those high in fiber and low in fat (and most likely rich in certain micronutrients), are beneficial.

More research is needed to assess the effects of long-term dietary fat and fiber intake with respect to incidence and progression of DR. The small number of published studies of dietary fiber and fat have a number of limitations: cross-sectional designs [255, 256], dietary assessment methods, which do not accurately assess long-term dietary intake [255, 256, 260], prevalent disease [254, 256], lack of adjustment for confounding factors [254–256, 260], and small sample sizes [255, 256].

D. Antioxidants

Hyperglycemia is thought to increase oxidative stress as a result of several biochemical mechanisms: nonenzymatic protein glycosylation, auto-oxidative glycosylation, increased polyol pathway activity, tissue damage from ischemia reperfusion, and inflammatory responses [265–267]. Oxidative damage is hypothesized to promote diabetic vascular complications such as DR [267]. The increased production of reactive oxygen species (ROS) among persons with diabetes may intensify damage to the PUFA-rich endothelial cells of the retinal microvasculature, an area sensitive to oxidative damage [266, 267], and contribute to the early signs of endothelial cell vascular lesions in the pathology of DR.

Cell culture and animals studies of induced diabetes support the idea that hyperglycemia enhances oxidative stress and contributes to disturbances in the retinal vasculature. Studies of cultured endothelial cells incubated in high levels of glucose have increased markers of retinal lipid peroxidation [268], increased permeability [269], and decreased rates of proliferation [270] compared to cells incubated in lower glucose concentrations. Studies of diabetes-induced rodents also show increased markers of retinal lipid peroxidation [268], decreased activities of retinal antioxidant defense enzymes [271, 272], and increased production of retinal superoxide levels compared to rodents without induced diabetes [272]. Increased levels of lipid peroxidation have been shown to be positively associated with the development of damage to the structure of the retina in Streptozotocin (STZ)-induced diabetic rats [273]. Moreover, in rodents with experimental diabetes, supplementation with antioxidants prevents death of retinal microvascular cells [274] and animal model markers of early-stage DR [275].

Persons with diabetes have also been shown to have lower levels of antioxidants within plasma and serum [276–281] as well as lower carotenoids in the retina [17?], perhaps as a result of increased oxidative stress. The observed increase in markers of oxidative stress [278] and decrease in antioxidant status [277] were positively correlated with the level of diabetic complications in some studies, and markers of oxidative stress were significantly correlated to the level of glycosylated hemoglobin among persons with diabetes [282].

The intake of vitamins C (ascorbic acid (AA)), the most abundant aqueous phase blood antioxidant, and E (α-tocopherol), the most abundant lipid phase blood antioxidant, have been hypothesized to protect against DR. Both nutrients are thought to decrease nonenzymatic glycosylation of proteins; vitamin C, by competing with glucose for binding to proteins via a Schiff base [283], and vitamin E, by scavenging free radicals which may promote protein glycosylation, although the actual mechanism is unknown [284].

It has been hypothesized that lower tissue AA concentrations detected in persons with diabetes [285], perhaps due to competition between glucose and AA for tissue uptake [286, 287], contribute to capillary fragility and microvascular lesions associated with DR [288]. Vitamin C has also been hypothesized to act as an aldose reductase inhibitor (ARI) because of its chemical similarities to glucose. ARIs have been shown to reduce retinal microaneurysms in rats [289]. High blood glucose leads to increased aldose reductase activity in the diabetic individuals [290], leading to changes in retinal blood flow and the stimulation of growth factors in the diabetic retina [291, 292].

α-Tocopherol, through increasing the ratio of reduced glutathione to oxidized glutathione in the plasma [293–295], is hypothesized to improve the beta-cell's response to glucose [266, 297]. It has been shown that hyperglycemia induces production of diacylglycerol (DAG) [298] and subsequently protein kinase C (PKC) [299, 300]. α-Tocopherol is thought to inactivate specific cellular PKC isoforms either by changing their phosphorylation state or by enhancing DAG kinase activity [301, 302]. PKC is hypothesized to cause growth factor activation [302–305] and decreased retinal blood flow seen in persons with diabetes prior to development of DR [298, 303, 306].

Despite the strong evidence for protective effects of antioxidants in studies of cells or experimental animals with diabetes, discussed perviously, there are insufficient data in humans to conclude that diets or supplements high in antioxidants prevent or slow DR. Only three population-based studies have investigated the relationship between antioxidant micronutrients and DR [257–259]. Levels of vitamin C or E in the serum [258] or diet (assessed 6 years earlier) [307] were unrelated to prevalent DR in two large samples of people with type 2 diabetes. In a smaller cross-sectional study, the prevalence of DR was actually higher in people with type 2 diabetes who had high intakes of vitamin C and in subsamples of people who had high intake of beta-carotene or vitamin E [257].

However, the associations of prevalent DR to short-term diet and blood nutrient levels in the two cross-sectional studies [257, 258] may be biased by the recent use of popular antioxidant supplements of vitamins C and E and beta-carotene, particularly in people who may be experiencing more severe symptoms of diabetes. In one of these studies [258], the direct associations of blood levels of vitamin C (nonsignificant) and vitamin E to DR were reversed or attenuated after persons who reported use of supplements were removed from the analyses. Consistent with the idea that longer-term use of supplements may have benefit, lower prevalence of DR was observed in people who reported using vitamin C or E supplements 3 or more years before DR was assessed in one of these studies [259]. Currently, no clinical trials have studied the effect of antioxidant therapy on the development of DR. Moreover, there are no prospective cohort studies of the influence of a wider range of dietary antioxidants on the onset or worsening of DR. Given the strong evidence for protection by antioxidants in animal studies, and suggestion of benefit of long-term antioxidant use in one large observational study, additional studies are needed.

E. Other Dietary Factors

Hypomagnesemia is common in persons with diabetes, and it is hypothesized to be associated with decreased glycemic control and insulin sensitivity [308, 309]. Some [310–314], but not all [315, 316], previous small studies have shown that low levels of serum magnesium are associated with the presence, severity, or progression of DR. These observations have yet to be investigated in large prospective epidemiological studies.

Deficiencies in B vitamins (B_{12}, B_6, and folate) can lead to hyperhomocysteinemia. Hyperhomocysteinemia is thought to increase risk for cardiovascular disease via a number of proposed mechanisms, including increased oxidative stress [317], and among persons with type 1 and 2 diabetes, hyperhomocysteinemia has been shown to be associated with DR in some [318–325], but not all [326–331], small studies. Interestingly, some studies suggest that persons homozygous for the C677T polymorphism of the methylenetetrahydrofolate reductase (MTHFR) enzyme, responsible for homocysteine metabolism, may be at increased risk for PDR [319, 320, 332]. This polymorphism is associated with decreased activity of MTHFR and is thought to lead to greater accumulation of serum homocysteine levels.

Vitamin D has anti-inflammatory and antiangiogenic properties, as previously discussed, which may also help prevent DR development and progression. Aksoy *et al.* observed lower serum $1,25(OH)_2D_3$ levels in persons with more severe DR compared to persons without DR among 66 patients with diabetes [333]. However, serum 1,25-dihydroxyvitamin D is not the preferred marker of vitamin D status due to its short half-life of only hours [334]. New studies have suggested associations between different polymorphisms of the vitamin D receptor and risk of DR [335, 336], whereas others have not [337, 338]. Relationships of serum vitamin D status, assessed with the preferred biomarker of 25(OH)D, to risk of DR in large epidemiological studies have not been investigated.

Additional research is needed in population studies of individuals with diabetes to determine whether the protective effects observed in experimental animals and short-term clinical trials can be generalized to people over the long term. Research is also needed to evaluate the importance of diets rich in different nutrients on the prevalence and progression of DR in the general population. Given the broad aspects of diet that could protect against DR, studies of diet patterns related to lower risk for DR will particularly assist in making public health recommendations.

V. SUMMARY

Research of nutrition's effects on the development and progression of the three common causes of vision loss: cataract, AMD, and DR, has taken place largely since the 1980s. The evidence thus far supports the importance of diet in maintaining eye health with age. Limited research on the relationship of diet patterns to cataract and AMD suggests that the overall impact of nutrient-rich diets on risk for age-related cataract and AMD is likely to be greater than is estimated by examining associations of single aspects of diet with these conditions. Single nutrient associations with these eye diseases are also unlikely to account for the interrelatedness of several aspects of diet and the modification by host factors. Also, genetic predisposition for these age-related eye diseases most likely magnifies or weakens the impact of diet, and this has not been studied. Additionally, measurement error and bias must also be considered when studying diseases that manifest over decades. Unavoidable error in measuring diet, particularly over a long period of time, and failing to take into consideration bias in diet and disease associations that occur as a result of selective mortality, may result in an underestimation of the impact of diet.

What is clear is that there are many similarities in the pathogenic processes of age-related eye diseases with those of hypertension, stroke, heart disease, diabetes, and cancer. Oxidative stress, high blood pressure, and inflammation contribute to these other chronic diseases and are observed, increasingly, to be related to cataract, AMD, and DR. Research has shown that dietary patterns such as those high in plant foods [339], low in refined carbohydrates, and low or moderate in fat, such the DASH diet or the Mediterranean diet pattern, are related to reduced occurrence of many chronic diseases of aging [340–343]. Therefore, such nutrient-rich diet patterns are also likely to provide benefit in slowing the development of age-related cataract, AMD, and DR.

References

1. West, K. P., and Jr., McLaren, D. (2003). The epidemiology of vitamin A deficiency disorders (VADD). *In* "The Epidemiology of Eye Disease" (G. J. Johnson, D. C. Minassian, R. A. Weale, and S. K. West, Eds.), pp. 240–260. Arnold Publishing, London.
2. Semba, R. (2007). "Handbook of Nutrition and Ophthalmology." Human Press, Totowa, NJ.
3. Hartong, D. T., Berson, E. L., and Dryja, T. P. (2006). Retinitis pigmentosa. *Lancet* **368**, 1795–1809.
4. Williams, C., Birch, E. E., Emmett, P. M., and Northstone, K. (2001). Stereoacuity at age 3.5 y in children born full-term is associated with prenatal and postnatal dietary factors: a report from a population-based cohort study. *Am. J. Clin. Nutr.* **73**, 316–322.
5. Singhal, A., Morley, R., Cole, T. J., *et al.* (2007). Infant nutrition and stereoacuity at age 4–6 y. *Am. J. Clin. Nutr.* **85**, 152–159.
6. Fleith, M., and Clandinin, M. T. (2005). Dietary PUFA for preterm and term infants: review of clinical studies. *Crit. Rev. Food Sci. Nutr.* **45**, 205–229.
7. Ness, A. R., Maynard, M., Frankel, S., *et al.* (2005). Diet in childhood and adult cardiovascular and all cause mortality: the Boyd Orr cohort. *Heart* **91**, 894–898.
8. Resnikoff, S., Pascolini, D., Etya'aale, D., *et al.* (2004). Global data on visual impairment in the year 2002. *Bulletin of the World Health Organization* **82**, 844–851.

9. Congdon, N., Vingerling, J. R., Klein, B. E., *et al.* (2004). Prevalence of cataract and pseudophakia/aphakia among adults in the United States. *Arch. Opthalmol.* **122**, 487–494, April.
10. Steinberg, E. P., Javitt, J. C., Sharkey, P. D., *et al.* (1993). The content and cost of cataract surgery. *Arch. Ophthalmol.* **111**, 1041–1049.
11. Ellwein, L. B., and Urato, C. J. (2002). Use of eye care and associated charges among the medicare population: 1991–1998. *Arch. Ophthalmol.* **120**, 804–811.
12. Kupfer, C. (1985). Bowman lecture. The conquest of cataract: a global challenge. *Trans. Ophthalmol. Soc. UK* **104**(Pt 1), 1–10.
13. Klein, B. E., Klein, R., and Linton, K. L. (1992). Prevalence of age-related lens opacities in a population: the Beaver Dam Eye Study. *Ophthalmology* **99**, 546–552.
14. Klein, B. E., Klein, R., and Lee, K. E. (2002). Incidence of age-related cataract over a 10-year interval: the Beaver Dam Eye Study. *Ophthalmology* **109**, 2052–2057.
15. West, S. K., Munoz, B., Schein, O. D., Duncan, D. D., and Rubin, G. S. (1998). Racial differences in lens opacities: the Salisbury Eye Evaluation (SEE) project. *Am. J. Epidemiol.* **148**, 1033–1039.
16. Leske, M. C., Wu, S. Y., Connell, A. M., Hyman, L., and Schachat, A. P. (1997). Lens opacities, demographic factors and nutritional supplements in the Barbados Eye Study. *Intl. J. Epidemiol.* **26**, 1314–1322.
17. Klein, B. E., Klein, R., and Moss, S. E. (1997). Incident cataract surgery: the Beaver Dam eye study. *Ophthalmology* **104**, 573–580.
18. Klein, B. E., Klein, R., and Lee, K. E. (1997). Cardiovascular disease, selected cardiovascular disease risk factors, and age-related cataracts: the Beaver Dam Eye Study. *Am. J. Ophthalmol.* **123**, 338–346.
19. Klein, B. E., Klein, R., Lee, K. E., and Meuer, S. M. (2003). Socioeconomic and lifestyle factors and the 10-year incidence of age-related cataracts. *Am. J. Ophthalmol.* **135**, 506–512.
20. Christen, W. G., Glynn, R. J., Manson, J. E., Ajani, U. A., and Buring, J. E. (1996). A prospective study of cigarette smoking and risk of age-related macular degeneration in men. *JAMA* **276**, 1147–1151.
21. Weintraub, J. M., Willett, W. C., Rosner, B., Colditz, G. A., Seddon, J. M., and Hankinson, S. E. (2002). Smoking cessation and risk of cataract extraction among US women and men. *Am. J. Epidemiol.* **155**, 72–79.
22. West, S., Munoz, B., Schein, O. D., *et al.* (1995). Cigarette smoking and risk for progression of nuclear opacities. *Arch. Ophthalmol.* **113**, 1377–1380.
23. Age-Related Eye Disease Study Research Group. (2001). Risk factors associated with age-related nuclear and cortical cataract: a case-control study in the Age-Related Eye Disease Study, AREDS Report No. 5. *Ophthalmology* **108**, 1400–1408.
24. Smith, W., Mitchell, P., and Leeder, S. R. (1996). Smoking and age-related maculopathy: the Blue Mountains Eye Study. *Arch. Ophthalmol.* **114**, 1518–1523.
25. Tsai, S. Y., Hsu, W. M., Cheng, C. Y., Liu, J. H., and Chou, P. (2003). Epidemiologic study of age-related cataracts among an elderly Chinese population in Shih-Pai, Taiwan. *Ophthalmology* **110**(6), 1089–1095.
26. Mukesh, B. N., Le, A., Dimitrov, P. N., Ahmed, S., Taylor, H. R., and and McCarty, C. A. (2006). Development of cataract and associated risk factors: the Visual Impairment Project. *Arch. Ophthalmol.* **124**, 79–85.
27. West, S. K., Duncan, D. D., Munoz, B., *et al.* (1998). Sunlight exposure and risk of lens opacities in a population-based study: the Salisbury Eye Evaluation project. *JAMA* **280**, 714–718.
28. Taylor, H. R., West, S. K., Rosenthal, F. S., *et al.* (1988). Effect of ultraviolet radiation on cataract formation. *N. Engl. J. Med.* **319**, 1429–1433.
29. McCarty, C. A., and Taylor, H. R. (2002). A review of the epidemiologic evidence linking ultraviolet radiation and cataracts. *Dev. Ophthalmol.* **35**, 21–31.
30. McCarty, C. A. (2002). Cataract in the 21st century: lessons from previous epidemiological research. *Clin. Exp. Optom.* **85**, 91–96.
31. Rowe, N. G., Mitchell, P. G., Cumming, R. G., and Wans, J. J. (2000). Diabetes, fasting blood glucose and age-related cataract: the Blue Mountains Eye Study. *Ophthalmic Epidemiol.* **7**, 103–114.
32. Leske, M. C., Wu, S. Y., Hennis, A., Connell, A. M., Hyman, L., and Schachat, A. (1999). Diabetes, hypertension, and central obesity as cataract risk factors in a black population: the Barbados Eye Study. *Ophthalmology* **106**, 35–41.
33. McCarty, C. A., Mukesh, B. N., Fu, C. L., and Taylor, H. R. (1999). The epidemiology of cataract in Australia. *Am. J. Ophthalmol.* **128**, 446–465.
34. Klein, B. E., Klein, R., Wang, Q., and Moss, S. E. (1995). Older-onset diabetes and lens opacities: the Beaver Dam Eye Study. *Ophthalmic Epidemiol.* **2**, 49–55.
35. Caulfield, L. E., West, S. K., Barron, Y., and Cid-Ruzafa, J. (1999). Anthropometric status and cataract: the Salisbury Eye Evaluation Project. *Am. J. Clin. Nutrition* **69**, 237–242.
36. Glynn, R. J., Christen, W. G., Manson, J. E., Bernheimer, J., and Hennekens, C. H. (1995). Body mass index: an independent predictor of cataract. *Arch. Ophthalmol.* **113**, 1131–1137.
37. Hiller, R., Podgor, M. J., Sperduto, R. D., *et al.* (1998). A longitudinal study of body mass index and lens opacities: the Framingham Studies. *Ophthalmology* **105**, 1244–1250.
38. Morris, M. S., Jacques, P. F., Hankinson, S. E., Chylack, L. T. Jr., Willett, W. C., and Taylor, A. (2004). Moderate alcoholic beverage intake and early nuclear and cortical lens opacities. *Ophthalmic Epidemiol.* **11**(1), 53–65.
39. Klein, B. E., Klein, R. E., and Lee, K. E. (1999). Incident cataract after a five-year interval and lifestyle factors: the Beaver Dam Eye Study. *Ophthalmic Epidemiol.* **6**, 247–255.
40. Cumming, R. G., and Mitchell, P. (1997). Alcohol, smoking, and cataracts: the Blue Mountains Eye Study. *Arch. Ophthalmol.* **115**, 1296–1303.
41. Klein, B. E., Klein, R., Lee, K. E., Knudtson, M. D., and Tsai, M. Y. (2006). Markers of inflammation, vascular endothelial dysfunction, and age-related cataract. *Am. J. Ophthalmol.* **141**, 116–122.
42. Klein, B. E., Klein, R., Lee, K. E., and Grady, L. M. (2006). Statin use and incident nuclear cataract. *JAMA* **295**, 2752–2758.
43. Worzala, K., Hiller, R., Sperduto, R. D., *et al.* (2001). Postmenopausal estrogen use, type of menopause, and lens

opacities: the Framingham Studies. *Arch. Intern. Med* **161**, 1448–1454.
44. Freeman, E. E., Munoz, B., Schein, O. D., and West, S. K. (2001). Hormone replacement therapy and lens opacities: the Salisbury Eye Evaluation Project. *Arch. Ophthalmol.* **119**(11), 1687–1692.
45. Bunce, G. E., Kinoshita, J., and Horwitz, J. (1990). Nutritional factors in cataract. *Annu. Rev. Nutr.* **10**, 233–254.
46. Bunce, G. E. (1999). Animal studies on cataract. *In* "Nutritional and Environmental Influences on the Eye" (A. Taylor, Ed.), pp. 105–115. CRC Press, Boca Raton, FL.
47. Chatterjee, A., Milton, R. C., and Thyle, S. (1982). Prevalence and aetiology of cataract in Punjab. *Br. J. Ophthalmol.* **66**, 35–42.
48. Ojofeitimi, E. O., Adelekan, D. A., Adeoye, A., Ogungbe, T. G., Imoru, A. O., and Oduah, E. C. (1999). Dietary and lifestyle patterns in the aetiology of cataracts in Nigerian patients. *Nutr. Health* **13**, 61–68.
49. Mohan, M., Sperduto, R. D., Angra, S. K., *et al.* (1989). India-US case-control study of age-related cataracts: India-US Case-Control Study Group. *Arch. Ophthalmol.* **107**, 670–676.
50. Sperduto, R. D., Hu, T. S., Milton, R. C., *et al.* (1993). The Linxian cataract studies: two nutrition intervention trials. *Arch. Ophthalmol.* **111**, 1246–1253.
51. Fontana, L., and Klein, S. (2007). Aging, adiposity, and calorie restriction. *JAMA* **297**, 986–994.
52. Taylor, A., Lipman, R. D., Jahngen-Hodge, J., *et al.* (1995). Dietary calorie restriction in the Emory mouse: effects on lifespan, eye lens cataract prevalence and progression, levels of ascorbate, glutathione, glucose, and glycohemoglobin, tail collagen breaktime, DNA and RNA oxidation, skin integrity, fecundity, and cancer. *Mech. Aging Dev.* **79**, 33–57.
53. van Boekel, M. A., and Hoenders, H. J. (1992). Glycation of crystallins in lenses from aging and diabetic individuals. *FEBS Lett.* **314**, 1–4.
54. Swamy-Mruthinti, S., Shaw, S. M., Zhao, H. R., Green, K., and Abraham, E. C. (1999). Evidence of a glycemic threshold for the development of cataracts in diabetic rats. *Curr. Eye Res.* **18**, 423–429.
55. Di Benedetto, A., Aragona, P., Romano, G., *et al.* (1999). Age and metabolic control influence lens opacity in type I, insulin-dependent diabetic patients. *J. Diabetes Complications* **13**, 159–162.
56. Chiu, C.-J., Morris, M. S., Rogers, G., *et al.* (2005). Carbohydrate intake and glycemic index in relation to the odds of early cortical and nuclear lens opacities. *Am. J. Clin. Nutr.* **81**, 1411–1416.
57. Schaumberg, D. A., Liu, S., Seddon, J. M., Willett, W. C., and Hankinson, S. E. (2004). Dietary glycemic load and risk of age-related cataract. *Am. J. Clin. Nutr.* **80**, 489–495.
58. Jacques, P. (1997). Nutritional antioxidants and prevention of age-related eye disease. *In* "Antioxidants and Disease Prevention" (H. Garewal, Ed.), pp. 149–177. CRC Press, Boca Raton, FL.
59. Christen, W. G. (1999). Antioxidant vitamins and age-related eye disease. *Proc. Assoc. Am. Physicians* **111**, 16–21.
60. Chiu, C.-J., and Taylor, A. (2007). Nutritional antioxidants and age-related cataract and maculopathy. *Exp. Eye Res.* **84**, 229–245.
61. Taylor, A., Jacques, P. F., Nowell, T., *et al.* (1997). Vitamin C in human and guinea pig aqueous, lens and plasma in relation to intake. *Curr. Eye Res.* **16**, 857–864.
62. Yeum, K. J., Shang, F. M., Schalch, W. M., Russell, R. M., and Taylor, A. (1999). Fat-soluble nutrient concentrations in different layers of human cataractous lens. *Curr. Eye Res.* **19**, 502–505.
63. Chitchumroonchokchai, C., Bomser, J. A., Glamm, J. E., and Failla, M. L. (2004). Xanthophylls and alpha-tocopherol decrease UVB-induced lipid peroxidation and stress signaling in human lens epithelial cells. *J. Nutr.* **134**, 3225–3232.
64. Mares-Perlman, J. A., Brady, W. E., Klein, B. E., *et al.* (1995). Diet and nuclear lens opacities. *Am. J. Epidemiol.* **141**, 322–334.
65. Lyle, B. J., Mares-Perlman, J. A., Klein, B. E., Klein, R., and Greger J. L. (1999). Antioxidant intake and risk of incident age-related nuclear cataracts in the Beaver Dam Eye Study. *Am. J. Epidemiol.* **149**, 801–809.
66. Chasan-Taber, L., Willett, W. C., Seddon, J. M., *et al.* (1999). A prospective study of carotenoid and vitamin A intakes and risk of cataract extraction in US women. *Am. J. Clin. Nutr.* **70**, 509–516.
67. Brown, L., Rimm, E. B., Seddon, J. M., *et al.* (1999). A prospective study of carotenoid intake and risk of cataract extraction in US men [comment]. *Am. J. Clin. Nutr.* **70**, 517–524.
68. Jacques, P. F., Chylack, L. T., Jr., Hankinson, S. E., *et al.* (2001). Long-term nutrient intake and early age-related nuclear lens opacities. *Arch. Ophthalmol.* **119**, 1009–1019.
69. Vu, H. T. V., Robman, L., Hodge, A., McCarty, C. A., and Taylor, H. R. (2006). Lutein and zeaxanthin and the risk of cataract: the Melbourne Visual Impairment Project. *Invest. Ophthalmol. Vis. Sci.* **47**, 3783–3786.
70. Ferrigno, L., Aldigeri, R., Rosmini, F., Sperduto, R. D., and Maraini, G. (2005). Associations between plasma levels of vitamins and cataract in the Italian-American Clinical Trial of Nutritional Supplements and Age-Related Cataract (CTNS): CTNS Report #2. *Ophthalmic Epidemiol.* **12**, 71–80.
71. Jacques, P. F., Chylack, L. T., Jr., McGandy, R. B., and Hartz, S. C. (1988). Antioxidant status in persons with and without senile cataract. *Arch. Ophthalmol.* **106**, 337–340.
72. Mares-Perlman, J. A., Brady, W. E., Klein, B. E., *et al.* (1995). Serum carotenoids and tocopherols and severity of nuclear and cortical opacities. *Invest. Ophthalmol. Vis. Sci.* **36**, 276–288.
73. Mares-Perlman, J. A., Brady, W. E., Klein, B. E., *et al.* (1995). Diet and nuclear lens opacities. *Am. J. Epidemioi.* **141**, 322–334.
74. Christen, W. G., Manson, J. E., Glynn, R. J., *et al.* (2003). A randomized trial of beta carotene and age-related cataract in US physicians. *Arch. Ophthalmol.* **121**, 372–378.
75. McNeil, J. J., Robman, L., Tikellis, G., Sinclair, M. I., McCarty, C. A., and Taylor, H. R. (2004). Vitamin E supplementation and cataract: randomized controlled trial. *Ophthalmology* **111**, 75–84.
76. Christen, W., Glynn, R., Sperduto, R., Chew, E., and Buring, J. (2004). Age-related cataract in a randomized trial of beta-carotene in women. *Ophthalmic Epidemiol.* **11**, 401–412.

77. Schaumberg, D. A., Mendes, F., Balaram, M., Dana, M. R., Sparrow, D., and Hu, H. (2004). Accumulated lead exposure and risk of age-related cataract in men. *JAMA* **292**, 2750–2754.
78. Moeller, S. M., Taylor, A., Tucker, K. L., *et al.* (2004). Overall adherence to the dietary guidelines for Americans is associated with reduced prevalence of early age-related nuclear lens opacities in women. *J. Nutr.* **134**, 1812–1819.
79. Seddon, J. M. (2007). Multivitamin-multimineral supplements and eye disease: age-related macular degeneration and cataract. *Am. J. Clin. Nutr.* **85**, 304S–307S.
80. Mares, J. A. (2004). High-dose antioxidant supplementation and cataract risk. *Nutr. Rev.* **62**, 28–32.
81. Mares, J. A., La Rowe, T. L., and Blodi, B. A. (2004). Doctor, what vitamins should I take for my eyes? *Arch. Ophthalmol.* **122**, 628–635.
82. Nadalin, G., Robman, L. D., McCarty, C. A., Garrett, S. K., McNeil, J. J., and Taylor, H. R. (1999). The role of past intake of vitamin E in early cataract changes. *Ophthalmic Epidemiol.* **6**, 105–112.
83. Mares-Perlman, J. A., Lyle, B. J., Klein, R., *et al.* (2000). Vitamin supplement use and incident cataracts in a population-based study. *Arch. Ophthalmol.* **118**, 1556–1563.
84. Leske, M. C., ChylackL. T., Jr., and Wu, S. Y. (1991). The Lens Opacities Case-Control Study: risk factors for cataract [comment]. *Arch. Ophthalmol.* **109**, 244–251.
85. Taylor, A., Jacques, P. F., Chylack, L. T., Jr., *et al.* (2002). Long-term intake of vitamins and carotenoids and odds of early age-related cortical and posterior subcapsular lens opacities. *Am. J. Clin. Nutr.* **75**, 540–549.
86. Seddon, J. M., Christen, W. G., Manson, J. E., *et al.* (1994). The use of vitamin supplements and the risk of cataract among US male physicians. *Am. J. Public Health* **84**, 788–792.
87. Kuzniarz, M., Mitchell, P., Cumming, R. G., and Flood, V. M. (2001). Use of vitamin supplements and cataract: the Blue Mountains Eye Study. *Am. J. Ophthal.* **132**, 19–26.
88. Milton, RC, Sperduto, RD, Clemons, TE, and Ferris FL, 3rd. (2006). Centrum use and progression of age-related cataract in the Age-Related Eye Disease Study: a propensity score approach. AREDS report No. 21. *Ophthalmology* **113**, 1264–1270.
89. Chasan-Taber, L., Willett, W. C., Seddon, J. M., *et al.* (1999). A prospective study of vitamin supplement intake and cataract extraction among U.S. women. *Epidemiology* **10**, 679–684.
90. The Italian-American Cataract Study Group. (1991). Risk factors for age-related cortical, nuclear, and posterior subcapsular cataracts. *Am. J. Epidemiol.* **133**, 541–553.
91. Lyle, B. J., Mares-Perlman, J. A., Klein, B. E., Klein, R., and Greger, J. L. (1998). Supplement users differ from nonusers in demographic, lifestyle, dietary and health characteristics. *J. Nutr.* **128**, 2355–2362.
92. The Italian-American Clinical Trial of Nutritional Supplements and Age-Related Cataract (CTNS): Design Implications. (2003). CTNS report no. 1. *Control Clin. Trials* **24**, 815–829.
93. Friedman, D. S., O'Colmain, B. J., Munoz, B., *et al.* (2004). Prevalence of age-related macular degeneration in the United States. *Arch. Ophthalmol.* **122**, 564–572.
94. Beatty, S., Koh, H., Phil, M., Henson, D., and Boulton, M. (2000). The role of oxidative stress in the pathogenesis of age-related macular degeneration. *Surv. Ophthalmol.* **45**, 115–134.
95. Anderson, D. H., Mullins, R. F., Hageman, G. S., and Johnson, L. V. (2002). A role for local inflammation in the formation of drusen in the aging eye. *Am. J. Ophthalmol.* **134**, 411–431.
96. Hagemen, G., Luthert, P., Victor-Chong, N., Johnson, L., Anderson, D., and Mullins, R. (2001). An integrated hypothesis that considers drusen as biomarkers of immune-medicated processes at the RPE-Brunch's membrane interface in aging and age-related macular degeneration. *Prog. Retin. Eye Res.* **20**, 705–732.
97. Hageman, G. S., Anderson, D. H., Johnson, L. V., *et al.* (2005). A common haplotype in the complement regulatory gene factor H (HF1/CFH) predisposes individuals to age-related macular degeneration. *Proc. Natl. Acad. Sci. USA* **102**, 7227–7232.
98. Shaban, H., and Richter, C. (2002). A2E and blue light in the retina: the paradigm of age-related macular degeneration. *Biol. Chem.* **383**, 537–545.
99. Klein, B. E., and Klein, R. (1996). Smoke gets in your eyes too. *JAMA* **276**, 1179–1180.
100. Leeuwen, V. R., Ikram, M. K., Vingerling, J. R., Witteman, J. C., Hofman, A., and de Jong, P. T. (2003). Blood pressure, atherosclerosis, and the incidence of age-related maculopathy: the Rotterdam Study. *Invest. Ophthalmol. Vis. Sci.* **44**, 3771–3777.
101. Vingerling, J. R., Dielemans, I., Hofman, A., and Bots, M. (1995). Age-related macular degeneration is associated with atherosclerosis. *Am. J. Epidemiol.* **142**, 404–409.
102. Snow, K. K., and Seddon, J. M. (1999). Do age-related macular degeneration and cardiovascular disease share common antecedents? *Ophthalmic Epidemiol.* **6**, 125–143.
103. Gorin, M. B. (2007). A clinician's view of the molecular genetics of age-related maculopathy. *Arch. Ophthalmol.* **125**, 21–29.
104. Seddon, J. M., Gensler, G., Milton, R. C., Klein, M. L., and Rifai, N. (2004). Association between C-reactive protein and age-related macular degeneration. *JAMA* **291**, 704–710.
105. Seddon, J. M., George, S., Rosner, B., and Rifai, N. (2005). Progression of age-related macular degeneration: prospective assessment of C-reactive protein, interleukin 6, and other cardiovascular biomarkers. *Arch. Ophthalmol.* **123**, 774–782.
106. Klein, R., Klein, B. E., Tomany, S. C., and Cruickshanks, K. J. (2003). Association of emphysema, gout, and inflammatory markers with long-term incidence of age-related maculopathy. *Arch. Ophthalmol.* **121**, 674–678.
107. Clemons, T. E., Milton, R. C., Klein, R., Seddon, J. M., and Ferris, F. L. 3rd. (2005). Risk factors for the incidence of advanced age-related macular degeneration in the Age-Related Eye Disease Study (AREDS). AREDS report no. 19. *Ophthalmology* **112**, 533–539.
108. Nielsen, J. C., Naash, M. I., and Anderson, R. E. (1988). The regional distribution of vitamins E and C in mature and premature human retinas. *Invest. Ophthalmol. Vis. Sci.* **29**, 22–26.

109. Bone, R. A., Landrum, J. T., Fernandez, L., and Tarsis, S. L. (1988). Analysis of the macular pigment by HPLC: retinal distribution and age study. *Invest. Ophthalmol. Vis. Sci.* **29**, 843–849.
110. Bernstein, P. S., Khachik, F., Carvalho, L. S., Muir, G. J., Zhao, D. Y., and Katz, N. B. (2001). Identification and quantitation of carotenoids and their metabolites in the tissues of the human eye. *Exp. Eye Res.* **72**, 215–223.
111. Karcioglu, Z. A. (1982). Zinc in the eye. *Surv. Ophthalmol.* **27**, 114–122.
112. Mares-Perlman, J. A., and Klein, R., (1999). Diet and age-related macular degeneration. *In* "Nutritional and Environmental Influences on the Eye" (A. Taylor, Ed.), pp. 181–214. CRC Press, Boca Raton, FL.
113. Handelman, G. J., and Dratz, E. A. (1986). The role of antioxidants in the retina and retinal pigment epithelium and the nature of pro-oxidant damage. *Adv. Free Radic. Biol. Med.* **2**, 1–89.
114. Malinow, M. R., Feeney-Burns, L., Peterson, L. H., Klein, M. L., and Neuringer, M. (1980). Diet-related macular anomalies in monkeys. *Invest. Ophthalmol. Vis. Sci.* **19**, 857–863.
115. Teikari, J. M., Laatikainen, L., Virtamo, J., *et al.* (1998). Six-year supplementation with alpha-tocopherol and beta-carotene and age-related maculopathy. *Acta Ophthalmol. Scand.* **76**, 224–229.
116. Taylor, H. R., Tikellis, G., Robman, L. D., McCarty, C. A., and McNeil, J. J. (2002). Vitamin E supplementation and macular degeneration: randomised controlled trial. *BMJ* **325**, 11.
117. Stahl, W., Junghans, A., de Boer, B., Driomina, E. S., Briviba, K., and Sies, H. (1998). Carotenoid mixtures protect multilamellar liposomes against oxidative damage: synergistic effects of lycopene and lutein. *FEBS Lett.* **427**, 305–308.
118. Wrona, M., Korytowski, W., Rozanowska, M., Sarna, T., and Truscott, T. G. (2003). Cooperation of antioxidants in protection against photosensitized oxidation. *Free Radic. Biol. Med.* **35**, 1319–1329.
119. Wrona, M., Rozanowska, M., and Sarna, T. (2004). Zeaxanthin in combination with ascorbic acid or alpha-tocopherol protects ARPE-19 cells against photosensitized peroxidation of lipids. *Free Radic. Biol. Med.* **36**, 1094–1101.
120. Eye Disease Case-Control Study Group. (1993). Antioxidant status and neovascular age-related macular degeneration [erratum appears in. *Arch. Ophthalmol.*, November 1993, **111**(11), 1499].
121. Moriarty-Craige, S. E., Adkison, J., Lynn, M., *et al.* (2005). Antioxidant supplements prevent oxidation of cysteine/cystine redox in patients with age-related macular degeneration. *Am. J. Ophthalmol.* **140**, 1020–1026.
122. Richer, S., Stiles, W., Statkute, L., *et al.* (2004). Double-masked, placebo-controlled, randomized trial of lutein and antioxidant supplementation in the intervention of atrophic age-related macular degeneration: the Veterans LAST study (Lutein Antioxidant Supplementation Trial). *Optometry* **75**, 216–230.
123. West, S., Vitale, S., Hallfrisch, J., *et al.* (1994). Are antioxidants or supplements protective for age-related macular degeneration? [see comment]. *Arch. Ophthalmol.* **112**, 222–227.
124. VandenLangenberg, G. M., Mares-Perlman, J. A., Klein, R., Klein, B. E., Brady, W. E., and Palta, M. (1998). Associations between antioxidant and zinc intake and the 5-year incidence of early age-related maculopathy in the Beaver Dam Eye Study. *Am. J. Epidemiol.* **148**, 204–214.
125. van Leeuwen, R., Boekhoorn, S., Vingerling, J. R., *et al.* (2005). Dietary intake of antioxidants and risk of age-related macular degeneration. *JAMA* **294**, 3101–3107.
126. Mares, J. (2003). Carotenoids and eye disease: epidemiologic evidence. *In* "Carotenoids in Health and Disease" (N. I. Krinsky, and S. Mayne S., Eds.), Chapter 19. Marcel Dekker, New York.
127. Delcourt, C., Carriere, I., Delage, M., Barberger-Gateau, P., and Schalch, W., (2006). Plasma lutein and zeaxanthin and other carotenoids as modifiable risk factors for age-related maculopathy and cataract: the POLA Study. *Invest. Ophthalmol. Vis. Sci.* **47**, 2329–2335.
128. Cardinault, N., Abalain, J.-H., Sairafi, B., *et al.* (2005). Lycopene but not lutein nor zeaxanthin decreases in serum and lipoproteins in age-related macular degeneration patients. *Clinica Chimica Acta* **357**, 34–42.
129. Gale, C. R., Hall, N. F., Phillips, D. I., and Martyn, C. N. (2003). Lutein and zeaxanthin status and risk of age-related macular degeneration. *Invest. Ophthalmol. Vis. Sci.* **44**, 2461–2465.
130. Moeller, S. M., Parekh, N., Tinker, L., *et al.* (2006). Associations between intermediate age-related macular degeneration and lutein and zeaxanthin in the Carotenoids in Age-related Eye Disease Study (CAREDS): ancillary study of the Women's Health Initiative. *Arch. Ophthalmol.* **124**, 1151–1162.
131. Mares-Perlman, J. A., Klein, R., Klein, B. E., *et al.* (1996). Association of zinc and antioxidant nutrients with age-related maculopathy. *Arch. Ophthalmol.* **114**, 991–997.
132. Moeller, S. M., Mehta, N. R., Tinker, L. F., *et al.* (2006). Associations between intermediate age-related macular degeneration and lutein and zeaxanthin in the Carotenoids in Age-Related Eye Disease Study (CAREDS), an ancillary study of the Women's Health Initiative. *Arch. Ophthalmol.* **124**, 1–24.
133. Mares, J. A. (2006). Potential value of antioxidant-rich foods in slowing age-related macular degeneration. *Arch. Ophthalmol.* **124**, 1339–1340.
134. Fowke, J. H., Morrow, J. D., Motley, S., Bostick, R. M., and Ness, R. M. (2006). Brassica vegetable consumption reduces urinary F2-isoprostane levels independent of micronutrient intake. *Carcinogenesis,* **27**, 2096–2102.
135. Yemelyanov, A. Y., Katz, N. B., and Bernstein, P. S. (2001). Ligand-binding characterization of xanthophyll carotenoids to solubilized membrane proteins derived from human retina. *Exp. Eye Res.* **72**, 381–392.
136. Bhosale, P., Larson, A. J., Frederick, J. M., Southwick, K., Thulin, C. D., and Bernstein, P. S. (2004). Identification and characterization of a Pi isoform of glutathione S-transferase (GSTP1) as a zeaxanthin-binding protein in the macula of the human eye. *J. Biol. Chem.* **279**, 49447–49454.
137. Handelman, G. J., Nightingale, Z. D., Lichtenstein, A. H., Schaefer, E. J., and Blumberg, J. B. (1999). Lutein and zeaxanthin concentrations in plasma after dietary supplementation with egg yolk. *Am. J. Clin. Nutr.* **70**, 247–251.

138. U.S. Department of Agriculture Agricultural Research Service. (2004). *USDA* National Nutrient Database for Standard Reference, Release 16-1.
139. Mares-Perlman, J. A., Fisher, A. I., Klein, R., *et al.* (2001). Lutein and zeaxanthin in the diet and serum and their relation to age-related maculopathy in the third National Health and Nutrition Examination Survey. *Am. J. Epidemiol.* **153**, 424–432.
140. Snodderly, D. M., Brown, P. K., Delori, F. C., and Auran, J. D. (1984). The macular pigment. I. Absorbance spectra, localization, and discrimination from other yellow pigments in primate retinas. *Invest. Ophthalmol. Vis. Sci.* **25**, 660–673.
141. Junghans, A., Sies, H., and Stahl, W. (2001). Macular pigments lutein and zeaxanthin as blue light filters studied in liposomes. *Arch. Biochem. Biophys.* **391**, 160–164.
142. Hammond, J. B. R., Wooten, B. R., and Curran-Celentano, J. (2001). Carotenoids in the retina and lens: possible acute and chronic effects on human visual performance. *Arch. Biochem. Biophys.* **385**, 41–46.
143. Sparrow, J. R., Nakanishi, K., and Parish, C. A. (2000). The lipofuscin fluorophore A2E mediates blue light-induced damage to retinal pigmented epithelial cells. *Invest. Ophthalmol. Vis. Sci.* **41**, 1981–1989.
144. Sparrow, J. R., Vollmer-Snarr, H. R., Zhou, J., *et al.* (2003). A2E-epoxides damage DNA in retinal pigment epithelial cells: vitamin E and other antioxidants inhibit A2E-epoxide formation. *J. Biol. Chem.* **278**, 18207–18213.
145. Bone, R. A., Landrum, J. T., Mayne, S. T., Gomez, C. M., Tibor, S. E., and Twaroska, E. E. (2001). Macular pigment in donor eyes with and without AMD: a case-control study. *Invest. Ophthalmol. Vis. Sci.* **42**, 235–240.
146. Khachik, F., Bernstein, P. S., and Garland, D. L. (1997). Identification of lutein and zeaxanthin oxidation products in human and monkey retinas. *Invest. Ophthalmol. Vis. Sci.* **38**, 1802–1811.
147. Sujak, A., Gabrielska, J., Grudzinski, W., Borc, R., Mazurek, P., and Gruszecki, W. I. (1999). Lutein and zeaxanthin as protectors of lipid membranes against oxidative damage: the structural aspects. *Arch. Biochem. Biophys.* **371**, 301–307.
148. Leung, I. Y., Sandstrom, M. M., Zucker, C. L., Neuringer, M., and Snodderly, D. M. (2004). Nutritional manipulation of primate retinas, II: effects of age, n-3 fatty acids, lutein, and zeaxanthin on retinal pigment epithelium. *Invest. Ophthalmol. Vis. Sci.* **45**, 3244–3256.
149. Feeney-Burns, L., Neuringer, M., and Gao, C. L. (1989). Macular pathology in monkeys fed semipurified diets. *Prog. Clin. Biol. Res.* **314**, 601–622.
150. Thomson, L. R., Toyoda, Y., Langner, A., *et al.* (2002). Elevated retinal zeaxanthin and prevention of light-induced photoreceptor cell death in quail. *Invest. Ophthal. Vis. Sci.* **43**, 3538–3549.
151. Jahn, C., Wastemeyer, H., Brinkmann, C., Trautmann, S., Mayner, A., and Wolf, S. (2005). Macular pigment density in age-related maculopathy. *Graefe's Archive for Clin. Exp. Ophthalmol.* **243**, 222–227.
152. Beatty, S., Murray, I. J., Henson, D. B., Carden, D., Koh, H., and Boulton, M. E. (2001). Macular pigment and risk for age-related macular degeneration in subjects from a Northern European population. *Invest. Ophthalmol. Vis. Sci.* **42**, 439–446.
153. Bernstein, P. S., Zhao, D. Y., Wintch, S. W., Ermakov, I. V., McClane, R. W., and Gellermann, W. (2002). Resonance Raman measurement of macular carotenoids in normal subjects and in age-related macular degeneration patients. *Ophthalmology* **109**, 1780–1787.
154. La Rowe, T. L., Mares, J., Snodderly, D. M., Klein, M. L., Wooten, B. R., and Chappell, R. (2007). Cross-sectional relationships between macular pigment density and age-related maculopathy in women participating in the Carotenoids in Age-Related Eye Disease Study: an ancillary study of the Women's Health Initiative. *Ophthalmology,* (in press).
155. Snellen, E. L., Verbeek, A. L., Van Den Hoogen, G. W., Cruysberg, J. R., and Hoyng, C. B. (2002). Neovascular age-related macular degeneration and its relationship to antioxidant intake. *Acta Ophthalmologica Scand.* **80**, 368–371.
156. Seddon, J., AUSR, *et al.* (1994). Dietary carotenoids, vitamins A, C, and E, and advanced age-related macular degeneration. *JAMA* **272**, 1413–1420.
157. Cho, E., Seddon, J. M., Rosner, B., Willett, W. C., and Hankinson, S. E. (2004). Prospective study of intake of fruits, vegetables, vitamins, and carotenoids and risk of age-related maculopathy. *Arch. Ophthalmol.* **122**, 883–892.
158. Flood, V., Smith, W., Wang, J. J., Manzi, F., Webb, K., and Mitchell, P. (2002). Dietary antioxidant intake and incidence of early age-related maculopathy: the Blue Mountains Eye Study. *Ophthalmology* **109**, 2272–2278.
159. Flood, V., Rochtchina, E., Wang, J. J., Mitchell, P., and Smith, W. (2006). Lutein and zeaxanthin dietary intake and age related macular degeneration. *Br. J. Ophthalmol.* **90**, 927–928.
160. Mares-Perlman, J. A., Klein, R., Klein, B. E., *et al.* (1996). Association of zinc and antioxidant nutrients with age-related maculopathy. *Arch. Ophthalmol.* **114**, 991–997.
161. Mares-Perlman, J. A., Brady, W. E., Klein, R., *et al.* (1995). Serum antioxidants and age-related macular degeneration in a population-based case-control study. *Arch. Ophthalmol.* **113**, 1518–1523.
162. Vaicaitiene, R., Luksiene, D. K., Paunksnis, A., Cerniauskiene, L. R., Domarkiene, S., and Cimbalas, A. (2003). Age-related maculopathy and consumption of fresh vegetables and fruits in urban elderly. *Medicina (Kaunas)* **39**, 1231–1236.
163. Goldberg, J., Flowerdew, G., Smith, E., Brody, J. A., and Tso, M. O. (1988). Factors associated with age-related macular degeneration: an analysis of data from the first National Health and Nutrition Examination Survey. *Am. J. Epidemiol.* **128**, 700–710.
164. Cangemi, F. E. (2007). TOZAL Study: an open case control study of an oral antioxidant and omega-3 supplement for dry AMD. *BMC Ophthalmol.* **7**, 3.
165. Kvansakul, J., Rodriguez-Carmona, M., Edgar, D., *et al.* (2006). Supplementation with the carotenoids lutein or zeaxanthin improves human visual performance. *Ophthalmic Physiol. Opt.* **26**, 362–371.
166. Bartlett, H. E., and Eperjesi, F. (2007). Effect of lutein and antioxidant dietary supplementation on contrast sensitivity in age-related macular disease: a randomized controlled trial. *Eur. J. Clin. Nutr.* **61**, 1121–1127.

167. Coleman, H., and Chew, E. (2007). Nutritional supplementation in age-related macular degeneration. *Curr. Opin. Ophthalmol.* **18**, 220–223.
168. Hammond, B. R. Jr., Johnson, E. J., Russell, R. M., *et al.* (1997). Dietary modification of human macular pigment density. *Invest. Ophthalmol. Vis. Sci.* **38**, 1795–1801.
169. Bone, R. A., Landrum, J. T., Guerra, L. H., and Ruiz, C. A. (2003). Lutein and zeaxanthin dietary supplements raise macular pigment density and serum concentrations of these carotenoids in humans. *J. Nutr.* **133**, 992–998.
170. Aleman, T. S., Duncan, J. L., Bieber, M. L., *et al.* (2001). Macular pigment and lutein supplementation in retinitis pigmentosa and Usher syndrome. *Invest. Ophthalmol. Vis. Sci.* **42**, 1873–1881.
171. Bowen, P. E., Garg, V., Stacewicz-Sapuntzakis, M., Yelton, L., and Schreiner, R. S. (1993). Variability of serum carotenoids in response to controlled diets containing six servings of fruits and vegetables per day. *Ann. NY Acad. Sci.* **691**, 241–243.
172. Bowen, P. E., Herbst-Espinosa, S. M., Hussain, E. A., and Stacewicz-Sapuntzakis, M. (2002). Esterification does not impair lutein bioavailability in humans. *J. Nutr.* **132**, 3668–3673.
173. Mares, J. A., LaRowe, T. L., Snodderly, D. M., *et al.* (2006). Predictors of optical density of lutein and zeaxanthin in retinas of older women in the Carotenoids in Age-Related Eye Disease Study, an ancillary study of the Women's Health Initiative. *Am. J. Clin. Nutr.* **84**, 1107–1122.
174. Food and Nutrition Board IoM. (2000). "Scientific Evaluation of Dietary Reference Intake, Food and Nutrition Board, Institute of Medicine; Dietary Reference Intakes (DRI) for Vitamin A, Vitamin K, Arsenic, Boron, Chromium, Copper, Iodine, Iron, Manganese, Molybdenum, Nickel, Silicon, Vanadium, and Zinc." National Academy Press, Washington, DC.
175. Tate, D. J. Jr., Miceli, M. V., and Newsome, D. A. (1999). Zinc protects against oxidative damage in cultured human retinal pigment epithelial cells. *Free Radic. Biol. Med.* **26**, 704–713.
176. Mocchegiani, E., Muzzioli, M., and Giacconi, R. (2000). Zinc, metallothioneins, immune responses, survival and ageing. *Biogerontology* **1**, 133–143.
177. Shankar, A. H., and Prasad, A. S. (1998). Zinc and immune function: the biological basis of altered resistance to infection. *Am. J. Clin. Nutr.* **68**, 447S–463S.
178. Hyun, H. J., Sohn, J. H., Ha, D. W., Ahn, Y. H., Koh, J.-Y., and Yoon, Y. H. (2001). Depletion of intracellular zinc and copper with TPEN results in apoptosis of cultured human retinal pigment epithelial cells. *Invest. Ophthalmol. Vis. Sci.* **42**, 460–465.
179. Wood, J. P. M., and Osborne, N. N. (2003). Zinc and energy requirements in induction of oxidative stress to retinal pigmented epithelial cells. *Neurochem. Res.* **28**, 1525–1533.
180. Cho, E., Stampfer, M. J., Seddon, J. M., *et al.* (2001). Prospective study of zinc intake and the risk of age-related macular degeneration. *Ann. Epidemiol.* **11**, 328–336.
181. A randomized, placebo-controlled, clinical trial of high-dose supplementation with vitamins C and E, beta carotene, and zinc for age-related macular degeneration and vision loss: AREDS report no. 8. (2001). *Arch. Ophthalmol.* **119**, 1417–1436.
182. Newsome, D. A., Swartz, M., Leone, N. C., Elston, R. C., and Miller, E. (1988). Oral zinc in macular degeneration [comment]. *Arch. Ophthalmol.* **106**, 192–198.
183. Stur, M., Tittl, M., Reitner, A., and Meisinger, V. (1996). Oral zinc and the second eye in age-related macular degeneration. *Invest. Ophthalmol. Vis. Sci.* **37**, 1225–1235.
184. Clemons, T. E., Kurinij, N., and Sperduto, R. D. (2004). Associations of mortality with ocular disorders and an intervention of high-dose antioxidants and zinc in the Age-Related Eye Disease Study: AREDS Report No. 13. *Arch. Ophthalmol.* **122**, 716–726.
185. Seddon, J. M., Rosner, B., Sperduto, R. D., *et al.* (2001). Dietary fat and risk for advanced age-related macular degeneration. *Arch. Ophthalmol.* **119**, 1191–1199.
186. Schaumberg, D. A., Christen, W. G., Hankinson, S. E., and Glynn, R. J. (2001). Body mass index and the incidence of visually significant age-related maculopathy in men. *Arch. Ophthalmol.* **119**, 1259–1265.
187. Seddon, J. M., Cote, J., Davis, N., and Rosner, B. (2003). Progression of age-related macular degeneration: association with body mass index, waist circumference, and waist-hip ratio. *Arch. Ophthalmol.* **121**, 785–792.
188. Miceli, M. V., Newsome, D. A., Tate D. J. and Jr., Sarphie, T. G. (2000). Pathologic changes in the retinal pigment epithelium and Bruch's membrane of fat-fed atherogenic mice. *Curr. Eye Res.* **20**, 8–16.
189. Kliffen, M., Lutgens, E., Daemen, M. J., de Muinck, E. D., Mooy, C. M., and de Jong, P. T. (2000). The APO(*)E3-Leiden mouse as an animal model for basal laminar deposit. *Br. J. Ophthalmol.* **84**, 1415–1419.
190. Espinosa-Heidmann, D. G., Sall, J., Hernandez, E. P., and Cousins, S. W. (2004). Basal laminar deposit formation in APO B100 transgenic mice: complex interactions between dietary fat, blue light, and vitamin E. *Invest. Ophthalmol. Vis. Sci.* **45**, 260–266.
191. Dithmar, S., Sharara, N. A., Curcio, C. A., *et al.* (2001). Murine high-fat diet and laser photochemical model of basal deposits in Bruch membrane. *Arch. Ophthalmol.* **119**, 1643–1649.
192. Heuberger, R. A., Mares-Perlman, J. A., Klein, R., Klein, B. E., Millen, A. E., and Palta, M. (2001). Relationship of dietary fat to age-related maculopathy in the Third National Health and Nutrition Examination Survey. *Arch. Ophthalmol.* **119**, 1833–1838.
193. Mares-Perlman, J. A., Brady, W. E., Klein, R., VandenLangenberg, G. M., Klein, B. E., and Palta, M. (1995). Dietary fat and age-related maculopathy. *Arch. Ophthalmol.* **113**, 743–748.
194. Smith, W., Mitchell, P., and Leeder, S. R. (2000). Dietary fat and fish intake and age-related maculopathy. *Arch. Ophthalmol.* **118**, 401–404.
195. Cho, E., Hung, S., Willett, W. C., *et al.* (2001). Prospective study of dietary fat and the risk of age-related macular degeneration. *Am. J. Clin. Nutr.* **73**, 209–218.
196. Chua, B., Flood, V., Rochtchina, E., Wang, J. J., Smith, W., and Mitchell, P. (2006). Dietary fatty acids and the 5-year incidence of age-related maculopathy. *Arch. Ophthalmol.* **124**, 981–986.
197. Seddon, J. M., Cote, J., and Rosner, B. (2003). Progression of age-related macular degeneration: association with dietary

fat, transunsaturated fat, nuts, and fish intake. *Arch. Ophthalmol.* **121**, 1728–1737.

198. Age-Related Eye Disease Study Research Group. (2007). The relationship of dietary lipid intake and age-related macular degeneration in a case-control study: AREDS Report No. 20. *Arch. Ophthalmol.* **125**, 671–679.
199. Parekh, N., Chappell, R., Millen, A., Albert, D. M., and Mares, J. A. (2007). Association between vitamin D and age-related macular degeneration in the Third National Health and Nutrition Examination Survey 1988-94. *Arch. Ophthalmol.* **125**, 661–669.
200. Seddon, J. M., George, S., and Rosner, B. (2006). Cigarette smoking, fish consumption, omega-3 fatty acid intake, and associations with age-related macular degeneration: the U.S. twin study of age-related macular degeneration. *Arch. Ophthalmol.* **124**, 995–1001.
201. Arnarsson, A., Sverrisson, T., Stefansson, E., *et al.* (2006). Risk factors for five-year incident age-related macular degeneration: the Reykjavik Eye Study. *Am. J. Ophthal.* **142**, 419.
202. Jonasson, F., Arnarsson, A., Peto, T., Sasaki, H., Sasaki, K., and Bird, A. C. (2005). 5-year incidence of age-related maculopathy in the Reykjavik Eye Study. *Ophthalmology* **112**, 132–138
203. Jonasson, F., Arnarsson, A., Sasaki, H., Peto, T., Sasaki, K. and Bird, A. C. (2003). The prevalence of age-related maculopathy in Iceland: Reykjavik Eye Study. *Arch. Ophthalmol.* **121**, 379–385.
204. Larsson, S. C., Kumlin, M., Ingelman-Sundberg, M., and Wolk, A. (2004). Dietary long-chain n-3 fatty acids for the prevention of cancer: a review of potential mechanisms. *Am. J. Clin. Nutr.* **79**, 935–945.
205. Breslow, J. L. (2006). n-3 Fatty acids and cardiovascular disease. *Am. J. Clin. Nutr.* **83**, S1477–S1482.
206. Bazan, N. G., and Scott, B. L. (1990). Dietary omega-3 fatty acids and accumulation of docosahexaenoic acid in rod photoreceptor cells of the retina and at synapses. *Ups. J. Med. Sci. Suppl.* **48**, 97–107.
207. Fliesler, S. J., and Anderson, R. E. (1983). Chemistry and metabolism of lipids in the vertebrate retina. *Prog. Lipid Res.* **22**, 79–131.
208. Arterburn, L. M., Hall, E. B., and Oken, H. (2006). Distribution, interconversion, and dose response of n-3 fatty acids in humans. *Am. J. Clin. Nutr.* **83**, S1467–S1476.
209. Litman, B. J., Niu, S. L., Polozova, A., and and Mitchell, D. C. (2001). The role of docosahexaenoic acid containing phospholipids in modulating G protein-coupled signaling pathways: visual transduction. *J. Mol. Neurosci.* **16**, 237–242; discussion 279–284.
210. Rotstein, N. P., Politi, L. E., German, O. L., and Girotti, R. (2003). Protective effect of docosahexaenoic acid on oxidative stress-induced apoptosis of retina photoreceptors. *Invest. Ophthalmol. Vis. Sci.* **44**, 2252–2259.
211. Mukherjee, P. K., Marcheselli, V. L., Serhan, C. N., and Bazan, N. G. (2004). Neuroprotectin D1: a docosahexaenoic acid-derived docosatriene protects human retinal pigment epithelial cells from oxidative stress. *Proc. Natl. Acad. Sci. USA* **101**, 8491–8496.
212. SanGiovanni, J. P., and Chew, E. Y. (2005). The role of omega-3 long-chain polyunsaturated fatty acids in health and disease of the retina. *Prog. Retin. Eye Res.* **24**, 87–138.
213. Klein, R. J., Zeiss, C., Chew, E. Y., *et al.* (2005). Complement factor H polymorphism in age-related macular degeneration. *Science* **308**, 385–389.
214. Haines, J. L., Hauser, M. A., Schmidt, S., *et al.* (2005). Complement factor H variant increases the risk of age-related macular degeneration. *Science* **308**, 419–421.
215. Edwards, A. O., Ritter, R. 3rd, Abel, K. J., Manning, A., Panhuysen, C., and Farrer, L. A. (2005). Complement factor H polymorphism and age-related macular degeneration. *Science* **308**, 421–424.
216. Anderson, D. H., Mullins, R. F., Hageman, G. S., and Johnson, R. V. (2002). A role for local inflammation in the formation of drusen in the aging eye. *Am. J. Ophthalmol.* **134**, 411–431.
217. Klein, R., Klein, B. E., Marino, E. K., *et al.* (2003). Early age-related maculopathy in the cardiovascular health study. *Ophthalmology* **110**, 25–33.
218. Klein, R., Klein, B. E., Knudtson, M. D., Wong, T. Y., Shankar, A., and Tsai, M. Y. (2005). Systemic markers of inflammation, endothelial dysfunction, and age-related maculopathy. *Am. J. Ophthalmol.* **140**, 35–44.
219. Holick, M. F. (2004). Vitamin D: importance in the prevention of cancers, type 1 diabetes, heart disease, and osteoporosis. *Am. J. Clin. Nutr.* **79**, 362–371.
220. Mark, B. L., and Carson, J. A. (2006). Vitamin D and autoimmune disease: implications for practice from the multiple sclerosis literature. *J. Am. Diet. Assoc.* **106**, 418–424.
221. Grant, W. B. (2006). Epidemiology of disease risks in relation to vitamin D insufficiency. *Prog. Biophys. Mol. Biol.* **92**, 65–79.
222. Provvedini, D. M., Rulot, C. M., Sobol, R. E., Tsoukas, C. D., and Manolagas, S. C. (1987). 1 alpha,25-Dihydroxyvitamin D3 receptors in human thymic and tonsillar lymphocytes. *J. Bone. Miner. Res.* **2**, 239–247.
223. Veldman, C. M., Cantorna, M. T., and DeLuca, H. F. (2000). Expression of 1,25-dihydroxyvitamin D(3) receptor in the immune system. *Arch. Biochem. Biophys.* **374**, 334–338.
224. Cantorna, M. T., and Mahon, B. D. (2004). Mounting evidence for vitamin D as an environmental factor affecting autoimmune disease prevalence. *Exp. Biol. Med. (Maywood).* **229**, 1136–1142.
225. Lemire, J. M., Archer, D. C., Beck, L., and Spiegelberg, H. L. (1995). Immunosuppressive actions of 1,25-dihydroxyvitamin D3: preferential inhibition of Th1 functions. *J. Nutr.* **125**, 1704S–1708S.
226. Iseki, K., Tatsuta, M., Uehara, H., *et al.* (1999). Inhibition of angiogenesis as a mechanism for inhibition by 1alpha-hydroxyvitamin D3 and 1,25-dihydroxyvitamin D3 of colon carcinogenesis induced by azoxymethane in Wistar rats. *Int. J. Cancer* **81**, 730–733.
227. Shokravi, M. T., Marcus, D. M., Alroy, J., Egan, K., Saornil, M. A., and Albert, D. M. (1995). Vitamin D inhibits angiogenesis in transgenic murine retinoblastoma. *Invest Ophthalmol. Vis. Sci.* **36**, 83–87.
228. Looker, A. C., Dawson-Hughes, B., Calvo, M. S., Gunter, E. W., and Sahyoun, N. R. (2002). Serum 25-hydroxyvitamin D status of adolescents and adults in two seasonal subpopulations from NHANES III. *Bone* **30**, 771–777.
229. Hanley, D. A., and Davison, K. S. (2005). Vitamin D insufficiency in North America. *J. Nutr.* **135**, 332–337.

230. Holick, M. F., Matsuoka, L. Y., and Wortsman, J. (1989). Age, vitamin D, and solar ultraviolet. *Lancet* **2**, 1104–1105.
231. Heaney, R. P. (2006). Barriers to optimizing vitamin D3 intake for the elderly. *J. Nutr.* **136**, 1123–1125.
232. Chiu, C. J., Hubbard, L. D., Armstrong, J., *et al.* (2006). Dietary glycemic index and carbohydrate in relation to early age-related macular degeneration. *Am. J. Clin. Nutr.* **83**, 880–886.
233. West, A. L., Oren, G. A., and Moroi, S. E. (2006). Evidence for the use of nutritional supplements and herbal medicines in common eye diseases. *Am. J. Ophthal.* **141**, 157–166.
234. Klein, R., and Klein, B. (1995). Vision disorders in diabetes, *In* "Diabetes in America" (M. I. Harris, C. C. Cowie, M. P. Stern, E.J. Boyko, G. E, Reiber, and P. H. Bennett, Eds.), pp. 293–338. NIH, National Institutes of Diabetes and Digestive and Kidney Disease (NIH Publ. No. 95–1468), U. S. Government Printing Office, Bethesda, MD.
235. National Institute of Diabetes and Digestive and Kidney Diseases. National Diabetes Statistics fact sheet: general information and national estimates on diabetes in the United States, 2005. (2005). U.S. Department of Health and Human Services, National Institute of Health, Bethesda, MD.
236. Narayan, K. M., Boyle, J. P., Geiss, L. S., Saaddine, J. B., and Thompson, T. J. (2006). Impact of recent increase in incidence on future diabetes burden: U.S., 2005-2050. *Diabetes Care* **29**, 2114–2116.
237. Ogden, C.L, Carroll, M. D., Curtin, L. R., McDowell, M. A., Tabak, C. J., and Flegal, K. M. (2006). Prevalence of overweight and obesity in the United States, 1999–2004. *JAMA* **295**, 1549–1555.
238. Cowie, C. C., Rust, K. F., Byrd-Holt, D. D., *et al.* (2006). Prevalence of diabetes and impaired fasting glucose in adults in the U.S. population: National Health And Nutrition Examination Survey 1999–2002. *Diabetes Care* **29**, 1263–1268.
239. Bennett, P. H. (1971). Diabetes mellitus in Pima Indians. *Lancet* **2**, 488–489.
240. Knowler, W. C., Pettitt, D. J., Saad, M. F., and Bennett, P. H. (1990). Diabetes mellitus in the Pima Indians: incidence, risk factors and pathogenesis. *Diabetes Metab. Rev.* **6**, 1–27.
241. Porta, M., and Allione, A. (2004). Current approaches and perspectives in the medical treatment of diabetic retinopathy. *Pharmacol. Therapeut.* **103**, 167–177.
242. Feman, S. (1992). "Ocular Problems in Diabetes mellitus." Blackwell Scientific, Boston.
243. Aiello, L. P., Gardner, T. W., King, G. L., *et al.* (1998). Diabetic retinopathy. *Diabetes Care* **21**, 143–156.
244. Javitt, J. C., Aiello, L. P., Chiang, Y., Ferris, F. L. 3rd, Canner, J. K., and Greenfield, S. (1994). Preventive eye care in people with diabetes is cost-saving to the federal government: implications for health-care reform. *Diabetes Care* **17**, 909–917.
245. The effect of intensive treatment of diabetes on the development and progression of long-term complications in insulin-dependent diabetes mellitus. (1993). The Diabetes Control and Complications Trial Research Group. *N. Engl. J. Med.* **329**, 977–986.
246. Intensive blood-glucose control with sulphonylureas or insulin compared with conventional treatment and risk of complications in patients with type 2 diabetes (UKPDS 33). (1998). UK Prospective Diabetes Study (UKPDS) Group. *Lancet* **352**, 837–853.
247. Shichiri, M., Kishikawa, H., Ohkubo, Y., and Wake, N. (2000). Long-term results of the Kumamoto study on optimal diabetes control in type 2 diabetic patients. *Diabetes Care* **23** (suppl 2), B21–B29.
248. Klein, R., Klein, B. E., Moss, S. E., Davis, M. D., and DeMets, D. L. (1984). The Wisconsin epidemiologic study of diabetic retinopathy. III. Prevalence and risk of diabetic retinopathy when age at diagnosis is 30 or more years. *Arch. Ophthalmol.* **102**, 527–532.
249. Matthews, D. R., Stratton, I. M., Aldington, S. J., Holman, R. R., and Kohner, E. M. (2004). Risks of progression of retinopathy and vision loss related to tight blood pressure control in type 2 diabetes mellitus: UKPDS 69. *Arch. Ophthalmol.* **122**, 1631–1640.
250. Ferris, F. L. 3rd, Chew, E. Y., and Hoogwerf, B. J. (1996). Serum lipids and diabetic retinopathy: Early Treatment Diabetic Retinopathy Study Research Group. *Diabetes Care* **19**, 1291–1293.
251. Davis, M. D., Fisher, M. R., Gangnon, R. E., *et al.* (1998). Risk factors for high-risk proliferative diabetic retinopathy and severe visual loss: Early Treatment Diabetic Retinopathy Study Report #18. *Invest. Ophthalmol. Vis. Sci.* **39**, 233–252.
252. Cusick, M., Chew, E. Y., Chan, C. C., Kruth, H. S., Murphy, R. P., and Ferris, F. L. 3rd. (2003). Histopathology and regression of retinal hard exudates in diabetic retinopathy after reduction of elevated serum lipid levels. *Ophthalmology* **110**, 2126–2133.
253. Leal, E. C., Santiago, A. R., and Ambrosio, A. F. (2005). Old and new drug targets in diabetic retinopathy: from biochemical changes to inflammation and neurodegeneration. *Curr. Drug Targets CNS Neurol. Disord.* **4**, 421–434.
254. Paisey, R. B., Arredondo, G., Villalobos, A., Lozano, O., Guevara, L., and Kelly, S. (1984). Association of differing dietary, metabolic, and clinical risk factors with microvascular complications of diabetes: a prevalence study of 503 Mexican type II diabetic subjects. II. *Diabetes Care* **7**, 428–433.
255. Raheja, B., Modi, K., Barua, J., Jain, S., Shahani, V., and Koppikar, G. (1987). Proliferative diabetic retinopathy in NIDDM and Indian diet. *J. Med. Assoc. Thai.* **70** (suppl 2), 139–143.
256. Roy, M. S., Stables, G., Collier, B., Roy, A., and Bou, E. (1989). Nutritional factors in diabetics with and without retinopathy. *Am. J. Clin. Nutr.* **50**, 728–730.
257. Mayer-Davis, E. J., Bell, R. A., Reboussin, B. A., Rushing, J., Marshall, J. A., and Hamman, R. F. (1998). Antioxidant nutrient intake and diabetic retinopathy: the San Luis Valley Diabetes Study. *Ophthalmology* **105**, 2264–2270.
258. Millen, A. E., Gruber, M., Klein, R., Klein, B. E., Palta, M., and Mares, J. A. (2003). Relations of serum ascorbic acid and alpha-tocopherol to diabetic retinopathy in the Third National Health and Nutrition Examination Survey. *Am. J. Epidemiol.* **158**, 225–233.
259. Millen, A. E., Klein, R., Folsom, A. R., Stevens, J., Palta, M., and Mares, J. A. (2004). Relation between intake of vitamins C and E and risk of diabetic retinopathy in the Atherosclerosis Risk in Communities Study. *Am. J. Clin. Nutr.* **79**, 865–873.

260. Cundiff, D. K., and Nigg, C. R. (2005). Diet and diabetic retinopathy: insights from the Diabetes Control and Complications Trial (DCCT). *MedGenMed* **7**, 3.
261. Houtsmuller, A. J., van Hal-Ferwerda, J., Zahn, K. J., and Henkes, H. E. (1980). Influence of different diets on the progression of diabetic retinopathy. *Prog. Food Nutr. Sci.* **4**, 41–46.
262. Howard-Williams, J., Patel, P., Jelfs, R., *et al.* (1985). Polyunsaturated fatty acids and diabetic retinopathy. *Br. J. Ophthalmol.* **69**, 15–18.
263. Toeller, M., Buyken, A. E., Heitkamp, G., Berg, G., and Scherbaum, W. A. (1999). Prevalence of chronic complications, metabolic control and nutritional intake in type 1 diabetes: comparison between different European regions. EURODIAB Complications Study Group. *Horm. Metab Res.* **31**, 680–685.
264. Chandalia, M., Garg, A., Lutjohann, D., von Bergmann, K., Grundy, S. M., and Brinkley, L. J. (2000). Beneficial effects of high dietary fiber intake in patients with type 2 diabetes mellitus. *N. Engl. J. Med.* **342**, 1392–1398.
265. Giugliano, D., Ceriello, A., and Paolisso, G. (1996). Oxidative stress and diabetic vascular complications. *Diabetes Care* **19**, 257–267.
266. van Reyk, D. M., Gillies, M. C., and Davies, M. J. (2003). The retina: oxidative stress and diabetes. *Redox. Rep.* **8**, 187–192.
267. Kowluru, RA, and Chan P. Oxidative stress and diabetic retinopathy 2007. Article ID 43603, 12 pages, 2007. doi:10.115/2007/43603.
268. Kowluru, R. A. (2001). Diabetes-induced elevations in retinal oxidative stress, protein kinase C and nitric oxide are interrelated. *Acta Diabetol.* **38**, 179–185.
269. Gillies, M. C., Su, T., Stayt, J., Simpson, J. M., Naidoo, D., and Salonikas, C. (1997). Effect of high glucose on permeability of retinal capillary endothelium in vitro. *Invest. Ophthalmol. Vis. Sci.* **38**, 635–642.
270. Curcio, F., and Ceriello, A. (1992). Decreased cultured endothelial cell proliferation in high glucose medium is reversed by antioxidants: new insights on the pathophysiological mechanisms of diabetic vascular complications. *In Vitro Cell Dev. Biol.* **28A**, 787–790.
271. Kowluru, R. A., Kern, T. S., and Engerman, R. L. (1997). Abnormalities of retinal metabolism in diabetes or experimental galactosemia. IV. Antioxidant defense system. *Free Radic. Biol. Med.* **22**, 587–592.
272. Kowluru, R. A., Kowluru, V., Xiong, Y., and Ho, Y. S. (2006). Overexpression of mitochondrial superoxide dismutase in mice protects the retina from diabetes-induced oxidative stress. *Free Radic. Biol. Med.* **41**, 1191–1196.
273. Armstrong, D., and al-Awadi, F. (1991). Lipid peroxidation and retinopathy in streptozotocin-induced diabetes. *Free Radic. Biol. Med.* **11**, 433–436.
274. Yatoh, S., Mizutani, M., Yokoo, T., *et al.* (2006). Antioxidants and an inhibitor of advanced glycation ameliorate death of retinal microvascular cells in diabetic retinopathy. *Diabetes Metab. Res. Rev.* **22**, 38–45.
275. Kowluru, R. A., Tang, J., and Kern, T. S. (2001). Abnormalities of retinal metabolism in diabetes and experimental galactosemia. VII. Effect of long-term administration of antioxidants on the development of retinopathy. *Diabetes* **50**, 1938–1942.
276. Nourooz-Zadeh, J., Rahimi, A., Tajaddini-Sarmadi, J., *et al.* (1997). Relationships between plasma measures of oxidative stress and metabolic control in NIDDM. *Diabetologia* **40**, 647–653.
277. Maxwell, S. R., Thomason, H., Sandler, D., *et al.* (1997). Antioxidant status in patients with uncomplicated insulin-dependent and non-insulin-dependent diabetes mellitus. *European J. Clin. Invest.* **27**, 484–490.
278. Sundaram, R. K., Bhaskar, A., Vijayalingam, S., Viswanathan, M., Mohan, R., and Shanmugasundaram, K. R. (1996). Antioxidant status and lipid peroxidation in type II diabetes mellitus with and without complications. *Clin. Sci. (Lond).* **90**, 255–260.
279. Jennings, P. E., Chirico, S., Jones, A. F., Lunec, J., and Barnett, A. H. (1987). Vitamin C metabolites and microangiopathy in diabetes mellitus. *Diabetes Res.* **6**, 151–154.
280. Sinclair, A. J., Girling, A. J., Gray, L., Le Guen, C., Lunec, J., and Barnett, A. H. (1991). Disturbed handling of ascorbic acid in diabetic patients with and without microangiopathy during high dose ascorbate supplementation. *Diabetologia* **34**, 171–175.
281. Sinclair, A. J., Girling, A. J., Gray, L., Lunec, J., and Barnett, A. H. (1992). An investigation of the relationship between free radical activity and vitamin C metabolism in elderly diabetic subjects with retinopathy. *Gerontology* **38**, 268–274.
282. Jain, S. K., McVie, R., Duett, J., and Herbst, J. J. (1989). Erythrocyte membrane lipid peroxidation and glycosylated hemoglobin in diabetes. *Diabetes* **38**, 1539–1543.
283. Davie, S. J., Gould, B. J., and Yudkin, J. S. (1992). Effect of vitamin C on glycosylation of proteins. *Diabetes* **41**, 167–173.
284. Jain, S. K., and Palmer, M. (1997). The effect of oxygen radicals, metabolites and vitamin E on glycosylation of proteins. *Free Radic. Biol. Med.* **22**, 593–596.
285. Will, J. C., and Byers, T. (1996). Does diabetes mellitus increase the requirement for vitamin C? *Nutr. Rev.* **54**, 193–202.
286. Mann, G. V. (1974). Hypothesis: the role of vitamin C in diabetic angiopathy. *Perspect. Biol. Med.* **17**, 210–217.
287. Cunningham, J. J. (1988). Altered vitamin C transport in diabetes mellitus. *Med. Hypotheses* **26**, 263–265.
288. Cox, B. D., and Butterfield, W. J. (1975). Vitamin C supplements and diabetic cutaneous capillary fragility. *Br. Med. J.* **3**, 205.
289. Robinson, W. G. Jr., Laver, N. M., Jacot, J. L, *et al.* (1996). Diabetic-like retinopathy ameliorated with the aldose reductase inhibitor WAY-121,509. *Invest. Ophthalmol. Vis. Sci.* **37**, 1149–1156.
290. Gabbay, K. H. (1973). The sorbitol pathway and the complications of diabetes. *N. Engl. J. Med.* **288**, 831–836.
291. Williamson, J. R., Chang, K., Frangos, M., *et al.* (1993). Hyperglycemic pseudohypoxia and diabetic complications. *Diabetes* **42**, 801–813.
292. Van den Enden, M. K., Nyengaard, J. R., Ostrow, E., Burgan, J. H., and Williamson, J. R. (1995). Elevated glucose levels increase retinal glycolysis and sorbitol pathway

metabolism: implications for diabetic retinopathy. *Invest. Ophthalmol. Vis. Sci.* **36**, 1675–1685.

293. Paolisso, G., D'Amore, A., Galzerano, D., *et al.* (1993). Daily vitamin E supplements improve metabolic control but not insulin secretion in elderly type II diabetic patients. *Diabetes Care* **16**, 1433–1437.
294. Paolisso, G., D'Amore, A., Giugliano, D., Ceriello, A., Varricchio, M., and D'Onofrio, F. (1993). Pharmacologic doses of vitamin E improve insulin action in healthy subjects and non-insulin-dependent diabetic patients. *Am. J. Clin. Nutr.* **57**, 650–656.
295. Paolisso, G., Di Maro, G., Galzerano, D., *et al.* (1994). Pharmacological doses of vitamin E and insulin action in elderly subjects. *Am. J. Clin. Nutr.* **59**, 1291–1296.
296. Ammon, H. P., Klumpp, S., Fuss, A., *et al.* (1989). A possible role of plasma glutathione in glucose-mediated insulin secretion: in vitro and in vivo studies in rats. *Diabetologia* **32**, 797–800.
297. Paolisso, G., Giugliano, D., Pizza, G., *et al.* (1992). Glutathione infusion potentiates glucose-induced insulin secretion in aged patients with impaired glucose tolerance. *Diabetes Care* **15**, 1–7.
298. Xia, P., Inoguchi, T., Kern, T. S., Engerman, R. L., Oates, P. J., and King, G. L. (1994). Characterization of the mechanism for the chronic activation of diacylglycerol-protein kinase C pathway in diabetes and hypergalactosemia. *Diabetes* **43**, 1122–1129.
299. Kunisaki, M., Bursell, S. E., Clermont, A. C., *et al.* (1995). Vitamin E prevents diabetes-induced abnormal retinal blood flow via the diacylglycerol-protein kinase C pathway. *Am. J. Physiol.* **269**, E239–E246.
300. Kunisaki, M., Fumio, U., Nawata, H., and King, G. L. (1996). Vitamin E normalizes diacylglycerol-protein kinase C activation induced by hyperglycemia in rat vascular tissues. *Diabetes* **45**(suppl 3), S117–S119.
301. Ricciarelli, R., Tasinato, A., Clement, S., Ozer, N. K., Boscoboinik, D., and Azzi, A. (1998). alpha-Tocopherol specifically inactivates cellular protein kinase C alpha by changing its phosphorylation state. *Biochem. J.* **334** (Pt 1), 243–249.
302. Lee, I. K., Koya, D., Ishi, H., Kanoh, H., and King, G. L. (1999). d-Alpha-tocopherol prevents the hyperglycemia induced activation of diacylglycerol (DAG)-protein kinase C (PKC) pathway in vascular smooth muscle cell by an increase of DAG kinase activity. *Diabetes Res. Clin. Pract.* **45**, 183–190.
303. King, G. L., Kunisaki, M., Nishio, Y., Inoguchi, T., Shiba, T., and Xia, P. (1996). Biochemical and molecular mechanisms in the development of diabetic vascular complications. *Diabetes* **45** (Suppl 3), S105–S108.
304. Lynch, J. J., Ferro, T. J., Blumenstock, F. A., Brockenauer, A. M., and Malik, A. B. (1990). Increased endothelial albumin permeability mediated by protein kinase C activation. *J. Clin. Invest.* **85**, 1991–1998.
305. Rasmussen, H., Forder, J., Kojima, I., and Scriabine, A. (1984). TPA-induced contraction of isolated rabbit vascular smooth muscle. *Biochem. Biophys. Res. Commun.* **122**, 776–784.
306. Bursell, S. E., Clermont, A. C., Kinsley, B. T., Simonson, D. C., Aiello, L. M., and Wolpert, H. A. (1996). Retinal blood flow changes in patients with insulin-dependent diabetes mellitus and no diabetic retinopathy. *Invest. Ophthalmol. Vis. Sci.* **37**, 886–897.
307. Millen, A. E., Klein, R., Folsom, A. R., Stevens, J., Palta, M., and Mares, J. A. (2004). Relation between intake of vitamins C and E and risk of diabetic retinopathy in the Atherosclerosis Risk in Communities Study. *Am. J. Clin. Nutr.* **79**, 865–873.
308. Sales, C. H., and Pedrosa, Lde F. (2006). Magnesium and diabetes mellitus: their relation. *Clin. Nutr.* **25**, 554–562.
309. Barbagallo, M., and Dominguez, L. J. (2007). Magnesium metabolism in type 2 diabetes mellitus, metabolic syndrome and insulin resistance. *Arch. Biochem. Biophys.* **458**, 40–47.
310. McNair, P., Christiansen, C., Madsbad, S., *et al.* (1978). Hypomagnesemia, a risk factor in diabetic retinopathy. *Diabetes* **27**, 1075–1077.
311. Ceriello, A., Giugliano, D., Dello Russo, P., and Passariello, N. (1982). Hypomagnesemia in relation to diabetic retinopathy. *Diabetes Care* **5**, 558–559.
312. Hatwal, A., Gujral, A. S., Bhatia, R. P., Agrawal, J. K., and Bajpai, H. S. (1989). Association of hypomagnesemia with diabetic retinopathy. *Acta Ophthalmol. (Copenh).* **67**, 714–716.
313. de Valk, H. W., Hardus, P. L., van Rijn, H. J., and Erkelens, D. W. (1999). Plasma magnesium concentration and progression of retinopathy. *Diabetes Care* **22**, 864–865.
314. Mahaba, H. M., Ben Thabet, E. A., el-Ghazali, S., el-Ebiari, H., and Abd el-Aziz, H. F. (2000). Magnesium deficiency and other risk factors for diabetic retinopathy. *J. Egypt Public Health Assoc.* **75**, 323–333.
315. Erasmus, R. T., Olukoga, A. O., Alanamu, R. A., Adewoye, H. O., and Bojuwoye, B. (1989). Plasma magnesium and retinopathy in black African diabetics. *Tropical Geograph. Med.* **41**, 234–237.
316. Walter, R. M. Jr., Uriu-Hare, J. Y., Olin, K.L. *et al.* (1991). Copper, zinc, manganese, and magnesium status and complications of diabetes mellitus. *Diabetes Care* **14**, 1050–1056.
317. Herrmann, W., Herrmann, M., and Obeid, R. (2007). Hyperhomocysteinaemia: a critical review of old and new aspects. *Curr. Drug Metab.* **8**, 17–31.
318. Hoogeveen, E. K., Kostense, P. J., Eysink, P. E., *et al.* (2000). Hyperhomocysteinemia is associated with the presence of retinopathy in type 2 diabetes mellitus: the Hoorn study. *Arch. Intern. Med.* **160**, 2984–2990.
319. Vaccaro, O., Perna, A. F., Mancini, F. P., *et al.* (2000). Plasma homocysteine and microvascular complications in type 1 diabetes. *Nutr. Metab. Cardiovasc. Dis.* **10**, 297–304.
320. Vaccaro, O., Perna, A. F., Mancini, F. P., *et al.* (2000). Plasma homocysteine and its determinants in diabetic retinopathy. *Diabetes Care* **23**, 1026–1027.
321. Agullo-Ortuno, M. T., Albaladejo, M. D., Parra, S., *et al.* (2002). Plasmatic homocysteine concentration and its relationship with complications associated to diabetes mellitus. *Clin. Chim. Acta* **326**, 105–112.
322. Looker, H. C., Fagot-Campagna, A., Gunter, E. W., *et al.* (2003). Homocysteine as a risk factor for nephropathy and retinopathy in Type 2 diabetes. *Diabetologia* **46**, 766–772.
323. Goldstein, M., Leibovitch, I., Yeffimov, I., Gavendo, S., Sela, B. A., and Loewenstein, A. (2004). Hyperhomocysteinemia in

patients with diabetes mellitus with and without diabetic retinopathy. *Eye* **18**, 460–465.

324. Yucel, I., Yucel, G., and Muftuoglu, F. (2004). Plasma homocysteine levels in noninsulin-dependent diabetes mellitus with retinopathy and neovascular glaucoma. *Int. Ophthalmol.* **25**, 201–205.
325. Huang, E. J., Kuo, W. W., Chen, Y. J., *et al.* (2006). Homocysteine and other biochemical parameters in type 2 diabetes mellitus with different diabetic duration or diabetic retinopathy. *Clin. Chim. Acta* **366**, 293–298.
326. Agardh, C. D., Agardh, E., Andersson, A., and Hultberg, B. (1994). Lack of association between plasma homocysteine levels and microangiopathy in type 1 diabetes mellitus. *Scand. J. Clin. Lab. Invest.* **54**, 637–641.
327. Agardh, E., Hultberg, B., and Agardh, C. D. (2000). Severe retinopathy in type 1 diabetic patients is not related to the level of plasma homocysteine. *Scand. J. Clin. Lab. Invest.* **60**, 169–174.
328. Buysschaert, M., Jamart, J., Dramais, A. S., Wallemacq, P., and Hermans, M. P. (2001). Micro- and macrovascular complications and hyperhomocysteinaemia in type 1 diabetic patients. *Diabetes Metab.* **27**, 655–659.
329. Abdella, N. A., Mojiminiyi, O. A., Akanji, A. O., and Moussa, M. A. (2002). Associations of plasma homocysteine concentration in subjects with type 2 diabetes mellitus. *Acta Diabetol.* **39**, 183–190.
330. Saeed, B. O., Nixon, S. J., White, A. J., Summerfield, G. P., Skillen, A. W., and Weaver, J. U. (2004). Fasting homocysteine levels in adults with type 1 diabetes and retinopathy. *Clin. Chim. Acta* **341**, 27–32.
331. Soedamah-Muthu, S. S., Chaturvedi, N., Teerlink, T., Idzior-Walus, B., Fuller, J. H., and Stehouwer, C. D. (2005). Plasma homocysteine and microvascular and macrovascular complications in type 1 diabetes: a cross-sectional nested case-control study. *J. Intern. Med.* **258**, 450–459.
332. Sun, J., Xu, Y., Zhu, Y., *et al.* (2003). The relationship between MTHFR gene polymorphisms, plasma homocysteine levels and diabetic retinopathy in type 2 diabetes mellitus. *Chin. Med. J. (Engl.)* **116**, 145–147.
333. Aksoy, H., Akcay, F., Kurtul, N., Baykal, O., and Avci, B. (2000). Serum 1,25 dihydroxy vitamin D (1,25(OH)2D3), 25 hydroxy vitamin D (25(OH)D) and parathormone levels in diabetic retinopathy. *Clin. Biochem.* **33**, 47–51.
334. Jones, G., Strugnell, S. A., and DeLuca, H. F. (1998). Current understanding of the molecular actions of vitamin D. *Physiol. Rev.* **78**, 1193–1231.
335. Taverna, M. J., Sola, A., Guyot-Argenton, C., *et al.* (2002). Taq I polymorphism of the vitamin D receptor and risk of severe diabetic retinopathy. *Diabetologia* **45**, 436–442.
336. Taverna, M. J., Selam, J. L., and Slama, G. (2005). Association between a protein polymorphism in the start codon of the vitamin D receptor gene and severe diabetic retinopathy in C-peptide-negative type 1 diabetes. *J. Clin. Endocrinol. Metab.* **90**, 4803–4808.
337. Capoluongo, E., Pitocco, D., Concolino, P., *et al.* (2006). Slight association between type 1 diabetes and "ff" VDR FokI genotype in patients from the Italian Lazio region. Lack of association with diabetes complications. *Clin. Biochem.* **39**, 888–892.
338. Cyganek, K., Mirkiewicz-Sieradzka, B., Malecki, M. T., *et al.* (2006). Clinical risk factors and the role of VDR gene polymorphisms in diabetic retinopathy in Polish type 2 diabetes patients. *Acta Diabetol.* **43**, 114–119.
339. Hu, F. B. (2003). Plant-based foods and prevention of cardiovascular disease: an overview. *Am. J. Clin. Nutr.* **78**, 544S–551S.
340. McCullough, M. L., Feskanich, D., Stampfer, M. J., *et al.* (2002). Diet quality and major chronic disease risk in men and women: moving toward improved dietary guidance. *Am. J. Clin. Nutr.* **76**, 1261–1271.
341. Hoffmann, K., Zyriax, B.-C., Boeing, H., and Windler, E. (2004). A dietary pattern derived to explain biomarker variation is strongly associated with the risk of coronary artery disease. *Am. J. Clin. Nutr.* **80**, 633–640.
342. Appel, L. J., Moore, T. J., Obarzanek, E., *et al.* (1997). A clinical trial of the effects of dietary patterns on blood pressure. DASH Collaborative Research Group. *N. Engl. J. Med.* **336**, 1117–1124.
343. Tucker, K. L., Chen, H., Hannan, M. T., *et al.* (2002). Bone mineral density and dietary patterns in older adults: the Framingham Osteoporosis Study. *Am. J. Clin. Nutr.* **76**, 245–252.

CHAPTER **19**

Nutrition Requirements for Athletes

ASKER E. JEUKENDRUP
University of Birmingham, Edgbaston, Birmingham, United Kingdom

Contents

I. INTRODUCTION

The combination of poor diet and poor physical activity has been recognized as the number two cause of death in the United States [1]. In the next few years it is predicted to overtake smoking as the number one killer. The effects of physical activity and diet on health and performance are intrinsically related. For example, when physical activity levels drop (as they have done since the 1950s), energy requirements are reduced, and in order to maintain a healthy body weight, energy intake must be reduced as well. On the other hand when performing regular exercise, hard physical labor, or exercise training, requirements of macro- and micronutrients increase.

The relationship between nutrition and physical performance has fascinated people for a long time. There are reports from ancient Greece that describe special nutrition regimes for athletes preparing for the Olympic Games [2]. It has become clear that different types of exercise and different sports have different energy and nutrient requirements, and therefore food intake must be adjusted accordingly. It has clearly been shown that certain nutritional strategies can enhance performance, improve recovery, and result in more profound training adaptations. More recently, special drinks and energy bars have been developed, and these have been marketed as sports foods. There is also a considerable amount of quackery in sports nutrition, and large numbers of nutrition supplements are on the market that claim to improve performance and recovery. This chapter reviews the nutritional demands of exercise in relation to the physiological demands. It then examines strategies to improve exercise performance. Where possible, practical implications and detailed guidelines have been provided.

II. ENERGY REQUIREMENTS FOR ATHLETES

During exercise, the energy expenditure increases severalfold mostly as a result of skeletal muscle contraction. The thermic effect of exercise (TEE) or energy expenditure for activity (EEA) is by far the most variable component of the daily energy expenditure. It includes all energy expended above the resting energy expenditure (REE), and diet-induced thermogenesis (DIT) and can contribute from virtually nothing to over 80% of energy expenditure (EE). In highly trained, very active individuals, the TEE can amount to up to 32 MJ (8000 kcal) per day. In sedentary people, the thermic effect of exercise may be as low as 400 kJ (100 kcal) per day. Energy expenditure during physical activity ranges from 20 kJ/min (5 kcal/min) for very light activities to up to 100 kJ/min (25 kcal/min) for very high-intensity exercise (Table 1).

In some situations, the energy provision can become critical and continuation of the exercise is dependent on the availability of energy reserves. Most of these reserves must be obtained through nutrition. In endurance athletes, for example, energy depletion (carbohydrate depletion) is one of the most common causes of fatigue. Carbohydrate intake is essential to prevent early fatigue as a result of carbohydrate depletion (see the next section).

A. Energy Balance

The energy balance is usually calculated over longer periods of time (days or weeks) and represents the difference between energy intake and energy expenditure. When the energy intake exceeds the energy expenditure, there is a positive energy balance, which results in weight gain. When the energy intake is below the energy expenditure, there is a negative energy balance and weight loss results. Over the

TABLE 1 Rough Estimation of Energy Expenditure in a Variety of Sports (energy expenditure depends on body mass, the intensity, and the duration of rest periods)

Activity Level	kJ/min	Kcal/min	Examples
Resting	4	1	Sleeping, watching TV
Very light activities	12–20	3–5	Sitting and standing activities, driving, cooking, card playing, cooking, desk work, typing
Light activities	20–28	5–7	Walking (slowly), baseball, bowling, horseback riding, cycling (very slowly), gymnastics, golf
Moderate activities	28–36	7–9	Jogging, cycling (at a moderate pace), basketball, badminton, soccer, tennis, volleyball, brisk walking, swimming (at an easy pace)
Strenuous activities	36–52	9–13	Running (10–13 km/h), cross-country skiing, boxing, cycling (30–35 km/h), swimming, judo
Very strenuous activities	>52	>13	Running (>14 km/h), cycling (>35 km/h)

Source: McArdle, Katch, and Katch (2006). *Exercise Physiology: Energy, Nutrition and Human Performance*, 5th Ed. Philadelphia: Lippincott, Williams & Wilkins.

long term, energy balance is maintained in weight-stable individuals, even though on a day-to-day basis this balance may sometimes be positive and sometimes negative. For individuals who want to lose weight, it is important to increase the energy expenditure relative to the energy intake. In many sports where body composition or body weight is believed to be important (gymnastics, dancing, body building, and weight category sports like judo and boxing), athletes often try to maintain a negative energy balance in order to lose weight [3]. It is known that the energy intakes in these sports can be very low [4, 5].

At the other extreme are the endurance sports in which we can find extremely high energy expenditures and energy intakes. These sports include triathlon, cycling, cross-country skiing, and ultraendurance running. In these sports, it is crucial for performance to maintain energy balance on a day-to-day basis. Both the upper and the lower limits of energy expenditure will be discussed later.

Table 2 presents daily energy intakes from various sports. As is obvious from the data, females generally have lower energy intakes than males. For females, energy intake ranged between 5.1 MJ (1600 kcal) to 10.2 MJ (3200 kcal). For males, energy intake ranged from 12.1 MJ (2900 kcal) to 24.7 MJ (10,500 kcal).

These differences in energy intake may be related to body size and weight, body composition, and the training volume. Team sport athletes have a moderate energy intake, whereas some of the endurance sports have been characterized by very high energy intakes. In fact, energy intakes in excess of 12.6 MJ (3000 kcal) for females and 16.7 MJ (4000 kcal) for males were reported only in endurance sports.

B. Energy Balance in Different Activities

Some physical activities require higher energy outputs than others, as shown in Table 1. Tennis, for example, has relatively low energy expenditure if played recreationally. Nevertheless, during a game the exercise can be intense, and energy expenditure during that short burst of exercise can be very high. However, because this is typically followed by a longer period of relatively low intensity (walking) or even

TABLE 2 Typical Daily (24-hour) Energy Expenditure in a Variety of Sports (m = male, f = female; data are estimations based on a variety of sources)

		MJ	kcal
Ultraendurance running (600-mile race)	m	44.0	10,500
Ironman triathlon	m	37.0	9500
Cross-country skiing	m	36.0	8500
Tour de France (cycling)	m	24.6	6000
Tour de l'Avenir (cycling)	m	23.3	5500
Triathlon	m	19.1	4600
Cycling amateur	m	18.3	4400
Water polo	m	16.6	4000
Swimming	m	16.1	3800
Marathon speed skating	m	16.1	3800
Rowing	m	14.6	3500
Soccer	m	14.3	3400
Body building	m	13.7	3300
Field hockey	m	13.6	3200
Running	m	13.3	3200
Rowing	f	13.0	3101
Judo	m	13.0	3100
Weight lifting	m	12.8	3000
Judo	m	12.2	2900
Cycling amateur	f	10.8	2600
Volley	f	9.2	2200
Hockey	f	9.0	2200
Handball	f	9.0	2100
Running	f	8.8	2100
Subtop swimming	f	8.7	2100
Subtop gymnastics	f	8.2	2000
Top gymnastics	f	7.4	1800
Body building	f	6.2	1500

standing, the average energy expenditure for this activity is relatively low. On the other hand, in continuous sports such as cycling and running, where there is usually no recovery during the activity, energy expenditures can be relatively high.

C. Lower Limits of Energy Expenditure

Female gymnasts, ballet dancers, and ice dancers often have very low energy intakes [4, 5]. Energy intakes have been reported between 6 MJ (1500 kcal) and 8 MJ (2000 kcal). In some cases, this is only 1.2 to 1.4 times the RMR. This is lower than that for sedentary people who on average expend 1.4 to 1.6 times the RMR despite the fact that these athletes may be involved in 3 to 4 hours of training per day. Although not all the time gymnasts and dancers spend in the gym involves high-intensity training, we expect an increased metabolic rate compared to the average sedentary individual. It has been argued that the food records in this group of athletes who are thriving for low body weight may not be accurate and may be an underestimation of the true energy intake. However, even when the reported intakes are corrected for this underestimation, the energy intakes are still very low.

The lower limits of energy expenditure are determined by the sum of RMR, DIT, and a minimum of physical activity. The DIT is directly affected by the amount of food consumed. Reducing food intake decreases the DIT and may also indirectly influence the RMR.

One of the potential problems associated with reducing energy intake to very low levels is marginal nutrition in terms of essential nutrients. Especially the intake of fat-soluble vitamins, calcium, iron, and essential fatty acids may suffer from a maintained low energy intake.

D. Upper Limits of Energy Expenditure

Well-trained endurance athletes can expend more than 4 MJ (1000 kcal) per hour for prolonged periods of time, resulting in very high daily energy expenditures. In some sports, energy expenditures of 16 MJ (4000 kcal) per day or more are common and may be seen on a daily basis. To maintain performance, energy stores need to be replenished and energy balance has to be restored. This means that these athletes have to eat very large amounts in periods of heavy training or competition.

The Tour de France is a 3-week, 20-stage cycling event where the cyclists cover approximately 3500 km, including various mountain stages. On some days, the cyclists spend up to 7 or 8 hours on their bicycles. Average energy expenditure in the Tour de France reached values over 24 MJ (6000 kcal) when measured in weekly intervals. The highest recorded average energy intake during the entire 3 weeks of the Tour de France was 32 MJ (8000 kcal) per day. Thus, maintaining energy balance is not an easy task, even more so because time to eat is limited as the cyclists spend 3 to 7 hours on the bike every day, hunger feelings may be suppressed for several hours after strenuous exercise, and gastrointestinal problems are often reported during the last part of the 3-week race. Nevertheless, there are hardly any changes in body weight during the course of the Tour de France, indicating that these cyclists are indeed able to maintain energy balance [6]. It is also possible that the cyclists who do not maintain their body weight are likely to drop out.

Although it is difficult to eat large amounts during the race, energy intake in the form of carbohydrate solutions has been shown to be crucial. When cyclists during 2 days of high-intensity cycling and 24-MJ (6000-kcal) energy expenditure were not supplemented with carbohydrate during the rides, cyclists were not able to maintain energy balance (5 to 10 MJ [1200 to 2400 kcal] negative energy balance) [7]. When the cyclists were given a 20% carbohydrate solution during exercise from which they could drink as much as they liked, energy balance was maintained.

In some sports, even higher 24-hour energy expenditures have been recorded. Using doubly-labeled water, similar estimates of energy expenditure were made in Norwegian cross-country skiers during training [8]. Energy expenditure amounted to up to 36 MJ (8600 kcal) per day. A case report indicated that an ultraendurance runner expended 134 MJ (55,970 kcal) during a 1000-km (600-mile) running race [9]. The runner covered the distance in 5 days and 5 hours during which he averaged an energy expenditure of 45 MJ (10,750 kcal) per day. Interestingly, in both cases the athletes' food intake was extremely high and almost matched the energy expenditure.

E. Body Composition

Athletes often want to change their body weight or body mass in an attempt to improve performance. Physical characteristics, such as height, body mass, muscle mass, and body fat levels, can all play a role in the performance of sport. Inherited characteristics, as well as the conditioning effects of the athlete's training program and diet, determine an athlete's physique. Often, "ideal" physiques for individual sports are set, based on a rigid set of characteristics of successful athletes. However, this process fails to take into account that there is considerable variability in the physical characteristics of sports people, even among elite athletes in the same sport. Therefore, it is dangerous to establish rigid prescriptions for individuals, particularly with regard to body composition. Instead it is preferable to nominate a range of acceptable values for body fat and body weight within each sport, and then monitor the health and performance of individual athletes within this range. These values may change over the athlete's career.

Many sports dietitians note that losing body weight or, more precisely, losing body fat is the most common reason for an athlete to seek nutrition counseling. A small body size is useful to reduce the energy cost of activity, to improve temperature regulation in hot conditions, and to allow mobility to undertake twists and turns in a confined space. This physique is characteristic of athletes such as gymnasts, divers, ski jumpers, and marathon runners. A low level of body fat enhances the "power-to-weight" ratio over a range of body sizes and is a desirable characteristic of many sports that require weight-bearing movement, particularly against gravity (e.g., distance running, mountain bike and uphill cycling, jumps, and hurdles). However, low body fat levels are also important for aesthetic sports such as diving, gymnastics, and figure skating where judging involves appearance as well as skill.

However, many athletes in weight/fat-conscious sports strive to achieve very low body fat levels or reduce body fat levels to below what seems their "natural" or "healthy" level. Although weight-loss efforts often produce a short-term improvement in performance, this must be balanced against the disadvantages related to having very low body fat stores or following unsafe weight-loss methods. Excessive training, chronic low energy and nutrient intake, and psychological distress are often involved in fat-loss strategies and may cause long-term damage to health, happiness, or performance.

It is suggested that there is a higher risk of eating disorders, or disordered eating behaviors and body perceptions, among athletes in weight-division sports or sports in which success is associated with low body fat levels than might be expected in the general community. Females seem at greater risk than males, reflecting the general dissatisfaction of females in the community with their body shape, as well as the biological predisposition for female athletes to have higher body fat levels than male athletes, despite undertaking the same training program. Even where clinical eating disorders do not exist, many athletes appear to be "restrained eaters," reporting energy intakes that are considerably less than their expected energy requirements. An adequate intake of energy is a prerequisite for many of the goals of sports nutrition.

The "female athlete triad"—the coexistence of disordered eating, disturbed menstrual function, and suboptimal bone density—has received considerable publicity [10–12]. Expert advice from sports medicine professionals, including dietitians, psychologists, and physicians, is important in the early detection and management of problems related to body composition and nutrition.

III. CARBOHYDRATE REQUIREMENTS FOR ATHLETES

Carbohydrate fuel plays a major role in the performance of many types of exercise and sport. The depletion of body carbohydrate stores is a cause of fatigue or performance impairments during exercise, particularly during prolonged (>90 min) sessions of submaximal or intermittent high-intensity activity. This fatigue may be seen both in the muscle (peripheral fatigue) and in the central nervous system (central fatigue) [13, 14]. Unfortunately, total body carbohydrate stores are limited and are often substantially less than the fuel requirements of the training and competition sessions undertaken by athletes. Therefore, strategies for athletes include consuming carbohydrate before, during, and in the recovery period between prolonged exercise bouts.

Many studies show that exercise is improved by strategies that enhance or maintain carbohydrate status during exercise [15–17] (for detailed reviews see [18–20]). Studies have provided preliminary evidence that there are other benefits from high-carbohydrate eating strategies apart from increasing fuel stores for exercise. For example, a growing number of investigations have reported performance benefits when carbohydrate is consumed before and during high-intensity exercise of ~1 hour [21–25]. In these situations, the athlete's body carbohydrate stores should already be sufficient to fuel the event, so it is not clear why additional carbohydrate would provide an advantage. It is possible that carbohydrate intake improves the function of the brain and central nervous system, reducing the perception of effort during the exercise task. In line with this, infusion of glucose did not affect performance [26] but a mouth rinse with a carbohydrate solution (without swallowing the solution) [27] did.

Finally, carbohydrate intake during and after exercise appears to assist the immune response to exercise [28–31]. Cellular immune parameters are often reduced or compromised after a prolonged workout. However, some studies have shown that carbohydrate intake can decrease or prevent this outcome. Whether this actually leads to an improvement in the immune status and health of athletes—for example, fewer sick days—remains to be seen and would require a sophisticated long-term study.

A. Fuelling up before Exercise

Carbohydrate stores in the muscle and liver should be maximized before exercise, particularly in the competition setting where the athlete wants to perform at his or her best. The key factors in glycogen storage are dietary carbohydrate intake and, in the case of muscle stores, tapered exercise or rest. In the absence of muscle damage, muscle glycogen stores can be returned to normal resting levels (to 350 to 500 mmol/kg dry weight muscle) with 24 to 36 hours of rest and an adequate carbohydrate intake (7 to 10 g per kilogram body weight per day). Normalized stores appear adequate for the fuel needs of events of less than 60 to 90 minutes in duration—for example, a soccer game, a half-marathon, or a basketball game.

Carbohydrate loading is a special practice that aims to maximize or "supercompensate" muscle glycogen stores up to twice the normal resting level (e.g., ~500 to 900 mmol/kg dry weight). The first protocol was devised in the late 1960s by Scandinavian exercise physiologists who found, using the muscle biopsy technique, that the size of preexercise muscle glycogen stores affected submaximal exercise capacity. Several days of a low-carbohydrate diet resulted in depleted muscle glycogen stores and reduced endurance capacity compared with a mixed diet. However, high carbohydrate intake for several days caused a "supercompensation" of muscle glycogen stores and prolonged the cycling time to exhaustion. These pioneering studies produced the "classical" 7-day model of carbohydrate loading. This model consists of a 3- to 4-day "depletion" phase of hard training and low carbohydrate intake, finishing with a 3- to 4-day "loading" phase of high carbohydrate intake and exercise taper (i.e., decreased amounts of training) [32]. Early field studies of prolonged running events showed that carbohydrate loading enhanced performance, not by allowing the athlete to run faster but by prolonging the time that the athlete can maintain the race pace.

Further studies undertaken on trained subjects have produced a "modified" carbohydrate loading strategy [33]. The muscle of well-trained athletes has been found to be able to supercompensate its glycogen stores without a prior depletion or "glycogen stripping" phase. For well-trained athletes at least, carbohydrate loading may be seen as an extension of "fuelling up"—involving rest/taper and high carbohydrate intake over 3 to 4 days. The modified carbohydrate loading protocol offers a more practical strategy for competition preparation by avoiding the fatigue and complexity of the extreme diet and training protocols associated with the previous depletion phase. Typically, carbohydrate loading postpones fatigue and extends the duration of steady-state exercise by ~20% and improves performance over a set distance or workload by 2% to 3% [34].

B. Preevent Meal

Food and fluids consumed in the 4 hours before an event may help to continue to fill muscle glycogen stores if they have not fully restored after the last exercise session, to restore liver glycogen levels, especially for events undertaken in the morning where liver stores are depleted from an overnight fast, to ensure that the athlete is well hydrated, to prevent hunger feelings, and to include foods and practices that are important to the athlete's psychology or superstitions.

Consuming carbohydrate-rich foods and drinks in the preevent meal is especially important in situations where body carbohydrate stores have not been fully recovered or where the event is of sufficient duration and intensity to deplete these stores. The intake of a substantial amount of carbohydrate (~200–300 g) in the 3 to 4 hours before exercise has been shown to enhance various measures of exercise performance compared to performance undertaken after an overnight fast [35–37].

However, it has been suggested that carbohydrate intake before exercise may have *negative* consequences for performance, especially when it is consumed in the hour before exercise. Carbohydrate intake causes a rise in plasma insulin concentrations, which in turn suppresses the availability and oxidation of fat as an exercise fuel. The final result is an increased reliance on carbohydrate oxidation at the onset of exercise, leading to faster depletion of muscle glycogen stores and a decline in plasma glucose concentration (rebound hypoglycemia) [38]. There has been considerable publicity surrounding one study from the 1970s, which found that subjects performed *worse* after consuming carbohydrate in the hour before exercise than when they cycled without consuming anything [39]. This has led to warnings that carbohydrate should not be consumed in the hour before exercise. However, a far greater number of studies have shown that any metabolic disturbances following preexercise carbohydrate feedings are short lived or unimportant [40]. These studies show that carbohydrate intake in the hour before exercise is associated with a neutral effect or a beneficial performance outcome [41–43].

Nevertheless, there is a small subgroup of athletes who experience a true fatigue, associated with a decline in blood glucose levels [44], if they start to exercise within the hour after consuming a carbohydrate snack. This problem can be avoided or diminished by a number of dietary strategies:

- Consume carbohydrate 5 to 10 minutes before the start of the exercise or incorporate this into a warmup. By the time insulin starts to rise, the exercise has already started and insulin release will be suppressed by catecholamines.
- Consume a substantial amount of carbohydrate (more than 75 g) rather than a small amount, so that the additional carbohydrate more than compensates for the increased rate of carbohydrate oxidation during the exercise.
- Choose a carbohydrate-rich food or drink that produces a low glycemic index response (that is, a low blood glucose and insulin response) rather than a carbohydrate source that produces a high glycemic index (that is, a large and rapid blood glucose and insulin response).
- Consume carbohydrate throughout the exercise session.

The type, timing, and quantity of preevent meals should be chosen according to the athlete's individual circumstances and experiences. Foods with a low-fat, low-fiber, and low–moderate protein content are the preferred choice for the preevent menu because they are less likely to cause gastrointestinal upsets.

C. Carbohydrate Intake during Exercise

Numerous studies show that the intake of carbohydrate during prolonged sessions of moderate-intensity or intermittent high-intensity exercise can improve endurance (i.e., prolong time to exhaustion) and performance. Although there is some evidence that increasing carbohydrate availability causes glycogen sparing in slow-twitch muscle fibers during running [45], the major mechanisms to explain the benefits of carbohydrate feedings during prolonged exercise are the maintenance of plasma glucose concentration (sustaining brain function) and the provision of an additional carbohydrate supply to allow the muscle to continue high rates of carbohydrate oxidation [20, 46].

Most carbohydrates (glucose, sucrose, maltose, maltodextrins, amylopectin) are oxidized at relatively high rates, whereas other carbohydrates (i.e., fructose, galactose, trehalose) are oxidized at slightly lower rates [47]. Carbohydrate consumed during exercise is oxidized in small amounts during the first hour of exercise (~20 g) and thereafter reaches a peak rate of ~ 60 g/h. Even ingestion of very large amounts of carbohydrate will not result in higher oxidation rates [19, 20]. In general a carbohydrate intake of 60 g/h is recommended with carbohydrate feedings starting well in advance of fatigue/depletion of body carbohydrate stores. Ingestion of more than 60 g/h will not have an additive effect and may even cause gastrointestinal distress.

It was discovered that exogenous carbohydrate oxidation is most likely limited by gastrointestinal absorption (for a review and detailed discussions, see [20]). Thus, strategies to improve absorption could provide more energy to the working muscle. It is likely feeding of a single carbohydrate source (e.g., glucose or maltodextrin) at high rates results in sodium-dependent glucose transporters (SGLT1) becoming saturated. Once these transporters are saturated, feeding more of that carbohydrate will not result in greater absorption and increased oxidation rates. Intestinal perfusion studies have suggested that the ingestion of carbohydrates that use different transporters might increase total carbohydrate absorption [48] (Fig. 1). In dual tracer studies, the oxidation of glucose and fructose mixtures during exercise was investigated. In the first study, subjects ingested a drink containing glucose and fructose. Glucose was ingested at a rate of 1.2 g/min and fructose at a rate of 0.8 g/min. In the control trials, the subjects ingested glucose at a rate of 1.2 g/min and 1.8 g/min (matching glucose intake or energy intake). It was found that the ingestion of glucose at a rate of 1.2 g/min resulted in oxidation rates around 0.8 g/min. Ingesting more glucose did not increase the oxidation. However, ingesting glucose + fructose increased the total exogenous carbohydrate oxidation rate to 1.26 g/min, an increase in oxidation of 45% compared with a similar amount of glucose!

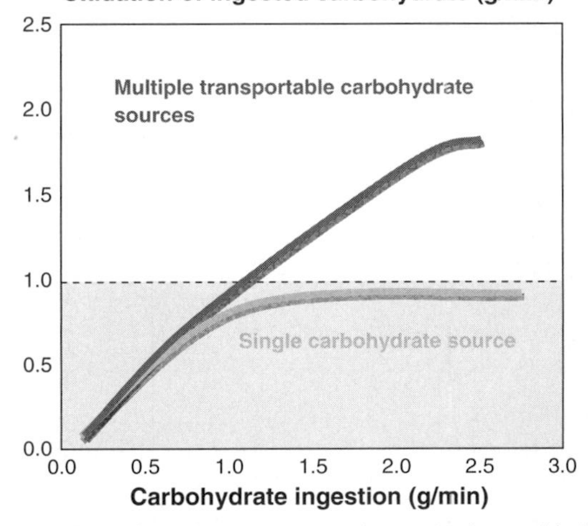

FIGURE 1 Schematic representation of the oxidation of ingested carbohydrate as a function of intake (both are averaged and expressed in grams per minute). The data are based on a large number of studies reviewed in Jeukendrup [20]. As the ingestion rate increases, the oxidation of the carbohydrate increases. However, if only a single carbohydrate is ingested, oxidation rates seem to level off at 1 g/min. By selecting multiple transportable carbohydrates and ingesting these at high rates (>1.2 g/min), exogenous carbohydrate oxidation can be increased significantly compared with a single carbohydrate.

In the following years, different combinations and different amounts of carbohydrates were tested in an attempt to find out what the maximal contribution could be of exogenous carbohydrate [49–57]. It was observed that very high oxidation rates were reached with combinations of glucose + fructose, maltodextrin + fructose, and glucose + sucrose + fructose. The highest rates were observed with a mixture of glucose and fructose ingested at a rate of 2.4 g/min. With this feeding regimen, exogenous CHO oxidation peaked at 1.75 g/min (see Fig. 1)! This is 75% higher than what was previously thought to be the absolute maximum.

The increased oxidation of ingested carbohydrate has been suggested to be beneficial, but concrete evidence for this has not yet been published. From a laboratory study in which subjects cycled for 5 hours with water, glucose, or glucose + fructose, there are some indications that drinks with multiple carbohydrates could improve performance [56]. In this study, carbohydrate was ingested at a rate of 90 g/h. The first indication of improved performance was that ratings of perceived exertion (RPE) tended to be lower with glucose + fructose compared with glucose, which in turn was lower than water placebo. In fact, with water not all participants were able to complete the 5 hours at 50%. In addition, the self-selected cadence dropped significantly with water, which is generally seen as an indication of developing fatigue. With glucose this was somewhat prevented, but with glucose + fructose cadence was highest and remained almost unchanged from the beginning of exercise [56]. One study confirmed the beneficial effects of glucose + fructose drinks compared with glucose on prolonged exercise performance (Currell *et al.*, Vol. 40, No. 2).

Another important observation is that the oxidation efficiency of drinks containing carbohydrates that use different transporters for intestinal absorption was higher than it was for drinks with a single carbohydrate. The amount of carbohydrate remaining in the intestine is therefore smaller, and osmotic shifts and malabsorption may be reduced. This likely means that drinks with multiple transportable carbohydrates are less likely to cause gastrointestinal distress. Interestingly, this is a consistent finding in studies that have attempted to register gastrointestinal discomfort during exercise [49–57]. Subjects tended to feel less bloated with the glucose + fructose drinks versus glucose drinks. A larger-scale study into the effects of drinks with different types of carbohydrates on gastrointestinal discomfort has not yet been conducted.

In the world of sport, athletes consume carbohydrate during exercise using a variety of foods and drinks and a variety of feeding schedules. Sports drinks (commercial solutions providing 4% to 8% carbohydrate—4 to 8 g carbohydrate/100 ml—electrolytes and palatable flavors) are particularly valuable because these allow athletes to replace their fluid and carbohydrate needs simultaneously. Each sport or exercise activity offers particular opportunities for fluid and carbohydrate to be consumed throughout the session, whether from aid stations, supplies carried by the athletes, or at formal stoppages in play such as time-outs or half-time breaks. Athletes should be creative in making use of these opportunities.

D. Postexercise Refueling

Restoration of muscle glycogen concentrations is an important component of postexercise recovery and is challenging for athletes who train or compete more than once each day. The main dietary issue in glycogen synthesis is the amount of carbohydrate consumed. The optimal carbohydrate intake for glycogen storage is 7 to 10 g/kg body weight/day [58]. However, there is some evidence that moderate and high glycemic index (GI) carbohydrate-rich foods and drinks may be more favorable for glycogen storage than some low GI food choices [58]. Glycogen storage may occur at a slightly faster rate during the first couple of hours after exercise; however, the main reason for encouraging an athlete to consume CHO-rich meals or snacks soon after exercise is that effective refueling does not start until a substantial amount of carbohydrate (~1 g/kg body weight) is consumed. When there is limited time between workouts or events (e.g., 1 hour or less) it makes sense to turn every minute into effective recovery time by consuming carbohydrate as soon as possible after the first session. However, when recovery time is longer, immediate carbohydrate intake after exercise is unnecessary and the athlete can afford to follow his or her preferred and practical eating schedule as long as goals for total carbohydrate intake are met over the day.

Certain amino acids have a potent effect on the secretion of insulin, which is a stimulator of glycogen resynthesis [59]. For this reason, the effects of adding amino acids and proteins to a carbohydrate solution have been investigated. One study compared glycogen resynthesis rates after ingestion of carbohydrate, protein, or carbohydrate plus protein [60]. As expected, very little glycogen was stored when protein alone is ingested, and glycogen storage increased when carbohydrate is ingested. But most interesting, glycogen storage further increased when carbohydrate was ingested together with protein. However, other studies have shown that if the amount of ingested carbohydrate is high, addition of protein or amino acids has no further effect [61–63]. Nevertheless, recovery goals also include attention to the immune system, muscle building, and injury repair. Therefore, it may be useful to eat nutrient-rich forms of carbohydrate foods and drinks during the recovery period to provide a range of valuable nutrients.

E. Everyday Eating for Recovery

Although strategies to promote carbohydrate availability have been shown to enhance performance and recovery after a *single* bout of exercise, it has been difficult to demonstrate that a *long-term* period of high carbohydrate eating will promote better training adaptations and long-term performance than a moderate carbohydrate diet. Theoretically, inadequate carbohydrate intake during repeated days of exercise leads to a gradual depletion of muscle glycogen stores and impairment of exercise endurance. More recently, investigators have suggested that training in a low glycogen state may have some beneficial effects on the adaptation of the muscle. This is further discussed in Section VII on nutrition and training adaptations.

Dietary guidelines for the general population make recommendations for carbohydrate intake as a percentage of dietary energy (for e.g., to increase carbohydrate to greater than 55% of total energy intake). Some sports nutrition experts have followed this terminology. However, athletes undertaking strenuous exercise have carbohydrate requirements based principally on muscle fuel needs, which are quantifiable according to the size of the athlete and the duration of the exercise program. Therefore, it makes sense, and is more practical, to be consistent, in describing carbohydrate goals in terms of grams per kg of the athlete's body mass. This allows athletes to quickly work out their carbohydrate requirement for a given situation—for example, 7 g/kg for a 70-kg athlete = 490 g. Meals and menus can then be constructed using information on food labels or in food composition tables to achieve this carbohydrate target.

General eating patterns that help to achieve a high carbohydrate intake are summarized later.

IV. PROTEIN REQUIREMENTS FOR ATHLETES

There is still considerable debate about how much dietary protein is required for optimal athletic performance. The interest in protein (meat) probably dates back to ancient Greece. There are reports that athletes in ancient Greece, in preparation of Olympic games, consumed large amounts of meat. This belief stems partly from the fact that muscle contains a large proportion of the total protein in a human body (about 40%). Muscle also accounts for 30% to 50% of all protein turnover in the body. Both the structural proteins that make up the myofibrillar proteins and the proteins that act as enzymes within a muscle cell change as an adaptation to exercise training. Indeed, muscle mass, muscle protein composition, and muscle protein content will change in response to training. Therefore, it is not surprising that meat has been popular as a protein source for athletes (especially strength athletes).

In muscle the majority of amino acids are incorporated into tissue proteins, with a small pool of free amino acids. This pool undergoes turnover receiving free amino acids from the breakdown of protein and contributing amino acids for protein synthesis. Protein breakdown in skeletal muscle serves two main purposes:

1. To provide essential amino acids when individual amino acids are converted to acetyl CO A or TCA cycle intermediates
2. To provide individual amino acids that can be used elsewhere in the body for the synthesis of neurotransmitters, hormones, glucose, and proteins

Clearly, if protein degradation rates are greater than the rates of synthesis, there will be a reduction of protein content; conversely, muscle protein content can only increase if the rate of synthesis exceeds that of degradation.

A. Increased Protein Requirements

Exercise (especially endurance exercise) results in increased oxidation of the branched chain amino acids (BCAA), which are essential amino acids and cannot be synthesized within the body. Therefore, increased oxidation would imply that the dietary protein requirements are increased. Some studies in which the nitrogen balance technique was used showed that the dietary protein requirements for athletes involved in prolonged endurance training were higher than those for sedentary individuals. Whether requirements are really higher remains somewhat controversial though (for review, see [64, 65]).

It has been estimated that protein may contribute up to about 15% to energy expenditure in resting conditions. During exercise, this relative contribution is likely to decrease because energy expenditure is increased and most of this energy is provided by carbohydrate and fat. During very prolonged exercise when carbohydrate availability becomes limited, the contribution of protein to energy expenditure may amount to about 10% of total energy expenditure. Thus, although protein oxidation is increased during endurance exercise, the relative contribution of protein to energy expenditure remains small. Protein requirements may increase somewhat, but this increased need may be met easily by a moderate protein intake. The research groups that advocate an increased protein intake for endurance athletes usually recommend a daily intake of 1.2 to 1.8 g/kg body mass. This is about twice the level of protein intake that is recommended for sedentary populations.

There are reports of increased protein breakdown after resistance exercise. The suggested increased dietary protein requirements with resistance training are related to increased muscle bulk (hypertrophy) rather than increased oxidation of amino acids. Muscle protein breakdown increases after resistance training, but to a smaller degree than muscle protein synthesis. The elevations in protein degradation and synthesis are transient. Protein breakdown and synthesis after exercise are elevated at 3 and 24 hours after exercise but return to baseline levels after 48 hours. These results seem to apply to resistance exercise and high-intensity dynamic exercise.

There is controversy as to whether strength athletes really need to eat large amounts of protein. The nitrogen balance studies conducted on such athletes have been criticized because they generally have been of short duration and a steady-state situation may not be established [65]. The recommendation for protein intakes for strength athletes is therefore generally 1.6 to 1.7 g/kg body mass per day. Again, this seems to be met easily with a normal diet, and no extra attention to protein intake is needed. Protein supplements are often used but are not necessary to meet the recommended protein intake. There is also no evidence that supplements would be more effective than normal foods.

B. Amino Acid versus Protein Intake

In the past, the amino acid needs of the body were primarily met by ingestion of whole proteins in the diet. However, the supplementation of individual amino acids has become increasingly popular. This is the result of technological advances that have made it possible to manufacture food-grade ultrapure amino acids, but it also reflects the general interest in the pharmacological and metabolic interactions of free-form amino acids in various areas of clinical nutrition. Here, individual amino acids are used to reduce nitrogen losses and improve organ functions in traumatized and critically ill patients. The results of these studies have been applied to populations of athletes and healthy individuals, where intake of separate amino acids is claimed to improve exercise performance, stimulate hormone release and improve immune function, among a variety of other positive effects.

However, amino acid metabolism is complex. One amino acid can be converted into another, and amino acids may influence nerve impulse transmission as well as hormone secretion. Composition of specific amino acid mixtures or even high protein diets may lead to nutritional imbalances because overload with one amino acid may reduce the absorption of other amino acids.

V. MICRONUTRIENT REQUIREMENTS FOR ATHLETES

The daily requirement for at least some vitamins and minerals is increased beyond population levels in people undertaking a strenuous exercise program. The potential reasons for this increased requirement are increased loss through sweat, urine, and perhaps feces and through increased production of free radicals. Unfortunately, at present we are unable to quantify the additional micronutrient requirements of athletes. Key factors ensuring an adequate intake of vitamins and minerals are a moderate to high-energy intake and a varied diet based on nutritious foods. Dietary surveys show that most athletes report dietary practices that easily supply vitamins and minerals in excess of nutrient requirements and are likely to meet any increases in micronutrient demand caused by training. However, not all athletes eat varied diets of adequate energy intake and may need help to improve both the quality and quantity of their food selections.

Studies of the micronutrient status of athletes have not revealed any significant differences between indices in athletes and sedentary controls [66]. The results suggest that athletic training per se does not lead to micronutrient deficiency. These data should, however, be interpreted carefully since most indices are not sensitive enough to detect marginal deficiencies. Overall, generalized vitamin and mineral supplementation for all athletes is not justified. Furthermore, studies do not support an increase in performance with such supplementation except in the case where a preexisting deficiency was corrected.

The best management for athletes with a high risk of suboptimal intake of micronutrients is to provide nutrition education to improve their food intake. However, a low-dose, broad-range multivitamin/mineral supplement may be useful when the athlete is unwilling or unable to make dietary changes or when the athlete is traveling to places with an uncertain food supply and eating schedule.

A. Antioxidant Vitamins

Exercise has been linked with an increased production of free oxygen radical species capable of causing cellular damage. A sudden increase in training stress (such as an increase in volume or intensity) or a stressful environment (training in hot conditions or at altitude) is believed to increase the production of these reactive oxygen species leading to an increase in markers of cellular damage. Supplementation with antioxidant vitamins such as vitamin C or vitamin E is often suggested to increase antioxidant status and provide protection against this damage.

However, the literature on the effects of antioxidant supplementation on antioxidant status, cellular damage, and performance is complex and confusing. Some, but not all, studies show that acute supplementation during periods of increased stress may provide bridging protection until the athlete is able to adapt his or her own antioxidant status to meet this stress. It is possible that subtle benefits occur at cellular levels, which are too small to translate into detectable performance benefits. Whether ongoing supplementation is necessary or even desirable for optimal training adaptations and the competition performance of athletes is also unknown. It has been suggested that the increase in free radical production during a period of nondamaging exercise may act as signal for adaptation, and therefore ingesting excessive amounts of antioxidant vitamins may actually interfere with the adaptation [67].

B. Iron

Minerals are the micronutrients at most risk of inadequate intake in the diets of athletes. Inadequate iron status can reduce exercise performance via suboptimal levels of hemoglobin and perhaps iron-related muscle enzymes. Reductions in the hemoglobin levels of distance runners first alerted sports scientists to the issue of iron status of athletes. However, more recent research has raised the problem of distinguishing true iron deficiency from alterations in iron status measures that are caused by exercise itself. Low iron status in athletes is overdiagnosed from single measures of low hemoglobin and ferritin levels. A major problem is the failure to recognize that the increase in blood volume that accompanies training causes a dilution of all the blood contents. This hemodilution, often termed "sports anemia," does not impair exercise performance.

Nevertheless, some athletes are at true risk of becoming iron deficient. The causes are essentially the same as for members of the general community: a lower than desirable intake of bioavailable iron or increased iron requirements or losses. Iron requirements may be increased in some athletes because of growth needs, or to increased losses of blood and red blood cell destruction. However, the most common risk factor among athletes is a low-energy or low-iron diet, with females, "restricted" eaters, vegetarians, and athletes eating high-carbohydrate/low-meat diets being likely targets.

Iron is found in a range of plant and animal food sources in two forms. Heme iron is found only in flesh- or blood-containing animal foods; organic iron is found both in animal foods and plant foods. Whereas heme iron is relatively well absorbed from single foods and mixed meals (15% to 35% bioavailability), the absorption of nonheme iron from single plant sources is low and variable (2% to 8%). The bioavailability of nonheme iron and, to a lesser extent, heme iron is affected by other foods consumed in the same meal. Factors that enhance iron absorption include vitamin C, peptides from fish/meat/chicken, alcohol, and food acids, whereas factors that inhibit absorption include phytates, polyphenols, calcium, and peptides from plant sources such as soy protein. The absorption of both heme and nonheme iron is increased as an adaptive response in people who are iron deficient or have increased iron requirements. Although the iron bioavailability studies from which these observations have been made have not been undertaken on special groups such as athletes, it is generally assumed that the results can be applied across populations of healthy people.

The assessment of total dietary iron intake of athletes is not necessarily a good predictor of their iron status; the mixing and matching of foods at meals plays an important role by determining the bioavailability of dietary iron intake. For example, in two groups of female runners who reported similar intakes of total dietary iron, the group who reported regular intake of meat was estimated to have a greater intake of absorbable iron and showed higher iron status than a matched group of runners who were semivegetarian.

Low iron status, indicated by serum ferritin levels lower than 20 ng/ml, should be considered for further assessment and treatment. Present evidence does not support the theory that low iron status without anemia reduces exercise performance. However, many athletes with such low iron stores, or a sudden drop in iron status, frequently complain of fatigue and inability to recover after heavy training. Many of these respond to strategies that improve iron status or prevent a further decrease in iron stores.

A sports medicine expert should evaluate and manage iron status on an individual basis. Prevention and treatment of iron deficiency may include iron supplementation, with a recommended therapeutic dose of 100 mg/day of elemental iron for 2 to 3 months. However, the management plan should include dietary counseling to increase the intake of bioavailable iron and appropriate strategies to reduce any unwarranted iron loss. Many athletes self-prescribe iron supplements; indeed, mass supplementation of athletes with iron has been fashionable at various times. However, these practices do not provide the athlete with the opportunity for adequate assessment of iron losses and expert dietary counseling from a sports dietitian. Dietary guidelines for increasing iron intake should be integrated with the athlete's other nutritional goals such as a need for high carbohydrate intake or reduced energy intake.

C. Calcium

Weight-bearing exercise is considered to be one of the best protectors of bone health. Therefore, it is puzzling to find reports of low body density in some female athletes, notably distance runners. However, a serious outcome of menstrual disturbances in female athletes is the high risk of either direct loss of bone density or failure to optimize the gaining of peak bone mass during early adulthood. Individually or in combination, these problems, referred to as the female athlete triad (disordered eating, menstrual dysfunction, and reduced bone status), can directly impair athletic performance. Significantly, they reduce the athlete's career span by increasing her risk of illness and injury, including stress fractures. Long-term problems may include an increased risk of osteoporosis in later life.

Optimal nutrition is important to correct factors that underpin the menstrual dysfunction, as well as those that contribute to suboptimal bone density. Adequate energy intake and the reversal of disordered eating or inadequate nutrient intake are important. A team approach involving sports physician, sports dietitian, psychologist and/or psychiatrist, coach, and family may be needed to treat the athlete with disordered eating or eating disorders.

Adequate calcium intake is important for bone health, and requirements may be increased to 1200 to 1500 mg/day in athletes with impaired menstrual function. Again, strategies to meet calcium needs must be integrated into the total nutrition goals of the athlete. Where adequate calcium intake cannot be met through dietary means, usually through use of low-fat dairy foods or calcium-enriched soy alternatives, a calcium supplement may be considered.

VI. FLUID REQUIREMENTS FOR ATHLETES

Water has many important functions in the human body. The total water content of the human body is between 30 and 50 liters, which is approximately 55% to 60% of our body mass. Each day, water is excreted in the form of sweat, urine, and evaporative losses and matched by usual water intake. Water turnover can be very high in some conditions, but the total body water content is remarkably constant and rarely exceeds variations of 1 liter. To maintain fluid balance, water intake may vary from 1 liter to up to about 12 liters per day. However, during exercise and especially during exercise in hot conditions, sweat rates (and thus water losses) may increase dramatically and dehydration may occur (i.e., the body is in negative fluid balance). Dehydration can have an enormous impact on physical and mental function and increased risk of heat illness. Even mild dehydration can result in reduced exercise capacity.

Exercise (muscle contraction) causes an increase in heat production in the body. Muscle contraction during most activities is only about 15% to 20% efficient. This means that of all the energy produced, only about 15% to 20% is used for the actual movement and the remainder is lost as heat. For every liter of oxygen consumed during exercise, approximately 16 kJ (4 kcal) of heat is produced and only 4 kJ (1 kcal) is actually used to perform mechanical work. If this heat were not dissipated, we would soon overheat.

When a well-trained individual is exercising at 80% to 90% VO_2max, the body's heat production may be more than 1000 W (i.e., 3.6 MJ or 900 kcal per hour). This could potentially cause body core temperature to increase by 1°C every 5 to 8 minutes if no heat could be dissipated. As a result, body core body temperature could approach dangerous levels in less than 20 minutes.

There are several mechanisms to dissipate this heat and to maintain body core temperature in a relatively narrow range: 36 to 38°C in resting conditions and 38 to 40°C during exercise and hot conditions. The most important cooling mechanism of the body is sweating, although radiation and convection can also contribute. Sweat must evaporate from the body surface in order to exert a cooling effect. Evaporation of 1 liter of water from the skin will remove 2.4 MJ of heat from the body. Although sweating is an effective way to dissipate heat, it may cause dehydration if sweat losses are not replenished. This may cause further problems for the athlete. Progressive dehydration impairs the ability to sweat and therefore to regulate body temperature. Body temperature rises faster in the dehydrated state and is commonly accompanied by a higher heart rate during exercise. The most dramatic consequence of dehydration-induced hyperthermia during exercise is a 25% to 30% reduction in stroke volume that is not generally met with a proportional increase in heart rate; this results in a decline in cardiac output and in arterial blood pressure [68].

Fluid losses are mainly dependent on three factors:

1. Ambient environmental conditions (temperature, humidity)
2. Exercise intensity
3. Duration of exercise and duration of the heat exposure

The environmental heat stress is determined by the ambient temperature, relative humidity, wind velocity, and solar radiation. The relative humidity is the most important of these factors, because a high humidity severely compromises the evaporative loss of sweat. Often sweat will drip off the skin in such conditions, rather than evaporate. This means that heat loss via this route will be less effective.

It is important to note that problems of hyperthermia and heat injury are not restricted to prolonged exercise in a hot environment. Heat production is directly proportional to exercise intensity, so that very strenuous exercise—even in a cool environment—can cause a substantial rise in body temperature.

To maintain water balance, fluid intake must compensate for the fluid loss that occurs during exercise. Fluid intake is usually dependent on thirst feelings, but thirst (or the lack of thirst) can also be overridden by conscious control. It is important to note, however, that thirst is a poor indicator of fluid requirements or the degree of dehydration. In general, the sensation of feeling thirsty is not perceived until a person has lost at least 2% of body mass. As already mentioned, even this mild degree of dehydration is sufficient to impair exercise performance [69]. It has also been shown that athletes tend to drink too little even when sufficient fluid is available. Runners, for instance, seldom drink more than 0.5 L/h, although sweat rates can be much higher than this.

A. Effects of Dehydration

As the body becomes progressively dehydrated, a reduction in skin blood flow and sweat rate may occur. A high humidity may limit evaporative sweat loss, which leads to further rises in core temperature, resulting in fatigue and possible heat injury to body tissues. The latter is potentially fatal.

1. Effect of Dehydration on Exercise Performance

Several studies have shown that mild dehydration, equivalent to the loss of only 2% body weight, is sufficient to significantly impair exercise performance [69, 70]. In addition, it is often reported that greater losses result in greater reductions in performance. Even very low-intensity exercise (i.e., walking) is affected by dehydration. The capacity to perform high-intensity exercise that results in exhaustion within only a few minutes has been shown to be reduced by as much as 45% by prior dehydration (2.5% of body weight) [71]. Although there is little opportunity for sweat loss during such short-duration, high-intensity events, athletes who travel to compete in hot climates are likely to experience acute dehydration, which can persist for several days and can be of sufficient magnitude to have a detrimental effect on performance in competition. Although dehydration has detrimental effects especially on performance in hot conditions, such effects can also be observed in cool conditions. Both decreases in maximal aerobic power (VO_2max) and decreases in endurance capacity have been reported with dehydration in temperate conditions [72], although not all studies found such an effect [73, 74].

There are several reasons why dehydration decreases exercise performance. First of all, a fall in plasma volume, a decreased blood volume, and increased blood viscosity and a lower central venous pressure can result in a reduced stroke volume and maximal cardiac output. In addition, during exercise in the heat, the dilation of the skin blood vessels reduces the proportion of the cardiac output that is devoted to perfusion of the working muscles. Dehydration also impairs the ability of the body to lose heat. Both sweat rate and skin blood flow are lower at the same core temperature for the dehydrated compared with the euhydrated state. This means that body temperature rises faster during exercise when the body is dehydrated. Finally, the larger rise in core temperature during exercise in the dehydrated state is associated with an increased rate of muscle glycogen breakdown. Depletion of these stores could also result in premature fatigue in prolonged exercise. In addition to the effects of dehydration on endurance, there are reported negative effects on coordination and cognitive functioning. This is likely to impact all sports where skill and decision making are involved.

2. Heat Illness

Dehydration also poses a serious health risk in that it increases the risk of cramps, heat exhaustion, and life-threatening heat stroke. Early symptoms of heat injury are excessive sweating, headache, nausea, dizziness, a reduced consciousness, and mental function. When the core temperature rises to over 40°C, heat stroke may develop, characterized by hot dry skin, confusion, and loss of consciousness. There are several anecdotal reports of athletes and army recruits who died because of heat stroke. Most of these deaths have been explained by exercise in hot conditions, often with insufficient fluid intake. These problems not only affect highly trained athletes but also less well-trained people participating in sport. Although well-trained individuals generally exercise at higher intensities and therefore produce more heat, less well-trained individuals have less effective thermoregulation during exercise and work less economically. Especially overweight, unacclimated and ill individuals are likely to develop heat stroke.

B. Hyponatremia

Sweat loss results not only in the loss of water but also in a loss of electrolytes. Sodium is the most important ion lost in sweat. An electrolyte imbalance commonly referred to as "water intoxication" that results from *hyponatremia* (low plasma sodium) because of excessive water consumption has occasionally been reported in endurance athletes. Hyponatremia is the dilution of serum sodium from normal levels of 135 to 145 mEq/L to levels below 130 mEq/L. This may result in intracellular swelling, which in turn can alter the function of the central nervous system. During prolonged exercise, serum sodium can be diluted by either excessive fluid intake or excessive sodium losses in sweat or both. The symptoms of hyponatremia are almost identical to those of dehydration, exertional heat exhaustion, and exertional heat stroke, and they may include nausea, confusion, disorientation, headache, vomiting, aphasia, impaired coordination, muscle cramps, and muscle weakness. Severe hyponatremia can result in complications like cerebral and pulmonary edema that can result in seizure, coma, and cardiorespiratory arrest.

Hyponatremia appears to be most common among slow runners in marathon and ultramarathon races and probably arises because of the loss of sodium in sweat coupled with very high intakes of water. This means that there can be a danger of misdiagnosis of this condition when it occurs in individuals participating in endurance races. The usual treatment for dehydration is administration of fluid intravenously and orally. If this treatment were to be given to a hyponatremic individual, the consequences could be fatal.

C. Fluid Intake Strategies

Fluid intake during exercise can help maintain plasma volume and prevent the adverse effects of dehydration on muscle strength, endurance, and coordination. Elevating blood volume just before exercise by various hyperhydration (greater than normal body water content) strategies has also been suggested to be effective in enhancing exercise performance. When there is only little time in between two exercise bouts, rapid rehydration is crucial and drinking regimens need to be employed to optimize fluid delivery. Strategies for fluid replacement before, during, and after exercise are discussed in the following sections.

1. Preexercise Hyperhydration

Because even mild dehydration has been shown to result in reduced exercise capacity, it has been hypothesized that hyperhydration could improve heat dissipation and exercise performance in the heat by expanding blood volume and reducing plasma osmolarity. Although some studies have reported higher sweating rates, lower core temperatures, and lower heart rates during exercise after hyperhydration, some studies also reported improvements in exercise performance. These results must be interpreted with caution. Some of these studies used a dehydrated state as a control condition and therefore it is impossible to conclude from these studies that hyperhydration improves thermoregulation and performance. However, the findings generally support the notion that hyperhydration reduces the thermal and cardiovascular strain of exercise.

The majority of studies have induced temporary hyperhydration by having subjects drink large volumes of water or water-electrolyte solutions for 1 to 3 hours before exercise. However, much of the fluid overload is rapidly excreted and so expansions of the body water and blood volume are only transient. Greater fluid retention can be achieved if glycerol is added to fluids consumed before exercise. This will be discussed in the nutrition supplements section.

2. Fluid Intake during Exercise

To avoid dehydration during prolonged exercise, fluids must be consumed to match the sweat losses. By regularly measuring body weight before and after a training session it is possible to get a good indication of fluid loss. Ideally the weight loss is compensated for by an equal amount of fluid intake. However, it may not always be possible to prevent dehydration completely.

Sweat rates during strenuous exercise in the heat can amount to up to 2 to 3 liters per hour. Such large volumes of fluid are difficult if not impossible to ingest, and even 1 liter may feel quite uncomfortable in the stomach. Therefore, it is often not practically possible to achieve fluid intakes that match sweat losses during exercise. Another factor that can make the ingestion of large amounts of fluid difficult is the fact that the rules or practicalities of some sports or disciplines may limit the opportunities for drinking during exercise.

Fluid intake may be useful during exercise longer than 30 to 60 minutes, but there is no advantage during strenuous exercise of less than 30 minutes' duration. During such high-intensity exercise, gastric emptying is inhibited, and the drink may cause gastrointestinal distress with no performance benefit.

3. Practice Drinking during Training

Although it is often difficult to tolerate the volumes of fluid needed to prevent dehydration, the volume of fluid that is tolerable is trainable and can be increased with frequent drinking in training. Often this is neglected during training. Training to drink will accustom athletes to the feeling of exercising with fluid in the stomach. It also gives the opportunity to experiment with different volumes and flavorings to determine how much fluid intake they can tolerate and which formulations suit them best.

D. Composition of Sports Drinks

Numerous studies have shown that regular water intake during prolonged exercise is effective in improving performance. Fluid intake during prolonged exercise, of course offers the opportunity to provide some fuel (carbohydrate) as well. The addition of some carbohydrate to drinks consumed during exercise has been shown to have an additive independent effect in improving exercise performance. The ideal drink for fluid and energy replacement during exercise is one that tastes good to the athlete, does not causes gastrointestinal discomfort when consumed in large volumes, is rapidly emptied from the stomach and absorbed in the intestine, and provides energy in the form of carbohydrate. Sports drinks typically have three main ingredients: water, carbohydrate, and sodium. The water and carbohydrate provide fluid and energy, respectively, and sodium is included to aid water absorption and retention.

Although carbohydrate is important, a too concentrated carbohydrate solution may provide more fuel for the working muscles but decreases the amount of water that can be absorbed because of a slowing of gastric emptying. Water is absorbed into the body primarily through the small intestine, but the absorption of water decreases if the concentration of dissolved carbohydrate (or other substances) in the drink is too high. In this situation, water is actually drawn out of the interstitial fluid and plasma into the lumen of the small intestine by osmosis. So long as the fluid remains hypotonic with respect to plasma, the uptake of water from the small intestine is not adversely affected. In fact, the presence of small amounts of glucose and sodium tend to slightly increase the rate of water absorption compared with pure water. It must be emphasized here that the addition of sodium and other electrolytes to sports drinks is to increase palatability, maintain thirst (and therefore promote drinking), prevent hyponatremia, and increase the rate of water uptake, rather than to replace the electrolyte losses through sweating. Replacement of the electrolytes lost in sweat can normally wait until the postexercise recovery period.

1. Rehydration after Exercise

When there is little time for recovery in between exercise bouts, the replacement of fluid and electrolytes in the postexercise recovery period is of crucial importance. In the limited time available the athlete should strive to maximize rehydration. The main factors influencing the effectiveness of postexercise dehydration are the volume and composition of the fluid consumed. Plain water is not the ideal postexercise rehydration beverage when rapid and complete restoration of body fluid balance is necessary and where all intake is in liquid form. Ingestion of water alone in the postexercise period results in a rapid fall in the plasma sodium concentration and the plasma osmolarity. These changes have the effect of reducing the stimulation to drink (thirst) and increasing the urine output, both of which will delay the rehydration process. Plasma volume is more rapidly and completely restored in the postexercise period if sodium chloride (77 mmol/L or 450 mg/L) is added to the water consumed. This sodium concentration is similar to the upper limit of the sodium concentration found in sweat, but it is considerably higher than the sodium concentration of many commercially available sports drinks, which usually contain 10 to 25 mmol/L (60 to 150 mg/L).

Ingesting a beverage containing sodium not only promotes rapid fluid absorption in the small intestine but also allows the plasma sodium concentration to remain elevated during the rehydration period and helps to maintain thirst while delaying stimulation of urine production. The inclusion of potassium in the beverage consumed after exercise would be expected to enhance the replacement of intracellular water and thus promote rehydration, but currently there is little experimental evidence to support this theory. The rehydration drink should also contain carbohydrate (glucose or glucose polymers) because the presence of glucose will also stimulate fluid absorption in the gut and improve beverage taste. Following exercise, the uptake of glucose into the muscle for glycogen resynthesis should also promote intracellular rehydration.

It has become clear that to restore fluid balance after exercise, much more than the fluid lost as sweat has to be consumed. This is because some of the ingested fluid will be excreted in urine. Studies indicate that ingestion of 150% or more of weight loss is required to achieve normal hydration within 6 hours following exercise [75].

VII. NUTRITION AND TRAINING ADAPTATIONS

Athletes spend hours and hours training in order to optimize their performance or change their physique. The training adaptations are the result of an accumulation of specific proteins. Each training session causes disruptions in cellular homeostasis, which signals the muscle to increase protein synthesis. These disruptions and the responses are highly specific to the sport and the type of training, and therefore different types of training result in vastly different proteins to be synthesized and different phenotypes [76, 77]. For example, strength training increases muscle mass and strength, whereas endurance training increases mitochondrial mass and endurance. Hence, the different types of training result in distinctly different phenotypes.

It is thought that chronic training adaptations are the result of the cumulative effect of transient events that occur during and after acute bouts of exercise. Evidence is emerging that certain nutrients can be potent modulators of these events that occur during exercise and in the recovery phase. Therefore, nutrition is a major determinant of the adaptations that occur in response to a training program.

At the onset of exercise, there are rapid changes in the muscle Ca^{2+} content, the ATP/ADP ratio, Na^{+}/K^{+} ATPase activity, pH, and other factors that can stimulate certain signaling proteins (kinases and phosphatases). The most studied signaling proteins are 5′-adenosine monophosphate-activated protein kinase (AMPK), mitogen-activated protein kinase (MAPK), and the mammalian target of rapamycin (mTOR). AMPK has been described as a "fuel sensing" molecule involved in many metabolic responses in skeletal muscle during exercise. AMPK is activated by disturbances in energy status and will stimulate various processes geared toward ATP production while inhibiting ATP consuming processes [78]. The effects of AMPK are both acute by switching on and off catabolic and anabolic processes (enzymes) and chronic by altering gene expression (e.g., AMPK increases GLUT4 protein expression and increases mitochondrial density) [79]. The MAPK signal transduction pathway has been suggested to be the system responsible for translating contraction-induced biochemical perturbations into appropriate intracellular responses [80]. Exercise activates the MAPK pathway and plays a role in the transcriptional regulation of protein synthesis as well as stimulating the translational stage of protein synthesis. The mTOR pathway plays a role especially in muscle growth. Both insulin and amino acids are potent stimulators of the mTOR signaling cascade [81]. Activation of AMPK, however, inhibits mTOR. This has important practical implications because training may increase mitochondrial biosynthesis but may not result in muscle hypertrophy.

Changes in diet can alter the concentrations of nutrients and hormones in blood and can change body energy stores. These factors in turn can have an impact on the initial signaling cascade and various stages of protein synthesis. The availability of macronutrients has been shown to have an impact on gene expression after exercise. Muscle glycogen content, for example, is a potent modulator of the resting and contraction-induced AMPK and MAPK responses. At rest, AMPK activity was 2.5-fold higher in a low-glycogen versus a high-glycogen state. Similarly, increasing glucose availability by feeding carbohydrate during exercise has been shown to blunt the effects on gene expression [82], and it has been proposed that training in a low-glycogen state [83] without carbohydrate feeding [82] may result in superior training adaptations. In one study, untrained participants were trained for 10 weeks [83]. One leg was trained in a low-glycogen state half of the time, whereas the other leg always trained with high glycogen. Both legs performed the same training, and it was observed that the low-glycogen leg had a more pronounced increase in resting muscle glycogen and in citrate synthase activity. However, there are several limitations in these studies that make it impossible to translate the findings into practical advice. For example, it is unlikely that in a free-living situation the same training would be performed in a low-glycogen versus a high-glycogen state. In fact, training with low glycogen in some studies, especially when training programs were strenuous, resulted in a reduced training adaptation [84] and in more symptoms of overreaching when very hard daily training was performed [85]. Also, one study showed no differences in training adaptation when carbohydrate was ingested during training sessions compared with fasted sessions [86]. So there is evidence that training in a low-glycogen state and with reduced carbohydrate availability enhances signaling and gene expression. However, more research is needed, and the practical implications of this are far from clear.

Diets high in fat have been shown to increase the expression of mRNA-encoding proteins that are relevant to fatty acid metabolism [87] within 5 to 7 days. It has therefore been suggested that high-fat diets might give similar adaptations to endurance training. However, Stellingwerff *et al.* [88] fed their subjects a high-fat diet for 5 days followed by 1 day of carbohydrate loading. They observed that the high-fat diet actually reduced the activity of pyruvate dehydrogenase (PDH), a key enzyme in glycolysis, and could therefore actually impair carbohydrate metabolism. The authors suggested that the observed glycogen sparing effect with high-fat diets is probably more a result of impaired carbohydrate metabolism than of increased fat metabolism.

In addition to insulin and exercise, amino acids stimulate the phosphotidylinositol 3-kinase (PI3K) mTOR signaling pathway. Acute protein ingestion near the time of exercise appears to have the greatest potential on training adaptations. Ingesting a mixture of carbohydrate and amino acid before or immediately after a training session increases amino acid availability and transport into the muscle, increases protein synthesis, and reduces protein breakdown, resulting in a net positive protein balance [89, 90]. It seems that the effect of postexercise amino acid ingestion is enhanced by the coingestion of carbohydrate [91] most likely because of the elevated insulin concentrations. The amount of protein required to elicit this effect is very small (~6 grams) [89, 92], but these small amounts are far more effective in increasing protein synthesis than the ingestion of carbohydrate only [93]. Little is known about the effects of the types of amino acids or the forms in which they are ingested, and the timing of intake on long-term training adaptations and future research will focus on these questions.

In summary, it is thought that chronic training adaptations are the result of the cumulative effect of transient events that occur during and after acute bouts of exercise. Evidence is emerging that certain nutrients can be potent modulators of these events that occur during exercise and in the recovery phase. Little is known about the effects of various nutrients and exercise on signaling responses, gene expression, and protein content. This is an area that is likely to develop rapidly over the next few years as athletes try to increase the effectiveness of their training.

References

1. Mokdad, A. H., Marks, J. S., Stroup, D. F., and Gerberding, J. L. (2004). Actual causes of death in the United States, 2000. *JAMA* **291**, 1238–1245.
2. Applegate, E. A., and Grivetti, L. E. (1997). Search for the competitive edge: a history of dietary fads and supplements. *J. Nutr.* **127** (5 Suppl): 8605–8685.
3. Loucks, A. B. (2004). Energy balance and body composition in sports and exercise. *Journal of sports sciences* **22**, 1–14.
4. Erp van-Baart, A. M. J., Saris, W. H. M., Binkhorst, R. A., Vos, J. A., and Elvers, J. W. H. (1989). Nationwide survey on nutritional habits in elite athletes. Part II: Mineral and vitamin intake. *Int. J. Sports Med.* **10**, S11–S16.
5. Dahlstrom, M., Jansson, E., Nordevang, E., and Kaijser, L. (1990). Discrepancy between estimated energy intake and requirement in female dancers. *Clin. Physiol.* **10**, 11–25.
6. Jeukendrup, A. E., Craig, N. P., and Hawley, J. A. (2000). The bioenergetics of world class cycling. *J. Sci. Med. Sport.* **3**, 414–433.
7. Brouns, F., Saris, W. H. M., Stroecken, J., Beckers, E., Thijssen, R., Rehrer, N. J., and ten Hoor, F. (1989). Eating, drinking, and cycling: a controlled Tour de France simulation study, part I. *Int. J. Sports Med.* **10**, S32–S40.
8. Sjodin, A. M., Andersson, A. B., Hogberg, J. M., and Westerterp, K. R. (1994). Energy balance in cross-country skiers: a study using doubly labeled water. *Med. Sci. Sports Exerc.* **26**, 720–724.
9. Rontoyannis, G. P., Skoulis, T., and Pavlou, K. N. (1989). Energy balance in ultramarathon running. *Am. J. Clin. Nutr.* **49**, 976–979.
10. Beals, K. .A, and Meyer, N. L. (2007). Female athlete triad update. *Clin. Sports Med.* **26**, 69–89.
11. Birch, K. (2005). Female athlete triad. *BMJ* **330**, 244–246.
12. Loucks, A. B. (2003). Introduction to menstrual disturbances in athletes. *Med. Sci. Sports Exerc.* **35**, 1551–1552.
13. Nybo, L. (2003). CNS fatigue and prolonged exercise: effect of glucose supplementation. *Med. Sci. Sports Exerc.* **35**, 589–594.
14. Noakes, T. D. (2000). Physiological models to understand exercise fatigue and the adaptations that predict or enhance athletic performance. *Scand. J. Med. Sci. Sports* **10**, 123–145.
15. Bergström, J., and Hultman, E. (1967). A study of glycogen metabolism during exercise in man. *Scand. J. Clin. Invest.* **19**, 218–228.
16. Bergström, J., Hermansen, L., Hultman, E., and Saltin, B. (1967). Diet, muscle glycogen and physical performance. *Acta Physiol. Scand.* **71**, 140–150.
17. Hultman, E. (1967). Physiological role of muscle glycogen in man, with special reference to exercise. *Circ. Res.* **XX** (Suppl. 1), I99–I112.
18. Hawley, J. A., Schabort, E. J., Noakes, T. D., and Dennis, S. C. (1997). Carbohydrate loading and exercise performance: an update. *Sports Med.* **24**, 73–81.
19. Jeukendrup, A. E., and Jentjens, R. (2000). Oxidation of carbohydrate feedings during prolonged exercise: current thoughts, guidelines and directions for future research. *Sports Med.* **29**, 407–424.
20. Jeukendrup, A. E. (2004). Carbohydrate intake during exercise and performance. *Nutrition* **20**, 669–677.
21. Jeukendrup, A. E., Brouns, F., Wagenmakers, A. J. M., and Saris, W. H. M. (1997). Carbohydrate feedings improve 1 h time trial cycling performance. *Int. J. Sports Med.* **18**, 125–129.
22. Anantaraman, R., Carmines, A. A., Gaesser, G. A., and Weltman, A. (1995). Effects of carbohydrate supplementation on performance during 1 h of high intensity exercise. *Int. J. Sports Med.* **16**, 461–465.
23. Below, P. R., Mora-Rodríguez, R., Gonzáles Alonso, J., and Coyle, E. F. (1995). Fluid and carbohydrate ingestion independently improve performance during 1 h of intense exercise. *Med. Sci. Sports Exerc.* **27**, 200–210.
24. Carter, J., Jeukendrup, A. E., Mundel, T., and Jones, D. A. (2003). Carbohydrate supplementation improves moderate and high-intensity exercise in the heat. *Pflugers Arch.* **446**, 211–219.

25. Carter, J., Jeukendrup, A. E., and Jones, D. A. (2005). The effect of sweetness on the efficacy of carbohydrate supplementation during exercise in the heat. *Can. J. Appl. Physiol.* **30**, 379–391.
26. Carter, J. M., Jeukendrup, A. E., Mann, C. H., and Jones, D. A. (2004). The effect of glucose infusion on glucose kinetics during a 1-h time trial. *Med. Sci. Sports Exerc.* **36**, 1543–1550.
27. Carter, J. M., Jeukendrup, A. E., and Jones, D. A. (2004). The effect of carbohydrate mouth rinse on 1-h cycle time trial performance. *Med. Sci. Sports Exerc.* **36**, 2107–2111.
28. Nieman, D. C., and Bishop, N. C. (2006). Nutritional strategies to counter stress to the immune system in athletes, with special reference to football. *J. Sports Sci.* **24**, 763–772.
29. Nieman, D. C. (2007). Marathon training and immune function. *Sports Med.* **37**, 412–415.
30. Gleeson, M. (2000). The scientific basis of practical strategies to maintain immunocompetence in elite athletes. *Exerc. Immunol. Rev.* **6**, 75–101.
31. Gleeson, M. (2006). Can nutrition limit exercise-induced immunodepression? *Nutr. Rev.* **64**, 119–131.
32. Sherman, W. (1983). Carbohydrates, muscle glycogen, and muscle glycogen supercompensation. *In* "Ergogenic Aids in Sports" (M. H. Williams Ed.), pp. 1–25. Human, Kinetics, Champaign, IL.
33. Sherman, W. M., Costill, D. L., Fink, W. J., and Miller, J. M. (1981). Effect of exercise-diet manipulation on muscle glycogen and its subsequent utilisation during performance. *Int. J. Sports Med.* **2**, 114–118.
34. Hawley, J. A., Schabort, E. J., Noakes, T. D., and Dennis, S. C. (1997). Carbohydrate loading and exercise performance. *Sports Med.* **24**, 1–10.
35. Sherman, W. M., Brodowicz, G., Wright, D. A., Allen, W. K., Simonsen, J., and Dernbach, A. (1989). Effects of 4 h preexercise carbohydrate feedings on cycling performance. *Med. Sci. Sports Exerc.* **21**, 598–604.
36. Wright, D. A., Sherman, W. M., and Dernbach, A. R. (1991). Carbohydrate feedings before, during, or in combination improve cycling endurance performance. *J. Appl. Physiol.* **71**, 1082–1088.
37. Schabort, E. J., Bosch, A. N., Weltan, S. M., and Noakes, T. D. (1999). The effect of a preexercise meal on time to fatigue during prolonged cycling exercise. *Med. Sci. Sports Exerc.* **31**, 464–471.
38. Koivisto, V. A., Karonen, S. L., and Nikkila, E. A. (1981). Carbohydrate ingestion before exercise: comparison of glucose, fructose, and sweet placebo. *J. Appl. Physiol.* **51**, 783–787.
39. Foster, C., Costill, D. L., and Fink, W. J. (1979). Effects of preexercise feedings on endurance performance. *Med. Sci. Sports* **11**, 1–5.
40. Hargreaves, M., Hawley, J. A., and Jeukendrup, A. E. (2004). Pre-exercsie carbohydrate and fat ingestion: effects on metabolism and performance. *J. Sports Sci.* **22**, 31–38.
41. Jentjens, R. L., Cale, C., Gutch, C., and Jeukendrup, A. E. (2003). Effects of pre-exercise ingestion of differing amounts of carbohydrate on subsequent metabolism and cycling performance. *Eur. J. Appl. Physiol.* **88**, 444–452.
42. Jentjens, R. L., and Jeukendrup, A. E. (2003). Effects of pre-exercise ingestion of trehalose, galactose and glucose on subsequent metabolism and cycling performance. *Eur. J. Appl. Physiol.* **88**, 459–465.
43. Moseley, L., Lancaster, G. I., and Jeukendrup, A. E. (2003). Effects of timing of pre-exercise ingestion of carbohydrate on subsequent metabolism and cycling performance. *Eur. J. Appl. Physiol.* **88**, 453–458.
44. Jentjens, R. L., and Jeukendrup, A. E. (2002). Prevalence of hypoglycemia following pre-exercise carbohydrate ingestion is not accompanied by higher insulin sensitivity. *Int. J. Sport Nutr. Exerc. Metab.* **12**, 398–413.
45. Tsintzas, O. K., Williams, C., Boobis, L., and Greenhaff, P. (1995). Carbohydrate ingestion and glycogen utilisation in different muscle fibre types in man. *J. Physiol.* **489**, 243–250.
46. Coyle, E. F., Coggan, A. R., Hemmert, M. K., and Ivy, J. L. (1986). Muscle glycogen utilization during prolonged strenuous exercise when fed carbohydrate. *J. Appl. Physiol.* **61**, 165–172.
47. Leijssen, D. P., Saris, W. H., Jeukendrup, A. E., and Wagenmakers, A. J. (1995). Oxidation of exogenous [^{13}C]galactose and [^{13}C]glucose during exercise. *J. Appl. Physiol.* **79**, 720–725.
48. Shi, X., Summers, R. W., Schedl, H. P., Flanagan, S. W., Chang, R., and Gisolfi, C. V. (1995). Effects of carbohydrate type and concentration and solution osmolality on water absorption. *Med. Sci. Sports Exerc.* **27**, 1607–1615.
49. Jentjens, R. L., Venables, M. C., and Jeukendrup, A. E. (2004). Oxidation of exogenous glucose, sucrose and maltose during prolonged cycling exercise. *J. Appl. Physiol.* **96(4)**, 1285–1291.
50. Jentjens, R. L. P. G., Achten, J., and Jeukendrup, A. E. (2004). High oxidation rates from a mixture of glucose, sucrose and fructose ingested during prolonged exercise. *Med. Sci. Sport Exerc.* **5**, **36(9)**, 1551–1558.
51. Jentjens, R. L., Venables, M. C., and Jeukendrup, A. E. (2004). Oxidation of exogenous glucose, sucrose, and maltose during prolonged cycling exercise. *J. Appl. Physiol.* **96**, 1285–1291.
52. Jentjens, R. L., Moseley, L., Waring, R. H., Harding, L. K., and Jeukendrup, A. E. (2004). Oxidation of combined ingestion of glucose and fructose during exercise. *J. Appl. Physiol.* **96**, 1277–1284.
53. Jentjens, R. L., and Jeukendrup, A. E. (2005). High rates of exogenous carbohydrate oxidation from a mixture of glucose and fructose ingested during prolonged cycling exercise. *Br. J. Nutr.* **93**, 485–492.
54. Jentjens, R. L., Shaw, C., Birtles, T., Waring, R. H., Harding, L. K., and Jeukendrup, A. E. (2005). Oxidation of combined ingestion of glucose and sucrose during exercise. *Metabolism* **54**, 610–618.
55. Jentjens, R. L., Underwood, K., Achten, J., Currell, K., Mann, C. H., and Jeukendrup, A. E. (2006). Exogenous carbohydrate oxidation rates are elevated after combined ingestion of glucose and fructose during exercise in the heat. *J. Appl. Physiol.* **100**, 807–816.
56. Jeukendrup, A. E., Moseley, L., Mainwaring, G. I., Samuels, S., Perry, S., and Mann, C. H. (2006). Exogenous carbohydrate oxidation during ultraendurance exercise. *J. Appl. Physiol.* **100**, 1134–1141.
57. Wallis, G. A., Yeo, S. E., Blannin, A. K., and Jeukendrup, A. E. (2007). Dose-response effects of ingested carbohydrate on

exercise metabolism in women. *Med. Sci. Sports Exerc.* **39**, 131–138.

58. Jentjens, R., and Jeukendrup, A. (2003). Determinants of post-exercise glycogen synthesis during short-term recovery. *Sports Med.* **33**, 117–144.
59. van Loon, L. J., Saris, W. H., Verhagen, H., and Wagenmakers, A. J. (2000). Plasma insulin responses after ingestion of different amino acid or protein mixtures with carbohydrate. *Am. J. Clin. Nutr.* **72**, 96–105.
60. Zawadzki, K. M., Yaspelkis B. B. III, and Ivy, J. L. (1992) Carbohydrate-protein complex increased the rate of muscle glycogen storage after exercise. *J. Appl. Physiol.* **72**, 1854–1859.
61. Jentjens, R. L., van Loon, L. J., Mann, C. H., Wagenmakers, A. J., and Jeukendrup, A. E. (2001). Addition of protein and amino acids to carbohydrates does not enhance postexercise muscle glycogen synthesis. *J. Appl. Physiol.* **91**, 839–846.
62. van Loon, L. J., Saris, W. H., Kruijshoop, M., and Wagenmakers, A. J. (2000). Maximizing postexercise muscle glycogen synthesis: carbohydrate supplementation and the application of amino acid or protein hydrolysate mixtures. *Am. J. Clin. Nutr.* **72**, 106–111.
63. van Hall, G., Shirreffs, S. M., and Calbet, J. A. (2000). Muscle glycogen resynthesis during recovery from cycle exercise: no effect of additional protein ingestion. *J. Appl. Physiol.* **88**, 1631–1636.
64. Phillips, S. M. (2006). Dietary protein for athletes: from requirements to metabolic advantage. *Appl. Physiol. Nutr. Metab.* **31**, 647–654.
65. Tipton, K. D., and Witard, O. C. (2007). Protein requirements and recommendations for athletes: relevance of ivory tower arguments for practical recommendations. *Clin. Sports Med.* **26**, 17–36.
66. Volpe, S. L. (2007). Micronutrient requirements for athletes. *Clin. Sports Med.* **26**, 119–130.
67. McArdle, A., and Jackson, M. J. (2000). Exercise, oxidative stress and ageing. *J. Anat.* **197** (Pt 4), 539–541.
68. Gonzalez-Alonso, J., Mora-Rodriguez, R., Below, P. R., and Coyle, E. F. (1995). Dehydration reduces cardiac output and increases systemic and cutaneous vascular resistance during exercise. *J. Appl. Physiol.* **79**, 1487–1496.
69. Cheuvront, S. N., Carter, R. 3rd, and Sawka, M. N. (2003). Fluid balance and endurance exercise performance. *Curr. Sports Med. Rep.* **2**, 202–208.
70. Sawka, M. N., Burke, L. M., Eichner, E. R., Maughan, R. J., Montain, S. J., and Stachenfeld, N. S. (2007). American College of Sports Medicine position stand. Exercise and fluid replacement. *Med. Sci. Sports Exerc.* **39**, 377–390.
71. Walsh, R. M., Noakes, T. D., Hawley, J. A., and Dennis, S. C. (1994). Impaired high-intensity cycling performance time at low levels of dehydration. *Int. J. Sports Med.* **15**, 392–398.
72. McConell, G. K., Burge, C. M., Skinner, S. L., and Hargreaves, M. (1997). Influence of ingested fluid volume on physiological responses during prolonged exercise. *Acta Physiol. Scand.* **160**, 149–156.
73. Robinson, T. A., Hawley, J. A., Palmer, G. S., Wilson, G. R., Gray, D. A., Noakes, T. D., and Dennis, S. C. (1995). Water ingestion does not improve 1-h cycling performance in moderate ambient temperatures. *Eur. J. Appl. Physiol. Occup. Physiol.* **71**, 153–160.
74. Maughan, R. J., Fenn, C. E., and Leiper, J. B. (1989). Effects of fluid, electrolyte and substrate ingestion on endurance capacity. *Eur. J. Appl. Physiol. Occup. Physiol.* **58**, 481–486.
75. Shirreffs, S. M., Taylor, A. J., Leiper, J. B., and Maughan, R. J. (1996). Post-exercise rehydration in man: effects of volume consumed and drink sodium content. *Med. Sci. Sports Exerc.* **28**, 1260–1271.
76. Nader, G. A., and Esser, K. A. (2001). Intracellular signaling specificity in skeletal muscle in response to different modes of exercise. *J. Appl. Physiol.* **90**, 1936–1942.
77. Hildebrandt, A. L., Pilegaard, H., and Neufer, P. D. (2003). Differential transcriptional activation of select metabolic genes in response to variations in exercise intensity and duration. *Am. J. Physiol. Endocrinol. Metab.* **285**, E1021–E1027.
78. Hardie, D. G., and Hawley, S. A. (2001). AMP-activated protein kinase: the energy charge hypothesis revisited. *Bioessays* **23**, 1112–1119.
79. Aschenbach, W. G., Sakamoto, K., and Goodyear, L. J. (2004). 5′ adenosine monophosphate-activated protein kinase, metabolism and exercise. *Sports Med.* **34**, 91–103.
80. Widegren, U., Ryder, J. W., and Zierath, J. R. (2001). Mitogen-activated protein kinase signal transduction in skeletal muscle: effects of exercise and muscle contraction. *Acta Physiol. Scand.* **172**, 227–238.
81. Deldicque, L., Theisen, D., and Francaux, M. (2005). Regulation of mTOR by amino acids and resistance exercise in skeletal muscle. *Eur. J. Appl. Physiol.* **94**, 1–10.
82. Civitarese, A. E., Hesselink, M. K., Russell, A. P., Ravussin, E., and Schrauwen, P. (2005). Glucose ingestion during exercise blunts exercise-induced gene expression of skeletal muscle fat oxidative genes. *Am. J. Physiol. Endocrinol. Metab.* **289**, E1023–E1029.
83. Hansen, A. K., Fischer, C. P., Plomgaard, P., Andersen, J. L., Saltin, B., and Pedersen, B. K. (2005). Skeletal muscle adaptation: training twice every second day vs. training once daily. *J. Appl. Physiol.* **98**, 93–99.
84. Simonsen, J. C., Sherman, W. M., Lamb, D. R., Dernbach, A. R., Doyle, J. A., and Strauss, R. (1991). Dietary carbohydrate, muscle glycogen, and power output during rowing training. *J. Appl. Physiol.* **70**, 1500–1505.
85. Achten, J., Halson, S. L., Moseley, L., Rayson, M. P., Casey, A., and Jeukendrup, A. E. (2004). Higher dietary carbohydrate content during intensified running training results in better maintenance of performance and mood state. *J. Appl. Physiol.* **96(4)**, 1331–1340.
86. Akerstrom, T. C., Birk, J. B., Klein, D. K., Erikstrup, C., Plomgaard, P., Pedersen, B. K., and Wojtaszewski, J. (2006). Oral glucose ingestion attenuates exercise-induced activation of 5′-AMP-activated protein kinase in human skeletal muscle. *Biochem. Biophys. Res. Commun.* **342**, 949–955.
87. Cameron-Smith, D., Burke, L. M., Angus, D. J., Tunstall, R. J., Cox, G. R., Bonen, A., Hawley, J. A., and Hargreaves, M. (2003). A short-term, high-fat diet up-regulates lipid metabolism and gene expression in human skeletal muscle. *Am. J. Clin. Nutr.* **77**, 313–318.
88. Stellingwerff, T., Spriet, L. L., Watt, M. J., Kimber, N. E., Hargreaves, M., Hawley, J. A., and Burke, L. M. (2006).

Decreased PDH activation and glycogenolysis during exercise following fat adaptation with carbohydrate restoration. *Am. J. Physiol. Endocrinol. Metab.* **290**, E380–E388.

89. Tipton, K. D., Rasmussen, B. B., Miller, S. L., Wolf, S. E., Owens-Stovall, S. K., Petrini, B. E., and Wolfe, R. R. (2001). Timing of amino acid-carbohydrate ingestion alters anabolic response of muscle to resistance exercise. *Am. J. Physiol. Endocrinol. Metab.* **281**, E197–E206.
90. Biolo, G., Tipton, K. D., Klein, S., and Wolfe, R. R. (1997). An abundant supply of amino acids enhances the metabolic effect of exercise on muscle protein. *Am. J. Physiol.* **273**, E122–E129.
91. Miller, S. L., Tipton, K. D., Chinkes, D. L., Wolf, S. E., and Wolfe, R. R. (2003). Independent and combined effects of amino acids and glucose after resistance exercise. *Med. Sci. Sports Exerc.* **35**, 449–455.
92. Tipton, K. D., Ferrando, A. A., Phillips, S. M., Doyle, D. Jr., and Wolfe, R. R. (1999). Postexercise net protein synthesis in human muscle from orally administered amino acids. *Am. J. Physiol.* **276**, E628–E634.
93. Borsheim, E., Aarsland, A., and Wolfe, R. R. (2004). Effect of an amino acid, protein, and carbohydrate mixture on net muscle protein balance after resistance exercise. *Int. J. Sport Nutr. Exerc. Metab.* **14**, 255–271.

CHAPTER **20**

Nutrition for Children with Special Health Care Needs

ANNE BRADFORD HARRIS,[1] MARION TAYLOR BAER,[2] CRISTINE M. TRAHMS,[3] AND JENNIFER WEBB[4]

[1]*Waisman Center, Department of Pediatrics, University of Wisconsin-Madison*

[2]*Department of Community Health Sciences, School of Public Health, University of California, Los Angeles, and USC UCEDD, Department of Pediatrics, University of Southern California, Los Angeles*

[3]*Center on Human Development and Disability, Department of Pediatrics, University of Washington, Seattle*

[4]*Children's Hospital, Los Angeles, CA*

Contents

I. INTRODUCTION

Approximately 13% of children in the United States have special conditions that increase their need for health care services and educational supports beyond those of children without special needs [1, 2]. Children with special health care needs include those with specific diagnoses related to genetic disorders (e.g., Trisomy 21, a.k.a. Down syndrome, and phenylketonuria, a.k.a. PKU), developmental disorders (e.g., autism, intellectual disabilities), and neuromotor disorders (e.g., cerebral palsy, spina bifida). Including children who are born prematurely or with a low birth weight, considered to be at high risk for developing a special health care need, as well as children with some common chronic health concerns (e.g., asthma) raises the overall prevalence to an estimated 30% [3].

As many as 40% to 60% of children with special health care needs are at risk for one or more problems with nutritional status, ranging from slow growth and poor feeding to more severe gastrointestinal problems and metabolic disorders [4] (also Baer and Herman, 1997, unpublished data). Table 1 shows a list of special health care needs, including the sequelae of prematurity, the more common genetic disorders, and other developmental disabilities, and the associated functional problems with nutrition. Because of changing public perspectives related to individuals with intellectual and other developmental disabilities since the 1950s, and legislation favoring their inclusion in all aspects of society, nutrition professionals will encounter children with special needs in their practices and should be able to anticipate their most commonly seen nutrition problems [5].

Nutrition screening should be conducted for all children with special needs in multiple settings to ensure that those at high risk are identified and referred for appropriate services [6]. Intrauterine nutrition insults occurring prenatally can predispose children to developmental disabilities and poor growth, as well as to problems with obesity and chronic illnesses later in life. Evidence is mounting establishing the cost effectiveness of appropriate nutrition interventions, especially preventive interventions, for high-risk populations of mothers and children [4, 7]. Mothers who receive adequate nutrition during pregnancy have better birth outcomes, and children who have their nutritional risks identified and receive appropriate interventions demonstrate better growth, health, and developmental outcomes than those who do not.

The approach for this chapter is to identify and assess nutrition risk for mothers, and for children identified with a special need, with the intent to prevent disability through good nutrition, to provide secondary prevention of further disability, and to preserve and promote function in order to enable the child to reach his or her potential in terms of growth and development.

Section II addresses primary and secondary prevention of nutrition issues for pregnant women and children with special needs, especially where there is evidence to support specific nutrition interventions.

Section III outlines the functional approach to identifying nutritional needs—how growth, diet, and medical issues are assessed, including issues specific to children with special needs in each of these areas.

Section IV covers selected conditions in more depth—inborn errors of metabolism (IEM), Prader-Willi syndrome, feeding problems, and autism spectrum disorders (ASD)—including a description of the diagnosis, assessment of

TABLE 1 A Comparison of Specific Developmental Disabilities and Chronic Conditions with Types of Nutrition Risks

	Growth			Diet		Medical		
Disorder	**Underweight**	**Overweight**	**Short Stature**	**Low Energy Needs**	**High Energy Needs**	**Feeding Issues/ Special Diet**	**Constipation**	**Chronic Medications**
Autism	X					X		X
Prematurity/ BPD	X		X		X	X		X
Cerebral Palsy	X	X	X	X	X	X	X	X
Down Syndrome		X	X	X		X		
Fetal Alcohol Spectrum Disorder	X		X					
Inborn Errors of Metabolism						X		X
Prader-Willi Syndrome		X	X	X				
Seizure Disorder								X
Spina Bifida	X	X	X	X		X	X	X

nutritional status, and ongoing intervention, where there is evidence for nutrition interventions.

II. THE ROLE OF NUTRITION IN PREVENTING DEVELOPMENTAL PROBLEMS

A. Nutrition and the Primary Prevention of Developmental Disabilities

Although the focus of this chapter is on screening, assessment, and treatment of nutrition problems in children with special health care needs, it should be noted that nutrition also plays an important role in the primary prevention of disability, a role that has been underplayed in the past but that is becoming increasingly recognized. This section outlines developments in our understanding of this relationship, the extent of which emphasizes the necessity of promoting nutritionally adequate diets for women of childbearing years, before pregnancy and during interconceptional intervals.

B. Risk for Developmental Disability Related to Nutritional Disorders

An adequate supply of nutrients, both macro and micro, may be the most important environmental factor in determining the optimal growth and development of the fetus. Low birth weight (LBW), resulting from preterm birth or intrauterine growth retardation, has long been established as a major risk factor for developmental disabilities (discussed next), but maternal nutritional status has not always been considered as a causal factor. The following brief review of research findings will illustrate that associations between nutrition and pregnancy outcome are becoming more apparent.

1. Low Birth Weight, Preterm Birth

Risk of preterm birth (before 37 weeks' gestation), LBW (<2500 g), or being small for gestational age (SGA) increases by about 50% with short interpregnancy intervals (<18 mo) and <2 years postmenarche in U.S. women compared to those who are adults or who have an interpregnancy interval of 18 to 23 months; this cannot be entirely explained by social or behavioral factors [8]. Evidence is accumulating to indicate that if the pregnant woman is still growing herself or has not had sufficient time since her last pregnancy to replace depleted stores, the physiological changes that normally favor improved nutrient utilization during gestation may not be adequate to meet the needs of both mother and fetus. In such cases, it appears that nutrients are partitioned between the mother and fetus depending on the initial nutritional status of the mother [8]. In one study, the incidence of preterm birth was lowered in women given a diet of fish, oils, low-fat meats and dairy, whole grains, fruits, vegetables, and legumes compared to women who continued their usual diet. Although the study goal was to lower cholesterol in the mothers, it is likely that this diet also improved their nutritional status [9]. Other studies have shown that zinc [10], folate [11], and iron [12] deficiencies all increase the risk of preterm birth and LBW. In addition, low serum selenium early in pregnancy has been shown to predict lower birth weight [13].

2. Other Risks

Beyond the risk of developmental disabilities subsequent to preterm delivery or LBW, intellectual disability (ID) in children has long been linked to known maternal nutrition-related conditions such as iodine deficiency. Recently, ID of unknown cause, including autism, has been found to be higher in children of mothers with diabetes, epilepsy (possibly due to drug-nutrient interactions), and anemia [14]. This large population-based study in Western Australia used linked databases of children born between 1983 and 1992 and studied in 2002 to identify 2865 children with ID of unknown cause (grouped into mild to moderate $n = 2462$; severe to profound $n = 212$; and ASD with ID $n = 181$) and 236,964 children without ID. Mothers with anemia ($n = 1101$) were five times more likely (OR = 5.26) to have a child with severe ID (IQ <35 to 40), even after socioeconomic variables were introduced into the stepwise logistic regression model (OR = 4.93).

In another large population-based study in Hungary of children with ($n = 781$) and without ($n = 22{,}843$) Down syndrome, the investigators found a protective effect for maternal iron supplementation (150 to 300 mg/d of ferrous sulfate) during the first month of gestation [15]. It should be noted that iron and folate, two nutrients related to LBW, are both depleted in pregnancy and must be replaced [8].

The toxic effects of mercury on the developing fetus are well known [16], and the U.S. Department of Health and Human Services (DHHS) has recommended limiting seafood consumption by pregnant women because of its high mercury content. However, a large study (11,875 women) in the United Kingdom [17] has suggested that restricting seafood intake to 340 g/day results in adverse neurodevelopmental outcomes in the children, whereas higher intakes had no ill effects. The assumption is that this is due to a deficiency of omega-3 fatty acids, as few women took fish oil supplements. These fatty acids have been shown to have positive effects on neurodevelopment [18, 19], as have other nutrients found in fish, such as choline, zinc, and copper [20]. Although protective effects of other dietary components were not considered in the Australian study, there is evidence to suggest that selenium plays a protective role by reducing the bioavailability of mercury [21].

3. Spina Bifida and Other Birth Defects

The recognition of a relationship between nutrition and the etiology of certain developmental disabilities has been most fully realized in the case of spina bifida and other neural tube defects (NTD). Although the link between diet and NTD was suspected as early as the 1970s [22] and demonstrated repeatedly in studies that implicated vitamins and minerals in general [23] and gradually focused on folate in particular [24], there was no public health campaign in the United States to increase folate intake until the mid-1990s after the publication of the Medical Research Council's definitive study [25]. Then, rather than directing efforts toward increasing women's intake of folate-containing foods, the emphasis was on folate supplementation for pregnant women followed by the passage of the mandatory fortification of cereal grain products which went into effect in January 1998. Since then, the Centers for Disease Control (CDC), using data from 23 population-based surveillance systems, has estimated an approximate decrease of 26% in the reported prevalence of NTD, comparing data from 1995–1996 with those from 1999–2000, the year following the mandate [26]. The report from the Center on Birth Defects and Developmental Disabilities noted that the decrease is less than predicted from research trials [27], which led to the Healthy People 2010 goal of 50%. The CDC recommends increased folic acid intake (supplements or fortified breakfast cereals) and charges health care professionals to recommend these to women of child-bearing age [26]. There is no mention of the desirability of improving the general diet of these women, although the editor acknowledges that there may be other causes of NTD besides folate insufficiency. Meanwhile, a group of investigators in the Netherlands has reported significantly lower dietary intake of iron, magnesium, niacin, fiber, plant protein, and polysaccharides in mothers of infants with spina bifida ($n = 106$) compared to controls ($n = 181$) independent of periconceptional folic acid use [28]. They also found increased risk related to dietary intake of mono- and disaccharides, which they suggest supports the importance of optimal carbohydrate intake during neural tube formation.

Another food-related factor that has been implicated in the high incidence of NTD among Mexican-Americans are fumonisins, a class of mycotoxins produced by mold, which interferes with folate metabolism. In response to an "outbreak" of NTD along the Texas–Mexico border in 1990–1991, an epidemiological study linked it to high levels of fumonisins in the corn used to make the tortillas, and other corn products, which form a large part of the diet of this population [29]. Other risk factors, such as folate intake (dietary plus multivitamin), were not confounders. In fact, B_{12} levels have been shown to be a more important predictor of risk in this population than folate levels [30], and low maternal B_{12} status has been shown to be an independent risk factor in a folic acid–fortified population in Canada [31].

Although orofacial clefts (cleft lip or palate or both) are not considered a developmental disability per se, there is evidence of an association between their occurrence and maternal nutrition as well as a rationale for the potential contribution of several nutrients (folate, niacin, thiamine, B_6, B_{12}, riboflavin, zinc, amino acids, and carbohydrates) to their etiology. In a large multistate case-control study using 1997–2000 data from the National Birth Defects Prevention Study, Shaw and colleagues [32], via phone interviews and food frequency questionnaires, found an association between iron and riboflavin intakes and a reduction in risk. Other studies [33] point to the contribution of vitamins B_1, B_6, and niacin.

4. Fetal Alcohol Syndrome

Fetal alcohol syndrome (FAS) or, as proposed by the American Academy of Pediatrics (AAP), alcohol-related neurodevelopmental disorder (ARND) and alcohol-related birth defects (ARBD) is defined by pre- and postnatal growth disturbances, dysmorphic facial features, and mental retardation; behavioral abnormalities are common, as are motor problems, facial anomalies, and cardiac defects [34]. FAS represents an example of a developmental disability where prenatal nutrition likely plays a major role in the etiology. This is due to (1) the dietary energy provided by alcohol (7.1 kcal/g or 29.7 kJ), which replaces foods containing essential micronutrients, thereby putting the mother and fetus at risk for nutrient deficiencies, and (2) the interference of alcohol with the metabolism of all fat and water-soluble vitamins [35], especially folate, and zinc. These nutrients are key to fetal development because of their involvement in multiple systems: enzyme, gene transcription, and transport. Folate insufficiency, mentioned earlier in the etiology of spina bifida and cleft lip and palate, has also been implicated in the etiology of limb and heart abnormalities and Down syndrome [36]. Severe zinc deficiency has been linked to intrauterine growth retardation and teratogenesis; mild to moderate deficiency has been associated with congenital malformations, LBW, and preterm delivery [37].

Because of the wide array of nutrients that have been implicated in promoting optimal birth outcomes, it would seem of utmost importance that clinicians and public health professionals urge women of childbearing age to eat a well-balanced diet rather than relying primarily on vitamin and mineral supplements or population-based fortification of foods or food ingredients as has been suggested by some [38].

C. Prematurity/Low Birth Weight as a Risk Factor for Nutrition and Feeding Problems

The number of preterm infants and infants of LBW continues to increase steadily, and in 2004, more than one-half million infants, or 12.5% of births in the United States, were preterm [39]. Though there have been dramatic improvements in mortality rates for infants who are born preterm or LBW, preterm birth remains a leading cause of infant mortality and morbidity, accounting for nearly half of all congenital neurological defects; however, the causes and best management of preterm labor are not well understood [39]. Infants who are born preterm are vulnerable to many complications leading to nutritional deficits and delayed feeding development, including congenital and neurological defects such as cerebral palsy, cognitive and learning disabilities, behavioral disabilities and developmental delays, sensory impairment, cardiovascular and respiratory disorders, and growth impairment [40, 41].

With increased survival rates of premature and medically fragile infants, there is an increased population of children who are at risk for, or who demonstrate, feeding difficulties with multifactorial origins, depending on the degree of prematurity or LBW. For example, gastroesophageal reflux (GER), a physiological and developmental consequence of prematurity, may occur three to five times per hour [42]. Poor coordination of the suck-swallow and suck-respiration are often predictive of feeding abnormalities [43]. Nutritional status and feeding development are of primary concern in the neonatal intensive care unit (NICU) [40]. Feeding skill development and transitions from parenteral to enteral nutrition in the medically fragile newborn are facilitated by an interdisciplinary team of health care providers, which includes the parents. Breastfeeding, or the feeding of breast milk, is encouraged as much as is possible for the newborn and his or her mother, as an important means of nourishment, developing attachment, and for feeding skills to develop at the appropriate gestational age. Fortified human milk is often indicated in preterm infants to support optimal growth and bone mineralization [44]. Many infants are discharged from the NICU only after achieving at least a portion of their nutritional intake through oral routes, though the use of nasogastric, orogastric, and gastrostomy tube feedings has increased. However, to prevent later feeding problems, it is best to minimize the use of tube feedings or to encourage developmentally appropriate oral feedings as soon as the child is medically stable and physically able. Early introduction of feedings, as early as 33 weeks gestational age, has been shown to accelerate the rate at which preterm infants are able to attain full oral feedings [45].

D. Newborn Screening

Population-based newborn screening (NBS) has been established as a preventive public health measure available to all neonates. If infants with specific metabolic, endocrine, and other disorders are identified by NBS in the first few days of life, the diagnostic process can be completed and treatment started before physical and neurological damage occurs. Many of the disorders screened by NBS programs are autosomal recessive. This means that each parent carries the affected gene and the incidence of an affected infant is 1 out of 4 for each pregnancy. Thus, affected infants identified by NBS usually do not have a positive family history of the disorder.

Successful NBS of infants depends on reliable, valid, and timely laboratory results. Many states have established a central NBS laboratory, which supports a rigid quality control process. The initial NBS report provides a presumptive positive result that must be confirmed by mandatory laboratory confirmation of the diagnosis.

The American Academy of Pediatrics [46] recommends that an NBS blood sample be collected before the infant is discharged from the hospital nursery. If the blood sample is collected before the infant is 24 hours of age, it is essential that a second screening sample be obtained. Generally accepted NBS criteria are shown in Table 2. NBS programs require a small sample of dried blood, which is submitted on filter paper; this type of sample is easily transported to the central laboratory, and many tests can be completed on a single blood sample. Tandem mass spectrometry technology has increased the number of disorders that can be effectively screened [47].

An effective NBS program involves timely community-based collection of initial blood samples, laboratory

TABLE 2 General Timeline for Collecting Newborn Screening (NBS) Samples

Between Day 1 (birth) and Discharge
- Inform parents of NBS and their right to refuse for specific statutory reasons.
- Send sample to state lab within 24 hours of collection.

Or
- If parents refuse screening (e.g., based on religious beliefs) obtain refusal signature before discharge.

No Later Than Day 5 of the Infant's Life
- Draw NBS sample for infants remaining in hospital and for those born out of hospital.
- Send sample to state lab within 24 hours of collection.

Between Day 7 and 14 of the Infant's Life
- Repeat newborn screening sample is recommended for all infants.
- Send sample to state lab within 24 hours of collection.

This is a generic newborn screening timeline; individual states have their own protocol. Timing of the newborn screening sample is critical to establishing the proper diagnosis and offering the earliest possible intervention before the infant is irreparably damaged.

confirmation, education of primary care providers and families, as well as specific disorder follow-up, management, and evaluation components [48]. Each state has developed individual legislative NBS mandates. However, the March of Dimes [49], the American Academy of Pediatrics [46], and the American Society Human of Genetics [50] have developed recommendations and guidelines for expanded NBS, which, it is hoped, will be universally adopted. The general criteria for inclusion of a disorder in an NBS panel are shown in Table 3. Resources for up-to-date information on expanded NBS programs are shown in Table 4.

TABLE 3 Generally Accepted Criteria for Newborn Screening for a Specific Disorder

Symptoms are usually absent in newborns.
Disease results in developmental impairment, serious illness, or death.
Sensitive, specific laboratory tests are available on a mass population basis.
Disease occurs frequently enough to warrant screening.
Successful treatment procedures are available.
Benefits of screening justify the cost.
Follow-up and treatment programs are available.

Adapted from Newborn Screening Fact Sheets, 2006 [46].

TABLE 4 Newborn Screening Informational Resources

Resource	URL
ACMG Model ACT Sheets and Screening Algorithms	www.acmg.net/resources/policies/ACT/condition-analyte-links.htm
ACTion (ACT) sheets and algorithms developed by experts of the American College of Human Genetics (ACMG) involved in newborn screening for endocrine, hematological, genetic, and metabolic diseases. The material describes the interrelationships between the conditions screened in newborn screening laboratories and the markers (analytes) used for screening. For each marker(s), there is (1) an ACT sheet that describes the short-term actions a health professional should follow in communicating with the family and determining the appropriate steps in the follow-up of the infant who has screened positive and (2) an algorithm that presents an overview of the basic steps involved in determining the final diagnosis in the infant.	
The National Newborn Screening and Genetics Resource Center (NNSGRC)	http://genes-r-us.uthscsa.edu
Provides information and resources in the area of newborn screening and genetics to benefit health professionals, the public health community, consumers, and government officials; describes the newborn screening program in each state.	
March of Dimes (MOD)	www.marchofdimes.com/professionals/14332_15455.asp
The MOD recommends that all babies in all states receive newborn screening for 29 disorders for which effective treatment is available. The 29 disorders can be grouped into five categories: (1) amino acid metabolism disorders, (2) organic acid metabolism disorders, (3) fatty acid oxidation disorders, (4) the hemoglobinopathies, and (5) others; the basic description and treatment for each disorder are provided.	
Kaye CI and the Committee on Genetics. Newborn Screening Fact Sheets; Pediatrics 2006, 118: 934–963	http://genes-r-us.uthscsa.edu/NBS_%20Fact%20Sheets.pdf
Fact sheets from the American Academy of Pediatrics that describe each disorder, frequency, treatment, and outcome.	

Expanded NBS programs provide the opportunity for earlier presumptive positive identification. Recent laboratory and clinical developments provide an increased specificity in diagnosis and treatment modalities. Thus, the long-term outcome for persons with inborn errors of metabolism is brighter than in the past. The contrast between outcome without early treatment and expected outcomes with early identification and treatment are shown in Table 5.

III. THE FUNCTIONAL APPROACH TO NUTRITION ASSESSMENT FOR CHILDREN WITH SPECIAL NEEDS

There are five functional areas that should be assessed when measuring a child's nutritional status. These include growth, dietary intake, elimination patterns, medical status and medications, and feeding behaviors and skills [51]. For children with special health care needs, these five areas may all contain specific and relevant information that relates to nutritional status. Special needs vary widely, from conditions that may only mildly affect nutritional status, such as ID, to those with severe and chronic nutritional implications, such as some neuromotor or metabolic disorders. However, these five functional areas can provide a framework for the nutritional assessment regardless of the type of disability.

Levels of nutrition assessment can be categorized from screening (level 1), to more comprehensive but still generalized (level 2), to specialized assessment as part of an interdisciplinary team (level 3) [6]. All three levels of nutrition screening and assessment should cover the five functional areas, but the amount and type of information gathered is different depending on the purpose of the interaction and the training of the health care provider. The purpose of screening is to identify children at risk for nutrition problems, to refer children with more severe problems to the next level of care, and to provide anticipatory guidance and educational materials to children and their families regarding prevention of nutrition problems [6]. Although screening can be completed with information from the caregiver working with health care providers from other disciplines [52], more in-depth nutrition assessment and intervention usually require a registered dietitian (RD), often in conjunction with a specialized interdisciplinary team. For many children with special health care needs, their nutritional status will be assessed, monitored, and intervention provided on an ongoing basis by a pediatric RD (preferably master's prepared) who has been trained in a particular area, such as pulmonary disease, metabolic disorders, or feeding skill development. These dietitians, working with the health care team, utilize their specialized skills to optimize a child's nutritional status in the context of the child's developmental level and medical conditions, as well as the other therapies being provided.

A. Measuring Growth

Growth information is used in nutritional assessment to compare an individual child's growth to that of his or her peers and to evaluate the growth pattern over time. For the information to be valuable, anthropometric measurements must be accurate, the data must be plotted correctly, and the appropriate reference data should be used for comparison [53]. At a minimum, weight, height (or length for children who cannot stand independently), and head circumference (especially for children up to 3 years of age) should be measured. Ideally, some form of body composition measurement is also used, the simplest and most well referenced being skinfold measurements obtained with calipers.

To determine growth rate in children, weight and height for age can be plotted on the CDC growth charts [54]. The CDC charts provide a useful reference to monitor weight in relation to stature, including body mass index (BMI) for children above age 2. In most cases, BMI-for-age charts are not available for special conditions and have not been validated for use with children whose body composition often differs from that of typical children. Triceps and subscapular skinfolds along with arm circumference measurements can further define the growth and body composition data for the child. All of these anthropometric data used together can give a general picture of the nutritional status of the child, whether the child is underweight or overfat, and how linear growth is progressing.

While the CDC growth charts, based on the growth of large numbers of healthy children, can be used to plot growth, some physical conditions affect growth potential. There are specialized charts that may be considered for use with children affected by these conditions. The rationale for using specialized growth charts is clear for genetic disorders where there is evidence that linear growth potential is different. However, the development and use of growth charts for nongenetic conditions, such as cerebral palsy, is complicated by the fact that the underlying cause of the disorder does not change the genetic growth potential. The CDC training module, "Children with Special Health Care Needs," is available at www.cdc.gov for a full discussion of these charts. While specialized growth charts may serve as useful references, they have significant limitations. Generally, they are developed from relatively small, homogeneous samples of children with unknown nutritional status, and data used to develop the charts may have been obtained using inconsistent measuring techniques. One recommendation is to plot the growth patterns of children on both the specialized charts, when appropriate, and the CDC growth charts. This will allow comparisons of growth to the general population of children and to the references for children identified with a similar condition. Assessing serial growth measurements and using disorder-specific growth charts for genetic disorders (e.g., Down syndrome, Prader-Willi syndrome)

TABLE 5 Selected Metabolic Disorders That Require Medical Nutrition Therapy (MNT)

Disorder	Pathway Affected	Outcome without Treatment	Outcome with Early Identification and Treatment	MNT	Supplements/ Medications	Biochemical Parameters Monitored
Disorders of Amino Acid Metabolism						
Phenylketonuria (PKU)	Phenylalanine hydroxylase	Mental retardation ($IQ < 40$)	Normal IQ (requires lifelong treatment)	Low phenylalanine, supplemented tyrosine	Potentially, BH4 for some patients	Plasma phenylalanine, tyrosine
Maple syrup urine disease (MSUD)	Branched-chain ketoacid dehydrogenase complex	Encephalopathy → death	Variable; may be cognitively compromised	Low leucine, isoleucine, valine	L-carnitine	Plasma leucine, isoleucine, valine, alloisoleucine
Glutaric acidemia, type 1	Glutaryl-CoA dehydrogenase	Impaired movement, dystonia, vomiting, seizures, coma	No long-term data	Low lysine, tryptophan	L-carnitine; riboflavin	Electrolytes, blood glucose, plasma amino acids
Homocystinuria (HCY)	Cystathionine-β-synthase	Cardiac problems, organ damage, psychiatric disturbances, death	Variable; may be physically and cognitively compromised	Low methionine, supplement cystine	Folate, Betaine	Plasma methionine, total and free homocystine
Isovaleric acidemia	Isovaleryl-CoA dehydrogenase	Metabolic acidosis → coma → death	Variable; may be cognitively compromised	Low leucine	L-carnitine, glycine	Plasma amino acids (isovaleryl carnitine), urine organic acids (isovaleryl glycine)
Tyrosinemia, type 1	Fumaryl-acetoacetate hydrolase	Liver failure → death	Prevention of neurologic crisis, renal and hepatic failure, rickets	Low tyrosine, phenylalanine	Orfadin® (nitisinone)	Plasma tyrosine, phenylalanine, methionine, succinate, alpha-feto-protein
Disorders of Carbohydrate Metabolism						
Hereditary fructose intolerance (HFI)	Fructose-1-phosphate aldolase B	Hypoglycemia, liver cancer → death	Typical growth, development	Restrict fructose, sucrose		Liver function enzymes
Galactosemia	Galactose-1-phosphate uridyl transferase	Sepsis → severe delays, death	Often learning disabilities	Restrict galactose, use soy formula		Galactose-1-phosphate
Glycogen storage disease, type Ia	Glucose-6-phosphatase	Glycogen stored in liver, severe hypoglycemia, liver cancer → death	Normalization of glucose levels, moderation in liver size, decreased risk of liver cancer	Low lactose, fructose, sucrose, low fat, high complex carbohydrates	Raw cornstarch, iron and calcium supplements	Cholesterol, triglycerides, uric acid, liver function (AST, ALT, GGT), blood glucose levels
Disorders of Fatty Acid Oxidation						
Long-chain Acyl-CoA dehydrogenase deficiency (LCAD)	Long-chain acyl-CoA dehydrogenase	Cardiomyopathy, hepatomegaly, encephalopathy, hypotonia	Variable	Low-fat, low long-chain fatty acids, avoid fasting	L-carnitine, medium-chain triglycerides, thiamine, DHA, glycine	Plasma acylcarnitines, CK, uric acid, liver function studies, CBC, iron status

(continues)

TABLE 5 *(continued)*

Disorder	Pathway Affected	Outcome without Treatment	Outcome with Early Identification and Treatment	MNT	Supplements/ Medications	Biochemical Parameters Monitored
Medium-chain acyl-CoA dehydrogenase deficiency (MCAD)	Medium-chain acyl-CoA dehydrogenase	Encephalopathy, hepatomegaly, disability due to severe episodes	No sequelae if no episodes of severe hypoglycemia	Moderate fat, low medium-chain fatty acids, avoid fasting	L-carnitine	If well, plasma carnitine; if ill, electrolytes, blood glucose
Disorders of Organic Acid Metabolism						
Methylmalonic aciduria (MMA)	Methylmalonyl-CoA mutase or cobalamin cofactor synthesis	Metabolic acidosis → coma → death	Variable; may be cognitively compromised	Low protein, isoleucine, methionine, threonine, valine	L-carnitine, sodium benzoate+, bicitra	Methylmalonic acid, electrolytes, kidney function (BUN, creatinine), carnitine
Propionic acidemia (PA)	Propionyl-CoA carboxylase	Metabolic acidosis → coma → death	Variable; may be cognitively compromised	Low protein, isoleucine, methionine, threonine, valine, long-chain unsaturated fatty acids	L-carnitine, sodium benzoate, biotin, bicitra	Urine organic acids, plasma electrolytes, L-carnitine, kidney function
Disorders of Urea Cycle Metabolism						
Ornithine transcarbamylase deficiency (OTC)	Ornithine transcarbamylase	Hyperammonemia → severe delays or death	Variable; may be cognitively compromised	Low protein, supplement essential amino acids, increased energy	L-carnitine, sodium benzoate, sodium phenylbutyrate, L-arginine	Plasma amino acids, ammonia, electrolytes, L-carnitine, citrulline, arginine
Citrullinemia (CIT)	Argininosuccinate synthetase	Hyperammonemia → severe delays or death	Variable; may be cognitively compromised	Low protein, supplement essential amino acids, increased energy	L-carnitine, sodium benzoate, sodium phenylbutyrate, L-citrulline	Plasma amino acids, ammonia, electrolytes, L-carnitine, citrulline, arginine
Carbamyl phosphate synthetase deficiency (CPS)	Carbamyl phosphate synthetase	Hyperammonemia → severe delays or death	Variable; may be cognitively compromised	Low protein, supplement essential amino acids, increased energy	L-carnitine, sodium benzoate, sodium phenylbutyrate, L-arginine	Plasma amino acids, ammonia, electrolytes, L-carnitine, citrulline, arginine
Argininosuccinic aciduria (ASA)	Argininosuccinate lyase	Hyperammonemia → severe delays or death	Variable; may be cognitively compromised	Low protein, supplement essential amino acids, increased energy	L-carnitine, sodium benzoate, sodium phenylbutyrate	Plasma amino acids, ammonia, electrolytes, L-carnitine, arginine

+Sodium benzoate and sodium phenylbutyrate are chemicals administered to enhance waste ammonia excretion; other compounds producing the same effect are also used.

will help differentiate between normal growth and alterations in growth rate resulting from poor nutrition [6]. To interpret the growth data and understand the causes for unusual growth patterns, further information is needed.

B. Assessing Dietary Intake

Analysis of a child's dietary pattern can predict and prevent nutrient deficiencies and, as with anthropometric data, requires accurate information to be useful. Data collection techniques may range from screening based on food groups and frequency of consumption to conducting a 24-hour recall to computer-assisted analysis of food records kept for 3 or more days. The assessment method should depend on the type and accuracy of information needed. Nutrient adequacy is usually compared to the daily recommended intakes (DRI) for a child's age and size, unless there are known high-risk nutrients (e.g., vitamin D with chronic use of some anticonvulsant medications). Dietary analysis will reveal if there are any specific nutrients or groups of nutrients of concern. In addition, energy intake must be assessed and compared to estimated needs for each child [6]. If anthropometric measurement data provide evidence that the child is overweight or underweight (indicating chronic or current inappropriate energy intake) or has altered lean or adipose tissue patterns, this information should be considered along with the dietary assessment in determining a child's energy needs. For children with short stature or who are overweight or underweight, it is often best to calculate energy intake recommendations based on current height (kcal/cm) rather than weight (kcal/kg).

Information relating to other functional areas—elimination patterns, the use of medications (and nutrition supplements), and feeding skills and behaviors—can be included as part of the dietary assessment. When conducting a 24-hour dietary recall or analyzing a food record, the dietitian should ask questions about (1) food textures; (2) feeding frequency, methods, caregiver and child behaviors and interactions; and (3) type, amount, and timing of supplement and medication use. Along with the nutrient analysis, dietary assessment should include checking for possible drug-nutrient interactions. Also, fluid intake and output and the frequency and texture of bowel movements should be assessed to check for chronic constipation or diarrhea, which may compromise feeding behaviors and nutritional status. Children with special needs may have significant feeding skill delays or feeding patterns that fit their developmental level rather than their chronological age, which will alter the type, texture, and quantity of foods they are able to consume. This may pose a challenge to meeting the nutrient and energy needs of the child in the volume of food consumed. Food preferences and refusals, which may be related to past or current medical conditions, need to be taken into account when designing dietary interventions. Often, working with a team that includes the parent/caregiver, a feeding therapist, and physician can aid in determining the most appropriate type and texture of foods to use to facilitate the child's optimal feeding skill development as well as his or her nutritional intake.

C. Considering Medical Conditions

Medical conditions can have an impact on nutritional status in a number of ways (shown in Table 1). Some conditions actually alter a child's energy and other nutrient needs, whereas others interfere with adequate nutrient intake or utilization. Examples of conditions that increase energy demands are cardiac and pulmonary complications of prematurity, and cerebral palsy where there are significant hypertonia/spasticity or athetoid movements. Other chronic illnesses associated with high energy needs are HIV/AIDS, cystic fibrosis, and mitochondrial disorders. Some genetic disorders are associated with lowered metabolic rates resulting in decreased energy needs (e.g., Down syndrome and Prader-Willi syndrome). Conditions associated with reduced or low muscle mass (as in muscular dystrophies) also result in decreased energy needs. In genetic disorders where short stature occurs, overall food, nutrient, and energy needs may be lower than age peers throughout adolescence and adulthood. If a child has reduced activity levels because of immobility, energy demands will also be lower.

With a few exceptions, such as copper deficiency in Wilson's disease and some metabolic disorders, most medical conditions do not have a primary effect on an individual's micronutrient needs [53]. There is not enough evidence to suggest that nutrients associated with energy metabolism be adjusted even with very high or low energy intakes. Therefore, the DRIs are still the most appropriate guideline for nutrient requirements for children with special needs [53]. In the self-study curriculum "Nutrition for Children with Special Health Care Needs," available online at www.pacificwestmch.org, the following questions are suggested to assess whether and how a medical condition will affect nutritional status:

1. Does the condition (or medication used to treat the condition) have an effect on the child's nutrient needs?
2. Does the condition change the types of foods the child can eat?
3. Does the condition alter the amount of food that the child can reasonably be expected to consume?
4. Does the condition affect the amount of time the child can spend at the table (eating)? Does this make a smaller intake likely?
5. Does the medication or therapy schedule interfere with scheduled meal or snack times?

Consideration of these issues and the impact on overall nutrition intake and nutritional status should be the final component of the nutrition assessment for children with special needs.

IV. EVIDENCE-BASED INTERVENTIONS FOR SELECTED CONDITIONS

A. Inborn Errors of Metabolism

Inborn errors of metabolism (IEM) are the classic example of the secondary prevention of a developmental disability through nutrition therapy. The basic concepts, principles, and strategies of treatment of selected disorders of protein, carbohydrate, and fat metabolism are presented here. A complete discussion of diagnosis and management of the array of metabolic disorders can be found elsewhere [55, 56]. Table 6 outlines the essential components of medical nutrition therapy (MNT) for treatment of metabolic disorders. Although the clinical and biochemical presentation of each metabolic disorder is unique and the disorders present a range from mild to life-threatening illness, metabolic disorders can be thought of as a group in which the absence or inactivity of a specific enzyme or cofactor causes the buildup of the substrate and deficiency of the product. The goal of treatment for inborn errors of metabolism is to strive for correction of the biochemical abnormality. The outcome of treatment for these disorders is variable and depends on early diagnosis and intensive and continuous intervention. Medical nutrition therapy is the primary mode of treating metabolic disorders.

1. Principles of Medical Nutrition Therapy

The two major principles of MNT for IEM are (1) to mitigate the effect of the altered enzyme by modifying components of dietary intake to adjust the environment at the cellular level and (2) to provide protein, energy, and nutrients to support growth and development.

For many disorders, the treatment is determined by the identification of the missing or inactive enzyme. In an effort to modify the detrimental effect of the decreased or absent enzyme activity, the paradigm of "working around" the enzyme is used. In many cases, decreasing the substrate available for the reaction and supplementing the product to promote "normal" blood levels prevents or decreases the deleterious effects of the disorder. For example, in the treatment of phenylketonuria (PKU), phenylalanine (substrate) is restricted because of the absence or inactivity of phenylalanine hydroxylase (enzyme), and tyrosine (product) is supplemented.

TABLE 6 Components of Medical Nutrition Therapy for Metabolic Disorders

Identify precise modifications required for treatment of the disorder.
Provide nutritional surveillance to ensure that MNT is adequate.
Provide a mechanism for follow-up of child for symptoms of nutritional deficiency or toxicity.
Provide emotional and educational support for child and family.

In some disorders the absent or inactive enzyme is further down the amino acid degradation pathway and may affect the metabolism of two or more amino acids, for example, leucine, isoleucine, and valine in maple syrup urine disease (MSUD). The affected amino acids in most disorders of amino acid metabolism are "essential"—that is, they cannot be synthesized by the body and therefore must be provided in the diet. These critical essential amino acids (EAA) must be provided at a level that promotes growth and development but is restricted enough to prevent toxic buildup of amino acid(s) that cannot be metabolized.

In some disorders, the additional step of enhancing enzyme activity by supplying its cofactor can be helpful. An example of this is providing pharmacological doses of biotin in biotinidase deficiency [57] or vitamin B_6 on which cystathionine β-synthase is dependent in some forms of homocystinuria [58].

In metabolic disorders of carbohydrate metabolism, such as galactosemia, the nutrient of concern (galactose) is not essential. Therefore, the goal of effective MNT is to eliminate as much of the exogenous component as possible from the diet. There is no established requirement for galactose because it is produced endogenously [59]. Other sources of nourishment need to be provided to compensate for the omitted foods and their nutrients.

Disorders of fatty acid metabolism require a source of energy other than fat, as fat cannot be metabolized to meet energy needs. Fatty acids of specific carbon lengths are often minimized or eliminated, depending on whether or not the fatty acid is essential for growth and development. For example, in long-chain acyl-CoA dehydrogenase (LCAD) or very long-chain acyl-CoA dehydrogenase (VLCAD), shorter-chain fats can be metabolized and so are often supplemented; for example, medium-chain triglycerides are provided as MCT oil [55]. However, in medium-chain acyl-CoA dehydrogenase deficiency (MCAD) sources of medium-chain fat should be modestly restricted. Other essential components of treatment are (1) supplementation of L-carnitine, an amino acid that functions in the transport of fatty acid acyl-coA esters during mitochondrial beta-oxidation, and (2) avoidance of fasting because of the accumulation of partially oxidized metabolites associated with impaired energy production [60].

2. Providing Medical Nutrition Therapy

The general principles of protein and energy management that support general nourishment as well as the issues of biochemical control specific to the disorder must be addressed [61, 62]. The infant or child should be provided adequate amino acids, total protein, nitrogen, and energy to support growth. Energy needs may be increased when L-amino acids provide the protein equivalent, and maintaining an adequate energy intake is essential in preventing catabolism. It must also be noted that suppression of destructive metabolites can produce striking biochemical and clinical improvement.

3. Providing Adequate Nourishment

For some disorders, total protein is not restricted, but the composition of the protein may be adjusted. For example, the treatment of PKU requires the restriction of phenylalanine intake but not total protein. This is accomplished by using a specially designed semisynthetic medical formula. Infants and children with IEM must obtain most of their protein from specialized metabolic formulas. Natural foods seldom provide more than 25% (often 10%) of the protein requirements of infants and children with amino acid, organic acid, or urea cycle disorders.

Protein in the specialized formulas is provided as individual L-amino acids (excluding those amino acids which are contraindicated for each condition). L-amino acids are more readily oxidized than intact protein; thus, the requirement for protein when L-amino acids are the protein source is greater than usual. Adequate protein and energy are required to maintain an anabolic state. If adequate formula is not prescribed or consumed, a catabolic state will develop, causing both high plasma amino acids levels and clinical problems.

The urea cycle disorders require the restriction of overall protein intake because excess nitrogen from any source can be neurotoxic. However, it is also imperative to provide enough protein and energy to support growth. The formulas for the treatment of urea cycle disorders have a high concentration of amino acids to assure ready incorporation into protein. Most children with these disorders also require medications to enhance the excretion of excess nitrogen through secondary pathways other than the urea cycle. For infants and children with urea cycle disorders, catabolism from excess protein, weight loss, illness, or infection is a danger. Hyperammonemia can occur rapidly (in several hours) and be life threatening. The total amount of protein tolerated by the individual child depends on residual enzyme activity, age, and growth rate.

4. Supplying Restricted Amino Acids

Small amounts of the restricted essential amino acids (EAA) specific to each disorder are required for growth and development. These EAA are provided to the infant by including small amounts of proprietary infant formula in the metabolic formula mixture. As the child grows and matures, adding small amounts of cow or soymilk or fruits, vegetables, and grains can provide these essential amino acids.

5. Breastfeeding

Some infants with metabolic disorders are able to maintain low and stable plasma amino acid levels with a combination of breast and metabolic formula feedings. Many infants are too metabolically fragile to maintain appropriate plasma levels with breastfeeding; for example, some infants with PKU are able to tolerate partial breastfeeding and maintain low plasma phenylalanine levels. Breastfeeding is overtly contraindicated for infants with some disorders such as galactosemia.

6. Feeding

Poor appetite is not uncommon for children with IEM, especially urea cycle disorders, MSUD, and the organic acidemias. Reasons for poor appetite and poor feeding may be organic, behavioral, or both. Many children with IEM who have had acute episodes may also have a history of vomiting which may lead to feeding avoidance.

7. Growth

Poor growth is often a reflection of several factors: (1) infants who have endured a severe neonatal illness may require an extended time of appropriate nourishment before they catch up in growth, (2) frequent febrile illness may interfere with achieving expected physical growth, or (3) poor metabolic control or an inadequate protein or energy intake may interfere with growth. Many children with IEM who have little or no enzyme activity are medically fragile, and it is difficult to maintain metabolic balance and support typical growth patterns.

8. Monitoring for Children with Metabolic Disorders

The frequency of monitoring biochemical parameters depends on the age and health status of the child. Maintaining metabolic balance for children with IEM requires frequent and intensive monitoring of biochemical parameters specific to the disorder as well as those that reflect general nutritional status. The goal of treatment for all disorders is to achieve biochemical levels at or near the normal range. Laboratory parameters that are frequently monitored include plasma amino acids, urine organic acids, hematological status, protein status, electrolytes, blood lipids, and ammonia levels. A general plan to guide biochemical assessment is shown in Table 5.

9. Specialized Metabolic Team

As a group, children with IEM comprise a small percentage of the pediatric population and even the most common of these disorders is rare in the general population. However, the health care needs of these children are specific and urgent. The American Academy of Pediatrics has recommended that a team experienced in management supervise the therapy of these children [46]. Effective treatment generally requires the expertise of a geneticist, dietitian, nurse, genetic counselor, psychologist, and neurologist. The complex nutritional and medical management of these children cannot occur effectively without the follow-up and support of the community teams. Communication between the team at the tertiary center, the community, and the family is crucial for supporting the best possible medical, nutritional, and intellectual outcome for these children.

TABLE 7 Feeding Problems in Children with Special Healthcare Needs

Feeding Difficulty	Frequent Causes	Associated Conditions
Delayed or slow development	Hypertonia	Developmental delay
Posturing or seating difficulties	Hypotonia	Cerebral palsy
Persistence of primitive reflexes	Hypersensitivity	Cleft lip and cleft palate
Craniofacial anatomic problems	Developmental immaturity (prematurity)	Down syndrome (trisomy 21)
Uncoordinated sucking, chewing, swallowing	Gastroesophageal reflux	Prader-Willi syndrome
Behavioral refusals	Unpleasant intrusions into the oral cavity	Rett syndrome
Decreased appetite	Unpleasant feeding experiences (past or present)	Muscular dystrophy
	Constipation, increased secretions, decreased gastric motility	Williams syndrome
	Medications	Myelomeningocele with accompanying Arnold-Chiari malformation
		Autism
		Chronic respiratory diseases (cystic fibrosis, BPD, RDS)

B. Prader-Willi Syndrome

Prader-Willi syndrome (PWS) is caused by a microdeletion on paternally inherited genes on chromosome 15. PWS is associated with early failure to thrive, hypotonia, abnormal body composition, hypogonadism, short stature, and behavioral and learning issues [63]. Hyperphagia, which usually begins toward the end of the first year and is thought to be due to a hypothalamic abnormality, leads to a lack of satiety. Thus, delay in identification and treatment of PWS can lead to the onset of gross obesity after infancy.

Abnormal food-seeking behavior is characteristic of PWS and includes hoarding and foraging for food as well as eating items generally considered to be inedible such as garbage, pet food, or uncooked frozen food. Infants and children with PWS demonstrate a low metabolic rate and thus decreased energy needs. An intake as low as 1000 to 1200 kcal/day is often required to maintain a stable weight. Strict and consistent behavioral limits and the establishment of regular routines may be helpful behavioral management strategies for food intake and behavioral concerns for children with PWS [64].

Recently, controlled trials have shown that children with PWS benefit from growth hormone therapy from infancy through childhood and into adulthood [65].

C. Dietary Interventions for Children with Feeding Problems

Among children with developmental delays and special health care needs, the estimated incidence of feeding disorders ranges from 33% up to 80%, depending on the definitions of feeding disorders [4, 66]. Problems with feeding occur when the components of normal feeding development are missing or delayed, there are significant behavioral refusals due to multiple etiologies, there are medical interventions and medications that interfere with appetite and regulation, or there are physiological problems related to the condition or disability (Table 7). An interdisciplinary team that can determine which of these factors are interacting as causes of the feeding difficulties is helpful in determining and coordinating appropriate interventions. A detailed description of feeding problems, developmental conditions leading to feeding problems, and suggested feeding and nutrition assessment strategies is included in the self-study curriculum referred to earlier, "Nutrition for Children with Special Health Care Needs," Module 3: Feeding Skills (www.pacificwestmch.org).

The challenge for the nutritionist on the interdisciplinary feeding team is to assess nutritional status and dietary intake and determine how best the child can meet his or her nutritional needs using the most developmentally appropriate feeding methods and forms of foods whenever possible. Use of dietary supplements and tube feedings, including nutritionally complete formulas, is sometimes necessary to supply the necessary energy, protein, and other nutrients needed. It is often best to use the child's size and developmental stage (e.g., pubertal, prepubertal) to estimate nutrient needs, rather than the child's chronological age. When estimating energy needs for children with feeding problems who are smaller, shorter, or heavier than their same-aged peer, it is better to use the child's height or length, rather than weight or ideal body weight or age, to determine the recommended caloric intake. As always, the goal of nutrition intervention is to provide adequate nutrients to support optimal growth and development while meeting the identified feeding goals for the child.

D. Autism Spectrum Disorders

Autism spectrum disorders (ASD) is a group of disorders that includes pervasive developmental disorder–not otherwise specified (PDD-NOS), Asperger syndrome, and autism. All diagnostic instruments rely on criteria outlined by the DSM-IV [67], which include stereotyped behaviors, interactions, and activities, as well as limited social and communication abilities. The level of cognitive impairment can range from mild to profound. Although the reasons are not clear, the prevalence of autism appears to be increasing in the United States; recent surveillance data (from only 14 counties) published by the CDC indicate that 1/150 American children has a potential diagnosis of autism [68]. For the purposes of this discussion, the broad category of ASD is used to encompass all nutritional issues that may be seen in this population.

Although the diagnosis itself need not trigger an automatic referral for nutrition assessment, some behaviors seen with ASD can lead to feeding or eating disorders. For example, a child with ASD may have difficulty eating school lunch with a group of peers because of other distractions in the cafeteria. If the child eats a broad range of foods that adequately meet protein, energy, and micronutrient needs, then basic nutrition education is appropriate, focusing on strategies for working with picky eaters in order to prevent development of a limited food repertoire. If the clinician determines that the child is at moderate to high risk, then the child can be referred for further assessment; in some cases an interdisciplinary care team may be required.

1. Anthropometric

The growth patterns of children with ASD are thought to be normal. However, studies have emerged showing an increased rate of head growth in children with ASD [69]. Others have shown an increased rate of linear growth, weight gain, and body mass index [70]. Anthropometric data for this population should still consist of length, weight, weight/length, and head circumference. When the child is found to have a large head size, that should be considered when making weight/height comparisons, as they could be deceptively increased. In children with ASD, it is not unusual to see a very low or very high BMI as a result of feeding issues.

2. Biochemical

Because of an increased incidence of a limited food repertoire and resultant limited nutrient intake, children with ASD are at particular risk for anemia. Similarly, an increased prevalence of pica may suggest the need to measure hemoglobin and lead levels to rule out lead toxicity [71].

3. Clinical

Comorbidities often seen in the ASD population, which may affect feeding or nutritional status, include fragile X syndrome, seizure disorder, and immune system dysregulation [72, 73]. Since the 1970s, researchers have consistently found a high rate of gastrointestinal (GI) complaints in this population, reporting a prevalence ranging from 9% to 84%; this wide variation is due to the use of small sample sizes and differing definitions of ASD. The most common GI complaints are lactose intolerance, gastroesophageal reflux, constipation, diarrhea, liver problems, feeding problems, and gastric-tube feedings [74]. More in-depth studies indicate an increased incidence of lymphoid nodular hyperplasia and lymphoid follicular growth in the GI tract and increased permeability of the bowel [71].

There are currently no drugs used specifically to treat ASD, but drug-nutrient interactions should be assessed if children are prescribed drugs to treat comorbidities such as seizures or psychiatric symptoms. The most common drugs used in this population that have significant drug/nutrient interactions are lithium, Depakote, and monoamine oxidase inhibitors (MAOIs). The use of secretin as an experimental therapy in the ASD population is currently being investigated and may have GI side effects [71].

4. Dietary/Feeding Issues

Generally, the diet of children with ASD meets or exceeds the RDA for protein, carbohydrates, and fat [71, 72]. However, pica may be more common than in the general population [71], and nutrition education and intervention may need to address this behavior. Rigid mealtime patterns or limited food repertoires are other common symptoms [71, 74, 75]. For example, the child may insist on only red foods or demand a particular plate. The child may have a preference for only salty or only sweet foods or only foods of a certain texture; these behaviors may make family meals difficult, and severe food jags can lead to nutrient deficiency. If entire food groups are missing, or if there is excessive intake of a particular nutrient, a vitamin/mineral supplement may be appropriate. Therapies to expand food choices vary and can be hard to implement in this population. Families should be encouraged to offer new items with preferred foods [76] and to offer them repeatedly, possibly over a period of months. Interdisciplinary intervention, including a behavior specialist in such cases, or an occupational or speech therapist where texture or oral issues present as feeding disorders can be helpful.

The child with ASD may also have difficulties with changes in the environment [74], which may lead to problems with school mealtimes where a loud and boisterous cafeteria may be overwhelming. The child with ASD may benefit from taking meals in the classroom with an aide, in order to minimize distracting stimuli.

5. Alternative Nutrition Therapies

The fact that the etiology of ASD is not well understood makes the diagnosis difficult for many families to accept and they may try multiple interventions to improve their child's symptomatology. Highly publicized case studies often lead families to choose an alternate diet for their child. Although there are many alternative therapies currently promoted as treatments, the most widely investigated is the gluten/casein free diet. Those who promote the diet theorize that metabolites of these foods build up in the bloodstream, penetrate the blood-brain barrier, and stimulate opioid receptors, causing the behaviors seen in autism [72]. An alternate theory is that children with ASD have an increased incidence of GI disorders such as gluten intolerance [71] because of autoimmune responses [77], as histopathologies of children with ASD have shown an irregular immune response [78]. Inflammation of the GI tract resulting from exposure to irritants is uncomfortable or even painful. Because of a limited ability to communicate, the child with ASD may not be able to make known his or her discomfort, which may worsen behavioral symptoms. Because most studies use parent or caregiver report of behavior changes in children after implementing the diet, a significant potential confounder is the parental placebo effect, as parents seeking to find any treatment for their child may be susceptible to suggestion. Also, most studies have not been of a prospective, randomized, blinded design [79]; for example, they begin with a population of children with ASD who already have GI issues [80]. This makes it difficult to generalize findings to the entire ASD population. It remains unclear whether the gluten/casein-free diet is a valid treatment for ASD or simply a treatment for a comorbid gluten/casein sensitivity in a subgroup of these children. Therefore, there are now no recommended diet therapies for ASD. As long as the diet provides sound nutrition for growth and development, it is the dietitian's role to support the family in its choice and offer current nutrition information.

V. CONCLUSION

In this brief review, we have provided examples of nutrition issues related to the role of nutrition in the primary (maternal nutrition) and secondary (MNT for IEM) prevention of developmental disabilities and in the secondary prevention of nutrition-related disorders that are commonly seen in children with special needs. Emphasis is given to several important concepts. First is rapidly emerging evidence regarding the contributory role of nutrition in the etiology of preterm birth, low birth weight, and some birth defects, as well as its essential role in early postnatal development. Because of the growing numbers of nutrients shown to be involved in normal development, the importance of whole foods versus supplements is also becoming apparent. The second major point, which follows from the first, is that early nutrition intervention is crucial to the individual child's attaining his or her highest potential. Third, we have presented a functional approach to assessment, early detection, and treatment of nutrition problems in these children, stressing the point that although diagnoses may differ, the nutrition issues are often similar. Finally, although we have not covered the importance of training dietetics professionals in this area or of the need to improve families' access to nutrition services for children with special needs, we refer the reader to the latest position paper of the American Dietetic Association for an excellent review of the legislative history and a presentation of the challenges that remain for the profession [5].

Acknowledgments

The authors would like to acknowledge Jennifer Webb, MS, RD, who contributed information for the section on autism.

References

1. McPherson, M., Arango, P., Fox, H., Lauver, C., McManus, M., Newacheck, P. W., Perrin, J. M., Shonkogg, J. P., and Strickland, B. (1998). A new definition of children with special health care needs. *Pediatrics* **102**, 137–140.
2. van Dyck, P., Kogan, M. D., McPherson, M. G., Weissman, G. R., and Newacheck, P. W. (2004). Prevalence and characteristics of children with special health care needs. *Arch. Pediar. Adolesc. Med.* **158**, 884–890.
3. Ireys, H. T., and Katz, S. (1997). The demography of disability and chronic illness among children. *In* "Mosby's Resource Guide to Children with Disabilities and Chronic Illnesses." Mosby, St. Louis.
4. Lucas, B., and Nardella, M. (1998). "Cost Considerations: The Benefits of Nutrition Services for a Case Series of Children with Special Health Care Needs in Washington State." Washington State Department of Health, Olympia, WA.
5. American Dietetic Association. (2004). Providing nutrition services for infants, children and adults with developmental disabilities and special health care needs. *JADA* **104**(1), 97–107.
6. Baer, M. T., and Harris, A. B. (1997). Pediatric nutrition assessment: identifying children at risk. *J. Am. Diet. Assoc.* **97** (S2), S107–S115.
7. GAO. (1992). "Early Intervention: Federal Investments Like WIC Can Produce Savings," GAO publication HRD-92-18. U.S. General Accounting Office, Gaithersburg, MD.
8. King, J. C. (2003). The risk of maternal nutritional depletion and poor outcomes increases in early or closely spaced pregnancies. *J. Nutr.* **133**,1732S–1736S.
9. Khoury, J., Henriksen, T., Christophersen, B., and Tonstad, S. (2005). Effect of a cholesterol-lowering diet on maternal, cord and neonatal lipids, and pregnancy outcome: a randomized clinical trial. *Am. J. Obstet. Gynecol.* **193**(4), 1292–1301.
10. Scholl, T. O., Hediger, M. L., Schall, J. I., Fischer, R. L., and Khoo, C. S. (1993). Low zinc intake during pregnancy: its association with preterm and very preterm delivery. *Am. J. Epidemiol.* **137**, 155–124.
11. Scholl, T. O., Hediger, M. L., Schall, J. I., Khoo, C. S., and Fischer, R. L. (1996). Dietary and serum folate: their influence on the outcome of pregnancy. *Am. J. Clin. Nutr.* **63**, 520–525.
12. Scholl, T. O. (1998). High third trimester ferritin concentration: association with very preterm delivery, infection, and maternal nutritional status. *Obstet. Gynecol.* **92**, 161–166.

13. Bogden, J. D., Kemp, F. W., Chen, X., Stagnaro-Green, A., Stein, T. P., and Scholl, T. O. (2006). Low-normal serum selenium early in human pregnancy predicts lower birth weight. *Nutr. Res.* **29**(10), 497–502.
14. Leonard, H., de Keirk, N., Bourke, J., and Bower, C. (2006). Maternal health in pregnancy and intellectual disability in the offspring: a population-based study. *Ann. Epidemiol.* **16**(6), 448–454.
15. Czeizel, A. E., and Puho, E. (2005). Maternal use of nutritional supplements during the first month of pregnancy and decreased risk of Down's syndrome: case-control study. *Intl J. Appl. Basic Nutr. Sci.* **21**, 698–704.
16. Jedrychowski, W., Jankowski, J., Flak, E., Skarupa, A., Mroz, E., Sochack-Tatara, E., Lisowska-Miszczyk, I., Szpanowska-Wohn, A., Rauh, V., Skolicki, Z., Kaim, I., and Perera, F. (2006). Effects of prenatal exposure to mercury on cognitive and psychomotor function in one-year-old infants: epidemiologic cohort study in Poland. *Ann. Epidemiol.* **16**(6), 439–447.
17. Hibbeln, J. R., Davis, J. M., Steer, C., Emmett, P., Rogers, I., Williams, C., and Golding, J. (2007). Maternal seafood consumption in pregnancy and neurodevelopmental outcomes in childhood (ALSPAC study): an observational cohort study. *Lancet* **369**, 578–585.
18. Colombo, J., Dannass, K. N., Shaddy, D. J., Kundurthi, S., Maidranz, J. M., Anderson, C. J., Blaga, O. M., and Carlson, S. E. (2004). Maternal DHA and the development of attention in infancy and toddlerhood. *Child Dev.* **75**, 1254–1267.
19. Willatts, R., Forsyth, J. S., DiModugno, M. K., Varma, S., and Colvin, M. (1998). Effect of long-chain polyunsaturated fatty acids in infant formula on infant cognitive function. *Lipids* **33**, 973–980.
20. Strain, J. J., Bonham, M. P., Duffy, E. M., Wallace, J. M. W., Robson, P. J., Clarkson, T. W., and Shamlaye, C. (2004). Nutrition and neurodevelopment: the search for candidate nutrients in the Seychelles Child Development Nutrition Study. *Seychelles Med. Dent. J.* **7**(1), 77.
21. Raymond, L. J., and Ralston, N. V. C. (2004). Mercury: selenium interactions and health implications. *Seychelles Med. Dent. J.* **7**(1), 72–77.
22. Fedrick, J. (1970). Anencephalus: variation with maternal age, parity, social class and region in England, Wales and Scotland. *Ann. Hum. Genet.* **34**, 31–38.
23. Smithells, R. W., Sheppard, S., and Schorah. (1976). Vitamin deficiencies and neural tube defects. *Arch. Dis. Child.* **51**, 944–950.
24. Laurence, K. M., James, N., Miller, M. H., Tennant, G. B., and Campbell, H. (1981). Double blind randomized controlled trial of folate treatment before conception to prevent recurrence of neural tube defects. *Fr. Med. J.* **282**, 1509–1511.
25. MRC Vitamin Study Research Group. (1991). The epidemiology of neural tube defects: a review of the Medical Research Council Vitamin Study. *Lancet* **338**, 131–137.
26. CDC. (2004). Spina bifida and anencephaly before and after folic acid mandate—United States, 1995–1996 and 1999–2000. *MMWR* **53**(17), 362–365.
27. CDC. (1992). Recommendations for the use of folic acid to reduce the number of cases of spina bifida and other neural tube defects. *MMWR* **41**, RR-14.
28. Groenen, P. M. W., van Rooij, I. A. L. M., Peer, P. G. M., Ocke, M. C., Zielhuis, G. A., and Steegers-Theunissen, R. P. M. (2004). Low maternal dietary intakes of iron, magnesium, and niacin are associated with spina bifida in the offspring. *J. Nutr.* **134**, 1516–1522.
29. Missmer, S. A., Suarez, L., Felkner, M., Wang, D., Merrill, A. H., Jr, Rothman, K. J., and Hendricks, K. A. (2006). Exposure to fumonisins and the occurrence of neural tube defects along the Texas-Mexico border. *Environ. Health Perspect.* **114**(2), 237–241.
30. Suarez, L., Hendricks, K., Cooper, S. P., Sweeney, A. M., Hardy, R. J., and Larsen, R. D. (2003). Maternal serum B-12 levels and risk for neural tube defects in a Texas-Mexico border population. *Ann. Epidemiol.* **13**, 81–88.
31. Ray, J. G., Wyatt, P. R., Thompson, M. D., Vermeulen, M. J., Meier, C., Wong, P.-Y., Farrell, S. A., and Cole, D. E. C. (2007). Vitamin B-12 and the risk of neural tube defects in a folic-acid fortified population. *Epidemiology* **18**(3), 362–366.
32. Shaw, G. M., Carmichael, S. L., Laurent, C., and Rasmussen, S. A. (2006). Maternal nutrient intakes and risk of orofacial clefts. *Epidemiology* **17**, 285–291.
33. Krapels, I. P. C., van Rooij, I. A. L. M., Ocke, M. C., van Cleef, B. A., Duijpers-Jagtman, A. M. M., and Steegers-Theunissen, R. P. M. (2004). Maternal dietary B vitamin intake, other than folate, and the association with orofacial cleft in the offspring. *Eur. J. Nutr.* **43**, 7–14.
34. American Academy of Pediatrics (APA), Committee on Substance Abuse and Committee on Children with Disabilities. (2000). Fetal alcohol syndrome and alcohol-related neurodevelopmental disorders: policy statement. *Pediatrics* **196**(2), 359–361.
35. Suter, P. M. (2001). Alcohol: its role in health and nutrition. *In* "Present Knowledge in Nutrition" (B. A. Bowman and R. M. Russell, Eds.), 8th ed. ILSI Press, Washington, DC.
36. Bailey, L. B., Moyers, S., and Gregory, J. F. (2001). Folate. *In* "Present Knowledge in Nutrition" (B. A. Bowman and R. M. Russell Eds.), 8th ed. ILSI Press, Washington, DC.
37. Dibley, M. J. Zinc. (2001). *In* "Present Knowledge in Nutrition"(B. A. Bowman and R. M. Russell, Eds.), 8th ed. ILSI Press Washington, DC.
38. Oakley, G. P. (2007). When will we eliminate folic acid-preventable spina bifida? *Epidemiology* **1**(3), 367–368.
39. National Vital Statistics Reports. (2006). Vol. 55, No. 1, September 29, 2006.
40. Anderson, D. M. (1999). Nutrition for premature infants. *In* "Handbook of Pediatric Nutrition" (P. Q. Samour, K. K. Helm, and C. E. Lang, Eds.), 2nd ed., pp. 43–63. Aspen, Gaithersburg, MD.
41. Wilson-Costello, D., Friedman, H., Minich, N., Fanaroff, A. A., and Hack, M. (2005). Improved survival rates with increased neurodevelopmental disability for extremely low birth weight infants in the 1990s. *Pediatrics* **116**, 997–1003.
42. Poets, C. F. (2004). Gastroesophageal reflux: a critical review of its role in preterm infants. *Pediatrics* **113**, e128–e132.
43. Gewolb, I. H., and Vice, F. L. (2006). Abnormalities in the coordination of respiration and swallow in preterm infants with bronchopulmonary dysplasia. *Dev. Med. Child Neurol.* **48**(7), 595–599.

44. Heiman, H., and Schanler, R. J. (2007). Enteral nutrition for premature infants: the role of human milk. *Sem. Fetal Neonatal Med.* **1291**, 26–34.
45. Simpson, C., Schanler, R., and Lau, C. (2002). Early introduction of oral feeding in preterm infants. *Pediatrics* (Sep) **110**(3), 517–522.
46. Kaye, C. I. and the Committee on Genetics. (2006). Newborn screening fact sheets. *Pediatrics* **118**, 2006 934–963.
47. Cipriano, L. E., Rupar, C. A., and Zaric, G. S. (2007). The cost-effectiveness of expanding newborn screening for up to 21 inherited metabolic disorders using tandem mass spectrometry: results from a decision-analytic model. *Value in Health* **10**, 83–97.
47a. Newborn screening: toward a uniform screening panel and system—executive summary. (2006). *Pediatrics* **117**, S296–S307.
48. Watson, M. S., Mann, M. Y., Lloye-Puryear, M. A., Rinaldo, P., and Howell, R. R. (2006). American College of Medical Genetics Newborn Screening Expert Group. *Pediatrics* **117**, S296–S307.
49. March of Dimes. (2007) www.marchofdimes.com/pnhec/298_834.asp.
50. American Society of Human Genetics. (2007). www.acmg.net/resources/policies/ACT/condition-analyte-links.htm. Accessed August 2, 2007.
51. Lucas, B. L., Feucht, S. A., and Grieger, L. Eds. (2004). "Children with Special Health Care Needs: Nutrition Care Handbook." Pediatric Nutrition Practice Group and Dietetics in Developmental and Psychiatric Disorders, American Dietetic Association.
52. Bujold, C. R. (2006). Validation of a Parent Completed Nutrition Screening Questionnaire for Children with Special Health Care Needs (thesis), University of Washington. (1994).
53. Pacific West MCH Distance Learning Network. (2005). Self-study modules: "Nutrition for Children with Special Health Care Needs," Module 3: Feeding Skills, and Group-study modules: "Nutrition for Children with Special Health Care Needs"(www.pacificwestmch.org).
54. National Center for Health Statistics (NCHS). (2000). www.cdc.gov/growthcharts.
55. Scriver, C. R., Beaud, A. L., Sly, W. S., and Valle, D., Eds. (1995). "The Metabolic and Molecular Basis of Inherited Disease." McGraw Hill, New York.
56. Online Mendelian Inheritance in Man, OMIM (TM). (2000). McKusick-Nathans Institute for Genetic Medicine, Johns Hopkins University (Baltimore, MD) and National Center for Biotechnology Information, National Library of Medicine, Bethesda, MD, www.ncbi.nlm.nih.gov/omim. Accessed July 25, 2007.
57. Wolf, B. Biotinidase deficiency. (2007). *GeneReviews*, www.geneclinics.org. Accessed July 29, 2007.
58. Picker, J. D., and Levy, H. L. (2007). Homocystinuria caused by cystathionine β-synthase deficiency. *GeneReviews*, www.geneclinics.org. Accessed July 29, 2007.
59. Bosch, A. M. (2006). Classical galactosemia revisited. *J. Inherit. Metab. Dis.* **29**, 516–525.
60. Matern, D., and Rinaldo, P. (2007). Medium chain acyl-Coenzyne A dehydrogenase deficiency. *GeneReviews*, www.geneclinics.org. Accessed July 29, 2007.
61. Acosta, P. B., and Yanicelli, S. (1993a). Nutrition support of inherited disorders of amino acid metabolism: part 1. *Top. Clin. Nutr.* **9**, 65–82.
62. Acosta, P. B., and Yanicelli, S. (1993b). Nutrition support of inherited disorders of amino acid metabolism: part 2. *Top. Clin. Nutr.* **9**, 48–72.
63. Chen, C., Visootsak, J., Dills, S., and Graham, J. M. (2007). Prader-Willi syndrome: an update and review for the primary pediatrician. *Clin. Pediatr. Online,* May 23, 2007.
64. Benarroch, F., Hirsch, H. J., Genstil, L., Landau, M. A., and Gross-Tsur, V. (2007). Prader-Willi syndrome: medical prevention and behavioral challenges. *Child Adol. Psych. Clin. N. Amer* **16**, 695–708.
65. Angulo, M. A., Castro-Magana, M., Lamerson, M., Arguello, R., Accacha, S., and Khan, A. (2007). Final adult height in children with Prader-Willi syndrome with and without human growth hormone treatment. *Am. J. Med. Genet.,* Part A **143A**, 1456–1461.
66. Linsheid, T., Budd, K., and Rasnake, L. (2003). Pediatric feeding problems. *In* "Handbook of Pediatric Psychology" (M. Roberts, Ed.), 3rd ed., pp. 481–498. Guilford Press, New York.
67. American Psychiatric Association. (1994). "Diagnostic and Statistical Manual of Mental Disorders," 4th ed. American Psychiatric Association, Washington, DC.
68. Kuehn, B. M. (2007). CDC: autism spectrum disorders common. *JAMA* **297**(9), 940.
69. Dissanayake, C., Bui, Q. M., Huggins, R., and Loesch, D. Z. (2006). Growth in stature and head circumference in high-functioning autism and Asperger disorder during the first 3 years of life. *Dev. Psychopathology* **18**, 381–393.
70. Mills, J. L., Hediger, M. L., Molloy, C. A., Chrousos, G. P., Manning-Courtney, P., Yu, K. F., Brasington, M., and England, L. J. (2007). Elevated levels of growth-related hormones in autism and autism spectrum disorder. *Clin. Endocrinology* Epub June 4, 2007.
71. Erickson, C. A., Stigler, K. A., Corkins, M. R., Posey, D. J., Fitzgerald, J. F., and McDougle, J. (2005). Gastrointestinal factors in autistic disorder: a critical review. *J. Autism Dev. Disord.,* **35**, 713–727.
72. Levy, S. E., Souders, M. C., Ittenbach, R. F., Giarelli, E., Mulberg, A. E., and Pinto-Martin, J. A. (2007). Relationship of dietary intake to gastrointestinal symptoms in children with autistic spectrum disorders. *Biol. Psychiatry* **61**, 492–497.
73. Newschaffer, C. J., Fallin, D., and Lee, N. L. (2002). Heritable and nonheritable risk factors for autism spectrum disorders. *Epidemiol. Rev.* **24**, 137–153.
74. Schreck, K. A., and Williams, K. (2006). Food preferences and factors influencing food selectivity for children with autism spectrum disorders. *Res. Dev. Disabil.* **27**, 353–363.
75. Schreck, K. A., Williams, K., and Smith, A. F. (2004). A comparison of eating behaviors between children with and without autism. *J. Autism Dev. Disord.* **34**, 433–438.
76. Ahearn, W. H. (2003). Using simultaneous presentation to increase vegetable consumption in a mildly selective child with autism. *J. Appl. Behav. Anal.* **26**, 361–365.
77. Jyonouchi, H., Geng, L., Ruby, A., Reddy, C., and Zimmerman-Bier, B. (2005). Evaluation of an association between gastrointestinal symptoms and cytokine production against common

dietary proteins in children with autism spectrum disorders. *J. Pediatr.* **146**, 605–610.

78. Jyonouchi, H., Geng, L., Ruby, A., and Zimmerman-Bier, B. (2005). Dysregulation innate immune responses in young children with autism spectrum disorders: their relationship to gastrointestinal symptoms and dietary intervention. *Neuropsychobiology,* **51**, 77–85.

79. Wong, H. H., and Smith, R. G. (2006). Patterns of complementary and alternative medical therapy use in children diagnosed with autism spectrum disorders. *J. Autism Dev. Disord.* **36**, 902–909.

80. Elder, J. H., Shankar, M., Shuster, J., Theriaque, D., Burns, S., and Sherrill, L. (2006). The gluten-free, casein free diet in autism: results of a preliminary double blind clinical trial. *J. Autism Dev. Disord.* **36**, 413–420.

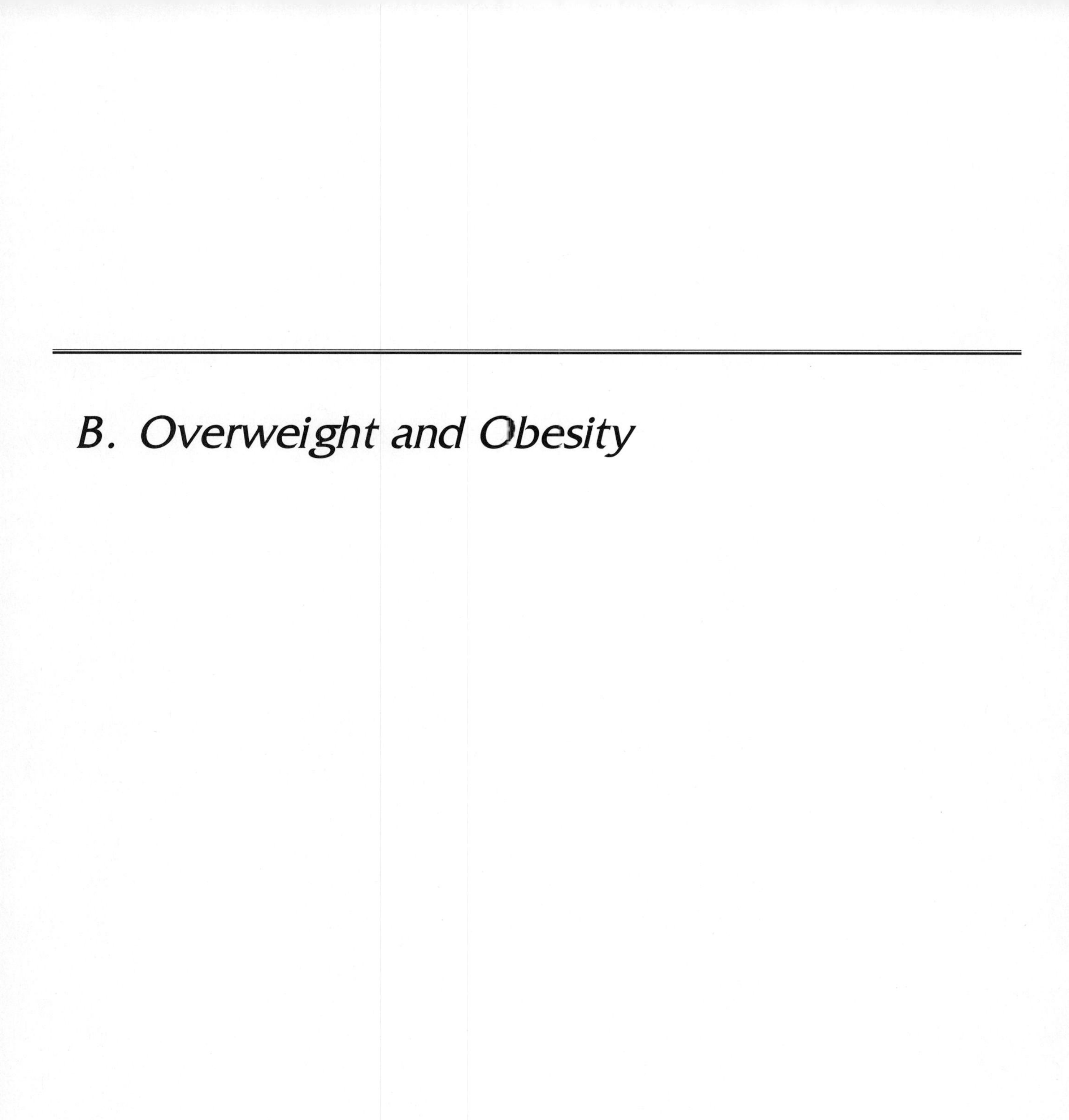

B. Overweight and Obesity

CHAPTER 21

Genetics of Human Obesity

JANIS S. FISLER[1] AND NANCY A. SCHONFELD-WARDEN[2]
[1]*Department of Nutrition, University of California, Davis*
[2]*Department of Pediatrics, University of California, Davis*

Contents

I. INTRODUCTION

Complex and incompletely defined interactions between environment and genetics determine each individual's height and weight, as well as other human quantitative traits.[1] The result is a population in which individuals vary widely for height and weight, but no one factor can be identified as controlling either trait. In humans, long-term adult weight is relatively stable, as evidenced by the difficulty of sustaining intentional weight loss and the automatic return to previous weight following brief periods of overeating. This drive to constancy of body weight is due to both behavioral and physiological alterations that accompany weight change. Convincing evidence of the biological basis of the regulation of body fat stores comes from the identification of more than 50 single-gene mutations and Mendelian syndromes that result in spontaneous massive obesity or in adipose tissue atrophy.

Most human obesity, however, is not due to mutations in single genes but is inherited as a complex, polygenic, quantitative trait influenced by many genetic and environmental variables. There are also likely to be interactions among genes and between genes and environmental factors such that some alleles of one gene will not cause obesity unless specific alleles of another gene or some environmental pressure are present. Genetic heterogeneity[2] and incomplete penetrance[3] of the trait also make dissection of complex phenotypes difficult. Expression of an obesity gene may also be age or gender dependent. Thus, identification of all the genes promoting human obesity is not a trivial task.

Genetics is a rapidly progressing field, and knowledge of the genetic basis for obesity is expanding exponentially. Therefore, the reader should use this chapter only as the starting point for an understanding of this exciting body of knowledge.

II. GENETIC EPIDEMIOLOGY OF HUMAN OBESITY

Genetic epidemiology of human obesity is the study of the relationships of the various factors determining the frequency and distribution of obesity in the population. Such studies of obesity are limited in that they do not examine DNA and rarely directly measure the amount or location of body fat. However, genetic epidemiology studies do provide information as to whether there is a genetic basis for the trait, whether a major gene is involved in the population, whether inheritance is maternal or paternal, and whether expression of the trait is gender or age dependent.

Genetic epidemiology studies of human obesity employ a variety of designs and statistical methods, each giving somewhat different heritability[4] estimates for obesity. For a discussion of genetic epidemiology methods employed in the study of obesity, see Bouchard *et al.* [1].

[1] A quantitative trait is one that varies over a continuous range, such as body weight and height, and is controlled by multiple genes.

[2] Genetic heterogeneity refers to similar phenotypes caused by mutations in more than one gene.

[3] Incomplete penetrance means that not all individuals who have the gene mutation develop the phenotype.

[4] Genetic heritability is the percentage of interindividual variation in a trait that is explained by genetic factors.

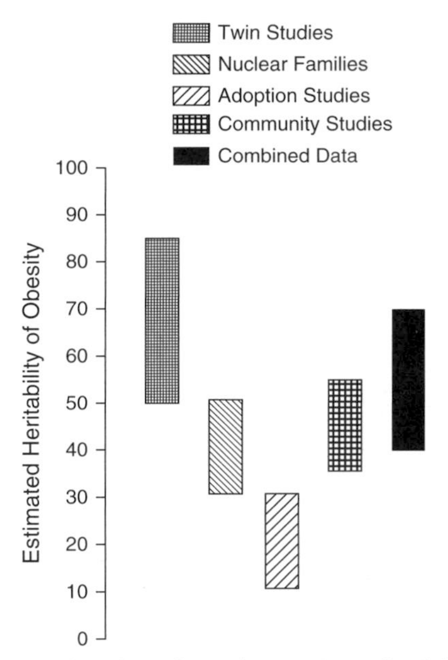

FIGURE 1 Heritability of obesity as determined by different study types. (Data for studies of twins, nuclear families, and adoption studies are taken from [1]. Data for community-based studies are taken from [102]. Range of heritability estimated from all study types is taken from [5].)

The heritability estimates for human obesity are derived from a large number of studies of adoptees, twins, families, or communities. Population or family studies tend to have lower, and twin studies to have higher, heritability for body mass index[5] (BMI). Heritability of BMI has been estimated from adoption studies to be as low as 10% and from twin studies to be as high as 85% [1, 2] (Fig. 1). In a pediatric twin study, genetic influences contributed 75% to 80% in the percentage of body fat [3]. The heritability estimate for BMI in a study of childhood obesity in an Hispanic population was 40% and, in that study, heritability of diet and physical activity phenotypes ranged from 32% to 69% [4]. By using data from all types of studies, it is estimated that 40% to 70% of the within population variation in obesity is due to genetic variation [5] (Fig. 1). Most studies indicate that familial environment has only a minor impact on obesity.

Genes implicated in obesity can have major, minor, or polygenic effects. A major gene is a single gene that has a large effect on the phenotype. Mode of transmission[6] may be recessive, codominant, or dominant. Polygenic effects are due to many genes, each with a small effect on the phenotype. Segregation analyses[7] indicate that the percentage of minor gene transmission ranges from 25% to 42% and that there is a single major gene for high BMI segregating from the parents to their offspring. These data do not mean that there is only one major gene contributing to obesity in humans. Rather, the specific obesity gene may vary among population subgroups, such that there may be several major obesity genes in the entire population. Recently, a purported major gene was identified in European populations [6].

III. GENE-ENVIRONMENT INTERACTIONS

Why some people in modern societies become obese, despite considerable effort and expense to avoid this condition, whereas others stay lean without such effort, appears to have a genetic basis [7]. The chronic overfeeding studies by Sims and colleagues beginning in the 1960s showed interindividual differences in weight gain [8, 9]. More recently, Bouchard and colleagues determined the response to changes in energy balance by submitting pairs of monozygotic twins either to positive energy balance induced by overeating or to negative energy balance induced by exercise training. During 100 days of overfeeding by 1000 kcal/day, significant intrapair resemblance was observed for changes in body composition and was particularly striking for changes in regional fat distribution and amount of visceral fat, with six times as much variance among as within twin pairs [10]. During long-term energy deficit induced by exercise training, intrapair resemblance was observed for changes in body weight, fat mass, percentage fat, and abdominal visceral fat [11]. One explanation for these differences is that some twin pairs were found to be better oxidizers of lipid, as evidenced by reduced respiratory quotient, during submaximal work than were other twin pairs [11].

An important component of the interindividual difference in response to overeating may be individual differences in spontaneous physical activity or "fidgeting" [12, 13]. A large portion of the variability in total daily energy expenditure, independent of lean body mass, is due to fidgeting, which varies by more the sevenfold among subjects [13], is a familial trait, and is a predictor of future weight gain [14]. In a very elegant study of the fate of excess energy during overfeeding, Levine *et al.* [12] again demonstrated the considerable interindividual variation in susceptibility to weight gain. Two-thirds of the increases in total daily energy expenditure in nonobese subjects overfed by 1000 kcal/day for 8 weeks was due to increased spontaneous physical activity

[5] Body mass index (BMI) is calculated as body weight divided by height squared and is a good surrogate for percentage body fat.

[6] Mode of transmission can be recessive, co-dominant, or dominant. If a mutation has a recessive transmission, both copies of the gene must be defective to produce the phenotype. Codominant is inheritance in which two alleles of a pair of genes both have full phenotype expression causing individuals with two defective genes to be more obese than those with only one defective gene. With a dominant mode of transmission, only one defective gene need be present to express the phenotype.

[7] Segregation analysis is the determination of the number of progeny that have inherited distinct and mutually exclusive phenotypes.

associated with fidgeting, maintenance of posture, and other daily activities of life independent of volitional exercise [12].

The genetic contributions to physical performance and to voluntary food intake and eating behavior are less well known than for obesity. Nevertheless, the genes contributing to these phenotypes are being increasingly identified [15, 16].

IV. GENE-GENE INTERACTIONS

Gene-gene interactions are likely a universal phenomenon in common human diseases and may be more important in determining the phenotype than the independent main effects of any one susceptibility gene [17, 18]. A classic example of the effect of gene-gene interactions on a complex trait comes from mouse studies. The severity of diabetes in both *Lep[ob]* (leptin) and *Lepr[db]* (leptin receptor) mutant mice is determined by the genetic background upon which the mutation is expressed. Both *Lep[ob]* and *Lepr[db]* mutations in C57BL/6 mice result in hyperinsulinemia and obesity, whereas these mutations in C57BLKs mice result in severe diabetes and early death [19].

Because gene-gene interactions are difficult to determine using traditional genetic studies in humans, currently little is known of these significant contributions to phenotypic variation in obesity. However, gene-gene interaction effects have been shown on extreme obesity [20] and on immune dysfunction in obesity [21]. Gene-gene interactions among variants of the β-adrenergic receptor genes (*ADRB1*, *ADRB2*, and *ADRB3*) contribute to longitudinal weight changes in African and Caucasian American subjects [22]. A three-way gene-gene interaction has been reported to affect abdominal fat [23]. The new availability of whole genome maps at reasonable cost [24, 25] and new statistical approaches [26] will speed the identification of genes and gene-gene interactions contributing to obesity.

V. THE OBESITY GENE MAP

A genetic map is a representation of the distribution of a set of genetic loci or markers. (For a discussion of genetic maps, see [27].) The three types of genetic maps are linkage, chromosomal, and physical. Genetic maps provide many kinds of information, from overall chromosomal views to more detailed molecular information, but all genetic maps place items (usually genes or clones) in an order, from top to bottom or left to right.

The Human Obesity Gene Map [16], established by Dr. Claude Bouchard and his colleagues at Laval University, Quebec, incorporates information from all three types of maps: linkage, chromosomal, and physical. The map and its associated summary provide an overview of the data reported in peer-reviewed journals on all markers, genes, and mutations associated or linked with obesity phenotypes. The map includes many obesity-related phenotypes, including BMI, percentage body fat, fat mass, skinfolds, abdominal fat, macronutrient intake, metabolic rate, energy expenditure, and fat-free mass. The current compilation of obesity gene information, covering all 22 autosomal chromosomes as well as the X and Y chromosomes, is available in the Obesity Gene Map Database available online at http://obesitygene.pbrc.edu.

The Obesity Gene Map, as with all genetic knowledge, has expanded dramatically since its initial publication in 1996 [28] (Fig. 2). The current version of the Obesity Gene Map [16] reports that as of October 2005 there were 176 human obesity cases caused by single gene mutations in 11 genes (plus one identified after the most recent map update), 50 chromosomal loci related to Mendelian syndromes associated with obesity, and 127 candidate genes and 253 quantitative trait loci[8] (QTLs) associated with obesity related phenotypes in humans (Fig. 2).

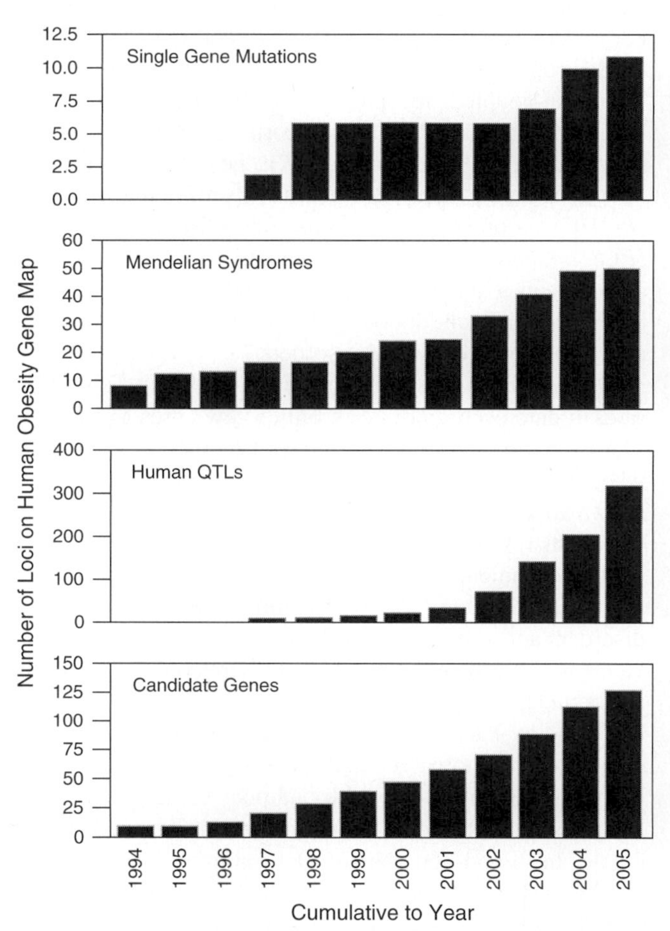

FIGURE 2 Increase from 1994 to 2005 in the number of identified genes contributing to obesity phenotypes. (Data adapted from [16].)

[8] A quantitative trait locus (QTL) is a chromosomal region in which alleles are linked to variations of a quantitative trait.

VI. SINGLE-GENE OBESITY IN HUMANS

Cloning of the mouse obesity genes,[9] *Lep*^{ob}, *Lepr*^{db}, *Cpe*^{fat}, *Tub*, and A^y, from naturally occurring mutant models between 1992 and 1996 led to an explosion of knowledge of the genetic causes of obesity. Obesity in these rodent models exhibits Mendelian segregation, indicating that the obesity is inherited as a single-gene mutation. With one exception, these mutations result in the loss of gene or protein function and are expressed only when both copies of the gene in an individual are defective. Therefore, obesity caused by these mutations in humans would not be common. Nevertheless, these genes are of great interest, because subtle mutations may contribute to common forms of obesity. Study of these genes led to the identification of the leptin/melanocortin pathway, the central regulator of energy homeostasis.

As of early 2007, 12 genes have been shown to cause spontaneous Mendelian obesity in humans (Table 1): brain-derived neurotrophic factor (gene abbreviation *BDNF*), cortiotropin-releasing hormone receptor 1 (*CRHR1*), corticotropin-releasing hormone receptor 2 (*CRHR2*), G-protein-coupled receptor 24 (*GPR24*), leptin (*LEP*), leptin receptor (*LEPR*), melanocortin 3 receptor (*MC3R*), melanocortin 4 receptor (*MC4R*), neurotrophic tyrosine kinase receptor type 2 (*NTRK2*), proopiomelanocortin (*POMC*), proprotein convertase subtilisin/kexin type 1 (*PCSK1*), and single-minded homolog 1 (*SIM1*). Mutations in these genes in humans are associated with severe obesity, beginning in childhood, along with developmental, endocrine, and behavioral disorders.

Most of the single gene obesities reported in Table 1 are rare, to date each reported in only a few cases or families. The exceptions are *MC4R* [29] and perhaps *NTRK2* [30], where for each gene more than 100 obese cases have been shown to harbor a mutation. Generally, these single gene mutations result in severe obesity and as more such individuals are studied, additional cases will surely be identified. (For a description of the most common single gene obesity disorders and syndromes, see [31].)

Most of the genes listed in Table 1 are components of the leptin/melanocortin pathway that functions to control whole body energy balance (Fig. 3). The mutations of this pathway that are shown to cause obesity are highlighted in Figure 3. Leptin is secreted by adipocytes, and its concentration in blood is therefore proportional to fat mass. Leptin crosses the blood-brain barrier and activates leptin receptors on the surface of neurons in the arcuate nucleus of the hypothalamus. This activation stimulates proprotein convertase (product of *PCSK1*) to cleave proopiomelanocortin (product of *POMC*) into the melanocortins including α-MSH, the primary ligand for melanocortin receptors and activation of downstream signaling to regulate energy balance. (For reviews of the leptin/melanocortin pathway and downstream signaling in obesity, see [2, 32–35].)

A. Leptin and Leptin Receptor Deficiencies

Cloning and characterization of the mouse *Lep*^{ob} gene identified its protein product, leptin, a hormone that is secreted from adipose tissue [36]. Leptin circulates in the blood [37], crosses the blood-brain barrier [38], and binds to its receptor in the hypothalamus to regulate food intake and energy expenditure [39]. Thus, leptin functions as an afferent signal in a negative feedback loop to maintain constancy of body fat stores.

Leptin clearly has a broader physiological role than just the regulation of body fat stores. Leptin deficiency results in many of the abnormalities seen in starvation, including reduced body temperature, reduced activity, decreased immune function, and infertility. (For reviews of the physiological role of leptin, see [39, 40].)

Known mutations in leptin causing spontaneous massive obesity in humans are autosomal recessive and are rare in the population. However, two highly consanguineous[10] families were identified that carry mutations in *LEP* (reviewed in [2, 29]).

Leptin acts through the leptin receptor, a single-transmembrane-domain receptor of the cytokine-receptor family [41]. The leptin receptor is found in many tissues in several alternatively spliced forms, raising the possibility that leptin affects many tissues in addition to the hypothalamus. (For additional discussion of the leptin receptor, see [40].) As with the *LEP* gene, an autosomal recessive mutation in the human leptin receptor gene (*LEPR*) that results in a truncated leptin receptor was discovered in homozygosity in a consanguineous family [29]. In a study of 300 subjects with severe early onset obesity (before age 10) [42], eight families (two of which were nonconsanguineous) were identified where severe obesity segregated with mutations in *LEPR*. Individuals homozygous for these mutations were hyperphagic, had delayed or no pubertal development, had defects in immune function, and were hyperinsulinemic and hyperleptinemic consistent with the degree of obesity. This phenotype is similar to that seen in individuals with mutation of the leptin gene. Heterozygotes for *LEP* and *LEPR* mutations have increased fat mass but are not obese.

---

[9] The designations *Lep*^{ob}, *Lepr*^{db}, and so on represent mutations of the genes coding for leptin and the leptin receptor that occur in the *obese* mouse and the *diabetes* mouse, and so on, respectively. The homologous genes in humans are designated LEP and LEPR.

[10] If an allele is rare and if two copies (homozygocity) of a mutation are required for the phenotype to be expressed, then homozygotes are usually found only in highly consanguineous (inbred) families.

TABLE 1 Single Gene Mutations Causing Uncomplicated (Nonsyndromic) Obesity in Humans

Gene Name	Chromosomal Locus	Function of Gene Product Relative to Obesity
Brain-derived neurotrophic factor (*BDNF*)[a]	11p13	Regulation of eating behavior. *BDNF* and its receptor *NTRK2* are downstream components of the *MC4R*-mediated control of energy balance
Corticotropin-releasing hormone receptor 1 (*CRHR1*)[b]	17q12–q22	Receptor for CRH which is principal neuroregulator of the hypothamic-pituitary-adrenocortical (HPA) axis
Corticotropin-releasing hormone receptor 2 (*CRHR2*)[b]	7p21–p15	Receptor for CRH involved in HPA axis drive
G-protein-coupled receptor 24 (*GPR24*)[b]	22q13.3	Receptor for melanin-concentrating hormone (MCH), which plays a role in energy homeostasis through actions on locomotor activity, appetite, and neuroendocrine function
Leptin (*LEP*)[b]	7q31.3	Hormone secreted from adipocytes that plays a critical role in regulation of body weight by inhibiting food intake and stimulating energy expenditure
Leptin receptor (*LEPR*)[b]	1p31	Receptor for leptin
Melanocortin 3 receptor (*MC3R*)[b]	20q13.2	Receptor for MC[c] as autoreceptor to suppress activity of POMC neurons
Melanocortin 4 receptor (*MC4R*)[b]	18q22	Receptor for MC *MC4R* mutations affect appetite, cause severe obesity, and increased lean mass, linear growth, and bone mass
Neurotrophic tyrosine kinase receptor type 2 (*NTRK2*)[b]	9q22.1	Receptor for brain-derived neurotrophic factor (BDNF), affects appetite and causes severe early onset obesity and developmental delay
Proopiomelanocortin (*POMC*)[b]	2p23.3	Precursor molecule of adrenocorticotropin, MC, β-isotropin, β-endorphin
Proprotein convertase subtilisin/kexin type 1 (*PCSK1*)[b]	5q15–q21	Neuroendocrine convertase that cleaves POMC (also cleaves proinsulin to insulin)
Single-minded homolog 1 (*SIM1*)[b]	6q16.3–q21	Transcription factor that is essential for the development of the paraventricular nucleus of the hypothalamus

[a]Data from [50, 101].
[b]Data from [16, 50].
[c]MC, α-MSH, β-MSH, and δ-MSH.

B. Mutations in the Melanocortin System

Sequential cleavage of the precursor protein proopiomelanocortin (product of *POMC*) generates the melanocortin peptides adreno-corticotrophin (ACTH), the melanocyte-stimulating hormones (α-, β-, and δ-MSH), and the opioid-receptor ligand β-endorphin. α-MSH plays a central role in the regulation of food intake by the activation of the brain melanocortin 4 receptor (product of *MC4R*). The dual role of α-MSH in regulating food intake and influencing hair pigmentation predicts that the phenotype associated with a defect in *POMC* function would include obesity, alteration in pigmentation (for example, red hair and pale skin in Caucasians), and ACTH deficiency. The observations of these symptoms in two probands[11] led to the identification of three separate mutations within their *POMC* genes [43]. Another *POMC* variant in a region encoding β-MSH results in severe early onset obesity, hyperphagia, and increased linear growth, a phenotype much like seen with mutations in *MC4R* [44]. Heterozygosity[12] for a *POMC* mutation having subtle effects on proopiomelanocortin expression and function was shown to influence susceptibility to obesity in a large family of Turkish origin [33].

A wide variety of hormones, enzymes, and receptors are initially synthesized as large inactive precursors. To release the active hormone, enzyme, or receptor, these precursors must undergo limited proteolysis by specific convertases. An example is the clipping of proopiomelanocortin by proprotein convertase subtilisin/kexin type 1, also known as prohormone convertase-1 (product of *PCSK1*). A recessive mutation of the gene producing carboxypeptidase E, a homologous enzyme active in the processing and sorting of prohormones, causes obesity in the Cpe^{fat} mouse. Mutations in the human homolog *PCSK1* were found in individuals with extreme childhood obesity and elevated proinsulin and proopiomelanocortin concentrations but very low insulin levels (for review see [29]). Because the human cases and the Cpe^{fat} mouse share similar phenotypes, it can be inferred that molecular defects in prohormone conversion represent a generic mechanism for obesity.

The agouti protein, identified in the yellow obese (A^y) mouse, inhibits binding of α-MSH to melanocortin receptors. Obesity and yellow coat color in the A^y mouse result from expressing agouti in all tissues, not just in skin, which

[11] A proband is the index case, the person through whom the pedigree (family) was ascertained.

[12] Heterozygosity refers to having different alleles of a gene on the two chromosomes.

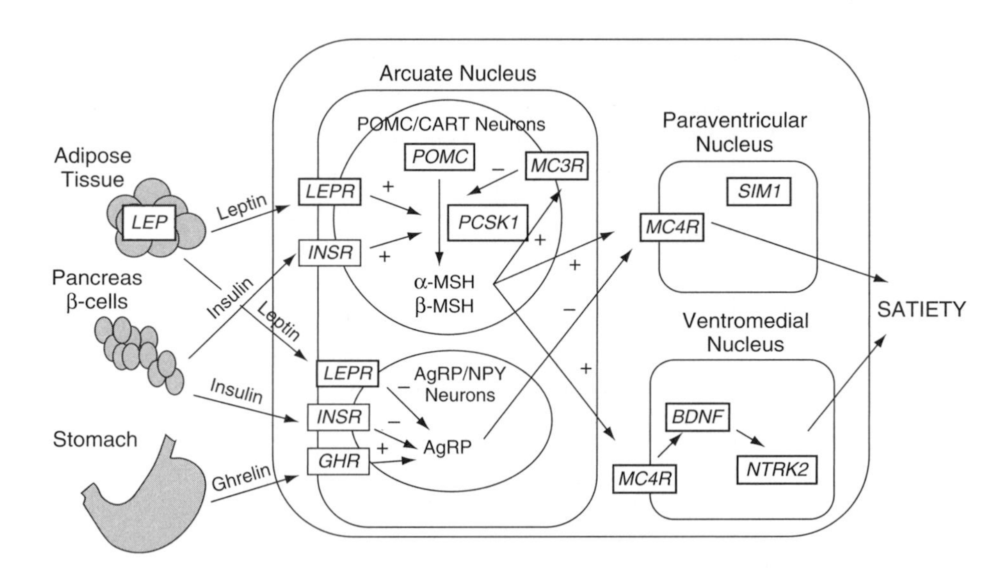

FIGURE 3 Simplified schematic of the leptin/melanocortin pathway in the hypothalamus. Mutations of genes are known to cause monogenic obesity in humans. This pathway is an essential component of the central control of energy homeostasis, propagating the signals that result in satiety. Leptin is secreted from adipocytes, crosses the blood-brain barrier, and activates leptin receptors. This activation stimulates POMC neurons producing melanocortins that activate melanocortin 4 receptors (MC4R) in the paraventricular and ventromedial nuclei, resulting in satiety. Melanocortin 3 receptor (MC3R) is an autoregulator of the POMC system. MC3R also affects obesity independently of its autoregulation of the POMC system, but whether it acts through the same pathway as MC4R is not known. SIM1 is essential for development of the paraventricular nucleus. Brain-derived neurotrophic factor and its receptor, neurotrophic tyrosine kinase receptor, type 2, are part of the MC4R cascade leading to satiety. When stimulated by ghrelin receptors in the arcuate nucleus, agouti-related protein inhibits MC4R activity. Abbreviations: AgRP, agouti-related protein; BDNF, brain-derived neurotrophic factor; CART, cocaine- and amphetamine-related transcript; GHR, ghrelin receptor; INSR, insulin receptor; LEP, leptin; LEPR, leptin receptor; MC3R, melanocortin 3 receptor; MC4R, melanocortin 4 receptor; α-MSH, α-melanocyte stimulating hormone; β-MSH, β-melanocyte stimulating hormone; NPY, neuropeptide Y; NTRK2, neurotrophic tyrosine kinase receptor, type 2; PCSK1, proprotein convertase subtilisin/kexin-type 1; POMC, proopiomelanocortin; SIM1, single-minded, Drosophila, homolog of, 1. (Figure adapted from data from [34, 103].)

is the normal condition. The melanocortin 4 receptor, a G-protein-coupled receptor, is highly expressed in the hypothalamus, a region of the brain intimately involved in appetite regulation. It is a receptor for α-MSH, a product of the *POMC* gene, which inhibits feeding. Inactivation of *Mc4r* by gene targeting in mice results in a maturity-onset obesity syndrome associated with hyperphagia and impaired glucose tolerance [45]. *Mc4r*-deficient mice do not respond to an α-MSH-like agonist, suggesting that α-MSH inhibits feeding primarily by activating melanocortin 4 receptors [46].

Mutations in *MC4R* are found in various ethnic groups and cause the most common form of monogenic obesity in humans. The global presence of obesity-specific *MC4R* mutations is estimated to be 2% to 3% [47]. *MC4R*-linked obesity in humans is dominantly inherited with incomplete penetrance. Adrenal function is not impaired but severe hyperinsulinemia is present in the *MC4R*-deficient subjects. Sexual development and fertility are normal. Affected subjects are hyperphagic and have increased linear growth, similar to what occurs in heterozygous *Mc4r*-deficient mice. *MC4R*-deficient humans also have increased lean mass and bone mineral density and mild central hypothyroidism. Female haploinsufficiency[13] carriers are heavier then male carriers in their families, a pattern also seen in *Mc4r*-deficient mice. These data are strong evidence for dominantly inherited obesity, not associated with infertility, because of haploinsufficiency mutations in *MC4R*.

Mutation of a second melanocortin receptor, *MC3R*, is also shown to cause single gene obesity [16]. *MC3R* acts as an autoreceptor indicating the tight regulation of the melanocortin system in energy balance. Two variants of the *MC3R* gene interact with diet to affect weight loss success in an Italian clinic treating severe childhood obesity [48].

Three genes important in downstream signaling of the melanocortin system have also been shown to cause single gene obesities. Single-minded homolog 1 (*SIM1*) is a regulator of neurogenesis and is essential to the development of the paraventricular nucleus of the hypothalamus [16, 49].

[13] Haploinsufficiency occurs when a gene, or a group of genes, is present in too few or too many copies. In autosomes (all but the X and Y chromosomes), one copy of each gene is inherited from each parent. The presence of extra chromosomal material or lack of chromosomal material alters gene dosage, causing abnormalities in gene function.

Brain-derived neurotrophic factor (*BDNF*) and its receptor, neurotrophic tyrosine kinase receptor type 2 (*NTRK2*), are involved in signaling in the ventromedial nucleus of the hypothalamus [30].

VII. RARE GENETIC SYNDROMES WITH OBESITY AS A PROMINENT FEATURE

Currently, 50 identified genes are associated with rare Mendelian syndromes in which obesity is a prominent clinical feature. These are listed in the Obesity Gene Map [16] and described in the Online Mendelian Inheritance in Man (OMIM) database[14] [50]. Among the better known of these syndromes are the Prader-Willi syndrome, the Bardet-Biedl syndromes, and Alstrom syndrome. Currently, the pathophysiological mechanisms leading to obesity in these syndromes are largely unknown. One gene, *SIM1*, causes single gene obesity (see Table 1) and is also involved in a Prader-Willi–like syndrome. The characteristic features of these syndromes were reviewed [31, 51].

With an incidence of about 1 in 25,000 births, the most common of the Mendelian syndromes is the autosomal dominant Prader-Willi syndrome, which results from a microdeletion of paternal chromosome 15q11–q13 or, more rarely, as a result of maternal disomy[15] of chromosome 15. In addition to obesity, the Prader-Willi syndrome is characterized by hypotonic musculature, mental retardation, hypogonadism, short stature, and small hands and feet (for review, see [52, 53]). Aberrant behavior, including hyperphagia and aggressive food seeking, makes management of these patients difficult.

The better known of the autosomal recessive Mendelian obesity syndromes are the Bardet-Biedl syndromes (BBS1–BBS8) and the Alstrom syndrome (ALMS1). The eight Bardet-Biedl syndromes are associated with variants in genes on eight separate chromosomes but are all characterized by mental retardation, pigmentary retinopathy, polydactyly, obesity, and hypogonadism. It is, therefore, apparent that mutation in at least eight separate genes can result in the same phenotype. The gene coding for BBS2 may predispose males carrying only one copy of the gene to obesity and may explain approximately 3% of severely overweight males [54]. Of interest, BBS2 parents of both sexes were significantly taller than U.S. individuals of comparable age [54].

[14] The Online Mendelian Inheritance in Man (OMIM) database is available at www.ncbi.nlm.nih.gov/omim. This database, a catalog of human genes and genetic disorders, is authored and edited by Dr. Victor A. McKusick and his colleagues at Johns Hopkins University and elsewhere, and it was developed for the World Wide Web by the National Center for Biotechnology information (NCBI).

[15] Maternal disomy is inheritance of an extra maternal chromosome or part thereof.

In addition to obesity, Alstrom syndrome (causative gene *ALMS1*) is characterized by retinitis pigmentosa leading to blindness, insulin resistance, diabetes, and deafness, but it does not involve mental retardation or polydactyly (for review, see [55]). Onset of obesity is usually between 2 and 10 years of age and can range from mild to severe.

VIII. EVIDENCE FROM LINKAGE STUDIES OF OBESITY PHENOTYPES

A. Mapping of Loci in Animals

Animal models have been very important in the dissection of complex traits [56]. Quantitative trait locus (QTL) mapping is a method for mapping Mendelian factors that underlie complex traits, in virtually any animal model, by using genetic linkage maps[16] (for discussion, see [27, 56]). As of October 2005, 408 animal QTLs were linked variously to body weight, body fat, energy expenditure, food intake, leptin levels, or weight gain [16]. A number of these QTLs identified in separate crossbreeding experiments of different strains are overlapping, and it is likely that the same underlying gene is responsible for these overlapping QTLs. Many of these QTLs have pleiotropic effects.[17]

B. From Mouse to Human

QTLs are valuable for identifying candidate genes to be further evaluated by gene targeting experiments in mice or by linkage studies or association studies of the candidate genes in humans. Because of the evolutionary relationship between mice and humans, many ancestral chromosomal segments have been retained where the same genes occur in the same order within discrete regions of chromosome. These regions of homology[18] may include many hundreds to thousands of genes in the same orders, although some regions of homology are made more complex by chromosomal rearrangements within the region of homology. Because regions of homology between mouse and human chromosomes are well defined, the identification of a gene in the mouse frequently gives the chromosomal location of the same gene in the human. For example, to test whether an obesity QTL on mouse chromosome 2 contributes to human obesity, linkage analysis between markers located within the homologous region on chromosome 20 and

[16] A locus is any segment of DNA that is measurable in genetic analysis. A locus may be within a gene or may be a DNA sequence of no known function. A linkage map represents a set of loci on a single chromosome in which all members of the set are linked either directly or indirectly with all other members of the set. Linkage in humans refers to the cosegregation of a genetic marker and a trait together in families.

[17] Pleiotropy means that one gene has a primary effect on more than one phenotype.

[18] Links to homology and other maps are available at www.ncbi.nlm.nih.gov/Homology.

measures of obesity was performed in more than 150 French Canadian families; a locus on 20q13 that contributes to body fat and fasting insulin was found [57]. This locus was later confirmed in a second population group [58].

C. Linkage Studies in Humans

Linkage studies in humans are conducted with large extended families or with nuclear families. A conceptually simple and practical method is the nonparametric sib-pair linkage method that provides statistical evidence of linkage between a quantitative phenotype and a genetic marker [1, 59, 60]. The method is based on the concept that siblings who share a greater number of alleles (1 or 2) identical by descent[19] at a linked marker locus should also share more alleles at the phenotypic locus of interest and should be phenotypically more similar than siblings who share fewer marker alleles (0 or 1). The method has been expanded to use data from multiple markers, allowing higher resolution mapping. Linkage studies do not identify any specific gene but are useful in identifying candidate genes for further study.

Whole genome scans and linkage studies covering smaller chromosomal regions, published as of October 2005, identified 317 QTLs for various measures of adiposity, respiratory quotient, metabolic rate, and plasma leptin levels in humans (for details, see [16, 61]). Many of the reported linkages are based on small samples and do not have strong statistical support. Moreover, many loci are broad and contain far too many genes to be tested by previous methods. However, these chromosomal loci do contain candidate genes for obesity, including genes known to cause single-gene obesity. Using newly practical whole genome scanning approaches in humans [24, 25], it should soon be possible to test all genes within a quantitative trait locus for association with obesity or any other complex trait of interest.

IX. STUDIES OF CANDIDATE GENES FOR OBESITY AND RELATED PHENOTYPES

A. Insertion/Deletion Studies in Rodents

Expression of genes can be altered in the mouse by targeted insertion or deletion in the whole animal or by newer strategies that cause gene deficiencies in only those tissues of interest [62, 63]. Insertion/deletion studies in rodents to date have identified approximately 250 genes that affect obesity or leanness [16]. The development of obesity models with under- or overexpression of a gene not only confirms that the gene contributes to obesity but also provides a model to investigate the resulting phenotype including energy balance, feeding and activity behaviors, and fat and carbohydrate metabolism.

Several confounding factors are involved in gene insertion/deletion studies. A confounding factor in many gene deletion studies is the redundancy of gene function. For example, deletion of the β1-adrenergic receptor only shows an obesity phenotype if there are also mutations in both β2- and β3-adrenergic receptors [64]. Other confounders are gene-gene interactions, requiring mutations in two or even three independent genes to have the phenotype expressed and gene-environment interactions such as the necessity of a high-fat diet for obesity to develop. Also, in comparing gene-targeted models to wild type, it must not be overlooked that the source of DNA surrounding the altered gene generally began as a 129-strain embryonic stem cell. The deletion model retains 129-strain alleles linked to the altered gene no matter how many generations the mice are crossed to a background strain. Because there are many obesity genes and because the average included 129-strain region is about 20 cM, it is possible that these gene-targeted models also include 129-strain alleles of obesity genes within the retained donor region. Despite these confounding factors, mouse insertion/deletion models have been a valuable tool in the dissection of the genetic causes of obesity.

B. Association Studies of Candidate Genes in Humans

Association studies examine the correlation of a genetic variant (polymorphism) within a gene with the phenotype of interest. It is assumed that variants within a gene's coding region alter gene function, although proof of that requires gene-targeting experiments in cell or animal models. Association studies are generally carried out in unrelated individuals and are frequently designed as case-control studies. Although case-control studies remain a powerful tool in some areas, they are less powerful for genetic studies because of methodological issues that complicate analyses in complex populations such as those found in the United States. Therefore, most association studies are now conducted in isolated populations, such as occur in Finland and Quebec. A positional candidate gene is identified both by its location in a chromosomal region that has significant linkage to obesity in family studies and because its biological functions are generally consistent with a role in body weight regulation. As of October 2005, 127 candidate genes have been associated with varying degrees of confidence with obesity phenotypes [16]. Table 2 lists 18 of these genes, each of which has been confirmed by at least five positive independent population samples. (For an example and discussion of difficulties with replication in association studies, see [65]).

[19] Identical by descent is in contrast to identical by state. Two siblings sharing the same allele are identical by descent if you know that it is the same allele inherited from the same parent. They are identical by state if they have the same allele, but you do not know if they are derived from the same parental haplotype.

TABLE 2 Association of Candidate Genes with Obesity Phenotypes: Only Candidates with a Minimum of Five Positive Studies (Range 5–30) Are Included[a]

Gene Name	Chromosomal Locus	Phenotypes Associated with Different Gene Variants
Angiotensin 1-converting enzyme (*ACE*)	17q23	Obesity, BMI, percentage body fat
Adipocyte, C1Q (adiponectin) (*ADIPOQ*)	3q27	Obesity, BMI, abdominal fat
β2-adrenergic receptor (*ADRB2*)	5q32–34	Mild obesity, severe obesity with high fat diet, BMI, abdominal fat, lipolysis, leptin
β3-adrenergic receptor (*ADRB3*)	8p12–p11.2	Mild obesity, severe obesity with high fat diet, BMI, percentage body fat, abdominal fat, weight change
Dopamine receptor (*DRD2*)	11q23	Obesity, energy expenditure
Glucocorticoid receptor (*GCCR* or *NR3C1*)	5q31	BMI, abdominal obesity, leptin
Guanine nucleotide-binding protein, β-3 (*GNB3*)	12p13	Obesity, BMI, percentage body fat, lipolysis, weight gain
5-hydroxytryptamine receptor 2C (*HTR2C*)	Xq24	Maturity onset obesity, BMI, weight change
Interleukin 6 (*IL6*)	7p21	Maturity onset obesity, BMI, abdominal fat, weight change, energy expenditure
Insulin (*INS*)	11p15.5	Obesity, BMI, abdominal fat
Leptin (*LEP*)	7q31.3	Obesity, BMI, weight change, leptin
Leptin receptor (*LEPR*)	1p31	Obesity, BMI, percentage body fat, abdominal fat, weight change, energy expenditure, leptin
Lipase, Hormone-sensitive (*LIPE*)	19q13.1–q13.2	Obesity, BMI, percentage body fat, abdominal fat, lipolysis
Melanocortin 4 receptor (*MC4R*)	18q22	Obesity, BMI, percentage body fat, weight change, energy expenditure
Peroxisome-proliferator-activated receptor δ (*PPARG*)	3p25	Obesity, BMI, percentage body fat, abdominal fat, weight change, leptin, lipid oxidation
Tumor necrosis factor (*TNF*)	6p21.3	Obesity, BMI, percentage body fat, abdominal fat
Uncoupling protein 1 (*UCP1*)	4q31	Obesity, BMI, percentage body fat, abdominal fat, weight change, energy expenditure
Uncoupling protein 2 (*UCP2*)	11q13	Obesity, BMI, percentage body fat, energy expenditure, glucose and lipid oxidation
Uncoupling protein 3 (*UCP3*)	11q13	BMI, percentage body fat, energy expenditure, lipid oxidation, leptin

[a]Data from [16, 50].

Newer and more cost-efficient technologies have allowed a different approach to association studies, one in which whole-genome maps are used to identify genes influencing traits anywhere in the genome [24, 25]. This approach has the same advantage as linkage studies, in that no prior assumptions about the identity of the underlying genes need to be made. Also, the overall data can be evaluated to identify the most significant results, including gene-gene interactions. However, whole genome association studies also have several disadvantages. Whole genome studies necessarily restrict the number of markers to genotype per gene, whereas a more limited association study can investigate a greater number of markers within target genes. In addition, there are greater statistical problems with whole genome studies. However, we can expect to see more data coming from whole genome association studies in the future.

1. Role of Single-Gene Obesity in Common Forms of Human Obesity

Eight of the 12 genes causing single gene obesity in humans have been implicated in polygenic obesity using association studies [16]. The best studied are *LEP, LEPR*, and *MC4R* genes (see Table 2). *LEP* in 11 studies is associated with body weight, weight loss, or leptin levels. The *LEPR* gene is associated with obesity in 16 studies, including one of severe obesity in children [66]. Populations from the Pacific Island of Nauru have some of the highest rates of obesity and non-insulin-dependent diabetes mellitus (NIDDM) in the world. In Nauruan males, specific combinations of alleles in the *LEP* and *LEPR* genes are associated with increased risk for development of insulin resistance [67]. *MC4R* is associated with BMI, percentage fat, weight change, and energy expenditure in eight studies. These data suggest that mutations of the genes causing Mendelian obesity also contribute to common (polygenic) obesity in humans.

2. Candidate Genes with Variants Causing Altered Function

A number of other genes, with variants that may alter gene product function, are believed to contribute to common obesity and its comorbidities. Obesity frequently clusters with insulin resistance, hyperlipidemia, and hypertension [68]. This clustering might arise by any of several mechanisms: obesity might promote comorbidities, comorbidities might promote obesity, or some genes might promote development of both obesity and its comorbidities. Several association studies have reported evidence for genes that influence obesity and one or more comorbidities. However, the demonstration that these variants result in single-gene obesity in any case remains to be established. The more promising candidates are described in Table 2, and several are discussed later to emphasize the difficulty in interpreting such studies.

Adiponectin, like leptin, is a hormone secreted by adipocytes. Adiponectin (product of *ADIPOQ*) regulates energy balance and glucose and lipid metabolism. *ADIPOQ* has been associated with obesity, BMI, abdominal fat, and metabolic rate in Hispanic, Japanese, and Caucasian populations [16, 69].

The β2-adrenergic receptor (product of *ADRB2*) is a major lipolytic receptor in human adipose tissue and thus plays a significant role in lipid mobilization. Several polymorphisms have been identified and their frequencies compared between lean and obese subjects. These variants were associated with body mass index and blood triglycerides in both Swedish and Japanese populations [70–72]. Swedish women with two copies of a common polymorphism, Glu27, were about 10 times more likely to be obese than those with the wild-type *ADRB2* gene, with approximately 20 kg excess body fat and a 50% increase in fat cell size. Another variant of the *ADRB2* gene, Gly16, was associated with improved adipocyte β2-adrenergic receptor function. Thus, genetic variability in the human *ADRB2* gene may be a significant contributor to human obesity.

The β3-adrenergic receptor (product of *ADRB3*) is expressed in adipose tissue and is involved in the regulation of lipolysis and thermogenesis. Disruption of the *Adrb3* gene in mice results in moderate obesity [73]. This potential relevance to human obesity led to an initial positive report in 1995 of an association between *ADRB3* and obesity followed by numerous studies with conflicting results (for review, see [74]). A paired sibling design that aimed to detect effects of the variant by accounting for background genes examined 45 nondiabetic sibling-pairs discordant for the variant who were identical by descent at another marker that is known to be associated with obesity in this population. Presence of the variant was significantly associated with increases in BMI, fat mass, and waist circumference [75]. However, a meta-analysis[20] combining 23 studies and 7399 subjects concluded that the *ADRB3* gene variant is not significantly associated with BMI [74]. The possible association of the variant with diabetes phenotypes was not examined in the meta-analysis. Thus, whether the *ADRB3* gene contributes to obesity or diabetes phenotypes is still subject to debate.

Peroxisome-proliferator-activated receptor (PPAR) γ (product of *PPARG*) is a member of the nuclear hormone receptor subfamily of transcription factors that includes T3 and vitamin D_3 receptors. PPARs regulate expression of genes involved in, among other things, lipid metabolism and energy balance (for review, see [76]). Because of the combined effects on both fat and muscle, activation of PPARγ improves insulin sensitivity and glucose metabolism and decreases blood triglyceride levels. Three association studies have implicated *PPAR*γ in obesity and insulin-resistance phenotypes. Four of 121 obese German subjects had a missense mutation in the *PPAR*γ gene compared to none in 237 normal-weight individuals [77]. All of the subjects with the mutant allele were severely obese. The mutant gene was then overexpressed in mouse fibroblasts, which led to the accelerated differentiation of the cells into adipocytes with greater accumulation of triglyceride than seen with the wild-type *PPAR*γ gene. A different mutation of the *PPAR*γ gene was identified in Finnish populations and was found to be associated with lower BMI, lower fasting insulin levels, and greater insulin sensitivity [78].

Uncoupling proteins 2 and 3 (products of *UCP2* and *UCP3*) are structurally related to uncoupling protein 1, a mitochondrial protein found in brown fat that plays an important role in generating heat and burning calories without the production of adenosine triphosphate. Uncoupling protein 1 is critical in the maintenance of body temperature of newborn humans but is unlikely to be significantly involved in weight regulation because brown fat is normally atrophied in adult humans. *UCP2* and *UCP3* are recently identified genes [79, 80] located very near each other on chromosome 11 that, like *UCP1*, encode mitochondrial transmembrane carrier proteins. *UCP2* is widely expressed in human tissues, whereas *UCP3* is expressed only in skeletal muscle. In a large French Canadian study, *UCP2* and *UCP3* were linked with resting metabolic rate, BMI, percentage of body fat, and fat mass [81]. Several groups have, therefore, examined polymorphisms within the coding regions of both *UCP2* and *UCP3*. Polymorphisms in *UCP2* were associated with metabolic rate during sleep in older, but not younger, Pima Indians [82]. A stronger association between a *UCP2* exon variant and BMI was found in South Indian subjects [83]. The same variant was not associated with obesity in a British population but was

[20] Meta-analysis is a statistical tool for pooling data from many studies into a single analysis, thus greatly increasing the statistical power of the analysis.

correlated with fasting serum leptin concentrations in the presence of extreme obesity [83]. Two other studies failed to find a relationship between *UCP2* variants and energy expenditure, obesity, or insulin resistance [84, 85]. To determine whether *UCP3* mutations could contribute to human obesity, the nucleotide sequence of coding exons was determined in obese or diabetic Africans, African Americans, and Caucasians. A mutation and two missense polymorphisms in the *UCP3* gene were identified in two severely obese probands of African descent [86]. The gene variants were not found in the Caucasian population. The variants were transmitted in a Mendelian fashion; however, they were not consistently associated with obesity in other family members. Individuals who carried one copy of the exon 6-splice polymorphism were found to have only 50% the capacity to oxidize fat and had elevated respiratory quotients (RQs), even though they were not obese. These data indicate that *UCP3* could alter the availability in the cell of fatty acids for oxidation, promoting fat storage. High RQ and low fat oxidation were previously identified risk factors for future weight gain in Pima Indians [87] and African Americans [88]. Thus, *UCP3* is a potentially important obesity gene in certain population groups.

X. CLINICAL IMPLICATIONS OF THE DISCOVERY OF OBESITY GENES

The discoveries of human obesity genes will ultimately have broad implications for clinical practice. The best known of the monogenic, nonsyndromic obesities are those involving *LEP* and *LEPR* mutations, both extremely rare, and *MC4R* mutations, the most common known cause of nonsyndromic, monogenic obesity. For the more infrequent obesity gene mutations, there is little information on the physiological impacts of these mutations except for obesity. Thus, methods for their diagnosis and the implications of their discovery for treatment have not been discussed in depth.

A. Diagnosis of Obesity Disorders 2007

Until recently, only the rare Mendelian, syndromic mutations, such as Prader-Willi and Bardet-Biedl syndromes, were known to cause heritable obesity [50]. These disorders are easily recognized, both by a wide spectrum of phenotypes [31, 51] and by the use of cytogenetics assays that are widely available. However, the Mendelian, nonsyndromic obesity disorders are not so easily diagnosed, because obesity is often the only apparent phenotype, and clinical assays for known obesity gene mutations are currently not practical. It is estimated that 2% to 7% of morbidly obese patients have mutations in *MC4R* [89–92], 3% have mutations in *PPAR*γ [77, 93], and an unknown but smaller percentage have mutations in other obesity genes, including *POMC* [94] and *NTRK2* [30]. Thus, only about 1 in 10 morbidly obese patients has a known mutation that explains the obesity, and molecular assays for the currently known Mendelian obesities would be negative in the majority of morbidly obese patients. Also, there are several known distinct mutations in each of these genes. Thus, no clinical laboratories yet provide diagnosis of these mutations; rather they have only been diagnosed by research laboratories, which are not licensed to provide patient information. However, the inability to make a specific molecular diagnosis does not mean that one cannot identify people with increased risk for genetic obesity, and this may influence choices or approaches to treatment.

Several criteria can be used to estimate individual risks for genetic obesity (Table 3). At the present time, because of the lack of data, these estimates do not produce any quantitative values revealing individual risk that obesity is monogenic.

A strong family history of obesity is consistent with the presence of an obesity gene shared among family members.

The earlier the age of onset and the more extreme the obesity, the more likely that there is a genetic basis for the obesity. Extreme trait values are more likely to be genetic for many complex diseases, simply because extremes tend to result from the actions of severe mutations or from mutations in genes that have larger effects [95]. Children with single-gene obesity are normal weight at birth but severe early hyperphagia, often associated with aggressive food-seeking behavior, results in rapid weight gain, usually beginning in the first year of life. Severe obesity in children has been variously defined as a standard deviation score for BMI of more than 2.5 [96] or 3 [42] relative to the appropriate reference population.

At present, few diagnostic tools are available for the medical evaluation of patients suspected of having monogenic obesity. The only screening tests available are for endocrine abnormalities. Leptin should be measured. Very low or very high serum leptin levels indicate mutation in *LEP* or *LEPR*, respectively. However, lack of very high leptin levels cannot rule out homozygous mutations in *LEPR* [42]. A subset of obese individuals has inappropriately low leptin levels for their fat mass, suggesting a less severe defect in leptin regulation [97]. ACTH and proinsulin should be measured to indicate defects in *POMC* or in prohormone processing. Insulin should be measured to evaluate the appropriateness of the degree of hyperinsulinemia, as this may indicate an *MC4R* mutation.

Physical appearance provides evidence of *POMC* mutations or the syndromic obesities. *POMC* defects can cause red hair and obesity [43], although most red hair results from mutations in melanocortin 1 receptor (*MC1R*) [98], which does not influence obesity. Thus, red hair is only informative when red hair and obesity cosegregate within a family. Prader-Willi, Bardet-Biedl, and other syndromic

TABLE 3 Evaluation of Suspected Monogenic Etiology for Severe Obesity[a,b]

Characteristic	Phenotype Indicative of Genetic Etiology
Family history	Having first-degree relatives with severe obesity
Age of onset	Normal birth weight but age of onset of obesity before age 10
Hyperphagia	Hyperphagia developing within first year of life and aggressive food-seeking behavior
Low ACTH or high proinsulin levels	Mutation in *POMC* or *PCSK1*
Frequent infections	Mutation in *POMC* (ACTH), *LEP*, or *LEPR*
Very low leptin levels	Homozygous mutations in *LEP*
Delayed puberty, lack of growth spurt	Homozygous mutations in *LEP* or *LEPR*
Red hair segregating with obesity	Mutation in *POMC*
Severe hyperinsulinemia, acanthosis nigricans	Mutation in *MC4R*
Accelerated linear growth, increased bone mass and age	Mutation in *MC4R*
Delayed language skills, impaired short-term memory	Mutation in *NTRK2* or *BDNF*
Developmental delay, mental retardation	Prader-Willi, Bardet-Biedl, or other genetic syndrome
Visual and hearing impairments	Bardet-Biedl or other genetic syndrome
Polydactyly or small hands and feet	Bardet-Biedl or Prader-Willi syndromes, respectively

[a]Data adapted from [31, 42, 96, 100].
[b]For a complete algorithm for the assessment of a severely obese individual, see [31].
Abbreviations: BDNF, brain derived neurotrophic factor; LEP, leptin; LEPR, leptin receptor; MC4R, melanocortin 4 receptor; NTRK2, neurotrophic tyrosine kinase receptor, type 2; PCSK1, proprotein convertase subtilisin/kexin-type 1; POMC, proopiomelanocortin.

obesities can be diagnosed by a variety of characteristic phenotypes, such as small hands and feet, polydactyly, and mental retardation as well as by cytogenetic assays. Thus, one should rule out these diagnoses by phenotype determination and by absence of characteristic chromosomal abnormalities.

B. Implications of Obesity Genes for Obesity Treatment

Patients with monogenic obesity will probably be more difficult to treat than those with polygenic obesity, because individuals with monogenic obesity will likely have strong food-seeking behavior and may have physiological resistance to fat loss.

The discovery of human obesity genes may have future implications for diet, behavioral, and drug therapy of genetically obese humans, and perhaps of all obese people. But as yet, no specific treatments are available, with the exception of leptin, for most genetic obesities. The identification and characterization of gene products associated with obesity have provided novel pathways that can be targeted for drug intervention. Certainly, the primary therapy for individuals with documented leptin deficiency is recombinant leptin. A potential drug target identified by the cloning of obesity genes is the melanocortin 4 receptor. Development of safe and effective drugs, such as a small-molecule, α-MSH-like, agonist for the melanocortin receptor to increase satiety, or stimulators of the expression of *UCP2* or *UCP3* to enhance energy expenditure, are certainly goals of the pharmaceutical industry [99]. Specific pharmaceutical treatment of other single-gene obesities will have to await development of drugs targeted further along the pathway of the mutated gene.

Lifestyle changes that may promote weight loss and improve metabolic fitness and quality of life should be recommended [100]. Drugs should also be considered, but whether currently available drug therapies, such as the appetite suppressant, sibutramine, or an inhibitor of fat absorption, orlistat, are more or less effective in individuals with single-gene obesity than those with polygenic obesity is unknown. However, as with any severe obesity, when lifestyle changes and pharmaceutical approaches are inadequate to ameliorate morbidity, surgical treatment of the obesity may be considered. Unfortunately, individuals with aggressive food-seeking behavior, as frequently seen in monogenic obesities, are poor candidates for bariatric surgery. The effect of obesity gene variants on treatment (diet, drug, or surgery) outcomes is an important area for future research.

References

1. Bouchard, C., Perusse, L., Rice, T., and Rao, D. C. (2004). Genetics of human obesity. *In* "Handbook of Obesity" (G. A. Bray and C. Bouchard, Eds.), pp. 157–200. Marcel Dekker, New York.
2. O'Rahilly, S., and Farooqi, S. (2006). Genetics of obesity. In *Phil. Trans. R. Soc. B.* 1–11.
3. Faith, M. S., Pietrobelli, A., Nunez, C., Heo, M., Heymsfield, S. B., and Allison, D. B. (1999). Evidence for independent genetic influences on fat mass and body mass index in a pediatric twin sample. *Pediatrics* **104**, 61–67.

4. Butte, N. F., Cai, G., Cole, S. A., and Comuzzie, A. G. (2006). Viva la Familia Study: genetic and environmental contributions to childhood obesity and its comorbidities in the Hispanic population. *Am. J. Clin. Nutr.* **84**, 646–654; quiz, 673–644.
5. Comuzzie, A. G., and Allison, D. B. (1998). The search for human obesity genes. *Science* **280**, 1374–1377.
6. Frayling, T. M., Timpson, N. J., Weedom, M. N., Zeggini, E., Freathy, R. M., Lindgren, C. M., Perry, J. R. B., Elliott, K. S., Lango, H., Rayner, N. W., Shields, B., Harries, L. W., Barrett, J. C., Ellard, S., Groves, C. J., Knight, B., Patch, A.-M., Ness, A. R., Ebrahim, S., Lawlor, D. A., Ring, S. M., Ben-Shlomo, Y., Jarvelin, M.-R., Sovio, U., Bennett, A. J., Melzer, D., Furrucci, L., Loos, R. J. F., Barroso, I., Wareham, N. J., Karpe, F., Owen, K. R., Cardon, L. R., Walker, M., Hitman, G. A., Palmer, C. N. A., Doney, A. S., Morris, A. D., Davey-Smith, G., Consortium, T. W. T. C.C., Hattersley, A. T., and McCarthy, M. I. (2007). A common variant in the FTO gene is associated with body mass index and predisposes to childhood and adult obesity. *Science* **316**, 889–894.
7. Ravussin, E., and Danforth, E. Jr., (1999). Beyond sloth physical activity and weight gain. *Science* **283**, 184–185.
8. Sims, E. A., Danforth, E. Jr., Horton, E. S., Bray, G. A., Glennon, J. A., and Salans, L. B. (1973). Endocrine and metabolic effects of experimental obesity in man. *Recent Prog. Horm. Res.* **29**, 457–496.
9. Sims, E. A., Goldman, R. F., Gluck, C. M., Horton, E. S., Kelleher, P. C., and Rowe, D. W. (1968). Experimental obesity in man. *Trans. Assoc. Am. Physicians* **81**, 153–170.
10. Bouchard, C., Tremblay, A., Despres, J. P., Nadeau, A., Lupien, P. J., Theriault, G., Dussault, J., Moorjani, S., Pinault, S., and Fournier, G. (1990). The response to long-term overfeeding in identical twins. *N. Engl. J. Med.* **322**, 1477–1482.
11. Bouchard, C., Tremblay, A., Despres, J. P., Theriault, G., Nadeau, A., Lupien, P. J., Moorjani, S., Prudhomme, D., and Fournier, G. (1994). The response to exercise with constant energy intake in identical twins. *Obes. Res.* **2**, 400–410.
12. Levine, J. A., Eberhardt, N. L., and Jensen, M. D. (1999). Role of nonexercise activity thermogenesis in resistance to fat gain in humans. *Science* **283**, 212–214.
13. Ravussin, E., Lillioja, S., Anderson, T. E., Christin, L., and Bogardus, C. (1986). Determinants of 24-hour energy expenditure in man. Methods and results using a respiratory chamber. *J. Clin. Invest.* **78**, 1568–1578.
14. Zurlo, F., Ferraro, R. T., Fontvielle, A. M., Rising, R., Bogardus, C., and Ravussin, E. (1992). Spontaneous physical activity and obesity: cross-sectional and longitudinal studies in Pima Indians. *Am. J. Physiol.* **263**, E296–E300.
15. Rankinen, T., Bray, M. S., Hagberg, J. M., Perusse, L., Roth S. M., Wolfarth, B., and Bouchard, C. (2006). The human gene map for performance and health-related fitness phenotypes: the 2005 update. *Med. Sci. Sports Exerc.* **38**, 1863–1888.
16. Rankinen, T., Zuberi, A., Chagnon, Y. C., Weisnagel, S. J., Argyropoulos, G., Walts, B., Perusse, L., and Bouchard, C. (2006). The human obesity gene map: the 2005 update. *Obesity (Silver Spring)* **14**, 529–644.
17. Moore, J. H. (2003). The ubiquitous nature of epistasis in determining susceptibility to common human diseases. *Hum. Hered.* **56**, 73–82.
18. Warden, C. H., Yi, N., and Fisler, J. (2004). Epistasis among genes is a universal phenomenon in obesity: evidence from rodent models. *Nutrition* **20**, 74–77.
19. Hummel, K. P., Coleman, D. L., and Lane, P. W. (1972). The influence of genetic background on expression of mutations at the diabetes locus in the mouse. I. C57BL-KsJ and C57BL-6J strains. *Biochem. Genet.* **7**, 1–13.
20. Dong, C., Wang, S., Li, W. D., Li, D., Zhao, H., and Price, R. A. (2003). Interacting genetic loci on chromosomes 20 and 10 influence extreme human obesity. *Am. J. Hum. Genet.* **72**, 115–124.
21. Skibola, C. F., Holly, E. A., Forrest, M. S., Hubbard, A., Bracci, P. M., Skibola, D. R., Hegedus, C., and Smith, M. T. (2004). Body mass index, leptin and leptin receptor polymorphisms, and non-Hodgkin lymphoma. *Cancer Epidemiol. Biomarkers Prev.* **13**, 779–786.
22. Ellsworth, D. L., Coady, S. A., Chen, W., Srinivasan, S. R., Boerwinkle, E., and Berenson, G. S. (2005). Interactive effects between polymorphisms in the beta-adrenergic receptors and longitudinal changes in obesity. *Obes. Res.* **13**, 519–526.
23. Ukkola, O., Perusse, L., Chagnon, Y. C., Despres, J. P., and Bouchard, C. (2001). Interactions among the glucocorticoid receptor, lipoprotein lipase and adrenergic receptor genes and abdominal fat in the Quebec Family Study. *Int. J Obes. Relat. Metab. Disord.* **25**, 1332–1339.
24. Carlson, C. S., Eberle, M. A., Kruglyak, L., and Nickerson, D. A. (2004). Mapping complex disease loci in whole-genome association studies. *Nature* **429**, 446–452.
25. Lawrence, R. W., Evans, D. M., and Cardon, L. R. (2005). Prospects and pitfalls in whole genome association studies. *Philos. Trans. R. Soc. Lond. B. Biol. Sci.* **360**, 1589–1595.
26. Foulkes, A. S., Reilly, M., Zhou, L., Wolfe, M., and Rader, D. J. (2005). Mixed modelling to characterize genotype-phenotype associations. *Stat. Med.* **24**, 775–789.
27. Silver, L. M. (1995). "Mouse Genetics." Oxford University Press, New York.
28. Bouchard, C., and Perusse, L. (1996). Current status of the human obesity gene map. *Obes. Res.* **4**, 81–90.
29. Farooki, I., and O'Rahilly, S. (2006). Genetics of obesity in humans. *Endocr. Rev.* **27**, 710–718.
30. Gray, J., Yeo, G., Hung, C., Keogh, J., Clayton, P., Banerjee, K., McAulay, A., O'Rahilly, S., and Farooqi, I. S. (2007). Functional characterization of human NTRK2 mutations identified in patients with severe early-onset obesity. *Int. J. Obes. (Lond.)* **31**, 359–364.
31. Farooqi, I. S. (2005). Genetic and hereditary aspects of childhood obesity. *Best Pract. Res. Clin. Endocrinol. Metab.* **19**, 359–374.
32. Adan, R. A., Tiesjema, B., Hillebrand, J. J., la Fleur, S. E., Kas, M. J., and de Krom, M. (2006). The MC4 receptor and control of appetite. *Br. J. Pharmacol.* **149**, 815–827.
33. Farooqi, I. S., Drop, S., Clements, A., Keogh, J. M., Biernacka, J., Lowenbein, S., Challis, B. G., and O'Rahilly, S. (2006). Heterozygosity for a POMC-null mutation and increased obesity risk in humans. *Diabetes* **55**, 2549–2553.

34. Mutch, D. M., and Clement, K. (2006). Unraveling the genetics of human obesity. *PLoS Genet.* **2**, e188.
35. Zigman, J. M., and Elmquist, J. K. (2003). Minireview: From anorexia to obesity—the yin and yang of body weight control. *Endocrinology* **144**, 3749–3756.
36. Zhang, Y., Proenca, R., Maffei, M., Barone, M., Leopold, L., and Friedman, J. M. (1994). Positional cloning of the mouse obese gene and its human homologue. *Nature* **372**, 425–432.
37. Halaas, J. L., Gajiwala, K. S., Maffei, M., Cohen, S. L., Chait, B. T., Rabinowitz, D., Lallone, R. L., Burley, S. K., and Friedman, J. M. (1995). Weight-reducing effects of the plasma protein encoded by the obese gene. *Science* **269**, 543–546.
38. Golden, P. L., Maccagnan, T. J., and Pardridge, W. M. (1997). Human blood-brain barrier leptin receptor. Binding and endocytosis in isolated human brain microvessels. *J. Clin. Invest.* **99**, 14–18.
39. Cohen, M. M. Jr. (2006). Role of leptin in regulating appetite, neuroendocrine function, and bone remodeling. *Am. J. Med. Genet. A.* **140**, 515–524.
40. Friedman, J. M., and Halaas, J. L. (1998). Leptin and the regulation of body weight in mammals. *Nature* **395**, 763–770.
41. Tartaglia, L. A., Dembski, M., Weng, X., Deng, N., Culpepper, J., Devos, R., Richards, G. J., Campfield, L. A., Clark, F. T., Deeds, J., Muir, C., Sanker, S., Moriarty, A., Moore, K. J., Smutko, J. S., Mays, G. G., Wool, E. A., Monroe, C. A., and Tepper, R. I. (1995). Identification and expression cloning of a leptin receptor, OB-R. *Cell* **83**, 1263–1271.
42. Farooqi, I. S., Wangensteen, T., Collins, S., Kimber, W., Matarese, G., Keogh, J. M., Lank, E., Bottomley, B., Lopez-Fernandez, J., Ferraz-Amaro, I., Dattani, M. T., Ercan, O., Myhre, A. G., Retterstol, L., Stanhope, R., Edge, J. A., McKenzie, S., Lessan, N., Ghodsi, M., De Rosa, V., Perna, F., Fontana, S., Barroso, I., Undlien, D. E., and O'Rahilly, S. (2007). Clinical and molecular genetic spectrum of congenital deficiency of the leptin receptor. *N. Engl. J. Med.* **356**, 237–247.
43. Krude, H., Biebermann, H., Luck, W., Horn, R., Brabant, G., and Gruters, A. (1998). Severe early-onset obesity, adrenal insufficiency and red hair pigmentation caused by POMC mutations in humans. *Nat. Genet.* **19**, 155–157.
44. Lee, Y. S., Challis, B. G., Thompson, D. A., Yeo, G. S., Keogh, J. M., Madonna, M. E., Wraight, V., Sims, M., Vatin, V., Meyre, D., Shield, J., Burren, C., Ibrahim, Z., Cheetham, T., Swift, P., Blackwood, A., Hung, C. C., Wareham, N. J., Froguel, P., Millhauser, G. L., O'Rahilly, S., and Farooqi, I. S. (2006). A POMC variant implicates beta-melanocyte-stimulating hormone in the control of human energy balance. *Cell Metab.* **3**, 135–140.
45. Huszar, D., Lynch, C. A., Fairchild-Huntress, V., Dunmore, J. H., Fang, Q., Berkemeier, L. R., Gu, W., Kesterson, R. A., Boston, B. A., Cone, R. D., Smith, F. J., Campfield, L. A., Burn, P., and Lee, F. (1997). Targeted disruption of the melanocortin-4 receptor results in obesity in mice. *Cell* **88**, 131–141.
46. Marsh, D. J., Hollopeter, G., Huszar, D., Laufer, R., Yagaloff, K. A., Fisher, S. L., Burn, P., and Palmiter, R. D. (1999). Response of melanocortin-4 receptor-deficient mice to anorectic and orexigenic peptides. *Nat. Genet.* **21**, 119–122.
47. Lubrano-Berthelier, C., Dubern, B., Lacorte, J. M., Picard, F., Shapiro, A., Zhang, S., Bertrais, S., Hercberg, S., Basdevant, A., Clement, K., and Vaisse, C. (2006). Melanocortin 4 receptor mutations in a large cohort of severely obese adults: prevalence, functional classification, genotype-phenotype relationship, and lack of association with binge eating. *J. Clin. Endocrinol. Metab.* **91**, 1811–1818.
48. Santoro, N., Perrone, L., Cirillo, G., Raimondo, P., Amato, A., Brienza, C., and Del Giudice, E. M. (2007). Effect of the melanocortin-3 receptor C17A and G241A variants on weight loss in childhood obesity. *Am. J. Clin. Nutr.* **85**, 950–953.
49. Hung, C. C., Luan, J., Sims, M., Keogh, J. M., Hall, C., Wareham, N. J., O'Rahilly, S., and Farooqi, I. S. (2007). Studies of the SIM1 gene in relation to human obesity and obesity-related traits. *Int. J. Obes. (Lond.)* **31**, 429–434.
50. Online Mendelian Inheritance in Man, O. T. (2007). McKusick-Nathans Institute of Genetic Medicine, Johns Hopkins University (Baltimore, MD) and National Center for Biotechnology Information, National Library of Medicine (Bethesda, MD), www.ncbi.nlm.hih.gov/omim.
51. Bray, G. A. (1989). Classification and evaluation of the overweight patient. *In* "Handbook of Obesity" (G. A. Bray, C. Bouchard, and W. P. T. James, Ed.), pp. 831–854. Marcel Dekker, New York.
52. Couper, R. (1999). Prader-Willi syndrome. *J. Paediatr. Child Health* **35**, 331–334.
53. Khan, N. L., and Wood, N. W. (1999). Prader-Willi and Angelman syndromes: update on genetic mechanisms and diagnostic complexities. *Curr. Opin. Neurol.* **12**, 149–154.
54. Croft, J. B., Morrell, D., Chase, C. L., and Swift, M. (1995). Obesity in heterozygous carriers of the gene for the Bardet-Biedl syndrome. *Am. J. Med. Genet.* **55**, 12–15.
55. Marshall, J. D., Bronson, R. T., Collin, G. B., Nordstrom, A. D., Maffei, P., Paisey, R. B., Carey, C., Macdermott, S., Russell-Eggitt, I., Shea, S. E., Davis, J., Beck, S., Shatirishvili, G., Mihai, C. M., Hoeltzenbein, M., Pozzan, G. B., Hopkinson, I., Sicolo, N., Naggert, J. K., and Nishina, P. M. (2005). New Alstrom syndrome phenotypes based on the evaluation of 182 cases. *Arch. Intern. Med.* **165**, 675–683.
56. Warden, C. H., and Fisler, J. S. (1998). Molecular genetics of obesity. *In* "Handbook of Obesity" (G. A. Bray, C. Bouchard, and W. P. T. James, Eds.), pp. 223–242. Marcel Dekker, New York.
57. Lembertas, A. V., Perusse, L., Chagnon, Y. C., Fisler, J. S., Warden, C. H., Purcell-Huynh, D. A., Dionne, F. T., Gagnon, J., Nadeau, A., Lusis, A. J., and Bouchard, C. (1997). Identification of an obesity quantitative trait locus on mouse chromosome 2 and evidence of linkage to body fat and insulin on the human homologous region 20q. *J. Clin. Invest.* **100**, 1240–1247.
58. Lee, J. H., Reed, D. R., Li, W. D., Xu, W., Joo, E. J., Kilker, R. L., Nanthakumar, E., North, M., Sakul, H., Bell, C., and Price, R. A. (1999). Genome scan for human obesity and linkage to markers in 20q13. *Am. J. Hum. Genet.* **64**, 196–209.
59. Haseman, J. K., and Elston, R. C. (1972). The investigation of linkage between a quantitative trait and a marker locus. *Behav. Genet.* **2**, 3–19.
60. Kruglyak, L., and Lander, E. S. (1995). High-resolution genetic mapping of complex traits. *Am. J. Hum. Genet.* **56**, 1212–1223.
61. Jacobson, P., Rankinen, T., Tremblay, A., Perusse, L., Chagnon, Y. C., and Bouchard, C. (2006). Resting metabolic rate and respiratory quotient: results from a genome-wide scan in the Quebec Family Study. *Am. J. Clin. Nutr.* **84**, 1527–1533.

62. Glaser, S., Anastassiadis, K., and Stewart, A. F. (2005). Current issues in mouse genome engineering. *Nat. Genet.* **37**, 1187–1193.
63. McMinn, J. E., Liu, S. M., Liu, H., Dragatsis, I., Dietrich, P., Ludwig, T., Boozer, C. N., and Chua, S. C. Jr. (2005). Neuronal deletion of Lepr elicits diabesity in mice without affecting cold tolerance or fertility. *Am. J. Physiol. Endocrinol. Metab.* **289**, E403–411.
64. Bachman, E. S., Dhillon, H., Zhang, C. Y., Cinti, S., Bianco A. C., Kobilka, B. K., and Lowell, B. B. (2002). betaAR signaling required for diet-induced thermogenesis and obesity resistance. *Science* **297**, 843–845.
65. Herbert, A., Gerry, N. P., McQueen, M. B., Heid, I. M. Pfeufer, A., Illig, T., Wichmann, H.-E., Meitinger, T., Hunter D., Hu, F. B., Colditz, G., Hinney, A., Hebebrand, J., Koberwitz, K., Zhu, X., Cooper, R., Ardlie, K., Lyon, H., Hirschhorn, J. N., Laird, N. M., Lenburg, M. E., Lange, C., and Christman, M. F. (2007). Response to comments on "A common genetic variant is associated with adult and childhood obesity." *Science* **315**, 187e.
66. Roth, H., Korn, T., Rosenkranz, K., Hinney, A., Ziegler, A., Kunz, J., Siegfried, W., Mayer, H., Hebebrand, J., and Grzeschik, K. H. (1998). Transmission disequilibrium and sequence variants at the leptin receptor gene in extremely obese German children and adolescents. *Hum. Genet.* **103**, 540–546.
67. de Silva, A. M., Walder, K. R., Aitman, T. J., Gotoda, T., Goldstone, A. P., Hodge, A. M., de Courten, M. P., Zimmet, P. Z., and Collier, G. R. (1999). Combination of polymorphisms in OB-R and the OB gene associated with insulin resistance in Nauruan males. *Int. J. Obes. Relat. Metab. Disord.* **23**, 816–822.
68. Must, A., Spadano, J., Coakley, E. H., Field, A. E., Colditz, G., and Dietz, W. H. (1999). The disease burden associated with overweight and obesity. *JAMA* **282**, 1523–1529.
69. Loos, R. J., Ruchat, S., Rankinen, T., Tremblay, A., Perusse, L., and Bouchard, C. (2007). Adiponectin and adiponectin receptor gene variants in relation to resting metabolic rate, respiratory quotient, and adiposity-related phenotypes in the Quebec Family Study. *Am. J. Clin. Nutr.* **85**, 26–34.
70. Ishiyama-Shigemoto, S., Yamada, K., Yuan, X., Ichikawa, F., and Nonaka, K. (1999). Association of polymorphisms in the beta2-adrenergic receptor gene with obesity, hypertriglyceridaemia, and diabetes mellitus. *Diabetologia* **42**, 98–101.
71. Large, V., Hellstrom, L., Reynisdottir, S., Lonnqvist, F., Eriksson, P., Lannfelt, L., and Arner, P. (1997). Human beta-2 adrenoceptor gene polymorphisms are highly frequent in obesity and associate with altered adipocyte beta-2 adrenoceptor function. *J. Clin. Invest.* **100**, 3005–3013.
72. Mori, Y., Kim-Motoyama, H., Ito, Y., Katakura, T., Yasuda, K., Ishiyama-Shigemoto, S., Yamada, K., Akanuma, Y., Ohashi, Y., Kimura, S., Yazaki, Y., and Kadowaki, T. (1999). The Gln27Glu beta2-adrenergic receptor variant is associated with obesity due to subcutaneous fat accumulation in Japanese men. *Biochem. Biophys. Res. Commun.* **258**, 138–140.
73. Susulic, V. S., Frederich, R. C., Lawitts, J., Tozzo, E., Kahn, B. B., Harper, M. E., Himms-Hagen, J., Flier, J. S., and Lowell, B. B. (1995). Targeted disruption of the beta 3-adrenergic receptor gene. *J. Biol. Chem.* **270**, 29483–29492.
74. Allison, D. B., Heo, M., Faith, M. S., and Pietrobelli, A. (1998). Meta-analysis of the association of the Trp64Arg polymorphism in the beta3 adrenergic receptor with body mass index. *Int. J. Obes. Relat. Metab. Disord.* **22**, 559–566.
75. Mitchell, B. D., Blangero, J., Comuzzie, A. G., Almasy, L. A., Shuldiner, A. R., Silver, K., Stern, M. P., MacCluer, J. W., and Hixson, J. E. (1998). A paired sibling analysis of the beta-3 adrenergic receptor and obesity in Mexican Americans. *J. Clin. Invest.* **101**, 584–587.
76. Clarke, S. D., Thuillier, P., Baillie, R. A., and Sha, X. (1999). Peroxisome proliferator-activated receptors: a family of lipid-activated transcription factors. *Am. J. Clin. Nutr.* **70**, 566–571.
77. Ristow, M., Muller-Wieland, D., Pfeiffer, A., Krone, W., and Kahn, C. R. (1998). Obesity associated with a mutation in a genetic regulator of adipocyte differentiation. *N. Engl. J. Med.* **339**, 953–959.
78. Deeb, S. S., Fajas, L., Nemoto, M., Pihlajamaki, J., Mykkanen, L., Kuusisto, J., Laakso, M., Fujimoto, W., and Auwerx, J. (1998). A Pro12Ala substitution in PPARgamma2 associated with decreased receptor activity, lower body mass index and improved insulin sensitivity. *Nat. Genet.* **20**, 284–287.
79. Boss, O., Samec, S., Paoloni-Giacobino, A., Rossier, C., Dulloo, A., Seydoux, J., Muzzin, P., and Giacobino, J. P. (1997). Uncoupling protein-3: a new member of the mitochondrial carrier family with tissue-specific expression. *FEBS Lett.* **408**, 39–42.
80. Fleury, C., Neverova, M., Collins, S., Raimbault, S., Champigny, O., Levi-Meyrueis, C., Bouillaud, F., Seldin, M. F., Surwit, R. S., Ricquier, D., and Warden, C. H. (1997). Uncoupling protein-2: a novel gene linked to obesity and hyperinsulinemia. *Nat. Genet.* **15**, 269–272.
81. Bouchard, C., Perusse, L., Chagnon, Y. C., Warden, C., and Ricquier, D. (1997). Linkage between markers in the vicinity of the uncoupling protein 2 gene and resting metabolic rate in humans. *Hum. Mol. Genet.* **6**, 1887–1889.
82. Walder, K., Norman, R. A., Hanson, R. L., Schrauwen, P., Neverova, M., Jenkinson, C. P., Easlick, J., Warden, C. H., Pecqueur, C., Raimbault, S., Ricquier, D., Silver, M. H., Shuldiner, A. R., Solanes, G., Lowell, B. B., Chung, W. K., Leibel, R. L., Pratley, R., and Ravussin, E. (1998). Association between uncoupling protein polymorphisms (UCP2-UCP3) and energy metabolism/obesity in Pima Indians. *Hum. Mol. Genet.* **7**, 1431–1435.
83. Cassell, P. G., Neverova, M., Janmohamed, S., Uwakwe, N., Qureshi, A., McCarthy, M. I., Saker, P. J., Albon, L., Kopelman, P., Noonan, K., Easlick, J., Ramachandran, A., Snehalatha, C., Pecqueur, C., Ricquier, D., Warden, C., and Hitman, G. A. (1999). An uncoupling protein 2 gene variant is associated with a raised body mass index but not Type II diabetes. *Diabetologia* **42**, 688–692.
84. Klannemark, M., Orho, M., and Groop, L. (1998). No relationship between identified variants in the uncoupling protein 2 gene and energy expenditure. *Eur. J. Endocrinol.* **139**, 217–223.
85. Urhammer, S. A., Dalgaard, L. T., Sorensen, T. I., Moller, A. M., Andersen, T., Tybjaerg-Hansen, A., Hansen, T., Clausen, J. O., Vestergaard, H., and Pedersen, O. (1997). Mutational analysis of the coding region of the uncoupling

protein 2 gene in obese NIDDM patients: impact of a common amino acid polymorphism on juvenile and maturity onset forms of obesity and insulin resistance. *Diabetologia* **40**, 1227–1230.
86. Argyropoulos, G., Brown, A. M., Willi, S. M., Zhu, J., He, Y., Reitman, M., Gevao, S. M., Spruill, I., and Garvey, W. T. (1998). Effects of mutations in the human uncoupling protein 3 gene on the respiratory quotient and fat oxidation in severe obesity and type 2 diabetes. *J. Clin. Invest.* **102**, 1345–1351.
87. Ravussin, E. (1995). *Metabolic differences and the development of obesity. Metabolism* **44**, 12–14.
88. Jakicic, J. M., and Wing, R. R. (1998). Differences in resting energy expenditure in African-American vs Caucasian overweight females. *Int. J. Obes. Relat. Metab. Disord.* **22**, 236–242.
89. Hinney, A., Schmidt, A., Nottebom, K., Heibult, O., Becker, I., Ziegler, A., Gerber, G., Sina, M., Gorg, T., Mayer, H., Siegfried, W., Fichter, M., Remschmidt, H., and Hebebrand, J. (1999). Several mutations in the melanocortin-4 receptor gene including a nonsense and a frameshift mutation associated with dominantly inherited obesity in humans. *J. Clin. Endocrinol. Metab.* **84**, 1483–1486.
90. Sina, M., Hinney, A., Ziegler, A., Neupert, T., Mayer, H., Siegfried, W., Blum, W. F., Remschmidt, H., and Hebebrand, J. (1999). Phenotypes in three pedigrees with autosomal dominant obesity caused by haploinsufficiency mutations in the melanocortin-4 receptor gene. *Am. J. Hum. Genet.* **65**, 1501–1507.
91. Vaisse, C., Clement, K., Guy-Grand, B., and Froguel, P. (1998). A frameshift mutation in human MC4R is associated with a dominant form of obesity. *Nat Genet* **20**, 113–114.
92. Yeo, G. S., Farooqi, I. S., Aminian, S., Halsall, D. J., Stanhope, R. G., and O'Rahilly, S. (1998). A frameshift mutation in MC4R associated with dominantly inherited human obesity. *Nat. Genet.* **20**, 111–112.
93. Valve, R., Sivenius, K., Miettinen, R., Pihlajamaki, J., Rissanen, A., Deeb, S. S., Auwerx, J., Uusitupa, M., and Laakso, M. (1999). Two polymorphisms in the peroxisome proliferator-activated receptor-gamma gene are associated with severe overweight among obese women. *J. Clin. Endocrinol. Metab.* **84**, 3708–3712.
94. Hixson, J. E., Almasy, L., Cole, S., Birnbaum, S., Mitchell, B. D., Mahaney, M. C., Stern, M. P., MacCluer, J. W., Blangero, J., and Comuzzie, A. G. (1999). Normal variation in leptin levels in associated with polymorphisms in the proopiomelanocortin gene, POMC. *J. Clin. Endocrinol. Metab.* **84**, 3187–3191.
95. Lander, E. S., and Schork, N. J. (1994). Genetic dissection of complex traits. *Science* **265**, 2037–2048.
96. Farooqi, I. S., and O'Rahilly, S. (2005). New advances in the genetics of early onset obesity. *Int. J. Obes. (Lond.)* **29**, 1149–1152.
97. Hager, J., Clement, K., Francke, S., Dina, C., Raison, J., Lahlou, N., Rich, N., Pelloux, V., Basdevant, A., Guy-Grand, B., North, M., and Froguel, P. (1998). A polymorphism in the 5′ untranslated region of the human ob gene is associated with low leptin levels. *Int. J. Obes. Relat. Metab. Disord.* **22**, 200–205.
98. Palmer, J. S., Duffy, D. L., Box, N. F., Aitken, J. F., O'Gorman, L. E., Green, A. C., Hayward, N. K., Martin, N. G., and Sturm, R. A. (2000). Melanocortin-1 receptor polymorphisms and risk of melanoma: is the association explained solely by pigmentation phenotype? *Am. J. Hum. Genet.* **66**, 176–186.
99. Campfield, L. A., Smith, F. J., and Burn, P. (1998). Strategies and potential molecular targets for obesity treatment. *Science* **280**, 1383–1387.
100. Dietz, W. H., and Robinson, T. N. (2005). Clinical practice. Overweight children and adolescents. *N. Engl. J. Med.* **352**, 2100–2109.
101. Gray, J., Yeo, G. S., Cox, J. J., Morton, J., Adlam, A. L., Keogh, J. M., Yanovski, J. A., El Gharbawy, A., Han, J. C., Tung, Y. C., Hodges, J. R., Raymond, F. L., O'Rahilly, S., and Farooqi, I. S. (2006). Hyperphagia, severe obesity, impaired cognitive function, and hyperactivity associated with functional loss of one copy of the brain-derived neurotrophic factor (BDNF) gene. *Diabetes* **55**, 3366–3371.
102. Herbert, A., Gerry, N. P., McQueen, M. B., Heid, I. M., Pfeufer, A., Illig, T., Wichmann, H. E., Meitinger, T., Hunter, D., Hu, F. B., Colditz, G., Hinney, A., Hebebrand, J., Koberwitz, K., Zhu, X., Cooper, R., Ardlie, K., Lyon, H., Hirschhorn, J. N., Laird, N. M., Lenburg, M. E., Lange, C., and Christman, M. F. (2006). A common genetic variant is associated with adult and childhood obesity. *Science* **312**, 279–283.
103. Xu, B., Goulding, E. H., Zang, K., Cepoi, D., Cone, R. D., Jones, K. R., Tecott, L. H., and Reichardt, L. F. (2003). Brain-derived neurotrophic factor regulates energy balance downstream of melanocortin-4 receptor. *Nat. Neurosci.* **6**, 736–742.

CHAPTER 22

Obesity: Overview of Treatments and Interventions

HELEN M. SEAGLE,[1] HOLLY R. WYATT,[2] AND JAMES O. HILL[3]

[1]*Kaiser Permanente, Denver, Colorado*

[2]*University of Colorado at Denver and Health Sciences Center, Denver, Colorado*

[3]*University of Colorado at Denver and Health Sciences Center, Denver, Colorado*

Contents

I. INTRODUCTION

Obesity is a pervasive disease in our world. Currently an estimated 1.6 billion of the global population are overweight, and at least 400 million adults are obese [1]. In the United States, almost two-thirds of the population is either overweight or obese, and this disease affects all age, sex, and race-ethnic groups [2]. The prevalence of overweight and obesity made a dramatic leap upward toward the end of the 20th century; in 1960, only about one-third of the U.S. population was overweight [2]. Because excess weight is associated with a higher prevalence of diseases such as diabetes, cardiovascular diseases, and osteoarthritis [3], the burden of obesity is high. Health care providers and policymakers alike are faced with the dilemma of managing this obesity epidemic and preventing this situation from worsening.

In this chapter, we give a broad overview of weight-management interventions. We focus first on the clinical assessment and relevance of excess fat accumulation (i.e., total body fat assessed by body mass index as well as visceral adiposity assessed by waist circumference). Next we review specific interventions ranging from relatively low risk (e.g., lifestyle modifications) to pharmaceutical and surgical interventions that confer successively higher risk but potentially a higher benefit for those individuals more compromised by their higher body weights. Ultimately, we encourage the reader to think about the treatment of obesity not just from an acute weight-loss perspective but also from the perspective of both prevention of weight gain as well as prevention of weight regain following the acute weight-loss phase. It is from this broadened perspective that the current obesity epidemic is most likely to be curtailed.

II. ASSESSMENT OF OVERWEIGHT AND OBESITY

The primary care physician is often the first person to assess an individual's need for obesity treatment. Ideally, the physician would then refer the client to a comprehensive weight-management program or obesity specialist, or recommend a commercial weight-loss program and provide follow-up care to monitor progress. If the primary care physician decides to treat the client, referrals should be made for additional services (e.g., the services of a psychologist, an exercise physiologist, or a registered dietitian [4]). Determining the appropriate treatment approach for an individual requires a careful assessment of an individual's needs [5]. An obese person with two or three comorbidities clearly requires a more aggressive approach than an overweight individual suffering no medical impact from excess body fat. The use of a tailored approach to obesity treatment will ensure that more people become successful in their weight-management efforts. Obesity treatment is one arena where a one-size-fits-all approach does not ensure success.

Obesity is characterized by the accumulation of excess body fat. Although accurate methods to measure body fat do exist, they are expensive and impractical for general use.

Body weight has traditionally been used as a surrogate measure of excess body fat. In the past, an obese or overweight body weight was based on ideal body weight tables established by the Metropolitan Life Insurance Company [6]. These insurance tables estimated an "ideal" weight for a given height, frame size, and gender based on collected mortality data. Overweight and obesity were then defined as some percentage above the estimated ideal body weight. Although widely used, these tables were criticized for being derived from populations with body fat contents that did not reflect those of the general public, for using frame size (an arbitrary assessment), and for being based on mortality outcomes alone without evaluating morbidity data [5].

A. Clinical Assessment of Body Fat

1. Body Mass Index

The use of body mass index (BMI) has replaced the insurance tables and has become the recommended method to estimate body fat and to define both overweight and obesity in a clinical setting [1, 5]. BMI is determined by weight in kilograms divided by height in meters squared ($BMI = kg/m^2$). The formula—weight (pounds)/height $(inches)^2 \times 704.5$—can be used to convert height in inches and weight in pounds directly into BMI units [7]. Figure 1 converts measures of height and weight into BMI units.

BMI is a better estimate of body fat than body weight [5, 8] and has advantages over the ideal body weight estimation that preceded it. Unlike the ideal body weight tables that were based on mortality data alone, BMI correlates with morbidity. The relationship between a given BMI and the risk of both mortality and morbidity has been assessed in several large epidemiological studies [9–11]. BMI does not require a subjective assessment for frame size and the same formula is used for both men and women (unlike the insurance tables, which were specific for gender). However, there are limitations to the usefulness of the BMI. BMI may overestimate total body fat in persons who are very muscular (such as elite athletes) and may underestimate

ft/in lbs.	4'10"	4'11"	5'0"	5'1"	5'2"	5'3"	5'4"	5'5"	5'6"	5'7"	5'8"	5'9"	5'10"	5'11"	6'0"	6'1"	6'2"	6'3"	6'4"
120	25	24	23	23	22	21	21	20	19	19	18	18	17	17	16	16	15	15	15
125	26	25	24	24	23	22	22	21	20	20	19	18	18	17	17	17	16	16	15
130	27	26	25	25	24	23	22	22	21	20	20	19	19	18	18	17	17	16	16
135	28	27	26	26	25	24	23	23	22	21	21	20	19	19	18	18	17	17	16
140	29	28	27	27	26	25	24	23	23	22	21	21	20	20	19	19	13	18	17
145	30	29	28	27	27	26	25	24	23	23	22	21	21	20	20	19	19	18	18
150	31	30	29	28	27	27	26	25	24	24	23	22	22	21	20	20	19	19	18
155	32	31	30	29	28	28	27	26	25	24	24	23	22	22	21	20	20	19	19
160	34	32	31	30	29	28	28	27	26	25	24	24	23	22	22	21	21	20	19
165	35	33	32	31	30	29	28	28	27	26	25	24	24	23	22	22	21	21	20
170	36	34	33	32	31	30	29	28	27	27	26	25	24	24	23	22	22	21	21
175	37	35	34	33	32	31	30	29	28	27	27	26	25	24	24	23	23	22	21
180	38	36	35	34	33	32	31	30	29	28	27	27	26	25	24	24	23	22	22
185	39	37	36	35	34	33	32	31	30	29	28	27	27	26	25	24	24	23	22
190	40	38	37	36	35	34	33	32	31	30	29	28	27	27	26	25	24	24	23
195	41	39	38	37	36	35	34	33	32	31	30	29	28	27	27	26	25	24	24
200	42	40	39	38	37	36	34	33	32	31	30	30	29	28	27	26	26	25	24
205	43	41	40	39	38	36	35	34	33	32	31	30	29	29	28	27	26	26	25
210	44	43	41	40	38	37	36	35	34	33	32	31	30	29	29	28	27	26	26
215	45	44	42	41	39	38	37	36	35	34	33	32	31	30	29	28	28	27	26
220	46	45	43	42	40	39	38	37	36	35	34	33	32	31	30	29	28	27	27
225	47	46	44	43	41	40	39	38	36	35	34	33	32	31	31	30	29	28	27
230	48	47	45	44	42	41	40	38	37	36	35	34	33	32	31	30	30	29	28
235	49	48	46	44	43	42	40	39	38	37	36	35	34	33	32	31	30	29	29
240	50	49	47	45	44	43	41	40	39	38	37	36	35	34	33	32	31	30	29
245	51	50	48	46	45	43	42	41	40	38	37	36	35	34	33	32	32	31	30
250	52	51	49	47	46	44	43	42	40	39	38	37	36	35	34	33	32	31	30
255	53	52	50	48	47	45	44	43	41	40	39	38	37	36	35	34	33	32	31
260	54	53	51	49	48	46	45	43	42	41	40	38	37	36	35	34	33	33	32

FIGURE 1 Body mass index using height in feet and inches and weight in pounds. (Note: From the first edition.)

body fat in persons who have lost muscle mass (such as the elderly). Additionally, BMI inaccurately reflects body fat in edematous states or in individuals who are less than 5 feet tall. Clinical judgment must always be used in the interpretation of BMI on an individual case basis.

Despite these limitations, the correlation between BMI and excess body fat in the general population is good [12]. Both the National Institutes of Health and the World Health Organization have defined overweight as a BMI of 25.0 to 29.9 kg/m^2 and obesity as a BMI of 30 kg/m^2 or greater [1, 5]. These BMI cutoffs were determined using studies evaluating the relationship between BMI and mortality and morbidity risk [9, 10]. In general, morbidity and mortality risk increases as BMI rises, but this relationship is curvilinear. Increases in BMI between 20 and 25 kg/m^2 alter the morbidity and mortality risk less than increases in BMI above 25 kg/m^2. For example, in the Nurses' Health Study, relative to a woman with a BMI $<$ 21 kg/m^2, heart disease risk was 1.8 times greater in women with a BMI between 25 and 29 kg/m^2 but 3.3 times greater for women with a BMI greater than 29 kg/m^2 [9, 13]. Gender does not alter this relationship; therefore, the same cutoff points are used to define obesity (BMI $\geq$ 30 kg/m^2) and overweight (BMI 25.0–29.9 kg/m^2) in both men and women [14].

2. Waist Circumference

The waist circumference as a measure of visceral adiposity can complement the BMI for assessing disease risk. Excess fat located in the upper abdominal region (visceral fat) is associated with a greater disease risk than fat located in other areas [15]. Abdominal fatness is an independent risk factor (even when BMI is not increased) and is predictive of comorbidities and mortality [15, 16]. A high BMI alerts the provider that a client is carrying too much body fat, whereas a high waist circumference signals that a significant amount of the excess fat is visceral fat. High risk is defined by a waist circumference >40 inches (102 cm) for men and >35 inches (88 cm) for women [5]. The power of waist circumference to predict disease risk may vary by ethnicity and age [12] as waist circumference is a better disease risk indicator than BMI in Asian Americans and in older individuals. For this reason, waist circumference cutoffs may need to be adjusted in the future based on age and ethnicity. Sex-specific cutoffs for waist circumference can be used for adults with a BMI less than 35 kg/m^2. In individuals with a BMI above 35 kg/m^2, a waist circumference does not confer additional disease risk and therefore it is not necessary to measure waist circumference in clients with BMI >35 kg/m^2. Figure 2 illustrates an appropriate technique to measure waist circumference.

3. Weight Gain

Several long-term cohort studies have shown that the average U.S. adult gains about 1 pound each year or about 10 pounds each decade [17]. In addition to BMI and the waist circumference measurement, providers should assess increases in a client's weight over time. Even at a normal or healthy BMI, clients with a weight gain of more than 2.2 pounds (1 kg) per year or more than 22 pounds (10 kg) overall have an increased disease risk [9]. Providers should identify clients who are gaining weight from one year to the next and provide appropriate interventions to prevent further gain.

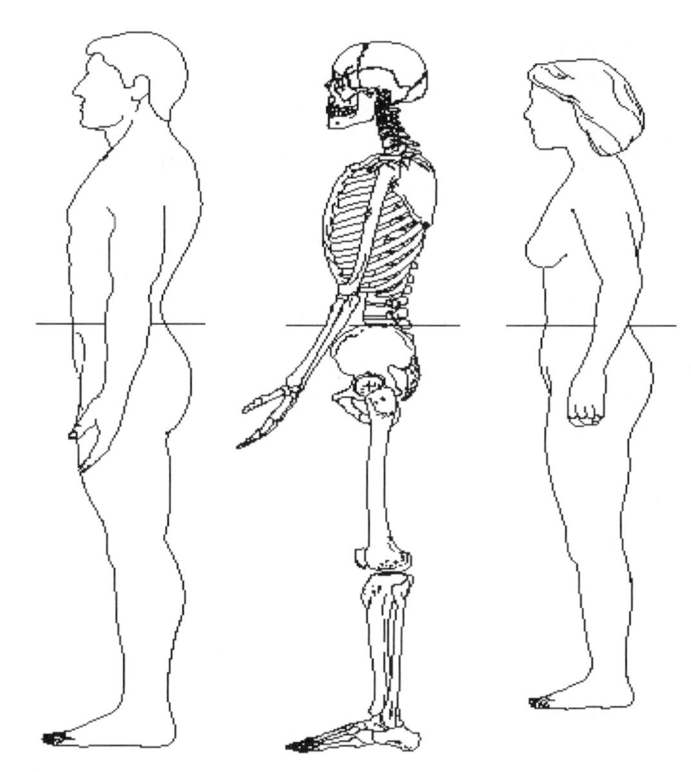

FIGURE 2 Measuring tape position for waist circumference. (Adapted from [22].)

B. Assessment of Comorbidities and Risk Factors

Patients should be assessed for the presence of both obesity-related comorbidities and concomitant cardiovascular risk factors. This is important because some obesity-associated diseases and cardiovascular risk factors will place the patient in a very high risk category and the aggressiveness of the obesity treatment should be increased. Patients are considered to be at *very high risk* if they have existing comorbidities such as coronary heart disease, atherosclerotic diseases, type 2 diabetes mellitus, or sleep apnea. This would include patients with a history of myocardial infarction, angina pectoris, heart surgery, or angioplasty. Many obese patients may have "silent" comorbidities such as hyperlipidemia or hypertension. Obese patients should always have a blood pressure

TABLE 1 Criteria for Clinical Diagnosis of Metabolic Syndrome by NCEP (Any Three of the Five Risk Factors Constitute Diagnosis of Metabolic Syndrome)

Risk Factor	Defining Level
Abdominal obesity	Waist circumference
Men	>102 cm (>40 inches)
Women	>88 cm (>35 inches)
Triglycerides	≥150 mg/dL or On drug treatment for ↑ triglycerides
HDL cholesterol	
Men	< 40 mg/dL
Women	< 50 mg/dL or On drug treatment for ↓ HDL
Blood pressure	≥ 130 mmHg systolic or ≥ 85 mmHg diastolic or On drug treatment for history of high blood pressure
Fasting glucose	≥ 100 mg/dL or On drug treatment for ↑ glucose

measurement with an appropriate-sized cuff and complete cardiovascular exam. Physical examination signs of high cholesterol such as xanthomas, arcus cornelius, and xanthelasma and should be evaluated for along with retinopathy, neuropathy, and other signs of type 2 diabetes.

A key part of the health risk assessment during an office visit is determining the patient's risk of developing type 2 diabetes and cardiovascular disease. The National Cholesterol Education Program (NCEP) Adult Treatment Panel III encourages the identification of a constellation of risk factors (Table 1) that, when found in the same individual, can be diagnosed as the metabolic syndrome [18]. The purpose of identifying these risk factors is to prevent cardiovascular disease in a population group thought to be at high risk for developing the diseases in the future. Patients who have three or more of the listed cardiovascular risk factors can be classified as having the metabolic syndrome by NCEP Guidelines. Weight reduction, healthy diet, and physical activity are critical components of a treatment plan for these individuals [18].

C. Assessment of Readiness

After assessing the patient's need for weight reduction, the health care provider must assess the patient's readiness to participate in treatment. Even when a patient is seriously overweight, he or she may not be ready to make a commitment to weight reduction [4]. Providers should assess if the patient recognizes the need for weight loss and is willing (and able) to sustain a weight-loss effort. Brownell developed a series of questions that can be used to help assess a patient's readiness to accept and participate in a long-term treatment plan [19, 20]. Patients who are not ready and feel they are unable to make the commitments to treatment should not begin a weight-loss program. Goals for these patients may be simply encouraging them to think about what it would take in their lives to make a commitment. Preventing further weight gain rather than an immediate weight loss may be more appropriate for these patients.

D. Selecting Treatment Options

Obesity is a complex disease of multifactorial origins. However, there is simplicity in the underlying model of body weight change. The energy balance equation dictates that in order for body weight to change, there must be an energy imbalance. Either a change in energy intake or a change in energy output must occur so that body stores of energy are altered, causing a change in total body weight [21]. Therefore, treatments of obesity must either focus on diminishing energy intake (e.g., diet, medications, surgery), increasing energy output (e.g., physical activity), or a combination of both (e.g., behavior modification addressing changes in dietary intake and physical activity). However, it is imperative that any intervention for overweight and obesity address both weight gain prevention and weight maintenance as well as body weight reduction [5].

Many options are available for treating overweight and obese individuals. For each patient, the risks of each treatment option must be weighed against the benefit of the potential weight loss produced by that treatment. This risk/benefit assessment must take into account a patient's BMI, waist circumference, and the presence of comorbidities and cardiovascular risk factors. Patients at a higher BMI or with existing obesity-related diseases are at more risk from their excess weight, and, therefore, more aggressive treatments such as pharmacotherapy and surgery become appropriate options. For each patient, there is a level of obesity where the risk of the treatment is outweighed by the benefit the patient would receive from a long-term reduction in weight. Each treatment plan must be tailored to meet the BMI and risk/benefit assessment for each patient. Table 2 shows recommended treatment options based on BMI and the presence or absence of a serious health complication [22].

E. Appropriate Goal Setting

The evidence-based NIH clinical guidelines for obesity treatment set the following general goals for weight loss and management: (1) to prevent further weight gain, (2) to reduce body weight, and (3) to maintain a lower body weight long term [5]. Traditionally, the goal of obesity treatment was to achieve an ideal body weight, and for

TABLE 2 Selecting Treatment Options Based on BMI and Comorbidities

Body Mass Index	Comorbidities[a] Present	Diet	Exercise	Behavioral Therapy	Pharmacotherapy	Surgery
25–26.9	No	–[b]	–[b]	–[b]	–	–
	Yes	+	+	+	–	–
27–29.9	No	–[b]	–[b]	–[b]	–	–
	Yes	+	+	+	+	–
30–34.9	No	+	+	+	+	–
	Yes	+	+	+	+	–
35–39.9	No	+	+	+	+	–
	Yes	+	+	+	+	+
≥40	No	+	+	+	+	+
	Yes	+	+	+	+	+

Source: Based on "NIH Clinical Guidelines on the Identification, Evaluation, and Treatment of Overweight and Obesity in Adults: The Evidence Report" (1998). National Institutes of Health, Bethesda, MD.

+ indicates appropriate treatment option; – indicates inappropriate treatment option.

[a]Comorbidities include hypertension, sleep apnea, dyslipidemia, coronary heart disease, and type 2 diabetes.

[b]Prevention of weight gain with diet, exercise, and behavioral therapy is indicated.

many people this meant losing extremely large amounts of weight. However, a reduction to ideal body weight is not necessary for health improvement and risk reduction. Clinical studies indicate that moderate weight reduction (i.e., 5% to 10% of the initial body weight) can correct or ameliorate many of the metabolic abnormalities associated with obesity and that small weight losses are associated with improvements in hypertension, dyslipidemia, and type 2 diabetes mellitus [23–25]. Prescribing a weight-loss goal of 5% to 10% sets a reasonable and achievable goal that may be more easily maintained. Unfortunately, many patients are not satisfied with weight reduction in this range, and the provider must work closely with the patient to help set realistic expectations and provide guidance in this area [26].

III. LIFESTYLE MODIFICATION

Lifestyle modification—in particular, the modification of a person's diet and physical activity—is the cornerstone of most obesity treatment programs. For most people, this treatment incurs no side effects, has minimal cost, and, if the lifestyle changes can be maintained, has great potential for long-term effectiveness [27]. In addition to creating changes in diet and physical activity patterns, the lifestyle modification component of obesity interventions usually also includes some form of behavioral treatment to enhance the long-term effectiveness of the program.

A. Dietary Modification

National surveys indicate that about 30% of U.S. adults are currently trying to lose weight [28] and about 50% report having tried to lose weight in the past year [29]. Most people report modifying their diet in some way to achieve their goals [28, 29]. Because the prevalence of overweight and obesity is also at an all-time high, this interest in dieting and attempts to modify diet appear to be contradictory. This contradiction underscores the difficulty people face in making seemingly simple changes to their dietary intake in an environment that encourages easy overconsumption of energy.

1. Creating an Energy Deficit

The universal component of dietary interventions for weight loss is a creation of an energy deficit [1, 5]. Most recommendations encourage a slow rate of weight loss through an energy deficit (energy output minus energy intake) of 500 to 1000 kcal/day [5]. This recommendation is based on the energy cost of burning 1 pound of excess body weight per week. Typically, the composition of the weight loss is about 25% fat-free mass and 75% fat mass. Metabolically, the more obese person can handle a greater energy deficit, as demonstrated by a lower protein oxidation rate during fasting, than a lean person [30]. It is important to monitor the rate of weight loss during the active weight-loss phase. Initially, particularly at the greater energy deficits, diuresis may occur and weight will drop quickly. However, after this initial drop in weight, the rate of weight loss will slow and should not be greater than 1% body weight per week [31]. An energy deficit of 500 to 1000 kcal/day should produce about a 10% body weight reduction over 6 months [5].

Many health care providers prescribe a standard weight-loss diet of a preset calorie level (e.g., 1200 or 1500 kcal/day). This approach has the advantage of allowing the health care provider to give out preprinted diet plans already designed to achieve the calculated energy level. The disadvantages of this approach include inappropriately low energy intakes (i.e., >1000 kcal/day deficits) in very large individuals with high energy requirements, and diet plans that are not tailored to an individual's lifestyle. To avoid inappropriate energy deficits, it makes more sense to estimate a person's energy expenditure and subtract 500 to 1000 kcal/day (the greater energy deficit should be reserved for the heavier individuals). Estimating a person's energy expenditure based on sex, body weight, and age [32] is preferable to using self-reported food intakes, which are notoriously unreliable, particularly among obese subjects [33, 34]. Alternatively, as a general rule, an appropriate energy level can be determined by assigning 12 kilocalories per pound of current body weight and then subtracting 500 to 1000 calories to create an energy deficit. The NIH Clinical Guidelines recommend 1000 to 1200 kcal/day for women and 1200 to 1500 kcal/day for men [5]. However, they emphasize the need for dietary education and tailoring the diet plan to accommodate individual food preferences.

Decreasing one's food consumption without ensuring an intake of a variety of foods may compromise a person's nutrient intake, particularly of calcium, iron, and vitamin E. At energy intakes below 1200 kcal/day, it is difficult to consume an adequate intake of essential vitamins and minerals. Multivitamin and mineral supplements are recommended for intakes below 1200 kcal/day and for individuals whose food choices limit their abilities to consume a satisfactory nutrient intake.

2. Very Low Calorie/Energy Diets

Some dietary regimens, such as very low calorie diets (VLCDs), establish a greater than 1000 kcal/day energy deficit. A VLCD is typically a liquid formulation that contains up to 800 kcal/day (3350 kJ/day) [35, 36]. VLCDs are enriched in protein of high biological value (0.8 to 1.5 g/kg of ideal body weight per day) and are supplemented with essential vitamins, minerals, electrolytes, and fatty acids. A typical VLCD program lasts for 12 to 16 weeks, and the liquid formula completely replaces all usual foods. A structured period of refeeding usually occurs after a VLCD, with solid foods slowly being reintroduced into the patient's diet.

The purpose of VLCDs are to quickly achieve large weight loss while providing adequate nutrition and preserving lean body mass. The mean weight loss for a 12- to 16-week program (including the long-term weight loss of dropouts) is about 20 kg [37, 38]. Weight losses on VLCDs are greater for men than for women [37], and heavier people lose more than lighter people [39]. It has been hypothesized that, beyond the greater than usual energy deficit, the form of these diets is an important factor in the effectiveness of a VLCD to produce a weight loss [37, 39]. That is, the structured feeding regimen of a VLCD encourages excellent adherence to the low-calorie plan; a patient does not have to make food choices and refrains from making impulsive high-calorie selections.

Although effective at producing a quick weight loss, VLCDs have not proven to be effective for long-term weight-loss maintenance. Now, many programs combine VLCD with behavior modification, and this combination has slightly improved the long-term weight maintenance outcomes [40]. Other disadvantages of VLCDs include the expense of the programs and the side effects of the quick weight loss. The most serious side effects include hyperuricemia, gout, gallstones, and cardiac complications. These serious side effects underscore the need for (1) appropriate medical assessment for patients entering a program; (2) use of VLCDs in only obese patients (i.e., BMI > 30), especially those individuals with comorbid conditions that would be responsive to weight loss (e.g., diabetes, sleep apnea, or presurgery); and (3) ongoing medical supervision during the course of the program [35]. The NIH Clinical Guidelines do not recommend VLCDs for use in obesity treatment [5].

3. Meal Replacements

Meal replacement drinks are often used in weight-loss programs. These differ from VLCDs in that they are designed to replace only one or two of the day's meals rather than the whole day's intake, and they are sold over the counter. The advantage of a meal replacement drink is that it can replace a person's most problematic meal of the day with a drink of a known energy and macronutrient content, thus helping to achieve a targeted energy intake goal. Few studies have specifically evaluated the efficacy of meal replacement drinks for weight loss [41]. A 27-month study [42] showed a significantly greater weight loss in people following a low-calorie plan utilizing two meal replacement drinks per day versus people following a conventional low-calorie plan ($11.3 \pm 6.8\%$ vs. $5.9 \pm 5.0\%$ of initial body weight, $p < 0.0001$). Additionally, meal replacement drinks may be useful for weight maintenance or weight gain prevention. A study conducted in rural Wisconsin evaluated the efficacy of meal replacement drink treatment against the backdrop of weight gain experienced by matched controls over a 5-year period [43]. A total of 134 men and women participated in a self-management meal replacement drink program that included a 3-month active weight-loss phase (two meal replacement drinks per day with weekly weigh-ins) and a maintenance phase (self-monitoring of weight and use of a meal replacement drink if weight gain occurred). During the same 5-year period, 86% of the meal replacement drink participants were *at least* weight stable ($\leq$0.8 kg weight gain), whereas only 25% of the matched controls had prevented weight gain [43]. However, further research is required to verify the efficacy of this weight-loss treatment.

4. Macronutrient Composition

A negative energy balance is the most important factor affecting the amount, rate, and composition of weight loss. The macronutrient composition of the diet has no significant additive effect on these weight-loss parameters during the course of a usual obesity treatment program [44] (See Chapter 26 for further discussion on this topic of diet composition and its relationship to body weight.) However, it is apparent that the macronutrient composition of a weight-loss diet may have an important effect on behavioral compliance with a dietary regimen.

In feeding experiments, it has been shown that people typically eat a constant daily weight of food regardless of the type of diet consumed [45]. Because fat contains more kilocalories per gram than either protein or carbohydrate (9 versus 4 kcal/g), more energy is likely to be consumed when high-fat versus low-fat foods are eaten [45]. Additionally, fat has been proposed to have a weak effect on both satiation (process controlling meal size) and satiety (process controlling subsequent hunger and eating) [46] (see Chapter 27). "Passive overconsumption" is a term used to describe the likelihood of increased energy intake that occurs during high-fat feeding because of the combination of these two factors: high energy value of fat and its weak effect on both satiation and satiety [46]. For these reasons, a low-fat diet (less than 30% energy from fat) is currently the typical recommended macronutrient guideline for weight loss [5, 47]. Because diabetes and cardiovascular disease are frequent comorbidities of obesity, it is also recommended to modify the fat content of the diet to be low in saturated and *trans*-fatty acids, and higher in monounsaturated fats [48]. The effectiveness of low-fat, low-calorie diets in combination with lifestyle counseling and activity has been demonstrated in multicenter clinical trials where, in addition to 5% to 10% weight loss, the reduction or prevention of comorbidities such as diabetes or hypertension has also occurred [49–52].

Energy density (energy/food weight) of foods and diets has been evaluated as a potential dietary attribute that could be manipulated to influence energy intake [53], and the Dietary Guidelines for Americans 2005 [47] suggest consuming low energy-dense foods as a strategy to reduce energy intake. Laboratory studies have demonstrated that people consume fewer calories when presented with low energy-dense foods versus higher energy-dense foods [54]. Initial clinical studies indicate this strategy impacts body weight regulation [55].

5. Popular Diets

Many people become frustrated with their perceived inability to change their body weight in a timely manner and are enticed by quick and painless weight loss promised by popular diet programs and books. These diets tend to be limiting in food choices (e.g., grapefruit diet), restrictive of at least one macronutrient (e.g., Atkins diet), or rely on novel food combinations (e.g., the Zone diet). The initial dramatic weight loss that occurs on many of these diets is often due to the diuresis that occurs as glycogen stores are depleted in response to a low carbohydrate intake. This weight loss is temporary and if carbohydrate intake is increased, the glycogen stores are repleted and body weight increases accordingly. On very low carbohydrate diets, the body goes through a period of ketogenesis to provide fuel that can be utilized by the brain cells. Ketones are potent appetite depressants, and lack of appetite may be one reason why individuals, at least temporarily, follow such restrictive diets.

It has only been recently that carefully controlled trials have evaluated the efficacy of these types of diets for weight loss. Typically, more weight loss occurs with low-carbohydrate diets in the first 6 months when compared with low-fat diets, although this weight-loss difference does not persist by 12 months [56–60]. Concerns regarding an increase in cardiovascular risks with these low-carbohydrate diets do not appear to be as problematic as first thought [58]. These studies have underscored the need to better understand factors that impact study attrition rates as well as individual adherence to any type of weight-loss diet as these two parameters impacted the interpretation of these trial outcomes [58, 59, 61].

6. The Nondiet Approach

There has been a movement toward replacing restrictive diet approaches and unrealistic weight goals with promotion of healthful food choices and size acceptance. Proponents cite two rationales for adopting this approach: (1) there is no long-term effective strategy for treating obesity and (2) pressure to be thin impairs the psychological and social well-being of both people who are overweight and those who are not [62]. The focus of the nondiet approach is to encourage people to improve their self-acceptance (self-image) regardless of their current weight and to adopt healthy practices to promote physical well-being (e.g., promoting fitness and healthful food choices). Few studies have carefully evaluated the result of the nondiet approaches on weight loss or the reduction of comorbidities associated with obesity [63]. Repeated weight loss (also termed "weight cycling") does not appear to be associated with psychopathology or changes in weight- and eating-related constructs [64]. However, it is apparent that more research is needed to conclusively determine the psychological effect of repeated episodes of dieting and weight loss.

B. Physical Activity Modification

Physical activity is important for improving health-related outcomes across many disease states including heart disease, cancer, and diabetes (i.e., frequent comorbidities to obesity) [65]. In general, physical activity has been used as a key component of obesity treatment. However, studies looking at physical activity per se for weight loss have found only modest reductions in body weight using this strategy [66]. Weight losses in the range of 0.09 to 0.1 kg/week have been reported when exercise is used alone compared to a no-treatment control group [66, 67]. Exceptions to this trend have been reported with extreme levels of exercise as is seen in military-type training. Combining exercise with dietary restriction produces only a slight increase in weight loss over dietary restriction alone [66]. In general, groups using diet restriction plus exercise lost more weight than the diet-alone condition, but the magnitude of the difference was not significant in most studies.

A possible explanation for the relatively modest effect of physical activity on weight loss is that the energy cost of exercise is minimal compared to potential changes in energy intake. A person who exercises for 30 minutes 5 days a week may only burn 1000 kcal more per week depending on the individual's size, his or her fitness level, and the intensity of the exercise. In comparison, a person may consume an extra 1000 kcal in one or two unplanned snacks and easily negate the energy expended in exercise for the entire week.

Although its impact on weight loss may be minimal, physical activity appears to have a crucial role in the long-term maintenance of a weight loss (i.e., the prevention of weight regain). Many correlation studies show a strong association between self-reported exercise at follow-up and maintenance of a weight loss [66]. Studies using doubly-labeled water suggest physical activity in the range of 11 to 12 kcal/kg/day may be necessary to prevent weight regain following a weight loss [68].

Data from the National Weight Control Registry (NWCR) also support the concept that high levels of physical activity are crucial in preventing weight regain following a weight loss. The NWCR is a registry of more than 2000 individuals who have maintained a minimum of a 30-pound weight loss for at least 1 year. These individuals report using a variety of methods to lose weight initially, but more than 90% report exercise as a key element in maintaining the loss long term. They report expending, on average, 2682 kcal/week in physical activity [27]. This is approximately the equivalent of walking 4 miles 7 days a week, and many report much higher levels. This suggests that while physical activity may not have been essential for weight loss in these subjects, they believe it to be essential in prevention of weight regain.

Pedometers and step counters are relatively recent devices used to promote physical activity and are often included in weight-loss interventions. Currently 10,000 steps per day are recommended as this level is consistent with 30 minutes per day to improve health benefits [69]. These devices promote awareness of a person's current level of activity and may be used to give a goal for improvement. Progression to higher levels of steps would be necessary to promote prevention of weight gain.

The physical activity levels recommended in the 2005 U.S. Dietary Guidelines [47] include three categories that relate to weight-management goals: (1) 30 minutes per day (moderate intensity) for health benefits; (2) 60 minutes per day (moderate to vigorous intensity) for prevention of weight gain; and (3) 90 minutes per day (moderate intensity) for the maintenance of weight loss in formerly obese individuals. A complete review of the role of physical activity in obesity intervention appears in Chapter 23.

C. Behavior Modification

Behavior modification is included in many obesity interventions in order to help patients develop a skill set to achieve a healthier weight. That skill set builds on the specific information about *what* to change and helps people on the *how* to change. Behavioral modification treatment is goal directed (e.g., clear, measurable goals based on behaviors such as "walk for 20 minutes daily"), process-oriented (e.g., patients are encouraged to identify strategies to overcome anticipated barriers as well as learn from setbacks), and advocates small changes so a person can build on successive successful experiences with change [70]. Group behavioral programs, conducted on a weekly basis in university or hospital clinics, have been reported to produce average reductions of 8% to 10% of initial body weight over 16 to 26 weeks [4]. The key behavioral modification components utilized in obesity treatment include self-monitoring, stimulus control, and relapse-prevention strategies. Self-monitoring commonly includes the systematic recording of food intake, exercise activities, or weight change. The available scientific literature on obesity suggests that consistent self-monitoring, particularly of food intake, is associated with improved treatment outcome [71, 72], even during high-risk periods such as holidays [73]. Stimulus control involves the identification and modification of environmental cues associated with overeating and sedentary activity and is widely accepted as clinically effective [74]. Relapse-prevention strategies involve training patients to prepare for lapses in the weight-loss process and to utilize coping strategies to prevent complete relapse of behavior change efforts [75].

Finally, group-based behavior modification training is used in a variety of obesity treatment formats, including university-based programs, commercial weight-management programs (e.g., Weight Watchers, Jenny Craig), and self-help programs (e.g., Take Off Pounds Sensibly, Overeaters Anonymous). Although preliminary research indicates that group-based interventions may be equally as effective as individual interventions for a variety of problem behaviors, there has been no systematic assessment to date on the group process specifically for obesity intervention [76]. As such, a variety of important questions including optimal number of group participants, frequency and length of meetings, critical group leader skills, and strategies to identify the optimal candidates for group-based interventions remain unanswered [61]. Most behavioral modification techniques utilized in a group setting can be effectively incorporated into weight-loss counseling provided to an individual [4, 77].

IV. PHARMACEUTICAL INTERVENTION

A. Background

Although diet, physical activity, and behavior modification remain the cornerstone of obesity management, pharmaceutical intervention has become a legitimate therapeutic option for obesity treatment since the mid-1990s. The increasing interest in pharmaceutical interventions has primarily been fueled by the realization that obesity in the United States has reached epidemic proportions with no sign of slowing down. Additionally, the old perception that obesity is caused by a lack of willpower or gluttony has been replaced with a better scientific understanding of the genetics and biology that predispose certain individuals to gain weight in our current environment [78].

The discovery of leptin in the mid-1990s established the existence of a genetically controlled complex biological system for food intake and weight regulation. The view of obesity has shifted from that of a behavioral problem occurring in people lacking in self-control and discipline to that of a chronic disease model with genetic, biological, and behavioral roots. This critical shift in thinking also changed the view of drug therapy for obesity from that of a short-term quick fix for a social problem to a long-term treatment for a chronic disease. It is now acknowledged that obesity, like other chronic diseases such as hypertension or diabetes, will require long-term treatment.

Weight-loss medications were previously viewed as unsuccessful because weight was always regained following the withdrawal of the drug. Now it is recognized that medications need to be given long term to be effective in obesity treatment [79]. In 1992, Weintraub and coworkers published a landmark long-term weight-loss trial using the combination of phentermine and fenfluramine [80]. This trial showed these medications' efficacy in maintenance of a weight loss over 4 years. This long-term pharmaceutical treatment data, along with the view of obesity as a chronic disease, marked the acceptance of pharmacotherapy as legitimate intervention in obesity treatment.

In the mid-1990s, dexfenfluramine alone and fenfluramine in combination with phentermine were used for long-term treatment of obesity. In September 1997, reported concerns about serious unacceptable side effects, such as valvular heart lesions, led to the withdrawal of dexfenfluramine and fenfluramine from the market. This left no long-term medications for obesity treatment that had the approval of the Food and Drug Administration (FDA). Since that time, the FDA has approved two new medications for long-term use in the treatment of obesity. Neither of these drugs (to date) has been associated with heart valve lesions. All other weight-loss medications approved by the FDA for weight loss are approved only for short-term use (Table 3). Because obesity is a chronic disease and must be managed for long periods of time, if not a lifetime, only medications approved for long-term use are of real interest in obesity treatment.

B. Medications Approved for Short-Term Use in Weight Loss

The FDA has approved several medications for short-term use (less than 3 months). These drugs include mazindol, phentermine, and diethylpropion. These are noradrenergic or sympathomimetic drugs that either stimulate release or block reuptake of norepinephrine. In general, these medications have produced more weight loss than placebo in most short-term clinical trials, but the magnitude of the weight loss was variable [81, 82]. Large clinical trials evaluating these drugs for efficacy and safety in long-term obesity

TABLE 3 Pharmaceutical Interventions Used in Obesity Treatment

Drug	Mechanism of Action	Administration	Daily Dose Range (mg)	FDA Approval for Use
Benzphetamine	Stimulates NE release	Start with 25 mg qd; max dose 25–50 mg po tid	25–150	Short term
Phendimetrazine	Stimulates NE release	35 mg po tid before meals or 105 mg SR qd	35–105	Short term
Diethylpropion	Stimulates NE release	25 mg po tid or 75 mg SR qd	25–75	Short term
Mazindol	Blocks NE reuptake	Start with 1 mg po qd; max dose 1 mg po tid with meals	1–3	Short term
Phentermine	Stimulates NE release	15, 30, or 37.5 mg po qd in the AM	15–37.5	Short term
Sibutramine	Serotonin and NE reuptake inhibitor	Start with 10 mg po qd, may increase or decrease to 5–15 mg po qd	5–15	Long term
Orlistat	Lipase inhibitor	120 mg taken with each meal po tid	120–360	Long term
		60 mg taken with meal po tid	60–180	OTC
Rimonabant	CB1 antagonist	20 mg taken before breakfast po	20	Pending approval in the US

Key: po = by mouth; qd = once a day; tid = three times a day; SR = slow release capsule; NE = norepinephrine.

treatment are scarce, and the FDA has not approved their use for longer than 3 months.

C. Medications with No Approval for Use to Promote Weight Loss

Fluoxetine and bupropion (FDA-approved antidepressants) have been evaluated in several weight-loss trials [24, 83, 84]. Patients taking fluoxetine in doses greater than 60 mg/day have shown more weight loss than those using placebo in clinical trials. However, the long-term efficacy of fluoxetine is questionable, because weight regain, while *continuing* the medication, was also demonstrated [24, 83]. Bupropion is approved for the treatment of depression and for smoking cessation. Several clinical trials have evaluated its weight-loss potential. One study randomized 327 subjects to bupropion 300 mg/day, buproprion 400 mg/day, or placebo [84]. Weight loss at 6 months was $5 \pm 1\%$, $7.2 \pm 1\%$, and $10.1 \pm 1\%$ for the placebo, 300 mg, and 400 mg groups, respectively ($p < 0.0001$). Although bupropion is not approved for weight loss if a person is clinically depressed, it may be an appropriate antidepressant (if medication is indicated), especially in the overweight person already trying to lose weight. Neither fluoxetine nor bupropion have FDA approval for weight loss.

Zonisamide and topiramate (FDA-approved antiepileptic drugs) have demonstrated weight loss in clinical trials designed for the treatment of epilepsy. The FDA has not approved either drug for weight loss. Zonisamide is an antiepileptic drug that has serotonergic and dopaminergic activity in addition to inhibiting sodium and calcium channels. Zonisamide has been studied in one 16-week trial in obese subjects for weight loss with favorable results [85]. Topiramate has been evaluated for weight loss, binge-eating treatment, and the Prader-Willi syndrome [86–91]. Initial studies reported weight loss in the range of 6% to 10%. Adverse events in these trials included paresthesias, somnolence, difficulty concentrating, and difficulty with memory. Several large studies were terminated early because a time-release formula was developed and it was hoped this new preparation would minimize the adverse event profile of the drug. At this point in time, there is no indication that the time-release preparation succeeded in alleviating the side effect profile, and the development of the program to pursue the indication from the FDA for obesity treatment was terminated by the sponsor in December of 2004. This decision was probably related to the associated adverse events. Topiramate, however, is still available as an antiepileptic drug.

D. Medications Approved for Long-Term Use

1. Sibutramine

Sibutramine was approved in 1997 for long-term treatment of obesity. Sibutramine is a combination serotonin and norepinephrine reuptake inhibitor. Unlike the fenfluramine–phentermine combination, sibutramine does not stimulate the release of either of these neurotransmitters, and this may be why valvular heart lesions have not been associated with the use of this drug [92]. Sibutramine's main mechanism of action is a reduction in food and energy intake. In laboratory animals, sibutramine also increases total metabolic rate by an increase in thermogenesis, but this effect remains in question in humans and is small in magnitude if present [93–95].

The long-term efficacy of sibutramine has been evaluated in several 6- and 12-month double-blind randomized studies [96–98]. Sibutramine-treated patients lose significantly more weight, on average, than placebo-treated patients in these studies. For example, a greater than 10% weight loss was achieved by 30% to 39% of the sibutramine-treated patients (on 10 and 15 mg/day, respectively) versus only 8% of the placebo group achieving this level of weight loss in a 12-month study [96].

Side effects for sibutramine include headache, dry mouth, constipation, and insomnia. Increases in heart rate and blood pressure have been reported and, therefore, blood pressure should be monitored before and regularly after starting the medication. A few patients (<5%) may have a significant increase in blood pressure and, therefore, uncontrolled hypertension, a history of heart disease, heart failure, stroke, and arrhythmia are contraindications to the use of this drug. Sibutramine is also contraindicated with monoamine oxidase inhibitors and other serotonin uptake inhibitors, which include medications for depression and migraines. Sibutramine is available in 5-, 10-, and 15-mg tablets and is taken once a day. Weight loss is dose related with the 10- and 15-mg dose. Most patients should be started on the 10-mg dose and adjustments made as necessary.

2. Orlistat

The FDA approved the use of orlistat for the long-term treatment of obesity in 1999. Orlistat is a minimally absorbable agent (<1%) that works in the gastrointestinal tract by blocking gastrointestinal lipases and reducing the subsequent absorption of ingested fat by approximately 30% [99]. Orlistat has been studied in 1- and 2-year clinical trials [100–102]. In general, the orlistat-treated groups lost more weight and had a higher percentage of subjects able to achieve a 10% weight loss than subjects in the placebo-treated group [103]. In a large study evaluating more than 600 patients, 38.8% of the patients in the orlistat group achieved a 10% or greater weight loss compared to only 17.7% in the placebo group [104]. Orlistat's use for weight-loss maintenance has been evaluated in 2-year studies. In these studies, a hypocaloric diet in combination with orlistat was used for weight loss in the first year, and then a eucaloric diet with orlistat or placebo was used in the second year to evaluate weight-loss maintenance. In a large study by Davidson and coworkers [101], almost twice as many subjects (34.1%) in the orlistat group were able to maintain a ≥10% weight loss for the 2-year period compared to the placebo group (17%) [101]. Similar results indicate that mean body weight regain is less with orlistat treatment than placebo (32.4 versus 56.0%; orlistat and placebo, respectively) [100]. In these clinical trials, orlistat was used in combination with either a low-fat hypocaloric diet for weight loss or a eucaloric diet for weight maintenance, and a portion of the benefit seen in these trials must be attributable to that diet.

The optimal dosing of orlistat is 120-mg po tid with meals that contain fat. Higher doses do not have any additional efficacy in weight loss and may result in more adverse events [105]. Orlistat is minimally absorbed; therefore, any systemic adverse events would be expected to be negligible. Orlistat should be used with a diet that is less than 30% energy from fat to prevent adverse side effects that include oily stools, oily spotting, flatus with discharge, fecal urgency, and fecal incontinence [105]. These events are due to the drug inhibiting fat absorption rather than a direct effect of the drug itself. Patients should be advised to maintain a low-fat diet while using the medication, because these side effects increase with diets that have over 30% energy from fat. In general, these events tend to decrease over time.

Mean plasma levels of vitamins A, D, E, and β-carotene were monitored during the trials. In general, plasma levels of these vitamins decreased but remained in the reference ranges [104, 105]; however, in the United States, a multivitamin supplement is recommended for patients prescribed orlistat and should be taken 2 hours before or after the dose of orlistat.

Over-the-counter (OTC) low-dose orlistat is the newest preparation available to help individuals manage weight. Over-the-counter orlistat (sold as alli in the United States) is the identical molecular compound of the prescription strength orlistat; however, it is provided in a 60-mg capsule rather than the 120-mg capsule. Dosing is one 60-mg capsule three times a day with meals. OTC orlistat represents a change in the scheduling of orlistat from a 120-mg prescription compound requiring a medical physician to write a prescription to an over-the-counter status that will allow patients to buy the medication without a medical consultation or oversight. Its OTC indication promotes weight loss in overweight adults when used along with a reduced-energy and low-fat diet. The OTC 60-mg dose of orlistat inhibits absorption of approximately 25% of ingested fats [99]. In contrast, the prescription dose of 120 mg blocks the absorption of about 30% of the fat consumed in a meal that contains approximately 30% fat. Similar to the 120-mg orlistat studies, significantly greater weight loss was observed with an orlistat dosage of 60 mg plus diet compared to placebo plus diet in overweight and obese patients [106–108]. Weight losses in the range of 4.8 to 9.7% were reported with the 60-mg OTC dose at 1 year.

E. Medications in Phase III Clinical Trials with No Approval

1. Rimonabant XX

The newly discovered endocannabinoid system and cannabinoid receptor, with their reported roles in the regulation of energy balance and body composition, offer a new target to induce weight loss and improve the metabolism of carbohydrates and lipids. Rimonabant, a selective cannabinoid-1 receptor blocker, is the compound closest to achieving future FDA approval. It has been shown in four phase III trials to reduce body weight and improve cardiovascular risk factors in obese patients [109–112]. The most common side effects are nausea, dizziness, arthralgia, and diarrhea. Rare but significant psychiatric side effects including depression and suicide have been noted in some of the clinical trials. Further evaluation of this type of adverse event and who is at risk will be necessary before the FDA approves this drug for general use. If Rimonabant achieves FDA approval, it will represent a new class of obesity drugs available for health care providers to manage body weight. Rimonabant is approved for use in Europe and Canada, but the FDA has not approved it for use in the United States.

F. Risk / Benefit Ratio

1. Selecting Patients for Pharmacotherapy

After the degree of obesity has been assessed and the presence of comorbidities determined, the potential risks and benefits of a particular pharmacotherapy should be determined for each patient. The risk of the potential side effects of the medication must be carefully weighed against the benefits of the weight loss as a result of the treatment. The risk of not losing weight or even gaining weight must also be considered in the decision-making process. Additionally, many clinicians require that patients attempt weight loss in a structured program of diet, exercise, and behavioral modification before being considered for pharmacotherapy [4]. Severe obesity (BMI $>35\ kg/m^2$) carries a high risk of morbidity and mortality [113], and intensive therapeutic approaches are indicated for most individuals. The lower BMI threshold for pharmaceutical intervention is not as easily defined, and other factors such as a high waist circumference, recent weight gain, family history, and the presence of comorbidities become important considerations in the decision to treat with weight-loss medications. Cosmetic weight-loss attempts for patients who want to lose a few pounds are not appropriate. The risk of the medication in this case is not outweighed by the benefit because a reduction in weight at a lean weight (BMI $\leq 25\ kg/m^2$) is not associated with significant improvements in health risks. Currently, the evidence and the NHLBI guidelines justify the use of a weight-loss medication in patients with BMIs $\geq 30\,kg/m^2$ or $\geq 27\,kg/m^2$ if comorbidities such as hypertension or diabetes are present [5].

2. Predictors of Efficacy

Not every patient responds to drug therapy. Therefore, it is important to monitor patients on weight-loss medications, not only for potential side effects but also for the efficacy of the medication itself. Clinical trials have shown that initial responders tend to continue to lose weight, whereas initial nonresponders continue to be nonresponders. The initial rate of weight loss has been frequently noted to predict subsequent weight loss. If the patient does not lose weight or maintain a previous weight loss with the medication, the medication should be discontinued. In this case, the risk of the medication is not outweighed by the weight loss, because there has not been a reduction in weight. As a general guideline, patients who do not lose 1% of body weight during their first month of treatment should discontinue treatment [93].

G. Medications Combined with Lifestyle Modification

It is important to emphasize that weight-loss medications should be prescribed in combination with a diet, exercise, and behavioral modification program. Medications are not a substitute for, but an adjunct to, lifestyle intervention, providing additional benefit by helping patients adhere to the necessary diet and exercise changes. For some individuals, diet and exercise alone may be enough to produce a 10% weight loss and long-term maintenance. For others, medications may provide a necessary additional intervention to allow weight-loss success.

V. SURGICAL TREATMENT

Although the prevalence of obesity (BMI > 30) has increased by 24% from 2000 to 2005, the prevalence of a BMI > 40 and of a BMI > 50 increased two and three times faster, respectively, over the same period [114]. For people in these BMI categories, surgical intervention is considered the most effective treatment for weight loss and long-term weight maintenance. Increasingly, more people are considering surgery to reduce body weight: from 1998 to 2004, the total number of bariatric surgeries increased nine-fold in the United States from 13,386 to 121,055 [115]. Surgery, with its inherent permanence, clearly has an advantage in long-term success [116]. It is reserved for patients with severe disease who have failed less invasive interventions and are at a very high risk for obesity-related morbidity and mortality. In the past, practitioners have used a rough guide of an excess weight of 100 pounds (45.5 kg) as an indication to consider a surgical intervention. In 1991, The National Institutes of Health Consensus Development Conference on gastrointestinal surgery for severe obesity set the patient selection criterion for surgery as a BMI exceeding $40\,kg/m^2$ [117]. In certain instances, an obese patient with a BMI between 35 and $40\,kg/m^2$ may be considered for a surgical procedure if a severe comorbidity is present [117]. These life-threatening comorbidities commonly include sleep apnea, hypertension, diabetes, heart failure, and vertebral disk herniation [118].

Similar to pharmacotherapy, the use of a surgical intervention in obesity requires a case-by-case risk/benefit analysis. Patients with a low probability of success with nonsurgical interventions and who meet BMI criteria may be appropriate candidates for surgery. Patients who have been determined to have acceptable operative risks should be well informed about the procedure, the benefits, and the risks, as well as the potential impact on their lifestyles. A commitment to long-term follow-up is also essential. It is recommended that surgeries are performed in high-volume clinics, as the prevention of postoperative complications depends on not only the technical expertise of the surgeon but also on the careful preoperative assessment, screening, and minimization of preoperative risk [119].

A. Surgical Procedures

Bariatric (i.e., weight-loss) surgery procedures reduce energy intake by modifying the gastrointestinal tract to either restrict the stomach volume or bypass a portion of the small intestine to decrease nutrient absorption. A number of different procedures have been performed in the past, but Roux-en-Y gastric bypass is the most prevalent procedure currently performed in the United States [120]. A Roux-en-Y gastric bypass procedure involves constructing a small gastric pouch whose outlet is a Y-shaped limb of small bowel. The procedure results in ingested food bypassing the majority of the stomach and variable lengths of the small intestine. Gastric bypass is thus both a restrictive and malabsorptive procedure. Recently, adjustable gastric banding has become a more frequently used procedure. This restrictive-only procedure involves wrapping an inflatable band around the stomach to create a small pouch with a narrow outlet. The band is connected to a small subcutaneous reservoir so that the degree of gastric restriction can be adjusted postoperatively by injecting or aspirating saline from the device. Both procedures are typically performed laparoscopically.

1. Amount of Weight Loss

In general, malabsorptive procedures tend to result in greater weight loss than restrictive procedures. Gastric bypass produces a slightly higher degree of weight loss (61.6% excess weight loss[1]) than does gastric banding (47.4% excess weight

[1] Percentage excess weight loss = weight loss/(preoperative weight − ideal weight) × 100.

loss) [121]. This increase in weight loss has to be balanced against a higher risk of nutritional deficiencies associated with the gastric bypass procedure. Weight loss tends to occur over a period of 18 to 24 months and is typically well maintained for as long as 14 years [122].

2. Reduction in Obesity-Related Comorbid Conditions

Most patients experience substantial improvements in obesity-related comorbidities such as sleep apnea, glycemic control, and hypertension [121, 122]. Studies looking at quality of life also report improvements following the surgical procedure and weight loss [123, 124]. The Swedish Obesity Study, a large prospective randomized study, has shown that weight reduction following surgery reduced the 2-year incidence of diabetes, hypertension, and other health risks and that these benefits were still sustained, although to a lesser degree, at 10 years [125].

3. Complications and Side Effects

For centers that specialize in obesity surgery, the immediate operative mortality for both adjustable banding and gastric bypass procedures is relatively low and usually in the range of 0.5% to 1% [121]. These postoperative complications include wound infections, wound dehiscence, leaks from the staple line breakdown, stomal stenosis, marginal ulcers, pulmonary problems, and deep venous thrombosis. Other problems may arise in the later postoperative period. These include pouch and distal esophageal dilation, incisional hernias, strictures, persistent vomiting, cholecystitis, and diarrhea.

Nutritional deficiencies are more likely in the gastric bypass versus the gastric banding procedures. In addition to the type of procedure, the risk and severity of postoperative nutritional deficiency is dependent on factors such as preoperative nutrition status, occurrence of postoperative complications, the ability to modify eating behavior, and compliance with prescribed vitamin and mineral supplementation [126]. Micronutrient deficiencies of thiamin, vitamin B_{12}, folate, and iron are common and require treatment. Long-term vitamin and mineral supplementation is necessary as is ongoing nutritional monitoring. Nutritional deficiencies are of particular concern in women of childbearing years because nutrient deficiencies carry a high risk of fetal damage.

"Dumping syndrome," a side effect of a gastric bypass procedure, is characterized by tachycardia, palpitations, diaphoresis, and nausea. Ingestion of energy-dense high-carbohydrate food, which is rapidly emptied into the small intestine, causes the release of vasoactive gastrointestinal polypeptides that may lead to the syndrome. Up to 70% of patients undergoing a gastric bypass procedure will experience some degree of a "dumping syndrome."

Vomiting and intolerance to solid food are the most common side effects of a gastric restrictive procedure. Vomiting following a restrictive procedure is frequently due to behavioral origin (e.g., inadequate chewing, quick consumption of fluid leading to overdistention of gastric pouch, large-volume meals); however, stricture, stenosis, marginal ulcers, and intestinal obstruction must be ruled out [126].

VI. SPECIAL ISSUES IN THE TREATMENT OF PEDIATRIC OBESITY

The assessment for weight-related health risk is different for children than for adults. BMI is still calculated in children but requires adjustments for age and sex as a child's body fat level changes with age and is different between girls and boys. BMI must be plotted on BMI-for-age growth charts specific for either girls or boys [127]. These Centers for Disease Control (CDC) growth charts show weight status using statistical percentiles—the term "overweight" when applied to children refers to a BMI-for-age greater than the 95th percentile and "at-risk-for-overweight" refers to a BMI-for-age that is between the 85th and 95th percentiles.

Data from national surveys indicate that overweight prevalence in children and adolescents continues to rise [128]. Most recently, in NHANES III [127] overweight (i.e., $\geq$95th percentile BMI-for-age) prevalence among 6-to-11 year olds and among 12-to-19 year olds is 18.8% and 17.4%, respectively. In the NHANES I (1971 to 1974), the levels of overweight in these age groups were 4% and 6%, respectively [127]. Because obese children tend to become obese adults and face an increased risk of chronic diseases (e.g., diabetes and cardiovascular diseases), pediatric obesity is considered a major public health problem [129–130]. However, even during childhood and adolescence, disorders such as hyperlipidemia, hypertension, and abnormal glucose tolerance occur with increased frequency in overweight and obese children [131]. In addition to the metabolic sequelae, significant negative psychosocial consequences of obesity are observed in this population [131].

In 2005, the Institute of Medicine (IOM) committee for the prevention of childhood obesity acknowledged the lack of strong evidence regarding the best ways to prevent childhood obesity. The committee stated that the urgent need to control childhood overweight must be met by acting on the best available evidence rather than waiting on the best possible evidence to emerge [129]. A number of professional associations [130, 132–134] concur with this view but have made specific recommendations for assessment and treatment. These recommendations include at least a yearly assessment of weight status (i.e., BMI-for-age) in all children as well as a routine assessment of eating and activity patterns with anticipatory guidance where necessary [130, 133, 134]. In children identified as at-risk-for-overweight or overweight, assessment for obesity-related comorbid conditions is also recommended (e.g., family history, blood pressure, fasting lipids, glucose and liver enzymes, blood urea nitrogen, and creatinine) [134].

Treatment recommendation goals for pediatric obesity are different than those for adult obesity. In general, because children are still growing with increases in lean body mass, treatment focuses on preventing weight gain rather than the weight-loss focus of adult treatment. However, in the most recent recommendations [134], specific weight maintenance/loss targets do include recommendations for small weight losses (no more than an average of 2 lbs per week) only for children with very high levels of obesity (i.e., ≥99th percentile BMI-for-age). Additionally, any treatment aimed at regulating body weight and body fat must also provide adequate nutrition for the growth and development of the child. It is also important that the health care provider evaluate the physiological and psychological impact of an obesity treatment for children and adolescents.

Any dietary intervention to regulate a child's body weight must provide sufficient energy for growth and development. In particular, the diet must adequately provide micronutrients such as iron and calcium. Dietary advice and modification are most likely to be effective if there is parental involvement, because parental attitudes, purchase, and presentation of food as well as modeling of eating behavior can impact a child's intake [132]. Some child-feeding practices can have negative (and unintended) effects on a child's food preferences and ability to control food intake [135]. For example, stringent parental control can increase a child's preference for high-energy-dense foods and limit a child's acceptance of a wide variety of foods [135]. Parental support involves being a positive role model for healthful feeding behavior as well as providing a wide array of healthful food choices in a supportive eating context [136]. Finally, because the potential for emergence of eating disorders is high in children and adolescents, dietary modifications must occur in the context of the promotion of realistic body weight goals, positive self-esteem, and body image satisfaction. Specific dietary and physical activity strategies are listed in Table 4.

It has been stated that our current environment facilitates underactivity through increased opportunity for sedentary activities [138]. This statement is just as true for children as it is for adults with sedentary activities (e.g., television viewing, video game playing) frequently taking prominence in children's lives. Physical activity interventions in the treatment of pediatric obesity often include decreasing these sedentary behaviors [139, 140]. Studies have shown this approach to be effective both in increasing physical activity and producing relative changes in BMI [139, 140]. In addition, because children are more likely to continue being active if they are able to choose their own activity, providing a choice of activities appears to be superior to providing a specific exercise prescription [139].

TABLE 4 Strategies to Prevent or Control Overweight in Children and Adolescents

Increased physical activity
—accumulate at least 60 minutes per day
Increased breastfeeding
Reduced television viewing
Limit to screen time to < 2 hours per day
No TV in child's bedroom
Reduced intake of sugar-sweetened beverages
Increased consumption of fruit and vegetables
—5 or more servings per day
Family meals eaten together where possible
Limit number of meals eaten outside the home

Adapted from [130], [133], [134], and [137].

The fact that children routinely attend school, an environment that presents continuous and intensive contact with children, creates a unique opportunity for pediatric obesity intervention. Schools have the necessary resources to promote physical activity (e.g., gym equipment, playing fields), they provide at least one meal a day where children can be exposed to healthful food choices, and many schools have access to school nurses or health clinics, which could potentially provide services to overweight children [141]. The effectiveness of school-based programs for the treatment of obesity has been modest, but the results are encouraging and are worthy of more research [132].

VII. ACUTE WEIGHT LOSS VERSUS MAINTAINING LONG-TERM WEIGHT LOSS

Effective weight management comprises both a weight-loss phase and a weight-loss maintenance phase. Most people are relatively successful at achieving a short-term weight loss, but few people can sustain that weight loss over long periods of time. Although experts now acknowledge that obesity is not an acute disorder that is "cured" by a single short-term intervention, the public is still focused much more on achieving weight loss than on maintaining the weight loss. This is evidenced by the use of popular diets that advocate dietary changes that produce weight loss but that are difficult to maintain over time. Although a large number of resources are available to help people lose weight, there are few to help people keep the weight off. Considering weight loss as a different process than weight-loss maintenance may help us develop different strategies and tools for each phase.

If weight loss and weight-loss maintenance are considered as separate processes, we can consider how strategies for each phase might vary. Many different dietary strategies (e.g., following low-fat, low-carbohydrate, or low-energy-dense diets) can produce weight loss [57]. Further, physical activity added to food restriction does not add much to amount of weight loss [66]. High levels of physical activity seem to be critical for most people to maintain a substantial weight loss [27, 68]. The composition of the diet may be less important, although much data favor a low-fat diet for weight-loss maintenance [27]. It is not surprising that physical activity is important for weight-loss maintenance. Weight loss is accompanied by a decline in energy requirements, so that the amount of energy needed to achieve energy balance is lower after weight loss than before. Keeping weight off by diet alone means continuous food restriction for most people, and this seems to be difficult. Increasing physical activity increases total energy requirements and can allow energy intake to be sufficiently high to avoid chronic hunger.

The real challenge in weight management lies in the *prevention* of weight gain after weight loss, not in the accomplishment of weight loss itself. Thus, a treatment plan for obesity cannot just involve a plan for producing negative energy balance but must also involve a plan for achieving and maintaining energy balance at a new, lower body weight. Treatments that produce the largest degree of negative energy balance will produce the most weight loss, but these treatments may not be easy to maintain chronically. Similarly, the treatments that most effectively maintain a weight loss may not be effective in producing weight loss. The best overall treatment for an obese individual may incorporate treatments for acute weight loss paired with a different treatment in the weight maintenance phase. Potentially, weight-loss medications may have their greatest role in helping obese individuals maintain a weight loss achieved by different interventions (such as diet and exercise) that may be difficult to maintain at the necessary level for long periods of time.

VIII. THE FUTURE OF WEIGHT MANAGEMENT

Preventing weight gain is the true future of weight management. The most efficient way to address obesity is in preventing excess weight gain from ever occurring, therefore health care professionals should consider addressing prevention of weight gain in all patients. Because the entire U.S. population appears to be in a gradual weight gain state, a first goal for all patients should be not to gain additional weight. This would apply to patients whether they are currently at a healthy weight, overweight, or obese. In children, this goal can be translated to preventing excessive weight gain. Hill *et al.* [143] suggested that modifying energy balance by only 100 kcal/day could prevent weight gain in 90% of adults. Wang *et al.* [144] found that 100 to 150 kcal/day could prevent excessive weight gain in most children.

Once individuals are obese, the real challenge is not in producing weight loss but in preventing weight gain after weight loss. It seems that larger lifestyle changes are required to prevent weight gain after weight loss than to prevent weight gain from occurring in the first place. There has been surprisingly little research into obesity prevention and, to date, prevention efforts have met with modest success at best. Although obesity prevention can be targeted as a high priority, there is much to learn about how to do it effectively.

It is becoming clear that preventing weight gain requires attention to individual behavior change but also to the larger environmental and societal factors that ultimately influence an individual's behavior. The current environment in the United States promotes weight gain and obesity by encouraging excessive food consumption and discouraging physical activity. We live in an environment that has an abundance of cheap, good-tasting, energy-dense foods but requires little physical exertion for day-to-day living. It is difficult for individuals in this environment to consistently maintain behaviors that would support a healthy body weight.

There is a need for a concerted, integrated effort from all sectors of society to create an environment that is less obesity conducive. Despite the daunting nature of this task, the rapidity of the increase in obesity underscores the urgency to address the issue.

References

1. World Health Organization. Obesity and overweight fact sheet. www.who.int/mediacentre/factsheets/fs311/en/print.html. Accessed May 2007.
2. Ogden, C. L., Carroll, M. D., Curtin, L. R., McDowell, M. A., Tabak, C. J., and Flegal, K. M., (2006). Prevalence of overweight and obesity in the United States: 1999–2004. *JAMA* **295**, 1549–1555.
3. Must, A., Spadano, J., Coakley, E. H., Field, A. E., Colditz, G., and Dietz, W. H., (1999). The disease burden associated with overweight and obesity. *JAMA* **282**, 1523–1529.
4. Anderson, D., and Wadden, T. (1999). Treating the obese patient: Suggestions for primary care practice. *Arch. Fam. Med.* **8**, 156–167.
5. National Heart, Lungs, and Blood Institute, Obesity Education Initiative Expert Panel (1998). Clinical guidelines on the identification, evaluation, and treatment of overweight and obesity in adults: the evidence report. *Obes. Res.* **6** (suppl. 2), 51S–209S, www.nhlbi.nih.gov/guidelines/obesity/ob_home.htm.
6. Metropolitan Life Insurance Company. (1983). Metropolitan height and weight tables. *Stat. Bull. Met. Life Ins. Co.* **64**, 2.
7. Stensland, S., and Margolis, S. (1990). Simplifying the calculation of body mass index for quick reference. *J. Am. Diet. Assoc.* **90**, 856.
8. Heymsfield, S. B., Allison, D. B., Heshka, S., and Pierson, R. N. (1995). "Handbook of Assessment Methods for Eating Behavior and Weight Related Problems: Measures, Theory, and Research," pp. 515–560. Sage, Thousand Oaks, CA.

9. Manson, J. E., Willett, W. C., Stampfer, M. J., Colditz, G. A., Hunter, D. J., Hankinson, S. E., Hennekens, C. H., and Speizer, F. E. (1995). Body weight and mortality among women. *N. Engl. J. Med.* **333**, 677–685.
10. Lew, E. A., and Garfinkel, L. (1979). Variations in mortality by weight among 750,000 men and women. *J. Chronic Dis.* **32**, 563–576.
11. Gordon, T., and Doyle, J. T. (1988). Weight and mortality in men: The Albany Study. *Int. J. Epidemiol.* **17**, 77–81.
12. Gallagher, D., Visser, M., Sepulveda, D., Pierson, R. N., Harris, T., and Heymsfield, S. B. (1996). How useful is body mass index for comparison of body fatness across age, sex, and ethnic groups? *Am. J. Epidemiol.* **143**, 228–239.
13. Bray, G. A. (1996). Health hazards of obesity. *Endocrin. Metab. Clin. No. Am.* **4**, 907–919.
14. Seidel, J. (1998). Epidemiology: definition and classification of obesity. *In* "Clinical Obesity" (P. G. Kopleman and M. J. Stock, Eds.), pp. 1–17. Blackwell Science, Malden, MA.
15. Despres, J. P., Moorjani, S., Lupien, P. J., Tremblay, A., Nadeau, A., and Bouchard, C. (1990). Regional distribution of body fat, plasma lipoproteins, and cardiovascular disease. *Arteriosclerosis* **10**, 497–511.
16. Lemieux, S., Prud'homme, D., Bouchard, C., Tremblay, A., and Despres, J. P. (1996). A single threshold value of waist girth identifies normal-weight and overweight subjects with excess visceral adipose tissue. *Am. J. Clin. Nutr.* **64**, 685–693.
17. Lewis, C. E., Jacobs, D. R., McCreath, H., Kiefe, C. I., Schreiner, P. J., Smith, D. E., and Williams, O. D. (2000). Weight gain continues in the 1990s: 10 year trends in weight and overweight from the CARDIA study. *Am. J. Epidemiol.* **151**, 1172–81.
18. Expert Panel on Detection, Evaluation and Treatment of High Blood Cholesterol in Adults. (2001). Executive Summary of the Third Report of the National Cholesterol Education Program (NCEP) Expert Panel on the Detection, Evaluation and Treatment of High Blood Cholesterol in Adults (Adult Treatment Panel III). *JAMA* **285**, 2486–2497.
19. Brownell, K. D. (1990). Dieting readiness. *Weight Control Digest* **1**, 1–9.
20. Bray, G. A. (1998). The dieting readiness test. *In* "Contemporary Diagnosis and Management of Obesity," pp. A1–A7. Handbooks in Health Care, Newtown, PA.
21. Hill, J. O. (1997). Energy metabolism and obesity. *In* "Clinical Research in Diabetes and Obesity, Vol. II: Diabetes and Obesity" (B. Drazin and R. Rizza, Eds.), pp. 3–12. Humana Press, Totowa, NJ.
22. North American Association for the Study of Obesity. (1998). "The Practical Guide to the Identification, Evaluation, and Treatment of Overweight and Obese Adults." National Institutes of Health, National Heart, Lung, and Blood Institute, Bethesda, MD.
23. Wing, R., Koeske, R., Epstein, Z. L., Norwalk, M. P., Gooding, W., and Becker, D. (1987). Long-term effects of modest weight loss in type II diabetic patients. *Arch. Intern. Med.* **147**, 1749–1753.
24. Goldstein, D. J., Rampey, A. H., Roback, P. J., Wilson, M. G., Hamilton, S. H., Sayler, M. E., and Tollefson, G. D. (1995). Efficacy and safety of long-term fluoxetine treatment of obesity: maximizing success. *Obes. Res.* **3** (suppl. 4), 481S–490S.
25. Dattilo, A. M., and Kris-Etherton, P. M. (1992). Effects of weight reduction on blood lipids and lipoproteins: a meta-analysis. *Am. J. Clin. Nutr.* **56**, 320–328.
26. Foster, G. D., and Makris, A. (2004). Behavioral treatment: Part B. Practical applications. *In* "Managing Obesity: A Clinical Guide" (G. D. Foster and C. A. Nonas, Eds.), pp. 76–90. American Dietetic Association, Chicago.
27. Klem, M. L., Wing, R. R., McGuire, M. T., Seagle, H. M., and Hill, J. O. (1997). A descriptive study of individuals successful at long-term maintenance of substantial weight loss. *Am. J. Clin. Nutr.* **66**, 239–246.
28. Kruger, J., Galuska, D. A., Serdula, M. K., and Jones, D. A. (2004). Attempting to lose weight: specific practices among U.S. adults. *Am. J. Prev. Med.* **26** (5), 402–406.
29. Weiss, E. C., Galuska, D. A., Kettel Khan, L., and Serdula, M. K. (2006). Weight control practices among U.S. adults, 2001–2002. *Am. J. Prev. Med.* **31** (1), 18–24.
30. Elia, M., Stubbs, R. J., and Henry, C. J. K. (1999). Differences in fat, carbohydrate, and protein metabolism between lean and obese subjects undergoing total starvation. *Obes. Res.* **7**, 597–604.
31. VanItallie, T. B. (1999). Treatment of obesity: can it become a science? *Obes. Res.* **7**, 605–606.
32. Food and Agricultural Organization/World Health Organization/United Nations University. (1985). "Report of a Joint Expert Consultation: Energy and Protein Requirement of Adults." Technical Report No. 724. World Health Organization, Geneva.
33. Schoeller, D. A., Bandini, L. G., and Dietz, W. H. (1989). Inaccuracies in self-reported intake identified by comparison with the doubly labeled water method. *Can. J. Physiol. Pharmacol.* **68**, 941–949.
34. Lichtman, S. W., Pisarska, K., Berman, E. R., Pestone, M., Dowling, H., Offenbacher, E., Weisel, H., Heshka, S., Matthews, D. E., and Heymsfield, S. B. (1992). Discrepancy between self-reported and actual caloric intake and exercise in obese subjects. *N. Engl. J. Med.* **327**, 1893–1898.
35. National Task Force on the Prevention and Treatment of Obesity. (1993). Very low-calorie diets. *JAMA* **270**, 967–974.
36. Saris, W. H. M. (2001). Very-low-calorie diets and sustained weight loss. *Obes. Res.* **9**, 295S–301S.
37. Wadden, T. A. (1993). Treatment of obesity by moderate and severe caloric restriction. Results of clinical research trials. *Ann. Intern. Med.* **119**, 688–693.
38. Anderson, J. W., Konz, E. C., Frederich, R. C., and Wood, C. L. (2001). Long-term weight-loss maintenance: a meta-analysis of U.S. studies. *Am. J. Clin. Nutr.* **74**, 579–584.
39. Wing, R. R. (1992). Don't throw out the baby with the bathwater: a commentary on very-low-calorie diets. *Diabetes Care* **15**, 293–296.
40. Wadden, T. A., Sternberg, J. A., Letizia, K. A., Stunkard, A. J., and Foster, G. D. (1989). Treatment of obesity by very low calorie diet, behavior therapy, and their combination: A five-year perspective. *Int. J. Obes.* **13** (suppl. 2), 39–46.
41. Heber, D., Ashley, J. M., Wang, H. J., and Elashoff, R. M. (1994). Clinical evaluation of a minimal intervention meal

replacement regimen for weight reduction. *J. Am. College. Nutr.* **13**, 608–614.
42. Ditschuneit, H. H., Flechtner-Mors, M., Johnson, T. D., and Adler, G. (1999). Metabolic and weight-loss effects of a long-term dietary intervention in obese patients. *Am. J. Clin. Nutr.* **69**, 198–204.
43. Quinn Rothacker, D. (2000). Five-year self-management of weight using meal replacements: Comparison with matched controls in rural Wisconsin. *Nutrition* **16**, 344–348.
44. Hill, J. O., Drougas, H., and Peters, J. C. (1993). Obesity treatment: Can diet composition play a role? *Ann. Intern Med.* **119**, 694–697.
45. Lissner, L., Levitsky, D. A., Strupp, B. J., Kalkwarf, H. J., and Roe, D. A. (1987). Dietary fat and the regulation of energy intake in human subjects. *Am. J. Clin. Nutr.* **46**, 886–892.
46. Blundell, J. E., and Stubbs, R. J. (1999). High and low carbohydrate and fat intakes: Limits imposed by appetite and palatability and their implications for energy balance. *Eur. J. Clin. Nutr.* **53**, S148–S165.
47. U.S. Department of Health and Human Services and U.S. Department of Agriculture. Dietary Guidelines for Americans. 6th ed. (2005, January). U.S. Government Printing Office, Washington, DC.
48. Lichtenstein, A. H., Appel, L. J., Brands, M., Carnetho, M., Daniels, S., Franch, H. A., Franklin, B., Kris-Etherton, P., Harris, W. S., Howard, B., Karanja, N., Lefevre, M., Rudel, L., Sacks, F., Van Horn, L., Winston, M., and Wylie-Rosett, J. (2006). Diet and lifestyle recommendations revision 2006: a scientific statement from the American Heart Association Nutrition Committee. *Circulation.* **114**, 82–96.
49. Finnish Diabetes Prevention Study Group. (2006). Sustained reduction in the incidence of type 2 diabetes by lifestyle intervention: follow-up of the Finnish Diabetes Prevention Study. *Lancet* **368**, 1673–1679.
50. Diabetes Prevention Program Research Group. (2002). Reduction in the incidence of type 2 diabetes with lifestyle intervention or metformin. *N. Engl. J. Med.* **346**, 393–403.
51. The Look AHEAD Research Group. (2007). Reduction in weight and cardiovascular disease risk factors in individuals with type 2 diabetes: one-year results of the Look AHEAD trial. *Diab. Care* **30**, 1374–1383.
52. Appel, L. J., Champagne, C. M., Harsha, D. W., Cooper, L. S., Obarzanek, E., Elmer, P. J., Stevens, V. J., Vollmer, W. M., Lin, P. H., Svetkey, L. P., Stedman, S. W., and Young, D. R. Writing Group of the PREMIER Collaborative Research Group (2003). Effects of the comprehensive lifestyle modification on blood pressure control: main results of the PREMIER clinical trial. *JAMA* **289**, 2083–2093.
53. Bell, E. A., Castellanos, V. A., Pelkman, C. L., Thorwart, M. L., and Rolls, B. J. (1998). Energy density of foods affects energy intake in normal-weight women. *Am. J. Clin. Nutr.* **67**, 412–420.
54. Rolls, B. J., Roe, L. S., and Meengs, J. S. (2006). Reductions in portion size and energy density of foods are additive and lead to sustained decreases in energy intake. *Am. J. Clin. Nutr.* **83**, 11–17.
55. Ledikwe, J. H., Rolls, B. J., Smiciklas-Wright, H., Mitchell, D. C., Ard, J. D., Champagne, C., Karanja, N., Lin, P., Stevens, V. J., and Appel, L. J. (2007). Reductions in dietary energy density are associated with weight loss in overweight and obese participants in the PREMIER trial. *Am. J. Clin. Nutr.* **85**, 1212–1221.
56. Brehm, B. J., Seeley, R. J., Daniels, S. R., and D'Alessio, D. A. (2003). A randomized trial comparing a very low carbohydrate diet and a calorie-restricted low fat diet on body weight and cardiovascular risk factors in healthy women. *J. Clin. Endocrinol. Metab.* **88**, 1617–1623.
57. Foster, G. D., Wyatt, H. R., Hill, J. O., McGuckin, B. G., Brill, C., Mohammed, S., Szapary, P. O., Rader, D. J., Edman, J. S., and Klein, S. (2003). A multi-center, randomized, controlled clinical trial of the Atkins diet. *N. Engl. J. Med.* **348**, 282–290.
58. Nordmann, A. J., Nordmann, A., Briel, M., Keller, U., Yancy, W. S., Brehm, B. J., and Bucher, H. C. (2006). Effects of low-carbohydrate vs. low-fat diets on weight loss and cardiovascular risk factors: a meta-analysis of randomized controlled trials. *Arch. Intern. Med.* **166**, 285–293.
59. Dansinger, M. L., Gleason, J. A., Griffith, J. L., Selker, H. P., and Schaefer, E. J. (2005). Comparison of the Atkins, Ornish, Weight Watchers, and Zone diets for weight loss and heart disease risk reduction: a randomized trial. *JAMA* **293**, 43–53.
60. Gardner, C. D., Kiazand, A, Alhassan, S., Kim, S., Stafford, R. S., Balise, R. R., Kraemer, H. C., and King, A. C. (2007). Comparison of the Atkins, Zone, Ornish and LEARN diet for change in weight and related risk factors among overweight premenopausal women. *JAMA* **297**, 969–977.
61. Arterburn, D. (2006). The BBC diet trials: reality television and academic researchers jointly tackle the weight loss industry. *BMJ* **332**, 1284–1285.
62. Ikeda, J. P., Hayes, D., Satter, E., Parham, E. S., Kratina, K., Woolsey, M., Lowey, M., and Tribole, E. (1999). A commentary on the new obesity guidelines from NIH. *J. Am. Diet. Assoc.* **99**, 918–919.
63. Higgins, L., and Gray, W. (1999). What do anti-dieting programs achieve? A review of research. *Austral. J. Nutr. Diet.* **56** 128–136.
64. Foster, G. D., Sarwer, D. B., and Wadden, T. A. (1997). Psychological effects of weight cycling in obese persons: a review and research agenda. *Obes. Res.* **5**, 474–488.
65. U.S. DHHS. (1996). Physical activity and health: a report of the surgeon general. Centers for Disease Control, Atlanta, GA.
66. Wing, R. R. (1999). Physical activity in the treatment of the adulthood overweight and obesity: current evidence and research issues. *Med. Sci. Sports Exerc.* **30**(11), S547–S552.
67. Zachwieja, J. J. (1996). Exercise as treatment for obesity. *Endocrinol. Metab. Clin. N. Am.* **25**(4), 965–988.
68. Schoeller, D. A., Shay, K., and Kushner, R. F. (1997). How much physical activity is needed to minimize weight gain in previously obese women? *Am. J. Clin. Nutr.* **66**, 551–556.
69. Jakicic, J. M., and Otto, A. D. (2006). Treatment and prevention of obesity: what is the role of exercise? *Nutr. Rev.* **64**, S57–S61.
70. Foster, G. D., Makris, A. P., and Bailer, B. A. (2005). Behavioral treatment of obesity. *Am. J. Clin. Nutr.* **82**, 230s–235s.
71. Boutelle, K. N., and Kirshenbaum, D. S. (1998). Further support for consistent self-monitoring as a vital component of successful weight control. *Obes. Res.* **6**, 219–224.

72. Streit, K. J., Stevens, N. H., Stevens, V. J., and Rossner, J. (1991). Food records: A predictor and modifier of weight change in a long-term weight loss program. *J. Am. Diet. Assoc.* **91**, 213–216.
73. Boutelle, K. N., Kirshenbaum, D. S., Baker, R. C., and Mitchell, M. E. (1999). How can obese weight controllers minimize weight gain during high risk holiday season? By self-monitoring very consistently. *Health Pysch.* **18**, 364–368.
74. Foreyt, J. P., and Goodrick, G. K. (1993). Evidence for success of behavior modification in weight loss and control. *Ann. Intern. Med.* **119**, 698–701.
75. Foreyt, J. P., and Poston, W. S., II (1998). The role of the behavioral counselor in obesity treatment. *J. Am. Diet. Assoc.* **98**, S27–S30.
76. Hayaki, J., and Brownell, K. D. (1996). Behaviour change in practice: Group approaches. *Int. J. Obes.* **20**, S27–S30.
77. Frank, A. (1998). A multidisciplinary approach to obesity management: the physician's role and team care alternatives. *J. Am. Diet. Assoc.* **98**, S44–S48.
78. Comuzzie, A. G., and Allison, D. B. (1998). The search for human obesity genes. *Science* **280**, 1374–1377.
79. Bray, G. A. (1998). Drug treatment of overweight. *In* "Contemporary Diagnosis and Management of Obesity," pp. 246–273. Handbooks in Health Care, Newtown, PA.
80. Weintraub, M., Sundaresan, P. R., and Cox, C. (1992). Long-term weight control study VI. *Clin. Pharmacol. Ther.* **51**, 619–633.
81. Ryan, D. H. (1996). Medicating the obese patient. *Endocrinol. Metab. Clin. N. Am.* **25**, 989–1004.
82. Bray, G. A., and Greenway, F. L. (1999). Current and potential drugs for treatment of obesity. *Endocrinol. Rev.* **20**(6), 805–875.
83. Darga, L. L., Carroll-Michals, L., Botsford, S. J., and Lucas, C. P. (1991). Fluoxetine's effects on weight loss in obese subjects. *Am. J. Clin. Nutr.* **54**, 321–325.
84. Anderson, J. W., Greenway, F. L., Fujioka, K., Gadde, K. M., McKenney, J., and O'Neil, P. M. (2002). Bupropion SR enhances weight loss: a 48-week double blinded, placebo controlled trial. *Obes. Res.* **10**, 633–641.
85. Gadde, K. M., Franciscy, D. M., Wagner, H. R. 2nd, and Krishnan, K. R. (2003). Zonisamide for weight loss in obese adults: a randomized controlled trial. *JAMA* **289**, 1820–1825.
86. Bray, G. A., Hollander, P., Klein, S., Kushner, R., Levy, B., Fitchet, M., and Perry, B. H. (2003). A 6 month randomized placebo controlled dose ranging study of topiramate for weight loss in obesity. *Obes. Res.* **11**, 722–733.
87. Wilding, J., Van Gaal, L. F., Rissanen, A., Vercruysse, F., and Fitchet, M. (2004). A randomized double-blinded placebo-controlled study of the long-term efficacy and safety of topiramate in the treatment of obese subjects. *Int. J. Obes. Rel. Metab. Disord.* **28**, 1399–1410.
88. Astrup, A., Caterson, I., Zelissen, P., Guy-Grand, B, Carruba, M., Levy, B., Sun, X., and Fitchet, M. (2004). Topiramate: long-term maintenance of weight loss induced by a low calorie diet in obese subjects. *Obes. Res.* **12**, 1658–1669.
89. Shapira, N. A., Goldsmith, T. D., and McElroy, S. L. (2000). Treatment of binge eating disorder with topiramate: a clinical case series. *J. Clin. Psychiatry.* **61**, 368–372.
90. McElroy, S. L., Arnold, L. M., Shapira, N. A., Keck, P. E., Jr., Rosenthal, N. R., Karim, M. R., Kamin, M., and Hudson, J. I. (2003). Topiramate in the treatment of binge eating disorder associated with obesity: a randomized placebo controlled trial. *Am. J. Psychiatry.* **160**, 255–261.
91. Smathers, S. A., Wilson, J. G., and Nigro, M. A. (2003). Topiramate effectiveness in Prader-Willi syndrome. *Pediatr. Neurol.* **28**, 130–133.
92. Heal, D. J., Aspley, S., Prow, M. R., Jackson, H. C., Martin, K. F., and Cheetham, S. C. (1998). Sibutramine: A novel anti-obesity drug: a review of the pharmacological evidence to differentiate it from amphetamine and d-fenfluramine. *Int. J. Obes.* **22**(suppl. 1), S18–S28.
93. Astrup, A., Hansen, H. L., Lundsgaard, C., and Toubro, S. (1998). Sibutramine and energy balance. *Int. J. Obes.* **22** (suppl. 1), S30–S35.
94. Seagle, H. M., Bessesen, D. H., and Hill, J. O. (1998). Effects of sibutramine on resting metabolic rate and weight loss in overweight women. *Obes. Res.* **6**, 115–121.
95. Hansen, D. L., Toubro, S., Macdonald, I., Stock, M. J., and Astrup, A. (1997). Thermogenic properties of sibutramine in humans. *Int. J. Obes.* **22**(suppl. 2), 102A.
96. Jones, S. P., Smith, I. G., Kelly, F., and Gray, J. A. (1995). Long-term weight loss with sibutramine. *Int. J. Obes.* **19** (suppl. 2), 41.
97. Apfelbaum, M., Vague, P., Ziegler, O., Hanotin, C., Thomas, F., and Leutenegger, E. (1999). Long-term maintenance of weight loss after a very-low calorie diet: a randomized blinded trial of the efficacy and tolerability of sibutramine. *Am. J. Med.* **106**, 179–184.
98. Bray, G. A. (1996). Health hazards of obesity. *Endocrinol. Metab. Clin. N. Am.* **25**, 907–919.
99. Zhi, J., Melia, A. T., Guerciolini, R., Chung, J., Kingberg, J., and Hauptman, J. R. (1994). Retrospective population-based analysis of the dose-response (fecal fat excretion) relationship of orlistat in normal and obese volunteers. *Clin. Pharmacol. Ther.* **56**, 82–85.
100. Hill, J. O., Hauptman, J., Anderson, J. W., Fujioka, K., O'Neil, P. M., Smith, D. K., Zavoral, J. H., and Aronne, L. J. (1999). Orlistat, a lipase inhibitor, for weight maintenance after conventional dieting: a 1-y study. *Am. J. Clin. Nutr.* **69**, 1108–1116.
101. Davidson, M. H., Hauptman, J., Di Girolamo, M., Foreyt, J. P., Halsted, C. H., Heber, D., Heimburger, D. C., Lucas, C. P., Robbins, D. C., Chung, J., and Heymsfield, S. B. (1999). Weight control and risk factor reduction in obese subjects treated with orlistat: a randomized, controlled trial. *JAMA* **281**, 235–242.
102. James, W. P. T., Avenell, A., Broom, J., and Whitehead, J. (1997). A one-year trial to assess the value of orlistat in the management of obesity. *Int. J. Obes. Metab. Disord.* **21**(suppl.), 24–30.
103. Hill, J. O., and Wyatt, H. R. (1999). The efficacy of orlistat (xenical) in promoting weight loss and preventing weight regain. *Curr. Prac. Med.* **2**(11), 228–231.
104. Sjostrom, L. M., Rissanen, A., Andersen, T., Boldrin, M., Golay, A., Koppeschaar, H. P., and Krempf, M., for the European Multi-Center Orlistat Study Group (1998). Randomised placebo-controlled trial of orlistat for weight loss and prevention of weight regain in obese patients. *Lancet* **352**, 167–172.

105. Van Gaal, L. F., Broom, J. I., Enzi, G., and Toplak, H., for the Orlistat Dose-Ranging Group. (1998). Efficacy and tolerability of orlistat in the treatment of obesity: a 6-month dose ranging study. *Eur. J. Clin. Pharmacol.* **54**, 125–132.
106. Rossner, S., Sjostrom, L., Noack, R., Meinders, E., and Noseda G. (2000). Weight loss, weight maintenance, and improved cardiovascular risk factors after 2 years treatment with orlistat for obesity. *Obes. Res.* **8**, 49–61.
107. Hauptman, J., Lucas, C., Boldrin, M. N., Collins, H., and Segal, K., for the Orlistat Primary Care Study. (2000). Orlistat in the long-term treatment of obesity in the primary care setting.. *Arch. Fam. Med.* **9**, 160–167.
108. Anderson, J., Schwartz, S., Hauptman, J., Boldrin, M., Rossi, M., Bansal, V., and Hale, C. (2006). Low-dose orlistat effects on body weight on mildly to moderately overweight individuals: a 16-week, double-blind, placebo-controlled trial. *Ann Pharmaco.* **40**, 1717–1723.
109. Van Gaal, L. F., Rissanen, A. M., Scheen, A. J., Ziegler, O., and Rossner, S. (2005). Effects of the cannabinoid-1 receptor blocker rimonabant on weight reduction and cardiovascular risk factors in overweight patients: 1 year experience from the RIO_Europe Study. *Lancet* **365**, 1389–1397.
110. Scheen, A., Finer, N., Hollander, P., Jensen, M., Van Gaal, L., and Rio Diabetes study Group. (2006). Efficacy and tolerability of rimonabant in overweight or obese patients with type 2 diabetes: a randomised controlled study. *Lancet.* **368**, 1660–1672.
111. The RIO-North America Study Group. (2006). Effect of rimonabant, a cannabinoid-1 receptor blocker, on weight and cardiometabolic risk factors in overweight or obese patients. *JAMA* **295**, 761–775.
112. Despres, J. P., Golay, A., and Sjostrom, L., Rimonabant in Obesity-Lipids Study Group. (2005). Effects of rimonabant on metabolic risk factors in overweight patients with dyslipidemia. *N. Engl. J. Med.* **353**, 2121–2134.
113. Sjostrom, L. V. (1992). Mortality of severely obese subjects. *Am. J. Clin. Nutr.* **55**, 516S–523S.
114. Sturm, R. (2007). Increases in morbid obesity in the USA: 2000–2005. *Public Health.* doi:10.1016/j.puhe.2007.01.006.
115. Zhao, Y., and Encinosa, W. (2007). Bariatric surgery utilization and outcomes in 1998 and 2004. Statistical brief#23. Agency for Healthcare Research and Quality, Rockville, MD. www.hcup-us.ahrq.gov/reports/statbriefs/sb23.pdf.
116. Greenway, F. L. (1996). Surgery for obesity. *Endocrinol. Metab. Clin. N. Am.* **25**(4), 1005–1027.
117. National Institutes of Health, Consensus Development Conference. (1992). Gastrointestinal surgery for severe obesity. *Am. J. Clin. Nutr.* **55**, 487S–619S.
118. Kral, J. (1998). Surgical treatment of obesity. *In* "Handbook of Obesity" (G. A. Bray, C. Bouchard, and W. P. James, Eds.), pp. 977–993. Marcel Dekker, New York.
119. Kuruba, R., Koche, L. S., and Murr, M. M. (2007). Preoperative assessment and perioperative care of patients undergoing bariatric surgery. *Med. Clin. N. Am* **91**, 339–351.
120. Santry, H. P., Gillen, D. L., and Lauderdale, D. S. (2005). Trends in bariatric surgical procedures. *JAMA* **294**, 1909–1917.
121. Buchwald, H., Avidor, Y., Braunwald, E., Jensen, M. D., Pories, W., Fahrbach, K., and Schoelles, K. (2004). Bariatric surgery: a systematic review and meta-analysis. *JAMA* **292**, 1724–1737.
122. Poiries, W. J., Swanson, M. S., MacDonald, K. G., Long, S. B., Morris, P. G., Brown, B. M., Barakat, H. A., de Ramon, R. A., Israel, G., Dolezal, J. M., and Dohm, L. (1995). Who would have thought it? An operation proves to be the most effective therapy for adult onset diabetes mellitus. *Ann. Surg.* **222**, 339–352.
123. Castelnuovo-Tedesco, P. (1986). Psychiatric complications of surgery for superobesity: a review and reappraisal. *Clin. Nutr.* **5**(suppl.), 163–166.
124. Rand, S. W., and Macgregor, M. C. (1991). Successful weight loss following obesity surgery and the perceived liability of morbid obesity. *Int. J. Obes.* **15**, 577–579.
125. Sjostrom, L., Lindroos, A.K, Peltonen, M., Torgerson, J., Bouchard, C., Carlsson, B., Dahlgren, S., Larsson, B., Narbro, K., Sjostrom, C. D., Sullivan, M., Wedel, H., and Swedish Obese Subjects Study Scientific, Group (2004). Lifestyle, diabetes, and cardiovascular risk factors 10 years after bariatric surgery. *N. Engl. J. Med.* **351**, 2683–2693.
126. Tucker, O. N., Szomstein, S., and Rosenthal, R. J. (2007). Nutritional consequences of weight-loss surgery. *Med. Clin. N. Am.* **91**, 499–514.
127. Prevalence of overweight among children and adolescents: United States 2003–2004. Centers for Disease Control and Prevention, National Center for Chronic Disease Prevention and Health Promotion, Division of Nutrition and Physical Activity web site. Available at www.cdc.gov/nchs/products/pubs/pubd/hestats/obese03_04/overwght_child_03.htm. Accessed May 2007.
128. Ogden, C. L., Flegal, K. M., Carroll, M. D., and Johnson, C. L. (2002). Prevalence and trends in overweight among U.S. children and adolescents, 1999–2000. *JAMA* **288**, 1728–1732.
129. Institute of Medicine. (2006). Progress in preventing childhood obesity: how do we measure up? Report Brief GPO, Washington, DC.
130. Daniels, S. R., Arnett, D. K., Eckel, R. H., Gidding, S. S., Hayman, L. L., Kumanyika, S., Robinson, T. N., Scott, B. J., St. Jeor, S., and Williams, C. L. (2005). Overweight in children and adolescents: pathophysiology, consequences, prevention, and treatment. *Circulation* **111**, 1999–2012.
131. Dietz, W. H. (1998). Health consequences of obesity in youth: childhood predictors of adult disease. *Pediatrics* **101**, 518–525.
132. Ritchie, L. D., Crawford, P. B., Hoelscher, D. M., and Sothern, M. S. (2006). Position of the American Dietetic Association: individual-, family-, school-, and community-based interventions for pediatric overweight. *J. Am. Diet. Assoc.* **106**, 925–945.
133. Committee on Nutrition, American Academy of Pediatrics. (2003). Policy statement: prevention of pediatric overweight and obesity. *Pediatrics* **112**, 424–430.
134. Appendix: Expert committe recommendations on the assessment, prevention, and treatment of child and adolescent overweight and obesity, www.ama-assn.org/ama/pub/category/11759.html. Accessed June 2007.
135. Birch, L. L., and Fisher, J. O. (1998). Development of eating behaviors among children and adolescents. *Pediatrics* **101**, 539–549.
136. Satter, E. M. (1996). Internal regulation and the evolution of normal growth as the basis for prevention of obesity in children. *J. Am. Diet. Assoc.* **96**, 860–864.

137. Dietz, W., Lee, J., Wechsler, H., Malepati, and Sherry, B. (2007). Health plans' role in preventing overweight in children and adolescents. *Health Affairs* **26**, 430–440.
138. Hill, J. O., and Peters, J. C. (1998). Environmental contributions to the obesity epidemic. *Science* **280**, 1371–1374.
139. Epstein, L. H., Paluch, R. A., Gordy, C. C., and Dorn, J. (2000). Decreasing sedentary behaviors in treating pediatric obesity. *Arch. Pediatr. Adolesc. Med.* **154**, 220–226.
140. Robinson, T. N. (1999). Reducing children's television viewing to prevent obesity. *JAMA* **282**, 1561–1567.
141. Story, M. (1999). School-based approaches for preventing and treating obesity. *Int. J. Obes.* **22**, S43–S51.
142. Veugelers, P. J., and Fitzgerald, A. L. (2005). Effectiveness of school programs in preventing childhood obesity: a multilevel comparison. *Am. J. Pub. Health.* **95**, 432–435.
143. Hill, J. O., Wyatt, H. R., Reed, G. W., and Peters, J. C. (2003). Obesity and the environment: where do we go from here? *Science* **299**, 853–855.
144. Wang, Y. C., Gortmaker, S. L., Sobol, A. M., and Kuntz, K. M. (2006). Estimating the energy gap among U.S. children: a counterfactual approach. *Pediatrics* **118**, 1721–1733.

CHAPTER 23

Obesity: The Role of Physical Activity in Adults

MARCIA L. STEFANICK
Stanford University, Stanford, California

Contents

I. INTRODUCTION

The prevalence of obesity, currently defined as having a body mass index (BMI, i.e., height / weight^2) greater than or equal to 30 kg/m^2, has increased markedly in the United States in the past three decades, with the percent of obese adults aged 20–74 years increasing from 15.0% in the second National Health and Nutrition Examination Survey (NHANES II; 1976–1980), to 23.3% in NHANES III (1988–1994), and to 30.5% in the 1999–2000 NHANES, and the percent of extremely obese (BMI $\geq$ 40 kg/m^2) adults increasing from 2.9% in NHANES III to 4.7% in 1999–2000 [1]. The corresponding prevalence of overweight plus obese (BMI $\geq$ 25 kg/m^2) increased from 55.9% in NHANES III to 64.5% in NHANES 1999–2000, with increases occurring in both men and women, in all age groups, and in each racial/ethnic group surveyed (i.e., non-Hispanic white, non-Hispanic black, and Mexican-American) [1], and with minority and low-socioeconomic groups being disproportionately affected at all ages [2].

The public health significance of this obesity "epidemic" prompted the National Heart, Lung, and Blood Institute (NHLBI) to convene the Obesity Education Initiative (OEI) Expert Panel to review the literature on obesity and health risks and produce evidence-based "Clinical Guidelines on the Identification, Evaluation, and Treatment of Overweight and Obesity in Adults" in 1998 [3]. The OEI Expert Panel identified many obesity-related comorbidities, including coronary heart disease and major cardiovascular disease risk factors, e.g., type 2 diabetes, hypertension, and dyslipoproteinemias (in particular, low high-density lipoprotein [HDL] cholesterol and elevated plasma triglycerides), as well as gallbladder disease, respiratory disease, certain cancers (colorectal and prostate, endometrial, cervical, and breast), and osteoarthritis [3]. The guidelines specified ***three general goals of weight loss and management***: *to prevent further weight gain, to reduce body weight, and to maintain a lower body weight over the long term* and included the recommendation that physical activity be "part of a comprehensive weight loss therapy and weight control program, because it: modestly contributes to weight loss in overweight and obese adults, may decrease abdominal fat, increases cardiorespiratory fitness, and may help with maintenance of weight loss" [3].

The 1998 OEI exercise prescription that "moderate levels of physical activity for 30–45 minutes, 3–5 days a week should be encouraged initially, with a long-term goal of accumulating at least 30 minutes or more of moderate-intensity physical activity on most, and preferably all, days of the week" [3] was based on the level of physical activity recommended in 1995 by the Centers for Disease Control and Prevention and the American College of Sports Medicine (ACSM) to promote and maintain health [4]. This prescription was endorsed by the 1996 NIH Consensus Development Panel on Physical Activity and Cardiovascular Health [5] and incorporated into the 1996 ***Physical Activity and Health: Report of the Surgeon General*** [6]. Although there was substantial evidence that this level of physical activity would limit health risks for a number of chronic diseases [4–6], its role in weight control was unclear, thus prompting the ACSM to sponsor a scientific roundtable on the role of physical activity in the prevention and treatment of obesity and its comorbidities [7].

Setting randomized controlled trial (RCT) data as the highest level of evidence, the consensus of the ACSM roundtable was that the evidence for the role of physical activity in achieving the three OEI weight control goals was weak, as no RCT had addressed whether physical activity could prevent weight gain in the general population, exercise alone had been shown to produce only modest weight loss (generally 1–2 kg), the addition of exercise to reduced-energy diets had generally failed to produce significantly more weight loss than the diet alone, and few RCTs had been extended to examine whether the addition of exercise to diet would provide benefit over diet alone in maintaining weight loss [7, 8]. When available RCTs of at least 3 months' duration, completed by at least 20 participants, were separated by obesity status in a review of this literature for the first edition of ***Nutrition in the Prevention and Treatment of Disease***, the evidence that the recommended levels of physical activity produced or maintained weight loss in overweight or obese adults was particularly weak for women [9].

Considerable scientific discourse on the role of physical activity in weight control has occurred since the release of the 1998 OEI guidelines and the 1999 ACSM roundtable, and new evidence and perspectives are presented here. This review does not, however, address weight loss in elderly adults, which may be accompanied by a substantial loss of muscle mass associated with aging, or weight maintenance in elderly adults, which may therefore be accompanied by increasing adiposity. (Note also that recently updated physical activity recommendations for older adults, i.e., men and women aged 65 years and over [10], differ from those for younger adults [11].) Furthermore, although the prevalence of overweight among children and adolescents has also increased in the United States (to 15.5% among 12- to 18-year-olds, 15.3% among 6- to 11-year-olds, and 10.4% among 2- to 5-year-olds in 1999–2000, from 10.5%, 11.3%. and 7.2%, respectively, in 1988–1994) [12]), as there is no standard definition of "obesity" in children, this review will be restricted to adults.

II. CURRENT PHYSICAL ACTIVITY RECOMMENDATIONS FOR WEIGHT LOSS AND PREVENTION OF WEIGHT REGAIN

Acknowledging that BMI may misclassify the health risk of very active and/or lean individuals, a 2001 ACSM Position Stand recommended that overweight (BMI = 25–29.9 kg/m^2) and obese (BMI $\geq$ 30 kg/m^2) individuals reduce their body weight by a minimum of 5–10% by decreasing energy intake by 500–1000 kcal/day, reducing dietary fat to less than 30% of total energy intake, and progressively increasing moderate intensity exercise to 200–300 minutes (3.3–5 hours) per week [13]. This recommendation was supported by data derived primarily from 32 women who had previously lost weight, utilizing the doubly-labeled water (DLW) method, a relatively precise objective measure of total energy expenditure (TEE), which suggested a threshold-like relation between physical activity and weight gain corresponding to an average of 80 min/day of moderate activity or 35 min/day of vigorous activity for weight maintenance [14]. The ACSM also pronounced that there was no scientific evidence to suggest that resistance exercise will attenuate the loss of fat free mass (FFM) typically observed with reductions in total energy intake and loss of body weight; thus, resistance exercise was promoted to improve muscular strength and function in overweight adults, but not for weight loss or maintenance [13].

In 2002, consensus reached at the first Stock Conference of the International Association for the Study of Obesity (IASO) deemed the 1995 physical activity guideline for health insufficient to prevent unhealthy weight gain and recommended 45–60 minutes of moderate-intensity activity per day to prevent the transition to overweight or obesity, with even more activity recommended for children, and 60–90 minutes for preventing weight gain in formerly obese individuals [15]. The same year, new dietary recommendations published by the Institute of Medicine (IOM) consisted of guidelines to meet the body's daily energy and nutrient needs while minimizing the risk of weight gain [16]. Targeting a "healthy" BMI range of 18.5 to $<$ 25.0 kg/m^{2}and estimating energy requirements at sedentary, low active, active, and very active levels of energy expenditure, the IOM Panel produced total daily energy expenditure (TEE) results, based on doubly labeled water studies, in units of physical activity level (PAL), defined as the ratio of TEE to basal rate of energy expenditure (BEE) extrapolated to 24 hours [16, 17]. Studies with DLW, TEE, or both in titles, abstracts, or keywords were identified and used to compile age- and sex-specific data, which suggested that most (66%) adults who maintained a "healthy" BMI had PAL values greater than 1.6, prompting the IOM to recommend the equivalent of 60 or more minutes of physical activity of moderate intensity each day for adults [16, 17], which was reasonably consistent with the IASO consensus that 1.7 PAL, or 1.7 times one's basal metabolic rate, is required to prevent the transition to overweight or obesity [15].

Concluding that dietary and physical activity recommendations for healthful living are inextricably intertwined, the IOM presented physical activity as the primary means by which a person can vary energy expenditure to balance dietary energy intake once BEE and obligatory thermogenesis (see later discussion) in the healthy population are covered [16, 17]. Subsequently, both the American Heart Association (AHA) [18] and the American Cancer Society (ACS) [19] released combined nutrition and physical activity guidelines in 2006, emphasized a "healthy" body weight among the primary aims of diet and lifestyle recommendations, and endorsed more than 30 minutes of physical activity per day, with the AHA specifically recommending

60 minutes most days of the week for adults who are attempting to lose weight or maintain weight loss [18].

Recently, the ACSM and AHA jointly released updated physical activity recommendations, which generally reinforce the 1995 guideline, but clarify that "to promote and maintain health, all healthy adults aged 18 to 65 years need moderate-intensity aerobic (endurance) physical activity for a minimum of 30 minutes on five days each week or vigorous-intensity activity for a minimum of 20 minutes on three days each week" and that combinations of moderate- and vigorous-intensity activity can be performed to meet this recommendation [11]. (The accumulation of physical activity in "intermittent bouts" totaling 30 minutes or more is retained in the updated recommendations; and the minimum length of the "short bout" is defined as 10 minutes.) The updated ACSM/AHA recommendations acknowledge the complex set of cultural, psychological, and biological factors that influence maintenance of a healthy weight and the difficulty of accurately identifying the primary cause of obesity for any individual, and promote as a high priority for physical activity and nutrition professionals the development of integrated "calorie balance" guidelines and specific strategies on how to implement them effectively [11]. Thus, rather than promoting a higher volume of activity for overweight and obese adults, it is stated that "to help prevent unhealthy weight gain, some adults will need to exceed minimum recommended amounts of physical activity to a point that is individually effective in achieving energy balance, while considering their food intake and other factors that affect body weight" [11]. The recommendations specifically promote adoption of the minimum physical activity guidelines by adults, regardless of body size or shape, because of the obesity-independent benefits of physical activity, but add: "For individuals who achieve this level of activity, but remain overweight, an increase in physical activity is a reasonable component of any strategy to lose weight" [11]

III. RELATIONSHIP OF PHYSICAL ACTIVITY TO OBESITY IN POPULATIONS

Physical activity data from the 1996 U.S. Behavioral Risk Factor Surveillance System (BRFSS) showed that the prevalence of inactivity was high in both normal-weight (BMI of 18.5 to < 25.0 kg/m^2) men (26.8%) and women (27.8%) and in overweight (BMI of 25.0 to < 30.0 kg/m^2) men (25.6%) and women (31.7%), but was considerably higher in obese (BMI ≥ 30.0 kg/m^2) men (32.7%) and women (40.9%) [20], and that participation in recommended physical activity levels was low in both normal-weight men (29.3%) and women (30.7%) and in overweight men (29.2%) and women (26.0%), but was substantially lower among obese men (23.5%) and women (18.9%) [20]. Though one cannot assume causality, it is likely that inactivity promotes weight gain in some individuals and that excess weight deters some individuals from exercising, but also, some overweight adults increase activity in the hope of losing weight. In fact, the prevalence of reported regular leisure-time physical activity in adults aged 18 years and older in NHANES 1999–2002 was 28.3% overall (29.7% for men; 26.9% for women), but it was slightly higher, 32.6%, for those who reported trying to lose weight (45.6% of a sample of 9496 adults surveyed) and for 37.9% of those trying to maintain weight (9.5% of the sample), compared to 21.8% of those not trying to lose or maintain weight (44.9% of the sample) [21]. Those trying to lose weight were almost three times more likely to be regularly active (versus inactive), and those trying to maintain weight were over three times more likely to be regularly active (versus inactive), compared to those not trying to lose or maintain weight [21].

Several large-scale observational studies have reported a relationship between higher physical activity at baseline or improvements in physical activity and either attenuated weight gain or a lower odds of significant weight gain [22]. Conversely, a decrease in physical activity has been inferred in several studies that have reported secular decreases in energy intake concurrent with increases in weight and/or fatness; and in one study that had both energy intake and physical activity data, secular trends in obesity were accompanied by changes in transport to work, work-related activity, and/or changes in leisure-time activity [23]. One systematic review of observational cohort studies on physical activity and weight gain in adults concluded, however, that there was inconsistent evidence of the predictive effect of baseline physical activity on subsequent weight gain, though there was a stronger, albeit modest, association between weight gain and change in activity or activity at follow-up [24].

Several large longitudinal, prospective cohort studies have reported an inverse relationship between physical activity and weight change, including a study of more than 12,500 Finnish adults, aged 25–64 years followed over 4 to 7 years (median 5.7 years) [25]; a study of 2564 men and 2695 women, aged 19–63 years old and followed for 10 years in three municipalities in Finland, which reported that clinically significant body mass gain, defined as 5 kg or more, had an odds ratio of 2.59 in men and 2.67 in women with no regular physical activity, compared to the most active groups [26]; a 4-year follow-up of about 10,000 U.S. male health professionals aged 45–64 years [27]; a 2-year follow-up of approximately 3500 men and women in Minnesota, of a mean age of 38 years [28]; the Coronary Artery Risk Development in Young Adults (CARDIA) study of 5115 black and white men and women aged 18–30 years, followed over a 10-year period, which reported that the change in physical activity was inversely associated with change in body weight within all four race and sex subgroups ($p < 0.005$) [29]; and in 1319 subjects aged 9–18 years at baseline and followed over a 21-year period in the Cardiovascular Risk in Young Finns Study, both decreasingly active and persistently inactive subjects (11.5% of men and 7.4% of women) were more likely to be obese as adults compared with persistently active subjects (33.1% of men and 32.0% of women) [30].

In 1004 premenopausal white women who participated in NHANES III, mean BMI, percent body fat, and waist-to-hip ratio were significantly less ($p < 0.001$) at each higher physical activity level of four levels based on ACSM fitness criteria, except between level 0 and 1 [31]. In 3064 racially/ethnically diverse women aged 42–52 years at baseline of the Study of Women Across the Nation (SWAN), an observational study over the menopausal transition, a one-unit increase in reported level of sports/exercise (on a scale of 1–5) over 3 years of follow-up was longitudinally related to decreases of 0.32 kg in weight ($p < 0.0001$) and similar inverse relations were observed for daily routine physical activity (biking and walking for transportation and less television watching) [32]. And, in the Australian Longitudinal Study on Women's Health (ALSWH) of 8071 women aged 45–55 years, there were independent relationships between the odds of gaining more than 5 kg over 5 years and lower levels of habitual physical activity, as well as more time spent sitting, energy intake (in women with BMI > 25 kg/m^2), menopause transition, and hysterectomy [33].

Maintaining a given physical activity level over time may not be adequate to maintain a given weight, however. A study of 2501 healthy men in the Aerobics Center Longitudinal Study (ACLS), aged 20–55 years and followed over an average of 5 years, indicated weight gain among those who maintained the same daily PAL, whereas weight gain accelerated among men who decreased their activity [34]. Recent analyses from the National Runner's Health Survey of 6119 male and 2221 female runners whose running distances had changed less than 5 km/week between baseline and follow-up surveys 7 years later (19% of the surviving cohort) found that those who maintained modest (0–23 km/wk), intermediate (24–47 km/wk), or prolonged running distances ($\geq$48 km/wk) gained weight, through age 64 in men and in all age groups, from age 18 to 74 years, in women; however, those who maintained $\geq$48 km/wk had half the average annual weight gain of those who maintained <24 km/wk [35].

On the other hand, results from more than 9000 adults responding to the NHANES I Epidemiologic Follow-up Study (1971–1975 to 1982–1984) suggested that even relative to people who stay very active over time, sedentary people who increase activity can minimize the risk of major weight gain compared with people who decrease their activity, thus supporting the notion that increasing physical activity over time reduces the risk of becoming overweight [36]. In CARDIA, an increase in physical activity during the initial 2–3 years of follow-up was associated with an attenuation of weight gain that was sustained through 5-year follow-up, whether or not the physical activity increase was maintained in later years [29].

Twin studies, designed to partially control for genetic liability and childhood environment study, have also addressed the relationship of physical activity and obesity. A study of nearly 1000 female twins showed that current physical activity was a stronger predictor of total-body and central adiposity of healthy middle-aged women than dietary intake or smoking; furthermore, participants with a predisposition to adiposity did not show a lesser effect of physical activity on body mass [37]. In a 30-year follow-up study in Finland, 42 twin pairs discordant for physical activity had a mean weight gain of 5.4 kg less in the active compared with the inactive co-twins and mean waist circumference was 8.4 cm less in the active versus inactive co-twin; furthermore, trends were similar in both monozygotic and digyzotic twins, whereas pairwise differences in weight gain and waist circumference were not seen in 47 twin pairs who were not consistently discordant for physical activity [38]. Persistent participation in leisure-time activity was thus associated with decreased rate of weight gain and with smaller waist circumference to a clinically significant extent.

Supporting the alternative hypothesis that obesity may lead to inactivity, the odds ratio of later physical inactivity among adults with high BMI was significant in both 3653 women (1.91) and 2626 men (1.50), aged 20–78, randomly selected within age-sex strata from the general population of Copenhagen and followed over a 5-year interval, compared to median BMI, whereas there was no significant relationship between medium and high level of activity, compared to inactivity, and development of obesity [39]. BMI was also highly significantly associated with physical activity (PA), measured by accelerometry, in 85 severely obese outpatients (mean BMI of 42.7 kg/m^2) in a Swedish study, whereas the association between BMI and PA was weak in 193 nonobese control subjects (mean BMI of 24.0 kg/m^2) [40].

In the ALSWH study of middle-aged Australian women, the average weight gain was 0.5 kg/year, which, based on a food frequency questionnaire estimate of energy intake, was equated to an energy imbalance of only 10 kcal or 40 kJ per day [33]. Using available NHANES and CARDIA data, Hill *et al.* [41] estimated the average weight gain of Americans (20 to 40 years of age) to be 1.8 to 2.0 pounds (0.82–0.90 kg) per year, at least over the past decade, and further estimated that 90% of the U.S. population is gaining weight because of $\leq$100 kcal/day (i.e., only 10% is gaining weight at a rate greater than could be explained with an extra 100 kcal/day). Hill *et al.* proceeded to suggest that increasing physical activity by 100 kcal/day could prevent weight gain in most of the population and noted that this is comparable to about 1 mile (1.6 km) of walking, or an additional 2000 steps each day [41].

Because most researchers do not have the capability to measure either caloric intake or energy expenditure accurately enough to assess the independent contribution of either side of the equation or of the individual components of energy expenditure discussed below, accumulating evidence regarding the validity of this estimate, for the general population, study cohorts, or even individuals, poses a major challenge.

IV. ENERGY EXPENDITURE AND ETIOLOGY OF OBESITY

Underlying the recommendation to integrate physical activity guidelines with calorie intake guidelines [11] is the generally accepted principle that the development of obesity in any given individual requires that energy intake exceed energy expenditure over a period of time. To lose weight, one must expend more energy than is consumed. To prevent weight gain or maintain reduced weight over time, one must balance energy intake and energy expenditure long term.

Energy expenditure includes the energy expended to maintain normal body functions and homeostasis at rest, in the postabsorptive state, with the rate of this expenditure generally referred to as basal metabolic rate (BMR), which is commonly extrapolated to 24 hours as BEE, (expressed as kcal/24 hr); the energy costs of absorbing, metabolizing, and storing food, referred to as the thermic effect of food (TEF, which is determined primarily by the amount and composition of the foods consumed); and the energy expended above Resting Metabolic Rate (RMR) and TEF for both voluntary exercise and involuntary activity such as shivering, fidgeting, and postural control, i.e., the energy to regulate body temperature within narrow limits (a process that rarely affects TEE to an appreciable extent because most people adjust their clothing and environment to maintain comfort), and the energy to perform "physical activity" and more vigorous "exercise," a term often used for an activity level that improves physical fitness [16]. (Energy is also expended for growth, pregnancy, and lactation, but a discussion of individual components of energy expenditure with age, sex/gender, and reproductive status is beyond the scope of this review.)

Considerable research has been conducted during the past several decades on adaptations of muscle and fat tissue to exercise training, body composition changes associated with exercise-induced weight loss, differences between lipolytic responses of subcutaneous and intra-abdominal (visceral) fat in response to norepinephrine, which may influence differential mobilization of fat stores during exercise, and interactions between exercise and diet (both energy intake and relative contributions from fat, carbohydrate, and protein) during weight loss [42], including gender differences [16]. A clear physiological rationale supports the notion that exercise could bring about loss of excess body fat and/or serves to maintain weight loss and/or prevent accumulation of excess adipose tissue over time. Metabolic rate during exercise can increase up to 20 times above resting levels; therefore, it seems obvious that if a person were to engage in exercise regularly without increasing energy intake appreciably, weight loss should occur.

In general, RMR accounts for about 60–75% of daily energy expenditure, TEF for 5–10%, and physical activity for 15% in sedentary individuals and 30% or more in highly active individuals. Whereas all cells and tissues contribute to what may be considered obligatory energy expenditure, i.e., RMR and TEF, other than the participation of brown adipose tissue in cold-induced and diet-induced nonshivering thermogenesis, skeletal muscle expends the remaining energy, i.e., that for exercise-induced and cold-induced shivering thermogenesis. Skeletal muscle, which comprises about 40% of body weight in nonobese persons, has also long been held to be an important determinant of RMR [43]. In response to energy restriction (i.e., dieting), a modest suppression of RMR has been reported to range between 5% and 25%, with an average maximum change of about 15% [44], largely due to a decrease in fat free mass, which often accompanies diet-induced weight loss. Although small perturbations in RMR can have a substantial impact on the regulation of body weight in a given individual, there is no indication that RMR has declined in the population during the past few decades [45]. Similarly, there is no evidence that small differences in TEF between lean and obese subjects play a role in the development of obesity, nor is there any evidence that differences by BMI in the cost of weight-bearing physical activity have played a major causal role; however, a continued decline in daily energy expenditure, primarily due to decreases in activities of daily living and work-related physical activity during the past several decades, may be a primary factor contributing to the increased prevalence of obesity [45]. In contrast, leisure-time physical activity, which is the primary target of the exercise guidelines, has not decreased appreciably, but has been consistently low, with 23.7% of adults reporting no leisure-time activity in 2005 [11]. It is well known, of course, that within any given environment, there is considerable variation in body fatness among people who are in energy balance, due to genetic factors, which have been estimated to contribute anywhere from 25% to 70% of the variability in adiposity [46–49]. However, the U.S. genotype has not changed substantially enough during the past few decades to explain the increased prevalence of obesity; therefore, it seems that a focus on increasing physical activity is appropriate.

Adopting the exercise prescription presented in the OEI guidelines [3] would result in an energy expenditure of about 750 to 1000 kcal (3138–4184 kJ) per week. Noting that weight loss and preservation of lean body mass should be the goals of a weight loss program, the facts that 1 pound of fat is estimated to supply about 3500 kcal and that 1 mile of walking or jogging requires about 100 kcal mean that a pound of fat supplies enough energy to support about 35 miles of walking or jogging, or comparable activities (i.e., over 100 km per kg of fat). This makes it obvious that fat loss will be a slow process. Nonetheless, activity equivalent to a 1- to 1.5-mile (1.6- to 2.4-km) walk or jog (a 20- to 30-minute effort) five times a week, as recommended in the OEI guidelines, would be expected to bring about a loss of 6.5 kg of fat over the course of a year, if one did not consume excess energy or expend less energy in the remaining 23.5 hours of the day. If fat loss were to be achieved solely through the direct energy cost associated with increased physical exercise, a person must be committed to expending the energy over a reasonable period of time, with respect to his or her weight loss goal.

V. ROLE OF PHYSICAL ACTIVITY IN TREATMENT (WEIGHT LOSS) OF OVERWEIGHT AND OBESITY: EVIDENCE FROM RANDOMIZED CONTROLLED TRIALS

Both the 1998 OEI Panel [3] and the 1999 ACSM Roundtable panel [7] considered data arising from randomized controlled trials, involving a substantial number of participants, as providing the highest quality evidence; however, relatively few such trials specifically addressed the effect of physical activity in overweight and obese adults. In fact, early trials often reported baseline weight, without height, and did not report BMI, thereby making it difficult to determine the initial obesity status of the participants, especially when data are not separated by gender [9]. Many studies reported significant weight changes within groups, that is, baseline versus posttreatment weights, without specifying whether there were significant differences between groups, thereby making poor use of the control group. In studies with more than two treatment groups, pairwise comparisons were often made without an initial analysis of variance for all randomized participants; therefore, the significance of differences between groups at baseline is not always clear. Furthermore, retention of randomized participants is often low in physical activity and weight loss trials, and final results may be biased by overrepresentation of the most successful exercisers or weight losers. Finally, regardless of the physical activity prescription, adherence to the prescriptions is generally not reported, so it is often unclear what proportion of participants randomized to aerobic exercise or resistance-training programs actually achieve the expected levels of activity.

A. Role of Physical Activity in the Weight Loss

1. Aerobic Exercise Only versus Control

In an earlier review of RCTs of aerobic exercise (without diet restriction) of at least 3 months' duration and completed by at least 20 participants [9], only two [50, 51] of five [50–54] trials identified of overweight (baseline mean) men and one [55] of two [55, 56] trials of obese men reported significant weight reduction with aerobic exercise versus control; however, the aim of the second trial of obese men was to compare effects of weight loss (by diet only) with exercise (without weight loss), and thus the exercise-only group was expected to consume more calories to maintain weight [56]. In the trials of overweight men that reported weight loss with exercise only, Hellenius *et al.* [50] reported BMI changes for each of four randomly assigned groups (158 men total), with the exercise only group starting with a BMI of 26.1, which was reduced 0.3 over the 6 months of the study, and the control starting at BMI of 24.5, i.e., normal weight, and increasing 0.3 units. Pritchard *et al.* [51] reported a 2.6-kg loss over 12 months in the exercise only group, a 6.3-kg loss in a diet-restricted only group, and a 0.9-kg gain in the control (66 men total).

The one-year Diet and Exercise for Elevated Risk (DEER) trial [54] was not designed as a weight loss trial; however, overweight and obese participants among 197 men recruited to have low HDL cholesterol combined with elevated low-density lipoprotein (LDL) cholesterol received individual counseling on weight reduction if assigned to a group-based dietary intervention, with or without exercise, but not if assigned to exercise only or control, both of which were instructed not to make dietary changes. Nonetheless, energy intake in the exercise-only group, determined through five unannounced 24-hour dietary recall telephone interviews, was nonsignificantly increased by about 100 kcal per day, compared to control, which would have been enough to counteract the 700–1000 kcal/week energy expenditure achieved by most exercisers. In the trial of obese men that did report significant weight loss in the exercise-only group versus control, the first Stanford Weight Control Project (SWCP-I) [55], the exercise prescription of approximately 45 minutes of supervised walking or jogging three times per week was integrated into a goal of reducing body fat, rather than total weight, by one-third over a 9- to 12-month intervention period. Energy intake, assessed by 7-day food records, did not differ between exercisers and controls at either 7 months or 1 year, and the design goal of exercise-induced weight loss (4.6 kg versus control) was successfully achieved [55]. It is worth noting that the OEI Panel [3] unknowingly included two additional publications from SWCP-I among its set of randomized controlled trials that reported successful weight loss by aerobic exercise only, thereby (inappropriately) strengthening the data in support of the role of aerobic exercise in producing weight loss.

Meta-analyses reviewed by the panel may have made similar errors.

An earlier metabolic study by Sopko *et al.* [57], designed to tease apart the effects of diet- and exercise-induced weight loss on lipid changes, involved 40 initially sedentary men, aged 19–44, who were slightly overweight (110% of standard weight). Men were randomly assigned for 12 weeks to control, weight loss by energy restriction, weight loss by exercise, and exercise with weight maintenance. Meals were prepared in a metabolic kitchen for all groups and were fixed at 40% energy from fat. Energy intake was adjusted to maintain weight every 3 days, according to group assignment. Exercisers engaged in supervised treadmill walking at a pace of 5.6–6.4 km/hr for most subjects, with an energy expenditure of 700 kcal per session, 5 times a week, to total 3500 kcal per week. Significant weight was lost by men assigned to weight loss by exercise (−6.2 kg), but not to control (−0.5 kg) or exercise with weight maintenance (−0.5 kg). Together with the SWCP-I [55] and Katzel *et al.* [56] trials, these data highlight the role of energy intake in determining the effectiveness of exercise on weight loss in overweight and obese men.

Neither of two reasonably large 1-year trials of overweight (baseline mean), postmenopausal women ($N = 160$ and 180, respectively, and both completed by more than 90% of randomized subjects) reported weight loss by exercise only, despite significant increases in VO_{2max} [52, 54]; however, a 6-month trial of 40 borderline obese postmenopausal women, completed by 25 women, reported a modest weight reduction (−1.3 kg) in women assigned to walking (average walking speed increased from 5.5 to 6.0 km/hr with an estimated weekly energy expenditure of 6367 kJ) versus control [58]. Analysis of 3-day food records indicated that there were no significant changes in energy intake or diet composition throughout the program and walkers lost a small but significant amount of fat mass (−1.7 kg) compared with controls ($p < 0.05$) [58]. Two similarly sized studies that mixed obese men and women with type 2 diabetes found no effect of exercise only on weight [59, 60].

In summary, the randomized controlled trials available for review at the time of the OEI and ACSM consensus meetings, none of which exceeded 1 year, showed only modest (or no) weight loss with aerobic exercise programs that met the recommended exercise prescription, despite improved cardiorespiratory fitness. There were too few trials of women to provide much evidence regarding sex differences and it was clear that attention to energy intake was needed.

Since the OEI and ACSM consensus meetings and the review of this literature for the first edition of this text [9], several additional trials have been published, some of which have included information on body composition, beyond skinfolds. Magnetic resonance imaging (MRI) was used in a 3-month trial of 52 obese men assigned to diet-induced weight loss, exercise-induced weight loss, exercise without weight loss, or control, to provide evidence that total body fat, abdominal subcutaneous and visceral fat, and visceral fat-to-subcutaneous fat ratios were substantially reduced by both increased daily physical activity without caloric restriction and caloric restriction without exercise, compared to control, with both weight loss groups decreasing body weight by 7.5 kg (8%); however, total fat was decreased more (1.3 kg) in the exercise-induced weight loss group than in the diet-induced group ($p = 0.03$), whereas reductions in abdominal subcutaneous, visceral, and visceral fat-to-subcutaneous ratios were similar between weight loss groups [61]. Abdominal and visceral fat were also decreased in the exercise without weight loss group ($p = 0.001$), versus control, despite no significant change in body weight [61]. Dual emission absorptiometry (DXA) and computed tomography (CT) were employed in a 12-month randomized trial of 173 sedentary, overweight (BMI $\geq 24.0\ kg/m^2$) postmenopausal women, aged 50–75 years, to demonstrate significantly greater reductions in total body fat (−1.0%) by DXA and intra-abdominal fat ($-8.6\ g/cm^2$) and subcutaneous abdominal fat ($-28.8\ g/cm^2$) by CT, in women assigned to an intervention consisting of exercise facility and home-based moderate-intensity exercise compared to a stretching control group, with exercisers losing significant, but modest, body weight (−1.4 kg); furthermore, a significant dose response was observed for greater body fat loss by DXA and intra-abdominal body fat loss by CT with increasing duration of exercise [62].

The Midwest Exercise Trial for overweight and obese (BMI $25.0–34.9\ kg/m^2$) young adults, aged 17–35 years was completed by only 74 (43 women; 31 men) of 132 participants randomized to a 16-month program of verified exercise (primarily walking on motor-driven treadmills) or control at an approximately 2:1 ratio (65% to exercise, $N = 87$, 41 of whom completed the trial; 35% to control, $N = 44$, 33 of whom completed the trial). The study concluded that moderate-intensity exercise sustained for 16 months is effective for weight management in young adults [63]. Though retention for men versus women was not presented, it was reported that exercise prevented weight gain and increases in fat mass (by hydrostatic weighing) in women, compared to increases in control women of 2.9 ± 5.5 kg and 2.1 kg, respectively, whereas in men, exercise produced weight loss of 5.2 ± 4.7 kg and reduction of fat mass of 4.9 kg, versus control. However, no significant changes were seen in fat-free mass in either sex and there were no significant between-group differences for visceral, subcutaneous, or total fat measured by CT, despite significant reductions from baseline for men (but not for women) assigned to exercise [63].

The Studies of Targeted Risk Reduction Interventions through Defined Exercise (STRRIDE) reported findings from 120 (65 men, 55 postmenopausal women) sedentary, overweight adults, aged 40–65 years, who completed the trial of 182 who had been randomly assigned to control or one of three 8-month exercise intervention groups: high amount/vigorous intensity (calorically equivalent to approximately 20 miles [32 km] of jogging per week at 65–80% peak oxygen consumption); low amount/vigorous intensity (equivalent to approximately 12 miles [19.2 km] of jogging per week at 65–80%); and low amount/moderate intensity (12 miles [19.2 km] at 40–55%), all of whom were counseled not to change their diet and were encouraged to maintain body weight. However (noting that data were not presented by gender), the controls gained weight, both low-amount exercise groups lost weight and fat (via sum of skinfolds) versus control, but did not differ from each other, and the high-amount exercise group lost more weight and fat, compared to both low-amount exercise group and controls, and in a dose-response manner [64].

2. Diet plus Aerobic Exercise Only versus Diet Alone

The OEI Expert Panel [3] stated that a combination of a reduced-energy diet and increased physical activity produces greater weight loss than diet alone or exercise alone, based on results from 12 studies [50, 53, 54, 65–73]. Seven additional trials that met the criteria of the OEI Panel were identified for the previous review for the first edition of this book [9; 74–80]. The studies included fairly equal numbers of overweight men and women, but nearly twice as many obese women as men, including women who were considerably more obese than those who were included in studies of exercise only. None of the five trials of overweight individuals, which included more than 100 premenopausal women, 300 postmenopausal women, and about 550 men, reported greater weight loss in men or women assigned to aerobic exercise plus diet versus those assigned to diet only [50, 53, 54, 65, 66]; however, both treatments resulted in significant weight loss compared to controls in all five trials. Only one, the second Stanford Weight Control Project (SWCP-II) [65], of the nine trials [65, 67–69, 74–79] of obese class 1 (BMI $= 30.0–34.9\,kg/m^2$) individuals, which included more than 300 women, mostly premenopausal, and approximately 300 men, reported greater weight loss in men assigned to the exercise plus diet group versus the diet-only group [65], noting that the women in this trial were overweight, per mean baseline BMI, and did not show greater weight loss in diet plus exercise versus exercise only [65].

Only one [70] of five trials [70–73, 80] in obese class 2 (BMI $= 35–39.9\,kg/m^2$) individuals, which also included more than 300 women but only about 25 men, reported greater weight loss in the exercise plus diet group versus the diet-only group, and this was a small study that mixed nine men and 21 women and did not have a (no weight loss treatment) control group [70].

In summary, only two of 19 studies involving about 2000 overweight or obese women and men showed a significant benefit of the addition of aerobic exercise to a weight-reducing diet for weight loss, providing little support for the contention that the addition of exercise to an energy-restricted diet will result in significant additional weight loss in a majority of overweight or obese individuals; however, the majority of these studies included fewer than 20 participants per treatment assignment, whereas the SWCP-II trial [65] included more than twice that number.

3. Comparisons of Different Aerobic Exercise Programs Combined with Diet

Following the release of several supportive physical activity guidelines in the mid-to-late 1990s [3–6], research was conducted on various components of the recommendations. Jakicic *et al.* [81, 82] reported trial-based effects of doing exercise in multiple short bouts (three 10-minute bouts) versus a long bout (one 30-minute bout) on weight changes produced predominantly by a behavioral weight loss program consisting of energy-restricted diet combined with a specific exercise prescription. In the first [81], obese women, aged 25–50 years, were randomly assigned to a behavioral weight loss program consisting of an energy-restricted diet combined with 5 days per week of either single aerobic exercise bouts per day, starting as 20-minute bouts (weeks 1–4), increasing to 30-minute bouts (weeks 5–8) and to 40-minute bouts (weeks 9–20); or multiple 10-minute bouts per day, starting as 2 bouts per day, increasing to 3, then 4, respectively. Both groups reduced energy intake and percent energy from fat significantly. Women performing multiple short bouts lost 8.9 kg in the 20-week period, while those exercising in single long bouts lost 6.4 kg (not significant). It is difficult to determine the independent contribution of the exercise to the weight loss, as the dietary changes (energy restriction) contributed substantially to the weight loss; however, exercising in multiple short bouts was shown to improve adherence to exercise and to result in significantly greater improvement in aerobic capacity, as well as the trend for greater weight loss. In the second trial [82], obese women were assigned to one of three treatment groups, all of which included an energy-restricted diet: long-bout exercise, multiple short-bout exercise, or multiple short-bout exercise with home exercise equipment using a treadmill. Total and fat weight loss was significantly greater in the short-bout exercise with home exercise equipment group compared to multiple short-bout exercise, but did not differ between long-bout exercise and either multiple short-bout exercise or short-bout exercise with home exercise equipment.

The effect of accumulating activity over the course of the day through "lifestyle activity" versus structured aerobic exercise on weight control was studied in a 16-week randomized trial of 40 obese women, all of whom received a low-fat energy-restricted diet intervention, with the finding that weight loss was similar between groups; however, the aerobic group lost less fat-free mass [83]. During a 1-year follow-up, the aerobic group regained 1.6 kg, while the lifestyle group regained 0.08 kg, suggesting that lifestyle intervention may be better for weight maintenance [83].

A randomized, controlled trial of 88 mixed-race (50% Hispanic, 25% Caucasian, 21% African-American, 2% Indian, 2% Asian) premenopausal, overweight (BMI $\geq$ 25 kg/m^2) women participating in a 12-week weight loss intervention with a 5.0–5.8 Megajoule (MJ) diet daily compared the effects of walking 30 minutes, 5 days per week, versus 60 minutes, 5 times per week, versus diet only and found that all groups had similar and significant ($p < 0.001$) declines in body weight, fat mass, fat-free mass, and percentage body fat, by hydrodensitometry, BMI, and waist-to-hip circumference ratio, with no evidence of a dose-response effect and no baseline differences between the 56 completers and 32 dropouts [84].

Also, Jakicic *et al.* [85] found no differences in weight loss after 12 months among groups of overweight and obese (BMI = 27–40 kg/m^2) sedentary women, ages 21–45, enrolled in a standard behavioral weight loss intervention and receiving instruction on reducing energy intake to 1200–1500 kcal/d and assigned to one of four groups: vigorous intensity/high duration (2000 kcal/wk); moderate intensity/high duration (2000 kcal/wk); moderate intensity/moderate duration (1000 kcal/wk); and vigorous intensity/moderate duration (1000 kcal/wk), with all groups losing weight (−6.3 to −8.9 kg).

4. Comparisons of Different Aerobic Exercise Programs without Dietary Change

The effect of accumulating activity over the course of the day through "lifestyle activity," without a dietary intervention, was as effective as a structured exercise program in improving physical activity, cardiorespiratory fitness, and blood pressure in a moderately overweight cohort of men and women; however, neither group changed weight [86].

Whether a higher dose of exercise will produce greater weight loss than a lower dose was studied in 202 (58% female) overweight (14–32 kg overweight, according to actuarial norms) men and women, aged 25–50 years, who were randomly assigned to either a standard behavioral therapy (SBT) treatment with an exercise goal of 1000 kcal/wk or a high physical activity (HPA) treatment with a goal of 2500 kcal which included SBT plus encouragement to recruit 1–3 exercise partners and small-group counseling with an exercise coach [87]. The HPA group achieved greater exercise levels than the SBT group at 6 months, 12 and 18 months and had significantly greater weight losses versus SBT at 12 months (8.5 ± 7.9 versus 6.1 ± 8.8 kg) and 18 months (6.7 ± 8.1 versus 4.1 ± 7.3 kg), noting that 82% of HPA and 79% of SBT who completed 12 month weights and 87% of HPA and 80% of SBT who completed 18-month follow-up, with no differences reported for baseline characteristics between the 168 study completers and 34 dropouts [87].

In a small, randomized trial of 22 overweight subjects, Donnelly *et al.* [88] compared the effects of 18 months of continuous exercise (30 minutes at 60–75% maximal aerobic capacity, three times per week, with supervision in a Human Performance Laboratory) to intermittent exercise (walking briskly, 2 times per day, 15 minutes per session 5 days per week, at home or at work) and reported that continuous exercisers lost significant, but modest, weight (2.1%) from baseline, while the intermittent group did not; however, there were no significant differences between groups.

5. Resistance Exercise Only

ACSM recommendations for the quantity and quality of exercise to be achieved by adults includes a progressive resistance training component that provides a stimulus to all major muscle groups, with the plan to complete 10–15 repetitions of a set of 8–10 exercises, 2–3 days/week [89]. Although these guidelines do not address obesity management, some research has assessed the potential role of resistance exercise or strength training in bringing about weight loss and maintaining lean body mass. In the previous review of this literature, two trials of obese class 1 women [69, 76] and three trials of obese class 2 women [71, 73, 80] were identified which compared aerobic exercise with resistance exercise or strength training, none of which showed a difference in weight loss between these forms of exercise when combined with the diets. In a trial that included measurement of adipose tissue by MRI [90], 24 women with BMI over 27 kg/m^2 and waist-to-hip ratio greater than 0.85 were randomly assigned to aerobic exercise or resistance exercise for 16 weeks, with both groups instructed on an energy-restricted diet. The aerobic exercise group, which started at a mean BMI of 34.4 kg/m^2, lost 10.9 kg, while the resistance exercise group started at mean BMI of 31.8 kg/m^2 and lost 10.1 kg. These differences were not significant among groups, and both groups reduced their volume ratio of visceral to subcutaneous fat, with no differences between groups [90].

In a 12-week strength training study of 22 obese women (mean BMI of 31.4 kg/m^2 for the resistance exercise group and 32.8 kg/m^2 for the controls), total weight was increased 1.4 kg in resistance exercise and 0.4 kg in controls, but there were no significant differences between groups [91]. The fact that weight was increased, rather than decreased, emphasizes the need to separate possible lean body mass gain and fat weight loss in such studies.

The 2001 ACSM Position Stand on intervention strategies for weight loss stated that there was no scientific evidence to suggest that resistance exercise will attenuate the loss of FFM typically observed with reductions in total energy intake and loss of body weight and promoted resistance exercise to improve muscular strength and function in overweight adults, not for weight loss or maintenance [13].

B. Role of Physical Activity in the Maintenance of Weight Loss

The 1998 OEI Panel stated that physical activity may prevent weight regain; however, there were few randomized controlled trials that included long-term follow-up at the time of the report. A 1997 meta-analysis [92] noted that 1-year follow-up patients in the diet-only groups maintained a weight loss of 6.6 kg, whereas those in the diet plus exercise groups maintained a weight loss of 8.6 kg; however, neither the overall weight loss nor the percent of weight loss retained differed significantly between conditions. A 2001 meta-analysis, restricted to studies that provided follow-up data for at least 2 years in a structured weight loss program, reported that of 29 studies identified, only six provided information related to exercise on weight-loss maintenance with the finding that initial body weights and weight losses did not differ significantly between lower and higher exercise groups; however, groups with higher amounts of exercise were significantly more successful in maintaining their weight loss for periods 2–3.3 years [93].

King *et al.* [94] focused on adherence issues surrounding group versus home-based exercise as well as optimal intensity of exercise and did not identify a better approach to weight loss during the initial year or during a second year of follow-up; however, the higher intensity, group-based men and women increased BMI slightly (0.1 and 0.2 kg/m^2, respectively) from baseline to 24 months, whereas the higher intensity, home-based men and women decreased BMI slightly (−0.1 kg/m^2 each), as did the lower intensity, home-based men and women (−0.2 and −0.4 kg/m^2, respectively).

Wing *et al.* [70] published two small trials and their follow-up studies in one report. The 2-year weight change data for a study with aerobic exercise plus diet (−7.9 kg) versus diet only (−3.8 kg) were similar to the data for a study of aerobic exercise plus diet (−7.8 kg) versus a low-intensity, flexibility ("placebo") exercise plus diet group (−4.0 kg); however, the differences among groups reached significance in the first study, but not the second.

Svendsen *et al.* [95] published a 6-month follow-up to a 12-week weight loss trial [66], which reported that both the exercise plus diet and diet-only women gained about 2 kg during the period, and although weight loss was still significant versus control, it remained not significant between the combined and diet-only women. Wadden *et al.* [96] reported 1-year follow-up data for the study of obese class 2 individuals [73]; all groups, regardless of assignment to aerobic exercise, strength training, a combination, or no exercise gained back a substantial portion of the weight they had lost in the initial year and there were no differences among groups.

Skender *et al.* [97] published a one-year follow-up to a trial in which obese men and women had been randomly assigned to a Help Your Heart Eating Plan (HYHEP) or to aerobic exercise, primarily walking, for three to five 45-minute periods per week, at an intensity that subjectively was perceived as "vigorous" but not "strenuous," without changing diet; or to exercise plus the eating plan. At 1 year, the exercise-only group (−2.9 kg) had lost less weight than the diet-only (−6.8 kg) or exercise plus diet group (−8.9 kg), ANOVA $p < 0.09$; however, both the diet-only and exercise plus diet groups regained weight during the second year, while the exercise-only group maintained its weight loss. Weight change from baseline to year 2 for 86 (of an original 127) women who returned was diet only (+0.9), exercise only (−2.7 kg), and diet plus exercise (−2.2 kg). These few studies provided only minimal support for the notion that the addition of exercise to diet is effective for maintenance of weight loss over time; however, the data were clearly very limited.

Subsequently, Fogelholm *et al.* [98] reported results after a 2-year follow-up of 74 of 82 obese (mean BMI of 34.0 kg/m^2), premenopausal women who participated in a 12-week weight reduction by mostly a low-energy diet and relapse prevention and a 40-week maintenance program randomized in three groups, each of which had counseling on diet and relapse prevention: control with no increase in habitual exercise; a walking program targeted to expend 4.2 MJ/wk; and a walking program targeted to expend 8.4 MJ/wk, with follow up for 2 years after the intervention, with the finding that weight regain (after a mean weight loss of 13.1 kg) was 3.5 kg less and waist circumference regain was 3.8 cm less in the 4.2-MJ walking group versus control, whereas the higher training target (8.4 MJ/wk) group did not maintain weight [98].

Tate *et al.* [99] recently reported 30-month follow-up data for an 18-month trial in which overweight men and women aged 25–50 years were randomly assigned to either an SBT treatment with an exercise goal of 1000 kcal/wk or the HPA treatment with a goal of 2500 kcal presented earlier [87], and reported that there were no longer differences between SBT and HPA for either exercise levels or weight loss, which were 2.86 ± 8.6 kg for HPA and 0.90 ± 8.9 kg for SBT ($p = 0.16$).

Although the randomized controlled trial data are mixed and may not provide strong evidence, there is considerable anecdotal support for the concept that exercise is important in weight maintenance after losses of substantial weight. In a study involving interviews of 44 obese women who regained weight after successful weight reduction (relapsers), 30 formerly obese, average-weight women who maintained weight loss (maintainers), and 34 women who had always remained at the same average, nonobese weight ($n = 34$), 90% of maintainers and 82% of controls reported exercising regularly, compared to only 34% of relapsers [100].

The National Weight Control Registry, a large study of successful long-term maintainers of weight loss, includes 629 women and 155 men who lost an average of 30 kg and maintained a required minimum weight loss of 13.6 kg for 5 years [101]. These men and women report using both diet and exercise to maintain weight loss and report a very high activity level of approximately 11,830 kJ being expended through physical activity per week. This exceeds the ACSM recommendation of 2000 kcal/week (8368 kJ/week) energy expenditure as optimal physical activity, a goal that was achieved by 52% of the registry sample (50% of women, 62% of men) [101]. Proposing that successful long-term weight loss maintenance be defined as intentionally losing at least 10% of initial body weight and keeping it off for 1 year, Wing and Hill [102] suggested that eating a diet low in fat, frequent self-monitoring of body weight and food intake, and high levels of regular physical activity (such as 1 hour of brisk walking per day) does improve weight maintenance, which may get easier over time.

A review by Ross and Janssen [103] of the literature on exercise-induced weight loss on total and/or abdominal fat subdivided studies into 20 short-term ($\leq$16 weeks) and 11 long-term ($\geq$26 weeks) studies and reported that short-term studies were characterized by exercise programs that had increased energy expenditure by values double that of long-term studies (2200 versus 1100 kcal/week). Reductions in body weight (−0.18 versus −0.06 kg/week) and total fat (−0.21 versus 0.06 kg/week) were threefold higher in short-term studies than those reported for long-term studies [103]. Short-term studies suggested that exercise-induced weight loss was positively related to reductions in total fat in a dose-response manner, whereas no such relationship was observed in long-term studies [103].

C. Demonstration of Successful Diet plus Exercise Weight Loss Intervention in Preventing Type 2 Diabetes

The success of the Diabetes Prevention Program Trial [104] of 3234 nondiabetic persons, at least 25 years of age, with elevated fasting and postload plasma glucose concentrations and BMI of 24 kg/m^2 or higher (BMI of 22 in Asians) in achieving substantial weight loss with dietary intervention combined with increased physical activity has promoted the clinical value of a combined diet and exercise weight loss intervention. Fifty-one percent of participants in the lifestyle group achieved 7% weight loss at 24 weeks and 38% had 7% weight loss by the most recent visit (mean 2.8 years of follow-up) and, in turn significantly reduced the incidence of type 2 diabetes (with one case being prevented per seven persons treated for 3 years). This clearly demonstrates the clinical (and population health) value of a combined diet and exercise weight loss intervention. The potential role of a similar weight loss program in overweight people with type 2 diabetes (designed to achieve a mean loss of >7% of initial weight and to increase participants' physical activity to >175 min/week) in preventing cardiovascular disease is currently being investigated in the NIH Look AHEAD (Action for Health in Diabetes) study [105], which will undoubtedly also produce valuable information on the role of physical activity in managing overweight and obesity.

Its seems unlikely that an exercise-only recommendation will be built into future guidelines, particularly without a better understanding of how to get overweight and obese persons to increase their physical activity levels without increasing caloric intake or making adjustments in other components of their overall energy balance.

VI. CONCLUSION

Epidemiological studies suggest that physical inactivity is related to obesity, and that active adults are less likely to be obese and that they gain less weight as they age than do inactive adults; however, a causal relationship between a sedentary lifestyle and obesity has not been established. Despite ample biological rationale to support the contention that being more physically active will facilitate weight control, the evidence from clinical trials to support a strong role for adopting a more active lifestyle to bring about weight loss and weight maintenance in overweight or obese adults remains weak, primarily due to a dearth of randomized trials with adequate retention rates and long-term follow-up. In general, neither aerobic exercise nor resistance exercise seems to be powerful enough to cause substantial weight loss, without the addition of restriction of energy intake. Noting that diet-induced weight loss alone is generally not associated with long-term success and that the goals of healthy weight control should emphasize the loss of body fat, while maintaining lean body mass, it is important to acknowledge that weight loss will likely be a slow process. Recently, the American College of Sports Medicine and the American Heart Association jointly released physical activity recommendations which promote the development of integrated "calorie balance" guidelines and specific strategies on how to implement them, rather than stressing the need for even higher volumes of physical activity for overweight and obese adults than what has been promoted for limiting health risks for chronic diseases, many of which are obesity-related comorbidities. Both physical-activity and nutritional professionals are encouraged to make this a high priority, because the steady increase in the prevalence of obesity in adults and children in the United States and other developed countries is likely to have an increasingly high adverse impact of public health.

References

1. Flegal, K. M., Carroll, M. D., Ogden, C. L., and Johnson, C. L. (2002). Prevalence and trends in obesity among U.S. adults, 1999–2000. *JAMA* **288**, 1723–1727.
2. Wang, Y., and Beydoun, M. A. (2007). The obesity epidemic in the United States—gender, age, socioeconomic, racial/ethnic, and geographic characteristics: a systematic review and meta-regression analysis. *Epidemiol. Rev.* **29**, 6–28.
3. Expert Panel on the Identification, Evaluation, and Treatment of Overweight in Adults (1998). Clinical guidelines on the identification, evaluation, and treatment of overweight and obesity in adults—The evidence report. *Obes. Res.* **6**(Suppl. 2), 51S–209S.
4. Pate, R. R., Pratt, M., Blair, S. N., Haskell, W. L., Macera, C. A., Bouchard, C., Buchner, D., Ettinger, W., Heath, G. W., King, A. C., Kriska, A., Leon, A. S., Marcus, B. H., Morris, J., Paffenbarger, R. S., Jr., Patrick, K., Pollock, M. L., Rippe, J. M., Sallis, J., and Wilmore, J. H. (1995). Physical activity and public health: A recommendation from the Centers for Disease Control and Prevention and the American College of Sports Medicine. *JAMA* **273**, 402–407.
5. NIH Consensus Development Panel on Physical Activity and Cardiovascular Health (1996). Physical activity and cardiovascular health. *JAMA* **276**, 241–246.
6. U.S. Department of Health and Human Services (1996). "Physical Activity and Health: A Report of the Surgeon General." U.S. Department of Health and Human Services, Centers for Disease Control and Prevention, National Center for Chronic Disease Prevention and Health Promotion, Atlanta, GA.
7. Grundy, S. M., Blackburn, G., Higgins, M., Lauer, R., Perri, M. B., and Ryan, D. (1999). Physical activity in the prevention and treatment of obesity and its comorbidities: Evidence report of independent panel to assess the role of physical activity in the treatment of obesity and its comorbidities. *Med. Sci. Sports Exerc.* **31**, S502–S507.
8. Wing, R. R. (1999). Physical activity in the treatment of adulthood overweight and obesity: Current evidence and research issues. *Med. Sci. Sports Exerc.* **31**, S447–S552.
9. Stefanick, M. L. (2001). Obesity: role of physical activity. *In* "Nutrition in the Prevention and Treatment of Disease" (A.M. Coulston, C. L. Rock C, and E. R. Monsen, Eds.), pp. 481–497. Academic Press, San Diego, CA.
10. Nelson, M. E., Rejeski, W. J., Blair, S. N., Duncan, P. W., Judge, J. O., King, A. C., Macera, C. A., and Castaneda-Sceppa, C. (2007). Physical activity and public health in older adults: recommendations from the American College of Sports Medicine and the American Heart Association. *Med. Sci. Sports Exerc.* **39**, 1435–1445.
11. Haskell, W. L., Lee, I-M. L., Pate, R. R., Powell, K. E., Blair, S. N., Franklin, B. A., Macera, C. A., Heath, G. W., Thomson, P. D., Bauman, A., and Whitehead, J. R. (2007). Physical activity and public health: Updated recommendation for adults for the American College of Sports Medicine and the American Heart Association. *Med. Sci. Sports Exerc.* **39**, 1423–1434.
12. Ogden, C. L., Flegal, K. M., Carroll, M. D., and Johnson, C. L. (2002). Prevalence and trends in overweight among U.S. children and adults, 1999–2000. *JAMA* **288**, 1728–1732.
13. American College of Sports Medicine. (2001). Position Stand. Appropriate intervention strategies for weight loss and prevention of weight regain for adults. *Med. Sci. Sports Exerc.* **33**, 2145–2156.
14. Schoeller, D. A., Shay, K., and Kushner, R. F. (1997). How much physical activity is needed to minimize weight gain in previously obese women? *Am. J. Clin. Nutr.* **66**, 551–556.
15. Saris, W. J. M., Blair, S. N., van Baak, M. A., Eaton, S. G., Davies, P. S. W., Di Pietro, L., Fogelholm, M., Rissanen, A., Schoeller, D., Swinburn, B., Tremblay, A., Westerterp, K. R., and Wyatt, H. (2003). How much physical activity is enough to prevent unhealthy weight gain? Outcome of the IASO 1st Stock Conference and consensus statement. *Obesity Rev* **4**, 101–114.
16. Institute of Medicine. (2002). "Dietary Reference Intakes for Energy, Carbohydrate, Fiber, Fat, Fatty Acids, Cholesterol, Protein and Amino Acids," pp. 880–935 (http://www.nap.edu/books/0309085373/html, accessed August 12, 2007). National Academies Press, Washington, D.C.
17. Brooks, G. A., Butte, N. F., Rand, W. M., Flatt, J-P., and Cabellero, B. (2004). Chronicle of the Institute of Medicine physical activity recommendation: how a physical activity recommendation came to be among dietary recommendations. *Am. J. Clin. Nutr.* **79**(suppl), 921S–930S.
18. Lichtenstein, A. H., Appel, L. J., Brands, M., Carnethon, M., Daniels, S., Franch, H. A., Franklin, B., Kris-Etheron, P., Harris, W. S., Howard, B., Karanja, N., Lefevre, M., Rudel, L., Sacks, F., van Horn, L., Winston, M., and Wylie-Rosett, J. (2006). Diet and lifestyle recommendations revision 2006: a scientific statement from the American Heart Association Nutrition Committee. *Circulation* **114**, 82–96.
19. Kushi, L. H., Byers, T., Doyle, C., Bandera, E. V., McCulloug, M., Gansler, T., Andrews, K. S., and Thun, M. J. and the American Cancer Society 2006 Nutrition and Physical Activity Guidelines Advisory Committee. (2006). American Cancer Society guidelines on nutrition and physical activity for cancer prevention: Reducing the risk of cancer with healthy food choices and physical activity. *CA. Cancer J. Clin.* **56**, 254–281.
20. Pratt, M., Macera, D. A., and Blanton, C. (1999). Levels of physical activity and inactivity in children and adults in the United States: Current evidence and research issues. *Med. Sci. Sports Exerc.* **31**, S526–S533.
21. Kruger, J., Yore, M. M., and Kohl III, H. W. (2007). Leisure-time physical activity patterns by weight control status: 1999–2002 NHANES. *Med. Sci. Sports Exerc.* **39**, 788–795.
22. Di Pietro, L. (1999). Physical activity in the prevention of obesity: current evidence and research issues. *Med. Sci. Sports Exerc.* **31**, S542–S546.
23. Jebb, S. A., and Moore, M. S. (1999). Contribution of a sedentary lifestyle and inactivity to the etiology of overweight and obesity: Current evidence and research issues. *Med. Sci. Sports Exerc.* **31**, S534–S541.
24. Fogelholm, M., and Kukkonen-Harjula, K. (2000). Does physical activity prevent weight gain – a systematic review. *Obes. Rev.* **1**, 95–111.
25. Rissanen, A. M., Heliovaara, M., Knekt, P., Reunanen, A., and Aroma, A. (1991). Determinants of weight gain and overweight in adult Finns. *Eur. J. Clin. Nutr.* **45**, 419–430.

26. Haapanen, N., Miilunpalo, S., Pasanen, M., Oja, P., and Vuori, I. (1997). Association between leisure time physical activity and 10-year body mass change among working-aged men and women. *Int. J. Obes.* **21**, 288–296.
27. Coakley, E. H., Rimm, E. B., Colditz, G., Kawachi, I., and Willett, W. (1998). Predictors of weight change in men: Results from the Health Professionals Follow-Up Study. *Int. J. Obes.* **22**, 89–96.
28. French, S. A., Jeffery, R. W., Forster, J. L., McGovern, P. G., Kelder, S. H., and Baxter, J. E. (1994). Predictors of weight change over two years among a population of working adults: The Healthy Worker Project. *Int. J. Obes.* **18**, 145–154.
29. Schmitz, K. H., Jacobs, D. R., Leon, A. S., Schreiner, P. J., and Sternfeld, B. (2000). Physical activity and body weight: Associations over ten years in the CARDIA study. *Int. J. Obes.* **24**, 1475–1487.
30. Yang, X., Telama, R., Viikari, J., and Raitakari, O. T. (2006). Risk of obesity in relation to physical activity from youth to adulthood. *Med. Sci. Sports Exerc.* **38**, 919–925.
31. Holcomb, C. A., Heim, D. L., and Loughin, T. M. (2004). Physical activity minimizes the association of body fatness with abdominal obesity in white, premenopausal women: Results from the Third National Health and Nutrition Examination Survey. *J. Am. Diet. Assoc.* **104**, 1859–1862.
32. Sternfeld, B., Wang, H., Quesenberry, C. P., Jr., Abrams, B., Everson-Rose, S. A., Greendale, G. A., Matthews, K. A., Torrens, J. I., and Sowers, M. F. (2004). Physical activity and changes in weight and waist circumference in midlife women: findings from the Study of Women Across the Nation. *Am. J. Epidemiol.* **160**, 912–922.
33. Brown, W. J., Williams, L., Ford, H. H., Ball, K., and Dobson, A. J. (2005). Identifying the energy gap: magnitude and determinants of 5-year weight gain in midage women. *Obes. Res* **13**, 1431–1441.
34. DiPetro, L., Dziura, J., and Blair, S. N. (2004). Estimated change in physical activity level (PAL) and prediction of 5-year weight change in men: The Aerobics Center Longitudinal Study. *Int. J. Obes.* **28**, 1541–1547.
35. Williams, P. T. (2007). Maintaining vigorous activity attenuates 7-yr weight gain in 8340 runners. *Med. Sci. Sports Exerc.* **39**, 801–809.
36. Williamson, D. R., Kahn, H. S., Remington, P. L., and Anda, R. F. (1990). The 10-year incidence of overweight and major weight gain in U.S. adults. *Arch. Intern. Med.* **150**, 665–672.
37. Samaras, K., Kelly, P. J., Chiano, M. N., Spector, T. D., and Campbell, L. V. (1999). Genetic and environmental influences on total-body and central abdominal fat: The effect of physical activity in female twins. *Ann. Intern. Med.* **130**, 873–882.
38. Waller, K., Kaprio, J., and Kujala, U. M. (2007). Associations between long-term physical activity, waist circumference and weight gain: A 30-year longitudinal twin study. Int. *J. Obes.* advance online publication 24 July 2007; doi: 10.1038/sj.ijo.0803692.
39. Petersen, L., Schnohr, P., and Sorensen, T. I. A. (2004). Longitudinal study of the long-term relation between physical activity and obesity in adults. *Int. J. Obes.* **28**, 105–112.
40. Hemmingsson, E., and Ekelund, U. (2007). Is the association between physical activity and body mass index obesity dependent? *Int. J. Obes.* **31**, 663–668.
41. Hill, J. O., Wyatt, H. R., Reed, G. W., and Peters, J. C. (2003). Obesity and the environment: where do we go from here? *Science* **299**, 853–855.
42. Stefanick, M. L. (1993). Exercise and weight control. *Exerc. Sports Sci. Rev.* **21**, 363–396.
43. Zurlo, F., Larson, K., Bogardus, C., and Ravvussin, E. (1990). Skeletal muscle metabolism is a major determinant of resting energy expenditure. *J. Clin. Invest.* **86**, 1423–1427.
44. Prentice, A. M., Goldberg, F. R., Jebb, S. A., Black, A. E., Murgatroyd, P. R., and Diaz, E. O. (1991). Physiological responses to slimming. *Proc. Nutr. Soc.* **50**, 441–458.
45. Hill, J. O., and Melanson, E. L. (1999). Overview of the determinants of overweight and obesity: Current evidence and research issues. *Med. Sci. Sports Exerc.* **31**, S515–S521.
46. Stunkard, A. J., Foch, T. T., and Hrubec, Z. (1986). A twin study of human obesity. *JAMA* **256**, 51–54.
47. Stunkard, A. J., Sorenson, T. I., and Hanis, C. (1986). An adoption study of human obesity. (1986). *N. Engl. J. Med.* **314**, 193–198.
48. Bouchard, C., and Tremblay, A. (1990). Genetic effects in human energy expenditure components. *Int. J. Obes.* **14**, 49–55.
49. Cardon, L. R., Carmelli, R. D., Fabsitz, R. R., and Reed, T. (1994). Genetic and environmental correlations between obesity and body fat distribution in adult male twins. *Hum. Biol.* **66**, 465–479.
50. Hellenius, M. L., de Faire, U. H., Berglund, B. H., Hamsten, A., and Krakau, I. (1993). Diet and exercise are equally effective in reducing risk for cardiovascular disease. Results of a randomized controlled study in men with slightly to moderately raised cardiovascular risk factors. *Atherosclerosis* **103**, 81–91.
51. Pritchard, J. E., Nowson, C. A., and Wark, J. D. (1997). A worksite program for overweight middle-aged men achieves lesser weight loss with exercise than with dietary change. *J. Am. Diet. Assoc.* **97**, 37–42.
52. King, A. C., Haskell, W. L., Taylor, C. B., Kraemer, H. C., and DeBusk, R. F. (1991). Group- vs home-based exercise training in healthy older men and women: A community-based clinical trial. *JAMA* **266**, 1535–1542.
53. Anderssen, S., Holme, I., Urdal, P., and Hjermann, I. (1995). Diet and exercise intervention have favourable effects on blood pressure in mild hypertensives: The Oslo Diet and Exercise Study (ODES). *Blood Press.* **4**, 343–349.
54. Stefanick, M. L., Mackey, S., Sheehan, M., Ellsworth, N., Haskell, W. L., and Wood, P. D. (1998). Effects of diet and exercise in men and postmenopausal women with low levels of HDL cholesterol and high levels of LDL cholesterol. *N. Engl. J. Med.* **339**, 12–20.
55. Wood, P. D., Stefanick, M. L., Dreon, D. M., Frey-Hewit, B., Garay, S. C., Williams, P. T., Superko, H. R., Fortmann, S. P., Albers, J. J., Vranizan, K. M., Ellsworth, N. M., Terry, R. B., and Haskell, W. L. (1988). Changes in plasma lipids and lipoproteins in overweight men during weight loss through dieting as compared with exercise. *N. Engl. J. Med.* **319**, 1173–1179.
56. Katzel, L. I., Bleecker, E. T., Colman, E. G., Rogus, E. M., Sorking, J. D., and Goldberg, A. P. (1995). Effects of weight loss vs aerobic exercise training on risk factors for

coronary disease in healthy, obese, middle-aged and older men: A randomized controlled trial. *JAMA* **274**, 1915–1921.

57. Sopko, G., Leon, A. S., Jacobs, D. R., Foster, N., Moy, J., Kuba, K., Anderson, J. T., Casal, D., McNally, C., and Frantz, I. (1985). The effects of exercise and weight loss on plasma lipids in young obese men. *Metabolism* **34**, 227–236.
58. Ready, A. E., Drinkwater, D. T., Ducas, J., Fitzpatrick, D. W., Brereton, D. G., and Oades, S. C. (1995). Walking program reduces elevated cholesterol in women postmenopause. *Can. J. Cardiol.* **11**, 905–912.
59. Ronnemaa, T., Mattila, K., Lehtonen, A., and Kallio, V. (1986). A controlled randomized study on the effect of long-term physical exercise on the metabolic control in type 2 diabetic patients. *Acta Med. Scand.* **220**, 219–224.
60. Raz, I., Hauser, E., and Bursztyn, M. (1994). Moderate exercise improves glucose metabolism in uncontrolled elderly patients with non-insulin-dependent diabetes mellitus. *Isr. J. Med. Sci.* **30**, 766–770.
61. Ross, R., Dagnone, D., Jones, P. J. H., Smith, H., Paddags, A., Hudson, R., and Janssen, I. (2000). Reduction in obesity and related comorbid conditions after diet-induced weight loss or exercise-induced weight loss in men—a randomized controlled trial. *Ann Intern Med,* **133**, 92–103.
62. Irwin, M. L., Yasui, Y., Ulrich, C. M., Bowen, D., Rudolph, R. E., Schwartz, R. S., Yukawa, M., Aiello, E., Potter, J. D., and McTiernan, A. (2003). Effect of exercise on total and intra-abdominal body fat in postmenopausal women—a randomized controlled trial. *JAMA* **289**, 323–330.
63. Donnelly, J. E., Hill, J. O., Jacobsen, D. J., Potteiger, J., Sullivan, D. K., Johnson, S. L., Heelan, K., Hise, M., Fennessey, P. V., Sonko, B., Sharp, T., Jakicic, J. M., Blair, S. N., Tran, Z. V., Mayo, M., Gibson, C., and Washburn, R. A. (2003). Effects of a 16-month randomized controlled exercise trial on body weight and composition in young, overweight men and women. *Arch. Intern. Med.* **163**, 1343–1350.
64. Slentz, D. A., Duscha, B. D., Johnson, J. L., Ketchum, K., Atken, L. B., Samsa, G. P., Houmard, J. A., Bales, C. W., and Kraus, W. E. (2004). Effects of the amount of exercise on body weight, body composition, and measures of central obesity: STRRIDE—A randomized controlled study. *Arch. Intern. Med.* **164**, 31–39.
65. Wood, P. D., Stefanick, M. L., Williams, P. T., and Haskell, W. L. (1991). The effects on plasma lipoproteins of a prudent weight-reducing diet, with or without exercise, in overweight men and women. *N. Engl. J. Med.* **325**, 461–466.
66. Svendsen, O. L., Hassager, C., and Christiansen, C. (1993). Effect of an energy-restrictive diet with or without exercise on lean tissue, resting metabolic rate, cardiovascular risk factors, and bone in overweight postmenopausal women. *Am. J. Med.* **95**, 131–140.
67. Hammer, R. L., Barrier, C. A., Roundy, E. S., Bradford, J. M., and Fisher, A. G. (1989). Calorie-restricted low-fat diet and exercise in obese women. *Am. J. Clin. Nutr.* **49**, 77–85.
68. Blonk, M. C., Jacobs, M. A. J. M., Biesheuvel, E. H. E., Weeda-Mannak, W. L., and Heine, R. J. (1994). Influences on weight loss in type 2 diabetic patients: Little long-term benefit from group behaviour therapy and exercise training. *Diab. Med.* **11**, 449–457.
69. Marks, B. L., Ward, A., Morris, D. H., Castellani, J., and Rippe, J. M. (1995). Fat-free mass is maintained in women following a moderate diet and exercise program. *Med. Sci. Sports Exerc.* **27**, 1243–1251.
70. Wing, R. R., Epstein, L. H., Paternostro-Bayles, M., Nowalk, M. P., and Gooding, W. (1988). Exercise in a behavioural weight control programme for obese patients with Type 2 (noninsulin-dependent) diabetes. *Diabetologia* **31**, 902–909.
71. Andersen, R. E., Wadden, T. A., Bartlett, S. J., Vogt, R. A., and Weinstock, R. S. (1995). Relation of weight loss to changes in serum lipids and lipoproteins in obese women. *Am. J. Clin. Nutr.* **62**, 350–357.
72. Gordon, N. F., Scott, C. B., and Levine, B. D. (1997). Comparison of single versus multiple lifestyle interventions: Are the antihypertensive effects of exercise training and diet-induced weight loss additive? *Am. J. Cardiol.* **79**, 763–767.
73. Wadden, T. A., Vogt, R. A., Andersen, R. E., Bartlett, S. J., and Foster, G. D. (1997). Exercise in the treatment of obesity: Effects of four interventions on body composition, resting energy expenditure, appetite, and mood. *J. Consult. Clin. Pschyol.* **65**, 269–277.
74. Hill, J. O., Schlundt, D. G., Sbrocco, T., Sharp, T., Pope-Cordle, J., Stetson, B., Kaler, M., and Heim, C. (1989). Evaluation of an alternating-calorie diet with and without exercise in the treatment of obesity. *Am. J. Clin. Nutr.* **50**, 248–254.
75. Bertram, S. R., Venter, I., and Stewart, R. I. (1990). Weight loss in obese women—exercise v. dietary education. *S. Afr. Med. J.* **78**, 15–18.
76. Sweeney, M. E., Hill, J. O., Heller, P. A., Baney, R., and Di-Girolamo, M. (1993). Severe vs moderate energy restriction with and without exercise in the treatment of obesity: Efficiency of weight loss. *Am. J. Clin. Nutr.* **57**, 127–134.
77. Dengel, D. R., Hagberg, J. M., Coon, P. J., Drinkwater, D. T., and Goldberg, A. P. (1994). Effects of weight loss by diet alone or combined with aerobic exercise on body composition in older obese men. *Metabolism* **43**, 867–871.
78. Fox, A. A., Thompson, J. L., Butterfield, G. E., Gylfadottir, U., Moynihan, S., and Spiller, G. (1996). Effects of diet and exercise on common cardiovascular disease risk factors in moderately obese older women. *Am. J. Clin. Nutr.* **63**, 225–233.
79. Evans, E. M., Saunders, M. J., Spano, M. A., Arngrimsson, S. A., Lewis, R. D., and Cureton, K. J. (1999). Effects of diet and exercise on the density and composition of the fat-free mass in obese women. *Med. Sci. Sports Exerc.* **31**, 1778–1787.
80. Donnelly, J. E., Pronk, N. P., Jacobsen, D. J., Pronk, S. J., and Jakicic, J. M. (1991). Effects of a very-low-calorie diet and physical-training regimens on body composition and resting metabolic rate in obese females. *Am. J. Clin. Nutr.* **54**, 56–61.
81. Jakicic, J. M., Wing, R. R., Butler, B. A., and Robertson, R. J. (1995). Prescribing exercise in multiple short bouts versus one continuous bout: Effects on adherence, cardiorespiratory fitness, and weight loss in overweight women. *Int. J. Obes.* **19**, 893–901.
82. Jakicic, J. M., Winters, C., Lang, W., and Wing, R. R. (1999). Effects of intermittent exercise and use of home exercise equipment on adherence, weight loss, and fitness in over weight women: A randomized trial. *JAMA* **282**, 1554–1560.
83. Andersen, R. E., Wadden, T. A., Bartlett, S. J., Zemel, B., Verde, T. J., and Franckowiak, S. C. (1999). Effects of

lifestyle activity vs structured aerobic exercise in obese women: A randomized trial. *JAMA* **281**, 335–340.

84. Bond Brill, J., Perry, A. C., Parker, L., Robinson, A., and Burnett, K. (2002). Dose-response effect of walking exercise on weight loss. How much is enough? *Int. J. Obes. Relat. Metab. Disord.* **26**, 1484–1493.
85. Jakicic, J. M., Marcus, B. H., Gallagher, K. I., Napolitano, M., and Lang, W. (2003). Effects of exercise duration and intensity on weight loss in overweight, sedentary women: A randomized trial. *JAMA* **290**, 1323–1330.
86. Dunn, A. L., Marcus, B. H., Kampert, J. B., Garcia, M. E., Kohl, H. W., III, and Blair, S. N. (1999). Comparison of lifestyle and structured interventions to increase physical activity and cardiorespiratory fitness: A randomized trial. *JAMA* **281**, 327–334.
87. Jeffery, R. W., Wing, R. R., Sherwood, N. E., and Tate, D. F. (2003). Physical activity and weight loss: Does prescribing higher physical activity goals improve outcome? *Am. J. Clin. Nutr.* **78**, 684–689.
88. Donnelly, J. E., Jacobsen, D. J., Snyder Heelan, K., Seip, R., and Smith, S. (2000). The effects of 18 months of intermittent vs continuous exercise on aerobic capacity, body weight and composition, and metabolic fitness in previously sedentary, moderately obese females. *Int. J. Obes.* **24**, 566–572.
89. American College of Sports Medicine. (1998). The recommended quantity and quality of exercise for developing and maintaining cardiorespiratory and muscular fitness, and flexibility in healthy adults. *Med. Sci. Sports Exerc.* **30**, 975–991.
90. Ross, R., and Rissanen, J. (1994). Mobilization of visceral and subcutaneous adipose tissue in response to energy restriction and exercise. *Am. J. Clin. Nutr.* **60**, 695–703.
91. Manning, J. M., Dooly-Manning, C. R., White, K., Kampa, L, Silas, S., Kesselhaut, M., and Ruoff, M. (1991). Effects of a resistive training program on lipoprotein-lipid levels in obese women. *Med. Sci. Sports Exerc.* **23**, 1222–1226.
92. Miller, W. C., Koceja, D. M., and Hamilton, E. J. (1997). A meta-analysis of the past 25 years of weight loss research using diet, exercise or diet plus exercise intervention. *Int. J. Obes.* **21**, 941–947.
93. Anderson, J. W., Konz, E. C., Frederich, R. C., and Wood, C. L. (2001). Long-term weight-loss maintenance: a meta-analysis of US studies. *Am. J. Clin. Nutr.* **74**, 579–584.
94. King, A. C., Haskell, W. L., Young, D. R., Oka, R. K., and Stefanick, M. L. (1995). Long-term effects of varying intensities and formats of physical activity on participation rates fitness, and lipoproteins in men and women aged 50–65 years. *Circulation* **91**, 2596–2604.
95. Svendsen, O. L., Hassager, C., and Christiansen, C. (1994). Six months' follow-up on exercise added to a short-term diet in overweight postmenopausal women—Effects on body composition, resting metabolic rate, cardiovascular risk factors and bone. *Int. J. Obes.* **18**, 692–698.
96. Wadden, T. A., Vogt, R. A., Foster, G. D., and Anderson, D. A. (1998). Exercise and the maintenance of weight loss: 1-year follow-up of a controlled clinical trial. *J. Consult. Clin. Pschyol.* **66**, 429–433.
97. Skender, M. L., Goodrick, G. K., Del Junco, D. J., Reeves, R. S., Darnell, L., Gotto, A. M., and Foreyt, J. P. (1996). Comparison of 2-year weight loss trends in behavioral treatments of obesity, diet, exercise, and combination interventions. *J. Am. Diet. Assoc.* **96**, 342–346.
98. Fogelholm, M., Kukkonen-Harjula, K., Nenonen, A., and Pasanen, M. (2000). Effects of walking training on weight maintenance after a very-low-energy diet in premenopausal obese women—a randomized controlled trial. *Arch. Intern. Med.* **160**, 2177–2184.
99. Tate, D. F., Jeffery, R. W., Sherwood, N. E., and Wing, R. R. (2007). Long-term weight losses associated with prescription of higher physical activity goals. Are higher levels of physical activity protective against weigh regain? *Am. J. Clin. Nutr.* **85**, 954–959.
100. Kayman, S., Bruvol, W., and Stern, J. S. (1990). Maintenance and relapse after weight loss in women: Behavioral aspects. *Am. J. Clin. Nutr.* **52**, 800–807.
101. Klem, M. L., Wing, R. R., McGuire, M. T., Seagle, H. M., and Hill, J. O. (1997). A descriptive study of individuals successful at long-term maintenance of substantial weight loss. *Am. J. Clin. Nutr.* **66**, 239–246.
102. Wing, R. R., and Hill, J. O. (2001). Successful weight loss maintenance. *Annu. Rev. Nutr.* **21**, 323–341.
103. Ross, R., and Janssen, I. (2001). Physical activity, total and regional adiposity: dose-response considerations. *Med. Sci. Sports Exerc.* **33** (6 Suppl). S521–S527.
104. Diabetes Prevention Program Research Group. (2002). Reduction in the incidence of type 2 diabetes with lifestyle intervention or metformin. *N. Engl. J. Med.* **346**, 393–403.
105. The Look AHEAD Research Group. (2006). The Look AHEAD Study: A description of the lifestyle intervention and the evidence supporting it. *Obesity* **14**, 737–752.

CHAPTER **24**

Macronutrient Intake and the Control of Body Weight

DAVID A. LEVITSKY
Cornell University, Ithaca, New York

Contents

I. INTRODUCTION

Scholars of the control of food intake and the regulation of body weight have given little attention to the role played by macronutrients until fairly recently. Most researchers in this field were convinced by the dictum issued by the famous physiologist, Adolph [1] , who declared that "Within limits, rats eat for calories," a conclusion readily transferred to humans as well. Although Mayer [2] and Mellinkoff *et al.* [3] tried to argue particular roles for their pet nutrients (glucose and amino acids, respectively), the domination of the field by the set-point theories of body weight [4–17] shifted the focus of research on eating behavior from dietary variables to physiology in the search for those mechanisms that controlled body weight through the control of food intake.

Meanwhile, the public was being fed (sold) great dreams of easy weight reduction and magical cures by eating certain macronutrients and avoiding others. In 1972, Robert Atkins was one of the first of a series of diet gurus to pontificate on the virtues of eating a high-protein, low-carbohydrate, high-fat diet as a means of losing weight and curing obesity. He was followed by a rash of other "non-nutritionist" experts who pushed their own versions of the single macronutrient theory (carbohydrate is the culprit) and made lots of money and converts of the public on the way to the bank.

Fortunately, in the past 30 years the medical/nutritional establishment has begun to listen to the public's questions and has generated a considerable amount of research on the role macronutrients play in the control of food intake and body weight. In some areas, there is almost complete agreement. In others, the answers are split almost down the middle. The purpose of this chapter is to review this literature and reveal what we do and do not know about the role played by macronutrients in determining energy intake and, ultimately, body weight.

II. FAT CHANCE

The macronutrient that has received the most attention in the scientific appetite and body weight literature is dietary fat. The relationship between dietary fat and body weight has been examined at almost every level of analysis from molecular to epidemiological. Using country as the unit of analysis and national food disappearance data as a surrogate for energy intake, Bray and Popkin [18] observed a fairly strong relationship between percent of the population who are overweight and the dietary fat expressed as a percent of the total energy as shown in Figure 1. A similar function was found by West and Kalbfleisch [19] in which 11 underdeveloped countries were examined and earlier by Keys [20] in his famous seven-country study. However, no relationship was observed in a study of percent calories as fat and body weight in males in 18 European countries and a negative relationship was observed in females [21]. One possible reason for this discrepancy in findings is that the range of fat values for the European studies was almost all at the high end of distribution ranging from 27% to 46% energy as fat with the majority of studies having fat intakes greater than 35% calories as fat, whereas the majority of countries analyzed by Bray and Popkin [18], by West and Kalbfleisch, [19] and by Keys [20] consumed a diet

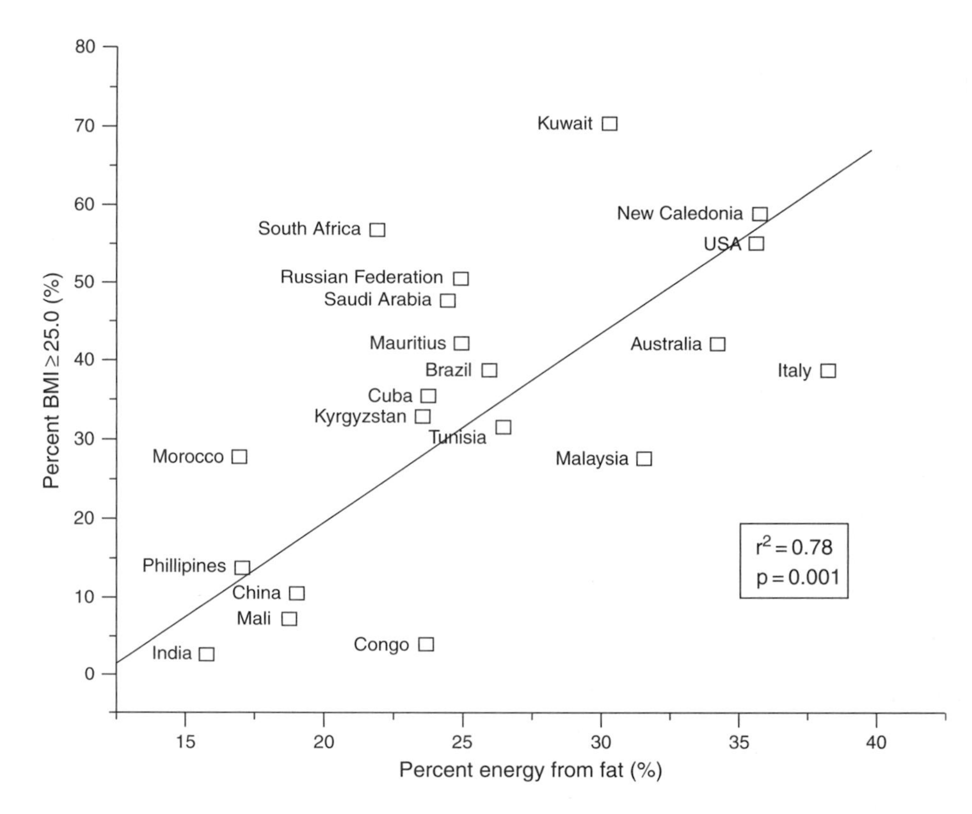

FIGURE 1 Cross-country depiction of the percent daily calories consumed as fat and body weight. (Adapted from Bray, G. A., and Popkin, B. M. (1998). Dietary fat intake does affect obesity! *Am. J. Clin. Nutr.* **68,** 1157–1173.)

consisting of less than 35% calories as fat. Moreover, the five countries that contributed most to the significant negative correlation among women were Eastern European, and the accuracy of the food disappearance data on which the dietary intake is based is questionable. Indeed when these five points are removed from the data, the correlation becomes no longer significant.

The epidemiological literature on the relationship between fat intake and body weight using individuals as the unit of analysis was critically reviewed by Lissner and Heitmann [22]. Although the data are not entirely consistent, they concluded that the greater the amount of dietary fat humans consume, the greater their body weight. Although the preponderance of the data indicates a direct relationship between fat consumption and body weight, they probably *underestimated* the association between dietary fat and body weight for two reasons. First, poor measures of food consumption lead to an underestimation of the magnitude of the effect, and most measures of food consumption in the home environment are quite poor. Even more serious, though, is the problem that the underestimation of fat intake and energy is directly related to body weight. The larger the person, the more greatly he or she underestimates his or her energy intake [23–29], particularly fat intake [23, 30]. Expressing fat intake as a percent of energy ingested reduces this problem somewhat, but not entirely. Despite these limitations, the literature supports the conclusions reached by Lissner and Heitmann [22]. As depicted in Figure 2 the preponderance of studies that have addressed this issue have found that a positive and significant relationship exists between dietary fat and body weight and/or fat composition. The greater the amount of fat consumed, the greater the amount of body fat.

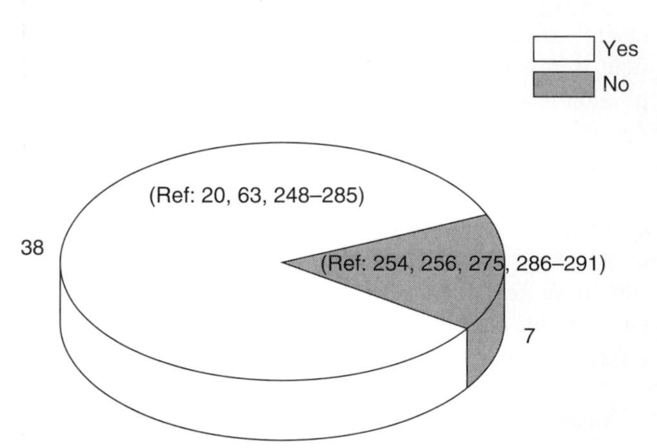

FIGURE 2 Distribution of ecological studies relating body composition or change in body composition to dietary fat. Note: Howarth *et al.* [224] observed the correlation in young people, but not older ones. Jackson *et al.* [225] and Guillaume *et al.* [226] found correlation in males, but not females.

III. ESTABLISHING CAUSAL LINKS

However, because the epidemiological data just reported are correlational, they do not prove that eating a diet rich in fat causes an increase in body weight. It is possible that having a body composed of a high percent of fat causes an increased preference for dietary fat. Establishing a causal link between dietary fat and body weight requires an experimental design where the amount of fat that people consume must be experimentally manipulated and its effects on food intake or body weight (fat) measured. Two kinds of studies have been used to examine the causal links between macronutrient intake and body weight: laboratory studies and clinical studies.

A. Laboratory Studies

Unfortunately, studies of the effects of macronutrients in the laboratory rarely are of long enough duration to observe changes in body weight or fat composition. The dependent measure in most of these studies is the amount of food consumed. In many ways, this is a far more accurate measure of energy balance than body weight or even body composition measures, particularly if the design utilizes each subject as his or her own control. Any inaccuracy in estimating the energy or nutrient content of the food is nullified because each subject is tested under all conditions.

Figure 3 illustrates the results of studies on the effects of alterations in dietary fat on energy consumption. Although the weight of the evidence appears to support the epidemiological studies (people consume fewer calories when the diet contains less fat) important and interesting methodological differences exist that are cause for concern.

Most of the studies that failed to find that humans change the amount of food they consume (compensate) for changes in dietary fat had altered the fat content of all the foods that were accessible to the subjects. For most of these studies, all foods offered to the subjects were approximately of the same relative fat content. Therefore, when the fat content was reduced, the fat content of all foods was decreased. Consequently, the only way subjects could avoid reducing their daily energy intake when faced with a low-fat diet was to increase the amount of lower fat foods that they consumed.

Most of those studies that found subjects do compensate for alterations in dietary fat manipulated the fat content of only a single meal such as lunch or changed the fat content of only particular foods in the diet. Poppitt and Swann [31] provided a direct comparison of the two methodologies within a single study. Their results are replotted in Figure 4. Under conditions where the fat content of only the lunch was manipulated (left panel), subjects appear to energetically compensate by changing the amount of food they consumed. However, as the right panel indicates, when the fat content of all the food is altered, humans do not compensate for the reduced energy content and, therefore, energy intake is diminished.

This dependence of finding energetic compensation on specific methodology is quite important, but not understood. More importantly, if energy compensation for low-fat foods does exist when subjects have free access to foods of different fat composition, as suggested by these studies, then eating low-fat foods in the "real world" will never succeed as a strategy to chronically suppress body weight, unless one eats only low-fat foods. The examination of clinical studies, however, indicates that people do lose weight when they use some low-fat foods, suggesting that the laboratory conditions or procedures may produce an artifact making it difficult to extrapolate the results to the

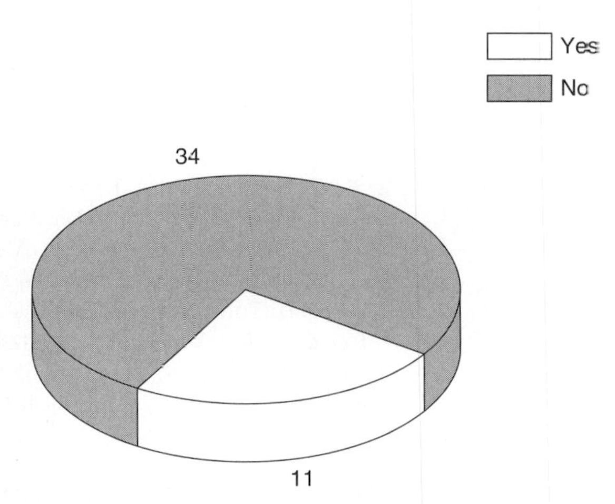

FIGURE 3 Distribution of experimental studies on the effects of dietary fat on food intake.

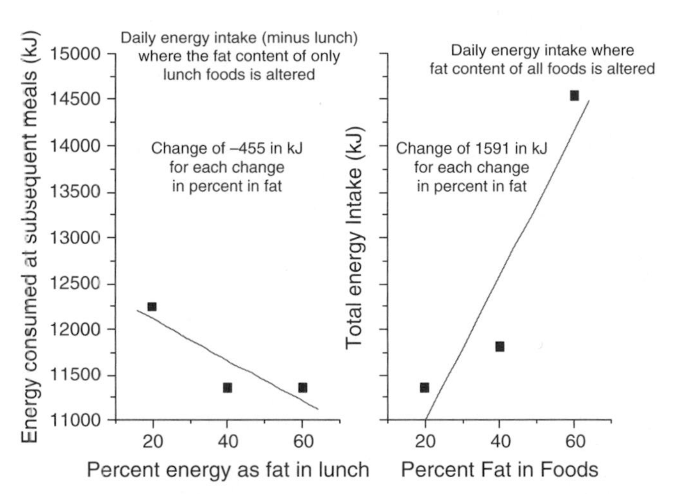

FIGURE 4 Energy intake as a function of level of dietary fat. (Left) Only the fat content of the lunch was varied. (Right) Fat content of all foods were varied. (Adapted from Poppitt, S. D., and Swann, D. L. (1998). Dietary manipulation and energy compensation. *Int. J. Obes. Relat. Metab. Disord.* **22,** 1024–1031.)

"real world." It is also possible that if conditions are sufficiently well controlled in the laboratory, people will demonstrate physiological regulation, but that the external variables associated with eating behavior in humans are so powerful as to easily obscure its demonstration.

B. Clinical Studies

Clinical studies are more realistic than laboratory studies. However, it is usually quite difficult to obtain weighed food intake measures from subjects in a clinical study as a measure of energy intake. Therefore, such studies require subjects to prepare their own meals and record what they eat. Such records, unfortunately, involve large measurement errors. As a consequence, clinical studies rely on physiological measures to corroborate their energy intake measures such as body weight or blood lipid concentrations.

Figure 5 shows the results of 34 published clinical studies in which the fat content of the diet was changed and body weights of subjects in the community were measured at the beginning and the end of the study. This figure clearly illustrates that changes in dietary fat in the "real world" are not totally compensated for by an accurate adjustment of energy intake. A reduction of 1% in the fat composition of the diet results in an average weight loss of 0.26 kg, a value slightly lower than that derived from a meta-analysis of dietary interventions performed by Astrup *et al.* (0.37 kg weight loss for every 1% decrease in dietary fat) [32] but very close to a value of weight loss of 0.28 kg for every percent decrease in total calories as fat derived from an independent meta-analysis by Yu-Poth *et al.* [33].

It appears, therefore, that body weight is directly related to the amount of dietary fat consumed in the diet. But if that is correct, then why does the "American paradox" exist? The American paradox, or the fat paradox, refers to the often-quoted statement (uttered mostly by proponents of low carbohydrate diet; see later discussion) that despite the fact that we are consuming less dietary fat, we are still getting fatter [34]. The panel of Figure 6 illustrates this paradox. The data are taken from two sources. Intake is approximated from the USDA Food Availability Tables and represents the total fat produced in the United States per capita from 1909 until 2004 expressed as a percent to total daily intake. The bars represent adjusted body weights obtained from NHANES survey at four time points. Mean body weight started to increase in the early 1980s at a time when fat intake, expressed as a percent of total daily energy intake, began to decline. The middle panel in Figure 6 shows the same mean adjusted body weights but now superimposed on total daily energy intake. This figure demonstrates very clearly that the reason the population is gaining weight is directly related to the increase in energy intake. Moreover, absolute fat intake never declined during the 1980s and 1990s, as can be seen in the bottom panel of Figure 6. The reason for the decline in percent calories consumed as fat is that the amount of carbohydrate consumed during this period increased (see Figure 12, later) sufficiently to cause the total energy intake to increase as illustrated in the middle panel. What appeared to be paradoxical (the decrease in fat intake) was really a result of expressing fat intake as a percent of total calories consumed. It should be noted from the bottom panel of Figure 6 that the

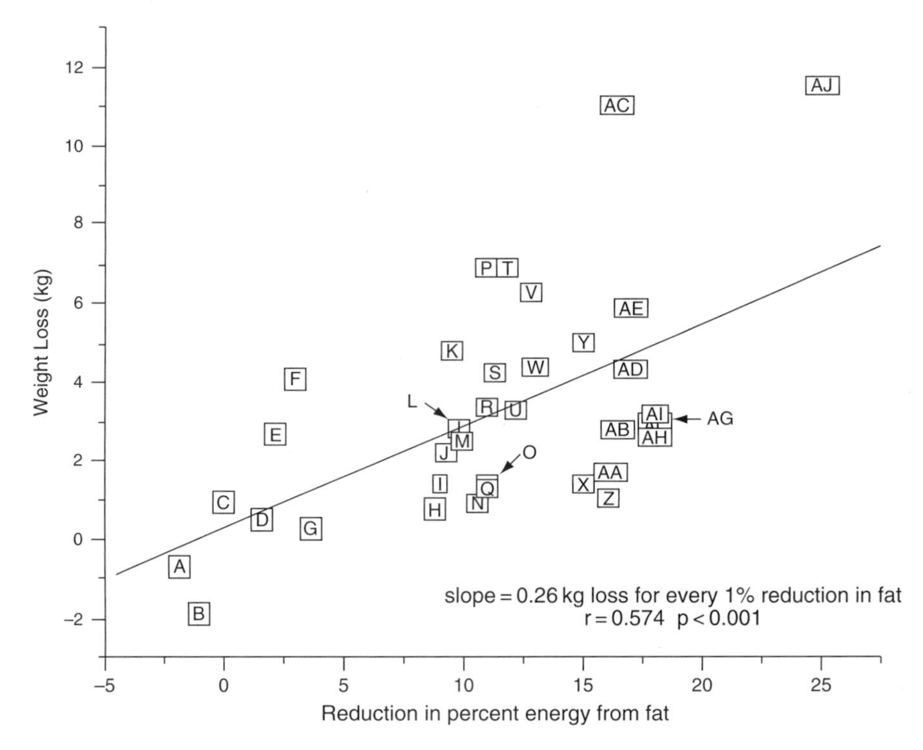

FIGURE 5 Maximum weight loss as a function of percent calories from fat. A, Archer *et al.* [227]; B, Black *et al.* [228]; C, Bloemberg *et al.* [229]; D, Boyd *et al.* [230]; E, Buzzard *et al.* [231]; F, Chlebowski *et al.* [232]; G, Djuric *et al.* [233]; H, Howard *et al.* [234]; I, Hunninghake *et al.* [235]; J, Insull *et al.* [236]; K, Jeffery *et al.* [237]; L, Kasim *et al.* [238]; M, Kendall *et al.* [79]; N, Knopp *et al.* [239]; O, Lee-Han *et al.* [240]; P, Marckmann [241]; Q, McManus *et al.* [242]; R, Ornish *et al.* [243]; S, Peterson *et al.* [244]; T, Poppitt *et al.* [245]; U, Pritchard *et al.* [246]; V, Raben *et al.* [247]; W, Rock *et al.* [248]; X, Saris *et al.* [96]; Y, Schlundt *et al.* [249]; Z, Segal-Isaacson *et al.* [250]; AA, Shah *et al.* [251]; AB, Sheppard *et al.* [252]; AC, Siggaard *et al.* [253]; AD, Simon *et al.* [254]; AE, Singh *et al.* [255]; AF, Sorensen *et al.* [256]; AG, Stefanick *et al.* [257]; AH, Thompson *et al.* [258]; AI, Toubro and Astrup [259]; AJ, Winters *et al.* [260].

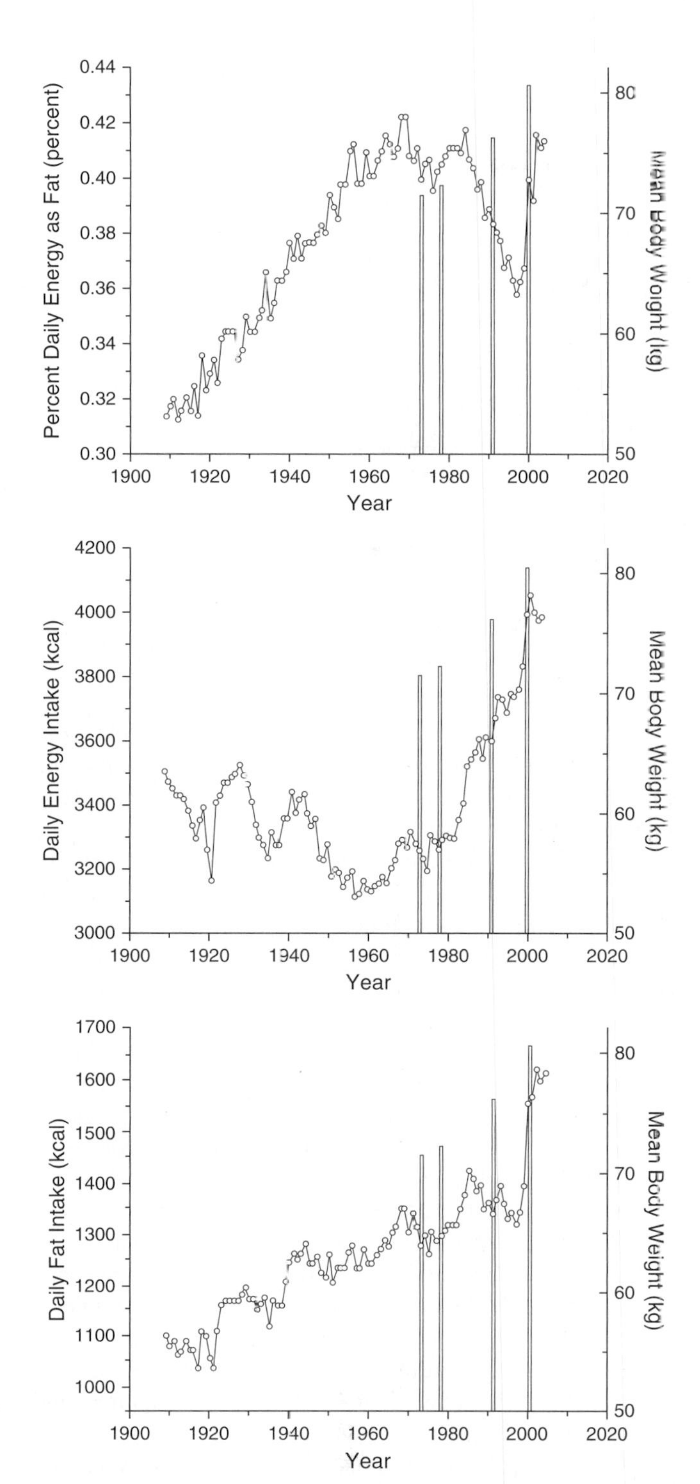

FIGURE 6 (Top panel) Daily fat consumption per person per day expressed as a percent of total daily intake from 1907 until 2004 (from USDA Food Availability Data [261]) and the age-adjusted mean body weight of the population taken from NHANES surveys (from Zhang and Wang [262]). (Middle panel) Daily energy intake per person per day expressed as kcal from 1907 until 2004 (from USDA Food Availability Data [261]) and the mean body weight of the population (from Zhang and Wang [262]). (Bottom panel) Daily fat intake per person per day expressed as kcal from 1907 until 2004 (from USDA Food Availability Data [261]) and the mean body weight of the population (from Zhang and Wang [262]).

amount of fat consumed per capita has been increasing every year since records began and increased remarkably from the year 2000 until 2004, when the last data point was published. This fact plus the observation that current surveillance has not detected any decrease in body weight [35] or energy intake (see Figure 6, middle panel) suggests that the prevalence of overweight and obesity cannot be expected to decrease in the near future.

The food production data are remarkably consistent with the majority of laboratory and epidemiological studies indicating that the amount of calories consumed is directly related to the fat content of the foods humans consume. There does not appear to be any physiological mechanism operating in humans to prevent overconsuming energy when confronted with a diet high in fat.

Despite the high degree of congruent data relating the consumption of dietary fat to body weight and adiposity, there are several perplexing issues that must be solved before we suggest to the public that the only thing they have to do to lose weight is to reduce their intake of dietary fat. A number of studies have examined the correlation between the amount of dietary fat consumed and subsequent weight gain over long periods of time. These studies are shown graphically in Figure 7. The literature is evenly split—half find increased weight gain with increasing amounts of dietary fat and half do not. If the previous data concerning dietary fat and body weight are correct, why would you not expect to find weight gain to be related to the amount of food consumed?

There are a number of possible explanations. First, it is possible that when tested at the first point in time, the individual consuming a habitually high-fat diet could be in energy balance. Therefore, one would not expect that the continued consumption of high-fat foods would

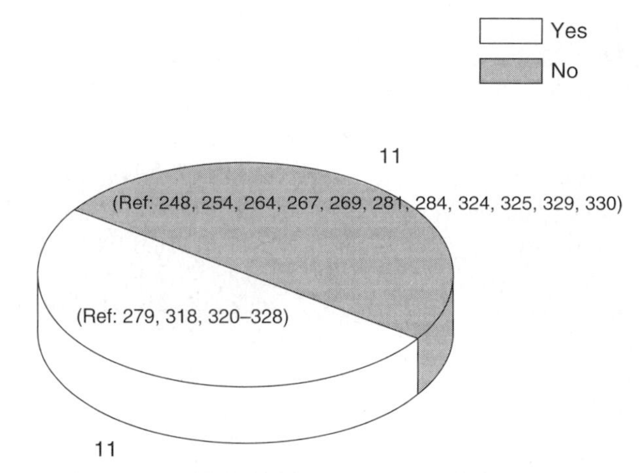

FIGURE 7 Distribution of longitudinal studies relating dietary fat intake to weight gain. Kant *et al.* [263] found that fat intake predicted weight gain in males, but not in females.

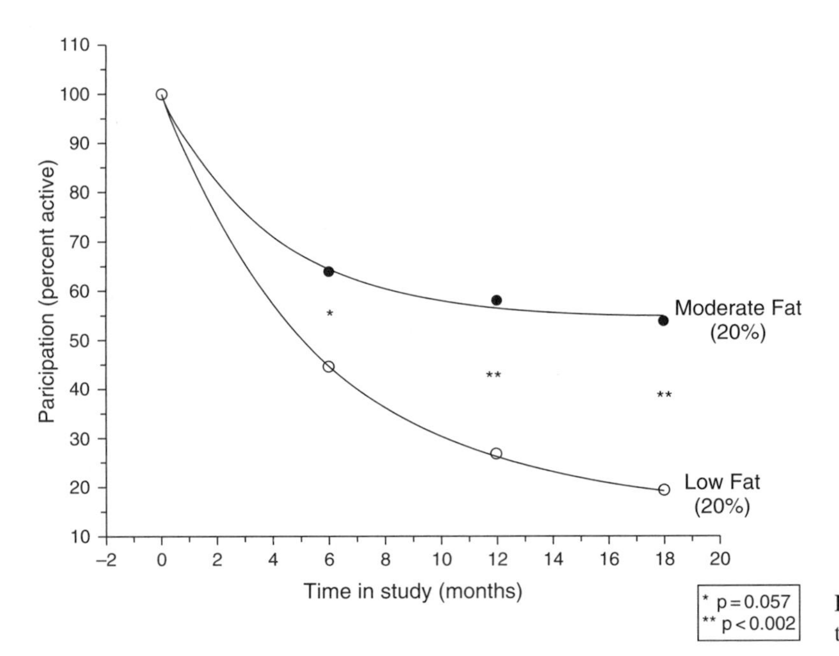

FIGURE 8 Participation rates as a function of time in the study. (Adapted from McManus *et al.* [242].)

necessarily promote any increase in body weight unless the individual increased fat consumption even further. Second, the problem of underreporting fat looms over such studies that do not find an effect of dietary fat on weight gain.

Third, and most disappointing, is the fact that people do not seem able to continue to eat a low-fat diet for long periods of time. Although short-term studies of changing fat content clearly demonstrate an effect on body weight, long-term studies of weight loss have been disappointingly small. The average body weight of subjects consuming a low-fat diet for 12 months or greater is about 1 kg less than when they started [21]. One reason for such small effects may be in the data provided by an 18-month study of the effect of a low-fat diet on weight loss in overweight adults. Figure 8 shows their results. The rate of participation of the subjects in their study decreased exponentially with the duration of the study. The rate of participation of the low-fat group, however, decreased twice as rapidly as the moderate-fat group, dropping to 20% of the original sample by 18 months.

What is responsible for the low rate of participation for subjects on the low-fat diet is not clear. Even more puzzling, however, is that maintaining a reduced fat intake is a common characteristic of people who have lost weight and have sustained that weight loss for at least 1 year [36–39].

IV. IS IT FAT OR ENERGY DENSITY?

The fact that high-fat diets cause an increase in energy intake and obesity in animals has been evident for a long time in the animal feeding literature (see West and York [40] for a thorough review). Whereas this effect was thought to occur because of some unique property relating to the chemical structure of fat, Ramirez and Friedman [41] performed a series of interesting studies demonstrating that the excessive energy intake was due to an increase in the energy density of the diet rather than to the fat content per se.

In one of the few human studies that failed to observe an increase in energy intake with increasing dietary fat, van Stratum *et al.* [42] examined the effects of introducing liquid diets that varied in the amount of fat and carbohydrate they contained, but were of equal energy density. They failed to find a difference in total energy intake between the two liquid diets either in the amount of liquid diet consumed or the amount of solid food eaten during the rest of the day. Stubbs *et al.* [43] reported similar results using a more elegant experimental design in which the amount of dietary fat was varied (20%, 40%, and 59%), but where the energy density of the three diets was maintained constant. A 3-day menu rotation was used for 14 days and was repeated three times; each time all the foods contained one of the three amounts of dietary fat. Subjects were free to eat as much or as little as they desired. The results of this extraordinary study are shown in Figure 9 and are very consistent with the work of Ramirez and Friedman [41] in animals and with van Stratum *et al.* [42] in humans. Total energy intake did not differ between any condition despite the large differences between fat and carbohydrate. The study quite clearly demonstrates that the reason humans (and animals) overeat on a high-fat diet is because of the high energy density of fat and not because of any unique metabolic or physical property of high dietary fat.

In the laboratory, the idea that energy density is the cause of overeating on a high-fat diet has been firmly confirmed and broadened to include changes in energy density created by diluting with water, air, and fiber

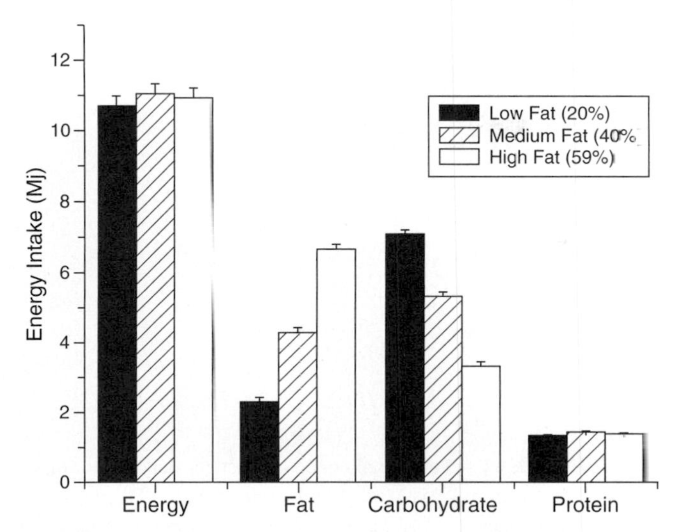

FIGURE 9 Intake of energy and macronutrients as a function of dietary fat when the energy density of the diet is held constant. (Adapted from Stubbs, R. J., Harbron, C. G., and Prentice, A. M. (1996). Covert manipulation of the dietary fat to carbohydrate ratio of isoenergetically dense diets. *Int. J. Obes. Relat. Metab. Disord.* **20,** 651–660.)

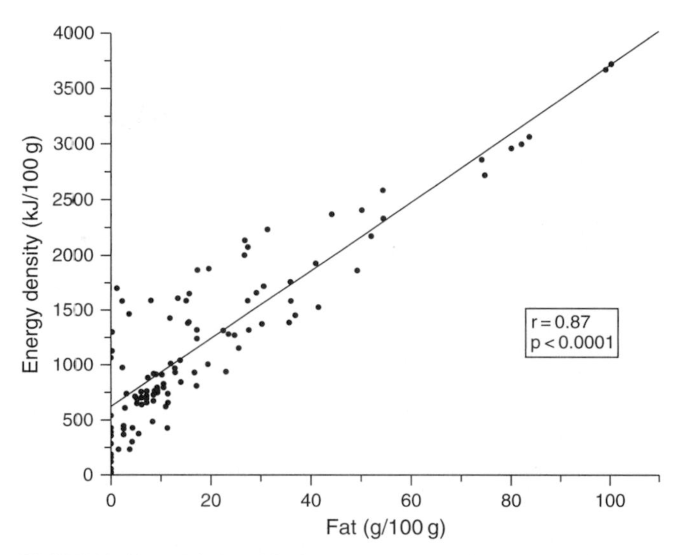

FIGURE 10 Relationship between energy density and fat concentration of 125 randomly selected foods. (Data from USDA (1996). "Continuing Survey of Food Intakes by Individuals (CSFII): Diet and Health Knowledge Survey, 1994" Department of Agriculture, Beltsville Human Nutrition Research Center/ARS, Beltsville, MD.)

[43–50]. The direct relationship between the amount of energy consumed and the energy density of foods eaten is one of the most consistently reported functions in the eating literature [51–74]. The one exception may be small children who respond to reduced energy density by increasing their intake [61], although others have been unable to confirm this effect [57]. Unfortunately epidemiological studies that have examined the relationship between energy density of foods people consume and their body weight are split: half have observed a significant correlation [58, 63, 69] while half have not [55, 75, 76].

Although this finding is of major theoretical and applied significance, in reality, the major determinant of energy density is the amount of fat in food. Figure 10 shows the relationship between the fat content and energy density for 125 randomly selected foods. The foods were consumed by subjects in the 1994 Continuing Survey of Food Intakes of Individuals conducted by the U.S. Department of Agriculture (USDA) [77]; similar functions have been reported by others [76, 78].

The most straightforward explanation for the effect of fat and energy density on human food intake is that humans tend to eat a constant volume of food [49, 79]. Implicit in this explanation is a notion that humans do not possess (or use) physiological–behavioral mechanisms that would be necessary for the precise regulation of energy balance [80]. Consequently, if the fat or energy density of the food supply is increased, the population has no choice but to gain weight. On the other hand, the scientific data are abundantly clear—one effective means to reduce energy intake is to eat foods low in fat content.

V. ARE CARBOHYDRATES THE CULPRIT RESPONSIBLE FOR OVERWEIGHT?

Although Figure 10 clearly indicates that energy density is determined primarily by dietary fat, there is a popular notion that carbohydrates are mainly responsible for overweight and obesity. Carbohydrates have been blamed as the culprit for obesity in such best-selling books as *Dine Out and Lose Weight: The French Way to Culinary "Savoir Vivre"* by Michel Montignac [81], *Sugar Busters! Cutting Sugar to Trim Fat* by H. Leighton Steward *et al.* [82], *Dr. Atkins' New Diet Revolution* by Robert C. Atkins [83] actually an evolution from his previous book, *Dr. Atkins' Diet Revolution: The High Calorie Way to Stay Thin Forever* [84], *The Zone* by Barry Sears and Bill Lawren [85], *The Carbohydrate Addict's Diet: The Lifelong Solution to Yo-Yo Dieting* by Rachael and Richard Heller [86], and *Protein Power: The High-Protein/Low-Carbohydrate Way to Lose Weight, Feel Fit, and Boost Your Health in Just Weeks* by Rachael and Michael Eades [87].

The basic premise behind all of these money-making books is that despite the fact that Americans are more concerned about fat than any other aspect of food, as is depicted in Figure 11 and have been reducing their consumption of fat (at least at home) [34], Americans are still getting fatter [35]. Therefore, these nutrition gurus argue that it is not dietary fat that is causing the overweight and obesity; it is the carbohydrate content that is the culprit.

The gurus were partially right; see Figure 12 for estimates of carbohydrate intake over time using the USDA

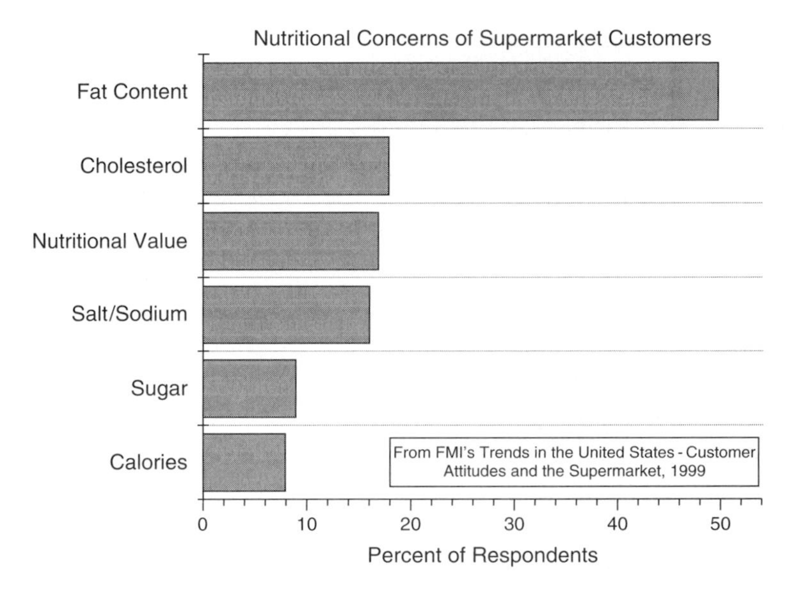

FIGURE 11 Nutritional concerns of supermarket customers. (Adapted from Food Marketing Institute (1999). "Trends in the United States—Consumer attitudes and the supermarket." Food Marketing Institute, Washington, DC.)

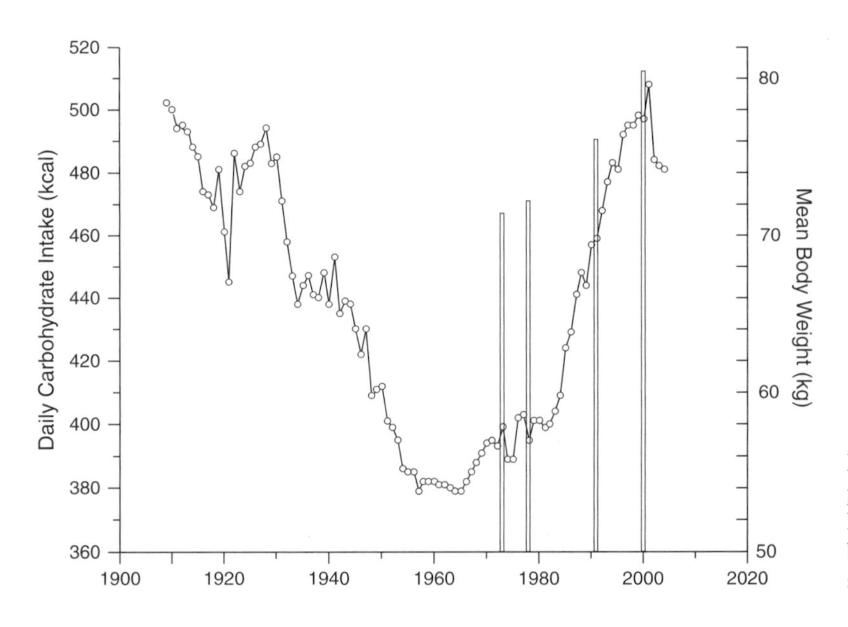

FIGURE 12 Daily carbohydrate consumption per person per day from 1907 until 2004 (from USDA Food Availability Data [261]) and age-adjusted mean body weight of the population taken from NHANES surveys (from Zhang and Wang [262]).

food availability data. It is clear that the amount of carbohydrate consumed per capita has been increasing since the early 1980s. However, what is rarely pointed out by these carbohydrate haters is that (a) absolute fat intake also has been rising during this period (see lower panel of Figure 6) and (b) carbohydrate intake initially decreased at the beginning of the twentieth century until about 1960 at a time where there is no evidence that body weight also decreased. Interestingly, in the most recent four years there has been a steady decrease in carbohydrate intake, but there is no sign of any decrease in body weight [35].

Regardless of these dietary trend data, does eating a diet high in carbohydrate cause overeating and obesity? Figure 13 shows the number of studies that have examined the relationship between total carbohydrate intake and body weight and/or fat composition. Unfortunately, the literature is somewhat split between studies that have found a positive relationship between energy consumed as carbohydrate and studies that have not. In fact, more studies find a negative correlation between carbohydrate consumption and body weight than do not.

This literature becomes a little clearer when the form of the carbohydrate consumed is considered. If simple carbohydrates, such as glucose or fructose, are examined as a function of body weight, then as Figure 14 demonstrates, the great majority of studies indicate that the greater the amount of sugar consumed, the *smaller* the body weight. On the other hand, if the amount of sugar in sweetened drinks is used, then the opposite effect is found—body weight increases as a function of the amount of sweetened drinks consumed, as illustrated in Figure 15. How can this apparent contradiction in findings be reconciled?

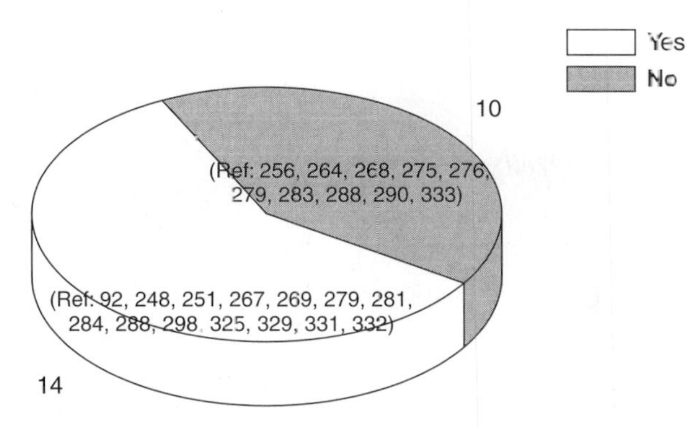

FIGURE 13 Distribution of studies demonstrating a negative correlation between carbohydrate consumption and body weight or fat composition. Jackson *et al.* [264] observed negative correlation in males, but not if females. Maskarinec *et al.* [265] and Toeller *et al.* [266] observed negative correlation in females, but not males.

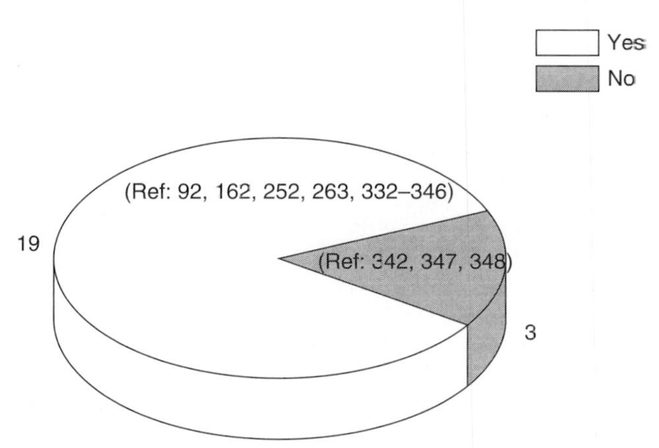

FIGURE 14 Distribution of studies demonstrating a negative correlation between sugar consumption and body weight. Ma *et al.* [267] observed a negative correlation with high-glycemic foods, but not with low-glycemic foods.

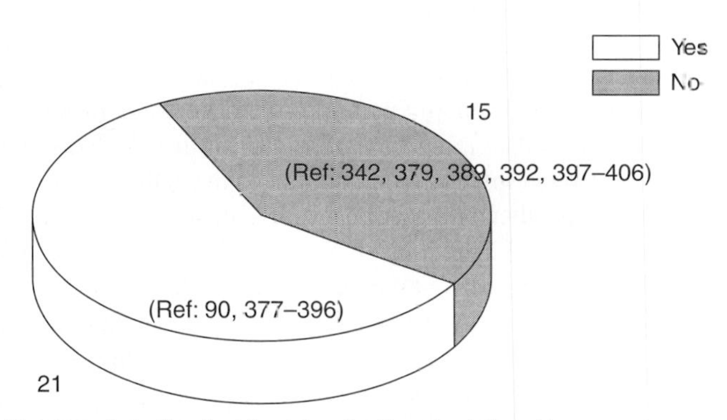

FIGURE 15 Distribution of studies demonstrating a significant relationship between sugar beverage consumption and body weight.

There are two explanations. First, "liquid calories" result in less energy compensation when they are used as a preload (consumed within a half hour of eating) than when solid foods are used [88–91]. Consequently, total energy intake will be greater if the energy is consumed as a liquid just prior to eating than if the food were in solid form. Second, consuming food in any form an hour before eating does not suppress subsequent consumption. As a result if a caloric beverage or high-carbohydrate snack is consumed as a midmorning or midafternoon snack, little reduction in energy intake will result at the subsequent meal [80]. If continued over time, the increase in total daily energy intake must result in an increase in body mass.

Contrary to sugars being consumed with liquids, if they are eaten as part of meals, they tend to displace foods that are higher in fat as indicated by an inverse relationship between the amount of sugar and the amount of fat consumed, known as the "sugar–fat see-saw" [92–94]. Because the volume of daily food consumed remains fairly constant [78, 79, 95], increasing the amount of sugar in the diet displaces dietary fat, resulting in significant weight loss [94, 96–99].

Further evidence that the consumption of carbohydrate, particularly complex carbohydrate, leads to a smaller body weight rather than increased weight emerges from literature on vegetarians. Along with eating a diet greater in the amount of complex carbohydrate, vegetarians consume significantly more fiber than non-vegetarians [100] and, as indicated in Figure 16 consistently exhibit smaller body weight than non-vegetarians [101–117].

No review of the relationship between carbohydrates and obesity would be complete without a discussion of the relationship between the glycemic index of foods and body weight. The "glycemic" theory of overeating and obesity emerged from a study of obese, adolescent males by Ludwig *et al.* [118] in which they tested the effect of

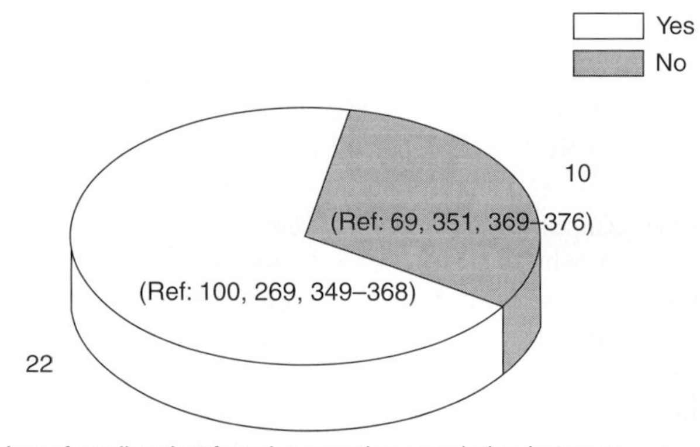

FIGURE 16 Distribution of studies demonstrating a significant negative correlation between dietary fiber and body weight. Howarth *et al.* [268] observed the negative correlation in women, but not men.

high-, medium-, or low-glycemic foods given at breakfast and lunch on food intake over the next 5 hours after lunch. They observed a significant reduction in food intake after feeding the low-glycemic meals versus the high-glycemic meals. Studies of the relationship between the glycemic index of foods and measurements of appetite clearly support the finding: Foods that have a high glycemic index are less satiating than foods having a lower glycemic index [118–137], although not every study observed such a relationship [123, 138–146]. Unfortunately, the evidence that foods having a high glycemic index cause a significant increase in energy intake is far less clear, with seven studies observing such an effect [118, 134, 135, 137, 147–149] while seven other studies have not [128, 137, 139, 140, 142, 144, 150]. Casting even more doubt on this hypothesis is a recent study by Reid *et al.* [151], who gave subjects either sugar drinks or artificially sweetened drinks to consume daily for a 4-week period and observed that feeding the sugars caused a significant suppression of daily energy intake, rather than an increase, a finding consistent with the majority of studies relating carbohydrate composition of the diet and body weight. It appears more likely that foods varying in glycemic index cause changes in subsequent food intake through mechanisms other than the insulin stimulating action of dietary carbohydrate, possibly through the action of dietary fiber to slow gastric emptying [152].

In summary, it appears that the gurus are wrong. Carbohydrates, as a macronutrient, do not lead to increased energy intake and are not a cause of increased body weight. They may, however, play a role as a major constituent of snack foods. Because humans do not precisely compensate for foods eaten between meals [80], the increased consumption of snacks can lead to an increase in total daily intake and a significant weight gain. However, it is the eating of snacks rather than the unique role played by carbohydrates as a macronutrient that is responsible for the weight gain.

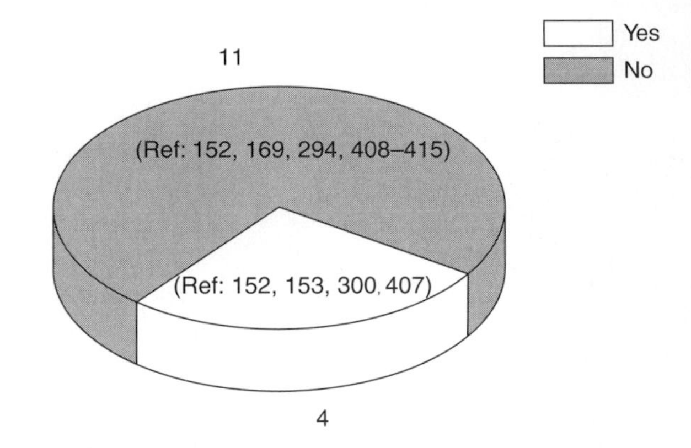

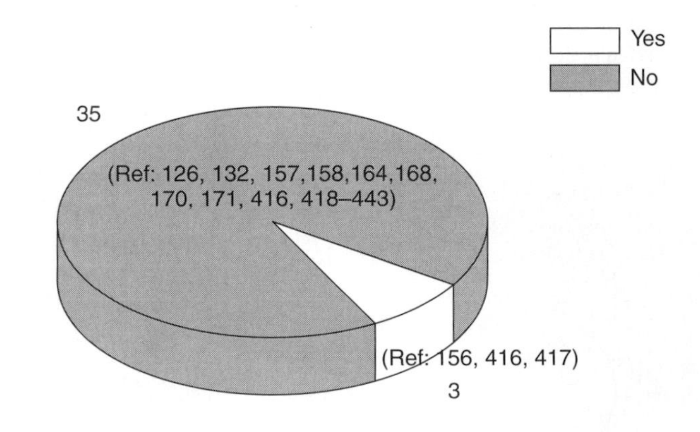

FIGURE 17 (Top panel) Distribution of experimental studies that failed to find a significant increase in energy intake in response to fat substitutes. (Bottom panel) Distribution of experimental studies that failed to find a significant increase in energy intake in response to sugar substitutes.

VI. ENERGY COMPENSATION FOR FAT AND SUGAR SUBSTITUTES

If we can generalize from the majority of studies just cited above that humans appear to demonstrate very little energy compensation for reductions in either dietary fat or carbohydrate, then using palatable, low-calorie fat or sugar substitutes should be an effective way to reduce daily energy intake. Indeed, in many respects, the use of fat substitutes is a better test of the role of dietary fat in energy balance than merely reducing the amount of fat, because it replaces the food with a product of similar properties. The results of studies that have examined the relationship between fat and sugar substitutes and energy intake are presented in the two panels in Figure 17. Similar to the studies on the effect of decreasing fat intake, the vast majority of published studies demonstrate that humans do not accurately compensate for the reduced energy when the fat in the diet is replaced by a fat substitute, as can be seen in top panel of Figure 17. The study by Cotton *et al.* [153] is particularly important because although they found about 72% energy compensation, the level of dietary fat was reduced to about 20% of energy from 34%. This high degree of fat restriction is substantially more than what was observed in other studies. The study by Kelly *et al.* [154] failed to show a difference in energy intake between a group receiving sucrose polyester in place of dietary fat for a 3-month trial. However, it was very possible that the measure of energy intake (dietary records) was too variable to allow accurate assessment. This interpretation is supported by the fact that the dietary change was sufficient to cause a statistically significant reduction in blood lipid and in body weight of the group receiving the fat substitute. A reduction in energy may have occurred in this group as a function of eating the fat substitutes, but the intake measure was not sufficiently sensitive to detect it.

The experimental studies on the effects of sugar substitutes on energy intake are even clearer than those for fat substitutes. As is evident from bottom panel of Figure 17, the vast majority of studies fail to find energy compensation for the energy lost when the sugar in foods is replaced by non-caloric sweeteners regardless of the particular food in which the sweetener was added, for example, water, soft drinks, yogurt, and cheese. With such a preponderance of studies showing lack of energy compensation when noncaloric sweeteners are substituted for sugar in the diet, one would expect it to be easy to demonstrate that the use of these sweeteners should facilitate weight loss. Unfortunately, the data are not clear and are riddled with controversy.

Two papers that fueled the controversy appeared in 1986 and caused a stir in the popular press and in the scientific community. In a fascinating, but brief, letter to *The Lancet*, Blundell and Hill [155] reported that the ingestion of aspartame caused an increase in hunger ratings in human subjects, and in another paper, Stellman and Garfinkel [156] reported that women who consumed the sweetener saccharin were more likely to gain body weight than nonusers of saccharin. Rogers and Blundell [157] extended their finding by demonstrating that following the ingestion of saccharin, not only are subjects hungrier, but they increased their food intake. Thus, rather than being an aid to weight reduction, these studies were suggesting that artificial sweeteners may actually cause a gain in body weight.

Conflicts in science usually resolve themselves with replication and increased scrutiny. As evidence in bottom panel of Figure 17 indicates, the effect observed by Rogers and Blundell [157] was never replicated by any other investigator and may be attributed to a type 2 statistical error. Critics of the Stellman and Garfinkel [156] paper revealed sufficient weaknesses to cause considerable doubt as to its conclusion: (1) The data were based on the subject's memory of their body weight, not measured body weight, and (2) anyone who actively tried to lose body weight was eliminated from the analysis [158, 159]. Either of these conditions would have led to the *false* conclusion that weight gain is *related* to sweetener use.

Other published papers, however, also shed doubt on the idea that low-calorie sweeteners could be a significant aid to weight control. No difference in body weight was observed between low-calorie sweetener users and nonusers [160, 161]. Colditz *et al.* [162] reported a *positive* relationship between the use of the sweetener saccharin and a gain in body weight. These results would suggest that sweeteners are not effective in reducing body weight. However, it might be argued that those subjects with the largest body weight had the greatest need for sweeteners to reduce their caloric intake, but apparently, this is not the case. Richardson [163] found no difference in sweetener use between people who restrict their sugar intake compared to those who do not. Consequently, the "need" to use sweeteners for weight reduction cannot explain these results.

These studies clearly raise serious doubts as to whether humans use low-calorie sweeteners as substitutes for sugar as originally intended. Even more disturbing from an energetic perspective is that if consumers did substitute low-calorie sweeteners for sugar, then the proportion of their total energy intake derived from fat would increase. As is evident from the data presented earlier, increasing the percentage of energy from fat may result in an increase in body weight, not a decrease.

One of the few studies that examined the issue of how consumers were using sugar substitutes was provided by Chen and Parham [164]. They found that consumers, college students, were not using the sugar substitutes to substitute for carbohydrates, but rather they were consuming foods containing the sweeteners in addition to the sugars consumed in their diet. This pattern of behavior was not seen when low-calorie sweeteners were substituted covertly for sugar in the diet [165]. It seems that the knowledge of the contents of foods drastically changes the decision to consume it or not, an effect well documented in the laboratory [166–168], although not universally [169]. Miller *et al.* [170] found that information about the caloric content of potato chips significantly affected restrained eaters, but had no effect on unrestrained eaters.

The effectiveness of artificial sweeteners as a weight reduction aid remains tenuous at best, particularly in light of the data presented in Figure 13. It is possible that artificial sweeteners may play a greater role in weight maintenance following a weight loss than in actually aiding the weight loss. Blackburn and his associates [171] performed one of the only long-term studies of the effectiveness of sugar substitutes in a weight reduction program. Their results can be seen in Figure 18. All subjects in this study were prescribed a 4180-kJ diet for 16 weeks, and then observed for 2 years. The left panel illustrates that the weight loss in the group who used aspartame was not different from the group that did not use aspartame. However, the right panel shows the typical weight recovery that follows diet-induced weight loss programs. The group that had used aspartame during the dieting phase continued to use it during the follow-up phase and showed less rapid relapse than the group that did not use aspartame. Unfortunately, linear regression analysis of the weight loss at 77 weeks indicated that the sustained weight loss was unrelated to aspartame use. Most probably, the subjects' measures of aspartame use were too inaccurate to allow such a relationship to be observed. Nevertheless, at present no clear demonstration exists in the literature that sustained use of low-calorie sweeteners produces a significant weight loss.

In summary, fat and sugar substitutes take advantage of the human inability to accurately regulate calorie intake. As

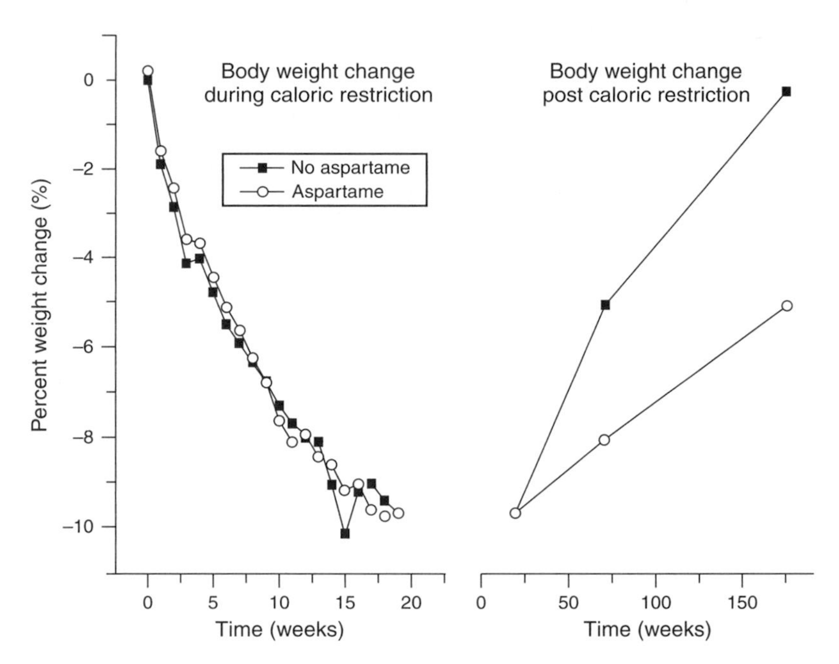

FIGURE 18 Effect of aspartame on the change in body weight during weight reduction and recovery. (Adapted from Blackburn *et al.* [171].)

a result they may be effective at producing a reduction in energy intake and a sustained loss in body weight.

VII. PROTEIN PARADOX

Although all the popular books cited earlier blame carbohydrates in our diet as being the major culprit responsible for our expanding waistlines, the solution they propose is a diet that is not only low in carbohydrate, but high in protein (30% of energy or greater). Although the scientific community was originally skeptical of the claims that a high-protein, high-fat diet could cause a weight loss, controlled studies clearly confirmed the effectiveness of a diet high in protein and low in carbohydrate to produce a reduction in body weight [99, 172–185].

The skepticism by the more orthodox nutritionists stemmed from four arguments. First, the deleterious health effects of eating a diet high in fat, particularly high in saturated fat, are well known. Second, foods containing large amounts of protein are usually accompanied by large amounts of fat. Figure 19 shows the relationship between the protein and fat content of 500 randomly selected foods from the USDA food consumption survey [77]. As discussed earlier, the evidence overwhelmingly indicates that a diet high in fat is associated with an increase, not a decrease, in body weight (see Figures 2 and 3). Third, as indicated in Figure 20 and consistent with the last argument, people who consume greater amounts of total protein have higher body mass index (BMI) than people who eat less protein. Finally, vegetarians, who habitually consume considerably less total protein than non-vegetarians [186], have a lower body weight than non-vegetarians (see earlier discussion).

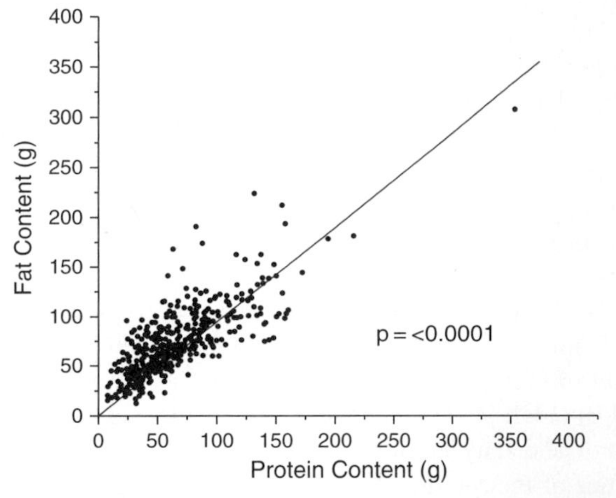

FIGURE 19 Relationship between protein content and fat content of 500 randomly selected foods. (Data from USDA (1996). "Continuing Survey of Food Intakes by Individuals (CSFII): Diet and Health Knowledge Survey, 1994." Department of Agriculture, Beltsville Human *Nutr. Res.* Center/ARS, Beltsville, MD.)

One possible resolution to this apparent paradox may lie in the increase in energy expenditure produced by the protein. It is well accepted that the thermogenic effect of protein is considerably higher than those of carbohydrate and fat [187]. A number of studies have examined the energetic effect of consuming equal quantities of a high-versus a low-protein diet and have found that the total energy expenditure is higher after eating meals consisting of high-protein food [187–196]. Eisenstein and her collaborators [197] used the data from these studies to estimate the total increase in daily energy expenditure after consuming a high-protein diet (30% of total calories consumed)

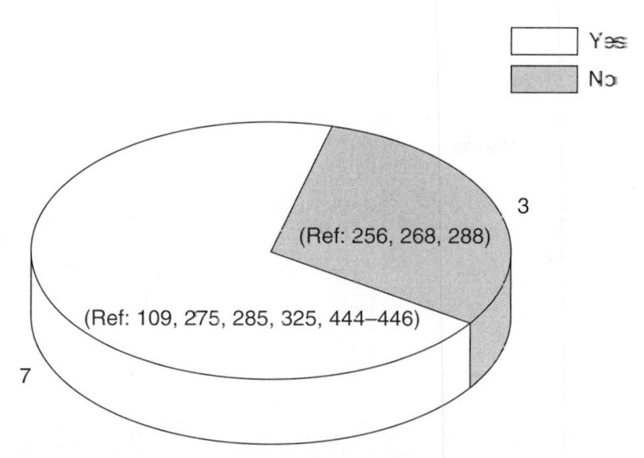

FIGURE 20 Distribution of studies that found a significant correlation between protein consumption and body weight. Jackson *et al.* [225] observed the correlation in women, but not men.

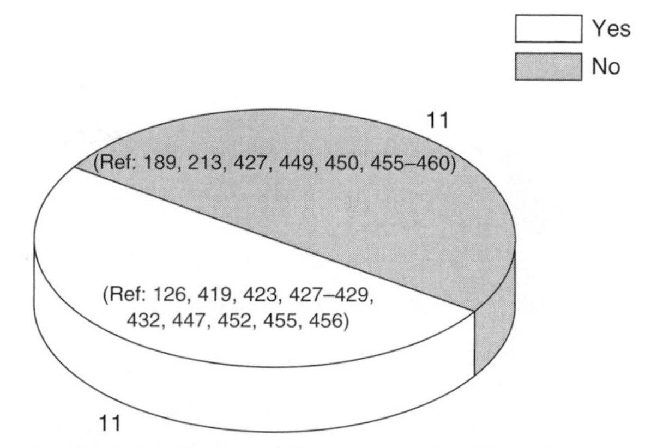

FIGURE 21 Distribution of studies that found the consumption of protein produces a significant suppression in food intake. Porrini *et al.* [269] observed suppression only when water was removed from the test diet.

compared to a "normal" level of protein (15% of total calories consumed) for 1 day and arrived at a value of 23 kcal, hardly a value of energetic significance for anyone who desires to lose weight.

The determination of such a trivial effect of high-protein diets on energy expenditure is consistent with studies of weight loss comparing high- and low-protein diets when energy intake is maintained constant for the two diets [172, 189, 198–202]. Of these seven studies, only two found a statistically ($p < 0.05$) greater loss on the high-protein diet [172, 199]. Both studies are questionable. The DeHaven *et al.* study [199] used an unnatural, 100% protein diet for comparison, raising the issue of generalizability to the population at large, while the authors of the Baba *et al.* study [172] admit that because the entire weight loss difference between the high- and low-protein groups occurred during the first week of the 4-week study, and not thereafter, it was probably due to "greater water losses" rather than changes in body tissue. Eisenstein *et al.* [203] calculated the statistical difference in actual weight loss in the seven studies and found it to be statistically insignificant ($p = 0.904$).

If high-protein diets do not cause an increase in metabolism yet people lose weight when they consume high protein diets *ad libitum*, then the only logical path out of the protein paradox is that the consumption of protein must cause a reduction in energy intake. Indeed, it has been known for almost a century that feeding high-protein diets to animals produces a very significant reduction in spontaneous food intake [204–210].

The data in humans are not as consistent, as shown in Figure 21. About half the studies that have compared the food-inhibiting effect of protein versus carbohydrate or fat have found greater suppression of intake, whereas half have not. Part of the discrepancy in findings can be explained by the kind of protein examined. Anderson and Moore observed a significant suppression in intake at a subsequent meal when whey and soy protein were consumed but not when egg albumin was the source of protein [211]. Bowen *et al.* [212] observed a significant suppression in subsequent intake when the premeal load consisted of gluten, but not when the protein source was soy and whey. Although it is clear that eating a diet high in protein reduces food intake and causes a weight loss, the mechanism through which protein inhibits food intake remains ambiguous.

An alternative explanation as to why the consumption of a high-protein diet inhibits food intake was suggested almost 50 years ago by Yudkin and Carey [213]. Their subjects maintained an exhaustive, weighed dietary records of everything they ate during a 2-week control period, then again during the next 2 weeks, during which time they were instructed to minimize their consumption of carbohydrate and eat as much protein and fat as they desired. The results for all six of their study subjects are shown in Figure 22. There are several remarkable features about this work that should give us insight into why the population is buying into the high-protein, high-fat diets, the most important of which is that all the subjects lost weight. The reason for the weight loss is shown in the top left panel of this figure. Although asked to eat more protein and fat and less carbohydrate, the subjects did not increase their intake of protein and fat (lower two panels) but rather reduced the amount of carbohydrate (top panel) they ate. The average reduction in energy intake was quite large, approximately 37%. In a subsequent replication, Stock and Yudkin [214] examined 11 more subjects in a very similar experimental paradigm. They observed a 33% decrease in energy intake and, again, no change was observed in the amount of protein or fat consumed, only a reduction in carbohydrate intake.

These two older studies may be criticized because they had subjects record their own data and did not use a

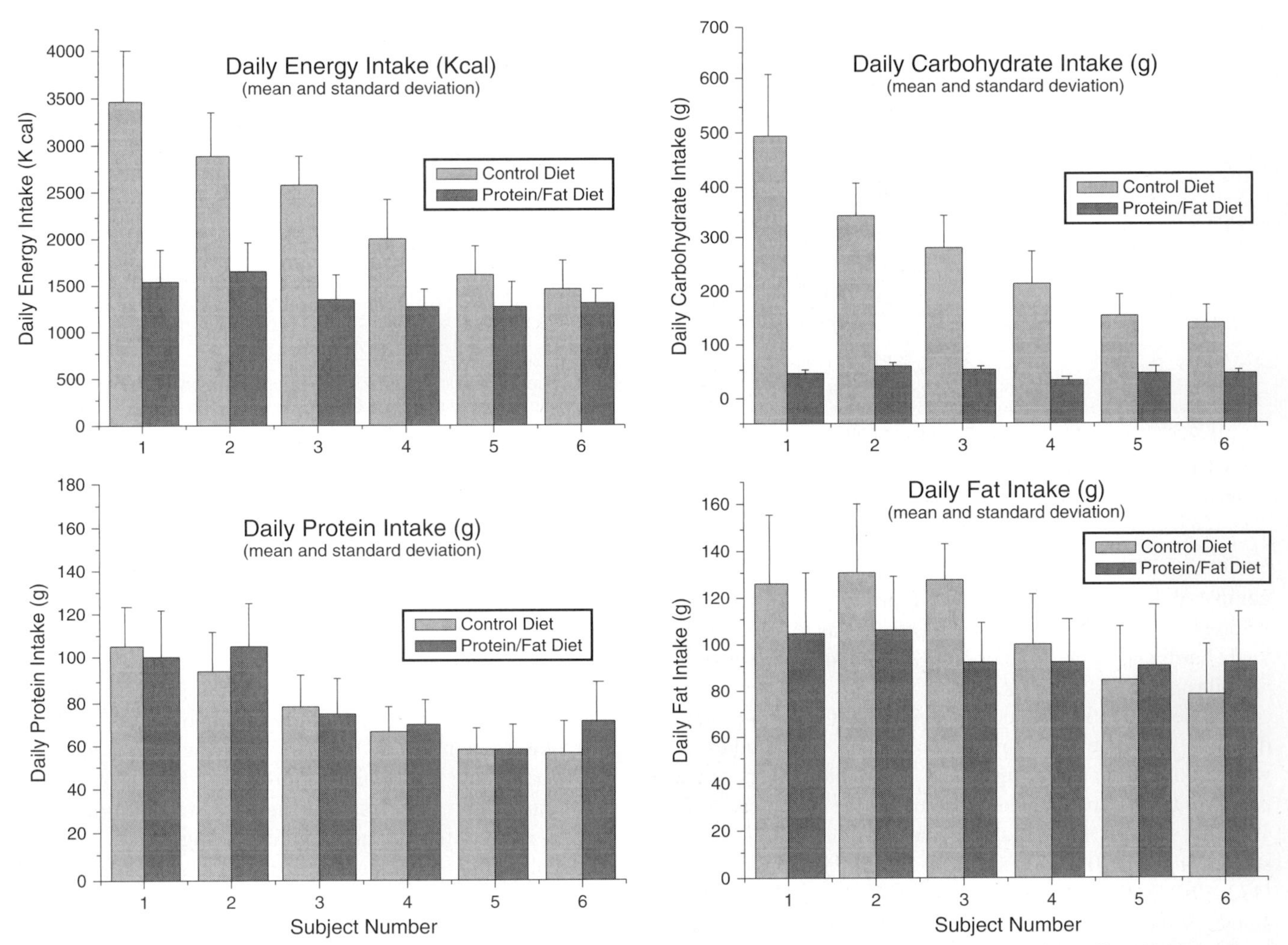

FIGURE 22 Daily caloric and macronutrient intake of six subjects consuming a high-protein, low-carbohydrate diet. (Adapted from Yudkin, J., and Carey, M. (1960). The treatment of obesity by the "high-fat" diet: The inevitability of calories. *Lancet* **2,** 939–941.)

crossover design. However, similar observations were made by Skov *et al.* [182], who also observed a reduction in energy intake on a low-carbohydrate, high-protein diet using a more experimentally sophisticated design, more subjects, and measured over a longer period of time (26 weeks). They observed an approximately 18% reduction in energy intake in a group of overweight and obese subjects who ate *ad libitum* from a menu composed of high-protein (22%), low-carbohydrate (46%) foods prepared by their staff compared to a matched group of subjects who consumed a high-carbohydrate (59%), low-protein (12%) diet.

One reason why the Skov *et al.* study [182] observed a smaller difference in energy intake than the Yudkin [214] group may be due to the greater degree of experimental control. Skov *et al.* controlled the composition of the diet consumed by their subjects. The reduction in energy derived from dietary carbohydrate was offset by an increase in the amount of energy derived from protein. In the Yudkin studies, the subjects determined the composition of the diet and did not increase the amount of protein in response to the decrease in carbohydrate. This control allowed Skov and his associates to rule out palatability and energy density as factors contributing to their results.

These studies indicate that when people are asked to increase their intake of protein and reduce their consumption of carbohydrates, they can more easily restrict their intake of carbohydrates and they do not compensate for the reduced energy by increasing their intake of protein or fat. As a result, their total energy intake is reduced, and they lose weight. Why don't they compensate? One reason may lie in the "anatomy" of eating macronutrients, rather than in metabolic properties of micronutrients. Yudkin and Carey [213] suggested that one effect of restricting carbohydrates is that it reduces the variety of foods that can be eaten. There is a considerable amount of evidence indicating that reducing the variety of foods available reduces total food consumption [80, 215–222], at least in the short term. Perhaps the weight loss achieved by the high-protein, low-carbohydrate diet, or any other diet for that matter, is simply due to voluntarily reducing the variety of food that can be consumed, and because there is little evidence that energy compensation occurs [80], energy intake is diminished. The problem, of course, is that people may not be able to withstand the reduction in food variety for very long and quickly return to their original intake and previous body weight.

VIII. SUMMARY AND IMPLICATIONS OF THE RESEARCH ON MACRONUTRIENTS AND INTAKE

In viewing the long list of studies in macronutrients and food intake, it is apparent that humans living in contemporary society do not eat for energy. In fact, it appears that humans possess very poor mechanisms to adjust the volume of food they consume in response to alterations in energy density of the foods they consume [80]. If these observations are true, then the use of "artificial" sweeteners and fats should successfully cause a sustained reduction in the body weight in our population as long they are substituted for "real" carbohydrates and fat. Why sweeteners and fat substitutes do not cause a greater weight loss than they do requires more research as to how consumers use these products. If energy density is the major determinant of human energy intake, as suggested by this research, then the major thrust of any program aimed at weight reduction or obesity prevention should concentrate on increasing the consumption of energy-dilute foods such as soups, salads, and casseroles. Using this approach, Rolls has produced some promising data [223].

From a more theoretical perspective, the plasticity of body weight, rather than the constancy of body weight, that occurs when macronutrients are manipulated should evoke a reevaluation of the theory that energy intake in humans is well controlled and that body weight is well regulated. It is quite possible that humans maintain a vestigial process that "controls" our behavior of eating, but that system is easily dominated by more powerful environmental determinants. If this is the case, then it is not too optimistic to believe that by understanding more about these environmental determinants, such as macronutrient composition, it may be possible to not only understand why we are getting fatter, but also and most important, be able to reduce and even prevent some of the overweight and obesity from occurring.

References

1. Adolph, E. (1947). Urges to eat and drink in rats. *Am. J. Physiol.* **151**, 110–125.
2. Mayer, J. (1955). Regulation of energy intake and the body weight: The glucostatic theory and the lipostatic hypothesis. *Ann. N. Y. Acad. Sci.* **63**, 15–43.
3. Mellinkoff, S. M., Frankland, M., Boyle, D., and Greipel, M. (1956). Relationship between serum amino acid concentration and fluctuations in appetite. *J. Appl. Physiol.* **8**, 535–538.
4. Michel, C., and Cabanac, M. (1999). Effects of dexamethasone on the body weight set point of rats. *Physiol. Behav.* **68**, 145–150.
5. Cabanac, M., and Morrissette, J. (1992). Acute, but not chronic, exercise lowers the body weight set-point in male rats. *Physiol. Behav.* **52**, 1173–1177.
6. Cabanac, M. (1991). Role of set-point theory in body weight. *FASEB J.* **5**, 2105–2106.
7. Garn, S. M. (1992). Role of set-point theory in regulation of body weight. *FASEB J.* **6**, 794.
8. Bernardis, L. L., McEwen, G., and Kodis, M. (1986). Body weight set point studies in weanling rats with dorsomedial hypothalamic lesions (DMNL rats). *Brain Res. Bull.* **17**, 451–460.
9. Fantino, M., and Brinnel, H. (1986). Body weight set-point changes during the ovarian cycle: experimental study of rats using hoarding behavior. *Physiol. Behav.* **36**, 991–996.
10. Fantino, M., Faion, F., and Rolland, Y. (1986). Effect of dexfenfluramine on body weight set-point: study in the rat with hoarding behaviour. *Appetite* **7**(Suppl), 115–126.
11. Harris, R. B., and Martin, R. J. (1984). Recovery of body weight from below "set point" in mature female rats. *J. Nutr.* **114**, 1143–1150.
12. Keesey, R. E., and Corbett, S. W. (1984). Metabolic defense of the body weight set-point. *Res. Publ. Assoc. Res. Nerv. Ment. Dis.* **62**, 87–96.
13. Stunkard, A. J. (1982). Anorectic agents lower a body weight set point. *Life Sci.* **30**, 2043–2055.
14. DeCastro, J. M., Paullin, S. K., and Delugas, G. M. (1978). Insulin and glucagon as determinants of body-weight set point and microregulation in rats. *J. Comp. Physiol. Psych.* **92**, 571–579.
15. Garrow, J. S., and Stalley, S. (1975). Is there a "set point" for human body-weight? *Proc. Nutr. Soc.* **34**, 84A–85A.
16. Barnes, D. S., and Mrosovsky, N. (1974). Body weight regulation in ground squirrels and hypothalamically lesioned rats: Slow and sudden set point changes. *Physiol. Behav.* **12**, 251–258.
17. Myers, R. D., and Martin, G. E. (1973). 6-OHDA lesions of the hypothalamus: interaction of aphagia, food palatability, set-point for weight regulation, and recovery of feeding. *Pharmacol. Biochem. Behav.* **1**, 329–345.
18. Bray, G. A., Paeratakul, S., and Popkin, B. M. (2004). Dietary fat and obesity: a review of animal, clinical and epidemiological studies. *Physiol. Behav.* **83**, 549–555.
19. West, K. M., and Kalbfleisch, J. M. (1971). Influence of nutritional factors on prevalence of diabetes. *Diabetes* **20**, 99–108.
20. Keys, A. (1970). Coronary heart disease in 7 countries. *Circulation* **41**, 1–211.
21. Willett, W. C. (1998). Is dietary fat a major determinant of body fat? *Am. J. Clin. Nutr.* **67**, 556S–562S.
22. Lissner, L., and Heitmann, B. L. (1995). Dietary fat and obesity: evidence from epidemiology. *Eur. J. Clin. Nutr.* **49**, 79–90.
23. Goris, A. H., Westerterp-Plantenga, M. S., and Westerterp, K. R. (2000). Undereating and underrecording of habitual food intake in obese men: Selective underreporting of fat intake. *Am. J. Clin. Nutr.* **71**, 130–134.

24. Black, A. E., Goldberg, G. R., Jebb, S. A., Livingstone, M. B., Cole, T. J., and Prentice, A. M. (1991). Critical evaluation of energy intake data using fundamental principles of energy physiology: 2. Evaluating the results of published surveys. *Eur. J. Clin. Nutr.* **45**, 583–599.
25. Heitmann, B. L. (1993). The influence of fatness, weight change, slimming history and other lifestyle variables on diet reporting in Danish men and women aged 35–65 years. *Int. J. Obes. Relat. Metab. Disord.* **17**, 329–336.
26. Heitmann, B. L., and Lissner, L. (1995). Dietary underreporting by obese individuals—is it specific or non-specific? *BMJ* **311**, 986–989.
27. Heitmann, B. L., Lissner, L., Sorensen, T. I., and Bengtsson, C. (1995). Dietary fat intake and weight gain in women genetically predisposed for obesity. *Am. J. Clin. Nutr.* **61**, 1213–1217.
28. Heitmann, B. L., Lissner, L., and Osler, M. (2000). Do we eat less fat, or just report so? *Int. J. Obes. Relat. Metab. Disord.* **24**, 435–442.
29. Prentice, A. M., Black, A. E., Coward, W. A., *et al.* (1986). High levels of energy expenditure in obese women. *Br. Med. J. (Clin. Res. Ed.)* **292**, 983–987.
30. Lafay, L., Mennen, L., Basdevant, A., *et al.* (2000). Does energy intake underreporting involve all kinds of food or only specific food items? Results from the Fleurbaix Laventie Ville Sante (FLVS). study. *Int. J. Obes. Relat. Metab. Disord.* **24**, 1500–1506.
31. Poppitt, S. D., and Swann, D. L. (1998). Dietary manipulation and energy compensation: does the intermittent use of low-fat items in the diet reduce total energy intake in free-feeding lean men? *Int. J. Obes. Relat. Metab. Disord.* **22**, 1024–1031.
32. Astrup, A., Grunwald, G. K., Melanson, E. L., Saris, W. H., and Hill, J. O. (2000). The role of low-fat diets in body weight control: a meta-analysis of ad libitum dietary intervention studies. *Int. J. Obes. Relat. Metab. Disord.* **24**, 1545–1552.
33. Yu-Poth, S., Zhao, G., Etherton, T., Naglak, M., Jonnalagadda, S., and Kris-Etherton, P. M. (1999). Effects of the National Cholesterol Education Program's Step I and Step II dietary intervention programs on cardiovascular disease risk factors: a meta-analysis. *Am. J. Clin. Nutr.* **69**, 632–646.
34. Heini, A. F., and Weinsier, R. L. (1997). Divergent trends in obesity and fat intake patterns: The American paradox. *Am. J. Med.* **102**, 259–264.
35. Ogden, C. L., Carroll, M. D., Curtin, L. R., McDowell, M. A., Tabak, C. J., and Flegal, K. M. (2006). Prevalence of overweight and obesity in the United States, 1999–2004. *JAMA* **295**, 1549–1555.
36. Wing, R. R., and Hill, J. O. (2001). Successful weight loss maintenance. *Annu. Rev. Nutr.* **21**, 323–341.
37. Astrup, A., Toubro, S., Raben, A., and Skov, A. R. (1997). The role of low-fat diets and fat substitutes in body weight management: What have we learned from clinical studies? *J. A. Diet. Assoc.* **97**, S82–S87.
38. Jeffery, R. W., and French, S. A. (1997). Preventing weight gain in adults: Design, methods and one year results from the Pound of Prevention study. *Int. J. Obes. Relat. Metab. Disord.* **21**, 457–464.
39. Leser, M. S., Yanovski, S. Z., and Yanovski, J. A. (2002). A low-fat intake and greater activity level are associated with lower weight regain 3 years after completing a very-low-calorie diet. *J. Am. Diet. Assoc.* **102**, 1252–1256.
40. West, D. B., and York, B. (1998). Dietary fat, genetic predisposition, and obesity: Lessons from animal models. *Am. J. Clin. Nutr.* **67**, 505S–512S.
41. Ramirez, I., and Friedman, M. I. (1990). Dietary hyperphagia in rats: Role of fat, carbohydrate, and energy content. *Physiol. Behav.* **47**, 1157–1163.
42. van Stratum, P., Lussenburg, R. N., van Wezel, L. A., Vergroesen, A. J., and Cremer, H. D. (1978). The effect of dietary carbohydrate:fat ratio on energy intake by adult women. *Am. J. Clin. Nutr.* **31**, 206–212.
43. Stubbs, R. J., Harbron, C. G., and Prentice, A. M. (1996). Covert manipulation of the dietary fat to carbohydrate ratio of isoenergetically dense diets: Effect on food intake in feeding men ad libitum. *Int. J. Obes.* **20**, 651–660.
44. Stubbs, R. J., Johnstone, A. M., O'Reilly, L. M., Barton, K., and Reid, C. (1998). The effect of covertly manipulating the energy density of mixed diets on ad libitum food intake in "pseudo free-living" humans. *Int. J. Obes.* **22**, 980–987.
45. Saltzman, E., Dallal, G. E., and Roberts, S. B. (1997). Effect of high-fat and low-fat diets on voluntary energy intake and substrate oxidation: Studies in identical twins consuming diets matched for energy density, fiber, and palatability. *Am. J. Clin. Nutr.* **66**, 1332–1339.
46. Bell, E. A., Castellanos, V. H., Pelkman, C. L., Thorwart, M. L., and Rolls, B. J. (1998). Energy density of foods affects energy intake in normal-weight women. *Am. J. Clin. Nutr.* **37**, 412–420.
47. Rolls, B. J., and Bell, E. A. (1999). Intake of fat and carbohydrate: role of energy density. *Eur. J. Clin. Nutr.* **53**, S166–S173.
48. Rolls, B. J., Bell, E. A., and Thorwart, M. L. (1999). Water incorporated into a food but not served with a food decreases energy intake in lean women. *Am. J. Clin. Nutr.* **70**, 448–455.
49. Rolls, B. J., Bell, E. A., Castellanos, V. H., Chow, M., Pelkman, C. L., and Thorwart, M. L. (1999). Energy density but not fat content of foods affected energy intake in lean and obese women. *Am. J. Clin. Nutr.* **69**, 863–871.
50. Chapelot, D., Fumeron, F., and Fricker, J. (1998). Dietary fat, energy density and BMI: A case of a missing flower? *Int. J. Obes.* **22**, 1032–1033.
51. Bell, E. A., and Rolls, B. J. (2001). Energy density of foods affects energy intake across multiple levels of fat content in lean and obese women. *Am. J. Clin. Nutr.* **73**, 1010–1018.
52. Crowe, T. C., Fontaine, H. L., Gibbons, C. J., Cameron-Smith, D., and Swinburn, B. A. (2004). Energy density of foods and beverages in the Australian food supply: influence of macronutrients and comparison to dietary intake. *Eur. J. Clin. Nutr.* **58**, 1485–1491.
53. de Castro, J. M. (2006). Macronutrient and dietary energy density influences on the intake of free-living humans. *Appetite* **46**, 1–5.
54. de Castro, J. M. (2005). Stomach filling may mediate the influence of dietary energy density on the food intake of free-living humans. *Physiol. Behav.* **86**, 32–45.
55. de Castro, J. M. (2004). Dietary energy density is associated with increased intake in free-living humans. *J. Nutr.* **134**, 335–341.

56. Devitt, A. A., and Mattes, R. D. (2004). Effects of food unit size and energy density on intake in humans. *Appetite* **42**, 213–220.
57. Fisher, J. O., Liu, Y., Birch, L. L., and Rolls, B. J. (2007). Effects of portion size and energy density on young children's intake at a meal. *Am. J. Clin. Nutr.* **86**, 174–179.
58. Kant, A. K., and Graubard, B. I. (2005). Energy density of diets reported by American adults: Association with food group intake, nutrient intake, and body weight. *Int. J. Obes. (Lond.)* **29**, 950–956.
59. Kral, T. V., Roe, L. S., and Rolls, B. J. (2002). Does nutrition information about the energy density of meals affect food intake in normal-weight women? *Appetite* **39**, 137–145.
60. Kral, T. V., and Rolls, B. J. (2004). Energy density and portion size: Their independent and combined effects on energy intake. *Physiol. Behav.* **82**, 131–138.
61. Kral, T. V., Stunkard, A. J., Berkowitz, R. I., Stallings, V. A., Brown, D. D., and Faith, M. S. (2007). Daily food intake in relation to dietary energy density in the free-living environment: A prospective analysis of children born at different risk of obesity. *Am. J. Clin. Nutr.* **86**, 41–47.
62. Kral, T. V. E., Roe, L. S., and Rolls, B. J. (2004). Combined effects of energy density and portion size on energy intake in women. *Am. J. Clin. Nutr.* **79**, 962–968.
63. Ledikwe, J. H., Blanck, H. M., Khan, L. K., *et al.* (2006). Dietary energy density is associated with energy intake and weight status in US adults. *Am. J. Clin. Nutr.* **83**, 1362–1368.
64. McCrory, M. A., Saltzman, E., Rolls, B. J., and Roberts, S. B. (2006). A twin study of the effects of energy density and palatability on energy intake of individual foods. *Physiol. Behav.* **87**, 451–459.
65. Poppitt, S. D. (1995). Energy density of diets and obesity. *Int. J. Obes.* **19**, S20–S26.
66. Prentice, A. M., and Poppitt, S. D. (1996). Importance of energy density and macronutrients in the regulation of energy intake. *Int. J. Obes.* **20**, S18–S23.
67. Rolls, B. J., Roe, L. S., and Meengs, J. S. (2006). Reductions in portion size and energy density of foods are additive and lead to sustained decreases in energy intake. *Am. J. Clin. Nutr.* **83**, 11–17.
68. Rolls, B. J., Roe, L. S., and Meengs, J. S. (2004). Salad and satiety: Energy density and portion size of a first-course salad affect energy intake at lunch. *J. Am. Diet. Assoc.* **104**, 1570–1576.
69. Stookey, J. D. (2001). Energy density, energy intake and weight status in a large free-living sample of Chinese adults: Exploring the underlying roles of fat, protein, carbohydrate, fiber and water intakes. *Eur. J. Clin. Nutr.* **55**, 349–359.
70. Stubbs, R. J., and Whybrow, S. (2004). Energy density, diet composition and palatability: Influences on overall food energy intake in humans. *Physiol. Behav.* **81**, 755–764.
71. Wallengren, O., Lundholm, K., and Bosaeus, I. (2005). Diet energy density and energy intake in palliative care cancer patients. *Clin. Nutr.* **24**, 266–273.
72. Westerterp-Plantenga, M. S. (2004). Effects of energy density of daily food intake on long-term energy intake. *Physiol Behav.* **81**, 765–771.
73. Westerterp-Plantenga, M. S. (2004). Modulatory factors in the effect of energy density on energy intake. *Br. J. Nutr.* **92** (Suppl 1), S35–S39.
74. Yeomans M.R. Weinberg L. James S.(2005) Effects of palatability and learned satiety on energy density influences on breakfast intake in humans. *Physiol. Behav.*
75. Cuco, G., Arija, V., Marti-Henneberg, C., and Fernandez-Ballart, J. (2001). Food and nutritional profile of high energy density consumers in an adult Mediterranean population. *Eur. J. Clin. Nutr.* **55**, 192–199.
76. Drewnowski, A., Almiron–Roig, E., Marmonier, C., and Lluch, A. (2004). Dietary energy density and body weight: Is there a relationship? *Nutr. Rev.* **62**, 403–413.
77. USDA. (1996). Continuing Survey of Food Intakes by Individuals (CSFII): Diet and health knowledge survey, 1994. Beltsville, MD: Department of Agriculture, Beltsville Human Nutrition Research Center.
78. Poppitt, S. D., and Prentice, A. M. (1996). Energy density and its role in the control of food intake: Evidence from metabolic and community studies. *Appetite* **26**, 153–174.
79. Kendall, A., Levitsky, D. A., Strupp, B. J., and Lissner, L. (1991). Weight loss on a low-fat diet: Consequence of the imprecision of the control of food intake in humans. *Am. J. Clin. Nutr.* **53**, 1124–1129.
80. Levitsky, D. A. (2005). The non-regulation of food intake in humans: Hope for reversing the epidemic of obesity. *Physiol. Behav.* **86**, 623–632.
81. Montignac M. (1989). "Dine Out and Lose Weight: The French Way to Culinary 'Savoir Vivre.'" Paris: Editions Artulen.
82. Steward, H., Bethea, M., Andrews, S., Balart, L., and Brennan, E. (1995). "Sugar Busters! Cut Sugar to Trim Fat." Westminster, MD: Ballantine.
83. Atkins, R. (1999). "Dr. Atkins' Age–Defying Diet Revolution." St. Martin's, New York.
84. Atkins, R. (1972). "Dr. Atkins' Diet Revolution: The High Calorie Way to Stay Thin Forever." MacKay, New York.
85. Sears, B., and Lawren, B. (1996). "The Zone: A Dietary Road Map to Lose Weight Permanently: Reset Your Genetic Code: Prevent Disease: Achieve Maximum Physical Performance." Harper Collins, New York.
86. Heller, R., and Heller, R. (1992). "The Carbohydrate Addict's Diet: The Carbohydrate Addict's Program for Success." Signet, New York.
87. Eades, M., and Eades, M. (1999). "The Protein Power Lifeplan." Warner, New York.
88. Tieken, S. M., Leidy, H. J., Stull, A. J., Mattes, R. D., Schuster, R. A., and Campbell, W. W. (2007). Effects of solid versus liquid meal-replacement products of similar energy content on hunger, satiety, and appetite-regulating hormones in older adults. *Horm. Metab. Res.* **39**, 389–394.
89. Mattes, R. (2006). Fluid calories and energy balance: The good, the bad, and the uncertain. *Physiol. Behav.* **89**, 66–70.
90. Mattes, R. D. (2006). Fluid energy—Where's the problem? *J. Am. Diet. Assoc.* **106**, 1956–1961.
91. DiMeglio, D. P., and Mattes, R. D. (2000). Liquid versus solid carbohydrate: Effects on food intake and body weight. *Int. J. Obes. Relat. Metab. Disord.* **24**, 794–800.
92. Gibney, M. (1990). Dietary guidelines: A critical appraisal. *J. Hum. Nutr. Diet.* **6**, 13–22.
93. Bolton-Smith, C., and Woodward, M. (1994). Dietary-composition and fat to sugar ratios in relation to obesity. *Int. J. Obes.* **18**, 820–828.

94. Kirk, T. R. (2000). Role of dietary carbohydrate and frequent eating in body-weight control. *Proc. Nutr. Soc.* **59**, 349–358.
95. Rolls, B. J., Castellanos, V. H., Halford, J. C., *et al.* (1998). Volume of food consumed affects satiety in men. *Am. J. Clin. Nutr.* **67**, 1170–1177.
96. Saris, W. H., Astrup, A., Prentice, A. M., *et al.* (2000). Randomized controlled trial of changes in dietary carbohydrate/fat ratio and simple vs complex carbohydrates on body weight and blood lipids: The CARMEN study. The Carbohydrate Ratio Management in European National diets. *Int. J. Obes. Relat. Metab. Disord.* **24**, 1310–1318.
97. Kirk, T., Crombie, N., and Cursiter, M. (2000). Promotion of dietary carbohydrate as an approach to weight maintenance after initial weight loss: A pilot study. *J. Hum. Nutr. Diet.* **13**, 277–285.
98. Kirk, T. R., Burkill, S., and Cursiter, M. (1997). Dietary fat reduction achieved by increasing consumption of a starchy food—An intervention study. *Eur. J. Clin. Nutr.* **51**, 455–461.
99. West, J. A., and de Looy, A. E. (2001). Weight loss in overweight subjects following low-sucrose or sucrose-containing diets. *Int. J. Obes.* **25**, 1122–1128.
100. Davey, G. K., Spencer, E. A., Appleby, P. N., Allen, N. E., Knox, K. H., and Key, T. J. (2003). EPIC-Oxford: Lifestyle characteristics and nutrient intakes in a cohort of 33,883 meat-eaters and 31,546 non meat-eaters in the UK. *Public Health Nutr.* **6**, 259–269.
101. Appleby, P. N., Thorogood, M., Mann, J. I., and Key, T. J. (1998). Low body mass index in non-meat eaters: The possible roles of animal fat, dietary fibre and alcohol. *Int. J. Obes. Relat. Metab. Disord.* **22**, 454–460.
102. Chang-Claude, J., Frentzel-Beyme, R., and Eilber, U. (1992). Mortality pattern of German vegetarians after 11 years of follow-up. *Epidemiology* **3**, 395–401.
103. Famodu, A. A., Osilesi, O., Makinde, Y. O., and Osonuga, O. A. (1998). Blood pressure and blood lipid levels among vegetarian, semi-vegetarian, and non-vegetarian native Africans. *Clin. Biochem.* **31**, 545–549.
104. Fraser, G. E. (1999). Associations between diet and cancer, ischemic heart disease, and all-cause mortality in non-Hispanic white California Seventh-day Adventists. *Am. J. Clin. Nutr.* **70**, 532S–538S.
105. Key, T., and Davey, G. (1996). Prevalence of obesity is low in people who do not eat meat. *BMJ* **313**, 816–817.
106. Key, T. J., Fraser, G. E., Thorogood, M., *et al.* (1998). Mortality in vegetarians and non-vegetarians: A collaborative analysis of 8300 deaths among 76,000 men and women in five prospective studies. *Public Health Nutr.* **1**, 33–41.
107. Li, D., Sinclair, A., Mann, N., *et al.* (1999). The association of diet and thrombotic risk factors in healthy male vegetarians and meat–eaters. *Eur. J. Clin. Nutr.* **53**, 612–619.
108. Lu, S. C., Wu, W. H., Lee, C. A., Chou, H. F., Lee, H. R., and Huang, P. C. (2000). LDL of Taiwanese vegetarians are less oxidizable than those of omnivores. *J. Nutr.* **130**, 1591–1596.
109. Melby, C. L., Goldflies, D. G., and Toohey, M. L. (1993). Blood-pressure differences in older Black and White long-term vegetarians and nonvegetarians. *J. Am. Coll. Nutr.* **12**, 262–269.
110. Newby, P. K., Tucker, K. L., and Wolk, A. (2005). Risk of overweight and obesity among semivegetarian, lactovegetarian, and vegan women. *Am. J. Clin. Nutr.* **81**, 1267–1274.
111. Rosell, M., Appleby, P., Spencer, E., and Key, T. (2006). Weight gain over 5 years in 21,966 meat-eating, fish-eating, vegetarian, and vegan men and women in EPIC-Oxford. *Int. J. Obes. (Lond.)* **30**, 1389–1396.
112. Sanders, T. A., Ellis, F. R., and Dickerson, J. W. (1978). Studies of vegans: The fatty acid composition of plasma choline phosphoglycerides, erythrocytes, adipose tissue, and breast milk, and some indicators of susceptibility to ischemic heart disease in vegans and omnivore controls. *Am. J. Clin. Nutr.* **31**, 805–813.
113. Slattery, M. L., Jacobs, D. R., Jr., Hilner, J. E., *et al.* (1991). Meat consumption and its associations with other diet and health factors in young adults: The CARDIA study. *Am. J. Clin. Nutr.* **54**, 930–935.
114. Snowdon, D. A., and Phillips, R. L. (1985). Does a vegetarian diet reduce the occurrence of diabetes? *Am. J. Public Health* **75**, 507–512.
115. Spencer, E. A., Appleby, P. N., Davey, G. K., and Key, T. J. (2003). Diet and body mass index in 38,000 EPIC-Oxford meat-eaters, fish-eaters, vegetarians and vegans. *Int. J. Obes. Relat. Metab. Disord.* **27**, 728–734.
116. Thorogood, M., Mann, J., Appleby, P., and McPherson, K. (1994). Risk of death from cancer and ischaemic heart disease in meat and non-meat eaters. *BMJ* **308**, 1667–1670.
117. Toohey, M. L., Harris, M. A., DeWitt, W., Foster, G., Schmidt, W. D., and Melby, C. L. (1998). Cardiovascular disease risk factors are lower in African-American vegans compared to lacto-ovo-vegetarians. *J. Am. Coll. Nutr.* **17**, 425–434.
118. Ludwig, D. S., Majzoub, J. A., Al Zahrani, A., Dallal, G. E., Blanco, I., and Roberts, S. B. (1999). High glycemic index foods, overeating, and obesity. *Pediatrics* **103**, E26.
119. Benini, L., Castellani, G., Brighenti, F., *et al.* (1995). Gastric emptying of a solid meal is accelerated by the removal of dietary fibre naturally present in food. *Gut* **36**, 825–830.
120. Haber, G. B., Heaton, K. W., Murphy, D., and Burroughs, L. F. (1977). Depletion and disruption of dietary fibre. Effects on satiety, plasma-glucose, and serum-insulin. *Lancet* **2**, 679–682.
121. Holt, S., Brand, J., Soveny, C., and Hansky, J. (1992). Relationship of satiety to postprandial glycaemic, insulin and cholecystokinin responses. *Appetite* **18**, 129–141.
122. Jimenez-Cruz, A., Gutierrez-Gonzalez, A. N., and Bacardi-Gascon, M. (2005). Low glycemic index lunch on satiety in overweight and obese people with type 2 diabetes. *Nutr. Hosp.* **20**, 348–350.
123. Granfeldt, Y., Liljeberg, H., Drews, A., Newman, R., and Bjorck, I. (1994). Glucose and insulin responses to barley products: Influence of food structure and amylose–amylopectin ratio. *Am. J. Clin. Nutr.* **59**, 1075–1082.
124. Gustafsson, K., Asp, NG, Hagander, B., and Nyman, M. (1995). Satiety effects of spinach in mixed meals: Comparison with other vegetables. *Int. J. Food Sci. Nutr.* **46**, 327–334.
125. Holt, S. H. A., Delargy, H. J., Lawton, C. L., and Blundell, J. E. (1999). The effects of high-carbohydrate vs high-fat

breakfasts on feelings of fullness and alertness, and subsequent food intake. *Int. J. Food Sci. Nutr.* **50**, 13–28.
126. Krotkiewski, M. (1984). Effect of guar gum on body-weight, hunger ratings and metabolism in obese subjects. *Br. J. Nutr.* **52**, 97–105.
127. Latner, J. D., and Schwartz, M. (1999). The effects of a high-carbohydrate, high-protein or balanced lunch upon later food intake and hunger ratings. *Appetite* **33**, 119–128.
128. Lavin, J. H., and Read, N. W. (1995). The effect on hunger and satiety of slowing the absorption of glucose: Relationship with gastric emptying and postprandial blood glucose and insulin responses. *Appetite* **25**, 89–96.
129. Leathwood, P., and Pollet, P. (1988). Effects of slow release carbohydrates in the form of bean flakes on the evolution of hunger and satiety in man. *Appetite* **10**, 1–11.
130. Liljeberg, H. G., and Bjorck, I. M. (1996). Delayed gastric emptying rate as a potential mechanism for lowered glycemia after eating sourdough bread: Studies in humans and rats using test products with added organic acids or an organic salt. *Am. J. Clin. Nutr.* **64**, 886–893.
131. Raben, A., Andersen, K., Karberg, M. A., Holst, J. J., and Astrup, A. (1997). Acetylation of or beta-cyclodextrin addition to potato beneficial effect on glucose metabolism and appetite sensations. *Am. J. Clin. Nutr.* **66**, 304–314.
132. Rigaud, D., Paycha, F., Meulemans, A., Merrouche, M., and Mignon, M. (1998). Effect of psyllium on gastric emptying, hunger feeling, and food intake in normal volunteers: A double blind study. *Eur. J. Clin. Nutr.* **52**, 239–245.
133. Rodin, J. (1990). Comparative effects of fructose, aspartame, glucose, and water preloads on calorie and macronutrient intake. *Am. J. Clin. Nutr.* **51**, 428–435.
134. Spitzer, L., and Rodin, J. (1987). Effects of fructose and glucose preloads on subsequent food intake. *Appetite* **8**, 135–145.
135. Rodin, J. (1991). Effects of pure sugar vs. mixed starch fructose loads on food intake. *Appetite* **17**, 213–219.
136. van Amelsvoort, J. M., and Weststrate, J. A. (1992). Amylose–amylopectin ratio in a meal affects postprandial variables in male volunteers. *Am. J. Clin. Nutr.* **55**, 712–718.
137. Rodin, J., Reed, D., and Jamner, L. (1988). Metabolic effects of fructose and glucose: Implications for food intake. *Am. J. Clin. Nutr.* **47**, 683–689.
138. Alfenas, R. C., and Mattes, R. D. (2005). Influence of glycemic index/load on glycemic response, appetite, and food intake in healthy humans. *Diabetes Care* **28**, 2123–2129.
139. Anderson, G. H., Catherine, N. L., Woodend, D. M., and Wolever, T. M. (2002). Inverse association between the effect of carbohydrates on blood glucose and subsequent short-term food intake in young men. *Am. J. Clin. Nutr.* **76**, 1023–1030.
140. Guss, J. L., Kissileff, H. R., and Pi-Sunyer, F. X. (1994). Effects of glucose and fructose solutions on food intake and gastric emptying in nonobese women. *Am. J. Physiol.* **267**, R1537–1544.
141. Holm, J., and Bjorck, I. (1992). Bioavailability of starch in various wheat-based bread products: Evaluation of metabolic responses in healthy subjects and rate and extent of in vitro starch digestion. *Am. J. Clin. Nutr.* **55**, 420–429.
142. Holt, S. H., and Miller, J. B. (1995). Increased insulin responses to ingested foods are associated with lessened satiety. *Appetite* **24**, 43–54.
143. Raben, A., Tagliabue, A., Christensen, N. J., Madsen, J., Holst, J. J., and Astrup, A. (1994). Resistant starch—The effect on postprandial glycemia, hormonal response, and satiety. *Am. J. Clin. Nutr.* **60**, 544–551.
144. Raben, A., and Astrup, A. (2000). Leptin is influenced both by predisposition to obesity and diet composition. *Int. J. Obes. Relat. Metab. Disord.* **24**, 450–459.
145. Liljeberg, H., Granfeldt, Y., and Bjorck, I. (1992). Metabolic responses to starch in bread containing intact kernels versus milled flour. *Eur. J. Clin. Nutr.* **46**, 561–575.
146. Weststrate, J. A., and van Amelsvoort, J. M. (1993). Effects of the amylose content of breakfast and lunch on postprandial variables in male volunteers. *Am. J. Clin. Nutr.* **58**, 180–186.
147. Agus, M. S., Swain, J. F., Larson, C. L., Eckert, E. A., and Ludwig, D. S. (2000). Dietary composition and physiologic adaptations to energy restriction. *Am. J. Clin. Nutr.* **71**, 901–907.
148. Dumesnil, J. G., Turgeon, J., Tremblay, A., *et al.* (2001). Effect of a low-glycaemic index–low-fat–high protein diet on the atherogenic metabolic risk profile of abdominally obese men. *Br J. Nutr.* **86**, 557–568.
149. Warren, J. M., Henry, C. J., and Simonite, V. (2003). Low glycemic index breakfasts and reduced food intake in preadolescent children. *Pediatrics* **112**, e414.
150. Barkeling, B., Granfelt, Y., Bjorck, I., and Rossner, S. (1995). Effects of carbohydrates in the form of pasta and bread on food-intake and satiety in man. *Nutr. Res.* **15**, 467–476.
151. Reid, M., Hammersley, R., Hill, A. J., and Skidmore, P. (2007). Long-term dietary compensation for added sugar: Effects of supplementary sucrose drinks over a 4-week period. *Br. J. Nutr.* **97**, 193–203.
152. Gonlachanvit, S., Hsu, C. W., Boden, G. H., *et al.* (2003). Effect of altering gastric emptying on postprandial plasma glucose concentrations following a physiologic meal in type-II diabetic patients. *Dig. Dis. Sci.* **48**, 488–497.
153. Cotton, J. R., Weststrate, J. A., and Blundell, J. E. (1996). Replacement of dietary fat with sucrose polyester: Effects on energy intake and appetite control in nonobese males. *Am. J. Clin. Nutr.* **63**, 891–896.
154. Kelly, S. M., Shorthouse, M., Cotterell, J. C., *et al.* (1998). A 3-month, double-blind, controlled trial of feeding with sucrose polyester in human volunteers. *Br. J. Nutr.* **80**, 41–49.
155. Blundell, J. E., and Hill, A. J. (1986). Paradoxical effects of an intense sweetener (aspartame) on appetite. *Lancet* **1**, 1092–1093.
156. Stellman, S. D., and Garfinkel, L. (1986). Artificial sweetener use and one-year weight change among women. *Prev. Med.* **15**, 195–202.
157. Rogers, P. J., and Blundell, J. E. (1989). Separating the actions of sweetness and calories—Effects of saccharin and carbohydrates on hunger and food-intake in human subjects. *Physiol. Behav.* **45**, 1093–1099.
158. Lavin, P. T., Sanders, P. G., Mackey, M. A., and Kotsonis, F. N. (1994). Intense sweeteners use and weight change among women—A critique of the Stellman and Garfinkel study. *J. Am. Coll. Nutr.* **13**, 102–105.

159. Renwick, A. G. (1994). Intense sweeteners, food-intake, and the weight of a body of evidence. *Physiol. Behav.* **55**, 139–143.
160. McCann, M. B., Trulson, M. F., and Stulb, S. C. (1956). Non-caloric sweeteners and weight reduction. *J. Am. Diet. Assoc.* **32**, 327–330.
161. Parham, E. S., and Parham, A. R. Jr. (1980). Saccharin use and sugar intake by college students. *J. Am. Diet. Assoc.* **76**, 560–563.
162. Colditz, G. A., Willett, W. C., Stampfer, M. J., London, S. J., Segal, M. R., and Speizer, F. E. (1990). Patterns of weight change and their relation to diet in a cohort of healthy women. *Am. J. Clin. Nutr.* **51**, 1100–1105.
163. Richardson, J. F. (1972). The sugar intake of businessmen and its inverse relationship with relative weight. *Br. J. Nutr.* **27**, 449–460.
164. Chen, L. N. A., and Parham, E. S. (1991). College students' use of high-intensity sweeteners is not consistently associated with sugar consumption. *J. Am. Diet. Assoc.* **91**, 686–689.
165. Naismith, D. J., and Rhodes, C. (1995). Adjustment in energy-intake following the covert removal of sugar from the diet. *J. Hum. Nutr. Diet* **8**, 167–175.
166. Wooley, S. C. (1972). Physiologic versus cognitive factors in short term food regulation in the obese and nonobese. *Psychosom. Med.* **34**, 62–68.
167. Caputo, F. A., and Mattes, R. D. (1993). Human dietary responses to perceived manipulation of fat content in a mid-day meal. *Int. J. Obes. Relat. Metab. Disord.* **17**, 237–240.
168. Shide, D. J., and Rolls, B. J. (1995). Information about the fat content of preloads influences energy intake in healthy women. *J. Am. Diet. Assoc.* **95**, 993–998.
169. Rolls, B. J., Laster, L. J., and Summerfelt, A. (1989). Hunger and food intake following consumption of low-calorie foods. *Appetite* **13**, 115–127.
170. Miller, D. L., Castellanos, V. H., Shide, D. J., Peters, J. C., and Rolls, B. J. (1998). Effect of fat-free potato chips with and without nutrition labels on fat and energy intakes. *Am. J. Clin. Nutr.* **68**, 282–290.
171. Blackburn, G. L., Kanders, B. S., Lavin, P. T., Keller, S. D., and Whatley, J. (1997). The effect of aspartame as part of a multidisciplinary weight-control program on short- and long-term control of body weight. *Am. J. Clin. Nutr.* **65**, 409–418.
172. Baba, N. H., Sawaya, S., Torbay, N., Habbal, Z., Azar, S., and Hashim, S. A. (1999). High protein vs high carbohydrate hypoenergetic diet for the treatment of obese hyperinsulinemic subjects. *Int. J. Obes. Relat. Metab. Disord.* **23**, 1202–1206.
173. Boden, G., Sargrad, K., Homko, C., Mozzoli, M., and Stein, T. P. (2005). Effect of a low-carbohydrate diet on appetite, blood glucose levels, and insulin resistance in obese patients with type 2 diabetes. *Ann. Intern. Med.* **142**, 403–411.
174. Brehm, B. J., Seeley, R. J., Daniels, S. R., and D'Alessio, D. A. (2003). A randomized trial comparing a very low carbohydrate diet and a calorie-restricted low fat diet on body weight and cardiovascular risk factors in healthy women. *J. Clin. Endocrinol. Metab.* **88**, 1617–1623.
175. Dansinger, M. L., Gleason, J. A., Griffith, J. L., Selker, H. P., and Schaefer, E. J. (2005). Comparison of the Atkins, Ornish, Weight Watchers, and Zone diets for weight loss and heart disease risk reduction: A randomized trial. *JAMA* **293**, 43–53.
176. Foster, G. D., Wyatt, H. R., Hill, J. O., *et al.* (2003). A randomized trial of a low-carbohydrate diet for obesity. *N. Engl. J. Med.* **348**, 2082–2090.
177. Gately, P. J., King, N. A., Greatwood, H. C., *et al.* (2007). Does a high-protein diet improve weight loss in overweight and obese children? *Obesity* **15**, 1527–1534.
178. Meckling, K. A., Gauthier, M., Grubb, R., and Sanford, J. (2002). Effects of a hypocaloric, low-carbohydrate diet on weight loss, blood lipids, blood pressure, glucose tolerance, and body composition in free-living overweight women. *Can. J. Physiol. Pharmacol.* **80**, 1095–1105.
179. Meckling, K. A., O'Sullivan, C., and Saari, D. (2004). Comparison of a low-fat diet to a low-carbohydrate diet on weight loss, body composition, and risk factors for diabetes and cardiovascular disease in free-living, overweight men and women. *J. Clin. Endocrinol. Metab.* **89**, 2717–2723.
180. Nickols-Richardson, S. M., Coleman, MD, Volpe, J. J., and Hosig, KW. (2005). Perceived hunger is lower and weight loss is greater in overweight premenopausal women consuming a low-carbohydrate/high-protein vs high-carbohydrate/low-fat diet. *J. Am. Diet. Assoc.* **105**, 1433–1437.
181. Samaha, F. F., Iqbal, N., Seshadri, P., *et al.* (2003). A low-carbohydrate as compared with a low-fat diet in severe obesity. *N. Engl. J. Med.* **348**, 2074–2081.
182. Skov, A. R., Toubro, S., Ronn, B., Holm, L., and Astrup, A. (1999). Randomized trial on protein vs carbohydrate in ad libitum fat reduced diet for the treatment of obesity. *Int. J. Obes. Relat. Metab. Disord.* **23**, 528–536.
183. Stern, L., Iqbal, N., Seshadri, P., *et al.* (2004). The effects of low-carbohydrate versus conventional weight loss diets in severely obese adults: one–year follow-up of a randomized trial. *Ann. Intern. Med.* **140**, 778–785.
184. Worthington, B. S., and Taylor, L. E. (1974). Balanced low-calorie vs. high-protein–low-carbohydrate reducing diets. I. Weight loss, nutrient intake, and subjective evaluation. *J. Am. Diet. Assoc.* **64**, 47–51.
185. Yancy, W. S. Jr., Olsen, M. K., Guyton, J. R., Bakst, R. P., and Westman, E. C. (2004). A low-carbohydrate, ketogenic diet versus a low-fat diet to treat obesity and hyperlipidemia: A randomized, controlled trial. *Ann. Intern. Med.* **140**, 769–777.
186. Fontana, L., Meyer, T. E., Klein, S., and Holloszy, J. O. (2007). Long-term low-calorie low-protein vegan diet and endurance exercise are associated with low cardiometabolic risk. *Rejuv. Res.* **10**, 225–234.
187. Westerterp, K. R., Wilson, S. A., and Rolland, V. (1999). Diet induced thermogenesis measured over 24h in a respiration chamber: effect of diet composition. *Int. J. Obes. Relat. Metab. Disord.* **23**, 287–292.
188. Crovetti, R., Porrini, M., Santangelo, A., and Testolin, G. (1998). The influence of thermic effect of food on satiety. *Eur. J. Clin. Nutr.* **52**, 482–488.
189. Hendler, R., and Bonde, A. A., 3rd. (1988). Very-low-calorie diets with high and low protein content: Impact on triiodothyronine, energy expenditure, and nitrogen balance. *Am. J. Clin. Nutr.* **48**, 1239–1247.

190. Karst, H., Steiniger, J., Noack, R., and Steglich, H. D. (1984). Diet-induced thermogenesis in man: Thermic effects of single proteins, carbohydrates and fats depending on their energy amount. *Ann. Nutr. Metab.* **28**, 245–252.
191. Nair, K. S., Halliday, D., and Garrow, J. S. (1983). Thermic response to isoenergetic protein, carbohydrate or fat meals in lean and obese subjects. *Clin. Sci. (Lond.)* **65**, 307–312.
192. Robinson, S. M., Jaccard, C., Persaud, C., Jackson, A. A., Jequier, E., and Schutz, Y. (1990). Protein turnover and thermogenesis in response to high-protein and high-carbohydrate feeding in men. *Am. J. Clin. Nutr.* **52**, 72–80.
193. Schutz, Y., Bray, G., and Margen, S. (1987). Postprandial thermogenesis at rest and during exercise in elderly men ingesting two levels of protein. *J. Am. Coll. Nutr.* **6**, 497–506.
194. Steiniger, J., Karst, H., Noack, R., and Steglich, H. D. (1987). Diet-induced thermogenesis in man: Thermic effects of single protein and carbohydrate test meals in lean and obese subjects. *Ann. Nutr. Metab.* **31**, 117–125.
195. Swaminathan, R., King, R. F., Holmfield, J., Siwek, R. A., Baker, M., and Wales, J. K. (1985). Thermic effect of feeding carbohydrate, fat, protein and mixed meal in lean and obese subjects. *Am. J. Clin. Nutr.* **42**, 177–181.
196. Welle, S., Lilavivat, U., and Campbell, R. G. (1981). Thermic effect of feeding in man: increased plasma norepinephrine levels following glucose but not protein or fat consumption. *Metabolism* **30**, 953–958.
197. Eisenstein, E. L., Shaw, L. K., Nelson, C. L., Anstrom, K. J., Hakim, Z., and Mark, D. B. (2002). Obesity and long-term clinical and economic outcomes in coronary artery disease patients. *Obes. Res.* **10**, 83–91.
198. Alford, B. B., Blankenship, A. C., and Hagen, R. D. (1990). The effects of variations in carbohydrate, protein, and fat content of the diet upon weight loss, blood values, and nutrient intake of adult obese women. *J. Am. Diet. Assoc.* **90**, 534–540.
199. DeHaven, J., Sherwin, R., Hendler, R., and Felig, P. (1980). Nitrogen and sodium balance and sympathetic-nervous-system activity in obese subjects treated with a low-calorie protein or mixed diet. *N. Engl. J. Med.* **302**, 477–482.
200. Piatti, P. M., Monti, F., Fermo, I., *et al.* (1994). Hypocaloric high-protein diet improves glucose oxidation and spares lean body mass: comparison to hypocaloric high-carbohydrate diet. *Metabolism* **43**, 1481–1487.
201. Vazquez, J. A., Kazi, U., and Madani, N. (1995). Protein metabolism during weight reduction with very-low-energy diets: Evaluation of the independent effects of protein and carbohydrate on protein sparing. *Am. J. Clin. Nutr.* **62**, 93–103.
202. Yang, M. U., and van Itallie, T. B. (1984). Variability in body protein loss during protracted, severe caloric restriction: role of triiodothyronine and other possible determinants. *Am. J. Clin. Nutr.* **40**, 611–622.
203. Eisenstein, J., Roberts, S. B., Dallal, G., and Saltzman, E. (2002). High-protein weight–loss diets: Are they safe and do they work? A review of the experimental and epidemiologic data. *Nutr. Rev.* **60**, 189–200.
204. Krauss, R. M., and Mayer, J. (1965). Influence of protein and amino acids on food intake in rat. *Am. J. Physiol.* **209**, 479–483.
205. Peng, Y. S., Meliza, L. L., Vavich, M. G., and Kemmerer, A. R. (1974). Changes in food-intake and nitrogen-metabolism of rats while adapting to a low or high protein diet. *J. Nutr.* **104**, 1008–1017.
206. Semon, B. A., Leung, P. M. B., Rogers, Q. R., and Gietzen, D. W. (1987). Effect of type of protein on food–intake of rats fed high protein diets. *Physiol. Behav.* **41**, 451–458.
207. Semon, B. A., Leung, P. M., Rogers, Q. R., and Gietzen, D. W. (1988). Increase in plasma ammonia and amino acids when rats are fed a 44% casein diet. *Physiol. Behav* **43**, 631–636.
208. Harper, A. E. (1976). Protein and amino acids in the regulation of food intake. In "Hunger: Basic Mechanisms and Clinical Implications" (D. Novin, W. Wyrwicka, and G. A. Bray, Eds.), pp. 103–113. New York: Raven.
209. Hannah, J. S., Dubey, A. K., and Hansen, B. C. (1990). Postingestional effects of a high-protein diet on the regulation of food intake in monkeys. *Am. J. Clin. Nutr.* **52**, 320–325.
210. McArthur, L. H., Kelly, W. F., Gietzen, D. W., and Rogers, Q. R. (1993). The role of palatability in the food intake response of rats fed high-protein diets. *Appetite* **20**, 181–196.
211. Anderson, G. H., and Moore, S. E. (2004). Dietary proteins in the regulation of food intake and body weight in humans. *J. Nutr.* **134**, 974S–979S.
212. Bowen, J., Noakes, M., and Clifton, P.M. (2007). Appetite hormones and energy intake in obese men after consumption of fructose, glucose and whey protein beverages. *Int. J. Obes.* **31**, 1696–1703.
213. Yudkin, J., and Carey, M. (1960). The treatment of obesity by the "highfat" diet. The inevitability of calories. *Lancet* **2**, 939–941.
214. Stock, A. L., and Yudkin, J. (1970). Nutrient intake of subjects on low carbohydrate diet used in treatment of obesity. *Am. J. Clin. Nutr.* **23**, 948–952.
215. Kennedy, E. (2004). Dietary diversity, diet quality, and body weight regulation. *Nutr. Rev.* **62**, S78–S81.
216. McCrory, M. A., Hajduk, C. L., and Roberts, S. B. (2002). Dietary variety within food groups, not dietary fat, predicts BMI: An analysis using U.S. national survey data. *FASEB J.* **16**, A368–A369.
217. Norton, G. N. M., Anderson, A. S., and Hetherington, M. M. (2006). Volume and variety: Relative effects on food intake. *Physiol. Behav.* **87**, 714–722.
218. Raynor, H. A., and Epstein, L. H. (2001). Dietary variety, energy regulation, and obesity. *Psychol. Bull.* **127**, 325–341.
219. Raynor, H. A., Jeffery, R. W., Phelan, S., Hill, J. O., and Wing, R. R. (2005). Amount of food group variety consumed in the diet and long-term weight loss maintenance. *Obes. Res.* **13**, 883–890.
220. Roberts, S. B., Hajduk, C. L., Howarth, N. C., Russell, R., and McCrory, M. A. (2005). Dietary variety predicts low body mass index and inadequate macronutrient and micronutrient intakes in community–dwelling older adults. *J. Gerontol. A Biol. Sci. Med. Sci.* **60**, 613–621.
221. Rolls, B. J., Rowe, E. A., Rolls, E. T., Kingston, B., Megson, A., and Gunary, R. (1981). Variety in a meal enhances food intake in man. *Physiol. Behav.* **26**, 215–221.
222. Rozin, P., and Markwith, M. (1991). Cross-domain variety seeking in human food choice. *Appetite* **16**, 57–59.

223. Ledikwe, J. H., Rolls, B. J., Smiciklas–Wright, H., *et al.* (2007). Reductions in dietary energy density are associated with weight loss in overweight and obese participants in the PREMIER trial. *Am. J. Clin. Nutr.* **85**, 1212–1221.
224. Howarth, N. C., Huang, T. T., Roberts, S. B., Lin, B. H., and McCrory, M. A. (2007). Eating patterns and dietary composition in relation to BMI in younger and older adults. *Int. J. Obes. (Lond.)* **31**, 675–684.
225. Jackson, M., Walker, S., Forrester, T., Cruickshank, J. K., and Wilks, R. (2003). Social and dietary determinants of body mass index of adult Jamaicans of African origin. *Eur. J. Clin. Nutr.* **57**, 621–627.
226. Guillaume, M., Lapidus, L., and Lambert, A. (1998). Obesity and nutrition in children. The Belgian Luxembourg Child Study IV. *Eur. J. Clin. Nutr.* **52**, 323–328.
227. Archer, W. R., Lamarche, B., Deriaz, O., *et al.* (2003). Variations in body composition and plasma lipids in response to a high-carbohydrate diet. *Obes. Res.* **11**, 978–986.
228. Black, H. S., Herd, J. A., Goldberg, L. H., *et al.* (1994). Effect of a low-fat diet on the incidence of actinic keratosis. *N. Engl. J. Med.* **330**, 1272–1275.
229. Bloemberg, B. P. M., Kromhout, D., Goddijn, H. E., Jansen, A., and Obermanndeboer, G. L. (1991). The impact of the Guidelines for a Healthy Diet of the Netherlands Nutrition Council on total and high-density-lipoprotein cholesterol in hypercholesterolemic free-living men. *Am. J. Epidemiol.* **134**, 39–48.
230. Boyd, N. F., McGuire, V., Shannon, P., *et al.* (1988). Effect of a low-fat high-carbohydrate diet on symptoms of cyclical mastopathy. *Lancet* **2**, 128–132.
231. Buzzard, I. M., Asp, E. H., Chlebowski, R. T., *et al.* (1990). Diet intervention methods to reduce fat intake: Nutrient and food group composition of self-selected low-fat diets. *J. Am. Diet. Assoc.* **90**, 42–53.
232. Chlebowski, R. T., Blackburn, G. L., Thomson, C. A., *et al.* (2006). Dietary fat reduction and breast cancer outcome: Interim efficacy results from the Women's Intervention Nutrition Study. *J. Natl. Cancer Inst.* **98**, 1767–1776.
233. Djuric, Z., Poore, K. M., Depper, J. B., *et al.* (2002). Methods to increase fruit and vegetable intake with and without a decrease in fat intake: Compliance and effects on body weight in the nutrition and breast health study. *Nutr. Cancer* **43**, 141–151.
234. Howard, B. V., Manson, J. E., Stefanick, M. L., *et al.* (2006). Low-fat dietary pattern and weight change over 7 years: The Women's Health Initiative Dietary Modification Trial. *JAMA* **295**, 39–49.
235. Hunninghake, D. B., Stein, E. A., Dujovne, C. A., *et al.* (1993). The efficacy of intensive dietary therapy alone or combined with lovastatin in outpatients with hypercholesterolemia. *N. Engl. J. Med.* **328**, 1213–1219.
236. Insull, W., Henderson, M. M., Prentice, R. L., *et al.* (1990). Results of a randomized feasibility study of a low-fat diet. *Arch. Int. Med.* **150**, 421–427.
237. Jeffery, R. W., Hellerstedt, W. L., French, S. A., and Baxter, J. E. (1995). A randomized trial of counseling for fat restriction versus calorie restriction in the treatment of obesity. *Int. J. Obes. Relat. Metab. Disord.* **19**, 132–137.
238. Kasim, S. E., Martino, S., Kim, P. N., *et al.* (1993). Dietary and anthropometric determinants of plasma lipoproteins during a long-term low-fat diet in healthy women. *Am. J. Clin. Nutr.* **57**, 146–153.
239. Knopp, R. H., Walden, C. E., Retzlaff, B. M., *et al.* (1997). Long-term cholesterol-lowering effects of 4 fat-restricted diets in hypercholesterolemic and combined hyperlipidemic men. The Dietary Alternatives Study. *JAMA* **278**, 1509–1515.
240. Lee-Han, H., Cousins, M., Beaton, M., *et al.* (1988). Compliance in a randomized clinical trial of dietary fat reduction in patients with breast dysplasia. *Am. J. Clin. Nutr.* **48**, 575–586.
241. Marckmann, P. (1994). Fat from fish-oil. *Int. J. Obes.* **18**, 185.
242. McManus, K., Antinoro, L., and Sacks, F. (2001). A randomized controlled trial of a moderate-fat, low-energy diet compared with a low fat, low-energy diet for weight loss in overweight adults. *Int. J. Obes. Relat. Metab. Disord.* **25**, 1503–1511.
243. Ornish, D., Brown, S. E., Scherwitz, L. W., *et al.* (1990). Can lifestyle changes reverse coronary heart disease? The Lifestyle Heart Trial. *Lancet* **336**, 129–133.
244. Petersen, M., Taylor, M. A., Saris, W. H., *et al.* (2006). Randomized, multi-center trial of two hypo-energetic diets in obese subjects: High- versus low-fat content. *Int. J. Obes. (Lond.)* **30**, 552–560.
245. Poppitt, S. D., Keogh, G. F., Prentice, A. M., *et al.* (2002). Long-term effects of ad libitum low-fat, high-carbohydrate diets on body weight and serum lipids in overweight subjects with metabolic syndrome. *Am. J. Clin. Nutr.* **75**, 11–20.
246. Pritchard, J. E., Nowson, C. A., and Wark, J. D. (1996). Bone loss accompanying diet-induced or exercise-induced weight loss: A randomised controlled study. *Int. J. Obes.* **20**, 513–520.
247. Raben, A., Jensen, N. D., Marckmann, P., Sandstrom, B., and Astrup, A. (1995). Spontaneous weight loss during 11 weeks' ad libitum intake of a low fat/high fiber diet in young, normal weight subjects. *Int. J. Obes. Relat. Metab. Disord.* **19**, 916–923.
248. Rock, C. L., Thomson, C., Caan, B. J., *et al.* (2001). Reduction in fat intake is not associated with weight loss in most women after breast cancer diagnosis: Evidence from a randomized controlled trial. *Cancer* **91**, 25–34.
249. Schlundt, D. G., Hill, J. O., Pope-Cordle, J., Arnold, D., Virts, K. L., and Katahn, M. (1993). Randomized evaluation of a low fat ad libitum carbohydrate diet for weight reduction. *Int. J. Obes. Relat. Metab. Disord.* **17**, 623–629.
250. Segal-Isaacson, C. J., Johnson, S., Tomuta, V., Cowell, B., and Stein, D. T. (2004). A randomized trial comparing low-fat and low-carbohydrate diets matched for energy and protein. *Obes. Res.* **12**(Suppl 2), 130S–140S.
251. Shah, M., McGovern, P., French, S., and Baxter, J. (1994). Comparison of a low-fat, ad libitum complex-carbohydrate diet with a low-energy diet in moderately obese women. *Am. J. Clin. Nutr.* **59**, 980–984.
252. Sheppard, L., Kristal, A. R., and Kushi, L. H. (1991). Weight loss in women participating in a randomized trial of low-fat diets. *Am. J. Clin. Nutr.* **54**, 821–828.

253. Siggaard, R., Raben, A., and Astrup, A. (1996). Weight loss during 12 weeks' ad libitum carbohydrate-rich diet in overweight and normal-weight subjects at a Danish work site. *Obes. Res.* **4**, 347–356.
254. Simon, M. S., Heilbrun, L. K., Boomer, A., *et al.* (1997). A randomized trial of a low-fat dietary intervention in women at high risk for breast cancer. *Nutr. Cancer* **27**, 136–142.
255. Singh, R. B., Rastogi, S. S., Verma, R., *et al.* (1992). Randomised controlled trial of cardioprotective diet in patients with recent acute myocardial infarction: Results of one year follow up. *BMJ* **304**, 1015–1019.
256. Sorensen, T. I., Boutin, P., Taylor, M. A., *et al.* (2006). Genetic polymorphisms and weight loss in obesity: A randomised trial of hypo-energetic high- versus low-fat diets. *PLoS Clin. Trials* **1**, e12.
257. Stefanick, M. L., Mackey, S., Sheehan, M., Ellsworth, N., Haskell, W. L., and Wood, P. D. (1998). Effects of diet and exercise in men and postmenopausal women with low levels of HDL cholesterol and high levels of LDL cholesterol. *N. Engl. J. Med.* **339**, 12–20.
258. Thomson, C. A., Rock, C. L., Giuliano, A. R., *et al.* (2005). Longitudinal changes in body weight and body composition among women previously treated for breast cancer consuming a high-vegetable, fruit and fiber, low-fat diet. *Eur J. Nutr.* **44**, 18–25.
259. Toubro, S., and Astrup, A. (1997). Randomised comparison of diets for maintaining obese subjects' weight after major weight loss: Ad lib, low fat, high carbohydrate diet v fixed energy intake. *BMJ* **314**, 29–34.
260. Winters, B. L., Mitchell, D. C., Smiciklas-Wright, H., Grosvenor, M. B., Liu, W., and Blackburn, G. L. (2004). Dietary patterns in women treated for breast cancer who successfully reduce fat intake: The Women's Intervention Nutrition Study (WINS). *J. Am. Diet. Assoc.* **104**, 551–559.
261. USDA (2007). U.S. food supply: Nutrients and other food components, 1909 to 2004. Center for Nutrition Policy and Promotion, United States Department of Agriculture, **57**, 1–48.
262. Zhang, Q., and Wang, Y. (2004). Trends in the association between obesity and socioeconomic status in U.S. adults: 1971 to 2000. *Obes. Res.* **12**, 1622–1632.
263. Kant, A. K., Schatzkin, A., Graubard, B. I., and Ballard-Barbash, R. (1995). Frequency of eating occasions and weight change in the NHANES I Epidemiologic Follow-up Study. *Int. J. Obes. Relat. Metab. Disord.* **19**, 468–474.
264. Jackson, M., Walker, S., Cruickshank, JK, *et al.* (2007). Diet and overweight and obesity in populations of African origin: Cameroon, Jamaica and the UK. *Public Health Nutr.* **10**, 122–130.
265. Maskarinec, G., Takata, Y., Pagano, I., *et al.* (2006). Trends and dietary determinants of overweight and obesity in a multiethnic population. *Obes. (Silver Spring)* **14**, 717–726.
266. Toeller, M., Buyken, A. E., Heitkamp, G., Cathelineau, G., Ferriss, B., and Michel, G. (2001). Nutrient intakes as predictors of body weight in European people with type 1 diabetes. *Int. J. Obes. Relat. Metab. Disord.* **25**, 1815–1822.
267. Ma, Y., Olendzki, B., Chiriboga, D., *et al.* (2005). Association between dietary carbohydrates and body weight. *Am. J. Epidemiol.* **161**, 359–367.
268. Howarth, N. C., Saltzman, E., McCrory, M. A., *et al.* (2003). Fermentable and nonfermentable fiber supplements did not alter hunger, satiety or body weight in a pilot study of men and women consuming self-selected diets. *J. Nutr.* **133**, 3141–3144.
269. Porrini, M., Santangelo, A., Crovetti, R., Riso, P. T., and Blundell, J. E. (1997). Weight, protein, fat, and timing of preloads affect food intake. *Physiol. Behav.* **62**, 563–570.

CHAPTER 25

Behavioral Risk Factors for Overweight and Obesity: Diet and Physical Activity

NANCY E. SHERWOOD,[1,2] MARY STORY,[2] DIANNE NEUMARK-SZTAINER,[2] AND KATE BAUER[2]
[1]HealthPartners Research Foundation, Minneapolis, Minnesota
[2]University of Minnesota, Minneapolis, Minnesota

Contents

I. INTRODUCTION

Obesity is a significant problem that affects children, adolescents, and adults across gender, race, and socioeconomic strata [1, 2]. Dramatic increases in the prevalence of obesity in recent years have focused attention on this important public health problem. Comprehensive data of trends since the 1960s in the prevalence of obesity provided by the National Health and Nutrition Examination Surveys (NHANES) show that the percentage of obese adults has increased over time, particularly during the past two decades (Figure 1) [3]. The percentage of adults classified as obese, that is, those with a body mass index (BMI) $\geq$ 30 kg/m^2, increased from 15% in NHANES II (1976–80) to 33% in the most recent NHANES data (2003–04). An additional 33% of adults in NHANES 1999–2000 were classified as overweight according to the BMI $\geq$ 25 kg/m^2 standard. According to these national data, an estimated 66% of the adult population is classified as overweight or obese. Prevalence rates of obesity and overweight among children and adolescents have also increased dramatically since the mid-1960s and appear to be on the rise. NHANES data indicate that the prevalence of overweight (BMI $\geq$ 95th percentile) among youth doubled from 1976–1980 to 1988–1994, increasing from 6.5% to 11.3% for 6- to 11-year-olds and from 5% to 12% for 12- to 17-year-olds (Figure 2) [4]. The prevalence of overweight among preschool children (2- to 5-year-olds) doubled from 1976–80 to 1999–2000, increasing from 5% to 10.3%. The prevalence of overweight continued to rise with 19% of 6- to 11-year-olds and 17% of 12- to 17-year-olds categorized as overweight in 2003–04. Overall, most recent estimates suggest that 17% of U.S. children and adolescents are overweight, and an additional 17% have a BMI between the 85th and 95th percentiles, indicating risk of overweight [5]. The high prevalence of obesity among children is of particular concern given that childhood-onset obesity often tracks with adult obesity [6, 7].

The alarming increase in the prevalence of obesity during the past few decades has raised concerns about associated health risks for children, adolescents, and adults. Persistence of this trend could lead to substantial increases in the number of people affected by obesity-related health conditions and

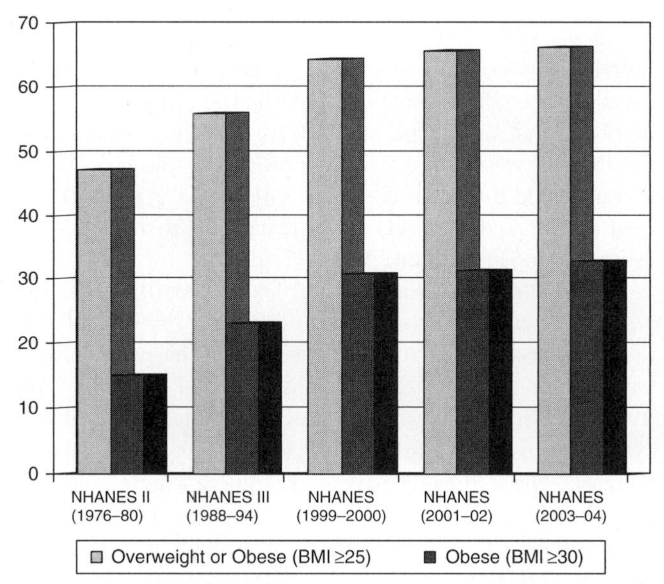

*Age-adjusted by the direct method to the year 2000 U.S. Bureau of the Census estimates using the age groups 20–39, 40–59, and 60–74 years.
**NHANES II did not include individuals over 74 years of age, thus trend estimates are based on age 20–74 years.

FIGURE 1 Age-adjusted* prevalence of overweight and obesity among U.S. adults, age 20–74 years.**

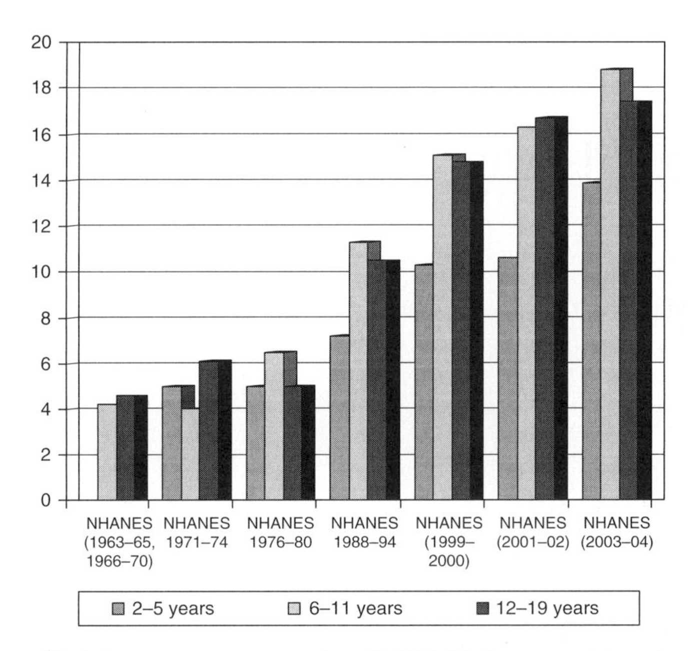

[1]Excludes pregnant women starting with 1971–74. Pregnancy status not available for 1963–65 and 1966–70.

[2]Data for 1963–65 are for children 6–11 years of age; data for 1966–70 are for 12–17 years of age, not 12–19 years.

FIGURE 2 Prevalence of overweight among children and adolescents age 6–19 years.[1,2]

premature mortality. The health risks associated with obesity are numerous and include hypertension, type 2 diabetes mellitus, dyslipidemia, stroke, gallbladder disease, osteoarthritis, sleep apnea, respiratory problems, and certain cancers (e.g., endometrial, breast, prostate, and colon) [8–16]. Obesity is also associated with psychosocial problems such as binge eating disorders and depression for some individuals [17]. Individuals who are obese are also adversely impacted by social bias and discrimination [18–20]. The economic burden of obesity is sizable because of its impact on individual health, costs to society due to lost productivity, and premature mortality and treatment costs [21–28]. In the year 2000, the total (direct and indirect) cost attributable to obesity was estimated at $117 billion [29].

Pediatric obesity is increasingly concurrently associated with significant health problems. Between 1979 and 1999, rates of obesity and associated hospital discharge diagnoses tripled among 6- to 17-year-olds [30]. Two disturbing trends are recognition of type 2 diabetes and the metabolic syndrome in youth. Referred to as adult-onset diabetes in the past, type 2 diabetes is becoming increasingly common in children, accounting for up to 45% of all new cases of diabetes [31]. The prevalence of metabolic syndrome among overweight youth has also been recently documented and implications for cardiovascular risk are concerning [32, 33]. Additional medical and psychosocial concerns associated with childhood overweight include asthma, sleep-disordered breathing, fatty liver disease, depression, low self-esteem, and social stigma [34, 35].

Obesity and overweight are multidetermined chronic problems resulting from complex interactions between genes and an environment characterized by energy imbalance due to sedentary lifestyles and ready access to an abundance of food [36]. Research suggests that obesity runs in families and that some individuals are more vulnerable than others to weight gain and developing obesity [37]. Various mechanisms through which genetic susceptibility to weight gain have been proposed include low resting metabolic rate, low level of lipid oxidation rate, low fat-free mass, and poor appetite control [36]. Genetic research holds considerable promise for understanding the development of obesity and identifying those at risk for obesity. However, the rapid increase in rates of obesity has occurred over too brief a time period for there to have been significant genetic changes in the population. Although body weight is primarily regulated by a series of physiological processes, it is also influenced by behavioral and environmental factors [36]. Recent epidemiological trends in obesity have been linked to behavioral and environmental changes that have occurred in recent years. The higher proportion of fat and the higher energy density of the diet in combination with reductions in physical activity levels and increases in sedentary behavior have been implicated as significant contributors to the obesity epidemic [36, 38]. It is important to note that these dietary and activity behavioral risk factors are modifiable and can be targets for change in obesity prevention and treatment efforts.

Understanding the determinants of obesity and developing appropriate prevention and treatment strategies require an in-depth examination of behavioral risk factors for obesity. The goal of this chapter is to review available data regarding behavioral and environmental determinants of dietary intake and physical activity and to discuss implications for future public health research and intervention. The review will focus on behavioral risk factors for obesity in both children and adults. Figure 3 presents a conceptual model for understanding how behavior risk factors influence weight regulation and obesity.

II. PHYSICAL ACTIVITY

This section examines the role of physical activity in the development of obesity. Multiple factors that influence physical activity levels are discussed, including sedentary behavior, psychological variables such as self-efficacy and social support, and environmental and societal influences.

Prominent among the health benefits associated with a physically active lifestyle is the protective effect of physical activity on obesity. An abundance of cross-sectional research shows that lighter individuals are more active than heavier individuals and prospective research indicates that changes in physical activity level are associated with changes in body weight in the direction predicted by the

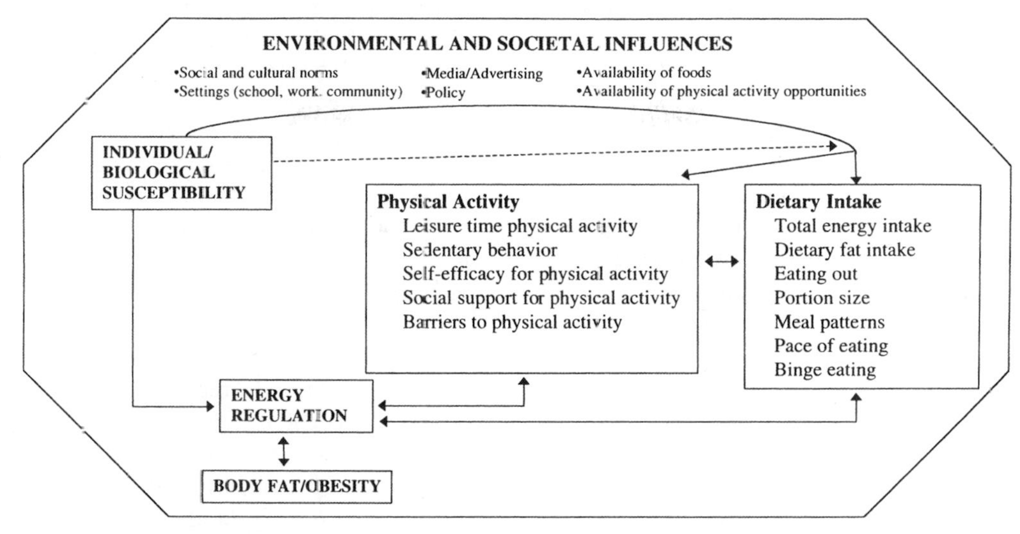

FIGURE 3 Influences on energy balance and obesity. This model shows the direct and indirect influence of biological susceptibility on energy regulation and obesity. This model also shows that both biological susceptibility and environmental factors influence dietary intake and physical activity, which, in turn, directly affect energy regulation and obesity. [Adapted with permission from World Health Organization (1997). "Obesity: Preventing and Managing the Global Epidemic," Report of a WHO Consultation on Obesity, Geneva, June 3–5, 1997. World Health Organization, Geneva.]

energy balance equation [39–48]. The majority of studies conducted in children also find that physical activity levels and body weight are negatively associated [49, 50]. Exercise has also been shown to improve short- and long-term weight loss in experimental studies in both children and adults [45, 50] and is a key factor in successful weight loss maintenance [51, 52]

A. Prevalence of Leisure-Time Physical Activity in Adults

Despite the benefits of physical activity for body weight regulation and health, we are in the midst of a sedentary behavior epidemic. Overall, less than half of adults (48% of men and 43% of women) were active at recommended levels (defined as at least 30 minutes 5 or more days per week in moderate-intensity activities, equivalent to brisk walking, or at least 20 minutes 3 or more days per week in vigorous activities, equivalent to running, heavy yard work, or aerobic dance) [53]. Less than 16% of adults (15% of men and 17% of women) reported no moderate or vigorous activity in a usual week. Demographic differences in physical activity levels have also been observed, with men more likely to engage in vigorous physical activity than women [53]. Physical activity also declines with age, with women experiencing a greater decline in older age groups than men [54]. African-American and Hispanic adults are less physically active than non-Hispanic white adults [55, 56]. Education and income are also both positively associated with physical activity level [57–60].

B. Prevalence of Leisure-Time Physical Activity in Youth

Although estimates of physical activity levels among youth are slightly higher than self-reports by adults, the prevalence of regular physical activity is still surprisingly low. According to data from the CDC's Youth Media Campaign Longitudinal Survey (YMCLS), a nationally representative survey of children aged 9–13 years and their parents, 61.5% of children aged 9–13 years do not participate in any organized physical activity during their non-school hours. Examination of free-time physical activity levels, defined as self-reported engagement in a free-time physical activity, showed that 77.4% of children aged 9–13 reported being engaged in some form of free-time activity during the previous 7 days [61]. Similar to activity patterns with adults, girls tend to be less active than boys and declines with age are more striking for girls compared to boys [55]. African-American girls, in particular, appear to be at greater risk for inactivity [55]. Representative survey data on the physical activity patterns of young children are not available, in part because of methodological difficulties in collecting such data from children. Young children are limited in their ability to accurately recall their activity patterns, and the unplanned, unstructured nature of children's physical activity patterns does not lend itself well to the self-report format employed in large-scale surveys.

C. Sedentary Behavior

Although low levels of leisure-time physical activity likely contribute to the epidemic of obesity, it is noteworthy that among adults, leisure-time activity has remained stable or increased since the mid-1980s, the period during which the prevalence of obesity increased [55]. The past century, however, has produced dramatic changes in physical activity patterns in the United States. Machines with motors have replaced human labor in virtually every aspect of life, so that the energy expenditure now required for daily life is a fraction of what it was a generation or two ago. The consequences of this dramatic change are far reaching and only now are beginning to be carefully studied. It is likely that increases in sedentary activities such as television watching and computer use and decreases in lifestyle, household, and occupational activity that have been less carefully measured have contributed to reductions in overall energy expenditure at the population level. Television viewing is a major source of inactivity and has received considerable attention as a risk factor for obesity [62–64]. In addition to potentially contributing to lower energy expenditure by displacing time potentially spent in more active pursuits, television viewing has been hypothesized to contribute to excess energy intake. Television watching can serve as a cue for eating given the numerous references to food and commercials for food, often high-fat, high-energy foods, on television [64, 65].

According to data provided by A. C. Nielsen Company, the average household television set is turned on for more than 7 hours per day [66]. Survey data estimating the frequency of television watching are necessary, however, because television viewing is not necessarily the primary activity when the television set is turned on. Data from the Americans' Use of Time study show that free time spent watching television increased from about 10.4 hours per week in 1965 to about 15.1 hours in 1985 [67]. Survey data from adults also indicate that time spent watching television averages about 2–3 hours a day [68–71].

The high frequency of television viewing and other "screen time" among youth is disturbing. Dietz and Gortmaker [64] report that children in the United States spend, on average, as much time watching television each year as they do attending school. Data from the 1999 YRBS showed that almost two-thirds of students age 14–18 reported watching 2 or more hours of television per day [72]. Other national data show that the average amount of television viewing is about 5 hours per day among 10- to 15-year-old children [62]. About one-third of these children watched more than 5 hours per day and only 1% watched 0–2 hours per day.

Examination of media use data in younger children, including toddlers and infants, is also concerning. The prevalence of daily media use in a large sample of children younger than 11 (mean age: 5.1 years) was as follows: television (1.45 hours; SD, 1.5); videos (1.1 hours; SD, 1.30); and computer games (0.54 hours; SD, 0.96) [73]. Data from the National Longitudinal Survey of Youth, 1990 to 1998, were used to analyze reported television viewing at 0 to 35 months of age and evaluate adherence to the American Academy of Pediatrics (AAP) recommendations that children 2 years and older limit their time with entertainment media (television, video games, the Internet) to 2 hours per day and that children younger than 2 watch no television [74]. These data showed that 17% of 0- to 11-month-olds, 48% of 12- to 23-month-olds, and 41% of 24- to 35-month-olds were reported by their mothers to watch more television than the AAP recommends. Mothers with less education reported that their children viewed more television compared to mothers with higher levels of education. For example, 29% of mothers with less than 12 years of education reported that their infants watched television, compared with only 14% of college graduates.

Cross-sectional research has shown that there is a consistently strong positive relationship between television watching and obesity in children [62, 64, 72, 75–90] and adults [68, 69, 91, 92]. Prospective relationships between television watching and weight are unclear with some studies showing that frequency of television watching is weakly or moderately associated with future weight gain [64, 93, 94], and other studies not detecting an association [68]. Although it has been hypothesized that television watching influences obesity by replacing time that could otherwise be spent engaging in more active pursuits, only some studies have shown that children and adults who watch more television are less physically active [75, 80, 84, 91, 92]. Additional hypotheses regarding associations between television watching and obesity implicate increases in energy intake linked with television watching. The 2005 Institute of Medicine report focused on food marketing to children concluded that there is strong evidence that television advertising influences children's food preferences and requests, short-term food consumption patterns, and possibly usual dietary intake, and that exposure to advertising is associated with adiposity in children [95]. Television watching is associated with increases in energy intake and repeated episodes of eating while watching television may result in television's becoming a trigger for eating [96]. Crawford *et al.* [68] caution that the link between obesity and television watching appears to be complex and that targeting television viewing alone would not be an adequate public health strategy for addressing the obesity epidemic.

Despite the fact that relationships between sedentary behavior and obesity are somewhat unclear, intervention research with children and adolescents focused on reducing television watching shows promise as an obesity prevention and treatment strategy [97]. The work of Epstein *et al.* [65, 98–100] has figured prominently in the literature on decreasing sedentary activity as a strategy for promoting higher levels of physical activity. According to Epstein, the principles of behavioral economics or behavioral choice theory can be applied to sedentary individuals who, given the opportunity to choose between sedentary and physically active alternatives, will consistently choose the sedentary alternative. Choice of a given alternative, in this case, sedentary behavior, depends on the behavioral "cost" of that choice. Epstein argues and has demonstrated empirically in the laboratory that reducing the accessibility of sedentary behaviors or increasing the cost of being sedentary are both methods for reducing sedentary behavior. Epstein and colleagues [98] have also demonstrated that obese children participating in family-based weight control programs show the best changes when they are reinforced for being less sedentary as opposed to being reinforced for being more active.

School-based research targeting change in sedentary behaviors also shows promise. Robinson [101] conducted a small randomized trial in which one elementary school received an 18-lesson, 6-month classroom curriculum to reduce television, videotape, and video game use, and one school served as the control group. At follow-up, children in the intervention school had lower BMIs, tricep skinfold thicknesses, waist circumferences, and waist-to-hip ratios relative to children in the control school. Significant differences were also observed in children's reported television viewing and meals eaten in front of the television, although no significant differences were observed for moderate-to-vigorous physical activity levels, cardiorespiratory fitness, or high-fat food intake. The Planet Health Program, another school-based obesity prevention trial, found that the decreasing television component of their multicomponent program appeared to contribute most to the decrease in prevalence of obesity observed among girls [102]. The effectiveness of reducing sedentary behaviors in preventing and treating obesity clearly deserves further exploration. One reason for the success of such strategies may be the simplicity or clarity of the intervention message.

D. Self-Efficacy for Physical Activity

Exercise self-efficacy is one of the strongest and most consistent predictors of exercise behavior. Self-efficacy predicts both exercise intention and several forms of exercise behavior [44, 103–113]. Self-efficacy is an individual's belief in his or her ability to successfully engage in a given behavior. It is theorized to influence the activities that individuals choose to approach, the effort expended on such activities, and the degree of persistence demonstrated in the face of failure or aversive stimuli [114]. Exercise self-efficacy is the degree of confidence an individual has in his or her ability to be physically active under a number of specific/different circumstances or, in other words, efficacy to overcome barriers to exercise [105]. Among adults, self-efficacy is thought to be particularly important in the early stages of exercise [109]. In the early stage of an exercise program, exercise frequency is related to one's general beliefs regarding physical abilities and one's confidence that continuing to exercise in the face of barriers will pay off. Self-efficacy has also been shown to be highly related to physical activity in youth. Children with higher perceived self-efficacy for physical activity and for overcoming barriers to physical activity were more active [115–120]. Gender differences in self-efficacy have also been observed, with boys reporting higher self-efficacy than girls [118].

Given that self-efficacy for exercise is such a strong predictor of exercise behavior, enhancing self-efficacy should figure prominently in physical activity intervention programs [121]. Strategies for enhancing self-efficacy could include teaching individuals how to overcome barriers to exercise, creating opportunities for positive experiences with physical activity, and providing positive feedback regarding exercise performance [110, 121, 122]. Self-monitoring can also be used as a tool for those who exercise on their own to monitor performance and provide evidence regarding physical activity accomplishments [121, 123]. Emphasis should also be placed on increasing girls' self-efficacy for exercise, given observed gender differences in exercise self-efficacy and physical activity levels.

E. Exercise History

Prior history of physical activity should positively influence future physical activity behavior by promoting and shaping self-efficacy for exercise and by developing physical activity skills. The observed relationship between exercise history and exercise behavior varies, however, depending on how exercise history is defined and the time period over which physical activity behavior is "tracked." Physical activity has been shown to track in early childhood [84] and from childhood to adolescence [124]. Recent exercise history is generally predictive of future exercise behavior [125]. Childhood exercise history, however, is inconsistently related to physical activity in adulthood [106, 126], although one recent study that tracked participants over a 21-year period showed that a high level of physical activity at ages 9 to 18, especially when continuous, significantly predicted a high level of adult physical activity. Additionally, a history of sport participation during childhood was associated with higher total and sport activity levels in a sample of non-Hispanic white and African-American women [127]. The perception of the exercise experience as a child may be as important as amount of childhood exercise. One recent study found that recalling being forced to exercise as a child was associated with lower levels of physical activity in adulthood [128]. A child's enjoyment of physical activity [116, 118] and enjoyment of physical education experiences [129] are significant predictors of physical activity levels. Creating positive environments for physical activity for youth is likely a key factor in promoting higher levels of physical activity as a lifestyle habit.

F. Social Support for Physical Activity

Social support is another strong correlate of physical activity for both youth and adults. Adults who engage in regular exercise report more support for activity from people in their homes and work environments [44, 106, 130, 131]. Adults who are initiating exercise programs are more likely to perceive their families as being supportive of their desire to maintain good health [132]. Additionally, individuals who joined a fitness program with their spouse had higher rates of adherence at 12 months compared to those who joined without a spouse [133]. Carron *et al.* [134] examined six major sources of social influence on physical activity, including important others such as physicians or work colleagues, family member, exercise instructors or other in-class professionals, coexercisers, and members of exercise groups, in a comprehensive review. The authors concluded that social influence generally has a small to moderate effect on exercise behavior. Moderate to large effect sizes were found for family support and attitudes about exercise, important others and attitudes about exercise, and family support and compliance behavior.

Positive family environments and family support for physical activity is a robust correlate of physical activity for both boys and girls [115–117, 129, 135–138]. Parental prompts for children to play outdoors instead of watching television or playing video games have been shown to positively influence children's activity levels [139]. Parental prompts to be active have also been shown to be related to young children's activity levels in some [140–142], but not all studies [143, 144]. Taylor *et al.* [145] argue that age of the child needs to be taken into account when understanding the impact of parental behavior on children's activity levels. Specifically, as children make the transition to school and spend more time in broader social contexts, other social influences may become more prominent. Not surprisingly, social support for physical activity from peers is also positively correlated with physical activity levels [117, 146]. Taylor *et al.* [145] also suggest that attention should be paid to children's perception of parents' social influence and potential adverse forms of social influence such as nagging, discouragement, or excessive pressure to be physically active.

G. Barriers to Physical Activity

1. Time

Among adults, time constraints are the most frequent barriers to exercise and are reported by both sedentary and active individuals [130, 147]. Even among regular exercisers, scheduling efficacy remains an important and significant predictor of adherence [105]. Therefore, to maintain exercise adherence, regular exercisers have to become adept at dealing with time as a barrier. The time barrier may be a particular problem for certain population subgroups. For example, Schmitz *et al.* [148] reported that becoming a parent is associated with reductions in physical activity for mothers. Time spent caring for children may make it difficult for parents to maintain a regular physical activity program.

Several physical activity intervention approaches geared toward addressing the time barrier have been developed in recent years. Although group exercise programs can provide support and structure for participants, these advantages may be outweighed by the long-term costs involved in traveling to exercise sites at specific times for physical activity involvement [149]. Home-based physical activity programs appear to be a positive option for many adults [130, 150–154]. In the context of a weight loss program, Perri and colleagues [151] compared the effects of two exercise regimens, a group-based program versus a home-based program, on exercise participation, adherence, and fitness. Their results indicated that participants in the home-based program demonstrated better adherence and exercise performance data, particularly at 12-month follow-up. Two recent studies, which compared structured exercise program to a lifestyle approach, reported that the lifestyle program was as effective as the structured exercise program in improving physical activity and health outcomes including body weight and cardiorespiratory fitness [153, 154]. The recent focus on the health benefits of short bouts versus long bouts of activity given the public health recommendation to accumulate 30 minutes of physical activity on most days of the week also has implications for addressing the problem of time as a barrier to activity [155–159]. Short bouts of activity may be easier for people to incorporate into their schedules and may be particularly suitable for those initiating physical activity. Research suggests that multiple short bouts of exercise have been effective at promoting adherence to exercise programs [153, 154, 158, 159]. Interestingly, Epstein and colleagues [100, 160] have also reported that a lifestyle approach to exercise is associated with better adherence and greater weight loss in children.

2. Access and Environmental Factors and the "Built Environment"

Another barrier that has received increasing attention in the determinants literature is access to exercise facilities and safe and attractive places to walk and play outside. Distance between individuals' homes and exercise facilities has been shown to be negatively correlated with exercise behavior in adults [161]. Depending on an individual's activity preference, access to exercise facilities may or may not be related to exercise levels. For those individuals who prefer exercises such as walking or running, which can be done anywhere, access to facilities may be less relevant. Additionally, for those who exercise with home equipment, which could include stationary bikes, treadmills, and even exercise videos, access to facilities may also not affect exercise adherence. Regardless, the extent to which environments are conducive to physical activity (e.g., walking/biking paths, safe streets) likely has a strong impact on population activity levels. One recent study examining the association between neighborhood safety and sedentary behavior in a population-based sample found that there was a higher prevalence of physical inactivity among persons who perceived their neighborhoods as unsafe [162]. Better measurement of environmental resources for physical activity and strategies for improving access to physical activity facilities are needed.

Physical activity among youth appears to be particularly strongly influenced by environmental factors. First of all, the amount of time children spend playing outside has been shown to be a strong correlate of physical activity levels [143, 163]. Clearly, children who live in neighborhoods where play spaces are not adequate are going to have more difficulty achieving recommended levels of physical activity. Inequality in availability of physical activity facilities and access to safe play spaces may contribute to ethnic and socioeconomic disparities in physical activity and overweight patterns among youth. Use of after-school time for sports and physical activity [129], access to community sports activities [118], and frequency of parents transporting children to activity locations have all been shown to be correlates of physical activity in boys and girls [144, 164]. The extent to which families have time and resources to support their children in physical activity pursuits will also have a strong impact on children's activity levels. Anecdotal reports suggest that children spend less time in unstructured physical activities (e.g., neighborhood pick-up games, hide and seek, tag) than in previous years. In contrast, there appears to have been an increase in community-organized sports (e.g., traveling soccer and basketball teams) that require increased parental time, involvement, and financial resources. These factors may potentially contribute to decreases in physical activity and increased socioeconomic differences in physical activity and obesity risk among youth and is an area worthy of further exploration.

3. Overweight and/or Discomfort with Physical Activity

Clearly, body weight and physical activity are inextricably linked. Although it is clear that increasing physical activity is an important factor in regulating body weight, weight status may serve as a barrier to physical activity. Sedentary individuals may be heavier than those who initiate exercise programs [132]. This is due in part to physical activity being less pleasurable (e.g., it is uncomfortable for people to exercise when they are heavier), and in part because of embarrassment (e.g., individuals report feeling embarrassed about being seen in public in exercise clothes, at gyms, due to weight status and societal reactions toward overweight individuals). However, weight status can also be a motivator for initiating exercise. One of the most common reasons adults give for exercising is weight control. Also, dieting to control weight is positively associated with frequency of participation in both high- and moderate-intensity physical activity [42]. Physical activity promotion programs, however, need to be modified to address the needs of overweight youth and adults. Longitudinal data on adolescents indicate that body dissatisfaction predicts lower levels of physical activity, suggesting a need for physical activity interventions that help individuals feel more comfortable with their bodies, regardless of their weight status.

Another important aspect of physical activity promotion for weight control is how much exercise to recommend. Recognition that more exercise may be necessary for promoting long-term weight control after weight loss has emerged during the past decade. Data from the National Weight Control Registry and several other sources suggest that considerably higher physical activity levels (e.g., 2500 kcal per week compared to 1000 kcal per week typically recommended in many weight loss programs) than may be achievable by many overweight people are more beneficial for long-term weight loss [165–167]. This higher level of activity may be perceived as intimidating and difficult to achieve for those who have a history of sedentary behavior and are overweight. Strategies and support for helping individuals gradually increase their physical activity level while remaining injury free are needed.

III. DIETARY FACTORS

This section examines the role of dietary factors in the development of overweight and obesity. Multiple factors that influence food intake will be discussed, including the macronutrient composition of the diet, environmental and societal influences, and specific eating patterns.

A. Macronutrient Composition

1. Energy Intake

Laboratory experiments in animals and human clinical studies have repeatedly shown that the level of fat and energy intake in the diet is strongly and positively related to excess body weight. Population-based surveys of diet and obesity have reported inconsistent results [36]; however, the most recent examination of secular trends in self-reported overall eating frequency and energy density of diets suggests that increases in food consumption roughly parallel the pattern of obesity increase observed in the United States over the past 30 years [168]. Examination of trends in the frequency of eating episodes, meal and snack consumption, quantity of food consumed, and the energy density of foods reported by U.S. adults using data from four consecutive National Health and Nutrition Examination Surveys (NHANES) showed that the quantity of foods and their energy density increased beginning in NHANES III (1988–94), although the increases were not large. Between 1971 and 2000, average energy intake increased from 2450 kcal to 2618 kcal ($p < 0.01$) among men, and from 1542 kcal to 1877 kcal among women ($p < 0.01$). Additionally, among men, the percentage of kcal from carbohydrate increased between 1971–1974 and 1999–2000, from 42.4% to 49.0% ($p < 0.01$), and among women, from 45.4% to 51.6% ($p < 0.01$). Correspondingly, the percentage of kcal from total fat decreased among both men and women.

Secular trends in energy intakes of youth aged 2–19 years have also been examined. Data from NHANES found that mean energy intake changed little from the 1970s to 1988–1994, except for an increase among adolescent girls [169]. Between NHANES II and NHANES III, mean energy intake increased 1–4% among most age groups under age 20; mean intakes declined 3% for ages 6–11 years and increased 16% in females ages 16–19 years [170]. Among adolescent girls, energy intake increased by 225 calories. The increase between surveys for black females aged 12–19 years was slightly larger at 249 calories. Mean energy intakes from the USDA's national food consumption surveys showed little change for young children and slightly lower mean intakes for adolescents in 1989–1991 compared to 1977–1978 [171]. More recent data examining national trends in total energy intake among youth have not been published.

The inconsistencies in secular-trend surveys have been attributed to a number of factors, including weaknesses in the study design, methodological flaws, confounders, and random or systematic measurement error in the dietary data [36]. For example, the procedural changes between NHANES II and III in dietary survey methodologies, survey food coding, and nutrient composition databases make comparisons between the two surveys difficult [172]. Some evidence suggests that people participating in nutrition surveys underreport the food they eat, either by completely omitting food items or by inaccurately estimating the amount eaten [173]. Underreporting of food intake is discussed later.

The mixed dietary results of national surveys and the paradoxical findings that energy and fat intakes have decreased or stayed about the same, despite an increase in overweight prevalence, have led some researchers to conclude that the observed trend in obesity is not related to a shift in energy and fat intake but rather to a decline in physical activity. Secular trends in energy intake suggest that increased intake over time is not the major contributor to the increased prevalence of overweight among Americans and that decreased physical activity may play a more important role in overweight population prevalence [169]. However, given the problems inherent in measuring dietary intakes, and methodological survey issues, it should be noted that under more controlled laboratory conditions consistent findings show a strong association between dietary factors and obesity [36] (see Chapter 4, Energy Requirement Methodology).

In understanding the development of obesity, an important question is whether obese individuals eat more than leaner individuals. In most studies with both adults and children, energy intake has not been found to correlate with the degree of obesity [55]. However, this may be due, in part, to greater underreporting of food intake by the obese.

2. Underreporting of Food Intake

In contrast to measures of body weight, dietary intake is difficult to measure accurately. Underreporting must be considered when interpreting dietary survey data. Studies have documented that food consumption is underreported by about 20–25% by people participating in dietary studies and occurs more often in women, overweight persons, and weight-conscious persons [172, 174, 175]. Discrepancies between reported energy intakes and measured energy expenditures (with the doubly-labeled water method) of 20–50% have been described in overweight individuals [174, 176]. The systematic bias of underreporting in both overweight and nonoverweight individuals may be due to socially desirable responses, poor memory for foods consumed, lack of awareness of food consumed, difficulty with portion size estimation, or undereating (consuming less food than usual because of the requirement to record food intake). A recent study of underreporting of habitual food intake in obese men found that about 70% of the total underreporting was due to a diminished intake of food over the reporting period; that is, subjects changed their food patterns during the recording period [176]. Selective underreporting of fat intake was also found. The magnitude of underestimations observed in various studies indicates the considerable error in dietary intake data and highlights the need for improved techniques of data collection [177] (see discussion in Chapter 1).

3. Fat Intake

Dietary fat has a higher energy density than either protein or carbohydrates. Controversy remains about whether the percentage of dietary fat plays an important role in development of obesity and in its treatment once it has developed [178, 179]. There is evidence that consumption of high-fat diets increases total energy intake and that excess dietary fat is stored with a greater efficiency than similar excesses of dietary carbohydrate or protein [178, 180]. On the other hand, it has been argued that in short-term studies only a modest reduction in body weight is typically seen in individuals assigned to diets with a lower percentage of dietary fat, which suggests that dietary fat does not play a role in the development of obesity [179]. Results of studies in laboratory animals and metabolic studies clearly and strongly show that a high percentage of fat in the diet relative to other macronutrients contributes to the development of obesity. These kinds of research, however, do not prove causality in humans [178, 180, 181]. It has also been pointed out that the focus on dietary fat may have been overemphasized at the expense of total energy intake. Total energy balance is what matters most and the focus on dietary fat intake must be viewed through its effects on total energy intake [180]. Reduction of dietary fat is one of the most practical ways to reduce energy density of the diet.

National data show that Americans have dramatically lowered the percent of energy intake from total fat during the past three decades. The reduction is from about 45% of energy in 1965 to about 34% in 1995 [182]. Levels, though, continue to be higher than the 30% recommended. Interestingly, the percent of energy from fat continued to decrease from 1990 to 1995 even as the daily grams of fat intake remained steady or increased. The explanation for this paradox is that although daily fat consumption was increasing or remained unchanged, the total energy intake was increasing at a faster pace. A higher intake of energy will reduce the percent of energy from fat even when there is no decrease in total fat consumption. Therefore, the decrease in percent of energy from fat observed recently may be a result of increased total energy intake and not necessarily due to decreased fat consumption [182].

4. Sugar Intake

Americans are consuming record high amounts of caloric sweeteners, mainly sucrose and corn sweeteners. Per-capita consumption increased 45 pounds or 41% between 1950–1959 and 1997 [183]. In 1997, Americans consumed on average 154 pounds of caloric sweeteners; when adjusted for losses this number translates into 33 teaspoons of added sugar per person per day [184]. Regular (nondiet) soft drinks are the major contributor of added sweeteners in the American diet and account for one-third of the intake of added sweeteners [185]. There has been a 47% increase in annual per-capita consumption of regular carbonated soft drinks, from 28 gallons per person in 1986 to 41 gallons in 1997.

Guthrie and Morton [185] found increased intake of regular soft drinks to be one of the major changes in children's diets between 1989–1991 and 1994–1995. High soft-drink consumption may lead to excessive energy intake (a 12-ounce soft drink contains approximately 150 kcal), which theoretically may contribute to obesity. As consumption of sweetened beverages has been increasing among children and adolescents, milk consumption has been decreasing. Using nationally representative data from the 1977–1978 Nationwide Food Consumption Survey, the 1989–1991 and 1994–96 Continuing Survey of Food Intake by Individuals (SCFII), and 1999–2001 NHANES data, Nielsen and Popkin [185a] reported that energy intake from sweetened beverages increased 135% and that from milk was reduced by 38%, resulting in an overall energy intake increase of 278 calories. Harnack and colleagues [186] found that energy intake was positively associated with consumption of regular soft drinks. Energy intake was higher for those in the highest soft drink consumption category. In this study the association of obesity and soft drink consumption was not assessed. However, using NHANES III data, Troiano *et al.* [169] reported that soft drink energy contribution was higher among overweight children and adolescents.

There is no consensus regarding the role of sugar intake on body weight regulation. A preference for sweet–fat foods has been observed in obese individuals, which may be a factor in promoting excess energy consumption [36]. However, the notion that a sensory "sweet tooth," that is, a heightened preference for sweet taste, is a direct cause of obesity is not well supported [187].

B. Environmental Influences

1. U.S. Food Supply

Data on dietary levels of individuals provide key information on energy and nutrient intakes. However, food supply data can also provide a measure of changes in food consumption over time and estimated nutrient content of the food supply. Food supply data, also known as food disappearance data, reflect the amount of food commodities entering the market, regardless of final use. USDA's Economic Research Service (ERS) has developed methods to adjust the food supply for spoilage, plate waste, and other losses. USDA's Center for Nutrition Policy and Promotion provides food supply nutrient estimates derived from the ERS data [184]. These data are used as a proxy to estimate human consumption. Adjusted food supply data suggest that average daily energy intake increased 14.7%, or about 340 calories, between 1984 and 1994.

USDA food supply data indicate that Americans are consuming record amounts of some high-fat dairy products (e.g., cheese) and caloric sweeteners and near record amounts of added fats, including salad and cooking oils and baking fats. The hefty increase in grain consumption reflects higher consumption of mostly refined rather than high-fiber, whole-grain products [183]. On the positive side, fruit and vegetable consumption continues to rise. Americans consumed about a fifth (22%) more fruit and vegetables in 1997 than in the 1970s [184]. Supermarket produce departments carry more than 400 produce items today, up from 250 in the late 1980s and 150 in the mid-1970s. Also, the number of ethnic and natural food stores that offer fresh produce continue to increase [184].

The food industry has responded to consumer demand for lower fat products. For example, 2076 new food products introduced in 1996 claimed to be reduced in fat or fat free—nearly 16% of all new food products introduced that year, and more than twice the number just 3 years earlier. The number dropped in 1997 and it is unclear whether that represents a backlash to health concerns [188]. The Calorie

Control Council reports a notable rise in the percentage of the U.S. population consuming low-calorie products: 19% of the population in 1978, 29% in 1984, and 76% in 1991 [189]. Still, the use of these products has not prevented the progression of obesity in the population. It should be pointed out that many commercially available low-fat or fat-free foods are not lower in energy density than their full-fat counterparts [190].

2. Eating Out

During the past 20 years, one of the most noticeable changes in eating patterns of Americans has been the increased popularity of eating out [191, 192]. The proportion of meals and snacks eaten away from home increased by more than two-thirds between 1977–1978 and 1995, rising from 16% of all meals and snacks in 1977–1978 to 27% in 1995 [193]. Almost half of all adults (46%) were restaurant patrons on a typical day during 1998 [194]. In 1999–2000, U.S. adults reported consuming an average of almost three commercially prepared meals each week, an 11% increase compared to the number of commercially prepared meals reported in 1987 and 1992 [192]. Currently, almost half (47%) of a family's food budget is spent on away-from-home food [195] compared to only 26% in 1970 [193]. Food away from home includes foods obtained at restaurants, fast-food places, school cafeterias, and vending machines. A number of factors account for the increasing trend in eating out, including a growing number of working women (75% of women 25–50 years old are in the workforce), more two-earner households, higher incomes, a desire for convenience foods because of busy lifestyles and little time for preparing meals, more fast-food outlets offering affordable food, smaller families, and increased advertising and promotion by large food-service chains and fast-food outlets [191].

The trend toward eating away from home more frequently has also been observed among adolescents and young adults [196]. Data from the 1977–78 Nationwide Food Consumption Survey and the 1989–91 and 1994–96 Continuing Surveys of Food Intake by Individuals document a decrease in the percentage of energy obtained from foods consumed at home and school and an increase in the percentage of energy obtained from foods consumed at fast food establishments and restaurants among adolescents and young adults.

The trend in eating out may be related to the observed increase in energy intake among Americans, because food away from home is generally higher in energy and fat than food consumed at home [184, 193]. Many table-service restaurants provide 1000–2000 calories per meal, amounts equivalent to 35–100% of a full day's energy requirement for most adults [197]. Data from USDA's food intake surveys conducted during the past 20 years have been analyzed to compare the nutritional qualities of at-home and away-from-home foods and changes over time. In 1995, the average total fat and saturated fat content of away-from-home foods, expressed as a percentage of energy, was 38% and 13%, respectively, compared with 32% and 11% for at-home foods [193]. Foods eaten away from home provided 34% of total food energy consumption in 1995, up from 19% in 1977–1978. The 1995 data also suggest that, when eating out, people tend to eat larger amounts, eat higher energy foods, or both [184]. Lin and colleagues [191] at the USDA calculated that if away-from-home food had the same average nutrient densities as food at home in 1995, Americans would have consumed 197 fewer calories per day, and reduced their fat intake to 31.5% of energy from fat (instead of 33.6%).

Consumers may view food differently when eating out than when eating at home. Consumers may view eating away from home as an exception to their usual dietary patterns, regardless of how frequently it occurs, and an opportunity to "splurge" [198]. Consumers also may not be aware of the fat or energy content of prepared foods because nutrition information is generally not provided in restaurants and other eating places. Restaurants generally do not provide detailed nutritional profiles of foods served, although most fast-food restaurants have this information available on request. New restaurant regulations mandate that menu items labeled as low-fat must comply with defined standards. However, consumers by and large must rely on their own knowledge to identify healthful menu options [198]. Because the trend of eating out is expected to increase even more in the next decade, nutrition interventions to improve the nutritional quality of food choices made away from home are needed. Such interventions should include both environmental changes (e.g., nutrition information on foods) and efforts to change consumer attitudes toward eating out and increase motivation to make healthier choices [193, 198].

3. Fast Foods

Fast food has become a significant part of the American diet. In the United States, more than 200 people are served a hamburger every second of the day [36]. The number of fast-food outlets in the United States has risen steadily during the past 25 years, increasing from roughly 75,000 outlets in 1972 to almost 200,000 in 1997 [199]. Fast-food sales in the United States rose 56% to $102,387 million between 1988 and 1998 [200]. Recent estimates show that in 2001, there were about 222,000 fast-food locations in the United States, generating sales of more than $125 billion. Many items on fast-food menus are high in fat. The fat density of fast foods is about 40% of total energy. Several fast-food restaurants offer large-size burgers that are exceptionally high in fat and energy, even after public pressure to move away from "super sizing." For example, McDonald's Double Quarter Pounder® with Cheese has 740 calories and 42 grams of fat; Burger King's Triple Whopper with Cheese burger has 1130 calories and 74 grams of fat; and Hardee's Double Thickburger has 1240 calories with

90 grams of fat. In addition, large soft drinks containing substantial amounts of sugar are often consumed as part of the fast-food meal and contribute substantially to total energy intake [36, 193].

Several fast-food chains have tried to introduce reduced-fat entrees, but later withdrew them because of slow sales. For example, in 1991, McDonald's introduced the McLean Deluxe, which used a 91% fat-free beef patty, but due to slow sales and poor public acceptance it was taken off the market after a few years. Taco Bell introduced a line of low-fat menu items in 1994, called Border Lights, but these were also largely removed because of sluggish sales [199]. Many fast-food chains offer other low-fat items, such as grilled chicken sandwiches, wraps, and salads. A recent trend in response to growing concern about the obesity epidemic has been the inclusion of healthier items on children's menus at fast-food establishments, including offering fruit choices as an alternative to French fries and low-fat milk as an alternative to soda. However, observations that fast-food establishments continue to offer menu items with excessive portion sizes despite recommendations to reduce portion sizes and public assertions by these companies that they will make changes, voluntary efforts to reduce portion sizes are unlikely to be effective [201].

There is increasing evidence that increased fast-food consumption is associated with obesity and weight gain. One study examined the relationship between fast-food eating and body mass index in 1059 adult men and women over the course of 1 year. The researchers found that the number of meals eaten at fast-food restaurants per week was positively associated with energy intake and body mass index in women but not in men. The strongest relationship was observed among low-income women [93]. Another recent study assessed energy intake and the frequency of consuming food from seven fast-food and table-service restaurants in 73 adults [202]. The researchers found that the frequency of consuming restaurant food was positively associated with body fatness, total daily energy intake, and fat intake. Two recent studies [203, 204] show that higher levels of fast-food consumption are positively associated with BMI cross-sectionally and with greater increases in BMI over time. More such studies are warranted to assess whether frequency of fast-food consumption and eating out is related to the secular trends in obesity prevalence in the United States and globally.

Fast-food outlets are receiving increasing competition from supermarkets and other establishments offering fully or partially prepared entrees or multicourse meals for eat-in or carry-out. On an average day in 1998, 21% of U.S. households used some form of takeout or delivery [194]. Home meal replacements, as they are called, are intended to be easily reheated in the oven or microwave and are designed to eliminate the need to cook at home by providing a wide variety of higher quality foods that are as convenient and affordable as fast food [199]. The widespread adoption of microwave ovens by U.S. households (now in nearly 90% of homes) contributes to the convenience of home meal replacements for takeout [199]. Driven by consumer demand for convenience due to hectic schedules, the market for home meal replacements will continue to rise. It is speculated that miniaturized outlets offering hot fast-food meals might one day be as common in public buildings as soft drink machines are today [199]. This speaks to the need for consumer education and nutrition labeling of these products.

4. Changing Portion Sizes

It has been suggested that food portion sizes in food service establishments have become larger and thereby have increased energy intake, which may lead to obesity. However, empirical data to support this association have only just begun to emerge. It is noticeable that many restaurants, especially fast-food restaurants, in recent years have been offering large and extra-large portion sizes of products and meals at low cost. A comparison of food service portion sizes during the past 30 years is remarkable. Putnam [184] reports that the typical fast-food outlet's hamburger in 1957 contained a little more than 1 ounce of cooked meat, compared with up to 6 ounces in 1997. Soda pop was 8 ounces in 1957, compared with 32 ounces to 64 ounces in 1997. A theater serving of popcorn was 3 cups in 1957, compared with 16 cups (medium size popcorn) in 1997. A muffin was less than 1.5 ounces in 1957, compared with 5–8 ounces in 1997.

Smiciklas-Wright *et al.* examined data from the Continuing Survey of Food Intakes by Individuals (CSFII) in 1989–1991 and 1994–1996 to compare quantities of a variety of foods consumed per eating occasion. In 1994–1996, all persons aged 2 years and over reported consuming larger amounts of soft drinks, tea, coffee and ready-to-eat cereal and smaller amounts of margarine, mayonnaise, chicken, macaroni and cheese, and pizza. Nielsen and Popkin used data from the Nationwide Food Consumption Survey (1977–1978) and the Continuing Survey of Food Intake by Individuals (1989–1991, 1994–1996, and 1998) to examine trends in average portion sizes consumed from specific food items (salty snacks, desserts, soft drinks, fruit drinks, French fries, hamburgers, cheeseburgers, pizza, and Mexican food) and whether portion sizes consumed varied by eating location (home, restaurant, or fast food). Food portion sizes increased both inside and outside the home for all categories except pizza between 1977 and 1996. The energy intake and portion size of salty snacks, soft drinks, hamburgers, French fries, and Mexican food increased by 93 kcal, 49 kcal, 97 kcal, 68 kcal, and 133 kcal, respectively, with portion sizes the largest at fast-food restaurants. Experimental data show that increasing the portion size of foods and beverages is associated with a significant increase in energy intake sustained over 2 days, supporting the hypothesis that large portions are associated with excess energy intake and potential for increased body weight [205].

5. Food Marketing

Although multiple factors influence eating behaviors, one potent force is food marketing. Over the past few decades, U.S. children and adolescents have increasingly been targeted with intensive and aggressive forms of food marketing and advertising practices [206, 207]. Multiple techniques and channels are used to reach youth, beginning when they are toddlers, to foster brand building and influence product purchase behavior. Recently the Kaiser Family Foundation [208] released the largest study conducted on TV food advertising to children. The study found that children ages 8–12 years see the most food ads on TV, an average of 21 ads a day or more than 7600 per year. The majority of the ads were for candy, snacks, sugared cereals, and fast foods; none of the 8854 ads reviewed was for fruits and vegetables. Food marketing to children now extends beyond television and is widely prevalent on the Internet [209]; it is expanding rapidly into a ubiquitous digital media culture of new techniques including cell phones, instant messaging, video games, and three-dimensional virtual worlds, often under the radar of parents [210].

The Institute of Medicine report on Food Marketing to Children and Youth [207] conducted a systematic review of the evidence and concluded that food and beverage marketing practices geared to children and youth are out of balance with recommended healthful diets and contribute to an environment that puts youth's health at risk [207]. The report set forth recommendations for different segments of society to guide the development of effective marketing strategies that promote healthier food, beverages, and meals for children and youth. Among the major recommendations for the food, beverage, and restaurant industries was that industry should shift their advertising and marketing emphasis to child- and youth-oriented foods and beverages that are healthier. If voluntary efforts related to children's television programming are unsuccessful in shifting the emphasis away from high-energy and low-nutrient foods and beverages to healthful foods and beverages, then Congress should enact legislation mandating the shift. Advocacy and public health groups are also calling on the Federal Trade Commission, the Federal Communications Commission, and Congress to work together with industry to develop a new set of rules governing the marketing of food and beverages to children—rules that take into account the full spectrum of advertising and marketing practices across all media and apply to all children, including adolescents [210]. Marketing efforts need to serve the health of children rather than undermine health [210].

C. Eating and Dietary Practices

The majority of research examining the potential role of diet in the etiology of obesity has focused on associations between obesity and dietary intake (e.g., intake of energy, macronutrients, and food groups). As previously discussed, findings from this large body of research leave many questions unanswered. Fewer studies have examined associations between obesity and eating practices such as the pace of eating, meal patterns, dieting, and binge eating. In this section, we highlight some of the research that examines associations between specific eating practices and obesity; identify questions that remain unanswered; and make some recommendations based on this body of research for future studies and for interventions aimed at obesity prevention/treatment.

1. Pace of Eating

Eating practices aimed at eating at a slower pace are often encouraged in obesity treatment programs. Efforts to slow down the pace of eating may assist certain individuals to consume smaller amounts of food. However, research findings comparing the pace of eating among overweight and nonoverweight individuals have not been consistent. Some studies have shown that overweight individuals do have a rapid eating pace [211–213]. However, other studies have not found a characteristic eating style among overweight individuals [214, 215]. Based on findings from their work and the work of others, Terri and Beck [215] have concluded that differences in eating behaviors between overweight and nonoverweight individuals are "extremely variable and cannot be presumed." Therefore, they suggest that individual assessment is required before embarking on a treatment program that attempts to modify eating behaviors in overweight individuals [215]. For overweight individuals who tend to eat large amounts of food at a quick pace, behavioral strategies aimed at slowing the pace of eating may be helpful, whereas for others these strategies may not be suitable.

2. Meal Patterns

Concerns about skipping meals exist in that meals are important for ensuring an adequate nutrient intake, for socializing (if eaten with family and/or friends), and for avoiding hunger, which may then lead to binge-eating episodes. Meal skipping has been found to be higher among overweight adolescents than among their nonoverweight peers, with breakfast being the most common meal skipped. In a cross-sectional study of more than 8000 adolescents, usual breakfast consumption was reported by 53% of non-overweight youth, 48% of moderately overweight youth (85th–95th percentile), and 43% of very overweight youth (BMI $>$ 95th percentile) [216]. Because of the cross-sectional nature of these findings, it is not clear whether breakfast skipping leads to obesity (e.g., in that it may be associated with higher energy intake at later times in the day) or rather that breakfast skipping was a consequence of obesity (e.g., meals are being skipped for weight-control purposes). A recent review of the literature summarized the results of 47 studies examining the association of breakfast consumption with nutritional adequacy (nine studies), body weight (16 studies), and academic performance (22 studies) in children and adolescents and concluded that children who reported eating breakfast on a consistent basis tended to have nutritional profiles superior to those of their breakfast-skipping peers [217]. Breakfast eaters generally consumed more daily energy yet were less likely to be overweight. The quality of the breakfast meal is a key consideration, with consumption of ready-to-eat or cooked cereal typically associated with better nutrition and weight outcomes [218, 219].

Meal skipping is frequently used as a weight control method. Among adolescents and adults trying to control their weight, Neumark-Sztainer *et al.* [220] found that skipping meals was commonly reported; 18.6% of adult males and females, 22.8% of adolescent females, and 14.1% of adolescent males trying to control their weight reported skipping meals. An important question relates to the impact of skipping meals on overall energy and nutrient intake. In a study of women participating in a weight gain prevention study, skipping meals for weight-control purposes was not associated with overall energy intake [221]. However, meal skippers reported higher percentages of total energy intake from fat and from sweets, lower percentages of total energy intake from carbohydrates, and lower fiber intakes than women who did not report meal skipping [221].

In summary, existing research suggests that meal skipping is associated with a poorer nutrient intake and with obesity status [216, 221, 222]. However, it is not clear whether meal skipping plays an etiological role in the onset of obesity or rather is a consequence of obesity. Additional prospective studies and evaluations of interventions aimed at decreasing meal skipping are needed to assess causality. However, in light of the inverse associations between meal skipping and nutrient intake, and the potential for leading to uncontrolled eating due to hunger, meal skipping should not be recommended as a weight-control strategy. Rather, careful planning of meals with nutrient-dense foods that are low in fat and energy should be encouraged.

3. Dieting Behaviors and Dietary Restraint

Research findings clearly indicate that overweight individuals are more likely to report engaging in dieting and other weight-control behaviors than nonoverweight individuals. For example, in a large cross-sectional study of adolescents, dieting behaviors were reported by 17.5% of underweight girls (BMI $\leq$ 15th percentile), 37.9% of average weight girls (BMI 15th–85th percentile), 49.3% of moderately overweight girls (BMI 85th–95th percentile), and 52.1% of very overweight girls (BMI $\geq$ 95th percentile) [223]. In a 5-year follow-up study on the same population, dieters were found to be at increased risk for weight gain and for being overweight, even after adjusting for baseline weight status. Adolescents using unhealthy weight-control behaviors were at approximately three times the risk for being overweight 5 years later, suggesting that behaviors such as skipping meals, taking diet pills, and imposing strict dietary restrictions may be counterproductive to long-term weight management [224]. Similarly, in a prospective study on adolescent girls by Stice and his colleagues [225], baseline dieting behaviors and dietary restraint were found to be associated with obesity onset 4 years later. After controlling for baseline BMI values, the hazard for obesity onset over the 4-year study period was 324% greater for baseline dieters than for baseline nondieters. For each unit increase on the restraint scale, there was a corresponding 192% increase in the hazard for obesity onset [225].

These findings suggest that for some individuals self-reported dieting may be associated with a higher energy intake, and not a lower energy intake as intended. One explanation for this is that self-reported "dieting" may represent a temporary change in eating behaviors, which may be alternated with longer term eating behaviors that are not conducive to weight control. Another explanation is that self-reported dieting and dietary restraint may be associated with increased binge-eating episodes resulting from excessive restraint, control, and hunger. Indeed, Stice and his colleagues did report positive, albeit modest, associations between binge eating and both dieting behaviors ($r = 0.20$) and dietary restraint ($r = 0.20$). In an analysis aimed at addressing the somewhat perplexing question as to why dieting leads to weight gain, Neumark-Sztainer and her colleagues found that binge eating was an important mediating variable between dieting and weight gain over time in adolescents [224]. Other researchers have also suggested that dietary restraint may lead to binge-eating behaviors [226, 227], thereby placing individuals at risk for weight gain, rather than the intended weight loss or maintenance [228].

It is noteworthy that retrospective studies of adults who have been successful in losing weight and in maintaining their weight loss over extended periods of time suggest that modifications in eating and physical behaviors are helpful strategies. The National Weight Control Registry is a large study of individuals who have been successful at long-term maintenance of weight loss [165]. The most common dietary strategy used by these successful weight-loss maintainers was limiting intake of certain types or classes of foods. Other commonly used dietary strategies included limiting quantities of food eaten, limiting the percentage of daily energy from fat, and counting calories. Meal patterns tended to be regular; on average, subjects reported eating nearly five times per day and only a small proportion ate less than two times per day. Also, most members of the registry reported increasing physical activity levels as part of their weight-loss effort [165].

4. Binge Eating

Overweight individuals are more likely to engage in binge-eating behaviors than their nonoverweight counterparts

[17]. In a nonclinical sample of adult women enrolled in a weight gain prevention program, binge eating was reported by 9% of nonoverweight women and 21% of overweight women [229]. Furthermore, binge eating tends to be more prevalent among overweight individuals seeking treatment for weight loss. Based on a review of the literature, Devlin *et al.* [230] have estimated that between 25% and 50% of overweight people seeking treatment for weight loss engage in binge eating. Factors contributing to the higher rates of binge eating among overweight individuals, as compared to nonoverweight individuals, may include the following: greater appetites brought on by higher physiological needs of a larger body size, greater emotional disturbances (e.g., depressive symptoms) or different responses to stressful situations, greater exposure to stressful situations (e.g., related to weight stigmatization), increased weight preoccupation and dieting behaviors, and stronger dietary restraint.

In working with overweight individuals within health care and other settings, it is essential to be sensitive to the daily struggles overweight clients may face within thin-oriented societies. Overweight clients may be reluctant to share their binge-eating experiences; therefore a nonjudgmental attitude on the part of the health care provider is critical. For some individuals, hunger resulting from dieting or meal skipping may be a major cause of binge eating while for others binge eating may be a response to stress. Some individuals may be experiencing cyclical patterns; for example, emotional stress leads to binge eating, which leads to further emotional stress, which leads to further binge eating. Strategies for avoiding binge eating should be linked to factors that appear to be leading to binge eating for each individual.

5. Family Influences on Dietary Intake and Eating Practices

Research has demonstrated that familial factors contribute to the etiology of obesity via genetic and shared environmental factors [231, 232]. A considerable amount of research has been devoted to the role of the family in the etiology, prevention, and treatment of obesity [233–235]. With regard to dietary intake and eating practices, questions arise as to how the family environment influences individual family members' eating behaviors and what the family can do to improve eating behaviors of its members. The aim is clearly to provide for an environment in which healthful food is available, eaten in an enjoyable manner, and consumed in appropriate amounts. Most of the research in this area has focused on the influence of parents on their children's eating behaviors.

Parents/caretakers may influence their children's dietary intake and eating practices via numerous channels. Some of the key channels include food availability within the home setting (including food purchasing, food preparation, and food accessibility), family meal patterns, infant and child feeding practices, role modeling of eating behaviors and body image attitudes, and verbal encouragement of specific eating practices.

In focus group discussions, adolescents reported that their parents influence their food choices in different ways: parental eating and cooking behaviors, parental food purchasing patterns, parental concern about foods their children eat, family meal patterns, overall parent–child relations, and family cultural/religious practices [223]. Adolescents also perceived that they were more likely to eat healthier foods when eating meals with their families than when eating in other situations [236, 237].

Research in the arena of adolescent health indicates that general family context variables are strongly associated with the emotional well-being of adolescents and with eating and other health-related behaviors [238]. For example, Neumark-Sztainer and her colleagues [239, 240] found that low family connectedness (i.e., perceived level of caring and communication within the family) placed adolescents at increased risk for inadequate intake of fruit, vegetables, and dairy foods. Furthermore, among overweight youth, Mellin and colleagues [241] found that high family connectedness was associated with more regular breakfast eating, increased fruit and vegetable consumption, and lower rates of unhealthy dieting behaviors. These findings indicate the importance of a positive familial environment for adolescents and, in particular, for overweight youth who may be experiencing social stigmatization and need additional support.

An important question relates to how involved parents should be in their children's eating practices; that is, what is an appropriate parental role? Birch and her colleagues [242, 243] have examined associations between child-feeding practices and children's ability to regulate their energy intake. Birch and Fisher [242] have suggested that individual differences in self-regulation of energy intake are associated with differences in child-feeding practices and with children's adiposity. They state that "initial evidence indicates that imposition of stringent parental controls can potentiate preferences for high-fat, energy-dense foods, limit children's acceptance of a variety of foods, and disrupt children's regulation of energy intake by altering children's responsiveness to internal cues of hunger and satiety." In a study of 77 children between 3 and 5 years of age, Johnson and Birch [243] found that children with greater body fat stores were less able to regulate energy intake accurately. The strongest predictor of children's ability to regulate energy intake was parental control in the feeding situation; mothers who were more controlling of their children's food intake had children who showed less ability to self-regulate energy intake ($r = -0.67$). They concluded from this study that "the optimal environment for children's development of self-control of energy intakes is that in which parents provide healthy food choices but allow children to assume control of how much they consume."

Parents of overweight children and adolescents are often in a difficult situation in that they want to be supportive of their children, yet also want to help them to modify behavioral patterns that may increase their risk of obesity [244]. Furthermore, they may feel as though they are being blamed for their child's obesity. Existing research suggests that parental involvement and support is important [235, 241], but efforts to control a child's intake may be counterproductive [242, 243]. It may be most helpful for parents to provide an environment that makes it easy for children to adopt healthy eating and activity patterns (e.g., increasing the availability of fruits and vegetables at meals or encouraging physical activity), but includes less talking about weight per se. Indeed, longitudinal research with adolescents has found that negative weight comments by family members predicts unhealthy weight control behaviors, binge eating, and obesity [245]. Further studies are needed to explore how parents can best help their overweight children develop a positive self-image and healthy eating and physical activity behaviors.

IV. SUMMARY AND PUBLIC HEALTH RECOMMENDATIONS

We have reviewed the literature on key physical activity and diet-related risk factors for obesity in children and adults. Highlights of the review from the physical activity domain include (1) the importance of addressing the influence of both leisure-time physical activity and sedentary behavior to total energy expenditure, (2) the importance of fostering both self-efficacy and social support for physical activity to promote higher levels of physical activity, and (3) the influence of environmental factors on physical activity levels. Highlights related to dietary intake include (1) recognizing the contribution of both dietary fat intake and total energy intake to energy regulation; (2) the importance of accurately assessing portion size; (3) the influence of eating practices such as eating out, breakfast skipping, restrained eating, and binge eating on obesity; and (4) social and environmental factors that promote excess energy intake.

To effectively combat the public health problem of obesity, interventions and policies that target change in dietary intake and physical activity are necessary. Intervention efforts must take into account that dietary intake and physical activity are complex, multidetermined behaviors influenced by individual, social, and environmental factors. Although obesity intervention approaches have traditionally focused primarily on an individual's change, both individual- and population-level approaches are essential. Community-based interventions including school-based programs, after-school programs, and work-site programs as well as clinic-based programs focused on diet, physical activity, and weight management all have the potential to effectively address the problem of obesity. Environmental and policy interventions to prevent obesity have also begun to receive greater attention [149, 197, 246].

A. School and Community-Based Youth Programs

The increasing prevalence of obesity in adults and children, coupled with difficulty in successfully treating adult obesity, highlights the urgent need for prevention approaches geared toward children. Table 1 provides an overview of essential components for obesity prevention programs in schools. School-based programs must encompass both educational and environmental strategies to promote healthy eating and activity levels among students. A number of school-based programs targeting physical activity and eating behaviors have shown that it is possible to modify school environments and show improvements in diet and physical activity levels [102, 247–250].

Although the majority of health promotion programs for children are provided in school settings [251], community-based programs have considerable potential for helping children acquire positive health behaviors. The school environment confers many advantages including the reduction of barriers of cost and transportation and the provision of access to a large, already-assembled population. Community-based after-school programs have the potential to complement health promotion and education efforts made by schools for several reasons. Although children spend a large proportion of time in school each day, after-school hours constitute a substantial amount of time each week. Often children do not have opportunities to spend this time constructively, particularly in at-risk communities. According to the National Education Longitudinal Survey, the average eighth grader spends between 2 and 3 hours a day at home alone after school [252]. These hours have been shown to be those when many youth engage in high-risk behaviors because of lack of supervision. When young people are provided with safe and healthy activities in which to participate during critical gap periods (e.g., after school, on weekends), they are less likely to have time to participate in the high-risk, unhealthy activities that can delay or derail positive development [253–255].

Community-based after-school programs provide multiple opportunities for teaching and reinforcing healthy patterns of physical activity and eating. Through their involvement in such programs, children can be educated about healthful eating habits and active lifestyles, but also have ample opportunity to practice these new habits and skills. Classroom-based eating habit programs that emphasize the behavioral skills needed for planning,

TABLE 1 Strategies for School Programs to Promote Healthy Eating and Physical Activity

Healthy Eating	Physical Activity
• Supply health education regarding nutrition and healthy weight management as part of required curriculum for health education. • Train staff involved in nutrition education and offer work-site health promotion opportunities for staff. • Educate children on the influence of food advertising on eating habits. • Ban required watching of commercials for foods high in calories, fat, or sugar on school television programs. • Conduct school-wide media campaigns to promote healthy eating (e.g., the "5 a day message"). • Emphasize dietary changes that can be implemented on a long-term basis and discourage the use of unhealthy weight control behaviors, which can be counterproductive to weight management and increase risk for eating disorders. • Modify cafeteria staff food preparation practices (e.g., low-fat food preparation methods). • Increase availability of fresh fruits, vegetables, and low-fat milk and low-fat choices in cafeterias. • Reduce the availability of vending machines with soft drinks and high-fat, high-sugar snack foods. • Involve families and communities in supporting and reinforcing healthy eating patterns.	• Require and fund daily physical education and sports programs in primary and secondary schools. • Provide adequate training for physical education teachers to promote higher levels of physical activity during classes. • Ensure the provision of adequate space and necessary athletic/ sports equipment. • Provide a positive environment regarding physical activity to promote the development of self-efficacy and enjoyment of physical activity for all students. • Provide after-school physical education/sports opportunities for students. • Utilize approaches that help participants feel comfortable with their bodies in order to promote physical activity. • Incorporate lifestyle activities that students can continue to implement on a long-term basis. • Incorporate traditional and "nontraditional" (e.g., dance) types of activities in physical education classes. • Incorporate strategies to promote decreases in sedentary behavior. • Involve families and communities in supporting and reinforcing higher levels of physical activity. • Provide links to community programs and resources for physical activity.

preparing, and selecting healthy foods can be easily adapted for this setting. Children may have the opportunity to practice new skills in such programs by engaging in activities such as planning and preparing healthy snacks. According to Sallis and McKenzie [256], physical activity programs for children should include (1) activities and skills that have the potential for carryover into adult life, (2) moderate-intensity activities, and (3) a focus on maximizing the participation of all children. Community based after-school programs are ideal for achieving such goals.

B. Policy Recommendations

Table 2 provides a thorough overview of policy recommendations for obesity prevention across a number of domains. These policy recommendations acknowledge the multiple levels at which change must occur from city, state, and federal tax programs to fund campaigns to promote healthy eating and physical activity and curb unhealthy patterns to developing better walking/biking paths for the purposes of both recreation and transportation to work [197]. Recommendations for work-site programs are also made including environmental changes in work setting such as promoting healthy eating by increasing the availability of healthy food choices in cafeterias, instituting work-site campaigns to promote physical activity and healthy eating, and providing tax incentives to employers for providing weight management programs.

V. CONCLUSION

The goals of Healthy People 2010 are to reduce the prevalence of obesity among adults from 23% to 15% and to reduce the prevalence of obesity among children and adolescents from 11% to 5% [257, 258]. The etiology of obesity is complex and encompasses a wide variety of social, behavioral, cultural, environmental, physiological, and genetic factors. To achieve these ambitious goals, considerable effort must be focused on helping individuals at the population level modify their diets and increase their physical activity levels, key behaviors involved in the regulation of body weight. Educational and environmental interventions that support diet and exercise patterns associated with healthy body weight must be developed and evaluated. Prevention of obesity should begin early in life and involve the development and maintenance of healthy eating and physical activity patterns. These patterns need to be reinforced at home, in schools, and throughout the community. Public health agencies, communities, government, health organizations, the media, and the food and health industry must form alliances if we are to combat obesity.

TABLE 2 Reducing the Prevalence of Obesity: Policy Recommendations

Education

- Provide federal funding to state public health department for mass media health promotion campaigns that emphasize healthful eating and physical activity patterns.
- Require instruction in nutrition and weight management as part of the school curriculum for future health education teachers.
- Make a plant-based diet the focus of dietary guidance.
- Ban required watching of commercials for foods high in energy, fat, or sugar on school television programs (for example, Channel One).
- Declare and organize an annual National "No-TV" Week.
- Require and fund daily physical education and sports programs in primary and secondary schools, extending the school day if necessary.
- Develop culturally relevant obesity prevention campaigns for high-risk and low-income Americans.
- Promote healthy eating in government cafeterias, Veterans Administration medical centers, military installations, prisons, and other venues.
- Institute campaigns to promote healthy eating and activity patterns among federal and state employees in all departments.

Food labeling and advertising

- Require chain restaurants to provide information about calorie content on menus or menu boards and nutrition labeling on wrappers.
- Require that containers for soft drinks and snacks sold in movie theaters, convenience stores, and other venues bear information about calories, fat, or sugar content.
- Require nutrition labeling on fresh meat and poultry products.
- Restrict advertising of high-calorie, low-nutrient foods on television shows commonly watched by children or require broadcasters to provide equal time for messages promoting healthy eating and physical activity.
- Require print advertisements to disclose the caloric content of the foods being marketed.

Food assistance programs

- Protect school food programs by eliminating the sale of soft drinks, candy bars, and foods high in calories, fat, or sugar in school buildings.
- Require that any foods that compete with school meals be consistent with federal recommendations for fat, saturated fat, cholesterol, sugar, and sodium content.
- Develop an incentive system to encourage food stamp recipients to purchase fruits, vegetables, whole grains, and other healthful foods, such as by earmarking increases in food stamp benefits for the purchase of those foods.

Health Care and Training

- Require curricula for medical, nursing, and other health professions to teach the principles and benefits of healthful diets and exercise patterns.
- Require health care providers to learn about behavioral risks for obesity and how to counsel patients about health-promoting behavior change.
- Develop and fund a research agenda focused on behavioral as well as metabolic determinants of weight gain and maintenance, and on the most cost-effective methods for promoting healthful diet and activity patterns.
- Revise Medicaid and Medicare regulations to provide incentives to health care providers for nutrition and obesity counseling and other interventions that meet specified standards of cost and effectiveness.

Transportation and urban development

- Provide funding and other incentives for bicycle paths, recreation centers, swimming pools, parks, and sidewalks.
- Develop and provide guides for cities, zoning authorities, and urban planners on ways to modify residential neighborhoods, workplaces, and shopping centers to promote physical activity.

Taxes

- Levy city, state, or federal taxes on soft drinks and other foods high in calories, fat, or sugar to fund campaigns to promote good nutrition and physical activity.
- Subsidize the prices of low-calorie nutritious foods, perhaps by raising the prices of selected high-calorie, low-nutrient foods.
- Remove sales taxes on, or provide other incentives for, purchase of exercise equipment.
- Provide tax incentives to encourage employers to provide weight management programs.

Policy development

- Use the National Nutrition Summit to develop a national campaign to prevent obesity.
- Produce a *Surgeon General's Report on Obesity Prevention.*
- Expand the scope of the President's Council on Physical Fitness and Sports to include nutrition and to emphasize obesity prevention.
- Develop a coordinated federal implementation plan for the Healthy People 2010 nutrition and physical activity objectives.

Source: Nestle, M., and Jacobson, M. F. (2000). Halting the obesity epidemic: A public health policy approach. *Public Health Rep.* **115**, 12–24.

References

1. Flegal, K. M., Carroll, M. D., Kuczmarski, R. J., and Johnson, C. L. (1998). Overweight and obesity in the United States: Prevalence and trends, 1960–1994. *Int. J. Obes. Relat. Metab. Disord.* **22**, 39–47.
2. Troiano, R. P., and Flegal, K. M. (1998). Overweight children and adolescents: Description, epidemiology, and demographics. *Pediatrics* **101**(Suppl. 1), 497–504.
3. http://www.cdc.gov/nchs/products/pubs/pubd/hestats/overweight/overwght_adult_03.htm. Access date: July 18, 2007.
4. http://www.cdc.gov/nchs/products/pubs/pubd/hestats/overweight/overwght_child_03.htm#Table%201. Access date: July 18, 2007.
5. Ogden, C. L., Carroll, M. D., Curtin, L. R., McDowell, M. A., Tabak, C. J., and Flegal, K. M. (2006). Prevalence of overweight and obesity in the United States, 1999–2004. *JAMA* **295**, 1549–1555.
6. Whitaker, R. C., Wright, J. A., Pepe, M. S., Seidel, K. D., and Dietz, W. H. (1997). Predicting obesity in young adulthood from childhood and parental obesity. *N. Engl. J. Med.* **337**, 869–873.
7. Nader, P. R., O'Brien, M., Houts, R., Bradley, R., Belsky, J., Crosnoe, R., Friedman, S., Mei, Z., and Susman, E. J., National Institute of Child Health and Human Development Early Child Care Research Network. (2006). Identifying risk for obesity in early childhood. *Pediatrics* **118**, e594–e601.
8. Manson, J. E., and Bassuk, S. S. (2003). Obesity in the United States: A fresh look at its high toll. *JAMA* **289**, 229–230.
9. Fontaine, K. R., Redden, D. T., Wang, C., Westfall, A. O., and Allison, D. B. (2003). Years of life lost due to obesity. *JAMA* **289**, 187–193.
10. Wilson, P. W., D'Agostino, R. B., Sullivan, L., Parise, H., and Kannel, W. B. (2002). Overweight and obesity as determinants of cardiovascular risk: The Framingham experience. *Arch. Intern. Med.* **162**, 1867–1872.
11. Field, A. E., Coakley, E. H., Must, A., Spadano, J. L., Laird, N., Dietz, W. H., Rimm, E., and Colditz, G. A. (2001). Impact of overweight on the risk of developing common chronic diseases during a 10-year period. *Arch. Intern. Med.* **161**, 1581–1586.
12. Chang, S. C., and Lacey, J. V. Jr., Brinton, L. A., Hartge, P., Adams, K., Mouw, T., Carroll, L., Hollenbeck, A., Schatzkin, A., and Leitzmann, M. F. (2007). Lifetime weight history and endometrial cancer risk by type of menopausal hormone use in the NIH-AARP diet and health study. *Cancer Epidemiol. Biomarkers Prev.* **16**, 723–730.
13. Wenten, M., Gilliland, F. D., Baumgartner, K., and Samet, J. M. (2002). Associations of weight, weight change, and body mass with breast cancer risk in Hispanic and non-Hispanic white women. *Ann. Epidemiol.* **12**, 435–444.
14. Venkat Narayan, K. M., Boyle, J. P., Thompson, T. J., Gregg, E. W., and Williamson, D. F. (2007). Effect of body mass index on lifetime risk for diabetes mellitus in the United States. *Diabetes Care* **30**, 1562–1566.
15. Allison, D. B., Fontaine, K. R., Manson, J. E., Stevens, J., and VanItallie, T. B. (1999). Annual deaths attributable to obesity in the United States. *JAMA* **282**, 1530–1538.
16. National Institutes of Health, National Heart, Lung, and Blood Institute (1999). "Clinical Guidelines on the Identification, Evaluation, and Treatment of Overweight and Obesity in Adults." National Institutes of Health, National Heart, Lung, and Blood Institute, Bethesda, MD.
17. Marcus, M. (1993). Binge eating in obesity. *In* "Binge Eating: Nature, Assessment, and Treatment" (C. G. Fairburn and G. T. Wilson, Eds.), pp. 77–96. Guilford Press, New York.
18. Falkner, N. H., French, S. A., Jeffery, R. W., Neumark-Sztainer, D., Sherwood, N. E., and Morton, N. (1999). Mistreatment due to weight: Prevalence and sources of perceived mistreatment in women and men. *Obes. Res.* **7**, 572–576.
19. Wadden, T. A., and Stunkard, A. J. (1985). Social and psychological consequences of obesity. *Ann. Intern. Med.* **103**, 1062–1067.
20. Gortmaker, S. L., Must, A., Perrin, J. M., Sobol, A. M., and Dietz, W. H. (1993). Social and economic consequences of overweight in adolescence and young adulthood [see comments]. *N. Engl. J. Med.* **329**, 1008–1012.
21. Oster, G., Thompson, D., Edelsberg, J., Bird, A. P., and Colditz, G. A. (1999). Lifetime health and economic benefits of weight loss among obese persons. *Am. J. Public Health* **89**, 1536–1542.
22. Thompson, D., Edelsberg, J., Kinsey, K. L., and Oster, G. (1998). Estimated economic costs of obesity to U.S. business. *Am. J. Health Promot.* **13**, 120–127.
23. Wolf, A. M., and Colditz, G. A. (1996). Social and economic effects of body weight in the United States. *Am. J. Clin. Nutr.* **63**, 466S–469S.
24. Wolf, A. M. (2002). Economic outcomes of the obese patient. *Obes. Res.* **10**(Suppl 1), 58S–62S.
25. Thompson, D., and Wolf, A. M. (2001). The medical-care cost burden of obesity. *Obes, Rev.* **2**, 189–197.
26. Finkelstein, E. A., Fiebelkorn, I. C., and Wang, G. (2003). National medical spending attributable to overweight and obesity: How much, and who's paying? *Health Aff. (Millwood)* **W3**, 219–226.
27. Finkelstein, E. A., Brown, D. S., Trogdon, J. G., Segel, J. E., and Ben-Joseph, R. H. (2007). Age-specific impact of obesity on prevalence and costs of diabetes and dyslipidemia. *Value Health* **10**, S45–S51.
28. Wang, F., Schultz, A. B., Musich, S., McDonald, T., Hirschland, D., and Edington, D. W. (2003). The relationship between National Heart, Lung, and Blood Institute Weight Guidelines and concurrent medical costs in a manufacturing population. *Am. J. Health Promot.* **17**, 183–189.
29. Finkelstein, E. A., Ruhm, C. J., and Kosa, K. M. (2005). Economic causes and consequences of obesity. *Annu. Rev. Public Health,* **26**, 239–257.
30. Wang, G., and Dietz, W. H. (2002). Economic burden of obesity in youths aged 6 to 17 years: 1979–1999. *Pediatrics* **109**, E81.
31. American Dietetic Association (2000). Type 2 diabetes in children and adolescents. *Pediatrics,* **105**, 671–680.
32. de Ferranti, S. D., Gauvreau, K., Ludwig, D. S., Neufeld, E. J., Newburger, J. W., and Rifai, N. (2004). Prevalence of the metabolic syndrome in American adolescents: Findings from

the Third National Health and Nutrition Examination Survey. *Circulation,* **110**, 2494–2497.
33. Weiss, R., Dziura, J., Burgert, T. S., Tamborlane, W. V., Taksali, S. E., Yeckel, C. W., Allen, K., Lopes, M., Savoye, M., Morrison, J., Sherwin, R. S., and Caprio, S. (2004). Obesity and the metabolic syndrome in children and adolescents. *N. Engl. J. Med.* **350**, 2362–2374.
34. Lobstein, T., Baur, L., and Uauy, R. (2004). Obesity in children and young people: A crisis in public health. *Obes. Rev.* **5**(Suppl. 1), 4–104.
35. Williams, J., Wake, M., Hesketh, K., Maher, E., and Waters, E. (2005). Health-related quality of life of overweight and obese children. *JAMA* **293**, 70–76.
36. World Health Organization (1997). "Obesity: Preventing and Managing the Global Epidemic," Report of a WHO Consultation on Obesity, Geneva, June 3–5, 1997. World Health Organization, Geneva.
37. Bouchard, C., and Perusse, L. (1988). Heredity and body fat. *Annu. Rev. Nutr.* **8**, 259–277.
38. National Institutes of Health, National Heart, Lung, and Blood Institute (1999). "Clinical Guidelines on the Identification, Evaluation, and Treatment of Overweight and Obesity in Adults." National Institutes of Health, National Heart, Lung, and Blood Institute, Bethesda, MD.
39. Dannenberg, A. L., Keller, J. B., Wilson, P. W., and Castelli, W. P. (1989). Leisure time physical activity in the Framingham Offspring Study. Description, seasonal variation, and risk factor correlates. *Am. J. Epidemiol.* **129**, 76–88.
40. DiPietro, L., Williamson, D. F., Caspersen, C. J., and Eaker, E. (1993). The descriptive epidemiology of selected physical activities and body weight among adults trying to lose weight: The Behavioral Risk Factor Surveillance System Survey, 1989. *Int. J. Obes. Relat. Metab. Disord.* **17**, 69–76.
41. Folsom, A. R., Caspersen, C. J., Taylor, H. L., Jacobs, D. R., Jr., Luepker, R. V., Gomez-Marin, O., Gillum, R. F., and Blackburn, H. (1985). Leisure time physical activity and its relationship to coronary risk factors in a population-based sample. The Minnesota Heart Survey. *Am. J. Epidemiol.* **121**, 570–579.
42. French, S. A., Jeffery, R. W., Forster, J. L., McGovern, P. G., Kelder, S. H., and Baxter, J. E. (1994). Predictors of weight change over two years among a population of working adults: The Healthy Worker Project. *Int. J. Obes. Relat. Metab. Disord.* **18**, 145–154.
43. Gibbons, L. W., Blair, S. N., Cooper, K. H., and Smith, M. (1983). Association between coronary heart disease risk factors and physical fitness in healthy adult women. *Circulation* **67**, 977–983.
44. Hovell, M., Sallis, J., Hofstetter, R., Barrington, E., Hackley, M., Elder, J., Castro, F., and Kilbourne, K. (1991). Identification of correlates of physical activity among Latino adults. *J. Community Health* **16**, 23–36.
45. King, A. C., and Tribble, D. L. (1991). The role of exercise in weight regulation in nonathletes. *Sports Med.* **11**, 331–349.
46. Slattery, M. L., McDonald, A., Bild, D. E., Caan, B. J., Hilner, J. E., and Jacobs, D. R., Jr. and Liu, K. (1992). Associations of body fat and its distribution with dietary intake, physical activity, alcohol, and smoking in blacks and whites. *Am. J. Clin. Nutr.* **55** 943–949.
47. Voorrips, L. E., Lemmink, K. A., van Heuvelen, M. J., Bult, P., and van Staveren, W. A. (1993). The physical condition of elderly women differing in habitual physical activity. *Med. Sci. Sports Exerc.* **25**, 1152–1157.
48. Voorrips, L. E., van Staveren, W. A., and Hautvast, J. G. (1991). Are physically active elderly women in a better nutritional condition than their sedentary peers? *Eur. J. Clin. Nutr.* **45**, 545–552.
49. Bovet, P., Auguste, R., and Burdette, H. (2007). Strong inverse association between physical fitness and overweight in adolescents: a large school-based survey. *Int. J. Behav. Nutr. Phys. Act.* **4**, 24.
50. Epstein, L. H. (1995). Exercise in the treatment of childhood obesity. *Int. J. Obes. Relat. Metab. Disord.* **19**(Suppl. 4), S117–S121.
51. Jeffery, R. W., Drewnowski, A., Epstein, L. H., Stunkard, A. J., Wilson, G. T., and Wing, R.R, Hill, D. R. (2000). Long-term maintenance of weight loss: current status. *Health Psychol.* **19**(1 Suppl), 5–16. Review.
52. Tate, D. F., Jeffery, R. W., Sherwood, N. E., and Wing, R. R. (2007). Long-term weight losses associated with prescription of higher physical activity goals. Are higher levels of physical activity protective against weight regain? *Am. J. Clin. Nutr.* **85**(4), 954–959.
53. Macera, C. A., Ham, S. A., Yore, M. M., Jones, D. A., Ainsworth, B. E., and Kimsey C.D., Kohl, H. W. 3rd. (2005). Prevalence of physical activity in the United States: Behavioral Risk Factor Surveillance System, 2001. *Prev. Chronic Dis.* **2**, A17, Epub March 15, 2005.
54. Caspersen, C. J., Merritt, R. K., Health, G. W., and Yeager, K. K. (1990). Physical activity patterns of adults aged 60 years and older. *Med. Sci. Sports Exerc.* **22**, S79.
55. U.S. Department of Health and Human Services (1988). "Surgeon General's Report on Nutrition and Health." U.S. Department of Health and Human Services, Public Health Service, Washington, DC.
56. Caspersen, C. J., and Merritt, R. K. (1995). Physical activity trends among 26 states, 1986–1990. *Med. Sci. Sports Exerc.* **27**, 713–720.
57. Caspersen, C. J., and Merritt, R. K. (1992). Trends in physical activity patterns among older adults: The Behavioral Risk Factor Surveillance System, 1986–1990. *Med. Sci. Sports Exerc.* **24**, 526.
58. Bauman, A., Owen, N., and Rushworth, R. L. (1990). Recent trends and socio-demographic determinants of exercise participation in Australia. *Community Health Studies* **14**, 19–26.
59. Folsom, A. R., Cook, T. C., Sprafka, J. M., Burke, G. L., Norsted, S. W., and Jacobs, D. R., Jr. (1991). Differences in leisure-time physical activity levels between blacks and whites in population-based samples: The Minnesota Heart Survey. *J. Behav. Med.* **14**, 1–9.
60. King, A. C., Blair, S. N., Bild, D. E., Dishman, R. K., Dubbert, P. M., Marcus, B. H., Oldridge, N. B., Paffenbarger, R. S. J., Powell, K. E., and Yeager, K. K. (1992). Determinants of physical activity and interventions in adults. *Med. Sci. Sports Exerc.* **24**, S221–S236.
61. Centers for Disease Control and Prevention (CDC). (2003). Physical activity levels among children aged 9–13

years—United States, 2002. *MMWR Morb. Mortal. Wkly Rep.* **52**, 785–788.
62. Gortmaker, S. L., Must, A., Sobol, A. M., Peterson, K., Colditz, G. A., and Dietz, W. H. (1996). Television viewing as a cause of increasing obesity among children in the United States, 1986–1990. *Arch. Pediatr. Adolesc. Med.* **150**, 356–362.
63. Dietz, W. H., and Strasburger, V. C. (1991). Children, adolescents, and television. *Curr. Probl. Pediatr.* **21**, 8–31; discussion 32.
64. Dietz, W. H., Jr., and Gortmaker, S. L. (1985). Do we fatten our children at the television set? Obesity and television viewing in children and adolescents. *Pediatrics* **75**, 807–812.
65. Epstein, L. H. (1992). Exercise and obesity in children. *J. Appl. Sport Psychol.* **4**, 120–133.
66. A. C. Nielsen Company. (2000). "2000 Report on Television. The First 50 Years." Nielsen Media Research, New York.
67. Robinson, J., and Godbey, G. (1997). "Time for Life." Pennsylvania State University Press, University Park.
68. Crawford, D. A., Jeffery, R. W., and French, S. A. (1999). Television viewing, physical inactivity and obesity. *Int. J. Obes. Relat. Metab. Disord.* **23**, 437–440.
69. Sidney, S., Sternfeld, B., Haskell, W. L., Jacobs, D. R., Jr., Chesney, M. A., and Hulley, S. B. (1996). Television viewing and cardiovascular risk factors in young adults: The CARDIA study. *Ann. Epidemiol.* **6**, 154–159.
70. Coakley, E. H., Rimm, E. B., Colditz, G., Kawachi, I., and Willett, W. (1998). Predictors of weight change in men: Results from the Health Professionals Follow-up Study. *Int. J. Obes. Relat. Metab. Disord.* **22**, 89–96.
71. Bowman, S. A. (2006). Television-viewing characteristics of adults: Correlations to eating practices and overweight and health status. *Prev. Chronic Dis.* **3**, 38. Epub March 15, 2006.
72. Eisenmann, J. C., Bartee, R. T., and Wang, M. Q. (2002). Physical activity, TV viewing, and weight in U.S. youth: 1999 Youth Risk Behavior Survey. *Obes. Res.* **10**(5), 379–385.
73. Christakis, D. A., Ebel, B. E., Rivara, F. P., and Zimmerman, F. J. (2004). Television, video, and computer game usage in children under 11 years of age. *J. Pediatr.* **145**, 652–656.
74. Certain, L. K., and Kahn, R. S. (2002). Prevalence, correlates, and trajectory of television viewing among infants and toddlers. *Pediatrics* **109**, 634–642.
75. Pate, R. R., and Ross, J. G. (1987). The national children and youth fitness study II: Factors associated with health-related fitness. *J. Phys. Educ. Recreat. Dance* **58**, 93–95.
76. Obarzanek, E., Schreiber, G. B., Crawford, P. B., Goldman, S. R., Barrier, P. M., Frederick, M. M., and Lakatos, E. (1994). Energy intake and physical activity in relation to indexes of body fat: The National Heart, Lung, and Blood Institute Growth and Health Study. *Am. J. Clin. Nutr.* **60**, 15–22.
77. Shannon, B., Peacock, J., and Brown, M. J. (1991). Body fatness, television viewing and calorie-intake of a sample of Pennsylvania sixth grade children. *J. Nutr. Educ.* **23**, 262–268.
78. Locard, E., Mamelle, N., Billette, A., Miginiac, M., Munoz, F., and Rey, S. (1992). Risk factors of obesity in a five year old population. Parental versus environmental factors. *Int. J. Obes. Relat. Metab. Disord.* **16**, 721–729.
79. Andersen, R. E., Crespo, C. J., Bartlett, S. J., Cheskin, L. J., and Pratt, M. (1998). Relationship of physical activity and television watching with body weight and level of fatness among children: Results from the Third National Health and Nutrition Examination Survey [see comments]. *JAMA* **279**, 938–942.
80. Robinson, T. N., Hammer, L. D., Killen, J. D., Kraemer, H. C., Wilson, D. M., Hayward, C., and Taylor, C. B. (1993). Does television viewing increase obesity and reduce physical activity? Cross-sectional and longitudinal analyses among adolescent girls [see comments]. *Pediatrics* **91**, 273–280.
81. Robinson, T. N., and Killen, J. D. (1995). Ethnic and gender differences in the relationships between television viewing and obesity, physical activity and dietary fat intake. *J. Health Educ.* **26**, S91–S98.
82. Tucker, L. A. (1986). The relationship of television viewing to physical fitness and obesity. *Adolescence* **21**, 797–806.
83. Wolf, A. M., Gortmaker, S. L., Cheung, L., Gray, H. M., Herzog, D. B., and Colditz, G. A. (1993). Activity, inactivity, and obesity: Racial, ethnic, and age differences among schoolgirls. *Am. J. Public Health* **83**, 1625–1627.
84. DuRant, R. H., Baranowski, T., Johnson, M., and Thompson, W. O. (1994). The relationship among television watching, physical activity, and body composition of young children. *Pediatrics* **94**, 449–455.
85. DuRant, R. H., Thompson, W. O., Johnson, M., and Baranowski, T. (1996). The relationship among television watching, physical activity, and body composition of 5- or 6-year-old children. *Pediatr. Exerc. Sci.* **8**, 15–26.
86. Dwyer, J. T., Stone, E. J., Yang, M., Feldman, H., Webber, L. S., Must, A., Perry, C. L., Nader, P. R., and Parcel, G. S. (1998). Predictors of overweight and overfatness in a multiethnic pediatric population. Child and Adolescent Trial for Cardiovascular Health Collaborative Research Group. *Am. J. Clin. Nutr.* **67**, 602–610.
87. Armstrong, C. A., Sallis, J. F., Alcaraz, J. E., Kolody, B., McKenzie, T. L., and Hovell, M. F. (1998). Children's television viewing, body fat, and physical fitness. *Am. J. Health Promot.* **12**, 363–368.
88. O'Brien, M., Nader, P. R., Houts, R. M., Bradley, R., Friedman, S. L., Belsky, J., and Susman, E. (2007). The ecology of childhood overweight: A 12-year longitudinal analysis. *Int. J. Obes. (Lond.).* **31(9)**: 1469–1478.
89. Gable, S., Chang, Y., and Krull, J. L. (2007). Television watching and frequency of family meals are predictive of overweight onset and persistence in a national sample of school-aged children. *J. Am. Diet. Assoc.* **107**(1), 53–61.
90. Utter, J., Neumark-Sztainer, D., Jeffery, R., and Story, M. (2003). Couch potatoes or French fries: Are sedentary behaviors associated with body mass index, physical activity, and dietary behaviors among adolescents? *J. Am. Diet. Assoc.* **103**(10), 1298–1305.
91. Tucker, L. A., and Friedman, G. M. (1989). Television viewing and obesity in adult males. *Am. J. Public Health* **79**, 516–518.

92. Tucker, L. A., and Bagwell, M. (1991). Television viewing and obesity in adult females. *Am. J. Public Health* **81**, 908–911.
93. Jeffery, R. W., and French, S. A. (1998). Epidemic obesity in the United States: Are fast foods and television viewing contributing? *Am. J. Public Health* **88**, 277–280.
94. Ching, P. L., Willett, W. C., Rimm, E. B., Colditz, G. A., Gortmaker, S. L., and Stampfer, M. J. (1996). Activity level and risk of overweight in male health professionals. *Am. J. Public Health* **86**, 25–30.
95. Committee on Food Marketing and the Diets of Children and Youth, J. Michael McGinnis, Jennifer Appleton Gootman, Vivica I. Kraak, Editors (2006). Food Marketing to Children and Youth: Threat or Opportunity? The National Academies Press.
96. Temple, J. L., Giacomelli, A. M., Kent, K. M., Roemmich, J. N., and Epstein, L. H. (2007). Television watching increases motivated responding for food and energy intake in children. *Am. J. Clin. Nutr.* **85**, 355–361.
97. DeMattia, L., Lemont, L., and Meurer, L. (2007). Do interventions to limit sedentary behaviours change behaviour and reduce childhood obesity? A critical review of the literature. *Obes. Rev.* **8**(1), 69–81.
98. Epstein, L. H., Valoski, A. M., Vara, L. S., McCurley, J., Wisniewski, L., Kalarchian, M. A., Klein, K. R., and Shrager, L. R. (1995). Effects of decreasing sedentary behavior and increasing activity on weight change in obese children. *Health Psychol.* **14**, 109–115.
99. Epstein, L. H., Saelens, B. E., Myers, M. D., and Vito, D. (1997). Effects of decreasing sedentary behaviors on activity choice in obese children. *Health Psychol.* **16**, 107–113.
100. Epstein, L. H. (1998). Integrating theoretical approaches to promote physical activity. *Am. J. Prevent. Med.* **15**, 257–265.
101. Robinson, T. N. (1999). Reducing children's television viewing to prevent obesity: A randomized controlled trial. *JAMA* **282**, 1561–1567.
102. Gortmaker, S. L., Peterson, K., Wiecha, J., Sobol, A. M., Dixit, S., Fox, M. K., and Laird, N. (1999). Reducing obesity via a school-based interdisciplinary intervention among youth: Planet Health. *Arch. Pediatr. Adolesc. Med.* **153**, 409–418.
103. Brawley, L. R., and Rodgers, W. M. (1993). Social psychological aspects of fitness promotion. *In* "Exercise Psychology" (P. Seraganian, Ed.), pp. 254–298. Wiley, New York.
104. Courneya, K. S., and McAuley, E. (1993). Can short-range intentions predict physical activity participation? *Percept. Mot. Skills* **77**, 115–122.
105. DuCharme, K. A., and Brawley, L. R. (1995). Predicting the intentions and behavior of exercise initiates using two forms of self-efficacy. *J. Behav. Med.* **18**, 479–497.
106. Hovell, M. F., Sallis, J. F., Hofstetter, C. R., Spry, V. M., Faucher, P., and Caspersen, C. J. (1989). Identifying correlates of walking for exercise: An epidemiologic prerequisite for physical activity promotion. *Prevent. Med.* **18**, 856–866.
107. Marcus, B. H., Pinto, B. M., Simkin, L. R., Audrain, J. E., and Taylor, E. R. (1994). Application of theoretical models to exercise behavior among employed women. *Am. J. Health Promot.* **9**, 49–55.
108. Marcus, B. H., Selby, V. C., Niaura, R. S., and Rossi, J. S. (1992). Self-efficacy and the stages of exercise behavior change. *Res. Q. Exerc. Sport.* **63**, 60–66.
109. McAuley, E. (1992). The role of efficacy cognitions in the prediction of exercise behavior in middle-aged adults. *J. Behav. Med.* **15**, 65–88.
110. McAuley, E., and Jacobson, L. (1991). Self-efficacy and exercise participation in sedentary adult females. *Am. J. Health Promot.* **5**, 185–191.
111. Poag, K., and McAuley, E. (1992). Goal setting, self-efficacy, and exercise bahavior. *J. Sport Exerc. Psychol.* **14**, 352–360.
112. Poag-DuCharme, K. A., and Brawley, L. R. (1993). Self-efficacy theory: Use in the prediction of exercise behavior in the community setting. *J. Appl. Sport Pscyhol.* **5**, 178–194.
113. Rodgers, W. M., and Brawley, L. R. (1993). Using both self-efficacy theory and the theory of planned behavior to discriminate adherers and dropouts from stuctured programs. *J. Appl. Sport Psychol.* **5**, 195–206.
114. Bandura, A. (1986). "Social Foundations of Thought and Action: A Social Cognitive Theory." Prentice Hall, Englewood Cliffs, NJ.
115. O'Loughlin, J., Paradis, G., Kishchuk, N., Barnett, T., and Renaud, L. (1999). Prevalence and correlates of physical activity behaviors among elementary schoolchildren in multiethnic, low income, inner-city neighborhoods in Montreal, Canada. *Ann. Epidemiol.* **9**, 397–407.
116. DiLorenzo, T. M., Stucky-Ropp, R. C., Vander Wal, J. S., and Gotham, H. J. (1998). Determinants of exercise among children. II. A longitudinal analysis. *Prevent. Med.* **27**, 470–477.
117. Simons-Morton, B. G., McKenzie, T. J., Stone, E., Mitchell, P., Osganian, V., Strikmiller, P. K., Ehlinger, S., Cribb, P., and Nader, P. R. (1997). Physical activity in a multiethnic population of third graders in four states. *Am. J. Public Health* **87**, 45–50.
118. Sallis, S. G., Pate, R. R., Saunders, R., Ward, D. S., Dowda, M., and Felton, G. (1997). A prospective study of the determinants of physical activity in rural fifth-grade children. *Prevent. Med.* **26**, 257–263.
119. Strauss, R. S., Rodzilsky, D., Burack, G., and Colin, M. (2001). Psychosocial correlates of physical activity in healthy children. *Arch. Pediatr. Adolesc. Med.* **155**, 897–902.
120. Dishman, R. K., Motl, R. W., Sallis, J. F., Dunn, A. L., Birnbaum, A. S., Welk, G. J., Bedimo-Rung, A. L., Voorhees, C. C., and Jobe, J. B. (2005). Self-management strategies mediate self-efficacy and physical activity. *Am. J. Prev. Med.* **29**(1), 10–18.
121. McAuley, E., Lox, C., and Duncan, T. E. (1993). Long-term maintenance of exercise, self-efficacy, and physiological change in older adults. *J. Gerontol.* **48**, 218–224.
122. McAuley, E., Courneya, K. S., and Lettunich, J. (1991). Effects of acute and long-term exercise on self-efficacy responses in sedentary, middle-aged males and females. *Gerontologist* **31**, 534–542.
123. Williams, P., and Lord, S. R. (1995). Predictors of adherence to a structured exercise program for older women. *Psychol. Aging* **10**, 617–624.

124. Kristensen, P. L., Moller, N. C., Korsholm, L., Wedderkopp, N., Andersen, L. B., and Froberg, K. (2007). Tracking of objectively measured physical activity from childhood to adolescence: The European youth heart study. *Scand. J. Med. Sci. Sports.* doi: 10.1111/j.1600-0838.2006.00622.x.
125. Dishman, R. K., and Sallis, J. F. (1994). Determinants and interventions for physical activity and exercise. *In* "Physical Activity, Fitness, and Health: International Proceedings and Consensus Statement" (C. Bouchard, R. J. Shephard, and T. Stephens, Eds.), pp. 214–238. Hum. Kinet, Champaign.
126. Hoftstetter, C. R., Hovell, M. F., and Sallis, J. F. (1990). Social learning correlates of exercise self-efficacy: Early experiences with physical activity. *Soc. Sci. Med.* **31**, 1169–1176.
127. Alfano, C. M., Klesges, R. C., Murray, D. M., Beech, B. M., and McClanahan, B. S. (2002). History of sport participation in relation to obesity and related health behaviors in women. *Prev. Med.* **34**(1), 82–89.
128. Taylor, W. C., Blair, S. N., Cummings, S. S., Wun, C. C., and Malina, R. M. (1999). Childhood and adolescent physical activity patterns and adult physical activity. *Med. Sci. Sports Exerc.* **31**, 118–123.
129. Sallis, J. F., Prochaska, J. J., Taylor, W. C., Hill, J. O., and Geraci, J. C. (1999). Correlates of physical activity in a national sample of girls and boys in grades 4 through 12. *Health Psychol.* **18**, 410–415.
130. King, A. C., Taylor, C. B., Haskell, W. L., and DeBusk, R. F. (1990). Identifying strategies for increasing employee physical activity levels: Findings from the Stanford/Lockheed Exercise Survey. *Health Educ. Q.* **17**, 269–285.
131. Treiber, F. A., Baranowski, T., Braden, D. S., Strong, W. B., Levy, M., and Knox, W. (1991). Social support for exercise: Relationship to physical activity in young adults. *Prevent. Med.* **20**, 737–750 [erratum **21**(3), 392].
132. Hooper, J. M., and Veneziano, L. (1995). Distinguishing starters from nonstarters in an employee physical activity incentive program. *Health Educ. Q.* **22**, 49–60.
133. Wallace, J. P., Raglin, J. S., and Jastremski, C. A. (1995). Twelve month adherence of adults who joined a fitness program with a spouse vs without a spouse. *J. Sports Med. Phys. Fitness* **35**, 206–213.
134. Carron, A. V., Hausenblaus, H. A., and Mack, D. (1996). Social influence and exercise: A meta-analysis. *J. Sports Exerc. Psychol.* **18**, 1–16.
135. Kuo, J., Voorhees, C. C., Haythornthwaite, J. A., and Young, D. R. (2007). Associations between family support, family intimacy, and neighborhood violence and physical activity in urban adolescent girls. *Am. J. Public Health* **97**(1), 101–103.
136. Ornelas, I. J., Perreira, K. M., and Ayala, G. X. (2007). Parental influences on adolescent physical activity: A longitudinal study. *Int. J. Behav. Nutr. Phys. Act.* **4**, 3.
137. Dowda, M., Dishman, R. K., Pfeiffer, K. A., and Pate, R. R. (2007). Family support for physical activity in girls from 8th to 12th grade in South Carolina. *Prev. Med.* **44**(2), 153–159.
138. Heitzler, C. D., Martin, S. L., Duke, J., and Huhman, M. (2006). Correlates of physical activity in a national sample of children aged 9–13 years. *Prev. Med.* **42**(4), 254–260.
139. Epstein, L. H., Smith, J. A., Vara, L. S., and Rodefer, J. S. (1991). Behavioral economic analysis of activity choice in obese children. *Health Psychol.* **10**, 311–316.
140. Klesges, R. C., Costes, T. J., Moldenhauer-Klesges, L. M., Holzer, B., Gustavson, J., and Barnes, J. (1984). The FATS: An observational system for assessing physical activity in children and associated parent behavior. *Behav. Assess.* **6**, 333–345.
141. Klesges, R. C., Malott, J. M., Buschee, P. F., and Weber, J. M. (1986). The effects of parental influences on children's food intake, physical activity and relative weight. *Int. J. Eat. Disord.* **5**, 335–346.
142. McKenzie, T. L., Sallis, J. F., Nader, P. R., Patterson, T. L., Elder, J. P., Berry, C. C., Rupp, J. W., Atkins, C. J., Buono, M. J., and Nelson, J. A. (1991). BEACHES: An observational system for assessing children's eating and physical activity behaviors and associated events. *J. Appl. Behav. Anal.* **24**, 141–151.
143. Klesges, R. C., Eck, L. H., Hanson, C. L., Haddock, C. K., and Klesges, L. M. (1990). Effects of obesity, social interactions, and physical environment on physical activity in preschoolers. *Health Psychol.* **9**, 435–449.
144. Sallis, J. F., Alcaraz, J. E., McKenzie, T. L., Hovell, M. F., Kolody, B., and Nader, P. R. (1992). Parental behavior in relation to physical activity and fitness in 9-year-old children. *Am. J. Dis. Child.* **146**, 1383–1388.
145. Ross, J. G., and Gilbert, G. G. (1985). The National Children and Youth Fitness Study. A summary of findings. *J. Phys. Educ. Recreat. Dance* **56**, 45–50.
146. Taylor, W. C., Baranowski, T., and Sallis, J. F. (1994). Family determinants of childhood physical activity. *In* "Advances in Exercise Adherence" (R. Dishman, Ed.). Human Kinetics, Champaign, IL.
147. Dishman, R. K., Sallis, J. F., and Orenstein, D. R. (1985). The determinants of physical activity and exercise. *Public Health Rep.* **100**, 158–171.
148. Schmitz, M. K. H., Jacobs, D. R., French, S., Lewis, C. E., Caspersen, C. J., and Sternfeld, B. (1999). The impact of becoming a parent on physical activity: The CARDIA Study. *Circulation* **99**, 1108.
149. King, A. C. (1994). Community and public health approaches to the promotion of physical activity. *Med. Sci. Sports Exerc.* **26**, 1405–1412.
150. Sherwood, N. E., Morton, N., Jeffery, R. W., French, S. A., Neumark-Sztainer, D., and Falkner, N. H. (1998). Consumer preferences in format and type of community-based weight control programs. *Am. J. Health Promot.* **13**, 12–18.
151. Perri, M. G., Martin, A. D., Leermakers, E. A., Sears, S. F., and Notelovitz, M. (1997). Effects of group- versus home-based exercise in the treatment of obesity. *J. Consult. Clin. Psychol.* **65**, 278–285.
152. Craighead, L. W., and Blum, M. D. (1989). Supervised exercise in behavioral treatment for moderate obesity. *J. Behav. Med.* **20**, 49–60.
153. Andersen, R. E., Wadden, T. A., Bartlett, S. J., Zemel, B., Verde, T. J., and Franckowiak, S. C. (1999). Effects of lifestyle activity vs structured aerobic exercise in obese women: A randomized trial. *JAMA* **281**, 335–340.

154. Dunn, A. L., Marcus, B. H., Kampert, J. B., Garcia, M. E., Kohl, H. W., III, and Blair, S. N. (1999). Comparison of lifestyle and structured interventions to increase physical activity and cardiorespiratory fitness: A randomized trial. *JAMA* **281**, 327–334.
155. Murphy, M. H., and Hardman, A. E. (1998). Training effects of short and long bouts of brisk walking in sedentary women. *Med. Sci. Sports Exerc.* **30**, 152–157.
156. DeBusk, R. F., Stenestrand, U., Sheehan, M., and Haskell, W. L. (1990). Training effects of long versus short bouts of exercise in healthy subjects. *Am. J. Cardiol.* **65**, 1010–1013.
157. Ebisu, T. (1985). Splitting the distances of endurance training: On cardiovascular endurance and blood lipids. *Jpn. J. Phys. Educ.* **32**, 37–43.
158. Jakicic, J. M., Wing, R. R., Butler, B. A., and Robertson, R. J. (1995). Prescribing exercise in multiple short bouts versus one continuous bout: Effects on adherence, cardiorespiratory fitness, and weight loss in overweight women. *Int. J. Obes. Relat. Metab. Disord.* **19**, 893–901.
159. Jakicic, J. M., Winters, C., Lang, W., and Wing, R. R. (1999). Effects of intermittent exercise and use of home exercise equipment on adherence, weight loss, and fitness in overweight women: A randomized trial. *JAMA* **282**, 1554–1560.
160. Epstein, L. H., Valoski, A., Wing, R. R., and McCurley, J. (1994). Ten-year outcomes of behavioral family-based treatment for childhood obesity [see comments]. *Health Psychol.* **13**, 373–383.
161. Sallis, J. F., Hovell, M. F., Hofstetter, C. R., Elder, J. P., Hackley, M., Caspersen, C. J., and Powell, K. E. (1990). Distance between homes and exercise facilities related to frequency of exercise among San Diego residents. *Public Health Rep.* **105**, 179–185.
162. Centers for Disease Control and Prevention. (1999). Neighborhood safety and the prevalence of physical inactivity—Selected states, 1996. *MMWR Morbid. Mortal. Wkly. Rep.* **48**, 143–146.
163. Baranowski, T., Thompson, W. O., DuRant, R. H., Baranowski, J., and Puhl, J. (1993). Observations on physical activity in physical locations: Age, gender, ethnicity, and month effects. *Res. Q. Exerc. Sport.* **64**, 127–133.
164. Sallis, J. F., Alcaraz, J. E., McKenzie, T. L., and Hovell, M. F. (1999). Predictors of change in children's physical activity over 20 months. Variations by gender and level of adiposity. *Am. J. Prevent. Med.* **16**, 222–229.
165. Klem, M. L., Wing, R. R., McGuire, M. T., Seagle, H. M., and Hill, J. O. (1997). A descriptive study of individuals successful at long-term maintenance of substantial weight loss. *Am. J. Clin. Nutr.* **66**, 239–246.
166. McGuire, M. T., Wing, R. R., Klem, M. L., and Hill, J. O. (1999). Behavioral strategies of individuals who have maintained long-term weight losses. *Obes. Res.* **7**, 334–341.
167. Tate, D. F., Jeffery, R. W., Sherwood, N. E., and Wing, R. R. (2007). Long-term weight losses associated with prescription of higher physical activity goals. Are higher levels of physical activity protective against weight regain? *Am. J. Clin. Nutr.* **85**(4), 954–959.
168. Kant, A. K., and Graubard, B. I. (2006). Secular trends in patterns of self-reported food consumption of adult Americans: NHANES 1971–1975 to NHANES 1999–2002. *Am. J. Clin. Nutr.* **84**(5), 1215–1223.
169. Troiano, R. P., Briefel, R. R., Carroll, M. D., and Bialostosky, K. (2000). Energy and fat intake of children and adolescents in the United States. Data from the National Health and Nutrition Examination Surveys. *Am. J. Clin. Nutr.* **72**, 1343S–1354S.
170. Briefel, R. R., McDowell, M. A., Alaimo, K., Caughman, C. R., Bischof, A. L., Carroll, M. D., and Johnson, C. L. (1995). Total energy intake of the U.S. population: The third National Health and Nutrition Examination Survey, 1988–1991. *Am. J. Clin. Nutr.* **62**, 1072S–1080S.
171. Federation of Associated Societies for Experimental Biology (1995). "Third Report on Nutrition Monitoring in the United States." U.S. Government Printing Office, Washington, DC.
172. McDowell, M. A., Briefel, R. R., Alaimo, K., Bischof, A. M., Caughman, C. R., Carroll, M. D., Loria, C. M., and Johnson, C. L. (1994). Energy and macronutrient intakes of persons ages 2 months and over in the United States: Third National Health and Nutrition Examination Survey, Phase 1, 1988–91. *Adv. Data* **255**, 1–24.
173. Tippett, K. S., and Cleveland, L. E. (1999). How current diets stack up: Comparison with dietary guidelines. *In* "America's Eating Habits: Changes and Consequences" (E. Frazao, Ed.), pp. 51–70. USDA, Washington, DC.
174. Schoeller, D. A. (1990). How accurate is self-reported dietary intake? *Nutr. Rev.* **48**, 373–379.
175. Bingham, S. A. (1987). The dietary assessment of individuals; Methods, accuracy, new techniques and recommendations. *Nutr. Abstr. Rev.* **57**, 705–742.
176. Goris, A. H., Westerterp-Plantenga, M. S., and Westerterp, K. R. (2000). Undereating and underrecording of habitual food intake in obese men: Selective underreporting of fat intake. *Am. J. Clin. Nutr.* **71**, 130–134.
177. Jonnalagadda, S. S., Mitchell, D. C., Smiciklas-Wright, H., Meaker, K. B., Heel, N. V., Karmally, W., Ershow, A. G., and Kris-Etherton, P. M. (2000). Accuracy of energy intake data estimated by a multiple-pass, 24-hour dietary recall technique. *J. Am. Diet. Assoc.* **100**, 303–308; quiz 309–311.
178. Bray, G. A., and Popkin, B. M. (1998). Dietary fat intake does affect obesity! *Am. J. Clin. Nutr.* **68**, 1157–1173.
179. Willett, W. C. (1998). Is dietary fat a major determinant of body fat? *Am. J. Clin. Nutr.* **67**, 556S–562S.
180. Hill, J. O., Melanson, E. L., and Wyatt, H. T. (2000). Dietary fat intake and regulation of energy balance: Implications for obesity. *J. Nutr.* **130**, 284S–288S.
181. West, D. B., and York, B. (1998). Dietary fat, genetic predisposition, and obesity: Lessons from animal models. *Am. J. Clin. Nutr.* **67**, 505S–512S.
182. Anand, R. S., and Basiotis, P. P. (1998). Is total fat consumption really decreasing? *Family Econom. Nutr. Rev.* **11**, 58–60.
183. Putnam, J., and Gerrior, S. (1999). Trends in the U.S. food supply, 1970–1997. *In* "America's Eating Habits: Changes and Consequences" (E. Frazao, Ed.), pp. 133–160. USDA, Washington, DC.
184. Putnam, J. (1999). U.S. food supply providing more food and calories. *FoodReview* **22**, 2–12.

185. Guthrie, J. F., and Morton, J. F. (2000). Food sources of added sweeteners in the diets of Americans. *J. Am. Diet. Assoc.* **100**, 43–51, quiz 49–50.
185a. Nielsen, S. J., and Popkin, B. M. (2004). Changes in beverage intake between 1997 and 2001. *Am. J. Prev. Med.* **27**, 205–210.
186. Harnack, L., Stang, J., and Story, M. (1999). Soft drink consumption among U.S. children and adolescents: Nutritional consequences. *J. Am. Diet. Assoc.* **99**, 436–441.
187. Drewnowski, A. (1997). Taste preferences and food intake. *Annu. Rev. Nutr.* **17**, 237–253.
188. Weimer, J. (1999). Accelerating the trend toward healthy eating. *In* "America's Eating Habits: Changes and Consequences" (E. Frazao, Ed.), pp. 385–401. USDA, Washington, DC.
189. Heini, A. F., and Weinsier, R. L. (1997). Divergent trends in obesity and fat intake patterns: The American paradox. *Am. J. Med.* **102**, 259–264.
190. Bell, E. A., Castellanos, V. H., Pelkman, C. L., Thorwart, M. L., and Rolls, B. J. (1998). Energy density of foods affects energy intake in normal-weight women. *Am. J. Clin. Nutr.* **67**, 412–420.
191. Lin, B. H., Guthrie, J., and Frazao, E. (1999). Nutrient contribution of food away from home. *In* "America's Eating Habits: Changes and Consequences" (E. Frazao, Ed.), pp. 213–242. USDA, Washington, DC.
192. Kant, A. K., and Graubard, B. I. (2004). Eating out in America, 1987–2000: Trends and nutritional correlates. *Prev. Med.* **38**, 243–249.
193. Lin, B. H., Guthrie, J., and Frazao, E. (1998). Popularity of dining out presents barrier to dietary improvements. *Food Rev.* **21**, 2–10.
194. National Restaurant Association (2000). "Restaurant Industry Pocket Factbook." National Restaurant Association, Washington, DC.
195. Clauson, A. (1999). Share of food spending for eating out reaches 47%. *FoodReview* **22**, 20–33.
196. Nielsen, S. J., Siega-Riz, A. M., and Popkin, B. M. (2002). Trends in food locations and sources among adolescents and young adults. *Prev. Med.* **35,** 107–113.
197. Nestle, M., and Jacobson, M. F. (2000). Halting the obesity epidemic: A public health policy approach. *Public Health Rep.* **115**, 12–24.
198. Guthrie, J. F., Derby, B. M., and Levy, A. S. (1999). What people know and do not know about nutrition. *In* "America's Eating Habits: Changes and Consequences" (E. Frazao, Ed.), pp. 243–280. USDA, Washington, DC.
199. Jekanowski, M. D. (1999). Causes and consequences of fast food sales growth. *Food Review* **22**, 11–16.
200. Anonymous (1999). Food away from home sales at a glance 1988–98. *FoodReview* **22**, 26.
201. Young, L. R., and Nestle, M. (2007). Portion sizes and obesity: Responses of fast-food companies. *J. Public Health Pol.* **28**(2), 238–248.
202. McCrory, M. A., Fuss, P. J., Hays, N. P., Vinken, A. G., Greenberg, A. S., and Roberts, S. B. (1999). Overeating in America: Association between restaurant food consumption and body fatness in healthy adult men and women ages 19 to 80. *Obes. Res.* **7**, 564–571.
203. Taveras, E. M., Berkey, C. S., Rifas-Shiman, S. L., Ludwig, D. S., Rockett, H. R., Field, A. E., Colditz, G. A., and Gillman, M. W. (2005). Association of consumption of fried food away from home with body mass index and diet quality in older children and adolescents. *Pediatrics* **116**(4), e518–e524.
204. Thompson, O. M., Ballew, C., Resnicow, K., Must, A., Bandini, L. G., Cyr, H., and Dietz, W. H. (2004). Food purchased away from home as a predictor of change in BMI z-score among girls. *Int. J. Obes. Relat. Metab. Disord.* **28**(2), 282–289.
204a. Smiciklas-Wright, H., Michell, D., Michle, S., Goldman, J., and Cook, A. (2003). Foods commonly eaten in the United States, 1989–1991 and 1994–1996: are portion sizes changing? *J. Am. Diet. Assoc.* **103**, 41–47.
205. Rolls, B. J., Roe, L. S., and Meengs, J. S. (2006). Larger portion sizes lead to a sustained increase in energy intake over 2 days. *J. Am. Diet. Assoc.* **106**(4), 543–549.
206. Story, M., and French, S. (2004). Food advertising and marketing directed at children and adolescents in the U.S. *Int. J. Behav. Nutr. Phys. Act.* **1**, 3.
207. Institute of Medicine (U.S.), Committee on Food Marketing and the Diets of Children and Youth, McGinnis, J. M., Gootman, J., and Kraak, V. I. (2006). "Food Marketing to Children: Threat or Opportunity?" National Academies Press, Washington, DC.
208. Kaiser Family Foundation. (2007). "Food for thought: Television food advertising to children in the United States." Available at: http://www.kff.org.
209. Kaiser Family Foundation. (2006). "It's child's play: Advergaming and the online marketing of food to children." Available at: http://www.kff.org.
210. Chester, J., and Montgomery, K. (2007). "Interactive Food and Beverage Marketing: Targeting Children and Youth in the Digital Age." Berkeley Media Studies Group, Berkeley, CA.
211. Keane, T. M., Geller, S. E., and Scheirer, C. J. (1981). A parametric investigation of eating styles in obese and nonobese children. *Behav. Ther.* **12**, 280–286.
212. LeBow, M. D., Goldberg, P. S., and Collins, A. (1977). Eating behavior of overweight and nonoverweight persons in the natural environment. *J. Consult. Clin. Psychol.* **45**, 1204–1205.
213. Barkeling, B., Ekman, S., and Rossner, S. (1992). Eating behaviour in obese and normal weight 11-year-old children. *Int. J. Obes. Relat. Metab. Disord.* **16**, 355–360.
214. Adams, N., Ferguson, J., Stunkard, A. J., and Agras, S. (1978). The eating behavior of obese and nonobese women. *Behav. Res. Ther.* **16**, 225–232.
215. Terry, K., and Beck, S. (1985). Eating style and food storage habits in the home. Assessment of obese and nonobese families. *Behav. Modif.* **9**, 242–261.
216. Boutelle, K., Neumark-Sztainer, D., Story, M., and Resnick, M. (2002). Weight control behaviors among obese, overweight, and non overweight adolescents. *J. Pediatr. Psychol.* **27**, 531–540.
217. Rampersaud, G. C., Pereira, M. A., Girard, B. L., Adams, J., and Metzl, J. D. (2005). Breakfast habits, nutritional status, body weight, and academic performance in children and

adolescents. *J. Am. Diet. Assoc.* **105**(5), 743–760: quiz 761–762. Review.
218. Cho, S., Dietrich, M., Brown, C. J., Clark, C. A., and Block, G. (2003). The effect of breakfast type on total daily energy intake and body mass index: Results from the Third National Health and Nutrition Examination Survey (NHANES III). *J. Am. Coll. Nutr.* **22**(4), 296–302.
219. Barton, B. A., Eldridge, A. L., Thompson, D., Affenito, S. G., Striegel-Moore, R. H., Franko, D. L., Albertson, A. M., and Crockett, S. J. (2005). The relationship of breakfast and cereal consumption to nutrient intake and body mass index: The National Heart, Lung, and Blood Institute Growth and Health Study. *J. Am. Diet. Assoc.* **105**(9), 1383–1389.
220. Neumark-Sztainer, D., Rock, C., Thornquist, M., Cheskin, L., Neuhouser, M., and Barnett, M. (2000). Weight control behaviors among adults and adolescents. *Prevent. Med.* **30**, 381–391.
221. Neumark-Sztainer, D., French, S. A., and Jeffery, R. W. (1996). Dieting for weight loss: Associations with nutrient intake among women. *J. Am. Diet. Assoc.* **96**, 1172–1175.
222. Nicklas, T. A., Bao, W., Webber, L. S., and Berenson, G. S. (1993). Breakfast consumption affects adequacy of total daily intake in children. *J. Am. Diet. Assoc.* **93**, 886–891.
223. Neumark-Sztainer, D., Story, M., Falkner, N. H., Behuring, T., and Resnick, M. D. (1999). Sociodemographic and personal characteristics of adolescents engaged in weight loss and weight/muscle gain behaviors: Who is doing what? *Prevent. Med.* **28**, 4–5.
224. Neumark-Sztainer, D., Wall, M., Guo, J., Story, M., Haines, J., and Eisenberg, M. (2006). Obesity, disordered eating, and eating disorders in a longitudinal study of adolescents: How do dieters fare five years later? *J. Am. Diet. Assoc.* **106**, 559–568.
225. Stice, E., Cameron, R. P., Killen, J. D., Hayward, C., and Taylor, C. B. (1999). Naturalistic weight-reduction efforts prospectively predict growth in relative weight and onset of obesity among female adolescents. *J. Consult. Clin. Psychol.* **67**, 967–974.
226. Wardle, J., and Beinart, H. (1981). Binge eating: A theoretical review. *Br. J. Clin. Psychol.* **20**, 97–109.
227. Polivy, J., and Herman, C. P. (1985). Dieting and binging. A causal analysis. *Am. Psychol.* **40**, 193–201.
228. Neumark-Sztainer, D., Wall, M., Haines, J., Story, M., and Eisenberg, M. E. (2007). Why does dieting predict weight gain in adolescents? Findings from Project EAT-II: A five-year longitudinal study. *J. Am. Diet. Assoc.* **107**, 448–455.
229. French, S. A., Jeffery, R. W., Sherwood, N. E., and Neumark-Sztainer, D. (1999). Prevalence and correlates of binge eating in a nonclinical sample of women enrolled in a weight gain prevention program. *Int. J. Obes. Relat. Metab. Disord.* **23**, 576–585.
230. Devlin, M. J., Walsh, B. T., Spitzer, R. L., and Hasin, D. (1992). Is there another binge eating disorder? A review of the literature on overeating in the absence of bulimia nervosa. *Int. J. Eating Disord.* **11**, 333–340.
231. Teasdale, T. W., Sorensen, T. I., and Stunkard, A. J. (1990). Genetic and early environmental components in sociodemographic influences on adult body fatness. *Br. Med. J.* **300**, 1615–1618.
232. Bouchard, C. (1989). Genetic factors in obesity. *Med. Clin. N. Am.* **73**, 67–81.
233. Nader, P. R. (1993). The role of the family in obesity prevention and treatment. *Ann. NYAcad. Sci.* **699**, 147–153.
234. Golan, M., Fainaru, M., and Weizman, A. (1998). Role of behaviour modification in the treatment of childhood obesity with the parents as the exclusive agents of change. *Int. J. Obes. Relat. Metab. Disord.* **22**, 1217–1224.
235. Epstein, L. H. (1985). Family-based treatment for pre- adolescent obesity. *Adv. Devel. Behav. Pediatr.* **6**, 1–39.
236. Neumark-Sztainer, D., Story, M., Ackard, D., Moe, J., and Perry, C. (2000). The "family meal": Views of adolescents. *J. Nutr. Educ.* **32**, 329–354.
237. Neumark-Sztainer, D., Story, M., Ackard, D., Moe, J., and Perry, C. (2000). Family meals among adolescents: Findings from a pilot study. *J. Nutr. Educ.* **32**, 335–340.
238. Resnick, M. D., Bearman, P. S., Blum, R. W., Bauman, K. E., Harris, K. M., Jones, J., Tabor, J., Beuhring, T., Sieving, R. E., Shew, M., Ireland, M., Bearinger, L. H., and Udry, J. R. (1997). Protecting adolescents from harm. Findings from the National Longitudinal Study on Adolescent Health. *JAMA* **278**, 823–832.
239. Neumark-Sztainer, D., Story, M., Resnick, M. D., and Blum, R. W. (1996). Correlates of inadequate fruit and vegetable consumption among adolescents. *Prevent. Med.* **25**, 497–505.
240. Neumark-Sztainer, D., Story, M., Dixon, L. B., Resnick, M., and Blum, R. (1997). Correlates of inadequate consumption of dairy products among adolescents. *J. Nutr. Educ.* **29**, 12–20.
241. Mellin, A., Neumark-Sztainer, D., Story, M., Ireland, M., and Resnick, M. (submitted). Unhealthy behaviors and psychosocial difficulties among overweight youth: The potential impact of familial factors.
242. Birch, L. L., and Fisher, J. O. (1998). Development of eating behaviors among children and adolescents. *Pediatrics* **101**, 539–549.
243. Johnson, S. L., and Birch, L. L. (1994). Parents' and children's adiposity and eating style. *Pediatrics* **94**, 653–661.
244. Neumark-Sztainer, D. (2005). "'I'm, like, SO fat!': Helping Your Teen Make Healthy Choices about Eating and Exercise in a Weight Obsessed World." Guilford, New York.
245. Neumark-Sztainer, D., Wall, M., Haines, J., Story, M., Sherwood, N. E., and van den Berg, P. (2007). Shared risk and protective factors for overweight and disordered eating in adolescents. *Am. J., Prev. Med.*, **33**, 359–369.
246. Sallis, J. F., Bauman, A., and Pratt, M. (1998). Environmental and policy interventions to promote physical activity. *Am. J. Prevent. Med.* **15**, 379–397.
247. Nader, P. R., Stone, E. J., Lytle, L. A., Perry, C. L., Osganian, S. K., Kelder, S., Webber, L. S., Elder, J. P., Montgomery, D., Feldman, H. A., Wu, M., Johnson, C., Parcel, G. S., and Luepker, R. V. (1999). Three-year maintenance of improved diet and physical activity: The CATCH cohort. Child and Adolescent Trial for Cardiovascular Health. *Arch. Pediatr. Adolesc. Med.* **153**, 695–704.

248. Lytle, L. A., Stone, E. J., Nichaman, M. Z., Perry, C. L., Montgomery, D. H., Nicklas, T. A., Zive, M. M., Mitchell, P., Dwyer, J. T., Ebzery, M. K., Evans, M. A., and Galati, T. P. (1996). Changes in nutrient intakes of elementary school children following a school-based intervention: Results from the CATCH Study. *Prevent. Med.* **25**, 465–477.
249. McKenzie, T. L., Nader, P. R., Strikmiller, P. K., Yang, M., Stone, E. J., Perry, C. L., Taylor, W.C, Epping, J. N., Feldman, H. A., Luepker, R. V., and Kelder, S. H. (1996). School physical education: Effect of the Child and Adolescent Trial for Cardiovascular Health. *Prevent. Med.* **25**, 423–431.
250. Sallis, J. F., McKenzie, T. L., Alcaraz, J. E., Kolody, B., Faucette, N., and Hovell, M. F. (1997). The effects of a 2-year physical education program (SPARK) on physical activity and fitness in elementary school students. Sports, Play and Active Recreation for Kids. *Am. J. Public Health* **87**, 1328–1334.
251. Contento, L., Blach, G., Bronner, Y., Lytle, L., Maloney, S., Olson, C., and Swadener, S. (1995). The effectiveness of nutrition education and implications for nutrition education policy, programs and research: A review of research. *J. Nutr. Educ.* **27**, 284–418.
252. U.S. Department of Education, Office of Education Research and Improvement NCFES. (1990). "A Profile of the American Eighth Grader: NEL: 88 Description Summary." U.S. Government Printing Office, Washington, DC.
253. Carnegie Council on Adolescent Development. (1992). "Matter of Time: Risk and Opportunity in the Nonschool Hours." Carnegie Council on Adolescent Development, Waldorf, MD.
254. Carnegie Council of Adolescent Development. (1995). "Great Transitions: Preparing Adolescents for a New Century." Carnegie Corporation, New York.
255. Mulhall, P. F., Stone, D., and Stone, B. (1996). Home alone: Is it a risk factor for middle school youth and drug use? *J. Drug Educ.* **26**, 39–48.
256. Sallis, J. F., and McKenzie, T. L. (1991). Physical education's role in public health. *Res. Q. Exerc. Sport.* **62**, 124–137.
257. U.S. Department of Health and Human Services. (2000). "Healthy People 2010: Conference Edition, Vol I." U.S. Department of Health and Human Services, Washington, DC.
258. U.S. Department of Health and Human Services. (2000). "Healthy People 2010: Conference Edition, Vol II." U.S. Department of Health and Human Services, Washington, DC.

CHAPTER 26

Dietary Approaches to Exploit Energy Balance Utilities for Body Weight Control

RICHARD D. MATTES
Department of Foods and Nutrition, Purdue University, W. Lafayette, Indiana

Contents

I. INTRODUCTION

While the prevalence of obesity has increased dramatically over the past 30 years [1], most believe the etiology is subtle. Indeed, until the application of doubly-labeled water to dietary assessment revealed greater under reporting among the obese, the field was stymied by the inability to document greater energy intake by obese individuals compared to their lean counterparts [2, 3]. The prevailing current view is that the obesity problem reflects a small systematic positive energy balance that is manifest through a slow, but continuous small weight gain over time. One aim of this review is to explore the veracity and implications of this concept. Is this the only dietary pattern that can account for the observed change of weight? What is the theoretical weight gain associated with this pattern of intake? If slow sustained weight gain is the pattern responsible for the trends in body weight, are only small sustained changes of behavior or diet required to remedy the problem? The balance of this review will critically evaluate the leveraging potential of selected dietary components reported to moderate weight gain. The goal is to minimally change the diet while reducing its energy contribution and thereby improve dietary adherence and weight loss.

II. DOES A SMALL POSITIVE ENERGY BALANCE LEAD TO SUBSTANTIVE WEIGHT GAIN OVER TIME?

To demonstrate the importance of a small sustained positive energy balance on body weight and health, an arbitrary small increment in intake, often exemplified by the potential addition of a small quantity of some popular food, is commonly factored into standard equations for estimation of energy needs. When the calculated excess ingested energy is converted to the energy content of body fat, marked gains of body weight/fat are predicted. An example will highlight the limitations of this argument. An equation for estimating the energy needs of a young adult female proposed by Schofield *et al.* is as follows:

$$EER = (0.034\,WT(kg) + 0.006\,HT(m) + 3.530) \times 239 \times 1.55$$

(EER = estimated energy requirement; PA = physical activity factor; WT = weight; HT = height). Thus a sedentary, 1.58-m, 35-year-old female weighing 68 kg would need an estimated 2168 kcal/d to maintain body weight. If she were to adopt a diet that included just 100 additional kcal/d, with no compensatory response, it could be predicted that this would result in a gain of 23.7 kg over 5 years as follows:

$$100 \text{ extra kcal/d} \times 365\text{d/yr} \times 5 \text{ years} = 182{,}500 \text{ extra kcal};$$

assuming 7700 kcal/kg body fat; 182,500/7700 = 23.7 kg.

This is depicted by the linear solid line in Figure 1. However, there are several reasons to challenge this calculation. First, it assumes a 100-kcal surplus every day and the absence of any compensatory dietary, metabolic, or lifestyle change. Thus, it ignores the evidence of compensatory mechanisms at each of these levels. More important, it fails to account for the cumulative effects the weight gain will

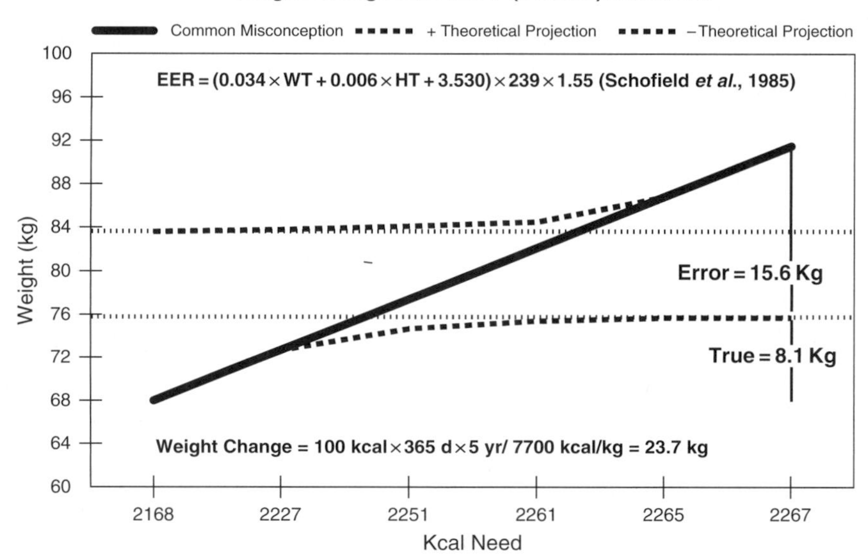

FIGURE 1 Theoretical changes of body weight associated with small changes (100 kcal) in energy intake.

have on the impact of the energy supplement. As weight increases, energy requirements rise, so the 100-kcal supplement will have a diminishing incremental effect. If it is assumed that the 100-kcal supplement led to the theoretical gain of 4.7 kg over the first year, the estimated energy needs of the test individual will have increased to 2227 kcal. So, the 100-kcal excess is now only a 59-kcal excess relative to need. This diminishing effect will continue as described by the lower curvilinear function that falls away from the original linear projection. It begins to asymptote around 8.1 kg, rather than reaching 23.7 kg. After 5 years, energy needs estimated from linear increases will be 2466 kcal for this female. So, the original 100-kcal supplemented diet would now represent a 200 kcal hypocaloric diet and would result in weight loss. The only way to sustain the linear gain would be to consume 100 kcal more energy than is needed every day. Accounting for the cumulative weight gain, this would amount to a nontrivial 300-kcal increment relative to maintenance needs 5 years previously. So, the question remains, does a small systematic positive error in energy intake result in significant weight gain? These calculations indicate it is not as great as often claimed, but may still be substantive. If the latter holds health significance, efforts to offset the energy imbalance are warranted. How can this be achieved?

Before addressing this issue, one additional point requires clarification: What is the pattern of weight gain in the population? The scenario just described is predicated on a continuous incremental gain due to a small, but chronic, positive energy balance. The rise is about 0.2 to 0.8 kg/year [4, 5]. This is consistent with the difficulty of detecting the underlying dietary contribution and lack of overt gluttony. However, episodic gains due to acute marked excess intake could also be difficult to detect if dietary assessments are not conducted at the times of these imbalances. If they occur in conjunction with holidays and special events, they may be times when efforts to collect dietary information are curtailed in the knowledge that they are not reflective of customary behavior. However, if these episodes lead to gains that are not fully offset by subsequent losses, they, too, could lead to apparent incremental weight gain. Indeed, there is evidence supporting this pattern. In one trial, 195 adults were weighed four times at intervals of 6 to 8 weeks before, during, and following the winter holidays as well as approximately 1 year from the first weighing [6]. Body weight increased during the holiday period, but not the pre- or post-holiday periods, and the increment of approximately 0.48 kg was retained at the 1-year follow-up. In a more tightly controlled feeding trial, six nonobese males were fed 20%, 40%, and 60% more energy than required for three 21-day periods with 1-week ad libitum intakes between [7]. As expected, weight gain was noted during the intervention periods, but the washout periods were not characterized by consistent compensation. Thus, weight gain ensued. A study of 94 college students noted a small (0.5-kg), but significant, increase of body weight over the Thanksgiving period [8].

Despite the small magnitude of these increases, they may be of clinical significance because small uncorrected errors over time will lead to weight gain. Furthermore, the group mean obscures significant weight gain in a subset of the population [8–11]. Hull *et al.* [8] reported fifteen percent of a college study sample gained 2.0 kg or more over the course of the holiday period. Approximately ten percent of the sample observed by Yanovski *et al.* [6] gained 2.3 kg or more during the 6-week holiday period. Specifically, significantly different patterns appear to exist across body mass index (BMI) categories, as the greater weight gain was predominately reported in those who were overweight or obese at the outset of the holiday period [6, 8]. No significant increase in body weight was observed in normal-BMI participants [8]. This suggests that holiday periods are a critical period of weight gain for overweight and obese individuals.

However, in a study of college students [9], it was notable that despite the normal-weight individuals losing a significant amount of body weight by the end of the holiday period (post–New Year), they gained a significant amount of total body fat ($p < 0.05$), particularly in the leg and trunk areas. In consideration of the increased risks of cardiovascular disease, diabetes, and hyperlipidemia associated with excess fat in the abdominal region, redistribution or increases in total body fat are undesirable, even if accompanied by maintenance of body weight. Thus, the holiday period may be a critical period for overall health risk for all weight categories.

Possible factors promoting weight gain and alterations in fat patterning during holiday periods include increased availability of palatable food, excessive portion sizes, increased variety, increased eating frequency and duration, social facilitation, and increased stress. The overall disruption in normal routine and heightened focus on enjoyment and relaxation may also increase sedentary behaviors. Studies to date have failed to measure food intake, physical activity, and weight changes simultaneously during these potentially critical periods. Consequently, the factors that are particularly conducive to weight gain during these periods have not been identified. A study assessing dietary intake 2 days pre-, 4 days during, and 2 days post-Thanksgiving reported a significant increase in energy intake during the holiday period relative to both pre- and post-holiday periods (356 kcal more, $p < 0.05$) [12]. However, overweight participants reported eating significantly less than normal-weight participants ($p < 0.05$). A large longitudinal study ($n = 593$) reported that peak caloric intake is in November, and the lowest physical activity level is in December [13], the 2 months during which the two main holiday periods occur. Nonetheless, causal relationships cannot be inferred or specific effects on health determined.

Improved understanding of the patterns of weight gain should yield insights useful for development of weight management recommendations. If the gain is largely due to episodic, event (holiday)-related changes of behavior, interventions may only need to be short-term and highly specific. However, if the gain is attributable to an incremental, sustained positive energy balance, an ongoing but modest intervention will be required. Assuming the latter and focusing on energy intake, rather than expenditure, there are several potential approaches. These are outlined in Table 1 with several examples of each.

TABLE 1 Dietary Approaches to Reduce Energy Balance

Enhanced Satiety	Inefficient Absorption	Increased Energy Expenditure
Protein	Fiber	Protein
Fiber	Nuts	Nuts
Nuts	Dairy	Dairy
Soup	Physical Processing	Tea
Rheology		Irritants
Glycemic index		
Irritants		
Meal pattern		

III. MECHANISMS FOR DIETARY APPROACHES TO MODERATE ENERGY BALANCE

One popular approach is to maximize the satiation/satiety value of a food or diet. Satiation is the set of sensations that determine meal size, typically measured in weight or energy. Satiety is the set of sensations that determine eating frequency, commonly measured in the time interval between eating occurrences or the amount eaten subsequent to a given eating occurrence. The concept is that a food with high satiation or satiety value will result in fewer eating occurrences and/or smaller sizes of subsequent eating occurrences. This will offset some portion of the energy contributed by the food itself. This is referred to as dietary compensation. Importantly, this property may hold for foods of any energy content or density. So, a high-energy food with a high satiation/satiety value that actually elicits a subsequent reduction of energy intake equal to or greater than the energy it contributes is not problematic for energy balance. Alternatively, a food of lower energy value or density that elicits only a weak compensatory dietary response may contribute to weight gain.

A second approach that may ameliorate the positive impact of a food on energy balance is to take advantage of foods that naturally, or through processing, have a structure or composition that limits the bioaccessibility or bioavailability of its energy sources. Bioaccessibility refers to the degree to which the energy is released from a food matrix, whereas bioavailability is determined by the processes that influence the ability to absorb and utilize constituents. The fat in a nut may not be bioaccessible unless the cell wall in which it is located is ruptured during mastication or digestion [14], and free fatty acids may not be fully bioavailable if they bind with dietary fiber from other plant sources. This approach is currently exploited pharmacologically. Orlistat, the active ingredient in the drug Xenical, aids weight management by inhibiting lipase-mediated hydrolysis of triacylglycerol, which reduces the bioavailability of dietary fat.

Third, energy balance may be shifted downward through the action of food constituents that stimulate or enhance energy expenditure. Theoretically, this would occur if a food constituent, whole food, or diet augmented thermogenesis, resting energy expenditure, and/or physical activity. The digestion of protein, fat, and carbohydrate entails the generation of heat that is unavailable to anabolic processes. This is termed thermogenesis. Thus, the instrumentally measured energy content of a food is not equal to the energy available to the body. More chronic effects on metabolism are also possible that result in energy loss as measured by a rise of resting energy expenditure (REE). REE may also be augmented by increased accretion of lean body mass or reduced loss of lean mass when dieting. Direct links between diet and physical activity are less clear, but associations have been reported. For example there are reports of increases associated with ingestion of alcohol [15–17]. The increment in expenditure would offset a portion of the energy provided by the alcohol.

Each of these processes may act alone or in combination. Three examples are provided to highlight the potential to exploit them for therapeutic or preventative purposes. Data on the impact of protein (a macronutrient), fiber (a nonnutritive food constituent), and capsaicin (a sensory stimulus) are critically reviewed to identify the magnitude of advantage that may be achieved with their manipulation versus of the cost or change of customary diet to realize the benefit.

A. Protein

Protein has been ascribed special status among the macronutrients with respect to feeding and energy balance. It is reported to have the strongest satiety effect and may thereby moderate energy intake. In addition, it has the highest thermogenic effects and so should augment energy expenditure. The combination then would be predicted to help curb weight gain. The validity and functional significance of these properties warrant critical evaluation to better interpret the potential role of protein in energy balance.

Numerous studies have explored the satiety value of protein [18, 19]. Most commonly, this is undertaken with a preload study. This is an approach where individuals who have fasted (over some length of time that presumably standardizes appetitive sensations) are provided a known food challenge (typically fixed in energy or weight with a specified nutrient content) and appetitive sensations are monitored over a period of time. This may be followed by a challenge meal and the amount consumed is monitored. Table 2 is a summary of findings from one review of protein [19]. Most, but not all, trials report stronger satiation/satiety effects for protein loads. However, from an energy balance perspective, the most relevant outcome is reduced energy intake. By this criterion, only the first four trials listed in Table 2 noted that the higher protein load elicited stronger satiation effects and a reduction of energy intake. In contrast, the next three trials observed stronger satiety but no impact on intake. The next three entries failed to find a differential protein effect on satiety and two also detected no change of intake. The remaining entries measured only one of these outcomes.

TABLE 2 Effects of Protein on Appetite and Energy Intake

Reference	Duration	Effect on Satiety	Effect on Food Intake
Johnson, 1993 [20]	90 m	Increase	Decrease
Poppitt, 1998 [21]	90 m	Increase	Decrease
Porrini, 1995 [22]	2 h	Increase	Decrease
Rolls, 1988 [23]	2 h	Increase	Decrease
Stubbs, 1996 [24]	5 h	Increase	NS
Stubbs, 1999 [25]	24 h	Increase	NS
Johnstone, 1996 [26]	15 d	Increase	NS
Barkling, 1990 [27]	4 h	NS	Decrease
DeGraaf, 1992 [28]	day	NS	NS
Geliebter, 1978 [29]	70 m	NS	NS
Long, 2000 [30]	13 h/13 d	I (acute) D (chronic)	–
Hill, 1986 [31]	1 h	Increase	–
Holt, 1995 [32]	2 h	Increase	–
Vanderwater, 1994 [33]	2 m	Increase	–
Westerterp-Plantenga, 1999 [34]	24 h	Increase	–
Booth, 1970 [35]	2.5 h	–	Decrease
Ludwig, 1999 [36]	5 h	–	Decrease
Araya, 2000 [37]	3.5 h	–	Decrease
Porrini, 1997 [38]	2 h	–	NS
Teff, 1989 [39]	3 h	–	NS

A number of methodological issues help to explain this diversity of outcomes. First, the studies reporting a consistent association were all trials lasting 2 hours or less, whereas the studies noting that stronger satiation effects were not associated with reduced intake were of longer duration. The 2-hour interval between the protein load and challenge meal is less than most intermeal intervals reported in free-living individuals. Thus, test participants were being required to eat at a time they would not ordinarily choose to do so. Consequently, the relevance of these findings is uncertain. If, instead of requiring individuals to eat at an experimentally fixed short time interval, they are allowed to eat upon request after a preload, they wait longer and the protein effect may be lost [40]. One possible explanation is that when meal requests are spontaneous, they elicit a consumption pattern more in line with the individual's custom, thus overriding a transient or subtle preload effect.

Three additional methodological issues warrant consideration in this context. First, the format in which a protein load is presented may alter the magnitude of the satiety effect it elicits. As shown in Table 3, for trials where the protein is presented in a solid food form, the effects are relatively robust whereas this is not the case for responses

TABLE 3 Effects of Food Matrix on the Appetitive and Dietary Properties of Proteins

Reference	Form	Effect on Hunger	Effect on Food Intake
Booth *et al.* [35]	Solid		Decrease
Hill & Blundell [31]	Solid	Decrease	
Hill & Blundell [31]	Solid	Decrease	Decrease
Rolls *et al.* [23]	Solid	Decrease	Decrease
Barkeling *et al.* [2]	Solid	=	Decrease
Stubbs *et al.* [24]	Solid	Decrease	=
Vanderwater & Vickers [33]	Solid	Decrease	
Johnson & Vickers [20]	Solid	Decrease	Decrease
Poppitt *et al.* [21]	Solid	Decrease	Decrease
Lambert *et al.* [49]	Liquid	Decrease	Decrease
Latner & Schwarts [50]	Liquid	Decrease	Decrease
Driver [51]	Liquid	=	
de Graaf *et al.* [28]	Liquid	=	=
Stockley *et al.* [52]	Liquid		=
Sunkin & Garrow [53]	Liquid		=
Geilebter [29]	Liquid	=	=
Rumpler *et al.* [41]	Liquid (34 g)	=	

to protein-containing beverages. One issue with this comparison is that protein concentrations in beverages tend to be lower than concentrations in solid foods. However, this does not appear to fully explain the findings because a recent trial incorporated 34 g of protein into a beverage and still reported no greater effect on intake (satiety was not monitored) relative to an isocaloric load high in carbohydrate [41]. Second, there is preliminary evidence that one's customary exposure to protein may modulate the satiety effect it elicits. Individuals accustomed to a high-protein diet may develop a tolerance to this property [42]. Third, the form of protein used as a challenge may be important. For example, there is some evidence that whey protein is more satiating than casein [43]. The former is more readily absorbed so the postprandial rise of plasma amino acids, the presumed satiety signal, is greater. However, this finding is not robust [43, 44]. There is also evidence that dairy protein may have stronger satiety effects because of its high concentration of casinomacropeptide (CMP) [45, 46]. CMP is a potent stimulus for the release of gut peptides associated with satiety such as cholecystokinin (CCK). However, other data do not confirm a high satiety value for dairy [47] generally, or CMP [48] in particular.

Taken together, the preponderance of evidence supports a higher satiety property for protein than for other macronutrients. However, the translation of this sensation to altered feeding or energy balance is less well defined. Thus, it is presently not possible to quantify the independent effect of protein-induced satiety on intake.

The data are somewhat clearer for the effects of protein on metabolism. A review of the thermogenic effects of protein concluded it was higher than the response to ingestion of carbohydrate or fat. These authors developed a predictive equation for the thermogenic effect of manipulations of the macronutrient composition of foods: TEF = 0.132 Protein (g) + 0.051 Fat (g) + 0.054 Carbohydrate (g) [18]. Assuming ingestion of a 2000-kcal diet where the proportion of energy derived from protein, fat, and carbohydrate is 15%, 30%, and 55%, respectively, the predicted loss of energy due to thermogenesis is 130 kcal. If the mix is changed such that 30% of energy is derived from protein at the expense of carbohydrate, the predicted value is 153 kcal, or an increase of 23 kcal. Although 23 kcal/day may seen trivial, under the assumption that small, sustained changes can exert substantive effects over time and assuming no other forms of compensation, it would account for an approximately 1.7-kg weight loss over 3 years, when its effect will be lost because of the lower body weight. However, the caveat is that this level of advantage comes at a cost of doubling the contribution of protein from 15% of energy to 30%. Experience to date in longer-term trials of high-protein diets that sought such a shift indicates this is not easily sustained. In one recent trial contrasting popular diets with marked differences in macronutrient content, the high-protein group achieved an intake of 27% energy from protein at 2 months, but despite the reinforcement of marked weight loss in a motivated group, the protein contribution had dropped to only 22% by 6 months [54]. Thus, whether this 23-kcal advantage is realistically achievable is uncertain.

Higher protein diets may also facilitate energy loss during energy restriction by reducing the loss of lean body mass (LBM). A number of studies document that LBM is better maintained by higher protein diets [55], but the expected benefit with respect to energy balance is less widely considered. Interindividual lean body mass may vary widely. Young healthy males may have 30–55 kg while sedentary elderly females may have less than 13 kg [56]. It is estimated that the metabolic contribution of 10 kg of lean mass equals approximately 100 kcal/d [56]. Thus, to achieve a 100 kcal/d increment in energy expenditure would require almost doubling the lean body mass of the elderly female and increasing that of the young male by 18–33%. Given recent evidence that adherence to a 750-kcal energy deficit diet containing 30% energy from protein versus 18% of energy for 9 weeks results in a loss of about 8–10 kg of body weight with retention of about 1.5 kg more lean body mass with the former diet, the benefit may be about 15 kcal. Again, this is an increment in the desired direction and could lead to about a 1.1-kg weight loss before the advantage is lost, but it is only achieved by a marked change of diet, that is, roughly doubling protein intake.

B. Fiber

Fiber has been recognized as a satiety factor for some time, but it also decreases energy intake by reducing the efficiency of energy absorption. Its satiety effects may be attributable to greater demand for oral processing [57], greater gastric distention [58], altered gut hormone secretion or efficacy [59], and, in the case of soluble fiber, decreased gastric transit [60]. Quantifying fiber's satiety effect is complicated because it is consumed in many forms and nearly always as a component of a food with other constituents that may reduce or augment the fiber's actions. A relatively direct association has been noted between the fiber content of foods and self-reported satiety [61]. However, test foods varied in other properties such as weight, nutrient content, and palatability. So, it was not possible to isolate the independent effect of the fiber. It is noteworthy that, in this trial, only one food, a high-fiber cereal, produced an effect greater than eggs and bacon, which contain no fiber. Thus, it would appear that the fiber effect is modest. Many trials have sought to demonstrate the satiety value of fiber by contrasting responses between a control food and the item supplemented with a recommended daily fiber portion (e.g., 20–35 g). For individuals in many Western countries who are accustomed to a daily intake of only half this amount, the challenge is substantive and may result in gastric distress. Indeed, in some trials this has been acknowledged [62]. The practical interpretation of such trials is uncertain.

From the current perspective, the most relevant outcome measure for a fiber supplement may be altered feeding and body weight. This has been the topic of a number of reviews [63, 64]. One in particular provides useful insights [63]. These authors segregated 38 trials that explored the effects of soluble, insoluble, or mixed fiber challenges. All trials had treatment and control conditions matched on energy and fat and measured satiety by scaled self-report. No substantive differences were observed between fiber types. All resulted in approximately 10% reductions of energy intake. By dividing the trials into those that used a fixed intake or *ad libitum* test paradigm, it is possible to estimate the role of satiety on body weight change. Presumably, in a fixed feeding protocol, satiety effects are eliminated and any increment in weight loss is attributable to physical processes (discussed later). In contrast, any greater weight loss noted in *ad libitum* protocols, where satiety effects can be manifest, may serve as an approximation of the satiety contribution. For the studies included in the review, the fixed feeding approach led to a 20 g/d loss of body weight whereas the *ad libitum* trials yielded a 24 g/day reduction. This would suggest a 4 g/d benefit and could translate into a 31 kcal/d reduction of energy intake. If true, a maximum weight loss of approximately 2.3 kg would be achieved over the 3 years this would promote a negative energy balance. However, the *ad libitum* trials used a mean supplement of 12 g/d of fiber whereas the fixed protocols used only 10. Thus, the weight loss expressed per gram of fiber was 2 g/d in both cases, indicating no independent effect of the non-physical properties of dietary fiber on body weight. This is supported by a further analysis of the data showing there is no dose-response relationship between the fiber challenge and percent decrease of energy intake (Table 4). Although a direct satiety effect on weight loss may be limited, the value of the satiety effect cannot be dismissed. If it improves the sense of well-being and thereby promotes better compliance with an energy-restricted diet, greater benefit may still be realized.

Fiber was long branded an anti-nutrient because it binds with various nutrients in the gastrointestinal tract rendering them less bioaccessible or bioavailable. Although this may be undesirable for some nutrients, it holds potential benefit as well with respect to lower cholesterol and energy absorption. There is an inverse association between the fiber content of foods/diets and measured metabolizable energy [65, 66]. In one trial, the metabolizable energy of diets varying in fruits and vegetables was measured. The lower fiber condition provided 16 g of total dietary fiber and the higher provided 37 g. The metabolizable energy value of the lower and higher fiber diets were 4.6% and 6.0% less than values listed in food tables [65]. If the 1.4% difference is applied to a 2000-kcal diet, the estimated increment in lost energy would be about 28 kcal/d or a maximum 1.7-kg weight reduction over 2 years. However, to achieve the

TABLE 4 Percentage Decrease in Energy Intake During Consumption of Higher-fiber Diet Compared with a Lower-fiber Control Diet*

Fiber Amount (g/day)	% of Energy Consumed on High versus Control Fiber Challenges
5	98+
6 (obese)	105
6 (nonobese)	59+
7	76+
7.3	99
8	94
8	82+
9	80
9.5	99
10	90
10	90
14	78+
17	91+
19	98
19	107
19	95+
19	93+
19	97+
20	87
29	96+
39	89+
40	81+

*Data obtained from Howarth, NC, *et al.* Dietary Fiver and Weight Regulation. Nutrition Reviews. 2001;59/5:129–139.

benefit requires the ingestion of foods with 2.3-fold higher fiber content. The feasibility of such a change is uncertain as it would require a substantive dietary modification. Other published estimates of fiber's reduction of available energy have been lower. In one case, increasing consumption of fruits, vegetables, and potatoes from 505 to 1235 g/d resulted in fiber concentrations of 18.8 and 52 g/d. The benefit was calculated to be 0.7 kcal/g of added fiber. Assuming a linear relationship, changing fiber intake from 15 g/d to 30 g/d, more in line with current recommendations, while holding energy intake constant would result in the loss of about 10 kcal/d more on the higher fiber diet (~0.5 kg maximal weight loss). Using these two estimates, roughly doubling the fiber content in the diet would have the potential to decrease the available energy by 10–25 kcal/d. This would lead to an associated loss of 0.7–1.7 kg body weight (BW) over 3 years, assuming stability of other energy sources and losses.

Thus, taken together, the data indicate that fiber does contribute to satiety, but the effects are quite modest and may not result in altered energy intake. There are no clear differences according to fiber type. Fiber does bind with energy-yielding compounds in foods, reducing the metabolizable energy of the food. This increases directly with fiber content. The estimated gain is small but robust and may produce weight loss or reduced weight gain. A reasonable estimate may be about 2 g BW/d with a doubling of fiber intake from 15 g to 30 g/d assuming the diet otherwise meets energy needs.

C. Irritancy (Capsaicin)

Interest in the association between irritancy and body weight stems from observations in animal and human trials that capsaicin, the primary compound responsible for the burn of red chili peppers, elicits responses potentially linked to energy balance. The human trials alone suggest that ingestion of the compound stimulates the sympathetic nervous system resulting in elevated thermogenesis [67]; enhances fat [68, 69] and carbohydrate [70, 71] oxidation; promotes satiety [72]; and reduces postprandial insulin concentrations [73], hunger [67], and energy intake [67]. However, there are conflicting findings as well, necessitating a critical review of the literature. This is focused on the effects of capsaicin on satiety and thermogenesis.

Two studies permit an estimation of the contribution of oral irritancy to satiety because they tested the same dose of capsaicin administered orally and intragastrically. The difference between the two may be attributed to the sensory role. In one trial, 12 male and 12 female adults were provided a dose of 0.0025 g of capsaicin in either tomato juice (oral stimulation) or capsules with tomato juice (gastric stimulation) 30 minutes prior to 4 test meals over 2 days [72]. Hunger and satiety were monitored. The reduction of energy intake associated with the gastric load of capsaicin relative to placebo was 1.1 MJ/d (263 kcal/d) for both males and females. This was a mean reduction of 9.6% and 11.7% in total daily energy intake. With oral exposure, energy intakes were 1.6 MJ/d (382 kcal/d) and 2.0 MJ/d (478 kcal/d) for males and females, respectively. These reductions were 13.9% and 21.3% of daily intake. There was a 0.5 MJ/d (119 kcal/d) and 0.4 MJ/d (96 kcal/d) greater loss in the oral versus gastric load conditions among males and females. These 31% and 20% reductions of intake may be estimated to reflect the oral contribution. They constitute 4.3% reductions of total daily energy intake. This was predominantly due to a reduction of fat intake. Satiety was significantly increased by 9.9% and 13.2%. Hunger was not significantly changed. Only in the oral condition was the change in satiety significantly correlated with the change of intake ($r = 0.66$ and $r = 0.69$ in males and females, respectively). Based on a 2000-kcal diet, capsaicin-induced irritancy would account for an approximately 85 kcal/d reduction of energy intake. This would correspond to an approximately 6.5-kg loss of body weight over the roughly 4 years the capsaicin addition would pose a challenge.

In contrast, another study involving 16 Japanese males observed no significant difference of energy intake between capsaicin administered orally or by capsule [74]. Intakes during both interventions were lower than the placebo treatment by 550 kJ and 514 kJ (8.1–8.4%), but this difference was not statistically significant. The trend for lower energy intake was largely attributable to reduced fat intake in both groups, and this was significant for the capsule group. The dose used in the trial equaled 0.0036 g, which was the maximal tolerable concentration determined in a pretest.

In several other trials, the oral and gastric effects cannot be isolated [67]. In one trial with 13 Japanese females, the addition of capsaicin (0.027g) to a breakfast meal resulted in lower desire to eat and hunger ratings, but no change of satiety (absolute values could not be discerned from the published figures). There was no significant reduction of energy intake, although protein and fat were reduced at a subsequent test meal by 20% and 6%, and by 17% and 11% when added to high-fat or high-carbohydrate breakfasts, respectively. A second trial with 10 non-Hispanic white males entailed adding 0.018 g of capsaicin to a prelunch appetizer. This led to a significant reduction of energy (11%) and carbohydrate (18%) intake at the subsequent lunch and snack.

Taken together, these studies fail to provide a convincing effect of capsaicin on satiety due either to orosensory or gastric stimulation. Most reported reduced energy intake, but the limited sample sizes led to equivocal outcomes, precluding a definitive conclusion. The trials used concentrations of capsaicin that approach or equal the maximal tolerated dose, so there is no option to increase the dose to document effects, nor is there an ecological justification to do so. As noted later, even these doses may not be well tolerated on a chronic basis. Testing of more moderate doses in larger study samples may still reveal modest benefits with respect to satiety and reduced energy intake.

Although the sensory properties of foods are important determinants of food choice, they also elicit a cascade of physiological responses, termed cephalic phase responses, that are believed to optimize the digestion of foods and the absorption and utilization of their nutrients by activating many, if not all, normal ingestive and postingestive processes [75–77]. Cephalic phase responses are typically small and transient, but by modulating postingestive responses, their impact may be magnified. Generally, there is a hierarchy of effective stimuli such that responses build with the addition of cognitive + vision + olfactory + taste + mastication + swallowing inputs. Little is known about the role of irritancy in eliciting cephalic phase responses, but there has been interest in a contribution of capsaicin to a cephalic phase thermogenic response.

Interest in this property stemmed from animal studies documenting effects on sympathetic nervous system activity and adiposity [78] as well as suggestive human evidence. For example, administration of 3g of chili and mustard sauces to the meals of 12 individuals resulted in a 25% increment of thermogenesis over a 180-minute period [79]. This was noted in several subsequent human trials focusing specifically on capsaicin. In one instance, the ingestion of 10 g of red pepper (~30 mg capsaicin) led to an increase of thermogenesis for the subsequent 30 minutes, but this returned to baseline for the subsequent 120 minutes of the study. [71]. This effect was blocked by treatment with propranolol, indicating it was attributable to beta-adrenergic stimulation. Additional studies have also revealed that the sympathetic nervous system (SNS) activation is transient [67]. Other work noted transient significant elevations of epinephrine (81%) and norepinephrine (76%) 30 minutes after capsaicin ingestion (30 mg) relative to control treatment, but this was not associated with an additional effect on thermogenesis [70].

Where a thermogenic effect is noted, the magnitude of response appears to be related to the capsaicin concentration rather than total amount ingested, that is, a lower but more concentrated dose is a better stimulus than a larger but more diluted exposure [67]. However, preexposure and high exposures can lead to desensitization and loss of the effect [80]. The obese may also be less responsive to the thermogenic effects of capsaicin [81]. Diet composition is a factor as well. The thermogenic effect is greater (about 45% estimated from a published figure) with the addition of 10 g of red pepper to a higher fat (45% energy from fat) compared to a lower fat (25% energy from fat) (about 17% estimated from a published figure) diet. Indeed, the addition of red pepper to the higher fat diet led to a thermogenic effect that was comparable to that noted with the unsupplemented lower fat diet, suggesting that the addition of capsaicin can offset the lower thermogenic effect of a high-fat diet [68]. When diets contained both capsaicin (12 g red pepper; ~0.36 mg capsaicin) and caffeine (200 mg), both SNS stimulants, a nonsignificant 320 kJ/d (76 kcal/d) increase in energy expenditure was observed. Compared to baseline, total energy balance was actually elevated in the intervention and control groups, but was lower with the supplemented group by approximately 4000 kJ/d (955 kcal/d) [82]. Given the multitude of methodological factors that may influence the thermogenic response to capsaicin and failure of any study to control them all, it is not possible to quantitatively estimate the magnitude of irritancy-induced thermogenesis. At present it appears as though there is a phenomenon, but it is limited in size and duration. Its health significance would best be determined by a long-term trial of energy balance.

There are only limited data on the effects of chronic capsaicin exposure on body weight. In one randomized, double-blind, placebo-controlled trial, the administration of 135 mg of capsaicin per day for 9 weeks (as a supplement) led to no differential response of appetite, serum glucose, insulin, energy expenditure, or improved weight loss maintenance [69]. Weight regain was about 33% in the treatment group versus 19% in the control group. However, fat oxidation was significantly elevated with the capsaicin treatment. Ten participants were not able to tolerate the intervention and were reduced to half the dose, but their outcome did not differ from the compliant group. In a crossover design trial with 36 participants, 33 mg/d of capsaicin was added to the diet for one 4-week period or omitted for another 4 weeks [73]. No significant changes of energy expenditure or body weight were observed. At the end of the trial, the postprandial insulin response was blunted. It is noteworthy that the reduced postprandial insulin response was more marked in the participants with greater BMI. It was hypothesized that this may be causally related to insulin's influence on SNS activity and thermogenesis. In another trial, capsaicin was only a component of a combination preparation including chromium picolinate, inulin, and L-phenylalanine that was administered daily in conjunction with a 4-week diet and exercise weight-loss program [83]. The dose of capsaicin could not be determined from the report, but no effects of the supplement were observed on energy intake or body weight over the 4-week trial.

Thus, the longer term trials offer little support for a measurable effect of capsaicin exposure on energy balance. However, none was adequately powered to detect a realistic effect. Interestingly, there is another branch of literature suggesting that the addition of spices, and capsaicin in particular, may augment appetite [84], digestive processes [85, 86], and growth efficiency [87]. This work largely stems from cultures accustomed to high levels of spices in the diet and suggests a possible confounding effect of palatability on outcome variables. Generally, palatability enhances intake [88] and cephalic phase responses such as thermogenesis [89, 90]. In one trial [68], the addition of capsaicin to a high-fat diet increased the thermogenic response, but lowered the diet's palatability. Thus, the degree to which capsaicin could augment thermogenesis if added in a way that had a neutral or positive effect on palatability or among those with a preference for higher irritancy requires further consideration.

IV. CONCLUSION

Considerable experience indicates that few individuals attain marked and sustained weight reduction with radical changes of diet. More favorable outcomes may be expected if a diet is designed to fit an individual's lifestyle rather than expecting the individual to change his/her lifestyle to accommodate a new dietary prescription. This review critically assessed the evidence related to selected food components often ascribed a role in moderating appetite, reducing the efficiency of energy absorption and/or promoting energy expenditure. Although their effects are modest and self-limiting, there are data supporting a role for each. Whether combinations of food constituents, ingredients, or whole foods can lead to additive or synergistic effects has not been adequately evaluated. In one food, nuts, this appears to hold. Although they are high-fat and energy dense, the evidence indicates they do not promote weight gain, reportedly because of the combination of factors identified here [47]. Whether the advantages of manipulating these aspects of the diet outweigh the costs of implementing them will be determined on an individual basis. Small positive errors in energy balance may have led to the current overweight/obesity problem and, with patience, small changes may be able to reverse the trend. Health care providers should provide realistic estimates of the benefits consumers may realize from shifts of diet composition, estimates that require objective critical appraisals of the literature.

References

1. Flegal, K. M., Carroll, M. D., Ogden, C. L., and Johnson, C. L. (2002). Prevalence and trends in obesity among U.S. adults, 1999–2000. *JAMA* **288**, 1723–1727.
2. Schoeller, D. A., Bandini, L. G., and Dietz, W. H. (1990). Inaccuracies in self-reported intake identified by comparison with the doubly labeled water method. *Can. J. Physiol. Pharmacol.* **68**, 941–949.
3. Hill, R. J., and Davies, P. S. W. (2001). The validity of self-reported energy intake as determined using the doubly labelled water technique. *Br. J. Nutr.* **85**, 415–430.
4. Williamson, D. F. (1993). Descriptive epidemiology of body weight and weight change in U.S. adults. *Ann. Intern. Med.* **119**, 646–649.
5. Jeffrey, R. W., and French, S. A. (1997). Preventing weight gain in adults: design, methods and one year results from the Pound of Prevention study. *Int. J. Obes. Relat. Metab. Disord.* **21**, 457–464.
6. Yanovski, J. A., Yanovski, S. Z., Sovik, K. N., Nguyen, T. T., O'Neil, P. M., and Sebring, N. G. (2000). A prospective study of holiday weight gain. *N. Engl. J. Med.* **342**, 861–867.
7. Jebb, S. A., Siervo, M., Fruhbeck, G., Goldberg, G. R., Murgatroyd, P. R., and Prentice, A. M. (2006). Variability of appetite control mechanisms in response to 9 weeks of progressive overfeeding in humans. *Int. J. Obes.* **30**, 1160–1162.
8. Hull, H. R., Radley, D., Dinger, M. K., and Fields, D. A. (2006). The effect of the Thanksgiving Holiday on weight gain. *Nutr. J.* **5**, 1–6.
9. Hull, H. R., Hester, C. N., and Fields, D. A. (2006). The effect of the Holiday season on body weight and composition in college students. *Nutr. Metab.* **3**, 44–50.
10. Van Staveren, W. A., Deurenberg, P., Burema, J., DeGroot, L. C., and Hautvast, J. G. (1986). Seasonal variation in food intake, pattern of physical activity and change in body weight in a group of young adult Dutch women consuming self-selected diets. *Int. J. Obes.* **10**, 133–145.
11. Reid, R., and Hackett, A. F. (1999). Changes in nutritional status in adults over Christmas 1998. *J. Hum. Nutr. Diet.* **12**, 133–145.
12. Klesges, R. C., Klem, M. L., and Bene, C. R. (1989). Effects of dietary restraint, obesity, and gender on holiday eating behavior and weight gain. *J. Abnormal Psychol.* **98**, 499–503.
13. Ma, Y., Olendzki, B. C., Li, W., Hafner, A. R., Chiriboga, D., Hebert, J. R., Campbell, M., Sarnie, M., and Ockene, I. S. (2006). Seasonal variation in food intake, physical activity, and body weight in a predominantly overweight population. *Eur. J. Clin. Nutr.* **60**, 519–528.
14. Ellis, P. R., Kendall, C. W. C., Ren, Y., Parker, C., Pacy, J. F., Waldron, K. W., and Jenkins, D. J. A. (2004). Role of cell walls in the bioaccessibility of lipids in almond seeds. *Am. J. Clin. Nutr.* **80**, 604–613.
15. Eisen, S. A., Lyons, M. J., Goldberg, J., and True, W. R. (1993). The impact of cigarette and alcohol consumption on weight and obesity: an analysis of 1911 monozygotic male twin pairs. *Arch. Intern. Med.* **153**, 2457–2463.
16. Fagrell, B., Defaire, U., Bondy, S., Criqui, M., Gasiano, M., Gronbaek, M., Jackson, R., Klatsky, A., Salonen, J., and Shaper, A. G. (1999). The effects of light to moderate drinking on cardiovascular diseases. *J. Intern. Med.* **246**, 331–340.
17. Westerterp, K. R., Meijer, E. P., Goris, A. H. C., and Kester, A. D. M. (2004). Alcohol energy intake and habitual physical activity in older adults. *Br. J. Nutr.* **91**, 149–152.
18. Eisenstein, J., Roberts, S. B., Dallal, G., and Saltzman, E. (2002). High-protein weight-loss diets: Are they safe and do they work? A review of the experimental and epidemiologic data. *Nutr. Rev.* **60**, 189–200.
19. Halton, T. L., and Hu, F. B. (2004). The effects of high protein diets on thermogenesis, satiety and weight loss: A critical review. *J. Am. Coll. Nutr.* **23**, 373–385.
20. Johnson, J., and Vickers, Z. (1993). Effects of flavor and macronutrient composition of food servings on liking, hunger and subsequent intake. *Appetite* **21**, 25–39.
21. Poppitt, S. D., McCormack, D., and Buffenstein, R. (1998). Short term effects of macronutrient preloads on appetite and energy intake in lean women. *J. Am. Coll. Nutr.* **6**, 497–506.
22. Porrini, M., Crovetti, R., and Testolin, G. (1995). Evaluation of satiety sensations and food intake after different preloads. *Appetite* **25**, 17–30.
23. Rolls, B. J., Hetherington, M., and Burley, V. J. (1988). The specificity of satiety: The influence of foods of different macronutrient content on the development of satiety. *Physiol. Behav.* **43**, 145–153.

24. Stubbs, R. J., Van Wyk, M. C. W., Johnstone, A. M., and Harbron, C. G. (1996). Breakfasts high in protein, fat or carbohydrate: effect on within-day appetite and energy balance. *Eur. J. Clin. Nutr.* **50**, 409–417.
25. Stubbs, R. J., O'Reilly, L. M., Johnstone, A. M., Harrison, C. L. S., Clark, H., and Franklin, M. F. (1999). Description and evaluation of an experimental model to examine changes in selection between high protein, high carbohydrate and high fat foods in humans. *Eur. J. Clin. Nutr.* **53**, 13–21.
26. Johnstone, A. M., Stubbs, R. J., and Harbron, C. G. (1996). Effect of overfeeding macronutrients on day to day food intake in man. *Eur. J. Clin. Nutr.* **50**, 418–430.
27. Barkeling, B., Rossner, S., and Bjorvell, H. (1990). Effects of a high protein meal and a high carbohydrate meal on satiety measured by automated computerized monitoring of subsequent food intake, motivation to eat and food preferences. *Int. J. Obes.* **14**, 743–751.
28. DeGraaf, C., Hulshof, T., Weststrate, J. A., and Jas, P. (1992). Short Term effects of different amounts of protein, fats and carbohydrate on satiety. *Am. J. Clin. Nutr.* **55**, 33–38.
29. Geliebter, A. (1979). Effects of equicaloric loads of protein, fat, and carbohydrate on food intake in the rat and man. *Physiol. Behav.* **22**, 267–273.
30. Long, S. J., Jeffcoat, A. R., and Millward, D. J. (2000). Effect of habitual dietary protein intake on appetite and satiety. *Appetite* **35**, 78–88.
31. Hill, A. J., and Blundell, J. E. (1986). Macronutrients and satiety: The effects of a high protein or high carbohydrate meal on subjective motivation to eat and food preferences. *Nutr. Behav.* **3**, 133–144.
32. Holt, S. H. A., Brand Miller, J. C., Petocz, P., and Farmakalidis, E. (1995). A satiety index of common foods. *Eur. J. Clin. Nutr* **49**, 675–690.
33. Vanderwater, K., and Vickers, Z. (1996). Higher protein foods produce greater sensory specific satiety. *Physiol. Behav.* **59**, 579–583.
34. Westerterp-Plantenga, M. S., Rolland, V., Wilson, S. A. J., and Westerterp, K. R. (1999). Satiety related to 24h diet-induced thermogenesis during high protein/carbohydrate vs high fat diets measured in a respiration chamber. *Eur. J. Clin. Nutr.* **53**, 495–502.
35. Booth, D. A., Chase, A., and Campbell, A. T. (1970). Relative effectiveness of protein in the late stages of appetite suppression in man. *Physiol. Behav.* **5**, 1299–1302.
36. Ludwig, D. S., Majzoub, J. A., Al-Zahrani, A., Dallal, G. E., Blanco, I., and Roberts, S. B. (1999). High glycemic index foods, overeating and obesity. *Pediatrics* **E261**, 261–266.
37. Araya, H., Hills, J., Alvina, M., and Vera, G. (2000). Short-term satiety in preschool children: a comparison between high protein meal and a high complex carbohydrate meal. *Int. J. Food Sci. Nutr.* **51**, 119–124.
38. Porrini, M., Santangelo, A., Crovetti, R., Riso, P., Testolin, G., and Blundell, J. E. (1997). Weight, protein, fat and timing of preloads affect food intake. *Physiol Behav.* **65**, 563–570.
39. Teff, K. L., Young, S. N., and Blundell, J. E. (1989). The effect of protein or carbohydrate breakfasts on subsequent plasma amino acid levels, satiety and nutrient selection in normal males. *Pharmacol. Biochem. Behav.* **34**, 829–837.
40. Marmonier, C., Chapelot, D., and Louis-Sylvestre, J. (2000). Effects of macronutrient content and energy density of snacks consumed in a satiety state on the onset of the next meal. *Appetite* **34**, 161–168.
41. Rumpler, W. V., Kramer, M., Rhodes, D. G., and Paul, D. R. (2006). The impact of the covert manipulation of macronutrient intake on energy intake and the variability in daily food intake in nonobese men. *Int. J. Obes.* **30**, 774–781.
42. Long, S. J., Jeffcoat, A. R., and Millward, D. J. (2000). Effect of habitual dietary protein intake on appetite and satiety. *Appetite* **35**, 79–88.
43. Hall, W. L., Millward, D. J., Long, S. J., and Morgan, L. M. (2003). Casein and whey exert different effects on plasma amino acid profiles, gastrointestinal hormone secretion and appetite. *Br. J. Nutr.* **89**, 239–248.
44. Bowen, J, Noakes, M., Trenerry, C., and Clifton, P. M. (2006). Energy intake, ghrelin, and cholecystokinin after different carbohydrate and protein preloads in overweight men. *J. Clin. Endocrinol. Metab.* **91**, 1477–1483.
45. Pedersen, N. L., Nagain-Domaine, C., Mahe, S., Chariot, J., Roze, C., and Tome, D. (2000). Caseinomacropeptide specifically stimulates exocrine pancreatic secretion in the anesthetized rat. *Peptides* **21**, 1527–1535.
46. Schneeman, B. O., Burton-Freeman, B., and Davis, P. (2003). Incorporating dairy foods into low and high fat diets increases the postprandial cholecystokinin response in men and women. *J. Nutr.* **133**, 4124–4128.
47. Hollis J, Mattes RD. (2007). Effect of chronic consumption of almonds on body weight in healthy humans. *Br. J. Nutr.* **98**, 651–656.
48. Gustafson, D. R., McMahon, D. J., Morrey, J., and Nan, R. (2001). Appetite is not influenced by a unique milk peptide: Caseinomacropeptide (CMP). *Appetite* **36**, 157–163.
49. Lambert, T. C., Hill, A. J., and Blundell, J. E. (1989). Investigating the satiating effect of protein with disguised liquid preloads. *Appetite* **12**, 220.
50. Latner, J. D., and Schwartz, M. (1999). The effects of a high-carbohydrate, high-protein or balanced lunch upon later food intake and hunger ratings. *Appetite* **33**, 119–128.
51. Driver, C. J. I. (1988). The effect of meal composition on the degree of satiation following a test meal and possible mechanisms involved. *Br. J. Nutr.* **60**, 441–449.
52. Stockley, L, Jones, F. A., and Broadhurst, A. J. (1984). The effects of moderate protein or energy supplements on subsequent nutrient intake in man. *Appetite* **5**, 209–219.
53. Sunkin, S., and Garrow, J. S. (1982). The satiety value of protein. *Hum. Nutr. Appl. Nutr.* **36A**, 197–201.
54. Gardner, C. D., Kiazand, A., Alhassan, S., Kim, S., Stafford, R. S., Balise, R. R., Kraemer, H. C., and King, A. C. (2007). Comparison of the Atkins, Zone, Ornish, and LEARN Diets for change in weight and related risk factors among overweight premenopausal women. The A to Z Weight Loss Study: A randomized trial. *JAMA* **297**, 969–977.
55. Leidy, H. J., Carnell, N. S., Mattes, R. D., and Campbell, W. W. (2007). Higher protein intake preserves lean mass and satiety with weight loss in pre-obese and obese women. *Obesity* **15**, 421–429.
56. Wolfe, R. R. (2006). The underappreciated role of muscle in health and disease. *Am. J. Clin. Nutr.* **84**, 475–482.

57. Fujise, T., Yoshimatsu, H., Kurokawa, M., Oohara, A., Kang, M., Nakata, M., and Sakata, T. (1998). Satiation and masticatory function modulated by brain histamine in rats. *Proc. Soc. Exp. Biol. Med.* **217**, 228–234.
58. Delgado-Aros, S., Cremonini, F., Castillo, J. E., Chial, H. J., Burton, D. D., Ferber, I., and Camilleri, M. (2004). Independent influences of body mass and gastric volumes on satiation in humans. *Gastroenterology* **126**, 432–440.
59. Kissileff, H. R., Carretta, J. C., Gelibter, A., and Pi-Synyer, F. (2003). Cholecystokinin and stomach distension combine to reduce food intake in humans. *Am. J. Physiol. Regul. Integr. Comp. Physiol.* **285**, R992–R998.
60. Clarkston, W. K., Pantano, M. M., Morley, J. E., Horowitz, M., Littlefield, J. M., and Burton, F. R. (1997). Evidence for the anorexia of aging: Gastrointestinal transit and hunger in healthy elderly vs. young adults. *Am. J. Physiol.* **272**, R243–R248.
61. Holt, S., Brand, J., Soveny, C., and Hansky, J. (1992). Relationship of satiety to postprandial glycaemic, insulin and cholecystokinin response. *Appetite* **18**, 129–141.
62. Tiwary, C. M., Ward, J. A., and Jackson, B. A. (1997). Effect of pectin on satiety in healthy U.S. army adults. *J. Am. Coll. Nutr.* **16**, 423–428.
63. Howarth, N. C., Saltzman, E., and Roberts, S. B. (2001). Dietary fiber and weight regulation. *Nutr. Rev.* **59**, 129–139.
64. Burton-Freeman, B. (2000). Dietary fiber and energy regulation. *J. Nutr.* **130**, 272S–275S.
65. Miles, C. W., Kelsay, J. L., and Wong, N. P. (1988). Effect of dietary fiber on the metabolizable energy of human diets. *J. Nutr.* **118**, 1075–1081.
66. Wisker, E., and Feldheim, W. (1990). Metabolizable energy of diets low or high in dietary fiber from fruits and vegetables when consumed by humans. *J. Nutr.* **120**, 1331–1337.
67. Yoshioka, M., St-Pierre, S., Drapeau, V., Dionne, I., Doucet, E., Suzuki, M., and Tremblay, A. (1999). Effects of red pepper on appetite and energy intake. *Br. J. Nutr.* **82**, 115–123.
68. Yoshioka, M., St-Pierre, S., Suzuki, M., and Tremblay, A. (1998). Effects of red pepper added to high-fat and high-carbohydrate meals on energy metabolism and substrate utilization in Japanese women. *Br. J. Nutr.* **80**, 503–510.
69. Lejeune, M. P. G. M., Kovacs, E. M. R., and Westerterp-Plantenga, M. S. (2003). Effect of capsaicin on substrate oxidation and weight maintenance after modest body-weight loss in human subjects. *Br. J. Nutr.* **90**, 651–659.
70. Lim, K., Yoshioka, M., Kikuzato, S., Kiyonaga, A., Tanaka, H., Shindo, M., and Suzuki, M. (1997). Dietary red pepper ingestion increases carbohydrate oxidation at rest and during exercise in runners. *Med. Sci. Sport Exerc.* **29**, 355–361.
71. Yoshioka, M., Lim, K., Kikuzato, S., Kiyonaga, A., Tanaka, H., Shindo, M., and Suzuki, M. (1995). Effects of red-pepper diet on the energy metabolism in men. *J. Nutr. Sci. Vitaminol. (Tokyo)* **41**, 647–656.
72. Westerterp-Plantenga, M. S., Smeets, A., and Lejeune, M. P. G. (2005). Sensory and gastrointestinal satiety effects of capsaicin on food intake. *Int. J. Obes.* **29**, 682–688.
73. Ahuja, K. D. K., Robertson, I. K., Geraghty, D. P., and Ball, M. J. (2006). Effects of chili consumption on postprandial glucose, insulin, and energy metabolism. *Am. J. Clin. Nutr.* **84**, 63–69.
74. Yoshioka, M., Imanaga, M., Ueyama, H., Yamane, M., Kubo, Y., Boivin, A., St-Amand, J., Tanaka, H., and Kiyonaga, A. (2004). Maximum tolerable dose of red pepper decreases fat intake independently of spicy sensation in the mouth. *Br. J. Nutr.* **91**, 991–995.
75. Mattes, R. D. (1997). Physiological responses to sensory stimulation by food: Nutritional implications. *J. Am. Diet. Assoc.* **97**, 406–413.
76. Zafra, M. A., Molina, F., and Peurto, A. (2006). The neural/cephalic phase reflexes in the physiology of nutrition. *Neurosci. Biobehav. Rev.* **30**, 1032–1044.
77. LeBlanc, J., and Soucy, J. (1996). Interactions between postprandial thermogenesis, sensory stimulation of feeding, and hunger. *Am. J. Physiol. Regul. Integr. Comp. Physiol.* **271**, R936–R940.
78. Kawada, T., Sakabe, S., Watanabe, T., Yamamoto, M., and Iwai, K. (1988). Some pungent principles of spices cause the adrenal medulla to secrete catecholamine in anesthetized rats. *Proc. Soc. Exp. Biol. Med.* **188**, 229–233.
79. Henry, C. J. K., and Emery, B. (1986). Effect of spiced food on metabolic rate. *Hum. Nutr.* **40C**, 165–168.
80. Watanabe, T., Kawada, T., Kurosawa, M., Sato, A., and Iwai, K. (1991). Thermogenic action of capsaicin and its analogs. *In* "Obesity: Dietary Factors and Control" (M. Suzuki, Ed.), pp. 67–77. Japan Scientific Societies Press and S. Karger, AG, Tokyo and Basel.
81. Matsumoto, T., Miyawaki, C., Ue, H., Miyatsuji, A., and Moritani, T. (2000). Effects of capsaicin-containing yellow curry sauce on sympathetic nervous system activity and diet induced thermogenesis in lean and obese young women. *J. Nutr. Sci. Vitaminol. (Tokyo).* **46**, 309–315.
82. Yoshioka, M., Doucet, E., Drapeau, V., Dionne, I., and Tremblay, A. (2001). Combined effects of red pepper and caffeine consumption on 24h energy balance in subjects given free access to foods. *Br. J. Nutr.* **85**, 203–211.
83. Hoeger, W. W. K., Harris, C., and Hopkins, D. R. (1998). Four-week supplementation with a natural dietary compound produces favorable changes in body composition. *Adv. Ther.* **15**, 305–314.
84. Pradeep, K. U., Geervani, P., and Eggum, B. O. (1991). Influence of spices on utilization of sorghum and chickpea protein. *Plant Food Hum. Nutr.* **41**, 269–276.
85. Desai, H. G., Venugopalan, K., Philipose, M., Zaveri, M. P., Kalro, R. H., and Antia, F. P. (1977). Effect of red chili powder on gastric mucosal barrier and acid secretion. *Indian J. Med.* **66**, 440–448.
86. Limlomwongse, L., Chaitauchawong, C., and Tongyai, S. (1979). Effect of capsaicin on gastric acid secretion and mucosal blood flow in the rat. *J. Nutr.* **109**, 773–777.
87. Nopanitaya, W. (1973). Long term effects of capsaicin on fat absorption and the growth of the rat. *Growth* **37**, 269–279.
88. Sorensen, L. B., Moller, P., Flint, A., Martens, M., and Raben, A. (2003). Effect of sensory perception of foods on appetite and food intake: a review of studies on humans. *Int. J. Obes.* **27**, 1152–1166.
89. Raben, A., Christensen, N. J., Madsen, J., Holst, J. J., and Astrup, A. (1994). Decreased postprandial thermogenesis and fat oxidation but increased fullness after a high-fiber meal compared with a low-fiber meal. *Am. J. Clin. Nutr.* **80**, 503–510.
90. LeBlanc, J., and Brondel, L. (1985). Role of palatability on meal-induced thermogenesis in human subjects. *Am. J. Physiol. Endocrinol. Metab.* **248**, E333–E336.

CHAPTER 27

Properties of Foods and Beverages That Influence Energy Intake and Body Weight

JENNY H. LEDIKWE AND BARBARA J. ROLLS
The Department of Nutritional Sciences, The Pennsylvania State University, University Park

Contents

I. INTRODUCTION

Although genetic disposition is a risk factor of obesity, the recent increase in the prevalence of overweight and obesity suggests that environmental factors play an important role. Understanding these factors is critical in the development of strategies to help people achieve and maintain a healthy body weight.

The fundamental rule of weight management is that people gain weight when they consume more energy than they expend. Although we recognize the importance of energy expenditure in weight management [1], energy intake is the focus of this chapter.

Eating behavior is influenced by a number of environmental factors, such as food marketing, cost, and availability, as well as various properties of foods. This review summarizes how energy intake is influenced by three properties of foods and beverages: variety, energy density, and portion size. Research has shown that these properties influence energy intake and can be modified to reduce energy consumption. Foods and beverages are considered separately because they may affect energy balance differently.

Studies have shown that energy intake is increased when the variety, energy density, or portion size of foods is increased. Similar effects are likely to be seen for beverages, though they have not been as extensively studied. The influence of variety, energy density, and portion size on energy intake is mostly unrecognized by consumers; therefore, these factors are particularly amenable to changes that will help create a more healthful eating environment. This review presents information on the influence of variety, energy density, and portion size on body weight, along with strategies for utilizing these food-related factors to aid in weight management.

II. VARIETY

A. Variety and Energy Intake

1. Food Variety

The variety of foods available can affect energy intake. Studies have shown that people eat more during a meal consisting of a variety of foods than during a meal consisting of just one food [2, 3]. In a laboratory-based study reported by Rolls *et al.* [3], participants were provided with a four-course lunch on two different days that consisted of either a different food in each course or the same food in each course. Energy intake was elevated by 60% when participants were offered a meal consisting of four different foods compared to when they were offered a single food. Even similar foods presented in a variety of flavors can enhance intake. A study found that offering sandwiches with three different flavors of fillings (salty, curried, and sweet) increased energy intake by 15% compared to when the participants were offered sandwiches with only their favorite filling [2]. Other laboratory-based studies have shown similar results [4–7].

The influence of variety on energy intake is related to changes in the palatability of a food as it is consumed. Palatability is a primary determinant of the foods people choose to eat. When queried about factors affecting food selection, a nationally representative sample of U.S. adults indicated that taste was the most important influence on their food choices [8]. Laboratory studies have confirmed that palatability influences the type and amount of food that people consume [9]. In general, for a given food, the greater the palatability, the greater is the intake of that food at a meal [9]. However, the palatability of a particular food is not constant; it declines as that food is consumed [10–14]. This decline contributes to the termination of the consumption of that food and foods with similar sensory properties [2, 3]. As the palatability of a particular food declines with consumption, the palatability of other foods with different sensory properties remains relatively unchanged [15]. The decrease in the palatability of a food as it is eaten, relative to other foods, is called sensory-specific satiety [4]. Sensory-specific satiety is involved in decreasing consumption of a particular food at a meal and promoting the consumption of other foods, which explains why increasing the variety of foods served at a meal increases energy intake.

2. Beverage Variety

Just as palatability influences food intake, the palatability of beverages influences fluid intake. Studies have shown that adding flavoring and sweeteners to beverages to make them taste better can increase consumption [16–18]. Sensory-specific satiety has also been demonstrated for beverages. For example, a study found that the pleasantness ratings of a milk-based drink declined as participants consumed greater amounts of the drink [2, 19].

Beverage variety has been shown to increase the amount people drink. This was demonstrated in a study in which participants were presented with three drinks successively, with a 10-minute period to consume as much of each beverage as they wanted. Participants consumed a greater amount when they were presented with a different flavor during each 10-minute period as opposed to when they were presented with water or the same flavor during each period [20]. The effect of beverage variety on intake has also been demonstrated during a meal [21]. When four fruit-flavored beverages were available with lunch, participants consumed 17% more beverage than when only one flavor, the participant's favorite, was available during the meal.

B. Variety and Body Weight

While having a variety of foods and beverages is important for ensuring that a balance of nutrients is consumed [22], the increase in energy intake associated with variety may affect body weight. This was tested in a laboratory-based study in which participants had access to 5, 10, or 15 different food items during each of three week-long study periods [7]. Researchers found that increasing the variety of foods available to study participants led not only to increased energy intake, but also to increased body weight. Additional support linking variety and body weight is provided by a recent study with individuals successful at losing weight. Researchers found that restricting variety was important in helping individuals maintain weight loss [23].

The relationship between food variety and body fatness may, however, depend on the types of foods being consumed [24]. Roberts and colleagues [25] found that consuming a variety of foods with a high energy content was associated with higher body mass index (BMI) values among a representative sample of U.S. adults. In another study examining food consumption patterns of adults, there was a direct correlation between intake of relatively high energy foods and adiposity [26]. As depicted in Figure 1, body fatness was positively correlated with the variety of sweets, snacks, condiments, entrées, and carbohydrates consumed by study participants. Conversely, consuming a greater variety of vegetables, which provide relatively less energy, was associated with less body fat. This is supported by data from an 18-month clinical trial in which greater weight loss was associated with increases in the variety of vegetables consumed, and with decreases in the variety of high-fat foods and sweets consumed [27].

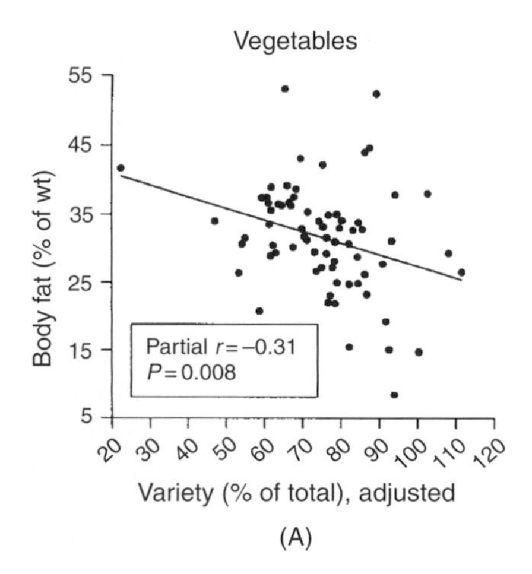

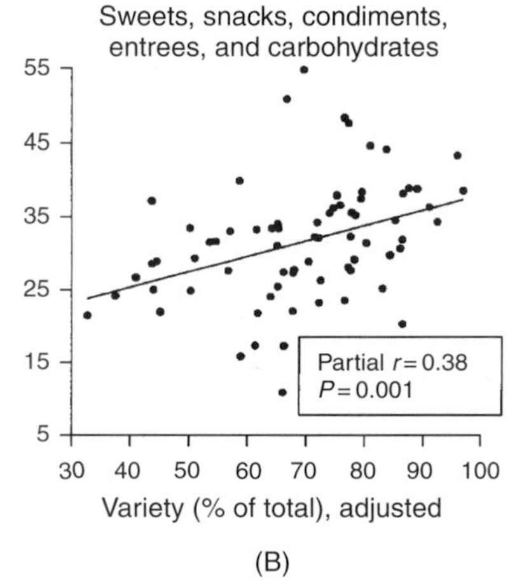

FIGURE 1 A cross-sectional study of 71 adults found that the variety of vegetables consumed was inversely associated with body fatness. Conversely, the variety of sweets, snacks, condiments, entrees, and carbohydrates consumed was positively associated with body fatness. Reproduced with permission from the *American Journal of Clinical Nutrition* [26].

C. Summary

While food and beverage consumption is influenced by the sensory attributes or perceived pleasantness of these items, eating and drinking behaviors are complex, with many factors influencing consumption patterns. The availability of a variety of foods and beverages can contribute to the overconsumption of energy. Studies indicate that the availability of a wide variety of foods with a high energy content increases energy intake and could be contributing to the increased incidence of overweight and obesity. Research also suggests that an increased variety of foods with low energy content, such as vegetables, may help to moderate energy intake. Given that the association between adiposity and food variety differed depending on the energy content of the foods consumed, it is likely that dietary energy density (discussed later) plays a role in the regulation of food intake and body weight.

III. ENERGY DENSITY

Energy density refers to the amount of energy in a given weight of food (kcal/g). Energy density values, which are influenced by the macronutrient composition and moisture content of foods and beverages, range from 0 kcal/g to 9 kcal/g (Fig. 2). Because of its high energy content, fat (9 kcal/g) influences energy density values more than carbohydrate or protein (4 kcal/g). Foods and beverages with a high fat content generally have a relatively high energy density, but a high moisture content lowers the energy density of foods, even those high in fat. Water has an energy density of 0 kcal/g because it contributes weight but not energy to foods and beverages. Because beverages are primarily water, they tend to have a lower energy density than most foods.

Foods and beverages with a lower energy density provide less energy per gram than items with a higher energy density. Figure 3 shows the amount of energy provided by 100-gram portions of foods differing in energy density. A 100-gram portion of spaghetti with tomato sauce (1 kcal/g) provides 100 kcal whereas 100 grams of higher-energy-density pretzels (4 kcal/g) provide 400 kcal. As outlined in the studies discussed hereafter, people tend to eat a consistent amount of food on a day-to-day basis; therefore intakes of energy vary directly with changes in the energy density of foods consumed.

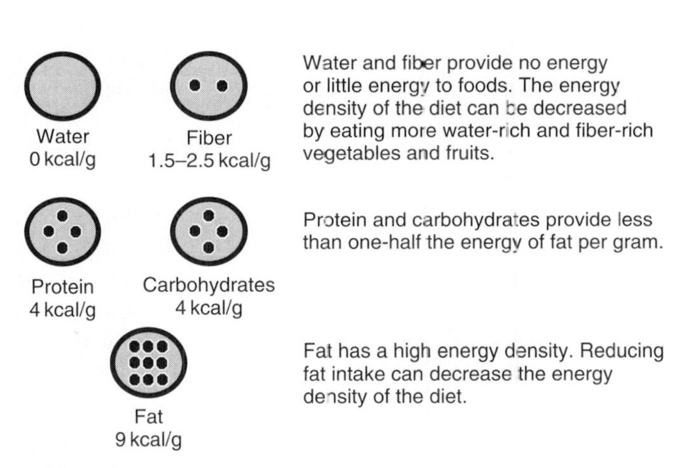

FIGURE 2 This diagram illustrates the energy density of the macronutrients that contribute energy and weight to foods and beverages. Each oval represents 1 gram of a macronutrient, and the number of dots indicates the amount of energy provided by each gram. The energy density of foods ranges from 0 to 9 kcal/g depending on the macronutrient composition.

A. Energy Density and Energy Intake

1. Food Energy Density

The importance of food energy density in the determination of energy intake has been demonstrated in a number of laboratory studies [28–32]. In these studies, when individuals were offered *ad libitum* access to the same foods on different occasions, they tended to eat a consistent weight of food. As a result, energy intake varied directly with changes in the energy density of the foods.

One method of altering the energy density of a food is through the use of fat substitutes. In a study reported by Miller and colleagues [28], participants came to the laboratory every afternoon for 10 successive weekdays. During one 10-day period, they were offered a snack of regular potato chips, and during another they were offered equally palatable fat-free potato chips made with a fat replacement (olestra). Subjects ate similar amounts of the regular and nonfat chips in each of the 10-day periods. Because the nonfat chips were reduced in energy density, participants consumed less energy when eating these as a snack. Other trials have found that reductions in energy density through the use of fat substitutes led to reduced energy consumption not only at single test meals [33, 34], but also when the diet was manipulated during long-term studies [35–37].

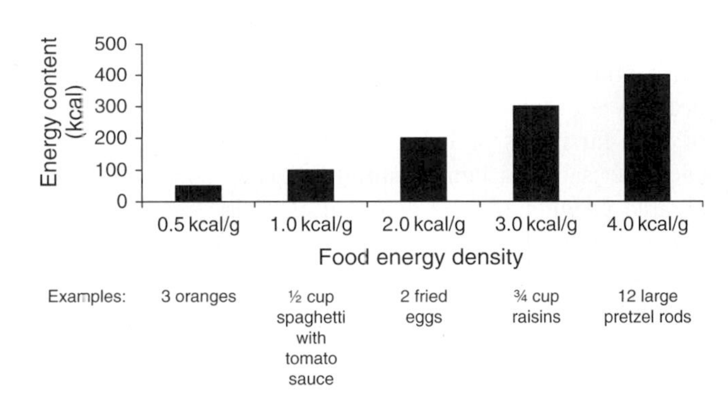

FIGURE 3 The energy content for 100-gram portions of foods differing in energy density is depicted. A 100-gram portion of a food with a relatively low energy density will provide less energy than a food with a higher energy density.

Reducing the total fat content of foods, without the addition of fat substitutes, is another strategy for reducing energy density. Changes in energy density achieved through reducing the absolute fat content of foods have also been shown to be effective at reducing energy consumption [38–41]. This was demonstrated in a study in which participants were provided with meals varying in energy density and fat content over 2-week study periods [41]. Participants consumed a constant weight of food across each study period; therefore energy intake was greatest on the high-fat/high-energy-density diet.

Reducing the total fat content of the diet generally reduces energy density, but fat and energy density can be manipulated independently. However, lowering the fat content of the diet has not been shown to influence energy intake if the energy density of the diet is unchanged [42–45]. In one study, participants were provided with diets over 2-week periods that were matched for energy density, but varied in fat content (20–60% of energy) [44]. Participants consumed similar amounts of food in each condition; as a result, energy intake was similar, regardless of the fat content of the diet. Variations in dietary fat have only been shown to affect energy consumption when accompanied by variations in energy density.

Energy density can be reduced not only by decreasing the fat content of foods, but also by increasing the water content, for example, by adding fruits and vegetables. Bell and colleagues [29] used this strategy to alter the energy density of 2-day diets provided to subjects on three different occasions. Although they could eat as much or as little as they liked, the study participants ate similar amounts of food, by weight, over the 2-day sessions. Consequently, reducing the energy density of the diet by 30% through the addition of extra vegetables led to a 30% reduction in energy intake. Despite the substantial reduction in energy intake, subjects rated themselves equally full. Studies manipulating the energy density of the diet through *both* reductions in fat and increases in water-rich vegetables have found similar results [30–32, 38]. Duncan and colleagues [38] provided participants with meals over 5 days on two separate occasions. On one occasion a lower-fat diet with substantial amounts of fruits, vegetables, whole grains, and beans was provided; the other diet included large amounts of high-fat meats and desserts. Participants reported that each diet satisfied hunger similarly. They consumed comparable weights of food during each 5-day session, which resulted in a 50% reduction in energy intake on the lower energy-density diet.

Epidemiological studies examining the foods people habitually consume provide further evidence that energy density is related to energy intake. These studies have investigated the energy density of the entire diet. A representative study of U.S. adults found that men and women who reported eating a diet based on low-energy-density foods had lower daily energy intakes than persons with a higher-energy-density diet [46, 47]. Studies among free-living Mediterranean [48], Chinese [49], and French [50] adults have also found that people with a diet low in energy density consume less energy.

In addition to energy intake, population-based studies have found the energy density of the diet to be associated with body weight. Normal-weight adults have been shown to consume diets with a lower energy density than obese individuals [47]. Data have also shown that the prevalence of obesity was lowest among those individuals with a high intake of fruits and vegetables; this was found even among individuals with a diet relatively high in fat (>30% of kcal) [47]. This highlights the potential of diets rich in low-energy-density foods to be useful for weight management.

2. Beverage Energy Density

The energy density of foods is emerging as an important influence on energy intake, but little is known about the effect of the energy density of beverages on energy balance.

Studies comparing beverages sweetened with regular and intense sweeteners have provided an initial indication that beverage energy density influences energy intake. Several laboratory-based studies suggest that substituting a beverage having a lower energy density for one with a higher energy density may be an effective strategy for reducing energy consumption at a meal [51–54]. In a study reported by DellaValle and colleagues [51], participants were provided with a compulsory caloric or noncaloric beverage along with a standard lunch on different occasions. The beverages included water and diet cola, which have an energy density of 0 kcal/g, and regular cola, orange juice, and 1% milk, which each have an energy density of 0.4 kcal/g. Food intake was similar regardless of the beverage served with the meal. Therefore total energy intake at the lunch meal was lower, by approximately 100 kcal, when the lower energy-density beverages (water and diet soda) were consumed. Subjects' ratings of fullness after lunch did not differ among the beverage conditions, which suggests that having a low-energy-density beverage with a meal can help reduce energy intake without significantly affecting satiety ratings.

Given that laboratory-based studies suggest that reductions in beverage energy density may be an effective strategy for reducing energy consumption, it is of interest to examine relationships between daily beverage energy density and energy intake among free-living individuals. Relatively little is known about the energy density of the beverages typically consumed by individuals on a day-to-day basis. Until recently it has been difficult to determine the overall daily energy density of the beverages consumed by free-living individuals because water intake has generally not been measured. As methodological improvements allow for accurate assessments of beverage intakes in large epidemiological studies, more will be learned about relationships between beverage energy density and energy consumption.

B. Energy Density and Body Weight

1. Food Energy Density

Strategies typically used to reduce energy intake for weight loss include limitations in portion sizes, food groups, or certain macronutrients. These sorts of restrictive approaches may result in feelings of hunger or dissatisfaction, which can limit their acceptability, sustainability, and long-term effectiveness [55–57]. Conversely, consuming a diet low in energy density allows energy intake to be reduced without strictly limiting food portions. Figure 4 depicts the total amount of food that can be consumed for 1400 kcal at three different energy density levels. The energy density values in the figure correspond to a low- (1.4 kcal/g), medium- (1.9 kcal/g), and high-energy-density (2.2 kcal/g) diet, as defined by the average dietary energy density of a representative group of U.S. adults [47]. When consuming a diet with an energy density of 1.4 kcal/g, rich in low-energy-density foods, 1000 grams of food can be consumed for 1400 kcal. Only 636 grams can be consumed on the high-energy-density, 2.2 kcal/g diet. At any energy level, the lower the energy density of the diet, the greater the amount of food that can be consumed.

Few longitudinal studies have specifically assessed the role of energy density in weight management. Nevertheless, it is likely that weight-loss interventions advocating fat reduction along with increased fruit and vegetable consumption operate at least in part by decreasing dietary energy density. In a short-term study, Shintani and colleagues [58, 59] provided participants with a low-fat, fruit- and vegetable-rich diet that was considerably lower in energy density than the participants' habitual diet. Even though they could eat as much as they liked, the participants consumed their usual amount of food over the 3-week test period, which resulted in a reduction in energy intake as well as body weight (mean weight loss = 17.2 lb). Despite the reduction in energy intake, the subjects reported the diet to be moderately to highly satiating.

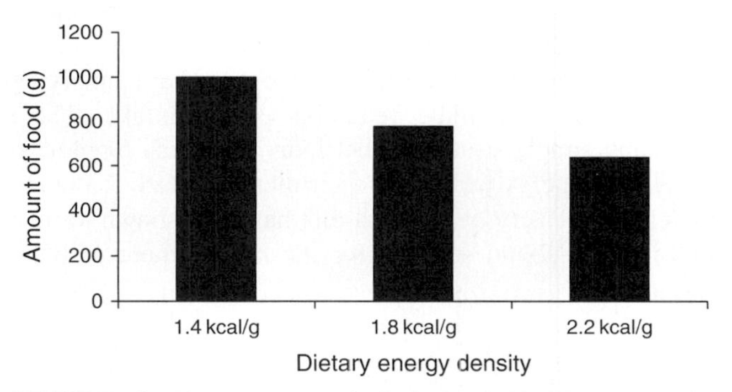

FIGURE 4 The amount of food a person consumes for 1400 kcal varies depending on the energy density of the diet. At a given energy level, a greater amount of food is consumed on a low-energy-density diet. The energy-density values in this figure were based on values corresponding to a low-, medium-, and high-energy-density diet, as defined using food intake data from a representative group of U.S. adults [46, 47].

Stronger evidence regarding the influence of energy-density reduction on body weight is provided by longer-term clinical trials. Rolls and colleagues [60] found that the incorporation of a single low-energy-density food into a reduced-energy diet was sufficient to reduce the overall energy density of the diet and increase weight loss. In this year-long study with overweight and obese individuals, weight loss among those consuming two servings of low-energy-density soup a day was 50% greater (16 versus 11 lb) than among those consuming two servings of high-energy-density dry snacks daily.

Taking a broader dietary approach, Ello-Martin and colleagues [61] found that advice to reduce the energy density of the diet was effective in achieving weight loss. In this study, one group of obese women was counseled to decrease the energy density of their diet by increasing consumption of water-rich foods, such as fruits and vegetables, and choosing reduced-fat foods. A comparison group was counseled only to choose reduced-fat foods. Participants in both groups lost weight without receiving specific goals for intake of energy or fat. After 12 months, the group counseled to eat more fruits and vegetables had a greater reduction in the energy density of their diet and lost more weight than the group instructed only to choose reduced-fat foods (17.4 versus 14.1 lb). Over the course of the year, participants who ate the lower-energy-density diet (higher in fruits and vegetables) reported consuming an average of 25% more food and reported less hunger than those in the comparison group. This study indicates that advice leading to a reduction in dietary energy density is effective for weight loss, particularly when guidance to increase fruit and vegetable consumption is coupled with instruction in decreasing fat intake.

Weight loss is not the only beneficial change associated with reducing dietary energy density. Ledikwe and colleagues [62] recently used data from a large intervention study to examine the relationship between changes in dietary energy density after 6 months and changes in body weight and diet quality. In this study, participants received one of three lifestyle interventions to reduce blood pressure that included information on physical activity, diet, and weight loss. Because each intervention group experienced a decline in energy density and body weight, analyses were conducted by classifying participants into tertiles based on the magnitude of change in energy density after 6 months. Participants with a large reduction in energy density lost more weight (13 lb) than those with a modest reduction (9 lb) or those with a slight reduction or increase in energy density (5 lb). In addition to weight loss, reductions in energy density were associated with improved diet quality. Both large and modest reductions in energy density were associated with increased intakes of fruits, vegetables, fiber, vitamins, and minerals, indicating that reductions in dietary energy density are a healthy weight management strategy. Furthermore, participants with both large and modest decreases in energy density increased the amount of food they consumed, which may contribute to the long-term acceptability of a low-energy-density eating plan (Fig. 5).

Relationships between dietary energy density and the *maintenance* of lost weight have not been extensively investigated. Greene and colleagues [63] examined energy density values 2 years after participation in a weight-loss program that encouraged consumption of low-energy-density foods. The researchers found that individuals who maintained weight loss had a lower-energy-density diet than those who regained 5% or more of their body weight. Although additional long-term studies are required to understand the impact of reduced-energy-density diets on weight maintenance, these initial findings are promising.

2. Beverage Energy Density

Research specifically examining the influence of beverage energy density on body weight has not been conducted. An examination of studies comparing beverages sweetened with intense sweeteners with those containing higher-energy-density sweeteners (i.e., sucrose, high-fructose corn syrup) provides some information about beverage energy density. Three investigations with intense sweeteners provide an indication that beverage energy density is related to body weight. In one study, normal-weight adults supplemented their daily diets with soft drinks for two 3-week periods [64]. They were provided with regular soft drinks during one study period and diet soft drinks during the other. Body weight increased for males and females when consuming the higher-energy-density regular soft drinks. Conversely, body weight decreased for males, but not females, when consuming the diet beverages. In another study, one group of overweight adults was provided with beverages and foods sweetened with intense sweeteners to incorporate into their diet over a 10-week period. A comparison group was provided with drinks and foods sweetened with sucrose [65]. An increase in body weight was experienced by the group consuming the higher-energy-density beverages and foods. Body weight decreased when consuming the items containing intense sweeteners. Additionally, a recent pilot study with adolescents tested the effect of replacing caloric beverages with lower-energy-density beverages containing intense sweeteners [66]. Over 25 weeks, an intervention group received home delivery of diet beverages, while a control group continued their usual beverage consumption. There was no overall effect of the intervention on BMI, but analysis based on participants with relatively high BMI values found that participants in the intervention group lost weight, and those in the control group gained weight.

Even though these studies have not purposely focused on energy density, they support the hypothesis that beverage energy density is related to body weight. These data suggest that individuals may want to choose beverages containing intense sweeteners rather than regular sweeteners as a weight management strategy. Nevertheless, further studies are needed to understand the specific influence of beverage energy density on energy balance.

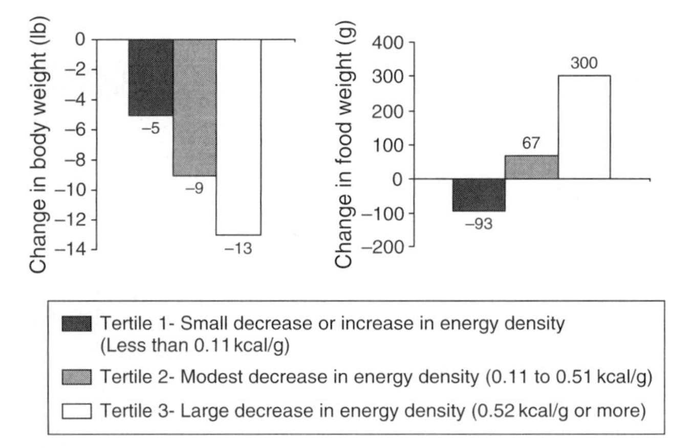

FIGURE 5 Participants in the PREMIER trial ($n = 658$) were classified into tertiles based on the magnitude of change in dietary energy density after 6 months of intervention. Those with a large reduction in energy density lost more weight (13 lb) than those with a modest reduction (9 lb) or those with a slight reduction or increase in energy density (5 lb). Additionally, participants with either a large or modest decrease in energy density increased the amount of food they consumed, which may contribute to the long-term acceptability of a low-energy-density eating plan [62].

C. Summary

Studies have consistently shown that people consume more energy when presented with foods having a higher energy density than with similar foods having a lower energy density. This effect has been shown for snack foods [28] and meal components [45, 67] as well as for all meals provided over multiple days [29]. The consumption of diets low in energy density allows people to have satisfying amounts of food while reducing energy intake. There are many simple strategies that individuals can employ to reduce the energy density of their diets [68, 69]. Reducing the energy density of the total diet has been shown to be a nutritionally sound strategy for the management of body weight.

IV. PORTION SIZE

A. Portion Size and Energy Intake

Since the 1970s the portion sizes of many foods and beverages have been increasing [70], a trend that has been observed in a variety of settings including restaurants, supermarkets, and homes [71–73]. Increasing portion size has occurred in parallel with the rise in the prevalence of obesity [74]. This suggests that portion size could play a role in body weight. The first step in establishing a relationship between portion size and body weight is to investigate the influence of portion size on energy intake.

1. Food Portion Size

A variety of studies have been carried out in which subjects were presented with different amounts of food on different occasions to determine the influence of portion size on the amount consumed. In a study by Rolls *et al.* [75], adults were served four different portions of macaroni and cheese on different days. As shown in Figure 6, the bigger the portion, the more the participants ate. They consumed 30% more energy when offered the largest portion compared to the smallest portion. This effect was seen both when the portion on the plate was determined by the investigator and when the participants served themselves from bowls containing different portions. The portion size of foods with clearly defined shapes or units has also been shown to influence energy intake. On different days, Rolls and colleagues [76] offered participants submarine sandwiches that varied in size and found there was a systematic and significant effect of portion size on intake. When served a 12-inch sandwich, compared with the 8-inch sandwich, females consumed 12% more energy and males consumed 23% more energy. In both of these studies, participants reported similar ratings of hunger and fullness at the end of their meals, despite consuming 12–30% more when offered larger portions.

Studies in more natural eating environments confirm that food portions influence energy intake. A study in a cafeteria-style restaurant tested whether increasing the portion size of a pasta entrée by 50%, while keeping the price the same, would affect intake [77]. This increase in the portion size of the main entrée was associated with a 43% increase in energy intake for the pasta and a 25% increase for the entire meal. A customer survey showed no difference in ratings of the appropriateness of the two portions. Another study [78], also conducted in a natural setting, found that patrons at a movie theater ate a significantly greater amount of popcorn when provided with a large-sized bucket, compared to a medium-sized bucket.

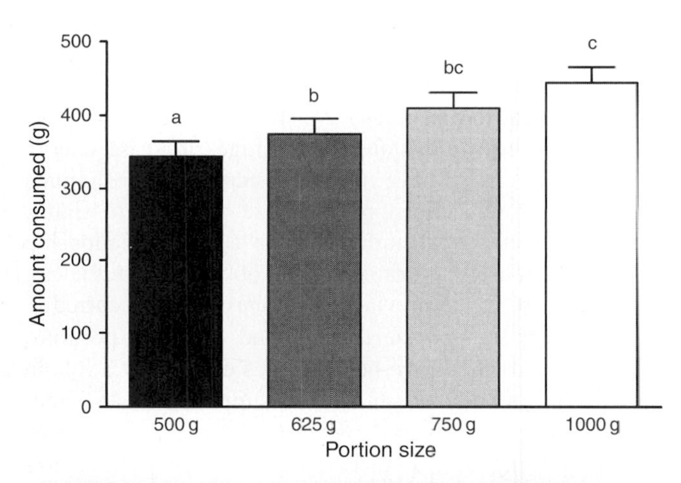

FIGURE 6 When adults ($n = 51$) were provided with four different portions of macaroni and cheese served on different days, researchers found that the bigger the portion, the more participants ate. They consumed 30% more energy (162 kcal) when offered the largest portion (1000 g) compared to the smallest portion (500 g). Of particular interest was the finding that despite the difference in intake, participants reported similar ratings of hunger and fullness after eating [75].

2. Beverage Portion Size

The rise in portion size is not limited to food; beverage portion sizes have also increased in recent years. For example, between 1977 and 1996, the average soft drink portion increased significantly from 13.1 fl oz to 19.9 fl oz, an increase in energy content from approximately 144 kcal to 193 kcal [72].

Few studies have investigated the influence of beverage portion size on food and energy intake. In a recent study reported by Flood and colleagues [54], participants were provided with cola, diet cola, or water in one of two portion sizes along with a standard lunch on six different occasions. As depicted in Figure 7, food intake did not differ, regardless

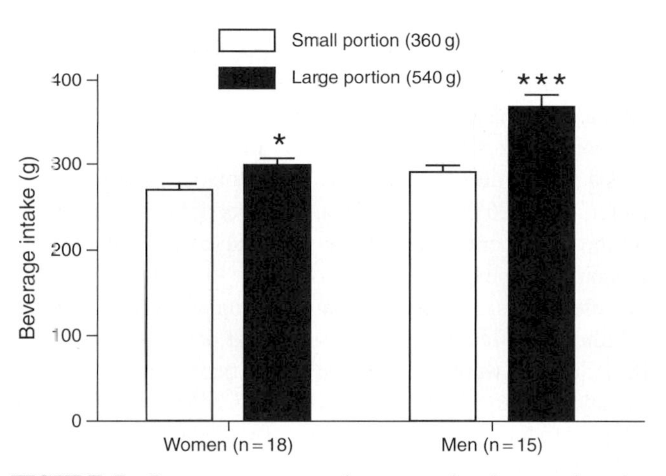

FIGURE 7 Beverage consumption was related to portion size when participants ($n = 33$) were provided with a caloric or noncaloric beverage in one of two portion sizes along with a standard lunch on six different occasions. Increasing beverage portion size increased the weight of beverage consumed. For the caloric beverage, energy intake from the beverage increased by 10% for women and 26% for men when there was a 50% increase in the portion served. Reproduced with permission from the *Journal of the American Dietetic Association* [54].

of the type of beverage that accompanied the meal or the portion size of the beverage. Beverage intake did, however, vary with portion size. Participants consumed more of a beverage when it was presented in the larger portion size. As a consequence, total energy intake increased when the meals were accompanied by an energy-containing beverage, particularly when it was offered in the larger portion size. Even though they consumed more energy, the study participants did not report feeling more full or less hungry after the lunch meals with the energy-containing beverages. These results suggest that large portions of caloric beverages can increase energy intake from beverages as well as energy intake at a meal.

3. Compensation for Large Portions

Studies have shown that in laboratory and in more naturalistic situations, the portion size of foods can affect the amount people consume at a particular eating occasion. An essential question is whether compensatory mechanisms will limit intake at subsequent eating occasions after a bout of overeating stimulated by large portions.

In a laboratory-based study reported by Rolls *et al.* [79], subjects were provided with different package sizes of potato chips as an afternoon snack, and were later served a standard dinner. As in other studies, portion size had a significant effect on snack intake. When served a 170-g package, women ate 18% more and men ate 37% more than when served an 85-g package. As subjects increased their snack intake with increasing package size, they also reported feeling more full, but there was little adjustment of intake at the subsequent dinner meal to compensate for the increased energy intake and fullness.

To investigate further whether portion size has an impact on intake beyond a single eating occasion, Rolls and colleagues conducted a study in which the portion size of all foods served over 2 days was increased [80]. It was again found that increasing portion sizes led to significantly increased energy intake. When the portions of all foods were doubled, energy intake increased by 26% on both days. Subjects did not compensate for the excess energy eaten over the course of the first day by reducing their intake on the second day. In a follow-up study, the researchers extended the manipulation period to determine whether the effect of portion size would persist over 11 days [81]. Subjects were provided with all of their foods during two 11-day periods, which were separated by a 2-week interval. In addition to foods, participants were provided with orange juice at breakfast, regular cola and iced tea at lunch, and milk at dinner. During one period, standard portions of all foods and beverages were served, and during the other, all portions were increased by 50%. Analysis showed that for both men and women the 50% increase in portion sizes resulted in a 16% increase in mean daily energy intake. The effect of portion size on energy intake was significant for both foods and beverages. Furthermore, this effect was sustained for 11 days and did not decline significantly over time (Fig. 8). The 50% increase in portion sizes resulted in a mean increase in energy intake of 423 kcal/day, with a mean cumulative increase in intake of 4636 kcal in 11 days. These data demonstrate that the effects of portion size can persist over multiple days, resulting in substantial increases in energy intake.

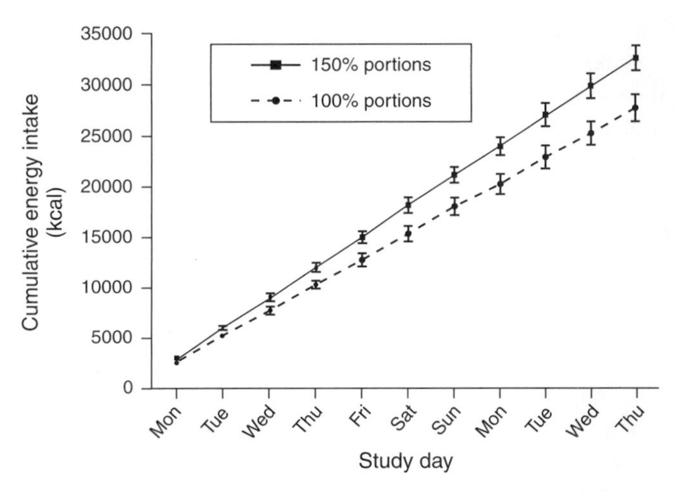

FIGURE 8 Mean cumulative energy intakes (13 men and 10 women) when the portion size of all foods and beverages served over 11 days was increased. The 50% increase in portion sizes resulted in a mean increase in energy intake of 423 kcal/day, with a mean cumulative increase in intake of 4636 kcal in 11 days. The effect of portion size on energy intake was significant and did not decline significantly over time [81].

B. Portion Size and Obesity

Although short-term studies have shown that large portions of food and energy-containing beverages increase energy intake, investigators have not yet systematically examined the relationship between portion size and weight status. There are no empirical studies to show a causal relationship between increased portion sizes and obesity. Nevertheless, the recent rise in the prevalence of obesity has occurred in parallel with an increase in food and beverage portions, which suggests that large portion sizes could play a role in the increase in body weight. This is supported by an epidemiological study reported by McConahy and colleagues [82] suggesting that a relationship between portion size and BMI is evident at an early age. This evaluation of reported food intake and BMI in a representative sample of children aged 2 to 5 years found that children with a higher BMI consumed portions of many foods that were as much as 100% larger than those consumed by children with a lower BMI. Thus, these data suggest that an association between body weight and portion size starts at an early age.

C. Summary

The experimental evidence demonstrates that portion size has a significant effect on food intake not only in the short term, but for periods of up to 11 days. The effect of portion size has been demonstrated in both the laboratory and in more natural settings. Although there is little empirical evidence indicating a causal relationship between increased portion sizes and obesity, studies have shown that controlling portion sizes helps limit energy intake.

V. THE COMPLEX EATING ENVIRONMENT

There are likely a multitude of environmental factors that influence eating behaviors. For example, eating in a convivial atmosphere with friends may increase energy intake. Even though many environmental factors have not been well studied, there is a substantial body of literature indicating that properties of foods and beverages affect consumption. Studies have shown that increases in energy intake can result from increases in food and beverage variety, increases in food energy density, and increases in the portion size of foods and beverages. These effects have been demonstrated among individuals with a wide range of subject characteristics. In the current eating environment, snacks and meals vary simultaneously in variety, energy density, and portion size. Examining how these factors interact to affect energy intake is critical in developing practical strategies to aid in weight management.

A. Simultaneous Changes in Variety, Energy Density, and Portion Size

To date, studies investigating the combined influence of variety, energy density, and portion size have not been reported in the literature. However, there have been several studies investigating the combined effects of food energy density and portion size on energy intake.

The combined effects of food energy density and portion size on energy intake at a single meal were investigated in a controlled laboratory study reported by Kral *et al.* [32]. On six different days, study participants were served a lunch entrée at one of two levels of energy density in one of three portion sizes. Subjects consumed 56% more energy when served the largest portion of the high-energy-density entrée, compared to when they were served the smallest portion of the low-energy-density entrée. Despite the 56% difference in energy intake between these two entrées, participant ratings of hunger and fullness did not differ after the meals. Interestingly, there was no statistical interaction between energy density and portion size, which indicated that these two factors were acting independently to influence energy intake.

To determine whether the combined effect of energy density and portion size persists beyond a single meal, Rolls and colleagues [30] provided study participants with a variety of popular, commercially available foods over two consecutive days on four different occasions. The same menus were served during each 2-day session, but all foods were varied in both energy density and portion size between a standard level and a reduced level (75% of the standard). Again, the effects of energy density and portion size combined so that the participants consumed the greatest amount of energy when provided with the larger portions of the higher-energy-density foods, and the least amount of energy when provided with the reduced portions of the lower-energy-density foods (Fig. 9). Reducing the energy density of foods by 25% led to a 24% decrease in daily energy intake. Reducing the portion size of foods by 25% led to a 10% decrease in daily energy intake. Thus, when smaller portions of lower-energy-density foods were served, daily energy intake was 32% less on both days than when larger portions of higher-energy-density foods were served. As in the previously discussed study, there were no significant differences in the ratings of hunger and fullness across conditions over the 2 days of the study. This study shows that the effects of energy density and portion size on energy intake are additive and persist over at least 2 days without differentially affecting ratings of satiety. Additional studies are needed to determine whether these effects persist for longer periods.

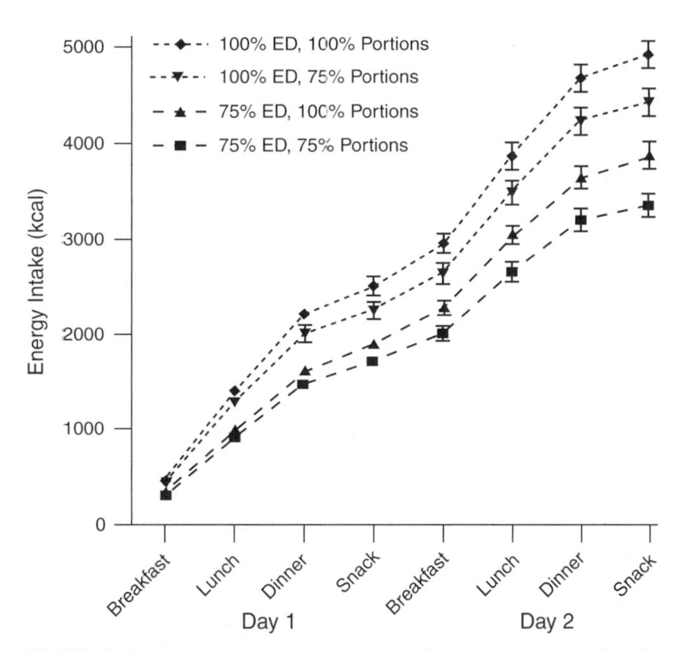

FIGURE 9 Mean cumulative energy intakes by meal for 24 women who were served 2-day menus that varied in portion size and energy density. Reducing the portion size and energy density of all foods led to significant and independent decreases in energy intake over 2 d. A 25% decrease in portion size led to a 10% decrease in energy intake (231 kcal/d), and a 25% decrease in energy density led to a 24% decrease in energy intake (575 kcal/d). Reproduced with permission from the *American Journal of Clinical Nutrition* [30].

B. Strategies for Managing Intake

The studies discussed in the previous section suggest that habitual exposure to large portions of energy-dense foods is likely to be problematic for managing energy consumption and perhaps body weight. It is interesting to note that the consumption of smaller portions of lower-energy-density foods was associated with lower energy consumption, without increases in hunger. This suggests that modest reductions in the energy density and portion size of meals and snacks may be an effective strategy to manage energy intake.

Two other studies suggest an additional strategy for using energy density and portion size to moderate energy intake at a meal: consuming a relatively large portion of a low-energy-density food as a first course [31, 67]. In one study, subjects were required to consume a first-course salad, which was varied in energy density and portion size on different days; this was followed by a main course of pasta consumed *ad libitum* [31]. Compared to having no first course, consuming a low-energy-density salad as a first course led to a decrease in total energy intake at the meal. This reduction in energy intake was greater when subjects consumed the larger rather than the smaller low-energy-density salad. Consuming either portion of the high-energy-density salad increased energy intake at the meal. The energy content of the low-energy-density salads, which helped reduce energy intake at the meal, was fairly low, less than 150 kcal. Consumption of a low-energy-density soup at the start of a meal has also been shown to reduce overall meal intake [67]. This indicates that when choosing a first course, overall energy intake can be reduced by selecting a large portion of a food low in energy density.

An additional strategy for managing energy intake is to consume a variety of foods with a low energy density. Studies have found that consuming a variety of high-energy-density foods was associated with adiposity [25, 26]. Conversely, consuming a greater variety of vegetables, which generally have a low energy density, was associated with less body fat [26] and weight loss [27]. Studies are needed that systematically investigate the combined influences of variety and energy density on energy intake; nevertheless, these initial findings are promising.

C. Moving from Research to Practice

As obesity rates continue to rise, it is important to understand the influence of environmental factors on energy intake and weight status. Studies have shown that increases in food and beverage variety, food energy density, and the portion size of foods and beverages can increase energy intake and may influence body weight. Changes in these factors are relatively unrecognized by consumers. Therefore, minor changes in variety, energy density, and portion size can be made to help moderate energy consumption and produce a more healthful eating environment. There remains a need for further research exploring the effect of simultaneous modifications of these and other environmental factors on food intake and body weight. Nevertheless, the available research suggests several practical approaches that may help create an environment in which consumers are better able to maintain a healthy weight.

One of the keys to consuming a healthful diet is to structure the environment so that a variety of palatable foods low in energy density is readily available. Data suggest that increasing the availability of fruits and vegetables will increase consumption of these items [83, 84]. Individual consumers will be more likely to make better dietary choices if they ensure that a variety of fruits, vegetables, and other low-energy-density foods are readily available at home. A recent systematic review of studies examining associations between fruit and vegetable availability and consumption suggests that small changes in availability may yield changes in consumption patterns that could provide important health benefits [85].

Restaurants, worksites, educational institutions, health care facilities, and faith-based organizations should ensure access to healthful foods and beverages when these items are sold or are made available for consumption. Minor modifications to foods could be made without affecting taste and acceptability. For example, the addition of water-rich vegetables along with a decrease in fat could reduce the energy density of many popular foods such as burgers, sandwiches, and pizza [86]. Providing foods in a range of portion sizes may make it easier for people to choose items that are appropriate for their energy needs. Pricing strategies could also be used to promote the selection of healthier options [87, 88].

In addition to changes in our food environment, it is important for consumers to recognize the influence of various environmental factors on their food choices. Ultimately, consumers must understand and accept the value to their health of food and beverage choices. Policy makers need to organize a well-funded campaign using sophisticated marketing and psychological techniques to help consumers equip themselves with the knowledge and skills to make the most appropriate food choices. Educational messages that encompass interrelationships between these environmental factors are likely to be more effective than messages focusing on a single aspect of our eating environment. Educational initiatives should encourage the consumption of a variety of foods low in energy density, such as fruits and vegetables, while promoting more moderate portions of foods high in energy density.

An imbalance in energy intake compared to expenditure is a primary cause of the recent obesity epidemic. This may be attributed in part to the ready availability of a wide variety of large portions of energy-dense foods. The portion size, energy density, and variety of foods served at a meal are readily modifiable factors that can be utilized in the development of effective dietary strategies for weight management as well as in the promotion of a healthier eating environment.

Acknowledgments

Supported by NIH grants R37DK039177 and R01DK059853.

References

1. U.S. Department of Health and Human Services, and U.S. Department of Agriculture. (2005). "Dietary guidelines for Americans 2005" (6th ed.). Washington, DC.
2. Rolls, B. J. (2000). Sensory-specific satiety and variety in the meal. *In* "Dimensions of the Meal" (H.L. Meiselman, Eds.), pp. 107–116. Aspen, Gaithersburg, MD.
3. Rolls, B. J., Van Duijvenvoorde, P. M., and Rolls, E. T. (1984). Pleasantness changes and food intake in a varied four-course meal. *Appetite* **5**, 337–348.
4. Rolls, B. J. (1986). Sensory-specific satiety. *Nutr. Rev.* **44**, 93–101.
5. Norton, G. N., Anderson, A. S., and Hetherington, M. M. (2006). Volume and variety: relative effects on food intake. *Physiol. Behav.* **87**, 714–722.
6. Rolls, B. J., Rowe, E. A., Rolls, E. T., Kingston, B., Megson, A., and Gunary, R. (1981). Variety in a meal enhances food intake in man. *Physiol. Behav.* **26**, 215–221.
7. Stubbs, R. J., Johnstone, A. M., Mazlan, N., Mbaiwa, S. E., and Ferris, S. (2001). Effect of altering the variety of sensorially distinct foods, of the same macronutrient content, on food intake and body weight in men. *Eur. J. Clin. Nutr.* **55**, 19–28.
8. Glanz, K., Basil, M., Maibach, E., Goldberg, J., and Snyder, D. (1998). Why Americans eat what they do: taste, nutrition, cost, convenience, and weight control as influences on food consumption. *J. Am. Diet. Assoc.* **98**, 1118–1126.
9. Sorensen, L. B., Moller, P., Flint, A., Martens, M., and Raben, A. (2003). Effect of sensory perception of foods on appetite and food intake: a review of studies on humans. *Int. J. Obes.* **27**, 1152–1166.
10. Rolls, B. J., Rowe, E. A., and Rolls, E. T. (1982). How sensory properties of foods affect human feeding behavior. *Physiol. Behav.* **29**, 409–417.
11. Bell, E. A., Roe, L. S., and Rolls, B. J. (2003). Sensory-specific satiety is affected more by volume than by energy content of a liquid food. *Physiol. Behav.* **78**, 593–600.
12. Miller, D. L., Bell, E. A., Pelkman, C. L., Peters, J. C., and Rolls, B. J. (2000). Effects of dietary fat, nutrition labels, and repeated consumption on sensory-specific satiety. *Physiol. Behav.* **71**, 153–158.
13. Rolls, B. J., Hetherington, M., and Burley, V. J. (1988). The specificity of satiety: the influence of foods of different macronutrient content on the development of satiety. *Physiol. Behav.* **43**, 145–153.
14. Johnson, J., and Vickers, Z. (1993). Effects of flavor and macronutrient composition of food servings on liking, hunger and subsequent intake. *Appetite* **21**, 25–39.
15. Rolls, B. J., Rolls, E. T., Rowe, E. A., and Sweeney, K. (1981). Sensory specific satiety in man. *Physiol. Behav.* **27**, 137–142.
16. Engell, D., and Hirsch, E. (1991). Environmental and sensory modulation of fluid intake in humans. *In* "Thirst – Physiological and Psychological Aspects" (D.J. Ramsay and D.A. Booth, Eds.), pp. 382–390. Springer-Verlag, London.
17. Tuorila, H. (1991). Individual and cultural factors in the consumption of beverages. *In* "Thirst – Physiological and Psychological Aspects" (D.J. Ramsay and D.A. Booth, Eds.), pp. 354–364. Springer-Verlag, London.
18. Rolls, B. J. (1998). Homeostatic and non-homeostatic controls of drinking in humans. *In* "Hydration Throughout Life" (M.J. Arnaud, Eds.), pp.19–28. John Libbey Eurotext.
19. Bell, E. A., Roe, L. S., and Rolls, B. J. (2003). Sensory-specific satiety is affected more by volume than by energy content of a liquid food. *Physiol. Behav.* **78**, 593–600.
20. Rolls, B. J., Wood, R. J., and Rolls, E. T. (1980). Thirst: The initiation, maintenance, and termination of drinking. *Prog. Psycho. Physiol.* **9**, 263–321.
21. Engell, D. B., Maller, O., Sawka, M. N., Francesconi, R. N., Drolet, L., and Young, A. J. (1987). Thirst and fluid intake following graded hypohydration levels in humans. *Physiol. Behav.* **40**, 229–236.
22. Foote, J. A., Murphy, S. P., Wilkens, L. R., Basiotis, P. P., and Carlson, A. (2004). Dietary variety increases the probability of nutrient adequacy among adults. *J. Nutr.* **134**, 1779–1785.
23. Raynor, H. A., Jeffery, R. W., Phelan, S., Hill, J. O., and Wing, R. R. (2005). Amount of food group variety consumed in the diet and long-term weight loss maintenance. *Obes. Res.* **13**, 883–890.
24. Sea, M. M., Woo, J., Tong, P. C., Chow, C. C., and Chan, J. C. (2004). Associations between food variety and body fatness in Hong Kong Chinese adults. *J. Am. Coll. Nutr.* **23**, 404–413.
25. Roberts, S. B., Hajduk, C. L., Howarth, N. C., Russell, R., and McCrory, M. A. (2005). Dietary variety predicts low body mass index and inadequate macronutrient and micronutrient intakes in community-dwelling older adults. *J. Gerontol.* **60**, 613–621.
26. McCrory, M. A., Fuss, P. J., McCallum, J. E., Yao, M., Vinken, A. G., Hays, N. P., *et al.* (1999). Dietary variety within food groups: association with energy intake and body fatness in men and woman. *Am. J. Clin. Nutr.* **69**, 440–447.
27. Raynor, H. A., Jeffery, R. W., Tate, D. F., and Wing, R. R. (2004). Relationship between changes in food group variety, dietary intake, and weight during obesity treatment. *Int. J. Obes.* **28**, 813–820.
28. Miller, D. L., Castellanos, V. H., Shide, D. J., Peters, J. C., and Rolls, B. J. (1998). Effect of fat-free potato chips with and without nutrition labels on fat and energy intakes. *Am. J. Clin. Nutr.* **68**, 282–290.
29. Bell, E. A., Castellanos, V. H., Pelkman, C. L., Thorwart, M. L., and Rolls, B. J. (1998). Energy density of foods affects energy intake in normal-weight women. *Am. J. Clin. Nutr.* **67**, 412–420.
30. Rolls, B. J., Roe, L. S., and Meengs, J. S. (2006). Reductions in portion size and energy density of foods are additive and lead to sustained decreases in energy intake. *Am. J. Clin. Nutr.* **83**, 11–17.
31. Rolls, B. J., Roe, L. S., and Meengs, J. S. (2004). Salad and satiety: energy density and portion size of a first course salad

affect energy intake at lunch. *J. Am. Diet. Assoc.* **104**, 1570–1576.

32. Kral, T. V. E., Roe, L. S., and Rolls, B. J. (2004). Combined effects of energy density and portion size on energy intake in women. *Am. J. Clin. Nutr.* **79**, 962–968.
33. Rolls, B. J., Pirraglia, P. A., Jones, M. B., and Peters, J. C. (1992). Effects of olestra, a noncaloric fat substitute, on daily energy and fat intakes in lean men. *Am. J. Clin. Nutr.* **56**, 84–92.
34. Cotton, J. R., Burley, V. J., Weststrate, J. A., and Blundell, J. E. (1996). Fat substitution and food intake: effect of replacing fat with sucrose polyester at lunch or evening meals. *British J. Nutr.* **75**, 545–556.
35. Hill, J. O., Seagle, H. M., Johnson, S. L., Smith, S., Reed, G. W., Tran, Z. V., *et al.* (1998). Effects of 14 d of covert substitution of olestra for conventional fat on spontaneous food intake. *Am. J. Clin. Nutr.* **67**, 1178–1185.
36. Roy, H. J., Most, M. M., Sparti, A., Lovejoy, J. C., Volaufova, J., Peters, J. C., *et al.* (2002). Effect on body weight of replacing dietary fat with olestra for two or ten weeks in healthy men and women. *J. Am. Coll. Nutr.* **21**, 259–267.
37. Bray, G. A., Lovejoy, J. C., Most-Windhauser, M., Smith, S. R., Volaufova, J., Denkins, Y., *et al.* (2002). A 9-mo randomized clinical trial comparing fat-substituted and fat-reduced diets in healthy obese men: the Ole Study. *Am. J. Clin. Nutr.* **76**, 928–934.
38. Duncan, K. H., Bacon, J. A., and Weinsier, R. L. (1983). The effects of high and low energy density diets on satiety, energy intake, and eating time of obese and nonobese subjects. *Am. J. Clin. Nutr.* **37**, 763–767.
39. Lissner, L., Levitsky, D. A., Strupp, B. J., Kalkwarf, H. J., and Roe, D. A. (1987). Dietary fat and the regulation of energy intake in human subjects. *Am. J. Clin. Nutr.* **46**, 886–892.
40. Kendall, A., Levitsky, D. A., Strupp, B. J., and Lissner, L. (1991). Weight loss on a low-fat diet: consequence of the imprecision of the control of food intake in humans. *Am. J. Clin. Nutr.* **53**, 1124–1129.
41. Stubbs, R. J., Ritz, P., Coward, W. A., and Prentice, A. M. (1995). Covert manipulation of the ratio of dietary fat to carbohydrate and energy density: effect on food intake and energy balance in free-living men eating *ad libitum*. *Am. J. Clin. Nutr.* **62**, 330–337.
42. van Stratum, P., Lussenburg, R. N., van Wezel, L. A., Vergroesen, A. J., and Cremer, H. D. (1978). The effect of dietary carbohydrate:fat ratio on energy intake by adult women. *Am. J. Clin. Nutr.* **31**, 206–212.
43. Saltzman, E., Dallal, G. E., and Roberts, S. B. (1997). Effect of high-fat and low-fat diets on voluntary energy intake and substrate oxidation: studies in identical twins consuming diets matched for energy density, fiber and palatability. *Am. J. Clin. Nutr.* **66**, 1332–1339.
44. Stubbs, R. J., Harbron, C. G., and Prentice, A. M. (1996). Covert manipulation of the dietary fat to carbohydrate ratio of isoenergetically dense diets: effect on food intake in feeding men ad libitum. *Int. J. Obes.* **20**, 651–660.
45. Rolls, B. J., Bell, E. A., Castellanos, V. H., Chow, M., Pelkman, C. L., and Thorwart, M. L. (1999). Energy density but not fat content of foods affected energy intake in lean and obese women. *Am. J. Clin. Nutr.* **69**, 863–871.
46. Ledikwe, J. H., Blanck, H. M., Khan, L. K., Serdula, M. K., Seymour, J. D., Tohill, B. C., *et al.* (2006). Low-energy-density diets are associated with high diet quality in adults in the United States. *J. Am. Diet. Assoc.* **106**, 1172–1180.
47. Ledikwe, J. H., Blanck, H. M., Kettel Khan, L., Serdula, M. K., Seymour, J. D., Tohill, B. C., *et al.* (2006). Dietary energy density is associated with energy intake and weight status in U.S. adults. *Am. J. Clin. Nutr.* **83**, 1362–1368.
48. Cuco, G., Arija, V., Marti-Henneberg, C., and Fernandez-Ballart, J. (2001). Food and nutritional profile of high energy density consumers in an adult Mediterranean population. *Eur. J. Clin. Nutr.* **55**, 192–199.
49. Stookey, J. D. (2001). Energy density, energy intake and weight status in a large free-living sample of Chinese adults: Exploring the underlying roles of fat, protein, carbohydrate, fiber and water intakes. *Eur. J. Clin. Nutr.* **55**, 349–359.
50. Drewnowski, A., Almiron-Roig, E., Marmonier, C., and Lluch, A. (2004). Dietary energy density and body weight: Is there a relationship? *Nutr. Rev.* **62**, 403–413.
51. DellaValle, D. M., Roe, L. S., and Rolls, B. J. (2005). Does the consumption of caloric and non-caloric beverages with a meal affect energy intake? *Appetite* **44**, 187–193.
52. Rolls, B. J., Kim, S., and Fedoroff, I. C. (1990). Effects of drinks sweetened with sucrose or aspartame on hunger, thirst and food intake in men. *Physiol. Behav.* **48**, 19–26.
53. Almiron-Roig, E., and Drewnowski, A. (2003). Hunger, thirst, and energy intakes following consumption of caloric beverages. *Physiol. Behav.* **79**, 767–773.
54. Flood, J. E., Roe, L. S., and Rolls, B. J. (2006). The effect of increased beverage portion size on energy intake at a meal. *J. Am. Diet. Assoc.* **106**, 1984–1990.
55. Cuntz, U., Leibbrand, R., Ehrig, C., Shaw, R., and Fichter, M. M. (2001). Predictors of post-treatment weight reduction after in-patient behavioral therapy. *Int. J. Obes.* S99–S101.
56. Pasman, W. J., Saris, W. H., and Westerterp-Plantenga, M. S. (1999). Predictors of weight maintenance. *Obes. Res.* **7**, 43–50.
57. Elfhag, K., and Rossner, S. (2005). Who succeeds in maintaining weight loss? A conceptual review of factors associated with weight loss maintenance and weight regain. *Obes. Rev.* **6**, 67–85.
58. Shintani, T. T., Beckham, S., Brown, A. C., and O'Connor, H. K. (2001). The Hawaii Diet: *ad libitum* high carbohydrate, low fat multi-cultural diet for the reduction of chronic disease risk factors: obesity, hypertension, hypercholesterolemia, and hyperglycemia. *Hawaii Med. J.* **60**, 69–73.
59. Shintani, T. T., Hughes, C. K., Beckham, S., and O'Connor, H. K. (1991). Obesity and cardiovascular risk intervention through the *ad libitum* feeding of traditional Hawaiian diet. *Am. J. Clin. Nutr.* **53**, 1647S–1651S.
60. Rolls, B. J., Roe, L. S., Beach, A. M., and Kris-Etherton, P. M. (2005). Provision of foods differing in energy density affects long-term weight loss. *Obes. Res.* **13**, 1052–1060.
61. Ello-Martin, J. A., Roe, L. S., Ledikwe, J. H., Beach, A. M., and Rolls, B. J. (2007). Dietary energy density in the treatment of obesity: A year-long trial comparing two weight-loss diets. *Am. J. Clin. Nutr.* **85**, 1465–1477.
62. Ledikwe, J. H., Rolls, B. J., Smiciklas-Wright, H., Mitchell, D. C., Ard, J. D., Champagne, C., *et al.* (2007). Reductions in

dietary energy density are associated with weight loss in overweight and obese participants in the PREMIER trial. *Am. J. Clin. Nutr.* **85**, 1212–1221.
63. Greene, L. F., Malpede, C. Z., Henson, C. S., Hubbert, K. A., Heimburger, D. C., and Ard, J. D. (2006). Weight maintenance 2 years after participation in a weight loss program promoting low-energy density foods. *Obesity (Silver Spring)*, **14**, 1795–1801.
64. Tordoff, M. G., and Alleva, A. M. (1990). Effect of drinking soda sweetened with aspartame or high-fructose corn syrup on food intake and body weight. *Am. J. Clin. Nutr.* **51**, 963–969.
65. Raben, A., Vasilaras, T. H., Moller, A. C., and Astrup, A. (2002). Sucrose compared with artificial sweeteners: different effects on *ad libitum* food intake and body weight after 10 wk of supplementation in overweight subjects. *Am. J. Clin. Nutr.* **76**, 721–729.
66. Van Wymelbeke, V., Beridot-Therond, M. E., de La Gueronniere, V., and Fantino, M. (2004). Influence of repeated consumption of beverages containing sucrose or intense sweeteners on food intake. *Eur. J. Clin. Nutr.* **58**, 154–161.
67. Rolls, B. J., Bell, E. A., and Thorwart, M. L. (1999). Water incorporated into a food but not served with a food decreases energy intake in lean women. *Am. J. Clin. Nutr.* **70**, 448–455.
68. Rolls, B. J., and Barnett, R. A. (2003). "The Volumetrics Weight-Control Plan: feel full on fewer calories." HarperTorch, New York.
69. Rolls, B. J. (2007). "The Volumetrics eating plan." HarperCollins, New York.
70. Young, L. R., and Nestle, M. (2002). The contribution of expanding portion sizes to the U.S. obesity epidemic. *Am. J. Public Health* **92**, 246–249.
71. Young, L. R., and Nestle, M. (2003). Expanding portion sizes in the U.S. marketplace: implications for nutrition counseling. *J. Am. Diet. Assoc.* **103**, 231–234.
72. Nielsen, S. J., and Popkin, B. M. (2003). Patterns and trends in food portion sizes, 1977–1998. *JAMA* **289**, 450–453.
73. Smiciklas-Wright, H., Mitchell, D. C., Mickle, S. J., Goldman, J. D., and Cook, A. (2003). Foods commonly eaten in the United States, 1989–1991 and 1994–1996: are the portion sizes changing? *J. Am. Diet. Assoc.* **103**, 41–47.
74. Ledikwe, J. H., Ello-Martin, J. A., and Rolls, B. J. (2005). Portion sizes and the obesity epidemic. *J. Nutr.* **135**, 905–909.
75. Rolls, B. J., Morris, E. L., and Roe, L. S. (2002). Portion size of food affects energy intake in normal-weight and overweight men and women. *Am. J. Clin. Nutr.* **76**, 1207–1213.
76. Rolls, B. J., Roe, L. S., Meengs, J. S., and Wall, D. E. (2004). Increasing the portion size of a sandwich increases energy intake. *J. Am. Diet. Assoc.* **104**, 367–372.
77. Diliberti, N., Bordi, P., Conklin, M. T., Roe, L. S., and Rolls, B. J. (2004). Increased portion size leads to increased energy intake in a restaurant meal. *Obes. Res.* **12**, 562–568.
78. Wansink, B., and Park, S. B. (2001). At the movies: how external cues and perceived taste impact consumption volume. *Food Qual. Prefer.* **12**, 69–74.
79. Rolls, B. J., Roe, L. S., Kral, T. V. E., Meengs, J. S., and Wall, D. E. (2004). Increasing the portion size of a packaged snack increases energy intake in men and women. *Appetite* **42**, 63–69.
80. Rolls, B. J., Roe, L. S., and Meengs, J. S. (2006). Larger portion sizes lead to a sustained increase in energy intake over 2 days. *J. Am. Diet. Assoc.* **106**, 543–549.
81. Rolls, B. J., Roe, L. S., and Meengs, J. S. (2007). The effect of large portion sizes on energy intake is sustained for 11 days. *Obesity.* **15**, 1535–1543.
82. McConahy, K. L., Smiciklas-Wright, H., Birch, L. L., Mitchell, D. C., and Picciano, M. F. (2002). Food portions are positively related to energy intake and body weight in early childhood. *J. Pediatr.* **140**, 340–347.
83. Cullen, K. W., Baranowski, T., Owens, E., Marsh, T., Rittenberry, L., and de Moor, C. (2003). Availability, accessibility, and preferences for fruit, 100% fruit juice, and vegetables influence children's dietary behavior. *Health Educ. Behav.* **30**, 615–626.
84. Reinaerts, E., de Nooijer, J., Candel, M., and de Vries, N. (2007). Explaining school children's fruit and vegetable consumption: the contributions of availability, accessibility, exposure, parental consumption and habit in addition to psychosocial factors. *Appetite* **48**, 248–258.
85. Jago, R., Baranowski, T., and Baranowski, J. C. (2007). Fruit and vegetable availability: a micro environmental mediating variable? *Public Health Nutr.* 1–9.
86. Rolls, B. J., Drewnowski, A., and Ledikwe, J. H. (2005). Changing the energy density of the diet as a strategy for weight management. *J. Am. Diet. Assoc.* **105**, 98–103.
87. French, S. A. (2003). Pricing effects on food choices. *J. Nutr.* **133**, 841S–843S.
88. Horgen, K. B., and Brownell, K. D. (2002). Comparison of price change and health message interventions in promoting healthy food choices. *Health Psychol.* **21**, 505–512.

C. Cardiovascular Disease

CHAPTER **28**

Genetic Influences on Blood Lipids and Cardiovascular Disease Risk

JOSE M. ORDOVAS
USDA-HNRCA, Tufts University, Boston, Massachusetts

Contents

I. INTRODUCTION

The major public health concerns in the developed world (i.e., cardiovascular disease, cancer, and diabetes) have both genetic and environmental causes. The interface between public health and genetics consists of working toward an understanding of how genes and the environment act together to cause these diseases and how the environment (e.g., diet), rather than genes, might be manipulated to help prevent or delay the onset of disease.

Cardiovascular disease (CVD), the leading cause of mortality in most industrialized countries, is a multifactorial disease that is associated with nonmodifiable risk factors, such as age, gender, and genetic background, and with modifiable risk factors, including elevated total and low-density lipoprotein (LDL) cholesterol levels, as well as reduced high-density lipoprotein (HDL) cholesterol levels. Heritability estimates for blood lipids are high, including approximately 40–60% for high-density lipoprotein cholesterol (HDL-C), approximately 40–50% for low-density lipoprotein cholesterol (LDL-C), and approximately 35–48% for triglycerides (TG) [1].

Traditionally, the major emphasis has been on lowering the concentration of serum cholesterol, which has long been recognized as an important risk factor for the development and progression of atherosclerotic vascular disease; multiple clinical studies for the last 30 years have demonstrated the benefits of this approach. In view of these findings, the National Heart, Lung, and Blood Institute (NHLBI) convened the National Cholesterol Education Program-Adult Treatment Panel I (NCEP ATP I) [2]. This panel and similar ones around the world have served to set the standards for lipid lowering in clinical practice. Subsequent revisions of these standards (NCEP ATP II and III) [3, 4] have led to greater focus being placed on LDL, and targets for lowering LDL levels being based on patients' risk of subsequent coronary disease events. Since the publication of the NCEP ATP III guidelines, several large-scale clinical trials of cholesterol lowering have been conducted, the findings of which have the potential to further refine current clinical practice standards [5, 6]. Although most of the beneficial evidence from lowering serum LDL cholesterol values in reducing CVD morbidity and mortality comes from pharmacological interventions, the NCEP has emphasized that therapeutic lifestyle change (TLC) should be the primary treatment for lowering cholesterol values, with drug therapies reserved for cases in which lifestyle modification is ineffective. The modifications advocated include dietary changes, increased physical activity, and weight management. The recommended dietary changes include restriction in the amount of saturated fat to less than 7% of calories and cholesterol to less than 200 mg/day; increased viscous fiber (10–24 g/day); and plant stanols/sterols (2 g/day) to enhance LDL lowering [4]. However, it is not known how many individuals can achieve the recommended levels of serum lipids using this approach, one of the major reasons being our current inability to predict individual plasma lipid response to dietary changes [7].

In addition to cholesterol lowering, other pharmacological approaches to CVD risk reduction have been investigated. These include the lowering of serum TG and the raising of HDL-C concentrations. The most commonly used drugs to lower TG concentrations are the fibrates. These drugs also raise HDL-C and improve small dense LDL, so would be expected to have large beneficial effects. Their use in the management of lipoprotein disorders has a history dating back to the mid-1960s. However, their prominence has lessened over the years because of unimpressive results in major clinical trials, safety concerns, and the emergence of 3-hydroxy-3-methylglutaryl coenzyme A reductase inhibitors (statin drugs). Moreover, the general trial results with these agents have been confusing, with varying cardiovascular benefits [8–10].

The epidemiologic, experimental, and other circumstantial evidence implicating low HDL-C concentrations as a major CVD risk has driven considerable attention to this lipoprotein fraction as a potential therapeutic target to further reduce CVD risk beyond what can be achieved with the use of statin drugs. However, doubts on the clinical benefit achievable with treatments enhancing plasma HDL-C levels have been raised by the premature termination of a large phase III trial with torcetrapib, the most potent and furthest developed HDL-C raising compound, because of excess mortality in patients receiving the drug [11]. The causes of torcetrapib failure are presently unknown and may be related to the drug mode of action, to off-target toxic effects of the drug, or to a mixture of these. Torcetrapib failure does not mean that the concept of targeting HDL in CVD prevention is definitely dead. Other HDL-C raising therapies, which act through disparate molecular mechanisms, are in various stages of preclinical and clinical development.

The considerable interindividual variation observed following these therapies, in terms of lipid response, cardiovascular event response, and adverse events is bringing considerable attention to the concept of more targeted therapies based on genetic information. Pharmacogenomics (pharmacogenetics) involves the search and identification of genetic variants that influence response to drug therapy. Over the past decade, some progress has been made in our understanding of the variability associated with statin therapy [12]. Important limitations and issues have been raised, however, which need to be resolved before its clinical application.

Regarding the connection between diet and plasma cholesterol concentrations, several studies demonstrated, during the first half of the twentieth century, that serum cholesterol could be modified by the composition of dietary fat [13, 14], and studies by Keys *et al.* [15] and Hegsted *et al.* [16] provided the first quantitative estimates of the relative effects of the various classes of dietary fatty acids and amount of cholesterol on serum cholesterol changes. Other predictive algorithms have been developed during the ensuing years, including predictions of response for LDL-C and HDL-C [17–19]. These relationships between dietary changes and serum lipid changes are well founded and predictable for groups; however, a striking variability in the response of serum cholesterol to diet between subjects was reported as early as 1933 [20], and this variability has been the subject of multiple reports. In some individuals, plasma total and LDL-C levels dramatically decrease following consumption of lipid-lowering diets, whereas they remain unchanged in others [18, 21–25]. More recently, multiple studies in experimental animal models and in humans have shown that the serum lipoprotein response to dietary manipulation has a significant genetic component [26–31]. Such genetic variability could have a significant impact on the success of public health policies and individual therapeutic interventions. Moreover, it could be partially responsible for the apparent lack of hard end point benefits shown by many dietary studies aimed to decrease CVD [32, 33].

As indicated earlier, we have traditionally measured the success of CVD risk-reducing strategies based on their effect on plasma lipids and, more specifically, lipoprotein levels. Lipoproteins are macromolecular complexes of lipids and proteins that originate mainly from the liver and intestine and are involved in the transport and redistribution of lipids in the body. Lipid and lipoprotein metabolism can be viewed as a complex biological pathway containing multiple steps. Lipid homeostasis is achieved by the coordinated action of a large number of nuclear factors, binding proteins, apolipoproteins, enzymes, and receptors involving hundreds of genes. Lipid metabolism is also closely linked with energy metabolism and is subjected to many hormonal controls that are essential for adjustment to environmental and internal conditions. Genetic variability at candidate genes involved in lipoprotein metabolism has been profusely associated with abnormal lipid metabolism and plasma lipoprotein profiles that may contribute to the pathogenesis of atherosclerosis. This complex picture can be dissected into three pathways (Fig. 1). The exogenous lipoprotein pathway describes the metabolism of lipoproteins synthesized in the intestine following dietary fat intake; the endogenous pathway deals with the metabolism of lipoproteins involved in the transport of liver lipids to peripheral tissues; and finally the reverse cholesterol transport depicts the process by which excess peripheral lipids (primarily cholesterol) are transported to the liver for catabolism. Our current knowledge about how variants at candidate genes involved in each of these three interrelated processes affect dietary response and cardiovascular risk is presented next. Therefore, only those candidate genes with reported gene–diet interactions are discussed in more detail.

II. EXOGENOUS LIPOPROTEIN PATHWAY

The exogenous lipoprotein pathway has its origin in the enterocyte with the synthesis of chylomicron particles. Dietary fats absorbed in the intestine are packaged into large, triglyceride-rich chylomicrons for delivery to sites of lipid metabolism or storage. During their transit to the liver, these particles interact with lipoprotein lipase (LPL) and undergo partial lipolysis to form chylomicron remnants. These chylomicron remnants pick up

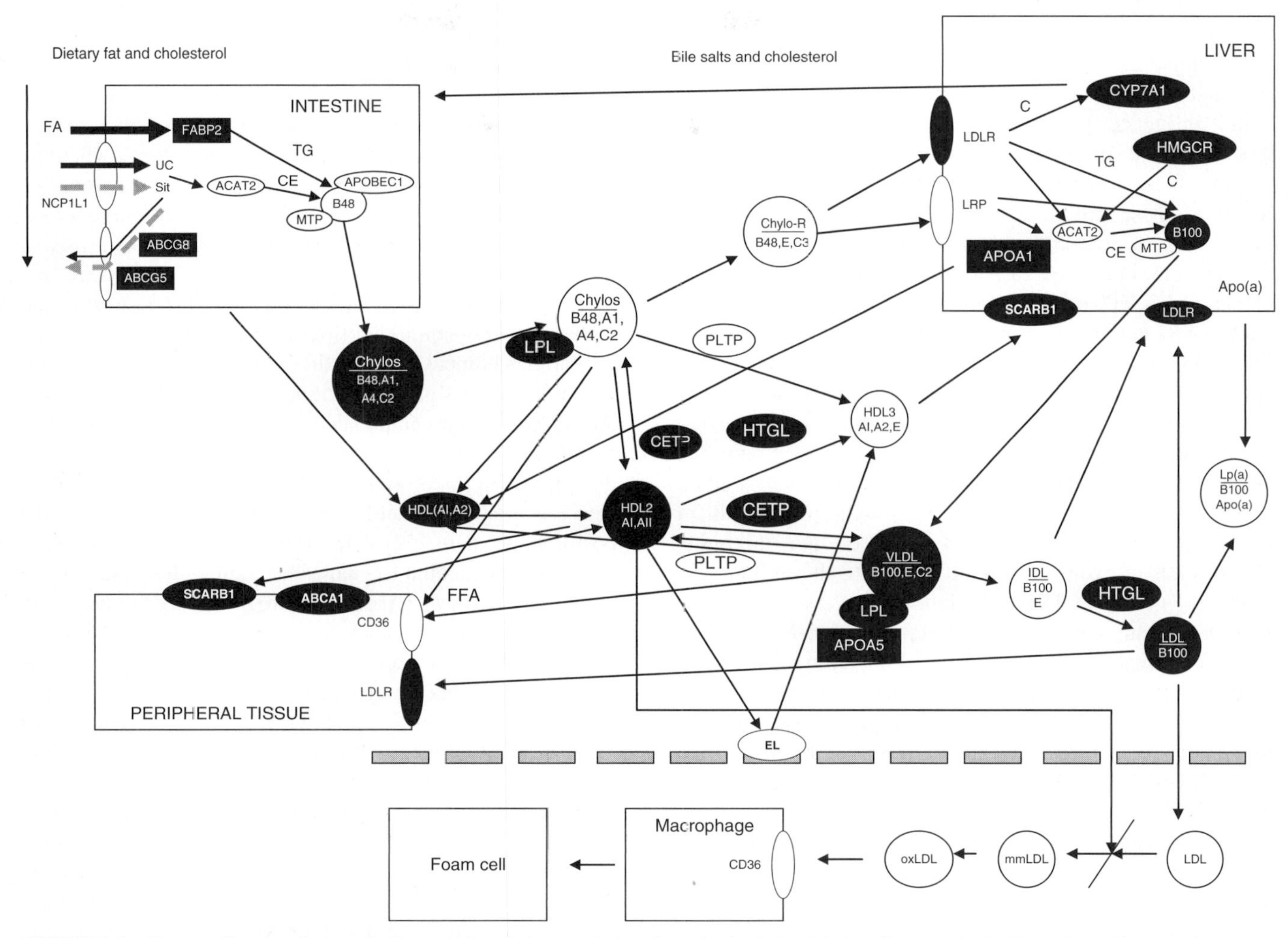

FIGURE 1 Human lipoprotein metabolism. Abbreviations: A1 (apolipoprotein A-I); A2 (apolipoprotein A-II); A4 (apolipoprotein A-IV); A5 (apolipoprotein A-V); ABCA1 (ATP-binding cassette, subfamily A, member 1); ABCG5 (ATP-binding cassette, subfamily G, member 5); ABCG8 (ATP-binding cassette, subfamily G, member 8); ACAT2 (cytosolic acetoacetyl-CoA thiolase); apo (apolipoprotein); APOBEC1 (apolipoprotein B mRNA editing enzyme); B100 (apolipoprotein B-100); B48 (apolipoprotein B-48); C2 (apolipoprotein C-II); C3 (apolipoprotein C-III); CE (cholesteryl esters); CD36 (CD36 antigen) CETP (cholesteryl ester transfer protein); Chylos (chylomicron); Chilo-R (chylomicron remnants); CYP7A1 (cholesterol-7-alpha-hydroxylase); E (apolipoprotein E); FA (fatty acids); FABP2 (intestinal fatty acid binding protein); FFA (free fatty acids); LDL (low-density lipoprotein); HMGCR (3-hydroxy-3-methylglutaryl-CoA reductase); LDLR (LDL receptor); LPL (lipoprotein lipase); NPC1L1 (Niemann-Pick C1-like 1); TG (triglycerides); HDL (high-density lipoprotein); HTGL (hepatic triglyceride lipase); IDL (intermediate density lipoproteins); mmLDL (minimally modified LDL); Lp(a) (lipoprotein (a)); MTP (microsomal triglyceride transfer protein); oxLDL (oxidized LDL); PLTP (phospholipid transfer protein); VLDL (very-low-density lipoprotein); SCARB1 (scavenger receptor type I B); Sit (sitosterol); UC (unesterified cholesterol). Those components of the lipoprotein pathway highlighted in the chapter are indicated with white font over black background. (Adapted from [30].)

apolipoprotein E (APOE) and cholesteryl ester from HDL and are taken up by the liver via a process mediated by the interaction of APOE with hepatic receptors. For most people, this is a fast process and chylomicrons are not usually present in the blood after a prolonged fasting period. However, dramatic individual variability is seen in postprandial lipoprotein metabolism, which is in part determined by genetic factors, and this variability could be highly relevant to achieving a more precise definition of individualized CVD risk [34]. The most relevant candidate genes involved in this metabolic pathway and their known associations with plasma lipid levels and dietary response are described next.

A. Apolipoprotein B (APOB)

APOB is an essential protein component of intestinal chylomicron particles. In humans, the form of APOB synthesized by the intestine is APOB-48, produced by an mRNA editing mechanism [35]. However, some APOB-100 synthesis may occur also in the intestine [36].

The *APOB* gene is on chromosome 2p24–p23. Because APOB is the major protein of chylomicrons, as well as of LDL and very low-density lipoprotein (VLDL), it is reasonable to expect that genetic variation at this locus could influence plasma cholesterol and/or triglyceride levels. A total of 284 polymorphisms have been reported in this locus (http://www.ncbi.nlm.nih.gov/sites/entrez). Some of these polymorphic sites have been utilized as markers in population or case-control studies in an attempt to correlate individual alleles or haplotypes with lipid levels and CVD risk. In general, the outcome of these studies has not been unanimous [37]; however, some of the polymorphisms have been functionally implicated in severe forms of familial hypercholesterolemia and hypobetalipoproteinemia [38].

A silent mutation causing a cytosine to thymidine (ACC→ACT) change characterizes the well-studied *APOB XbaI* restriction fragment length polymorphism (RFLP). It involves the third base of the threonine codon 2488 in exon 26 without changing the amino acid sequence [39–43]. The association between this polymorphism and variability in dietary response has been studied by several investigators and the results are quite controversial. Some reports show that subjects carrying the X+X+ or X+X– genotypes respond to a low-fat, low-cholesterol diet with greater reductions in plasma total cholesterol, LDL-C, APOB [44, 45], and, surprisingly, HDL-C levels than subjects with the X–X– genotype [44]. However, a meta-analysis suggests that this polymorphism plays a minor role in determining individual variability in response to dietary intervention [46]. In addition, the *APOB XbaI* polymorphism has been investigated for potential associations with the interindividual variability observed during postprandial lipemia [47, 48]. We carried out an oral fat load study involving 51 healthy young male volunteers [20 X–/X– (X–), and 31 X+/X– or X+/X+(X+)], homozygous for the apolipoprotein E APOE3 allele [48]. Our data showed that subjects with the X– genotype had significantly greater retinyl palmitate (RP) and APOB-48 postprandial responses on both the large and the small triglyceride-rich lipoprotein (TRL) fractions compared with X+ subjects, suggesting that the X–/X– genotype is associated with greater postprandial response as compared with the X–/X+ or the X+/X+ genotypes. These differences observed in postprandial lipoprotein metabolism could explain some reported associations of this polymorphism to coronary artery disease risk. As indicated earlier, this mutation does not result in an amino acid change at the affected codon and it cannot have a direct functional effect. However, it is in strong linkage disequilibrium with the *APOB* Val591→Ala polymorphism (Ag al/d), which may be the functional sequence change. An alternative hypothesis relates to the position of this mutation near the APOB-48 editing site at position 2153.

Another potentially interesting polymorphism is the three-codon (leu-ala-leu) I/D polymorphism within the APOB signal peptide [49]. Xu *et al.* [50] reported that subjects with the I/I genotype had the highest TG levels and D/D subjects had the lowest, while consuming a high-fat, high-cholesterol diet. This effect disappeared when the subjects were consuming a low-fat, low-cholesterol diet. These results were not confirmed by Boerwinkle *et al.* [51] in a study in which subjects received two levels of dietary cholesterol without modification of dietary fat. In a more recent study, the D allele was found to be associated with reduced postprandial lipid response as compared with individuals homozygous for the I allele, suggesting that this mutation in the signal peptide may affect APOB secretion during the postprandial state [52]. Moreover, this polymorphism has been reported to be involved with postprandial lipoparticle responses [53] and with the association between free fatty acid concentrations and TRL in the postprandial state [52]. The *Msp*I polymorphism in exon 26 causes an adenine to guanosine change that results in an amino acid change (Arg^{3611}→Gln) [54]. We found a significant association between the less common allele (M2) and the presence of premature CVD [43], with this allele being nearly twice as frequent in CVD (0.105) as in the control population (0.057). However, no associations of this allele with alterations in plasma APOB or LDL cholesterol levels in subjects with coronary artery disease were noted [43]. The association between this RFLP and variability in dietary response has been examined and the results of a recent meta-analysis suggest that this polymorphism may play a minor role as a determinant of dietary response [46].

A comprehensive analysis of these polymorphisms and the Bsp 1261I RFLP in exon 4 (the second base in codon 71, causing a cytosine to thymidine change and changing threonine to isoleucine) was reported by Rantala *et al.* [46]. Moreover, the authors conducted a meta-analysis on the available data. A controlled dietary intervention study was conducted in 44 healthy, middle-aged subjects, consisting of a 3-month baseline, a 1-month fat-controlled, a 1-month high-fat, and a 1-month habitual diet period. In this dietary study, the *APOB XbaI* restriction-site polymorphism affected the responsiveness to diet of the plasma LDL-C concentration, especially during the high-fat diet. The X–/X– subjects had a greater increase in LDL-C ($44 \pm 5\%$) than X+/X+ ($27 \pm 7\%$) or X+/X– ($40 \pm 5\%$) subjects. The high-fat diet also induced a larger increase in plasma LDL-C in subjects with the R–/R– genotype ($59 \pm 10\%$) than in those with the R+/R– ($39 \pm 6\%$) or R+/R+ ($36 \pm 4\%$) genotypes. M+/M+ subjects were also more responsive ($41 \pm 3\%$ increase in LDL-C) than the M+/M– subjects ($27 \pm 10\%$ increase). Their meta-analysis supported the finding of the significant role of the *Eco*RI and *Msp*I polymorphisms, but not that of the *Xba*I polymorphism [46]. The results of the meta-analysis support the association between the *APOB Eco* RI and *Msp*I genotypes and responsiveness to diet. However, the percent variability in response due to this locus is very small and this information does not yet provide enough information to be used as a clinical tool in dietary counseling.

The availability of the dbSNP and the HAPMAP databases has dramatically increased the number of genetic polymorphisms available for study at each gene locus. In the case of APOB, recent research has focused on several other single-nucleotide polymorphisms (SNPs). Thus, a common functional polymorphism in the promoter of the *APOB* gene, consisting of a C to T change at position –516, increases the transcription rate of APOB, resulting in elevated circulating levels of LDL-C [55], and it is independently associated with the presence of carotid atherosclerotic disease [56]. Our own data show that this polymorphism is associated with differential lipid postprandial response and insulin sensitivity in healthy males during the consumption of diets with different fat content, but it does not have a significant effect on the lipid profile [57–59].

B. Apolipoprotein A-IV (APOA4)

In humans, APOA4 is synthesized primarily in the intestine as a 46-kDa glycoprotein. The experimental evidence from familial apolipoprotein A1/C3/A4 deficiency [60] suggests that it plays a role in dietary fat absorption and chylomicron synthesis. *In vitro* studies have shown that the activation of lipoprotein lipase by apolipoprotein C-II (APOC2) is mediated by APOA4 [61] and that APOA4 can serve as an activator of lecithin cholesterol acyltransferase (LCAT) [62]. APOA4-containing lipoproteins promote cholesterol efflux from cultured fibroblasts and adipose cells *in vitro*, and there is evidence that APOA4 may be one of the ligands for the HDL receptor [63, 64]. Overall, the data suggest that APOA4 plays a role in fat absorption and reverse cholesterol transport.

Multiple genetic variants have been examined within the *APOA4* locus in relation to several lipid phenotypes as well as association with disease risk. However, these studies have been complicated by the close proximity of other candidate genes such as *APOA1*, *APOC3*, and more recently *APOA5*. The most common isoform detected using isoelectric focusing is APOA4-1, with an allele frequency in Caucasians ranging from 88% to 95%. APOA4-2 (Gln^{360}→His) is the second most common isoform with an allele frequency in the range of 5–12% in Caucasians [65, 66]. Additional variation within these isoforms was detected using the polymerase chain reaction (PCR). A relatively common mutation (Thr^{347}→Ser) has been documented within subjects with the APOA4-1 isoform. The effect of APOA4 genetic variation on plasma lipid levels has been studied in several populations [66–71]. In some Caucasian populations, the APOA4-2 allele has been associated with higher levels of HDL-C and/or lower TG levels [65, 71–74], but no associations have been observed in other studies [69, 75–79].

The effect of genetic variation at this locus on dietary response of fasting plasma lipids has been examined [80–83]. Our own data show that the APOA4-2 isoform was associated with hyporesponsiveness of LDL-C to dietary therapy consisting of reduction in total fat and cholesterol intakes. McCombs *et al.* [81] also demonstrated that this effect may be due to the reduction in dietary cholesterol. We also observed that subjects with the APOA4-2 allele tended to have greater decreases in HDL-C concentrations following a low-fat and low-cholesterol diet. To follow up on this observation, we have examined the effect of this polymorphism on HDL-C response in 41 healthy males [83]. They were fed three consecutive diets (high-saturated-fat, low-fat, and high-monounsaturated-fat diets) for 4 weeks each. After consuming the high-saturated-fat diet, carriers of the APOA4-2 allele had a greater decrease in HDL-C and APOA1. In these subjects, replacement of a high-carbohydrate diet by monounsaturated fat resulted in a greater increase in HDL-C and APOA1 as compared with homozygotes for the APOA4-1 allele. These data suggest that in APOA4-2 subjects, a high-carbohydrate diet may induce an apparently increased atherogenic lipid profile (LDL-C does not decrease, but HDL-C does decrease). Therefore, these subjects may benefit specially from a diet relatively high in monounsaturated fat.

The APOA4-2 allele has been also studied in relation to the changes in CVD risk factors associated with urbanization in developing countries. We demonstrated in a population-based study in Costa Rica that lifestyles associated with an urban environment, such as increased smoking and saturated fat intake, elicit a more adverse plasma lipoprotein profile among subject carriers of the APOA4-2 allele than in APOA4-1 homozygotes, which could make them more susceptible to CVD [84]. These results may be difficult to reconcile with those from the dietary metabolic studies. However, note that the changes associated with "modernization" (i.e., increased fat and cholesterol intakes, smoking, and decreased physical activity) are more complex gene–environment interactions than those taking place during a well-controlled dietary protocol carried out in a metabolic unit.

We examined the association between the APOA4 (Thr^{347}→Ser) variant observed within the APOA4-1 allele and the LDL-C response to dietary intervention [83]. Our data show that carriers of the less common Ser^{347} allele have a greater decrease in total cholesterol, LDL-C, and APOB concentrations when they are switched from a low-fat and -cholesterol diet to a high-monounsaturated-fat diet as compared with homozygous carriers of the Thr^{347} allele.

Similar studies in subjects heterozygous for familial hypercholesterolemia [85] show that the APOA4-2 allele was associated with lower LDL-C and APOB levels independent of diet effects. No differences in total cholesterol,

LDL-C, HDL-C, and APOB levels were observed between subjects homozygous for the APOA4 Thr^{347} allele and those carriers of the APOA4 Ser^{347} allele. After dietary intervention, Ser/Ser subjects showed significant reductions in plasma TG and VLDL-C levels, but no changes were found in carriers of the Ser allele.

The combined information for the Thr^{347}→Ser and the Gln^{360}→His suggests that the responsiveness of LDL-C to changes on dietary fat could follow this pattern: 347Ser/360Gln > 347Thr/360Gln > 347Thr/360His. The mechanisms by which these mutations may exert the observed effects are still unknown. The APOA4-2 allele binds to lipoproteins with higher affinity than APOA4-1, which may result in delayed hepatic clearance of chylomicron remnants as shown in metabolic studies. The substitution of Ser for Thr at position 347 induces changes in the secondary structure and a slight increase in hydrophilic profile at this position, which could result in a decrease in its affinity for lipids on the TRL particles. This could facilitate the exchange with APOC2, thereby increasing LPL activity over those particles, which would in turn accelerate clearance of remnants. The increased influx of dietary cholesterol would down-regulate the LDL receptors, with consequent increases in LDL-C concentrations. Therefore, consumption of fat-rich diets would produce a greater increase in LDL-C in Ser^{347} carriers.

After several decades of research, it is becoming evident that postprandial lipemia is a major determinant of blood lipoprotein concentrations and CVD risk [86, 87]. The postprandial response is highly heterogeneous, and multiple factors such as age, exercise levels, body weight, fasting lipid levels, diet, and genetics have been noted to be responsible for this variability. We have also examined whether the *APOA4* locus could have an effect on the interindividual variability observed during postprandial lipemia [88, 89]. In this regard, we examined the *APOA4* Gln^{360}→His polymorphism in 51 healthy male volunteers (42 homozygous for the APOA4 360Gln allele (Gln/Gln) and nine carriers of the A4-360His allele), homozygous for the *APOE3* allele. These individuals were subjected to a vitamin A-fat load test consisting of 1 g of fat/kg body weight and 60,000 IU of vitamin A. Blood was drawn at time 0 and every hour for 11 hours. Plasma cholesterol, TG, and cholesterol, TG, APOB100, APOB48, APOA4, and retinyl palmitate (RP) were determined in lipoprotein fractions. Data on postprandial lipemia revealed that subjects with the *APOA4* 360His allele had significantly greater postprandial levels in small TRL-C ($p < 0.02$), small TRL-TG ($p < 0.01$), and large TRL-TG ($p < 0.05$) than *APOA4* 360Gln/Gln subjects.

Similar analysis for the *APOA4*-347Ser polymorphism [89] revealed that carriers of the A4-347Ser allele ($n = 14$) had a lower postprandial response in total TG ($p < 0.025$), large TRL-TG ($p < 0.02$), and small TRL-TG levels ($p < 0.007$), and a higher postprandial response in large TRL-APOA4 ($p < 0.006$) and APOB100 ($p < 0.041$) levels than subjects homozygous for A4-347Thr ($n = 36$). These data suggest that the modifications observed in postprandial lipoprotein metabolism in subjects with *APOA4* polymorphisms could be involved in the different LDL-C responses observed in these subjects following a diet rich in cholesterol and saturated fats.

C. Apolipoprotein E (APOE)

APOE in serum is associated with chylomicrons, VLDL, and HDL and serves as a ligand for the LDL receptor and the LDL receptor-related protein [90–93]. When APOE deficiency is present, there is marked accumulation of cholesterol-enriched lipoproteins of density less than 1.006 g/ml containing APOB48 and APOA4, as well as APOB100 [94]. Moreover, in this disorder there is delayed clearance of both APOB100 and APOB48 within TRL. Genetic variation at the *APOE* locus results from three common alleles in the population, E4, E3, and E2, with frequencies in Caucasian populations of approximately 15, 77, and 8%, respectively [93]. Population studies have shown that plasma cholesterol, LDL-C, and APOB levels are highest in subjects carrying the *APOE4* isoform, intermediate in those with the *APOE3* isoform, and lowest in those with the *APOE2* isoform [95, 96]. It has been suggested that *APOE* allelic variation may account for up to 7% of the variation in total and LDL-C levels in the general population [97]. This relationship between LDL-C levels and *APOE* genetic variation is not independent of environmental and ethnic factors. Thus, it appears that the association of the *APOE4* isoform with elevated serum cholesterol levels is greater in populations consuming diets rich in saturated fat and cholesterol than in other populations. These data indicate that the higher LDL cholesterol levels observed in subjects carrying the *APOE4* isoform are manifested primarily in the presence of an atherogenic diet characteristic of certain societies, and that the response to dietary saturated fat and cholesterol may differ among individuals with different APOE phenotypes. Many studies have been conducted to prove this hypothesis [30]. Some investigators reported greater plasma lipid responses in subjects carrying the *APOE4* allele, whereas others failed to find significant associations between *APOE* genotype and plasma lipid response [30, 98]. Important differences exist among these studies that could account for some of the discrepancies observed. Studies differed in subject gender, age, and baseline lipid levels, and all of these variables are known to play an important role in the variability of dietary response.

The *APOE* gene has been implicated as one of the genetic factors mediating variability in postprandial lipemia response. The *APOE2* isoform is considered to decrease remnant clearance because of impaired affinity for the receptors. Conversely, the APOE4 isoform should induce a faster clearance. However, studies that have compared postprandial triglyceride responses across different *APOE* genotypes have produced conflicting results, especially regarding the effects associated with the *APOE4* allele [99–105]. Postprandial response was examined at 4 and 8 hours by Boerwinkle *et al.* [106] in a sample of individuals taking part in the Atherosclerosis Risk in Communities Study (ARIC), following a single high-fat meal containing vitamin A (used as a marker for intestinal lipoprotein synthesis). Postprandial plasma retinyl palmitate response was significantly different among *APOE* genotypes, with delayed clearance in subjects carrying the *APOE2* allele compared with E3/3 and E3/4 subjects; however, measurements of other lipid variables, such as triglyceride concentration and triglyceride-rich lipoprotein, were not sensitive enough to detect these effects. Another study by Nikkilä *et al.* [102], carried out in CVD cases and controls, showed that in CVD patients with the APOE2/3 phenotype triglyceride levels were highest and still increasing after 7 hours, reflecting delayed chylomicron remnant clearance. The same effect was observed also in normotriglyceridemic non-insulin-dependent diabetic patients [107] and in nondiabetic normolipemic subjects [108], although in this report, the delayed chylomicron remnant was observed only in E2/2 individuals. The findings associated with the *APOE4* allele have been more discordant. In an earlier report, heterozygosity for this allele was associated with lower lipemic response in relation to other phenotype groups [101]; however, in a more recent study the E4 allele was associated with prolonged postprandial responses of lipids and apolipoproteins in TRL [109].

Several mechanisms have been proposed to explain these APOE-related differences in individual response to dietary therapy. Some studies have shown that intestinal cholesterol absorption is related to *APOE* phenotype, with *APOE4* carriers absorbing more cholesterol than non-*APOE4* carriers. Other mechanisms, such as different distribution of APOE on the lipoprotein fractions, LDL APOB production, bile acid and cholesterol synthesis, and postprandial lipoprotein clearance, may also be involved.

More recently, other genetic variants within the *APOE* locus have been investigated in relation to association with lipid phenotypes and response to dietary intervention. The current evidence suggests that variability in the APOE promoter region is associated with plasma lipid concentrations [110], dietary response [111–113], and CVD risk [114], independently of the traditional E2, E3, and E4 alleles.

D. Apolipoprotein C-III (APOC3)

Plasma APOC3 is a component of chylomicrons, VLDL, and HDL. This protein is synthesized primarily in the liver and to a lesser extent in the intestine [115]. *In vitro*, APOC-III inhibits LPL activity [116] and the binding of APOE-containing lipoproteins to the LDL receptor, but not to the LDL receptor-related protein. In agreement with the observations *in vitro*, the overexpression of the human *APOC3* gene in transgenic mice resulted in severe hypertriglyceridemia [117]. The *APOC3* gene is closely linked to the *APOA1*, *APOA4*, and *APOA5* genes [118]. Several SNPs have been described at this locus. The S2 allele of the SstI SNP (C3238G), 3′ to the APOC3 gene, has been associated in some studies with hypertriglyceridemia and increased coronary artery disease risk [118]. A PvuII RFLP located in the first intron of the APOC3 gene has also been associated with variation in HDL-C levels. In addition, several polymorphisms have been identified in the promoter region of the gene (C^{-641}→A, G^{630}→A, T^{-625}→deletion, C^{-482}→T, and T^{-455}→C) [119]. These mutations are in linkage disequilibrium with the *SstI* site in the 3′ untranslated region [119]. An insulin response element has been mapped to a 42-nucleotide-long fragment located between –490 and –449 relative to the transcription start site, and *in vitro* studies demonstrated that transcriptional activity of the *APOC3* gene was down-regulated by insulin only in the construct bearing the wild-type promoter, but not in those constructs containing the C^{-482}→T, and T^{-455} →C variants [120]. These results may provide the molecular basis to understand the increased levels of APOC3 found in subjects carrying the S2 allele and its association with hypertriglyceridemia. Our own studies show that following an increase in dietary monounsaturated fat intake, S1S1 subjects responded with an increase in LDL-C levels, whereas S1S2 subjects experienced a significant decrease [121], suggesting that the *APOC3* locus is involved in LDL-C responsiveness to dietary fat. This interaction could begin to explain the different effects associated with this polymorphism that have been reported in the literature. Along those lines, Waterworth *et al.* [122] examined the association of the S2 allele and several other polymorphisms at this locus on postprandial lipid levels following an oral fat load test. The population consisted of young, healthy male offspring whose fathers had had a myocardial infarction before the age of 55 years, compared with age-matched controls. The *APOC3* variations examined were C3238G (SstI) in the 3′-UTR, C1 100T in exon 3, C-482T in the insulin response element, and T-2854G in the APOC3–APOA4 intergenic region. The postprandial response was regulated by variation at the T-2854G and C3238G sites. After the oral fat tolerance test, carriers of the rare alleles had significantly delayed clearance of TG levels; G-2854 carriers showed the largest effect on triglycerides, and G3238 carriers showed a lesser response. However, after adjustment for fasting TG, only the effect with the T-2854G remained significant.

The reported association of the S2 allele with elevated TG and cholesterol concentrations and with high blood pressure, all of which are characteristic of an insulin-resistant state, and the presence of an insulin response element in the promoter region of the *APOC3* gene suggest that this polymorphism could also be involved in dysfunctional glucose metabolism. Therefore, we have examined the effect of the C3238G polymorphism on carbohydrate metabolism in healthy subjects. We gave 41 males three consecutive diets [123]. The first was rich in saturated fat, the second was an NCEP Step 1 diet, and the last was rich in monounsaturated fat. At the end of each dietary period, subjects received an oral glucose tolerance test. *APOC3* genotype significantly affected basal glucose concentrations and insulin concentrations after the test. Carriers of the S2 allele ($n = 13$) had higher insulin concentrations after the test than S1/S1 subjects ($n = 28$) in the three periods. Multiple regression analysis showed that this polymorphism predicted the insulin response to the oral glucose tolerance test and the difference between basal insulin concentrations and insulin concentrations in response to the saturated fat-rich diet.

An association between the *APOC3* locus and carbohydrate metabolism was also reported by Waterworth *et al.* [122] on the EARSII study. Variation at the C-482T insulin response element determined response to the oral glucose tolerance test, with carriers of the rare T-482 having significantly elevated glucose and insulin concentrations.

These data suggest that specific genetic variants at the *APOC3* gene locus differentially affect postprandial lipids [122] and response to oral glucose [122, 123], which could result in reduced sensitivity to insulin, especially when persons consume diets rich in saturated fat.

Lipoprotein phenotypes have been associated with CVD risk but in addition, they may be involved in longevity [124]. A recent report [125] has demonstrated that homozygosity for the *APOC3*-641C allele is associated with a favorable lipoprotein profile, cardiovascular health, insulin sensitivity, and longevity. Because modulation of lipoproteins is also seen in genetically altered longevity models, it may be a common pathway influencing lifespan from nematodes to humans.

E. Apolipoprotein A5 (APOA5)

A new candidate gene for lipid metabolism was identified in 2001 using computational biology; the gene was named *APOA5* [126]. One of the reasons why it had been hiding for so many years was the very low concentrations in which it is found in plasma. However, the current evidence suggests that *APOA5* and its genetic variability may parallel for the metabolism of TRL the impact that *APOE* has had for the metabolism of LDL.

Five common *APOA5* SNPs have been reported in several populations: −1131T > C, −3A > G, 56C > G, IVS3 + 476G > A, and 1259T > C. These *APOA5* gene variants have been associated with increased TG concentrations [127–132], whereas other SNPs have been reported so far in one ethnic group [133, 134]. With the exception of the 56C > G SNP, the SNPs are reported to be in significant linkage disequilibrium [135, 136]. Despite the relatively good consistency with plasma TG concentrations, the association between *APOA5* SNPs (or haplotypes) and CVD risk is still controversial and the scientific literature is beginning to be populated by controversial reports regarding the contribution of *APOA5* variants to disease risk [132, 137–144]. Learning from the experience accumulated with *APOE*, it is reasonable to implicate gene–environment interactions modulating the association with CVD risk factors as well as with the disease risk itself. Consistent with this hypothesis, several intriguing interactions are beginning to emerge. In the Framingham Heart Study, we have reported a gene–diet interaction between the *APOA5* gene variation and the polyunsaturated fatty acids (PUFAs) in relation to plasma lipid concentrations and lipoprotein particle size [145]. Furthermore, we have demonstrated that obesity modulates the effect of the *APOA5* gene variation in carotid intimal medial thickness (IMT), a surrogate measure of atherosclerosis burden. This association remained significant even after the adjustment for triglycerides [146]. More recently, we have found a gene–diet interaction between the *APOA5* gene and total fat intake in relation to body mass index (BMI), overweight, and obesity risk [147]. The homogeneity of this interaction was detected in both men and women and followed a dose-response relationship. In comparison with wild-type subjects, the carriers of the variant C allele at the –1131T > C SNP appeared to be protected from an increase in body weight when consuming more dietary fat. Interactions between BMI and *APOA5* had been previously reported by Aberle *et al.* [148].

Behind these headlines, some interesting and potentially revealing peculiarities are emerging that deserve further examination, which could lead to a better understanding of the mechanisms of action of *APOA5*. One relates to the specificity of the interactions in terms of the location of the tagging SNPs, one in the promoter region (–1131T > C) and the other affecting the primary structure of the protein (56C > G). We have reported that the complex interactions among APOA5 SNPs, dietary intake, BMI, and IMT appear to be related to the promoter polymorphism, whereas the *APOA5* 56C > G SNP appears to associate with the different traits independently of environmental triggers [145–147]. The second aspect to consider is the specificity of the interactions in terms of dietary fat, being different for saturated, monounsaturated, and polyunsaturated n-6 and n-3. We investigated the interaction between these APOA5 SNPs and dietary fat in determining plasma fasting TGs, remnant-like particle (RLP) concentrations, and lipoprotein particle size in Framingham Heart Study participants. Significant gene–diet interactions between the –1131T > C SNP and PUFA intake were found in determining fasting TGs, RLP concentrations, and lipoprotein particle size, but these interactions were not found for the 56C > G polymorphism [145]. The –1131C allele was associated with higher fasting TGs and RLP concentrations in only the subjects consuming a high-PUFA diet (> 6% of total energy) regardless of gender. Interestingly, when we further analyzed the effects of n-6 and n-3 fatty acids, we found that the PUFA-*APOA5* interactions were specific for dietary n-6 fatty acids. Thus, higher n-6 (but not n-3) PUFA intake increased fasting TGs, RLP concentrations, and VLDL size and decreased LDL size in *APOA5*-1131C carriers, suggesting that n-6 PUFA-rich diets are related to a more atherogenic lipid profile in these subjects. This could shed some light on the data coming from postprandial studies [149–153] showing slightly different outcomes regarding the interaction between postprandial metabolism and the *APOA5* SNPs and directed us to hypothesize that an impaired postprandial response in carriers of the C allele at the –1131T > C SNP could be triggered by PUFA n-6 rather than by saturated fatty acids. Similar specificities in terms of SNP, exclusive for the –1131T > C, and dietary fat, primarily monounsaturated fat, were found in relation to the interaction between fat intake and BMI within the Framingham Study [147].

These observations suggest some new hypothesis regarding some of the observed geographical differences in CVD risk as well as the changes in risk observed following westernization of lifestyles. In this regard, it is important to pay attention to the differences in allele frequencies of the *APOA5* SNPs identified so far. These frequencies appear to be higher in Asian as compared to White populations [131, 154]. If the interactions observed in Framingham apply to other geographical areas, we should expect that an increased fraction of the Asian population will be placed at higher CVD risk because of the current trend toward higher dietary fat intakes, especially if they are driven by increases in PUFA n-6. On the other hand, some of the benefits of the Mediterranean diet, rich in monounsaturated fatty acids from olive oil (as well as the high fish consumption providing adequate levels of PUFA n-3) may derive from the intriguing finding in the Framingham study supporting the notion that some individuals (carriers of the C allele at the –1131T > C SNP) may consume a high percentage of fat, especially MUFA, in their habitual diet without adversely affecting their BMI and their risk of overweight and obesity [147].

F. Lipoprotein Lipase (LPL)

LPL is a heparin-releasable enzyme, bound to glycosaminoglycan components of the capillary endothelium. It plays a key role in lipoprotein metabolism by catalyzing the hydrolysis of 1,3 ester bonds of TG in chylomicrons and VLDL. The active form of LPL is constituted by two identical subunits, each of approximately 60,000 daltons, which require APOC2 as a cofactor, whereas APOC3 acts as an inhibitor. LPL is synthesized in the adipose and muscle tissue, as well as in macrophages [155, 156]. The gene for LPL is located on the short arm of chromosome 8 (region 8p22). Hundreds of polymorphisms have been reported at this locus [157, 158], some of which cause a loss of enzymatic function. In contrast to this, the LPL(S447X) variant has been associated with increased lipolytic function and can therefore be regarded as a gain-of-function mutation [159]. The consensus of the literature for this polymorphism shows that carriers of the 447X mutation have lower TG levels and increased HDL-C levels with a concomitant lower incidence of CVD compared with noncarriers. Most studies in humans indicate that the beneficial effects are associated with enhanced TG-lowering capacity mainly attributed to increased lipolytic function [159]. Moreover, we have demonstrated that some of these benefits may occur through the modulation of the postprandial response to fat-rich meals [160].

Regarding those mutations reducing LPL activity, the Asn291Ser is the most common; the Caucasian population has 2–5% heterozygote carriers. Less common mutations are Asp9Asn (1.5%) and Gly188Glu (0.06%). The French Canadian population has an especially high rate of LPL mutation carriers, up to 17% (Pro207Leu, Gly188Glu, and Asp250Asn). In the Japanese population, the frameshift mutation G916 deletion (first position of Ala221) is most common. The promoter variant T-93G was found in 76.4% of South African Blacks and in only 1.7% of Caucasians. This promoter variant is in near-complete linkage disequilibrium with the Asp9Asn mutation in Caucasians, but not in South African Blacks [161]. The common missense mutation (Asn291Ser) in exon 6 appears to be enriched in familial combined hyperlipidemia and low HDL cholesterol. *In vivo* and *in vitro* measurements of LPL activity indicate that this mutation is associated with approximately 50–70% of normal lipase catalytic activity. A recent systematic review summarized the associations between the LPL Asn291Ser variant and dyslipidemia, the risk of type 2 diabetes mellitus (T2DM), obesity, hypertension, and coronary heart disease (CHD) [162]. Twenty-one articles, including 19,246 White subjects, were included in this meta-analysis. The summary standardized mean difference (SMD) of plasma TG for carriers compared with noncarriers of the Asn291Ser variant was 3.23 ($p < 0.00001$). The summary SMD of plasma HDL-C for carriers compared with noncarriers of the Asn291Ser variant was –3.42 ($p < 0.0001$). The summary SMD of the association of the Asn291Ser variant with plasma TG increased with increasing age and weight gain. Significant interactions between the LPL Asn291Ser variant and fasting glucose, T2DM, and CHD were seen ($p = 0.02$, 0.04, and 0.01, respectively). No significant interactions were seen between the LPL Asn291Ser variant and BMI, waist–hip ratio, and blood pressure ($p > 0.05$). In addition, there is evidence supporting the conclusion that normolipidemic Asn291Ser carriers exhibited a more pronounced postprandial response compared with noncarriers as shown by higher chylomicron TG and retinyl palmitate peaks and alters postprandial HDL composition compared to matched noncarriers [163, 164]. It is possible that carriers of this mutation may be unable to respond to a high-fat diet by an increase in their LPL activity as normal subjects do. Thus, an oral fat challenge may unmask a hidden defect in lipolysis in these subjects that may not be generally evident in the fasting state. Therefore, the current evidence indicates that the Asn291Ser variant in the *LPL* gene is a risk factor for dyslipidemia, characterized by hypertriglyceridemia and low HDL-C levels, which worsens with increasing age and weight gain. Moreover, this gene variant may predispose to T2DM and CHD.

In addition to the *LPL* polymorphisms just described, some other less frequent mutations have been associated with LPL deficiency. Although a direct association has not been found, mutations resulting in LPL deficiencies have been linked to unfavorable lipid profiles. Although homozygous or compound heterozygous mutations in *LPL* resulting in complete loss of catalytical activity are rare, heterozygous mutations are quite common worldwide. It is when these moderate mutations are compounded by other risk factors that a significant increase in risk of cardiovascular disease is observed. Therefore, the *LPL* locus may be considered as a major player in the genetic component of CVD risk [165].

G. Intestinal Fatty Acid Binding Protein (IFABP)

The IFABP is a member of a family of small (14–15 kDa) intracellular lipid binding proteins. Besides the IFABP, this family also includes the heart, epidermal, brain, and liver fatty acid binding proteins, testicular, myelin, adipocyte, and ileal lipid binding proteins, and cellular retinol and retinoic acid binding proteins [166, 167]. The IFBP gene (*FABP2*) is located in 4q28–q31 and has the conserved four exons and three introns characteristic of this family of genes; its expression is under nutritional control [168].

IFABP plays important roles in several steps of fat absorption and transport, including uptake and trafficking of saturated and unsaturated long-chain fatty acids, targeting free fatty acids toward different metabolic pathways, protecting the cytosol from the cytotoxic effects of free fatty acids, and modulating enzyme activity involved in lipid metabolism [169]. Besides free fatty acids, the IFABP may bind other ligands, such as phenolic antioxidants. The IFABP is abundant in the enterocyte and represents 2–3% of an enterocyte's cytoplasmic mass [170]. Sacchettini *et al.* [171] found that the fatty acid binding proteins each have 10 antiparallel β strands, which form two orthogonal β sheets. There are two short α helices, which are connected to the first two β strands. This structure has been characterized as a β clam because it resembles a clamshell. It is hypothesized that fatty acids enter the fatty acid binding protein by passing a portal that includes three structures; the portal is comprised of one of the α helices, and two sharp turns between β strands, which include residues 54–55 and 73–74 [171]. A mutation in the IFABP at any of these amino acid residues that surround the portal may alter fatty acid entry into the protein. This may result in abnormal binding of fatty acids, which can then influence lipid profiles and consequently disease risk.

Baier *et al.* [172] initially reported a G→A mutation, which results in an amino acid substitution in IFABP at residue 54, alanine54 (wild-type)→threonine (mutant-type). This polymorphism is very common, with a thr54 allelic frequency of around 29% in most populations. This amino acid substitution was found to be associated with elevated fasting insulin levels, insulin resistance, and increased fatty acid binding in Pima Indians [172, 173]. Associations of this polymorphism with several biochemical and anthropometric variables have been subsequently carried out in Japanese, Mexican-American, Native American, and Caucasian populations [166]. In general, the presence of the thr54 allele has been associated in some, but not all, studies with higher fasting insulin concentrations and insulin resistance. Conversely, a study of Keewatin Inuit (Eskimos) found the thr54 allele to be associated with lower 2-hour glucose concentrations [174]. It is well known that Eskimos have large dietary intakes of omega-3 fatty acids from fish. This finding in Eskimos suggests that differences in the type of fatty acid consumed interact with the functional differences in the gene products to produce phenotypic differences.

Several studies have utilized the euglycemic, hyperinsulinemic clamp for determining insulin resistance. Using this approach, the *FABP2* mutation was associated with a lower mean insulin-stimulated glucose uptake rate in Pima Indians. However, no significant findings were noted in subjects with familial combined hyperlipidemia [175] and in overweight Finnish subjects [176]. In a separate study, Stern *et al.* [177] found that the IFABP was not significantly linked to the major gene for the age of onset for type 2 diabetes.

Other studies have found an association between the presence of the thr54 mutation and higher mean fat oxidation rate [172, 176], greater fasting plasma triglycerides [178], and internal carotid artery (ICA) stenosis [179].

Regarding differences in dietary response, Hegele *et al.* [180] found that the ala54thr mutation was associated with variation in the response of plasma lipoproteins to dietary fiber. Subjects with the mutant thr54 allele, compared to those homozygous for the ala54 allele, had significantly greater decreases in plasma total and LDL-C and APOB when consuming a high-soluble-fiber diet than when consuming a high-insoluble-fiber diet. Moreover, Agren *et al.* [181] have shown that the TG response to an oral fat load test was greater in Thr homozygotes. In addition, the classical correlation between fasting and postprandial triglyceride levels was present in Ala carriers, but not so in Thr individuals, suggesting that the delayed triglyceride-rich lipoprotein clearance may affect insulin action (or vice versa). Finally, Marin *et al.* [182] found that insulin sensitivity decreased in subjects with the Thr54 allele of the FABP2 polymorphism when SFAs were replaced by MUFAs and carbohydrates.

In vitro studies have found that the thr54 IFABP differs from the ala54 IFABP, and these studies offer a biologically plausible mechanism for thr54's role in altered lipid metabolism and disease states. Baier *et al.* [172] found, through titration microcalorimetry, that the thr54 IFABP had a twofold greater affinity for long-chain fatty acids than the ala54 IFABP. In addition, in another study, Baier's group created permanently transfected Caco-2 cells that express the ala54 or thr54 IFABP [183]. The cells expressing the mutant thr54 protein were found to have greater transport of long-chain fatty acids and greater secretion of triglycerides than the ala54 protein. A recent study with mouse L-cells transfected with wild-type IFABP found that the increase in secretion of triglycerides over nontransfected cells was due to increased synthesis in the cell [184]. The results of these studies support the findings of Baier *et al.* in which the thr54 protein has a higher affinity for long-chain fatty acids than the ala54 protein [172, 183]. The three-dimensional structure of the wild-type IFABP has been studied by nuclear magnetic resonance (NMR), and it was shown that amino acid 54 in the IFABP is important in the stabilization of the portal region, where the fatty acid enters the protein [185, 186]. Amino acid 54 interacts with α-helix II and helps to stabilize the portal region. Hodsdon and Cistola [187] hypothesize that since threonine can bond hydrogen and alanine cannot, then thr54 will be better able to interact with α-helix II and stabilize the portal, which may lead to increased binding of fatty acids [185].

In summary, the physiological function of the protein coded by this gene and the findings suggesting functionality for the Ala54Thr polymorphism warrant further research to assess the full impact of this locus on the interindividual variability in CVD risk and dietary response.

III. ENDOGENOUS LIPOPROTEIN METABOLISM

The hepatocyte synthesizes and secretes triglyceride-rich VLDL, which may be converted first to intermediate-density lipoproteins and then to LDL through lipolysis by a mechanism involving LPL, similar to that described for the exogenous lipoprotein pathway. The excess surface components are usually transferred to HDL. Some of these remnants may be taken up by the liver, whereas others are further lipolyzed to become LDL, which, in humans, contains APOB as their only apolipoprotein. This pathway shares many of the genes described already for the exogenous pathway. This section describes briefly some of the studies related to the LDL receptor gene, which has major responsibility for catabolism of the LDL particles in the liver and peripheral tissues, and the HMGCoA reductase gene, coding for a key enzyme in the synthesis *de novo* of cholesterol.

A. LDL Receptor (LDLR)

The LDL receptor gene has been localized to the 19p13.1 region. Mutations at this locus have been found to be responsible for familial hypercholesterolemia, an autosomal dominant disorder characterized by elevation of LDL cholesterol and premature CVD. The frequency of homozygosity is estimated to be one in a million, and the frequency of heterozygotes approximately 1 in 500. Several hundred different mutations have been already described in the LDL receptor gene and many of them have been reviewed in detail [188]. In addition to those rare mutations associated with familial hypercholesterolemia, several common SNPs have been studied in this locus, and their results suggest that this locus may be also involved in plasma lipid concentrations in the general population.

The role of common genetic variations in the 3′UTR of the LDLR in relation to plasma cholesterol has been recently examined [189]. Six SNPs, G44243A, G44332A, C44506G, G44695A, C44857T, and A44964G, within the 5′ region of the 3′UTR fall into three common haplotypes, GGCGCA, AGCACG, and GGCGTA, occurring at frequencies of 0.45, 0.31 and 0.17, respectively, in Caucasians and 0.13, 0.13 and 0.38, respectively, in African Americans. In a tissue culture–based system, expression of a reporter gene carrying a 3′UTR that includes the 1-kb nucleotide sequences corresponding to the AGCACG or GGCGTA was 70% or 63%, respectively, of the same sequence with GGCGCA. Genotyping of two "haplotype tagging" SNPs, C44857T and A44964G, in the Atherosclerosis Risk in Communities (ARIC) study population showed that in Caucasians, but not in African Americans, the inferred TA haplotype had a significant LDL-C lowering effect. The adjusted LDL-C levels in the TA/TA diplotypes were significantly lower, by 6.10 mg/dl in men and by 4.63 mg/dl in women, than in individuals with other diplotypes. Caucasian men homozygous for CA, in contrast, showed significantly higher LDL-C, lower HDL-C, and higher LDL/HDL ratios. These data show that 3′UTR sequences that cause higher reporter gene expression *in vitro* are associated in Caucasians with plasma lipid profiles indicative of higher cardiovascular risk. Another single common amino acid–changing polymorphism (A370T) has been investigated, and some investigators found an association between this polymorphism and stroke, independent of lipid levels [190]. However, another study did not show any association between this SNP and plasma lipids [191]. No reports of gene–diet interactions have been reported for this locus despite its pivotal role in lipoprotein metabolism.

B. 3-Hydroxy-3-Methylglutaryl-Coenzyme A Reductase

3-Hydroxy-3-methylglutaryl-coenzyme A reductase (HMG-CoA) catalyzes a rate-limiting step in cholesterol biosynthesis and also participates in the production of a variety of other compounds. This enzyme has been an extremely successful drug target in the form of HMG-CoA reductase inhibitors. The gene is located on 5q13.3–q14.

Similarly to the LDLR locus and despite its important biological role and the fact that it has been a very successful drug target, this gene has been barely examined in terms of association with plasma lipid levels and related phenotypes. Some studies have investigated this locus regarding its potential pharmacogenetic value [192, 193].

IV. REVERSE CHOLESTEROL TRANSPORT

Both the liver and the intestine synthesize HDL. Its precursor form is discoidal in shape and matures in circulation as it picks up unesterified cholesterol from cell membranes (see the later discussion of the ATP binding cassette-1), other lipids (phospholipids and TG), and proteins from TRL as these particles undergo lipolysis. The cholesterol is esterified by the action of LCAT, and the small HDL3 particle becomes a larger HDL2 particle. The esterified cholesterol is either delivered to the liver or transferred by the action of the cholesterol ester transfer protein (CETP) to TRL in exchange for TG. The liver may then take up this cholesterol via receptors specific for these lipoproteins, or it can be delivered again to the peripheral tissues. The TG received by the HDL2 is hydrolyzed by hepatic lipase and the particle is converted back to HDL3, completing the HDL3 cycle in plasma. In the liver, cholesterol can be excreted directly into bile, converted to bile acids, or reutilized in lipoprotein production.

The study of the genetics of reverse cholesterol transport has been impaired by the intrinsic complexity of this process and the difficulty of measuring precisely some of the most relevant phenotypes (i.e., flux). Some of our most important advances have come out from the study of rare HDL deficiencies. Taking advantage of this approach Kiss *et al.* [194] have used a multitiered approach to identify genetic and cellular contributors to HDL deficiency in 124 human subjects. These investigators resequenced four candidate genes for HDL regulation and identified several functional nonsynonymous mutations including two in APOA1, four in lecithin:cholesterol acyltransferase (LCAT), one in phospholipid transfer protein (PLTP), and seven in the ATP-binding cassette transporter ABCA1. Interestingly, in 88% [110/124) of HDL-deficient subjects these genes did not provide any information about the molecular basis of their genetic defect. Cholesterol efflux assays performed using cholesterol-loaded monocyte-derived macrophages from the 124 low-HDL subjects and 48 control subjects revealed that 33% (41/124) of low-HDL subjects had low efflux, despite the fact that the majority of these subjects (34/41) were not carriers of dysfunctional ABCA1 alleles. In contrast, only 2% of control subjects presented with low efflux (1/48). In three families without ABCA1 mutations, efflux defects were found to cosegregate with low HDL. The information from this comprehensive approach drives us to the following conclusions: Efflux defects are frequent in low-HDL syndromes, but the majority of HDL-deficient subjects with cellular cholesterol efflux defects do not harbor ABCA1 mutations, suggesting that unidentified pathways are responsible for their phenotype.

The understanding of the molecular basis for HDL deficiency is crucial to using this information for CVD prevention and the development of new therapeutic targets. It is important to keep in mind that although observational data support an inverse relationship between HDL-C and CHD, genetic HDL-deficiency states often do not correlate with premature CHD. The work by Miller *et al.* [195] has investigated this apparent paradox more in detail. These investigators obtained carotid intimal medial thickness (cIMT) measurements in cases comprising 10 different mutations in LCAT, ABCA1, and APOA1 to further evaluate the relationship between low HDL resulting from genetic variation and early atherosclerosis. Their results show that in a 1:2 case-control study of sex- and age-related subjects ($n = 114$), cIMT was nearly identical between cases (0.66 ± 0.17 cm) and controls (0.65 ± 0.18 cm) despite significantly lower HDL-C (0.67 versus 1.58 mmol/l) and APOA1 levels (96.7 versus 151.4 mg/dl; $p < 0.05$), concluding that genetic HDL deficiency may be insufficient to promote early carotid atherosclerosis.

The following section provides an overview of the current status of the genetics of reverse cholesterol transport as it relates to common polymorphisms within some of the most relevant candidate genes and their potential interactions with dietary components.

A. Apolipoprotein A-I (APOA1)

APOA1, the major protein of HDL, plays several crucial roles in the metabolism of these particles as (1) structural component, (2) activator of the enzyme LCAT, and (3) key component of the reverse cholesterol transport process [196–198]. The gene for APOA1 is located on the long arm of human chromosome 11 [118]. This DNA region has been extensively analyzed since the early 1980s, resulting in the identification of scores of SNPs that have been examined in relation to variability in plasma lipids and CVD risk with mixed results [199]. One of the difficulties associated with the study of the APOA1 locus is the close proximity of the other candidate genes in the cluster (APOC3, APOA4, and

APOA5) and the complex linkage disequilibrium present within SNPs in this genomic region [118].

The common variant most intensively studied is the one resulting from an adenine (A) to guanine (G) transition (G/A), 75 bp upstream from the APOA1 gene transcription start site. Several studies have reported that individuals with the A allele, which occurs at a frequency of 0.15–0.20 in Caucasian populations, have higher levels of HDL-C than those subjects homozygous for the most common G allele. The magnitude and the gender distribution of this effect have differed among studies. These data were summarized in meta-analysis [200]. The reasons for these discrepancies are not known. Our own data [201, 202] support the notion that in well-controlled dietary studies performed in normolipidemic subjects, the A allele of this G/A polymorphism appears to be associated with hyperresponse to changes in the amount and saturation of dietary fat. Moreover, other investigators have shown that the significance of the associations may depend on cigarette smoking. Therefore, these data suggest an interaction between this polymorphism and environmental factors such as dietary and smoking habits [201–203].

It is not clear whether the putative effect of this variant on HDL-C levels is due to the G to A substitution per se, or to linkage disequilibrium between the A locus and a distinct and as yet unidentified effector locus. *In vitro* analysis of the effects of this polymorphism on transcription has yielded conflicting results. Smith *et al.* [204] reported that the A allele decreased *in vitro* transcription by 30%, consistent with their own *in vivo* turnover studies that showed decreased APOA1 synthetic rates in individuals with the A allele, although plasma HDL-C did not differ between GG and GA individuals. Tuteja *et al.* [205] reported that substitution of A for G decreased transcription about twofold, and Jeenah *et al.* [206] reported a fourfold increase in transcription. Angotti *et al.* [207] reported a five- to sevenfold increase in transcription associated with the A allele and demonstrated that this may be due to reduced binding affinity of a nuclear factor to the A allele that results in increased transcription efficiency of the *APOA1* promoter. More recent research has demonstrated that the induction of *APOA1* mRNA level in response to gemfibrozil is mediated at the transcriptional level and this effect is mediated by two copies of the "drug-responsive element" in the *APOA1* promoter region in the same area as the –75 G/A polymorphism. These data suggest that this *APOA1* variant may be a functional mutation responsible for some of the individual variability in HDL cholesterol response to gemfibrozil therapy [208].

In summary, the mechanisms responsible for the observed effect are still unknown. This mutation may have a direct effect on liver and/or intestinal *APOA1* gene expression as suggested in previous studies or it may be in linkage disequilibrium with a functional mutation in either of the neighboring genes (*APOC3* and *APOA4*). Further studies are needed to clarify these results.

In addition to the common variants, many different rare genetic abnormalities have been reported within this locus, and some have been associated with severe HDL deficiency and premature coronary atherosclerosis; however, as indicated earlier, many others have not been associated with atherosclerotic risk. At least one of these mutations shows a protective effect despite its association with low HDL [209], and its use as a therapeutic agent is being investigated [210].

B. Apolipoprotein A-II (APOA2)

APOA2 is the second most abundant protein of HDL particles. Therefore, it constitutes an obvious candidate gene for the study of reverse cholesterol transport. However, its function remains largely unknown [211]. Some initial studies reported an inverse relationship between plasma APOA2 concentrations and CVD risk [212, 213]; subsequent studies did not find significant associations or even suggested a proatherogenic role [214, 215]. Overexpression of murine *APOA2* in mice resulted in greater development of aortic fatty streak lesions [216]. There are marked differences between murine and human *APOA2*, however, and overexpression of human *APOA2* in mice does not exert effects similar to those of murine *APOA2* on atherosclerosis [217–219]. Relatively few studies have examined the association between *APOA2* polymorphisms and phenotypic traits [220–223]. Moreover, few genetic variants have been identified in the *APOA2* gene [224]. Interestingly, a T > C transition at position –265 affecting element D of the *APOA2* promoter has been reported to be functional in two independent studies, both demonstrating an approximately 30% drop in basal transcription activity [225, 226]. In one of these studies, the –265T > C polymorphism was associated with waist circumference in men [225]. Another study [227] reported an association between this polymorphism and abdominal fat depots in women. Although Castellani *et al.* [228] found increased body weight in mice overexpressing murine *APOA2*, the mechanism by which *APOA2* may influence body weight is largely unknown. *APOA2* is a member of the apolipoprotein multigene superfamily, which includes genes encoding soluble apolipoproteins (e.g., APOA1 and APOA4) that share genomic structure and several functions. Although all these apolipoprotein genes have been related to obesity in at least one epidemiological study [229], only APOA4 has been subscribed in regulation of food intake, acting as a satiety signal [230].

We studied the association between a functional *APOA2* promoter polymorphism (–265T > C) and plasma lipids (fasting and postprandial), anthropometric variables, and food intake in 514 men and 564 women who participated in the Genetics of Lipid Lowering Drugs and Diet Network (GOLDN) Study [231]. We obtained fasting and postprandial (after consuming a high-fat meal) measures. We measured lipoprotein particle concentrations by proton nuclear magnetic resonance spectroscopy and estimated dietary intake by use of a validated questionnaire. We observed recessive effects for this polymorphism that were homogeneous by sex. Individuals homozygous for the –265C allele had statistically higher body mass index (BMI) than did carriers of the T allele. Consistently, after multivariate adjustment, the odds ratio for obesity in CC individuals compared with T allele carriers was 1.70 ($p = 0.039$). Interestingly, total energy intake in CC individuals was statistically higher (mean [SE] 9371 [497] versus 8456 [413] kJ/d, $p = 0.005$) than in T allele carriers. Likewise, total fat and protein intakes (expressed in grams per day) were statistically higher in CC individuals ($p = 0.002$ and $p = 0.005$, respectively). After adjustment for energy, percent of carbohydrate intake was statistically lower in CC individuals. These associations remained statistically significant even after adjustment for BMI. We found no associations with fasting lipids and only some associations with HDL subfraction distribution in the postprandial state. We concluded that the –265T > C polymorphism is consistently associated with food consumption and obesity, suggesting a new role for APOA2 in regulating dietary intake.

C. Cholesteryl Ester Transfer Protein

CETP mediates the exchange of neutral lipid core constituents (cholesteryl ester and triglycerides) between plasma lipoproteins. The facilitation of the transfer of cholesteryl ester from HDL to TRL results in a reduction in HDL-C levels, but CETP may also promote the reverse cholesterol transport. Therefore, the overall effect of CETP expression on atherogenesis is uncertain, as has been recently seen with the results obtained with torcetrapib [11, 232]. The gene has been located on chromosome 16 adjacent to the *LCAT* gene (16q21). The most commonly studied SNP at this locus is the so-called *Taq*IB located within intron 1 [233]. The minor allele, referred to as B2, is associated with lower lipid transfer activity and higher HDL-C concentrations [233]. In addition, several reports have found significant interactions with alcohol intake [234, 235], cigarette smoking [236], physical activity [237], diet [238], and drug therapy [239–244]. The evidence for significant interactions between the CETP-*TaqIB* and the environmental factors just listed is stronger for smoking and drug therapy; however, considering the potential confounding effect of the several other common polymorphisms at this locus [245], these findings need to be interpreted with caution, especially taking into consideration the small effects reported by those studies examining particular interactions.

Given the consistent association between polymorphisms at this locus and CVD risk factors, especially HDL-C concentrations, several studies have reported significant associations between CETP SNPs and CVD risk [246], and again the most consistent data relate to the *Taq1B* SNP, with the minor allele associated with lower CVD risk.

In view of the associations with HDL-C and CVD risk, this locus has been examined as a potential candidate gene for extreme longevity, and data from both the general population and from elderly cohorts suggest that this gene could be involved in human longevity [247, 248].

D. Hepatic Lipase (LIPC)

Hepatic lipase (HL) is a lipolytic enzyme that is synthesized in the hepatocytes, secreted, and bound extracellularly to the liver. Hepatic lipase participates in the metabolism of intermediate-density lipoproteins and large LDL to form smaller, denser LDL particles and in the conversion of HDL_2 to HDL_3. In addition, hepatic lipase can mediate the unloading of cholesterol from HDL to the plasma membrane in the liver. It has also been suggested that hepatic lipase may act as a ligand protein with cell surface proteoglycans in the uptake of lipoproteins by cell surface receptors. The lipid profile of individuals with complete hepatic lipase deficiency is characterized by elevated plasma cholesterol and triglyceride levels, triglyceride enrichment of lipoprotein fractions with a density greater than 1.006 g/ml, the presence of a β-VLDL, and an impaired metabolism of postprandial triglyceride-rich lipoproteins [249].

The hepatic lipase gene (*LIPC*) maps to chromosome 15q21, spans 35 kbp, and is composed of nine exons. Four polymorphisms in complete linkage disequilibrium (–250 G to A, –514 C to T, –710 T to C, –763 A to G) have been identified in the 780-bp region upstream of the transcription initiation site. The presence of the –514TT genotype was initially found to be far more common among men with high HDL-C than among men with low HDL-C, suggesting that the –514T allele could be a marker for elevated HDL-C levels. Subsequent studies demonstrated that the –514T allele was also found to be associated with increased plasma HDL-C and apolipoprotein AI concentrations in the Finnish participants of the EARS study [250] and in Dutch coronary artery disease patients from the REGRESS study [251]. Tahvanainen *et al.* [252] reported that the –514T allele was associated with TG enrichment of the intermediate-density lipoproteins (IDL), LDL, and HDL, and an increase in IDL (but not HDL) cholesterol in 395 male Finnish patients with CHD. Therefore, the association between the –514T allele and increased HDL-C concentrations observed in previous studies was not apparent in this cohort. Likewise, Hegele *et al.* [253] did not find any association between this polymorphism and lipoprotein levels in three Canadian populations. Our own association data in the Framingham study showed a gender-specific and mild association with HDL-C levels that became stronger after examination of more specific markers of HDL metabolism such as HDL subfractions and HDL size [254]. This was also demonstrated by other investigators for HDL subfractions as well as for LDL and TRL subfractions [255].

Despite this lack of consensus on the outcome from association studies, association between the –514T allele and plasma HDL-C concentrations does not appear to be an artifact [256]. Several studies have shown that the –514T allele is associated with a 35–45% decrease in HL activity in White men and women [251, 252, 257]. However, the molecular mechanism responsible for the decreased HL activity associated with the –514T allele has not been elucidated. Because the coding region of the –514T allele is identical to that of the –514C allele [258], the decrease in HL activity associated with the –514T allele presumably reflects a decrease in *LIPC* gene expression. In this regard, Deeb and Peng [259] have investigated whether these polymorphisms have any effect on transcriptional activity of the proximal promoter (–639 to +29) in transient transfection assays. They found that the promoter with T at position –514 had approximately 30% lower activity than one with C at the same position ($P < 0.0005$) regardless of the genotype at position –250. Moreover, it has been suggested that the presence of the T-allele abolishes the insulin sensitivity of *LIPC* expression, meaning that the phenotypic expression of the polymorphism may be largest in conditions with high plasma insulin. This has also been demonstrated by Hadzopoulou-Cladaras and Cardot [260], who showed that deletion of the region –483 to +129 in the *LIPC* gene leads to a 60% drop in promoter activity, suggesting the presence of important stimulatory elements in this region. In addition, disruption of the upstream stimulatory factor binding site, which mediates insulin stimulation of protein expression by the C-514T substitution (change of E-box motif CACGTG to CATGTG), may be of importance. Conversely, van't Hooft *et al.* [261] found no consistent evidence for a significant contribution of any of these polymorphisms to the basal rate of transcription of the *LIPC* gene. According to the latter results, the four polymorphisms in the promoter region of the *LIPC* gene may be in linkage disequilibrium with one or more as-yet-unknown functional polymorphisms in the *LIPC* locus with a significant effect on HL metabolism and/or enzymatic activity. Alternatively, this result may be a consequence of the specific *in vitro* conditions used by these investigators. In addition to those polymorphisms present in several ethnic groups, other common SNPs have been shown in Chinese populations that have shown highly significant associations with promoter activity, HDL-C concentrations and CVD risk [262]. These polymorphisms may confound the results of the –514(C/T) polymorphisms and we hypothesize that the same could occur in other ethnic groups.

Other studies have focused on the potential role of this *LIPC* polymorphism as modulated by postprandial lipid response. This was investigated by Gomez *et al.* [263], who demonstrated that the T allele of the –514C/T polymorphism was associated with lower postprandial lipemic response. However, these results were not consistent with those previously reported by others [264].

In addition to the gene–diet interactions that we have reported [265, 266], other investigators have examined the interaction between the same *LIPC* allele and other behavioral risk factors. Hokanson *et al.* [267] used a population-based prospective study in San Luis Valley, Colorado, to investigate the role of the *LIPC* –480C > T polymorphism (also known as –514C > T) in predicting clinical CHD and the modifying effect of physical activity. Hispanics and non-Hispanic Whites ($n = 966$) were followed for 14 years (1984–1998); 91 CHD events occurred. Despite the association reported by some between the *LIPC* –480 TT genotype and increased HDL-C concentration, this genotype predicted an increase in CHD in both ethnic groups. Most important, physical activity altered this relation. In the full cohort, the rate of CHD was significantly elevated among subjects with the high-risk genotype and normal levels of physical activity (hazard ratio [HR] = 2.58, 95% confidence interval: 1.39, 4.77) but was not elevated among those with the high-risk genotype who participated in vigorous physical activity (HR = 0.52, 95% confidence interval: 0.12, 2.19) (reference group: *LIPC* –480 CC/CT, normal physical activity). Thus, in this study, the *LIPC* –480 TT genotype increased susceptibility to CHD only in those subjects with normal levels of physical activity, not in those with the high-risk genotype who were vigorously active. This information suggests that TT subjects may benefit more from a traditional lifestyle, in the case presented by Hokanson *et al.* [267] represented by physical activity, whereas in our case represented by a low-fat diet [265, 266].

Obesity has been shown to modulate the associations observed for the *LIPC* –514C > T polymorphism. This was investigated by St-Pierre *et al.* [268] in a population of French-Canadian men from the greater Quebec City area. In that study, the highest values of BMI, waist circumference, and accumulation of visceral adipose tissue (VAT) were observed among TT homozygotes ($p < 0.05$). After adjustment for age and BMI, TT homozygotes still showed higher plasma apolipoprotein APOA1 and HDL-TG concentrations than the two other groups ($p < 0.05$). When the two genotype groups (CC versus CT/TT) were further divided on the basis of VAT accumulation using a cutoff point of 130 cm (high versus low) it appears that, irrespective of the genotype, subjects with low VAT had higher HDL2-C concentrations ($p < 0.0001$). However, lean carriers of the T allele had higher plasma HDL2-C levels than did lean CC homozygotes. The beneficial effect of the T allele on plasma HDL2-C levels was abolished in the presence of visceral obesity (VAT > 130 cm). In summary, the presence of visceral obesity attenuates the impact of the *LIPC* –514C > T polymorphism on plasma HDL2-C levels. Again, this study shows that the TT genotype carries a potential benefit under optimal conditions, in the example here, low VAT [268]. Additional support for the idea that –514 TT carriers benefit only under optimal conditions is further supported by studies examining insulin resistance (IR) and diabetes. In population-based studies, dyslipidemia related to IR (high TG and low HDL-C levels) is a risk factor for type 2 diabetes (T2DM). Therefore, considering the association of the *LIPC* gene with HDL-C and TRL concentrations, Todorova *et al.* [269] investigated whether the *LIPC* G-250A polymorphism predicted the conversion from impaired glucose tolerance (IGT) to T2DM in the Finnish Diabetes Prevention Study. This study randomized subjects to either the intervention group (lifestyle modification aimed at weight loss, such as changes in diet and increased physical exercise) or the control group. In the entire study population, the conversion rate to T2DM was 17.8% among subjects with the –250 GG genotype and 10.7% among subjects with the –250A allele ($p = 0.03$). In multivariate logistic regression analysis, the –250GG genotype predicted the conversion to T2DM independently of the study group (control or intervention), gender, weight, waist circumference at baseline, and change in weight and waist circumference. In the intervention group, 13.0% of subjects with the –250GG genotype and 1.0% of the subjects with the –250A allele converted to diabetes ($p = 0.001$). The authors concluded that the –250GG genotype of the *LIPC* gene is a risk factor for T2DM. Therefore, genes regulating lipid and lipoprotein metabolism may be potential candidate genes for T2DM. But even more interesting was the fact that carriers of the –250A allele benefited more from the intervention than did those who were homozygotes for the G allele.

E. Cholesterol 7 Alpha-Hydroxylase (CYP7)

The first reaction of the catabolic pathway of cholesterol is catalyzed by *CYP7* and serves as the rate-limiting step and major site of regulation of bile acid synthesis in the liver. The cholesterol catabolic pathway, exclusive to the liver, comprises several enzymes, which convert cholesterol into bile acids. Previous studies show that *CYP7* is regulated by bile acid feedback, cholesterol, and hormonal factors [270]. In addition, dietary fats modulate the regulatory potential of dietary cholesterol on *CYP7* gene expression [271]. The human *CYP7* has been localized to chromosome 8q11–q12. A common A-to-C substitution at position –204 of the promoter of the *CYP7* gene is associated with variations in plasma LDL-C [272]. Moreover, the study of the molecular mechanisms of the transcriptional regulation of *CYP7* by sterols and bile acids has revealed that the promoter region between –432 and –220 contains several cell-specific enhancer elements whose activity is controlled, in part, by hepatocyte nuclear factor 3 [273]. Therefore, it is conceivable that the A-204C polymorphism might modulate transcription of the *CYP7* gene and, consequently, the rate of cholesterol catabolism.

The effect of the A278-C promoter polymorphism in the CYP7A1 gene on responses of plasma lipids to an increased intake in dietary cholesterol, cafestol, saturated fat, and *trans* fat was investigated in 496 normolipidemic subjects [274]. These responses were measured in 26 previously published dietary trials. After adjustment for the APOE genotype effect, AA subjects consuming a cholesterol-rich diet had a significantly smaller increase in plasma HDL cholesterol than did CC subjects. On intake of cafestol, AA subjects had a significantly smaller increase in plasma total cholesterol than did CC subjects. No effects of the polymorphism were found in the saturated and *trans* fat interventions. Therefore, the CYP7A1 promoter polymorphism had a small but significant effect on the increase in plasma HDL-C and plasma total cholesterol after an increased intake of dietary cholesterol and cafestol, respectively.

Another study revisited this issue in a smaller number of selected subjects [275]. The concentrations of total cholesterol and LDL-C were measured in 11 healthy men who were homozygous for either the –204A or –204C allele, after 3 weeks on a low-fat (LF) diet and 3 weeks on a high-fat (HF) diet. During both dietary regimens, the isocaloric amount of food was provided to volunteers. In six subjects homozygous for the –204C allele, the concentrations of cholesterol and LDL-C were significantly higher on HF than on LF diet (cholesterol: 4.62 versus 4.00 mmol/l, $p < 0.05$; LDL-C: 2.15 versus 1.63 mmol/l, $p < 0.01$, respectively); no significant change was observed in five subjects homozygous for the –204A allele. Therefore, this preliminary finding suggests that the polymorphism in the cholesterol 7-alpha-hydroxylase promoter region seems to be involved in the determination of cholesterol and LDL-C responsiveness to a dietary fat challenge.

Other studies have investigated the involvement of this locus on the plasma cholesterol response to statin therapy [276–278] and the current, limited evidence supports a statistically significant but still minor role of this locus in explaining the variability to drug response.

F. Scavenger Receptor B Type I (SCARB1)

Scavenger receptor B type I (SRBI) is a multilipoprotein receptor found in the liver and steroidogenic glands of mice and humans. The cDNA for human SRBI (also known as LIMPII analogous 1 [CLA-1] and SCARB1) was originally cloned by homology to human CD36 and rat lysosomal integral membrane protein II (LIMPII), members of a family of transmembrane proteins. Another study identified the hamster homolog by its ability to mediate the binding of modified LDL, and it was also shown to bind native LDL [279]. Subsequently, murine SRBI was shown to mediate the uptake of lipid, but not apoprotein, from HDL into cells [280], a process described as selective uptake. This finding established SRBI as the first HDL transmembrane receptor to be identified and cloned. Further studies of the human homolog demonstrated that it also is a multilipoprotein receptor that binds HDL, LDL, and VLDL [281]. Preliminary evidence from our studies indicated that the *SCARB1* gene plays a significant role in determining lipid concentrations and body mass index [282]. This has been further demonstrated by other investigators [283]. Moreover, the observation that SRBI mediates absorption of dietary cholesterol in the intestine suggests that it may also play a role in the postprandial response and insulin metabolism [284]. In this regard, we have examined the relation of common genetic variants at the SCARB1 locus with fasting and postprandial lipid responses [59, 285–287], and our results support previous findings in experimental models in relation to postprandial lipids and insulin metabolism.

G. ATP-Binding Cassette Transporters (ABCA1, ABCG5, ABCG8)

For several decades the elucidation of the reverse pathway involving the cholesterol efflux from cells and its transport and internalization by the liver has been intensively studied. However, these mechanisms have been poorly understood. The identification of SRBI as an HDL receptor was a step forward in our understanding of reverse cholesterol transport, but the role of this receptor within the scheme of human lipoprotein metabolism is still unknown. We may have reached a breaking point in our understanding of reverse cholesterol transport with the elucidation of a biological mystery that began almost four decades ago on the small Tangier Island located in Chesapeake Bay in the United States. Two brothers living on the island presented with orange tonsils, peripheral neuropathy, and, more interesting for the present focus, deficiency of HDL cholesterol. Scores of investigators have studied families affected with this rare disease with the notion that a better understanding of this disease could provide the key to the molecular basis of reverse cholesterol transport, therefore providing new mechanistic tools to prevent the development of atherosclerosis and the treatment of CVD [288, 289].

Patients with Tangier disease are characterized by their almost complete absence of HDL. They accumulate cholesteryl esters in tissues, resulting in enlarged orange tonsils, hepatosplenomegaly, peripheral neuropathy, and deposits in the rectal mucosa. After 40 years of research, several groups have reported, almost simultaneously, the gene locus and several specific mutations responsible for Tangier disease and for some forms of familial hypoalphalipoproteinemia [288, 289]. These investigations clearly demonstrate that mutations in the ATP-binding cassette transporter 1 (ABCA1) gene are responsible for the Tangier phenotype in several kindred from a variety of ethnic backgrounds. ABC1 belongs to the ABC family of genes. These genes encode for proteins involved in vectorial movement of substrates across biological membranes. The ABCA1 transporter consists of two symmetric halves, each including six membrane-spanning domains and a nucleotide-binding fold. Moreover, a long charged region and a highly hydrophobic segment link these two halves of the molecule. The transporter activity depends on its interaction with ATP. This specific transporter appears to be involved in transport of free cholesterol (and perhaps phospholipids), and the alteration of this important cellular function results in impaired efflux of free cellular cholesterol onto HDL particles, thus preventing the formation of normal nascent HDL particles. Consequently, the lipid-poor HDL is quickly removed from the circulation without being able to accomplish its function. This abnormality clearly prevents normal reverse cholesterol transport, and patients with Tangier disease tend to develop premature CVD, despite relatively low levels of LDL cholesterol [290].

It is important to remark that all these reports have found different mutations in the different kindreds studied. These data support the concept that Tangier disease may be very heterogeneous, a situation similar to that of familial hypercholesterolemia, where several hundred mutations have been reported for the LDL receptor locus. This may also explain the heterogeneous phenotypic expression of this disease, with pedigrees showing a high incidence of premature CVD, whereas in other families the risk does not appear to be much higher than in normal subjects.

These findings point toward a new physiological mechanism controlling cellular cholesterol metabolism and its efflux to HDL. However, given the rarity of Tangier disease, mutations at the ABC 1 gene could have very limited impact in modulating HDL levels in the population at large. In this regard, the findings initially reported by Marcil *et al.* [291] were of great relevance, because they demonstrated that the ABC1 locus was also responsible for a much more common HDL deficiency known as familial hypoalphalipoproteinemia that shares with Tangier's disease the low levels of HDL-C but not some of the other phenotypic characteristics (such as neuropathy and orange tonsils). Familial hypoalphalipoproteinemia is one of the most common genetic abnormalities in subjects with premature CVD.

Since the initial discovery of ABCA1 as a key player in lipoprotein metabolism, other members of the ATP-binding cassette (ABC) transporter family have been shown to play similar roles. Like ABCA1, ABCG1 exports excess cellular cholesterol into the HDL pathway and reduces cholesterol accumulation in macrophages. ABCG5 and ABCG8 form heterodimers that limit absorption of dietary sterols in the intestine and promote cholesterol elimination from the body through hepatobiliary secretion. All four transporters are induced by the same sterol-sensing nuclear receptor system. Unlike the mutations in ABCA1 that can cause severe HDL deficiency, those in the ABCG5 or ABCG8 genes can cause sitosterolemia, in which patients accumulate cholesterol and plant sterols in the circulation and develop premature CVD. Experimental animal models have shown that the disruption of ABCA1 or ABCG1 in mice promotes accumulation of excess cholesterol in macrophages, and manipulating mouse macrophage ABCA1 expression affects atherogenesis. Overexpressing ABCG5 and ABCG8 in mice attenuates diet-induced atherosclerosis in association with reduced circulating and liver cholesterol. Metabolites elevated in individuals with the metabolic syndrome and diabetes destabilize ABCA1 protein and inhibit transcription of all four transporters. Thus, impaired ABC cholesterol transporters might contribute to the enhanced atherogenesis associated with common inflammatory and metabolic disorders. Their beneficial effects on cholesterol homeostasis have made these transporters important new therapeutic targets for preventing and reversing CVD [292].

ABCA1 has not been properly investigated in terms of dietary response; however, several studies have examined the contribution of ABCG5 polymorphisms to the variability in dietary response. We examined differences in the plasma lipid response to consumption of cholesterol and carotenoids from eggs [293]. For this purpose, genotyping was conducted for 40 men and 51 premenopausal women who were randomly assigned to consume an egg (EGG, 640 mg/d additional dietary cholesterol and 600 μg lutein + zeaxanthin) or placebo (SUB, 0 mg/d cholesterol, 0 μg lutein + zeaxanthin) diet for 30 days. The two arms of the dietary intervention were separated by a 3-week washout period. Plasma concentrations of total cholesterol, LDL cholesterol (LDL-C), and HDL cholesterol were determined. Because eggs are an excellent source of lutein and zeaxanthin, the plasma levels of these carotenoids were also measured in a subset of subjects to determine whether the response to carotenoid intake was similar to that seen for dietary cholesterol and to evaluate the contribution of ABCG5 polymorphism to both responses. Individuals possessing the C/C genotype experienced a greater increase in both LDL-C ($p < 0.05$) and a trend for lutein ($p = 0.08$) during the EGG period compared with those individuals with the C/G (heterozygote) or G/G genotypes (homozygotes). These results, although obtained from a small number of subjects, suggest that the ABCG5 polymorphism may play a role in the plasma response to dietary cholesterol and carotenoids.

V. CONCLUSION

The mechanisms involved in the regulation of plasma lipoprotein levels by dietary factors, such as intakes of cholesterol and fatty acids, have been partially elucidated during the past century, thanks to the research efforts of scores of investigators. Genetic research has also brought up a number of candidate genes involved in the regulation of the homeostasis of blood lipids. Considering the significant effect of diet on plasma lipids, genetic variation at those loci is expected to explain some of the dramatic variations in lipoprotein response to dietary change that have been shown to exist among individuals. In fact, the current evidence supports the importance of gene–diet interactions in humans. Multiple candidate genes have been examined under different experimental conditions and the major findings have been highlighted in this chapter. However, because of conflicting results, more studies will be required to increase our predictive capacity and to reconcile the multiple discordances found in the literature.

To this regard, we have to keep in mind that lipoprotein response to dietary factors is extremely complex, as illustrated by the multiple interrelated pathways and genes discussed herein. Therefore, the effects of individual gene variants can be difficult to identify. In fact, the concerted action of differences in "gene families" may be required to elicit significant interindividual differences in responsiveness to diet. This systems biology approach has been facilitated by the availability of the public information generated by the Human Genome Project and the HAPMAP [294].

We should also be cautious concerning the interpretation of studies of association between allelic variants and common phenotypes. We should direct attention to the population admixture, which can cause an artificial association if a study includes genetically distinct subpopulations, one of which coincidentally displays a higher frequency of disease and allelic variants. Consideration of the ethnic backgrounds of subjects and the use of multiple, independent populations can help avoid this problem.

Another source of concern is multiple-hypothesis testing, which is aggravated by publication bias. Authors who test a single genetic variant for an association with a single phenotype base statistical thresholds for significance on a single hypothesis. But many laboratories search for associations using different variants. Each test represents an independent hypothesis, but only positive results are reported, leading to an overestimate of the significance of any positive associations. Statistical correction for multiple testing is possible, but the application of such thresholds results in loss of statistical power.

An additional caveat about the published literature relates to the fact that most studies were not initially designed to examine gene–diet interactions, but they are reanalyses of previously obtained data using new information from genetic analysis carried out *a posteriori*. Future studies need to be carefully designed in terms of sample size, taking into consideration the frequencies of the alleles examined. Moreover, we do not really know yet the specific dietary factors responsible for most of the effects already reported. Therefore, baseline and intervention diets should be carefully controlled in terms of dietary cholesterol, individual fatty acids, and levels of fat, as well as of fiber and other minor components of the diet such as phytosterols. It is also important to emphasize that some allele effects may be apparent primarily during situations of metabolic "stress," such as the postprandial state. Therefore, studies should be designed to test gene–diet interactions, in both the fasting and fed states. As indicated earlier, the most plausible scenario is that multiple genes will determine the response to dietary manipulation. Consequently, attention should be paid to gene–gene interactions. However, the large number of study subjects required and subsequent costs involved may make such studies infeasible. Two alternatives to examine these complex interactions in humans are possible: One consists of selecting study participants based on their genetic variants; the second needs to make better use of the large cohort studies for which dietary information has been or will be collected. Both approaches combined will enable us to take the concept of gene–diet interactions beyond the research laboratory into the real world.

Acknowledgments

This work was supported by NIH grants HL54776 and DK75030 and contract 58-1950-9-001 from the U.S. Department of Agriculture Research Service.

References

1. Weiss, L. A., Pan, L., Abney, M., and Ober, C. (2006). The sex-specific genetic architecture of quantitative traits in humans. *Nat. Genet.* **38**, 218–222.
2. Cleeman, J. I., and Lenfant, C. (1987). New guidelines for the treatment of high blood cholesterol in adults from the National Cholesterol Education Program. From controversy to consensus. *Circulation* **76**, 960–962.
3. (1994). National Cholesterol Education Program. Second report of the Expert Panel on Detection, Evaluation, and Treatment of High Blood Cholesterol in Adults (Adult Treatment Panel II). *Circulation* **89**, 1333–1445.
4. (2002). Third report of the National Cholesterol Education Program (NCEP). Expert Panel on Detection, Evaluation, and Treatment of High Blood Cholesterol in Adults (Adult Treatment Panel III). Final report. *Circulation* **106**, 3143–3421.
5. Grundy, S. M., Cleeman, J. I., Merz, C. N., *et al.* (2004). A summary of implications of recent clinical trials for the National Cholesterol Education Program Adult Treatment Panel III guidelines. *Arterioscler. Thromb. Vasc. Biol.* **24**, 1329–1330.
6. Stone, N. J., Bilek, S., and Rosenbaum, S. (2005). Recent National Cholesterol Education Program Adult Treatment Panel III update: adjustments and options. *Am. J. Cardiol.* **96**, 53E–59E.
7. Corella, D., and Ordovas, J. M. (2005). Integration of environment and disease into "omics" analysis. *Curr. Opin. Mol. Ther.* **7**, 569–576.
8. Backes, J. M., Gibson, CA, Ruisinger, J. F., and Moriarty, P. M. (2007). Fibrates: What have we learned in the past 40 years? *Pharmacotherapy* **27**, 412–424.
9. Brown, W. V. (2007). Expert commentary: The safety of fibrates in lipid-lowering therapy. *Am. J. Cardiol.* **99**, 19C–21C.
10. Wierzbicki, A. S. (2006). Fibrates after the FIELD study: Some answers, more questions. *Diabetes Vasc. Dis. Res.* **3**, 166–171.
11. Bots, M. L., Visseren, F. L., Evans, G. W., *et al.* (2007). Torcetrapib and carotid intima-media thickness in mixed dyslipidaemia (RADIANCE 2 study): A randomised, double-blind trial. *Lancet* **370**, 153–160.
12. Schmitz, G., Schmitz-Madry, A., and Ugocsai, P. (2007). Pharmacogenetics and pharmacogenomics of cholesterol-lowering therapy. *Curr. Opin. Lipidol.* **18**, 164–173.
13. Anderson, J. T., Grande, F., and Keys, A. (1957). Essential fatty acids, degree of unsaturation, and effect of corn (maize) oil on the serum-cholesterol level in man. *Lancet* **272**, 66–68.
14. Ahrens, E. H. Jr., (1957). Seminar on atherosclerosis: Nutritional factors and serum lipid levels. *Am. J. Med.* **23**, 928–952.
15. Keys, A., Anderson, J. T., and Grande, F. (1957). Prediction of serum-cholesterol responses of man to changes in fats in the diet. *Lancet* **273**, 959–966.
16. Hegsted, D. M., McGandy, R. B., Myers, M. L., and Stare, F.J. (1965). Quantitative effects of dietary fat on serum cholesterol in man. *Am. J. Clin. Nutr.* **17**, 281–295.
17. Mensink, R. P., and Katan, M. B. (1992). Effect of dietary fatty acids on serum lipids and lipoproteins. A meta-analysis of 27 trials. *Arterioscler. Thromb.* **12**, 911–919.
18. Cobb, M. M., and Teitlebaum, H. (1994). Determinants of plasma cholesterol responsiveness to diet. *Br. J. Nutr.* **71**, 271–282.
19. Hegsted, D. M., Ausman, L. M., Johnson, J. A., and Dallal, G. E. (1993). Dietary fat and serum lipids: An evaluation of the experimental data. *Am. J. Clin. Nutr.* **57**, 875–883.
20. Okey, R. (1933). Dietary blood and cholesterol in normal women. *J. Biol. Chem.* **99**, 717–727.
21. Katan, M. B., Beynen, A. C., de Vries, J. H., and Nobels, A. (1986). Existence of consistent hypo- and hyperresponders to dietary cholesterol in man. *Am. J. Epidemiol.* **123**, 221–234.
22. Jacobs, D.R, Jr., Anderson, J. T., Hannan, P., Keys, A., and Blackburn, H. (1983). Variability in individual serum cholesterol response to change in diet. *Arteriosclerosis (Dallas, Tex.)* **3**, 349–356.
23. O'Hanesian, M. A., Rosner, B., Bishop, L. M., and Sacks, F. M. (1996). Effects of inherent responsiveness to diet and day-to-day diet variation on plasma lipoprotein concentrations. *Am. J. Clin. Nutr.* **64**, 53–59.
24. Cobb, M. M., and Risch, N. (1993). Low-density lipoprotein cholesterol responsiveness to diet in normolipidemic subjects. *Metab. Clin. Exp.* **42**, 7–13.
25. Lefevre, M., Champagne, C. M., Tulley, R. T., Rood, J. C., and Most, M. M. (2005). Individual variability in cardiovascular disease risk factor responses to low-fat and low-saturated-fat diets in men: Body mass index, adiposity, and insulin resistance predict changes in LDL cholesterol. *Am. J. Clin. Nutr.* **82**, 957–963; quiz 1145–1146.
26. Mahaney, M. C., Blangero, J., Rainwater, D. L., *et al.* (1999). Pleiotropy and genotype by diet interaction in a baboon model for atherosclerosis: A multivariate quantitative genetic analysis of HDL subfractions in two dietary environments. *Arterioscler. Thromb. Vasc. Biol.* **19**, 1134–1141.

27. Rainwater, D. L., Kammerer, C. M., Carey, K. D., *et al.* (2002). Genetic determination of HDL variation and response to diet in baboons. *Atherosclerosis* **161**, 335–343.
28. Rainwater, D. L., Kammerer, C. M., Cox, L. A., *et al.* (2002). A major gene influences variation in large HDL particles and their response to diet in baboons. *Atherosclerosis* **163**, 241–248.
29. Rainwater, D. L., Kammerer, C. M., and VandeBerg, J. L. (1999). Evidence that multiple genes influence baseline concentrations and diet response of Lp(a) in baboons. *Arterioscler. Thromb. Vasc. Biol.* **19**, 2696–2700.
30. Corella, D., and Ordovas, J. M. (2005). Single nucleotide polymorphisms that influence lipid metabolism: Interaction with dietary factors. *Annu. Rev. Nutr.* **25**, 341–390.
31. Ordovas, J. M., and Corella, D. (2005). Genetic variation and lipid metabolism: Modulation by dietary factors. *Curr. Cardiol. Rep.* **7**, 480–486.
32. Brunner, E. J., Thorogood, M., Rees, K., and Hewitt, G. (2005). Dietary advice for reducing cardiovascular risk. *Cochrane database of systematic reviews (Online)*, CD002128.
33. Hooper, L., Summerbell, C. D., Higgins, J. P., *et al.* (2001). Reduced or modified dietary fat for preventing cardiovascular disease. *Cochrane database of systematic reviews (Online)*, CD002137.
34. Ordovas, J. M. (2001). Genetics, postprandial lipemia and obesity. *Nutr. Metab. Cardiovasc. Dis.* **11**, 118–133.
35. Davidson, N. O., and Shelness, G. S. (2000). Apolipoprotein B:mRNA editing, lipoprotein assembly, and presecretory degradation. *Annu. Rev. Nutr.* **20**, 169–193.
36. Lopez-Miranda, J., Kam, N., Osada, J., *et al.* (1994). Effect of fat feeding on human intestinal apolipoprotein B mRNA levels and editing. *Biochim. Biophys. Acta* **1214**, 143–147.
37. Boekholdt, S. M., Peters, R. J., Fountoulaki, K., Kastelein, J. J., and Sijbrands, E. J. (2003). Molecular variation at the apolipoprotein B gene locus in relation to lipids and cardiovascular disease: A systematic meta-analysis. *Hum. Genet.* **113**, 417–425.
38. Whitfield, A. J., Barrett, P. H., van Bockxmeer, F. M., and Burnett, J. R. (2004). Lipid disorders and mutations in the APOB gene. *Clin. Chem.* **50**, 1725–1732.
39. Aalto-Setala, K., Tikkanen, M. J., Taskinen, M. R., Nieminen, M., Holmberg, P., and Kontula, K. (1988). XbaI and c/g polymorphisms of the apolipoprotein B gene locus are associated with serum cholesterol and LDL-cholesterol levels in Finland. *Atherosclerosis* **74**, 47–54.
40. Talmud, P. J., Barni, N., Kessling, A. M., *et al.* (1987). Apolipoprotein B gene variants are involved in the determination of serum cholesterol levels: A study in normo– and hyperlipidaemic individuals. *Atherosclerosis* **67**, 81–89.
41. Series, J., Cameron, I., Caslake, M., Gaffney, D., Packard, C. J., and Shepherd, J. (1989). The Xba1 polymorphism of the apolipoprotein B gene influences the degradation of low density lipoprotein *in vitro*. *Biochim. Biophys. Acta* **1003**, 183–188.
42. Wiklund, O., Darnfors, C., Bjursell, G., *et al.* (1989). XbaI restriction fragment length polymorphism of apolipoprotein B in Swedish myocardial infarction patients. *Eur. J. Clin. Invest.* **19**, 255–258.
43. Genest, J. J. Jr., Ordovas, J. M., McNamara, J. R., *et al.* (1990). DNA polymorphisms of the apolipoprotein B gene in patients with premature coronary artery disease. *Atherosclerosis* **82**, 7–17.
44. Tikkanen, M. J., Xu, C. F., Hamalainen, T., *et al.* (1990). XbaI polymorphism of the apolipoprotein B gene influences plasma lipid response to diet intervention. *Clin. Genet.* **37**, 327–334.
45. Talmud, P. J., Boerwinkle, E., Xu, C. F., *et al.* (1992). Dietary intake and gene variation influence the response of plasma lipids to dietary intervention. *Genet. Epidemiol.* **9**, 249–260.
46. Rantala, M., Rantala, T. T., Savolainen, M. J., Friedlander, Y., and Kesaniemi, Y. A. (2000). Apolipoprotein B gene polymorphisms and serum lipids: Meta-analysis of the role of genetic variation in responsiveness to diet. *Am. J. Clin. Nutr.* **71**, 713–724.
47. Syvanne, M., Talmud, P. J., Humphries, S. E., *et al.* (1997). Determinants of postprandial lipemia in men with coronary artery disease and low levels of HDL cholesterol. *J. Lipid Res.* **38**, 1463–1472.
48. Lopez-Miranda, J., Ordovas, J. M., Ostos, M. A., *et al.* (1997). Dietary fat clearance in normal subjects is modulated by genetic variation at the apolipoprotein B gene locus. *Arterioscler. Thromb. Vasc. Biol.* **17**, 1765–1773.
49. Boerwinkle, E., and Chan, L. (1989). A three codon insertion/deletion polymorphism in the signal peptide region of the human apolipoprotein B (APOB) gene directly typed by the polymerase chain reaction. *Nucleic Acids Res.* **17**, 4003.
50. Xu, C. F., Tikkanen, M. J., Huttunen, J. K., *et al.* (1990). Apolipoprotein B signal peptide insertion/deletion polymorphism is associated with Ag epitopes and involved in the determination of serum triglyceride levels. *J. Lipid Res.* **31**, 1255–1261.
51. Boerwinkle, E., Brown, S. A., Rohrbach, K., Gotto, A. M., Jr., and Patsch, W. (1991). Role of apolipoprotein E and B gene variation in determining response of lipid, lipoprotein, and apolipoprotein levels to increased dietary cholesterol. *Am. J. Hum. Genet.* **49**, 1145–1154.
52. Byrne, C. D., Wareham, N. J., Mistry, P. K., *et al.* (1996). The association between free fatty acid concentrations and triglyceride–rich lipoproteins in the post-prandial state is altered by a common deletion polymorphism of the apo B signal peptide. *Atherosclerosis* **127**, 35–42.
53. Regis-Bailly, A., Visvikis, S., Steinmetz, J., Fournier, B., Gueguen, R., and Siest, G. (1996). Effects of apo B and apo E gene polymorphisms on lipid and apolipoprotein concentrations after a test meal. *Clin. Chim. Acta* **253**, 127–143.
54. Huang, L. S., de Graaf, J., and Breslow, J. L. (1988). ApoB gene MspI RFLP in exon 26 changes amino acid 3611 from Arg to Gln. *J. Lipid Res.* **29**, 63–67.
55. van't Hooft, F. M., Jormsjo, S., Lundahl, B., Tornvall, P., Eriksson, P., and Hamsten, A. (1999). A functional polymorphism in the apolipoprotein B promoter that influences the level of plasma low density lipoprotein. *J. Lipid Res.* **40**, 1686–1694.
56. Sposito, A. C., Gonbert, S., Turpin, G., Chapman, M. J., and Thillet, J. (2004). Common promoter C516T polymorphism in the ApoB gene is an independent predictor of carotid atherosclerotic disease in subjects presenting a broad range of

plasma cholesterol levels. *Arterioscler. Thromb. Vasc. Biol.* **24**, 2192–2195.

57. Perez-Martinez, P., Perez-Jimenez, F., Ordovas, J. M., *et al.* (2007). Postprandial lipemia is modified by the presence of the APOB–516C/T polymorphism in a healthy Caucasian population. *Lipids* **42**, 143–150.
58. Perez-Martinez, P., Perez-Jimenez, F., Ordovas, J. M., *et al.* (2007). The APOB –516C/T polymorphism is associated with differences in insulin sensitivity in healthy males during the consumption of diets with different fat content. *Br. J. Nutr.* **97**, 622–627.
59. Perez-Martinez, P., Lopez-Miranda, J., Ordovas, J. M., *et al.* (2004). Postprandial lipemia is modified by the presence of the polymorphism present in the exon 1 variant at the SR-BI gene locus. *J. Mol. Endocrinol.* **32**, 237–245.
60. Ordovas, J. M., Cassidy, D. K., Civeira, F., Bisgaier, C. L., and Schaefer, E. J. (1989). Familial apolipoprotein A-I, C-III, and A-IV deficiency and premature atherosclerosis due to deletion of a gene complex on chromosome 11. *J. Biol. Chem.* **264**, 16339–16342.
61. Goldberg, I. J., Scheraldi, C. A., Yacoub, L. K., Saxena, U., and Bisgaier, C. L. (1990). Lipoprotein ApoC-II activation of lipoprotein lipase. Modulation by apolipoprotein A-IV. *J. Biol. Chem.* **265**, 4266–4272.
62. Steinmetz, A., and Utermann, G. (1985). Activation of lecithin: Cholesterol acyltransferase by human apolipoprotein A-IV. *J. Biol. Chem.* **260**, 2258–2264.
63. Stein, O., Stein, Y., Lefevre, M., and Roheim, P. S. (1986). The role of apolipoprotein A-IV in reverse cholesterol transport studied with cultured cells and liposomes derived from an ether analog of phosphatidylcholine. *Biochim. Biophys. Acta* **878**, 7–13.
64. Steinmetz, A., Barbaras, R., Ghalim, N., Clavey, V., Fruchart, J. C., and Ailhaud, G. (1990). Human apolipoprotein A-IV binds to apolipoprotein A-I/A-II receptor sites and promotes cholesterol efflux from adipose cells. *J. Biol. Chem.* **265**, 7859–7863.
65. Menzel, H. J., Sigurdsson, G., Boerwinkle, E., Schrangl-Will, S., Dieplinger, H., and Utermann, G. (1990). Frequency and effect of human apolipoprotein A-IV polymorphism on lipid and lipoprotein levels in an Icelandic population. *Hum. Genet.* **84**, 344–346.
66. de Knijff, P., Johansen, L. G., Rosseneu, M., Frants, R. R., Jespersen, J., and Havekes. L. M. (1992). Lipoprotein profile of a Greenland Inuit population. Influence of anthropometric variables, Apo E and A4 polymorphism, and lifestyle. *Arterioscler. Thromb.* **12**, 1371–1379.
67. von Eckardstein, A., Funke, H., Schulte, M., Erren, M., Schulte, H., and Assmann, G. (1992). Nonsynonymous polymorphic sites in the apolipoprotein (apo) A-IV gene are associated with changes in the concentration of apo B- and apo A-I-containing lipoproteins in a normal population. *Am. J. Hum. Genet.* **50**, 1115–1128.
68. Kaprio, J., Ferrell, R. E., Kottke, B. A., Kamboh, M. I., and Sing, C. F. (1991). Effects of polymorphisms in apolipoproteins E, A-IV, and H on quantitative traits related to risk for cardiovascular disease. *Arterioscler. Thromb.* **11**, 1330–1348.
69. Kamboh, M. I., Hamman, R. F., and Ferrell, R. E. (1992). Two common polymorphisms in the APO A-IV coding gene: Their evolution and linkage disequilibrium. *Genet. Epidemiol.* **9**, 305–315.
70. Kamboh, M. I., Hamman, R. F., Iyengar, S., Aston, C. E., and Ferrell, R. E. (1991). Apolipoprotein A-IV polymorphism, and its role in determining variation in lipoprotein-lipid, glucose and insulin levels in normal and non-insulin-dependent diabetic individuals. *Atherosclerosis* **91**, 25–34.
71. Eichner, J. E., Kuller, L. H., Ferrell, R. E., and Kamboh, M. I. (1989). Phenotypic effects of apolipoprotein structural variation on lipid profiles: II. Apolipoprotein A-IV and quantitative lipid measures in the healthy women study. *Genet. Epidemiol.* **6**, 493–499.
72. Menzel, H. J., Boerwinkle, E., Schrangl-Will, S., and Utermann, G. (1988). Human apolipoprotein A-IV polymorphism: Frequency and effect on lipid and lipoprotein levels. *Hum. Genet.* **79**, 368–372.
73. Menzel, H. J., Dieplinger, H., Sandholzer, C., Karadi, I., Utermann, G., and Csaszar, A. (1995). Apolipoprotein A-IV polymorphism in the Hungarian population: Gene frequencies, effect on lipid levels, and sequence of two new variants. *Hum. Mutat.* **5**, 58–65.
74. Ganan, A., Corella, D., Guillen, M., Ordovas, J. M., and Pocovi, M. (2004). Frequencies of apolipoprotein A4 gene polymorphisms and association with serum lipid concentrations in two healthy Spanish populations. *Hum. Biol.* **76**, 253–266.
75. Bai, H., Saku, K., Liu, R., Funke, H., von Eckardstein, A., and Arakawa, K. (1993). Polymorphic site study at codon 347 of apolipoprotein A-IV in a Japanese population. *Biochim. Biophys. Acta* **1174**, 279–281.
76. de Knijff, P., Rosseneu, M., Beisiegel, U., de Keersgieter, W., Frants, R. R., and Havekes, L. M. (1988). Apolipoprotein A-IV polymorphism and its effect on plasma lipid and apolipoprotein concentrations. *J. Lipid Res.* **29**, 1621–1627.
77. Crews, D. E., Kamboh, M. I., Mancilha-Carvalho, J. J., and Kottke, B. (1993). Population genetics of apolipoprotein A-4, E, and H polymorphisms in Yanomami Indians of northwestern Brazil: Associations with lipids, lipoproteins, and carbohydrate metabolism. *Hum. Biol.* **65**, 211–224.
78. Hanis, C. L., Douglas, T. C., and Hewett-Emmett, D. (1991). Apolipoprotein A-IV protein polymorphism: Frequency and effects on lipids, lipoproteins, and apolipoproteins among Mexican-Americans in Starr County, Texas. *Hum. Genet.* **86**, 323–325.
79. Zaiou, M., Visvikis, S., Gueguen, R., Parra, H. J., Fruchart, J. C., and Siest, G. (1994). DNA polymorphisms of human apolipoprotein A-IV gene: Frequency and effects on lipid, lipoprotein and apolipoprotein levels in a French population. *Clin. Genet.* **46**, 248–254.
80. Mata, P., Ordovas, J. M., Lopez-Miranda, J., *et al.* (1994). ApoA-IV phenotype affects diet-induced plasma LDL cholesterol lowering. *Arterioscler. Thromb.* **14**, 884–891.
81. McCombs, R. J., Marcadis, D. E., Ellis, J., and Weinberg, R. B. (1994). Attenuated hypercholesterolemic response to a high–cholesterol diet in subjects heterozygous for the apolipoprotein A-IV-2 allele. *N. Engl. J. Med.* **331**, 706–710.
82. Hubacek, J. A., Bohuslavova, R., Skodova, Z., Pitha, J., Bobkova, D., and Poledne, R. (2007). Polymorphisms in the APOA1/C3/A4/A5 gene cluster and cholesterol responsiveness to dietary change. *Clin. Chem. Lab. Med.* **45**, 316–320.

83. Jansen, S., Lopez-Miranda, J., Salas, J., *et al.* (1997). Effect of 347-serine mutation in apoprotein A-IV on plasma LDL cholesterol response to dietary fat. *Arterioscler. Thromb. Vasc. Biol.* **17**, 1532–1538.
84. Campos, H., Lopez-Miranda, J., Rodriguez, C., Albajar, M., Schaefer, E. J., and Ordovas, J. M. (1997). Urbanization elicits a more atherogenic lipoprotein profile in carriers of the apolipoprotein A-IV-2 allele than in A-IV-1 homozygotes. *Arterioscler. Thromb. Vasc. Biol.* **17**, 1074–1081.
85. Carmena—Ramon, R., Ascaso, J. F., Real, J. T., Ordovas, J.M., and Carmena, R. (1998). Genetic variation at the apoA-IV gene locus and response to diet in familial hypercholesterolemia. *Arterioscler. Thromb. Vasc. Biol.* **18**, 1266–1274.
86. Huff, M. W. (2003). Dietary cholesterol, cholesterol absorption, postprandial lipemia and atherosclerosis. *Can. J. Clin. Pharmacol.* **10**(Suppl), 26A–32A.
87. Hyson, D., Rutledge, J. C., and Berglund, L. (2003). Postprandial lipemia and cardiovascular disease. *Curr. Atheroscler. Rep.* **5**, 437–444.
88. Ostos, M. A., Lopez-Miranda, J., Marin, C., *et al.* (2000). The apolipoprotein A-IV-360His polymorphism determines the dietary fat clearance in normal subjects. *Atherosclerosis* **153**, 209–217.
89. Ostos, M. A., Lopez-Miranda, J., Ordovas, J. M., *et al.* (1998). Dietary fat clearance is modulated by genetic variation in apolipoprotein A-IV gene locus. *J. Lipid Res.* **39**, 2493–2500.
90. Mahley, R. W. (1988). Apolipoprotein E: Cholesterol transport protein with expanding role in cell biology. *Science* **240**, 622–630.
91. Beisiegel, U., Weber, W., Ihrke, G., Herz, J., and Stanley, K. K. (1989). The LDL-receptor-related protein, LRP, is an apolipoprotein E-binding protein. *Nature* **341**, 162–164.
92. Hatters, D. M., Peters-Libeu, C. A., and Weisgraber, K. H. (2006). Apolipoprotein E structure: insights into function. *Trends Biochem. Sci.* **31**, 445–454.
93. Eichner, J. E., Dunn, S. T., Perveen, G., Thompson, D. M., Stewart, K. E., and Stroehla, B. C. (2002). Apolipoprotein E polymorphism and cardiovascular disease: A HuGE review. *Am. J. Epidemiol.* **155**, 487–495.
94. Schaefer, E. J., Gregg, R. E., Ghiselli, G., *et al.* (1986). Familial apolipoprotein E deficiency. *J. Clin. Invest.* **78**, 1206–1219.
95. Ordovas, J. M., Litwack-Klein, L., Wilson, P. W., Schaefer, M. M., and Schaefer, E. J. (1987). Apolipoprotein E isoform phenotyping methodology and population frequency with identification of apoE1 and apoE5 isoforms. *J. Lipid Res.* **28**, 371–380.
96. Schaefer, E. J., Lamon-Fava, S., Johnson, S., *et al.* (1994). Effects of gender and menopausal status on the association of apolipoprotein E phenotype with plasma lipoprotein levels. Results from the Framingham Offspring Study. *Arterioscler. Thromb.* **14**, 1105–1113.
97. Davignon, J., Gregg, R. E., and Sing, C. F. (1988). Apolipoprotein E polymorphism and atherosclerosis. *Arteriosclerosis (Dallas, Tex.)* **8**, 1–21.
98. Masson, L. F., and McNeill, G. (2005). The effect of genetic variation on the lipid response to dietary change: Recent findings. *Curr. Opin. Lipidol.* **16**, 61–67.
99. Brown, A. J., and Roberts, D. C. (1991). The effect of fasting triacylglyceride concentration and apolipoprotein E polymorphism on postprandial lipemia. *Arterioscler. Thromb.* **11**, 1737–1744.
100. Kesaniemi, Y. A., Ehnholm, C., and Miettinen, T. A. (1987). Intestinal cholesterol absorption efficiency in man is related to apoprotein E phenotype. *J. Clin. Invest.* **80**, 578–581.
101. Weintraub, M. S., Eisenberg, S., and Breslow, J. L. (1987). Dietary fat clearance in normal subjects is regulated by genetic variation in apolipoprotein E. *J. Clin. Invest.* **80**, 1571–1577.
102. Nikkilä, M., Solakivi, T., Lehtimaki, T., Koivula, T., Laippala, P., and Astrom, B. (1994). Postprandial plasma lipoprotein changes in relation to apolipoprotein E phenotypes and low density lipoprotein size in men with and without coronary artery disease. *Atherosclerosis* **106**, 149–157.
103. Superko, H. R., and Haskell, W. L. (1991). The effect of apolipoprotein E isoform difference on postprandial lipoprotein in patients matched for triglycerides, LDL-cholesterol, and HDL-cholesterol. *Artery* **18**, 315–325.
104. Talmud, P. J., Waterworth, D. M., and Humphries, S. E. (2001). Effect of genetic variation on the postprandial response. Results from the European Atherosclerosis Research Study II. *World Rev. Nutr. Diet.* **89**, 53–60.
105. Cardona, F., Morcillo, S., Gonzalo-Marin, M., and Tinahones, F. J. (2005). The apolipoprotein E genotype predicts postprandial hypertriglyceridemia in patients with the metabolic syndrome. *J. Clin. Endocrinol. Metab.* **90**, 2972–2975.
106. Boerwinkle, E., Brown, S., Sharrett, AR, Heiss, G., and Patsch, W. (1994). Apolipoprotein E polymorphism influences postprandial retinyl palmitate but not triglyceride concentrations. *Am. J. Hum. Genet.* **54**, 341–360.
107. Reznik, Y., Pousse, P., Herrou, M., *et al.* (1996). Postprandial lipoprotein metabolism in normotriglyceridemic non-insulin-dependent diabetic patients: Influence of apolipoprotein E polymorphism. *Metab. Clin. Exp.* **45**, 63–71.
108. Orth, M., Wahl, S., Hanisch, M., Friedrich, I., Wieland, H., and Luley, C. (1996). Clearance of postprandial lipoproteins in normolipemics: Role of the apolipoprotein E phenotype. *Biochim. Biophys. Acta* **1303**, 22–30.
109. Bergeron, N., and Havel, R. J. (1996). Prolonged postprandial responses of lipids and apolipoproteins in triglyceride-rich lipoproteins of individuals expressing an apolipoprotein epsilon 4 allele. *J. Clin. Invest.* **97**, 65–72.
110. Viiri, L. E., Loimaala, A., Nenonen, A., *et al.* (2005). The association of the apolipoprotein E gene promoter polymorphisms and haplotypes with serum lipid and lipoprotein concentrations. *Atherosclerosis* **179**, 161–167.
111. Moreno, J. A., Lopez-Miranda, J., Marin, C., *et al.* (2003). The influence of the apolipoprotein E gene promoter (–219G/T) polymorphism on postprandial lipoprotein metabolism in young normolipemic males. *J. Lipid Res.* **44**, 2059–2064.
112. Moreno, J. A., Perez-Jimenez, F., Marin, C., *et al.* (2004). Apolipoprotein E gene promoter –219G→T polymorphism increases LDL-cholesterol concentrations and susceptibility to oxidation in response to a diet rich in saturated fat. *Am. J. Clin. Nutr.* **80**, 1404–1409.
113. Moreno, J. A., Perez-Jimenez, F., Marin, C., *et al.* (2005). The apolipoprotein E gene promoter (–219G/→T)

polymorphism determines insulin sensitivity in response to dietary fat in healthy young adults. *J. Nutr.* **135**, 2535–2540.

114. Stengard, J. H., Frikke-Schmidt, R., Tybjaerg-Hansen, A., Nordestgaard, B. G., and Sing, C. F. (2007). Variation in 5′ promoter region of the APOE gene contributes to predicting ischemic heart disease (IHD) in the population at large: The Copenhagen City Heart Study. *Ann. Hum. Genet.* **71**, 762–771.
115. Zannis, V. I., Cole, F. S., Jackson, C. L., Kurnit, D. M., and Karathanasis, S. K. (1985). Distribution of apolipoprotein A-I, C-II, C-III, and E mRNA in fetal human tissues. Time-dependent induction of apolipoprotein E mRNA by cultures of human monocyte-macrophages. *Biochemistry* **24**, 4450–4455.
116. Wang, C. S., McConathy, W. J., Kloer, H. U., and Alaupovic, P. (1985). Modulation of lipoprotein lipase activity by apolipoproteins. Effect of apolipoprotein C-III. *J. Clin. Invest.* **75**, 384–390.
117. Ito, Y., Azrolan, N., O'Connell, A., Walsh, A., and Breslow, J. L. (1990). Hypertriglyceridemia as a result of human apo CIII gene expression in transgenic mice. *Science* **249**, 790–793.
118. Lai, C. Q., Parnell, L. D., and Ordovas, J. M. (2005). The APOA1/C3/A4/A5 gene cluster, lipid metabolism and cardiovascular disease risk. *Curr. Opin. Lipidol.* **16**, 153–166.
119. Dammerman, M., Sandkuijl, L. A., Halaas, J. L., Chung, W., and Breslow, J. L. (1993). An apolipoprotein CIII haplotype protective against hypertriglyceridemia is specified by promoter and 3′ untranslated region polymorphisms. *Proc. Natl. Acad. Sci. USA* **90**, 4562–4566.
120. Li, W. W., Dammerman, M. M., Smith, J. D., Metzger, S., Breslow, J. L., and Leff, T. (1995). Common genetic variation in the promoter of the human apo CIII gene abolishes regulation by insulin and may contribute to hypertriglyceridemia. *J. Clin. Invest.* **96**, 2601–2605.
121. Lopez-Miranda, J., Jansen, S., Ordovas, J. M., *et al.* (1997). Influence of the SstI polymorphism at the apolipoprotein C-III gene locus on the plasma low-density-lipoprotein-cholesterol response to dietary monounsaturated fat. *Am. J. Clin. Nutr.* **66**, 97–103.
122. Waterworth, D. M., Ribalta, J., Nicaud, V., Dallongeville, J., Humphries, S. E., and Talmud, P. (1999). ApoCIII gene variants modulate postprandial response to both glucose and fat tolerance tests. *Circulation* **99**, 1872–1877.
123. Salas, J., Jansen, S., Lopez-Miranda, J., *et al.* (1998). The SstI polymorphism of the apolipoprotein C-III gene determines the insulin response to an oral-glucose-tolerance test after consumption of a diet rich in saturated fats. *Am. J. Clin. Nutr.* **68**, 396–401.
124. Ordovas, J. M., and Mooser, V. (2005). Genes, lipids and aging: Is it all accounted for by cardiovascular disease risk? [editorial review]. *Curr. Opin. Lipidol.* **16**, 121–126.
125. Atzmon, G., Rincon, M., Schechter, C. B., *et al.* (2006). Lipoprotein genotype and conserved pathway for exceptional longevity in humans. *PLoS Biol.* **4**, e113.
126. Pennacchio, L. A., Olivier, M., Hubacek, J. A., *et al.* (2001). An apolipoprotein influencing triglycerides in humans and mice revealed by comparative sequencing. *Science* **294**, 169–173.
127. Talmud, P. J., Hawe, E., Martin, S., *et al.* (2002). Relative contribution of variation within the APOC3/A4/A5 gene cluster in determining plasma triglycerides. *Hum. Mol. Genet.* **11**, 3039–3046.
128. Baum, L., Tomlinson, B., and Thomas, G. N. (2003). APOA5-1131T > C polymorphism is associated with triglyceride levels in Chinese men. *Clin. Genet.* **63**, 377–379.
129. Evans, D., Buchwald, A., and Beil, F. U. (2003). The single nucleotide polymorphism –1131T > C in the apolipoprotein A5 (APOA5). gene is associated with elevated triglycerides in patients with hyperlipidemia. *J. Mol. Med.* **81**, 645–654.
130. Kao, J. T., Wen, H. C., Chien, K. L., Hsu, H. C., and Lin, S. W. (2003). A novel genetic variant in the apolipoprotein A5 gene is associated with hypertriglyceridemia. *Hum. Mol. Genet.* **12**, 2533–2539.
131. Lai, C. Q., Tai, E. S., Tan, C. E., *et al.* (2003). The APOA5 locus is a strong determinant of plasma triglyceride concentrations across ethnic groups in Singapore. *J. Lipid Res.* **44**, 2365–2373.
132. Lai, C. Q., Demissie, S., Cupples, L. A., *et al.* (2004). Influence of the APOA5 locus on plasma triglyceride, lipoprotein subclasses, and CVD risk in the Framingham Heart Study. *J. Lipid Res.* **45**, 2096–2105.
133. Priore Oliva, C., Tarugi, P., Calandra, S., *et al.* (2006). A novel sequence variant in APOA5 gene found in patients with severe hypertriglyceridemia. *Atherosclerosis* **188**, 215–217.
134. Talmud, P. J. (2007). Rare APOA5 mutations – Clinical consequences, metabolic and functional effects. An ENID review. *Atherosclerosis* **194**, 287–292.
135. Pennacchio, L. A., Olivier, M., Hubacek, J. A., Krauss, R. M., Rubin, E. M., and Cohen, J. C. (2002). Two independent apolipoprotein A5 haplotypes influence human plasma triglyceride levels. *Hum. Mol. Genet.* **11**, 3031–3038.
136. Wang, Q. F., Liu, X., O'Connell, J., *et al.* (2004). Haplotypes in the APOA1-C3-A4-A5 gene cluster affect plasma lipids in both humans and baboons. *Hum. Mol. Genet.* **13**, 1049–1056.
137. Bi, N., Yan, S. K., Li, G. P., Yin, Z. N., and Chen, B. S. (2004). A single nucleotide polymorphism –1131T > C in the apolipoprotein A5 gene is associated with an increased risk of coronary artery disease and alters triglyceride metabolism in Chinese. *Mol. Genet. Metab.* **83**, 280–286.
138. Lee, K. W., Ayyobi, A. F., Frohlich, J. J., and Hill, J. S. (2004). APOA5 gene polymorphism modulates levels of triglyceride, HDL cholesterol and FERHDL but is not a risk factor for coronary artery disease. *Atherosclerosis* **176**, 165–172.
139. Szalai, C., Keszei, M., Duba, J., *et al.* (2004). Polymorphism in the promoter region of the apolipoprotein A5 gene is associated with an increased susceptibility for coronary artery disease. *Atherosclerosis* **173**, 109–114.
140. Talmud, P. J., Martin, S., Taskinen, M. R., *et al.* (2004). APOA5 gene variants, lipoprotein particle distribution, and progression of coronary heart disease: Results from the LOCAT study. *J. Lipid Res.* **45**, 750–756.
141. Liu, H., Zhang, S., Lin, J., *et al.* (2005). Association between DNA variant sites in the apolipoprotein A5 gene and coronary heart disease in Chinese. *Metab. Clin. Exp.* **54**, 568–572.
142. Ruiz–Narvaez, E. A., Yang, Y., Nakanishi, Y., Kirchdorfer, J., and Campos, H. (2005). APOC3/A5 haplotypes, lipid levels, and risk of myocardial infarction in the Central Valley of Costa Rica. *J. Lipid Res.* **46**, 2605–2613.
143. Tang, Y., Sun, P., Guo, D., *et al.* (2006). A genetic variant c.553G > T in the apolipoprotein A5 gene is associated with

an increased risk of coronary artery disease and altered triglyceride levels in a Chinese population. *Atherosclerosis* **185**, 433–437.

144. Hsu, L. A., Ko, Y. L., Chang, C. J., *et al.* (2006). Genetic variations of apolipoprotein A5 gene is associated with the risk of coronary artery disease among Chinese in Taiwan. *Atherosclerosis* **185**, 143–149.
145. Lai, C. Q., Corella, D., Demissie, S., *et al.* (2006). Dietary intake of n-6 fatty acids modulates effect of apolipoprotein A5 gene on plasma fasting triglycerides, remnant lipoprotein concentrations, and lipoprotein particle size: The Framingham Heart Study. *Circulation* **113**, 2062–2070.
146. Elosua, R., Ordovas, J. M., Cupples, L. A., *et al.* (2006). Variants at the APOA5 locus, association with carotid atherosclerosis, and modification by obesity: The Framingham Study. *J. Lipid Res.* **47**, 990–996.
147. Corella, D., Lai, C. Q., Demissie, S., *et al.* (2007). APOA5 gene variation modulates the effects of dietary fat intake on body mass index and obesity risk in the Framingham Heart Study. *J. Mol. Med.* **85**, 119–128.
148. Aberle, J., Evans, D., Beil, F. U., and Seedorf, U. (2005). A polymorphism in the apolipoprotein A5 gene is associated with weight loss after short-term diet. *Clin. Genet.* **68**, 152–154.
149. Martin, S., Nicaud, V., Humphries, S. E., and Talmud, P. J. (2003). Contribution of APOA5 gene variants to plasma triglyceride determination and to the response to both fat and glucose tolerance challenges. *Biochim. Biophys. Acta* **1637**, 217–225.
150. Masana, L., Ribalta, J., Salazar, J., Fernandez-Ballart, J., Joven, J., and Cabezas, M. C. (2003). The apolipoprotein AV gene and diurnal triglyceridaemia in normolipidaemic subjects. *Clin. Chem. Lab. Med.* **41**, 517–521.
151. Jang, Y., Kim, J. Y., Kim, O. Y., *et al.* (2004). The –1131T→C polymorphism in the apolipoprotein A5 gene is associated with postprandial hypertriacylglycerolemia; elevated small, dense LDL concentrations; and oxidative stress in nonobese Korean men. *Am. J. Clin. Nutr.* **80**, 832–840.
152. Kim, J. Y., Kim, O. Y., Koh, S. J., *et al.* (2006). Comparison of low-fat meal and high-fat meal on postprandial lipemic response in non-obese men according to the –1131T > C polymorphism of the apolipoprotein A5 (APOA5) gene (randomized cross-over design). *J. Am. Coll. Nutr.* **25**, 340–347.
153. Moreno, R., Perez-Jimenez, F., Marin, C., *et al.* (2006). A single nucleotide polymorphism of the apolipoprotein A-V gene –1131T > C modulates postprandial lipoprotein metabolism. *Atherosclerosis* **189**, 163–168.
154. Chandak, G. R., Ward, K. J., Yajnik, C. S., *et al.* (2006). Triglyceride associated polymorphisms of the APOA5 gene have very different allele frequencies in Pune, India compared to Europeans. *BMC Med. Genet.* **7**, 76.
155. Karpe, F., Bickerton, A. S., Hodson, L., Fielding, B. A., Tan, G. D., and Frayn, K. N. (2007). Removal of triacylglycerols from chylomicrons and VLDL by capillary beds: The basis of lipoprotein remnant formation. *Biochem. Soc. Trans.* **35**, 472–476.
156. Tsutsumi, K. (2003). Lipoprotein lipase and atherosclerosis. *Curr. Vasc. Pharmacol.* **1**, 11–17.
157. Templeton, A. R., Clark, A. G., Weiss, K. M., Nickerson, D. A., Boerwinkle, E., and Sing, C. F. (2000). Recombinational and mutational hotspots within the human lipoprotein lipase gene. *Am. J. Hum. Genet.* **66**, 69–83.
158. Templeton, A. R., Weiss, K. M., Nickerson, D. A., Boerwinkle, E., and Sing, C. F. (2000). Cladistic structure within the human lipoprotein lipase gene and its implications for phenotypic association studies. *Genetics* **156**, 1259–1275.
159. Rip, J., Nierman, M. C., Ross, C. J., *et al.* (2006). Lipoprotein lipase S447X: A naturally occurring gain-of-function mutation. *Arterioscler. Thromb. Vasc. Biol.* **26**, 1236–1245.
160. Lopez-Miranda, J., Cruz, G., Gomez, P., *et al.* (2004). The influence of lipoprotein lipase gene variation on postprandial lipoprotein metabolism. *J. Clin. Endocrinol. Metab.* **89**, 4721–4728.
161. Merkel, M., Eckel, R. H., and Goldberg, I. J. (2002). Lipoprotein lipase: Genetics, lipid uptake, and regulation. *J. Lipid Res.* **43**, 1997–2006.
162. Hu, Y., Liu, W., Huang, R., and Zhang, X. (2006). A systematic review and meta-analysis of the relationship between lipoprotein lipase Asn291Ser variant and diseases. *J. Lipid Res.* **47**, 1908–1914.
163. Pimstone, S. N., Gagne, S. E., Gagne, C., *et al.* (1995). Mutations in the gene for lipoprotein lipase. A cause for low HDL cholesterol levels in individuals heterozygous for familial hypercholesterolemia. *Arterioscler. Thromb. Vasc. Biol.* **15**, 1704–1712.
164. Mero, N., Suurinkeroinen, L., Syvanne, M., Knudsen, P., Yki–Jarvinen, H., and Taskinen, M. R. (1999). Delayed clearance of postprandial large TG-rich particles in normolipidemic carriers of LPL Asn291Ser gene variant. *J. Lipid Res.* **40**, 1663–1670.
165. Evans, V., and Kastelein, J. J. (2002). Lipoprotein lipase deficiency—rare or common? *Cardiovasc. Drugs Ther.* **16**, 283–287.
166. Chmurzynska, A. (2006). The multigene family of fatty acid-binding proteins (FABPs): Function, structure and polymorphism. *J. Appl. Genet.* **47**, 39–48.
167. Ono, T. (2005). Studies of the FABP family: A retrospective. *Mol. Cell. Biochem.* **277**, 1–6.
168. Clarke, S. D., and Armstrong, M. K. (1989). Cellular lipid binding proteins: Expression, function, and nutritional regulation. *FASEB J.* **3**, 2480–2487.
169. Tso, P., Nauli, A., and Lo, C. M. (2004). Enterocyte fatty acid uptake and intestinal fatty acid-binding protein. *Biochem. Soc. Trans.* **32**, 75–78.
170. Bass, N. M., Manning, J. A., Ockner, R. K., Gordon, J. I., Seetharam, S., and Alpers, D. H. (1985). Regulation of the biosynthesis of two distinct fatty acid-binding proteins in rat liver and intestine. Influences of sex difference and of clofibrate. *J. Biol. Chem.* **260**, 1432–1436.
171. Sacchettini, J. C., Gordon, J. I., and Banaszak, L. J. (1989). Crystal structure of rat intestinal fatty-acid-binding protein. Refinement and analysis of the *Escherichia coli*-derived protein with bound palmitate. *J. Mol. Biol.* **208**, 327–339.
172. Baier, L. J., Sacchettini, J. C., Knowler, W. C., *et al.* (1995). An amino acid substitution in the human intestinal fatty acid

binding protein is associated with increased fatty acid binding, increased fat oxidation, and insulin resistance. *J. Clin. Invest.* **95**, 1281–1287.

173. Tataranni, P. A., Baier, L. J., Paolisso, G., Howard, B. V., and Ravussin, E. (1996). Role of lipids in development of noninsulin-dependent diabetes mellitus: Lessons learned from Pima Indians. *Lipids* **31**(Suppl), S267–270.
174. Hegele, R. A., Young, T. K., and Connelly, P. W. (1997). Are Canadian Inuit at increased genetic risk for coronary heart disease? *J. Mol. Med.* **75**, 364–370.
175. Pihlajamaki, J., Rissanen, J., Heikkinen, S., Karjalainen, L., and Laakso, M. (1997). Codon 54 polymorphism of the human intestinal fatty acid binding protein 2 gene is associated with dyslipidemias but not with insulin resistance in patients with familial combined hyperlipidemia. *Arterioscler. Thromb. Vasc. Biol.* **17**, 1039–1044.
176. Rissanen, J., Pihlajamaki, J., Heikkinen, S., Kekalainen, P., Kuusisto, J., and Laakso, M. (1997). The Ala54Thr polymorphism of the fatty acid binding protein 2 gene does not influence insulin sensitivity in Finnish nondiabetic and NIDDM subjects. *Diabetes* **46**, 711–712.
177. Stern, M. P., Mitchell, B. D., Blangero, J., *et al.* (1996). Evidence for a major gene for type II diabetes and linkage analyses with selected candidate genes in Mexican-Americans. *Diabetes* **45**, 563–568.
178. Hegele, R. A., Connelly, P. W., Hanley, A. J., Sun, F., Harris, S. B., and Zinman, B. (1997). Common genomic variants associated with variation in plasma lipoproteins in young aboriginal Canadians. *Arterioscler. Thromb. Vasc. Biol.* **17**, 1060–1066.
179. Wanby, P., Palmquist, P., Brudin, L., and Carlsson, M. (2005). Genetic variation of the intestinal fatty acid–binding protein 2 gene in carotid atherosclerosis. *Vasc. Med.* **10**, 103–108.
180. Hegele, R. A., Wolever, T. M., Story, J. A., Connelly, P. W., and Jenkins, D. J. (1997). Intestinal fatty acid-binding protein variation associated with variation in the response of plasma lipoproteins to dietary fibre. *Eur. J. Clin. Invest.* **27**, 857–862.
181. Agren, J. J., Valve, R., Vidgren, H., Laakso, M., and Uusitupa, M. (1998). Postprandial lipemic response is modified by the polymorphism at codon 54 of the fatty acid-binding protein 2 gene. *Arterioscler. Thromb. Vasc. Biol.* **18**, 1606–1610.
182. Marin, C., Perez-Jimenez, F., Gomez, P., *et al.* (2005). The Ala54Thr polymorphism of the fatty acid-binding protein 2 gene is associated with a change in insulin sensitivity after a change in the type of dietary fat. *Am. J. Clin. Nutr.* **82**, 196–200.
183. Baier, L. J., Bogardus, C., and Sacchettini, J. C. (1996). A polymorphism in the human intestinal fatty acid binding protein alters fatty acid transport across Caco-2 cells. *J. Biol. Chem.* **271**, 10892–10896.
184. Prows, D.R, Murphy, E. J., Moncecchi, D., and Schroeder, F. (1996). Intestinal fatty acid-binding protein expression stimulates fibroblast fatty acid esterification. *Chem. Phys. Lipids* **84**, 47–56.
185. Hodsdon, M. E., and Cistola, D. P. (1997). Discrete backbone disorder in the nuclear magnetic resonance structure of apo intestinal fatty acid-binding protein: Implications for the mechanism of ligand entry. *Biochemistry* **36**, 1450–1460.
186. Zhang, F., Lucke, C., Baier, L. J., Sacchettini, J. C., and Hamilton, J. A. (1997). Solution structure of human intestinal fatty acid binding protein: Implications for ligand entry and exit. *J. Biomol. NMR* **9**, 213–228.
187. Hodsdon, M. E., and Cistola, D. P. (1997). Ligand binding alters the backbone mobility of intestinal fatty acid-binding protein as monitored by ^{15}N NMR relaxation and ^{1}H exchange. *Biochemistry* **36**, 2278–2290.
188. Soutar, A. K., and Naoumova, R. P. (2007). Mechanisms of disease: Genetic causes of familial hypercholesterolemia. *Nat. Clin. Pract.* **4**, 214–225.
189. Muallem, H., North, K. E., Kakoki, M., *et al.* (2007). Quantitative effects of common genetic variations in the 3′UTR of the human LDL-receptor gene and their associations with plasma lipid levels in the Atherosclerosis Risk in Communities study. *Hum. Genet.* **121**, 421–431.
190. Frikke-Schmidt, R., Nordestgaard, B. G., Schnohr, P., and Tybjaerg-Hansen, A. (2004). Single nucleotide polymorphism in the low-density lipoprotein receptor is associated with a threefold risk of stroke. A case-control and prospective study. *Eur. Heart J.* **25**, 943–951.
191. Vieira, J. R., Whittall, R. A., Cooper, J. A., Miller, G. J., and Humphries, S. E. (2006). The A370T variant (StuI polymorphism). in the LDL receptor gene is not associated with plasma lipid levels or cardiovascular risk in UK men. *Ann. Hum. Genet.* **70**, 697–704.
192. Chasman, D. I., Posada, D., Subrahmanyan, L., Cook, N. R., Stanton, V.P. Ridker, P. M. (2004). Pharmacogenetic study of statin therapy and cholesterol reduction. *JAMA* **291**, 2821–2827.
193. Plat, J., and Mensink, R. P. (2002). Relationship of genetic variation in genes encoding apolipoprotein A-IV, scavenger receptor BI, HMG-CoA reductase, CETP and apolipoprotein E with cholesterol metabolism and the response to plant stanol ester consumption. *Eur. J. Clin. Invest.* **32**, 242–250.
194. Kiss, R. S., Kavaslar, N., Okuhira, K., *et al.* (2007). Genetic etiology of isolated low HDL syndrome: Incidence and heterogeneity of efflux defects. *Arterioscler. Thromb. Vasc. Biol.* **27**, 1139–1145.
195. Miller, M., Rhyne, J., Hong, S. H., Friel, G., Dolinar, C., and Riley, W. (2007). Do mutations causing low HDL-C promote increased carotid intima-media thickness? *Clin. Chim. Acta* **377**, 273–275.
196. Krimbou, L., Marcil, M., and Genest, J. (2006). New insights into the biogenesis of human high-density lipoproteins. *Curr. Opin. Lipidol.* **17**, 258–267.
197. Sviridov, D., and Nestel, P. J. (2007). Genetic factors affecting HDL levels, structure, metabolism and function. *Curr. Opin. Lipidol.* **18**, 157–163.
198. Lewis, G. F. (2006). Determinants of plasma HDL concentrations and reverse cholesterol transport. *Curr. Opin. Cardiol.* **21**, 345–352.
199. Dastani, Z., Engert, J. C., Genest, J., and Marcil, M. (2006). Genetics of high-density lipoproteins. *Curr. Opin. Cardiol.* **21**, 329–335.
200. Juo, S. H., Wyszynski, D. F., Beaty, T. H., Huang, H. Y., and Bailey-Wilson, J. E. (1999). Mild association between the A/G polymorphism in the promoter of the apolipoprotein A-I

gene and apolipoprotein A-I levels: A meta-analysis. *Am. J. Med. Genet.* **82**, 235–241.

201. Lopez-Miranda, J., Ordovas, J. M., Espino, A., *et al.* (1994). Influence of mutation in human apolipoprotein A-1 gene promoter on plasma LDL cholesterol response to dietary fat. *Lancet* **343**, 1246–1249.
202. Mata, P., Lopez-Miranda, J., Pocovi, M., *et al.* (1998). Human apolipoprotein A-I gene promoter mutation influences plasma low density lipoprotein cholesterol response to dietary fat saturation. *Atherosclerosis* **137**, 367–376.
203. Sigurdsson, G., Jr., Gudnason, V., Sigurdsson, G., and Humphries, S. E. (1992). Interaction between a polymorphism of the apo A-I promoter region and smoking determines plasma levels of HDL and apo A-I. *Arterioscler. Thromb.* **12**, 1017–1022.
204. Smith, J. D., Brinton, E. A., and Breslow, J. L. (1992). Polymorphism in the human apolipoprotein A-I gene promoter region. Association of the minor allele with decreased production rate *in vivo* and promoter activity *in vitro*. *J. Clin. Invest.* **89**, 1796–1800.
205. Tuteja, R., Tuteja, N., Melo, C., Casari, G., and Baralle, F. E. (1992). Transcription efficiency of human apolipoprotein A-I promoter varies with naturally occurring A to G transition. *FEBS Lett.* **304**, 98–101.
206. Jeenah, M., Kessling, A., Miller, N., and Humphries, S. (1990). G to A substitution in the promoter region of the apolipoprotein AI gene is associated with elevated serum apolipoprotein AI and high density lipoprotein cholesterol concentrations. *Mol. Biol. Med.* **7**, 233–241.
207. Angotti, E., Mele, E., Costanzo, F., and Avvedimento, E. V. (1994). A polymorphism (G→A transition) in the −78 position of the apolipoprotein A-I promoter increases transcription efficiency. *J. Biol. Chem.* **269**, 17371–17374.
208. Zhang, X., Chen, Z. Q., Wang, Z., Mohan, W., and Tam, S. P. (1996). Protein–DNA interactions at a drug-responsive element of the human apolipoprotein A-I gene. *J. Biol. Chem.* **271**, 27152–27160.
209. Chiesa, G., and Sirtori, C. R. (2003). Apolipoprotein A-I(Milano): Current perspectives. *Curr. Opin. Lipidol.* **14**, 159–163.
210. Calabresi, L., Sirtori, C. R., Paoletti, R., and Franceschini, G. (2006). Recombinant apolipoprotein A-IMilano for the treatment of cardiovascular diseases. *Curr. Atheroscler. Rep.* **8**, 163–167.
211. Blanco-Vaca, F., Escola-Gil, J. C., Martin-Campos, J. M., and Julve, J. (2001). Role of apoA-II in lipid metabolism and atherosclerosis: Advances in the study of an enigmatic protein. *J. Lipid Res.* **42**, 1727–1739.
212. Fager, G., Wiklund, O., Olofsson, S. O., Wilhelmsen, L., and Bondjers, G. (1981). Multivariate analyses of serum apolipoproteins and risk factors in relation to acute myocardial infarction. *Arteriosclerosis (Dallas, Tex.)* **1**, 273–279.
213. Buring, J. E., O'Connor, G. T., Goldhaber, S. Z., *et al.* (1992). Decreased HDL2 and HDL3 cholesterol, Apo A-I and Apo A-II, and increased risk of myocardial infarction. *Circulation* **85**, 22–29.
214. Ridker, P. M., Rifai, N., Cook, N. R., Bradwin, G., and Buring, J. E. (2005). Non-HDL cholesterol, apolipoproteins A-I and B100, standard lipid measures, lipid ratios, and CRP as risk factors for cardiovascular disease in women. *JAMA* **294**, 326–333.
215. Tailleux, A., Duriez, P., Fruchart, J. C., and Clavey, V. (2002). Apolipoprotein A-II, HDL metabolism and atherosclerosis. *Atherosclerosis* **164**, 1–13.
216. Warden, C. H., Hedrick, C. C., Qiao, J. H., Castellani, L. W., and Lusis, A. J. (1993). Atherosclerosis in transgenic mice overexpressing apolipoprotein A-II. *Science* **261**, 469–472.
217. Schultz, J. R., Gong, E. L., McCall, M. R., Nichols, A. V., Clift, S. M., and Rubin, E. M. (1992). Expression of human apolipoprotein A-II and its effect on high density lipoproteins in transgenic mice. *J. Biol. Chem.* **267**, 21630–21636.
218. Castellani, L. W., Navab, M., Van Lenten, B. J., *et al.* (1997). Overexpression of apolipoprotein AII in transgenic mice converts high density lipoproteins to proinflammatory particles. *J. Clin. Invest.* **100**, 464–474.
219. Tailleux, A., Bouly, M., Luc, G., *et al.* (2000). Decreased susceptibility to diet-induced atherosclerosis in human apolipoprotein A-II transgenic mice. *Arterioscler. Thromb. Vasc. Biol.* **20**, 2453–2458.
220. Scott, J., Knott, T. J., Priestley, L. M., *et al.* (1985). High-density lipoprotein composition is altered by a common DNA polymorphism adjacent to apoprotein AII gene in man. *Lancet* **1**, 771–773.
221. Ferns, G. A., Shelley, C. S., Stocks, J., *et al.* (1986). A DNA polymorphism of the apoprotein AII gene in hypertriglyceridaemia. *Hum. Genet.* **74**, 302–306.
222. Vohl, M. C., Lamarche, B., Bergeron, J., *et al.* (1997). The MspI polymorphism of the apolipoprotein A–II gene as a modulator of the dyslipidemic state found in visceral obesity. *Atherosclerosis* **128**, 183–190.
223. Martin-–Campos, J. M., Escola-Gil, J. C., Ribas, V., and Blanco-Vaca, F. (2004). Apolipoprotein A-II, genetic variation on chromosome 1q21–q24, and disease susceptibility. *Curr. Opin. Lipidol.* **15**, 247–253.
224. Fullerton, S. M., Clark, A. G., Weiss, K. M., *et al.* (2002). Sequence polymorphism at the human apolipoprotein AII gene (APOA2): Unexpected deficit of variation in an African-American sample. *Hum. Genet.* **111**, 75–87.
225. van't Hooft, F. M., Ruotolo, G., Boquist, S., de Faire, U., Eggertsen, G., and Hamsten, A. (2001). Human evidence that the apolipoprotein a-II gene is implicated in visceral fat accumulation and metabolism of triglyceride-rich lipoproteins. *Circulation* **104**, 1223–1228.
226. Takada, D., Emi, M., Ezura, Y., *et al.* (2002). Interaction between the LDL-receptor gene bearing a novel mutation and a variant in the apolipoprotein A-II promoter: molecular study in a 1135-member familial hypercholesterolemia kindred. *J. Hum. Genet.* **47**, 656–664.
227. Lara-Castro, C., Hunter, G. R., Lovejoy, J. C., Gower, B. A., and Fernandez, J. R. (2005). Apolipoprotein A-II polymorphism and visceral adiposity in African-American and white women. *Obes. Res.* **13**, 507–512.
228. Castellani, L. W., Goto, A. M., and Lusis, A. J. (2001). Studies with apolipoprotein A-II transgenic mice indicate a

role for HDLs in adiposity and insulin resistance. *Diabetes* **50**, 643–651.

229. Rankinen, T., Zuberi, A., Chagnon, Y. C., *et al.* (2006). The human obesity gene map: The 2005 update. *Obesity (Silver Spring, Md.)* **14**, 529–644.
230. Tso, P., Sun, W., and Liu, M. (2004). Gastrointestinal satiety signals IV. Apolipoprotein A-IV. *Am. J. Physiol.* **286**, G885–G890.
231. Corella, D., Arnett, D. K., Tsai, M. Y., *et al.* (2007). The –256T > C polymorphism in the apolipoprotein A-II gene promoter is associated with body mass index and food intake in the genetics of lipid lowering drugs and diet network study. *Clin. Chem.* **53**, 1144–1152.
232. Kastelein, J. J., van Leuven, S. I., Burgess, L., *et al.* (2007). Effect of torcetrapib on carotid atherosclerosis in familial hypercholesterolemia. *N. Engl. J. Med.* **356**, 1620–1630.
233. Boekholdt, S. M., and Thompson, J. F. (2003). Natural genetic variation as a tool in understanding the role of CETP in lipid levels and disease. *J. Lipid Res.* **44**, 1080–1093.
234. Fumeron, F., Betoulle, D., Luc, G., *et al.* (1995). Alcohol intake modulates the effect of a polymorphism of the cholesteryl ester transfer protein gene on plasma high density lipoprotein and the risk of myocardial infarction. *J. Clin. Invest.* **96**, 1664–1671.
235. Tsujita, Y., Nakamura, Y., Zhang, Q., *et al.* (2007). The association between high-density lipoprotein cholesterol level and cholesteryl ester transfer protein TaqIB gene polymorphism is influenced by alcohol drinking in a population-based sample. *Atherosclerosis* **191**, 199–205.
236. Hodoglugil, U., Williamson, D. W., Huang, Y., and Mahley, R. W. (2005). An interaction between the TaqIB polymorphism of cholesterol ester transfer protein and smoking is associated with changes in plasma high-density lipoprotein cholesterol levels in Turks. *Clin. Genet.* **68**, 118–127.
237. Mukherjee, M., and Shetty, K. R. (2004). Variations in high-density lipoprotein cholesterol in relation to physical activity and Taq 1B polymorphism of the cholesteryl ester transfer protein gene. *Clin. Genet.* **65**, 412–418.
238. Dullaart, R. P., Hoogenberg, K., Riemens, S. C., *et al.* (1997). Cholesteryl ester transfer protein gene polymorphism is a determinant of HDL cholesterol and of the lipoprotein response to a lipid-lowering diet in type 1 diabetes. *Diabetes* **46**, 2082–2087.
239. Kuivenhoven, J. A., Jukema, J. W., Zwinderman, A. H., *et al.* (1998). The role of a common variant of the cholesteryl ester transfer protein gene in the progression of coronary atherosclerosis. The Regression Growth Evaluation Statin Study Group. *N. Engl. J. Med.* **338**, 86–93.
240. Carlquist, J. F., Muhlestein, J. B., Horne, B. D., *et al.* (2003). The cholesteryl ester transfer protein Taq1B gene polymorphism predicts clinical benefit of statin therapy in patients with significant coronary artery disease. *Am. Heart J.* **146**, 1007–1014.
241. van Venrooij, F. V., Stolk, R. P., Banga, J. D., *et al.* (2003). Common cholesteryl ester transfer protein gene polymorphisms and the effect of atorvastatin therapy in type 2 diabetes. *Diabetes Care* **26**, 1216–1223.
242. Freeman, D. J., Samani, N. J., Wilson, V., *et al.* (2003). A polymorphism of the cholesteryl ester transfer protein gene predicts cardiovascular events in non-smokers in the West of Scotland Coronary Prevention Study. *Eur. Heart J.* **24**, 1833–1842.
243. de Grooth, G. J., Smilde, T. J., Van Wissen, S., *et al.* (2004). The relationship between cholesteryl ester transfer protein levels and risk factor profile in patients with familial hypercholesterolemia. *Atherosclerosis* **173**, 261–267.
244. Klerkx, A. H., de Grooth, G. J., Zwinderman, A. H., Jukema, J. W., Kuivenhoven, J. A., and Kastelein, J. J. (2004). Cholesteryl ester transfer protein concentration is associated with progression of atherosclerosis and response to pravastatin in men with coronary artery disease (REGRESS). *Eur. J. Clin. Invest.* **34**, 21–28.
245. Thompson, J. F., Wood, L. S., Pickering, E. H., Dechairo, B., and Hyde, C. L. (2007). High-density genotyping and functional SNP localization in the CETP gene. *J. Lipid Res.* **48**, 434–443.
246. Barter, P. J., Brewer, H. B. Jr., Chapman, M. J., Hennekens, C. H., Rader, D. J., and Tall, A. R. (2003). Cholesteryl ester transfer protein: A novel target for raising HDL and inhibiting atherosclerosis. *Arterioscler. Thromb. Vasc. Biol.* **23**, 160–167.
247. Barzilai, N., Atzmon, G., Schechter, C., *et al.* (2003). Unique lipoprotein phenotype and genotype associated with exceptional longevity. *JAMA* **290**, 2030–2040.
248. Cellini, E., Nacmias, B., Olivieri, F., *et al.* (2005). Cholesteryl ester transfer protein (CETP) I405V polymorphism and longevity in Italian centenarians. *Mech. Ageing Dev.* **126**, 826–828.
249. Santamarina-Fojo, S., Gonzalez-Navarro, H., Freeman, L., Wagner, E., and Nong, Z. (2004). Hepatic lipase, lipoprotein metabolism, and atherogenesis. *Arterioscler. Thromb. Vasc. Biol.* **24**, 1750–1754.
250. Murtomaki, S., Tahvanainen, E., Antikainen, M., *et al.* (1997). Hepatic lipase gene polymorphisms influence plasma HDL levels. Results from Finnish EARS participants. European Atherosclerosis Research Study. *Arterioscler. Thromb. Vasc. Biol.* **17**, 1879–1884.
251. Jansen, H., Verhoeven, A. J., Weeks, L., *et al.* (1997). Common C-to-T substitution at position –480 of the hepatic lipase promoter associated with a lowered lipase activity in coronary artery disease patients. *Arterioscler. Thromb. Vasc. Biol.* **17**, 2837–2842.
252. Tahvanainen, E., Syvanne, M., Frick, M. H., *et al.* (1998). Association of variation in hepatic lipase activity with promoter variation in the hepatic lipase gene. The LOCAT Study Invsestigators. *J. Clin. Invest.* **101**, 956–960.
253. Hegele, R. A., Harris, S. B., Brunt, J. H., *et al.* (1999). Absence of association between genetic variation in the LIPC gene promoter and plasma lipoproteins in three Canadian populations. *Atherosclerosis* **146**, 153–160.
254. Couture, P., Otvos, J. D., Cupples, L. A., *et al.* (2000). Association of the C-514T polymorphism in the hepatic lipase gene with variations in lipoprotein subclass profiles: The Framingham Offspring Study. *Arterioscler. Thromb. Vasc. Biol.* **20**, 815–822.
255. Deeb, S. S., Zambon, A., Carr, M. C., Ayyobi, A. F., and Brunzell, J. D. (2003). Hepatic lipase and dyslipidemia:

Interactions among genetic variants, obesity, gender, and diet. *J. Lipid Res.* **44**, 1279–1286.

256. Isaacs, A., Sayed-Tabatabaei, F. A., Njajou, O. T., Witteman, J. C., and van Duijn, C. M. (2004). The –514 C→T hepatic lipase promoter region polymorphism and plasma lipids: A meta-analysis. *J. Clin. Endocrinol. Metab.* **89**, 3858–3863.
257. Vega, G. L., Clark, L. T., Tang, A., Marcovina, S., Grundy, S. M., and Cohen, J. C. (1998). Hepatic lipase activity is lower in African American men than in white American men: Effects of 5′ flanking polymorphism in the hepatic lipase gene (LIPC). *J. Lipid Res.* **39**, 228–232.
258. Guerra, R., Wang, J., Grundy, S. M., and Cohen, J. C. (1997). A hepatic lipase (LIPC) allele associated with high plasma concentrations of high density lipoprotein cholesterol. *Proc. Natl. Acad. Sci. USA* **94**, 4532–4537.
259. Deeb, S. S., and Peng, R. (2000). The C-514T polymorphism in the human hepatic lipase gene promoter diminishes its activity. *J. Lipid Res.* **41**, 155–158.
260. Hadzopoulou-Cladaras, M., and Cardot, P. (1993). Identification of a cis-acting negative DNA element which modulates human hepatic triglyceride lipase gene expression. *Biochemistry* **32**, 9657–9667.
261. van't Hooft, F. M., Lundahl, B., Ragogna, F., Karpe, F., Olivecrona, G., and Hamsten, A. (2000). Functional characterization of 4 polymorphisms in promoter region of hepatic lipase gene. *Arterioscler. Thromb. Vasc. Biol.* **20**, 1335–1339.
262. Su, Z., Zhang, S., Nebert, D. W., *et al.* (2002). A novel allele in the promoter of the hepatic lipase is associated with increased concentration of HDL-C and decreased promoter activity. *J. Lipid Res.* **43**, 1595–1601.
263. Gomez, P., Miranda, J. L., Marin, C., *et al.* (2004). Influence of the –514C/T polymorphism in the promoter of the hepatic lipase gene on postprandial lipoprotein metabolism. *Atherosclerosis* **174**, 73–79.
264. Jackson, K. G., Zampelas, A., Knapper, J. M., *et al.* (2000). Differences in glucose-dependent insulinotrophic polypeptide hormone and hepatic lipase in subjects of southern and northern Europe: Implications for postprandial lipemia. *Am. J. Clin. Nutr.* **71**, 13–20.
265. Ordovas, J. M., Corella, D., Demissie, S., *et al.* (2002). Dietary fat intake determines the effect of a common polymorphism in the hepatic lipase gene promoter on high-density lipoprotein metabolism: Evidence of a strong dose effect in this gene–nutrient interaction in the Framingham Study. *Circulation* **106**, 2315–2321.
266. Tai, E. S., Corella, D., Deurenberg-Yap, M., *et al.* (2003). Dietary fat interacts with the –514C > T polymorphism in the hepatic lipase gene promoter on plasma lipid profiles in a multiethnic Asian population: The 1998 Singapore National Health Survey. *J. Nutr.* **133**, 3399–3408.
267. Hokanson, J. E., Kamboh, M. I., Scarboro, S., Eckel, R. H., and Hamman, R. F. (2003). Effects of the hepatic lipase gene and physical activity on coronary heart disease risk. *Am. J. Epidemiol.* **158**, 836–843.
268. St-Pierre, J., Miller-Felix, I., Paradis, M. E., *et al.* (2003). Visceral obesity attenuates the effect of the hepatic lipase –514C > T polymorphism on plasma HDL-cholesterol levels in French-Canadian men. *Mol. Genet. Metab.* **78**, 31–36.
269. Todorova, B., Kubaszek, A., Pihlajamaki, J., *et al.* (2004). The G-250A promoter polymorphism of the hepatic lipase gene predicts the conversion from impaired glucose tolerance to type 2 diabetes mellitus: The Finnish Diabetes Prevention Study. *J. Clin. Endocrinol. Metab.* **89**, 2019–2023.
270. Norlin, M., and Wikvall, K. (2007). Enzymes in the conversion of cholesterol into bile acids. *Curr. Mol. Med.* **7**, 199–218.
271. Cheema, S. K., Cikaluk, D., and Agellon, L. B. (1997). Dietary fats modulate the regulatory potential of dietary cholesterol on cholesterol 7 alpha-hydroxylase gene expression. *J. Lipid Res.* **38**, 315–323.
272. Hubacek, J. A., and Bobkova, D. (2006). Role of cholesterol 7alpha-hydroxylase (CYP7A1) in nutrigenetics and pharmacogenetics of cholesterol lowering. *Mol. Diagn. Ther.* **10**, 93–100.
273. Pikuleva, I. A. (2006). Cholesterol-metabolizing cytochromes P450. *Drug Metab. Dispos.* **34**, 513–520.
274. Hofman, M. K., Weggemans, R. M., Zock, P. L., Schouten, E. G., Katan, M. B., and Princen, H. M. (2004). CYP7A1 A-278C polymorphism affects the response of plasma lipids after dietary cholesterol or cafestol interventions in humans. *J. Nutr.* **134**, 2200–2204.
275. Kovar, J., Suchanek, P., Hubacek, J. A., and Poledne, R. (2004). The A-204C polymorphism in the cholesterol 7alpha-hydroxylase (CYP7A1) gene determines the cholesterolemia responsiveness to a high-fat diet. *Physiol. Res.* **53**, 565–568.
276. Takane, H., Miyata, M., Burioka, N., *et al.* (2006). Pharmacogenetic determinants of variability in lipid-lowering response to pravastatin therapy. *J. Hum. Genet.* **51**, 822–826.
277. Kajinami, K., Brousseau, M. E., Ordovas, J. M., and Schaefer, E. J. (2004). Interactions between common genetic polymorphisms in ABCG5/G8 and CYP7A1 on LDL cholesterol-lowering response to atorvastatin. *Atherosclerosis* **175**, 287–293.
278. Kajinami, K., Brousseau, M. E., Ordovas, J. M., and Schaefer, E. J. (2005). A promoter polymorphism in cholesterol 7alpha-hydroxylase interacts with apolipoprotein E genotype in the LDL-lowering response to atorvastatin. *Atherosclerosis* **180**, 407–415.
279. Acton, S. L., Scherer, P. E., Lodish, H. F., and Krieger, M. (1994). Expression cloning of SR-BI, a CD36-related class B scavenger receptor. *J. Biol. Chem.* **269**, 21003–21009.
280. Acton, S., Rigotti, A., Landschulz, K. T., Xu, S., Hobbs, H. H., and Krieger, M. (1996). Identification of scavenger receptor SR-BI as a high density lipoprotein receptor. *Science* **271**, 518–520.
281. Calvo, D., Gomez–Coronado, D., Lasuncion, M. A., and Vega, M. A. (1997). CLA-1 is an 85-kD plasma membrane glycoprotein that acts as a high-affinity receptor for both native (HDL, LDL, and VLDL) and modified (OxLDL and AcLDL) lipoproteins. *Arterioscler. Thromb. Vasc. Biol.* **17**, 2341–2349.
282. Acton, S., Osgood, D., Donoghue, M., *et al.* (1999). Association of polymorphisms at the SR-BI gene locus with plasma lipid levels and body mass index in a white population. *Arterioscler. Thromb. Vasc. Biol.* **19**, 1734–1743.

283. Miller, M., Rhyne, J., Hamlette, S., Birnbaum, J., and Rodriguez, A. (2003). Genetics of HDL regulation in humans. *Curr. Opin. Lipidol.* **14**, 273–279.
284. Van Eck, M., Pennings, M., Hoekstra, M., Out, R., and Van Berkel, T. J. (2005). Scavenger receptor BI and ATP-binding cassette transporter A1 in reverse cholesterol transport and atherosclerosis. *Curr. Opin. Lipidol.* **16**, 307–315.
285. Perez-Martinez, P., Ordovas, J. M., Lopez-Miranda, J., *et al.* (2003). Polymorphism exon 1 variant at the locus of the scavenger receptor class B type I gene: Influence on plasma LDL cholesterol in healthy subjects during the consumption of diets with different fat contents. *Am. J. Clin. Nutr.* **77**, 809–813.
286. Perez-Martinez, P., Perez-Jimenez, F., Bellido, C., *et al.* (2005). A polymorphism exon 1 variant at the locus of the scavenger receptor class B type I (SCARB1) gene is associated with differences in insulin sensitivity in healthy people during the consumption of an olive oil-rich diet. *J. Clin. Endocrinol. Metab.* **90**, 2297–2300.
287. Tanaka, T., Delgado-Lista, J., Lopez-Miranda, J., *et al.* (2007). Scavenger receptor class B type I (SCARB1) c.1119C > T polymorphism affects postprandial triglyceride metabolism in men. *J. Nutr.* **137**, 578–582.
288. Kolovou, G. D., Mikhailidis, D. P., Anagnostopoulou, K. K., Daskalopoulou, S. S., and Cokkinos, D. V. (2006). Tangier disease four decades of research: A reflection of the importance of HDL. *Curr. Med. Chem.* **13**, 771–782.
289. Brunham, L. R., Singaraja, R. R., and Hayden, M. R. (2006). Variations on a gene: Rare and common variants in ABCA1 and their impact on HDL cholesterol levels and atherosclerosis. *Annu. Rev. Nutr.* **26**, 105–129.
290. Kaminski, W. E., Piehler, A., and Wenzel, J. J. (2006). ABC A-subfamily transporters: Structure, function and disease. *Biochim. Biophys. Acta* **1762**, 510–524.
291. Marcil, M., Brooks-Wilson, A., Clee, S. M., *et al.* (1999). Mutations in the ABC1 gene in familial HDL deficiency with defective cholesterol efflux. *Lancet* **354**, 1341–1346.
292. Oram, J. F., and Vaughan, A. M. (2006). ATP-binding cassette cholesterol transporters and cardiovascular disease. *Circ. Res.* **99**, 1031–1043.
293. Herron, K. L., McGrane, M. M., Waters, D., *et al.* (2006). The ABCG5 polymorphism contributes to individual responses to dietary cholesterol and carotenoids in eggs. *J. Nutr.* **136**, 1161–1165.
294. Frodsham, A. J., and Higgins, J. P. (2007). Online genetic databases informing human genome epidemiology. *BMC Med. Res. Methodol.* **7**, 31.

CHAPTER 29

The Role of Diet in the Prevention and Treatment of Cardiovascular Disease

MICHAEL ROUSSELL, JESSICA GRIEGER, AND PENNY M. KRIS-ETHERTON
Pennsylvania State University, University Park, Pennsylvania

Contents

I. INTRODUCTION

Cardiovascular disease (CVD) is the leading cause of death in the United States, accounting for more deaths per year than all other causes. In the past 25 years, there has been progress in reducing the number of deaths from CVD. For example, CVD mortality rate has declined by 41% since the early 1980s [1]. Despite this, medical treatments for cardiac conditions have increased. In addition, hospital discharges related to CVD are at an all-time high (>6,000,000 per year). The American College of Cardiology recently predicted that by 2050 the number of Americans diagnosed with CVD will double to 25 million [2]. Thus, despite the progress that has been made in reducing death rates from CVD, much remains to be achieved to decrease CVD risk, which in turn will reduce onset and progression of CVD on a population basis.

Prevention of CVD is a major public health goal. Accordingly, intensive efforts are ongoing to decrease CVD risk through the reduction of CVD risk factors in all population groups. Risk factors are classified as nonmodifiable or modifiable [3]. Nonmodifiable risk factors include age and family history. Major modifiable CVD risk factors include elevated total cholesterol, low-density lipoprotein cholesterol (LDL-C) and triglyceride (TG) levels, reduced high-density lipoprotein cholesterol (HDL-C) levels, hypertension, diabetes mellitus, and overweight and obesity. This chapter focuses primarily on lipids and lipoproteins, a major modifiable CVD risk factor, and the role that diet can play in reducing CVD risk via modifications in lipids and lipoproteins. Other major CVD risk factors are mentioned in this chapter; however, they are discussed in greater detail elsewhere in this book. Other increasingly important emerging CVD risk factors that can be modified by diet include elevated levels of lipoprotein (a); insulin; altered hemostatic factors; C-reactive protein; cytokines and inflammatory mediators (e.g., interleukin [IL]-6, IL-7, IL-8, IL-18); and small, dense LDL particles, among others. Thus, there is great potential for further decreasing CVD by favorably modifying multiple risk factors.

The importance of reducing major risk factors is illustrated by results from three large prospective studies (Chicago Heart Association Detection Project in Industry, Multiple Risk Factor Intervention Trial, and the Framingham Heart Study), which reported that 87–100% of men and women (ages 18–59 years) with at least one major risk factor died from coronary heart disease (CHD) [4]. In addition, another analysis of 112,458 patients with CHD reported that 80–90% of the participants had one or more major CVD factors [5]. It is important to note, however, that other risk factors also contribute to the development of CHD. This is best illustrated by the evidence that about 35% of CHD occurs in individuals with a total cholesterol (TC) less than 200 mg/dl [6]. Thus, modifying as many CVD risk factors as possible will have the greatest impact on decreasing CVD risk.

The impact of lowering major coronary heart disease risk factors has been reported in a recent analysis demonstrating that approximately one-half of the decrease in CHD in the United States between 1980 and 2000 can be attributed to reductions in total cholesterol, blood pressure, and body mass index (BMI) [1]. The reduction in these major risk factors is due to lifestyle and behavioral interventions, as well as pharmacotherapy. Thus, diet can have a significant impact on CVD risk factors, and, consequently, healthy diet and lifestyle practices can markedly decrease the risk for CHD.

Diet has been a cornerstone in the management of heart disease risk factors for more than 50 years. The American Heart Association (AHA) published their first dietary recommendations for CVD risk reduction in 1957 [7]. The AHA updates dietary recommendations routinely as new science emerges. Other groups such as the U.S. Department of Agriculture (USDA), Health and Human Services (HHS), National Cholesterol Education Program (NCEP), American Dietetic Association (ADA), and the American Diabetes Association continually update and publish diet and lifestyle recommendations to reduce risk of chronic diseases, including (or specifically focusing on) CVD. Traditionally these organizations have made dietary recommendations based on targeted nutrient levels (e.g., <7 to <10% of energy from saturated fatty acids [SFA]). Recently a more consumer-friendly approach has evolved and food-based dietary recommendations have been made (e.g., "Consume 3 cups per day of fat-free or low fat milk or equivalent products" [8]). This change is most notable in development of the MyPyramid.gov, the new graphic for communicating the 2005 U.S. Dietary Guidelines [9, 10]. These food-based recommendations are based on macronutrient and micronutrient recommendations made by the National Academies (Dietary Reference Intakes), as well as other organizations (e.g., NCEP). The American Dietetic Association Evidence Analysis Library on Disorders of Lipid Metabolism is an excellent summary of the literature about the role of diet on lipid and lipoprotein risk factors, including dietary recommendations for the management of CVD risk factors [11]. A food-based approach that integrates all nutrient recommendations is encouraged because it targets multiple CVD risk factors, as well as many other chronic diseases. It is important to note that this approach encompasses all dietary recommendations and translates to a greater health benefit for the population.

Historically dietary recommendations have focused on modifying the type and amount of fat. However, modifying the type and amount of carbohydrate and protein to lower CVD risk factors has attracted recent attention. The reduction and replacement of SFA and *trans* fatty acids (TFA) with unsaturated fat or carbohydrates is currently a widely accepted approach for decreasing major CVD risk factors. In addition, dietary protein is being considered as a substitute for SFA and TFA. The effects of protein on CVD is becoming a more well-defined area as an increasing number of studies are being performed that examine both type and amount of protein. In the past decade, dietary recommendations have been made for other nutrients based on the emerging evidence. For example, the cardioprotective benefits of a diet rich in omega-3 fatty acids, both marine- and plant-based, have been intensively studied, leading to specific dietary recommendations that have been made by certain organizations. With respect to dietary carbohydrates, studies are being conducted to evaluate dietary fiber, the glycemic index of carbohydrate-rich foods, and how the glycemic load of the diet affects CVD risk factors. Likewise, studies are ongoing to evaluate the physiological effects of amount and type of animal protein and plant protein on CVD risk.

This is an exciting era for gaining a better understanding of how macronutrients, micronutrients, and other dietary factors affect CVD risk, and thus it is not unreasonable to speculate that more effective dietary approaches for reducing CVD risk will be identified in the future. This chapter reviews the present understanding of how diet affects CVD risk status via changes in plasma lipid and lipoproteins, emerging physiological risk factors, and overall CVD-related morbidity and mortality.

II. DIETARY FAT

A. Total Fat

A diet low in SFA (<7% of energy from SFA) continues to be a major focus in the prevention and treatment of CVD. Typically, reducing total fat has been a major strategy recommended for decreasing SFA. Efforts are ongoing to determine the optimal quantity of total dietary fat. Currently the Dietary Reference Intakes (DRI) for Macronutrients from the National Academies recommend 20–35% of energy from fat for adults (>19 years old) [12]. The NCEP recommends 25–35% of energy from fat, an amount intended for the management of dyslipidemia (specifically high TG and low HDL-C levels) [13].

The upper and lower ranges for total fat intake were defined on the basis of achieving nutrient adequacy, as well as the range considered optimal for health. The upper range for total fat was set because SFA and energy intakes have been shown to increase beyond recommended levels when total fat exceeds 35% of energy [14–19]. Within the context of the amount of total fat in the diet that meets current recommendations, SFA, TFA, and cholesterol should be reduced as much as possible and nutrient adequacy should be met.

The lower range of the total fat recommendation was set to achieve nutrient adequacy and meet essential fatty acid requirements as well as promote compliance with the total fat recommendation. Adherence to very low-fat diets (<20% of energy) is problematic [20]. The decrease in compliance commonly seen with reduced-fat diets was most recently observed in the Women's Health Initiative Dietary Modification Trial (WHI) [21]. In this study, which enrolled almost 49,000 women, the total fat intake goal of 20% of energy was not achieved. However, the low-fat intervention group was able to decrease their total fat intake by at least 9 percentage points (37.8% down to 28.8% of energy), an average that was achieved after 6 years. Some women were able to reduce total fat further. The WHI trial demonstrates that the total fat recommendations set by the National Academies and NCEP are achievable.

B. Saturated Fatty Acids

Dietary recommendations have been made by numerous government agencies and health care organizations for SFA. All recommendations consistently advocate reductions in SFA. In 2005, the DRI Report [12] recommended that SFA be a low as possible within the context of a nutritionally adequate diet. Specific targets have been recommended by the Dietary Guidelines for Americans, 2005 [22], the National Cholesterol Education Program (NCEP) ATP III, and the American Heart Association (AHA) in 2006. The Dietary Guidelines Report recommends that SFA be less than 10% of energy. Further reductions are recommended for individuals with high TC and LDL-C. Both the NCEP and the AHA recommend less than 7% of energy from SFA for the treatment of high LDL-C levels and for the prevention of CVD, respectively.

The Seven Countries Study [23], a landmark epidemiologic investigation, demonstrated that SFA intake (as a percent of energy) was positively correlated with serum cholesterol levels as well as with 5-year incidence of CHD. Many well-controlled clinical studies followed, resulting in the development of blood cholesterol predictive equations for estimating the changes in total cholesterol in response to changes in type of fat and amount of dietary cholesterol consumed. The original equations developed by Keys *et al.* [24] and Hegsted *et al.* [25] demonstrated that SFA was twice as potent in raising blood cholesterol levels as polyunsaturated fat (PUFA) was in lowering cholesterol levels. Monounsaturated fat (MUFA) was shown to have a neutral effect and dietary cholesterol raised the blood cholesterol level but less so than SFA. Other predictive equations have been developed for LDL-C and HDL-C [26–29]. The LDL-C response to fatty acid classes tracks with that reported for TC.

Figure 1 reports the results of a meta-analysis of 60 studies in which carbohydrates were isoenergetically replaced with different fatty acid classes (SFA, MUFA, PUFA, TFA) and shows the relative effects on LDL-C, HDL-C, and the TC:HDL-C ratio. SFA and TFA increase LDL-C, and MUFA and PUFA decrease LDL-C, the latter more so. All fatty acid classes except TFA increase HDL-C; SFA is the most potent, PUFA is least potent, and MUFA has an intermediate effect. Because SFA increases both LDL-C and HDL-C, there is a slight increase in the TC:HDL-C ratio. The TC:HDL-C ratio is most favorably affected by unsaturated fatty acids, and PUFA more so than MUFA because PUFA has a greater TC-lowering effect than MUFA [30].

A positive dose-response relationship between SFA intake, TC (Fig. 2), and LDL-C has also been reported in meta-analyses by several investigators [12, 26]. Figure 2 includes data from 395 studies and displays the linear relationship between SFA intake and total cholesterol concentrations.

There is a modest dose-response effect of SFA on the LDL:HDL-C ratio [31]. As shown in Figure 3, the increase in the LDL:HDL-C ratio elicited by SFA is less than that observed for TFA.

Recent studies also have evaluated the effects of individual SFA on plasma lipids and lipoproteins (Fig. 4) [32]. The cholesterolemic effects of the individual SFA vary. Lauric acid (C:12:0) has the most potent LDL-C and TC raising effect. Because lauric acid increases HDL-C proportionally more than it does LDL-C, a lowering of the TC:HDL-C ratio is observed [32]. Stearic acid has a neutral effect on LDL-C. However, results from the Nurses' Health Study have reported a high correlation between stearic acid and other SFA in the diet, and therefore, distinguishing

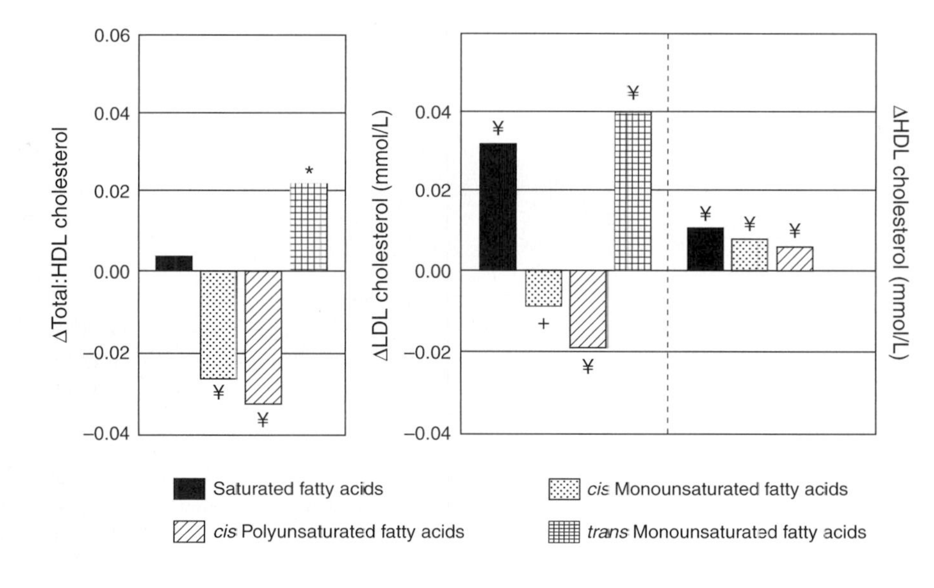

FIGURE 1 Predicted changes (Δ) in the ratio of serum total to HDL cholesterol and in LDL- and HDL-cholesterol concentrations when carbohydrates constituting 1% of energy are replaced isoenergetically with saturated, *cis* monounsaturated, *cis* polyunsaturated, or *trans* monounsaturated fatty acids. $*p < 0.05$. $+p < 0.01$. $¥p < 0.001$ [32]. Reprinted with permission from the American Society of Nutritional Sciences.

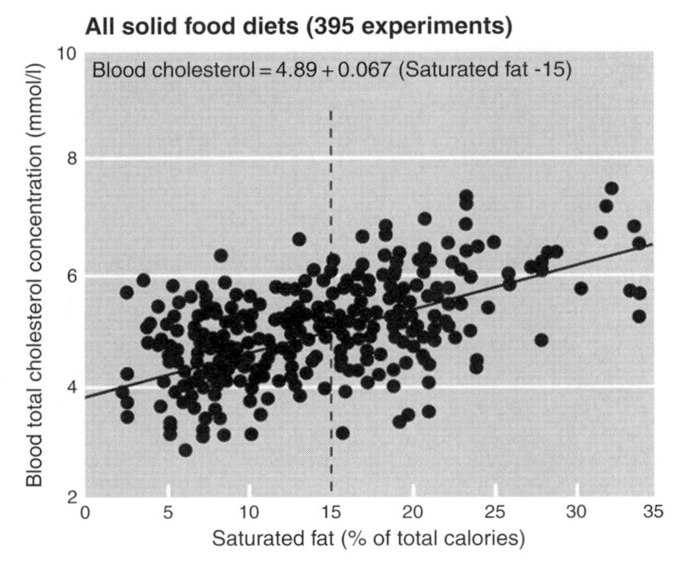

FIGURE 2 Univariate multilevel regression slopes of blood total cholesterol versus dietary saturated fat in metabolic ward experiments. (The equations take account of which experiments were part of the same study.) The dotted line represents a typical intake of saturated fat (15% of dietary energy), and this is used as an intercept for the regression equations. No adjustments are made for total energy intake or any other aspects of experimental diets [26]. Reprinted with permission from the *British Journal of Medicine*.

between individual SFA should not be a major dietary intervention in the treatment and prevention of CVD [33].

SFA also has been shown to have adverse effects on various emerging CVD risk factors such as flow-mediated dilation (FMD) of the brachial artery, cellular adhesion molecules, and hemostatic factors. FMD is a measurement of vascular reactivity commonly used to assess vascular endothelial function. Decreases in FMD are indicative of decreased vascular elasticity, which increases risk for CVD. Both acute and chronic consumption of SFA have been shown to adversely affect FMD. A single meal high in SFA has been shown to significantly decrease FMD for up to 6 hours [34]. With respect to chronic consumption of a diet high in SFA, a recent free-living crossover (3 weeks each diet period) study compared the effects of a diet that provided varying levels of energy from SFA (19% of energy), MUFA (19% of energy), PUFA (15% of energy), or carbohydrates (68% of energy) on FMD and found that the high-SFA diet significantly decreased FMD, indicating an adverse effect compared to the other diets, which had no appreciable effect [35].

Other important markers of vascular function are cellular adhesion molecules. Cellular adhesion molecules (ICAM-1, sICAM-1, VCAM-1, P-selectin, E-selectin) are compounds found on the luminal blood vessel epithelium that bind leukocytes and initiate the leukocyte-endothelial cell adhesion cascade [36]. Increased levels of adhesion molecules increase risk of myocardial infarction (MI) [37] and CVD [37, 38]. The effects of SFA on cellular adhesion molecules also have been examined in a postprandial setting. Subjects were given a high-fat milk shake (1 g fat/kg body weight) containing mainly SFA (89.6% SFA) or unsaturated fat (8.8% SFA). Increased consumption of SFA at this single meal significantly elevated the postprandial expression of the adhesion molecules ICAM-1 and VCAM-1 for up to 6 hours [34]. Levels of another adhesion molecule, P-selectin, which have been positively correlated with CVD, also have been shown to increase following a diet high in SFA [35]. Compared to individuals on a Mediterranean-style diet, the LDL-C particles from individuals on a high-SFA diet (20% of energy) induce increased expression of VCAM-1 and E-selectin [39]. Small reductions (–1.8% of total energy) in dietary SFA also significantly decreased sICAM-1 in hypercholesterolemic men and women [40].

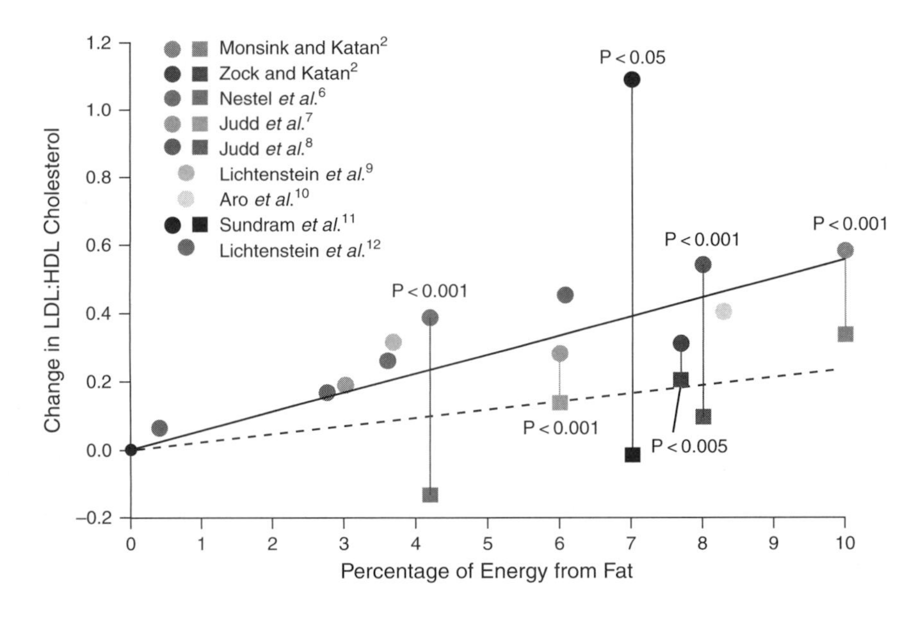

FIGURE 3 Results of randomized studies of the effects of a diet high in *trans* fatty acids (circles) or saturated fatty acids (squares) on the ratio of LDL cholesterol to HDL cholesterol. A diet with isoenergetic amounts of *cis* fatty acids was used as the comparison group. The solid line indicates the best-fit regression for *trans* fatty acids. The dashed line indicates the best-fit regression for saturated fatty acids [31].

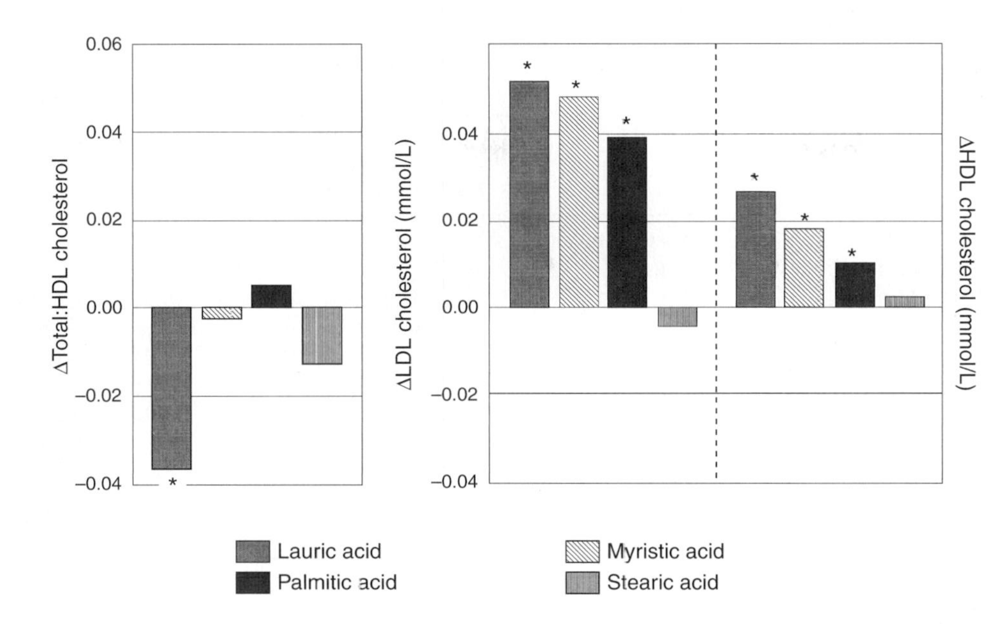

FIGURE 4 Predicted changes (Δ) in the ratio of serum total to HDL cholesterol and in LDL- and HDL-cholesterol concentrations when carbohydrates constituting 1% of energy are replaced isoenergetically with lauric acid (12:0), myristic acid (14:0), palmitic acid (16:0), or stearic acid (18:0). $^*p < 0.001$ [32]. Reprinted with permission from the American Society of Nutritional Sciences.

Epidemiologic evidence from the Atherosclerosis Risk in Communities Study [41] showed that a high intake of total fat, SFA, and cholesterol was associated with higher levels of factor VII and fibrinogen, two hemostatic factors that play a role in blood clot formation and are CVD risk factors. Likewise, in the Dietary Effects on Lipoproteins and Thrombogenic Activity (DELTA) Study, a well-controlled multicenter feeding study, investigators reported that reductions in SFA led to decreased factor VII levels [42]. Replacing SFA with MUFA has been shown to significantly decrease factor VII in mildly hypercholesterolemic subjects [43]. In a review on the topic of fatty acids and hemostasis, SFA (namely, C12–C16 SFA) was shown to be the main determinant of factor VII activation over time [44]. Collectively, the results from studies conducted to date provide a strong rationale for current dietary guidance to decrease SFA. The database for the recommendation to decrease TFA also is convincing. Decreasing these two fatty acid classes can be achieved in a variety of ways.

Strategies for reducing saturated fat that can be implemented singly or in combination are as follows:

1. *Replace energy from SFA and TFA with carbohydrate energy sources*—This is a common approach for reducing total fat. The potential downside to this approach is that replacing SFA with carbohydrate leads to a decrease in HDL-C and often an increase in TG [45–47], both of which increase risk of CVD. The increases in TG can be attenuated when whole grain carbohydrates are consumed, and when dietary fiber is increased [48].
2. *Replace energy from SFA and TFA with unsaturated fat energy sources*—The addition of PUFA and MUFA to the diet at the expense of SFA and TFA is an increasingly common approach that is consistent with a Mediterranean-style diet. When MUFA replace SFA, plasma TG is decreased and HDL-C remains unchanged or is decreased less compared with a high-carbohydrate, reduced-fat diet [47, 49, 50].
3. *Replace energy from SFA and TFA with protein energy sources*—There is a recent interest in replacing SFA and TFA energy with dietary protein to prevent some of the adverse effects that have been reported when dietary carbohydrate replaces SFA and TFA [51].
4. *Decrease energy from SFA and TFA*—Decreasing SFA and TFA without replacing energy with other macronutrients results in a reduction in total fat and energy, yielding a diet that is reduced in total fat.

The evidence is compelling that SFA has a potent and dose-response effect on TC and LDL-C. The individual SFA elicit different effects, with C12:0–C16:0 being hypercholesterolemic and stearic acid having a neutral effect. However, because these fatty acids track together in foods, the most prudent advice is to decrease total SFA to recommended levels. SFA intakes should meet current dietary recommendations within the context of a diet that achieves nutritional adequacy and is consistent with recommendations for total fat.

C. Unsaturated Fatty Acids

1. Monounsaturated Fatty Acids

MUFA along with PUFA provide great flexibility in diet planning because they are a vehicle for increasing total fat and can be used to replace energy from SFA, TFA, or carbohydrate. The current dietary recommendations from the National Academies do not include specific recommendations for MUFA, whereas the NCEP ATP III recommends that MUFA can make up to 20% of total energy

[13]. The average MUFA intake among the adult population provides about 15% of energy, whereas a high-MUFA diet typically provides about 20–22% of energy. During the past decade there has been a surge in research that examined the effects of using MUFA as a substitute for dietary carbohydrate and SFA because of beneficial effects: A moderate fat diet mediated CVD risk factors [47, 49]. MUFA have a slight LDL-C lowering effect, however, as noted in the meta-analysis conducted by Mensink *et al.* [32]; when MUFA are substituted for SFA and TFA, the expected decrease in TC and LDL-C will be significant. Replacing dietary carbohydrate with MUFA will decrease TG and increase HDL-C [32, 51]. Figure 5 shows an inverse-linear relationship between MUFA intake and TC:HDL-C levels [8].

Inclusion of MUFA in the diet is commonly achieved through the replacement of energy from SFA and/or carbohydrate. A meta-analysis of 60 controlled trials showed that the replacement of carbohydrate with MUFA resulted in significant decreases in TC:HDL-C ratio as well as LDL-C. This was accompanied by an increase in HDL-C [32]. These effects were examined more recently in the OmniHeart study. The OmniHeart study was a three-way crossover ($n = 164$), controlled-feeding trial that compared the effect of blood cholesterol-lowering diets that were either high in carbohydrate, protein, or unsaturated fats mostly MUFA on serum lipids and blood pressure (Table 1) [51, 52].

The high-unsaturated-fat (predominately MUFA) diet group experienced greater reductions in TG compared to the high-carbohydrate group (–9.3 mg/dl versus 0.1 mg/dl), LDL-C (–13.1 mg/dl versus –11.6 mg/dl), and TC (–15.4 mg/dl versus –12.4 mg/dl). The high-unsaturated-fat diet group also was the only diet group in which subjects did not experience a reduction in HDL-C [51].

TABLE 1 Macronutrient Comparisons of Three OmniHeart Diets [51]

	Diets		
	Carbohydrate	**Protein**	**Unsaturated Fat**
Diet composition, kcal%			
Fat	27	27	37
Saturated	6	6	6
Monounsaturated	13	13	21
Polyunsaturated	8	8	10
Carbohydrate	58	48	48
Protein	15	25	15
Meat	5.5	9	5.5
Dairy	4	4	4
Plant	5.5	12	5.5

Evidence also suggests that MUFA may decrease susceptibility of LDL particles to oxidative modification, an important initiating event in the development of atherosclerosis, thereby reducing the atherogenic potential of LDL [53, 54]. High-MUFA diets have been shown to increase the resistance of LDL to oxidation when compared to a high-PUFA diet, low-fat/high-carbohydrate diet, the average American diet, or the NCEP Step 1 diet (energy distributed as 55% carbohydrate, 30% fat, and 15% protein) [55, 56]. Moreover, a diet higher in total fat (and higher in MUFA) versus a lower-fat diet has been shown to maintain a larger LDL particle diameter [57, 58], which is important because small, dense LDL has been identified as a risk factor for CVD [59, 60].

There are conflicting views on the benefits of MUFA in the prevention and treatment of cardiovascular disease

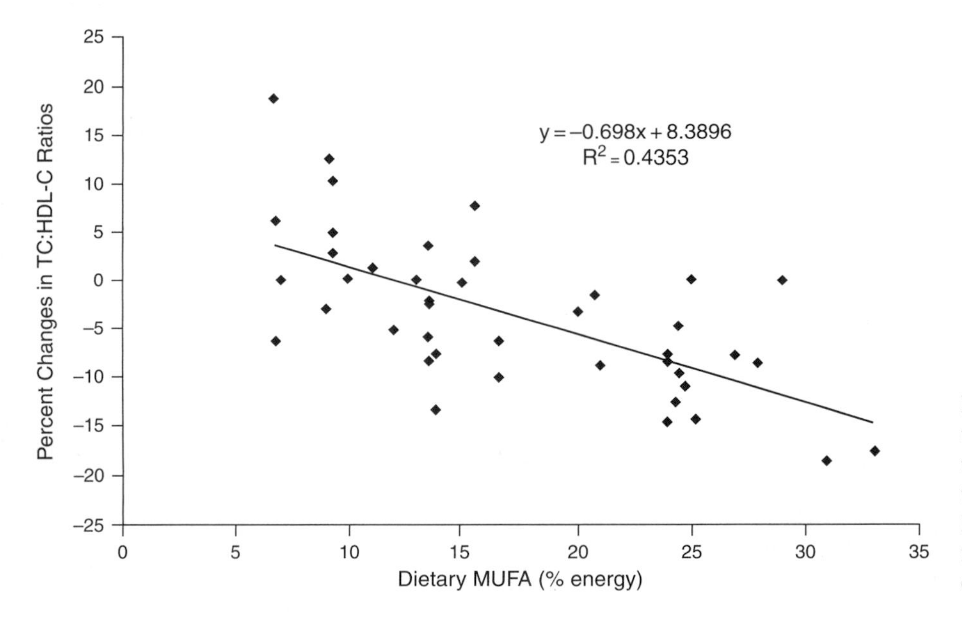

FIGURE 5 Relationship between monounsaturated fatty acid (MUFA) intake and total cholesterol (TC): high-density lipoprotein cholesterol (HDL-C) ratio. Weighted least-squares regression analyses were performed using the mixed procedure to test for differences in lipid concentrations [12].

based on animal studies conducted by Rudel *et al.* [61]. Nonhuman primates were fed diets high in PUFA, MUFA, or SFA. The lipid and lipoprotein responses were similar to those reported for humans when diets high in PUFA, MUFA, or SFA were consumed. On the high-MUFA diet, nonhuman primates developed atherosclerosis similar to that observed for a high-SFA diet. The high-PUFA diet showed decreased atherosclerosis compared to the other two diet groups. More recently it was demonstrated in the mouse model that a diet high in MUFA leads to a greater formation of oleoyl-CoA and an increase in ACAT2-derived cholesteryl oleate. These results suggest that the potential atherogenic effects of a MUFA-rich diet are due to an increase in monounsaturated cholesterol esters in LDL particles [62].

In summary, the incorporation of MUFA into a heart-healthy diet appears to have favorable effects on some important risk factors for heart disease, including lipids and lipoproteins, oxidized lipids, and blood pressure in healthy individuals and persons with diabetes. On the other hand, results from some animal studies suggesting adverse effects indicate that more research comparing the effects of MUFA and PUFA is needed to determine the optimal ratio of these important unsaturated fats in relation to the total fat content of the diet to maximally reduce CVD risk.

2. *Trans* Fatty Acids

TFA are unsaturated fatty acids (mono- or poly- in the case of conjugated linoleic acid) with *trans* stereochemistry configuration of the double bonds. TFA have two main origins: ruminant animals and industrial production. The ruminant-produced TFA vaccenic acid (C18:1 Δ11t) is a precursor to conjugated linoleic acid (CLA). Because of the substantive evidence base, industrially produced TFA and their effects on CVD risk will be emphasized. The DRI report recommended that TFA intake be kept as low as possible as any increase in TFA intake will negatively impact CVD risk [12]. NCEP ATP III also recommends that TFA consumption be kept as low as possible while the AHA recommends that TFA intake be kept at 1% of energy or below [63].

Industrially produced TFA is made by the hydrogenation of unsaturated vegetable oils. The use of TFA became very popular in the mid-twentieth century when public health recommendations encouraged people to decrease their animal fat and tropical oil intakes. However, research that started in the early 1990s demonstrated numerous adverse effects of TFA [64, 65].

The adverse cardiovascular effects of TFA have been demonstrated in epidemiologic and clinical trials. In the Seven Countries Study, higher CHD mortality rates were observed in countries with greater TFA consumption (northern European countries versus Japan and Mediterranean countries) [66]. Other epidemiologic studies such as the Nurses' Health Study [67], a population-based case-control study [68], and Zutphen Elderly Study [69] also showed an increased risk of CHD with increased intake of TFA.

An analysis that included 140,000 subjects, pooled from several studies, concluded that a 2% increase in energy from TFA was associated with 23% increase in incidence of CHD [70]. One large case-control study found an association between erythrocyte TFA levels and sudden death (odds ratio = 1.47) [68].

TFA elicit both similar and different lipid and lipoprotein responses compared to SFA. SFA has consistently increased both LDL-C and HDL-C, whereas TFA increased LDL-C but decreased HDL-C, compared to SFA [65] (Fig. 4). Thus, SFA increase the LDL:HDL-C ratio modestly, whereas TFA increase the ratio appreciably, that is, towards a more atherogenic profile. Figure 1, from a meta-analysis of 60 controlled trials, displays the potent effect TFAs has on increasing LDL-C [32].

TFA intakes as low as 1–3% of total energy (current consumption is 2.6% of energy) adversely affect many CVD risk factors beyond increasing LDL-C [71] such as increasing inflammatory markers and decreasing endothelial function [70]. Thus, as a result of the evidence base, it is clear that TFA should be reduced in the diet. Because of the adverse effects that TFA has on CVD risk factors and positive association with CVD morbidity and mortality, the AHA has actively disseminated information about the health effects of TFA. Central to this is sharing information about the development of alternative fats and oils devoid of TFA that have acceptable functional properties and sensory characteristics and do not increase CVD risk [64].

D. Polyunsaturated Fatty Acids

Polyunsaturated fats are long-chain fatty acids that contain more than one double bond. There are two major classes of PUFA that are defined by the position of the first double bond relative to the methyl terminus; omega-6 (n-6 FA) and omega-3 (n-3 FA) fatty acids. There is a large evidence base demonstrating benefits of PUFA on CVD risk and risk factors.

1. Omega-6 Fatty Acids

Linoleic acid (LA) is an essential fatty acid. The major source of LA in the diet is vegetable oil. Intakes of LA have almost doubled from the 1930s (3% of energy) to the present (5–6% of energy) in the United States and Canada [72]. The adequate intake (AI) for LA is 17 g/day and 12 g/day for men and women, respectively, between 19 and 50 years of age [12]. The Acceptable Macronutrient Distribution Range (AMDR) for n-6 FA (LA) is 5–10% of total energy. The lower range of the AMDR for LA is the AI. The upper range of PUFA intake set by the DRI Committee and Dietary Guidelines [8] because it represents the upper range of PUFA consumption in the United States. The

NCEP ATP III guidelines also recommend up to 10% of energy from PUFA [13].

Predictive equations developed by Keys *et al.* [23] and Hegsted *et al.* [25] showed that a 1% increase in energy from PUFA resulted in a 0.9 mg/dl decrease in TC. Replacing of carbohydrate calories with PUFA increases HDL-C [32], but less than was observed for MUFA. PUFA intake also decreases the TC:HDL-C ratio. Three early controlled trials verified the cardioprotective effects of PUFA as observed by Keys *et al.* and Hegsted *et al.* [73]. The Oslo Heart Study [74], the Finnish Mental Hospital Study [75], and the Wadsworth Hospital and Veterans Administration Center in Los Angeles Study [76] all observed marked hypocholesterolemic effects of diets very high in PUFA from vegetable oils. Importantly, in two of these studies [75, 76], the cholesterol-lowering response was associated with a reduction in the incidence of CVD (16–34%).

The Nurses' Health Study reported a dose-response relationship regarding CVD risk and PUFA intake with the highest quintile of intake (6.4% of energy) conferring approximately a 30% reduction in risk [77]. A review [78] of the influence of PUFA on CVD concluded that there are beneficial effects of PUFA intakes above the recommended intake range (14–21% of energy); thus more research is needed to determine the optimal level of PUFA in the diet.

There is some concern that n-6 fatty acids may have adverse health effects because of their increased susceptibility to oxidation and the impact that high intakes have on n-3 fatty acid status and inflammation [79]. However, many long-term clinical studies have shown that increased PUFA intakes have a beneficial effect on CVD risk, and thus the impact of increased susceptibility to oxidation is unlikely if the diet provides ample antioxidants [73, 75, 80].

Concern also has been expressed because of the possibility that higher consumption of n-6 fatty acids will inhibit metabolism of n-3 fatty acids. This concern stems from n-3 and n-6 fatty acids sharing common enzymatic and metabolic pathways en route to eicosanoid synthesis (Fig. 6).

The impact of higher intakes of n-6 fatty acids on metabolism of n-3 fatty acids is hypothesized to occur in two areas. First, the increased LA levels can interfere with the formation of eicosapentaenoic acid from the alpha-linolenic acid because LA uses the same desaturase and elongase enzymes to form arachidonic acid. The second area is the direct formation of eicosanoids. Arachidonic acid, eicosapentaenoic acid, and docosahexaenoic acid all occupy space in the lipid bilayers of cellular membranes. It is hypothesized that increased levels of n-6 fatty acids and decreased levels of n-3 fatty acids will lead to more n-6 fatty acids being used for eicosanoid synthesis, leading to an increased production of proinflammatory molecules. However, it has been argued that the ratio of n-6 to n-3 fatty acids is not as important as the total amount of n-3 fatty acids [81]. A review of the results from the Health Professionals Follow-up Study based on information from 45,722 men concluded that irrespective of background n-6 fatty acid intake, increased consumption of n-3 fatty acids reduced risk of CVD [82].

n-6 fatty acids have been shown to have great benefit in reducing CVD risk factors and CVD events. The public health message regarding n-6 PUFA consumption should not be to decrease intake, but rather to achieve the recommended intake (5–10% of energy). More research is needed to examine the health effects and safety of PUFA intake above 10% of energy.

2. Omega-3 Fatty Acids

The two main sources of n-3 fatty acids are either plant derived or marine derived and have unique benefits with respect to CVD risk factors. Alpha-linolenic acid (ALA,

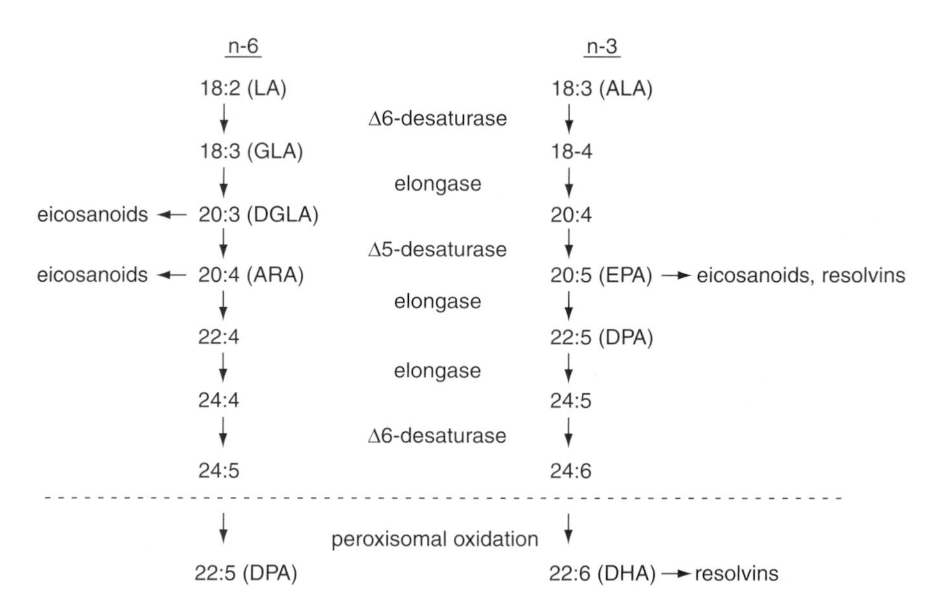

FIGURE 6 Biochemical pathway for the interconversion of n-6 and n-3 fatty acids. ALA, (alpha)-linolenic acid; ARA, arachidonic acid; DGLA, dihomo-γ-linolenic acid; DHA, docosahexaenoic acid; DPA, docosapentaenoic acid; EPA, eicosapentaenoic acid; GLA, (gamma)-linolenic acid; LA, linoleic acid [85]. Reprinted with permission from the American Society of Nutritional Sciences.

C18:3) is the major plant-derived n-3 fatty acid. The two major marine-derived omega-3 fatty acids are eicosapentaenoic acid (EPA, C20:5) and docosahexaenoic acid (DHA, C22:6). Fatty fish are the main source of EPA and DHA in the diet; these fatty acids are synthesized by cold water algae which are part of the food chain consumed by fish [83, 84]. ALA, commonly found in flax/flaxseed oil, canola oil, walnuts/walnut oil, and soybean oil, can undergo a series of elongations and desaturations by the body to yield both EPA and DHA (Fig. 6); however, these conversion rates are low, especially for DHA [85–88]. ALA is an essential fatty acid. The AI for individuals aged between 19 and 50 years for linolenic acid is 1.6 g/day and 1.1 g/day for men and women, respectively. The AMDR for ALA is 0.6–1.2% of total energy. It is recommended that up to 10% of the AMDR for ALA can be consumed as EPA and/or DHA [8]. Recommendations for EPA and DHA for the primary prevention of CVD from the United Kingdom Scientific Advisory Committee [89], Dietary Guidelines Advisory Committee [8], and the National Heart Foundation of Australia [90] are to consume 450–500 mg EPA and DHA per day.

The AHA has both food-based and nutrient-based recommendations (Table 2) [91] regarding n-3 FA, and specifically EPA and DHA. In a Science Advisory published in 2003, the AHA made recommendations for n-3 fatty acids for individuals without heart disease, patients with documented CHD, and patients with hypertriglyceridemia. The AHA Diet and Lifestyle Recommendations Revision 2006 recommend that people eat two servings (4 oz-each) of fatty fish weekly. This is the equivalent of 500 mg/day of EPA and DHA. The NCEP ATP III did not make specific n-3 fatty acid recommendations but endorsed those made by the AHA in 2002 [91].

TABLE 2 Summary of AHA Recommendations for Omega-3 Fatty Acid Intake [91]

Population	Recommendation
Patients without documented coronary heart disease (CHD)	Eat a variety of (preferably fatty) fish at least twice a week. Include oils and foods rich in alpha-linolenic acid (flaxseed, canola and soybean oils; flaxseed and walnuts).
Patients with documented CHD	Consume about 1 g of EPA+DHA per day, preferably from fatty fish. EPA+DHA in capsule form could be considered in consultation with the physician.
Patients who need to lower triglycerides	2 to 4 g of EPA+DHA per day provided as capsules under a physician's care.

2.1. Alpha-Linolenic Acid

Observational studies have shown cardioprotective effects of ALA on risk of coronary morbidity and mortality. The Nurses' Health Study reported a 30% reduction in the relative risk of fatal coronary heart disease in individuals who consumed more than 1 gram of ALA per day [92]. In a subset of participants from the Cardiovascular Health Study, consumption of tuna or other broiled or baked fish assessed with a food frequency was found to correlate with plasma phospholipid long-chain n-3 fatty acid levels. Among the entire cohort of 4775 adults 65 years or older followed for 12 years, tuna and other fish consumption was associated with a 27% lower risk of ischemic stroke with intake of one to four times per week compared with an intake of less than once per month [93]. In the Iowa Women's Health Study, the highest tertile of ALA intake was associated with a 15% reduction in total mortality [94]. Finally, in a secondary analysis of the 24-hour recall data from the Multiple Risk Factor Intervention Trial (MRFIT), a primary prevention study that examined the effects of reducing elevated serum cholesterol and diastolic blood pressure along with smoking cessation on CHD mortality, observed that the highest quintile of ALA intake, 2.81 g/day, yielded a multivariate-adjusted relative risk for all-cause mortality of 0.67 [95].

The Lyon Diet Heart Study is the largest clinical trial to examine the effects of ALA on CVD [96, 97]. In this randomized secondary prevention trial, an AHA Step 1 Mediterranean dietary pattern (high in ALA) reduced cardiac death and nonfatal MI by approximately 70%, and all coronary events by about 50% despite no improvement in lipids and lipoproteins. The authors attributed the benefits to the 68% increase in ALA intake (~1.7 g/day) compared to the control group. On the other hand, the AHA Science Advisory suggested that other differences between the two diet groups could have played a role in the reduction in CVD risk observed in the Lyon Diet Heart Study [98]. For example, in the Lyon Diet Heart Study, subjects in the experimental group were instructed to adopt a Mediterranean-type diet that contained more bread, root vegetables, and green vegetables, fish, fruit at least once daily, less red meat (replaced with poultry), and margarine supplied by the Study to replace butter and cream.

Despite the beneficial effects of ALA on coronary disease risk seen in observational studies and in the Lyon Diet Heart Study, a recent review questioned the cardiovascular benefits of plant-derived n-3 fatty acids, stating that there "was no high quality evidence to support the beneficial effects of ALA" with respect to eliciting reductions in all-cause mortality, cardiac and sudden death, and stroke [99]. The majority of evidence comes from epidemiologic data, and the one controlled clinical study conducted to date did not specifically evaluate ALA effects. More high-quality clinical trials are needed that examine the effects of ALA on CVD outcomes before definite conclusions can be drawn regarding its role in the treatment and prevention of heart disease.

2.2. Eicosapentaenoic Acid and Docosahexaenoic Acid

The 1970s marked the beginning of an extensive scientific evaluation of the role of n-3 fatty acids in the development of CVD. The seminal studies of Dyerberg *et al.* [100] noted that coronary atherosclerotic disease was rare in Greenland Eskimos and prevalent in a Danish population. These scientists attributed this difference in the incidence of CHD to the high intake of marine oils by the Eskimos and, in particular, EPA and DHA. During the past 30 years numerous studies have demonstrated that these fatty acids may confer cardioprotective effects via multiple mechanisms of action. As shown in Figure 7, the effects of EPA and DHA on clinical outcomes (e.g., antiarrhythmic, TG-lowering, BP-lowering) occur in a time-dependent manner. Of importance is that the antiarrhythmic effect is achieved at relatively low doses of EPA and DHA. This is clinically important as arrhythmias are the cause of sudden cardiac death, the leading cause of cardiac death in the United States [101]. The effects of EPA and DHA on lowering TG, heart rate, and blood pressure all occur within months to years at doses that are consistent with current dietary recommendations. Fish and/or fish oil consumption has consistently been shown to reduce CHD death (~35%), CHD sudden death (~50%), and ischemic stroke (~30%). Modest benefits of fish and/or fish oil consumption also have been observed in regard to nonfatal MI, delayed progression of atherosclerosis, recurrent ventricular tachyarrhythmias, and postangioplasty restenosis [102].

Several large epidemiologic studies have demonstrated EPA and DHA intakes of 250–500 mg/day yield significant reductions in CHD mortality and sudden death [102]. Researchers have suggested that there is a threshold of effect where intakes above 900 mg/day do not elicit a greater decrease in risk [93, 102, 103]. The beneficial effects of increased EPA and DHA intake in the prevention of CVD have been examined in both primary and secondary prevention populations.

The results from the MRFIT trial found that individuals in the highest quintile of EPA and DHA had a 40% reduction in risk from cardiac death [104]; Data from the Nurses' Health Study showed that women in the highest quintile of EPA and DHA intake had a 31% lower risk of heart attack than those in the lowest quintile [105]. Participants in the Physicians Health Study who consumed fatty fish at least once a month had a ~50% reduction in risk of sudden death from MI; however, no associations between fish intake and reduction in incidence of MI were found [106]. Not all primary prevention studies have found a benefit of an increased intake of EPA and DHA. A 6-year prospective study from Finland, which included 21,930 men, found no benefit to EPA and DHA or ALA in reducing cardiac death [107]. The researchers questioned the external validity of these findings because the subjects were mainly middle-aged, smoking men, with high intakes of dietary fat. Also, this study was conducted in an area in

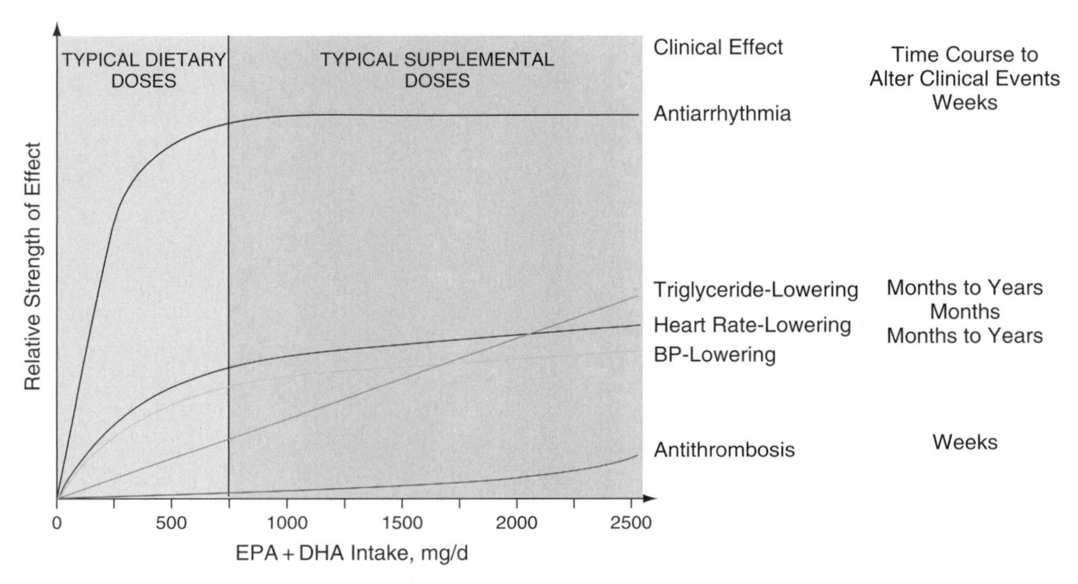

FIGURE 7 Schema of potential dose responses and time courses for altering clinical events of physiologic effects of fish or fish oil intake [102]. Reprinted with permission from the American Medical Association.

which mercury intake from fish is known to be high. Mercury levels can increase CVD risk and were not controlled for in the analysis.

There have been three large secondary prevention intervention studies where patients with CHD were given dietary advice to consume at least two servings of fatty fish a week (200–400 grams of fish) or given supplemental fish oil capsules (850 mg EPA and DHA and 1800 mg EPA). These interventions resulted in a 21–29% reduction in all-cause mortality [108], 45% reduction in sudden death from MI [108, 109], and 19% reduction in all coronary events [110].

Despite the significant evidence base reporting benefits of fish consumption in many populations, other studies have revealed that not all populations benefit from increased EPA and DHA consumption [111–114]. The Diet and Reinfarction Trial 2 (DART-2), a randomized clinical trial of patients with angina, was conducted to determine if fish or fish oil consumption would reduce risk of cardiac and sudden-death end points. The DART-2 Trial showed significantly higher mortality rates from sudden and cardiac death. The findings have been questioned because it is not clear how carefully subjects adhered to advice regarding increased fish consumption. In this trial, serum EPA levels were only measured in a small subset ($n = 39$) of the subjects at 6 months into the study [78].

The benefits of fish oil supplementation have been questioned in patients with implanted cardioverter/defibrillators (ICDs). Three fish-oil supplementation trials have been conducted in this population with mixed results. One trial found a significant trend favoring the use of fish-oil supplementation to prevent fatal ventricular arrhythmias; however, there was not enough statistical power to yield a significant result [115]. Another trial found that fish-oil supplementation did not reduce the risk of ventricular fibrillation/ventricular tachycardia and a potential adverse effect was seen in individuals with an ejection fraction less than 40% [114]. Finally, in the most recent trial, the Study on Omega-3 Fatty Acids and Ventricular Arrhythmia (SOFA), a protective effect attributed to fish-oil supplementation was not found in patients with ICDs [116]. Because of the discordant findings regarding increased EPA and DHA intake and patients with angina or ICDs, further studies are warranted.

Fish oil also has a marked hypotriglyceridemic effect in individuals with normal or elevated TG levels (≥ 2 mmol/l). A review of 21 studies that examined the effects of fish or fish oil on lipids and lipoproteins concluded that while the effects of EPA and DHA (0.1–5.4 g/day) on TC, LDL-C, and HDL-C are modest, TGs were reduced an average of 15%. TG were reduced with fish intakes as low as 0.9 servings per day, as high as 5.4 g/day, and with the greatest effect occurring when intake of EPA and DHA was greater than or equal to 2.6 g/day [117]. Also from this analysis, a dose-response relationship was reported. For every 1 g/day of fish oil, TG levels decreased 8 mg/dl. Individuals with elevated TG are more responsive to the TG-lowering effects of fish oil [118]; for every 10 mg/dl increase in baseline TG levels there was an additional 1.6 mg/dl decrease in TG with the consumption of EPA and DHA.

The effects of n-3 FA consumption on LDL-C levels were evaluated by the Agency for Healthcare Research and Quality of the U.S. Department of Health and Human Services. Data from 15 randomized clinical trials showed that n-3 FA consumption (ranging from 45 mg to 5.4 g of fish oil) led to a net increase of 10 mg/dl in LDL-C [119]. The increase in LDL-C levels may be due partly to an increase in LDL-C size as some [121, 122], but not all [120] studies have shown an increase in LDL-C particle size. It has been proposed that the change in LDL-C particle size relates to a patient's starting LDL-C particle size. Individuals with a Pattern B phenotype, defined as having a higher amount of small, dense LDL-C particles, will respond to fish oil supplementation with an increase LDL-C particle size, whereas individuals without this phenotype will not [120]. Fish-oil supplements can be an effective treatment for patients with hypertriglyceridemia, although monitoring by a physician is essential to ensure that LDL-C levels are closely observed [91].

One area of concern when recommending increased fish intake pertains to the issue of environmental contaminants such as mercury (Fig. 8). A majority of the fatty fish consumed are smaller with a shorter lifespan and thus have lower mercury concentrations than larger, longer-lived fish such as shark and swordfish [102]. It is also important to note that high consumption of certain lean fish high in mercury plays an important role in increased mercury accumulation in humans [123]. Therefore, emphasis on high omega-3 fatty acid fish (that is low in mercury) is recommended.

There have been two studies that observed higher risk of CVD with increased mercury intake; however, fish consumption in these situations still was cardioprotective [124, 125]. Moreover, recent reports describe the benefits of fish consumption far outweigh the risks [102, 124, 125]. Another strategy for increasing EPA and DHA consumption is fish-oil supplements as they are cost effective, convenient, and contain negligible quantities of environmental contaminants [126]. This is a strategy recommended for individuals who do not eat fish or those who need high doses of marine-derived omega-3 fatty acids. The best guidance is to follow FDA guidelines as well as state and local advisories for safe fish consumption.

In conclusion, long-chain n-3 fatty acids have been shown to have significant cardioprotective benefits in observational studies and secondary prevention trials. This research supports current dietary recommendations for two fish meals per week, preferably fatty fish. The efficacy of EPA and DHA in primary prevention along with the

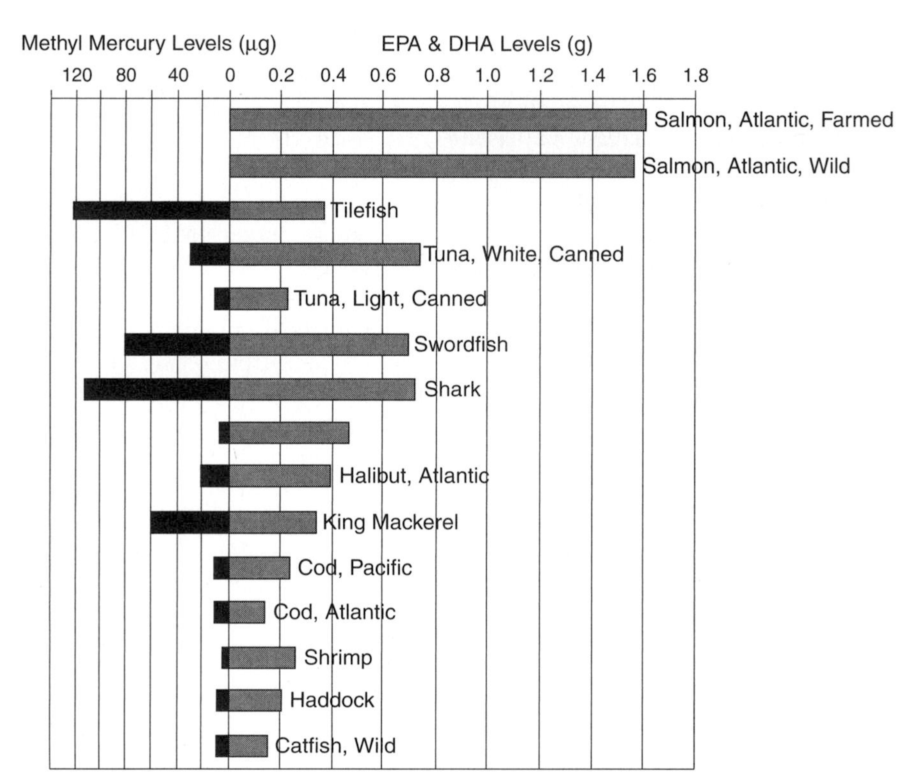

FIGURE 8 Comparison of EPA and DHA content of fish with methyl mercury levels (3-oz serving) [10, 288].

cardioprotective effects of ALA need to be clarified and explored further via randomized clinical trials.

III. DIETARY CARBOHYDRATE

The RDA for glucose is 130 g/day. This level of consumption is enough to prevent ketosis and provide enough glucose to the brain for proper function. The AMDR for carbohydrate is 45–65% of total energy, with no more than 25% of total energy coming from added sugar [12]. The 2005 Dietary Guidelines for Americans do not recommend specific intakes for added sugar, but rather recommend choosing and preparing foods and beverages with little added sugar; the Guidelines also suggest a discretionary allowance for various energy levels (e.g., 32 g/day [8 tsp] suggested for a 2000-kcal diet) [22]. The NCEP ATP III [13] recommends that 50–60% of energy come from carbohydrate.

When discussing different types of carbohydrates, scientists historically used the terms *complex* versus *simple* carbohydrates. We have some understanding of the role of some carbohydrate food sources in CVD risk; thus this simple terminology needs further explanation. Many foods that are included in a high-carbohydrate diet, such as fruits and vegetables, breads and cereals, and legumes, contain multiple compounds that could favorably affect CVD risk. In this section, a few of the active areas of scientific investigation associated with carbohydrate and reduction of CVD risk are covered, including low-carbohydrate diets and the role of glycemic index, dietary fiber, and soluble fiber.

A. High-Carbohydrate, Low-Fat Diets

Replacing energy from SFA with carbohydrate is one strategy for reducing SFA to decrease LDL-C. The resulting high-carbohydrate, low-fat diet, when not accompanied by weight loss, decreases HDL-C and typically increases TG, especially on low-fiber diets [127]. This is particularly problematic for people with insulin resistance and accompanying dyslipidemia characterized by low HDL-C and elevated TG [128]. A high-carbohydrate, low-fat diet also has been shown to increase plasma glucose and insulin in individuals with type 2 diabetes [129] and healthy women [130]. On the other hand, it can be argued that a high-carbohydrate, low-fat diet facilitates a reduction in energy intake [131] and promotes weight loss [132]. In the Women's Health Initiative Dietary Modification Trial [21], subjects in the intervention group who were instructed on a diet that provided 20% of energy from fat maintained a lower weight (1.9 kg, $p < 0.001$) during an average 7.5 years follow-up period than did women in the control group. In this study, it is important to point out that the women did not achieve the 20% energy as fat target and consumed 29.8% of energy from fat. However, they did decrease total fat intake from 38.8% of energy. An extreme reduction in fat (10% energy) and thus a much higher carbohydrate (70–80%) diet has been advocated by some [134]. This latter approach has been shown to be effective when combined with other intensive lifestyle changes (e.g., stress management and aerobic exercise), in achieving athero-regression as measured by percent change in diameter stenosis [135]. Unfortunately, very low fat diets can be very hard to follow in free-living situations, and compliance with the diet significantly decreases over time (by 50%) [20].

A higher carbohydrate diet does not always increase TG levels. The OmniHeart Trial [51] and Dietary Approaches to Stop Hypertension (DASH) Diet studies [136, 137] both provided diets relatively high in carbohydrate (58% of energy), and no significant increase in TG was observed. This is most likely due to the type of carbohydrate consumed as well as dietary fiber. The diets tested in the OmniHeart Trial and DASH Diet studies were high in fruits, vegetables, and whole grains, which yielded diets containing at least 30 grams of fiber/day (2100-kcaldiet) [48].

B. Low-Carbohydrate Diets

Although low-carbohydrate diets have been around for decades, it has only been recently that their popularity has increased, principally for both weight loss and, in turn, prevention of cardiovascular disease. There is no standard definition of a low-carbohydrate diet, but it is generally accepted when a diet contains less than 50 grams of carbohydrate or less than 10% of energy from carbohydrate, it is considered a low-carbohydrate diet [138]. When compared to low-fat diets, the low-carbohydrate diets have been shown to elicit greater weight loss at 6 months (–7.3 lb versus −3.1 lb) but at 12 months weight loss is similar between diets [20, 139]. The mechanism by which low-carbohydrate diets promote greater weight loss at 6 months is unclear however several factors have been implicated. These include improved insulin/glucose control, increased thermic effect of food, enhanced maintenance of basal metabolic rate, ketosis-induced appetite suppression, and increased water loss [140]. The reduced weight loss in subjects on low-carbohydrate diets between months 6 and 12 is likely due to decreased dietary compliance [20].

Low-carbohydrate diets have been shown to reduce triglyceride levels and increase HDL-C [139]. The debate surrounding the effectiveness of low-carbohydrate diets in the treatment and prevention of cardiovascular disease centers on the effect of these diets on total cholesterol and LDL-C due to the increased intake of SFA that accompanies low-carbohydrate diets. Recent work by Krauss *et al.* [141] found that in overweight or obese men, reductions in carbohydrate intake (26% of total energy) versus 54% energy from carbohydrate along with controlled weight loss reduces the proportion of small, dense LDL-C particles. Importantly, this effect is primarily driven by changes in dietary carbohydrate. Of note also is that although there is an effect of dietary carbohydrate on LDL pattern B phenotype, this is attenuated after weight loss. Another dietary factor that affects the lipid and lipoprotein responses to a low-carbohydrate diet is soluble fiber. Wood *et al.* [142] found that the addition of 3 grams of supplemental soluble fiber (Konjac-mannan) to a low-carbohydrate diet reduced LDL-C by 14% while still reducing triglycerides and increasing HDL-C. Potentially one of the most important findings regarding a low-carbohydrate diet is the improved clearance of triglyceride-rich chylomicrons [138]. Chylomicron remnants increase risk for coronary disease [143]. Thus, low-carbohydrate diets appear to be of benefit for short-term weight loss (6 months). In addition, they beneficially affect triglycerides and HDL-C compared with higher carbohydrate diets, as well as postprandial TG clearance.

C. Glycemic Index and Glycemic Load

Carbohydrates differ in terms of their effects on glucose metabolism [144]. Carbohydrates can be classified according to their blood glucose-raising effects using the glycemic index (GI). Low-GI foods, such as mature beans and green vegetables, elicit less of a glycemic response than high-GI foods, such as potatoes and ready-to-eat cereals. Independent of weight loss, diets composed of low-GI foods increase insulin sensitivity [145, 146] and decrease total serum cholesterol [146] and LDL-C [145] in people with type 2 diabetes. LDL-C decreased 6% more in subjects following a low-GI diet compared to a high-GI diet [145]. These findings illustrate the importance of considering not only the macronutrient composition of the diet, but also the type of carbohydrate consumed within the context of a high-carbohydrate, low-fat diet.

A review of 13 intervention studies that examined the effects of glycemic index on triglycerides (5–20% reduction), LDL-C (>5% reduction), and total:HDL-C ratio (>5% reduction) consistently showed the benefit of a lower GI diet [147]. In general, foods high in soluble fiber have a low GI; however, this is an oversimplification because food preparation and consumption of a specific food in a mixed meal can alter the GI. Thus, a mixed meal can have a low GI with the selection of certain foods that elicit a high glycemic response when tested individually. Because of the complexity of implementing the GI, it likely will be difficult for consumers to adopt at the present time with currently available foods and contemporary lifestyle practices. Furthermore, the concept of glycemic index must also address the role of other nutrients in a way that is consistent with current dietary recommendations. For example, some low-GI foods are high in total fat, SFA, and sugar and therefore should be limited. This example elucidates the importance of not just considering glycemic index but also considering other nutrient and dietary factors. Thus, it is apparent that many questions remain about implementation of glycemic index in practice.

The glycemic load is a measure of both type and amount of carbohydrates while the glycemic index is a measure of only carbohydrate type. Glycemic load is defined as the mathematical product of the carbohydrate amount and glycemic index. This has been shown to be a more effective classification method for carbohydrates. In a prospective analysis of carbohydrate intake using data from the Nurses' Health Study the glycemic load of the diet was directly associated with CHD. The upper two quintiles for glycemic load had relative risk ratings of 1.51 and 1.98. This relationship was most evident among women with a BMI above 23 [148]. An analysis [149] of 244 healthy women enrolled in the Women's Health Study (WHS) showed a strong correlation between high-sensitivity C-reactive protein (CRP) levels and glycemic load. This again suggests a relationship between type and amount of carbohydrate consumed and cardiovascular disease, specifically in middle-aged women. In contrast, the Zutphen Elderly Study [150] found no relationship between a high-GI diet and CHD in elderly men.

In a year randomized clinical trial, subjects with elevated insulin levels following a large oral dose of glucose lost more weight after 18 months on a low-glycemic-load diet compared with a high-glycemic-load weight-loss diet (12.76 lb versus 2.64 lb). Subjects with normal insulin levels lost similar amounts of weight with either low-glycemic-load or low-fat diets [151]. Changes in cardiovascular risk factors were diet specific. The low-glycemic-load diet resulted in greater reductions in TG (–21.2 mg/dl versus –4.0 mg/dl) and an increase in HDL-C (1.6 mg/dl versus –4.4 mg/dl), whereas the low-fat diet group experienced a much greater reduction in LDL-C (–5.8 mg/dl versus –16.3 mg/dl). However, another randomized clinical trial [152] that compared the effects of a low-glycemic-load diet versus a low-fat diet on lipids and lipoproteins found the only difference in lipid changes between diet groups to be TG, which decreased more in the low-glycemic-load group while LDL-C and HDL-C changes did not differ between diet groups. In this study, the low-glycemic-load group also experienced a 50% reduction in CRP, whereas CRP remained unchanged in the low-fat diet group. A 12-month randomized trial that compared low-glycemic-load compared to high-glycemic-load diets found no difference between diets with respect to TG, TC, LDL-C, and HDL-C [153]. Thus, further work is needed before conclusions are made regarding the importance of glycemic load for the treatment and prevention of CVD. Nonetheless, research to date suggests that a low-glycemic-load diet is more efficacious in individuals with insulin resistance.

D. Dietary Fiber

An abundance of evidence supports a beneficial association between dietary fiber intake and risk of CVD [154]. Dietary fiber is found naturally in fruits, vegetables, whole-grain cereals, and legumes. Fiber-fortified foods and supplements also are available that are intended to increase dietary fiber [155, 156]. The DRI Report on Macronutrients set the AI for dietary fiber at 38 g/day and 25 g/day for men and women, respectively [12]. This translates into 14 grams of fiber per 1000 kcal of energy consumed. Numerous epidemiologic studies support the cardioprotective effect of dietary fiber. A group of researchers recently pooled data from 10 prospective cohort studies in the United States and Europe and reported that each 10 g/day increment of dietary fiber yielded a 23% reduction in risk of coronary mortality and 14% reduction in all coronary events [157]. The association was strongest with cereal fiber compared to fruit/vegetable fiber. Increased fiber intake via whole grains also has been associated with improved insulin sensitivity [158].

E. Soluble Fiber

Soluble fiber, including oat bran, psyllium, guar gum, and pectin, has been shown to reduce CVD risk through its action on lipids and lipoproteins and glucose metabolism. Soluble fiber has numerous properties that mediate its cholesterol-lowering effects, such as binding bile acids, increasing gastrointestinal tract viscosity, and inhibiting cholesterol synthesis following fermentation in the colon [159, 160]. NCEP ATP III guidelines recommend 10–25 grams of soluble fiber each day using evidence that a dietary intake of 5–10 grams of soluble fiber per day has been shown to lower LDL-C levels by 5% [13]. Observational data from the Los Angeles Atherosclerosis Study showed that increased fiber intake (namely pectin) had a strong inverse relationship with intima-media thickness progression in otherwise healthy individuals [161].

A meta-analysis of 67 controlled human trials [162] determined that various soluble fibers (2–10 g/day) modestly reduced total and LDL-C and did not affect HDL-C cholesterol and TG levels. A randomized trial examining the effects of increased fiber in normolipidemic participants found that 3.5 g/day soluble fiber (29.5 g/day total fiber) yielded a 12.8% decrease in LDL-C. This effect was attributed to the differences in total and soluble fiber intake because no differences in intake of other macronutrients were found [163].

The effects of different food sources of soluble fiber on lipids and lipoproteins have been studied. Incorporating 15–115 grams of beans (navy or pinto) has been shown to reduce both total and LDL-C by 15–23% and 13–24%, respectively [160]. Arguably the most commercialized source of soluble fiber, oats (rolled or bran), are known for their cholesterol-lowering properties [164]. Oats contain β-glucan, which has been shown to reduce total cholesterol along with fasting glucose and insulin [165]. In addition, soluble fiber has been shown to lower glucose and insulin levels in healthy individuals [166] and favorably affect insulin sensitivity in individuals with diabetes [167] and moderate hypercholesterolemia [168].

A diet that meets current recommendations for dietary fiber that is high in both total and soluble fiber can be achieved by consuming recommended amounts of fruits (four servings), vegetables (five servings), and whole grains (at least 3-oz, but preferably 6 oz equivalent servings). This level of dietary fiber has been shown in observational studies to reduce cardiovascular events and in clinical trials to reduce LDL-C.

IV. DIETARY PROTEIN

Epidemiologic and controlled clinical studies have shown benefits of dietary protein on CVD risk, although some studies suggest an adverse relationship. Data from the Nurses' Health Study found high protein intakes (up to 24% of total energy intake), which included animal and plant protein, were associated with a significantly reduced risk of CVD (RR = 0.75; 95% CI: 0.61, 0.92) [92]. Likewise, in the largest randomized controlled clinical trial to date, the OmniHeart Trial, found that a high-protein diet (25% of energy intake) reduced the estimated 10-year risk (5.8% lower) for CHD compared with a high-carbohydrate diet (58% of energy intake) [51]. However, in a 12-year follow-up study of middle-aged Swedish women participating in the Women's Lifestyle and Health Cohort, a diet low in carbohydrate and high in protein (assessed from protein and carbohydrate intakes according to deciles of individual energy intake) was associated with a higher total and CVD mortality [169]. In addition, in the Greek component of the European Prospective Investigation into Cancer and Nutrition, prolonged consumption of a low-carbohydrate, high-protein diet but not a high-protein diet alone was associated with an increase in total mortality [170]. This study suggests that marked reductions in carbohydrate together with an increase in dietary protein are problematic.

A. Animal Protein

There is uncertainty whether animal protein, particularly from meat, affects CVD risk. In a prospective study that examined whether overall dietary patterns derived from a food frequency questionnaire (FFQ) predicted risk of CHD in men found a dietary pattern including high intakes of red meat and processed meat was associated with a higher risk for CVD compared lower intakes (RR adjusted for lifestyle variables and fat intake: 1.43 [1.01, 2.01]) [171]. In contrast, the 2-year randomized Cholesterol Lowering Atherosclerosis Study found a reduction in new coronary artery lesions among those with increased dietary protein (lean meat and low-fat dairy) compared with subjects who decreased their protein intakes [172]. The early work that reported associations between animal protein and CVD mortality [173] likely was confounded by correlations between protein intake and dietary SFA and cholesterol [174].

The dietary guidance that recommends a reduction in SFA and cholesterol has been interpreted by some to restrict red meat consumption. Several studies, however, have been conducted to evaluate the effect of blood cholesterol-lowering diets containing lean red meat on lipids and lipoproteins. In three randomized controlled trials [287–289] conducted with hypercholesterolemic subjects, lean red meat, fish or lean poultry elicited similar effects on TC, LDL-C, and TG or HDL-C (Table 3). In addition, a small, recent, 8-week parallel study conducted in Australia found that markers of oxidative stress and inflammation were not elevated when energy from carbohydrate was partially replaced with 200 g/day of lean red meat [175]. These data suggest that lean red meat can be included in a blood cholesterol-lowering diet. Moreover, a diet that includes lean red meat also must meet current dietary recommendations.

B. Soy (Vegetable) Protein

Studies conducted in the 1970s and 1980s found that a diet rich in soy decreased TC by 20%–25% in hyperlipidemic patients [176–178]. In 1995, a meta-analysis of clinical trials of soy intakes (ranging between 31 and 47 g/day) reduced TC by 23.2 mg/dl (9.3%), LDL-C by 21.7 mg/dl (12.9%), and TG by 13.3 mg/dl (10.5%), with the greatest changes in serum TC and LDL-C related to initial serum cholesterol concentrations. There was a trend toward a 1.2 mg/dl (2.4%) increase in HDL-C [179]. Results of the 6 week randomized controlled OmniHeart trial, found that a high protein diet (25% energy (12 servings of plant protein/day)) reduced HDL-C (–2.6 mg/dal) over a six week period compared to a high carbohydrate (58% energy (5.5 servings of plant protein/day)) diet (–1.4 mg/dl) and a high unsaturated fat (31% energy (5.5 servings of plant protein/day)) diet (–0.3 mg/dl)) [51].

A reevaluation of the prior scientific evidence of soy protein effects on lipids and lipoproteins, as well as soy isoflavones, was conducted by Sacks *et al.* and published as an AHA Science Advisory [180]. Twenty-two randomized trials (conducted between 1998 and 2005) were included that evaluated isolated soy protein with isoflavones compared with casein or milk protein, wheat protein, or mixed animal protein. In these studies, soy protein intake ranged between 25 and 135 g/day with the range of soy isoflavones between 40 and 318 mg. A very modest 3% reduction in LDL-C was found due to soy proteins; soy isoflavones generally had no effect on LDL-C [180], despite large

TABLE 3 Dietary Cholesterol Changes Following Consumption of Varying Protein Food Sources

Author	Year	Subjects	Design		Results TC	LDC-C	TG	HDL-C
Beauchesne-Rondeau, E. *et al.* [287]	2003	Hypercholesterolemic men (TC: >5.2 mmol/l; LDL-C: >3.4 mmol/l)	RCT: 26 days ($n = 17$ each; 50 years) of lean meat (~180 g) versus lean poultry (~180 g) versus fish (~270 g). 6-week washout period in between each diet. 69% animal protein and no milk products allowed: given 600 mg Ca + 125 IU vitamin D.	Lean beef	↓ 18%	↓ 7%	↓ 19%	
				Lean poultry	↓ 8%	↓ 9%	↓ 25%	
				Fish	↓ 5%	↓ 5%	↓ 20%	
Davidson, M. *et al.* [288]	1999	Hypercholesterolemic men and women (LDL-C: 3.37–4.92 mmol/l; TG: <3.96 mmol/l)	RCT: 36 days lean white meat (n = 102, 55 years) or lean red meat (n = 89, 57 years). 170 g (6 oz) for 5–7 days/week.	Lean red meat	NS	↓ 1.7%	—	↑ 2.3%
				Lean white meat	↓ 1.8%	↓ 3%	↓ 0.5%	↑ 2.4%
Hunninghake, D. *et al.* [289]	2000	Hypercholesterolemic men and women (LDL-C: 3.37–4.92 mmol/l; TG: <3.96 mmol/l)	RCT: 36 weeks lean red meat (n = 72, 57 years) versus lean white meat (n = 73, 56 years).	Lean red meat	↓ 0.9%	↓ 1.9%	NS	↑ 2.8%
				Lean white meat	↓ 1.2%	↓ 2%	NS	↑ 2.2%

RCT, randomized controlled trial.

increases in blood isoflavone concentrations. A recent report from Agency for Healthcare Research and Quality of the soy protein literature demonstrated a decrease in TC of 6 mg/dl (2.5%), with a median decrease in LDL-C of 5 mg/dl (3%). There was no effect of isoflavones [181]. Figure 9 presents the changes in LDL-C in response to soy protein and soy isoflavones as a function of baseline LDL-C. It can be seen in Figure 9 that those with high baseline LDL-C have a greater LDL-C lowering response, and there is a dose-response relationship between soy protein intake and LDL-C lowering.

A subsequent meta-analysis was published [182], including 11 trials on soy protein containing isoflavones, in addition to the effects of soy protein containing enriched and depleted isoflavones. Compared with animal protein (containing no isoflavones), soy protein (containing no isoflavones) decreased TC by 7.7 mg/dl (3.6%) and decreased LDL-C by 3.9 mg/dl (2.8%); soy protein (plus enriched isoflavones) lowered LDL-C by 7.0 mg/dl (5%) and increased HDL-C by 1.6 mg/dl (3%) [182]. The effects of isoflavones (after controlling for soy protein intake) were found to reduce TC by 3.9 mg/dl (1.8%) and decrease LDL-C by 5 mg/dl (3.6%) compared with soy protein (containing no isoflavones). Reductions in LDL-C were greatest in subjects with high versus normal cholesterol levels; however, there was no association between the changes in LDL-C levels and isoflavone intake. Although it was concluded that 102 mg ingested soy-derived isoflavones (independent of ingested soy protein) can lower TC and LDL-C, this intake is approximately double current intakes in Japan. Consuming 42 mg soy (which included 6 mg isoflavones—a very low content) also can improve LDL-C and HDL-C; however, this level of soy protein intake is quite large, representing at least 50% of the average daily total protein intake in the United States [182]. Moreover, ingestion of large amounts of isoflavones is necessary to identify any benefits on the lipid profile.

In addition to the effects on lipids, a recent 8-week crossover trial was conducted among postmenopausal women with metabolic syndrome comparing the DASH diet (17% energy from protein) versus soy protein (replaced one serving of red meat with soy protein—30 g) versus soy nut (replaced red meat with soy nuts—30 g) [183]. The soy nut and soy protein diets decreased the inflammatory markers E-selectin and CRP, compared with the DASH diet; IL-18 was lower on the soy nut diet, and nitric oxide production improved. In contrast, a recent randomized controlled trial in hypercholesterolemic patients found varying diets containing soybeans, soy flour, or soy milk (15% of energy as protein—7.5% of energy as experimental protein; 37.5 g/day) had no effect on blood pressure, vascular endothelial function, or CRP concentrations [184]. Previous

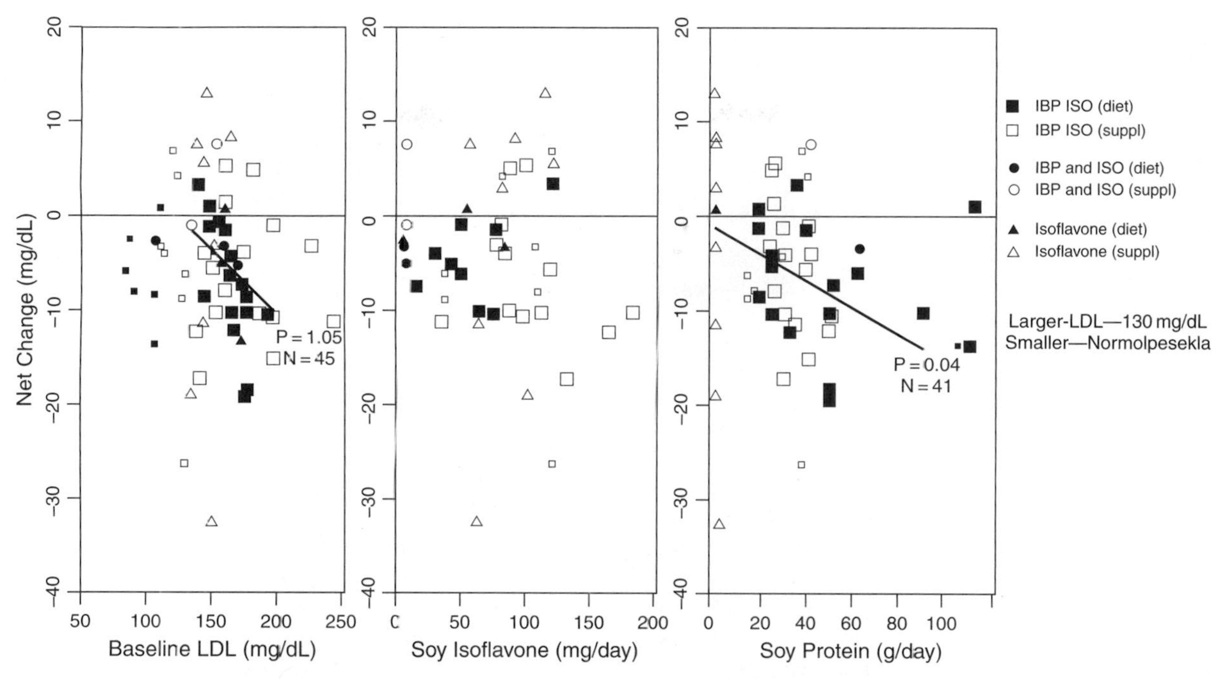

FIGURE 9 LDL changes after soy products consumption. (1) Net change of LDL-C with soy product consumption compared to control, by baseline level, isoflavone content, and soy protein content. Studies without non-soy control are not included. Studies without data on isoflavone or protein content are omitted from relevant graphs. (2) IBP w/ Iso, soy protein with isoflavones; IBP w/o Iso, soy protein without isoflavones; suppl, supplement. (3) Dashed lines represent adjusted regressions for studies with sufficient data for regression. Regression lines are drawn only within the range of independent variable (*x*-axis) data examined. *P*-values and number of studies included in regressions are shown. Both regression lines drawn are for all studies with abnormal baseline LDL. Reprinted with permission from [181].

controlled studies also have reported that soy intake inhibited LDL oxidative susceptibility [185, 186] and improved arterial compliance [187].

The evidence is conflicting about the role of soy protein in CVD risk reduction, and consumption of large amounts of isoflavones may be required to provide further benefits. Further trials are required investigating the isoflavone component on CVD mortality.

In summary, these studies highlight the favorable effect that both animal and plant proteins have on CVD risk factors. Of specific interest are their effects on TG and HDL-C, as well as inflammatory markers and blood pressure. A reduced-fat, high-protein (18–25% energy from protein) diet may be an alternative approach to the traditional reduced-fat, high-carbohydrate diets. With respect to soy protein, large intakes of soy protein (more than half the daily protein intake) may lower LDL-C by a few percent when it replaces dairy protein or a mixture of animal proteins [180]. Moreover, soy proteins with isoflavones do not appear to further benefit CVD risk reduction and therefore isoflavone supplement use is not recommended. In addition, soy protein also may be used to increase total dietary protein intake and to reduce carbohydrate and fat intake [180]. Soy products (e.g., tofu, soy butter, and soy nuts) may be beneficial to health because of their high unsaturated fat content, fiber, vitamins, minerals, and low content of saturated fat [180]. Further studies are required investigating the effect of high-protein diets on CVD risk factors. Studies are needed with both plant and animal protein sources to determine whether their effects on CVD risk and risk factors differ. This information will be important for future dietary guidance about protein.

V. ALCOHOL

There is accumulating evidence demonstrating that moderate alcohol consumption reduces CHD risk, and the pattern of drinking has important implications for various CVD risk factors [188–190], clinical outcomes [191–194], and mortality [195–197]. Consumption of one to three drinks a day has been shown to lower risk of CHD by 10–40% compared with those who abstain; the majority of studies have identified a lower risk with up to three alcoholic drinks a day [198–201]. The epidemiologic evidence suggests a J- or U-shaped relationship between alcohol and CHD [202–204], such that moderate alcohol consumption appears more beneficial than no alcohol or excessive alcohol consumption.

The mechanisms underlying alcohol and CVD risk have been a result of some beneficial effects in hemostatic factors [205, 206]; however, the putative benefits

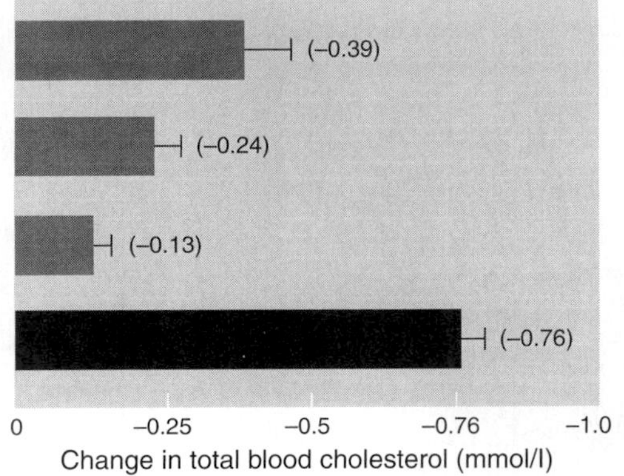

FIGURE 10 Mean (SE) changes in blood total cholesterol concentration associated with replacing dietary saturated fat by polyunsaturated and monounsaturated fats and with reducing dietary cholesterol [26]. Reprinted with permission from the *British Journal of Medicine*.

have been mainly credited to an increase in HDL-C [207–209], thus inhibiting LDL oxidation during the atherogenic process. In a meta-analysis of 42 trials investigating alcohol consumption and HDL-C, a 24.7% reduction in CHD was found following consumption of 30 g alcohol/day, directly attributable to increased HDL-C levels (3.99 mg/dl) [210].

With respect to specific beverage type, early studies have suggested that in wine-drinking countries, there was less coronary artery disease (CAD) incidence compared to beer- or liquor-drinking countries [211]; with later international comparison studies confirming this [212, 213]. Particularly in France, the markedly low incidence of CHD, despite intake of a high-fat diet, has been attributed to the consumption of red wine containing high levels of polyphenolic compounds. This was termed the "French paradox," referring to the observation that France has the highest wine intake, the highest total alcohol intake, yet the second lowest CHD mortality rate [212, 213]. The mechanisms underlying this lower CHD incidence have been associated with antioxidant phenolic compounds or antithrombotic substances in red wine [213–215].

A recent review, however, reported that prospective population studies provide no consensus that wine has additional benefits, and various studies show benefit for all three major beverage types (wine, beer, and liquor) [216]. In contrast to this, controlled trials have found moderate consumption of red wine to inhibit oxidative modification of LDL-C [217, 218], improve coronary blood flow [219], and improve markers of inflammation [220], as well as reverse the impaired endothelial function caused by cigarette smoking [221].

Therefore, further randomized studies are required identifying benefits of other alcoholic beverages on CVD risk factors, and to assess whether moderate wine consumption is more protective than these beverages against cardiovascular risk.

Recommendations for alcohol consumption by the NCEP panel have stated that no more than two drinks per day for men and no more than one drink per day for women should be consumed. A drink is defined as 5 oz of wine, 12 oz of beer, or $1^1/_2$ oz of 80-proof whiskey. Persons who do not drink should not be encouraged to initiate regular alcohol consumption [13].

VI. DIETARY CHOLESTEROL

Cholesterol plays an important role in steroid hormone and bile acid biosynthesis and serves as an integral component of cell membranes. There is no biological requirement for dietary cholesterol because all tissues synthesize a sufficient amount of cholesterol to meet metabolic and structural needs. The DRI Report for Macronutrients recommended that cholesterol intake be as low as possible because of the dose-response relationship between dietary cholesterol and total and LDL-C [12, 26]. The Dietary Guidelines for Americans, 2005, as well as AHA, recommend less than 300 mg/day for normocholesterolemic individuals. NCEP ATP III recommends less than 200 mg/day for those with CHD or elevated LDL-C [13, 22].

There is some epidemiologic evidence demonstrating a relationship between dietary cholesterol and CVD risk [222]. Among Japanese women, one to two eggs per week was associated with a 12% lower all-cause death rate, with a trend toward a lower mortality due to stroke, ischemic heart disease (IHD), and cancer, compared with women consuming 1 egg per day [222]. In contrast, among 117,933 subjects in the United States, no relationship was found between consumption of 1 egg or fewer per day and the risk of IHD or stroke [33]. The Framingham Heart Study reported that egg consumption was not related to CHD [223] and was not related to serum cholesterol levels [224].

In normolipidemic healthy young men, subjects were randomized to consume either a low-cholesterol diet (n = 15, 3 egg whites: total intake of cholesterol 174 mg/day) or a high-cholesterol diet (n = 12, 3 whole eggs: total intake of cholesterol 804 mg/day) for 15 days. Compared with the low-cholesterol diet group, the high-cholesterol diet group had higher serum TC (135 mg/dl versus 181 mg/dl), LDL-C (85 mg/dl versus 117 mg/dl), and HDL-C levels (35 mg/dl versus 50 mg/dl). Triglyceride and Lp(a) concentration did not differ [225].

In a meta-analysis of 395 metabolic ward trials investigating the importance of dietary fatty acids and dietary cholesterol in serum TC, LDL-C, and HDL-C, it was found that isoenergetic replacement of saturated fats (10% of dietary energy) by complex carbohydrates was associated with a decrease in TC by 20 mg/dl; replacing carbohydrates by PUFA (5% energy) would further reduce TC by 5 mg/dl. A reduction of 200 mg/day in dietary cholesterol was associated with a further reduction in TC of 5 mg/dl. The sum of these isoenergetic changes is displayed in Figure 1, indicating a reduction in TC by 29 mg/dl (99% CI: 0.67, 0.85) and LDL-C by 24 mg/dl and an increase in HDL-C by 4 mg/dl [26]. On the basis of this meta-analysis, dietary cholesterol elicits a small hypercholesterolemic effect that is less than that of saturated fat.

In another meta-analysis of 17 trials, consumption of dietary cholesterol increased the ratio of TC:HDL-C concentrations by 0.02 (95% CI: 0.010, 0.030). Furthermore, consuming approximately 200 mg/day of cholesterol (equivalent to one egg per day) increased the ratio of TC:HDL-C by 0.04 units, TC by 4.3 mg/dl, LDL-C by 3.9 mg/dl, and HDL-C by 0.6 mg/dl [226]. This effect was independent of other dietary factors such as type and amount of fat.

Collectively, the evidence shows that dietary cholesterol plays a role in modifying CVD risk; however, the effects on total and LDL-C are relatively modest. Nonetheless, decreasing dietary cholesterol would be expected to reduce the risk of CHD, especially in subgroups of individuals who are responsive to changes in cholesterol intake [227].

VII. PLANT STEROLS/STANOLS

Phytosterols are found in seed oil (e.g., sunflower oil and maize oil), fruits, vegetables, legumes, cereals, and some nuts. Although more than 40 phytosterols have been identified, β-sitosterol, campesterol, and stigmasterol are the most abundant and are the most effective at reducing cholesterol levels. Stanols are saturated molecules produced by the hydrogenation of sterols but are less abundant in nature [228]. The mechanisms by which plant sterols and stanols lower cholesterol levels involve the displacement of cholesterol from micelles, thus reducing intestinal cholesterol absorption and increasing fecal cholesterol excretion [229, 230].

Several nested case-control studies have found conflicting results between plasma sitosterol concentrations and CHD [231, 232]. Because nested case-control studies involve subsamples of participants in prospective studies, biological samples collected prior to onset of disease can be analyzed as a measure of exposure prior to onset of disease. Likewise, biological samples are analyzed in a matched group of disease-free controls. Thus, in these three nested case-control studies, the plasma for determination of sitosterol was collected prior to onset of disease for the cases. In the Prospective Cardiovascular Munster (PROCAM) study, men who suffered a coronary event within the previous 10 years had higher plasma sitosterol concentrations compared with those who had no event (0.19 mg/dl versus 0.17 mg/dl), with a 1.8-fold increased risk of coronary events in subjects with sitosterol levels above 0.21 mg/dl [231]. In contrast the Longitudinal Aging Study Amsterdam (LASA), plasma concentrations of all plant sterols were lower in those with CHD versus those free of CHD (e.g., sitosterol: 0.29 mg/dl versus 0.34 mg/dl); yet higher plasma concentrations of sitosterol were associated with a 28% reduction in CHD risk. The odds ratio with a twofold increase in sitosterol was associated with a 22% decrease in CHD risk [232]. On the other hand, the EPIC-Norfolk Population study, which compared 373 CAD cases with 758 controls, found that there was no difference in sitosterol between CAD cases and controls (0.21 mg/dl versus 0.21 mg/dl). However, there was a 21% lower risk (NS) for future CAD in the highest tertile of sitosterol concentration after adjusting for traditional risk factors [233]. Because these plasma sitosterol concentrations are similar to what was found in the PROCAM men, it appears that higher sitosterol concentrations, such as those found in the LASA study (approximately >0.34 mg/dl), may be necessary to achieve any sort of reduction in CHD risk in subjects initially free of the disease.

Numerous clinical studies conducted with hypercholesterolemic subjects, individuals with type 2 diabetes, and healthy adults and children have shown that consumption of 2–3 g/day of plant sterols or stanols lowers plasma LDL-C by 6–15% [13, 234]. Studies that examined sterol-enriched foods (spreads, low-fat yogurt, bakery products) in healthy subjects found that consumption of 1.6–3.2 g/day of sterols for 4 weeks to 1 year reduced TC 4–8.9% and LDL-C 6–14.7% [32, 235–238].

Some [236, 239], but not all [237, 240, 241], studies have found a decrease in serum antioxidant levels (corrected for TC and TG levels) (range of decrease ~6–14%) with sterol/stanol intakes ranging between 1.8 g and 3.4 g/day. Noakes *et al.*, however, observed that the variations in plasma carotenoids are within observed seasonal and individual variations [239]. Increased consumption of one serving of carotene-rich fruit and vegetables has been effective in preventing this decline, after consumption of a sterol-rich spread [240].

Consumption of 2 g/day of plant sterols and stanols is recommended by the NCEP ATP III as a therapeutic option for maximal LDL-C lowering (6–15%) [242]. The usual dietary intake of plant sterols and stanols ranges between 200 and 450 mg/day [228], but diets based on high intakes of vegetable fats, whole grains, fruits, and vegetables provide higher amounts but still fall far short of meeting the 2 g/day recommendation. Thus, fortified foods and/or supplements are needed to attain this level of intake. Plant stanols and sterols are now included in many foods such as margarine, low-fat milk, yogurt, cereal bars, and orange juice, as well as gel-capped sterol and stanol supplements to assist in increasing consumption.

VIII. SUPPLEMENTS

The dietary supplement market is one of the fastest growing industries [243]. Data from the Third National Health and Nutrition Examination Survey (NHANES III) and the more recent 1999–2002 NHANES found that approximately 55% of U.S. men and women aged 40 or older reported the use of some dietary supplement. Multivitamin use was most popular (38%) followed by single or combined antioxidant use (21%) and B vitamins (8%) [244]. The primary reason for using dietary supplements was to achieve self-care goals, as a means of ensuring good health, alleviating depression, and for "medicinal" purposes to treat and prevent various illnesses, including the common cold and flu [243].

Epidemiologic and clinical studies that have evaluated dietary supplements in the prevention or treatment of disease are inconsistent, showing little or no benefit. Nevertheless, the few beneficial results obtained in some studies likely influence the public's attitude toward the use of dietary supplements.

A. Niacin

High-dose niacin therapy is used in clinical practice mainly to increase HDL-C [245], and also favorably affect TC, LDL-C, and TG levels [246–248]. In men with a previous MI participating in the Coronary Drug Project, 5-year supplementation with 3 g/day niacin reduced incidence of nonfatal MI and cerebrovascular end points (stroke or transient ischemic attack [TIA]) by 26% and 24%, respectively, compared to placebo, with a reduction in TC and TG by 10% and 26% [249]. After 9 years of follow-up, all-cause mortality was 11% lower in the niacin group, primarily due to a reduction in CHD death [250]. Since then, numerous randomized studies have reported that niacin in doses ranging between 1 and 3 g/day decrease TC (8%), LDL-C (6–21%), and TG (16–29%) and increase HDL-C (17–30%) [246–248]. Several randomized controlled trials reported a reduced frequency of cardiovascular events and an improved lipid profile when niacin was taken alone or in combination with statins (3-hydroxy-3-methylglutaryl coenzyme A reductase inhibitors) [251–255]. Of note in the study by Brown *et al.*, however, was the inclusion of the "antioxidant cocktail" (800 IU vitamin E, 1000 mg vitamin C, 25 mg β-carotene, and 100 μg selenium) with drug therapy. The antioxidants blunted the effect of simvastatin/niacin treatment and did not reduce stenosis progression, unlike the simvastatin/niacin treatment.

It is recommended that niacin (1–3 g/day) be used to increase HDL-C levels, typically in patients with low HDL-C levels; they are also often prescribed with statins to treat patients with hypercholesterolemia or dyslipidemia. Although the use of niacin in clinical practice has been limited because of adverse side effects such as cutaneous flushing and hepatic toxicity [256], recently, dyslipidemia has been treated through the development of new niacin formulations that elicit less flushing and hepatic toxicity [257, 258].

B. Fish Oil

As mentioned previously, two meta-analyses have demonstrated benefits of EPA and DHA in the prevention of CHD mortality and stroke [259, 260] and improvements in the lipid profile (mainly TG) [261]. With respect to fish-oil supplements, however, a review published in 2006 reported increased consumption of omega-3 fatty acids from dietary fish intake or fish-oil supplements reduced the rates of all-cause mortality, cardiac and sudden death, and possibly stroke [99]. The evidence for the benefits of fish oil appeared stronger in secondary rather than primary-prevention settings, and adverse effects appeared to be minor [99].

Two servings per week of fatty fish are recommended (equivalent to 500 mg EPA+DHA) for primary prevention of CVD. For secondary prevention, 1 g/day is recommended; and for TG lowering, 2–4 g/day is recommended under a physician's supervision. For patients unable to meet these recommendations by consuming oily fish, fish-oil supplements are advised.

C. Fiber

Several cross-sectional and prospective studies have reported that an increased intake of dietary fiber is associated with a reduced risk of CHD and CVD [262–267]. In terms of supplemental fiber, however, in a randomized controlled trial, subjects with mild to moderate hypercholesterolemia were treated with a Step I diet for 12 weeks before receiving placebo or 3.4 g of psyllium (equivalent to 1 teaspoon) three times per day for eight weeks. Compared with placebo, psyllium decreased TC by a further 4.8%, LDL-C by 8.2%, and apolipoprotein (apo) B concentration by 8.8% [268]. A meta-analysis of eight trials also found supplementation with 10.2 g/day of psyllium (Metamucil), adjunct to a low-fat diet in men and women with hypercholesterolemia, lowered serum TC by 4%, LDL-C by 7%, and the ratio of apo B to apo A-I by 6%, relative to placebo, with no effect on serum HDL or triacylglycerol concentrations [155]. Soluble fiber (psyllium—given as Metamucil supplement) appears to have an additive cholesterol-lowering effect when combined with statins. In a randomized clinical trial of hyperlipidemic subjects, 10 mg simvastatin plus 15 g psyllium decreased TC by 66 mg/dl (26%) and decreased LDL-C by 63 mg/dl (36%) [269].

The NCEP ATP III recommends that dietary supplementation with viscous soluble fiber is an effective therapeutic option to lower LDL-C. On average, an increase in viscous fiber of 5–10 g/day is accompanied by an approximate 5% reduction in LDL-C [270, 271]. To achieve the upper end of the ATP III soluble fiber recommendation, there must be a major emphasis on fruits, vegetables, cereal grains, and legumes. It is challenging to consume 25 g of soluble fiber each day, but a lower intake (5–10 g/day) will still reduce LDL-C by approximately 3% to 5%. Despite the availability of soluble fiber supplements, it is important to note that some provide energy and many do not deliver the same variety of nutrients that soluble-fiber-rich foods do.

D. Antioxidants

There has been controversy about whether antioxidant vitamins decrease CVD risk. In general, epidemiologic studies demonstrate a very modest beneficial association of antioxidants on CVD risk, whereas clinical studies have consistently not reported benefits. In a recent prospective report, the pooled analysis of 10 trials ($n = 227{,}443$) investigating dietary antioxidant intakes found that higher intakes of vitamin E (median intake in highest quartile: 8.2 mg) and β-carotene (median intake in highest quartile: 523 μg) were inversely associated with incidence of all major CHD events (vitamin E RR: 0.77 [0.64, 0.92]; β-carotene: 0.84 [0.74, 0.95]); whereas vitamin C was not (median intake highest quartile: 152 mg; 1.03 [0.91, 1.16]) [272]. When the intake of antioxidant supplements was combined with antioxidant intakes from food, no additional benefits were reported [272].

Relatively few [273] clinical studies have found beneficial effects on CVD reduction following antioxidant supplementation (ranging between 10 and 5000 IU vitamin E, 60 and 200 mg vitamin C, 15 and 50 mg β-carotene), compared with the majority that have not demonstrated beneficial effects [273–275]. Moreover, some have even found adverse effects [274–276].

The use of antioxidant supplements as a means to reduce CVD mortality is inconsistent and generally ineffective. The American Heart Association currently recommends consumption of antioxidant-rich foods such as fruits, vegetables, whole grains, and nuts [277, 278] to achieve recommended intakes, and supplementation is not recommended for CVD risk reduction. The increases in mortality rates following some supplementation studies further support not using antioxidant supplements to prevent or reduce CVD risk factors. Further studies are required assessing whether long-term dietary intakes of antioxidants are beneficial for CVD risk reduction, and whether excessive consumption of dietary intakes (from food and/or supplements) adversely affect health in the long term. Antioxidant supplements are therefore not recommended for the treatment or prevention of CVD.

E. B-Vitamins

The B-vitamins, in particular folate or folic acid, vitamin B_6, and vitamin B_{12}, decrease serum homocysteine levels [279–283], which is a risk factor for CVD. In the Norwegian Vitamin Trial (NORVIT), supplementation with folic acid 0.8 mg/day, vitamin B_{12} at a dosage of 0.4 mg/day, and vitamin B_6 at a dosage of 40 mg/day lowered plasma homocysteine levels by 27%, yet did not result in any reduction in the risk of MI or stroke in patients who have already had an MI [280]. Despite other studies also finding reductions in homocysteine, there was no substantial evidence indicating that B vitamins lowered CVD risk [281, 284]. Furthermore, supplemental doses that have been used (ranging between 20 μg and 2.5 mg folate; 200 μg and 50 mg vitamin B_6; and 0.4 mg and 1 mg vitamin B_{12}) are greater than what can be achieved through dietary sources alone and are far higher than current recommended intakes. Presently, the AHA does not recommend use of folic acid and B vitamin supplements to reduce the risk of CVD [63, 278], but rather recommends consumption of a healthy dietary pattern consisting of vegetables, fruits, legumes, nuts, lean meats, poultry, fatty fish, whole grains, and cereals to meet current recommendations for all nutrients.

IX. FOOD-BASED GUIDANCE

Food-based dietary guidance has been issued that translates energy and nutrient recommendations into healthful dietary patterns that can be implemented by individuals. There is great support for a food-based approach to meet nutrient needs, because certain dietary patterns deliver multiple, rather than single, nutrients to target CVD risk reduction, as well as risk of other chronic diseases, and to promote health and well-being. There are several models for designing dietary patterns. Although there are some subtle differences among them, they all recommend a diet that is high in fruits and vegetables, whole grains, low-fat and skim milk dairy products, lean meats, poultry and fish, legumes, and food sources of unsaturated fats including liquid vegetable oils, and nuts and seeds. All dietary patterns are low in SFA, TFA, and dietary cholesterol and high in dietary fiber. In addition, they all meet current recommendations for sodium

TABLE 4 USDA Food Guide

Daily Amount of Food from Each Group (vegetable subgroup amounts are per week)

Energy Level (kcal)	1000	1200	1400	1600	1800	2000	2200	2400	2600	2800	3000	3200
Fruits	1 c (2 srv)	1 c (2 srv)	1.5 c (3 srv)	1.5 c (3 srv)	1.5 c (3 srv)	2 c (4 srv)	2 c (4 srv)	2 c (4 srv)	2 c (4 srv)	2.5 c (5 srv)	2.5 c (5 srv)	2.5 c (5 srv)
Vegetables	1 c (2 srv)	1.5 c (3 srv)	1.5 c (3 srv)	2 c (4 srv)	2.5 c (5 srv)	2.5 c (5= srv)	3 c (6 srv)	3 c (6 srv)	3.5 c (7 srv)	3.5 c (7 srv)	4 c (8 srv)	4 c (8 srv)
Dark green veg.	1 c/wk	1.5c c/wk	1.5 c/wk	2 c/wk	3 c/wk	3 c/wk	3 c/wk	3 c/wk	3 c/wk	3 c/wk	3 c/wk	3 c/wk
Orange veg.	0.5 c/wk	1 c/wk	1 c/wk	1.5 c/wk	2 c/wk	2 c/wk	2 c/wk	2 c/wk	2.5 c/wk	2.5 c/wk	2.5 c/wk	2.5 c/wk
Legumes	0.5 c/wk	1 c/wk	1 c/wk	2.5 c/wk	3 c/wk	3 c/wk	3 c/wk	3 c/wk	3.5 c/wk	3.5 c/wk	3.5 c/wk	3.5 c/wk
Starchy veg.	1.5 c/wk	2.5 c/wk	2.5 c/wk	2.5 c/wk	3 c/wk	3 c/wk	6 c/wk	6 c/wk	7 c/wk	7 c/wk	9 c/wk	9 c/wk
Other veg.	3.5 c/wk	4.5 c/wk	4.5 c/wk	5.5 c/wk	6.5 c/wk	6.5 c/wk	7 c/wk	7 c/wk	8.5 c/wk	8.5 c/wk	10 c/wk	10 c/wk
Grains	3 oz eq	4 oz eq	5 oz eq	5 oz eq	6 oz eq	6 oz eq	7 oz eq	8 oz eq	9 oz eq	10 oz eq	10 oz eq	10 oz eq
Whole grains	1.5	2	2.5	3	3	3	3.5	4	4.5	5	5	5
Other grains	1.5	2	2.5	2	3	3	3.5	4	4.5	5	5	5
Lean meat and beans	2 oz eq	3 oz eq	4 oz eq	5 oz eq	5 oz eq	5.5 oz eq	6 oz eq	6.5 oz eq	6.5 oz eq	7 oz eq	7 oz eq	7 oz eq
Milk	2 c	2 c	2 c	3 c	3 c	3 c	3 c	3 c	3 c	3 c	3 c	3 c
Oils (g)	15 g	17 g	17 g	22 g	24 g	27 g	29 g	31 g	34 g	36 g	44 g	51 g
Discretionary energy allowance	165	171	171	132	195	267	290	362	410	426	512	648

Food group amounts shown in cup (c) or ounce equivalents (oz eq), with number of servings (srv) in parentheses when it differs from the other units.
srv, serving; c,cup; wk, week; oz eq, ounce equivalent; g, gram.

TABLE 5 Discretionary Energy Allowances for Various Age Groups as kcal

	Estimated Total Energy Need (kcal)[a]	Estimated Discretionary Energy Allowance (kcal)[a]	Estimated Total Energy Need (kcal)[b]	Estimated Discretionary Energy Allowance (kcal)[b]
Children 2–3 years old	1000	165[c]	1000–1400	165–170
Children 4–8 years old	1200–1400	170[c]	1400–1800	170–195
Girls 9–13 years old	1600	130	1600–2200	130–290
Boys 9–13 years old	1800	195	1800–2600	195–410
Girls 14–18 years old	1800	195	2000–2400	265–360
Boys 14–18 years old	2200	290	2400–3200	360–650
Females 19–30 years old	2000	265	2000–2400	265–360
Males 19–30 years old	2400	360	2600–3000	410–510
Females 31–50 years old	1800	195	2000–2200	265–290
Males 31–50 years old	2200	290	2400–3000	360–510
Females 51+ years old	1600	130	1800–2200	195–290
Males 51+ years old	2000	265	2200–2800	290–425

[a]These amounts are appropriate for individuals who get less than 30 minutes of moderate physical activity most days.

[b]These amounts are appropriate for individuals who get at least 30 minutes (lower energy level) to at least 60 minutes (higher energy level) of moderate physical activity most days.

[c]The level of discretionary energy is higher for children 8 and younger than it is for older children or adults consuming the same amount of energy, because younger children's nutrient needs are lower.

(<2300 mg/day). The specific food-based dietary recommendations that have been made are presented next.

A. USDA Food Guide

To complement the Dietary Guidelines, two examples of eating patterns have been developed, which include the U.S. Department of Agriculture (USDA) Food Guide and the Dietary Approaches to Stop Hypertension (DASH) Eating Plan (see below). These two similar eating patterns are designed to integrate dietary recommendations into a healthy way to eat and are presented in the Dietary Guidelines to provide examples of how nutrient-focused recommendations can be expressed in terms of food choices. (The Food Guide is available at http://www.health.gov/dietaryguidelines/dga2005/document/pdf/DGA2005.pdf.)

Table 4 displays food-based dietary recommendations from each of five food groups and liquid vegetable oil at 13 different energy levels. At each energy level, there is a discretionary energy allowance (Table 5) that can be included as solid fats, sugar, and/or both within a specified energy level. The discretionary energy allowance is based on estimated energy needs by age/sex group. The discretionary energy allowance is part of total estimated energy needs, not in addition to total energy needs. The chart gives a general guide (http://www.mypyramid.gov/pyramid/discretionary_calories_amount_table.html).

MyPyramid.gov (http://www.mypyramid.gov/) is the new online Food Guide that was developed to help Americans implement the Dietary Guidelines for Americans, 2005, and plan a healthful dietary pattern. The MyPyramid Plan offers a personal eating plan with the foods and amounts that are right for individuals. Using the MyPyramid Plan helps individuals make smart choices from every food group; find the appropriate balance between food and physical activity; achieve nutrient adequacy within energy needs; and identify daily energy needs.

B. Dietary Approaches to Stop Hypertension (DASH) Dietary Pattern

The DASH dietary pattern is rich in fruits, vegetables, and low-fat dairy foods. Whole grains, poultry, fish, and nuts also are emphasized and sweets are reduced. Table 6 displays the DASH eating plan with the recommended daily number of servings in a food group depending on energy needs.

C. Theraputic Lifestyle Changes (TLC) Diet

The TLC diet is a low SFA, TFA, and cholesterol diet aimed to reduce LDL-C. The TLC also includes two therapeutic diet options: plant stanol/sterol (add 2 g/day) and soluble fiber (add 10 to 25 g/day).

Table 7 details an example of the expected LDL-C lowering response to each dietary recommendation of the TLC diet, as well as the additive effects of all dietary strategies. The TLC diet recommendations are consistent with the AHA Diet and Lifestyle Recommendations (revised 2006). AHA recommendations designed specifically for daily energy needs, recommended range of total fat intake, and limits for SFA and TFA can be found at http://www.myfatstranslator.com/.

TABLE 6 The DASH Eating Plan at 1600, 2000, 2600, and 3100 kcal Levels[a]

Food Groups	1600 kcal	2000 kcal	2600 kcal	3100 kcal	Serving Sizes	Examples and Notes	Significance of Each Food Group to the DASH Eating Plan
Grains[b]	6 servings	7–8 servings	10–11 servings	12–13 servings	1 slice bread 1, oz dry cereal, $^{1}/_{2}$ cup cooked rice, pasta, or cereal[c]	Whole wheat bread, English muffin, pita bread, bagel, cereals, grits, oatmeal, crackers, unsalted pretzels, and popcorn	Major sources of energy and fiber
Vegetables	3–4 servings	4–5 servings	5–6 servings	6 servings	1 cup raw leafy vegetable, $^{1}/_{2}$ cup cooked vegetable, 6 oz vegetable juice	Tomatoes, potatoes, carrots, green peas, squash, broccoli, turnip greens, collards, kale, spinach, artichokes, green beans, lima beans, sweet potatoes	Rich source of potassium, magnesium, and fiber
Fruits	4 servings	4–5 servings	5–6 servings	6 servings	6 oz fruit juice, 1 medium fruit, $^{1}/_{4}$ cup dried fruit, $^{1}/_{2}$ cup fresh, frozen, or canned fruit	Apricots, bananas, dates, grapes, oranges, orange juice, grapefruit, grapefruit juice, mangoes, melons, peaches, pineapples, prunes, raisins, strawberries, tangerines	Important source of potassium, magnesium, and fiber
Low-fat or fat-free dairy foods	2–3 servings	2–3 servings	3 servings	3–4 servings	8 oz milk, 1 cup yogurt, $1^{1}/_{2}$ oz cheese	Fat-free or low-fat milk, fat-free or low-fat buttermilk, fat-free or low-fat regular or frozen yogurt, low-fat and fat-free cheese	Major sources of calcium and protein
Meat, poultry, fish	1–2 servings	2 or fewer servings	2 servings	2–3 servings	3 oz cooked meats, poultry, or fish	Select only lean; trim away visible fats; broil, roast, or boil instead of frying; remove skin from poultry	Rich sources of protein and magnesium
Nuts, seeds, legumes	3–4 servings/ week	4–5 servings/ week	1 serving	1 serving	$^{1}/_{3}$ cup or $1^{1}/_{2}$ oz nuts, 2 Tbsp or $^{1}/_{2}$ oz seeds, $^{1}/_{2}$ cup cooked dry beans or peas	Almonds, filberts, mixed nuts, peanuts, walnuts, sunflower seeds, kidney beans, lentils, pistachios	Rich sources of energy, magnesium, potassium, protein, and fiber
Fat and oils[d]	2 servings	2–3 servings	3 servings	4 servings	1 tsp soft margarine, 1 Tbsp low-fat mayonnaise, 2 Tbsp light salad dressing, 1 tsp vegetable oil	Soft margarine, low-fat mayonnaise, light salad dressing, vegetable oil (such as olive, corn, canola, or safflower)	DASH has 27% of energy as fat (low in saturated fat), including fat in or added to foods
Sweets	0 servings	5 servings/ week	2 servings	2 servings	1 Tbsp sugar, 1 Tbsp jelly or jam, $^{1}/_{2}$ oz jelly beans, 8 oz lemonade	Maple syrup, sugar, jelly, jam, fruit-flavored gelatin, jelly beans, hard candy, fruit punch sorbet, ices	Sweets should be low in fat

[a]NIH publication No. 034082; Karanja NM *et al. JADA* 8:S19–27, 1999.

[b]Whole grains are recommended for most servings to meet fiber recommendations.

[c]Equals $^{1}/_{2}$ to $1^{1}/_{4}$ cups, depending on cereal type. Check the product's Nutrition Facts Label.

[d]Fat content changes serving counts for fats and oils: For example, 1 Tbsp of regular salad dressing equals 1 serving; 1 Tbsp of a low-fat dressing equals $^{1}/_{2}$ serving; 1 Tbsp of a fat-free dressing equals 0 servings. Need to put reference number in. . . .

TABLE 7 Therapeutic Lifestyle Diet, Eating Pattern

Eating Pattern	TLC	Serving Sizes
Grains	7 servings/day[a]	1 slice bread; 1 oz dry cereal[b]; $^{1}/_{2}$ C cooked rice, pasta, or cereal
Vegetables	5 servings/day[a]	1 C raw leafy vegetable, $^{1}/_{2}$ C cut-up raw or cooked vegetable, $^{1}/_{2}$ C vegetable juice
Fruits	4 servings/day	1 medium fruit; $^{1}/_{4}$ C dried fruit; $^{1}/_{2}$ C fresh, frozen, or canned fruit; $^{1}/_{2}$ C fruit juice
Fat-free or low-fat milk and milk products	2–3 servings/day	1 C milk, 1 C yogurt, $1^{1}/_{2}$ oz cheese
Lean[c] meats, poultry, and fish	≤5 oz/day	
Nuts, seeds, and legumes	Counted in vegetable servings	$^{1}/_{3}$ C ($1^{1}/_{2}$ oz), 2 Tbsp peanut butter, 2 Tbsp or $^{1}/_{2}$ oz seeds, $^{1}/_{2}$ C dry beans or peas
Fats and oils	Amount depends on daily energy level	1 tsp soft margarine, 1 Tbsp mayonnaise, 2 Tbsp salad dressing, 1 tsp vegetable oil
Sweets and added sugars	No recommendation	1 Tbsp sugar, 1 Tbsp jelly or jam, $^{1}/_{2}$ C sorbet and ices, 1 C lemonade

[a]This number can be less or more depending on other food choices to meet 2000 kcal.

[b]Equals $^{1}/_{2}$ to $1^{1}/_{4}$ C, depending on cereal type. Check the product's Nutrition Facts Label.

[c]Lean cuts include sirloin tip, round steak, and rump roast; extra lean hamburger; and cold cuts made with lean meat or soy protein. Lean cuts of pork are center-cut ham, loin chops, and pork tenderloin. More information can be found at http://www.nhlbi.nih.gov/cgi-bin/chd/step2intro.cgi.

TABLE 8 Approximate and Cumulative LDL Cholesterol Reduction Achievable by Dietary Modification [13]

Dietary Component	Dietary Change	Approx. LDL-C Reduction
Major		
Saturated fat	<7% of energy	8–10%
Dietary cholesterol	<200 mg/d	3–5%
Weight reduction	Lose 10 lb	5–8%
Other LDL-Lowering Options		
Viscous fiber	5–10 g/d	3–5%
Plant sterol/stanol esters	2 g/d	6–15%
Cumulative estimate		**20–30%**

X. SUMMARY/CONCLUSION

Two very recent reviews have reinforced the importance of a dietary pattern that is low in SFA, TFA and cholesterol, and high in viscous fiber that beneficially affects lipids and lipoproteins, as well as other CVD risk factors that collectively decrease CVD risk [285, 286]. A dietary pattern that is low in SFA, TFA, and cholesterol, and high in viscous fiber beneficially affects lipids and lipoproteins, and other CVD risk factors that collectively decrease CVD risk. Table 8 shows the impact of these interventions, singly and collectively, on LDL-C lowering. Taken together, along with modest weight loss and inclusion of plant sterol/stanols, they can have a big impact on LDL-C lowering and, in turn, CVD risk reduction. Other dietary factors that beneficially affect CVD risk are omega-3 fatty acids, including fish/fish oil (to provide EPA and DHA), and ALA. These fatty acids act primarily through non-lipid-mediated mechanisms. A food-based dietary pattern that meets current nutrient goals is recommended because it delivers the full complement of nutrients and dietary factors that target a broad array of health benefits. This dietary pattern promotes consumption of fruits, vegetables, whole grains, low-fat/nonfat dairy products, lean meats, poultry and fish, and liquid vegetable oils, nuts, and seeds. Implementation of this dietary pattern will affect multiple, major CVD risk factors, including lipids and lipoproteins, blood pressure, and body weight, which can markedly decrease CVD risk. Thus, a healthful dietary pattern is an important tool for combating heart disease, and implementation of the food-based dietary recommendations that have been made by many organizations can have an important public health benefit.

References

1. Ford, E. S., Ajani, U.A, Croft, J.B, *et al.* (2007). Explaining the decrease in U.S. deaths from coronary disease, 1980–2000. *N. Engl. J. Med.* **356**, 2388–2398.
2. Foot, D. K., Lewis, R. P., Pearson, T. A., and Beller, G. A. (2000). Demographics and cardiology, 1950-2050. *J. Am. Coll. Cardiol.* **35**, 66B–80B.
3. Pearson, T. A. (2007). The prevention of cardiovascular disease: Have we really made progress? *Health Aff.* **26**, 49–60.
4. Greenland, P., Knoll, M. D., Stamler, J, *et al.* (2003). Major risk factors as antecedents of fatal and nonfatal coronary heart disease events. *JAMA* **290**, 891–897.
5. Khot, U. N., Khot, M. B., Bajzer, C. T., *et al.* (2003). Prevalence of conventional risk factors in patients with coronary heart disease. *JAMA* **290**, 898–904.
6. Castelli, W. P. (1996). Lipids, risk factors and ischaemic heart disease. *Atherosclerosis* **124**, S1–S9.

7. Page, I. H., Stare, F. J., Corcoran, A. C., Pollack, H., and Wilkinson, C. F. Jr, (1957). Atherosclerosis and the fat content of the diet. *JAMA* **164**, 2048–2051.
8. Dietary Guidelines Advisory Committee (2005). "Dietary Guidelines for Americans." Retrieved July 2007 from http://www.health.gov/dietaryguidelines/.
9. Kris-Etherton, P. M. (2007). The Minnesota Heart Survey: Nutrition successes pave the way for strategic opportunities for food and nutrition professionals. *J. Am. Diet. Assoc.* **107**, 209–212.
10. United States Department of Agriculture (2007). USDA National Nutrient Database for Standard Reference. Retrieved July 2007 from http://www.nal.usda.gov/fnic/foodcomp/search/.
11. American Dietetic Association (2007). "Disorders of Lipid Metabolism—Evidence Based Nutrition Practice Guideline." Retrieved August 2007 from http://www.adaevidencelibrary.com/topic.cfm?cat=2651&library=EBG.
12. OR Panel on Dietary Reference Intakes for Macronutrients. (2005). "Dietary Reference Intakes for Energy, Carbohydrate, Fiber, Fat, Fatty Acids, Cholesterol, Protein, and Amino Acids (Macronutrients)." Retrieved July 2007 from http://www.nal.usda.gov/fnic/DRI//DRI_Energy/energy_full_report.pdf, The National Academies Press, Washington, DC.
13. National Cholesterol Education Program (NCEP) Expert Panel on Detection, Evaluation, and Treatment of High Blood Cholesterol in Adults (Adult Treatment Panel III). (2002). Final report. *Circulation* **106**, 3143.
14. Bell, E. A., Castellanos, V. H., Pelkman, C. L., Thorwart, M. L., and Rolls, B. J. (1998). Energy density of foods affects energy intake in normal-weight women. *Am. J. Clin. Nutr.* **67**, 412–420.
15. Kendall, A., Levitsky, D. A., Strupp, B. J., and Lissner, L. (1991). Weight loss on a low-fat diet: Consequence of the imprecision of the control of food intake in humans. *Am. J. Clin. Nutr.* **53**, 1124–1129.
16. Kennedy, E., and Bowman, S. (2001). Assessment of the effect of fat-modified foods on diet quality in adults, 19 to 50 years, using data from the continuing survey of food intake by individuals. *J. Am. Diet. Assoc.* **101**, 455–460.
17. Lissner, L., Levitsky, D. A., Strupp, B. J., Kalkwarf, H. J., and Roe, D. A. (1987). Dietary fat and the regulation of energy intake in human subjects. *Am. J. Clin. Nutr.* **46**, 886–892.
18. Stubbs, R. J., Harbron, C. G., and Prentice, A. M. (1996). Covert manipulation of the dietary fat to carbohydrate ratio of isoenergetically dense diets: Effect on food intake in feeding men *ad libitum*. *Int. J. Obes. Relat. Metab. Disord.* **20**, 651–660.
19. van Stratum, P., Lussenburg, R. N., van Wezel, L. A., Vergroesen, A. J., and Cremer, H. D. (1978). The effect of dietary carbohydrate:fat ratio on energy intake by adult women. *Am. J. Clin. Nutr.* **31**, 206–212.
20. Dansinger, M. L., Gleason, J. A., Griffith, J. L., Selker, H. P., and Schaefer, E. J. (2005). Comparison of the Atkins, Ornish, Weight Watchers, and Zone diets for weight loss and heart disease risk reduction: A randomized trial. *JAMA* **293**, 43–53.
21. Howard, B. V., Van Horn, L., Hsia, J., *et al.* (2006). Low-fat dietary pattern and risk of cardiovascular disease: The Women's Health Initiative randomized controlled dietary modification trial. *JAMA* **295**, 655–666.
22. Dietary Guidelines Advisory Committee Report. (2005). Retrieved July 2007 from http://www.health.gov/dietaryguidelines/dga2005/report/ Or U.S. Department of Health and Human Services and U.S. Department of Agriculture (2005). "Dietary Guidelines for Americans" Washington, DC. Retrieved June 2007 from www.healthierus.gov/dietaryguidelines.
23. Keys, A. (1970). Coronary heart disease in seven countries—American Heart Association monograph No. 29. *Circulation* **41–42**, I-1–I-211.
24. Keys, A., Anderson, J. T., and Grande, F. (1965). Serum cholesterol response to changes in the diets. IV. Particular saturated fatty acids in the diet. *Metabolism* **14**, 776–787.
25. Hegsted, D. M., McGandy, R. B., Myers, M. L., and Stare, F. J. (1965). Quantitative effects of dietary fat on serum cholesterol in man. *Am. J. Clin. Nutr.* **17**, 281–295.
26. Clarke, R., Frost, C., Collins, R., Appleby, P., and Peto, R. (1997). Dietary lipids and blood cholesterol: Quantitative meta-analysis of metabolic ward studies. *BMJ* **314**, 112–117.
27. Hegsted, D. M., Ausman, L. M., Johnson, J. A., and Dallal, G E. (1993). Dietary fat and serum lipids: An evaluation of the experimental data. *Am. J. Clin. Nutr.* **57**, 875–883.
28. Mensink, R. P., and Katan, M. B. (1992). Effect of dietary fatty acids on serum lipids and lipoproteins. A meta-analysis of 27 trials. *Arterioscler. Thromb.* **12**, 911–919.
29. Yu, S., Derr, J., Etherton, T. D., and Kris-Etherton, P. M. (1995). Plasma cholesterol-predictive equations demonstrate that stearic acid is neutral and monounsaturated fatty acids are hypocholesterolemic. *Am. J. Clin. Nutr.* **61**, 1129–1139.
30. Muller, H., Lindman, A. S., Brantsaeter, A. L., and Pedersen, J. I. (2003). The serum LDL/HDL cholesterol ratio is influenced more favorably by exchanging saturated with unsaturated fat than by reducing saturated fat in the diet of women. *J. Nutr.* **133**, 78–83.
31. Ascherio, A., Katan, M. B., Zock, P. L., Stampfer, M. J., and Willett, W. C. (1999). Trans fatty acids and coronary heart disease. *N. Engl. J. Med.* **340**, 1994–1998.
32. Mensink, R. P., Zock, P. L., Kester, A. D., and Katan, M. B. (2003). Effects of dietary fatty acids and carbohydrates on the ratio of serum total to HDL cholesterol and on serum lipids and apolipoproteins: A meta-analysis of 60 controlled trials. *Am. J. Clin. Nutr.* **77**, 1146–1155.
33. Hu, F. B., Stampfer, M. J., Manson, J. E., *et al.* (1999). Dietary saturated fats and their food sources in relation to the risk of coronary heart disease in women. *Am. J. Clin. Nutr.* **70**, 1001–1008.
34. Nicholls, S. J., Lundman, P., Harmer, J. A., *et al.* (2006). Consumption of saturated fat impairs the anti-inflammatory properties of high-density lipoproteins and endothelial function. *J. Am. Coll. Cardiol.* **48**, 715–720.
35. Keogh, J. B., Grieger, J. A., Noakes, M., and Clifton, P. M. (2005). Flow-mediated dilatation is impaired by a high-saturated fat diet but not by a high-carbohydrate diet. *Arterioscler. Thromb. Vasc. Biol.* **25**, 1274–1279.
36. Albelda, S. M., Smith, C. W., and Ward, P. A. (1994). Adhesion molecules and inflammatory injury. *FASEB J.* **8**, 504–512.
37. Ray, K. K., Morrow, D. A., Shui, A., Rifai, N., and Cannon, C. P. (2006). Relation between soluble intercellular adhesion molecule-1, statin therapy, and long-term risk of clinical cardiovascular events in patients with previous acute coronary

syndrome (from PROVE IT-TIMI 22). *Am. J. Cardiol.* **98**, 861–865.

38. Shai, I., Pischon, T., Hu, F. B., Ascherio, A., Rifai, N., and Rimm, E. B. (2006). Soluble intercellular adhesion molecules, soluble vascular cell adhesion molecules, and risk of coronary heart disease. *Obesity* **14**, 2099–2106.
39. Bellido, C., Lopez-Miranda, J., Perez-Martinez, P., *et al.* (2006). The Mediterranean and CHO diets decrease VCAM-1 and E-selectin expression induced by modified low-density lipoprotein in HUVECs. *Nutr. Metab. Cardiovasc. Dis.* **16**, 524–530.
40. Bemelmans, W. J. E., Lefrandt, J. D., Feskens, E. J. M., *et al.* (2002). Change in saturated fat intake is associated with progression of carotid and femoral intima-media thickness, and with levels of soluble intercellular adhesion molecule-1. *Atherosclerosis* **163**, 113–120.
41. Shahar, E., Folsom, A. R., Wu, K. K., *et al.* (1993). Associations of fish intake and dietary n-3 polyunsaturated fatty acids with a hypocoagulable profile. The Atherosclerosis Risk in Communities (ARIC) Study. *Arterioscler. Thromb.* **13**, 1205–1212.
42. Ripsin, C. M., Keenan, J. M., Jacobs, D. R., Jr., *et al.* (1992). Oat products and lipid lowering. A meta-analysis. *JAMA* **267**, 3317–3325.
43. Allman-Farinelli, M. A., Gomes, K., Favaloro, E. J., and Petocz, P. (2005). A diet rich in high-oleic-acid sunflower oil favorably alters low-density lipoprotein cholesterol, triglycerides, and factor VII coagulant activity. *J. Am. Diet. Assoc.* **105**, 1071–1079.
44. Miller, G. J. (2005). Dietary fatty acids and the haemostatic system. *Atherosclerosis* **179**, 213–227.
45. Ginsberg, H. N., Kris-Etherton, P., Dennis, B., *et al.* (1998). Effects of reducing dietary saturated fatty acids on plasma lipids and lipoproteins in healthy subjects: The Delta Study, protocol 1. *Arterioscler. Thromb. Vasc. Biol.* **18**, 441–449.
46. Kris-Etherton, P. M., Pearson, T. A., Wan, Y., *et al.* (1999). High-monounsaturated fatty acid diets lower both plasma cholesterol and triacylglycerol concentrations. *Am. J. Clin. Nutr.* **70**, 1009–1015.
47. Kris-Etherton, P. M., Zhao, G., Pelkman, C. L., Fishell, V. K., and Coval, S. (2000). Beneficial effects of a diet high in mono-unsaturated fatty acids on risk factors for cardiovascular disease. *Nutr. Clin. Care* **3**, 153–162.
48. Griel, A. E., Ruder, E. H., and Kris-Etherton, P. M. (2006). The changing roles of dietary carbohydrates: From simple to complex. *Arterioscler. Thromb. Vasc. Biol.* **26**, 1958–1965.
49. Kris-Etherton, P. M. (1999). Monounsaturated fatty acids and risk of cardiovascular disease. *Circulation* **100**, 1253–1258.
50. Mensink, R. P., and Katan, M. B. (1987). Effect of monounsaturated fatty acids versus complex carbohydrates on high-density lipoproteins in healthy men and women. *Lancet* **1**, 122–125.
51. Appel, L. J., Sacks, F. M., Carey, V. J., *et al.* (2005). Effects of protein, monounsaturated fat, and carbohydrate intake on blood pressure and serum lipids: Results of the OmniHeart randomized trial. *JAMA* **294**, 2455–2464.
52. Carey, V. J., Bishop, L., Charleston, J., *et al.* (2005). Rationale and design of the Optimal Macro-Nutrient Intake Heart trial to prevent heart disease (OMNI-Heart). *Clin. Trials* **2**, 529–537.
53. Gumbiner, B., Low, C. C., and Reaven, P. D. (1998). Effects of a monounsaturated fatty acid-enriched hypocaloric diet on cardiovascular risk factors in obese patients with type 2 diabetes. *Diabetes Care* **21**, 9–15.
54. Reaven, P., Parthasarathy, S., Grasse, B. J., Miller, E., Steinberg, D., and Witztum, J. L. (1993). Effects of oleate-rich and linoleate-rich diets on the susceptibility of low density lipoprotein to oxidative modification in mildly hypercholesterolemic subjects. *J. Clin. Invest.* **91**, 668–676.
55. Lapointe, A., Couillard, C., and Lemieux, S. (2006). Effects of dietary factors on oxidation of low-density lipoprotein particles. *J. Nutr. Biochem.* **17**, 645–658.
56. Perez-Jimenez, F., Lopez-Miranda, J., and Mata, P. (2002). Protective effect of dietary monounsaturated fat on arteriosclerosis: Beyond cholesterol. *Atherosclerosis* **163**, 385–398.
57. Krauss, R. M., and Dreon, D. M. (1995). Low-density-lipoprotein subclasses and response to a low-fat diet in healthy men. *Am. J. Clin. Nutr.* **62**, 478S–487S.
58. Reaven, P. D., Grasse, B. J., and Tribble, D. L. (1994). Effects of linoleate-enriched and oleate-enriched diets in combination with alpha-tocopherol on the susceptibility of LDL and LDL subfractions to oxidative modification in humans. *Arterioscler. Thromb.* **14**, 557–566.
59. Austin, M. A. (1994). Small, dense low-density lipoprotein as a risk factor for coronary heart disease. *Int. J. Clin. Lab. Res.* **24**, 187–192.
60. Chait, A., Brazg, R. L., Tribble, D. L., and Krauss, R. M. (1993). Susceptibility of small, dense, low-density lipoproteins to oxidative modification in subjects with the atherogenic lipoprotein phenotype, pattern B. *Am. J. Med.* **94**, 350–356.
61. Rudel, L. L., Parks, J. S., and Sawyer, J. K. (1995). Compared with dietary monounsaturated and saturated fat, polyunsaturated fat protects African green monkeys from coronary artery atherosclerosis. *Arterioscler. Thromb. Vasc. Biol.* **15**, 2101–2110.
62. Bell, T. A., Wilson, M. D., Kelley, K., Sawyer, J. K., and Rudel, L. L. (2007). Monounsaturated fatty acyl-coenzyme A is predictive of atherosclerosis in human apoB-100 transgenic, LDLr-/- mice. *J. Lipid Res.* **48**, 1122–1131.
63. Lichtenstein, A. H., Appel, L. J., Brands, M., *et al.* (2006). Diet and lifestyle recommendations revision 2006: A scientific statement from the American Heart Association nutrition committee. *Circulation* **114**, 82–96.
64. Eckel, R. H., Borra, S., Lichtenstein, A. H., and Yin-Piazza, S. Y. (2007). Understanding the complexity of trans fatty acid reduction in the American diet: American Heart Association trans fat conference 2006: Report of the trans fat conference planning group. *Circulation* **115**, 2231–2246.
65. Mensink, R. P., and Katan, M. B. (1990). Effect of dietary trans fatty acids on high-density and low-density lipoprotein cholesterol levels in healthy subjects. *N. Engl. J. Med.* **323**, 439–445.
66. Kromhout, D., Menotti, A., Bloemberg, B., *et al.* (1995). Dietary saturated and trans fatty acids and cholesterol and 25-year mortality from coronary heart disease: The Seven Countries Study. *Prev. Med.* **24**, 308–315.

67. Willett, W. C., Stampfer, M. J., Manson, J. E., *et al.* (1993). Intake of trans fatty acids and risk of coronary heart disease among women. *Lancet* **341**, 581–585.
68. Lemaitre, R. N., King, I. B., Raghunathan, T. E., *et al.* (2002). Cell membrane trans-fatty acids and the risk of primary cardiac arrest. *Circulation* **105**, 697–701.
69. Oomen, C. M., Ocke, M. C., Feskens, E. J., van Erp-Baart, M. A., Kok, F. J., and Kromhout, D. (2001). Association between trans fatty acid intake and 10-year risk of coronary heart disease in the Zutphen Elderly Study: A prospective population-based study. *Lancet* **357**, 746–751.
70. Mozaffarian, D., Katan, M. B., Ascherio, A., Stampfer, M. J., and Willett, W. C. (2006). Trans fatty acids and cardiovascular disease. *N. Engl. J. Med.* **354**, 1601–1613.
71. Ascherio, A. (2006). Trans fatty acids and blood lipids. *Atheroscler. Suppl.* **7**, 25–27.
72. Loss, K., and Chow, C. K. (2000). "Fatty Acids in Foods and Their Health Implications" 2nd ed. Marcel Dekker, New York.
73. Sacks, F. M., and Katan, M. (2002). Randomized clinical trials on the effects of dietary fat and carbohydrate on plasma lipoproteins and cardiovascular disease. *Am. J. Med.* **113**(Suppl 9B), 13S–24S.
74. Leren, P. (1966). The effect of plasma cholesterol lowering diet in male survivors of myocardial infarction. A controlled clinical trial. *Acta Med. Scand. Suppl.* **466**, 1–92.
75. Turpeinen, O., Karvonen, M. J., Pekkarinen, M., Miettinen, M., Elosuo, R., and Paavilainen, E. (1979). Dietary prevention of coronary heart disease: The Finnish Mental Hospital Study. *Int. J. Epidemiol.* **8**, 99–118.
76. Dayton, S., Pearce, M. L., Hashimoto, S., Fakler, L. J., Hiscock, E., and Dixon, W. J. (1962). A controlled clinical trial of a diet high in unsaturated fat. Preliminary observations. *N. Engl. J. Med.* **266**, 1017–1023.
77. Hu, F. B., Stampfer, M. J., Manson, J. E., *et al.* (1997). Dietary fat intake and the risk of coronary heart disease in women. *N. Engl. J. Med.* **337**, 1491–1499.
78. Kris-Etherton, P. M., and Harris, W. S. (2004). Adverse effect of fish oils in patients with angina? *Curr. Atheroscler. Rep.* **6**, 413–414.
79. Grundy, S. M. (1997). What is the desirable ratio of saturated, polyunsaturated, and monounsaturated fatty acids in the diet? *Am. J. Clin. Nutr.* **66**, 988S–990S.
80. Research, Committee (1968). Controlled trial of soya-bean oil in myocardial infarction. *Lancet* **2**, 693–699.
81. Harris, W. S., Assaad, B., and Poston, W. C. (2006). Tissue omega-6/omega-3 fatty acid ratio and risk for coronary artery disease. *Am. J. Cardiol.* **98**, 19i–26i.
82. Mozaffarian, D. D. (2005). Does alpha-linolenic acid intake reduce the risk of coronary heart disease? A review of the evidence. *Altern. Ther. Health Med.* **11**, 24–30.
83. Kris-Etherton, P. M., Taylor, D. S., Yu-Poth, S., *et al.* (2000). Polyunsaturated fatty acids in the food chain in the United States. *Am. J. Clin. Nutr.* **71**, 179S–188S.
84. Simopoulos, A. P. (1991). Omega-3 fatty acids in health and disease and in growth and development. *Am. J. Clin. Nutr.* **54**, 438–463.
85. Arterburn, L. M., Hall, E. B., and Oken, H. (2006). Distribution, interconversion, and dose response of n-3 fatty acids in humans. *Am. J. Clin. Nutr.* **83**, S1467–S1476.
86. Burdge, G. C., Finnegan, Y. E., Minihane, A. M., Williams, C.M., and Wootton, S. A. (2003). Effect of altered dietary n-3 fatty acid intake upon plasma lipid fatty acid composition, conversion of [^{13}C]alpha-linolenic acid to longer-chain fatty acids and partitioning towards beta-oxidation in older men. *Br. J. Nutr.* **90**, 311–321.
87. Burdge, G. C., Jones, A. E., and Wootton, S. A. (2002). Eicosapentaenoic and docosapentaenoic acids are the principal products of alpha-linolenic acid metabolism in young men. *Br. J. Nutr.* **88**, 355–363.
88. Burdge, G. C., and Wootton, S. A. (2002). Conversion of alpha-linolenic acid to eicosapentaenoic, docosapentaenoic and docosahexaenoic acids in young women. *Br. J. Nutr.* **88**, 411–420.
89. U.K. Scientific Advisory Committee on Nutrition. (2004). Paper FICS/04/02. Food Standards Agency, London. Retrieved August 2007.
90. National Heart Foundation of Australia. (1999). A review of the relationship between dietary fat and cardiovascular disease. *Aust. J. Nutr. Diet.* **56**, S5–S22.
91. Kris-Etherton, P. M., Harris, W. S., and Appel, L. J., American Heart Association Nutrition Committee (2002). Fish consumption, fish oil, omega-3 fatty acids, and cardiovascular disease. *Circulation.* **21**, 2747–2757.
92. Hu, F. B., Stampfer, M. J., Manson, J. E., *et al.* (1999). Dietary protein and risk of ischemic heart disease in women. *Am. J. Clin. Nutr.* **70**, 221–227.
93. Mozaffarian, D., Longstreth, W. T., Jr., Lemaitre, R. N., Manolio, T. A., Kuller, L. H., Burke, G. L., and Siscovick, D. S. (2005). Fish consumption and stroke risk in elderly individuals: The Cardiovascular Health Study. *Arch. Intern. Med.* **165**, 200–206.
94. Folsom, A. R., and Demissie, Z. (2004). Fish intake, marine omega-3 fatty acids, and mortality in a cohort of postmenopausal women. *Am. J. Epidemiol.* **160**, 1005–1010.
95. Dolecek, T., and Granditis, G. (1991). Dietary polyunsaturated fatty acids and mortality in the Multiple Risk Factor Intervention Trial (MRFIT). *World Rev. Nutr. Diet.* **66**, 205–216.
96. de Lorgeril, M., Renaud, S., Mamelle, N., *et al.* (1994). Mediterranean alpha-linolenic acid-rich diet in secondary prevention of coronary heart disease. *Lancet* **343**, 1454–1459.
97. de Lorgeril, M., Salen, P., Martin, J.-L., Monjaud, I., Delaye, J., and Mamelle, N. (1999). Mediterranean diet, traditional risk factors, and the rate of cardiovascular complications after myocardial infarction: Final report of the Lyon Diet Heart Study. *Circulation* **99**, 779–785.
98. Kris-Etherton, P., Eckel, R. H., Howard, B. V., St. Jeor, S., and Bazzarre, T. L. (2001). Lyon Diet Heart Study: Benefits of a Mediterranean-style, National Cholesterol Education Program/American Heart Association Step I dietary pattern on cardiovascular disease. *Circulation* **103**, 1823–1825.
99. Wang, C., Harris, W. S., Chung, M., *et al.* (2006). n-3 fatty acids from fish or fish-oil supplements, but not alpha-linolenic acid, benefit cardiovascular disease outcomes in primary- and secondary-prevention studies: A systematic review. *Am. J. Clin. Nutr.* **84**, 5–17.

100. Dyerberg, J., Bang, H. O., and Hjorne, N. (1975). Fatty acid composition of the plasma lipids in Greenland Eskimos. *Am. J. Clin. Nutr.* **28**, 958–966.
101. Richter, S., Duray, G., Gronefeld, G., *et al.* (2005). Prevention of sudden cardiac death: Lessons from recent controlled trials. *Circ. J.* **69**, 625–629.
102. Mozaffarian, D., and Rimm, E. B. (2006). Fish intake, contaminants, and human health: Evaluating the risks and the benefits. *JAMA* **296**, 1885–1899.
103. Siscovick, D. S., Lemaitre, R. N., and Mozaffarian, D. (2003). The fish story: A diet-heart hypothesis with clinical implications: n-3 polyunsaturated fatty acids, myocardial vulnerability, and sudden death. *Circulation* **107**, 2632–2634.
104. Multiple Risk Factor Intervention Trial Research Group. (1982). Multiple risk factor intervention trial. Risk factor changes and mortality results. *JAMA* **248**, 1465–1477.
105. Hu, F. B., Bronner, L., Willett, W. C., *et al.* (2002). Fish and omega-3 fatty acid intake and risk of coronary heart disease in women. *JAMA* **287**, 1815–1821.
106. Morris, M. C., Manson, J. E., Rosner, B., Buring, J. E., Willett, W. C., and Hennekens, C. H. (1995). Fish consumption and cardiovascular disease in the Physicians' Health Study: A prospective study. *Am. J. Epidemiol.* **142**, 166–175.
107. Pietinen, P., Ascherio, A., Korhonen, P., *et al.* (1997). Intake of fatty acids and risk of coronary heart disease in a cohort of Finnish men: The alpha-tocopherol, beta-carotene cancer prevention study. *Am. J. Epidemiol.* **145**, 876–887.
108. Gruppo Italiano per lo Studio della Sopravvivenza nell'Infarto miocardico. (1999). Dietary supplementation with n-3 polyunsaturated fatty acids and vitamin E after myocardial infarction: Results of the GISSI-Prevenzione trial. *Lancet* **354**, 447–455.
109. Burr, M. L., Fehily, A. M., Gilbert, JF, *et al.* (1989). Effects of changes in fat, fish, and fibre intakes on death and myocardial reinfarction: Diet and Reinfarction Trial (DART). *Lancet* **2**, 757–761.
110. Yokoyama, M., Origasa, H., Matsuzaki, M., *et al.* (2007). Effects of eicosapentaenoic acid on major coronary events in hypercholesterolaemic patients (JELIS): A randomised open-label, blinded endpoint analysis. *Lancet* **369**, 1090–1098.
111. Brouwer, I. A., Heeringa, J., Geleijnse, J. M., Zock, P. L., and Witteman, J. C. (2006). Intake of very long-chain n-3 fatty acids from fish and incidence of atrial fibrillation. The Rotterdam Study. *Am. Heart J.* **151**, 857–862.
112. Burr, M. L., Ashfield-Watt, P. A., Dunstan, F. D., *et al.* (2003). Lack of benefit of dietary advice to men with angina: Results of a controlled trial. *Eur. J. Clin. Nutr.* **57**, 193–200.
113. Leaf, A., Albert, C. M., Josephson, M., *et al.* (2005). Prevention of fatal arrhythmias in high-risk subjects by fish oil n-3 fatty acid intake. *Circulation* **112**, 2762–2768.
114. Raitt, M. H., Connor, W. E., Morris, C., *et al.* (2005). Fish oil supplementation and risk of ventricular tachycardia and ventricular fibrillation in patients with implantable defibrillators: A randomized controlled trial. *JAMA* **293**, 2884–2891.
115. Leaf, A., Xiao, Y. F., Kang, J. X., and Billman, G. E. (2005). Membrane effects of the n-3 fish oil fatty acids, which prevent fatal ventricular arrhythmias. *J. Membr. Biol.* **206**, 129–139.
116. Brouwer, I. A., Zock, P. L., Camm, A. J., *et al.* (2006). Effect of fish oil on ventricular tachyarrhythmia and death in patients with implantable cardioverter defibrillators: The Study on Omega-3 Fatty Acids and Ventricular Arrhythmia (SOFA) randomized trial. *JAMA* **295**, 2613–2619.
117. Balk, E. M., Lichtenstein, A. H., Chung, M., Kupelnick, B., Chew, P., and Lau, J. (2006). Effects of omega-3 fatty acids on serum markers of cardiovascular disease risk: A systematic review. *Atherosclerosis* **189**, 19–30.
118. Harris, W. S. (1997). N-3 fatty acids and serum lipoproteins: Human studies. *Am. J. Clin. Nutr.* **65**, 1645S–1654S.
119. Balk, E., Chung, M., Lichtenstein, A., Chew, P., Kupelnick, B., Lawrence, A., DeVine, D., and Lau, J. (2004). Effects of omega-3 fatty acids on cardiovascular risk factors and intermediate markers of cardiovascular disease. Evidence Report/Technology Assessment No. 93. AHRQ Publication No. 04-E010-2. Agency for Healthcare Research and Quality, Rockville, MD. March 2004. Retrieved August 2007 from http://www.ahrq.gov/downloads/pub/evidence/pdf/o3cardrisk/o3cardrisk.pdf.
120. Rivellese, A. A., Maffettone, A., Vessby, B., *et al.* (2003). Effects of dietary saturated, monounsaturated and n-3 fatty acids on fasting lipoproteins, LDL size and post-prandial lipid metabolism in healthy subjects. *Atherosclerosis* **167**, 149–158.
121. Minihane, A. M., Khan, S., Leigh-Firbank, EC, *et al.* (2000). ApoE polymorphism and fish oil supplementation in subjects with an atherogenic lipoprotein phenotype. *Arterioscler. Thromb.* **20**, 1990–1997.
122. Mori, T. A., Burke, V., Puddey, I. B., *et al.* (2000). Purified eicosapentaenoic and docosahexaenoic acids have differential effects on serum lipids and lipoproteins, LDL particle size, glucose, and insulin in mildly hyperlipidemic men. *Am. J. Clin. Nutr.* **71**, 1085–1094.
123. Salonen, J. T., Seppanen, K., Nyyssonen, K., *et al.* (1995). Intake of mercury from fish, lipid peroxidation, and the risk of myocardial infarction and coronary, cardiovascular, and any death in eastern Finnish men. *Circulation* **91**, 645–655.
124. Guillar, E., Sanz-Gallardo, M. I., van't Veer, P., *et al.* (2002). Mercury, fish oils, and the risk of myocardial infarction. *N. Engl. J. Med.* **347**, 1747–1754.
125. Virtanen, J. K., Voutilainen, S., Rissanen, T. H., *et al.* (2005). Mercury, fish oils, and risk of acute coronary events and cardiovascular disease, coronary heart disease, and all-cause mortality in men in eastern Finland. *Arterioscler. Thromb.* **25**, 228–233.
126. Foran, S. E., Flood, J. G., and Lewandrowski, K. B. (2003). Measurement of mercury levels in concentrated over-the-counter fish oil preparations: Is fish oil healthier than fish? *Arch. Pathol. Lab. Med.* **127**, 1603–1605.
127. Yu-Poth, S., Zhao, G., Etherton, T., Naglak, M., Jonnalagadda, S., and Kris-Etherton, P. M. (1999). Effects of the National Cholesterol Education Program's Step I and Step II dietary intervention programs on cardiovascular disease risk factors: A meta-analysis. *Am. J. Clin. Nutr.* **69**, 632–646.
128. DeFronzo, R. A., and Ferrannini, E. (1991). Insulin resistance. A multifaceted syndrome responsible for NIDDM,

obesity, hypertension, dyslipidemia, and atherosclerotic cardiovascular disease. *Diabetes Care* **14**, 173–194.

129. Parillo, M., Rivellese, A. A., Ciardullo, A. V., *et al.* (1992). A high-monounsaturated-fat/low-carbohydrate diet improves peripheral insulin sensitivity in non-insulin-dependent diabetic patients. *Metabolism* **41**, 1373–1378.
130. Jeppesen, J., Schaaf, P., Jones, C., Zhou, M. Y., Chen, Y. D., and Reaven, G. M. (1997). Effects of low-fat, high-carbohydrate diets on risk factors for ischemic heart disease in postmenopausal women. *Am. J. Clin. Nutr.* **65**, 1027–1033.
131. Hammer, V., and Rolls, B. (1997). Diet composition and the regulation of body weight. *In* "Obesity and Weight Management: The Health Professional's Guide to Understanding and Treatment" (S. Dalton, Ed.). Aspen, Gaithersburg, MD.
132. Bray, G. A., and Popkin, B. M. (1998). Dietary fat intake does affect obesity! *Am. J. Clin. Nutr.* **68**, 1157–1173.
133. Willett, W. C. (2002). Dietary fat plays a major role in obesity: No. *Obes. Rev.* **3**, 59–68.
134. Lichtenstein, A. H., and Van Horn, L. (1998). Very low fat diets. *Circulation* **98**, 935–939.
135. Ornish, D., Scherwitz, L. W., Billings, J. H., *et al.* (1998). Intensive lifestyle changes for reversal of coronary heart disease. *JAMA* **280**, 2001–2007.
136. Harsha, D. W., Sacks, F. M., Obarzanek, E., *et al.* (2004). Effect of dietary sodium intake on blood lipids: Results from the DASH-sodium trial. *Hypertension* **43**, 393–398.
137. Obarzanek, E., Sacks, F. M., Vollmer, W. M., *et al.* (2001). Effects on blood lipids of a blood pressure-lowering diet: The Dietary Approaches to Stop Hypertension (DASH) Trial. *Am. J. Clin. Nutr.* **74**, 80–89.
138. Volek, J. S., Sharman, M. J., and Forsythe, C. E. (2005). Modification of lipoproteins by very low-carbohydrate diets. *J. Nutr.* **135**, 1339–1342.
139. Nordmann, A. J., Nordmann, A., Briel, M., *et al.* (2006). Effects of low-carbohydrate vs low-fat diets on weight loss and cardiovascular risk factors: A meta-analysis of randomized controlled trials. *Arch. Intern. Med.* **166**, 285–293.
140. St. Jeor, S. T., Howard, B. V., Prewitt, T. E., Bovee, V., Bazzarre, T., and Eckel, R. H. (2001). Dietary protein and weight reduction: A statement for healthcare professionals from the nutrition committee of the council on nutrition, physical activity, and metabolism of the American Heart Association. *Circulation* **104**, 1869–1874.
141. Krauss, R. M., Blanche, P. J., Rawlings, R. S., Fernstrom, H. S., and Williams, P. T. (2006). Separate effects of reduced carbohydrate intake and weight loss on atherogenic dyslipidemia. *Am. J. Clin. Nutr.* **83**, 1025–1031.
142. Wood, R. J., Fernandez, M. L., Sharman, M. J., *et al.* (2007). Effects of a carbohydrate-restricted diet with and without supplemental soluble fiber on plasma low-density lipoprotein cholesterol and other clinical markers of cardiovascular risk. *Metabolism* **56**, 58–67.
143. Ginsberg, H. N. (2002). New perspectives on atherogenesis: Role of abnormal triglyceride-rich lipoprotein metabolism. *Circulation* **106**, 2137–2142.
144. Wolever, T. M. S. (1999). Dietary recommendations for diabetes: High carbohydrate or high monounsaturated fat. *Nutr. Today* **34**, 73–77.
145. Jarvi, A. E., Karlstrom, B. E., Granfeldt, Y. E., Bjorck, I. E., Asp, N. G., and Vessby, B. O. (1999). Improved glycemic control and lipid profile and normalized fibrinolytic activity on a low-glycemic index diet in type 2 diabetic patients. *Diabetes Care* **22**, 10–18.
146. Wolever, T. M., Jenkins, D. J., Vuksan, V., Jenkins, A. L., Wong, G. S., and Josse, R. G. (1992). Beneficial effect of low-glycemic index diet in overweight NIDDM subjects. *Diabetes Care* **15**, 562–564.
147. Ludwig, D. S. (2002). The glycemic index: Physiological mechanisms relating to obesity, diabetes, and cardiovascular disease. *JAMA* **287**, 2414–2423.
148. Liu, S., Manson, J. E., Stampfer, M. J., *et al.* (2000). A prospective study of whole-grain intake and risk of type 2 diabetes mellitus in U.S. women. *Am. J. Public Health* **90**, 1409–1415.
149. Liu, S., Manson, J. E., Buring, J. E., Stampfer, M. J., Willett, W. C., and Ridker, P. M. (2002). Relation between a diet with a high glycemic load and plasma concentrations of high-sensitivity C-reactive protein in middle-aged women. *Am. J. Clin. Nutr.* **75**, 492–498.
150. van Dam, R. M., Visscher, A. W., Feskens, E. J., Verhoef, P., and Kromhout, D. (2000). Dietary glycemic index in relation to metabolic risk factors and incidence of coronary heart disease: The Zutphen Elderly Study. *Eur. J. Clin. Nutr.* **54**, 726–731.
151. Ebbeling, C. B., Leidig, M. M., Feldman, H. A., Lovesky, M. M., and Ludwig, D. S. (2007). Effects of a low-glycemic load vs low-fat diet in obese young adults: A randomized trial. *JAMA* **297**, 2092–2102.
152. Pereira, M. A., Swain, J., Goldfine, A. B., Rifai, N., and Ludwig, D. S. (2004). Effects of a low-glycemic load diet on resting energy expenditure and heart disease risk factors during weight loss. *JAMA* **292**, 2482–2490.
153. Das, S. K., Gilhooly, C. H., Golden, J. K., *et al.* (2007). Long-term effects of 2 energy-restricted diets differing in glycemic load on dietary adherence, body composition, and metabolism in CALERIE: A 1-y randomized controlled trial. *Am. J. Clin. Nutr.* **85**, 1023–1030.
154. Anderson, J. W., and Hanna, T. J. (1999). Impact of nondigestible carbohydrates on serum lipoproteins and risk for cardiovascular disease. *J. Nutr.* **129**, 1457.
155. Anderson, J. W., Davidson, M. H., Blonde, L., *et al.* (2000). Long-term cholesterol-lowering effects of psyllium as an adjunct to diet therapy in the treatment of hypercholesterolemia. *Am. J. Clin. Nutr.* **71**, 1433–1438.
156. Davidson, M. H., Maki, K. C., Kong, J. C., *et al.* (1998). Long-term effects of consuming foods containing psyllium seed husk on serum lipids in subjects with hypercholesterolemia. *Am. J. Clin. Nutr.* **67**, 367–376.
157. Pereira, M. A., O'Reilly, E., Augustsson, K., *et al.* (2004). Dietary fiber and risk of coronary heart disease: A pooled analysis of cohort studies. *Arch. Intern. Med.* **164**, 370–376.
158. Liese, A. D., Roach, A. K., Sparks, K. C., Marquart, L., D'Agostino, R. B., and Jr., Mayer-Davis, E. J. (2003). Whole-grain intake and insulin sensitivity: the insulin resistance atherosclerosis study. *Am. J. Clin. Nutr.* **78**, 965–971.

159. Fernandez, M. L. (2001). Soluble fiber and nondigestible carbohydrate effects on plasma lipids and cardiovascular risk. *Curr. Opin. Lipidol.* **12**, 35–40.
160. Glore, S. R., Van Treeck, D., Knehans, A. W., and Guild, M. (1994). Soluble fiber and serum lipids: A literature review. *J. Am. Diet. Assoc.* **94**, 425–436.
161. Wu, H., Dwyer, K. M., Fan, Z., Shircore, A., Fan, J., and Dwyer, J. H. (2003). Dietary fiber and progression of atherosclerosis: the Los Angeles Atherosclerosis Study. *Am. J. Clin. Nutr.* **78**, 1085–1091.
162. Brown, L., Rosner, B., Willett, W. W., and Sacks, F. M. (1999). Cholesterol-lowering effects of dietary fiber: A meta-analysis. *Am. J. Clin. Nutr.* **69**, 30–42.
163. Aller, R., de Luis, D. A., Izaola, O., *et al.* (2004). Effect of soluble fiber intake in lipid and glucose levels in healthy subjects: A randomized clinical trial. *Diabetes Res. Clin. Pract.* **65**, 7–11.
164. de Groot, A., Luyken, R., and Pikaar, N. A. (1963). Cholesterol-lowering effect of rolled oats. *Lancet* **2**, 303–304.
165. Biorklund, M., van Rees, A., Mensink, R. P., and Onning, G. (2005). Changes in serum lipids and postprandial glucose and insulin concentrations after consumption of beverages with beta-glucans from oats or barley: A randomised dose-controlled trial. *Eur. J. Clin. Nutr.* **59**, 1272–1281.
166. Fukagawa, N. K., Anderson, J. W., Hageman, G., Young, V. R., and Minaker, K. L. (1990). High-carbohydrate, high-fiber diets increase peripheral insulin sensitivity in healthy young and old adults. *Am. J. Clin. Nutr.* **52**, 524–528.
167. Anderson, J. W., Zeigler, J. A., Deakins, D. A., *et al.* (1991). Metabolic effects of high-carbohydrate, high-fiber diets for insulin- dependent diabetic individuals. *Am. J. Clin. Nutr.* **54**, 936–943.
168. Hallfrisch, J., Scholfield, D. J., and Behall, K. M. (1995). Diets containing soluble oat extracts improve glucose and insulin responses of moderately hypercholesterolemic men and women. *Am. J. Clin. Nutr.* **61**, 379–384.
169. Lagiou, P., Sandin, S., Weiderpass, E., *et al.* (2007). Low carbohydrate-high protein diet and mortality in a cohort of Swedish women. *J. Intern. Med.* **261**, 366–374.
170. Trichopoulou, A., Psaltopoulou, T., Orfanos, P., Hsieh, C. C., and Trichopoulos, D. (2007). Low-carbohydrate-high-protein diet and long-term survival in a general population cohort. *Eur. J. Clin. Nutr.* **61**, 575–581.
171. Hu, F. B., Rimm, E. B., Stampfer, M. J., Ascherio, A., Spiegelman, D., and Willett, W. C. (2000). Prospective study of major dietary patterns and risk of coronary heart disease in men. *Am. J. Clin. Nutr.* **72**, 912–921.
172. Blankenhorn, D. H., Johnson, R. L., Mack, W. J., el Zein, H. A., and Vailas, L. I. (1990). The influence of diet on the appearance of new lesions in human coronary arteries. *JAMA* **263**, 1646–1652.
173. Terpstra, A. H., Hermus, R. J., and West, C. E. (1983). The role of dietary protein in cholesterol metabolism. *World Rev. Nutr. Diet.* **42**, 1–55.
174. Stamler, J., (1979). Population studies. *In* "Nutrition, Lipids and Coronary Heart Disease," Raven, New York.
175. Hodgson, J. M., Ward, N. C., Burke, V., Beilin, L. J., and Puddey, I. B. (2007). Increased lean red meat intake does not elevate markers of oxidative stress and inflammation in humans. *J. Nutr.* **137**, 363–367.
176. Descovich, G. C., Ceredi, C., Gaddi, A., *et al.* (1980). Multicentre study of soybean protein diet for outpatient hypercholesterolaemic patients. *Lancet* **2**, 709–712.
177. Sirtori, C. R., Gatti, E., Mantero, O., *et al.* (1979). Clinical experience with the soybean protein diet in the treatment of hypercholesterolemia. *Am. J. Clin. Nutr.* **32**, 1645–1658.
178. Sirtori, C. R., Zucchi-Dentone, C., Sirtori, M., *et al.* (1985). Cholesterol-lowering and HDL-raising properties of lecithinated soy proteins in type II hyperlipidemic patients. *Ann. Nutr. Metab.* **29**, 348–357.
179. Anderson, J. W., Johnstone, B. M., and Cook-Newell, M. E. (1995). Meta-analysis of the effects of soy protein intake on serum lipids. *N. Engl. J. Med.* **333**, 276–282.
180. Sacks, F. M., Lichtenstein, A., Van Horn, L., Harris, W., Kris-Etherton, P., and Winston, M. (2006). Soy protein, isoflavones, and cardiovascular health: An American Heart Association Science Advisory for professionals from the Nutrition Committee. *Circulation* **113**, 1034–1044.
181. Balk, E., Chung, M., Chew, P., Ip, S., Raman, G., Kupelnick, B., Tatsioni, A., Sun, Y., Wolk, B., DeVine, D., and Lau, J. (2005). Effects of soy on health outcomes. Evidence Report/Technology Assessment No. 126. AHRQ Publication No. 05-E024-2. Agency for Healthcare Research and Quality, Rockville, MD. Retrieved August 2007 from http://www.ahrq.gov/downloads/pub/evidence/pdf/soyeffects/soy.pdf
182. Taku, K., Umegaki, K., Sato, Y., Taki, Y., Endoh, K., and Watanabe, S. (2007). Soy isoflavones lower serum total and LDL cholesterol in humans: A meta-analysis of 11 randomized controlled trials. *Am. J. Clin. Nutr.* **85**, 1148–1156.
183. Azadbakht, L., Kimiagar, M., Mehrabi, Y., Esmaillzadeh, A., Hu, F. B., and Willett, W. C. (2007). Soy consumption, markers of inflammation, and endothelial function: A crossover study in postmenopausal women with the metabolic syndrome. *Diabetes Care* **30**, 967–973.
184. Matthan, N. R., Jalbert, S. M., Ausman, L. M., Kuvin, J. T., Karas, R. H., and Lichtenstein, A. H. (2007). Effect of soy protein from differently processed products on cardiovascular disease risk factors and vascular endothelial function in hypercholesterolemic subjects. *Am. J. Clin. Nutr.* **85**, 960–966.
185. Tikkanen, M. J., Wahala, K., Ojala, S., Vihma, V., and Adlercreutz, H. (1998). Effect of soybean phytoestrogen intake on low density lipoprotein oxidation resistance. *Proc. Natl. Acad. Sci. USA* **95**, 3106–3110.
186. Wiseman, H., O'Reilly, J. D., Adlercreutz, H., *et al.* (2000). Isoflavone phytoestrogens consumed in soy decrease F(2)-isoprostane concentrations and increase resistance of low-density lipoprotein to oxidation in humans. *Am. J. Clin. Nutr.* **72**, 395–400.
187. Nestel, P. J., Yamashita, T., Sasahara, T., *et al.* (1997). Soy isoflavones improve systemic arterial compliance but not plasma lipids in menopausal and perimenopausal women. *Arterioscler. Thromb. Vasc. Biol.* **17**, 3392–3398.
188. Russell, M., Cooper, M. L., Frone, M. R., and Welte, J. W. (1991). Alcohol drinking patterns and blood pressure. *Am. J. Public Health* **81**, 452–457.

189. Stranges, S., Wu, T., Dorn, J. M., *et al.* (2004). Relationship of alcohol drinking pattern to risk of hypertension: a population-based study. *Hypertension* **44**, 813–819.
190. Trevisan, M., Krogh, V., Farinaro, E., Panico, S., and Mancini, M. (1987). Alcohol consumption, drinking pattern and blood pressure: Analysis of data from the Italian National Research Council Study. *Int. J. Epidemiol.* **16**, 520–527.
191. Dorn, J. M., Hovey, K., Williams, B. A., *et al.* (2007). Alcohol drinking pattern and non-fatal myocardial infarction in women. *Addiction* **102**, 730–739.
192. Kabagambe, E. K., Baylin, A., Ruiz-Narvaez, E., Rimm, E. B., and Campos, H. (2005). Alcohol intake, drinking patterns, and risk of nonfatal acute myocardial infarction in Costa Rica. *Am. J. Clin. Nutr.* **82**, 1336–1345.
193. Mukamal, K. J., Jensen, M. K., Gronbaek, M., *et al.* (2005). Drinking frequency, mediating biomarkers, and risk of myocardial infarction in women and men. *Circulation* **112**, 1406–1413.
194. Tolstrup, J., Jensen, M. K., Tjonneland, A., Overvad, K., Mukamal, K. J., and Gronbaek, M. (2006). Prospective study of alcohol drinking patterns and coronary heart disease in women and men. *BMJ* **332**, 1244–1248.
195. Jackson, R., Scragg, R., and Beaglehole, R. (1991). Alcohol consumption and risk of coronary heart disease. *BMJ* **303**, 211–216.
196. Tolstrup, J. S., Jensen, M. K., Tjonneland, A., Overvad, K., and Gronbaek, M. (2004). Drinking pattern and mortality in middle-aged men and women. *Addiction* **99**, 323–330.
197. Trevisan, M., Schisterman, E., Mennotti, A., Farchi, G., and Conti, S. (2001). Drinking pattern and mortality: The Italian Risk Factor and Life Expectancy Pooling Project. *Ann. Epidemiol.* **11**, 312–319.
198. Doll, R., Peto, R., Hall, E., Wheatley, K., and Gray, R. (1994). Mortality in relation to consumption of alcohol: 13 years' observations on male British doctors. *BMJ* **309**, 911–918.
199. Fuchs, C. S., Stampfer, M. J., Colditz, G. A., *et al.* (1995). Alcohol consumption and mortality among women. *N. Engl. J. Med.* **332**, 1245–1250.
200. Rimm, E. B., Giovannucci, E. L., Willett, W. C., *et al.* (1991). Prospective study of alcohol consumption and risk of coronary disease in men. *Lancet* **338**, 464–468.
201. Thun, M. J., Peto, R., Lopez, A. D., *et al.* (1997). Alcohol consumption and mortality among middle-aged and elderly U.S. adults. *N. Engl. J. Med.* **337**, 1705–1714.
202. Agarwal, D. P. (2002). Cardioprotective effects of light-moderate consumption of alcohol: A review of putative mechanisms. *Alcohol Alcohol.* **37**, 409–415.
203. Marmot, M. G., Rose, G., Shipley, M. J., and Thomas, B. J. (1981). Alcohol and mortality: A U-shaped curve. *Lancet* **1**, 580–583.
204. Shaper, A. G., Wannamethee, G., and Walker, M. (1988). Alcohol and mortality in British men: Explaining the U-shaped curve. *Lancet* **2**, 1267–1273.
205. Renaud, S. C., and Ruf, J. C. (1996). Effects of alcohol on platelet functions. *Clin. Chim. Acta* **246**, 77–89.
206. Ridker, P. M., Vaughan, D. E., Stampfer, M. J., Glynn, R. J., and Hennekens, C. H. (1994). Association of moderate alcohol consumption and plasma concentration of endogenous tissue-type plasminogen activator. *JAMA* **272**, 929–933.
207. De Oliveira, E., Silva, E. R., Foster, D., McGee Harper, M., Seidman, C. E., Smith, J. D., Breslow, J. L., and Brinton, E. (2000). Alcohol consumption raises HDL cholesterol levels by increasing the transport rate of apolipoproteins A-I and A-II. *Circulation* **102**, 2347–2352.
208. Srivastava, L. M., Vasisht, S., Agarwal, D. P., and Goedde, H. W. (1994). Relation between alcohol intake, lipoproteins and coronary heart disease: The interest continues. *Alcohol Alcohol.* **29**, 11–24.
209. Roussell, M., and Kris-Etherton, P. (2007). Effects of lifestyle interventions on high-density lipoprotein cholesterol levels. *J. Clin. Lipid.* **1**, 65–73.
210. Rimm, E. B., Williams, P., Fosher, K., Criqui, M., and Stampfer, M. J. (1999). Moderate alcohol intake and lower risk of coronary heart disease: Meta-analysis of effects on lipids and haemostatic factors. *BMJ* **319**, 1523–1528.
211. St. Leger, A. S., Cochrane, A. L., and Moore, F. (1979). Factors associated with cardiac mortality in developed countries with particular reference to the consumption of wine. *Lancet* **1**, 1018–1020.
212. Criqui, M. H., and Ringel, B. L. (1994). Does diet or alcohol explain the French paradox? *Lancet* **344**, 1719–1723.
213. Renaud, S., and de Lorgeril, M. (1992). Wine, alcohol, platelets, and the French paradox for coronary heart disease. *Lancet* **339**, 1523–1526.
214. Booyse, F. M., and Parks, D. A. (2001). Moderate wine and alcohol consumption: Beneficial effects on cardiovascular disease. *Thromb. Haemost.* **86**, 517–528.
215. Frankel, E. N., Kanner, J., German, J. B., Parks, E., and Kinsella, J. E. (1993). Inhibition of oxidation of human low-density lipoprotein by phenolic substances in red wine. *Lancet* **341**, 454–457.
216. Klatsky, A. L. (2007). Alcohol, cardiovascular diseases and diabetes mellitus. *Pharmacol. Res.* **55**, 237–247.
217. Gorelik, S., Ligumsky, M., Kohen, R., and Kanner, J. (2007). A novel function of red wine polyphenols in humans: Prevention of absorption of cytotoxic lipid peroxidation products. *FASEB J.* (in press).
218. Pignatelli, P., Ghiselli, A., Buchetti, B., *et al.* (2006). Polyphenols synergistically inhibit oxidative stress in subjects given red and white wine. *Atherosclerosis* **188**, 77–83.
219. Kiviniemi, T. O., Saraste, A., Toikka, J. O., *et al.* (2007). A moderate dose of red wine, but not de-alcoholized red wine increases coronary flow reserve. *Atherosclerosis* (in press).
220. Avellone, G., Di Garbo, V., Campisi, D., *et al.* (2006). Effects of moderate Sicilian red wine consumption on inflammatory biomarkers of atherosclerosis. *Eur. J. Clin. Nutr.* **60**, 41–47.
221. Karatzi, K., Papamichael, C., Karatzis, E., *et al.* (2007). Acute smoking induces endothelial dysfunction in healthy smokers. Is this reversible by red wine's antioxidant constituents? *J. Am. Coll. Nutr.* **26**, 10–15.
222. Nakamura, Y., Okamura, T., Tamaki, S., *et al.* (2004). Egg consumption, serum cholesterol, and cause-specific and all-cause mortality: The National Integrated Project for Prospective Observation of Non-communicable Disease and Its Trends in the Aged (1980, NIPPON DATA80). *Am. J. Clin. Nutr.* **80**, 58–63.

223. Dawber, T. R., Nickerson, R. J., Brand, F. N., and Pool, J. (1982). Eggs, serum cholesterol, and coronary heart disease. *Am. J. Clin. Nutr.* **36**, 617–625.
224. Kannel, W. B., and Gordon, T. (1970). The Framingham Diet Study: Diet and the regulations of serum cholesterol (Section 24). Department of Health, Education and Welfare, Washington, DC.
225. Cesar, T. B., Oliveira, M. R., Mesquita, C. H., and Maranhao, R. C. (2006). High cholesterol intake modifies chylomicron metabolism in normolipidemic young men. *J. Nutr.* **136**, 971–976.
226. Weggemans, R. M., Zock, P. L., and Katan, M. B. (2001). Dietary cholesterol from eggs increases the ratio of total cholesterol to high-density lipoprotein cholesterol in humans: A meta-analysis. *Am. J. Clin. Nutr.* **73**, 885–891.
227. Kratz, M. (2005). Dietary cholesterol, atherosclerosis and coronary heart disease. *Handb. Exp. Pharmacol.* **2005**, 195–213.
228. Perisee, D. M. (2005). Food fortification with plant sterol/stanol for hyperlipidemia: Management in free-living populations. *J. Am. Diet. Assoc.* **105**, 52–53.
229. Gylling, H., Puska, P., Vartiainen, E., and Miettinen, T. A. (1999). Serum sterols during stanol ester feeding in a mildly hypercholesterolemic population. *J. Lipid Res.* **40**, 593–600.
230. Miettinen, T. A., Vuoristo, M., Nissinen, M., Jarvinen, H. J., and Gylling, H. (2000). Serum, biliary, and fecal cholesterol and plant sterols in colectomized patients before and during consumption of stanol ester margarine. *Am. J. Clin. Nutr.* **71**, 1095–1102.
231. Assmann, G., Cullen, P., Erbey, J., Ramey, D. R., Kannenberg, F., and Schulte, H. (2006). Plasma sitosterol elevations are associated with an increased incidence of coronary events in men: results of a nested case-control analysis of the Prospective Cardiovascular Munster (PROCAM) study. *Nutr. Metab. Cardiovasc. Dis.* **16**, 13–21.
232. Fassbender, K., Lutjohann, D., Dik, M. G., *et al.* (2006). Moderately elevated plant sterol levels are associated with reduced cardiovascular risk—The LASA study. *Atherosclerosis* (in press).
233. Pinedo, S., Vissers, M. N., von Bergmann, K., *et al.* (2007). Plasma levels of plant sterols and the risk of coronary artery disease: The prospective EPIC-Norfolk Population Study. *J. Lipid Res.* **48**, 139–144.
234. Devaraj, S., and Jialal, I. (2006). The role of dietary supplementation with plant sterols and stanols in the prevention of cardiovascular disease. *Nutr. Rev.* **64**, 348–354.
235. Hendriks, H. F., Brink, E. J., Meijer, G. W., Princen, H. M., and Ntanios, F. Y. (2003). Safety of long-term consumption of plant sterol esters-enriched spread. *Eur. J. Clin. Nutr.* **57**, 681–692.
236. Homma, Y., Ikeda, I., Ishikawa, T., Tateno, M., Sugano, M., and Nakamura, H. (2003). Decrease in plasma low-density lipoprotein cholesterol, apolipoprotein B, cholesteryl ester transfer protein, and oxidized low-density lipoprotein by plant stanol ester-containing spread: A randomized, placebo-controlled trial. *Nutrition* **19**, 369–374.
237. Ntanios, F. Y., Homma, Y., and Ushiro, S. (2002). A spread enriched with plant sterol-esters lowers blood cholesterol and lipoproteins without affecting vitamins A and E in normal and hypercholesterolemic Japanese men and women. *J. Nutr.* **132**, 3650–3655.
238. Quilez, J., Rafecas, M., Brufau, G., *et al.* (2003). Bakery products enriched with phytosterol esters, alpha-tocopherol and beta-carotene decrease plasma LDL-cholesterol and maintain plasma beta-carotene concentrations in normocholesterolemic men and women. *J. Nutr.* **133**, 3103–3109.
239. Noakes, M., Clifton, P. M., Doornbos, A. M., and Trautwein, E. A. (2005). Plant sterol ester-enriched milk and yoghurt effectively reduce serum cholesterol in modestly hypercholesterolemic subjects. *Eur J. Nutr.* **44**, 214–222.
240. Noakes, M., Clifton, P., Ntanios, F., Shrapnel, W., Record, I., and McInerney, J. (2002). An increase in dietary carotenoids when consuming plant sterols or stanols is effective in maintaining plasma carotenoid concentrations. *Am. J. Clin. Nutr.* **75**, 79–86.
241. Plat, J., van Onselen, E. N., van Heugten, M. M., and Mensink, R. P. (2000). Effects on serum lipids, lipoproteins and fat soluble antioxidant concentrations of consumption frequency of margarines and shortenings enriched with plant stanol esters. *Eur. J. Clin. Nutr.* **54**, 671–677.
242. Grundy, S. M. (2005). Stanol esters as a component of maximal dietary therapy in the National Cholesterol Education Program Adult Treatment Panel III report. *Am. J. Cardiol.* **96**, 47D–50D.
243. Levitt, J. A. (2001). Center for Food Safety and Applied Nutrition, Food and Drug Administration. Statement to the U.S. House of Representatives Committee on Government Reform, U.S. Food and Drug Administration. Accessed August 2007. http://www.cfsan.fda.gov/~lrd/st010320.html/
244. Buettner, C., Phillips, R. S., Davis, R. B., Gardiner, P., and Mittleman, M. A. (2007). Use of dietary supplements among United States adults with coronary artery disease and atherosclerotic risks. *Am. J. Cardiol.* **99**, 661–666.
245. Carlson, L. A. (2005). Nicotinic acid: The broad-spectrum lipid drug. A 50th anniversary review. *J. Intern Med.* **258**, 94–114.
246. Goldberg, A., Alagona, P., Jr., Capuzzi, D. M., *et al.* (2000). Multiple-dose efficacy and safety of an extended-release form of niacin in the management of hyperlipidemia. *Am. J. Cardiol.* **85**, 1100–1105.
247. Knopp, R. H., Alagona, P., Davidson, M., *et al.* (1998). Equivalent efficacy of a time-release form of niacin (Niaspan) given once-a-night versus plain niacin in the management of hyperlipidemia. *Metabolism* **47**, 1097–1104.
248. Morgan, J. M., Capuzzi, D. M., Guyton, J. R., *et al.* (1996). Treatment effect of Niaspan, a controlled-release niacin, in patients with hypercholesterolemia: A placebo-controlled trial. *J. Cardiovasc. Pharmacol. Ther.* **1**, 195–202.
249. The Coronary Drug Project Research Group. (1975). Clofibrate and niacin in coronary heart disease. *JAMA* **231**, 360–381.
250. Canner, P. L., Berge, K. G., Wenger, N. K., *et al.* (1986). Fifteen year mortality in coronary drug project patients: Long-term benefit with niacin. *J. Am. Coll. Cardiol.* **8**, 245–1255.
251. Birjmohun, R. S., Kastelein, J. J., Poldermans, D., Stroes, E. S., Hostalek, U., and Assmann, G. (2007). Safety and

tolerability of prolonged-release nicotinic acid in statin-treated patients. *Curr. Med. Res. Opin.* **23**, 1707–1713.
252. Blankenhorn, D. H., Nessim, S. A., Johnson, R. L., Sanmarco, M. E., Azen, S. P., and Cashin-Hemphill, L. (1987). Beneficial effects of combined colestipol-niacin therapy on coronary atherosclerosis and coronary venous bypass grafts. *JAMA* **257**, 3233–3240.
253. Brown, B. G., Hillger, L., Zhao, X. Q., Poulin, D., and Albers, J. J. (1995). Types of change in coronary stenosis severity and their relative importance in overall progression and regression of coronary disease. Observations from the FATS Trial. Familial Atherosclerosis Treatment Study. *Ann. N. Y. Acad. Sci.* **748**, 407–417, discussion 417–418.
254. Brown, B. G., Zhao, X. Q., Chait, A., *et al.* (2001). Simvastatin and niacin, antioxidant vitamins, or the combination for the prevention of coronary disease. *N. Engl. J. Med.* **345**, 1583–1592.
255. Hunninghake, D. B., McGovern, M. E., Koren, M., *et al.* (2003). A dose-ranging study of a new, once-daily, dual-component drug product containing niacin extended-release and lovastatin. *Clin. Cardiol.* **26**, 112–118.
256. Berra, K. (2004). Clinical update on the use of niacin for the treatment of dyslipidemia. *J. Am. Acad. Nurse Pract.* **16**, 526–534.
257. Brown, B. G. (2007). Expert commentary: Niacin safety. *Am. J. Cardiol.* **99**, 32C–34C.
258. Tavintharan, S., and Kashyap, M. L. (2001). The benefits of niacin in atherosclerosis. *Curr. Atheroscler. Rep.* **3**, 74–82.
259. He, K., Song, Y., Daviglus, M. L., *et al.* (2004). Accumulated evidence on fish consumption and coronary heart disease mortality: a meta-analysis of cohort studies. *Circulation* **109**, 2705–2711.
260. He, K., Song, Y., Daviglus, M. L., *et al.* (2004). Fish consumption and incidence of stroke: A meta-analysis of cohort studies. *Stroke* **35**, 1538–1542.
261. Panagiotakos, D. B., Zeimbekis, A., Boutziouka, V., *et al.* (2007). Long-term fish intake is associated with better lipid profile, arterial blood pressure, and blood glucose levels in elderly people from Mediterranean islands (MEDIS epidemiological study). *Med. Sci. Monit.* **13**, CR307–CR312.
262. Bazzano, L. A., He, J., Ogden, L. G., Loria, C. M., and Whelton, P. K. (2003). Dietary fiber intake and reduced risk of coronary heart disease in US men and women: The National Health and Nutrition Examination Survey I Epidemiologic Follow-up Study. *Arch. Intern. Med.* **163**, 1897–1904.
263. Liu, S., Buring, J. E., Sesso, H. D., Rimm, E. B., Willett, W. C., and Manson, J. E. (2002). A prospective study of dietary fiber intake and risk of cardiovascular disease among women. *J. Am. Coll. Cardiol.* **39**, 49–56.
264. Merchant, A. T., Hu, F. B., Spiegelman, D., Willett, W. C., Rimm, E. B., and Ascherio, A. (2003). Dietary fiber reduces peripheral arterial disease risk in men. *J. Nutr.* **133**, 3658–3663.
265. Mozaffarian, D., Kumanyika, S. K., Lemaitre, R. N., Olson, J. L., Burke, G. L., and Siscovick, D. S. (2003). Cereal, fruit, and vegetable fiber intake and the risk of cardiovascular disease in elderly individuals. *JAMA* **289**, 1659–1666.
266. Rimm, E. B., Ascherio, A., Giovannucci, E., Spiegelman, D., Stampfer, M. J., and Willett, W. C. (1996). Vegetable, fruit, and cereal fiber intake and risk of coronary heart disease among men. *JAMA* **275**, 447–451.
267. Wolk, A., Manson, J. E., Stampfer, M. J., *et al.* (1999). Long-term intake of dietary fiber and decreased risk of coronary heart disease among women. *JAMA* **281**, 1998–2004.
268. Bell, L. P., Hectorne, K., Reynolds, H., Balm, T. K., and Hunninghake, D. B. (1989). Cholesterol-lowering effects of psyllium hydrophilic mucilloid. Adjunct therapy to a prudent diet for patients with mild to moderate hypercholesterolemia. *JAMA* **261**, 3419–3423.
269. Moreyra, A. E., Wilson, A. C., and Koraym, A. (2005). Effect of combining psyllium fiber with simvastatin in lowering cholesterol. *Arch. Intern. Med.* **165**, 1161–1166.
270. U.S. Department of Health and Human Services. Food and Drug Administration. (1997). Food labeling: health claims. Soluble fiber from certain foods and coronary heart disease: Proposed rule. *Fed. Register* **62**, 28234–28245.
271. U.S. Department of Health and Human Services. Food and Drug Administration. (1998). Food labeling: health claims. Soluble fiber from certain foods and coronary heart disease: Final rule. *Fed. Register* **63**, 8103–8121.
272. Knekt, P., Ritz, J., Pereira, M. A., *et al.* (2004). Antioxidant vitamins and coronary heart disease risk: A pooled analysis of 9 cohorts. *Am. J. Clin. Nutr.* **80**, 1508–1520.
273. Virtamo, J., Rapola, J. M., Ripatti, S., *et al.* (1998). Effect of vitamin E and beta carotene on the incidence of primary nonfatal myocardial infarction and fatal coronary heart disease. *Arch. Intern. Med.* **158**, 668–675.
274. Bjelakovic, G., Nikolova, D., Gluud, L. L., Simonetti, R. G., and Gluud, C. (2007). Mortality in randomized trials of antioxidant supplements for primary and secondary prevention: Systematic review and meta-analysis. *JAMA* **297**, 842–857.
275. Vivekananthan, D. P., Penn, M. S., Sapp, S. K., Hsu, A., and Topol, E. J. (2003). Use of antioxidant vitamins for the prevention of cardiovascular disease: Meta-analysis of randomised trials. *Lancet* **361**, 2017–2023.
276. Cook, N. R. (2007). A randomized factorial trial of vitamins C and E and beta carotene in the secondary prevention of cardiovascular events in women: Results from the Women's Antioxidant Cardiovascular Study. *Arch. Intern. Med.* **167**, 1610–1618.
277. Lichtenstein, A. H., Appel, L. J., Brands, M., *et al.* (2006). Summary of the American Heart Association diet and lifestyle recommendations revision 2006. *Arterioscler. Thromb. Vasc. Biol.* **26**, 2186–2191.
278. Mosca, L., Banka, C. L., Benjamin, E. J., *et al.* (2007). Evidence-based guidelines for cardiovascular disease prevention in women: 2007 update. *J. Am. Coll. Cardiol.* **49**, 1230–1250.
279. Homocysteine Lowering Trialists' Collaboration (1998). Lowering blood homocysteine with folic acid based supplements: Meta-analysis of randomised trials. *BMJ* **316**, 894–898.
280. Bonaa, K. H., Njolstad, I., Ueland, P. M., *et al.* (2006). Homocysteine lowering and cardiovascular events after acute myocardial infarction. *N. Engl. J. Med.* **354**, 1578–1588.

281. Toole, J. F., Malinow, M. R., Chambless, L. E., *et al.* (2004). Lowering homocysteine in patients with ischemic stroke to prevent recurrent stroke, myocardial infarction, and death: The Vitamin Intervention for Stroke Prevention (VISP) randomized controlled trial. *JAMA* **291**, 565–575.
282. Wang, X., Demirtas, H., and Xu, X. (2006). Homocysteine, B vitamins, and cardiovascular disease. *N. Engl. J. Med.* **355**, 207–209.
283. Zee, RY, Mora, S., Cheng, S., *et al.* (2007). Homocysteine, 5,10-methylenetetrahydrofolate reductase 677C > T polymorphism, nutrient intake, and incident cardiovascular disease in 24,968 initially healthy women. *Clin. Chem.* **53**, 845–851.
284. Bazzano, L. A., Reynolds, K., Holder, K. N., and He, J. (2006). Effect of folic acid supplementation on risk of cardiovascular diseases: A meta-analysis of randomized controlled trials. *JAMA* **296**, 2720–2726.
285. Kris-Etherton, P. M., Innis, S., American Dietetic Association, and Dietitians of Canada. (2007). Position of the American Dietetic Association and Dietitians of Canada: dietary fatty acids. *J. Am. Diet. Assoc.* **9**, 1599–1611.
286. Van Horn, L. V., McCoin, M., Kris-Etherton, P. M., *et al.* The evidence for dietary prevention and treatment of cardiovascular disease. *J. Am. Diet Assoc.* (in press).
287. Beauchesne-Rondeau, E., Gascon, A., and Bergeron, J., Jacques, H. (2003). Plasma lipids and lipoproteins in hypercholesterolemic men fed a lipid-lowering diet containing lean beef, lean fish, or poultry. *Am. J. Clin. Nutr.* **3**, 587–593.
288. Davidson, M. H., Hunninghake, D., Maki, K. C., Kwiterovich, P. O. Jr, and Kafonek, S. (1999). Comparison of the effects of lean red meat vs. lean white meat on serum lipid levels among free-living persons with hypercholesterolemia: a long-term, randomized clinical trial. *Arch Intern Med.* **12**, 1331–1338.
289. Hunninghake, D. B., Maki, K. C., Kwiterovich, P. O. Jr, Davidson, M. H., Dicklin, M. R., and Kafonek, S. D. (2000). Incorporation of lean red meat into a National Cholesterol Education Program Step I diet: a long-term, randomized clinical trial in free-living persons with hypercholesterolemia. *J. Am. Coll. Nutr.* **3**, 351–60.

CHAPTER 30

Nutrition, Lifestyle, and Hypertension

PAO-HWA LIN[1], BRYAN C. BATCH[2], AND LAURA P. SVETKEY[3]

[1]*Department of Medicine, Nephrology Division, Duke University Medical Center, Durham, North Carolina*
[2]*Division of Endocrinology, Metabolism and Nutrition, Duke University Medical Center, Durham, North Carolina*
[3]*Department of Medicine, Nephrology Division, Duke Hypertension Center, Duke University Medical Center, Durham, North Carolina*

Contents

I. INTRODUCTION

Approximately 30% of U.S. adults have hypertension,[1] a major risk factor for coronary heart disease, stroke, and premature death [1–3]. Further, approximately 28% of U.S. adults have prehypertension,[2] which is also associated with a graded, increased risk of cardiovascular disease (CVD) and progression to hypertension [2–4]. Studies have shown that for every 20-mmHg increase in systolic blood pressure (SBP) or 10-mmHg increase in diastolic blood pressure (DBP) there is a doubling of mortality from ischemic heart disease (IHD) and stroke (Table 1) [2]. In industrialized societies, blood pressure (BP) increases with age: More than 50% of Americans ages 60–69 and more than three-fourths of those ages 70 and above are affected [5]. Hypertension is more common in African Americans [1]. Although the cause of hypertension is largely unknown, 1–5% of hypertension cases are due to a secondary underlying correctable condition.

In contrast to the prevalence in the United States, many non-Westernized, remote populations have a low prevalence of hypertension and do not experience an increase in BP with age [6, 7]. Their protection from hypertension is often attributed to a very low salt intake [8–10], a rich potassium intake [9, 11], being physically active [6], low alcohol consumption [9, 12], and generally high plant food and fish consumption. Migration studies of indigenous populations also report increasing prevalence of hypertension with urbanization, providing additional evidence for a role of environmental factors [13, 14]. With urbanization, access to processed food increases, and fresh foods that were previously readily available become less affordable. In addition to other lifestyle changes, increases in body weight, sodium intake, dietary fat, and the ratio of urinary sodium to potassium have been observed during the process of acculturation [12–18]. Taken all together, these observations support an important role of diet and lifestyle in BP and inspired much of the later research in this area.

In this chapter, we provide an overview of epidemiologic and clinical evidence for established and potential dietary factors for hypertension prevention and control. Because a comprehensive review of all individual trials in this area is beyond the scope of this chapter, when applicable, we review meta-analyses. Although meta-analyses are useful for evaluating consistency in the literature, they tend to weigh large studies more heavily [19, 20] and it is rarely feasible to conduct subanalyses on potentially important modifying factors or to account for variable dietary adherence among studies.

The last two sections of this chapter are devoted to the review of several large-scale intervention trials. These trials include the whole diet–based controlled feeding trials such as the Dietary Approaches to Stop Hypertension (DASH) trial [21], DASH-Sodium [22], and OmniHeart [23] trials. Additionally, large-scale multi-lifestyle intervention trials including Trials of Hypertension Prevention (TOHP) I [24], TOHP II [25], and PREMIER [26] are also reviewed. Issues related to implementation of the current national guidelines are discussed. We conclude with a summary of qualitative

[1] The operational definition of hypertension is a systolic blood pressure (SBP) of 140 mmHg or greater, a diastolic blood pressure (DBP) of 90 mmHg or greater, or current use of an antihypertensive medication. *Source*: The Seventh Report of the Joint National Committee on prevention, detection, evaluation, and treatment of high blood pressure. National Institutes of Health. A complete list of the classifications of high BP, used to guide treatment, is included in Appendix 1.

[2] Defined as an SBP of 120–139 mmHg or a DBP of 80–89 mmHg.

TABLE 1 Baseline Systolic BP and Age-Adjusted 10-Year Mortality from Cardiovascular Disease from the Multiple Risk Factor Intervention Trial

Systolic BP (mmHg)	*n*	Deaths	Rate per 1000	Relative Risk	Excess Deaths	% of All Excess Deaths
<110	21,379	202	10.5	1.0	0.0	0.0
110–119	66,080	658	11.0	1.0	33.0	1.0
120–129	98,834	1324	14.3	1.4	375.6	11.5
130–139	79,308	1576	19.8	1.9	737.6	22.6
140–149	44,388	1310	27.3	2.6	745.7	22.8
150–159	21,477	946	38.1	3.6	592.8	18.2
160–169	9308	488	44.8	4.3	319.3	9.8
170–179	4013	302	65.5	6.2	220.7	6.8
≥180	3191	335	85.5	8.1	239.3	7.3

Source: Reprinted with permission from Stamler, J. (1991). BP and high BP: Aspects of risk. *Hypertension* **18** (Suppl 1), I95–I107.
Note: Men free of history of myocardial infarction at baseline (N = 347,978); Multiple Risk Factor Intervention Trial primary screenees [4].

and quantitative recommendations on dietary and lifestyle changes for the prevention and treatment of hypertension.

II. INDIVIDUAL NUTRIENTS AND BLOOD PRESSURE

A. Micronutrients

Micronutrients associated with BP tend to be highly correlated with each other because of similar food sources [27], thus limiting the interpretability of observational studies. For example, foods high in magnesium (e.g., nuts) are also high in fiber and potassium, making it virtually impossible to attribute associations with BP to the effects of a single nutrient. To isolate and test the effect of individual nutrients, intervention studies typically use dietary supplements. The use of supplements allows for any changes in BP seen as a result of the intervention to be attributed solely to the nutrient being examined. However, supplements may not be absorbed as well or have the same physiologic effects as when they are consumed in natural form. Varying levels of other dietary components may also modify the effectiveness of supplements.

1. Sodium

The mechanism by which a high sodium intake may affect BP is generally accepted to be related to sodium retention. Little research has been conducted on alternative mechanisms such as effects on vascular reactivity. It is clear that the association is influenced by genetics [28] and other dietary components [29, 30]. The potential for sodium reduction to lower BP is supported by a broad range of data. Whether salt reduction should be broadly recommended to lower BP in individuals without hypertension has been the subject of much controversy [31, 32]. Advocates propose that population-wide sodium reduction to 50–100 mmol/day (1150–2300 mg/day) would substantially lower the incidence of CVD in the general population [33, 34]. Some express concern that sodium reduction may raise vasoconstrictive hormones and lipid levels [35] and increase BP in certain individuals [36, 37]. However, the preponderance of evidence supports the safety and efficacy of moderate sodium reduction [38], which is part of the national recommendations for preventing and treating high BP.

Dietary sodium intake is not easily measured by standard dietary assessment methods, because salt added at the table and during cooking is difficult to quantify, and because processed foods vary widely in sodium content. Most often, 24-hour urine collections are used to assess daily sodium intake. Under stable conditions (e.g., adequate health, hydration, no excessive sweating), 90–95% of dietary sodium is excreted in the urine [39]. Wide variations in day-to-day sodium excretion within individuals will weaken the correlation of single 24-hour urinary sodium levels with BP. This weakness may be minimized by collecting multiple samples in individuals or by increasing sample size in group analyses. Several investigators now employ statistical methods to correct for this source of error by using data from repeated collections in a subset of the study population [40, 41]. Improperly collected urine samples and varying geographic conditions (e.g., climate) among populations [6] can introduce additional error. Despite these methodologic limitations, a relationship between dietary sodium intake and BP is well established.

a. Observational Studies. Observations of a direct relationship between sodium intake and BP across populations support a causal role for sodium in hypertension [42, 43]. The INTERSALT study measured the relationship between 24-hour urinary sodium excretion and BP in 10,079 men and women from 52 centers around the world [44]. Mean urinary sodium by center was positively associated with BP (Fig. 1a, solid line). When excluding data from the four isolated, traditional populations in whom other unmeasured potentially relevant factors were of particular concern,

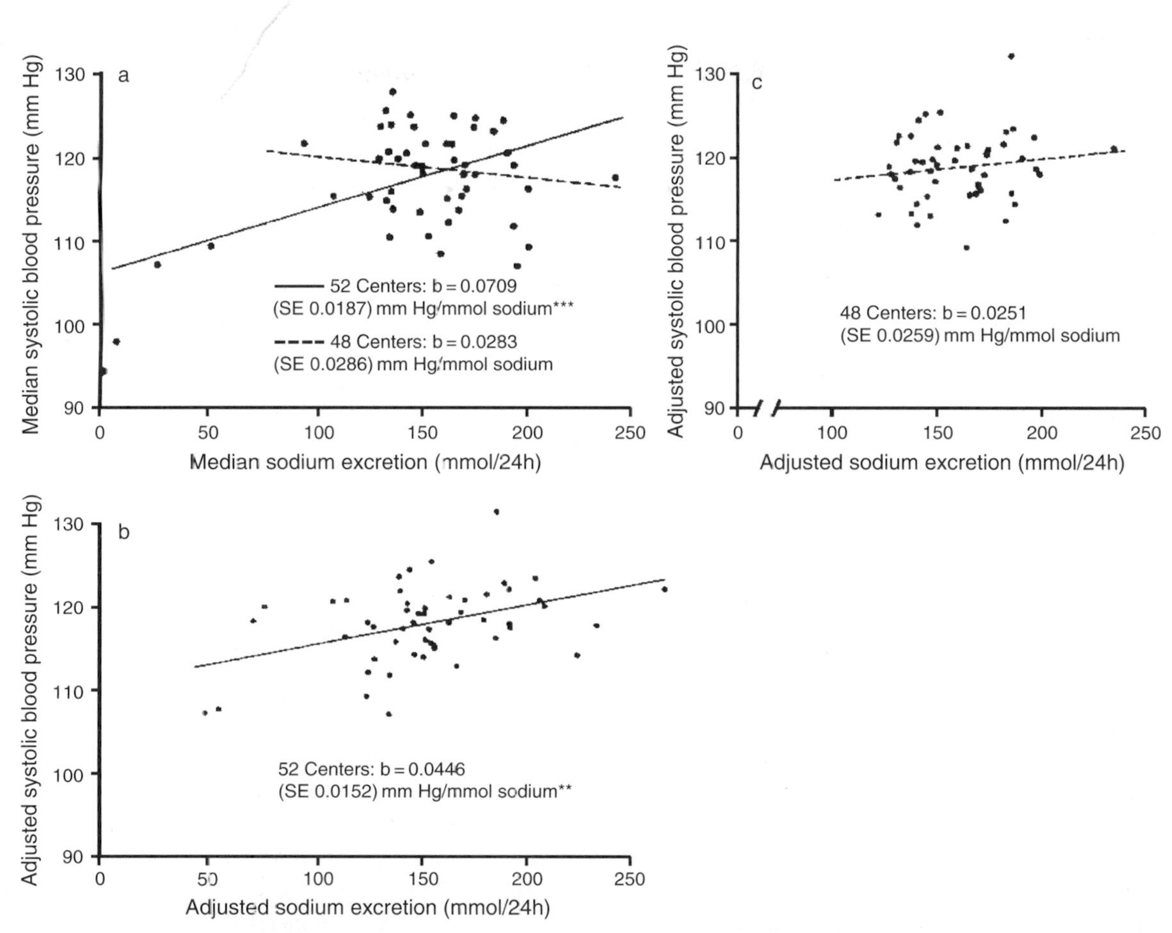

FIGURE 1 Intersalt figure panel. Relationship between age and sex-standardized 24-hour urinary sodium excretion and systolic blood pressure from the INTERSALT study. (a) with and without including four remote populations, (b) also adjusted for alcohol intake and body mass index (52 centers), and (c) also adjusted for alcohol intake and body mass index (48 centers).
$*p < 0.05$; $**p < 0.01$; $***p < 0.001$ [44].

the association disappeared (Fig. 1a, dashed line). After adjusting for alcohol and body mass index (Figs. 1b and 1c), a slightly positive relationship was again noted. The investigators also reported a strong positive relationship between sodium intake and the slope of BP increase with age across populations [44], suggesting a role for sodium in age-related BP increase. In a reanalysis of the original INTERSALT data, corrected for measurement error due to use of single 24-hour urine collections, results were stronger: A 100 mmol/day (2300 mg sodium) increase in urinary sodium was associated with an increase in SBP of 3–6 mmHg and DBP of 0–3 mmHg [45]. In a meta-analysis of observational studies, Law *et al.* [46] reported somewhat stronger findings than INTERSALT, especially in the elderly and those with higher baseline BP, but diet and other confounders were not assessed in a standard manner across studies.

b. Interventional Studies. The early observation of a large BP reduction in patients with severe hypertension consuming the Kempner rice diet [47] is often attributed to its low sodium content (7 mmol or 150 mg/day), though the diet was also rich in fruit, low in fat and protein, and supplemented with vitamins. The many trials of sodium reduction conducted since then have varied in design, population age, BP status, race and gender composition, amount and form of sodium provided, quantity and control of other nutrients in the diet, trial procedures (inpatient versus outpatient), and degree of adherence to the diets. Several meta-analyses of clinical trials of sodium reduction have been published in the past 16 years [33–35, 48, 49]. In a meta-analysis of 28 trials published by Midgley *et al.* [48], a reduction in urinary sodium by 100 mmol/day (2300 mg/day) was associated with a 3.7/0.9-mmHg lower SBP/DBP in individuals with hypertension and 1.0/0.1-mmHg BP reduction in nonhypertensives. A meta-analysis by Cutler and others [33] in 1997 of 78 published trials reported stronger results. In people with hypertension, a 100-mmol/day (2300 mg/day) reduction in sodium intake was associated with a reduction in SBP/DBP of 5.8/2.5 mmHg; corresponding reductions were 2.3/1.4 mmHg for nonhypertensives. Statistical approaches used to quantify the relationship between sodium and BP reduction may partly explain differing conclusions of these investigators.

The recent meta-analysis by Graudal and others [35] included 58 trials in individuals with hypertension and 56 trials in nonhypertensives. Despite greater mean reductions in sodium intake used to estimate the BP-lowering effect (a 118-mmol [2714-mg] sodium reduction in individuals with hypertension, and 160-mmol [3680-mg] reduction in nonhypertensives), their findings of a 3.9/1.9-mmHg and 1.2/0.26-mmHg BP reduction, respectively, were closer to those of Midgley *et al.* [48]. Graudal *et al.* [35] and Midgley *et al.* [48] included studies of short-term acute sodium reduction (which stimulates BP regulating hormones), while Cutler *et al.* [33] excluded these studies. In two more recent meta-analyses, salt reduction was evaluated among trials of longer term [49, 50]. Hooper reviewed only trials longer than 6 months, but the salt reduction achieved was smaller (on average by 87 mmol/day or 2000 mg/day). Thus, they found a smaller fall in BP and did not observe a dose-response to salt reduction. In one of the analyses [49], 20 salt reduction trials in hypertensives and 11 trials in nonhypertensives that were 4 weeks or longer were examined. The median reduction in urinary sodium was 78 mmol/day (1794 mg/day) and the mean reduction in SBP/DBP was 5.06/2.7 mmHg among hypertensives. Corresponding findings among nonhypertensives were a reduction of urinary sodium of 74 mmol/day (1702 mg/day) and 2.03/0.99 mmHg in SBP/DBP. There was a significant relationship between sodium reduction and BP. Despite variations in estimated effect size, all meta-analyses reported a direct relationship between sodium and BP, which was consistently significant in hypertensives. The estimate of the magnitude of the reduction varied at least partly because of statistical methods used, and the choice of studies included in the meta-analyses.

The DASH-Sodium multicenter trial [22] provides more precise estimates of the magnitude of the effect of sodium reduction on BP. The study, funded by the National Heart, Lung, and Blood Institute, used a controlled feeding design to compare the effect of three sodium levels (resulting in urinary sodium excretion of 65, 107, and 142 mmol/day or 1495, 2461, and 3266 mg/day) on BP in adults with prehypertension or stage 1 hypertension. The sodium intervention was provided in conjunction with a typical U.S. (control) diet or the DASH dietary pattern (Appendix 2), and intake was precisely regulated—participants ate foods prepared in a metabolic kitchen. In the context of a typical American diet, reducing sodium intake to 107 mmol/day (2461 mg/day) reduced SBP/DBP significantly by 2.1/1.1 mmHg. Reducing sodium further to 65 mmol/day (1495 mg/day) caused additional significant SBP/DBP reduction by 4.6/2.4 mmHg. The effects of sodium were observed in participants with and in those without hypertension, African Americans and those of other races, and women and men. In addition, as noted later, the lowest sodium intake (65 mmol/day or 1495 mg/day) superimposed on the already effective DASH dietary pattern provided the most effective BP-lowering combination. Compared to the control diet with the highest sodium level, the DASH dietary pattern with the lowest sodium level reduced SBP significantly by 8.9 and 7.1 mmHg, respectively, among participants with and without hypertension at baseline.

In addition to lowering BP, there is evidence that sodium reduction prevents hypertension incidence and cardiovascular events. Several long-term, randomized clinical trials have provided evidence that moderate sodium reduction with or without weight loss reduces the incidence of hypertension and cardiovascular events, especially in overweight participants [51]. In the Trial of Nonpharmacologic Interventions in the Elderly (TONE) [52], a 10-pound weight loss and dietary sodium reduction of 40 mmol/day (920 mg/day) were independently associated with approximately a 40% reduced risk of hypertension or cardiovascular event after medication withdrawal, compared with those receiving standard care. In the obese participants, the combined intervention was associated with a 53% reduced risk of hypertension or CVD event. In the Trials of Hypertension Prevention (TOHP) Phase II [53], overweight adults who were counseled to reduce sodium for 6 months achieved a 2.9/1.6-mmHg BP reduction from sodium reduction alone, and 4.0/2.8 mmHg when coupled with weight loss. Although effects on average BP declined over time with recidivism, 20% reductions in hypertension incidence remained after 48 months of follow-up in each intervention group. Thus, combining sodium reduction with other lifestyle modifications such as weight loss is feasible and can increase effectiveness of BP control.

c. Salt Sensitivity. Approximately 50% of individuals with hypertension and 25% of nonhypertensives are considered by some to be salt sensitive [54–56], defined arbitrarily as a mean arterial pressure reduction of at least 10 mmHg or 10% with salt restriction. It is more common in the elderly and in African Americans [57]. Various protocols have been employed to diagnose salt sensitivity, including feeding low- and high-sodium diets, and a more rigorous protocol of saline infusion (to expand blood volume), followed by volume contraction with a low-sodium diet and administration of a diuretic [54]. The concept of salt sensitivity has been challenged [58]. Using the data collected in the DASH-Sodium trial, the variability and consistency of individual SBP responses to changes in sodium intake was examined. A total of 188 participants consumed a run-in (typical) diet at the higher sodium level and continued to eat the same diet at three sodium levels (higher: 142, intermediate: 107, lower: 65 mmol/day) sequentially in random order for 30 days each. Changes in SBP from run-in to higher sodium (no sodium level change) ranged from −24 to +25 mmHg and 8.0% of participants decreased 10 mmHg or more. Comparing the higher versus lower sodium levels (78-mmol sodium difference), the range of SBP change was −32 to +17 mmHg, and 33.5% decreased 10 mmHg or more. In addition, SBP change with run-in versus lower sodium was modestly correlated with SBP change with higher versus intermediate sodium;

SBP change with higher versus lower sodium was similarly correlated with run-in versus intermediate sodium (combined Spearman $r = 0.27, p = 0.002$). These results show low-order consistency of response and confirm that identifying individuals as sodium responders or salt sensitive is difficult.

Even if the concept of a dichotomous trait in salt sensitivity is not useful, it is clear that there is a wide range of individual response to changes in sodium intake. The reasons why certain individuals respond differently to a sodium load are unclear but may be due to differences in the ability of the vascular system to adjust to a change in circulating blood volume and to blunted sodium excretion [59], abnormalities in kidney function, or the renin-angiotensin system. Physiologic abnormalities, such as low renin hypertension and abnormal modulation ("nonmodulation") of these hormones, predispose to salt sensitivity [60]. Sodium chloride–induced increases in BP may also be enhanced by a low calcium or potassium intake [30, 61, 62]. Genetic factors modulate the degree of salt sensitivity [63, 64], but much work remains to explain the magnitude of heterogeneity observed. Ideally, future studies will identify markers of susceptibility to aid in targeting interventions.

Sodium is not the only factor that affects BP, as it appears to interact with body weight and levels of other nutrients in the diet. However, sodium reduction is likely to benefit most people. Because sodium is ubiquitous in the U.S. food supply, large reductions in intake are not easily attainable. For example, the TOHP Phase I research group reported a mean decrease in sodium excretion of 44 mmol/day (1012 mg/day) after 18 months of intensive dietary counseling in a free-living population [65]. Only 20% of the population met the study goal of 60 mmol/day (1380 mg/day), but 56% were able to reduce sodium to less than 100 mmol/day (2300 mg/day). Counseling was least effective in men and in African Americans [65]. In the PREMIER clinical trial [26], a 6-month multi-lifestyle intensive intervention increased the proportion of participants who met the goal of less than 100 mmol/day in urinary sodium by about 18–28% compared to baseline. The control group, who received only a 30-minute advice session, also increased the corresponding figure by about 8%. All participants returned toward their baseline sodium intake levels at 18 months of follow-up. Thus, like all other lifestyle modifications, reducing sodium intake is no exception to the challenge of long-term maintenance. Seventy-five percent of sodium intake is derived from processed foods [66]; therefore, significant reductions in sodium intake will not be feasible on a population-wide basis unless sodium is reduced in processed and restaurant foods.

2. Potassium

The evidence for a role of potassium in lowering BP is relatively consistent across study types and is biologically plausible [67]. Potassium may lower BP through a direct vasodilatory role, alterations in the renin-angiotensin system [68], nitric oxide production [69], renal sodium handling, and/or natriuretic effects [70, 71] (Fig. 2).

a. Observational Studies. Epidemiologic observations support a role of dietary potassium in BP control [72]. Both potassium alone (inversely) and the sodium-to-potassium ratio (directly) have been associated with BP in cross-cultural studies [44]. Ophir *et al.* [73] noted that urinary potassium was the strongest discriminating feature related to low BP in vegetarian Seventh Day Adventists in Australia, compared to nonvegetarians.

b. Interventional Studies. Several meta-analyses have examined the effect of potassium supplementation on BP. In 1997, Whelton *et al.* [74] reviewed 33 randomized

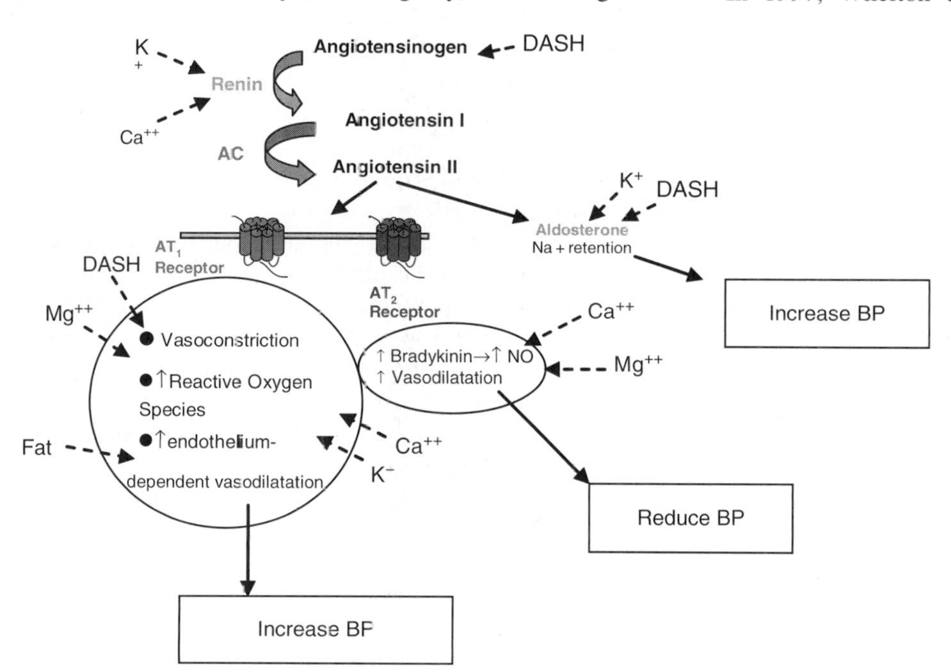

FIGURE 2 The potential effect of various dietary factors and DASH dietary pattern on BP regulation.

clinical trials of potassium and BP. These studies included mostly individuals with hypertension, some of whom were receiving antihypertensive medication. In all studies, potassium was provided as a supplement (median 75 mmol [2925 mg]), either superimposed on a controlled research diet or added to participants' usual diets. After excluding one outlier study, a high potassium intake was associated with 3.11/1.97-mmHg reduction in SBP/DBP. Interestingly, greater BP reductions occurred in those with progressively higher urinary sodium excretion during follow-up (measured at the end of the study). This suggests that potassium is more effective at higher levels of sodium intake. Other research evidence also points to the interaction of potassium and sodium in BP [67]. In addition, results were significantly stronger in studies that included a high proportion of African Americans. This meta-analysis reached qualitatively similar conclusions to those of a 1995 analysis by Cappuccio *et al.* [75] with roughly a 50% overlap in studies included.

In a large intervention trial of U.S. female nurses (Nurses Health Study II), Sacks and colleagues [76] administered supplemental potassium (40 mmol [1560 mg]), calcium (30 mmol [1200 mg]), magnesium (14 mmol [336 mg]), all three minerals, or placebo to women who reported habitually low intakes of these nutrients, for a 6-month period. Potassium, administered alone, was the only intervention that reduced BP [76]. The mild yet significant reduction occurred even though the women were nonhypertensive (mean BP: 116/73 mmHg).

Furthermore, two recent meta-analyses provide supporting evidence for an effect of potassium supplement on BP. Geleijnse *et al.* [77] examined 27 randomized trials with at least 2 weeks of follow-up among both hypertensives and nonhypertensives. They found that a median potassium supplementation of 44 mmol/day (1716 mg/day) was associated with a reduction of SBP/DBP by 2.42/1.57 mmHg. This reduction was even greater among those who were hypertensive (3.51/2.51 mmHg), but it was only borderline significant (~ 0.08). Dickinson *et al.* [78] reviewed four of the 27 trials in the Geleijnse study plus two other trials, all of which included at least 8 weeks follow-up and only hypertensives. They found a large reduction in BP by potassium supplementation (11.2/5.0 mmHg), but the reduction was not statistically significant, perhaps partially because of the heterogeneity of the trials. The authors stated that the heterogeneity of the trials was not explained by the varying dosage of potassium supplementation, study quality, or the baseline BP.

Strong support for the effect of dietary potassium on BP comes from the DASH trial as well. A fruit and vegetable diet, which contained increased amounts of potassium, magnesium, and fiber but was otherwise similar to the typical American control diet, significantly reduced SBP/DBP by 2.8/1.1 mmHg [21]. Further, the BP reduction was much greater among the African Americans (3.5/1.4 mmHg) than among the non–African Americans (0.7/0.3 mmHg), suggesting that African Americans may be more responsive to the BP effect of potassium.

Thus, research strongly supports the current JNC recommendation of increasing potassium intake to 4700 mg/day for the prevention and control of high BP. Individuals should be encouraged to increase potassium intake by including more whole grains, fruits, vegetables, and low-fat dairy products.

3. Calcium

Although a modest BP-lowering effect of calcium is noted in some observational studies, results from intervention trials have been inconsistent. Studies showing the greatest effect have tended to use dietary sources of calcium (e.g., dairy products), in which several potentially confounding dietary factors also change [21]. In addition, the BP-lowering effect of calcium may be greater among those with a low habitual intake of calcium. Potential mechanisms by which calcium may affect BP include effects on plasma renin activity, endothelial function, or the production of nitric oxide (Fig. 2).

a. Observational Studies. The higher calcium and magnesium content of "hard" water, and its inverse relation to cardiovascular mortality [79] initially sparked epidemiologic investigation into the relation of both minerals to BP. Cutler and Brittain [80] reviewed 25 observational studies and showed only modest associations between calcium and BP. Because most studies used 24-hour recall methods to assess diet, random day-to-day variation may have obscured any relation with BP [80]. It is also worth noting that some low-BP populations have minimal calcium intakes [11]—more counterevidence to the link between calcium and BP. Cappuccio and others [81] later conducted a meta-analysis of 23 observational studies, and found negligible associations for calcium with both SBP and DBP, though a reanalysis of this data resulted in a somewhat stronger inverse association [19]. More recently, a cross-sectional analysis of the NHANES III data showed that increasing calcium intake is significantly associated with a lower rate of age-related increase in SBP and pulse pressure [82]. Other cross-sectional studies also have shown a significant and inverse association between calcium intake [83, 84] or dairy intake [85–87] and BP.

b. Interventional Studies. In 1924, Addison and Clark [88] recorded the BP response to repeated administration and discontinuation of oral calcium supplements in a convenience sample of hospital outpatients and inpatients. They noted that BP decreased after 2 weeks of calcium supplementation and immediately rose with calcium discontinuation. It was not until much later in the century that the calcium hypothesis was tested in a more methodologically rigid fashion in a number of intervention studies.

Several meta-analyses of calcium-intervention trials have been published since 1989 [89–94], all showing only a slight or negligible BP reduction, primarily of SBP, with calcium supplementation of about 1000 mg. Intervention studies using calcium from food sources have sometimes [95], but not always [96], been more effective, though these studies also involve changes in several other nutrients.

A recent randomized controlled trial in 732 postmenopausal women showed that calcium supplementation of 1 g per day significantly reduced SBP by 1–2 mmHg at 6 months; however, this effect disappeared at 30 months despite continual calcium supplementation [97]. Two other small randomized trials did not find any significant effect of calcium supplementation on BP [98, 99]. An interaction between calcium and other minerals in their effect on BP has been observed [100] such that calcium supplementation may prevent a salt-induced rise in BP in susceptible individuals. Inability to either control for sodium intake or stratify by level of sodium intake in most meta-analyses may obscure a relationship. Despite weak evidence for a role of calcium in BP, it is advisable for the public to consume adequate dietary calcium for the benefit of bone health.

4. Magnesium

Adequate magnesium is required for the Na/K-ATPase pump, which regulates intracellular calcium—one of the critical determinants of vascular smooth-muscle contraction [101] (Fig. 2). In animal models, magnesium deficiency has been shown repeatedly to be associated with hypertension [102–104]. In humans, magnesium deficiency, although recognized rarely, is seen in severe malnutrition, in chronic alcoholism, and in association with malabsorption [101].

a. Observational Studies. As with calcium, one of the early suggestions for a role of magnesium in hypertension came from reports that water hardness (increased calcium and magnesium) was associated with lower cardiovascular mortality [79]. This finding was corroborated by Yang and Chin [105]. Several cross-sectional [106] and prospective observational analyses [107, 108] have found higher-magnesium diets to be associated with lower BP. However, interpretation of these results is limited by the fact that a high-magnesium diet tends to be high in other beneficial dietary factors as well. In a recent prospective study, magnesium intake was found to be inversely associated with risk for hypertension after adjustment for known risk factors among 28,349 participants of the Women's Health Study [109].

b. Interventional Studies. Many intervention trials of magnesium supplementation have been conducted [24, 76, 110–120] and some have shown a beneficial effect on BP [110, 111, 116, 117, 119]. Patients were magnesium depleted (due to diuretic treatment) in two of the studies [110, 111], and in one study an effect was only seen in those with a low baseline intake of dietary magnesium [120]. Kawano *et al.* [119] found the greatest BP reduction with magnesium supplementation in older men on antihypertensive medications. More than half of these studies, however, found no BP-lowering effect with magnesium supplementation. In a recent meta-analysis of 12 randomized trials, magnesium supplement significantly reduced DBP only (by 2.2 mmHg) but had no significant effect on SBP [121]. Overall, the quality of the included trials were poor, results varied, and thus the evidence was weak. In another meta-analysis of nine randomized trials with diabetic participants, a median level of 360 mg/day magnesium supplementation did not have any significant effect on BP [122]. Thus, despite the potential role for magnesium in BP control, evidence from intervention trials to date has not been supportive.

B. Macronutrients

Studies of macronutrients and BP are often subject to various design limitations. Thus, results can be difficult to interpret. For example, when a study is designed to examine the effect of the amount of fat intake on BP, alteration in fat intake under isocaloric conditions inevitably will change intake of protein and/or carbohydrate and may change the intake of other nutrients as well. As a result, it may be difficult to attribute the effect on BP to change in fat intake alone. In addition, the impact of macronutrients on BP potentially involves aspects of the absolute quantity and type of macronutrients consumed. Both aspects can affect BP independently, but they are not always distinguishable in research designs.

1. Protein

Despite evidence that high protein intake may promote renal injury, especially in those with existing kidney disease, the overall effect of protein on BP appears to be salutary. Although the exact mechanism(s) linking amount or type of protein to BP is still unclear, possible explanations exist. First, an increase in protein intake may induce increases in renal plasma flow, renal size, glomerular filtration rate, and sodium excretion [113, 123–125]. Second, the amino acid arginine may act as a vasodilator and contribute to BP lowering [126].

a. Observational Studies. Many observational studies have shown an inverse relationship between total dietary protein intake and BP [127]. Nevertheless, results have been mixed when studies examined animal and plant proteins separately. Cross-sectional studies conducted in rural Japanese and Chinese populations found only an inverse relationship between animal protein and BP [128, 129]. In contrast, cohort studies conducted in the United States have shown an inverse association between the increase in SBP with aging and baseline intakes of plant protein but not animal protein [130, 131]. Thus, there may be differential effects of animal and plant protein on BP, but the exact relationship is not clear.

b. Interventional Studies. Recent evidence from randomized controlled trials suggests that an increased intake of protein, from both plant and animal sources, may lower BP. In a 6-week randomized three-period crossover trial in subjects with diagnosed hypertension, substitution of protein for carbohydrate led to a decrease in SBP of 3.5 mmHg [23]. Recently, Hodgson *et al.* [132] demonstrated in a small trial that replacing carbohydrate with lean red meat while keeping total calorie and fat intake constant reduced SBP but not DBP. Although the reduction in SBP was statistically significant, the magnitude within the active intervention participants was quite small (–1.9 mmHg using clinic BP and –0.6 mmHg using ambulatory 24-hour BP). In contrast, using data collected from the PREMIER clinical trial, we found that dietary plant protein intake was inversely associated with both SBP and DBP in cross-sectional analyses at the 6-month follow-up ($p = 0.0009$ and $p = 0.0126$, respectively) [133]. An increase in plant protein intake from baseline to month 6 was marginally associated with a reduction of both SBP and DBP from baseline to month 6 ($p = 0.0516$ and 0.082, respectively), independent of change in body weight. Animal protein was not associated with BP or change in BP in PREMIER. Furthermore, among those with prehypertension, increased intake of plant protein was significantly associated with a lower risk of developing hypertension at 6 months. It should be noted that many intervention trials employ different types or amounts of protein, and the delivery mechanism may differ as well, all of which contribute to variation in results. Most trials use protein supplements that include soy, vegetable, or animal protein and do not control for other nutrients.

Soy protein was hypothesized to reduce BP because it is rich in arginine, a potential vasodilator and a precursor for the vasodilator nitric oxide [134]. The results of numerous randomized controlled trials that supplemented soy protein have been mixed. In a 12-week randomized double-blind controlled trial, He *et al.* [135] demonstrated a decrease in SBP/DBP of 4.31/2.76 mmHg soybean protein supplements as compared to the placebo group. Further, Rivas *et al.* [136] demonstrated that soy milk supplement significantly reduced SBP by 18.4 mmHg in a 3-month randomized trial. In contrast, three randomized controlled trials [137–139] showed no difference in SBP or DBP between the soy protein–supplemented group and the placebo group. It should be noted that some studies used milk supplements, which provide not only protein but also other nutrients that may affect BP. Thus, the BP responses cannot be attributed to protein alone.

Considering the evidence available, it is not advisable to use protein supplements as a dietary intervention to treat hypertension. Instead, ensuring an adequate protein intake on the basis of a healthy eating pattern may likely benefit BP.

2. Dietary Fat

Numerous studies have investigated the relationship between dietary fat and BP. However, because of differences in study design, lack of adequate sample size, and other design limitations, the issue remains controversial. Both the absolute total intake of dietary fat and the relative fatty acid composition may independently relate to BP control. Research has suggested that increased intake of total fat and saturated fats may impair endothelial function, which may subsequently affect BP (Fig. 2). Conversely, diets rich in omega-3 fatty acids may improve endothelial function and therefore may lower BP.

a. Observational Studies. Most observational studies have not found an association between total fat intake and BP [27, 140–143]. However, two large studies [140, 142, 143] but not others [27, 144] show a positive relationship between saturated fatty acids and BP. In addition, Hajjar and Kotchen [145] examined the NHANES III data and reported that the southern region of the United States, which consumed the highest amounts of monounsaturated and polyunsaturated fats, also had the highest SBP and DBP compared to other regions.

b. Interventional Studies. Many studies have examined the impact of the amount or type of fat on BP. However, as discussed previously in this chapter, any change in total fat intake often introduces changes in other dietary factors as well, leading to potential confounding. Thus, the BP responses may not be attributed solely to the change in fat intake. In a crossover randomized study, consumption of a single high-fat meal (42 g) was found to increase both SBP and DBP significantly more than a low-fat meal did (1 g) [146]. In 2000, Ferrara *et al.* [147] investigated the effect of monounsaturated versus polyunsaturated fat on BP in a 6-month double-blind randomized crossover study. The monounsaturated fat diet reduced SBP and DBP significantly more than the polyunsaturated fat diet. Recently, Appel *et al.* [23] reported that substituting monounsaturated fats for saturated fats significantly reduced SBP by 2.9 mmHg in a group of 191 subjects with prehypertension or stage 1 hypertension. This study finding suggests that a high fat intake (37% energy) mainly provided as monounsaturated fat, in combination with other beneficial dietary factors, can lower BP effectively.

Many short-term intervention trials have been undertaken to determine whether supplementation of either fish or fish oil lowers BP. Because of variations in research design, participant criteria, dosage and type of supplements, and length of intervention, the results have been inconsistent. Recently, a meta-analysis of randomized controlled trials (36 of which included fish oil) explored the impact of fish-oil intake on hypertension prevalence in five populations (Finland, Italy, the Netherlands, the United Kingdom, and the United States) [148]. Pooled meta-regression of data from these trials showed that a 4.1-g supplement of fish oil is associated with a 2.1/1.6-mmHg decrease in SBP/DBP. Long-term studies are required to confirm the benefit of fish oil on BP lowering. Until such information is available, it is more advisable to encourage greater fish consumption as part of a healthy diet than taking fish-oil supplementation. Further, it remains unclear how fat intake and various fatty acids affect BP and whether an interaction exists among these factors.

3. Carbohydrates

Although not well understood, carbohydrate may contribute to the development of essential hypertension through its glycemic effect. Kopp [149] suggests that consumption of a high-glycemic-index diet may create a chronic state of postprandial hyperinsulinemia, sympathetic nervous system overactivity, and vascular remodeling of renal vessels leading to chronic activation of the renin-angiotensin-aldosterone system and development of essential hypertension. Although logical, the relationship of a high carbohydrate diet to high BP has not been consistent in all studies. Very few studies have been designed specifically to investigate the impact of the quantity or type of carbohydrate intake on BP. Nevertheless, studies of the relationship between fat intake and BP often alter intakes of both fat and carbohydrate while keeping protein intake constant. Thus, interpretation of the effect of fat intake on BP is potentially confounded by the effects of carbohydrate intake.

a. Observational Studies. Very few observational studies have specifically reported the relationship between carbohydrate intake and BP. Inevitably, examination of carbohydrate intake is often confounded by other nutrients or dietary factors. Using the NHANES III data, quintile of carbohydrate intake was not associated with the risk for increased SBP (≥140 mmHg) after adjustment for potential confounders and total sugar intake [150].

b. Interventional Studies. No research study has been designed specifically to examine the impact of the amount of carbohydrate alone on BP. It is difficult to manipulate the amount of carbohydrate without changing intake of other nutrients if total energy is to be kept constant. However, in the context of whole dietary pattern change, both low- and high-carbohydrate intakes in combination with other dietary changes have been associated with BP lowering [21–23, 151].

Some human studies have examined the impact of the type of carbohydrate on BP, but the research has been limited in number and yielded inconsistent results. In one study [152], SBP rose significantly 1 hour after ingestion of a sucrose solution. However, in an earlier study of patients with coronary artery disease, both SBP and DBP decreased after 4-days of a sucrose load at 4 g/kg/day [153]. In another study that was designed to examine the metabolic effect of sucrose in a group of overweight women, BP was not changed over the 6 weeks after consuming a hypocaloric diet with sucrose as the main source of carbohydrate [154]. In the study by Palumbo *et al.* [153], fructose loading reduced BP significantly, but glucose loading had no effect. Similar hypotensive effects of fructose were observed in other studies [155, 156], yet the response to oral glucose loading was found to be either hypertensive [152, 155] or hypotensive [157].

Thus, it is unclear if the amount of carbohydrate alone or the type of carbohydrate affect BP. Future research designed specifically to examine these questions is needed to help clarify the roles of carbohydrate in BP control.

4. Fiber

Fiber may indirectly lower BP through reduction of insulin levels [158]. Hyperinsulinemia is often associated with obesity, impaired glucose tolerance, and hypertension.

a. Observational Studies. Both cross-sectional [27, 158, 159] and prospective analyses [107] have demonstrated inverse associations between fiber and BP but have also noted a high correlation of fiber with other nutrients that can affect BP in a salutary manner. In a prospective 8-year study of 12,741 subjects, the highest quintile of total dietary fiber and nonsoluble dietary fiber intake was associated with an 11.6% lower risk of hypertension as compared to that of the lowest quintile [160].

b. Interventional Studies. Several interventional studies have examined the effect of fiber on BP, with most adding cereal fiber to the diet. These studies suggest that an average supplementation of 14 g fiber reduces SBP/DBP by about 1.6/2.0 mmHg, respectively [158]. Similarly, a meta-analysis of 25 randomized controlled trials between 1966 and 2003 demonstrates that fiber supplementation (average 11.5 g/day) leads to a reduction in SBP/DBP of 1.13/1.26 mmHg [161].

Thus, this evidence suggests a consistent small beneficial effect of fiber on BP. Individuals should be encouraged to increase fiber intake to the current recommended level not only for BP control but possibly for other benefits to cardiovascular health. Such recommendation should be achieved by increasing fruits, vegetables, and whole grains based on the foundation of a healthy eating pattern, rather than using a supplement.

5. Alcohol

The exact mechanism for an alcohol–BP association is not clear, but possibilities include stimulation of the sympathetic nervous system, inhibition of vascular relaxing substances, calcium or magnesium depletion, and increased intracellular calcium in vascular smooth muscle [162–164].

a. Observational Studies. Excessive alcohol consumption is associated with higher BP and higher prevalence of hypertension in observational studies [163]. Men who consume three or more drinks per day [165], and women who consume two or more drinks per day [166] may be at higher risk, but levels below this are not associated with increased risk. There is no consistent relationship between BP and type of alcohol consumed. Chronic, habitual intake may be more related to BP than recent intake [165]. In one study, men who had quit alcohol drinking during an 18-year follow-up period experienced less age-related increase in BP than those who did not [167].

Recently, Fuchs *et al.* examined incident hypertension in a cohort study of 8334 participants from the Atherosclerosis Risk in Communities (ARIC) study [168]. Participants were free of hypertension at baseline and were followed for 6 years. Risk of incident hypertension was increased in those who consumed greater than or equal to 210 g alcohol (78 oz wine, 191 oz beer, or 21 oz liquor) per week as compared to those who did not consume alcohol.

b. Interventional Studies. The relatively few intervention studies of alcohol and BP have tended to be small and of short duration, and are reviewed by Cushman *et al.* [164]. In nine of 10 studies examined, SBP was significantly reduced after a reduction of one to six alcoholic beverages per day. The Prevention and Treatment of Hypertension Study (PATHS) [169] was designed to evaluate the long-term BP-lowering effect of reducing alcohol consumption in nondependent moderate drinkers (those who consumed more than three drinks per day). The goal of intervention was either two or fewer drinks daily or a 50% reduction in intake (whichever was less). After 6 months, the intervention group experienced a 1.2/0.7 mmHg greater reduction in BP than the control group (NS), and among hypertensives, this reduction was more modest. In this study, the intervention group reduced their intake by two alcoholic drinks per day, but the control group also lowered their alcohol intake during intervention, so that the difference in intake between the groups was only 1.3 drinks/day. This small difference between the two groups may have limited the interpretation of the true BP effect of the intervention group. In addition, perhaps a greater reduction in alcohol consumption is necessary to see a significant effect on BP. However, this level of reduction appears realistic in moderate alcohol drinkers and is similar to the absolute reduction achieved in an earlier study [170].

In another study [171, 172] of mainly heavier drinkers (> 5 drinks/day), replacing alcohol with low-alcoholic substitutes resulted in a reduction of approximately five drinks per week and a greater reduction in BP (–4.8/3.3 mmHg). Importantly, this intervention also reduced body weight of the participants by an average of 2.1 kg, which may explain the larger BP reduction than that observed in the PATHS trial. However, in a recent meta-analysis [173] of 15 randomized controlled trials including a total of 2234 participants who drank an average of 3–6 drinks per day at baseline, the effect of alcohol reduction on BP was explored. Reduced alcohol consumption, varying from 29% to 100% reduction from baseline, was associated with a significant reduction in mean SBP (3.31 mmHg) and DBP (2.04 mmHg), whereas the body weight was minimally changed (mean change: –0.56 kg). There was a dose-response relationship between the mean percentage of alcohol reduction and mean percentage of BP reduction.

Thus, limiting alcohol consumption to the current recommendation of two or fewer drinks per day for men and one or fewer for women is supported by most research evidence and will likely improve BP. There is no need to recommend total abstinence; indeed, moderate alcohol consumption (as compared to no alcohol intake) also has well-known benefits for overall CVD risk [174, 175].

III. OTHER DIETARY AND LIFESTYLE MODIFICATIONS

A. Dietary Patterns

The previous sections on micro- and macronutrients highlight extensive confounding due to simultaneous changes in multiple nutrients. It is difficult to study (and change) one component of the diet without affecting others. Thus, it may be more appropriate to focus on dietary pattern rather than on individual micro- and macronutrients, and indeed there is extensive evidence that dietary pattern affects BP. For example, vegetarian groups in the United States and abroad have been observed to have lower BP than their nonvegetarian counterparts in many [176, 177] but not all studies [178]. The term *vegetarian* comprises several heterogeneous groups [179], but in general, the diet tends to be high in whole grains, beans, vegetables, and sometimes fish, dairy products, eggs, and fruit [177]. Aspects of the vegetarian diet suggested to benefit BP include ample amount of plant foods, a low intake of animal products [177], and a high potassium, magnesium, fiber, (sometimes) calcium [179, 180], a high ratio of polyunsaturated to saturated fat, and often a low sodium intake. However, as outlined in previous sections of this chapter, studies on individual nutrients have shown inconsistent results. Explanations for such inconsistencies may include the following: (1) The effect of individual nutrients may be too small to be detected, particularly when trials contain insufficient sample size and thus statistical power; (2) most intervention studies employed supplements of nutrients, which may function differently from nutrients in foods; (3) other dietary factors naturally occurring in foods that are not hypothesized to affect BP may also have an impact on BP; and (4) nutrients occurring in foods simultaneously may exert synergistic effects on BP. Clearly, differences in physical activity, stress, alcohol consumption, and other unmeasured factors may also contribute to a lower BP among vegetarians. But when research participants were counseled to follow the vegetarian diet pattern in intervention studies, significant reductions in BP in both nonhypertensive [181] and mildly hypertensive [180] participants were reported.

Despite the clear BP effect of a vegetarian diet, it is not realistic to expect a wide-scale adoption of such dietary pattern. In addition, a vegetarian diet does not include all dietary factors associated with lower BP. Thus, the Dietary Approaches to Stop Hypertension (DASH) multicenter trials were designed to test the impact of whole dietary patterns on BP simultaneously while controlling for multiple nutrients, weight, and dietary factors [182, 183].

1. The Dietary Approaches to Stop Hypertension (DASH) Dietary Pattern

The original DASH trial [21] was an 11-week randomized controlled feeding trial of 459 individuals with prehypertension or stage 1 hypertension. Three dietary patterns varying in amounts of fruits, vegetables, dairy products, meats, sweets, nuts and seeds and thus fats, cholesterol, fiber, calcium, potassium, and magnesium were tested (Table 2). In brief, the dietary patterns were (1) the control diet, which mimicked what most Americans were

TABLE 2 Nutrient Target for the Three Dietary Patterns Tested in the DASH Trial

Item	Control Diet Nutrient Target	Fruits-and-Vegetables Diet Nutrient Target	Combination Diet Nutrient Target
Nutrients			
Fat (% kcal)	37	37	27
Saturated	16	16	6
Monounsaturated	13	13	13
Polyunsaturated	8	8	8
Carbohydrates (% kcal)	48	48	55
Protein (% kcal)	15	15	18
Cholesterol (mg/day)	300	300	150
Fiber (g/day)	9	31	31
Potassium (mg/day)	1700	4700	4700
Magnesium (mg/day)	165	500	500
Calcium (mg/day)	450	450	1240
Sodium (mg/day)	3000	3000	3000
Food groups (servings/day)			
Fruits and juices	1.6	5.2	5.2
Vegetables	2.0	3.3	4.4
Grains	8.2	6.9	7.5
Low-fat dairy	0.1	0.0	2.0
Regular-fat dairy	0.4	0.3	0.7
Nuts, seeds, and legumes	0.0	0.6	0.7
Beef, pork, and ham	1.5	1.8	0.5
Poultry	0.8	0.4	0.6
Fish	0.2	0.3	0.5
Fat, oils, and salad dressing	5.8	5.3	2.5
Snacks and sweets	4.1	1.4	0.7

consuming at the time the trial was conducted; (2) a fruits and vegetables diet, which contained a macronutrient profile similar to that of the control diet except with a higher amount of fruits and vegetables; and (3) the DASH dietary pattern, which was higher in fruits, vegetables, and low-fat dairy products; lower in total and saturated fats and cholesterol; and rich in fiber, potassium, magnesium, and calcium. Sodium intake, body weight, and alcohol consumption were kept constant throughout the intervention.

The DASH dietary pattern reduced BP by 5.5/3.0 mmHg more than the control group (both SBP and DBP $p < 0.001$). The fruits and vegetables diet reduced BP by 2.8/1.1 mmHg more than the control diet ($p < 0.001$ for SBP and $p = 0.07$ for DBP). The reductions in BP were significant after participants consumed the diets for 2 weeks and were sustained for the following 6 weeks (Fig. 3). In addition, BP lowering was similarly effective in men and women and in younger and older persons, and it was particularly effective among African Americans and those who had high BP. These reductions occurred while body weight, sodium intake, alcohol consumption, and exercise patterns remained stable. Of note, sodium intake was not reduced and was identical in all treatment groups (3000 mg/day). Among the 133 participants with hypertension (SBP $\geq$ 140 mmHg and/or DBP $\geq$ 90 mmHg), the DASH dietary pattern lowered SBP and DBP by 11.4 and 5.5 mmHg, respectively. These effects in hypertensives are similar to reductions seen with single drug therapy [184] and more effective than most of the other lifestyle modifications for BP reduction (Table 3).

Even though the DASH trial was not designed to identify specific nutrient(s) responsible for the BP lowering effect, data from the fruits and vegetables group support the hypothesis that increasing potassium, magnesium, and dietary fiber intake reduces BP. In addition, by further lowering total and saturated fat and cholesterol, and increasing low-fat dairy products in the DASH dietary pattern, BP reduction was nearly doubled compared to the fruits and vegetables diet. Because whole food items rather than single nutrients were manipulated in this trial, other nutrients that were not controlled for in the study or other beneficial factors as yet unrecognized may also have contributed to the BP responses. Further research is needed to analyze the specific nutrients or factors responsible for the effect. Further details on DASH can be found in Appendix 2 and on the following web site: http://www.nhlbi.nih.gov.

2. Variations of DASH Dietary Pattern

Even though the DASH trial provided strong evidence of the efficacy of the DASH dietary pattern in reducing BP, this trial alone was not able to test all hypotheses related to dietary pattern and BP. As discussed in the previous sections, research suggests that a high unsaturated fat intake and high protein intake may benefit BP. Thus, the Omni-Heart study [23] was designed to further understand the impact on BP of macronutrient variations of the DASH

TABLE 3 Effect of Different Lifestyle Modification on Blood Pressure Reduction

Modification	Approximate SBP Reduction (range)
Dietary sodium reduction	2–8 mm Hg
Moderation of alcohol consumption	2–4 mm Hg
Adopt DASH eating plan	8–14 mm Hg
Weight reduction	5–20 mm Hg/10-kg weight loss
Physical activity	4–9 mm Hg

Source: The Seventh Report of the Joint National Committee on Prevention, Detection, Evaluation and Treatment of High BP, National Institutes of Health, NHLBI, NIH Publication No. 03–5233, December 2003.

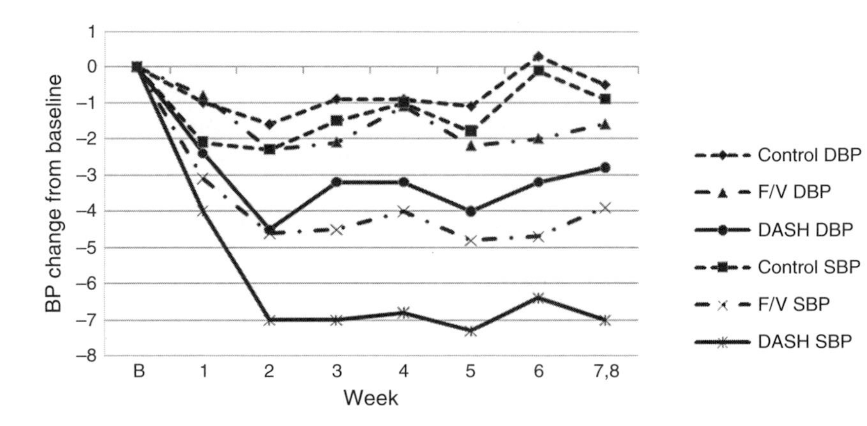

FIGURE 3 Mean changes in systolic and diastolic blood pressures from base line during each intervention week, according to diet, for participants in the DASH study.

dietary pattern. The dietary patterns tested were (1) the DASH dietary pattern with a slight reduction in total protein (called the carbohydrate diet, with 15% kcal protein instead of 18% tested in the original DASH trial); (2) the DASH dietary pattern with 10% of the carbohydrate energy replaced with unsaturated fats (called the unsaturated fat diet with mainly monounsaturated fats); and (3) the DASH dietary pattern with 10% of the carbohydrate energy replaced with protein (called the protein diet). A total of 164 adults with prehypertension or stage 1 hypertension were randomized into the three diets in a crossover fashion for 6 weeks each. All three diets lowered BP significantly but the protein and unsaturated fat diets significantly reduced SBP by 1.4 and 1.3 mmHg more than the carbohydrate diet. The further reductions in BP were even greater among those who were hypertensive at baseline.

Thus, these studies have vigorously and consistently proven that whole dietary patterns such as the DASH dietary pattern or the two modified DASH patterns in the OmniHeart study are effective strategies for BP control. As noted earlier, DASH in combination with reduced sodium intake lowers BP more than either intervention alone [22]. In addition, both the DASH and OmniHeart studies demonstrate that benefits of adopting a whole dietary approach extend beyond BP to other health indicators such as lipids [23, 185]. Thus a nutritional approach to BP control that involves changes in overall dietary pattern appears to be superior to approaches that manipulate only a small number of nutritional factors.

B. Weight Reduction and Multi-Lifestyle Modification

Weight loss alone is an effective strategy for BP reduction. Potential mechanisms for the effect of weight loss on BP include suppression of sympathetic nervous system activity, lowered insulin resistance, normalization of BP regulating hormones [186], decreased body sodium stores, decreased blood volume and cardiac output, and reduction of salt sensitivity [187–189].

a. Observational Studies. Several observational studies have reported a positive relationship between several indices of body weight or body fatness with BP [44, 190–194].

b. Intervention Studies. Several large-scale clinical trials have been conducted to evaluate different lifestyle modification programs on BP, several of which provide evidence for a BP-lowering effect of weight loss. For example, in the Trials of Hypertension Prevention Phase I (TOHP Phase I) [195], weight loss was found to be the most successful intervention in lowering BP compared to sodium reduction, stress management, or nutritional supplements (calcium, magnesium, or potassium). At the 6-month follow-up, men and women in the intervention group had lost 6.5 and 3.7 kg, respectively. This level of weight loss was achieved with a fairly rigorous counseling approach aimed at simultaneously reducing energy intake and increasing exercise [196]. At study termination, BP fell an average of 2.9/2.3 mmHg overall (after subtracting the BP change in the control group). In this study, some recidivism occurred, and at 18 months, men had maintained a 4.7-kg reduction and women, 1.6 kg. After 7 years' follow-up in a subset of study participants, the odds of developing hypertension were reduced by 77% in the weight-loss group [197], even though their long-term weight loss was nearly identical to that of the control group (4.9 kg and 4.5 kg, respectively).

As a follow-up to TOHP Phase I, TOHP Phase II [25] was conducted to examine the effects of weight loss and dietary sodium reduction alone, and in combination, on BP in overweight adults. The intervention lasted 3 years and included individual and group counseling meetings focusing on diet, exercise, and social support. The weight loss intervention group ($n = 595$) achieved a mean reduction in weight from baseline to 6 months of 4.4 kg, 2.0 kg at 18 months, and 0.2 kg at 36 months. BP was significantly lower at all time points mentioned and the greater the weight loss, the greater the BP reduction. At 36 months, every kilogram of weight loss was associated with a reduction of SBP/DBP of 0.35/0.45 mmHg. Further, this weight loss intervention was associated with a 42% reduction in incident hypertension.

In a meta-analysis of 25 randomized, controlled trials, weight loss of 1 kg was associated with approximately 1 mmHg reduction in both SBP and DBP in individuals with prehypertension [198]. The largest of the trials included in this analysis, Trials of Hypertension Prevention [195], demonstrated a larger effect. In that trial, a behavioral weight loss intervention in adults with prehypertension led to an average reduction in body weight of 4.4 pounds at 6 months, associated with an average reduction in BP of 3.7/2.7 mmHg [25]. Although point estimates of this effect vary [199–202], evidence of an effect of weight loss on BP is strong and consistent. Thus, several studies demonstrate the BP-lowering effect of weight loss alone. However, weight loss often involves change in dietary pattern, energy restriction, and increased physical activity. Implementing multiple simultaneous lifestyle changes may be an effective strategy for lowering BP.

In the more recent PREMIER clinical trial [26], the effects on BP of two multicomponent lifestyle interventions compared to an advice-only control group were tested for 18 months. The two behavioral interventions were designed to stimulate adoption of what were at the time the well-established lifestyle guidelines for BP control (EST) or the well-established guidelines plus the recently reported DASH dietary pattern (EST+DASH). The well-established guidelines included weight loss if overweight (95% of participants), reducing sodium intake to less than 2400 mg/day, increasing physical activity to 180 minutes of moderate activity/week, and alcohol consumption not exceeding 2 drinks/day for men and 1 drink/day for women. A total of 810 individuals were randomized and completed the study. Participants in both intervention groups significantly reduced weight, improved fitness, and lowered sodium intake, and the EST+DASH group also increased fruit, vegetable, and dairy intake. Mean reduction in SBP, net of the control group, was 3.7 mmHg ($p < 0.001$) in EST group and 4.3 mmHg ($p < 0.001$) in EST+DASH group [26]. Each individual lifestyle modification was independently and significantly associated with SBP reduction at 6 and 18 months [203].

Overall, these studies demonstrate that individuals can make multiple dietary and lifestyle changes to reduce BP. Although the beneficial effect of lifestyle changes on BP may last beyond the intervention period, recidivism is often observed. Thus, it is important to develop and test strategies to help individuals maintain these changes long term.

IV. CURRENT RECOMMENDATIONS AND IMPLEMENTATION

The Joint National Committee on Prevention, Detection, Evaluation and Treatment of High BP was appointed by the National Heart Lung and Blood Institute to provide evidenced-based clinical guidelines for the prevention and management of hypertension [5]. Current lifestyle guidelines include salt reduction, moderate alcohol consumption, weight loss if overweight, increased potassium intake, aerobic exercise, and following the DASH eating pattern. These lifestyle modifications are all recommended as part of the first-line therapy for low-risk individuals, defined as those without diabetes or CVD and with SBP less than 160 mmHg and DBP less than 100 mmHg. They are also recommended in combination with pharmacotherapy for high-risk individuals. These recommendations also apply to individuals with prehypertension in order to prevent the development of high BP.

Although much evidence indicates that diet and lifestyle modification can prevent and treat hypertension, recent reports suggest that 64.9% of hypertensives are not well controlled (SBP/DBP < 140/90 mmHg) [5, 204]. The full potential of diet and lifestyle modification for treatment has likely not yet been realized, both because dietary causes of hypertension have not been completely identified, and because of poor adherence of both the clinicians and the public alike to the established medical and dietary guidelines. Despite data linking overweight and obesity to numerous adverse health outcomes including hypertension, clear guidelines for weight control, and a plethora of weight loss programs, the obesity epidemic continues to grow. Similarly, despite clear and long-standing recommendations to reduce sodium intake to 2400 mg/day or less, the average intake remains above 3000 mg/day. Furthermore, according to the 1999–2000 NHANES survey [204, 205], Americans are consuming far less than the nationally recommended amounts of fruit, vegetable, fiber, calcium, magnesium, and potassium—all of which are key components of the DASH dietary pattern.

Clearly, implementing dietary and lifestyle modifications is challenging. Lessons learned from past research indicate that behavioral intervention programs that result in successful behavior change are generally rooted in social cognitive theory [207] and techniques of behavioral self-management [208]. They are ideally constructed using the transtheoretical or stages-of-change model [209, 210] and use motivational enhancement approaches [211, 212]. These approaches emphasize the importance of the individual's ability to regulate behavior by setting goals, developing specific behavior change plans, monitoring progress toward the goals, and attaining skills necessary to reach the goals. Self-efficacy (one's confidence in performing a given behavior) and outcome expectancies (one's expectations concerning the outcome of that behavior) are critical mediators of behavior change [207, 213]. The transtheoretical model recognizes that behavior change is a dynamic process of moving through different motivational stages of readiness for change. Different behavioral strategies may need to be emphasized at different times, depending on the individual's stage of change.

In addition, behavioral interventions conducted with small groups can take advantage of the economy of scale and the social support provided by a group of peers. In the PREMIER trial of lifestyle intervention for lowering BP, the behavioral strategies discussed earlier were incorporated into an intervention consisting of frequent group sessions conducted by a trained interventionist. It is also important that any intervention program make efforts to create a culturally appropriate behavioral intervention. Such effort may include but is not limited to (1) having intervention encounters take place at a location in the community; (2) employing staff from the same cultural background as participants; (3) selecting foods, music, or examples from within the culture; and (4) involving staff of the same cultural background in program design and in consultation with the potential participants [140]. Despite considerable understanding of the theory and practice of behavior change, additional research is needed to develop effective strategies for sustaining dietary and lifestyle change long term.

V. SUMMARY

The evidence that diet modification can prevent and treat hypertension is strong, and recommendations are summarized in Table 4. In some cases, the effective intervention strategy and mechanisms involved are still being clarified. Because of various design limitations, inadequate statistical power, and measurement issues, studies of single nutrients, with the exception of sodium and potassium, have generally provided inconsistent results. However, when multiple nutrients or dietary factors are combined in a whole dietary strategy, as seen in the DASH and OmniHeart studies, BP is significantly and effectively reduced. Nutrients may have additive or interactive effects when provided together in whole foods. Thus, the current national guideline of lifestyle modification for BP control includes the DASH pattern, sodium reduction, weight loss, increased physical activity, and consumption of alcohol in moderation. Concurrent adherence to several recommendations is likely to hold the greatest promise for preventing and treating hypertension and has been shown to be feasible. In addition to addressing unresolved nutritional hypotheses, future research should focus on strategies to motivate and maintain lifestyle changes for BP control. At both the population and individual levels, success in dietary and lifestyle intervention relies on multiple levels of support ranging from clinicians to government agencies to private institutes and industries. In particular, partnering with industry to improve the nutritional quality of the food supply, such as reducing sodium and fat content of processed foods, and promoting foods and nutrients consistent with the DASH dietary pattern will play a critical role in implementing dietary and lifestyle modifications. Consistent efforts to educate and promote adherence to dietary and lifestyle guidelines by dietetic and other health care professionals are also instrumental to the prevention and management of hypertension.

APPENDIX 1.

Classification of BP and Lifestyle Modification for Adults Ages 18 and Over[a,b]

Category	Systolic (mmHg)	Diastolic (mmHg)	Lifestyle/ Nutrition Modification
Normal	<120	And <80	Encourage
Prehypertension	120–139	Or 80–89	Yes
Stage 1 hypertension	140–159	Or 90–99	Yes
Stage 2 hypertension	≥160	Or ≥100	Yes

[a]Based on the average of two or more properly measured, seated BP readings on each of two or more office visits.

[b]*Source*: The Seventh Report of the Joint National Committee on Prevention, Detection, Evaluation and Treatment of High BP, National Institutes of Health, NHLBI, NIH Publication No. 03–5233, December 2003.

APPENDIX 2. NATIONAL INSTITUTES OF HEALTH

The DASH Dietary Pattern

This eating plan is from the "Dietary Approaches to Stop Hypertension" (DASH) clinical study. The research was funded by the National Heart, Lung, and Blood Institute (NHLBI), with additional support by the National Center for Research Resources and the Office of Research on Minority Health, all units of the National Institutes of Health. The final results of the DASH study appear in the April 17, 1997, issue of the *New England Journal of Medicine*. The results show that the DASH "combination diet" lowered BP and so may help prevent and control high BP. The DASH dietary pattern is rich in fruits, vegetables, and low-fat dairy foods and low in saturated and total fat. It also is low in cholesterol; high in dietary fiber, potassium, calcium, and magnesium; and moderately high in protein.

The DASH eating plan shown below is based on 2000 calories a day. Depending on your caloric needs, your number of daily servings in a food group may vary from those listed. It should be noted that the servings listed here may be slightly different from those published previously due to the changes in serving sizes. For example, 1 serving of meat was defined previously as 3 oz and the new definition is 1 oz.

FOLLOWING THE DASH EATING PLAN

The DASH eating plan shown in Table 5 is based on 2000 calories a day. The number of daily servings in a food group may vary from those listed depending on your caloric needs. Use Table 6 and Table 7 to help you plan your menus or take them with you when you go to the store.

TABLE 4 Summary of Evidence Relating Dietary Factors with BP

Dietary Factors	Strength of Relationship with BP[a]	Direction of Association	Potential Mechanisms	Recommendations	Those Most Likely to Benefit
Sodium	1A	Direct	Changes in blood volume and BP regulating hormones	<65–100 mmol/day	Hypertensives, salt-sensitive individuals, African Americans, elderly, and those consuming typical American diet
Potassium	1A	Inverse	Vasodilatory; natriuretic	Increase potassium-rich foods	Those on high-salt diet
Calcium	2A	Inverse	Regulation of parathyroid hormone (PTH) and intracellular calcium; natriuretic	2–3 reduced-fat dairy products/day	Possibly salt-sensitive individuals
Magnesium	2B	–	Modification of Na-K/ATPase activity	Maintain adequate magnesium intake	Those depleted in magnesium, e.g., from diuretics
Protein	2A	Inverse	BP-related amino acid may act as vasodilator	Consume adequate protein (particularly plant sources); limit intake of high-fat animal protein	
Fat	1C	Direct for saturated fat, inverse for monounsaturated fats	Vasodilatory action of prostaglandins (PGE)	Moderate total fat (include more monounsaturated fat sources; reduce saturated fat	
Carbohydrates	2B	–	Unclear	Consume more whole grains and less sugar	
Fiber	2B	–	Unclear, possible reduction of insulin levels	Increase fiber-rich foods for overall health benefits	Those who have insulin resistance or impaired glucose tolerance
Alcohol	1B	Direct	Unclear, possible role of sympathetic nervous system, inhibition of vascular relaxing substances, calcium and magnesium depletion	Moderate consumption: ≤ 2 drinks/day for men, ≤ 1 drink/day for women	Those consuming >2–3 drinks/day
Body weight	1A	Direct	Lowering of blood volume, cardiac output, and insulin resistance, and raising salt sensitivity	Maintain healthy weight; lose weight if overweight	Overweight individuals
Dietary patterns	1A	Depends on factor	Multiple mechanisms, as above	DASH dietary pattern (see Appendix 2)	Those at risk for, or with, hypertension

[a] 1A, clear, consistent, and strong randomized controlled trial (RCT); 1B, clear RCT but inconsistent results; 1C+, clear, no RCT but strong observational studies; 1C, clear, observational studies; 2A, unclear, consistent RCT—intermediate strength; 2B, unclear, RCT with inconsistent results; 2C, unclear, observational studies.

TABLE 5 The DASH Eating Plan

Food Group	Daily Servings	Serving Sizes	Examples and Notes	Significance of Each Food Group to the DASH Eating Plan
Grains[a]	6–8	1 slice bread 1 oz dry cereal ½ cup cooked rice, pasta, or cereal[b]	Whole wheat bread and rolls, whole wheat pasta, English muffin, pita bread, bagel, cereals, grits, oatmeal, brown rice, unsalted pretzels and popcorn	Major sources of energy and fiber
Vegetables	4–5	1 cup raw leafy vegetable ½ cup cut-up raw or cooked vegetable ½ cup vegetable juice	Broccoli, carrots, collards, green beans, green peas, kale, lima beans, potatoes, spinach, squash, sweet potatoes, tomatoes	Rich source of potassium, magnesium, and fiber
Fruits	4–5	1 medium fruit ¼ cup dried fruit ½ cup fresh, frozen, or canned fruit ½ cup fruit juice	Apples, apricots, bananas, dates, grapes, oranges, grapefruit, grapefruit juice, mangoes, melons, peaches, pineapples, raisins, strawberries, tangerines	Important sources of potassium, magnesium, and fiber
Fat-free or low-fat milk and milk products	2–3	1 cup fat-free milk or yogurt 1½ oz cheese	Fat-free (skim) or low-fat (1%) milk, fat-free, low-fat, or reduced-fat cheese, fat-free or low-fat regular or frozen yogurt	Major sources of calcium and protein
Lean meats, poultry, fish	≤6	1 oz cooked meats, poultry, or fish 1 egg[c]	Select only lean; trim away visible fats; broil, roast, or poach; remove skin from poultry	Rich sources of protein and magnesium
Nuts, seeds, and legumes	4–5 per week	⅓ cup or 1½ oz nuts 2 Tbsp peanut butter 2 Tbsp or ½ oz seeds ½ cup cooked dry beans or peas	Almonds, filberts, mixed nuts, peanuts, walnuts, sunflower seeds, kidney beans, lentils, split peas	Rich sources of energy, fiber, magnesium, potassium, and protein
Fats and oils[d]	2–3	1 tsp soft margarine 1 tsp vegetable oil 1 Tbsp mayonnaise 2 Tbsp salad dressing	Soft margarine, vegetable oil (such as olive, corn, canola, or safflower), low-fat mayonnaise, light salad dressing	The DASH study had 27% of calories as fat, including fat in or added to foods
Sweets and added sugars	≤5 per week	1 Tbsp sugar 1 Tbsp jelly or jam ½ cup sorbet, gelatin 1 cup lemonade	Fruit-flavored gelatin, fruit punch, hard candy, jelly, maple syrup, sorbet and ices, sugar	Sweets should be low in fat

[a]Whole grains are recommended for most grain servings.
[b]Serving sizes vary from ½ to 1¼ cups, depending on cereal type. Check the product's Nutrition Facts label.
[c]Because eggs are high in cholesterol, limit egg yolk intake to no more than 4 per week; 1 egg white has the same protein content as 1 oz of meat.
[d]Fat content changes serving counts for fats and oils. For example, 1 Tbsp of regular salad dressing equals 1 serving; 1 Tbsp of a low-fat dressing equals ½ serving; 1 Tbsp of a fat-free dressing equals 0 servings.

TABLE 6 The DASH Dietary Pattern: Sample Menu (Based on 2000 Calories and 2300 mg Sodium per Day)

Food	Amount	Servings Provided	Substitution to Reduce Sodium to 1500 mg
Breakfast			
Orange juice	1 C	2 fruits	
Low-fat milk	1 C	1 dairy	
Bran flake cereal	¾ C	1 grain	2 cups puffed wheat cereal
Banana	1 medium	1 fruit	
Whole wheat bread	1 slice	1 grain	
Soft margarine	1 tsp	1 fats and oils	1 tsp unsalted soft (tub) margarine

(continues)

TABLE 6 *(continued)*

Food	Amount	Servings Provided	Substitution to Reduce Sodium to 1500 mg
Lunch			
Beef barbeque sandwich:			
Beef, eye of round, cooked	2 oz	2 meats	
Barbeque sauce	1 Tbsp		
Natural cheddar cheese, reduced fat	2 slices (1½ oz)	1 dairy	1½ oz natural cheddar cheese, reduced fat, low sodium
Hamburger bun	1 ea	2 grains	
Romaine lettuce	1 large leaf	¼ vegetable	
Tomato	2 slices	½ vegetable	
New potato salad	1 C	2 vegetables	
Orange	1 medium	1 fruit	
Dinner			
Herbed baked cod	3 oz	3 meats	
Brown rice	½ C	1 grain	
Spinach, sauteed from frozen with:	1 C	2 vegetables	
Canola oil	1 tsp	1 fats and oils	
Almonds, slivered	1 Tbsp	¼ nuts and seeds	
Cornbread muffin, made with oil	1 small	1 grain	
Soft margarine (tub)	1 tsp	1 fats and oils	1 tsp unsalted soft margarine
Snacks			
Fruit yogurt, fat free, no added sugar	1 C	1 dairy	
Sunflower seeds, unsalted	1 Tbsp	½ nuts and seeds	
Graham cracker rectangles	2 large	1 grain	
Peanut butter	1 Tbsp	½ nuts and seeds	

TABLE 7 Total Number of Servings in 2000 Calories/Day Menu

Food Group	Servings
Grains	= 7
Vegetables	= 4¾
Fruits	= 4
Dairy foods	= 3
Meats, poultry, and fish	= 5
Nuts, seeds, and legumes	= 1¼
Fats and oils	= 3

DASH DIETARY PATTERN AT OTHER CALORIE LEVELS

The DASH eating plan used in the studies calls for a certain number of daily servings from various food groups. The number of servings you require may vary, depending on your caloric need—Table 8 gives the servings for 1600, 2600, and 3100 calories.

TABLE 8 DASH Eating Plan: Number of Daily Servings for Other Calorie Levels

	Servings/Day		
Food Group	1600 Calories/Day	2600 Calories/Day	3100 Calories/Day
Grains[a]	6	10–11	12–13
Vegetables	3–4	5–6	6
Fruits	4	5–6	6
Fat-free or low-fat milk and milk products	2–3	3	3–4
Lean meats, poultry, and fish	3–4	6	6–7
Nuts, seeds, and legumes	3/week	1	1
Fats and oils	2	3	4
Sweets and added sugars	0	≤2	≤2

[a]Whole grains are recommended for most grain servings as a good source of fiber and other nutrients.

GETTING STARTED WITH DASH

It's easy to adopt the DASH eating plan. Here are some ways to get started:

Change gradually

- If you now eat one or two vegetables a day, add a serving at lunch and another at dinner.
  - If you don't eat fruit now or have only juice at breakfast, add a serving to your meals or have it as a snack.
  - Gradually increase your use of fat-free and low-fat milk and milk products to three servings a day. For example, drink milk with lunch or dinner, instead of soda, sugar-sweetened tea, or alcohol. Choose fat-free (skim) or low-fat (1 percent) milk and milk products to reduce your intake of saturated fat, total fat, cholesterol, and calories and increase your calcium.
  - Read Nutrition Facts label on margarines and salad dressings to choose those lowest in saturated fat. Some margarines are now *trans*-fat free.

Treat meats as one part of the whole meal, instead of the focus

- Limit lean meats to 6 ounces a day—all that's needed. Have only three ounces at a meal, which is about the size of a deck of cards.
- If you now eat large portions of meats, cut them back gradually—by a half or a third at each meal.
- Include two or more vegetarian-style (meatless) meals each week.
- Increase servings of vegetables, brown rice, whole wheat pasta, and cooked dry beans in meals. Try casseroles, whole wheat pasta, and stir-fry dishes, which have less meat and more vegetables, grains, and dry beans.

Use fruits or other foods low in saturated fat, *trans* fat, cholesterol, sodium, sugar, and calories as desserts and snacks

- Fruits and other lower-fat foods offer great taste and variety. Use fruits canned in their own juice or packed in water. Fresh fruits require little or no preparation. Dried fruits are a good choice to carry with you or to have ready in the car.
- Try these snack ideas: unsalted rice cakes or nuts mixed with raisins; graham crackers; fat-free and low-fat yogurt and frozen yogurt; popcorn with no salt or butter added; raw vegetables.

Other tips

- Choose whole grain foods for most grain servings to get added nutrients, such as minerals and fiber. For example, choose whole wheat bread or whole grain cereals.
- If you have trouble digesting milk and milk products, try taking lactase enzyme pills (available at drugstores and groceries) with the milk products. Or, buy lactose-free milk, which has the lactase enzyme already added to it.
- If you are allergic to nuts, use seeds or legumes (cooked dried beans or peas).
- Use fresh, frozen, or no-salt-added canned vegetables and fruits.

To learn more about high BP, call 1-800-575-WELL or visit the NHLBI web site at www.nhlbi.nih.gov/nhlbi/nhlbi.htm. DASH is also online at http://dash.bwh.harvard. edu.

Source: "Your guide to lowering your blood pressure with DASH," U.S. Department of Health and Human Services, National Institutes of Health, National Heart, Lung, and Blood Institute, NIH Publication No. 06–4082, revised April 2006.

References

1. Burt, V. L., *et al.* (1995). Prevalence of hypertension in the U.S. adult population. Results from the Third National Health and Nutrition Examination Survey, 1988–1991. *Hypertension* **25**, 305–313.
2. Stamler, J. (1991). Blood pressure and high blood pressure. Aspects of risk. *Hypertension* **18**(Suppl I), I-95–I-107.
3. U.S. National Center for Health Statistics. (2006). "Health, United States, 2006, with Chartbook on Trends in the Health of Americans." CDC, Hyattsville, MD.
4. Stamler, J., Neaton, J. D., and Wentworth, D. N. (1989). Blood pressure (systolic and diastolic) and risk of fatal coronary heart disease. *Hypertension* **13**(Suppl I), I-2–I-12.
5. The Seventh Report of the Joint National Committee on Prevention, Detection, Evaluation, and Treatment of High Blood Pressure. (2004). National Heart, Lung, and Blood Institute, Bethesda, MD.
6. James, G. D., and Baker, P. T. (1990). Human population biology and hypertension. Evolutionary and ecological aspects of blood pressure. *In* "Hypertension: Pathophysiology, Diagnosis and Management" (J. H. Laragh and B. M. Brenner, Eds.), pp. 137–145. Raven, New York.
7. Lowenstein, F. W. (1961). Blood-pressure in relation to age and sex in the tropics and subtropics. *Lancet* **1**, 389–392.
8. Page, L. B., Damon, A., and Moellering, R. C. (1974). Antecedents of cardiovascular disease in six Solomon Islands societies. *Circulation* **49**, 1132–1146.
9. Carvalho, J. J. M., *et al.* (1989). Blood pressure in four remote populations in the INTERSALT study. *Hypertension* **14**, 238–246.
10. Oliver, W. J., Cohen, E. L., and Neel, J. V. (1975). Blood pressure, sodium intake, and sodium related hormones in the

Yanomamo Indians, a "no-salt" culture. *Circulation* **52**, 146–151.
11. Hollenberg, N. K., *et al.* (1997). Aging, acculturation, salt intake, and hypertension in the Kuna of Panama. *Hypertension* **29**(2), 171–176.
12. Poulter, N. R., *et al.* (1990). The Kenyan Luo migration study: observations on the initiation of a rise in blood pressure. *BMJ* **300**, 967–972.
13. He, J., *et al.* (1991). Effect of migration on blood pressure: the Yi People study. *Epidemiology* **2**, 88–97.
14. Prior, I. A. M. (1974). Cardiovascular epidemiology in New Zealand and the Pacific. *N. Z. Med. J.* **80**, 245–252.
15. Eason, R. J., *et al.* (1987). Changing patterns of hypertension, diabetes, obesity, and diet among Melanesians and Micronesians in the Solomon Islands. *Med. J. Aust.* **146**, 465–473.
16. Hanna, J. M., Pelletier, D. L., and Brown, V. J. (1986). The diet and nutrition of contemporary Samoans. *In* "The Changing Samoans: Behavior and Health in Transition" (P. T. Baker, J. M. Hanna, and T. S. Baker, Eds.), pp. 275–296. Oxford University Press, Oxford.
17. Zimmet, P., Jackson, L., and Whitehouse, S. (1980). Blood pressure studies in two Pacific populations with varying degrees of modernisation. *N. Z. Med. J.* **91**, 249–252.
18. Kaufman, J. S., *et al.* (1996). Determinants of hypertension in West Africa: contribution of anthropometric and dietary factors to urban–rural and socioeconomic gradients. *Am. J. Epidemiol.* **143**(12), 1203–1218.
19. Birkett, N. J. (1998). Comments on a meta-analysis of the relation between dietary calcium intake and blood pressure. *Am. J. Epidemiol.* **148**(3), 223–228.
20. Stoto, M. A. (1998). Invited commentary on meta-analysis of epidemiologic data: the case of calcium intake and blood pressure. *Am. J. Epidemiol.* **148**(3), 229–230.
21. Appel, L., *et al.* (1997). A clinical trial of the effects of dietary patterns on blood pressure. *N. Engl. J. Med.* **336**, 1117–1124.
22. Sacks, F. M., *et al.* (2001). Effects on blood pressure of reduced dietary sodium and the Dietary Approaches to Stop Hypertension (DASH) diet. DASH-Sodium Collaborative Research Group. *N. Engl. J. Med.* **344**(1), 3–10.
23. Appel, L. J., *et al.* (2005). Effects of protein, monounsaturated fat, and carbohydrate intake on blood pressure and serum lipids: results of the OmniHeart randomized trial. *JAMA* **294**(19), 2455–2464.
24. Yamamato, M. E., *et al.* (1995). Lack of blood pressure effect with calcium and magnesium supplementation in adults with high-normal blood pressure. Results from Phase I of the Trials of Hypertension Prevention (TOHP). *Ann. Epidemiol.* **5**, 96–107.
25. Stevens, V. J., *et al.* (2001). Long-term weight loss and changes in blood pressure: results of the Trials of Hypertension Prevention, Phase II. *Ann. Intern. Med.* **134**(1), 1–11.
26. Appel, L. J., Harsha, D. W., Cooper, L. S., Obarzanek, E., Elmer, P. J., Stevens, V. J., Vollmer, W. M., Lin, P. H., Svetkey, L. P., Stedman, S. W., and Young, D. R. Writing Group of the PREMIER Collaborative Research Group. (2003). Effects of comprehensive lifestyle modification on blood pressure control: main results of the PREMIER clinical trial. *JAMA* **289**(16), 2083–2093.
27. Reed, D., *et al.* (1985). Diet, blood pressure, and multicollinearity. *Hypertension* **7**, 405–410.
28. Luft, F. C., *et al.* (1988). Genetic influences on the response to dietary salt reduction, acute salt loading, or salt depletion in humans. *J. Cardiovasc. Pharmacol.* **12**(Suppl. 3), S49–S55.
29. Sowers, J. R., *et al.* (1991). Calcium metabolism and dietary calcium in salt sensitive hypertension. *Am. J. Hypertens.* **4**, 557–563.
30. Cutler, J. A. (1999). The effects of reducing sodium and increasing potassium intake for control of hypertension and improving health. *Clin. Exp. Hypertens.* **21**(5&6), 769–783.
31. McCarron, D. A. (2000). The dietary guideline for sodium: should we shake it up? Yes. *Am. J. Clin. Nutr.* **71**, 1013–1019.
32. Kaplan, N. M. (2000). The dietary guideline for sodium: should we shake it up? No. *Am. J. Clin. Nutr.* **71**, 1020–1026.
33. Cutler, J. A., Follmann, D., and Allender, P. S. (1997). Randomized trials of sodium reduction: an overview. *Am. J. Clin. Nutr.* **65**, 643S–651S.
34. Law, M. R., Frost, C. D., and Wald, N. J. III (1991). Analysis of data from trials of salt reduction. *BMJ* **302**, 819–824.
35. Graudal, N. A., Galloe, A. M., and Garred, P. (1998). Effects of sodium restriction on blood pressure, renin, aldosterone, catecholamines, cholesterols, and triglyceride. A meta-analysis. *JAMA* **279**(17), 1383–1391.
36. Alderman, M. H., and Lamport, B. (1990). Moderate sodium restriction: do the benefits justify the hazards? *Am. J. Hypertens.* **3**, 499–504.
37. Luft, F. C. (1988). Sodium: complexities in a simple relationship. *Hosp. Pract.* 73–80.
38. Kumanyika, S. K., and Cutler, J. A. (1997). Dietary sodium reduction: Is there cause for concern? *J. Am. Coll. Nutr.* **16**(3), 192–203.
39. Pietinen, P., and Tuomilehto, J. (1980). Estimating sodium intake in epidemiological studies. *In* "Epidemiology of Arterial Blood Pressure" (H. Kestleloot and J. V. Joossens, Eds.), pp. 29–44. Springer-Kerlog, New York.
40. Dyer, A. R., Elliott, P., and Shipley, M. (1994). Urinary electrolyte excretion in 24-hours and blood pressure in the INTERSALT study. II. Estimates of electrolyte–blood pressure associations corrected for regression dilution bias. *Am. J. Epidemiol.* **139**, 940–951.
41. Frost, C. D., Law, M. R., and Wald, N. J. (1991). Analysis of observational data within populations. *BMJ* **302**, 815–818.
42. Dahl, L. (1960). Possible role of salt intake in the development of hypertension. *In* "Essential Hypertension: An International Symposium." Springer-Verlag, Berlin.
43. Elliott, P. (1991). Observational studies of salt and blood pressure. *Hypertension* **17**(suppl), I-3–I-8.
44. Intersalt Cooperative Research Group. (1988). Intersalt: an international study of electrolyte excretion and blood pressure: results for 24-hour urinary sodium and potassium excretion. *BMJ* **297**, 319–328.
45. Stamler, J. (1997). The INTERSALT study: background, methods, findings and implications. *Am. J. Clin. Nutr.* **65**(suppl), 626S–642S.
46. Law, M. R., Frost, C. D., and Wald, N. J. (1991). By how much does dietary salt reduction lower blood pressure? I—Analysis of observational data among populations. *BMJ* **302**, 811–814.

47. Kempner, W. (1948). Treatment of hypertensive vascular disease with rice diet. *Am. J. Med.* **4**, 545–577.
48. Midgley, J. P., *et al.* (1996). Effect of reduced dietary sodium on blood pressure. *JAMA* **275**(20), 1590–1597.
49. He, F. J., and MacGregor, G. A. (2004). Effect of longer-term modest salt reduction on blood pressure. *Cochrane Database Syst. Rev.* **2004**(3), CD004937.
50. Hooper, L., *et al.* (2002). Systematic review of long term effects of advice to reduce dietary salt in adults. *BMJ* **325**, 628–632.
51. Chobanian, A. V., and Hill, M. (2000). National Heart, Lung, and Blood Institute workshop on sodium and blood pressure. A critical review of current scientific evidence. *Hypertension* **35**, 858–863.
52. Whelton, P. K., *et al.* (1998). Sodium reduction and weight loss in the treatment of hypertension in older persons: a randomized controlled trial of nonpharmacologic interventions in the elderly (TONE). TONE Collaborative Research Group. *JAMA* **279**(11), 839–846.
53. The Trials of Hypertension Prevention Collaborative Research Group. (1997). Effects of weight loss and sodium reduction intervention on BP and hypertension incidence in overweight people with high-normal blood pressure: the Trials of Hypertension Prevention, Phase II. *Arch. Intern. Med.* **157**(6), 657–667.
54. Weinberger, M. H., *et al.* (1986). Definitions and characteristics of sodium sensitivity and blood pressure resistance. *Hypertension* **8**(Supp II), II-127–II-134.
55. Kawasaki, T., *et al.* (1978). The effect of high-sodium and low-sodium intakes on blood pressure and other related variables in human subjects with idiopathic hypertension. *Am. J. Med.* **64**, 193–198.
56. Fujita, T., *et al.* (1980). Factors influencing blood pressure in salt-sensitive patients with hypertension. *Am. J. Med.* **69**, 334–344.
57. Luft, F. C., and Weinberger, M. H. (1997). Heterogeneous responses to changes in dietary salt intake: the salt-sensitivity paradigm. *Am. J. Clin. Nutr.* **65**(suppl), 612S–617S.
58. Obarzanek, E., *et al.* (2003). Individual blood pressure responses to changes in salt intake: results from the DASH-Sodium trial. *Hypertension* **42**(4), 459–467.
59. Falkner, B. (1988). Sodium sensitivity: a determinant of essential hypertension. *J. Am. Coll. Nutr.* **7**(1), 35–41.
60. Williams, G. H., and Hollenberg, N. K. (1985). Sodium-sensitive essential hypertension. Emerging insights into pathogenesis and therapeutic implications. *In* "Contemporary Nephrology" (S. Klahr and S. G. Massry, Eds.), pp. 303–331. Plenum, New York.
61. Zemel, M. B., *et al.* (1988). Erythrocyte cation metabolism in salt-sensitive hypertensive blacks as affected by dietary sodium and calcium. *Am. J. Hypertens.* **1**, 386–92.
62. Kotchen, T. A., and Kotchen, J. M. (1997). Dietary sodium and blood pressure: interactions with other nutrients. *Am. J. Clin. Nutr.* **65**, 708S–711S.
63. Svetkey, L. P., *et al.* (2001). Angiotensinogen genotype and blood pressure response in the Dietary Approaches to Stop Hypertension (DASH) study. *J. Hypertens.* **19**(11), 1949–1956.
64. Hunt, S. C., *et al.* (1998). Angiotensinogen genotype, sodium reduction, weight loss, and prevention of hypertension. Trials of Hypertension Prevention, Phase II. *Hypertension* **32**, 393–401.
65. Kumanyika, S. K., *et al.* (1993). Feasibility and efficacy of sodium reduction in the Trials of Hypertension Prevention, Phase I. *Hypertension* **22**, 502–512.
66. National Heart, Lung, and Blood Institute (1996). "Implementing Recommendations for Dietary Salt Reduction," pp. 1–28. National Institutes of Health, Bethesda, MD.
67. Ray, P. E., *et al.* (2001). Chronic potassium depletion induces renal injury, salt sensitivity, and hypertension in young rats. *Kidney Int.* **59**(5), 1850–1858.
68. Zhou, M. S., Kosaka, H., and Yoneyama, H. (2000). Potassium augments vascular relaxation mediated by nitric oxide in the carotid arteries of hypertensive Dahl rats. *Am. J. Hypertens.* **13**(6 Pt 1), 666–672.
69. Luft, F. C., *et al.* (1986). Effects of volume expansion and contraction on potassium homeostasis in normal and hypertensive humans. *J. Am. Coll. Nutr.* **5**, 357–369.
70. Pamnani, M. B., *et al.* (2000). Mechanism of antihypertensive effect of dietary potassium in experimental volume expanded hypertension in rats. *Clin. Exp. Hypertens.* **22**(6), 555–569.
71. Hajjar, I. M., *et al.* (2001). Impact of diet on blood pressure and age-related changes in blood pressure in the U.S. population: analysis of NHANES III. *Arch. Intern. Med.* **161**(4), 589–593.
72. Ophir, O., *et al.* (1983). Low blood pressure in vegetarians: the possible role of potassium. *Am. J. Clin. Nutr.* **37**, 755–762.
73. Whelton, P. K., *et al.* (1997). Effects of oral potassium on blood pressure. Meta-analysis of randomized controlled clinical trials. *JAMA* **277**(20), 1624–1632.
74. Cappuccio, F. P., and MacGregor, G. A. (1991). Does potassium supplementation lower blood pressure? A meta-analysis of published trials. *J. Hypertens.* **9**, 465–473.
75. Sacks, F. M., *et al.* (1998). Effect on blood pressure of potassium, calcium, and magnesium in women with low habitual intake. *Hypertension* **31**(1), 131–138.
76. Geleijnse, J. M., Kok, F. J., and Grobbee, D. E. (2003). Blood pressure response to changes in sodium and potassium intake: a metaregression analysis of randomised trials. *J. Hum. Hypertens.* **17**(7), 471–480.
77. Dickinson, H. O., *et al.* (2006). Potassium supplementation for the management of primary hypertension in adults. *Cochrane Database Syst. Rev.* **3**, CD004641.
78. Crawford, M. D., Gardner, M. J., and Morris, J. N. (1968). Mortality and hardness of local water supplies. *Lancet* **1**, 827–831.
79. Cutler, J. A., and Brittain, E. (1990). Calcium and blood pressure. An epidemiologic perspective. *Am. J. Hypertens.* **3**, 137S–146S.
80. Cappuccio, F. P., *et al.* (1995). Epidemiologic association between dietary calcium intake and blood pressure: a meta-analysis of published data. *Am. J. Epidemiol.* **142**(9), 935–945.
81. Hajjar, I. M., Grim, C. E., and Kotchen, T. A. (2003). Dietary calcium lowers the age-related rise in blood pressure in the United States: the NHANES III survey. *J. Clin. Hypertens. (Greenwich)* **5**(2), 122–126.
82. Schroder, H., Schmelz, E., and Marrugat, J. (2002). Relationship between diet and blood pressure in a representative Mediterranean population. *Eur. J. Nutr.* **41**(4), 161–167.

83. Morikawa, Y., *et al.* (2002). A cross-sectional study on association of calcium intake with blood pressure in Japanese population. *J. Hum. Hypertens.* **16**(2), 105–110.
84. Djousse, L., *et al.* (2006). Influence of saturated fat and linolenic acid on the association between intake of dairy products and blood pressure. *Hypertension* **48**(2), 335–341.
85. Jorde, R., and Bonaa, K. H. (2000). Calcium from dairy products, vitamin D intake, and blood pressure: the Tromso Study. *Am. J. Clin. Nutr.* **71**(6), 1530–1535.
86. Alonso, A., Ruiz-Gutierrez, V., and Martinez-Gonzalez, M. A. (2006). Monounsaturated fatty acids, olive oil and blood pressure: epidemiological, clinical and experimental evidence. *Public Health Nutr.* **9**(2), 251–257.
87. Addison, W. L. T., and Clark, H. G. (1924). Calcium and potassium chlorides in the treatment of arterial hypertension. *Can. Med. Assoc. J.* **15**, 913–915.
88. Cappuccio, F. P., Siani, A., and Strazzullo, P. (1989). Oral calcium supplementation and blood pressure: an overview of randomized controlled trials. *J. Hypertens.* **7**, 941–946.
89. Bucher, H. C., *et al.* (1986). Effects of dietary calcium supplementation on blood pressure. *JAMA* **275**(13), 1016–1022.
90. Allender, P. S., *et al.* (1996). Dietary calcium and blood pressure: a meta-analysis of randomized clinical trials. *Ann. Intern. Med.* **124**, 825–831.
91. Griffith, L. E., *et al.* (1999). The influence of dietary and nondietary calcium supplementation on blood pressure. An updated metaanalysis of randomized controlled trials. *Am. J. Hypertens.* **12**, 84–92.
92. van Mierlo, L. A., *et al.* (2006). Blood pressure response to calcium supplementation: a meta-analysis of randomized controlled trials. *J. Hum. Hypertens.* **20**(8), 571–580.
93. Dickinson, H. O., *et al.* (2006). Calcium supplementation for the management of primary hypertension in adults. *Cochrane Database Syst. Rev.* **2006**(2), CD004639.
94. Cappuccio, F. P. (1999). The "calcium antihypertension theory." *Am. J. Hypertens.* **12**, 93–95.
95. Kynast-Gales, S. A., and Massey, L. K. (1992). Effects of dietary calcium from dairy products on ambulatory blood pressure in hypertensive men. *J. Am. Diet. Assoc.* **92**, 1497–1501.
96. Reid, I. R., *et al.* (2005). Effects of calcium supplementation on body weight and blood pressure in normal older women: a randomized controlled trial. *J. Clin. Endocrinol. Metab.* **90**(7), 3824–3829.
97. Nowson, C. A., *et al.* (2004). Blood pressure response to dietary modifications in free-living individuals. *J. Nutr.* **134**(9), 2322–2329, 2004.
98. Jorde, R., *et al.* (2002). The effects of calcium supplementation to patients with primary hyperparathyroidism and a low calcium intake. *Eur. J. Nutr.* **41**(6), 258–263.
99. Elmarsafawy, S. F., *et al.* (2006). Dietary calcium as a potential modifier of the relationship of lead burden to blood pressure. *Epidemiology* **17**(5), 531–637.
100. Moore, T. J. (1989). The role of dietary electrolytes in hypertension. *J. Am. Coll. Nutr.* **8**(S), 1–12.
101. Touyz, R. M., *et al.* (2002). Effects of low dietary magnesium intake on development of hypertension in stroke-prone spontaneously hypertensive rats: role of reactive oxygen species. *J. Hypertens.* **20**(11), 221–2232.
102. Blache, D., *et al.* (2006). Long-term moderate magnesium-deficient diet shows relationships between blood pressure, inflammation and oxidant stress defense in aging rats. *Free Radic. Biol. Med.* **41**(2), 277–284.
103. Kisters, K., *et al.* (2004). Early-onset increased calcium and decreased magnesium concentrations and an increased calcium/magnesium ratio in SHR versus WKY. *Magnes. Res.* **17**(4), 264–269.
104. Yang, C. Y., and Chin, H. F. (1999). Calcium and magnesium in drinking water and risk of death from hypertension. *Am. J. Hypertens.* **12**(9), 894–899.
105. Joffres, M. R., Reed, D. M., and Yano, K. (1987). Relationship of magnesium intake and other dietary factors to blood pressure: the Honolulu heart study. *Am. J. Clin. Nutr.* **45**, 469–475.
106. Ascherio, A., *et al.* (1992). A prospective study of nutritional factors and hypertension among U.S. men. *Circulation* **86**, 1475–1484.
107. Witteman, J. C. M., *et al.* (1989). A prospective study of nutritional factors and hypertension among U.S. women. *Circulation* **80**, 1320–1327.
108. Song, Y., *et al.* (2006). Dietary magnesium intake and risk of incident hypertension among middle-aged and older U.S. women in a 10-year follow-up study. *Am. J. Cardiol.* **98**(12), 1616–21, 2006.
109. Dyckner, T., and Wester, P. O. (1983). Effect of magnesium on blood pressure. *BMJ* **286**, 1847–1849.
110. Reyes, A. J., *et al.* (1984). Magnesium supplementation in hypertension treated with hydrochlorothiazide. *Curr. Ther. Res.* **36**(2), 332–340.
111. Cappuccio, F. P., *et al.* (1985). Lack of effect of oral magnesium on high blood pressure: a double blind study. *BMJ* **291**, 235–238.
112. Henderson, D. G., Schierup, J., and Schodt, T. (1986). Effect of magnesium supplementation on blood pressure and electrolyte concentrations in hypertensive patients receiving long term diuretic treatment. *BMJ* **293**, 664–665.
113. Nowson, C. A., and Morgan, T. O. (1989). Magnesium supplementation in mild hypertensive patients on a moderately low sodium diet. *Clin. Exp. Pharm. Physiol.* **16**, 299–302.
114. Zemel, P. C., *et al.* (1990). Metabolic and hemodynamic effects of magnesium supplementation in patients with essential hypertension. *Am. J. Clin. Nutr.* **51**, 665–669.
115. Wirell, M. P., Wester, P. O., and Stegmayr, B. G. (1994). Nutritional dose of magnesium in hypertensive patients on beta blockers lowers systolic blood pressure: a double-blind, cross-over study. *J. Int. Med.* **236**, 189–195.
116. Witteman, J. C. M., *et al.* (1994). Reduction of blood pressure with oral magnesium supplementation in women with mild to moderate hypertension. *Am. J. Clin. Nutr.* **60**, 129–135.
117. Sacks, F. M., *et al.* (1995). Combinations of potassium, calcium, and magnesium supplements in hypertension. *Hypertension* **26**(1), 950–956.
118. Kawano, Y., *et al.* (1998). Effects of magnesium supplementation in hypertensive patients. Assessment by office, home, and ambulatory blood pressures. *Hypertension* **32**, 260–265.
119. Lind, L., *et al.* (1991). Blood pressure response during long-term treatment with magnesium is dependent on magnesium

status. A double-blind, placebo-controlled study in essential hypertension and in subjects with high-normal blood pressure. *Am. J. Hypertens.* **4**, 674–679.
120. Dickinson, H. O., *et al.* (2006). Magnesium supplementation for the management of essential hypertension in adults. *Cochrane Database Syst. Rev.* **2006**(3), CD004640.
121. Song, Y., *et al.* (2006). Effects of oral magnesium supplementation on glycaemic control in type 2 diabetes: a meta-analysis of randomized double-blind controlled trials. *Diabetes Med.* **23**(10), 1050–1056.
122. Yamori, Y., *et al.* (1981). Genetics of hypertensive diseases: experimental studies on pathogenesis, detection of predisposition and prevention. *Adv. Nephrol.* **10**, 51–74.
123. Yamori, Y., *et al.* (1984). Dietary prevention of hypertension in animal models and its applicability to humans. *Ann. Clin. Res.* **43**, 28–31.
124. He, J., *et al.* (1995). Dietary macronutrients and blood pressure in southwestern China. *J. Hypertens.* **13**(11), 1267–1274.
125. Palloshi, A., *et al.* (2004). Effect of oral L-arginine on blood pressure and symptoms and endothelial function in patients with systemic hypertension, positive exercise tests, and normal coronary arteries. *Am. J. Cardiol.* **93**(7), 933–935.
126. Obarzanek, E., Velletri, P., and Cutler, J. (1996). Dietary protein and blood pressure. *JAMA* **275**(20), 1598–1603.
127. Yamori, Y., *et al.* (1981). Hypertension and diet: multiple regression analysis in a Japanese farming community. *Lancet* **1**(8231), 1204–1205.
128. Zhou, B. F., *et al.* (1989). Dietary patterns in 10 groups and the relationship with blood pressure. Collaborative Study Group for Cardiovascular Diseases and Their Risk Factors. *Chin. Med. J. (Engl.)* **102**(4), 257–261.
129. Stamler, J., *et al.* (2002). Eight-year blood pressure change in middle-aged men: relationship to multiple nutrients. *Hypertension* **39**(5), 1000–1006.
130. Stamler, J., Caggiula, A. W., and Grandits, G. A. (1997). Relation of body mass and alcohol, nutrient, fiber, and caffeine intakes to blood pressure in the special intervention and usual care groups in the Multiple Risk Factor Intervention Trial. *Am. J. Clin. Nutr.* **65**(1 Suppl), 338S–365S.
131. Hodgson, (2006). Partial substitution of carbohydrate intake with protein intake from lean red meat lowers blood pressure in hypertensive persons. *Am. J. Clin. Nutr.* **83**, 780–787.
132. Wang, Y., *et al.* The relationship between dietary protein intake and blood pressure results from the PREMIER study. Submitted for publication.
133. Moncada, S., and Higgs, A. (1993). The L-arginine-nitric oxide pathway. *N. Engl. J. Med.* **329**, 2002–2012.
134. Fisher, N. D., and Hollenberg, N. K. (2005). Flavanols for cardiovascular health: the science behind the sweetness. *J. Hypertens.* **23**(8), 1453–1459.
135. Rivas, M., *et al.* (2002). Soy milk lowers blood pressure in men and women with mild to moderate essential hypertension. *J. Nutr.* **132**(7), 1900–1902.
136. Teede, H. J., *et al.* (2006). Randomised, controlled, crossover trial of soy protein with isoflavones on blood pressure and arterial function in hypertensive subjects. *J. Am. Coll. Nutr.* **25**(6), 533–540.
137. Kreijkamp-Kaspers, S., *et al.* (2005). Randomized controlled trial of the effects of soy protein containing isoflavones on vascular function in postmenopausal women. *Am. J. Clin. Nutr.* **81**(1), 189–195.
138. Sagara, M., *et al.* (2003). Effects of dietary intake of soy protein and isoflavones on cardiovascular disease risk factors in high risk, middle-aged men in Scotland. *J. Am. Coll. Nutr.* **23**(1), 85–91.
139. Stamler, J., *et al.* (1996). Inverse relation of dietary protein markers with blood pressure. Findings for 10,020 men and women in the INTERSALT Study. *Circulation* **94**, 1629–1634.
140. Elliott, P., *et al.* (1987). Diet, alcohol, body mass, and social factors in relation to blood pressure: the Caerphilly Heart Study. *J. Epidemiol. Commun. Health* **41**, 37–43.
141. Salonen, J., Tuomilehto, J., and Tanskanen, A. (1983). Relation of blood pressure to reported intake of salt, saturated fats, and alcohol in healthy middle-aged population. *J. Epidemiol. Commun. Health* **37**, 32–37.
142. Salonen, J., *et al.* (1988). Blood pressure, dietary fats, and antioxidants. *Am. J. Clin. Nutr.* **48**, 1226–1232.
143. Gruchow, H., Sobocinski, K., and Barboriak, J. (1985). Alcohol, nutrient intake and hypertension in U.S. adults. *JAMA* **253**, 1567–1570.
144. Hajjar, I., and Kotchen, T. (2003). Regional variations of blood pressure in the United States are associated with regional variations in dietary intakes: The NHANES-III data. *J. Nutr.* **133**(1), 211–214.
145. Jakulj, F., *et al.* (2007). A high-fat meal increases cardiovascular reactivity to psychological stress in healthy young adults. *J. Nutr.* **137**(4), 935–939.
146. Ferrara, L. A., *et al.* (2000). Olive oil and reduced need for antihypertensive medications. *Arch. Intern. Med.* **160**(6), 837–842.
147. Geleijnse, J. M., Grobbee, D. E., and Kok, F. J. (2005). Impact of dietary and lifestyle factors on the prevalence of hypertension in Western populations. *J. Hum. Hypertens.* **19**(Suppl 3), S1–4.
148. Kopp, W. (2005). Pathogenesis and etiology of essential hypertension: role of dietary carbohydrate. *Med. Hypotheses* **64**(4), 782–787.
149. Yang, E., *et al.* (2003). Carbohydrate intake is associated with diet quality and risk factors for cardiovascular disease in U.S. adults: NHANES III. *J. Am. Coll. Nutr.* **22**(1), 71–79.
150. Wood, R. J., *et al.* (2007). Effects of a carbohydrate-restricted diet with and without supplemental soluble fiber on plasma low-density lipoprotein cholesterol and other clinical markers of cardiovascular risk. *Metabolism* **56**(1), 58–67.
151. Hodges, R., and Rebello, T. (1983). Carbohydrates and blood pressure. *Ann. Intern. Med.* **98**(Part 2), 838–841.
152. Palumbo, P. J., *et al.* (1977). Sucrose sensitivity of patients with coronary-artery disease. *Am. J. Clin. Nutr.* **30**, 394–401.
153. Surwit, R., *et al.* (1997). Metabolic and behavioral effects of a high-sucrose diet during weight loss. *Am. J. Clin. Nutr.* **65**, 908–915.
154. Koh, E., Ard, N., and Mendoza, F. (1988). Effects of fructose feeding on blood parameters and blood pressure in impaired glucose-tolerant subjects. *J. Am. Diet. Assoc.* **88**, 932–938.

155. Hallfrisch, J., Reiser, S., and Prather, E. (1983). Blood lipid distribution of hyperinsulinemic men consuming three levels of fructose. *Am. J. Clin. Nutr.* **37**, 740–748.
156. Jansen, R., *et al.* (1987). Blood pressure reduction after oral glucose loading and its relation to age, blood pressure and insulin. *Am. J. Cardiol.* **60**, 1087–1091.
157. He, J., and Whelton, P. K. (1999). Effect of dietary fiber and protein intake on blood pressure: a review of epidemiologic evidence. *Clin. Exp. Hypertens.* **21**(5&6), 785–796.
158. Ascerio, A., *et al.* (1991). Nutrient intakes and blood pressure in normotensive males. *Int. J. Epidemiol.* **20**(4), 886–891.
159. Lairon, D., *et al.* (2005). Dietary fiber intake and risk factors for cardiovascular disease in French adults. *Am. J. Clin. Nutr.* **82**(6), 1185–1194.
160. Streppel, M. T., *et al.* (2005). Dietary fiber and blood pressure: a meta-analysis of randomized placebo-controlled trials. *Arch. Intern. Med.* **165**(2), 150–156.
161. Piano, M. R. (2005). The cardiovascular effects of alcohol: the good and the bad: how low-risk drinking differs from high-risk drinking. *Am. J. Nurs.* **105**(7), 89–91, 87.
162. MacMahon, S. (1987). Alcohol consumption and hypertension. *Hypertension* **9**, 111–121.
163. Cushman, W. C., *et al.* (1994). Prevention and treatment of hypertension study (PATHS). Rationale and design. *Am. J. Hypertens.* **7**, 814–823.
164. Klatsky, A. L., Friedman, G. D., and Armstrong, M. A. (1986). The relationships between alcoholic beverage use and other traits to blood pressure: a new Kaiser-Permanente study. *Circulation* **73**, 628–636.
165. Witteman, J. C., *et al.* (1990). Relation of moderate alcohol consumption and risk of systemic hypertension in women. *Am. J. Cardiol.* **65**(9), 633–637.
166. Gordon, T., and Doyle, J. T. (1986). Alcohol consumption and its relationship to smoking, weight, blood pressure, and blood lipids. *Arch. Int. Med.* **146**, 262–265.
167. Fuchs, F., *et al.* (2001). Alcohol consumption and the incidence of hypertension: the Atherosclerosis Risk in Communities Study. *Hypertension* **37**(5), 1242–1250.
168. Cushman, W. C., *et al.* (1998). Prevention and treatment of hypertension study (PATHS): Effects of an alcohol treatment program on blood pressure. *Arch. Intern. Med.* **158**, 1197–1207.
169. Wallace, P., Cutler, S., and Haines, A. (1988). Randomised controlled trial of general practitioner intervention in patients with excessive alcohol consumption. *BMJ* **297**, 663–668.
170. Puddey, I. B., Beilin, L. J., and Vandongen, R. (1987). Regular alcohol use raises blood pressure in treated hypertensive subjects. *Lancet*, 647–651.
171. Puddey, I. B., *et al.* (1992). Effects of alcohol and caloric restrictions on blood pressure and serum lipids in overweight men. *Hypertension* **20**, 533–541.
172. Xin, X., *et al.* (2001). Effects of alcohol reduction on blood pressure: a meta-analysis of randomized controlled trials. *Hypertension* **38**(5), 1112–1117.
173. Rimm, E. B., *et al.* (1991). A prospective study of alcohol consumption and the risk of coronary disease in men. *Lancet* **338**, 464–468.
174. Beilin, L. J. (1995). Alcohol, hypertension and cardiovascular disease. *J. Hypertens.* **13**, 939–942.
175. Armstrong, B., Van Merwyk, A. J., and Coates, H. (1977). Blood pressure in Seventh-Day Adventist vegetarians. *Am. J. Epidemiol.* **105**(5), 444–449.
176. Sacks, F. M., Rosner, B., and Kass, E. H. (1974). Blood pressure in vegetarians. *Am. J. Epidemiol.* **100**(5), 390–398.
177. Burr, M. L., *et al.* (1981). Plasma cholesterol and blood pressure in vegetarians. *J. Hum. Nutr.* **35**, 437–441.
178. Beilin, L. J., *et al.* (1988). Vegetarian diet and blood pressure levels: incidental or causal association? *Am. J. Clin. Nutr.* **48**, 806–810.
179. Margetts, B. M., *et al.* (1986). Vegetarian diet in mild hypertension: a randomised controlled trial. *BMJ* **293**, 1468–1471.
180. Rouse, I. L., *et al.* (1983). Blood-pressure-lowering effect of a vegetarian diet: controlled trial in normotensive subjects. *Lancet* i: 5–10.
181. Sacks, F., *et al.* (1995). Rationale and design of the Dietary Approaches to Stop Hypertension trial. *Ann. Epidemiol.* **5**, 108–118.
182. Vogt, T., *et al.* (1999). Dietary Approaches to Stop Hypertension: Rationale, design, and methods. *J. Am. Diet. Assoc.* **99**(suppl), S12–S18.
183. Materson, B. J., *et al.* (1993). Single-drug therapy for hypertension in men. A comparison of six antihypertensive agents with placebo. The Department of Veterans Affairs Cooperative Study Group on Antihypertensive Agents. *N. Engl. J. Med.* **328**(13), 914–921.
184. Obarzanek, E., *et al.* (2001). Effects on blood lipids of a blood pressure-lowering diet: the Dietary Approaches to Stop Hypertension (DASH) trial. *Am. J. Clin. Nutr.* **74**(1), 80–89.
185. Dustan, H. P., and Weinsier, R. L. (1991). Treatment of obesity-associated hypertension. *Ann. Epidemiol.* **1**, 371–379.
186. He, J., *et al.* (1999). Dietary sodium intake and subsequent risk of cardiovascular disease in overweight adults. *JAMA* **282**(21), 2027–2034.
187. Rocchini, A. P., *et al.* (1989). The effect of weight loss on the sensitivity of blood pressure to sodium in obese adolescents. *N. Engl. J. Med.* **321**(9), 580–585.
188. McKnight, J. A., and Moore, T. J. (1994). The effects of dietary factors on blood pressure. *Comp. Ther.* **20**(9), 511–517.
189. Spiegelman, D., *et al.* (1992). Absolute fat mass, percent body fat, and body-fat distribution: which is the real determinant of blood pressure and serum glucose? *Am. J. Clin. Nutr.* **55**, 1033–1044.
190. Stamler, J. (1991). Epidemiologic findings on weight and blood pressure in adults. *Ann. Epidemiol.* **1**, 347–362.
191. Harlan, W. R., *et al.* (1984). Blood pressure and nutrition in adults. The National Health and Nutrition Examination Survey. *Am. J. Epidemiol.* **120**(1), 17–28.
192. Ford, E. S., and Cooper, R. S. (1991). Risk factors for hypertension in a national cohort study. *Hypertension* **18**(5), 598–606.
193. Okosun, I. S., Prewitt, T. E., and Cooper, R. S. (1999). Abdominal obesity in the United States: prevalence and attributable risk of hypertension. *J. Hum. Hypertens.* **13**(7), 425–430.

194. The Trials of Hypertension Prevention Collaborative Research Group. (1992). The effects of nonpharmacologic interventions on blood pressure of persons with high normal levels. Results of the Trials of Hypertension, Phase I. *JAMA* **267**, 1213–1220.
195. Stevens, V. J., *et al.* (1993). Weight loss intervention in Phase I of the trials of hypertension prevention. *Arch. Intern. Med.* **153**, 849–858.
196. He, J., *et al.* (2000). Long-term effects of weight loss and dietary sodium reduction on incidence of hypertension. *Hypertension* **35**(2), 544–549.
197. Neter, J. E., *et al.* (2003). Influence of weight reduction on blood pressure: a meta-analysis of randomized controlled trials. *Hypertension* **42**(5), 878–884.
198. Staessen, J., Fagard, R., and Amery, A. (1988). The relationship between body weight and blood pressure. *J. Hum. Hypertens.* **2**(4), 207–217.
199. Ebrahim, S. (1998). Detection, adherence and control of hypertension for the prevention of stroke: a systematic review. *Health Technol. Assess.* **2**(11), i–iv, 1–78.
200. Mulrow, C. D., *et al.* (2000). Dieting to reduce body weight for controlling hypertension in adults. *Cochrane Database Syst. Rev.* **2000**(2), CD000484.
201. Cutler, J. (1991). Randomized clinical trials of weight reduction in nonhypertensive persons. *Ann. Epidemiol.* **1**, 363–370.
202. Elmer, P., *et al.* (2006). Effects of comprehensive lifestyle modification on diet, weight, physical fitness, and blood pressure control: 18 month results of a randomized trial. *Ann. Intern. Med.* **144**(7), 485–495.
203. Rosamond, W., *et al.* (2007). Heart disease and stroke statistics—2007 update: a report from the American Heart Association Statistics Committee and Stroke Statistics Subcommittee. *Circulation* **115**(5), e69–171.
204. Casaqrande, S., *et al.* (2007). Have Americans increased their fruit and vegetable intake? The trends between 1988 and 2002. *Am. J. Prev. Med.* **32**(4), 257–263.
205. U.S. Department of Agriculture. (2001). Nutrient intakes: mean amount consumed per individual, one day, Retrieved May 4, 2007, from http://www.ars.usda.gov/SP2UserFiles/Place/12355000/pdf/Table_1_BIA.pdf.
206. Bandura, A. (1986). "Social Foundations of Thought and Action: a Social Cognitive Theory." Prentice-Hall, Englewood Cliffs, NJ.
207. Watson, D., and Tharp, R. (2002). "Self-Directed Behavior: Self-Modification for Personal Adjustment" (8th ed). Wadsworth, Belmont, CA.
208. Prochaska, J., and DiClemente, C. (1983). Stages and processes of self-change of smoking: toward an integrative model of change. *J. Consult. Clin. Psychol.* **51**, 390–395.
209. Prochaska, J., *et al.* (1994). Stages of change and decisional balance for 12 problem behaviors. *Health Psychol.* **13**(1), 39–46.
210. Miller, W., and Rollnick, S. (1991). "Motivational Interviewing: preparing People to Change Addictive Behavior." Guilford, New York.
211. Rollnick, S., Mason, P., and Butler, C. (1999). "Health Behavior Change: a Guide for Practitioners." Churchill Livingstone, London.
212. Bandura, A. (1997). The anatomy of stages of change. *Am. J. Health Promot.* **12**(1), 8–10.

D. Diabetes Mellitus

CHAPTER **31**

Obesity and the Risk for Diabetes

REJEANNE GOUGEON
Royal Victoria Hospital, Crabtree Laboratory, Montreal, Canada

Contents

I. INTRODUCTION

One of the medical consequences of obesity is development of type 2 diabetes mellitus. Epidemiological studies, both cross-sectional [1–12] and prospective [8, 13–24], show a positive relationship between degree of obesity, notably of central adiposity [7, 11, 15, 16, 23, 24], and the risk for diabetes. Two large prospective studies have examined the impact of obesity on the incidence of diabetes in women [25] and in men [22] and calculated that 77% of new cases in women and 64% in men could be prevented if body mass index [BMI, weight (kg)/height (m^2)] was maintained below 25 kg/m^2. A follow-up study in a middle-aged Japanese population found that increases in BMI of 1, even within nonobese levels, increased the risk for diabetes by about 25% [26]. In the EPIC-Potsdam study, weight gain between ages 25 and 40 was associated with higher risk of diabetes than weight gain in later life [27]. Although being obese and inactive contributed independently to the development of diabetes, the magnitude of risk associated with adiposity was much greater than with lack of physical activity [24]. An increase in the prevalence of obesity in certain countries may well explain the concurrent and projected increase in diabetes prevalence [28–30] and coining the term *diabesity pandemic* [31]. Furthermore, body fat mass has also been shown to be associated with an increase in the risk of prediabetic conditions such as glucose intolerance and insulin resistance [6].

Despite these strong associations with diabetes, obesity does not appear to be an essential condition for type 2 diabetes to express itself in a genetically predisposed person. Indeed, 20–25% of type 2 diabetic persons are not obese [25, 32, 33] and 80% of individuals with elevated BMI and indicators of high intra-abdominal adiposity do not develop diabetes [25]. Still, the delaying of diabetes or improvement of its control with weight loss [33–39] indicate that obesity has an impact on diabetes, its prevention, and its management.

This chapter (1) provides the current definitions and diagnostic criteria of obesity and diabetes, (2) examines some of the evidence that suggests a major role for excess adiposity in the etiology of diabetes and its complications, (3) examines some of the mechanisms that relate increased adiposity to diabetes, and (4) describes the contributions of weight loss and energy restriction in the treatment of the obese person at risk for diabetes, particularly through lifestyle interventions [40].

II. DEFINITIONS AND CLASSIFICATIONS OF OBESITY AND DIABETES

A. Obesity

Obesity has been explained as the result of an imbalance between the intake of energy substrates and energy utilization [41]. This imbalance promotes the shunting of substrates into anabolic pathways for synthesis and storage of fat [41]. Obesity is referred to as a condition when fat accumulation is excessive to an extent that it increases risk of ill health [42], especially if it is stored in the abdominal region [43]. The relationship between BMI and adverse health outcomes exists in all age groups, including those older than 75 years [44]. A World Health Organization (WHO) expert committee has proposed cutoff points for the classification of overweight and obesity [45] such that it would be possible to identify individuals or groups at risk and to compare weight status among populations. A classification can also provide a basis for the evaluation of interventions. BMI has been used to classify obesity in populations because it correlates with percentage of body fat and with mortality and morbidity. Furthermore, the values are the same for both sexes because the relationship between BMI and mortality is similar in men and women. Because women have a higher percentage of body fat for

TABLE 1 Cutoff Points for Body Mass Index [45]

BMI (kg/m^2)	WHO Classification of Overweight in Adults According to BMI
<18.5	Underweight
18.5–24.9	Normal range
25.0–29.9	Overweight, preobese at increased risk
30.0–34.9	Obese class I at high risk
35.0–39.9	Obese class II at very high risk
≥40.0	Obese class III at extremely high risk

a comparable weight than men, this indicates that at least premenopausal women can carry fat better than men. It has been suggested that they do so because their excess fat is mainly subcutaneous and peripherally distributed (thighs, buttocks, breasts) compared with men in whom fat is stored in the abdominal region [46].

The BMI classification from WHO (Table 1) is based on the relationship between BMI and mortality. BMIs between 18.5 and 24.9 kg/m^2 are referred to as normal, between 30 and 39 as obese, and above 40 as morbidly obese [45, 46]]

A BMI of 30 kg/m^2 does not always correspond to excess adiposity. A muscle builder may have a BMI of 30 that is associated with a large muscle mass. Ethnic groups with deviating body proportions, such as being very tall and thin, have healthy BMI ranging from 17 to 22 and have excessive fat mass at a BMI of 25 kg/m^2 [47].

Cutoff points have also been defined for waist circumference as a surrogate marker of abdominal fat mass [48], a better predictor of intra-abdominal adipose tissue than BMI and a strong predictor of metabolic comorbidities [49]. Waist circumference of 94 cm or more in men (37 inches) and 80 cm (32 inches) or more in women indicates a need for concern [50] even in persons with a normal BMI, and 102 cm (40 inches) or more in men and 88 cm (35 inches) or more in women a need for action and intervention [48].

The waist circumference is measured at the narrowest part of the torso, at mid-point between the lower border of the rib cage and the iliac crest, with a non-stretchable tape measure, placed horizontal to the floor. It is considered a useful tool for initial screening and follow-up assessment of change. Wang *et al.* [23] have found that waist circumferences of 94 cm in men already predicted type 2 diabetes and suggested that a lower cutoff of 94 cm for critical level would be more appropriate than 102 cm. However, increased waist girths have been seen to relate to excess abdominal subcutaneous fat rather than intra-abdominal fat in very obese persons with a normal risk-factor profile [51]. More recently, it has been suggested that accumulation of intra-abdominal fat may be a marker for a relative inability of subcutaneous fat to store excess energy intake; the overflow would also find its way in ectopic sites such as the liver and skeletal muscles [52]. Indeed, studies have shown an inverse association with hip circumference, a marker for subcutaneous fat, and prevalence of undiagnosed diabetes in a large population-based survey [53]. These findings suggest that larger hips may have a protective effect possibly linked to greater muscle mass or femoral fat mass. Femoral fat has been shown to have greater lipoprotein lipase activity and lower rates of lipolysis than intra-abdominal fat, favoring fat uptake and storage, particularly in women [54]. In 2484 men and women, hip circumference was negatively associated with markers of glucose intolerance, independent of the waist circumference, which was positively associated with glucose intolerance [55].

B. Diabetes Mellitus

Diabetes mellitus is a metabolic disorder characterized by elevated blood glucose concentrations and disturbances of carbohydrate, fat, and protein metabolism secondary to defective insulin secretion and/or action [56, 57].

Asymptomatic diabetes, particularly type 2 diabetes mellitus, may be diagnosed when abnormal blood or urine glucose levels are found during routine testing. Exposure to chronic hyperglycemia is associated with pathologic and functional changes in organs such as the eyes, blood vessels, heart, kidneys, and nerves [57].

The classification of diabetes mellitus is based primarily on its clinical description and comprises four major types: type 1, type 2, other specific types, and gestational diabetes mellitus. All differ in their etiology. For example, type 1 diabetes, which is prone to ketoacidosis, is due to autoimmune or idiopathic destruction of the β cells of the pancreas, destruction that results in a deficiency in insulin secretion; type 2 diabetes is due to metabolic abnormalities leading to a diminished response to the action of insulin, along with a defect in insulin secretion. Other types relate to genetic defects or pancreatopathy or are first diagnosed during pregnancy (such as gestational diabetes); these types may or may not subsequently develop into diabetes after parturition [58].

The classification also includes prediabetes, a term used for impaired glucose tolerance and impaired fasting glucose, which place people at risk of developing diabetes and its complications [59]. Although diabetes is characterized by alterations of fat [60] and of protein metabolism [61, 62], glucose impairment remains the hallmark of diabetes diagnosis and control.

The American Diabetes Association (ADA) diagnostic criteria for diabetes mellitus [63] are summarized in Table 2 and those for prediabetes in Table 3. Criteria for the metabolic syndrome are given in Table 4.

The diagnosis of diabetes is made in (1) a person who shows symptoms of diabetes (polyuria, polydipsia, unexplained weight loss) and has a plasma glucose concentration, taken at

TABLE 2 Criteria for the Diagnosis of Diabetes [57]

1. Symptoms of diabetes and a casual plasma glucose ≥ 200 mg/dl (11.1 mmol/L). Casual is defined as any time of day without regard to time since last meal. The classic symptoms of diabetes include polyuria, polydipsia, and unexplained weight loss.

OR

2. Fasting plasma glucose (FPG) ≥ 126 mg/dl (7.0 mmol/L). Fasting is defined as no caloric intake for at least 8 hours.

OR

3. 2-hour plasma glucose ≥ 200 mg/dl (11.1 mmol/L) during an OGTT. The test should be performed as described by the WHO, using a glucose load containing the equivalent of 75 g anhydrous glucose dissolved in water.

any time of day, above 11.1 mmol/L (≥ 200 mg/dl); (2) a person who has a fasting plasma glucose concentration (FPG) of 7.0 mmol/L (≥ 126 mg/dl) or higher, fasting meaning at least 8 hours after food consumption; or (3) a person with a plasma glucose concentration of 11.1 mmol/L (≥ 200 mg/dl) or higher 2 hours after an oral challenge of 75 g glucose—called an oral glucose tolerance test (OGTT). Those values are chosen because they best predict the development of microvascular diseases, such as retinopathy and nephropathy [57], although the thresholds for increased risk of macrovascular diseases also correspond with those values.

Fasting plasma glucose (FPG) values have been established to define normality, FPG less than 5.6 mmol/L (< 100 mg/dl), and to define an intermediate category that is too high to be considered normal but too low for a diagnosis of diabetes. That category is called impaired fasting glucose (IFG) and corresponds to glucose values between 5.6 and 6.9 mmol/L (100 and 125 mg/dl).

The results of an OGTT are classified within one of three categories:

1. One of normal glucose tolerance when the 2-hour plasma glucose is below 7.8 mmol/L (< 140 mg/dl).
2. One of impaired glucose tolerance (IGT) when the 2-hour plasma glucose is at or above 7.8 mmol/L (≥140 mg/dl) but below 11.1 mmol/L (<200 mg/dl).
3. One of a provisional diagnosis of diabetes, which needs to be confirmed on a subsequent day, when the 2-hour plasma glucose is at or above 11.1 mmol/L (≥ 200 mg/dl).

TABLE 4 Clinical Identification of the Metabolic Syndrome using NCEP-ATP III Criteria [65]

Risk Factor	Defining Level[a]	
FPG	≥5.6 mmol/L	100 mg/dl
BP	≥130/85 mm Hg	
TG	≥1.7 mmol/L	>150 mg/dl
HDL-C		
Men	<1.03 mmol/L	<40 mg/dl
Women	<1.29 mmol/L	<50 mg/dl
Abdominal obesity (waist circumference)		
Men	>102 cm	
Women	>88 cm	

[a]A diagnosis of metabolic syndrome is made when three or more of the five disorders are present.

BP, blood pressure; FPG, fasting plasma glucose; HDL-C, high-density lipoprotein cholesterol; NCEP-ATP III, National Cholesterol Education Program Adult Treatment Panel III; TG, triglyceride.

The ADA criteria recommend that testing should be done in individuals 45 years of age and older and, when results are normal, should be repeated every 3 years thereafter. However, testing should be done at a younger age and more frequently in individuals with a BMI above 25 kg/m^2, because obese individuals are considered more at risk of having undiagnosed diabetes. The early detection and treatment of this disease could decrease mortality and minimize complications, especially those related to renal disease, peripheral vascular disease, and cardiovascular disease [56]. Thus testing individuals at high risk becomes highly cost effective and has implications in the prevention of

TABLE 3 Plasma Glucose Levels for Diagnosis of IFG, IGT, and Diabetes [57]

	FPG			2hPG in the 75-g OGTT	
	(mmol/liter)	(mg/dl)		(mmol/liter)	(mg/dl)
IFG	5.6–6.9	100–125		NA	
IFG (isolated)	5.6–6.9	100–125	and	<7.8	<140
IGT (isolated)	<5.6	<100	and	7.8–11.0	140–200
IFG and IGT	5.6–6.9	100–125	and	7.8–11.0	140–200
Diabetes	≥7.0	≥126	or	≥11.1	≥200

2hPG, 2-hour plasma glucose; FPG, fasting plasma glucose; IFG, impaired fasting glucose; IGT, impaired glucose tolerance; NA, not applicable; OGTT, oral glucose tolerance test.

diabetes [56]. Obese individuals, particularly those with IFG, IGT, or the metabolic syndrome, a condition characterized by a constellation of abnormalities that predicts diabetes beyond IGT alone [64] (see Table 4), are such individuals at high risk for diabetes and cardiovascular disease [57].

III. WHY ARE THE OBESE AT RISK?

A. Epidemiological Evidence

Obesity has been implicated as a risk factor for diabetes in cross-sectional [66–68] and longitudinal studies [69–77]. Population-based studies have shown strong associations between central adiposity, assessed by measurements of skinfolds on the trunk, and type 2 diabetes [75, 78, 79]. Ratios of waist-to-hip circumferences have been reported to be highly predictive of not only abnormal blood lipids and lipoproteins but of glucose intolerance as well [1–21, 80–84]. A meta-analysis of prospective studies has provided evidence that, as upper-body adiposity increased and criteria of the metabolic syndrome were present, the risk of developing diabetes increased [85]. Elevated glucose [17, 72, 86] and insulin [17] concentrations were likewise associated with a greater risk for diabetes.

Hyperinsulinemia, an indirect indicator of insulin resistance, mediated the development of diabetes in subjects characterized by an unfavorable body fat distribution [87]. However, in other population studies carried out in whites [72], Nauruans [88], and Japanese [89], obesity remained an independent predictive factor after adjusting for glucose and insulin, indicating that it may act through other pathways than insulin action and secretion. Even if insulin and glucose concentrations are better predictors of diabetes and their screening identifies well the subjects who could benefit from intervention, it remains true that more obese than lean persons are at risk of being resistant to insulin and are candidates who can benefit from weight loss and improved physical fitness. Oral glucose tolerance tests were used to assess type 2 diabetes mellitus and IGT in a large multiethnic population in Mauritius. In that population, age, family history, overall body mass, abdominal fat, and low physical activity scores were independent risk factors for diabetes [5]. However, there were ethnic differences in the relative importance of some factors. For example, physical inactivity had a lower impact in Chinese people of both sexes, and so had the waist-to-hip ratio in Muslim women. In other ethnic groups, the waist-to-hip ratio was a stronger predictor for glucose intolerance than BMI in women compared to men, for whom it was the reverse [5].

The data analyzed from a large cohort of 51,529 U.S. male health professionals, 40–75 years of age, and followed during 5 years, showed that as BMI increased beyond 24 kg/m^2, the risk for diabetes increased. It was 77 times greater in men whose BMI was above 35 kg/m^2 than in lean men. This study provides strong evidence that weight gain during adult life increases the risk of developing diabetes. In men who had gained more than 11 kg, their relative risk was amplified according to their BMI at age 21 years by 6.3 with a BMI below 22, by 9.1 with a BMI of 22–23, and by 21.1 with a BMI above 24 kg/m^2 [22].

The data also suggested that waist circumference was a better predictor of diabetes than waist-hip ratio [22]. The reliability of waist and hip measurements is limited by their difficulty and the fact that they are taken at different sites of the body in various studies. Nevertheless, population data report strong associations between waist-to-hip ratio and glucose intolerance [8], accounting for the associations with BMI. A high prevalence of diabetes has been found in populations that have a tendency for central adiposity and have been exposed to the Westernization of their lifestyle and its consequence, weight gain [90].

B. Fetal Origins of Type 2 Diabetes and Obesity

Some individuals may be more prone to insulin resistance and diabetes with weight gain if they have been malnourished *in utero*. Malnutrition during pregnancy forces the fetus to adapt during its development to such an extent that permanent changes occur in the structure and the physiology of its body [91]. These changes seen in low-birth-weight infants have been identified as contributing factors in adult life to chronic diseases such as type 2 diabetes, coronary heart disease, stroke, and hypertension [91–93]. It was first reported in England at the beginning of the 1990s that middle-aged adults who had low weights at birth and during infancy were at greater risk of type 2 diabetes and insulin resistance [93]. This association was confirmed in later studies in Europe and the United States [94–96] suggesting that endocrine and metabolic adaptations induced by malnutrition could explain insulin resistance in skeletal muscle.

These associations only applied to infants born of mothers without gestational diabetes. The infants of the latter, by contrast, tended to be of high birth weight (macrosomia), but also at risk for developing type 2 diabetes as adults [96].

Prenatal exposure to the Dutch famine in 1944 was associated with glucose intolerance and insulin resistance in the offspring once adults, even if the famine had little effect on their birth weights [97]. These associations support a role for prenatal nutrition in type 2 diabetes.

In the Nurses' Health Study [96], the inverse association between birth weights and diabetes remained significant even after adjustment for adult adiposity. The increased relative risk was seen among lean, moderate, and obese women, indicating that *in utero* growth had independent effects from adult body weight on the risk for developing type 2 diabetes. The greatest risk remained, however, in women of low birth weight who developed obesity as adults [96].

Data obtained from rat studies confirm human observations. Pregnant rats fed isoenergetic protein-restricted diets gave birth to offspring with low birth weights, reduced pancreatic β-cell mass and islet vascularization, and an impaired insulin response [98, 99], conditions that were not restored by normal nutrition after birth [99]. These offspring experienced diabetic pregnancies, exposing their fetus to hyperglycemia and increasing their risk of becoming diabetic adults. These observations indicated that *in utero* environment affects endocrine function and when deficient, contributes to insulin resistance and β-cell dysfunction that may lead to type 2 diabetes, especially in the presence of obesity [90].

Most prospective studies of prediabetic individuals demonstrate that they are hyperinsulinemic [100], or have impaired insulin secretion, subtle abnormalities of β-cell function that may be present before overt diabetes develops. Absence of rapid oscillations of insulin has been reported in first-degree relatives of patients with type 2 diabetes [101]. Other abnormalities include a greater proportion of proinsulin in plasma of patients with type 2 diabetes [102].

C. Maternal Obesity and Gestational Weight Gain

Maternal obesity has also been shown to alter gestational metabolic adjustments and fetal growth and development as reviewed by King [103]. It is associated with greater incidence of larger placentas and bigger babies [104], significant predictors of type 2 diabetes in the offspring [96, 105]. Considering that one-third of pregnant women in the United States are obese [106], metabolic complications, preeclampsia, fetal anomalies, and poor pregnancy outcomes have become more prevalent [107]. Infants large or small for gestational age are more often seen in obese compared with normal-weight mothers and are at increased risk for diabetes and hypertension later in life [108]. Some of the factors implicated in maternal obesity, beyond maternal overnutrition, are (1) maternal hypertension and its influence on placental size, (2) a subclinical inflammatory state and elevated cytokine levels [109], and (3) an enhanced insulin resistance that exaggerates the pregnancy-associated increase in circulating plasma glucose, lipids, and amino acids, exposing the fetus to an excess of all fuel sources [110].

Although the 1990 gestational weight gain guidelines from the Institute of Medicine (IOM) recommend smaller gains in obese mothers [111], excessive gains (> 19 kg) have become the common trend and account for a growing proportion of cesarean deliveries [112] and postpartum weight retention and obesity in the mother [113]. Oken *et al.* [106] have recently found, in a sample of 1044 mother–child pairs, a fourfold increased risk of having an overweight child at age 3 years in mothers whose gestational weight gain was adequate or excessive compared to mothers whose weight gain was inadequate based on the 1990 IOM guidelines. The increased risk was independent of parental BMI, maternal glucose tolerance, breastfeeding duration, fetal and infant growth, and child behaviors such as frequency of TV watching or fast-food consumption. These results pointed to the urgency of recommending more moderate weight gain during pregnancy, considering the current obesity and diabetes pandemic. However, weight control strategies to prevent metabolic complications should exclude strict energy restriction and its associated ketosis, which may have negative effects on fetal cognitive development [103]. The dietary approach in treating maternal obesity must attempt to improve insulin sensitivity in the mother without reducing fetal glucose levels and compromising fetal growth. This is achieved by the consumption of dark green and orange vegetables, fruits, legumes, whole grain breads and cereals, carbohydrates with a low glycemic index [114], nuts, fish, and low-fat meat and dairy products, and the use of oils such as canola oil for cooking and/or olive oil for salad dressing. It is recommended to limit added fat, sugar, and salt. Because motivation for dietary changes may be high during pregnancy, it is an opportune time to establish healthier eating habits and influence those of other family members, father included.

D. Visceral Obesity, Inflammation and Insulin Resistance: Major Predictors of Diabetes

Type 2 diabetes is a syndrome of diseases with different causes. For a minority of patients, they include mutations in some genes such as the insulin receptor gene that appear to cause insulin resistance, or in other genes that impair β-cell functions, compromise the glucose-sensing mechanism in the β-cell, and appear to be a mild form of the autoimmune disease [115, 116]; however, not all individuals with identified mutations express the diabetic phenotype [117]. Four genes that predispose individuals to developing type 2 diabetes have been recently identified. Two are involved in the development or function of insulin-secreting cells and one is involved in the transport of zinc, a mineral required in regulation of insulin secretion [118]. The knowledge of specific genetic mutations associated with type 2 diabetes could help in the early identification of people with a genetic risk and in the development of better treatments. It has long been recognized that family history plays an important part in determining whether one will develop diabetes. Impaired insulin secretion may reflect a genetic susceptibility to impaired glucose tolerance and type 2 diabetes [119, 120], but for the majority of patients, the best predictor is insulin resistance. Insulin resistance, characterized by hyperinsulinemia, is closely related to abdominal adiposity, and the inflammatory process is most marked in the visceral fat depot (Figure 1) [121–125].

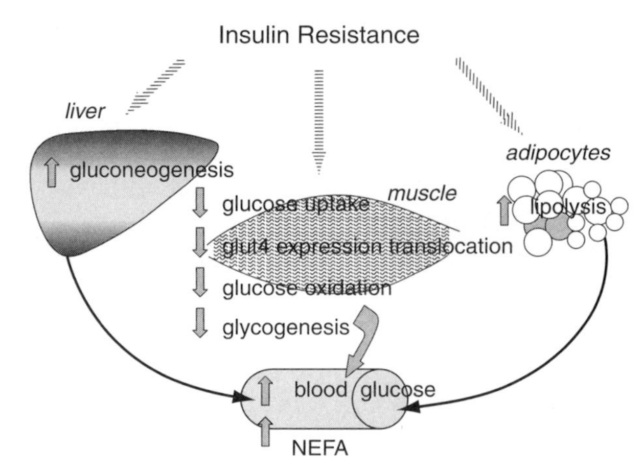

FIGURE 1 As fat mass and abdominal obesity increase, insulin sensitivity falls in skeletal muscle, leading to a decrease in glucose uptake; glucose oxidation, especially postprandially; and glycogen stores. In liver, insulin clearance is contributing to hyperinsulinemia, and gluconeogenesis increases as does glucose output. Enlarged visceral adipose tissue is associated with less response to the antilipolytic action of insulin and more response to the lipolytic action of catecholamines associated with increased nonesterified fatty acids (NEFA) in circulation.

A prospective study [126] that followed second- and third-generation Japanese Americans for up to 10 years confirms that the amount of intra-abdominal fat plays an important role in the development of diabetes. In this study, visceral adiposity, measured by computed tomography, was predictive of diabetes incidence, regardless of age, sex, family history of diabetes, fasting insulin, insulin secretion, glycemia, or total and regional adiposity. By increasing the demand for insulin, insulin resistance becomes a risk factor for diabetes, causing glucose intolerance in subjects who have impaired insulin secretory capacity and a reduction in the glucose potentiation of insulin secretion [127]. Incremental insulin response to an oral glucose challenge, an assessment of insulin secretion, was depressed in an older generation, suggesting that a failure in β-cell function preceded the onset of diabetes [128]. Insulin resistance was also associated with an impaired suppression of glucagon secretion by glucose in impaired glucose tolerance, suggesting that β-cell dysfunction and local insulinopenia may exaggerate glucagon secretion because the α-cell becomes less sensitive to glucose [127].

E. Insulin Resistance in Nonobese with Visceral Adiposity

The effects of nutrition and being sedentary on body composition and metabolic fitness are becoming the burden of nonobese as well. Normal weight individuals are displaying a cluster of characteristics that predispose them to type 2 diabetes. Thirteen young women were identified in a cohort of 71 healthy nonobese women, for being insulin insensitive to glucose [129]. The same women had a higher body fat percentage and higher visceral adiposity compared to the insulin-sensitive group. The energy expended in physical activity, measured by doubly-labeled water methodology and indirect calorimetry, was less (2.7 ± 0.9 versus 4.4 ± 1.5 MJ/d, $p = 0.01$) than in the insulin sensitive group. This study indicates that body fat and inactivity override body weight in determining glucose metabolism. Furthermore, there is evidence that there is a high prevalence of these individuals in the general population [130]. Although assessed from self-reported questionnaires, physical activity has been shown to relate negatively with the incidence of type 2 diabetes [131].

F. Metabolic Alterations in Obesity That Predispose to Type 2 Diabetes Mellitus

Total fatness and a body fat distribution that reflects visceral fat cell hypertrophy [81, 82] alter the metabolism of glucose: Both are associated with more insulin resistance than observed in obese individuals with lower-body adiposity [132, 133], greater breakdown of the stored fat in adipose cells [134] and elevated plasma nonesterified free fatty acid concentrations, which can lead to impaired glucose tolerance [135–138]. See Table 5.

1. Visceral Fat Alterations in Obesity

The extent of fat storage as triacylglycerol in adipocytes will depend on the balance between its formation and its mobilization. Both processes are under hormonal and nervous system regulations. Insulin plays a major role in suppressing the breakdown of triglycerides or lipolysis, by lowering cyclic AMP concentrations [139], which causes a dephosphorylation of hormone-sensitive lipase and, especially after food intake, by activating the enzyme lipoprotein

TABLE 5 Metabolic Alterations in Obesity Related to Diabetes

1. α_2 adrenoceptor activity in abdominal tissue of men that favors greater visceral adiposity
2. β_3-adrenoceptor sensitivity in visceral fat that increases the lipolytic response to catecholamine
3. Strong relationship between insulin resistance and visceral adiposity
4. Inflammatory cytokines by enlarged adipocytes, altering insulin action
5. Reduction of hepatic insulin clearance in upper-body obesity leading to hyperinsulinemia
6. Defects in intracellular glucose transporters
7. Less suppression of fat mobilization by insulin in visceral fat
8. Hypercortisolism in upper body obesity that favors lipolysis
9. Metabolic inflexibility of oxidation fuel selection in skeletal muscle
10. Endothelial dysfunction

lipase. Lipoprotein lipase releases fatty acids from chylomicrons and very low-density lipoproteins (VLDL) in the circulation. Half of those are taken up for storage [135]. Catecholamines can suppress lipolysis [140] via their action on α_2-adrenoceptors, but they mostly have the opposite effect of insulin and, by acting on β-adrenoreceptors, increase cyclic AMP and the phosphorylation of hormone-sensitive lipase, stimulating lipolysis and the release of fatty acids in the circulation [140].

Body fat distribution may be influenced by differences in lipoprotein lipase activity. In premenopausal women, lipoprotein lipase activity is higher in femoral and gluteal compared with abdominal regions, where fat cells are larger. These differences are not seen in men or in postmenopausal women [141]. Furthermore, it has been suggested that a greater α_2 activity in the abdominal tissue of men may explain greater adiposity in that location [142].

Body fat distribution also affects the lipolytic response to the catecholamine norepinephrine. It has been reported that in men and women, the response is greater in abdominal compared with gluteal and femoral adipose tissues, abdominal adipocytes having greater β_3-adrenoceptor sensitivity. Receptor numbers are reported to be increased in obese subjects [143]. Lipolysis being relatively more elevated in visceral fat cells, subjects with upper-body obesity would be exposed to a greater release of free fatty acids into the portal system.

2. Insulin Sensitivity in Obesity

Studies measuring insulin sensitivity showed a positive correlation between the degree of upper body adiposity and the steady-state plasma glucose, an index of the capacity a subject has to dispose of a glucose challenge under insulin stimulus [81]. The correlation between abdominal fat and insulin resistance has been shown to be independent of total body fatness [81, 144, 145]. Overall glucose disposal by skeletal muscle in premenopausal women was lower with greater upper body fatness [146]. Insulin resistance of glucose in skeletal muscle is seen in both obesity and type 2 diabetes mellitus. The decreased efficiency in glucose disposal has been related to the reduction in insulin-stimulated activity of the glucose-6-phosphate independent form of glycogen synthase.

3. Inflammation and Insulin Resistance

Other factors that may explain why insulin resistance worsens in some obese individuals and not in others as their fat mass increases are the overproduction by specific adipocytes of pro-inflammatory cytokines such as tumor necrosis factor α (TNF-α) and interleukin-6 (IL-6). These cytokines block the effect of insulin on glucose transport in skeletal muscle by producing nitric oxide in excess via the induction of the expression of inducible nitric oxide synthase [147, 148]. TNF-α mRNA expression in adipose tissue and skeletal muscle correlates with BMI and plasma insulin levels. It inhibits the tyrosine kinase activity of the insulin receptor, altering its action and possibly leading to insulin resistance [149]. In visceral obesity, plasma levels of C-reactive protein, a marker of inflammation, are elevated [150]; C-reactive protein, along with leptin, is produced only in the adipocyte. By contrast, adiponectin concentrations are reduced in visceral obesity [151]. Adiponectin improves insulin sensitivity by inhibiting the action of TNF-α on the 1-κ B-kinase B/nuclear factor κ B [152]. Visceral fat cells appear more pathogenic. In fat tissue and liver, lipid accumulation, due to excess dietary intake and obesity, activates inflammatory signaling pathways, such that proinflammatory cytokines (TNF-α, IL-6, leptin, resistin, chemokines) are produced and induce local insulin resistance. Systemic proinflammatory mediators cause insulin resistance in skeletal muscle and other tissues such as the kidney. Thus chronic subacute inflammation modulates metabolism in obesity, increasing the risk for insulin resistance and eventually type 2 diabetes (Figure 2).

Fractional hepatic clearance of insulin, postabsorptive and during stimulation by intravenous glucose or an oral glucose load, is reduced in upper-body obesity compared

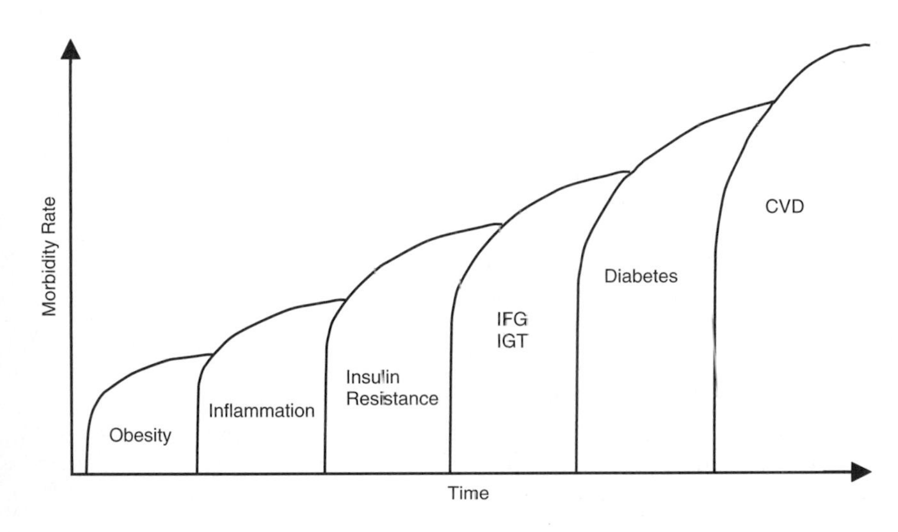

FIGURE 2 Probable progression from obesity, particularly visceral obesity, to cardiovascular disease, via the inflammatory process, insulin resistance, impaired fasting glucose, impaired glucose tolerance, and type 2 diabetes mellitus. IFG, impaired fasting glucose; IGT, impaired glucose tolerance; CVD, cardiovascular disease.

with lower-body obesity, despite indications that the portal plasma insulin levels do not differ [153].

Forty percent of the insulin secreted by the pancreas is removed by the liver. Less hepatic insulin extraction leads to greater peripheral insulin concentrations, a downregulation of insulin receptors, and insulin resistance [154]. This effect may be a consequence of elevated levels of free testosterone and decreased sex-hormone-binding globulin [155] that characterize upper-body obesity [83].

The transport of glucose into the adipose cell was reduced in obesity and more so in type 2 diabetes because of a depletion of intracellular glucose transporters and of carriers for their recruitment to the plasma membrane of the cell [156]. Although the content in intracellular glucose transporters did not differ in the skeletal muscle of subjects who were obese compared with controls, the defect associated with impaired responsiveness to insulin was a loss in the functional activity of the transporters or a decrease in their translocation to the cell surface. The latter may be a consequence of long-term exposure to hyperglycemia in type 2 diabetes [157].

Visceral adipose tissue shows a greater response to lipolytic stimuli but is less sensitive to suppression of fat mobilization by insulin than is subcutaneous adipose tissue. The elevated concentrations in insulin in upper-body obesity will have a greater inhibitory effect on lipolysis in subcutaneous adipose fat, and proportionally a greater amount of nonesterified fatty acids will be mobilized from visceral fat [158]. It is conceivable that more portal nonesterified fatty acids will be conducive to greater hepatic and skeletal muscle insulin resistance [159].

Upper-body obesity is associated with hypercortisolism, characterized by increased degradation and clearance rates of cortisol that are compensated by increased production rates. Cortisol favors lipolysis by inhibiting the action of insulin and permitting that of catecholamines [160]. Because visceral fat shows a higher density of glucocorticoid receptors compared to subcutaneous fat, hypercortisolism in obesity may contribute to more portal nonesterified fatty acids and hepatic and skeletal muscle insulin resistance.

Circulating nonesterified fatty acid concentrations increase as fat mass increases [159]. Furthermore, as insulin action diminishes with obesity, suppression of nonesterified acids is decreased and their concentrations in plasma increased [154]. The larger supply of fatty acids to the liver, particularly if they are not totally suppressed after meals, is associated with accumulation of acetyl CoA and inhibition of pyruvate carboxylase, altering glucose utilization and stimulating gluconeogenesis and inappropriate hepatic glucose production. In muscle, maintaining or increasing plasma nonesterified fatty acid concentrations decreased insulin-stimulated glucose uptake. Using the euglycemic-hyperinsulinemic clamp and indirect calorimetry, lipid infusions produced a decrease in insulin-stimulated glucose oxidation and a greater decrease in glycogen synthesis [161]. These results suggested that in insulin-resistant states with increased lipid availability and oxidation, glucose oxidation was reduced. However, by contrast, in fasting conditions, hyperglycemia in type 2 diabetes was shown to override the effect of increased lipid availability and to be associated with increased glucose oxidation in leg muscle. Normalization of the leg muscle hyperglycemia with a low-dose of insulin increased fat oxidation [162]. There is evidence (reviewed in [163]) to suggest that, postabsorptively, skeletal muscle fat oxidation in insulin-resistant states is decreased. This decrease is explained by the presence in muscle of carbohydrate derived malonyl CoA, which inhibits carnitine palmitoyl transferase blocking the entry of free fatty acids into the mitochondria [164]. Carnitine palmitoyl transferase activity has been shown to be reduced in the vastus lateralis muscle of insulin-resistant obese individuals [165]. The excess free fatty acids may increase long-chain acyl CoA concentrations and diacylglycerol, leading to the accumulation of lipids in muscle, an effect conducive to alterations of insulin signaling and insulin action.

There are human studies that suggest that muscle lipid content in obesity could be a determinant of insulin insensitivity [166, 167], independent of visceral fat [166]. Lipid accumulation in muscle is observed in trained athletes who are insulin sensitive but who have a great metabolic capacity for lipid utilization, which is not the case in muscles of sedentary individuals who are obese. Obese subjects appear to display what has been called metabolic inflexibility of oxidative fuel selection [163]. Compared with insulin sensitive lean individuals, obese subjects have shown lower fat oxidation after an overnight fast, as indicated by higher leg respiratory quotient (RQ), and less capacity to switch to glucose in insulin-stimulated conditions, as indicated by an absence of increase in leg RQ. Furthermore, high levels of fatty acids may suppress insulin secretion and their effect may be toxic to the β cells of the pancreas, becoming an important contributor to the pathogenesis of obesity-dependent type 2 diabetes [59]. Excess nonesterified fatty acid reduces hepatic insulin extraction, further increasing its serum concentration [159].

The prevalence of visceral fat measured by computer tomography correlated with endothelial dysfunction, independently of BMI, in otherwise healthy women who are obese [168]. Endothelial dysfunction was also closely associated with a marker of insulin sensitivity. The authors attributed these early alterations involved in the atherosclerotic process to high nonesterified fatty acid levels.

As long as the pancreas maintains sufficient insulin secretion to compensate for insulin resistance, glycemia remains within the normal range. Once the β-cells fail to compensate, insulin response is decreased and glucose uptake by muscle diminished. Then, the liver preferentially oxidizes fatty acids and gluconeogenesis is stimulated. The consequences are a conversion from normal to impaired glucose tolerance and, in genetically susceptible individuals, to diabetes [169].

G. Associations among Psychologic Stress, Diabetes, and Visceral Adiposity

Psychologic stress can be associated with high cortisol and low sex steroid concentrations. These antagonize insulin action and cause visceral adiposity, which may contribute to insulin resistance and the onset of type 2 diabetes mellitus. Psychologic stress is known to aggravate glycemia in diabetes [170]. A greater prevalence of previously undetected diabetes was reported with more stressful events; the age- and sex-adjusted association between major life changes and diabetes was independent of family history of diabetes, physical activity, heavy alcohol use, or low level of education. There was no association between work-related stressful events and type 2 diabetes in this cross-sectional study. A weak positive association was found with waist-to-hip ratio, but visceral fat was not the main mediating factor between stress and diabetes [171]. It has been suggested that the link between stress and diabetes may be through other factors such as the chronic stimulation of the autonomic nervous system and its resulting hyperglycemia [171, 172]. More recently, the inflammatory consequences of psychologic stress have been proposed as the link to insulin resistance and diabetes [173]. Psychologic stress via stimulation of the major stress hormones, norepinephrine and cortisol, the renin-angiotensin system, the proinflammatory cytokines, and free fatty acid fluxes produces an inflammatory response by activation of nuclear factor κB in macrophages, visceral fat, and endothelial cells. This process leads to insulin resistance and, if maintained, to increased risk for diabetes.

H. Improvement in Insulin Sensitivity with Weight Loss and Energy Intake Restriction

In obesity characterized by diminished insulin action, any treatment associated with weight loss improves insulin sensitivity and fasting plasma glucose; these improvements are related to losses of abdominal fat [174, 175]. Fujioka *et al.* [174] measured the changes in body fat distribution and those in metabolic disorders after an 8-week energy intake restricted diet (800 kcal/day) in 40 women aged 38 ± 9 (mean $\pm$ SD) years with uncomplicated obesity. Body fat was determined by the computed tomographic (CT) scan method, and 14 women were characterized as having visceral fat obesity rather than subcutaneous fat obesity. They found that at week 8 of diet, the decrease in visceral fat volume was more sharply correlated with the changes in plasma glucose and in lipid metabolism than the decrease in body weight, total fat, or subcutaneous fat volume, and that these correlations were independent of total fat loss. The reduction in visceral fat was more pronounced than the reduction in subcutaneous fat.

Another study [175] reported that the improvements in fasting plasma glucose and insulin sensitivity with weight loss related to losses of abdominal fat. That study was designed to distinguish the effects of energy intake restriction before substantial weight loss from those of a weight loss of 6.3 ± 0.4 kg on glycemia, hepatic glucose production, and insulin action and secretion in 20 overweight subjects with or without type 2 diabetes. At day 4 of reducing energy intake by 50%, a significant decrease was observed in basal hepatic glucose output with a greater insulin suppression of endogenous glucose production in all subjects, measured during a hyperinsulinemic euglycemic clamp. Although the decrease in glucose output was counteracted by a decrease in metabolic clearance rate of glucose, it resulted in a fall in fasting plasma glucose. This fall related to the decrease in carbohydrate intake and possibly an associated reduction in hepatic glycogen content. At day 4 of energy intake restriction, fat oxidation was increased as were the nonesterified fatty acid levels. With substantial weight loss (at day 28 of the energy intake restricted diet), plasma glucose was further decreased in the diabetic subjects only, a decrease associated with increased metabolic clearance rates. Fasting insulin concentrations were reduced compared with those preceding weight loss, and insulin sensitivity was improved in both groups. These improvements were associated with a reduction in the abdominal fat depot. These data support a role for central adiposity in the alterations of glucose metabolism observed in obesity.

Loss of weight and visceral fat also improved endothelial dysfunction and vascular inflammatory markers when combined with exercise, or secondary to gastric bypass surgery [176, 177]. Dietary factors modulate endothelial function; not only have postprandial hyperglycemia and hypertriglyceridemia been shown to generate oxidative stress and endothelial dysfunction [178], but meals high in advanced glycation end products, secondary to frying or broiling and cooking at high temperatures for long durations, have been reported to induce acute impairment of vascular function as well [179]. These results provide evidence for the wisdom of using cooking techniques such as steaming, boiling, poaching, or stewing to minimize dietary sources of advanced glycation end products.

I. Obesity and Treatment of Diabetes

The conclusions of the U.K. Prospective Diabetes Study are that [180] optimal diabetic control aiming at a glycosylated hemoglobin below 7.0% must be achieved if morbidity is to be significantly prevented. To do so, intensive therapy using insulin alone or combined with oral antihyperglycemic agents may be required. Results from the Finnish Multicenter Insulin Therapy Study, in which 100 insulin-treated type 2 diabetic patients were followed for 12 months, showed that good glycemic control started to deteriorate after 3 months and more so in the obese subjects, the latter being attributed to their greater insulin resistance. Glycemic control was best achieved in the nonobese subjects whether with insulin alone or in combination with other therapeutic agents. However, the combination therapy was associated with less weight gain [181].

Obesity in persons with type 1 diabetic persons can affect their insulin requirements by aggravating their insulin resistance. The consequences of increasing the insulin dosages can be further weight gain associated with the lipogenic effects of insulin and the additional eating in reaction to the hypoglycemic events often seen with intensive therapy [182]. Weight gain with insulin therapy is often associated with higher waist-to-hip ratios [180]. Given the associations between obesity and dyslipidemia, atherosclerosis and hypertension, intensive therapy of diabetes should aim at not producing weight gain [180].

IV. CONCLUSION

Family history of diabetes and age are recognized risk factors for diabetes [18], but they cannot be controlled. Recommendations for prevention of diabetes address factors that can be controlled. Thus it is crucial to screen early for persons at risk of developing diabetes using the recommended criteria for BMI, waist circumference, metabolic syndrome, and inflammatory markers. Tight control of glycemia in the person with diabetes is necessary to prevent microvascular complications [180], but fewer than half achieve such control [183]. Furthermore, the increased cardiovascular risk associated with the disease requires that therapeutic strategies be devised to correct the factors related to that risk. These include elevated blood pressure, smoking, abnormal lipid profile, low physical activity, and obesity. Current data support that to reduce the incidence of diabetes it is advisable to achieve a healthy weight when young and maintain it throughout life [22]. Prevention of overall obesity should be the result of a lifestyle that includes healthy eating habits favoring foods with a low energy density [184] and levels of physical activity associated with the maintenance of optimal fitness for age. Innovative ways are especially needed to promote and achieve proficient physical activity and exercise habits in those who are obese and have diabetes [185].

References

1. Haffner, S. M., Stern, M. P., Hazuda, H. P., Pugh, J., and Patterson, J. K. (1987). Do upper-body and centralized adiposity measure different aspects of regional body-fat distribution? Relationship to non-insulin-dependent diabetes mellitus, lipids, and lipoproteins. *Diabetes* **36**, 43–51.
2. van Noord, P. A., Seidell, J. C., den Tonkelaar, I., Baanders-van Halewijn, E. A., and Ouwehand, I. J. (1990). The relationship between fat distribution and some chronic diseases in 11,825 women participating in the DOM-project. *Int. J. Epidemiol.* **19**, 564–570.
3. Skarfors, E. T., Selinus, K. I., and Lithell, H. O. (1991). Risk factors for developing non-insulin dependent diabetes: A 10 year follow up of men in Uppsala. *BMJ* **303**, 755–760.
4. Shelgikar, K. M., Hockaday, T. D., and Yajnik, C. S. (1991). Central rather than generalized obesity is related to hyperglycaemia in Asian Indian subjects. *Diabetes Med.* **8**, 712–717.
5. Dowse, G. K., Zimmet, P. Z., Gareeboo, H., George, K., Alberti, M. M., Tuomilehto, J., Finch, C. F., Chitson, P., and Tulsidas, H. (1991). Abdominal obesity and physical inactivity as risk factors for NIDDM and impaired glucose tolerance in Indian, Creole, and Chinese Mauritians. *Diabetes Care* **14**, 271–282.
6. Tai, T. Y., Chuang, L. M., Wu, H. P., and Chen, C. J. (1992). Association of body build with non-insulin-dependent diabetes mellitus and hypertension among Chinese adults: A 4-year follow-up study. *Int. J. Epidemiol.* **21**, 511–517.
7. Schmidt, M. I., Duncan, B. B., Canani, L. H., Karohl, C., and Chambless, L. (1992). Association of waist-hip ratio with diabetes mellitus. Strength and possible modifiers. *Diabetes Care* **15**, 912–914.
8. McKeigue, P. M., Pierpoint, T., Ferrie, J. E., and Marmot, M. G. (1992). Relationship of glucose intolerance and hyperinsulinaemia to body fat pattern in south Asians and Europeans. *Diabetologia* **35**, 785–791.
9. Shaten, B. J., Smith, G. D., Kuller, L. H., and Neaton, J. D. (1993). Risk factors for the development of type II diabetes among men enrolled in the usual care group of the Multiple Risk Factor Intervention Trial. *Diabetes Care* **16**, 1331–1339.
10. Marshall, J. A., Hamman, R. F., Baxter, J., Mayer, E. J., Fulton, D. L., Orleans, M., Rewers, M.J, and Jones, R.H (1993). Ethnic differences in risk factors associated with the prevalence of non-insulin-dependent diabetes mellitus. The San Luis Valley Diabetes Study. *Am. J. Epidemiol.* **137**, 706–718.
11. Collins, V. R., Dowse, G. K., Toelupe, P. M., Imo, T. T., Aloaina, F. L., Spark, R. A., and Zimmet, P. Z. (1994). Increasing prevalence of NIDDM in the Pacific island population of Western Samoa over a 13-year period. *Diabetes Care* **17**, 288–296.
12. Chou, P., Liao, M. J., and Tsai, S. T. (1994). Associated risk factors of diabetes in Kin-Hu, Kinmen. *Diabetes Res. Clin. Pract.* **26**, 229–235.
13. Ohlson, L. O., Larsson, B., Svardsudd, K., Welin, L., Eriksson, H., Wilhelmsen, L., Bjorntorp, P., and Tibblin, G. (1985). The influence of body fat distribution on the incidence of diabetes mellitus. 13.5 years of follow-up of the participants in the study of men born in 1913. *Diabetes* **34**, 1055–1058.
14. Modan, M., Karasik, A., Halkin, H., Fuchs, Z., Lusky, A., Shitrit, A., and Modan, B. (1986). Effect of past and concurrent body mass index on prevalence of glucose intolerance and type 2 (non-insulin-dependent) diabetes and on insulin response. The Israel study of glucose intolerance, obesity and hypertension. *Diabetologia* **29**, 82–89.
15. Lundgren, H., Bengtsson, C., Blohme, G., Lapidus, L., and Sjostrom, L. (1989). Adiposity and adipose tissue distribution in relation to incidence of diabetes in women: Results

from a prospective population study in Gothenburg, Sweden. *Int. J. Obes.* **13**, 413–423.

16. Lemieux, S., Prud'homme, D., Nadeau, A., Tremblay, A., Bouchard, C., and Despres, J. P. (1996). Seven-year changes in body fat and visceral adipose tissue in women. Association with indexes of plasma glucose-insulin homeostasis. *Diabetes Care* **19**, 983–991.
17. Haffner, S. M., Stern, M. P., Mitchell, B. D., Hazuda, H. P., and Patterson, J. K. (1990). Incidence of type II diabetes in Mexican Americans predicted by fasting insulin and glucose levels, obesity, and body-fat distribution. *Diabetes* **39**, 283–288.
18. Colditz, G. A., Willett, W. C., Stampfer, M. J., Manson, J. E., Hennekens, C. H., Arky, R. A., and Speizer, F. E. (1990). Weight as a risk factor for clinical diabetes in women. *Am. J. Epidemiol.* **132**, 501–513.
19. Knowler, W. C., Pettitt, D. J., Saad, M. F., Charles, M. A., Nelson, R. G., Howard, B. V., Bogardus, C., and Bennett, P. H. (1991). Obesity in the Pima Indians: Its magnitude and relationship with diabetes. *Am. J. Clin. Nutr.* **53**, 1543S–1551S.
20. Charles, M. A., Fontbonne, A., Thibult, N., Warnet, J. M., Rosselin, G. E., and Eschwege, E. (1991). Risk factors for NIDDM in white population. Paris prospective study. *Diabetes* **40**, 796–799.
21. Cassano, P. A., Rosner, B., Vokonas, P. S., and Weiss, S. T. (1992). Obesity and body fat distribution in relation to the incidence of non-insulin-dependent diabetes mellitus. A prospective cohort study of men in the normative aging study. *Am. J. Epidemiol.* **136**, 1474–1486.
22. Chan, J. M., Rimm, E. B., Colditz, G. A., Stampfer, M. J., and Willett, W. C. (1994). Obesity, fat distribution, and weight gain as risk factors for clinical diabetes in men. *Diabetes Care* **17**, 961–969.
23. Wang, Y., Rimm, E. B., Stampfer, M. J., Willett, W. C., and Hu, F. B. (2005). Comparison of abdominal adiposity and overall obesity in predicting risk of type 2 diabetes among men. *Am. J. Clin. Nutr.* **81**, 555–563.
24. Rana, J. S., Li, T. Y., Manson, J. E., and Hu, F. B. (2007). Adiposity compared with physical inactivity and risk of type 2 diabetes in women. *Diabetes Care* **30**, 53–58.
25. Colditz, G. A., Willett, W. C., Rotnitzky, A., and Manson, J. E. (1995). Weight gain as a risk factor for clinical diabetes mellitus in women. *Ann. Intern. Med.* **122**, 481–486.
26. Nagaya, T., Yoshida, H., Takahashi, H., and Kawai, M. (2005). Increases in body mass index, even within non-obese levels, raise the risk for Type 2 diabetes mellitus: A follow-up study in a Japanese population. *Diabetes Med.* **22**, 1107–1111.
27. Schienkiewitz, A., Schulze, M. B., Hoffmann, K., Kroke, A., and Boeing, H. (2006). Body mass index history and risk of type 2 diabetes: Results from the European Prospective Investigation into Cancer and Nutrition (EPIC)-Potsdam Study. *Am. J. Clin. Nutr.* **84**, 427–433.
28. Ruwaard, D., Gijsen, R., Bartelds, A. I., Hirasing, R. A., Verkleij, H., and Kromhout, D. (1996). Is the incidence of diabetes increasing in all age-groups in The Netherlands? Results of the second study in the Dutch Sentinel Practice Network. *Diabetes Care* **19**, 214–218.
29. Midthjell, K., Kruger, O., Holmen, J., Tverdal, A., Claudi, T., Bjorndal, A., and Magnus, P. (1999). Rapid changes in the prevalence of obesity and known diabetes in an adult Norwegian population. The Nord-Trondelag Health Surveys: 1984–1986 and 1995–1997. *Diabetes Care* **22**, 1813–1820.
30. Mainous, A. G., 3rd, Baker, R., Koopman, R. J., Saxena, S., Diaz, V. A., Everett, C. J., and Majeed, A. (2007). Impact of the population at risk of diabetes on projections of diabetes burden in the United States: An epidemic on the way. *Diabetologia* **50**, 934–940.
31. Astrup, A., and Finer, N. (2000). Redefining type 2 diabetes: "Diabesity" or "obesity dependent diabetes mellitus"? *Obes. Rev.* **1**, 57–59.
32. Hadden, D. R., Montgomery, D. A., Skelly, R. J., Trimble, E. R., Weaver, J. A., Wilson, E. A., and Buchanan, K. D. (1975). Maturity onset diabetes mellitus: Response to intensive dietary management. *BMJ* **3**, 276–278.
33. Lean, M. E., Powrie, J. K., Anderson, A. S., and Garthwaite, P. H. (1990). Obesity, weight loss and prognosis in type 2 diabetes. *Diabetes Med.* **7**, 228–233.
34. UKPDS (1990). UK Prospective Diabetes Study 7. Response of fasting plasma glucose to diet therapy in newly presenting type II diabetic patients. *Metabolism* **39**, 909–912.
35. Vessby, B., Boberg, M., Karlstrom, B., Lithell, H., and Werner, I. (1984). Improved metabolic control after supplemented fasting in overweight type II diabetic patients. *Acta Med. Scand.* **216**, 67–74.
36. Williamson, D. F., Pamuk, E., Thun, M., Flanders, D., Byers, T., and Heath, C. (1995). Prospective study of intentional weight loss and mortality in never-smoking overweight U.S. white women aged 40–64 years. *Am. J. Epidemiol.* **141**, 1128–1141.
37. Tuomilehto, J., Lindstrom, J., Eriksson, J. G., Valle, T. T., Hamalainen, H., Ilanne-Parikka, P., Keinanen-Kiukaanniemi, S., Laakso, M., Louheranta, A., Rastas, M., Salminen, V., and Uusitupa, M. (2001). Prevention of type 2 diabetes mellitus by changes in lifestyle among subjects with impaired glucose tolerance. *N. Engl. J. Med.* **344**, 1343–1350.
38. Knowler, W. C., Barrett-Connor, E., Fowler, S. E., Hamman, R. F., Lachin, J. M., Walker, E. A., and Nathan, D. M. (2002). Reduction in the incidence of type 2 diabetes with lifestyle intervention or metformin. *N. Engl. J. Med.* **346**, 393–403.
39. Gillies, C. L., Abrams, K. R., Lambert, P. C., Cooper, N. J., Sutton, A. J., Hsu, R. T., and Khunti, K. (2007). Pharmacological and lifestyle interventions to prevent or delay type 2 diabetes in people with impaired glucose tolerance: Systematic review and meta-analysis. *BMJ* **334**, 299.
40. Laakso, M. (2005). Prevention of type 2 diabetes. *Curr. Mol. Med.* **5**, 365–374.
41. Rosenbaum, M., Leibel, R. L., and Hirsch, J. (1997). Obesity. *N. Engl. J. Med.* **337**, 396–407.
42. Garrow, J. S. (1988). Health implications of obesity. *In* "Obesity Related Diseases," pp. 1–16. Churchill Livingstone, London.
43. Vague, J. (1956). The degree of masculine differentiation of obesities: A factor determining predisposition to diabetes, atherosclerosis, gout, and uric calculous disease. *Am. J. Clin. Nutr.* **4**, 20–34.
44. Stevens, J., Cai, J., Pamuk, E. R., Williamson, D. F., Thun, M. J., and Wood, J. L. (1998). The effect of age on the association between body-mass index and mortality. *N. Engl. J. Med.* **338**, 1–7.

45. (1995). Physical status: The use and interpretation of anthropometry. Report of a WHO Expert Committee. *World Health Organ Tech. Rep. Ser.* **854**, 1–452.
46. Seidell, J. C., and Flegal, K. M. (1997). Assessing obesity: classification and epidemiology. *Br. Med. Bull.* **53**, 238–252.
47. Norgan, N. G. J., and Jones, P. R. (1995). The effect of standardising the body mass index for relative sitting height. *Int. J. Obes. Relat. Metab. Disord.* **19**, 206–208.
48. Lemieux, S., Prud'homme, D., Bouchard, C., Tremblay, A., and Despres, J. P. (1996). A single threshold value of waist girth identifies normal-weight and overweight subjects with excess visceral adipose tissue. *Am. J. Clin. Nutr.* **64**, 685–693.
49. Klein, S., Allison, D. B., Heymsfield, S. B., Kelley, D. E., Leibel, R. L., Nonas, C., and Kahn, R. (2007). Waist circumference and cardiometabolic risk: A consensus statement from Shaping America's Health: Association for Weight Management and Obesity Prevention; NAASO, The Obesity Society; the American Society for Nutrition; and the American Diabetes Association. *Am. J. Clin. Nutr.* **85**, 1197–1202.
50. Lean, M. E., Han, T. S., and Morrison, C. E. (1995). Waist circumference as a measure for indicating need for weight management. *BMJ* **311**, 158–161.
51. Lemieux, I., Drapeau, V., Richard, D., Bergeron, J., Marceau, P., Biron, S., and Mauriege, P. (2006). Waist girth does not predict metabolic complications in severely obese men. *Diabetes Care* **29**, 1417–1419.
52. Miranda, P. J., DeFronzo, R. A., Califf, R. M., and Guyton, J. R. (2005). Metabolic syndrome: Definition, pathophysiology, and mechanisms. *Am. Heart J.* **149**, 33–45.
53. Snijder, M. B., Zimmet, P. Z., Visser, M., Dekker, J. M., Seidell, J. C., and Shaw, J. E. (2004). Independent and opposite associations of waist and hip circumferences with diabetes, hypertension and dyslipidemia: The AusDiab Study. *Int. J. Obes. Relat. Metab. Disord.* **28**, 402–409.
54. Rebuffe-Scrive, M., Enk, L., Crona, N., Lonnroth, P., Abrahamsson, L., Smith, U., and Bjorntorp, P. (1985). Fat cell metabolism in different regions in women. Effect of menstrual cycle, pregnancy, and lactation. *J. Clin. Invest.* **75**, 1973–1976.
55. Snijder, M. B., Dekker, J. M., Visser, M., Yudkin, J. S., Stehouwer, C. D., Bouter, L. M., Heine, R. J., Nijpels, G., and Seidell, J. C. (2003). Larger thigh and hip circumferences are associated with better glucose tolerance: The Hoorn study. *Obes. Res.* **11**, 104–111.
56. (1994). Prevention of diabetes mellitus. Report of a WHO Study Group. *World Health Organ Tech. Rep. Ser.* **844**, 1–100.
57. (2000). Report of the Expert Committee on the Diagnosis and Classification of Diabetes Mellitus. *Diabetes Care* **23**(Suppl 1), S4–S19.
58. Stern, M. P. (1988). Type II diabetes mellitus. Interface between clinical and epidemiological investigation. *Diabetes Care* **11**, 119–126.
59. Fuller, J. H., Shipley, M. J., Rose, G., Jarrett, R. J., and Keen, H. (1980). Coronary-heart-disease risk and impaired glucose tolerance. The Whitehall study. *Lancet* **1**, 1373–1376.
60. Unger, R. H. (1995). Lipotoxicity in the pathogenesis of obesity-dependent NIDDM. Genetic and clinical implications. *Diabetes* **44**, 863–870.
61. Nair, K. S., Garrow, J. S., Ford, C., Mahler, R. F., and Halliday, D. (1983). Effect of poor diabetic control and obesity on whole body protein metabolism in man. *Diabetologia* **25**, 400–403.
62. Gougeon, R., Styhler, K., Morais, J. A., Jones, P. J., and Marliss, E. B. (2000). Effects of oral hypoglycemic agents and diet on protein metabolism in type 2 diabetes. *Diabetes Care* **23**, 1–8.
63. (2007). Diagnosis and classification of diabetes mellitus. *Diabetes Care* **30**(Suppl 1), S42–47.
64. Lorenzo, C., Williams, K., Hunt, K. J., and Haffner, S. M. (2007). The National Cholesterol Education Program—Adult Treatment Panel III, International Diabetes Federation, and World Health Organization definitions of the metabolic syndrome as predictors of incident cardiovascular disease and diabetes. *Diabetes Care* **30**, 8–13.
65. Grundy, S. M., Cleeman, J. I., Daniels, S. R., Donato, K. A., Eckel, R. H., Franklin, B. A., Gordon, D. J., Krauss, R. M., Savage, P. J., Smith, S. C., Jr, , Spertus, J. A., and Costa, F. (2005). Diagnosis and management of the metabolic syndrome: An American Heart Association/National Heart, Lung, and Blood Institute Scientific Statement. *Circulation* **112**, 2735–2752.
66. King, H., Zimmet, P., Raper, L. R., and Balkau, B. (1984). Risk factors for diabetes in three Pacific populations. *Am. J. Epidemiol.* **119**, 396–409.
67. King, H., Taylor, R., Koteka, G., Nemaia, H., Zimmet, P., Bennett, P. H., and Raper, L. R. (1986). Glucose tolerance in Polynesia. Population-based surveys in Rarotonga and Niue. *Med. J. Aust.* **145**, 505–510.
68. McLarty, D. G., Swai, A. B., Kitange, H. M., Masuki, G., Mtinangi, B. L., Kilima, P. M., Makene, W. J., Chuwa, L. M., and Alberti, K. G. (1989). Prevalence of diabetes and impaired glucose tolerance in rural Tanzania. *Lancet* **1**, 871–875.
69. Medalie, J. H., Papier, C. M., Goldbourt, U., and Herman, J. B. (1975). Major factors in the development of diabetes mellitus in 10,000 men. *Arch. Intern. Med.* **135**, 811–817.
70. Stanhope, J. M., and Prior, I. A. (1980). The Tokelau island migrant study: Prevalence and incidence of diabetes mellitus. *N. Z. Med. J.* **92**, 417–421.
71. Knowler, W. C., Pettitt, D. J., Savage, P. J., and Bennett, P. H. (1981). Diabetes incidence in Pima Indians: Contributions of obesity and parental diabetes. *Am. J. Epidemiol.* **113**, 144–156.
72. Keen, H., Jarrett, R. J., and McCartney, P. (1982). The ten-year follow-up of the Bedford survey (1962–1972): Glucose tolerance and diabetes. *Diabetologia* **22**, 73–78.
73. Butler, W. J., Ostrander, L. D., Jr., Carman, W. J., and Lamphiear, D. E. (1982). Diabetes mellitus in Tecumseh, Michigan. Prevalence, incidence, and associated conditions. *Am. J. Epidemiol.* **116**, 971–980.
74. Balkau, B., King, H., Zimmet, P., and Raper, L. R. (1985). Factors associated with the development of diabetes in the Micronesian population of Nauru. *Am. J. Epidemiol.* **122**, 594–605.
75. Haffner, S. M., Stern, M. P., Hazuda, H. P., Rosenthal, M., Knapp, J. A., and Malina, R. M. (1986). Role of obesity and fat distribution in non-insulin-dependent diabetes mellitus in Mexican Americans and non-Hispanic whites. *Diabetes Care* **9**, 153–161.

76. Ohlson, L. O., Larsson, B., Bjorntorp, P., Eriksson, H., Svardsudd, K., Welin, L., Tibblin, G., and Wilhelmsen, L. (1988). Risk factors for type 2 (non-insulin-dependent) diabetes mellitus. Thirteen and one-half years of follow-up of the participants in a study of Swedish men born in 1913. *Diabetologia* **31**, 798–805.
77. Schranz, A. G. (1989). Abnormal glucose tolerance in the Maltese. A population-based longitudinal study of the natural history of NIDDM and IGT in Malta. *Diabetes Res. Clin. Pract.* **7**, 7–16.
78. Feldman, R., Sender, A. J., and Siegelaub, A. B. (1969). Difference in diabetic and nondiabetic fat distribution patterns by skinfold measurements. *Diabetes* **18**, 478–486.
79. Joos, S. K., Mueller, W. H., Hanis, C. L., and Schull, W. J. (1984). Diabetes alert study: Weight history and upper body obesity in diabetic and non-diabetic Mexican American adults. *Ann. Hum. Biol.* **11**, 167–171.
80. Kissebah, A. H., Vydelingum, N., Murray, R., Evans, D. J., Hartz, A. J., Kalkhoff, R. K., and Adams, P. W. (1982). Relation of body fat distribution to metabolic complications of obesity. *J. Clin. Endocrinol. Metab.* **54**, 254–260.
81. Krotkiewski, M., Bjorntorp, P., Sjostrom, L., and Smith, U. (1983). Impact of obesity on metabolism in men and women. Importance of regional adipose tissue distribution. *J. Clin. Invest.* **72**, 1150–1162.
82. Evans, D. J., Hoffmann, R. G., Kalkhoff, R. K., and Kissebah, A. H. (1983). Relationship of androgenic activity to body fat topography, fat cell morphology, and metabolic aberrations in premenopausal women. *J. Clin. Endocrinol. Metab.* **57**, 304–310.
83. Hartz, A. J., Rupley, D. C., and Rimm, A. A. (1984). The association of girth measurements with disease in 32,856 women. *Am. J. Epidemiol.* **119**, 71–80.
84. Evans, D. J., Hoffmann, R. G., Kalkhoff, R. K., and Kissebah, A. H. (1984). Relationship of body fat topography to insulin sensitivity and metabolic profiles in premenopausal women. *Metabolism* **33**, 68–75.
85. Galassi, A., Reynolds, K., and He, J. (2006). Metabolic syndrome and risk of cardiovascular disease: A meta-analysis. *Am. J. Med.* **119**, 812–819.
86. Saad, M. F., Knowler, W. C., Pettitt, D. J., Nelson, R. G., Mott, D. M., and Bennett, P. H. (1988). The natural history of impaired glucose tolerance in the Pima Indians. *N. Engl. J. Med.* **319**, 1500–1506.
87. Diehl, A. K., and Stern, M. P. (1989). Special health problems of Mexican-Americans: obesity, gallbladder disease, diabetes mellitus, and cardiovascular disease. *Adv. Intern. Med.* **34**, 73–96.
88. Sicree, R. A., Zimmet, P. Z., King, H. O., and Coventry, J. S. (1987). Plasma insulin response among Nauruans. Prediction of deterioration in glucose tolerance over 6 yr. *Diabetes* **36**, 179–186.
89. Kadowaki, T., Miyake, Y., Hagura, R., Akanuma, Y., Kajinuma, H., Kuzuya, N., Takaku, F., and Kosaka, K. (1984). Risk factors for worsening to diabetes in subjects with impaired glucose tolerance. *Diabetologia* **26**, 44–49.
90. Wilding, J., and Williams, G. (1998). Diabetes and Obesity. *In* "Clinical Obesity" (P. G. Kopelman, and M. J. Stock, Eds.), pp. 309–349. Blackwell Science, Cambridge, U.K.
91. Barker, D. J. P. (1998). "Mothers, Babies and Health in Later Life" (2nd ed.). Churchill Livingstone, Edinburgh.
92. Barker, D. J. (1995). Fetal origins of coronary heart disease. *BMJ* **311**, 171–174.
93. Hales, C. N., Barker, D. J., Clark, P. M., Cox, L. J., Fall, C., Osmond, C., and Winter, P. D. (1991). Fetal and infant growth and impaired glucose tolerance at age 64. *BMJ* **303**, 1019–1022.
94. Lithell, H. O., McKeigue, P. M., Berglund, L., Mohsen, R., Lithell, U. B., and Leon, D. A. (1996). Relation of size at birth to non-insulin dependent diabetes and insulin concentrations in men aged 50–60 years. *BMJ* **312**, 406–410.
95. McCance, D. R., Pettitt, D. J., Hanson, R. L., Jacobsson, L. T., Knowler, W. C., and Bennett, P. H. (1994). Birth weight and non-insulin dependent diabetes: Thrifty genotype, thrifty phenotype, or surviving small baby genotype? *BMJ* **308**, 942–945.
96. Rich-Edwards, J. W., Colditz, G. A., Stampfer, M. J., Willett, W. C., Gillman, M. W., Hennekens, C. H., Speizer, F. E., and Manson, J. E. (1999). Birthweight and the risk for type 2 diabetes mellitus in adult women. *Ann. Intern. Med.* **130**, 278–284.
97. Ravelli, A. C., van der Meulen, J. H., Michels, R. P., Osmond, C., Barker, D. J., Hales, C. N., and Bleker, O. P. (1998). Glucose tolerance in adults after prenatal exposure to famine. *Lancet* **351**, 173–177.
98. Snoeck, A., Remacle, C., Reusens, B., and Hoet, J. J. (1990). Effect of a low protein diet during pregnancy on the fetal rat endocrine pancreas. *Biol. Neonate* **57**, 107–118.
99. Dahri, S., Snoeck, A., Reusens-Billen, B., Remacle, C., and Hoet, J. J. (1991). Islet function in offspring of mothers on low-protein diet during gestation. *Diabetes* **40**(Suppl 2), 115–120.
100. Martin, B. C., Warram, J. H., Krolewski, A. S., Bergman, R. N., Soeldner, J. S., and Kahn, C. R. (1992). Role of glucose and insulin resistance in development of type 2 diabetes mellitus: Results of a 25-year follow-up study. *Lancet* **340**, 925–929.
101. O'Rahilly, S., Turner, R. C., and Matthews, D. R. (1988). Impaired pulsatile secretion of insulin in relatives of patients with non-insulin-dependent diabetes. *N. Engl. J. Med.* **318**, 1225–1230.
102. Porte, D., Jr, and Kahn, S. E. (1989). Hyperproinsulinemia and amyloid in NIDDM. Clues to etiology of islet beta-cell dysfunction? *Diabetes* **38**, 1333–1336.
103. King, J. C. (2006). Maternal obesity, metabolism, and pregnancy outcomes. *Annu. Rev. Nutr.* **26**, 271–291.
104. Williams, L. A., Evans, S. F., and Newnham, J. P. (1997). Prospective cohort study of factors influencing the relative weights of the placenta and the newborn infant. *BMJ* **314**, 1864–1868.
105. Barker, D. J. (1999). The fetal origins of type 2 diabetes mellitus. *Ann. Intern. Med.* **130**, 322–324.
106. Oken, E., Taveras, E. M., Kleinman, K. P., Rich-Edwards, J. W., and Gillman, M. W. (2007). Gestational weight gain and child adiposity at age 3 years. *Am. J. Obstet. Gynecol.* **196**, 322, e321–328.
107. Galtier-Dereure, F., Boegner, C., and Bringer, J. (2000). Obesity and pregnancy: Complications and cost. *Am. J. Clin. Nutr.* **71**, 1242S–1248S.

108. Whitaker, R. C., and Dietz, W. H. (1998). Role of the prenatal environment in the development of obesity. *J. Pediatr.* **132**, 768–776.
109. Greenberg, A. S., and Obin, M. S. (2006). Obesity and the role of adipose tissue in inflammation and metabolism. *Am. J. Clin. Nutr.* **83**, 461S–465S.
110. Butte, N. F. (2000). Carbohydrate and lipid metabolism in pregnancy: Normal compared with gestational diabetes mellitus. *Am. J. Clin. Nutr.* **71**, 1256S–1261S.
111. Board, F. A. N. (1990). "Nutrition During Pregnancy" National Academies Press, Washington, DC.
112. Rhodes, J. C., Schoendorf, K. C., and Parker, J. D. (2003). Contribution of excess weight gain during pregnancy and macrosomia to the cesarean delivery rate, 1990–2000. *Pediatrics* **111**, 1181–1185.
113. Siega-Riz, A. M., Evenson, K. R., and Dole, N. (2004). Pregnancy-related weight gain—a link to obesity? *Nutr. Rev.* **62**, S105–111.
114. Wolever, T. M., and Mehling, C. (2003). Long-term effect of varying the source or amount of dietary carbohydrate on postprandial plasma glucose, insulin, triacylglycerol, and free fatty acid concentrations in subjects with impaired glucose tolerance. *Am. J. Clin. Nutr.* **77**, 612–621.
115. Steiner, D. F., Tager, H. S., Chan, S. J., Nanjo, K., Sanke, T., and Rubenstein, A. H. (1990). Lessons learned from molecular biology of insulin-gene mutations. *Diabetes Care* **13**, 600–609.
116. Froguel, P., Zouali, H., Vionnet, N., Velho, G., Vaxillaire, M., Sun, F., Lesage, S., Stoffel, M., Takeda, J., Passa, P. *et al.* (1993). Familial hyperglycemia due to mutations in glucokinase. Definition of a subtype of diabetes mellitus. *N. Engl. J. Med.* **328**, 697–702.
117. Taylor, S. I., Accili, D., and Imai, Y. (1994). Insulin resistance or insulin deficiency. Which is the primary cause of NIDDM? *Diabetes* **43**, 735–740.
118. Sladek, R., Rocheleau, G., Rung, J., Dina, C., Shen, L., Serre, D., Boutin, P., Vincent, D., Belisle, A., Hadjadj, S., Balkau, B., Heude, B. (2007). A genome-wide association study identifies novel risk loci for type 2 diabetes. *Nature* **445**, 881–885.
119. Gerich, J. E. (1998). The genetic basis of type 2 diabetes mellitus: Impaired insulin secretion versus impaired insulin sensitivity. *Endocr. Rev.* **19**, 491–503.
120. Vauhkonen, I., Niskanen, L., Vanninen, E., Kainulainen, S., Uusitupa, M., and Laakso, M. (1998). Defects in insulin secretion and insulin action in non-insulin-dependent diabetes mellitus are inherited. Metabolic studies on offspring of diabetic probands. *J. Clin. Invest.* **101**, 86–96.
121. Kohrt, W. M., Kirwan, J. P., Staten, M. A., Bourey, R. E., King, D. S., and Holloszy, J. O. (1993). Insulin resistance in aging is related to abdominal obesity. *Diabetes* **42**, 273–281.
122. DeFronzo, R. A., and Ferrannini, E. (1991). Insulin resistance. A multifaceted syndrome responsible for NIDDM, obesity, hypertension, dyslipidemia, and atherosclerotic cardiovascular disease. *Diabetes Care* **14**, 173–194.
123. Beard, J. C., Ward, W. K., Halter, J. B., Wallum, B. J., Porte, D., Jr (1987). Relationship of islet function to insulin action in human obesity. *J. Clin. Endocrinol. Metab.* **65**, 59–64.
124. Prager, R., Wallace, P., and Olefsky, J. M. (1986). In vivo kinetics of insulin action on peripheral glucose disposal and hepatic glucose output in normal and obese subjects. *J. Clin. Invest.* **78**, 472–481.
125. Wisse, B. E. (2004). The inflammatory syndrome: The role of adipose tissue cytokines in metabolic disorders linked to obesity. *J. Am. Soc. Nephrol.* **15**, 2792–2800.
126. Boyko, E. J., Fujimoto, W. Y., Leonetti, D. L., and Newell-Morris, L. (2000). Visceral adiposity and risk of type 2 diabetes: A prospective study among Japanese Americans. *Diabetes Care* **23**, 465–471.
127. Larsson, H., and Ahren, B. (2000). Islet dysfunction in insulin resistance involves impaired insulin secretion and increased glucagon secretion in postmenopausal women with impaired glucose tolerance. *Diabetes Care* **23**, 650–657.
128. Samaras, K., and Campbell, L. V. (2000). Increasing incidence of type 2 diabetes in the third millennium: Is abdominal fat the central issue? *Diabetes Care* **23**, 441–442.
129. Dvorak, R. V., DeNino, W. F., Ades, P. A., and Poehlman, E. T. (1999). Phenotypic characteristics associated with insulin resistance in metabolically obese but normal-weight young women. *Diabetes* **48**, 2210–2214.
130. Ruderman, N., Chisholm, D., Pi-Sunyer, X., and Schneider, S. (1998). The metabolically obese, normal-weight individual revisited. *Diabetes* **47**, 699–713.
131. Helmrich, S. P., Ragland, D. R., Leung, R. W., Paffenbarger, R. S., Jr (1991). Physical activity and reduced occurrence of non-insulin-dependent diabetes mellitus. *N. Engl. J. Med.* **325**, 147–152.
132. Salans, L. B., Knittle, J. L., and Hirsch, J. (1968). The role of adipose cell size and adipose tissue insulin sensitivity in the carbohydrate intolerance of human obesity. *J. Clin. Invest.* **47**, 153–165.
133. Olefsky, J. M. (1976). The insulin receptor: its role in insulin resistance of obesity and diabetes. *Diabetes* **25**, 1154–1162.
134. Goldrick, R. B., and McLoughlin, G. M. (1970). Lipolysis and lipogenesis from glucose in human fat cells of different sizes. Effects of insulin, epinephrine, and theophylline. *J. Clin. Invest.* **49**, 1213–1223.
135. Bjorntorp, P., Bengtsson, C., Blohme, G., Jonsson, A., Sjostrom, L., Tibblin, E., Tibblin, G., and Wilhelmsen, L. (1971). Adipose tissue fat cell size and number in relation to metabolism in randomly selected middle-aged men and women. *Metabolism* **20**, 927–935.
136. Bjorntorp, P. (1984). Hazards in subgroups of human obesity. *Eur J. Clin. Invest.* **14**, 239–241.
137. Randle, P. J., Garland, P. B., Hales, C. N., and Newsholme, E. A. (1963). The glucose fatty-acid cycle. Its role in insulin sensitivity and the metabolic disturbances of diabetes mellitus. *Lancet* **1**, 785–789.
138. Stern, M. P., Olefsky, J., Farquhar, J. W., and Reaven, G. M. (1973). Relationship between fasting plasma lipid levels and adipose tissue morphology. *Metabolism* **22**, 1311–1317.
139. Smith, C. J., Vasta, V., Degerman, E., Belfrage, P., and Manganiello, V. C. (1991). Hormone-sensitive cyclic GMP-inhibited cyclic AMP phosphodiesterase in rat adipocytes. Regulation of insulin- and cAMP-dependent activation by phosphorylation. *J. Biol. Chem.* **266**, 13385–13390.
140. Lafontan, M., and Berlan, M. (1993). Fat cell adrenergic receptors and the control of white and brown fat cell function. *J. Lipid. Res.* **34**, 1057–1091.

141. Rebuffe-Scrive, M., and Bjorntorp, P. (1985). Regional adipose tissue metabolism in man. *In* "Metabolic Complications of Human Obesity," pp. 149–159. Excerpta Medicine, Amsterdam.
142. Lafontan, M., Dang-Tran, L., and Berlan, M. (1979). Alpha-adrenergic antilipolytic effect of adrenaline in human fat cells of the thigh: Comparison with adrenaline responsiveness of different fat deposits. *Eur J. Clin. Invest.* **9**, 261–266.
143. Lonnqvist, F., Thome, A., Nilsell, K., Hoffstedt, J., and Arner, P. (1995). A pathogenic role of visceral fat beta 3-adrenoceptors in obesity. *J. Clin. Invest.* **95**, 1109–1116.
144. Carey, D. G., Jenkins, A. B., Campbell, L. V., Freund, J., and Chisholm, D. J. (1996). Abdominal fat and insulin resistance in normal and overweight women: Direct measurements reveal a strong relationship in subjects at both low and high risk of NIDDM. *Diabetes* **45**, 633–638.
145. Despres, J. P. (1993). Abdominal obesity as important component of insulin-resistance syndrome. *Nutrition* **9**, 452–459.
146. Evans, D. J., Murray, R., and Kissebah, A. H. (1984). Relationship between skeletal muscle insulin resistance, insulin-mediated glucose disposal, and insulin binding. Effects of obesity and body fat topography. *J. Clin. Invest.* **74**, 1515–1525.
147. Bedard, S., Marcotte, B., and Marette, A. (1997). Cytokines modulate glucose transport in skeletal muscle by inducing the expression of inducible nitric oxide synthase. *Biochem. J.* **325**(Pt 2), 487–493.
148. Kapur, S., Bedard, S., Marcotte, B., Cote, C. H., and Marette, A. (1997). Expression of nitric oxide synthase in skeletal muscle: A novel role for nitric oxide as a modulator of insulin action. *Diabetes* **46**, 1691–1700.
149. Hotamisligil, G. S., and Spiegelman, B. M. (1994). Tumor necrosis factor alpha: A key component of the obesity–diabetes link. *Diabetes* **43**, 1271–1278.
150. Lemieux, I., Pascot, A., Prud'homme, D., Almeras, N., Bogaty, P., Nadeau, A., Bergeron, J., and Despres, J. P. (2001). Elevated C-reactive protein: Another component of the atherothrombotic profile of abdominal obesity. *Arterioscler. Thromb. Vasc. Biol.* **21**, 961–967.
151. Cote, M., Mauriege, P., Bergeron, J., Almeras, N., Tremblay, A., Lemieux, I., and Despres, J. P. (2005). Adiponectinemia in visceral obesity: Impact on glucose tolerance and plasma lipoprotein and lipid levels in men. *J. Clin. Endocrinol. Metab.* **90**, 1434–1439.
152. Shoelson, S. E., Lee, J., and Goldfine, A. B. (2006). Inflammation and insulin resistance. *J. Clin. Invest.* **116**, 1793–1801.
153. Peiris, A. N., Mueller, R. A., Smith, G. A., Struve, M. F., and Kissebah, A. H. (1986). Splanchnic insulin metabolism in obesity. Influence of body fat distribution. *J. Clin. Invest.* **78**, 1648–1657.
154. Campbell, P. J., Carlson, M. G., and Nurjhan, N. (1994). Fat metabolism in human obesity. *Am. J. Physiol.* **266**, E600–605.
155. Kissebah, A. H., Evans, D. J., Peiris, A. N., and Wilson, C. R. (1985). Endocrine characteristics in regional obesities: Role of sex steroids. *In* "Metabolic Complications for Human Obesities" (J. Vague, P. Bjorntorp, B. Guy-Grand, M. Rebuffe-Scrive, and P. Vague, Eds.), pp. 115–130. Elsevier, Amsterdam.
156. Garvey, W. T., Maianu, L., Huecksteadt, T. P., Birnbaum, M. J., Molina, J. M., and Ciaraldi, T. P. (1991). Pretranslational suppression of a glucose transporter protein causes insulin resistance in adipocytes from patients with non-insulin-dependent diabetes mellitus and obesity. *J. Clin. Invest.* **87**, 1072–1081.
157. Garvey, W. T., Olefsky, J. M., Matthaei, S., and Marshall, S. (1987). Glucose and insulin co-regulate the glucose transport system in primary cultured adipocytes. A new mechanism of insulin resistance. *J. Biol. Chem.* **262**, 189–197.
158. Rebuffe-Scrive, M., Andersson, B., Olbe, L., and Bjorntorp, P. (1989). Metabolism of adipose tissue in intraabdominal depots of nonobese men and women. *Metabolism* **38**, 453–458.
159. Boden, G. (1997). Role of fatty acids in the pathogenesis of insulin resistance and NIDDM. *Diabetes* **46**, 3–10.
160. Cigolini, M., and Smith, U. (1979). Human adipose tissue in culture. VIII. Studies on the insulin-antagonistic effect of glucocorticoids. *Metabolism* **28**, 502–510.
161. Kelley, D. E., Mokan, M., Simoneau, J. A., and Mandarino, L. J. (1993). Interaction between glucose and free fatty acid metabolism in human skeletal muscle. *J. Clin. Invest.* **92**, 91–98.
162. Kelley, D. E., and Mandarino, L. J. (1990). Hyperglycemia normalizes insulin-stimulated skeletal muscle glucose oxidation and storage in noninsulin-dependent diabetes mellitus. *J. Clin. Invest.* **86**, 1999–2007.
163. Kelley, D. E., and Mandarino, L. J. (2000). Fuel selection in human skeletal muscle in insulin resistance: A reexamination. *Diabetes* **49**, 677–683.
164. Winder, W. W., Arogyasami, J., Elayan, I. M., and Cartmill, D. (1990). Time course of exercise-induced decline in malonyl-CoA in different muscle types. *Am. J. Physiol.* **259**, E266–E271.
165. Simoneau, J. A., Veerkamp, J. H., Turcotte, L. P., and Kelley, D. E. (1999). Markers of capacity to utilize fatty acids in human skeletal muscle: Relation to insulin resistance and obesity and effects of weight loss. *Faseb J* **13**, 2051–2060.
166. Pan, D. A., Lillioja, S., Kriketos, A. D., Milner, M. R., Baur, L. A., Bogardus, C., Jenkins, A. B., and Storlien, L. H. (1997). Skeletal muscle triglyceride levels are inversely related to insulin action. *Diabetes* **46**, 983–988.
167. Goodpaster, B. H., Thaete, F. L., Simoneau, J. A., and Kelley, D. E. (1997). Subcutaneous abdominal fat and thigh muscle composition predict insulin sensitivity independently of visceral fat. *Diabetes* **46**, 1579–1585.
168. Arcaro, G., Zamboni, M., Rossi, L., Turcato, E., Covi, G., Armellini, F., Bosello, O., and Lechi, A. (1999). Body fat distribution predicts the degree of endothelial dysfunction in uncomplicated obesity. *Int. J. Obes. Relat. Metab. Disord.* **23**, 936–942.
169. Reaven, G. M. (1995). The fourth musketeer—from Alexandre Dumas to Claude Bernard. *Diabetologia* **38**, 3–13.
170. Surwit, R. S., Schneider, M. S., and Feinglos, M. N. (1992). Stress and diabetes mellitus. *Diabetes Care* **15**, 1413–1422.
171. Mooy, J. M., de Vries, H., Grootenhuis, P. A., Bouter, L. M., and Heine, R. J. (2000). Major stressful life events in relation to prevalence of undetected type 2 diabetes: The Hoorn Study. *Diabetes Care* **23**, 197–201.
172. Surwit, R. S., and Feinglos, M. N. (1988). Stress and autonomic nervous system in type II diabetes. A hypothesis. *Diabetes Care* **11**, 83–85.

173. Black, P. H. (2006). The inflammatory consequences of psychologic stress: Relationship to insulin resistance, obesity, atherosclerosis and diabetes mellitus, type II. *Med. Hypotheses* **67**, 879–891.
174. Fujioka, S., Matsuzawa, Y., Tokunaga, K., Kawamoto, T., Kobatake, T., Keno, Y., Kotani, K., Yoshida, S., and Tarui, S. (1991). Improvement of glucose and lipid metabolism associated with selective reduction of intra-abdominal visceral fat in premenopausal women with visceral fat obesity. *Int. J. Obes.* **15**, 853–859.
175. Markovic, T. P., Jenkins, A. B., Campbell, L. V., Furler, S. M., Kraegen, E. W., and Chisholm, D. J. (1998). The determinants of glycemic responses to diet restriction and weight loss in obesity and NIDDM. *Diabetes Care* **21**, 687–694.
176. Esposito, K., Pontillo, A., Di Palo, C., Giugliano, G., Masella, M., Marfella, R., and Giugliano, D. (2003). Effect of weight loss and lifestyle changes on vascular inflammatory markers in obese women: A randomized trial. *Jama* **289**, 1799–1804.
177. Ramos, E. J., Xu, Y., Romanova, I., Middleton, F., Chen, C., Quinn, R., Inui, A., Das, U., and Meguid, M. M. (2003). Is obesity an inflammatory disease? *Surgery* **134**, 329–335.
178. Ceriello, A., Taboga, C., Tonutti, L., Quagliaro, L., Piconi, L., Bais, B., Da Ros, R., and Motz, E. (2002). Evidence for an independent and cumulative effect of postprandial hypertriglyceridemia and hyperglycemia on endothelial dysfunction and oxidative stress generation: Effects of short- and long-term simvastatin treatment. *Circulation* **106**, 1211–1218.
179. Negrean, M., Stirban, A., Stratmann, B., Gawlowski, T., Horstmann, T., Gotting, C., Kleesiek, K., Mueller-Roesel, M., Koschinsky, T., Uribarri, J., Vlassara, H., and Tschoepe, D. (2007). Effects of low- and high-advanced glycation endproduct meals on macro- and microvascular endothelial function and oxidative stress in patients with type 2 diabetes mellitus. *Am. J. Clin. Nutr.* **85**, 1236–1243.
180. (1998). Intensive blood-glucose control with sulphonylureas or insulin compared with conventional treatment and risk of complications in patients with type 2 diabetes (UKPDS 33). UK Prospective Diabetes Study (UKPDS) Group. *Lancet* **352**, 837–853.
181. Yki-Jarvinen, H., Ryysy, L., Kauppila, M., Kujansuu, E., Lahti, J., Marjanen, T., Niskanen, L., Rajala, S., Salo, S., Seppala, P., Tulokas, T., and Viikari, J. (1997). Effect of obesity on the response to insulin therapy in noninsulin-dependent diabetes mellitus. *J. Clin. Endocrinol. Metab.* **82**, 4037–4043.
182. (1993). The effect of intensive treatment of diabetes on the development and progression of long-term complications in insulin-dependent diabetes mellitus. The Diabetes Control and Complications Trial Research Group. *N. Engl. J. Med.* **329**, 977–986.
183. Harris, M. I., Eastman, R. C., Cowie, C. C., Flegal, K. M., and Eberhardt, M. S. (1999). Racial and ethnic differences in glycemic control of adults with type 2 diabetes. *Diabetes Care* **22**, 403–408.
184. Ledikwe, J. H., Rolls, B. J., Smiciklas-Wright, H., Mitchell, D. C., Ard, J. D., Champagne, C., Karanja, N., Lin, P. H., Stevens, V. J., and Appel, L. J. (2007). Reductions in dietary energy density are associated with weight loss in overweight and obese participants in the PREMIER trial. *Am. J. Clin. Nutr.* **85**, 1212–1221.
185. Pronk, N. P., and Wing, R. R. (1994). Physical activity and long-term maintenance of weight loss. *Obes. Res.* **2**, 587–599.

CHAPTER **32**

Nutrition Management of Diabetes Mellitus

ANN M. COULSTON
Mountain View, California

Contents

I. INTRODUCTION

Diabetes mellitus is a group of metabolic disorders characterized by hyperglycemia resulting from defects in insulin secretion, insulin action, or both. The cause of diabetes continues to be a mystery, although both genetic and environmental factors such as obesity and lack of exercise play a role. The hyperglycemia of diabetes increases the risk of a variety of complications including cardiovascular disease, stroke, visual impairment and blindness, nephropathy leading to renal failure, and neuropathy. In the United States, the annual economic cost of diabetes in 2002 was estimated at $132 billion, or about one of every 10 health care dollars spent [1]. Diabetes is a chronic illness that requires continuing medical care and patient self-management to prevent acute and long-term complications.

Broadly, diabetes is classified into two major forms. Type 1 diabetes is characterized by a complete inability of the beta cells of the pancreas to produce insulin. It most commonly occurs during childhood and young adulthood and accounts for about 10% of all persons diagnosed with diabetes. Most of Americans who are diagnosed with diabetes have type 2, characterized by a combination of a defect in insulin secretion and insulin resistance at the site of insulin action in the muscle, liver, and adipose tissue [2]. Often people with type 2 diabetes are obese, and obesity itself causes some degree of insulin resistance. Insulin secretion improves with weight loss but hyperglycemia seldom returns to normal. The increase in diabetes parallels the rising rates of obesity observed over the past decade [3]. Type 2 diabetes has a gradual onset, usually beginning with an impairment of glucose tolerance that frequently goes undiagnosed in the early stages. It is estimated that, at diagnosis, most adults with type 2 diabetes have had the disease for an average of 7 years and the ravages of chronic hyperglycemia may already be damaging vital body tissues [4]. Type 2 diabetes is now epidemic. In the United States there has been a 61% increase in incidence between 1990 and 2001 [5]. There are 1.5 million new cases of diabetes per year and the prevalence in 2005 was almost 21 million [6]. We are witnessing a trend toward an increase in younger individuals developing type 2 diabetes in association with an increase in childhood and adolescent obesity.

Results of the United Kingdom Prospective Diabetes Study (UKPDS) [7] demonstrated that the hyperglycemia in type 2 diabetes is no less deadly than that in type 1. Advances in medical management and improved techniques for self-monitoring of blood glucose have resulted in more aggressive management of all types of diabetes.

The medical management of diabetes is complex. Patients require a combination of medications and lifestyle changes to achieve blood glucose control [8]. Many are obese or overweight and frequently have associated disease states that require medical management, such as hypertension. Once diabetes is diagnosed, time continues to be a factor as beta-cell exhaustion progresses (in type 2) and eventually necessitates the use of insulin secretagogues or exogenous insulin therapy. Early detection followed by early intervention might slow this progress and reduce complications [4]. Reduced energy intake, moderate weight loss, and moderate activity reduce the incidence of diabetes in those with impaired fasting glucose (100–125 mg/dl) [9–11]. Glycemic improvement, as a result of energy restriction, is due to the combined effects of reduced energy and carbohydrate intake.

Medical nutrition therapy is important in preventing diabetes, managing diagnosed diabetes, and slowing the rate of the development of complications. Current nutrition recommendations to achieve and maintain glucose, lipid, and blood pressure goals are simple to state, but difficult to initiate and even more difficult to maintain. Nutrition therapy can be summarized as follows: lose weight if overweight (more than 25 kg/m^2), restrict saturated and *trans*-fat intake, and divide total nutrient intake throughout the day [12]. As with treatment regimens for many chronic diseases, behavior modification and lifestyle change are essential. Achieving nutrition goals requires a coordinated effort including the person with diabetes. Medical nutrition therapy for treatment of diabetes takes into account the best available scientific evidence, treatment goals, strategies to attain such goals, and changes individuals with diabetes are willing and able to make.

The traditional approach to medical management of type 2 diabetes, consisting of monotherapy, a weight-loss or "no sugar" diet, and advice to "get some exercise," does not yield desired medical outcomes. Current recommendations are for persons with diabetes to have targeted blood glucose levels, made possible by feedback from daily self-monitoring of blood glucose and routine laboratory evaluations. Diabetes self-management education is an important part of overall medical management [13]. Ongoing training programs for educators of patients with diabetes are essential [14]. Table 1 defines glycemic and lipid concentration treatment goals for adults with diabetes.

This chapter addresses the nutritional management of persons with diabetes. As mentioned earlier, persons with type 2 diabetes tend to be overweight, so a discussion of diabetes management relative to weight loss is included. People with type 1 diabetes usually have normal body weight. Remember that for optimal management of plasma glucose, people with type 2 diabetes may require insulin therapy. Nutrition issues for people with gestational diabetes are discussed in Chapter 33 in this volume. Throughout the discussion of energy and nutrient intake, the focus is to achieve the goals of medical nutrition therapy of persons with diabetes. These are to achieve and maintain targeted blood glucose levels, a lipid profile that reduces cardiovascular disease, and blood pressure in the normal range. Table 2 summarizes types of diabetes.

TABLE 1 Recommendations for Adults with Diabetes

Glycemic Control	
HbAlc	<7.0% (normal 4.0–6.0%)
Preprandial capillary plasma glucose	90–130 mg/dl (5.0–7.2 mmol/liter)
Peak postprandial capillary plasma glucose (1–2 hours after start of meal)	<180 mg/dl (<10.0 mmol/liter)
Blood pressure	<130/80 mmHg
Lipids	
LDL cholesterol	<100 mg/dl (<2.6 mmol/liter)
Triglycerides	<150 mg/dl (<1.7 mmol/liter)
HDL cholesterol	>40 mg/dl (>1.0 mmol/liter)

From [12].
HDL, high-density lipoprotein; LDL, low-density lipoprotein.

TABLE 2 Classification Criteria

Diabetes Type	Description
Type 1	Results from pancreatic beta-cell destruction, usually leading to absolute insulin deficiency
Type 2	Results from a progressive insulin secretory defect in conjunction with insulin resistance
Gestational diabetes mellitus	Diagnosed during pregnancy
Diabetes due to other causes	Genetic defects in beta-cell function or insulin action
	Diseases of the exocrine pancrease, i.e., cystic fibrosis
	Drug- or chemical-induced

II. ENERGY INTAKE AND BODY WEIGHT MANAGEMENT

Approximately 90% of people with type 2 diabetes are overweight [4]. Exercise and restriction of energy intake can improve glucose tolerance, diminish insulin resistance, and improve coronary risk factors in many patients with established type 2 diabetes. On the other hand, the therapeutic efficacy of diet and exercise in such patients has been limited because most individuals with this disorder are more than 40 years old at the time of diagnosis and many tend to be resistant to lifestyle changes [1].

Obesity, upper-body obesity in particular, aging, and a sedentary lifestyle are independent environmental factors that contribute to insulin resistance [15]. Weight loss improves blood glucose control by decreasing insulin resistance, which improves glucose uptake and reduces hepatic glucose output [16]. Studies have evaluated blood glucose control during weight loss and demonstrate that metabolic changes occur with as little as a 10-kg weight loss [16]. Blood glucose levels and insulin sensitivity continue to improve as weight loss progresses on an energy-restricted constant diet [17]. Patients with hyperinsulinemia respond most dramatically to weight loss; thus early intervention before beta-cell exhaustion occurs provides the best possibility for improving blood glucose control with weight loss. Because we know that 30–40% of patients diagnosed with type 2 diabetes already have clinically significant ischemic heart disease, microvascular disease, and neuropathy at the time of diagnosis [4], diet and exercise may be more beneficial if instituted earlier in life, before the onset of hyperglycemia [9–11]. Glycemic control begins to improve within 24 hours of initiating a hypocaloric diet, even before any weight is lost [18]. In fact, within 10 days of a controlled hypocaloric diet 87% of the eventual drop in blood glucose occurs. These metabolic changes are temporary and require continuation of an energy intake that maintains a decreased body weight. Three landmark studies showed that moderate weight loss and increased physical activity can prevent or delay the development of type 2 diabetes in high-risk groups [9–11]. Data from the Diabetes Prevention Program (DPP) demonstrated that weight loss of 7% in the first year and increased physical activity of walking 150 minutes per week reduced the 4-year incidence of type 2 diabetes by 58% in men and women with impaired glucose tolerance [10].

Many suggestions have been proposed for the most effective macronutrient composition to achieve weight loss in patients with diabetes. Traditionally, weight-loss diets are low in fat (25–30% of energy); however, persons following these diets tend to have higher carbohydrate intakes. Because of the carbohydrate intolerance of patients with diabetes, scientists have questioned the wisdom of this approach especially for people with diabetes. At the same time there is concern that higher fat diets might promote obesity. Two reports in the literature have shown equal amounts of weight loss on hypocaloric low-fat or higher fat diets [19, 20]. When metabolic parameters were examined during weight loss, investigators found that energy restriction and weight loss improved glycemic control, lipid profiles, and blood pressure independent of diet composition. In another report, the cardiovascular risk profile was improved when monounsaturated fatty acids or carbohydrate replaced saturated fatty acids on weight loss diets [21]. More recently several studies have been conducted on low-carbohydrate, high-protein diets designed to promote weight loss. Subjects on the low-carbohydrate, high-protein/high-fat diet achieved greater short-term weight loss than those on the low-fat/high-carbohydrate diets, but by 12 months there was no difference in the two approaches [22–24]. Very-low-calorie diet therapy consisting of 400–800 kcal/day has been shown to be safe in obese individuals with diabetes [25]. However, this is a weight reduction method that can only be used for 3–4 months and then weight maintenance or moderately restricted diets must be initiated. To determine how this technique might work for the long term, Wing and colleagues [25, 26] studied short periods of energy restriction, ranging from 1 day/week to 1 week/month, to enhance weight loss efforts. Wing and colleagues in review of behavioral interventions conclude that although weight regain is common, about 66% of weight lost is maintained at 1 year [27]. One characteristic that all current weight loss methods have in common is that long-term weight loss is difficult and results are not dramatically different between weight loss regimens. The complexity of weight management has led clinicians to adopt the philosophy that absence of weight gain is itself a reputable goal.

Long-term maintenance of weight loss is more challenging than initial weight loss. Some strategies that are associated with success include eating a low-calorie diet (~1400 kcal/day) and low fat (<30% of total intake), daily monitoring of body weight, and regular physical activity. Other contributing factors include reduced portion sizes and snacking, eating breakfast, reduction in meals eaten away from home to fewer than four times per week, and watching television less than 3 hours per week on average [28].

To help rein in total energy intake, many in the United States select beverages and carbohydrate-rich foods sweetened with nonnutritive sweeteners. These substances are under surveillance of the FDA. To date, five nonnutritive sweeteners have FDA approval: acesulfame potassium, aspartame, neotame, saccharin, and sucralose. Prior to approval, all of these compounds have been tested in humans including people with diabetes and women during pregnancy [12].

III. MACRONUTRIENT INTAKE

A. Protein

Historically, the protein content, both type and amount, of the diet in patients with type 2 diabetes has played a secondary role to carbohydrate and fat. General concerns for dietary protein adequacy are to maintain lean body mass and nitrogen balance whether people have diabetes or not.

Body proteins are continuously being synthesized and degraded. The estimated turnover is about 280 g/day [29]. The amino acids resulting from protein degradation can be recycled, but this process is not well understood. Data indicate that the efficiency with which amino acids are recycled may be regulated by the amount of protein in the diet [29]. The lower the protein content of the diet, the more efficiently amino acids are utilized. The current recommended amount, 0.8 g/kg body weight of good-quality protein, represents an intake of about 10–11% of energy from protein [30]. Good-quality protein sources are defined as having high protein digestibility scores and providing all nine indispensable amino acids, such as meat, poultry, fish, eggs, milk, cheese, and soy. By comparison, protein sources with lower quality scores include cereals, grains, nuts, and vegetables. Thus, in meal planning, protein intake will be higher than the 0.8 gm/kg body weight to account for the mixed protein quality in foods. The estimated amount of protein ingested by the general population in the United States represents 15–20% of energy intake or 1.1–1.4 g/kg body weight based on 2000 kcal/day. This is considerably more than the minimum protein necessary even in people with diabetes.

Specific dietary protein recommendations for people with diabetes have not been established. An increased loss of body protein with severe insulin deficiency has been known for years [31]. This is especially striking in patients with type 1 diabetes in whom withdrawal of insulin results in a marked increase in protein loss [32]. Abnormalities in protein metabolism may be caused by insulin deficiency and insulin resistance in type 2 diabetes and in both cases are corrected by blood glucose control [31]. Protein malnutrition is uncommon in people with diabetes, probably because the amount of dietary protein ingested by the average adult regularly exceeds the required amount.

In people with established renal insufficiency, with or without diabetes, a restriction in dietary protein has been considered desirable to modify the progression of the disease. This is controversial and there is concern that frank protein deficiency may result [33, 34]. At the present time, there is no evidence that restricting protein in the diet of people with diabetes will prevent or delay the onset of renal insufficiency [12].

From reports in the literature, we know that the minimum amount of protein necessary to replace body stores is relatively small, and the amount of protein that can be tolerated without toxic effects is high [35]. Data to support beneficial effects of protein in the diet lower or higher than the typical Western diet (approximately 15–20% of energy from protein) are lacking. The position of the American Diabetes Association recommends 10–20% of energy from protein, with the lower amount being recommended for patients with overt nephropathy [12].

In the liver, amino acids that are not required to replace body proteins, particularly nonessential amino acids, are deaminated [36]. The amino group is condensed with carbon dioxide to form urea, which is then carried to the kidneys and excreted in the urine. The amount of urea excreted per day is an index of the amount of protein metabolized, although about 14% of newly synthesized urea is utilized by bacteria in the colon [37]. Small amounts of amino acids are also metabolized (deaminated) in the kidney [38]. The carbon skeletons remaining after deamination can be converted to glucose. The resulting glucose may contribute to plasma glucose concentration. Despite the conversion of amino acids to glucose, controlled feeding studies of known amounts of protein do not result in the predicted (calculated) increase in peripheral glucose concentration in normal or type 2 subjects [39, 40]. Further studies designed to elucidate these findings are described next.

It is known that the carbon skeleton of all amino acids derived from protein digestion, with the exception of leucine, can be used to synthesize glucose. Theoretically, ingestion of 100 g protein can yield 50–80 g glucose depending on the amino acid composition of the protein. Using isotope dilution techniques combined with determination of urea formation rates, it was calculated that ingestion of 50 g of cottage cheese protein would result in 34 g being deaminated over the 8-hour period of the study in normal subjects. However, the amount of glucose entering the circulation was only 9.7 g [41]. Thus, the amount of glucose produced was considerably less than the amount theorized by about 25 g. A similar technique, applied to patients with type 2 diabetes following the ingestion of beef protein, found that of the 50 g protein ingested only ~2 g could be accounted for by the appearance of glucose in the circulation over an 8-hour period [42]. Thus, in patients with diabetes, the amount of glucose appearing in the circulation was even less than in normal subjects. From the beef protein load, 28 g glucose was predicted. The fate of the remaining carbon skeletons is unknown.

In healthy people, protein is a weak insulin stimulator compared with glucose; however, in patients with type 2 diabetes, protein and glucose are equipotent in stimulating insulin secretion. After the ingestion of protein, the circulating insulin concentration is increased in both healthy subjects and patients with type 2 diabetes [43, 44]. When protein is ingested with glucose, healthy subjects have an additive insulin response [45], whereas obese subjects with type 2 diabetes exhibit a synergistic insulin response [46].

Protein-containing foods are generally classified as plant or animal based. Studies comparing animal to plant proteins are confounded by the fact that ingestion of plant protein diets is accompanied by additional dietary fiber, whereas diets based on animal protein are associated with an increased ingestion of dietary fat. Animal studies have demonstrated an atherogenicity of animal protein diets as compared to plant- or soy protein-based diets [47]. However, it is not possible with the data currently available to draw conclusions as to the benefits or hazards of either plant or animal proteins in the diets of people with diabetes.

B. Carbohydrate

Fasting blood glucose is determined by the overproduction of glucose from the liver and the body's ability to remove glucose from the bloodstream. Blood glucose concentration following a meal is determined by the rate of appearance of glucose into the bloodstream and its clearance or disappearance from the circulation [48]. Patients with diabetes require from 3 to 4 hours for blood glucose to return to fasting or premeal levels after eating. However, gastric emptying rate, intestinal motility, and factors that affect glucose removal from the circulation, such as insulin response or insulin resistance, modify the absolute plasma glucose response. Thus, plasma glucose concentration at any given time is the result of the action of medication, glucose absorption from meals, and endogenous production of glucose mainly from the liver.

Prior to the availability of insulin and other pharmaceutical agents, both glucose overproduction and the impaired ability to metabolize absorbed glucose were shown to be treatable by dietary manipulations. In the early 1900s semistarvation or low-calorie diets controlled overproduction of glucose. The amount of carbohydrate in the diet depended on the severity of diabetes and was tailored to the individual utilization rate [49]. Dietary patterns were low in carbohydrate and very high in fat as an energy fuel. Adhering to these diets was difficult, and semistarvation could only be followed for a limited time.

Following the availability of insulin and oral agents for the treatment of diabetes, food energy restriction and the carbohydrate content of the diet have been greatly relaxed. In fact, there has been a universal trend toward advocating a higher and higher carbohydrate content in the diet, not only for people with diabetes, but also for the general population [50]. An increase in carbohydrate intake allows a decrease in fat intake to meet overall energy needs. The protein content of the diet remains relatively constant. This high-carbohydrate dietary recommendation has been driven by the concern that dietary fat, especially saturated fat, in the diet is responsible for the increasing incidence of cardiovascular disease.

Because cardiovascular disease is more common in people with diabetes and accounts for the majority of deaths in patients with diabetes [51], dietary recommendations have followed the American Heart Association and recommended a high-carbohydrate, low-fat diet aimed at lowering plasma cholesterol concentration. About 40% of patients with type 2 diabetes have an elevation in LDL cholesterol [52]; the most frequent lipid abnormalities in patients with diabetes are hypertriglyceridemia, increased very low density lipoprotein (VLDL) cholesterol, and reduced HDL cholesterol [53]. The hypertriglyceridemia of type 2 diabetes is believed to be due in part to increased hepatic production of triglyceride-rich VLDL particles induced by increased dietary carbohydrate intake. An increase in small, dense LDL particles along with decreased HDL cholesterol and hypertriglyceridemia appear to be sequelae of insulin resistance, although a complete understanding of the mechanism remains to be elucidated [53, 54]. The insulin resistance that accompanies the development of type 2 diabetes results in hyperinsulinemia in the face of hyperglycemia. Epidemiological studies have demonstrated that, in addition to the dyslipidemia of type 2 diabetes, hyperglycemia and hyperinsulinemia also contribute to an increased risk for cardiovascular disease in people with diabetes [55, 56].

Dietary intervention studies of high-carbohydrate, low-fat diets as compared to moderate-carbohydrate, higher fat diets have indicated that the risk factors for cardiovascular disease in people with type 2 diabetes are exacerbated with high-carbohydrate, low-fat diets [57, 58]. Diets with relatively high fat (40–45% of energy from fat) in which monounsaturated and polyunsaturated fatty acids predominate demonstrate an improved lipid profile, and the increase in dietary fat does not hinder glucose disposal [59]. As a result of studies that demonstrate that higher intakes of carbohydrate accentuate plasma glucose, insulin, and triglyceride response, general nutrition recommendations currently are to individualize the amount of carbohydrate in the diet to optimize patient blood glucose and lipid goals [12].

Concern for the most appropriate type and amount of dietary carbohydrate for people with diabetes remains. A recent statement from the American Diabetes Association has addressed these issues [60]. For many years, it was felt that refined sucrose (i.e., table sugar) should be eliminated from the diets of people with diabetes despite the lack of convincing scientific data. Now clinical studies indicate that the amount of sucrose typically found in the American diet does not have an adverse effect on blood glucose control [61, 62]. Current dietary recommendations from the American Diabetes Association conclude that sucrose and sucrose-containing foods do not need to be restricted because of concern for aggravating hyperglycemia. Sucrose can be substituted for other carbohydrates in the meal plan. However, the intake of other nutrients, such as fat, that frequently accompany sucrose-containing foods needs to be taken into account so as to avoid excess energy intake [12]. Rather, the focus is on the amount of dietary carbohydrate at each meal and for the total day because the amount of carbohydrate at meals and throughout the day has a major impact on day-long blood glucose control.

Dietary fiber recommended intake mirrors guidelines for the general public, 14 g/1000 kcal [30]. Increasing consumption of soluble fiber has been shown to reduce serum cholesterol and improve colonic function [63]. Whether patients with diabetes can achieve improved glycemic control by consuming diets high in dietary fiber, and especially soluble fiber, has been debated. Studies have shown that mixing viscous, nonstarch polysaccharides with carbohydrate-rich foods reduces the postprandial plasma glucose response, presumably by slowing the intestinal absorption of glucose [64, 65]. Studies indicate that about 50 g fiber per day reduces glycemia and improves lipid concentrations in subjects with diabetes [66]. Palatability and gastrointestinal side effects limit the ability to achieve such high fiber intakes on a regular basis. Recommendations to choose a variety of fiber-containing foods such as legumes, whole grain breads and cereals, fruits, and vegetables daily seem to be the most appropriate guideline.

During the past century, investigators have reported differences in the plasma glucose response to the ingestion of various carbohydrate-containing foods despite the foods being matched for total carbohydrate content [67, 68]. Dietary carbohydrate foods are composed of a variety of mono-, di-, and polysaccharide structures. These sugars and starches are hydrolyzed into constituent glucose molecules by a combination of pancreatic amylase and intestinal mucosal glucosidase and maltase enzymes. These enzymes are present in excess so the rate-limiting process is not related to hydrolysis but rather to the physical state of the starch. Uncooked starch in general is poorly digested and thus is not well absorbed. Another hydrolysis factor is how readily the starch granules, which differ in size and structure depending on the plant source, undergo gelatinization. As a result, starch in many foods is not completely hydrolyzed or digestible.

Sugars, on the other hand, are readily digested and absorbed in the intestine. However, the monosaccharide fructose does not cause an elevation in blood glucose in normoglycemic patients or in patients treated for diabetes. Gannon and colleagues [69] verified this concept by comparing the day-long plasma glucose and insulin response to identical carbohydrate content meals in patients with type 2 diabetes. One day the meals were high in starch foods and the other day in sugars from fruit, vegetables, and dairy products. On the day of the higher sugars diet, plasma glucose and insulin were significantly lower following each meal. Although these findings are interesting and help explain the glycemic impact of foods, dietary patterns are a combination of starches and sugars. Consequently, when individual carbohydrate-containing foods are tested for their glycemic response, foods differ.

Jenkins and colleagues [70, 71] tested a large number of foods and compared them to the plasma glucose response for pure glucose or white bread and developed a "glycemic index." The glycemic index of a food is the increase of blood glucose urea above fasting over 2 hours following ingestion of a specific carbohydrate load, usually 50 g, from a specific food divided by the response to a reference food, glucose or white bread. Using this index the glycemic load of food portions, meals, or total diets can be calculated by multiplying the glycemic index of the constituent foods by the amounts of carbohydrate in each food and then totaling the values for all foods. Potential methodological problems with the glycemic index have been noted [72].

Of concern for the nutritional management of patients with diabetes is to determine whether or not these differences in glycemic response have clinical relevance. Recent reports from the Health Professionals Follow-up Study cohort and the Nurses' Health Study cohort indicate that over time, diets high in glycemic load and low in cereal fiber significantly increase the risk of type 2 diabetes [73, 74]. These observational studies need to be addressed in research settings with careful controls on potential confounding factors and in settings where the mechanisms of disease can be studied. Several clinical trials have reported that low-glycemic-index diets reduce glycemia in people with diabetes, but others have not confirmed this effect [60]. A recent meta-analysis of low-glycemic-index diet trials showed an improvement of 0.4% in glycosylated hemoglobin when compared with high-glycemic index diets [75]. Clinical trials that have examined eating patterns of subjects with diabetes conclude that most people are consuming diets of a moderate glycemic index level [76]. Thus, the modest impact of low- versus high-glycemic index foods relies on food choices to the extreme low or high indices.

As mentioned earlier, regulation of blood glucose to achieve near-normal levels is a primary goal in the management of diabetes. To the extent that individual people with diabetes can improve blood glucose control and estimate insulin and insulin secretagogue doses with the use of dietary techniques, reducing postprandial hyperglycemia can be important in limiting the complications of diabetes. Much of this control is achieved through personal experience in addition to food tables of carbohydrate amounts and glycemic indices.

C. Fat

Cardiovascular disease (CVD) is a major cause of mortality for people with diabetes [13]. Patients with type 2 diabetes have an increased prevalence of lipid abnormalities, which contribute to higher rates of CVD. Because diabetes is associated with a marked increase in coronary artery disease, there is a strong focus on dietary fatty acid intake. Target lipid concentrations are shown in Table 1. Several studies have shown that increased cardiovascular risk factors, including hypertension, precede the onset of type 2 diabetes [77]. Nutritional intervention coupled with medication therapy aimed at lowering LDL cholesterol, raising HDL cholesterol, and lowering triglyceride concentrations have been shown to reduce macrovascular disease and mortality in patients with type 2 diabetes, particularly in those who have had prior cardiovascular events [78, 79]. The National Cholesterol Education Program (NCEP) recognized that the atherogenic dyslipidemia associated with type 2 diabetes designates diabetes as a CVD risk equivalent [80]. Consequently, although the dyslipidemia of diabetes is associated with hypertriglyceridemia, LDL cholesterol is the primary target of therapy. Several studies have indicated effective results from pharmacological therapy to reduce LDL cholesterol and cardiovascular risk [78, 79].

Adequate glycemic control improves plasma lipid concentrations. In patients with very high triglycerides and poor glycemic control, glucose lowering may be necessary to control hypertriglyceridemia. People with type 2 diabetes have a dyslipidemia characterized by increased triglyceride concentrations and decreased HDL cholesterol levels, and small dense LDL cholesterol particles. This dyslipidemia is strongly associated with increased central (visceral) obesity, insulin resistance, and cardiovascular disease [54, 81, 82]. Much has been written about the association of hyperinsulinemia and/or insulin resistance with cardiovascular disease [83]. A study by Despres and colleagues [84] provides strong evidence that hyperinsulinemia is associated prospectively with the development of cardiovascular disease. LDL cholesterol is usually not different in people with and without diabetes, but a number of metabolic and compositional changes in LDL particles have been described. A preponderance of smaller denser LDL particles (subclass pattern B) has been identified as a risk for cardiovascular disease [82]. Although this has not been studied extensively in people with type 2 diabetes, small, dense LDL particles are associated with increased triglyceride and decreased HDL cholesterol levels, male sex, hyperinsulinemia, and insulin resistance [54]. Recently, a study aimed at lowering LDL cholesterol with statin therapy in people with diabetes over 40 years of age showed that a 30% decrease in LDL cholesterol was associated with a 25% reduction in first event rate for major coronary artery events independent of baseline LDL, preexisting vascular disease, type or duration of diabetes, or adequacy of glycemic control [78]. Similarly, a study in people with type 2 diabetes had a significant reduction in cardiovascular events including stroke [79].

Because of a lack of specific studies of dietary manipulation with CVD end points in patients with diabetes, the American Diabetes Association recommends that dietary goals be the same as for individuals with preexisting CVD, because the two groups appear to have equivalent CVD risk [80]. The dietary recommendations are less than 7% of total energy from saturated fatty acids, minimal intake of *trans*-fatty acids, and cholesterol intake less than 200 mg/day [12].

People with type 2 diabetes and hypertriglyceridemia usually have both overproduction and impaired catabolism of VLDL triglyceride. Lipoprotein lipase (LPL) activity can be reduced in patients with poor glucose control and profound hypertriglyceridemia [85]. Decreased HDL cholesterol levels may be due to increased catabolism, or to decreased production of HDL due to impaired catabolism of VLDL and decreased LPL activity [86]. For additional information, Ginsberg [87] has reviewed the pathophysiology of dyslipidemia in diabetes, and Haffner [88] has reviewed the effectiveness of medical therapy for the dyslipidemia of adults with type 2 diabetes.

Controversy has arisen over recommending a low-fat, high-carbohydrate diet to patients with diabetes, especially type 2, because of concern that this may worsen triglyceride and HDL abnormalities and ultimately increase cardiovascular risk [89]. Current American Diabetes Association nutrition guidelines permit either a high-carbohydrate, low-fat diet, or a moderate carbohydrate, higher fat diet enriched with polyunsaturated or monounsaturated fat. The issue of whether a high-monounsaturated-fat diet is preferable to a high-carbohydrate diet remains controversial [90]. In short-term studies, a high-carbohydrate diet has been associated with higher triglyceride levels and lower HDL levels than a higher fat diet [91, 92].

It has been demonstrated in numerous studies that glycemic control and insulin action are improved—or at least not made worse—when the diet is restricted in saturated fatty acids [93–95]. In clinical studies where energy intake and weight are held constant, diets low in saturated fatty acids and high in either carbohydrate or monounsaturated fatty acids lowered plasma LDL cholesterol equivalently [58, 66]. High (~55%) carbohydrate diets increase postprandial plasma glucose, insulin, and triglycerides when compared with high monounsaturated fat diets. However, high-monounsaturated fat diets have not been shown to improve glycosylated hemoglobin values. In other studies when energy intake was reduced, the adverse effects of high-carbohydrate diets were not observed [96, 97]. Individual variability in response to higher carbohydrate diets suggests that the plasma triglyceride response to dietary modification should be monitored carefully, particularly in the absence of weight loss.

Diets high in polyunsaturated fatty acids appear to have effects on plasma lipid concentrations similar to those of diets high in monounsaturated fatty acids [98, 99]. Very long-chain n-3 polyunsaturated fatty acid supplements have been shown to lower plasma triglyceride levels in people with type 2 diabetes who are hypertriglyceridemic. Although the accompanying small rise in plasma LDL cholesterol is of concern, an increase in HDL cholesterol may offset this concern [100]. A lower total glucose load keeps postprandial glucose and insulin concentrations reduced, low dietary saturated fat keeps LDL cholesterol in check, and the higher monounsaturated and polyunsaturated content of the diet prevents HDL cholesterol decrease and triglyceride concentration increase. With attention to food choices, it is possible to keep the saturated fat content of the diet low without decreasing the total dietary fat intake. The addition of unsaturated fatty acids to the diet to maintain energy needs appears to be the optimal approach to nutrition management, especially for patients with the typical dyslipidemia of type 2 diabetes and with insulin resistance.

Nutrition intervention should be tailored for each patient based on age, type of diabetes, pharmacological treatment, lipid levels, and other medical conditions. Dietary alterations focus on reduction of saturated fat, cholesterol, and *trans*-fat intake.

IV. SELECTED MICRONUTRIENTS

A. Magnesium

Magnesium levels remain remarkably constant in people without diabetes, because of regulatory mechanisms. However, those with diabetes appear to be prone to low serum magnesium levels [101, 102]. The clinical significance of this finding remains undefined, because only about 0.3% of total body magnesium is in the blood. Low serum levels may be related to increased urinary magnesium losses secondary to glycosuria-induced renal wasting [103], although short-term improvement in glycemic control has not been shown to restore serum magnesium levels [104].

Magnesium is intimately involved in a number of important biochemical reactions, particularly processes that involve the formation and use of high-energy phosphate bonds. As a cofactor in more than 300 enzyme reactions, it modulates glucose transport through membranes and is a cofactor in several enzymatic systems involving glucose oxidation [105]. Some believe that magnesium deficiency may increase or cause insulin resistance [106].

Low levels of magnesium have been associated with hypertension, cardiac arrhythmia, congestive heart failure, retinopathy, and insulin resistance [107–109]. Magnesium depletion in a few studies has been shown to result in insulin resistance as well as impaired insulin secretion and thereby may worsen control of diabetes [109]. Nadler and colleagues [110] conducted a magnesium depletion study. Men were fed 12 mg magnesium daily for 3 weeks. Intravenous glucose tolerance tests performed at the beginning and end of the 3-week period revealed a significant increase in insulin sensitivity. These findings raise the possibility that insulin resistance and abnormal glucose tolerance might be due to inadequate magnesium. In fact, Paolisso and colleagues [111] have shown magnesium supplementation to improve glucose tolerance in nondiabetic elderly and improve insulin response in elderly patients with type 2 diabetes. It is possible that decreased magnesium levels may represent a marker rather than a cause of the disease. In a recent report on the role of magnesium deficiency in the pathogenesis of type 2 diabetes from the large (12,218 adults) Atherosclerosis Risk in Communities (ARIC) Study, the authors demonstrated in white, but not black, adults a strong relationship between type 2 diabetes and serum magnesium levels [112]. Findings in this study suggest that modification of magnesium intake by dietary means alone is insufficient to achieve an effect to prevent type 2 diabetes.

Two interesting prospective studies suggested that an increased magnesium intake from foods could have a protective role in reducing the risk of type 2 diabetes. Data from the Nurses Health Study and the Health Professional's Follow-up Study indicated a significant reduction in the relative risk of type 2 diabetes in both men and women in the highest quintiles of magnesium intake [113]. Similarly, a cohort study of U.S. women in the Women's Health Study found a statistically significant inverse relationship between magnesium intake and risk of developing type 2 diabetes [114].

Until magnesium depletion studies conducted in normal individuals can relate specific dietary intake levels with abnormal glucose tolerance testing or other indicators of glucose metabolism, it is premature to consider the prevalence of diabetes mellitus as a functional indicator of adequacy for magnesium [115]. There is compelling evidence to justify support for a randomized prospective clinical trial to test the effect of consuming major food sources of magnesium, such as whole grains, nuts, and leafy green vegetables, on the development of type 2 diabetes in a high-risk population [116]. In the most recent Dietary Reference Intakes (DRIs) for magnesium, the Recommended Dietary Allowance (RDA) value for magnesium for women over 50 years of age is 320 mg/day and 420 mg/day for men in the same age group [115]. A review of the literature by the Standing Committee resulted in a gradual increase in the RDA for magnesium with age. Thus, it makes sense, at the least, to assess the dietary intake of older patients with type 2 diabetes for magnesium adequacy.

B. Chromium

In laboratory animals chromium deficiency is associated with an increase in blood glucose, cholesterol, and triglyceride concentrations. This observation has led to periodic and ongoing investigation of chromium status in humans with glucose intolerance. The biologically active elemental chromium product is called glucose tolerance factor [117]. Glucose tolerance factor is composed of nicotinic acid, elemental chromium, and the amino acids glutamic acid, glycine, and cysteine. It acts as a cofactor for insulin and may facilitate insulin–membrane receptor interaction. However, glucose tolerance factor lowers plasma glucose only in the presence of insulin (postprandial state) and not in overnight fasted animals [118].

Most chromium supplements are poorly absorbed, but when combined with picolinate, absorption is improved. Because no method exists to determine chromium deficiency, the prevalence of a deficiency is unknown. Rabinowitz and colleagues [119] studied chromium in hair, serum, urine, and red blood cells and could not identify a deficiency of chromium in people with or without diabetes. In three double-blind crossover studies in patients with diabetes, chromium supplementation failed to improve glucose and lipid levels [120–122].

Evidence from China suggests a role for chromium supplementation in people with type 2 diabetes. Individuals with type 2 diabetes were randomly divided into three groups and supplemented with 1000 μg/day or 2000 μg/day chromium or placebo. In all three groups, fasting, 2-hour glucose concentrations and glycosylated hemoglobin values decreased but the decreases in the subjects receiving supplemental chromium were much larger. Fasting and 2-hour insulin concentrations were also lowered in the two supplemented groups. Total plasma cholesterol concentration decreased in the group receiving supplements, but there was no impact of supplementation on HDL cholesterol, triglyceride concentration, or body weight [123].

An extensive review of the relationship of chromium status and diabetes has been published by Cefalu and Hu [124]. Several studies have explored chromium supplementation in people with type 1 and 2 diabetes and with gestational and steroid-induced diabetes. Most of these studies are limited by small size, short term, nonrandomized design, and different doses of chromium supplementation, all of which contribute to the high variability of findings.

Before chromium supplementation can be recommended, double-blind crossover studies of the effect of chromium supplementation in people with diabetes with known dietary intake of chromium need to be conducted. Until proven otherwise, it is assumed that chromium functions as a nutrient, not as a therapeutic agent, and that it benefits only individuals with marginal ... glucose intolerance whose signs and symptoms are due ... marginal or overt chromium deficiency.

...idants

... diabetes is ... a state of increased ...ative stress, ... interest in antioxidant ...erapy. Altho... nutrients appear to play a role in redu... and possibly in insulin sensitivity ... but evidence at present to warran... recommendations about use of th... management [12]. Two large cl... a positive impact of antioxidants ... including patients with diabetes ... studies have been conducted in ... examining antioxidants and inflam... on the use of antioxidant supple... E, carotene, and other supplements, ... positive results for improved glycemic ...7]. Foods with high antioxidant poten... and cinnamon, have received increased ... diabetes literature over the past ...

...consumption has been associated with improved ...cose tolerance and lower risk of type 2 diabetes in diverse populations from Europe, United States, and Japan [128–130]. When the data from the Nurses' Health Study was examined, higher coffee consumption was associated with a lower risk of type 2 diabetes [131]. In this study, both caffeinated and decaffeinated coffee, but not tea, consumption were associated with lower risk for diabetes. They also looked at components of coffee and found that caffeine intake was associated with a lower risk of type 2 diabetes. Although not explored in this study, the authors discussed the possibility that beneficial effects of coffee components other than caffeine might contribute. Coffee has strong antioxidant properties *in vivo* [132]. These observations are interesting, but coffee consumption cannot be recommended as a preventive measure for type 2 diabetes.

Several botanical products and a number of medicinal/culinary herbs have been reported to improve glucose metabolism and alter plasma lipid concentrations. For example, cinnamon and its ability to improve glucose and lipid concentrations in patients with type 2 diabetes has been reported and picked up by popular publications [133]. However, a follow-up study in adolescents with type 1 diabetes failed to show improvement in glycosylated hemoglobin concentrations [134].

V. CONCLUSION

We are witnessing a surge in public and media attention to type 2 diabetes mellitus. This is due in part to the increasing worldwide incidence of type 2 diabetes and to the reports of disease management studies that continue to show marked improvement in disease outcome with improved glycemic control. Nutrition intervention remains a key component of type 2 diabetes management.

In this chapter, nutrient components of the diet are discussed in relation to the impact they have on diabetes treatment. The entire disease management process has been enhanced because of the advances in oral medications for glucose and lipid control. In addition, with an increased focus on near-normal glycemic control in this population, all health care team members are more closely involved in assisting the patients with disease management.

The use of herbal products (commonly viewed as a component of nutrition therapy) in disease prevention and treatment is gaining in popularity. Despite the tremendous recent growth in popularity, carefully controlled studies are few. For these reasons, no discussion of this topic has been included here. However, with the development of guidelines for clinical investigation, this is an area that will expand in the future.

The public health burden of type 2 diabetes causes us to take notice not only for treatment following diagnosis, but more important, disease prevention for those genetically susceptible and attention to the current epidemic of obesity in adults and children. The role for nutrition is expanding as our knowledge and treatment options increase.

References

1. American Diabetes Association. (2007). Diabetes statistics. Retrieved March 27, 2007, from http://www.diabetes.org/diabetes-statistics.jsp.
2. DeFronzo, R. (1988). Lilly lecture 1987. The triumvirate: Beta-cell, muscle, liver. *Diabetes* **37**, 667–687.
3. Must, A., Spando, J., Coakley, E. H., Field, A. E., Colditz, G., and Dietz, W. H. (1999). The disease burden associated with overweight and obesity. *JAMA* **282**, 1523–1529.
4. American Diabetes Association. (2007). Diagnosis and classification of diabetes mellitus [committee report]. *Diabetes Care* **30**(Suppl. 1), S42–S47.
5. Mokdad, A. H., Ford, E. S., Bowman, B. A., Dietz, W. H., Vinicor, F., Bales, V. S., and Marks, J. S. (2003). Prevalence of obesity, diabetes, and obesity-related health risk factors, 2001. *JAMA* **289**, 76–169.
6. Centers for Disease Control and Prevention (CDC). (2005). National diabetes fact sheet, 2005. Retrieved March 27, 2007, from http://www.cdc.gov/diabetes/pubs/factsheet05.htm.
7. U.K. Prospective Diabetes Study Group. (1998). Intensive blood glucose control with sulfonylureas or insulin compared with conventional treatment and risk of complications in patients with type 2 diabetes (UKPDS 33). *Lancet* **352**, 837–853.

8. Turner, R. C., Cull, C. A., Frighi, V., and Holman, R. R. for the U.K. Prospective Diabetes Study (UKPDS) Group. (1999). Glycemic control with diet, sulfonylurea, metformin, or insulin in patients with type 2 diabetes mellitus. *JAMA* **281**, 2005–2012.
9. Tuomilehto, J., Lindstrom, J., Eriksson, J. G., Valle, T. T., Hamalainen, H., Ilanne-Parikka, P., Keinanen-Iukaanniemi, S., Laakso, M., Louheranta, A., Rastas, M., Salminen, V., and Uusitupa, M. (2001). Prevention of type 2 diabetes mellitus by changes in lifestyle among subjects with impaired glucose tolerance. *N. Engl. J. Med.* **344**, 1343–1350.
10. Knowler, W. C., Barrett-Connor, E., Fowler, S. E., Hamman, R. F., Lachin, J. M., Walker, E. A., and Nathan, D. M. (2002). Reduction in the incidence of type 2 diabetes with lifestyle intervention or metformin. *N. Engl. J. Med.* **346**, 393–403.
11. Pan, S. R., Li, G. W., Hu, Y. H., Wang, J. X., Yang, W. Y., An, Z. S., Hu, Z. X., Lin, J., Ziao, J. Z., Cao, H. B., Liu, P. A., Jang, S. G., Jiang, Y. Y., Wang, J. P., Aheng, H., Ahang, H., Bennett, P. H., and Howard, B. V. (1997). Effects of diet and exercise in preventing NIDDM in people with impaired glucose tolerance: The DaQing IGT and Diabetes Study. *Diabetes Care* **20**, 537–544.
12. American Diabetes Association. (2006). Nutrition recommendations and interventions for diabetes 2006: A position statement of the American Diabetes Association. *Diabetes Care* **29**, 2140–2157.
13. American Diabetes Association. (2007). Standards of medical care in diabetes—2007. *Diabetes Care* **30**, S4–S41.
14. Lorenz, R. A., Gregory, R. P., Davis, D. L., Schlundt, D. G., and Wermager, J. (2000). Diabetes training for dietitians: Needs assessment, program description and effects on knowledge and problem solving. *J. Am. Diet. Assoc.* **100**, 225–228.
15. Eriksson, K. F., and Lindgarde, F. (1996). Poor physical fitness and impaired early insulin response but late hyperinsulinemia, as predictors of NIDDM in middle-aged Swedish men. *Diabetologia* **39**, 573–579.
16. Williams, K. V., and Kelley, D. E. (2000). Metabolic consequences of weight loss on glucose metabolism and insulin actionin type 2 diabetes. *Diabetes Obes. Metab.* **2**, 121–129.
17. Wing, R. R., Koeske, R., Epstein, L. H., Nowarlk, M. P., Gooding, W., and Becker, D. (1987). Long-term effects of modest weight loss in type 2 diabetic patients. *Arch. Intern. Med.* **147**, 1749–1753.
18. Henry, R. R., Schaefer, L., and Olesfsky, J. M. (1985). Glycemic effects of intensive caloric restriction and isocaloric refeeding in noninsulin dependent diabetes mellitus. *J. Clin. Endocrinol. Metab.* **61**, 917–925.
19. Low, C. C., Grossman, E. B., and Gumbiner, B. (1995). Potentiation of effects of weight loss by monounsaturated fatty acids in obese NIDDM patients. *Diabetes* **45**, 569–571.
20. Golay, A., Allaz, A. F., Morel, Y., de Tonnac, N., Tankova, S., and Reaven, G. (1996). Similar weight loss with low or high carbohydrate diets. *Am. J. Clin. Nutr.* **63**, 174–178.
21. Heilbronn, L. K., Noakes, M., and Clifton, P. M. (1999). Effect of energy restriction, weight loss and diet composition on plasma lipids and glucose in patients with type 2 diabetes. *Diabetes Care* **22**, 889–895.
22. Foster, G. D., Wyatt, H. R., Hill, J. O., McGuckin, B. G., Jill, C., Mohammed, B. S., Szapary, P. O., Rader, D. J., Edman, J. S., and Klein, S. (2003). A randomized trial of a low-carbohydrate diet for obesity. *N. Engl. J. Med.* **348**, 2082–2090.
23. Brehm, B. J., Seeley, R. J., Daniles, S. R., and D'Alessio, D. A. (2003). A randomized trial comparing a very low carbohydrate diet and a ca[...]ed low fat diet on body weight and cardiovascul[...] in healthy women. *J. Clin. Endocrinol. Metab.* [...]
24. Yancy, W. S., Jr., O[...] Westman, E. C. (200[...] versus a low-fat diet t[...] randomized controlled tr[...]
25. Wing, R. R., Blair, E., a[...] weight loss treatment for o[...] Does inclusion of an inter[...] improve control? *Am. J. Med.* [...]
26. Williams, K., Mullen, M., Kelly, [...]R., Bakst, R. P., and [...] effect of short periods of caloric re[...] ketogenic diet [...]demia: A glycemic control in type 2 diabetes. [...] 777
27. Wing, R. R. (2003). Behavioral in[...] Recognizing our progress and future [...] **11**(Suppl), 3S–6S.
28. Klein, S., Sheard, N. F., Pi-Sunyer, X., [...] Rosett, J., Kulkarni, K., and Clark, N. G[...] management through lifestyle modification for the p[...] and management of type 2 diabetes: Rationale and str[...] A statement of the American Diabetes Association, the [...] American Association for the Study of Obesity, and the Am[...]ican Society for Clinical Nutrition. *Diabetes Care* **2**[...] 2067–2073.
29. Newby, F. D., and Price, S. R. (1998). Determinants of protein turn-over in health and disease. *Mineral Electrolyte Metab.* **24**, 6–12.
30. Institute of Medicine. (2002). "Dietary Reference Intakes: Energy, Carbohydrate, Fiber, Fat, Fatty Acids, Cholesterol, Protein and Amino Acids" pp. 589–768. National Academies Press, Washington, DC.
31. Gougeon, R., Styhler, K., Morais, J. A., Jones, P. J., and Marliss, E. B. (2000). Effects of oral hypoglycemic agents and diet on protein metabolism in type 2 diabetes. *Diabetes Care* **23**, 1–8.
32. Nair, K. S., Garrow, J. S., Ford, C., Mahler, R. F., and Halliday, D. (1983). Effect of poor diabetic control and obesity on the whole body protein metabolism in man. *Diabetologia* **25**, 400–403.
33. Henry, R. R. (1994). Protein content of the diabetic diet. *Diabetes Care* **17**, 1502–1513.
34. Maroni, B. J., and Mitch, W. E. (1997). Role of nutrition in prevention of the progression of renal disease. *Annu. Rev. Nutr.* **17**, 435–455.
35. Gannon, M. C., and Nuttall, F. Q. (1999). Protein and diabetes. *In* "American Diabetes Association Guide to Medical Nutrition Therapy for Diabetes" (M. J. Franz, and J. P. Bantle, Eds.), pp. 107–125. American Diabetes Association, Alexandria, VA.
36. Wahren, J., Felig, P., and Hagenfeldt, L. (1976). Effect of protein ingestion on splanchnic and leg metabolism in normal men and in patients with diabetes mellitus. *J. Clin. Invest.* **57**, 987–999.

37. Vilstrup, H. (1980). Synthesis of urea after stimulation with amino acids: Relation to liver function. *Gut* **21**, 990–995.
38. Ganong, W. F. (1979). "Review of Medical Physiology." Lange Medical Publications, Los Altos, CA.
39. Nuttal, F. Q., and Gannon, M. C. (1991). Plasma glucose and insulin response to macronutrients in nondiabetic and NIDDM subjects. *Diabetes Care* **14**, 824–838.
40. Gannon, M. C., Nuttal, F. Z., Lane, J. T., and Burmeister, L. A. (1992). Metabolic response to cottage cheese or egg white protein, with or without glucose in type 2 diabetic subjects. *Metabolism* **41**, 1137–1145.
41. Kahn, M. A., Gannon, M. C., and Nuttall, F. Q. (1992). Glucose appearance rate following protein ingestion in normal subjects. *J. Am. College Nutr.* **11**, 701–706.
42. Gannon, M. C., Nuttal, J. A., Damberg, G., Gupta, V., and Nuttall, F. Q. (2001). Effect of protein ingestion on the glucose appearance rate in people with type 2 diabetes. *J. Clin. Endocrinol. Metab.* **86**, 1040–1047.
43. Floyd, J. C., Fajans, S. S., Conn, J. W., Knopf, R. F., and Rull, J. (1996). Insulin secretion in response to protein ingestion. *J. Clin. Invest.* **45**, 1479–1486.
44. Rabinowitz, D., Merimee, T. J., Maffezzoli, R., and Burggess, J. A. (1966). Patterns of hormonal release after glucose, protein, and glucose plus protein. *Lancet* **2**, 454–457.
45. Krezowski, P. A., Nuttall, F. Q., Gannon, M. C., and Bartosh, N. H. (1986). The effect of protein ingestion on the metabolic response to oral glucose in normal individuals. *Am. J. Clin. Nutr.* **44**, 847–856.
46. Gannon, M. C., Nuttall, F. Q., Neil, B. J., and Westphal, S. A. (1988). The insulin and glucose responses to meals of glucose plus various proteins in type 2 diabetic subjects. *Metabolism* **37**, 1081–1088.
47. Carroll, K. K., and Kurowska, E. M. (1995). Soy consumption and cholesterol reduction: Review of animal and human studies. *J. Nutr.* **125**, 594S–597S.
48. Schenk, S., Davidson, C. J., Zderic, T. W., Byerly, L. O., and Coyle, E. F. (2003). Different glycemic indexes are not due to glucose entry into blood but to glucose removal by tissue. *Am. J. Clin. Nutr.* **78**, 742–748.
49. Nuttall, F. Q., and Gannon, M. C. (1999). Carbohydrate and diabetes. *In* "American Diabetes Association Guide to Medical Nutrition Therapy for Diabetes" (M. J. Franz, and J. P. Bantle, Eds.), pp. 85–106. American Diabetes Association, Alexandria, VA.
50. Expert Panel on Detection, Evaluation, and Treatment of High Blood Cholesterol in Adults. (1993). Summary of the Second Report of the National Cholesterol Education Program (NCEP) Expert Panel on Detection, Evaluation, and Treatment of High Blood Cholesterol in Adults (Adult Treatment Panel II). *JAMA* **269**, 3015–3023.
51. Stamler, J., Vaccaro, O., Neaton, J. D., and Wentworth, D. for the Multiple Risk Factor Intervention Trial Research Group. (1993). Diabetes, other risk factors, and 12-yr cardiovascular mortality for men screened in the Multiple Risk Factor Intervention Trial. *Diabetes Care* **16**, 434–444.
52. Stern, M. P., Patterson, J. K., Haffner, S. M., Hazuda, H. P., and Mitchell, B. D. (1989). Lack of awareness and treatment of hyperlipidemia in type 2 diabetes in a community survey. *JAMA* **262**, 360–364.
53. Reaven, G. M. (1992). The role of insulin resistance and hyperinsulinemia in coronary heart disease. *Metabolism* **41**, 16–19.
54. Reaven, G.M, Chen, Y. D., Jeppesen, J., Maheux, P., and Krause, R. (1993). Insulin resistance and hyperinsulinemia in individuals with small, dense, low density lipoprotein particles. *J. Clin. Invest.* **92**, 141–146.
55. Fontbonne, A., Eschwege, E., Cambien, F., Richard, J. L., Ducimetiere, P., Thibult, N., Warnet, J. M., Claude, J. R., and Rosselin, G. E. (1989). Hypertriglyceridemia as a risk factor of coronary heart disease mortality in subjects with impaired glucose tolerance or diabetes. Results from the 11-year follow-up of the Paris Prospective Study. *Diabetes Care* **3**, 300–304.
56. Fontbonne, A. M., and Eschwege, E. M. (1991). Insulin and cardiovascular disease: Paris Prospective Study. *Diabetes Care* **14**, 461–469.
57. Coulston, A. M., Hollenbeck, C. B., Swislocki, A. L. M., Chen, Y.-D. I., and Reaven, G. M. (1987). Deleterious metabolic effects of high-carbohydrate, sucrose-containing diets in patients with non-insulin-dependent diabetes mellitus. *Am. J. Med.* **82**, 213–220.
58. Garg, A., Bantle, J. P., Henry, R. R., Coulston, A. M., Griver, K. A., Raatz, S. K., Brinkley, L., Chen, Y.-D. I., Grundy, S. M., Huet, B. A., and Reaven, G. M. (1994). Effects of varying carbohydrate content of the diet in patients with non-insulin-dependent diabetes mellitus. *JAMA* **271**, 1421–1428.
59. Reaven, G. M. (1997). Do high carbohydrate diets prevent the development or attenuate the manifestations (or both) of syndrome X? A viewpoint strongly against. *Curr. Opin. Lipidol.* **8**, 23–27.
60. Sheard, N. F., Clark, N. G., Brand-Miller, J. C., Franz, M. J., Pi-Sunyer, F. X., Mayer-Davis, E., Kulkarni, K., and Geil, P. (2004). Dietary carbohydrate (amount and type) in the prevention and management of diabetes: A statement of the American Diabetes Association. *Diabetes Care* **27**, 2266–2271.
61. Coulston, A. M., Hollenbeck, C. B., Donner, C. C., Williams, R., Chiou, Y.-A. M., and Reaven, G. M. (1985). Metabolic effects of added dietary sucrose in individuals with non-insulin-dependent diabetes mellitus (NIDDM). *Metabolism* **34**, 962–966.
62. Bantle, J. P., Swanson, J. E., Thomas, W., and Laine, D. C. (1993). Metabolic effects of dietary sucrose in type 2 diabetic subjects. *Diabetes Care* **16**, 1301–1305.
63. Bruce, B., Spiller, G. A., Klevay, L. M., and Gallagher, S. K. (2000). A diet high in whole and unrefined foods favorably alters lipids, antioxidant defenses, and colonic function. *J. Am. College Nutr.* **19**, 61–67.
64. Nuttall, F. Q. (1997). Dietary fiber in the management of diabetes. *Diabetes* **42**, 503–508.
65. Wursch, P., and Pi-Sunyer, F. X. (1997). The role of viscous soluble fiber in the metabolic control of diabetes: A review with special emphasis on cereals rich in beta-glucan. *Diabetes Care* **20**, 1774–1780.
66. Franz, M. J., Bantle, J. P., Beebe, C. A., Brunzell, J. D., Chiasson, J. L., Garg, A., Holzmeister, L. A., Hoogwerf, B., Mayer-Davis, E., Mooradian, A. D., Purnell, J. Q., and Wheeler, M. (2002). Evidence-based nutrition principles and

recommendations for the treatment and prevention of diabetes and related complications. *Diabetes Care* **25**, 148–198.

67. Labbe, M. (1907). Tolerame comparee des divers hydrates de carbone part l'organisme des diabetiques. *Bull. Mem. Soc. Med. Hosp.* **24**, 221–234.
68. Otto, H., Bleyer, G., and Pehhartz, M. (1983). Kehlenhydrataustausch nach biologischen Äquivalenten. *In* "Diatetik bei Diabetes Mellitus," pp. 41–51. Verlag Has Huber, Bern, Switzerland.
69. Gannon, M. C., Nuttall, F. Q., Westphal, S. A., Fang, S., and Ercan-Fang, N. (1998). Acute metabolic response to high-carbohydrate, high-starch meals compared with moderate-carbohydrate, low-starch meals in subjects with type 2 diabetes. *Diabetes Care* **21**, 1619–1626.
70. Jenkins, D. J. A., Wolever, T. M. S., and Taylor, R. H. (1981). Glycemic index of foods: A physiological basis for carbohydrate exchange. *Am. J. Clin. Nutr.* **34**, 362–366.
71. Wolever, T. M. S., Katzman-Relle, L., Jenkins, A. L., Vuksan, V., Josse, R. G., and Jenkins, D. J. A. (1994). Glycemic index of 102 complex carbohydrate foods in patients with diabetes. *Nutr. Res.* **14**, 651–669.
72. Mayer-Davis, E. J., Dhawan, A., Liese, A. D., Teff, K., and Schulz, M. (2006). Towards understanding of glycaemic index and glycaemic load in habitual diet: Associations with measures of glycaemia in the Insulin Resistance Atherosclerosis Study. *Br. J. Nutr.* **95**, 397–405.
73. Salmeron, J., Mason, J. E., Stampfer, M. J., Colditz, G. A., Wing, A. L., and Willett, W. C. (1997). Dietary fiber, glycemic load, and risk of non-insulin-dependent diabetes mellitus in women. *JAMA* **277**, 472–477.
74. Salmeron, J., Ascherio, A., Rimm, E. B., Colditz, G. A., Spiegelman, D., Jenkins, D. J., Stamfer, M. J., Wing, A. L., and Willett, W. C. (1997). Dietary fiber, glycemic load, and risk of NIDDM in men. *Diabetes Care* **20**, 545–550.
75. Brand-Miller, J., Hayne, S., Petocz, P., and Colagiuri, S. (2003). Low-glycemic index diets in the management of diabetes: A meta-analysis of randomized controlled trials. *Diabetes Care* **26**, 2261–2267.
76. Liese, A. D., Schulz, M., Fang, F., Wolever, T.M, D'Agostino, R. B., Jr., Sparks, K. C., and Mayer-Davis, E. J. (2005). Dietary glycemic index and glycemic load, carbohydrate and fiber intake, and measures of insulin sensitivity, secretion, and adiposity in the Insulin Resistance Atherosclerosis Study. *Diabetes Care* **28**, 2832–2838.
77. Haffner, S. M., Stern, M. P., Hazuda, H. P., Mitchell, B. D., and Patterson, J. K. (1990). Cardiovascular risk factors in confirmed prediabetic individuals: Does the clock for coronary heart disease start ticking before the onset of clinical diabetes? *JAMA* **263**, 2893–2898.
78. Heart Prevention Study Collaborative Group. (2003). RC/BHF Heart Protection Study of cholesterol-lowering with simvastatin in 5963 people with diabetes: A randomized placebo-controlled trial. *Lancet* **361**, 2005–2016.
79. Colhoun, H. M., Betteridge, D. J., Durrington, P. N., Hitman, G. A., Neil, H. A., Livingstone, S. J., Thomason, M. J., Mackness, M. I., Chariton-Menys, V., and Fuller, J. H. (2004). Primary prevention of cardiovascular disease with atorvastatin in type 2 diabetes in the Collaborative Atorvastatin Diabetes Study (CARDS): Multicenter randomized placebo-controlled trial. *Lancet* **364**, 685–696.
80. National Cholesterol Education Program Expert Panel on Detection, Evaluation, and Treatment of High Blood Cholesterol in Adults. (2001). Executive Summary of the Third Report of the National Cholesterol Education Program (NCEP) Expert Panel on Detection, Evaluation, and Treatment of High Blood Cholesterol in Adults (Adult Treatment Panel III). *JAMA* **285**, 2486–2497.
81. Gardner, C. D., Fortmann, S. P., and Krauss, R. M. (1996). Association of small low-density lipoprotein particles with the incidence of coronary artery disease in men and women. *JAMA* **276**, 875–881.
82. Stampfer, M. J., Krauss, R. M., Ma, J., Blanche, P. H., Holl, L. G., Sacks, F. M., and Hennekens, C. H. (1996). A prospective study of triglyceride level, low-density lipoprotein particle diameter, and risk of myocardial infarction. *JAMA* **276**, 882–888.
83. Reaven, G. M. (1988). 1988 Banting Lecture: Role of insulin resistance in human disease. *Diabetes* **37**, 1595–1607.
84. Despres, J. P., Lamarch, B., Mauriege, P., Cantin, B., Dagenais, G. R., Moorjani, S., and Lupien, P. J. (1996). Hyperinsulinemia as an independent risk factor for ischemic heart disease. *N. Engl. J. Med.* **334**, 952–957.
85. Taskinen, M.-R., Beltz, W. F., Harper, I., Fields, R. M., Schonfeld, G., Grundy, S. M., and Howard, B. V. (1986). Effects of NIDDM on very-low-density lipoprotein triglyceride and apolipoprotein B metabolism: Studies before and after sulfonylurea therapy. *Diabetes* **35**, 1268–1277.
86. Golay, A., Zech, L., Shi, M. Z., Chiou, Y. A., Reaven, G. M., and Chen, Y. D. (1987). High-density lipoprotein (HDL) metabolism in non-insulin dependent diabetes mellitus: Measurement of HDL turnover using tritiated HDL. *J. Clin. Endocrinol. Metab.* **65**, 512–518.
87. Ginsberg, H. (1991). Lipoprotein physiology in non-diabetic and diabetic states: Relationship to atherogenesis. *Diabetes Care* **14**, 839–855.
88. Haffner, S. M. (1998). Management of dyslipidemia in adults with diabetes (a technical review). *Diabetes Care* **21**, 160–178.
89. Garg, A. (1994). High-monounsaturated fat diet for diabetic patients: Is it time to change the current dietary recommendations? *Diabetes Care* **17**, 242–246.
90. Berry, E. M. (1997). Dietary fatty acids in the management of diabetes mellitus. *Am. J. Clin. Nutr.* **66**, 991S–997S.
91. Garg, A., Bonanome, A., Grundy, S. M., Zhang, A. J., and Unger, R. H. (1988). Comparison of high-carbohydrate diet with high monounsaturated-fat diet in patients with non-insulin-dependent diabetes mellitus. *N. Engl. J. Med.* **391**, 829–834.
92. Coulston, A. M., Hollenbeck, C. B., Swislocki, A. L. M., and Reaven, G. M. (1989). Persistence of hypertriglyceridemic effect of low-fat high-carbohydrate diets in NIDDM patients. *Diabetes Care* **12**, 94–101.
93. Stone, D. B., and Conner, W. E. (1963). The prolonged effects of a low cholesterol, high carbohydrate diet upon the serum lipids in diabetic patients. *Diabetes* **12**, 127–132.
94. Anderson, J. W., and Ward, K. (1979). High-carbohydrate, high fiber diets for insulin-treated men with diabetes mellitus. *Am. J. Clin. Nutr.* **32**, 2312–2321.
95. Kolterman, O. G., Greenfield, M., Reaven, G. M., Saekow, M., and Olefsky, J. M. (1979). Effect of a high carbohydrate diet

on insulin binding to adipocytes and on insulin action *in vivo* in man. *Diabetes* **28**, 731–736.
96. Heilbronn, L. K., Noakes, M., and Clifton, P. M. (1999). Effect of energy restriction, weight loss, and diet composition on plasma lipids and glucose in patients with type 2 diabetes. *Diabetes Care* **22**, 889–895.
97. Parker, B., Noakes, M., Luscombe, N., and Clifton, P. (2002). Effect of a high-protein, high-monounsaturated fat weight loss diet on glycemic control and lipid levels in type 2 diabetes. *Diabetes Care* **25**, 425–430.
98. Salmeron, J., Hu, F. B., Manson, J. E., Stampfer, M. J., Colditz, G. A., Rimm, E. B., and Willett, W. C. (2001). Dietary fat intake and risk of type 2 diabetes in women. *Am. J. Clin. Nutr.* **73**, 1019–1026.
99. Tapsell, L. C., Gillen, L. J., Patch, C. S., Batterham, M., Owen, A., Bare, M., and Kennedy, M. (2004). Including walnuts in a low-fat/modified fat diet improves HDL cholesterol-to-total cholesterol ratios in patients with type 2 diabetes. *Diabetes Care* **27**, 2777–2783.
100. West, S. G., Hecker, K. D., Mustad, V. A., Nicholson, S., Schoemer, S. L., Wagner, P., Hinderliter, A. L., Ulbrecht, J., Ruey, P., and Kris-Etherton, P. M. (2005). Acute effects of monounsaturated fatty acids with and without omega-3 fatty acids on vascular reactivity in individuals with type 2 diabetes. *Diabetologia* **48**, 113–122.
101. Nadler, J. L., Malayan, S., Luong, H., Shaw, S., Natarajan, R. D., and Rude, R. K. (1992). Intracellular free magnesium deficiency plays a key role in increased platelet reactivity in type II diabetes mellitus. *Diabetes Care* **15**, 835–841.
102. Lima, J. D. L., Cruz, T., Pousada, J. C., Rodrigues, L. E., Barbosa, K., and Cangucu, V. (1990). The effect of magnesium supplementation in increasing doses on the control of type 2 diabetes. *Diabetes Care* **21**, 682–686.
103. Rude, R. K. (1993). Magnesium metabolism and deficiency. *Endocrinol. Metab. Clin. North Am.* **22**, 377–395.
104. American Diabetes Association (1992). Magnesium supplementation in the treatment of diabetes (consensus statement). *Diabetes Care* **15**, 1065–1067.
105. Mooradian, A. D., Faila, M., Hoogwerf, B., Isaac, R., Maryniuk, M., and Wylie-Rosett, J. (1994). Selected vitamins and minerals in diabetes mellitus [technical review]. *Diabetes Care* **17**, 464–479.
106. Alzaid, A., Dinneen, S., Moyer, T., and Rizza, R. (1995). Effects of insulin on plasma magnesium in non-insulin dependent diabetes mellitus: Evidence for insulin resistance. *J. Clin. Endocrinol. Metab.* **80**, 1376–1381.
107. Whelton, P. K., and Klag, M. J. (1989). Magnesium and blood pressure: Review of the epidemiologic and clinical trial experience. *Am. J. Cardiol.* **63**, 26G–30G.
108. Shattock, M. J., Hearse, D. J., and Fry, C. H. (1987). The ionic basis of anti-ischemic and anti-arrhythmic properties of magnesium in the heart. *J. Am. College Nutr.* **6**, 27–33.
109. Paolisso, G., Scheen, A., D'Onofrio, E., and Lefebvre, P. (1990). Magnesium and glucose homeostasis. *Diabetologia* **33**, 511–514.
110. Nadler, J. L., Buchanan, T., Natarajan, R., Antonipillai, I., Bergman, R., and Rude, R. K. (1993). Magnesium deficiency produces insulin resistance and increased thromboxane synthesis. *Hypertension* **21**, 1024–1029.
111. Paolisso, G., Sgambato, S., Gambardella, A., Pizza, G., Tesauro, P., Varricchio, M., and D'Onofrio, F. (1992). Daily magnesium supplements improve glucose handling in elderly subjects. *Am. J. Clin. Nutr.* **55**, 1161–1167.
112. Kao, W. H. L., Folsom, A. R., Nieto, F. J., Mo, J.-P., Watson, R. S., and Brancati, F. L. (1999). Serum and dietary magnesium and the risk for type 2 diabetes mellitus: The Atherosclerosis Risk in Communities Study. *Arch. Intern. Med.* **159**, 2151–2159.
113. Lopez-Ridaura, R., Willett, W. C., Rimm, E. B., Liu, S., Stampfer, M. J., Manson, J. E., and Hu, F. B. (2003). Magnesium intake and risk of type 2 diabetes in men and women. *Diabetes Care* **27**, 134–140.
114. Song, Y., Manson, M. E., Buring, J. E., and Liu, S. (2003). Dietary magnesium intake in relation to plasma insulin levels and risk of type 2 diabetes in women. *Diabetes Care* **27**, 59–65.
115. Food Nutrition Board, Institute of Medicine, National Academy of Sciences. (1997). "Dietary Reference Intakes for Calcium, Phosphorus, Magnesium, Vitamin D, and Fluoride." National Academies Press, Washington, DC.
116. Nadler, J. L. (2004). A new dietary approach to reduce the risk of type 2 diabetes? *Diabetes Care* **27**, 270–271.
117. Schwarz, K., and Mertz, W. (1957). A glucose tolerance factor and its differentiation from factor 3. *Arch. Biochem. Biophys.* **72**, 515–518.
118. Truman, R. W., and Doisy, R. J. (1977). Metabolic effects of the glucose tolerance factor (GTF) in normal and genetically diabetic mice. *Diabetes* **26**, 820–826.
119. Rabinowitz, M. B., Levin, S. R., and Gonick, J. E. (1980). Comparisons of chromium status in diabetic and normal men. *Metabolism* **29**, 355–364.
120. Rabinowitz, M. B., Gonick, H. C., Levin, S. R., and Davidson, M. B. (1983). Effect of chromium and yeast supplements on carbohydrate and lipid metabolism. *Diabetes Care* **6**, 319–327.
121. Sherman, L., Glennon, J. A., Brech, W. J., Klomberg, G. H., and Gordon, E. S. (1968). Failure of trivalent chromium to improve hyperglycemia in diabetes mellitus. *Metabolism* **17**, 439–442.
122. Usitupa, M. I. J., Kumpulainen, J. T., and Voutilainen, E. (1983). Effect of inorganic chromium supplementation on glucose intolerance, insulin response and serum lipids in non-insulin-dependent diabetics. *Am. J. Clin. Nutr.* **38**, 404–410.
123. Anderson, R. A., Cheng, N., Bryden, N. A., Polansky, M. M., Cheng, N., Chi, J., and Feng, J. (1997). Elevated intakes of supplemental chromium improve glucose and insulin variables in individuals with type 2 diabetes. *Diabetes* **46**, 1786–1791.
124. Cefalu, W. T., and Hu, F. B. (2004). Role of chromium in human health and in diabetes. *Diabetes Care* **27**, 2741–2751.
125. HOPE and HOPE-TOO Trial Investigators. (2005). Effects of long-term vitamin E supplementation on cardiovascular events and cancer: A randomized controlled trial. *JAMA* **293**, 1338–1347.
126. Toole, J., Malinow, M., Chambliss, L., Spence, J. D., Pettigrew, L. C., Howard, V. J., Sides, E. G., Wang, C.-H., and Stampfer, M. (2004). Lowering homocysteine in patients with ischemic stroke to prevent recurrent stroke, myocardial infarction, and

heath: The Vitamin Intervention for Stroke Prevention (VISP) randomized controlled trial. *JAMA* **291**, 565–575.

127. Hasanain, B., and Mooradian, A. D. (2002). Antioxidant vitamins and their influence in diabetes mellitus. *Curr. Diabetes Rep.* **2**, 448–456.
128. Van Dam, R. M., and Hu, F. B. (2005). Coffee consumption and risk of type 2 diabetes: A systemic review. *JAMA* **294**, 97–104.
129. Faerch, K., Lau, C., Tetens, I., Pedersen, O. B., Jorgensen, T., Borch-Johnsen, K., and Glulmer, C. (2005). A statistical approach based on substitution of macronutrients provides additional information to models analyzing single dietary factor in the relation to type 2 diabetes in Danish adults: The Inter99 study. *J. Nutr.* **135**, 1177–1182.
130. Greenberg, J. A., Axen, K. V., Schnoll, R., and Boozer, C. N. (2005). Coffee, tea and diabetes: The role of weight loss and caffeine. *Int. J. Obes. Relat. Metab. Disord.* **29**, 1121–1129.
131. Van Dam, R. M., Willett, W. C., Manson, J. E., and Hu, F. B. (2006). Coffee, caffeine, and risk of type 2 diabetes. *Diabetes Care* **29**, 398–403.
132. Svilaas, A., Salkhi, A. K., Anderson, L. F., Svilaas, T., Strom, E. C., Jacobs, D. R., Jr., Ose, L., and Blomhoff, R. (2004). Intakes of antioxidants in coffee, wine, and vegetables are correlated with plasma carotenoids in humans. *J. Nutr.* **134**, 562–567.
133. Kahn, A., Safdar, M., Kahn, M. M. S., Khattak, K. N., and Anderson, R. A. (2003). Cinnamon improves glucose and lipids of people with type 2 diabetes. *Diabetes Care* **26**, 3215–3218.
134. Altschuler, J. A., Casella, S. J., MacKenzie, T. A., and Curtis, K. M. (2007). The effect of cinnamon on A1C among adolescents with type 1 diabetes. *Diabetes Care* **30**, 813–816.

CHAPTER **33**

Nutrition Management for Gestational Diabetes

MARIA DUARTE-GARDEA
University of Texas at El Paso, El Paso, Texas

Contents

I. INTRODUCTION

Gestational diabetes mellitus (GDM) is defined as glucose intolerance with onset or first recognition during pregnancy [1, 2]. Exposure to maternal hyperglycemia conveys a high risk for obesity and type 2 diabetes in the offspring, in addition to genetic predisposition, and regardless of maternal diabetes type [3–6].

Nutrition management is an essential component in the overall medical care of the patient diagnosed with GDM and it plays an important role in controlling blood glucose levels throughout the pregnancy. Ideally, an interdisciplinary medical team that includes endocrinologists, obstetricians, diabetes educators, nurses, social workers, nutritionists among other professionals, work in collaboration with the pregnant woman toward a common goal to achieve normoglycemia from diagnosis and throughout delivery. Nutrition management consists of the implementation of a meal plan that provides adequate nutrients and promotes appropriate weight gain, normoglycemia, and the absence of ketonuria [7]. The meal plan will provide both additional calories to sustain weight gain, and a specific amount and type of carbohydrates distributed throughout the day to maximize blood glucose control [8].

A nutrition assessment is initially completed with the purpose of obtaining medical and obstetrical history, anthropometric and laboratory data, lifestyle, and food preferences. This information is useful in the design of an individualized and culturally appropriate meal plan that contributes to the ultimate goal of delivering a healthy infant. The nutrition educational materials should be appropriate for the patient's educational level. During intervention the nutritionist emphasizes lifestyle modification aimed at reducing the possibility of developing diabetes later in life. Frequent follow-up visits are required to monitor nutrient intake and parameters such as weight gain, blood glucose control, prevention of ketosis, and compliance.

Pregnancy complicated with diabetes offers a great opportunity for the nutritionist and other health care providers to instill lifestyle patterns beneficial for the woman and her family.

A. Prevalence

The prevalence of GDM varies worldwide and among different racial/ethnic groups within a country. Approximately 135,000 cases of GDM, representing on average 3–8% of all pregnancies, are diagnosed annually in the United States [1, 9]. There is variability of GDM among members of certain racial/ethnic groups (African American, Native American, Hispanic, South or East Asian, Pacific Islands, or Indigenous Australian) [10]. The prevalence among Zuni Indians is 14.3% [11]. The prevalence of GDM among women with diverse racial/ethnic backgrounds was examined between 1994 and 2002. The prevalence of GDM in United States increased from 1.7% to 3.1% among non-Hispanic whites; from 2.8 to 5.4% in Hispanics, and from 2.9 to 5.4% in African Americans [12]. It has been reported that the prevalence of GDM among the Hispanic population in the United States ranges between 5 and 15% depending on the geographical location. The prevalence of GDM in Hispanic women of Mexican origin living in border communities ranges from 10 to 15% [13–15]. A similar trend has been reported for Canada's aboriginal population. The prevalence in the inland communities was twice as high as that in the coastal communities (18.0% versus 9.3%, $p = 0.002$) [16]. The increase in GDM prevalence may represent a major determinant of the recent increase in obesity among women of childbearing age, which may lead to further increases in GDM [12].

B. Risk Factors

Risk factors associated with the development of GDM include age, obesity, family history of diabetes, and being a member of a high-prevalence racial/ethnic group. The risk can be classified as low, average, or high according to the criteria described later and should be determined during the first prenatal visit. Table 1 describes characteristics associated with risk for developing GDM. A patient identified as low risk for developing GDM does not require a routine screening. Patients with one or more characteristics in the average risk category should be screened for GDM between the 24th and 28th weeks of gestation. If any characteristic is present from the high-risk category, screening is performed as soon as possible. If GDM is not diagnosed, blood glucose testing should be repeated at 24–28 weeks or at any time the patient has symptoms or signs that are suggestive of hyperglycemia [10].

C. Dietary Risk Factors

Several predisposing dietary factors have been identified with the development of GDM. Ying and Wang [17] examined the effects of dietary fat on GDM in nulliparous pregnant (NP) women with GDM ($n = 85$) compared with normal pregnant women ($n = 159$). Dietary assessments were conducted and evaluated in both groups. The analysis focused on carbohydrate, protein, and fat, especially on dietary fat and fat subtypes. Fat intake in the GDM group was significantly higher than that in the NP group, whereas carbohydrate and protein intake in the GDM group was not significantly different compared with the NP group. Polyunsaturated fatty acid intake in the GDM group was lower than that in the NP group, but saturated fatty acid intake was higher than that in the NP group. Although monounsaturated fatty acid intake in the GDM group was higher than in the NP group, there was no significant difference between the two groups. The authors concluded that total dietary fat, high saturated fatty acids, and low polyunsaturated fatty acids may be a risk factor for GDM [17].

In a large cohort prospective study of 13,110 U.S. women of reproductive age (The Nurses' Health Study II), the authors examined whether pre-gravid dietary fiber consumption from cereal, fruit, and vegetable sources, dietary glycemic index, and glycemic load were related to GDM risk. They concluded that total dietary fiber, in particular cereal and fruit fiber, was inversely associated with GDM risk. A multivariate analysis indicated that for each 10 g/day increment in total fiber, the risk of GDM was reduced by 26%. No significant associations were observed for vegetable fiber. Dietary glycemic load was significantly and positively associated with GDM risk after adjustment for nondietary and dietary covariates. The association between dietary glycemic index alone and GDM risk was not statistically significant. Women with low cereal fiber intake and high glycemic load had a 2.15-fold (95% CI 1.04–4.29) higher risk for GDM [18].

TABLE 1 Risk Categories for Screening for Gestational Diabetes

Risk Category	Characteristics
Low	• Not a member of an ethnic group with a higher prevalence of GDM (see Prevalence section, this chapter) • No known diabetes in first-degree relatives (parents, siblings) • Normal pre-pregnant body mass index • No previous history of abnormal glucose tolerance • No history of adverse pregnancy outcome associated with GDM (e.g., unexplained loss, macrosomia or excessive-sized infant)
Average	• 25 years of age • $<$25 years of age and obese (i.e., 20% over desired body weight or body mass index $>27\,kg/m^2$) • Family history of diabetes in first-degree relatives • Member of an ethnic/racial group of high prevalence (e.g., Hispanic American, Native American, Asian American, African American, or Pacific Islander)
High	• Significant obesity • Family history of diabetes • GDM in previous pregnancy • History of glucose intolerance • Glucosuria • History of adverse pregnancy outcome associated with GDM

In summary it has been reported that a diet high in total fat containing high saturated and low polyunsaturated fatty acids, as well as low cereal fiber and high glycemic load, has been associated with the risk of developing GDM [17, 18].

II. SCREENING AND DIAGNOSIS

The World Health Organization (WHO) and the American Diabetes Association use different criteria to diagnose GDM. The criteria for the diagnosis of GDM are based on an oral glucose tolerance test (OGTT).

According to the World Health Organization (WHO), women with high risk for GDM (older women, from high-risk ethnic groups, personal history of glucose intolerance, personal history of large for gestational age offspring, and those presenting elevated fasting or casual blood glucose levels) should be screened during the first trimester of pregnancy. Systematic testing of all pregnant women should be performed between 24 and 28 weeks of pregnancy with a 75-g OGTT. Pregnant women who meet the WHO criteria for diabetes mellitus [fasting glucose ≥7.0 mmol/liter (≥126 mg/dl) or 2-hour glucose ≥11.1 mmol/liter (≥200 mg/dl)] or impaired glucose tolerance (IGT) [fasting glucose ≥7.0 mmol/liter (≥126 mg/dl) or 2-hour post glucose load ≥7.8 mmol/liter (≥140 mg/dl) and <11.1 mmol/liter (<200 mg/dl)] are classified as having GDM [19].

The criteria used by the American Dietetic Association indicate that a fasting plasma glucose level above 126 mg/dl or a casual plasma glucose above 200 mg/dl meets the threshold for the diagnosis of diabetes. In the absence of this degree of hyperglycemia, evaluation of GDM in women with average or high-risk characteristics should follow either a one- or two-step approach to screening and diagnosis during the first trimester.

In the one-step approach, the woman is given a 75-g OGTT without a prior plasma or serum glucose screen. The one-step approach may be cost-effective in high-risk patients or populations (e.g., some Native American groups).

The two-step approach uses a glucose challenge test and the OGTT [10].

Glucose Challenge Test

1. A 50-g oral glucose load, administered between the 24th and 28th week, and without regard to time of day or time of the last meal, to all pregnant women who have not been identified as having glucose intolerance before the 24th week.
2. Venous plasma glucose is measured 1 hour later. A value greater than 130 mg/dl in venous plasma indicates the need for a full diagnostic OGTT.

Oral Glucose Tolerance Test

1. 100-g glucose load, administered in the morning after overnight fast of at least 8 but not more than 14 hours, and after at least 3 days of unrestricted diet (≥ 150 g carbohydrate) and physical activity.
2. Venous plasma glucose is measured fasting and at 1, 2, and 3 hours. Women should remain seated without smoking throughout the test. Two or more of the following venous plasma concentrations must be met or exceeded for a positive diagnosis of GDM:

Fasting, ≥95 mg/dl or 5.3 mmol/liter

1 hour, ≥180 mg/dl or 10.0 mmol/liter

2 hours, ≥155 mg/dl or 8.6 mmol/liter

3 hours, ≥140 mg/dl or 7.8 mmol/liter

One abnormal value on the 100-g glucose load as well as significantly elevated (>185 mg/dl) glucose screen despite normal results has been associated with fetal macrosomia and may warrant management. If one abnormal value is seen during the 100-g 3-hour OGTT, it is recommended that the test be repeated approximately 1 month later. There is growing evidence that one abnormal value is sufficient to have an impact on the health of the fetus and now most clinicians initiate treatment using this criterion. When the two-step approach is employed, a glucose threshold value above 140 mg/dl (7.8 mmol/liter) identifies approximately 80% of women with GDM, and the yield is further increased to 90% by using a cutoff of greater than 130 mg/dl (7.2 mmol/liter) [20, 21].

A fasting glucose level above 126 mg/dl or a casual plasma glucose above 200 mg/dl that is confirmed on a subsequent day precludes the need for any glucose challenge.

III. COMPLICATIONS

Perinatal morbidity and mortality in GDM are associated with uncontrolled blood glucose. When managed intensively, the risk of intrauterine fetal death is similar to that in normal pregnancies [22]. The severity of risk is associated with the degree of maternal hyperglycemia. Even mild blood glucose elevations, however, may cause fetal macrosomia. Whereas maternal glucose readily crosses the placenta, insulin does not. Elevated maternal blood glucose levels produce fetal hyperglycemia, which results in fetal hyperinsulinemia, lipogenesis, increased glycogen and protein synthesis, and subsequent macrosomia. In a prospective study that included 115 untreated women with borderline GDM, it was found that compared with normoglycemic controls, the untreated borderline GDM group had increased rates of macrosomia (28.7% versus 13.7%, $p < 0.001$) and cesarean delivery

(29.6% versus 20.2%, $p = 0.02$) [23]. In a retrospective case control study of 970 women, infant and maternal morbidity was assessed in 114 mother–child pairs with an infant birth weight over the 90th percentile [24]. The incidence of pre-eclampsia was 8.8% versus 2.7% in the peer group ($p = 0.002$), and neonatal jaundice was 16.7% in the undiagnosed GDM group versus 4.5% in the peer group ($p < 0.0001$). Cord-blood insulin levels were significantly elevated in comparison to the peer group of mothers without metabolic disorders and who had borne eutrophic infants (8.4 mU/liter [3.0–100.0] versus 5.3 mU/liter [3.0–30.7], $p = 0.01$). Cord-blood insulin levels of 11.4% of all macrosomic infants were above the normal range.

Major complications of uncontrolled GDM have been associated with traumatic birth, shoulder dystocia, clavicular fracture, brachial palsy or other traumatic injury, congenital anomalies, neonatal metabolic disorders, cardiomyopathy, neurological symptoms, renal vein thrombosis, hyaline membrane disease, neonatal respiratory distress syndrome, asphyxia, resuscitation at delivery, and transient tachypnea [25]. In a retrospective case control study of 472 pregnant women with GDM the relationship among unrecognized gestational diabetes and infant birth weight, delivery mode, and perinatal complications was evaluated in 297 women treated with diet, 76 treated with diet and insulin and 16 with unrecognized GDM [26]. Controlled confounding variables were maternal body mass index (BMI [kg/m^2]), weight gain, age, parity, and gestational age. It was found that shoulder dystocia occurred in 19% of the unrecognized group compared with 3% in the control group ($p < 0.05$) and 3% in the diet group ($p < 0.02$). The incidence of birth trauma in the unrecognized group (25%) was significantly greater than in control group (0%), diet group (0.3%), and insulin group (1%). There was a significantly higher rate of hyperbilirubinemia, hypocalcemia, polycythemia, and neonatal ICU admission in the unrecognized group when compared to the control group (13% versus 0%).

In a case control study of 555 women with untreated GDM, one or more of five indicators were measured (stillbirth, neonatal macrosomia, neonatal hypoglycemia, erythrocytosis, and hyperbilirubinemia). It was reported that for every 10 mg/dl increment in fasting plasma glucose there was a 15% increase in adverse outcome [27].

IV. NUTRITION MANAGEMENT

Nutrition management represents the cornerstone of treatment during GDM for blood glucose control. This was recently documented when the results of a randomized clinical trial reported a reduced risk of serious perinatal outcomes in a group of women who received individualized dietary advice by a qualified nutritionist, blood glucose monitoring, and insulin therapy as needed. Infants born to women in the intervention group had birth weights significantly lower than did their counterparts [28]. Another study looking at the impact of nutrition practice guidelines implemented by registered dietitians on pregnancy outcomes showed lower insulin use, reduced glycated hemoglobin, and lower birth weight [29].

During nutrition management of GDM, specific caloric and nutrient recommendations are determined and subsequently modified as needed based on individual assessment and self-monitoring of blood glucose. Because of the continuous fetal draw of glucose from the mother, maintaining consistency of schedule and amount of food is important to avoid hypoglycemia. Plasma glucose monitoring and daily food records provide valuable information for insulin and meal plan adjustments [8].

According to the American Diabetes Association, all women with GDM should receive nutritional counseling, by a registered dietitian when possible. Individualization of medical nutrition therapy (MNT) depending on maternal weight and height is recommended. MNT should include the provision of adequate calories and nutrients to meet the needs of pregnancy and should be consistent with the maternal blood glucose goals that have been established [20].

A. Weight Gain Recommendations

Total weight gain during pregnancy varies widely among women. The current weight gain recommendations according to the Institute of Medicine are based on pre-pregnancy BMI. Table 2 displays recommendations for total weight gain for each BMI category and the recommended total weight for the first trimester and weekly weight gain for the second and third trimesters [30].

TABLE 2 Recommended Total Weight Gain and Mean Rate of Weight Gain for Each Body Mass Index (BMI) Category

Pre-pregnancy BMI Category	Total Weight Gain (kg)	First Trimester Total (kg)	Second Trimester (kg/week)	Third Trimester (kg/week)
Low (BMI < 18.5)	12.5–18	2.3	0.49	0.49
Normal (BMI 18.5–24.9)	11.5–16	1.6	0.44	0.44
High (BMI 25–29.9)	7–11.5	0.9	0.30	0.30
Obese (BMI > 29)	>7			

Source: Institute of Medicine. (1990). "Nutrition during Pregnancy. Part 1: Weight Gain. Part 2: Nutrient Supplements." National Academy of Sciences, Washington, DC.

B. Caloric Requirements

The energy allowance for pregnancy may be estimated by dividing the gross energy cost (80,000 kcal) by the approximate duration (250 days following the first month), yielding an average value of 300 kcal per day in addition to the allowance of nonpregnant women [31, 32].

The caloric requirements are different for each trimester. During the first trimester the energy needs are the same as for a nonpregnant woman. An initial guideline for the first trimester is to use 30 kcal/kg of ideal pre-pregnant body weight. The energy requirements for the second and third trimesters can be calculated using a rough method by multiplying actual maternal weight in kilograms by the recommended kcal/kg correspondent to each pre-pregnant BMI category (see Table 2). Thomas-Dobersen recommends using 36 kcal/kg ideal body weight for the second trimester and 38 kcal/kg of ideal body weight for the third trimester [33]. The following studies have reported using different caloric recommendations based on actual body weight or pre-pregnant body weight. For example, Jovanic-Peterson recommends using a euglycemic diet (40% of calories from carbohydrate) with 30 kcal/kg of current weight recommended for normal pre-pregnant weight, 24 kcal/kg of current weight for overweight, and 12 to 15 kcal/kg for obese pregnant women during their second and third trimesters. In their hands, this provides sufficient energy without causing excessive weight gain or hyperglycemia [34]. It has been documented that when women with GDM followed the euglycemic diet, normal glycemia is achieved in 75% to 80% of cases. The diet reduces caloric intake to just above the ketonuric threshold [35]. Langer *et al.* studied the relationship between pre-pregnant weight, treatment modality (diet and insulin), levels of glycemic control and pregnancy outcome [36]. Normal-weight women were prescribed 35 kcal/kg for actual maternal weight and obese women 25 kcal/kg. They concluded that well-controlled gestational diabetes was associated with good outcomes in all maternal weight groups. In a study of Mexican American women with GDM who received culturally appropriate nutrition therapy, energy intake was prescribed at 37 ± 3 kcal/kg for normal pre-pregnant weight ($n = 42$). Overweight ($n = 46$) and obese ($n = 47$) pregnant women were prescribed an average of 23 ± 3 kcal/kg. At delivery neonatal weight of infants from normal-weight, overweight, and obese women were 3313 g $\pm$ 671, 3413 g $\pm$ 727, and 3448 g $\pm$ 654, respectively [37]. In this study, the caloric prescription was based on pre-pregnant body weight for all BMI categories, whereas the majority of studies base their requirements on current weight.

The energy requirements for pregnant adolescents (14–18 years) and adult women (19–50 years) are determined by using the estimated energy requirements (EER) formula [32]. This method is not applicable to overweight and obese women. Available guidelines suggest that caloric requirement for overweight and obese women can be obtained using adjusted body weight in the Harris Benedict formula and adding 150–300 kcal for the second and third trimesters [8]. Table 3 displays the EER formulas to determine energy requirements for normal pre-pregnant BMI category for the second and third trimesters. It also displays formulas to determine energy requirements for high pre-pregnant BMI category [8, 32].

Independent of the method used to estimate caloric level, appetite, weight gain pattern, and testing for ketones will indicate whether adjustments of caloric recommendations are necessary.

C. Carbohydrate

Carbohydrate is the primary nutrient affecting postprandial glucose levels, and postprandial blood glucose concentration is the most important role in macrosomia [38]. The amount and kind of carbohydrate eaten at meals and snacks are essential to maintaining optimal blood glucose and reducing the need for insulin while controlling maternal weight gain and infant birth weight [8]. In summary, it is recommended that carbohydrates be provided from a variety of food sources such as fruits, vegetables, low-fat milk, and whole grains [7].

1. Total Carbohydrate Intake

To ensure provision of glucose to the fetal brain, which needs about 33 g/day, and supply the glucose fuel requirement for the mother's brain independent of utilization of ketoacids (or other substrates), a minimum amount of 175 g of carbohydrates is necessary to prevent ketosis [32]. Carbohydrate recommendations vary from 40 to 45% of total daily calories [8, 36, 39, 40]. An estimated amount of carbohydrates generally comprises between 200 and 225 g of carbohydrates for a 2000-calorie meal plan. On the euglycemic diet, the amount of carbohydrate consumption is restricted to less than 40% of total calories, allowing 40% or more of calories to be fat and the remaining 20% to be protein [35]. Low-carbohydrate diets (<130 g/day) are not recommended [7].

2. Carbohydrate Distribution

The distribution of energy and carbohydrate in the meal plan is based on woman's eating habits and plasma glucose responses. A better glucose response is achieved when the total amount of carbohydrate is distributed throughout the day and limited at breakfast [7]. This was demonstrated by Peterson *et al.*, in 14 women prescribed a diet with 24 kcal/kg of current pregnant weight/day (12.5% of calories at breakfast, 28% of calories at lunch and dinner, with the remaining calories divided among three snacks) [41]. Blood glucose response was measured four times a day. The glycemic response to a meal was highly correlated with the carbohydrate content of the meal. The glycemic response was more consistent for dinner ($r = 0.95$, $p < 0.001$), with greater variability at breakfast ($r = 0.75$, $p = 0.002$) and lunch ($r = 0.86$, $p = 0.001$). They found that the percentage of

TABLE 3 Energy Requirements for Pregnancy

Name ________________________________ **Age (A)** ____________

Gestational age __________ **wks**

Pre-pregnancy Wt ________ **kg**

Height ________ **m × 100 cm/m =** __________ **cm**

BMI ____________ **kg/m²**

Pre-pregnancy BMI Category
Low (BMI < 18.5)
Normal (BMI 18.5 – 24.9)
High (BMI 25 – 29.9)
Obese (BMI > 29)

Estimated Energy Requirements for Normal Pre-pregnant BMI

$\text{Energy}_{\text{First Trimester}} = \text{EER} + 0$	$\text{EER} = 354 - (6.91 \times \text{A}) + \text{PA} \times (9.36 \times \text{Wt kg} + 726 \times \text{Ht m})$
$\text{Energy}_{\text{Second Trimester}} = \text{EER} + 340$	$\text{EER} = 354 - (6.91 \times \text{A}) + \text{PA} \times (9.36 \times \text{Wt kg} + 726 \times \text{Ht m}) + 340$
$\text{Energy}_{\text{Third Trimester}} = \text{EER} + 452$	$354 - (6.91 \times \text{A}) + \text{PA} \times (9.36 \times \text{Wt kg} + 726 \times \text{Ht m}) + 452$

Source: Data from [31].

Physical Activity Coefficient (PA)

Sedentary	Low Activity	Active	Very Active
1.0	1.12	1.27	1.45

Energy Requirements for High and Obese Pre-pregnant BMI for Second and Third Trimesters

$$\text{Energy} = [655 + (9.6 \times \text{AdBWt kg}) + (1.8 \times \text{Ht cm}) - (4.7 \times \text{A})] \times \text{PA} + \text{X}$$

X = 150 – 300 kcal
Adjusted Body Weight (AdBWt)
$\text{AdBWt} = [(\text{Actual Body Wt} - \text{Desirable Body Wt}) \times 0.25\,] + \text{Desirable Body Wt}$

carbohydrate in each meal should be 33% for breakfast, 45% for lunch, and 40% for dinner to keep the postprandial blood glucose below 6.7 mmol/liter (120 mg/dl).

Carbohydrate tolerance is often lower in the morning because high cortisol levels interfere with glucose clearance in the morning and are particularly high during pregnancy [35]. Postbreakfast hyperglycemia can be prevented by limiting the total carbohydrate to less than 30 g and minimizing or avoiding intake of fruit, fruit juice, and refined grain products [42]. The total amount of carbohydrate is generally distributed into three meals and two to four snacks. The amount of carbohydrate recommended for snacks is smaller than the amount recommended for meals [8, 33, 41, 42]. An intake of 15–30 g of carbohydrates during breakfast is recommended based on the production of higher levels of plasma cortisol and growth hormone in the morning affecting glucose levels [41]. This carbohydrate distribution throughout the day helps in the prevention of hunger, ketonuria, heartburn, and nausea [8]. A consistent amount of carbohydrate at meals and snacks will result in better glucose control and will facilitate insulin adjustments [43].

An acceptable caloric distribution would be 10% of calories at breakfast, 30% at both lunch and dinner, and 30% distributed in between meals as snacks [41].

3. Type of Carbohydrate

Gestational diabetes is associated with increased carbohydrate sensitivity. To address this sensitivity, it is advised to use the glycemic index to customize the food plan. The applicability of the glycemic index concept in pregnancy was studied by Lock *et al.* [44]. They studied glycemic responses to foods in women with GDM. The postprandial curves of bread, raisins, dates, sweet corn, bananas, oranges, spaghetti, and green peas were measured in 28 gestational diabetic subjects and compared to glycemic indices reported for nonpregnant women. Each food test contained 50 g of carbohydrate with the exception of peas, which contained 25 g. Glucose (50 and 25 g) was the reference food. After bread ingestion, the glycemic excursion was similar to glucose in form but slightly blunted. Compared to bread, other test foods had earlier glucose peaks and more blunted postprandial glucose curves. The glycemic index of bread (77.7 ± 5.0) was significantly higher than those of corn (51.8 ± 6.8), bananas (49.0 ± 5.5), oranges (46.6 ± 4.7), spaghetti (42.2 ± 7.3), and peas (35.1 ± 4.2). The glycemic indices of dates (61.6 ± 3.5) and raisins (65.7 ± 5.8) were also significantly higher than that of peas. There was a weak positive correlation between individual glycemic indices and the means of the fasting glucose levels used in their calculation. It was concluded that during GDM the glycemic indices were uniform after the ingestion of the foods and therefore pregnancy does not appear to alter the glycemic indices of the foods tested in nonpregnant subjects.

A study of pregnant (nondiabetic) women found that glycemic index of the diet was positively and significantly related to maternal glycosylated hemoglobin and plasma glucose. It was also reported that pregnant women with low dietary glycemic index had reduced infant birth weight and approximately twofold increased risk of a small-for-gestational-age baby [45].

The glycemic response does not solely depend on dietary carbohydrates, but is also influenced by other meal components. The glycemic index can be used as an adjunct for the fine tuning of postprandial blood glucose responses [46]. Because many clinicians find that certain foods lead to a higher glycemic response causing a higher elevation in postprandial blood glucose [8], it is the general recommendation to restrict high-glycemic-index foods (e.g., highly processed breakfast cereals, instant potatoes, instant noodles, sugar, honey, molasses, corn syrup, candy, sweetened beverages) and limit intake of fruit, fruit juice, and milk. Low-glycemic-index foods (e.g., whole-wheat bread, old-fashioned oatmeal, bran cereal, nuts, legumes, and lentils) and protein sources are emphasized [44].

4. Consistency of Carbohydrates

Consistency of carbohydrates is encouraged, specifically to prevent undereating and to assist with postprandial blood glucose evaluation. A consistent amount and type of carbohydrates at meals and snacks are recommended for better glucose control. Consistency of carbohydrates will facilitate insulin adjustments if necessary [8, 42].

5. Dietary Fiber

Dietary fiber such as pectin and guar gum and those present in oatmeal products and beans have been found to produce a reduction to glycemic response and high insulin sensitivity. Specifically, it has been reported that vegetable fibers with a high viscosity reduce the levels of basal and postprandial glycemia in both normal and diabetic subjects. The value of glucomannan and guar in the treatment of excessive weight gain in pregnancy and GDM was studied. Thirty-four patients were included in the study, of whom 13 received glucomannan (3 g/day) and 21 guar (10 g/day). An OGTT was performed in all patients before and after therapy. A reduction of basal and post-OGTT glycemia values was observed in all subjects [47].

In a study by Reece *et al.*, non-insulin-requiring GDM patients were placed in three different fiber dose groups (moderate-fiber dose, 40 to 60 g; high-fiber dose, 70 to 80 g; and a normal fiber dose, 20 g or less of fiber). No significant difference was observed in the mean blood glucose and postprandial glucose levels among the three groups. It was concluded that a high amount of fiber intake (40–80 g/day) in GDM is not associated with a better glucose response [48].

Women with diabetes are advised to ingest a variety of fiber-containing foods. Higher fiber intake (>50 g/day) does not provide additional benefits [7]. An adequate intake of dietary fiber for pregnant women is 28 grams per day. This amount of fiber should be enough to slow gastric emptying time, resulting in a significant reduction in postprandial glucose [32].

6. Simple Carbohydrates and Nutritive Sweeteners

Limiting foods high in total carbohydrates such as regular sodas, other sweetened beverages, candies, and desserts may help in attaining glucose levels [8]. Although it has been demonstrated that the consumption of sucrose does not compromise blood glucose when total carbohydrate content remains the same [49]. However because of its low nutrient density, the consumption of foods containing high amounts of sucrose is not recommended. Other carbohydrate containing foods, such as whole grains, are more nutrient dense. Foods that contain large amounts of sucrose, such as sweet rolls and cookies, often contain large amounts of fat [8]. Sugar alcohols and non-nutritive sweeteners should be consumed in accordance to the parameters established by FDA. Fructose intake is not advised except as a food constituent. [7].

Sorbitol, mannitol, and xylitol are sugar alcohols supplying 4 kcal/g. They have less influence on blood glucose or insulin levels than sugar because they are absorbed more slowly from the gastrointestinal tract. The slow, passive absorption of sugar alcohols can produce osmotic diarrhea, malabsorption, and abdominal discomfort, and so they should be avoided or used in limited amounts [49].

D. Fat

Fat content in the diet of a pregnant woman usually ranges from 30% to 40% of total calories; however, large amounts of fat beyond total caloric needs should be avoided to prevent excessive weight gain, which in turn can result in further insulin resistance [50, 51]. The acceptable macronutrient distribution range for fat is 20–35% of total energy, the same as for nonpregnant women [32].

The type of fat in the diet may play an important role in the glucose response to a meal, independent of the carbohydrate content. In a study by Ilic *et al.* [52], 10 GDM women who were well controlled on diet alone were randomized to receive a meal after overnight fast containing saturated fat (SF) or monounsaturated fat (MUFA). Blood was drawn at 0, 60, 120, and 180 min for plasma glucose, insulin, lipid profile, and free fatty acid concentration. After 2 weeks, each patient received the other type of meal. The test meal was composed of 20% of the total daily caloric needs based on ideal body weight. The area under the curve showed a significantly lower glucose concentration for SF meal ($p = 0.001$). Serum insulin concentrations followed the glucose response with the peak at the 60-minute time point and a significantly lower concentration at the 180-minute time point in the SF than in the MUFA group. That study demonstrated that the addition of SF to the meal resulted in lower postprandial glucose and insulin than when the meal contained MUFA. Therefore, SF may be useful in controlling postprandial glucose.

In addition to total fat recommendations, dietary reference intakes have been formulated for essential polyunsaturated fatty acids (PUFA). There is evidence that docosahexaenoic acid modulates insulin resistance and that it is vital for fetal neurovisual development [53]. The consumption of omega-3 fatty acids in pregnant women continues to be below the recommendation. In a study by Thomas *et al.*, the consumption of omega-3 fatty acids and fiber was increased in women with GDM after nutrition intervention [54]. Adequate Intake for omega-6 (PUFA) linoleic acid is 13 g/day and omega-3 (PUFA) α-linolenic acid is 1.4 g/day [32].

Within that fat allowance, no more than 7% to 10% of total energy intake should come from saturated fat, *trans* fat should be reduced, and cholesterol intake should be less than 200 mg/day [7].

E. Protein

The diet of well-nourished women in the preconception period and throughout most of pregnancy has a significant effect on birth weight; and proteins are the macronutrient that has the greatest impact. Cuco *et al.* demonstrated that a 1-g increase in maternal protein intake during preconception and in the 10th, 26th, and 38th weeks of pregnancy leads to a significant increase of 7.8–11.4 g in birth weight [55]. On the other hand, high intakes of protein and fat during pregnancy may impair development of the fetal pancreatic beta cells and lead to insulin deficiency in the offspring [51]. In a case case-control study of 2341 women with singleton pregnancy, three different levels of protein in the diet were associated with birth weight. It was found that birth weight was 77 g lower ($p = 0.021$) in the low-protein group, and 71 g lower ($p = 0.009$) in the high-protein group, compared with the intermediate-protein group. Birth weight increased with protein levels up to 69.5 g/day and declined with higher protein intake. A high average prenatal protein consumption results in a significant depression of birth weight, in fact, a protein intake of more than 84 g/day on average is more detrimental than is low protein. It appears that moderate protein intake is optimal during pregnancy [56].

The inclusion of protein in meals and snacks does not significantly affect blood glucose excursions and thus can be added for additional calories in place of carbohydrate foods [8]. The RDA for pregnancy for all age groups is 1.1 g/kg/day or 25 grams additional protein [32]. This recommendation is easily met in meal plans for GDM when the carbohydrate is controlled to 40–45% of total energy intake. Protein generally comprises 20–25% of total daily calories [8].

F. Nonnutritive Sweeteners

According to the American Dietetic Association Position Statement on the use of nonnutritive sweeteners in pregnancy must be based on well-designed and approved clinical investigation to ensure healthy pregnancy outcomes. Although no specific recommendations have been made regarding their use during pregnancy, moderation seems appropriate [49].

Aspartame (Equal or NutraSweet) does not cross the placenta when consumed in usual amounts. It is useful in uncooked foods but unstable when heated. Women who have phenylketonuria (PKU) must restrict their phenylalanine intake. Because the amino acid phenylalanine is one of the metabolites of aspartame, it should be avoided by women with PKU. Maternal plasma levels of phenylalanine after ingestion of aspartame-containing foods, in normal amounts, are no higher than after the ingestion of protein-containing meals [57]. Use of aspartame within FDA guidelines appears to be safe during pregnancy [49]. Acesulfame potassium (acesulfame-K; Sweet One, Sunette) is considered to be safe during pregnancy. Sucralose (Splenda), derived from sucrose, is the most recently FDA-approved high-intensity sweetener. The FDA concluded that this sweetener does not pose carcinogenic, reproductive, or neurological risk to humans. Saccharin (Sweet'N Low) crosses the placenta and may remain in fetal tissue because of slow fetal clearance. The American Medical Association and the American Dietetic Association suggest careful use of saccharin in pregnancy. Many practitioners suggest complete avoidance of saccharin in pregnancy [49].

In summary, medical nutrition therapy remains the cornerstone of treatment for GDM and is best prescribed by a registered dietitian or a qualified individual with experience in the management of GDM. Nutrition recommendations for GDM, including gestational weight gain, calorie intake, and macronutrient composition and distribution, are based on limited scientific evidence [58]. Currently, nutrition practice guidelines for GDM recommend a carbohydrate controlled meal plan with adequate nutrient content aimed to sustain maternal needs and fetal growth. Besides practice guidelines, research findings associating macronutrients and caloric prescriptions to birth outcomes are also available. Using the current nutrition practice guidelines in combination with research findings can help the nutritionist to individualize a meal plan that will contribute to the delivery of a healthy infant. Table 3 displays equations to calculate estimated energy requirements for normal weight, high weight and obese pre-pregnant women. In addition, Table 4 summarizes the Institute of Medicine micronutrient recommendations for pregnant women [32]. Refer to the Appendix at the end of this book for complete Dietary Reference Intakes for vitamins, minerals, and macronutrients.

TABLE 4 Recommended Dietary Allowances (RDA) and Adequate Intakes (AI) for Micronutrients

Micronutrient	≤18 Years	19–30 Years	31–50 Years
Thiamin (mg/day)[a]	1.4	1.4	1.4
Riboflavin (mg/day)[a,c]	1.4	1.4	1.4
Niacin (mg/day)[a]	18	18	18
Biotin (μg/day)[b]	30	30	30
Pantothenic acid (mg/day)[b]	6	6	6
Vitamin B_6 (mg/day)[a]	1.9	1.9	1.9
Folate (μg/day)[a,d]	600	600	600
Vitamin B_{12} (μg/day)[a]	2.6	2.6	2.6
Choline (mg/day)[b]	450	450	450
Vitamin C (mg/day)[a]	80	85	85
Vitamin A (μg/day)[a,e]	750	770	770
Vitamin D (μg/day)[b,f]	5	5	5
Vitamin E (mg/day)[a,g]	15	15	15
Vitamin K (μg/day)[b]	75	90	90
Sodium (mg/day)[b]	1500	1500	1500
Chloride (mg/day)[b]	2300	2300	2300
Potassium (mg/day)[b]	4700	4700	4700
Calcium (mg/day)[b]	1300	1000	1000
Phosphorus (mg/day)[a]	1250	700	700
Magnesium (mg/day)[a]	400	350	360
Iron (mg/day)[a]	27	27	27
Zinc (mg/day)[a]	12	11	11
Iodine (μg/day)[a]	220	220	220
Selenium (μg/day)[a]	60	60	60
Copper (μg/day)[a]	1000	1000	1000
Manganese (mg/day)[b]	2.0	2.0	2.0
Fluoride (mg/day)[b]	3	3	3
Chromium (μg/day)[b]	29	30	30
Molybdenum (μg/day)[a]	50	50	50

[a]Recommended Dietary Allowance (RDA).
[b]Adequate Intake (AI).
[c]Niacin recommendations are expressed as niacin equivalents (NE).
[d]Folate recommendations are expressed as dietary folate equivalents (DFE).
[e]Vitamin A recommendations are expressed as retinol activity equivalents (RAE).
[f]Vitamin D recommendations are expressed as cholecalciferol and assume absence of adequate exposure to sunlight.
[g]Vitamin E recommendations are expressed as α-tocopherol.

V. CLINICAL OUTCOMES

Nutrition management in combination with intensive monitoring can contribute to the delivery of a healthy infant. Frequent follow-up visits are needed to determine whether the patient is achieving the clinical outcomes originally established by the medical team. The information obtained from blood-glucose monitoring, food and physical activity records, urinary ketone testing, glycosylated hemoglobin, and weight gain patterns, as well as the patient's ability to understand and follow a meal plan, will allow the practitioner to evaluate glycemic control. This information assists in meal plan and insulin therapy adjustment.

A. Blood Glucose Monitoring

The use of self-monitoring blood glucose (SMBG) allows the medical team to objectively evaluate the nutrition management recommended and reinforces selection of appropriate amounts and types of foods that are likely to produce normoglycemia. Blood glucose data permit evaluation of the glycemic response to meals and snacks, and to determine whether changes in macronutrient amount or distribution are needed [59].

Maternal metabolic surveillance should be directed at detecting hyperglycemia severe enough to increase risks to the fetus. Daily SMBG appears to be superior to intermittent office monitoring of plasma glucose. For women treated with insulin, limited evidence indicates that postprandial monitoring is superior to preprandial monitoring. However, the success of either approach depends on the glycemic targets that are set and achieved [10].

The initial daily testing schedule is five times per day: fasting; 1 hour after meals (breakfast, lunch, and dinner); and at bedtime. The frequency may be reduced from daily to every third to fourth day once control of blood glucose is established. Increased surveillance for pregnancies at risk for fetal demise is appropriate, particularly when fasting glucose levels exceed 105 mg/dl (5.8 mmol/liter) or pregnancy progresses past term [10]. The SMBG levels recommended by the American Diabetes Association are as follows: fasting plasma glucose less than 105 mg/dl (5.8 mmol/liter); 1-hour postprandial plasma glucose less than or equal to 155 mg/dl (8.6 mmol/liter); 2-hour postprandial plasma glucose less than or equal to 130 mg/dl (7.2 mmol/liter). Many obstetricians prefer lower glucose targets: fasting less than 95 mg/dl; 1-hour postprandial 120 mg/dl; 2-hour postprandial less than 120 mg/dl.

B. Food Records

Detailed information recorded in food records is useful to make adjustments to the meal plan or to the insulin regimen. By keeping food records the patient becomes aware of individual blood glucose response to particular types and amounts of foods. Food intake records also indicate the patient's level of understanding and compliance with the recommended meal plan.

C. Urinary Ketone Testing

Urinary ketone tested on the first morning voided specimen may be useful in detecting insufficient caloric or carbohydrate intake in women treated with calorie restriction [10].

D. Glycosylated Hemoglobin

Periodic glycosylated hemoglobin determinations evaluate glycemic control over the previous 4–6 weeks because the attachment of glucose molecules to hemoglobin occurs in relation to the concentration of plasma glucose over the lifespan of the red blood cells. Levels above 7% (varies with individual laboratory) indicate suboptimal blood glucose control [60].

E. Weight Gain Patterns

A weight gain grid in the medical records helps to evaluate weight gain progress. It is helpful to identify inadequate weight gain that may be related to restriction of food intake. Guidelines for weight gain recommendations are given in Table 2 [30].

VI. PHARMACOLOGICAL AGENTS

Pharmacological therapy is added to nutrition therapy when women with GDM are unable to achieve or maintain target blood glucose levels and/or when fetal growth rate is above normal [8, 61].

A. Insulin Therapy

Insulin therapy is recommended when nutrition therapy fails to maintain target blood glucose concentrations. The American Diabetes Association recommends the initiation of insulin therapy when unable to maintain a fasting blood glucose level of 105 mg/dl (5.8 mmol/liter), a 1-hour postprandial glucose level at or below 155 mg/dl (8.6 mmol/liter), or a 2-hour postprandial glucose level at or below 130 mg/dl (7.2 mmol/liter) [10]. The amount of insulin to be administered for GDM in the first trimester is 0.6 to 0.8 U/kg body weight; in the second trimester the insulin requirement is 1.0 U/kg body weight; and in the third trimester, the requirement is 1.2 U/kg body weight [62]. The total amount of insulin is distributed as follows: two-thirds in the morning and one-third in the evening. The morning dose is administered as two-thirds intermediate-acting insulin and one-third short-acting or regular insulin. The evening dose is divided into half short-acting or regular insulin and half intermediate-acting insulin [63].

If the postprandial glucose concentration is elevated, premeal rapid-acting insulin is usually prescribed, beginning with a dose of 1 U per 10 g carbohydrate in the meal. If both the fasting and postprandial glucose levels are elevated, or if a women's postprandial glucose levels can only be blunted if starvation ketosis occurs, a four-injections-per-day regimen should be prescribed. The latter can be based on combinations of intermediate insulin and regular insulin, timed to provide basal and meal-related insulin boluses.

The total daily insulin dose for the four-injections regimen should be adjusted according to pregnant body weight and gestational week (0.7–1 U/kg/day). Doses may need to be increased for the morbidity obese or when there is twin gestation [64].

B. Oral Hypoglycemic Agents

There is some evidence that oral hypoglycemic agents may be beneficial in GDM and that they can be used as alternative for insulin. It has been demonstrated that the use of glyburide in GDM is effective in lowering fasting glucose [65, 66]. Langer *et al.* reported no significant differences between the use of glyburide and insulin in two groups of GDM in regards to maternal outcomes such as glycosylated hemoglobin, rate of preeclampsia, cesarean section, neonatal complications including neonatal intensive care admissions, and risk or neonatal hypoglycemia [67, 68]. However, another study reported that women with GDM using glyburide had a higher risk of preeclampsia and infants had an increased risk of phototherapy [69]. Although the use of glyburide has been suggested as a cost-effective, patient-friendly alternative modality to insulin therapy, more studies of maternal and neonatal safety are needed.

VII. PHYSICAL ACTIVITY

Moderate exercise can be an important adjunctive therapy in the management of diabetes in pregnancy, particularly in GDM. Regular physical activity decreases the common discomforts associated with pregnancy without a negative effect on maternal or neonatal outcome. Regular physical activity is also beneficial in reducing insulin resistance, postprandial hyperglycemia, and excessive weight gain. In combination with dietary management, planned physical activity of 30 minutes per day is recommended for all individuals capable of participating [10]. Exercise is not recommended if blood glucose levels are poorly controlled [70]. To prevent hypoglycemia, exercise should be avoided in the fasting state and during periods of peak insulin activity. Energy and carbohydrate intake should be adequate prior to exercise and a rapidly absorbable form of carbohydrate readily available during exercise. Blood glucose should be monitored before and after exercise.

VIII. POSTPARTUM FOLLOW-UP

Glucose tolerance testing should be delayed until 6–12 weeks after delivery in women with GDM who do not have diabetes immediately postpartum. A 2-hour, 75-g glucose tolerance test can be coordinated with the postpartum visit. If glucose levels are normal postpartum, reassessment of glycemia should be undertaken at a minimum of 3-year intervals. Women with impaired fasting glucose (IFG: < 140 mg/dl) or impaired glucose tolerance (IGT: ≥ 140 to < 200 mg/dl) in the postpartum periods should be tested for diabetes annually. Because of the high risk for diabetes, these patients should receive medical nutrition therapy and be placed on an individualized exercise program [10].

IX. PREVENTION

The risk for developing diabetes after GDM has been reported to be higher than 70%. Current recommendations for women with GDM to prevent the onset of type 2 diabetes include diet counseling, awareness of the benefits of weight reduction or weight maintenance, and participation in regular physical activity. In addition to lifestyle modification, pharmacological therapies (metformin, troglitazone, and pioglitazone) have been shown to reduce diabetes development by 50% or more [71].

It has been reported that dietary modifications of lowering fat and carbohydrate intake in combination with physical activity can reduce the risk of developing GDM in subsequent pregnancies [72]. An intervention study designed to increase physical activity and healthy behaviors through dietary changes improved body weight and glucose tolerance over a period of 2–3 years [73]. Another study documented that a diagnosis of diabetes could be prevented when diet and intense exercise reduced fasting glucose and body weight over a 2-month period [74]. In addition to the implementation of dietary modifications and increased physical activity, other studies have demonstrated that oral hypoglycemic agents can be used to prevent or delay a positive diagnosis of type 2 diabetes [75]. On the other hand, there is new evidence indicating that a prepregnancy diet, in particular a diet with low fiber and high glycemic load, has been associated with an increased risk of gestational diabetes [18].

X. CONCLUSION

Currently, nutrition practice guidelines for GDM recommend a carbohydrate controlled food plan designed to promote adequate nutrition, appropriate weight gain, normoglycemia, and the absence of ketonuria. The food plan should be individualized, culturally appropriate, and adapted to the patient's food preferences and lifestyle. Monitoring of blood glucose concentration, urinary ketones, food intake, and physical activity is crucial in evaluating the nutrition intervention. Proper nutrition management in combination with intensive monitoring can contribute to the delivery of a healthy infant.

A diagnosis of GDM initiates a need for continued intervention and monitoring to minimize the risk of developing diabetes or to diagnose it as early as possible [71]. Pregnancy complicated with diabetes represents a window of opportunity for the implementation of nutrition education and a physical activity program aimed at diabetes prevention and healthier lifestyle patterns for the woman and her family.

References

1. Metzger, B. E., and Constan, D. R. (1998). Summary and recommendations of the Fourth International Workshop-Conference on Gestational Diabetes Mellitus: The Organizing Committee. *Diabetes Care* **21**, B161–B167.
2. Buchanan, T. A., Xiang, A., Kjos, S. L., and Watanabe, R. (2007). What is gestational diabetes? *Diabetes Care* **30**, S105–S111.
3. Dabelea, D. (2007). The predisposition to obesity and diabetes in offspring of diabetic mothers. *Diabetes Care* **30**, S169–S174.
4. Dabelea, D., Knowler, W. C., and Pettitt, D. J. (2000). Effect of diabetes in pregnancy on offspring: Follow-up research in the Pima Indians. *J. Matern. Fetal Med.* **9**, 83–88.
5. Pettitt, D. J., and Knowler, W. C. (1988). Diabetes and obesity in the Pima Indians: A crossgenerational vicious cycle. *J. Obes. Weight Regul.* **7**, 61–65.
6. Silverman, B. L., Rizzo, T., Green, O. C., Cho, N. H., Winter, R. J., Ogata, E. S., Richards, G. E., and Metzger, B. E. (1991). Long-term prospective evaluation of offspring of diabetic mothers. *Diabetes* **40**, S121–S125.
7. American Diabetes Association. (2007). Nutrition recommendations and interventions for diabetes. A position statement of the American Diabetes Association. *Diabetes Care* **30**, S48–S65.
8. American Dietetic Association. (2001). "Medical Nutrition Therapy Evidence-Based Guide for Practice: Nutrition Practice Guidelines for Gestational Diabetes Mellitus" [CD-ROM]. American Dietetic Association, Chicago.
9. Beischer, N. A., Oats, J. N., Henry, O. A., Cedí, M. T., and Walstab, J. E. (1991). Incidence and severity of gestational diabetes mellitus according to country of birth in women living in Australia. *Diabetes* **40**, S35–S38.
10. American Diabetes Association. (2004). Gestational diabetes mellitus. *Diabetes Care* **27**, S88–S90.
11. King, H. (1998). Epidemiology of glucose intolerance and gestational diabetes in women of childbearing age. *Diabetes Care* **21**, B9–B13.
12. Dabelea, D., Snell-Bergeon, J. K., Hartsfield, C. L., Bischoff, K. J., Hamman, R. F., and McDuffie, R. S. (2005). Increasing prevalence of gestational diabetes mellitus (GDM) over time and by birth cohort. *Diabetes Care* **28**, 579–584.
13. O Brian, M. E., and Gilson, G. (1987). Detection and management of gestational diabetes in out-of-hospital birth center. *J. Nurse-Midwifery* **32**, 79–84.
14. Mestman, J. (1980). Outcome of diabetes screening in pregnancy and perinatal morbidity in infants of mothers with mild impairment in glucose intolerance. *Diabetes Care* **3**, 447–452.
15. Duarte-Gardea, M., and Gonzalez, J. L. (1999). Screening and diagnosis scheme for glucose intolerance and gestational diabetes in Hispanic women. American Diabetes Association 59th Scientific Sessions. Abstract 2071.
16. Rodríguez, S., Robinson, E., and Gary-Donald, K. (1999). Prevalence of gestational diabetes mellitus among James Bay Cree women in northern Quebec. *Can. Med. Assoc. J.* **160**, 1293–1297.
17. Ying, H., and Wang, D. F. (2006). Effects of dietary fat on onset of gestational diabetes mellitus [in Chinese]. *Zhonghua Fu Chan Ke Za Zhi* **41**, 729–731.
18. Zhang, C., Liu, S., Solomon, C. G., and Hu, F. B. (2006). Dietary fiber intake, dietary glycemic load, and the risk for gestational diabetes mellitus. *Diabetes Care* **29**, 2223–2230.
19. World Health Organization. (1999). Definition, diagnosis and classification of Diabetes Mellitus and its complications: Report of a WHO Consultation. Part 1. Diagnosis and classification of diabetes mellitas. Retrieved January 5, 2007 from http://www.staff.ncl.ac.uk/philip.home/who_dmc.htm#DiagGDM.
20. American Diabetes Association. (2004). Gestational diabetes. *Diabetes Care* **27**, S88–S90.
21. Metzger, B. E., and Coustan, D. R. (1998). Summary and recommendations of the Fourth International Workshop-Conference on Gestational Diabetes Mellitus. *Diabetes Care* **21**, B161–B167.
22. American Diabetes Association. (1999). Position statement: Gestational diabetes mellitus. *Diabetes Care* **22**, S74–S76.
23. Naylor, C. H., Sermer, M., Chen, E., and Sykora, K. (1996). Cesarean delivery in relation to birth weight and gestational glucose tolerance: Pathophysiology or practice style? Toronto Trihospital Gestational Diabetes Investigators. *JAMA* **275**, 1165–1179.
24. Hunger-Dathe, W., Volk, K., Braun, A., Sammann, A., Muller, U. A., Peiker, G., and Huller, M. (2005). Perinatal morbidity in women with undiagnosed gestational diabetes in northern Thuringia in Germany. *Exp. Clin. Endocrinol. Diabetes* **113**, 160–166.
25. American Diabetes Association. (1998). Proceedings of the Fourth International workshop-conference on gestational diabetes mellitus. *Diabetes Care* **21**, B79.
26. Adams, K. M., Li, H., Nelson, R. L., Ogburn, P. L. Jr., and Danilenko-Dixon, D. R. (1998). Sequelae of unrecognized gestational diabetes. *Am. J. Obstet. Gynecol.* **178**, 1321–1332.
27. Langer, O., Yogev, Y., Most, O., and Xenakis, E. M. J. (2005). Gestational diabetes: The consequences of not treating. *Am. J. Obstet. Gynecol.* **192**, 989–997.
28. Crowther, C. A., Hiller, J. E., Moss, J. R., Jeffries, W. S., and Robinson, J. S. (2005). Australian carbohydrate intolerance study in pregnant women (ACHOIS). Effect of treatment of gestational diabetes mellitus on pregnancy outcomes. *N. Engl. J. Med.* **16**, 2477–2486.
29. Reader, D., Splett, P., and Gunderson, E. P. (2006). Diabetes care and education practice group. Impact of gestational diabetes mellitus nutrition practice guidelines implemented by registered dietitians on pregnancy outcome. *J. Am. Diet. Assoc.* **106**, 1426–1433.
30. Institute of Medicine. (1990). "Nutrition during Pregnancy. Part 1: Weight Gain. Part 2: Nutrient Supplements." National Academy of Sciences, Washington, DC.

31. Institute of Medicine. (1989). "Recommended Dietary Allowances"(10th ed.). National Academies Press, Washington, DC.
32. Institute of Medicine. (2002). "Dietary Reference Intakes for Energy, Carbohydrates, Fiber, Fat, Fatty Acids, Cholesterol, Protein and Amino Acids (Macronutrients)." National Academies Press, Washington, DC. Retrieved March 6, 2007, from http://www.iom.edu.
33. Thomas-Dobersen, D. (1999). Nutritional management of gestational diabetes and nutritional management of women with a history of gestational diabetes: Two different therapies or the same? *Clin. Diabetes* **17**. Retrieved March 20, 2007, from http://journal.diabetes.org/clinicaldiabetes/V17N41999/pg170.htm.
34. Jovanic, L. (1998). American Diabetes Association's Fourth International Workshop-Conference on gestational diabetes mellitus. Summary and discussion. Therapeutic interventions. *Diabetes Care* **21**, B131–B137.
35. Jovanovic, L. (2000). Controversies in the diagnosis and treatment of gestational diabetes. *Clev. Clin. J. Med.* **67**, 481–488.
36. Langer, O., Yogev, Y., Xenakis, E., and Brustman, L. (2005). Overweight and obese in gestational diabetes: The impact of pregnancy outcome. *Am. J. Obst. Gynecol.* **192**, 1768–1776.
37. Duarte-Gardea, M., and Gonzalez, J. (2003). Outcomes of nutrition therapy in Mexican-American women with gestational diabetes. Overcoming Diabetes Health Disparities Conference Nashville, TN, November 2003. Abstract.
38. Jovanovic, L., and Nakai, Y. (2006). Successful pregnancy in women with type 1 diabetes: From preconception through postpartum care. *Endocrinol. Metab. Clin. North Am.* **35**, 79–97.
39. Romon, M., Nuttens, M. C., Vambergue, A., Verier-Mine, O., Biausque, S., Lemaire, C., Fontaine, P., Salomez, J. L., and Beuscart, R. (2001). Higher carbohydrate intake is associated with decreased incidence of newborn macrosomia in women with gestational diabetes. *J. Am. Diet. Assoc.* **101**, 897–902.
40. Thomas, A. M., and Gutierrez, Y. M. (2005). "American Dietetic Association Guide to Gestational Diabetes Mellitus." American Dietetic Association, Chicago.
41. Peterson, C. M., Jovanovic-Peterson, L., and Nakai, Y. (1991). Percentage of carbohydrate and glycemic response to breakfast, lunch, and dinner in women with gestational diabetes. *Diabetes* **40**, 172–174.
42. Gunderson, E. (1997). Intensive nutrition therapy for gestational diabetes. *Diabetes Care* **20**, 221–226.
43. Gunderson, E. P. (2004). Gestational diabetes and nutritional recommendations. *Curr. Diabetes Rep.* **4**, 377–386.
44. Lock, D. R., Bar-Eyal, A., Voet, H., and Madar, Z. (2000). Glycemic indices of various foods given to pregnant diabetic subjects. *Obstet. Gynecol.* **71**, 180–183.
45. Scholl, T. O., Chen, X., Khoo, C. S., and Lenders, C. (2004). The dietary glycemic index during pregnancy: Influence on infant birth weight, fetal growth, and biomarkers of carbohydrate metabolism. *Am. J. Epidemiol.* **159**, 467–474.
46. Franz, M. J. (2006). The argument against glycemic index: What are the other options? *Nestle Nutr. Workshop Ser. Clin. Perform. Programme* **11**, 57–68.
47. Cesa, F., Mariani, S., Fava, A., Rauseo, R., and Zanetti, H. (1990). [The use of vegetable fiber in the treatment of pregnacy diabetes and/or excessive weight gain during pregnancy]. *Minerva Ginecol.* **42**, 271–274.
48. Reece, E. A., Hagay, Z., Caseria, D., Gay, L. J., and DeGennaro, N. (1993). Do fiber-enriched diabetic diets have glucose-lowering effects in pregnancy? *Am. J. Perinatol.* **10**, 272–274.
49. Position of the American Dietetic Association. (1998). Use of nutritive and nonnutritive sweeteners. *J. Am. Diet. Assoc.* **104**, 255–275.
50. American Diabetes Association. (2006). Nutrition recommendations and interventions for diabetes—2006. A position statement of the American Diabetes Association. *Diabetes Care* **29**, 2140–2157.
51. Shiell, A. W., Campbell, D. M., Hall, M. H., and Barker, D. J. (2000). Diet in late pregnancy and glucose-insulin metabolism of the offspring 40 years later. *Br. J. Obstet. Gynecol.* **107**, 890–895.
52. Ilic, S., Jovanovic, L., and Pettitt, D. J. (1999). Comparison of the effect of saturated and monounsaturated fat on postprandial plasma glucose and insulin concentration in women with gestational diabetes mellitus. *Am. J. Perinatol.* **16**, 489–495.
53. Uauy, R., and Dangour, A. R. (2006). Nutrition in brain development and aging: Role of essential fatty acids. *Nutr. Rev.* **64**, S24–S33.
54. Thomas, B., Ghebremeskel, K., Lowy, C., Crawford, M., and Offley-Shore, B. (2006). Nutrient intake of women with and without gestational diabetes with a specific focus on fatty acids. *Nutrition* **22**, 230–236.
55. Cuco, G., Arija, V., Iranzo, R., Vila, J., Prieto, M. T., and Fernandez-Ballart, J. (2006). Association of maternal protein intake before conception and throughout pregnancy with birth weight. *Acta Obstet. Gynecol. Scand.* **85**, 413–421.
56. Sloan, N. L., Lederman, S. A., Leighton, J., Himes, J. H., and Rush, D. (2001). The effect of prenatal dietary protein intake on birth weight. *Nutr. Res.* **21**, 129–139.
57. Pitkin, R. M. (1984). Aspartame ingestion during pregnancy. In "Aspartame" (L. D. Stegink, and J. F. Filer, Eds.). Marcel Dekker, New York.
58. Reader, D. M. (2007). Medical nutrition therapy and lifestyle interventions. *Diabetes Care* **30**, S188–S193.
59. American Diabetes Association. (2001). Position statement. Gestational Diabetes. *Diabetes Care* **24**, S77.
60. Powers, M. (1996). "Handbook of Diabetes Medical Nutrition Therapy." Aspen, Gaithersburg, MD.
61. Langer, O. (1998). Maternal glycemic criteria for insulin therapy in gestational diabetes. *Diabetes Care* **21**, S91–S98.
62. Gabbe, S. G., and Graves, C. R. (2003). Management of diabetes mellitus complicating pregnancy. *Obstet. Gynecol* **102**, 857–868.
63. Jones, M. W., and Stone, L. C. (1998). Management of the women with gestational diabetes mellitus. *J. Perinat. Neonat. Nurs.* **11**, 13–24.
64. Jovanic, L. (2004). Achieving euglycaemia in women with gestational diabetes mellitus: Current options for screening, diagnosis and treatment. *Drugs* **64**, 1401–1417.
65. Chamit, S., Denise, T., and Moore, T. (2004). Prospective observational study to establish predictors of glyburide success in women with gestational diabetes mellitus. *J. Perinatol.* **24**, 617–622.

66. Conway, D. L., Gonzalez, O., and Skiver, D. (2004). Use of glyburide for the treatment of gestational diabetes: The San Antonio experience. *J. Matern. Fetal Neonat. Med.* **15**, 51–55.
67. Langer, O., Conway, D. L., Berkus, M. D., Xenakis, E. M., and Gonzalez, O. (2003). A comparison of glyburide and insulin in women with gestational diabetes mellitus. *N. Engl. J. Med* **343**, 1134–1138.
68. Ramos, G. A., Jacobson, G. F., Kirby, R. S., Ching, J. Y., and Field, D. R. (2007). Comparison of glyburide and insulin for the management of gestational diabetics with markedly elevated oral glucose challenge test and fasting hyperglycemia. *J. Perinatol.* **27**, 262–267.
69. Jacobson, G. F., Ramos, G. A., Ching, J. Y., Kirby, R. S., Ferrara, A., and Field, D. R. (2005). Comparison of glyburide and insulin for the management of gestational diabetes in a large managed care organization. *Am. J. Obstet. Gynecol.* **193**, 118–124.
70. Maternal and Child Health Branch, Department of Health and Human Services, State of California. (1998). "Sweet Success Guidelines for Care." California Diabetes and Pregnancy Program.
71. Ratner, R. E. (2007). Prevention of type 2 diabetes in women with previous gestational diabetes. *Diabetes Care* **30**, S242–S245.
72. Moses, R. G., Shand, J. L., and Tapsell, L. C. (1997). The recurrence of gestational diabetes: Could dietary differences in fat intake be an explanation? *Diabetes Care* **20**, 1647–1650.
73. Swinburn, B. A., Metcalf, P. A., and Lay, S. J. (2001). Long-term (5-year) effects of a reduced-fat diet intervention in individuals with glucose intolerance. *Diabetes Care* **24**, 619–624.
74. Duarte-Gardea, M. (2004). Case study: The prevention of diabetes through diet and intense exercise. *Clin. Diabetes* **22**, 45–46.
75. Knowler, W. C., Barret-Connor, E., Fowler, S. E., Hamman, R. F., Lachin, J. M., Walter, E. A., Nathan, D. M., the Diabetes Prevention Program Research Group. (2002). Reduction in the incidence of type 2 diabetes with lifestyle intervention or metformin. *N. Engl. J. Med.* **346**, 393–403.

E. Cancer

CHAPTER **34**

Interaction of Nutrition and Genes in Carcinogenesis

JO L. FREUDENHEIM
School of Public Health and Health Professions, University at Buffalo, State University of New York, Buffalo, New York

Contents

I. INTRODUCTION

Cancer leads to the death of about 6.7 million people in the world each year with more than 10.8 million new cases of cancer each year. In the United States alone, there are about 1.4 million new cases annually [1]. In the United States, about 65% of cancer patients survive for 5 years after diagnosis [2]. Therapies to cure cancer continue to improve, but prevention is clearly an essential strategy in ameliorating cancer-related morbidity and mortality.

Given differences in rates of cancer worldwide and given changes in rates of cancer among populations who migrate from one environment to another, it is generally agreed that most common cancers are caused by environmental factors with the potential to be controlled [3, 4]. An understanding of how dietary factors contribute to cancer etiology is an important part of understanding cancer prevention. There is accumulating evidence that this is a complicated process involving the interactions of an individual's endogenous milieu, including their genetic makeup, with diet and other exogenous exposures of an individual.

In conceptualizing what causes cancer, it is crucial to understand that a cancer may have more than one cause. For example, both occupational exposures and diet could contribute to the etiology of one kind of cancer and even to the cancer of one individual, and either a change in the occupational exposure or a change in diet might prevent that cancer or reduce the risk. It is estimated that nutritional factors contribute to somewhere between 35% and 75% of tumors [5]. Genetic factors likely play a role in many if not all cancers. Even when an exposure is clearly required for a particular kind of cancer, genetic factors often play a role in determining who, among exposed individuals, develops that cancer. In particular for dietary risk factors, evidence is now accumulating that both genetic and dietary factors and their interactions are important in explaining part of the etiology of the disease.

II. BACKGROUND AND DEFINITIONS

Cancer is characterized by a number of changes in cell. Tumor cells have their own growth signals, they do not respond to signals inhibiting growth, they can replicate indefinitely without going into programmed cell death (apoptosis), they can invade other tissues (metastasis), and they can develop a blood supply for those metastases (angiogenesis) [6]. Cancer is characterized by the accumulation of changes in the structure of the genes in a cell such that there are these significant changes in the functioning of the cell.

Genes are the basic unit of heredity with each gene having its own location in a particular chromosome [7]. A *mutation* is a structural change in the base pair sequence of DNA, the chromosomal material that provides the code for gene expression, DNA [7]. In tumor cells, there are generally large numbers of mutations to the DNA. In addition, there may also be *epigenetic* changes. These are changes that are other than changes to the DNA base pair sequences. Epigenetic changes affect gene structure, function, and expression. One such change is *methylation,* the addition or loss of methyl groups on the DNA, which can affect gene expression. *Hypermethylation*, addition of methyl groups to the promoter region of tumor suppressor genes, is common in tumors and may be as important a mechanism as mutation in carcinogenesis. Other epigenetic changes include changes in the *histones*, small proteins that appear to be important in the packing of the chromosomes [7].

Genotype refers to an individual's genetic structure based on his or her DNA [7]; it is the nucleic acid sequence of the DNA. Each individual has two copies of each gene, one inherited from each parent. From knowledge of the genotype, it is possible to infer the amino acid sequence of the protein encoded by a gene. Differences in genotype may be used to infer differences in protein activity and to determine, for example, whether an individual has the gene for the faster or slower version of a particular enzyme. Other factors besides genotype will also influence actual enzyme activity, including the available concentration of substrate and the rates of synthesis and degradation of the enzyme. These factors as a group, the observable characteristics and the expression of the genotype with interactions with other genes and the environment, are referred to as *phenotype* [7].

Agreement between genotype and phenotype often is not perfect. Phenotype is generally not dependent on a single gene but is the result of the genotype of a number of different genes in a pathway combined with the impact of external exposures. In many instances there are gene–gene interactions as well as gene–environment interactions. For example, a person may have the genotype for lower activity for a particular enzyme, but other exposures including dietary factors may induce a high level of expression of that enzyme. A measurement of phenotype in that individual might be based on the rate of metabolism of a particular compound, and therefore phenotype might not be different for high activity as a result of genotype and that as a result of enzyme induction.

Penetrance of a genetic factor refers to the likelihood that those with a specific gene will exhibit a particular phenotype under given environmental circumstances [7]. A gene with high penetrance would be one for which virtually everybody with the gene has the expressed trait, whereas for a gene with low penetrance there is a lower likelihood of expression of the gene. The measurement of phenotype is of interest because it gives an indication as to the true level of exposure, whereas the measurement of genotype is of interest because it gives some indication of exposure over the lifetime and is not biased by factors such as disease state and recent exposures.

Some individuals are at greater risk for certain types of cancers. This increased risk can be related to differences in exposure to cancer-inducing or -protecting agents, and/or to the individual's genetic makeup, or to a combination of these factors. Genetic factors vary widely in the magnitude of their impact. Some inherited mutations greatly increase the risk of cancer; others cause smaller increases in risk. Some mutations may only increase risk in the presence of a particular exposure. Examples of inherited mutations with high penetrance are mutations identified in the *BRCA1* gene. Particular mutations in this gene have been shown to be strongly related to risk of breast and ovarian cancer. Current estimates are that carriers of this genetic mutation have a 65% risk of breast cancer by age 70; that is, many but not all women with this mutation will have cancer by age 70 [8]. However, relatively few individuals in the population carry this factor (estimates range between 1 in 2000 and 1 in 500). Therefore, although those with the gene are at high risk, only about 1–5% of women with breast cancer have this mutation [9–11].

There is a considerable range in the impact of genetic differences on phenotype. Other genetic variants that are more common and that have weaker effects on risk are often referred to as *genetic polymorphisms.* Genetic polymorphism means that the gene has more than one form; its structure can vary among individuals. Many polymorphisms are silent, not affecting phenotype. Other polymorphisms may alter the effectiveness of a protein and/or may affect its interaction with other compounds including nutrients. One or more polymorphisms are found in most genes. The frequency of the occurrence of a genetic polymorphism among individuals within a population can vary greatly. Generally polymorphisms have less impact on cancer risk than highly penetrant mutations such as *BRCA1*. However, when they are common, they may still have a significant effect on the rate of disease in a population. In general, common polymorphisms affect the response to an exposure so that their effect on risk is evident only when that exposure or exposures are present, referred to as a gene–environment interaction. For these types of gene–environment interactions, an understanding of both the genetic factor and the environmental exposure is important in order to understand disease etiology and prevention.

There are several terms that have been developed more recently to describe the interaction of food components with genes. *Nutrigenomics* is the broadest term, describing the interaction of dietary components with genes. Included in this term is *nutrigenetics*, the study of the impact of genetic variation on the response to dietary components; *nutritional epigenetics*, the study of the effect of food-derived compounds on DNA structure other than base pairs, that is, on methylation and on chromatin structure; and *nutritional*

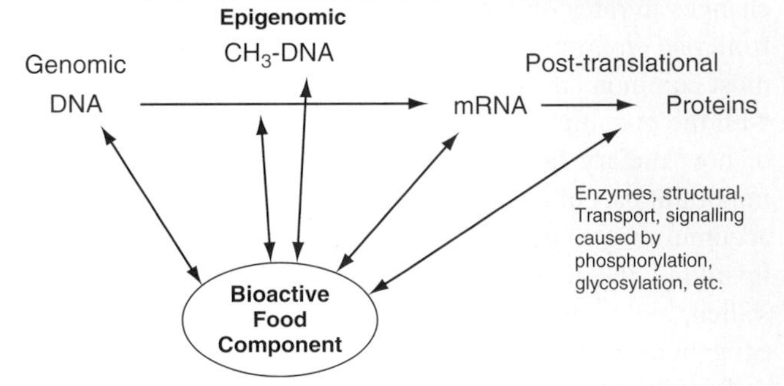

FIGURE 1 Bioactive food components may interact with genes and their products in a number of different ways, including interactions with the coding DNA, with other parts of the chromosome structure, with the messenger RNA, and with the protein product after translation. (From Milner, J. A."Molecular Targets for Bioactive Food Components." Copyright The American Society of Nutrition, Reprinted by permission from *The Journal of Nutrition*, volume 134, pp. 2492S–8S, 2004.)

transcriptomics, the study of the impact of dietary components on gene transcription. In addition, dietary components may affect protein structure post-translation; the study of these interactions is *nutritional proteomics* [12] (Figure 1 [13]).

III. MECHANISMS OF DIET–GENE INTERACTIONS

A. Overview of Interactions

In the exploration of the role of dietary factors in the process of cancer, both those compounds that have been identified as essential, that is, nutrients, as well as other compounds with biological activity may be important. These interactions may operate in two directions. Genes can affect the action of these compounds of dietary origin, and these compounds can affect the action of genes. To clearly understand these processes, several layers of complexity need to be taken into account. Individual bioactive food components may have more than one action, in some cases interacting with other food components and with more than one gene. Foods can differ significantly with respect to content of the compounds of interest, depending on growing conditions, genetic differences in the foodstuff, storage and cooking, and other food processing [14]. After consumption, there can be interactions of a particular bioactive compound both with other food constituents, other environmental exposures (e.g., from air and water) and with genes. Further, genes interact with each other. The resulting metabolic environment is the result of these numerous interactions of factors both of endogenous and exogenous origin. In understanding the role of a particular food component, examination of its interactions with other food components and with one or more metabolic pathways may be necessary, as well as possible consideration of differing actions depending on dose or on tissue site.

Most of this complexity is at this time only partially understood for most of the bioactive compounds which have been studied. Up until now, mostly single food components or at most single foods have been studied in relation to single genes for cancer. Although studies of this kind are essential first steps, there is much to be learned as the more complex interacting systems are examined. Presented here is information first about the impact of genetic variation on the availability and utilization of dietary compounds in relation to carcinogenesis and then the impact of diet on genetic factors in relation to carcinogenesis.

B. Genetic Variation in Relation to Metabolism of Food Components

1. Carcinogen Metabolism

Much of the cancer research on the impact of genetic variation on metabolism has been of genes that modulate the metabolism of carcinogens. These carcinogens include both compounds occurring naturally in foods and others contaminating foods. If an individual has a genetic variant that results in slower metabolism of a carcinogen, that individual will be more affected by a given dose of that carcinogen than would someone who can metabolize and excrete the carcinogen more rapidly. Conversely, if an individual has a gene that results in more rapid metabolism of a carcinogen, then he or she will be less affected by the same dose than someone who does not have that variant.

The enzymes related to carcinogen metabolism and excretion are divided into two groups: phase I and phase II. Phase I enzymes activate the compound and phase II enzymes attach polar groups to the activated compound so that it will be more water soluble and can be excreted in the urine. Many of the phase I enzymes are in the cytochrome P-450 (CYP) family. Phase II enzymes include glutathione-*S*-transferases (GSTs) and *N*-acetyltransferase (NAT). In some cases, phase I activation leads to the production of compounds with greater carcinogenic potential. Often rather than carcinogens, foods contain procarcinogens, compounds that are carcinogenic after metabolic activation. The process of metabolic activation may vary depending on genetic factors. For example, heterocyclic amines are a group of compounds found in meat cooked at high temperatures, primarily "well-done" meat and fish. There is some, although inconsistent, evidence that the heterocyclic amines are associated with cancer. For example, although there is not strong evidence of an association of intake of these compounds from meat with risk of breast cancer [15], there may be some increased risk with lifetime intake and postmenopausal breast cancer [16]. This association may depend on genetic factors. Metabolism and excretion of heterocyclic amines involve CYP1A2, *N*-acetyltransferases 1 and 2 (NAT1 and 2), and UDP-glucuronosyltransferases (UGTs) [17]. In a study examining cells from the breast that had been sloughed into breast milk, the amount of DNA adducts among women varied with NAT1 or NAT2 genotype [18]. With regard to colorectal cancer, there is some, although inconsistent, evidence that individuals who eat well-cooked meat and who have the rapid variant of both *CYP1A2* and *NAT2* are at increased risk [17]. Figure 2 illustrates the relationships between these phase I and II enzyme systems [19, 20].

2. Metabolism of Nutrients and Other Food Components

There can be significant genetic variation among individuals in the absorption, metabolism, transport, and excretion of nutrients and other bioactive food components [13]. All of these can alter the effect of intake of the same amount of a food component by individuals with different genes. Further, genetic factors may alter food preferences and therefore food consumption [21].

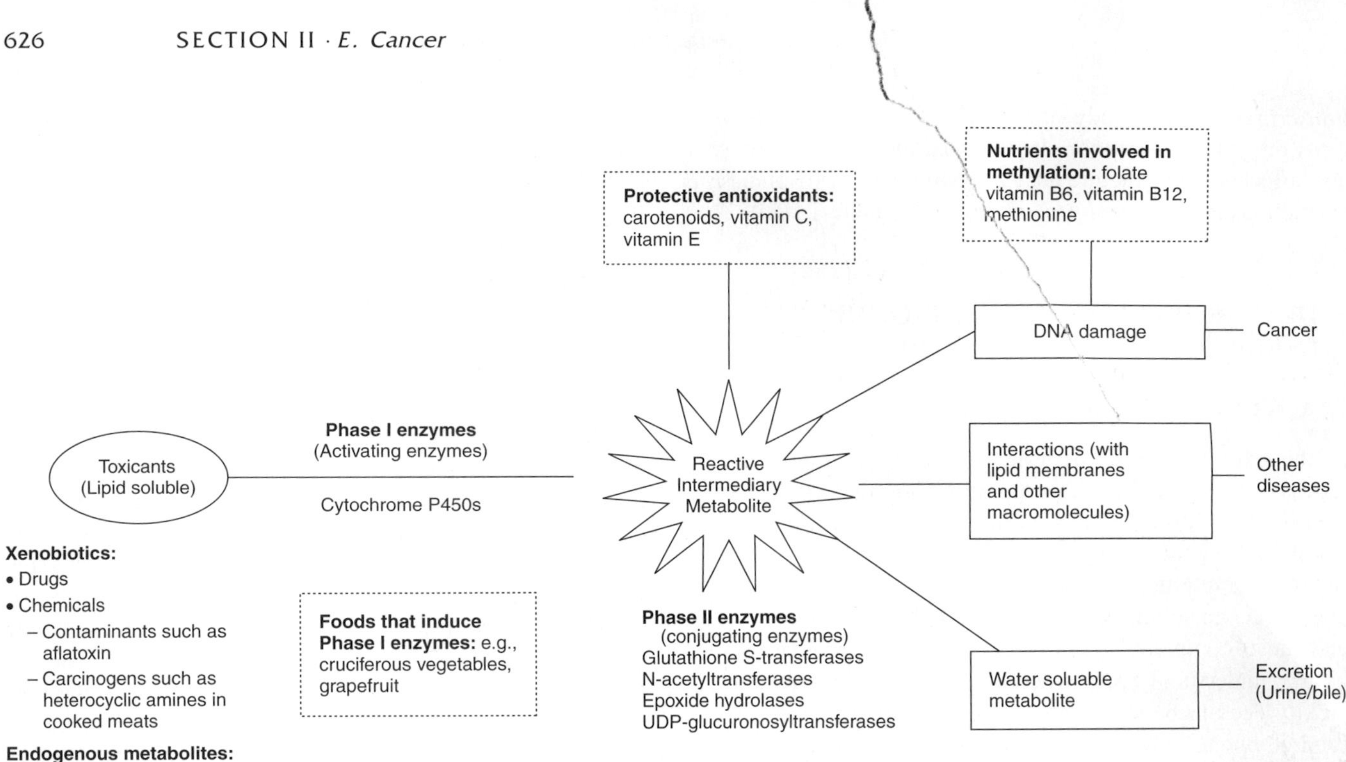

FIGURE 2 Interrelationship between the biotransformation enzyme systems. (From Patterson, R. E., Eaton, D. L., and Potter, J. D. The genetic revolution: Change and challenge for the dietetics profession. Copyright The American Dietetic Association. Reprinted by permission from *Journal of the American Dietetic Association*, vol. 99, pp. 1412–1420, 1999.)

The impact of a particular genetic polymorphism may differ for different compounds and may even protect against cancer for one compound and increase risk for another. For example, the GSTs are phase II enzymes that catalyze transformation of carcinogens such as polycyclic aromatic hydrocarbons (PAHs) for excretion; for those with higher exposure to PAHs, including PAHs of dietary origin, the more rapid metabolism would be advantageous. Some of the GSTs also deactivate other phytochemicals that are thought to exert protective effects. Included in the latter group would be the isothiocyanates from cruciferous vegetables, which are metabolized by GST, also resulting in their excretion [21].

Genetic variation in the pathway for a vitamin can affect the physiologic impact of intake of that vitamin. If a vitamin plays a role in cancer prevention, differences in the receptor for the vitamin could also affect the susceptibility of an individual to cancer. For example, there are commonly occurring genetic differences in the vitamin D receptor that may have functional significance. There have been a number of studies of the association of those variants with risk of several cancers. For example, for breast cancer, there is some, although inconsistent, evidence that genetic variation in the vitamin D receptor is associated with risk [22]. There has been considerable exploration of genetic variants in the one-carbon metabolism pathway. Genetic variation in this pathway affects blood folate concentrations and folate transport and utilization. This genetic variation may also affect the synthesis of purines and pyrimidines for DNA synthesis and may affect gene methylation and thus the expression of genes. There is some, although not consistent, evidence of the association of variants in genes in this pathway and risk of several cancers [12]. Another example of a gene that affects nutrient utilization includes the *HFE* gene; a variant that is common among Europeans affects iron storage and is part of the etiology of hemochromatosis [23].

There is genetic variation in several of the enzymes involved in alcohol metabolism. For example, one variant in the gene for aldehyde dehydrogenase (ALDH2), carried by about 50% of Asians, results in greatly reduced clearance of acetaldehyde, a likely carcinogen. Those who are either heterozygous or homozygous for the variant experience an adverse reaction to alcohol consumption; there is a characteristic flushing associated with drinking. This variant has been shown to be associated with both decreased alcohol consumption and alcoholism [24].

C. Interactions of Bioactive Food Components with Gene Functions in Relation to Carcinogenesis

1. Mutations and Metastasis

Nutrients and other bioactive food components appear to contribute both in positive and negative ways at every stage of the carcinogenic process including DNA mutations and

Genotypes are then determined, and the gene–environment interactions analyzed for this subset of the whole cohort. Environmental exposures are estimated from interview data or from other, more direct measures of exposure. In a clinical trial, study participants are randomly allocated to exposure category; in this case, a dietary intervention would be compared to the lack of that intervention. To examine diet–gene interactions, the effect of the intervention would be compared by genotype. It is possible to determine genotype in advance of randomization in order to block a genotype, to ensure that the number of participants with a particular genotype is the same in the intervention and control groups.

For both case-control and nested case-control studies, the examination of gene–diet interactions generally involves a few steps. These include (1) the examination of the association between risk of a particular cancer and the genetic factor(s) alone—genetic factors can be a single polymorphism with known functional significance, haplotypes within a gene, or a group of genes in one or more pathways; (2) the examination of the association between risk of that cancer and the dietary factor(s) alone—dietary factors can be nutrients, single food components, foods, groups of bioactive agents, or dietary patterns; and (3) the examination of the gene–diet interactions. In some cases, there will be further examination of interaction with other exposures (e.g., exposure to hormone replacement therapy or body weight).

Although this is a relatively new area of research, in some clinical trials, the interaction of genes and environment is studied. The effect of a dietary intervention might be examined within groups defined by strata to determine if one group is more susceptible to the intervention.

For all of these study designs, there are some important methodological considerations. Many of these are the same considerations as for any epidemiologic study of diet and cancer [36]. A significant one is that, as in all epidemiology, conclusions about causality need to come from a synthesis of epidemiologic findings in a number of populations, animal research, and metabolic studies. No single epidemiologic study can be considered definitive and used to establish causality. In case-control and cohort studies, there is always concern that the findings may not be causal but rather the result of confounding. That is, another factor may be correlated with the exposure under study and with the disease and may be the causative agent. If the investigators are unaware of this confounding factor or they are unable to control for it sufficiently well, it might appear that the exposure under study is associated with disease simply because it is correlated with the second factor. The confounding factor can be an exogenous exposure or another genetic factor. When genes tend to be inherited together and therefore can confound an investigation of each other, it is called linkage disequilibrium.

An example of confounding would be in a study of diet and lung cancer: If individuals who smoke are also more likely to drink alcohol, unless smoking is measured well and controlled for in analysis, one might incorrectly assess the relation of alcohol to lung cancer because of the correlation between these two behaviors. However, when studies are done in different populations in different cultures, the likelihood of confounding is diminished; it is less likely that the same correlations exist between behaviors. Additionally, other sources of error may differ among studies making a consistent finding more believable. Only with a randomized trial can a causal link be identified with certainty because of the randomization to intervention or control group. However, even among clinical trials, differences can result because of differences in participant populations, so results need to be carefully interpreted. Clearly it is not possible to randomize genotypes, and linkage disequilibrium can make determination difficult as to which gene or genes are important in the process being studied.

Because the availability of the technology to measure genotypes in large studies is relatively new, there are few hypotheses related to the interaction of diet and genetics in cancer etiology that have been examined in numerous studies. Many questions have been examined in only a single study. This is a rapidly expanding field. With further development of the field, it will be possible to begin to identify the most consistent and important associations. At all times, it is important to examine findings critically and to look for consistent findings from well-conducted studies.

A major concern in evaluating diet and genetic interactions in relation to risk of cancer is that studies need to have large numbers of participants in order to have sufficient power to examine risk within strata defined by genotype or by diet. Because of the need to examine interactions, large studies are required to provide a sufficient number of individuals within the strata of interest. Even if the results are statistically significant, such findings can be unstable. That is, if the study were redone, findings may not be consistent. In particular, for genes with low frequencies, the number of participants required can be very large. Further, to examine the interactions of more than one gene and more than one dietary or other exposure, the number of participants required may be exceedingly large. In evaluation of any study, it is important to look at the number of participants in each cell to make a determination of the likely stability of the findings.

At this time, most of the research regarding interactions of diet and genetics has focused on a single dietary factor and a single nutrient. Such a focus is of interest. Nonetheless, cancer is a multifactorial disease, and it is likely that several genetic factors are of importance even within a single causative pathway. Analysis of a single polymorphism might not capture the total picture of variation in risk. Similarly, several nutrients are likely to be important even within a single causal pathway. As noted earlier, the examination of gene–gene and gene–environment interactions can be seriously hampered by the required sample size.

Diet too is a complex set of exposures. A single food may include different factors with both carcinogenic and anti-carcinogenic properties. Intakes of different nutrients are generally highly correlated, so that an association attributed to intake of one nutrient may in fact be the result of a causal relationship with a different nutrient. Finally, when there is evidence that a particular nutrient is related to a decrease in risk, it cannot be assumed that larger quantities of that nutrient would be even more protective. For example, whereas vitamin C has antioxidant properties at the level found in most foods, at higher intake levels, it may have pro-oxidant properties [37].

A final challenge in this field is in the measurement of dietary exposure. Much literature is available on the problems of measurement of diet for epidemiologic purposes [38]. Beyond those concerns, the study of gene–diet interactions in relation to cancer risk has led to an interest in a number of new dietary factors. For many of these compounds, the dietary instruments used in the past may not provide sufficient detail for assessment of intake of that compound. For example, the study of heterocyclic amines has led to the development of a questionnaire that specifically addresses the sources of these compounds, because the information regarding intake on the existing questionnaires did not include the necessary detail to assess intake of heterocyclic amines [39]. Further, nutrient composition databases may be limited and may need work in order to determine the composition for compounds that may relate to disease risk but that have not been previously analyzed.

V. DIET–GENE INTERACTIONS AND CANCER

As described in the section on definitions of terms, there are a number of properties that distinguish a tumor. These properties contribute to the ability of a tumor to break loose from normal controls on growth and to invade tissues. Tumor cells have their own growth signals, they do not respond to signals inhibiting growth, they can replicate indefinitely and they do not go into programmed cell death (apoptosis), they can invade other tissues (metastasis), and they can develop a blood supply for those metastases (angiogenesis) [6]. All of these pathways may be appropriate targets for cancer prevention. Given the nature of tumor biology, it is unlikely that there will be one food or compound that would prevent all or even of one kind of tumor; rather the cumulative effect of a number of compounds may be important in prevention [40]. There are a number of processes that are important in carcinogenesis: carcinogen metabolism, hormone regulation, cell differentiation, DNA repair, apoptosis, cell growth cycle, and inflammatory response. Nutrients and other bioactive food components may play a role in each of these processes (Figure 3 [41]). In this section, breast cancer is used as an example and recent data regarding diet and gene interactions for some of these processes for breast cancer are described. This area of research is very rapidly changing and our understanding of these relationships is likely to shift with further investigation. This overview is provided to give some sense of the developing understanding.

A. Carcinogen Metabolism

There is quite consistent epidemiologic evidence that alcohol consumption is associated with a moderate increase in breast cancer risk [42]. While there are several possible mechanisms to explain these observations, there is also some evidence that the mechanism of this association involves gene–environment interactions in carcinogen metabolism. Alcohol is metabolized to acetaldehyde in the liver and in other tissues including breast tissue by a group of enzymes called alcohol dehydrogenases. Acetaldehyde has been identified as a likely carcinogen [43]. It may be that the exposure of breast tissue to acetaldehyde accounts for the increased risk associated with alcohol consumption. There are several studies that have examined the interaction of alcohol consumption and a common genetic variant in one isozyme of alcohol dehydrogenase in relation to risk of breast cancer. This variant affects how quickly acetaldehyde is formed from ingested ethanol. Some [44, 45] but not all [46, 47] studies have found that there was more risk associated with alcohol consumption among women who also had the variant resulting in more rapid metabolism to acetaldehyde.

B. Regulation of Hormones

There is considerable evidence that estrogen exposure is a strong risk factor for breast cancer [48]. There is considerable interest in whether compounds of plant origins with estrogenic properties, called phytoestrogens, affect breast cancer risk. Among these is indole 3-carbinol, found in cruciferous vegetables after they are cooked or crushed. There is evidence from studies of breast cells in culture that this compound can affect the estrogen receptor, ER-α. It can alter the estrogen and the DNA binding regions of the ER, leading to decreased response to estrogen in estrogen-responsive cells [49]. Thus the interaction of this dietary compound with DNA may affect breast cancer risk because of the dampening of response to estrogen.

C. Cellular Differentiation

Genistein, a polyphenol found in soy, has been found to decrease mammary tumors in animal models. There is evidence that when genistein is ingested in the prepubertal rat, it can help to regulate mammary development and differentiation. An increase in differentiation of mammary tissue is observed, as well as differences in cellular

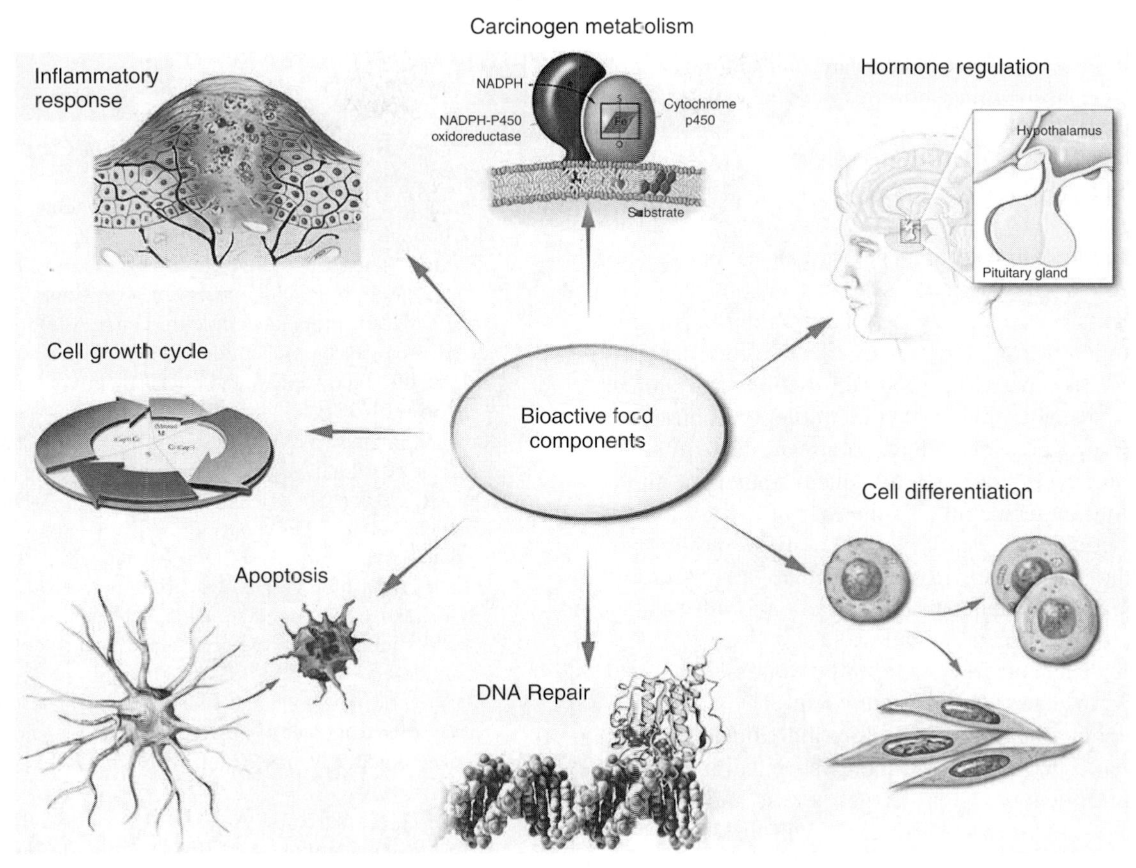

FIGURE 3 Interactions of bioactive food components with genes appear to be important mechanisms for carcinogenesis. (From Trujillo, E., Davis, C., Milner, J. "Nutrigenomics, Proteomics, Metabalomics and the Practice of Dietetics." Copyright The American Dietetic Association. Reprinted by permission from *Journal of The American Dietetic Association*, vol. 106, pp. 403–413, 2006.) See color plate.

proliferation, apoptosis, and tumor suppressor expression. The effects of genistein on mammary development appear to be different and most beneficial with administration of this bioactive compound in young animals and may not have the same impact in older ones [50].

D. DNA Repair

Repair of DNA errors resulting from exogenous exposures or from errors in replication is an important function of the cell. Errors in DNA repair may be important in the carcinogenic process. There are several different processes to maintain the integrity of the DNA. In one epidemiologic study, there was evidence of interaction of a variant in a gene in the excision repair pathway with intake of fruits and vegetables and risk of breast cancer. Among women with at least one copy of the variant that appears to be protective and with higher intake of fruits and vegetables, risk of breast cancer was lower than for women with lower intake or without the variant [51]. Another food compound that may also affect DNA repair is indole 3-carbinol. There is evidence that it may block DNA strand breaks. It can upregulate the *BRCA1* gene, which plays a role in DNA repair [49].

E. Apoptosis

Programmed cell death or apoptosis is an important function in the maintenance of the integrity of tissues. Dietary factors may also affect apoptosis and may interact with genes that control it. For example, there is evidence that vitamin D can influence genes controlling apoptosis [52]. Other dietary factors may also play a role. In one study, overweight or obese postmenopausal women were put on a low-fat, high-fiber diet and an exercise program. A comparison was made of the growth-promoting properties of the women's serum taken before and after the intervention on three different estrogen receptor positive breast cancer cell lines. Among other changes, there was an increase in apoptosis for several different cell lines when the postintervention serum was used as the growth medium [53]. Other dietary factors that may also affect the expression of genes important in apoptosis include organosulfur compounds from foods from the allium family, including garlic, polyphenols from green tea, chocolate, and chili peppers, and isothiocyanates from cruciferous vegetables [54].

VI. FUTURE DIRECTIONS

Understanding of the role of genes in cancer and in particular the role of the interactions of diet with genes in this disease process is a relatively new and rapidly expanding field of inquiry. Most of the available evidence remains fairly preliminary, based only on cell cultures and animal studies or on a small number of epidemiologic studies. Our understanding of gene–diet interactions and how to study them is advancing rapidly with tremendous potential implications for our understanding of the role of dietary factors in cancer etiology. Clearly, as the field progresses, extremely large studies will be needed to allow for the examination of multiple levels of interaction to fully understand the relevant etiological pathways.

In the future, genotype is likely to be examined in metabolic studies to determine the short-term effects on intermediate outcome measures in healthy individuals. A great deal remains to be understood about the relation between genotype and phenotype for the genes that appear to be important in terms of risk. Further, we will need to understand the factors—both dietary and from other sources—that affect gene induction and the etiological pathways. More work is going to have to be done to identify the relevant genes that are likely to be both rate limiting and also vary frequently enough.

There is little doubt that a large portion of the etiology of cancer can be attributed to environmental factors. Twin studies indicate that genetic factors alone are not sufficient to explain who will get the disease and who will not [46]. Important among the likely environmental factors are dietary factors. With increased understanding of genetic susceptibility as additional risk factors, it may eventually become possible to identify individuals with higher or lower requirements for particular nutrients or to identify individuals with greater sensitivity to agents such as alcohol. Further, the elucidation of genetic factors in relation to diet will help us to truly understand the natural history of this disease.

References

1. Ferlay, J., Bray, P., Pisani, P., and Parkin, D. M. (2004). Globocan 2002: Cancer incidence, mortality and prevalence worldwide. IARC CancerBase No. 5. version 2.0. IARCPress, Lyon, available at http://www-dep.iarc.fr.
2. American Cancer Society (2006). "Cancer Facts and Figures 2006." American Cancer Society, Atlanta, GA.
3. World Cancer Research Fund in association with American Institute for Cancer Research (1997). "Food, Nutrition and the Prevention of Cancer: A Global Perspective." World Cancer Research Fund in association with American Institute for Cancer Research,Washington, DC.
4. Weinberg, R. A. (2007). "The Biology of Cancer." Garland Science, New York.
5. Kushi, L. H., Byers, T., Doyle, C., Bandera, E. V., McCullough, M., Gansler, T., Andrews, K. S., Thun, M. J., and the American Cancer Society 2006. Nutrition and Physical Activity Guidelines Advisory Committee (2006). American Cancer Society guidelines on nutrition and physical activity for cancer prevention: Reducing the risk of cancer with healthy food choices and physical activity. *CA Cancer J. Clin.* **56**, 254–281.
6. Hanahan, D., and Weinberg, R. A. (2000). The hallmarks of cancer. *Cell* **100**, 57–70.
7. King, R. C., and Stansfield, W. D. (1997). "A Dictionary of Genetics." Oxford University Press, New York.
8. Antoniou, A., Pharoah, P. D. P., Narod, S., *et al.* (2003). Average risks of breast and ovarian cancer associated with BRCA1 and BRCA2 mutations detected in case series unselected for family history: A combined analysis of 22 studies. *Am. J. Hum. Genet.* **72**, 1117–1130.
9. Easton, D. F., Ford, D., and Bishop, D. T. (1995). Breast and ovarian cancer incidence in BRCA1-mutation carriers. Breast Cancer Linkage Consortium. *Am. J. Hum. Genet.* **56**, 265–271.
10. Easton, D. F., Narod, S. A., Ford, D., and Steel, M. (1994). The genetic epidemiology of BRCA1. Breast Cancer Linkage Consortium [letter]. *Lancet* **344**, 761.
11. Newman, B., Mu, H., Butler, L., Milliken, R. C., Moorman, P. G., and King, M.-C. (1998). Frequency of breast cancer attributable to BRCA1 in a population based series of American women. *JAMA* **279**, 915–921.
12. Davis, C. D., and Hord, N. G. (2005). Nutritional "omics" technologies for elucidating the role(s) of bioactive food components in colon cancer prevention. *J. Nutr.* **135**, 2694–2697.
13. Milner, J. A. (2004). Molecular targets for bioactive food components. *J. Nutr.* **134**, 2492S–2498S.
14. Kaput, J. (2007). Nutrigenomics—2006 update. *Clin. Chem. Lab. Med.* **45**, 279–287.
15. Ochs-Balcom, H. M., Wiesner, G., and Elston, R. C. (2007). A meta-analysis of the association of *N-acetyltransferase* gene (NAT2) variants with breast cancer. *Am. J. Epidemiol.* **166**, 246–254.
16. Steck, S. E., Gaudet, M. M., Eng, S. M., Britton, J. A., Teitlebaum, S. L., Neugat, A. I., Santella, R. M., and Gammon, M. D. (2007). Cooked meat and risk of breast cancer—lifetime versus recent dietary intake. *Epidemiology* **18**, 373–382.
17. Cross, A. J., and Sinha, R. (2004). Meat related mutagens/carcinogens in the etiology of colorectal cancer. *Environ Mol Mutagen* **44**, 44–55.
18. Ambrosone, C. B., Abrams, S. M., Gorlewska-Roberts, K., and Kadlubar, F. F. (2007). Hair dye use, meat intake and tobacco exposure and presence of DNA-carcinogen adducts in exfoliated breast ductal epithelial cells. *Arch. Biochem. Biophys.* **464**, 169–175.

19. Patterson, R. E., Eaton, D. L., and Potter, J. D. (1999). The genetic revolution: Change and challenge for the dietetics profession. *J. Am. Diet. Assoc.* **99**, 1412–1420.
20. Rock, C. L., Lampe, J. W., and Patterson, R. E. (2000). Nutrition, genetics, and risks and cancer. *Ann. Rev. Public Health* **21**, 47–64.
21. Lampe, J. W. (2007). Diet, genetic polymorphisms, detoxification and health risks. *Altern. Ther. Health Med.* **13**, S108–S111.
22. Cui, Y., and Rohan, T. E. (2006). Vitamin D, calcium and breast cancer risk: A review. *Cancer Epidemiol. Biomarkers Prev.* **15**, 1427–1437.
23. Stover, P. J. (2006). Influence of human genetic variation on nutritional requirements. *Am. J. Clin. Nutr.* **83**(suppl), 436S–442S.
24. Agarwal, D. P. (2001). Genetic polymorphisms of alcohol metabolizing enzymes. *Pathol. Biol.* **49**, 703–709.
25. Mooney, L. A., Bell, D. A., Santella, R. M., Van Bennekum, A. M., Ottman, R., Paik, M., Blaner, W. S., Lucier, G. W., Covey, L., Young, T. L., Cooper, T. B., Glassman, A. H., and Perera, F. P. (1997). Contribution of genetic and nutritional factors to DNA damage in heavy smokers. *Carcinogenesis* **18**, 503–509.
26. Ambrosone, C. B., Freudenheim, J. L., Thompson, P. A., Bowman, E., Vena, J. E., Marshall, J. R., Graham, S., Laughlin, R., Nemoto, T., and Shields, P. G. (1999). Manganese superoxide dismutase (MnSOD) genetic polymorphisms, dietary antioxidants and risk of breast cancer. *Cancer Res.* **59**, 602–606.
27. Herbert, V., and Das, K. (1994). Folic acid and vitamin B_{12}. *In* "Modern Nutrition in Health and Disease" (M. E. Shils, J. A. Olson, and M. Shike, Eds.) 8th ed., pp. 402–425. Lea and Febiger, Philadelphia.
28. Jang, H., Mason, J. B., and Choi, S.-W. (2005). Genetic and epigenetic interactions between folate and aging in carcinogenesis. *J. Nutr.* **135**, 2967S–2971S.
29. Pogribny, I. P., Basnakian, A. G., Miller, B. J., Lopatina, N. G., Poirier, L. A., and James, S. J. (1995). Breaks in genomic DNA and within the p53 gene are associated with hypomethylation in livers of folate/methyl-deficient rats. *Cancer Res.* **55**, 1894–1901.
30. Mariappen, D., Winkler, J., Parthiban, V., Doss, M. X., Hescheler, J., and Sachinidis, A. (2006). Dietary small molecules and large-scale gene expression studies: An experimental approach for understanding their beneficial effects on the development of malignant and non-malignant proliferative diseases. *Curr. Med. Chem.* **13**, 1481–1489.
31. Sinha, R., and Caporaso, N. (1999). Diet, genetic susceptibility and human cancer etiology. *J. Nutr.* **129**, 556S–559S.
32. Bouillon, R., Eelen, G., Verlinden, L., Mathieu, C., Carmeliet, G., and Vestuyf, A. (2006). Vitamin D and cancer. *J. Steroid Biochem. Mol. Biol.* **102**, 156–162.
33. Mariman, E. C. M. (2006). Nutrigenomics and nutrigenetics: The "omics" revolution in nutritional science. *Biotechnol. Appl. Biochem.* **44**, 119–128.
34. Ross, S. A. (2003). Diet and DNA methylation interactions in cancer prevention. *Ann. N. Y. Acad. Sci.* **983**, 197–207.
35. Ip, C. (1993). Controversial issues of dietary fat and experimental mammary carcinogenesis. *Prev. Med.* **22**, 728–737.
36. Freudenheim, J. L. (1999). Study design and hypothesis testing: Issues in the evaluation of evidence from nutritional epidemiology. *Am. J. Clin. Nutr.* **69**, 1315S–1321S.
37. Panel on Dietary Antioxidants and Related Compounds, Food and Nutrition Board, Institute of Medicine (2000). "Dietary Reference Intakes for Vitamin C, Vitamin E, Selenium and Carotenoids." National Academy Press, Washington, DC.
38. Willett, W. C. (1998). "Nutritional Epidemiology,"2nd ed. Oxford University Press, New York.
39. Sinha, R., and Rothman, N. (1997). Exposure assessment of heterocyclic amines (HCAs) in epidemiologic studies. *Mutat. Res.* **376**, 195–202.
40. Davis, C. D. (2007). Nutritional interactions: Credentialing of molecular targets for cancer prevention. *Exp. Biol. Med.* **232**, 176–183.
41. Trujillo, E., Davis, C., and Milner, J. (2006). Nutrigenomics, proteomics, metabalomics and the practice of dietetics. *J. Am. Diet. Assoc.* **106**, 403–413.
42. Smith-Warner, S. A., Spiegelman, D., Yaun, S. S., van den Brandt, P. A., Folsom, A. R., Goldbohm, R. A., Graham, S., Holmberg, L., Howe, G. R., Marshall, J. R., Miller, A. B., Potter, J. D., Speizer, F. E., Willett, W. C., Wolk, A., and Hunter, D. J. (1998). Alcohol and breast cancer in women: A pooled analysis of cohort studies. *JAMA* **279**, 535–540.
43. __________Report on Carcinogens, Eleventh Edition, United States Department of Health and Human Services, Public Health Service, National Toxicology Program. Available at http://ntp.niehs.nih.gov/ntp/roc/toc11.html.
44. Freudenheim, J. L., Ambrosone, C.B, Moysich, K. B., Vena, J. E., Graham, S., Marshall, J. R., Muti, P., Laughlin, R., Nemoto, T., Harty, L. C., Crits, G. A., Chan, A. W. K., and Shields, P. G. (1999). Alcohol dehydrogenase 3 genotype modification of the association of alcohol consumption with breast cancer. *Cancer Causes Control* **10**, 369–377.
45. Terry, M. B., Gammon, M. D., Zang, F. F., Knight, J. A., Wang, Q., Britton, J. A., Teitelbaum, S. L., Neugat, A. I., and Santella, R. M. (2006). ADH3 genotype, alcohol intake and breast cancer risk. *Carcinogenesis* **27**, 840–847.
46. Hines, L. M., Hankinson, S. E., Smith-Warner, S. A., Spiegleman, D., Kelsey, K. T., Colditz, G. A., Willett, W. C., and Hunter, D. J. (2000). A prospective study of the effect of alcohol consumption and ADH3 genotype on plasma steroid hormone levels and breast cancer risk. *Cancer Epidemiol. Biomarkers Prev.* **9**, 1099–1105.
47. Vishvanathan, K., Crum, R. M., Strickland, P. T., You, X., Ruscinski, I., Berndt, S. I., Alberg, A. J., Hoffman, S. C., Comstock, G. W., Bell, D. A., and Helzlsouer, K. J. (2007). Alcohol dehydrogenase polymorphisms, low-to-moderate alcohol consumption and risk of breast cancer. *Alcohol Clin. Exp. Res.* **31**, 46–76.
48. Key, T. J., Appleby, P. N., Reeves, G. K., *et al.* (2003). Body mass index, serum sex hormones, and breast cancer risk in postmenopausal women. *J. Natl. Cancer Inst.* **95**, 1218–1226.
49. Kim, Y. S., and Milner, J. A. (2005). Targets for indole-3-carbinol in cancer prevention. *J. Nutr. Biochem.* **16**, 65–73.
50. Witsett, T. G., and Lamartiniere, L. A. (2006). Genistein and reservatrol: Mammary cancer chemoprevention and

mechanisms of action in the rat. *Expert Rev. Anticancer Ther.* **6**, 1699–1706.

51. Shen, J., Gammon, M. D., Terry, M. B., Wang, L., Wang, Q., Zhang, F., Teitlebaum, S. L., Eng, S. M., Sagiv, S. K., Gaudet, M. M., Neugat, A. I., and Santella, R. M. (2005). Polymorphisms in *XRCC1* modify the association between aromatic hydrocarbon-DNA adducts, cigarette smoking, dietary antioxidants and breast cancer. *Cancer Epidemiol. Biomarkers Prev.* **14**, 336–342.
52. Holick, M. F. (2007). Vitamin D deficiency. *N. Engl. J. Med.* **357**, 266–281.
53. Barnard, R. J., Gonzalez, J. H., Liva, M. E., and Ngo, T. H. (2006). Effects of a low-fat, high-fiber diet and exercise on breast cancer risk factors *in vivo* and tumor cell growth and apoptosis *in vitro*. *Nutr. Cancer* **55**, 28–34.
54. Martin, K. R. (2006). Targeting apoptosis with dietary bioactive agents. *Exp. Biol. Med.* **231**, 117–129.

CHAPTER **35**

Nutrition and Cancers of the Breast, Endometrium, and Ovary

CHERYL L. ROCK[1] AND WENDY DEMARK-WAHNEFRIED[2]

[1]*Department of Family and Preventive Medicine, and Cancer Prevention and Control Program, University of California, San Diego, California*

[2]*Department of Behavioral Science, The University of Texas—MD Anderson Cancer Center, Houston, Texas*

Contents

I. INTRODUCTION

Carcinomas of the breast, endometrium, and ovary are hormone-related cancers that have biologic similarities. Among U.S. women, breast cancer is far more common than endometrial or ovarian cancers. In 2007, it is estimated that approximately 178,480 women will be diagnosed with breast cancer, thus comprising roughly 26% of the incident cancers among females [1, 2]. In contrast, cancers of the endometrium (uterine corpus) and ovary are estimated to comprise approximately 6% and 3% of female cancers, respectively. Because of differential effectiveness of treatments, the mortality estimates for these cancers vary a great deal [1, 3]. Breast cancer will be the cause of death for approximately 40,460 women during 2007 (approximately 15% of cancer-related deaths among U.S. females), whereas endometrial and ovarian cancers account for approximately 3% and 6% of cancer-related mortality among women, respectively.

Breast cancer occurs infrequently in men, although the incidence is rising [4]. In 2007, an estimated 2030 men will be diagnosed with breast cancer, and 450 men will die of this disease [1], accounting for less than 1% of all malignancies and cancer deaths in men. Although male breast cancer is uncommon, men tend to have a less favorable outcome than women. Advanced stage at diagnosis is largely responsible for these gender differences, given that over 40% of these patients present with stage III or IV disease [5]. However, anatomical differences between male and female breasts also may contribute to the more aggressive clinical behavior of mammary carcinoma in men [4].

Estrogens are thought to play an important role in breast, endometrial, and ovarian cancers [6, 7]. Normal cell proliferation and differentiation in these tissues is highly responsive to estrogens and the other gonadal hormones, as well as cellular factors and mitogens that affect growth regulation and apoptosis and thus influence carcinogenesis across these cell types. In addition to the ovarian steroids, other growth factors and mitogens influenced by nutritional factors and dietary patterns also appear to play an important role in the initiation and promotion of breast cancer. Two factors that are currently of intense scientific interest are insulin and insulin-like growth factor 1 (IGF-1), and the interactions of these factors with adiposity and weight gain [8–12].

Inherited variations in other biochemical or metabolic pathways relevant to mammary, endometrial, and ovarian cell biology, such as those involved in estrogen metabolism or the growth factor axis, are currently under study [13, 14], and their contributions to genetic risk of hormone-related cancers are not yet known. Interactions between genetic and dietary factors also are likely to be among the determinants of risk for these cancers, and these interactions are currently the focus of intensive research [15–18].

Diet and/or nutritional status is presumed to play a major role in the risk and progression of these hormone-related cancers, either through influence on the hormonal milieu or gene expression or via direct effects. The incidence of these cancers varies widely by geographical location and with migration, so environmental factors appear to contribute substantially to risk. Over the past several decades, there has been a considerable amount of research exploring the possible interrelationships between various nutritional factors and breast cancer in women. In contrast, only a limited amount of effort has been focused specifically on nutritional factors and breast cancer in males [19]. However, female and male breast cancer share several clinical and biological features, and the available epidemiologic evidence is similar; thus, findings from studies focused on nutrition and breast cancer in women are presumed to be relevant to breast cancer in men. Compared to the amount of research on breast cancer, far fewer studies have examined the relationship between nutritional factors and risk and/or progression of endometrial and ovarian cancers. However, significant associations are suggested by the extant data.

This chapter reviews and summarizes evidence on the relationships between nutritional factors and breast, endometrial, and ovarian cancers. The emphasis is on recent clinical and epidemiological studies, with the goal of identifying clinically useful strategies for prevention and patient management.

II. NUTRITIONAL FACTORS AND BREAST CANCER

A. Nutritional Factors and Primary Breast Cancer Risk

1. Height, Weight, and Body Fat Distribution

The association between increased height and risk of breast cancer was first reported by de Waard in 1975 [20]. Over the past four decades, there have been several other studies [21–26] that corroborate these findings, including recent studies confirming a strong association for both pre- and postmenopausal disease [27], as well as consistent observations across a variety of populations, that is, Japanese and Nigerian women [28–31]. Furthermore, height also has been associated with increased mammographic density, which is considered a strong risk factor for breast cancer [32]. A pooled analysis of seven prospective cohort studies (4385 incident invasive breast cancer cases within a total sample of 337,819 women) identified a relative risk (RR) of 1.03 (95% confidence interval [CI] 0.96, 1.10) per 5-cm increment in premenopausal women and an RR of 1.07 (95% CI 1.03, 1.12) in postmenopausal women [33]. The majority of studies conducted in Western cultures have used heights of 160 cm or less as a referent and have defined heights of 175 cm or above as "tall." It is hypothesized that growth factors that drive skeletal development also may stimulate the proliferation of mammary stem cells, which are also in turn modified by the hormonal environment [23, 31]. Most recently, studies have focused on length or height attainment throughout the life cycle to determine whether acceleration of growth is linked with increased breast cancer risk. Some studies have investigated height attainment during puberty and selected periods during childhood in relation to risk [34, 35]. Others, such as those by Lagiou *et al.* and Vatten *et al.* [36, 37], have concentrated on *in utero* exposure and have found positive associations between increased risk and birth length, as well as maternal height. Such studies increase our understanding of the etiology of breast cancer and ultimately may uncover mechanisms and critical periods for delivering future interventions aimed at prevention. Although maternal height and early height attainment may hold promise for elucidating these pathways, such exploration is in its infancy and much more study is needed [30, 31].

Other skeletal indices, such as increased elbow breadth (a frame size marker), femur and trunk length, and bone density also have been associated with increased risk [22, 38, 39]. Data in these areas, however, are far too preliminary to permit any firm conclusions to be drawn.

A clear discrepancy exists with regard to adiposity and its apparent effect on risk for pre- versus postmenopausal disease. Leanness is an acknowledged risk factor for premenopausal breast cancer, whereas obesity serves as a strong risk factor for postmenopausal disease [40]. In the pooled analysis noted earlier, women with a body mass index (BMI) above 31 kg/m^2 had an RR of 0.54 (95% CI 0.34, 0.85) for premenopausal breast cancer, whereas women whose BMI was above 28 kg/m^2 had an RR of 1.26 (95% CI 1.09, 1.46) for postmenopausal disease [33]. Similar associations have been found by a myriad of studies completed over the past half century, with more recent findings from the Swedish Twin Registry and the French 3N women's cohort study confirming these results [41, 42]. Furthermore, these associations have been corroborated in a variety of populations (e.g., Caucasians, Asians, and African Americans, and young and old) [43–47]; however, these risks may be modified by the following factors: (1) a positive family history (which may exacerbate the impact of obesity) [48, 49]; (2) relative chronicity within the broad categories of pre- versus postmenopause (i.e., younger obese postmenopausal women may be at most risk) [50]; (3) the use of hormone-replacement therapy (which may mask associations between obesity and postmenopausal disease) [51]; and (4) various genetic polymorphisms [12, 52]. Recently reported data also suggest that the effect of obesity may be age-independent among Hispanic women; that is, obesity may increase risk for both pre- and postmenopausal disease [53]. Clearly, these results represent an exception to the rule, and more research is needed. Recent studies also suggest that obesity is a strong risk factor only for estrogen and progesterone receptor positive breast cancer, which tends to be more prevalent in some populations [29, 54, 55], such as older women—a finding that may help explain why weight status plays a complex role in this disease.

Two observational studies that focused specifically on male breast cancer indicate that the relationship between obesity and risk for breast cancer in men is similar to the relationship observed in postmenopausal breast cancer in women. In a case-control study of 178 men who died of breast cancer and 512 men who died of other causes, Hsing *et al.* [56] found increased risk for men who were described by their next of kin as being very overweight (odds ratio

[OR] 2.3, 95% CI 1.1, 5.0). Ewertz *et al.* [57] found obesity 10 years before diagnosis to be associated with increased risk for breast cancer in men (OR 3.3, 95% CI 1.0, 4.5) in a population-based case-control study of 156 incident cases of male breast cancer and 468 matched controls. As reviewed recently by Lorincz and Sukumar [58], there are several hypothesized mechanisms by which overweight and obesity confer risk for postmenopausal breast cancer. A classic hypothesis involves the increased peripheral aromatization of androgens within the adipose tissue, thus yielding increased levels of circulating estrogens. Studies by McTiernan *et al.* [59] and Rinaldi *et al.* [60] show that obese women have significantly higher circulating levels of total testosterone, estrone, and estradiol, as well as higher free levels of these hormones, thus providing support to this premise. A second hypothesis is that obesity results in increased circulating levels of insulin and insulin-like growth factors, which act as mitogens to transform normal cells into those that are neoplastic. Although this hypothesis fits well into theories surrounding metabolic syndrome, supporting data are mixed and suggest a complex pathway that requires further study to be fully elucidated [58, 61, 62]. For premenopausal breast cancer in which increased skeletal structure is associated with risk, the IGF-1 pathway also has been implicated as a potentially viable mechanism [63]. Finally, a newer hypothesis posits that obesity should be considered an endocrine tumor that affects cancer by secreting various cytokines, polypeptides, and hormone-like molecules [64]. Studies currently are ongoing to determine the role of leptin, adiponectin, interleukin-6, tumor necrosis factor, and other potential mediating factors that may have an impact on carcinogenesis [58, 65–67].

Because body weight fluctuates throughout life, it is conceivable that risk may be modified by body weight status at differing ages. A handful of studies have investigated birth weight in relation to risk and have found mixed results [34, 37, 38, 68–70]. Other studies that have explored obesity during childhood suggest that it may be protective [71–74], with one study suggesting that this relationship is particularly significant among females with a positive family history of breast cancer [75]. This relationship appears logical for premenopausal breast cancer, in which obesity is an acknowledged protective factor, and indeed, a 2006 study by Michels and colleagues [76] suggests that an increased BMI at age 18 years may be the strongest protective factor for premenopausal disease. However, given the relationship between increased weight and early menarche, as well as increased weight and postmenopausal breast cancer, it is difficult to reason why early obesity would be protective for postmenopausal disease. Recent studies, however, indicate that obese girls have significantly fewer ovulatory cycles and therefore have lower circulating levels of both estrogen and progesterone—hormones associated with mammary proliferation [77, 78]. On attainment of adulthood, there appears to be a consistent finding of weight gain and increased body weight being highly associated with increased risk for postmenopausal cancer [21, 26, 33, 79–85], although risk appears affected by use of hormone therapy and ethnicity [86, 87]. Data from the European Prospective Investigation into Cancer (EPIC) Study indicate an 8% increase in risk for each 5-kg gain among non-hormone-therapy users [86]. This increased risk may be largely explained by the fact that as women age, their circulating levels of estrogen become more influenced by estrogens produced by adipose tissue rather than those produced by the ovary [58, 83, 88]. In contrast, studies that have investigated weight loss have found a protective effect [82, 89, 90]. For example, Parker and Folsom [90] found a risk reduction for postmenopausal breast cancer of roughly 19% among women who intentionally lost 20 pounds or more during adulthood. A study by Kotsopoulous *et al.* [89] suggests that this protective effect may be further amplified among BRCA 1 and 2 carriers during critical periods in time, because they found a risk reduction of roughly 50% for weight loss of 10 or more pounds reported during ages 18–30 among high-risk women. Of note, weight gain during adulthood also is more likely to be deposited in an android versus gynoid pattern and hence may promote insulin resistance and the increased production of insulin and insulin-like growth factors that may act synergistically with estrogen to confer risk [21, 80, 85].

Although several previous studies have found that central or visceral obesity (primarily assessed via waist: hip ratio) is an additional risk factor for breast cancer [21, 24, 91–94], results of a 2003 systematic review and meta-analysis by Harvie and colleagues [95] found that these associations disappear when analyses are controlled for BMI. The results of subsequent studies that have used similar methods of analysis largely parallel these findings; however, associations between central obesity and increased risk may still exist for select populations, such as Asians [60, 96–99].

2. Dietary Composition

In ecologic studies, a fivefold difference in breast cancer mortality rates has been observed across countries, and dietary patterns are a major aspect of the environmental exposures that differ across these countries [100, 101]. Also, risk for breast cancer increases on relocation from low-risk countries to high-risk countries, concurrent with the adoption of the dietary and lifestyle patterns of the new locale [101–104]. These findings from hypothesis-generating ecologic and migration studies have inspired much research effort aimed toward identifying the specific dietary factors that increase risk for breast cancer.

The vast majority of the observational studies that have examined the link between dietary intake and breast cancer risk have relied on self-reported dietary intake data, and only a limited number of studies have utilized serologic biomarkers of diet. Self-reported dietary intakes should be interpreted as estimates that may allow ranking rather than producing absolute values, even when the best-developed methodologies are used. Thus, a high risk for reporting bias and misclassification of subjects is inherent in this type of research.

The possible link between dietary fat intake and risk for breast cancer has historically received the most attention. Several comprehensive reviews and discussions of the evidence reported to date, and the challenges in interpreting these data, have been published [105–111]. A major issue in the interpretation of the data suggesting an adverse effect of dietary fat on breast cancer risk is that fat intake, total energy consumption, and adiposity are typically inextricably linked; therefore, demonstrating an independent effect of total dietary fat per se is a challenge.

Despite a strong and direct link between fat intake and breast cancer risk in ecologic studies, this relationship has generally not been as evident in observational studies conducted within populations, particularly in cohort studies. For example, in a pooled analysis of a U.S. cohort of 337,819 U.S. women with 4980 incident cases [112], the multivariate RR of breast cancer was 1.05 (95% CI 0.94, 1.16) when comparing women in the highest quintile of energy-adjusted total fat intake to women in the lowest quintile. Relative risks for intakes of saturated, monounsaturated, and polyunsaturated fat also were nonsignificant. In a single large cohort study of 88,795 women, of whom 2956 were diagnosed with breast cancer over a follow-up period of 14 years, those consuming 30–35% energy intake from fat had a relative risk of 1.15 (95% CI 0.73, 1.80) compared to women consuming 20% or less energy intake from fat [113]. In that study, specific types of fat also were not found to be associated with risk of breast cancer, although a small but significant increase in risk was associated with intake of omega-3 fatty acids from fish (RR 1.09, 95% CI 1.03, 1.16), and *trans*-fatty acid intake had a slightly protective effect (RR 0.92, 95% CI 0.86, 0.98). Disparate findings on the association between fat intake and breast cancer risk in ecologic versus within-population cohort studies have been attributed to the limited range of fat intake that is typically reported by the homogeneous within-country populations and also to the underreporting of high-fat foods, which may limit the capability to detect an effect and also increases risk of classification errors [111].

Laboratory animal studies quite consistently demonstrate that except for a specific tumor-promoting effect of linoleic acid, the effects of dietary fat on tumor growth is attributable to the high energy content of a high-fat diet, rather than the content of fat [114]. In laboratory animal models, omega-6 fatty acids (i.e., linoleic acid) have been shown to promote tumor development, particularly in comparison to omega-3 fatty acid–rich diets [107, 110, 115], theoretically by influencing tumor eicosanoid production and possibly other cellular mechanisms (e.g., protein kinase C, membrane permeability). However, epidemiological studies have not consistently revealed a protective pattern of specific fatty acids, so this issue remains unresolved.

Given that a reduction in fat intake has been hypothesized to influence breast cancer risk by reducing serum estrogen concentrations, various feeding studies and small diet intervention trials have examined the effect of dietary fat reduction on serum estrogen levels. Low-fat diets were associated with an average 13% reduction in serum estradiol concentration in a meta-analysis of several small feeding studies [116]. However, significant weight loss occurred in the majority of the studies in which serum estradiol was significantly reduced, and dietary fiber was concurrently increased in eight of the 13 studies included in the analysis. Thus, an energy deficit, weight loss, or increased fiber intake could be the more important factors promoting a reduction in hormone levels, and an independent effect of fat intake cannot be assumed to be the primary factor modulating serum estrogens.

Several epidemiological studies have examined the association between vegetable and fruit intake and risk for primary breast cancer. In the past few years, two large observational studies addressed this relationship using combined and pooled data. These pooled studies were based on combinations of published observational studies with different study designs, and they produced somewhat divergent results. In a meta-analysis of 26 studies (21 case-control and five cohort studies) published from 1982 to 1997, the relationships between risk for breast cancer and intakes of vegetables, fruit, beta-carotene, and vitamin C were examined [117]. High (versus low) consumption of vegetables exhibited the strongest protective effect (RR 0.75, 95% CI 0.66, 0.85), whereas the relationship with fruit consumption was not significant (RR 0.94, 95% CI 0.74, 1.11) [117]. Data from 11 of these studies allowed analysis of beta-carotene intake, which was significantly inversely associated with risk (RR 0.82, 95% CI 0.76, 0.91 for approximately >7000 versus $<1000\,\mu$g/day). Results from this meta-analysis are very similar to a summary based on 19 case-control and three cohort studies reported in 1997 [118], which found at least a 25% reduction in risk in the majority of the studies and greater consistency for vegetable compared to fruit intake. In a subsequent pooled analysis of 7377 incident breast cancer cases from women enrolled in eight prospective cohort studies, the protective effect of total fruit and vegetable intake was found to be small and nonsignificant (RR 0.93, 95% CI 0.86, 1.00 for highest versus lowest quintiles) [119]. Differences in the study designs and in the approaches used to estimate intake likely contribute to these inconsistent results. The pooled analysis that relied exclusively on data from cohort studies utilizes a better study design, because the dietary data were collected prior to diagnosis. However, the number of vegetable and fruit questions on the various instruments used to collect the dietary data across these cohort studies varied by a factor of more than 4, and combining widely disparate ranges of servings and types of vegetables and foods may have limited the capability of identifying a relationship with risk. Although the period of follow-up is rather limited (median 5.3 years), the relationship between vegetable and fruit intake and risk of breast cancer also was not significant in the large EPIC study ($n = 285{,}526$ women) [120].

To date, only one study has explored the association between intake of selected foods and male breast cancer [56]. In that study, nonsignificant protective associations were observed for intakes of seven or more times/week versus less than one time/week of vegetables (OR 0.5, 95% CI 0.2, 1.7), and to a lesser extent, intake of fruit (OR 0.8, 95% CI 0.4, 1.3 for 7+ times/week versus <1 times/day). Point estimates of red meat consumption from that study also suggested a nonsignificant direct association with risk for breast cancer in these men (OR 1.8, 95% CI 0.6, 4.9, for 7+ servings per week versus <1 serving/week) [56].

Fewer studies in which tissue concentrations of carotenoids (a marker of vegetable and fruit intake) have been quantified and analyzed in relation to risk for breast cancer have been reported. These have often (but not always) suggested a protective effect of higher concentrations of these compounds, indicative of a diet high in vegetables and fruit [121–125]. In the most recent prospective cohort study that examined this relationship [123], the OR for the lowest versus highest quartile of total serum carotenoids was 2.31 (95% CI 1.35, 3.96), with serum concentrations of beta-carotene (OR 2.21, 95% CI 1.29, 3.79), alpha-carotene (OR 1.99, 95% CI 1.18, 3.34), and lutein (OR 2.08, 95% CI 1.11, 3.90) inversely associated with risk.

Several biologically feasible mechanisms and supportive laboratory evidence for constituents of vegetables and fruits have been demonstrated in cell culture. For example, carotenoids have retinoid-like effects on cellular differentiation and also exhibit inhibitory effects on mammary cell growth [126–129]. Antioxidants, such as vitamin C, may reduce the risk for breast cancer by protecting against DNA damage and other free-radical–induced cellular changes associated with neoplasia. Vegetables of the *Brassica* genus, such as broccoli, may favorably alter estrogen metabolism via the induction of cytochromes P-450 [130].

Alcohol intake has been consistently and positively associated with risk for breast cancer in epidemiological studies. Pooled analysis of data from cohort studies ($n = 22{,}647$) indicates that alcohol intake exhibits a dose-response relationship with risk for breast cancer, at least up to 60 g/day [131]. The multivariate adjusted relative risk for total alcohol intake of 30–60 g/day (approximately 2–5 drinks/day), compared to no alcohol, was 1.41 (95% CI 1.18, 1.69). The prevailing theory is that alcohol intake promotes increased serum estrogen levels, which would theoretically increase risk for breast cancer. This effect of alcohol has been observed in studies of premenopausal women and in postmenopausal women administered hormone therapy but not among postmenopausal women who do not use estrogen therapy [132].

Although an association with intake of folate per se and breast cancer risk has not been consistently observed in epidemiological studies, results from several recent cohort studies suggest that dietary folate intake may influence breast cancer risk through an interaction with alcohol [133–135]. Alcohol has well-known effects on folate status by interfering with folate metabolism. In a large prospective U.S. cohort study involving 88,818 women with 3483 incident breast cancer cases over a 16-year follow-up period [135], folate intake *of* 600 μg/day or more compared to 150–299 μg/day was associated with a multivariate adjusted RR of 0.55 (95% CI 0.39, 0.76, $p = 0.01$ for trend) among women who consumed at least 15 g/day of alcohol. A similar relationship was identified in a cohort of 41,836 postmenopausal women, in whom the relative risk of breast cancer associated with low folate intake was 1.08 (95% CI 0.78, 1.49) among nondrinkers, 1.33 (95% CI 0.86, 2.05) among drinkers of no more than 4 g alcohol/day, and 1.59 (95% CI 1.05, 2.41) among drinkers of more than 4 g alcohol/day [134].

As previously noted, dietary fiber also may exert a protective effect against breast cancer. Postulated mechanisms include the effect of fiber on circulating estrogens, because fiber has been shown to bind to estrogen in the enterohepatic circulation and hinder reabsorption [136]. However, case-control and cohort studies that have examined the relationship between fiber intake and risk for breast cancer have generally not found a significant protective effect [105, 109].

Evidence from large U.S. cohort and other studies have generally not found significant associations between vitamin E intake and risk for breast cancer [109, 137–142]. Similarly, caffeine has not been identified as influencing risk for either breast cancer or fibrocystic breast condition in women [143]. Observational studies that addressed the relationship between meat and/or protein consumption and breast cancer risk in women have produced inconsistent results and generally do not support this relationship [118, 144].

Countries that consume greater amounts of soy and soy products, mainly native Japanese and Chinese populations, have historically exhibited the lowest breast cancer mortality rates, when compared with the United States and most European countries [145]. A dietary pattern characterized by higher intakes of soy and vegetables, compared to a dietary pattern emphasizing animal foods and sweets, was associated with reduced risk for breast cancer in the Shanghai Breast Cancer Study [146]. Soy is a rich source of phytoestrogens, and soy isoflavones have been shown to exert hormonal effects in cell culture systems and laboratory animal models [147, 148]. Thus, a role in breast cancer prevention for soy and soy isoflavones, as well as other phytoestrogens, has been investigated in numerous laboratory and epidemiological studies. However, phytoestrogens can act as estrogen agonists as well as antagonists, so it is possible that phytoestrogens could potentially promote mammary neoplasia. This would be most relevant for postmenopausal women, who do not have high levels of endogenous estrogens in the circulation that would compete for estrogen receptor binding sites. Recent results from several types of studies, including cell culture, clinical, and laboratory animal studies, have produced evidence suggesting the possibility of an adverse effect of soy phytoestrogens on risk and progression of breast cancer, when amounts greater than that provided by the typical Asian diet are consumed, that is, three servings per day [149].

In the Women's Health Initiative (WHI) randomized controlled dietary modification trial, the impact of a low-fat diet (intervention goal of <20% of energy intake) was tested in comparison to a usual care control group as a potential means of reducing risk of breast cancer in 48,835 postmenopausal women during an 8-year follow-up period [150]. Adherence was an issue in both study arms, with those in the intervention arm having higher fat intakes than planned, and those in the control arm reporting reductions in intake such that the difference in fat intake between the two study arms was not as great as forecasted (i.e., a difference of only 10.7% in year 1 and 8.1% in year 2). Results indicate a nonsignificant impact on risk, with a hazard ratio (HR) of 0.91 (95% CI 0.83, 1.01). Given the borderline nature of this HR and an increasing separation in mortality rates in the two arms in the study's latter years, the WHI trial is currently being extended to determine whether the low-fat diet may make an impact given a longer period of study. The ongoing Canadian Diet and Breast Cancer Prevention Trial, a multicenter randomized controlled study, is testing whether a low-fat (15% of energy), high-carbohydrate diet intervention can reduce the incidence of breast cancer over a 10-year period among women aged 30–65 years who have increased mammographic density [151]. Table 1 summarizes the nature of the evidence for a relationship with breast cancer and nutrients and other dietary constituents.

B. Nutritional Factors and Recurrence, Progression, and Survival

Breast cancer mortality has been declining in recent years, a trend that has been attributed to earlier diagnosis and improvements in initial treatments [2]. A majority of all

TABLE 1 A Summary of the Nature of the Evidence for a Relationship with Breast Cancer and Nutrients and Other Dietary Constituents

Strength of Current Evidence	Nutritional Status, Nutrient, Food or Dietary Constituent	Proposed Direction of Association	Comments
The consensus of case-control and cohort studies indicate that an increased adult height (defined as 175 cm or more in Western populations) is associated with risk for both pre- and postmenopausal disease.	Height	Increases risk	Current studies are now focused on investigating rates of growth in childhood and maternal height as possible risk factors and postulate that increased production of growth factors as contributory mechanisms.
A large body of case-control and cohort studies conducted in several populations throughout the world provide consensus for these associations, thus making weight status the strongest nutritional parameter associated with risk.	Weight, body mass index	Decreases risk for premenopausal disease Increases risk for postmenopausal disease	It is generally accepted that obesity is a protective factor for premenopausal breast cancer since it interrupts ovulatory function. In contrast, obesity is a risk factor for postmenopausal disease, because of a host of contributory factors, i.e., increased exposure to bioavailable estrogens, growth factors, and adipokines. Current studies are now focused on investigating underlying mechanisms, and the impact of weight status throughout the life cycle on risk (including the impact of intentional weight loss).

(continues)

TABLE 1 *(continued)*

Strength of Current Evidence	Nutritional Status, Nutrient, Food or Dietary Constituent	Proposed Direction of Association	Comments
Several studies and one systematic review and meta-analysis have explored the potential contribution of central adiposity to risk.	Waist-to-hip ratio	Increases risk	Currently, it is unclear whether central adiposity is a risk factor for breast cancer independent of weight status.
Somewhat consistent evidence from epidemiologic analytical research, and biological mechanism(s) are plausible.	Vegetable and fruit intake	Decreases risk	Several potential contributing mechanisms, attributable to numerous dietary constituents
	Alcohol intake	Increases risk	Possibly affects risk through hormonal effects, although clinical studies suggest a complicated relationship; possibly linked through effect on folate status
Inconsistent evidence from epidemiologic analytical research, but biological mechanism(s) are plausible and/ or demonstrable in laboratory studies	Dietary fiber intake	Decreases risk	Effects on hormonal factors demonstrable in biological systems
	Folate intake	Decreases risk	May contribute to beneficial effects of vegetables and fruits; possibly interacts with alcohol intake
	Carotenoid and vitamin A intake	Decreases risk	May contribute to beneficial effects of vegetables and fruits; strongly supportive evidence from cell culture studies
Evidence from epidemiologic analytical research and limited clinical trial data generally not supportive, but biological mechanism(s) may be plausible or demonstrable	Dietary fat	Decreases risk	Possible effects on hormonal factors, but clinical studies are difficult to interpret due to confounding variables
	Specific fatty acids	Variable effects	Cell culture studies show differential effects of omega-6 and omega-3 fatty acids, but relevance to dietary patterns is not established or evident
	Vitamin E, vitamin C, and selenium	Decreases risk	Antioxidant effects possible, hypothesized to interact with genetic polymorphisms (e.g., glutathione-*S*-transferase)
Insufficient evidence from epidemiologic analytical research, but biological mechanism(s) may be plausible or demonstrable	Well-done meat	Increases risk	Linked to carcinogens such as heterocyclic amines, hypothesized to interact with genetic polymorphisms (e.g., *N*-acetyl transferase)
	Soy and soy isoflavones, flaxseed	Variable effects	Although inverse associations suggested by ecologic studies, clinical studies indicate possibility of adverse effects as well
Evidence from epidemiologic analytical research generally not supportive, and biological mechanism(s) unknown or not demonstrable	Caffeine	–	No association with increased risk found consistently in case-control and cohort studies
	Protein	–	Associated in some case-control but not cohort studies, may be confounded by other characteristics of the diet

breast cancers are now diagnosed at a localized stage, with 98% 5-year relative survival rates [2]. As a result, there are increasing numbers of women in the population who are breast cancer survivors and are at risk for breast cancer recurrence. Recurrence is an important issue in the management of these patients, because the yearly rate of secondary cancer events, even for women who have been diagnosed with very early stage cancers, does not return to the level of similarly aged women who have not been diagnosed with breast cancer [2]. Perhaps even more important, women who have been diagnosed with breast cancer have higher rates of mortality from other causes, such as diabetes and cardiovascular disease, comorbid factors in which diet and nutritional status play an important role [152].

1. Obesity and Overweight

The nutritional factor associated most consistently with reduced likelihood of survival is increased weight. Four critical reviews published within the past 5 years conclude that increased body weight either at prediagnosis or at the time of diagnosis exerts a negative prognostic effect [153–155]. Although several studies included within these reviews and those reported afterward suggest that overweight and obesity may be linked to recurrence or to progressive disease [153, 155–159], findings are somewhat mixed with regard to these outcomes [160–162]. However, because overweight is associated with several comorbid conditions, data are fairly consistent regarding overall survival, for which normal-weight women demonstrate a significant advantage [163–165]. Recent findings by Pierce and colleagues [166] suggest that physical activity may be more important than weight in conferring a survival advantage; however, further study is needed. It should be noted that during breast cancer treatment, normal-weight women, as compared to those who are overweight, manifest significantly fewer surgical complications [167, 168]; less lymphedema [169, 170]; and fewer thromboembolic events while on hormonal therapy [171]. Studies exploring body fat distribution in relation to survival or comorbid conditions have yielded mixed findings [165, 172–174].

Weight gain after the diagnosis of breast cancer is a common occurrence and may be undesirable for several reasons [40]. First, gains in weight may negatively affect quality of life, and previous studies indicate that a majority of patients with breast cancer find weight gain distressing and report it as a major concern [147, 175–179]. Second, as mentioned previously, weight gain may predispose women to other chronic disorders, such as hypertension, cardiovascular disease, gallbladder disease, orthopedic disturbances, and diabetes mellitus [152, 180–182]. Finally, gathering evidence suggests that weight gain may adversely affect disease-free survival. In one of the initial studies, Camoriano *et al.* [175] followed 646 patients with breast cancer for a median of 6.6 years and found that premenopausal patients who gained more than the median amount of weight (5.9 kg) were 1.5 times more likely to relapse and 1.6 times more likely to die of their breast cancer. More recently, and in the largest study conducted to date (N = 5204), Kroenke and colleagues [183] found that breast cancer survivors who increased their BMI by 0.5 to 2 units had an RR of recurrence of 1.40 (95% CI 1.02–1.92) and those who gained more than 2.0 BMI units had an RR of 1.53 (95% CI 1.54–2.34); both groups also experienced significantly higher all-cause mortality. In contrast, Caan and colleagues [160] found no evidence of increased mortality with weight gain in a cohort of 3215 early-stage breast cancer patients. Given the pandemic of obesity occurring worldwide, more study is needed in this area, as well as determining the impact of weight loss interventions on survival and other short- and long-term outcomes.

Recent reports suggest that mean weight gain during chemotherapy ranges from 1 to 5 kg; however, roughly 20% of women gain 10–20 kg [40, 184, 185]. Weight gain is most prevalent among premenopausal patients receiving adjuvant chemotherapy, and may vary by treatment regimen, with anthracycline agents associated with greater gains [40, 184, 186]. To date, only four reported studies have examined weight loss interventions among women with breast cancer [187–190]. Differential effects were found for individual diet counseling by a dietitian, with one study showing significantly favorable effects on body weight status [187], and the other showing no effect [190]. A more recent study by Djuric and colleagues [188] found that counseling by a dietitian was most effective if combined with a structured Weight Watchers program that included exercise, with weight change at 12 months being $+0.85 \pm 6.0$ kg versus -8.0 ± 5.5 kg or -9.4 ± 8.6 kg in the control versus dietitian or dietitian + Weight Watchers program, respectively. Multiple behavioral interventions that utilize a comprehensive approach to energy balance, and that include both diet and exercise components, may have the potential to be more effective than interventions relying on either component alone. In their evaluation of a diet and exercise intervention among early-stage breast cancer patients that was begun during the time of treatment and extended throughout the year following diagnosis, Goodwin *et al.* [189] found that exercise was the strongest predictor of weight loss. Given evidence that sarcopenic obesity (gain of adipose tissue at the expense of lean body mass) is a documented side effect of both chemotherapy and hormonal therapy, exercise (especially strength training exercise) may be of particular importance for cancer survivors because it is considered the cornerstone of treatment for this condition [40, 152, 191].

2. Dietary Composition

The relationship between dietary composition and survival or recurrence has been examined in 18 studies involving cohorts of women who had been diagnosed with breast cancer. In most cases, the dietary data were collected at the time of diagnosis or soon thereafter. Most of these studies examined the effect of intake of dietary fat or selected high-fat foods. Results of the analysis of the relationship between fat intake and survival are generally inconsistent, with some studies finding inverse associations between survival and dietary fat intake at diagnosis, either adjusted [192–196] or unadjusted [93, 128, 129] for energy, whereas others [197–201] have not. One study ($n = 472$)

found stage- and age-adjusted risk for recurrence to be associated with intakes of butter, margarine, and lard with a RR of 1.30 (95% CI 1.03, 1.64) for each time these foods were consumed per day [202]. In a prospective study by Holmes *et al.* ($n = 1982$), in which dietary data were collected preceding and after diagnosis, no apparent association between energy-adjusted fat intake and mortality was observed [203]. However, protein intake, mainly linked to poultry, fish, and dairy food sources, was found to exert a protective effect in that cohort (RR 0.65, 95% CI 0.47, 0.88, for highest versus lowest quintile).

Protective effects of vegetables and fruits and the micronutrients provided by these foods (e.g., vitamin C., carotenoids) have been observed in several of these follow-up cohort studies, with findings somewhat more consistent than those for dietary fat, although the strength of the association is modest. Of the 10 studies that examined these dietary factors, four found a significant inverse association with risk of death [194–196, 204], two found that risk of dying was nonsignificantly decreased in association with frequent vegetable consumption [197, 205], and one found a significant inverse association in women with node-negative disease, who comprised 62% of that cohort (but not in the total group that included women at all stages of invasive breast cancer) [203]. In the studies that found an inverse relationship with survival and intakes of vegetables, fruit, and related nutrients (beta-carotene, vitamin C), the magnitude of the protective effect was a 20–90% reduction in risk for death. In a cohort study involving 1511 women previously diagnosed and treated for breast cancer, who were followed for an average of 7 years, women in the highest quartile of plasma total carotenoid concentration had an estimated 43% reduction in risk for a new breast cancer event (recurrence or new primary) compared to the lowest quintile [206]. In the same cohort, breast cancer survivors who reported consuming five or more daily servings of vegetables and fruit and exercised at least at a moderate level 30 minutes nearly every day had a 50% reduction in risk associated with these healthy lifestyles [166]. An overall dietary pattern characterized by higher intakes of vegetables, fruit, whole grains, and low-fat dairy products was not related to all-cause or breast cancer mortality but was related to a significantly lower risk of mortality from other causes during a 20-year follow period in another cohort of 2619 women [207].

Although alcohol intake has been identified as a risk factor for primary breast cancer, no significant relationships between alcohol intake and survival or recurrence were found in the eight studies that examined this relationship, although one of these studies reported that risk of dying was slightly (but not significantly) increased in association with frequent alcohol consumption [197]. In a study that examined specific foods [202], beer intake was directly related to risk for recurrence (but not survival), and wine and hard liquor intake were unrelated to risk for either outcome. Likewise, significant associations between dietary fiber intake and likelihood of survival have generally not been observed in these cohort studies.

Two multicenter randomized controlled trials have tested whether diet modification can influence the risk for recurrence and overall survival following the diagnosis of early-stage breast cancer. In the Women's Intervention Nutrition Study (WINS), which involved 2437 postmenopausal women randomized within 12 months of primary surgery, the primary dietary goal was a reduction in dietary fat intake ($\leq$15% energy from fat) [208]. Women in the intervention arm in the WINS study reported a reduction in fat intake (33.3 g/day versus 51.3 g/d in the control group), which was associated with an average 6-pound weight loss. The HR of relapse events in the intervention group compared with the control group was 0.76 (95% CI 0.60, 0.98) after approximately 5 years follow-up [209]. Results of secondary analysis suggest a greater protective effect in women with hormone receptor-negative breast cancers compared with women whose cancers were hormone receptor-positive. In the Women's Healthy Eating and Living (WHEL) Study, the target population consisted of 3109 pre- and postmenopausal women who had been diagnosed with breast cancer within the preceding 4 years and who had completed initial therapies, prior to randomization. The primary emphasis of the WHEL Study diet intervention was on increased vegetable and fruit intake, with daily dietary goals of five vegetable servings, 16 ounces of vegetable juice, three fruit servings, 15–20% energy intake from fat, and 30 g dietary fiber. Feasibility study reports and trial data from this study indicated excellent adherence [210–212] and increased intake of vegetables and fruit in the intensive intervention group was validated by plasma carotenoid concentrations. Overall findings indicate no significant differences in risks for recurrence or survival in the intensive intervention versus the control group over a mean 7.3-year follow-up. Notably, the study participants reported at baseline an average daily consumption of 7.4 servings of vegetables and fruit, which actually meets or exceeds current recommendations, so the WHEL Study results mainly indicate that an extraordinarily high intake of vegetables and fruit does not appear to further reduce risk for recurrence.

III. ENDOMETRIAL CANCER

Cancer of the endometrium (uterine corpus) is the most common invasive gynecologic cancer [2]. Similar to breast and ovarian cancers, endometrial cancer is most common after menopause and is more prevalent among whites, although blacks have higher rates of mortality. Relative 5-year survival rates are 96%, 67%, and 23% for cancers diagnosed at local, regional, and distant stages, respectively [2]. High cumulative exposure to estrogen is the major risk factor for endometrial cancer, whereas exposure to progesterone is protective [2]. Hence, early menarche, late menopause, and nulliparity are risk factors for endometrial cancer, just as they are for breast and ovarian cancers [2]. However, unlike these cancers, risk factors for endometrial cancer also include tamoxifen use and factors associated with unopposed estrogens [213]. Given the key role that unopposed estrogens play in this disease, it is not surprising that obesity is one of the major nutrition-related risk factors for this cancer [214]. Polycystic ovarian syndrome, which also is more prevalent among overweight and obese women, also is a risk factor for this disease [215].

The link between obesity and endometrial cancer is well established [213, 214, 216, 217]. A 2001 review and meta-analysis by Bergstrom *et al.* [79] found an RR of 1.10 (95% CI 1.07, 1.14) for each unit increase in BMI, equating to a relative risk for overweight ($25\,kg/m^2 < BMI < 30\,kg/m^2$) of 1.59 and a relative risk for obesity ($BMI > 30\,kg/m^2$) of 2.52. When compared to parallel meta-analyses conducted for cancers at other sites (i.e., cancers of the colon, gallbladder, kidney, prostate, and postmenopausal breast), Bergstrom *et al.* [79] concluded that the strongest association was between excess weight and cancer of the endometrium. This exceptionally strong association between excess body weight and endometrial cancer also was reported in a 2003 prospective cohort study (n = 495,477) by Calle and colleagues [216] who investigated cancer-related mortality of various sites in relation to body weight status (Figure 1). Their findings also indicate that compared to other cancers, the strongest association between obesity and cancer-related mortality exists for endometrial cancer, where the RR for a BMI of $40\,kg/m^2$ or more is 6.25 (95% CI 3.75, 10.42), as compared to cancers of the breast (postmenopausal) and ovary, whereas the point estimates are still significant, but much lower, that is, 2.12 (for BMI of $40+\,kg/m^2$) and 1.51 (for BMI of $35+\,kg/m^2$), respectively. Over the past decade, such findings have been confirmed in a broad array of case-control and cohort studies conducted in a wide variety of populations [218–233]. More recent studies have focused on determining the contribution of obesity in relation to other factors that are weight-related, such as diabetes, physical activity, and energy intake. In general, study findings support that

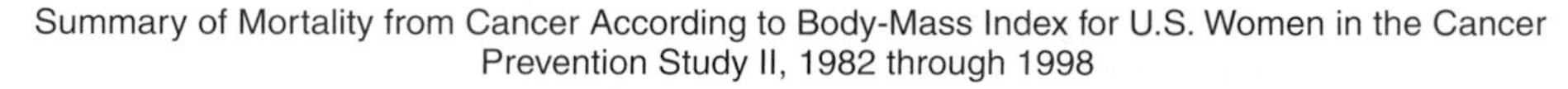

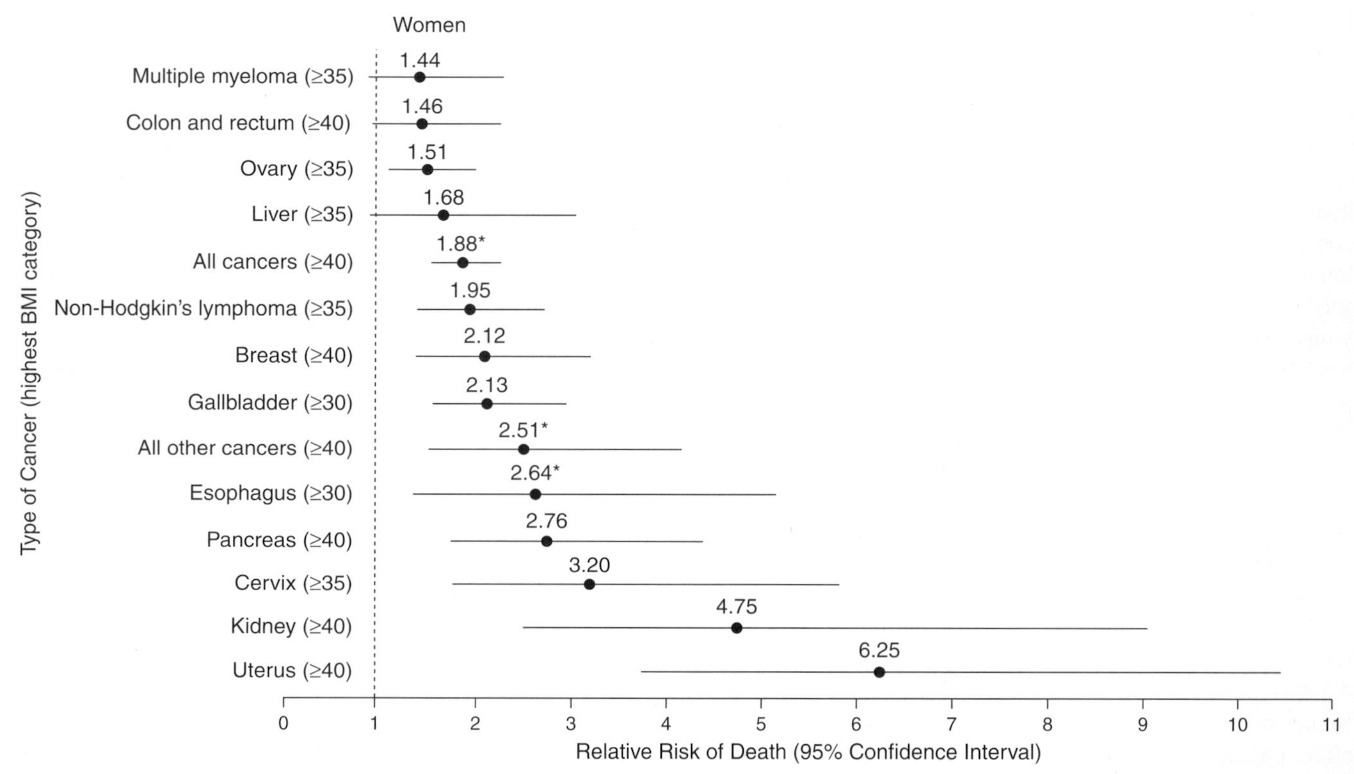

FIGURE 1 Summary of mortality from cancer according to body-mass index for U.S. women in the Cancer Prevention Study I. I., 1982 through 1998. For each relative risk, the comparison was between women in the highest body-mass-index (BMI) category (indicated in parentheses) and women in the reference category (body-mass index, 18.5 to 24.9). Asterisks indicate relative risks for women who never smoked. Results of the linear test for trend were significant ($p \leq 0.05$) for all cancer sites. [Frome Calle *et al.* (2003). *N. Engl. J. Med.* **348**, 1625–1638.]

concept that each of these factors plays a significant, yet independent, role in the etiology of endometrial carcinoma [221, 231, 233–236]. Beginning research also suggests that biomarkers associated with adiposity, such as higher circulating levels of leptin and insulin-like growth factors, and lower levels of insulin-like binding proteins and adiponectin, also may be independently linked with increased endometrial cancer risk; however, further research is needed to confirm these findings [10, 237–239]. In addition, data continue to increase regarding associations between fat patterning and weight gain throughout the life cycle and risk for endometrial cancer [224, 240–250]. Recent data from two large cohort studies (EPIC; $n = 223{,}008$] and the AARP Diet and Health Study [$n = 103{,}882$]) and one large case-control study ($n = 1678$) suggest that waist circumferences over 88 cm and increased waist-to-hip ratio are associated with increased endometrial cancer risk, independent of body weight status [219, 220, 232]. Findings from the two cohort studies of Chang *et al.* [219] and Freidenreich *et al.* [220] also suggest that compared to women who are weight stable, those who gain at least 20 kg during adulthood have an increased risk that is roughly 2–3 times higher. Although further data are needed to draw a consensus regarding these associations, the balance of studies in these two arenas now favors adult weight gain and upper body obesity as probable independent risk factors for endometrial carcinoma. Given the strong associations between endometrial cancer and obesity and other markers of adiposity, it is somewhat surprising that intentional weight loss of 20 kg or more was found to be associated with a nonsignificant risk reduction of only 4% among women participating in the Iowa Women's Health Cohort [90]. Although these data suggest that the consequences of obesity may be far-reaching and permanent, such views remain speculative until more research is completed.

A dearth of data exists regarding endometrial cancer and anthropometric factors that are unrelated to adiposity. To date, there is only one report that describes risk in relation to body height. In the Netherlands Cohort Study (226 cases within a cohort of 62,573), an RR of 2.57 (95% CI 1.32, 4.99) was found for women who were at least 175 cm tall compared to those measuring 160 cm or less [236]. This association is similar and of the same magnitude as found between height and breast cancer and points to the potential role of growth factors and early nutritional status in the etiology of this disease.

To date, 24 reported studies have explored various dietary factors in relation to endometrial carcinoma. Eight of these studies were hospital-based case-control studies [251–257], eight were population-based case-control studies [258–264], and eight were cohort studies [221, 223, 265–270]. The cohort studies found relatively few significant associations, with investigations by Larsson *et al.* [267], Silvera *et al.* [268], and Folsom *et al.* [265] focusing on carbohydrate intake and glycemic indexes or load and finding no associations with carbohydrate intake and either an absence or very weak associations between glycemic index or load and endometrial cancer risk. Data suggest, however, that these factors may behave differently in obese versus normal weight women or in women with or without diabetes [265, 267, 268]. In contrast, a cohort study by Terry *et al.* [263] found significant associations between endometrial cancer risk and very low intakes of vegetables and fruit. Fruit consumption was found to be a protective factor in three of 16 studies [252, 253, 261]; vegetable intake (or vegetable protein) also was found to be protective in five of these studies [225, 251, 253, 262, 271], with odds ratios suggesting an average 40% reduction in risk when intakes in the fourth versus the first quartiles are compared. Data largely show few associations for other plant constituents, such as dietary lycopene or dietary fiber; however, either these elements were not analyzed in several studies (as in the case of lycopene) or analysis did not take into account various forms (as in the case of fiber). A 2003 study by Horn-Ross and colleagues [258], however, suggests that isoflavones and lignins may be protective given RRs of 0.59 (95% CI 0.37, 0.93) and 0.68 (95% CI 0.44, 1.1) in highest versus lowest quartiles, respectively. In this study, obesity was found to interact with phytoestrogen intake, suggesting a sevenfold increase in risk among obese women with the lowest intakes of phytoestrogens. These findings, as well as sporadic reports of decreased risk with intakes of specific plant foods (e.g., brassica vegetables [263], allium vegetables [262], legumes [263], dark-yellow or green vegetables [262], pulses, nuts, and seeds [255]) are of interest, but must be confirmed by other studies before consensus can be achieved.

Prior to 1995, dietary fat was associated with risk in a majority of studies of associations between dietary intake and risk for endometrial cancer. However, in recent years, and especially in the cohort studies in which odds ratios have been adjusted for BMI, total energy intake, and other confounding factors, dietary fat intake is less often found to be an independent risk factor for endometrial cancer. A similar situation is true for energy intake, and it is unclear whether excess energy intake independently affects risk, or whether risk is conferred largely by obesity [272]. Although a 2003 cohort study by Furberg and Thune [221] found a significant association between energy intake and endometrial cancer risk, such findings may need to be replicated using collection methods other than food frequency questionnaires to assess energy intake.

Finally, although reports exist that suggest increased risk with some foods, such as meat [221] and animal protein [271], as well as protective benefit of various foods, such as coffee [263] or fatty fish [273], or dietary constituents such as vitamin C [271], lactose [256], calcium [256, 263], vitamin D [256], and other vitamins [255], further research is necessary to corroborate these findings.

Alcohol use has been shown to be associated with elevated circulating estrogen levels and reduced progesterone;

however, studies conducted to date have generally found no association or modest inverse relationships between alcohol and risk for endometrial cancer [274]. Small sample sizes, limited range of alcohol intake, and confounding factors (e.g., possible interaction with exogenous estrogens or factors such as age) may have limited the ability to detect associations [274, 275]. Thus, more research is needed to define the potential role of alcohol in the etiology of this cancer.

Comparatively little is known regarding nutritional issues after the diagnosis of endometrial cancer, though research on this topic is increasing. Five years ago, studies were largely limited to reports that chronicled the increased prevalence of treatment-related complications among obese patients [276–279]. Although data from more recent reports confirm these findings [280, 281], data regarding recurrence, disease-free survival, and overall survival are conflicting. For example, in three moderate-sized studies with case numbers ranging from 380 to 492, obesity was found to be unassociated with recurrence in one study [281], but inversely associated with recurrence in two others [280, 282]. Furthermore, in two moderate-sized studies conducted in the United States ($n = 380$ and $n = 745$), obesity was associated with increased overall mortality [281, 283], whereas in another smaller study ($n = 121$) conducted in Poland, it was found to be protective [284]. Thus, the only consensus regarding obesity and survivorship of endometrial cancer relates to comorbid conditions, in which obesity was found to be associated with the increased prevalence of diabetes, hypertension, and pulmonary disease [280, 283] and also was found to be consistently associated with poorer quality of life [285, 286], although the number of studies are few. To date, no studies have investigated other nutritional issues beyond weight status in endometrial cancer survivors.

IV. OVARIAN CANCER

Ovarian cancer is the most lethal of these cancers, being the fourth most common cause of cancer death among U.S. women despite having a relatively low incidence rate [1]. Women diagnosed with ovarian cancer typically present at an advanced stage of the disease, because the early stages are asymptomatic. The established risk factors for ovarian cancer are older age ($\geq$50 years), low parity, never having used oral contraceptives, and having a family history of breast or ovarian cancer [287, 288]. Although the etiology is still a focus of intense investigation, a current theory is that increased ovulation or hormonal stimulation of ovarian epithelial cells plays a role in the development of ovarian cancer [289]. Current prevention strategies for women at high risk include oral contraceptive use, tubal ligation, and prophylactic oophorectomy [290]. Relatively few studies on the relationship between nutritional factors and risk for ovarian cancer have been reported, and there is a dearth of research that examines how these factors may influence progression or survival after the diagnosis of ovarian cancer.

Several case-control and three prospective studies have examined the relationship between adiposity and risk for ovarian cancer, and the results are mixed. In two case-control studies [291, 292] and one prospective study [293], obesity was found to be significantly protective against ovarian cancer. In the most recent of these reports, Lukanova *et al.* [293] found in their multicenter prospective study that women in the highest versus lowest quartile of BMI had a 54% lower risk of ovarian cancer (OR 0.23, 95% CI 0.23, 0.92). In contrast, one population-based case-control study found women in the highest versus lowest category of BMI to have an increased risk with an odds ratio of 1.7 (95% CI 1.1, 2.7) [294], and two case-control studies [295, 296] and one prospective study [243] found no association between degree of obesity and risk for ovarian cancer. The relationship between body fat distribution and risk for ovarian cancer was examined in one prospective study involving postmenopausal women [297], in which a direct relationship between risk and waist:hip ratio was suggested (RRs for the upper three quintiles compared with the lowest quintile were 2.0, 1.6, and 2.3). Based on the mixed findings of these studies, it appears that the relationship between obesity and ovarian cancer risk clearly differs from the relationships observed between obesity and risk for either postmenopausal breast cancer or endometrial cancer.

Similar to the evidence for breast cancer, international comparisons of incidence of ovarian cancer indicate a strong inverse relationship with per capita dietary fat consumption [102], and the hypothesis that dietary fat intake increases circulating estrogen concentrations (discussed earlier) has stimulated several observational studies of the relationship between dietary fat intake and risk for ovarian cancer. A meta-analysis of pooled data from all available observational studies was conducted to examine the relationship between dietary fat intake and risk of invasive ovarian cancer [298]. In eight observational studies (one prospective and seven case-control studies) involving 6689 subjects, the RR for total fat intake was 1.24 (95% CI 1.07, 1.43) for higher versus lower intakes, suggesting that fat intake imparts a modestly increased risk [298]. For saturated fat, the RR was 1.20 (95% CI 1.04, 1.39) and for animal fat, the RR was 1.70 (95% CI 1.43, 2.03). However, the data on fat intake were energy-adjusted in only four of these eight studies, so whether fat intake exerts an independent effect on risk for ovarian cancer remains unresolved. In a few studies, food sources of fat also have been quantified for analysis. Saturated fatty acids or sources of these fatty acids (i.e., red meat) have been associated with increased risk in a few studies; for example, Bosetti *et al.* [299] found an OR of 1.53 (95% CI 1.13, 2.05) for the highest compared to the lowest quintile of red meat consumption. Intakes of fish [299, 300] and monounsaturated fatty acids [301] have been identified as being protective. However, the data to support relationships between risk and intakes of these fat subtypes are still limited, and replication in additional populations is needed.

Since 1987, 14 observational studies (two cohort and 12 case-control studies) have examined the relationships between intakes of antioxidant micronutrients, selected food sources, and risk for ovarian cancer. The most consistent relationship that has been observed is an inverse association between intake of vegetables and fruit and risk for ovarian cancer, which was found in seven of these studies [289, 300, 302–306], with intakes of green leafy vegetables, carrots, and tomato sauce the vegetables specifically associated with reduced risk. Additionally, one of the cohort studies found that women who consumed at least 2.5 servings of vegetables and fruit as adolescents had a 46% reduction in risk for ovarian cancer [307], although adult vegetable and fruit intake was unrelated to risk in that study. Data on the relationship between risk and specific nutrients examined in these studies are inconsistent. For example, four case-control studies found a protective effect of carotenoids [302, 308–310], whereas two case-control studies [301, 305] and two prospective studies [303, 307] did not find a protective effect of carotenoid intake. Dietary vitamin C intake was examined in five studies [291, 301, 303, 308, 310] and was found to be inversely related to risk in two of these studies [301, 310]. Vitamin E intake from food sources has not been found to relate to risk [291, 303]. One study found vitamin C and vitamin E intake from dietary supplements, but not from food sources, to be associated with reduced risk of ovarian cancer [291]; however, there was no indication of a dose-response effect in that study, and the lifestyle and behavioral characteristics of dietary supplement users are known to differ from those of nonusers. Results of analysis of relationships between ovarian cancer risk and intakes of other micronutrients and dietary constituents, such as vitamin A and alcohol, are similarly inconsistent.

One small prospective cohort study (35 cases, 37 controls) examined associations between serum concentrations of micronutrients and risk for ovarian cancer [311], using sera collected prior to diagnosis. They found no relationship between risk and serum retinol, beta-carotene, lycopene, and lipid-adjusted alpha-tocopherol and gamma-tocopherol concentrations, but serum selenium concentration was inversely associated with risk of ovarian cancer among cases diagnosed 4 or more years after blood collections (p for trend $= 0.02$) [311]. In contrast to these findings linking serum selenium to risk, another prospective study did not find an association between toenail selenium level and ovarian cancer risk [312]. Several earlier case-control studies compared serum nutrient concentrations of patients with ovarian cancer to those of matched controls, but when blood samples are collected following diagnosis, the effect of the diagnosis and treatment cannot be ruled out.

A possible role for intake of lactose (or more specifically, galactose, a unique monosaccharide constituent of lactose) in the development of ovarian cancer has been the focus of some research. Several years ago, it was suggested that increased exposure to galactose, due to high intakes of dairy foods or altered galactose metabolism, might be associated with the development of ovarian cancer. This suggestion was based on animal studies showing that a diet very high in galactose is toxic to oocytes and evidence that the genetic disorder of galactosemia causes premature ovarian failure in women. Indeed, an early case-control study found intake of dairy foods to be directly associated with risk for ovarian cancer [309]. Since that time, however, several case-control and cohort studies have investigated this relationship, and these more recent studies have not found significant associations between intakes of lactose or dairy foods and risk for ovarian cancer [302, 303, 313–315]. Thus, the majority of the evidence suggests that adult consumption of galactose, lactose, or dairy foods is not associated with increased risk for ovarian cancer. Recent findings from studies on the possible role for galactose in the development of ovarian cancer indicate that women with ovarian cancer may be more likely to have genetic polymorphisms that affect the metabolism of galactose at very high concentrations [313, 315]. The effects of these genetic differences would be more relevant to fetal galactose, which is influenced by both maternal and fetal genotypes, than to intakes of lactose and dairy foods in adulthood [313], and the clinical importance of intrauterine galactose exposure to later ovarian cancer risk is unknown.

To our knowledge, studies that have examined the influence of nutritional factors on survival after the diagnosis of ovarian cancer have not been conducted or reported to date.

V. SUMMARY AND CONCLUSION

Although a considerable amount of research has been devoted to understanding the effects of nutritional factors on the risk for breast cancer, much remains to be learned. More research on the relationships between these factors and risk for endometrial and ovarian cancers is sorely needed, and to date, few studies have examined how these factors may influence overall survival in women who have been diagnosed with hormone-related cancers. At this time, guidelines from the American Cancer Society [181, 316] (summarized in Table 2) and the American Institute for Cancer Research [118] form the basis of current dietary recommendations, with encouragement toward more plant-based diets that promote healthy weight control serving as the underlying theme behind these recommendations.

TABLE 2 American Cancer Society Guidelines on Nutrition and Physical Activity for Cancer Prevention: ACS Recommendations for Individual Choices

Maintain a healthy weight throughout life

- Balance caloric intake with physical activity.
- Avoid excessive weight gain throughout the life cycle.
- Achieve and maintain a healthy weight if currently overweight or obese.

Adopt a physically active lifestyle

- Adults: engage in at least 30 minutes of moderate-to-vigorous physical activity, above usual activities, on five or more days of the week (45–60 minutes of intentional physical activity are preferable).
- Children and adolescents: engage in at least 60 minutes/day of moderate-to-vigorous physical activity at least 5 days per week.

Consume a healthy diet, with an emphasis on plant sources

- Choose foods and beverages in amounts that achieve and maintain a healthy weight.
- Eat five or more servings of a variety of vegetables and fruits each day.
- Choose whole grains in preference to processed (refined) grains.
- Limit consumption of processed and red meats.

Limit consumption of alcoholic beverages

- Drink no more than one drink per day for women or two drinks per day for men.

Used with permission from Kushi LH *et al.* [316].

The risk for morbidity and mortality from causes other than breast, endometrial, and ovarian cancer also should be considered in dietary recommendations for women and men at risk for cancer and cancer survivors, especially those diagnosed with early-stage cancers [157]. For example, even though evidence to support a link between fat intake and breast cancer risk and prognosis is inconsistent, limiting saturated fat intake is an established strategy to reduce risk for cardiovascular disease. Similarly, eating a diet with adequate dietary fiber has been associated with decreased risk of coronary heart disease and may contribute to overall health [317], irrespective of a specific link between fiber and hormone-related cancers. Diets that emphasize vegetables, fruit, whole grains, low-fat dairy foods, and lean meats and poultry have been associated with decreased risk of all-cause mortality [318].

References

1. Jemal, A., Siegel, R., Ward, E., Murray, T., Xu, J., and Thun, M. J. (2007). Cancer statistics, 2007. *CA Cancer J. Clin.* **57**, 43–66.
2. American Cancer Society. (2007). "Cancer Facts & Figures 2007." American Cancer Society, Atlanta, GA.
3. Bast, R. C., Jr., Brewer, M., Zou, C., *et al.* (2007). Prevention and early detection of ovarian cancer: mission impossible? *Recent Results Cancer Res.* **174**, 91–100.
4. Nahleh, Z., and Girnius, S. (2006). Male breast cancer: A gender issue. *Nat. Clin. Pract. Oncol.* **3**, 428–437.
5. Fentiman, I. S., Fourquet, A., and Hortobagyi, G. N. (2006). Male breast cancer. *Lancet* **367**, 595–604.
6. Persson, I. (2000). Estrogens in the causation of breast, endometrial and ovarian cancers—evidence and hypotheses from epidemiological findings. *J Steroid Biochem. Mol. Biol.* **74**, 357–364.
7. Zografos, G. C., Panou, M., and Panou, N. (2004). Common risk factors of breast and ovarian cancer: Recent view. *Int. J. Gynecol. Cancer* **14**, 721–740.
8. Druckmann, R., and Rohr, U. D. (2002). IGF-1 in gynaecology and obstetrics: Update 2002. *Maturitas* **41**(Suppl 1), S65–83.
9. Peeters, P. H., Lukanova, A., Allen, N., *et al.* (2007). Serum IGF-I, its major binding protein (IGFBP-3) and epithelial ovarian cancer risk: The European Prospective Investigation into Cancer and Nutrition (EPIC). *Endocr. Relat. Cancer* **14**, 81–90.
10. Cust, A. E., Allen, N. E., Rinaldi, S., *et al.* (2007). Serum levels of C-peptide, IGFBP-1 and IGFBP-2 and endometrial cancer risk; results from the European prospective investigation into cancer and nutrition. *Int. J. Cancer* **120**, 2656–2664.
11. Slattery, M. L., Baumgartner, K. B., Byers, T., *et al.* (2005). Genetic, anthropometric, and lifestyle factors associated with IGF-1 and IGFBP-3 levels in Hispanic and non-Hispanic white women. *Cancer Causes Control* **16**, 1147–1157.
12. Wasserman, L., Flatt, S. W., Natarajan, L., *et al.* (2004). Correlates of obesity in postmenopausal women with breast cancer: Comparison of genetic, demographic, disease-related, life history and dietary factors. *Int. J. Obes. Relat. Metab. Disord.* **28**, 49–56.
13. Low, Y. L., Taylor, J. I., Grace, P. B., *et al.* (2005). Phytoestrogen exposure correlation with plasma estradiol in postmenopausal women in European Prospective Investigation of Cancer and Nutrition—Norfolk may involve diet–gene interactions. *Cancer Epidemiol. Biomarkers Prev.* **14**, 213–220.
14. Le Marchand, L., Haiman, C. A., Wilkens, L. R., Kolonel, L. N., and Henderson, B. E. (2004). MTHFR polymorphisms, diet, HRT, and breast cancer risk: The multiethnic cohort study. *Cancer Epidemiol. Biomarkers Prev.* **13**, 2071–2077.

15. Ahn, J., Gammon, M. D., Santella, R. M., *et al.* (2006). Effects of glutathione *S*-transferase A1 (GSTA1) genotype and potential modifiers on breast cancer risk. *Carcinogenesis* **27**, 1876–1882.
16. Xu, X., Gammon, M. D., Wetmur, J. G., *et al.* (2007). A functional 19–base pair deletion polymorphism of dihydrofolate reductase (DHFR) and risk of breast cancer in multivitamin users. *Am. J. Clin. Nutr.* **85**, 1098–1102.
17. McCullough, M. L., Stevens, V. L., Diver, W. R., *et al.* (2007). Vitamin D pathway gene polymorphisms, diet, and risk of postmenopausal breast cancer: A nested case-control study. *Breast Cancer Res.* **9**, R9.
18. Low, Y. L., Dunning, A. M., Dowsett, M., *et al.* (2006). Implications of gene–environment interaction in studies of gene variants in breast cancer: An example of dietary isoflavones and the D356N polymorphism in the sex hormone-binding globulin gene. *Cancer Res.* **66**, 8980–8983.
19. Weiss, J. R., Moysich, K. B., and Swede, H. (2005). Epidemiology of male breast cancer. *Cancer Epidemiol. Biomarkers Prev.* **14**, 20–26.
20. de Waard, F. (1975). Breast cancer incidence and nutritional status with particular reference to body weight and height. *Cancer Res.* **35**, 3351–3356.
21. Ballard-Barbash, R. (1994). Anthropometry and breast cancer. Body size—a moving target. *Cancer* **74**, 1090–1100.
22. Brinton, L. A., and Swanson, C. A. (1992). Height and weight at various ages and risk of breast cancer. *Ann. Epidemiol.* **2**, 597–609.
23. Clinton, S. K. (1997). Diet, anthropometry and breast cancer: Integration of experimental and epidemiologic approaches. *J. Nutr.* **127**, 916S–920S.
24. Ng, E. H., Gao, F., Ji, C. Y., Ho, G. H., and Soo, K. C. (1997). Risk factors for breast carcinoma in Singaporean Chinese women: The role of central obesity. *Cancer* **80**, 725–731.
25. Swanson, C. A., Jones, D. Y., Schatzkin, A., Brinton, L. A., and Ziegler, R. G. (1988). Breast cancer risk assessed by anthropometry in the NHANES I epidemiological follow-up study. *Cancer Res.* **48**, 5363–5367.
26. Ziegler, R. G. (1997). Anthropometry and breast cancer. *J. Nutr.* **127**, 924S–928S.
27. Baer, H. J., Rich-Edwards, J. W., Colditz, G. A., Hunter, D. J., Willett, W. C., and Michels, K. B. (2006). Adult height, age at attained height, and incidence of breast cancer in premenopausal women. *Int. J. Cancer* **119**, 2231–2235.
28. Adebamowo, C. A., Ogundiran, T. O., Adenipekun, A. A., *et al.* (2003). Obesity and height in urban Nigerian women with breast cancer. *Ann. Epidemiol.* **13**, 455–461.
29. Iwasaki, M., Otani, T., Inoue, M., Sasazuki, S., and Tsugane, S. (2007). Body size and risk for breast cancer in relation to estrogen and progesterone receptor status in Japan. *Ann. Epidemiol.* **17**, 304–312.
30. Okasha, M., McCarron, P., Gunnell, D., and Smith, G. D. (2003). Exposures in childhood, adolescence and early adulthood and breast cancer risk: A systematic review of the literature. *Breast Cancer Res. Treat.* **78**, 223–276.
31. Velie, E. M., Nechuta, S., and Osuch, JR. (2005). Lifetime reproductive and anthropometric risk factors for breast cancer in postmenopausal women. *Breast Dis.* **24**, 17–35.
32. Heng, D., Gao, F., Jong, R., *et al.* (2004). Risk factors for breast cancer associated with mammographic features in Singaporean chinese women. *Cancer Epidemiol. Biomarkers Prev.* **13**, 1751–1758.
33. van den Brandt, P. A., Spiegelman, D., Yaun, S. S., *et al.* (2000). Pooled analysis of prospective cohort studies on height, weight, and breast cancer risk. *Am. J. Epidemiol.* **152**, 514–527.
34. Ahlgren, M., Melbye, M., Wohlfahrt, J., and Sorensen, T. I. (2006). Growth patterns and the risk of breast cancer in women. *Int. J. Gynecol. Cancer* **16**(Suppl 2), 569–575.
35. De Stavola, B. L., dos Santos Silva, I., McCormack, V., Hardy, R. J., Kuh, D. J., and Wadsworth, M. E. (2004). Childhood growth and breast cancer. *Am. J. Epidemiol.* **159**, 671–682.
36. Lagiou, P., Samoli, E., Lagiou, A., Hsieh, C. C., Adami, H. O., and Trichopoulos, D. (2005). Maternal height, pregnancy estriol and birth weight in reference to breast cancer risk in Boston and Shanghai. *Int. J. Cancer* **117**, 494–498.
37. Vatten, L. J., Nilsen, T. I., Tretli, S., Trichopoulos, D., and Romundstad, P. R. (2005). Size at birth and risk of breast cancer: Prospective population-based study. *Int. J. Cancer* **114**, 461–464.
38. Lawlor, D. A., Okasha, M., Gunnell, D., Smith, G. D., and Ebrahim, S. (2003). Associations of adult measures of childhood growth with breast cancer: Findings from the British Women's Heart and Health Study. *Br. J. Cancer* **89**, 81–87.
39. Mondina, R., Borsellino, G., Poma, S., Baroni, M., Di Nubila, B., and Sacchi, P. (1992). Breast carcinoma and skeletal formation. *Eur. J. Cancer* **28A**, 1068–1070.
40. Rock, C. L., and Demark-Wahnefried, W. (2002). Can lifestyle modification increase survival in women diagnosed with breast cancer? *J. Nutr.* **132**, 3504S–3507S.
41. Jonsson, F., Wolk, A., Pedersen, N. L., *et al.* (2003). Obesity and hormone-dependent tumors: Cohort and co-twin control studies based on the Swedish Twin Registry. *Int. J. Cancer* **106**, 594–599.
42. Tehard, B., Lahmann, P. H., Riboli, E., and Clavel-Chapelon, F. (2004). Anthropometry, breast cancer and menopausal status: Use of repeated measurements over 10 years of follow-up results of the French E3N women's cohort study. *Int. J. Cancer* **111**, 264–269.
43. Chow, L. W., Lui, K. L., Chan, J. C., *et al.* (2005). Association between body mass index and risk of formation of breast cancer in Chinese women. *Asian J. Surg.* **28**, 179–184.
44. Cui, Y., Whiteman, M. K., Langenberg, P., *et al.* (2002). Can obesity explain the racial difference in stage of breast cancer at diagnosis between black and white women? *J. Women's Health Gend. Based Med.* **11**, 527–536.
45. Krebs, E. E., Taylor, B. C., Cauley, J. A., Stone, K. L., Bowman, P. J., and Ensrud, K. E. (2006). Measures of adiposity and risk of breast cancer in older postmenopausal women. *J. Am. Geriatr. Soc.* **54**, 63–69.
46. McCullough, M. L., Feigelson, H. S., Diver, W. R., Patel, A. V., Thun, M. J., and Calle, E. E. (2005). Risk factors for fatal breast cancer in African-American women and White women in a large U.S. prospective cohort. *Am. J. Epidemiol.* **162**, 734–742.
47. Sweeney, C., Blair, C. K., Anderson, K. E., Lazovich, D., and Folsom, A. R. (2004). Risk factors for breast cancer in elderly women. *Am. J. Epidemiol.* **160**, 868–875.
48. Carpenter, C. L., Ross, R. K., Paganini-Hill, A., and Bernstein, L. (2003). Effect of family history, obesity and exercise on

breast cancer risk among postmenopausal women. *Int. J. Cancer* **106**, 96–102.
49. Marsden, D. E., Friedlander, M., and Hacker, N. F. (2000). Current management of epithelial ovarian carcinoma: A review. *Semin. Surg. Oncol.* **19**, 11–19.
50. Morimoto, L. M., White, E., Chen, Z., *et al.* (2002). Obesity, body size, and risk of postmenopausal breast cancer: The Women's Health Initiative (United States). *Cancer Causes Control* **13**, 741–751.
51. Modugno, F., Kip, K. E., Cochrane, B., *et al.* (2006). Obesity, hormone therapy, estrogen metabolism and risk of postmenopausal breast cancer. *Int. J. Cancer* **118**, 1292–1301.
52. Lee, S. A., Lee, K. M., Park, W. Y., *et al.* (2005). Obesity and genetic polymorphism of ERCC2 and ERCC4 as modifiers of risk of breast cancer. *Exp. Mol. Med.* **37**, 86–90.
53. Wenten, M., Gilliland, F. D., Baumgartner, K., and Samet, J. M. (2002). Associations of weight, weight change, and body mass with breast cancer risk in Hispanic and non-Hispanic white women. *Ann. Epidemiol.* **12**, 435–434.
54. Colditz, G. A., Rosner, B. A., Chen, W. Y., Holmes, M. D., and Hankinson, S. E. (2004). Risk factors for breast cancer according to estrogen and progesterone receptor status. *J. Natl. Cancer Inst.* **96**, 218–228.
55. Suzuki, R., Rylander-Rudqvist, T., Ye, W., Saji, S., and Wolk, A. (2006). Body weight and postmenopausal breast cancer risk defined by estrogen and progesterone receptor status among Swedish women: A prospective cohort study. *Int. J. Cancer* **119**, 1683–1689.
56. Hsing, A. W., McLaughlin, J. K., Cocco, P., Co Chien, H. T., and Fraumeni, J. F. Jr. (1998). Risk factors for male breast cancer (United States). *Cancer Causes Control* **9**, 269–275.
57. Ewertz, M., Holmberg, L., Tretli, S., Pedersen, B. V., and Kristensen, A. (2001). Risk factors for male breast cancer—a case-control study from Scandinavia. *Acta Oncol.* **40**, 467–471.
58. Lorincz, A. M., and Sukumar, S. (2006). Molecular links between obesity and breast cancer. *Endocr. Relat. Cancer* **13**, 279–292.
59. McTiernan, A., Rajan, K. B., Tworoger, S. S., *et al.* (2003). Adiposity and sex hormones in postmenopausal breast cancer survivors. *J. Clin. Oncol.* **21**, 1961–1966.
60. Rinaldi, S., Key, T. J., Peeters, P. H., *et al.* (2006). Anthropometric measures, endogenous sex steroids and breast cancer risk in postmenopausal women: A study within the EPIC cohort. *Int. J. Cancer* **118**, 2832–2839.
61. Falk, R. T., Brinton, L. A., Madigan, M. P., *et al.* (2006). Interrelationships between serum leptin, IGF-1, IGFBP3, C-peptide and prolactin and breast cancer risk in young women. *Breast Cancer Res. Treat.* **98**, 157–165.
62. Stephenson, G. D., and Rose, D. P. (2003). Breast cancer and obesity: An update. *Nutr. Cancer* **45**, 1–16.
63. Schernhammer, E. S., Holly, J. M., Pollak, M. N., and Hankinson, S. E. (2005). Circulating levels of insulin-like growth factors, their binding proteins, and breast cancer risk. *Cancer Epidemiol. Biomarkers Prev.* **14**, 699–704.
64. Dizdar, O., and Alyamac, E. (2004). Obesity: An endocrine tumor? *Med. Hypotheses* **63**, 790–792.
65. Goodwin, P. J., Ennis, M., Fantus, I. G., *et al.* (2005). Is leptin a mediator of adverse prognostic effects of obesity in breast cancer? *J. Clin. Oncol.* **23**, 6037–6042.
66. Rose, D. P., Gilhooly, E. M., and Nixon, D. W. (2002). Adverse effects of obesity on breast cancer prognosis, and the biological actions of leptin (review). *Int. J. Oncol.* **21**, 1285–1292.
67. Rose, D. P., Komninou, D., and Stephenson, G. D. (2004). Obesity, adipocytokines, and insulin resistance in breast cancer. *Obes. Rev.* **5**, 153–165.
68. dos Santos Silva, I., De Stavola, B. L., Hardy, R. J., Kuh, D. J., McCormack, V. A., and Wadsworth, M. E. (2004). Is the association of birth weight with premenopausal breast cancer risk mediated through childhood growth? *Br. J. Cancer* **91**, 519–524.
69. Sanderson, M., Shu, X. O., Jin, F., *et al.* (2002). Weight at birth and adolescence and premenopausal breast cancer risk in a low-risk population. *Br. J. Cancer* **86**, 84–88.
70. Michels, K. B., Xue, F., Terry, K. L., and Willett, W. C. (2006). Longitudinal study of birthweight and the incidence of breast cancer in adulthood. *Carcinogenesis* **27**, 2464–2468.
71. Baer, H. J., Colditz, G. A., Rosner, B., *et al.* (2005). Body fatness during childhood and adolescence and incidence of breast cancer in premenopausal women: A prospective cohort study. *Breast Cancer Res.* **7**, R314–325.
72. Le Marchand, L., Kolonel, L. N., Earle, M. E., and Mi, M. P. (1988). Body size at different periods of life and breast cancer risk. *Am. J. Epidemiol.* **128**, 137–152.
73. Magnusson, C. M., and Roddam, A. W. (2005). Breast cancer and childhood anthropometry: Emerging hypotheses? *Breast Cancer Res.* **7**, 83.
74. Radimer, K., Siskind, V., Bain, C., and Schofield, F. (1993). Relation between anthropometric indicators and risk of breast cancer among Australian women. *Am. J. Epidemiol.* **138**, 77–89.
75. Weiderpass, E., Braaten, T., Magnusson, C., *et al.* (2004). A prospective study of body size in different periods of life and risk of premenopausal breast cancer. *Cancer Epidemiol. Biomarkers Prev.* **13**, 1121–1127.
76. Michels, K. B., Terry, K. L., and Willett, W. C. (2006). Longitudinal study on the role of body size in premenopausal breast cancer. *Arch. Intern. Med.* **166**, 2395–2402.
77. Apter, D. (1996). Hormonal events during female puberty in relation to breast cancer risk. *Eur. J. Cancer Prev* **5**, 476–482.
78. Stoll, B. A. (1997). Impaired ovulation and breast cancer risk. *Eur. J. Cancer* **33**, 1532–1535.
79. Bergstrom, A., Pisani, P., Tenet, V., Wolk, A., and Adami, H. O. (2001). Overweight as an avoidable cause of cancer in Europe. *Int. J. Cancer* **91**, 421–430.
80. Carroll, K. K. (1998). Obesity as a risk factor for certain types of cancer. *Lipids* **33**, 1055–1059.
81. Friedenreich, C. M., Courneya, K. S., and Bryant, H. E. (2002). Case-control study of anthropometric measures and breast cancer risk. *Int. J. Cancer* **99**, 445–452.
82. Harvie, M., Howell, A., Vierkant, R. A., *et al.* (2005). Association of gain and loss of weight before and after menopause with risk of postmenopausal breast cancer in the Iowa women's health study. *Cancer Epidemiol. Biomarkers Prev.* **14**, 656–661.
83. Kumar, N. B., Lyman, G. H., Allen, K., Cox, C. E., and Schapira, D. V. (1995). Timing of weight gain and breast cancer risk. *Cancer* **76**, 243–249.

84. Radimer, K. L., Ballard-Barbash, R., Miller, J. S., *et al.* (2004). Weight change and the risk of late-onset breast cancer in the original Framingham cohort. *Nutr. Cancer* **49**, 7–13.
85. Stoll, B. A. (1999). Western nutrition and the insulin resistance syndrome: A link to breast cancer. *Eur. J. Clin. Nutr.* **53**, 83–87.
86. Lahmann, P. H., Schulz, M., Hoffmann, K., *et al.* (2005). Long-term weight change and breast cancer risk: The European prospective investigation into cancer and nutrition (EPIC). *Br. J. Cancer* **93**, 582–589.
87. Slattery, M. L., Sweeney, C., Edwards, S., *et al.* (2007). Body size, weight change, fat distribution and breast cancer risk in Hispanic and non-Hispanic white women. *Breast Cancer Res. Treat.* **102**, 85–101.
88. Vihko, R., and Apter, D. (1989). Endogenous steroids in the pathophysiology of breast cancer. *Crit. Rev. Oncol. Hematol.* **9**, 1–16.
89. Kotsopoulos, J., Olopado, O. I., Ghadirian, P., *et al.* (2005). Changes in body weight and the risk of breast cancer in BRCA1 and BRCA2 mutation carriers. *Breast Cancer Res.* **7**, R833–843.
90. Parker, E. D., and Folsom, A. R. (2003). Intentional weight loss and incidence of obesity-related cancers: The Iowa Women's Health Study. *Int. J. Obes. Relat. Metab. Disord.* **27**, 1447–1452.
91. London, S. J., Colditz, G. A., Stampfer, M. J., Willett, W. C., Rosner, B., and Speizer, F. E. (1989). Prospective study of relative weight, height, and risk of breast cancer. *JAMA* **262**, 2853–2858.
92. Schapira, D. V., Clark, R. A., Wolff, P. A., Jarrett, A. R., Kumar, N. B., and Aziz, N. M. (1994). Visceral obesity and breast cancer risk. *Cancer* **74**, 632–639.
93. Sellers, T. A., Drinkard, C., Rich, S. S., *et al.* (1994). Familial aggregation and heritability of waist-to-hip ratio in adult women: the Iowa Women's Health Study. *Int. J. Obes. Relat. Metab. Disord.* **18**, 607–613.
94. Sonnenschein, E., Toniolo, P., Terry, M. B., *et al.* (1999). Body fat distribution and obesity in pre- and postmenopausal breast cancer. *Int. J. Epidemiol.* **28**, 1026–1031.
95. Harvie, M., Hooper, L., and Howell, A. H. (2003). Central obesity and breast cancer risk: A systematic review. *Obes. Rev.* **4**, 157–173.
96. Delort, L., Kwiatkowski, F., Chalabi, N., Satih, S., Bignon, Y. J., and Bernard-Gallon, D. J. (2007). Risk factors for early age at breast cancer onset—the "COSA program" population-based study. *Anticancer Res.* **27**, 1087–1094.
97. Tehard, B., and Clavel-Chapelon, F. (2006). Several anthropometric measurements and breast cancer risk: Results of the E3N cohort study. *Int. J. Obes. (Lond.)* **30**, 156–163.
98. Wu, A. H., Yu, M. C., Tseng, C. C., and Pike, M. C. (2007). Body size, hormone therapy and risk of breast cancer in Asian-American women. *Int. J. Cancer* **120**, 844–852.
99. Wu, M. H., Chou, Y. C., Yu, J. C., *et al.* (2006). Hormonal and body-size factors in relation to breast cancer risk: A prospective study of 11,889 women in a low-incidence area. *Ann. Epidemiol.* **16**, 223–229.
100. Harris, J. R., Lippman, M. E., Veronesi, U., and Willett, W. (1992). Breast cancer (1). *N. Engl. J. Med.* **327**, 319–328.
101. Kelsey, J. L., and Horn-Ross, P. L. (1993). Breast cancer: Magnitude of the problem and descriptive epidemiology. *Epidemiol. Rev.* **15**, 7–16.
102. Armstrong, B., and Doll, R. (1975). Environmental factors and cancer incidence and mortality in different countries, with special reference to dietary practices. *Int. J. Cancer* **15**, 617–631.
103. McMichael, A. J., and Giles, G. G. (1988). Cancer in migrants to Australia: Extending the descriptive epidemiological data. *Cancer Res.* **48**, 751–756.
104. Ziegler, R. G., Hoover, R. N., Pike, M. C., *et al.* (1993). Migration patterns and breast cancer risk in Asian-American women. *J. Natl. Cancer Inst.* **85**, 1819–1827.
105. Clavel-Chapelon, F., Niravong, M., and Joseph, R. R. (1997). Diet and breast cancer: Review of the epidemiologic literature. *Cancer Detect. Prev.* **21**, 426–440.
106. Greenwald, P. (1999). Role of dietary fat in the causation of breast cancer: Point. *Cancer Epidemiol. Biomarkers Prev.* **8**, 3–7.
107. Guthrie, N., and Carroll, K. K. (1999). Specific versus non-specific effects of dietary fat on carcinogenesis. *Prog. Lipid Res.* **38**, 261–271.
108. Hunter, D. J. (1999). Role of dietary fat in the causation of breast cancer: Counterpoint. *Cancer Epidemiol. Biomarkers Prev.* **8**, 9–13.
109. Hunter, D. J., and Willett, W. C. (1996). Nutrition and breast cancer. *Cancer Causes Control* **7**, 56–68.
110. Rose, D. P. (1997). Dietary fatty acids and cancer. *Am. J. Clin. Nutr.* **66**, 998S–1003S.
111. Wynder, E. L., Cohen, L. A., Muscat, J. E., Winters, B., Dwyer, J. T., and Blackburn, G. (1997). Breast cancer: Weighing the evidence for a promoting role of dietary fat. *J. Natl. Cancer Inst.* **89**, 766–775.
112. Hunter, D. J., Spiegelman, D., Adami, H. O., *et al.* (1996). Cohort studies of fat intake and the risk of breast cancer—a pooled analysis. *N. Engl. J. Med.* **334**, 356–361.
113. Holmes, M. D., Hunter, D. J., Colditz, G. A., *et al.* (1999). Association of dietary intake of fat and fatty acids with risk of breast cancer. *JAMA* **281**, 914–920.
114. Klurfeld, D. M., Welch, C. B., Davis, M. J., and Kritchevsky, D. (1989). Determination of degree of energy restriction necessary to reduce DMBA-induced mammary tumorigenesis in rats during the promotion phase. *J. Nutr.* **119**, 286–291.
115. Bartsch, H., Nair, J., and Owen, R. W. (1999). Dietary polyunsaturated fatty acids and cancers of the breast and colorectum: Emerging evidence for their role as risk modifiers. *Carcinogenesis* **20**, 2209–2218.
116. Wu, A. H., Pike, M. C., and Stram, D. O. (1999). Meta-analysis: Dietary fat intake, serum estrogen levels, and the risk of breast cancer. *J. Natl. Cancer Inst.* **91**, 529–534.
117. Gandini, S., Merzenich, H., Robertson, C., and Boyle, P. (2000). Meta-analysis of studies on breast cancer risk and diet: The role of fruit and vegetable consumption and the intake of associated micronutrients. *Eur. J. Cancer* **36**, 636–646.
118. World Cancer Research Fund, American Institute for Cancer Research. (1997). "Food, Nutrition and the Prevention of Cancer: A Global Perspective." American Institute for Cancer Research, Washington, DC.

119. Smith-Warner, S. A., Spiegelman, D., Yaun, S. S., *et al.* (2001). Intake of fruits and vegetables and risk of breast cancer: A pooled analysis of cohort studies. *JAMA* **285**, 769–776.
120. van Gils, C. H., and Peeters, P. H., Bueno-de-Mesquita, H. B. *et al.* (2005). Consumption of vegetables and fruits and risk of breast cancer. *JAMA* **293**, 183–193.
121. Ching, S., Ingram, D., Hahnel, R., Beilby, J., and Rossi, E. (2002). Serum levels of micronutrients, antioxidants and total antioxidant status predict risk of breast cancer in a case control study. *J. Nutr.* **132**, 303–306.
122. Potischman, N., McCulloch, C. E., Byers, T., *et al.* (1990). Breast cancer and dietary and plasma concentrations of carotenoids and vitamin A. *Am. J. Clin. Nutr.* **52**, 909–915.
123. Toniolo, P., Van Kappel, A. L., Akhmedkhanov, A., *et al.* (2001). Serum carotenoids and breast cancer. *Am. J. Epidemiol.* **153**, 1142–1147.
124. Yeum, K. J., Ahn, S. H., Rupp de Paiva, S. A., Lee-Kim, Y. C., Krinsky, N. I., and Russell, R. M. (1998). Correlation between carotenoid concentrations in serum and normal breast adipose tissue of women with benign breast tumor or breast cancer. *J. Nutr.* **128**, 1920–1926.
125. Zhang, S., Tang, G., Russell, R. M., *et al.* (1997). Measurement of retinoids and carotenoids in breast adipose tissue and a comparison of concentrations in breast cancer cases and control subjects. *Am. J. Clin. Nutr.* **66**, 626–632.
126. Dawson, M. I., Chao, W. R., Pine, P., *et al.* (1995). Correlation of retinoid binding affinity to retinoic acid receptor alpha with retinoid inhibition of growth of estrogen receptor-positive MCF-7 mammary carcinoma cells. *Cancer Res.* **55**, 4446–4451.
127. Prakash, P., Krinsky, N. I., and Russell, R. M. (2000). Retinoids, carotenoids, and human breast cancer cell cultures: A review of differential effects. *Nutr. Rev.* **58**, 170–176.
128. Rock, C. L., Kusluski, R. A., Galvez, M. M., and Ethier, S. P. (1995). Carotenoids induce morphological changes in human mammary epithelial cell cultures. *Nutr. Cancer* **23**, 319–333.
129. Sumantran, V. N., Zhang, R., Lee, D. S., and Wicha, M. S. (2000). Differential regulation of apoptosis in normal versus transformed mammary epithelium by lutein and retinoic acid. *Cancer Epidemiol. Biomarkers Prev.* **9**, 257–263.
130. Fowke, J. H., Longcope, C., and Hebert, J. R. (2000). Brassica vegetable consumption shifts estrogen metabolism in healthy postmenopausal women. *Cancer Epidemiol. Biomarkers Prev.* **9**, 773–779.
131. Smith-Warner, S. A., Spiegelman, D., Yaun, S. S., *et al.* (1998). Alcohol and breast cancer in women: A pooled analysis of cohort studies. *JAMA* **279**, 535–540.
132. Ginsburg, E. S. (1999). Estrogen, alcohol and breast cancer risk. *J. Steroid Biochem. Mol. Biol.* **69**, 299–306.
133. Rohan, T. E., Jain, M. G., Howe, G. R., and Miller, A. B. (2000). Dietary folate consumption and breast cancer risk. *J. Natl. Cancer Inst.* **92**, 266–269.
134. Sellers, T. A., Kushi, L. H., Cerhan, J. R., *et al.* (2001). Dietary folate intake, alcohol, and risk of breast cancer in a prospective study of postmenopausal women. *Epidemiology* **12**, 420–428.
135. Zhang, S., Hunter, D. J., Hankinson, S. E., *et al.* (1999). A prospective study of folate intake and the risk of breast cancer. *JAMA* **281**, 1632–1637.
136. Arts, C. J., Govers, C. A., van den Berg, H., Wolters, M. G., van Leeuwen, P., and Thijssen, J. H. (1991). In vitro binding of estrogens by dietary fiber and the in vivo apparent digestibility tested in pigs. *J. Steroid Biochem. Mol. Biol.* **38**, 621–628.
137. Dorgan, J. F., Sowell, A., Swanson, C. A., *et al.* (1998). Relationships of serum carotenoids, retinol, alpha-tocopherol, and selenium with breast cancer risk: results from a prospective study in Columbia, Missouri (United States). *Cancer Causes Control* **9**, 89–97.
138. Graham, S., Zielezny, M., Marshall, J., *et al.* (1992). Diet in the epidemiology of postmenopausal breast cancer in the New York State Cohort. *Am. J. Epidemiol.* **136**, 1327–1337.
139. Hunter, D. J., Manson, J. E., Colditz, G. A., *et al.* (1993). A prospective study of the intake of vitamins, C., E., and A and the risk of breast cancer. *N. Engl. J. Med.* **329**, 234–240.
140. Kushi, L. H., Fee, R. M., Sellers, T. A., Zheng, W., and Folsom, A. R. (1996). Intake of vitamins A, C, and E and postmenopausal breast cancer. The Iowa Women's Health Study. *Am. J. Epidemiol.* **144**, 165–174.
141. Stoll, B. A. (1998). Breast cancer and the Western diet: Role of fatty acids and antioxidant vitamins. *Eur. J. Cancer* **34**, 1852–1856.
142. van 't Veer, P., Strain, J. J., Fernandez-Crehuet, J., *et al.* (1996). Tissue antioxidants and postmenopausal breast cancer: The European Community Multicentre Study on Antioxidants, Myocardial Infarction, and Cancer of the Breast (EURAMIC). *Cancer Epidemiol. Biomarkers Prev.* **5**, 441–447.
143. Horner, N. K., and Lampe, J. W. (2000). Potential mechanisms of diet therapy for fibrocystic breast conditions show inadequate evidence of effectiveness. *J. Am. Diet. Assoc.* **100**, 1368–1380.
144. Ambrosone, C. B., Freudenheim, J. L., Sinha, R., *et al.* (1998). Breast cancer risk, meat consumption and *N*-acetyltransferase (NAT2) genetic polymorphisms. *Int. J. Cancer* **75**, 825–830.
145. Henderson, B. E., and Bernstein, L. (1991). The international variation in breast cancer rates: An epidemiological assessment. *Breast Cancer Res. Treat.* **18**(Suppl 1), S11–17.
146. Cui, X., Dai, Q., Tseng, M., Shu, X. O., Gao, Y. T., and Zheng, W. (2007). Dietary patterns and breast cancer risk in the Shanghai Breast Cancer Study. *Cancer Epidemiol. Biomarkers Prev.* **16**, 1443–1448.
147. Adlercreutz, C. H., Goldin, B. R., Gorbach, S. L., *et al.* (1995). Soybean phytoestrogen intake and cancer risk. *J. Nutr.* **125**, 757S–770S.
148. Kurzer, M. S. (2000). Hormonal effects of soy isoflavones: Studies in premenopausal and postmenopausal women. *J. Nutr.* **130**, 660S–661S.
149. Messina, M. J., and Loprinzi, C. L. (2001). Soy for breast cancer survivors: A critical review of the literature. *J. Nutr.* **131**, 3095S–3108S.
150. Prentice, R. L., Caan, B., Chlebowski, R. T., *et al.* (2006). Low-fat dietary pattern and risk of invasive breast cancer: The Women's Health Initiative Randomized Controlled Dietary Modification Trial. *JAMA* **295**, 629–642.
151. Boyd, N. F., Fishell, E., Jong, R., *et al.* (1995). Mammographic densities as a criterion for entry to a clinical trial of breast cancer prevention. *Br. J. Cancer* **72**, 476–479.

152. Jones, L. W., and Demark-Wahnefried, W. (2006). Diet, exercise, and complementary therapies after primary treatment for cancer. *Lancet Oncol.* **7**, 1017–1026.
153. Carmichael, A. R. (2006). Obesity and prognosis of breast cancer. *Obes. Rev.* **7**, 333–340.
154. Chlebowski, R. T., Aiello, E., and McTiernan, A. (2002). Weight loss in breast cancer patient management. *J. Clin. Oncol.* **20**, 1128–1143.
155. Demark-Wahnefried, W., and Rock, C. L. (2003). Nutrition-related issues for the breast cancer survivor. *Semin. Oncol.* **30**, 789–798.
156. Loi, S., Milne, R. L., Friedlander, M. L., *et al.* (2005). Obesity and outcomes in premenopausal and postmenopausal breast cancer. *Cancer Epidemiol. Biomarkers Prev.* **14**, 1686–1691.
157. Rock, C. L., Demark-Wahnefried W. (in press). Nutrition and survival after the diagnosis of breast cancer: A review of the evidence. *J. Clin. Oncol.*
158. Whiteman, M. K., Hillis, S. D., Curtis, K. M., McDonald, J. A., Wingo, P. A., and Marchbanks, PA. (2005). Body mass and mortality after breast cancer diagnosis. *Cancer Epidemiol. Biomarkers Prev.* **14**, 2009–2014.
159. Enger, S. M., Greif, J. M., Polikoff, J., and Press, M. (2004). Body weight correlates with mortality in early-stage breast cancer. *Arch. Surg.* **139**, 954–958, discussion 958–960.
160. Caan, B. J., Emond, J. A., Natarajan, L., *et al.* (2006). Postdiagnosis weight gain and breast cancer recurrence in women with early stage breast cancer. *Breast Cancer Res. Treat.* **99**, 47–57.
161. Enger, S. M., and Bernstein, L. (2004). Exercise activity, body size and premenopausal breast cancer survival. *Br. J. Cancer* **90**, 2138–2141.
162. Polednak, A. P. (2004). Racial differences in mortality from obesity-related chronic diseases in U.S. women diagnosed with breast cancer. *Ethn Dis* **14**, 463–468.
163. Dignam, J. J., Wieand, K., Johnson, K. A., Fisher, B., Xu, L., and Mamounas, E. P. (2003). Obesity, tamoxifen use, and outcomes in women with estrogen receptor-positive early-stage breast cancer. *J. Natl. Cancer Inst.* **95**, 1467–1476.
164. Dignam, J. J., Wieand, K., Johnson, K. A., *et al.* (2006). Effects of obesity and race on prognosis in lymph node-negative, estrogen receptor-negative breast cancer. *Breast Cancer Res. Treat.* **97**, 245–254.
165. Tao, M. H., Shu, X. O., Ruan, Z. X., Gao, Y. T., and Zheng, W. (2006). Association of overweight with breast cancer survival. *Am. J. Epidemiol.* **163**, 101–107.
166. Pierce, J. P., Stefanick, M. L., Flatt, S. W., *et al.* (2007). Greater survival after breast cancer in physically active women with high vegetable-fruit intake regardless of obesity. *J. Clin. Oncol.* **25**, 2345–2351.
167. Pinsolle, V., Grinfeder, C., Mathoulin-Pelissier, S., and Faucher, A. (2006). Complications analysis of 266 immediate breast reconstructions. *J. Plast. Reconstr. Aesthet. Surg.* **59**, 1017–1024.
168. Spear, S. L., Ducic, I., Cuoco, F., and Taylor, N. (2007). Effect of obesity on flap and donor-site complications in pedicled TRAM flap breast reconstruction. *Plast. Reconstr. Surg.* **119**, 788–795.
169. Paskett, E. D., Naughton, M. J., McCoy, T. P., Case, L. D., and Abbott, J. M. (2007). The epidemiology of arm and hand swelling in premenopausal breast cancer survivors. *Cancer Epidemiol. Biomarkers Prev.* **16**, 775–782.
170. Shaw, C., Mortimer, P., and Judd, P. A. (2007). Randomized controlled trial comparing a low-fat diet with a weight-reduction diet in breast cancer-related lymphedema. *Cancer* **109**, 1949–1956.
171. Decensi, A., Maisonneuve, P., Rotmensz, N., *et al.* (2005). Effect of tamoxifen on venous thromboembolic events in a breast cancer prevention trial. *Circulation* **111**, 650–656.
172. Abrahamson, P. E., Gammon, M. D., Lund, M. J., *et al.* (2006). General and abdominal obesity and survival among young women with breast cancer. *Cancer Epidemiol. Biomarkers Prev.* **15**, 1871–1877.
173. Borugian, M. J., Sheps, S. B., Kim-Sing, C., *et al.* (2003). Waist-to-hip ratio and breast cancer mortality. *Am. J. Epidemiol.* **158**, 963–968.
174. Kumar, N. B., Cantor, A., Allen, K., and Cox, C. E. (2000). Android obesity at diagnosis and breast carcinoma survival: Evaluation of the effects of anthropometric variables at diagnosis, including body composition and body fat distribution and weight gain during life span, and survival from breast carcinoma. *Cancer* **88**, 2751–2757.
175. Camoriano, J. K., Loprinzi, C. L., Ingle, J. N., Therneau, T. M., Krook, J. E., and Veeder, M. H. (1990). Weight change in women treated with adjuvant therapy or observed following mastectomy for node-positive breast cancer. *J. Clin. Oncol.* **8**, 1327–1334.
176. Ganz, P. A., Schag, C. C., Polinsky, M. L., Heinrich, R. L., and Flack, V. F. (1987). Rehabilitation needs and breast cancer: The first month after primary therapy. *Breast Cancer Res. Treat.* **10**, 243–253.
177. Goodwin, P. J., Ennis, M., Pritchard, K. I., *et al.* (1999). Adjuvant treatment and onset of menopause predict weight gain after breast cancer diagnosis. *J. Clin. Oncol.* **17**, 120–129.
178. Kornblith, A. B., Hollis, D. R., Zuckerman, E., *et al.* (1993). Effect of megestrol acetate on quality of life in a dose-response trial in women with advanced breast cancer. The Cancer and Leukemia Group B. *J. Clin. Oncol.* **11**, 2081–2089.
179. Monnin, S., Schiller, M. R., Sachs, L., and Smith, A. M. (1993). Nutritional concerns of women with breast cancer. *J. Cancer Educ.* **8**, 63–69.
180. Brown, B. W., Brauner, C., and Minnotte, M. C. (1993). Noncancer deaths in white adult cancer patients. *J. Natl. Cancer Inst.* **85**, 979–987.
181. Doyle, C., Kushi, L. H., Byers, T., *et al.* (2006). Nutrition and physical activity during and after cancer treatment: An American Cancer Society guide for informed choices. *CA Cancer J. Clin.* **56**, 323–353.
182. Li, F. P., and Stovall, E. L. (1998). Long-term survivors of cancer. *Cancer Epidemiol. Biomarkers Prev.* **7**, 269–270.
183. Kroenke, C. H., Chen, W. Y., Rosner, B., and Holmes, M. D. (2005). Weight, weight gain, and survival after breast cancer diagnosis. *J. Clin. Oncol.* **23**, 1370–1378.
184. Ingram, C., and Brown, J. K. (2004). Patterns of weight and body composition change in premenopausal women with early stage breast cancer: Has weight gain been overestimated? *Cancer Nurs.* **27**, 483–490.

185. Irwin, M. L., McTiernan, A., Bernstein, L., *et al.* (2005). Relationship of obesity and physical activity with C-peptide, leptin, and insulin-like growth factors in breast cancer survivors. *Cancer Epidemiol. Biomarkers Prev.* **14**, 2881–2888.
186. Makari-Judson, G., Judson, C. H., and Mertens, W. C. (2007). Longitudinal patterns of weight gain after breast cancer diagnosis: Observations beyond the first year. *Breast J.* **13**, 258–265.
187. de Waard, F., Ramlau, R., Mulders, Y., de Vries, T., and van Waveren, S. (1993). A feasibility study on weight reduction in obese postmenopausal breast cancer patients. *Eur. J. Cancer Prev,* **2**, 233–238.
188. Djuric, Z., DiLaura, N. M., Jenkins, I., *et al.* (2002). Combining weight-loss counseling with the Weight Watchers plan for obese breast cancer survivors. *Obes. Res.* **10**, 657–665.
189. Goodwin, P., Esplen, M. J., Butler, K., *et al.* (1998). Multidisciplinary weight management in locoregional breast cancer: Results of a phase II study. *Breast Cancer Res. Treat.* **48**, 53–64.
190. Loprinzi, C. L., Athmann, L. M., Kardinal, C. G., *et al.* (1996). Randomized trial of dietician counseling to try to prevent weight gain associated with breast cancer adjuvant chemotherapy. *Oncology* **53**, 228–232.
191. Demark-Wahnefried, W., Peterson, B. L., Winer, E. P., *et al.* (2001). Changes in weight, body composition, and factors influencing energy balance among premenopausal breast cancer patients receiving adjuvant chemotherapy. *J. Clin. Oncol.* **19**, 2381–2389.
192. Borugian, M. J., Sheps, S. B., Kim-Sing, C., *et al.* (2004). Insulin, macronutrient intake, and physical activity: Are potential indicators of insulin resistance associated with mortality from breast cancer? *Cancer Epidemiol. Biomarkers Prev.* **13**, 1163–1172.
193. Holm, L. E., Nordevang, E., Hjalmar, M. L., Lidbrink, E., Callmer, E., and Nilsson, B. (1993). Treatment failure and dietary habits in women with breast cancer. *J. Natl. Cancer Inst.* **85**, 32–36.
194. Jain, M., and Miller, AB. (1994). Pre-morbid body size and the prognosis of women with breast cancer. *Int. J. Cancer* **59**, 363–368.
195. McEligot, A. J., Largent, J., Ziogas, A., Peel, D., and Anton-Culver, H. (2006). Dietary fat, fiber, vegetable, and micronutrients are associated with overall survival in postmenopausal women diagnosed with breast cancer. *Nutr. Cancer* **55**, 132–140.
196. Rohan, T. E., Hiller, J. E., and McMichael, A. J. (1993). Dietary factors and survival from breast cancer. *Nutr. Cancer* **20**, 167–177.
197. Ewertz, M., Gillanders, S., Meyer, L., and Zedeler, K. (1991). Survival of breast cancer patients in relation to factors which affect the risk of developing breast cancer. *Int. J. Cancer* **49**, 526–530.
198. Kyogoku, S., Hirohata, T., Nomura, Y., Shigematsu, T., Takeshita, S., and Hirohata, I. (1992). Diet and prognosis of breast cancer. *Nutr. Cancer* **17**, 271–277.
199. Newman, S. C., Miller, A. B., and Howe, G. R. (1986). A study of the effect of weight and dietary fat on breast cancer survival time. *Am. J. Epidemiol.* **123**, 767–774.
200. Saxe, G. A., Rock, C. L., Wicha, M. S., and Schottenfeld, D. (1999). Diet and risk for breast cancer recurrence and survival. *Breast Cancer Res. Treat.* **53**, 241–253.
201. Goodwin, P. J., Ennis, M., Pritchard, K. I., Koo, J., Trudeau, M. E., and Hood, N. (2003). Diet and breast cancer: Evidence that extremes in diet are associated with poor survival. *J. Clin. Oncol.* **21**, 2500–2507.
202. Hebert, J. R., Hurley, T. G., and Ma, Y. (1998). The effect of dietary exposures on recurrence and mortality in early stage breast cancer. *Breast Cancer Res. Treat.* **51**, 17–28.
203. Holmes, M. D., Stampfer, M. J., Colditz, G. A., Rosner, B., Hunter, D. J., and Willett, W. C. (1999). Dietary factors and the survival of women with breast carcinoma. *Cancer* **86**, 826–835.
204. Ingram, D. (1994). Diet and subsequent survival in women with breast cancer. *Br. J. Cancer* **69**, 592–595.
205. Fink, B. N., Gaudet, M. M., Britton, J. A., *et al.* (2006). Fruits, vegetables, and micronutrient intake in relation to breast cancer survival. *Breast Cancer Res. Treat.* **98**, 199–208.
206. Rock, C. L., Flatt, S. W., Natarajan, L., *et al.* (2005). Plasma carotenoids and recurrence-free survival in women with a history of breast cancer. *J. Clin. Oncol.* **23**, 6631–6638.
207. Kroenke, C. H., Fung, T. T., Hu, F. B., and Holmes, M. D. (2005). Dietary patterns and survival after breast cancer diagnosis. *J. Clin. Oncol.* **23**, 9295–9303.
208. Chlebowski, R. T., Blackburn, G. L., Buzzard, I. M., *et al.* (1993). Adherence to a dietary fat intake reduction program in postmenopausal women receiving therapy for early breast cancer. The Women's Intervention Nutrition Study. *J. Clin. Oncol.* **11**, 2072–2080.
209. Chlebowski, R. T., Blackburn, G. L., Thomson, C. A., *et al.* (2006). Dietary fat reduction and breast cancer outcome: Interim efficacy results from the Women's Intervention Nutrition Study. *J. Natl. Cancer Inst.* **98**, 1767–1776.
210. Pierce, J. P., Faerber, S., Wright, F. A., *et al.* (1997). Feasibility of a randomized trial of a high-vegetable diet to prevent breast cancer recurrence. *Nutr. Cancer* **28**, 282–288.
211. Rock, C. L., Flatt, S. W., Wright, F. A., *et al.* (1997). Responsiveness of carotenoids to a high vegetable diet intervention designed to prevent breast cancer recurrence. *Cancer Epidemiol. Biomarkers Prev.* **6**, 617–623.
212. Rock, C. L., Thomson, C., Caan, B. J., *et al.* (2001). Reduction in fat intake is not associated with weight loss in most women after breast cancer diagnosis: Evidence from a randomized controlled trial. *Cancer* **91**, 25–34.
213. Hale, G. E., Hughes, C. L., and Cline, J. M. (2002). Endometrial cancer: Hormonal factors, the perimenopausal "window of risk," and isoflavones. *J. Clin. Endocrinol. Metab.* **87**, 3–15.
214. Kaaks, R., Lukanova, A., and Kurzer, M. S. (2002). Obesity, endogenous hormones, and endometrial cancer risk: A synthetic review. *Cancer Epidemiol. Biomarkers Prev.* **11**, 1531–1543.
215. Affenito, S., Lambert-Lagace, L., Kerstetter, J., and Demark-Wahnefried, W. (2004). Nutrition and women's health: Position of the American Dietetic Association and Dietitians of Canada. *Can. J. Diet. Pract. Res.* **65**, 85–89.
216. Calle, E. E., Rodriguez, C., Walker-Thurmond, K., and Thun, M. J. (2003). Overweight, obesity, and mortality

from cancer in a prospectively studied cohort of U.S. adults. *N. Engl. J. Med.* **348**, 1625–1638.

217. Modesitt, S. C., and van Nagell, J. R. Jr. (2005). The impact of obesity on the incidence and treatment of gynecologic cancers: A review. *Obstet. Gynecol. Surv* **60**, 683–692.
218. Anderson, K. E., Anderson, E., Mink, P. J., *et al.* (2001). Diabetes and endometrial cancer in the Iowa women's health study. *Cancer Epidemiol. Biomarkers Prev.* **10**, 611–616.
219. Chang, S. C., Lacey, J. V., Jr., Brinton, L. A., *et al.* (2007). Lifetime weight history and endometrial cancer risk by type of menopausal hormone use in the NIH-AARP Diet and Health Study. *Cancer Epidemiol. Biomarkers Prev.* **16**, 723–730.
220. Friedenreich, C., Cust, A., Lahmann, P. H., *et al.* (2007). Anthropometric factors and risk of endometrial cancer: The European prospective investigation into cancer and nutrition. *Cancer Causes Control* **18**, 399–413.
221. Furberg, A. S., and Thune, I. (2003). Metabolic abnormalities (hypertension, hyperglycemia and overweight), lifestyle (high energy intake and physical inactivity) and endometrial cancer risk in a Norwegian cohort. *Int. J. Cancer* **104**, 669–676.
222. Iatrakis, G., Zervoudis, S., Saviolakis, A., *et al.* (2006). Women younger than 50 years with endometrial cancer. *Eur. J. Gynaecol. Oncol.* **27**, 399–400.
223. Jain, M. G., Rohan, T. E., Howe, G. R., and Miller, A. B. (2000). A cohort study of nutritional factors and endometrial cancer. *Eur. J. Epidemiol.* **16**, 899–905.
224. Levi, F., La Vecchia, C., Negri, E., Parazzini, F., and Franceschi, S. (1992). Body mass at different ages and subsequent endometrial cancer risk. *Int. J. Cancer* **50**, 567–571.
225. McCann, S. E., Freudenheim, J. L., Marshall, J. R., Brasure, J. R., Swanson, M. K., and Graham, S. (2000). Diet in the epidemiology of endometrial cancer in western New York (United States). *Cancer Causes Control* **11**, 965–974.
226. Salazar-Martinez, E., Lazcano-Ponce, E. C., Lira-Lira, G. G., *et al.* (2000). Case-control study of diabetes, obesity, physical activity and risk of endometrial cancer among Mexican women. *Cancer Causes Control* **11**, 707–711.
227. Shields, T. S., Weiss, N. S., Voigt, L. F., and Beresford, S. A. (1999). The additional risk of endometrial cancer associated with unopposed estrogen use in women with other risk factors. *Epidemiology* **10**, 733–738.
228. Shoff, S. M., and Newcomb, P. A. (1998). Diabetes, body size, and risk of endometrial cancer. *Am. J. Epidemiol.* **148**, 234–240.
229. Soliman, P. T., Oh, J. C., Schmeler, K. M., *et al.* (2005). Risk factors for young premenopausal women with endometrial cancer. *Obstet. Gynecol.* **105**, 575–580.
230. Weiderpass, E., Persson, I., Adami, H. O., Magnusson, C., Lindgren, A., and Baron, J. A. (2000). Body size in different periods of life, diabetes mellitus, hypertension, and risk of postmenopausal endometrial cancer (Sweden). *Cancer Causes Control* **11**, 185–192.
231. Weiss, J. M., Saltzman, B. S., Doherty, J. A., *et al.* (2006). Risk factors for the incidence of endometrial cancer according to the aggressiveness of disease. *Am. J. Epidemiol.* **164**, 56–62.
232. Xu, W., Dai, Q., Ruan, Z., Cheng, J., Jin, F., and Shu, X. (2002). Obesity at different ages and endometrial cancer risk factors in urban Shanghai, China. *Zhonghua Liu Xing Bing Xue Za Zhi* **23**, 347–351.
233. Yamazawa, K., Matsui, H., Seki, K., and Sekiya, S. (2003). A case-control study of endometrial cancer after antipsychotics exposure in premenopausal women. *Oncology* **64**, 116–123.
234. Friberg, E., Mantzoros, C. S., and Wolk, A. (2007). Diabetes and risk of endometrial cancer: A population-based prospective cohort study. *Cancer Epidemiol. Biomarkers Prev.* **16**, 276–280.
235. Papanas, N., Giatromanolaki, A., Galazios, G., Maltezos, E., and Sivridis, E. (2006). Endometrial carcinoma and diabetes revisited. *Eur. J. Gynaecol. Oncol.* **27**, 505–508.
236. Schouten, L. J., Goldbohm, R. A., and van den Brandt, P. A. (2004). Anthropometry, physical activity, and endometrial cancer risk: Results from the Netherlands Cohort Study. *J. Natl. Cancer Inst.* **96**, 1635–1638.
237. Cust, A. E., Kaaks, R., Friedenreich, C., *et al.* (2007). Plasma adiponectin levels and endometrial cancer risk in pre- and postmenopausal women. *J. Clin. Endocrinol. Metab.* **92**, 255–263.
238. Petridou, E., Belechri, M., Dessypris, N., *et al.* (2002). Leptin and body mass index in relation to endometrial cancer risk. *Ann Nutr Metab* **46**, 147–151.
239. Sharma, D., Saxena, N. K., Vertino, P. M., and Anania, F. A. (2006). Leptin promotes the proliferative response and invasiveness in human endometrial cancer cells by activating multiple signal-transduction pathways. *Endocr. Relat. Cancer* **13**, 629–640.
240. Austin, H., Drews, C., and Partridge, E. E. (1993). A case-control study of endometrial cancer in relation to cigarette smoking, serum estrogen levels, and alcohol use. *Am. J. Obstet. Gynecol.* **169**, 1086–1091.
241. Elliott, E. A., Matanoski, G. M., Rosenshein, N. B., Grumbine, F. C., and Diamond, E. L. (1990). Body fat patterning in women with endometrial cancer. *Gynecol. Oncol.* **39**, 253–258.
242. Le Marchand, L., Wilkens, L. R., and Mi, M. P. (1991). Early-age body size, adult weight gain and endometrial cancer risk. *Int. J. Cancer* **48**, 807–811.
243. Tornberg, S. A., and Carstensen, J. M. (1994). Relationship between Quetelet's index and cancer of breast and female genital tract in 47,000 women followed for 25 years. *Br. J. Cancer* **69**, 358–361.
244. de Waard, F., de Ridder, C. M., Baanders-van Halewyn, E. A., and Slotboom, B. J. (1996). Endometrial cancer in a cohort screened for breast cancer. *Eur. J. Cancer Prev.* **5**, 99–104.
245. Folsom, A. R., Kaye, S. A., Potter, J. D., and Prineas, R. J. (1989). Association of incident carcinoma of the endometrium with body weight and fat distribution in older women: Early findings of the Iowa Women's Health Study. *Cancer Res.* **49**, 6828–6831.
246. Goodman, M. T., Hankin, J. H., Wilkens, L. R., *et al.* (1997). Diet, body size, physical activity, and the risk of endometrial cancer. *Cancer Res.* **57**, 5077–5085.
247. Lapidus, L., Helgesson, O., Merck, C., and Bjorntorp, P. (1988). Adipose tissue distribution and female carcinomas.

A 12–year follow-up of participants in the population study of women in Gothenburg, Sweden. *Int. J. Obes.* **12**, 361–368.

248. Schapira, D. V., Kumar, N. B., Lyman, G. H., Cavanagh, D., Roberts, W. S., and LaPolla, J. (1991). Upper-body fat distribution and endometrial cancer risk. *JAMA* **266**, 1808–1811.
249. Shu, X. O., Brinton, L. A., Zheng, W., *et al.* (1992). Relation of obesity and body fat distribution to endometrial cancer in Shanghai, China. *Cancer Res.* **52**, 3865–3870.
250. Swanson, C. A., Potischman, N., Wilbanks, G. D., *et al.* (1993). Relation of endometrial cancer risk to past and contemporary body size and body fat distribution. *Cancer Epidemiol. Biomarkers Prev.* **2**, 321–327.
251. Barbone, F., Austin, H., and Partridge, E. E. (1993). Diet and endometrial cancer: A case-control study. *Am. J. Epidemiol.* **137**, 393–403.
252. La Vecchia, C., Decarli, A., Fasoli, M., and Gentile, A. (1986). Nutrition and diet in the etiology of endometrial cancer. *Cancer* **57**, 1248–1253.
253. Levi, F., Franceschi, S., Negri, E., and La Vecchia, C. (1993). Dietary factors and the risk of endometrial cancer. *Cancer* **71**, 3575–3581.
254. Negri, E., La Vecchia, C., Franceschi, S., Levi, F., and Parazzini, F. (1996). Intake of selected micronutrients and the risk of endometrial carcinoma. *Cancer* **77**, 917–923.
255. Petridou, E., Kedikoglou, S., Koukoulomatis, P., Dessypris, N., and Trichopoulos, D. (2002). Diet in relation to endometrial cancer risk: A case-control study in Greece. *Nutr. Cancer* **44**, 16–22.
256. Salazar-Martinez, E., Lazcano-Ponce, E., Sanchez-Zamorano, L. M., Gonzalez-Lira, G., Escudero, D. E. L. R. P., and Hernandez-Avila, M. (2005). Dietary factors and endometrial cancer risk. Results of a case-control study in Mexico. *Int. J. Gynecol. Cancer* **15**, 938–945.
257. Villani, C., Pucci, G., Pietrangeli, D., Pace, S., and Tomao, S. (1986). Role of diet in endometrial cancer patients. *Eur. J. Gynaecol. Oncol.* **7**, 139–143.
258. Horn-Ross, P. L., John, E. M., Canchola, A. J., Stewart, S. L., and Lee, M. M. (2003). Phytoestrogen intake and endometrial cancer risk. *J. Natl. Cancer Inst.* **95**, 1158–1164.
259. Obermair, A., Kurz, C., Hanzal, E., *et al.* (1995). The influence of obesity on the disease-free survival in primary breast cancer. *Anticancer Res.* **15**, 2265–2269.
260. Potischman, N., Swanson, C. A., Brinton, L. A., *et al.* (1993). Dietary associations in a case-control study of endometrial cancer. *Cancer Causes Control* **4**, 239–250.
261. Shu, X. O., Zheng, W., Potischman, N., *et al.* (1993). A population-based case-control study of dietary factors and endometrial cancer in Shanghai, People's Republic of China. *Am. J. Epidemiol.* **137**, 155–165.
262. Tao, M. H., Xu, W. H., Zheng, W., *et al.* (2005). A case-control study in Shanghai of fruit and vegetable intake and endometrial cancer. *Br. J. Cancer* **92**, 2059–2064.
263. Terry, P., Vainio, H., Wolk, A., and Weiderpass, E. (2002). Dietary factors in relation to endometrial cancer: A nationwide case-control study in Sweden. *Nutr. Cancer* **42**, 25–32.
264. Xu, W. H., Dai, Q., Xiang, Y. B., *et al.* (2006). Animal food intake and cooking methods in relation to endometrial cancer risk in Shanghai. *Br. J. Cancer* **95**, 1586–1592.
265. Folsom, A. R., Demissie, Z., and Harnack, L. (2003). Glycemic index, glycemic load, and incidence of endometrial cancer: The Iowa Women's Health Study. *Nutr. Cancer* **46**, 119–124.
266. Hirose, K., Tajima, K., Hamajima, N., *et al.* (1996). Subsite (cervix/endometrium)-specific risk and protective factors in uterus cancer. *Jpn. J. Cancer Res.* **87**, 1001–1009.
267. Larsson, S. C., Friberg, E., and Wolk, A. (2007). Carbohydrate intake, glycemic index and glycemic load in relation to risk of endometrial cancer: A prospective study of Swedish women. *Int. J. Cancer* **120**, 1103–1107.
268. Silvera, S. A., Rohan, T. E., Jain, M., Terry, P. D., Howe, G. R., and Miller, A. B. (2005). Glycaemic index, glycaemic load and risk of endometrial cancer: A prospective cohort study. *Public Health Nutr.* **8**, 912–919.
269. Terry, P., Baron, J. A., Weiderpass, E., Yuen, J., Lichtenstein, P., and Nyren, O. (1999). Lifestyle and endometrial cancer risk: A cohort study from the Swedish Twin Registry. *Int. J. Cancer* **82**, 38–42.
270. Zheng, W., Kushi, L. H., Potter, J. D., *et al.* (1995). Dietary intake of energy and animal foods and endometrial cancer incidence. The Iowa Women's Health Study. *Am. J. Epidemiol.* **142**, 388–394.
271. Xu, W. H., Dai, Q., Xiang, Y. B., *et al.* (2007). Nutritional factors in relation to endometrial cancer: A report from a population-based case-control study in Shanghai, China. *Int. J. Cancer* **120**, 1776–1781.
272. Boyle, P., Maisonneuve, P., and Autier, P. (2000). Update on cancer control in women. *Int. J. Gynaecol. Obstet.* **70**, 263–303.
273. Terry, P., Wolk, A., Vainio, H., and Weiderpass, E. (2002). Fatty fish consumption lowers the risk of endometrial cancer: A nationwide case-control study in Sweden. *Cancer Epidemiol. Biomarkers Prev.* **11**, 143–145.
274. Bandera, E. V., Kushi, L. H., Olson, S. H., Chen, W. Y., and Muti, P. (2003). Alcohol consumption and endometrial cancer: Some unresolved issues. *Nutr. Cancer* **45**, 24–29.
275. Weiderpass, E., and Baron, J. A. (2001). Cigarette smoking, alcohol consumption, and endometrial cancer risk: A population-based study in Sweden. *Cancer Causes Control* **12**, 239–247.
276. Eltabbakh, G. H., Shamonki, M. I., Moody, J. M., and Garafano, L. L. (2000). Hysterectomy for obese women with endometrial cancer: Laparoscopy or laparotomy? *Gynecol. Oncol.* **78**, 329–335.
277. Holub, Z., Bartos, P., Jabor, A., Eim, J., Fischlova, D., and Kliment, L. (2000). Laparoscopic surgery in obese women with endometrial cancer. *J. Am. Assoc. Gynecol. Laparosc.* **7**, 83–88.
278. Scribner, D. R., Jr., Walker, J. L., Johnson, G. A., McMeekin, D. S., Gold, M. A., and Mannel, R. S. (2002). Laparoscopic pelvic and paraaortic lymph node dissection in the obese. *Gynecol. Oncol.* **84**, 426–430.
279. Thomadsen, B. R., Paliwal, B. R., Petereit, D. G., and Ranallo, F. N. (2000). Radiation injury from x-ray exposure during brachytherapy localization. *Med, Phys,* **27**, 1681–1684.
280. Everett, E., Tamimi, H., Greer, B., *et al.* (2003). The effect of body mass index on clinical/pathologic features, surgical

morbidity, and outcome in patients with endometrial cancer. *Gynecol. Oncol.* **90**, 150–157.
281. von Gruenigen, V. E., Tian, C., Frasure, H., Waggoner, S., Keys, H., and Barakat, R. R. (2006). Treatment effects, disease recurrence, and survival in obese women with early endometrial carcinoma: A Gynecologic Oncology Group study. *Cancer* **107**, 2786–2791.
282. Anderson, B., Connor, J. P., Andrews, J. I., *et al.* (1996). Obesity and prognosis in endometrial cancer. *Am. J. Obstet. Gynecol.* **174**, 1171–1178, discussion 1178–1179.
283. Chia, V. M., Newcomb, P. A., Trentham-Dietz, A., and Hampton, J. M. (2007). Obesity, diabetes, and other factors in relation to survival after endometrial cancer diagnosis. *Int. J. Gynecol. Cancer* **17**, 441–446.
284. Studzijnski, Z., and Zajewski, W. (2003). Factors affecting the survival of 121 patients treated for endometrial carcinoma at a Polish hospital. *Arch. Gynecol. Obstet.* **267**, 145–147.
285. Courneya, K. S., Karvinen, K. H., Campbell, K. L., *et al.* (2005). Associations among exercise, body weight, and quality of life in a population-based sample of endometrial cancer survivors. *Gynecol. Oncol.* **97**, 422–430.
286. von Gruenigen, V. E., Gil, K. M., Frasure, H. E., Jenison, E. L., and Hopkins, M. P. (2005). The impact of obesity and age on quality of life in gynecologic surgery. *Am. J. Obstet. Gynecol.* **193**, 1369–1375.
287. Edmondson, R. J., and Monaghan, J. M. (2001). The epidemiology of ovarian cancer. *Int. J. Gynecol. Cancer* **11**, 423–429.
288. Runnebaum, I. B., and Stickeler, E. (2001). Epidemiological and molecular aspects of ovarian cancer risk. *J. Cancer Res. Clin. Oncol.* **127**, 73–79.
289. Risch, H. A. (1998). Hormonal etiology of epithelial ovarian cancer, with a hypothesis concerning the role of androgens and progesterone. *J. Natl. Cancer Inst.* **90**, 1774–1786.
290. Narod, S. A., and Boyd, J. (2002). Current understanding of the epidemiology and clinical implications of BRCA1 and BRCA2 mutations for ovarian cancer. *Curr. Opin. Obstet. Gynecol.* **14**, 19–26.
291. Fleischauer, A. T., Olson, S. H., Mignone, L., Simonsen, N., Caputo, T. A., and Harlap, S. (2001). Dietary antioxidants, supplements, and risk of epithelial ovarian cancer. *Nutr. Cancer* **40**, 92–98.
292. Parazzini, F., Moroni, S., La Vecchia, C., Negri, E., dal Pino, D., and Bolis, G. (1997). Ovarian cancer risk and history of selected medical conditions linked with female hormones. *Eur. J. Cancer* **33**, 1634–1637.
293. Lukanova, A., Toniolo, P., Lundin, E., *et al.* (2002). Body mass index in relation to ovarian cancer: A multi-centre nested case-control study. *Int. J. Cancer* **99**, 603–608.
294. Farrow, D. C., Weiss, N. S., Lyon, J. L., and Daling, J. R. (1989). Association of obesity and ovarian cancer in a case-control study. *Am. J. Epidemiol.* **129**, 1300–1304.
295. Hartge, P., Schiffman, M. H., Hoover, R., McGowan, L., Lesher, L., and Norris, H. J. (1989). A case-control study of epithelial ovarian cancer. *Am. J. Obstet. Gynecol.* **161**, 10–16.
296. Mori, M., Nishida, T., Sugiyama, T., *et al.* (1998). Anthropometric and other risk factors for ovarian cancer in a case-control study. *Jpn. J. Cancer Res.* **89**, 246–253.
297. Mink, P. J., Folsom, A. R., Sellers, T. A., and Kushi, L. H. (1996). Physical activity, waist-to-hip ratio, and other risk factors for ovarian cancer: A follow-up study of older women. *Epidemiology* **7**, 38–45.
298. Huncharek, M., and Kupelnick, B. (2001). Dietary fat intake and risk of epithelial ovarian cancer: A meta-analysis of 6,689 subjects from 8 observational studies. *Nutr. Cancer* **40**, 87–91.
299. Bosetti, C., Negri, E., Franceschi, S., *et al.* (2001). Diet and ovarian cancer risk: A case-control study in Italy. *Int. J. Cancer* **93**, 911–915.
300. Fernandez, E., Chatenoud, L., La Vecchia, C., Negri, E., and Franceschi, S. (1999). Fish consumption and cancer risk. *Am. J. Clin. Nutr.* **70**, 85–90.
301. Tzonou, A., Hsieh, C. C., Polychronopoulou, A., *et al.* (1993). Diet and ovarian cancer: A case-control study in Greece. *Int. J. Cancer* **55**, 411–414.
302. Engle, A., Muscat, J. E., and Harris, R. E. (1991). Nutritional risk factors and ovarian cancer. *Nutr. Cancer* **15**, 239–247.
303. Kushi, L. H., Mink, P. J., Folsom, A. R., *et al.* (1999). Prospective study of diet and ovarian cancer. *Am. J. Epidemiol.* **149**, 21–31.
304. La Vecchia, C., Decarli, A., Negri, E., *et al.* (1987). Dietary factors and the risk of epithelial ovarian cancer. *J. Natl. Cancer Inst.* **79**, 663–669.
305. Shu, X. O., Gao, Y. T., Yuan, J. M., Ziegler, R. G., and Brinton, LA. (1989). Dietary factors and epithelial ovarian cancer. *Br. J. Cancer* **59**, 92–96.
306. Parazzini, F., Chatenoud, L., Chiantera, V., Benzi, G., Surace, M., and La Vecchia, C. (2000). Population attributable risk for ovarian cancer. *Eur. J. Cancer* **36**, 520–524.
307. Fairfield, K. M., Hankinson, S. E., Rosner, B. A., Hunter, D. J., Colditz, G. A., and Willett, W. C. (2001). Risk of ovarian carcinoma and consumption of vitamins A, C, and E and specific carotenoids: A prospective analysis. *Cancer* **92**, 2318–2326.
308. Byers, T., Marshall, J., Graham, S., Mettlin, C., and Swanson, M. (1983). A case-control study of dietary and nondietary factors in ovarian cancer. *J. Natl. Cancer Inst.* **71**, 681–686.
309 Cramer, D. W., Kuper, H., Harlow, B. L., and Titus-Ernstoff, L. (2001). Carotenoids, antioxidants and ovarian cancer risk in pre- and postmenopausal women. *Int. J. Cancer* **94**, 128–134.
310. Slattery, M. L., Schuman, K. L., West, D. W., French, T. K., and Robison, L. M. (1989). Nutrient intake and ovarian cancer. *Am. J. Epidemiol.* **130**, 497–502.
311. Helzlsouer, K. J., Alberg, A. J., Norkus, E. P., Morris, J. S., Hoffman, S. C., and Comstock, G. W. (1996). Prospective study of serum micronutrients and ovarian cancer. *J. Natl. Cancer Inst.* **88**, 32–37.
312. Garland, M., Morris, J. S., Stampfer, M. J., *et al.* (1995). Prospective study of toenail selenium levels and cancer among women. *J. Natl. Cancer Inst.* **87**, 497–505.
313. Cramer, D W., Greenberg, E. R., Titus-Ernstoff, L., *et al.* (2000). A case-control study of galactose consumption and metabolism in relation to ovarian cancer. *Cancer Epidemiol. Biomarkers Prev.* **9**, 95–101.

314. Mettlin, C. J., and Piver, M. S. (1990). A case-control study of milk-drinking and ovarian cancer risk. *Am. J. Epidemiol.* **132**, 871–876.
315. Webb, P. M., Bain, C. J., Purdie, D. M., Harvey, P. W., and Green, A. (1998). Milk consumption, galactose metabolism and ovarian cancer (Australia). *Cancer Causes Control* **9**, 637–644.
316. Kushi, L. H., Byers, T., Doyle, C., *et al.* (2006). American Cancer Society Guidelines on Nutrition and Physical Activity for cancer prevention: Reducing the risk of cancer with healthy food choices and physical activity. *CA Cancer J. Clin.* **56**, 254–281.
317. Wolk, A., Manson, J. E., Stampfer, M. J., *et al.* (1999). Long-term intake of dietary fiber and decreased risk of coronary heart disease among women. *JAMA* **281**, 1998–2004.
318. Kant, A. K., Schatzkin, A., Graubard, B. I., and Schairer, C. (2000). A prospective study of diet quality and mortality in women. *JAMA* **283**, 2109–2115.

CHAPTER 36

Nutrition and Prostate Cancer

LAURENCE N. KOLONEL AND SONG-YI PARK
University of Hawaii, Honolulu, Hawaii

Contents

I. INTRODUCTION

In this chapter, we discuss the epidemiologic evidence for associations of dietary factors with prostate cancer risk, and the potential for diet to play a role in prostate cancer prevention. We begin with some general background on the disease and its diagnosis, followed by a description of incidence patterns and risk factors for prostate cancer other than diet. The relationship of nutrition to prostate cancer includes foods and dietary constituents that have been associated with increased risk of the disease, as well as those that have been associated with decreased risk. Some findings from animal and *in vitro* studies, as well as possible mechanisms for the carcinogenic effects, are presented in support of the epidemiologic findings. We conclude with a brief review of the genetics of prostate cancer and of ongoing nutritional intervention trials.

A. Normal Prostate Anatomy and Function

The normal adult prostate gland is a walnut-sized organ that surrounds the urethra and the neck of the bladder. The gland is composed of three distinct zones: peripheral, central, and transition. The peripheral zone is comprised of left and right lobes that can be palpated during digital rectal examination. The transition zone is the region that enlarges in benign prostatic hyperplasia, which is common in older men [1]. The prostate gland is a male secondary sex organ that secretes one fluid component of semen. Prostatic fluid is essential for male fertility.

Normal growth and activity of the prostate gland are under the control of androgenic hormones. Circulating testosterone, primarily produced in the testes, diffuses into the prostate where it is irreversibly converted by the enzyme steroid 5α-reductase type II to dihydrotestosterone (DHT), a metabolically more active form of the hormone. Dihydrotestosterone binds to the androgen receptor, and this complex then translocates to the cell nucleus where it activates selected genes [2].

B. Pathology and Diagnosis of Prostate Cancer

Almost all prostate tumors are classified as adenocarcinomas (i.e., they arise from the glandular epithelial cells) and occur most commonly in the peripheral zone of the gland. Accordingly, they can often be felt by the physician during digital rectal examination. A unique feature of human prostate cancer is the high frequency of small, latent tumors in older men. A clear relationship between these occult tumors and those that become clinically apparent has not been established, although it is commonly assumed that the latter evolve from the former as a consequence of additional genetic mutations.

Generally, prostate cancer in its early stages is asymptomatic. Enlargement of the prostate gland (benign prostatic hyperplasia or BPH) commonly begins after the age of 45, ultimately leading to urinary tract symptoms (difficult and frequent urination). Many cases of prostate cancer are diagnosed as a result of digital rectal examination performed when a man visits his physician for relief of these symptoms. (Suspicious lesions on examination may be confirmed by transrectal ultrasound, followed by a biopsy of the gland.) In recent years, the prostate-specific antigen (PSA) test has come into widespread use. This test is not specific for prostate cancer, however, and gives an abnormal result if there is any increased tissue growth in the gland, such as occurs in BPH. Because of its sensitivity, the PSA test can lead to the diagnosis of very early, microscopic tumors. Although such lesions might never progress to clinical disease, surgical removal carries a risk of major complications (notably incontinence and/or impotence), leading to controversy regarding the proper use of PSA as a screening test for early prostate cancer [3, 4].

II. DESCRIPTIVE EPIDEMIOLOGY OF PROSTATE CANCER

A. Incidence and Mortality Trends

Prostate cancer is a common cancer among men in many Western countries and is the leading male incident cancer in the United States, where 218,890 new cases are projected for the year 2007 [5, 6]. Incidence trends in the United States show a rather slow increase over most of the past 50 years, with a rather striking increase between 1989 and 1992, attributable in large measure to the widespread adoption of the PSA screening test, which first became available in the early 1980s [7, 8]. Since 1992, the incidence has declined, reflecting an end to the surge in cases due to the introduction of this new screening procedure [9, 10], as well as, perhaps, to the more judicious application of PSA screening. Moreover, mortality from prostate cancer is low relative to its incidence. This is because prostate cancer is generally well controlled by treatment (surgery, radiation, and androgen ablation) and occurs at relatively late ages, so that even men who are not cured of the disease often die from other causes. Interestingly, a parallel increase in prostate cancer mortality did not occur during the period 1989–1992, presumably because most of the additional cases diagnosed would not otherwise have led to fatal outcomes.

B. Risk Factors for Prostate Cancer

Few risk factors for prostate cancer have been established. Proposed factors are listed in Table 1. Age is the strongest risk factor. Prostate cancer incidence increases more sharply with age than does any other cancer; more than 50% of cases in the United States are diagnosed in men above the age of 70 [5, 11].

TABLE 1 Proposed Risk Factors for Prostate Cancer

Category	Characteristic or Exposure
Demographic	Age, ethnicity, geography
Genetic	Family history (father, brothers); high-penetrance gene; susceptibility genes
Occupational	Cadmium products, rubber industry, agricultural chemicals
Hormonal	Androgens (testosterone, dihydrotestosterone)
Lifestyle	Sexually transmitted agents, smoking, alcohol, vasectomy, physical activity, diet

Race/ethnicity is a second risk factor for prostate cancer. In the United States, the lowest incidence rates are seen among Korean and Vietnamese men, both relatively recent immigrant groups from Asia; the rates are somewhat higher among Chinese, American Indian, Alaska Native, and Native Hawaiian men. Caucasian men have very high rates, but by far, the highest incidence of this cancer is among African American men [12, 13].

The incidence of prostate cancer varies nearly 50-fold in populations around the world (Figure 1). Indeed, of all common cancers, this site shows the widest variation

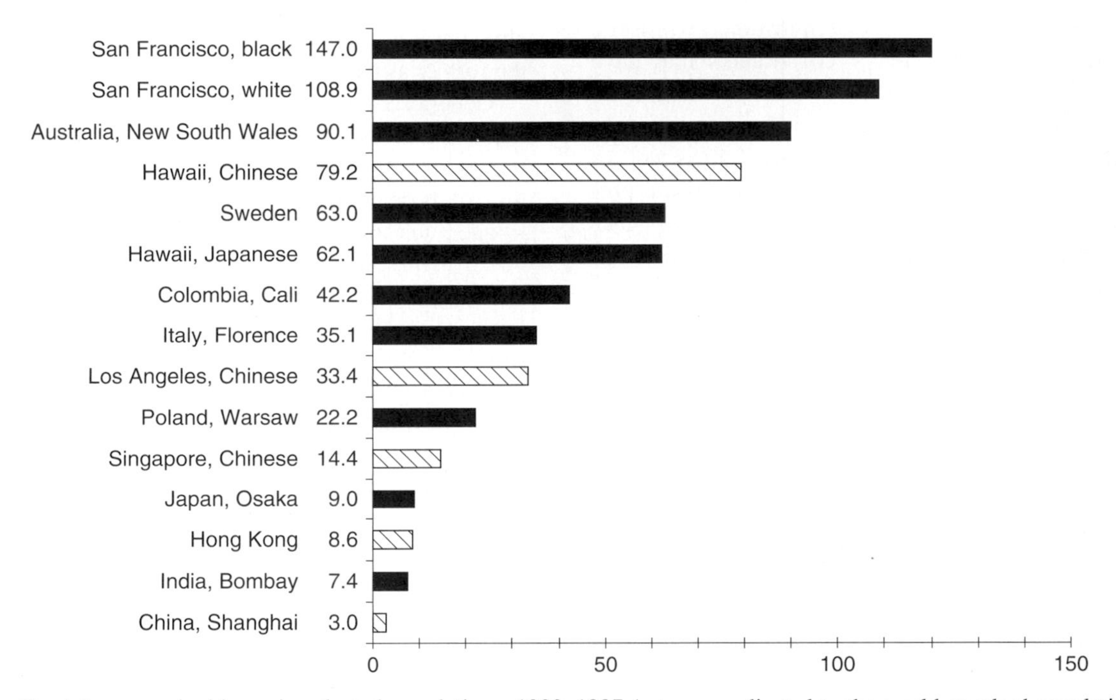

FIGURE 1 Prostate cancer incidence in selected populations, 1993–1997 (rates age adjusted to the world standard population).

between low- and high-risk countries or populations. High rates are seen in developed, especially Western, countries, including the United States, Canada, parts of Europe, and Australasia. Low rates tend to occur in Asia, particularly China [5, 14]. The highest reported rates in the world are among African Americans, whereas the lowest reported rates are among Chinese men in Shanghai. Interestingly, Chinese men in more developed areas of Asia (Singapore and Hong Kong) and Chinese men in the United States have much higher incidence rates than men in mainland China (see cross-hatched populations in Figure 1). Furthermore, immigrants from Japan to Brazil and the United States have higher rates than do men in Japan [11]. Although the incidence of prostate cancer in Japan increased about threefold between 1970 and 1990, which was similar to the rate of increase in the United States during the same period, the actual incidence in Japan remains very low.

Men with a first-degree male relative who has had prostate cancer are at a two- to threefold increased risk; whether this reflects an inherited predisposition for the disease or a shared environmental exposure has not been confirmed [15–17]. The search for high-penetrance, rare genes for prostate cancer has identified some candidates, though none has yet been confirmed; of even greater interest is the potential role in this disease of low-penetrance, highly prevalent susceptibility genes (discussed later).

Apart from these few established risk factors, the etiology of prostate cancer is unknown. Among the several potential causal agents, apart from diet, that have been proposed are (1) occupational exposures (rubber industry; manufacture of products containing cadmium, such as paints and batteries; use of agriculture chemicals); (2) sexually transmitted agents (e.g., cytomegalovirus); (3) smoking; (4) alcohol use; (5) vasectomy; and (6) physical activity [15, 18–21]. However, the evidence is not yet convincing for any of these exposures.

Although it is suspected that most exogenous factors affecting prostate cancer risk exert their influence by altering endogenous androgen levels [22, 23], epidemiologic studies have not yet clearly established the role of androgens in prostate cancer. Studies of prediagnostic circulating levels of individual androgens generally showed no clear association with prostate cancer [15, 24–29]. However, two studies showed weak positive relationships between the ratio of testosterone to dihydrotestosterone and prostate cancer [30, 31]. A third study showed a strong positive association of plasma testosterone with prostate cancer only after adjustment for the sex hormone binding globulin level [32]. The latter finding suggests that the level of free (unbound) testosterone may be most relevant.

The most promising area of research, apart from genetics, on the etiology of prostate cancer pertains to diet.

III. STUDIES OF DIET IN RELATION TO PROSTATE CANCER

A. Origin of the Diet–Prostate Cancer Hypothesis

The descriptive patterns of prostate cancer, especially data showing very different rates of the disease in the same ethnic/racial group living in different geographic settings, as well as changing rates in migrants and their offspring [11], prompted investigators to seek environmental risk factors for this cancer. Diet became an important focus of this research because (1) geographic variations in food and nutrient intakes are known to be large [14]; and (2) components of the diet can influence the levels of circulating androgens [33–36], which, as noted earlier, are thought to play a role in prostate cancer risk. Many different dietary factors, including both foods and particular constituents of foods, have been proposed and studied. Some of these appear to increase risk, while others are possibly protective. These factors are listed in Table 2, and the supporting evidence is discussed in the following sections of this chapter.

B. Dietary Factors That Increase Risk

1. Foods and Beverages

a. Red Meat. Many epidemiologic investigations of different designs, including ecologic [37–41], case-control [42–47], and cohort [48–54] studies, have reported positive associations between the consumption of meat, especially red meat, and prostate cancer. However, not all studies reproduced this finding [55–68].

TABLE 2 Proposed Dietary Risk Factors for Prostate Cancer

Increasing Risk	Decreasing Risk
Foods and Beverages	
Red meat	Vegetables
Dairy products	Fruits
Alcohol	Legumes
	Tea
Food Components	
Total energy	Vitamin D
Fat	Vitamin E
Calcium	Carotenoids
Zinc	Fructose
Cadmium	Selenium
	Isoflavonoids
Diet-Associated Factors	
Obesity	Physical activity

Explaining the association with meat is not straightforward. Initially, the finding was thought to reflect a high exposure to dietary fat, especially saturated fat, because meat and dairy products are the major contributors to fat intake in the Western diet. However, because the findings on dietary fat per se and prostate cancer are equivocal (discussed later), other explanations for the association should be considered. There are several possibilities: (1) In the American diet, red meat is a major source of zinc, which is essential for testosterone synthesis and may have other effects in the prostate (discussed later). (2) Diets high in meat and other animal products may be relatively deficient in certain anticarcinogenic constituents found primarily in plant foods. (3) Red meat contains high levels of heme iron, which is a source for free radical formation and oxidative damage to tissues [69]. (4) Most intriguingly, many meats are cooked at high temperatures, such as by pan-frying, grilling, or barbecuing. Cooking meats at high temperatures can result in the formation of heterocyclic amines, which are potent carcinogens in animals, including the rat prostate [70, 71]. Furthermore, when meats are cooked on charcoal grills, rendered fat is pyrolyzed by the coals, leading to the deposition of polycyclic aromatic hydrocarbons, which are also carcinogenic in animals, on the outer surface of the meat [72]. Few epidemiologic studies have been able to examine the relationship of such exposures to prostate cancer risk, because their levels in the diets of individuals cannot be easily and precisely assessed. In one study, a positive association between prostate cancer and estimated intakes of very well done meat and of a particular heterocyclic amine (PhIP) was reported [67]. However, another study that estimated heterocyclic amine intake from cooked meat and risk of prostate cancer did not lead to a clear result [73].

b. Dairy Products. Several case-control [42, 45, 46, 59, 74, 75] and cohort [48, 51, 52, 68, 76–79] studies found positive associations between the consumption of milk and other dairy products and the risk of prostate cancer. Nevertheless, other studies did not find this association [47, 55, 58, 60–62, 64–66, 80–85]. Despite these mixed results, two meta-analyses of many of these studies reached the conclusion that milk or total dairy consumption was associated with an increased risk of prostate cancer [86, 87]. One of these reports, a meta-analysis of 10 prospective studies [87], found that high dairy product intake was associated with an 11% increase in prostate cancer risk (Figure 2). One explanation for this positive association could be an adverse effect on the prostate of the high fat, especially saturated fat, content of dairy products. Another prominent constituent of these foods is calcium, which has also been proposed as a risk factor for prostate cancer (discussed later).

c. Alcoholic Beverages. Most case-control studies showed no association of prostate cancer either with total alcohol intake or with the intake of specific types of alcoholic beverages [44, 55, 56, 61, 81, 88–93], although a few studies in North America found a positive association for total alcohol intake [94–96]. Whereas a few cohort studies, including one in Japan [97], two in the United States [64, 98], and one in Europe [99] found positive associations with alcohol in general or with specific beverage types, most cohort studies reported null results for alcohol and prostate cancer risk [48, 51, 66, 100–106]. Furthermore, a meta-analysis of six cohort and 27 case-control studies found no overall association between alcohol consumption and prostate cancer risk [107]. Some general mechanisms by which alcohol might enhance carcinogenesis have been proposed, including the activation of environmental nitrosamines, production of carcinogenic metabolites (acetaldehyde), immune suppression, and secondary nutritional deficiencies [94, 99, 108].

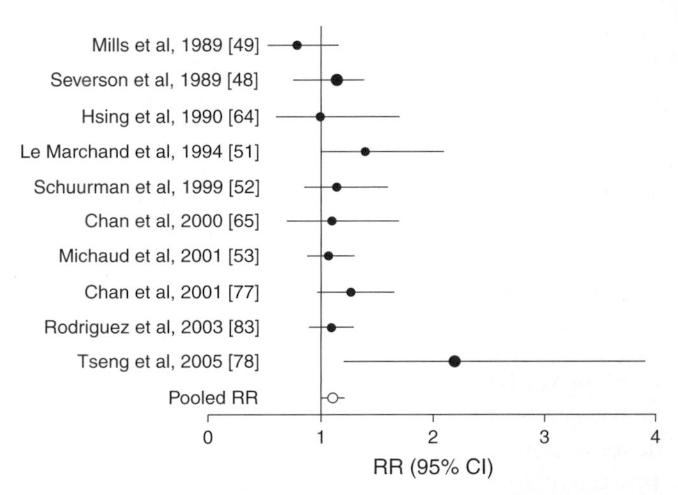

FIGURE 2 Dairy product intake and prostate cancer (relative risks, RR, compare highest with lowest intake categories; error bars indicate 95% confidence intervals; reference numbers are in square brackets, [ ]). (Adapted, with permission of Oxford University Press, from Gao, X., LaValley, M. P., and Tucker, K. L. (2005). Prospective studies of dairy product and calcium intakes and prostate cancer risk: A meta-analysis. *J. Natl. Cancer Inst.* **97**, 1768–1777 [87].)

2. Nutrients and Other Food Constituents

a. Energy. Total energy intake was not reported in early epidemiologic studies of diet and prostate cancer, because the dietary assessment methods used at the time were incomplete. However, several studies in recent years have examined this variable, particularly in relation to the effects of dietary fat intake. The findings have been very inconsistent. Some studies reported a positive association for energy intake and no independent association with dietary fat [109–111]. Others reported an effect of dietary fat, but no independent effect of energy [46, 80, 112, 113]. Two studies reported a positive association with both total energy and total fat [47, 114], and five studies found no association with either total energy or fat [56, 62, 65, 115, 116]. An experimental study in rodents (rats and mice) found that energy restriction reduced prostate tumor growth, possibly by inhibiting tumor angiogenesis [117].

b. Fat. Dietary fat has been the most studied nutrient with regard to effects on prostate cancer risk. Detailed reviews on this topic have been published [118–120]. Total fat intake has been associated with the risk of prostate cancer in several case-control studies [44, 88, 114, 121–123], although most found the association to be strongest for saturated or animal fat [43, 112, 121–124]. In two cohort studies that examined total and saturated fat, one [115] found no evidence of an effect on prostate cancer, and the other [50] found an elevated risk for total, but not saturated, fat. When limited to studies that controlled for energy intake, however, the findings from both case-control and cohort studies are more equivocal, with many investigations showing no independent effect of fat [56, 62, 109–111, 113, 115]. In some studies, the association with total and saturated fat was stronger for the advanced cases [50, 112, 123]. A meta-analysis found a positive association for total fat intake, whether based on all studies included (15) or whether limited to studies that adjusted for energy (10), but did not find an association for saturated fat intake [125].

Some epidemiologic studies examined intakes of monounsaturated and polyunsaturated fat as well. Although many of these studies found no association of either of these classes of fat with prostate cancer [60, 62, 115, 116, 118], several case-control studies reported positive associations [58, 80, 114, 123]. One study reported a positive association with monounsaturated fat and an inverse association with polyunsaturated fat [126]. A meta-analysis of published studies reported no association with either monounsaturated or polyunsaturated fat [125].

Some studies also examined specific fatty acids (including several omega-3 and omega-6 polyunsaturated fatty acids), based either on dietary intake data or biochemical measurements in blood or adipose tissue [115, 118]. A few studies suggested that consumption of fatty fish containing abundant long-chain polyunsaturated fatty acids may be protective [127–130]. This conclusion was supported by studies showing that eicosapentaenoic acid (EPA) and docosahexaenoic acid (DHA) were inversely associated with prostate cancer [131, 132], though other studies did not reproduce this finding [133, 134]. The effect of α-linolenic acid (ALA) intake was examined in several studies, with some showing a positive association [132, 133, 135] and some no association [134, 136]. However, a meta-analysis involving five cohort and four case-control studies found that ALA intake or blood concentrations was positively associated with prostate cancer [137]. Finally, one study reported that prediagnostic serum concentration of trans-fatty acids was associated with increased risk of prostate cancer [138]. Overall, these reports are not very consistent, and no conclusions as to the role of specific fatty acids in the risk of prostate cancer can be reached on the basis of current data.

Some animal experiments have tested the fat–prostate cancer hypothesis. For example, a high-fat diet increased prostate cancer incidence and shortened the latency period in Lobund Wistar rats treated with exogenous testosterone to induce the tumors [139]. Conversely, prostate tumor growth rate was reduced by a fat-free diet in Dunning rats [140] or by lowering dietary fat intake in athymic nude mice injected with LNCaP cells (a human prostate cancer cell line) [141, 142]. With regard to specific types of fat, fish oils containing high levels of omega-3 fatty acids, such as EPA and DHA, generally suppressed prostate tumor growth in rodents, whereas omega-6 polyunsaturated fatty acids, such as linoleic acid, promoted tumor growth [130, 143, 144]. However, because most animal studies have been conducted in rodents, whose prostate glands differ anatomically from that of the human, extrapolation of these findings to humans is particularly tenuous.

A number of plausible mechanisms by which dietary fat could increase cancer risk have been proposed. These include the formation of lipid radicals and hydroperoxides that can produce DNA damage, increased circulating androgen levels, decreased gap-junctional communication between cells, altered activity of signal transduction molecules, effects on eicosanoid metabolism, and decreased immune responsiveness [118].

c. Calcium. As noted earlier, a number of studies that have examined the relationship of dairy product consumption to prostate cancer risk found a positive association. Dairy products could be a marker of exposure to calcium, though this food group is also a major source of saturated or animal fat in the Western diet. Two case-control studies [75, 114] and several cohort studies [77–79, 83, 84] reported statistically significant positive associations for this nutrient, especially with advanced or metastatic cancer. A meta-analysis of six prospective studies confirmed this finding [87]. Nevertheless, several other case-control [46, 60, 116, 145, 146] and cohort [52, 65, 68, 82, 85] studies did not show an effect of calcium on prostate cancer risk. Furthermore, a clinical trial that entailed daily supplementation with 1200 mg of calcium for 4 years found no effect on prostate cancer risk [147].

A mechanism for an adverse effect of calcium on prostate carcinogenesis has been proposed, based on the observation that a high intake of calcium decreases the circulating levels of 1,25-dihydroxyvitamin D [$1,25(OH)_2D$], which may inhibit cell proliferation and promote differentiation in prostatic tissue [148, 149]. The role of vitamin D in prostatic carcinogenesis is discussed later.

d. Zinc and Cadmium. The trace elements zinc and cadmium are considered together because they act as antagonists in biological systems. Few epidemiologic studies have examined dietary zinc in relationship to prostatic cancer. Most of the studies found no effect of zinc, especially after energy adjustment [56, 80, 109, 123], although a positive association was suggested in three reports [122, 150, 151] and an inverse association was found in one report [152]. The frequent association of prostate cancer risk with high intake of red meat (discussed earlier) could also be explained by a higher intake of zinc, rather than animal fat, because meat, especially red meat, is an important source of zinc in the American diet (other sources are shellfish, whole-grain cereals, nuts, and legumes) [153]. Reports based on zinc levels in serum or prostatic tissue of patients with cancer and controls have not been consistent [154, 155], but such studies are unreliable, because the levels of zinc measured after diagnosis in the cases may reflect physiologic changes in the prostate as a result of the cancer. One study measured prediagnostic toenail zinc in prostate cancer patients and in controls; no association was found [156].

As a major constituent of prostatic fluid [157–159], zinc is essential for normal prostate function. Zinc is also essential for normal testicular function, and high levels of zinc have been proposed to increase the production of testosterone, leading to enhanced tissue growth in the prostate. Plasma levels of zinc have been positively correlated with testosterone and dihydrotestosterone levels in men [160, 161]. Furthermore, in the rat prostate, zinc has been shown to increase 5α-reductase activity [162] and to potentiate androgen receptor binding [163]. Thus, one might speculate that higher intake of zinc could partially offset the normal decline in testosterone levels with age [164–166], thereby contributing to prostate cancer risk.

Epidemiologic evidence for cadmium as a risk factor for prostate cancer is also limited [167]. Only two studies attempted to assess dietary exposure [168, 169], and neither found very convincing evidence for an effect of diet alone, although both studies found some evidence for a positive association when combined cadmium exposure from multiple sources (diet, cigarette smoking, occupation) was considered. Most studies of cadmium and prostate cancer have been among exposed workers in industries that utilize cadmium, and the findings of those investigations are only weakly suggestive of an adverse effect [15]. A study of toenails and prostate cancer [156], mentioned earlier with regard to zinc, also examined cadmium and found no association with prostate cancer risk. Cadmium is a competitive inhibitor of zinc in enzyme systems and accumulates in the body throughout life, because no mechanism exists for excreting it. Thus, the hypothesis that cadmium may be carcinogenic for the prostate has biologic plausibility. This hypothesis is further supported by studies showing that cadmium is carcinogenic in animals and that the effect can be blocked by simultaneous injection of zinc [170, 171].

3. Diet-Associated Risk Factors

a. Obesity, Weight, and Height. Prostate cancer is sometimes considered a male counterpart to breast cancer in women, for which there is clear evidence of a positive association with obesity, especially in postmenopausal cases. However, evidence for a similar association of adult obesity with prostate cancer is much less clear. Although several epidemiological studies reported a significant positive association [42, 76, 95, 172–179], the majority of such studies found no clear relationship of measures of obesity to prostate cancer risk [49, 51, 81, 88, 110, 112, 121, 123, 180–193]. Indeed, several case-control [194, 195] and prospective cohort studies [196–200] even reported inverse associations. Two studies reported reversals of effect: In one, obesity increased the risk of high-grade but decreased the risk of low-grade prostate cancer [201]; in the other, obesity was associated with decreased prostate cancer incidence but with increased prostate cancer mortality [202]. Although some studies suggested that obesity in childhood [198] or before age 30 [203] may be protective against adult prostate cancer, another study found that obesity at age 20 was associated with an increased risk [196].

A few studies reported that height was associated with an increased risk of prostate cancer [177, 186, 201], but the evidence is limited.

A recent meta-analysis considered 31 cohort and 25 case-control studies and found a weak direct association for obesity, as well as for height and weight; the relationship was stronger among the cohort studies compared with the case-control studies [204].

The basis for an association between obesity and prostate cancer could involve endocrine factors, because adult obesity in men has been associated with decreased circulating levels of testosterone and increased levels of estrogen [205, 206]. However, this mechanism would suggest an inverse rather than a direct association between obesity and this cancer.

C. Dietary Factors That Decrease Risk

1. Foods and Beverages

a. Vegetables. Intake of vegetables has been inversely associated with cancer risk at many sites. This has led to strong recommendations to consume significant quantities of these foods as part of a healthful diet. However, the evidence for a beneficial effect of vegetables on prostate cancer risk is not overwhelming. Although several case-control studies showed inverse associations with selected vegetables or vegetable groupings, including green and yellow-orange vegetables, cruciferous vegetables, and allium vegetables [44, 47, 59, 60, 62, 207–211], many others showed no significant associations [42, 45, 46, 57, 58, 61, 212, 213]. The findings from prospective cohort studies have been mostly null [49, 64–66, 76, 214–219]. The findings for legumes, a vegetable subgroup, are considered separately (discussed later), and the findings for tomatoes are included in the later discussion of carotenoids.

Because vegetables contain numerous compounds that can act through a variety of mechanisms to inhibit carcinogenesis [220, 221], an inverse association between vegetables and prostate cancer is plausible. Some of these mechanisms are discussed with respect to specific food constituents hereafter.

b. Fruits. The epidemiologic data on fruits and prostate cancer are also inconsistent, but interestingly, several studies, both case-control [57, 122, 207, 208] and cohort [48, 216], showed a direct (positive) association. Other case-control [45–47, 58–61, 80, 209, 210] and cohort [49, 64–66, 76, 214, 215, 218, 219] studies showed either no association or a statistically nonsignificant inverse relationship to prostate cancer risk. In one cohort study [222], a statistically significant inverse association was found for fruit intake and advanced prostate cancer; this finding did not persist after adjustment for total fructose intake.

Why intake of fruits might have an adverse effect on the prostate is not clear. Fruits contain many of the same compounds with anticarcinogenic properties that are found in vegetables, such as various carotenoids and vitamin C [220, 223]. Although this finding is not yet established firmly, it does appear that fruit intake has no particular benefit with regard to the risk of prostate cancer.

c. Legumes, Including Soy Products. Prostate cancer rates have traditionally been low in populations such as those of Japan and China, where the intake of soy products is relatively high. Several case-control [56, 57, 61, 208, 209, 224] and cohort [48, 49, 216, 225] studies have reported inverse associations between intake of legumes and prostate cancer, including soy foods specifically [48, 57, 61, 209, 224, 225]. When a meta-analysis of two cohort and six case-control studies was conducted, it showed an inverse association between soy food consumption and prostate cancer risk [226]. At present, the data are suggestive of a beneficial effect of legumes in general, not limited to soy products specifically. Additional research may clarify this issue.

In the past, legumes were of interest in nutritional epidemiology primarily because of their important contribution to fiber intake. However, these foods also contain phytoestrogens, plant constituents that have mild estrogenic properties. Because estrogens are associated with lower risk of prostate cancer and are used in prostate cancer therapy, there is a good rationale for the hypothesis that phytoestrogen intake can protect against prostate cancer. Soybeans and many products made from soy, such as tofu, are rich in a class of phytoestrogens known as isoflavones (other classes of phytoestrogens include the coumestans and lignans). The main isoflavones found in soy include genistein, daidzein, and glycitein [227, 228]. Although a few epidemiologic studies assessed the intake of dietary phytoestrogens, particularly isoflavones, and found inverse associations with prostate cancer risk [62, 224, 229, 230], analytical epidemiologic data on this topic are limited [231]. The mechanism for a protective effect of soy products on prostate carcinogenesis could entail the estrogenic effects of isoflavones, though other actions of these compounds, such as inhibition of protein tyrosine phosphorylation, induction of apoptosis, and suppression of angiogenesis, have also been proposed [232, 233]. Laboratory data based on human tissue as well as animal models offer support for the hypothesis that soy products may protect against prostate cancer [232, 234–236].

Although soy products and isoflavones are of particular interest, legumes contain other bioactive microconstituents, including saponins, protease inhibitors, inositol hexaphosphate, γ-tocopherol, and phytosterols; mechanisms by which each of these compounds can inhibit carcinogenesis have been proposed [220, 232, 233, 237–239].

d. Tea. A few studies have examined the relationship between tea consumption and prostate cancer risk, and the findings have been inconsistent. A cohort study in Hawaii [240] and a case-control study in China [241] showed an inverse relationship between daily tea consumption and prostate cancer risk, but several other cohort studies [48, 66, 102, 242, 243] as well as case-control studies [56, 57, 61, 244] did not reproduce this finding.

Tea contains polyphenols that are potentially anticarcinogenic because of their antioxidant properties, effects on signal transduction pathways, inhibition of cell proliferation, and other actions in the body [245].

2. Nutrients and Other Food Constituents

a. Vitamin D. Evidence for a protective effect of vitamin D against prostate cancer is not yet convincing. In five cohort studies [65, 78, 82, 83, 222] and five case-control studies [47, 75, 114, 145, 146], estimated dietary intake of vitamin D, whether from foods or supplements, was not associated with risk. However, since circulating vitamin D levels are substantially determined by the conversion of 7-dehydrocholesterol in the skin in response to solar UVB radiation, studies based on dietary vitamin D intake alone may be misleading [246, 247].

Although $1{,}25(OH)_2D$ is the biologically active form of the vitamin, it is homeostatically controlled over a narrow range in the blood. The most abundant circulating form of the vitamin is 25-hydroxyvitamin D [25(OH)D], which more closely reflects solar exposure and dietary intake. Eight cohort studies that examined the relationship of prediagnostic serum levels of 25(OH)D to subsequent development of prostate cancer produced discrepant results. One study found a clear inverse association [248], but the other seven studies did not reproduce this finding [249–255].

Vitamin D reduces cell proliferation in the prostate (and other tissues) and enhances cell differentiation, both of which would be expected to lower the risk of cancer [148]. In a mouse model, an analog of $1{,}25(OH)_2D$, the hormone produced from 25(OH)D, inhibited the growth of prostate tumors [256].

b. Vitamin E. The intake of vitamin E is not adequately assessed with dietary histories, because much of the vitamin is obtained from fats and oils added during food preparation; estimating the amounts of added fats and oils is especially difficult. Nevertheless, a few epidemiologic studies attempted to determine vitamin E intake, with variable results. Whereas some case-control [60, 109, 110, 116] and cohort [219, 257] studies found no association with prostate cancer risk, other case-control studies found an inverse association [58, 62] and one cohort study reported a positive association [258]. Use of vitamin E supplements was not associated with prostate cancer risk overall in two U.S. cohort studies [259, 260], although a reduced risk for advanced cancers was seen among current smokers and recent ex-smokers in one of them [259]. Some cohort studies reported on findings for vitamin E and prostate cancer based on prediagnostic serum levels with mixed results, some showing an inverse association [261–265] and others no association [266–268]. In an intervention trial among male heavy smokers in Finland, an incidental finding was a reduced risk of prostate cancer associated with intake of vitamin E supplements [269–271]. Because the study was not specifically designed to test hypotheses related to prostate cancer, these results need confirmation. Furthermore, neither dietary nor serum vitamin E level at baseline was associated with subsequent development of prostate cancer in the group without vitamin E supplementation [272].

Vitamin E inhibits the growth of human prostate tumors in nude mice [273, 274]. As a powerful antioxidant, one mechanism for a protective effect of vitamin E against carcinogenesis in the prostate could be inhibition of lipid peroxidation [275].

c. Carotenoids (β-Carotene and Lycopene). The epidemiologic evidence related to carotenoids and prostate cancer is inconsistent. With regard to β-carotene intake, several case-control studies reported an inverse association [62, 74, 88, 123, 207, 276, 277], though more of them offered no support for a protective effect [47, 56, 60, 80, 109, 110, 113, 116, 122, 210, 213]. Furthermore, the findings often differed between younger and older men. Some cohort studies reported that dietary intake of β-carotene decreased risk [64] or had no effect on risk [214, 215, 219, 257], but, as in the case-control studies, the findings sometimes differed between younger and older men [64]. One of these studies also showed no effect of α-carotene, β-cryptoxanthin, or lutein on the risk of prostate cancer [215], whereas another showed a positive association for β-cryptoxanthin but not other carotenoids [258]. Studies based on prediagnostic serum have reported both an increased risk [278] and no association [264, 266, 279] with elevated β-carotene levels. It is of interest that circulating levels of β-carotene were shown to be significantly correlated with the levels in prostatic tissue [280]. Finally, an intervention trial in the United States found no effect of β-carotene supplementation on prostate cancer risk [281], though a similar trial in Finland found a decreased risk among nondrinkers of alcohol but an increased risk among drinkers [269].

A carotenoid of particular interest with regard to prostate cancer is lycopene, found primarily in tomatoes and tomato products (other food sources include watermelon, grapefruit, and guava). Although a few case-control studies showed an inverse association between tomato consumption, particularly cooked tomatoes, and prostate cancer [46, 58, 208, 277], most such studies found no association [56, 57, 61, 209, 210, 212, 213]. One study [46] found a weak inverse association with raw tomatoes but not cooked tomatoes (which is surprising, because the lycopene should be more bioavailable in the cooked tomatoes because of enhanced absorption as a result of heat processing and the presence of lipids [282]). Of the studies that estimated intake of lycopene per se, most failed to show a clear inverse relationship to prostate cancer risk [47, 56, 208, 213], though two studies found an inverse association [81, 283]. The findings from cohort studies are also inconsistent: Two cohorts found a significant inverse association with lycopene intake [49, 215, 284], whereas another two found no association [216, 219]. A meta-analysis of 11 case-control and 10 cohort studies found a protective effect of tomato (raw and cooked) and lycopene intake against prostate cancer, but the effect was restricted to high levels of intake [285]. The results of investigations based on prediagnostic circulating levels of lycopene are also inconsistent. Although three studies showed an inverse relationship [279, 286, 287], three others [263, 266, 288], one of which [288] was able to exclude screening bias, found no association.

β-Carotene, lycopene, and other carotenoids are widely distributed in human tissues, including the prostate [280, 289], where, as potent antioxidants, they help protect cell membranes, DNA, and other macromolecules from damage by reactive oxygen species. Other biological activities of carotenoids, such as the upregulation of gap-junctional communication [290], may also contribute to their anticarcinogenic effects. In three human prostate cancer cell lines (PC-3, DU 145, and LNCaP), β-carotene significantly inhibited *in vitro* growth rates [291].

d. Fructose. Intake of fructose was inversely associated with the risk of prostate cancer in one cohort study [222], which found a similar relationship for the intake of fruit, a major source of fructose in the diet. However, two other cohort studies found no association [65, 82]. Although many epidemiologic studies have not assessed fructose intake per se, several, as noted earlier, examined the relationship of fruit intake to prostate cancer and most did not find an inverse association. However, the metabolism of calcium, phosphorus, fructose, and vitamin D are interrelated, and unless all components are considered simultaneously in an analysis, their individual effects could be missed [222].

The hypothesis for the protective effect of fructose is that it reduces plasma phosphate levels, resulting in increased levels of circulating $1,25(OH)_2D$, which in turn may reduce the risk of prostate cancer [222, 292] (discussed earlier).

e. Selenium. Several case-control studies [123, 208, 293–296] found no association between estimated selenium intake and prostate cancer risk, and baseline selenium intake was not associated with subsequent development of prostate cancer in a cohort study [219] and an intervention trial [272]. Prediagnostic serum selenium was inversely related to prostate cancer in several [262, 297–299], though not all [300–302], prospective cohort studies. Prediagnostic selenium levels in toenails were also inversely associated with advanced or total prostate cancer [303, 304]. A meta-analysis of 11 cohort and five case-control studies that mostly measured selenium concentrations in blood or toenails as a surrogate for selenium intake concluded that selenium intake reduces the risk of prostate cancer (by 28% based on cohort studies and 26% based on case-control studies) [305] (Figure 3).

Selenium is a component of glutathione peroxidase, an important enzyme in certain antioxidative pathways. In *in vitro* experiments, selenium was shown to inhibit the growth of human (DU-145) prostate carcinoma cells [306, 307]. Selenium may exert its anticancer effects through any of several proposed mechanisms, such as antioxidation, enhanced immune function, inhibition of cell proliferation, and induction of apoptosis [308].

f. Isoflavonoids. The potential role of isoflavonoids in prostate carcinogenesis was discussed in the section on legumes.

3. Diet-Associated Protective Factors

a. Physical Activity. The role of physical activity in human prostate carcinogenesis is at present quite unclear. Several epidemiologic studies have examined this relationship, but the findings have been inconsistent. Although some studies showed an inverse association with prostate cancer risk [309–312], others showed no association [18, 81, 189, 313–316]. These discrepancies may be resolved if future studies distinguish better between different types of physical activities and can establish the time of life (e.g., young adulthood versus older ages) that may be most relevant. Indeed, one study found inconsistent associations depending on the type of activity (occupational versus recreational), age period of life, and intensity of physical activity [317].

Because exercise influences androgen levels in the body, an effect of physical activity on prostate cancer risk is biologically plausible. Exercise lowers testosterone in the blood and also raises the level of sex hormone binding globulin, which reduces the circulating free testosterone levels; both effects would be expected to lower prostate cancer risk [18, 318, 319].

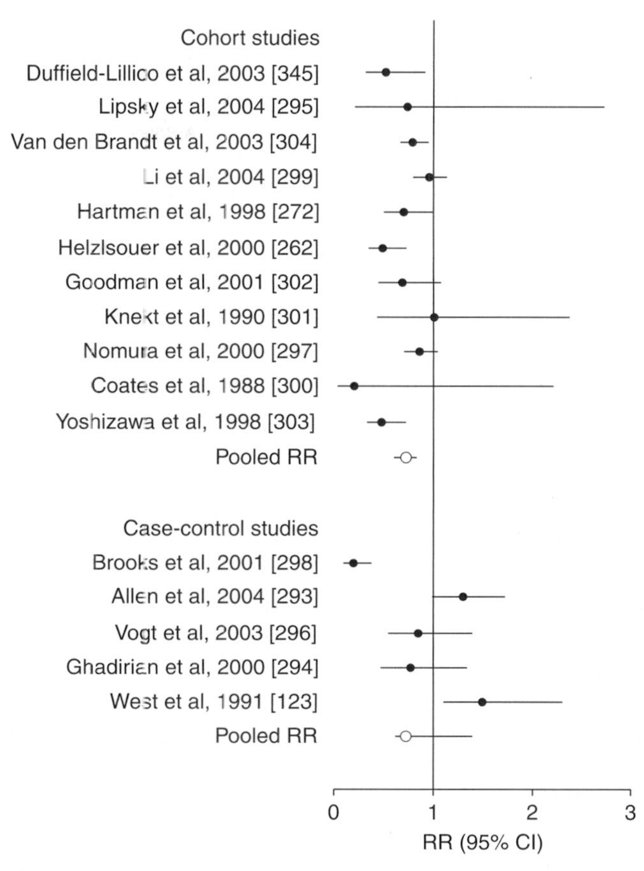

FIGURE 3 Selenium intake and prostate cancer (relative risks, RR, comparing moderate with low intake; error bars indicate 95% confidence intervals; reference numbers are in square brackets, [ ]). (Adapted, with permission of Springer Science + Business Media, from Etminan, M., FitzGerald, J. M., Gleave, M., and Chambers, K. (2005). Intake of selenium in the prevention of prostate cancer: A systematic review and meta-analysis. *Cancer Causes Control* **16**, 1125–1131 [305].)

IV. GENETICS AND GENE–ENVIRONMENT INTERACTIONS

As noted earlier, prostate cancer shows a familial association (i.e., men whose fathers or brothers have had prostate cancer are at a two- to threefold increased risk of getting the disease compared to men without such a family history). The search for one or more highly penetrant genes that predispose to prostate cancer and that may explain at least part of this familial association has not yet been successful. Several candidate genes have been identified thus far, including a promising locus at q24–25 on chromosome 1 [320–322].

However, genetic predisposition may have a more indirect relationship to prostate cancer. Cell growth and differentiation in the prostate are regulated by androgens and various growth factors. It has been shown that several genes related to these cellular constituents are polymorphic (i.e., exist in variant forms) [323, 324]. Some of these variants may have functional effects, leading to altered activity of the gene product. For example, a variant form, A49T, of the gene *SRD5A2* (steroid 5α-reductase type II) that encodes the enzyme responsible for converting testosterone to dihydrotestosterone in the prostate, was found to have lower activity in an *in vitro* assay [325, 326]. Such functional consequences of genetic polymorphisms in constitutional DNA could lead to differing susceptibilities to prostate cancer. Variant forms of the androgen receptor gene (*AR*), the vitamin D receptor gene (*VDR*), and the insulin-like growth factor 1 gene (*IGF-1*) have been examined in relation to prostate cancer risk [327–335]. Such polymorphisms may affect not only the risk for prostate cancer, but also the pathological characteristics of the tumors and hence prognosis [336]. A region of 8q24 was recently found to be associated with an increased risk of prostate cancer [337]. Identification of this susceptibility locus has been consistently reproduced by several other investigators [338–343]. Although neither a specific gene nor a mechanism for the association has yet been proposed, this region appears to be of major importance, carrying a significant population-attributable risk (10–20%).

Because diet (and other behaviors, such as physical activity [344]) can influence androgen, vitamin D, and IGF-1 levels in the body, interactions may occur between dietary exposures and inherited susceptibilities in determining actual risk for prostate cancer. Such gene–environment interactions offer considerable potential for elucidating the etiology of prostate cancer.

V. DIETARY INTERVENTION TRIALS

A randomized intervention trial is the closest method to an experimental design that can be implemented in human subjects. Such a trial should be conducted only when there is considerable evidence for a beneficial effect, without apparent harm, based on observational studies and supported by animal and *in vitro* studies. As the preceding review indicates, the data in support of most dietary factors are either limited or inconsistent and would not justify the expense and risks of an intervention trial specifically for prostate cancer. However, two randomized intervention trials that were conducted for other purposes have shown unexpected reductions in prostate cancer incidence in the intervention groups. One of these studies [345] was designed to test the potential of a daily selenium supplement (200 μg) to reduce the occurrence of basal and squamous cell skin cancers in men with a prior history of such lesions. The second trial [269–271] tested the effects of daily β-carotene (20 mg) and/or vitamin E (50 mg *dl*-α-tocopherol acetate) on the risk of lung cancer in a group of male smokers. In both trials, the incidence of prostate cancer was significantly lower in the men who received the intervention compared with the placebo groups. In the lung cancer trial, the reduced incidence was in the men who received vitamin E rather than β-carotene. However, because protection against prostate cancer was not a prespecified hypothesis in either trial, the results cannot be taken as definitive. Based on these findings, and the supportive evidence from other epidemiologic and laboratory research, a double-blind randomized trial to test the potential benefit of supplemental selenium (200 μg/day) and vitamin E (400 mg/day), alone and in combination, was begun in the United States [346, 347]. However, the results will not be available for several years.

Intervention trials in prostate cancer patients, using specific nutrients or dietary patterns to reduce the rate of progression or recurrence of the disease, have been proposed [348–351]. However, no results from such randomized trials have been reported.

VI. CONCLUSION AND IMPLICATIONS FOR PREVENTION AND TREATMENT

Considering the combined evidence from descriptive epidemiologic studies (especially the remarkable changes in migrant populations), analytic epidemiologic studies in widely varying populations, experimental studies in animals, and *in vitro* studies, the likelihood that certain dietary components or general patterns of eating influence the risk of prostate cancer remains high. However, at the present time, no specific relationships have been established conclusively. Continued research on this topic should be a high priority because diet is a modifiable risk factor and because prostate cancer incidence is extremely high in many populations. In addition, further research on genetic polymorphisms that affect susceptibility to prostate cancer should contribute to better identification of high-risk groups of men who can be targeted for future preventive dietary interventions.

Currently, the primary treatment modalities for prostate cancer consist of surgery, radiation, and hormonal therapy. The fact that the findings for many dietary factors (e.g., saturated fat) were often stronger in advanced or metastatic cases of prostate cancer implies that dietary effects can occur very late in the disease process. This suggests that dietary interventions have the potential not only to reduce the incidence, but also to improve the survival rates of the disease, providing another possible treatment modality. Indeed, one study showed significantly worse survival for prostate cancer patients in the upper tertile of saturated fat intake (>13.2% of energy) compared with men in the lower tertile (<10.8% of energy) [352]. A possible mechanism to explain this result is reduced circulating androgen levels in the men with lower fat intake [33, 34, 36], since prostate tumors are androgen sensitive, at least early in their clinical course.

Based on current knowledge, it is not prudent to make very specific dietary recommendations to prevent or treat prostate cancer. However, taken as a whole, the evidence offers reasonably strong support for a diet that emphasizes vegetables, including legumes, and is moderate or low in the consumption of meat, especially red meat, and dairy products.

References

1. Bostwick, D. G., and Amin, M. B. (1996). Prostate and seminal vesicles. *In* "Anderson's Pathology," 10th ed. (I. Damjanov and J. Linder, Eds.), Vol. 2, pp. 2197–2230. Mosby, St. Louis, MO.
2. Partin, A. W., and Coffey, D. S. (1998). The molecular biology, endocrinology and physiology of the prostate and seminal vesicles. *In* "Campbell's Urology," 7th ed. (P. C. Walsh, A. B. Retik, E. D. Vaughan Jr., and A. J. Wein, Eds.), pp. 1381–1428. W. B. Saunders, Philadelphia.
3. Barry, M. J. (1998). PSA screening for prostate cancer: The current controversy—a viewpoint. Patient Outcomes Research Team for Prostatic Diseases. *Ann. Oncol.* **9**, 1279–1282.
4. Chodak, G. (2006). Prostate cancer: Epidemiology, screening, and biomarkers. *Rev. Urol.* **8**, S3–S8.
5. Parkin, D. M., Whelan, S. L., Ferlay, J., Teppo, L., and Thomas, D. B. (2002). "Cancer Incidence in Five Continents," Vol VIII. IARC Sci. Pub., No.155, IARC Lyon, France.
6. American Cancer Society. (2007). "Cancer Facts and Figures 2007." American Cancer Society, Inc., Atlanta, GA.
7. Hankey, B. F., Feuer, E. J., Clegg, L. X., Hayes, R. B., Legler, J. M., Prorok, P. C., Ries, L. A., Merrill, R. M., and Kaplan, R. S. (1999). Cancer surveillance series: Interpreting trends in prostate cancer—part I: Evidence of the effects of screening in recent prostate cancer incidence, mortality, and survival rates. *J. Natl. Cancer Inst.* **91**, 1017–1024.
8. Sarma, A. V., and Schottenfeld, D. (2002). Prostate cancer incidence, mortality, and survival trends in the United States: 1981–2001. *Semin. Urol. Oncol.* **20**, 3–9.
9. Legler, J. M., Feuer, E. J., Potosky, A. L., Merrill, R. M., and Kramer, B. S. (1998). The role of prostate-specific antigen (PSA) testing patterns in the recent prostate cancer incidence decline in the United States. *Cancer Causes Control* **9**, 519–527.
10. Etzioni, R., Penson, D. F., Legler, J. M., di Tommaso, D., Boer, R., Gann, P. H., and Feuer, E. J. (2002). Overdiagnosis due to prostate-specific antigen screening: Lessons from U.S. prostate cancer incidence trends. *J. Natl. Cancer Inst.* **94**, 981–990.
11. Kolonel, L. N. (1997). Racial and geographic variations in prostate cancer and the effect of migration. *In* "Accomplishments in Cancer Research 1996" (J. G. Fortner and P. A. Sharp Eds.), pp. 221–230. Lippincott-Raven, Philadelphia.
12. Miller, B. A., Kolonel, L. N., Bernstein, L., Young, J. L. J., Swanson, G. M., West, D. W., Key, C. R., Liff, J. M., Glover,C. S., and Alexander, G. A. (Eds.). (1996). "Racial/Ethnic Patterns of Cancer in the United States, 1988–1992." NIH Pub. No. 96–4104. National Cancer Institute, Bethesda, MD.
13. Powell, I. J. (2007). Epidemiology and pathophysiology of prostate cancer in African-American men. *J. Urol.* **177**, 444–449.
14. World Cancer Research Fund/American Institute for Cancer Research. (1997). "Food, Nutrition and the Prevention of Cancer: A Global Perspective," pp. 20–52. American Institute for Cancer Research, Washington, DC.
15. Nomura, A. M., and Kolonel, L. N. (1991). Prostate cancer: A current perspective. *Epidemiol. Rev.* **13**, 200–227.
16. Whittemore, A. S., Wu, A. H., Kolonel, L. N., John, E. M., Gallagher, R. P., Howe, G. R., West, D. W., Teh, C. Z., and Stamey, T. (1995). Family history and prostate cancer risk in black, white, and Asian men in the United States and Canada. *Am. J. Epidemiol.* **141**, 732–740.
17. Johns, L. E., and Houlston, R. S. (2003). A systematic review and meta-analysis of familial prostate cancer risk. *BJU Int.* **91**, 789–794.
18. McTiernan, A., Ulrich, C., Slate, S., and Potter, J. (1998). Physical activity and cancer etiology: Associations and mechanisms. *Cancer Causes Control* **9**, 487–509.
19. Parker, A. S., Cerhan, J. R., Putnam, S. D., Cantor, K. P., and Lynch, C. F. (1999). A cohort study of farming and risk of prostate cancer in Iowa. *Epidemiology* **10**, 452–455.
20. John, E. M., Whittemore, A. S., Wu, A. H., Kolonel, L. N., Hislop, T. G., Howe, G. R., West, D. W., Hankin, J., Dreon, D. M., Teh, C. Z., *et al.* (1995). Vasectomy and prostate cancer: Results from a multiethnic case-control study. *J. Natl. Cancer Inst.* **87**, 662–669.
21. Bostwick, D. G., Burke, H. B., Djakiew, D., Euling, S., Ho, S. M., Landolph, J., Morrison, H., Sonawane, B., Shifflett, T., Waters, D. J., and Timms, B. (2004). Human prostate cancer risk factors. *Cancer* **101**, 2371–2490.
22. Wilding, G. (1995). Endocrine control of prostate cancer. *Cancer Surv.* **23**, 43–62.
23. Gann, P. H. (2002). Androgenic hormones and prostate cancer risk: Status and prospects. *IARC Sci. Publ.* **156**, 283–288.
24. Nomura, A. M., Stemmermann, G. N., Chyou, P. H., Henderson, B. E., and Stanczyk, F. Z. (1996). Serum androgens and prostate cancer. *Cancer Epidemiol. Biomarkers Prev.* **5**, 621–625.
25. Guess, H. A., Friedman, G. D., Sadler, M. C., Stanczyk, F. Z., Vogelman. J. H., Imperato-McGinley, J., Lobo, R. A., and Orentreich, N. (1997). 5-Alpha-reductase activity and prostate cancer: A case-control study using stored sera. *Cancer Epidemiol. Biomarkers Prev.* **6**, 21–24.
26. Vatten, L. J., Ursin, G., Ross, R. K., Stanczyk, F. Z., Lobo, R. A., Harvei, S., and Jellum, E. (1997). Androgens in serum and the risk of prostate cancer: A nested case-control study from the Janus serum bank in Norway. *Cancer Epidemiol. Biomarkers Prev.* **6**, 967–969.
27. Mohr, B. A., Feldman, H. A., Kalish, L. A., Longcope, C., and McKinlay, J. B. (2001). Are serum hormones associated with the risk of prostate cancer? Prospective results from the Massachusetts Male Aging Study. *Urology* **57**, 930–935.
28. Chen, C., Weiss, N. S., Stanczyk, F. Z., Lewis, S. K., DiTommaso, D., Etzioni, R., Barnett, M. J., and Goodman, G. E. (2003). Endogenous sex hormones and prostate cancer risk: A case-control study nested within the Carotene and Retinol Efficacy Trial. *Cancer Epidemiol. Biomarkers Prev.* **12**, 1410–1416.

29. Tsai, C. J., Cohn, B. A., Cirillo, P. M., Feldman, D., Stanczyk, F. Z., and Whittemore, A. S. (2006). Sex steroid hormones in young manhood and the risk of subsequent prostate cancer: A longitudinal study in African-Americans and Caucasians (United States). *Cancer Causes Control* **17**, 1237–1244.
30. Nomura, A., Heilbrun, L. K., Stemmermann, G. N., and Judd, H. L. (1988). Prediagnostic serum hormones and the risk of prostate cancer. *Cancer Res.* **48**, 3515–3517.
31. Hsing, A. W., and Comstock, G. W. (1993). Serological precursors of cancer: Serum hormones and risk of subsequent prostate cancer. *Cancer Epidemiol. Biomarkers Prev.* **2**, 27–32.
32. Gann, P. H., Hennekens, C. H., Ma, J., Longcope, C., and Stampfer, M. J. (1996). Prospective study of sex hormone levels and risk of prostate cancer. *J. Natl. Cancer Inst.* **88**, 1118–1126.
33. Hamalainen, E., Adlercreutz, H., Puska, P., and Pietinen, P. (1984). Diet and serum sex hormones in healthy men. *J. Steroid Biochem.* **20**, 459–464.
34. Hill, P., Wynder, E. L., Garbaczewski, L., Garnes, H., and Walker, A. R. (1979). Diet and urinary steroids in black and white North American men and black South African men. *Cancer Res.* **39**, 5101–5105.
35. Berrino, F., Bellati, C., Secreto, G., Camerini, E., Pala, V., Panico, S., Allegro, G., and Kaaks, R. (2001). Reducing bioavailable sex hormones through a comprehensive change in diet: The diet and androgens (DIANA) randomized trial. *Cancer Epidemiol. Biomarkers Prev.* **10**, 25–33.
36. Wang, C., Catlin, D. H., Starcevic, B., Heber, D., Ambler, C., Berman, N., Lucas, G., Leung, A., Schramm, K., Lee, P. W., Hull, L., and Swerdloff, R. S. (2005). Low-fat high-fiber diet decreased serum and urine androgens in men. *J. Clin. Endocrinol. Metab.* **90**, 3550–3559.
37. Howell, M. A. (1974). Factor analysis of international cancer mortality data and per capita food consumption. *Br. J. Cancer.* **29**, 328–336.
38. Armstrong, B., and Doll, R. (1975). Environmental factors and cancer incidence and mortality in different countries, with special reference to dietary practices. *Int. J. Cancer.* **15**, 617–631.
39. Koo, L. C., Mang, O. W., and Ho, J. H. (1997). An ecological study of trends in cancer incidence and dietary changes in Hong Kong. *Nutr. Cancer.* **28**, 289–301.
40. Ferguson, L. R. (2002). Meat consumption, cancer risk and population groups within New Zealand. *Mutat. Res.* **506**, 215–224.
41. Colli, J. L., and Colli, A. (2005). Comparisons of prostate cancer mortality rates with dietary practices in the United States. *Urol. Oncol.* **23**, 390–398.
42. Talamini, R., La Vecchia, C., Decarli, A., Negri, E., and Franceschi, S. (1986). Nutrition, social factors and prostatic cancer in a Northern Italian population. *Br. J. Cancer* **53**, 817–821.
43. Bravo, M. P., Castellanos, E., and del Rey Calero, J. (1991). Dietary factors and prostatic cancer. *Urol. Int.* **46**, 163–166.
44. Walker, A. R., Walker, B. F., Tsotetsi, N. G., Sebitso, C., Siwedi, D., and Walker, A. J. (1992). Case-control study of prostate cancer in black patients in Soweto, South Africa. *Br. J. Cancer* **65**, 438–441.
45. Talamini, R., Franceschi, S., La Vecchia, C., Serraino, D., Barra, S., and Negri, E. (1992). Diet and prostatic cancer: A case-control study in northern Italy. *Nutr. Cancer* **18**, 277–286.
46. Hayes, R. B., Ziegler, R. G., Gridley, G., Swanson, C., Greenberg, R. S., Swanson, G. M., Schoenberg, J. B., Silverman, D. T., Brown, L. M., Pottern, L. M., Liff, J., Schwartz, A. G., Fraumeni, J. F., Jr., and Hoover, R. N. (1999). Dietary factors and risks for prostate cancer among blacks and whites in the United States. *Cancer Epidemiol. Biomarkers Prev.* **8**, 25–34.
47. Deneo-Pellegrini, H., De Stefani, E., Ronco, A., and Mendilaharsu, M. (1999). Foods, nutrients and prostate cancer: A case-control study in Uruguay. *Br. J. Cancer* **80**, 591–597.
48. Severson, R. K., Nomura, A. M., Grove, J. S., and Stemmermann, G. N. (1989). A prospective study of demographics, diet, and prostate cancer among men of Japanese ancestry in Hawaii. *Cancer Res.* **49**, 1857–1860.
49. Mills, P. K., Beeson, W. L., Phillips, R. L., and Fraser, G. E. (1989). Cohort study of diet, lifestyle, and prostate cancer in Adventist men. *Cancer* **64**, 598–604.
50. Giovannucci, E., Rimm, E. B., Colditz, G. A., Stampfer, M. J., Ascherio, A., Chute, C. C., and Willett, W. C. (1993). A prospective study of dietary fat and risk of prostate cancer. *J. Natl. Cancer Inst.* **85**, 1571–1579.
51. Le Marchand, L., Kolonel, L. N., Wilkens, L. R., Myers, B. C., and Hirohata, T. (1994). Animal fat consumption and prostate cancer: A prospective study in Hawaii. *Epidemiology* **5**, 276–282.
52. Schuurman, A. G., van den Brandt, P. A., Dorant, E., and Goldbohm, R. A. (1999). Animal products, calcium and protein and prostate cancer risk in The Netherlands Cohort Study. *Br. J. Cancer* **80**, 1107–1113.
53. Michaud, D. S., Augustsson, K., Rimm, E. B., Stampfer, M. J., Willet, W. C., and Giovannucci, E. (2001). A prospective study on intake of animal products and risk of prostate cancer. *Cancer Causes Control* **12**, 557–567.
54. Rodriguez, C., McCullough, M. L., Mondul, A. M., Jacobs, E. J., Chao, A., Patel, A. V., Thun, M. J., and Calle, E. E. (2006). Meat consumption among Black and White men and risk of prostate cancer in the Cancer Prevention Study II Nutrition Cohort. *Cancer Epidemiol. Biomarkers Prev.* **15**, 211–216.
55. Gronberg, H., Damber, L., and Damber, J. E. (1996). Total food consumption and body mass index in relation to prostate cancer risk: A case-control study in Sweden with prospectively collected exposure data. *J. Urol.* **155**, 969–974.
56. Key, T. J., Silcocks, P. B., Davey, G. K., Appleby, P. N., and Bishop, D. T. (1997). A case-control study of diet and prostate cancer. *Br. J. Cancer* **76**, 678–687.
57. Villeneuve, P. J., Johnson, K. C., Kreiger, N., and Mao, Y. (1999). Risk factors for prostate cancer: Results from the Canadian National Enhanced Cancer Surveillance System. The Canadian Cancer Registries Epidemiology Research Group. *Cancer Causes Control* **10**, 355–367.
58. Tzonou, A., Signorello, L. B., Lagiou, P., Wuu, J., Trichopoulos, D., and Trichopoulou, A. (1999). Diet and cancer of the prostate: A case-control study in Greece. *Int. J. Cancer* **80**, 704–708.
59. Bosetti, C., Micelotta, S., Dal Maso, L., Talamini, R., Montella, M., Negri, E., Conti, E., Franceschi, S., and La Vecchia, C.

(2004). Food groups and risk of prostate cancer in Italy. *Int. J. Cancer* **110**, 424–428.

60. Hodge, A. M., English, D. R., McCredie, M. R., Severi, G., Boyle, P., Hopper, J. L., and Giles, G. G. (2004). Foods, nutrients and prostate cancer. *Cancer Causes Control* **15**, 11–20.
61. Sonoda, T., Nagata, Y., Mori, M., Miyanaga, N., Takashima, N., Okumura, K., Goto, K., Naito, S., Fujimoto, K., Hirao, Y., Takahashi, A., Tsukamoto, T., Fujioka, T., and Akaza, H. (2004). A case-control study of diet and prostate cancer in Japan: Possible protective effect of traditional Japanese diet. *Cancer Sci.* **95**, 238–242.
62. McCann, S. E., Ambrosone, C. B., Moysich, K. B., Brasure, J., Marshall, J. R., Freudenheim, J. L., Wilkinson, G. S., and Graham, S. (2005). Intakes of selected nutrients, foods, and phytochemicals and prostate cancer risk in western New York. *Nutr. Cancer* **53**, 33–41.
63. Hirayama, T. (1979). "Epidemiology of Prostate Cancer with Special Reference to the Role of Diet." U.S. Government Printing Office, Washington, DC.
64. Hsing, A. W., McLaughlin, J. K., Schuman, L. M., Bjelke, E., Gridley, G., Wacholder, S., Chien, H. T., and Blot, W. J. (1990). Diet, tobacco use, and fatal prostate cancer: Results from the Lutheran Brotherhood Cohort Study. *Cancer Res.* **50**, 6836–6840.
65. Chan, J. M., Pietinen, P., Virtanen, M., Malila, N., Tangrea, J., Albanes, D., and Virtamo, J. (2000). Diet and prostate cancer risk in a cohort of smokers, with a specific focus on calcium and phosphorus (Finland). *Cancer Causes Control* **11**, 859–867.
66. Allen, N. E., Sauvaget, C., Roddam, A. W., Appleby, P., Nagano, J., Suzuki, G., Key, T. J., and Koyama, K. (2004). A prospective study of diet and prostate cancer in Japanese men. *Cancer Causes Control* **15**, 911–920.
67. Cross, A. J., Peters, U., Kirsh, V. A., Andriole, G. L., Reding, D., Hayes, R. B., and Sinha, R. (2005). A prospective study of meat and meat mutagens and prostate cancer risk. *Cancer Res.* **65**, 11779–11784.
68. Rohrmann, S., Platz, E. A., Kavanaugh, C. J., Thuita, L., Hoffman, S. C., and Helzlsouer, K. J. (2007). Meat and dairy consumption and subsequent risk of prostate cancer in a U.S. cohort study. *Cancer Causes Control.* **18**, 41–50.
69. Tappel, A. (2007). Heme of consumed red meat can act as a catalyst of oxidative damage and could initiate colon, breast and prostate cancers, heart disease and other diseases. *Med. Hypotheses* **68**, 562–564.
70. Shirai, T., Sano, M., Tamano, S., Takahashi, S., Hirose, M., Futakuchi, M., Hasegawa, R., Imaida, K., Matsumoto, K., Wakabayashi, K., Sugimura, T., and Ito, N. (1997). The prostate: A target for carcinogenicity of 2-amino-1-methyl-6-phenylimidazo[4,5-*b*]pyridine (PhIP) derived from cooked foods. *Cancer Res.* **57**, 195–198.
71. Sugimura, T., Wakabayashi, K., Nakagama, H., and Nagao, M. (2004). Heterocyclic amines: Mutagens/carcinogens produced during cooking of meat and fish. *Cancer Sci.* **95**, 290–299.
72. Lijinsky, W., and Shubik, P. (1964). Benzo(*a*)pyrene and other polynuclear hydrocarbons in charcoal-broiled meat. *Science* **145**, 53–55.
73. Norrish, A. E., Ferguson, L. R., Knize, M. G., Felton, J. S., Sharpe, S. J., and Jackson, R. T. (1999). Heterocyclic amine content of cooked meat and risk of prostate cancer. *J. Natl. Cancer Inst.* **91**, 2038–2044.
74. Mettlin, C., Selenskas, S., Natarajan, N., and Huben, R. (1989). Beta-carotene and animal fats and their relationship to prostate cancer risk. A case-control study. *Cancer* **64**, 605–612.
75. Chan, J. M., Giovannucci, E., Andersson, S. O., Yuen, J., Adami, H. O., and Wolk, A. (1998). Dairy products, calcium, phosphorus, vitamin D, and risk of prostate cancer (Sweden). *Cancer Causes Control* **9**, 559–566.
76. Snowdon, D. A., Phillips, R. L., and Choi, W. (1984). Diet, obesity, and risk of fatal prostate cancer. *Am. J. Epidemiol.* **120**, 244–250.
77. Chan, J. M., Stampfer, M. J., Ma, J., Gann, P. H., Gaziano, J. M., and Giovannucci, E. L. (2001). Dairy products, calcium, and prostate cancer risk in the Physicians' Health Study. *Am. J. Clin. Nutr.* **74**, 549–554.
78. Tseng, M., Breslow, R. A., Graubard, B. I., and Ziegler, R. G. (2005). Dairy, calcium, and vitamin D intakes and prostate cancer risk in the National Health and Nutrition Examination Epidemiologic Follow-up Study cohort. *Am. J. Clin. Nutr.* **81**, 1147–1154.
79. Kesse, E., Bertrais, S., Astorg, P., Jaouen, A., Arnault, N., Galan, P., and Hercberg, S. (2006). Dairy products, calcium and phosphorus intake, and the risk of prostate cancer: Results of the French prospective SU.VI.MAX (Supplementation en Vitamines et Mineraux Antioxydants) study. *Br. J. Nutr.* **95**, 539–545.
80. Lee, M. M., Wang, R. T., Hsing, A. W., Gu, F. L., Wang, T., and Spitz, M. (1998). Case-control study of diet and prostate cancer in China. *Cancer Causes Control* **9**, 545–552.
81. Sanderson, M., Coker, A. L., Logan, P., Zheng, W., and Fadden, M. K. (2004). Lifestyle and prostate cancer among older African-American and Caucasian men in South Carolina. *Cancer Causes Control* **15**, 647–655.
82. Berndt, S. I., Carter, H. B., Landis, P. K., Tucker, K. L., Hsieh, L. J., Metter, E. J., and Platz, E. A. (2002). Calcium intake and prostate cancer risk in a long-term aging study: The Baltimore Longitudinal Study of Aging. *Urology* **60**, 1118–1123.
83. Rodriguez, C., McCullough, M. L., Mondul, A. M., Jacobs, E. J., Fakhrabadi-Shokoohi, D., Giovannucci, E. L., Thun, M. J., and Calle, E. E. (2003). Calcium, dairy products, and risk of prostate cancer in a prospective cohort of United States men. *Cancer Epidemiol. Biomarkers Prev.* **12**, 597–603.
84. Giovannucci, E., Liu, Y., Stampfer, M. J., and Willett, W. C. (2006). A prospective study of calcium intake and incident and fatal prostate cancer. *Cancer Epidemiol. Biomarkers Prev.* **15**, 203–210.
85. Koh, K. A., Sesso, H. D., Paffenbarger, R. S., Jr., and Lee, I. M. (2006). Dairy products, calcium and prostate cancer risk. *Br. J. Cancer* **95**, 1582–1585.
86. Qin, L. Q., Xu, J. Y., Wang, P. Y., Kaneko, T., Hoshi, K., and Sato, A. (2004). Milk consumption is a risk factor for prostate cancer: Meta-analysis of case-control studies. *Nutr. Cancer* **48**, 22–27.
87. Gao, X., LaValley, M. P., and Tucker, K. L. (2005). Prospective studies of dairy product and calcium intakes and prostate

cancer risk: A meta-analysis. *J. Natl. Cancer Inst.* **97**, 1768–1777.
88. Ross, R. K., Shimizu, H., Paganini-Hill, A., Honda, G., and Henderson, B. E. (1987). Case-control studies of prostate cancer in blacks and whites in southern California. *J. Natl. Cancer Inst.* **78**, 869–874.
89. Yu, H., Harris, R. E., and Wynder, E. L. (1988). Case-control study of prostate cancer and socioeconomic factors. *Prostate* **13**, 317–325.
90. Tavani, A., Negri, E., Franceschi, S., Talamini, R., and La Vecchia, C. (1994). Alcohol consumption and risk of prostate cancer. *Nutr. Cancer* **21**, 24–31.
91. Crispo, A., Talamini, R., Gallus, S., Negri, E., Gallo, A., Bosetti, C., La Vecchia, C., Dal Maso, L., and Montella, M. (2004). Alcohol and the risk of prostate cancer and benign prostatic hyperplasia. *Urology* **64**, 717–722.
92. Chang, E. T., Hedelin, M., Adami, H. O., Gronberg, H., and Balter, K. A. (2005). Alcohol drinking and risk of localized versus advanced and sporadic versus familial prostate cancer in Sweden. *Cancer Causes Control* **16**, 275–284.
93. Schoonen, W. M., Salinas, C. A., Kiemeney, L. A., and Stanford, J. L. (2005). Alcohol consumption and risk of prostate cancer in middle-aged men. *Int. J. Cancer* **113**, 133–140.
94. Hayes, R. B., Brown, L. M., Schoenberg, J. B., Greenberg, R. S., Silverman, D. T., Schwartz, A. G., Swanson, G. M., Benichou, J., Liff, J. M., Hoover, R. N., and Pottern, L. M. (1996). Alcohol use and prostate cancer risk in U.S. blacks and whites. *Am. J. Epidemiol.* **143**, 692–697.
95. Putnam, S. D., Cerhan, J. R., Parker, A. S., Bianchi, G. D., Wallace, R. B., Cantor, K. P., and Lynch, C. F. (2000). Lifestyle and anthropometric risk factors for prostate cancer in a cohort of Iowa men. *Ann. Epidemiol.* **10**, 361–369.
96. Sharpe, C. R., and Siemiatycki, J. (2001). Case-control study of alcohol consumption and prostate cancer risk in Montreal, Canada. *Cancer Causes Control* **12**, 589–598.
97. Hirayama, T. (1992). "Life-Style and Cancer: From Epidemiological Evidence to Public Behavior Change to Mortality Reduction of Target Cancers." U.S. Government Printing Office, Washington, DC.
98. Sesso, H. D., Paffenbarger, R. S., Jr., and Lee, I. M. (2001). Alcohol consumption and risk of prostate cancer: The Harvard Alumni Health Study. *Int. J. Epidemiol.* **30**, 749–755.
99. Schuurman, A. G., Goldbohm, R. A., and van den Brandt, P. A. (1999). A prospective cohort study on consumption of alcoholic beverages in relation to prostate cancer incidence (The Netherlands). *Cancer Causes Control* **10**, 597–605.
100. Hiatt, R. A., Armstrong, M. A., Klatsky, A. L., and Sidney, S. (1994). Alcohol consumption, smoking, and other risk factors and prostate cancer in a large health plan cohort in California (United States). *Cancer Causes Control* **5**, 66–72.
101. Breslow, R. A., Wideroff, L., Graubard, B. I., Erwin, D., Reichman, M. E., Ziegler, R. G., and Ballard-Barbash, R. (1999). Alcohol and prostate cancer in the NHANES I epidemiologic follow-up study. First National Health and Nutrition Examination Survey of the United States. *Ann. Epidemiol.* **9**, 254–261.
102. Ellison, L. F. (2000). Tea and other beverage consumption and prostate cancer risk: A Canadian retrospective cohort study. *Eur. J. Cancer Prev.* **9**, 125–130.
103. Albertsen, K., and Gronbaek, M. (2002). Does amount or type of alcohol influence the risk of prostate cancer? *Prostate* **52**, 297–304.
104. Platz, E. A., Leitzmann, M. F., Rimm, E. B., Willett, W. C., and Giovannucci, E. (2004). Alcohol intake, drinking patterns, and risk of prostate cancer in a large prospective cohort study. *Am. J. Epidemiol.* **159**, 444–453.
105. Baglietto, L., Severi, G., English, D. R., Hopper, J. L., and Giles, G. G. (2006). Alcohol consumption and prostate cancer risk: Results from the Melbourne Collaborative Cohort Study. *Int. J. Cancer.* **119**, 1501–1504.
106. Weinstein, S. J., Stolzenberg-Solomon, R., Pietinen, P., Taylor, P. R., Virtamo, J., and Albanes, D. (2006). Dietary factors of one-carbon metabolism and prostate cancer risk. *Am. J. Clin. Nutr.* **84**, 929–935.
107. Dennis, L. K. (2000). Meta-analysis for combining relative risks of alcohol consumption and prostate cancer. *Prostate* **42**, 56–66.
108. Boffetta, P., and Hashibe, M. (2006). Alcohol and cancer. *Lancet Oncol.* **7**, 149–156.
109. Andersson, S. O., Wolk, A., Bergstrom, R., Giovannucci, E., Lindgren, C., Baron, J., and Adami, H. O. (1996). Energy, nutrient intake and prostate cancer risk: A population-based case-control study in Sweden. *Int. J. Cancer* **68**, 716–722.
110. Rohan, T. E., Howe, G. R., Burch, J. D., and Jain, M. (1995). Dietary factors and risk of prostate cancer: A case-control study in Ontario, Canada. *Cancer Causes Control* **6**, 145–154.
111. Hsieh, L. J., Carter, H. B., Landis, P. K., Tucker, K. L., Metter, E. J., Newschaffer, C. J., and Platz, E. A. (2003). Association of energy intake with prostate cancer in a long-term aging study: Baltimore Longitudinal Study of Aging (United States). *Urology* **61**, 297–301.
112. Whittemore, A. S., Kolonel, L. N., Wu, A. H., John, E. M., Gallagher, R. P., Howe, G. R., Burch, J. D., Hankin, J., Dreon, D. M., West, D. W., and *et al.* (1995). Prostate cancer in relation to diet, physical activity, and body size in blacks, whites, and Asians in the United States and Canada. *J. Natl. Cancer Inst.* **87**, 652–661.
113. Ghadirian, P., Lacroix, A., Maisonneuve, P., Perret, C., Drouin, G., Perrault, J. P., Beland, G., Rohan, T. E., and Howe, G. R. (1996). Nutritional factors and prostate cancer: A case-control study of French Canadians in Montreal, Canada. *Cancer Causes Control* **7**, 428–436.
114. Kristal, A. R., Cohen, J. H., Qu, P., and Stanford, J. L. (2002). Associations of energy, fat, calcium, and vitamin D with prostate cancer risk. *Cancer Epidemiol. Biomarkers Prev.* **11**, 719–725.
115. Schuurman, A. G., van den Brandt, P. A., Dorant, E., Brants, H. A., and Goldbohm, R. A. (1999). Association of energy and fat intake with prostate carcinoma risk: Results from The Netherlands Cohort Study. *Cancer* **86**, 1019–1027.
116. Ramon, J. M., Bou, R., Romea, S., Alkiza, M. E., Jacas, M., Ribes, J., and Oromi, J. (2000). Dietary fat intake and

prostate cancer risk: A case-control study in Spain. *Cancer Causes Control* **11**, 679–685.

117. Mukherjee, P., Sotnikov, A. V., Mangian, H. J., Zhou, J. R., Visek, W. J., and Clinton, S. K. (1999). Energy intake and prostate tumor growth, angiogenesis, and vascular endothelial growth factor expression. *J. Natl. Cancer Inst.* **91**, 512–523.
118. Kolonel, L. N., Nomura, A. M., and Cooney, R. V. (1999). Dietary fat and prostate cancer: Current status. *J. Natl. Cancer Inst.* **91**, 414–428.
119. Kolonel, L. N. (2001). Fat, meat, and prostate cancer. *Epidemiol. Rev.* **23**, 72–81.
120. Kushi, L., and Giovannucci, E. (2002). Dietary fat and cancer. *Am. J. Med.* **113**, 63S–70S.
121. Graham, S., Haughey, B., Marshall, J., Priore, R., Byers, T., Rzepka, T., Mettlin, C., and Pontes, J. E. (1983). Diet in the epidemiology of carcinoma of the prostate gland. *J. Natl. Cancer Inst.* **70**, 687–692.
122. Kolonel, L. N., Yoshizawa, C. N., and Hankin, J. H. (1988). Diet and prostatic cancer: A case-control study in Hawaii. *Am. J. Epidemiol.* **127**, 999–1012.
123. West, D. W., Slattery, M. L., Robison, L. M., French, T. K., and Mahoney, A. W. (1991). Adult dietary intake and prostate cancer risk in Utah: A case-control study with special emphasis on aggressive tumors. *Cancer Causes Control* **2**, 85–94.
124. Harvei, S., Bjerve, K. S., Tretli, S., Jellum, E., Robsahm, T. E., and Vatten, L. (1997). Prediagnostic level of fatty acids in serum phospholipids: Omega-3 and omega-6 fatty acids and the risk of prostate cancer. *Int. J. Cancer* **71**, 545–551.
125. Dennis, L. K., Snetselaar, L. G., Smith, B. J., Stewart, R. E., and Robbins, M. E. (2004). Problems with the assessment of dietary fat in prostate cancer studies. *Am. J. Epidemiol.* **160**, 436–444.
126. Bidoli, E., Talamini, R., Bosetti, C., Negri, E., Maruzzi, D., Montella, M., Franceschi, S., and La Vecchia, C. (2005). Macronutrients, fatty acids, cholesterol and prostate cancer risk. *Ann. Oncol.* **16**, 152–157.
127. Terry, P., Lichtenstein, P., Feychting, M., Ahlbom, A., and Wolk, A. (2001). Fatty fish consumption and risk of prostate cancer. *Lancet* **357**, 1764–1766.
128. Augustsson, K., Michaud, D. S., Rimm, E. B., Leitzmann, M. F., Stampfer, M. J., Willett, W. C., and Giovannucci, E. (2003). A prospective study of intake of fish and marine fatty acids and prostate cancer. *Cancer Epidemiol. Biomarkers Prev.* **12**, 64–67.
129. Terry, P. D., Terry, J. B., and Rohan, T. E. (2004). Long-chain (n-3) fatty acid intake and risk of cancers of the breast and the prostate: Recent epidemiological studies, biological mechanisms, and directions for future research. *J. Nutr.* **134**, 3412S–3420S.
130. Astorg, P. (2004). Dietary N-6 and N-3 polyunsaturated fatty acids and prostate cancer risk: A review of epidemiological and experimental evidence. *Cancer Causes Control* **15**, 367–386.
131. Norrish, A. E., Skeaff, C. M., Arribas, G. L., Sharpe, S. J., and Jackson, R. T. (1999). Prostate cancer risk and consumption of fish oils: A dietary biomarker-based case-control study. *Br. J. Cancer* **81**, 1238–1242.
132. Leitzmann, M. F., Stampfer, M. J., Michaud, D. S., Augustsson, K., Colditz, G. C., Willett, W. C., and Giovannucci, E. L. (2004). Dietary intake of n-3 and n-6 fatty acids and the risk of prostate cancer. *Am. J. Clin. Nutr.* **80**, 204–216.
133. Newcomer, L. M., King, I. B., Wicklund, K. G., and Stanford, J. L. (2001). The association of fatty acids with prostate cancer risk. *Prostate* **47**, 262–268.
134. Mannisto, S., Pietinen, P., Virtanen, M. J., Salminen, I., Albanes, D., Giovannucci, E., and Virtamo, J. (2003). Fatty acids and risk of prostate cancer in a nested case-control study in male smokers. *Cancer Epidemiol. Biomarkers Prev.* **12**, 1422–1428.
135. De Stefani, E., Deneo-Pellegrini, H., Boffetta, P., Ronco, A., and Mendilaharsu, M. (2000). Alpha-linolenic acid and risk of prostate cancer: A case-control study in Uruguay. *Cancer Epidemiol. Biomarkers Prev.* **9**, 335–338.
136. Koralek, D. O., Peters, U., Andriole, G., Reding, D., Kirsh, V., Subar, A., Schatzkin, A., Hayes, R., and Leitzmann, M. F. (2006). A prospective study of dietary alpha-linolenic acid and the risk of prostate cancer (United States). *Cancer Causes Control* **17**, 783–791.
137. Brouwer, I. A., Katan, M. B., and Zock, P. L. (2004). Dietary alpha-linolenic acid is associated with reduced risk of fatal coronary heart disease, but increased prostate cancer risk: A meta-analysis. *J. Nutr.* **134**, 919–922.
138. King, I. B., Kristal, A. R., Schaffer, S., Thornquist, M., and Goodman, G. E. (2005). Serum trans-fatty acids are associated with risk of prostate cancer in beta-Carotene and Retinol Efficacy Trial. *Cancer Epidemiol. Biomarkers Prev.* **14**, 988–992.
139. Pollard, M., and Luckert, P. H. (1986). Promotional effects of testosterone and high fat diet on the development of autochthonous prostate cancer in rats. *Cancer Lett.* **32**, 223–227.
140. Clinton, S. K., Palmer, S. S., Spriggs, C. E., and Visek, W. J. (1988). Growth of Dunning transplantable prostate adenocarcinomas in rats fed diets with various fat contents. *J. Nutr.* **118**, 908–914.
141. Wang, Y., Corr, J. G., Thaler, H. T., Tao, Y., Fair, W. R., and Heston, W. D. (1995). Decreased growth of established human prostate LNCaP tumors in nude mice fed a low-fat diet. *J. Natl. Cancer Inst.* **87**, 1456–1462.
142. Ngo, T. H., Barnard, R. J., Anton, T., Tran, C., Elashoff, D., Heber, D., Freedland, S. J., and Aronson, W. J. (2004). Effect of isocaloric low-fat diet on prostate cancer xenograft progression to androgen independence. *Cancer Res.* **64**, 1252–1254.
143. Pandalai, P. K., Pilat, M. J., Yamazaki, K., Naik, H., and Pienta, K. J. (1996). The effects of omega-3 and omega-6 fatty acids on *in vitro* prostate cancer growth. *Anticancer Res.* **16**, 815–820.
144. Karmali, R. A., Reichel, P., Cohen, L. A., Terano, T., Hirai, A., Tamura, Y., and Yoshida, S. (1987). The effects of dietary omega-3 fatty acids on the DU-145 transplantable human prostatic tumor. *Anticancer Res.* **7**, 1173–1179.
145. Tavani, A., Gallus, S., Franceschi, S., and La Vecchia, C. (2001). Calcium, dairy products, and the risk of prostate cancer. *Prostate* **48**, 118–121.

146. Tavani, A., Bertuccio, P., Bosetti, C., Talamini, R., Negri, E., Franceschi, S., Montella, M., and La Vecchia, C. (2005). Dietary intake of calcium, vitamin D, phosphorus and the risk of prostate cancer. *Eur. Urol.* **48**, 27–33.
147. Baron, J. A., Beach, M., Wallace, K., Grau, M. V., Sandler, R. S., Mandel, J. S., Heber, D., and Greenberg, E. R. (2005). Risk of prostate cancer in a randomized clinical trial of calcium supplementation. *Cancer Epidemiol. Biomarkers Prev.* **14**, 586–589.
148. Feldman, D., Zhao, X. Y., and Krishnan, A. V. (2000). Vitamin D and prostate cancer. *Endocrinology* **141**, 5–9.
149. Bonjour, J. P., Chevalley, T., and Fardellone, P. (2007). Calcium intake and vitamin D metabolism and action, in healthy conditions and in prostate cancer. *Br. J. Nutr.* **97**, 611–616.
150. Schrauzer, G. N., White, D. A., and Schneider, C. J. (1977). Cancer mortality correlation studies—III: Statistical associations with dietary selenium intakes. *Bioinorg. Chem.* **7**, 23–31.
151. Leitzmann, M. F., Stampfer, M. J., Wu, K., Colditz, G. A., Willett, W. C., and Giovannucci, E. L. (2003). Zinc supplement use and risk of prostate cancer. *J. Natl. Cancer Inst.* **95**, 1004–1007.
152. Kristal, A. R., Stanford, J. L., Cohen, J. H., Wicklund, K., and Patterson, R. E. (1999). Vitamin and mineral supplement use is associated with reduced risk of prostate cancer. *Cancer Epidemiol. Biomarkers Prev.* **8**, 887–892.
153. Shils, M. E., Olson, J. A., and Shike, M. Eds. (1994). "Modern Nutrition in Health and Disease," 8th ed. Lea & Febiger, Philadelphia.
154. Feustel, A., and Wennrich, R. (1986). Zinc and cadmium plasma and erythrocyte levels in prostatic carcinoma, BPH, urological malignancies, and inflammations. *Prostate* **8**, 75–79.
155. Whelan, P., Walker, B. E., and Kelleher, J. (1983). Zinc, vitamin A and prostatic cancer. *Br. J. Urol.* **55**, 525–528.
156. Platz, E. A., Helzlsouer, K. J., Hoffman, S. C., Morris, J. S., Baskett, C. K., and Comstock, G. W. (2002). Prediagnostic toenail cadmium and zinc and subsequent prostate cancer risk. *Prostate* **52**, 288–296.
157. Tisell, L. E., Fjelkegard, B., and Leissner, K. H. (1982). Zinc concentration and content of the dorsal, lateral and medical prostatic lobes and of periurethral adenomas in man. *J. Urol.* **128**, 403–405.
158. Feustel, A., and Wennrich, R. (1984). Determination of the distribution of zinc and cadmium in cellular fractions of BPH, normal prostate and prostatic cancers of different histologies by atomic and laser absorption spectrometry in tissue slices. *Urol. Res.* **12**, 253–256.
159. Costello, L. C., Franklin, R. B., Feng, P., Tan, M., and Bagasra, O. (2005). Zinc and prostate cancer: A critical scientific, medical, and public interest issue (United States). *Cancer Causes Control* **16**, 901–915.
160. Habib, F. K., Mason, M. K., Smith, P. H., and Stitch, S. R. (1979). Cancer of the prostate: Early diagnosis by zinc and hormone analysis? *Br. J. Cancer* **39**, 700–704.
161. Hartoma, T. R., Nahoul, K., and Netter, A. (1977). Zinc, plasma androgens and male sterility. *Lancet* **2**, 1125–1126.
162. Om, A. S., and Chung, K. W. (1996). Dietary zinc deficiency alters 5 alpha-reduction and aromatization of testosterone and androgen and estrogen receptors in rat liver. *J. Nutr.* **126**, 842–848.
163. Colvard, D. S., and Wilson, E. M. (1984). Zinc potentiation of androgen receptor binding to nuclei in vitro. *Biochemistry (Mosc.)* **23**, 3471–3478.
164. Vermeulen, A., Kaufman, J. M., and Giagulli, V. A. (1996). Influence of some biological indexes on sex hormone-binding globulin and androgen levels in aging or obese males. *J. Clin. Endocrinol. Metab.* **81**, 1821–1826.
165. Wu, A. H., Whittemore, A. S., Kolonel, L. N., John, E. M., Gallagher, R. P., West, D. W., Hankin, J., Teh, C. Z., Dreon, D. M., and Paffenbarger, R. S. Jr. (1995). Serum androgens and sex hormone-binding globulins in relation to lifestyle factors in older African-American, white, and Asian men in the United States and Canada. *Cancer Epidemiol. Biomarkers Prev.* **4**, 735–741.
166. Gray, A., Feldman, H. A., McKinlay, J. B., and Longcope, C. (1991). Age, disease, and changing sex hormone levels in middle-aged men: Results of the Massachusetts Male Aging Study. *J. Clin. Endocrinol. Metab.* **73**, 1016–1025.
167. Sahmoun, A. E., Case, L. D., Jackson, S. A., and Schwartz, G. G. (2005). Cadmium and prostate cancer: A critical epidemiologic analysis. *Cancer Invest.* **23**, 256–263.
168. Kolonel, L., and Winkelstein, W. Jr. (1977). Cadmium and prostatic carcinoma. *Lancet* **2**, 566–567.
169. Elghany, N. A., Schumacher, M. C., Slattery, M. L., West, D. W., and Lee, J. S. (1990). Occupation, cadmium exposure, and prostate cancer. *Epidemiology* **1**, 107–115.
170. Haddow, A., Roe, F. J., Dukes, C. E., and Mitchley, B. C. (1964). Cadmium neoplasia: Sarcomata at the site of injection of cadmium sulphate in rats and mice. *Br. J. Cancer* **18**, 667–673.
171. Gunn, S. A., Gould, T. C., and Anderson, W. A. (1964). Effect of zinc on cancerogenesis by cadmium. *Proc. Soc. Exp. Biol. Med.* **115**, 653–657.
172. Lew, E. A., and Garfinkel, L. (1979). Variations in mortality by weight among 750,000 men and women. *J. Chronic Dis.* **32**, 563–576.
173. Hsing, A. W., Deng, J., Sesterhenn, I. A., Mostofi, F. K., Stanczyk, F. Z., Benichou, J., Xie, T., and Gao, Y. T. (2000). Body size and prostate cancer: A population-based case-control study in China. *Cancer Epidemiol. Biomarkers Prev.* **9**, 1335–1341.
174. von Hafe, P., Pina, F., Perez, A., Tavares, M., and Barros, H. (2004). Visceral fat accumulation as a risk factor for prostate cancer. *Obes. Res.* **12**, 1930–1935.
175. Freedland, S. J., Terris, M. K., Platz, E. A., and Presti, J. C. Jr., (2005). Body mass index as a predictor of prostate cancer: Development versus detection on biopsy. *Urology* **66**, 108–113.
176. Rodriguez, C., Patel, A. V., Calle, E. E., Jacobs, E. J., Chao, A., and Thun, M. J. (2001). Body mass index, height, and prostate cancer mortality in two large cohorts of adult men in the United States. *Cancer Epidemiol. Biomarkers Prev.* **10**, 345–353.
177. Engeland, A., Tretli, S., and Bjorge, T. (2003). Height, body mass index, and prostate cancer: A follow-up of 950,000 Norwegian men. *Br. J. Cancer* **89**, 1237–1242

178. Calle, E. E., Rodriguez, C., Walker-Thurmond, K., and Thun, M. J. (2003). Overweight, obesity, and mortality from cancer in a prospectively studied cohort of U.S. adults. *N. Engl. J. Med.* **348**, 1625–1638.
179. Bassett, W. W., Cooperberg, M. R., Sadetsky, N., Silva, S., DuChane, J., Pasta, D. J., Chan, J. M., Anast, J. W., Carroll, P. R., and Kane, C. J. (2005). Impact of obesity on prostate cancer recurrence after radical prostatectomy: Data from CaPSURE. *Urology* **66**, 1060–1065.
180. Nomura, A., Heilbrun, L. K., and Stemmermann, G. N. (1985). Body mass index as a predictor of cancer in men. *J. Natl. Cancer Inst.* **74**, 319–323.
181. Thompson, M. M., Garland, C., Barrett-Connor, E., Khaw, K. T., Friedlander, N. J., and Wingard, D. L. (1989). Heart disease risk factors, diabetes, and prostatic cancer in an adult community. *Am. J. Epidemiol.* **129**, 511–517.
182. Nilsen, T. I., and Vatten, L. J. (1999). Anthropometry and prostate cancer risk: A prospective study of 22,248 Norwegian men. *Cancer Causes Control* **10**, 269–275.
183. Boland, L. L., Mink, P. J., Bushhouse, S. A., and Folsom, A. R. (2003). Weight and length at birth and risk of early-onset prostate cancer (United States). *Cancer Causes Control* **14**, 335–338.
184. Friedenreich, C. M., McGregor, S. E., Courneya, K. S., Angyalfi, S. J., and Elliott, F. G. (2004). Case-control study of anthropometric measures and prostate cancer risk. *Int. J. Cancer* **110**, 278–283.
185. Liu, X., Rybicki, B. A., Casey, G., and Witte, J. S. (2005). Relationship between body size and prostate cancer in a sibling based case-control study. *J. Urol.* **174**, 2169–2173.
186. Cox, B., Sneyd, M. J., Paul, C., and Skegg, D. C. (2006). Risk factors for prostate cancer: A national case-control study. *Int. J. Cancer* **119**, 1690–1694.
187. Baillargeon, J., Platz, E. A., Rose, D. P., Pollock, B. H., Ankerst, D. P., Haffner, S., Higgins, B., Lokshin, A., Troyer, D., Hernandez, J., Lynch, S., Leach, R. J., and Thompson, I. M. (2006). Obesity, adipokines, and prostate cancer in a prospective population-based study. *Cancer Epidemiol. Biomarkers Prev.* **15**, 1331–1335.
188. Habel, L. A., Van Den Eeden, S. K., and Friedman, G. D. (2000). Body size, age at shaving initiation, and prostate cancer in a large, multiracial cohort. *Prostate* **43**, 136–143.
189. Lee, I. M., Sesso, H. D., and Paffenbarger, R. S. Jr. (2001). A prospective cohort study of physical activity and body size in relation to prostate cancer risk (United States). *Cancer Causes Control* **12**, 187–193.
190. MacInnis, R. J., English, D. R., Gertig, D. M., Hopper, J. L., and Giles, G. G. (2003). Body size and composition and prostate cancer risk. *Cancer Epidemiol. Biomarkers Prev.* **12**, 1417–1421.
191. Jonsson, F., Wolk, A., Pedersen, N. L., Lichtenstein, P., Terry, P., Ahlbom, A., and Feychting, M. (2003). Obesity and hormone-dependent tumors: Cohort and co-twin control studies based on the Swedish Twin Registry. *Int. J. Cancer* **106**, 594–599.
192. Hubbard, J. S., Rohrmann, S., Landis, P. K., Metter, E. J., Muller, D. C., Andres, R., Carter, H. B., and Platz, E. A. (2004). Association of prostate cancer risk with insulin, glucose, and anthropometry in the Baltimore longitudinal study of aging. *Urology* **63**, 253–258.
193. Kurahashi, N., Iwasaki, M., Sasazuki, S., Otani, T., Inoue, M., and Tsugane, S. (2006). Association of body mass index and height with risk of prostate cancer among middle-aged Japanese men. *Br. J. Cancer* **94**, 740–742.
194. Porter, M. P., and Stanford, J. L. (2005). Obesity and the risk of prostate cancer. *Prostate* **62**, 316–321.
195. Bradbury, B. D., Wilk, J. B., and Kaye, J. A. (2005). Obesity and the risk of prostate cancer (United States). *Cancer Causes Control* **16**, 637–641.
196. Schuurman, A. G., Goldbohm, R. A., Dorant, E., and van den Brandt, P. A. (2000). Anthropometry in relation to prostate cancer risk in the Netherlands Cohort Study. *Am. J. Epidemiol.* **151**, 541–549.
197. Cerhan, J. R., Torner, J. C., Lynch, C. F., Rubenstein, L. M., Lemke, J. H., Cohen, M. B., Lubaroff, D. M., and Wallace, R. B. (1997). Association of smoking, body mass, and physical activity with risk of prostate cancer in the Iowa 65+ Rural Health Study (United States). *Cancer Causes Control* **8**, 229–238.
198. Giovannucci, E., Rimm, E. B., Stampfer, M. J., Colditz, G. A., and Willett, W. C. (1997). Height, body weight, and risk of prostate cancer. *Cancer Epidemiol. Biomarkers Prev.* **6**, 557–563.
199. Andersson, S. O., Wolk, A., Bergstrom, R., Adami, H. O., Engholm, G., Englund, A., and Nyren, O. (1997). Body size and prostate cancer: A 20–year follow-up study among 135,006 Swedish construction workers. *J. Natl. Cancer Inst.* **89**, 385–389.
200. Giovannucci, E., Rimm, E. B., Liu, Y., Leitzmann, M., Wu, K., Stampfer, M. J., and Willett, W. C. (2003). Body mass index and risk of prostate cancer in U.S. health professionals. *J. Natl. Cancer Inst.* **95**, 1240–1244.
201. Gong, Z., Neuhouser, M. L., Goodman, P. J., Albanes, D., Chi, C., Hsing, A. W., Lippman, S. M., Platz, E. A., Pollak, M. N., Thompson, I. M., and Kristal, A. R. (2006). Obesity, diabetes, and risk of prostate cancer: Results from the prostate cancer prevention trial. *Cancer Epidemiol. Biomarkers Prev.* **15**, 1977–1983.
202. Wright, M. E., Chang, S. C., Schatzkin, A., Albanes, D., Kipnis, V., Mouw, T., Hurwitz, P., Hollenbeck, A., and Leitzmann, M. F. (2007). Prospective study of adiposity and weight change in relation to prostate cancer incidence and mortality. *Cancer* **109**, 675–684.
203. Robinson, W. R., Stevens, J., Gammon, M. D., and John, E. M. (2005). Obesity before age 30 years and risk of advanced prostate cancer. *Am. J. Epidemiol.* **161**, 1107–1114.
204. MacInnis, R. J., and English, D. R. (2006). Body size and composition and prostate cancer risk: Systematic review and meta-regression analysis. *Cancer Causes Control* **17**, 989–1003.
205. Pasquali, R., Casimirri, F., Cantobelli, S., Melchionda, N., Morselli Labate, A. M., Fabbri, R., Capelli, M., and Bortoluzzi, L. (1991). Effect of obesity and body fat distribution on sex hormones and insulin in men. *Metabolism* **40**, 101–104.
206. Amling, C. L. (2005). Relationship between obesity and prostate cancer. *Curr Opin Urol.* **15**, 167–171.

207. Ohno, Y., Yoshida, O., Oishi, K., Okada, K., Yamabe, H., and Schroeder, F. H. (1988). Dietary beta-carotene and cancer of the prostate: A case-control study in Kyoto, Japan. *Cancer Res.* **48**, 1331–1336.
208. Jain, M. G., Hislop, G. T., Howe, G. R., and Ghadirian, P. (1999). Plant foods, antioxidants, and prostate cancer risk: Findings from case-control studies in Canada. *Nutr. Cancer* **34**, 173–184.
209. Kolonel, L. N., Hankin, J. H., Whittemore, A. S., Wu, A. H., Gallagher, R. P., Wilkens, L. R., John, E. M., Howe, G. R., Dreon, D. M., West, D. W., and Paffenbarger, R. S. Jr. (2000). Vegetables, fruits, legumes and prostate cancer: A multiethnic case-control study. *Cancer Epidemiol. Biomarkers Prev.* **9**, 795–804.
210. Cohen, J. H., Kristal, A. R., and Stanford, J. L. (2000). Fruit and vegetable intakes and prostate cancer risk. *J. Natl. Cancer Inst.* **92**, 61–68.
211. Hsing, A. W., Chokkalingam, A. P., Gao, Y. T., Madigan, M. P., Deng, J., Gridley, G., and Fraumeni, J. F. Jr. (2002). Allium vegetables and risk of prostate cancer: A population-based study. *J. Natl. Cancer Inst.* **94**, 1648–1651.
212. Le Marchand, L., Hankin, J. H., Kolonel, L. N., and Wilkens, L. R. (1991). Vegetable and fruit consumption in relation to prostate cancer risk in Hawaii: A reevaluation of the effect of dietary beta-carotene. *Am. J. Epidemiol.* **133**, 215–219.
213. Norrish, A. E., Jackson, R. T., Sharpe, S. J., and Skeaff, C. M. (2000). Prostate cancer and dietary carotenoids. *Am. J. Epidemiol.* **151**, 119–123.
214. Shibata, A., Paganini-Hill, A., Ross, R. K., and Henderson, B. E. (1992). Intake of vegetables, fruits, beta-carotene, vitamin C and vitamin supplements and cancer incidence among the elderly: A prospective study. *Br. J. Cancer* **66**, 673–679.
215. Giovannucci, E., Ascherio, A., Rimm, E. B., Stampfer, M. J., Colditz, G. A., and Willett, W. C. (1995). Intake of carotenoids and retinol in relation to risk of prostate cancer. *J. Natl. Cancer Inst.* **87**, 1767–1776.
216. Schuurman, A. G., Goldbohm, R. A., Dorant, E., and van den Brandt, P. A. (1998). Vegetable and fruit consumption and prostate cancer risk: A cohort study in The Netherlands. *Cancer Epidemiol. Biomarkers Prev.* **7**, 673–680.
217. Giovannucci, E., Rimm, E. B., Liu, Y., Stampfer, M. J., and Willett, W. C. (2003). A prospective study of cruciferous vegetables and prostate cancer. *Cancer Epidemiol. Biomarkers Prev.* **12**, 1403–1409.
218. Key, T. J., Allen, N., Appleby, P., Overvad, K., Tjonneland, A., Miller, A., Boeing, H., Karalis, D., Psaltopoulou, T., Berrino, F., Palli, D., Panico, S., Tumino, R., Vineis, P., Bueno-De-Mesquita, H. B., Kiemeney, L., Peeters, P. H., Martinez, C., Dorronsoro, M., Gonzalez, C. A., Chirlaque, M. D., Quiros, J. R., Ardanaz, E., Berglund, G., Egevad, L., Hallmans, G., Stattin, P., Bingham, S., Day, N., Gann, P., Kaaks, R., Ferrari, P., and Riboli, E. (2004). Fruits and vegetables and prostate cancer: No association among 1104 cases in a prospective study of 130,544 men in the European Prospective Investigation into Cancer and Nutrition (EPIC). *Int. J. Cancer* **109**, 119–124.
219. Stram, D. O., Hankin, J. H., Wilkens, L. R., Park, S., Henderson, B. E., Nomura, A. M., Pike, M. C., and Kolonel, L. N. (2006). Prostate cancer incidence and intake of fruits, vegetables and related micronutrients: The multiethnic cohort study (United States). *Cancer Causes Control* **17**, 1193–1207.
220. Steinmetz, K. A., and Potter, J. D. (1991). Vegetables, fruit, and cancer. II. Mechanisms. *Cancer Causes Control* **2**, 427–442.
221. Terry, P., Terry, J. B., and Wolk, A. (2001). Fruit and vegetable consumption in the prevention of cancer: An update. *J. Intern. Med.* **250**, 280–290.
222. Giovannucci, E., Rimm, E. B., Wolk, A., Ascherio, A., Stampfer, M. J., Colditz, G. A., and Willett, W. C. (1998). Calcium and fructose intake in relation to risk of prostate cancer. *Cancer Res.* **58**, 442–447.
223. Chan, J. M., and Giovannucci, E. L. (2001). Vegetables, fruits, associated micronutrients, and risk of prostate cancer. *Epidemiol. Rev.* **23**, 82–86.
224. Lee, M. M., Gomez, S. L., Chang, J. S., Wey, M., Wang, R. T., and Hsing, A. W. (2003). Soy and isoflavone consumption in relation to prostate cancer risk in China. *Cancer Epidemiol. Biomarkers Prev.* **12**, 665–668.
225. Jacobsen, B. K., Knutsen, S. F., and Fraser, G. E. (1998). Does high soy milk intake reduce prostate cancer incidence? The Adventist Health Study (United States). *Cancer Causes Control* **9**, 553–557.
226. Yan, L., and Spitznagel, E. L. (2005). Meta-analysis of soy food and risk of prostate cancer in men. *Int. J. Cancer* **117**, 667–669.
227. Wang, H., and Murphy, P. A. (1994). Isoflavone content in commercial soybean foods. *J. Agric. Food Chem.* **42**, 1666–1673.
228. Franke, A. A., Custer, L. J., Wang, W., and Shi, C. Y. (1998). HPLC analysis of isoflavonoids and other phenolic agents from foods and from human fluids. *Proc. Soc. Exp. Biol. Med.* **217**, 263–273.
229. Strom, S. S., Yamamura, Y., Duphorne, C. M., Spitz, M. R., Babaian, R. J., Pillow, P. C., and Hursting, S. D. (1999). Phytoestrogen intake and prostate cancer: A case-control study using a new database. *Nutr. Cancer* **33**, 20–25.
230. Kurahashi, N., Iwasaki, M., Sasazuki, S., Otani, T., Inoue, M., and Tsugane, S. (2007). Soy product and isoflavone consumption in relation to prostate cancer in Japanese men. *Cancer Epidemiol. Biomarkers Prev.* **16**, 538–545.
231. Ganry, O. (2005). Phytoestrogens and prostate cancer risk. *Prev. Med.* **41**, 1–6.
232. Fournier, D. B., Erdman, J. W., Jr., and Gordon, G. B. (1998). Soy, its components, and cancer prevention: A review of the *in vitro,* animal, and human data. *Cancer Epidemiol. Biomarkers Prev.* **7**, 1055–1065.
233. Messina, M. J. (2003). Emerging evidence on the role of soy in reducing prostate cancer risk. *Nutr. Rev.* **61**, 117–131.
234. Aronson, W. J., Tymchuk, C. N., Elashoff, R. M., McBride, W. H., McLean, C., Wang, H., and Heber, D. (1999). Decreased growth of human prostate LNCaP tumors in SCID mice fed a low-fat, soy protein diet with isoflavones. *Nutr. Cancer* **35**, 130–136.
235. Suzuki, K., Koike, H., Matsui, H., Ono, Y., Hasumi, M., Nakazato, H., Okugi, H., Sekine, Y., Oki, K., Ito, K., Yamamoto, T., Fukabori, Y., Kurokawa, K., and Yamanaka, H. (2002). Genistein, a soy isoflavone, induces glutathione

peroxidase in the human prostate cancer cell lines LNCaP and PC-3. *Int. J. Cancer* **99**, 846–852.

236. Rice, L., Handayani, R., Cui, Y., Medrano, T., Samedi, V., Baker, H., Szabo, N. J., Rosser, C. J., Goodison, S., and Shiverick, K. T. (2007). Soy isoflavones exert differential effects on androgen responsive genes in LNCaP human prostate cancer cells. *J. Nutr.* **137**, 964–972.
237. Rao, A. V., and Sung, M. K. (1995). Saponins as anticarcinogens. *J. Nutr.* **125**, 717S–724S.
238. Kennedy, A. R. (1998). The Bowman-Birk inhibitor from soybeans as an anticarcinogenic agent. *Am. J. Clin. Nutr.* **68**, 1406S–1412S.
239. Wyatt, C. J., Carballido, S. P., and Mendez, R. O. (1998). Alpha- and gamma-tocopherol content of selected foods in the Mexican diet: Effect of cooking losses. *J. Agric. Food Chem.* **46**, 4657–4661.
240. Heilbrun, L. K., Nomura, A., and Stemmermann, G. N. (1986). Black tea consumption and cancer risk: A prospective study. *Br. J. Cancer* **54**, 677–683.
241. Jian, L., Xie, L. P., Lee, A. H., and Binns, C. W. (2004). Protective effect of green tea against prostate cancer: A case-control study in southeast China. *Int. J. Cancer.* **108**, 130–135.
242. Kinlen, L. J., Willows, A. N., Goldblatt, P., and Yudkin, J. (1988). Tea consumption and cancer. *Br. J. Cancer* **58**, 397–401.
243. Kikuchi, N., Ohmori, K., Shimazu, T., Nakaya, N., Kuriyama, S., Nishino, Y., Tsubono, Y., and Tsuji, I. (2006). No association between green tea and prostate cancer risk in Japanese men: The Ohsaki Cohort Study. *Br. J. Cancer.* **95**, 371–373.
244. Sharpe, C. R., and Siemiatycki, J. (2002). Consumption of non-alcoholic beverages and prostate cancer risk. *Eur. J. Cancer Prev.* **11**, 497–501.
245. Yang, C. S., Chung, J. Y., Yang, G., Chhabra, S. K., and Lee, M. J. (2000). Tea and tea polyphenols in cancer prevention. *J. Nutr.* **130**, 472S–478S.
246. Holick, M. F. (2005). The vitamin D epidemic and its health consequences. *J. Nutr.* **135**, 2739S–2748S.
247. Giovannucci, E. (2005). The epidemiology of vitamin D and cancer incidence and mortality: A review (United States). *Cancer Causes Control* **16**, 83–95.
248. Ahonen, M. H., Tenkanen, L., Teppo, L., Hakama, M., and Tuohimaa, P. (2000). Prostate cancer risk and prediagnostic serum 25-hydroxyvitamin D levels (Finland). *Cancer Causes Control* **11**, 847–852.
249. Corder, E. H., Guess, H. A., Hulka, B. S., Friedman, G. D., Sadler, M., Vollmer, R. T., Lobaugh, B., Drezner, M. K., Vogelman, J. H., and Orentreich, N. (1993). Vitamin D and prostate cancer: A prediagnostic study with stored sera. *Cancer Epidemiol. Biomarkers Prev.* **2**, 467–472.
250. Gann, P. H., Ma, J., Hennekens, C. H., Hollis, B. W., Haddad, J. G., and Stampfer, M. J. (1996). Circulating vitamin D metabolites in relation to subsequent development of prostate cancer. *Cancer Epidemiol. Biomarkers Prev.* **5**, 121–126.
251. Braun, M. M., Helzlsouer, K. J., Hollis, B. W., and Comstock, G. W. (1995). Prostate cancer and prediagnostic levels of serum vitamin D metabolites (Maryland, United States). *Cancer Causes Control* **6**, 235–239.
252. Nomura, A. M., Stemmermann, G. N., Lee, J., Kolonel, L. N., Chen, T. C., Turner, A., and Holick, M. F. (1998). Serum vitamin D metabolite levels and the subsequent development of prostate cancer (Hawaii, United States). *Cancer Causes Control* **9**, 425–432.
253. Platz, E A., Leitzmann, M. F., Hollis, B. W., Willett, W. C., and Giovannucci, E. (2004). Plasma 1,25-dihydroxy- and 25-hydroxyvitamin D and subsequent risk of prostate cancer. *Cancer Causes Control* **15**, 255–265.
254. Tuohimaa, P., Tenkanen, L., Ahonen, M., Lumme, S., Jellum, E., Hallmans, G., Stattin, P., Harvei, S., Hakulinen, T., Luostarinen, T., Dillner, J., Lehtinen, M., and Hakama, M. (2004). Both high and low levels of blood vitamin D are associated with a higher prostate cancer risk: A longitudinal, nested case-control study in the Nordic countries. *Int. J. Cancer* **108**, 104–108.
255. Jacobs, E. T., Giuliano, A. R., Martinez, M. E., Hollis, B. W., Reid, M. E., and Marshall, J. R. (2004). Plasma levels of 25-hydroxyvitamin D, 1,25-dihydroxyvitamin D and the risk of prostate cancer. *J. Steroid Biochem. Mol. Biol.* **89–90**, 533–537.
256. Schwartz, G G., Hill, C. C., Oeler, T. A., Becich, M. J., and Bahnson, R. R. (1995). 1,25-Dihydroxy-16-ene-23-yne-vitamin D_3 and prostate cancer cell proliferation in vivo. *Urology* **46**, 365–369.
257. Kirsh, V. A., Hayes, R. B., Mayne, S. T., Chatterjee, N., Subar, A. F., Dixon, L. B., Albanes, D., Andriole, G. L., Urban, D. A., and Peters, U. (2006). Supplemental and dietary vitamin E, beta-carotene, and vitamin C intakes and prostate cancer risk. *J. Natl. Cancer Inst.* **98**, 245–254.
258. Schuurman, A. G., Goldbohm, R. A., Brants, H. A., and van den Brandt, P. A. (2002). A prospective cohort study on intake of retinol, vitamins C and E, and carotenoids and prostate cancer risk (Netherlands). *Cancer Causes Control.* **13**, 573–582.
259. Chan, J. M., Stampfer, M. J., Ma, J., Rimm, E. B., Willett, W. C., and Giovannucci, E. L. (1999). Supplemental vitamin E intake and prostate cancer risk in a large cohort of men in the United States. *Cancer Epidemiol. Biomarkers Prev.* **8**, 893–899.
260. Rodriguez, C., Jacobs, E. J., Mondul, A. M., Calle, E. E., McCullough, M. L., and Thun, M. J. (2004). Vitamin E supplements and risk of prostate cancer in U.S. men. *Cancer Epidemiol. Biomarkers Prev.* **13**, 378–382.
261. Eichholzer, M., Stahelin, H. B., Gey, K. F., Ludin, E., and Bernasconi, F. (1996). Prediction of male cancer mortality by plasma levels of interacting vitamins: 17-year follow-up of the prospective Basel study. *Int. J. Cancer.* **66**, 145–150.
262. Helzlsouer, K. J., Huang, H. Y., Alberg, A. J., Hoffman, S., Burke, A., Norkus, E. P., Morris, J. S., and Comstock, G. W. (2000). Association between alpha-tocopherol, gamma-tocopherol, selenium, and subsequent prostate cancer. *J. Natl. Cancer Inst.* **92**, 2018–2023.
263. Huang, H. Y., Alberg, A. J., Norkus, E. P., Hoffman, S. C., Comstock, G. W., and Helzlsouer, K. J. (2003). Prospective study of antioxidant micronutrients in the blood and the risk of developing prostate cancer. *Am. J. Epidemiol.* **157**, 335–344.

264. Goodman, G. E., Schaffer, S., Omenn, G. S., Chen, C., and King, I. (2003). The association between lung and prostate cancer risk, and serum micronutrients: Results and lessons learned from beta-carotene and retinol efficacy trial. *Cancer Epidemiol. Biomarkers Prev.* **12**, 518–526.
265. Weinstein, S. J., Wright, M. E., Pietinen, P., King, I., Tan, C., Taylor, P. R., Virtamo, J., and Albanes, D. (2005). Serum alpha-tocopherol and gamma-tocopherol in relation to prostate cancer risk in a prospective study. *J Nat Cancer Inst.* **97**, 396–399.
266. Nomura, A. M., Stemmermann, G. N., Lee, J., and Craft, N. E. (1997). Serum micronutrients and prostate cancer in Japanese Americans in Hawaii. *Cancer Epidemiol. Biomarkers Prev.* **6**, 487–491.
267. Comstock, G. W., Bush, T. L., and Helzlsouer, K. (1992). Serum retinol, beta-carotene, vitamin E, and selenium as related to subsequent cancer of specific sites. *Am. J. Epidemiol.* **135**, 115–121.
268. Knekt, P., Aromaa, A., Maatela, J., Aaran, R. K., Nikkari, T., Hakama, M., Hakulinen, T., Peto, R., Saxen, E., and Teppo, L. (1988). Serum vitamin E and risk of cancer among Finnish men during a 10-year follow-up. *Am. J. Epidemiol.* **127**, 28–41.
269. Heinonen, O. P., Albanes, D., Virtamo, J., Taylor, P. R., Huttunen, J. K., Hartman, A. M., Haapakoski, J., Malila, N., Rautalahti, M., Ripatti, S., Maenpaa, H., Teerenhovi, L., Koss, L., Virolainen, M., and Edwards, B. K. (1998). Prostate cancer and supplementation with alpha-tocopherol and beta-carotene: Incidence and mortality in a controlled trial. *J. Natl. Cancer Inst.* **90**, 440–446.
270. The Alpha-Tocopherol, Beta Carotene Cancer Prevention Study Group. (1994). The effect of vitamin E and beta carotene on the incidence of lung cancer and other cancers in male smokers. *N. Engl. J. Med.* **330**, 1029–1035.
271. Virtamo, J., Pietinen, P., Huttunen, J. K., Korhonen, P., Malila, N., Virtanen, M. J., Albanes, D., Taylor, P. R., and Albert, P. (2003). Incidence of cancer and mortality following alpha-tocopherol and beta-carotene supplementation: A postintervention follow-up. *JAMA.* **290**, 476–485.
272. Hartman, T. J., Albanes, D., Pietinen, P., Hartman, A. M., Rautalahti, M., Tangrea, J. A., and Taylor, P. R. (1998). The association between baseline vitamin E, selenium, and prostate cancer in the alpha-tocopherol, beta-carotene cancer prevention study. *Cancer Epidemiol. Biomarkers Prev.* **7**, 335–340.
273. Fleshner, N., Fair, W. R., Huryk, R., and Heston, W. D. (1999). Vitamin E inhibits the high-fat diet promoted growth of established human prostate LNCaP tumors in nude mice. *J. Urol.* **161**, 1651–1654.
274. Basu, A., and Imrhan, V. (2005). Vitamin E and prostate cancer: Is vitamin E succinate a superior chemopreventive agent? *Nutr. Rev.* **63**, 247–251.
275. Burton, G. W., and Ingold, K. U. (1981). Autoxidation of biological molecules. 1. Antioxidant activity of vitamin E and related chain-breaking phenolic antioxidants *in vitro*. *J. Am. Chem. Soc.* **103**, 6472–6477.
276. Bosetti, C., Talamini, R., Montella, M., Negri, E., Conti, E., Franceschi, S., and La Vecchia, C. (2004). Retinol, carotenoids and the risk of prostate cancer: A case-control study from Italy. *Int. J. Cancer* **112**, 689–692.
277. Jian, L., Du, C. J., Lee, A. H., and Binns, C. W. (2005). Do dietary lycopene and other carotenoids protect against prostate cancer? *Int. J. Cancer* **113**, 1010–1014.
278. Knekt, P., Aromaa, A., Maatela, J., Aaran, R. K., Nikkari, T., Hakama, M., Hakulinen, T., Peto, R., and Teppo, L. (1990). Serum vitamin A and subsequent risk of cancer: Cancer incidence follow-up of the Finnish Mobile Clinic Health Examination Survey. *Am. J. Epidemiol.* **132**, 857–870.
279. Hsing, A. W., Comstock, G. W., Abbey, H., and Polk, B. F. (1990). Serologic precursors of cancer. Retinol, carotenoids, and tocopherol and risk of prostate cancer. *J. Natl. Cancer Inst.* **82**, 941–946.
280. Freeman, V. L., Meydani, M., Yong, S., Pyle, J., Wan, Y., Arvizu-Durazo, R., and Liao, Y. (2000). Prostatic levels of tocopherols, carotenoids, and retinol in relation to plasma levels and self-reported usual dietary intake. *Am. J. Epidemiol.* **151**, 109–118.
281. Cook, N. R., Le, I. M., Manson, J. E., Buring, J. E., and Hennekens, C. H. (2000). Effects of beta-carotene supplementation on cancer incidence by baseline characteristics in the Physicians' Health Study (United States). *Cancer Causes Control* **11**, 617–626.
282. Sies, H., and Stahl, W. (1998). Lycopene: Antioxidant and biological effects and its bioavailability in the human. *Proc. Soc. Exp. Biol. Med.* **218**, 121–124.
283. Goodman, M., Bostick, R. M., Ward, K. C., Terry, P. D., van Gils, C. H., Taylor, J. A., and Mandel, J. S. (2006). Lycopene intake and prostate cancer risk: Effect modification by plasma antioxidants and the XRCC1 genotype. *Nutr. Cancer.* **55**, 13–20.
284. Giovannucci, E., Rimm, E. B., Liu, Y., Stampfer, M. J., and Willett, W. C. (2002). A prospective study of tomato products, lycopene, and prostate cancer risk. *J. Natl. Cancer Inst.* **94**, 391–398.
285. Etminan, M., Takkouche, B., and Caamano-Isorna, F. (2004). The role of tomato products and lycopene in the prevention of prostate cancer: A meta-analysis of observational studies. *Cancer Epidemiol. Biomarkers Prev.* **13**, 340–345.
286. Gann, P. H., Ma, J., Giovannucci, E., Willett, W., Sacks, F. M., Hennekens, C. H., and Stampfer, M. J. (1999). Lower prostate cancer risk in men with elevated plasma lycopene levels: Results of a prospective analysis. *Cancer Res.* **59**, 1225–1230.
287. Wu, K., Erdman, J. W., Jr., Schwartz, S. J., Platz, E. A., Leitzmann, M., Clinton, S. K., DeGroff, V., Willett, W. C., and Giovannucci, E. (2004). Plasma and dietary carotenoids, and the risk of prostate cancer: A nested case-control study. *Cancer Epidemiol. Biomarkers Prev.* **13**, 260–269.
288. Peters, U., Leitzmann, M. F., Chatterjee, N., Wang, Y., Albanes, D., Gelmann, E. P., Friesen, M. D., Riboli, E., and Hayes, R. B. (2007). Serum lycopene, other carotenoids, and prostate cancer risk: A nested case-control study in the Prostate, Lung, Colorectal, and Ovarian Cancer Screening Trial. *Cancer Epidemiol. Biomarkers Prev.* **16**, 962–968.
289. Clinton, S. K., Emenhiser, C., Schwartz, S. J., Bostwick, D. G., Williams, A. W., Moore, B. J., and Erdman, J. W. Jr. (1996). cis-trans lycopene isomers, carotenoids, and retinol in the human prostate. *Cancer Epidemiol. Biomarkers Prev.* **5**, 823–833.

290. Bertram, J. S. (1999). Carotenoids and gene regulation. *Nutr. Rev.* **57**, 182–191.
291. Williams, A. W., Boileau, T. W., Zhou, J. R., Clinton, S. K., and Erdman, J. W. Jr. (2000). Beta-carotene modulates human prostate cancer cell growth and may undergo intracellular metabolism to retinol. *J. Nutr.* **130**, 728–732.
292. Giovannucci, E. (1998). Dietary influences of 1,25(OH)$_2$ vitamin D in relation to prostate cancer: A hypothesis. *Cancer Causes Control.* **9**, 567–582.
293. Allen, N. E., Morris, J. S., Ngwenyama, R. A., and Key, T. J. (2004). A case-control study of selenium in nails and prostate cancer risk in British men. *Br. J. Cancer* **90**, 1392–1396.
294. Ghadirian, P., Maisonneuve, P., Perret, C., Kennedy, G., Boyle, P., Krewski, D., and Lacroix, A. (2000). A case-control study of toenail selenium and cancer of the breast, colon, and prostate. *Cancer Detect. Prev.* **24**, 305–313.
295. Lipsky, K., Zigeuner, R., Zischka, M., Schips, L., Pummer, K., Rehak, P., and Hubmer, G. (2004). Selenium levels of patients with newly diagnosed prostate cancer compared with control group. *Urology* **63**, 912–916.
296. Vogt, T. M., Ziegler, R. G., Graubard, B. I., Swanson, C. A., Greenberg, R. S., Schoenberg, J. B., Swanson, G. M., Hayes, R. B., and Mayne, S. T. (2003). Serum selenium and risk of prostate cancer in U.S. blacks and whites. *Int. J. Cancer* **103**, 664–670.
297. Nomura, A. M., Lee, J., Stemmermann, G. N., and Combs, G. F. Jr. (2000). Serum selenium and subsequent risk of prostate cancer. *Cancer Epidemiol. Biomarkers Prev.* **9**, 883–887.
298. Brooks, J. D., Metter, E. J., Chan, D. W., Sokoll, L. J., Landis, P., Nelson, W. G., Muller, D., Andres, R., and Carter, H. B. (2001). Plasma selenium level before diagnosis and the risk of prostate cancer development. *J. Urol.* **166**, 2034–2038.
299. Li, H., Stampfer, M. J., Giovannucci, E. L., Morris, J. S., Willett, W. C., Gaziano, J. M., and Ma, J. (2004). A prospective study of plasma selenium levels and prostate cancer risk. *J. Natl. Cancer Inst.* **96**, 696–703.
300. Coates, R. J., Weiss, N. S., Daling, J. R., Morris, J. S., and Labbe, R. F. (1988). Serum levels of selenium and retinol and the subsequent risk of cancer. *Am. J. Epidemiol.* **128**, 515–523.
301. Knekt, P., Aromaa, A., Maatela, J., Alfthan, G., Aaran, R. K., Hakama, M., Hakulinen, T., Peto, R., and Teppo, L. (1990). Serum selenium and subsequent risk of cancer among Finnish men and women. *J. Natl. Cancer Inst.* **82**, 864–868.
302. Goodman, G. E., Schaffer, S., Bankson, D. D., Hughes, M. P., and Omenn, G. S. (2001). Predictors of serum selenium in cigarette smokers and the lack of association with lung and prostate cancer risk. *Cancer Epidemiol. Biomarkers Prev.* **10**, 1069–1076.
303. Yoshizawa, K., Willett, W. C., Morris, S. J., Stampfer, M. J., Spiegelman, D., Rimm, E. B., and Giovannucci, E. (1998). Study of prediagnostic selenium level in toenails and the risk of advanced prostate cancer. *J. Natl. Cancer Inst.* **90**, 1219–1224.
304. van den Brandt, P. A., Zeegers, M. P., Bode, P., and Goldbohm, R. A. (2003). Toenail selenium levels and the subsequent risk of prostate cancer: A prospective cohort study. *Cancer Epidemiol. Biomarkers Prev.* **12**, 866–871.
305. Etminan, M., FitzGerald, J. M., Gleave, M., and Chambers, K. (2005). Intake of selenium in the prevention of prostate cancer: A systematic review and meta-analysis. *Cancer Causes Control* **16**, 1125–1131.
306. Webber, M. M., Perez-Ripoll, E. A., and James, G. T. (1985). Inhibitory effects of selenium on the growth of DU-145 human prostate carcinoma cells *in vitro*. *Biochem. Biophys. Res. Commun.* **130**, 603–609.
307. Menter, D. G., Sabichi, A. L., and Lippman, S. M. (2000). Selenium effects on prostate cell growth. *Cancer Epidemiol. Biomarkers Prev.* **9**, 1171–1182.
308. Medina, D. (1986). Mechanisms of selenium inhibition of tumorigenesis. *Adv. Exp. Med. Biol.* **206**, 465–472.
309. Giovannucci, E., Leitzmann, M., Spiegelman, D., Rimm, E. B., Colditz, G. A., Stampfer, M. J., and Willett, W. C. (1998). A prospective study of physical activity and prostate cancer in male health professionals. *Cancer Res.* **58**, 5117–5122.
310. Bairati, I., Larouche, R., Meyer, F., Moore, L., and Fradet, Y. (2000). Lifetime occupational physical activity and incidental prostate cancer (Canada). *Cancer Causes Control* **11**, 759–764.
311. Jian, L., Shen, Z. J., Lee, A. H., and Binns, C. W. (2005). Moderate physical activity and prostate cancer risk: A case-control study in China. *Eur. J. Epidemiol.* **20**, 155–160.
312. Nilsen, T. I., Romundstad, P. R., and Vatten, L. J. (2006). Recreational physical activity and risk of prostate cancer: A prospective population-based study in Norway (the HUNT study). *Int. J. Cancer* **119**, 2943–2947.
313. Liu, S., Lee, I. M., Linson, P., Ajani, U., Buring, J. E., and Hennekens, C. H. (2000). A prospective study of physical activity and risk of prostate cancer in U.S. physicians. *Int. J. Epidemiol.* **29**, 29–35.
314. Zeegers, M. P., Dirx, M. J., and van den Brandt, P. A. (2005). Physical activity and the risk of prostate cancer in the Netherlands cohort study, results after 9.3 years of follow-up. *Cancer Epidemiol. Biomarkers Prev.* **14**, 1490–1495.
315. Patel, A. V., Rodriguez, C., Jacobs, E. J., Solomon, L., Thun, M. J., and Calle, E. E. (2005). Recreational physical activity and risk of prostate cancer in a large cohort of U.S. men. *Cancer Epidemiol. Biomarkers Prev.* **14**, 275–279.
316. Littman, A. J., Kristal, A. R., and White, E. (2006). Recreational physical activity and prostate cancer risk (United States). *Cancer Causes Control* **17**, 831–841.
317. Friedenreich, C. M., McGregor, S. E., Courneya, K. S., Angyalfi, S. J., and Elliott, F. G. (2004). Case-control study of lifetime total physical activity and prostate cancer risk. *Am. J. Epidemiol.* **159**, 740–749.
318. Hackney, A. C., Fahrner, C. L., and Gulledge, T. P. (1998). Basal reproductive hormonal profiles are altered in endurance trained men. *J. Sports Med. Phys. Fitness* **38**, 138–141.
319. Friedenreich, C. M., and Orenstein, M. R. (2002). Physical activity and cancer prevention: Etiologic evidence and biological mechanisms. *J. Nutr.* **132**, 3456S–3464S.
320. Goode, E. L., Stanford, J. L., Chakrabarti, L., Gibbs, M., Kolb, S., McIndoe, R. A., Buckley, V. A., Schuster, E. F., Neal, C. L., Miller, E. L., Brandzel, S., Hood, L., Ostrander, E. A., and

Jarvik, G. P. (2000). Linkage analysis of 150 high-risk prostate cancer families at 1q24–25. *Genet. Epidemiol.* **18**, 251–275.

321. Xu, J., Meyers, D., Freije, D., Isaacs, S., Wiley, K., Nusskern, D., Ewing, C., Wilkens, E., Bujnovszky, P., Bova, G. S., Walsh, P., Isaacs, W., Schleutker, J., Matikainen, M., Tammela, T., Visakorpi, T., Kallioniemi, O. P., Berry, R., Schaid, D., French, A., McDonnell, S., Schroeder, J., Blute, M., Thibodeau, S., Gronberg, H., Emanuelsson, M., Damber, J. E., Bergh, A., Jonsson, B. A., Smith, J., Bailey-Wilson, J., Carpten, J., Stephan, D., Gillanders, E., Amundson, I., Kainu, T., Freas-Lutz, D., Baffoe-Bonnie, A., Van Aucken, A., Sood, R., Collins, F., Brownstein, M., and Trent, J. (1998). Evidence for a prostate cancer susceptibility locus on the X chromosome. *Nat. Genet.* **20**, 175–179.
322. Xu, J. (2000). Combined analysis of hereditary prostate cancer linkage to 1q24–25: Results from 772 hereditary prostate cancer families from the International Consortium for Prostate Cancer Genetics. *Am. J. Hum. Genet.* **66**, 945–957.
323. Ross, R. K., Pike, M. C., Coetzee, G. A., Reichardt, J. K., Yu, M. C., Feigelson, H., Stanczyk, F. Z., Kolonel, L. N., and Henderson, B. E. (1998). Androgen metabolism and prostate cancer: Establishing a model of genetic susceptibility. *Cancer Res.* **58**, 4497–4504.
324. Coughlin, S. S., and Hall, I. J. (2002). A review of genetic polymorphisms and prostate cancer risk. *Ann. Epidemiol.* **12**, 182–196.
325. Makridakis, N. M., Ross, R. K., Pike, M. C., Crocitto, L. E., Kolonel, L. N., Pearce, C. L., Henderson, B. E., and Reichardt, J. K. (1999). Association of mis-sense substitution in SRD5A2 gene with prostate cancer in African-American and Hispanic men in Los Angeles, USA. *Lancet* **354**, 975–978.
326. Ntais, C., Polycarpou, A., and Ioannidis, J. P. (2003). SRD5A2 gene polymorphisms and the risk of prostate cancer: A meta-analysis. *Cancer Epidemiol. Biomarkers Prev.* **12**, 618–624.
327. Platz, E. A., Giovannucci, E., Dahl, D. M., Krithivas, K., Hennekens, C. H., Brown, M., Stampfer, M. J., and Kantoff, P. W. (1998). The androgen receptor gene GGN microsatellite and prostate cancer risk. *Cancer Epidemiol. Biomarkers Prev.* **7**, 379–384.
328. Stanford, J. L., Just, J. J., Gibbs, M., Wicklund, K. G., Neal, C. L., Blumenstein, B. A., and Ostrander, E. A. (1997). Polymorphic repeats in the androgen receptor gene: Molecular markers of prostate cancer risk. *Cancer Res.* **57**, 1194–1198.
329. Habuchi, T., Suzuki, T., Sasaki, R., Wang, L., Sato, K., Satoh, S., Akao, T., Tsuchiya, N., Shimoda, N., Wada, Y., Koizumi, A., Chihara, J., Ogawa, O., and Kato, T. (2000). Association of vitamin D receptor gene polymorphism with prostate cancer and benign prostatic hyperplasia in a Japanese population. *Cancer Res.* **60**, 305–308.
330. Ingles, S. A., Ross, R. K., Yu, M. C., Irvine, R. A., La Pera, G., Haile, R. W., and Coetzee, G. A. (1997). Association of prostate cancer risk with genetic polymorphisms in vitamin D receptor and androgen receptor. *J. Natl. Cancer Inst.* **89**, 166–170.
331. Takacs, I., Koller, D. L., Peacock, M., Christian, J. C., Hui, S. L., Conneally, P. M., Johnston, C. C., Jr., Foroud, T., and Econs, M. J. (1999). Sibling pair linkage and association studies between bone mineral density and the insulin-like growth factor I gene locus. *J. Clin. Endocrinol. Metab.* **84**, 4467–4471.
332. Berndt, S. I., Dodson, J. L., Huang, W. Y., and Nicodemus, K. K. (2006). A systematic review of vitamin D receptor gene polymorphisms and prostate cancer risk. *J. Urol.* **175**, 1613–1623.
333. Ntais, C., Polycarpou, A., and Ioannidis, J. P. (2003). Vitamin D receptor gene polymorphisms and risk of prostate cancer: A meta-analysis. *Cancer Epidemiol. Biomarkers Prev.* **12**, 1395–1402.
334. Zeegers, M. P., Kiemeney, L. A., Nieder, A. M., and Ostrer, H. (2004). How strong is the association between CAG and GGN repeat length polymorphisms in the androgen receptor gene and prostate cancer risk? *Cancer Epidemiol. Biomarkers Prev.* **13**, 1765–1771.
335. Tsuchiya, N., Wang, L., Horikawa, Y., Inoue, T., Kakinuma, H., Matsuura, S., Sato, K., Ogawa, O., Kato, T., and Habuchi, T. (2005). CA repeat polymorphism in the insulin-like growth factor-I gene is associated with increased risk of prostate cancer and benign prostatic hyperplasia. *Int. J. Oncol.* **26**, 225–231.
336. Jaffe, J. M., Malkowicz, S. B., Walker, A. H., MacBride, S., Peschel, R., Tomaszewski, J., Van Arsdalen, K., Wein, A. J., and Rebbeck, T. R. (2000). Association of SRD5A2 genotype and pathological characteristics of prostate tumors. *Cancer Res.* **60**, 1626–1630.
337. Amundadottir, L. T., Sulem, P., Gudmundsson, J., Helgason, A., Baker, A., Agnarsson, B. A., Sigurdsson, A., Benediktsdottir, K. R., Cazier, J. B., Sainz, J., Jakobsdottir, M., Kostic, J., Magnusdottir, D. N., Ghosh, S., Agnarsson, K., Birgisdottir, B., Le Roux, L., Olafsdottir, A., Blondal, T., Andresdottir, M., Gretarsdottir, O. S., Bergthorsson, J. T., Gudbjartsson, D., Gylfason, A., Thorleifsson, G., Manolescu, A., Kristjansson, K., Geirsson, G., Isaksson, H., Douglas, J., Johansson, J. E., Balter, K., Wiklund, F., Montie, J. E., Yu, X., Suarez, B. K., Ober, C., Cooney, K. A., Gronberg, H., Catalona, W. J., Einarsson, G. V., Barkardottir, R. B., Gulcher, J. R., Kong, A., Thorsteinsdottir, U., and Stefansson, K. (2006). A common variant associated with prostate cancer in European and African populations. *Nat. Genet.* **38**, 652–658.
338. Yeager, M., Orr, N., Hayes, R. B., Jacobs, K. B., Kraft, P., Wacholder, S., Minichiello, M. J., Fearnhead, P., Yu, K., Chatterjee, N., Wang, Z., Welch, R., Staats, B. J., Calle, E. E., Feigelson, H. S., Thun, M. J., Rodriguez, C., Albanes, D., Virtamo, J., Weinstein, S., Schumacher, F. R., Giovannucci, E., Willett, W. C., Cancel-Tassin, G., Cussenot, O., Valeri, A., Andriole, G. L., Gelmann, E. P., Tucker, M., Gerhard, D. S., Fraumeni, J. F., Jr., Hoover, R., Hunter, D. J., Chanock, S. J., and Thomas, G. (2007). Genome-wide association study of prostate cancer identifies a second risk locus at 8q24. *Nat. Genet.* **39**, 645–649.
339. Wang, L., McDonnell, S. K., Slusser, J. P., Hebbring, S. J., Cunningham, J. M., Jacobsen, S. J., Cerhan, J. R., Blute, M. L., Schaid, D. J., and Thibodeau, S. N. (2007). Two common

chromosome 8q24 variants are associated with increased risk for prostate cancer. *Cancer Res.* **67**, 2944–2950.

340. Suuriniemi, M., Agalliu, I., Schaid, D. J., Johanneson, B., McDonnell, S. K., Iwasaki, L., Stanford, J. L., and Ostrander, E. A. (2007). Confirmation of a positive association between prostate cancer risk and a locus at chromosome 8q24. *Cancer Epidemiol. Biomarkers Prev.* **16**, 809–814.
341. Schumacher, F. R., Feigelson, H. S., Cox, D. G., Haiman, C. A., Albanes, D., Buring, J., Calle, E. E., Chanock, S. J., Colditz, G. A., Diver, W. R., Dunning, A. M., Freedman, M. L., Gaziano, J. M., Giovannucci, E., Hankinson, S. E., Hayes, R. B., Henderson, B. E., Hoover, R. N., Kaaks, R., Key, T., Kolonel, L. N., Kraft, P., Le Marchand, L., Ma, J., Pike, M. C., Riboli, E., Stampfer, M. J., Stram, D. O., Thomas, G., Thun, M. J., Travis, R., Virtamo, J., Andriole, G., Gelmann, E., Willett, W. C., and Hunter, D. J. (2007). A common 8q24 variant in prostate and breast cancer from a large nested case-control study. *Cancer Res.* **67**, 2951–2956.
342. Haiman, C. A., Patterson, N., Freedman, M. L., Myers, S. R., Pike, M. C., Waliszewska, A., Neubauer, J., Tandon, A., Schirmer, C., McDonald, G. J., Greenway, S. C., Stram, D. O., Le Marchand, L., Kolonel, L. N., Frasco, M., Wong, D., Pooler, L. C., Ardlie, K., Oakley-Girvan, I., Whittemore, A. S., Cooney, K. A., John, E. M., Ingles, S. A., Altshuler, D., Henderson, B. E., and Reich, D. (2007). Multiple regions within 8q24 independently affect risk for prostate cancer. *Nat. Genet.* **39**, 638–644.
343. Freedman, M. L., Haiman, C. A., Patterson, N., McDonald, G. J., Tandon, A., Waliszewska, A., Penney, K., Steen, R. G., Ardlie, K., John, E. M., Oakley-Girvan, I., Whittemore, A. S., Cooney, K. A., Ingles, S. A., Altshuler, D., Henderson, B. E., and Reich, D. (2006). Admixture mapping identifies 8q24 as a prostate cancer risk locus in African-American men. *Proc. Natl. Acad. Sci. USA* **103**, 14068–14073.
344. Tymchuk, C. N., Tessler, S. B., Aronson, W. J., and Barnard, R. J. (1998). Effects of diet and exercise on insulin, sex hormone-binding globulin, and prostate-specific antigen. *Nutr. Cancer* **31**, 127–131.
345. Duffield-Lillico, A. J., Dalkin, B. L., Reid, M. E., Turnbull, B. W., Slate, E. H., Jacobs, E. T., Marshall, J. R., and Clark, L. C. (2003). Selenium supplementation, baseline plasma selenium status and incidence of prostate cancer: An analysis of the complete treatment period of the Nutritional Prevention of Cancer Trial. *BJU Int.* **91**, 608–612.
346. Klein, E. A., Thompson, I. M., Lippman, S. M., Goodman, P. J., Albanes, D., Taylor, P. R., and Coltman, C. (2000). SELECT: The Selenium and Vitamin E Cancer Prevention Trial: Rationale and design. *Prostate Cancer Prostatic Dis.* **3**, 145–151.
347. Lippman, S. M., Goodman, P. J., Klein, E. A., Parnes, H. L., Thompson, I. M., Jr., Kristal, A. R., Santella, R. M., Probstfield, J. L., Moinpour, C. M., Albanes, D., Taylor, P. R., Minasian, L. M., Hoque, A., Thomas, S. M., Crowley, J. J., Gaziano, J. M., Stanford, J. L., Cook, E. D., Fleshner, N. E., Lieber, M. M., Walther, P. J., Khuri, F. R., Karp, D. D., Schwartz, G. G., Ford, L. G., and Coltman, C. A. Jr. (2005). Designing the Selenium and Vitamin E Cancer Prevention Trial (SELECT). *J. Natl. Cancer Inst.* **97**, 94–102.
348. Ornish, D. M., Lee, K. L., Fair, W. R., Pettengill, E. B., and Carroll, P. R. (2001). Dietary trial in prostate cancer: Early experience and implications for clinical trial design. *Urology* **57**, 200–201.
349. Bowen, P., Chen, L., Stacewicz-Sapuntzakis, M., Duncan, C., Sharifi, R., Ghosh, L., Kim, H. S., Christov-Tzelkov, K., and van Breemen, R. (2002). Tomato sauce supplementation and prostate cancer: Lycopene accumulation and modulation of biomarkers of carcinogenesis. *Exp. Bio. Med.* **227**, 886–893.
350. Demark-Wahnefried, W., Morey, M. C., Clipp, E. C., Pieper, C. F., Snyder, D. C., Sloane, R., and Cohen, H. J. (2003). Leading the Way in Exercise and Diet (Project LEAD): Intervening to improve function among older breast and prostate cancer survivors. *Control. Clin. Trials* **24**, 206–223.
351. Li, Z., Aronson, W. J., Arteaga, J. R., Hong, K., Thames, G., Henning, S. M., Liu, W., Elashoff, R., Ashley, J. M., and Heber, D. (2007). Feasibility of a low-fat/high-fiber diet intervention with soy supplementation in prostate cancer patients after prostatectomy. *Eur. J. Clin. Nutr.* [Epub ahead of print].
352. Meyer, F., Bairati, I., Shadmani, R., Fradet, Y., and Moore, L. (1999). Dietary fat and prostate cancer survival. *Cancer Causes Control* **10**, 245–251.

CHAPTER 37

Nutrition and Colon Cancer

MAUREEN A. MURTAUGH,[1] MARTHA L. SLATTERY,[1] AND BETTE J. CAAN[2]
[1]*Division of Epidemiology, Department of Medicine, University of Utah, Salt Lake City, Utah*
[2]*Kaiser Foundation Research Institute, Oakland, California*

Contents

I. INTRODUCTION

In the United States, colon cancer is the third most common cancer and second leading cause of death. There are more than 55,000 deaths attributed to colorectal cancer and 148,000 new cases of colon cancer every year in the United States [1]. Worldwide, there were 655,000 deaths from colorectal cancer in 2005, accounting for 8.5% of all incident cancers [2]. There is wide variation in reported incidence of colorectal cancer around the world. Developed countries have high incidence rates and developing countries have much lower incidence rates. Migrant populations who move from countries of low to high incidence adopt the rates of the host country [3]. These facts suggest that changes in diet and lifestyle associated with development or Westernization are associated with colorectal cancer. Much research attention has been focused on diet—with the expectation that identification of specific dietary factors that contribute to changing incidence rates will provide avenues for prevention of colorectal cancer.

Ecological studies, correlating differences in colorectal cancer mortality with differences in population intakes of nutrients such as fat, fiber, and calcium [4–6], suggest that populations with diets low in fat and high in fiber or calcium have lower mortality from colorectal cancer. Analytic epidemiological studies, however, have produced inconsistent findings regarding the relationship between many dietary factors and colorectal cancer.

Case-control studies comparing recalled diet in those with and without cancer and cohort studies linking reported diet to disease development have also singled out many suspect dietary components, but findings are often inconsistent. Varying results can stem from many sources, including study design, methods used to collect and analyze data, age and sex of the population being studied, range of dietary exposures captured from the questionnaire, the referent period for which the dietary data were collected, and possibly the tumor site itself. In the case-control study, participants are asked to recall past dietary intake; it is possible that recalled dietary intake might be influenced by changes in diet as a result of disease. Secular changes in food supply and diet trends may also contribute to inconsistent findings.

Cohort studies frequently have limited ranges of dietary exposure, because they are often based on a more select population than are population-based case-control studies. If dietary factors are only linked at high levels of intake, associations might be missed in cohort studies because of a truncated range of dietary intake. Data are usually obtained from self-administered questionnaires in cohort studies; therefore, less detailed information on diet and other possible exposures is obtained than with the interviewer-administered questionnaires usually used in case-control studies.

Results from controlled clinical trials studying the relation between diet and colonic polyps (adenomas) are providing credence to associations for some dietary factors. Although polyps are precursor lesions to colon tumors, not all polyps progress to tumors. Therefore, some differences in dietary associations could be expected for studies that focus on adenomas rather than adenocarcinomas.

There are several major hypotheses regarding the nutrition-related mechanisms involved in the etiology of adenocarcinoma of the colon (Table 1). These hypotheses are overlapping and complementary. They stress the complexity of colon cancer and the difficulty of identifying a single mechanism, and also elucidate the diverse approaches being used to examine the links between nutrition and colon cancer.

In this chapter we discuss five different models describing mechanisms by which diet may either promote or prevent colon cancer and review the supporting epidemiologic data for each (Table 2).

TABLE 1 Summary of Selected Hypothesized Mechanisms Related to a Diet and Colon Cancer Relationship

Model	Hypotheses
Bile acid/volatile fatty acid	Bile acid metabolism is central to colon cancer. A high-fat diet increases levels of bile acids, and bile acids damage cells in the colon and induce proliferation.
Fat and fiber	International correlation studies suggest that dietary fat increases risk of colon cancer; fat is associated with bile acid metabolism. Colon cancer was rare among Africans who ate a diet high in whole grains; early case-control studies supported this.
Meats and vegetables	A variation of the fat and fiber hypothesis. Data suggest variability in cancer risk that correlates with the spectrum of meat and vegetables in the diet [43].
Cooked foods	Meat prepared at high temperatures increases heterocyclic amines and polycyclic aromatic hydrocarbons. These carcinogens increase risk of colon cancer.
Insulin resistance	Insulin resistance hypothesis links many of the previously identified risk factors for colon cancer. Increased insulin levels influence cell proliferation and stimulate growth of colon tumors.
DNA methylation	Dietary factors such as folate, vitamin B_6, vitamin B_{12}, methionine, and alcohol are involved in DNA methylation pathways. Levels of intake of these nutrients that lead to availability of low levels of methyl groups cause hypomethylation of DNA. DNA damage leads to colon cancer.
Growth medium	Dietary factors act as initiators and promoters of the transition from normal to malignant cell. Dietary factors are involved in apoptosis and cell growth and regulation.

TABLE 2 Summary of Epidemiological Studies of Diet and Colon Cancer Associations

Dietary Factor	Risk Estimates	Supporting Evidence	Comments
Model 1: Bile acids, fat/fiber, meats/vegetables (plant foods)			
Fat	0.8–3.0	Older case-control studies and those that did not adjust for total energy more likely to report associations.	Specific types of fat or uses of fat may be important. Association with fat may be confounded by total energy.
Meats	0.7–2.5	Many cohort and case-control studies observed no increased risk.	Associations may depend on amount of meat eaten at a given time, other components of the diet, method of preparation, age of the population, or other factors that could modify these effects.
Fiber	0.6–1.0	Most case-control and cohort studies report reduced risk at high levels of fiber intake.	Association may be dependent on level of fiber intake.
Vegetables	0.5–1.2	Older case-control studies observed risk; newer cohort studies are less positive.	Some vegetables (leafy greens, garlic, onion, and legumes) may be more important than others. Studies examining cruciferous vegetables are inconclusive in their findings.
Fruits	0.8–1.5	No consistent association	Specific types of fruits may be important, such as apples and apricots.
Whole grains	0.6–1.0	Many studies support inverse association.	Different definitions of whole grains are used. Some studies have not adjusted for dietary fiber.
Refined grains	1.1–2.0	Most studies suggest an increase in risk.	Different definitions of refined grains are used.
Beta-carotene	~1.0	Inconsistent associations	Some studies show an increased risk with higher levels of intake among smokers.
Lutein	0.6–0.8	Report from one study	Associations appear to be stronger for younger people and those with proximal tumors. Lutein is a good marker for green vegetable intake.
Lycopene	0.8–1.0	Inconclusive	Although protective effect not observed, high intake of tomatoes has been reported as protective in three studies that examined the association.

(continues)

TABLE 2 *(continued)*

Dietary Factor	Risk Estimates	Supporting Evidence	Comments
Vitamin E	0.6–1.0	Inconclusive	Strong protective effect reported from a cohort study of older women; other studies have not identified similar protective effect. Slightly more positive association for rectal cancer than colon cancer.
Vitamin C	0.6–1.0	Inconclusive	Although plausible mechanisms exist, studies have generally not detected associations.
Model 2: Cooked foods			
Fried foods	Null–6.0	Inconclusive	Many studies show no association; three show strong associations for highly browned or processed food.
Model 3: Insulin resistance			
Energy and energy balance	RR: 0.8–1.8	Case-control studies report increased risk, with associations that are generally not observed in cohort studies.	Possible recall bias in case-control studies; cohort studies have narrower range of exposures. Some studies suggest that risk associated with energy intake is related to level of energy expenditure.
Sucrose	1.0–2.0	Inconsistent	Many studies have examined sugar-containing foods rather than sucrose. Sucrose-to-dietary-fiber ratio better indicator of risk than sucrose alone in one study. Age of participants and secular trends may contribute to disparate results.
Glycemic index and glycemic load	1.0–1.7 1.0–1.8	Inconsistent	Inconsistent results; glycemic index or load may be more important in obese or physically inactive populations.
Model 4: DNA methylation			
Folate	0.3–1.1	Inconsistent	Some data support a protective effect in conjunction with other related factors; however, few studies show significant reduction in risk for dietary folate. Folate supplements generally have little effect except for a possible reduced risk if taken for 15 or more years.
Vitamin B_6	0.6–0.9	Suggestive of protective effect	May reduce risk of colon cancer.
Methionine	0.7–1.1	Inconsistent	May be associated with risk in conjunction with other related dietary factors, although not all studies that examined the combined effect of these DNA-related factors observed an association.
Alcohol	0.7–7.0	Inconsistent	Some studies suggest stronger associations for distal and rectal tumors. Cigarette smoking and gender may confound associations.
Model 5: Cell growth regulators			
Calcium	0.5–1.6	Consistent among more recent studies	Studies with a wide range of calcium intake generally show a protective effect.
Vitamin D	0.5–1.0	Consistent	Most show a reduction in risk from diet or supplemental vitamin D reduces risk. May be an important factor.

Note: Nutrients fit into multiple models. They are listed with the model with which they have been most aligned.

II. MODEL 1: BILE ACIDS, DIETARY COMPONENTS

Perhaps the oldest theory driving epidemiologic studies on diet and colon cancer is bile acids [7]. In theory, high fat consumption stimulates bile acid output and higher concentrations of conjugated bile acid in the colonic contents. In the colon, bile acids are deconjugated and dehydroxylated to form secondary bile acids that, in turn, damage cells and induce proliferation. In more recent years, it has been recognized that fiber binds bile acids and may modify the association of fat to colon cancer.

Dietary fiber also reduces transit time, increases stool bulk, and helps ferment volatile fatty acids—all factors that may reduce the conversion of primary to secondary bile acids in the colon [8]. The products of fermentable volatile fatty acids include butyrate, which, in addition to being a major colonic epithelial cell fuel, may play a role in apoptosis and cell replication [9]. Pectin, a water-soluble fiber, reduces the rate of glucose absorption and decreases the rate of absorption and/or availability of lipids.

Meat and vegetable intake is also relevant to this hypothesis [7]. Meat contains high levels of fat and protein that may be harmful to the colon, whereas vegetables are rich sources of vitamins, minerals, fiber, and phytochemicals (plant substances) that may protect against the development of colon cancer.

A. Dietary Fat

Ecological studies have reported that fat in diets varies directly with rates of colon cancer deaths [6]. Whereas many early case-control and cohort studies detected an increased risk of colon cancer with increasing intake of dietary fat [4], more recent cohort [10], large case-control [11], and a pooled analysis of 13 case-control studies [12] have failed to find an association between dietary fat and colon cancer or even reported an inverse association [13]. The Women's Health Initiative (WHI), a randomized trial of a low-fat dietary pattern in more than 46,000 postmenopausal women, confirms these findings; they also found no association between a low-fat diet and colorectal cancer [14]. The Nurses' Health Study, a cohort study, is one of the few recent studies reporting an association between dietary fat and colon cancer [15].

Studies examining types of fat or fat intake patterns, rather than total fat content, in women's diets suggest a positive association. Cohort studies of women report an increased risk of colon cancer associated with high intakes of animal fat [10, 15] and fried foods eaten away from home [16]. A large case-control study reported increases in risk among women with a high use of fats in food preparation [11]. It is unclear whether these associations exist only among women or whether men report these aspects of food consumption differently, leading to a missed association.

Studies have investigated the association of the subtypes of fat with risk of colorectal cancer following the paradigm that fatty acids of different lengths have different associations with other heath outcomes such as heart disease, with inconclusive results. Several studies identified no increased risk associated with types of dietary fat [14, 17–20], whereas two case-control studies reported inverse associations with monounsaturated and polyunsaturated fatty acids [11, 21]. The WHI found no association with saturated fat as a percent of energy intake and colorectal cancer risk [14]. A U.S.-based study found the increased risk was most evident among women with family history of colorectal cancer and who had diets high in both monounsaturated and polyunsaturated fatty acids and among younger men with high intakes of monounsaturated fatty acids, linoleic acid, and 20-carbon polyunsaturated fatty acids [11]. An Italian study found no association with monounsaturated fat, but did report increased risk with high polyunsaturated fat intake and risk of colon cancer, particularly in the right colon [19].

The potential for modification of the association of dietary fat and risk of colorectal adenomas by nonsteroidal anti-inflammatory drugs (NSAIDs) has also been reported [22]. Increased risk has been reported with high *trans*-fatty acid intake among older (≥67 years of age) men and women, among those who did not use NSAIDs, and among postmenopausal women who were not using exogenous estrogen [23]. These results suggest that differing population characteristics may contribute to disparate associations of total fat and individual fatty acids with risk of colorectal cancer.

Although the data only weakly support an association between dietary fat and colon cancer, there are plausible biological explanations in addition to its role in bile acid production that support an association between specific types of fat and colon cancer. At a low concentration butyrate promotes intestinal barrier function, but at high concentrations may be associated with cytotoxicity due to apoptosis [24]. Eicosanoids derived from 20-carbon polyunsaturated fatty acids have been shown to regulate cell proliferation and immune response [25]. Eicosanoids (20:4, n-6) and docosahexaenoic fatty acids (22:6, n-3) appear to have beneficial effects on apoptosis [26]. Linoleic acid has been shown to be involved in prostaglandin synthesis as well as acting synergistically with other growth factors [27, 28].

Early studies linking dietary fats to colon cancer could have been confounded by the effect of total energy, because fats are major contributors to variability in total energy in the diet, and statistical methods to sort these factors were often not employed [29–31]. Studies that have tried to separate the effects have generally not shown an association for dietary fat [32]. It is possible that fat is only important at high levels of intake, and the range of fat consumption observed in population studies is not wide enough to show associations or does not encompass the range of fat where effects are measurable. It is also possible that other factors that modify the harmful or helpful effects of dietary fat are changing in the population.

B. Meat

It is possible that red meat, a major source of animal fat, may actually be the agent responsible for the increased risk attributed to dietary fat. The American Cancer Society considers the evidence of an association of red and processed meat with risk of colorectal cancer great enough to make the recommendation to "limit intake of processed and red meats" [33]. However, although several cohort studies found increased risk with higher consumption of red meat, processed meat, or other types of meat [10, 15, 34], 10 other cohort studies did not find an association [35]. In a Norwegian cohort there was an increased risk associated with sausage intake as a main meal in women, but not in men [36]. Some case-control studies detected associations [37–39], whereas others did not [40–42]. Two studies suggest that the association of meat with risk of colon cancer is modified by other components of the diet, specifically vegetables [43] and legumes [44].

C. Dietary Fiber

Fiber has been associated with reduced risk of colon cancer in numerous studies [4, 45–51]. A recent pooled analysis of cohort studies suggested an inverse association of fiber and colon and rectal cancer risk; however, the association was attenuated after adjustment for other risk factors [52]. Two studies reported no differences in risk from soluble and insoluble fiber [50, 53]. One of these [53] observed that the strongest associations were for older people and those with proximal tumors. Some research based on a wide range of intake shows protection only at high levels of fiber intake [53]; some shows increased risk at very low levels of fiber intake (<10 g/day) [52]; still other studies may fail to identify the association because of truncated levels of fiber intake [54].

Despite the relatively consistent association between dietary fiber and colorectal cancer in case-control and cohort studies, randomized intervention studies of fiber have failed to detect reduced polyp recurrence. In the Polyp Prevention Trial [55], a randomized study of 2079 men and women with recently detected colorectal adenomas, an increase in dietary fiber intake (from approximately 10 to 17 g/1000 kcal) combined with reduction in dietary fat (from 35% to 24% of energy) and increased consumption of fruits and vegetables did not reduce the likelihood of having recurrent adenomatous polyps during the 4-year follow-up. Likewise, no increase in recurrence of polyps was found at 3-year follow-up with dietary supplementation of 13.5 g wheat bran fiber/day in a randomized study of 1303 men and women [56]. These clinical trials are critically limited by the use of polyps as the end point. Polyps are generally considered precursors to cancer, but the majority of polyps does not advance to cancer. The critical time for an influence of fiber may be before the development of polyps. Additionally, participants in clinical trials often change other behaviors that may influence the effect of the dietary intervention or the outcome itself. Thus, the interpretation of these nutrition intervention studies is not straightforward.

D. Plant Foods: Vegetables, Fruits, Legumes, and Grains

Plant foods such as vegetables, fruits, legumes, whole grains, and refined grains are the custodians of numerous dietary constituents, including fiber, vitamins, minerals, and other potentially anticarcinogenic factors. Of these plant foods, vegetables had been one of the most consistently identified factors associated with a reduced risk of colon cancer [4]. Most studies through the 1990s reported a 30–40% reduction in risk in those with the highest level of vegetable intake relative to those with the lowest level of intake [45, 53, 57–60]. In 1997, the World Cancer Research Fund reported, "Evidence that diets rich in vegetables protect against cancer of the colon and rectum is convincing" [61].

Recently published studies are less convincing. Cohort studies from Japan [62], the Netherlands [63], and the United States [64, 65] show no association of total vegetable intake with risk of colorectal cancer. Biologic plausibility of finding associations with subtypes of fruits and vegetables is supported by growing documentation of the anticarcinogenic potential of individual food components from fruits and/or vegetables with colorectal cancer from animal models and *in vitro* studies (Table 3). This shift in association could be related to secular changes in fruit and vegetable intake, characteristics of the fruits and vegetables, overall diet, or other characteristics (e.g., rates of obesity) of the populations studied [66]. Several recent cohort studies with colorectal cancers did find associations in subgroups of the populations and/or with specific subtypes of vegetables [60, 63]. Accordingly, fruit and vegetable intake was associated with a decreased risk of colon cancer in the distal colon, but not the proximal colon, in women [63] and older [51–70] adults [60]. Cruciferous vegetables were associated with a decreased risk of colon cancer, but an increased risk of rectal cancer in women from the Netherlands [63]. Cruciferous vegetables are the most widely studied type of vegetable, with inconsistent results associated with high levels of intake [4].

Other subtypes of vegetables are also reported to be associated with colon cancer. Leafy vegetables [60, 63], dark yellow vegetables [odds ratio (OR) 0.7; 95% confidence interval (CI) 0.6–1.0], and tomatoes (OR 0.6; 95% CI 0.4–0.8) [53] are reported to be associated with reduced risk of colon cancer. European case-control studies have reported a significant inverse association between garlic [37, 67] and onion [67] consumption and colorectal cancer. Studies are supportive of a protective effect of legumes and nuts. Two studies report reduced adenomas with increased legume consumption [68, 69]. Reports also suggested increased legume fiber [65] or peanuts are associated with reduced colorectal cancer.

Whereas some studies report inverse associations of fruit with colorectal cancer [57], others report null associations [60], or even a slight increase in risk for canned fruits and juices (OR 1.2; 95% CI 0.9–1.5) [53]. Levi and colleagues [37] observed a halving of risk of colorectal cancer associated with high levels of citrus fruit intake (OR 0.5; 95% CI 0.3–0.8). These inconsistencies would make sense if different types of fruits have different influences or juices are included as fruit.

TABLE 3 Food Components Associated with Anticarcinogenic Effects

Food Component	Effect	Food Source	Type of Study/ Evidence
Falcarinol	Pro-proliferative at low concentration, apoptotic at high concentration [260]	Carrots	Rats
Anthocyanin	Inhibition of cell growth, induction of apoptosis [261, 262]	Grapes, blueberries	
Pterostilbene	Inhibit aberrant crypt foci formation [263]	Blueberries	Rats
Sulforaphane	Inhibition of histone deacetylase (HDAC) [264]	Cruciferous vegetables	Mice and *in vitro*
	Induce phase II detoxification enzymes [265] Inhibit angiogenesis [266] Apoptosis [265]		
Allyl sulfides	Suppress cell growth and induce apoptosis [267]	Garlic	*In vitro*
Indole-3-carbinol	Induces apoptosis [268]	Cruciferous vegetables	*In vitro*
Ferulic acid	Inhibit cell growth, COX-1, and COX-2 and inhibit lipid peroxidation [269]	Ubiquitous in plants, abundant in fruits and vegetables	*In vitro*
Caffeic acid	Inhibit cell growth, COX-1, and COX-2 and inhibit lipid peroxidation [269], reduction of oxidative DNA damage [270]	Ubiquitous in plants, abundant in fruits and vegetables	*In vitro*
Flavone	Inhibit aberrant crypt foci formation [271]	Vegetables	Mice
Quercetin	Inhibit angiogenesis [272] and potent antioxidant [270]	Onions, apples, green and black tea	*In vitro*
Salicylates	Inhibit COX-2 gene transcription [273]	Strawberries and other fruits, organic vegetable soups [274]	
Chlorophyll	Prevent formation of cytotoxic haem metabolites [275, 276]	Green vegetables	Rats
Puniclagin, ellagitannin, and ellagic acid	Antiproliferative, apoptotic	Pomegranate	*In vitro* [277]
Resversatol, catechin, epicatechin, and ellagic acid	Antioxidant activity [278]	Grapes	
Fisetin	Inhibition of cell-cycle progression [279]	Vegetables, fruits, and wine	
Spearmint	Inhibit carcinogen activation in the cytochrome P-450 system [280]	Spearmint tea	*In vitro*
Green tea extract	Inhibit activation of heterocyclic amines and polycyclic aromatic hydrocarbons [281]	Green tea	*In vitro*
Green, black, and decaffeinated tea	Antimutagenic, prevent activation of mutagens [282]	Teas	*In vitro*

Whole grain and refined grain products have also been evaluated. Some studies show that whole-grain products [46, 47, 53, 70, 71] reduce risk of colon cancer and refined grains are associated with an increased risk [37, 40, 72]. A recent study of Swedish women suggested reduction in risk of colon cancer with consumption of whole grains, but no association with rectal cancer [71]. It is interesting that the evidence for whole and refined grains is more consistent than the evidence for an association with fiber. Residual confounding could come from behaviors associated with whole grain consumption or other nutrients present in whole grains (e.g., fiber) as compared with refined grains. Similarly, studies that have tried to ferret out the association between the different compounds present in plant foods and colon cancer have shown that, even when controlling for dietary fiber, trace minerals, antioxidants, and other nutrients found in plant foods, vegetables still remain significantly inversely associated with colon cancer [53], implying that other unmeasured compounds contained in plant foods could be protective. Biological plausibility is based on abundant evidence that components in fruits and vegetables have potential influence on carcinogenesis (Table 3). However, many of these bioactive compounds are unavailable in standard nutrient databases, leaving assessment of foods that contain these compounds as a surrogate for their intake.

E. Dietary Antioxidants

Vitamins with antioxidant properties, including beta-carotene, vitamin E, and vitamin C, are associated with decreased colon cancer risk in some but not all studies [4, 58, 73–76]. Enger and colleagues [77] observed that beta-carotene intake was inversely associated with polyp development. However, several clinical trials [74, 78–80] have found that high-dose beta-carotene supplementation increases risk of colorectal cancer and adenomas among smokers. In addition, in the Australian Polyp Study [74] higher levels of beta-carotene were associated with greater recurrence of large polyps. Increased vitamin C, on the other hand [17], was associated with a marked decrease in colon cancer risk (OR 0.5; 95% CI 0.3–0.9).

Carotenoids, long recognized for their antioxidant properties, are of increasing interest in relation to cancer because of their effect on regulation of cell growth and modulation of gene expression, as well as their possible effect on immune response [81]. However, evidence from cohort studies conducted in North America and Europe [82] and from the Shanghai Women's Health Study [83] does not strongly support an association of dietary intake of carotenoids with colorectal cancer. Associations of serum carotenoids with risk of colorectal cancer are also equivocal. At the highest level of serum carotenoids, colorectal cancer risk is reported as not associated in men [84] and increased in women [85]. Data from case-control studies suggest an inverse relationship between both lutein (40% to 10% lower risk depending on subgroup) and lycopene-rich tomato intake [53, 86]. A report of lycopene itself revealed no association with colon cancer [87], but an inverse association was reported with rectal cancer in women, particularly women over 60 years of age [88]. Unfortunately, no definitive conclusions are possible from these mixed results.

Vitamin E also been inversely associated with colon cancer in some but not all studies [4, 58, 73, 83, 89, 90]. This vitamin represents a group of tocol and tocotrienol derivatives that may have specific biological mechanisms of action [91]. Alpha-tocopherol is the most biologically active tocopherol and has been the major tocopherol considered in dietary calculations of vitamin E intake. Despite the lower biological activity of other tocopherols, such as beta-, gamma-, and delta-tocopherol, their presence in the diet in amounts two to four times that of alpha-tocopherol makes them potentially important chemopreventive agents [91]. The biological import of the mixture of tocopherols is supported by the observation that mixed tocopherols inhibited aberrant crypt foci in rats [92].

The association between colorectal cancer and vitamin E, mainly in the form of alpha-tocopherol, is mixed among analytic epidemiologic studies [89, 90, 93, 94] and clinical trials [95, 95]. The work of Bostick and colleagues [93] provides the strongest evidence of an association between vitamin E and colon cancer. In their cohort of women living in Iowa, a strong protective effect was seen in women categorized in the upper quintile of vitamin E, after adjusting for other confounding factors. Women younger than 59 years old were the most protected by vitamin E, and supplemental vitamin E appeared to be more protective than vitamin E from dietary sources. On the other hand, clinical trials have not supported the observation that vitamin E supplements decrease risk of colorectal polyps, a precursor to colon cancer [95, 96], nor have data from a large case-control study of colon cancer, after adjusting for related dietary and lifestyle factors [97] or examining different forms of vitamin E [97].

The evidence for an association between vitamin E intake and rectal cancer is slightly more positive. A large U.S.-based case-control study of rectal cancer found an inverse association of dietary alpha-, delta-, and gamma-tocopherols with rectal cancer in women; the association was stronger among women over 60 years [88]. An inverse dose-response association was reported for vitamins E and C with rectal cancer [73], although the association with vitamin E was no longer significant after adjustment for serum cholesterol [94]. Whether the serum level of vitamin E is not relevant to tissue levels in the rectum or whether there are other reasons for differences in results remains to be determined.

III. MODEL 2: COOKED FOODS

A second hypothesis is that the methods used to cook meat and other foods may be more important than the foods themselves [98–100]. Diets high in fat and protein—high-meat diets, for example—contain greater amounts of heterocyclic amines when cooked at high temperatures. Greater exposure to these carcinogens may increase cancer risk. Cooked sucrose also be harmful in that it has been shown to promote aberrant crypt foci in rodents [101]. Using data from a case-control study, Gerhardsson de Verdier *et al.* [102] reported an increased risk of colon cancer (OR 2.7; 95% CI 1.4–5.9), comparing most frequent consumers of fried meat with a heavily browned surface to those who did not use these cooking methods; associations in this study were higher for rectal cancer than for colon cancer (OR 6.0; 95% CI 2.9–12.6). Schiffman and Felton [103] reported more than a threefold increase in risk associated with high levels of consumption of well-done meat. Other studies failed to detect significant associations or reported much weaker associations with colon cancer (OR 1.3; 95% CI 1.0–1.7) [42, 104] or identified associations with rectal cancer only among men [105] when using a mutagen index that was based on both frequency of consumption and temperature of cooking.

IV. MODEL 3: INSULIN RESISTANCE

McKeown-Eyssen [106] first proposed the theory that nutrition relates to colon cancer through dietary contributions to insulin resistance, and Giovannucci [107] provided additional support in a literature review. The insulin resistance hypothesis pulls together many risk factors into a central biological mechanism. Long-term leisure physical activity is protective for colon cancer [97] and inversely associated with development of insulin resistance independently of body size [108]. Obesity is directly associated with increasing glucose and insulin levels as well as with colon cancer [109, 110]; diets high in fiber and low in sucrose are inversely associated with glucose levels and colon cancer [111]. Moreover, increased levels of blood glucose and/or triglycerides can result in increased insulin levels that may influence cell proliferation and stimulate growth of colon tumors. The relevance of the insulin resistance model or pathway to nutrition and colon cancer is the association of diet with hyperinsulinemia and, therefore, the potential for modifying risk [112].

Metabolic syndrome is associated with a higher risk of adenomas [113] and proximal and synchronous colorectal cancer [113–115], and some evidence suggests that the association is stronger among men than women [116, 117]. Mechanistically, *in vivo* animal models suggest that hyperinsulinemia rather than hyperglycemia was associated with colorectal epithelial proliferation [118]. Data from three large cohort studies support this idea [117, 119, 120]. Risk was increased in the highest versus the lowest quintile of C-peptide, a marker of insulin secretion, in all studies [117, 119, 120] and was greater for colon cancer than rectal cancer [117, 120]. Associations were similar for men and women from the European cohort [120] and significant only for men in the Japanese cohort [117].

A. Energy and Energy Balance

Energy intake, energy expenditure, and body size have been repeatedly studied in relation to colon cancer. High levels of energy intake are associated with increased risk of colon cancer in many case-control studies, although cohort studies generally have not observed the association [4, 61]. In contrast, high levels of physical activity are consistently associated with reduced risk of colon cancer (this association has not been detected in studies of rectal cancer) [4]. Body mass index (kg/m^2; BMI) is directly associated with colon cancer, but not rectal cancer [121, 122]. Some studies suggest increased risk of colon cancer with increasing BMI only in men [110, 121] and others in both men and women less than 67 years old [122]. Indicators of central adiposity are significantly associated in both genders [121]. High energy intake and high body mass both become highly significantly associated with colon cancer when levels of physical activity are low [123]. Similarly, being physically active is most critical when energy intake is high. Thus, the underlying activity level of the population studied may be important to detect associations between colon cancer and energy intake.

B. Sugar and Glycemic Index

Levels of simple sugar consumption also vary from country to country, and high intake may typify a Western-style diet. High consumption of simple sugars may result in increased triglyceride and plasma glucose levels, especially among those who are insulin resistant [124]. Findings regarding sucrose itself, however, have been mixed [125–131]. Of the studies reported, only three found a significant association between sucrose and colon or colorectal cancer [8, 126, 131]. Bostick and colleagues [132] observed that sucrose-containing foods showed stronger associations with colon cancer than did sucrose itself, with the strongest associations being observed when dairy foods that were high in sugar, such as ice cream, were excluded. Others have demonstrated links to specific foods [133, 134], although the high sugar content by itself may not be driving the association [133, 135]. Desserts or dairy products with high sugar content, for instance, have other components, such as fat, that may account for the observed associations with colon cancer. Findings reported by Slattery and colleagues [125] and Bostick and colleagues [132] suggest that the strongest associations are among older women. Some of these discrepancies in findings could be accounted for by age of study participants or secular changes in sucrose intake.

The variation in reported risk estimates between studies for high-sugar foods and/or sugar may be the result of different metabolic responses to specific foods, commonly reported as glycemic index or glycemic load [136, 137]. Glycemic index, however, does not correlate with simple sugar intake. Increased risk of colon cancer has been observed with both dietary glycemic index [125, 138] and glycemic load [131, 138] in U.S. studies. No association was noted of glycemic index or glycemic load with colon or rectal cancer in a cohort of Swedish women [139]. In the U.S. studies, increased risks were most evident in the obese [131, 138] or physically inactive [125].

V. MODEL 4: DNA METHYLATION

Methylation of DNA is one step in the regulation of gene activity [140–142]. Disturbances in DNA methylation are thought to result in abnormal expression of oncogenes and tumor suppressor genes [142–144]. In the case of colorectal tumors, for example, both generalized genomic hypomethylation and hypermethylation of usually unmethylated sites occur frequently [143–145]. Several dietary components, including folate, methionine, vitamin B_{12}, and vitamin B_6, are involved either directly or indirectly in DNA methylation [145, 146]. Alcohol may also alter DNA methylation patterns indirectly by affecting the intestinal absorption, hepatobiliary metabolism, and renal excretion of folate [147, 148]. Dietary involvement in the DNA methylation process has become a focus of research into the relationship of nutrition and colon cancer, although a recent study did not support a role of dietary folate, alcohol, methionine, or vitamins B_6 or B_{12} in the CpG island methylator phenotype (CIMP) [149]; however, diets with combined low folate and methionine and high alcohol in conjunction with *MTHFR* genotype were associated with CIMP+ tumors [150].

A. Folate, Vitamin B_6, and Methionine

Freudenheim and colleagues [151] originally observed that high intakes of dietary folate in women [relative risk (RR) 0.69; 95% CI 0.36–1.30], but not men (RR 1.03; 95% CI 0.56–1.89), were inversely associated with colon cancer, with stronger associations observed as the tumor site became more distal. Other studies confirm the original findings [146, 152]; however, the association may not be linear and may be confounded by other dietary or environmental factors. In the Health Professionals Follow-up Study [153], men in the highest quintile of dietary folate intake compared to the lowest were not at reduced risk (RR 0.85; 95% CI 0.54–1.39). Confounding of the association by dietary fiber is controversial. Adjustment for dietary fiber attenuated the association in a large case-control study [146], although it did not in other studies [151, 154]. Modification of the association of folate with colorectal cancer by smoking has been reported [155], and polymorphisms in folic acid metabolism genes [156, 157], although their effects appeared to be independent in a report focusing on rectal cancer [158], and different associations with tumor characteristics are reported [159, 160].

Inclusion of vitamin supplement data had little effect in findings from several large cohort and case-control studies [146, 153, 154]. One study noted that people taking folate supplements for 15 or more years were at reduced risk of colon cancer, but there was no effect from folate supplements taken for less than 15 years [153]. It is not clear if this finding is one of chance, if it points to a time when folate supplementation may have had important protective influences, or if it points to other characteristics that may be linked to reduced colon cancer risk in a small subset of the population.

Other nutrients involved in this pathway have been examined less extensively than dietary folate. Methionine was not associated with risk of colon cancer in a large case-control study [12], although other cohort and case-control studies did find inverse associations [73, 145]. Vitamin B_6 intake has also been shown by some investigators to reduce risk of colon cancer [73, 147] as have serum levels of pyridoxal 5′-phosphate (PLP), the main circulating form of vitamin B_6 [161].

B. Alcohol

A recent meta-analysis of cohort studies suggests a linear increase in risk for both colon and rectal cancer with alcohol consumption (approximately 15% increase in risk with an increment of 5–7 drinks/week), with similar increases for colon and rectal cancer and for men and women [162]. Geographical area of study was a significant source of heterogeneity. A systematic review of the literature suggests an increase in risk of colon cancer that is stronger than for rectal cancer in Japanese studies [163]. Independent reports have been less consistent with some reporting a positive association [4, 164–166] and the rest showing no association [167–169]. Evidence linking alcohol and colorectal cancer has usually been more consistent and stronger for more distal or rectal tumors, and some investigators have pointed to cigarette smoking [166, 169] and gender [170] as confounding or effect modifying variables.

A synergy of the nutrients involved in DNA methylation is needed for disturbance to occur. Such a synergy is exemplified by a report that men who both consumed large amounts of alcohol and had low intakes of dietary folate were those at highest risk for rectal cancer [151]. Giovannucci *et al.* [145] also found the greatest increase in risk of colon cancer when a combination of factors was present: high alcohol and low folate or low methionine. However, a study by Slattery and colleagues, in which a large sample size provided the most power to detect an association, found no significant increase in risk of colon cancer with a dietary profile that could be considered "high risk" with respect to alcohol and folate intakes [146]. Because several genetic variants have been identified that may affect this pathway, these potential interactions are being investigated and are discussed later in this chapter.

VI. MODEL 5: CELL GROWTH REGULATORS

Dietary factors act as initiators and promoters of tumors that advance along the continuum from polyp, the common precursor lesion, to colon cancer. This model of cell growth regulation encompasses several dietary factors because the proliferative activity of colon mucosa is modified by different nutrients [171, 172]. Newmark and colleagues [173] have suggested that calcium may induce saponification of free fatty acids and bile acids in the gut, thereby diminishing the proliferative stimulus of these compounds on the colon mucosa. Calcium might also directly influence proliferation by inducing cell differentiation [174]. Animal and human experimental studies suggest inhibition of cell proliferation by vitamin D, either directly or by way of its effect on calcium absorption [173–175].

A. Calcium

Support for an association between calcium and colon cancer also comes from controlled studies which show that epithelial proliferation of the colon mucosa, a presumed intermediate marker of colorectal carcinogenesis, is diminished by calcium supplementation in humans [176]. Additionally, clinical trials of calcium in sporadic adenoma patients showed that, although the overall proliferation rate was unchanged, the distribution of proliferating cells was favorably altered [177] and that polyp recurrence was reduced by 15–20% [178].

Although a meta-analysis conducted in 1996 was not supportive of a protective role of calcium in colorectal cancer etiology [179], some newer evidence does support a protective role [180–186]. A cohort study in Finland, for example, found a significant reduction in risk at high levels of calcium intake [18], as did a large multicenter case-control study from the southwest United States [180] and case-control studies conducted in Wisconsin [181], Japan [187, 188], and Shanghai [83], as well as a pooled analysis of 10 cohort studies from five countries [186]. A prospective study among Iowa women, including 241 colon cancer cases after 10 years of follow-up, suggested an inverse association between calcium and colon cancer only among those without a family history of colorectal cancer among first-degree relatives [185]. However, the WHI, a large randomized clinical trial in more than 36,000 postmenopausal women, found no effect of calcium/vitamin D supplementation on colorectal cancer [189].

B. Vitamin D

Vitamin D is recognized by the American Cancer Society as a factor in prevention of colon cancer. It may contribute to colorectal cancer prevention through enabling apoptosis [190]. A recent review of epidemiological data on vitamin D and colorectal cancer suggests protective effects of vitamin D on colon cancer risk [191]. In large studies—both case-control and cohort—the inverse association between dietary vitamin D and colorectal cancer ranges between 0.5 and 0.9, although not always reaching statistical significance after multivariate adjustment [180–184, 192–195], and supplemental vitamin D is also related to lower risk of colon cancer [181].

VII. FOOD INTAKE RELATIONSHIPS

A. Dietary Patterns

Diet is complex and comprised of foods that are usually consumed together and clusters of nutrients stemming from foods. Therefore, although individual foods and nutrients are studied, it is not clear that associations can be restricted to the individual food or nutrient. Whereas the association of colorectal cancer with individual food items is not consistent between studies, the data are more unified on the more broadly focused eating patterns.

Slattery and colleagues first identified two major dietary patterns using a large case-control study [196]. Although the descriptive labels were arbitrarily given, foods clustering together represent what could be described as the "Western diet" and the "Prudent diet." The Western diet represented an eating pattern characterized by high intakes of red meat, fast foods, high-fat dairy foods, refined grains, and foods with high sugar content. The Prudent diet, on the other hand, was typified by high intakes of fruits, vegetables, whole grains, and fish and poultry. Subsequent to the initial study, these same dietary patterns were identified and validated using data from a large cohort study even though the diet questionnaire was different [197]. The Western diet is consistently associated with increased risk of colon cancer [198–201], although in one study it was associated only with colon cancer in women [200], and in others the risk was greater among people with a family history of colorectal cancer [198, 201]. The association of the Prudent diet, however, is less consistent. It was associated with decreased risk among younger (<67 years) people in a large U.S.-based case-control study [198], but not associated with risk of colorectal cancers in other studies [199, 200].

B. Dietary and Genetic Interactions

It is estimated that 15% of the people who develop colon cancer have familial disease; however, only 3–5% of colon cancers are associated with high-risk inherited syndromes [202, 203] such as the adenomatous polyposis coli (*APC*) gene [204], or mismatch repair genes leading to hereditary nonpolyposis colon cancer (*HPNCC*) [205]. Low-penetrance genes carry a much lower independent risk than high-penetrance genes, but they may be associated with a higher attributable population risk than high-penetrance genes, especially when coupled with dietary or environmental factors that may serve as modulators. To date, however, no low-penetrance gene or gene–environment combination has been identified universally as an established risk factor for colorectal cancer.

Carcinogens in meat prepared at high temperatures contain heterocyclic amines and polycyclic aromatic hydrocarbons that are metabolized by enzymes such as *N*-acetyltransferase-1 and -2 (NAT1, -2) and glutathione *S*-transferase (GST) [206]. An ecological study suggests that international variability in colorectal cancer incidence may be a result of the combination of meat intake and genetic susceptibility to heterocyclic amines [207]. Refined grains and fats also contain polycyclic aromatic hydrocarbons and heterocyclic amines and may interact with variants of GST and NAT

[208, 209]. Actual DNA exposure in tissues to heterocyclic amines (HCA), polycyclic aromatic hydrocarbons (PAH), and *N*-nitroso compounds seem to be particularly important for those who are rapid metabolizers of *N*-acetyltransferase. Case-control comparisons confirm increased risk with high intake of red meats or smoking among NAT1 or NAT2 fast acetylators, although interactions reflect generally modest increases in risk [42, 210, 211] and suggest that combinations of these polymorphisms may be necessary to significantly increase risk [212, 213].

Several constituents of cruciferous vegetables, including isothiocyanates and indoles, have been hypothesized to reduce risk of cancer through activation of GST [214, 215] or though modulating gene expression of GST or UDP-glucuronosyltransferases (UGTs) [216]. Sulforaphane, an isothiocyanate compound found predominately in broccoli, is one of the most potent inducers of GST [217–219]. Studies suggest a lower risk of colon adenomas [220] or cancer [221] among individuals with the *GST*μ*-1* null genotype; the reduced risk of cancer was among younger (<50 years) individuals who smoked cigarettes [221]. The combined *GST*μ*-1* and *GST T-1* null genotypes and high cruciferous vegetable intake [222] were associated with a 57% decrease in risk, supporting the idea that combinations of risk factors are needed in order to detect changes in risk.

Methylenetetrahydrofolate reductase (MTHFR) is a key enzyme in the conversion to 5-methyltetrahydrofolate, the major circulating form of folate in the body and the primary methyl donor for the methylation of homocysteine to methionine. This enzyme is also key in the methylation process of DNA. A polymorphism of the human *MTHFR* gene that leads to reduced *MTHFR* activity resulting in elevated plasma homocysteine levels has been described [223]. One large case-control study observed weak differences in risk between levels of dietary intake of folate, vitamin B_{12}, vitamin B_6, alcohol, and *MTHFR* 677 polymorphism [156]. However, reduced colon cancer risk was observed with at least one variant allele in the thymidylate synthase gene and low folate consumption [224]. A combination of the *MTHFR* 677CC/1298AA genotypes and higher folate, vitamins B_2, B_6, and B_{12}, and methionine reduced colon cancer risk in women from the same study population [157]. Several other studies indicate reduced risk of colorectal cancer among individuals with the *MTHFR* 677TT genotype [225, 226], particularly among those with high levels of folate intake [225] and low alcohol consumption [225, 227]. The MTHFR 1298CC genotype and high intake of nonfried vegetables were associated with a reduction in risk of rectal cancer [228]. Other genes involved in methylation processes, such as methionine synthase [229] and alcohol dehydrogenase, also may be influenced by dietary factors. However, one study evaluating methionine synthase did not find variation in risk by genotype in conjunction with either alcohol or folate intake [230]. Further study with more combinations of genes in the methylation process should be informative.

Other dietary factors may influence colon cancer risk by interacting with genetic variants. In a large U.S. case-control study, specific sources of energy appear to be more related to colon cancer risk in the presence of specific insulin receptor substrate 2 (*IRS2*) and insulin-like growth factor 1 (*IGF1*) genotypes. A high sucrose-to-fiber ratio increased risk of colon cancer in men who had the *IRS2* DD genotype and among men who did not have the 192/192 *IGF1* genotype [231]. Studies examining the vitamin D receptor (*VDR*) with calcium and vitamin D have not found an association with colorectal cancer [232, 233]; however, the *VDR* genotype appeared to modify the association of BMI and physical activity with colorectal cancer [234]. Risk of colon cancer was decreased with a low sucrose-to-fiber ratio and the Ff/ff *FOKI* genotype. Rectal cancer risk, on the other hand, decreased with greater consumption of dairy products and increased with red or processed meat consumption and the FF *FOKI* genotype [235]. Among older people (>64 years), associations of BMI (positive association) and Prudent diet (inverse association) were stronger among those without an *e3* apolipoprotein E (*apoE*) [236, 237]. Although the influence of peroxisome proliferator-activated receptor gamma (*PPAR*γ) variants on colon and rectal cancer did not appear to operate through energy balance [238], high lutein intake or Prudent diet score and low refined grain intake were associated with reduced risk of colon cancer among individuals with the PA/AA *PPAR*γ genotype [239] (Table 4).

Potentially important diet and genetic interactions may occur for each hypothesized model of the relationship between diet and colon cancer [240]. There is both promise and hope for improved understanding of the combined effects of diet and genetics on colon and rectal cancer. Newer studies are utilizing more complex approaches such as examining effects of multiple genes and their interactions with dietary components in etiologic pathways. More advanced techniques, such as microarray analysis and whole-genome scans, are also being utilized to understand the complexities of the epigenetics of colorectal cancer.

C. Diet and Specific Mutations in Tumors

A spectrum of mutations occurs in colon cancer tumors, implying that multiple pathways to disease exist. The primary alterations observed in colon tumors are in the adenomatous polyposis coli (*APC*), *Ki-ras*, and *p53* genes, and microsatellite instability [241]. Studies have reported approximately 85–90% of colon tumors have an *APC* mutation, although recent work by Samowitz *et al.* suggest that the percentage with *APC* mutations is closer to 60–70% [242]. Additionally, those with the homozygous variant genotype of the D1822V *APC* gene who consumed a low-fat diet had lower risk of colon cancer relative to those with

TABLE 4 Interrelationship between Diet and Genetic Factors

Disease Pathways	Genetic Factors[a]	Dietary Factors
Meat/fat/cooking	*NAT2/NAT1/CYP1A1/GST*	Meat, protein, fat, cruciferous vegetables, coffee
DNA methylation	*MTHFR*	Folate, vitamin B_{12}, vitamin B_6
	Methionine synthase	Methionine
	Alcohol dehydrogenase	Alcohol
Insulin	*IGF1, IRS1*	Sugar, fat, energy balance
Cell growth regulation	*VDR*	Calcium, vitamin D
	PPARγ	Lutein, prudent diet pattern, refined grain
	ApoE	Fat, sugar

[a]*NAT2*, *N*-acetyltransferase 2; *NAT1*, *N*-acetyltransferase 1; *CYP1A1*, cytochrome P-450 1A1; *GST*, glutathione-*S*-transferase; *MTHFR*, methylenetetrahydrofolate reductase; *IGF*, insulin-like growth factor; *IRS*, insulin receptor substrate; *VDR*, vitamin D receptor; *ApoE*, apolipoprotein E; *PPAR*γ peroxisome proliferator-activated receptor gamma.

the wild-type genotype who consumed a high-fat diet [243]. Most population-based studies estimate that 30–40% of colon tumors have *Ki-ras* mutations [244, 245], whereas microsatellite instability has been found to occur in 14–18% of cases [246]. The prevalence of *p53* mutations, the most commonly mutated gene in most cancers including colon cancer, varies by the method used to detect the abnormality [247, 248]. Processed meat [248, 249] and meat in general [159] have been reported to be associated with microsatellite instability of colon and rectal tumors [250]. Liquor consumption was associated with an increase in risk of rectal cancer in men with the wild-type *K-ras* [251], monounsaturated fat [252], and low levels of nutrients associated with DNA methylation [44] or vitamins B_1 or A and iron intake [253], whereas total dietary fat [254] and linoleic acid [244] and linoleic acid [255] have been reported to be associated with greater likelihood of tumors with *Ki-ras* mutations [244].

Evidence is accumulating to suggest that diet influences the likelihood of *p53* mutation [249]. Fat intake was associated with a greater likelihood of a *p53* mutation, especially transversions. A large U.S.-based case-control study suggested an increase in *p53* mutations with a diet with a high sugar index [256], or consuming a Western diet, high glycemic load, and high red meat, fast food, and *trans*-fatty acid consumption [257], suggesting an insulin-related mechanism. In one dissenting study, beef intake was more strongly associated with *p53* negative cases than *p53* positive cases [258].

Associations between dietary factors and microsatellite instability (MSI) are beginning to appear in the literature. A small North Carolina based case-control study suggests strong inverse associations with MSI among those with adequate folate intake (>400 μg) and combined *MTHFR* 677 CC and 1298AC/CC the *MTHFR* 677 CT/TT genotype; no association was observed with lower folate intake [259]. Other studies have suggested that high levels of dietary alcohol may increase risk of MSI (OR for MSI+ versus MSI– tumors for alcohol 1.6, 95% CI 1.0–2.5; OR for liquor 1.6, 95% CI 1.1–2.4) [23].

VIII. PREVENTION OF COLON CANCER

The link between diet and colon cancer is clearly complex. It is understandable that attempts to identify consistent associations between dietary factors and colon cancer are difficult because of the interactions of individual nutrients and overall dietary composition with genetic predisposition and other environmental influences. However, there is now adequate evidence to support putting forth several important public health recommendations that may aid in the prevention of colorectal cancer and will promote lifestyle patterns consistent with healthy living. The American Cancer Society has specific advice for prevention of colon cancer and in the "Complete Guide—Nutrition and Physical Activity" recommends the following [33]:

Colorectal Cancer

The risk of colorectal cancer is higher for those with a family history of colorectal cancer. Risk is also increased by long-term tobacco use and possibly excessive alcohol use. Risk may be decreased by use of aspirin or other non-steroidal anti-inflammatory drugs (NSAIDs) and postmenopausal hormone replacement therapy. But neither aspirin-like drugs nor postmenopausal hormones are currently recommended to prevent colorectal cancer because of their potential side effects.

Some studies show a lower risk of colon cancer among those who are moderately active on a regular basis, and more vigorous activity may even further reduce the risk of colon cancer. Obesity raises the risk of colon cancer in both men and women, but the link seems to be stronger in men. Diets high in vegetables and fruits have been linked with lower risk, and diets high in processed and/or red meats have been linked with a higher risk of colon cancer.

Several studies have found that calcium, vitamin D, or a combination of the two may help protect against colorectal cancer. But because of the possible increased risk of prostate cancer with high calcium intake, it may be wise for men to limit their daily calcium intake to less than 1500 mg until further studies are done.

The best advice to reduce the risk of colon cancer is to:

- Increase the intensity and duration of physical activity
- Limit intake of processed and red meats
- Get the recommended levels of calcium
- Eat more vegetables and fruits
- Avoid obesity
- Avoid excess alcohol

In addition, it is very important to follow the ACS guidelines for regular colorectal screening because finding and removing polyps in the colon can prevent colorectal cancer.

Future research will focus on the interactions between diet and genetics. As our understanding increases we should be able, within subsets of the population with specified susceptibilities, to identify stronger and more consistent associations as well as develop targeted interventions with specific dietary recommendations.

References

1. American Cancer Society. (2006). "Colorectal Cancer." ACS, Atlanta, GA.
2. World Health Organization. (2007). "Cancer." WHO, Geneva, Switzerland.
3. Haenszel, W., and Kurihara, M. (1968). Studies of Japanese migrants. I. Mortality from cancer and other diseases among Japanese in the United States. *J. Natl. Cancer Inst.* **40**, 43–68.
4. Potter, J. D., Slattery, M. L. Bostick, R. M., and Gapstur, S. M. (1993). Colon cancer: A review of the epidemiology. *Epidemiol. Rev.* **15**, 499–545.
5. Sorenson, A. W., Slattery, M. L., and Ford, M. H. (1988). Calcium and colon cancer: A review. *Nutr. Cancer* **11**, 135–145.
6. Adlercreutz, H. (1990). Western diet and Western diseases: Some hormonal and biochemical mechanisms and associations. *Scand. J. Lab. Invest.* **50**, 3–23.
7. Potter, J. D. (1992). Reconciling the epidemiology, physiology, and molecular biology of colon cancer. *JAMA* **268**, 1573–1577.
8. Hill, M. J. (1998). Cereals, cereal fibre and colorectal cancer risk: A review of the epidemiological literature. *Eur. J. Cancer Prev.* **7**(Suppl 2), S5–10.
9. Hague, A., Elder, D. J., Hicks, D. J., and Paraskeva, C. (1995). Apoptosis in colorectal tumour cells: Induction by the short chain fatty acids butyrate, propionate and acetate and by the bile salt deoxycholate. *Int. J. Cancer* **60**, 400–406.
10. Giovannucci, E., Rimm, E. B., Stampfer, M. J., Colditz, G. A., Ascherio, A., and Willett, W. C. (1994). Intake of fat, meat, and fiber in relation to risk of colon cancer in men. *Cancer Res.* **54**, 2390–2397.
11. Slattery, M. L., Potter, J. D., Duncan, D. M., and Berry, T. D. (1997). Dietary fats and colon cancer: Assessment of risk associated with specific fatty acids. *Int. J. Cancer* **73**, 670–677.
12. Howe, G. R., Aronson, K. J., Benito, E., *et al.* (1997). The relationship between dietary fat intake and risk of colorectal cancer: Evidence from the combined analysis of 13 case-control studies. *Cancer Causes Control* **8**, 215–228.
13. Le Marchand, L., Wilkens, L. R., Hankin, J. H., Kolonel, L. N., and Lyu, L. C. (1997). A case-control study of diet and colorectal cancer in a multiethnic population in Hawaii (United States): Lipids and foods of animal origin. *Cancer Causes Control* **8**, 637–648.
14. Beresford, S. A., Johnson, K. C., Ritenbaugh, C., *et al.* (2006). Low-fat dietary pattern and risk of colorectal cancer: The Women's Health Initiative Randomized Controlled Dietary Modification Trial. *JAMA* **295**, 643–654.
15. Willett, W. C., Stampfer, M. J., Colditz, G. A., Rosner, B. A., and Speizer, F. E. (1990). Relation of meat, fat, and fiber intake to the risk of colon cancer in a prospective study among women. *N. Engl. J. Med.* **323**, 1664–1672.
16. Lin, J., Zhang, S. M., Cook, N. R., Lee, I. M., and Buring, J. E. (2004). Dietary fat and fatty acids and risk of colorectal cancer in women. *Am. J. Epidemiol.* **160**, 1011–1022.
17. Van den Brandt, P. A., Goldbohm, R. A., van't Veer, P., Volovics, A., Hermus, R. J., and Sturmans, F. (1990). A large-scale prospective cohort study on diet and cancer in The Netherlands. *J. Clin. Epidemiol.* **43**, 285–295.
18. Pietinen, P., Malila, N., Virtanen, M., *et al.* (1999). Diet and risk of colorectal cancer in a cohort of Finnish men. *Cancer Causes Control* **10**, 387–396.
19. Franceschi, S., La Vecchia, C., Russo, A., *et al.* (1998). Macronutrient intake and risk of colorectal cancer in Italy. *Int. J. Cancer* **76**, 321–324.
20. Oh, K., Willett, W. C., Fuchs, C. S., and Giovannucci, E. (2005). Dietary marine n-3 fatty acids in relation to risk of distal colorectal adenoma in women. *Cancer Epidemiol. Biomarkers Prev.* **14**, 835–841.
21. Levi, F., Pasche, C., Lucchini, F., and La Vecchia, C. (2002). Macronutrients and colorectal cancer: A Swiss case-control study. *Ann. Oncol.* **13**, 369–373.
22. Hartman, T. J., Yu, B., Albert, P. S., *et al.* (2005). Does nonsteroidal anti-inflammatory drug use modify the effect of a low-fat, high-fiber diet on recurrence of colorectal adenomas? *Cancer Epidemiol. Biomarkers Prev.* **14**, 2359–2365.
23. Slattery, M. L., Benson, J., Ma, K. N., Schaffer, D., and Potter, J. D. (2001). Trans-fatty acids and colon cancer. *Nutr. Cancer* **39**, 170–175.
24. Peng, L., He, Z., Chen, W., Holzman, I. R., and Lin, J. (2007). Effects of butyrate on intestinal barrier function in a Caco-2 cell monolayer model of intestinal barrier. *Pediatr. Res.* **61**, 37–41.

25. Sears, B. (1993). Essential fatty acids, eicosanoids, and cancer. *In* "Adjuvant Nutrition in Cancer Treatment" (Quillin, P., Williams, R. M., Ed.). Cancer Treatment and Research Foundation, Arlington Heights, IL.
26. Hofmanova, J., Vaculova, A., Lojek, A., and Kozubik, A. (2005). Interaction of polyunsaturated fatty acids and sodium butyrate during apoptosis in HT-29 human colon adenocarcinoma cells. *Eur. J. Nutr.* **44**, 40–51.
27. Hsiao, W. L., Pai, H. L., Matsui, M. S., and Weinstein, I. B. (1990). Effects of specific fatty acids on cell transformation induced by an activated c-H-ras oncogene. *Oncogene* **5**, 417–421.
28. Rose, D. P., and Connolly, J. M. (1990). Effects of fatty acids and inhibitors of eicosanoid synthesis on the growth of a human breast cancer cell line in culture. *Cancer Res.* **50**, 7139–7144.
29. Willett, W., and Stampfer, M. J. (1986). Total energy intake: Implications for epidemiologic analyses. *Am. J. Epidemiol.* **124**, 17–27.
30. Brown, C. C., Kipnis, V., Freedman, L. S., Hartman, A. M., Schatzkin, A., and Wacholder, S. (1994). Energy adjustment methods for nutritional epidemiology: The effect of categorization. *Am. J. Epidemiol.* **139**, 323–338.
31. Wacholder, S., Schatzkin, A., Freedman, L. S., Kipnis, V., Hartman, A., and Brown, C. C. (1994). Can energy adjustment separate the effects of energy from those of specific macronutrients? *Am. J. Epidemiol.* **140**, 848–855.
32. Slattery, M. L., Caan, B. J., Potter, J. D., *et al.* (1997). Dietary energy sources and colon cancer risk. *Am. J. Epidemiol.* **145**, 199–210.
33. Kushi, L. H., Byers, T., Doyle, C., *et al.* (2006). American Cancer Society Guidelines on Nutrition and Physical Activity for cancer prevention: Reducing the risk of cancer with healthy food choices and physical activity. *CA Cancer J. Clin.* **56**, 254–281; quiz 313–254.
34. Robertson, D. J., Sandler, R. S., Haile, R., *et al.* (2005). Fat, fiber, meat and the risk of colorectal adenomas. *Am. J. Gastroenterol.* **100**, 2789–2795.
35. Hill, M. J. (1999). Meat and colo-rectal cancer. *Proc. Nutr. Soc.* **58**, 261–264.
36. Gaard, M., Tretli, S., and Loken, E. B. (1996). Dietary factors and risk of colon cancer: A prospective study of 50,535 young Norwegian men and women. *Eur. J. Cancer Prev.* **5**, 445–454.
37. Levi, F., Pasche, C., La Vecchia, C., Lucchini, F., and Franceschi, S. (1999). Food groups and colorectal cancer risk. *Br. J. Cancer* **79**, 1283–1287.
38. Norat, T., Bingham, S., Ferrari, P., *et al.* (2005). Meat, fish, and colorectal cancer risk: The European Prospective Investigation into cancer and nutrition. *J. Natl. Cancer Inst.* **97**, 906–916.
39. Kampman, E., Verhoeven, D., Sloots, L., and van't Veer, P. (1995). Vegetable and animal products as determinants of colon cancer risk in Dutch men and women. *Cancer Causes Control* **6**, 225–234.
40. Boutron-Ruault, M. C., Senesse, P., Faivre, J., Chatelain, N., Belghiti, C., and Meance, S. (1999). Foods as risk factors for colorectal cancer: A case-control study in Burgundy (France). *Eur. J. Cancer Prev.* **8**, 229–235.
41. Kimura, Y., Kono, S., Toyomura, K., *et al.* (2007). Meat, fish and fat intake in relation to subsite-specific risk of colorectal cancer: The Fukuoka Colorectal Cancer Study. *Cancer Sci.* **98**, 590–597.
42. Kampman, E., Slattery, M. L., Bigler, J., *et al.* (1999). Meat consumption, genetic susceptibility, and colon cancer risk: A United States multicenter case-control study. *Cancer Epidemiol. Biomarkers Prev.* **8**, 15–24.
43. Manousos, O., Day, N. E., Trichopoulos, D., Gerovassilis, F., Tzonou, A., and Polychronopoulou, A. (1983). Diet and colorectal cancer: A case-control study in Greece. *Int. J. Cancer* **32**, 1–5.
44. Singh, P. N., and Fraser, G. E. (1998). Dietary risk factors for colon cancer in a low-risk population. *Am. J. Epidemiol.* **148**, 761–774.
45. Howe, G. R., Benito, E., Castelleto, R., *et al.* (1992). Dietary intake of fiber and decreased risk of cancers of the colon and rectum: Evidence from the combined analysis of 13 case-control studies. *J. Natl. Cancer Inst.* **84**, 1887–1896.
46. Trock, B., Lanza, E., and Greenwald, P. (1990). Dietary fiber, vegetables, and colon cancer: Critical review and meta-analyses of the epidemiologic evidence. *J. Natl. Cancer Inst.* **82**, 650–661.
47. Lipkin, M., Reddy, B., Newmark, H., and Lamprecht, S. A. (1999). Dietary factors in human colorectal cancer. *Annu. Rev. Nutr.* **19**, 545–586.
48. Negri, E., Franceschi, S., Parpinel, M., and La Vecchia, C. (1998). Fiber intake and risk of colorectal cancer. *Cancer Epidemiol. Biomarkers Prev.* **7**, 667–671.
49. Jansen, M. C., Bueno-de-Mesquita, H. B., Buzina, R., *et al.* (1999). Dietary fiber and plant foods in relation to colorectal cancer mortality: The Seven Countries Study. *Int. J. Cancer* **81**, 174–179.
50. Wakai, K., Date, C., Fukui, M., *et al.* (2007). Dietary fiber and risk of colorectal cancer in the Japan collaborative cohort study. *Cancer Epidemiol. Biomarkers Prev.* **16**, 668–675.
51. Bingham, S. A., Day, N. E., Luben, R., *et al.* (2003). Dietary fibre in food and protection against colorectal cancer in the European Prospective Investigation into Cancer and Nutrition (EPIC): An observational study. *Lancet* **361**, 1496–1501.
52. Park, Y., Hunter, D. J., Spiegelman, D., *et al.* (2005). Dietary fiber intake and risk of colorectal cancer: A pooled analysis of prospective cohort studies. *JAMA* **294**, 2849–2857.
53. Slattery, M. L., Potter, J. D., Coates, A., *et al.* (1997). Plant foods and colon cancer: An assessment of specific foods and their related nutrients (United States). *Cancer Causes Control* **8**, 575–590.
54. Fuchs, C. S., Giovannucci, E. L., Colditz, G. A., *et al.* (1999). Dietary fiber and the risk of colorectal cancer and adenoma in women. *N. Engl. J. Med.* **340**, 169–176.
55. Schatzkin, A., Lanza, E., Corle, D., *et al.* (2000). Lack of effect of a low-fat, high-fiber diet on the recurrence of colorectal adenomas. Polyp Prevention Trial Study Group. *N. Engl. J. Med.* **342**, 1149–1155.
56. Alberts, D. S., Martinez, M. E., Roe, D. J., *et al.* (2000). Lack of effect of a high-fiber cereal supplement on the recurrence of colorectal adenomas. Phoenix Colon Cancer Prevention Physicians' Network. *N. Engl. J. Med.* **342**, 1156–1162.
57. Shannon, J., White, E., Shattuck, A. L., and Potter, J. D. (1996). Relationship of food groups and water intake to colon cancer risk. *Cancer Epidemiol. Biomarkers Prev.* **5**, 495–502.

58. Steinmetz, K. A., and Potter, J. D. (1991). Vegetables, fruit, and cancer. I. Epidemiology. *Cancer Causes Control* **2**, 325–357.
59. Terry, P., Giovannucci, E., Michels, K. B., *et al.* (2001). Fruit, vegetables, dietary fiber, and risk of colorectal cancer. *J. Natl. Cancer Inst.* **93**, 525–533.
60. Park, Y., Subar, A. F., Kipnis, V., *et al.* (2007). Fruit and vegetable intakes and risk of colorectal cancer in the NIH-AARP Diet and Health Study. *Am J. Epidemiol.* **166**, 170–180.
61. World Cancer Research Fund. (1997). "Food, Nutrition and the Prevention of Cancer: A Global Perspective." American Institute for Cancer Research, Washington, DC.
62. Sato, Y., Tsubono, Y., Nakaya, N., *et al.* (2005). Fruit and vegetable consumption and risk of colorectal cancer in Japan: The Miyagi Cohort Study. *Public Health Nutr.* **8**, 309–314.
63. Voorrips, L. E., Goldbohm, R. A., van Poppel, G., Sturmans, F., Hermus, R. J., and van den Brandt, P. A. (2000). Vegetable and fruit consumption and risks of colon and rectal cancer in a prospective cohort study: The Netherlands Cohort Study on Diet and Cancer. *Am. J. Epidemiol.* **152**, 1081–1092.
64. Flood, A., Velie, E. M., Chaterjee, N., *et al.* (2002). Fruit and vegetable intakes and the risk of colorectal cancer in the Breast Cancer Detection Demonstration Project follow-up cohort. *Am. J. Clin. Nutr.* **75**, 936–943.
65. Lin, J., Zhang, S. M., Cook, N. R., *et al.* (2005). Dietary intakes of fruit, vegetables, and fiber, and risk of colorectal cancer in a prospective cohort of women (United States). *Cancer Causes Control* **16**, 225–233.
66. Potter, J. D. (2005). Vegetables, fruit, and cancer. *Lancet* **366**, 527–530.
67. Galeone, C., Pelucchi, C., Levi, F., *et al.* (2006). Onion and garlic use and human cancer. *Am. J. Clin. Nutr.* **84**, 1027–1032.
68. Michels, K. B., Giovannucci, E., Chan, A. T., Singhania, R., Fuchs, C. S., and Willett, W. C. (2006). Fruit and vegetable consumption and colorectal adenomas in the Nurses' Health Study. *Cancer Res.* **66**, 3942–3953.
69. Lanza, E., Hartman, T. J., Albert, P. S., *et al.* (2006). High dry bean intake and reduced risk of advanced colorectal adenoma recurrence among participants in the polyp prevention trial. *J. Nutr.* **136**, 1896–1903.
70. Jacobs, D. R. Jr., Slavin, J., and Marquart, L. (1995). Whole grain intake and cancer: A review of the literature. *Nutr. Cancer* **24**, 221–229.
71. Larsson, S. C., Giovannucci, E., Bergkvist, L., and Wolk, A. (2005). Whole grain consumption and risk of colorectal cancer: A population-based cohort of 60,000 women. *Br. J. Cancer* **92**, 1803–1807.
72. Chatenoud, L., La Vecchia, C., Franceschi, S., *et al.* (1999). Refined-cereal intake and risk of selected cancers in Italy. *Am. J. Clin. Nutr.* **70**, 1107–1110.
73. Kune, G., and Watson, L. (2006). Colorectal cancer protective effects and the dietary micronutrients folate, methionine, vitamins B_6, B_{12}, C, E, selenium, and lycopene. *Nutr. Cancer* **56**, 11–21.
74. MacLennan, R., Macrae, F., Bain, C., *et al.* (1995). Randomized trial of intake of fat, fiber, and beta carotene to prevent colorectal adenomas. *J. Natl. Cancer Inst.* **87**, 1760–1766.
75. Dorgan, J. F., and Schatzkin, A. (1991). Antioxidant micronutrients in cancer prevention. *Hematol. Oncol. Clin. North Am.* **5**, 43–68.
76. Hennekens, C. H. (1994). Antioxidant vitamins and cancer. *Am. J. Med.* **97**, 2S–4S; discussion 22S–28S.
77. Enger, S. M., Longnecker, M. P., Chen, M. J., *et al.* (1996). Dietary intake of specific carotenoids and vitamins A, C, and E and prevalence of colorectal adenomas. *Cancer Epidemiol. Biomarkers Prev.* **5**, 147–153.
78. Omenn, G. S., Goodman, G. E., Thornquist, M. D., *et al.* (1996). Risk factors for lung cancer and for intervention effects in CARET, the Beta-Carotene and Retinol Efficacy Trial. *J. Natl. Cancer Inst.* **88**, 1550–1559.
79. Alpha Tocopherol Beta Carotene Cancer Prevention Study Group. (1994). The effect of vitamin E and beta carotene on the incidence of lung cancer and other cancers in male smokers. The Alpha-Tocopherol, Beta Carotene Cancer Prevention Study Group. *N. Engl. J. Med.* **330**, 1029–1035.
80. Hennekens, C. H., Buring, J. E., Manson, J. E., *et al.* (1996). Lack of effect of long-term supplementation with beta carotene on the incidence of malignant neoplasms and cardiovascular disease. *N. Engl. J. Med.* **334**, 1145–1149.
81. Rock, C. L. (1997). Carotenoids: Biology and treatment. *Pharmacol. Ther.* **75**, 185–197.
82. Mannisto, S., Yaun, S. S., Hunter, D. J., *et al.* (2007). Dietary carotenoids and risk of colorectal cancer in a pooled analysis of 11 cohort studies. *Am. J. Epidemiol.* **165**, 246–255.
83. Shin, A., Li, H., Shu, X. O., Yang, G., Gao, Y. T., and Zheng, W. (2006). Dietary intake of calcium, fiber and other micronutrients in relation to colorectal cancer risk: Results from the Shanghai Women's Health Study. *Int. J. Cancer* **119**, 2938–2942.
84. Malila, N., Virtamo, J., Virtanen, M., Pietinen, P., Albanes, D., and Teppo, L. (2002). Dietary and serum alpha-tocopherol, beta-carotene and retinol, and risk for colorectal cancer in male smokers. *Eur. J. Clin. Nutr.* **56**, 615–621.
85. Wakai, K., Suzuki, K., Ito, Y., *et al.* (2005). Serum carotenoids, retinol, and tocopherols, and colorectal cancer risk in a Japanese cohort: Effect modification by sex for carotenoids. *Nutr. Cancer* **51**, 13–24.
86. Le Marchand, L., Franke, A. A., Custer, L., Wilkens, L. R., and Cooney, R. V. (1997). Lifestyle and nutritional correlates of cytochrome CYP1A2 activity: Inverse associations with plasma lutein and alpha-tocopherol. *Pharmacogenetics* **7**, 11–19.
87. Slattery, M. L., Benson, J., Curtin, K., Ma, K. N., Schaeffer, D., and Potter, J. D. (2000). Carotenoids and colon cancer. *Am. J. Clin. Nutr.* **71**, 575–582.
88. Murtaugh, M. A., Ma, K. N., Benson, J., Curtin, K., Caan, B., and Slattery, M. L. (2004). Antioxidants, carotenoids, and risk of rectal cancer. *Am. J. Epidemiol.* **159**, 32–41.
89. Flagg, E. W., Coates, R. J., and Greenberg, R. S. (1995). Epidemiologic studies of antioxidants and cancer in humans. *J. Am. Coll. Nutr.* **14**, 419–427.
90. Byers, T., and Guerrero, N. (1995). Epidemiologic evidence for vitamin C and vitamin E in cancer prevention. *Am. J. Clin. Nutr.* **62**, 1385S–1392S.
91. Stone, W. L., and Papas, A. M. (1997). Tocopherols and the etiology of colon cancer. *J. Natl. Cancer Inst.* **89**, 1006–1014.
92. Newmark, H. L., Huang, M. T., and Reddy, B. S. (2006). Mixed tocopherols inhibit azoxymethane-induced aberrant crypt foci in rats. *Nutr. Cancer* **56**, 82–85.

93. Bostick, R. M., Potter, J. D., McKenzie, D. R., *et al.* (1993). Reduced risk of colon cancer with high intake of vitamin E: The Iowa Women's Health Study. *Cancer Res.* **53**, 4230–4237.
94. Longnecker, M. P., Martin-Moreno, J. M., Knekt, P., *et al.* (1992). Serum alpha-tocopherol concentration in relation to subsequent colorectal cancer: Pooled data from five cohorts. *J. Natl. Cancer Inst.* **84**, 430–435.
95. McKeown-Eyssen, G., Holloway, C., Jazmaji, V., Bright-See, E., Dion, P., and Bruce, W. R. (1988). A randomized trial of vitamins C and E in the prevention of recurrence of colorectal polyps. *Cancer Res.* **48**, 4701–4705.
96. Greenberg, E. R., Baron, J. A., Stukel, T. A., *et al.* (1990). A clinical trial of beta carotene to prevent basal-cell and squamous-cell cancers of the skin. The Skin Cancer Prevention Study Group. *N. Engl. J. Med.* **323**, 789–795.
97. Slattery, M. L., Edwards, S. L., Anderson, K., and Caan, B. (1998). Vitamin E and colon cancer: Is there an association? *Nutr. Cancer* **30**, 201–206.
98. Sinha, R., Rothman, N., Brown, E. D., *et al.* (1995). High concentrations of the carcinogen 2-amino-1-methyl-6-phenylimidazo[4,5-*b*]pyridine (PhIP) occur in chicken but are dependent on the cooking method. *Cancer Res.* **55**, 4516–4519.
99. Hayatsu, H., Hayatsu, T., Wataya, Y., and Mower, H. F. (1985). Fecal mutagenicity arising from ingestion of fried ground beef in the human. *Mutat. Res.* **143**, 207–211.
100. Skog, K., Steineck, G., Augustsson, K., and Jagerstad, M. (1995). Effect of cooking temperature on the formation of heterocyclic amines in fried meat products and pan residues. *Carcinogenesis* **16**, 861–867.
101. Corpet, D. E., Stamp, D., Medline, A., Minkin, S., Archer, M. C., and Bruce, W. R. (1990). Promotion of colonic microadenoma growth in mice and rats fed cooked sugar or cooked casein and fat. *Cancer Res.* **50**, 6955–6958.
102. Gerhardsson de Verdier, M., Hagman, U., Peters, R. K., Steineck, G., and Overvik, E. (1991). Meat, cooking methods and colorectal cancer: A case-referent study in Stockholm. *Int. J. Cancer* **49**, 520–525.
103. Schiffman, M. H., and Felton, J. S. (1990). Re: Fried foods and the risk of colon cancer. *Am. J. Epidemiol.* **131**, 376–378.
104. Wu, K., Giovannucci, E., Byrne, C., *et al.* (2006). Meat mutagens and risk of distal colon adenoma in a cohort of U.S. men. *Cancer Epidemiol. Biomarkers Prev.* **15**, 1120–1125.
105. Murtaugh, M. A., Ma, K. N., Sweeney, C., Caan, B. J., and Slattery, M. L. (2004). Meat consumption patterns and preparation, genetic variants of metabolic enzymes, and their association with rectal cancer in men and women. *J. Nutr.* **134**, 776–784.
106. McKeown-Eyssen, G. (1994). Epidemiology of colorectal cancer revisited: Are serum triglycerides and/or plasma glucose associated with risk? *Cancer Epidemiol. Biomarkers Prev.* **3**, 687–695.
107. Giovannucci, E. (1995). Insulin and colon cancer. *Cancer Causes Control* **6**, 164–179.
108. Berentzen, T., Petersen, L., Pedersen, O., Black, E., Astrup, A., and Sorensen, T. I. (2007). Long-term effects of leisure time physical activity on risk of insulin resistance and impaired glucose tolerance, allowing for body weight history, in Danish men. *Diabetes Med.* **24**, 63–72.
109. Krotkiewski, M., Bjorntorp, P., Sjostrom, L., and Smith, U. (1983). Impact of obesity on metabolism in men and women. Importance of regional adipose tissue distribution. *J. Clin. Invest.* **72**, 1150–1162.
110. Caan, B. J., Coates, A. O., Slattery, M. L., Potter, J. D., Quesenberry C. P. Jr., and Edwards, S. M. (1998). Body size and the risk of colon cancer in a large case-control study. *Int. J. Obes. Relat. Metab. Disord.* **22**, 178–184.
111. Coulston, A. M., Liu, G. C., and Reaven, G. M. (1983). Plasma glucose, insulin and lipid responses to high-carbohydrate low-fat diets in normal humans. *Metab. Clin. Exp.* **32**, 52–56.
112. Giovannucci, E. (2002). Modifiable risk factors for colon cancer. *Gastroenterol. Clin. North Am.* **31**, 925–943.
113. Morita, T., Tabata, S., Mineshita, M., Mizoue, T., Moore, M. A., and Kono, S. (2005). The metabolic syndrome is associated with increased risk of colorectal adenoma development: The Self-Defense Forces health study. *Asian Pac. J. Cancer Prev.* **6**, 485–489.
114. Chiu, H. M., Lin, J. T., Shun, C. T., *et al.* (2007). Association of metabolic syndrome with proximal and synchronous colorectal neoplasm. *Clin. Gastroenterol. Hepatol.* **5**, 221–229, quiz 141.
115. Bowers, K., Albanes, D., Limburg, P., *et al.* (2006). A prospective study of anthropometric and clinical measurements associated with insulin resistance syndrome and colorectal cancer in male smokers. *Am. J. Epidemiol.* **164**, 652–664.
116. Ahmed, R. L., Schmitz, K. H., Anderson, K. E., Rosamond, W. D., and Folsom, A. R. (2006). The metabolic syndrome and risk of incident colorectal cancer. *Cancer* **107**, 28–36.
117. Otani, T., Iwasaki, M., Sasazuki, S., Inoue, M., and Tsugane, S. (2007). Plasma C-peptide, insulin-like growth factor-I, insulin-like growth factor binding proteins and risk of colorectal cancer in a nested case-control study: The Japan Public Health Center-Based Prospective Study. *Int. J. Cancer* **120**, 2007–2012.
118. Tran, T. T., Naigamwalla, D., Oprescu, A. I., *et al.* (2006). Hyperinsulinemia, but not other factors associated with insulin resistance, acutely enhances colorectal epithelial proliferation *in vivo*. *Endocrinology* **147**, 1830–1837.
119. Ma, J., Giovannucci, E., Pollak, M., *et al.* (2004). A prospective study of plasma C-peptide and colorectal cancer risk in men. *J. Natl. Cancer Inst.* **96**, 546–553.
120. Jenab, M., Riboli, E., Cleveland, R. J., et al. (2007). Serum C-peptide, IGFBP-1 and IGFBP-2 and risk of colon and rectal cancers in the European Prospective Investigation into Cancer and Nutrition. *Int. J. Cancer.* **121**, 368–376.
121. Pischon, T., Lahmann, P. H., Boeing, H., *et al.* (2006). Body size and risk of colon and rectal cancer in the European Prospective Investigation Into Cancer and Nutrition (EPIC). *J. Natl. Cancer Inst.* **98**, 920–931.
122. Adams, K. F., Leitzmann, M. F., Albanes, D., et al. (2007). Body mass and colorectal cancer risk in the NIH-AARP Cohort. *Am. J. Epidemiol.* **166**, 36–45.
123. Slattery, M. L., Potter, J., Caan, B., *et al.* (1997). Energy balance and colon cancer—beyond physical activity. *Cancer Res.* **57**, 75–80.

124. Kim, Y. I. (1998). Diet, lifestyle, and colorectal cancer: Is hyperinsulinemia the missing link? *Nutr. Rev.* **56**, 275–279.
125. Slattery, M. L., Benson, J., Berry, T. D., *et al.* (1997). Dietary sugar and colon cancer. *Cancer Epidemiol. Biomarkers Prev.* **6**, 677–685.
126. Bristol, J. B., Emmett, P. M., Heaton, K. W., and Williamson, R. C. (1985). Sugar, fat, and the risk of colorectal cancer. *BMJ* **291**, 1467–1470.
127. Macquart-Moulin, G., Riboli, E., Cornee, J., Charnay, B., Berthezene, P., and Day, N. (1986). Case-control study on colorectal cancer and diet in Marseilles. *Int. J. Cancer* **38**, 183–191.
128. La Vecchia, C., Negri, E., Decarli, A., *et al.* (1988). A case-control study of diet and colo-rectal cancer in northern Italy. *Int. J. Cancer* **41**, 492–498.
129. Bidoli, E., Franceschi, S., Talamini, R., Barra, S., and La Vecchia, C. (1992). Food consumption and cancer of the colon and rectum in north-eastern Italy. *Int. J. Cancer* **50**, 223–229.
130. Peters, R. K., Pike, M. C., Garabrant, D., and Mack, T. M. (1992). Diet and colon cancer in Los Angeles County, California. *Cancer Causes Control* **3**, 457–473.
131. Michaud, D. S., Fuchs, C. S., Liu, S., Willett, W. C., Colditz, G. A., and Giovannucci, E. (2005). Dietary glycemic load, carbohydrate, sugar, and colorectal cancer risk in men and women. *Cancer Epidemiol. Biomarkers Prev.* **14**, 138–147.
132. Bostick, R. M., Potter, J. D., Kushi, L. H., *et al.* (1994). Sugar, meat, and fat intake, and non-dietary risk factors for colon cancer incidence in Iowa women (United States). *Cancer Causes Control* **5**, 38–52.
133. Miller, A. B., Howe, G. R., Jain, M., Craib, K. J., and Harrison, L. (1983). Food items and food groups as risk factors in a case-control study of diet and colo-rectal cancer. *Int. J. Cancer* **32**, 155–161.
134. Tuyns, A. J., Kaaks, R., and Haelterman, M. (1988). Colorectal cancer and the consumption of foods: A case-control study in Belgium. *Nutr. Cancer* **11**, 189–204.
135. Pickle, L. W., Greene, M. H., Ziegler, R. G., *et al.* (1984). Colorectal cancer in rural Nebraska. *Cancer Res.* **44**, 363–369.
136. Crapo, P. A., Reaven, G., and Olefsky, J. (1977). Postprandial plasma-glucose and -insulin responses to different complex carbohydrates. *Diabetes* **26**, 1178–1183.
137. Vaaler, S., Hanssen, K. F., and Aagenaes, O. (1980). Plasma glucose and insulin responses to orally administered carbohydrate-rich foodstuffs. *Nutr. Metab.* **24**, 168–175.
138. McCarl, M., Harnack, L., Limburg, P. J., Anderson, K. E., and Folsom, A. R. (2006). Incidence of colorectal cancer in relation to glycemic index and load in a cohort of women. *Cancer Epidemiol. Biomarkers Prev.* **15**, 892–896.
139. Larsson, S. C., Giovannucci, E., and Wolk, A. (2007). Dietary carbohydrate, glycemic index, and glycemic load in relation to risk of colorectal cancer in women. *Am. J. Epidemiol.* **165**, 256–261.
140. Laird, P. W., and Jaenisch, R. (1994). DNA methylation and cancer. *Hum. Mol. Genet.* **3**(Spec No), 1487–1495.
141. Issa, J. P., Vertino, P. M., Wu, J., *et al.* (1993). Increased cytosine DNA-methyltransferase activity during colon cancer progression. *J. Natl. Cancer Inst.* **85**, 1235–1240.
142. Goelz, S. E., Vogelstein, B., Hamilton, S. R., and Feinberg, A. P. (1985). Hypomethylation of DNA from benign and malignant human colon neoplasms. *Science* **228**, 187–190.
143. Makos, M., Nelkin, B. D., Lerman, M. I., Latif, F., Zbar, B., and Baylin, S. B. (1992). Distinct hypermethylation patterns occur at altered chromosome loci in human lung and colon cancer. *Proc. Natl. Acad. Sci. USA* **89**, 1929–1933.
144. Baylin, S. B., Makos, M., Wu, J. J., *et al.* (1991). Abnormal patterns of DNA methylation in human neoplasia: Potential consequences for tumor progression. *Cancer Cells* **3**, 383–390.
145. Giovannucci, E., Stampfer, M. J., Colditz, G. A., *et al.* (1993). Folate, methionine, and alcohol intake and risk of colorectal adenoma. *J. Natl. Cancer Inst.* **85**, 875–884.
146. Slattery, M. L., Schaffer, D., Edwards, S. L., Ma, K. N., and Potter, J. D. (1997). Are dietary factors involved in DNA methylation associated with colon cancer? *Nutr. Cancer* **28**, 52–62.
147. Shaw, S., Jayatilleke, E., Herbert, V., and Colman, N. (1989). Cleavage of folates during ethanol metabolism. Role of acetaldehyde/xanthine oxidase-generated superoxide. *Biochem. J.* **257**, 277–280.
148. Seitz, H. K., and Simanowski, U. A. (1988). Alcohol and carcinogenesis. *Annu. Rev. Nutr.* **8**, 99–119.
149. Slattery, M. L., Curtin, K., Sweeney, C., *et al.* (2007). Diet and lifestyle factor associations with CpG island methylator phenotype and BRAF mutations in colon cancer. *Int. J. Cancer* **120**, 656–663.
150. Curtin, K., Slattery, M. L., Ulrich, C. M., Bigler, J., Levin, T. R., Wolff, R. K., Albertsen, H., Potter, J. D., and Samowitz, W. S. (2007). Genetic polymorphisms in one-carbon metabolism: Associations with CpG island methylator phenotype (CIMP) in colon cancer and the modifying effects of diet. *Carcinogenesis.* **28**, 1672–1679.
151. Freudenheim, J. L., Graham, S., Marshall, J. R., Haughey, B. P., Cholewinski, S., and Wilkinson, G. (1991). Folate intake and carcinogenesis of the colon and rectum. *Int. J. Epidemiol.* **20**, 368–374.
152. Jiang, Q., Chen, K., Ma, X., *et al.* (2005). Diets, polymorphisms of methylenetetrahydrofolate reductase, and the susceptibility of colon cancer and rectal cancer. *Cancer Detect. Prev.* **29**, 146–154.
153. Giovannucci, E., Stampfer, M. J., Colditz, G. A., *et al.* (1998). Multivitamin use, folate, and colon cancer in women in the Nurses' Health Study. *Ann. Intern. Med.* **129**, 517–524.
154. Bingham, S. (2006). The fibre-folate debate in colo-rectal cancer. *Proc. Nutr. Soc.* **65**, 19–23.
155. Larsson, S. C., Giovannucci, E., and Wolk, A. (2005). A prospective study of dietary folate intake and risk of colorectal cancer: Modification by caffeine intake and cigarette smoking. *Cancer Epidemiol. Biomarkers Prev.* **14**, 740–743.
156. Slattery, M. L., Potter, J. D., Samowitz, W., Schaffer, D., and Leppert, M. (1999). Methylenetetrahydrofolate reductase, diet, and risk of colon cancer. *Cancer Epidemiol. Biomarkers Prev.* **8**, 513–518.
157. Curtin, K., Bigler, J., Slattery, M. L., Caan, B., Potter, J. D., and Ulrich, C. M. (2004). MTHFR C677T and A1298C

polymorphisms: Diet, estrogen, and risk of colon cancer. *Cancer Epidemiol. Biomarkers Prev.* **13**, 285–292.

158. Murtaugh, M. A., Curtin, K., Sweeney, C., *et al.* (2007). Dietary intake of folate and co-factors in folate metabolism, MTHFR polymorphisms, and reduced rectal cancer. *Cancer Causes Control* **18**, 153–163.
159. de Vogel, S., van Engeland, M., Luchtenborg, M., *et al.* (2006). Dietary folate and APC mutations in sporadic colorectal cancer. *J. Nutr.* **136**, 3015–3021.
160. Brink, M., Weijenberg, M. P., de Goeij, A. F., *et al.* (2005). Dietary folate intake and k-ras mutations in sporadic colon and rectal cancer in The Netherlands Cohort Study. *Int. J. Cancer* **114**, 824–830.
161. Wei, E. K., Giovannucci, E., Selhub, J., Fuchs, C. S., Hankinson, S. E., and Ma, J. (2005). Plasma vitamin B_6 and the risk of colorectal cancer and adenoma in women. *J. Natl. Cancer Inst.* **97**, 684–692.
162. Moskal, A., Norat, T., Ferrari, P., and Riboli, E. (2007). Alcohol intake and colorectal cancer risk: A dose-response meta-analysis of published cohort studies. *Int. J. Cancer* **120**, 664–671.
163. Mizoue, T., Tanaka, K., Tsuji, I., *et al.* (2006). Alcohol drinking and colorectal cancer risk: An evaluation based on a systematic review of epidemiologic evidence among the Japanese population. *Jpn. J. Clin. Oncol.* **36**, 582–597.
164. Munoz, S. E., Navarro, A., Lantieri, M. J., *et al.* (1998). Alcohol, methylxanthine-containing beverages, and colorectal cancer in Cordoba, Argentina. *Eur. J. Cancer Prev.* **7**, 207–213.
165. Akhter, M., Kuriyama, S., Nakaya, N., *et al.* (2007). Alcohol consumption is associated with an increased risk of distal colon and rectal cancer in Japanese men: The Miyagi Cohort Study. *Eur. J. Cancer* **43**, 383–390.
166. Tuyns, A. J., Kaaks, R., Haelterman, M., and Riboli, E. (1992). Diet and gastric cancer. A case-control study in Belgium. *Int. J. Cancer* **51**, 1–6.
167. Kune, G. A., and Vitetta, L. (1992). Alcohol consumption and the etiology of colorectal cancer: A review of the scientific evidence from 1957 to 1991. *Nutr. Cancer* **18**, 97–111.
168. Hsing, A. W., McLaughlin, J. K., Chow, W. H., *et al.* (1998). Risk factors for colorectal cancer in a prospective study among U.S. white men. *Int. J. Cancer* **77**, 549–553.
169. Slattery, M. L., West, D. W., Robison, L. M., *et al.* (1990). Tobacco, alcohol, coffee, and caffeine as risk factors for colon cancer in a low-risk population. *Epidemiology* **1**, 141–145.
170. Gapstur, S. M., Potter, J. D., and Folsom, A. R. (1994). Alcohol consumption and colon and rectal cancer in postmenopausal women. *Int. J. Epidemiol.* **23**, 50–57.
171. Risio, M., Lipkin, M., Newmark, H., *et al.* (1996). Apoptosis, cell replication, and Western-style diet-induced tumorigenesis in mouse colon. *Cancer Res.* **56**, 4910–4916.
172. Stamp, D., Zhang, X. M., Medline, A., Bruce, W. R., and Archer, M. C. (1993). Sucrose enhancement of the early steps of colon carcinogenesis in mice. *Carcinogenesis* **14**, 777–779.
173. Newmark, H. L., Wargovich, M. J., and Bruce, W. R. (1984). Colon cancer and dietary fat, phosphate, and calcium: A hypothesis. *J. Natl. Cancer Inst.* **72**, 1323–1325.
174. Buset, M., Lipkin, M., Winawer, S., Swaroop, S., and Friedman, E. (1986). Inhibition of human colonic epithelial cell proliferation *in vivo* and *in vitro* by calcium. *Cancer Res.* **46**, 5426–5430.
175. Shabahang, M., Buras, R. R., Davoodi, F., *et al.* (1994). Growth inhibition of HT-29 human colon cancer cells by analogues of 1,25-dihydroxyvitamin D_3. *Cancer Res.* **54**, 4057–4064.
176. Bostick, R. M. (1997). Human studies of calcium supplementation and colorectal epithelial cell proliferation. *Cancer Epidemiol. Biomarkers Prev.* **6**, 971–980.
177. Bostick, R. M., Boldt, M., Darif, M., Wood, J. R., Overn, P., and Potter, J. D. (1997). Calcium and colorectal epithelial cell proliferation in ulcerative colitis. *Cancer Epidemiol. Biomarkers Prev.* **6**, 1021–1027.
178. Baron, J. A., Beach, M., Mandel, J. S., *et al.* (1999). Calcium supplements for the prevention of colorectal adenomas. Calcium Polyp Prevention Study Group. *N. Engl. J. Med.* **340**, 101–107.
179. Bergsma-Kadijk, J. A., van't Veer, P., Kampman, E., and Burema, J. (1996). Calcium does not protect against colorectal neoplasia. *Epidemiology* **7**, 590–597.
180. Kampman, E., Slattery, M. L., Caan, B., and Potter, J. D. (2000). Calcium, vitamin D, sunshine exposure, dairy products and colon cancer risk (United States). *Cancer Causes Control* **11**, 459–466.
181. Marcus, P. M., and Newcomb, P. A. (1998). The association of calcium and vitamin D, and colon and rectal cancer in Wisconsin women. *Int. J. Epidemiol.* **27**, 788–793.
182. Pritchard, R. S., Baron, J. A., and Gerhardsson de Verdier, M. (1996). Dietary calcium, vitamin D, and the risk of colorectal cancer in Stockholm, Sweden. *Cancer Epidemiol. Biomarkers Prev.* **5**, 897–900.
183. Kearney, J., Giovannucci, E., Rimm, E. B., *et al.* (1996). Calcium, vitamin D, and dairy foods and the occurrence of colon cancer in men. *Am. J. Epidemiol.* **143**, 907–917.
184. Zheng, W., Anderson, K. E., Kushi, L. H., *et al.* (1998). A prospective cohort study of intake of calcium, vitamin D, and other micronutrients in relation to incidence of rectal cancer among postmenopausal women. *Cancer Epidemiol. Biomarkers Prev.* **7**, 221–225.
185. Sellers, T. A., Bazyk, A. E., Bostick, R. M., *et al.* (1998). Diet and risk of colon cancer in a large prospective study of older women: An analysis stratified on family history (Iowa, United States). *Cancer Causes Control* **9**, 357–367.
186. Cho, E., Smith-Warner, S. A., Spiegelman, D., *et al.* (2004). Dairy foods, calcium, and colorectal cancer: A pooled analysis of 10 cohort studies. *J. Natl. Cancer Inst.* **96**, 1015–1022.
187. Tsong, W. H., Koh, W. P., Yuan, J. M., Wang, R., Sun, C. L., and Yu, M. C. (2007). Cigarettes and alcohol in relation to colorectal cancer: The Singapore Chinese Health Study. *Br. J. Cancer* **96**, 821–827.
188. Wakai, K., Hirose, K., Matsuo, K., *et al.* (2006). Dietary risk factors for colon and rectal cancers: A comparative case-control study. *J. Epidemiol.* **16**, 125–135.
189. Wactawski-Wende, J., Kotchen, J. M., Anderson, G. L., *et al.* (2006). Calcium plus vitamin D supplementation and the risk of colorectal cancer. *N. Engl. J. Med.* **354**, 684–696.
190. Miller, E. A., Keku, T. O., Satia, J. A., Martin, C. F., Galanko, J. A., and Sandler, R. S. (2005). Calcium, vitamin D, and

apoptosis in the rectal epithelium. *Cancer Epidemiol. Biomarkers Prev.* **14**, 525–528.

191. Martinez, M. E. (2005). Primary prevention of colorectal cancer: Lifestyle, nutrition, exercise. *Recent Results Cancer Res.* **166**, 177–211.
192. Martinez, M. E., Giovannucci, E. L., Colditz, G. A., *et al.* (1996). Calcium, vitamin D, and the occurrence of colorectal cancer among women. *J. Natl. Cancer Inst.* **88**, 1375–1382.
193. Bostick, R. M., Kushi, L. H., Wu, Y., Meyer, K. A., Sellers, T. A., and Folsom, A. R. (1999). Relation of calcium, vitamin D, and dairy food intake to ischemic heart disease mortality among postmenopausal women. *Am. J. Epidemiol.* **149**, 151–161.
194. Kampman, E., Goldbohm, R. A., van den Brandt, P. A., and van't Veer, P. (1994). Fermented dairy products, calcium, and colorectal cancer in The Netherlands Cohort Study. *Cancer Res.* **54**, 3186–3190.
195. Slattery, M. L., Sweeney, C., Murtaugh, M., *et al.* (2006). Associations between vitamin D, vitamin D receptor gene and the androgen receptor gene with colon and rectal cancer. *Int. J. Cancer* **118**, 3140–3146.
196. Slattery, M. L., Boucher, K. M., Caan, B. J., Potter, J. D., and Ma, K. N. (1998). Eating patterns and risk of colon cancer. *Am. J. Epidemiol.* **148**, 4–16.
197. Hu, F. B., Rimm, E., Smith-Warner, S. A., *et al.* (1999). Reproducibility and validity of dietary patterns assessed with a food-frequency questionnaire. *Am. J. Clin. Nutr.* **69**, 243–249.
198. Slattery, M. L., Levin, T. R., Ma, K., Goldgar, D., Holubkov, R., and Edwards, S. (2003). Family history and colorectal cancer: Predictors of risk. *Cancer Causes Control* **14**, 879–887.
199. Wu, K., Hu, F. B., Fuchs, C., Rimm, E. B., Willett, W. C., and Giovannucci, E. (2004). Dietary patterns and risk of colon cancer and adenoma in a cohort of men (United States). *Cancer Causes Control* **15**, 853–862.
200. Kim, M. K., Sasaki, S., Otani, T., and Tsugane, S. (2005). Dietary patterns and subsequent colorectal cancer risk by subsite: A prospective cohort study. *Int. J. Cancer* **115**, 790–798.
201. Le Marchand, L., Wilkens, L. R., Hankin, J. H., Kolonel, L. N., and Lyu, L. C. (1999). Independent and joint effects of family history and lifestyle on colorectal cancer risk: Implications for prevention. *Cancer Epidemiol. Biomarkers Prev.* **8**, 45–51.
202. Kaz, A. M., and Brentnall, T. A. (2006). Genetic testing for colon cancer. *Nat. Clin. Pract.* **3**, 670–679.
203. Kwak, E. L., and Chung, D. C. (2007). Hereditary colorectal cancer syndromes: An overview. *Clin. Colorect. Cancer* **6**, 340–344.
204. Groden, J., Thliveris, A., Samowitz, W., *et al.* (1991). Identification and characterization of the familial adenomatous polyposis coli gene. *Cell* **66**, 589–600.
205. Toniolo, P., Boffetta, P., Shuker, D., Rothman, N., Hulka, B., and Pearce, N. (1997). "Application of Biomarkers in Cancer Epidemiology." International Association for Research in Cancer, Oxford, U.K.
206. Dennis, M. J., Massey, R. C., Cripps, G., Venn, I., Howarth, N., and Lee, G. (1991). Factors affecting the polycyclic aromatic hydrocarbon content of cereals, fats and other food products. *Food Addit. Contam.* **8**, 517–530.
207. Ognjanovic, S., Yamamoto, J., Maskarinec, G., and Le Marchand, L. (2006). NAT2, meat consumption and colorectal cancer incidence: An ecological study among 27 countries. *Cancer Causes Control* **17**, 1175–1182.
208. Takeuchi, M., Hara, M., Inoue, T., and Kada, T. (1988). Adsorption of mutagens by refined corn bran. *Mutat. Res.* **204**, 263–267.
209. Vineis, P., and McMichael, A. (1996). Interplay between heterocyclic amines in cooked meat and metabolic phenotype in the etiology of colon cancer. *Cancer Causes Control* **7**, 479–486.
210. Goode, E. L., Potter, J. D., Bamlet, W. R., Rider, D. N., and Bigler, J. (2007). Inherited variation in carcinogen-metabolizing enzymes and risk of colorectal polyps. *Carcinogenesis* **28**, 328–341.
211. Chan, A. T., Tranah, G. J., Giovannucci, E. L., Willett, W. C., Hunter, D. J., and Fuchs, C. S. (2005). Prospective study of *N*-acetyltransferase-2 genotypes, meat intake, smoking and risk of colorectal cancer. *Int. J. Cancer* **115**, 648–652.
212. Murtaugh, M. A., Sweeney, C., Ma, K. N., Caan, B. J., and Slattery, M. L. (2005). The CYP1A1 genotype may alter the association of meat consumption patterns and preparation with the risk of colorectal cancer in men and women. *J. Nutr.* **135**, 179–186.
213. Lilla, C., Verla-Tebit, E., Risch, A., *et al.* (2006). Effect of NAT1 and NAT2 genetic polymorphisms on colorectal cancer risk associated with exposure to tobacco smoke and meat consumption. *Cancer Epidemiol. Biomarkers Prev.* **15**, 99–107.
214. Wattenberg, L. W. (1990). Inhibition of carcinogenesis by minor nutrient constituents of the diet. *Proc. Nutr. Soc.* **49**, 173–183.
215. Zhang, Y., and Talalay, P. (1994). Anticarcinogenic activities of organic isothiocyanates: Chemistry and mechanisms. *Cancer Res.* **54**, 1976s–1981s.
216. Svehlikova, V., Wang, S., Jakubikova, J., Williamson, G., Mithen, R., and Bao, Y. (2004). Interactions between sulforaphane and apigenin in the induction of UGT1A1 and GSTA1 in CaCo-2 cells. *Carcinogenesis* **25**, 1629–1637.
217. Gerhauser, C., You, M., Liu, J., *et al.* (1997). Cancer chemopreventive potential of sulforamate, a novel analogue of sulforaphane that induces phase 2 drug-metabolizing enzymes. *Cancer Res.* **57**, 272–278.
218. Fahey, J. W., Zhang, Y., and Talalay, P. (1997). Broccoli sprouts: An exceptionally rich source of inducers of enzymes that protect against chemical carcinogens. *Proc. Natl. Acad. Sci. USA* **94**, 10367–10372.
219. Sreerama, L., Hedge, M. W., and Sladek, N. E. (1995). Identification of a class 3 aldehyde dehydrogenase in human saliva and increased levels of this enzyme, glutathione *S*-transferases, and DT-diaphorase in the saliva of subjects who continually ingest large quantities of coffee or broccoli. *Clin. Cancer Res.* **1**, 1153–1163.
220. Lin, H. J., Probst-Hensch, N. M., Louie, A. D., *et al.* (1998). Glutathione transferase null genotype, broccoli, and lower prevalence of colorectal adenomas. *Cancer Epidemiol. Biomarkers Prev.* **7**, 647–652.
221. Slattery, M. L., Kampman, E., Samowitz, W., Caan, B. J., and Potter, J. D. (2000). Interplay between dietary inducers

of GST and the GSTM-1 genotype in colon cancer. *Int. J. Cancer* **87**, 728–733.

222. Seow, A., Yuan, J. M., Sun, C. L., Van Den Berg, D., Lee, H. P., and Yu, M. C. (2002). Dietary isothiocyanates, glutathione *S*-transferase polymorphisms and colorectal cancer risk in the Singapore Chinese Health Study. *Carcinogenesis* **23**, 2055–2061.
223. Frosst, P., Blom, H. J., Milos, R., *et al.* (1995). A candidate genetic risk factor for vascular disease: A common mutation in methylenetetrahydrofolate reductase. *Nat. Genet.* **10**, 111–113.
224. Ulrich, C. M., Curtin, K., Potter, J. D., Bigler, J., Caan, B., and Slattery, M. L. (2005). Polymorphisms in the reduced folate carrier, thymidylate synthase, or methionine synthase and risk of colon cancer. *Cancer Epidemiol. Biomarkers Prev.* **14**, 2509–2516.
225. Le Marchand, L., Wilkens, L. R., Kolonel, L. N., and Henderson, B. E. (2005). The MTHFR C677T polymorphism and colorectal cancer: The multiethnic cohort study. *Cancer Epidemiol. Biomarkers Prev.* **14**, 1198–1203.
226. Sharp, L., and Little, J. (2004). Polymorphisms in genes involved in folate metabolism and colorectal neoplasia: A HuGE review. *Am. J. Epidemiol.* **159**, 423–443.
227. Weisberg, I., Tran, P., Christensen, B., Sibani, S., and Rozen, R. (1998). A second genetic polymorphism in methylenetetrahydrofolate reductase (MTHFR) associated with decreased enzyme activity. *Mol. Genet. Metab.* **64**, 169–172.
228. Wang, J., Gajalakshmi, V., Jiang, J., *et al.* (2006). Associations between 5,10–methylenetetrahydrofolate reductase codon 677 and 1298 genetic polymorphisms and environmental factors with reference to susceptibility to colorectal cancer: A case-control study in an Indian population. *Int. J. Cancer* **118**, 991–997.
229. Chen, L. H., Liu, M. L., Hwang, H. Y., Chen, L. S., Korenberg, J., and Shane, B. (1997). Human methionine synthase. cDNA cloning, gene localization, and expression. *J. Biol. Chem.* **272**, 3628–3634.
230. Ma, J., Stampfer, M. J., Christensen, B., *et al.* (1999). A polymorphism of the methionine synthase gene: Association with plasma folate, vitamin B_{12}, homocyst(e)ine, and colorectal cancer risk. *Cancer Epidemiol. Biomarkers Prev.* **8**, 825–829.
231. Slattery, M. L., Murtaugh, M., Caan, B., Ma, K. N., Neuhausen, S., and Samowitz, W. (2005). Energy balance, insulin-related genes and risk of colon and rectal cancer. *Int. J. Cancer* **115**, 148–154.
232. Peters, U., McGlynn, K. A., Chatterjee, N., *et al.* (2001). Vitamin D, calcium, and vitamin D receptor polymorphism in colorectal adenomas. *Cancer Epidemiol. Biomarkers Prev.* **10**, 1267–1274.
233. Grau, M. V., Baron, J. A., Sandler, R. S., *et al.* (2003). Vitamin D, calcium supplementation, and colorectal adenomas: Results of a randomized trial. *J. Natl. Cancer Inst.* **95**, 1765–1771.
234. Slattery, M. L., Murtaugh, M., Caan, B., Ma, K. N., Wolff, R., and Samowitz, W. (2004). Associations between BMI, energy intake, energy expenditure, VDR genotype and colon and rectal cancers (United States). *Cancer Causes Control* **15**, 863–872.
235. Murtaugh, M. A., Sweeney, C., Ma, K. N., *et al.* (2006). Vitamin D receptor gene polymorphisms, dietary promotion of insulin resistance, and colon and rectal cancer. *Nutr. Cancer* **55**, 35–43.
236. Davidson, N. O. (1996). Apolipoprotein E polymorphism: Another player in the genetics of colon cancer susceptibility? *Gastroenterology* **110**, 2006–2009.
237. Slattery, M. L., Sweeney, C., Murtaugh, M., *et al.* (2005). Associations between apoE genotype and colon and rectal cancer. *Carcinogenesis* **26**, 1422–1429.
238. Slattery, M. L., Murtaugh, M. A., Sweeney, C., *et al.* (2005). PPARgamma, energy balance, and associations with colon and rectal cancer. *Nutr. Cancer* **51**, 155–161.
239. Murtaugh, M. A., Ma, K. N., Caan, B. J., *et al.* (2005). Interactions of peroxisome proliferator-activated receptor γ and diet in etiology of colorectal cancer. *Cancer Epidemiol. Biomarkers Prev.* **14**, 1224–1229.
240. Gertig, D. M., and Hunter, D. J. (1998). Genes and environment in the etiology of colorectal cancer. *Semin. Cancer Biol.* **8**, 285–298.
241. Vogelstein, B., Fearon, E. R., Hamilton, S. R., *et al.* (1988). Genetic alterations during colorectal-tumor development. *N. Engl. J. Med.* **319**, 525–532.
242. Samowitz, W. S., Slattery, M. L., Sweeney, C., Herrick, J., Wolff, R. K., and Albertsen, H. (2007). APC mutations and other genetic and epigenetic changes in colon cancer. *Mol. Cancer Res.* **5**, 165–170.
243. Slattery, M. L., Samowitz, W., Ballard, L., Schaffer, D., Leppert, M., and Potter, J. D. (2001). A molecular variant of the APC gene at codon 1822: Its association with diet, lifestyle, and risk of colon cancer. *Cancer Res.* **61**, 1000–1004.
244. Slattery, M. L., Curtin, K., Anderson, K., *et al.* (2000). Associations between dietary intake and Ki-ras mutations in colon tumors: A population-based study. *Cancer Res.* **60**, 6935–6941.
245. Martinez, M. E., Maltzman, T., Marshall, J. R., *et al.* (1999). Risk factors for Ki-ras protooncogene mutation in sporadic colorectal adenomas. *Cancer Res.* **59**, 5181–5185.
246. Thibodeau, S. N., Bren, G., and Schaid, D. (1993). Microsatellite instability in cancer of the proximal colon. *Science* **260**, 816–819.
247. Baker, S. J., Fearon, E. R., Nigro, J. M., *et al.* (1989). Chromosome 17 deletions and p53 gene mutations in colorectal carcinomas. *Science* **244**, 217–221.
248. Fearon, E. R. (1994). Molecular genetic studies of the adenoma-carcinoma sequence. *Adv. Intern. Med.* **39**, 123–147.
249. Voskuil, D. W., Kampman, E., van Kraats, A. A., *et al.* (1999). p53 over-expression and p53 mutations in colon carcinomas: Relation to dietary risk factors. *Int. J. Cancer* **81**, 675–681.
250. Luchtenborg, M., Weijenberg, M. P., de Goeij, A. F., *et al.* (2005). Meat and fish consumption, APC gene mutations and hMLH1 expression in colon and rectal cancer: A prospective cohort study (The Netherlands). *Cancer Causes Control* **16**, 1041–1054.
251. Bongaerts, B. W., de Goeij, A. F., van den Brandt, P. A., and Weijenberg, M. P. (2006). Alcohol and the risk of colon and rectal cancer with mutations in the K-ras gene. *Alcohol* **38**, 147–154.
252. Bongaerts, B. W., de Goeij, A. F., de Vogel, S., van den Brandt, P. A., Goldbohm, R. A., and Weijenberg, M. P. (2007). Alcohol consumption and distinct molecular pathways to colorectal cancer. *Br. J. Nutr.* **97**, 430–434.

253. Laso, N., Mas, S., Jose Lafuente, M., Casterad, X., Trias, M., Ballesta, A., Molina, R., Salas, J., Ascaso, C., Zheng, S., Wiencke, J. K., and Lafuente, A. (2004). Decrease in specific micronutrient intake in colorectal cancer patients with tumors presenting *Ki-ras* mutation. **24**, 2011–2020.
254. Bautista, D., Obrador, A., Moreno, V., *et al.* (1997). Ki-ras mutation modifies the protective effect of dietary monounsaturated fat and calcium on sporadic colorectal cancer. *Cancer Epidemiol. Biomarkers Prev.* **6**, 57–61.
255. Brink, M., Weijenberg, M. P., De Goeij, A. F., *et al.* (2004). Fat and K-ras mutations in sporadic colorectal cancer in The Netherlands Cohort Study. *Carcinogenesis* **25**, 1619–1628.
256. Slattery, M. L., Ballard-Barbash, R., Potter, J. D., *et al.* (2004). Sex-specific differences in colon cancer associated with p53 mutations. *Nutr. Cancer* **49**, 41–48.
257. Slattery, M. L., Curtin, K., Ma, K., *et al.* (2002). Diet activity, and lifestyle associations with p53 mutations in colon tumors. *Cancer Epidemiol. Biomarkers Prev.* **11**, 541–548.
258. Freedman, A. N., Michalek, A. M., Marshall, J. R., *et al.* (1996). Familial and nutritional risk factors for p53 overexpression in colorectal cancer. *Cancer Epidemiol. Biomarkers Prev.* **5**, 285–291.
259. Eaton, A. M., Sandler, R., Carethers, J. M., Millikan, R. C., Galanko, J., and Keku, T. O. (2005). 5,10-Methylenetetrahydrofolate reductase 677 and 1298 polymorphisms, folate intake, and microsatellite instability in colon cancer. *Cancer Epidemiol. Biomarkers Prev.* **14**, 2023–2029.
260. Young, J. F., Duthie, S. J., Milne, L., Christensen, L. P., Duthie, G. G., and Bestwick, C. S. (2007). Biphasic effect of falcarinol on caco-2 cell proliferation, DNA damage, and apoptosis. *J. Agric. Food Chem.* **55**, 618–623.
261. Yi, W., Fischer, J., and Akoh, C. C. (2005). Study of anticancer activities of muscadine grape phenolics *in vitro*. *J. Agric. Food Chem.* **53**, 8804–8812.
262. Yi, W., Fischer, J., Krewer, G., and Akoh, C. C. (2005). Phenolic compounds from blueberries can inhibit colon cancer cell proliferation and induce apoptosis. *J. Agric. Food Chem.* **53**, 7320–7329.
263. Suh, N., Paul, S., Hao, X., *et al.* (2007). Pterostilbene, an active constituent of blueberries, suppresses aberrant crypt foci formation in the azoxymethane-induced colon carcinogenesis model in rats. *Clin. Cancer Res.* **13**, 350–355.
264. Myzak, M. C., Tong, P., Dashwood, W. M., Dashwood, R. H., and Ho, E. (2007). Sulforaphane retards the growth of human PC-3 xenografts and inhibits HDAC activity in human subjects. *Exp. Biol. Med.* **232**, 227–234.
265. Gamet-Payrastre, L., Li, P., Lumeau, S., *et al.* (2000). Sulforaphane, a naturally occurring isothiocyanate, induces cell cycle arrest and apoptosis in HT29 human colon cancer cells. *Cancer Res.* **60**, 1426–1433.
266. Jackson, S. J., Singletary, K. W., and Venema, RC. (2007). Sulforaphane suppresses angiogenesis and disrupts endothelial mitotic progression and microtubule polymerization. *Vasc. Pharmacol.* **46**, 77–84.
267. Hosono, T., Fukao, T., Ogihara, J., *et al.* (2005). Diallyl trisulfide suppresses the proliferation and induces apoptosis of human colon cancer cells through oxidative modification of beta-tubulin. *J. Biol. Chem.* **280**, 41487–41493.
268. Kim, E. J., Park, S. Y., Shin, H. K., Kwon, D. Y., Surh, Y. J., and Park, J. H. (2007). Activation of caspase-8 contributes to 3,3′-diindolylmethane-induced apoptosis in colon cancer cells. *J. Nutr.* **137**, 31–36.
269. Jayaprakasam, B., Vanisree, M., Zhang, Y., Dewitt, D. L., and Nair, M. G. (2006). Impact of alkyl esters of caffeic and ferulic acids on tumor cell proliferation, cyclooxygenase enzyme, and lipid peroxidation. *J. Agric. Food Chem.* **54**, 5375–5381.
270. Schaefer, S., Baum, M., Eisenbrand, G., Dietrich, H., Will, F., and Janzowski, C. (2006). Polyphenolic apple juice extracts and their major constituents reduce oxidative damage in human colon cell lines. *Mol. Nutr. Food Res.* **50**, 24–33.
271. Winkelmann, I., Diehl, D., Oesterle, D., Daniel, H., and Wenzel, U. (2007). The suppression of aberrant crypt multiplicity in colonic tissue of 1,2-dimethylhydrazine-treated C57BL/6J mice by dietary flavone is associated with an increased expression of Krebs-cycle enzymes. *Carcinogenesis.* **28**, 1446–1454.
272. Jackson, S. J., and Venema, R. C. (2006). Quercetin inhibits eNOS, microtubule polymerization, and mitotic progression in bovine aortic endothelial cells. *J. Nutr.* **136**, 1178–1184.
273. Paterson, J. R., and Lawrence, J. R. (2001). Salicylic acid: A link between aspirin, diet and the prevention of colorectal cancer. *QJM* **94**, 445–448.
274. Baxter, G. J., Graham, A. B., Lawrence, J. R., Wiles, D., and Paterson, J. R. (2001). Salicylic acid in soups prepared from organically and non-organically grown vegetables. *Eur. J. Nutr.* **40**, 289–292.
275. de Vogel, J., Jonker-Termont, D. S., van Lieshout, E. M., Katan, M. B., and van der Meer, R. (2005). Green vegetables, red meat and colon cancer: Chlorophyll prevents the cytotoxic and hyperproliferative effects of haem in rat colon. *Carcinogenesis* **26**, 387–393.
276. de Vogel, J., Jonker-Termont, D. S., Katan, M. B., and van der Meer, R. (2005). Natural chlorophyll but not chlorophyllin prevents heme-induced cytotoxic and hyperproliferative effects in rat colon. *J. Nutr.* **135**, 1995–2000.
277. Seeram, N. P., Adams, L. S., Zhang, Y., *et al.* (2006). Blackberry, black raspberry, blueberry, cranberry, red raspberry, and strawberry extracts inhibit growth and stimulate apoptosis of human cancer cells *in vitro*. *J. Agric. Food Chem.* **54**, 9329–9339.
278. Yilmaz, Y., and Toledo, R. T. (2004). Major flavonoids in grape seeds and skins: Antioxidant capacity of catechin, epicatechin, and gallic acid. *J. Agric. Food Chem.* **52**, 255–260.
279. Lu, X., Jung, J., Cho, H. J., *et al.* (2005). Fisetin inhibits the activities of cyclin-dependent kinases leading to cell cycle arrest in HT-29 human colon cancer cells. *J. Nutr.* **135**, 2884–2890.
280. Yu, T. W., Xu, M., and Dashwood, R. H. (2004). Antimutagenic activity of spearmint. *Environ. Mol. Mutagen.* **44**, 387–393.
281. Bu-Abbas, A., Clifford, M. N., Walker, R., and Ioannides, C. (1994). Marked antimutagenic potential of aqueous green tea extracts: Mechanism of action. *Mutagenesis* **9**, 325–331.
282. Bu-Abbas, A., Nunez, X., Clifford, M. N., Walker, R., and Ioannides, C. (1996). A comparison of the antimutagenic potential of green, black and decaffeinated teas: Contribution of flavanols to the antimutagenic effect. *Mutagenesis* **11**, 597–603.

F. Gastrointestinal Health and Disease

CHAPTER 38

Intestinal Microflora and Diet in Health

MERLIN W. ARIEFDJOHAN AND DENNIS A. SAVAIANO
Department of Foods and Nutrition, Purdue University, West Lafayette, Indiana

Contents

I. INTRODUCTION

The human gastrointestinal (GI) tract contains bacterial communities that are diverse and complex. This intestinal microflora plays an important role in digestion and production of essential vitamins and protects the GI tract from pathogen colonization. Although the intestinal microflora appears to be relatively stable, it can be altered by environmental factors such as disease, antibiotics, and diet. Further, it has been postulated that dominant bacterial groups influence the well-being of their host by forming close interactions with the mammalian cells.

For the past three decades, there has been relatively little progress in understanding the GI microbe–host relationship, primarily because of methodological limitations. Consequently, the mechanisms by which the intestinal microflora influences human health and disease are not well understood. Traditional plating methods are both time consuming and insensitive. Recent advances in molecular techniques can overcome these limitations and pave the way for a better understanding of the complex GI tract ecosystem. Scientific breakthroughs in this field will permit researchers to move from observation to the prediction of disease using biomarkers based on the metabolic capabilities of intestinal microflora. Dietary intervention strategies including the consumption of prebiotics, probiotics, and synbiotics may also be developed to enhance overall health and reduce disease incidence. Therefore, understanding the intricate relationship between GI tract microflora and health is an important item on the research agenda.

This chapter serves to provide a comprehensive review of the following: (1) the concept of intestinal microflora, (2) the methodology used to investigate the intestinal microflora and their limitations, (3) the influence of diet on intestinal microflora, and (4) future directions in the field.

II. DISTRIBUTION AND DIVERSITY OF THE HUMAN INTESTINAL MICROFLORA

When the word *ecosystem* comes to mind, most people conjure up images of the rainforest with its myriads of animals, plants, and insects or the oceans teeming with fish, algae, and phytoplankton. Rarely do we think of the ecosystem on our body in the form of bacterial communities residing on the skin, oral cavity, genitals, and GI tract. On average, the human GI tract [i.e., small intestine and large intestine (colon)] is 27 feet long. Its surface area is greatly enhanced by the formation of microvilli on the surface of each intestinal villus (Figure 1). Because of this anatomical folding, the GI tract creates a large surface area for bacterial colonization (approximately 150–200 m^2 or 1614–2153 ft^2) [1]. As a comparison, the human skin covers only approximately 2 m^2 (21.5 ft^2). The adult GI tract is estimated to harbor up to 10^{14} bacteria/g of intestinal contents. This number easily exceeds the population size of other microbial communities associated with the human body. Remarkably, the microbial population in the GI tract is approximately 10 times greater than the total number of cells in the human body [2].

The intestinal microflora of the adult GI tract is composed of all three domains of life—*Bacteria*, *Archaea*, and *Eukarya*, with *Bacteria* having the highest cell densities [3]. The flora is distributed along the entire GI tract from the esophagus to the rectum (Figure 2). The intestinal microflora is also distributed in a vertical gradient within a specific part of the GI tract. Four microhabitats have been described (Figure 3): the intestinal lumen, the

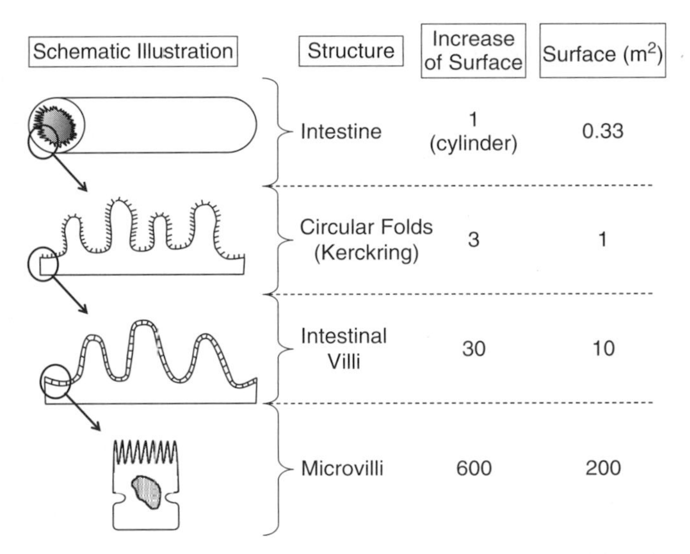

Schematic Illustration
Structure
Increase of Surface
Surface (m2)
Intestine
1 (cylinder)
0.33
Circular Folds (Kerckring)
3
1
Intestinal Villi
30
10
Microvilli
600
200

FIGURE 1 Folding on the intestinal mucosa significantly increased the surface area of the GI tract, providing a large surface area for bacterial colonization (modified from [222]).

unstirred water layer, the mucous layer at the surface of the mucosal epithelial cells, and the mucus layer in the intestinal crypts [4].

The population size and diversity of the intestinal microflora are influenced by intrinsic factors such as pH, secretion of intestinal fluids, and transit time. The stomach and duodenum create a harsh environment for bacterial colonization. The former has low pH (ranging from 2.5 to 3.5) due to secretions of intestinal fluids (e.g., hydrochloric acid) and the latter has a short transit time. Consequently, approximately only 10 to 10^3 bacteria/ml of intestinal content reside in the stomach and duodenum, with a majority of them being transient. As the environment becomes less acidic and transit time gradually increases toward the distal end of the GI tract, the microflora community flourishes both in number and diversity (Figure 2).

In comparison to the other parts of the GI tract, the coloni has the slowest cell turnover rate, the lowest redox potential, and the longest transit time. Hence, the colon harbors the most diverse and the highest number of bacteria (Figure 2). The colon is the major site for bacterial fermentation of nondigestible food components. Approximately 10^{10}–10^{12} bacteria/ml of intestinal contents reside in the colon. This microbial population includes more than 500 species belonging to more than 190 genera [5]. Because of the complexity of collecting samples from within the GI tract, most studies investigating the intestinal microflora have been based on fecal sample analyses. Thus, our current understanding of the population size may not accurately reflect "true" species abundance and their relative importance in metabolic processes [6].

A few major groups of strict anaerobes dominate the colonic microflora community including *Bacteroides* spp., *Eubacterium* spp., and *Bifidobacterium* spp. [7]. Facultative aerobes such as *Enterobacter* spp., *Streptococcus* spp., and *Lactobacillus* spp. are also present as subdominant flora. Minor groups of pathogenic and opportunistic microflora (e.g., *Clostridium* spp., *Vibrio* spp.) are also present in low numbers [8]. The metabolic functions of many of the dominant bacterial species in the GI tract are not well understood. However, several studies indicate that *Bifidobacterium* spp. and *Lactobacillus* spp. are intestinal bacterial species that directly contribute toward health [9] (Figure 4).

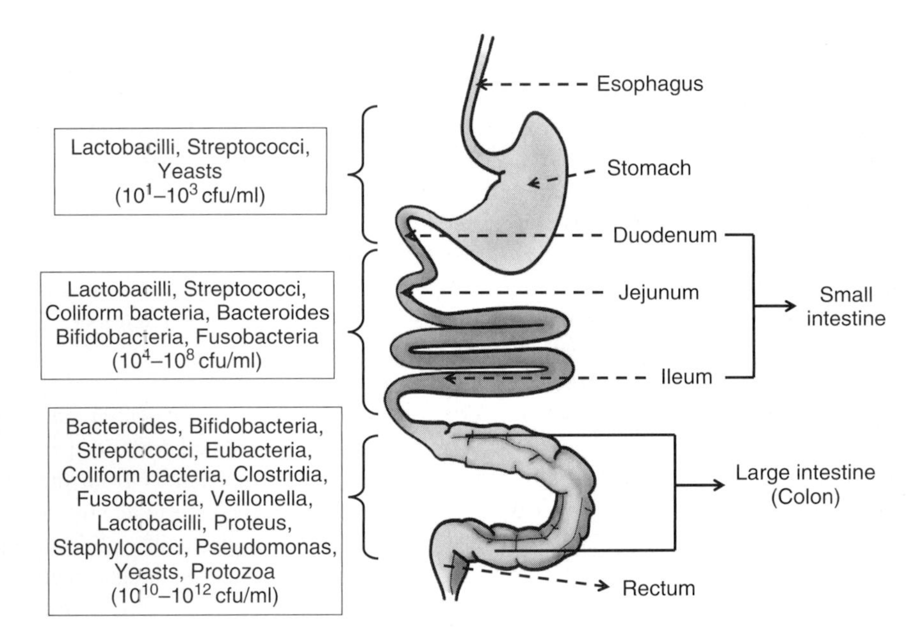

FIGURE 2 The human GI tract and the distribution of the intestinal microflora (modified from [223]).

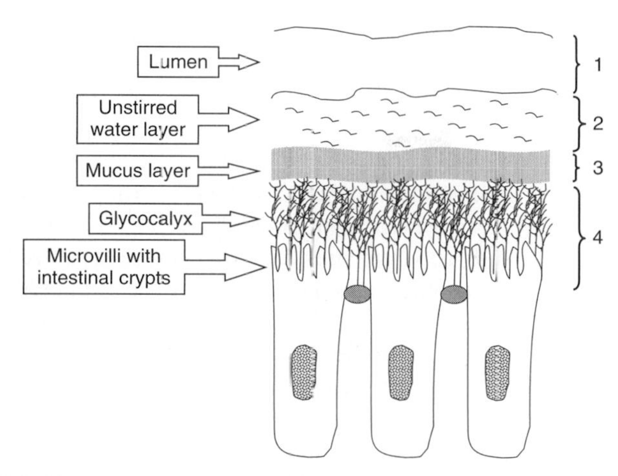

FIGURE 3 The four microhabitats within the GI tract representing the vertical distribution of the intestinal microflora: (1) the intestinal lumen, (2) the unstirred water layer, (3) the mucus layer at the surface of the mucosal epithelial cells, and (4) the mucus layer in the intestinal crypts.

III. BACTERIAL COLONIZATION, SUCCESSION, AND METABOLISM

A. Colonization and Succession

The GI tract of a fetus is apparently sterile [10, 11]. Bacterial colonization occurs immediately at delivery and gradually becomes more extensive as the newborn is introduced to the living environment and various foods [6, 10–14]. Bacterial colonization of the GI tract occurs in four phases [15, 16]. Phase I, the initial acquisition phase, occurs between birth and 1 to 2 weeks. Phase II is considered a transitional period that takes place during lactation with either breast milk or formula. This event typically starts at the end of the second week after birth and ends when supplementary feeding begins. Phase III is initiated with the introduction of other food sources, especially at weaning (e.g., when breast milk or formula is supplemented with solid foods). Phase IV follows

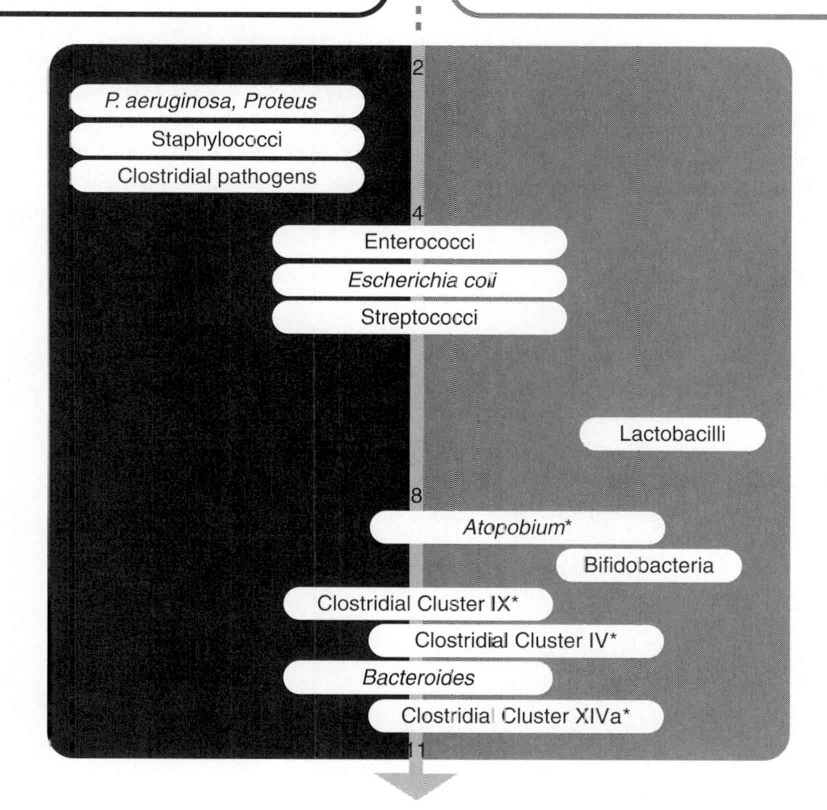

FIGURE 4 Dominant intestinal microflora as categorized into potentially harmful or health-promoting groups (reprinted with permission [224]).

once weaning is completed and the child is introduced to an adult diet.

In phase I, *Enterobacter* and *Streptococcus* appear to colonize the infant GI tract within 48 hours of birth. *Escherichia coli* follows soon after. These bacteria are thought to create a favorable environment for subsequent colonization by anaerobic bacteria including *Bacteroides* spp., *Bifidobacterium* spp., and *Clostridium* spp. This secondary colonization usually occurs within 4–7 days [6, 17, 18]. During later stages of phase I, a marked reduction in the levels of *E. coli* and *Streptococcus* was observed in the stools of exclusively breast-fed infants. This event was followed by a decrease in the levels of *Clostridium* and *Bacteroides* spp., while *Bifidobacterium* spp. gradually dominate. It appears that these trends are not as distinct in formula-fed infants [15].

In phase II, *Bifidobacterium* spp. remain at high levels and become the dominant bacterial group in the intestinal microflora of breast-fed infants [15]. Apparently, this is different from the intestinal microflora profile of formula-fed infants where relatively high numbers of *Bacteroides* spp., *Clostridium* spp., and *Streptococcus* spp. are present and *Bifidobacterium* spp. is no longer the dominant group [15].

The introduction of new foods to the infants in phase III results in major shifts in microbial succession [12, 13, 15, 18]. This shift is more significant in breast-fed infants. Following weaning, the intestinal microflora of breast-fed infants gradually changes to resemble the community found in formula-fed infants. *Clostridium* spp., *Streptococcus* spp. and *E. coli* reappear, followed by *Bacteroides* spp. and other anaerobic gram-positive cocci such as *Peptococcus* spp. and *Peptostreptococcus* spp. [18]. At the end of phase III, differences in the intestinal microflora of breast-fed and formula-fed infants are no longer observed.

Bacterial succession continues until weaning is completed (i.e., beginning of phase IV). This phase is denoted with a continued increase in *Bacteroides* spp. and anaerobic gram-positive cocci. Colonic levels of *Bifidobacterium* spp. continue to remain high. At this time, *Bifidobacterium* spp. is found in all individuals regardless of their starting diet [15]. *E. coli* and *Streptococcus* spp. gradually decline to a typical adult level (i.e., approximately 10^6–10^8 bacteria/g of feces) [19]. *Clostridium* spp. is also present [19]. At the end of phase IV, the infant intestinal microflora begins to resemble the bacterial community profile of an adult, which is typically diverse, but relatively stable [12, 13, 18] (Figure 5). Phase IV is usually attained by 2 years of age [12, 13, 18].

During adulthood, the intestinal microflora can be modified by external factors including diet, environment, and medication. Due to reasons that are still unclear, the number of bacterial species residing in the GI tract decline with age [20–22] (Figure 5). Typically, the population levels (in terms of $\log_{10}$ CFU1g) of *Bifidobacterium* spp. are markedly decreased and reduced to one or two dominant species, in particular, *B. adolescentis* and *B. longum* [20, 23]. In contrast, *Enterococcus* spp., *Lactobacillus* spp., *Bacteroides* spp., and *Clostridium* spp. are increased in GI tracts of the elderly [21, 22].

B. Metabolic Consequences of Bacterial Colonization and Succession

Changes in the intestinal microflora of infants due to bacterial colonization and succession are reflected metabolically, especially in terms of fecal short-chain fatty acid (SCFA) profiles [24]. During the period when *Bifidobacterium* spp. predominate in the GI tract of breast-fed infants, acetic acid is the major SCFA detected [25, 26]. Accumulation of acetic acid decreases the stool pH from values

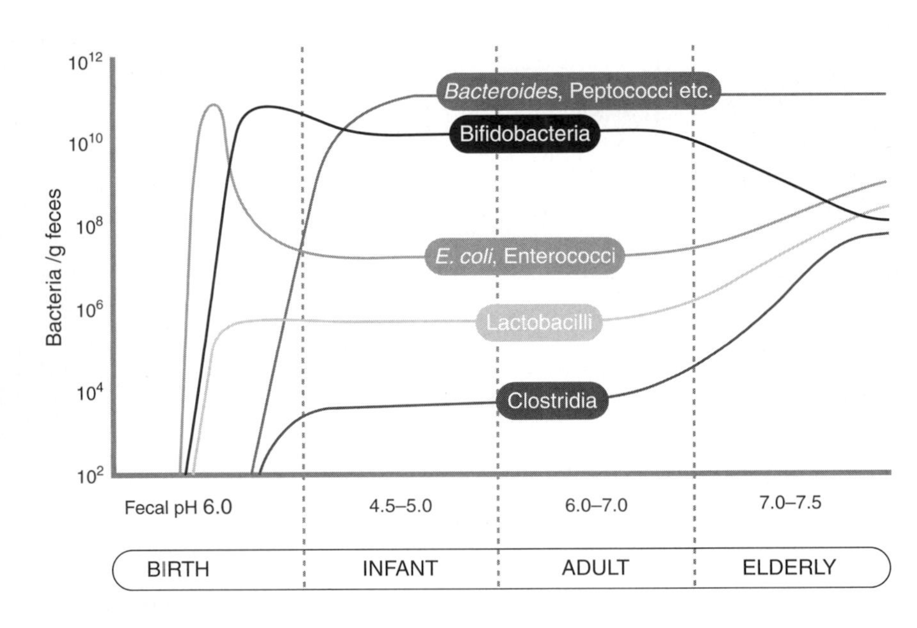

FIGURE 5 Bacterial succession throughout the lifetime (reprinted with permission [224]).

originally observed at birth. This trend is expected because *Bifidobacterium* spp. are a major producer of acetic acid. In formula-fed infants, however, lower levels of acetic acid are present and the stool pH tends to rise slightly. Therefore, changes in fecal SCFA profile of breast-fed infants appear to be more profound than that in formula-fed infants [15, 26]. Breast-fed infants are also observed to exhibit a gradual increase in total SCFA. Concurrently, lactic acid decreases while acetic and propionic acid increase. At later stages of weaning, butyric acid production gradually increases as well. In contrast, changes in the fecal SCFA profile of formula-fed infants are less profound [15, 26, 27]. This may be attributed to fewer fluctuations in the GI microflora in these infants.

The ability to ferment complex carbohydrates from diet may also reflect the variability in bacterial succession rate in breast-fed and formula-fed infants. Although variations exist among individual infants, fermentation capability appears to develop faster in formula-fed infants than in breast-fed infants [27]. This is expected because the colons of formula-fed infants are inhabited by more diverse bacterial strains and a greater population of gram-negative anaerobes, many of which are involved in the fermentation of complex carbohydrates. Unlike breast-fed infants, the colonic fermentation capacity of formula-fed infants does not vary significantly through weaning stages [27]. This observation suggests that the colonic microflora of formula-fed infants matures faster than that of breast-fed infants and does not experience major shifts in composition [26–28].

C. Factors Influencing Bacterial Colonization and Succession

In phase I of colonization, environmental factors introduce bacteria to the infant GI tract. With infants born by vaginal delivery, the length of the birthing process significantly influences the chance of detecting viable bacteria from the mouth and stomach of the newborn [29]. Infants born by cesarean section may also acquire the mother's microflora. However, initial exposure is most likely from the environment (e.g., hospital condition, nursing staffs, and other infants in the nursery) [30–33]. Besides mode of delivery, hygiene practices may influence the bacterial species introduced to the infant GI tract. For example, bacterial colonization of the GI tract of Pakistani infants born in poor areas with minimal sanitation occurred significantly earlier than in Swedish infants delivered in more sanitary conditions, regardless of delivery method [14].

There are numerous external and internal factors that shape the composition of the microbial community in the GI tract (Figure 6). External factors include mode of birth, composition of the diet, sanitation of the living environment, and the use of medicines, especially antibiotics. Internal factors include changes in the physiological condition of the host (e.g., stress, health status, and aging) and conditions in the GI tract (e.g., pH, substrate availability, redox potential, transit time, flow of enteric fluid, IgA secretions) [6]. A disruption of the "normal" intestinal bacteria may lead to the growth of those that are potentially harmful and physiologically manifest as a disease. Studies show that changes in diet, climate, aging, medication, illness, stress, and/or infection generally lead to an increase in anaerobes and *E. coli* in the small intestine and to an increase of *Enterobacter* and streptococci in the colon, with a concurrent decrease of *Bifidobacterium* spp. [34]. Although predicting the occurrence of a disease based on intestinal microflora profile is not possible at this time, it is hypothesized that a marked decline in intestinal bacteria that are beneficial to health, such as *Bifidobacterium* spp., will ultimately influence the health of the host.

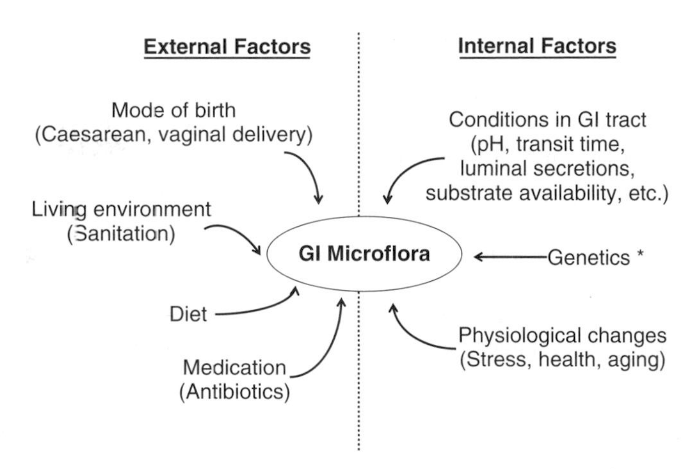

FIGURE 6 External and internal factors influencing bacterial succession in the GI tract. (*Genetics has been speculated to play a role in intestinal microflora succession, but the current body of literature has not provided strong evidence for it.)

IV. FUNCTIONS OF THE GI TRACT MICROFLORA

A. Production of Vitamin K

Prothrombin, which is involved in blood clotting, is activated by vitamin K. Food sources of vitamin K are liver, eggs, milk, and spinach. However, this vitamin can also be synthesized by bacteria in the GI tract. Administration of antibiotics eradicates intestinal bacteria and consequently may diminish *de novo* vitamin K production. Newborn infants, whose GI tracts are typically devoid of bacteria, may be given an injection of vitamin K at birth to counter deficiency until their vitamin K–producing bacteria are established [35]. Another vitamin that is exclusively synthesized by bacteria is vitamin B_{12} (cobalamin) [36], which is important for the formation of red blood cells.

B. Protection against Pathogens

Colonization of the GI tract by the intestinal microflora confers protection on the human host. A fully established GI tract microflora community can "out-compete" pathogens for carbon and other energy sources, as well as for adhesion sites on the intestinal mucosa. Consequently, pathogens are less able to establish themselves on the intestinal mucosa and thus are prevented from causing physiological damage to the host [37]. Some species of indigenous intestinal microflora may also produce bacteriocins that kill pathogens [38].

C. Enhanced Histological and Physiological Development of the GI Tract and the Immune System

Intestinal microflora plays a role in the development of intestinal mucosa and gut-associated lymphoid tissue (GALT). Evidence for this role was provided by comparative assessments of histological and physiological data from germ-free (GF) mice and conventionally raised (CONV-R) mice. GF mice are devoid of intestinal microflora because they are delivered by cesarean and raised in a sterile environment. On the other hand, CONV-R mice are delivered and raised in a standard environment in order to acquire a "normal" intestinal microflora. Histological data show that CONV-R mice have a higher epithelial cell turnover rate than GF mice (i.e., 2 days in CONV-R versus 4 days in GF). In addition, the secondary lymphatic organs of GF mice (e.g., GALT, spleen) are significantly less developed than those of CONV-R mice [3]. The intestinal microflora has also been found to stimulate the development of immune tissues in the GI tract such as the lamina propria, Peyer's patches, and mesenteric lymph nodes [39]. In the absence of the intestinal microflora, these tissues do not fully develop and become less "primed" to fight infection. The serum of GF mice contains lower concentrations of immunoglobulins than that of CONV-R mice; GF mice have a reduced immune function, are more susceptible to severe infection, and usually have poorer survival.

In humans, differences in the composition of intestinal microflora have also been suggested to influence immune function. Evidence for this is derived from comparing the prevalence of allergies and atopic diseases in infants [40]. GI tracts of infants born and raised in developing countries (i.e., assumed to have a low level of sanitation) appear to be colonized at an early stage by gram-negative bacteria and variable enterobacterial strains. On the other hand, infants in developed countries (i.e., assumed to have a high level of sanitation) acquired gram-negative bacteria later and more stable enterobacterial strains. Such differences in the intestinal microflora composition have been associated with a higher prevalence of allergies and atopic diseases in infants born and raised in developed countries than those in developing countries [40]. Another study that compared the intestinal microflora profiles of children in Europe observed that Swedish and Estonian toddlers with low counts of *Lactobacillus* spp., *Bifidobacterium* spp., *Bacteroides* spp., and *Enterococcus* spp., but having higher levels of clostridia and *Staphylococcus aureus*, were more prone to allergies than were healthy infants [41, 42]. Much remains to be learned about intestinal colonization, immune function, and disease in humans.

D. Production of Short-Chain Fatty Acids

Food components that are not digested in the small intestine travel to the colon. These nondigestible elements are substrates for fermentation by the intestinal microflora. By-products of fermentation include carbon dioxide, hydrogen, and methane gases. Nongaseous by-products include SCFA such as acetate, propionate, and butyrate (Figure 7). Butyrate is mainly produced by *Clostridium* spp. and *Eubacterium* spp. [43]. Recently, *Roseburia intestinalis*, *Eubacterium rectale*, and *Faecalibacterium prausnitzii* have also been identified as butyrate producers [44, 45]. Acetate is produced by *Lactobacillus* spp. and *Bifidobacterium* spp. [43].

Approximately 40–50% of the available energy from carbohydrate in the diet is converted into SCFA by the colonic microflora [46]. These volatile fatty acids provide energy for cellular maintenance and metabolism by being passively absorbed by the colonic epithelium (butyrate), liver (propionate), and muscle (acetate) [43, 47, 48]. Recent studies have suggested that butyrate plays an important role in cellular differentiation and proliferation in the colonic mucosa by inducing apoptosis and may confer protection

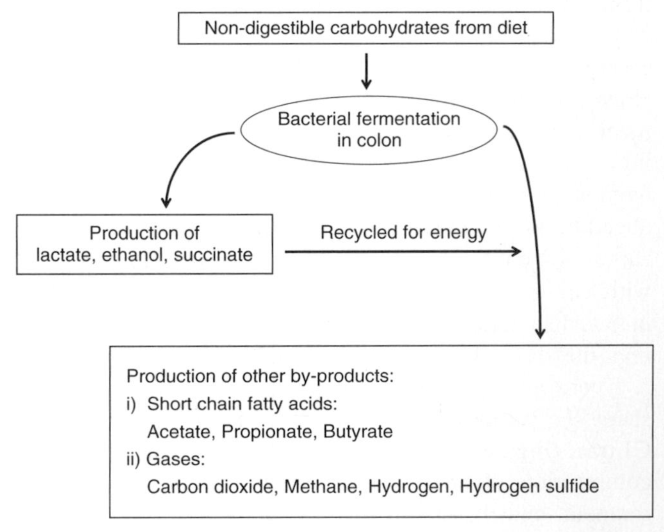

FIGURE 7 Bacterial fermentation of nondigestible food components.

against colitis and colorectal cancer by modulating oncogene expression [49, 50]. Production of SCFA also lowers intestinal pH, which increases the solubility of minerals such as calcium and magnesium [51, 52] and consequently enhances absorption [53]. Furthermore, reduction of colonic pH by accumulation of SCFA has been hypothesized to protect the intestinal mucosa from being colonized by pathogens that are less tolerant of an acidic environment (e.g., *Helicobacter pylori*) [54]. Additional benefits of SCFA are outlined in Figure 8.

E. Utilization of Nutrients

Absorption of nutrients from food is largely dependent on the action of various digestive enzymes. However, numerous food components cannot be digested. For example, stachyose (a tetrasaccharide) and raffinose (a trisaccharide) are long-chain carbohydrates (i.e., commonly termed as oligosaccharides) found in soy. The unique α-(1-6) galactose linkage present in stachyose and raffinose can be broken down by α-galactosidase, which is not secreted by the human intestinal mucosa. As a result, these oligosaccharides (and other nondigestible food components) pass to the large intestine where they become fermentation substrates for the colonic microflora. Fructo-oligosaccharides (FOS), transgalacto-oligosaccharides (TOS), and galacto-oligosaccharides (GOS) are other types of nondigestible oligosaccharides that can be fermented by the colonic microflora. Thus, the intestinal microflora salvages a significant amount of energy from an otherwise nonavailable source [43, 46–48].

Evidence to show that the intestinal microflora plays an important role in fermenting nondigestible foodstuffs is based on studies using GF and CONV-R mice [3, 55]. The latter were more efficient in nutrient absorption than GF mice because they had acquired "normal" intestinal microflora. CONV-R mice gained more body weight, grew faster, and had as much as 40% more body fat than their GF counterparts even though they were maintained on the same diet. This growth difference was present even though CONV-R mice consumed less chow per day [3]. In order for GF mice to achieve the body weight of the CONV-R counterparts, they had to consume approximately 30% more calories [55]. However, following inoculation of GF mice with the intestinal microflora of CONV-R mice, GF mice quickly gained body fat to levels equivalent to those of CONV-R mice without having to consume additional chow [55].

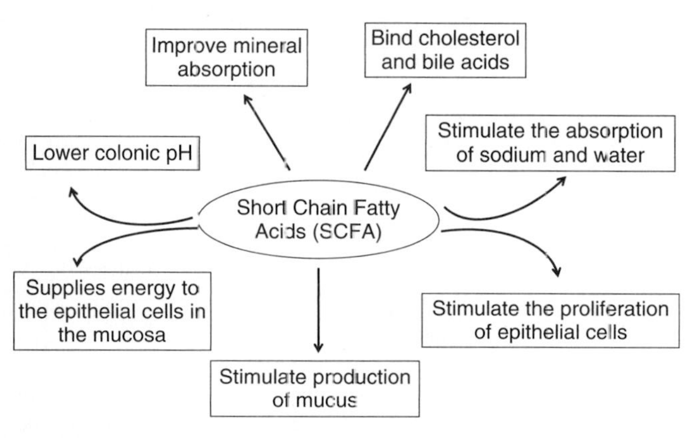

FIGURE 8 Benefits of SCFA.

F. Conversion of Isoflavones

Soy isoflavones, which may be classified as "other nonnutritive components," are phytochemicals found in soy beans. Currently, soy is the only recognized nutritionally relevant source of isoflavones. The primary isoflavones in soy are genistin and daidzin. Following ingestion, these glycosides are hydrolyzed by intestinal glucosidases and converted to the aglycone form of genistein and daidzein. These aglycones are further converted by certain intestinal microflora into specific metabolites, such as equol. Chemically, equol is similar to the hormone estradiol. Results of *in vitro* animal studies indicate that equol has a higher estrogenic effect than its precursor, daidzein [56]. Hence, equol has garnered much attention for its potential in the prevention and/or treatment of chronic diseases or conditions associated with estrogen levels (e.g., breast cancer, osteoporosis, and menopause) [57].

In humans, the conversion of soy isoflavones (genistein and daidzein) to the more potent metabolite (equol) appears to be dependent on intestinal microflora. Evidence for this conversion comes from animal and clinical studies. First, all rodents are equol producers, except those that are bred germ-free. Second, infants fed soy-based formulas do not form a substantial amount of equol for the first 4 months of life, coinciding with intestinal microflora development. In addition, individuals who are known to be equol producers have significantly lower equol excretion after antibiotic treatment [58]. It appears that the large interindividual variability in the intestinal microflora composition results in only 30–40% of individuals producing equol after soy consumption [56, 59, 60]. It is unclear which bacterial strain is involved in equol production [61, 62] and whether the ability to convert daidzein to equol can be induced in nonproducers [63].

V. METHODOLOGY FOR STUDYING INTESTINAL MICROFLORA

A. Conventional Methodology and Its Limitations

Our present knowledge of intestinal microflora is largely based on classical approaches of cultivation, direct microscopic observation, and biochemical analysis [64]. Results obtained using these conventional methodologies have improved our understanding of the intestinal microflora. However, many intestinal bacteria are difficult to culture because the media may not be specific (i.e., causing overestimation) or may be too selective (i.e., resulting in underestimation or absence of growth) for culturing the particular bacteria of interest [65]. According to recent estimates, only 20–40% [66–68] of the total intestinal microflora can be cultured and identified using conventional microbiological methods. Thus, evaluating intestinal microflora using conventional methods will bias our knowledge in favor of those genera that are most easily grown under laboratory conditions [69]. In addition, media preparation and biochemical analyses associated with these approaches are also time-consuming.

B. Molecular Analysis of the Human Intestinal Microflora

In the past two decades, novel analytical approaches based on the manipulation of 16S rDNA and other genetic materials have been developed to analyze bacterial communities in environmental samples (e.g., soil, lakes, oceans, hydrothermal vents). Recently, these methods have been adapted to evaluate bacterial communities in the GI tract [9, 70]. Molecular techniques do not require the presence of viable bacteria [9, 71]. Further, the use of genetic materials (e.g., DNA, RNA) allows species that cannot be cultured using current standard laboratory protocols to be detected. Thus, data derived from molecular techniques depict a more complete and real picture of the bacterial community. These "new" molecular methods have shown a great potential to overcome the limitations associated with conventional techniques and have become increasingly favored [71]. Many molecular techniques, especially after being optimized, are now being developed as rapid assays, allowing large-scale studies with high throughput analyses. Biological samples can now be frozen at the site of collection and then transported to laboratories for analyses. However, because RNA is degraded more easily than DNA, sample storage and handling largely depend on the choice of method and outcome measures.

Molecular techniques can be categorized as follows:

1. *Direct molecular detection and/or enumeration:* dot blot hybridization, fluorescent *in situ* hybridization (FISH), real-time polymerase chain reaction (RT-PCR)
2. *Molecular fingerprinting techniques to monitor changes in the composition of bacterial community:* terminal restriction fragment length polymorphism (T-RFLP), denaturing gradient gel electrophoresis (DGGE)
3. *Genotyping:* Rapid amplification polymorphic DNA–polymerase chain reaction (RAPD-PCR), enterobacterial repetitive intergenic consensus–polymerase chain reaction (ERIC-PCR)
4. *Other novel molecular methods under development:* microarray, magnetic-immuno PCR, *rec*A gene analysis

There are other molecular techniques that have been developed to study microbial communities in the environment. Figure 9 outlines several common methods of analysis.

C. Limitations Associated with Molecular Techniques

Despite their superiority to conventional microbiological approaches, molecular techniques do have limitations. The main drawback of molecular techniques, which significantly influences downstream analytical processes, is their reliance on cell lysis efficiency and the quality of DNA recovered from the environmental samples. DNA isolation methods that contribute to insufficient cell lysis or shearing of DNA bias PCR amplification [72–75]. Inhibitors in feces, such as bile salts and complex polysaccharides, will create similar problems [72, 76, 77]. Furthermore, special equipment and reagents are needed for sample processing and analyses (e.g., thermal cycler, RT-PCR unit, DGGE setup, bead-beating equipment for lysing cells, DNA or RNA extraction kits). Thus, initial laboratory setup may be costly, especially for novice researchers. However, research in this field has advanced rapidly in the past decade. Technological progress has significantly reduced costs, as well as improved quality and performance of equipments and reagents. It must

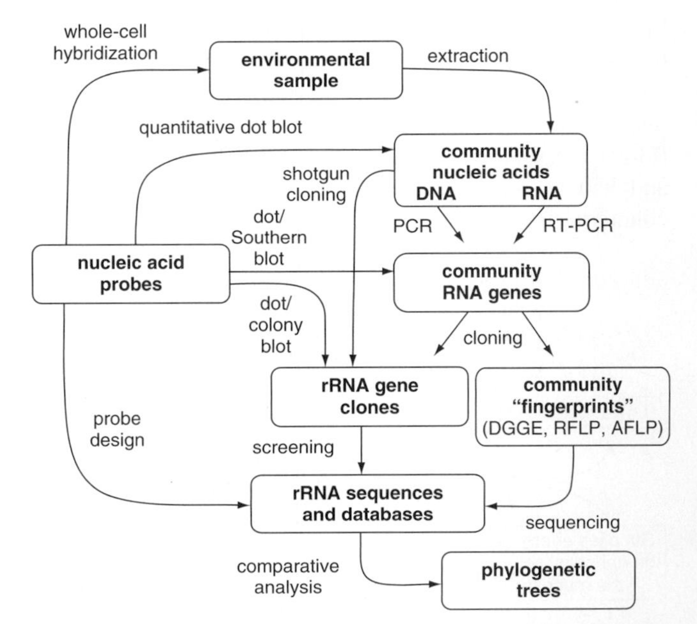

FIGURE 9 Various molecular methods and analytical approaches that are applicable for studying the intestinal microflora communities (reprinted with permission [225]).

be noted that some molecular techniques are semi-quantitative (DGGE, TGGE) and may lack sensitivity (e.g., dot blot hybridization, FISH). Primer bias may also occur in PCR reactions. In addition, almost all of these molecular techniques rely on primers and oligonucleotide probes to detect bacteria. Hence, the choice of primers and probes in analyses is crucial for detecting the bacteria of interest. Although primer and probe availability may be limited, their numbers are steadily increasing as interest in this field grows. Despite these limitations, molecular techniques are continually being developed and modified to increase efficiency and sensitivity in analyzing the diverse bacterial community in environmental samples, including those found in the GI tract and feces.

VI. INFLUENCE OF DIET ON INTESTINAL MICROFLORA

Several bacterial genera make up a large majority of the intestinal microflora (Figures 2 and 4). However, external and internal factors can modify the overall number and dominant species within the community (i.e., the concept of microbial succession; Figure 6). One of the most studied external factors that influences microbial succession is diet.

A. Can Diet Alter the Intestinal Microflora?

Comparative analyses of fecal samples from breast-fed and formula-fed infants have shown differences in the composition of their intestinal microflora as a response to the diet (Figures 10 and 11). The intestinal microflora of adults is more diverse and stable than that of infants; however, it can still be modified by diet. In 1974, Finegold *et al.* [78] compared the fecal flora of subjects who consumed a traditional Japanese diet to that of those who consumed a Western diet. In this study, a significantly higher number of *Clostridium* spp., *Eubacterium* spp., and *E. coli* were recovered in the feces of subjects on the traditional Japanese diet than in those on the Western diet. On the other hand, *Bacteroides* spp. (especially *B. infantis* and *B. putredinis*) and *Bifidobacterium* spp. were more prevalent in the fecal samples of subjects on the Western diet. In addition, differences in intestinal bacterial profiles were also found to vary by ethnicity (e.g., Asians, North Americans, and Europeans), implicating regional dietary habits [34]. Hence, from these observational studies, it appears that dietary patterns influence intestinal microflora communities.

Intervention studies provide a more direct evidence of the ability of diet to modify intestinal microflora. We have recently conducted unpublished experiments where diet was strictly controlled and modified from basal, free-living diet. Asian adolescents participated in two 3-week sessions of supervised clinical camp. The "camp" diet comprised food items such as peanut butter and jelly sandwiches, pasta, and pizza, but without the addition of probiotics, prebiotics, and other high-fiber food. Meals were given such that all subjects consumed the same types of food at any given meal time (i.e., controlled diet). Changes in intestinal microflora profile were assessed based on fecal samples using the DGGE method. It was observed that the bacterial community shifts to a different profile within 2–4 days of consuming the new diet and the profile is maintained as long as the diet remains largely unchanged. The intestinal microflora profiles of these subjects changed from "basal" profile to "camp" profile in the first session of camp, reverted to "basal" profile during the wash-out period, and

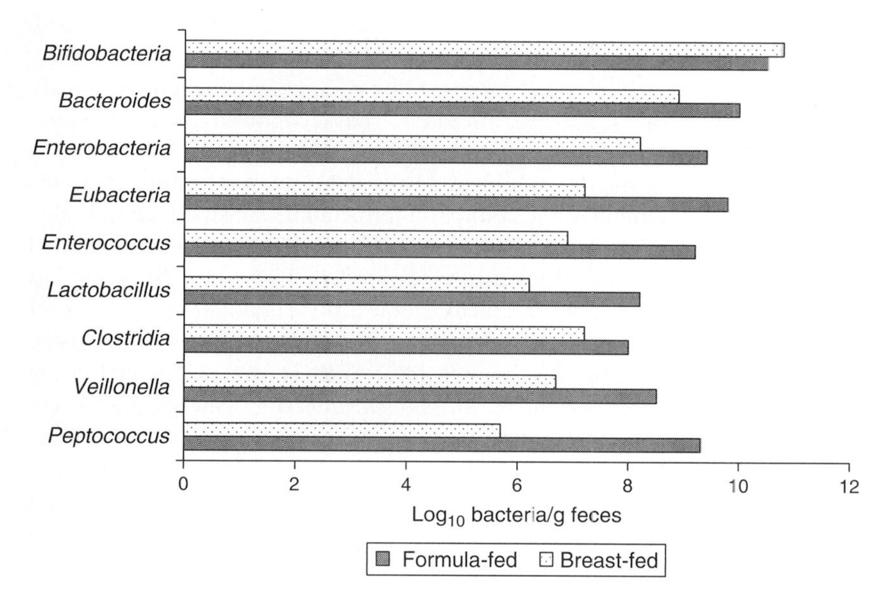

FIGURE 10 Comparison of bacterial populations in breast-fed versus formula-fed infants (data adapted from [226]).

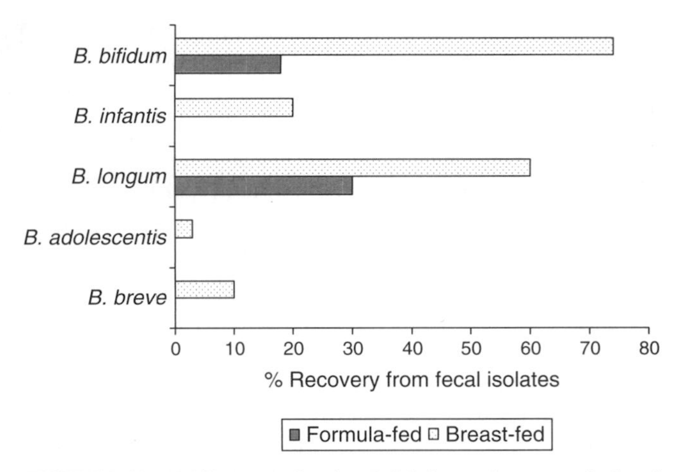

FIGURE 11 Differences in the *Bifidobacterium* spp. isolated from fecal samples of breast-fed versus formula-fed infants (data adapted from [227]).

then slipped back to "camp" profile in the second session of camp [79]. However, intersubject variability remained high, indicating that regardless of dietary changes each individual maintains his or her unique intestinal microflora profile [79, 80]. Intestinal microflora profile can also be altered when one type of food is consumed at a higher than usual intake [81, 82]. The number of *Bacteroides* spp. and *Staphylococcus* spp. as elevated during periods of high meat consumption, whereas the number of these bacteria decreased when the diet was devoid of meat. In addition, the number of *Eubacterium* spp., *Bifidobacterium* spp., and *Lactobacillus* spp. was much lower following a high-meat diet compared to a meat-free diet [81]. A high intake of cruciferous vegetables has also been shown to cause a major shift in the intestinal microflora composition, as compared to the intestinal microflora profile of subjects maintained on a fruit- and vegetable-free diet [82].

B. Altering the Intestinal Microflora using Food Supplements

1. Probiotics

Probiotics contain viable microorganisms that are proposed to improve the health of the host primarily by altering its intestinal microbial communities. Various probiotic bacterial strains are available, but the genera *Lactobacillus* spp., *Bifidobacterium* spp., and *Streptococcus* spp. are the most commonly used. Yogurt, kefir, and capsules containing freeze-dried probiotic bacteria are examples of commercially available probiotics.

The beneficial effects of consuming probiotics, which include improvement in lactose tolerance and a reduction in the risk and severity of diarrheal symptoms (i.e., as a side effect to antibiotics, traveler's diarrhea, or induced by rotavirus as in the case of gastroenteritis), have been observed in clinical studies. Inflammatory bowel disease (IBD), which is characterized by chronic or recurring intestinal inflammation, has been postulated to be caused by abnormal host immune responses to certain members of the intestinal microflora and/or from a defective mucosal barrier [83]. Some clinical studies have indicated that probiotics can treat IBD (e.g., ulcerative colitis, Crohn's disease, and pouchitis; Table 1). Probiotics may also be a treatment for intestinal infection caused by pathogens (e.g., *Clostridium difficile*, *Helicobacter pylori*) (Table 1). Furthermore, it is hypothesized that regular consumption of probiotics may reduce the levels of secondary bile acids and other mutagens that are involved in colon carcinogenesis [84]. Other studies outlining the efficacy of probiotics in various clinical applications are summarized in Table 1.

The efficacy of probiotics remains unproven. Mixed results raise concerns regarding the suitability of the strains, effective dosage, toxicity levels, and the viability of the bacteria in the product. Furthermore, mechanisms by which these bacteria colonize the intestinal tract have not been described [85–87]. Colonization success of the ingested probiotic bacteria is questionable because probiotic bacteria typically are not able to permanently colonize the GI tract. Hence, colonization and benefits are usually transient in nature [88, 89].

2. Prebiotics

Prebiotics are defined as nondigestible and fermentable food components that selectively stimulate the growth and/or activity of certain colonic bacteria that bring beneficial health effects to the host. Unlike probiotics, where one attempts to introduce beneficial bacteria into the GI tract, prebiotics attempt to stimulate the growth of endogenous beneficial bacteria by providing specific fermentable substrate(s). Examples of prebiotics are inulin, fructo-oligosaccharides (FOS), galacto-oligosaccharides (GOS), and lactulose.

It has been proposed that a food component has to satisfy the following criteria to be recognized as a prebiotic: (1) resistant to the acidic condition of the stomach, (2) unable to be broken down in the small intestine, (3) transferable to the colon as a fermentable substrate, and (4) able to stimulate the growth of a selective group of beneficial colonic bacteria (e.g., *Bifidobacterium* spp., *Lactobacillus* spp.) [90]. Clinical studies investigating the effects of prebiotics such as inulin, FOS, GOS, and lactulose are listed in Table 2. Although most of these studies resulted in the enhancement of *Bifidobacterium* spp., few of them were able to correlate bacterial data with health-related biomarkers. It is speculated that increases in the colonic levels of *Bifidobacterium* spp. modulate immune function, inhibit pathogen colonization, and augment the production of SCFA (Figure 12).

TABLE 1 Clinical Studies Investigating Various Applications of Probiotics

Intestinal Condition	Probiotic	Reference
Lactose intolerance	*Lactobacillus acidophilus*	[91]
	Lactobacillus acidophilus + *Lactobacillus casei*	[92]
	Lactobacillus acidophilus	[93]
	Bifidobacterium longum	[94]
	Lactobacillus bulgaricus + *Streptococcus thermophilus*	[95]
	Lactobacillus bulgaricus	[96]
	Lactobacillus bulgaricus	[97]
Diarrhea associated with the use of antibiotics	*Saccharomyces boulardii*	[98]
	Lactobacillus acidophilus + *Lactobacillus bulgaricus*	[99]
	Streptococcus faecium	[100]
	Lactobacillus acidophilus + *Lactobacillus bulgaricus*	[101]
	Bifidobacterium longum	[102]
	Saccharomyces boulardii	[103]
	Enterococcus faecium SF68	[104]
	Lactobacillus GG	[105]
	Bifidobacterium longum + *Lactobacillus acidophilus*	[106]
	Saccharomyces boulardii	[107]
	Lactobacillus acidophilus	[108]
	Lactobacillus GG	[109]
	Lactobacillus GG	[110]
	Lactobacillus GG	[111]
	Saccharomyces boulardii	[112]
Rotavirus-induced diarrhea and/or pediatric diarrhea, including gastroenteritis	*Enterococcus faecium* SF68	[113]
	Lactobacillus acidophilus	[114]
	Saccharomyces boulardii	[115]
	Lactobacillus GG	[116]
	Lactobacillus GG	[117]
	Bifidobacterium bifidum + *Streptococcus thermophilus*	[118]
	Lactobacillus spp.	[119]
	Lactobacillus GG	[120]
	Lactobacillus GG	[121]
	Lactobacillus GG	[122]
	Enterococcus faecium SF68	[123]
	Lactobacillus GG	[124]
	Lactobacillus GG	[125]
	Lactobacillus reuteri; *Lactobacillus* GG	[126, 127]
	Lactobacillus GG	[128]
	Lactobacillus casei DN-114001	[129]
	Bifidobacterium sp. Bb12 + *Streptococcus thermophilus*	[130]
	Lactobacillus GG	[131]
	Lactobacillus casei	[132]
	Lactobacillus GG	[133]
	Lactobacillus GG	[134]
	Lactobacillus sporogenes	[135]
	Lactobacillus rhamnosus 19070-2 + *Lactobacillus reuteri* DSM 12246	[136, 137]
	Lactobacillus casei	[138]
	Saccharomyces boulardii	[139]

(continues)

TABLE 1 *(continued)*

Intestinal Condition	Probiotic	Reference
Traveler's diarrhea	*Lactobacillus acidophilus* + *Lactobacillus bulgaricus*	[140]
	A mixture of Lactobacilli, Bifidobacteria, and Streptococci	[141]
	Lactobacillus GG	[142]
	Saccharomyces boulardii	[143]
	Lactobacillus GG	[144]
Diarrhea induced by tube feeding	*Saccharomyces boulardii*	[145]
	Saccharomyces boulardii	[146]
	Saccharomyces boulardii	[147]
	Lactobacillus sp.	[148]
Diarrhea in immunocompromised individuals	*Lactobacillus acidophilus*	[149]
	Bifidobacterium sp.	[150]
	Saccharomyces boulardii	[151]
	Lactobacillus reuteri	[152]
	Lactobacillus rhamnosus[a]	[153]
	VSL#3[b]	[154]
Small bowel bacterial overgrowth	*Lactobacillus acidophilus*	[155]
	Lactobacillus acidophilus	[156]
	Lactobacillus plantarum 299V + *Lactobacillus* GG	[157]
	Lactobacillus casei + *Lactobacillus acidophilus* Cerela	[158]
Allergic dermatitis	*Lactobacillus* GG	[159]
	Bifidobacterium lactis Bb-12 or *Lactobacillus* GG	[160]
	Lactobacillus GG	[161]
	Lactobacillus rhamnosus GG	[162]
Necrotizing enterocolitis	*Lactobacillus acidophilus* + *Bifidobacterium infantis*	[163]
Irritable bowel syndrome (IBS)	*Saccharomyces boulardii*	[164]
	Lactobacillus acidophilus	[165]
	Enterococcus faecium M74	[166]
	Lactobacillus acidophilus[c]	[167]
	Enterococcus faecium PRSS	[168]
	Lactobacillus plantarum	[169]
	Lactobacillus helveticus + *Lactobacillus acidophilus* + *Escherichia coli*[d]	[170]
	Lactobacillus plantarum	[171]
	Propionibacterium freudenreichii	[172]
	Lactobacillus plantarum	[173]
	VSL#3[b]	[174]
	VSL#3[b] + *Enterococcus faecium* SF68[e]	[175]
	Lactobacillus plantarum 299V	[176]
	Bifidobacterium animalis DN-173010	[177]
	Lactobacillus plantarum 299V	[178]
	VSL#3[b]	[179]
	Bifidobacterium infantis 35624	[180]
Inflammatory bowel disease (IBD) (e.g., ulcerative colitis, pouchitis, Crohn's disease)	*Saccharomyces boulardii*	[181]
	Lactobacillus GG	[182]
	Escherichia coli	[183]
	Escherichia coli	[184]
	Escherichia coli	[185]
	VSL#3[b]	[186]
	Saccharomyces boulardii + 5-aminosalicylic acid	[187]
	Lactobacillus GG	[188]

(continues)

TABLE 1 *(continued)*

Intestinal Condition	Probiotic	Reference
	Saccharomyces boulardii	[189]
	VSL#3[b]	[190]
	Lactobacillus GG	[191]
	Lactobacillus GG	[192]
	Escherichia coli Nissle 1917	[193]
	Lactobacillus GG	[194]
	Lactobacillus GG	[195]
Intestinal infection caused by pathogens: *Clostridium difficile*	*Lactobacillus* GG	[196]
	Saccharomyces boulardii	[197]
	Saccharomyces boulardii	[198]
	Saccharomyces boulardii	[199]
	Saccharomyces boulardii	[200]
	Lactobacillus GG	[201]
	Lactobacillus GG	[202]
	Lactobacillus plantarum 299v	[203]
Intestinal infection caused by pathogens: *Helicobacter pylori*	*Lactobacillus acidophilus* (*Johnsonii*) La1	[204]
	Lactobacillus acidophilus	[205]
	Lactobacillus gasseri 2716 (LG21)	[206]

[a]Antibiophilus.
[b]VSL#3 contains four strains of lactobacilli (*L. casei*, *L. plantarum*, *L. acidophilus*, and *L. delbruekii* subsp. *bulgaricus*), three strains of bifidobacteria (*B. longum*, *B. breve*, and *B. infantis*), and one strain of *Streptococcus salivarius* subsp. *thermophilus*.
[c]Lacteol Forte.
[d]Hylac N and Hylac N Forte.
[e]Bioflorin.

TABLE 2 Clinical Studies Investigating Alterations in Intestinal Microflora as a Response to Prebiotics

Prebiotic	Dose and Duration	Microflora Modulation	Reference
Inulin	8 g/day, 14 days	Increase in bifidobacteria, slight increase in clostridia	[208]
Inulin	Up to 34 g/day, 64 days	Increase in bifidobacteria	[209]
Inulin	15 g/day, 15 days	Increase in bifidobacteria	[210]
Inulin	20 g/day, 19 days	Increase in bifidobacteria, decrease in enterococci, and enterobacteria	[211]
FOS	15 g/day, 15 days	Increase in bifidobacteria, decrease in *Bacteroides,* clostridia and fusobacteria	[210]
FOS and PHGG	6.6 g/day FOS, 3.4 g/day PHGG, 21 days	Increase in bifidobacteria	[208]
FOS	0–20 g/day, 7 days	Increase in bifidobacteria	[212]
FOS	4 g/day, 42 days	Increase in bifidobacteria	[213]
FOS + GOS	0.04 g/l, 0.08 g/l, 28 days	Increase in bifidobacteria and lactobacilli	[214]
FOS + GOS	10 g/l, 28 days	Increase in bifidobacteria	[215]
Lactulose	10 g/day, 26 days	Increase in bifidobacteria	[216]
Lactulose	3 g/day, 14 days	Increase in bifidobacteria, decrease in lactobacilli	[217]
Lactulose	2 × 10 g/day, 4 weeks	Increase in bifidobacteria and lactobacilli	[218]
Lactulose	5 g/l and 10 g/l, 3 weeks	Increase in bifidobacteria, decrease in coliforms	[219]
GOS	0–10 g/day, 8 weeks	Increase in bifidobacteria and lactobacilli	[220]
GOS	2.5 g/day, 3 weeks	Increase in bifidobacteria, decrease in *Bacteroides* and clostridia	[221]

FOS, fructo-oligosaccharides; GOS, galacto-oligosaccharides; PHGG, partially hydrolyzed guar gum.
Reprinted with permission from [207].

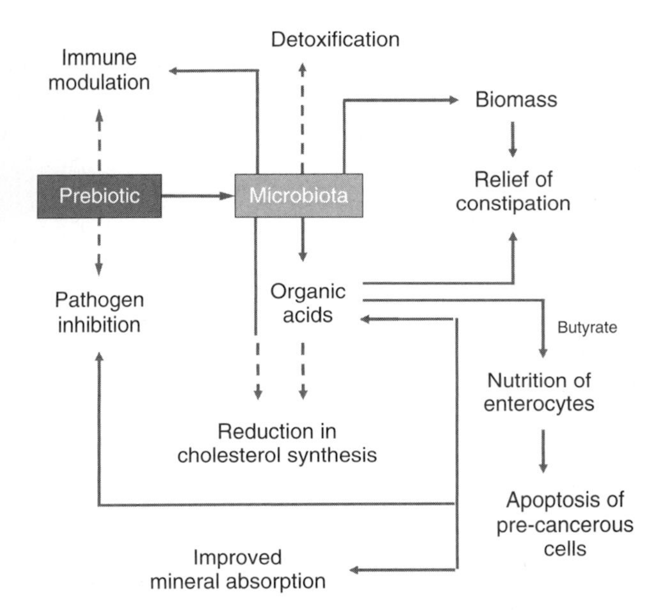

FIGURE 12 Possible mechanisms of prebiotic action. Solid lines indicate well-established modes of action and dotted lines indicate speculative mechanisms (modified from [228]).

VII. CHALLENGES IN THE FIELD

The study of the human intestinal microflora is a relatively new field. Interest in this field has been stimulated by the active marketing of probiotics and prebiotics for intestinal health. In addition, concerns about the effects of antibiotics (i.e., due to a medical treatment or residues in animal products) have focused research on the GI microflora.

Given the unknown role that the GI tract plays in maintaining overall health, efforts to advance our knowledge of the complex interaction between host and microflora are needed. As a first goal, the diversity of the intestinal microflora needs to be characterized. With better basic knowledge, we can move toward elucidating microbe–microbe and host–microbe interactions. Investigations of the effects of age, gender, host genotype, various components of diet, and the environment on intestinal microbial community are additional goals. Reliable methods to detect alterations in the composition of the intestinal microflora are required to meet these objectives. Molecular methods hold much promise in this regard [64, 65].

References

1. van Dijk, J. (1997). Morphology of the gut barrier. *Eur. J. Compar. Gastroenterol.* **2**, 23–27.
2. Luckey, T., and Floch, M. (1972). Introduction to intestinal microecology. *Am. J. Clin. Nutr.* **25**, 1291–1295.
3. Backhed, F., Ley, R. E., Sonnenburg, J. L., Peterson, D. A., and Gordon, J. I. (2005). Host-bacterial mutualism in the human intestine. *Science* **307**, 1915–1920.
4. Berg, R. D. (1996). The indigeneous gastrointestinal microflora. *Trends Microbiol.* **4**, 430–435.
5. Savage, D. S. (1977). Microbial ecology of the human large intestine. *Ann. Rev. Microbiol.* **31**, 107–133.
6. Mackie, R. I., Sghir, A., and Gaskins, H. R. (1999). Developmental microbial ecology of the neonatal gastrointestinal tract. *Am. J. Clin. Nutr.* **69**, 1035–1045.
7. Holzapfel, W. H., Haberer, P., Snel, J., Schillinger, U., and Huis in't Veld, J. H. J. (1998). Overview of gut flora and probiotics. *Int. J. Food Microbiol.* **41**, 85–101.
8. Gedek, B. (1993). Darmflora—Physiologie und Okologie. *Chemother. J.* Suppl. **1,** 2–6.
9. Tannock, G. W. (2002). Molecular methods for exploring the intestinal ecosystem. *Br. J. Nutr.* **87**, 199–201.
10. Tannock, G. W., Fuller, R., Smith, S., and Hall, M. (1990). Plasmid profiling of members of the family Enterobacteriaceae, lactobacilli and bifidobacteria to study the transmission of bacteria from mother to infant. *J. Clin. Microbiol.* **28**, 1225–1228.
11. Zetterstrom, R., Bennet, R., and Nord, K. (1994). Early infant feeding and microecology of the gut. *Acta Pediatr. Japonica* **36**, 562–571.
12. Edwards, C., and Parrett, A. (2002). Intestinal flora during the first months of life: New perspectives. *Br. J. Nutr.* **88**, S11–S18.
13. Favier, C. F., Vaughan, E. E., De Vos, W. M., and Akkermans, A. D. L. (2002). Molecular monitoring of succession of bacterial communities in human neonates. *Appl. Environ. Microbiol.* **68**, 219–226.
14. Adlerberth, I., Carlsson, B., de Man, P., Jalil, F., Khan, S. R., Larsson, P., Mellander, L., Svanborg, C., Wold, A. E., and Hanson, L. A. (1991). Intestinal colonization with Enterobacteriaceae in Pakistani and Swedish hospital-delivered infants. *Acta Paediatr. Scand.* **80**, 602–610.
15. Cooperstock, M. S., and Zedd, A. J. (1983). Intestinal flora of infants. *In* "Human Intestinal Microflora in Health and Disease" (D. J. Hentges, Ed.), pp. 79–99. Academic Press, New York.
16. Orrhage, K., and Nord, C. (1999). Factors controlling the bacterial colonization of the intestine in breastfed infants. *Acta Paediatr.* **88**, 47–57.
17. Dai, D., and Walker, W. A. (1999). Protective nutrients and bacterial colonization in the immature human gut. *Adv. Pediatr.* **46**, 353–382.
18. Stark, P. L., and Lee, A. (1982). The microbial ecology of the large bowel of breast-fed and formula-fed infants during the first year of life. *J. Med. Microbiol.* **15**, 189–203.
19. Ellis-Pegler, R., Crabtree, C., and Lambert, H. (1975). The faecal flora of children in the United Kingdom. *J. Hyg.* **75**, 135–142.
20. Gavini, F., Cayuela, C., Antoine, J.-M., Lecoq, C., Lefebvre, B., Membre, J.-M., and Neut, C. (2001). Differences in the distribution of bifidobacterial and enterobacterial species in human faecal microflora of three different (children, adults, elderly) age groups. *Microb. Ecol. Health Dis.* **13**, 40–45.
21. Hopkins, M. J., and Macfarlane, G. T. (2002). Changes in predominant bacterial populations in human faeces with age and with *Clostridium difficile* infection. *J. Med. Microbiol.* **51**, 448–454.

22. Woodmansey, E. J. (2007). Intestinal bacteria and ageing. *J. Appl. Microbiol.* **102**, 1178–1186.
23. He, F., Ouwehand, A., Isolauri, E., Hosoda, M., Benno, Y., and Salminen, S. (2001). Differences in composition and mucosal adhesion of bifidobacteria isolated from healthy adults and healthy seniors. *Curr. Microbiol.* **43**, 351–354.
24. Midtvedt, A. C., and Midtvedt, T. (1992). Production of short chain fatty acids by the intestinal microflora during the first two years of human life. *J. Pediatr. Gastroenterol. Nutr.* **18**, 321–326.
25. Bullen, C., Tearle, P., and Stewart, M. (1977). The effect of "humanised" milks and supplemented breast feeding on the faecal flora of infants. *J. Med. Microbiol.* **10**, 403–413.
26. Parrett, A., Farley, K., Fletcher, A., and Edwards, C. (2001). Comparison of faecal short chain fatty acids in breast-fed, formula-fed and mixed-fed neonates. *Proc. Nutr. Soc.* **60**, 48A.
27. Parrett, A. (2001). "Development of Colonic Fermentation in Early Life." Glasgow University.
28. Sakata, S., Tonooka, T., Ishizeki, S., Takada, M., Sakamoto, M., Fukuyama, M., and Benno, Y. (2005). Culture-independent analysis of fecal microbiota in infants, with special reference to *Bifidobacterium* species. *FEMS Microbiol. Lett.* **243**, 417–423.
29. Brook, I., Barett, C., Brinkman, C., Martin, W., and Finegold, S. (1979). Aerobic and anaerobic bacterial flora of the maternal cervix and newborn gastric fluid and conjunctiva: A prospective study. *Pediatrics* **63**, 451–455.
30. Lennox-King, S., O'Farrell, S., Bettelheim, K., and Shooter, R. (1976). Colonization of caesarean section babies by *Escherichia coli. Infection* **4**, 134–138.
31. Lennox-King, S., O'Farrol, S., Bettelheim, K., and Shooter, R. (1976). *Escherichia coli* isolated from babies delivered by caesarean section and their environment. *Infection* **4**, 439–445.
32. Gronlund, M., Lehtonen, O., Eerola, E., and Kero, P. (1999). Fecal microflora in healthy infants born by different methods of delivery: Permanent changes in intestinal flora after cesarean delivery. *J. Pediatr. Gastroenterol. Nutr.* **28**, 19–25.
33. Penders, J., Thijs, C., Vink, C., Stelma, F. F., Snijders, B., Kummeling, I., van den Brandt, P. A., and Stobberingh, E. E. (2006). Factors influencing the composition of the intestinal microbiota in early infancy. *Pediatrics* **118**, 511–521.
34. Mitsuoka, T. (1992). Intestinal flora and aging. *Nutr. Rev.* **50**, 438–446.
35. Whitney, E., and Rolfes, S. (2002). The fat soluble vitamins: A, D, E, and K. *In* "Understanding Nutrition," pp. 370–372. Wadsworth Thomson Learning.
36. Wardlaw, G., and Hampl, J. (2006). The water-soluble vitamins. *In* "Perspectives in Nutrition" (7th Ed.), pp. 335–378. McGraw-Hill Science/Engineering/Math.
37. Raibaud, P. (1992). Bacterial interactions in the gut. *In* "Probiotics" (R. Fuller, Ed.), pp. 9–28. Chapman & Hall, London.
38. Tannock, G. W. (1995). Invisible forces: The influence of the normal microflora on host characteristics. *In* "Normal Microflora: An Introduction to Microbes Inhabiting the Human Body," pp. 63–84. Chapman & Hall.
39. Gordon, H., and Pesti, L. (1971). The gnotobiotic animal as a tool in the study of host–microbial relationship. *Bacteriol. Rev.* **35**, 390–429.
40. Wills-Karp, M., Santeliz, J., and Karp, C. L. (2001). The germless theory of allergic disease: Revisiting the hygiene hypothesis. *Nat. Rev. Immunol.* **1**, 69–75.
41. Bjorksten, B., Naaber, P., Sepp, E., and Mikelaar, M. (1999). The intestinal microflora in allergic Estonian and Swedish 2-year-old children. *Clin. Exp. Allergy* **29**, 342–1246.
42. Bjorksten, B., Sepp, E., Julge, K., Voor, T., and Mikelsaar, M. (2001). Allergy development and the intestinal microflora during the first year of life. *J. Allergy Clin. Immunol.* **108**, 516–520.
43. Cummings, J. H., and Macfarlane, G. T. (1991). A review: The control and consequences of bacterial fermentation in the colon. *J. Appl. Microbiol.* **70**, 443–459.
44. Duncan, S., Hold, G., Barcenilla, A., Stewart, C., and Flint, H. (2002). *Roseburia intestinalis* spp. nov., a novel saccharolytic, butyrate-producing bacterium from human faeces. *Int. J. Syst. Evol. Microbiol.* **52**, 1–6.
45. Duncan, S. H., Louis, P., and Flint, H. J. (2004). Lactate-utilizing bacteria, isolated from human feces, that produce butyrate as a major fermentation product. *Appl. Environ. Microbiol.* **70**, 5810–5817.
46. Roediger, W. E. W. (1982). The effect of bacterial metabolites on nutrition and function of the colonic mucosa. Symbiosis between men and bacteria. *In* "Colon and Nutrition" (H. Kasper, and H. Goebell, Eds.), pp. 11–24. Plenum Press, Lancaster.
47. Hoverstad, T. (1989). The normal microflora and short-chain fatty acids. *In* "The Regulatory and Protective Role of the Normal Microflora" (R. Grubb, T. Midtvedt, and E. Norin, Eds.), pp. 89–108. Macmillan, London.
48. Pryde, S. E., Duncan, S. H., Hold, G. L., Stewart, C. S., and Flint, H. J. (2002). The microbiology of butyrate formation in the human colon. *FEMS Microbiol. Lett.* **217**, 133–139.
49. McIntyre, A., Gibson, P., and Young, G. P. (1993). Butyrate production from dietary fiber and protection against large bowel cancer in a rat model. *Gut* **34**, 386–391.
50. Demigne, C., Remesy, C., and Morand, C. (1999). Short chain fatty acids. *In* "Colonic Microbiota, Nutrition and Health" (G. R. Gibson, and M. B. Gibson, Eds.), pp. 55–70. Kluwer Academic Publishers, Dordrecht, The Netherlands.
51. Younes, H., Coudray, C., Bellanger, J., Demigne, C., Rayssiguier, Y., and Remesy, C. (2001). Effects of two fermentable carbohydrates (inulin and resistant starch) and their combination on calcium and magnesium balance in rats. *Br. J. Nutr.* **86**, 479–485.
52. Coudray, C., Tressol, J. C., Gueux, E., and Rayssiguier, Y. (2003). Effects of inulin-type fructans of different chain length and type of branching on intestinal absorption and balance of calcium and magnesium in rats. *Eur. J. Nutr.* **42**, 91–98.
53. van den Heuvel, E. G., Muys, T., van Dokkum, W., and Schaafsma, G. (1999). Oligofructose stimulates calcium absorption in adolescents. *Am. J. Clin. Nutr.* **69**, 544–548.
54. Cummings, J. H. (1998). Dietary carbohydrates and the colonic microflora. *Curr. Opin. Clin. Nutr. Metab. Care* **1**, 409–414.
55. Hooper, L. V., and Gordon, J. I. (2001). Commensal host–bacterial relationships in the gut. *Science* **292**, 1115–1118.
56. Setchell, K. D. R., Brown, N. M., and Lydeking-Olsen, E. (2002). The clinical importance of the metabolite equol—a

clue to the effectiveness of soy and its isoflavones. *J. Nutr.* **132**, 3577–3584.

57. Duncan, A. M., Merz-Demlow, B. E., Xu, X., Phipps, W. R., and Kurzer, M. S. (2000). Premenopausal equol excretors show plasma hormone profiles associated with lowered risk of breast cancer. *Cancer Epidemiol. Biomarkers Prev.* **9**, 581–586.
58. Setchell, K. D. R., and Cassidy, A. (1999). Dietary isoflavones: Biological effects and relevance to human health. *J. Nutr.* **129**, 758S–767S.
59. Cassidy, A., Brown, J. E., Hawdon, A., Faughnan, M. S., King, L. J., Millward, J., Zimmer-Nechemias, L., Wolfe, B., and Setchell, K. D. R. (2006). Factors affecting the bioavailability of soy isoflavones in humans after ingestion of physiologically relevant levels from different soy foods. *J. Nutr.* **136**, 45–51.
60. Karr, S., Lampe, J., Hutchins, A., and Slavin, J. (1997). Urinary isoflavonoid excretion in humans is dose dependent at low to moderate levels of soy-protein consumption. *Am. J. Clin. Nutr.* **66**, 46–51.
61. Atkinson, C., Berman, S., Humbert, O., and Lampe, J. W. (2004). *In vitro* incubation of human feces with daidzein and antibiotics suggests interindividual differences in the bacteria responsible for equol production. *J. Nutr.* **134**, 596–599.
62. Atkinson, C., Frankenfeld, C. L., and Lampe, J. W. (2005). Gut bacterial metabolism of the soy isoflavone daidzein: Exploring the relevance to human health. *Exp. Biol. Med.* **230**, 155–170.
63. Vedrine, N., Mathey, J., Morand, C., Brandolini, M., Davicco, M.-J., Guy, L., Remesy, C., Coxam, V., Manach, C. (2006). One-month exposure to soy isoflavones did not induce the ability to produce equol in postmenopausal women. *European J. Clin. Nutr.* **60**, 1039–1045.
64. O'Sullivan, D. (2000). Methods for analysis of the intestinal microflora. *Curr. Issues Intest. Microbiol.* **1**, 39–50.
65. Franks, A. H., Harmsen, H. J. M., Raangs, G. C., Jansen, G. J., Schut, F., and Welling, G. W. (1998). Variations of bacterial populations in human feces measured by fluorescent *in situ* hybridization with group-specific 16S rRNA-targeted oligonucleotide probes. *Appl. Environ. Microbiol.* **64**, 3336–3345.
66. Wilson, K., and Blitchington, R. (1996). Human colonic biota studied by ribosomal DNA sequence analysis. *Appl. Environ. Microbiol.* **62**, 2273–2278.
67. Sghir, A., Gramet, G., Suau, A., Rochet, V., Pochart, P., and Dore, J. (2000). Quantification of bacterial groups within human fecal flora by oligonucleotide probe hybridization. *Appl. Environ. Microbiol.* **66**, 2263–2266.
68. Apajalahti, J. H. A., Kettunen, A., Nurminen, P. H., Jatila, H., and Holben, W. E. (2003). Selective plating underestimates abundance and shows differential recovery of bifidobacterial species from human feces. *Appl. Environ. Microbiol.* **69**, 5731–5735.
69. Satokari, R. M., Vaughan, E. E., Akkermans, A. D. L., Saarela, M., and de Vos, W. M. (2001). Bifidobacterial diversity in human feces detected by genus-specific PCR and denaturing gradient gel electrophoresis. *Appl. Environ. Microbiol.* **67**, 504–513.
70. Blaut, M., Collins, M. D., Welling, G. W., Dore, J., van Loo, J., and de Vos, W. (2002). Molecular biological methods for studying the gut microbiota: The EU Human Gut Flora Project. *Br. J. Nutr.* **8**, S203–S211.
71. Vaughan, E., Schut, F., Heilig, H., Zoetendal, E., de Vos, W., and Akkermans, A. (2000). A molecular view of the intestinal ecosystem. *Curr. Issues Intest. Microbiol.* **1**, 1–12.
72. Holland, J. L., Louie, L., Simor, A. E., and Louie, M. (2000). PCR detection of *Escherichia coli* O157:H7 directly from stools: Evaluation of commercial extraction methods for purifying fecal DNA. *J. Clin. Microbiol.* **38**, 4108–4113.
73. Zoetendal, E., Ben-Amor, K., Akkermans, A., Abee, T., and de Vos, W. (2001). DNA isolation protocols affect the detection limit of PCR approaches of bacteria in samples from the human gastrointestinal tract. *Syst. Appl. Microbiol.* **24**, 405–410.
74. McOrist, A. L., Jackson, M., and Bird, A. R. (2002). A comparison of five methods for extraction of bacterial DNA from human faecal samples. *J. Microbiol. Methods* **50**, 131–139.
75. Li, M., Gong, J., Cottrill, M., Yu, H., de Lange, C., Burton, J., and Topp, E. (2003). Evaluation of QIAamp DNA Stool Mini Kit for ecological studies of gut microbiota. *J. Microbiol. Methods* **54**, 13–20.
76. Monteiro, L., Bonnemaison, D., Vekris, A., Petry, K. G., Bonnet, J., Vidal, R., Cabrita, J., and Megraud, F. (1997). Complex polysaccharides as PCR inhibitors in feces: *Helicobacter pylori* model. *J. Clin. Microbiol.* **35**, 995–998.
77. Cavallini, A., Notarnicola, M., Berloco, P., Lippolis, A., and Di Leo, A. (2000). Use of macroporous polypropylene filter to allow identification of bacteria by PCR in human fecal samples. *J. Microbiol. Methods* **39**, 265–270.
78. Finegold, S., Attebery, H., and Sutter, V. (1974). Effect of diet on human fecal flora: Comparison of Japanese and American diets. *Am. J. Clin. Nutr.* **27**, 1456–1469.
79. Ariefdjohan, M. W., Savaiano, D., and Nakatsu, C. H. (2006). Dietary changes may influence the stability of the bacterial community in the human gut. 106th General Meeting of the American Society for Microbiology, Orlando, Florida, Poster No. N–221 (Abstract 206-GM-A-1530-ASM).
80. Zoetendal, E. G., Akkermans, A. D. L., and De Vos, W. M. (1998). Temperature gradient gel electrophoresis analysis of 16S rRNA from human fecal samples reveals stable and host-specific communities of active bacteria. *Appl. Environ. Microbiol.* **64**, 3854–3859.
81. Maier, B. R., Flynn, M. A., Burton, G. C., Tsutakawa, R. K., and Hentges, D. J. (1974). Effects of a high-beef diet on bowel flora: A preliminary report. *Am. J. Clin. Nutr.* **27**, 1470–1474.
82. Li, F., Hullar, M. A. J., and Lampe, J. W. (2007). Human gut bacterial community change due to cruciferous vegetables in a controlled feeding study. 107th General Meeting of the American Society for Microbiology, Toronto, Canada, Poster No. N-16 (Session 264/N).
83. Shanahan, F. (2001). Inflammatory bowel disease: Immunodiagnostics, immunotherapeutics, and ecotherapeutics. *Gastroenterology* **120**, 622–635.
84. Wollowski, I., Rechkemmer, G., and Pool-Zobel, B. L. (2001). Protective role of probiotics and prebiotics in colon cancer. *Am. J. Clin. Nutr.* **73**, 451S–455S.

85. Sanders, M. E., and Huis in't Veld, J. (1999). Bringing a probiotic-containing functional food to the market: Microbiological, product, regulatory and labeling issues. *Antonie Van Leeuwenhoek* **76**, 293–315.
86. Fasoli, S., Marzotto, M., Rizzotti, L., Rossi, F., Dellaglio, F., and Torriani, S. (2003). Bacterial composition of commercial probiotic products as evaluated by PCR-DGGE analysis. *Int. J. Food Microbiol.* **82**, 59–70.
87. Saxelin, M., Tynkkynen, S., Mattila-Sandholm, T., and de Vos, W. M. (2005). Probiotic and other functional microbes: From markets to mechanisms. *Curr. Opin. Biotechnol.* **16**, 204–211.
88. Alander, M., Satokari, R., Korpela, R., Saxelin, M., Vilpponen-Salmela, T., Mattilla-Sandholm, T., and von Wright, A. (1999). Persistence of colonization of human colonic mucosa by a probiotic strain, *Lactobacillus rhamnosus* G. G., after oral consumption. *Appl. Environ. Microbiol.* **65**, 351–354.
89. Tannock, G. W., Munro, K., Harmsen, H. J. M., Welling, G. W., Smart, J., and Gopal, P. K. (2000). Analysis of the fecal microflora of human subjects consuming a probiotic product containing *Lactobacillus rhamnosus* DR20. *Appl. Environ. Microbiol.* **66**, 2578–2588.
90. Tuohy, K. M., Rouzaud, G. C. M., Bruck, W. M., and Gibson, G. R. (2005). Modulation of the human gut microflora towards improved health using prebiotics—assessment of efficacy. *Curr. Pharm. Des.* **11**, 75–90.
91. Savaiano, D., Abou ElAnouar, A., Smith, D., and Levitt, M. (1984). Lactose malabsorption from yogurt, pasteurized yogurt, sweet acidophilus milk, and cultured milk in lactase-deficient individuals. *Am. J. Clin. Nutr.* **40**, 1219–1223.
92. Gaon, D., Doweck, Y., Zavaglia, A., Holgado, A., and Oliver, G. (1995). Digestion de la lactosa por una leche fermentada con *Lactobacillus acidophilus* y *Lactobacillus casei* de origen humano. (Lactose digestion of a milk fermented by *Lactobacillus acidophilus* and *Lactobacillus casei* of human origin.). *Medicina (B. Aires)* **55**, 237–242, in Spanish.
93. Montes, R. G., Bayless, T. M., Saavedra, J. M., and Perman, J. A. (1995). Effect of milks inoculated with *Lactobacillus acidophilus* or a yogurt starter culture in lactose-maldigesting children. *J. Dairy Sci.* **78**, 1657–1664.
94. Jiang, T., Mustapha, A., and Savaiano, D. A. (1996). Improvement of lactose digestion in humans by ingestion of unfermented milk containing *Bifidobacterium longum*. *J. Dairy Sci.* **79**, 750–757.
95. Shermak, M., Saavedra, J., Jackson, T., Huang, S., Bayless, T., and Perman, J. (1995). Effect of yogurt on symptoms and kinetics of hydrogen production in lactose-malabsorbing children. *Am. J. Clin. Nutr.* **62**, 1003–1006.
96. Lin, M.-Y., Yen, C.-L., and Chen, S.-H. (1998). Management of lactose maldigestion by consuming milk containing Lactobacilli. *Dig. Dis. Sci.* **43**, 133–137.
97. de Vrese, M., Stegelmann, A., Richter, B., Fenselau, S., Laue, C., and Schrezenmeir, J. (2001). Probiotics—compensation for lactase insufficiency. *Am. J. Clin. Nutr.* **73**, 421S–429S.
98. Ligny, G. (1975). Le traitement par l'Ultralevure des troubles intestinaux secondaires à l'antibiothérapie. Etude en double aveugle et étude clinique simple. (*Saccharomyces boulardii* as a treatment for antibiotic associated disorders. A double blind study.). *Revue Française de Gastroentérologie* **114**, 45–50 in French.
99. Gotz, V., Romankiewicz, J., Moss, J., and Murray, H. (1979). Prophylaxis against ampicillin-associated diarrhea with a *Lactobacillus* preparation. *Am. J. Hosp. Pharm.* **36**, 754–757.
100. Borgia, M., Sepe, N., Brancato, V., *et al.* (1982). A controlled clinical study on *Streptococcus faecium* preparation for the prevention of side reactions during long-term antibiotic treatments. *Curr. Ther. Res.* **31**, 265–271.
101. Clements, M., Levine, M., Ristaino, P., Daya, V., and Hughes, T. (1983). Exogenous lactobacilli fed to man—their fate and ability to prevent diarrheal disease. *Prog. Food Nutr. Sci.* **7**, 29–37.
102. Colombel, J., Cortot, A., Neut, C., and Romond, C. (1987). Yoghurt with *Bifidobacterium longum* reduces erythromcyin-induced gastrointestinal effects. *Lancet* **2**, 43.
103. Surawicz, C., Elmer, G., Speelman, P., McFarland, L., Chinn, J., and van Belle, G. (1989). Prevention of antibiotic associated diarrhea by *Saccharomyces boulardii*: A prospective study. *Gastroenterology* **96**, 981–988.
104. Wunderlich, P., Braun, L., Fumagalli, I., D'Apuzzo, V., Heim, F., Karly, M., Lodi, R., Politta, G., Vonbank, F., and Zeltner, L. (1989). Double-blind report on the efficacy of lactic acid-producing *Enterococcus* SF68 in the prevention of antibiotic-associated diarrhoea and in the treatment of acute diarrhoea. *J. Int. Med. Res.* **17**, 333–338.
105. Siitonen, S., Vapaatalo, H., Salminen, S., Gordin, A., Saxelin, M., Wikberg, R., and Kikkola, A. L. (1990). Effect of *Lactobacillus* GG yoghurt in prevention of antibiotic associated diarrhoea. *Ann. Med.* **22**, 57–59.
106. Orrhage, K., Brismar, B., and Nord, C. (1994). Effects of supplements of *Bifidobacterium longum* and *Lactobacillus acidophilus* on the intestinal microbiota during administration of clindamycin. *Microb. Ecol. Health Dis.* **7**, 17–25.
107. McFarland, L. V., Surawicz, C. M., Greenberg, R. N., Elmer, G. W., Moyer, K. A., Melcher, S. A., Bowen, K. E., and Cox, J. L. (1995). Prevention of beta-lactam-associated diarrhea by *Saccharomyces boulardii* compared with placebo. *Am. J. Gastroenterol.* **90**, 439–448.
108. Witsell, D., Garrett, C., Yarbrough, W., Dorrestein, S., Drake, A., and Weissler, M. (1995). Effect of *Lactobacillus acidophilus* on antibiotic-associated gastrointestinal morbidity: A prospective randomized trial. *J. Otolaryngol.* **24**, 230–233.
109. Arvola, T., Laiho, K., Torkkeli, S., Mykk, H., Salminen, S., Maunula, L., and Isolauri, E. (1999). Prophylactic *Lactobacillus* GG reduces antibiotic-associated diarrhea in children with respiratory infections: A randomized study. *Pediatrics* **104**, 64–68.
110. Vanderhoof, J., Whitney, D., Antonson, D., Hanner, T., Lupo, J., and Young, R. (1999). *Lactobacillus* GG in the prevention of antibiotic-associated diarrhea in children. *J. Pediatr.* **135**, 564–568.
111. Armuzzi, A., Cremonini, F., Bartolozzi, F., Canducci, F., Candelli, M., Ojetti, V., Cammarota, G., Anti, M., DeLorenzo, A., Pola, P., Gasbarini, G., and Gasbarrini, A. (2001). The effect of oral administration of *Lactobacillus* GG on antibiotic-associated gastrointestinal side-effects

during *Helicobacter pylori* eradication therapy. *Aliment. Pharmacol. Ther.* **15**, 163–169.

112. Kotowska, M., Albrecht, P., and Szajewska, H. (2005). *Saccharomyces boulardii* in the prevention of antibiotic-associated diarrhoea in children: A randomized double-blind placebo-controlled trial. *Aliment. Pharmacol. Ther.* **21**, 583–590.
113. Bellomo, G., Mangiagle, A., Nicastro, L., and Frigerio, G. (1980). A controlled double blind study of SF68 strain as a new biological preparation for the treatment of diarrhea in pediatrics. *Curr. Ther. Res.* **28**, 927–936.
114. Bodilis, J. (1983). Etude controlée du Lactéol fort contre placebo et contre produit de référence dans les diarrhées aigues de l'adulte. (Lacteol versus placebo in acute adult diarrhea: A controlled study.). *Médecine Actuelle* **10**, 232–235 in French.
115. Chapoy, P. (1985). Traitement des diarrhées aigues infantiles: Essai contrôlé de *Saccharomyces boulardii*. (Treatment of acute diarrhea in children: A controlled trial with *Saccharomyces boulardii*.). *Ann. Pediatr. (Paris)* **32**.
116. Isolauri, E., Rautanen, T., Juntunen, M., Sillanaukee, P., and Koivula, T. (1991). A human *Lactobacillus* strain (*Lactobacillus casei* spp. strain GG) promotes recovery from acute diarrhea in children. *Pediatrics* **88**, 90–97.
117. Isolauri, E., Kaila, M., Mykkanen, H., Ling, W., and Salminen, S. (1994). Oral bacteriotherapy for viral gastroenteritis. *Dig. Dis. Sci.* **39**, 2595–2600.
118. Saavedra, J. M., Bauman, N. A., Oung, I., Perman, J. A., and Yolken, R. H. (1994). Feeding of *Bifidobacterium bifidum* and *Streptococcus thermophilus* to infants in hospital for prevention of diarrhea and shedding of rotavirus. *Lancet* **344**, 1046–1049.
119. Sugita, T., and Togawa, M. (1994). Efficacy of lactobacillus preparation biolactis powder in children with rotavirus enteritis. *Jpn. Pediatr.* **47**, 2755–2762 in Japanese.
120. Kaila, M., Isolauri, E., Saxelin, M., Arvilommi, H., and Vesikari, T. (1995). Viable versus inactivated lactobacillus strain GG in acute rotavirus diarrhoea. *Arch. Dis. Child.* **72**, 51–53.
121. Majamaa, H., Isolauri, E., Saxelin, M., and Vesikari, T. (1995). Lactic acid bacteria in the treatment of acute rotavirus gastroenteritis. *J. Pediatr. Gastroenterol. Nutr.* **20**, 333–338.
122. Raza, S., Graham, S., Allen, S., Sultana, S., Cuevas, L., and Hart, C. (1995). *Lactobacillus* GG promotes recovery from acute nonbloody diarrhea in Pakistan. *Pediatr. Infect. Dis. J.* **14**, 107–111.
123. Buydens, P., and Debeuckelaere, S. (1996). Efficacy of SF 68 in the treatment of acute diarrhea. A placebo-controlled trial. *Scand J Gastroenterol.* **31**, 887–891.
124. Pant, A. R., Graham, S. M., Allen, S. J., Harikul, S., Sabchareon, A., Cuevas, L., Med, M. T., and Hart, C. A. (1996). Lactobacillus GG and acute diarrhoea in young children in the tropics. *J. Trop. Pediatr.* **42**, 162–165.
125. Guarino, A., Canani, R., Spagnuolo, M., Albano, F., and Di Benedetto, L. (1997). Oral bacterial therapy reduces the duration of symptoms and of viral excretion in children with mild diarrhea. *J. Pediatr. Gastroenterol. Nutr.* **25**, 516–519.
126. Shornikova, A., Casas, I., Mykkanen, H., Salo, E., and Vesikari, T. (1997). Bacteriotherapy with *Lactobacillus reuteri* in rotavirus gastroenteritis. *Pediatr. Infect. Dis. J.* **16**, 1103–1107.
127. Shornikova, A., Isolauri, E., Burkanova, L., Lukovnikova, S., and Vesikari, T. (1997). A trial in the Karelian Republic of oral rehydration and *Lactobacillus* GG for treatment of acute diarrhoea. *Acta Paediatr.* **86**, 460–465.
128. Oberhelman, R., Gilman, R., Sheen, P., Taylor, D. N., Black, R. E., Cabrera, L., Lescano, A. G., Meza, R., and Madico, G. (1999). A placebo-controlled trial of *Lactobacillus* GG to prevent diarrhea in undernourished Peruvian children. *J. Pediatr.* **134**, 15–20.
129. Pedone, C., Bernabeu, A., Postaire, E., Bouley, C., and Reinert, P. (1999). The effect of supplementation with milk fermented by *Lactobacillus casei* (strain DN-114001) on acute diarrhoea in children attending day care centres. *Int. J. Clin. Pract.* **53**, 179–184.
130. Phuapradit, P., Varavithya, W., Vathanophas, K., Sangchai, R., Podhipak, A., Suthutvoravut, U., Nopchinda, S., Chantraruksa, V., and Haschke, F. (1999). Reduction of rotavirus infection in children receiving bifidobacteria-supplemented formula. *J. Med. Assoc. Thai.* **82**, S43–S48.
131. Guandalini, S., Pensabene, I., Zikri, M., Dias, J. A., Casali, L. G., Hoekstra, H., Kolacek, S., Massar, K., Micetic-Turk, D., Papadopoulou, A., de Sousa, J. S., Sandhu, B., Szajewska, H., and Weizman, Z. (2000). *Lactobacillus* GG administered in oral rehydration solution to children with acute diarrhea: A multicenter European trial. *J. Pediatr. Gastroenterol. Nutr.* **30**, 54–60.
132. Pedone, C., Arnaud, C., Postaire, E., Bouley, C., and Reinert, P. (2000). Multicentric study of the effect of milk fermented by *Lactobacillus casei* on the incidence of diarrhoea. *Int. J. Clin. Pract.* **54**, 568–571.
133. Hatakka, K., Savilahti, E., Ponka, A., Meurman, J. H., Poussa, T., Nase, L., Saxelin, M., and Korpela, R. (2001). Effect of long term consumption of probiotic milk on infections in children attending day care centres: Double blind, randomised. *BMJ* **322**, 1327–1330.
134. Szajewska, H., Kotowska, M., Mrukowics, J. Z., Armanska, M., and Mikolajczyk, W. (2001). Efficacy of *Lactobacillus* GG in prevention of nosocomial diarrhea in infants. *J. Pediatr.* **138**, 361–365.
135. Chandra, R. (2002). Effect of *Lactobacillus* on the incidence and severity of acute rotavirus diarrhea in infants. A prospective placebo-controlled double-blind study. *Nutr. Res.* **22**, 65–69.
136. Rosenfeldt, V., Michaelsen, K., Jakobsen, M., Larsen, C. N., Moller, P. L., Pedersen, P., Tvede, M., Weyrehter, H., Valerius, N. H., and Paerregaard, A. (2002). Effect of probiotic *Lactobacillus* strains in young children hospitalized with acute diarrhea. *Pediatr. Infect. Dis. J.* **21**, 411–416.
137. Rosenfeldt, V., Michaelsen, K., Jakobsen, M., Larsen, C. N., Moller, P. L., Tvede, M., Weyrehter, H., Valerius, N. H., and Paerregaard, A. (2002). Effect of probiotic *Lactobacillus* strains on acute diarrhea in a cohort of nonhospitalized children attending day-care centers. *Pediatr. Infect. Dis. J.* **21**, 417–419.

138. Pereg, D., Kimhi, O., Tirosh, A., Orr, N., Kayouf, R., and Lishner, M. (2005). The effect of fermented yogurt on the prevention of diarrhea in a healthy adult population. *Am. J. Infect. Control* **33**, 122–125.
139. Billoo, A., Memon, M., Khaskheli, S., Murtaza, G., Iqbal, K., Saeed Shekhani, M., and Siddiqi, A. Q. (2006). Role of a probiotic (*Saccharomyces boulardii*) in management and prevention of diarrhoea. *World J. Gastroenterol.* **12**, 4557–4560.
140. De Dios-Pozo-Alano, J., Warram, J., Gomez, R., and Cavazos, M. (1978). Effect of a Lactobacilli preparation on traveler's diarrhea: A randomized double-blind clinical trial. *Gastroenterology* **74**, 829–830.
141. Black, F., Anderson, P., Orskow, J., Orskow, F., Gaarslev, K., and Laudlund, S. (1989). Prophylactic efficacy of lactobacilli on travellers diarrhea. *In* "Travel Medicine Conference on International Travel Medicine" (R. Steffen, Ed.), pp. 333–335. Springer, Berlin.
142. Oksanen, P. J., Salminen, S., Saxelin, M., Hamalainen, P., Ihantolavormisto, A., Muurasniemiisoviita, L., Nikkari, S., Oksanen, T., Porsti, I., Salminen, E., Siitonen, S., Stuckey, H., Toppila, A., and Vapaatalo, H. (1990). Prevention of traveler's diarrhea by *Lactobacillus* GG. *Ann. Med.* **22**, 53–56.
143. Kollaritsch, v.H., Holst, H., Grobara, P., and Wiedermann, G. (1993). Prophylaxe des Reisediarrhöe mit *Saccharomyces boulardii*. (Prevention of travelers' diarrehea *by Saccharomyces boulardi.*). *Fortischritte der Medizin* **111**, 153–156.
144. Hilton, E., Kolakowski, P., Singer, C., and Smith, M. (1997). Efficacy of *Lactobacillus* GG as a diarrheal preventive in travelers. *J. Travel Med.* **4**, 41–43.
145. Tempé, J., Steidel, A., Bléhaut, H., Hasselmann, M., Lutun, P., and Maurier, F. (1983). Prévention par *Saccharomyces boulardii* des diarrhées de l'alimentation entérale à débit continu. (Prevention of tube feeding-induced diarrhea by *Saccharomyces boulardii.*). *Semaine des Hôpitaux de Paris* **59**, 1409–1412 in French.
146. Schlotterer, M., Bernasconi, P., Lebreton, F., and Wassermann, D. (1987). Intérêt de *Saccharomyces boulardii* dans la tolérance digestive de la nutrition entérale à débit continu chez le brûlé. (Effect of *Saccharomyces boulardii* on the digestive tolerance of enteral nutrition in burn.). *Nutrition Clinique et Métabolisme* **1**, 31–34 in French.
147. Bleichner, G., Bléhaut, H., Mentec, H., and Moyse, D. (1997). *Saccharomyces boulardii* prevents diarrhea in critically ill tube-fed patients. A multicenter, randomized, double-blind placebo-controlled trial. *Intens. Care Med.* **23**, 517–523.
148. Heimburger, D., Sockwell, D., and Geels, W. (1994). Diarrhea with enteral feeding: Prospective reappraisal of putative causes. *Nutrition* **10**, 392–1296.
149. Salminen, E., Elomaa, I., Minkkinen, J., Vapaatalo, H., and Salminen, S. (1988). Preservation of intestinal integrity during radiotherapy using live *Lactobacillus acidophilus* cultures. *Clin. Radiol.* **39**, 435–437.
150. Tomoda, T., Nakano, Y., and Kageyama, T. (1988). Intestinal *Candida* overgrowth and *Candida* infection in patients with leukemia: Effect of *Bifidobacterium* administration. *Bifidobacteria Microflora* **7**, 71–74.
151. Elmer, G., Moyer, K., Surawicz, C., Collier, A., Hooton, T., and McFarland, L. (1995). Evaluation of *Saccharomyces boulardii* for patients with HIV-related chronic diarrhoea and healthy volunteers receiving antifungals. *Microecol. Ther.* **25**, 23–31.
152. Wolf, B., Wheeler, K., Ataya, D., and Garleb, K. (1998). Safety and tolerance of *Lactobacillus reuteri* supplementation to a population infected with the human immunodeficiency virus. *Food Chem. Toxicol.* **36**, 1085–1094.
153. Urbancsek, H., Kazar, T., Mezes, I., and Neumann, K. (2001). Results of a double-blind, randomized study to evaluate the efficacy and safety of Antibiophilus in patients with radiation-induced diarrhoea. *Eur. J. Gastroenterol. Hepatol.* **13**, 391–396.
154. Delia, P., Sansotta, G., Donato, V., Messina, G., Frosina, P., Pergolizzi, S., De Renzis, C., and Famularo, G. (2002). Prevention of radiation-induced diarrhea with the use of VSL#3, a new high-potency probiotic preparation. *Am. J. Gastroenterol.* **97**, 2150–2152.
155. Simenhoff, M. L., Dunn, S. R., Zollner, G. P., Fitzpatrick, M. E., Emery, S. M., Sandine, W. E., and Ayres, J. W. (1996). Biomodulation of the toxic and nutritional effects of small bowel bacterial overgrowth in end-stage kidney disease using freeze-dried *Lactobacillus acidophilus*. *Miner. Electrolyte Metab.* **22**, 92–96.
156. Dunn, S., Simenhoff, M., Ahmed, K., Gaughan, W. J., Eltayeb, B. O., Fitzpatrick, M.-E. D., Emery, S. M., Ayres, J. W., and Holt, K. E. (1998). Effect of oral administration of freeze-dried *Lactobacillus acidophilus* on small bowel bacterial overgrowth in patients with end stage kidney disease: Reducing uremic toxins and improving nutrition. *Int. Dairy J.* **8**, 545–553.
157. Vanderhoof, J., Young, R., Murray, N., and Kaufman, S. (1998). Treatment strategies for small bowel bacterial overgrowth in short bowel syndrome. *J. Pediatr. Gastroenterol. Nutr.* **27**, 155–160.
158. Gaon, D., Garmendia, C., Murrielo, N., de Cucco Games, A., Cerchio, A., Quintas, R., Gonzalez, S. N., and Oliver, G. (2002). Effect of *Lactobacillus* strains (*L. casei* and *L. acidophilus* Strains cerela) on bacterial overgrowth-related chronic diarrhea. *Medicina (B. Aires)* **62**, 159–163.
159. Majamaa, H., Isolauri, E., Saxelin, M., and Vesikari, T. (1997). Probiotics: A novel approach in the management of food allergy. *J. Allergy Clin. Immunol.* **99**, 179–185.
160. Isolauri, E., Arvola, T., Sutas, Y., Moilanen, E., and Salminen, E. (2000). Probiotics in the management of atopic eczema. *Clin. Exp. Allergy* **30**, 1604–1610.
161. Kalliomaki, M., Salminen, S., Arvilommi, H., Kero, P., Koskinen, P., and Isolauri, E. (2001). Probiotics in primary prevention of atopic disease: A randomised placebo-controlled trial. *Lancet* **357**, 1076–1079.
162. Kalliomaki, M., Salminen, S., Poussa, T., Arvilommi, H., and Isolauri, E. (2003). Probiotics and prevention of atopic disease: 4-year follow-up of a randomised placebo-controlled trial. *Lancet* **361**, 1869–1871.
163. Hoyos, A. (1999). Reduced incidence of necrotizing enterocolitis associated with enteral administration of *Lactobacillus acidophilus* and *Bifidobacterium infantis* to neonates in an intensive care unit. *Int. J. Infect. Dis.* **3**, 197–202.

164. Maupas, J., Champemont, P., and Delforge, M. (1983). Traitement des colopathies fonctionnelles—Essai en double aveugle de l'ultra-levure. (Treatment of irritable bowel syndrome with *Saccharomyces boulardii*—a double-blind, placebo controlled study.). *Med. Chirurg. Dig.* **12**, 77–79, in French.
165. Newcomer, A., Park, H., O'Brien, P., and McGill, D. (1983). Response of patients with irritable bowel syndrome and lactase deficiency using unfermented acidophilus milk. *Am. J. Clin. Nutr.* **38**, 257–263.
166. Gade, J., and Thorn, P. (1989). Paraghurt for patients with irritable bowel syndrome: A controlled clinical investigation from general practice. *Scand. J. Prim. Health Care* **7**, 23–26.
167. Halpern, G., Prindiville, T., Blankenburg, M., Hsia, T., and Gershwin, M. (1996). Treatment of irritable bowel syndrome with Lacteol Fort: A randomized, double-blind, crossover trial. *Am. J. Gastroenterol.* **91**, 1579–1584.
168. Hunter, J., Lee, A., King, T., Barratt, M., Linggood, M., and Blades, J. (1996). *Enterococcus faecium* strain PR88—an effective probiotic. *Gut* **38**, A62, Supplement.
169. Young, R., and Vanderhoof, J. (1997). Successful probiotic therapy of chronic recurrent abdominal pain in children. *Gastroenterology* **112**, A856.
170. Hentschel, C., Bauer, J., and Dill, N. (1997). Complementary medicine in non-ulcer-dyspepsia: Is alternative medicine a real alternative? A randomised placebo-controlled double blind clinical trial with two probiotic agents (Hylac N and Hylac N Forte). *Gastroenterology* **112**, A146.
171. Kordecki, H., and Niedzielin, K. (1998). New possibility in the treatment of irritable bowel syndrome. Probiotics as a modification of the microflora of the colon. *Gastroenterology* **114**, A402.
172. Bougle, D., Roland, N., Lebeurrier, F., and Arhan, P. (1999). Effect of propionibacteria supplementation on fecal bifidobacteria and segmental colonic transit time in healthy human subjects. *Scand. J. Gastroenterol.* **34**, 144–148.
173. Nobaek, S., Johansson, M.-L., Molin, G., Ahrne, S., and Jeppsson, B. (2000). Alteration of intestinal microflora is associated with reduction in abdominal bloating and pain in patients with irritable bowel syndrome. *Am. J. Gastroenterol.* **95**, 1231–1238.
174. Brigidi, P., Vitali, B., Swennen, E., Bazzocchi, G., and Matteuzzi, D. (2001). Effects of probiotic administration upon the composition and enzymatic activity of human fecal microbiota in patients with irritable bowel syndrome or functional diarrhea. *Res. Microbiol.* **152**, 735–741.
175. De Simone, C., Famularo, G., Salvadori, B., Moretti, S., Marcellini, S., Trinchieri, V., and Santini, G. (2001). Treatment of irritable bowel syndrome (IBS) with the newer probiotic VSL#3: A multicenter trial. *Am. J. Clin. Nutr.* **73**, 491S.
176. Niedzielin, K., Kordecki, H., and Birkenfeld, B. (2001). A controlled, double-blind, randomized study on the efficacy of *Lactobacillus plantarum* 299V in patients with irritable bowel syndrome. *Eur. J. Gastroenterol. Hepatol.* **13**, 1143–1147.
177. Marteau, P., Cuillerier, E., Meance, S., Gerhardt, M. F., Myara, A., Bouvier, M., Bouley, C., Tondu, F., Bommelaer, G., and Grimaud, J. C. (2002). *Bifidobacterium animalis* strain DN-173 010 shortens the colonic transit time in healthy women: A double-blind, randomized, controlled study. *Aliment. Pharmacol. Ther.* **16**, 587–593.
178. Sen, S., Mullan, M., Parker, T., Woolner, J., Tarry, S., and Hunter, J. (2002). Effect of *Lactobacillus plantarum* 299v on colonic fermentation and symptoms of irritable bowel syndrome. *Dig. Liver Dis.* **47**, 2615–2620.
179. Kim, H., Vazquez Roque, M., Camilleri, M., Stephens, D., Burton, D., Baxter, K., Thomforde, G., and Zinsmeister, A. (2005). A randomized controlled trial of a probiotic combination VSL#3 and placebo in irritable bowel syndrome with bloating. *Neurogastroenterol. Motil.* **17**, 687–696.
180. Whorwell, P. J., Altringer, L., Morel, J., Bond, Y., Charbonneau, D., O'Mahony, L., Kiely, B., Shanahan, F., and Quigley, E. M. M. (2006). Efficacy of an encapsulated probiotic *Bifidobacterium infantis* 35624 in women with irritable bowel syndrome. *Am. J. Gastroenterol.* **101**, 1581–1590.
181. Plein, K., and Hotz, J. (1993). Therapeutic effects of *Saccharomyces boulardii* on mild residual symptoms in a stable phase of Crohn's disease with special respect to chronic diarrhea—a pilot study. *Z. Gastroenterol.* **31**, 129–134.
182. Malin, M., Suomalainen, H., Saxelin, M., and Isolauri, E. (1996). Promotion of IgA immune response in patients with Crohn's disease by oral bacteriotherapy with *Lactobacillus* GG. *Ann. Nutr. Metab.* **40**, 137–145.
183. Kruis, W., Schutz, E., Fric, P., Fixa, B., Judmaier, G., and Stolte, M. (1997). Double-blind comparison of an oral *Escherichia coli* preparation and mesalazine in maintaining remission of ulcerative colitis. *Aliment. Pharmacol. Ther.* **11**, 853–858.
184. Malchow, H. (1997). Crohn's disease and *Escherichia coli*. A new approach in therapy to maintain remission of colonic Crohn's disease? *J. Clin. Gastroenterol.* **25**, 653–658.
185. Rembacken, B., Snelling, A., Hawkey, P., Chalmers, D., and Axon, A. (1999). Non-pathogenic *Escherichia coli* versus mesalazine for the treatment of ulcerative colitis: A randomised trial. *Lancet* **354**, 635–639.
186. Campieri, M., Rizzello, F., Venturi, A., Poggioli, G., Ugolini, F., Helwig, U., Amadini, C., Romboli, E., and Gionchetti, P. (2000). Combination of antibiotic and probiotic treatment is efficacious in prophylaxis of post-operative recurrence of Crohn's disease: A randomized controlled study vs mesalamine (abstract). *Gastroenterology* **118**, G4179.
187. Copaci, I., Micu, L., Chira, C., and Rovinaru, I. (2000). Maintenance of remission of ulcerative colitis (UC): Mesalamine, dietary fiber, *S. boulardii* (abstract). *Gut* **47** (Suppl. III), A240–P929.
188. Gupta, P., Andrew, H., Kirschner, B., and Guandalini, S. (2000). Is *Lactobacillus* GG helpful in children with Crohn's disease? Results of a preliminary, open-label study. *J. Pediatr. Gastroenterol. Nutr.* **31**, 453–457.
189. Guslandi, M., Mezzi, G., Sorghi, M., and Testoni, P. A. (2000). *Saccharomyces boulardii* in maintenance treatment of Crohn's disease. *Dig. Dis. Sci.* **45**, 1462–1464.
190. Gionchetti, P., Rizzello, F., Venturi, A., Brigidi, P., Matteuzzi, D., Bazzochi, G., Poggioli, G., Miglioli, M., and Campieri, M. (2000). Oral bacteriotherapy as maintenance

treatment in patients with chronic pouchitis: A double-blind, placebo-controlled trial. *Gastroenterology* **119**, 305–309.

191. Guandalini, S. (2002). Use of *Lactobacillus*-GG in paediatric Crohn's disease. *Dig. Liver Dis.* **34**, S63–S65.
192. Prantera, C., Scribano, M. L., Falasco, G., Andreoli, A., and Luzi, C. (2002). Ineffectiveness of probiotics in preventing recurrence after curative resection for Crohn's disease: A randomised controlled trial with *Lactobacillus* GG. *Gut* **51**, 405–409.
193. Kruis, W., Fric, P., Pokrotnieks, J., Lukas, M., Fixa, B., Kascak, M., Kamm, M. A., Weismueller, J., Beglinger, C., Stolte, M., Wolff, C., and Schulze, J. (2004). Maintaining remission of ulcerative colitis with the probiotic *Escherichia coli* Nissle 1917 is as effective as with standard mesalazine. *Gut* **53**, 1617–1623.
194. Schultz, M., Timmer, A., Herfarth, H., Balfour Sartor, R., Vanderhoof, J., and Rath, H. (2004). *Lactobacillus* GG in inducing and maintaining remission of Crohn's disease. *BMC Gastroenterol.* **4**, 5–8.
195. Bousvaros, A., Guandalini, S., Baldassano, R., Botelho, C., Evans, J., Ferry, G. D., Goldin, B., Hartigan, L., Kugathasan, S., Levy, J., Murray, K. F., Oliva-Hemker, M., Rosh, J. R., Tolia, V., Zholudev, A., Vanderhoof, J. A., and Hibberd, P. L. (2005). A randomized, double-blind trial of *Lactobacillus* GG versus placebo in addition to standard maintenance therapy for children with Crohn's disease. *Inflamm. Bowel Dis.* **11**, 833–839.
196. Gorbach, S., Chang, T., and Goldin, B. (1987). Successful treatment of relapsing *Clostridium difficile* colitis with *Lactobacillus* GG. *Lancet* **2**, 1519.
197. Surawicz, C., McFarland, L., Elmer, G., and Chinn, J. (1989). Treatment of recurrent *Clostridium difficile* colitis with vancomycin and *Saccharomyces boulardii*. *Am. J. Gastroenterol.* **84**, 1285–1287.
198. Kimmey, M., Elmer, G., Surawicz, C., and McFarland, L. (1990). Prevention of further recurrence of *Clostridium difficile* colitis with *Saccharomyces boulardii*. *Dig. Dis. Sci.* **35**, 897–901.
199. Buts, J., Corthier, G., and Delmée, M. (1993). *Saccharomyces boulardii* for *Clostridium difficile* associated enteropathies in infants. *J. Pediatr. Gastroenterol. Nutr.* **16**, 419–425.
200. McFarland, L. V., Surawicz, C. M., Greenberg, R. N., Fekety, R., Elmer, G. W., Moyer, K. A., Melcher, S. A., Bowen, K. E., Cox, J. L., and Noorani, Z. (1994). A randomized placebo-controlled trial of *Saccharomyces boulardii* in combination with standard antibiotics for *Clostridium difficile* disease. *JAMA* **271**, 1913–1918.
201. Biller, J., Katz, A., Flores, A., Buie, T., and Gorbach, S. (1995). Treatment of recurrent *Clostridium difficile* colitis with *Lactobacillus* GG. *J. Pediatr. Gastroenterol. Nutr.* **21**, 224–226.
202. Bennet, R. G., Gorbach, S. L., Goldin, B. R., Chang, T. W., Laughon, B. E., Greenough, W. B., and III, Bartlett, J. G. (1996). Treatment of relapsing *C. difficile* diarrhea with *Lactobacillus* GG. ***31***(Suppl), 35S–38S.
203. Levy, J. (1997). Experience with live *Lactobacillus plantarum* 299v: A promising adjunct in the management of recurrent *Clostridium difficile* infection. *J. Gastroenterol.* **112**, A379, Abstract.
204. Michetti, P., Dorta, G., Wiesel, P., Brassart, D., Verdu, E., Herranz, M., Felley, C., Porta, N., Rouvet, M., Blum, A. L., and Corthésy-Theulaz, I. (1999). Effect of whey-based culture supernatant of *Lactobacillus acidophilus* (*johnsonii*) La1 on *Helicobacter pylori* infection in humans. *Digestion* **60**, 203–209.
205. Canducci, F., Armuzzi, A., Cremonini, F., Cammarota, G., Bartolozzi, F., Pola, P., Gasbarrini, G., and Gasbarrini, A. (2000). A lyophilized and inactivated culture of *Lactobacillus acidophilus* increases *Helicobacter pylori* eradication rates. *Aliment. Pharmacol. Ther.* **14**, 1625–1629.
206. Sakamoto, I., Igarashi, M., Kimura, S., Takagi, A., Miwa, T., and Koga, Y. (2001). Suppressive effect of *Lactobacillus gasseri* OLL 2716 (LG21) on *Helicobacter pylori* infection in humans. *J. Antimicrob. Chemother.* **47**, 709–710.
207. Tuohy, K. M., Rouzaud, G. C. M., Bruck, W. M., and Gibson, G. R. (2005). Modulation of the human gut microflora towards improved health using prebiotics—assessment of efficacy. *Curr. Pharm. Des.* **11**, 75–90.
208. Tuohy, K., Kolida, S., Lustenberger, A., and Gibson, G. (2001). The prebiotic effects of biscuits containing partially hydrolysed guar gum and fructo-oligosaccharides—a human volunteer study. *Br. J. Nutr.* **86**, 341–348.
209. Kruse, H., Kleesen, B., and Blaut, M. (1999). Effect of inulin on faecal bifidobacteria in human subjects. *Br. J. Nutr.* **82**, 375–382.
210. Gibson, G., Beatty, E., Wang, X., and Cummings, J. (1995). Selective stimulation of bifidobacteria in the human colon by oligofructose and inulin. *Gastroenterology* **108**, 975–982.
211. Kleessen, B., Sykura, B., Zunft, H., and Blaut, M. (1997). Effects of inulin and lactose on fecal microflora, microbial activity, and bowel habit in elderly constipated persons. *Am. J. Clin. Nutr.* **65**, 1397–1402.
212. Bouhnik, Y., Vahedi, K., Achour, L., Attar, A., Salfati, J., Pochart, P., Marteau, P., Flourie, B., Bornet, F., and Rambaud, J.-C. (1999). Short-chain fructo-oligosaccharide administration dose-dependently increases fecal bifidobacteria in healthy humans. *J. Nutr.* **129**, 113–116.
213. Buddington, R., Williams, C., Chen, S., and Witherly, S. (1996). Dietary supplement of neosugar alters the fecal flora and decreases activities of some reductive enzymes in human subjects. *Am. J. Clin. Nutr.* **63**, 709–716.
214. Moro, G., Minoli, I., Mosca, M., Fanaro, S., Jelinek, J., Stahl, B., and Boehm, G. (2002). Dosage-related bifidogenic effects of galacto- and fructooligosaccharides in formula-fed term infants. *J. Pediatr. Gastroenterol. Nutr.* **34**, 291–295.
215. Boehm, G., Lidestri, M., Casetta, P., Jelinek, J., Negretti, F., Stahl, B., and Marini, A. (2002). Supplementation of a bovine milk formula with an oligosaccharide mixture increases counts of faecal bifidobacteria in preterm infants. *Arch. Dis. Child. Fetal Neonatal* **86**, F178–F181.
216. Tuohy, K., Ziemer, C., Klinder, A. Y. K., Pool-Zobel, B. L., and Gibson, G. R. (2002). A human volunteer study to determine the prebiotic effects of lactulose powder on human colonic microbiota. *Microb. Ecol. Health Dis.* **14**, 165–173.
217. Terada, A., Hara, H., Kataoka, M., and Mitsuoka, T. (1992). Effect of lactulose on the composition and metabolic activity

of the human faecal flora. *Microb. Ecol. Health Dis.* **5**, 43–50.

218. Ballongue, J., Schumann, C., and Quignon, P. (1997). Effects of lactulose and lactitol on colonic microbiota and enzymatic activity. *Scand. J. Gastroenterol.* **32**, 41–44.
219. Nagendra, R., Viswanatha, S., and Arun Kumar, S. (1995). Effect of feeding milk formula containing lactulose to infants on faecal bifidobacterial flora. *Nutr. Res.* **15**, 15–24.
220. Ito, M., Deguchi, Y., Miyamori, A., Matsumoto, K., Kikuchi, H., Matsumoto, K., Kobayashi, Y., Yajima, T., and Kan, T. (1990). Effects of administration of galactooligosaccharides on the human faecal microflora, stool weight and abdominal sensation. *Microb. Ecol. Health Dis.* **3**, 285–292.
221. Ito, M., Deguchi, Y., Matsumoto, K., Kimura, M., Onodera, N., and Yajima, T. (1993). Influence of galactooligosaccharides on the human fecal microflora. *J. Nutr. Sci. Vitaminol.* **39**, 635–640.
222. Waldeck, F. (1990). Funktionen des magen-darm-kanals. *In* "Physiologie des Menschen" (R. Schmidt and G. Thews, Eds.), p. 24. Springer, Berlin.
223. Simon, G., and Gorbach, S. (1982). Intestinal microflora. *Med. Clin. North Am.* **66**, 557–574.
224. Thomas, L., and Flower, M. (2006). "A guide for healthcare professional: *Lactobacillus casei* Shirota and Yakult." Science for Health, Yakult UK.
225. Lawson, P. (1999). Taxonomy and systematics of predominant gut anaerobes. *In* "Colonic Microbiota, Nutrition and Health" (G. R. Gibson and M. B. Roberfroid, Eds.), pp. 149–166. Kluwer Academic, Dordrecht, The Netherlands.
226. Benno, Y., Sawada, K., and Mitsuoka, T. (1984). The intestinal microflora of infants: Composition of fecal flora in breast-fed and bottle-fed infants. *Microbiol. Immunol.* **28**, 975–986.
227. Beerens, H., Romond, C., and Neut, C. (1980). Influence of breast-feeding on the bifid flora of the newborn intestine. *Am. J. Clin. Nutr.* **33**, 2434–2439.
228. Ouwehand, A. C., Derrien, M., de Vos, W., Tiihonen, K., and Rautonen, N. (2005). Prebiotics and other microbial substrates for gut functionality. *Curr. Opin. Biotechnol.* **16**, 212–217.

CHAPTER 39

Nutrition in Inflammatory Bowel Disease and Short Bowel Syndrome

PETER L. BEYER
Department of Dietetics and Nutrition, University of Kansas Medical Center, Kansas City, Kansas

Contents

I. INTRODUCTION

The two primary forms of inflammatory bowel disease (IBD), Crohn's disease and ulcerative colitis, are characterized as chronic, relapsing inflammatory diseases of varying severity. The cause is not completely understood, but the strongest evidence supports the notion that something has altered the body's ability to maintain a normal relationship between the intestinal environment and the host's immune system. The underlying etiology of IBD appears to involve an abnormal innate and adaptive immune response to intestinal bacteria in genetically susceptible individuals [1–4]. Lifelong interaction of microbial, dietary, and other environmental antigens with the hosts' gastrointestinal (GI) immunoregulatory genome appears to influence the presentation and severity of the disease. Inflammatory bowel disease can significantly compromise the individual's health, well-being, and quality of life. From a nutrition perspective, the disease may result in decreased intake of foods, adverse responses to ingested foods, malabsorption, obstruction, malnutrition, increased nutrient requirements, and certainly decreased pleasure associated with eating. Medications and other therapies employed in treating inflammatory bowel disease can further compromise dietary intake and nutritional status.

Nutrition can have a significant role in the management of IBD. Dietary factors may enhance or temper the underlying inflammatory processes or may trigger, exacerbate, or reduce symptoms of the disease. Inflammatory bowel disease and its treatment may significantly compromise nutritional status. Surgical resections are not uncommon and, in some cases, may significantly affect the ability of the patient to maintain adequate nutritional status without specialized nutritional support. In the first section of the chapter, characteristics of inflammatory bowel disease, the nature of the inflammatory process, and the potential role of nutrition interactions are reviewed. In the second section, nutrition and short bowel syndrome, one of the potential complications of Crohn's disease and other gastrointestinal maladies, are discussed.

II. INFLAMMATORY BOWEL DISEASE

Inflammatory bowel disease (IBD) refers to any type of inflammatory disorder involving the small and/or large intestine including those caused by infectious and immunologic or toxic agents. Crohn's disease and ulcerative colitis are the two most common forms of idiopathic inflammatory bowel disease and will be the only conditions discussed here. Although Crohn's disease and ulcerative colitis share many clinical, genetic, and pathologic features, they also exhibit distinguishing characteristics.

A. Characteristics of Inflammatory Bowel Disease

Symptoms of both forms of IBD appear to result from a robust, cytokine-driven inflammation of the intestinal tract [1]. In Crohn's disease (CD) and ulcerative colitis (UC), weight loss, growth failure, anemias, diarrhea, fever, arthritic and dermatologic manifestations, and risk for colon cancer are commonly noted [2, 3]. Macro- and micronutrient deficiencies of malnutrition can occur in both disorders. Each disease can have a significant impact on quality of life, use of health care resources, and costs. See Table 1 for common disease characteristics.

TABLE 1 Characteristics of Crohn's Disease and Ulcerative Colitis

A. Features common to both Crohn's disease and ulcerative colitis
- Chronic episodes of relapse and remission
- Diarrhea
- Extraintestinal manifestations to varying degrees
- Fever
- Increased risk of colorectal cancer
- Weight loss, growth failure

B. Characteristics more common with Crohn's disease
- Any area of the GI tract but more commonly involves the distal small bowel and colon
- Segments of normal mucosa interrupted by diseased bowel
- Transmural involvement, thickening of the intestinal wall
- 70% have surgery in their lifetime; surgery does not guarantee a cure
- Abscesses, sinus tract, fistulas
- Strictures
- Hyperoxaluria, oxalate stones

C. Characteristics more common in ulcerative colitis
- More superficial mucosal disease involvement
- Confined to the colon
- Surgical resection of the colon eliminates the disease
- Normally originates in the distal colon and migrates proximally
- GI bleeding more common

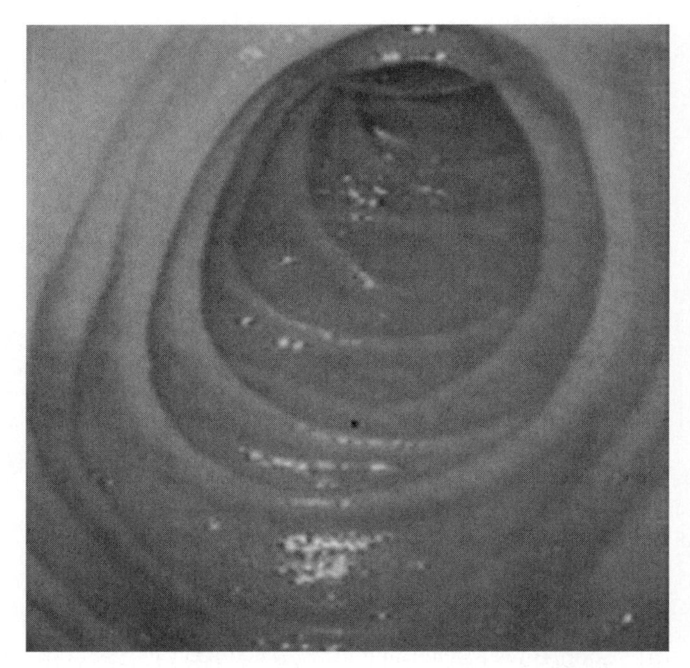

FIGURE 1 Normal small intestine. Photo courtesy of Dr. Gottumukkala S. Raju, Kansas University Medical Center, Division of Gastroenterology. See color plate.

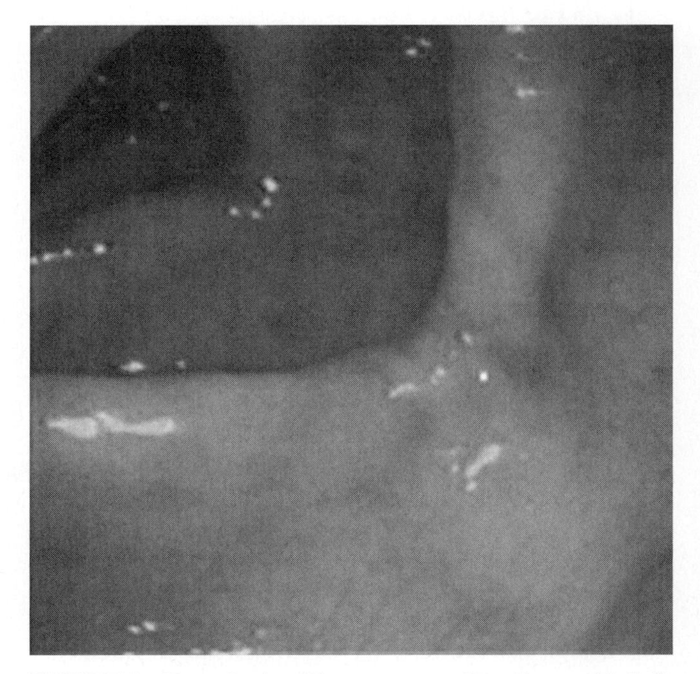

FIGURE 2 Normal colon. Photo courtesy of Dr. Gottumukkala S. Raju, The University of Texas Medical Branch at Galveston, Division of Gastroenterology. See color plate.

In CD, malabsorption, abdominal pain, mucosal thickening, strictures, obstruction, abscess and fistula formation, and nephrolithiasis are more common. (See Figs. 1 and 2 for representation of normal small intestine and colon, Fig. 3 for CD, and Fig. 4 for UC.) Crohn's disease may appear anywhere along the gastrointestinal tract. Although the location of the disease may change over time, at diagnosis it is most commonly found in the ileum (47%), in the colon alone (28%), and 21% of the time in the ileum and colon together [2]. Signs of CD are found in the upper GI tract in 3% of cases. Diseased bowel may be adjacent to "skipped" areas that appear normal. In contrast to ulcerative colitis, the disease is transmural rather than mucosal. Over the course of the disease about two-thirds of persons with CD will eventually have at least one surgery [5]. Surgery does not guarantee a cure: Many persons with CD have had numerous surgical procedures. Endoscopic evidence of the recurrence at 1 year is approximately 65–90%, although clinical relapse is closer to 20–25% [5, 6].

In UC only the colon is involved. In about 30% of patients, the disease remains confined to the rectum. In about 25–50% of those who present with proctitis, the disease tends to progress eventually to more extensive colonic involvement. The disease begins distally, extends proximally, and unlike CD, is continuous rather than normal segments interrupted by diseased bowel. Diarrhea with blood and mucus, cramps, and abdominal pain are common symptoms, and obstruction is rare [2, 3, 7]. Cigarette smoking tends to increase the likelihood of CD, whereas smoking

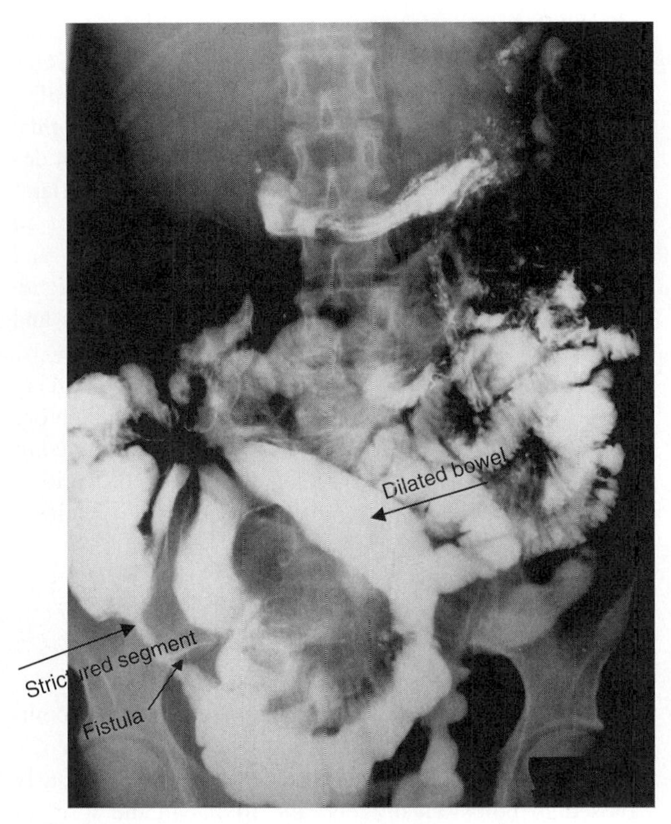

FIGURE 3 X-ray, patient with Crohn's disease with small bowel stricture, fistula, and loss of normal small bowel architecture. Distention, loss of haustra, narrowed lumen, fistula. See color plate.

is associated with a lower incidence of UC. One-third or fewer of persons with UC eventually require colectomy during their lifetime, and surgery is curative.

B. Incidence, Prevalence, and Disease Course

Inflammatory bowel disease may occur at any age but onset is most frequently at 15–35 years of age. About 25% of IBD cases begin in childhood [3]. Neither CD nor UC is very common but because of the frequency of clinic and hospital visits associated with the diseases, health care professionals tend to have a distorted sense of the prevalence. Worldwide, the incidence of CD ranges from less than 1 to 15 new cases per 100,000 annually and UC, from 2 to greater than 20 per 100,000 annually. The incidence is severalfold greater in relatives of persons with IBD, persons of Jewish descent, and Caucasians [8–10]. If siblings or both parents have CD, the risk increases 30-fold. The broad range may reflect the genetic and environmental influence and the sophistication of the health care and reporting mechanisms throughout the world. The disease tends to be greater in developed countries. In North America, the incidence of new cases of CD is about 5–15 per 100,000 and for UC, about 3–15 per 100,000 [9, 10]. The prevalence of IBD is greater than 300 cases per 100,000 persons [8–10]. Because of the differences in the prevalence, onset, severity, and course of the disease among races and populations, IBD is being viewed as a polygenic disorder.

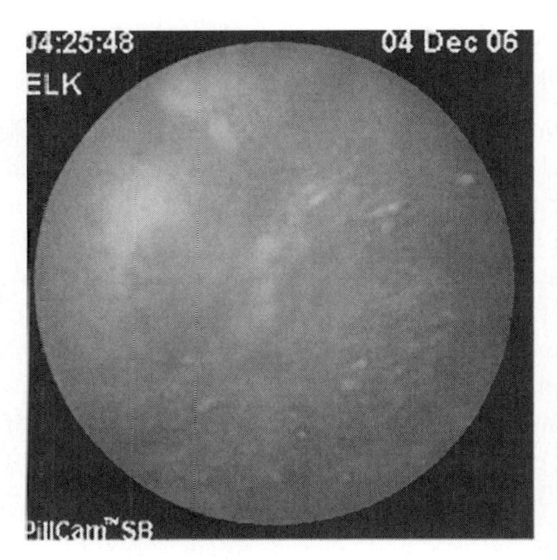

IBD Capsule 1

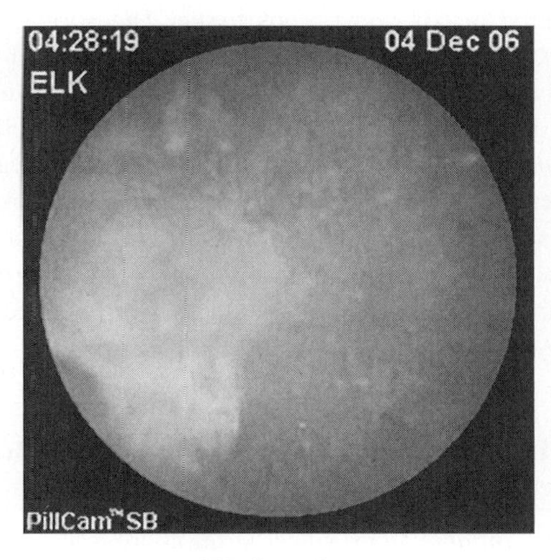

IBD Capsule 2

FIGURE 4 Ulcerative colitis. Photo from swallowed capsule. See color plate.

The course of IBD can vary from tranquil to severe and intermittent to unrelenting. In persons with more severe and active forms of disease, a number of factors can significantly affect quality of life including symptoms, health care requirements, dietary changes, malnutrition, pain, fecal continence, surgery, medical procedures, medications, and social embarrassment [11–13]. Underlying disease factors include genetic susceptibility, environmental factors, and therapeutic interventions. Taken together, these factors affect the severity, clinical course, treatment, and overall outcomes of the disease [1–3]. Diet and nutrition play a role in the symptoms, treatment, quality of life, and perhaps even the expression or remission of disease.

C. Normal Injury/Inflammatory Response and Pathogenesis of Inflammatory Bowel Disease

1. Normal Immune and Inflammatory Response

Because several dietary factors may be involved in the exacerbation, modulation, or resolution of the inflammatory state, a brief review of the normal immune and inflammatory response to pathologic events is appropriate. A discussion of the role of diet will follow. The normal immune and inflammatory process is typically initiated by microbial invasion, physical trauma, and exposure to toxic agents or autoimmune reactions. Subsequently a cascade of events occurs, eventually leading to resolution/repair [14, 15]. The events include recruitment and proliferation of leukocytes; release of proinflammatory and regulatory cytokines and eicosanoids; release of proteases and generation of oxygen radicals; activation of clotting mechanisms; and production of granular matrix and fibrous tissue. The normal inflammatory response is directed toward containment of the injury, destruction of invading microbes, inactivation of toxins, and repair of the injured tissue. Normally resolution is mediated by apoptosis, phagocytosis, and eradication of pathogens, anti-inflammatory cytokines, and eicosanoids. Overzealous or chronic inflammation can result in a continuous cycle of tissue damage and fibrosis, leading to mucosal thickening, obstruction, and dysfunction of involved tissues and organs.

D. Immune Response in Inflammatory Bowel Disease

The term *inflammatory bowel* is an appropriate descriptor for IBD because the classic events of acute and chronic inflammation are exhibited—but several components of immune response and its resolution may be defective. In IBD, the consequences of a prolonged and dysfunctional immune and inflammatory response account for most of the symptoms and the physical and pathological consequences described in the disorders. Changes in intestinal mucosal structure, fibrosis, strictures, bowel thickening, ulcerations, altered motility, and malabsorption are all examples of the consequences of the heightened inflammatory response [1, 14–17]. Symptoms of fatigue, intestinal and extraintestinal pain, bloating, diarrhea, steatorrhea, weight loss, and malnutrition are manifestations of increased local and circulating levels of cytokines and other products of inflammation. Proinflammatory cytokines dominate and a number of alterations in the control and regulation of immune defense mechanisms [1, 14–17]. Levels of peroxidation and generation of oxygen radicals are increased and circulating levels of antioxidants are decreased [18]. Consistent with the inflammatory process, platelets, prothrombin, fibrinogen, and collagen formation are increased, resulting in increased risk of systemic thromboembolic episodes, local fibrosis, and thickening of the bowel wall (most notably in CD).

Increased gut permeability, local edema, ulceration, and damage to the intestinal architecture may occur. Gastrointestinal secretion, digestion, absorption, and motility may be abnormal, especially in active stages, and the normal barrier to microbial fragments, toxins, and large peptides may be compromised [1, 16]. As in any significant inflammatory response, acute-phase proteins are found at increased levels and may serve, along with specific cytokine and leukotriene levels, as markers of the disease. Fatigue, malaise, anorexia, and disturbances in sleep patterns and mood can be seen in active disease.

A number of dietary factors are important in treatment and have the potential to influence the inflammatory response. These are discussed later. In addition, many complications of the disease may require dietary modifications, therapeutic supplementation of nutrients, or specialized nutrition support.

E. What Is the Underlying Cause of Inflammatory Bowel Disease?

The cause of CD and UC is unknown, but several hypotheses have been evaluated through the years (Table 2). Under normal circumstances, the GI tract is continuously exposed to potential antigens and toxins in the form of intestinal and food-borne microbes, foods, and medications. Physical mucosal barriers normally prevent excessive host interaction with the luminal environment and the host immune system is able to maintain a remarkable tolerance to thousands of food antigens and potential antigens from the highly concentrated normal (or commensal) microbes, numbering 400–500 species [1, 16]. At the same time, the host system is able to aggressively mount, when appropriate, a specific attack against pathogenic organisms. In the past, a common hypothesis was that IBD resulted from an inflammatory response to specific pathogens and/or environmental agents.

Specific bacteria and viruses have long been etiologic candidates because of (1) associations between the incidence of inflammatory bowel and community exposure to microbial infections; (2) antibodies in tissues to specific microbial fragments; or (3) similar GI damage provoked by infectious agents in human and animal models. Measles virus, strains of *Escherichia coli*, and *Mycobacterium paratuberculosis* have all been

TABLE 2 Possible Causes of Inflammatory Bowel Disease

- Genetic defect(s) in GI immune mechanisms
- Virus or bacterium
- Defective mucosal barrier functions with increased permeability
- Abnormal response to commensal bacteria, pathogens/antigens
- Imbalance of proinflammatory/anti-inflammatory cytokines
- Dietary: refined sugars, specific lipids, microparticles, sulfur compounds, dietary antigens

implicated, but not proven as causal factors [1, 3, 6]. Current investigations into the genetics of IBD point to loss of physiologic tolerance to commensal microflora in genetically susceptible individuals as the probable etiology [1, 4, 16, 17]. The potential abnormalities in the abnormal gastrointestinal immune and inflammatory responses (and its regulation) include at least pattern recognition receptors, defenses, adhesion molecules, toll-like receptors, chemokines, cytokines, eicosanoids, and roles of lymphocytes, macrophages, neutrophils, and fibroblasts. Abnormal epithelial cell functions and increased intestinal permeability [1, 16] may also be involved. The result is an upregulation of the inflammatory response related to an abnormal relationship with organisms that commonly reside in the intestinal tract. From recent genetic studies, mutations of NOD2/CARD15 and IL23R are examples of several important susceptibility genes for IBD (especially CD) [1, 3, 4, 16]. The CARD15 gene is involved in the several aspects of bacterial recognition and immune response. Other genetic susceptibility loci have been identified for both CD and UC, such as IBD3, an area that includes the HLA complex. Presence of single mutations by themselves, however, are not a guarantee that an individual will develop overt disease. Other genetic polymorphisms appear to affect the presentation and characteristics of an individual case of IBD including the onset, severity, clinical characteristics, presence of extraintestinal manifestations, and response to therapy.

Some of the therapeutic approaches and differences in response to various interventions, including medical, surgical, and nutritional, may be more accurately targeted toward the specific genetic and clinical profile for individual patients with IBD (Table 1).

1. What May Trigger the Onset or Exacerbations of Inflammatory Bowel Disease?

Inflammatory bowel disease is characteristically a disease of relapse and remission. Knowledge of environmental factors that trigger the onset or help maintain remission would be valuable in clinical practice. Identifying the initial trigger for the disease onset, however, is difficult. The genetic and/or environmental contributors may be different among populations or individuals. In addition, a number of years may elapse between histologic onset, symptoms, and its diagnosis. Microscopic or endoscopic evidence of the disease may occur long before clinical symptoms become apparent. The disease process and the inflammatory response result in pathologic changes such as increased gut permeability and antigen exposure, dysmotility, obstruction, or small intestinal bacterial overgrowth—all of which compound the problem of identifying whether the disease or the environmental trigger came first. Normal reactions to dietary indiscretions may cause GI symptoms (gas, bloating, pain, cramping, diarrhea) and may be confused with active disease. Patients with inflammatory bowel disease have sufficient reason to have as many or more dietary intolerances simply based on the structural and functional complications associated with the disease.

Nondietary factors associated with the onset or relapse of IBD include smoking (worsens CD and reduces UC risk), appendectomy (decreases risk of UC), nonsteroidal medications, measles virus, and modern lifestyle [4, 9, 16, 19]. According to the hygiene hypothesis, an ultraclean environment and reduced exposure to microbial antigens and parasites during childhood may impair the maturation of the GI immune system [4, 9, 19, 20].

Frequent use of antibiotics and other drugs has also been considered [20]. Dietary factors that been considered include high sugar intake, specific lipids, lack of dietary fiber (fruits and vegetables) [3, 4, 19–22], intake of microparticles [20, 23], abnormal metabolism of sulfur-containing amino acids as in hyperhomocysteinemia [20, 24], inadequate micronutrient intake (e.g., folate [24, 25] and vitamin D [26]), and individual food intolerances and food allergies [9, 16, 20–22]. Breast-feeding and consumption of fruits, vegetables, dietary fiber, and n-3 lipids have been considered protective [3, 16, 22, 27].

2. Dietary Patterns, Specific Foods, and IBD

Russel and colleagues [28] attempted to associate the rising incidence of IBD with changes in dietary consumption patterns. The authors showed a positive risk for IBD with increases in consumption of cola beverages, chocolate, and chewing gum, but negative risk for consumption of citrus fruits. The authors admitted that these nutritional items could be true risk factors or they could be coincidental dietary patterns adopted during the same time period. Joachim [29] reported subjective responses to foods of 60 patients with IBD. Patients were asked to report whether specific foods made them feel better or worse. The foods most commonly perceived to be associated with feeling worse included chocolate, dairy products, fats, and artificial sweeteners. No objective measures of illness were measured, and unfortunately only 122 foods were included on the questionnaire. The author did not compare the responses to foods in healthy controls or patients with other forms of gastrointestinal illness. Magee *et al.* [30] attempted to associate sigmoidoscopy disease activity scores with amounts of foods reported on diet records of 81 patients with UC. High food–sigmoidoscopy scores were related to amount of wine, soft drinks from concentrate, sausages, lager, and caffeinated coffee. Low activity scores were associated with consumption of several vegetables and fruits, pork, beef, and fish, breakfast cereals, potatoes, legumes, and milk and other dairy products.

A number of other investigators have attempted to associate foods or food types with the onset or relapse of IBD. The studies are often contradictory and the relationships between food and disease are often based on the patients' attempt to recall foods consumed prior to the onset of symptoms, and seldom related to markers of inflammation or histological changes. Both historical dietary recall and onset of symptoms are notoriously difficult to pinpoint. This is not to say that patients with IBD might not have food allergies or that certain foods/food types might not bring about symptoms and temper or heighten the immunologic activity.

3. Dietary Lipids

Several investigations have been centered on the lipid content of the diet and inflammation. High-fat diets and an increased ratio of n-6 to n-3 fatty acids have been associated with the increased occurrence of IBD and perhaps the duration of remission [31, 32]. Animal and human studies have demonstrated that consumption of predominantly n-6 lipid can increase several mediators of the inflammatory response by increasing production of leukotrienes and prostaglandins from arachidonic acid. The eicosanoids produced from n-6 fatty acids are considered more inflammatory and thrombotic than those from n-3 fatty acids. In animal models of IBD and a host of other human inflammatory disorders, increasing the n-3 fatty acid content of the diet decreases the production of proinflammatory eicosanoids, cytokines, adhesion molecules, and reactive oxygen species. Additional studies have shown beneficial effects on membrane functions, signal transduction pathways, and resolvin synthesis [33, 34].

Geerling and associates [32] examined the fatty acid intake and the fatty acid composition of plasma phospholipids and adipose tissue in 20 newly diagnosed CD patients, 32 patients with long-standing CD in remission, and matched controls. Fatty acid intake was not different between cases and controls, but a lower percentage of the sum of n-3 fatty acids and other changes in the lipid profile were seen in the patients. Kuroki and colleagues [35] measured serum fatty acid levels in 20 patients with CD without major resections and 18 healthy controls. In this population of Japanese patients and controls, neither n-6 nor n-3 fatty acids were deficient, but levels of total polyunsaturated fatty acids were lower in patients and total n-3 fatty acids was associated with a increased CD disease activity index. In separate analyses of studies using n-3 fatty acids to maintain remission or modify treatment, only a few of the studies actually measured remission/relapse rates or clinical outcome [36, 37]. The most consistent finding in the initial review [36] was that n-3 supplementation reduced corticosteroid requirement, and in the more recent review [37], prolonged remission was associated with use of with enteric-coated fish oil capsules. The authors concluded that larger controlled studies are needed with histologic and clinical evidence of effect. Other dietary lipids, including gamma linolenic acid and conjugated linoleic acid, may play anti-inflammatory roles by modifying lipid and nonlipid mediators of the inflammatory response [38–40], but additional studies are needed in IBD applications.

4. Sugar and a Refined Diet

In earlier reports, consumption of sucrose, refined cereals, fast foods, and fat was positively associated with either the incidence or onset of inflammatory bowel disease [22, 41, 42]. Intake of dietary fiber, fruits and vegetables, and specific nutrients, on the other hand, was negatively associated with the onset or severity of IBD [22, 41–44]. In other disorders with an inflammatory element, such as heart and respiratory disease and some cancers, consumption of fruits, vegetables, whole grains, and the phytochemicals contained therein was considered protective and anti-inflammatory. Phytochemicals in fruits, vegetables, whole grains, nuts, and legumes have been shown to inhibit COX-2 and nitric oxide synthase, block NF-kappa B activation, decrease cytokine expression, alter cell cycle activity, and decrease oxidative stress [45–48]. Large epidemiologic or cohort studies are needed that can effectively evaluate the roles of the consumption of fruit, vegetables, and whole grains and the risk of IBD. The anti-inflammatory effects of fibrous foods are commonly attributed to the short-chain fatty acids produced during colonic fermentation [49], but the harmful effect of a refined sugar diet may be more related to lack of immunoregulatory and anti-inflammatory micronutrients such as vitamins, phytochemicals, and antioxidants provided by fruits, vegetables, and whole grains. As mentioned, several dietary lipids can also affect the inflammatory process [22, 27, 34–40].

5. Food Allergies as Triggers

The lumen of the gastrointestinal tract is constantly exposed to antigens that have the potential to initiate at least part of the local immune and inflammatory processes. Loss of tolerance to some antigenic characteristics of commensal bacteria seems to be the driving force in IBD, but the role of dietary antigens as triggers or co-conspirators has not been established. Normally the antigenic response to the myriad of foods consumed is prevented or mitigated by acidic and enzymatic digestion of dietary proteins, mucous and epithelial cell barriers, peristalsis, and a number of secreted substances that allow oral tolerance [50–54]. Because persons with IBD may have several defects in gut barrier and immune regulatory functions plus malnutrition and maldigestion, they may be at increased risk for food allergies, food intolerances, and food-borne illness. However, food allergies have not been established as a major player in the cause of IBD. For example, Bichoff and associates [54] studied 375 adults with gastrointestinal disorders, including IBD, for objective evidence of intestinal allergic reactions. Thirty-two percent of the individuals complained of adverse reactions to foods as a cause of abdominal symptoms, 14% had indirect evidence of allergic manifestations, but only 3% of the cases could be confirmed by allergen provocation, elimination, and rechallenge testing. The authors concluded that food antigens could be a factor in a subgroup of persons with inflammatory and functional disease. The incidence of allergies in the gastrointestinal disorder patients (3.2%) was only slightly greater than the approximately 1.5% occurrence seen in the overall adult population [55].

In several earlier reports, milk allergy was suggested as a cause or trigger in the relapse of UC. Increased antibodies to milk protein have been reported [1], but their presence did not correlate with other markers of allergy or severity of symptoms. Withdrawal of milk or any other specific food does not consistently result in remission of active disease. Huber and colleagues [55] detected no food-specific IgE after studying the sera of patients with CD. Local allergic response to foods in the gastrointestinal tract does not appear to be a primary component of the pathology of IBD [55–57].

In general, associations made among the onset, prevalence, or symptoms of IBD and diet are complicated and may be explained by several potential relationships: (1) food intake patterns just prior to the initial onset of the disease, (2) entry and reaction to dietary antigens normally well tolerated due to the increased mucosal permeability resulting from the disease process, (3) food intolerances related to GI dysmotility, maldigestion, malabsorption, or changes in the intestinal flora, (4) protective effects of macronutrients, micronutrients, and/or phytochemicals, or (5) the effects of micro- or macronutrients on the GI flora. Intolerance to some foods, such as lactose, fructose, caffeine, fiber, or legumes, in patients with IBD is also often suspected because these foods have the same effects in normal individuals. In the latter case, the disease or its consequences (e.g., short bowel, small intestinal bacterial overgrowth, obstruction) may make the consequences of the food intolerance more profound.

F. Nutrition-Related Problems in Inflammatory Bowel Disease

Patients with IBD are at increased risk of several forms of malnutrition [13, 58–60]. Intake of macronutrients, vitamins, minerals, electrolytes, and phytochemicals may be compromised.

1. Macronutrient and Overall Nutrient Intake

Weight loss, growth failure, muscle wasting, and delayed maturation are some of the most common problems in adults and children with IBD, but several micronutrient deficiencies also occur. Anorexia, malaise, and malabsorption [61], concerns about potential adverse effects of eating, and perceived food intolerances may all contribute to decreased quality and quantity of food intake.

Patients may be fearful of eating specific foods or food types because they have heard that these foods may worsen symptoms or they experienced problems with specific foods. Food aversions and questionable associations of foods with subsequent symptoms are not uncommon. The pattern is not unlike that seen in cancer patients who undergo aggressive chemotherapy, become ill as a result of the side effects of therapy, and associate the illness with ingestion, taste, or smell of foods.

Inflammation, obstruction, dysmotility, bowel resections, and medications can alter nutrient needs and utilization, but most investigators and experienced clinicians attribute the nutritional inadequacy to lack of sufficient food intake [62–64]. Decreased levels of transport proteins such as albumin, transferrin, or prealbumin may be more related to the inflammatory state [61, 65] and a shift to synthesis of acute-phase proteins. Increased intestinal loss of nitrogen during relapse and decreased protein and total energy intake may contribute to growth failure, weight loss, muscle wasting, and reduced serum transport of protein concentrations.

2. Fluid and Electrolyte Losses

Diarrhea is one of the most common problems seen in IBD [2, 3]. In flares of IBD and after significant small bowel resections in CD, losses may be of sufficient quantities and duration to cause dehydration and electrolyte disturbances. Several dietary factors including specific carbohydrates and lipids, magnesium salts, and pre- and probiotics may worsen or attenuate diarrhea and fluid losses [66–69]. Incontinence, increased stool frequency, and the occurrence of anal burning may make the diarrhea even more unpleasant.

3. Micronutrient Deficiencies

Numerous reports of micronutrient deficiencies in persons with IBD have been published worldwide in the last two or three decades [60, 62, 71]. Despite the availability of micronutrients, some vitamin and mineral deficiencies nevertheless do occur. Vitamin B_{12}, folate, zinc, calcium, iron, magnesium, selenium, copper, and vitamin A, D, and E inadequacies have all been reported [61, 62, 70, 71]. Deficiencies of the same nutrients may explain the consequences, exacerbation, and perhaps even the susceptibility to IBD [58, 62, 72–74].

Several forms of anemia may occur as a result of inadequate intake, malabsorption, increased requirements, and drug–nutrient interactions. Iron deficiency anemia in ulcerative colitis is more commonly related to blood loss than poor iron intake or malabsorption. Vitamin B_{12} deficiency is more likely to be seen in Crohn's patients who have had resections of the terminal ileum, the site of active absorption of vitamin B_{12}. Anemia related to folate deficiency could be a result of poor intake, incomplete absorption, and/or drug–nutrient interactions [75, 76].

Fear of eating, food aversions, self-imposed or iatrogenic diet restrictions, surgical resections, and drug–nutrient interactions are likely to increase the risk of micronutrient deficiencies. Because the cells lining the gastrointestinal tract are especially metabolically active, macro- and micronutrient deficiencies have the potential to contribute to the pathology of the disease and compromise its resolution. Use of enteral nutrition

TABLE 3 Nutrition-Related Problems in Inflammatory Bowel Disease

- Nutrient inadequacy and/or limited food choices, anorexia
- Malabsorption, diarrhea, steatorrhea
- Symptoms related to ingestion of foods, intolerances, aversions
- Anemias (blood loss, B_{12} or folate deficiency)
- Weight loss, growth failure in children
- Protein, energy, vitamin mineral deficiencies
- Osteoporosis, osteomalacia

formulas, vitamin and mineral supplements, and fortified foods can reduce the risk of deficiencies. See Table 3 for a list of nutrition-related problems.

G. Nutrition Fears, Concerns, and Quality of Life Indicators Expressed by Persons with Inflammatory Bowel Disease

Patients often share concerns regarding their symptoms and potential outcomes, as outlined in Table 4. Although no studies have been reported on nutritional quality of life, IBD and its symptoms offers much to deal with in diet and nutrition. Total diet and specific foods are among the most common concerns expressed when patients visit health care providers and when they share their concerns in live and virtual online support groups. Patients with IBD may experience bloating, abdominal cramping and pain, nausea, and vomiting. Diarrhea and malabsorption may be common and in some patients, GI urgency and lack of bowel control may cause inconvenience or social embarrassment [11, 12]. Malnutrition may be a contributor to the overall symptoms of fatigue and malaise [13]. Whereas healthy persons normally endure minor abdominal pains and changes in bowel habit with little thought, patients with IBD may associate gastrointestinal signals with symptoms that previously preceded exacerbation of disease, surgery, diagnostic procedures, or prolonged hospitalizations. Persons may also instill fear about worst-case scenarios. Patients with CD who have already had significant small bowel resections may worry about their ability to absorb sufficient food and liquid, and persons with more severe forms of both CD and UC may worry about ostomy surgery and its sequelae. Because neither the cause nor the triggers for the onset of the disease are clear, frustration is common among patients attempting to resolve or temper their disease activity. Moreover, anxiety itself may worsen the manifestations of the disease [4, 76]. Appropriate dietary education and guidance may help ease symptoms and anxiety (Table 4).

TABLE 4 Concerns Expressed by Patients with Inflammatory Bowel Disease and persons with IBS in Internet Newsgroups, Crohn's Colitis Support Groups

- Need for procedures, surgery/ostomy/hospitalization
- Source or significance of pains and other symptoms
- Diarrhea, incontinence, embarrassment
- Occurrence of obstruction or fistulas
- The possibility that symptoms will lead to surgery
- Symptoms or appliances interfering with daily activity, quality of life, or personal interactions
- Concern whether nutrient intake, digestion, or absorption is adequate
- Side effects of medications
- Incomplete understanding of symptoms; no definite etiology for the disease
- Inconsistent advice and answers from health care professionals, scientists, and media

H. Treatment of Inflammatory Bowel Disease

Treatment of IBD usually involves medication, surgery, and nutrition and psychosocial support (see Table 5). The goals of treatment are essentially to bring about remission (or at least resolve symptoms), treat and prevent complications, restore and maintain nutritional status, and improve quality of life. Because the specific etiology of IBD is unknown and no single therapy is completely effective in accomplishing the overall goals, therapies are often used in combination and are based on individual disease profile and needs. As the underlying causes of CD and UC are better defined, more targeted therapies are likely to address individual patient phenotypes.

1. Medical Management of Inflammatory Bowel Disease

Over the course of a lifetime, patients with IBD are likely to be treated with several types of medications. Drugs may be used for inducing or maintaining remission, and treating specific complications and symptoms of the disease. Medications may be used in different combinations depending on the severity and presentation of the disease, and newer information about the genetic polymorphisms of CD is allowing more tailored treatment. General categories of medications commonly used to treat the primary events of IBD include anti-inflammatory agents, immunomodulator/

TABLE 5 Treatment Options for Inflammatory Bowel Disease

- Medications—Anti-inflammatory agents, immune modulators, antibiotics
- Surgery—Repair strictures, fistulas, resections of small and large bowel
- Psychosocial—Counseling, education, support
- Nutrition—Restore deficiencies, reduce symptoms, prevent relapse, possible primary therapy

suppressive drugs, biologic agents, and antibiotics. Several other medications, however, may also be used specifically to treat symptoms at various stages of the disease, such as antidiarrheal or antiemetic medications and specific antibiotics, for example, to treat opportunistic infections or bacterial overgrowth. More comprehensive reviews are available [2, 3, 5, 7, 78–80].

Although most of the same drugs are used for both CD and UC, some differences in application and effectiveness are apparent. Drugs may be required at high doses and for prolonged periods of time to bring on remission, but even at high doses and in combination, medications may fail to induce and maintain remission. Even when considered effective, medications bring additional costs, potential short-term and long-term side effects, and drug–drug and drug–nutrient interactions, and they affect quality of life. Side effects of medications that may affect nutrition include bone disease, glucose intolerance, folate depletion, altered taste, nausea, abdominal pain, and diarrhea. They invariably alter daily life patterns. See Table 6 for a listing of commonly used medications.

Drugs containing mesalamine have been considered first-line therapy in UC to bring on remission in mild to moderate disease and for maintenance. Typically the drug is started topically (suppository) for distal disease and by oral dosing for more proximal disease. Corticosteroids are generally used in CD and more severe forms of UC. They may be used in combination with other classes of drugs. Because of the metabolic, cosmetic, and psychogenic complications associated with use of some steroids, however, large doses and prolonged treatment are avoided when possible. Steroids are not considered a valuable alternative in maintaining remission in either form of inflammatory bowel disease [2, 5, 7, 78–80]. Infliximab, a monoclonal antibody to TNF alpha, has been used primarily in patients with fistulizing CD but has more recently been used for the management of both forms of IBD resistant to other drugs. Antibiotics, probiotics, and infliximab, among other agents, have been used in the treatment of pouchitis, a complication in patients who have had ileorectal pouches created as fecal reservoirs after colectomy.

TABLE 6 Medical Management of Inflammatory Bowel Disease

Anti-inflammatory drugs
Corticosteroids, sulfasalazine, 5-aminosalicylic acid (5-ASA or mesalamine)
Immune modulators or suppressants
Methotrexate, azothioprine, 6-mercaptopurine, cyclosporine, monoclonal antibody to TNF alpha (Remicade or infliximab)
Antibiotics
Metronidazole, ciprofloxacin

A number of new agents are being evaluated for treatment of IBD, especially with advancing knowledge of the GI immune system, including use of inhibitors of other proinflammatory cytokines, anti-inflammatory cytokines, inhibitors of inflammatory leukotrienes, adhesion molecule inhibitors, and T-cell inhibitors [2, 80, 81]. Leukocyte apheresis has been performed in a growing number of patients with both CD and UC with some reported benefit [81–82, 83].

2. Surgical Treatment of Inflammatory Bowel Disease

Indications for surgery typically include severe, unrelenting disease, strictures, obstruction, hemorrhage, increasing risk of cancer, repair of fistulas, and failure of medical therapy [3, 5, 7, 79]. In UC, colectomy "cures" the disease in that the disease does not subsequently occur elsewhere in the gastrointestinal tract. Approximately 25–40% of patients with UC have surgery during their lifetime [79, 84]. The most common form of surgery is colectomy with the creation of an ileoanal pouch (J-pouch, W-pouch), using folds of ileum pulled into the rectal canal and anastomosed to the ileum. The pouch develops a flora and serves to some degree as a colonic/rectal reservoir. Another form is the continent ileostomy or Kock pouch in which folds of ileum are use to create an internal reservoir that is drained with a catheter several times per day [85, 86]. No appliance is worn. The traditional colectomy involves the creation of an ileostomy and the use of an external stool collection appliance. Patient problems or concerns associated with the ileostomy may include stomal irritation, irritation of the skin surrounding the ostomy, leakage or dislodgment of the ostomy bag, odors, and social stigmata associated with the use of an external appliance. Although quality of life and bowel habit are not normal with either an ostomy or an ileoanal pouch, most patients feel these procedures are an improvement compared to the disease state [87, 88].

With CD, surgical resection of severely involved segments of small or large bowel does not, unfortunately, bring resolution of the disease. Endoscopic evidence of disease recurrence is seen in half or more of patients within a year after surgery. Duration of clinical remission varies greatly. In some patients relapse may occur in months, and in some patients, apparent remission may last for years. Approximately 70–80% of patients with CD eventually have at least one surgery during their lifetime [84, 85].

Strictureplasty has been used in relatively uncomplicated cases of CD to relieve narrowed segments of bowel [86]. In some cases, patients with prolonged severe courses of CD have multiple resections resulting in short bowel syndrome and the attendant nutritional and medical consequences.

3. Nutritional Treatment of Inflammatory Bowel Disease: Potential Roles

Nutrition and dietary patterns may play several roles in the management of IBD. Nutritional rehabilitation may be required after acute or prolonged reduction in the quantity

TABLE 7 Possible Roles of Nutrition Interventions

- Help patient understand roles of diet/foods and digestive and absorptive functions; sort out food associations and intolerances
- Treat symptoms/complications
- Improve nutritional status of the host/gut
- Treat disease/initiate and maintain remission
- Temper, "detoxify" the immune response
- Compensate for drug–nutrient interactions

or quality of dietary intake. Dietary modifications may be required to increase the tolerance of foods during exacerbation of the active disease. Special diets and supplements may be needed to provide adequate nourishment with complications such as short bowel syndrome, strictures, or fistulas. Refeeding protocols may be appropriate after severe malnutrition, after severe bouts of the disease, or after surgical procedures. Nutrition may also have some role in the regulation of the inflammatory response and inducing or maintaining remission (Table 7).

For a number of reasons, patients with IBD are at increased risk for vitamin and mineral deficiencies and protein and energy malnutrition [58–65, 88, 89]. During active disease, patients may avoid foods to prevent adverse symptoms such as diarrhea, abdominal pain, nausea, and bloating. Patients may associate adverse symptoms with consumption of specific foods or food types and may be on restricted diets imposed by health professionals, themselves, or the advice of self-proclaimed nutrition "experts." The reason for food restrictions may be to prevent symptoms, maintain remission, or treat the disease. As a result of extensive resections and severe stages of the disease, malabsorption of macro- and micronutrients could occur. Medications may also limit appetite and produce gastrointestinal-related symptoms, resulting in decreased intake, decreased absorption, and/or increased requirement for nutrients. Restoration of nutritional status may include a carefully selected oral diet, vitamin and mineral supplements, or special enteral and/or parenteral nutrition.

I. Nutrition Support as a Primary Therapy

As discussed earlier in the chapter, although the cause of IBD is unknown, current information suggests that it results from a genetic predisposition to an abnormal GI immune response toward microbial and perhaps other environmental triggers. The primary environmental factors to which the gastrointestinal tract is exposed are foods and commensal microbes. Several individual organisms and dietary factors have been scrutinized but no specific or combinations of etiologic agent(s) have been identified as yet. The GI flora is composed of approximately 500 species of bacteria and a large percent of the organisms have yet to be identified. The human diet may be a significant source of antigens, inflammatory or regulatory agents, antioxidants, and nutrients that alter the normal function of the GI tract and the immune system (see Chapter 38). Components of the diet can also serve as major sources of fuels for cells lining the GI tract and microbes residing there. It has also become increasingly clear that diet and nutrients not only can alter the host cell membranes and the way the GI tract functions but may also change the characteristics of the luminal microflora. The question, then, is to what degree can diet treat or prevent relapses of IBD?

1. Can Parenteral, Enteral, or Special Diets Induce/Prolong Remission Compared to Other Therapies?

a. Parenteral Nutrition

Parenteral nutrition has long been considered an appropriate approach in some patients because nutritional status can be restored and sufficient energy and nutrients can be provided for growth and weight gain. In the earlier days of parenteral nutrition, some patients, especially with CD, appeared to enter remission while being provided parenteral support. These earlier observations gave hope that parenteral nutrition would serve as a primary or at least adjunctive therapy [91–93]. Theoretically withdrawal of oral nutrition and use of parenteral nutrition might "rest" the bowel, reduce the antigenic load from foods, reduce microbial populations, and feed the gut and host. Use of parenteral nutrition as a primary therapy, however, has not been as successful as anticipated [62, 94]. Although parenteral nutrition may result in short-term remission [91–94], its primary role is for support of patients with complications such as gastrointestinal obstruction, complicated fistulas, perforation, toxic megacolon, and short bowel syndrome that preclude adequate enteral nutrition. Withholding oral/enteral nutrition may also be considered undesirable because it may compromise gut integrity or contribute to biliary stasis. In addition, parenteral nutrition does not contain the phytonutrients and wide range of lipids and microelements found in a normal diet. Additional study with other supportive additives such as specific amino acids, lipids, and selected micronutrients may provide more effective parenteral solutions.

2. Enteral Nutrition as Primary Therapy for IBD

The effectiveness of various forms of enteral feedings as a primary treatment for IBD has been evaluated in several reviews, meta-analyses, and individual studies [94–101, 104]. In almost all studies, CD appeared to be more amenable to enteral treatment than UC although different forms of nutrition interventions (other than standard enteral formulas) may be of value in UC. Generally enteral nutrition appeared to be more effective than parenteral nutrition, but the degree to which enteral nutrition can replace current medication protocols is not clear. Guidelines from the European Society for Parenteral and Enteral Nutrition [102] cite that enteral feeding is first-line therapy in CD and should be used as sole therapy in adults when treatment with steroids is not feasible and that undernourished patients with CD or UC should receive oral or tube-fed supplements. In North America, the position on enteral nutrition is less clear for children and adults.

Many difficulties are inherent in performing and interpreting studies for determining efficacy of enteral feedings. Numerous criteria have been used to describe the severity of disease both at initiation of therapy and when remission was achieved. Initial disease activity may have ranged from mild to severe. Severity of disease and/or resolution of disease have been measured using clinical measures such as diarrhea, weight change, and ratings of well-being. In other studies, more objective indicators such as endoscopic and histologic evidence, levels of inflammatory mediators, or measures of gastrointestinal structure/permeability were used. Duration of the patient's disease prior to initiating enteral or medical therapy may or may not have been reported. In some of the study designs, diet therapy may have been evaluated with and without various medications to determine whether diet improved the overall management. Dose and route of administration of medications varied greatly among studies. The diets have been described in great detail in some studies and in general terms in others. Duration of treatment with enteral therapy may have ranged from 2 weeks to 8 weeks or longer and the diet may have been fed orally, by tube, or by combinations of routes. One of the problems identified by even the proponents of nutritional therapy was that some of the diets were unpalatable or it was difficult for some patients to take the majority of their diet in the form of an oral formula. Improved nutritional status was considered one of the measures of success in many of the studies but nutritional status may not have been well described or was described in a variety of ways. In addition to differences in the approaches to comparisons with diet and medications, it should be noted that the natural course of IBD includes unexplained exacerbations and remission. In the history of therapeutic interventions with medications, 20–40% of patients may improve with placebo alone, so studies must be designed with comparisons against standard therapies or placebo [103, 104].

In the most recent and most rigorous reviews of randomized and quasi-randomized controlled trials of CD [99–101], the general consensus was that (1) corticosteroid treatment was more effective than enteral nutrition for active disease, (2) one form of enteral diet (elemental versus polymeric) was not more effective than the other, (3) exclusive enteral feeding was considerably more effective than partial use of enteral diet, and (4) lower fat content tended to be favorable compared to larger amounts. Claimed remission rates in the individual studies ranged from 20% to 84%. The total number of subjects from all studies meeting the criteria for inclusion in the comparisons was typically fewer than 200, and all authors noted the need for additional study, especially for sufficient power to evaluate subgroups. More data are needed to be able to determine whether age, duration of the disease, location of disease, or different lipid formulations would favor use of different types and combinations of enteral feedings and medications. Enteral diets have the advantage of providing/restoring nutritional status and reducing the risks of medications. Most reports state that mucosal healing is more likely to occur with enteral feeding than with steroids. On the other hand, enteral diets required that they be the sole source of nutrition for 4–8 weeks, and palatability and inadequate compliance for some patients required that they be tube fed. Relapse was not uncommon within a year after discontinuation of the exclusive enteral diet.

Explanations for the effectiveness of enteral nutrition include decrease in antigen load, changes in the GI flora, and improved structure and function of the GI mucosa and immune system as a result of nutritional restoration. Most enteral feedings are based on one or two protein sources as compared to hundreds of potential proteins consumed during even a simple daily diet. Enteral feedings are typically sterile and well fortified with vitamins, minerals, and trace elements. They are normally relatively low in fat and some may contain at least token amounts of n-3 fatty acids. On the other hand, most have no fiber and lack the phytochemicals provided by a normal diet. Some study of enteral products tailored for use in IBD are under way and additional study may be ultimately performed for the various subtypes of inflammatory bowel disease [99–102]. Subsequent results may better help clarify the role, effectiveness, and mechanisms of action for nutrition therapies.

Probiotics are a relatively new and promising application in the management of IBD. If IBD is a result of an aberrant response to luminal bacteria, the attractiveness of probiotics is understandable. It has been shown in animal models of IBD that bacteria must be present for the expression of the disease. The specific species or combinations of GI organisms that trigger the inflammatory state have not been identified. In both animals and humans, use of probiotics and prebiotics can change the intestinal flora, which in turn can change the relationship with the mucosal immune system and the levels of inflammatory and regulatory mediators in the gut [105–107]. A tremendous resurgence has occurred in interest in the interaction of the host microflora and their role in antigen tolerance, antigen presentation, levels of cytokines, adhesion molecules resulting from addition of probiotics, changes in dietary components, and medications [104–108]. Results of the human intervention trials with probiotics for treating CD have not been very successful, but results are more promising for UC [107, 108]. Some combinations of probiotics, notably VSL#3, have considerable evidence of effectiveness in pouchitis, the inflammatory state that occurs in the "pouch" surgically created from distal loops of ileum after colectomy [107–109]. As knowledge increases about the composition of the microflora and the roles of commensal and therapeutic probiotics and prebiotics, their roles in treatment and maintenance may improve.

In UC, therapeutic trials with fermentable fibers, short-chain and medium-chain fatty acids, n-3 fatty acids, and probiotics may have as much success as use of standard parenteral and enteral formulas. Early reports of success with open trials of dietary fibers and short-chain fatty acids, the end products of carbohydrate/fiber fermentation, offered some promise that dietary fiber might offer some therapeutic effect [110–112]. Epidemiologic reports were consistent with increased fiber intake and reduced presentation of UC. Many nutrients and phytochemicals in high-fiber foods may temper inflammation, and dietary fiber and other carbohydrates may provide a prebiotic effect, favoring the growth of beneficial organisms. Short-chain fatty acids serve as fuel for colonocytes, enhance salt and water absorption, and influence several markers of mucosal function and immune response [112–115]. Germinated barley foodstuffs, oat bran, and *Plantago ovata* seeds are examples of fibrous foods that, when ingested, raise short-chain fatty acid production. They have shown some value in maintaining remission and or reducing medication requirements in patients with mild to moderate UC [112–115].

3. How Might Diets/Nutrients Work as Therapy?

Patients with IBD are often malnourished. Inadequate intake of protein, energy, and micronutrients by themselves may create changes in gastrointestinal barrier functions and digestive functions and may alter immune mechanisms. Individual dietary components may act as antioxidants, serve as precursors to inflammatory mediators, and alter the relative types of microbes existing in the gastrointestinal tract. Dietary excesses, such as osmolarity, particle size, or volume, may also bring about undue symptoms even though they may have nothing to do with the pathology of the disease. It stands to reason, then, that selected enteral feedings could play several roles in the management of the disease. Many of the formulations used in past studies did not include many of the newer individual nutrient additives or modifications, yet many appeared to provide some level of success in the management of IBD. Continued trials, for example, with lipids, amino acids, probiotics, short-chain fatty acid and fiber sources, antioxidant nutrients, and phytochemicals, may make enteral diets more effective in managing IBD (Table 8).

Regardless of the issue of effectiveness of nutrition support as a primary therapy, both parenteral and enteral feedings are valuable in restoring nutritional reserves and allowing growth and maturation in children. Enteral nutrition is typically the route of choice for nutrition support whenever possible. Appropriate selection and use of enteral nutrition has the potential to provide enterocytes and colonocytes with a direct source of preferred fuels, allow normal proliferation and differentiation of those cells, increase gut barrier functions, and suppress the proliferation of unhealthy microbes. If diet does serve as a source of antigens, inflammatory stimulants, or toxins, enteral formulas may be designed specifically to reduce the offending agent yet provide reasonable nutrition support. Sufficient success with nutrition as a sole therapy and in combination with other treatments warrants additional study in inducing and maintaining remission in both CD and UC.

TABLE 8 How Enteral Diets Might Work to Temper Inflammatory Bowel Disease

- Restore nutritional dependent GI/immune function
- Reduce antigen load
- Repair GI barrier functions
- Alter cytokine, eicosanoid patterns
- Alter GI microflora
- Alter inflammatory response
- Prevent symptoms related to food intolerance, obstruction, enzyme deficiency
- Provide conditionally essential nutrients

J. Nutrition Assessment in Inflammatory Bowel Disease

A comprehensive assessment is required when evaluating nutrition risk and considering nutritional interventions in inflammatory bowel disease. Nutrition assessment for the person with inflammatory bowel disease would likely include (1) consideration of several elements of the patient's medical and surgical history, such as duration and severity of the disease, presence of strictures, fistulas, resections, ostomies; (2) presence of nutrition-related symptoms such as diarrhea (stool volume, frequency, and duration) or malabsorption (increased fecal fat, very low serum cholesterol, abdominal cramping, bloating, or distention); (3) physical measures such as growth rate for age, body mass index, weight changes; (4) diet history that includes habitual food intake, quantity and quality of food choices, intolerance and aversions to various foods, perceived and documented food allergies; (5) use of herbal and nutritional supplements and alternative therapies, prescription and over-the-counter medications; and (6) pertinent laboratory data which reflect protein-calorie nutriture and micronutrient status. Oddly enough, assessment of the dietary intake patterns and duration of those patterns is often omitted in a nutrition assessment. The nutrition assessment should also include an evaluation of the patient's knowledge and understanding of his or her nutritional status, needs, problems, and therapeutic options.

Energy requirements of persons with IBD are not usually increased except with sepsis and fever, but protein requirements often are. Energy requirements may be increased to restore weight, to return to normal growth curves, or to compensate for malabsorption. The presence of active disease, by itself, however, does not appear to raise energy requirements appreciably [63, 64]. Refeeding severely depleted patients too quickly may result in more harm than good. Protein needs may be increased significantly because of gastrointestinal nitrogen losses, the inflammatory response, and the need for new tissue for weight gain and growth.

K. Dietary Practices in Remission and Self-Management of Symptoms

Patients certainly have a great deal of interest in the role of diet and nutrition in day-to-day management of their disease and may be taught to deal with nutrition-related symptoms such as abdominal bloating, gas and diarrhea, strictures, malabsorption, and nutritional deficiencies. Several lines of evidence also suggest that dietary practices might reduce the severity of the disease, increase the effectiveness of medications, and/or prolong remission.

1. Lipids

The amount and nature of the fatty acids consumed in the diet may play a role in the management of several inflammatory diseases, including IBD. The ratio of n-6 to n-3 polyunsaturated fatty acids in the Western diet greatly favors n-6 fatty acids. Estimates of the ratio of n-6 to n-3 fatty acids consumed in various parts of the world range from approximately 9–1 to 30–1 [116, 117]. Consumption of oils such as corn, cottonseed, safflower, and sunflower was originally encouraged for their cholesterol-lowering effect and as a source of linoleic acid, an essential n-6 fatty acids. Each of the oils is rich in n-6 fatty acid and lacking in n-3. In the past two or three decades, consumption of polyunsaturated n-6 fatty acids largely replaced a significant portion of saturated fats that had been habitually consumed. Consumption of marine and terrestrial sources of n-3 fatty acids is now receiving more attention because of their potential to affect several forms of inflammatory and chronic disease. Fairly recently the ratio of fatty acid consumption shifted slightly toward n-3 fatty acids primarily as a result of consumption of more soy and canola oils.

Dietary fatty acids not only serve as a fuel but become components of the phospholipids in virtually every cell membrane in the body. The nature of the fatty acids in the cell membranes affects the physical, chemical, and functional properties of the cell and cell wall [116–118]. The lipids also serve as precursors for potent mediators in many physiologic reactions including the immune and inflammatory response, and lipids themselves may alter several players in the inflammatory response [118, 119].

Consumption of a diet that contains predominantly n-6 fatty acids, for example, results in adipose lipid and cell membrane phospholipids that are high in n-6 fatty acyl groups. Membrane lipids are involved in the immune and inflammatory processes. The cell membranes of neutrophils, monocytes, eosinophils, lymphocytes, and platelets, for example, are among those that are altered with the lipid composition of the diet. When these cells are activated they release long-chain lipids that are rapidly transformed to eicosanoids such as prostaglandins and leukotrienes that participate in the regulation of the inflammatory response. Their release and synthesis can affect recruitment and proliferation of other leukocytes and fibroblasts, alter the permeability of tissues, increase adherence of leukocytes and platelets, and increase the generation of harmful oxygen metabolites.

Eicosanoids produced from the metabolism of polyunsaturated fatty acids of the n-6 family (arachidonic acid, linoleic acid) tend to be proinflammatory and thrombotic, whereas those from n-3 fatty acids in general tend to be anti-inflammatory and decrease cellular adherence. In animal and human studies, an increased ratio of n-6 fatty acids has been shown to increase the synthesis of potent proinflammatory cytokines such as TNF alpha, IL-6, and IL-1; enhance leukocyte chemotaxis, membrane permeability, and edema; and increase generation of reactive oxygen species [33, 34, 118, 119]. n-3 oils have been shown to decrease gene expression of inflammatory cytokines, increase anti-inflammatory resolvins, and alter signaling pathways in the inflammatory response [33, 34, 119].

In clinical trials the value of n-3 fatty acids in the management of IBD has been considered promising, but more as a complementary approach rather than single independent treatment [33–37]. Belluzzi *et al.* [116] reviewed 10 studies in which varying doses and forms of n-3 oils were used in patients with both UC and CD. At initiation of diet therapy, patients differed in stage and severity of disease and some received concurrent medical therapies. Two of the studies were uncontrolled and several types of placebos were incorporated. End points included measures of duration of histologic and symptomatic evidence of disease or remission, steroid-sparing effect, Crohn's Disease Activity Index, and levels of proinflammatory leukotrienes. In these and similar studies, choice of the placebo, subject compliance, level and form of n-3 fatty acids, whether n-6 fatty acids or total fats were altered, whether subjects were truly blinded to the taste/smell of the oils, duration of the washout period, and simultaneous use of antioxidants and other nutrients were among the variables that could have altered study outcome. More recent analyses demonstrate some effectiveness, especially with the use of enteric coated fish oil capsules in maintenance of remission and reducing corticosteroid requirements [36, 37].

In general, n-3 lipid content of the Western diet is lacking for general health maintenance and may be especially low in some patients with IBD. n-3 lipid supplements appear to be reasonably safe in the small doses required to alter membrane eicosanoids and may have value in either sparing other therapies or independently tempering the inflammatory process and its consequences [33–36, 117–119]. Because n-3 fatty acids are highly polyunsaturated and therefore subject to oxidation, caution must be used in the purchase, storage, and preparation of foods and supplements.

Gamma-linolenic acid (GLA, C18-3 n-6) and dihomo-gamma-linolenic acid (DGLA), the elongase-catalyzed product of GLA, have also been shown to alter membrane lipids of inflammatory cells [40, 119]. These n-6 fatty acids have been evaluated in animals and in humans in the attenuation of inflammatory and proliferative disorders, but not to the same extent as n-3 fatty acids. Reports have shown reduced levels of proinflammatory cytokines, reduced leukotriene synthesis, and modulated autoimmune activity. Examples of the sources of the fatty acids include borage oil and evening primrose oil.

Conjugated linoleic acids (CLA) are a group of isomers of linoleic acid found in bovine lipids and in other ruminant animals that have been shown to have antiproliferative properties against several cancer cell lines [121]. These lipids also tend to reduce cell adhesion, decrease platelet aggregation, and alter several chemical mediators in the immune and inflammatory process [37, 38, 119–121]. Although they have been evaluated to some degree in some inflammatory states, neither GLA nor CLA has been evaluated in treatment of IBD.

Medium-chain triglycerides may have little to do with the inflammatory process, but they may be valuable as an energy source in patients with lipid malabsorption. Absorption of long-chain fats may be compromised during prolonged intestinal disease as a result of bile salt malabsorption that comes with ileal resection. Medium-chain triglycerides are known to enter the portal circulation without the more extensive processing that long-chain fats undergo. Studies suggest that medium-chain triglycerides are absorbed in the colon and may serve as an energy substrate for colonocytes in the same manner as short-chain fatty acids [122].

2. Dietary Fiber, Resistant Starches, and Short-Chain Fatty Acids

Consumption of dietary fiber and certain carbohydrates that are incompletely digested and absorbed in the small intestine have the potential to alter colonic flora and affect several gastrointestinal functions. The malabsorbed polysaccharides and sugars serve as substrates for microbes in the colon, and the end products of microbial fermentation can in turn serve as a primary fuel for cells lining the colon and modify metabolic activity. Inulin, fructose oligosaccharides, pectin, banana, oat and soy fiber, and several resistant carbohydrates have been shown to alter microbial populations in the colon [110–112]. Several have been shown to increase, at least transiently, *Bifidobacterium* and *Lactobacillus* groups that are viewed as more healthful microbes. Various saccharides might also reduce the growth of potentially pathogenic microbial populations such as *Clostridium difficile* [123].

The end products of fiber and carbohydrate fermentation in the colon are short-chain fatty acids and gases. Short-chain fatty acids, and in particular butyrate, serve as primary and preferred fuel for colonocytes and appear to be semiessential for their normal function. In several animal and human studies, addition of various fermentable saccharides or short-chain fatty acids has been reported to increase colonic blood flow, maintain normal mucosal barrier function, enhance fluid and electrolyte absorption (at least at physiologic doses), prevent antibiotic-associated diarrhea, provide protection from toxins, reduce the potential for inflammatory activity, and maintain normal proliferation and differentiation of colonocytes [109–111]. In several trials, addition of dietary fiber sources or topical administration of short-chain fatty acids has been shown to have some value in treating mild forms of UC or in maintaining remission—either as single agents or when used in combination with more traditional therapy. The effect is modest but further work with fermentable saccharides and combinations of short-chain fatty acids may provide effective therapy [110, 114, 115]. Most of the work with dietary fiber has appropriately been focused toward UC, but use of fiber sources has been considered in the management of CD.

Dietary fiber intake in Western countries is only about half that recommended [124], and some patients with inflammatory bowel disease may consume even less. Fear of obstruction and concern about coarseness of fibrous foods or increased gas may result in very low intakes, making the patient more susceptible to the reported effects of insufficient colonic substrate. Fibrous foods in the diet are not consistently thoroughly reduced to small particles, and partial or complete obstruction is a reasonable concern in patients with strictures. Even boluses of powdered fiber sources have been shown to cause obstruction in the upper gastrointestinal tract. However, use of dietary fiber sources, even if in powdered or blenderized form, appears to be appropriate in the diets of persons with IBD and especially in UC. Whenever possible, it appears to be logical to use complete foods as the fiber source. Fruits, vegetables, and whole grains are excellent sources of fiber, vitamins, minerals, trace elements, and a legion of potentially protective phytochemicals that may favorably affect the health of the gastrointestinal tract and the individual. A number of nutrients and even more phytochemicals serve as anti-inflammatory agents and antioxidants [45–48].

3. Dietary Sugars

Consumption of refined foods, especially sugar, at least in some countries, has been linked to the preillness diet or rise in the incidence of IBD [41–43]. No causal association, however, has been shown [3, 41]. On the other hand, consumption of large amounts of sugars could displace foods that are more wholesome and perhaps protective in the diet. Certain sugars, such as lactose, fructose, and alcohol sugars, are not as well absorbed as other carbohydrates and may

TABLE 9 Dietary Recommendations for Healthy Diet in Inflammatory Bowel Disease

- When possible, adequate intake of fruits, vegetables, and whole grains breads and cereals
- Healthy ratio of n-3 to n-6 fatty acids
- Adequate intake of vitamins, minerals, and trace elements—compensate for influence of maldigestion, malabsorption, or drug–nutrient interactions
- Modest intake of dietary sugars
- Minimal intake of alcoholic beverages
- Probiotics as needed for symptom control as indicated

contribute to symptoms such as bloating, gas, and diarrhea. Malabsorption and fermentation of relatively modest amounts of low-molecular-weight sugars can result in osmotic diarrhea [125, 126].

4. Vitamins, Minerals, Antioxidant Nutrients

The dietary habits in Western populations lead to inadequate nutrient intake of nutrients such as zinc, magnesium, potassium, calcium, folate, vitamin A, vitamin D, and vitamin E, in several age and gender groups, primarily women and the elderly [127–129]. The American diet is lacking in fruits, vegetables, fiber, and calcium food sources while being excessive in energy, sugars, and fats. In patients with IBD, micronutrient and antioxidant status may be compromised by a combination of an incomplete diet, maldigestion, malabsorption, medications, and the inflammatory process. When healthy and able, persons with IBD have a more compelling reason to consume a nutritious diet, especially from whole foods. In addition to vitamins, minerals, and trace elements, a good diet is an excellent source of antioxidants and protective phytochemicals not found in nutrient supplements. If, however, the patient is unable or unwilling to consume an adequate diet, vitamin and mineral supplements and/or liquid meal "replacement" supplements may be appropriate. Numerous oral supplements are available with varying lipid mixtures, protein sources, fiber, and other additives. Some patients will have select nutrient malabsorption, such as B_{12} malabsorption due to ileal resection or calcium/vitamin D inadequacy due to GI surgeries or avoidance of dairy products and lactose. Specific supplementation of individual nutrients may be necessary with careful attention to excess doses and drug–nutrient and nutrient–nutrient interactions (Table 9).

III. SHORT BOWEL SYNDROME

A. Definition

Short bowel syndrome refers to the set of symptoms and complications, which occur after significant small bowel resection. Resection of the colon or stomach does not constitute or result in short bowel syndrome, but stomach or colon surgery can compromise the patient who has had a small intestinal resection. The syndrome and its consequences are primarily related to the inability to absorb digesta and liquids and reabsorb endogenous secretions. Malabsorption of macronutrients, micronutrients, fluids, and electrolytes helps define the disorder and the severity varies considerably. Weight loss, growth retardation, diarrhea, dehydration, electrolyte disturbances, bone loss, renal oxalate stones, gallstones, lactic acidosis, and bacterial overgrowth may occur depending on the extent and nature of the resection [130, 131]. The syndrome may be temporary with less drastic resections or permanent as is the case with severe shortening of the gastrointestinal tract. The symptoms appear immediately after the resection and continue until sufficient adaptation occurs and/or medical, nutritional, or surgical interventions are successful. Typically, maximal adaptation of the remaining intestine takes months or sometimes a year or more.

B. Causes

In adults, Crohn's disease is a primary cause of short bowel syndrome, but other causes include mesenteric infarct, radiation enteritis, and volvulus. In infants and children, atresia, volvulus, gastroschisis, and necrotizing enterocolitis are more common reasons for significant resections leading to short bowel syndrome [130, 131]. The incidence and prevalence are not known, but tens of thousands are supported on total or supplemental parenteral nutrition.

C. Predictors of the Severity and Duration of the Syndrome

The severity and duration of symptoms, the need for nutrition support, and the survival of the patient all depend on the length and function of the remaining GI tract and the health of the host. Patients who survive without parenteral nutrition tend to adapt more completely, have fewer adverse symptoms, and are at decreased nutritional risk. These are patients who (1) are very young; (2) have remaining distal ileum and ileocecal valve; (3) have retained colon; and (4) are otherwise healthy and well nourished. Alternatively, predictors of poor survival, dependence on parenteral nutrition support, increased nutritional risk, complications, and compromised quality of life include (1) loss of terminal ileum and ileocecal valve; (2) advanced age at resection; (3) loss of the colon in addition to small bowel; and (4) presence of residual gastrointestinal disease [130, 132].

D. Variants of Short Bowel Syndrome

Requirements for nutritional, medical, and surgical care of the patient after bowel resections can be predicted somewhat by the normal physiologic role and capacity of

TABLE 10 Predictors of Severity of Short Bowel Syndrome

- Loss of distal ileum or ileocecal valve
- Loss of colon in addition to small bowel
- Advanced age
- Residual disease
- Mesenteric infarct as etiology for resection

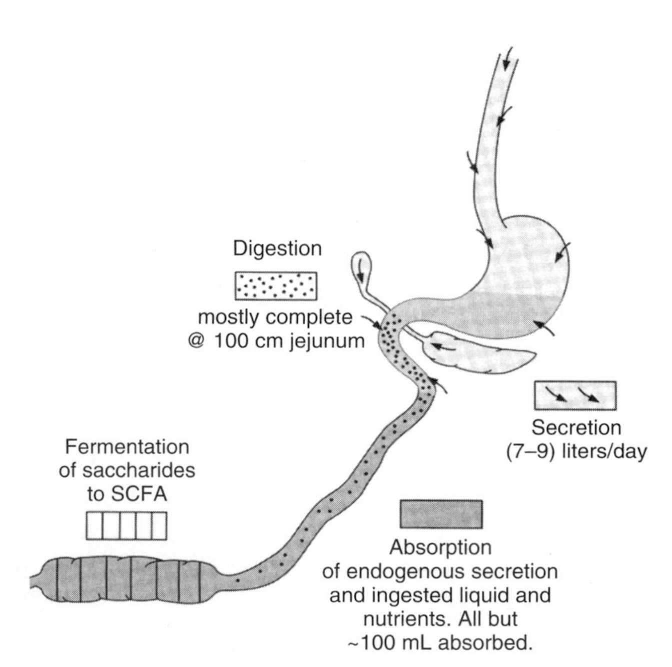

FIGURE 5 Normal secretion, digestion, and absorption.

segments removed and remaining. See Table 10 for predictors of severity of short bowel syndrome and nutrition risk and see Figure 5, which shows normal events in gastrointestinal secretion, digestion, absorption, and fermentation processes, for comparison with changes with bowel resections.

1. Jejunal Resections (with Intact Ileum and Colon)

Despite the fact that most gastrointestinal fluids are secreted into the proximal small intestine and the jejunum is normally the site of intense digestive and absorptive activity, loss of the entire jejunum in infants or adults typically results in only transient and relatively mild malabsorption as long as the patient and the remaining gastrointestinal tract are healthy. Figure 6 depicts jejunal resection.

Initially, the patient may require parenteral nutrition support but should and will be able to consume small, frequent, simple foods and beverages. Adaptation of the

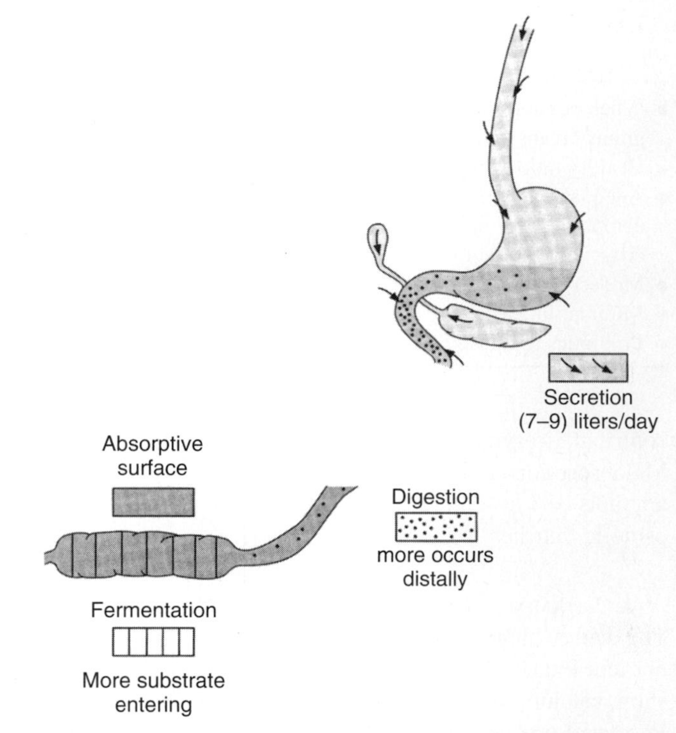

FIGURE 6 Jejunal resection, remaining ileum and colon.

remaining intestine will likely occur over the next several months. The patient will ultimately be able to consume regular foods with only limited restrictions and minimal compromise in digestive and absorptive capacity. The ileum has sufficient reserve to compensate for the decreased surface area. Because lactase is found primarily in the proximal small intestine, jejunal resection will result in lactose maldigestion and malabsorption. The shortened small intestine may also result in somewhat shortened transit time, so patients may not tolerate dietary abuses such as consumption of large quantities of hyperosmolar or highly concentrated foods and beverages. Net protein and energy intake and fluid and electrolyte utilization are not seriously threatened. The efficiency of absorption of certain nutrients such as iron, calcium, magnesium, and some lipid-soluble nutrients is often slightly compromised. Food selections and eating habits of many normal persons in the United States are hardly ideal; consequently patients with GI resections should be advised to pay greater attention to a high-quality diet and nutrient adequacy.

2. Ileal Resections

Major resections of the ileum create several problems in maintaining nutritional status [130–134]. Figure 7 depicts ileal resection, with the remaining jejunum and colon. Adults consume about 2 liters of fluid per day, and the proximal gastrointestinal tract secretes approximately 7–9 liters of fluid per day into the intestinal tract. In the normal GI tract, most of this fluid load is absorbed in the distal

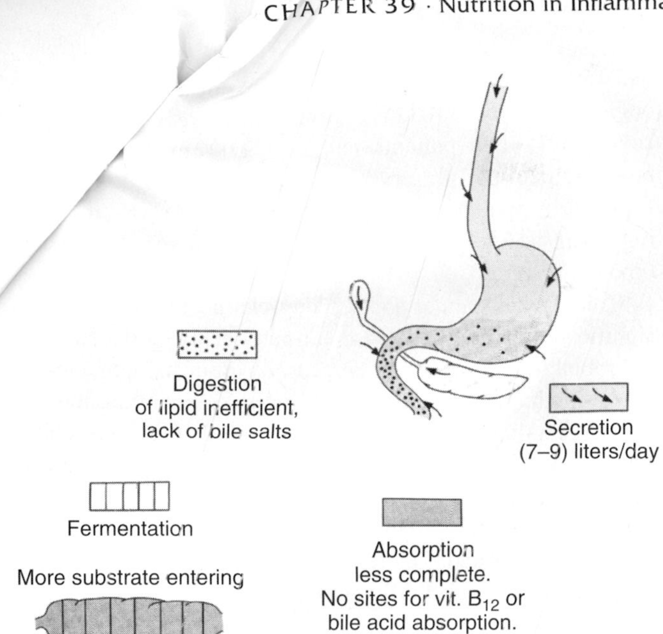

FIGURE 7 Ileal resection, remaining jejunum and colon.

small intestine and only 1–1.5 liters of liquid per day enters the colon. Normally about 100 ml of water is lost in the feces daily. When ileal resections occur, the jejunum and colon will be challenged to absorb dietary fluids and reabsorb endogenous secretions. If the colon is also resected, maintaining adequate hydration and electrolyte status is difficult and may become the determining factor for the requirement of parenteral nutrition. Table 11 summarizes the consequences of ileal resection.

The last meter of ileum is also the primary site for the absorption of vitamin B_{12} and is the only site for reabsorption of bile acids entering the intestine from the biliary tract. Normal adults usually have sufficient stores of B_{12}, but after ileal resection patients need to receive regular injections of B_{12}. More important, loss of the distal ileal segment where 90% or more of bile salts are reabsorbed results in a significant malabsorption cascade. The role of bile salts is to provide surfactant action to help reduce the particle size of lipid droplets and to participate in the formation of micelles. Micelles are the polar particles in which free fatty acids, monoglycerides, and fat-soluble vitamins are absorbed into cells of the small intestine. If the bile salts are no longer "recycled" in the distal ileum, *de novo* synthesis in the liver cannot supply enough bile salts for satisfactory absorption of lipids and lipid-soluble nutrients. In addition, unabsorbed bile acids entering the colon are considered "irritants" and serve as secretagogues at a time when colon is already burdened with the task of absorbing extraordinary amounts of fluid. Increased amounts of unabsorbed fatty acids in the lumen of the bowel can result in the formation of fatty acid soaps with calcium, magnesium, and zinc, which interfere with the absorption of these minerals. These malabsorbed fatty acids may also serve as secretagogues in the colon.

TABLE 11 Consequences of Ileal Resection, Especially with Loss of Ileocecal Valve

- Increased fluid loss
- Shorter transit
- Bile acid malabsorption; insufficient *de novo* bile salt synthesis
- Fat, fatty acid malabsorption
- Malabsorption of fat-soluble nutrients
- Decreased calcium, zinc, magnesium absorption
- In colon remains:
  —Increased colonic oxalate absorption
  —Increased secretion from malabsorbed bile acids and fatty acids
  —Increased risk of small bowel overgrowth

Dietary oxalate normally binds with calcium and other divalent cations to form insoluble complexes, but following surgery, free oxalate becomes available for absorption in the colon. In both ileal resection and long-term Crohn's disease involving the ileum, hyperoxaluria and renal stone formation is not uncommon. In an attempt to bind bile acids, improve colonic fluid and electrolyte absorption, and reduce absorption of oxalate and cholestyramine, supplemental calcium and modest dietary oxalate restriction are often prescribed [130, 134].

Another potential complication of distal ileal resections, especially involving loss of the ileocecal valve, is bacterial overgrowth in the small intestine. The small intestine is normally relatively sterile, but with loss of the one-way barrier provided by the ileocecal valve, microbes from the colon may mix with small bowel contents. Bacterial overgrowth is usually treated with antibiotics, but more recently use of small bowel irrigation and probiotics and prebiotic foods has been tried with some success [135].

3. Loss of Ileum and Colon

If both the ileum and colon are resected, maintenance of adequate hydration, electrolyte, magnesium, and energy balance becomes more difficult. Figure 8 depicts resection of both ileum and colon, with only jejunum remaining. Although only 1–1.5 liters of fluid normally enters the colon, a healthy colon is capable of absorbing 3–4 liters. With only a jejunum, patients have difficulty absorbing ingested fluid and endogenous gastrointestinal secretions. Fecal outputs from jejunostomies may exceed 3–5 liters daily and patients may lose 100–200 mEq of sodium in the effluent. In addition to increased risk of hypovolemia, patients are also more at risk for hypokalemia and hypomagnesemia and other forms of nutrient depletion [130, 133, 134].

The normal GI tract has a remarkable autoregulatory system for initiating, controlling, and slowing secretions, motility, absorption, and transport. The presence of foodstuffs or partially digested foods in the distal intestine normally stimulates the release of a host of neuropeptide

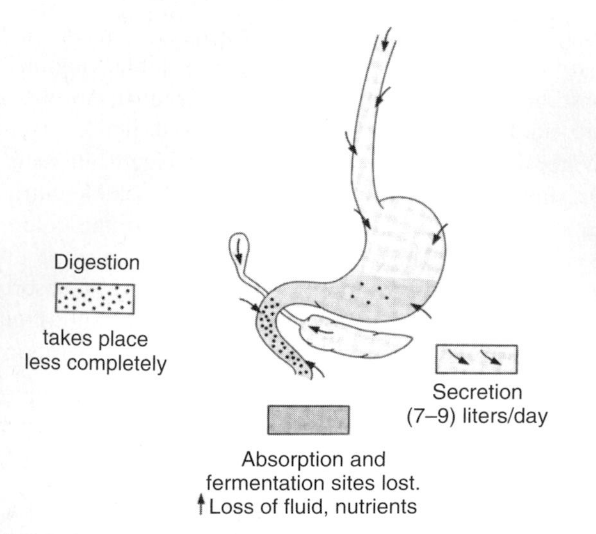

FIGURE 8 Resection of both ileum and colon, only jejunum remaining.

hormones that in turn affect gene expression, immunologic and neurologic activity, endocrine secretions, transport mechanisms, and muscular activity [74, 137]. Adaptation does not occur to any significant extent without luminal stimulation of foods and digesta. Examples of the hundred or so GI neuropeptides that have regulatory functions include gastrin, glucagon-like peptide-1 (GLP-1), glucagon-like-peptide 2 (GLP-2), neuropeptide-Y, epidermal growth factor-1 (EGF-1), and insulin-like growth factor-1 (IGF-1). The GI hormones and digested nutrients also trigger the synthesis of mucosal cells, immune cells, and the GI flora. Some of those autoregulatory mechanisms may be compromised with significant bowel resections, especially loss of the ileal braking effect, a feedback loop that slows proximal movement of the GI tract, and loss of the ileocecal valve. Use of supplemental or exogenous growth hormone, GLP-2, EGF, amino acids, and probiotics continues to be evaluated and shows promise in enhancement of adaptation following bowel resections [137–139]. For example, teduglutide is a GLP-2 analog that has been shown to enhance mucosal growth and fluid absorption and decrease stool weight and energy loss [140].

Presence of residual colon also serves to salvage energy from malabsorbed substrates. In humans, the role of the colon in extracting energy from foodstuffs is normally relatively minor compared to that in some animals. In short bowel syndrome, however, more carbohydrate, lipids, and nitrogenous compounds enter the colon than usual. The amount of energy the colon can extract after small bowel resections from malabsorbed substrates varies with the length of small intestine remaining. In normal individuals consuming a typical diet, only about 5–10% of energy is salvaged by colonic fermentation [140]. Nordgaard and colleagues [141] demonstrated fecal energy loses with and without the colon using bomb calorimetry. Patients who had 150–200 cm of small intestine witho[illegible] only 0.8 MJ/day (110 kcal) more than those w[illegible] colon. But when patients had only 100 cm or less [illegible] bowel remaining, preservation of the colon for energy sa[illegible]vage from malabsorbed substrates became much more important. Colectomy resulted in loss of 4.9 MJ/day (about 1150 kcal).

Colonocytes cannot absorb long-chain fatty acids to any significant degree, but short-chain fatty acids produced from microbial fermentation of fiber, carbohydrate, and, to a lesser extent, amino acids can be used by the colonocytes and the host. Recently, investigators have shown that medium-chain fatty acids can also be used by colonocytes and absorbed systemically [142]. Loss of the colon and ileum results not only in loss of the energy salvage mechanism, but also in decreased fluid and electrolyte absorption, rapid GI transit, and overall decreased efficiency of nutrient absorption.

4. Major Intestinal Resections in Infants

Volvulus, gastroschisis, or atresias often result in a small bowel length of only 100 cm or less. The same risk factors for increased morbidity apply in infants, including loss of ileum, ileocecal valve, and colon, except that infants appear to adapt significantly better after resections than do adults. In normal healthy infants and children, intestinal growth occurs from preterm until early adulthood. Weaver and associates [143] studied 1010 necropsy specimens from preterm fetuses to subjects up to 20 years of age. Small bowel length more than doubled from 20 to 40 weeks' gestation and almost doubled again from term (275 cm) to age 10 (500 cm). From 10 to 20 years intestinal length increased only an additional 75 cm. The extent to which lengthening and other forms of adaptation occur after resection (beyond normal growth) is not clear, but infants appear to function reasonably well with very short intestinal remnants. In infants, survival without parenteral nutrition has been reported with as little as approximately 30–60 cm of small bowel if the children have retained the ileocecal valve and as little as 30–100 cm without the ileocecal valve [143, 144]. When the ileocecal valve is lost, retained colon becomes more important for maintenance of electrolyte balance and adequate hydration and to salvage malabsorbed substrates. Adults typically require at least 100–200 cm of small intestine to avoid at least partial dependence on parenteral nutrition, and adaptation may be more prolonged and less complete [130, 132]. For children with extremely short small bowel lengths who cannot adapt, several options remain. They can remain on parenteral nutrition, undergo small bowel lengthening procedures or undergo small bowel transplantation. Each has its consequences but may be the only option for survival. Parenteral nutrition imposes on the quality of life for the patient and caretakers and carries increased risk of infection, cholestasis and hepatic failure, gallstone formation, and nutrient deficiencies.

5. Major Resections in Adults

The ability to tolerate major bowel resections in adults is considerably less than in children. When the patient has residual disease and has had other changes in gastrointestinal pathology because of radiation enteritis, atherosclerosis, diabetes, or inflammatory bowel disease, then bowel resections pose greater risk. In patients with radiation enteritis, even relatively small resections of the ileum, especially if the resection includes the ileocecal valve, can create short bowel syndrome.

Patients who suffer mesenteric infarct may lose almost the entire small intestine and most of the colon. Because mesenteric infarct typically occurs in older patients as a result of atherosclerotic vascular disease, adaptation of the remaining segment may not be sufficient. The patient may require lifelong support with at least supplemental parenteral nutrition. Maximal adaptation may require up to 2 years and may still not be adequate to maintain fluid, electrolyte, nutrient, and energy requirements. Reintroduction and advancement of foods should be gradual. Unfortunately, some patients may never be able to absorb sufficient amounts of nutrients by way of the gastrointestinal tract. Hyperphagia (compensatory overeating) can be a helpful response in many patients with short bowel, but overconsumption of refined carbohydrates, fluids, and large volumes of foods may result in net loss of endogenous fluid and electrolyte [130–134].

E. Principles of Nutrition Care and Feeding in Short Bowel Syndrome

One of the primary principles of managing patients with short bowel syndrome is to start using the gut as soon as possible. Foods and the gastrointestinal secretions are trophic to the small bowel and malnutrition compromises the adaptive process [131, 137]. In animals, the gastrointestinal tract atrophies markedly during starvation and parenteral nutrition. In humans, the degree of gut atrophy during parenteral nutrition may not be as striking as in animals but appears to be significant [74]. Biasco *et al.* [145] described the gastrointestinal response in a patient fed by parenteral and oral nutrition after a subtotal resection for mesenteric obstruction. The patient had 30 cm of proximal jejunum anastomosed to his descending colon. The patient was fed exclusively by parenteral nutrition for 30 days, then oral nutrition was gradually introduced over the next 30 days. Jejunal biopsies were taken during parenteral and oral periods and examined for cell kinetic studies. Hypoplasia was demonstrated soon after parenteral nutrition started. Hyperplasia occurred soon after starting oral feeding, and after 14 days crypt depth and villous height had increased by approximately 50%.

A significant portion of the energy supplied to the small intestine comes from sugars, keto acids, and amino acids (notably glutamine) from the lumen of the GI tract. An even greater percent of the energy for colonocytes come from substrates in the lumen, primarily short-chain fatty acid produced in fermentation of saccharides. Nutrition provided into the lumen of the GI tract provides fuels for the luminal cells, the host, and the microflora. Adequate nutrition in terms of macro- and micronutrients is also necessary for the normal GI function and for adaptation [74, 130, 137].

In animal studies with short bowel syndrome, the observed response to feeding after major bowel resection is vigorous hyperplasia and hypertrophy, increased length and diameter of the remaining small bowel. The response in humans is not as obvious as in animals. If oral or enteral feeding is not provided, some atrophy occurs. Even small amounts of foodstuffs provide sufficient stimulus for intestinal growth and improved function. Initial feedings after bowel resections may include small, frequent meals, sips of liquid supplements, or dilute enteral feedings to gradually restore gastrointestinal function and stimulate adaptation.

Adaptation in the GI tract is not as complete as that seen, for example, in regrowth of a lobe of liver after partial hepatic resection, and when adaptation is maximal, the patient may still have inadequate absorptive function. The extent to which adaptation takes place depends on the age of the individual, the health and vascular patency of the remaining intestine, and nutritional factors. Relatively few studies have been reported in which changes in intestinal length and diameter of the bowel following intestinal resection have been measured. Shin *et al.* [146] studied the adaptive response in dogs with massive resections after receiving nutrition support. Seven mongrel dogs had surgical resections leaving 20 cm of proximal jejunum anastomosed to the mid-transverse colon. The dogs were initially fed by parenteral nutrition, and after 4 days, oral food and water were reintroduced. Parenteral nutrition was continued in addition to the oral diet for 30 days and was then stopped. Parenteral nutrition proved to be essential initially because oral feeding induced massive stool outputs. Six weeks after the resection, the five dogs that survived were reexplored and the remaining small bowel was examined. The length and circumference of the jejunum increased by 30%, and the crypt depth, number of villi, and width of the villi had each increased by approximately 20%. Stool volume and frequency gradually decreased but were still abnormal.

Although marked changes occur in many animal models, humans have not been as well studied. Morphologic and functional changes occur, but may not occur in the same manner as in animals. In some cases, the functional changes do not correspond with morphologic findings [137]. Solhaug [147] studied five female patients who had undergone jejuno-ileal bypass procedures for treatment of obesity. Patients had end-to-end anastomosis of their proximal jejunum to their distal ileum and had undergone a second operation because of complications or unsuccessful weight loss. The remaining segments of small intestine increased during a 1- to 2-year period from a mean of 52 cm to 82 cm while the subjects consumed an *ad libitum* diet. The bypassed segments essentially remained unchanged. Mucosal thickness was approximately 20–40% greater than the bypassed segment, villous height was 25–50% greater, and small intestinal circumference was double that of the bypassed segment.

When starting feedings in patients with short bowel syndrome, the diet is normally provided in very small meals or in sips of dilute formula diets [133, 148, 149]. During the advancement of feeding, the patient is normally supported at least in part by parenteral feeding so that oral feeding proceeds gradually to allow the gut to adapt and to prevent diarrhea and malabsorption. If the patient is unable or unwilling to eat consistently in a controlled or scheduled manner, enteral feedings can be spaced in small meals or delivered by tube feeding. Tube feedings can be scheduled at night or between meals as needed, then decreased, as the patient is able to take more foods and beverages orally. Many patients are anorectic initially and may later become hyperphagic. Overfeeding foods and beverages, especially at the beginning of refeeding, can easily result in loss of more fluid than was originally consumed. Patients sometimes express extreme hunger and weakness. They may subsequently consume large volumes of food and beverages only to experience malabsorption of foodstuffs and large fluid effluent. Typically dietary modifications are included to ensure adequate nutrition without overwhelming digestive and absorptive capacity.

Principles of managing nutritional aspects of short bowel syndrome are generally based on the sites and capacities of the remaining digestive and absorptive function, adaptive processes, and the experience of practitioners familiar with diet and short bowel [133, 148–150]. The mainstays of the diet are complex carbohydrates (starch) and relatively lean animal proteins with modest amounts of fat. Patients with retained colon may do better with higher amounts of complex carbohydrate and lower amounts of fat (20–20% of calories) than those without. Sources of both n-6 and n-3 fatty acids should be consumed. Medium-chain triglyceride products can be helpful in providing a source of energy, especially in patients without the distal ileum. Poorly absorbed sugars (e.g., fructose and alcohol sugars) [126] and consumption of large volumes of hyperosmolar sugars should be limited to avoid diarrhea and reduce the risk of lactic acidosis. Lactase may be reduced in some patients, but reasonable amounts of lactose can be tolerated, especially when supplied in foods such as cheeses, yogurts, or mixed foods rather than milk as a beverage [151]. Commercially available lactase enzyme can also be used to improve the tolerance to foods containing lactose.

Some foods are discouraged simply because they provide little or no nutrient value and may displace nutrient-dense foods. Caffeinated and alcoholic beverages tend to increase gastrointestinal secretions, alter motility, or increase permeability and contribute little to nutritional value. The diet need not be devoid of fiber; in fact, modest amounts of dietary fiber may help to regulate gastric emptying. If the patient has a residual colon, fiber may be fermented to short-chain fatty acids to enhance gastrointestinal function and increase absorption of fluid and electrolytes. In patients with strictures, dietary fiber may need to be in the form of small particulate matter and added to juices or other foods.

A typical diet in the patient with short bowel may include small but frequent portions of foods such as eggs, meats, poultry, fishes, breads and cereals, rice, potatoes, pastas, vegetables, and vegetable juices. Foods that incorporate a mixture of ingredients such as casseroles, sandwiches, and low-sugar baked goods (e.g., banana muffin, pumpkin bread) may be used. Sauces, spices, and flavoring agents add appeal and variety and are not likely to alter absorption if used modestly. To meet energy targets, specially tailored oral electrolyte replacement fluids may be needed. Low-calorie/caffeine-free beverages and flavored waters can be used but are not recommended when fluid/electrolyte replacement is needed and hydration status is tenuous. Use of vitamin and mineral supplements should be tailored to the patients' dietary habits, tolerances and absorptive capacity. Particular attention is normally directed toward lipid-soluble vitamins, B_{12}, magnesium, and sodium and potassium. Because transit through the remaining GI tract may be rapid, some tablets may not disintegrate in time for maximal absorption. Supplements can be crushed or provided in liquid form. Adaptive hyperphagia occurs in most patients with short bowel syndrome and may be helpful in the adaptive process [152]. In patients with very little small bowel remaining, however, hyperphagia may result in net fluid and electrolyte loss. Appropriate attention should be provided to the trauma and life-altering challenges facing the patient. Nutrition and other forms of counseling may need to be provided in small doses, consistently and frequently. Follow-up evaluation of nutritional status is usually needed throughout the adaptive period and, in some patients, throughout life.

F. Nutrients, Growth Factors, Medical and Surgical Approaches to Restoring Gastrointestinal Function

Over the years several specific nutrients have been evaluated to hasten or enhance the adaptive process with or without neuroendocrine hormones. Examples of nutrients considered include supplemental vitamins, glutamine and other amino acids, fatty acids, nucleotides, fermentable fibers, and short-chain fatty acids [74, 137, 148].

Wilmore described the use of a combination of growth hormone, supplemental dietary glutamine, and a low-fat diet high in complex carbohydrate in managing more than 300 patients who were dependent on parenteral nutritional support or at significant nutrition risk [148]. Wilmore reported that the combined therapy was successful in increasing body weight and lean mass, reducing dependence on parenteral nutrition, and improving overall well-being of patients with short bowel syndrome. Other investigators [141, 142] were not as enthusiastic. Scolapio [153], for example, in a randomized, placebo-controlled, double-blind crossover trial, studied growth hormone and glutamine in eight patients who had been dependent on parenteral nutrition for 3–19 years. Scolapio explained that while receiving the treatment, patients gained fluid weight and lean mass but only as long as the therapy was continued. The patients did not demonstrate increased fluid and macronutrient absorption during a 21-day trial but admitted that results could have been different with a longer treatment period. Based on their study with animal models, Vanderhoof *et al.* [154] concluded that any adaptive response was more likely to be the result of enteral nutrition rather than the glutamine and growth hormone treatment. In other more recent reviews attempting to compare similar interventions among individual studies, the combination of growth hormone and glutamine, with or without a modified diet, was deemed at least likely to be beneficial [155, 156]. The individualized attention to diet, medications, surgery, treatment of any underlying disease, and education adds to the success of programs designed to improve the quality of life and independence of persons with short bowel.

Dietary fiber or short-chain fatty acids have been studied in the regulation of normal gastrointestinal function as a therapeutic agent in the management of gastrointestinal disease, and as a regulatory agent in short bowel syndrome. Dietary fiber may help regulate gastric emptying by: (1) increasing the viscosity of gastric contents, or (2) as a result of a feedback mechanism when dietary fiber and resistant starch are fermented to short-chain fatty acids in the colon. Short-chain fatty acids serve as a primary fuel for colonocytes, help maintain a healthy colonic flora, enhance the absorption of fluid and electrolytes, and may be trophic to intestinal cells [157, 158]. Short-chain fatty acids and certain dietary fibers may also regulate colonic cellular differentiation and serve to prevent the overgrowth of potentially pathogenic organisms [159].

Patients with short bowel syndrome may be also treated with several medications to improve symptoms, decrease GI secretions, delay gastric emptying, slow intestinal transit, or bind bile acids. Proton pump inhibitors or histamine-2 receptor blockers are used in the initial stages of gastric acid hypersecretion; loperamide, atropine, opiates, and octreotide are used for management of diarrhea; and cholestyramine is used to bind bile acids [130, 160]. Other agents are being studied in the management/prevention of complications of short bowel symptoms. Teglutide, a GLP-2 analog, is currently in clinical trials. Earlier studies have shown effectiveness in enhancing intestinal growth and function [130, 140].

Several surgical procedures have been considered in extreme short bowel and intestinal failure including intestinal tapering and lengthening procedures, intestinal loops, reversal of intestinal segments, insertion of colonic segments (which normally exhibit slower transit), insertion of reversed-segment colon, creation of intestinal loops, and valve-type passages [161]. Serial transverse enteroplasty (STEP) is a newer procedure that has shown to be effective in intestinal lengthening in children [160]. Small intestinal and "multivisceral" transplants have now become more successful for patients with intestinal failure and or complications of long-term parenteral nutrition [163, 164]. Small bowel transplant can include small intestinal segments alone, small intestine with colon, and intestine with liver and/or pancreatic transplant. Transplant procedures and immunosuppressive therapies continue to improve and with time may provide more viable option for more patients with severe short bowel syndrome.

IV. CONCLUSION

Both inflammatory bowel disease and short bowel syndrome are gastrointestinal disorders that may lead to serious nutrition-related consequences including diarrhea and malabsorption, weight loss, growth retardation, dehydration, micronutrient deficiencies, electrolyte disturbances, and bone loss. New therapeutic approaches and a renewed enthusiasm for the treatment of IBD are the result of the increased knowledge of interactions of commensal bacteria, with the immune system and individual genes. Nutrition therapy in the primary and complementary treatment of IBD and its complications will continue to be advanced. The role of diet is probably underappreciated and not well understood by health care practitioners and patients. Dietary manipulations can reduce the requirement for medications, decrease symptoms of the disease, and improve general well-being and quality of life. More information about the genetic and environmental contributors to IBD and additional nutrition studies will lead to effective and tailored medical and nutrition interventions.

Each disorder requires thorough evaluation of nutritional status and that nutrition intervention be tailored to the individual situation. Nutrition interventions typically take the form of dietary modifications, enteral and parenteral nutrition support, nutrition supplements, and education of the patient and family or caretaker. Newer medical, surgical, and nutritional approaches continue to improve the survival and quality of life in patients with inflammatory bowel disease and short bowel syndrome. However, despite the numerous investigations already performed related to nutrition in inflammatory bowel disease and short bowel syndrome, knowledge and practice of nutrition in treatment of these disorders are still in their infancy.

References

1. Strober, W., Fuss, I., and Mannon, P. (2007). The fundamental basis of inflammatory bowel disease. *J. Clin. Invest.* **117**, 514–521.
2. Baumgart, D., and Sandborn, W. J. (2007). Inflammatory bowel disease: Clinical aspects and established and evolving therapies. *Lancet* **369**, 1641–1657.
3. Beattie, R. M., Croft, N. M., Fell, J. M., Afzal, N. A., and Heuschkel, R. B. (2006). Inflammatory bowel disease. *Arch. Dis. Child.* **91**, 426–432.
4. Lukas, M., Bortlik, M., and Maratka, Z. (2006). What is the origin of ulcerative colitis? Still more questions than answers. *Postgrad. Med. J.* **82**, 620–625.
5. Caprilli, R., Gassal, M. A., Escher, J. C., Moser, G., Munkholm, P., Forbes, A., Hommes, D. W., Lochs, H., Angelucci, E., Cocco, A., Vucelic, B., Hildebrand, H., Kolacek, S., Riis, L., Lukas, M., de Franchis, R., Hamilton, M., Jantschek, G., Michetti, P., O'Morain, C., Anwar, M. M., Freitas, J. L., Mouzas, I. A., Faert, F., Mitchell, R., and Hawkey, C. J. (2006). European evidence based consensus on the diagnosis and management of Crohn's disease: Special situations. *Gut* **55**, 36–58.
6. Cottone, M., Orlando, A., and Modesto, I. (2006). Postoperative maintenance therapy for inflammatory bowel disease. *Curr. Opin. Gastroenterol.* **22**, 377–381.
7. Cima, R. R., and Pemberton, J. H. (2005). Medical and surgical management of chronic ulcerative colitis. *Arch. Surg.* **140**, 300–310.
8. Herrinton, L. J., Liu, L., Lafata, J. E., Allison, J. E., Andrade, S. E., Korner, E. J., Chan, K. A., Platt, R., Hiatt, D., and O'Connor, S. (2007). Estimation of the period of prevalence of inflammatory bowel disease among nine health plans using computerized diagnoses and outpatient pharmacy dispensings. *Inflamm. Bowel Dis.* **13**, 451–461.
9. Loftus, E. V. (2004). Clinical epidemiology of inflammatory bowel disease: Incidence, prevalence and environment. *Gastroenterology* **126**, 1504–1517.
10. Loftus, C. G., Loftus, E. V., Harmsen, W. S., Zinsmeister, A. R., Tremaine, W. J., Melton, L. J., and Sandborn, W. J. (2007). Update on the incidence and prevalence of Crohn's disease and ulcerative colitis in Olmsted County, Minnesota, 1940–2000. *Inflamm. Bowel Dis.* **13**, 354–361.
11. Graff, L. A., Walker, J. R., Lix, L., Clara, I., Rawsthorne, P., Rogala, I., Miller, N., Jakul, L., McPhail, C., Ediger, J., and Bernstein, C. N. (2006). The relationship of inflammatory bowel disease type and activity to psychological functioning and quality of life. *Clin. Gastroenterol. Hepatol.* **4**, 1491–1501.
12. Nicolas, D. B., Oatley, A., Smith, C., Avolio, J., Munk, M., and Griffiths, A. M. (2007). Challenges and strategies of children and adolescents with inflammatory bowel disease: A qualitative examination. *Health Qual. Life Outcomes* **5**, 28.
13. Norman, K., Kirchner, H., Lochs, H., and Pirlich, M. (2006). Malnutrition affects quality of life in gastroenterology patients. *World J. Gastroenterol.* **12**, 3380–3385.
14. Lawrence, T., and Gilroy, D. W. (2007). Chronic inflammation: A failure of resolution? *Int. J. Exp. Path.* **88**, 85–94.
15. Meneghin, A., and Hogaboam, C. M. (2007). Infectious disease, the innate immune response, and fibrosis. *J. Clin. Invest.* **117**, 530–538.
16. Baumgart, D. C., and Carding, S. R. (2007). Inflammatory bowel disease: Cause and immunobiology. *Lancet* **369**, 1627–1638.
17. Latinne, D., and Fiasse, R. (2006). New insights into the cellular immunology of the intestine in relation to the pathophysiology of inflammatory bowel diseases. *Acta Gastroenterol. Belg.* **69**, 393–405.
18. Thompson, A., Hemphill, D., and Jeejeebhooy, K. N. (1998). Oxidative stress and antioxidants in intestinal disease. *Dig. Dis.* **16**, 152–158.
19. Danese, S., and Fiocchi, C. (2006). Etiopathogenesis of inflammatory bowel diseases. *World J. Gastroenterol.* **12**, 4807–4812.
20. Korzenik, J. R. (2005). Past and current theories of etiology of IBD. Toothpaste, worms and refrigerators. *J. Clin. Gastroenterol.* **39**, S59–S65.
21. Hunter, J. O. (1998). Nutritional factors in inflammatory bowel disease (comment). *Eur. J. Gastroenterol. Hepatol.* **10**, 235–237.
22. Cashman, K. D., and Shanahan, F. (2003). Is nutrition an aetiological factor for inflammatory bowel disease? *Eur. J. Gastroenterol. Hepatol.* **15**, 607–613.
23. Schneider, J. C. (2007). Can microparticles contribute to inflammatory bowel disease: Innocuous or inflammatory? *Exp. Biol. Med. (Maywood)* **232**, 107–117.
24. Burrin, D. G., and Stoll, B. (2007). Emerging aspects of gut sulfur amino acid metabolism. *Curr. Opin. Nutr. Metab. Care* **10**, 63–68.
25. Danese, S., Sgambato, A., Papa, A., Scaldaferri, F., Pola, R., Sans, M., Lovecchio, M., Gasbarrini, G., Citadini, A., and Gasbarrini, A. (2005). Homocysteine triggers mucosal microvascular activations in inflammatory bowel disease. *Am. J. Gastroenterol.* **100**, 886–895.
26. Stio, M., Martinesi, M., Bruni, S., Treves, C., d'Albasio, G., Bagnoli, S., and Bonanomi, A. G. (2006). Interaction among vitamin D_3 analogue KH 1060, TNF-alpha, and vitamin D receptor protein in peripheral blood mononuclear cells of inflammatory bowel disease patients. *Int. Immunopharmacol.* **6**, 1083–1092.
27. Rose, D. J., DeMeo, M. T., Keshavarzian, A., and Hamaker, B. R. (2007). Influence of dietary fiber on inflammatory bowel disease and colon cancer: Importance of fermentation pattern. *Nutr. Rev.* **65**, 51–62.
28. Russel, M. G., Engels, L. G., Muris, J. W., Limonard, C. B., Volovics, A., Brummer, R. J., and Stockbrugger, R. W. (1998). Modern life in the epidemiology of inflammatory bowel disease: A case control study with special emphasis on nutritional factors. *Eur. J. Gastroenterol. Hepatol.* **10**, 243–249.
29. Joachim, G. (1999). The relationship between habits of food consumption and reported reactions to food in people with inflammatory bowel disease—testing the limits. *Nutr. Health* **13**, 69–83.
30. Magee, E. A., Edmond, L. M., Tasker, S. M., Kong, S. C., Curno, R., and Cummings, J. H. (2005). Associations between diet and disease activity in ulcerative colitis patients using a novel method of data analysis. *Nutr. J.* **4**, 7.
31. Shoda, R., Matsueda, K., Yamato, S., and Umeda, N. (1996). Epidemiologic analysis of Crohn disease in Japan: Increased

dietery intake of n-6 polyunsaturated fatty acids and animal protein relates to the increased incidence in Japan. *Am. J. Clin. Nutr.* **63**, 741–745.

32. Geerling, B. J., v-Houwelingen, A. C., Badart-Smook, A., Stockbrugger, R. W., and Burmmer, R. J. (1999). Fat intake and fatty acid profile in plasma phospholipids and adipose tissue in patients with Crohn's disease. *Am. J. Gastroenterol.* **94**, 410–417.
33. Calder, P. C. (2006). Polyunsaturated fatty acids and inflammation. *Prostaglandins Leukot. Essent. Fatty Acids* **96**, 93–99.
34. Shaikh, S. R., and Edidin, M. (2006). Polyunsaturated fatty acids, membrane organization, T cells and antigen presentation. *Am. J. Clin. Nutr.* **84**, 1277–1289.
35. MacLean, C. H., Mojica, W. A., Newberry, S. J., Pencharz, J., Garland, R. H., Tu, W., Hilton, L. G., Gralnek, I. M., Rhodes, S., Khanna, P., and Morton. S. C. (2005). Systematic review of the effects of n-3 fatty acids in inflammatory bowel disease. *Am. J. Clin. Nutr.* **82**, 611–619.
36. Kuroki, F., Iida, M., Matsumoto, T., Aoyagi, K., Kanamoto, K., and Fujishima, M. (1997). Serum N3 polyunsaturated fatty acids are depleted in Crohn's disease. *Dig. Dis. Sci.* **42**, 1137–1141.
37. Turner, D., Zlotkin, Sh., Shah, P., and Griffiths, A. Omega 3 fatty acids (fish oil) for maintenance of remission in Crohn's disease. Cochrane Database Syst. Rev. CDOO6320.
38. Badinga, L., and Greene, E. S. (2006). Physiological properties of conjugated linoleic acid and implications for human health. *Nutr. Clin. Prac.* **21**, 367–372.
39. O'Shea, M., Bassaganya-Riera, J., and Mohede, I. C. (2004). Immunomodulatory properties of conjugated linoleic acid. *Am. J. Clin. Nutr.* **79**, 1199S–1206S.
40. Kapoor, R., and Huang, Y. S. (2006). Gamma linolenic acid: An anti-inflammatory omega-6 fatty acid. *Curr. Pharm. Biotechnol.* **7**, 531–534.
41. Persson, P. G., Ahlbom, A., and Hellers, G. (1992). Diet and inflammatory bowel disease: A case-control study. *Epidemiology* **3**, 47–52.
42. Reif, S., Klein, I., Farbstein, M., Hallak, A., and Gilat, T. (1997). Pre-illness dietary factors in inflammatory bowel disease. *Gut* **40**, 754–760.
43. Jones, V. A., Dickinson, R. J., Workman, E., Wilson, A. J., Freeman, A. H., and Hunter, J. O. (1985). Crohn's disease: Maintenance of remission by diet. *Lancet* **2**, 177–180.
44. Hallert, C., Kaldma, M., and Petersson, B. G. (1991). Ispagula husk may relieve gastrointestinal symptoms in ulcerative colitis in remission. *Scan J. Gastroenterol.* **7**, 757–750.
45. Surh, Y. J., Chun, K. S., Cha, H. H., Han, S. S., Keum, Y. S., Park, K. K., and Lee, S. S. (2001). Molecular mechanisms underlying chemopreventive activities of anti-inflammatory phytochemicals: Down-regulation of COX-2 and iNOS through suppression of NF-kappa B activation. *Mutat. Res.* **480–482**, 243–268.
46. Heber, D. (2004). Vegetables, fruits and phytoestrogens in the prevention of diseases. *J. Postgrad. Med.* **50**, 145–149.
47. Jacobs, D. R., Anderson. L. F., and Blomhoff, R. (2007). Whole-grain consumption is associated with a reduced noncardiovascular, noncancer death attributed to inflammatory diseases in the Iowa Women's Health Study. *Am. J. Clin. Nutr.* **85**, 1606–1614.
48. Seaman, D. R. (2002). The diet-induced proinflammatory state: A cause of chronic pain and other degenerative diseases? *J. Manipulative Physiol. Ther.* **25**, 168–179.
49. Rose, D. J., DeMeo, M. T., Keshavarzian, A., and Hamaker, B. R. (2007). Influence of dietary fiber on inflammatory bowel disease and colon cancer: Importance of fermentation pattern. *Nutr. Rev.* **65**, 51–62.
50. Siebold, F. (2005). Food induced immune responses as origin of bowel disease? *Digestion* **71**, 251–260.
51. Macdonald, T. T., and Monteleone, G. (2005). Immunity, inflammation and allergy in the gut. *Science* **307**, 1920–1925.
52. Wittin, B. M., and Zeitz, M. (2003). The gut as an organ of immunology. *Int. J. Colorectal Dis.* **18**, 181–187.
53. Winkler, P., Ghadimi, D., Schrezenmeir, J., and Kraehenbuhl, J. P. (2007). Molecular and cellular basis of microflora–host interactions. *J. Nutr.* **137**, 756S–772S.
54. Bichoff, S. C., Herrmann, A., and Manns, M. P. (1996). Prevalence of adverse reactions to food in patients with gastrointestinal disease. *Allergy* **51**, 811–818.
55. Sampson, H. A. (1997). Food allergy. *JAMA* **278**, 1888–1894.
56. Huber, A., Genser, D., Spitzauer, S., Scheiner, O., and Jensen-Jarolim, E. (1998). *Int. Arch. Allergy Immunol.* **115**, 67–72.
57. Mowat, A. M., and Viney, J. L. (1997). The anatomical basis of intestinal immunity. *Immunol. Rev.* **156**, 145–166.
58. Gassull, M. A. (2004). Review article: Role of nutrition in the treatment of inflammatory bowel disease. *Aliment. Pharmacol. Ther.* **4**, 79S–83S.
59. McClave, S. A. (2007). Nutritional assessment in inflammatory bowel disease: Application of nutrition strategies to the management of the difficult Crohn's patient. *Am. J. Gastroenterol.* **102**, 88S–93S.
60. Wiskin, A. E., Wootton, S. A., and Beattie, R. M. (2007). Nutrition issues in pediatric Crohn's disease. *Nutr. Clin. Pract.* **22**, 214–222.
61. Kleinman, R. E., Baldassano, R. N., Caplan, A., Griffiths, A. M., Heyman, M. B., Issenman, R. M., and Lake, A. M. (2004). Nutrition support for pediatric patients with inflammatory bowel disease: A clinical report of the North American Society of Pediatric Gastroenterology, Hepatology and Nutrition. *J Pediatr. Gastroenterol. Nutr.* **39**, 15–27.
62. Goh, J., and O'Morain, C. A. (2003). Nutrition and adult inflammatory bowel disease. *Aliment. Pharmacol. Ther.* **17**, 317–320.
63. Rigaud, D., Angel, L. A., Carduner, M. C. M. J., Melchior, J. C., Sautier, C., Rene, E., Apfelbaum, M., and Mignon, M. (1994). Mechanisms of decreased food intake during weight loss in adult Crohn's patients without obvious malabsorption. *Am. J. Clin. Nutr.* **60**, 775–781.
64. Schneeweiss, B., Lochs, H., Zauner, C., Fischer, M., Wyatt, J., Maier-Dobersberger, T., and Schneider, B. (1999). Energy and substrate metabolism in patients with active Crohn's disease. *J. Nutr.* **129**, 844–848.
65. Plata-Salaman, C. R. (1998). Cytokine-induced anorexia. Behavorial cellular, and molecular mechanisms. *Ann. N. Y. Acad. Sci.* **856**, 160–170.
66. Schiller, L. R. (2006). Nutrition management of chronic diarrhea and malabsorption. *Nutr. Clin. Pract.* **21**, 34–39.
67. Sabol, V. K., and Carlson, K. K. (2007). Diarrhea. Applying research to bedside practice. *AACN Adv. Crit. Care.* **18**, 32–44.

68. Spiller, R. (2006). Role of motility in chronic diarrhea. *Neurogastroenterol. Motil.* **18**, 1045–1055.
69. Yan, F., and Polk, D. B. (2006). Probiotics as functional foods in the treatment of diarrhea. *Curr. Opin. Clin. Nutr. Metab. Care.* **9**, 717–712.
70. Solomons, N. W. (1983). Micronutrient deficiencies in inflammatory bowel disease. *Clin. Nutr.* **2**, 19–25.
71. Rath, H. C., Caesar, I., Roth, M., and Scholmerich, J. (1998). Nutritional deficiencies and complications in chronic inflammatory bowel disease. *Med. Klin.* **93**, 6–10.
72. Lim, W. C., Hanauer, S. B., and Li, Y. C. (2005). Mechanisms of disease: Vitamin D and inflammatory bowel disease. *Nat. Clin. Pract. Gastroenterol. Hepatol.* **2**, 308–315.
73. Reifen, R., Nut, T., Ghebermeskel, K., Zaiger, G., Urizky, R., and Pines, M. (2002). Vitamin A deficiency exacerbates inflammation in a rat model of colitis through activation of nuclear factor-κB and collagen formation. *J. Nutr.* **132**, 2743–2747.
74. Zeigler, T. R., Evans, M. E., Fernandez-Estivariz, C., and Jones, D. P. (2003). Trophic and cytoprotective nutrition for intestinal adaptation, mucosal repair, and barrier function. *Ann. Rev. Nutr.* **23**, 229–261.
75. Kulnigg, S., and Gasche, C. (2006). Systematic review: Managing anemia in Crohn's disease. *Aliment. Pharmacol. Ther.* **24**, 1507–1523.
76. Giannini, S., and Martes, C. (2006). Anemia in inflammatory bowel disease. *Minerva Gastroenterol. Dietol.* **52**, 275–291.
77. Irvine, E. J. (2004). Review article: Patients' fears and unmet needs in inflammatory bowel disease. *Aliment. Pharmacol. Ther.* **4**, 54S–59S.
78. Travis, S. P. L., Strange, E. F., Lemann, M., Oresland, T., Chowers, Y., Forbes, A., D'Haens, G., Kitis, G., Carlot, A., Prantera, C., Marteau, P., Colombel, J. F., Gionchetti, P., Bouhnik, Y., Tiret, E., Krosen, J., Starlinger, M., and Mortensen, N. J. (2006). European evidence based consensus on the diagnosis and management of Crohn's Disease: Current management. *Gut.* **55**, 16S–35S.
79. Collins, P., and Rhodes, J. (2006). Ulcerative colitis: Diagnosis and management. *BMJ* **333**, 340–343.
80. Nayar, M., and Rhodes, J. M. (2004). Management of inflammatory bowel disease. *Postgrad. Med. J.* **80**, 206–213.
81. Bamias, G., and Cominelli, F. (2006). Novel strategies to attenuate immune activation in Crohn's disease. *Curr. Opin. Pharmacol.* **6**, 401–407.
82. Kosaka, T., Sawada, K., Ohnishi, K., Egashira, A., Yamamura, M., Tanida, N., Satomi, M., and Shimoyama, T. (1999). Effect of leukocytapheresis therapy using a leukocyte removal filter in Crohn's disease. *Intern. Med.* **38**, 102–111.
83. Ljung, T., Thomsen, O. O., Vatn, M., Karlen, P., Karlsen, L. N., Tysk, C. Nilsson, S. U., Kilander, A., Gillberg, R., Grip, O., Lindgren, S., Befrits, R., and Lofberg, R. (2007). Granulocyte, monocyte/macrophage apheresis for inflammatory bowel disease: The first 100 patients treated in Scandinavia. *Scand. J. Gastroenterol.* 42, 221–227.
84. Hancock, L., Windsor, A. C., and Mortensen, N. J. (2006). Inflammatory bowel disease: The view of the surgeon. *Colorectal Dis.* **1**, 10S–14S.
85. O'Reilly Brown, M. (1999). Inflammatory bowel disease. *Primary Care* **26**, 141–171.
86. Becker, J. M. (1999). Surgical therapy for ulcerative colitis and Crohn's disease. *Gastroenterol Clin. North Am.* **28**, 371–390.
87. Cohen, R. D., Broadski, A. L., and Hanauer, S. B. (1999). A comparison of quality of life in patients with severe ulcerative colitis after total colectomy versus medical treatment with intravenous cyclosporine. *Inflamm. Bowel Dis.* **5**, 1–10.
88. Lichtenstein, G. R., Cohen, R., Yamashita, B., and Diamond, R. H. (2006). Quality of life after proctocolectomy with ileoanal anastomosis for patients with ulcerative colitis. *J. Clin. Gastroenterol.* **40**, 669–677.
89. Vagianos, K., Bector, S., McConnell, J., and Bernstein, C. N. (2007). Nutrition assessment of patients with inflammatory bowel disease. *JPEN J. Parenter. Enteral Nutr.* **31**, 311–319.
90. Geerling, B. J., Badart-Smook, A., Stockbrugger, R. W., and Brummer, R. J. M. (1998). Comprehensive nutritional status in patients with long-standing Crohn disease currently in remission. *Am. J. Clin. Nutr.* **67**, 919–926.
91. Fischer, J. E., Foster, G. S., Abel, R. M., Abbott, W. M., and Ryan, J. A. (1973). Hyperalimentation as primary therapy for inflammatory bowel disease. *Am. J. Surg.* **125**, 165–175.
92. Reilly, J., Ryan, J. A., Strole, W., and Fischer, J. E. (1976). Hyperalimentation in inflammatory bowel disease. *Am. J. Surg.* **131**, 192–200.
93. Zurita, V. F., Rawls, D. E., and Dyck, W. P. (1995). Nutritional support in inflammatory bowel disease. *Dig. Dis.* **13**, 92–107.
94. Wild, G. E., Drozdowski, L., Tartaglia, C., Clandinin, M. T., and Thompson, A. B. R. (2007). Nutritional modulation of the inflammatory response in inflammatory bowel disease—from the molecular to the integrative to the clinical. *World J. Gastroenterol.* **13**, 1–7.
95. Sitrin, M. D. (1992). Nutrition support in inflammatory bowel disease. *Nutr. Clin. Pract.* **7**, 53–60.
96. Messori, A., Trallori, G., D'Albasio, G., Milla, M., and Pacini, F. (1996). Defined-formula diets versus steroids in the treatment of active Crohn's disease: A meta-analysis. *Scand. J. Gastroenterol.* **31**, 267–272.
97. Heuschkel, R. B., and Walker-Smith, J. A. (1999). Enteral nutrition in inflammatory bowel disease of childhood. *JPEN J. Parenter. Enteral Nutr.* **23**, S29–S32.
98. Han, P. D., Burke, A., Baldassano, R. N., Rombeau, J. L., and Lichtenstein, G. R. (1999). Nutrition and inflammatory bowel disease. *Gastroenterol. Clin. North Am.* **28**, 423–443.
99. Razack, R., and Seidner, D. L. (2007). Nutrition in inflammatory bowel disease. *Curr. Opin. Gastroenterol.* **23**, 400–405.
100. Zachos, M., Tondeur, M., and Griffiths, A. M. (2007). Enteral nutritional therapy for induction of remission in Crohn's disease (review). *Cochrane Database of Systematic Reviews Issue 1*, art. CD00542.
101. Griffiths, A. M. (2006). Enteral feeding in inflammatory bowel disease. *Curr. Opin. Nutr. Metab. Care* **9**, 314–318.
102. Lochs, H., Dejong, C., Hammarqvist, F., Hebuterne, X., Leoj-Sanz, M., Schytz, T., van Gemert, W., van Gossum, A., Valentini, L., Lubke, H., Bischoff, S., Engelmann, N., and Thul, P. (2006). ESPEN guidelines on enteral nutrition: Gastroenterology. *Clin. Nutr.* **25**, 260–274.

103. Su, C., Lewis, J. D., Goldberg, B., Brensinger, C., and Lichtenstein, G. R. (2007). A meta-analysis of the placebo rates of remission and response in clinical trials of active ulcerative colitis. *Gastroenterology* **132**, 516–526.
104. Su, C., Lichtenstein, G. R., Krok, K., Lewis, J. D., Brensinger, C. M., and Goldberg, B. (2004). A meta-analysis of the placebo rates of remission and response in clinical trials of active Crohn's disease. *Gastroenterology* **125**, 1257–1269.
105. Ewaschuk, J. B., and Dieleman, L. A. (2006). Probiotics and prebiotics in chronic inflammatory bowel disease. *World J. Gastroenterol.* **12**, 5941–5950.
106. Bai, A. P., and Quyang, Q. (2006). Probiotics and inflammatory bowel disease. *Postgrad. Med. J.* **82**, 376–382.
107. Rioux, K. P., and Fedorak, R. N. (2006). Probiotics in the treatment of inflammatory bowel disease. *J. Clin. Gastroenterol.* **40**, 260–263.
108. Jones, J. L., and Foxx-Orenstein, A. E. (2007). The role of probiotics in inflammatory bowel disease. *Dig. Dis. Sci.* **106**, 607–611.
109. Chapnan, T. M., Plosker, G. L., and Figgitt, D. P. (2007). Spotlight on VSL#3 probiotic mixture in chronic inflammatory bowel diseases. *BioDrugs* **21**, 61–63.
110. Breuer, R. I., Soergel, K. H., Lashner, B. A., Christ, M. L., Hanauer, S. B., Vanagunas, A., Harig, J. M., Keshavarzian, A., Robinson, M., Sellen, J. M., Weinberg, D., Vidican, D. E., Flemal, K. L., and Rademaker, A. W. (1997). Short chain fatty acid rectal irrigation for left-sided ulcerative colitis: A randomized, placebo controlled trial. *Gut.* **40**, 485–491.
111. Andoh, A., Bamba, T., and Sasaki, M. (1999). Physiologic roles of dietary fiber and butyrate in intestinal functions. *JPEN J. Parenter. Enteral Nutr.* **23**, S70–S73.
112. Wong, J. M. W., De Souza, R., Kendall, C. W. C., Emam, A., and Jenkins, D. J. A. (2006). Colonic health: Fermentation and short chain fatty acids. *J. Clin. Gastroenterol.* **40**, 235–243.
113. Gassull, M. A. (2006). Review article: The intestinal lumen as a therapeutic target in inflammatory bowel disease. *Aliment. Pharmacol. Ther.* **24**, 90S–95S.
114. Hanai, H., Kanauchi, O., Mitsuyama, K., Andoh, A., Takeuchi, K., Takayuki, I., Araki, Y., Fujiyama, Y., Tononaga, A., Sata, M., Kojima, A., Fukuda, M., and Bamba, T. (2004). Germinated barley foodstuff prolongs remission in patients with ulcerative colitis.. *Int. J. Mol. Med.* **13**, 642–7.
115. Galvez, J., Rodriguez-Cabezas, M. E., and Zarzuelo, A. (2005). Effects of dietary fiber on inflammatory bowel disease. *Mol. Nutr. Food Res.* **49**, 601–608.
116. Belluzzi, A., Boschi, S.. Brignola, C., Munarini, A., Carinani, G., and Miglio, F. (2000). Polyunsaturated fatty acids and inflammatory bowel disease. *Am. J. Clin. Nutr.* **71**, 339S–342S.
117. Kris-Etherton, P. M., Taylor, D. S., Yu-Poth, S., Huth, P., Moriarty, K., Fishell, V., Harbrove, R. L., Zhao, G., and Etherton, T. D. (2000). Polyunsaturated fatty acids in the food chain in the United States. *Am. J. Clin. Nutr.* **71**, 179S–188S.
118. Simopoulos, A. P. (1999). Essential fatty acids in health and chronic disease. *Am. J. Clin. Nutr.* **70**(suppl), 560S–569S.
119. Das, U. N. (2006). Essential fatty acids: Biochemistry, physiology and pathology. *Biotechnol. J.* **1**, 420–439.
120. Mancuso, P., Whelan, J., DeMichele, S. J., Snider, C. C., Guszcza, J. A., and Karlstad, M. D. (1997). Dietary fish oil and fish and borage oil suppress intrapulmonary proinflammatory eicosanoid biosynthesis and attenuate pulmonary neutrophil accumulation in endotoxic rats. *Crit. Care. Med.* **25**, 1198–1206.
121. Sugano, M., Tsujita, A., Yamasaki, M., Noguchi, M., and Yamada, K. (1998). Conjugated linoleic acid modulates tissue levels of chemical mediators and immunoglobulins in rats. *Lipids.* **33**, 521–527.
122. Jeppeson, P. B., and Mortensen, P. B. (1999). Colonic digestion and absorption of energy from carbohydrates and medium-chain triglycerides in small bowel failure. *JPEN J. Parenter. Enteral Nutr.* **23**(Suppl), S101–S105.
123. Emery, E. A., Ahmad, S., Koethe, J. D., Skipper, A., Peerlmutter, S., and Paskin, D. L. (1997). Banana flakes control diarrhea in enterally fed patients. *Nutr. Clin. Pract.* **12**, 72–75.
124. Food and Nutrition Board. (2005). Dietary, functional and total fiber. *In* "Dietary Reference Intakes for Energy, Carbohydrate, Fiber, Fat, Fatty Acids, Cholesterol, Protein and Amino Acids (Macronutrients)." National Academies Press, Washington, DC.
125. Hyams, J. (1983). Sorbitol intolerance: An unappreciated cause of functional gastrointestinal complaints. *Gastroenterology* **84**, 30–33.
126. Beyer, P. L., Caviar, E. M., and McCallum, R. W. (2005). Fructose intake at current levels in the United States may cause gastrointestinal distress in normal adults. *J. Am. Diet. Assoc.* **105**, 1559–1566.
127. Wright, J. D., Wang, C. Y., Kennedy-Stephenson, J., and Ervin, R. B. (2003). Dietary intake of ten key nutrients for pubic health, United States: 1999–2000. Advance Data from Vital and Health Statistics, No. 334, National Center for Health Statistics, Washington, DC.
128. Ervin, R. B., Wright, J. D., Wang, C. Y., and Kennedy-Stephenson, J. (2004). Dietary intake of selected vitamins for the United States population, 1999–2000. Advance Data from Vital and Health Statistics, No. 339. National Center for Health Statistics, Washington, DC.
129. The Report of the Dietary Guidelines Advisory Committee on Dietary Guidelines for Americans (2005). Retrieved August 6, 2007, from http://www.health.gov/dietaryguidelines/dga2005/report/default.htm.
130. Nightengale, J., and Woodward, J. M. (2006). Guidelines for management of patients with a short bowel. *Gut* **55**, 1–12.
131. Gupte, G. L., Beath, S. V., Kelly, D. A., Millar, A. J. W., and Booth, I. W. (2006). Current issues in the management of intestinal failure. *Arch. Dis. Child.* **91**, 256–264.
132. Carbonnel, F., Cosnes, J., Chevret, S., Beaugerie, L., Ngo, Y., Malafosse, M., Rolland, P., Quintrec, Y., and Gendre, J. P. (1996). The role of anatomic factors in nutritional autonomy after extensive small bowel resection. *JPEN J. Parenter. Enteral Nutr.* **20**, 275–279.
133. Messing, B., Crenn, P., Beau, P., Boutron-Rualt, M. C., Rambaud, J. C., and Matchuansky, C. (1999). Long-term survival and parenteral nutrition dependence in adult patients with the short bowel syndrome. *Gastroenterology* **117**, 1043–1050.

134. Matarese, L. E., and Steiger, E. (2006). Dietary and medical management of short bowel syndrome in adult patients. *J. Clin. Gastroenterol* **40**, S85–S93.
135. Vanderhoof, J. A., Young, R. J., Murray, N., and Kaufman, S. S. (1998). Treatment strategies for small bowel bacterial overgrowth in short bowel syndrome. *J. Pediatr. Gastroenterol. Nutr.* **27**, 155–160.
136. Keller, J., Panter, H., and Layer, P. (2004). Management of the short bowel syndrome after extensive small bowel resection. *Best Pract. Res. Clin. Gastroenterol.* **18**, 977–992.
137. Drozdowski, L., and Thompson, ABR. (2006). Intestinal mucosal adaptation. *World J. Gastroenterol.* **12**, 4614–4627.
138. Pereira, P. M., and Bines, J. E. (2006). New growth factor therapies aimed at improving intestinal adaptation in short bowel syndrome. *J. Gastroenterol. Hepatol.* **21**, 932–940.
139. Yang, H., and Teitelbaum, D. H. (2006). Novel agents in the treatment of intestinal failure: Humoral factors. *Gastroenterology* **130**, S117–S121.
140. Jeppesen, P. B., Sanguinetti, E. L., Buchman, A., Howard, L., Scolapio, J. S., Ziegler, T. R., Gregory, J., Tappenden, K. A., Holst, J., and Mortensen, P. B. (2005). Tedglutide (ALX-0600), a dipeptidyl peptidase IV resistant glucagon-like peptide 2 analogue, improves intestinal function in short bowel syndrome. *Gut.* **54**, 1224–1231.
141. Nordgaard, I., Hansen, B. S., and Mortensen, P. B. (1996). Importance of colonic support for energy absorption as small-bowel failure proceeds. *Am. J. Clin. Nutr.* **64**, 222–231.
142. Jeppesen, P. B., and Mortensen, P. B. (1998). The influence of a preserved colon on the absorption of medium chain fat in patients with small bowel resection. *Gut.* **43**, 478–483.
143. Weaver, L. T., Austin, S., and Cole, T. J. (1991). Small intestinal length: A factor essential for gut adaptation. *Gut.* **32**, 1321–23.
144. Goulet, O. (1998). Short bowel syndrome in pediatric patients. *Nutrition* **14**, 784–787.
145. Biasco, G., Callegari, C., Lami, F., Minarini, A., Miglioli, M., and Barbara, L. (1984). Intestinal morphological changes during oral refeeding in a patient previously treated with total parenteral nutrition for small bowel resection. *Am. J. Gastroenterol.* **79**, 585–588.
146. Shin, C. S., Chaudhry, A. G., Khaddam, M. H., Penha, P. D., and Dooner, R. (1980). Early morphologic changes in the intestine following massive resection of the small intestine and parenteral nutrition therapy. *Surg. Gynecol. Obstet.* **151**, 246–250.
147. Solhaug, H. (1976). Morphometric studies of the small intestine following jejunoileal shunt operation. *Scand. J. Gastroent.* **11**, 155–160.
148. Wilmore, D. W. (1999). Growth factors and nutrients in the short bowel syndrome. *JPEN J. Parenter. Enteral Nutr.* **23**, S117–S120.
149. Beyer, P. L. Short-Bowel Syndrome. (1998). *In* "Dietitian's Handbook of Enteral and Parenteral Nutrition". (A. Skipper, Ed.), 2nd ed. Aspen, Gaithersburg, MD.
150. Dibaise, J. K., Matarese, L. E., Messing, B., and Steiger, E. (2006). Strategies for parenteral nutrition weaning in adult patients with short bowel syndrome. *J. Clin. Gastroenterol.* **40**, S94–S98.
151. Arrigoni, E., Marteau, P., Briet, F., Pochart, P., Rambaud, J. C., and Messing, B. (1994). Tolerance and absorption of lactose from milk and yogurt during short-bowel syndrome in humans. *Am. J. Clin. Nutr.* **60**, 926–929.
152. Crenn, P., Morin, M. C., Joly, F., Penven, S., Thuilleir, F., and Messing, B. (2004). Net digestive absorption and adaptive hyperphagia in adult short bowel patients. *Gut.* **53**, 1279–1286.
153. Scolapio, J. S. (1999). Effect of growth hormone, glutamine, and diet on body composition in short bowel syndrome: A randomized, controlled study. *JPEN J. Parenter. Enteral Nutr.* **23**, 309–313.
154. Vanderhoof, J. A., Kollman, K. A., Griffin, S., and Adrian, T. E. (1997). Growth hormone and glutamine do not stimulate intestinal adaptation following massive small bowel resection in the rat. *J. Pediatr. Gastroenterol. Nutr.* **25**, 327–331.
155. Messing, B., Blethen, S., Dibaise, J. K., Matarese, L. E., and Steiger, E. (2006). Treatment of adult short bowel patients with recombinant human growth hormone: A review of clinical studies. *J. Clin. Gastroenterol.* **40**, 75S–84S.
156. Matarese, L. E., Seidner, D. L., and Steiger, E. (2004). Growth hormone, glutamine and modified diet for intestinal adaptation. *J. Am. Diet. Assoc.* **104**, 1265–1272.
157. Bragg, L. E., Thompson, J. S., and Rikkers, L. F. (1991). Influence of nutrient delivery on gut structure and function. *Nutrition* **7**, 237–243.
158. Bengmark, S. (2000). Colonic food: Pre- and probiotics. *Am. J. Gastroenterol.* **95**(Suppl), S5–S7.
159. Scheppach, W. (1994). Effects of short chain fatty acids on gut morphology and function. *Gut.* **35**(Suppl), S35–S38.
160. Nightingale, J.M.D. (1999). Management of a patient with short bowel syndrome. *Nutrition* **15**, 633–637.
161. Thompson, J. S., and Langnas, A. N. (1999). Surgical approaches to improving intestinal function in the short-bowel syndrome. *Arch. Surg.* **134**, 706–709.
162. Chang, R. W., Javid, P. J., Oh, J. T., Andreoli, S., Kim, H. B., Fauza, D., and Jaksic, T. (2006). Serial transverse enteroplasty enhances intestinal function in a model of short bowel syndrome. *Ann. Surg.* **243**, 223–228.
163. DeLegge, M., Alsolaiman, M. M., Barbour, E., Bassas, S., Siddiqi, M. F., and Moore, N. W. (2007). Short bowel syndrome: Parenteral nutrition versus intestinal transplantation. Where are we today? *Dig. Dis. Sci.* **52**, 87–892.
164. Selvaggi, G., and Tzakis, A. G. (2007). Intestinal and multivisceral transplantation: Future perspectives. *Front. Biosci.* **12**, 4742–4754.

CHAPTER **40**

Nutrient Considerations in Lactose Intolerance

STEVE HERTZLER,[1] DENNIS A. SAVAIANO,[2] KARRY A. JACKSON,[2] AND FABRIZIS L. SUAREZ[1]

[1]*Abbort Laboratories, Ross Products Division, Columbus, Ohio*

[2]*Purdue University, West Lafayette, Indiana*

Contents

I. INTRODUCTION

Ingestion of a large single dose of lactose (e.g., 50 g, the quantity in a quart of milk) by lactose maldigesters commonly results in diarrhea, bloating, and flatulence [1]. The wide dissemination of this information has led some of the lay population and a fraction of the medical community to attribute common gastrointestinal symptoms to lactose intolerance, independent of the dose of lactose ingested. As a result, a segment of the population avoids dairy products in the belief that even trivial doses of lactose will induce diarrhea or gas. However, multiple factors affect the ability of lactose to induce perceptible symptoms. These factors include residual lactase activity [2], gastrointestinal transit time [3], lactose consumed with other foods [4], lactose load [5], and colonic fermentation [6]. The estimated 25% of adults in the United States who maldigest lactose are composed mainly of the Hispanic, Asian, and African American populations (Table 1). These race/ethnic groups are rapidly growing segments of the population. Thus, the overall number of lactose maldigesters will grow in the United States in coming years. A major challenge for diet therapy of lactose maldigesters is to ensure adequate intakes of calcium, vitamin D, and other nutrients found largely in dairy products while, at the same time, minimizing the occurrence of lactose intolerance symptoms that would tend to limit milk consumption.

This chapter (1) reviews the pathophysiology of lactose maldigestion, (2) attempts to correct common misconceptions concerning the frequency and severity of lactose intolerance symptoms, and (3) provides dietary strategies to minimize symptoms of intolerance.

II. LACTOSE IN THE DIET

Lactose is the primary disaccharide in virtually all mammalian milks. It is unique among the major dietary sugars because of the β-1→4 linkage between its component monosaccharides, galactose and glucose. Lactose production in nature is limited to the mammalian breast, which

TABLE 1 Prevalence of Lactase Nonpersistence in Various Populations

Group	Prevalence (%)
Northern European	2 to 15
American white	6 to 22
Central European	9 to 23
Indian (Indian subcontinent)	
Northern	20 to 30
Southern	60 to 70
Hispanic	50 to 80
Ashkenazi Jew	60 to 80
Black	60 to 80
American Indian	80 to 100
Asian	95 to 100

Source: Used with permission from Swagerty, D. L., Walling, A. D., and Klein, R. M. (2002). Lactose intolerance. *Am. Fam. Physician* **65**, 1845–1850.

contains the enzyme system (lactose synthase) necessary to create this linkage [7]. Human milk contains approximately 7% lactose by weight, which is among the highest lactose concentrations of all mammalian milks [5]. Cow's milk contains 4–5% lactose. Lactose, being water soluble, is associated with the whey portion of dairy foods. Thus, hard cheeses (with the whey removed from the curds) contain very little lactose compared with fluid milk (Table 2 shows the lactose content of selected foods).

In addition to food sources of lactose, small amounts of lactose are found in a wide variety of medications because of the excellent tablet-forming properties of lactose [5]. However, lactose is usually present in milligram, rather than gram, quantities in most medications and the amount is biologically insignificant for lactose maldigesters.

III. DIGESTION OF LACTOSE

The small intestine is normally impermeable to lactose. Lactose must first be hydrolyzed to glucose and galactose, which are subsequently absorbed. Inability to digest lactose is referred to as *lactose maldigestion.* Lactose digestion is dependent on the enzyme lactase-phlorizin hydrolase (LPH), a microvillar protein that has at least three enzyme activities: β-galactosidase, phlorizin hydrolase, and glycosylceramidase [8, 9]. Synthesis of LPH occurs in enterocytes, with the highest and most uniform synthesis being in the jejunum in humans [10]. The LPH gene is located on chromosome 2 and directs the synthesis of a pre-proLPH that is processed intracellularly (and possibly by pancreatic proteases) into the mature form that is anchored in the cell membrane at the brush border [11, 12]. Lactase activity develops late in gestation compared to other disaccharidases. Lactase activity in a fetus at 34 weeks is only 30% that of a full-term infant, rising to 70% of the full-term activity by 35–38 weeks [13].

TABLE 2 Lactose Content of Selected Foods

Product	Portion Size	Lactose Content (g/portion)
Milk, full fat	1 c. (244 g)	11
Milk, reduced fat (2%)	1 c. (244 g)	9–13
Milk, nonfat	1 c. (244 g)	12–14
Milk, chocolate	1 c. (244 g)	10–12
Buttermilk, fluid	1 c. (245 g)	9–11
Half and half	1 T. (15 g)	0.6
Yogurt, low fat	8 fl. oz. (227–258 g)	11–15
Cheese (blue, Camembert, cheddar, Colby, cream, Gouda, Limburger, grated Parmesan)	1 oz. (28 g)	0.1–0.8
Cheese, pasteurized processed (American, pimento, Swiss)	1 oz. (28 g)	0.4–1.7
Cottage cheese, whole	1 c. (210 g)	5–6
Cottage cheese, 2% fat	1 c. (226 g)	7–8
Butter	2 pats (10 g)	0.1
Ice cream, vanilla, regular	1 c. (133 g)	9
Ice milk, vanilla	1 c. (131 g)	10
Sherbet, orange	1 c. (193 g)	4

Source: Adapted with permission from Welsh, J. D. (1978). Diet therapy in adult lactose malabsorption: present practices. *Am. J. Clin. Nutr.* **31**, 592–596.

IV. LOSS OF LACTASE ACTIVITY

Full-term infants possess high lactase activity, except for *congenital lactase deficiency*, in which lactase is completely absent at birth. Holzel *et al.* [14] first described congenital lactase deficiency in 1959. It is a very rare condition, such that even in Finland, where it is most common, only 42 cases were diagnosed from 1966 to 1998 [11]. Lactase activity in jejunal biopsy specimens from infants with congenital lactase deficiency is reduced to 0–10 IU/g protein, and severe diarrhea results from unabsorbed lactose [11]. Treatment with a lactose-free formula eliminates symptoms and promotes normal growth and development [15].

Primary acquired hypolactasia, in which there is up to a 90–95% reduction in lactase activity, is much more common than congenital lactase deficiency (alactasia) [16]. The preferred term for this type of hypolactasia is *lactase non-persistence* (LNP). It is estimated that approximately 75% of the world's population are LNP (see Table 1), with the exception of Northern Europeans and a few pastoral tribes in Africa and the Middle East that maintain infantile levels of lactase throughout life [17]. Thus, LNP is not a "lactase deficiency" disease, but is the normal pattern in human physiology, similar to the physiology of other mammalian species. This permanent loss of lactase occurs sometime after 3–5 years of age [9, 18].

Lactase persistence is inherited as a highly penetrant, autosomal-dominant characteristic [17]. It has been hypothesized that individuals with a genetic mutation coding for lactase persistence would have gained a selective evolutionary advantage over LNP individuals in areas where dairy farming developed several thousand years ago [19, 20]. Under marginal nutritional conditions, the individual with lactase persistence would be able to comfortably consume dairy products, deriving greater nutritional benefit. A key question has been whether the lactase persistence mutation in humans occurred before (the reverse cause hypothesis) or after (the cultural-historical hypothesis) the advent of dairying. DNA studies on the skeletal remains of pre–dairy farming Neolithic Europeans [21] indicate that the most common allele for lactase persistence in Europeans (–13910*T) was not present, arguing for the cultural-historical hypothesis (this mutation is discussed in more detail in the section on diagnosis of lactose maldigestion). In addition, Tishkoff *et al.* [22] have identified three single-nucleotide polymorphisms (SNPs) for lactase persistence that developed in African populations as little as 3000 years ago. This date corresponds well with archeological data suggesting that cattle domestication came to different parts of Africa 3300 to 9000 years ago. These SNPs developed independently of the European mutation, providing striking evidence of both convergent evolution and the strong and relatively recent impact of a cultural practice such as dairy farming on the genome.

The genetic regulation of LPH has been studied extensively. Most evidence supports reduced levels of lactase mRNA in lactose maldigesters, suggesting that regulation is primarily at the level of transcription [23–26]. However, hypolactasia is sometimes present even when lactase mRNA is abundant, suggesting that posttranscriptional factors play a role [10, 27, 28]. One potential reason for conflicting results is the intestinal segment examined (duodenum versus jejunum). Lactase expression is higher and more uniform in the jejunum compared to the duodenum [29, 30]. Another potential discrepancy is the age of the subjects studied. A poor correlation between lactase mRNA and lactase activity was reported in intestinal biopsies from children, although the biopsy specimens in this study were from duodenal sites [28]. Lactase activity in the jejunal enterocytes is found in a "mosaic"-type pattern [31]. In hypolactasic individuals, some jejunal enterocytes produce high amounts of lactase while others, even those sharing the same villus, do not produce lactase [10]. Thus, rather than a uniform reduction in lactase production among all enterocytes, a hypolactasic individual may have a "patchy" distribution of lactose-producing enterocytes that are low in number relative to the non-lactase-producing enterocytes. In lactase-persistent individuals, all villus enterocytes may produce lactase. Current evidence suggests that the regulation of lactase is accomplished primarily at the level of transcription, although posttranscriptional factors (e.g., degradation of mRNA and posttranslational processing of the LPH protein) could be important in some individuals.

Secondary hypolactasia occurs as the result of damage to the enterocytes via disease, medications, surgery, or radiation to the gastrointestinal tract (Table 3) [5, 32, 33]. For example, the prevalence of microsporidiosis, which is associated with hypolactasia, can be as high as 50% in HIV-infected patients [34]. Seventy percent of HIV-infected patients showed evidence of lactose maldigestion compared to only 34% of controls [35]. In addition, the severity of lactose maldigestion increases in the more advanced stages of the disease. In general, secondary hypolactasia is reversible once the underlying cause is treated, but this reversal may require 6 months or more of diet therapy [5].

V. DIAGNOSIS OF LACTOSE MALDIGESTION

A. Genetic Testing

Historically, biochemical and/or symptom tests (described in parts B and C) have been used to diagnosis lactose maldigestion. In the past 5–10 years, however, the focus has been the development of genetic tests to identify markers of lactose maldigestion. These types of tests have

TABLE 3 Potential Causes of Secondary Hypolactasia

Diseases		
Small Bowel	**Multisystem**	**Iatrogenic**
HIV enteropathy	Carcinoid syndrome	Chemotherapy
Regional enteritis (e.g., Crohn's disease)	Cystic fibrosis	Radiation enteritis
Sprue (celiac and tropical)	Diabetic gastropathy	Surgical resection of intestine
Whipple's disease (intestinal lipodystrophy)	Protein energy malnutrition	Medications
Ascaris lumbricoides infection	Zollinger-Ellison syndrome	Colchicine (antigout)
Blind loop syndrome	Alcoholism	Neomycin (antibiotic)
Giardiasis	Iron deficiency	Kanamycin (antibiotic)
Infectious diarrhea		Aminosalicylic acid (antibiotic)
Short gut		

Sources: Adapted with permission from: Srinivasan, R., and Minocha, A. (1998). When to suspect lactose intolerance: symptomatic, ethnic, and laboratory issues. *Postgrad. Med.* **104**(3), 109–123; Scrimshaw, N. S., and Murray, E. B. (1998). The acceptability of milk and milk products in populations with a high prevalence of lactose intolerance. *Am. J. Clin. Nutr.* **48**, 1083–1159; and Savaiano, D. A., and Levitt, M. D. (1987). Milk intolerance and microbe-containing dairy foods. *J. Dairy Sci.* **70**, 397–406.

several advantages over present methods. First, the measurements can be done on buccal cell samples, which are rapidly and easily obtained and require little or no preparation on the part of the individual being tested. Second, because the symptoms of lactose intolerance often may be confused with other gastrointestinal disorders, such as irritable bowel syndrome or Crohn's disease, genetic testing would allow for the ready differentiation of lactose maldigestion (resulting from LNP) from other gastrointestinal conditions. Enattah *et al.* [36] have isolated two SNPs that are strongly associated with LNP in a primarily Finnish population. Both of these SNPs are located in a region adjacent to the lactase gene (LCT) on chromosome 2q21. The first, a substitution of cytosine for thymine 13,910 base pairs upstream of the 5′ end of LCT (termed C/T_{-13910}), was completely associated with biochemical evidence of LNP in 236 individuals. The second, g/A_{-22018}, was found in 229 of 236 cases. A subsequent study of intestinal biopsy samples showed that these SNPs coincided with low levels of mRNA for LPH consistent with transcriptional regulation [37]. It is likely that the C/T_{-13910} mutation impairs the binding of transcription factor Oct-1 [38]. The association between C/T_{-13910} and LNP was also observed in a study of children [39]. The presence of C/T_{-13910} had 100% specificity and 93% sensitivity in children older than 12 years when compared with intestinal biopsy. Although these findings caused a great deal of excitement over the possibility of widespread genetic testing for LNP, other researchers have argued that genetic testing for LNP, based on C/T_{-13910}, is premature. In two reports of genetic testing of LNP among sub-Saharan African subjects, the C/T_{-13910} variant was extremely rare and the authors suggested that other SNPs may be responsible for the LNP found in these populations [38, 40]. Currently, more research is required to determine the appropriate genetic markers of LNP in different populations before genetic testing becomes common.

B. Direct Assessment of Lactase Activity

Lactose digestion can be assessed directly or indirectly. The direct method involves obtaining a biopsy specimen of intestinal tissue and assaying for lactase activity or by intestinal perfusion studies [41]. Although these tests can accurately measure lactase activity, they are invasive and seldom used clinically.

C. Indirect Assessment Methods for Lactose Maldigestion

The metabolic basis for different indirect tests of lactose maldigestion is shown in Figure 1. Several indirect methods for assessing lactose digestion are available, including blood, urine, stool, and breath tests. Blood tests involve feeding a standard 50-g lactose dose and measurement of plasma glucose every 15–30 minutes over a period of 30 minutes to 2 hours. A rise in blood glucose of at least 25–30 mg/dl (1.5–1.7 mmol/l) is indicative of normal lactose digestion [41]. Unfortunately, blood glucose levels are subject to a variety of hormonal influences, reducing the reliability of this test. A blood test for galactose has been developed to correct this problem. The lactose dose is administered with a 500-mg/kg dose of ethanol (to prevent conversion of galactose to glucose in the liver) [41]. The galactose test is more reliable than the glucose test, but the ethanol exposure and somewhat invasive blood sampling are disadvantages.

A test has been devised to simultaneously measure intestinal lactose digestion (lactose digestion index, or

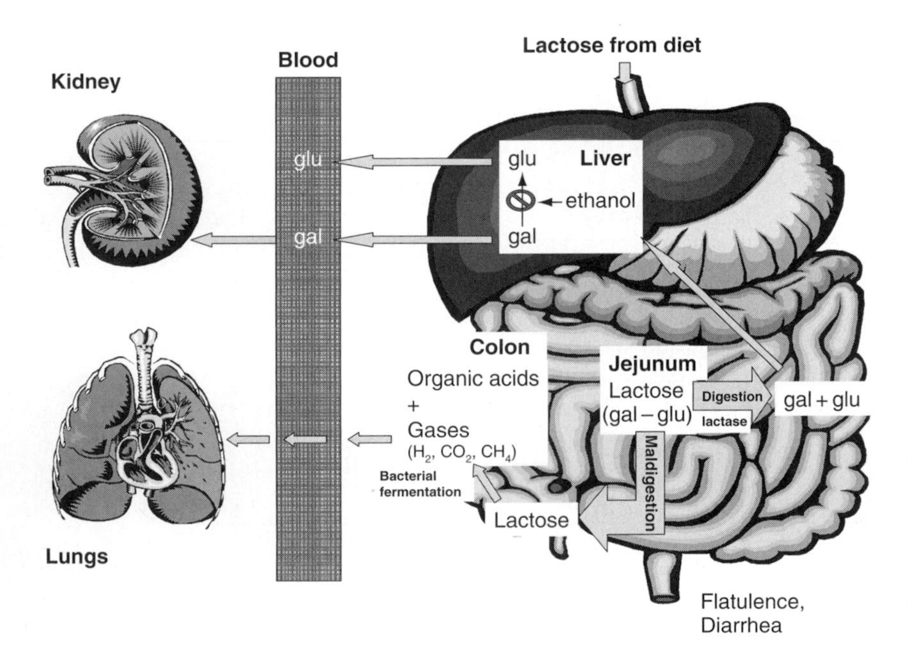

FIGURE 1 Metabolic background for understanding the diagnosis of lactose maldigestion and intolerance. Abbreviations: glu, glucose; gal, galactose. [Adapted with permission from: Arola, H. (1994). Diagnosis of hypolactasia and lactose malabsorption. *Scand. J. Gastroenterol.* **29**(Suppl 202), 26–35.] See color plate.

LDI) and intestinal permeability (sugar absorption test, or SAT) [42]. The LDI/SAT test consists of the oral administration of a 250-ml solution containing 25 g ^{13}C-lactose, 0.5 g ^{2}H-glucose, 5 g lactulose (a nonabsorbable disaccharide of galactose and fructose) and 1 g L-rhamnose. For the LDI test, the blood levels of ^{13}C-glucose and ^{2}H-glucose are measured before and at several time intervals after the LDI solution and a low ^{13}C glucose:^{2}H-glucose ratio (<0.60) indicates significant failure to hydrolyze lactose. Urine collections more than 10 hours after the solution are obtained for measurements of the lactulose:rhamnose ratio, with a higher ratio indicating increased intestinal permeability. The LDI test was shown to outperform other measures of lactose digestion in one study [43] and it is less invasive than an intestinal biopsy, but it still has a number of limitations. The test requires the use of expensive isotopes and analytical equipment, involves both blood and urine collection, and has not been evaluated in persons with mucosal damage that might increase intestinal permeability.

A sometimes-used urine test involves the measurement of galactose in the urine, rather than the blood, during the lactose tolerance test with ethanol. Another urine test is conducted by simultaneously administering lactose and lactulose [41]. Small amounts of lactose (up to 1% of the ingested dose) and lactulose diffuse unmediated across the intestinal mucosa and are excreted in the urine. The ratio of lactose to lactulose in the urine (collected over 10 hours) is determined by the hydrolysis of lactose. A value of less than 0.3 indicates normal lactose digestion and a ratio approaching 1.0 is observed in hypolactasia [41].

The measurement of stool pH and reducing substances in the stools has been used to assess lactose digestion in children. The analyses are easy to perform and convenient for the individuals being tested. However, stool pH has been shown to be unreliable in the diagnosis of hypolactasia in children and adults [41]. Furthermore, changes in gut motility and water excretion can alter the level of reducing substances in the stool. Thus, diagnosis of hypolactasia should not be based on stool tests alone [41].

Breath tests are the most widely used method for diagnosing carbohydrate maldigestion/malabsorption. The principle behind breath tests is that lactose, which escapes digestion in the small intestine, is fermented by bacteria in the colon, producing short-chain fatty acids and hydrogen, carbon dioxide, and methane (in some individuals). One breath test measures the amount of $^{13}CO_2$ excreted in the breath following administration of ^{13}C-lactose [41]. This stable isotope test has a safety advantage over older tests employing radioactive ^{14}C-lactose, but the high cost of the equipment prohibits widespread use of this method.

The current "gold standard" for diagnosis of carbohydrate maldigestion/malabsorption is the breath hydrogen test. Bacterial fermentation is the only source of hydrogen gas in the body. A portion of the hydrogen gas produced in the colon diffuses into the blood, with ultimate pulmonary excretion [44]. The hydrogen breath test is widely used because it is noninvasive and easy to perform. Typically, a subject is given an oral dose of lactose following an overnight (≥12 hours) fast. Breath samples are collected at regular intervals for a period of 3–8 hours. In early studies, 50 g of lactose was used as a challenge dose. Almost all lactose maldigesters will experience intolerance symptoms following a dose of lactose this large [32], and yet many will be able to tolerate smaller, more physiologic doses of lactose. Doses of lactose that are in the range of 1–2 cups (240–480 ml) of milk (12–24 g lactose) have been more frequently used in recent years [45]. The dose of lactose used in the breath hydrogen test influences the diagnostic criterion for lactose maldigestion. Early studies with 50 g lactose showed perfect separation of lactose digesters from maldigesters using a rise in breath hydrogen of greater than 20 parts per million (ppm) above the fasting level [46]. Strocchi *et al.* [47] evaluated different criteria for diagnosis of carbohydrate maldigestion, using small doses of carbohydrate (10 g lactulose). Using a cutpoint of a 10-ppm or greater rise in breath hydrogen above fasting over an 8-hour period resulted in improved sensitivity (93% versus 76%) and only a slight decrease in specificity (95% versus 100%) compared to the 20-ppm cutoff. Further, it was shown that using a sum of breath hydrogen values from hours 5, 6, and 7 and a 15-ppm or greater rise above fasting cutpoint resulted in 100% sensitivity and specificity.

Despite the advantages of breath hydrogen testing, care must be taken to ensure an accurate test. First, it is important to establish a low baseline breath hydrogen value, to which subsequent values are compared. This is accomplished by fasting before and after consumption of the lactose dose. In addition, it has been shown that a meal low in nondigestible carbohydrate (e.g., white rice and ground meat) the evening before the test results in lower baseline hydrogen [48]. Second, it is possible that some individuals may have a colonic microflora that is incapable of producing hydrogen. However, these individuals are rare and the possibility of a non-hydrogen-producing flora can be ruled out by the administration of lactulose [47]. Third, approximately 40% of adults harbor significant numbers of methane-producing bacteria in the colon [49]. Because methanogenic bacteria consume four parts of hydrogen to produce one part methane [49], some authors have suggested that simultaneous measurement of methane will improve the accuracy of breath hydrogen testing in methane-producing subjects [50]. The availability of gas chromatographs that can analyze both hydrogen and methane in breath samples eliminates this potential problem. Finally, a number of factors (sleep, antibiotics, smoking, bacterial overgrowth of the small intestine, and exercise) may complicate the interpretation of breath hydrogen tests [41]. Therefore, standardization of the breath test protocol and appropriate controls are important.

VI. LACTOSE MALDIGESTION AND INTOLERANCE SYMPTOMS

A positive breath hydrogen test is indicative of lactose maldigestion. However, reduced lactase levels do not necessarily lead to intolerance symptoms. Symptoms of intolerance occur when the amount of lactose consumed exceeds the ability of both the small intestine and colon to effectively metabolize the dose. Unhydrolyzed lactose passes from the small intestine to the large intestine where it is fermented by enteric bacteria, producing the gases that are partially responsible for causing intolerance symptoms. The intensity of symptoms varies with the amount of lactose consumed [33, 51–53], the degree of colonic adaptation [6, 54], and the physical form of the lactose-containing food [55].

The correlation between lactose maldigestion and reported intolerance symptoms is unclear. Most maldigesters can tolerate the amount of lactose in up to 1–2 cups of milk without experiencing severe symptoms. However, some lactose maldigesters believe that small amounts of lactose, such as the amount used with coffee or cereal, cause gastrointestinal distress [56]. Individual differences observed in symptom reporting may reflect learned behaviors, cultural attitudes, or other social issues.

Lactose maldigesters, unselected for their degree of lactose intolerance, tolerated a cup of milk without experiencing appreciable symptoms [57–59]. However, the results of these studies did not gain general acceptance, in part because of failure to enlist subjects with "severe" lactose intolerance. In 1995, Suarez *et al.* [56] conducted a study in 30 self-described "severely lactose intolerant individuals." Initial breath hydrogen test measurements indicated that approximately 30% (9 of 30) of the subjects claiming severe lactose intolerance were digesters and, thus, had no physiological basis for intolerance symptoms. These findings further demonstrate how strongly behavioral and psychological factors influence symptom reporting. Additional research is necessary to evaluate the psychological component of symptom reporting in lactose maldigesters.

VII. LACTOSE DIGESTION, CALCIUM, AND OSTEOPOROSIS

Individuals who are lactose intolerant can generally tolerate moderate amounts of lactose with minimal to no gastrointestinal discomfort [56, 60]. However, some lactose maldigesting individuals may unnecessarily restrict their intake of lactose-containing, calcium-rich dairy foods, thus compromising calcium intake. Milk and milk products contribute 73% of the calcium to the U.S. food supply [61]. Lactose maldigestion is associated with lower calcium intakes and is more frequent in osteoporotic cases than in controls [62–65]. For example, Newcomer *et al.* [62] found that 8 of 30 women with osteoporosis were lactose maldigesters compared to only 1 of 30 controls. In addition, calcium intakes of postmenopausal women positive for LNP in this study were significantly lower than in the lactase-persistent women (530 mg/day versus 811 mg/day). Interestingly, in this report, and another by Horowitz *et al.* [63], few of the LNP subjects reported a history of milk intolerance and yet they still restricted milk intake. The lower milk intakes in these subjects may have been due to factors other than lactose intolerance. However, it is also possible that these subjects restricted their milk intakes because of lactose intolerance in childhood, forgot that they had done so, and simply maintained that pattern of milk intake throughout life.

Another potential explanation for the increased prevalence of osteoporosis among lactose maldigesters is that maldigestion of lactose might decrease the absorption of calcium. Human and animal studies suggest that lactose stimulates the intestinal absorption of calcium [61]. However, there is considerable disagreement regarding the influence of lactose and lactose maldigestion on calcium absorption in adults. This disagreement results from a number of factors including the dose of lactose given, the choice of method for assessing calcium absorption (single isotope, double isotope, balance methods), prior calcium intake of the subjects, and the form in which the calcium is given (milk versus water).

Kocian *et al.* [66], using a single-isotope (^{47}Ca) method, demonstrated improved absorption of a 972-mg calcium dose from lactose-hydrolyzed milk as compared to milk containing lactose in lactose maldigesters. Conversely, the regular milk resulted in increased calcium absorption versus the lactose-hydrolyzed milk in lactose digesters. Another study, using dual-isotope methods, a 50-g lactose load, and 500 mg of calcium chloride in water found similar results [67]. Total fractional calcium absorption was decreased in maldigesters and increased in digesters with lactose feeding. However, the doses of lactose given in these studies (39–50 g or the equivalent of 3–4 cups of milk) were unphysiologic and may have resulted in more rapid intestinal transit than would be observed with more physiologic amounts of lactose.

Several studies have been conducted with physiologic doses of lactose. Griessen *et al.* [68], using dual-isotope methods, found that lactose maldigesters ($n = 7$) had a slightly, but not statistically significant, greater total fractional calcium absorption from 500 ml of milk compared to 500 ml of lactose-free milk. They also observed a nonsignificant decline in fractional calcium absorption in normal subjects ($n = 8$) when comparing lactose-free milk with regular milk. In another dual-isotope study, lactose maldigesters absorbed more calcium from a 240-ml dose of milk than did digesters (about 35% versus 25%), which was thought to be due to lower calcium intakes in the lactose maldigesting group [69]. Most important, however, no difference was observed in fractional calcium absorption between lactose-hydrolyzed and regular milk in either group of subjects. Calcium absorption from milk and yogurt, each containing 270 mg of calcium, was studied using a single-isotope method [70]. No significant differences were observed in calcium absorption between milk and yogurt in either the lactose maldigesting or digesting subjects. Interestingly, yogurt resulted in slightly, but significantly ($p < 0.05$), greater calcium absorption in lactose maldigesters when compared to lactose digesters.

Differences in study methodology (milk versus water, dose of lactose, and the choice of method for determining calcium absorption) may explain contrasting results. Physiologic doses of lactose (e.g., amounts provided by up to 2 cups of milk) are not likely to have a significant impact on calcium absorption. The increased prevalence of osteoporosis in lactose maldigesters is most likely related to inadequate calcium intake rather than impaired intestinal calcium absorption.

VIII. DIETARY MANAGEMENT FOR LACTOSE MALDIGESTION

It is difficult for lactose maldigesters to consume adequate amounts of calcium if dairy products are eliminated from the diet. Fortunately, lactose intolerance is easily managed. Dietary management approaches that effectively reduce or eliminate intolerance symptoms are discussed next and shown in Table 4.

A. Dose Response to Lactose

There is a clear relationship between the dose of lactose consumed and the symptomatic response. Small doses (up to 12 g of lactose) yield no symptoms [1, 56, 58–60], whereas high doses (> 20–50 g of lactose) produce appreciable symptoms in most individuals [1, 71–73]. In a well-controlled trial, Newcomer *et al.* [1] demonstrated that more than 85% of lactose maldigesters developed intolerance symptoms after consuming 50 g of lactose (the approximate amount of lactose in 1 quart of milk) as a single dose. The frequency of reported symptoms may be attributed to the nonphysiologic nature of the lactose dose and the physical form of lactose load administered. A lactose dose of 15–25 g produces appreciable symptoms in some subjects [30, 66]. The incidence of symptom reporting generally remains above 50% with intermediate doses. However, the frequency varies from less than 40% to greater than 90% [32]. In a double-blind protocol, Suarez *et al.* [56] demonstrated that feeding 12 g of lactose with a meal resulted in

TABLE 4 Dietary Strategies for Lactose Intolerance

Factors Affecting Lactose Digestion	Dietary Strategy	References
Dose of lactose	Consume a cup of milk or less at a time, containing up to 12 g lactose.	Suarez *et al.* [56] Hertzler *et al.* [6] Suarez *et al.* [60]
Intestinal transit	Consume milk with other foods, rather than alone, to slow the intestinal transit of lactose.	Solomons *et al.* [77] Martini and Savaiano [4] Dehkordi *et al.* [78]
Yogurt	Consume yogurt containing active bacteria cultures. One serving, or even two, should be well tolerated. Lactose in yogurts is better digested than the lactose in milk. Pasteurized yogurt does not improve lactose digestion; however, these products, when consumed, produce few to no symptoms.	Kolars *et al.* [85] Gilliland and Kim [91] Savaiano *et al.* [55] Shermak *et al.* [90] Savaiano *et al.* [55] Kolars *et al.* [85] Gilliland and Kim [91]
Digestive aids	Over-the-counter lactase supplements (pills, capsules, and drops) may be used when large doses of lactose (> 12 g) are consumed at once. Lactose-hydrolyzed milk is also well tolerated.	Moskovitz *et al.* [108] Lin *et al.* [103] Ramirez *et al.* [109] Nielsen *et al.* [115] Biller *et al.* [120] Rosado *et al.* [117] Brand and Holt [113]
Colon adaptation	Consume lactose-containing foods daily to increase the ability of the colonic bacteria to metabolize undigested lactose.	Perman *et al.* [127] Florent *et al.* [54] Hertzler *et al.* [6]

minimal to no symptoms in maldigesters. Interestingly, in unblinded studies [74, 75], lactose maldigesters more frequently reported intolerance symptoms after consuming lactose loads similar to those given by Suarez *et al.* Subsequently, Suarez *et al.* [60] provided further evidence that individuals who are lactose intolerant can consume lactose-containing foods without experiencing appreciable symptoms by feeding lactose maldigesters 2 cups of milk daily. One cup of milk was given with breakfast, and the second was given with the evening meal. The symptoms reported by maldigesters after consumption of 2 cups of milk a day were trivial.

Symptoms from excessive lactose in the intestine may increase out of proportion to dose, which raises the possibility that the absorption efficiency decreases with increased loads. Fractional lactose absorption is most likely influenced by dose, with more effective absorption of small loads and less effective utilization of larger doses. Hertzler *et al.* [52], using breath hydrogen as an indicator, suggested that 2 g of lactose is almost completely absorbed, whereas there was some degree of maldigestion when a 6-g load was ingested. The only study directly measuring the lactose absorption efficiency in lactose maldigesting subjects is that of Bond and Levitt [76], who intubated the terminal ileum and then fed the subjects ^{14}C-lactose mixed with polyethylene glycol, a nonabsorbable volumetric marker. Analysis of the ratio of ^{14}C-lactose to polyethylene glycol passing through the terminal ileum allowed researchers to calculate the percentage of lactose absorbed. On average, maldigesters absorbed about 40% of a 12.5-g lactose load, whereas the other 60% passed to the terminal ileum. However, sizable differences were seen in absorption efficiency among lactose maldigesters. These differences could represent differences in residual lactase efficiency and/or gastric emptying and intestinal transit time.

B. Factors Affecting Gastrointestinal Transit of Lactose

Consuming milk with other foods can minimize symptoms from lactose maldigestion [4, 77, 78]. A probable explanation for these findings is that the presence of additional foods slows the intestinal transit of lactose. Slowed transit allows more contact between ingested lactose and residual lactase in the small intestine, thus improving lactose digestion. It is also possible that additional foods may simply slow the rate at which lactose arrives in the colon, because a delay in peak breath hydrogen production, rather than a significant decrease in total hydrogen production, has been reported [4]. The slower fermentation of lactose might allow for more efficient disposal of fermentation gases, reducing the potential for symptoms.

The energy content, fat content, and added components such as chocolate may influence gastrointestinal transit of lactose and subsequent lactose digestion. Leichter [79] showed that 50 g of lactose from whole milk (1050 ml) resulted in fewer symptoms (abdominal discomfort, bloating, and flatulence) compared to 50 g of lactose from either skim milk (1050 ml) or an aqueous solution (330 ml). However, only blood glucose was measured to determine lactose digestion and no statistical evaluation of symptoms was done in this study. More recent studies have demonstrated that higher fat milk may slightly decrease breath hydrogen relative to skim milk [78], but not improve tolerance [78, 80, 81]. Further, increasing the energy content or viscosity of milk has not been effective in improving lactose digestion or tolerance [82, 83].

Chocolate milk has been recommended for individuals who are lactose intolerant. Apparently, chocolate milk empties from the stomach more slowly than unflavored milk, possibly because of its higher osmolality or energy content [3]. Two reports [78, 84] have demonstrated improved lactose digestion (i.e., reduced breath hydrogen) following consumption of chocolate milk. One of these studies [84] documented fewer symptoms in subjects.

Clearly, consumption of milk with other foods results in improved tolerance compared to milk alone. Therefore, consuming small amounts of milk routinely with meals is a recommended approach for individuals who are lactose intolerant to obtain sufficient calcium from dairy products. These individuals might also try chocolate milk to improve tolerance.

C. Yogurt

The lactose in yogurt with live cultures is digested better than lactose in milk and is well tolerated by those who are lactose intolerant [85]. Prior to fermentation, most commercially produced yogurt is nearly 6% lactose because of the addition of milk solids to milk during yogurt production. However, as the lactic acid bacteria (*Lactobacillus delbrueckii* subsp. *bulgaricus* and *Streptococcus salivarius* subsp. *thermophilus*) multiply to nearly 100 million organisms per milliliter, 20–30% of the lactose is utilized, decreasing the lactose content of yogurt to approximately 4% [86]. During fermentation, the activity of the β-galactosidase (lactase) enzyme substantially increases. Casein, calcium phosphate, and lactate in yogurt act as buffers in the acidic environment of the stomach, thus protecting a portion of the microbial lactase from degradation and allowing the delivery of intact cells to the small intestine [87, 88]. In the duodenum, once the intact bacterial cells interact with bile acids, they are disrupted, allowing substrate access to enzyme activity.

Yogurt consumption results in enhanced digestion of lactose and improved tolerance [55, 85, 89–91]. In 1984, Kolars *et al.* [85] and Gilliland and Kim [91] reported enhanced lactose digestion from yogurt in lactose maldigesters. In both studies, breath hydrogen excretion was significantly reduced with the consumption of live culture yogurt. Furthermore, Kolars *et al.* [85] found that an 18-g load of lactose in yogurt resulted in significantly fewer intolerance symptoms reported by subjects as compared to the other forms of lactose given. Also in 1984, Savaiano *et al.* [55] demonstrated that yogurt feeding resulted in one-third to one-fifth less hydrogen excretion as compared to other lactose-containing dairy foods with no symptoms. Shermak *et al.* [90] reported that a 12-g load of lactose in yogurt resulted in lower peak hydrogen in children with a delay in the time for breath hydrogen to rise when compared to a similar lactose load given in milk. These children experienced significantly fewer intolerance symptoms with yogurt consumption.

Yogurt pasteurization following fermentation has been somewhat controversial [91]. One advantage of pasteurizing yogurt is a longer shelf life. However, removing the active cultures that are partly responsible for improved lactose digestion may increase intolerance symptoms and cause lactose maldigesters to avoid yogurt products. Pasteurizing yogurt increases the maldigestion of lactose [55, 90, 91]. However, pasteurized yogurt is moderately well tolerated, producing minimal symptoms [55, 89, 90]. Because pasteurized yogurt is relatively well tolerated, other factors such as the physical form, or gelling, and the energy density of yogurt may play a role in tolerance. The level of the β-galactosidase enzyme in yogurt may not be the limiting factor for improving lactose digestion because not all yogurts have the same level of lactase activity [92]. Martini *et al.* [92] fed yogurts with varying levels of microbial β-galactosidase. The remaining characteristics of the test yogurts (pH, cell counts, and lactose concentrations) were similar. Despite the different levels of β-galactosidase activity, all yogurts improved lactose digestion and minimized intolerance symptoms.

D. Kefir

Kefir is a fermented dairy beverage that originated in Eastern Europe and has been made for centuries. Unlike yogurt, it is drinkable and the kefir grains used to culture the milk contain a wider variety of starter culture microorganisms. Hertzler and Clancy [93] investigated the effect of kefir ingestion on lactose maldigestion and intolerance symptoms in a group of lactose maldigesters. The kefir used in the study contained *Streptococcus lactis, Lactobacillus plantarum, Streptococcus cremoris, Lactobacillus casei, Streptococcus diacetylactis, Saccharomyces florentinus*, and *Leuconostoc cremoris* (as per label information) as the starter cultures. Feeding a 20-g lactose load as plain kefir reduced breath hydrogen excretion (8-hour area under the curve) by nearly threefold and decreased flatulence symptoms by 50% compared with the same amount of lactose from milk. The response to yogurt was similar to that observed with kefir.

E. Unfermented Acidophilus Milk

Individuals who are lactose maldigesters may consume unfermented milk containing cultures of *Lactobacillus acidophilus* in an effort to consume adequate amounts of calcium and avoid intolerance symptoms [94–96]. Various strains of *L. acidophilus* exist; however, strain NCFM has been most extensively studied and used in commercial products. Unfermented acidophilus milk tastes identical to unaltered milk because the NCFM strain does not multiply in the product, provided that the storage temperature is below 40°F (5°C) [94, 95, 97]. *Lactobacillus acidophilus* strain NCFM is derived from human fecal samples [87] and contains β-galactosidase. The effectiveness of acidophilus milk on improving lactose digestion and intolerance symptoms has been evaluated. Most evidence suggests that unfermented acidophilus milk does not enhance lactose digestion or reduce intolerance symptoms [55, 98–100], primarily because of the low concentration of the species in the milk. Improved lactose digestion has been observed by some [100]; however, the test milk in this study contained a much higher concentration of *L. acidophilus* than is normally used to produce commercial acidophilus milks.

Further, the microbial lactase from *L. acidophilus* may not be available to hydrolyze the lactose *in vivo* [92, 95, 99]. *Lactobacillus acidophilus* is not a bile-sensitive organism [87, 97]. Therefore, once the intact bacterial cells reach the small intestine, bile acids may not disrupt the cell membrane to allow the release of the microbial lactase. However, sonicated acidophilus milk improved lactose digestion by reducing breath hydrogen [101]. Thus, if less bile-resistant strains were developed and used in adequate amounts, these strains may allow the β-galactosidase to be released, possibly yielding an effective approach to the dietary management of lactose maldigestion.

Finally, fermented and unfermented milks containing bifidobacteria may also be useful in the management of lactose intolerance. Studies of milks treated with *Bifidobacterium bifidus* GD428 [92] and *B. longum* (strains B6 and 15708) [102] lower breath hydrogen excretion and flatulence symptoms compared with milk in lactose maldigesters. However, the beneficial effects, as with acidophilus milk, were still less than those observed with yogurt.

F. Lactase Supplements and Lactose-Reduced Milk

Lactase pills, capsules, and drops contain lactase derived from yeast (*Kluyveromyces lactis*) or fungal (*Aspergillus niger, A. oryzae*) sources. Common brands include Lactaid and Lacteeze. Dosages of lactase per pill or caplet vary from 4000 to 9000 FCC (Food Chemical Codex) units [103–105]. Since 1984, these over-the-counter preparations have been given GRAS (generally recognized as safe) status by the U.S. Food and Drug Administration [106].

Lactase pretreated milks (100% lactose hydrolysis) are available from Lactaid and Dairy Ease [104, 107].

A number of studies have evaluated the effectiveness of these products. Doses of 3000–6000 FCC units of lactase administered just prior to milk consumption decrease both breath hydrogen and symptom responses to lactose loads ranging from 17 to 20 g [103, 108, 109]. The decrease in breath hydrogen and symptoms is generally dose dependent. Doses up to 9900 FCC units may be needed for digestion of a large lactose load, such as 50 g of lactose [103, 110, 111].

Lactose-hydrolyzed milks also improve lactose intolerance symptoms in both children and adults [72, 73, 112–122]. A by-product of lactose-hydrolyzed milk is increased sweetness, due to the presence of free glucose [56]. This increased sweetness may increase its acceptability in children [115].

G. Colonic Fermentation and Colonic Bacterial Adaptation of Lactose

The colonic bacteria ferment undigested lactose and produce short-chain fatty acids (SCFA) and gases. Historically, this fermentation process was viewed negatively as a cause of lactose intolerance symptoms. However, it is now recognized that the fermentation of lactose, as well as other nonabsorbed carbohydrates, plays an important role in the health of the colon and affects the nutritional status of the individual.

The loss of intestinal lactase activity in lactose maldigesters is permanent. Studies from Israel, India, and Thailand have reported that feeding 50 g of lactose or more per day for periods of 1–14 months has no impact on jejunal lactase activity [17, 123, 124]. Despite this observation, milk has been used successfully in the treatment of malnourished children in areas of the world where lactose maldigestion is common. In Ethiopia, for example, 100 school children, aged 6–10 years, were fed 250 ml of milk per day for a period of 4 weeks [125]. Although the children initially experienced some degree of gastrointestinal symptoms, the symptoms rapidly abated and returned to pretrial levels within 4 weeks. Similar results were observed with school children in India [123]. Finally, a study of African American males and females aged 13–39 years who were lactose maldigesters and lactose intolerant showed that 77% of the subjects could ultimately tolerate 12 g or more of lactose if lactose was increased gradually and fed daily over a period of 6–12 weeks [126]. Approximately 80% of the subjects (18 of 22) had rises in breath hydrogen of at least 10 ppm above baseline at the maximum dose of lactose tolerated, suggesting that improved digestion of lactose in the small intestine was not responsible for the increased tolerance. Therefore, the authors proposed that colonic bacterial adaptation was a likely explanation for these findings.

Evidence for colonic bacterial adaptation to disaccharides (lactulose, lactose) is substantial. Perman *et al.* [127] fed adults 0.3 g/kg lactulose per day for 7 days and observed a decrease in fecal pH from 7.1 ± 0.3 to 5.8 ± 0.6. The breath hydrogen response to a challenge dose of lactose (0.3 g/kg) fell significantly after lactulose adaptation. Employing the same experimental design and doses of lactulose, Florent *et al.* [54] measured fecal β-galactosidase, colonic pH, breath hydrogen, fecal carbohydrates, SCFA, and ^{14}C-lactulose catabolism in subjects before and after the 7-day lactulose maintenance period. Fecal β-galactosidase was six times greater after lactulose feeding and breath hydrogen fell significantly. Breath $^{14}CO_2$ (indicating catabolism of ^{14}C-lactulose) increased and fecal outputs of lactulose and total hexose units were low after the lactulose feeding. Symptoms were not measured; however, a follow-up study showed that adaptation to lactulose (40 g/day for 8 days) reduced symptoms of diarrhea induced by a large dose (60 g) of lactulose [128]. Breath hydrogen decreased significantly and fecal β-galactosidase activity increased as in the previous study.

Two feeding trials directed to adapting lactose maldigesters to lactose have been reported. The first was a blinded, crossover study conducted at the University of Minnesota [6]. Feeding increasing doses of lactose (from 0.3 up to 1.0 g/kg/day) for 16 days resulted in a threefold increase in fecal β-galactosidase activity, which returned to baseline levels within 48 hours after substitution of dextrose for lactose. Further, 10 days of lactose feeding (from 0.6 up to 1.0 g/kg/day), compared to dextrose feeding, dramatically decreased the breath hydrogen response to a lactose challenge dose (0.35 g/kg; Fig. 2). After lactose adaptation, the subjects no longer appeared to be lactose maldigesters, based on a 20-ppm rise in breath hydrogen above fasting. The large doses of lactose fed during the adaptation period (averaging 42–70 g/day) resulted in only minor symptoms. Additionally, the severity and frequency of flatus symptoms in response to the lactose challenge dose were reduced by 50%.

The second study was a double-blind, placebo-controlled trial conducted in France with a group of 46 subjects who were lactose intolerant [129]. Following a baseline lactose challenge with 50 g of lactose, subjects were randomly assigned to either a lactose-fed group ($n = 24$) or a sucrose-fed control group ($n = 22$). Subjects were fed 34 g of either lactose or sucrose per day for 15 days. Fecal β-galactosidase increased and breath hydrogen decreased as the result of lactose feeding. Clinical symptoms (except diarrhea) were 50% less severe after lactose feeding. However, the sucrose-fed control group also experienced a comparable decrease in symptoms, despite no evidence of metabolic adaptation. Thus, these authors concluded that the improvements in symptoms resulted from familiarization with the test protocol rather than from metabolic adaptation.

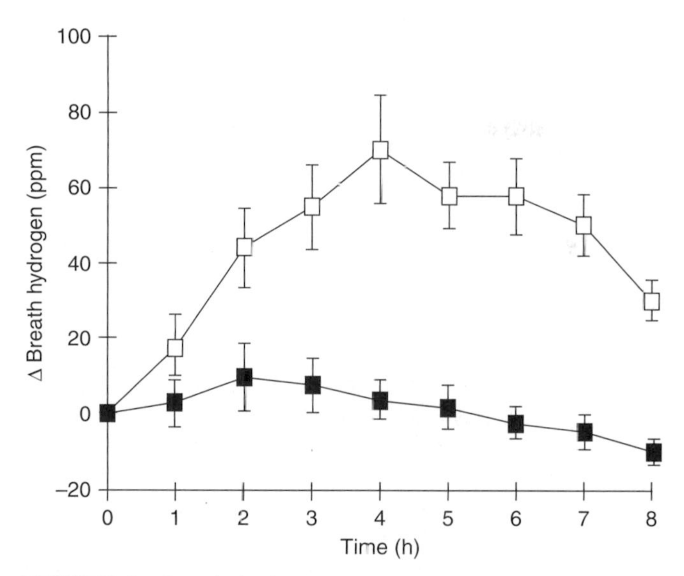

FIGURE 2 Breath hydrogen response to a lactose challenge after lactose (□) or dextrose (■) feeding periods. Data are the means ± SEM, $n = 20$. [Reprinted with permission from Hertzler, S. R., and Savaiano, D. A. (1996). Colonic adaptation to daily lactose feeding in lactose maldigesters reduces lactose intolerance. *Am. J. Clin. Nutr.* **64**, 232–236.

Colonic bacteria develop an increased ability to ferment lactose (indicated by increased fecal β-galactosidase) following prolonged lactose feeding. Because hydrogen gas is an end product of fermentation, it might be expected that the increased ability to ferment lactose would result in an increase, rather than the observed decrease, in breath hydrogen. However, breath hydrogen excretion represents the net of bacterial hydrogen production and consumption in the colon [49]. A decrease in net production of hydrogen could result from either decreased bacterial production or increased consumption. To examine the mechanism for decreased breath hydrogen after lactose adaptation, Hertzler *et al.* [130] employed metabolic inhibitors of bacterial hydrogen consumption (methanogenesis, sulfate reduction, and acetogenesis) to obtain measures of absolute hydrogen production. Subjects were fed increasing amounts of lactose or dextrose in a manner similar to previous studies. Fecal samples were assayed *in vitro* for absolute hydrogen production and hydrogen consumption. Absolute hydrogen production after 3 hours of incubation with lactose was threefold lower after lactose adaptation (242 ± 54 μl) compared to the dextrose feeding period (680 ± 79 μl, $p = 0.006$). Fecal hydrogen consumption was unaffected by either feeding period. These findings tend to support the hypothesis that prolonged lactose feeding favors the growth or metabolic activity of bacteria (e.g., bifidobacteria, lactic acid bacteria) that can ferment lactose without the production of hydrogen. Feeding lactose, lactulose, and nonabsorbable oligosaccharides stimulates the proliferation of lactic acid bacteria in the colon [131–133]. Additionally, high populations of bifidobacteria inhibit the growth of known hydrogen-producing organisms, such as clostridia or *Escherichia coli* [134].

Colonic bacterial adaptation to lactose does occur. Although the role of colonic adaptation in improving symptoms is not firmly established, it is clear that many individuals who are lactose intolerant can develop a tolerance to milk if they consume it regularly. This may represent a simpler and less expensive solution than the use of lactose digestive aids. Recently, a product called Lactagen has been marketed as a therapy for lactose intolerance, largely based on the colonic adaptation hypothesis. Lactagen is a powder containing lactose, *Lactobacillus acidophilus*, fructo-oligosaccharides, and small amounts of calcium and phosphorus. It is recommended by the manufacturer that the product be taken for a 38-day period, which is said to be sufficient to permanently "recondition" the intestinal microflora for lactose tolerance. One double-blind study of this product, published only as an abstract [135], reported that 80% of lactose-intolerant subjects had reductions in lactose intolerance symptoms, compared with 19% in the placebo group. However, objective evidence of improved lactose digestion or altered colonic bacterial fermentation of lactose (e.g., decreased breath hydrogen) was not obtained in this study. Further, longer-term follow-up studies to assess the permanence of the symptom reductions were not conducted.

IX. GENE THERAPY FOR LACTOSE INTOLERANCE

Although conventional dietary therapies for lactose intolerance exist, the possibility of gene therapy for lactase nonpersistence was examined by During *et al.* [136]. An adeno-associated virus vector was orally administered to hypolactasic rats to increase lactase mRNA. The adeno-associated virus vector is a defective, helper-dependent virus and the wild type is nonpathogenic in humans and other species. Following a single administration of a recombinant adeno-associated virus vector expressing β-galactosidase, all rats treated with this vector ($n = 4$) were positive for *lacZ* mRNA in the proximal intestine within 3 days. There was no lactase mRNA in the rats treated with the control vector. On day 7, following vector administration, the rats were challenged with a lactose solution. The treated rats had a rise in blood glucose from 114 ± 4 to 130 ± 3 mg/dl after 30 minutes, whereas the control rats had a flat blood glucose curve. Further, the treated rats still displayed similar lactase activity when challenged with lactose 6 months later. Thus, the potential of gene therapy for lactose intolerance is an interesting prospect, but this has not been studied in humans.

X. SUMMARY

A majority of the world's population and approximately 25% of the U.S. population are lactose maldigesters. Milk and milk products not only contain lactose, but also are important sources of calcium, riboflavin, and high-quality protein. Some maldigesters may avoid dairy products because of the perception that intolerance symptoms will inevitably follow dairy food consumption. Avoiding dairy products may limit calcium intake and bone density, thus increasing the risk for osteoporosis. Avoidance of milk and milk products is unnecessary because moderate lactose consumption does not usually produce a symptomatic response in maldigesters. Additionally, various dietary strategies effectively manage lactose intolerance by reducing or eliminating gastrointestinal symptoms. Dairy food consumption is possible for individuals who are lactose intolerant if simple dietary management strategies are incorporated into daily living.

References

1. Newcomer, A. D., McGill, D. B., Thomas, P. J., and Hofmann, A. F. (1978). Tolerance to lactose among lactase deficient American Indians. *Gastroenterology* **74**, 44–46.
2. Ravich, W. J., and Bayless, T. M. (1983). Carbohydrate absorption and malabsorption. *Clin. Gastroenterol.* **12**, 335–356.
3. Welsh, J. D., and Hall, W. H. (1977). Gastric emptying of lactose and milk in subjects with lactose malabsorption. *Am. J. Dig. Dis.* **22**, 1060–1063.
4. Martini, M. C., and Savaiano, D. A. (1988). Reduced intolerance symptoms from lactose consumed with a meal. *Am. J. Clin. Nutr.* **47**, 57–60.
5. Scrimshaw, N. S., and Murray, E. B. (1988). The acceptability of milk and milk products in populations with a high prevalence of lactose intolerance. *Am. J. Clin. Nutr.* **48**, 1083–1159.
6. Hertzler, S. R., and Savaiano, D. A. (1996). Colonic adaptation to the daily lactose feeding in lactose maldigesters reduces lactose tolerance. *Am. J. Clin. Nutr.* **64**, 1232–1236.
7. Kretchmer, N., and Sunshine, P. (1967). Intestinal disaccharidase deficiency in the sea lion. *Gastroenterology* **53**, 123–129.
8. Zeccha, L., Mesonero, J. E., Stutz, A., Poiree, J.-C., Giudicelli, J., Cursio, R., Gloor, S. M., and Semenza, G. (1998). Intestinal lactase-phlorizin hydrolase (LPH): The two catalytic sites; the role of the pancreas in pro-LPH maturation. *FEBS Lett.* **435**, 225–228.
9. Montgomery, R. K., Buller, H., Rings, E. H. H. M., and Grand, R. J. (1991). Lactose intolerance and the genetic regulation of intestinal lactase phlorizin hydrolase. *FASEB J.* **5**, 2824–2832.
10. Rossi, M., Maiuri, L., Fusco, M. I., Salvati, V. M., Fuccio, A., Aurrichio, S., Mantei, N., Zecca, L., Gloor, S. M., and Semenza, G. (1997). Lactase persistence versus decline in human adults: Multifactorial events are involved in downregulation after weaning. *Gastroenterology* **112**, 1506–1514.
11. Järvelä, I., Enattah, N. S., Kokkonen, J., Varilo, T., Sahvilahti, E., and Peltonen, L. (1998). Assignment of the locus for congenital lactase deficiency to 2q21, in the vicinity of but separate from the lactase-phlorizin hydrolase gene. *Am. J. Hum. Genet.* **63**, 1078–1085.
12. Sterchi, E. E., Mills, P. R., Fransen, J. A. M., Hauri, H.-P., Lentze, M. J., Naim, H. Y., Ginsel, L., and Bond, J. (1990). Biogenesis of intestinal lactase-phlorizin hydrolase in adults with lactose intolerance: Evidence for reduced biosynthesis and slowed-down maturation in enterocytes. *J. Clin. Invest.* **86**, 1329–1337.
13. Kien, C. L., Heitlinger, L. A., Li, U., and Murray, R. D. (1989). Digestion, absorption, and fermentation of carbohydrates. *Semin. Perinatol.* **13**, 78–87.
14. Holzel, A., Schwarz, V., and Sutcliffe, K. W. (1959). Defective lactose absorption causing malnutrition in infancy. *Lancet* **1**, 1126–1128.
15. Sahvilahti, E., Launiala, K., and Kuitunen, P. (1983). Congenital lactase deficiency. *Arch. Dis. Child.* **58**, 246–252.
16. Newcomer, A. D., and McGill, D. B. (1984). Clinical consequences of lactase deficiency. *Clin. Nutr.* **3**, 53–58.
17. Sahi, T. (1994). Genetics and epidemiology of adult-type hypolactasia. *Scand. J. Gastroenterol.* **29**, 7–20.
18. Gilat, T., Russo, S., Gelman-Malachi, E., and Aldor, T. A. M. (1972). Lactase in man: A nonadaptable enzyme. *Gastroenterology* **62**, 1125–1127.
19. Simoons, F. J. (1978). The geographic hypothesis and lactose malabsorption—A weighing of the evidence. *Am. J. Dig. Dis.* **23**, 963–980.
20. McCracken, R. D. (1970). Adult lactose tolerance [letter]. *JAMA* **213**, 2257–2260.
21. Burger, J., Kirchner, M., Bramanti, B., Haak, W., and Thomas, M. G. (2007). Absence of the lactose-persistence-associated allele in early Neolithic Europeans. *Proc. Natl. Acad. Sci. USA* **104**, 3736–3741.
22. Tishkoff, S. A., Reed, F. A., Ranciaro, A., Voight, B. F., Babbitt, C. C., Silverman, J. S., Powell, K., Mortensen, H. M., Hirbo, J. B., Osman, M., Ibrahim, M., Omar, S. A., Lema, G., Nyambo, T. B., GhoWri, J., Bumpstead, S., Pritchard, J. K., Wray, G. A., and Deloukas, P. (2007). Convergent adaptation of human lactase persistence in Africa and Europe. *Nat. Genet.* **39**, 31–40.
23. Lee, M.-F., and Krasinski, S. D. (1998). Human adult-onset lactase decline: An update. *Nutr. Rev.* **56**, 1–8.
24. Lloyd, M., Mevissen, G., Fischer, M., Olsen, W., Goodspeed, D., Genini, M., Boll, W., Semenza, G., and Mantei, N. (1992). Regulation of intestinal lactase in adult hypolactasia. *J. Clin. Invest.* **89**, 524–529.
25. Fajardo, O., Naim, H. Y., and Lacey, S. W. (1994). The polymorphic expression of lactase in adults is regulated at the messenger RNA level. *Gastroenterology* **106**, 1233–1241.
26. Eschler, J. C., de Koning, N., van Engen, C. G. J., Arora, S., Buller, H. A., Montgomery, R. K., and Grand, R. J. (1992). Molecular basis of lactase levels in adult humans. *J. Clin. Invest.* **89**, 480–483.
27. Sebastio, G., Villa, M., Sartorio, R., Guzetta, V., Poggi, V., Aurrichio, S., Boll, W., Mantei, N., and Semenza, G. (1989). Control of lactase in human adult-type hypolactasia and in weaning rabbits and rats. *Am. J. Hum. Genet.* **45**, 489–497.
28. Olsen, W. A., Li, B. U., Lloyd, M., and Korsmo, H. (1996). Heterogeneity of intestinal lactase activity in children: Relationship to lactase-phlorizin hydrolase messenger RNA abundance. *Pediatr. Res.* **39**, 877–881.

29. Newcomer, A. D., and McGill, D. B. (1966). Distribution of disaccharidase activity in the small bowel of normal and lactase-deficient subjects. *Gastroenterology* **51**, 481–488.
30. Triadou, N., Bataille, J., and Schmitz, J. (1983). Longitudinal study of the human intestinal brush border membrane proteins. *Gastroenterology* **85**, 1326–1332.
31. Maiuri, L., Rossi, M., Raia, V., Garipoli, V., Hughes, L. A., Swallow, D., Noren, O., Sjostrom, H., and Aurrichio, S. (1994). Mosaic regulation of lactase in human adult-type hypolactasia. *Gastroenterology* **107**, 54–60.
32. Savaiano, D. A., and Levitt, M. D. (1987). Milk intolerance and microbe-containing dairy foods. *J. Dairy Sci.* **70**, 397–406.
33. Srinivasan, R., and Minocha, A. (1998). When to suspect lactose intolerance: Symptomatic, ethnic, and laboratory clues. *Postgrad. Med.* **104**, 109–123.
34. Schmidt, W., Schneider, T., Heise, W., Schulze, J.-D., Weinke, T., Ignatius, R., Owen, R. L., Zeitz, M., Riecken, E.-O., and Ulrich, R. (1997). Mucosal abnormalities in microsporidiosis. *AIDS* **11**, 1589–1594.
35. Corazza, G. R., Ginaldi, L., Furia, N., Marani-Toro, G., Di Giammartino, D., and Quaglino, D. (1997). The impact of HIV infection on lactose absorptive capacity. *J. Infect.* **35**, 31–35.
36. Enattah, N. S., Sahi, T., Savilahti, E., Terwilliger, J. D., Peltonen, L., and Jarvella, I. (2002). Identification of a variant associated with adult-type hypolactasia. *Nat. Genet.* **30**, 233–237.
37. Kuokkanen, M., Enattah, N. S., Okasanen, A., Savilahti, E., Orpana, A., and Järvelä, I. (2003). Transcriptional regulation of the lactase-phlorizin hydrolase gene by polymorphisms associated with adult-type hypolactasia. *Gut* **52**, 647–652.
38. Ingram, C. J. E., Elamin, M. F., Mulcare, C. A., Weale, M. E., Tarekegn, A., Raga, T. O., Bekele, E., Elamin, F. M., Thomas, M. G., Bradman, N., and Swallow, D. M. (2007). A novel polymorphism associated with lactose tolerance in Africa: Multiple causes for lactase persistence? *Hum. Genet.* **120**, 779–788.
39. Rasinperä, H., Savilahti, E., Enattah, N. S., Kuokkanen, M., Tötterman, N., Lindahl, H., Järvelä, I., and Kolho, K.-L. (2004). A genetic test which can be used to diagnose adult-type hypolactasia in children. *Gut* **53**, 1571–1576.
40. Mulcare, C. A., Weale, M. E., Jones, A. L., Connell, B., Zeitlyn, D., Tarekegn, A., Swallow, D. M., Bradman, N., and Thomas, M. G. (2004). The T allele of a single-nucleotide polymorphism 13.9 kb upstream of the lactase gene (LCT)(C-13.9kbT) does not predict or cause the lactase-persistence phenotype in Africans. *Am. J. Hum. Genet.* **74**, 1102–1110.
41. Arola, H. (1994). Diagnosis of hypolactasia and lactose malabsorption. *Scand. J. Gastroenterol.* **29**, 26–35.
42. Koetse, H. A., Klaassen, D., van der Molen, A. R. H., Elzinga, H., Bijsterveld, K., Boverhof, R., and Stellaard, F. (2006). Combined LDI/SAT test to evaluate intestinal lactose digestion and mucosa permeability. *Eur. J. Clin. Invest.* **36**, 730–736.
43. Vonk, R. J., Stellarrd, F., Priebe, M. G., Koetse, H. A., Hagedoorn, R. E., de Bruijn, S., Elzinga, H., Lenoir-Wijnkoop, I., and Antoine, J.-M. (2001). The ^{13}Cy^{2}H-glucose test for determination of small intestinal lactase activity. *Eur. J. Clin. Invest.* **31**, 226–233.
44. Levitt, M. D., and Donaldson, R. M. (1970). Use of respiratory hydrogen (H_2) excretion to detect carbohydrate malabsorption. *J. Lab. Clin. Med.* **75**, 937–945.
45. Solomons, N. W. (1984). Evaluation of carbohydrate absorption: The hydrogen breath test in clinical practice. *Clin. Nutr.* **3**, 71–78.
46. Metz, G., Jenkins, D. J., Peters, J. J., Newman, A., and Blendis, L. M. (1975). Breath hydrogen as a diagnostic method for hypolactasia. *Lancet* **1**, 1155–1157.
47. Strocchi, A., Corazza, G., Ellis, C. J., Gasbarrini, G., and Levitt, M. D. (1993). Detection of malabsorption of low doses of carbohydrate: Accuracy of various breath hydrogen criteria. *Gastroenterology* **105**, 1404–1410.
48. Anderson, I. H., Levine, A. S., and Levitt, M. D. (1981). Incomplete absorption of the carbohydrate in all-purpose wheat flour. *N. Engl. J. Med.* **304**, 891–892.
49. Levitt, M. D., Gibson, G. R., and Christl, S. U. (1995). Gas metabolism in the large intestine. *In* "Human Colonic Bacteria: Role in Nutrition, Physiology, and Disease" (G. R. Gibson, and G. T. Macfarlane, Eds.), pp. 131–154. CRC Press, Boca Raton, FL.
50. Bjorneklett, A., and Jenssen, E. (1982). Relationships between hydrogen (H_2) and methane (CH_4) production in man. *Scand. J. Gastroenterol.* **17**, 985–992.
51. Gudmand Høyer, E. (1994). The clinical significance of disaccharide maldigestion. *Am. J. Clin. Nutr.* **59**, 735S–741S.
52. Hertzler, S. R., Huynh, B., and Savaiano, D. A. (1996). How much lactose is "low lactose"? *J. Am. Diet. Assoc.* **96**, 243–246.
53. Vesa, T. H., Korpela, R. A., and Sahi, T. (1996). Tolerance to small amounts of lactose in lactose maldigesters. *Am. J. Clin. Nutr.* **64**, 197–201.
54. Florent, C., Flourie, B., Leblond, A., Rautureau, M., Bernier, J.-J., and Rambaud, J.-C. (1985). Influence of chronic lactulose ingestion on the colonic metabolism of lactulose in man (an in vivo study). *J. Clin. Invest.* **75**, 608–613.
55. Savaiano, D. A., Abou El Anouar, A., Smith, D. E., and Levitt, M. D. (1984). Lactose malabsorption from yogurt, pasteurized yogurt, sweet acidophilus milk, and cultured milk in lactase-deficient individuals. *Am. J. Clin. Nutr.* **40**, 1219–1223.
56. Suarez, F. L., Savaiano, D. A., and Levitt, M. D. (1995). A comparison of symptoms after the consumption of milk or lactose-hydrolyzed milk by people with self-reported severe lactose intolerance. *N. Engl. J. Med.* **333**, 1–4.
57. Rorick, M. H., and Scrimshaw, N. S. (1979). Comparative tolerance of elderly from differing ethnic background to lactose-containing and lactose-free daily drinks: A double-blind study. *J. Gerontol.* **34**, 191–196.
58. Haverberg, L., Kwon, P. H., and Scrimshaw, N. S. (1980). Comparative tolerance of adolescents of differing ethnic backgrounds to lactose-containing and lactose-free dairy drinks. I. Initial experience with a double-blind procedure. *Am. J. Clin. Nutr.* **33**, 17–21.
59. Unger, M., and Scrimshaw, N. S. (1981). Comparative tolerance of adults of differing ethnic backgrounds to lactose-free and lactose-containing dairy drinks. *Nutr. Res.* **1**, 1227–1233.
60. Suarez, F. L., Savaiano, D. A., Arbisi, P., and Levitt, M. D. (1997). Tolerance to the daily ingestion of two cups of milk by individuals claiming lactose intolerance. *Am. J. Clin. Nutr.* **65**, 1502–1506.

61. Miller, G. D., Jarvis, J. K., and McBean, L. D. (2000). "Handbook of Dairy Foods and Nutrition," 2nd ed., pp. 311–354. CRC Press, Boca Raton, FL.
62. Newcomer, A. D., Hodgson, S. F., McGill, D. B., and Thomas, P. J. (1978). Lactase deficiency: Prevalence in osteoporosis. *Ann. Intern. Med.* **89**, 218–220.
63. Horowitz, M., Wishart, J., Mundy, L., and Nordin, B. E. C. (1987). Lactose and calcium absorption in postmenopausal osteoporosis. *Arch. Intern. Med.* **147**, 534–536.
64. Finkenstedt, G., Skrabal, F., Gasser, R. W., and Braunsteiner, H. (1986). Lactose absorption, milk consumption, and fasting blood glucose concentrations in women with idiopathic osteoporosis. *Br. Med. J.* **292**, 161–162.
65. Corazza, G. R., Benati, G., DiSario, A., Tarozzi, C., Strocchi, A., Passeri, M., and Gasbarrini, G. (1995). Lactose intolerance and bone mass in postmenopausal Italian women. *Br. J. Nutr.* **73**, 479–487.
66. Kocian, J., Skala, I., and Bakos, K. (1973). Calcium absorption from milk and lactose-free milk in healthy subjects and patients with lactose intolerance. *Digestion* **9**, 317–324.
67. Cochet, B., Jung, A., Griessen, M., Bartholdi, P., Schaller, P., and Donath, A. (1983). Effects of lactose on intestinal calcium absorption in normal and lactase-deficient subjects. *Gastroenterology* **84**, 935–940.
68. Griessen, M., Cochet, B., Infante, F., Jung, A., Bartholdi, P., Donath, A., Loizeau, E., and Courvoisier, B. (1989). Calcium absorption from milk in lactase-deficient subjects. *Am. J. Clin. Nutr.* **49**, 377–384.
69. Tremaine, W. J., Newcomer, A. D., Riggs, L., and McGill, D. B. (1986). Calcium absorption from milk in lactase-deficient and lactase-sufficient adults. *Dig. Dis. Sci.* **31**, 376–378.
70. Smith, T. M., Kolars, J. C., Savaiano, D. A., and Levitt, M. D. (1985). Absorption of calcium from milk and yogurt. *Am. J. Clin. Nutr.* **42**, 1197–1200.
71. Johnson, A. O., Semenya, J. G., Buchowski, M. S., Enwonwu, C. O., and Scrimshaw, N. S. (1993). Correlation of lactose maldigestion, lactose intolerance, and milk intolerance. *Am. J. Clin. Nutr.* **57**, 399–401.
72. Reasoner, J., Maculan, T. P., Rand, A. G., and Thayer, W. R. Jr. (1981). Clinical studies with low-lactose milk. *Am. J. Clin. Nutr.* **34**, 54–60.
73. Pedersen, E. R., Jensen, B. H., Jensen, H. J., Keldsbo, I. L., Moller, E. H., and Rasmussen, S. N. (1982). Lactose malabsorption and tolerance of lactose-hydrolyzed milk: A double-blind controlled crossover trial. *Scand. J. Gastroenterol.* **17**, 861–864.
74. Bayless, T. M., Rothfeld, B., Masser, C., Wise, L., Paige, D., and Bedine, M. S. (1975). Lactose and milk intolerance: Clinical implications. *N. Engl. J. Med.* **292**, 1156–1159.
75. Rosado, J. L., Gonzales, C., Valencia, M. E., Lopez, B., Mejia, L., and Del Carmen Baez, M. (1994). Lactose maldigestion and milk intolerance: A study in rural and urban Mexico using physiological doses of milk. *J. Nutr.* **124**, 1052–1059.
76. Bond, J. H., and Levitt, M. D. (1976). Quantitative measurement of lactose absorption. *Gastroenterology* **70**, 1058–1062.
77. Solomons, N. W., Guerrero, A.-M., and Torun, B. (1985). Dietary manipulation of postprandial colonic lactose fermentation: I. Effect of solid foods in a meal. *Am. J. Clin. Nutr.* **41**, 199–208.
78. Dehkordi, N., Rao, D. R., Warren, A. P., and Chawan, C. B. (1995). Lactose malabsorption as influenced by chocolate milk, skim milk, sucrose, whole milk, and lactic cultures. *J. Am. Diet. Assoc.* **95**, 484–486.
79. Leichter, J. L. (1973). Comparison of whole milk and skim milk with aqueous lactose solution in lactose tolerance testing. *Am. J. Clin. Nutr.* **26**, 393–396.
80. Vesa, T. H., Lember, M., and Korpela, R. (1997). Milk fat does not affect the symptoms of lactose intolerance. *Eur. J. Clin. Nutr.* **51**, 633–636.
81. Cavalli-Sforza, L. T., and Strata, A. (1986). Double-blind study on the tolerance of four types of milk in lactose malabsorbers and absorbers. *Hum. Nutr. Clin. Nutr.* **40C**, 19–30.
82. Ves, T. H., Marteau, P. R., Briet, F. B., Boutron-Ruault, M.-C., and Rambaud, J.-C. (1997). Raising milk energy content retards gastric emptying of lactose in lactose-intolerant humans with little effect on lactose digestion. *J. Nutr.* **127**, 2316–2320.
83. Vesa, T. H., Marteau, P. R., Briet, F. B., Flourie, B., Briend, A., and Rambaud, J.-C. (1997). Effects of milk viscosity on gastric emptying and lactose intolerance in lactose maldigesters. *Am. J. Clin. Nutr.* **66,** 123–126.
84. Lee, C. M., and Hardy, C. M. (1989). Cocoa feeding and human lactose intolerance. *Am. J. Clin. Nutr.* **49**, 840–844.
85. Kolars, J. C., Levitt, M. D., Aouji, M., and Savaiano, D. A. (1984). Yogurt—An autodigesting source of lactose. *N. Engl. J. Med.* **310**, 1–3.
86. Răsic, J., and Kurmans, J. A. (1978). The nutritional-physiological value of yoghurt. *In* "Yogurt; Scientific Grounds, Technology, Manufacture and Preparations," pp. 99–137. Tech. Dairy Pub. House, Copenhagen, Denmark.
87. Savaiano, D. A., and Kotz, C. (1988). Recent advances in the management of lactose intolerance. *Contemp. Nutr.* **13**, 9–10.
88. Pochart, P., Dewit, O., Desjeux, J.-H., and Bourlioux, P. (1989). Viable starter culture, β-galactosidase activity, and lactose in duodenum after yogurt ingestion in lactase-deficient humans. *Am. J. Clin. Nutr.* **49**, 828–831.
89. Lerebours, E., N'Djitoyap Ndam, C., Lavoine, A., Hellot, M. F., Antoine, J. M., and Colin, R. (1989). Yogurt and fermented-then-pasteurized milk: Effects of short-term and long-term ingestion on lactose absorption and mucosal lactase activity in lactase-deficient subjects. *Am. J. Clin. Nutr.* **49**, 823–827.
90. Shermak, M. A., Saavedra, J. M., Jackson, T. L., Huang, S. S., Bayless, T. M., and Perman, J. A. (1995). Effect of yogurt on symptoms and kinetics of hydrogen production in lactose-malabsorbing children. *Am. J. Clin. Nutr.* **62**, 1003–1006.
91. Gilliland, S. E., and Kim, H. S. (1984). Effect of viable starter culture bacteria in yogurt on lactose utilization in humans. *J. Dairy Sci.* **67**, 1–6.
92. Martini, M. C., Lerebours, E. C., Lin, W.-J., Harlander, S. K., Berrada, N. M., Antoine, J. M., and Savaiano, D. A. (1991). Strains and species of lactic acid bacteria in fermented milks (yogurts): Effect on *in vivo* lactose digestion. *Am. J. Clin. Nutr.* **54**, 1041–1046.
93. Hertzler, S. R., and Clancy, S. M. (2003). Kefir improves lactose digestion and tolerance in adults with lactose maldigestion. *J. Am. Diet. Assoc.* **103**, 582–587.
94. Lin, M.-Y., Savaiano, D., and Harlander, S. (1991). Influence of nonfermented dairy products containing bacterial starter

cultures on lactose maldigestion in humans. *J. Dairy Sci.* **74**, 87–95.
95. Hove, H., Nørgaard, H., and Mortensen, P. B. (1999). Lactic acid bacteria and the human gastrointestinal tract. *Eur. J. Clin. Nutr.* **53**, 339–350.
96. Gilliland, S. E. (1989). Acidophilus milk products: A review of potential benefits to consumers. *J. Dairy Sci.* **72**, 2483–2494.
97. Newcomer, A. D., Park, H. S., O'Brien, P. C., and McGill, D. B. (1983). Response of patients with irritable bowel syndrome and lactase deficiency using unfermented acidophilus milk. *Am. J. Clin. Nutr.* **38**, 257–263.
98. Payne, D. L., Welsh, J. D., Manion, C. V., Tsegaye, A., and Herd, L. D. (1981). Effectiveness of milk products in dietary management of lactose malabsorption. *Am. J. Clin. Nutr.* **34**, 2711–2715.
99. Onwulata, C. I., Rao, D. R., and Vankineni, P. (1989). Relative efficiency of yogurt, sweet acidophilus milk, hydrolyzed lactose milk, and a commercial lactase tablet in alleviating lactose maldigestion. *Am. J. Clin. Nutr.* **49**, 1233–1237.
100. Kim, H. S., and Gilliland, S. E. (1984). *Lactobacillus acidophilus* as a dietary adjunct for milk to aid lactose digestion in humans. *J. Dairy Sci.* **66**, 959–966.
101. Mcdonough, F. E., Hitchins, A. D., Wong, N. P., Wells, P., and Bodwell, C. E. (1987). Modification of sweet acidophilus milk to improve utilization by lactose-intolerant persons. *Am. J. Clin. Nutr.* **45**, 570–574.
102. Jiang., T., Mustapha, A., and Savaiano, D. A. (1996). Improvement of lactose digestion in humans by ingestion of unfermented milk containing *Bifidobacterium longum*. *J. Dairy Sci.* **79**, 750–757.
103. Lin, M.-Y., DiPalma, J. A., Martini, M. C., Gross, C. J., Harlander, S. K., and Savaiano, D. A. (1993). Comparative effects of exogenous lactase (β-galactosidase) preparations on in vivo lactose digestion. *Dig. Dis. Sci.* **38**, 2022–2027.
104. McNeil, P. P. C. "About Lactaid and Lactaid Ultra." Retrieved April 12, 2007, from http://www.lactaid.com.
105. digestMILK.com. "Lacteeze Enzyme Drops and Tablets." Retrieved April 12, 2007, from http://www.digestmilk.com.
106. Lactase preparation from *K. lactis* affirmed as GRAS. (1984). *Food Chem. News*, December 10, p. 30.
107. Land O' Lakes. "Dairy Ease Products." Retrieved April 12, 2007, from http://www.dairyease.com.
108. Moskovitz, M., Curtis, C., and Gavaler, J. (1987). Does oral enzyme replacement therapy reverse intestinal lactose malabsorption? *Am. J. Gastroenterol.* **82**, 632–635.
109. Ramirez, F. C., Lee, K., and Graham, D. Y. (1994). All lactase preparations are not the same: Results of a prospective, randomized, placebo-controlled trial. *Am. J. Gastroenterol.* **89**, 566–570.
110. DiPalma, J. A., and Collins, M. S. (1989). Enzyme replacement for lactose malabsorption using a beta-D-galactosidase. *J. Clin. Gastroenterol.* **11**, 290–293.
111. Sanders, S. W., Tolman, K. G., and Reitberg, D. P. (1992). Effect of a single dose of lactase on symptoms and expired hydrogen after lactose challenge in lactose-intolerant adults. *Clin. Pharm.* **11**, 533–538.
112. Cheng, A. H., Brunser, O., Espinoza, J., Fones, H. L., Monckeberg, F., Chichester, C. O., Rand, G., and Hourigan, A. G. (1979). Long-term acceptance of low-lactose milk. *Am. J. Clin. Nutr.* **32**, 1989–1993.
113. Brand, J. C., and Holt, S. (1991). Relative effectiveness of milks with reduced amounts of lactose in alleviating milk intolerance. *Am. J. Clin. Nutr.* **54**, 148–151.
114. Turner, S. J., Daly, T., Hourigan, J. A., Rand, A. J., and Thayer, W. R. Jr. (1976). Utilization of a low-lactose milk. *Am. J. Clin. Nutr.* **29**, 739–744.
115. Nielsen, O. H., Schiotz, P. O., Rasmussen, S. N., and Krasilnikoff, P. A. (1984). Calcium absorption and acceptance of low-lactose milk among children with lactase deficiency. *J. Pediatr. Gastroenterol.* **3**, 219–223.
116. Paige, D. M., Bayless, T. M., Mellits, E. D., Davis, L., and Dellinger, W. S., Jr., and Kreitner, M. (1979). Effects of age and lactose tolerance on blood glucose rise with whole cow and lactose-hydrolyzed milk. *J. Agric. Food Chem.* **27**, 677–680.
117. Rosado, J. L., Morales, M., and Pasquetti, A. (1989). Lactose digestion and clinical tolerance to milk, lactose-prehydrolyzed milk, and enzyme-added milk: A study in undernourished continuously enteral-fed patients. *J. Parenteral Enteral Nutr.* **13**, 157–161.
118. Nagy, L., Mozsik, G., Garamszegi, M., Sasreti, E., Ruzsa, C., and Javor, T. (1983). Lactose-poor milk in adult lactose intolerance. *Acta Med. Hung.* **40**, 239–245.
119. Payne, D. L., Welsh, J. D., Manion, C. V., Tsegaye, A., and Herd, L. D. (1981). Effectiveness of milk products in dietary management of lactose malabsorption. *Am. J. Clin. Nutr.* **34**, 2711–2715.
120. Biller, J. A., King, S., Rosenthal, A., and Grand, R. J. (1987). Efficacy of lactase-treated milk for lactose-intolerant pediatric patients. *J. Pediatr.* **111**, 91–94.
121. Rosado, J. L., Morales, M., Pasquetti, A., Nobara, R., and Hernandez, L. (1988). Nutritional evaluation of a lactose-hydrolyzed milk-based enteral formula diet: I. A comparative study of carbohydrate digestion and clinical tolerance. *Rev. Invest. Clin.* **40**, 141–147.
122. Payne-Bose, D., Welsh, J. D., Gearhart, H. L., and Morrison, R. D. (1977). Milk and lactose-hydrolyzed milk. *Am. J. Clin. Nutr.* **30**, 695–697.
123. Reddy, V., and Pershad, J. (1972). Lactase deficiency in Indians. *Am. J. Clin. Nutr.* **25**, 114–119.
124. Keusch, G. T., Troncale, F. J., Thavaramara, B., Prinyanot, P., Anderson, P. R., and Bhamarapravthi, N. (1969). Lactase deficiency in Thailand: Effect of prolonged lactose feeding. *Am. J. Clin. Nutr.* **22**, 638–641.
125. Habte, D., Sterky, G., and Hjalmarsson, B. (1973). Lactose malabsorption in Ethiopian children. *Acta Pediatr. Scand.* **62**, 649–654.
126. Johnson, A. O., Semenya, J. G., Buchowski, M. S., Enwonwu, C. O., and Scrimshaw, N. S. (1993). Adaptation of lactose maldigesters to continued milk intakes. *Am. J. Clin. Nutr.* **58**, 879–881.
127. Perman, J. A., Modler, S., and Olson, A. C. (1981). Role of pH in production of hydrogen from carbohydrates by colonic bacterial flora: Studies *in vivo* and *in vitro*. *J. Clin. Invest.* **67**, 643–650.
128. Flourie, B., Briet, F., Florent, C., Pellier, P., Maurel, M., and Rambaud, J.-C. (1993). Can diarrhea induced by lactulose be

reduced by prolonged ingestion of lactulose? *Am. J. Clin. Nutr.* **58**, 369–375.

129. Briet, F., Pochart, P., Marteau, P., Flourie, B., Arrigoni, E., and Rambaud, J. C. (1997). Improved clinical tolerance to chronic lactose ingestion in subjects with lactose intolerance: A placebo effect? *Gut* **41**, 632–635.
130. Hertzler, S. R., Savaiano, D. A., and Levitt, M. D. (1997). Fecal hydrogen production and consumption measurements: Response to daily lactose ingestion by lactose maldigesters. *Dig. Dis. Sci.* **42**, 348–353.
131. Terada, A., Hara, H., Kataoka, M., and Mitsuoka, T. (1992). Effect of lactulose on the composition and metabolic activity of the human fecal flora. *Microb. Ecol. Health Dis.* **5**, 43–50.
132. Gibson, G. R., Beatty, E. R., Wang, X., and Cummings, J. H. (1995). Selective stimulation of bifidobacteria in the human colon by oligofructose and insulin. *Gastroenterology* **108**, 975–982.
133. Ito, M., and Kimura, M. (1993). Influence of lactose on faecal microflora in lactose maldigesters. *Microb. Ecol. Health Dis.* **6**, 73–76.
134. Gibson, G. R., and Wang, X. (1994). Regulatory effects of bifidobacteria on the growth of other colonic bacteria. *J. Appl. Bacteriol.* **77**, 412–420.
135. Landon, C., Tran, T., and Connell, D. B. (2006). A randomized trial of a pre- and probiotic formula to reduce symptoms of dairy products in patients with dairy intolerance. *FASEB J.* **20**, abstract A1053.
136. During, M. J., Xu, R., Young, D., Kaplitt, M. G., Sherwin, R. S., and Leone, P. (1998). Peroral gene therapy of lactose intolerance using an adeno-associated virus vector. *Nat. Med.* **4**, 1131–1135.

CHAPTER 41

Nutritional Considerations in the Management of Celiac Disease

MICHELLE PIETZAK
University of Southern California Keck School of Medicine, Los Angeles, California

Contents

I. INTRODUCTION

Celiac disease (also spelled coeliac disease in the European literature) has been known by many names in the medical literature in the past, including gluten-sensitive enteropathy, gliadin-sensitive enteropathy, and celiac sprue (to differentiate it from tropical sprue). Celiac disease occurs in genetically predisposed individuals who develop a permanent immunologic reaction to gluten, found in wheat, rye, and barley. When individuals with celiac disease ingest gluten, the result is malabsorption of sugars, proteins, fats, vitamins, and other minerals. Individuals with celiac disease have a permanent intolerance to the gliadin fraction of wheat protein and related alcohol-soluble proteins (called prolamins) found in rye and barley. Susceptible individuals who ingest these proteins develop an immune-mediated enteropathy, which self-perpetuates as long as these gluten-containing grains are in the diet. Removal of gluten from the diet, in the majority of people with celiac disease, leads to both resolution of symptoms and improvement in the intestinal damage. However, in individuals with long-standing disease, there may not be a response to the gluten-free diet (called "refractory sprue"), and intense nutritional support and immune-suppressing medications may be required.

Classic symptoms of celiac disease in childhood include diarrhea, short stature, anemia, and obvious physical signs of protein-calorie malnutrition. In the United States, and worldwide, more individuals are being diagnosed with celiac disease as adults. They may present with gastrointestinal symptoms, such as gastroesophageal reflux, irritable bowel syndrome, and diarrhea or constipation. Adults may also present with symptoms outside the gastrointestinal tract; these may include complaints of the joints, skin, reproductive, hematologic, musculoskeletal, and neurologic systems. Individuals with celiac disease, and their family members, often have other autoimmune diseases, such as type 1 diabetes, inflammatory bowel disease, and thyroid disease. The diagnosis of celiac disease is often delayed for years because its symptoms are often attributed to these other autoimmune conditions.

Serum antibodies are useful as a screening method for celiac disease. However, a small-intestinal biopsy, done by endoscopy, and a patient's clinical response to a gluten-free diet, are used to confirm a diagnosis. Individuals with long-standing disease are at risk for complications, such as iron-deficiency anemia, vitamin deficiencies, osteoporosis, infertility, and gastrointestinal cancers. The long-term maintenance of a gluten-free lifestyle is challenging. Individuals diagnosed with celiac disease benefit most from a team approach, which includes regular supervision by a physician, nutritional counseling by a dietitian, and access to support groups knowledgeable about celiac disease and the gluten-free diet.

II. SYMPTOMS OF CELIAC DISEASE

The clinician must have a high index of suspicion for celiac disease, because its symptoms can affect almost every organ system of the body, and its clinical features can be highly variable. The disease may present at any age, with the "classic" presentation occurring after the introduction of gluten in the diet during infancy and the toddler years. However, many adolescents, young adults, and even the elderly are now being diagnosed. Likewise, the severity of the presentation of celiac disease can be highly variable, with some patients experiencing severe diarrhea and weight loss, and others having no gastrointestinal symptoms whatsoever. Because celiac disease is a multisystemic disease, it can affect not only the gastrointestinal tract, but also the neurologic, endocrine, orthopedic, reproductive, and hematologic systems. Thus, it is not uncommon that a patient with celiac disease may be seen by multiple physicians over many years before the disorder is correctly identified.

The presentations of celiac disease can be classified into the following six categories based on symptomatology: a classic gastrointestinal form, a late-onset gastrointestinal form, an extraintestinal form, a form presenting with associated conditions, an asymptomatic form, and a latent form. The patients diagnosed with celiac disease who have the "classic" gastrointestinal form in childhood are thought to represent the "tip of the iceberg," in that they are the obvious, visible patients [1]. The point in using this iceberg model is that the majority of patients with celiac disease are "submerged," that is, many patients are not presenting in childhood with "classic" disease. Therefore, what has historically been considered the "classic" presentation is now a misnomer [2]. Many individuals with undiagnosed celiac disease will fall into the other five classifications ("non-classic") and therefore may experience chronic ill health and multiple complications of the disease, without ever being correctly diagnosed.

A. Classic Gastrointestinal Form

The "classic" gastrointestinal form of celiac disease occurs in infants or toddlers usually between the ages of 6 and 18 months, after the introduction of gluten-containing foods into the diet. The onset is usually insidious, with "failure to thrive" and obvious physical signs of wasting over a period of weeks to months. The "classic" symptoms are those of diarrhea, weight loss, abdominal distention, gassiness, and foul-smelling stools due to carbohydrate (primarily lactose) and fat malabsorption. Parents will often report that the child becomes irritable after meals, and the child may also develop a secondary anorexia due to the pain associated with the diarrhea and abdominal distention. These children exhibit classic signs on physical exam of protein-calorie malnutrition: hypotonia, poor muscle bulk, peripheral edema, decreased subcutaneous fat stores, abdominal distention, and ascites in severe cases. Anemia is common. The anemia is usually microcytic, due to iron malabsorption. However, if the disease is long-standing, and small bowel involvement is severe, a macrocytic anemia can develop as a result of folate malabsorption. Radiographs can show osteopenia due to vitamin D and calcium malabsorption in prolonged, severe cases. As opposed to cases which present in adulthood, these children are usually referred to a gastroenterologist because of their poor nutritional status and the diagnosis is prompt, because of the "classic" presentation that most clinicians were taught to look for in medical school. However, as discussed later, most patients are now presenting outside of the "classic" age range with varied symptoms both within and outside the gastrointestinal tract, making the category of "classic" a misnomer.

B. Late-Onset Gastrointestinal Form

As in the "classic" gastrointestinal form, the late-onset gastrointestinal form also includes gastrointestinal symptoms, but at an age range outside of the infant and toddler years. School-age children, adolescents, young adults, and even the elderly often complain of mild to intermittent diarrhea, gassiness, urgency, and abdominal distention and cramping. A presenting patient may also complain of chronic abdominal pain; constipation rather than diarrhea; and upper gastrointestinal tract symptoms such as nausea, dyspepsia, indigestion, gastroesophageal reflux, and chronic vomiting. Rather than being examined for celiac disease, these patients often are labeled as having irritable bowel syndrome (IBS), lactose intolerance, recurrent abdominal pain of childhood, or even an inflammatory bowel disease such as Crohn's disease. Patients are therefore treated with lactose-free or high-fiber diets, medications to treat constipation or IBS, or medications to suppress gastric acid or the immune system. Not surprisingly, the celiac patient will not get relief from these treatments. Some of these therapies, such as a high-fiber diet, which often contains a great deal of gluten, may increase symptomatology. Many patients will have upper and lower endoscopies performed by a gastroenterologist, and the results may look visually normal without obvious ulcers or cancers; thus biopsies for celiac disease will not be taken. Because of this, the diagnosis of celiac disease is often missed.

C. Extraintestinal Form

Because celiac disease is a multisystemic immune-mediated disorder, it can present with signs and symptoms outside the gastrointestinal tract. Organ systems affected can include the musculoskeletal system, skin and mucous membranes, reproductive system, hematologic system, hepatic system, and central nervous system.

1. Musculoskeletal System

One of the most common presenting features in childhood in the musculoskeletal system is idiopathic short stature. In European studies [3–5], it is estimated that about one of 10 children whose height is far below his/her genetic predisposition for unclear reasons is short because of celiac disease. Short stature can be an isolated feature of this condition. The children often have delayed onset of puberty, and stimulation testing may reveal growth hormone deficiency [6]. The potential to achieve normal stature and bone mineralization is good if these patients are diagnosed with celiac disease before puberty and placed on a gluten-free diet in a timely fashion [7]. However, after the growth plates have closed, individuals will have short stature for life.

Dental enamel defects are common in children affected by celiac disease during the toddler years. They may experience caries at an early age and in atypical locations. One report has noted that up to 30% of older patients with celiac disease have enamel hypoplasia [8]. If gluten is in the diet, these enamel defects affect only the permanent (secondary) dentition that is forming before the age of 7 years. The primary dentition or "baby teeth" are formed *in utero*, which is a gluten-free environment, and thus the primary dentition is unaffected. The dental enamel defects are linear and occur symmetrically in all four quadrants [9]. The precise cause of these defects is unknown.

Joint pain is a common complaint among both adolescents and adults with undiagnosed celiac disease. The joints are often only painful, and not red, hot, or visibly swollen. The pain often resolves with the implementation of a gluten-free diet. Clubbing of the fingers and toes (broad digits with abnormally curved nails) has also been reported. Interestingly, and to confuse the issue further, celiac disease is also seen in higher incidence in patients with rheumatologic disorders such as systemic lupus erythematosus and rheumatoid arthritis [10–12]. In an adolescent, arthritis may be the only presenting symptom [13].

One of the most serious causes of morbidity in older individuals diagnosed with celiac disease is fractures associated with osteoporosis. The potential health implications are grave and include vertebral fractures, kyphosis, hip fractures, and Colles' fracture of the lower radius in the arm [14]. These problems can begin in childhood, when radiographs may reveal delayed bone age, rickets, or osteomalacia, which left untreated may lead to osteoporosis. It is believed that early interventional therapy with a gluten-free diet may prevent progression and may even reverse bone loss [15]. However, severe osteoporosis associated with celiac disease diagnosed very late in life will not revert on the glucose-free diet after a critical point in time [16].

2. Skin and Mucous Membranes

Often overlooked, the skin and mucous membranes can be obvious sites for the expression of symptoms from celiac disease. The classic skin manifestation of celiac disease is a skin rash called dermatitis herpetiformis, often abbreviated DH. It occurs primarily in adults and is rarely seen in children before puberty. This skin rash is classically pruritic (itchy) and symmetrical and does not respond to topical creams and medications. The rash begins as flat, red lesions, which then progress to erythematous, fluid-filled blisters. They usually occur on the face, elbows, back, buttocks, and knees. The lesions initially can be confused with urticaria (hives) or varicella (chickenpox). Patients will often scratch and pick at the lesions, so that the skin will then develop an eczematous appearance. A biopsy of the normal-appearing skin next to the affected area demonstrates the characteristic histology of granular IgA deposits [17]. DH is now known to be the pathognomonic skin rash for celiac disease, and once a skin biopsy has proven the diagnosis of DH, an intestinal biopsy for celiac disease is unnecessary. If a patient with DH does have a small-intestinal biopsy, the majority of the time there will be intestinal damage [18]. For complete resolution of DH, a gluten-free diet is required. Some patients with DH also require an anti-inflammatory antibiotic, in addition to a gluten-free diet, during acute onset of DH. The skin lesions, or just the pruritis of DH, may appear when gluten is intentionally or inadvertently ingested. This should alert the affected individual to scrutinize the diet for gluten contamination.

Celiac disease can have many other manifestations in the skin and mucous membranes including urticaria, psoriasis, and oral aphthous stomatitis [18–20]. Oral aphthous stomatitis ("fever sores" or "canker sores"), which can appear on the cheeks, tongue, gums, and lips, usually correlates with histologic changes in the small intestine. However, because aphthous ulcers are also seen with common viral infections (such as herpes viruses) and other types of autoimmune and inflammatory diseases of the gastrointestinal tract (in particular Crohn's disease), they are not pathognomonic for celiac disease.

3. Reproductive System

The reproductive systems of both women and men can be affected by celiac disease. Compared to the general population, women with undiagnosed celiac disease have greater difficulty becoming pregnant [21] and a higher risk for spontaneous abortions [22, 23]. Men with undiagnosed celiac disease may also experience infertility due to gonadal dysfunction [24]. Adult patients with undiagnosed celiac disease may often undergo exhaustive and expensive infertility studies, without discovery of an etiology. Many of these infertile women will also give a history of delayed menarche, anemia, and IBS symptoms, similar to the pediatric patients described previously with short stature and delayed puberty. Women with undiagnosed and diagnosed celiac disease have infants with higher rates of neural tube defects because of folic acid deficiency, as well as more infants with intrauterine growth retardation [25–27]. Whether these individuals experience reproductive challenges because of nutritional factors, or having an undiagnosed autoimmune disease, or a combination of both, is unclear.

4. Hematologic System

Iron-deficiency anemia is one of the most common micronutrient deficiencies in the undiagnosed celiac patient. In addition to anemia, leukopenia (low white blood cell count) and thrombocytopenia (low platelet count) have also been reported. The anemia in celiac disease is usually microcytic and hypochromic, due to iron deficiency [28]. However, a macrocytic anemia should warrant an investigation into gut malabsorption of vitamin B_{12} or folic acid. Because fat-soluble vitamins are also malabsorbed in this disorder, vitamin K deficiency may occur, resulting in an increased risk for bruising and bleeding. Vitamin E deficiency may also occur, resulting in a hemolytic anemia and jaundice.

5. Liver

In children and adults with celiac disease, diseases of the liver, such as autoimmune hepatitis and chronic transaminasemia (liver enzymes), have been reported [28–31]. In children initially diagnosed with celiac disease with elevated aspartate transaminase (AST) and alanine transaminase (ALT) liver enzymes at presentation, these biochemical markers of hepatocyte damage returned to normal with strict adherence to a gluten-free diet [32]. Autoimmune liver diseases, such as autoimmune hepatitis, sclerosing cholangitis, and primary biliary cirrhosis, are all more likely to occur in the individual with celiac disease [33]. Patients with celiac disease who have evidence of chronic liver disease that has not responded to a gluten-free diet require an evaluation for these autoimmune liver diseases.

6. Central Nervous System

There is evidence that patients with celiac disease have increased rates of both neurologic and psychiatric disorders. Their associations are not always clear: Some symptoms can be attributed to nutritional deficiencies; others perhaps related to autoimmunity; and many are hypothesized to be due to yet-unproven pathways by which gluten may cross the blood-brain barrier and interact with endogenous neurotransmitter receptors. It is unclear whether these patients truly have immune-mediated phenomena associated with celiac disease or suffer from a type of "gluten sensitivity." Neurological complications have been reported in association with established celiac disease for decades [34], and gluten sensitivity can present with neurological dysfunction as its sole symptom [35].

In the neurologic system, patients with celiac disease are 20 times more likely than the general population to have epilepsy and have been reported with associated cerebral and cerebellar calcifications imaged by both computed tomography (CT) and magnetic resonance imaging (MRI) [36]. In children with celiac disease, focal white-matter lesions in the brain have been reported and are thought to be either ischemic in origin as a result of vasculitis, or caused by inflammatory demyelination [37].

Gluten-associated ataxia is the most common neurological manifestation of gluten sensitivity [38] and presents as difficulty with speech, movement, and balance due to atrophy of the cerebellum. It rarely occurs before puberty, has a mean age of onset in the late 40s, and has a paucity of gastrointestinal symptoms [39]. Emerging evidence indicates that early initiation of a gluten-free diet is beneficial for treating the symptoms of both ataxia and peripheral neuropathy, even in the absence of intestinal damage [40].

In both treated and untreated persons with celiac disease, psychiatric comorbidities, such as depression, dementia, and schizophrenia, are common. In children, behavioral changes such as irritability, increased separation anxiety, emotional withdrawal, and autistic behaviors have improved, by parental report, on a gluten-free diet [41]. Although not scientifically validated, the gluten-free diet, along with a casein-free diet, is now also being advocated by several groups for children with autism [42–44]. Whether children with autism are at a higher risk for celiac disease, or celiac children have a higher incidence of autism, remains to be proven. However, children with Down syndrome, who often have autistic behaviors, are at higher risk for celiac disease [45]. It has been hypothesized that gluten may be broken down into small peptides that may cross the blood-brain barrier and interact with morphine receptors, leading to alterations in conduct and perceptions of reality [35]. It is important to note that only a proportion of patients presenting with neurological dysfunction associated with a gluten sensitivity will also have biopsy-proven intestinal damage [46].

D. Associated Conditions

There are several groups of patients who are at substantially increased risk of having celiac disease: those with another autoimmune disorder; those with certain syndromes; and relatives of biopsy-diagnosed patients. These associated conditions are further described in Table 1. Patients with other autoimmune diseases, such as type 1 diabetes, as well as relatives of persons with biopsy-diagnosed celiac disease, are more likely to carry the HLA (human leukocyte antigen) haplotypes that put them at risk for the disease. It has been demonstrated that patients diagnosed with celiac disease early in life, and treated early with a gluten-free diet, have a significantly lower risk of developing other autoimmune disorders than do individuals diagnosed later in life [47].

Persons with both Down and Turner syndromes classically have short stature and are at increased risk for skeletal and other autoimmune diseases such as diabetes mellitus, Crohn's disease, and thyroid disease [48], which makes the diagnosis in this population exceptionally difficult. A recent multicenter study from Italy strongly suggested the need for screening all children with Down syndrome, regardless of the presence or absence of gastrointestinal symptomatology [49]. It is unclear why individuals with certain syndromes are at higher risk for celiac disease. However, because several of these syndromes have as part of their constellation poor growth and short stature, anthropometric measures are poor screening tools for malabsorption in these populations. The NIH Consensus Development Conference on Celiac Disease, held in June 2004, recommended that patients with

TABLE 1 Conditions and Syndromes at Increased Risk for Celiac Disease (CD)

Condition	% Diagnosed with CD	Increased Risk	References
Arthritis	1.5–7.5%	3- to 10-fold	[10, 66, 113, 114]
Cardiomyopathy (idiopathic dilated)	5.7%		[115]
Dental enamel defects	19–30%	19-fold	[8, 9, 116]
Dermatitis herpetiformis	100%		[17, 78, 117]
Diabetes type 1	3.5–10%	4- to 10-fold	[118–122]
Down syndrome	4–20%	17 to 50-fold	[49, 66, 123, 124]
Epilepsy	2%		[36, 125]
IgA deficiency	7%	31-fold	[126]
Infertility (idiopathic)	6.3%		[66]
Iron-deficiency anemia	4.2%	16-fold	[28, 66]
Osteoporosis	2.6%		[66]
Primary biliary cirrhosis	6%		[33]
Relatives, first- or second-degree	2.6–20%	18-fold	[64–66]
Short stature (idiopathic)	4–10%	23-fold	[3–5, 66]
Sjögren's syndrome	2–3%		[66, 127]
Thyroid disease (autoimmune)	4%		[128]
Turner's syndrome	4–8%		[129–131]
Williams syndrome	9.5%		[132]

Down, Turner, or Williams syndrome should be offered screening for celiac disease at least once [50].

E. Asymptomatic Form

It can be debated whether or not there is truly an "asymptomatic" form of celiac disease, as historically, *asymptomatic* meant not fitting into the definition of the "classic" presentation, with severe diarrhea and protein-calorie malnutrition. As described in the previous sections, patients who present with gastrointestinal symptoms outside of the pediatric age range, with extraintestinal symptoms or associated conditions, may have previously been described as "asymptomatic." The medical literature describes patients lacking the signs and symptoms occurring in the gastrointestinal tract or extraintestinal systems as "asymptomatic." With a more detailed history, physical exam, and laboratory investigations, these "asymptomatics" may have revealed evidence of trace mineral deficiencies, anemia, short stature, low bone density, and low serum fat-soluble vitamin levels. These patients may be identified through mass serologic screenings and are often first- and second-degree relatives of a patient with biopsy-diagnosed disease. A family history of gastrointestinal cancers or other autoimmune disorders is often elicited [51–54].

F. Latent Form

Patients with the latent form of celiac disease have positive antibodies for the condition but an initial normal intestinal biopsy. It is believed that over time, with further ingestion of gluten, individuals with the latent form will develop celiac disease with subsequent abnormal histology of the small bowel. These small intestinal changes will then revert to normal with the initiation of a gluten-free diet [53–55].

III. DIAGNOSIS OF CELIAC DISEASE

A. Serologic and Genetic Screening Tests

The tests of choice to screen for celiac disease are serum immunological markers with high sensitivity and specificity. The classic screening tests that are used for other gastrointestinal malabsorption disorders, such as fecal fat, glucose tolerance tests, D-xylose uptake, serum carotene levels, and permeability tests (such as lactulose/mannitol ratios), are poorly specific for celiac disease and should not be used in place of the serologic markers that are discussed hereafter. The clinician must remember, however, that no screening test is perfect, and that the current "gold standard" to confirm the diagnosis of celiac disease remains a small-intestinal biopsy and the patient's clinical response to a gluten-free diet. Any patient with suggestive symptoms should have a small-bowel biopsy, even if the serology is negative.

There are three frequently used commercially available serologic tests (antibodies): anti-gliadin (AGA), anti-endomysial or anti-endomysium (EMA or AEA), and anti-tissue transglutaminase (tTG). Sensitivity, specificity, and positive and negative predictive values can vary widely for each antibody, depending on the age of the patient, the population being studied, the substrates and laboratory kits being used to run the assays, and the proficiency of the laboratory performing the test [56]. Conditions that may yield false negative antibody results include a patient who makes low levels of or no immunoglobulin A (IgA), young children who may not manufacture autoimmune antibodies, an inexperienced lab, and testing while the patient is already on a gluten-free diet. False positive tests can also be seen with these antibody tests, as delineated later, in normal individuals with other gastrointestinal disorders and in other autoimmune disorders.

1. Anti-Gliadin Antibodies

The anti-gliadin antibodies were the first serologic tests available to screen for celiac disease during the late 1970s and were the first step toward recognizing this condition as an immune-mediated disorder. The anti-gliadin antibodies IgG and IgA recognize a small antigenic portion of the gluten protein called gliadin [57]. AGA IgG has good sensitivity, whereas AGA IgA has good specificity, and therefore their combined use provided the first reliable screening test. Unfortunately, many normal individuals can have an elevated AGA IgG, causing much confusion among general practitioners. AGA IgG is useful in screening IgA-deficient individuals, because the other antibodies used for routine screening are usually of the IgA class. Whereas only 0.2–0.4% of the general population has selective IgA deficiency, as many as 2–3% of people with celiac disease are IgA deficient, complicating the screening procedure [58]. Other conditions under which an elevated AGA IgG can be seen include enteropathies where the gut is more permeable to gluten, such as parasitic infections, Crohn's disease, allergic gastroenteropathy, and autoimmune enteropathy. A strength of the AGA antibodies is that they are ELISA (enzyme-linked immunosorbent assay) tests, and the results are independent of observer variability.

2. Anti-Endomysium Antibodies

The serologic test currently commercially available with the highest sensitivity and specificity is the anti-endomysium IgA immunofluorescent antibody (EMA). The EMA was discovered in the early 1980s and rapidly gained use as part of a screening "celiac panel" by commercial labs in combination with AGA IgG and AGA IgA. False negative EMA can be seen in young children, those with IgA deficiency, and in the hands of an inexperienced laboratory because of the subjective nature of the test [56]. Also, the substrate for this antibody was initially monkey esophagus, making it expensive and unsuitable for screening large numbers of people. Human umbilical cord is now used as an alternative to monkey esophagus in most commercial laboratories [59].

3. Tissue Transglutaminase Antibodies

Tissue transglutaminase (tTG) was described in 1997 as the autoantigen of EMA [60]. The initial tTG ELISA was guinea pig IgA, with a lower sensitivity and specificity than EMA [61, 62]. However, most commercial labs now use human recombinant tTG, which has improved sensitivity and specificity and correlates better with EMA IgA and intestinal biopsy results [60–63]. The tTG IgA ELISA represents an improvement over the EMA IgA assay because it is less expensive, is less time-consuming, is not a subjective test, and can be performed on a single drop of blood using a dot-blot technique, making this an ideal test for mass serologic screenings. Positive tTG results may be seen in other autoimmune diseases, such as type 1 diabetes, autoimmune liver disease, autoimmune thyroid disease, and inflammatory bowel disease.

4. Genetic Testing

Although celiac disease is the only autoimmune disease for which we know the environmental trigger, gluten, we also know that there is a strong genetic influence. For example, as discussed under "associated conditions," in first- and second-degree relatives, there can be up to a 20% disease prevalence [64–66]. Also, identical twins have a 75% concordance rate for celiac disease (one of the highest rates reported for any disease), whereas nonidentical twins do not differ from siblings [67]. This again indicates a genetic, in addition to an environmental, component.

Celiac disease is a complex genetic disorder, but the actual "celiac genes" have not yet been identified. The strongest genetic determinant of risk for celiac disease appears to be the presence of certain HLA alleles. The presence of these HLA alleles, DQ2 and DQ8, is thought to account for up to 40% of the genetic load of the familial risk for celiac disease [68]. The HLA are markers that help identify cells as "self" versus "non-self." The HLA prevent the immune system from attacking "self." HLA DQ2 or DQ8 are found in 95% of celiac patients. It is extremely rare for a celiac patient to have neither of these genes. However, if an individual has these HLA alleles, it does not mean that the individual has celiac disease, because these alleles are found in 39.5% of the general population [69–71]. This is a great source of confusion for patients and physicians, because it is counterintuitive to other types of genetic testing, where the presence of the gene confirms the disease.

The value of genetic testing is that it has a high negative predictive value to rule out celiac disease for a patient's lifetime. Negativity for HLA DQ2 and DQ8 excludes the diagnosis of celiac disease with 99% confidence. However, positivity for DQ2 or DQ8 has limited diagnostic value, because of the high prevalence of these genotypes in the general population. The strength of genetic testing is that the patient does not need to be on a gluten-containing diet to be tested, because the presence of these genes is not affected by diet. Therefore, genetic testing can evaluate:

- Infants not yet exposed to gluten
- Young children who may not make all of the antibodies

- Patients who have self-imposed a gluten-free diet
- Patients with serology or biopsies that were not conclusive
- Relatives of biopsy-diagnosed individuals with celiac disease

5. Current Guidelines on the Use of Serologic and Genetic Testing to Screen for Celiac Disease

As described earlier, the IgA class human anti-tTG antibody, coupled with a determination of total serum IgA to rule out deficiency, currently seems to be the most cost-effective way to screen for celiac disease in an otherwise healthy adult. EMA should be used as a confirmatory, pre-biopsy test, whereas AGA determinations should be restricted to the diagnostic workup of younger children and patients with IgA deficiency. Given the high prevalence of the HLA haplotypes associated with celiac disease in the general population, genetic testing is not recommended as a routine screening test for celiac disease. The practitioner should remember that serologic tests are screens and that to confirm the diagnosis of celiac disease a small bowel biopsy must be performed, as discussed next.

B. The Intestinal Biopsy

Confirmation of either a clinical suspicion of celiac disease or a positive serologic screen requires a small intestinal biopsy. Before the advent of fiberoptic and chip technology for endoscopes, biopsies of the jejunum were obtained using a Crosby spring-loaded capsule, passed orally under fluoroscopic guidance. Most biopsies are performed today using a flexible endoscope, passed orally under either conscious sedation or general anesthesia. This has the advantages of allowing direct visualization of the mucosa with a camera to look for changes suggestive of small bowel damage, such as notching or scalloping of the small bowel folds or lymphonodular hyperplasia [72]. Endoscopy also allows the endoscopist to look for other lesions, such as ulcers, esophagitis, or gastritis, which may help explain the patient's symptomatology.

The detailed description of the characteristic small bowel changes seen in celiac disease that was given by Marsh in 1988 has become accepted as the standard [73]. The Marsh criteria describe four patterns of mucosal pathology: type 0 (pre-infiltrative) which is without detectable inflammation or changes in the crypt/villous architecture; type 1 (infiltrative) with an increase in the intraepithelial lymphocytes but without detectable changes in the crypt/villous architecture; type 2 (hyperplastic) with inflammation, villous blunting, and an increased crypt/villous height ratio; and type 3 (destructive) with severe inflammation, flat villi, and hyperplastic crypts. The clinician, however, needs to be aware that villous atrophy (shortening of the finger-like projections in the small bowel, which increase absorptive surface area) can be caused by a wide variety of gastrointestinal diseases and infections, and that correlation with serology and the patient's response to a gluten-free diet is imperative to confirm the diagnosis. In rare instances, a gluten challenge may be necessary to confirm that the villous atrophy was due to celiac disease, and not a concomitant gastrointestinal infection. In cases where serology is suggestive, the biopsy is confirmative, and the patient has had a clinical response to the gluten-free diet, a gluten challenge and repeat small bowel biopsy are no longer required [74].

IV. TREATMENT OF CELIAC DISEASE WITH A GLUTEN-FREE DIET

The only known treatment for celiac disease is a gluten-free diet. This was first discovered after World War II in children with celiac disease when the toxicity of wheat proteins was established after the bread shortages resolved in Europe [75, 76]. Gluten is important in baked goods, because it plays an important role in leavening, in forming the structure of the dough, and in holding the baked product together [57]. Removal of gluten from the diet of a biopsy-diagnosed person with celiac disease results in complete symptomatic and histologic resolution of the disease in the majority of patients. The identified agents responsible for the immune-mediated response and intestinal damage are prolamins, storage proteins located in the seeds of different grains. Gluten is the general name for the prolamins found in wheat (gliadin), rye (secalin), barley (hordein), and oats (avenin) [57].

The prolamin of oats, avenin, accounts for only 5–15% of the total seed protein, as opposed to gliadin, which comprises about 50% of wheat proteins [77]. This oat prolamin is thought not to elicit the same immune response as gliadin and is thought by some to be safe for patients with celiac disease to ingest [78]. The risk that oats are contaminated with wheat in the United States is great, because oats are often crop-rotated, harvested, and milled with wheat. A study in the United States in newly diagnosed children with celiac disease who were allowed to eat oats saw that these children had symptomatic and histologic resolution of the disease comparable to children who were denied oats [79]. Prolamins are also found in corn and rice, but they likewise do not elicit an immune reaction in the intestines of individuals with celiac disease [57].

Table 2 provides some basic dietary guidelines for persons following a gluten-free diet. Although not all-inclusive, it is meant to serve as a starting point for discussion between patients and health care practitioners. Many newly diagnosed individuals are not aware that "gluten-free" does not mean just eliminating bread and pastries from the diet, because gluten (especially wheat) can be identified on food labels and in restaurants by many other names. For example, triticale (a combination of wheat and rye), kamut, and spelt are all forms of wheat and are considered toxic [80]. Other forms of wheat, such as bulgur, couscous, einkorn, farina, and semolina (durum), are also not permitted on the gluten-free diet. Any food product that contains rye, barley, or malt (a partial hydrolysate of barley) has prolamins that are considered harmful [81]. In general, a food product that includes wheat in its name (such as cracked wheat, wheat bran, wheat grass, wheat germ, or whole wheat) or malt in its name (barley malt, malt extract, malt flavoring, or malt syrup) is considered to contain gluten. One notable exception, however, is buckwheat, which is not directly related to *Triticum* and is considered safe to consume. Distilled ingredients (such as vinegar and alcohol) are allowed, because gluten does not pass into the distillate. However, beverages made with barley (such as beer, ale, lager, and some rice and soy drinks) are not allowed [57].

Food labeling in the United States has recently undergone some changes. The Food Allergen Labeling and Consumer Protection Act was signed into law in August 2004. It requires food labels to clearly state if a product contains any of the top eight food allergens: milk, eggs, fish, crustacean shellfish, tree nuts, peanuts, soybeans, and wheat. All food products manufactured in the United States after January 1, 2006, are required to have updated labels declaring the presence of any of the top eight food allergens in the product. The Food Allergen Labeling and Consumer Protection Act of 2004 was primarily passed to benefit individuals with food allergies. However, it is also of tremendous value to those with celiac disease, because wheat is often hidden on ingredient labels as "starch," "flavorings," "seasonings," "couscous," "farro," "farina," or "hydrolyzed vegetable protein" (Table 2). Because wheat is the most commonly used grain in the United States, by clarifying the source of ingredients and identifying "wheat," about 90% of labeling concerns are resolved for celiac and gluten-sensitive patients. The new law also calls for the Food and Drug Administration (FDA) to issue rules, by 2008, detailing what it means when a product is labeled "gluten-free." Unlike Europe, Canada, and Australia, the United States does not have a defined standard for "gluten-free" foods, causing a great deal of confusion over

TABLE 2 Basic Dietary Guidelines for Individuals Following a Gluten-Free Diet

Not Allowed	Allowed	Questionable
Barley	Amaranth	Dextrin[a]
Bran	Beans	Flavorings[b]
Bulgur	Buckwheat	Hydrolyzed plant protein (HPP)
Cereal binding	Cheese[c]	Hydrolyzed vegetable protein (HVP)
Couscous	Corn (maize)	Modified food starch
Einkorn wheat	Egg	Oats[d]
Emmer wheat	Fish	Seasonings[e]
Farro	Fruit	Spices[f]
Farina	Kasha	Starch[g]
Filler	Meat[h]	
Graham flour	Milk[c]	
Kamut	Millet	
Malt	Nuts	
Rye	Potato	
Semolina	Peas	
Spelt	Quinoa	
Triticale	Rice, wild rice	
	Sorghum	
	Soybean	
	Teff	
	Tapioca	

Sources: Hardman *et al.* [78], Case [57], Forssell and Weiser [80], Ellis *et al.* [81].

[a]In North America, usually derived from corn or tapioca.

[b]In North America, gluten-containing grains are almost never used as flavorings, with the exception of barley malt, which is usually indicated on the label.

[c]Many individuals with celiac disease are lactose intolerant, and the coating of some cheeses may contain gluten.

[d]See text for detailed explanation on oats.

[e]A blend of flavoring agents, often using a carrier such as cereal flour or starch.

[f]Pure spices do not contain gluten, but imitation spices may have fillers.

[g]Corn, potato, tapioca, and rice are the usual sources of modified food starch in North America; however, food starch can be made from wheat.

[h]Without breading or gluten-containing seasonings.

what it means when an American manufacturer puts "gluten-free"' on a product's label. The FDA rules will establish a standard that will make it even easier for those with celiac disease to readily identify products that are safe.

Because of current food-labeling practices in the United States, it still may not be obvious whether or not an item contains gluten. Products to question include those labeled "wheat-free" (which may contain other harmful grains, such as rye or barley) and ingredients that do not state their sources (such as flavorings, spices, starch, or hydrolyzed vegetable protein). Gluten is often used as a flavoring in candy, sauces, seasonings, soups, and salad dressings and as a filler in vitamins and medications [57, 82]. Many times the only way for persons with celiac disease to be certain that a specific product is gluten-free is to call the manufacturer directly and often, because ingredients in products commonly change without warning. Toiletries, such as shampoos, conditioners, and skin care products, are thought not harmful as long as they are not ingested. Patients with open skin lesions, such as with dermatitis herpetiformis, may have reactions on the skin or systemically if gluten gets into an open wound.

The education of the person newly diagnosed with celiac disease should consist of a team approach between the patient (or parents); the gastroenterologist; the primary care physician; the dietitian; and local branches of national support groups. Medical management primarily consists of monitoring for compliance with the gluten-free diet and screening for the well-known complications to be discussed later. After the gastroenterologist who performed the biopsy confirms the diagnosis, the patient should be immediately referred to a knowledgeable dietitian for medical nutrition therapy [83]. Physicians and dietitians should encourage the patient to join local chapters of national support organizations, which can aid in finding local resources, such as supermarkets, food manufacturers, literature, and restaurants that are familiar with the gluten-free diet.

Lifelong compliance with the gluten-free diet is challenging. The most important factors in achieving compliance are patient education, close supervision by an interested physician, and regular nutritional counseling by a registered dietitian with expertise in this area [65, 83]. Compliance can be improved even in adolescents if they are seen by a physician on a regular basis [84, 85]. One of the best and least expensive markers for dietary compliance is assessment by a trained interviewer (either a physician or dietitian) because of the low cost and noninvasive nature of dietary assessment. There is a strong correlation between self-reported intake of foods containing gluten and intestinal damage [85].

V. MANAGEMENT OF THE COMPLICATIONS OF CELIAC DISEASE

Patients with undiagnosed and untreated celiac disease, as well as those diagnosed later in life, have increased morbidity and mortality due to associated conditions, including osteoporosis, nutritional deficiencies, other autoimmune diseases, and some cancers. These patients also incur increased health care costs because of being chronically ill, the need to see multiple subspecialists, and the tests performed on them until the correct diagnosis is obtained [86]. Corrao and other investigators of the Club del Tenue Study Group formed a prospective cohort study that included 1072 adults with diagnosed celiac disease and 3384 first-degree relatives. These individuals were followed for 32 years. The number of deaths between the two groups were compared and expressed as the standardized mortality ratio (SMR) and relative survival ratio. Two times the number of persons with celiac disease died compared to the relatives [SMR 2.0, 95% confidence interval (CI) 1.5–2.7]. The greatest excess of deaths occurred during the first 3 years after diagnosis. These results suggest that prompt and strict dietary treatment may decrease premature mortality among persons with celiac disease [87].

The primary reason for the increased mortality is the association with gastrointestinal malignancies, primarily intestinal lymphoma, which has been reported in up to 10–15% of adult patients who have been noncompliant with the diet [88]. The odds ratio overall for non-Hodgkin's lymphoma associated with celiac disease compared to first-degree relatives was reported to be 3.1, with odds ratios of 16.9 for gut lymphoma and 19.2 for T-cell lymphoma, respectively [89]. The good news is that the reported risk for lymphoma decreases to that of the general population on the gluten-free diet [90].

In a minority of patients with biopsy-diagnosed celiac disease, there will continue to be gastrointestinal symptoms and failure of normalization of intestinal damage, despite vigorous adherence to the gluten-free diet. Despite this, the majority of patients who are not better on the diet should not immediately be labeled as "refractory" to the diet. Given the challenges of living a gluten-free lifestyle, a thorough dietary history should first be taken to exclude inadvertent (or intentional) ingestion of gluten. Compliance can also be assessed by measuring the immunological markers (antibodies) for the disease. Also, in any celiac patient with persistent symptomatology, one should consider a repeat intestinal biopsy [91].

Histologic evidence of villous atrophy, despite rigorous adherence to the gluten-free diet for more than a year, should provoke a workup for gastrointestinal infections (such as viruses, bacteria, and parasites) and noninfectious diseases (such as other autoimmune and allergic diseases). Additional reasons for nonresponsiveness to the gluten-free diet include pancreatic insufficiency and T-cell lymphoma, both of which are complications of long-standing celiac disease [92, 93]. As many as 75% of adults with refractory sprue may have an aberrant clonal intraepithelial T-cell population associated with a condition classified as "cryptic enteropathy-associated T-cell lymphoma" [94].

These patients frequently require immunosuppressing medications, such as steroids, azathioprine, and cyclosporine [95–99] in addition to a gluten-free diet. For more information regarding refractory sprue, comprehensive reviews have been published [100, 101].

All patients with celiac disease, whether it be long- or short-standing, are at risk for nutritional deficiencies. Children should be examined for protein-calorie malnutrition, linear growth failure, and delayed puberty. All patients should be screened for the nutritional deficiencies that can accompany this malabsorptive disorder, such as iron deficiency anemia and fat-soluble vitamin deficiencies (vitamins A, E, 25-hydroxy-D, and a prothrombin time to check vitamin K status). Adult patients should also be monitored for the common extraintestinal complications, including osteoporosis, neurologic complaints, and the development of other autoimmune diseases, especially of the thyroid and liver [102, 103]. Bone density should be measured in the newly diagnosed celiac patient, because numerous studies have documented low bone density in both children and adults at the time of initial diagnosis of celiac disease. Osteopenia can improve with the gluten-free diet, and progression of osteoporosis can be halted with appropriate supplementation [104–107]. Osteopenic patients should be evaluated for deficient intake and absorption of vitamin D and calcium and the development of secondary hyperparathyroidism [108].

Special considerations need to be given to patients who have both celiac disease and type 1 diabetes, because many of the well-known complications of type 1 diabetes can be exacerbated by nutritional deficiencies. Nocturnal hypoglycemia with seizures and recurrent, unexplained hypoglycemia with a reduction in insulin requirements should prompt the physician to investigate for celiac disease [50, 109, 110]. In the young child, growth failure and delayed sexual maturation may be seen. Vitamin A deficiency may aggravate retinopathy; deficiencies of vitamins E and B_{12} may cause peripheral neuropathy; iron and folic acid deficiencies may lead to complications of fertility and pregnancy; and vitamin D can complicate dental disease, limit joint mobility, and cause osteopenia and osteoporosis. There is also an increased incidence of other autoimmune diseases in type 1 diabetics who have "silent" celiac disease [111, 112]. The gluten-free diet presents additional challenges to the diabetic patient, who may see acute hyperglycemia and a steady rise in hemoglobin A1c on initiation of this diet. This can be due to intestinal healing and better absorption, as well as gluten-free food substitutes, which can be corn-, rice-, or potato-based, and have a higher glycemic index.

Once the patient has undergone initial counseling, the primary care physician (or gastroenterologist) and dietitian should follow up with the patient in 3 to 6 months to discuss compliance with the diet and reinforce its importance. If the patient has been able to adjust to the lifestyle and has had no complications of the disorder, he or she can be seen annually. At the annual visit, a detailed dietary history should be elicited, and serum antibodies should be measured to gauge adherence. First- and second-degree relatives should be offered serologic screening. The primary care physician should perform a detailed history and physical aimed at screening for nutritional deficiencies and looking for signs and symptoms of other autoimmune disorders, gastrointestinal cancers, and refractory sprue. If the patient is doing well without clinical symptoms and has normal antibody titers, he or she should continue to be followed annually. If the patient is doing poorly, indicated by symptoms, nutritional deficiencies or elevated antibodies, more extensive medical nutritional therapy should be given by a knowledgeable dietitian [83]. This patient will also require closer monitoring for the development of the aforementioned nutritional, autoimmune and possibly malignant complications.

The reader is invited to review separate chapters in this book about additional nutritional considerations in colon cancer, type 1 diabetes, gastroesophageal reflux disease, diarrhea, constipation, lactose intolerance, liver disease, food allergy, osteomalacia, and osteoporosis, all of which can complicate celiac disease.

VI. SUMMARY

The astute clinician must have a high index of suspicion to make the diagnosis of celiac disease. Although this condition is very common, it is underdiagnosed because of its protean manifestations. Although classically thought to present in childhood with diarrhea and protein-calorie malnutrition, we now know this disease can present outside of the pediatric age range with a variety of symptoms ranging from joint pain to infertility to anemia. Serum antibodies are an excellent screening tool for this disease. However, confirmation requires a small bowel biopsy performed by a gastroenterologist via an upper endoscopy. Celiac disease is the only autoimmune disease for which we know the trigger: gluten. Removal of gluten from the diet results in a complete histologic and symptomatic recovery in the majority of patients. Not uncommon complications of celiac disease can include vitamin and mineral deficiencies, the development of other autoimmune diseases, and a higher risk for osteoporosis and gastrointestinal cancers. Living a gluten-free lifestyle is challenging, because the potential for contamination of foods by wheat, rye, and barley is great, and food labeling is not ideal. The patient benefits best from the involvement of a physician, a dietitian, and a support group who are up-to-date about the diet, as well as the latest literature and advances in the understanding of the complex interactions between gluten and the immune system.

References

1. Logan, R. F. (1991). Descriptive epidemiology of celiac disease. *In* "Gluten-Sensitive Enteropathy (Frontiers of Gastrointestinal Research)" (D. Branski, P. Rozen, and M. F. Kagnoff, Eds.), pp. 1–14. S. Karger, Basel, Switzerland.
2. Catassi, C., Fabiani, E., Ratsch, I. M., Coppa, G. V., Giorgi, P. L., Pierdomenico, R., Alessandrini, S., Iwanejko, G., Domenici, R., Mei, E., Miano, A., Marani, M., Bottaro, G., Spina, M., Dotti, M., Montanelli, A., Barbato, M., Viola, F., Lazzari, R., Vallini, M., Guariso, G., Plebani, M., Cataldo, F., Traverso, G., and Ventura, A. (1996). The coeliac iceberg in Italy. A multicentre antigliadin antibodies screening for coeliac disease in school-age subjects. *Acta Paediatr. Suppl.* **412**, 29–35.
3. Groll, A., Candy, D. C., Preece, M. A., Tanner, J. M., and Harries, J. T. (1980). Short stature as the primary manifestation of coeliac disease. *Lancet* **2**, 1097–1099.
4. Cacciari, E., Salardi, S., Lazzari, R., Cicignani, A., Collina, A., Pirazzoli, P., Tassoni, P., Biasco, G., Corazza, G. R., and Cassio, A. (1983). Short stature and celiac disease: A relationship to consider even in patients with no gastrointestinal tract symptoms. *J. Pediatr.* **103**, 708–711.
5. Stenhammar, L., Fallstrom, S. P., Jansson, G., Jansson, U., and Lindberg, T. (1986). Coeliac disease in children of short stature without gastrointestinal symptoms. *Eur. J. Pediatr.* **145**, 185–186.
6. Verkasalo, M., Kuitunen, P., Leisti, S., and Perheentupa, J. (1978). Growth failure from symptomless celiac disease. *Helv. Paediatr. Acta* **33**, 489–495.
7. Barr, D. G., Shmerling, D. H., and Prader, A. (1972). Catch-up growth in malnutrition, studied in celiac disease after institution of a gluten-free diet. *Pediatr. Res.* **6**, 521–527.
8. Smith, D. M., and Miller, J. (1979). Gastro-enteritis, coeliac disease and enamel hypoplasia. *Br. Dent. J.* **147**, 91–95.
9. Aine, L. (1986). Dental enamel defects and dental maturity in children and adolescents with celiac disease. *Proc. Finn. Dent. Soc.* **3**(Suppl), 1–71.
10. George, E. K., Hertzberger-ten Cate, R., van Suijekom-Smit, L. W., von Blomberg, B. M., Stapel, S. O., van Elburg, R. M., and Mearin, M. L. (1996). Juvenile chronic arthritis and coeliac disease in The Netherlands. *Clin. Exp. Rheumatol.* **14**, 571–575.
11. Lepore, L., Martelossi, S., Pennesi, M., Falcini, F., Ermini, M. L., Perticarari, S., Presani, G., Lucchesi, A., Lapin, M., and Ventura, A. (1996). Prevalence of celiac disease in patients with juvenile chronic arthritis. *J. Pediatr.* **129**, 111–113.
12. O'Farrelly, C., Marten, D., Melcher, D., McDougall, B., Price, R., Goldstein, A. J., Sherwood, R., and Fernandes, L. (1988). Association between villous atrophy in rheumatoid arthritis and rheumatoid factor and gliadin-specific IgG. *Lancet* **2**, 819–822.
13. Maki, M., Hallstrom, O., Verronen, P., Reunala, T., Lahdeaho, M. L., Holm, K., and Visokorpi, J. K. (1988). Reticulin antibody, arthritis, and coeliac disease in children. *Lancet* **1**, 479–480.
14. Valdimarsson, T., Lofmano, O., Toss, G., and Strom, M. (1996). Reversal of osteopenia with diet in adult coeliac disease. *Gut* **38**, 322–327.
15. Mora, S., Barera, G., Beccio, S., Proverbio, M. C., Weber, G., Bianchi, C., and Chiumello, G. (1999). Bone density and bone metabolism are normal after long-term gluten-free diet in young celiac patients. *Am. J. Gastroenterol.* **94**, 398–403.
16. Meyer, D., Stavropolous, S., Diamond, B., Shane, E., and Green, P. H. (2001). Osteoporosis in North American adult population with celiac disease. *Am. J. Gastroenterol.* **96**, 112–119.
17. Fry, L. (1995). Dermatitis herpetiformis. *Bailliere's Clin. Gastroenterol.* **9**, 371–393.
18. Zone, J. (2005). Skin manifestations of celiac disease. *Gastroenterology* **128**, S87–S91.
19. Scala, E., Giani, M., Pirrotta, L., Guerra, E. C., DePita, O., and Pudda, P. (1999). Urticaria and adult celiac disease. *Allergy* **54**, 1008–1009.
20. Michaelsson, G., Gerden, B., Hagforsen, E., Nilsson, B., Pihl-Ludin, I., Kraaz, W., Hjelmquist, G., and Loof, L. (2000). Psoriasis patients with antibodies to gliadin can be improved by gluten-free diet. *Br. J. Dermatol.* **142**, 44–51.
21. Auricchio, S., Greco, L., and Troncone, R. (1988). Gluten-sensitive enteropathy in childhood. *Pediatr. Clin. North Am.* **35**, 157–187.
22. Gasbarrini, A., Torre, E., Trivellini, C., DeCarolis, S., Carso, A., and Gasbarrini, G. (2000). Recurrent spontaneous abortion and intrauterine fetal growth retardation as symptoms of coeliac disease. *Lancet* **356**, 399–400.
23. Ciacci, C., Cirillo, M., Auriemma, G., DiDato, G., Sabbatini, F., and Mazzacca, G. (1996). Celiac disease and pregnancy outcome. *Obstet. Gynecol. Surv.* **51**, 643–644.
24. Farthing, M. J., Rees, L. H., Edwards, C. R., and Dawson, A. M. (1983). Male gonadal function in coeliac disease. 2. Sex hormones. *Gut* **24**, 127–135.
25. Collin, P., Vilska, S., Heinonen, P. K., Hallstrom, O., and Pikkarainen, P. (1996). Infertility and coeliac disease. *Gut* **39**, 382–384.
26. Kolho, K. L., Tiitinen, A., Tulppala, M., Unkila-kallio, L., and Savilahti, E. (1999). Screening for coeliac disease in women with a history of recurrent miscarriage or infertility. *Br. J. Obstet. Gynaecol.* **106**, 171–173.
27. Sher, K. S., and Mayberry, J. F. (1994). Female fertility, obstetric and gynaecological history in coeliac disease. *Digestion* **55**, 243–246.
28. Carroccio, A., Iannitto, E., Cavataio, F., Montalto, G., Tumminello, M., Campagna, P., Lipari, M. G., Notabartolo, A., and Iacono, G. (1998). Sideropenic anemia and celiac disease: One study, two points of view. *Dig. Dis. Sci.* **43**, 673–678.
29. Maggiore, G., De Giacomo, C., Scotta, M. S., and Sessa, F. (1985). Coeliac disease presenting as chronic hepatitis in a girl. *J. Pediatr. Gastroenterol. Nutr.* **5**, 501–503.
30. Loenardi, S., Bottaro, G., Patane, R., and Musumeci, S. (1990). Hypertransaminasemia as first symptom in infant coeliac disease. *J. Pediatr. Gastroenterol. Nutr.* **11**, 404–406.
31. Davison, S. (2002). Coeliac disease and liver dysfunction. *Arch. Dis. Child.* **87**, 293–296.
32. Fontanella, A., Vajro, P., Ardia, E., and Greco, L. (1987). Danno epatico in corso di malattia celiaca. Studio retrospettivo in 123 bambini. *Riv. Ital. Pediatr.* **5**, 80–85.
33. Kingham, J. G. C., and Parker, D. R. (1998). The association between biliary cirrhosis and coeliac disease: A study of relative prevalences. *Gut* **42**, 120–122.
34. Cooke, W. T., and Smith, W. T. (1966). Neurological disorders associated with adult coeliac disease. *Brain* **89**, 683–722.

35. Hadjivassiliou, M., Gibson, A., Davies-Jones, G. A., Lobo, A. J., Stephenson, T. J., and Milford-Ward, A. (1996). Does cryptic gluten sensitivity play a part in neurological illness? *Lancet* **347**, 371.
36. Gobbi, G., Bouquet, F., Greco, L., Lambertini, A., Tassinari, C. A., Ventura, A., and Zaniboni, M. G. (1992). Coeliac disease, epilepsy, and cerebral calcifications. *Lancet* **340**, 439–443.
37. Kieslich, M., Errazuriz, G., Posselt, H. G., Moeller-Hartmann, W., Zanella, F., and Boehles, H. (2001). Brain white-matter lesions in celiac disease: A prospecive study of 75 diet-treated patients. *Pediatrics* **108**, e21.
38. Hadjivassiliou, M., Grunewald, R. A., Chattopadhyay, A. K., Davies-Jones, G. A., Gibson, A., Jarratt, J. A., Kandler, R. H., Lobo, A., Powell, T., and Smith, C. M. (1998). Clinical, radiological, neurophysiological and neuropathological characteristics of gluten ataxia. *Lancet* **352**, 1582–1585.
39. Hadjivassiliou, M., Grunewald, R., Sharrack, B., Sanders, D., Lobo, A., Williamson, C., Woodroofe, N., Wood, N., and Davies-Jones, A. (2003). Gluten ataxia in perspective: Epidemiology, genetic susceptibility and clinical characteristics. *Brain* **126**, 685–691.
40. Hadjivassiliou, M., Grunewald, R. A., and Davies-Jones, G. A. (2002). Gluten sensitivity as a neurological illness. *J. Neurol. Neurosur. Psychiatry* **72**, 560–563.
41. Fabiani, E., Catassi, C., Villari, A., Gismondi, P., Pierdomenico, R., Ratsch, I. M., Coppa, G. V., and Giorgi, P. L. (1996). Dietary compliance in screening-detected coeliac disease adolescents. *Acta Paediatr. Acta Paediatr. Suppl.* **412**, 65–67.
42. Knivsberg, A. M., Reichelt, K. L., Nodland, M., and Torleiv, H. (1995). Autistic syndromes and diet: A follow-up study. *Scand. J. Educ. Res.* **39**, 223–236.
43. Autism Network for Dietary Intervention. (2007). Retrieved September 2007 from http://www.autismndi.com.
44. The gluten free casein free diet. (2007). Retrieved September 2007 from http://www.gfcfdiet.com.
45. Book, L., Hart, A., Black, J., Feolo, M., Zone, J. J., and Neuhausen, S. L. (2001). Prevalence and clinical characteristics of celiac disease in Down's syndrome in a U.S. study. *Am. J. Med. Genet.* **98**, 70–74.
46. Hadjivassiliou, M., Grunewald, R. A., and Davies-Jones, G. A. (1999). Gluten sensitivity: A many headed hydra. *BMJ* **318**, 1710–1711.
47. Ventura, A., Maazzu, G., and Greco, L. (1999). Duration of exposure to gluten and risk for autoimmune disorders in patients with celiac disease. *Gastroenterology* **117**, 297–303.
48. Smith, D. W. (1988). "Recognizable Patterns of Malformation," pp. 10, 12, 74, 75. Saunders, Philadelphia.
49. Bonamico, M., Mariani, P., Danesi, H. M., Crisogianni, M., Faill, P., Gemme, G., Quartino, A. R., Giannotti, A., Castro, M., Balli, F., Lecora, M., Andria, G., Guariso, G., Gabrielli, O., Catassi, C., Lazzari, R., Balocco, N. A., DeVirgilis, S., Culasso, F., and Romano, C. (2001). Prevalence and clinical picture of celiac disease in Italian Down syndrome patients: A multicenter study. *J. Pediatr. Gastroenterol. Nutr.* **33**, 139–143.
50. The NIH Consensus Development Conference on Celiac Disease. June 28–30, 2004. Retrieved September 2007 from http://consensus.nih.gov/2004/2004CeliacDisease118html.htm.
51. Auricchio, S., Mazzacca, G., Toi, R., Visakorpi, J., Maki, M., and Polanco, E. (1988). Coeliac disease as a familial condition: Identification of asymptomatic coeliac patients within family groups. *Gastroenterol. Int.* **1**, 25–31.
52. Hed, J., Leiden, G., Ottosson, E., Strom, M., Walan, A., Groth, O., Sjogren, F., and Franzen, L. (1986). IgA antigliadin antibodies and jejunal mucosal lesions in healthy blood donors. *Lancet* **2**, 215.
53. Ferguson, A., Arran, E., and O'Mahony, S. (1993). Clinical and pathological spectrum of coeliac disease—active, silent, latent, potential. *Gut* **34**, 150–151.
54. Troncone, R., Greco, L., Mayer, M., Paparo, F., Caputo, N., Micillo, M., Mugione, P., and Auricchio, S. (1996). Latent and potential coeliac disease. *Acta Paediatr. Acta Paediatr. Suppl.* **412** 10–14.
55. Collin, P., Kaukinen, K., and Maki, M. (1999). Clinical features of celiac disease today. *Dig. Dis.* **17**, 100–106.
56. Murray, J. A., Herlein, J., Mitros, F., and Goeken, J. A. (2000). Serologic testing for celiac disease in the United States: Results of a multilaboratory comparison study. *Clin. Diagn. Lab. Immunol.* **7**, 584–587.
57. Case, S. (2001). The gluten-free diet. *In* "Gluten-Free Diet, A Comprehensive Resource Guide," pp. 9–43. Centax Books, Saskatchewan, Canada.
58. Cataldo, F., Marino, V., Ventura, A., Botarro, G., and Corazza, G. R. (1998). Prevalence and clinical features of selective immunoglobulin A deficiency in coeliac disease: An Italian multicenter study. *Gut* **42**, 362–365.
59. Not, T., Horvath, K., Hill, I. D., Partanen, J., Hammed, A., Magazzu, G., and Fasano, A. (1997). Celiac disease risk in the USA: High prevalence of antiendomysium antibodies in healthy blood donors. *Scand. J. Gastroenterol.* **33**, 494–498.
60. Dieterich, W., Ehnis, T., Bauer, M., Donner, P., Volta, U., Riecken, E. O., and Schuppan, D. (1997). Identification of tissue transglutaminase as the autoantigen of celiac disease. *Nat. Med.* **3**, 797–801.
61. Sulkanen, S., Halttunen, T., Laurila, K., Kolho, K. L., Korponay-Szabo, I. R., Sarnesto, A., Savilawi, E., Collin, P., and Markku, M. (1998). Tissue transglutaminase autoantibody enzyme-linked immunosorbent assay in detecting celiac disease. *Gastroenterology* **115**, 1322–1328.
62. Troncone, R., Maurano, F., Rossi, M., Micillo, M., Greco, L., Auricchio, R., Salerno, G., Salvatore, F., and Sacchetti, L. (1999). IgA antibodies to tissue transglutaminase: An effective diagnostic test for celiac disease. *J. Pediatr.* **134**, 166–171.
63. Fasano, A. (1999). Tissue transglutaminase: The holy grail for the diagnosis of celiac disease, at last. *J. Pediatr.* **134**, 134–135.
64. Mylotte, M., Egan-Mitchel, B., Fottrell, P. F., McNicholl, B., and Carthy, C. F. (1975). Family studies in coeliac disease. *Gut* **16**, 598–602.
65. Stevens, F. M., Lloyd, R., Egan-Mitchel, B., Mylott, M. J., Fottrell, P. F., Write, R., McNicholl, B., and McCarthy, C. F. (1975). Reticulin antibodies in patients with coeliac disease and their relatives. *Gut* **16**, 598–602.
66. Fasano, A., Berti, I., Gerarduzzi, T., Not, T., Colletti, R. B., Drago, S., Elitsur, Y., Green, P. H. R., Guandalini, S., Hill, I.D., Pietzak, M., Ventura, A., Thorpe, M., Kryszak, D.,

Fornarolo, F., Wasserman, S. S., Murray, J. A., and Horvath, K. (2003). Prevalence of celiac disease in at-risk and not at-risk groups in the United States: A large multicenter study. *Arch. Intern. Med.* **163**, 286–292.

67. Greco, L., Romino, R., Coto, I., Di Cosmo, N., Percopo, S., Maglio, M., Para, F., Gasperi, V., Limongelli, M. G., Cotichini, R., D'Agate, C., Tinto, N., Sacchetti, L., Tosi, R., and Stazi, M. A. (2002). The first large population based twin study of coeliac disease. *Gut* **50**, 624–628.
68. Bevan, S., Popat, S., Braegger, C. P., Busch, A., O'Donoghue, D., Falth-Magnusson, K., Ferguson, A., Godkin, A., Hogberg, L., Holmes, G., Hosie, K. B., Howdle, P. D., Jenkins, H., Jewell, D., Johnston, S., Kennedy, N. P., Kerr, G., Kumar, P., Logan, R. F. A., Love, A. H. G., Marsh, M., Mulder, C. J. J., Sjoberg, K., Stenhammer, L., Walker-Smith, J., Marossy, A. M., and Houlston, R. S. (1999). Contribution of the MHC region to the familial risk of coeliac disease. *J. Med. Genet.* **36**, 687–690.
69. Hogberg, L., Falth-Magnusson, K., Grodzinsky, E., and Stenhammer, L. (2003). Familial prevalence of coeliac disease: A twenty-year follow-up study. *Scand. J. Gastroenterol.* **38**, 61–65.
70. Gudjonsdottir, A. H., Nilsson, S., Ek, J., Kristiansson, B., and Ascher, H. (2004). The risk of celiac disease in 107 families with at least two affected siblings. *J. Pediatr. Gastroenterol. Nutr.* **38**, 338–342.
71. Brook, L., Zone, J. J., and Neuhausen, S. L. (2003). Prevalence of celiac disease among relatives of sib pairs with celiac disease in U.S. families. *98* **2**, 381.
72. Jabbari, M., Wild, G., Goresky, A. C., Daly, D. S., Lough, J. O., Cleland, D. P., and Kinnear, D. G. (1988). Scalloped valvuae connivenetes: An endoscopic marker of celiac sprue. *Gastroenterology* **95**, 1518–1522.
73. Marsh, M. N. (1988). Studies on intestinal lymphoid tissue. XI. The immunopathology of the cell-mediated reactions in gluten sensitivity and other enteropathies. *Scanning Microsc.* **2**, 1663–1665.
74. Walker-Smith, J. A., Sandhu, B. K., Isolauri, E., Banchini, G., Van, C., Bertrand, M., Dias, J. A., Fasano, A., Guandalini, S., Hoekstra, J. H., Juntunen, M., Kolacek, S., Marx, D., Micenticc-Turk, D., Razenberg, M. C., Szajewska, H., Taminiau, J., Weizman, Z., Zanacca, C., and Zetterstrom, R. (1997). Guidelines prepared by the ESPGAN Working Group on Acute Diarrhoea. Recommendations for feeding in childhood gastroenteritis. European Society of Pediatric Gastroenterology and Nutrition. *J. Pediatr. Gastroenterol. Nutr.* **24**, 619–620.
75. Dicke, W. (1950). Coeliac disease: Investigation of harmful effects of certain types of cereal on patients with coeliac disease. University of Utrecht, The Netherlands.
76. Van de Kamer, J. H., Weijers, H. A., and Dicke, W. K. (1953). Coeliac disease. IV. An investigation into the injurious constituents of wheat in connection with their action in patients with coeliac disease. *Acta Paediatr.* **42**, 223–231.
77. Holmes, G., and Catassi, C. (2000). Pathophysiology, *In* "Coeliac Disease," pp. 18–20. Health Press, Oxford.
78. Hardman, C. M., Garioch, J. J., Leonard, J. N., Thomas, H. J. W., Walker, M. M., Lortan, J. E., Lister, A., and Fry, L. (1997). Absence of toxicity of oats in patients with dermatitis herpetiformis. *N. Engl. J. Med.* **337**, 1884–1887.
79. Hoffenberg, E. J., Haas, J., Drescher, A., Barnhurst, R., Osberg, I., Bao, F., and Eisenbarth, G. (2000). A trial of oats in children with newly diagnosed celiac disease. *J. Pediatr.* **137**, 361–366.
80. Forssell, F., and Weiser, H. (1995). Spelt wheat and celiac disease. *Z. Lebens. Untersuch. Forsch.* **201**, 35–39.
81. Ellis, H. J., Doyle, A. P., Day, P., Wieser, H., and Ciclitira, P. J. (1994). Demonstration of the presence of coeliac-activating gliadin-like epitopes in malted barley. *Int. Arch. Allergy Immunol.* **104**, 308–310.
82. Crowe, J. P., and Falini, N. P. (2001). Gluten in pharmaceutical products. *Am. J. Health Syst. Pharmacy* **58**, 396–401.
83. Pietzak, M. (2005). The follow-up of patients with celiac disease—Achieving compliance with treatment. *Gastroenterology* **128**, S135–S141.
84. Ljungman, G., and Myrdal, U. (1993). Compliance in teenagers with coeliac disease—A Swedish follow-up study. *Acta Paediatr.* **82**, 238.
85. Maki, M., Lahdeaho, M. L., Hallstrom, O., Viander, M., and Visokorpi, J. K. (1989). Postpubertal gluten challenge in coeliac disease. *Arch. Dis. Child.* **64**, 1604–1607.
86. Hankey, G. L., and Holmes, G. K. (1994). Coeliac disease in the elderly. *Gut* **35**, 65–67.
87. Corrao, G., Corazza, G. R., Bagnardi, V., Brusco, G., Ciacci, C., Cottone, M., Sategna Guidetti, C., Usai, P., Cesari, P., Pelli, M. A., Loperfido, S., Volta, U., Calabro, A., and Certo, M. Club del Tenue Study Group. (2001). Mortality in patients with coeliac disease and their relatives: A cohort study. *Lancet* **358**, 356–361.
88. Swinson, C. M., Slavin, G., Coles, E. C., and Booth, C. C. (1983). Coeliac disease and malignancy. *Lancet* **1**, 111–115.
89. Catassi, C., Fabiani, E., Corrao, G., Barbato, M., De Renzo, A., Carella, A. M., Gabrielli, A., Leoni, P., Carroccio, A., Baldassarre, M., Bertolani, P., Caramaschi, P., Sozzi, M., Guariso, G., Volta, U., and Corazza, G. (2002). Risk of non-Hodgkin lymphoma in celiac disease. *JAMA* **287**, 1413–1419.
90. Holmes, G. K., Prior, P., Lane, M. R., Pope, D., and Allan, R. N. (1989). Malignancy in coeliac disease: Effect on gluten-free diet. *Gut* **30**, 333–338.
91. Lee, S. K., Lo, W., Memeo, L., Rotterdam, H., and Green, P. H. (2003). Duodenal histology in patients with celiac disease after treatment with gluten-free diet. *Gastrointest. Endosc.* **57**, 187–191.
92. Carroccio, A., Iacono, G., Lerro, P., Cavataio, F., Malorgio, E., Soresi, M., Baldassarre, M., Nortabartolo, A., Ansaldi, N., and Montalto, G. (1997). Role of pancreatic impairment in growth recovery during gluten-free diet in childhood celiac disease. *Gastroenterology* **112**, 1839–1844.
93. Pink, I. J., and Creamer, B. (1967). Response to a gluten-free diet of patients with the coeliac syndrome. *Lancet* **1**, 300–304.
94. Cellier, C., Delabesse, E., Helmer, C., Patey, N., Matuchansky, C., Jalori, B., Macintyre, E., Cerf-Bensussan, M., and Brousse, N. (2000). Refractory sprue, coeliac disease, and enteropathy-associated T-cell lymphoma. *Lancet* **356**, 203–208.
95. Stuart, B. M., and Gent, A. E. (1998). Atrophy of the coeliac mucosa. *Eur. J. Gastroenterol. Hepatol.* **10**, 523–525.
96. Mitchison, H. C., al Mardini, H., Gillespie, S., Laker, M., Zaitoun, A., and Record, C. O. (1991). A pilot study of fluticasone propionate in untreated coeliac disease. *Gut* **32**, 260–265.

97. Viadya, A., Bolanos, J., and Berkelhammer, C. (1999). Azathioprine in refractory sprue. *Am. J. Gastroenterol.* **94**, 1967–1969.
98. Rolny, P., Sigurjonsdottir, H. A., Remotti, H., Nilsson, L. A., Ascher, H., Tlaskalova-Hogenova, H., and Tuckova, L. (1999). Role of immunosuppressive therapy in refractory sprue-like disease. *Am. J. Gastroenterol.* **94**, 219–225.
99. O'Mahony, S., Howdle, P. D., and Losowsky, S. M. (1996). Management of patients with non-responsive coeliac disease. *Aliment. Pharmacol. Ther.* **10**, 671–680.
100. Ryan, B. M., and Kelleher, D. (2000). Refractory celiac disease. *Gastroenterology* **119**, 243–251.
101. Daum, S., Cellier, C., and Mulder, C. (2005). Refractory coeliac disease. *Best Pract. Res. Clin. Gastroenterol.* **19**, 413–424.
102. Sategna-Guidetti, C., Volta, U., Ciacci, C., Usai, P., Carlino, A., Franceschi, L., Camera, A., Pelli, A., and Brossa, C. (2001). Prevalence of thyroid disorders in untreated adult celiac disease patients and effect of gluten withdrawal: An Italian multicenter study. *Am. J. Gastroenterol.* **96**, 751–757.
103. Kaukinen, K., Halme, L., Collin, P., Farkkila, M., Maki, M., Vehmanen, P., Partanen, P., and Hockerstedt, K. (2002). Celiac disease in patients with severe liver disease: Gluten-free diet may reverse hepatic failure. *Gastroenterology* **122**, 881–888.
104. Sategna-Guidetti, C., Grosso, S. B., Grosso, S., Mengozzi, G., Aimo, G., Zaccaria, T., Di Stefano, M., and Isaia, G. C. (2000). The effects of a 1-year gluten withdrawal on bone mass, bone metabolism and nutritional status in newly-diagnosed adult coeliac patients. *Aliment. Pharmacol. Ther.* **14**, 35–43.
105. Mora, S., Barera, G., Beccio, S., Menni, L., Proverbio, M. C., Bianchi, C., and Chiumello, G. (2001). A prospective, longitudinal study of the long-term effect of treatment on bone density in children with celiac disease. *J. Pediatr.* **139**, 516–521.
106. Kemppainen, T., Kroger, H., Janatuinen, E., Arnala, I., Lamberg-Allardt, C., Karkkainen, M., Kosma, V.-M., Julkunen, R., Jurvelin, J., Alhava, E., and Uusitupa, M. (1999). Bone recovery after a gluten-free diet: A 5-year follow-up study. *Bone* **25**, 355–360.
107. Kemppainen, T., Kroger, H., Janatuinen, E., Arnala, I., Kosma, V.-M., Pikkarainen, P., Julkunen, R., Jurvelin, J., Alhava, E., and Uusitupa, M. (1999). Osteoporosis in adult patients with celiac disease. *Bone* **24**, 249–255.
108. Valdimarsson, T., Toss, G., Lofman, O., and Strom, M. (2000). Three years' follow-up of bone density in adult coeliac disease: Significance of secondary hyperparathyroidism. *Scand. J. Gastroenterol.* **35**, 274–280.
109. Smith, C. M., Clarke, C. F., Porteous, L. E., Elsori, H., and Cameron, D. J. S. (2000). Prevalence of celiac disease and longitudinal follow-up of antigliadin antibody status in children and adolescents with type 1 diabetes mellitus. *Pediatr. Diabetes* **1**, 199–203.
110. Mohn, A., Cerruto, M., Iafusco, D., Prisco, F., Tumini, S., Stoppoloni, O., and Chiarelli, F. (2001). Celiac disease in children and adolescents with type 1 diabetes: Importance of hypoglycemia. *J. Pediatr. Gastroenterol. Nutr.* **32**, 37–40.
111. Not, T., Tommasini, A., Tonini, G., Buratti, E., Pocecco, M., Tortul, C., Valussi, M., Crichiutti, G., Berti, I., Trevisiol, C., Azzoni, E., Neri, E., Torre, G., Martelossi, S., Soban, M., Lenhardt, A., Cattin, L., and Ventura, A. (2001). Undiagnosed coeliac disease and risk of autoimmune disorders in subjects with type 1 diabetes mellitus. *Diabetologia* **44**, 151–155.
112. Jaeger, C., Hatziagelaki, E., Petzoldt, R., and Bretzel, R. (2001). Comparative analysis of organ-specific autoantibodies and celiac disease-associated antibodies in type 1 diabetic patients, their first-degree relatives, and healthy control subjects. *Diabetes Care* **24**, 27–32.
113. Lubrano, E., Ciacci, C., Ames, P. R. J., Mazzacca, G., Ordente, P., and Scarpa, R. (1996). The arthritis of coeliac disease: Prevalence and pattern in 200 adult patients. *Br. J. Rheumatol.* **35**, 1314–1318.
114. Paimela, L., Kurki, P., Leirisalo-Repo, M., and Piirainen, H. (1995). Gliadin immune reactivity in patients with rheumatoid arthritis. *Clin. Exp. Rheumatol.* **13**, 603–607.
115. Curione, M., Barbato, M., De Biase, L., Viola, F., LoRusso, L., and Cardi, E. (1999). Prevalence of coeliac disease in idiopathic dilated cardiomyopathy. *Lancet* **354**, 222–223.
116. Martelossi, S., Zanatta, E., Del Santo, E., Clarich, P., Radovich, P., and Ventura, A. (1996). Dental enamel defects and screening for coeliac disease. *Acta Paediatr. Acta Paediatr. Suppl.* **412**, 47–48.
117. Ruenala, T., and Collin, P. (1997). Diseases associated with dermatitis herpetiformis. *Br. J. Dermatol.* **136**, 315–318.
118. Collin, P., Reunala, T., Pukkala, E., Laippala, P., Keyrilainen, O., and Pasternak, A. (1994). Celiac disease-associated disorders and survival. *Gut* **35**, 1215–1218.
119. Thain, M. E., Hamilton, J. R., and Erlich, R. M. (1974). Coexistence of diabetes mellitus and celiac disease. *J. Pediatr.* **85**, 527–529.
120. Savilahti, E., Simell, O., Koskimies, S., Rilva, A., and Akerblom, H. K. (1986). Celiac disease in insulin-dependent diabetes mellitus. *J. Pediatr.* **108**, 690–693.
121. Collin, P., Salmi, J., Hallstrom, O., Oksa, H., Oksala, H., Maki, M., and Reunala, T. (1989). High frequency of coeliac disease in adult patients with type-I diabetes. *Scand. J. Gastroenterol.* **24**, 81–84.
122. Talal, A. H., Murray, J. A., Goeken, J. A., and Sivitz, W. F. (1997). Celiac disease in an adult population with insulin-dependent diabetes mellitus: Use of endomysial antibody testing. *Am. J. Gastroenterol.* **92**, 1280–1284.
123. Hill, I., Fasano, A., Schwartz, R., Counts, D., Glock, M., and Hovath, K. (2000). The prevalence of celiac disease in at-risk groups of children in the United States. *J. Pediatr.* **136**, 86–90.
124. Carlsson, A., Axelsson, I., Borulf, S., Bredberg, A., Forslund, M., Lindberg, B., Sjoberg, K., and Ivarsson, S. A. (1998). Prevalence of IgA-antigliadin antibodies and IgA-antiendomysium antibodies related to celiac disease in children with Down syndrome. *Pediatrics* **101**, 272–275.
125. Cronin, C. C., Jackson, L. M., Feighery, C., Shanahan, F., Abuzakouk, M., Ryder, D. Q., Welton, M. S., and Callahan, N. (1998). Coeliac disease and epilepsy. *Q. J. Med.* **71**, 359–369.
126. Collin, P., Maki, M., Keyrilainen, O., Hallstrom, O., Reunala, T., and Pasternak, A. (1992). Selective IgA deficiency and celiac disease. *Scand. J. Gastroenterol.* **27**, 367–371.
127. Collin, P., Korpela, M., Hallstrom, O., Viander, M., Keyrilainen, O., and Maki, M. (1992). Rheumatic complaints

as a presenting symptom in patients with celiac disease. *Scand. J. Rheumatol.* **21**, 20–23.

128. Collin, P., Salmi, J., Hallstrom, O., Reunala, T., and Pasternak, A. (1994). Autoimmune thyroid disorders and coeliac disease. *Eur. J. Endocrinol.* **130**, 140.
129. Rujner, J., Wisniewski, A., Gregorere, H., Wozniewicz, B., Mynarski, W., and Witas, H. W. (2001). Coeliac disease and HLA-DQ2 (DQA1*0501 and DQB1*0201) in patients with Turner syndrome. *J. Pediatr. Gastroenterol. Nutr.* **32**, 114–115.
130. Bonamico, M., Bottaro, G., Pasquino, A. M., Caruso-Nicoletti, M., Mariani, P., Gemme, G., Paradiso, E., Regusa, M. C., and Spina, M. (1998). Celiac disease and Turner syndrome. *J. Pediatr. Gastroenterol. Nutr.* **26**, 496–499.
131. Ivarsson, S. A., Carlsson, A., Bredberg, A., Alm, J., Aronsson, S., Gustafsson, J., Hagenas, L., Hager, A., Kriston, B., Marcus, C., Moelle, C., Nilsson, K. O., Turemo, T., Westphal, O., Albertson-Wikland, K., and Aman, J. (1999). Prevalence of coeliac disease in Turner syndrome. *Acta Paediatr.* **88**, 933–936.
132. Giannotti, A., Tiberio, G., Castro, M., Virgilii, F., Colistro, F., Ferretti, F., Digilio, M. C., Gambarara, M., and Dallapiccola, B. (2001). Celiac disease in Williams Syndrome. *Journal of Medical Genetics.* **38**, 767–768.

CHAPTER 42

Nutrition and Cystic Fibrosis

HUICHUAN J. LAI[1] AND PHILIP M. FARRELL[2]

[1]*Departments of Nutritional Sciences, Pediatrics, and Biostatistics and Medical Informatics, University of Wisconsin–Madison, Madison, Wisconsin*

[2]*Departments of Pediatrics and Population Health Sciences, University of Wisconsin–Madison, Madison, Wisconsin*

Contents

I. OVERVIEW OF CYSTIC FIBROSIS

Cystic fibrosis (CF) is the most common, life-threatening autosomal recessive disorder with estimated incidences of approximately 1 in 3000 white live births, 1 in 17,000 black live births, and 1 in 90,000 East Asian live births [1, 2]. CF was recognized as a distinct clinical entity in 1938. It is a generalized disease of the exocrine glands characterized by abnormal sodium and chloride transport, leading to elevated electrolyte levels in sweat [3, 4]. Dysfunction of the other exocrine glands occurs, producing viscid secretions of low water content. This results in pancreatic insufficiency (PI), which leads to malabsorption and failure to gain weight, as well as airway obstruction, which leads to increased susceptibility to recurrent bronchial infection, progressive lung damage, and eventual respiratory failure.

A. Clinical Presentation

There are three categories of major clinical abnormalities in CF: (1) gastrointestinal tract involvement, characterized by PI leading to malabsorption and malnutrition; (2) respiratory tract involvement, characterized by recurrent infections and chronic obstructive pulmonary disease; and (3) salt loss in sweat that can lead to severe hyponatremic dehydration. The pancreatic disturbance begins prenatally and can cause intestinal obstruction in newborns with CF (a problem referred to as meconium ileus). It has been estimated that 85–90% of CF patients have PI [5], and 15–20% have meconium ileus [6].

More recent studies have shown that PI develops during the first year of life such that about half the patients destined to have this problem show intestinal malabsorption by about 1 month of age, two-thirds by 6 months, and the remainder by 1 year [7]. Unlike PI, pulmonary status of patients with CF often appears normal at birth; however, it inevitably shows obstruction and infection. The onset and rate of progression of CF lung disease are not well understood but appear to vary widely among individuals [8–10]. Other complications may occur as the disease progresses. For example, glucose intolerance and diabetes mellitus may develop [11, 12], and about 15% of adults with CF develop diabetes requiring insulin therapy [13]. Because of focal biliary cirrhosis, up to 5% of CF patients develop overt liver disease in adolescence or adulthood [13, 14]. Because of an absent vas deferens at birth (presumably resulting from ductal obstruction with dehydrated secretions), infertility in males with CF is virtually universal [15].

B. Pathogenesis

On the basis of molecular genetics research [16], CF fundamentally can be attributed to mutations occurring in the long arm of chromosome 7. With cloning of the CF gene, it has been demonstrated that the most common mutation is a 3-base-pair (bp) deletion, which results in the loss of a phenylalanine residue at amino acid position 508 of the predicted gene product, namely the cystic fibrosis transmembrane regulator (CFTR) [17–19]. The 3-bp deletion mutant, commonly referred to as ΔF508 or more recently F508del, occurs in about 70% of the CF chromosomes [17], and more than 85% of CF patients in the United States have at least one F508del allele [20]. However, more than 1500 other DNA mutations in the CFTR gene have been identified. They are divided into five classes according to the molecular defect [4], as shown in Figure 1. Most mutations, especially classes I to III, are associated with PI, but patients with classes IV or V mutations tend to have pancreatic sufficiency [20].

The abnormal CFTR protein is the underlying pathogenic factor in the disease process because of its role in regulating ion transport across the apical membrane of epithelial cells, particularly chloride conductance, which is invariably defective in CF [2, 4]. This defect leads to

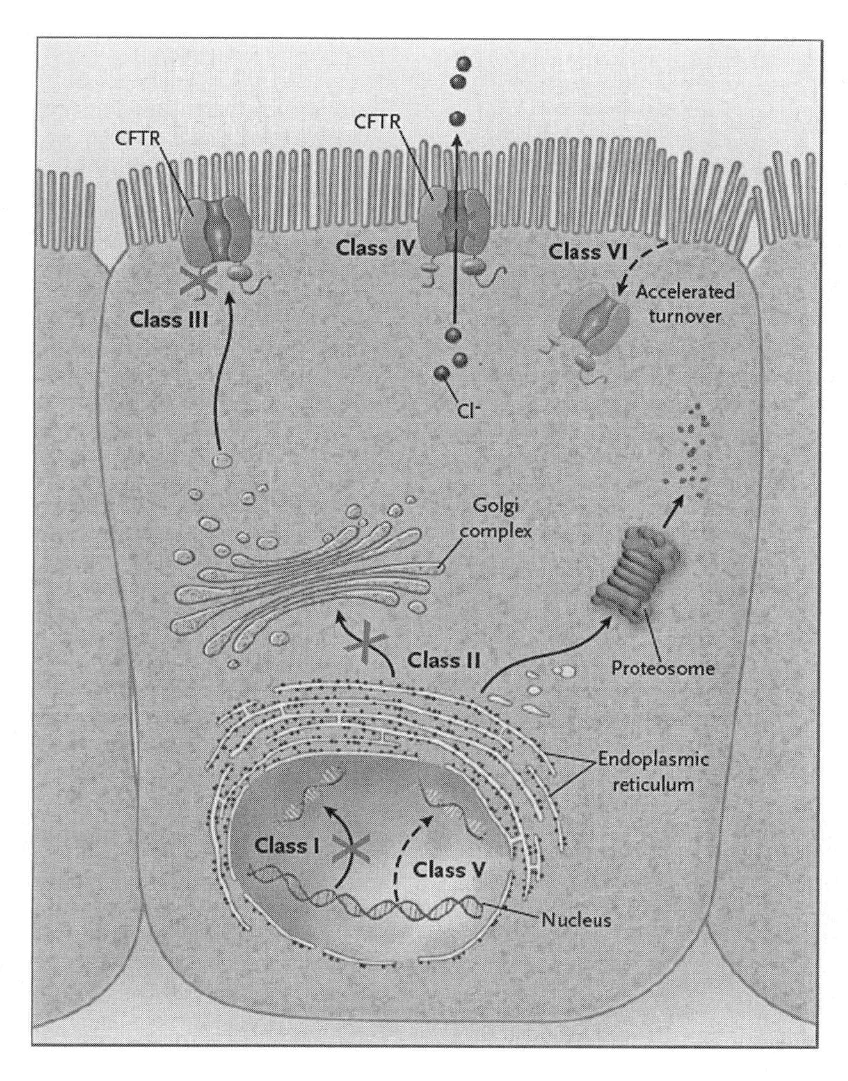

FIGURE 1 Categories of CFTR mutations: absence of synthesis (class I); defective protein maturation and premature degradation (class II); disordered regulation (class III); defective chloride conductance (class IV); reduced synthesis (class V); and accelerated turnover (class VI). Adapted from Rowe et al. [4]. Copyright © 2005, *Massachusetts Medical Society. All rights reserved.* See color plate.

abnormally high chloride concentration in the sweat, which constitutes the classical diagnostic test for CF [21]. More recent research suggests that the CFTR protein is a structural component of the chloride channel and may itself account for the channel core [4, 22].

C. Diagnosis and Treatment

Traditionally, the diagnosis of CF has been made because of (1) a positive family history, (2) the presence of meconium ileus, or (3) symptoms of malabsorption or pulmonary disease with infection, which occur at variable ages [23, 24]. Once the characteristic signs and symptoms become evident, the diagnosis of CF can be readily established by performing a sweat test using pilocarpine iontophoresis [21]. Traditional diagnosis by signs or symptoms of CF often leads to delays in diagnosis and referral to a CF center. These delays are associated with severe malnutrition in about half the patients [25], but this can be prevented with early diagnosis and treatment [26–28]. Consequently, there has been considerable interest in establishing newborn screening (NBS) methods for detection of presymptomatic cases for the purpose of instituting early treatment and preventing or ameliorating symptoms.

In 1979, Crossley *et al.* first described the use of dry-blood specimens obtained from the newborns to measure immunoreactive trypsinogen (IRT) level, which was shown to be highly elevated in patients with CF [29, 30]. The discovery of the CFTR gene in 1989 [16] has promoted the development of new screening methods where the IRT test is coupled with detection of the most common mutant allele (F508del) or with a multipanel CFTR mutation analysis [31, 32]. Various screening protocols are used to screen newborns for CF [33]. All protocols begin with a first-tier phenotypic test that measures IRT in dried blood spots. Infants who have an elevated IRT are then referred for further testing, either a repeat IRT test at 2 weeks of age or DNA analysis for CFTR mutations. Infants with a second positive screening result are referred to sweat testing to establish the diagnosis of CF. It's important to understand that a positive IRT test alone is not equivalent to a diagnosis of CF. In fact, only about 1% of newborns with positive IRT have CF [34]. In addition, not all CF cases are detected by screening for IRT; approximately 5% of CF patients show false-negative IRT in their dry-blood specimens obtained at birth [34].

In the United States, CF NBS began in Colorado [35] and Wisconsin [31, 32] in the mid-1980s. The potential benefits and risks associated with CF neonatal screening programs have been under investigation in various regions of the world [32, 36–38]. In many reports [27–32], clear evidence of nutritional benefits attributable to early diagnosis was demonstrated by anthropometric indexes. The most convincing evidence was obtained from a randomized clinical trial in Wisconsin after 10 years of investigation [26–28]. This trial and other more recent studies [41, 42] led the CDC to recommend universal screening in a MMWR published in 2004 [23]. Subsequently, during 2005–2007, the number of states screening newborns for CF has increased from 4 to 40 in the United States, which is an unprecedented rate of change in NBS. Most western European countries along with Australia and New Zealand have also implemented NBS for CF.

Clinical management of CF involves treatment programs with three principal objectives: (1) improve nutritional status, (2) promote clearance of respiratory secretions, and (3) control bronchopulmonary infections. Care programs for CF patients in North America and many European countries are organized in specialized regional centers. These centers have placed particular emphasis on enhancing nutrition and using aggressive strategies to prevent progressive pulmonary disease [43]. Although CF lung disease cannot be cured, treatment programs have been generally effective, as evidenced by the increasing longevity of CF patients in the United States from less than 20 years to approximately 36 years during the past two decades [44, 45]. The primary causes of death in patients with CF are cardiorespiratory complications, accounting for about 80% of deaths. For this reason, most CF Centers in the United States place a great deal of emphasis on respiratory management for patients with CF. In addition, with diagnosis through NBS, an increased emphasis has been placed on preventing malnutrition in the recent decade.

D. Consequences of Malnutrition

Evidence has accumulated in recent years from longitudinal studies that the consequences of malnutrition in CF are more severe than previously appreciated from clinic-based cross-sectional observations. Long-term adverse consequences of malnutrition include: (1) permanently stunted growth [26–28]; (2) cognitive dysfunction [46, 47]; and (3) greater susceptibility to lung disease [48–51]. More research is under way to address whether early nutritional intervention could alter long-term pulmonary disease progression.

II. MALNUTRITION IN CYSTIC FIBROSIS

CF is associated with an increased risk of protein-calorie malnutrition, as well as deficiencies in fat-soluble vitamins and other micronutrients. Malnutrition associated with CF is characterized by its early onset and is often present at the time of CF diagnosis. At the mild end of the malnutrition spectrum, CF patients may have depleted stores or low circulating concentrations of a given nutrient, but no associated signs or symptoms. More pronounced nutritional deficiencies lead to metabolic abnormalities, structural changes, functional disturbances, growth failure, developmental delay, and a variety of other characteristics of malnutrition. Malnutrition is most likely to occur during periods of rapid growth when nutritional requirements are high, during pulmonary exacerbations, and with increased severity of lung disease.

Growth impairment and abnormalities in the biochemical markers of nutritional status as well as clinical symptoms of malnutrition all have been reported in patients with CF. Historically, malnutrition in patients with CF was thought to represent either an inherent consequence of disease process or a physiologic adaptation to advanced pulmonary disease. However, it is now recognized that the causes of malnutrition in CF are multiple and can be attributed to three primary mechanisms [52–54]: increased energy and nutrient losses, increased energy expenditure as well as decreased energy and nutrient intakes. Table 1 lists the major risk factors for malnutrition associated with CF.

A. Causes of Malnutrition

1. Increased Losses

Loss of nutrients from maldigestion and malabsorption as a result of PI is the primary factor contributing to energy and

TABLE 1 Factors Contributing to Malnutrition in Cystic Fibrosis

Disease factors:
Presence of pancreatic insufficiency (PI)
Severity of PI (the degree of steatorrhea and azotorrhea)
Partial intestinal resection secondary to bowel obstruction (caused by meconium ileus)
Severity of respiratory disease
Loss of bile salts associated with steatorrhea
Cholestatic liver disease
Diabetes mellitus
Nutritional factors:
Growth rate (of particular concern in young children and adolescents with CF)
Energy and macronutrient intakes (e.g., the quantity and quality of food consumed)
Micronutrient deficiencies (e.g., fat-soluble vitamins)
Energy expenditure
Eating behaviors

nutrient deficiencies in patients with CF [55]. The ductular cells of the pancreas respond to stimulation with secretin by producing a high-volume, bicarbonate-rich secretion. This secretion functions to neutralize gastric acid, thus enabling the pancreatic digestive enzymes to function at their pH optimum. The abnormal chloride transport caused by the defective CFTR protein leads to thickened secretions that obstruct the pancreatic ducts and prevent the secretion of enzymes and bicarbonate. In CF, PI is defined by the presence of measurable steatorrhea. This does not occur until 1–2% of pancreatic enzymatic capacity remains [56]. Therefore, CF patients with PI have severe, irreversible loss of pancreatic function. It is also important to understand that CF patients with pancreatic sufficiency (PS) do not have normal pancreatic function [57]. They have decreased volume of bicarbonate-rich secretion but continue to produce enough pancreatic enzymes to avoid steatorrhea.

Maldigestion and malabsorption caused by PI can be attributed to three major abnormalities: lack of digestive enzymes, inadequate bicarbonate secretion, and loss of bile salts and bile acids. Inadequate bicarbonate secretion results in impaired capacity to neutralize gastric acid in the duodenum, and a lower intestinal pH until well into the jejunum, which often reduces the effectiveness of pancreatic enzyme replacement therapy [55–57] (see Section IV.B.1). Loss of bile salts and bile acids often exacerbates maldigestion and malabsorption. Bile acids are readily precipitated in an acid milieu, and duodenal bile acid concentration may fall below the critical micellar concentration, thereby exacerbating fat maldigestion. Precipitated bile salts also appear to be lost from the enterohepatic circulation in greater quantities, thus reducing the total bile acid pool and altering the glycocholate:taurocholate ratio. Oral taurine supplements have been reported to benefit some CF patients [58].

Other factors also contribute to energy and nutrient losses in CF. Patients presenting with meconium ileus, in particular those who have undergone intestinal resection, have further reduction in intestinal absorptive capabilities. Viscid, thick intestinal mucus, with altered physical properties, may affect the thickness of the intestinal unstirred layer, further limiting nutrient absorption. Diabetes mellitus may increase caloric losses due to glycosuria if not adequately controlled. Advanced liver disease and biliary cirrhosis may result in reduced bile salt synthesis and secretion, which may lead to severe fat malabsorption.

2. Increased Requirement

Energy requirements in patients with CF are highly variable. Several studies have reported that patients with CF are associated with increased energy expenditure compared with non-CF patients [59–62]. A variety of explanations have been proposed to explain the increased energy expenditure observed in CF patients. These include chronic lung infection, increase in work of breathing, genetic and cellular defects, and changes in body composition.

Chronic lung infections, particularly with *Pseudomonas aeruginosa*, have been shown to be associated with a 25–80% increase in metabolic rate and energy requirements [63]. The link between CF genotype and energy requirement was reported in a study by Tomezsco *et al.*, who demonstrated that energy expenditure was increased by 23% in CF patients with homozygous F508del mutations as compared with non-CF controls [62]. Therefore, a patient with advanced lung disease might not be able to ingest sufficient calories to meet energy needs.

The hypothesis that a basic cellular defect may increase energy requirement was supported by *in vitro* studies showing that mitochondria from cultured fibroblasts obtained from CF patients had higher rates of oxygen consumption compared with control tissues [64, 65]. When F508del was identified, it was proposed that the defective CFTR protein may affect cellular energy metabolism through its involvement in the regulation of ion transport across membranes, because CFTR is a cAMP-regulated chloride channel [4].

3. Decreased Consumption

The appetite or caloric intake of CF patients may be limited because of a variety of disease complications. Acute pulmonary exacerbations are a common cause of anorexia, and respiratory infections often give rise to nausea and vomiting, which may further reduce intake [52]. The biochemical causes of anorexia associated with acute infection are unclear, but elevated circulating levels of tumor necrosis factor may play a role [66].

In addition to pulmonary complications, a variety of gastrointestinal complications also contribute to anorexia and inadequate caloric intake [52]. Increased occurrence of gastroesophageal reflux disease and esophagitis is observed in patients with CF. Distal intestinal obstruction syndrome, a form of subacute or chronic partial bowel obstruction, usually occurs in older patients with PI. Large fecal masses, palpable in the abdomen, give rise to intermittent abdominal distention and cramping accompanied with reduced appetite. Constipation in the absence of distal intestinal obstruction syndrome is another cause of anorexia and abdominal discomfort in older patients with CF.

A number of dietary surveys indicate that CF patients often eat less than normal, particularly during the period of 1970s and early 1980s, when CF patients were commonly prescribed with fat-restricted diets on the assumption that a reduction in dietary fat intake might improve bowel symptoms [67]. The observation of better growth and survival in CF patients who received an unrestricted-fat, high-calorie diet in combination with pancreatic enzyme replacement therapy (PERT) compared with those who received a low-fat diet in the early 1980s has changed dietary practices in most CF centers [68, 69]. Energy intakes of 110% or greater than the recommended dietary allowances with 35–40% of energy from fat are now recommended for patients with CF [70, 71]. However, CF patients often fail to consume such high quantities of calories and/or fat because of their disease manifestation. In several cross-sectional or short-term studies [72–76] and a small prospective 3-year study of 25 patients [77], energy and fat intakes of CF patients were reported to be much lower than these recommendations. More recently, a longitudinal study evaluating dietary intake patterns in children with CF from the time of diagnosis to age 10 years revealed mean energy intake was at ~110% of RDA with fat consisting of ~37% of energy [27].

B. Common Nutritional Deficiencies

1. Energy and Macronutrients

As discussed earlier, patients with CF are at high risk of energy deficiency because of their increased requirement and decreased consumption. Protein poses less of a nutritional problem than does fat in the CF population. The major risk of protein deficiency in CF patients occurs during the first year of life, when the average requirement is at least three times as great as that in adulthood. Low serum markers of protein (e.g., albumin, prealbumin, and retinol binding protein) are commonly found in infants and young children with newly diagnosed CF. One-third to one-half of infants diagnosed through CF newborn screening were reported to be hypoalbuminemic [78–80]. Normalization of serum albumin level often occurs following comprehensive nutrition therapy [81].

The consequence of energy deficiency leads to impaired growth in children with CF. Weight retardation and linear growth failure are the most common observations documented in the CF clinics [82, 83], although its severity and prevalence vary greatly. Accurate estimates of the prevalence of malnutrition in the CF population have been difficult to obtain in the past, because of lack of sufficient data. In recent years, comprehensive national databases known as CF patient registries have been compiled by the U.S. and Canadian Cystic Fibrosis Foundations, as well as the European CF Society, making it possible to determine population estimates on the prevalence of malnutrition associated with CF. Analysis of 13,000 pediatric CF patients documented in the 1993 U.S. CF Patient Registry revealed that CF children grew substantially more slowly than normal at all ages [25]. Malnutrition was particularly prevalent in infants (47%) and adolescents (34%) as compared with children at other ages (22%), and in patients with newly diagnosed, untreated CF (44%). Underweight is also prevalent in adults with CF; approximately 35% of the 7200 adults with CF documented in the 1992–1994 U.S. and Canadian CF Patient Registries were found to be underweight [84]. During the past decade, the prevalence of malnutrition has decreased steadily from ~25% in 1995 to ~15% in 2005 in pediatric CF patients [45].

2. Essential Fatty Acids

Essential fatty acid deficiency (EFAD) has been known to occur in CF patients [78, 84–86]. In infancy, particularly before diagnosis, EFAD can occur with desquamating skin lesions, increased susceptibility to infection, poor wound healing, thrombocytopenia, and growth retardation. In patients who are adequately treated, clinical evidence of EFAD is rare, although biochemical abnormalities of EFA status remain common [88–90]. The major fatty acid abnormalities found in patients with CF are low level of linoleic acid and elevated levels of palmitoleic, oleic, and eicosatrienoic acid. Multiple hypotheses have been proposed to explain the underlying mechanisms of abnormal EFA status associated with CF. Fat malabsorption secondary to PI is the most common explanation for EFA. However, some investigators have postulated a primary metabolic defect in fatty acid metabolism [87–92].

Many studies have reported a clear association between better EFA status and better growth in children with CF. Plasma linoleic acid was shown to be correlated positively with growth in children whose CF was diagnosed before 3 months and followed up to 12 years of age [78]. Another study by van Egmond *et al.* showed better growth among CF infants who consumed a predigested formula that contained high linoleic acid (12% of energy) compared with those who consumed a comparable formula with lower linoleic acid (7% of energy), despite a lower total energy intake in the former group [93]. More recently, Shoff *et al.* demonstrated that, in children who experienced longer and more severe malnutrition due to delayed diagnosis, maintaining normal plasma linoleic acid (i.e., >26% of total plasma fatty acids) in addition to sustaining a high caloric intake (i.e., >120% of estimated energy requirement) is a critical determinant in promoting catch-up weight gain [94].

Despite the foregoing evidence, EFA supplementation for CF patients remains controversial for several reasons. First, not all patients respond to EFA supplementation; normalizing plasma linoleic acid is particularly difficult in patients with meconium ileus [54, 62]. Second, n-6 fatty acids, including linoleic acid and its metabolite arachidonic acid, have been proposed to play a role in CF inflammation in *in vitro* studies [95].

3. Fat-Soluble Vitamins

Deficiencies of fat-soluble vitamins in the CF population have been demonstrated in many studies [96–103]. Vitamins A and E are of the greatest concern, particularly in patients with severe malabsorption or liver disease. However, recent studies show that deficiencies in vitamin D or vitamin K are also common, especially in CF patients with advanced cholestatic liver disease. Abnormalities in fat-soluble vitamins are particularly prevalent in newly diagnosed infants with CF. Studies on infants diagnosed through NBS showed that 20–40% had low serum retinol, 35% had low serum 25-hydroxyvitamin D and 40–70% had low serum α-tocopherol [78, 79].

Vitamin A deficiency was the first micronutrient deficit demonstrated in patients with CF. Clinical symptoms of vitamin A deficiency reported in CF patients includes keratinizing metaplasia of the bronchial epithelium, xerophthalmia, and night blindness. Pancreatic lipase is required to digest retinyl esters prior to absorption. Other mechanisms for vitamin A deficiency in CF have been proposed, ranging from a defect in the mobilizing hepatic storage of vitamin A due to liver disease to low levels of retinol binding protein, which is responsible for transporting vitamin A in the circulation [96, 97].

Vitamin E deficiency in CF is most commonly evidenced by low plasma levels of α-tocopherol and has been associated with hemolytic anemia [98]. Low α-tocopherol levels are prevalent in infants with CF identified with NBS programs [78, 96]. Those with early, prolonged severe deficiency may also show cognitive dysfunction [46].

Recent evidence has also suggested that vitamin D deficiency is prevalent in both children and adults with CF [100–102]. Suboptimal vitamin D status has been directly linked to poor bone mineralization in CF patients, although the cause of CF bone disease is multifactorial and not completely understood [100].

Vitamin K deficiency has not been routinely demonstrated in patients with CF. However, vitamin K deficiency is likely to develop in CF patients with severe cholestatic liver disease, short-bowel syndrome, and lung disease requiring frequent antibiotic use [103–105].

4. Minerals

Minerals of concern in CF include sodium, calcium, and phosphorus. Sodium is of concern in CF patients because of its abnormally high content in the sweat. In hot climates, salt depletion can be catastrophic, leading to severe hyponatremic dehydration and shock [106, 107]. Therefore, sodium requirement may be considerably higher for CF patients than for normal individuals. Although routine sodium supplements may not be necessary because the average American diet contains an overabundance of sodium, sodium supplements are definitely needed in conditions that may cause prolonged sweat loss. In addition, there has been concern that marginal or low body sodium may limit the growth of children with CF [108].

Low plasma levels of zinc, calcium, magnesium, and iron in patients with CF have all been reported [109, 110]. In particular, iron-deficiency anemia with low serum ferritin has been shown to be quite frequent in CF patients with advanced pulmonary disease [110, 111].

III. NUTRITION ASSESSMENT

Frequent monitoring of the nutritional status for patients with CF is essential to ensure early detection of any deterioration and prompt initiation of nutrition intervention. Patients with CF are most vulnerable to experiencing malnutrition due to delayed diagnosis, during times of rapid growth (e.g., infancy and adolescence), and during pulmonary exacerbations. During these periods, close monitoring and intervention are critical to prevent nutritional decline. It should be emphasized that with comprehensive nutrition assessment and intervention, children with CF who are diagnosed early through NBS can achieve normal growth throughout childhood [28, 29], and adults with CF can maintain normal weight status. In addition, optimizing growth and nutritional status is critical for CF patients, because malnutrition worsens lung disease, affects the quality of life, and reduces survival [82, 112].

Assessment of nutritional status for patients with CF must include anthropometric, biochemical, and dietary assessments. The frequency with which the different indices of nutritional status monitoring should be measured is given in Table 2.

A. Anthropometric Assessment

Anthropometric assessment, with an emphasis on physical growth in children and body weight in adults, is an important component of nutritional assessment in patients with CF. For children with CF, accurate and sequential measurements of head circumference (age 0–3 years), recumbent length (age 0–2 years), height (age 2–20 years), weight (age 0–20 years), and body mass index (BMI) (age 2–20 years) should be obtained at each clinic visit using standardized techniques. For adults with CF, body weight should be measured at each clinic visit. Growth measurements should be plotted on the 2000 CDC growth charts [113] and converted to sex- and age-specific percentiles. Mid-arm circumference and triceps skinfold thickness measurements provide additional information about lean body mass and subcutaneous fat stores.

1. Evaluation of Height

In addition to evaluating whether an individual CF patient's height is appropriate for his or her age, it is useful to determine whether he or she is growing to genetic potential. Therefore, the genetic potential for height (i.e., target height) for each CF patient should be determined. This target height can be estimated by using mid-parental height, plus 6.5 cm for boys and minus 6.5 cm for girls, respectively [114]. The target height with a ±10 cm range should be noted on the patient's growth chart. According to the 2005 guidelines (Table 3), CF children who are below their genetic potential for height are considered at risk [70, 71].

2. Evaluation of Weight-for-Height

The 1992 and 2002 Consensus Reports [70, 115] recommend using two weight-for-height indexes to evaluate the relative proportion of weight for height, namely, percentage of ideal body weight (%IBW) based on the Moore method [116] and body mass index (BMI). However, recent studies and clinical applications demonstrated that use of the %IBW method is methodologically flawed [117–119]. Specifically, %IBW underestimates the severity of underweight in short children and overestimates it in tall children [118, 119]. In adults with CF, %IBW based on the Metropolitan Life

TABLE 2 Nutrition Assessment in Routine CF Care

	At Diagnosis	Every 3 Months (0–24 Months)	Every 3 Months (2–20 Years)	Every 3 Months (20+ Years)	Annually
Anthropometric assessment:					
Head circumference	x[a]	x			
Weight (to 0.1 kg)	x	x	x	x	
Length/height (to 0.1 cm)	x	x	x		
Body mass index	x		x	x	
Mid-arm circumference (MAC; to 0.1 cm)	x				x
Triceps skinfold (TSF; to 1.0 mm)	x[b]				x
Mid-arm muscle area, mm^2 (calculated from MAC and TSF)	x[b]				x
Mid-arm fat area, mm^2 (calculated from MAC and TSF)	x[b]				x
Biological mid-parental height[c]	x				
Pubertal status (Tanner staging)					x[d]
Biochemical assessment:[e]	x				x
Dietary assessment:					
24-hour diet recall					x
Nutritional supplement intake[f]					x
Eating behavior		x	x[g]		x

[a]If <24 months of age at diagnosis.
[b]Only in patients >1 year of age.
[c]Record in cm and gender-specific height percentile; note patient's target height percentile on all growth charts.
[d]Starting at age 9 years for girls and 10 years for boys until sexual maturation completes, annual pubertal self-assessment (patients, or parent and patient) or physician assessment using Tanner stage system [131, 132]; annual question as to menarchal status for girls.
[e]See Table 4 for details.
[f]A review of enzymes, vitamins, minerals, oral or enteral formulas, herbal, botanical, and other complementary and alternative medicine (CAM) products.
[g]Routine surveillance may be done informally by other team members, but the annual assessment and q3 monthly visits in the first 2 years of life and q3 monthly visits for patients at nutritional risk should be done by the center dietitian.

Insurance reference weights for medium/large frames overestimates the severity of underweight [118]. Consequently, use of %IBW leads to misclassification of nutritional status [118, 119]. The %IBW method has additional disadvantages in that its calculation is time-consuming and there is no readily available clinical resource to track its progress over time, making longitudinal monitoring difficult [120]. The Cystic Fibrosis Foundation (CFF) recognized the drawback associated with the use of %IBW and recommended the discontinuation of its use in 2005 [71].

Guidelines for classifying nutritional status based on height and weight-for-height indexes are summarized in Table 3. The U.S. CF Foundation made a major shift to broaden the screening of malnutrition from "nutritional failure" in the

TABLE 3 Classification of Height and Weight-for-Height Status in Patients with CF

	Children with CF		Adults with CF
	Linear Growth	Weight-for-Height	Weight-for-Height
U.S. Consensus Reports:[a]			
At risk	Height-for-age percentile not at genetic potential	Weight-for-height <50th percentile (age 0–2) BMI <50th percentile (age 2–20)	BMI <23 (males) BMI <22 (females)
European Consensus Report:[b]			
Malnutrition	Height <0.4 percentile or % Height <90%	%IBW <90%	BMI <18.5

[a]From Borowitz *et al.* [70], Stallings *et al.* [71], and Yankakas *et al.* [126].
[b]From Sinaasappel *et al.* [127].

2002 guidelines [70] to "nutritional risk" in the 2005 guidelines [71]. The new BMI indicators to identify nutritional risk, i.e., less than 50th percentile for children, less than 22nd for females, less than 23rd for males, were established based on their associations to lung function parameter forced expiratory volume in one second (FEV1) to replace the arbitrarily chosen 10th percentile cutoff that was recommended to define "nutritional failure" in the 2002 guidelines [70].

B. Biochemical Assessment

Monitoring biochemical indices of nutritional status is essential in patients with CF [70]. Current guidelines (Table 4) recommend routine, annual measurements of serum protein (albumin), vitamin A (retinol), vitamin D (25-hydroxycholecalciferol), vitamin E (α-tocopherol), and iron (hemoglobin, hematocrit).

Assessment of essential fatty acids is not routinely performed but only as indicated. However, recent findings on the relationships between abnormal essential fatty acid status and growth in children with CF [78, 79, 94], particularly those with meconium ileus, warrant the consideration of routine monitoring of essential fatty acid status, at least annually, in patients with CF. Similarly, routine measurements of vitamin K, calcium, zinc, and sodium are not regarded as necessary in the 2002 guidelines [70] but may be needed in individual patients.

C. Assessment of PI

Eighty-five to 90% of CF patients have PI. Pancreatic functional status not only has a direct influence on nutritional status but is also a strong predictor of long-term outcome [57, 121]. Data from the 1990–1995 U.S. CF Patient Registry demonstrated that patients with pancreatic sufficiency (PS) have an approximately 20-year longer lifespan than PI patients [57]. Specific mutations of the CFTR gene are associated with PS in a dominant fashion. Possessing an allele from this group (mostly "mild" mutations that belong to class IV or V; see Fig. 1) offers phenotypic protection even in combination with an allele normally associated with PI [57]. Therefore, the survival advantage associated with PS is likely due not only to better nutrient absorption but also to "mild" CF mutations.

The clinical signs and symptoms of PI include abdominal discomfort (bloating, flatus, pain), steatorrhea (frequent, malodorous, greasy stools), and the presence of meconium ileus or distal intestinal obstruction syndrome. Objective tests for PI include (1) duodenal measurement of pancreatic enzymes and bicarbonate; (2) 72-hour fecal fat balance study; and (3) fecal elastase in spot stool samples. Among these, 72-hour fecal fat balance study has been the gold standard. A high-fat diet is ingested for 72 hours, and stool is collected and analyzed for fat excreted. For the most precise results, oral dye markers are used to indicate the period of high-fat ingestion, and the stool that follows the first marker, up to and

TABLE 4 Laboratory Monitoring of Nutritional Status

	How Often to Monitor			
	At Diagnosis	Annually	Other	Tests
Pancreatic functional status	x		Check if maldigestion/malabsorption is present	72-hour fecal fat balance study or fecal elastase
Protein stores	x	x	Check if patient at nutritional risk	Albumin
Beta carotene			At physician's discretion	Serum levels
Vitamin A	x	x		Vitamin A (retinol)
Vitamin D	x	x		25-OH-D
Vitamin E	x	x		α-tocopherol
Vitamin K	x		If patient has hemoptysis or hematemesis; in patients with liver disease	Protein Induced by Vitamin K Absence-II (PIVKA-II) (preferably) or prothrombin time
Essential fatty acids			Consider checking in infants or those with failure to thrive	Triene:tetraene
Calcium/bone status			Age >8 years if risk factors are present (see text)	Calcium, phosphorus, ionized parathyroid hormone (PTH), dual-energy X-ray absorptiometry (DEXA)
Iron	x	x	Consider in-depth evaluation for patients with poor appetite	Hemoglobin, hematocrit; consider transferrin receptor levels
Zinc			Consider 6-month supplementation trial for patients with failure to thrive	No acceptable measurement
Sodium			Consider checking if exposed to heat stress and becomes dehydrated	Serum sodium; spot urine sodium if total body sodium depletion suspected

Adapted from Borowitz *et al.* [70].

including the second marker, represents the stool produced during the period of high fat intake. In clinical practice, diet is often measured for 3 days and stool collected simultaneously. A coefficient of fat absorption, that is, (fat intake – fecal fat loss)/fat intake × 100%, is calculated. In CF, PI is defined by a coefficient of fat absorption less than 93% [70].

Because 72-hour fecal fat balance study is cumbersome and not well accepted by CF patients and care providers, measurement of fecal elastase in a small stool sample has gained wide acceptance and is becoming the standard method of care in most CF centers [57]. Elastase is one of the more than 20 enzymes secreted by the pancreas. It has the physical property of being stable as it transits the intestinal tract, unlike other enzymes that may be degraded by intraluminal proteases. As water is withdrawn from the intestinal contents in the colon, elastase concentrations increase, making it easy to measure in stool. This protein is stable through a wide range of pH and temperature, making it ideal to collect at home. Levels greater than 100–200 μg fecal elastase per gram of stool generally indicate PS [57, 122].

It is important to assess pancreatic functional status as soon as CF diagnosis is made. Approximately half of CF patients whose initial tests indicate PS become PI later [4]. Therefore, PS patients should be reevaluated at least annually to determine if they have changed to PI, especially if genotype studies reveal mutations that are generally associated with PI. For patients diagnosed with PI, pancreatic enzyme replacement therapy (PERT) and vitamin supplementation should be started. It is important to understand that although PI can be treated with PERT, it cannot be corrected; many patients continue to have steatorrhea when they receive PERT [123]. In addition, response to PERT varies greatly among individual CF patients.

D. Dietary Assessment

Assessment of energy requirement and dietary intakes is an important way of determining whether the patient is at negative energy balance. Evaluation of dietary intakes is best performed by dietitians/nutritionists specializing in the care of patients with CF. For patients with good nutritional status, the dietitian may assess dietary habits and the quality of dietary intake using a 24-hour dietary recall. However, for patients with suboptimal nutritional status, a 3-day prospective food record is the best way to obtain quantitative estimates of energy and nutrient intakes. This assessment can then be used as the basis for initiating appropriate nutrition intervention.

Energy requirements for patients with CF are best determined by estimating the basal metabolic rate, the degree of malabsorption, and the severity of pulmonary disease. For children older than 2 years of age and adults, current CFF guidelines [71] recommend energy intakes at 110–200% of estimated energy requirement for the general population [124, 125] to support weight maintenance in adults and weight gain at an age-appropriate rate in children. Alternatively, the 2002 CFF guidelines [70] provide a method to calculate energy requirement for individual CF patients based on their pancreatic functional status and the severity of lung disease, as outlined in Table 5.

TABLE 5 Method for Estimating Energy Requirement for CF Patients

Step I: Estimate basal metabolic rate (BMR) by using the WHO equations [133]:

	Males	Females
0–3 years	60.9 × wt − 54	61.0 × wt − 51
3–10 years	22.7 × wt + 495	22.5 × wt + 499
10–18 years	17.5 × wt + 651	12.2 × wt + 476
18–30 years	15.3 × wt + 679	14.7 × wt + 496
>30 years	11.6 × wt + 879	8.7 × wt + 829

Step II: Estimate energy expenditure (EE) using the following equation:

EE = BMR (activity coefficient + disease coefficient)

Where activity coefficient = 1.3 (confined to bed)
1.5 (sedentary)
1.7 (active)

disease coefficient = 0 (normal lung function, i.e., $FEV_1 > 80\%$
0.2 (moderate lung disease, i.e., FEV_1 40–79%)
0.3 (severe lung disease, i.e., $FEV_1 < 40\%$)

Step III: Estimate total energy requirement (ER), taking into account pancreatic functional status:

a. For PS patients, i.e., coefficient of fat absorption (CFA) ≥ 93%:
ER = EE

b. For PI patients with a CFA < 93%:
ER = EE × (0.93 ÷ CFA)

c. For PI patients whose CFA has not been determined, use 0.85 as an approximate for CFA:
ER = EE × (0.93 ÷ 0.85)

Adapted from Borowitz *et al.* [70].

IV. NUTRITION MANAGEMENT

Nutrition management for patients with CF varies and depends on the stage of diagnosis (newly diagnosed CF versus routine management), patient's age (infancy, early childhood, adolescence, or adulthood), and disease severity. Nutrition management begins at the time of CF diagnosis. The first 6 months after the diagnosis of CF is a crucial period for establishing therapeutic interventions, dietary counseling, and nutritional education. Nutrition management for patients with stable CF focuses on maintaining optimal nutritional status and preventing malnutrition. Nutrition management for patients experiencing malnutrition focuses on achieving catch-up growth (for children) and weight gain (for adults). In addition, PERT and vitamin supplementation are essential for all categories of nutrition management.

The multidisciplinary CF care team should monitor growth, provide anticipatory counseling, and plan intervention strategies for each individual CF patient. Achieving and maintaining normal growth and nutritional status require management of gastrointestinal and pulmonary symptoms, dietary intakes and eating behaviors, and psychosocial and financial issues. Guidelines for nutritional assessment and intervention for patients with CF have been developed since the early 1990s by the U.S. and European Cystic Fibrosis Foundation [60, 70, 71, 115, 126, 127]. The following section describes the most recent updates published from 2002 to 2007 [70, 71, 126]. In addition, revisions of CFF guidelines for infants diagnosed through NBS are pending and will be published from a July 2007 Care Consensus Conference on "Management of Infants Diagnosed with CF by Newborn Screening."

A. At Diagnosis

All nutritional indices should be measured at the time of CF diagnosis. These include anthropometry, biochemical nutritional markers, pancreatic functional status, and dietary intake (Tables 2 and 4). If PI is diagnosed, PERT and nutrient supplementation (see following sections) should be initiated immediately. Extensive discussion of nutritional management with patients and/or parents is another important aspect of care during this period of time. By concentrating on nutrition during the family's first few visits to the CF center, CF care providers will be able to stress the importance of establishing and maintaining good nutrition.

The majority of CF patients (~70%) are diagnosed before age 2 years [41]. The first 2 years represent the phase of life with the highest growth rate and energy requirement. The severity of CF in infants and young children with newly diagnosed CF varies greatly depending on how they are diagnosed. For example, infants born with meconium ileus, in particular those who have bowel resection resulting from intestinal surgery, despite being diagnosed and treated during the neonatal period, are more likely to experience poor growth, essential fatty acid deficiency, and more severe lung disease [78, 89, 128]. Therefore, CF infants with meconium ileus may require enteral or parenteral nutrition support in addition to oral intakes to achieve optimal growth and correct nutrient deficiencies.

Infants and young children diagnosed with CF as a result of gastrointestinal and/or pulmonary symptoms are identified at variable ages. Half of these children present with growth failure at the time of CF diagnosis because of delayed diagnosis [25]. For these children, initial nutritional intervention focuses on reducing steatorrhea and promoting growth recovery.

Unlike infants presenting with relatively severe signs/symptoms of CF at the time of diagnosis, infants identified through NBS or a positive family history early in life often appear asymptomatic and show relatively good growth at the time of CF diagnosis [27, 41]. The goals of nutrition management for this group of infants are to maintain normal growth and to prevent the occurrence of malnutrition. Beginning in 2008, the majority of infants born with CF in the United States will be diagnosed through NBS programs. To standardize clinical care for these newborns with CF, the CFF convened a consensus conference in July of 2007; new recommendations are expected to be published in the near future.

For infants with CF, both the 2002 [70] and the proposed 2007 guidelines encourage breast-feeding for infants with CF. However, whether breast-feeding, in particular exclusive breast-feeding, can promote optimal growth in infants with CF is unclear, especially in infants who have more severe CF (e.g., meconium ileus, PI, growth failure). This is because human milk is relatively low in protein (7% of total calories) and sodium and may be low in essential fatty acids if maternal intake is low. Therefore, breast-feeding should be encouraged only in infants who show appropriate growth, and breast-fed infants should be closely monitored with regard to growth velocity, protein status, and electrolyte status. If growth faltering is observed, supplementation with cow's milk–based formula should be initiated, and a formula with caloric density greater than 20 kcal/oz may be needed. Soy-based formulas are more likely to be associated with hypoproteinemic edema and growth retardation and therefore are not recommended for infants with CF.

Another controversy in the nutritional care for newly diagnosed CF infants is the monitoring and supplementation of n-6 and n-3 long-chain polyunsaturated fatty acids such as linoleic acid and docosahexaenoic acid (DHA). The 2002 guidelines do not recommend routine monitoring of EFA status and supplementation of linoleic acid or DHA, but rather recommend it only in CF infants or children who are experiencing failure to thrive [70]. However, many studies demonstrated positive associations between EFA and growth in children with CF [87, 88, 93, 94]. If such evidence is combined with the high prevalence of poor growth and nutrient deficiencies at the time of diagnosis, assessing EFA in newly diagnosed CF infants is warranted. We believe that EFA monitoring should become routine in CF.

If an infant with CF is growing at a normal rate, transition to whole cow's milk is recommended beginning at 12 months of age, although formula feeding may be continued up to 24 months of age. Some families are surprised by the recommendation of whole milk when there is so much media and medical advocacy for low-fat milk. However, it is important for parents to ensure that two types of milk are available for themselves and their children because whole milk is arguably the best source of nutrients for the young child with CF.

Introduction of solid foods should be made according to the guidelines of the American Academy of Pediatrics, namely, between 4 and 6 months of age. For infants who have begun to consume solid foods but experience slower than normal growth rate, increasing the caloric density of solid foods, such as adding carbohydrate polymers (e.g., Polycose) and fats (e.g., vegetable oil, Microlipid, or MCT oil) to infant cereals and baby foods should be recommended. As infants transition to table foods, it is important to introduce families to the concept that children with CF should eat a balanced diet that is moderate to high in fat. They should be aware that this advice is counter to the usual dietary guidelines for children without CF. Parents should avoid giving their CF children low-fat or low-calorie foods.

B. Routine Management for Well-Nourished CF Patients

The primary goal of routine nutrition management is to optimize pancreatic enzyme replacement therapy, dietary intake, and nutrient supplementation to achieve and maintain growth and nutritional status. Nutrition management should also take into account the patient's pulmonary status. Assessment of nutritional status should be made at each routine clinic visit, that is, every 3 months according to the current CFF guidelines [70, 71]. Because nutrition requirement is influenced by age, specific recommendations from early childhood to adulthood are further described next.

1. Pancreatic Enzyme Replacement Therapy

PERT is used to improve maldigestion and malabsorption in CF patients with PI. Most enzyme products are capsules that contain enteric-coated microencapsulated enzymes. The enteric coating prevents inactivation of enzymes in the acidic gastric environment. All pancreatic enzyme products contain lipase, amylase, and protease. However, the potency of pancreatic enzyme is based on the content of lipase in each capsule, because fat is the macronutrient most often malabsorbed in PI.

Pancreatic enzymes should be taken with all foods and beverages, including breast milk and infant formula. The dosage for enzymes should be individualized based on the patient's diet, degree of PI, intestinal pH, and GI anatomy. In clinical practice, two approaches are used to calculate enzyme doses:

1. Based on the amount of fat in the diet, that is, units of lipase per gram of fat per meal. This method is more physiological than the weight-based method described next. Current CFF guidelines recommend 500–4000 lipase units per gram of fat per meal [70, 71]. A range of doses is recommended because there is great individual variation in response to enzymes.
2. Based on body weight, that is, units of lipase per kg body weight per meal. This method is less accurate, but it eliminates the need to estimate the amount of fat in the diet and thus is easier for patients and caretakers to understand and follow. However, it should be noted that the amount of fat ingested per kilogram of body fat generally decreases with age, so enzyme dose per kg of body weight should be adjusted as age increases. Current CFF guidelines recommend 500–2500 lipase units per kg body weight per meal. In addition, total enzyme dose should not exceed 10,000 units per kg body weight per day because of risk of developing fibrosing colonopathy [129, 130].

The total daily dose should reflect approximately three meals and two or three snacks. In general, half of the dose per meal is given with snacks. The adequacy of PERT is typically assessed clinically by following growth/weight status and stool patterns. If response to PERT is poor, adjunctive therapies such as bicarbonate or proton-pump inhibitor to decrease duodenal acidity may be helpful.

2. Energy Intake and Nutrient Supplementation

To obtain adequate energy intake and compensate for fat malabsorption, CF patients typically require a greater fat intake (35–40% of calories) than what is normally recommended for the general population (25–35%). Fat restriction is not recommended, because fat is the most energy-dense macronutrient and provides essential fatty acids. Medium-chain triglyceride (MCT) supplements may be utilized as a good source of fat because MCT requires less lipase activity, less bile salt for solubility, and can be transported as free fatty acids through the portal system.

In CF patients, vitamin supplementation is necessary to prevent the occurrence of deficiencies. A standard, age-appropriate multivitamin supplement should be given to all CF patients. Additional supplementation with fat-soluble vitamins is needed (Table 6).

CF patients are at risk of hyponatremia because of salt loss through the skin. Infants with CF should receive salt supplementation (1/8 tsp per day). Older children and adults are advised to consume a high-salt diet, especially during summer months and for those who live in hot climates.

TABLE 6 Recommendations for Daily Fat-Soluble Vitamin Supplementation

	Vitamin A (IU)	Vitamin E (IU)	Vitamin D (IU)	Vitamin K (mg)[a]
0–12 months	1,500	40–50	400	0.3–0.5
1–3 years	5,000	80–150	400–800	0.3–0.5
4–8 years	5,000–10,000	100–200	400–800	0.3–0.5
>8 years	10,000	200–400	400–800	0.3–0.5

[a]Currently, commercially available products do not have ideal doses for supplementation. Prothrombin time or, ideally, PIVKA-II levels should be checked in patients with liver disease, and vitamin K dose titrated as indicated.

Adapted from Borowitz *et al.* [70].

3. Age-Specific Recommendations

a. Preschool Age

Children in this age group have developed self-feeding skills, food preferences, and dietary habits. Food intake and physical activity vary from day to day. For these reasons, close monitoring of dietary habits, caloric intake, and growth velocity are important. Routinely adding calories to table foods may help with maintaining optimal growth at this stage. In addition to food intake, dietary counseling should include assessment of eating behaviors. The importance of serving calorie-dense foods (such as whole milk rather than low-fat milk) and establishing positive mealtime interactions should be emphasized.

b. School Age

Children in this age group are at risk for declining growth for various reasons. They typically participate in a variety of activities, leading to limited time for meals and snacks. They are also exposed to peer pressure and challenged to begin self-managing their disease. These may affect compliance with prescribed medications such as pancreatic enzymes and fat-soluble vitamins. In addition, acceptance and understanding by teachers and fellow students may be lacking, further stressing a child with CF. Encouraging children to help in meal planning and preparation may be helpful in improving food intake.

c. Adolescence

This stage represents another vulnerable period of developing malnutrition because of increased nutritional requirements associated with accelerated growth, endocrine development, and high levels of physical activity. In addition, pulmonary disease often becomes more severe in this period, increasing energy requirement. Puberty is often delayed in adolescents with CF; this usually is related to growth failure and poor nutritional status, rather than to a primary endocrine disorder. Assessment of puberty should be performed annually at age 9 in girls and age 10 in boys by a standardized self-assessment or physician examination [131, 132]. In addition to plotting growth on the (2000 CDC growth charts, evaluating height and weight velocity [131] in association with Tanner stages can be very useful in identifying delayed or attenuated pubertal growth.

Nutritional counseling should be directed toward the patient rather than the parents. Teenagers may be more receptive to efforts to improve muscular strength and body image as a justification for better nutrition than emphasis on weight gain and improved disease status.

d. Adulthood

CF patients reaching adulthood are usually responsible for the entire management of their disease, as well as for the financial burden of a chronic illness. While in college or working, adults with CF are constantly adapting to new schedules and stresses. The goal of nutrition management is to maintain optimal BMI and to prevent unintentional weight loss (Table 3). Nutritional counseling must be practical and pragmatic to help adults adjust to these changes.

e. Pregnancy and Lactation

Widespread experience in recent years has demonstrated that pregnancy and lactation can be accomplished successfully by some women with CF. Pregnant women with CF should follow the guidelines from the Dietary Reference Intakes (DRIs) for nutrient intakes (see Appendix). Special attention should be given to appropriate weight gain, particularly during the last trimester of the pregnancy. In addition to the usual multivitamin supplementation for CF, one prenatal vitamin should be consumed daily. During lactation, marked increase in caloric intake is necessary to meet the high energy requirement during this period.

C. Nutrition Intervention for Patients at Risk for or Experiencing Malnutrition

For CF patients who are at risk for or are experiencing malnutrition, nutritional intervention beyond of the level of routine management is required. First, the presence of comorbid medical conditions that are likely to affect growth and nutritional status, such as gastroesophageal reflux disease and CF-related diabetes, should be evaluated. Nutrition support is then delivered at various levels, ranging from behavioral intervention, dietary modification, or oral supplementation to enteral or parenteral supplementation.

TABLE 7 Maximizing Calories for Healthy Patients with CF

Adding Calories to Foods	High-Calorie Foods and Snacks[a]
• Add fats such as butter, gravy, cheese, or dressings to starches, fruits, and vegetables • Use whipped cream on fruits and desserts • Make "super" milk: ½ cup whole milk + ½ cup half-and-half • Flavor milk with syrups or powders (chocolate, strawberry, etc.) or add whole-milk yogurt to milk • Add eggs to hamburger meat or casseroles (never serve raw eggs) • Use extra salad dressing; avoid low-calorie or reduced-calorie dressings • Serve gravies and cheese sauces	• Full-fat ice cream, puddings • Cookies and milk • Cheese or peanut butter crackers • Muffins or bagels with cream cheese or butter • Cheese breadsticks • Chips and dip • French fries • Whole milk yogurt • Egg salad, tuna salad, cheese or avocado slices with crackers • Trail mixes, nuts, and granola (after the age of 2 years) • Cold cuts, pizza • Fresh vegetables with salad dressing or dip

[a]Assess age-appropriateness, especially with respect to choking risk in young children, before recommending.
Adapted from Borowitz *et al.* [70].

1. Behavioral Intervention

In an effort to increase dietary intakes, caregivers of young children with CF may be engaged in ineffective feeding practices such as coaxing, commanding, physical prompts, and parental feeding. Adolescents with CF may intentionally skip pancreatic enzymes in order to achieve a certain body image. An in-depth assessment of eating behavior, feeding patterns, and family interactions at mealtimes should be performed in CF patients at risk or experiencing malnutrition. If negative behaviors are present, behavioral intervention should be used in conjunction with dietary intervention to improve intake. For example, one behavioral strategy is to gradually increase calories by working on one meal at a time. Another strategy is to teach parents alternative ways of responding to their child who eats slowly or negotiates what he or she will eat. Referral for more in-depth behavioral therapy is also encouraged.

2. Dietary Intervention

a. Oral Supplements

Dietary intervention should begin with dietary modification to increase caloric density of the diet, that is, addition of high-calorie foods to the patient's regular diet without dramatically increasing the amount of food consumed. For example, margarine or butter may be added to many foods, and half-and-half can be used in place of skim milk or water when preparing canned soup. More examples of how to maximize the caloric density of the diet are given in Table 7. If dietary modification is ineffective, use of an energy supplement may be introduced. However, it is important to ensure that the energy supplement is not used as a substitute for normal food intake.

b. Enteral Feedings

Enteral feeding can be initiated when oral supplementation does not improve nutritional status significantly. Enteral feeding can be delivered via nasogastric tubes, gastrostomy tubes, and jejunostomy tubes. The choice of enterostomy tube and technique for its placement should be based on the expertise of the CF center. Nasogastric tubes are appropriate for short-term nutritional support in highly motivated patients. Gastrostomy tubes are more appropriate for patients who need long-term enteral nutrition. Jejunostomy tubes may be indicated in patients with severe GERD; use of predigested or elemental formula may be needed with jejunostomy feeding.

Standard enteral feeding formulas (complete protein, long-chain fat) are typically well tolerated. Calorie-dense formulas (1.5–2.0 kcal/ml) are usually required to provide adequate energy. Nocturnal infusion is encouraged to promote normal eating patterns during the day. Initially, 30–50% of estimated energy requirement may be provided overnight. Pancreatic enzymes should be given with enteral feeding; however, optimal dosing regimen is unclear with overnight feeding.

V. CONCLUSION

The clear associations between nutritional status and clinical outcomes in CF mandate careful nutritional assessment, management, and monitoring of all patients with CF. In recent years, with new knowledge arising from NBS research [33], there has been a shift away from the idea that malnutrition is inevitable for most CF patients toward the more optimistic view that normal nutrition and growth are possible if early diagnosis and aggressive nutritional monitoring and therapy are provided to each individual patient. This task is best accomplished by involving a multidisciplinary team that includes a dietitian in the care and management of CF patients. In this way, the goals of normal growth and prevention of malnutrition can be attained, which will improve the prognosis and quality of life for patients with CF.

Acknowledgments

We are grateful to our research colleagues in Nutritional Sciences (Suzanne Shoff and Zhumin Zhang) and Pediatrics (Anita Laxova and Michael Rock) at the University of Wisconsin—Madison and at the University of Wisconsin Hospital and Clinics (especially Lisa Davis and Mary Marcus). Dr. Lai is supported by NIH grant R01DK072126 and Dr. Farrell by NIH grants R01DK034108 and M01RR03186.

References

1. Kosorok, M. R., Wei, W. H., and Farrell, P. M. (1996). The incidence of cystic fibrosis. *Stat. Med.* **15**, 449–462.
2. Boat, T. F., Welsh, M. J., and Beaudet, A. L. (1989). Cystic fibrosis. "The Metabolic Basis of Inherited Disease" (C. R. Scriver, A. L. Beaudet, W. S. Sly, and D. Valle, Eds.), pp. 2649–2680. McGraw-Hill, New York.
3. di Sant'Agnese, P. A., and Davis, P. B. (1976). Research in cystic fibrosis. *N. Engl. J. Med.* **295**, 481.
4. Rowe, S. M., Miller, S., and Sorscher, E. J. (2005). Mechanisms of cystic fibrosis. *N. Engl. J. Med.* **352**, 1992–2001.
5. Waters, D. L., Dorney, S. F. A., Gaskin, K. J., Gruca, M. A., O'Halloran, M., and Wilcken, B. (1990). Pancreatic function in infants identified as having cystic fibrosis in a neonatal screening program. *N. Engl. J. Med.* **322**, 303–308.
6. Kerem, E., Corey, M., Kerem, B., Durie, P., Tsui L.-C., and Levison, H. (1989). Clinical and genetic comparisons of patients with cystic fibrosis, with or without meconium ileus. *J. Pediatr.* **114**, 767–73.
7. Bronstein, M. N., Sokol, R. J., Abman, S. H., Chatfield, B. A., Hammond, K. B., Hambidge, K. M., Stall, C. D., and Accurso, F. J. (1992). PI, growth, and nutrition in infants identified by newborn screening as having cystic fibrosis. *J. Pediatr.* **120**, 533–540.
8. Corey, M. L. (1980). Longitudinal studies in cystic fibrosis. *In* "Perspectives in Cystic Fibrosis, Proceedings of the 8th International Cystic Fibrosis Congress" (J. M. Sturgess, Ed.), pp. 246–261. Canadian Cystic Fibrosis Foundation, Toronto.
9. Katz, J. N., Horwitz, R. I., Dolan, T. F., and Shapiro, E. D. (1986). Clinical features as predictors of functional status in children with cystic fibrosis. *Pediatrics* **108**, 352–358.
10. Farrell, P. M., Li, Z., Kosorok, M. R., Laxova, A., Green, C. G., Collins, J., Lai, H. C., Rock, M. J., and Splaingard, M. L. (2003). Longitudinal evaluation of bronchopulmonary disease in children with cystic fibrosis. *Pediatr. Pulmonol.* **36**, 230–240.
11. Hardin, D. S., and Moran, A. (1999). Diabetes mellitus in cystic fibrosis. *Endocrinol. Metabol. Clin. North Am.* **28**, 787–800.
12. Mackie, A. D. R., Thornton, S. J., and Edenborough, F. P. (2003). Cystic fibrosis-related diabetes. *Diabet. Med.* **20**, 425–436.
13. Cystic Fibrosis Foundation (2001). National Cystic Fibrosis Patient Registry Annual Data Report, 2000 Bethesda, MD.
14. Scott-Jupp, R., Lana, M., and Tanner, M. S. (1991). Prevalence of liver disease in cystic fibrosis. *Arch. Dis. Child.* **66**, 698–701.
15. Sokol, R. Z. (2001). Infertility in men with cystic fibrosis. *Curr. Opin. Pulm. Med.* **7**, 421–426.
16. Kerem, B. S., Rommens, J. M., Buchanan, J. A., Markiewicz, D., Cox, T. K., Chakravarti, A., Buchwald, M., and Tsui, L. C. (1989). Identification of the cystic fibrosis gene: Genetic analysis. *Science* **245**, 1073–1080.
17. Collins, F. S. (1992). Cystic fibrosis: Molecular biology and therapeutic implications. *Science* **256**, 29–33.
18. Tsui, L. C., and Buchwald, M. (1991). Biochemical and molecular genetics of cystic fibrosis. *In* "Advances in Human Genetics" (H. Harris and K. Hirschhorn, Eds.), pp. 153–266. Plenum Press, New York.
19. Gregg, R. G., Simantel, A., and Farrell, P.M. *et al.* (1997). Newborn screening for cystic fibrosis in Wisconsin: Comparison of biochemical and molecular methods. *Pediatrics* **99**, 819–824.
20. The Cystic Fibrosis Genotype-Phenotype Consortium. (1993). Correlation between genotype and phenotype in patients with cystic fibrosis. *N. Engl. J. Med.* **329**, 1308–1313.
21. Gibson, L. E., Cooke, R. E. (1959). A test for the concentration of electrolytes in sweat in cystic fibrosis of the pancreas utilizing pilocarpine iontophoresis. *Pediatrics* **23**, 545–549.
22. Anderson, M. P., Rich, D. P., Gregory, R. J., Smith, A. E., and Welsh, M. J. (1991). Generation of cAMP-activated chloride currents by expression of CFTR. *Science* **251**, 679–682.
23. Blythe, S. A., and Farrell, P. M. (1984). Advances in the diagnosis and management of cystic fibrosis. *Clin. Biochem.* **17**, 277–283.
24. Rosenstein, B. J., and Cutting, G. R. (1998). The diagnosis of cystic fibrosis: A consensus statement. *J. Pediatr.* **132**, 589–595.
25. Lai, H. C., Kosorok, M. R., Sondel, S. A., Chen, S. T., FitzSimmons, S. C., Green, C., Shen, G., Walker, S., and Farrell, P. M. (1998). Growth status in children with cystic fibrosis based on National Cystic Fibrosis Patient Registry data: Evaluation of various criteria to identify malnutrition. *J. Pediatr.* **132**, 478–485.
26. Farrell, P. M., Kosorok, M. R., Laxova, A., *et al.* (1997). Nutritional benefits of neonatal screening for cystic fibrosis. *N. Engl. J. Med.* **337**, 963–969.
27. Farrell, P. M., Kosorok, M. R., Rock, M. J., Laxova, A., Zeng, L., Lai, H. C., Hoffman, G., Laessig, R. H., Splaingard, M. L., and the Wisconsin Cystic Fibrosis Neonatal Screening Study Group. (2001). Early diagnosis of cystic fibrosis through neonatal screening prevents severe malnutrition and improves long-term growth. *Pediatrics* **107**, 1–13.
28. Farrell, P. M., Lai, H. J., Li, Z., Kosorok, M. R., Laxova, A., Green, C. G., Collins, J., Hoffman, G., Laessig, R., Rock, M. J., and Splaingard, M. L. (2005). Evidence on improved outcomes with early diagnosis of cystic fibrosis through neonatal screening: Enough is enough! *J. Pediatr.* **147**, S30–36.
29. Crossley, J. R., Elliot, R. B., and Smith, R. A. (1979). Dried blood spot screening for cystic fibrosis in the newborn. *Lancet* **1**, 472–474.
30. Heely, A. F., Heely, M. E., King, D. N., Kuzemko, J. A., and Walsh, M. P. (1981). Screening for cystic fibrosis by dried blood spot trypsin assay. *Arch. Dis. Child.* **57**, 18.

31. Farrell, P. M., Mischler, E. H., Fost, N. C., Wilfond, B. S., Tluczek, A., Gregg, R. G., Bruns, W. T., Hassemer, D. J., and Laessig, R. H. (1991). Current issues in neonatal screening for cystic fibrosis and implications of the CF gene discovery. *Pediatr. Pulmonol.* **57**, 11–18.
32. Farrell, P. M., and Mischler, E. H. (1992). Newborn screening for cystic fibrosis. *Adv. Pediatr.* **39**, 31–64.
33. Centers for Disease Control and Prevention. (2004). Newborn screening for cystic fibrosis: Evaluation of benefits and risks and recommendations for state newborn screening programs. *MMWR Morb. Mortal. Wkly Rep.* **53**(No. RR-13), 1–36.
34. Rock, M. J., Hoffman, G., Laessig, R. H., Kopish, G. J., Litshem, T. J., and Farrell, P. M. (2005). Newborn screening for cystic fibrosis in Wisconsin: Nine-year experience with routine trypsinogen/DNA testing. *J. Pediatr.* **147**, S73–S77.
35. Hammond, K. B., Abman, S. H., Sokol, R. J., and Accurso, F. J. (1991). Efficacy of statewide neonatal screening for cystic fibrosis by assay of trypsinogen concentrations. *N. Engl. J. Med.* **325**, 769–774.
36. Mastella, G., Barlocco, E. G., Antonacci, B., Borgo, G., Braggion, C., Cazzola, G., Conforti, M., Doro, R., Faraguna, D., Giglio, L., Miano, A., Parmelli, C., and Riggio, S. (1988). Is neonatal screening for cystic fibrosis advantageous? The answer of a wide 15 years follow-up study. *In* "Mucoviscidose: Depistage Neonatal et Prise en Charge Precoce Caen," pp. 127–143. CHRU de Caen.
37. Wilcken, B. (1993). Newborn screening for cystic fibrosis: Its evolution and a review of the current situation. *Screening* **2**, 43–62.
38. Dankert-Roelse, J. E., te Meerman, G. J., Martin, A., ten Kate, L. P., and Knol, K. (1989). Survival and clinical outcome in patients with cystic fibrosis, with or without neonatal screening. *J. Pediatr.* **114**, 362–367.
39. Waters, D. L., Wilcken, B., Irwing, L., van Asperen, P., Mellis, C., Simpson, J. M., Brown, J., and Gaskin, K. J. (1999). Clinical outcomes of newborn screening for cystic fibrosis. *Arch. Dis. Child.* **80**, F1–7.
40. Ghosal, S., Taylor, C. J., Pickering, M., and McGaw, J. (1996). Head growth in cystic fibrosis following early diagnosis by neonatal screening. *Arch. Dis. Child.* **75**, 191–193.
41. Lai, H. J., Cheng, Y., and Farrell, P. M. The survival advantage of cystic fibrosis patients diagnosed through neonatal screening: Evidence from the U.S. Cystic Fibrosis Patient Registry data. *J. Pediatr.* (2005). **147**, S57–63.
42. Dankert-Roelse, J. E., Merelle, M. E. (2005). Review of outcomes of neonatal screening for cystic fibrosis versus non-screening in Europe. *J. Pediatr.* **147**, S15–S21.
43. Cystic Fibrosis Foundation. (1997). "Clinical Practice Guidelines for Cystic Fibrosis." Bethesda, MD.
44. FitzSimmons, S. C. (1993). The changing epidemiology of cystic fibrosis. *J. Pediatr.* **122**, 1–9.
45. Cystic Fibrosis Foundation. (2006). "National Cystic Fibrosis Patient Registry Annual Data Report, 2005." Bethesda, MD.
46. Koscik, R. L., Farrell, P. M., Kosorok, M. R., Zaremba, K. M., Laxova, A., Lai, H. C., Douglas, J. A., Rock, M. J., and Splaingard, M. L. (2004). Cognitive function of children with cystic fibrosis, deleterious effect of early malnutrition. *Pediatrics* **113**, 1549–1558.
47. Koscik, R. L., Lai, H. J., Laxova, A., Zaremba, K., Kosorok, M. R., Douglas, J. A., Rock, M. J., Splaingard, M. L., and Farrell, P. M. (2005). Preventing early, prolonged vitamin E deficiency: An opportunity for better cognitive outcomes via early diagnosis through neonatal screening. *J. Pediatr.* **147**, S51–56.
48. Rosenfeld, M. (2005). Overview of published evidence on outcomes with early diagnosis from large U.S. observational studies. *J. Pediatr.* **147**, S11–S14.
49. Sharma, R., Florea, V. G., Bolger, A. P., Doehner, Q., Florea, N. D., Coats, A. J. S., *et al.* (2001). Wasting as an independent predictor of mortality in patients with cystic fibrosis. *Thorax* **56**, 746–750.
50. Konstan, M. W., Butler, S. M., Wohl, M. E. B., Stoddard, M., Matouse, R., Wagner, J. S. *et al.* (2003). Growth and nutritional indexes in early life predict pulmonary function in cystic fibrosis. *J. Pediatr.* **142**, 624–630.
51. Zemel, B. S., Jawad, A. F., FitzSimmons, S., and Stallings, V. A. (2000). Longitudinal relationship among growth, nutritional status, and pulmonary function in children with cystic fibrosis: Analysis of the Cystic Fibrosis Foundation National CF Patient Registry. *J. Pediatr.* **137**, 374–380.
52. Durie, P. R., and Pencharz, P. B. (1992). Nutrition in cystic fibrosis. *Br. Med. Bull.* **48**, 823–847.
53. Pencharz, P. B., and Durie, P. R. (1993). Nutritional management of cystic fibrosis. *Annu. Rev. Nutr.* **13**, 111–136.
54. Stallings, V. A. (1994). Nutritional deficiencies in cystic fibrosis: causes and consequences. *New Insights Cystic Fibrosis* **2**, 1–5.
55. Pencharz, P. B., and Durie, P. R. (1993). Nutritional management of cystic fibrosis. *Annu. Rev. Nutr.* **13**, 111–136.
56. Gaskin, K. J., Durie, P. R., Lee, L., Hill, R., and Forstner, G. G. (1984). Colipase and lipase secretion in childhood-onset pancreatic insufficiency. Delineation of patients with steatorrhea secondary to relative colipase deficiency. *Gastroenterology* **86**, 1–7.
57. Borowitz, D. (2005). Update on the evaluation of pancreatic exocrine status in cystic fibrosis. *Curr. Opin. Pulm. Med.* **11**, 524–527.
58. Belli, D. C., Levy, E., Darling, P., Leroy, C., and Lepage, G. (1987). Taurine improves the absorption of a fat meal in patients with cystic fibrosis. *Pediatrics* **80**, 517–523.
59. Vaisman, N., Pencharz, P. B., Corey, M., Canny, G. J., and Hahn, E. (1987). Energy expenditure of patients with cystic fibrosis. *J. Pediatr.* **111**, 496–500.
60. Buchdahl, R. M., Cox, M., Fulleylove, C., Marchant, J. L., Tomkins, A. M., Brueton, M. J., Warner, J. O. (1988). Increased resting energy expenditure in cystic fibrosis. *J. Appl. Physiol.* **64**, 1810–1816.
61. Anthony, H., Bines, J., Phelan, P., and Paxton, S. (1998). Relation between dietary intake and nutritional status in cystic fibrosis. *Arch. Dis. Child.* **78**, 443–447.
62. Tomezsko, J. L., Stallings, V. A., Kawchak, D. A., Goin, J. E., Diamond, G., and Scanlin, T. F. (1994). Energy expenditure and genotype of children with cystic fibrosis. *Pediatr. Res.* **35**(4 Pt 1), 451–460.
63. Pencharz, P., Hill, R., Archibald, E., Levy, L., and Newth, C. (1984). Energy needs and nutritional rehabilitation in undernourished adolescents and young adult patients with cystic fibrosis. *J. Pediatr. Gastroenterol. Nutr.* **3**, S147–S153.
64. Feigal, R. J., and Shapiro, B. L. (1979). Mitochondrial calcium uptake and oxygen consumption in cystic fibrosis. *Nature* **278**, 276–277.

65. Stutts, M. J., Knowles, M. R., Gatzy, J. T., and Boucher, R. C. (1986). Oxygen consumption and ouabain binding sites in cystic fibrosis nasal epithelium. *Pediatr. Res.* **20**, 1316–1320.
66. Norman, D., Elborn, J. S., Cordon, S. M., Rayner, R. J., Wiseman, M. S., Hiller, E. J., and Shale, D. J. (1991). Plasma tumour necrosis factor alpha in cystic fibrosis. *Thorax* **46**, 91–95.
67. Dodge, J. A., and Yassa, J. G. (1980). Food intake and supplemental feeding programs. *In* "Perspectives in Cystic Fibrosis, Proceedings of the 8th International Cystic Fibrosis Congress" (J. M. Sturgess, Ed.), pp. 125–136. Canadian Cystic Fibrosis Foundation, Toronto.
68. Corey, M., McLaughlin F. J., Williams, M., *et al.* (1988). A comparison of survival, growth, and pulmonary function in patients with cystic fibrosis in Boston and Toronto. *J. Clin. Epidemiol.* **41**, 583–591.
69. Pencharz, P. B. (1983). Energy intakes and low-fat diets in children with cystic fibrosis. *J. Pediatr. Gastroenterol. Nutr.* **2**, 400–402.
70. Borowitz, D., Baker, R. D., and Stallings, V. (2002). Consensus report on nutrition for pediatric patients with cystic fibrosis. *J. Pediatr. Gastroenterol. Nutr.* **35**, 246–259.
71. Stallings, V. A., Stark L. S., Robinson, K. A., Feranchak, A. P., and Quinton, H. B. (2008). Evidence-based practice recommendations for nutrition-related management of children and adults with cystic fibrosis and pancreatic insufficiency: Results of a systematic review. *J. Am. Diet. Assoc.* **108**.
72. Bell, D., Durie, P., and Forstner, G. G. (1984). What do children with cystic fibrosis eat? *J. Paediatr. Gastroenterol. Nutr.* **3**(suppl 1), S137–S146.
73. Buchdahl, R. M., Fulleylov, C., Matchant, J. L., Warner, J. O., and Brueton, M. J. (1989). Energy and nutrient intakes in cystic fibrosis. *Arch. Dis. Child.* **64**, 373–378.
74. Hodges, P., Sauriol, D., Man S. F., Reichert, A., Grace, M., Talbot T. W., Brown, N., and Thomson, A. B. (1984). Nutrient intake of patients with cystic fibrosis. *J. Am. Diet. Assoc.* **84**, 664–669.
75. Lloyd-Still, J. D., Smith, A. E., and Wessel, H. U. (1989). Fat intake is low in cystic fibrosis despite unrestricted dietary practices. *JPEN J. Parenter. Enteral Nutr.* **13**, 296–298.
76. Tomezsko, J. L., Stallings, V. A., and Scanlin, T. F. (1992). Dietary intake of healthy children with cystic fibrosis compared with normal control children. *Pediatrics* **90**, 547–553.
77. Kawchak, D. A., Zhao, H., Scanlin, T. F., Tomezsko, J. L., Cnaan, A., and Stallings, V. A. (1996). Longitudinal, prospective analysis of dietary intake in children with cystic fibrosis. *J. Pediatr.* **129**, 119–129.
78. Lai, H. C., Kosorok, M. R., Laxova, A., Davis, L. A., FitzSimmons, S., and Farrell, P. M. (2000). Nutritional status of patients with cystic fibrosis with meconium ileus: A comparison with patients without meconium ileus and diagnosed early through neonatal screening. *Pediatrics* **105**, 53–61.
79. Sokol, R. J., Reardon, M. C., Accurso, F. J., Stall, C., Narkewicz, M., Abman, S. H., and Hammond, K. B. (1989). Fat-soluble vitamin status during the first year of life in infants with cystic fibrosis identified by screening of newborns. *Am. J. Clin. Nutr.* **50**, 1064–1071.
80. Benabdeslam, H., Garcia, I., Bellon, G., Gilly, R., and Revol, A. (1998). Biochemical assessment of the nutritional status of cystic fibrosis patients treated with pancreatic enzyme extracts. *Am. J. Clin. Nutr.* **67**, 912–918.
81. Marcus, M. S., Sondel, S. A., Farrell, P. M., Laxova, A., Carey, P. M., Langhough, R., and Mischler, E. H. (1991). Nutritional status of infants with cystic fibrosis associated with early diagnosis and intervention. *Am. J. Clin. Nutr.* **54**, 578–585.
82. Kraemer, R., Rudeberg, A., Hadorn, B., *et al.* (1978). Relative underweight in cystic fibrosis and its prognostic value. *Acta Paediatr. Scand.* **67**, 33–37.
83. Soutter, V. L., Kristidis, P., Gruca, M. A., *et al.* (1986). Chronic undernutrition/growth retardation in cystic fibrosis. *Clin. Gastroenterol.* **15**, 137–154.
84. Lai, H. C., Corey, M., FitzSimmons, S. C., Kosorok, M. R., and Farrell, P. M. (1999). Comparison of growth status in patients with cystic fibrosis in the United States and Canada. *Am. J. Clin. Nutr.* **69**, 531–538.
85. Farrell, P. M., Mischler, E. H., Engle, M. J., Brown, D. J., and Lau, S. (1985). Fatty acid abnormalities in cystic fibrosis. *Pediatr. Res.* **19**, 104–109.
86. Roulet, M., Frascarolo, P., Rappaz, I., and Pilet, M. (1997). Essential fatty acid deficiency in well nourished young cystic fibrosis patients. *Eur. J. Pediatr.* **156**, 952–6.
87. Strandvik, B. (2004). Fatty acid metabolism in cystic fibrosis. *N. Engl. J. Med.* **350**, 605–607.
88. Ozsoylu, S. (1998). Clinical importance of essential fatty acid deficiency. *Eur J. Pediatr.* **157**, 779.
89. Lloyd-Still, J. D., Bibus, D. M., Powers, C. A., Johnson, S. B., and Holman, R. T. (1996). Essential fatty acid deficiency and predisposition to lung disease in cystic fibrosis. *Acta Paediatr.* **85**, 1426–1432.
90. Hubbard, V. S., Dunn, D. G., and di Sant Agnese, P. A. (1977). Abnormal fatty acid composition of plasma lipids in cystic fibrosis: A primary or secondary effect? *Lancet* **2**, 1302–1304.
91. Bhura-Bandali, F. N., Suh, M., Man, S. F. P., and Clandinin, M. T. (2000). The F508del mutation in the CFTR alters control of essential fatty acid utilization in epithelial cells. *J. Nutr.* **130**, 2870–2875.
92. Strandvik, B., Gronowitz, E., Enlund, F., Martinsson, T., and Wahlstrom, J. (2001). Essential fatty acid deficiency in relation to genotype in patients with cystic fibrosis. *J. Pediatr.* **139**, 650–655.
93. van Egmond, A. W., Kosorok, M. R., Koscik, R., Laxova, R., and Farrell, P. M. (1996). Effect of linoleic acid intake on growth of infants with cystic fibrosis. *Am. J. Clin. Nutr.* **63**, 746–752.
94. Shoff, S. M., Ahn, H., Davis, L. A., and Lai, H. J., and the Wisconsin CF Neonatal Screening Group. (2006). Temporal associations among energy intake, plasma linoleic acid, and growth improvement in response to treatment initiation after diagnosis of cystic fibrosis. *Pediatrics* **117**, 391–400.
95. Freedman, S. D., Blanco, P. G., Zaman, M. M., *et al.* (2004). Association of cystic fibrosis with abnormalities in fatty acid metabolism. *N. Engl. J. Med.* **350**, 560–569.
96. Feranchak, A. P., Sontag, M. K., Wagener, J. S., Hammond, K. B., Accurso, F. J., and Sokol, R. J. (1999). Prospective, long-term study of fat-soluble vitamin status in children with cystic fibrosis identified by newborn screen. *J. Pediatr.* **1355**, 601–610.

97. Ahmed, F., Ellis, J., Murphy, J., Wootton, S., and Jackson, A. A. (1990). Excessive faecal losses of vitamin A (retinol) in cystic fibrosis. *Arch. Dis. Child.* **65**, 589–593.
98. Wilfond, B. S., Farrell, P. M., Laxova, A., and Mischler, E. (1994). Severe hemolytic anemia associated with vitamin E deficiency in infants with cystic fibrosis. Implications for neonatal screening. *Clin. Pediatr.* **33**, 2–7.
99. Lancellotti, L., D'Orazio, C., Mastella, G., Mazzi, G., and Lippi, U. (1996). Deficiency of vitamins E and A in cystic fibrosis is independent of pancreatic function and current enzyme and vitamin supplementation. *Eur. J. Pediatr.* **155**, 281–285.
100. Aris, R. M., Merkel, P. A., Bachrach, L. K., Borowitz, D. S., Boyle, M. P., Elkin, S. L., Guise, T. A., Hardin, D. S., Haworth, C. S., Holick, M. F., Joseph, P. M., O'Brien, K., Tullis, E., Watts, N. B., and White, T. B. (2005). Consensus statement: Guide to bone health and disease in cystic fibrosis. *J. Clin. Endocrinol. Metab.* **90**, 1888–1896.
101. Chavasse, R. J., Francis, J., Balfour-Lynn, I., Rosenthal, M., and Bush, A. (2004). Serum vitamin D levels in children with cystic fibrosis. *Pediatr. Pulmonol.* **38**, 119–122.
102. Stepheson, A., Brotherwood, M., Robert, R., Atenafu, E., Corey, M., and Tullis, E. (2007). Cholecalciferol significantly increases 25-hydroxyvitamin D concentrations in adults with cystic fibrosis. *Am. J. Clin. Nutr.* **85**, 1307–1311.
103. Durie, P. R. (1994). Vitamin K and the management of patients with cystic fibrosis. *CMAJ* **15**, 933–936.
104. van Hoorn, J. H., Hendriks, J. J., Vermeer, C., and Forget, P. P. (2003). Vitamin K supplementation in cystic fibrosis. *Arch. Dis. Child.* **88**, 974–975.
105. Conway, S. P., Wolfe, S. P., Brownlee, K. G., White, H., Oldroyd, B., Truscott, J. G., Harvey, J. M., and Shearer, M. J. (2005). Vitamin K status among children with cystic fibrosis and its relationship to bone mineral density and bone turnover. *Pediatrics* **115**, 1325–1331.
106. Sojo, A., Rodriguez-Soriano, J., Vitoria, J. C., Vazquez, C., Ariceta, G., and Villate, A. (1994). Chloride deficiency as a presentation or complication of cystic fibrosis. *Eur. J. Pediatr.* **153**, 825–828.
107. Corneli, H. M., Gormley, C. J., and Baker, R. C. (1985). Hyponatremia and seizures presenting in the first two years of life. *Pediatr. Emerg. Care* **1**, 190–193.
108. Ozcelik, U., Gocmen, A., Kiper, N., Coskun, T., Yilmaz, E., and Ozguc, M. (1994). Sodium chloride deficiency in cystic fibrosis patients. *Eur. J. Pediatr.* **153**, 829–831.
109. Vormann, J., Gunther, T., Magdorf, K., and Wahn, U. (1992). Mineral metabolism in erythrocytes from patients with cystic fibrosis. *Eur. J. Clin. Chem. Clin. Biochem.* **30**, 193–196.
110. Pond, M. N., Morton, A. M., and Conway, S. P. (1996). Functional iron deficiency in adults with cystic fibrosis. *Resp. Med.* **90**, 409–413.
111. Ater, J. L., Herbst, J. J., Landaw, S. A., and O'Brien, R. T. (1983). Relative anemia and iron deficiency in cystic fibrosis. *Pediatrics* **71**, 810–814.
112. Milla, C. E. (2004). Association of nutritional status and pulmonary function in children with cystic fibrosis. *Curr. Opin. Pulm. Med.* **10**, 505–509.
113. Kuczmarski, R. J., Ogden, C. L., Grummer-Strawn, L. M., *et al.* (2000). CDC growth charts: United States. Advanced Data from Vital and Health Statistics No. 314. National Center for Health Statistics, Hyattsville, MD.
114. Falkner, F., and Tanner, J. M., (Eds.). (1986). "Human Growth," 2nd ed., Vol. 3, pp. 104–107. Plenum Press, New York.
115. Ramsey, B. W., Farrell, P. M., Pencharz, P., The Consensus Committee. (1992). Nutritional assessment and management in cystic fibrosis: A consensus report. *Am. J. Clin. Nutr.* **55**, 108–116.
116. Moore, D. J., Durie, P. R., Forstner, G. G., and Pencharz, P. B. (1985). The assessment of nutritional status in children. *Nutr. Res.* **5**, 797–799.
117. Zhang, Z., and Lai, H. J. (2004). Comparison of the use of body mass index percentiles and percentage of ideal body weight to screen for malnutrition in children with cystic fibrosis. *Am. J. Clin. Nutr.* **80**, 982–991.
118. Lai, H. J. (2006). Classification of nutritional status in cystic fibrosis. *Curr. Opin. Pulm. Med.* **12**, 422–427.
119. Lai, H. J., and Shoff, S. M. (2008). Classification of malnutrition in cystic fibrosis: Implications for evaluating and benchmarking clinical practice performance. *AJCN* (in press).
120. Poustie, V. J., Watling, R. M., Ashby, D., and Smyth, R. L. (2000). Reliability of percentage ideal weight for height. *Arch. Dis. Child.* **83**, 183–184.
121. Gaskin, K., Gurwitz, D., Durie, P., *et al.* (1982). Improved respiratory prognosis in patients with cystic fibrosis with normal fat absorption. *J. Pediatr.* **100**, 857–862.
122. Kalnins, D., Durie, P. R., and Pencharz, P. (2007). Nutritional management of cystic fibrosis. *Curr. Opin. Clin. Nutr. Metab. Care* **10**, 348–354.
123. Mischler, E. J., Parrell, S., Farrell, P. M., and Odell, G. B. (1982). Effectiveness of enteric coated pancreatic enzymes compared to a conventional enzymes in males with cystic fibrosis. *Am. J. Dis. Child.* **136**, 1060–1063.
124. National Research Council. (1989). "Recommended Dietary Allowances" 10th ed. National Academies Press, Washington, DC.
125. National Academy of Sciences. (2002). "Dietary Reference Intakes for Energy, Carbohydrate, Fiber, Fat, Fatty Acids, Cholesterol, Protein, and Amino Acids (Macronutrients)." National Academies Press, Washington, DC.
126. Yankaskas, J. R., Marshall, B. C., Sufian, B., Simon, R. H., and Rodman, D. (2004). Cystic fibrosis adult care: Consensus conference report. *Chest* **125**, 1S–39S.
127. Sinaasappel, M., Stern, M., Littlewood, J., Wolfe, S., Steinkamp, G., Heijerman, H. G., Robberecht, E., and Doring, G. (2002). Nutrition in patients with cystic fibrosis: A European consensus. *J. Cystic Fibrosis* **1**, 51–75.
128. Li, Z., Lai, H. J., Kosorok, M. R., Laxova, A., Rock, M. J., Splaingard, M. L., and Farrell, P. M. (2004). Longitudinal pulmonary status in cystic fibrosis children presenting with melonium ileus. *Pediatr. Pulmonol.* **38**, 277–284.
129. Borowitz, D. S., Grand, R. J., Durie, P. R., Becker, L. T., *et al.* (1995). Use of pancreatic enzyme supplements for

patients with cystic fibrosis in the context of fibrosing colonopathy. *J. Pediatr.* **127**, 681–684.
130. Lloyd-Still, J. (2006). Gastrointestinal outcomes and confounders in cystic fibrosis. *J. Pediatr. Gastroenterol. Nutr.* **42**, 243–244.
131. Tanner, J. M., and Whitehouse, R. H. (1976). Clinical longitudinal standards for height, weight, height velocity, weight velocity, and stages of puberty. *Arch. Dis. Child.* **51**, 170–179.
132. Morris, N. M., and Udry, J. R. (1980). Validation of a self-administered instrument to assess stage of adolescent development. *J. Youth Adol.* **9**, 271–280.
133. World Health Organization. (1985). Energy and protein requirements. WHO Tech Rep Ser, No. 724, 924, 000.

G. Bone Health and Disease

CHAPTER 43

Current Understanding of Vitamin D Metabolism, Nutritional Status, and Role in Disease Prevention

SUSAN J. WHITING,[1] MONA S. CALVO,[2] AND CHARLES B. STEPHENSEN[3]

[1]*College of Pharmacy and Nutrition, University of Saskatchewan, Saskatoon, Saskatchewan, Canada*

[2]*Center for Food Safety and Applied Nutrition, U.S. Food and Drug Administration, College Park, Maryland*

[3]*USDA Western Human Nutrition Research Center, University of California, Davis, Davis, California*

Contents

I. INTRODUCTION

Vitamin D is a nutrient that, until recently, was neglected by the nutrition community. Although it was recognized in the early twentieth century as an essential nutrient, recommendations for intake were often qualified as being needed only in absence of sunlight. In theory (and in ancient times when early humans all lived closer to the equator), all vitamin D needs could be met by exposure to sunlight that provided ultraviolet (UV) B radiation, but only recently have we come to understand how UVB acts and what other factors—particularly environmental—mitigate cutaneous vitamin D synthesis. Studying vitamin D requirements is difficult. Previous dietary recommendations for vitamin D, such as the 1989 Recommended Dietary Allowance [1], indicated a "relative paucity of recent controlled studies [and] . . . lack of data" on which to base requirements. It further stated that "[c]linical osteomalacia appears to be rare in the United States." What is known today, however, is that vitamin D deficiency and insufficiency are more widespread [2] than imagined even a decade ago when the Dietary Reference Intakes (DRIs) were first published for vitamin D [3]. What has occurred in the past 10 years is a new understanding of vitamin D's roles in the body, especially for functions unrelated to calcium absorption, which has been the long-standing primary recognized function of vitamin D [4–7].

The role of vitamin D in preventing rickets was discovered early in the nineteenth century, but it was not until the 1970s that the sequence of steps from skin precursors to active metabolite was understood. Despite the interest generated in solving the puzzle of how vitamin D increased intestinal calcium absorption, there were several reasons why progress toward a better understanding of vitamin D requirements was not made. There were technical challenges in analyzing vitamin D and its metabolites. There was, beginning in the 1980s, a greater focus on dietary calcium as the major "bone" nutrient, leaving vitamin D with only a minor role in osteoporosis research. And finally, the important contribution of sun exposure to vitamin D status was not fully realized until recently. Indeed, estimations for dietary recommendations in the complete year-round absence of sun exposure give values that are 5 to 8 times higher than what is needed to maintain vitamin D status through the winter. It has been shown that globally there is greater prevalence of chronic diseases such as cancer and immune disorders at extremes of latitudes where sun exposure for skin synthesis of vitamin D is limited [4].

Vitamin D affects people starting with fetal development and continuing to old age, functioning at both the genomic and nongenomic level in the regulation of key protein synthesis or in the intracellular metabolic pathways in virtually all tissues [5–7]. Growth, development, and

The opinions expressed in this chapter are those of the authors and do not reflect those of the U.S. Food and Drug Administration.

maintenance of health are all affected, and in many regards, quality of life is as well. This chapter, although acknowledging vitamin D's contribution through the lifespan, focuses on vitamin D needs for maintenance of health, and on vitamin D's specific actions in selected clinical conditions. Because research is ongoing, the reader can expect to learn enough about vitamin D's roles to be able to understand and apply the research as it unfolds.

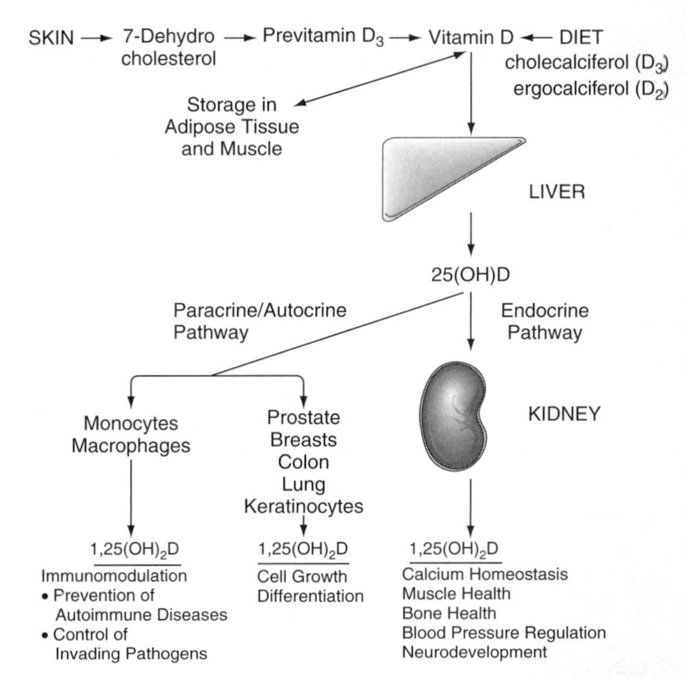

FIGURE 1 Overview of vitamin D metabolism, from synthesis (or provision in the diet) through to synthesis of active form. Once vitamin D_3 is made in skin, or provided from the diet (and some of this could be as vitamin D_2), it is converted in the liver to 25-hydroxyvitamin D [25(OH)D], which is the major circulating form of vitamin D. This 25(OH)D is now the substrate for production of the active form, 1,25-dihydroxyvitamin D, via two pathways. In the **endocrine pathway**, 1,25-dihydroxyvitamin D is made in the kidney under tight regulatory control; this 1,25-dihydroxyvitamin D circulates in the blood and acts to promote active calcium and phosphate absorption; working with parathyroid hormone it affects bone metabolism and kidney reabsorption of calcium. In the **paracrine/autocrine pathway**, 1,25-dihydroxyvitamin D is made and used locally by a variety of cells, including those of the immune system. Modified from Hollis, B. W., and Wagner, C. L. (2006). Nutritional vitamin D status during pregnancy: Reasons for concern. *CMAJ* **174**, 1287–1290.

II. METABOLISM OF VITAMIN D

A. Overview of Vitamin D Synthesis and Conversion to Its Active Metabolite

1. Vitamin D Is a Family of Compounds

To understand vitamin D is to appreciate the functions of the numerous metabolites that arise during its metabolism. Figure 1 shows how vitamin D is provided, either through skin synthesis or from diet, and undergoes successive hydroxylations to form the active metabolite, 1,25-dihydroxyvitamin D. There are several naturally occurring forms of vitamin D and many metabolites, and these are outlined in Table 1. The term *vitamin D* really represents all compounds having or potentially having the activity that we associate with the active metabolite of vitamin D; however, in nutrition we also refer to the precursor molecules provided in the diet or in supplements, which are cholecalciferol (for vitamin D_3) and ergocalciferol (for vitamin D_2), as "vitamin D." Generic use of the term *vitamin D* is a source of great confusion. Through this chapter, an attempt has been made to use the exact term for each metabolite in order to prevent confusion. Table 1 can be used as a guide for this purpose. When vitamin D_2 and D_3 can contribute to the same function, then "vitamin D" with no subscript is used.

2. Vitamin D_3 Synthesis in Skin

In the skin, there is 7-dehydrocholesterol (also called "provitamin D_3") in the epidermis and the dermis, which reacts when UVB radiation in the wavelength range of 280–315 nm passes through these skin layers to form previtamin D_3. Previtamin D_3 forms rapidly; however, skin pigmentation (melanin) competes with 7-dehydrocholesterol for the UVB photons, and therefore reduces the amount of UVB that can act on 7-dehydrocholesterol to form previtamin D_3. With prolonged exposure to UVB, inactive compounds are formed instead of previtamin D_3. Over a prolonged period of time, the previtamin D_3 that is formed is changed due to thermal isomerization to vitamin D_3 (more appropriately called "cholecalciferol" or, less commonly, "calciol"). The reaction to form provitamin D_3 takes little time, but the reaction converting previtamin D_3 to cholecalciferol takes hours to occur [7] and is a rate-limiting step. Should more UVB photons reach the epidermis and dermis, previtamin D_3 is converted to inactive compounds with no vitamin D activity (tachysterol and lumisterol). Thus excess exposure to UVB does not result in excess vitamin D production [7].

Besides the skin pigment melanin, other factors reduce skin synthesis of cholecalciferol. These include clothing (although some loosely woven clothing does permit UVB to pass through); window glass; sunscreens formulated to block UVB, particularly when SPF is over 8 [7]; being indoors; cloudy days; smog and light-blocking air pollution; and winter, when the sun does not rise far enough above the horizon to allow sufficient UVB irradiation to stimulate dermal vitamin D_3 synthesis. Thus the term *vitamin D winter* refers to the time of year when UVB radiation is not sufficient for cholecalciferol synthesis in the skin.

TABLE 1 A Glossary of Vitamin D Compounds and Metabolites

Vitamin D Metabolite	Alternate Name	Function	Clinical Utility of Measurement
"Vitamin D" (often used as synonym for vitamin D_3 or all dietary sources)	N/A	Term used to describe actions of the active metabolites of vitamins D_2 and D_3	N/A
7-Dehydrocholesterol	Provitamin D_3	Precursor to cholecalciferol found in skin; is acted on by UVB	Not measured
Previtamin D_3	NA	Intermediate in synthesis of cholecalciferol from 7-dehydrocholesterol	Not measured
Vitamin D_3	Cholecalciferol; calciol	Form of vitamin synthesized by animals in presence of UVB light	Provides information of recent sun exposure or ingestion of vitamin D_3
Vitamin D_2	Ergocalciferol	Form of vitamin synthesized by some plants in presence of UVB light	Provides information of ingestion of vitamin D_2
25-Hydroxyvitamin D_3 [25(OH)D_3]	25-Hydroxycholecalciferol; calcidiol	Circulating form of vitamin D_3, made from cholecalciferol	Measure of vitamin D_3 status
25-Hydroxyvitamin D_2 [25(OH)D_3]	25-Hydroxyergocalciferol	Circulating form of vitamin D_2, made from ergocalciferol	Measure of vitamin D_2 status
1,25-Dihydroxyvitamin D_3 [1,25-DHD]	1,25-Dihydroxycholecalciferol; calcitriol	Active metabolite of vitamin D_3	Measured to determine if active form can be made or to monitor treatment; not useful for vitamin D status
1,25-Dihydroxyvitamin D_2	1,25-Dihydroxyergocalciferol	Active metabolite of vitamin D_2	Measured to determine if active form can be made or to monitor treatment; not useful for vitamin D status
24,25-Dihydroxyvitamin D	24,25-Dihydroxy-cholecalciferol [or -ergocalciferol]	Inactive. Made instead of 1,25 metabolite if 1-hydroxylase not stimulated	Not measured
1,24,25-Trihydroxyvitamin D	Calcitroic acid	Inactive. Made from 1,25 metabolite as mechanism for inactivation	Not measured

Latitude is the major determining factor for intensity of UVB irradiation—whether in the southern or northern hemisphere. At the equator, vitamin D can be made year-round even in darker pigmented skin. Above latitude 37° there are 4 months of vitamin D winter; at latitude 42° there are 5 months, and close to the poles there would be no time during the year when vitamin D synthesis in the skin could occur [7, 8]. Yet not all analyses of vitamin D status and latitude show the expected relationship [9, 10]. For example, there is now considerable vitamin D deficiency and insufficiency in people living near the equator, in countries of the Middle East, southern Europe, and in India [10]. This suggests that many factors are determining skin synthesis, so lack of UVB (e.g., "vitamin D winter") affects some countries, whereas clothing and customs such as time spent outdoors determine vitamin D in other countries. Age is another factor decreasing skin synthesis of cholecalciferol, as described later.

3. Conversion of Cholecalciferol and Ergocalciferol to 25-Hydroxyvitamin D

The cholecalciferol made from skin synthesis is released from the epidermis into the blood, where it is bound to vitamin D binding protein (DBP). Cholecalciferol and ergocalciferol (vitamin D_2) in the diet are absorbed and carried to the liver in chylomicrons. Intestinal absorption is not a limiting factor except when there is fat malabsorption (e.g., cystic fibrosis, Crohn's disease). Generally fat-soluble vitamins are better absorbed with dietary fat, but vitamin D_3 added to orange juice is well absorbed [11]. Cholecalciferol and ergocalciferol circulate for only 1 to 2 days. This quick turnover is due to hepatic conversion and uptake by fat and muscle cells [7, 12].

There are two steps leading to the active form of vitamin D, which is 1,25-dihydroxyvitamin D (see Table 1). The first step is converting cholecalciferol or ergocalciferol to the major circulating form. This pathway involves four hepatic cytochrome P-450 enzymes (CYP2R1, CYP27A1,

CYP3A4, CYP2J3) [4] that hydroxylate cholecalciferol or ergocalciferol at carbon 25. The resulting metabolites, 25-hydroxyvitamin D_3 and 25-hydroxyvitamin D_2 (together denoted as 25(OH)D), are released into circulation. The amount of 25(OH)D that circulates is determined by the availability of its substrate, cholecalciferol or ergocalciferol [12, 13]. With increasing levels of cholecalciferol made in skin or of dietary cholecalciferol and ergocalciferol, levels of 25(OH)D increase [13]. It is important to appreciate that 25(OH)D is the key metabolite indicating vitamin D status. It is not the metabolically active form but it is the form that most accurately reflects deficiency or excess and is therefore used as a measure of vitamin D nutritional status.

Availability of serum 25(OH)D is required for vitamin D activity. Figure 2 illustrates the changes in 25(OH)D over a year, demonstrating how sun exposure affects vitamin D status in light-skinned individuals living at about 50°N latitude [14]. One can see the impact of "vitamin D winter" in these subjects. This figure also demonstrates how an oral supplement of cholecalciferol can maintain the summer level

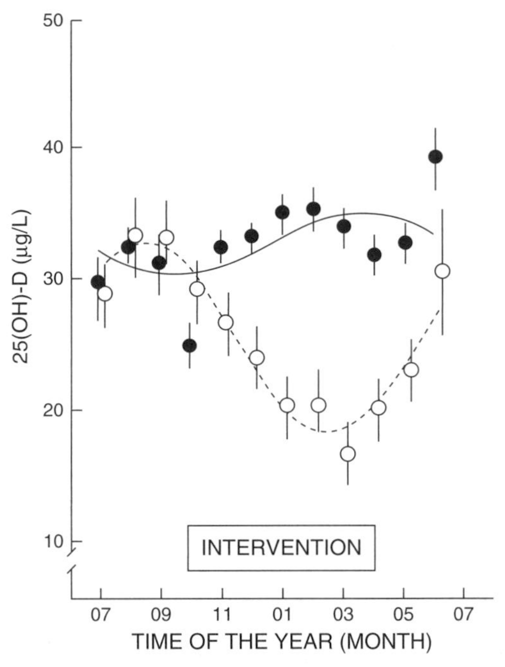

FIGURE 2 The data for this graph were derived from a randomized control trial conducted in Germany (latitude 50°N). Open circles (○) represent mean levels of serum 25(OH)D of subjects throughout the year. Closed circles (●) are mean levels of serum 25(OH) in subjects who received the treatment regimen of 12.5 µg (500 IU) cholecalciferol and 500 mg calcium. Note that the units for 25(OH)D are expressed as µg/liter (to convert to nmol/liter, multiply by 2.5). Reproduced with permission from Meier, C., Woitge, H. W., Witte, K., Lemmer, B., and Seibel, M. J. (2004). Supplementation with oral vitamin D_3 and calcium during winter prevents seasonal bone loss: A randomized controlled open-label prospective trial. *J. Bone Miner. Res.* **19**, 1221–1230.

of 25(OH)D through the winter months [14]. Studies indicate that whereas the molecule 25(OH)D has a half-life of 2 weeks [4], the amount in blood has an effective half-life of 2 months due to contribution from stores [15]. This means that in the absence of both sun and adequate dietary source of cholecalciferol, serum levels of 25(OH)D will decline throughout the winter months. Controversy surrounds the equivalency of efficacy of the two precursors to 25(OH)D, and this has been interpreted by some to reflect the inadequacy of vitamin D_2 as a dietary supplement or food fortificant [16]. When a single dose of either vitamin D_2 or vitamin D_3 was administered, $25(OH)D_2$ was shown to remain in blood for a much shorter time than the $25(OH)D_3$, declining after 1 week [17]. Further research is needed to determine the relative contribution of ergocalciferol to vitamin D status.

A recent study [12] sought to determine whether provision of the substrate, cholecalciferol, for synthesis of 25(OH)D was rate-limiting. They were able to obtain a wide range of substrate levels by giving subjects high doses of cholecalciferol or by investigating subjects living in a sunny near-equatorial region. As shown in Figure 3, high doses of cholecalciferol from supplements [as much as 160 µg (6400 IU)] in panel A or sun exposure in Hawaii in panel B affected serum 25(OH)D similarly. The relationship is not linear, indicating a controlled, saturable reaction. Unless cholecalciferol is provided in sufficient amounts, the production of 25(OH)D is limited by its substrate.

4. Renal Conversion to the Active Metabolite: Endocrine Pathway of 1,25-Dihydroxyvitamin D Synthesis and Use

There are two pathways for conversion of 25-hydroxyvitamin D to the active form of vitamin D, which is 1,25-dihydroxyvitamin D (Figure 1). The first to be described is the better known and understood pathway, now referred to as the endocrine pathway. In the endocrine pathway 1,25-dihydroxyvitamin D [$1,25(OH)_2D$] is synthesized in one tissue but acts elsewhere, thus fulfilling the definition of a "hormone."

In proximal renal epithelial cells, 1,25-dihydroxyvitamin D is made when the enzyme 1-α hydroxylase (also called CYP27B1) is stimulated by parathyroid hormone (which had previously been stimulated by a low circulating plasma calcium level) or by a fall in intracellular phosphate levels [7, 18]. This is a tightly controlled endocrine system and is considered to be the major contributor to the circulating levels of the active metabolite of vitamin D. Plasma levels of 1,25-dihydroxyvitamin D rise only when needed, as a result of synthesis in the kidney. The "need" for vitamin D action relates to its primary role (endocrine function) of providing the building blocks of bone—calcium and phosphate. When there is a need for calcium, this is expressed as hypocalcemia, which acts as the trigger for synthesis

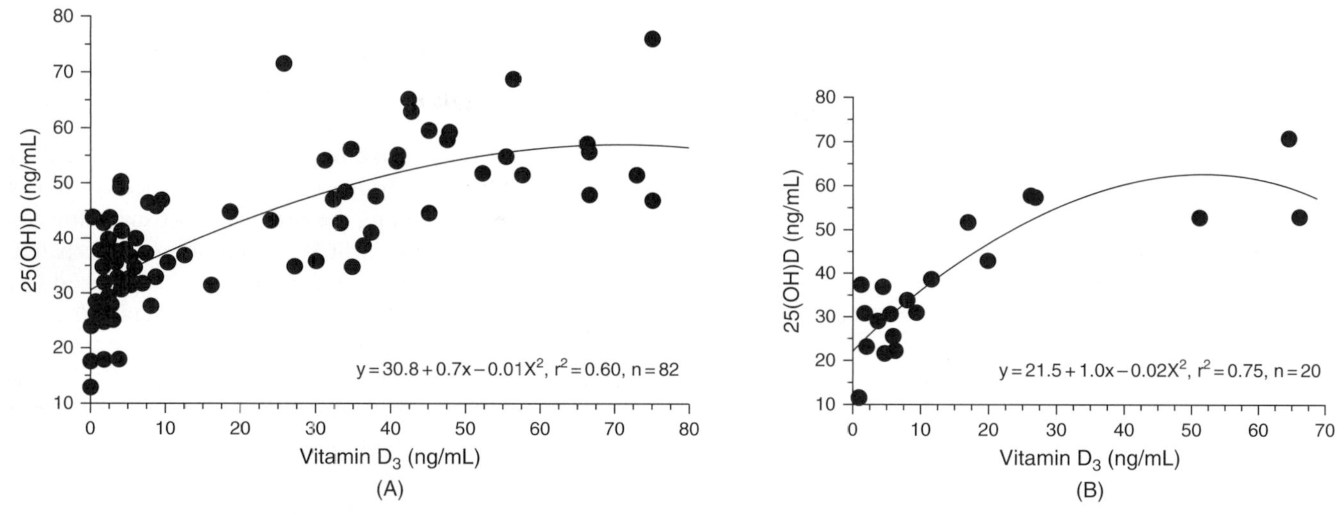

FIGURE 3 These graphs illustrate how either a high dose of cholecalciferol from a supplement of 160 μg (6400 IU) in panel A or sun exposure in Hawaii in panel B affect serum 25(OH)D. The authors of the study concluded that vitamin D status, as measured by serum 25(OH)D, is affected equally by diet or sun exposure. Further, unless cholecalciferol is provided in sufficient amounts, the production of 25(OH)D is limited by its substrate. Reproduced with permission from Hollis, B. W. (2007). Circulating vitamin D_3 and 25-hydroxyvitamin D in humans: An important tool to define adequate nutritional vitamin D status. *J. Steroid Biochem. Mol. Biol.* **103**, 631–634.

and release of parathyroid hormone (PTH). An increase in PTH has three actions relating to calcium metabolism: (1) increased resorption of bone in order to provide immediate calcium to the blood; (2) more efficient reabsorption of calcium in the renal tubule, in order to conserve blood calcium; and (3) increased activity of the 1-hydroxylase in the proximal renal tubule to increase levels of 1,25-dihydroxyvitamin D [19]. This third action, therefore, increases the concentration of 1,25-dihydroxyvitamin D in the blood that can travel to the small intestine and promote active absorption of calcium. The increase in plasma 1,25-dihydroxyvitamin D is also a way to provide this metabolite to other organs such as bone, parathyroid gland, and kidney, where it acts in concert with PTH to increase blood calcium levels. To complete the cycle, a rise in serum 1,25-dihydroxycholecalciferol acts to suppress parathyroid hormone secretion.

The main action of circulating 1,25-dihydroxyvitamin D is to increase calcium (and phosphate) absorption [20]. This occurs in the duodenum, in enterocytes, where active calcium (and phosphate) absorption is promoted through genomic and nongenomic actions. The genomic mechanism of action of 1,25-dihydroxyvitamin D operates through nuclear receptors in the cell, similar to that of other steroid hormones, while the nongenomic activity is thought to operate through plasma membrane receptors. In either case, there is binding to a specific receptor called the vitamin D receptor (VDR) at the cell membrane of the enterocyte. In the nucleus, this complex, along with other co-activators, binds to a vitamin D response element (VDRE) to promote transcription of specific gene products. It has been long assumed that one protein product, a calcium binding protein called calbindin 9K, was required for translocation of calcium through the enterocyte during active transport of calcium (otherwise calcium absorption is passive, via paracellular channels) [4]. However, recent studies with a knockout mouse having no gene for calbindin 9K showed that 1,25-dihydroxyvitamin D-mediated calcium absorption was normal in this mouse, thus leaving the exact gene products for active calcium translocation in doubt [21].

Without active absorption, the amount of calcium that can be absorbed is limited to about 15% of calcium intake [22]. The dependency of calcium absorption on vitamin D is illustrated in Figure 4. When levels of 25(OH)D are low, calcium absorption proceeds by passive absorption only, i.e., only 15% is absorbed. This occurs as paracellular absorption, dependent on the lumen concentration of calcium. As vitamin D status improves, as indicated by increasing 25(OH)D, calcium absorption rises and then reaches a threshold level. The data in the study described in Figure 4 were collected from healthy adults and 30% absorption of calcium is all one would expect in an adult who has no urgent need for calcium. In growth and pregnancy, calcium absorption can rise to levels of 80%, but only if vitamin D status is adequate; hormones, such as growth hormone and prolactin (respectively) are also important for this high absorption value [3, 4].

A second important action of circulating 1,25-dihydroxyvitamin D is to promote synthesis of mature osteoclasts [7]. Here, 1,25-dihydroxyvitamin D stimulates the osteoblasts to synthesize a specific receptor ligand, known as RANKL (receptor activator of the nuclear factor κ-B ligand, abbreviated as NF-κB). These cells have the

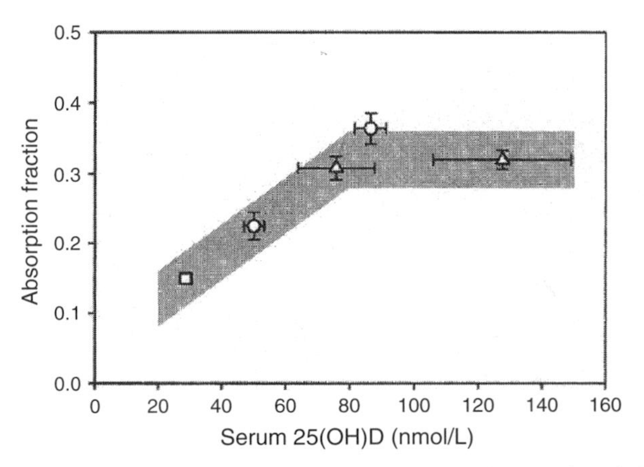

FIGURE 4 The dependency of calcium absorption on vitamin D status, as measured by serum 25(OH)D. Reproduced with permission from Heaney, R. P. (2007). The case for improving vitamin D status. *J. Steroid Biochem. Mol. Biol.* **103**, 635–641.

vitamin D receptors (VDR), and the action of 1,25-dihydroxyvitamin D is genomic. Once the RANKL is made, it exits the osteoblast and binds to RANK on preosteoclasts; this binding induces maturation of the cells into osteoclasts, which are responsible for bone resorption. The net result is release of calcium and phosphate into circulation. Thus, having been made in the kidney in response to a need for calcium or phosphate, 1,25-dihydroxyvitamin D's actions in the intestine and bone have resulted in an increase into the blood of both of these compounds.

5. Extrarenal Conversion to the Active Metabolite: Paracrine/Autocrine Pathway of 1,25-Dihydroxyvitamin D Synthesis and Use

The other pathway for conversion of 25(OH)D to $1,25(OH)_2D$ is called the paracrine/autocrine pathway because these terms denote that the molecule is used locally in adjacent cells or used in the same cell in which it is made, respectively (Figure 1). Less is known about this, but researchers found the 1-hydroxylase enzyme in many tissues other than the renal proximal tubule, and those working in cell culture systems were able to measure $1,25(OH)_2D$ production [23, 24]. Further, the vitamin D receptor has been identified in brain, prostate, breast, gonads, colon, pancreas, heart, monocytes, and lymphocytes. Extrarenal $1,25(OH)_2D$ production is not regulated by serum calcium, phosphate, or PTH. Many actions have been proposed, but the modulation of immune function through actions on lymphocytes and macrophages is the most developed.

Recent studies also demonstrate that cells of the immune system can produce $1,25(OH)_2D$ that acts in a paracrine manner to regulate the immune response. For example, macrophages are an important part of the host defense against a variety of pathogenic organisms, including *Mycobacterium tuberculosis*, the etiologic agent of tuberculosis. Macrophages are activated to kill such pathogens because pattern-recognition receptors, such as toll-like receptor (TLR) 1 and 2, recognize molecular patterns associated with microorganisms that are not found in mammalian cells. In the case of tuberculosis, macrophages are activated via recognition of *M. tuberculosis* lipopeptides by a TLR2/1 dimer. This interaction triggers antibacterial mechanisms of the macrophage, including the production of the antibacterial peptide cathelicidin, which can kill susceptible pathogens [25]. Cathelicidin transcription is enhanced by $1,25(OH)_2D$ acting via the VDR to increase gene transcription. This TLR2/1 interaction also stimulates the expression of the 1-hydroxylase enzyme that catalyzes $1,25(OH)_2D$ synthesis, suggesting that vitamin D deficiency may impair protection against tuberculosis by limiting $1,25(OH)_2D$ production by macrophages, although human intervention trials are needed to test this hypothesis [25].

In addition to vitamin D regulating aspects of innate immunity, adaptive immunity may also be compromised. Adaptive immunity refers to the acquired immune response that develops to infections (and vaccinations), preventing us from getting most infections a second time. T lymphocytes are an important component of adaptive immunity. Some T lymphocytes function as helper cells to promote antibody production by B cells. T lymphocytes also act as effector cells to kill virus-infected host cells or to promote macrophage-mediated responses, such as the one described in the previous paragraph against *M. tuberculosis*. In order to participate in such protective responses at specific sites in the body, such as the lung or intestine or skin, memory T lymphocytes must express homing receptors that allow recognition of specific factors expressed at these tissue sites. One such receptor is the chemokine receptor 10 (CCR10), which allows T lymphocytes to home to the skin by recognizing the chemokine CCL27 produced by the epidermis. T lymphocytes are activated by their initial contact with a foreign antigen (from a pathogen or a vaccine). The antigen is "presented" by an antigen-presenting cell (APC), which stimulates the naïve T lymphocyte to divide, expand, and differentiate into an effector T lymphocyte that will participate in the immune response to that antigen. Interaction with the APC-carrying antigen also can stimulate the expression of a homing receptor such as CCR10 that will bring the mature T lymphocyte back to the original site of infection (or immunization). Skin-derived APCs appear to be specifically activated to produce $1,25(OH)_2D$, which can act on a naïve T lymphocyte (through regulating gene expression via the VDR) to induce later expression of CCR10 [26]. Vitamin D deficiency may impair skin-specific homing of T lymphocytes and impair defense against skin infections, perhaps including cutaneous manifestations of tuberculosis.

6. Mechanism of Action of the Active Metabolite

The active form of vitamin D, 1,25-dihydroxyvitamin D ($1,25(OH)_2$ D), whether synthesized in the kidneys and released to the circulation or originating from extrarenal 1-alpha-hydroxylase activity, is thought to operate through two distinctly different mechanisms: (1) the classical genomic action, and (2) the rapid membrane-initiated action or nongenomic action [27, 28].

a. Genomic Action

The classical genomic action of $1,25(OH)_2D$ involves the binding of this steroid hormone to its nuclear receptor, which is a stereospecific interaction. The vitamin D nuclear receptor is a member of the superfamily of steroid hormone nuclear receptors. Ligand–nuclear receptor binding initiates the cell's transcriptional machinery to regulate gene transcription. In the currently accepted model of $1,25(OH)_2D$ and vitamin D receptor (VDR_{nuc}) activation of gene transcription, when the ligand or $1,25(OH)_2D$ binds the nuclear receptor, it forms a heterodimer with the nuclear retinoid-X receptor (RXR). This heterodimer–DNA complex then interacts with the appropriate vitamin D response element (VDRE) on the promoter genes of specific target cells that are up- or downregulated. The heterodimer–DNA complex then recruits necessary coactivator proteins to form a competent transcriptional complex capable of modulating mRNA production [28, 29]. The discovery of the presence of the VDR_{nuc} in more than 30 human tissues led to our new understanding of vitamin D's role in the regulation of B and T lymphocytes of the immune system, hair follicles, muscle, adipose tissue, bone marrow, and cancer cells through the mechanism of nuclear VDR regulation of gene transcription [29].

b. Nongenomic Action

More recently, a $1,25(OH)_2D$ mediated response that was observed to occur within minutes to an hour was discovered to be too rapid to be explained by nuclear VDR regulating gene transcription. Such rapid responses included secretion of insulin by pancreatic beta-cells, rapid migration of endothelial cells, Ca^{2+} influx in skeletal muscle cells as modulated by phospholipase C, protein kinase C, and tyrosine kinase, and activation of mitogen-activated kinase, to identify only a few [28, 29]. $1,25(OH)_2D$ can rapidly activate signal transduction pathways in addition to activation of the slower classical genomic mechanism of gene transcriptional regulation [30]. Studies show that $1,25(OH)_2D$ can rapidly stimulate ion fluxes and activate protein kinases after binding to a unique receptor in the caveolae of the plasma membrane [31]. The plasma membrane receptor, termed membrane-associated rapid response steroid binding protein (MARRS), does not behave like traditional membrane-spanning receptors and can be found in the endoplasmic reticulum or relocated in the nucleus [31]. Specific VDRs have been identified within the plasma membrane caveolae in a variety of different cell types, including intestine, kidney, lungs, leukemia, and osteoblast-like cells [29, 31, 32]. As emphasized by Fleet [31], many important questions remain to be resolved concerning membrane-initiated vitamin D action. Foremost among these unknowns is the need to better understand whether rapid vitamin D actions significantly influence physiologic processes unique to certain cell types.

B. Inactivation and Excretion of Vitamin D

There are turnover and excretion of vitamin D metabolites. When 25(OH)D is extracted by renal cells and there is no need for increasing calcium or phosphate levels in the body, 25(OH)D is hydroxylated at carbon 24 to form 24,25-dihydroxyvitamin D. This is catalyzed by 24-hydroxylase (also called CYP24) [18, 33]. This enzyme, located primarily in the kidney, plays an important role in preventing unwanted buildup of 1,25-dihydroxyvitamin D. The resulting metabolite, 24,25-dihydroxyvitamin D, is inactive. Additionally, 1,25-dihydroxyvitamin D can also be hydroxylated by the 24-hydroxylase enzyme to form 1,24,25-trihydroxyvitamin D (also called calcitroic acid), This compound is the first step in the inactivation of 1,25-dihydroxyvitamin D (Table 1). Many other hydroxylations of vitamin D metabolites have been observed [4], but the exact roles of these compounds are not known. Most vitamin D metabolites are excreted into the bile.

III. SOURCES OF VITAMIN D

A. Food Sources of Cholecalciferol and Ergocalciferol

There are only a few foods that naturally contain vitamin D as ergocalciferol and cholecalciferol [34] (Table 2). In the wild, fish are part of a food chain that allows for concentration of vitamin D in the flesh of fatty fish (e.g., salmon, sardines, mackerel), whereas in lean fish, vitamin D is concentrated in liver (e.g., cod liver oil). With fish farming, levels of cholecalciferol in fish raised in aquaculture cannot be assumed to be equivalent to those in wild species. Further, it is now recognized that levels in fish are much more variable than previously recognized [35]; therefore, caution must be taken in using data from Table 2. Other concerns about consumption of fish or fish oils include consumption of too much mercury and vitamin A. Land animals that are exposed to sunlight or have vitamin D in their feed may be a source of vitamin D, but the amount of vitamin D provided as meat is not well documented except for liver. Eggs are a natural source of vitamin D that can be increased when vitamin D is added to chicken feed, but the level is not usually significant. However, eggs processed to remove cholesterol and saturated fat in the United States have a restored vitamin D content of approximately 6% of the Daily Value. Presently, in the United States, there is no

TABLE 2 Predominant Food Sources of Vitamin D in the United States

Food	Source of Vitamin D	Vitamin D_3 μg (IU) per Serving
Fluid cow's milk, 250 ml (1 cup)	Fortification	2.5 (100)
Orange juice with added calcium and vitamin D, 125 ml (1/2 cup)	Fortification of selected brands	1.25 (50)
Yogurt 100 g	Fortification of selected brands	2.5 (100)
Margarine, 10 g (2 tsp)	Fortification of selected brands	0.84 (34)
Cereals, ready-to-eat, 1 serving (1/2 to 3/4 cup)	Fortification of selected brands	1.0–2.5 (40–100)
Salmon, canned, 85 g (3 ounces)	Naturally occurring but variable	9.9–16 (396–649)
Sun-dried shiitake mushrooms, 36 g (1/4 cup cooked)	Naturally occurring; exposed to sun (UVB)	2.8 (110) as vitamin D_2
Cod liver oil, 5 ml (1 tsp)	Naturally occurring	10 (400)

Sources: USDA, Agricultural Research Service. National Nutrient Database. www.nal.usda.gov/fnic/foodcomp/search/ and Johnson, M. A., and Kimlin, M. G. (2006). Vitamin D, aging, and the 2005 Dietary Guidelines for Americans. *Nutr. Rev.* **64**, 410–421.

mandatory labeling of the vitamin D content of foods on food labels [36].

Fortification of milk and other foods with vitamin D, such as selected cereals, margarines, juices, and a few selected brands of cheese, provide the majority (66–84% of the food sources) of the vitamin D intake of Americans [37]. Plant foods such as mushrooms that when briefly exposed to UVB produce significant amounts of vitamin D_2 [38] and some fortified foods may contain vitamin D_2. Such foods can appeal to vegetarians who may prefer to consume a plant-based form of vitamin D [39].

B. Supplement Sources of Vitamin D

Supplements provide another source of intake. Vitamin D (i.e., cholecalciferol or ergocalciferol) is usually (but not always) found in multivitamin preparations at 10 μg (400 IU) per tablet, some with only 5 μg (200 IU) but new ones being introduced with 25 μg (1000 IU). In addition to multivitamins, there are single vitamin D supplements largely available as cholecalciferol in 10 μg (400 IU), 25 μg (1000 IU), and 50 μg (2000 IU) dosages. Some calcium supplements contain various amounts of cholecalciferol or ergocalciferol, in the range 10 to 25 μg (400–1000 IU) per tablet. These supplements are intended for maintenance of status or for persons who have less-than-adequate vitamin D intakes. They are not intended for repletion of vitamin D deficiency. For that purpose, higher dosage forms are available through prescription, denoted in Table 3 as therapeutic preparations. In addition, vitamin D metabolites (primarily the active form 1,25-dihydroxyvitamin D) and analogs of this metabolite are available to treat clinical conditions, and these are also listed in Table 3.

The biological equivalency of the two forms of vitamin D, ergocalciferol and cholecalciferol, in humans, has recently been challenged [16, 17, 40]. Two studies have shown that vitamin D_2 is used less efficiently [17, 40]. Both forms of dietary vitamin D are converted to their corresponding 25(OH)D form equally well and remain at similar levels in plasma for a week. However, a week after a single dose of vitamin D_2, the resulting level of $25(OH)D_2$ begins to decline while after a dose of vitamin D_3, $25(OH)D_3$ does not [17]. In contrast, others [41] observed no difference in serum 25(OH)D concentrations with use of vitamin D_2 or vitamin D_3 containing supplements. There has been widespread use of vitamin D_2 supplements (e.g., Table 3), so there is no doubt that ergocalciferol-based supplements are efficacious. However, continuous use of these supplements, rather than intermittent dosing, may be required for efficacious use of vitamin D_2.

The positive aspect of the rapid disappearance of $25(OH)D_2$ is that it is a potentially safer form of the vitamin to use in higher doses on a daily basis. Evidence of its safe use at 125 and 250 μg per day (5000 and 10,000 IU/day, respectively) was recently demonstrated in postmenopausal women supplemented over 3 months [42]. Serum calcium levels did not exceed the upper limit of the normal range in any of the three treatment groups after 3 months and only one patient on the highest intake experienced hypercalciuria. Further research is needed to elucidate how to administer dietary or supplemental vitamin D when provided as vitamin D_2 versus D_3.

C. Dietary Intake of Vitamin D in the American Population

Only recently did vitamin D intakes of Americans become available [36, 37, 43]. Intake levels by various subgroups in the population as measured in 1999–2000 are shown in Table 4. Data shown are mean intake from food, and mean intake from food and supplements. For all age/sex groups, intake from food averaged only 4 to 6 μg (160–240 IU). Most of food-derived vitamin D was from fortified foods [44]. Supplement use contributed to vitamin D intake, adding 1 to 2 μg (40–80 IU) in each group except for women and men over 50 years who showed an increase

TABLE 3 Vitamin D and Related Compounds Found in Some Commonly Used Vitamin D Medications in the United States

Brand Name or Type of Product	Related Compound	Dosage Forms	
Vitamin D (supplemental)			
multivitamin	Cholecalciferol	tablet	5–25 μg (200–1000 IU)
Calcium supplement with vitamin D	Cholecalciferol	tablet	5–10 μg (200–400 IU) per 500–600 mg calcium
Vitamin D supplement	Cholecalciferol	tablet	10 μg (400 IU) 25 μg (1000 IU)
Vitamin D (therapeutic)			
Drisdol	Ergocalciferol	capsule	1250 μg (50,000 IU)
Drisdol Drops	Ergocalciferol	solution	207 μg/ml (8288 IU/ml)
1,25 Dihyroxyvitamin D (or similar acting analogs)			
Calcijex	Calcitriol	injection	1 μg/ml (40 IU)
Rocaltrol	Calcitriol	capsule solution	0.25 and 0.5 μg 1 μg/ml
Calderol	Calcifediol	capsule	20 μg, 50 μg
DHT	Dihydrotachysterol	tablet	125 μg, 200 μg, 400 μg
DHT Intensol	Dihydrotachysterol	solution	200 μg/ml
Hectorol	Doxercalciferol	capsule injection	2.5 μg (100 IU), 0.5 μg, 2 μg, 4 μg/ampoule
Hytakerol	Dihydrotachysterol	capsule	125 μg
Rocaltrol	Calcitriol	capsule solution	0.25 and 0.5 μg 1 μg/ml
Zemplar	Paricalcitol	capsule injection	1 μg, 2 μg, 4 μg 2 μg/ml, 5 μg/ml

Source: www.fda.gov/CDER/consumerinfo/druginfo.

TABLE 4 Estimated Mean Daily Intake of Vitamin D from Food and Supplements from NHANES 1999–2000

Sex	Age Group (years)	Total Intake Food, μg (IU)	Total Intake Food and Supplements, μg (IU)
Male, Female	1–8	5.9 (236)	8.2 (328)
Male	9–18	6.0 (240)	7.1 (284)
Male	19–50	5.4 (216)	7.5 (300)
Male	51+	5.3 (212)	8.8 (352)
Female	9–18	4.4 (176)	5.6 (224)
Female	19–50	4.2 (168)	7.1 (284)
Female	51+	4.7 (188)	9.5 (380)

Source: Moore, C. E., Murphy, M. M., and Holick, M. F. (2005). Vitamin D intakes by children and adults in the United States differ among ethnic groups. *J. Nutr.* **135**, 2478–2485.

of approximately 4–5 μg (160–200 IU). When considering total intakes, the mean intake of each age group met the current Dietary Reference Intake (DRI) recommendation except for adults over 50 years, where the Adequate Intakes (AI) of 10 μg (400 IU) and 15 μg (600 IU) were not met when food alone is considered; mean intakes of vitamin D do not meet the current DRI recommendation for females 9 years and over. Because many individuals are not consuming supplements, it is clear from these data that dietary intakes are well below current recommendations for most of the population. Considering that current DRIs are considered too low to maintain 25(OH)D at levels necessary for all functions, a need to improve dietary intakes is clear.

It is ironic that those individuals in greatest need of dietary sources of vitamin D because of aging or darker skin have the lowest intakes of vitamin D, thus further contributing to their low circulating 25(OH)D [36, 37]. African Americans consume less than the recommended intakes in every age group, with or without the inclusion of vitamin D from supplements as shown in Table 5 [36, 45–49]. Similar racial disparities in vitamin D intake were reported in more recent data from the NHANES 1999–2000 Survey [43]. Low vitamin D intakes among minority populations are largely attributed to lactose intolerance and limited consumption of milk, milk products, and ready-to-eat breakfast cereals [36, 43, 45].

In the NHANES III survey, vitamin supplementation was associated with a lower prevalence of vitamin D deficiency and higher serum 25(OH)D concentrations; however, at the current dose of vitamin D with routine use, a significant fraction of the general population was not able to

TABLE 5 Median Vitamin D Intakes among African Americans

Subject	Dietary Reference Intake, μg/day[a]	Intake from Food Alone, μg/day[b]	Intake from Food and Supplements, μg/day[b]
Females, years			
6–11	5	4.8	5.6
12–19	5	3.5	3.8
20–49	5	2.8	3.5
>50	10–15	3.3	4.0
Males, years			
6–11	5	5.5	6.1
12–19	5	4.7	4.9
20–49	5	3.7	4.2
>50	10–15	3.4	3.8

[a]Institute of Medicine. (1997). "Dietary Reference Intakes for Calcium, Phosphorus, Magnesium, Vitamin D and Fluoride." National Academies Press, Washington, DC.

[b]Calvo, M. S., Barton, C. N., and Whiting, S. J. (2004). Vitamin D fortification in the U.S. and Canada: Current status and data needs. *Am. J. Clin. Nutr.* **80**, 1710S–1716S; and Harris, S. S. (2006). Vitamin D and African Americans. *J. Nutr.* **136**, 1126–1129.

maintain adequate serum 25(OH)D levels. Table 6 demonstrates the clear racial, age, and gender disparity in the use of vitamin D supplements and the modest effect of supplementation on the prevalence of vitamin D deficiency [48]. There is an urgent need to reassess the adequacy of the vitamin D supplements currently available to meet individual needs, particularly in African Americans. This is illustrated by a study in African American women aged 15–49 who consumed 10 μg/day (400 IU) of supplemental vitamin D in addition to their usual diet. Serum 25(OH)D levels remained below 37.5 nmol≠1 in 11% of the African American women; thus a much higher percentage of black subjects could not attain the optimal serum range of 75–80 nmol≠1 at this modest level of vitamin D supplementation [43, 49].

D. Sun Exposure as a Source of Vitamin D

Sunlight contributes UVB and UVA radiation, but only UVB permits cholecalciferol synthesis [50]. UVB includes the wavelength range of 280–315 nm; conversion of 7-dehydrocholesterol to previtamin D_3 is better at 290–315 nm [4]. Exposure to sunlight can be quantified in erythemal doses, that is, the appearance of reddening of the skin. A minimal erythemal dose (1 MED) causes reddening, and further exposure results in more severe sunburning. Tanning (the induction of melanin synthesis in the skin) also occurs but takes longer to manifest and occurs with UVA as well as UVB exposure [51].

Many environmental factors affect synthesis of cholecalciferol, as listed in Table 7. One that affects a huge portion of the country is latitude. As the United States stretches from 20 °N (Puerto Rico) to over 70 °N (Alaska), it is important to recognize how latitude can impact on vitamin D status. As illustrated in Figure 2, there is a seasonal variation in 25(OH)D at latitudes close to 50° where presumably casual exposure to sun in the late spring, summer, and early autumn has resulted in levels of 25(OH)D that, on average, approach the level of 75–80 nmol/liter, which is closer to the level now recommended by many researchers [4, 14]. However, in winter, 25(OH)D levels fell to half of this level [14], presumably because dietary intake was not sufficient to provide enough vitamin D in the absence of UVB.

Studies have shown that one full-body (i.e., almost completely naked) minimal erythemal dose (1 MED) will synthesize as much as 500 μg (20,000 IU) of vitamin D_3 [48]. This intensity of sun exposure is not recommended because of concerns about skin phototoxicity, and an exposure less than 1 MED will provide maximal cholecalciferol production [51]. Accordingly, one can calculate that an exposure of one-fourth of an MED to 25% of body surface is sufficient for vitamin D_3 production of 25 μg (1000 IU) [8, 52]. To achieve 25% of body surface, one needs more than hands and face exposed, but also exposure of arms and legs; truncal exposure is also recommended [50]. Indeed, the "rule of nines" calculation for body surface area [53]

TABLE 6 Mean Serum Levels of 25(OH)D and Prevalence of Vitamin D Insufficiency (<70 nmol/liter) in Adult Participants in NHANES III

	No Vitamin Supplements		With Vitamin Supplements	
Population	25(OH)D (nmol/liter)	Prevalence (% below 70 nmol/liter)	25(OH)D (nmol/liter)	Prevalence (% below 70 nmol/liter)
Total	74.4	48	79.5	39
White	78.8	42	83.3	33
Black	48.0	86	54.6	77
Hispanic	62.7	66	66.2	60

Source: Taureen, N., Martins, D., Zadshir, A., Pan, D., and Norris, K. C. (2005). The impact of routine vitamin supplementation on serum levels of 25(OH)D3 among the general adult population and patients with chronic kidney disease. *Ethn. Dis.* **15**, S5-102–S5-106. with permission.

TABLE 7 Reasons for Low Sun (UVB) Exposure and Therefore Inability to Synthesize Cholecalciferol

Factor	Notes
Angle of sun in winter (November to March, inclusive at latitude 45°) not sufficient UVB	No synthesis
Being indoors and/or behind glass windows	No synthesis
Clothing, especially head-to-toe for cultural or environmental reasons	No synthesis if fabric blocks UVB. Some cloth is loosely woven and does permit UVB through.
Darkly pigmented skin	More sun exposure time is needed than person with little pigment
Impairment of skin synthesis of vitamin D with age	More sun exposure time needed compared to younger person
Sunscreen use (SPF 8 or greater)	Sun protection factor (SPF) indicates amount of UV blocked: SPF blocks at 1/SPF. E.g., SPF of 8 would allow 1/8 (~12%) of UV to penetrate.

Sources: Hollis, B. W. (2005). Circulating 25-hydroxyvitamin D levels indicative of vitamin D sufficiency: Implications for establishing a new effective dietary intake recommendation for vitamin D. *J. Nutr.* **135**, 317–322; Holick, M. F. (2006). Resurrection of vitamin D deficiency and rickets. *J. Clin. Invest.* **116**, 2062–2072; Holick, M. F. (2006). Vitamin D. *In* "Modern Nutrition in Health and Disease" (M. E. Shils, M. Shike, C. A. Ross, B. Caballero, R. J. Cousins, Eds.), 19th ed. Lippincott Williams & Wilkins. Accessed April 14, 2007, http://gateway.ut.ovid.com. Holick, M. F., and Jenkins, M. (2003). "The UV Advantage." Simon & Schuster, New York.

indicates that to achieve exposure of 25% of body surface area, one would expose both lower arms (9%) and lower legs (18%). Exposure of the head could contribute to an additional 9% of surface area. The time to achieve this exposure will vary greatly by season, latitude, and skin pigmentation.

One can measure skin type and the resulting amount of melanin that is produced in response to UV exposure. The Fitzpatrick skin type (also called skin phototypes [51, 54] was originally developed in the United States in 1975 to facilitate UV dosage for psoriasis photochemotherapy in subjects with "white" skin, and characterized for skin types (I through IV); it was later expanded to categories V and VI, as shown in Table 8 [54]. These skin types vary in ability to burn and tan. The time to burn (i.e., 1 MED) reflects melanin production and is an approximate indicator of the relative dose of UVB needed to synthesize equivalent amount of previtamin D_3 [55]. As shown in Table 8, skin type I needs only 40% of the time for 1 MED compared to skin type III, and skin type VI needs four times the exposure time of skin type III. The times given for ¼ MED are estimates based on exposure at 42 °N (i.e., Boston), at noon, in summer and may be used to calculate the exposure time to reach a vitamin D dose of 25 μg (1000 IU) when 25% of body surface is exposed [8]. Other factors related to timing, such as age, have not been considered.

Thus, there is no single answer to how much sun exposure is needed to achieve and maintain an adequate vitamin D status. Some dermatologists have advocated for no sun exposure [51]; however, the World Health organization (WHO) Report [56] on solar UV radiation indicates that it is not appropriate to strive for zero sun exposure as this would create a huge burden of skeletal disease from vitamin D deficiency. However, it is important to avoid excess exposure because this has been linked with skin cancers and skin photoaging [51]. A rational scheme to achieve skin synthesis of cholecalciferol without significant risk of overexposure has been published [8]. Although this subject remains a source of considerable controversy, the public now has access to information that will allow weighing of personal risk and benefits of sun exposure.

IV. VITAMIN D NUTRITIONAL STATUS ASSESSMENT

A. Indicators of Vitamin D Status

1. The Main Indicator of Vitamin D Status Is 25-Hydroxyvitamin D

Vitamin D status is evaluated by measuring the circulating levels of 25(OH)D, the level of which is related to the combined contributions of diet (including supplements) and skin synthesis. Measuring cholecalciferol and ergocalciferol is technically difficult and will only provide information on recent exposure (within the past 72 hours or less) [4]. In contrast, cholecalciferol and presumably ergocalciferol are stored in fatty tissues (adipose tissue, liver, muscle) and can be made available for conversion to 25(OH)D at a later time. However, these stores cannot maintain 25(OH)D levels through a period of dietary or UVB deprivation [12]. As well, there is some uncertainty as to whether cholecalciferol stored in adipose tissue is available to the body. Studies show that in obese individuals, serum levels of cholecalciferol, ergocalciferol, and 25(OH)D are lower than in the nonobese [57, 58]. This may relate to cholecalciferol and ergocalciferol being sequestered in adipose tissue without the ability to be released.

TABLE 8 Categorization of Skin Type Using a Traditional Dermatological System (Fitzpatrick Skin Type) and Association of These Categories with Vitamin D Synthetic Capacity of Skin

Fitzpatrick Skin Type	Skin Color	Common Geographic Origins	Melanin Index (M)[a]	Skin Response to Sun Exposure[b]	Relative MED[c]	Approximate Time for 1/4 MED[d]
I	White (blue eyes, freckled, albino)	Northern European	30 ± 3	Always burn, never tan	0.38	4 minutes
II	White (blond hair, blue or green eyes)	Northern European	–	Burn slightly, then tan slightly	0.75	6 minutes
III	White (brown eyes, darker complexion)	Southern European, Middle East	–	Rarely burn, tan moderately	1.0	7 minutes
IV	White ("Mediterranean")			Never burn, tan darkly	1.3	10 minutes
V	Brown	Asian, Native American, Pacific Islander	32 ± 2 (East Asian) 37 ± 4 (South Asian)	Never burn, tan darkly; Asian or Hispanic skin	2.0	13 minutes
VI	Black	African	57 ± 15	Never burn, tan darkly; black skin	3.8	21 minutes

Sources: Gilchrest, B. A. (2007). Sun protection and Vitamin D: Three dimensions of obfuscation. *J. Steroid Biochem. Mol. Biol.* **103**, 655–663; Webb, A. R., and Engelsen, O. (2006). Calculated ultraviolet exposure levels for a healthy vitamin D status. *Photochem. Photobiol.* **82**, 1697–1703; Fitzpatrick, T. B. (1988). The validity and practicality of sun-reactive skin types I through VI. *Arch. Dermatol.* **124**, 869–871; Holick, M. F., and Jenkins, M. (2003). "The UV Advantage." Simon & Schuster, New York; Shriver, M. D., and Parra, E. J. (2000). Comparison of narrow-band reflectance spectroscopy and tristimulus colorimetry for measurements of skin and hair color in persons of different biological ancestry. *Am. J. Phys. Anthropol.* **112**, 17–27.

[a]Narrow-band reflectance corrected for hemoglobin (red) to estimate melanin content (higher number indicates greater melanin content), from Shriver and Parra reference.

[b]Characteristics of previously unexposed skin after 30 minute direct exposure to sun, from Gilchrest reference.

[c]Relative minimal erythemal (MED) dose of UVB exposure derived from Fitzpatrick reference.

[d]Time estimated for sun exposure at 10:30 a.m. on June 21, to reach 1/4 MED at 11.5 °N, 29 °N, and 42.5 °N (values for 62.5 °N would be double) from Webb and Engelsen reference.

Measurement of 25(OH)D provides the best assessment of vitamin D "stores," that is, the form of vitamin D available to tissues for synthesis of 1,25-dihydroxyvitamin D by endocrine or paracrine/autocrine pathways (Figure 1). This is because 25(OH)D, although not the active form of vitamin D, rises with intake and/or sun exposure and declines with combined sun avoidance and low intake. Measurement of 1,25-dihydroxyvitamin D, the active vitamin D metabolite, is not the indicator of vitamin D status but rather a reflection, primarily, of the need for calcium and phosphate [4, 50], as described earlier. Further, the amount of 1,25-dihydroxyvitamin D in blood is influenced by renal function.

2. Classification of Vitamin D Status by Serum 25(OH)D

Until recently [59] there was a lack of understanding of tissue needs for 25(OH)D as a precursor for cellular synthesis of 1,25(OH)2D that functioned in the paracrine/autocrine pathways. Moreover, there was a lack of appreciation that vitamin D deficiency (low serum levels of 25(OH)D) may contribute to the risk of developing many chronic conditions beyond rickets and osteomalacia. The cutoff level determining vitamin D deficiency was serum 25(OH)D below 30 nmol/liter because, below this value, the patient has or will soon experience rickets (in children) or osteomalacia (in adults). The upper and lower values of normal for 25(OH)D, of 30–90 nmol/liter, were set using data from apparently healthy individuals (i.e., having neither rickets or osteomalacia) but as has been realized since, these apparently normal people were not adequate in vitamin D, being diet- or sun-deprived, or both [50].

There is disagreement as to what the normal range of 25(OH)D should be. Table 9 outlines the various cutoff points that have been used in the recent vitamin D literature. An analysis of several important physiologic endpoints was conducted to identify appropriate cutoffs for various physiologic functions [60]. These authors concluded that levels at or above 75–80 nmol/liter are needed to support most of the known endocrine and paracrine/autocrine functions. For example, vitamin D-dependent calcium absorption is maximal at a serum 25(OH)D level of 80 nmol/liter [22]. For many functions such as lower extremity strength and fracture risk, optimal 25(OH)D is

TABLE 9 Classification of Vitamin D Status by Serum 25(OH)D

Serum 25(OH)D	Category	Notes
<20 nmol/liter	Vitamin D deficiency	
<30 nmol/liter	Not at risk for clinical rickets or osteomalacia	Cutoff used for DRI recommendation
30 to 75 nmol/liter	Vitamin D insufficiency defined as at risk for chronic conditions except rickets and osteomalacia	Upper end of this range deemed lowest 25(OH)D for any health benefit
75 to 80 nmol/liter	Threshold for vitamin D-dependent calcium absorption	Strong evidence that 80 nmol/liter is required for calcium metabolism
90 to 100 nmol/liter	Optimal vitamin D status for some chronic conditions such as lower extremity function, periodontal disease, and fracture risk reduction; point at which substrate availability not limiting for 25(OH)D synthesis	
>220 nmol/liter	Potential adverse effects are seen above this level	Level of 25(OH)D when 250 μg is ingested chronically; no adverse effects were detected

Sources: Institute of Medicine. (1997). "Dietary Reference Intakes for Calcium. Phosphorus, Magnesium, Vitamin D and Fluoride." National Academies Press, Washington, DC; Heaney, R. P. (2007). The case for improving vitamin D status. *J. Steroid Biochem. Mol. Biol.* **103**, 635–641; Bischoff-Ferrari, H. A, Giovannucci, E., Willett, W. C., Dietrich, T., and Dawson-Hughes, B. (2006). Estimation of optimal serum concentrations of 25-hydroxyvitamin D for multiple health outcomes. *Am. J. Clin. Nutr.* **84**, 18–28; Hathcock, J. N., Shao, A., Vieth, R., and Heaney, R. P. (2007). Risk assessment for vitamin D. *Am. J. Clin. Nutr.* **85**, 6-18; Hollis, B. W. (2005). Circulating 25-hydroxyvitamin D levels indicative of vitamin D sufficiency: Implications for establishing a new effective dietary intake recommendation for vitamin D. *J. Nutr.* **135**, 317–322.

Based on references [3, 21].

90–100 nmol/liter [60]. As will be discussed later, excessive production of 25(OH)D can lead to toxicity. In adults, adverse effects are seen only at levels of 25(OH)D well above 250 nmol/liter [61]. The upper cutoff for children and infants may be lower and has not been reassessed from the original derivation in the DRI report [3].

Analysis of vitamin D status of Americans was performed using serum samples collected in the Third National Health and Nutrition Examination Survey (NHANES III) [62]. Using a cutoff of 50 nmol/liter, the prevalence of vitamin D insufficiency was determined for older adolescents and adults, and groups were further classified as being non-Hispanic white, non-Hispanic black, and Mexican American. As shown in Figure 5, the group with the highest prevalence of insufficiency was non-Hispanic black women. Risk for poor vitamin D status was related to skin color (although it was not directly measured), suggesting that sun exposure (and availability of UVB to make cholecalciferol) was an important determinant of status. As described earlier and in Table 5, African Americans tend to consume less dairy and therefore less dietary vitamin D.

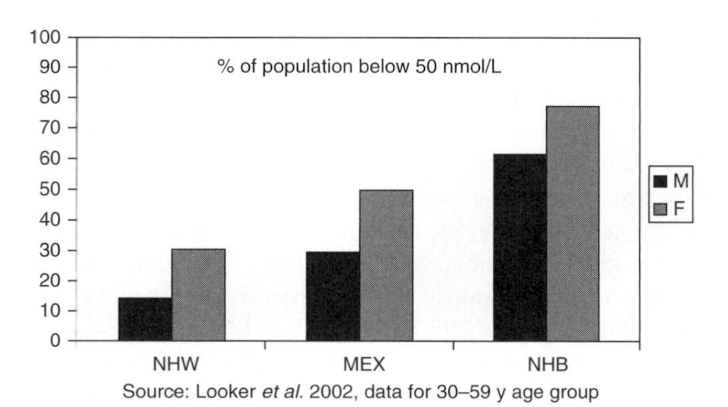

FIGURE 5 Prevalence of vitamin D insufficiency in the United States, as determined by the percent of the population below a serum 25(OH)D level of 50 nmol/liter. NHW (non-Hispanic white), MEX (Mexican), NHB (non-Hispanic black). Subjects were assessed during the Third National Health and Nutrition Examination Survey (NHANES III). Data are from Looker, A. C., Dawson-Hughes, B., Calvo, M. S., Gunter, E. W., and Sahyoun, N. R. (2002). Serum 25-hydroxyvitamin D status of adolescents and adults in two seasonal subpopulations from NHANES III. *Bone* **30**, 771–777.

Levels of 25(OH)D are influenced by exposure of skin to UVB and by dietary intakes of ergocalciferol or cholecalciferol. From its use in many research trials as well as in dosing studies [14], it can be shown that 1 μg (40 IU) of cholecalciferol daily raises serum 25(OH)D by 2 nmol/liter at basal levels, that is, in persons having serum levels of 25(OH)D less than 50 nmol/liter. Once past this level, 1 μg (40 IU) will raise 25(OH)D levels by 1 nmol/liter [60]. Sun exposure, as described earlier, will also affect serum levels of 25(OH)D. Effects of dietary vitamin D and skin synthesis of cholecalciferol are additive; both contribute to serum 25(OH)D [12]. Thus, it should be possible to predict vitamin D status in individuals. For example, if a background of moderate sun exposure during the summer achieves a serum 25(OH)D level of 50 nmol/liter, then a further 25 μg (1000 IU) per day of dietary cholecalciferol would be needed to reach a serum 25(OH)D level of 75 nmol/liter.

Most of the studies determining dietary or solar contributions to circulating levels of 25(OH)D have been conducted in light-skinned subjects. Our current understanding of dietary intakes needed to raise serum 25(OH)D is lacking for African Americans and others with darker skin, but evidence points to significant differences that should be taken into consideration when setting dietary requirements. Supplementation of African American women living above 40 °N latitude with up to 50 μg (2000 IU) of cholecalciferol for 3 years failed to increase serum 25(OH)D levels to the optimal goal of 80 nmol/liter in more than 40% of the subjects [63].

3. Technical Issues Related to 25-Hydroxyvitamin D Measurement

There are technical challenges to measurement of 25(OH)D, which have impeded the understanding of vitamin D deficiency and insufficiency. Although there are many methods for determining 25(OH)D, only two, a radioimmunoassay (RIA) and a competitive protein binding (CPB) assay, have been employed clinically. The RIA method has become automated and therefore widely available for routine analyses; the CPB assay is no longer widely used [64]. In the literature, both methods were widely reported. The CPB method yielded values that were approximately 30% higher than the RIA method due to nonspecific binding. The accepted standard is a relatively complex and time-consuming measurement by high performance liquid chromatography (HPLC) that is not practical for general clinical use. Recently, combination LC-mass spectrometry and a more rapid HPLC method have been promoted as more accurate than the commercially used immunoassay techniques. Fortunately, all of the widely used clinical laboratory methods perform reasonably well in identifying clinically important low levels. However, all clinical laboratories performing 25(OH)D testing should take part in external quality assessment programs, particularly because some of the commercially available RIA assays may perform poorly, failing to recognize and accurately measure $25(OH)D_2$.

Measurement of serum 25(OH)D, requires a fasting sample because lipid-rich plasma can interfere with the assay, and high-fat meals should be avoided at least 12 hours before the blood draw. Otherwise, blood may be taken at any time of day because there is no observed circadian variation in serum levels. Results are reported as nmol/liter or ng/ml. The conversion factor between these different units is the following: 1 ng/ml = 2.5 nmol/liter.

When repleting deficient patients, one expects the serum 25(OH)D response to high-dose supplementation with cholecalciferol or ergocalciferol to reach a plateau after 3 months of continuous treatment [14]. Therefore, if a patient's response to vitamin D supplements is to be monitored with blood levels, they should be checked no sooner than 3 months after treatment has been commenced. Monitoring of routine supplementation use is not necessary, but in situations where severe deficiency is being treated or there is reason to believe the patient may have impaired intestinal absorption, assessment of serum 25(OH)D would indicate the effectiveness of therapy.

4. Biomarkers of Vitamin Deficiency

Although it is well established that serum 25(OH)D is the best static indicator of vitamin D status, functional indicators provide information on how the deficiency/insufficiency is affecting the health of the person. They also confirm that a deficiency exists and may provide additional information for the diagnosis or monitoring of interventions. The most common responsive functional indicator examined to date is the suppression of secondary hyperparathyroidism, based on the endocrine function of 1,25-dihydroxyvitamin D in providing enough calcium through active intestinal absorption to suppress PTH levels that were raised with hypocalcemia [65]. Performance indicators such as the "get up and go" test vary in proportion to serum levels of 25(OH)D; they reflect muscle strength and balance that are negatively affected in vitamin D insufficiency [60, 66]. Pain is also associated with vitamin D deficiency, and minimal trauma fractures may also occur. These alone are not specific enough for diagnosis; however, they should be triggers to seek vitamin D testing.

B. Clinical Conditions Associated with Vitamin D Deficiency and Insufficiency

1. Rickets and Osteomalacia

Vitamin D deficiency results in rickets and osteomalacia, the former name used to describe the clinical signs observed in children, the latter name being those signs in adults. As outlined in Table 10, the physical signs of rickets in children result in severe bone, muscle, respiratory, and overall growth problems. Rickets is a disabling disease that usually begins in early childhood, before the age of 18 months [4]. Once the child can stand, gravity worsens the bone changes so that the classic picture of inward or outward bowed legs is observed. Osteomalacia is characterized by muscle atrophy (i.e., a waddling gait), by bone pain, and by fatigue; these signs are also seen in rickets.

The musculoskeletal effects of chronically low vitamin D, rickets in children and osteomalacia in adults, arise primarily because of a lack of calcium and phosphate for bone mineralization and muscle function. With chronic severe vitamin D deficiency, there is malabsorption of calcium and phosphate with a resultant hypocalcemia, hypophosphatemia, and secondary hyperparathyroidism. These lead to impaired growth plate development and bone mineralization [4, 67, 68]. Examples of different manifestations of rickets in African children can be viewed by accessing the Thachers website [67], which depicts rickets in Africa. Less severe rickets in African American children,

TABLE 10 Physical and Clinical Signs of Rickets

Body System Affected	Physical and Clinical Signs
Bone	• Poor mineralization • Hypertrophy of costochondral junctions • Rachitic rosary: involution of the ribs and protrusion of the sternum • Tibial and femoral bowing (inward or outward • Chest deformation • Delayed closing of fontanelles • Bone pain[a]
Teeth	• Delayed eruption • Enamel hypoplasia • Early dental caries
Muscle	• Delayed motor development • Toneless and flabby legs • Waddling gait[a]
Heart	• Myocardiopathy
Lungs	• Defective ventilation • Respiratory obstruction • Infections
Other	• Secondary hyperparathroidism • Low plasma phosphate • Tetany and seizures • Hypochromic anemia • Fatigue[a]

Source: Holick, M. F. (2006). Resurrection of vitamin D deficiency and rickets. *J. Clin. Invest.* **116**, 2062–2072.

[a]Symptoms of osteomalacia.

resulting from consumption of inappropriate milk alternatives, are at the website of the journal *Pediatrics* [68]. Not all rickets arise from a primary deficiency of cholecalciferol (or ergocalciferol). Although rare, there are many other types of rickets, as shown in Table 11; many of these rarer types are inherited forms of rickets due to mutations in important receptors (e.g., VDR) or enzymes (e.g., 1-hydroxylase) involved in vitamin D metabolism or action. Some forms of rickets are not due to problems with vitamin D (e.g., calcium-deficiency rickets).

There are drugs that cause osteomalacia. The most common are used in anticonvulsant therapy, particularly phenytoin, barbiturate derivatives, and carbamazepine. Although the exact mechanisms require further study, it is thought that these drugs result in greater inactivation of 1,25-dihydroxyvitamin D by inducing the enzyme 24-hydroxylase in the kidney [33].

2. Osteoporosis

The mild, secondary hyperparathyroidism that occurs with vitamin D insufficiency (i.e., levels of 25(OH)D below 80 nmol/liter) may cause increased bone turnover and bone loss—a clinical picture compatible with osteoporosis [59, 69]. Vitamin D insufficiency is very commonly found in patients with osteoporosis and contributes to the clinical presentation of osteoporosis (low bone density, fractures, falls), as well as the variety of conditions that have been found to have a higher incidence when associated with insufficient vitamin D.

TABLE 11 Types and Treatment of Rickets

Type of Rickets	Metabolic Abnormality Mechanism	Treatment
Vitamin D dependent	Insufficient cholecalciferol production from UVB exposure and lack of dietary cholecalciferol (or ergocalciferol)	Supplemental cholecalciferol (or ergocalciferol) or sun exposure (with appropriate UVB); aggressive therapy needed initially
Calcium-deficiency rickets	Low calcium intake	Provide adequate calcium and cholecalciferol (or ergocalciferol)
Fat malabsorption (e.g., cystic fibrosis)	Insufficient cholecalciferol production from UVB exposure and lack of dietary cholecalciferol (or ergocalciferol)	Subcutaneous or intramuscular injections of cholecalciferol recommended
Hereditary vitamin D-dependent rickets type 1	Inactive or absent renal 1-hydroxylase enzyme	Provide 1,25-dihydroxyvitamin D (or 1-hydroxyvitamin D)
Hereditary vitamin D-dependent rickets type 2	Mutations in VDR prevent normal actions of 1,25-dihydroxyvitamin D	Respond to high doses of either cholecalciferol or 1,25-dihydroxyvitamin D
Hereditary vitamin D-dependent rickets type 3	Abnormal hormone response element binding protein	Respond to high dose of either cholecalciferol or 1,25-dihydroxyvitamin D
Hypophosphatemic rickets	Phosphatemia (acquired and inherited) and decreased 1-hydroxylase enzyme	Intravenous phosphate; remove tumor
Tumor-induced osteomalacia	Tumor secretes phosphate factor that causes phosphatemia (acquired and inherited) and decreased 1-hydroxylase enzyme	

Source: Holick, M. F. (2006). Resurrection of vitamin D deficiency and rickets. *J. Clin. Invest.* **116**, 2062–2072.

Vitamin D insufficiency or deficiency should be considered in any patient with osteoporosis. Even after osteoporosis treatment has been initiated, a recent survey of patients attending osteoporosis clinics found approximately 50% of the patients had suboptimal serum 25(OH)D levels. Vitamin D insufficiency or deficiency should always be considered in patients with osteoporosis who do not appear to be responding to therapy (i.e., continuing to lose bone density, or continuing to suffer fragility fractures) [69].

Several pathological conditions contribute to secondary vitamin D deficiency that may lead to osteoporosis or osteomalacia. These include gastrointestinal disease, kidney disease, and drug-induced deficiency. Several common gastrointestinal disorders result in malabsorption [7], resulting in incomplete or absence of absorption of cholecalciferol/ergocalciferol. These include celiac disease, inflammatory bowel disease, gastrectomy, pancreatic insufficiency, and cystic fibrosis. As well, vitamin D deficiency can be precipitated by gastrointestinal surgeries such as bariatric surgery (intestinal bypass) or small-bowel resection (short gut syndrome).

In renal failure, several pathologies are related to vitamin D metabolism. The concentration of 1,25-dihydroxy vitamin D falls in chronic renal failure because of a reduced number of functional nephrons providing 1-hydroxylase for the endocrine pathway. A complicating factor, however, is the concomitant presence of secondary hyperparathyroidism, which can reduce renal tubular reabsorption of phosphorus and promotes renal synthesis of 1,25-dihydroxycholecalciferol [70]. However, parathyroid hormone adversely affects many organs and tissues and contributes to the uremic syndrome. In recognition of the low level of circulating 1,25-dihydroxyvitamin D in renal patients, management includes use of "active vitamin D," that is, 1,25-dihydroxyvitamin D.

3. Effects of Vitamin D Supplementation or Fortification in Osteoporosis

Randomized controlled trials (RCTs) in older adults have evaluated the effect of either vitamin D_3 or D_2 on bone mineral density (BMD). Most have found that supplementation with 12.5 μg (500 IU) to 20 μg (800 IU) of vitamin D_3 daily, combined with calcium (500–1000 mg), produces small but significant increases in lumbar spine and femoral neck BMD [14, 60]. A number of RCTs have evaluated the efficacy of vitamin D on fractures. A major limitation has been compliance rates for consuming the supplements of less than 80% and lack of assessment of vitamin D status (i.e., no 25(OH)D levels measured). Poor compliance with vitamin D supplementation was an issue noted in the large community-based trials. Most trials included nonvertebral or hip fractures as an outcome, although some trials assessed vertebral fractures. In a meta-analysis of the trials [71], which numbered almost 10,000 women, there was a 23% reduction in nonvertebral fractures. When the daily vitamin D dose was only 10 μg (400 IU), there was no reduction in fracture risk. This finding is not unexpected because that low amount of vitamin D_3 would result in only a small rise in serum 25(OH)D, approximately 7 to 10 nmol/liter [65], and the subjects receiving this might not have 25(OH)D levels above 50 nmol/liter.

Food fortification may prove to be as effective as supplements as a means of increasing serum 25(OH)D levels and reducing the risk of bone fracture for some populations. In a 2-year RCT in 167 men over 50, consumption of calcium and vitamin D-fortified milk, which provided an additional 1000 mg of calcium and 20 μg (800 IU) of vitamin D_3 daily, effectively raised serum 25(OH)D, lowered PTH, and stopped or slowed bone loss at several skeletal sites at risk for fracture with increasing age [72]. In this study the level of vitamin D fortification of the milk was markedly greater than the level allowed in the United States or Canada: 8.8 μg (352 IU) per 200 ml compared to 2.1 μg (85 IU) per 200 ml in North America [36, 72].

The risk of fracture is increased not only because of low bone density, but also because of falls. Studies show that vitamin D reduces fracture risk in part by reducing risk of falls [60, 73, 74]. Here, vitamin D is acting to increase muscle strength and may thus improve balance and reduce falls. In a study of adults 65 years and older, neuromuscular performance tests such as chair stands and walking tests were performed and related to vitamin D status (Figure 6). Performance improved as vitamin D status improved, up to a cutoff for serum 25(OH)D of 50–75 nmol/liter.

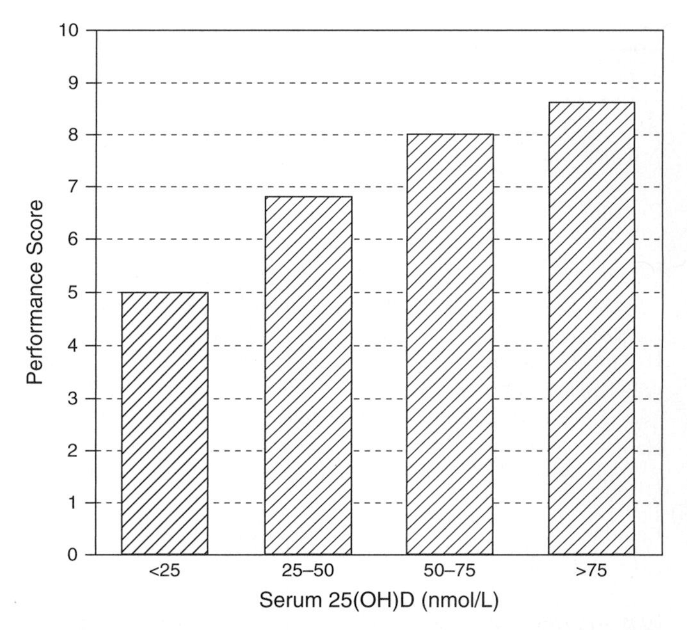

FIGURE 6 Improvement in physical performance of elderly residents who received graded amounts of vitamin D resulting in different serum levels of 25(OH)D. Reproduced with permission from Heaney, R. P. (2007). The case for improving vitamin D status. *J. Steroid Biochem. Mol. Biol.* **103**, 635–641.

C. Vitamin D Insufficiency as an Emerging Public Health Problem

Research is ongoing in determining how vitamin D status can impact on chronic disease. Both the endocrine and paracrine pathways are implicated in mechanisms for understanding how vitamin D is involved in chronic disease prevention. As shown in Figure 1, the endocrine pathway, which maintains calcium and phosphate homeostasis, affects bone, muscle, and cardiovascular health. The paracrine/autocrine pathways provide 1,25-dihydroxyvitamin D for cell functioning. Systems that are affected are those involving cell differentiation such as the immune system. These paracrine pathways may affect a variety of diseases under investigation such as cancer and autoimmune diseases [75]. A summary of disease conditions under investigation as being influenced by vitamin D status is given in Table 12.

TABLE 12 Health Outcomes Associated with Vitamin D Justify a Public Health Concern

Skeletal diseases
Rickets
Osteomalacia
Osteoporosis
Cancer
Breast
Prostate
Colon
Microbial infections
Tuberculosis
Autoimmune diseases
Multiple sclerosis
Diabetes (type 1 and 2)
Others, e.g., irritable bowel syndrome, rheumatoid arthritis, lupus
Cardiovascular disease
Hypertension
Arteriosclerosis
Malabsorption disorders
Crohn's
Celiac
Cystic fibrosis

Sources: Holick, M. F. (2004). Sunlight and vitamin D for bone health and prevention of autoimmune diseases, cancers, and cardiovascular disease. *Am. J. Clin. Nutr.* **80**, 1678s–1688s; Heaney, R. P. (2003). Long latency deficiency disease: Insights from calcium and vitamin D. *Am. J. Clin. Nutr.* **78**, 912–919; Bischoff-Ferrari, H. A. Giovannucci, E., Willett, W. C., Dietrich, T., and Dawson-Hughes, B. (2006). Estimation of optimal serum concentrations of 25-hydroxyvitamin D for multiple health outcomes. *Am. J. Clin. Nutr.* **84**, 18–28; Mathieu, C., Gysemans, C., Giulietti, A., and Bouillion, R. (2005). Vitamin D and diabetes. *Diabetologia* **48**, 1247–1257; Pittas, A., Lau, J., Hu, F., and Dawson-Hughes, B. (2007). The role of vitamin D and calcium in type 2 diabetes. A systematic review and meta-analysis. *J. Clin. Endocrinol. Metab.* First published ahead of print March 27, 2007, as DOI: 10.1210/jc.2007–0298; Pittas, A. G., Harris, S. S., Stark, P. C., and Dawson-Hughes, B. (2007). The effects of calcium and vitamin D supplementation on blood glucose and markers of inflammation in non-diabetic adults. *Diabetes Care* **30**, 980–986; Munger, K. L., Levin, L. I., Hollis, B. W., Howard, N. S., and Ascherio, A. (2006). Serum 25-hydroxyvitamin D levels and risk of multiple sclerosis. *JAMA* **296**, 2832–2838.

1. Vitamin D and Cancer Prevention

The connection between vitamin D status and cancer becomes clearer if one looks to the disruption of normal cell activities that characterizes cancers. Malignancies are characterized by dysregulation in cell growth and differentiation. Observational studies of sun exposure or vitamin D status in relation to cancer risk show protective relationships between sufficient vitamin D status and lower risk of cancer. A systematic review and meta-analysis were conducted for colorectal cancer by Gorham *et al.* [76]. Eighteen studies were included in the analysis, with four based on serum levels of 25(OH)D and 14 based on oral intake. Most of the studies showed that inadequate vitamin D status was significantly associated with higher risk of cancer of the colon and/or rectum. Individuals with 25 μg (1000 IU) per day or more of cholecalciferol intake or who had serum 25(OH)D concentrations greater than 81 nmol/liter had a 50% lower incidence of colorectal cancer compared to reference values. Breast cancer has also been studied [77]. Data pooled from two large studies indicated that women with serum 25(OH)D levels around 130 nmol had a 50% risk reduction for breast cancer compared to women with serum 25(OH)D levels of 30 nmol/liter.

A randomized control trial of vitamin D, originally intended as an osteoporosis study, provides compelling evidence of cancer prevention [78]. Almost 1200 white postmenopausal women living in Nebraska were enrolled in the 4-year study, which compared calcium alone (1400–1500 mg) with calcium plus vitamin D (27.5 μg [1100 IU]). Two-thirds of the women, at baseline, had 25(OH)D levels less than 80 nmol/liter. Vitamin D treatment increased levels in that group by an average of 24 nmol/liter, an expected amount (i.e., 1 μg vitamin D should raise 25(OH)D by 1 nmol/liter), considering compliance was 85%. As shown in Figure 7, the relative risk (RR) for developing cancer in these older women indicates that vitamin D supplemented women had only 40% of the risk of developing cancer compared to the placebo group; this risk fell to less than 25% when cancer incidence beyond the first year of the study was used in the analysis. Treatment with calcium alone had no significant effect on cancer incidence, but subject numbers may have been too low for the apparent 50% reduction by this treatment. There were no patterns in the types of cancer these women developed. Thus when vitamin D intakes were raised above an average level of 80 nmol/liter, a significant reduction in cancer incidence was observed, in agreement with the cohort studies described in the previous paragraph.

Further support of the role of vitamin D insufficiency in cancer development relates to the finding that in addition to their well-documented low 25(OH) D levels as a group, Black men in the United States have a 40% higher rate for total cancer mortality, whereas black women have a 20% higher mortality rate compared to their white counterparts [79]. Giovannucci *et al.* [79] demonstrated significantly higher risk of total cancer incidence and total cancer

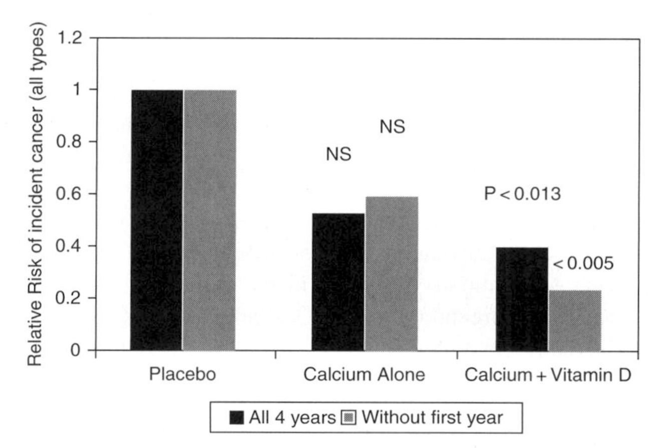

FIGURE 7 Relative risk (RR) of incident cancer plotted for the first randomized controlled trial of vitamin D and cancer risk. More than 1000 postmenopausal women in Nebraska were given one of three treatments (placebo, 1400–1500 mg calcium, or 1400–1500 mg calcium and 27.5 μg [1100 IU] vitamin D_3) and followed for 4 years. Data are taken from Lappe, J. M., Travers-Gustafson, D., Davies, K. M., Recker, R. R., and Heaney, R. P. (2007). Vitamin D and calcium supplementation reduces cancer risk: Results of a randomized trial. *Am. J. Clin. Nutr.* **85**, 1586–1591.

mortality among black men compared to whites in a study population that is relatively homogenous with respect to socioeconomic, education, lifestyle, and dietary factors.

There remains the debate as to whether vitamin D status should be improved by diet or sun exposure. In a study of persons with a diagnosis of nonmelanoma skin cancer, for whom sun exposure was the likely cause of this cancer, there were lower rates of second cancers (colon, gastric, and rectal cancers) [80]. Thus, there must be better understanding of the role of vitamin D, which for many people is through sun exposure. Although sun exposure increases the risk of developing skin cancer [51], it has the potential to reduce the risk of developing more severe forms of cancer [80].

2. Vitamin D and Autoimmune Diseases

Other chronic diseases have been implicated, especially in relation to immunity. Vitamin D deficiency may also predispose individuals to type 1 and type 2 diabetes. Several reviews and meta-analyses of the relationship between vitamin D status and development of diabetes are available [81–83]. Epidemiological studies have suggested that high circulating levels of vitamin D are associated with a lower risk of multiple sclerosis. Further support for this has emerged from a recent prospective case-control study of more than 7 million U.S. military personnel that identified 257 confirmed cases of multiple sclerosis cases [84]. Each case was matched with two controls and serum samples were collected before the date of initial symptoms and were assayed for 25(OH)D. The risk of developing multiple sclerosis significantly decreased with increasing levels of baseline 25(OH)D [84]. In conditions in which there is chronic pain and disability, such as fibromyalgia or chronic muscle fatigue, vitamin D deficiency should be investigated, because higher doses of vitamin D are known to promote muscle strength and balance in nursing home residents [85].

3. Vitamin D and Prevention of Chronic Kidney Disease and Cardiovascular Disease

Serum 25(OH)D levels are decreased in patients with varying stages of renal function [86–90]. Although this relationship was initially shown in small studies involving few subjects, more recent evidence from the general U.S. population using NHANES III data demonstrated significantly lower 25(OH)D levels only in those survey participants with estimated glomerular filtration rates (eGFR) below 29 ml/min/1.73 m^2 [91]. In contrast, patients in the general population with mild to moderate loss of renal function had "normal" levels of 25(OH)D. These findings support the hypothesis that vitamin D nutritional status may be involved in the pathophysiology of progressive renal disease. There are a number of mechanisms through which vitamin D status may affect the progression of renal failure, including control of cell proliferation and differentiation, changes in the loss of podocytes, modulation of inflammation such as the suppression of proinflammatory cytokines, regulation of the renin-angiotensin system to lower hypertension, and reduction of parathyroid hormone (PTH) levels [92]. Secondary hyperparathyroidism is a well-recognized hallmark of low circulating levels of 25(OH)D even in individuals with normal renal function; however, low serum 25(OH)D in chronic kidney disease (CKD) has been shown to aggravate the secondary hyperparathyroidism [92].

The U.S. National Kidney Foundation's 2003 Kidney Disease Outcome Quality Initiative (K/DOQI) guidelines recommend screening for hyperparathyroidism in patients with stage 3 and 4 CKD at least once per year, and if PTH concentrations are elevated, then measurement of serum 25(OH)D is recommended [86]. For CKD patients with hypovitaminosis D, therapy with ergocalciferol is recommended despite the fact that only a few studies have explored the efficacy of these guidelines to decrease serum iPTH and elevate 25(OH)D. To date, ergocalciferol treatment using a modified K/DOQI protocol corrects hypovitaminosis D, but treatment with vitamin D_2 effectively decreased serum iPTH by 13% over 7.4 months only in patients with CKD stage 3, and not in those with stage 4 who were followed over 6.8 months [90]. In another recent study, Deville *et al.* [89] examined ergocalciferol doses ranging from 800 IU/day to 100,000 IU/week over 90 days in patients with stages 3–5 CKD, but none on dialysis. Serum iPTH levels decreased among CKD stages 3 and 5, but reached statistically significant decreases only in patients with stage 4 CKD. As with the other vitamin D supplementation trial, there was a significant increase in serum levels of 25(OH)D in all the CKD patients. Whether or not cholecalciferol (vitamin D_3) treatment would be more effective in CKD remains to be determined; as well, the appropriate effective dose level of vitamin D and stage of CKD for initiation of treatment are not yet known.

Kidney dysfunction is also associated with insulin resistance and glucose intolerance. Chonchol and Scragg [91] in their analyses of kidney function and vitamin D status in the NHANES III survey also demonstrated that serum 25(OH)D and levels of kidney function (eGFR) were both inversely associated with homeostasis model assessment of insulin resistance (HOMA-IR) yet were independent of each other. Conversely, they found that survey participants with serum 25(OH)D levels in excess of 81 nmol/liter had lower HOMA-IR. These findings suggest that low serum 25(OH)D may be a risk factor in cardiovascular disease, because insulin resistance is an integral part of the putative path to cardiovascular disease development. Others have suggested that strategies to raise serum 25(OH)D levels, particularly in CKD patients, may decrease the risk of cardiovascular disease as well as possibly slow progression to renal failure [93–95]. Further evidence for a role of adequate vitamin D status in slowing the development of cardiovascular disease was demonstrated recently in a randomized, placebo-controlled, double-blind vitamin D supplementation trial where 50 μg/day (2000 IU) of cholecalciferol markedly improved the inflammatory cytokine profiles in patients with congestive heart failure [95]. There is clearly a need for more well-controlled intervention trials of longer duration and possibly higher doses of vitamin D to thoroughly establish safety and efficacy in slowing the progression of these chronic diseases.

D. Treatment of Vitamin D Deficiency

When an individual is depleted of vitamin D, aggressive therapy is needed. Table 3 provides a list of commonly used vitamin D supplements and analogs. New medications are updated continuously at the Food and Drug Administration website [96]. As indicated in Table 11, many forms of rickets require high doses of vitamin D. This vitamin D can be provided daily or in intermittent doses. It is not necessary to be concerned about exceeding the upper level (UL) for vitamin D as it is not intended as a barrier for treatment under a physician's care. In situations where rickets is determined to be a vitamin D-resistant form (Table 11), doses of the active metabolite 1,25-dihydroxyvitamin D, or an analog with similar properties, are often administered (Table 3).

Historically, sunlight was prescribed for vitamin D deficiency [4]. It was common practice to use UVB lamps for children with rickets, for adults with tuberculosis [4], and for some diseases where fat malabsorption precludes oral dosing. There is a growing interest in sunlamps to treat seasonal affective disorders. A new generation of lamps is being made available in which harmful UVC radiation is filtered [97]. A timing device is provided to prevent overexposure to UVB and to deliver only subthermal doses. The subject should receive exposure at different areas of the body to avoid tanning, which is an indicator of overexposure.

V. DIETARY REQUIREMENTS

A. Derivation of the 1997 Institute of Medicine Recommended Intake Values for Vitamin D

In 1997, the first report of the Dietary Reference Intakes (DRI) process included recommendations for vitamin D [3]. These recommendations are usually attributed to the Institute of Medicine (IOM) but may also be referred to as coming from the Food and Nutrition Board or the National Academy of Sciences. Table 13 lists the recommended values. These values for vitamin D are not Recommended Daily Allowances (RDAs) as were previous values; rather, they are Adequate Intake (AI) values, denoting the lack of scientific evidence needed to set an RDA [3]. In the DRI process, to determine an RDA, which can serve as an amount of a nutrient that is the goal for an individual, there must be an Estimated Average Requirement (EAR), which is derived from published data. An AI was set because there were not enough scientific studies defining a requirement for vitamin D at the time of deliberation. The AI was based on maintenance of serum 25(OH)D levels in the absence of sunlight (i.e., wintertime) at or above 27.5 nmol/liter for most age groups. Dietary intakes of a group of apparently healthy adults were taken as being the amount needed. It is not surprising that following over 10 years of new research on vitamin D, many authors argue for a revision (upward) for dietary vitamin D recommendations [e.g., 4, 12, 22, 44, 60].

B. Recommendations and Guidelines for Vitamin D Intake by Age Group

1. Recommendations for Adults

Since publication of the 1997 recommendations for vitamin D [3], much has been learned regarding vitamin D's metabolism. Most researchers believe that these recommendations (Table 13) are out-of-date with respect to recent evidence. Indeed, there is now sufficient information to set an EAR, one that more accurately reflects needs either with or without assurance of sun exposure. In the latter situation, the requirement and subsequent recommendation for vitamin D would be higher than one assuming sun exposure in the summer months. The advantage of an EAR in the absence of sun exposure is that it could be applied to specific populations such as shut-ins or persons choosing to wear clothing that permits no sun exposure. Concerns about adverse effects and the level of 25(OH)D associated with toxicity have been addressed [61].

TABLE 13 Dietary Recommendations and Expert Guidelines for Health Professionals for Vitamin D Intake

Organization and Date	Age Group	Recommendation	Notes
Institute of Medicine (IOM), 1997	0–1 year	5.0 μg (200 IU)	Many experts indicate a need for revision as these intakes will not maintain optimal 25(OH)D levels.
	1–50 years	5.0 μg (200 IU)	
	51–70 years	10 μg (400 IU)	
	71+ years	15 μg (600 IU)	
American College of Obstetricians and Gynecologists, 2003 www.acog.org	Women 51+ years	10–20 μg (400–800 IU)	Recommended for the prevention of osteoporosis in at-risk women (poor intake or sun exposure)
Dietary Guidelines for Americans, 2005 www.health.gov/dietaryguidelines	Men and women 51+ years	25 μg (1000 IU)	"Older adults, people with dark skin, and people exposed to insufficient ultraviolet band radiation (i.e. sunlight) should consume extra vitamin D from vitamin D-fortified foods and/or supplements"
North American Menopause Society (NAMS), 2006	Women 51+ years		In addition to IOM recommendations, intakes of 17.5–20 μg (700–800 IU) per day for women at deficiency risk because of inadequate sunlight exposure, such as older, frail, chronically ill, housebound, institutionalized, or those in northern latitudes. Doses as high as 50 μg (2000 IU) are safe.
American Academy of Pediatrics, 2003 www.aap.org	0–18 years	5 μg (200 IU)	"To prevent rickets and vitamin D deficiency in healthy infants and children and acknowledging that adequate sunlight exposure is difficult to determine, we reaffirm the adequate intake of 200 IU per day of vitamin D by the National Academy of Sciences and recommend a supplement of 200 IU per day for the following: 1. All breastfed infants unless they are weaned to at least 500 mL per day of vitamin D-fortified formula or milk. 2. All non breastfed infants who are ingesting less than 500 mL per day of vitamin D-fortified formula or milk. 3. Children and adolescents who do not get regular sunlight exposure, do not ingest at least 500 mL per day of vitamin D-fortified milk, or do not take a daily multivitamin supplement containing at least 200 IU of vitamin D."

No health professional group has made recommendations specific to adults. Speculation about the EAR has, however, been published. Using the data presented in Figure 2, it would appear that 12.5 μg (500 IU) could prevent the wintertime drop in 25(OH)D and maintain population levels of 25(OH)D close to 75–80 nmol/liter [14]. In a separate study [15] involving dosing with different levels of cholecalciferol, 12.5 μg (500 IU) was calculated to maintain 25(OH)D at 75–80 nmol/liter through several months in the winter. Taken together, these studies would support an EAR of 12.5 μg (500 IU) for adult Caucasians living at 35–45° latitude provided sun exposure occurred in the summer months. A vitamin D intake requirement set in absence of yearly summertime sun exposure or for individuals with poor cutaneous synthesis such as the elderly and dark-skinned individuals would be higher than 12.5 μg and would need to approach levels of intake that mimic vitamin D gained from sun exposure. To set an RDA from an EAR, the variation in the requirement estimate (standard deviation, SD) must be determined or estimated, and 2 SD are added to the EAR to obtain the RDA. The resulting RDA for vitamin D would be a value 20–40% greater than 12.5 μg (500 IU), that is, 15 μg (600 IU) to 17.5 μg (700 IU). For persons with no sun exposure providing skin synthesis of cholecalciferol, estimates of EARs are upwards of 25 to 50 μg (2000–4000 IU) [12, 22].

2. Recommendations for Older Adults

Older adults (age >50 years) have higher vitamin D needs for several reasons. Physiologically, there are two concerns. The enzyme responsible for synthesis of $1,25(OH)_2D$ in the kidney (the endocrine pathway) is resistant to PTH [4]. This means that the 1-hydroxylase is not increased by PTH when there is need for calcium, so there is prolonged secondary hyperparathyroidism leading to increased bone loss. A low level of 25(OH)D exacerbates this secondary hyperparathyroidism. Additionally, skin cells are less able to make cholecalciferol as there are fewer molecules of

7-dehydrocholesterol (provitamin D_3) in the epidermis. When young and old subjects are exposed to the same amount of UVB, elderly subjects produce only one-third the amount of cholecalciferol [4].

The special needs of the older and at-risk adults were recognized in the 2005 Dietary Guidelines for Americans [98, 99], as shown in Table 13. Men and women over 50 years were recommended to consume extra vitamin D from food and/or supplements. Those in high-risk groups, defined as being unable under usual circumstances to make cholecalciferol in skin, were advised to consume an additional 25 μg (1000 IU). This amount exceeds the IOM recommendations for adults over 50 years. A recommendation for postmenopausal women was made by the American Society of Obstetricians and Gynecologists [100] in 2003 to maintain an intake of vitamin D closer to 20 μg (800 IU). Similarly, the North American Menopause Society [101] made comparable recommendations in 2006.

3. Recommendations for Children and Infants

Few studies have addressed the specific needs of children, beyond describing low levels of circulating 25(OH)D that currently exist. For example, using NHANES III sera, levels of 25(OH)D of older adolescents was shown to be no better than that of adults in that survey [62]. When children have low levels of circulating 25(OH)D, as was the case in northern Europe when vitamin D fortification was not widespread, adolescents attained lower bone density than expected for their stage of development [102]. Promoting vitamin D supplementation generally improved serum 25(OH)D levels in winter time when sun exposure provided insufficient UVB. This is important as bone accrual during the adolescent growth spurt is compromised in vitamin D insufficiency [103].

Infants (age 0 to 12 months) represent a very vulnerable group. Infants do have the capacity for skin synthesis of vitamin D_3, however, dietary intakes are needed by almost all infants as they are usually kept indoors, swaddled, and/or may be born at times of the year when no skin synthesis is possible. For decades, researchers believed that vitamin D was present in only small amounts in breast milk, and therefore supplemental vitamin D is recommended for all breast-fed babies. An apparent reemergence of rickets in breast-fed babies has heightened this concern [4, 104, 105]. These babies were born in situations where the mothers' vitamin D status was poor because of poor dietary intake and limited sun exposure for mother or child (due to living conditions, season, skin pigmentation). Hollis and Wagner [106] have proposed that the problem of low vitamin D in breast milk should be solved by treating the lactating mother; thus, her milk provides additional vitamin D. In their preliminary studies, 100 μg (4000 IU) of vitamin D to vitamin D deficient lactating mothers normalized 25(OH)D levels in both mothers and infants where 10 μg (400 IU) alone did not. The effective level of cholecalciferol or ergocalciferol to provide sufficient vitamin D in breast milk remains to be determined.

The American Academy of Pediatrics (AAP) [107] in 2003 recommended that children up to age 18 years follow the DRI recommendations of 5 μg (200 IU) per day [3]. The AAP also recommends that the supplemental 5 μg (200 IU) is warranted when an infant or child is not ingesting fortified formula or milk, as described in Table 13. This recommendation for infants will prevent rickets but is suboptimal for preventing risk of chronic disease development [4].

VI. SAFETY OF VITAMIN D

A. The Tolerable Upper-Intake Level (UL) for Vitamin D

The daily tolerable upper-intake level (UL) for vitamin D of 50 μg (2000 IU) [3] was established in 1997 to discourage potentially dangerous self-medication. As discussed in Section V, recommendations for adults by health professional societies are as high as 25 μg (1000 IU), leaving a relatively small window between this value and the UL. The UL represents a safe intake (i.e., zero risk of adverse effects in an otherwise healthy person), and when a patient is undergoing therapeutic treatment under a health professional's care, this amount can be exceeded. However, in the absence of sun exposure, many researchers believe vitamin D intakes should approach 100 μg (4000 IU) [12, 22], a value greater than the UL. Therefore, it is important to consider whether the UL is too low to allow for ingestion of appropriate amounts of vitamin D to maintain optimal levels of 25(OH)D, particularly in the absence of sun exposure.

In a dosing study, Heaney *et al.* [15] gave cholecalciferol to subjects in doses up to 250 μg (10,000 IU), and found no adverse effects in men treated for 5 months. The primary adverse effect that is expected at very high levels of vitamin D is hypercalcemia, which can lead, over time, to calcification of soft tissues such as arteries (arteriosclerosis) and kidney (nephrocalcinosis) [7]. A less specific indicator is hypercalciuria, which may lead to increased risk of kidney stones [7].

A recent risk assessment for vitamin D used new data (post-1997) to derive a revised UL [61]. Studies indicated there was absence of any signs of toxicity when healthy adults were given over 250 μg (10,000 IU) daily. A comparison of this analysis with the UL is given in Table 14. It should be noted that at this time, revised data for a UL have not considered children or infants.

B. Vitamin D Intoxication

Accidental poisoning or uninformed supplementation with vitamin D can cause vitamin D intoxication [4, 61]. Levels reported in nine cases were in the range of 745 to 65,100 μg

TABLE 14 Comparison of Safety Assessments for Vitamin D

Reference	Tolerable Upper-Intake Level (UL)	Notes[a]
Institute of Medicine (IOM), 1997	0–1 year: 25 µg (1000 IU)	NOAEL = 45 µg (1800 IU)UF = 1.8 due to small sample sizes in studies
	1 year and above: 50 µg (2000 IU)	NOAEL = 60 µg (2400 IU)UF = 1.2 as potential for sensitive subjects to have been missed
Hathcock *et al.* (2007)	> 19 years: 250 µg (10,000 IU)	NOAEL = 250 µg (10,000 IU)UF = 1

Sources: Institute of Medicine. (1997). "Dietary Reference Intakes for Calcium, Phosphorus, Magnesium, Vitamin D and Fluoride." National Academies Press, Washington, DC; Hathcock, J. N., Shao, A., Vieth, R., and Heaney, R. P. (2007). Risk assessment for vitamin D. *Am. J. Clin. Nutr.* **85**, 6–18.

[a]NOAEL is the no-observed-adverse-effect-level, an intake level that has been shown to pose no risk of adverse effects. UF is uncertainty factor. UL = NOAEL/UF.

(approximately 30,000 to 2,600,000 IU), and durations were between days and decades [61]. One case involving a low dose of vitamin D was dismissed because no one has replicated toxicity on an intake of only 10 µg (400 IU) [61]. Generally, toxicity occurred in cases where intake was over 1250 µg (50,000 IU), well above the revised UL put forth by these authors. Levels of 25(OH)D associated with these high intakes ranged between 700 and 1600 nmol/liter, above the value of 250 nmol/liter considered high.

The signs and symptoms of vitamin D intoxication are associated with hypercalcemia [4]. These include constipation, lethargy, confusion, polyuria, and polydipsia. The resultant soft-tissue calcification can lead to calcific conjunctivitis, ectopic calcification, hypertension, and cardiac arrhythmias. Treatment for intoxication is elimination of all vitamin D supplements and avoidance of sun exposure. The imbalances in electrolyte and fluid balance can be corrected with appropriate hydration and diuresis. High-dose glucocorticoid therapy is not recommended [4].

VII. CONCLUSION

The high prevalence of vitamin D insufficiency in the United States stresses the need to address adequacy of vitamin D. Immediate action is needed to ensure adequate dietary intake and/or appropriate sun exposure. Vitamin D deficiency leads to rickets, osteomalacia, osteoporosis, and neuromuscular disabilities. Current understanding places importance on evaluation of the circulating levels of 25-hydroxyvitamin D, because this metabolite is the main indicator of vitamin D status. This intermediary metabolite must be available in sufficient quantities for synthesis of 1,25-dihydroxyvitamin D, the active metabolite, in both renal and nonrenal tissues: the former for calcium homeostasis and the latter for other, more fundamental purposes of cell growth and gene expression. Recent studies should allow more accurate dietary intake recommendations than the current 1997 Dietary Reference Intake values for vitamin D. Moreover, our understanding of vitamin D toxicity indicates that the current upper level is too low to provide assurance of safety for persons who would rely only on dietary sources. Better understanding of vitamin D metabolism and status will lead to its effective use in prevention and treatment of a variety of chronic diseases.

References

1. National Research Council. (1989). "*Recommended Dietary Allowances*" 10th ed. National Academies Press, Washington, DC.
2. Calvo, M. S., and Whiting, S. J. (2003). Prevalence of vitamin D insufficiency in Canada and the United States: Importance to health status and efficacy of current food fortification and dietary supplement use. *Nutr. Rev.* **61**, 107–113.
3. Institute of Medicine. (1997). "Dietary Reference Intakes for calcium phosphorus, magnesium, vitamin D and fluoride" National Academies Press, Washington DC.
4. Holick, M. F. (2006). Resurrection of vitamin D deficiency and rickets. *J. Clin. Invest.* **116**, 2062–2072.
5. Norman, A. (1998). Sunlight, season, skin pigment, vitamin D, and 25-hydroxy D: Integral components of the vitamin D endocrine system. *Am. J. Clin. Nutr.* **67**, 1108–1110.
6. Holick, M. F. (2004). Sunlight and vitamin D for bone health and prevention of autoimmune diseases, cancers, and cardiovascular disease. *Am. J. Clin. Nutr.* **80**, 1678s–1688s.
7. Holick, M. F. (2006). Vitamin D. *In* "Modern Nutrition in Health and Disease" (M. E. Shils, M. Shike, C. A. Ross, B. Caballero, and R. J. Cousins, Eds.), 19th ed. Lippincott Williams & Wilkins, Baltimore.
8. Webb, A. R., and Engelsen, O. (2006). Calculated ultraviolet exposure levels for a healthy vitamin D status. *Photochem. Photobiol.* **82**, 697–1703.
9. Kimlin, M. G., Olds, W. J., and Moore, M. R. (2007). Location and vitamin D synthesis: Is the hypothesis validated by geophysical data? *J. Photochem. Photobiol.* **86**, 234–239.
10. Lips, P. (2007). Vitamin D status and nutrition in Europe and Asia. *J. Steroid. Biochem. Mol. Biol.* **103**, 620–625.
11. Tangpricha, V., Koutkia, P., Rieke, S. M., Chen, T. C., Perez, A. A., and Holick, M. F. (2003). Fortification of orange juice

with vitamin D: A novel approach for enhancing vitamin D nutritional health. *Am. J. Clin. Nutr.* **77**, 1478–1483.

12. Hollis, B. W., Wagner, C. L., Drezner, M. K., and Binkley, N. C. (2007). Circulating vitamin D_3 and 25-hydroxyvitamin D in humans: An important tool to define adequate nutritional vitamin D status. *J. Steroid. Biochem. Mol. Biol.* **103**, 631–634.
13. Lips, P. (2006). Vitamin D physiology. *Prog. Biophys. Mol. Biol.* **92**, 4–8.
14. Meier, C., Woitge, H. W., Witte, K., Lemmer, B., and Seibel, M. J. (2004). Supplementation with oral vitamin D_3 and calcium during winter prevents seasonal bone loss: A randomized controlled open-label prospective trial. *J. Bone Miner. Res.* **19**, 1221–1230.
15. Heaney, R. P., Davies, K. M., Chen, T. C., Holick, M. F., and Barger-Lux, M. J. (2003). Human serum 25-hydroxycholecalciferol response to extended oral dosing with cholecalciferol. *Am. J. Clin. Nutr.* **77**, 204–210.
16. Houghton, L. A., and Vieth, R. (2006). The case against ergocalciferol (vitamin D_2) as a vitamin supplement. *Am. J. Clin. Nutr.* **84**, 694–697.
17. Armas, A. G., Hollis, B. W., and Heaney, R. P. (2004). Vitamin D_2 is much less effective than vitamin D_3 in humans. *J. Clin. Endocrinol. Metab.* **89**, 5387–5391.
18. Anderson, P. H., O'Loughlin, P. D., May, B. K., and Morris, H. A. (2003). Quantification of mRNA for the vitamin D metabolizing enzymes CYP27B1 and CYP24 and vitamin D receptor in kidney using real-time reverse transcriptase-polymerase chain reaction. *J. Mol. Endocrinol.* **31**, 123–132.
19. Dusso, A. S., Brown, A. J., and Slatopolsky, E. (2005). Vitamin D. *Am. J. Physiol. Renal Physiol.* **289**, F8–F28.
20. DeLuca, H. F. (2004). Overview of general physiologic features and functions of vitamin D. *Am. J. Clin. Nutr.* **80**, 1689s–1696s.
21. Akhter, S., Kutuzova, G. D., Christakos, S., and Deluca, H. F. (2007). Calbindin D(9k) is not required for 1, 25-dihydroxyvitamin D_3-mediated Ca^{2+} absorption in small intestine. *Arch. Biochem. Biophys.* **460**, 227–232.
22. Heaney, R. P. (2007). The case for improving vitamin D status. *J. Steroid. Biochem. Mol. Biol* **103**, 635–641.
23. Hewison, M., Burke, F., Evans, K. N., Lammas, D. A., Sansom, D. M., Liu, P., Modlin, R. L., and Adams, J. S. (2007). Extrarenal 25-hydroxyvitmain D_3-1α-hydroxylase in human health and disease. *J. Steroid. Biochem. Mol. Biol.* **103**, 316–321.
24. Hewison, M., Zehnder, D., Chakraverty, R., and Adams, J. S. (2004). Vitamin D and barrier function: A novel role for extrarenal 1 alpha-hydroxylase. *Mol. Cell. Endocrinol.* **215**, 31–38.
25. Liu, P. T., Stenger, S., Huiying, L., Wenzel, L., Tan, B. H., Krutzik, S. R., Ochoa, M. T., Schauber, K., Wu, K., Meinken, C., Kamen, D. L., Gallo, R. L., Eisenberg, D., Hewison, M., Hollis, B. W., Adams, J. S., Bloom, B. R., and Modlin, R. L. (2006). Toll-like receptor triggering of a vitamin D-mediated human antimicrobial response. *Science* **311**, 1770–1773.
26. Sigmundsdottir, H., Pan, J., Debes, G. F., Alt, C., Habtezion, A., Soler, D., and Butcher, E. C. (2007). DCs metabolize sunlight-induced vitamin D_3 to "program" T cell attraction to the epidermal chemokine CCL27. *Nat. Immunol.* **8**, 285–293.
27. Jones, G., Strugnell, S. A., and Deluca, D. F. (1998). Current understanding of the molecular actions of vitamin D. *Physiol. Rev.* **78**, 1193–1231.
28. Norman, A. W. (2001). Chapter 13. Vitamin D. In "Present Knowledge of Nutrition," 8th. ed., pp. 146–155. International Life Sciences Institute, Washington DC.
29. Norman, A. W. (2006). Minireview: Vitamin D receptor: New assignments for an already busy receptor. *Endocrinology* **147**, 5542–5548.
30. Fleet, J. C. (2004). Genomic and proteomic approaches for probing the role of vitamin D in health. *Am. J. Clin. Nutr.* **80**, 1730s–1734s.
31. Fleet, J. C. (2004). Rapid, membrane-initiated actions of 1, 25 dihydroxyvitamin D: What are they and what do they mean? *J. Nutr.* **134**, 3215–3218.
32. Norman, A. W., Okamura, W. H., Bishop, J. E., and Henry, H. L. (2002). Update on biological actions of 1-alpha, $25(OH)_2$-vitamin D_3 rapid effects and 24R,$25(OH)_2$-vitamin D_3. *Mol. Cell Endocrinol.* **197**, 1–13.
33. Zhou, C., Assem, M., Tay, J. C., Watkins, P. B., Blumberg, B., Schuetz, E. G., and Thummel, K. E. (2006). Steroid and xenobiotic receptor and vitamin D receptor cross-talk mediates CYP24 expression and drug-induced osteomalacia. *J. Clin. Invest.* **116**, 1703–1712.
34. USDA, Agricultural Research Service. (2007). "National Nutrient Database." Retrieved April 14, 2007, from http://www.nal.usda.gov/fnic/foodcomp/search/.
35. Lu, Z., Chen, T. C., Zhang, A., Persons, K. S., Kohn, N., Berkowiitz, R., Martinello, S., and Holick, M. F. (2007). An evaluation of the vitamin D_3 content in fish: Is the vitamin D content adequate to satisfy the dietary requirement for vitamin D. *J. Steroid. Biochem. Mol. Biol.* **103**, 642–644.
36. Calvo, M. S., Barton, C. N., and Whiting, S. J. (2004). Vitamin D fortification in the U.S. and Canada: Current status and data needs. *Am. J. Clin. Nutr.* **80**, 1710S–1716S.
37. Moore, C. E., Murphy, M. M., and Holick, M. F. (2005). Vitamin D intakes by children and adults in the United States differ among ethnic groups. *J. Nutr.* **135**, 2478–2485.
38. Calvo, M., Garthoff, L. H., Raybourne, R. B., Babu, U. S., Kelly, C., Lodder, S., Feeney, M. J., Minor, B., Beyer, D., Beelman, R., Pecchia, J., Paley, K., Chikthimmah, N., and Mattila, P. (2006). FDA's Center for Food Safety and Applied Nutrition and the Mushroom Council collaborate to optimize the natural vitamin D content of edible mushrooms and to examine their health benefits in different rodent models of innate immunity. Abstract of poster presented at FDA Science Forum. Retrieved May 25, 2007, from http://www.Accessdata.fda.gov/scripts/oc/scienceforum/sf2006/search/preview.cfm?keyword=mushrooms%20d&abstract_id=733&type=content&backto=searchmushrooms making vitamin D.
39. Calvo, M. S., and Whiting, S. J. (2006). Public health strategies to overcome barriers to optimal vitamin D status in populations with special needs. *J. Nutr.* **136**, 135–139.
40. Trang, H. M., Cole, D. E., Ribin, L. A., Pierratos, A., Siu, S., and Vieth, R. (1998). Evidence that vitamin D_3 increases serum 25-hydroxyvitamin D more efficiently than does vitamin D_2. *Am. J. Clin. Nutr.* **68**, 854–858.
41. Rapuri, P. B., Gallagher, J. C., and Haynatzki, G. (2004). Effect of vitamins D_2 and D_3 supplement use on serum 25OHD concentrations in elderly women in summer and winter. *Calcif. Tissue Int.* **74**, 150–156.

42. Mastaglia, S. R., Mautalen, C. A., Parisi, M. S., and Oliveri, B. (2006). Vitamin D_2 dose required to rapidly increase 25OHD levels in osteoporotic women. *Eur. J. Clin. Nutr.* **60**, 681–687.
43. Moore, C., Murphy, M. M., Keast, D. R., and Hollick, M. F. (2004). Vitamin D intake in the United States. *J. Am. Diet. Assoc.* **104**, 980–983.
44. Whiting, S. J., and Calvo, M. S. (2005). Dietary recommendations for vitamin D: A critical need for functional end points to establish an estimated average requirement. *J. Nutr.* **125**, 304–309.
45. Nesby-O'Dell, S., Scanlon, K. S., Cogswell, M. E., Gillespie, C., Hollis, B. W., Looker, A. C., Allen, C., Doughertly, C., Gunter, E. W., and Bowman, B. A. (2002). Hypovitaminosis D prevalence and determinants among African American and white women of reproductive age: Third National Health and Nutrition Examination Survey, 1988–1994. *Am. J. Clin Nutr.* **76**, 187–192.
46. Zadshir, A., Tareen, N., Pan, D., Norris, K., and Martins, D. (2005). The prevalence of hypovitaminosis D among US adults: Data from the NHANES III. *Ethn. Dis.* **15**, S5-97–S5-101.
47. Harris, S. S. (2006). Vitamin D and African Americans. *J. Nutr.* **136**, 1126–1129.
48. Taureen, N., Martins, D., Zadshir, A., Pan, D., and Norris, K. C. (2005). The impact of routine vitamin supplementation on serum levels of $25(OH)D_3$ among the general adult population and patients with chronic kidney disease. *Ethn. Dis.* **15**(S5), 102–106.
49. Harris, S. S., Soteriades, E., Coolidge, J. A., Mugal, S., and Dawson-Hughes, B. (2000). Vitamin D insufficiency and hyperparathyroidism in a low income, multiracial, elderly population. *J. Clin. Endocrinol. Metab.* **85**, 4125–4130.
50. Hollis, B. W. (2005). Circulating 25-hydroxyvitamin D levels indicative of vitamin D sufficiency: Implications for establishing a new effective dietary intake recommendation for vitamin D. *J. Nutr.* **135**, 317–322.
51. Gilchrest, B. A. (2007). Sun protection and vitamin D: Three dimensions of obfuscation. *J. Steroid Biochem. Mol. Biol.* **103**, 655–663.
52. Working Group of the Australian and New Zealand Bone and Mineral Society, Endocrine Society of Australia, Osteoporosis Australia. (2005). Vitamin D and adult bone health in Australia and New Zealand: A position statement. *Med. J. Austr.* **182**, 281–285.
53. Livingston, E. H., and Lee, S. (2000). Percentage of burned body surface area determination in obese and nonobese patients. *J. Surg. Res.* **91**, 106–110.
54. Fitzpatrick, T. B. (1988). The validity and practicality of sun-reactive skin types I through VI. *Arch. Dermatol.* **124**, 869–871.
55. Jablonski, N. G., and Chaplin, G. (2000). The evolution of human skin coloration. *J. Human Evol.* **39**, 57–106.
56. Lucas, R., McMichael, T., Smith, W., and Armstrong, B. (2006). Solar ultraviolet radiation: Global burden of disease from solar ultraviolet radiation. Environmental Burden of Disease series No. 13. World Health Organization.
57. Arunabh, S., Pollack, S., Yeh, J., and Aloia, J. F. (2003). Body fat content and 25-hydroxyvitamin D levels in healthy women. *J. Clin. Endocrinol. Metab.* **88**, 157–161.
58. Wortsman, J., Matsuoka, L. Y., Chen, T. C., Lu, Z., and Holick, M. F. (2000). Decreased bioavailability of vitamin D in obesity. *Am. J. Clin. Nutr.* **72**, 690–693.
59. Heaney, R. P. (2003). Long latency deficiency disease: insights from calcium and vitamin D. *Am. J. Clin. Nutr.* **78**, 912–919.
60. Bischoff-Ferrari, H. A., Giovannucci, E., Willett, W. C., Dietrich, T., and Dawson-Hughes, B. (2006). Estimation of optimal serum concentrations of 25-hydroxyvitamin D for multiple health outcomes. *Am. J. Clin. Nutr.* **84**, 18–28.
61. Hathcock, J. N., Shao, A., Vieth, R., and Heaney, R. P. (2007). Risk assessment for vitamin D. *Am. J. Clin. Nutr.* **85**, 6–18.
62. Looker, A. C., Dawson-Hughes, B., Calvo, M. S., Gunter, E. W., and Sahyoun, N. R. (2002). Serum 25-hydroxyvitamin D status of adolescents and adults in two seasonal subpopulations from NHANES III. *Bone* **30**, 771–777.
63. Aloia, J. F., Talwar, S. A., Pollack, S., and Yeh, J. (2005). A randomized controlled trial of vitamin D_3 supplementation in African American women. *Arch. Intern. Med.* **165**, 1618–1623.
64. Hollis, B. W., and Horst, R. L. (2007). The assessment of circulating 25(OH)D and 1, 25(OH)2D: Where are we and where are we going? *J. Steroid Biochem. Mol. Biol.* **103**, 473–476.
65. Lips, P., Duong, T., Oleksik, A., Black, D., Cummings, S., Cox, D., and Nickelsen, T. (2001). A global study of vitamin D status and parathyroid function in postmenopausal women with osteoporosis: Baseline data from the Multiple Outcomes of Raloxifene Evaluation clinical trial. *J. Clin. Endocrinol. Metab.* **86**, 1212–1221.
66. Pfeifer, M., Bergow, B., and Minne, H. (2002). Vitamin D and muscle function. *Osteoporosis Int.* **13**, 187–194.
67. African children with rickets. (2007). Retrieved April 28, 2007, from http://www.thachers.org/rickets_photos.htm.
68. Children with rickets. (2007). Retrieved April 28, 2007, from http://pediatrics. Aappublications.org/cgi/content/full/107/4/e46.
69. Holick, M. F., Siris, E. S., Binkley, N., Beard, M. K., Khan, A., Katzer, J. T., Petruschke, R. A., Chen, E., and dePapp, A. E. (2005). Prevalence of vitamin D inadequacy among postmenopausal North American women receiving osteoporosis therapy. *J. Clin. Endocrinol. Metab.* **90**, 3215–3224.
70. Kopple, J. (2006). Nutrition, diet, and the kidney. *In* "Modern Nutrition in Health and Disease" (M. E. Shils, M. Shike, C. A. Ross, B. Caballero, and R. J. Cousins, Eds.), 19th ed Lippincott Williams & Wilkins, Baltimore.
71. Bischoff-Ferrari, H. A., Willett, W. C., Wong, J. B., Giovannucci, E., Dietrich, T., and Dawson-Hughes, B. (2005). Fracture prevention with vitamin D supplementation: A meta-analysis of randomized controlled trials. *JAMA* **293**, 2257–2264.
72. Daly, R. M., Brown, M., Bass, S., Kukuljian, S., and Nowson, C. (2006). Calcium- and vitamin D_3-fortified milk reduces bone loss at clinically relevant skeletal sites in older men: A 2-year randomized controlled trial. *J. Bone Miner. Res.* **21**, 397–405.
73. Flicker, L., MacInnis, R. J., Stein, M. S., Scherer, S. C., Mead, K. E., Nowson, C. A., Thomas, J., Lowndes, C., Hopper, J. L., and Wark, J. D. (2005). Should older people in residential care receive vitamin D to prevent falls? Results of a randomized trial. *J. Am. Geriatr. Soc.* **53**, 1881–1888.
74. Jackson, C., Gaugris, S. S., and Hosking, D. (2007). The effect of cholecalciferol (vitamin D_3) on the risk of fall and fracture: A meta-analysis. *Q. J. Med.* **100**, 185–192.

75. Hollis, B. W., and Wagner, C. L. (2006). Nutritional vitamin D status during pregnancy: Reasons for concern. *CMAJ* **174**, 1287–1290.
76. Gorham, E. D., Garland, C. F., Garland, F. C., Grant, W. B., Mohr, S. B., Lipkin, M., Newmark, H. L., Giovannucci, E., Wei, M., and Holick, M. F. (2005). Vitamin D and prevention of colorectal cancer. *J. Steroid. Biochem. Mol. Biol.* **97**, 179–194.
77. Garland, C. F., Gorham, E. D., Mohr, S. B., Grant, W. B., Giovannucci, E. L., Lipkin, M., Newmark, H., Holick, M. F., and Garland, F. C. (2007). Vitamin D and prevention of breast cancer: Pooled analysis. *J. Steroid. Biochem. Mol. Biol.* **103**, 708–711.
78. Lappe, J. M., Travers-Gustafson, D., Davies, K. M., Recker, R. R., and Heaney, R. P. (2007). Vitamin D and calcium supplementation reduces cancer risk: Results of a randomized trial. *Am. J. Clin. Nutr.* **85**, 1586–1591.
79. Giovannucci, E., Liu, Y., and Willett, W. (2006). Cancer incidence and mortality and vitamin D in black and white male health professionals. *Cancer Epidemiol. Biomarkers Prev.* **15**, 2467–2472.
80. Grant, W. B. (2007). A meta-analysis of second cancers after a diagnosis of nonmelanoma skin cancer: Additional evidence that solar ultraviolet-B irradiance reduces the risk of internal cancers. *J. Steroid. Biochem. Mol. Biol.* **103**, 668–674.
81. Mathieu, C., Gysemans, C., Giulietti, A., and Bouillion, R. (2005). Vitamin D and diabetes. *Diabetologia* **48**, 1247–1257.
82. Pittas, A., Lau, J., Hu, F., and Dawson-Hughes, B. (2007). The role of vitamin D and calcium in type 2 diabetes. A systematic review and meta-analysis. *J. Clin. Endocrinol. Metab.* **92**, 2017–2029.
83. Pittas, A. G., Harris, S. S., Stark, P. C., and Dawson-Hughes, B. (2007). The effects of calcium and vitamin D supplementation on blood glucose and markers of inflammation in non-diabetic adults. *Diabetes Care* **30**, 980–986.
84. Munger, K. L., Levin, L. I., Hollis, B. W., Howard, N. S., and Ascherio, A. (2006). Serum 25–hydroxyvitamin D levels and risk of multiple sclerosis. *JAMA* **296**, 2832–2838.
85. Broe, K. E., Chen, T. C., Weinberg, J., Bischoff-Ferrari, H. A., Holick, M. J., and Kiel, D. P. (2007). A higher dose of vitamin D reduces the risk of falls in nursing home residents: A randomized, multiple-dose study. *J. Am. Geriatr. Soc.* **55**, 234–239.
86. National Kidney Foundation. (2003). K/DOQI clinical practice guidelines for bone metabolism and disease in chronic kidney disease. *Am. J. Kidney Dis.* **2**, S1–S201.
87. Gonzales, E. A., Sachdeva, A., Oliver, D. A., and Martin, K. J. (2004). Vitamin D insufficiency and deficiency in chronic kidney disease. A single center observation study. *Am. J. Nephrol.* **24**, 503–510.
88. LaClair, R. E., Hellman, R. N., Karp, S. L., Kraus, M., Ofner, S., Li, Q., Graves, K. L., and Moe, S. M. (2005). Prevalence of calcidiol deficiency in CKD: A cross-sectional study across latitudes in the United States. *Am. J. Kidney Dis* **45**, 1026–1033.
89. Deville, J., Thorp, M. L., Tobin, L., Gray, E., Johnson, E. S., and Smith, D. H. (2006). Effect of ergocalciferol supplementation on serum parathyroid hormone and serum 25-hydroxyvitamin D in chronic kidney disease. *Nephrology* **11**, 555–559.
90. Zisman, A. L., Hrisrova, M., Ho, L. T., and Sprague, S. M. (2007). Impact of ergocalciferol treatment of vitamin D deficiency on serum parathyroid hormone concentrations in chronic kidney disease. *Nephrology* **27**, 36–43.
91. Chonchol, M., and Scragg, R. (2007). 25-Hydroxyvitamin D, insulin resistance, and kidney function in the Third National Health and Nutrition Examination Survey. *Kidney Int.* **71**, 134–139.
92. Holick, M. F. (2005). Vitamin D for health and in chronic kidney disease. *Semin. Dial.* **18**, 266–275.
93. Zittermann, A. (2003). Vitamin D in preventive medicine: Are we ignoring the evidence? *Br. J. Nutr.* **89**, 552–572.
94. Zittermann, A., Schleithoff, S. S., and Koerfer, R. (2005). Putting cardiovascular disease and vitamin D insufficiency into perspective. *Br. J. Nutr.* **94**, 483–492.
95. Schleithoff, S. S., Zittermann, A., Tenderich, G., Berthold, H. K., Stehle, P, and Koerfer, R. (2006). Vitamin D supplementation improves cytokine profiles in patients with congestive heart failure: A double-blind, randomized, placebo-controlled trial. *Am. J. Clin. Nutr.* **83**, 754–759.
96. Food and Drug Administration. (2007). Centre for Drug Evaluation and Research. http://www.fda.gov/cder/index.html.
97. Sayre, R. M., Dowdy, J. C., and Shepherd, J. G. (2007). Reintroduction of a classic vitamin D ultraviolet source. *J. Steroid. Biochem. Mol. Biol.* **103**, 686–688.
98. Johnson, M. A., and Kimlin, M. G. (2006). Vitamin D, aging, and the 2005 Dietary Guidelines for Americans. *Nutr. Rev.* **64**, 410–421.
99. U.S. Department of Health Human Services U.S. Department of Agriculture. (2005). "Dietary Guidelines for Americans, 2005," 6th. ed. U.S. Government Printing Office, Washington, DC. Retrieved April 14, 2007 from http://www.healthierus.gov/dietaryguidelines.
100. American College of Obstetricians and Gynecologists. (2003). Osteoporosis. Retrieved April 14, 2007, from http://www.Acog.org/publications/patienteducation/bp048.cfm.
101. North American Menopause Society. (2006). Management of osteoporosis in postmenopausal women: 2006 position statement of the North American Menopause Society. *Menopause* **13**, 340–367.
102. Tylavsky, F. T., Cheng, S., Lyytikainen, A., Viljakainen, H., and Lamberg-Allardt, C. (2006). Strategies to improve vitamin D status in northern European children: Exploring the merits of vitamin D fortification and supplementation. *J. Nutr.* **136**, 1130–1134.
103. Outila, T. A., Karkkainen, M. U., and Lamberg-Allardt, C. J. (2001). Vitamin D status affects serum parathyroid hormone concentrations during winter in female adolescents: Associations with forearm bone mineral density. *Am. J. Clin. Nutr.* **74**, 206–210.
104. Weisberg, P., Scanlon, K. S., Li, R., and Cogswell, M. E. (2004). Nutritional rickets among children in the United

States: review of cases reported between 1986 and 2003. *Am. J. Clin. Nutr.* **80**, 1697S–1705S.
105. Carvalho, N. F., Kenney, R. D., Carrington, P. H., and Hall, D. E. (2001). Severe nutritional deficiencies in toddlers resulting from health food milk alternatives. *Pediatrics* **107**, 1–7.
106. Hollis, B. W., and Wagner, C. L. (2004). Assessment of vitamin D requirements during pregnancy and lactation. *Am. J. Clin. Nutr.* **79**, 717–726.
107. Gartner, L. M., and Greer, F. R. (2003). Prevention of rickets and vitamin D deficiency: New guidelines for vitamin D intake. *Pediatrics* **111**, 908–910.

CHAPTER **44**

Osteoporosis: The Early Years

CONNIE M. WEAVER
Purdue University, West Lafayette, Indiana

Contents

I. INTRODUCTION

The risk of developing osteoporosis is largely determined by the mass and size of bone acquired by adulthood known as peak bone mass [1]. The greater the skeletal mass at its peak and the stronger the geometry, the greater the amount of loss that can occur before entering the fracture risk zone. An interplay between heritable factors and environmental factors determine peak bone mass (Figure 1). Estimates for the contribution of genetics is 60–80% and the remaining 20–40% to lifestyle factors [2, 3]. Genes that control growth and development are thought largely to also determine bone acquisition [2].

Nutrition and physical activity are the primary lifestyle determinants of bone acquisition, and increasing evidence suggests that the interactions between these two are stronger determinants than either alone. Nutrition plays an important role in preventing osteoporosis through building optimal peak bone mass within one's genetic potential. This chapter reviews the influence of diet and nutrients on bone health and the basis of requirements for nutrients that are based on skeletal health. Nutrition likely also can play a role in fracture risk during childhood. Skeletal fragility in childhood occurs when bone mineral density (BMD) as assessed by dual energy x-ray absorptiometry (DXA) expressed as a Z-score is less than –2.0 [4]. Skeletal fragility can arise from a lag between peak height velocity and bone mineralization during the pubertal growth spurt or as a consequence of eating disorders or disease. Less is known about the role of diet in the context of pediatric disorders, and research is needed to address these gaps.

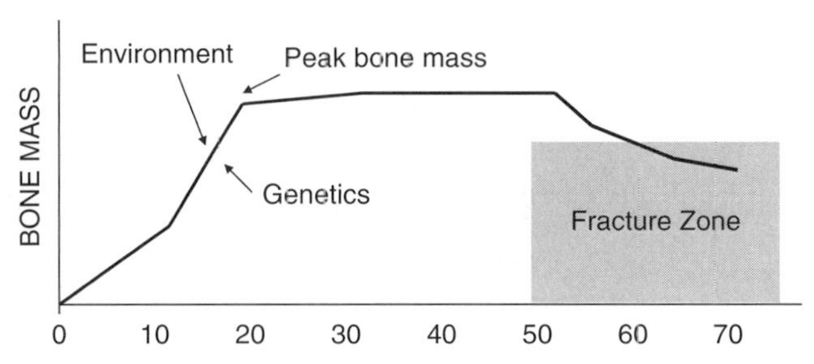

FIGURE 1 Bone mass throughout the lifespan. The influence of genetics and the environment is greatest during growth.

This chapter focuses on the development of peak bone mass and the role of nutrition in bone acquisition. The reader should consult the companion chapter by Marcus (Chapter 45) for a discussion of osteoporosis and bone qualities.

II. ACQUIRING PEAK BONE MASS AND BONE STRENGTH

Understanding the timing for an effective nutritional intervention can be as important as the nature of the intervention. Textbooks commonly cite that bone mass is acquired throughout the third decade of life and frequently quote a longitudinal study that averaged annual bone gains over a decade, ignoring the fact that the rate of accrual decreased over time and became trivial by age 30 years [5]. In a metabolic balance study, women over age 21 years were not in positive calcium retention on intakes of 1300 mg calcium per day [6]. Increase in BMD with age depends on the skeletal site. For total body BMD, 95% of adult peak bone mass was achieved by age 16.2 years and 99% was achieved by age 22.1 years [7]. Peak BMD in white women occurred at age 23 years for the spine, 18.5 years for the femoral neck, 14.2 years for the greater trochanter, and age 15.8 years for Wards Triangle [8] (Figure 2). Clearly, to markedly influence peak bone mass, lifestyle practices are more important prior to the end of puberty than post adolescence.

When there are transient periods when diet is inadequate, some catch-up growth is possible [9]. In a calcium-supplemented controlled trial from early puberty until

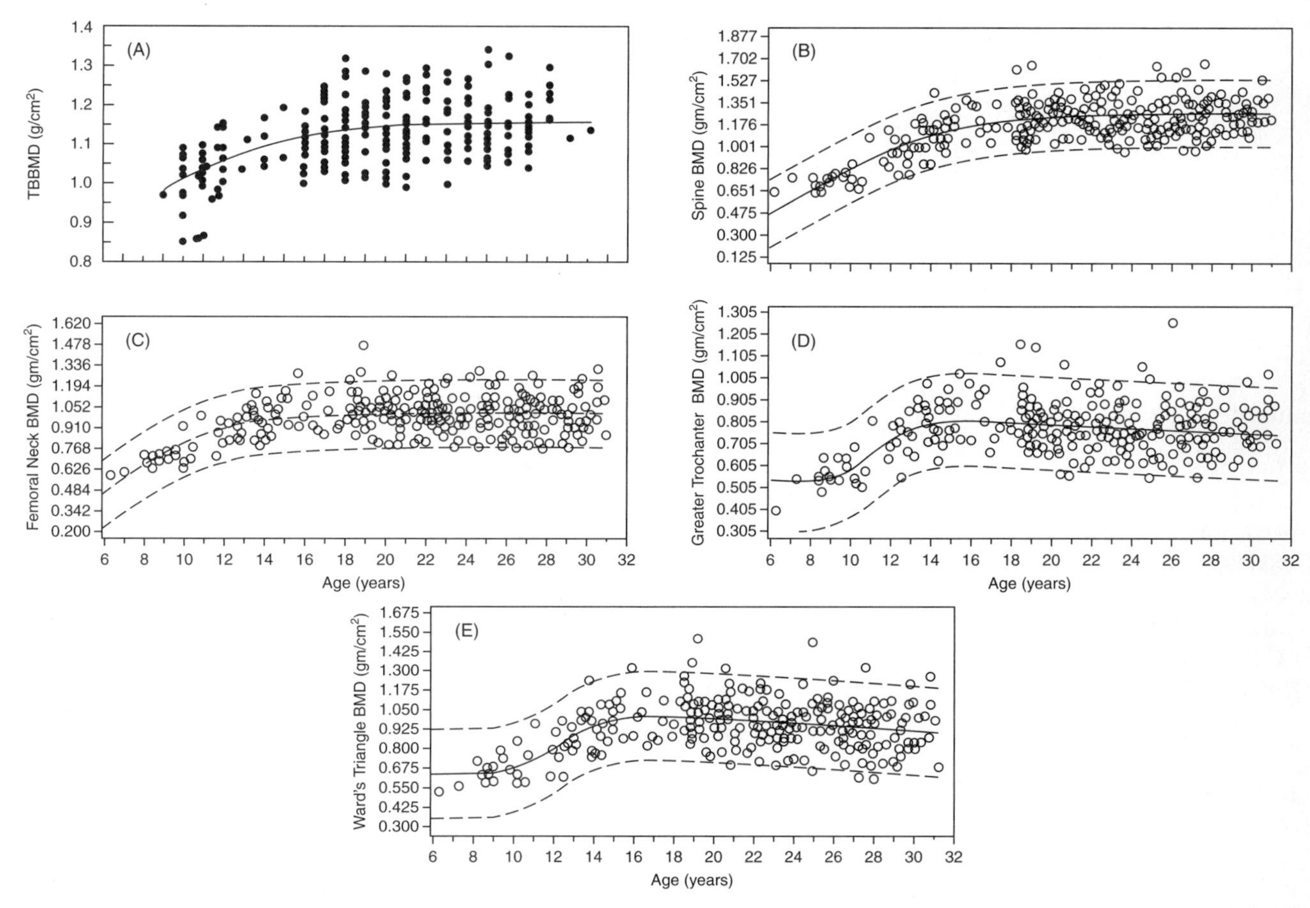

FIGURE 2 Bone mineral density accumulation with age in females. Gains in bone mineral density (BMD) vary with skeletal site. Almost 95% of adult peak total body BMD occurred by age 15.2 years (A) and the highest BMD of the spine occurs by age 23 years (B), by age 18.5 years for the femoral neck (C), by age 14.2 years for the greater trochanter (D), and by age 15.8 years for Ward's triangle (E). (Reproduced from Teegarden *et al.* [7] and Lin *et al.* [8].)

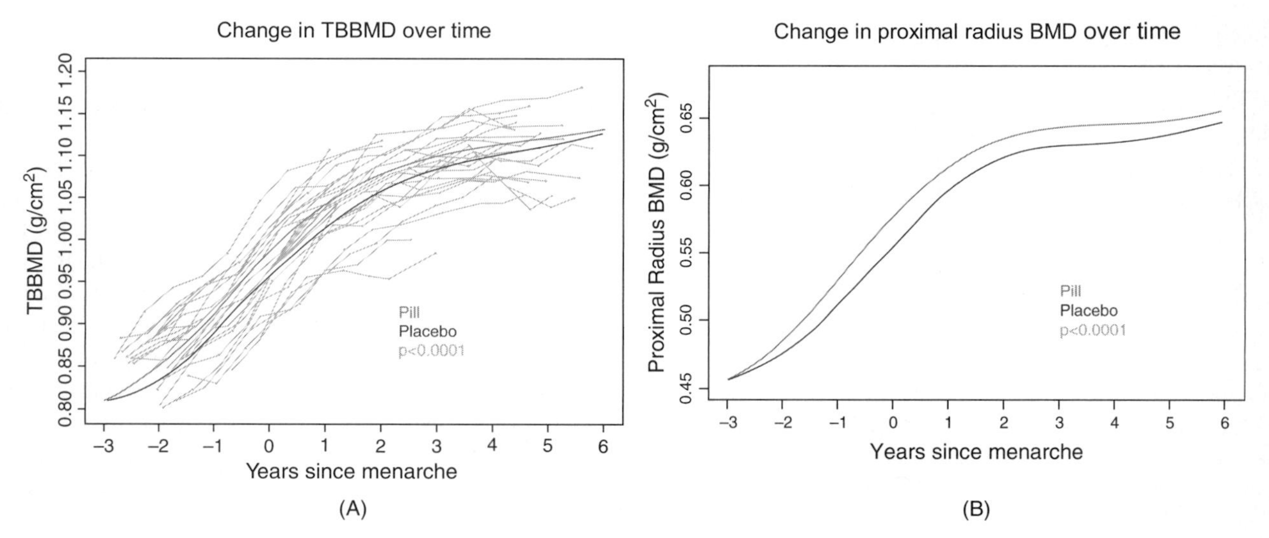

FIGURE 3 Total body BMD (A) and proximal radius (B) was significantly higher in prepubertal girls randomized to 1 g calcium supplement daily compared to those assigned to placebo from age 10 to 18 years [15].

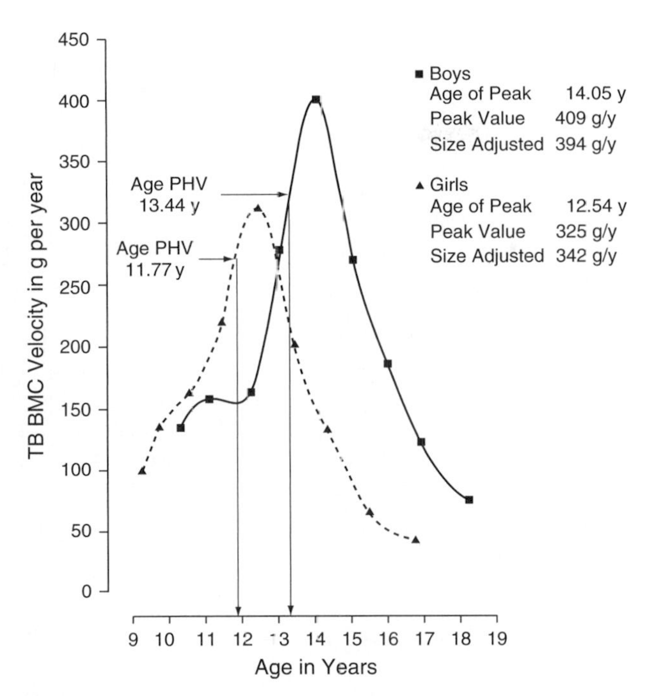

FIGURE 4 Peak height velocity (PHV) precedes total body (TB) peak bone mineral content (PBMCV) by approximately 0.77 years for girls and 0.61 years for boys. PHV occurs 1.67 years earlier in girls and PBMCV occurs 1.51 years earlier in girls than boys [13].

development of peak bone mass in girls habitually consuming on average 830 mg calcium per day, calcium supplementation increased accrual of BMD of total body (Figure 3) and radius through puberty over the placebo group [10]. But by age 18 years, the advantage was reduced. There was an interaction with size such that girls who were taller by age 18 did not fully catch up. The extent of catch-up growth undoubtedly depends on the degree of inadequacy of the diet, the timing and duration of the period of inadequacy, and the degree of repletion. Regardless, the period of inadequacy is itself a period of vulnerability, that is, in this example, a higher risk of fracture.

The impact of genetics on development of peak bone mass is evident in Figure 4. The light gray lines show tracking of total body BMD of 15 representative individuals. Those who began the trial with lower bone mineral density remained lower throughout development of peak bone mass. Modifications through lifestyle choices occur within one's genetic potential.

Bone geometry plays an additional role beyond bone mass in resistance to fracture. Bone geometry includes dimensions of bone such as length and cross-sectional area. The relationship of bone geometry to fracture risk in childhood and measurement by peripheral quantitative computer tomography (pQCT) has been described by Kalkwarf [11]. The effect of diet on bone geometry is a rather recent field of study because of the relative newness of this technique. The exercise intervention trials that have used this outcome measure have illustrated additional information gained about bone strength.

III. SKELETAL FRAGILITY IN CHILDREN

Fractures in children are associated with low bone mass and density for age [12]. If low bone mass in childhood leads to lower peak bone mass in adulthood, the risk of fracture increases later in life. Relative skeletal fragility can occur naturally with growth spurts. Beyond genetically programmed qualities, bone mass and density throughout life are influenced primarily by nutrition, physical activity, and hormones. Furthermore, eating disorders, smoking, alcohol abuse, and various drugs also influence bone mass. To a large extent, these factors work independently in that one cannot compensate for inadequacy of another. However, dietary calcium and physical activity have important interactions. Finally, some disorders are associated with low BMD.

A. Relatively Low BMD during Puberty

Puberty is a period of rapid skeletal growth that is genetically programmed and hormonally driven. The rate of total body bone mass accrual throughout adolescence was determined by Bailey *et al.* in a longitudinal study of white boys and girls [13]. Because they could plot the change in bone mineral content (BMC) over puberty for each individual and then determine the average for the cohort, they could more accurately determine how narrow and sharp the pubertal growth spurt is compared to previous cross-sectional studies (Figure 4).

From this study, we know that approximately 25% of adult peak bone mass is acquired over approximately 2 years—on average this occurs from age 12 to 14 years in girls and age 13 to 15 years in boys. Peak bone mineral content velocity (PBMCV) is higher and occurs later for boys than girls. The timing of bone mineral acquisition is more closely linked to pubertal development than to chronological age [2, 13].

During puberty bones first elongate and then mineralization, or bone consolidation, ensues. This is reflected in peak height velocity (PHV) preceding PBMCV in Figure 4. At the age of PHV, adolescents have acquired 90% of their adult height (or bone size), but only 60% of adult total body BMC. Thus, early puberty is a period of relatively low BMD, and, therefore, susceptibility to fracture not unlike that of age-related bone loss [14–16]. The higher incidence of fracture during this time of life is shown in Figure 5. Data collected by the Mayo Clinic [14] show that the incidence of childhood fracture is highest just prior to and during puberty and falls dramatically after adolescence. Approximately 51% of boys and 40% of girls experience fractures by age 18 years [16].

The dramatic increase in rates of childhood fracture over the past three decades is apparent in Figure 5. Fracture incidence increased 32% in males with the greatest increase at age 11–14 years and 56% in girls with the greatest increase at age 8–11 years. Increased rates of childhood fracture may

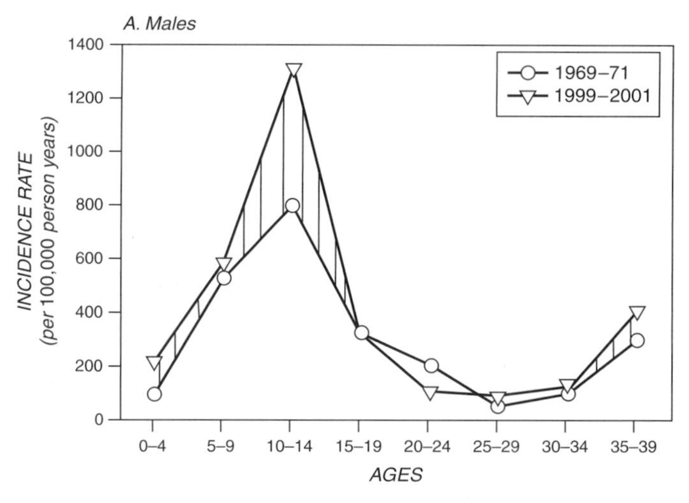

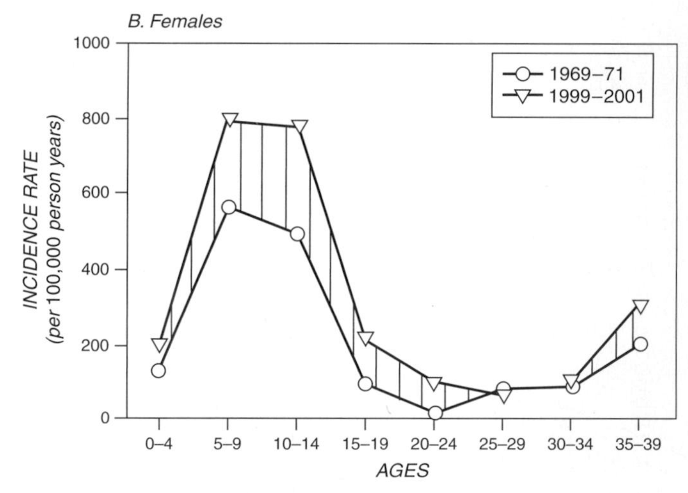

FIGURE 5 Childhood forearm fracture incidence in males (A) and females (B) from 1969–1971 (lower line) and 1999–2001 (upper line) in Rochester, Minnesota. (Reproduced from Heaney and Weaver [17] using data from Khosla *et al.* [14].)

relate to reduced consumption of milk, change in physical activity or recreational activities, and/or increased body weight over this time period. The prevalence of excessive adiposity in children and adolescents has nearly tripled while the incidence of fracture has increased [18, 19]. In adults, increased weight has been associated with increased bone mass [20], but overweight children and adolescents have higher rates of fracture [12, 21, 22]. The increased incidence in fracture with excessive body weight in children has been hypothesized to occur because of greater force being placed on bones such as the radius during falls, lower bone mass and bone strength with increasing body fat even when adjusted for total body weight, and impaired mobility [21, 23–28]. Changes in bone geometry that accompany increases in bone size throughout childhood include increases in cortical thickness and bone diameter [29]. Bone diameter and cortical thickness are less in children with excess body fat [28]. This emphasizes the need to maintain ideal body weight in children.

TABLE 1 Pediatric Conditions Associated with Low Bone Mass[a]

Amenorrhea
Anorexia nervosa
Brain tumors (craniospinal irradiation)
Celiac disease
Congenital hypothyroidism
Cystic fibrosis
Diabetes mellitus
Epilepsy
Hemophilia
HIV/antiretroviral therapy
Inflammatory bowel disease
Liver transplantation
Glucocorticoid-sensitive nephrotic syndrome
Osteogenesis imperfecta
Renal disease and transplantation
Rheumatologic disease
Turner's syndrome

[a]Derived from Gordon [30].

B. Disorders Associated with Low Bone Mass

A number of disorders have been associated with low bone mass in children (Table 1).

Skeletal health should be evaluated with diagnosis of these disorders. If skeletal fragility develops, bone fractures can occur with minimal or no trauma, a condition known as osteoporosis. The ability to reverse bone deficits with early diagnosis and treatment remains uncertain.

Amenorrhea most often results from excessive exercise or energy intakes too low to sustain physiologic levels of estrogen. Anorexia nervosa is an eating disorder characterized by intense fear of gaining weight or becoming fat. In this condition body weight is less than 86% of normal for height and age. Reduction in bone gains or acceleration of bone loss resulting in low BMD for age can occur with anorexia nervosa [31]. Osteoporosis that develops as a result of anorexia nervosa is more severe than with other causes of estrogen deficiency. Girls with anorexia nervosa had lower fat and energy intakes than girls without anorexia nervosa [31]. Severity of low bone mass, that is, osteopenia, is worse if the eating disorder is initiated in adolescence than in adulthood and worse for longer duration [32].

Little is known about specific diet therapies and improved bone health in the other disorders. Nutrient recommendations or dietary guidelines have not been established for any of these disorders for lowering risk of low bone mass. It is logical that recovery of low body weight would be helpful to bone. Some interesting associations suggest that future research may show a benefit for nutritional therapies. For example, vitamin K deficiency is prevalent among children with cystic fibrosis [33].

IV. NUTRITION AND DEVELOPMENT OF PEAK BONE MASS

The role of diet in development of peak bone mass is thought to have a great impact on risk of osteoporosis [1]. The formative years set the foundation for the skeletal reserves and for lifelong eating and exercise habits. In fact, osteoporosis has been called a pediatric disease. This chapter approaches the role of diet under two themes: diet patterns and individual nutrients. So much of the research base has focused on individual nutrients that supplementation with calcium and vitamin D has become a first-line strategy of prevention and therapy. Although calcium and vitamin D are the two nutrients important to bone most likely to be deficient, this reductionist approach to research and medical nutrition therapy aimed at building strong bones is woefully inadequate. Dietitians can play a critical role in assessing the overall diet and making recommendations that will have greater impact on body weight and health than simply to recommend supplements. To understand the quality of the evidence behind public health recommendations related to bone health, a discussion of the limitations of our research designs and measures is warranted.

A. Limitation in Methodology

Whether interpreting the effects of diet patterns or individual nutrients on bone, the limitations of investigation of nutritional effects on bone should be appreciated. Difficulty in assessing diet and individual nutrient consumption is not unique to studies of bone and are discussed in another chapter. To quantitatively determine the effect of intake of a nutrient or diet pattern is best accomplished through controlled feeding studies. But to control diets for sufficiently long periods for bone properties to change is not practical except for animal studies. Although changes in bone properties occur faster in pediatric studies than in studies of adults, it still requires a year or several years to evaluate the magnitude of the effect of interventions on bone. During the first 6 or more months following onset of an intervention, changes in bone reflect the bone remodeling transient [34]. This is a period of adaptation through changes in bone formation rate or bone resorption rate if the intervention is effective. Longer periods are required to determine the effect of the intervention on steady-state bone balance. In children, changes in bone are relatively rapid during periods of active growth. The biggest challenge is often to distinguish between the effects of intervention and growth [35].

The traditional outcome measures for assessing nutritional effects on bone properties, that is, BMC and BMD, may be altered by 2% or less per year as a result of a nutritional intervention. This magnitude of change is difficult to detect by DXA in short-term studies. Yet, this level of impact over many years can have a profound effect on risk of fracture. Other changes in bone that can indicate strength, such as bone geometry that can be assessed by three-dimensional imaging, also require substantial intervention periods and, to date, do not offer improved precision over DXA. It is preferable to use BMC rather than BMD as the primary outcome measure to assess efficacy of interventions in children. Effective dietary interventions could augment bone size, rather than merely affecting the mass of bone per unit volume. Thus, because BMD is calculated to adjust for size, it can miss most of the actual change in bone mass during growth. Yet, DXA only captures two-dimensional space, and consequently it is somewhat affected by size [36] and can correlate with changes in BMC. Interventions during growth that influence bone mass and bone size have a dramatic influence on bone strength [11]. A number of investigators have attempted to correct DXA measures for bone size such as calculating bone mineral apparent density, BMC for height, and BMC for bone area, but there is no consensus as to the best approach [37–39]. At present, it is recommended to use BMC as the primary outcome and helpful to also measure bone area and other indicators of bone size. Anthropometric measures can be good predictors of BMC in children [40].

Several shorter-term approaches for assessing perturbations in bone are available. Serum and urine biochemical markers of bone turnover can determine qualitative changes in bone turnover, but within-subject variation is high [41]. Furthermore, biochemical markers of bone turnover do not measure bone mineral, but rather measure fragments of collagen in connective tissue that leak into circulation when bone remodels. Thus, the units cannot be converted into units of bone.

Other short-term methods for analyzing perturbations in bone most often measure calcium as a surrogate for bone. Calcium is a good surrogate for bone because it is present in a fixed proportion of bone mineral. By multiplying bone mineral content determined from DXA by the fraction 0.31, one can derive calcium content in grams. Calcium levels in the blood are not a good indicator of calcium/bone status because they are tightly controlled within a narrow range. Short-term changes in bone balance can be estimated from calcium balance studies. During growth, bone balance is positive. The use of calcium balance studies to determine requirements and response to an intervention has been described [42]. In children, adaptation to a new calcium intake requires about 1 week before balance is determined. Calcium isotopic tracers can be used to measure all components of calcium metabolism [42–44]. Combining calcium balance studies and use of oral and intravenous tracers of calcium can provide the data for compartmental modeling whereby determination rates of transport of calcium in and out of pools and the mass of each pool is possible.

B. Dietary Patterns and Bone Health

When formulating recommendations for food patterns for different subgroups, the 2005 Dietary Guidelines Advisory Committee for Americans considered bone health [45]. The food patterns were designed to meet nutrient recommendations including bone-related nutrients. The evidence for a relationship between milk and milk products and bone health was also reviewed in determining quantities to recommend for optimal health. Beyond getting adequate nutrients and milk products, the committee considered dietary habits detrimental to health, especially with regard to energy excess, but also for bone health. Dietary patterns as they influence acid-base balance may also play a role in bone health.

1. Milk and Milk Products

The Dietary Guidelines for Americans include 2 cups of milk or milk product daily for children aged 2 to 8 years and 3 cups after age 8 [45]. The amount of milk was set to help meet requirements for several nutrients including calcium, magnesium, potassium, riboflavin, and vitamin D. Milk products provide approximately 50–79% of the calcium in the diets for children in the food patterns for age and gender recommended by the Dietary Guidelines Advisory Committee. The milk group also provided more than 10% of the nutrients in the pattern for riboflavin, vitamin B_{12}, vitamin A, thiamin, vitamin B_6, phosphorus, magnesium, zinc, potassium, protein, and carbohydrate. If milk products are excluded from the patterns, intake for calcium falls below 64% for all children and below 33% for young children, and below 88% for magnesium, but as low as 33% for some age groups [45]. Alternatives to milk products given in the Dietary Guidelines Advisory Committee report were low-lactose milk products [45]. Although some fortified foods, such as calcium-fortified soy milk, have nutrient profiles similar to that of milk, it is difficult to meet calcium and potassium requirements without milk. Milk product consumption has been associated with overall diet quality. Adequacy of milk intake has been associated with adequacy of calcium potassium, magnesium, zinc, iron, riboflavin, vitamin A, folate, and vitamin D for children [46].

Many calcium-fortified foods are on the market that could theoretically be used to provide the requirements for this nutrient. Gao *et al.* [47] evaluated the ability of dairy-free diets to meet calcium intake while meeting other nutrient requirements using diets in American children aged 9 to 18 years for those participants in NHANES 2001–2002 who reported no intake of dairy. Calcium requirements were not met without use of calcium-fortified foods and only one child accomplished this. Average calcium intakes without dairy products was 498 mg/day for girls and 480 mg/day for boys compared to 866 and 1070 mg/day, respectively, with dairy products. At calcium intakes of approximately 400 mg/day, calcium retention was only 131 mg/day compared to almost three times that much if the AI for calcium is met [48]. Milk intervention trials in children have shown increased bone mass over control groups [49, 50]. Interestingly, in trials in children using milk as the intervention, the positive effects of treatment were maintained after the intervention ceased [51, 52], in contrast to several trials using calcium supplements as the intervention [53, 54]. In a longitudinal study in 151 white girls, dairy product/calcium intake at age 9 years was associated with total body BMD gain from age 9 to 11 years [55]. Milk avoiders have increased risk for prepubertal bone fractures [56]. Retrospective studies have shown the incidence of postmenopausal fracture is inversely related to drinking milk in childhood [57, 58]. In the nationally representative NHANES data base, the incidence of hip fracture was twice as likely in those who consumed one glass of milk or less per week compared to those who consumed *at least* one serving per day during childhood [57].

Milk consumption in children has declined over time. In the Bogalusa Heart Study, over the two decades from 1972 to 1994 average milk consumption by 10-year-olds declined by an average of 64 g [59]. Fluid milk consumption was negatively correlated with soft drink consumption, which had a detrimental effect on bone gain in girls [60]. The displacement of milk with soft drinks removes a rich package of nutrients from the diet.

2. Plants versus Animal-Based Diets

The mix of animal- and plant-derived foods in the diet of an individual influences two postulated determinants of bone health, acid-base balance and amount of protein. Although dietary protein is a nutrient, because the type of protein influences acid-base balance, dietary protein is discussed in this section on dietary patterns. Intake of fruits and vegetables also influences the acid-base balance.

The role of type and amount of proteins in bone health has been studied for several decades, but little work has been done in children. Alkaline dietary salts contain the cations K^+, Ca^{2+}, and Mg^{2+}, which act as buffers for organic acids produced during metabolism and hepatic oxidation of S-containing amino acids that would otherwise lower blood pH. Increased metabolic acidosis has been associated with increased bone resorption in cell culture systems [61] and increased urinary calcium excretion in humans [62]. Bone is thought to serve as a reservoir of buffering capacity due to the carbonate and hydroxyapatite salts. Typical acid loads produced on a Western diet are on the order of 1 mEq of acid/kg/day. Investigators have attempted to estimate the renal acid load of diets as a measure of acid-base load by taking into account the mineral and protein composition of foods [63]. Thus, diets high in fruits and vegetables that contain potassium and produce an alkaline ash and those richer in plant proteins than animal proteins, which contribute more S-containing amino acids, have been promoted for better bone health through improving acid-base balance. A high ratio of dietary animal to vegetable protein intake has been associated with increased rates of bone loss and increased risk of fracture [64].

The hypothesis that increasing dietary protein or animal versus plant proteins or even acid-base balance influences bone has been challenged. Bonjour [65] argues that bone is unlikely to be the main source of buffering acid loads because bone mineral is not in direct contact with systemic circulation. Rather, buffering of acid loads is accomplished through elimination of carbon dioxide by the lungs and hydrogen ions by the kidney. The increased urinary calcium with increased protein intake generally, or S-containing amino acids specifically, is offset by, and in fact is due to, increased calcium absorption with no increase in bone resorption or net differences in calcium retention [66, 67]. Increased dietary potassium does reduce urinary calcium excretion, but this did not appear to affect calcium balance, because calcium absorption was also reduced as dietary potassium increased [68]. Intake of fruits, vegetables, and herbs does inhibit bone resorption, but this effect is independent of their alkali or potassium contributions [69].

There is some evidence that adequate dietary protein promotes bone accrual in children. In an 18-month randomized controlled trial (RCT) of a pint of milk a day in 12-year-old girls, an increase in total body BMD was associated with an increase in serum IGF-1 [70]. The relation between protein intake and bone gains in lumbar spine and femoral neck in 193 subjects aged 9 to 19 years was positive, and particularly in prepubertal children [71]. An interesting hypothesis has been put forth that aromatic amino acid intake may induce an increase in calcium absorption and serum IGF-1 compared with branched-chain amino acids through activating calcium sensor receptors in the gut to increase gastric acid production [72].

3. Salt

Dietary salt is the largest dietary predictor of urinary calcium excretion [73]. However, the response of adolescents to dietary salt is racially dependent. In controlled feeding studies using a crossover design with high (4 g, 172 mmol) and low (1.3 g, 5.7 mmol) sodium diets, white adolescent girls excreted more sodium on a high-salt diet than did black girls of matched weight and sexual maturity [74]. Because calcium is excreted with sodium through the shared transport proteins in the kidney, high-salt diets resulted in more calcium excretion and lower calcium retention in white girls than black girls in the same study [75]. Thus, a high intake of dietary salt is detrimental to growing bone, but the consequences to bone are greater for white than black individuals.

C. Individual Nutrients and Bone Health

Calcium is by far the most studied single nutrient related to bone health. Vitamin D is the focus of much current research. These two nutrients are the most likely ones of those important to bone health to be deficient. It should be understood that because bone is a living tissue, all essential nutrients are required for bone growth.

1. Calcium

a. Current Recommendations and Basis for Requirements

Current calcium recommendations from the Institute of Medicine's Food and Nutrition Board established in 1997 are in the form of Adequate Intakes (AI) [76]. AIs for children are given in Table 2.

TABLE 2 Dietary Reference Intakes for Bone-Related Nutrients in Children and Adolescents

	Nutrient							
	Calcium (mg/day)		Vitamin D (μg/day)		Phosphorus (mg/day)		Magnesium	
Life-Stage Group	AI	UL	AI	UL	RDA	UL	RDA	UL
(0–6 mo)	210	ND	5	25			30[a]	ND
(7–12 mo)	270	ND	5	25			75[a]	
(1–3 y)	500	2500	5	50	380	3000	80	65[b]
(4–8 y)	800	2500	5	50	405	3000	130	110[b]
9 through 13 y	1300	2500	5	50	1055	4000	240	350[b]
14 through 18 y	1300	2500	5	50	1055	4000		
(females)							360	350[b]
(males)							410	350[b]
Pregnancy ≤ 18 y	1300	2500	5	50	1055	3500	400	350[b]
Lactation ≤ 18 y	1300	2500	5	50	1055	4000	360	350[b]

[a]AI, Adequate Intake.
[b]Supplementary, not from food.

This was the first time calcium recommendations were not determined primarily on the basis of replacement of losses through urine, stools, and skin adjusted for absorption (the factorial method). The Panel on Calcium and Other Bone-Related Nutrients emphasized using the calcium intake associated with maximal calcium retention for population categories for which appropriate data were available. The panel reasoned that because bone mass is directly related to fracture risk and because calcium is a constant fraction of bone, it would be desirable to optimize calcium retention, that is, bone accretion during growth. This was a new approach so an AI was set rather than a Recommended Dietary Allowance (RDA). A group of experts, several of whom participated in the 1997 panel, suggested that it is time to move from AIs to RDAs for calcium [77].

Determining calcium intakes for maximal calcium retention requires studies on a range of calcium intakes. The RCTs described next studied only two levels of calcium intake, that of the self-selected diet or one's dietary intake plus the calcium intervention source. One can determine whether calcium supplementation is effective with this study design, but it is not possible to determine an optimal intake. It would be desirable to have bone accretion studies on a range of calcium intakes in different age groups, but this would require large, expensive studies. Instead, data on calcium retention at different calcium intakes are available from short-term controlled-feeding studies where composites of diet and complete urine and stool collections are analyzed for calcium so that calcium retention or balance (intake − loss from urine, stools, and sometimes sweat) can be calculated. As calcium intakes increase, calcium retention increases until a plateau intake when further calcium intakes are excreted. Data on intakes for maximal calcium retention were available for children aged 9–18 years and adolescents [6, 78, 79].

b. Newer Evidence

The data in adolescents for calcium intakes for maximal retention were in white girls [6]. Subsequently, calcium intakes for maximal retention have been reported in adolescent boys [80]. A comparison of calcium retention as a function of intake between boys and girls is shown in Figure 6 Although boys retained more calcium at a given level of calcium intake than girls, the calcium intake where retention was not further significantly increased was not different by gender. Similarly, black girls had higher calcium retention across a range in calcium intakes without different slopes [81]. Thus, boys are more efficient in using calcium than girls and blacks are more efficient than whites, which results in higher bone mass as adults, but the need for calcium to produce skeletons of greater bone mass is not detectably different over short periods. Asian girls are also more efficient than white girls in retaining calcium across a range of calcium intakes [82].

The relation between calcium intake and retention has been reported for children age 1–4 years [83]. Although a threshold intake was not found, expected growth was achieved at calcium intakes of about 470 mg/day, suggesting that the current AI of 500 mg/day is adequate.

c. Evidence for Relationship to Bone

There is quite a body of evidence from RCTs of calcium intake in children that have reported positive effects on one or more skeletal sites (Table 3). Intervention studies have used calcium salts, calcium-fortified foods, and milk with a variety of bone outcomes. Cheese was more effective at increasing tibial cortical thickness than calcium carbonate [95]. A meta-analysis of 19 RCTs reported that the effect of calcium supplementation on bone was modest and effective only on upper limbs [96]. However, this meta-analysis was

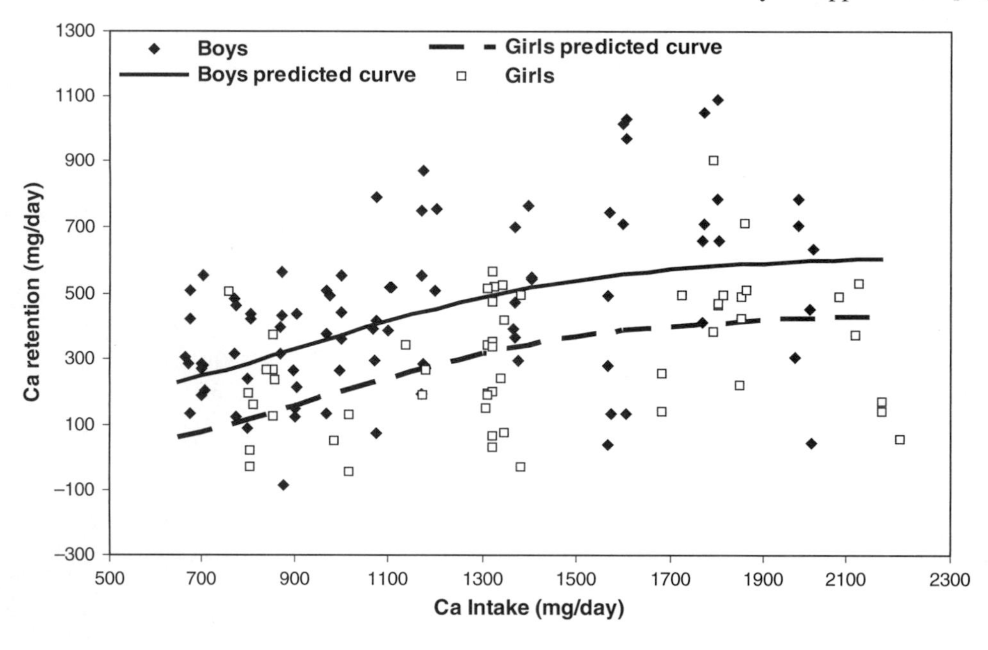

FIGURE 6 Calcium retention as a function of calcium intake in adolescent boys (upper curve) and girls (lower curve) [80].

TABLE 3 Differences in Mean Changes in Bone Mineral Content and Bone Mineral Density in Calcium Treated versus Placebo Groups in Randomized, Controlled Trials in Children

Source	Ref. No.	Subj. No.	Age (y)	Sex	Race/ Location	Length Study (mo)	Calcium Intake Controls (mg/day)	Calcium Intake Treatment (mg/day)	Site	Measure	Group Mean Increase	
											Treatment (T)	Placebo (P)
Johnson *et al.*, 1992	84	140 twins	6–14	F/M	White, IN	36	908	1612	Midshaft radius	BMD	17.7%	15.2%
									Distal radius	BMD	21.5%	18.2%
									Lumbar spine	BMD	20.1%	19.5%
									Femoral neck	BMD	15.3%	14.9%
									Ward's triangle	BMD	15.4%	14.2%
									Greater trochanter	BMD	18.1%	17.11%
Lloyd *et al.*, 1993	85	94	11.9 ± 0.5	F	White, PA	18	960	1314	Total body	BMC	9.6%	8.3%, $p = 0.003$
									Spine	BMC	2.9%, $p = 0.03$	4.7%, $p = 0.05$
Chan *et al.*, 1995	49	48	9–13	F	White, UT	12	728	1437	Total body	BMC	14.2 ± 7.0%	7.6 ± 6.0%, $p < 0.001$
									Lumbar spine	BMD	22.8 ± 6.9%	12.9 ± 8.3%, $p < 0.001$
Lee *et al.*, 1994	86	109	7	F/M	Asian, China	18	571	1363	Distal radius	BMC	15.92 (T) versus 14.95% (P) gain, $p = 0.53$	
										Area	7.74 (T) versus 6.00% (P) gain, $p = 0.081$	
									Lumbar spine	BMC	20.92 (T) versus 16.34% (P) gain, $p = 0.035$	
										Area	11.16 (T) versus 8.71% (P) gain, $p = 0.049$	
									Proximal femoral neck	BMC	24.19 (T) versus 23.42% (P) gain, $p = 0.37$	
Cadogan *et al.*, 1997	70	82	12.2	F	White, Sheffield UK	18	753	1125	Total body	BMD	9.6 (T) versus 8.5% (P), $p = 0.017$	
										BMC	27.0% (T) versus 24.1% (P) $p = 0.009$	
Bonjour *et al.*, 1997	50	149	7.9 ± 0.1	F	White, Geneva, CH	12	916	1723	Radial metaphysics	BMD	16 ± 3 g/cm^2 (T) versus 9 ± 2 g/cm^2 (P), $p < 0.08$	
Nowson *et al.*, 1997		87	10–17	F	White, Australia	18	692	>1600	Spine	BMD	1.62 ± 0.84% (TisP)	

(continues)

TABLE 3 *(continued)*

Source	Ref. No.	Subj. No.	Age (y)	Sex	Race/ Location	Length Study (mo)	Calcium Intake Controls (mg/day)	Calcium Intake Treatment (mg/day)	Site	Measure	Group Mean Increase	
											Treatment (T)	Placebo (P)
									Hip	BMD	NS	
									Femoral neck	BMD	22 ± 4 (T) versus 13 ± 4 g/cm^2 (P)	
									Femoral diaphysis	BMD	66 ± 3 (T) versus 54 ± 4 g/cm^2 (P) $p < 0.01$	
									Lumbar spine	BMD	25 ± 3 (T) versus 23 ± 3 g/cm^2 (P)	
Dibba *et al.*, 2000	88	160	8.3–11.9	F/M	Black, Gambia	12	342	1056	Midshaft radius	BMC	3.0 ± 1.4 (T-P), $p = 0.034$	
										BMD	4.5 ± 0.9 (T-P), $p < 0.0001$	
Moyer-Mileur *et al.*, 2003	89	71	12	F	White, UT	12	865	1524	pQCT of distal tibia	BMC	4.1(T) versus 1.6% (P), $p < 0.006$	
										vBMD	1.0 (T) versus –2.0% (P), $p < 0.006$	
Rozen *et al.*, 2003	90	100	14 ± 0.5	F	85 Jewish/26 Arab, Haifa, Israel	12	480	1110	Total body	BMC	4.63 (T) versus 4.65, NS	
										BMD	3.80 (T) versus 3.07, (P), $p < 0.05$	
									Lumbar spine	BMD	4.52 (T) versus 3.95 (P), NS	
										BMC	$3.66 \pm 0.35\%$ versus $3.00 \pm 0.43\%$, $p < 0.05$	
									Femoral neck	BMC	4.30 (T) versus 3.00, NS	
										BMD	2.00 (T) versus 1.39, (P), NS	
Stear *et al.*, 2003	91	144	17.3 ± 0.3	F	Cambridge, UK	15.5	938	+1 g Ca supple ment	Whole body	BMC	$0.8 \pm 0.5\%$ (T versus P), $p < 0.01$	
									Lumbar spine	BMC	$0.9 \pm 0.8\%$ (T versus P), $p < 0.001$	
									Hip	BMC	$5.3 \pm 1.0\%$ (T versus P), $p < 0.001$	
Cameron *et al.*, 2004	92	102	10.3 ± 1.5	F	White, Australia	2.4	715	+1200	Total body, lumbar spine, femoral neck, total hip	BMD	NS	

TABLE 3 *(continued)*

Source	Ref. No.	Subj. No.	Age (y)	Sex	Race/ Location	Length Study (mo)	Calcium Intake Controls (mg/day)	Calcium Intake Treatment (mg/day)	Site	Measure	Group Mean Increase	
											Treatment (T)	Placebo (P)
Mølgaard *et al.*, 2004	93	113	12–14	F	White, Denmark	12	2 groups, 1717/ 1320	2 groups 953/660	Whole body	BMC	0.8% (T versus P), $p = 0.049$	
											0.5% (T versus P), $p = 0.08$	
Prentice *et al.*, 2005	94	143	16–18	M	White, UK	13	1283	1858	Whole body	BMC	1.3% (T versus P), $p = 0.02$	
									Lumbar spine	BMC	2.5% (T versus P), $p = 0.004$	
									Total lumbar spine	Bone area	1.5% (T versus P), $p = 0.0003$	
									Total hip	BMC	23% (T versus P), $p = 0.01$	
									Femoral neck	BMC	2.4% (T versus P), $p = 0.01$	
									Intertrochanter	BMC	2.7% (T versus P), $p = 0.01$	
Matkovic *et al.*, 2005	10	354	10.9 ± 0.9	F	White, Ohio	48	830	1498	Whole body	BMD	0.215 versus 0.204 g/cm^2, $p < 0.0006$	
									Distal radius	BMD	0.106 versus 0.092 g/cm^2, $p < 0.0026$	
						84			Whole body	BMD	0.268 versus 0.263 g/cm^2, $p < 0.0006$	
									Distal radius	BMD	0.171 versus 0.165 g/cm^2, $p < 0.0006$	
Cheng *et al.*, 2005	95	195	10–12	F	White, Finland	24	1566	671	Whole body	BMC	34.7%	35.5%
									Femoral neck	BMC	24.0%	22.4%
									Total femur	BMC	33.6%	33.6%
									Spine	BMC	46.9%	47.0%
									Tibia cortical thickness	pQCT	31.7%	31.1%

criticized [97] for use of BMD as the outcome for evaluation rather than BMC for reasons already discussed. The authors rebutted that BMD correlates reasonably well with fracture risk in children [98].

The prudence of retaining a high calcium intake during growth for optimizing peak bone mass versus the risk of focusing too much on calcium, at least in children who have *at least* 800 mg/day, rather than on a more holistic evaluation of the many factors that affect bone, has been discussed [77]. The limitations of our methodologies contribute to lack of clarity, but several observations argue in support of the importance of adequate dietary calcium and bone development during childhood. The retrospective studies showing a benefit of childhood consumption of milk reducing fracture incidence later in life is one example. Also, adequate dietary calcium can have a positive interaction with exercise in young children and adolescents. In one study, the presence of both mechanical loading and adequate calcium was required for increased BMD [99]. In other studies, bone size was increased by calcium alone [50, 94], which indicates greater bone strength. The effects of dietary calcium on bone quality or interaction with physical activity were not usually considered in the RCTs evaluated using BMD in the meta-analysis by Winzenberg *et al.* [96].

A new theory for the influence of dietary calcium on bone through altering the timing of menarche was recently proposed [100]. Evidence for this possibility came from the RCT of Bonjour *et al.* [50]. The initial trial was a 1-year intervention of calcium-fortified foods in 7.9-year-old girls that showed a positive effect of calcium on mean BMD changes in six skeletal sites. The cohort was followed through age 16.4 years, which allowed determination of the age of menarche and subsequent changes in bone. Interestingly, those girls assigned to the calcium intervention for 1 year began menarche earlier than girls assigned to the placebo group [100]. The extra time of exposure to estrogen resulted in more bone gain in subsequent years even though the calcium intervention was discontinued.

d. Dietary Calcium

Calcium intake in most diets is predominantly from dairy products, that is, 75% of calcium intake in the early 1970s and 72% in 2000 [45]. The particular dairy product distribution varies somewhat with race and ethnicity [101]. The discrepancy between calcium intakes and recommendations is a large gap after age 11 years. This is a common problem. Looker [102] compiled calcium recommendations and intakes across the lifespan for 20 countries. The percent of adolescents meeting country-specific recommendations was just over 60% for males and approximately 50% for females. Most children over age 8 years in the United States have calcium intakes below recommended levels. Only 6% of girls and 28% of boys aged 9 to 13 years and 9% of girls and 31% of boys aged 14 to 18 years have calcium intakes above the recommended intakes [103].

It is important to reach children early to establish and maintain dietary habits that will ensure lifelong adequate calcium intakes. There is considerable tracking of calcium intake. In a 15-year longitudinal study begun as adolescents, tracking of calcium intake resulted in correlations of $r = 0.43$ for males and $r = 0.38$ for females [104]. Calcium intake is also a marker for intake of other nutrients [45]. This would not be true for individuals who consume calcium in the form of supplements.

Calcium absorption from dairy products is not dependent on the fat content or flavoring. A comparison of calcium bioavailability and the number of servings of various foods required to replace a cup of milk or yogurt on the basis of absorbed calcium is given in Table 4.

For children aged 4–8 years, the equivalent of 2.6 cups of milk would be required to meet the AI and for children above age 9 years, it would be 4.3 cups. Other foods typically provide the equivalent of at least two-thirds of a cup of milk, so milk is not expected to provide all of the calcium requirements.

e. Safety of High Intakes

The Upper Level for calcium is 2500 mg/day [76]. The primary concern for excessive calcium intake from supplements in adults is kidney stones. Interactions with trace mineral absorption are also a concern. Dose response safety data are not available in children aged 1 through 18.

2. Vitamin D

a. Recommendations, Dietary Sources and Intakes, and Status

The AI for children for vitamin D is given in Table 2. The American Academy of Pediatrics supports these recommendations, but recommends supplemental vitamin D of 5 µg/day to infants not ingesting fortified formula [106]. As discussed in Chapter 43 of this book, the 1997 Dietary Reference Intake values for vitamin D are undergoing revision. The amount of new evidence bearing on inputs for optimal health and safety has been remarkable. Yet, the strength of the new evidence is largely in adults and we have much to learn in children.

Vitamin D across the lifespan and dietary sources of vitamin D are described in the chapter devoted to vitamin D (Chapter 43). Dietary sources of vitamin D are limited and mostly consumed as fortified foods, especially milk. A summary of studies on vitamin D status in children and adolescents worldwide was recently compiled by El-Hajj Fuleihan [107]. The range in mean serum 25-hydroxyvitamin D levels (the best status indicator for vitamin D) was large: 13–142 nmol/liter. Until the optimal status of vitamin D is established, it is difficult to know how widespread vitamin D deficiency is, but nearly everyone would agree that means

TABLE 4 Comparing Sources for Absorbable Calcium

Source	Serving Size (g)	Calcium Content (mg/serving)	Estimated Absorption Efficiency (%)	Food Amount to Equal Calcium in 1 C milk
Milk	240	300	32.1	1.0 C
Beans, red	172	40.5	24.4	4.8 C
Beans, white	110	113	21.8	2.0 C
Bok choy	85	79	53.8	1.2 C
Broccoli	71	35	61.3	2.3 C
Cheddar cheese	42	303	32.1	1.5 oz
Cheese food	42	241	32.1	1.8 oz
Chinese cabbage flower leaves	85	239	39.6	0.5 C
Chinese mustard green	85	212	40.2	0.6 C
Chinese spinach	85	347	8.36	1.7 C
Kale	85	61	49.3	1.6 C
Orange juice w/Ca citrate maleate	240	300	36.3	0.9 C
Soy milk w/calcium phosphate	240	300	24.0	1.3 C
Spinach	85	115	5.1	8.1 C
Tofu, calcium set	126	258	31.0	0.6 C
Whole wheat bread	28	20	82.0	5.8 slices
Wheat bran cereal	28	20	38.0	12.8 oz
Yogurt	240	300	21.1	1.0 C

Source: Taken from [105]. With kind permission of Springer Science and Business Media.

less than 50 nmol/liter is suboptimal in 14 of the 20 studies, at least in winter months.

b. Evidence for a Relationship to Bone

Vitamin D deficiency in young children has long been associated with rickets. An intake of 2.5 μg (100 IU)/day is thought adequate to prevent rickets [108, 109]. This disorder is described by Whiting *et al.* in Chapter 43.

The vitamin D/PTH homeostatic regulatory axis helps regulate serum calcium levels in response to dietary calcium. Under conditions of low calcium intake, serum calcium levels fall, PTH is released, and vitamin D is converted to its active form, 1,25-dihydroxyvitamin D. Serum calcium levels rise as intestinal calcium absorption is upregulated, renal reabsorption of calcium increases, which conserves urinary excretion of calcium, and bone resorption increases. In adults, vitamin D status, measured as serum 25-hydroxyvitamin D, directly influences calcium absorption efficiency [110]. But there are no vitamin D intervention trials in children that have measured calcium absorption to determine what levels of vitamin D intake or status to recommend to optimize calcium absorption efficiency. Levels of oral vitamin D necessary for optimal calcium utilization would undoubtedly be influenced by inputs of vitamin D through both subcutaneous synthesis from sunlight exposure and dietary calcium levels. It is a difficult question to answer.

There have been three RCTs of vitamin D in children with bone outcomes (see Table 5). The results are quite varied among the studies. As with calcium absorption, vitamin D and calcium intakes would be interdependent. Additionally, response to vitamin D levels used in the RCT would only be effective if doses were adequate and beginning vitamin D status of the subjects sufficiently inadequate so that supplementation would result in a substantial change in vitamin D status. Short-term dose-response studies of vitamin D and calcium on calcium absorption and excretion and vitamin D status in children could inform RCTs with bone outcomes. One RCT of two doses of vitamin D equivalent to 200 IU and 2000 IU/day in 10- to 17-year-old Lebanese girls was able to improve vitamin D status on the higher dose and improve total hip BMC and area, but not other bone measures [111]. Dietary calcium intakes were low on that study. Lean mass was increased with vitamin D supplementation. Subjects in both the Cheng *et al.* [95] study and the Viljakainen et al. study [112] had adequate calcium together with vitamin D.

c. Safety of High Doses

Dose response studies with higher levels of longer duration than have been reported would be needed to truly determine upper levels of safety studies that risk intoxication are unmentioned. The current UL is given in Table 2.

Safety is assessed by observing and no change in serum or urinary calcium levels. The 5-μg dose of vitamin D was ineffective in the Cheng *et al.* [95] study, but effective for the femur in Viljakainen *et al.* [112] study, but only when a compliance-based analysis was used. That study showed a dose-response change in vitamin D. Single oral doses of vitamin D have been given with no change in mean serum or urinary Ca/Cr ratios [113].

TABLE 5 Differences in Mean Changes in Bone Mineral Content and Bone Mineral Density in Vitamin D versus Placebo in Randomized, Controlled Trials in Children

Source	Ref. No.	Subj. No.	Age (y)	Sex	Race/ Location	Length Study (mo)	Vitamin D Supplement (μg/ day)	Site	Measure	Group Mean % Increase	
										Treatment (T)	Placebo (P)
Cheng *et al.*, 2005	95	195	10–12	F	White, Finland	24	5	Whole body	BMC	34.7	35.0, NS
								Femoral neck	BMC	24.0	22.4, NS
								Total femur	BMC	33.6	33.6, NS
								Spine	BMC	46.9	47.0, NS
								Tibia cortical thickness	pQCT	31.7	31.1, NS
El-Hajj Fuleihan *et al.*, 2006	111	179	10–17	F	White, Lebanese	24	5	Total body	BMC	11.3	8.7, NS
								Total hip	BMC	11.2	7.8, NS
								Total hip	Area	4.0	2.4, $p = 0.05$
								Femoral neck	BMC	4.4	3.9, NS
								Femoral neck	Area	0.03	0.7, NS
								Trochanter	BMC	13.6	9.4, NS
								Trochanter	Area	6.8	4.7, NS
								Total body	BMC	12.0	8.7, NS
								Total hip	BMC	12.8	7.8, $p = 0.005$
								Total hip	Area	5.7	2.4, $p = 0.001$
								Femoral neck	BMC	5.2	3.9, NS
								Femoral neck	Area	0.8	0.7, NS
								Trochanter	BMC	14.2	9.4, NS
								Trochanter	Area	7.8	4.7, NS
Viljakainen *et al.*, 2006[a]	112	212	11.4 ± 0.4	F	White, Finland	12	5	Femur	BMC	14.3 (T versus P), $p = 0.012$	
								Lumbar spine	BMC	NS	
							10	Femur	BMC	17.2 (T versus P), $p = 0.012$	
								Lumbar spine	BMC	12.5 (T versus P), $p = 0.039$	

[a]Analysis included only those >80% compliant.

In the El Hajj Fuleihan study [111], adolescents given 14,000 IU (35 μg/week) of vitamin D_3 for 1 year showed no change in mean serum calcium level. This is equivalent on a daily dose to the UL. We do not know safety of much higher doses in children.

3. Other Nutrients

Although many other nutrients are necessary for growing bone, there is little evidence that current intakes compromise development of peak bone mass for most individuals. Two minerals besides calcium that comprise a substantial protein of bone mineral are phosphorus and magnesium. Bone mineral is calcium phosphate but 60% of the magnesium in the body resides in bone. RDAs and ULs for these nutrients for children are given in Table 2. The dietary Ca/P ratio affects bone mineralization and turnover through intestinal calcium and phosphorus transports [114]. Phosphorus is clearly essential for bone acquisition but deficiency has not been a concern for children. Excessive intakes of phosphorus in soft drinks have been a concern. However, as discussed earlier, the negative association of soft drink consumption and bone in girls is likely due to displacement of milk as a beverage.

Magnesium deficiency disrupts bone accretion. Rats fed 50% of their requirement for magnesium have structural changes that lead to reduced bone volume [115]. Obtaining recommended intakes of magnesium is not of concern for children. Iron deficiency also has a detrimental effect on bone morphology in growing animals that is exacerbated by calcium deficiency [116].

V. CONCLUSION

Making wise nutritional choices during growth is a window of opportunity to build optimal peak bone mass to reduce risk of fracture later in life. If the opportunity is neglected, the consequence can be fracture. During infancy, it is not as difficult to meet requirements through breast-feeding or infant formulas. The other accelerated growth period, puberty, is much more difficult. Diets vary widely in nutrient sufficiency; peers may have more influence than caregivers. Fracture incidence in childhood is highest during the pubertal growth spurt when bones elongate before they consolidate and bone mineral density is lower. The incidence of pubertal fractures is increasing, possibly related to the increase in obesity coupled with a decline in milk as the beverage of choice. Diet patterns may be as important in building strong bones as adequacy of individual nutrients. The Dietary Guidelines offer a good plan. Supplements may be useful for some individuals and in some conditions.

References

1. Heaney, R. P., Abrams, S., Dawson-Hughes, B., Looker, A., Marcus, R., Matkovic, V., and Weaver, C. M. (2000). Peak bone mass. *Osteoporosis Int.* **11**, 985–1009.
2. Bonjour, J.-P., and Chevalley, T. (2007). Pubertal timing, peak bone mass and fragility fracture risk. *Bone Key-Osteovision* **4**(2), 30–48.
3. Krall, E. A., and Dawson-Hughes, B. (1993). Heritable and lifestyle determinants of bone mineral density. *J. Bone Miner. Res.* **8**, 1–9.
4. Leib, E. S. (2004). Writing Group for the ISCD Position Development Conference Diagnosis of osteoporosis in men, premenopausal women and children. *J. Clin. Densitom.* **7**, 17–26.
5. Recker, R. R., Davies, M., Hinders, S. M., Heaney, R. P., Stegman, M. R., and Kimmel, D. B. (1992). Bone gain in young adult women. *JAMA* **268**, 2403–2408.
6. Jackman, L. A., Millane, S. S., Martin, B. R., Wood, O. B., McCabe, G. P., Peacock, M., and Weaver, C. M. (1997). Calcium retention in relation to calcium intake and postmenarcheal age in adolescent females. *Am. J. Clin. Nutr.* **66**, 327–333.
7. Teegarden, D., Proulx, W. R., Martin, B. R., Zhao, J., McCabe, G. P., Lyle, R. M., Peacock, M., Slemenda, C., Johnston, C. C., and Weaver, C. M. (1995). Peak bone mass in young women. *J. Bone Miner. Res.* **10**(5), 711–715.
8. Lin, Y.-C., Lyle, R. M., Weaver, C. M., McCabe, L. D., McCabe, G. P., Johnston, C. C., and Teegarden, D. (2003). Peak spine and femoral neck bone mass in young women. *Bone* **32**(5), 546–553.
9. Gafni, R. I., and Baron, J. (2000). Catch-up growth: Possible mechanisms. *Pediatr. Nephrol.* **14**, 616–619.
10. Matkovic, V., Goel, P. K., Badenkop-Stevens, N. E., *et al.* (2005). Calcium supplementation and bone mineral density in females from childhood to young adulthood: A randomized controlled trial. *Am. J. Clin. Nutr.* **81**, 175–188.
11. Kalkwarf, H. J. (2006). Forearm fractures in children and adolescents. *Nutr. Today* **41**(4), 171–177.
12. Goulding, A., Cannan, R., Williams, S. M., Gold, E. J., Taylor, R. W., and Lewis-Barned, N. J. (1998). Bone mineral density in girls with forearm fractures. *J. Bone Miner. Res.* **13**, 143–148.
13. Bailey, D. A., McKay, H. A., Mirwald, R. L., *et al.* (1999). A six-year longitudinal study of the relationship of physical activity to bone mineral accrual in growing children: The University of Saskatchewan Bone Mineral Accrual Study. *J. Bone Miner. Res.* **14**, 1672–1679.
14. Khosla, S., Melton, I. J., III, Delatoski, M. B., *et al.* (2003). Incidence of childhood distal forearm fractures over 30 years. *JAMA* **290**, 1479–1485.
15. Bailey, D. A., Martin, A. D., McKay, A. A., Whiting, S., and Miriwald, R. (2000). Calcium accretion in girls and boys during puberty: A longitudinal analysis. *J. Bone Miner. Res.* **15**, 2245–2250.
16. Jones, I. E., Williams, S. M., Dow, N., and Goulding, A. (2002). How many children remain fracture-free during growth? A longitudinal study of children and adolescents participating in Dunedin Multidisciplinary Health and Development Study. *Osteoporosis Int.* **13**, 990–995.

17. Heaney, R. P., and Weaver, C. M. (2005). Newer perspectives on calcium and bone quality. *J. Am. Coll. Nutr.* **24**(6), 574S–581S.
18. Flegal, K. M., Carroll, M. D., Ogden, C. L., and Johnson, C. L. (2002). Prevalence and trends in obesity among U.S. adults, 1999–2000. *JAMA* **288**, 1723–1727.
19. Ogden, C. L., Carroll, M. D., Curtin, L. R., McDowell, M. A., Tabak, C. J., and Flegal, K. M. (2006). Prevalence of overweight and obesity in the United States, 1999–2004. *JAMA* **295**, 1549–1555.
20. Leonard, M. B., Shults, J., Wilson, B. A., Tershakovec, A. M., and Zemel, B. S. (2004). Obesity during childhood and adolescence augments bone mass and bone dimensions. *Am. J. Clin. Nutr.* **80**, 514–523.
21. Skaggs, D. L., Loro, M. L., Pitukcheewanont, P., Tolo, V., and Gilsanz, V. (2001). Increased body weight and decreased radial cross-sectional dimensions in girls with forearm fractures. *J. Bone Miner. Res.* **16**, 1337–1342.
22. Goulding, A., Jones, I. E., Taylor, R. W., Williams, S. M., and Manning, P. J. (2001). Bone mineral density and body composition in boys with distal forearm fractures: A dual-energy x-ray absorptiometry study. *J. Pediatr.* **139**, 509–515.
23. Goulding, A., Taylor, R. W., Jones, I. E., McAuley, K. A., Manning, P. J., and Williams, S. M. (2000). Overweight and obese children have low bone mass and area for their weight. *Int. J. Obes. Relat. Metab. Disord.* **24**, 627–632.
24. Goulding, A., Jones, I. E., Taylor, R. W., Manning, P. J., and Williams, S. M. (2000). More broken bones: A 4-year double cohort study of young girls with and without distal forearm fractures. *J. Bone Miner. Res.* **15**, 2011–2018.
25. Goulding, A., Taylor, R. W., Jones, I. E., Manning, P. J., and Williams, S. M. (2002). Spinal overload: A concern for obese children and adolescents? *Osteporos. Int.* **13**, 835–840.
26. Goulding, A., Grant, A. M., and Williams, S. M. (2005). Bone and body composition of children and adolescents with repeated forearm fractures. *J. Bone Miner. Res.* **20**, 2090–2096.
27. Taylor, E. D., Theim, K. R., Mirch, M. C., Ghorbani, S., Tanofsky-Draff, M., Adler-Wailers, D. C., Brady, S., Reynolds, J. C., Cals, K. A., and Yanovski, J. A. (2006). Orthopedic complications of overweight in children and adolescents. *Pediatrics* **117**, 2167–2174.
28. Pollock, N. K., Laing, E. M., Baile, C. A., Hamrick, M. W., Hall, D. B., and Lewis, R. D. (2007). Is adiposity advantageous for bone strength? A peripheral quantitative computed tomography study in late adolescent females. *Am. J. Clin Nutr.* **86**, 1530–1538.
29. Seeman, E. (1997). From density to structure growing up and growing old on the surfaces of bone. *J. Bone Miner. Res.* **12**, 509–521.
30. Gordon, C. M. (2005). Evaluation of bone density in children. *Curr. Opin. Endocrinol. Diabetes* **12**, 444–451.
31. Misra, M., Tsai, P., Anderson, E. J., *et al.* (2006). Nutrient intake in community-dwelling adolescent girls with anorexia nervosa in healthy adolescents. *Am. J. Clin. Nutr.* **84**, 698–706.
32. Biller, B. M. K., Caughlin, J. F., Sake, V., Schoenfeld, D., Spratt, D. I., and Klitanski, A. (1991). Osteopenia in women with hypothalamic amenorrhea: A prospective study. *Obstet. Gynecol.* **78**, 996–1001.
33. Conway, S. P., Wolfe, S. P., and Brownie, K. G. (2005). Vitamin K status among children with cystic fibrosis and its relationship to bone mineral density and bone turnover. *Pediatrics* **115**, 1325–1331.
34. Heaney, R. P. (1994). The bone remodeling transient: Implications for the interpretation of clinical studies of bone mass change. *J. Bone Miner. Res.* **9**, 1515–1523.
35. Sawyer, A. J., Bachrach, L. K., and Fung, E. B. (Eds.). (2007). "Bone Densitometry in Growing Patients: Guidelines for Clinical Practices." Humana Press, Totowa, NJ.
36. Seeman, E. (1998). Editorial: Growth in bone mass and size—Are racial and gender differences in bone mineral density more apparent than real? *J. Clin. Endocrinol. Metab.* **83**, 1414–1419.
37. Katzman, D. K., Bachrach, L. K., Carter, D. R., and Marcus, R. (1991). Clinical and anthropometric correlates of bone mineral acquisition in healthy adolescent girls. *J. Clin. Endocrinol. Metab.* **73**, 1332–1339.
38. Carter, D. R., Bouxsein, M. L., and Marcus, R. (1992). New approaches for interpreting projected bone density data. *J. Bone Miner. Res.* **7**, 137–145.
39. Leonard, M. B., Shults, J., Elliott, D. M., Stallings, V. A., and Zemel, B. S. (2004). Interpretation of whole body dual energy x-ray absorptiometry measures in children: Comparison with peripheral quantitative computed tomography. *Bone* **34**, 1044–1052.
40. Weaver, C. M., McCabe, L. D., McCabe, G. P., Novotny, R., Van Loan, M., Going, S., Matkovic, V., Boushey, C., Savaiano, D. A., and the ACT research team. (2007). Bone mineral and predictors of bone mass in white, Hispanic, and Asian early pubertal girls. *Calcified Tissue Intl.* (in press).
41. Gundberg, C. M., Looker, A. C., Nieman, S. D., and Calvo, M. S. (2002). Patterns of osteocalcin and bone specific alkaline phosphatase by age, gender, and race on ethnicity. *Bone* **31**, 707–708.
42. Weaver, C. M. (2006). "Clinical approaches for studying calcium metabolism and its relationship to disease. *In* "Calcium in Human Health" (C. M. Weaver, and R. P. Heaney, Eds.), pp. 65–81. Humana Press, Totowa, NJ.
43. Wastney, M. E., Zhao, Y, and Weaver, C. M. (2006). Kinetic studies. *In* "Calcium in Human Health" (C. M. Weaver and R. P. Heaney, Eds.), pp. 83–93. Humana Press, Totowa, NJ.
44. Weaver, C. M., Rothwell, A. P., and Wood, K. V. (2006). Measuring calcium absorption and utilization in humans. *Curr. Opin. Clin. Nutr. Metab. Care* **9**, 568–574.
45. U.S. Department of Health and Human Services and U.S. Department of Agriculture. (2005). "Dietary Guidelines for Americans, 2005," 6th ed. U.S. Government Printing Office, Washington, DC. Retrieved September 12, 2006, from http://www.healthierus.gov/dietary guidelines
46. Ballow, C., Kuester, S., and Gillespie, C. (2000). Beverage choices affect adequacy of children's nutrient intakes. *Arch. Pediatr. Adolesc. Med.* **154**, 1148–1152.
47. Gao, X., Wilde, P. E., Lichtenstein, A. H., and Tucker, K. L. (2006). Meeting adequate intake for dietary calcium without dairy foods in adolescents aged 9 to 18 years (National Health and Nutrition Examination Survey 2001–2002). *J. Am. Diet. Assoc.* **106**, 1759–1765.
48. Abrams, S. A., Griffin, L. J., Hicks, P. D., and Gunn, S. K. (2004). Pubertal girls only partially adapt to low dietary calcium intakes. *J. Bone Miner. Res.* **19**, 759–763.

49. Chan, G. M., Hoffman, K., and McMurry, M. (1995). Effects of dairy products on bone and body composition in pubertal girls. *J. Pediatr.* **126**(4), 551–556.
50. Bonjour, J. P., Carrie, A. L., Ferrari, S., Clavien, H., Slosman, D., Thientz, G., and Rizzoli, R. (1997). Calcium-enriched foods and bone mass growth in prepubertal girls: A randomized, double-blind, placebo-controlled trial. *J. Clin. Invest.* **99**, 1287–1294.
51. Bonjour, J. P., Chevalley, T., Aminan, P., Slosman, D., and Rizzoli, R. (2001). Gain in bone mineral mass in prepubertal girls 3.5 y after discontinuation of calcium supplementation: A follow-up study. *Lancet* **358**, 1208–1213.
52. Ghatge, K. D., Lambert, H. L., Barker, M. E., and Eastell, R. (2001). Bone mineral gain following calcium supplementation in teenage girls is preserved two years after withdrawal of the supplement. *J. Bone Miner. Res.* **16**(S1), S173.
53. Lee, W. T. K., Leung, S. S. F., Leung, D. M. Y., and Cheng, J. C. Y. (1996). A follow-up study on the effect of calcium-supplement withdrawal and puberty on bone acquisition of children. *Am. J. Clin. Nutr.* **64**, 71–77.
54. Slemenda, C. W., Peacock, M., Hui, S., Zhou, L., and Johnston, C. C. (1997). Reduced rates of skeletal remodeling are associated with increased bone mineral density during the development of peak skeletal mass. *J. Bone Min. Res.* **12**, 676–682.
55. Fiorito, L. M., Mitchell, D. C., Smiciklas-Wright, H., and Berch, L. L. (2006). Girls' calcium intake is associated with bone mineral content during middle childhood. *J. Nutr.* **136**, 1281–1286.
56. Goulding, A., Rockell, J. E., Black, R. E., Grant, A. M., Jones, I. E., and Williams, S. M. (2004). Children who avoid drinking cow's milk are at increased risk for prepubertal bone fractures. *J. Am. Diet. Assoc.* **104**, 250–253.
57. Kalkwarf, H. J., Khoury, J. C., and Lanphear, B. P. (2003). Milk intake during childhood and adolescence, adult bone density, and osteoporotic fractures in U.S. women. *Am. J. Clin. Nutr.* **77**, 257–265.
58. Sandler, R. B., Slemenda, C. W., LaPorter, R. E., Cauley, J. A., Schramm, M. M., Barresi, M. I., and Kriska, A. M. (1985). Postmenopausal bone density and milk consumption in childhood and adolescence. *Am. J. Clin. Nutr.* **43**, 270–274.
59. Nicklas, T. A. (2003). Calcium intake trends and health consequences from childhood through adulthood. *J. Am. Coll. Nutr.* **22**, 340–356.
60. Whiting, S. J., Vatanparast, H., Baxter-Jones, A., Faulkner, R. A., Miriwald, R., and Bailey, D. A. (2004). Factors that affect bone mineral accrual in the adolescent growth spurt. *J. Nutr.* **134**, 696S–700S.
61. Barzel, U. S., and Massey, L. K. (1998). Excess dietary protein can adversely affect bone. *J. Nutr.* **128**, 1051–1053.
62. Kerstetter, J. E., and Allen, L. H. (1990). Dietary protein increases urinary calcium. *J. Nutr.* **120**, 134–136.
63. Remer, T., Dimitrios, T., and Manz, T. (2003). Dietary potential renal acid load and renal net acid excretion in healthy, free-living children and adolescents. *Am. J. Clin. Nutr.* **77**, 1255–1260.
64. Sellmeyer, D. E., Stone, K. L., Sebastian, A., and Cummings, S. R. (2001). A high ratio of dietary animal to vegetable protein increases the rate of bone loss and the risk of fracture in postmenopausal women. *Am. J. Clin. Nutr.* **73**, 118–122.
65. Bonjour, J. P. (2005). Dietary protein: An essential nutrient for bone health. *Am. J. Coll. Nutr.* **24**, 5265–5365.
66. Kerstetter, J. E., O'Brien, K. O., Caseria, D. M., Wall, D. E., and Insogna, K. L. (2005). The impact of dietary protein on calcium absorption and kinetic measures of bone turnover in women. *J. Clin. Endocrinol. Metab.* **90**, 26–31.
67. Roughead, Z. K., Johnson, L. K., Lykhen, G. I., and Hunt, J. R. (2003). Controlled high meat diets do not affect calcium retention or indices of bone status in healthy postmenopausal women. *J. Nutr.* **133**, 1020–1026.
68. Rafferty, K., Davies, K. M., and Heaney, R. P. (2005). Potassium intake and the calcium economy. *J. Am. Coll. Nutr.* **24**, 99–106.
69. Muhlbauer, R. C., Lozano, A., and Reiuli, A. (2002). Onion and a mixture of vegetables, salads, and herbs affect bone resorption in the rat by a mechanism independent of their base exceeds. *J. Bone Miner. Res.* **17**, 1230–1236.
70. Cadogan, J., Eastell, R., Jones, N., and Barker, M. E. (1997). Milk intake and bone mineral acquisition in adolescent girls: Randomized, controlled intervention trial. BMJ **315**, 1255–1260.
71. Bonjour, J.-P., Ammann, P., Chevalley, T., Ferrari, S., and Rizzoli, R. (2005). Nutritional aspects of bone growth: An overview. *In* "Nutritional Aspects of Bone Health" (S. A. Newand, J.-P. Bonjour, Eds.), pp. 111–127. The Royal Society of Chemistry, Cambridge, UK.
72. Dawson-Hughes, B. (2007). Protein and calcium absorption-potential role of the calcium sensor receptor. *In* "Nutritional Aspects of Osteoporosis" (P. Burckhardt, R. P. Heaney, and B. Dawson-Hughes, Eds.), Elsevier International Congress Series, Vol. 1297, pp. 217–227. Elsevier.
73. Matkovic, V., Ilich, J. Z., Andon, M. B., Hsieh, L. C., Tzagournis, M. A., Lagger, B. J., and Goel, P. K. (1995). Urinary calcium, sodium, and bone mass of young females. *Am. J. Clin. Nutr.* **62**, 417–425.
74. Palacios, C., Wigertz, K., Martin, B. R., Jackman, L., Pratt, J. H., Peacock, M., McCabe, G., and Weaver, C. M. (2004). Sodium retention in black and white female adolescents in response to salt intake. *J. Clin. Endocrin. Metab.* **89**(4), 1858–1863.
75. Wigertz, K., Palacios, C., Jackman, L. A., Martin, B. R., McCabe, L. D., McCabe, G. P., Peacock, M., Pratt, J. H., and Weaver, C. M. (2005). Racial differences in calcium retention in response to dietary salt in adolescent girls. *Am. J. Clin. Nutr.* **81**, 845–850.
76. Institute of Medicine. (1997). "Dietary Reference Intakes for Calcium, Phosphorus, Magnesium, Vitamin D, and Fluoride." National Academies Press, Washington, DC.
77. Atkinson, S., McCabe, G. P., Weaver, C. M., Abrams, S. A., and O'Brien, K. (2007). Are current calcium recommendations higher than needed to achieve optimal peak bone mass? The controversy. 2006 Experimental Biology Controversy Session, April 2, 2006, San Francisco, CA.
78. Greger, J. L., Baligai, P., Abernathy, R. P., Bennett, O. A., and Peterston, T. (1978). Calcium, magnesium, phosphorus, copper, and manganese balance in adolescent females. *Am. J. Clin. Nutr.* **31**, 117–121.

79. Matkovic, V., Fontana, D., Tominac, C., Goel, P., and Chestnut, C. H. III. (1990). Factors that influence peak bone mass formation: A study of calcium balance and the inheritance of bone mass in adult females. *Am. J. Clin. Nutr.* **52**, 878–888.
80. Braun, M. M., Martin, B. R., Kern, M., McCabe, G. P., Peacock, M., Jiang, Z., and Weaver, C. M. (2006). Calcium retention in adolescent boys on a range of controlled calcium intakes. *Am. J. Clin. Nutr.* **84**, 414–418.
81. Braun, M., Palacios, C., Wigertz, K., Jackman, L. A., Bryant, R. J., McCabe, L. D., Martin, B. R., McCabe, G. P., Peacock, M., and Weaver, C. M. (2007). Racial differences in skeletal calcium retention in adolescent girls on a range of controlled calcium intakes *Am. J. Clin. Nutr.* **85**, 1657–1663.
82. Wu, L., Martin, B. R., Braun, M., McCabe, G., McCabe, L., Kempa-Steczko, A., Dimeglio, L., Peacock, M., and Weaver, C. M. (2007). Calcium retention as a function of calcium intake in Asian adolescents. *FASEB J.* **21**(5), A354.
83. Lynch, M. F., Griffin, I. J., Hawthorne, K. M., Chen, Z., Hamzo, M., and Abrams, S. A. (2007). Calcium balance in 1–4 y old children. *Am. J. Clin. Nutr.* **85**, 750–754.
84. Johnston, C. C., Jr, Miller, J. Z., Slemenda, C. W., Reister, T. K., Hui, S., Christian, J. C., and Peacock, M. (1992). Calcium supplementation and increases in bone mineral density in children. *N. Engl. J. Med.* **327**, 82–87.
85. Lloyd, T., Andon, M. B., Rollings, N., Martel, J. K., Landis, J. R., Demers, C. M., Eggli, D. F., Kieselhorst, K., and Kulin, H. E. (1993). Calcium supplementation and bone mineral density in adolescent girls. *JAMA* **270**, 841–844.
86. Lee, W. T. K., Leung, S. S. F., Wang, S. H., Xu, Y. C., Zeng, W. P., Lau, J., Oppenheimer, S. J., and Cheng, J. C. (1994). Double-blind controlled calcium supplementation and bone mineral accretion in children accustomed to low calcium diet. *Am. J. Clin. Nutr.* **60**, 744–752.
87. Nowson, C. A., Green, R. M., Hopper, J. L., Sherwin, A. J., Young, D., Kaymakci, B., Guest, C. S., Smid, M., Larkins, R. G., and Wark, J. D. (1997). A co-twin study of the effect of calcium supplementation on bone density in adolescence. *Osteoporosis Int.* **7**, 219–225.
88. Dibba, B., Prentice, A., Ceesay, M., *et al.* (2000). Effect of calcium supplementation on bone mineral accretion in Gambian children accustomed to a low-calcium diet. *Am. J. Clin. Nutr.* **71**(2), 544–549.
89. Moyer-Mileur, L. J., Xie, B., Ball, S. D., and Pratt, T. (2003). Bone mass and density response to a 12-month trial of calcium and vitamin D supplementation preadolescent girls. *J. Musculoskel. Neuron. Interact.* **3**, 63–70.
90. Rozen, G. S., Rennert, G., Dodiuk-Gad, R., Rennert, H. S., Ish-Schalom, N., Diaf, G., Raz, B., and Ish-Shalom, S. (2003). Calcium supplementation provides an extended window of opportunity for bone mass accretion after menarche. *Am. J. Clin. Nutr.* **78**, 993–998.
91. Stear, S. J., Prentice, A., Jones, S. C., *et al.* (2003). Effect of a calcium and exercise intervention on the bone mineral status of 16–18 year-old adolescent girls. *Am. J. Clin. Nutr.* **77**(4), 985–992.
92. Cameron, M. A., Paton, L. M., Newson, C. A., Margerison, C., Frame, M., and Wark, J. D. (2004). The effect of calcium supplementation on bone density in premenarcheal females: A co-twin approach. *J. Clin. Endocrinol. Metab.* **89**, 4916–4922.
93. Mølgaard, C., Thomson, B. L., and Michaelson, K. F. (2004). Effect of habitual dietary calcium intake on calcium supplementation in 12–14 y old girls. *Am. J. Clin. Nutr.* **80**, 1422–1427.
94. Prentice, A., Gintz, F., Stear, S. J., Jones, S. C., Laskey, M. A., and Cole, T. J. (2005). Calcium supplementation increases stature and bone mineral mass of 16–18 year old boys. *J. Clin. Endocrinol. Metab.* **90**, 3153–3161.
95. Cheng, S., Lyytikäinen, A., Kröger, H., Lamberg-Allardt, C., Alén, M., Koisteinen, A., Wang, Q. J., Suuriniemi, M., Suominen, H., Mahonen, A., Nicholson, P. H. F., Ivaska, K. K., Korpela, R., Ohlsson, C., Väänänen, K. H., and Tylavsky, F. (2005). Effects of calcium, dairy product, and vitamin D supplementation on bone mass accrual and body composition in 10- to 12-y old girls: 12-y randomized trial. *Am. J. Clin. Nutr.* **82**, 1115–1126.
96. Winzenberg, T., Shaw, K., Fryer, J., and Jones, G. (2006). Effect of calcium supplementation on bone density in health children: Meta-analysis of randomized controlled trials. BMJ **333**, 775–781.
97. Heaney, R. P., and Weaver, C. M. (September 26, 2006). Letter to the editor. *BMJ.com Rapid Response.*
98. Jones, G., and Winzenberg, T. (October 3, 2006). In reply. *BMJ.com Rapid Response.*
99. Specker, B., Binkley, T., and Wermers, J. (2002). Randomized trial of physical activity and calcium supplementation BMC in 3–5 year old healthy children: The South Dakota Children's Health Study. *J. Bone Miner. Res.* **17**, S398.
100. Chevalley, T., Rizzoli, R., Hans, D., Ferrari, S., and Bonjour, J.-P. (2005). Interaction between calcium intake and menarcheal age on bone mass gain: An eight-year follow-up study from prepuberty to postmenarche. *J. Clin. Endocrinol. Metab* . **90**, 44–51.
101. Auld, G., Boushey, C. J., Bock, M. A., Bruhn, C., Gabel, K., Gustafson, D., Holmes, B., Misner, S., Novotony, R., Peck, L., Pelican, S., Pond-Smith, D., and Read, M. (2002). Perspectives on intake of calcium rich foods among Asian, Hispanic, and White preadolescent and adolescent females. *J. Nutr. Educ.* **34**, 242–251.
102. Looker, A. C. (2006). Dietary calcium: Recommendations and intakes around the world. *In* "Calcium in Human Health" (C. M. Weaver, and R. P. Heaney, Eds.), pp. 105–127. Humana Press, Totowa, NJ.
103. Moshfegh, A., Goldman, J., and Cleveland, L. (2005). "What We Eat in America, NHANES 2001–2002: Usual Nutrient Intakes from Food Compared to Dietary Reference Intakes." U.S. Department of Agriculture, Agricultural Research Service, www.ars.usda.gov/foodsurvey.
104. Lytle, L. A., Seifert, S., Greenstein, J., and McGovern, P. (2003). How do children's eating patterns and food choices change over time? Results from a cohort study. *Am. J. Health Promot.* **14**, 222–228.
105. Weaver, C. M., and Heaney, R. P. (Eds.). (2006). "Calcium in Human Health." Humana Press, Totowa, NJ.
106. Gartner, L. M., and Greer, F. R. (2003). Prevention of rickets and vitamin D deficiency: New guidelines for vitamin D intake. *Pediatrics* **111**, 908–910.

107. El-Hajj Fuleihan, G., and Vieth, R. (2007). Vitamin D insufficiency and musculoskeletal health in children and adolescents. *In* "Nutritional Aspects of Osteoporosis" (P. Burckhardt, R. P. Heaney, and B. Dawson-Hughes, Eds.), Elsevier International Congress Series, Vol. 1297, pp. 91–108. Elsevier.
108. Glazer, K., Parmelee, A. H., and Hoffman, W. S. (1949). Comparative efficacy of vitamin D preparations in prophylactic treatment of premature infants. *Am. J. Dis. Child.* **77**, 1–14.
109. Specker, B. L., Ho, M. L., Oesteich, A., Yin, T. A., Shui, Q. M., Chen, X. C., and Tsang, R. C. (1992). Prospective study of vitamin D supplementation and rickets in China. *J. Pediatr.* **120**, 733–739.
110. Heaney, R. P., Dowell, S., Hale, C. A., and Bendict, A. (2003). Calcium absorption varies within the reference range for serum 25-hydroxyvitamin D. *J. Am. Coll. Nutr.* **22**, 142–146.
111. El-Hajj Fuleihan, G., Nabulsi, M., Tamim, H., Maalouf, J., Salamoun, M., Khalife, H., Choucair, M., Arabi, A., and Veith, R. (2006). Effect of vitamin D replacement on musculoskeletal parameters in school children: A randomized controlled trial. *J. Clin. Endocrinol. Metab.* **91**, 405–412.
112. Vijakainen, H. T., Natri, A.-M., Kärkkäinen, M., Huttenen, M. M., Palssa, A., Jakobsen, J., Cashman, K. D., Mølgaard, C., and Lamberg-Allardt, C. (2006). A positive dose-response effect of vitamin D supplementation on site-specific bone mineral augmentation in adolescent girls: A double-blinded randomized placebo-controlled 1-year intervention. *J. Bone Miner. Res.* **21**, 836–844.
113. Oliveri, M. B., Ladizesky, M., Mautelen, C. A., Alouse, A., and Martinez, L. (1993). Seasonal variations of 25-hydroxyvitamin D and parathyroid hormone in Ushuaia (Argentina), the southernmost city of the world. *Bone Miner.* **20**, 99–108.
114. Masuyama, R., Nakaya, Y., Katsumata, S., Kajita, Y., Uehara, M., Tanaka, S., Sakai, A., Kato, S., Nahamma, T., and Suzuki, K. (2003). Dietary calcium and phosphorus ratio regulates bone mineralization and turnover in vitamin D receptor knockout mice by affecting intestinal calcium and phosphorus absorption. *J. Bone Miner. Res.* **18**, 1217–1226.
115. Rude, R. K., Gruber, H. E., Norton, H. J., Wei, L. Y., Fransto, A., and Kilburn, J. (2006). Reduction of dietary magnesium by only 50% in the rat disrupts bone and mineral metabolism. *Osteoporosis Int.* **17**, 1022–1032.
116. Medeiros, D. M., Plattner, A., Jennings, D., and Stoecker, B. (2002). Bone morphology, strength and density are comprised in iron-deficiency rats and exacerbated by calcium restriction. *J. Nutr.* **132**, 3135–3141.

CHAPTER **45**

Osteoporosis

ROBERT MARCUS
Professor Emeritus, Stanford University, Stanford, California

Contents

I. INTRODUCTION

Osteoporosis is a global skeletal disorder of decreased bone strength in which the only important consequence is an increased risk for fracture with minimal trauma. The term *porosis* means "spongelike," which aptly describes the appearance of that portion of the skeleton, the trabecular skeleton, which is most afflicted with this disease (Figure 1). Bone strength is a complex function integrating bone mineral density and bone quality. Bone mineral density, generally abbreviated BMD, is the concept most familiar to the public, to patients, and to physicians, but it is only one determinant of bone strength, all other contributions that are not captured by a BMD measurement falling under the rubric of "bone quality." The term *quality* should more properly be considered as the plural *qualities*, because it encompasses a wide variety of characteristics that are enumerated later in this chapter. Osteoporotic, or fragility, fractures traditionally have been grouped according to their location, either in the spine (vertebral) or nonspine, the latter including those of the distal radius (Colles' fractures) and the proximal femur (hip fracture). It is important to understand that because of its global nature, osteoporosis imposes an increased risk for virtually all fractures in affected individuals.

Already more common than any other generalized skeletal condition, osteoporosis continues to increase in prevalence. Based on data from the National Health and Nutrition Examination Survey III (NHANES) and from the 2000 National Census, the National Osteoporosis Foundation has estimated that, in 2002, 20% of postmenopausal white women in the United States had osteoporosis, and an additional 52% had low bone density at the hip [1]. In the whole population, about 8 million have osteoporosis, of whom about 1.5 million will fracture each year. One out of every two white women will experience an osteoporotic fracture at some point in her lifetime [1–3].

Although men have a lower prevalence of osteoporosis than women, perhaps 25% as great, fragility fractures certainly occur in men, and some of these, particularly at the hip, carry a less favorable prognosis in men than for women. The most common osteoporotic fractures occur as compression deformities of the thoracic or lumbar spine. Although two-thirds of these do not acutely produce symptoms, they must not be viewed in any sense as benign events. Even mild compression fractures aggravate by four- to fivefold the short-term risk for subsequent fractures. Approximately one-third of vertebral fractures produce symptoms when they occur and are referred to as "clinical" vertebral fractures. These are more likely to be of moderate or severe degree in deformity and are the fractures most likely to result in long-term pain, deformity, and disability. In the Study of Osteoporotic Fractures (SOF), a long-term observational study of thousands of older women, clinical vertebral fractures were associated with an eightfold excess of mortality similar to that observed with hip fracture [4]. The incidence of vertebral fractures in women begins to rise early in the sixth decade, corresponding in time to the menopausal loss of endogenous estrogen. The incidence continues to increase in succeeding decades (Figure 2) [3].

At nonvertebral sites, forearm fractures, particularly at the distal radius, also increase during the sixth decade but stabilize thereafter, at which time the incidence of hip fracture begins exponentially to increase [3]. Both forearm and hip fractures result directly from a fall. Whether an individual fractures the arm or the hip reflects the manner of falling. Younger women with normal locomotion generally fall while walking and break a fall by arm extension. Older and more frail women often fall while transferring from a seated to a standing position. If they fail to elevate their centers of gravity sufficiently to support an upright posture, they fall backwards or to the side, directly impacting the

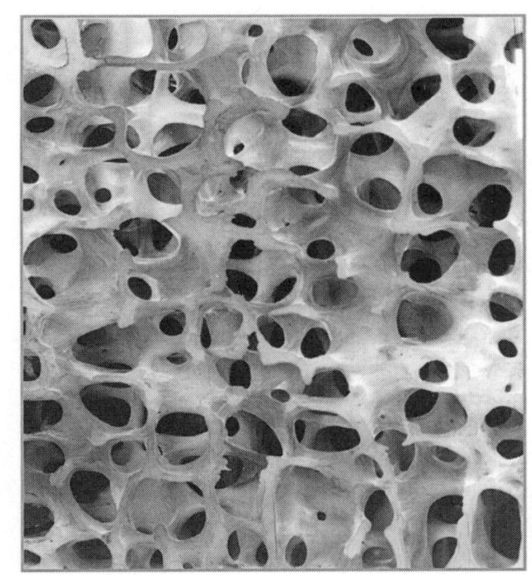

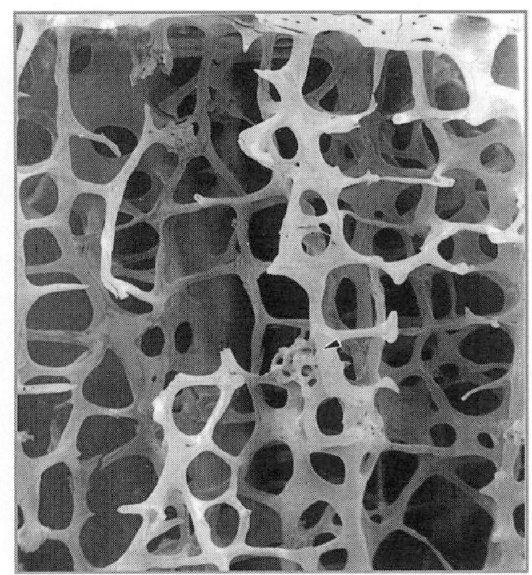

FIGURE 1 (A) Normal trabecular bone. Note the highly interconnected vertical and horizontal bars, fairly homogeneous size and shape of holes, and platelike appearance of many of the trabecular units. Courtesy Dr. David Dempster. Copyright David Dempster Ph.D. (B) Osteoporotic bone. Note substantial reduction in the amount of bone substance per unit volume compared to normal bone (A). Note the narrow rodlike appearance of vertical trabeculae compared to the normal platelike structures. Note the wide variation in the size of holes throughout the trabecular structure. In many regions trabecular struts are hanging in space without connection to neighboring structures. Courtesy Dr. David Dempster. Copyright David Dempster Ph.D.

femoral greater trochanter and possibly leading to hip fracture [5, 6].

This chapter focuses on the characteristics of a healthy skeleton, the underlying pathophysiology of osteoporosis, the characteristics of osteoporotic bone, approaches to conserving bone throughout adult life, and therapeutic approaches to treating skeletal fragility. A comprehensive

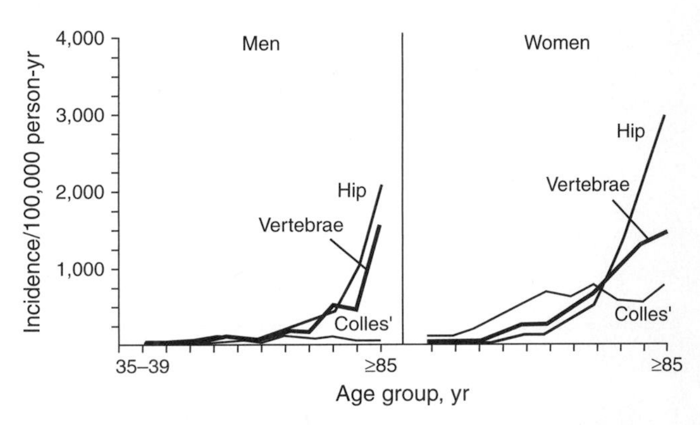

FIGURE 2 Age-specific incidence rates for hip, vertebral, and distal forearm fractures in men and women. Data derived from the population of Rochester, Minnesota. From Cooper, C., and Melton, L. J. III. (1992). Epidemiology of osteoporosis. *Trends Endocrinol. Metab.* **3,** 224–229, (1992).

discussion of the various pharmacologic agents available for patient management lies outside the scope of this chapter, and the critical role of bone acquisition during years of growth through adolescence appears in the companion chapter by Weaver (Chapter 44), which the reader is strongly urged to consult.

II. THE SKELETON

Bone is a complex cellular tissue that contains, by weight, about 30% organic constituents and 70% mineral. The most abundant protein in the organic compartment is type I collagen, a fibrillar structure consisting of three interweaving strands, normally two strands of alpha-1 collagen and one of alpha-2 collagen. Collagen represents 98% of the organic phase of bone and various noncollagen proteins account for the remainder [7].

The mineral phase of bone is about 95% hydroxyapatite, a highly organized crystal of calcium and phosphorus. Other minerals normally found in bone mineral include sodium (indeed, about 30% of total body sodium can be stored in bone crystal), magnesium, and fluoride. Incorporation of fluoride and strontium into bone crystal is of particular relevance because these compounds have seen use as therapy for osteoporosis.

A. Bone Cells

The processes of bone formation and breakdown (resorption) require cellular activity. Three major cell types reside in bone and conduct these processes, osteoblasts, osteocytes, and osteoclasts.

Osteoblasts are the primary bone-forming cells. They derive from stem cells in the bone marrow stroma. These stem cells are pluripotential, having the capacity to develop along multiple lineages, including fibroblasts, hematopoietic cells, smooth myocytes, adipocytes, chondrocytes, and osteoblasts. During linear growth, osteoblasts invade a temporary cartilaginous template to form primary lamellar bone. During remodeling (see later discussion), a wave of osteoblast precursors migrates to the base of a resorption cavity, acquires the characteristics of mature osteoblasts, and lays down new bone (Figure 3).

Osteocytes are osteoblasts that have become embedded within their own secreted matrix. Each osteocyte sits in its individual hole, or *lacuna*, connected to one another throughout the bone matrix by a highly developed network of channels, or *canaliculi*. Osteocytes appear to be the monitors and responders to a bone's mechanical environment (Figure 4).

Osteoclasts are multinucleated giant cells of macrophage lineage. They undertake the enzymatic destruction of bone during the resorption phase of remodeling (see later discussion). During this process, osteoclasts form a seal at the bone surface with the aid of anchoring proteins called integrins whose receptors exist in the bone matrix. This

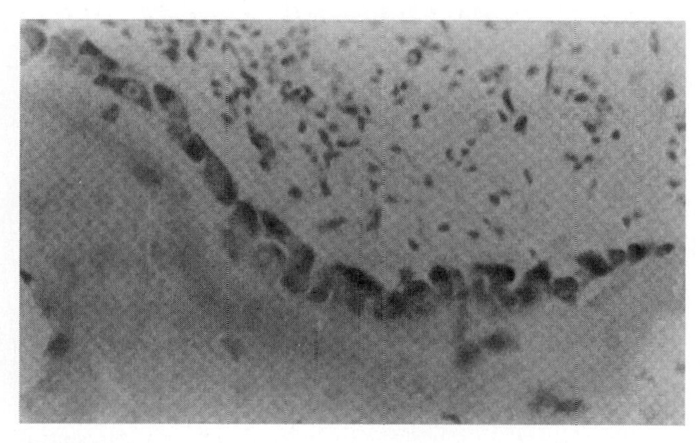

FIGURE 3 Low-power view of osteoblasts lining the bone surface. From [7], with permission.

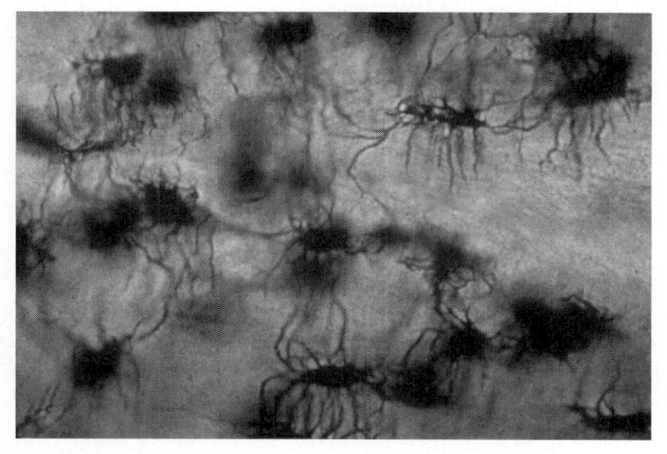

FIGURE 4 Osteocytes occupying individual lacunae with extensive canalicular interconnections. From [7], with permission.

seal creates a sequestered region underneath the osteoclast into which hydrogen ion is secreted using a carbonic anhydrase-dependent pump and resulting in a highly acidic local environment. In addition, the osteoclast secretes a variety of hydrolytic enzymes, such as cathepsins, which hydrolyze bone matrix (Figure 5).

A fourth cell is also observed in bone. So-called *lining cells* are seen as a syncytial layer of dormant cells that covers bone surfaces. This group of cells is thought to serve a surveillance function that responds to microscopic damage by locally stimulating new remodeling activity. Lining cells also originate from osteoblasts and, although dormant, retain the capacity under certain circumstances to convert into functional osteoblasts and lay down new bone. This appears to be one mechanism through which administration of parathyroid hormone achieves a rapid increase in bone formation (see section on Therapeutics below).

Eighty percent of the adult skeleton consists of compact bone. This is referred to as the *cortical* or *appendicular* skeleton and comprises mostly the long bones as well as the outer shells of the central, or *axial* skeleton, which includes the spine, pelvis, and the ends (metaphyses) of long bones. The axial skeleton has a heavy complement (perhaps 40% by weight or 80% by surface area) of a honeycomb-like series of vertical and horizontal bars, or *trabeculae*, and is therefore frequently called *trabecular* bone (in orthopedics this also may be called *cancellous* bone). In adults, the trabecular bone of the spine and pelvis constitutes the primary residence of red bone marrow. Because the cells responsible for conducting the processes underlying adult bone loss originate in the bone marrow, and because these processes occur on the surfaces of bone, it should be no surprise that trabecular bone, with its rich complement of bone marrow and extensive surface area, should be the bone compartment that experiences the earliest and most rapid loss of bone with aging.

At any time during adult life the amount of bone contained within the skeleton consists of that bone which was present at the end of growth, the so-called "peak bone mass," minus that which has been lost. One frequently encounters patients who report being told, following a bone density test, that they have "lost 30% of their skeleton." The problem with such conclusions is that one simply cannot determine from a single BMD measurement whether a deficit in bone mineral reflects bone *loss* or failure to achieve the peak bone mass that might have been predicted for that individual. In fact, the majority of young adults with low bone mass have not lost bone at all, but have age-related deficits related to poor acquisition of peak bone mass. (See Chapter 44 by Weaver.)

B. Physiologic Roles of the Skeleton

During vertebrate evolution, the skeleton acquired two fundamental but not necessarily compatible functions. By virtue of its dense mineralization, bone provides the *structural*

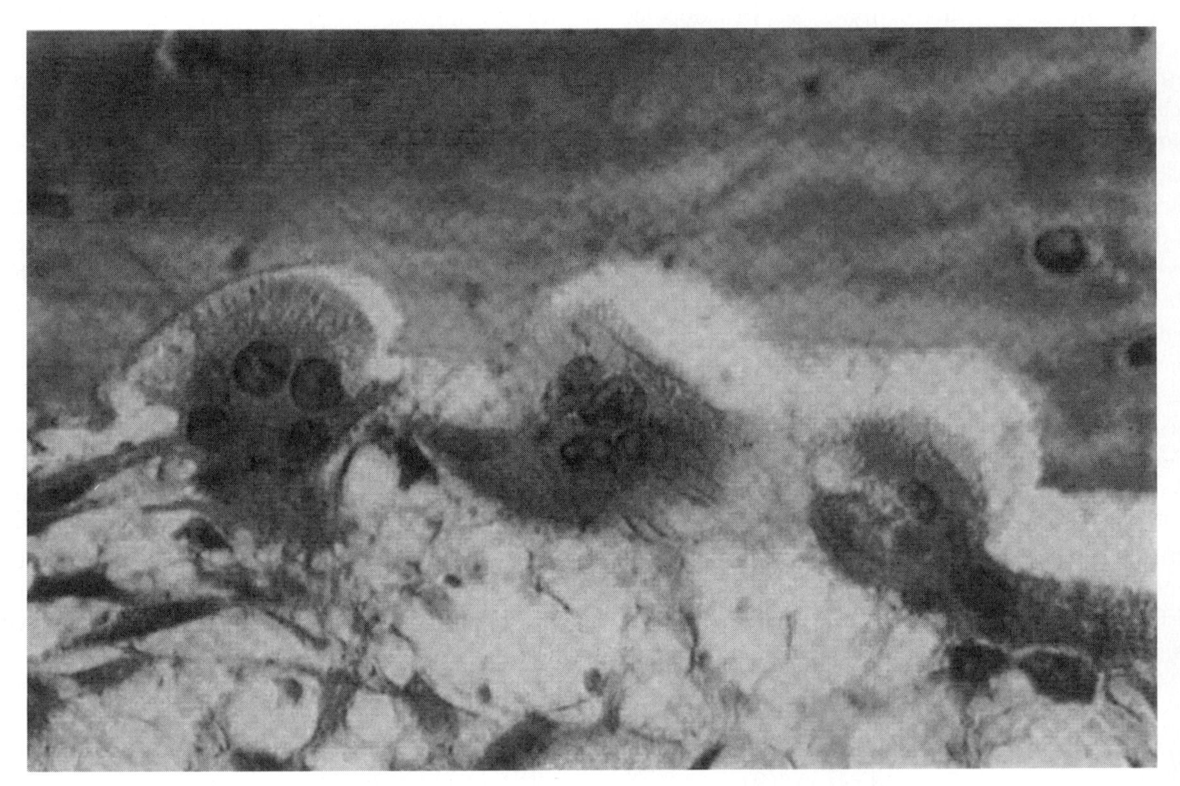

FIGURE 5 Low-power view of osteoclasts occupying resorption lacunae. From [7], with permission.

rigidity necessary to withstand the effect of gravity and support terrestrial locomotion. By adapting to region-specific differences in its mechanical environment, denser, stronger bone exists where it is needed without requiring a universal increase in skeletal weight to the point that mobility is jeopardized.

Bone also constitutes the *primary repository in the body for calcium*. Indeed, 99.5% of body calcium is contained within bone and can be mobilized to support the extracellular calcium concentration at times of need. For the great majority of vertebrates, the calcium environment is extremely high, reflecting its very high concentration in ocean water (~400 mg/liter). Facing the threat of calcium toxicity, ocean fish must be able to eliminate excess calcium from their bodies, which they accomplish through a calcium-dependent ATPase system in the gills. Progression of vertebrates onto land, with fresh water far more dilute in calcium, required mechanisms to promote calcium extraction from the environment and to conserve it within the body. Parathyroid hormone (PTH), the peptide secretory product of the parathyroid glands (first appearing in amphibia), serves this role. In response to minute-to-minute relatively mild reductions in extracellular calcium concentration, such as during the hours following a meal, PTH stimulates the kidney to conserve calcium by regulating renal tubular reabsorption efficiency. When calcium deficits become sustained or severe, such as in the face of chronically inadequate dietary calcium, PTH stimulates the renal production of $1,25(OH)_2$ vitamin D (calcitriol), the potent hormonal form of the parent vitamin, which in turn enhances intestinal calcium absorption. In this setting PTH also stimulates bone remodeling by initiating the formation of new remodeling units and resulting in delivery of calcium from the skeleton to the extracellular environment. Together, these actions restore plasma calcium concentrations to their normal level [8]. It must be understood that PTH action on the skeleton does *not* selectively remove calcium from the skeleton, but is accomplished by an increase in bone remodeling (see later discussion) so that the release of mineral to the plasma compartment is accompanied by a net loss of bone.

C. Remodeling: The Key to Understanding Age-Related Bone Loss

Many mammals, primates, and humans maintain skeletal integrity through a continuous process of breakdown and renewal known as *remodeling*. This process occurs lifelong, although during childhood and adolescence it is overshadowed by the events of linear growth (modeling). Once growth centers have fused and skeletal maturation is complete, remodeling becomes the dominant, indeed, with rare exception, the only mechanism through which bone is added to or removed from the skeleton. Each remodeling event is carried out by discrete bone multicellular units (BMUs) and consists of an initial phase of bone resorption that is coupled to a longer phase of bone formation (Figure 6). These are initiated when cells of macrophage lineage come from the bone marrow to points on the bone surface and fuse into multinucleated osteoclasts that dig into and remove bone. The cavity thus created reaches a depth of 60 microns within 6–8 weeks. In this manner both mineral and matrix constituents are returned to the circulating extracellular fluid. Released from the resorbed matrix is a rich assortment of cytokines and growth factors that then attract into the base of the cavity a wave of osteoblast precursor cells from the marrow stroma. These transform into functional osteoblasts and begin to lay down new bone matrix. Once the new bone reaches a thickness of about 20 microns it begins to accumulate mineral. By the end of about 6 months, bone formation is complete and the bone is restored almost to its basal state. However, like many biological processes, bone remodeling is not 100% efficient; that is, the amount of new bone formed does not completely make up for the older bone removed, so a small bone deficit remains as a consequence of each remodeling event, called the "remodeling imbalance." The cumulative effect of the hundreds of thousands of remodeling units in play at any one time is the readily observable phenomenon of age-related bone loss. Consequently, anything that promotes an overall increase in whole body bone remodeling aggravates the rate of bone loss and, by contrast, interventions that slow remodeling constrain bone loss. This forms the basis for using drugs that slow remodeling as a mainstay of osteoporosis treatment, as discussed at the end of this chapter.

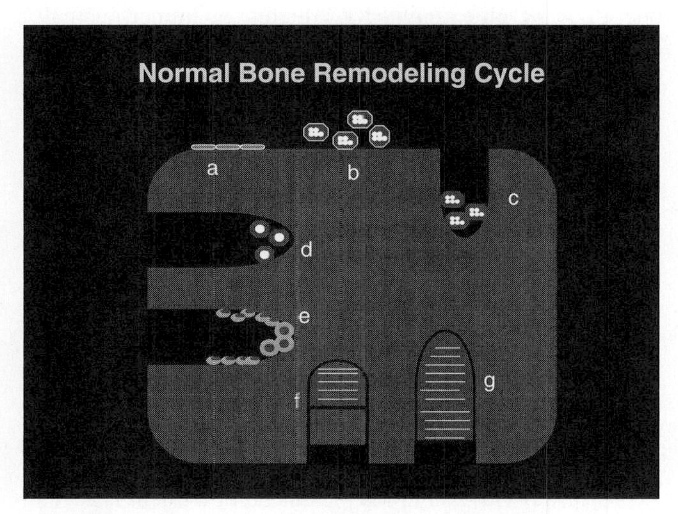

FIGURE 6 The bone remodeling cycle. This drawing represents a region of trabecular bone. All remodeling events occur on the bone surface. (a) 90% of bone surface generally covered by thin layer of dormant lining cells. (b) Coalescence of osteoclast precursors at a site on the bone surface with creation of multinucleated osteoclasts. (c) Osteoclasts remove a divot of bone, reaching 60 microns in depth by 6–8 weeks. (d) Soluble factors released by osteoclastic resorption recruit a new wave of cell proliferation (preosteoblasts) into the base of the resorption cavity. (e) Preosteoblasts acquire the osteoblast phenotype. (f) Osteoblasts secrete new bone matrix which begins to acquire mineral after a thickness of ~20 microns is achieved. (g) New mineralized bone almost fully replaces resorbed bone by about 6 months. Small deficits are left reflecting remodeling inefficiency and accounting for the process of age-related bone loss. Copyright Robert Marcus, M.D.

Another word should be said concerning matrix mineralization during the bone formation phase of remodeling. Mineral is rapidly laid down for the initial several weeks, but thereafter changes to a slow, linear rate. Mineral is never fully saturated in bone, so it continues to accumulate as long as that particular unit of bone survives. The only thing to terminate this process would be the initiation of a new wave of osteoclastic resorption to clear out this region of older bone. Thus, if overall remodeling slows, bone survives longer and becomes more densely mineralized. The consequences of this finding are also discussed later.

If, as described, bone remodeling leads to loss (and presumably weakening) of bone, one may ask why it has evolved and what its purpose may be. It appears that the cardinal role of remodeling is to serve a scavenger function, through which fatigued or damaged bone is cleared away and replaced (albeit at the long-term price of gradual bone loss). In addition, remodeling is the means by which PTH restores normocalcemia in response to hypocalcemia.

D. Intercellular Communication among Bone Cells: Triggers and Constraints on Remodeling

As initiation of each remodeling event begins with delivery of osteoclasts or their precursors to the bone surface, it is important to have passing familiarity with the signals that control this initial event [9, 10]. It turns out that the key cell for controlling osteoclast production is the osteoblast. This bone-forming cell elaborates two distinct proteins, a stimulator and a repressor, that regulate osteoclast production. The osteoclast stimulator was initially called "osteoclast differentiation factor," but now that its primary target is known, it has been given the less transparent name of RANK-ligand. RANK is an abbreviation for the "receptor that activates NF-kappa β (a gene present in osteoclasts and other cells of macrophage origin). Under stimulation by agents known to increase bone remodeling (e.g., PTH, L-thyroxine), the osteoblast synthesizes and extrudes RANK-ligand that binds to RANK located on osteoclast precursors, leading to production of new osteoclasts. The repressor molecule, also produced by osteoblasts, is a protein called osteoprotegerin (OPG). Production of this protein increases when the osteoblast binds agents that inhibit remodeling, such as estrogen. OPG exhibits high affinity for RANK-ligand and therefore acts as a false receptor, neutralizing the effect of any RANK-ligand with which it comes into contact.

The RANK–RANK-ligand–OPG complex acts in a push-pull manner to regulate osteoclast production and hence control the rate of bone remodeling. When stimuli favor greater remodeling, RANK-ligand production increases, OPG decreases, and RANK is activated. When remodeling is suppressed, RANK-ligand decreases, OPG increases, and RANK is constrained.

III. ADULT BONE MAINTENANCE

For many years scientific inquiry into the basis of adult bone loss and development of osteoporosis was highly parochial. Nutrition scientists focused on the diet, exercise physiologists and mechanical engineers focused on physical activity and the mechanical environment, and physicians whose responsibility it was to care for patients with osteoporosis focused largely on menopausal estrogen loss, It is now abundantly clear that acquisition and maintenance of a healthy skeleton is far more complex than can be explained by any of these individual spheres. One needs to view the skeleton as subject to diverse influences throughout life so that bone status at any particular time is the result of a stochastic process by which each individual insult or event over a lifetime has made its independent contribution.

A. Major Influences on Age-Related Bone Loss

Successful bone maintenance requires continued attention to the same "hygienic" factors that influenced bone acquisition: physical activity, diet, and reproductive status. Bone maintenance requires sufficiency in all areas, and deficiency in one is not compensated by the others. For example, amenorrheic athletes lose bone despite frequent high-intensity physical activity and supplemental calcium intake [11, 12]. Successful bone maintenance is also jeopardized by known toxic exposures such as smoking, alcohol excess, immobility, systemic illnesses, and many medications.

1. Habitual Physical Activity

The skeleton's mechanical function has been referred to earlier. To accomplish this role in a manner that optimizes bone strength while at the same time not unduly increasing its weight, bones accommodate the loads imposed on them by undergoing alterations in mass, in external geometry, and in internal microarchitecture. The first enunciation of this principle is credited to the German scientist Julius Wolff as "Wolff's law" [13]. As a consequence of such adaptation, steady-state bone mass should reflect the mechanical environment, a concept that applies when comparing bone mass among individuals, different bones within an individual, and even different regions within a single bone. A substantial body of research has addressed this prediction. These studies are of two general types: comparisons of bone mass of athletes to that of sedentary controls, and descriptions of associations between level of physical activity and bone mass within a general population. The first type of study generally considers only very active or sedentary individuals, and hence extreme differences in activity are represented. In the latter case, a broader range of physical activity is examined.

Considerable evidence indicates that elite athletes and chronic exercisers have higher BMD than age-matched, nonexercising subjects, a finding, not surprisingly, that applies primarily to sites that undergo loading during the exercise (reviewed in [14]). Activities associated with high load magnitude at low number of repetitions (cycles) are associated with substantial increases in bone mass. For example, world-class and recreational weight lifters have 10–35% greater lumbar spine BMD than sedentary age-matched controls. A special case is represented by comparing dominant to nondominant limbs in athletes whose sport involves unilateral loading. For example, increased BMD in the playing compared to the nonplaying arm of tennis players has been repeatedly observed. By contrast, swimming, a buoyant activity not associated with counteracting the effect of gravity, does not appear to increase BMD. In one study of elite university athletes, swimmers actually had lower bone mass than gymnasts or nonathletic controls, despite increased muscle bulk and regular weight training [15]. Young athletes who spend more than 20 hours each week in a buoyant environment for many years may simply not experience sufficient gravitational stress to promote fully the expected degree of bone acquisition.

These comparisons of athletes and control subjects must be interpreted with caution. Because no measurements of bone mass are made prior to initiating the exercise program, a *causal relationship* between exercise and bone cannot be proven. It may be that individuals with higher bone density are more apt to succeed in athletics, and therefore they enter the "athlete" and chronic exercising groups. Conversely, elite swimmers may have excelled in buoyant activity because of a lighter skeleton. In many studies important characteristics of the matched controls have been overlooked. Factors such as menstrual status, nutrient intake, and use of tobacco or alcohol may have confounded the results. Finally, skeletal status is most frequently expressed as the areal BMD (g/cm^2), a term that overestimates BMD in persons with large bones and underestimates it in smaller people. Thus, if exercisers and controls are not well matched for height, conclusions based on BMD may be spurious.

With respect to the impact of habitual physical activity within the general population, many studies now point to a significant skeletal effect of physical activity in children on the acquisition of bone during the second and third decades (see Chapter 44). The situation is less clear for moderately active adults, in whom no consistent relationship between current activity level and bone mass has been established. In our own work, strong relationships between estimates of daily energy expenditure and BMD were completely negated by normalizing the data for body weight or lean mass [14]. Several reports document positive relationships between lifelong physical activity and bone status.

Thus, cross-sectional studies generally support the notion that elite athletes and chronic exercisers have increased BMD, the magnitude of this difference likely depending on the type and intensity of exercise, age, sex, and hormonal status. However, data concerning moderate physical activity remain uncertain.

One approach to eliminating the selection bias of cross-sectional studies is a randomized controlled trial in which exercise is the intervention. A number of properly controlled studies of this sort have been reported. Although most indicate a positive effect of imposed exercise on BMD, the magnitude of response has been very disappointing to those who anticipated the large differences observed in cross-sectional studies. Rarely do the increases in BMD exceed 2% after 1–2 years of rigorous training, regardless of the type of exercise used (endurance versus resistance training) [16–19].

Understanding the meager response to exercise interventions is related to the fact that skeletal response to mechanical loads is curvilinear in nature. Complete immobilization, as seen with high-level spinal cord injury, leads rapidly to devastating bone loss, with deficits approaching 30–40% over several months. By contrast, imposition of even substantial training regimens on normally ambulatory people or animals increases bone mass by only a few percent over a similar period. This is illustrated in Figure 7, where the effect of walking on bone mass is schematized. As an individual goes from immobility to full ambulation, duration of time spent walking becomes progressively less efficient for

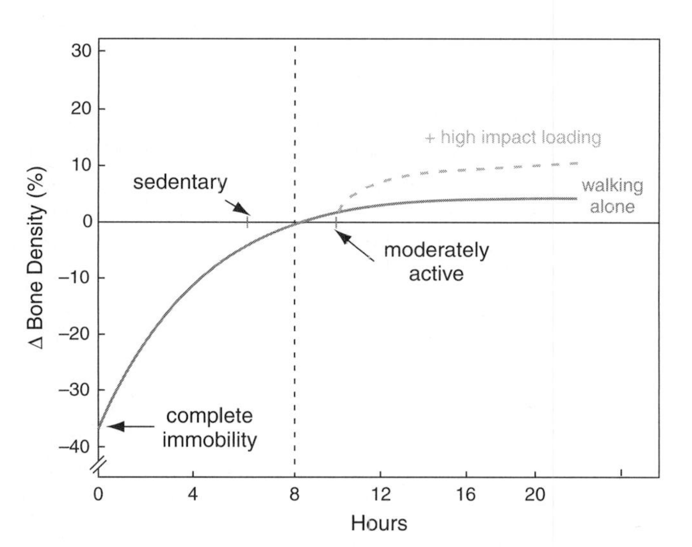

FIGURE 7 The curvilinear nature of skeletal response to mechanical loading. Copyright Robert Marcus, M.D.

increasing bone mass. A person who habitually walks 6 hours each day might require another 4–6 hours just to add a few more percent BMD. On the other hand, adding a more rigorous stimulus, such as high-impact loading, for even a few cycles would increase the response slope.

The worst thing that can happen to the skeleton is to be immobilized. For maintaining bone mass during adult life a certain degree of daily weight bearing is required, and this is achieved by the vast majority of even sedentary individuals. Small increments in BMD can be achieved by increasing one's daily exercise schedule, but, as a consequence of Wolff's law, these will remain only so long as the added activity continues. For most individuals, particularly if they are elderly or frail, walking provides the soundest and most prudent physical activity for skeletal maintenance.

2. Nutrition

a. Energy

Reflecting the major influence of weight-bearing activity on bone, bone mass is strongly related to body weight, so it should be no surprise that severe deficits in bone occur in states of profound malnutrition. Although frank starvation is extremely rare in developed societies, bone deficits are frequently encountered in medical conditions associated with extremely low body mass, such as anorexia nervosa, as well as various forms of intestinal malabsorption and cachexia. It may be difficult to assign responsibility for bone deficits in such patients to any specific nutrient, because patients generally show profound reductions in the consumption of many nutrients. In teenagers with anorexia nervosa, skeletal deficits appear very early, and their magnitude is exacerbated because bone is lost at a time when other girls of similar age are gaining bone at an accelerated rate. Bachrach *et al.* [20] observed that whole-body bone mass correlates linearly with body mass in normal teens, and that the bone mass of girls with anorexia nervosa lies exactly on this curve. In other words, skeletal deficits in young girls with anorexia nervosa primarily reflect their body mass. When these girls were observed over time, only weight rehabilitation (a gain of at least 5 kg) was associated with a gain in bone mass.

b. Calcium

The concept that osteoporosis is a disease of calcium deficiency was proposed more than a century ago, although a central role for calcium intake did not emerge into the scientific mainstream for many years. This largely reflected the overpowering influence of Professor Fuller Albright, who conceived of osteoporosis as a deficiency of bone matrix due to osteoblast failure, usually as a consequence of menopausal loss of estrogen [21].

Recent intensification of interest in calcium occurred with the publication of three independent reports. The first, by Matkovic *et al.* [22], demonstrated a difference in hip fracture incidence in two different regions of Croatia that were demographically very similar except for a substantial difference in the calcium content of local water supplies. Second was publication of the National Health and Nutrition Examination Survey II (NHANES), which showed that habitual calcium intakes of American girls and women failed to meet recommended daily standards (RDAs) as early as age 11 [23]. Third was a landmark paper by Chapuy *et al.* [24] that clearly demonstrated the ability of supplemental calcium and vitamin D to reduce the incidence of hip fracture in a highly vulnerable population.

When one considers that the skeleton is the repository for more than 99% of total body calcium, it is inconceivable that individuals whose intakes are below the amount necessary to maintain whole body calcium balance could be in negative balance without losing bone. Heaney has discussed the issues surrounding the concept of a Nutrient Requirement (see Chapter 14, this volume) and refers to calcium as a "threshold nutrient," by which he means that below a critical value, some physiological function (e.g., calcium balance) is dependent on intake, but above that value no further benefit accrues from additional intake. Because the mineral demands for growth, early adult, and older adult maintenance differ substantially, the "threshold" value, and therefore the intake requirements also differ by age [25, 26]. Table 1 shows that current recommendations for calcium intake have increased substantially from previous Recommended Dietary Allowances [27] and are closely allied to those derived from the literature of formal calcium balance studies [28].

Although a Consensus Conference on optimal calcium intake concluded that habitual calcium intakes by both men and women are inadequate for optimal bone health [25], and several clinical trials demonstrated the ability of calcium supplementation to constrain the rate of bone loss and even reduce fractures [24, 29–35], there remains in the medical and research communities some resistance to reaching consensus regarding the role of calcium inadequacy in the pathogenesis of osteoporosis. In part, this reflects considerable uncertainty in estimating the amount of calcium that people habitually consume. For example, dietary histories and food frequency questionnaires carry substantial imprecision. Also, it must be remembered that calcium is not absorbed in a vacuum, but in the course of eating or drinking other foods, and its availability from foods is highly influenced by other nutrients. For example, although the calcium content of spinach is quite good, it is rendered essentially nonabsorbable by the presence of oxalate and perhaps other anions to which it binds [36]. Further consideration of the roles of phosphorus, sodium, and protein appears later.

For adults, calcium intakes in the range of 1000–1500 mg/day are recommended. It should be noted that individuals in early to middle adult life, say 20–50 years of age, have calcium intakes within reasonable proximity to recommended values, and, by virtue of being young enough to undertake regular weight-bearing activity and to maintain normal reproductive function, generally have only modest stress on skeletal balance, as shown by very low rates of bone loss. After menopause (average age 51 years), and certainly at more advanced ages, declines in endogenous reproductive hormone concentrations and in the growth hormone–insulin-like growth factor (IGF-I) axis, coupled with overall trends toward less physical activity, make it less likely that dietary deficiency will be buffered and more likely that age-related bone loss will be aggravated.

c. Phosphorus

Various domestic animals respond to excessive dietary phosphorus by increasing the endogenous concentration of parathyroid hormone (PTH), resulting in negative calcium balance and bone loss. In such animals optimal dietary ratios of calcium to phosphorus approach 1.0. This has led to a popular theory that, because calcium:phosphorus ratios in human diets typically are well below 1.0, phosphorus overconsumption initiates a similar process in humans.

TABLE 1 Various Estimates of the Calcium Requirement (mg/day) in Women

Age (years)	1989 RDA[a]	NIH[b]	1997 AI[c]	Balance[d]
1–5	800	800	–	1100
6–10	800	800–1200	960	1100
11–24	1200	1200–1500	1560	1600
Pregnancy/lactation	1200	1200–1500	1200–1560	–
24–50/65	800	1000	1200	800–1000
65–	800	1500	1440	1500–1700

Adapted from Heaney [79].

[a]Reference [27].

[b]Recommendations for women as proposed by the Consensus Development Conference on Optimal Calcium Intake [25].

[c]The so-called adequate intakes of the new DRI values, multiplied by a factor of 1.2 to convert them into RDA format [26].

[d]Estimates derived from published balance studies [28].

Firm evidence to support a role for phosphorus excess in human osteoporosis has not been forthcoming. In contrast, some evidence indicates that intestinal calcium absorption and balance are fairly impervious to very wide variations in daily phosphorus consumption [37, 38]. Separate mention should be made concerning the role of phosphorus-containing soft drinks. Cola drinks contain phosphoric acid as their source of effervescence, and one reads frequently in the lay press that the high content of phosphorus in colas is an important contributor to developing osteoporosis. However, although excessive soft drink consumption may well contribute to poor bone status, this is so because these beverages have been substituted for milk, thus exchanging a calcium-rich drink for one that contains essentially no calcium [39].

d. Protein

For many years, a concept has circulated as a subtext in osteoporosis research that consumption of excess protein, particularly from animal sources, is an important contributor to the development of osteoporosis. This is considered the consequence of the fact that protein catabolism generates ammonium ion from ammonia and sulfate from sulfur-containing amino acids. When protein intake increases, citrate and carbonate ions are released from bone to neutralize these acids, and, as urinary calcium is closely linked to renal acid excretion, urinary calcium rises. However, plant proteins contain amounts of sulfur similar to those in eggs, milk, and meats, and therefore increased intake of protein from either animal or plant sources similarly increases urinary calcium. Further, the impact of protein on calcium balance depends to a major degree on other nutrients contained in the consumed food. For example, milk calcium compensates for urinary calcium losses generated by milk protein, potassium in such plants as legumes and grains decreases calcium excretion, and the high phosphorus content of meat offsets the calciuric effect of the protein [40].

Nutritional status assessments of patients with osteoporosis do not support a view that these patients have enjoyed a life of luxurious protein consumption. Barger-Lux *et al.* [41] reported that women with low calcium intakes generally consumed insufficient amounts of multiple nutrients, including protein. Sellmeyer *et al.* reported higher protein intakes to be associated with greater age-related bone loss in the Study of Osteoporotic Fractures cohort (SOF) [42]. However, a substantial body of evidence does not sustain this view. Protein consumption was reported to be an important positive predictor of bone mass in elderly women [43]. In the Framingham cohort, Hannan *et al.* [44] reported bone loss over time to be inversely related to protein intake, with rates of loss in the highest quartile of protein consumption less than one-third of those in the lowest quartile. Kerstetter *et al.* [45] conducted a series of controlled balance studies that showed that, although calciuria increased with increased protein intake, this did not come from bone but was rather a reflection of improved intestinal calcium absorption. In a randomized controlled trial, Dawson-Hughes and Harris [46] showed that the improvement in BMD in subjects supplemented with calcium and vitamin D was confined mainly to individuals whose protein intake was in the highest tertile. Delmi *et al.* [47] found that supplemental protein improved recovery from hip fracture, accelerated hospital discharge, and also slowed bone loss in the contralateral hip. Heaney has concluded that the weight of evidence shows high protein intake to be osteoprotective, but only if calcium intake is adequate, whereas the protective effect of calcium occurs only when protein intake is relatively high [48].

e. Sodium

Urinary excretion of calcium and that of sodium are highly linked, and increased sodium intake is known to promote calcium excretion. A similar effect occurs with most diuretic agents (with the single exception of thiazides, which uncouple the handling of these two cations). Thus, it is not unreasonable to expect that increased dietary sodium might be conducive to negative calcium balance and aggravate bone loss. In a 2-year prospective trial, Devine *et al.* [49] examined the influence of urinary sodium excretion and dietary calcium intake on bone density of postmenopausal women. Urinary sodium excretion (a robust indication of intake) correlated negatively with changes in BMD, and the data suggested that halving sodium intake had the same effect on BMD as increasing daily calcium intake by 891 mg [49].

f. Vitamin D

The importance of vitamin D to skeletal maintenance is thoroughly discussed in Chapter 43 by Whiting *et al.* (this volume). When severe, vitamin D deficiency is associated with significant undermineralization of bone matrix, a condition known as osteomalacia [50]. At milder levels of vitamin D inadequacy, impaired calcium absorption promotes compensatory secretion of PTH, which increases bone remodeling and aggravates bone loss. Maintaining vitamin D adequacy is an important and widespread issue for the general population, and for the vast majority of patients with osteoporosis. It appears that optimal vitamin D status is achieved at 25-hydroxyvitamin D concentrations of about 30 ng/ml (80 nmol/liter). To achieve this goal, previously recommended vitamin D consumption of 400–800 units/day is not adequate, and doses of 1000–2000 units/day are preferable.

3. Reproductive Hormonal Status

Decades of studies in animals and humans support the concept that achieving and maintaining normal gonadal function is a critical determinant of bone health during pubertal bone acquisition and throughout adult life in both women and men. Indeed, the impact of menopausal loss of endogenous estrogen on skeletal balance can be so profound that many investigators have focused solely on the contribution of estrogen withdrawal to osteoporosis without regard for the other contributions described earlier. We now understand that permanent loss of estrogen at menopause is not the only circumstance in which gonadal status has a skeletal impact. During earlier adult life, transient episodes of oligo- or amenorrhea may also be associated with at least transient bone loss [51].

The mechanisms by which estrogen withdrawal affects the skeleton involve multiple organs. The direct skeletal effect of the most potent circulating estrogen, 17-β estradiol, is to suppress the rate of bone remodeling. This is achieved by downregulating the formation of osteoclasts. In some rodents, this effect has been related to the suppression of an osteoclast-stimulating cytokine, interleukin-6 [52, 53]. In other animal models and in humans, strong evidence has been presented for the participation of other cytokines, including interleukins and transforming growth factor-β [54]. In addition, estrogen withdrawal reduces osteoblast production of osteoprotegerin, the decoy receptor for RANK-ligand discussed earlier [55]. These cytokine effects are reversed when estrogen is replaced. The use of estrogen as pharmacologic therapy for the prevention and treatment of osteoporosis will be described later.

TABLE 2 A Partial List of Risk Factors for Osteoporosis and Related Fractures

Major factors
- Personal history of fracture as an adult
- History of fragility fracture in a first-degree relative
- Low body weight (less than about 127 lb)
- Current smoking
- Use of oral corticosteroid therapy for more than 3 months

Additional risk factors
- Impaired vision
- Estrogen deficiency at an early age ($<$45 years)
- Thyroid hormone excess
- Intestinal malabsorption
- Some varieties of hypercalciuria
- Selected medications (cyclosporin and related agents, thiazolidinediones, possibly selective serotonin inhibitors [SSRIs])
- Dementia
- Poor health/frailty
- Recent falls
- Low calcium intake (lifelong)
- Immobilization or low physical activity
- Alcohol in amounts $>$2 drinks per day

Adapted from [1].

IV. DIAGNOSIS OF OSTEOPOROSIS

Because traditional radiographic techniques cannot distinguish osteoporosis until it is severe, diagnosis was, until recently, clinical, requiring a history of one or more low-trauma fractures. Although highly specifc, such a grossly insensitive diagnostic criterion offered no assistance to physicians who hope to identify and treat affected individuals who have been fortunate not yet to have sustained a fracture. The focus in this section is the measurement of bone mineral density. However, one must be cognizant of a wide variety of risk factors other than bone density that may profoundly influence an individual's risk for fracture. A partial list of these factors is presented in Table 2.

The introduction of accurate noninvasive bone mass measurements afforded the opportunity to estimate a person's fracture risk and to make an early diagnosis of osteoporosis. Large prospective studies have shown that a reduction in BMD of 1 standard deviation from the mean value for an age-specifc population confers a two- to threefold increase in long-term fracture risk [56–59]. In a manner similar to that by which serum cholesterol concentration predicts risk for heart attack or blood pressure predicts risk for stroke, BMD measurements can successfully identify subjects at risk of fracture and can help physicians select those individuals who will derive greatest benefit for initiation of therapy. Indeed, the gradient of risk associated with a 1 SD difference in BMD is even greater than those for cholesterol and heart attack.

Several factors limit the ability of BMD measurements to predict an individual's fracture risk with great accuracy. The normative data against which BMD comparisons are most often made have been determined for Caucasian men and women and do not necessarily apply to other ethnic groups. This problem is gradually being overcome as ethnic-specific BMD data have begun to penetrate the literature. BMD is clearly related to body weight, yet routine clinical bone mass assessments are not weight-adjusted. Various features of bone geometry that affect bone strength and fracture risk are not considered in the clinical interpretation of bone mass measurements. These include bone size, the distribution of bone mass around its bending axis (moments of inertia), and some derivative functions, such as the hip axis length [61]. Moreover, bone mass measurements cannot distinguish individuals with low mass and intact microarchitecture from those with equal mass who have trabecular disruption and cortical porosity. They also cannot distinguish other aspects of bone quality.

In 1994, a group of senior investigators in this field offered a working definition of osteoporosis based exclusively on bone mass [61]. The reasoning behind this proposal, made on behalf of the World Health Organization (WHO), was that the clinical significance of osteoporosis lies exclusively in the occurrence of fracture, that bone mass predicts long-term fracture risk, and that selection of rigorous diagnostic criteria would minimize the number of patients who are incorrectly diagnosed. The authors suggested a cutoff BMD value of 2.5 standard deviations below the average for healthy young adult women. Using this value, approximately 30% of postmenopausal women would be designated as osteoporotic, which gives a realistic projection of lifetime fracture rates. In addition, Kanis *et al.* [61] proposed that BMD values of 1–2 standard deviations below the young adult mean be designated as "osteopenic." Such values identify individuals at increased risk for fracture, but for whom a diagnosis of osteoporosis would not be justified because it would mislabel far more individuals than would actually be expected ever to fracture.

This approach has proven useful for clinical management, but has limitations. Its application to young people prior to their acquisition of peak bone mass would be inappropriate, and it remains unclear exactly what the best means to assess fracture risk in men may be. The BMD measurement is itself subject to several confounding factors, including bone size and geometry. As BMD correlations among skeletal sites are not strong, designating a person "normal" based on a single site, for example, the lumbar spine, necessarily overlooks individuals with low bone density elsewhere, such as the hip. It seems reasonable to suppose that adjustment of bone density readings for such factors as body size, bone geometry, and ethnic background might improve the accuracy of this technique. Finally, recent studies indicate that, although individuals with low BMD are at greater relative risk to fracture, many fractures in the population are experienced by individuals with normal bone mass [62–64]. Knowledge of a low bone density at a particular point in time offers no information regarding the adequacy of peak bone mass attained, the amount of bone that may have been lost, or the quality of bone that remains.

As of this writing, the World Health Organization has undertaken an initiative to give individual patients an estimate of absolute fracture risk. Because the goal of this initiative is to provide a useful document for all nations, it will be based primarily on osteoporosis risk factors rather than on BMD. However, for areas of the world in which BMD testing is widely available, it will be possible to add BMD scores as an additional predictive variable. At present, the WHO initiative has not been published, but it is anticipated that it will become available sometime during 2007.

A. Beyond BMD: The Question of Bone Quality

As stated in the introduction to this chapter, osteoporosis is a condition of decreased bone strength, where strength is a composite function of bone mineral density and bone quality. Several diverse qualitative characteristics have been described that directly influence bone strength. The more important factors, bone geometry, microarchitectural integrity, mineralization state, and remodeling rate, are briefly considered here. For more detailed information, the reader may consult Seeman and Delmas [65]. The first factor is overall bone geometry. A bone of greater diameter is better able to withstand either a compressive or a bending force than a smaller bone. One reason that men are relatively less susceptible to forearm fractures than are women is the fact that long bones of men are wider. A second feature, particularly in trabecular bone, is microarchitectural integrity. Normal trabecular bone is a honeycomb of highly connected vertical and horizontal trabeculae (Figure 1A), and the holes, or spaces between the trabeculae, are fairly uniform. By contrast, osteoporotic trabecular bone gives the appearance of a Swiss cheese (Figure 1B). The holes are not uniform because excessive bone resorption has perforated the trabeculae, leaving confluent, sometimes extremely large holes that represent the permanent loss of entire trabecular units. A third important qualitative factor is the state of bone remodeling. It will be remembered that each remodeling event begins with the removal by osteoclasts of a divot of bone from trabecular surface. At any one time, then, there are hundreds of thousands of resorption holes, or *lacunae*, on bone surfaces. These holes are an inherent point of weakness to the bone, permitting very modest mechanical loads to overload the structure and fracture it. In mechanical engineering parlance, such points of weakness are called "stress concentrators." Mechanical stresses that are generated across an area of bone divert from their normal path of stress transmission to focus on the area of weakening, thereby overloading it. Thus, individuals with higher levels of bone remodeling activity will have more stress-concentrating lacunae on their bone surfaces, giving them a greater chance for fracture than somebody in a low remodeling state. Finally, the degree of bone mineralization is also an important feature of bone quality. Roughly speaking, except in extreme cases, the more mineralized the bone matrix is, the greater its mechanical strength. Because a portion of bone continues to accumulate mineral as long as it exists, individuals with lower remodeling rates will have bone that is older, and therefore more mineralized and stronger, than individuals whose bone is remodeled more rapidly.

One message from this inquiry into bone quality is that a BMD measurement simply does not describe a sufficient amount of information about the nature of the skeleton ever to be a gold standard for the presence of osteoporosis. Unfortunately, at present, no suitable noninvasive measurements are available to assess bone quality in patients outside a research environment. Therefore, one is forced to rely on BMD measures to some degree. That being said, one also should not depend on changes in BMD over time as a true index of whether patients have improved or deteriorated in response to therapeutic interventions. One important indication of overall bone quality is whether a given person has already sustained a low-trauma fracture. Evidence has been published that individuals with such fractures have more serious disruption of qualitative measures seen on bone biopsies than individuals who have not fractured [66]. Therefore, the presence of a vertebral compression fracture on a radiograph of the spine may reasonably permit the physician to conclude that the patient has poor bone quality.

V. OSTEOPOROSIS PREVENTION AND TREATMENT

A. Hygienic Management

The term *hygienic* is used here in the sense of being nonpharmacologic. It refers to the appropriate attention to lifestyle factors that either protect or damage bone. These principles have universal application regardless of whether one wishes to forestall bone loss and development of osteoporosis or treat established disease.

1. Physical Activity

As described early in this chapter, the skeleton adapts to its mechanical environment so that the steady-state amount of bone reflects daily exposure to mechanical forces. Although the skeletal response to vigorous exercise, such as weight lifting (resistance activity) or running (endurance activity), may be greater than that observed with walking, the majority of middle-aged and older individuals are more likely to persist long term with a program of walking. A progressive schedule of walking 30 minutes or more several days each week provides a mechanical load to the legs and axial skeleton and should form the basis of skeletal maintenance for ambulatory persons in their middle and advanced years. Younger individuals should certainly be encouraged to pursue more vigorous activities, but it must be remembered that the effects of high-load environments persist only so long as the high loading schedule continues. If a person stops training, the bone perceives a reduction in mechanical environment and adaptive responses will lead to loss of bone until a new steady state is reached.

2. Dietary Factors

With respect to overall dietary intakes, substantial weight loss during adult life is itself a risk factor for fractures. Weight-loss programs are associated with measurable decreases in BMD, and the impact of so-called yo-yo weight fluctuations, although unclear, carries some skeletal risk. As discussed earlier, the nutrients of most relevance to skeletal maintenance are calcium and vitamin D. A daily calcium intake of 1200–1500 mg should be the goal for most healthy adults. This could be approached by consuming a quart of nonfat milk each day, but it is very unlikely that most adults, even dairy enthusiasts, will consistently accomplish that. Each quart of milk contains approximately 1100 mg calcium, although this may vary somewhat depending on its fat content. Because calcium is contained within the aqueous phase of milk, an equal volume of low-fat or skim milk has more water, and hence more calcium, than whole milk. In addition, some brands of reduced-fat milk are enriched by the addition of milk solids, which further increases calcium content.

Many segments of the population experience loss of the enzyme lactase, which is necessary for the hydrolysis of the major milk sugar lactose. Many afflicted individuals experience bloating, cramps, or other symptoms of intestinal distress if they consume lactose-containing dairy products. Several strategies can be recommended for such individuals. Lactose-free milk is readily available in markets. Lactose in this product has been prehydrolyzed to its constituent sugars, glucose and galactose, by incubation with commercial lactase. This milk has a slightly sweet taste due to the taste of its hexose sugars. For those who do not care for this taste, lactase tablets can be taken immediately prior to drinking milk. Yogurt and other products in which lactose is hydrolyzed provide an excellent substitute for milk. Many cheeses are rich in calcium (not cottage cheese, however), but have the complicating feature of high sodium content. The calcium:-sodium ratio of most liquid dairy products is about 1.0, but the ratio is well below 1 for hard cheeses. Given the relationship between sodium intake and calciuria [49], depending on cheese for calcium consumption may not be effective.

Other reasonably good food sources of calcium include small fish (anchovies, herrings, sardines), nuts, and some green vegetables (broccoli). However, it is quite difficult to attain calcium adequacy with these foods alone without also consuming dairy products. For example, it may require 6 cups of broccoli to provide as much calcium as would be consumed in a single 8-oz glass of milk.

For many individuals, a calcium supplement offers the most convenient approach to reaching target levels of calcium intake. Many calcium salts are available on an over-the-counter basis. Calcium carbonate has the advantage of having the highest percentage by weight in calcium (40% of calcium carbonate is calcium), so that the fewest number of pills need be taken. Calcium citrate is a reliable calcium supplement. It has been alleged that the citrate salt is better absorbed than the carbonate, particularly by individuals who have achlorhydria or are taking medications that reduce gastric acid secretion. However, it has been shown that when calcium carbonate is taken with food there is sufficient acidity in the food to permit normal calcium absorption [67]. Therefore my general recommendation for individuals who will likely not consume more than 1000 mg calcium per day through food alone is to take a 500-mg calcium supplement as calcium carbonate each morning with breakfast. For individuals with osteoporosis who are receiving pharmacological treatment, I increase that recommendation to 1000 mg calcium/day.

B. Pharmacologic Therapy

The variety of approved drugs for the prevention and treatment of osteoporosis has expanded enormously during the past decade (Table 3). These are grouped into two general categories, antiresorptive (sometimes called anticatabolic) and bone-forming (or anabolic). The first category contains the great majority of registered products. Although the fundamentals of their mechanism of action differ from one product to the next, they all act ultimately to inhibit

TABLE 3 Approved Medications for the Prevention and Treatment of Osteoporosis

Antiresorptive:
Calcium/vitamin D
Estrogens
Bisphosphonates:
• Alendronate
• Risedronate
• Ibandronate
• Zoledronate
Calcitonin
Strontium ranelate (not approved in United States)
Bone-forming:
Teriparatide
PTH(1-84) (not approved in United States)

either the development of osteoclasts from their precursors or the action of mature osteoclasts. Such action confers skeletal protection in two ways, both of which are predictable consequences of slowing the remodeling rate by 30–70%, as per the previous discussion of bone quality. First, and perhaps most important, at least during the first year or so of therapy, reducing the creation of new resorption lacunae results in a major reduction in the prevalence of stress concentrators on bone surfaces. Second, because any given point on the bone surface will survive longer when the remodeling rate is low, the bone continues to gain mineral for a longer period of time and ultimately becomes relatively hypermineralized.

By contrast, only one bone-forming agent is currently approved in the United States for the treatment of osteoporosis, and that is the 1-34 fragment of human parathyroid hormone (teriparatide; see later discussion). Its actions involve the direct stimulation of bone-forming osteoblasts.

Selection of appropriate therapies for individual patients lies beyond the scope of this chapter. The following material briefly describes each of these approved products.

1. Antiresorptive Agents

a. Calcium and Vitamin D

Strictly speaking, calcium and vitamin D have an antiresorptive activity on bone. By slightly elevating blood calcium concentrations, endogenous PTH secretion is reduced and overall bone remodeling decreases. The role of calcium and vitamin D in skeletal maintenance was described earlier. Calcium administration to osteoporotic women with previous vertebral fractures has been shown to decrease the risk for subsequent fracture [33] and, when combined with a modest amount of vitamin D, has been shown to reduce the incidence of hip fracture in elderly women [24]. Suffice it to say that calcium and vitamin D adequacy are essential components of all successful therapies, and the general approach described earlier must be pursued if any pharmacologic regimen is to succeed. Indeed, therapeutic failures are frequently attributable to inadequate vitamin D status. It should be noted that all major osteoporosis trials were designed to show an effect of drug on bone turnover, BMD, and fracture incidence superior to that of the control regimens which invariably consisted of calcium and vitamin D.

b. Estrogen

The use of estrogen to treat postmenopausal osteoporosis has been popular for at least 50 years. The era of the 1970s and 1980s saw the publication of many small clinical trials demonstrating the effect of various estrogen and estrogen/progestin combinations in early and postmenopausal women to improve calcium balance, to suppress markers of bone turnover, to increase BMD, and in small series to protect against vertebral compression fractures [68]. Because the available studies did not demonstrate a compelling case that estrogen administration to women with established osteoporosis led to a significant reduction in fracture incidence, FDA approval of estrogen was specified for prevention of osteoporosis, not for treatment. With recent publication of results from the Women's Health Initiative, we now have conclusive evidence for the nonvertebral fracture efficacy of estrogen, even in women who were not osteoporotic or at high risk for fracture [69]. Unfortunately, the Women's Health Initiative also established that estrogen, particularly when given in combination with progestin, increased the development of breast cancer, myocardial infarction, and other cardiovascular events and potentially contributed to cognitive decline. Thus, although estrogens remain highly effective when treating women with hot flashes, current opinion holds that they should be used at the lowest possible dose and for the shortest possible duration and are not recommended as long-term therapy for prevention or treatment of osteoporosis.

c. Selective Estradiol Receptor Modulators (SERMs)

These compounds are neither hormones nor estrogens, but molecules that interact with the estradiol receptor in multiple tissues of the body. For a comprehensive description of SERM physiology the reader is referred to the review of Siris and Muchmore [70]. Briefly summarized: Unlike estrogens, SERMs interact with the estradiol receptor and activate estrogen regulated genes in a manner that is tissue specific. For example, like estrogen, tamoxifen stimulates uterine hyperplasia and suppresses bone remodeling, but unlike estrogen it inhibits estrogen actions at the breast. On the other hand, raloxifene has no effect on the endometrium, acts like estrogen on bone, and antagonizes estrogen action at the breast. Although several such molecules were introduced into clinical medicine before their mechanisms of action were clarified (e.g., tamoxifen), recent years have seen the development of numerous molecules in this class for the intended purpose of treating osteoporosis. As of this writing, only one SERM, raloxifene (Evista), is FDA approved for prevention and treatment of osteoporosis. Raloxifene has been shown to prevent the development of frank osteoporosis in women with low bone mass and to offer substantial long-term protection against vertebral fracture in women with established osteoporosis [71].

d. Bisphosphonates

Bisphosphonates are analogs of the naturally occurring phosphate ester, pyrophosphate, in which the central oxygen bridge has been replaced by a carbon atom. This substitution renders the compounds nonsusceptible to hydrolysis by alkaline phosphatase, a ubiquitous enzyme that hydrolyzes pyrophosphate. Bisphosphonates are poorly absorbed from the gut, but once absorbed are taken up by bone. The presence of two phosphate groups permits the molecule to bind avidly to hydroxyapatite, so that the half-life of bisphosphonates in the skeleton may be a matter of many years. In theory, osteoclasts imbibe the bisphosphonate during the course of bone resorption, resulting in osteoclast death and a decrease in resorption. Differences in one of the two side chains underlie differences in action among agents of this class. The first generation of bisphosphonates acted to inhibit intermediary metabolism. More recent compounds that have amino groups on the side chain act in a manner similar to the statin class of antilipid drugs, that is, inhibiting the mevalonate synthesis pathway, but at a level (farnesyl diphosphate synthetase) that is further downstream, leading to interference with prenylation of plasma membrane lipids [72]. First developed and introduced for the treatment of Paget's disease of bone in the 1960s, several bisphosphonates have shown significant improvement in BMD and reductions in fracture. Three oral bisphosphonates are currently approved for prevention and treatment of osteoporosis: alendronate (Fosamax) [73], risedronate (Actonel) [74, 75], and ibandronate (Boniva) [76]. The last may also be administered by intravenous infusion. It is likely that another very potent bisphosphonate, zoledronic acid, will receive FDA approval for treatment of osteoporosis within the next year. This agent will be administered only once each year by intravenous infusion.

e. Calcitonin

This 32-amino-acid peptide is a natural hormone throughout the vertebrate phylum. Its primary physiological role appears to be to reduce the rate of bone remodeling by inhibiting the action of mature osteoclasts. Calcitonin obtained from salmon is considerably more potent than the human hormone and has been approved for treatment of osteoporosis [77]. Although calcitonin can be given by subcutaneous injection, most patients take the drug as a nasal spray (Miacalcin). In approved doses, calcitonin is a relatively weak antiresorptive drug with efficacy characteristics that are less pronounced than those of the other approved antiresorptive agents.

f. Strontium Ranelate

Strontium has recently received approval in Europe and Asia for treatment of osteoporosis. It becomes incorporated directly into the bone mineral, which creates an artifactual increase in BMD. However, strontium does appear to be an effective antiresorptive compound and has been shown to reduce fracture risk. Because no studies of this compound have been conducted in the United States, it does not appear that strontium can ever receive FDA approval and so it is unlikely to be introduced into the American market.

2. Bone Formation Therapy

a. Teriparatide (Forteo, Forsteo in Europe)

Teriparatide is the generic term for the 1-34 fragment of human PTH. It is approved as a single daily subcutaneous injection for up to 2 years at a dose of 20 μg/day. Teriparatide is the first bone anabolic agent approved for treatment of osteoporosis. (The full length PTH(1-84) molecule has not received approval in the United States, but is currently marketed in Europe. Its actions are qualitatively similar to those of teriparatide.) Teriparatide directly stimulates osteoblasts to form new bone and results in considerably greater increases in BMD than are observed with antiresorptive drugs. In addition, teriparatide uniquely repairs the disrupted microarchitecture of trabecular bone to a normal pattern and also increases the thickness of cortical bone. These effects result in substantial reduction in both vertebral and nonvertebral fracture [78]. Teriparatide given very long-term at high dose to Fischer 344 rats led to a high rate of osteosarcoma. Other animal models have shown no such effect, and although a relationship to human carcinogenesis seems very unlikely, one cannot be certain that no relationship exists. Because duration of therapy was a critical element for carcinogenesis in rats, teriparatide use is restricted to 2 years at the present time. Teriparatide is marketed for the treatment of men and postmenopausal women with osteoporosis whose physicians consider them at high risk for fracture. As opposed to several of the antiresorptive agents, it is not a drug for prevention of osteoporosis.

VI. CONCLUSION

At present, considerable efforts are under way to bring forth compounds of both the antiresorptive and bone-formation classes. A variety of new SERMS are in mid- to late-state human trials. Other developmental targets for antiresorptive agents include inhibitors of cathepsins, the osteoclast-derived enzymes that hydrolyze bone, and antibodies against RANK-ligand. With respect to bone-forming agents, new PTH analogs and modes of delivery for teriparatide are in early-phase human trials. Still largely in animal studies, investigation of means to activate a newly discovered anabolic pathway in bone, the Wnt pathway, has begun.

Thus, the future of pharmacologic therapy for osteoporosis seems reasonably bright. However, the overall outlook for this disease remains clouded. With aging of the population, trends toward increasingly sedentary life, and substandard intakes of critical nutrients, the worldwide burden of osteoporotic fractures is likely to increase dramatically. Even at the present time, it is difficult in the United States to get appropriate diagnosis and treatment for afflicted patients, even those with multiple fractures. With growing competition from other aspects of health care and with continued threats of onerous governmental and health insurance constraints on reimbursement for osteoporosis diagnosis and treatment, it is not at all certain that an explosive increase in fragility fractures and their consequences will be avoided.

References

1. National Osteoporosis Foundation. (1998), "Physician's Guide to Prevention and Treatment of Osteoporosis." Excerpta Medica, Bell Mead, NJ. Available at http://www.nof.org/physguide/inside_cover.htm.
2. Melton, L. J. III. (1995). How many women have osteoporosis now? *J. Bone Miner. Res.* **10**, 175–177.
3. Melton, L. J. III, and Cooper, C. (2001). Magnitude and impact of osteoporosis and fractures. *In* "Osteoporosis" (R. Marcus, D. Feldman, and J. Kelsey, Eds.), 2nd ed., pp. 557–567. Academic Press, San Diego, CA.
4. Cauley, J. A., Thompson, D. E., Ensrud, K. C., Scott, J. C., and Black, D. (2000). Risk of mortality following clinical fractures. *Osteoporos. Int.* **11**, 556–561.
5. Cummings, S. R., and Nevitt, M. C. (1989). A hypothesis: The causes of hip fractures. *J. Gerontol.* **44**, M107–M111.
6. Schwartz, A. V., Capezuti, E., and Grisso, J. A. (2001). *In* "Osteoporosis" (R. Marcus, D. Feldman, and J. Kelsey, Eds.), 2nd ed., pp. 795–807. Academic Press, San Diego, CA.
7. Lee, C. A., and Einhorn, T. A. (2001). The bone organ system, form and function. *In* "Osteoporosis" (R. Marcus, D. Feldman, and J. Kelsey, Eds.), 2nd ed., pp. 3–20. Academic Press, San Diego, CA.
8. Brown, E. M. (1994). Homeostatic mechanisms regulating extracellular and intracellular calcium metabolism. *In* "The Parathyroids" (J. P. Bilezikian, M. A. Levine, and R. Marcus, Eds.), pp. 15–54. Raven Press, New York.
9. Ross, F. P., and Teitelbaum, S. L. (2001). Osteoclast biology. *In* "Osteoporosis" (R. Marcus, D. Feldman, and J. Kelsey, Eds.), 2nd ed., pp. 73–105. Academic Press, San Diego, CA.
10. Asagiri, M., and Takayanagi, H. (2007). Review: The molecular understanding of osteoclast differentiation. *Bone* **40**, 251–264.
11. Drinkwater, B. L., Nilson, K., Chesnut, C. H., Bremner, W. J., Shainholtz, S., and Southworth, M. B. (1984). Bone mineral content of amenorrheic and eumenorrheic athletes. *N. Engl. J. Med.* **311**, 277–281.
12. Marcus, R., Cann, C., Madvig, P., Minkoff, J., Goddard, M., Bayer, M., Martin, M., Gaudiani, L., Haskell, W., and Genant, H. (1985). Menstrual function and bone mass in elite women distance runners. *Ann. Int. Med.* **102**, 158–163.
13. Wolff, J. (1892). "Das Gesetz der Transformation der Knochen." Hirschwald Verlag, Berlin.
14. Beck, B. R., Shaw, J., and Snow, C. M. (2001). Physical activity and osteoporosis. *In* "Osteoporosis" (R. Marcus, D. Feldman, and J. Kelsey, Eds.), 2nd ed., pp. 701–720. Academic Press, San Diego, CA.
15. Taaffe, D. R., Snow-Harter, C., Connolly, D. A., Robinson, T. L., and Marcus, R. (1995). Differential effects of swimming versus weight-bearing activity on bone mineral status of eumenorrheic athletes. *J. Bone Miner. Res.* **10**, 586–593.
16. Snow-Harter, C., Bouxsein, M., Lewis, B. T., Carter, D. R., and Marcus, R. (1992). Effects of resistance and endurance exercise on bone mineral status of young women: A randomized exercise intervention trial. *J. Bone Miner. Res.* **7**, 761–769.
17. Friedlander, A. L., Genant, H. K., Sadowsky, S., Byl, N. N., and Gluer, C. C. (1995). A two-year program of aerobics and weight-training enhances BMD of young women. *J. Bone Miner. Res.* **10**, 574–585.
18. Lohman, T., Going, S., Pamenter, R., Hall, M., Boyden, T., Houtkooper, L., Ritenbaugh, C., Bare, L., Hill, A., and Aickin, M. (1995). Effects of resistance training on regional and total bone mineral density in premenopausal women: A randomized prospective study. *J. Bone Miner. Res.* **10**, 1015–1024.
19. Heinonen, A., Sievanen, H., Kannus, P., Oja, P., and Vuori, I. (1996). Effects of unilateral strength training and detraining on bone mineral mass and estimated mechanical characteristics of the upper limb bones in young women. *J. Bone Miner. Res.* **11**, 490–501.
20. Bachrach, L. K., Guido, D., Katzman, D., Litt, I. F., and Marcus, R. (1990). Decreased bone density in adolescent girls with anorexia nervosa. *Pediatrics* **86**, 440–447.
21. Albright, F., and Reifenstein, E. C. Jr. (1948). "The Parathyroid Glands and Metabolic Bone Disease: Selected Studies," p. 145. Williams & Wilkins, Baltimore.
22. Matkovic, V., Kostial, K., Simonovic, I., Buzina, R., Brodarec, A., and Nordin, B. E. C. (1979). Bone status and fracture rates in two regions of Yugoslavia. *Am. J. Clin. Nutr.* **32**, 540–549.
23. Alaimo, K., McDowell, M. A., Briefel, R. R., Bischof, A. M., Caughman, C. R., Loria, C. M., and Johnson, C. L. (1994). Dietary intake of vitamins, minerals, and fiber of persons ages 2 months and over in the United States. Third National Health and Nutrition Examination Survey, Phase 1, 1988–1991, Advance Data from Vital and Health Statistics No. 258. National Center for Health Statistics, Hyattsville, MD.
24. Chapuy, M. C., Arlot, M. E., Duboeuf, F., Brun, J., Crouzet, B., Arnaud, S., Delmas, P. D., and Meunier, P. J. (1992). Vitamin D_3 and calcium to prevent hip fractures in the elderly women. *N. Engl. J. Med.* **327**, 1637–1642.
25. NIH Consensus Conference. (1994). Optimal calcium intake. *JAMA* **272**, 1942–1948.
26. Food and Nutrition Board, Institute of Medicine. (1997). "Dietary Reference Intakes for Calcium, Magnesium, Phosphorus, Vitamin D, and Fluoride." National Academies Press, Washington, DC.
27. "Recommended Dietary Allowances," 10th ed. (1989). National Academies Press, Washington, DC

28. Matkovic, V., and Heaney, R. P. (2002). Calcium balance during human growth. Evidence for threshold behavior. *Am. J. Clin. Nutr.* **55**, 992–996.
29. Dawson-Hughes, B., Dallal, G. E., Krall, E. A., Sadowski, L., Sahyoun, N., and Tannenbaum, S. (1990). A controlled trial of the effect of calcium supplementation on bone density in postmenopausal women. *N. Engl. J. Med.* **323**, 878–883.
30. Reid, I. R., Ames, R. W., Evans, M. C., Gamble, G. D., and Sharpe, S. J. (1993). Effect of calcium supplementation on bone loss in postmenopausal women. *N. Engl. J. Med.* **328**, 460–464.
31. Reid, I. R., Ames, R. W., Evans, M. C., Sharpe, S. J., and Gamble, G. D. (1994). Determinants of the rate of bone loss in normal postmenopausal women. *J. Clin. Endocrinol. Metab.* **79**, 950–954.
32. Chevalley, T., Rizzoli, R., Nydegger, V., Slosman, D., Rapin, C.-H., Michel, J.-P., Vasey, H., and Bonjour, J.-P. (1994). Effects of calcium supplements on femoral bone mineral density and vertebral fracture rate in vitamin D-replete elderly patients. *Osteoporos. Int.* **4**, 245–252.
33. Recker, R. R., Hinders, S., Davies, K. M., Heaney, R. P., Stegman, M. R., Kimmel, D. B., and Lappe, D. J. (1996). Correcting calcium nutritional deficiency prevents spine fractures in elderly women. *J. Bone Miner. Res.* **11**, 1961–1966.
34. Dawson-Hughes, B., Harris, S. S., Krall, E. A., and Dallal, G. E. (1997). Effect of calcium and vitamin D supplementation on bone density in men and women 65 years of age or older. *N. Engl. J. Med.* **337**, 670–676.
35. Aloia, J. F., Vaswani, A., Yeh, J. K., Ross, P. L., Flaster, E., and Dilmanian, F. A. (1994). Calcium supplementation with and without hormone replacement therapy to prevent postmenopausal bone loss. *Ann. Intern. Med.* **120**, 97–103.
36. Heaney, R. P., Weaver, C. M., and Recker, R. R. (1988). Calcium absorbability from spinach. *Am. J. Clin. Nutr.* **47**, 707–709.
37. Spencer, H., Kramer, L., Osis, D., and Norris, C. (1978). Effect of phosphorus on the absorption of calcium and on the calcium balance in man. *J Nutr* **108**, 447–457.
38. Heaney, R. P., and Recker, R. R. (1982). Effects of nitrogen, phosphorus, and caffeine on calcium balance in women. *J. Lab. Clin. Med.* **99**, 46–55.
39. Heaney, R. P., and Rafferty, K. (2001). Carbonated beverages and urinary calcium excretion. *Am. J. Clin. Nutr.* **74**, 343–347.
40. Massey, L. K. (2003). Dietary animal and plant protein and human bone health: A whole foods approach. *J. Nutr.* **133**, 862S–865S.
41. Barger-Lux, M. J., Heaney, R. P., Packard, P. T., Lappe, J. M., and Recker, R. R. (1992). Nutritional correlates of low calcium intake. *Clin. Appl. Nutr.* **2**, 39–44.
42. Sellmeyer, D. E., Stone, K. L., Sebastian, A., and Cummings, S. R. (2001). A high ratio of dietary animal to vegetable protein increases the rate of bone loss and the risk of fracture in postmenopausal women. Study of Osteoporotic Fractures Research Group. *Am. J. Clin. Nutr.* **73**, 118–122.
43 Devine, A., Dick, I. M., Islam, A. F., Dhaliwal, S. S., and Prince, R. L. (2005). Protein consumption is an important predictor of lower limb bone mass in elderly women. *Am. J. Clin. Nutr.* **81**, 1423–1428.
44. Hannan, M. T., Tucker, K. L., Dawson-Hughes, B., Cupples, L. A., Felson, D. T., and Kiel, D. P. (2000). Effect of dietary protein on bone loss in elderly men and women: The Framingham Osteoporosis Study. *J. Bone Miner. Res.* **15**, 2504–2512.
45. Kerstetter, J. E., O'Brien, K. O., Caseria, D. M., Wall, D. E., and Insogna, K. L. (2005). The impact of dietary protein on calcium absorption and kinetic measures of bone turnover in women. *J. Clin. Endocrinol. Metab.* **90**, 26–31.
46. Dawson-Hughes, B., and Harris, S. S. (2002). Calcium intake influences the association of protein intake with rates of bone loss in elderly men and women. *Am. J. Clin. Nutr.* **75**, 773–779.
47. Delmi, M., Rapin, C. H., Bengoa, J. M., Delmas, P. D., Vasey, H., and Bonjour, J. P. (1990). Dietary supplementation in elderly patients with fractured neck of the femur. *Lancet* **335**, 1013–1016.
48. Heaney, R. P. (2007). Effects of protein on the calcium economy." *In* "Nutritional Aspects of Osteoporosis 2006" (B. Dawson-Hughes, and R. P. Henry, Eds), pp. 191–197. Elsevier, Amsterdam.
49. Devine, A., Criddle, R. A., Dick, I. M., Kerr, D. A., and Prince, R. L. (1995). A longitudinal study of the effect of sodium and calcium intakes on regional bone density in postmenopausal women. *Am. J. Clin. Nutr.* **62**, 740–745.
50. Marcus, R. (2001). Osteomalacia. *In* "Nutrition in the Prevention and Treatment of Disease" (A. M. Coulston, C. L. Rock, and E. R. Monsen, Eds.), pp. 729–740. Academic Press, San Diego, CA.
51. Sowers, M. F. Premenopausal reproductive and hormonal characteristics and the risk for osteoporosis. *In* "Osteoporosis" (R. Marcus, D. Feldman, and J. Kelsey, Eds.), 2nd ed., pp. 721–740. Academic Press, San Diego, CA.
52. Jilka, R. L., Hangoc, G., Girasole, G., Passeri, G., Williams, D. C., Abrams, J. S., Boyce, B., Broxmeyer, H., and Manolagas, S. C. (1992). Increased osteoclast development after estrogen loss: mediation by interleukin-6. *Science* **3**, 88–91.
53. Girasole, G., Jilka, R. L., Passeri, G., Boswell, S., Boder, G., Williams, D. C., and Manolagas, S. C. (1992). 17-β-Estradiol inhibits interleukin-6 production by bone marrow-derived stromal cells and osteoblasts *in vitro*: A potential mechanism for the antiosteoporotic effect of estrogens. *J. Clin. Invest.* **89**, 883–891.
54. Weitzmann, M. N., and Pacifici, R. (2006). Estrogen deficiency and bone loss: An inflammatory tale. *J. Clin. Invest.* **116**, 1186–1194.
55. Zallone, A. (2006). Direct and indirect estrogen actions on osteoblasts and osteoclasts. *Ann. N. Y. Acad. Sci.* **1068**, 173–179.
56. Hui, S., Slemenda, C., and Johnston, C. J. (1989). Baseline measurement of bone mass predicts fracture in white women. *Ann. Intern. Med.* **111**, 355–361.
57. Melton, L., Atkinson, E., O'Fallon, W., Wahner, H., and Riggs, B. (1993). Long-term fracture prediction by bone mineral assessed at different skeletal sites. *J. Bone Miner. Res.* **8**, 1227–1233.
58. Cummings, S. R., Black, D. M., Nevitt, M. C., Browner, W., Cauley, J., Ensrud, K., Genant, H. K., Palermo, L., Scott, J., and Vogt, T. M. (1993). Bone density at various sites for prediction of hip fractures. The Study of Osteoporotic Fractures Research Group. *Lancet* **341**(8837), 72–75.
59. Johnell, O., Kanis, J. A., Oden, A., Johansson, H., De Laet, C., Delmas, P., Eisman, J. A., Fujiwara, S., Kroger, H.,

Mellstrom, D., Meunier, P. J., Melton, L. J. 3rd, O'Neill, T., Pols, H., Reeve, J., Silman, A., and Tenenhouse, A. (2005). Predictive value of BMD for hip and other fractures. *J. Bone Miner. Res.* **20**, 1185–1194.

60. Faulkner, K. G. (2003). Improving femoral bone density measurements. *J. Clin. Densitom.* **6**, 353–358.
61. Kanis, J. A., Melton, L. J. 3rd, Christiansen, C., Johnston, C. C., and Khaltaev, N. (1994). The diagnosis of osteoporosis. *J. Bone Miner. Res.* **9**, 1137–1141.
62. Siris, E. S., Chen, Y. T., Abbott, T. A., Barrett-Connor, E., Miller, P. D., Wehren, L. E., and Berger, M. L. (2004). Bone mineral density thresholds for pharmacological intervention to prevent fractures. *Arch. Intern. Med.* **164**, 1108–1112.
63. Sornay-Rendu, E., Munoz, F., Garnero, P., Duboeuf, F., and Delmas, P. D. (2005). Identification of osteopenic women at high risk of fracture: The OFELY study. *J. Bone Miner. Res.* **20**, 1813–1819.
64. Wainwright, S. A., Marshall, L. M., Ensrud, K. E., Cauley, J. A., Black, D. M., Hillier, T. A., Hochberg, M. C., Vogt, M. T., and Orwoll, E. S. (2005). Hip fracture in women without osteoporosis. *J. Clin. Endocrinol. Metab.* **90**, 2787–2793.
65. Seeman, E., and Delmas, P. D. (2006). Bone quality—the material and structural basis of bone strength and fragility. *N. Engl. J. Med.* **354**, 2250–2261.
66. Genant, H. K., Delmas, P. D., Chen, P., Jiang, Y., Eriksen, E. F., Dalsky, G. P., Marcus, R., and San Martin, J. (2007). Severity of vertebral fracture reflects deterioration of bone microarchitecture. *Osteoporos. Int.*, 69–76.
67. Heaney, R. P., Dowell, M. S., and Barger-Lux, M. J. (1999). Absorption of calcium as the carbonate and citrate salts, with some observations on method. *Osteoporos. Int.* **9**, 19–23.
68. Lindsay, R., Hart, D. M., Forrest, C., and Baird, C. (1980). Prevention of spinal osteoporosis in oophorectomized women. *Lancet* **2**, 1151–1153.
69. Jackson, R. D., LaCroix, A. Z., Gass, M., Wallace, R. B., Robbins, J., Lewis, C. E., Bassford, T., Beresford, S. A., Black, H. R., Blanchette, P., Bonds, D. E., Brunner, R. L., Brzyski, R. G., Caan, B., Cauley, J. A., Chlebowski, R. T., Cummings, S. R., Granek, I., Hays, J., Heiss, G., Hendrix, S. L., Howard, B. V., Hsia, J., Hubbell, F. A., Johnson, K. C., Judd, H., Kotchen, J. M., Kuller, L. H., Langer, R. D., Lasser, N. L., Limacher, M. C., Ludlam, S., Manson, J. E., Margolis, K. L., McGowan, J., Ockene, J. K., O'Sullivan, M. J., Phillips, L., Prentice, R. L., Sarto, G. E., Stefanick, M. L., Van Horn, L., Wactawski-Wende, J., Whitlock, E., Anderson, G. L., Assaf, A. R., and Barad, D. Women's Health Initiative Investigators. (2006) Calcium plus vitamin D supplementation and the risk of fractures. *N. Engl. J. Med.* **354**, 669–683 Erratum in: *N. Engl. J. Med.* (2006). **354,** 1102.
70. Siris, E. S., and Muchmore, D. B. (2001). Selective estrogen receptor modulators (SERMS). *In* "Osteoporosis" (R. Marcus, D. Feldman, and J. Kelsey, Eds.), 2nd ed., pp. 603–620. Academic Press, San Diego, CA.
71. Ettinger, B., Black, D. M., and Mitlak, B. (1999). Reduction of vertebral fracture risk in postmenopausal women with osteoporosis treated with raloxifene. Results from a 3-year randomized clinical trial. *JAMA* **282**, 637–645.
72. Kavanagh, K. I., Guo, K., Dunford, J. E., Wu, X., Knapp, S., and Ebetino, F. H. (2006). The molecular mechanism of nitrogen-containing bisphosphonates as antiosteoporosis drugs. *Proc. Natl. Acad. Sci. USA* **103**, 7829–7834.
73. Black, D. M., Cummings, S. R., Karpf, D. B., Cauley, J. A., Thompson, D. E., Nevitt, M. C., Bauer, D. C., Genant, H. K., Haskell, W. L., Marcus, R., Ott, S. M., Torner, J. C., Quandt, S. A., Reiss, T. F., and Ensrud, K. E. (1996). Randomised trial of effect of alendronate on risk of fracture in women with existing vertebral fractures. Fracture Intervention Trial Research Group. *Lancet* **348**, 1535–1541.
74. Harris, S. T., Watts, N. B., Genant, H. K., McKeever, C. D., Hangartner, T., Keller, M., Chesnut, C. H. 3rd, Brown, J., Eriksen, E. F., Hoseyni, M. S., Axelrod, D. W., and Miller, P. D. (1999). Effects of risedronate treatment on vertebral and nonvertebral fractures in women with postmenopausal osteoporosis: A randomized controlled trial. Vertebral Efficacy with Risedronate Therapy (VERT) Study Group. *JAMA* **282**, 1344–1352.
75. McClung, M. R., Geusens, P., Miller, P. D., Zippel, H., Bensen, W. G., Roux, C., Adami, S., Fogelman, I., Diamond, T., Eastell, R., Meunier, P. J., Reginster, J. Y.; Hip Intervention Program Study Group. (2001). Hip intervention program study group. Effect of risedronate on the risk of hip fracture in elderly women. *N. Engl. J. Med.* **344**, 333–340.
76. Delmas, P. D., Recker, R. R., Chesnut, C. H. 3rd, Skag, A., Stakkestad, J. A., Emkey, R., Gilbride, J., Schimmer, R. C., and Christiansen, C. (2004). Daily and intermittent oral ibandronate normalize bone turnover and provide significant reduction in vertebral fracture risk: Results from the BONE study. *Osteoporos. Int.* **15**, 792–798.
77. Chesnut, C. H., Silverman, S., Andriano, K., Genant, H., Gimona, A., Harris, S., Kiel, D., LeBoff, M., Maricic, M., Miller, P., Moniz, C., Peacock, M., Richardson, P., Watts, N., and Baylink, D. (2000). A randomized trial of nasal spray salmon calcitonin in postmenopausal women with established osteoporosis: The Prevent Recurrence of Osteoporotic Fractures Study. PROOF Study Group. *Am. J. Med.* **109**, 267–276.
78. Neer, R. M., Arnaud, C. D., Zanchetta, J. R., Prince, R., Gaich, G. A., Reginster, J. Y., Hodsman, A. B., Eriksen, E. F., Ish-Shalom, S., Genant, H. K., Wang, O., and Mitlak, B. H. (2001). Effect of parathyroid hormone (1-34) on fractures and bone mineral density in postmenopausal women with osteoporosis. *N. Engl. J. Med.* **344**, 1434–1441.
79. Heaney, R. P. (2007). Nutrition and risk for osteoporosis. *In* "Osteoporosis" (R. Marcus, D. Feldman, D. A. Nelson, and C. J. Rosen, Eds.), 3rd ed., pp. 799–828. Academic Press, San Diego, CA.

APPENDIX

DIETARY REFERENCE INTAKES (DRIs)

Dietary Reference Intakes (DRIs): Recommended Intakes for Individuals, Vitamins

Food and Nutrition Board, Institute of Medicine, National Academies

Life Stage Group	Vit A (μg/d)[a]	Vit C (mg/d)	Vit D (μg/d)[b,c]	Vit E (mg/d)[d]	Vit K (μg/d)	Thiamin (mg/d)	Riboflavin (mg/d)	Niacin (mg/d)[e]	Vit B_6 (mg/d)	Folate (μg/d)[f]	Vit B_{12} (μg/d)	Pantothenic Acid (mg/d)	Biotin (μg/d)	Choline[g] (mg/d)
Infants														
0–6 mo	400*	40*	5*	4*	2.0*	0.2*	0.3*	2*	0.1*	65*	0.4*	1.7*	5*	125*
7–12 mo	500*	50*	5*	5*	2.5*	0.3*	0.4*	4*	0.3*	80*	0.5*	1.8*	6*	150*
Children														
1–3 y	300	15	5*	6	30*	0.5	0.5	6	0.5	150	0.9	2*	8*	200*
4–8 y	400	25	5*	7	55*	0.6	0.6	8	0.6	200	1.2	3*	12*	250*
Males														
9–13 y	600	45	5*	11	60*	0.9	0.9	12	1.0	300	1.8	4*	20*	375*
14–18 y	900	75	5*	15	75*	1.2	1.3	16	1.3	400	2.4	5*	25*	550*
19–30 y	900	90	5*	15	120*	1.2	1.3	16	1.3	400	2.4	5*	30*	550*
31–50 y	900	90	5*	15	120*	1.2	1.3	16	1.3	400	2.4	5*	30*	550*
51–70 y	900	90	10*	15	120*	1.2	1.3	16	1.7	400	2.4[i]	5*	30*	550*
>70 y	900	90	15*	15	120*	1.2	1.3	16	1.7	400	2.4[i]	5*	30*	550*
Females														
9–13 y	600	45	5*	11	60*	0.9	0.9	12	1.0	300	1.8	4*	20*	375*
14–18 y	700	65	5*	15	75*	1.0	1.0	14	1.2	400[j]	2.4	5*	25*	400*
19–30 y	700	75	5*	15	90*	1.1	1.1	14	1.3	400[j]	2.4	5*	30*	425*
31–50 y	700	75	5*	15	90*	1.1	1.1	14	1.3	400[j]	2.4	5*	30*	425*
51–70 y	700	75	10*	15	90*	1.1	1.1	14	1.5	400	2.4[h]	5*	30*	425*
>70 y	700	75	15*	15	90*	1.1	1.1	14	1.5	400	2.4[h]	5*	30*	425*
Pregnancy														
14–18 y	750	80	5*	15	75*	1.4	1.4	18	1.9	600[j]	2.6	6*	30*	450*
19–30 y	770	85	5*	15	90*	1.4	1.4	18	1.9	600[j]	2.6	6*	30*	450*
31–50 y	770	85	5*	15	90*	1.4	1.4	18	1.9	600[j]	2.6	6*	30*	450*

(continues)

Dietary Reference Intakes (DRIs): Recommended Intakes for Individuals, Vitamins *(continued)*

Life Stage Group	Vit A (μg/d)[a]	Vit C (mg/d)	Vit D (μg/d)[b,c]	Vit E (mg/d)[d]	Vit K (μg/d)	Thiamin (mg/d)	Riboflavin (mg/d)	Niacin (mg/d)[e]	Vit B_6 (mg/d)	Folate (μg/d)[f]	Vit B_{12} (μg/d)	Pantothenic Acid (mg/d)	Biotin (μg/d)	Choline[g] (mg/d)
Lactation														
14–18 y	1200	115	5*	19	75*	1.4	1.6	17	2.0	500	2.8	7*	35*	550*
19–30 y	1300	120	5*	19	90*	1.4	1.6	17	2.0	500	2.8	7*	35*	550*
31–50 y	1300	120	5*	19	90*	1.4	1.6	17	2.0	500	2.8	7*	35*	550*

Note: This table (taken from the DRI reports, see www.nap.edu) presents Recommended Dietary Allowances (RDAs) and Adequate Intakes (AIs). AIs are followed by an asterisk (*). RDAs and AIs may both be used as goals for individual intake. RDAs are set to meet the needs of almost all (97 to 98 percent) individuals in a group. For healthy breastfed infants, the AI is the mean intake. The AI for other life stage and gender groups is believed to cover needs of all individuals in the group, but lack of data or uncertainty in the data prevent being able to specify with confidence the percentage of individuals covered by this intake.

[a]As retinol activity equivalents (RAEs). 1 RAE = 1 μg retinol, 12 μg β-carotene, 24 μg α-carotene, or 24 μg β-cryptoxanthin. The RAE for dietary provitamin A carotenoids is twofold greater than retinol equivalents (RE), whereas the RAE for preformed vitamin A is the same as RE.

[b]As cholecalciferol. 1 μg cholecalciferol = 40 IU vitamin D.

[c]In the absence of adequate exposure to sunlight.

[d]As α-tocopherol. α-Tocopherol includes *RRR*-α-tocopherol, the only form of α-tocopherol that occurs naturally in foods, and the 2*R*-stereoisomeric forms of α-tocopherol (*RRR*-, *RSR*-, *RRS*-, and *RSS*-α-tocopherol) that occur in fortified foods and supplements. It does not include the 2*S*-stereoisomeric forms of α-tocopherol (*SRR*-, *SSR*-, *SRS*-, and *SSS*-α-tocopherol), also found in fortified foods and supplements.

[e]As niacin equivalents (NE). 1 mg of niacin = 60 mg of tryptophan; 0–6 months = preformed niacin (not NE).

[f]As dietary folate equivalents (DFE). 1 DFE = 1 μg food folate = 0.6 μg of folic acid from fortified food or as a supplement consumed with food = 0.5 μg of a supplement taken on an empty stomach.

[g]Although AIs have been set for choline, there are few data to assess whether a dietary supply of choline is needed at all stages of the life cycle, and it may be that the choline requirement can be met by endogenous synthesis at some of these stages.

[h]Because 10 to 30 percent of older people may malabsorb food-bound B_{12}, it is advisable for those older than 50 years to meet their RDA mainly by consuming foods fortified with B_{12} or a supplement containing B_{12}.

[i]In view of evidence linking folate intake with neural tube defects in the fetus, it is recommended that all women capable of becoming pregnant consume 400 μg from supplements or fortified foods in addition to intake of food folate from a varied diet.

[j]It is assumed that women will continue consuming 400 μg from supplements or fortified food until their pregnancy is confirmed and they enter prenatal care, which ordinarily occurs after the end of the periconceptional period—the critical time for formation of the neural tube.

Sources: Dietary Reference Intakes for Calcium, Phosphorous, Magnesium, Vitamin D, and Fluoride (1997); Dietary Reference Intakes for Thiamin, Riboflavin, Niacin, Vitamin B_6, Folate, Vitamin B_{12}, Pantothenic Acid, Biotin, and Choline (1998); Dietary Reference Intakes for Vitamin C, Vitamin E, Selenium, and Carotenoids (2000); Dietary Reference Intakes for Vitamin A, Vitamin K, Arsenic, Boron, Chromium, Copper, Iodine, Iron, Manganese, Molybdenum, Nickel, Silicon, Vanadium, and Zinc (2001); and Dietary Reference Intakes for Water, Potassium, Sodium, Chloride, and Sulfate (2004). These reports may be accessed via http://www.nap.edu.

Dietary Reference Intakes (DRIs): Recommended Intakes for Individuals, Elements

Food and Nutrition Board, Institute of Medicine, National Academies

Life Stage Group	Calcium (mg/d)	Chromium (μg/d)	Copper (μg/d)	Fluoride (mg/d)	Iodine (μg/d)	Iron (mg/d)	Magnesium (mg/d)	Manganese (mg/d)	Molybdenum (μg/d)	Phosphorus (mg/d)	Selenium (μg/d)	Zinc (mg/d)	Potassium (g/d)	Sodium (g/d)	Chloride (g/d)
Infants															
0–6 mo	210*	0.2*	200*	0.01*	110*	0.27*	30*	0.003*	2*	100*	15*	2*	0.4*	0.12*	0.18*
7–12 mo	270*	5.5*	220*	0.5*	130*	11	75*	0.6*	3*	275*	20*	3	0.7*	0.37*	0.57*
Children															
1–3 y	500*	11*	340	0.7*	90	7	80	1.2*	17	460	20	3	3.0*	1.0*	1.5*
4–8 y	800*	15*	440	1*	90	10	130	1.5*	22	500	30	5	3.8*	1.2*	1.9*
Males															
9–13 y	1300*	25*	700	2*	120	8	240	1.9*	34	1250	40	8	4.5*	1.5*	2.3*
14–18 y	1300*	35*	890	3*	150	11	410	2.2*	43	1250	55	11	4.7*	1.5*	2.3*
19–30 y	1000*	35*	900	4*	150	8	400	2.3*	45	700	55	11	4.7*	1.5*	2.3*
31–50 y	1000*	35*	900	4*	150	8	420	2.3*	45	700	55	11	4.7*	1.5*	2.3*
51–70 y	1200*	30*	900	4*	150	8	420	2.3*	45	700	55	11	4.7*	1.3*	2.0*
>70 y	1200*	30*	900	4*	150	8	420	2.3*	45	700	55	11	4.7*	1.2*	1.8*
Females															
9–13 y	1300*	21*	700	2*	120	8	240	1.6*	34	1250	40	8	4.5*	1.5*	2.3*
14–18 y	1300*	24*	890	3*	150	15	360	1.6*	43	1250	55	9	4.7*	1.5*	2.3*
19–30 y	1000*	25*	900	3*	150	18	310	1.8*	45	700	55	8	4.7*	1.5*	2.3*
31–50 y	1000*	25*	900	3*	150	18	320	1.8*	45	700	55	8	4.7*	1.5*	2.3*
51–70 y	1200*	20*	900	3*	150	8	320	1.8*	45	700	55	8	4.7*	1.3*	2.0*
>70 y	1200*	20*	900	3*	150	8	320	1.8*	45	700	55	8	4.7*	1.2*	1.8*
Pregnancy															
14–18 y	1300*	29*	1000	3*	220	27	400	2.0*	50	1250	60	12	4.7*	1.5*	2.3*
19–30 y	1000*	30*	1000	3*	220	27	350	2.0*	50	700	60	11	4.7*	1.5*	2.3*
31–50 y	1000*	30*	1000	3*	220	27	360	2.0*	50	700	60	11	4.7*	1.5*	2.3*
Lactation															
14–18 y	1300*	44*	1300	3*	290	10	360	2.6*	50	1250	70	13	5.1*	1.5*	2.3*
19–30 y	1000*	45*	1300	3*	290	9	310	2.6*	50	700	70	12	5.1*	1.5*	2.3*
31–50 y	1000*	45*	1300	3*	290	9	320	2.6*	50	700	70	12	5.1*	1.5*	2.3*

Note: This table presents Recommended Dietary Allowances (RDAs) and Adequate Intakes (AIs). AIs are followed by an asterisk (*). RDAs and AIs may both be used as goals for individual intake. RDAs are set to meet the needs of almost all (97 to 98 percent) individuals in a group. For healthy breastfed infants, the AI is the mean intake. The AI for other life stage and gender groups is believed to cover needs of all individuals in the group, but lack of data or uncertainty in the data prevent being able to specify with confidence the percentage of individuals covered by this intake.

Sources: Dietary Reference Intakes for Calcium, Phosphorous, Magnesium, Vitamin D, and Fluoride (1997); Dietary Reference Intakes for Thiamin, Riboflavin, Niacin, Vitamin B_6, Folate, Vitamin B_{12}, Pantothenic Acid, Biotin, and Choline (1998); Dietary Reference Intakes for Vitamin C, Vitamin E, Selenium, and Carotenoids (2000); Dietary Reference Intakes for Vitamin A, Vitamin K, Arsenic, Boron, Chromium, Copper, Iodine, Iron, Manganese, Molybdenum, Nickel, Silicon, Vanadium, and Zinc (2001); and Dietary Reference Intakes for Water, Potassium, Sodium, Chloride, and Sulfate (2004). These reports may be accessed via http://www.nap.edu.

Dietary Reference Intakes (DRIs): Tolerable Upper Intake Levels (UL[a]), Vitamins

Food and Nutrition Board, Institute of Medicine, National Academies

Life Stage Group	Vit A (μg/d)[b]	Vit C (mg/d)	Vit D (μg/d)	Vit E (mg/d)[c,d]	Vit K	Thiamin	Riboflavin	Niacin (mg/d)[d]	Vit B_6 (mg/d)	Folate (μg/d)[f]	Vit B_{12}	Pantothenic Acid	Biotin	Choline (g/d)	Carotenoids[e]
Infants															
0–6 mo	600	ND[f]	25	ND	ND	ND	ND	ND	ND	ND	ND	ND	ND	ND	ND
7–12 mo	600	ND	25	ND	ND	ND	ND	ND	ND	ND	ND	ND	ND	ND	ND
Children															
1–3 y	600	400	50	200	ND	ND	ND	10	30	300	ND	ND	ND	1.0	ND
4–8 y	900	650	50	300	ND	ND	ND	15	40	400	ND	ND	ND	1.0	ND
Males															
Females															
9–13 y	1700	1200	50	600	ND	ND	ND	20	60	600	ND	ND	ND	2.0	ND
14–18 y	2800	1800	50	800	ND	ND	ND	30	80	800	ND	ND	ND	3.0	ND
19–70 y	3000	2000	50	1000	ND	ND	ND	35	100	1000	ND	ND	ND	3.5	ND
>70 y	3000	2000	50	1000	ND	ND	ND	35	100	1000	ND	ND	ND	3.5	ND
Pregnancy															
14–18 y	2800	1800	50	800	ND	ND	ND	30	80	800	ND	ND	ND	3.0	ND
19–50 y	3000	2000	50	1000	ND	ND	ND	35	100	1000	ND	ND	ND	3.5	ND
Lactation															
14–18 y	2800	1800	50	800	ND	ND	ND	30	80	800	ND	ND	ND	3.0	ND
19–50 y	3000	2000	50	1000	ND	ND	ND	35	100	1000	ND	ND	ND	3.5	ND

[a]UL = The maximum level of daily nutrient intake that is likely to pose no risk of adverse effects. Unless otherwise specified, the UL represents total intake from food, water, and supplements. Due to lack of suitable data, ULs could not be established for vitamin K, thiamin, riboflavin, vitamin B_{12}, pantothenic acid, biotin, carotenoids. In the absence of ULs, extra caution may be warranted in consuming levels above recommended intakes.

[b]As preformed vitamin A only.

[c]As α-tocopherol; applies to any form of supplemental α-tocopherol.

[d]The ULs for vitamin E, niacin, and folate apply to synthetic forms obtained from supplements, fortified foods, or a combination of the two.

[e]β-Carotene supplements are advised only to serve as a provitamin A source for individuals at risk of vitamin A deficiency.

[f]ND = Not determinable due to lack of data of adverse effects in this age group and concern with regard to lack of ability to handle excess amounts. Source of intake should be from food only to prevent high levels of intake.

Sources: Dietary Reference Intakes for Calcium, Phosphorous, Magnesium, Vitamin D, and Fluoride (1997); Dietary Reference Intakes for Thiamin, Riboflavin, Niacin, Vitamin B_6, Folate, Vitamin B_{12}, Pantothenic Acid, Biotin, and Choline (1998); Dietary Reference Intakes for Vitamin C, Vitamin E, Selenium, and Carotenoids (2000); and Dietary Reference Intakes for Vitamin A, Vitamin K, Arsenic, Boron, Chromium, Copper, Iodine, Iron, Manganese, Molybdenum, Nickel, Silicon, Vanadium, and Zinc (2001). These reports may be accessed via http://www.nap.edu.

Dietary Reference Intakes (DRIs): Tolerable Upper Intake Levels (UL[a]), Elements

Food and Nutrition Board, Institute of Medicine, National Academies

Life Stage Group	Arsenic[b]	Boron (mg/d)	Calcium (g/d)	Chromium	Copper (μg/d)	Fluoride (mg/d)	Iodine (μg/d)	Iron (mg/d)	Magnesium (mg/d)[c]	Manganese (mg/d)	Molybdenum (μg/d)	Nickel (mg/d)	Phosphorus (g/d)	Potassium	Selenium (μg/d)	Silicon[d]	Sulfate	Vanadium (mg/d)[e]	Zinc (mg/d)	Sodium (g/d)	Chloride (g/d)[f]
Infants																					
0–6 mo	ND	ND	ND	ND	ND	0.7	ND	40	ND	ND	ND	ND	ND	ND	45	ND	ND	ND	4	ND	ND
7–12 mo	ND	ND	ND	ND	ND	0.9	ND	40	ND	ND	ND	ND	ND	ND	60	ND	ND	ND	5	ND	ND
Children																					
1–3 y	ND	3	2.5	ND	1000	1.3	200	40	65	2	300	0.2	3	ND	90	ND	ND	ND	7	1.5	2.3
4–8 y	ND	6	2.5	ND	3000	2.2	300	40	110	3	600	0.3	3	ND	150	ND	ND	ND	12	1.9	2.9
Males																					
Females																					
9–13 y	ND	11	2.5	ND	5000	10	600	40	350	6	1100	0.6	4	ND	280	ND	ND	ND	23	2.2	3.4
14–18 y	ND	17	2.5	ND	8000	10	900	45	350	9	1700	1.0	4	ND	400	ND	ND	ND	34	2.3	3.6
19–70 y	ND	20	2.5	ND	10,000	10	1100	45	350	11	2000	1.0	4	ND	400	ND	ND	1.8	40	2.3	3.6
>70 y	ND	20	2.5	ND	10,000	10	1100	45	350	11	2000	1.0	3	ND	400	ND	ND	1.8	40	2.3	3.6
Pregnancy																					
14–18 y	ND	17	2.5	ND	8000	10	900	45	350	9	1700	1.0	3.5	ND	400	ND	ND	ND	34	2.3	3.6
19–50 y	ND	20	2.5	ND	10,000	10	1100	45	350	11	2000	1.0	3.5	ND	400	ND	ND	ND	40	2.3	3.6
Lactation																					
14–18 y	ND	17	2.5	ND	8000	10	900	45	350	9	1700	1.0	4	ND	400	ND	ND	ND	34	2.3	3.6
19–50 y	ND	20	2.5	ND	10,000	10	1100	45	350	11	2000	1.0	4	ND	400	ND	ND	ND	40	2.3	3.6

[a]UL = The maximum level of daily nutrient intake that is likely to pose no risk of adverse effects. Unless otherwise specified, the UL represents total intake from food, water, and supplements. Due to lack of suitable data, ULs could not be established for arsenic, chromium, silicon, potassium, and sulfate. In the absence of ULs, extra caution may be warranted in consuming levels above recommended intakes.

[b]Although the UL was not determined for arsenic, there is no justification for adding arsenic to food or supplements.

[c]The ULs for magnesium represent intake from a pharmacological agent only and do not include intake from food and water.

[d]Although silicon has not been shown to cause adverse effects in humans, there is no justification for adding silicon to supplements.

[e]Although vanadium in food has not been shown to cause adverse effects in humans, there is no justification for adding vanadium to food and vanadium supplements should be used with caution. The UL is based on adverse effects in laboratory animals and these data could be used to set a UL for adults but not children and adolescents.

[f]ND = Not determinable due to lack of data of adverse effects in this age group and concern with regard to lack of ability to handle excess amounts. Source of intake should be from food only to prevent high levels of intake.

Sources: Dietary Reference Intakes for Calcium, Phosphorous, Magnesium, Vitamin D, and Fluoride (1997); Dietary Reference Intakes for Thiamin, Riboflavin, Niacin, Vitamin B_6, Folate, Vitamin B_{12}, Pantothenic Acid, Biotin, and Choline(1998); Dietary Reference Intakes for Vitamin C, Vitamin E, Selenium, and Carotenoids (2000); Dietary Reference Intakes for Vitamin A, Vitamin K, Arsenic, Boron, Chromium, Copper, Iodine, Iron, Manganese, Molybdenum, Nickel, Silicon, Vanadium, and Zinc (2001); and Dietary Reference Intakes for Water, Potassium, Sodium, Chloride, and Sulfate (2004). These reports may be accessed via http://www.nap.edu.

Dietary Reference Intakes (DRIs): Estimated Energy Requirements (EER) for Men and Women 30 Years of Age[a]

Food and Nutrition Board, Institute of Medicine, National Academies

Height (m [in])	PAL[b]	Weight for BMI[c] of 18.5 kg/m^2 (kg [lb])	Weight for BMI of 24.99 kg/m^2 (kg [lb])	EER, Men[d] (kcal/day)		EER, Women[d] (kcal/day)	
				BMI of 18.5 kg/m^2	BMI of 24.99 kg/m^2	BMI of 18.5 kg/m^2	BMI of 24.99 kg/m^2
1.50 (59)	Sedentary	41.6 (92)	56.2 (124)	1848	2080	1625	1762
	Low active			2009	2267	1803	1956
	Active			2215	2506	2025	2198
	Very active			2554	2898	2291	2489
1.65 (65)	Sedentary	50.4 (111)	68.0 (150)	2068	2349	1816	1982
	Low active			2254	2566	2016	2202
	Active			2490	2842	2267	2477
	Very active			2880	3296	2567	2807
1.80 (71)	Sedentary	59.9 (132)	81.0 (178)	2301	2635	2015	2211
	Low active			2513	2884	2239	2459
	Active			2782	3200	2519	2769
	Very active			3225	3720	2855	3141

[a]For each year below 30, add 7 kcal/day for women and 10 kcal /day for men. For each year above 30, subtract 7 kcal/day for women and 10 kcal/day for men.

[b]PAL = physical activity level

[c]BMI = body mass index

[d]Derived from the following regression equations based on doubly-labeled water data:

Adult man: EER = 662 − 9.53 × age (y) + PA × (15.91 × wt [kg] + 539.6 × ht [m])

Adult woman: EER = 354 − 6.91 × age (y) + PA × (9.36 × wt [kg] + 726 × ht [m])

Where PA refers to coefficient for PAL

PAL = total energy expenditure ÷ basal energy expenditure

PA = 1.0 if PAL ≥ 1.0 < 1.4 (sedentary)

PA = 1.12 if PAL ≥ 1.4 < 1.6 (low active)

PA = 1.27 if PAL ≥ 1.6 < 1.9 (active)

PA = 1.45 if PAL ≥ 1.9 < 2.5 (very active)

Dietary Reference Intakes (DRIs): Acceptable Macronutrient Distribution Ranges

Food and Nutrition Board, Institute of Medicine, National Academies

Macronutrient	Range (percent of energy)		
	Children, 1–3 y	Children, 4–18 y	Adults
Fat	30–40	25–35	20–35
n-6 polyunsaturated fatty acids[a] (linoleic acid)	5–10	5–10	5–10
n-3 polyunsaturated fatty acids[a] (α-linoleic acid)	0.6–1.2	0.6–1.2	0.6–1.2
Carbohydrate	45–65	45–65	45–65
Protein	5–20	10–30	10–35

Source: Dietary Reference Intakes for Energy, Carbohydrate, Fiber, Fat, Fatty Acids, Cholesterol, Protein, and Amino Acids (2002).

[a]Approximately 10% of the total can come from longer-chain *n*-3 or *n*-6 fatty acids.

Index

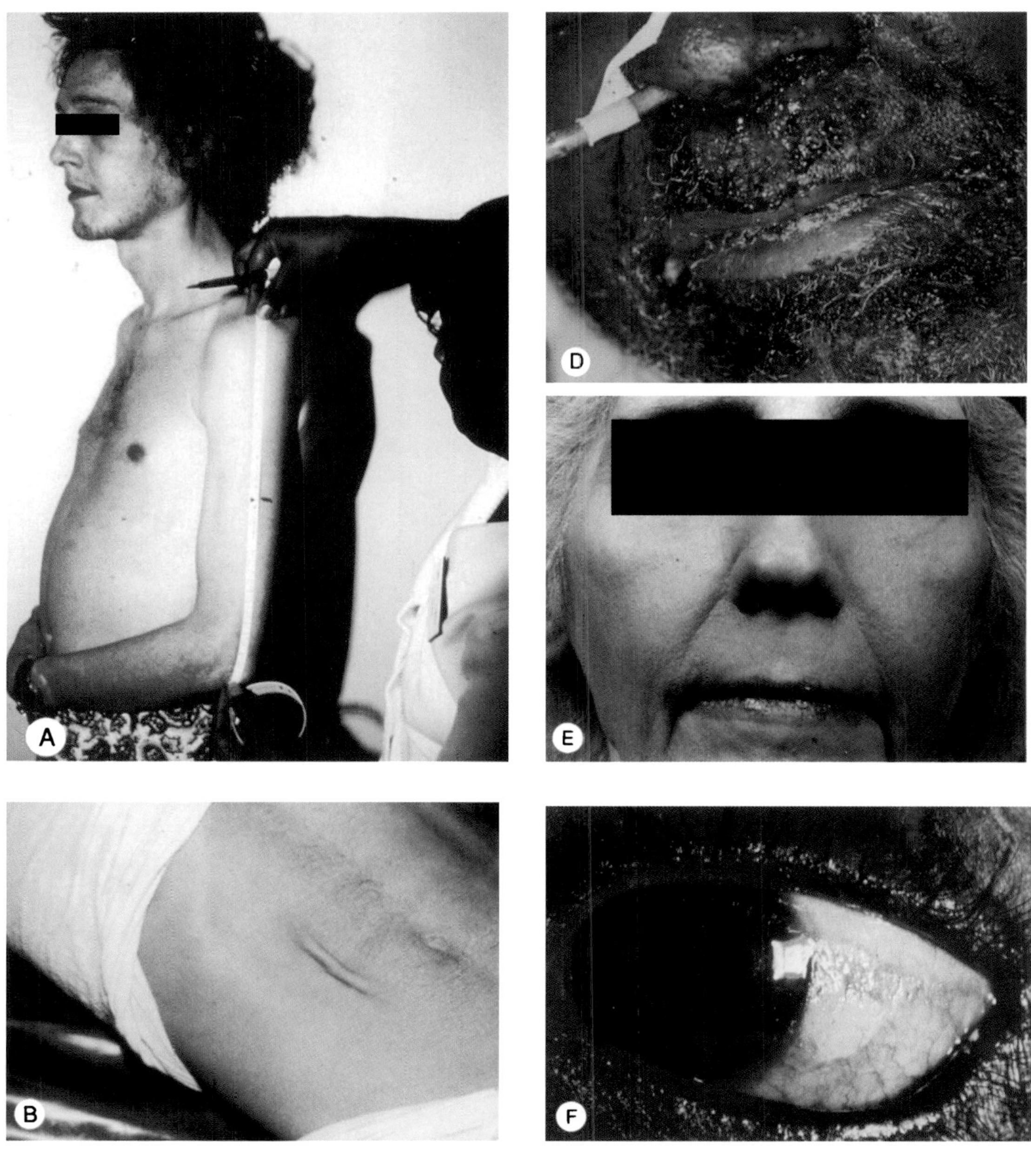

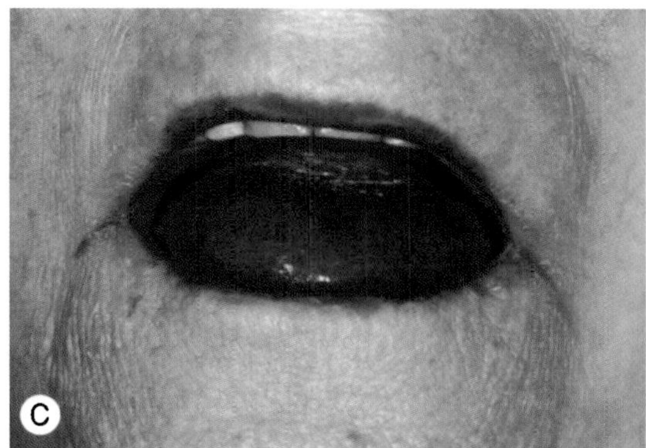

PLATE 1 Physical signs associated with nutrient deficiencies. (A) Muscle wasting in severe PEM. (B) Tenting of skin in dehydration; the skin retains the tented shape after being pinched. (C) Glossitis and angular stomatitis associated with multiple B vitamin deficiencies. (D) Dermatitis associated with zinc deficiency. (E) Cheilosis, or vertical fissuring of the lips, associated with multiple B vitamin deficiencies. (F) Bitot's spot accompanying vitamin A deficiency. (Photos courtesy of Dr. Robert Russell and Dr. Joel Mason.) (see Chapter 3, Figure 5).

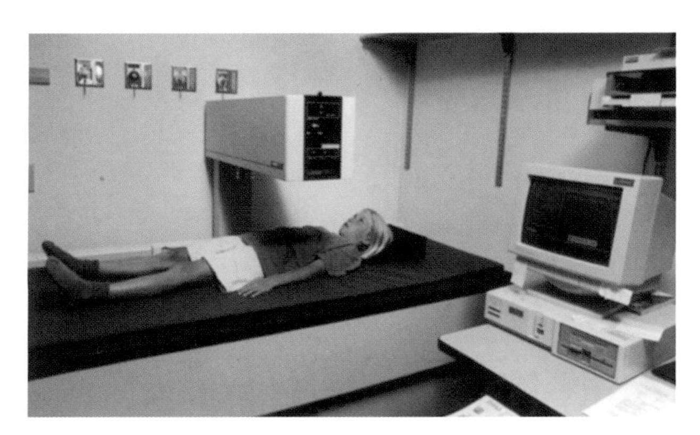

PLATE 2 Dual-energy x-ray absorptiometry is a scanning technique that accurately estimates bone mineral, fat, and fat-free soft tissue (see Chapter 4, Figure 2).

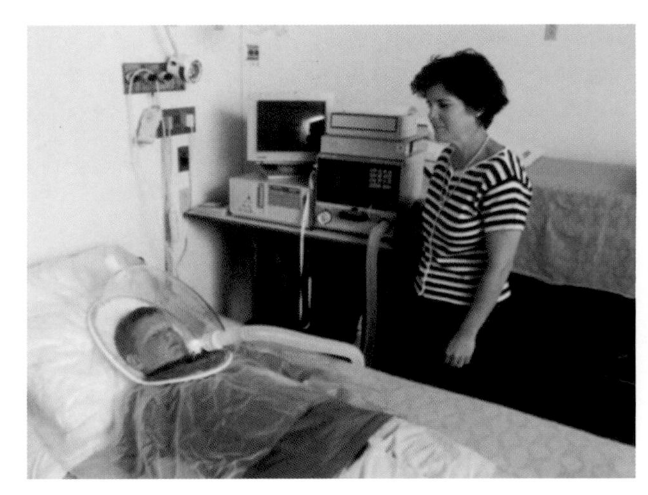

PLATE 3 Measurement of resting metabolic rate using indirect calorimetry (see Chapter 4, Figure 3).

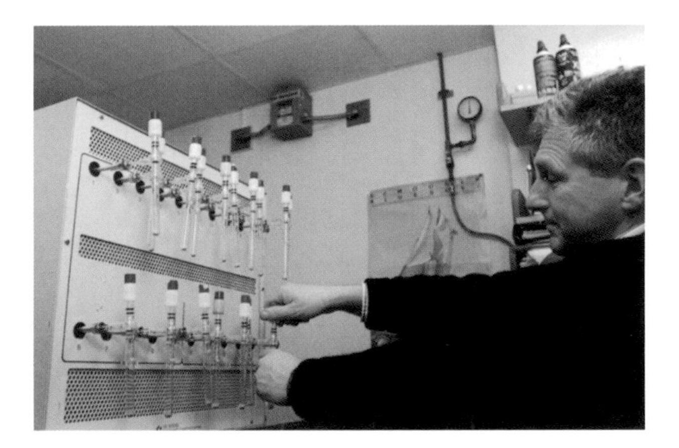

PLATE 4 Mass spectrometer used for the analysis of isotope enrichments in the DLW method (see Chapter 4, Figure 4).

GRAINS Make half your grains whole	VEGETABLES Vary your veggies	FRUITS Focus on fruits	MILK Get your calcium-rich foods	MEAT & BEANS Go lean with protein
Eat at least 3 oz. of whole-grain cereals, breads, crackers, rice, or pasta every day 1 oz. is about 1 slice of bread, about 1 cup of breakfast cereal, or ½ cup of cooked rice, cereal or pasta	Eat more dark-green veggies like broccoli, spinach, and other dark leafy greens Eat more orange vegetables like carrots and sweetpotatoes Eat more dry beans and peas like pinto beans, kidney beans, and lentils	Eat a variety of fruit Choose fresh, frozen, canned, or fried fruit Go easy on fruit juices	Go low-fat or fat-free when you choose milk, yogurt, and other milk products If you don't or can't consume milk, choose lactose-free products or other calcium sources such as fortified foods and beverages	Choose low-fat or lean meats and poultry Bake it, broil it, or grill it Vary your protein routine – choose more fish, beans, peas, nuts, and seeds
For a 2,000-calorie diet, you need the amounts below from each food group. To find the amounts that are right for you, go to MyPyramid.gov.				
Eat 6 oz. every day	Eat 2½ cups every day	Eat 2 cups every day	Get 3 cups every day; for kids aged 2 to 8, it's 2	Eat 5½ oz. every day

Find your balance between food and physical activity

- Be sure to stay within your daily calorie needs.
- Be physically active for at least 30 minutes most days of the week.
- About 60 minutes a day of physical activity may be needed to prevent weight gain.
- For sustaining weight loss, at least 60 to 90 minutes a day physical activity may be requires.
- Children and teenagers should be physically active for 60 minutes every day, or most days.

Know the limits on fats, sugars, and salt (sodium)

- Make most of your fat sources from fish, nuts, and vegetable oils.
- Limit solid fats like butter, margarine, shortening, and lard, as well as foods that contain these.
- Check the Nutrition Facts label to keep saturated fats, *trans* facts, and sodium low.
- Choose food and beverages low in added sugars. Added sugars contribute calories with few, if any, nutrients.

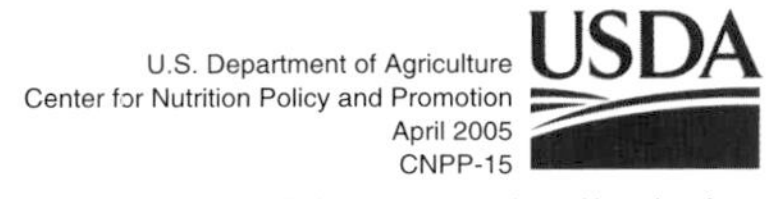

U.S. Department of Agriculture
Center for Nutrition Policy and Promotion
April 2005
CNPP-15
USDA

USDA is an equal opportunity provider and employer.

PLATE 5 MyPyramid: Steps to a healthier you (see Chapter 13, Figure 1).

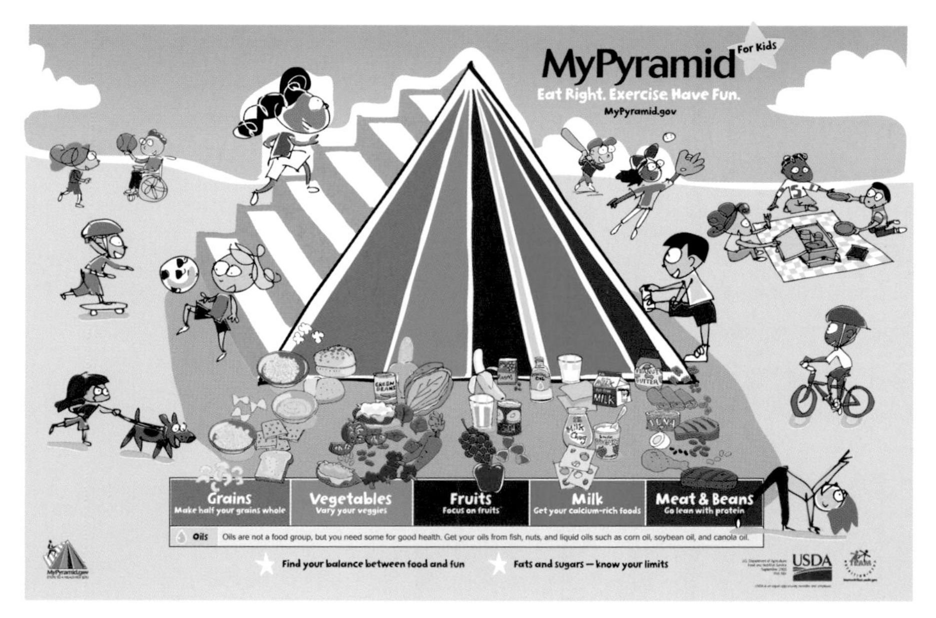

PLATE 6 MyPyramid for kids (see Chapter 13, Figure 4).

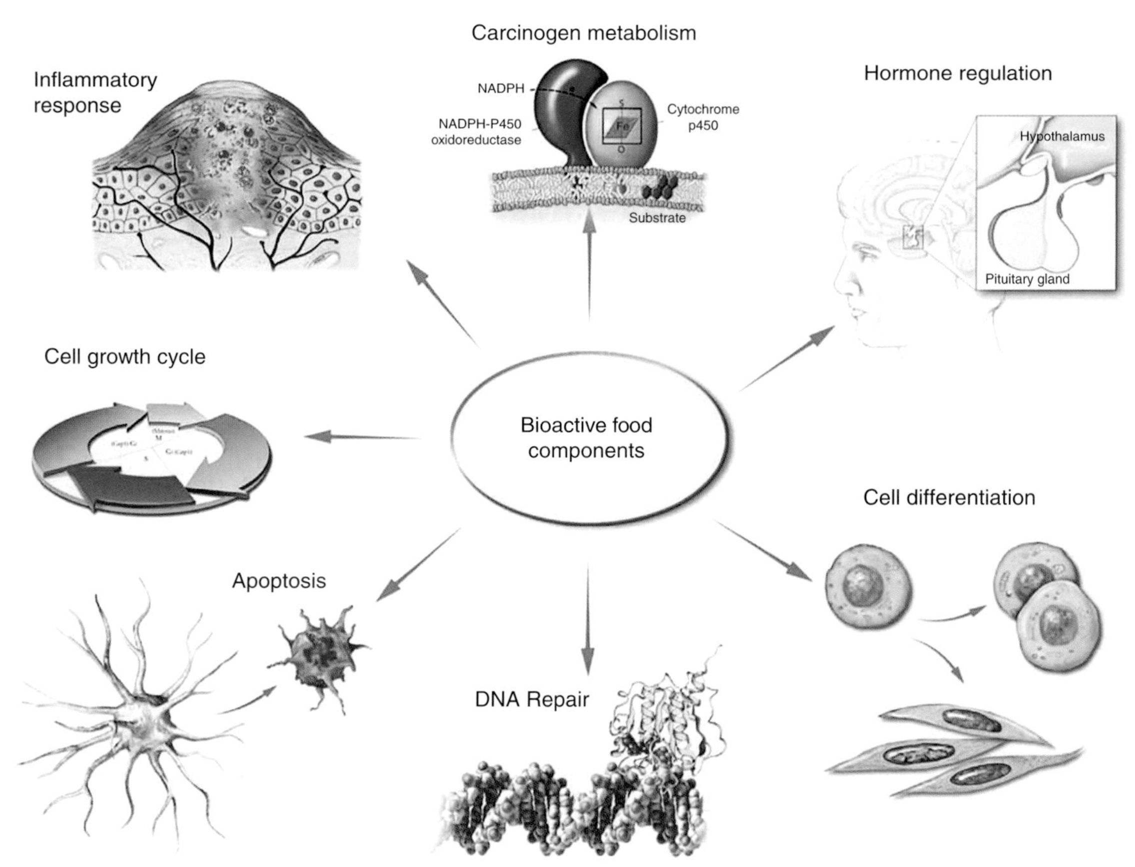

PLATE 7 Interactions of bioactive food components with genes appear to be important mechanisms for carcinogenesis. (From Trujillo, E., Davis, C., Milner, J. "Nutrigenomics, Proteomics, Metabalomics and the Practice of Dietetics." Copyright The American Dietetic Association. Reprinted by permission from *Journal of The American Dietetic Association*, vol. 106, pp. 403–413, 2006.) (see Chapter 34, Figure 3).

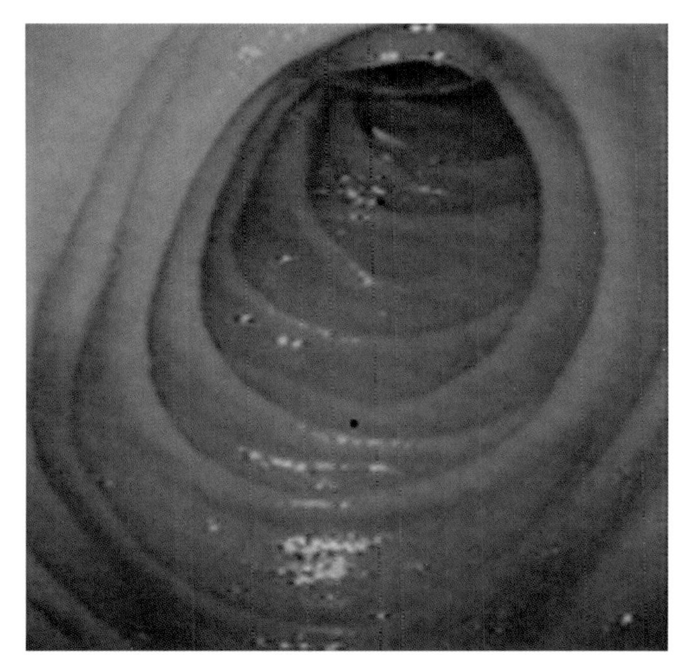

PLATE 8 Normal small intestine. Photo courtesy of Dr. Gottumukkala S. Raju, Kansas University Medical Center, Division of Gastroenterology (see Chapter 39, Figure 1).

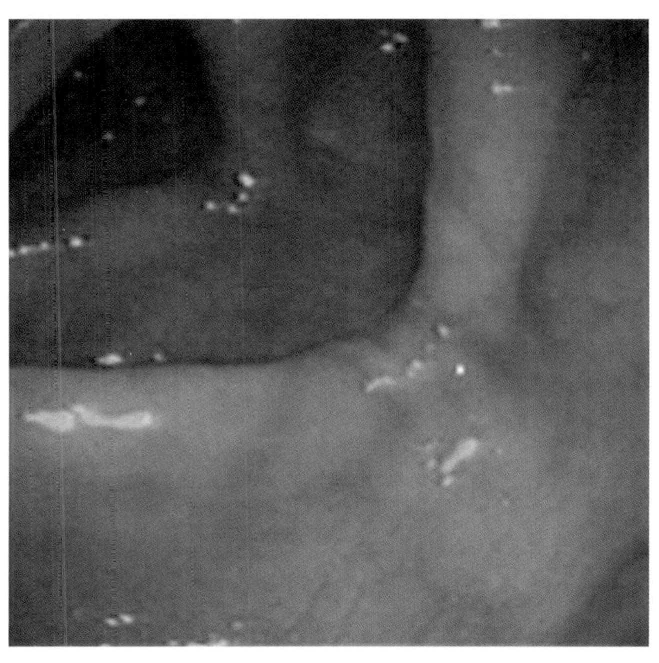

PLATE 9 Normal colon. Photo courtesy of Dr. Gottumukkala S. Raju, University of Texas Medical Branch at Galveston, Division of Gastroenterology (see Chapter 39, Figure 2).

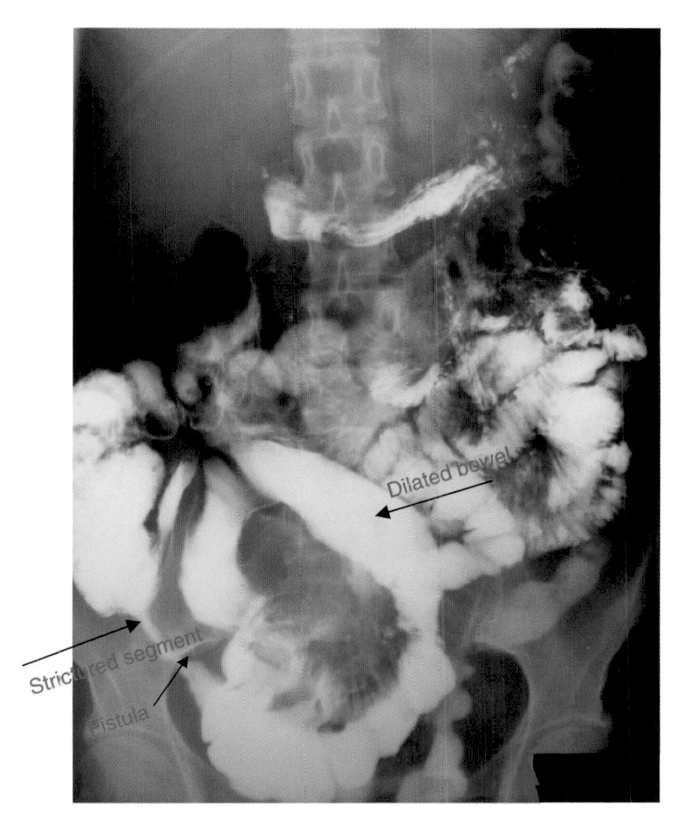

PLATE 10 X-ray, patient with Crohn's disease with small bowel stricture, fistula, and loss of normal small bowel architecture. Distension, loss of haustra, narrowed lumen, fistula (see Chapter 39, Figure 3).

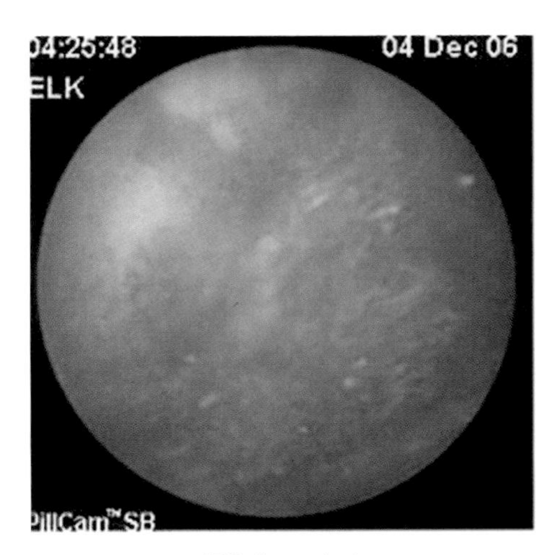

IBD Capsule 1

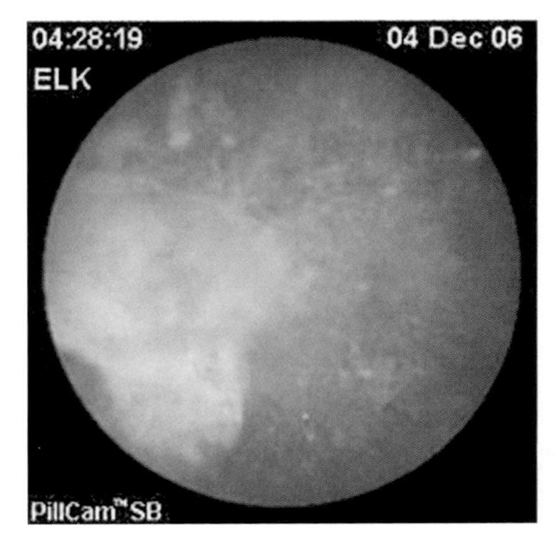

IBD Capsule 2

PLATE 11 Ulcerative colitis. Photo from swallowed capsule (see Chapter 39, Figure 4).

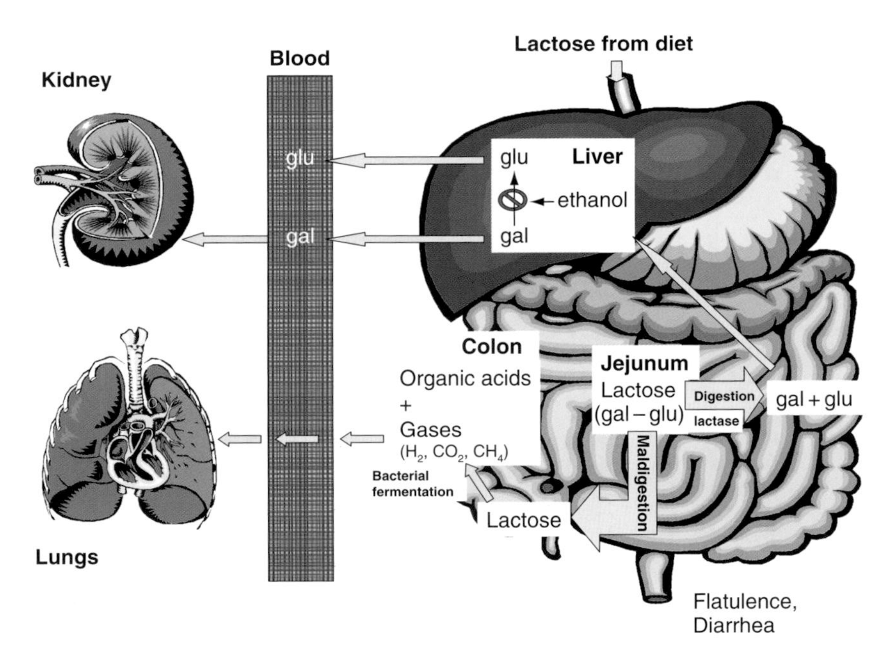

PLATE 12 Metabolic background for understanding the diagnosis of lactose maldigestion and intolerance. Abbreviations: glu, glucose; gal, galactose. [Adapted with permission from: Arola, H. (1994). Diagnosis of hypolactasia and lactose malabsorption. *Scand. J. Gastroenterol.* **29**(Suppl 202), 26–35.] (see Chapter 40, Figure 1).

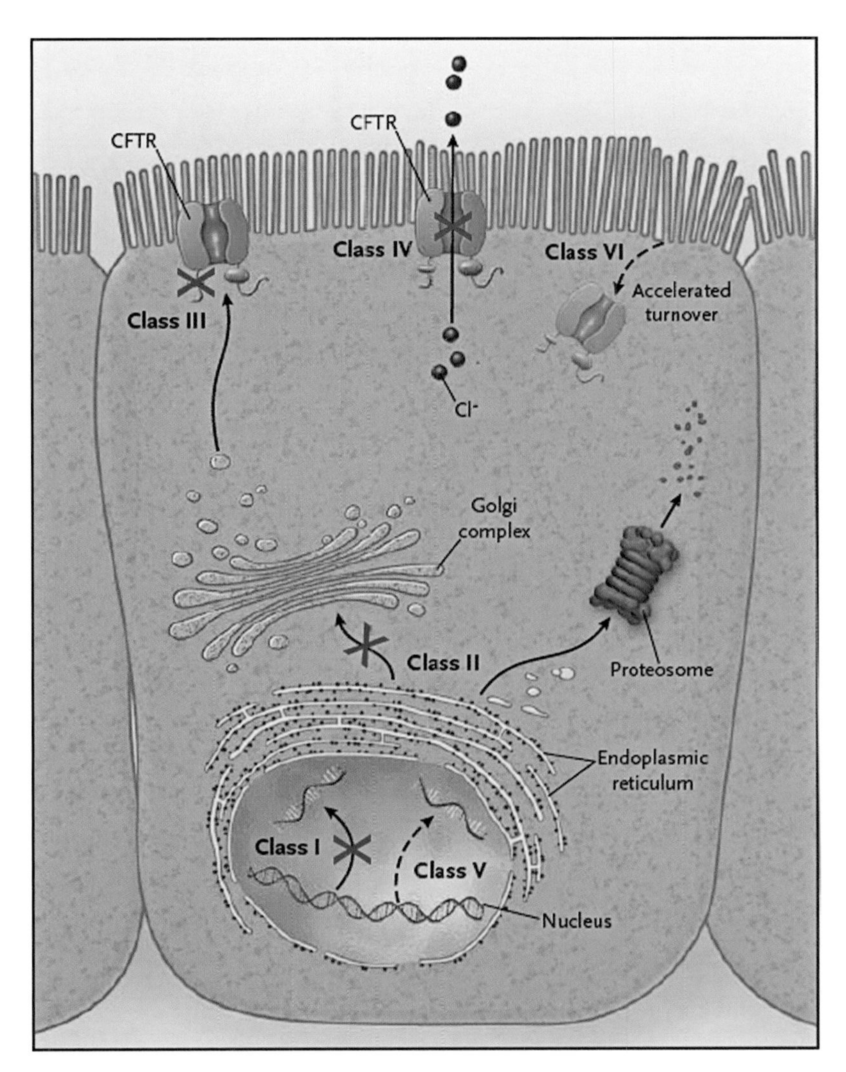

PLATE 13 Categories of CFTR mutations: absence of synthesis (class I); defective protein maturation and premature degradation (class II); disordered regulation (class III); defective chloride conductance (class IV); reduced synthesis (class V); and accelerated turnover (class VI). Adapted from Rowe et al. [4]. (see Chapter 42, Figure 1).